MODERN BIOLOGY

The Program of Choice Since 1921

For 80 years, *Modern Biology* has been setting the standard of excellence against which other biology textbooks are measured. The tradition continues with this newest edition of *Modern Biology,* the choice for a comprehensive, up-to-date approach to biology instruction.

SOUTH CAROLINA EDITION

correlated to

SOUTH CAROLINA Curriculum Standards

HOLT, RINEHART AND WINSTON

A Harcourt Education Company

Austin • New York • Orlando • Atlanta • San Francisco • Boston • Dallas • Toronto • London

SOUTH CAROLINA

Explanation of Correlation

*The following document is a correlation of **MODERN BIOLOGY** to the South Carolina Science Curriculum Standards. The format for this correlation follows the same basic format established by the Science Curriculum Standards, modified to accommodate the addition of page references. The correlation provides a cross-reference between the skills in the Science Curriculum Standards and representative page numbers where those skills are taught or assessed. Those references marked with an asterisk (*) represent pages which offer secondary support or where application of the required skill is implied.*

The references contained in this correlation reflect Holt, Rinehart and Winston's interpretation of the Science objectives outlined in the South Carolina curriculum.

KEY TO REFERENCES

Prefix	Explanation
SE	*Students's Edition*
ATE	*Annotated Teacher's Edition*
H	*History*
N	*Nature of Science*
P	*Science in Social and Personal Perspectives*
T	*Technology*

Printed in the United States of America

ISBN 0–03–068961–9

123 048 04 03 02

SOUTH CAROLINA

Detailed Correlation of *Modern Biology* to the South Carolina Curriculum Standards

I. Inquiry

Inquiry is not an isolated unit of instruction and should be embedded throughout the content areas. The nature of science and technology is incorporated within this area.

A. Identify Questions and Concepts that Guide Scientific Investigations

Experimental design should demonstrate logical connections between a knowledge base and conceptual understanding.

1.	Formulate a testable hypothesis based on literary research and previous knowledge.	**SE**	17, 92–93, 277, 317, 438, 462–463, 662–663, 740–741, 858–859, 952–953
		ATE	17
2.	Identify and select experimental variables (independent and dependent) and controlled conditions.	**SE**	17–18, 64–65, 142–143, 376–377, 462–463, 484–485, 504–505, 522–523, 662–663, 707, 740–741, 758–759, 858–859, 952–953, 1000–1001, 1044–1045
		ATE	17, 504

B. Design and Conduct Investigations

Prior knowledge about major concepts, laboratory apparatus, laboratory techniques, and safety should be used in designing and conducting a scientific investigation.

1.	Design a scientific investigation based on the major concepts in the area being studied.	SE	28–29, 46–47, 64–65, 92–93, 108–109, 124–125, 142–143, 160–161, 182–183, 200–201, 219–220, 236–237, 256–257, 276–277, 296–297, 316–317, 334–335, 354–355, 376–377, 394–395, 412–413, 438–439, 462–463, 484–485, 504–505, 522–523, 540–541, 556–557, 576–577, 596–597, 622–623, 644–645, 662–663, 690–691, 706–707, 722–723, 740–741, 758–759, 778–779, 794–795, 816–817, 836–837, 858–859, 878–879, 900–901, 928–929, 952–953, 974–975, 1000–1001, 1026–1027, 1044–1045, 1062–1063, 1080–1081
2.	Select and use appropriate instruments to make the observations necessary for the investigation, taking into consideration the limitations of the equipment.	SE	28–29, 64–65, 92–93, 142–143, 160–161, 256–257, 276–277, 376–377, 394–395, 412–413, 462–463, 522–523, 540–541, 556–557, 576–577, 596–597, 622–623, 644–645, 690–691, 706–707, 722–723, 778–779, 794–795, 878–879, 900–901, 928–929, 952–953, 974–975, 1000–1001, 1026–1027, 1044–1045, 1062–1063
3.	Identify technologies that could enhance the collection of data.	SE	152, 323, 447, 869, 921, 1080–1081
		ATE	152, 447, 921
4.	Select the appropriate safety equipment needed to conduct an investigation (e.g., goggles, aprons, etc.)	SE	28, 64, 92, 256, 276, 376, 394, 412, 462, 484, 504, 522, 556, 576, 644, 662, 690, 740, 778, 794, 928, 952, 974, 1000, 1026, 1044
5.	Suggest safety precautions that need to be implemented for the handling of materials and equipment used in an investigation.	SE	29, 64–65, 92, 200, 256–257, 276–277, 376–377, 394–395, 412–413, 462–463, 484–485, 504, 522–523, 556–557, 644–645, 662–663, 690–691, 706–707, 722–723, 740–741, 778–779, 794–795, 858, 928–929, 952–953, 974–975, 1000–1001, 1026–1027, 1044–1045
		ATE	64, 92, 142, 256, 276, 376, 394, 412, 462, 484, 504, 522, 556, 576, 622, 644, 662, 690, 722, 740, 778, 794, 858, 928, 952, 974, 1000, 1026, 1044
6.	Describe the proper response to emergency situations in the laboratory.	SE	256, 277, 376–377, 394, 413, 462–463, 794, 929, 952, 974, 1000, 1044
		ATE	256, 277, 376

SOUTH CAROLINA EDITION

SOUTH CAROLINA

7.	Conduct a laboratory investigation with repeated trials and systematic manipulation of variables.	**SE** **ATE**	17–18, 64–65, 142–143, 376–377, 462–463, 484–485, 504–505, 522–523, 662–663, 707, 740–741, 758–759, 858–859, 952–953, 1000–1001, 1044–1045 17, 504
8.	Identify possible sources of error inherent in an experimental design.	**SE**	29, 64–65, 161, 257, 277, 463, 484–485, 505, 557, 663, 859, 953
9.	Organize and display data in useable and efficient formats, such as tables, graphs, maps, and cross sections.	**SE**	28–29, 46–47, 64–65, 108–109, 124–125, 142–143, 160–161, 182–183, 236–237, 276–277, 296–297, 316–317, 354–355, 376–377, 394, 438, 462–463, 484–485, 504–505, 522–523, 540–541, 556–557, 576–577, 644–645, 662–663, 706–707, 722–723, 740–741, 758–759, 778–779, 794–795, 816–817, 836–837, 858–859, 878–879, 900–901, 928–929, 952–953, 974–975, 1000–1001, 1044–1045, 1080–1081
10.	Draw conclusions based on qualitative and quantitative data.	**SE**	28–29, 46–47, 64–65, 92–93, 108–109, 124–125, 142–143, 160–161, 182–183, 200–201, 219–220, 236–237, 256–257, 276–277, 296–297, 316–317, 334–335, 354–355, 376–377, 394–395, 412–413, 438–439, 462–463, 484–485, 504–505, 522–523, 540–541, 556–557, 576–577, 596–597, 622–623, 644–645, 662–663, 690–691, 706–707, 722–723, 740–741, 758–759, 778–779, 794–795, 816–817, 836–837, 858–859, 878–879, 900–901, 928–929, 952–953, 974–975, 1000–1001, 1026–1027, 1044–1045, 1062–1063, 1080–1081
11.	Discuss the impact of sources of error on experimental results.	**SE**	64–65, 277, 463, 484–485, 504–505, 557, 859, 953
12.	Communicate and defend the scientific thinking that resulted in conclusions.	**SE**	28–29, 46–47, 64–65, 92–93, 108–109, 124–125, 142–143, 160–161, 182–183, 200–201, 219–220, 236–237, 256–257, 276–277, 296–297, 316–317, 334–335, 354–355, 376–377, 394–395, 412–413, 438–439, 462–463, 484–485, 504–505, 522–523, 540–541, 556–557, 576–577, 596–597, 622–623, 644–645, 662–663, 690–691, 706–707, 722–723, 740–741, 758–759, 778–779, 794–795, 816–817, 836–837, 858–859, 878–879, 900–901, 928–929, 952–953, 974–975, 1000–1001, 1026–1027, 1044–1045, 1062–1063, 1080–1081

C. Use Technology and Mathematics to Improve Investigations and Communications

Scientific investigations can be improved through the use of technology and mathematics. While it is acknowledged that the SI system is the accepted measurement system in science, opportunities to use the English System are encouraged.

1.	Select and use appropriate technologies (e.g., computers, calculators, CBL's) to enhance the precision and accuracy of data collection, analysis, and display.	**SE** **ATE**	46–47, 108–109, 124–125, 152, 236–237, 322–323, 376–377, 816–817, 869, 872, 921, 1080–1081 46, 108, 124, 152, 236, 921
2.	Discriminate between data that may be valid or anomalous.	**SE**	29, 64–65, 161, 257, 277, 463, 484–485, 505, 557, 663, 859, 953
3.	Select and use mathematical formulas and calculations to extend the usefulness of laboratory measurements.	**SE** **ATE**	29, 47, 63, 124–125, 161, 182–183, 237, 256–257, 316–317, 394, 928–929, 975 194
4.	Draw a "best fit" curve through data points.	**SE**	47, 143, 741
5.	Calculate the slope of the curve and use correct units for the value of the slope for linear relationships.	**SE**	47, 143, 741
6.	Calculate interpolated and predict extrapolated data points.	**SE**	47, 143, 741
7.	Perform dimensional analysis calculations.	**SE**	47, 143, 741

D. Formulate and Revise Scientific Explanations and Models Using Logic and Evidence

Scientific explanations and models are developed and revised through discussion and debate.

1.	Construct experimental explanations or models through discussion, debate, logic, and experimental evidence.	**SE** **ATE**	18, 107, 182–183, 200–201, 218–219, 365, 367, 395–396, 510–511, 1000–1001 57, 137, 867
2.	Develop explanations and models that eliminate bias and demonstrate the use of ethical principles. (P)	**SE**	18, 182–183, 200–201, 365, 367384–385, 510–511
3.	Revise explanations or models after review.	**SE**	200–201, 365, 367, 384–385, 510–511

SOUTH CAROLINA EDITION

SOUTH CAROLINA

E. Recognize and Analyze Alternative Explanations and Models

Scientific criteria are used to discriminate among plausible explanations.

1.	Compare current scientific models with experimental results.	SE	200–201, 218–219, 365, 367, 384–385, 394–395, 1000–1001
2.	Select and defend, based on scientific criteria, the most plausible explanation or model.	SE	107, 200–201, 218–219, 322–323, 330, 365, 367, 384–385, 394–395, 510–511, 1000–1001
		ATE	510, 511

F. Communicate and Defend a Scientific Argument

Experimental processes, data, and conclusions should be communicated in a clear and logical manner.

1.	Develop a set of laboratory instructions that someone else can follow.	SE	27, 161, 377, 439, 463, 485, 505, 523, 556–557, 645, 663, 707, 722, 741, 759, 859, 953, 975, 1001
2.	Develop a presentation to communicate the process and conclusion of a scientific investigation.	SE	705, 899
		ATE	456

G. Understandings about Scientific and Technological Inquiry

Historical scientific knowledge, current research, technology, mathematics and logic should be the basis for conducting investigations and drawing conclusions.

1.	Analyze how science and technology explain and predict relationships.		
	a. Defend the idea that conceptual principles and knowledge guide scientific and technological inquiry.	SE	38–39, 60, 84–85, 152, 170, 247, 276–277, 282, 322–323, 340–341, 366–367, 447, 469, 510–511, 614, 680–681, 732, 921
		ATE	38, 60, 84, 152, 170, 247, 282, 322, 340, 341, 366, 367, 510, 511, 614, 680, 681, 921
	b. Explain how historical and current scientific knowledge influences the design, interpretation, and evaluations of investigations.	SE	38–39, 60, 84–85, 132, 141, 152, 170, 247, 276, 282, 303, 322–323, 340–341, 366–367, 447, 469, 510–511, 614, 680–681, 732, 869, 921, 938, 966
		ATE	38, 60, 84, 152, 170, 247, 282, 303, 322, 323, 340, 341, 366, 367, 469, 510, 511, 614, 680, 681, 938, 966
1.	Discuss the reasons scientists and engineers conduct investigations.	SE	18, 38–39, 60, 84–85, 152, 170, 199, 213, 247, 255, 282, 303, 322–323, 340–341, 366–367, 447, 469, 510–511, 568–569, 614, 680–681, 869, 921, 938, 966
		ATE	38, 60, 84, 152, 170, 213, 247, 282, 303, 322, 323, 340, 341, 366, 367, 469, 510, 511, 568, 569, 614, 680, 681, 921, 938, 966

2.	Defend the use of technology as a method for enhancing data collection, data manipulation, and advancing the fields of science and technology.	**SE**	39, 45, 60, 84–85, 91, 152, 170–171, 199, 213, 246–247, 249–251, 282, 322–323, 437, 447, 469, 510–511, 680–681, 732, 869, 921, 967
		ATE	85, 249, 250, 251, 265, 345, 447, 469, 921, 1007, 1018
3.	Explain how mathematics is important to scientific and technological inquiry.	**SE**	29, 124–125, 131, 137, 173, 181, 182–183, 224, 237, 256, 316–317, 365, 367, 383–384, 394–395, 595, 621
		ATE	194, 212, 224, 266, 299, 569, 683, 805
4.	Explain why scientific models and explanations need to be based on historical and current scientific knowledge.	**SE**	18, 60, 152, 276–277, 303, 322–323, 365, 367, 384–385, 447, 510–511, 732, 1000–1001
5.	Understand that scientific explanations must be logical, supported by the evidence, and open to revision.	**SE**	27, 38–39, 63, 107, 123, 125, 141, 159, 181, 199, 217, 235, 255, 257, 275, 282, 295, 303, 315, 322, 333, 353, 375, 393, 437, 461, 483, 503, 510–511, 521, 539, 555, 575, 595, 621, 643, 661, 689, 705, 721, 739, 757, 777, 793, 815, 835, 857, 877, 899, 927, 951, 973, 999, 1025, 1043, 1061, 1079

II. Life Science

A. The Cell

1. Cells have particular structures that underlie their function. Inside the cell is a concentrated mixture of thousands of different molecules which form a variety of specialized structures that carry out such cell functions as energy production, transport of molecules, waste disposal, synthesis of new molecules, and the storage of genetic material.

a.	Compare prokaryotic and eukaryotic cells.	**SE**	14, 72, 73–83, 91, 133, 148, 203, 206–208, 270, 510–511
		ATE	70, 71, 82, 87, 130, 133, 203, 271, 510
b.	Identify the cellular structures that are responsible for energy production, waste disposal, molecular synthesis, storage of genetic material, and cell movement.	**SE**	76–83, 132, 141
		ATE	77, 78, 132
c.	Trace the development of the cell theory. **(H)**	**SE**	69–70
d.	Discuss uses of technologies that enable in-depth studies of the cell such as microscopes, ultracentrifuge techniques, and radioscopy studies. **(T)**	**SE**	21–22, 27, 84–85, 91–93, 152, 170–171, 510–511
		ATE	21, 22, 78, 84, 85, 152, 171, 825

2. Most cell functions involve chemical reactions. Food molecules taken into the cell react to provide the chemical constituents needed to synthesize other molecules. Both breakdown and synthesis are made possible by a large set of protein catalysts, called enzymes. The breakdown of some of the food molecules enables the cell to store energy in specific chemicals that are used to carry out the many functions of the cell.

a.	Explain the role of enzymes in chemical reactions within the cell.	**SE**	36–37, 57, 986–988
		ATE	57, 58, 986
b.	Differentiate the functions of carbohydrates, proteins, lipids, and nucleic acids in the cell.	**SE**	55–59, 977–979, 986
		ATE	977, 978

SOUTH CAROLINA EDITION

SOUTH CAROLINA

3. Cells store and use information to guide their functions. The genetic information stored in DNA is used to direct the synthesis of the thousands of proteins each cell requires.

a.	Compare DNA and RNA.	**SE** **ATE**	81, 185–186, 190–191 191
b.	Explain the role of the triplet codon in protein synthesis.	**SE** **ATE**	193–196 194, 195
c.	Illustrate the steps of protein synthesis.	**SE** **ATE**	193–196 193, 194, 195

4. Cell functions are regulated. Regulation occurs through changes in the activity of the functions performed by proteins and by the selective expression of certain genes. This regulation allows cells to respond to their environment and to control and coordinate cell growth and division.

a.	Examine the importance of DNA and proteins in cell regulation.	**SE** **ATE**	209–210 210
b.	Discuss mishaps in cell regulation (e. g., tumors). **(P)**	**SE** **ATE**	211–214 211, 212, 213, 214

5. Cells can differentiate and complex multicellular organisms are formed as a highly organized arrangement of differentiated cells. In the development of these multicellular organisms, the progeny from a single cell form an embryo in which the cells multiply and differentiate to form the many specialized cells, tissues and organs that comprise the final organism. This differentiation is regulated through the expression of different genes.

a.	Illustrate the development of both an animal and a plant multicellular organism (cells, specialized cells, tissues, organs, organ systems, and organisms).	**SE** **ATE**	86–88, 209–210, 344, 625–628, 644–645, 667–668, 676–677, 679, 680–686, 768–769, 787, 905–908 87, 209, 625, 626, 627, 682, 683, 684, 685
b.	Describe how organs and systems in both plants and animals function. *[This concept has been taught at a previous grade level.]	**SE** **ATE**	86–87, 599–613, 615–618, 674–679, 693–698, 734–736, 748–750, 765–768, 785–787, 808–812, 824–829, 845–849, 864–868, 885–888, 907–908 87, 600, 601, 603, 604, 605, 606, 610, 611, 612, 617, 674, 675, 676, 677, 678, 810, 811, 826, 828, 845, 846, 847, 848, 866, 867, 907
c.	Recognize that a degenerative disease involves the deterioration of the organs or tissues.	**SE** **ATE**	228–230, 246 228, 230, 1011, 1016

B. The Molecular Basis of Heredity

1. In all organisms, the instructions for specifying the characteristics of the organism are carried in DNA, a large polymer formed from subunits of four kinds (A, T, G, and C). The chemical and structural properties of DNA explain how the genetic information that underlies heredity is both encoded in genes (as a string of molecular "letters") and replicated (by a templating mechanism). Each DNA molecule in a cell forms a single chromosome.

a.	Explain how DNA, genes and chromosomes are related.	**SE**	145–146, 169, 221–224
b.	Analyze the chemical structure of DNA.	**SE** **ATE**	185–187 186, 187, 189
c.	Explain how DNA replication occurs.	**SE** **ATE**	188–189, 199, 200–201 185, 188, 189
d.	Evaluate the impact of DNA technology on society (e.g., bioengineering, forensics, genome project, DNA fingerprinting). **(T, P)**	**SE** **ATE**	199, 243–252 243, 244, 245, 246, 247, 250, 251, 605, 608

2. Most of the cells in a human contain two copies of each of 22 different chromosomes. In addition, there is a pair of chromosomes that determines sex: a female contains two X–chromosomes and a male contains one X and one Y chromosome. Transmission of genetic information to offspring occurs through egg and sperm cells that contain only one representative from each chromosome pair. An egg and sperm unite to form a new individual. The fact that the human body is formed from cells that contain two copies of each chromosome—and therefore two copies of each gene—explains many features of human heredity, such as how variations that are hidden in one generation can be expressed in the next.

a.	Explain the process of meiosis.	**SE** **ATE**	153–156 153, 155, 287
b.	Make predictions concerning inheritance based on the laws of heredity.	**SE** **ATE**	168, 173–178, 221–224, 227–232 174, 175, 176, 221, 222, 223, 224, 227, 228, 229, 230, 231
c.	Discuss advancements in the study of heredity since Mendel including the chromosome theory. **(H)**	**SE** **ATE**	170, 221–226, 227–232 170

SOUTH CAROLINA EDITION

SOUTH CAROLINA

3. Changes in DNA (mutations) occur spontaneously at low rates. Some of these changes make no difference to the organism, whereas others can change cells and organisms. Only mutations in germ cells can create the variation that changes an organism's offspring.

a.	Discuss how both chromosomal and gene mutations might occur.	**SE** 217, 224–226 **ATE** 225
b.	Infer how mutations contribute to genetic diversity.	**SE** 217, 300, 304
c.	Discuss the characteristics and molecular basis of various genetic disorders, such as sickle cell anemia, Tay-Sachs, cystic fibrosis, and hemophilia. **(P)**	**SE** 228–231, 246 **ATE** 187, 228, 230, 246

C. Biological Evolution

1. Species evolve over time. Evolution is the consequence of the interactions of (1) the potential for a species to increase its numbers, (2) the genetic variability of offspring due to mutation and recombination of genes, (3) a finite supply of the resources required for life, and (4) the ensuing selection by the environment of those offspring better able to survive and leave offspring.

a.	Discuss evolution as a consequence of various interactions; such as the number of offspring, genetic variability, finite supply of resources, and environmental factors.	**SE** 284–288, 844
b.	Discuss the scientific evidence that illustrates change over time.	**SE** 289–292, 844 **ATE** 844

2. Natural selection and its evolutionary consequences provide a scientific explanation for the fossil record of ancient life forms, as well as for the striking molecular similarities observed among the diverse species of living organisms.

a.	Evaluate the process of natural selection and its consequences.	**SE** **ATE**	287–288, 296–297, 844 287, 396, 804, 844.
b.	Infer how the fossil record can reveal evolutionary changes over time.	**SE** **ATE**	279–282, 319–323, 819–820, 839–841, 844, 862–863, 882–884 279, 281, 285, 318, 319, 320, 322, 840, 841, 862, 863, 882, 883, 884
c.	Describe how carbon dating is utilized in the study of evolution. **(H, T)**	**SE** **ATE**	264–266 266
d.	Discuss Charles Darwin's contribution to the study of evolution. **(H)**	**SE** **ATE**	284–288, 401 286, 287

3. Biological classifications are based on how organisms are related.

a.	Investigate the modern kingdom classification system based on fossil record interpretation and similarities in structural and chemical make-up.	**SE** **ATE**	342–350, 508–509, 580–582, 669–673 342, 343, 344, 345, 347, 348, 349
b.	Analyze the complexity of classifying organisms based on structural adaptations, physiology, nutritional strategies, biochemical similarities, genetic similarities, embryological similarities, and methods of reproduction.	**SE** **ATE**	337–350, 467–468, 470–473, 526–527, 539, 546–549, 580–582, 669–673 336, 342, 343, 344, 345, 347, 348, 349, 467, 468, 470, 546, 872
c.	Justify why many scientists group viruses in a category separate from living things.	**SE**	487–488

SOUTH CAROLINA

D. Interdependence of Organisms

1. The atoms and molecules on the earth cycle among the living and nonliving components of the biosphere.

a.	Analyze how organisms interact with the biosphere as part of the geochemical cycles (carbon, nitrogen, phosphorous and water cycles).	**SE** **ATE**	420–423 420, 421, 422, 473
b.	Evaluate the importance of nutrient cycles in an ecosystem.	**SE** **ATE**	420–423 422, 423, 473

2. Energy flows through ecosystems in one direction, from photosynthetic organisms to herbivores to carnivores and decomposers.

a.	Trace the flow of energy through various trophic levels.	**SE** **ATE**	417–419, 437 418
b.	Assess the value of the carbon cycle to the flow of energy through the ecosystems.	**SE** **ATE**	111–112, 117–120, 138, 421–422 422, 423

3. Organisms both cooperate and compete in ecosystems. The interrelationships and interdependencies of these organisms may generate ecosystems that are stable for hundreds or thousands of years.

a.	Relate the terms of cooperation and competition to organisms within an ecosystem.	**SE** **ATE**	400–402 401, 617, 891
b.	Evaluate how interrelationships and interdependencies of living things contribute to the homeostasis of ecosystems.	**SE** **ATE**	359, 362–365 358, 359, 364, 430, 617

4. Living organisms have the capacity to produce populations of infinite size, but environments and resources are finite. This fundamental tension has profound effects on the interactions between organisms.

a.	Describe and give examples of demographic characteristics of populations (e.g., birth and death rates, age structure, and sex ratio).	**SE** **ATE**	381–382, 383–385, 393 381, 382, 383
b.	Give examples and explain how limiting factors such as water, food, oxygen, and living space play a role in the stability of ecosystems.	**SE** **ATE**	384–386, 393 384, 385, 386
c.	Predict how interactions among organisms, such as predation, competition, and parasitism affect population growth.	**SE** **ATE**	386, 397–402 386, 396, 398
d.	Discuss the effects of succession on terrestrial ecosystems.	**SE** **ATE**	406–408 406, 407
e.	Evaluate dynamic equilibrium as a result of checks and balances within populations, communities, and ecosystems.	**SE** **ATE**	363–367 366, 367

5. Human beings live within the world's ecosystems. Increasingly, humans modify ecosystems as a result of population growth, technology, and consumption. Human destruction of habitats through direct harvesting, pollution, atmospheric changes, and other factors is threatening current global stability, and if not addressed, ecosystems will be irreversibly affected.

a.	Identify events that lead to awareness of environmental concerns such as fish kills, destruction of the ozone layer, global warming, and decline of the bald eagle. **(H)**	**SE** **ATE**	115, 275, 361–363, 441–447, 452–458, 461, 539 360, 361, 362, 363, 445
b.	Discuss the conflicts that could occur between land developers and conservationists. **(P)**	**SE**	456–457
c.	Debate the consequences of extinction and introduction of species within ecosystems.	**SE** **ATE**	455–456, 614 614
d.	Assess the consequences of acid rain on ecosystems. **(P)**	**SE** **ATE**	41, 115 41
e.	Give examples of how technology has advanced the study of environmental science. **(T, P)**	**SE** **ATE**	444–445, 447 447

SOUTH CAROLINA

SOUTH CAROLINA EDITION

E. Matter, Energy and Organization in Living Systems

1. The energy for life primarily derives from the sun. Plants capture energy by absorbing light and using it to form strong (covalent) chemical bonds between the atoms of carbon–containing (organic) molecules. These molecules can be used to assemble larger molecules with biological activity (including proteins, DNA, sugars, and fats). In addition, the energy stored in bonds between the atoms (chemical energy) can be used as sources of energy for life processes.

a.	Summarize the basic process by which photosynthesis converts solar energy into chemical energy (food molecules).	**SE** **ATE**	111–116, 117–120 113, 114, 115, 117, 118, 119
b.	Summarize the basic aerobic and anaerobic processes by which cellular respiration breaks down food molecules into energy that can be used by cells.	**SE** **ATE**	127–131, 133–138 128, 129, 130, 133, 134, 135, 136, 137

2. The chemical bonds of food molecules contain energy. Energy is released when the bonds of food molecules are broken and new compounds with lower energy bonds are formed. Cells usually store this energy temporarily in phosphate bonds of a small high-energy compound called ATP.

a.	Analyze bond energy as it relates to food molecules.	**SE** **ATE**	33, 36, 54 33, 54, 55
b.	Discuss the importance of ATP and how it is cycled.	**SE** **ATE**	54, 116–118, 127–128, 130–131, 133–138 54, 128, 133, 135, 136, 137, 138

3. The complexity and organization of organisms accommodates the need for obtaining, transforming, transporting, releasing, and eliminating the matter and energy used to sustain the organism.

a.	Explain why energy is necessary for the development, growth, and maintenance of organisms.	**SE**	10, 35–37, 111–112, 130–131
b.	Explain homeostasis and predict the consequences of a lack of energy on homeostasis.	**SE** **ATE**	6–7, 101–104 36, 101, 102, 103

4. As matter and energy flow through different levels of organization of living systems (cells, organs, organisms, communities) and between living systems and the physical environment, chemical elements are recombined in different ways. Each recombination results in storage and dissipation of energy into the environment as heat. Matter and energy are conserved in each change.

a.	Discuss the dynamics of energy and entropy as they apply to biological systems.	**SE** 10, 35–37, 111–120, 127–131, 133–138 **ATE** 111, 112, 113, 114, 115, 128, 129, 130, 133, 135, 136, 137, 1012
b.	Analyze energy in biological systems in terms of transformation, conservation, and efficiency.	**SE** 35–37, 111–120, 127–138 **ATE** 111, 112, 113, 114, 115, 128, 129, 130, 133, 135, 136, 137

F. Behavior and Regulation

1. Multicellular animals have nervous systems that generate behavior. Nervous systems are formed from specialized cells that conduct signals rapidly through the long cell extensions that make up nerves. The nerve cells communicate with each other by secreting specific excitatory and inhibitory molecules. In sense organs, specialized cells detect light, sound, and specific chemicals and enable animals to monitor what is going on in the world around them.

a.	Describe how cells of multicellular animals communicate by signals conducted through a nervous system.	**SE** 698, 711, 735, 750, 767, 787, 811–812, 828, 847, 868, 887, 1012–1022 **ATE** 117, 118, 119, 120, 121, 1013, 1014, 1015
b.	Discuss the adaptive value of the reflexes such as blinking of the eye, opening/closing of the iris, responses to hot and cold, etc.	**SE** 1010–1011 **ATE** 1010
c.	Give examples of specialized cells, such as taste buds, touch receptors, and rods and cones, in sense organs that detect stimuli.	**SE** 1018–1022

SOUTH CAROLINA

SOUTH CAROLINA

2. Organisms have behavioral responses to internal change and external stimuli. Responses to external stimuli can result from interactions with the organism's own species and others, as well as environmental changes; these responses can be either innate or learned. The broad patterns of behavior exhibited by animals have evolved to ensure reproductive success. Animals often live in unpredictable environments, and so their behavior must be flexible enough to deal with uncertainty and change. Plants also respond to stimuli.

a.	Investigate how different organisms maintain homeostasis.	**SE** **ATE**	6–7, 12, 95–100, 101–104, 370, 802, 1038–1040 1038
b.	Give examples of feedback mechanisms.	**SE** **ATE**	1038–1040 1038
c.	Explain how organisms react to pathogens.	**SE** **ATE**	250, 483, 955–964, 966–967 959, 961, 962, 963
d.	Assess both the positive and negative effects of introducing chemical substances into the body. **(P)**	**SE** **ATE**	483, 966–967, 1073–1076 1073, 1074, 1075, 1076
e.	Give examples that illustrate innate behavior.	**SE** **ATE**	333, 771–774, 870 771, 772
f.	Give examples of learned behavior.	**SE** **ATE**	889 321, 889
g.	Discuss tropisms in plants as responses to external stimuli.	**SE** **ATE**	652–655 652, 653, 654

3. Like other aspects of an organism's biology, behaviors have evolved through natural selection. Behaviors often have an adaptive logic when viewed in terms of evolutionary principles.

a.	Give examples of common behavioral responses in organisms such as waggle dancing, courtship, and nesting behaviors that maximize their fitness and success.	**SE** 771–774, 870, 873 **ATE** 771, 772, 774, 852, 868
b.	Evaluate how computer technology has been instrumental in collecting and analyzing data in the study of animal behavior. **(T)**	**SE** 869

4. Behavioral biology has implications for humans, as it provides links to psychology, sociology, and anthropology.

a.	Describe classical studies of learned behavior such as B. F. Skinner, Jane Goodall, and Dian Fossey. **(H)**	**SE** 323, 889 **ATE** 889
b.	Give examples of how these classical studies relate to human behavior.	**SE*** 333 **ATE*** 319

SOUTH CAROLINA

SOUTH CAROLINA

SOUTH CAROLINA EDITION

Text Structures

Secret Key to Success on Standardized Testing

Whether students are being assessed in language arts, mathematics, science, or social studies, they will have a better chance to succeed if they are effective readers—those readers who can work to separate important ideas from the less important ones. These readers can then construct meaning around the important ideas they have read.

Why are effective readers able to separate important ideas from the less important ones? One reason is that they are aware of the **text structure** of the piece—that is, *the organization of the piece.* The passages students read on tests have structure; writers present their ideas with a structure in mind. The trick is to be able to identify that structure, and, thus, to see the major thought relationships that tie together the ideas contained in the passage.

FIVE OF THE MOST COMMON ORGANIZATIONS USED BY WRITERS	
Description provides information about a subject, concept, event, object, person, idea, and so on	**Important Questions** What is the central idea? What are its qualities?
Comparison-contrast shows similarities and differences between 2 things	**Important Questions** What is being compared? How are they similar? How are they different?
Sequence describes the stages of something, the steps in a process, or a sequence of events	**Important Questions** What is the beginning event? What are the effects or results? How does one lead to another? What is the final effect or result?
Problem-solution represents a problem, attempted solutions, and results	**Important Questions** What is the starting stage or step? What are the stages or steps? How does one lead to another? What is the end?
Cause-effect shows causal information	**Important Questions** What's the problem? Who had it? Why? What solution attempts were made and what succeeded?

☞ Help students to understand and remember these 5 common organizations by depicting them as graphic organizers.

❶ Description

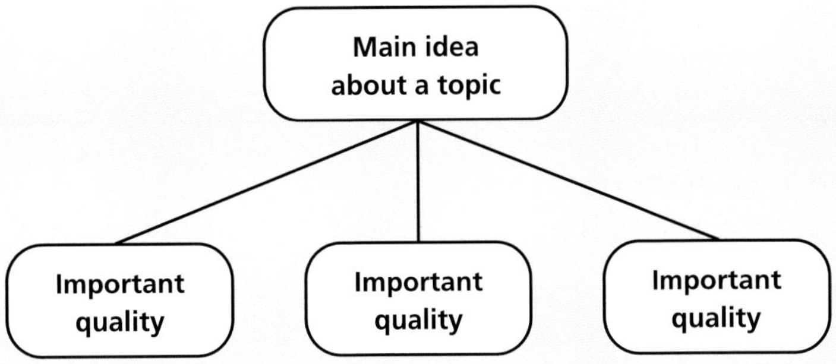

❷ Comparison-contrast

	Name 1	Name 2
Quality 1		
Quality 2		
Quality 3		

❸ Sequence

❹ Problem-solution

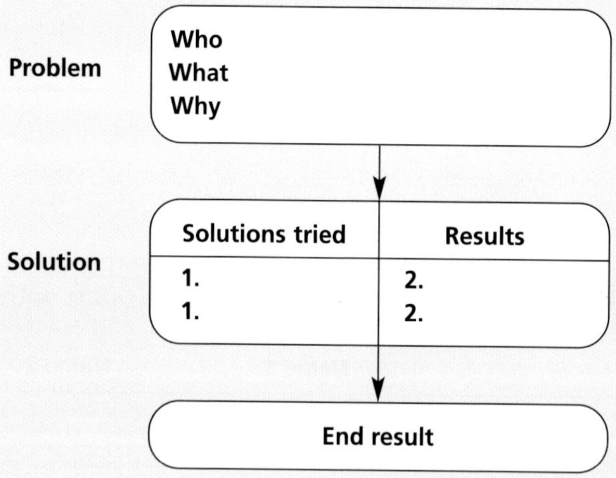

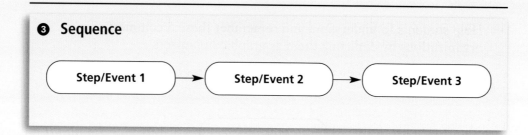

❺ Cause-effect

a. Cause-effect chain (Can be reversed to show causes instead.)

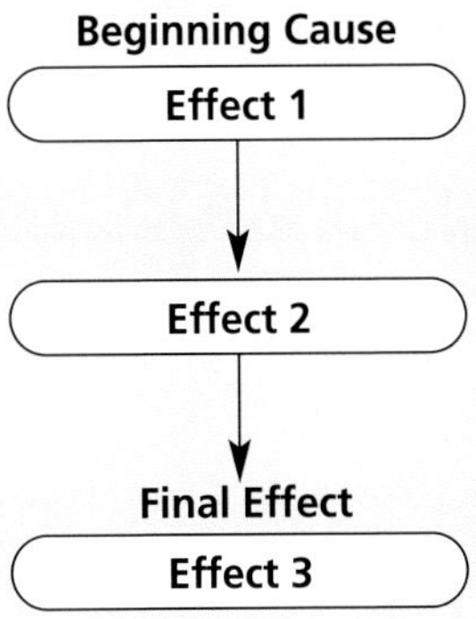

b. Cause-effect interaction (Can be reversed to start with result.)

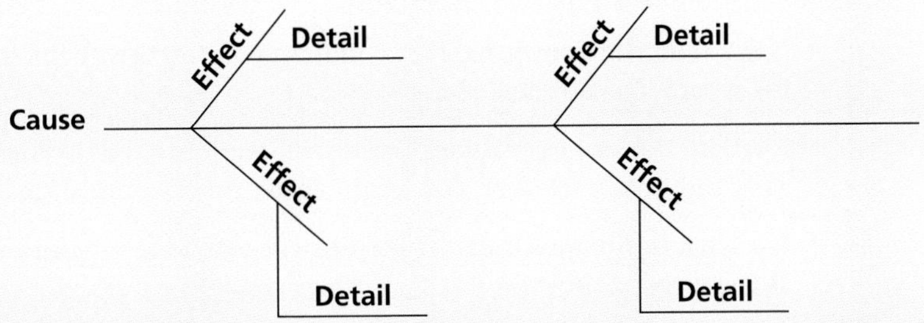

Test-taking Tips For Students

Every school year, there are students who are asked to take one or more standardized tests to demonstrate the content and skills they have learned. You can share the following test-taking tips with your students to help them as they prepare for these assessments.

Remind students, though, that the best way to prepare for any standardized test is to pay close attention in class and to take every opportunity to improve their science, reading, writing, and mathematical skills.

Tips for Standardized Tests: Reading Sections

The main goal of the reading sections of standardized tests is to determine your understanding of different aspects of a reading passage. Basically, if you can grasp the main idea and the author's purpose, and then pay attention to the details and vocabulary so that you are able to draw inferences and conclusions, you will do well on a test.

Here are some suggestions for answering questions based on reading passages:

- First, **read the passage as if you were not even taking a test.** Do this to get a general overview of both the topic and the tone of a passage.
- **Look at the big picture.** In other words, examine the most obvious features of the passage. To do this, ask yourself the following questions as you read:

  What is the title?

  What do the illustrations or pictures tell me?

  What is the main idea?

  What is the author's purpose? To inform? To entertain? To show how to do something?
- Next, **read the questions.** This will help you to know what information to look for when you re-read.
- Re-read the passage. **Underline the information** that relates to the questions. This will help you when you begin answering the questions.
- **Go back to the questions.** Try to answer each one in your mind before looking at the answer choices.
- Finally, **read *all* the answer choices and eliminate those that are obviously incorrect.** After this process, mark the best answer.

SOUTH CAROLINA EDITION

SOUTH CAROLINA

Types of Multiple-Choice Questions Based on a Passage

Realize that many multiple-choice questions fall into categories. The following categories are the most common.

1. **Main idea:** The main idea of a reading passage is the most important point expressed in the passage. The main idea must relate to the entire passage, not just to a portion of a selection. After reading a passage, locate and underline the main idea.
2. **Significant details:** You will most probably be asked to recall specific details from a reading passage. You will know what details to look for if you read the questions before reading the passage. Underline these details as you read. *Remember that correct answers do not always use the precise phrases or words that appear in the passage.*
3. **Reading graphic information:** These kinds of items test your ability to interpret information presented in a visual form or a graphic, such as a map, schedule, time line, or chart. If the question involves a graphic, follow these steps:
   a) Look at the title and major labels to figure out the focus or purpose of the graphic;
   b) Read the other headings or labels to find out what data is given and how it is organized;
   c) If the item includes a map, look at the map's legend or key, which will explain symbols, lines, and shadings in the map;
   d) Analyze the data in the graphic to determine quantities, relationships, intervals of time, directions, sequences, or other patterns
4. **Vocabulary:** Standardized tests will often ask you to determine the meaning of a word within the context of the passage. In many instances, an answer choice will include an actual meaning of the word that does not fit the context in which the word appears. To avoid choosing such an incorrect answer, read the answer choices and then plug them into the sentence to determine which answer fits the context of the passage.
5. **Conclusion and inference:** Standardized tests often ask you to draw conclusions or make inferences. There is often some idea within a passage that the author is trying to convey but does not state directly. Consider various parts of the passage together in order to determine what the author is implying. *If an answer choice refers to only one or two sentences or details within the passage, this is probably not the correct answer.*

If you do not understand a passage at first, keep reading. Many times you will find that you know more answers than you first thought. Once you understand the main idea of a passage, you can go from there to figure out the specific information.

Strategies for Answering Short-Answer Questions

- **Read the passage in its entirety,** paying close attention to the main ideas and details. Jot down information you think is important.
- **If you can't answer a question, skip it and come back later.**
- Words such as *compare, contrast, interpret, discuss,* and *summarize* appear often in short answer questions. **Be sure you have a complete understanding of each of these words.**
- For answers based on a passage, **return to the passage and skim the parts you underlined to find support.**
- **Organize your thoughts on a separate sheet of paper.** Write a general statement with which to begin. This will be your topic sentence.
- When writing your answer, **be precise but brief.** Be sure to refer to details in the passage in your answer. Try to get to the point quickly and never pad your answer.
- **Remember that how you use your time** on these kinds of questions is usually critical. Therefore, look over the questions in the beginning and divide your time based on your knowledge, required time for each item, and the scoring weight of the question.

Strategies for Answering Math Questions

- **Decide the goal of the question.** Read or study the problem carefully and determine what information must be found.
- Locate the factual information. **Decide what information represents key facts—the ones you must have to solve the problem.** You may also find facts you do not need to reach your solution. In some cases, you may determine that more information is needed to solve the problem. If so, ask yourself, "What assumptions can I make about this problem?" or "Do I need a formula to help solve this problem?"
- **Decide what strategies you might use to solve the problem, how you might use them, and what form your solution will be in.** For example, will you need to create a graph or chart? Will you need to solve an equation? Will your answer be in words or numbers? By knowing what type of solution you should reach, you may be able to eliminate some of the choices.
- **Apply your strategy** to solve the problem and compare your answer to the choices.
- **If the answer is still not clear, read the problem again.** If you had to make calculations to reach your answer, use estimation to see if your answer makes sense.

Other Tips

- **Adopt an acronym**—a word formed from the first letters of other words—that you will always use for analyzing a passage or a visual.

  HELPFUL ACRONYMS

  For a passage, use **SOAPS,** which stands for

  S subject
  O occasion
  A audience
  P purpose
  S speaker/author

  For a picture, cartoon, map, or other visual, use **OPTIC,** which stands for

  O overview
  P parts (labels for details of the visual)
  T title
  I interrelations (of how the parts work together)
  C conclusion (what the visual means)

- As mentioned before, **be sure to read all of the answer choices before choosing one.** Students often make the mistake of rushing through the multiple-choice questions and marking the first answer choice that seems correct.
- Also, keep in mind that the people who write standardized tests often create **incorrect answer choices that are designed to distract you from the right answer.** Such "distracters" include answer choices that are true but not relevant to the question, answer choices that relate to the wrong part of the passage, and answer choices that are too broad or too narrow.
- **Read the questions and the answers as carefully** as you would read the passage, and you should succeed on the reading sections of standardized tests.
- **Don't change your initial answer unless you have a good reason to do so.**

Tips for the Writing Portions of Tests

The primary goal of a writing portion of a standardized test is for you to demonstrate your ability to address and respond to a writing prompt or to an open-ended question. Writing also involves knowledge of the fundamentals of writing such as spelling, grammar, and punctuation through your response to multiple-choice and short-answer items.

Strategies for Responding to Composition Prompts and Open-Ended Questions

- First, **read the prompt or question carefully.** Be sure that you understand exactly what the prompt or question says.
- **Decide what kind of essay or response you are being asked to write.** You should ask yourself, "What is the purpose of my writing?" For example, are you trying to persuade your audience? Are you comparing and contrasting two works? Or issues? When you understand the type of writing you are being asked to produce, you will have a sense of the purpose of your essay or response.
- **Pay attention to key words.** There are certain words that appear again and again on the composition portions of standardized tests. The following is a list of these words and a general definition of each. It would be helpful if you looked up these words yourself and made sure you could use them in a sentence:

  To **compare** is to examine two or more subjects and note their similarities as well as their differences.

  To **contrast** is to stress the differences between two or more subjects.

  To **describe** something generally means to tell about it.

  To **classify** means to group together.

  To **explain** or **inform** is to clarify and develop an idea or thesis.

  To **persuade** is to **convince** others to accept a position you have taken on an issue or to take a specific action.
- Next, **organize your thoughts.** It is best to write down notes on a separate piece of paper before actually writing the essay.
  - Determine the main point of your essay or response. This sentence should include the general topic plus your idea or opinion about that topic. It should set the tone and catch the reader's attention. Most importantly, make sure it addresses the question or prompt. This will be the anchor to your response.
  - Come up with ideas to support your main idea sentence. If you wish, you can create your own graphic organizer to help you generate ideas and to organize them.
  - Be sure your ideas include the major points that you want to cover.

- Now you are ready to **begin writing your composition.** Write in complete sentences, and make sure your sentences and paragraphs flow smoothly. Sentences should come together smoothly to support your main idea and should be arranged in an order that makes sense to the reader. Be as specific as possible when stating your ideas. Make use of transitional words or phrases if necessary. Also, remember to write neatly.
- Finally, **proofread your essay or response.** Check for spelling and punctuation errors. Look for run-on sentences and sentence fragments. Look over verb tenses to see if you have used them correctly. Make the necessary edits as neat as possible.

Remember that practice makes perfect. Read and write as often as possible on whatever subjects you prefer, and you will see that writing will eventually come quite naturally.

Final Tips for Test Day

- Be sure that you are well rested.
- Be on time, and be sure that you have the necessary materials.
- Listen to the instructions given.
- Read directions and questions carefully. Many students are so anxious to finish the test that they don't read the directions. This is a costly mistake.
- Budget your time so that you can answer all parts of the test.
- If you have extra time, don't hand in your test. Check your answers, but do not dwell on questions you simply do not know how to answer.
- Remain calm and remember what you have learned in class.

Assessment and Higher-Order Thinking Skills

What Are Higher-Order Thinking Skills?

Higher-order thinking skills, sometimes called critical thinking skills, are not a new phenomenon on the education scene. In 1956, Benjamin Bloom published a book that listed a taxonomy of education objectives in the form of a pyramid similar to the one in the following illustration:

Bloom's Taxonomy of Educational Objectives

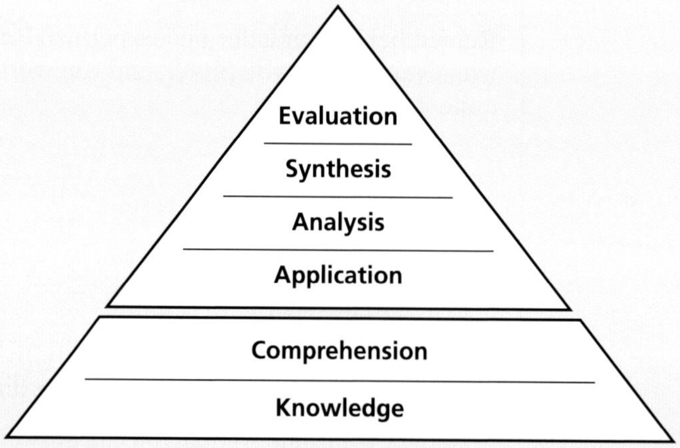

- **Knowledge** is the simplest level of education objectives and is not considered a higher-order thinking skill. It requires the learner to remember information without having to fully understand it. Tasks that students perform to demonstrate knowledge are recalling, identifying, recognizing, citing, labeling, listing, reciting, and stating.

  EXAMPLES

  1. *What is the capital of Minnesota?*

  2. *What is the French word for table?*

  3. *Label the parts of a plant.*

- **Comprehension** is not considered a higher-order thinking skill either. Learners demonstrate comprehension when they paraphrase, describe, summarize, illustrate, restate, or translate. Information isn't useful unless it's understood. Students can show they've understood by restating the information in their own words or by giving an example of the concept.

  EXAMPLES

  1. *Summarize the plot of the story in your own words.*
  2. *Interpret the information in the graph below.*
  3. *What were the underlying factors that contributed to the Revolutionary War?*

Many teachers tend to focus the most on knowledge and comprehension—and the tasks performed at these levels are important because they provide a solid foundation for the more complex tasks at the higher levels of Bloom's pyramid.

However, offering students the opportunity to perform at still higher cognitive levels provides them with more meaningful contexts in which to use the information and skills they have acquired, thus allowing them to more easily retain what they have learned.

When teachers incorporate **application, analysis, synthesis,** and **evaluation** as objectives, they allow students to utilize **higher-order thinking skills.**

- **Application** involves solving, transforming, determining, demonstrating, and preparing. Information becomes useful when students apply it to new situations—predicting outcomes, estimating answers—this is application.

  EXAMPLES

  1. *Organize the forms of pollution from most damaging to least damaging.*
  2. *Using the scale of 1 inch equals 200 miles, determine the point-to-point distance between Boston and Atlanta.*
  3. *Put the information below into graph form.*

- **Analysis** includes classifying, comparing, making associations, verifying, seeing cause-and-effect relationships, and determining sequences, patterns, and consequences. You can think of analysis as taking something apart in order to better understand it. Students must be able to think in categories in order to analyze.

  EXAMPLES

  1. *When it was written, how did the U. S. Constitution respond to the economic interests of certain classes of people?*
  2. *Using the vocabulary words from this unit, make a crossword puzzle.*
  3. *Analyze the literary elements in the following poem.*

- **Synthesis** requires generalizing, predicting, imagining, creating, making inferences, hypothesizing, making decisions, and drawing conclusions. Students create something which is new to them when they use synthesis. It's important to remember, though, that students can't create until they have the skills and information they have received in the comprehension through analysis levels.

  EXAMPLES

  1. *Create a newspaper using the details in this short story.*
  2. *Create a conversation that could have happened between General Grant and General Lee.*
  3. *Propose a plan for reorganizing your city's government.*

- **Evaluation** involves assessing, persuading, determining value, judging, validating, and solving problems. Evaluation is based on all the other levels. When students evaluate, they make judgments, but not judgments based on personal taste. These judgments must be based on criteria. It is important for students to evaluate because they learn to consider different points of view and to know how to validate their judgments.

  EXAMPLES

  1. *Justify the budget you created for your business.*
  2. *Explain and justify how this short story fulfills Edgar Allan Poe's requirements for a good story.*
  3. *Evaluate the methods used in your analysis of drinking water.*

MODERN BIOLOGY

The Program of Choice Since 1921

For 80 years, *Modern Biology* has been setting the standard of excellence against which other biology textbooks are measured. The tradition continues with this newest edition of *Modern Biology,* the choice for a comprehensive, up-to-date approach to biology instruction.

Comprehensive Content

Modern Biology offers complete coverage presented in a logical format. This content-rich approach provides you with all the tools you need to develop scientific literacy and prepare students for the future. See page 4T for a complete table of contents.

New Technology

Modern Biology brings your classroom into the new century with fully integrated technology. The *Holt Biology Interactive Tutor* uses video and interactive animations to present 42 key topics in biology. With *CNN Presents Science in the News: Biology Connections, Holt Biology Videodiscs,* and *Interactive Explorations in Biology,* your students can journey outside the boundaries of the classrooom. See pp. 36T–37T.

Internet Resources

NSTA *sci*LINKS, Smithsonian Institution Web resources, and www.go.hrw.com keep you up to date and extend your resources. See page 34T.

Accessible Coverage

Students can easily navigate their way through the text using the outline style headings and the Chapter Summary/Vocabulary that we've added to every chapter. They'll be guided through the Quick Labs and Chapter Investigations by sample data tables and helpful illustrations of laboratory techniques. See pp. 22T–27T.

Extraordinary Support

Holt BioSources Lab Program includes five lab manuals, including biotechnology labs and additional interactive explorations. The *Modern Biology One-Stop Planner CD-ROM with Test Generator* provides editable lesson plans, all *Modern Biology* worksheets, and hundreds of pages of additional worksheets, including science research papers, portfolio projects, supplemental reading guides, and occupational applications. The *Modern Biology Active Reading Guide Worksheets* help students learn the skills they need for reading success. See pp. 28T–35T to learn about these and other new teacher support materials.

HOLT, RINEHART AND WINSTON

A Harcourt Classroom Education Company

Austin • New York • Orlando • Atlanta • San Francisco • Boston • Dallas • Toronto • London

ACKNOWLEDGMENTS

STAFF CREDITS

Executive Editor
Susan Feldkamp

Managing Editor
Maureen Kilpatrick

Project Editors
Carolyn Biegert
Jennifer Cotner
Debra Hendricks
Joe Slapak

Copyeditors
Dawn Spinozza, *Copyediting Supervisor*
Brooke Fugitt

Editorial Staff
Jeanne Graham
Mary Helbling
Tanu'e White

Editorial Permissions
Ann B. Farrar

Production
Eddie Dawson, *Production Manager*

Electronic Publishing
Richard Chavez, *Project Coordinator*

Design
Richard Metzger, *Design Director*
David Hernandez, *Designer*

Image Acquisitions
Joe London, *Director*
Elaine Tate, *Art Buyer Supervisor*
David Knowles, *Art Buyer*
Jeannie Taylor, *Photo Research Supervisor*
Bob McClellan, *Senior Photo Researcher*
Elisabeth McCoy *Assistant Photo Researcher*

Photo Studio
Sam Dudgeon, *Senior Staff Photographer*
Victoria Smith *Staff Photographer*

Media
Kim Anderson
Adriana Bardin-Prestwood

Page design:
The Quarasan Group, Inc.
Cover image:
© Manfred Danegger/Okapai/Photo Researchers, Inc.

CONTRIBUTING WRITERS

I. Edward Alcamo, Ph.D.
Professor of Biology
State University of New York
Farmingdale, NY

Katy Z. Allen
Science Writer
Wayland, MA

David M. Armstrong, Ph.D.
Professor
University of Colorado
Boulder, CO

Barry Bogin, Ph.D.
Professor of Anthropology
Department of Behavioral Sciences
University of Michigan
Dearborn, MI

Curtis Chubb, Ph.D.
Blanco, TX

Linda Culp
Teacher
Thorndale High School
Thorndale, TX

Cara J. Cunningham
Teacher
Ursuline Academy of Dallas
Dallas, TX

Mary Pitt Davis
Teacher
Long Reach High School
Columbia, MD

William J. Ehmann, Ph.D.
Chair, Dept. of Environmental Science
Trinity College of Washington, D.C.
Washington, D.C.

Mark R. Feil, M.S.
Wilmington, DE

John Gallo
Science Curriculum Writer
Dallas,TX

Linda Gaul, Ph.D.
Senior Lecturer
The University of Texas at Austin
Austin, TX

Lisa A. Gloyna
Biology Teacher
James Bowie High School
Austin, TX

David R. Hershey, Ph.D.
Adjunct Professor, Biology/Horticulture Dept.
Prince George's Community College
Largo, MD

Debbie Hix
Science Teacher
Hays CISD
Kyle, TX

Mitchell Leslie
Science Writer and Editor
Albuquerque, NM

Karel F. Liem, Ph.D.
Professor
Harvard University
Cambridge, MA

Thomas Manerchia
Teacher
Archmere Academy
Wilmington, DE

James L. Middleton
Bear Creek High School
Lakewood, CO

Alyson Mike
Teacher
East Valley Middle School
East Helena, MT

Patricia Mokry
Teacher
Westlake High School
Austin, TX

C. Ford Morishita
Biology Teacher
Clackamas High School
Milwaukie, OR

Patricia Morrell, Ph.D.
Assistant Professor
University of Portland
Portland, OR

William Thwaites, Ph.D.
Professor Emeritus
Biology Department
San Diego State University
San Diego, CA

Salvatore Tocci
Science Writer
East Hampton, NY

Albert C. Wartski
Biology Teacher
Chapel Hill High School
Chapel Hill, NC

Lynn Marie Wartski
Science Writer
Hillsborough, NC

Karin E. Westerling, Ph.D.
Science Teacher
Moreno Valley High School
Moreno Valley, CA

Printed in the United States of America

ISBN 0-03-056542-1 3 4 5 6 7 8 048 06 05 04 03 02

REVIEWERS

Renato Aguilera, Ph.D.
Associate Professor
Molecular, Cell & Developmental Biology
UCLA
Los Angeles, CA

Robert Akeson
Boston Latin School
Boston, MA

David Armstrong, Ph.D.
University of Colorado
Boulder, CO

Allene C. Barans
Biology Teacher
Bishop England High School
Charleston, SC

Robert Baronak
High School Biology Teacher
Donegal High School
Mount Joy, PA

Carol Baskin, Ph.D.
Professor
School of Biological Sciences and Department of Agronomy
University of Kentucky
Lexington, KY

Lois Bergquist, Ph.D.
Professor—Microbiology
Biology Department
La Valley College
Valley Glen, CA

David Blinn
Biology Teacher
Wrenshall High School
Wrenshall, MN

Sonal S. D. Blumenthal, Ph.D.
The University of Texas at Austin
Austin, TX

Deborah Childs
Biology Teacher
Science Department
Gretha Public Schools
Gretha, NE

Mark Coyne
Associate Professor of Agronomy
University of Kentucky
Lexington, KY

Mary D. Coyne, Ph.D.
Professor of Biological Sciences
Wellesley College
Wellesley, MA

Joe W. Crim
Professor of Cellular Biology
University of Georgia
Athens, GA

Sandra Belden Curran
Biology Instructor
Science Department
Carl Junction High School
Carl Junction, MD

James Denbow, Ph.D.
Associate Professor
University of Texas at Austin
Austin, TX

Andrew A. Dewees, Ph.D.
Professor of Biology
Sam Houston State University
Huntsville, TX

Linda Gaul, Ph.D.
Senior Lecturer
The University of Texas at Austin
Austin, TX

Jan Gowin
Instructional Coordinator/ Biology Teacher
Jacksonville High School
Jacksonville, TX

Michael R. Harden
Chairperson of Science Department
Fall Creek High School
Fall Creek, WI

Y. Michelle Harman
Science Chairperson
Science Department
Northern Garrett High School
Accident, MD

Barry Harris
Biology Instructor
Ringgold High School
Monongahela, PA

Boyd Harrison
Science Coordinator
Muscatine High School
Muscatine, IA

John H. Honey
Fairfield High School
Fairfield, CT

John Hoover, Ph.D.
Associate Professor
Department of Biology
Millersville University
Millersville, PA

John J. Just, Ph.D.
Associate Professor
University of Kentucky
Lexington, KY

Roland Kays, Ph.D.
Mammal Curator
New York State Museum
Albany, NY

Clifford Keller
Institute of Neuroscience
University of Oregon
Eugene, OR

Jo Ann D. Lane
Science Department Chair
St. Ignatius High School
Cleveland, Ohio

Allen Lipke
Science Department Chairman
Hibbing High School
Hibbing, MN

V. Patteson Lombardi
Research Assistant Professor
Department of Biology
University of Oregon
Eugene, OR

Jeanne Olson, M.S. Med.
International Baccalaureate Program Teacher
Science
William Howard Taft High School
Chicago, IL

Nancy R. Parker, Ph.D.
Associate Professor of Biology
Southern Illinois University at Edwardsville
Edwardsville, IL

Kate Ruggless
Biology Teacher
Rochester High School
Rochester, IL

Michael Ryan, Ph.D.
Clark Hubbs Regents Professor in Zoology
The University of Texas at Austin
Austin, TX

John Richard Schrock
Director of Biology Education
Division of Biological Sciences
Emporia State University
Emporia, KS

Dale Simon
Biology Teacher
Biology Department
Central High School—Unit Dist. #3
Camp Point, IL

Marian Smith, Ph.D.
Professor of Biology
Southern Illinois University at Edwardsville
Edwardsville, IL

Joan E. Steffenhagen
Biology Teacher
Reitz Memorial High School
Evansville, IN

Richard D. Storey
Professor of Biology
Colorado College
Colorado Springs, CO

Gerald Summers
Associate Professor
University of Missouri
Columbia, MO

Cindy Tucci
Biology Teacher
Brookville High School
Brookville, OH

John P. Ulicney
Chairperson, Science Department
Ursuline High School
Youngstown, OH

Ben Vallejo
Mentor Science Teacher
Los Angeles Zoo Magnet at N. Hollywood H.S.
North Hollywood, CA

E. Peter Volpe, Ph.D.
Professor of Basic Medical Sciences
Mercer University School of Medicine
Macon, GA

Judith S. Weis, Ph.D.
Professor of Biology
Rutgers University
Newark, NJ

Mary E. White, M.S.
Biology Teacher
Round Rock High School
Round Rock, TX

Mary Wicksten, Ph.D.
Professor
Biology Department
Texas A&M University
College Station, TX

George C. Williams
Professor Emeritus
State University of New York
Stony Brook, NY

Kathy Yorks
Teacher/District Program Leader for Secondary Science
Science Department
Keystone Central School District
Lock Haven, PA

Michael Zimmerman
Dean, College of Letters and Sciences
Professor of Biology
University of Wisconsin–Oshkosh
Oshkosh, WI

CONTENTS

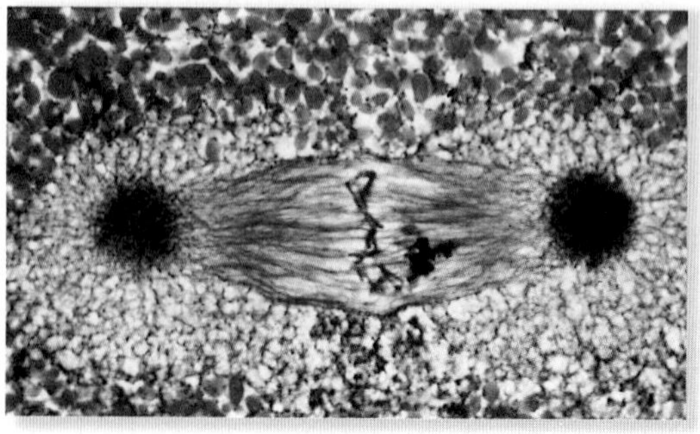

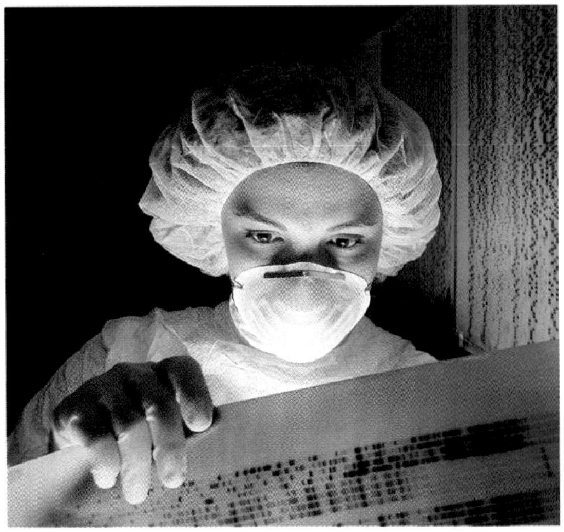

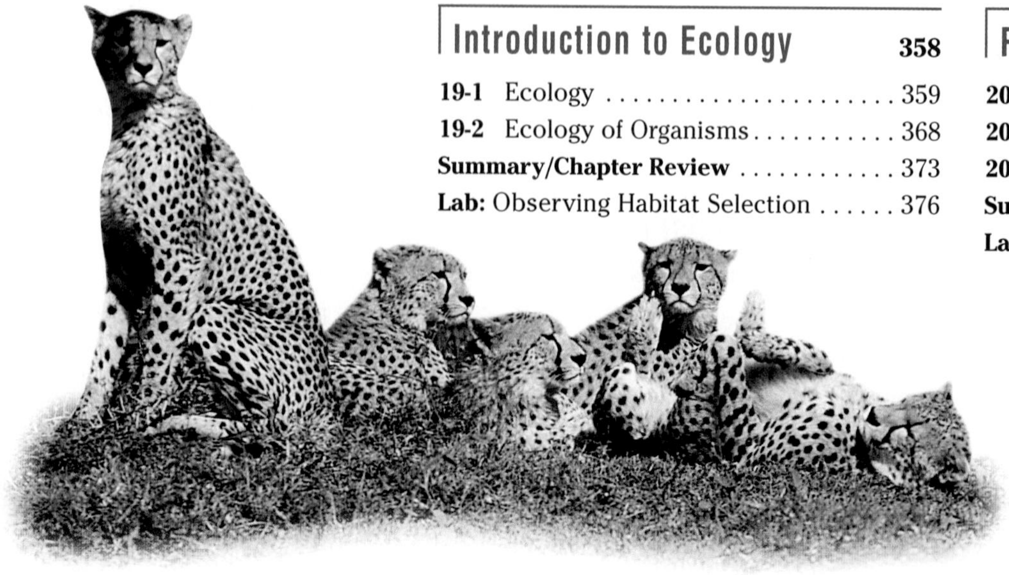

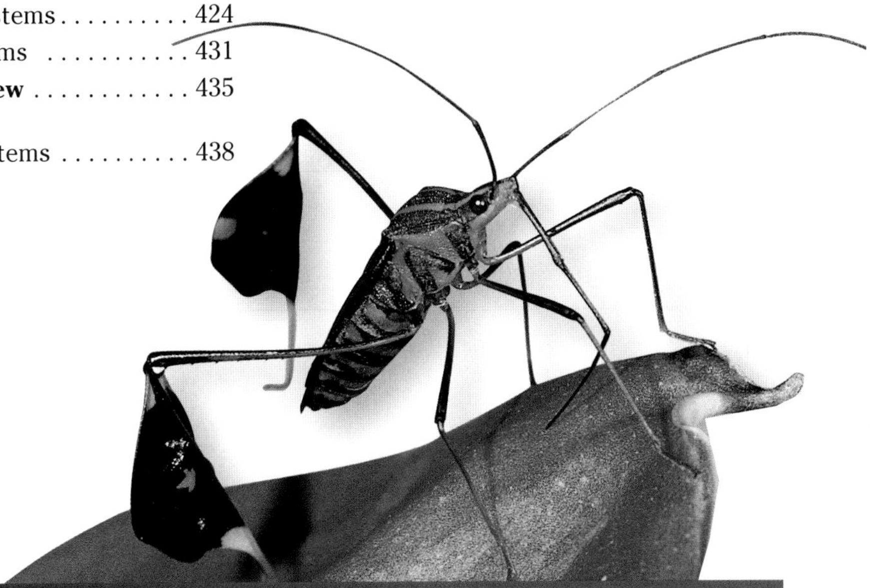

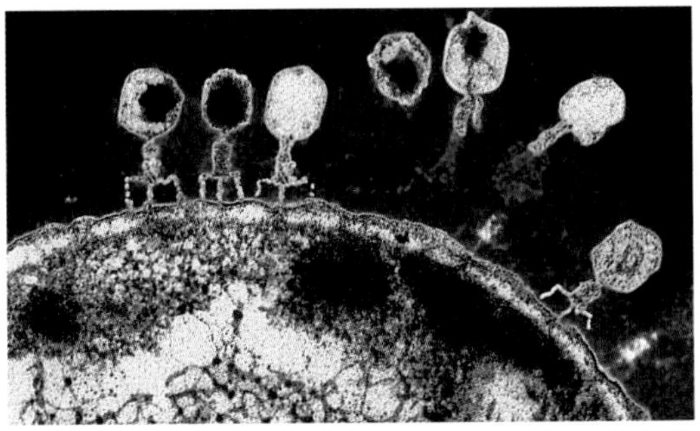

UNIT 10 HUMAN BIOLOGY (UNIT INTERLEAF 903A) 902

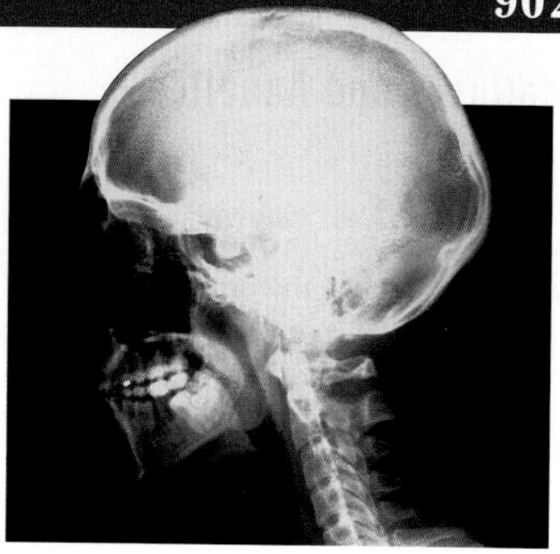

FEATURES

internetconnect

This textbook contains the following on-line resources to help you make the most of your science experience.

go.hrw.com

Visit **go.hrw.com** for extra help and study aids matched to your textbook. Just type in the keyword HM2 HOME.

sciLINKS NSTA

Visit **www.scilinks.org** to find resources specific to topics in your textbook. Keywords appear throughout your book to take you further.

Smithsonian Institution Internet Connections

Visit **www.si.edu/hrw** for specifically chosen on-line materials from one of our nation's premier science museums.

CNNfyi.com

Visit **www.cnnfyi.com** for late-breaking news and current events stories selected just for you.

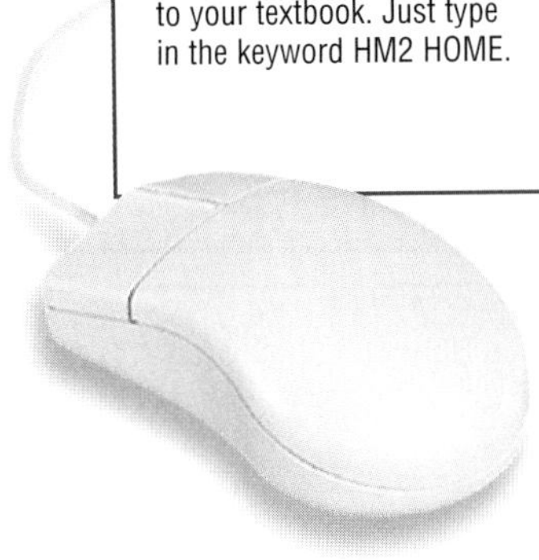

RESEARCH NOTES

These features present information about specific investigations, techniques, and scientists working in the field of biology.

LITERATURE & LIFE

Each Literature and Life feature includes an excerpt from the lay literature that is related to biology. Follow-up questions and additional readings help students digest and analyze each excerpt.

GREAT DISCOVERIES

Great Discoveries features provide students insight into the long and interesting history of inquiry upon which current biological principles and techniques are based. Each article describes a specific scientist's quest.

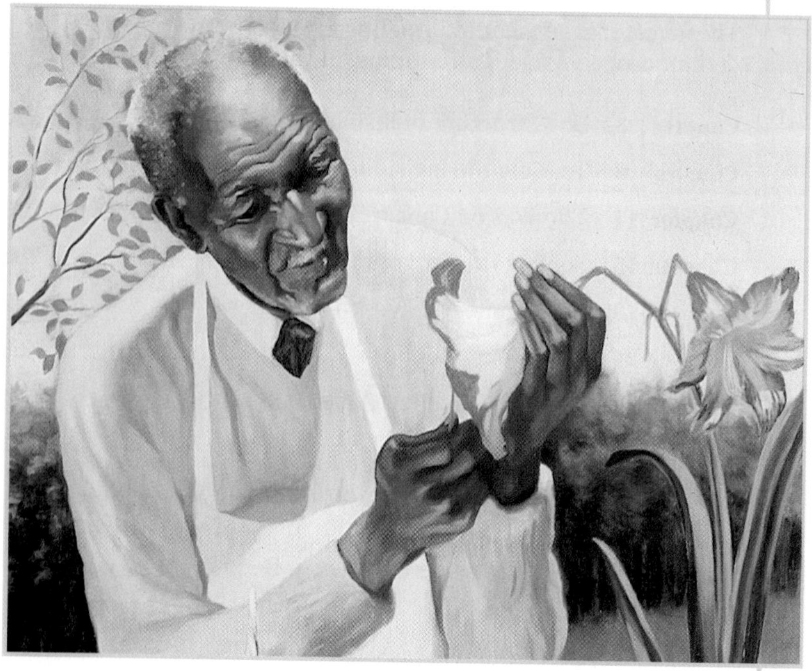

QUICK LABS AND CHAPTER INVESTIGATIONS

Quick Labs require few materials and are quick and easy to do. Chapter Investigations provide students a firsthand look at the scientific methods and data-collecting techniques used by biologists. A (CD icon) indicates an Interactive Exploration. See page 37T for details on how to use the Interactive Explorations.

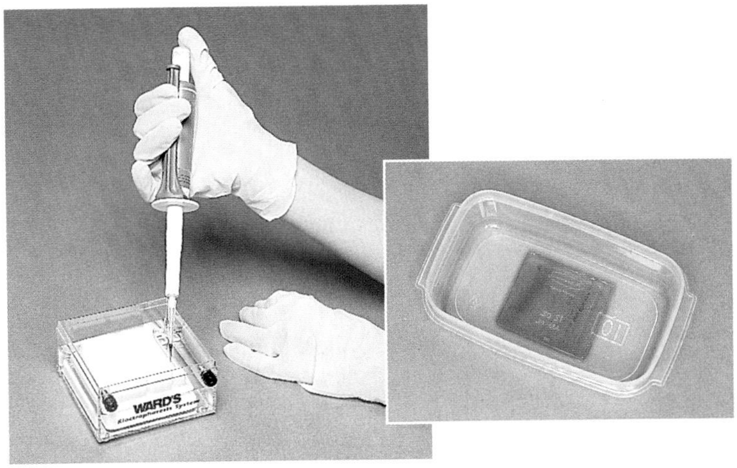

UNIT 1 *Biological Principles*

UNIT 2 *Cells*

UNIT 3 *Genetics*

UNIT 4 *Evolution*

UNIT 5 *Ecology*

UNIT 6 *Microorganisms*

UNIT 7 *Plants*

UNIT 8 *Invertebrates*

UNIT 9 *Vertebrates*

UNIT 10 *Human Biology*

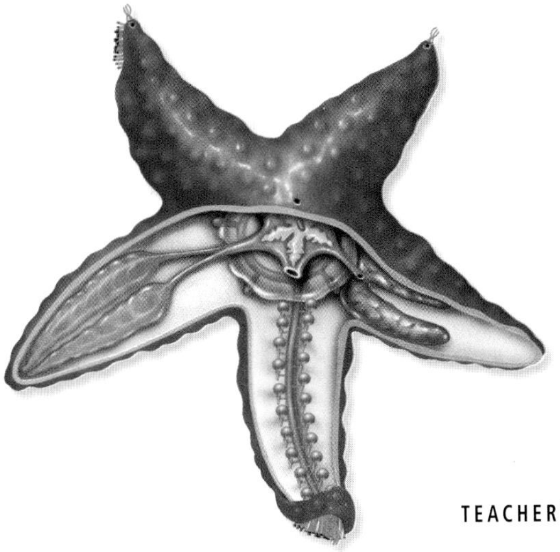

You Have High Expectations and *Modern Biology* Has High Standards

An Honored Tradition

Modern Biology has a proud tradition of excellence. For more than 75 years, teachers have used *Modern Biology* to develop students' understanding of biological concepts and terminology. Its classic distinction and reputation as the premier high school biology textbook is reinforced by this new edition.

Modern Biology has maintained high standards for content throughout the years. To meet the needs of parents and community leaders who are asking for more rigorous courses of study, this new edition of *Modern Biology* offers the quality and substance that ranks it first among high school biology textbooks.

Responding to an environment of rising educational standards, *Modern Biology* has been revised in accordance with the recommendations of the *National Science Education Standards* and *Benchmarks for Scientific Literacy*. Scientific literacy is an important goal for our society. *Modern Biology* is the one resource you can depend on to help students achieve scientific literacy.

> "Scientific literacy is of increasing importance in the workplace. More and more jobs demand advanced skills, requiring that people be able to learn, reason, think creatively, make decisions, and solve problems. An understanding of science and processes of science contributes in an essential way to these skills. Other countries are investing heavily to create scientifically and technically literate work forces. To keep pace in global markets, the United States needs to have an equally capable citizenry."
>
> From "National Science Education Standards: An Overview" from *National Science Education Standards,* 1996. Reprinted by permission of **National Academy Press.**

Up-to-Date and Comprehensive Content

Teachers have always counted on *Modern Biology* to deliver comprehensive coverage of biological concepts and topics. You can count on the most accurate, up-to-date biological science content. Careful reviews by content experts throughout the country have validated the accuracy of information presented in *Modern Biology*. NSTA *sci*LINKS Internet resources at point of use throughout the chapters provide content updates and extensions.

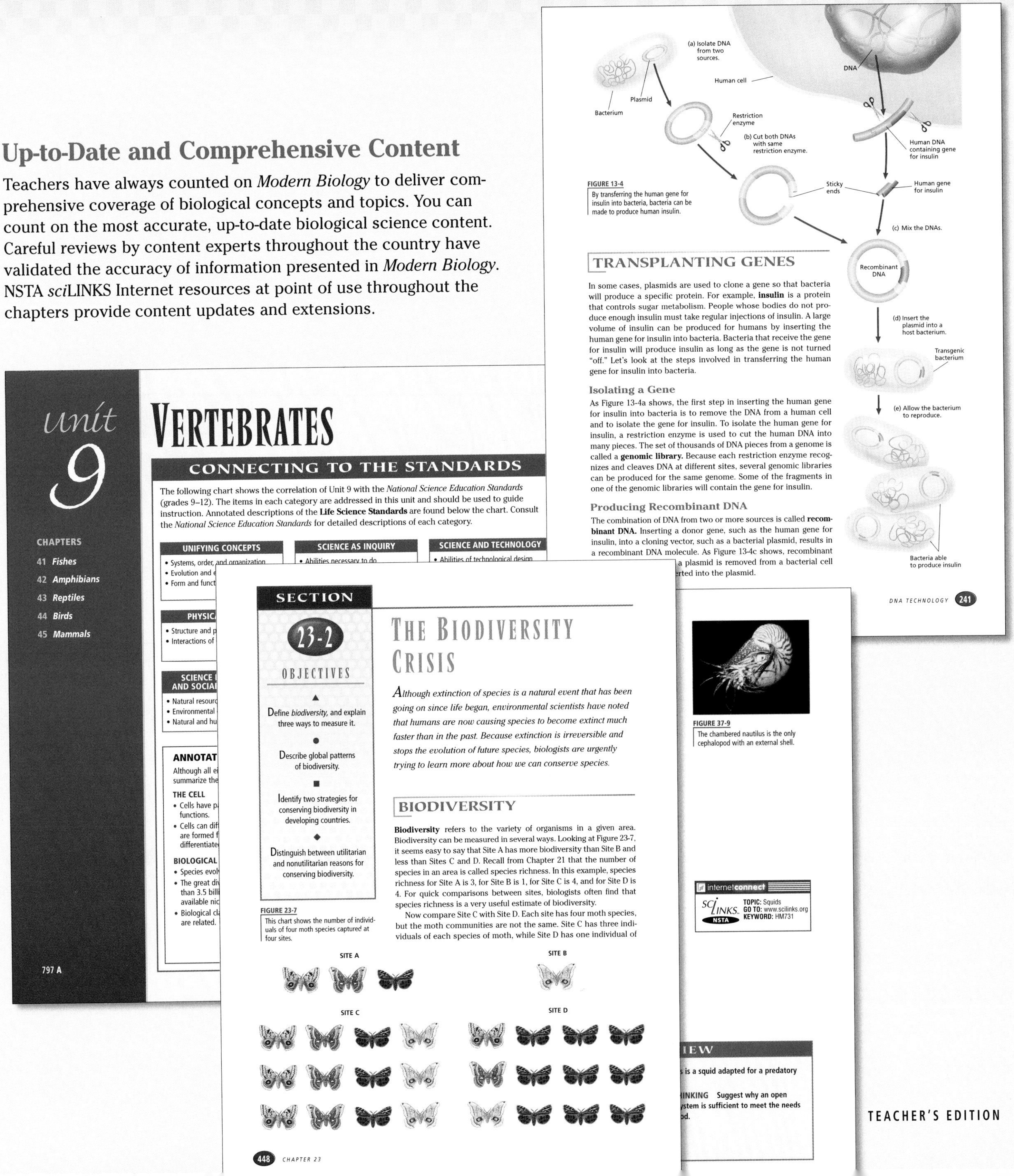

(a) Isolate DNA from two sources.
Human cell
DNA
Plasmid
Bacterium
Restriction enzyme
(b) Cut both DNAs with same restriction enzyme.
Human DNA containing gene for insulin
FIGURE 13-4
By transferring the human gene for insulin into bacteria, bacteria can be made to produce human insulin.
Sticky ends
Human gene for insulin
(c) Mix the DNAs.
TRANSPLANTING GENES
In some cases, plasmids are used to clone a gene so that bacteria will produce a specific protein. For example, insulin is a protein that controls sugar metabolism. People whose bodies do not produce enough insulin must take regular injections of insulin. A large volume of insulin can be produced for humans by inserting the human gene for insulin into bacteria. Bacteria that receive the gene for insulin will produce insulin as long as the gene is not turned "off." Let's look at the steps involved in transferring the human gene for insulin into bacteria.
Isolating a Gene
As Figure 13-4a shows, the first step in inserting the human gene for insulin into bacteria is to remove the DNA from a human cell and to isolate the gene for insulin. To isolate the human gene for insulin, a restriction enzyme is used to cut the human DNA into many pieces. The set of thousands of DNA pieces from a genome is called a genomic library. Because each restriction enzyme recognizes and cleaves DNA at different sites, several genomic libraries can be produced for the same genome. Some of the fragments in one of the genomic libraries will contain the gene for insulin.
Producing Recombinant DNA
The combination of DNA from two or more sources is called recombinant DNA. Inserting a donor gene, such as the human gene for insulin, into a cloning vector, such as a bacterial plasmid, results in a recombinant DNA molecule. As Figure 13-4c shows, recombinant
a plasmid is removed from a bacterial cell
rted into the plasmid.
Recombinant DNA
(d) Insert the plasmid into a host bacterium.
Transgenic bacterium
(e) Allow the bacterium to reproduce.
Bacteria able to produce insulin
DNA TECHNOLOGY 241
unit 9
VERTEBRATES
CHAPTERS
41 Fishes
42 Amphibians
43 Reptiles
44 Birds
45 Mammals
CONNECTING TO THE STANDARDS
The following chart shows the correlation of Unit 9 with the National Science Education Standards (grades 9–12). The items in each category are addressed in this unit and should be used to guide instruction. Annotated descriptions of the Life Science Standards are found below the chart. Consult the National Science Education Standards for detailed descriptions of each category.
UNIFYING CONCEPTS
SCIENCE AS INQUIRY
SCIENCE AND TECHNOLOGY
797 A
SECTION
23-2
OBJECTIVES
Define biodiversity, and explain three ways to measure it.
Describe global patterns of biodiversity.
Identify two strategies for conserving biodiversity in developing countries.
Distinguish between utilitarian and nonutilitarian reasons for conserving biodiversity.
THE BIODIVERSITY CRISIS
Although extinction of species is a natural event that has been going on since life began, environmental scientists have noted that humans are now causing species to become extinct much faster than in the past. Because extinction is irreversible and stops the evolution of future species, biologists are urgently trying to learn more about how we can conserve species.
BIODIVERSITY
Biodiversity refers to the variety of organisms in a given area. Biodiversity can be measured in several ways. Looking at Figure 23-7, it seems easy to say that Site A has more biodiversity than Site B and less than Sites C and D. Recall from Chapter 21 that the number of species in an area is called species richness. In this example, species richness for Site A is 3, for Site B is 1, for Site C is 4, and for Site D is 4. For quick comparisons between sites, biologists often find that species richness is a very useful estimate of biodiversity.
Now compare Site C with Site D. Each site has four moth species, but the moth communities are not the same. Site C has three individuals of each species of moth, while Site D has one individual of
FIGURE 23-7
This chart shows the number of individuals of four moth species captured at four sites.
SITE A
SITE B
SITE C
SITE D
448 CHAPTER 23
FIGURE 37-9
The chambered nautilus is the only cephalopod with an external shell.
internetconnect
SCILINKS NSTA
TOPIC: Squids
GO TO: www.scilinks.org
KEYWORD: HM731

Well-Crafted Lessons Promote Student Success

The logical organization and clear lesson design in *Modern Biology* make the relationships between ideas and their relative importance immediately apparent to students. The innovative visual program that accompanies the written text helps to clarify the content and amplify meaning. This solid, working relationship between the text and the visuals gives *Modern Biology* a distinctive, motivating approach to instruction.

Focus Concept
Every chapter begins with a Focus Concept that relates one of six recurring biological themes to the concepts found in each chapter.

Outline-Style Headings
Outline-style headings give students a clear, organized framework for the details of each chapter.

Objectives
The prominent objectives provide students with concise statements of the learning expectations.

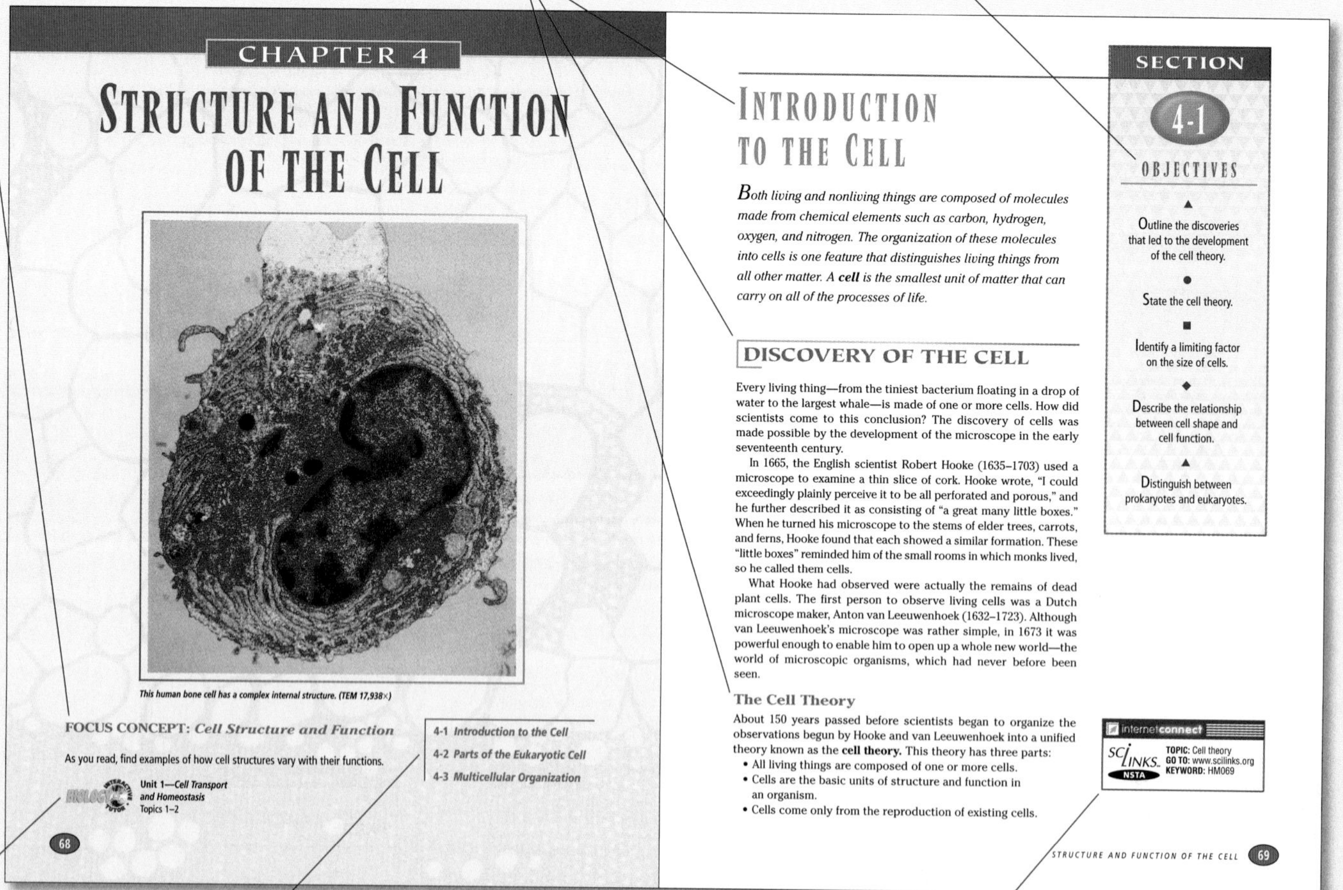

CHAPTER 4

STRUCTURE AND FUNCTION OF THE CELL

This human bone cell has a complex internal structure. (TEM 17,938×)

FOCUS CONCEPT: *Cell Structure and Function*

As you read, find examples of how cell structures vary with their functions.

BIOLOGY TUTOR Unit 1—*Cell Transport and Homeostasis* Topics 1–2

4-1 Introduction to the Cell
4-2 Parts of the Eukaryotic Cell
4-3 Multicellular Organization

68

INTRODUCTION TO THE CELL

Both living and nonliving things are composed of molecules made from chemical elements such as carbon, hydrogen, oxygen, and nitrogen. The organization of these molecules into cells is one feature that distinguishes living things from all other matter. A ***cell*** *is the smallest unit of matter that can carry on all of the processes of life.*

DISCOVERY OF THE CELL

Every living thing—from the tiniest bacterium floating in a drop of water to the largest whale—is made of one or more cells. How did scientists come to this conclusion? The discovery of cells was made possible by the development of the microscope in the early seventeenth century.

In 1665, the English scientist Robert Hooke (1635–1703) used a microscope to examine a thin slice of cork. Hooke wrote, "I could exceedingly plainly perceive it to be all perforated and porous," and he further described it as consisting of "a great many little boxes." When he turned his microscope to the stems of elder trees, carrots, and ferns, Hooke found that each showed a similar formation. These "little boxes" reminded him of the small rooms in which monks lived, so he called them cells.

What Hooke had observed were actually the remains of dead plant cells. The first person to observe living cells was a Dutch microscope maker, Anton van Leeuwenhoek (1632–1723). Although van Leeuwenhoek's microscope was rather simple, in 1673 it was powerful enough to enable him to open up a whole new world—the world of microscopic organisms, which had never before been seen.

The Cell Theory

About 150 years passed before scientists began to organize the observations begun by Hooke and van Leeuwenhoek into a unified theory known as the **cell theory.** This theory has three parts:
- All living things are composed of one or more cells.
- Cells are the basic units of structure and function in an organism.
- Cells come only from the reproduction of existing cells.

SECTION 4-1

OBJECTIVES

▲ Outline the discoveries that led to the development of the cell theory.

● State the cell theory.

■ Identify a limiting factor on the size of cells.

◆ Describe the relationship between cell shape and cell function.

▲ Distinguish between prokaryotes and eukaryotes.

internetconnect
sciLINKS NSTA
TOPIC: Cell theory
GO TO: www.scilinks.org
KEYWORD: HM069

STRUCTURE AND FUNCTION OF THE CELL 69

Biology Tutor
This shows the unit and the topics of the *Holt Biology Interactive Tutor* that students can use with this chapter.

Section Titles
A list of section titles alerts students and teachers to the chapter outline.

Internet Resources
NSTA *sci*LINKS Web site links show prescreened Web links at point of use.

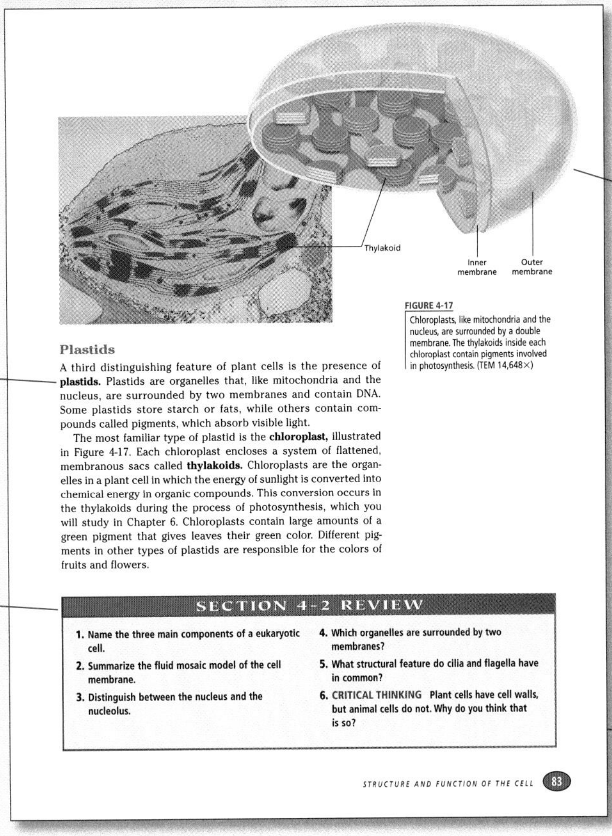

FIGURE 4-17
Chloroplasts, like mitochondria and the nucleus, are surrounded by a double membrane. The thylakoids inside each chloroplast contain pigments involved in photosynthesis. (TEM 14,648×)

Plastids

A third distinguishing feature of plant cells is the presence of **plastids.** Plastids are organelles that, like mitochondria and the nucleus, are surrounded by two membranes and contain DNA. Some plastids store starch or fats, while others contain compounds called pigments, which absorb visible light.

The most familiar type of plastid is the **chloroplast,** illustrated in Figure 4-17. Each chloroplast encloses a system of flattened, membranous sacs called **thylakoids.** Chloroplasts are the organelles in a plant cell in which the energy of sunlight is converted into chemical energy in organic compounds. This conversion occurs in the thylakoids during the process of photosynthesis, which you will study in Chapter 6. Chloroplasts contain large amounts of a green pigment that gives leaves their green color. Different pigments in other types of plastids are responsible for the colors of fruits and flowers.

SECTION 4-2 REVIEW

1. Name the three main components of a eukaryotic cell.
2. Summarize the fluid mosaic model of the cell membrane.
3. Distinguish between the nucleus and the nucleolus.
4. Which organelles are surrounded by two membranes?
5. What structural feature do cilia and flagella have in common?
6. CRITICAL THINKING Plant cells have cell walls, but animal cells do not. Why do you think that is so?

STRUCTURE AND FUNCTION OF THE CELL 83

Functional Illustrations
Beautiful photographs, detailed artwork, and meaningful tables bring life to the concepts, topics, and processes of biology.

Vocabulary
Vocabulary terms appear in boldfaced type and are clearly defined in context. Phonetic respellings help students pronounce difficult terms.

Section Reviews
Six questions at the end of each lesson cover all of the learning objectives.

Critical Thinking
Item 6 always challenges students to think critically about the main concepts of the lesson.

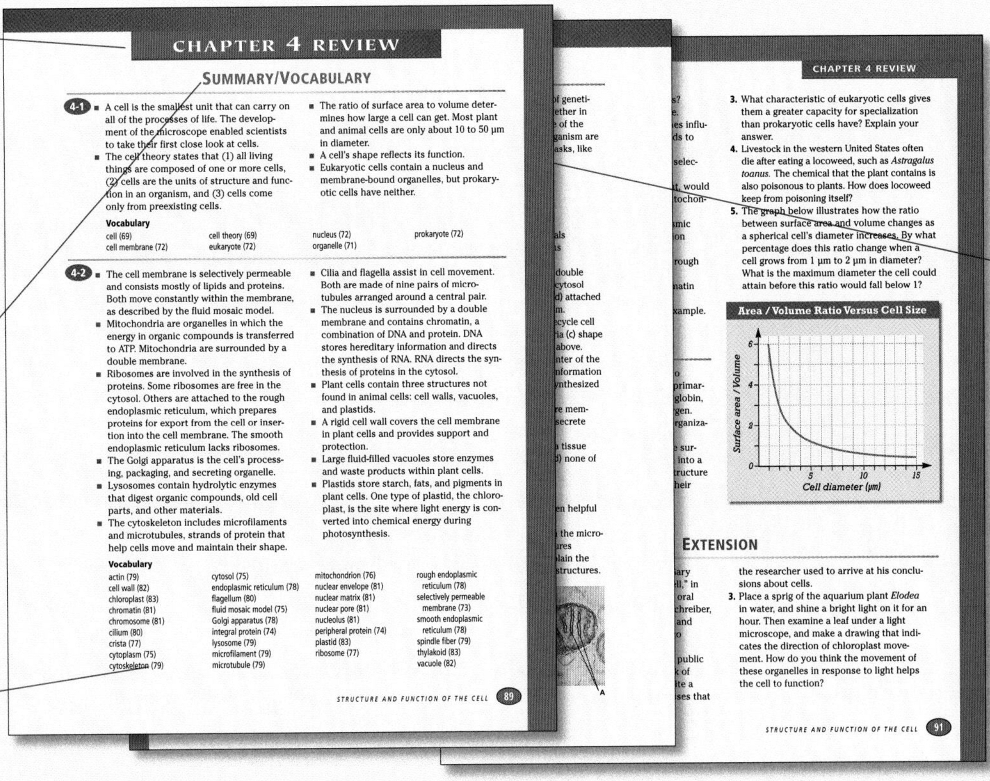

CHAPTER 4 REVIEW

SUMMARY/VOCABULARY

4-1

- A cell is the smallest unit that can carry on all of the processes of life. The development of the microscope enabled scientists to take their first close look at cells.
- The cell theory states that (1) all living things are composed of one or more cells, (2) cells are the units of structure and function in an organism, and (3) cells come only from preexisting cells.
- The ratio of surface area to volume determines how large a cell can get. Most plant and animal cells are only about 10 to 50 μm in diameter.
- A cell's shape reflects its function.
- Eukaryotic cells contain a nucleus and membrane-bound organelles, but prokaryotic cells have neither.

Vocabulary
cell (69), cell membrane (72), cell theory (69), eukaryote (72), nucleus (72), organelle (71), prokaryote (72)

4-2

- The cell membrane is selectively permeable and consists mostly of lipids and proteins. Both move constantly within the membrane, as described by the fluid mosaic model.
- Mitochondria are organelles in which the energy in organic compounds is transferred to ATP. Mitochondria are surrounded by a double membrane.
- Ribosomes are involved in the synthesis of proteins. Some ribosomes are free in the cytosol. Others are attached to the rough endoplasmic reticulum, which prepares proteins for export from the cell or insertion into the cell membrane. The smooth endoplasmic reticulum lacks ribosomes.
- The Golgi apparatus is the cell's processing, packaging, and secreting organelle.
- Lysosomes contain hydrolytic enzymes that digest organic compounds, old cell parts, and other materials.
- The cytoskeleton includes microfilaments and microtubules, strands of protein that help cells move and maintain their shape.
- Cilia and flagella assist in cell movement. Both are made of nine pairs of microtubules arranged around a central pair.
- The nucleus is surrounded by a double membrane and contains chromatin, a combination of DNA and protein. DNA stores hereditary information and directs the synthesis of RNA. RNA directs the synthesis of proteins in the cytosol.
- Plant cells contain three structures not found in animal cells: cell walls, vacuoles, and plastids.
- A rigid cell wall covers the cell membrane in plant cells and provides support and protection.
- Large fluid-filled vacuoles store enzymes and waste products within plant cells.
- Plastids store starch, fats, and pigments in plant cells. One type of plastid, the chloroplast, is the site where light energy is converted into chemical energy during photosynthesis.

Vocabulary
actin (79), cell wall (82), chloroplast (83), chromatin (81), chromosome (81), cilium (80), crista (77), cytoplasm (75), cytoskeleton (79), cytosol (75), endoplasmic reticulum (78), flagellum (80), fluid mosaic model (75), Golgi apparatus (78), integral protein (74), lysosome (79), microfilament (79), microtubule (79), mitochondrion (76), nuclear envelope (81), nuclear matrix (81), nuclear pore (81), nucleolus (81), peripheral protein (74), plastid (83), ribosome (77), rough endoplasmic reticulum (78), selectively permeable membrane (73), smooth endoplasmic reticulum (78), spindle fiber (79), thylakoid (83), vacuole (82)

STRUCTURE AND FUNCTION OF THE CELL 89

CHAPTER 4 REVIEW

3. What characteristic of eukaryotic cells gives them a greater capacity for specialization than prokaryotic cells have? Explain your answer.
4. Livestock in the western United States often die after eating a locoweed, such as *Astragalus toanus.* The chemical that the plant contains is also poisonous to plants. How does locoweed keep from poisoning itself?
5. The graph below illustrates how the ratio between surface area and volume changes as a spherical cell's diameter increases. By what percentage does this ratio change when a cell grows from 1 μm to 2 μm in diameter? What is the maximum diameter the cell could attain before this ratio would fall below 1?

EXTENSION

the researcher used to arrive at his conclusions about cells.

3. Place a sprig of the aquarium plant *Elodea* in water, and shine a bright light on it for an hour. Then examine a leaf under a light microscope, and make a drawing that indicates the direction of chloroplast movement. How do you think the movement of these organelles in response to light helps the cell to function?

STRUCTURE AND FUNCTION OF THE CELL 91

Chapter Reviews
Every chapter concludes with a three-page review of the major concepts.

Chapter Summary/ Vocabulary
Concise summary statements and a comprehensive vocabulary list are provided for each chapter. Teachers and students will find these features invaluable because they bring together all of the important information presented in the chapter.

Page References
The page on which each vocabulary term is defined is listed for student reference.

Summary Statements
The key concepts of each lesson are summarized by section.

A Comprehensive Program That Prepares You for a Diverse Classroom

Modern Biology Teaching Resources

- **Study Guide**
- **Study Guide Answer Key**
- **Chapter Tests with Answer Key**
- **Active Reading Guide with Answer Key**
- **Guided Reading Audio CD Program**
- **Assessment Item Listing**

One-Stop Planner

CD-ROM

Modern Biology One-Stop Planner CD-ROM with Test Generator
Contains Lesson Plans, *Modern Biology* Teaching Resources, Holt BioSources Lab Program and Teaching Transparencies, and more than 300 pages of blackline-master classroom-management tools and worksheets. Also contains Test Generator with more than 5,000 items for building custom tests

HOLT BIOSOURCES

Holt BioSources Lab Program
Five laboratory manuals, each with its own Teacher's Edition, in blackline-master form provide all the options you need for hands-on learning.

Holt BioSources Teaching Transparencies
238 color and 71 blackline-master illustrations

Holt BioSources Transparency Directory with Teacher's Notes
Teaching strategies for each transparency and transparency master

BIOLOGY TECHNOLOGY

Holt Biology Interactive Tutor
Models, videos, and interactive animations on CD-ROM help students learn the most challenging key topics in biology.

CNN Presents Science in the News: Biology Connections
Includes 35 short video clips that show students the newsworthy nature of biology. Includes **Teacher's Notes** and **Critical Thinking Worksheets** for each segment

Interactive Explorations in Biology: Human Biology
Eleven interactive investigations that enable students to explore biological systems on CD-ROM

Interactive Explorations in Biology: Cell Biology and Genetics
Eight additional interactive investigations on CD-ROM

Holt Biology Videodiscs
Contains four double-sided videodiscs with still images and video footage that are fully coordinated with *Modern Biology*

Concepts of Biology
Contains CD-ROMs with design-your-own multimedia presentations for *Modern Biology*

Thorough and Up-to-Date Coverage That Promotes Scientific Literacy

Nowhere has the explosion of knowledge and the impact of technology been more apparent than in the field of biology. *Modern Biology* provides students with the solid foundation they need to understand the expanding role of biology in society. *Modern Biology* also develops the skills students will need to excel in future science courses. A sample of the updated coverage you'll find in this edition of *Modern Biology* is shown below.

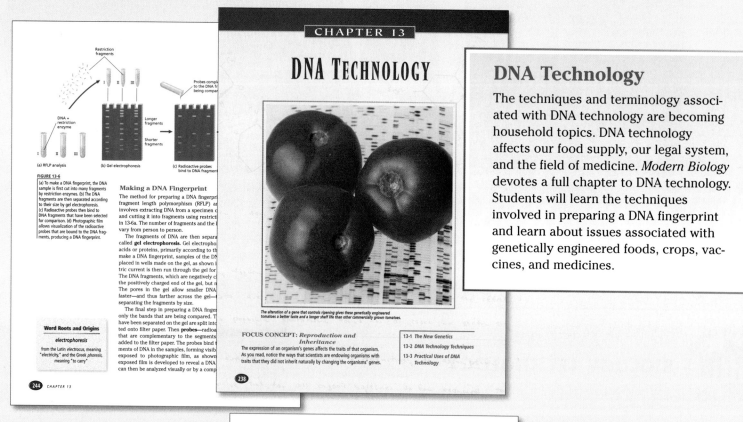

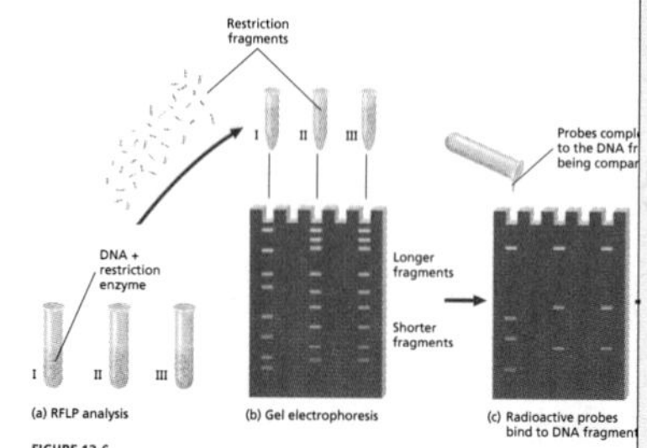

FIGURE 13-6
(a) To make a DNA fingerprint, the DNA sample is first cut into many fragments by restriction enzymes. (b) The DNA fragments are then separated according to their size by gel electrophoresis. (c) Radioactive probes then bind to DNA fragments that have been selected for comparison. (d) Photographic film allows visualization of the radioactive probes that are bound to the DNA fragments, producing a DNA fingerprint.

Making a DNA Fingerprint

The method for preparing a DNA fingerpri fragment length polymorphism (RFLP) ar involves extracting DNA from a specimen c and cutting it into fragments using restricti in 13-6a. The number of fragments and the l vary from person to person.

The fragments of DNA are then separa called **gel electrophoresis.** Gel electropho acids or proteins, primarily according to th make a DNA fingerprint, samples of the DN placed in wells made on the gel, as shown i tric current is then run through the gel for The DNA fragments, which are negatively c the positively charged end of the gel, but n The pores in the gel allow smaller DNA faster—and thus farther across the gel—t separating the fragments by size.

The final step in preparing a DNA finge only the bands that are being compared. T have been separated on the gel are split into ted onto filter paper. Then **probes**—radioa that are complementary to the segments added to the filter paper. The probes bind t ments of DNA in the samples, forming visib exposed to photographic film, as shown exposed film is developed to reveal a DNA can then be analyzed visually or by a comp

Word Roots and Origins

electrophoresis

from the Latin *electrocus,* meaning "electricity," and the Greek *phoresis,* meaning "to carry"

244 CHAPTER 13

CHAPTER 13

DNA TECHNOLOGY

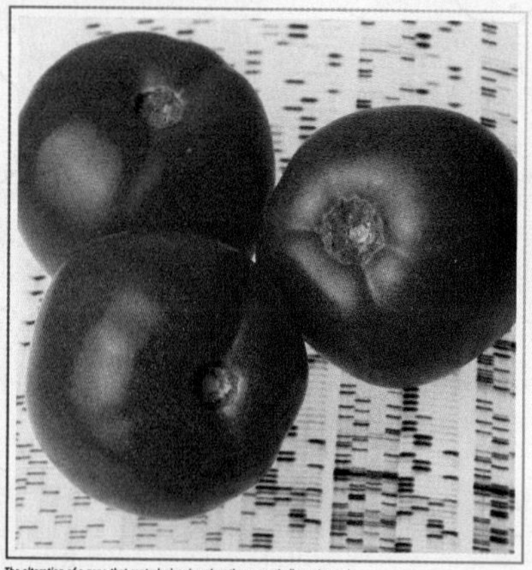

The alteration of a gene that controls ripening gives these genetically engineered tomatoes a better taste and a longer shelf life than other commercially grown tomatoes.

FOCUS CONCEPT: *Reproduction and Inheritance*

The expression of an organism's genes affects the traits of that organism. As you read, notice the ways that scientists are endowing organisms with traits that they did not inherit naturally by changing the organisms' genes.

13-1 *The New Genetics*
13-2 *DNA Technology Techniques*
13-3 *Practical Uses of DNA Technology*

238

DNA Technology

The techniques and terminology associated with DNA technology are becoming household topics. DNA technology affects our food supply, our legal system, and the field of medicine. *Modern Biology* devotes a full chapter to DNA technology. Students will learn the techniques involved in preparing a DNA fingerprint and learn about issues associated with genetically engineered foods, crops, vaccines, and medicines.

Classification

As new ways of investigating evolutionary relationships among organisms emerge, scientists are discovering new ways to classify organisms. This edition of *Modern Biology* includes the most recent and commonly accepted classification of organisms into six kingdoms. In addition, methods and new techniques used to establish evolutionary relationships are described.

TWO MODERN SYSTEMS OF CLASSIFICATION

Aristotle classified organisms as either plants or animals, but today we recognize that many forms of life are neither. In this section, you will read about two alternative classification systems that are in current use. But remember, organizational systems are imposed by humans and therefore may be flawed. As is true of everything in science, they are subject to change as new information arises.

SECTION 18-3

OBJECTIVES

- Describe the six-kingdom system of classification.
- List the characteristics that distinguish archaebacteria from eubacteria.
- Explain why the protists are grouped together in the six-kingdom system in spite of having differences that are greater than those between plants and animals.
- Describe the evidence that prompted the creation of the three-domain system of classification.
- Explain the principal difference between the six-kingdom system and the three-domain system of classification.

SIX-KINGDOM SYSTEM

A classification system based on five kingdoms of organisms was preferred by taxonomists for many years. But further studies of bacteria have shown that there are two important subtypes with very different morphologies and properties. Recognition of these two broad types of bacteria has driven the acceptance of a newer, six-kingdom system, illustrated in Table 18-2, which is used in this textbook.

TABLE 18-2 *Six Kingdoms of Life*

Kingdom	Cell type	Number of cells	Nutrition
Archaebacteria	prokaryotic	unicellular	autotrophy and heterotrophy
Eubacteria	prokaryotic	unicellular	autotrophy and heterotrophy
Protista	eukaryotic	unicellular and multicellular	autotrophy and heterotrophy
Fungi	eukaryotic	unicellular and multicellular	heterotrophy
Plantae	eukaryotic	multicellular	autotrophy and (rarely)

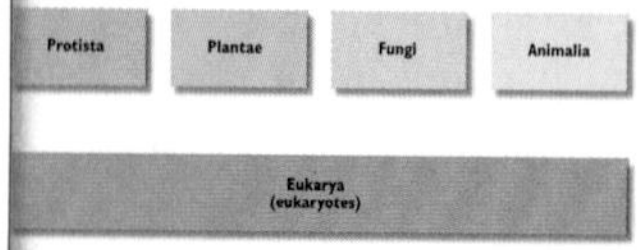

THREE-DOMAIN SYSTEM

The young science of molecular biology has led to an alternative to the six-kingdom system. By comparing sequences of ribosomal RNA in different organisms, molecular biologist Carl Woese (1928–), of the University of Illinois, has estimated how long ago pairs of different organisms shared a common ancestor. Because all organisms, even prokaryotes, have ribosomes, the ribosomal RNA molecule can be used to study the degree of relationship between any two living things. The phylogenetic tree drawn from these data shows that living things seem to fall naturally into three broad groups, or **domains.** Plants, animals, and fungi are one small twig of a large branch—a domain—that includes all the eukaryotes. These domains are shown in comparison with the six kingdoms in Figure 18-11.

Domain Archaea (ahr-KEE-uh) is known in the six-kingdom system as kingdom Archaebacteria. **Domain Bacteria** is known in the six-kingdom system as the kingdom Eubacteria.

Domain Eukarya (yoo-KAR-ee-uh) consist of the protists, the fungi, and the plants and animals. All eukaryotes have true nuclei with linear chromosomes and membrane-bound organelles. Most of the variation in this domain is among the protists. Surprisingly, when considered from the perspective of the complete diversity of life on Earth, the fungi, plants, and animals are quite similar to each other.

TION 18-3 REVIEW

ix-kingdom
haebacteria
ingdom in

5. Why do protists, fungi, plants, and animals share a domain in the six-kingdom system?
6. CRITICAL THINKING In the five-kingdom system, which is still used by some scientists, all species of bacteria are grouped in the kingdom Monera. Why might there have been only one bacteria kingdom recognized in the past?

Comparative Development

Modern Biology's in-depth coverage of the animal kingdom is preceded by an introduction to the animal kingdom. This chapter includes a comparison of invertebrate and vertebrate characteristics, phylogeny, and development, providing an excellent framework within which students can learn about each group of animals that follows.

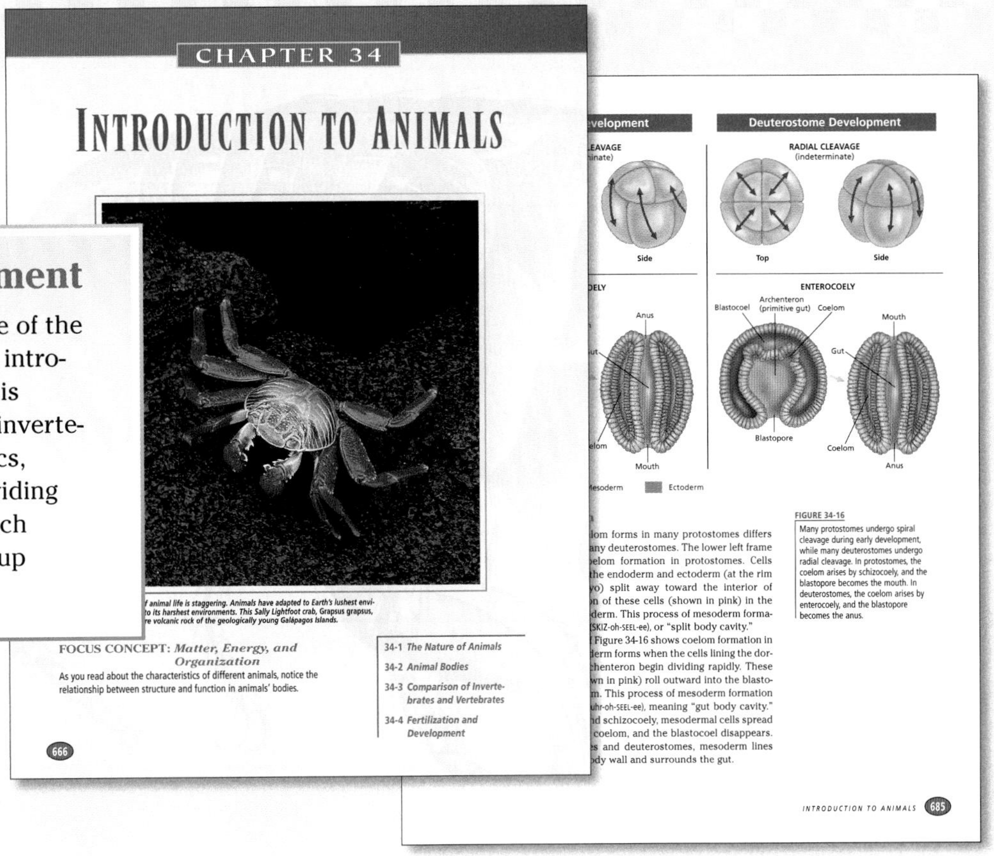

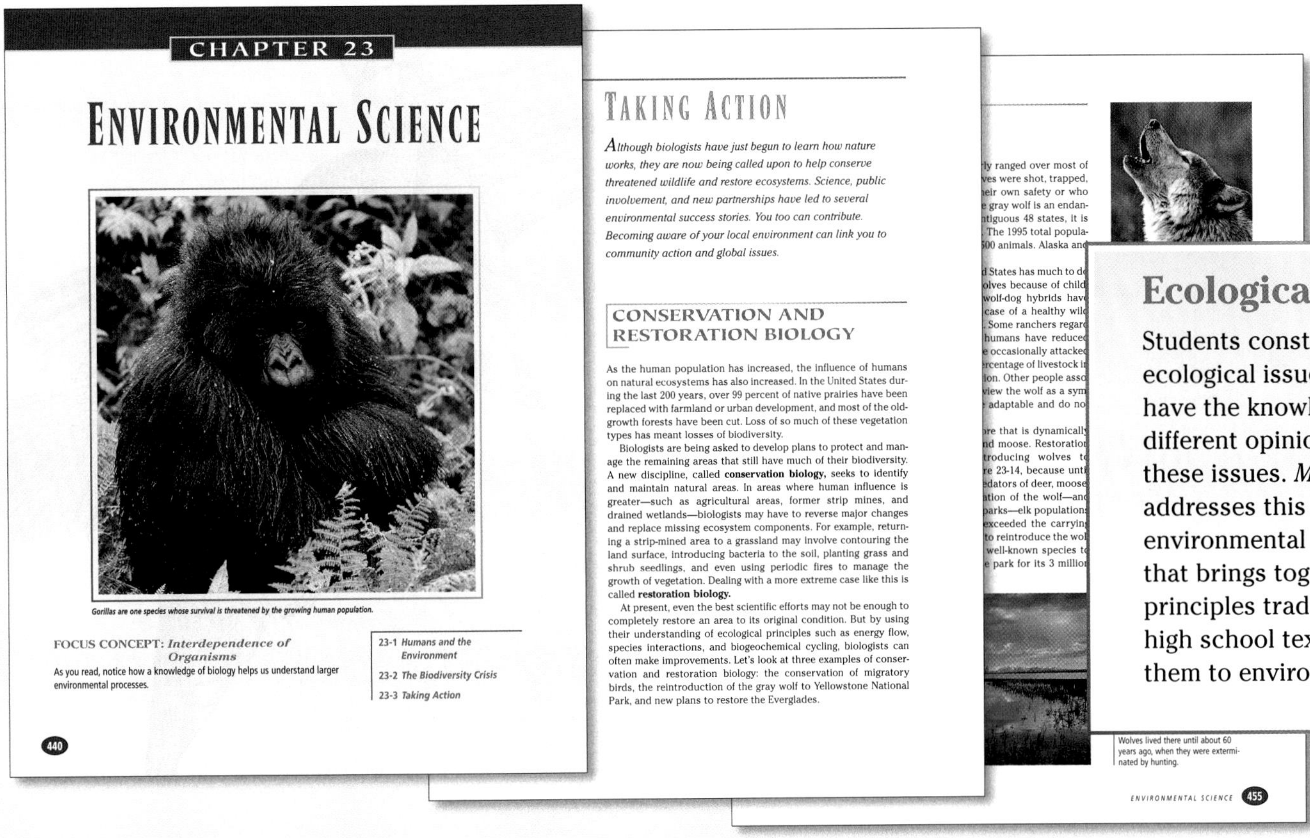

Ecological Awareness

Students constantly hear about ecological issues, but they don't have the knowledge to evaluate the different opinions expressed about these issues. *Modern Biology* addresses this need with an environmental science chapter that brings together the ecological principles traditionally presented in high school textbooks and applies them to environmental issues.

Accessible Coverage That Helps You Accommodate a Variety of Learning Styles

High-Quality Instructional Illustration Program
Flowcharts, summary tables, and diagrams help students understand the concepts and processes described in the text. Especially key to the illustration program is a pairing of photographs with diagrams, so that students can relate photographs of microscopic structures to explanatory diagrams.

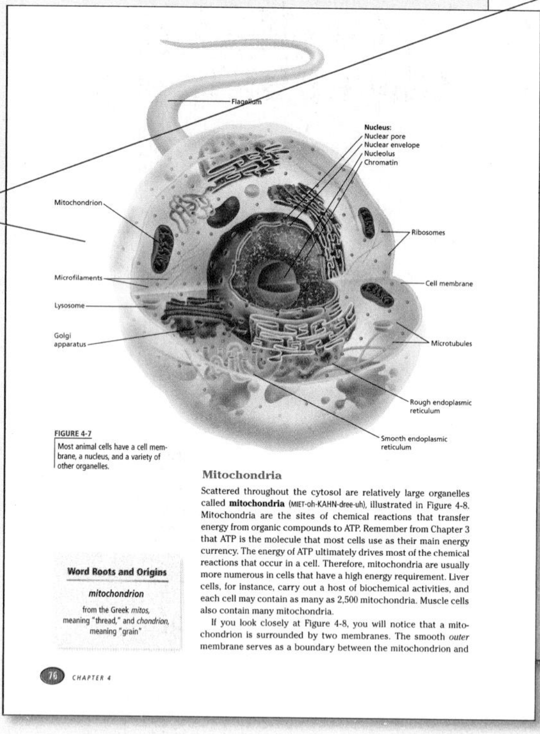

Flagellum
Nucleus:
Nuclear pore
Nuclear envelope
Nucleolus
Chromatin
Mitochondrion
Ribosomes
Microfilaments
Cell membrane
Lysosome
Golgi apparatus
Microtubules
Rough endoplasmic reticulum
Smooth endoplasmic reticulum

FIGURE 4-7
Most animal cells have a cell membrane, a nucleus, and a variety of other organelles.

Mitochondria

Scattered throughout the cytosol are relatively large organelles called **mitochondria** (MIET-oh-KAHN-dree-uh), illustrated in Figure 4-8. Mitochondria are the sites of chemical reactions that transfer energy from organic compounds to ATP. Remember from Chapter 3 that ATP is the molecule that most cells use as their main energy currency. The energy of ATP ultimately drives most of the chemical reactions that occur in a cell. Therefore, mitochondria are usually more numerous in cells that have a high energy requirement. Liver cells, for instance, carry out a host of biochemical activities, and each cell may contain as many as 2,500 mitochondria. Muscle cells also contain many mitochondria.

If you look closely at Figure 4-8, you will notice that a mitochondrion is surrounded by two membranes. The smooth *outer* membrane serves as a boundary between the mitochondrion and

Word Roots and Origins
mitochondrion
from the Greek *mitos*, meaning "thread," and *chondrion*, meaning "grain"

76 CHAPTER 4

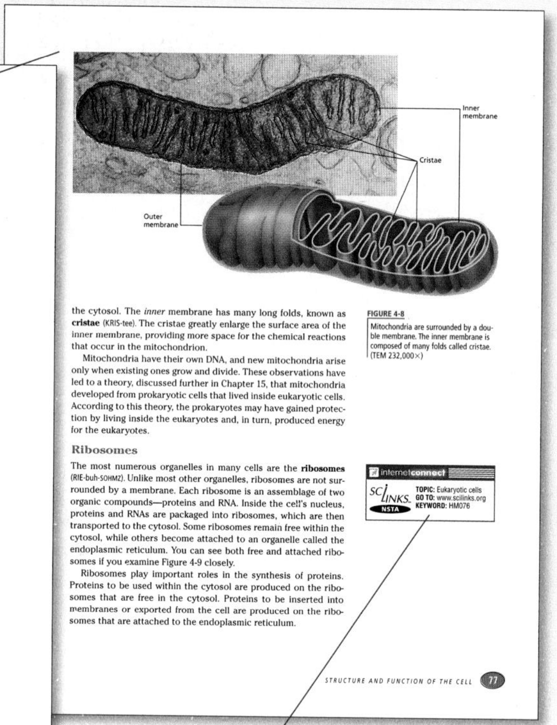

Inner membrane
Cristae
Outer membrane

the cytosol. The *inner* membrane has many long folds, known as **cristae** (KRIS-tee). The cristae greatly enlarge the surface area of the inner membrane, providing more space for the chemical reactions that occur in the mitochondrion.

Mitochondria have their own DNA, and new mitochondria arise only when existing ones grow and divide. These observations have led to a theory, discussed further in Chapter 15, that mitochondria developed from prokaryotic cells that lived inside eukaryotic cells. According to this theory, the prokaryotes may have gained protection by living inside the eukaryotes and, in turn, produced energy for the eukaryotes.

Ribosomes

The most numerous organelles in many cells are the **ribosomes** (RIE-buh-SOHMZ). Unlike most other organelles, ribosomes are not surrounded by a membrane. Each ribosome is an assemblage of two organic compounds—proteins and RNA. Inside the cell's nucleus, proteins and RNAs are packaged into ribosomes, which are then transported to the cytosol. Some ribosomes remain free within the cytosol, while others become attached to an organelle called the endoplasmic reticulum. You can see both free and attached ribosomes if you examine Figure 4-9 closely.

Ribosomes play important roles in the synthesis of proteins. Proteins to be used within the cytosol are produced on the ribosomes that are free in the cytosol. Proteins to be inserted into membranes or exported from the cell are produced on the ribosomes that are attached to the endoplasmic reticulum.

FIGURE 4-8
Mitochondria are surrounded by a double membrane. The inner membrane is composed of many folds called cristae. (TEM 232,000×)

internetconnect
SCILINKS. NSTA
TOPIC: Eukaryotic cells
GO TO: www.scilinks.org
KEYWORD: HM076

STRUCTURE AND FUNCTION OF THE CELL 77

Internet Resources
NSTA *sci*LINKS Web sites help today's students get excited about new developments in biology.

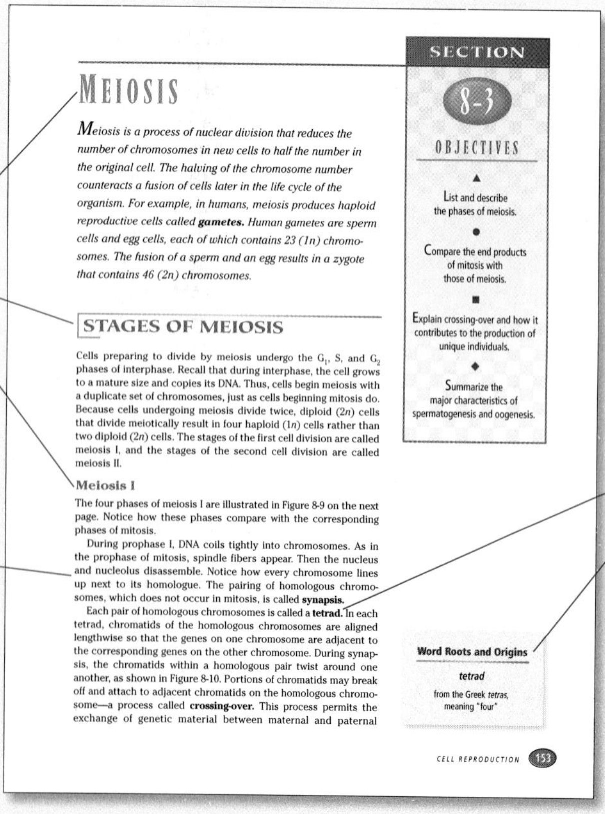

SECTION 8-3

MEIOSIS

Meiosis is a process of nuclear division that reduces the number of chromosomes in new cells to half the number in the original cell. The halving of the chromosome number counteracts a fusion of cells later in the life cycle of the organism. For example, in humans, meiosis produces haploid reproductive cells called ***gametes.*** *Human gametes are sperm cells and egg cells, each of which contains 23 (1n) chromosomes. The fusion of a sperm and an egg results in a zygote that contains 46 (2n) chromosomes.*

OBJECTIVES
▲ List and describe the phases of meiosis.
● Compare the end products of mitosis with those of meiosis.
■ Explain crossing-over and how it contributes to the production of unique individuals.
◆ Summarize the major characteristics of spermatogenesis and oogenesis.

STAGES OF MEIOSIS

Cells preparing to divide by meiosis undergo the G_1, S, and G_2 phases of interphase. Recall that during interphase, the cell grows to a mature size and copies its DNA. Thus, cells begin meiosis with a duplicate set of chromosomes, just as cells beginning mitosis do. Because cells undergoing meiosis divide twice, diploid (2*n*) cells that divide meiotically result in four haploid (1*n*) cells rather than two diploid (2*n*) cells. The stages of the first cell division are called meiosis I, and the stages of the second cell division are called meiosis II.

Meiosis I

The four phases of meiosis I are illustrated in Figure 8-9 on the next page. Notice how these phases compare with the corresponding phases of mitosis.

During prophase I, DNA coils tightly into chromosomes. As in the prophase of mitosis, spindle fibers appear. Then the nucleus and nucleolus disassemble. Notice how every chromosome lines up next to its homologue. The pairing of homologous chromosomes, which does not occur in mitosis, is called **synapsis.**

Each pair of homologous chromosomes is called a **tetrad.** In each tetrad, chromatids of the homologous chromosomes are aligned lengthwise so that the genes on one chromosome are adjacent to the corresponding genes on the other chromosome. During synapsis, the chromatids within a homologous pair twist around one another, as shown in Figure 8-10. Portions of chromatids may break off and attach to adjacent chromatids on the homologous chromosome—a process called **crossing-over.** This process permits the exchange of genetic material between maternal and paternal

Word Roots and Origins
tetrad
from the Greek *tetras*, meaning "four"

CELL REPRODUCTION 153

Outline-Style Headings
Students can easily glean the main ideas by following the headings in a chapter.

Concise Explanations
Every chapter is divided into well-circumscribed blocks of text that are geared for quick student comprehension.

Word Roots and Origins
These features help students relate Latin and Greek word parts to scientific terminology. These features also promote skills that students can use to decipher the meanings of unfamiliar terms as they read.

Chapter Summaries

Full-page Chapter Summaries are divided by section and include a list of the vocabulary terms.

Critical Thinking

Each Section Review concludes with a Critical Thinking question to help students put the main ideas of the lesson together. In addition, Chapter Reviews include a series of Critical Thinking questions.

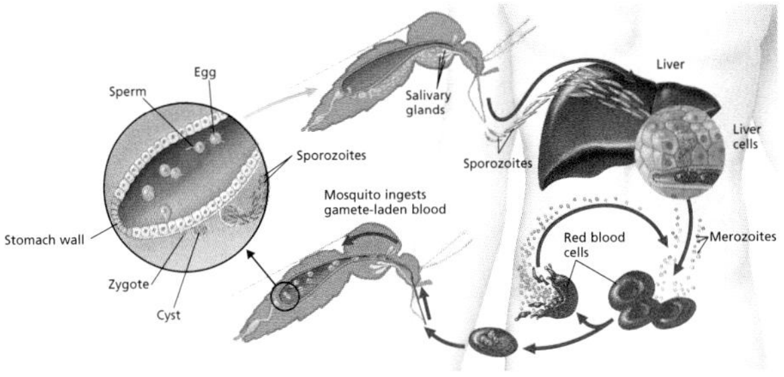

FIGURE 26-13

Malaria is caused by a sporozoan of the genus *Plasmodium*. *Plasmodium* is transmitted from host to host by female *Anopheles* mosquitoes. The host experiences attacks of chills and fever each time infected red blood cells burst and release the parasites that have multiplied within them. Some parasites develop into spores, which are picked up by uninfected *Anopheles* mosquitoes that bite the host.

body cavity and salivary glands of the mosquito. When the insect bites another person, the life cycle begins again. One effective way to reduce human deaths from malaria is to control mosquito populations. Without the mosquito hosts, the *Plasmodium* protozoans cannot complete their life cycle. The malaria life cycle is illustrated in Figure 26-13.

Malaria is usually cured with a drug derived from the cinchona tree, which is native to the Americas. This drug, quinine, has been used to treat malaria for over 500 years. As you learned in Chapter 24, bacteria can evolve resistance to antibiotics; similarly, sporozoans can evolve resistance to antimalarial drugs. Resistance to antimalarial drugs is a significant worldwide problem. Many more people will die from malaria despite the availability of drugs. Pharmaceutical companies have developed new drugs related to quinine, such as chloroquine and primaquine, but *Plasmodium* has evolved resistance to these newer quinine-related drugs. Although experimental malaria vaccines have not been successful, molecular techniques may offer new hope that a successful vaccine will be developed.

SECTION 26-2 REVIEW

1. What are pseudopodia? What function do they serve in sarcodines?
2. How have sarcodines built geological features of the environment?
3. What is conjugation? How is this process advantageous for ciliates, such as *Paramecium*?
4. What kinds of disease can zooflagellates cause in humans?
5. Describe the life cycle of *Plasmodium*, the sporozoan that causes malaria. What features typical of sporozoans does this life cycle exhibit?
6. CRITICAL THINKING How might health workers attempt to control diseases caused by protozoa?

518 CHAPTER 26

CHAPTER 8 REVIEW

SUMMARY/VOCABULARY

8-1

- Chromosomes are tightly coiled DNA molecules and associated proteins.
- In eukaryotes, histone proteins help maintain the compact structure of chromosomes.
- In dividing cells, chromosomes are composed of two identical chromatids constricted together at a centromere.
- Chromosomes are categorized as either sex chromosomes or autosomes.
- Homologous chromosomes consist of one autosomal chromosome from each parent.
- Diploid (2*n*) is the number of chromosomes in cells that have homologous pairs of autosomes and two sex chromosomes.
- Haploid cells (1*n*) have half the number of chromosomes that are present in diploid cells.

Vocabulary

autosome (146)
centromere (146)
chromatid (146)
diploid (147)
haploid (147)
histone (145)
homologous chromosome (146)
karyotype (147)
nonhistone (145)
sex chromosome (146)

8-2

- Cell division is the process by which cells reproduce themselves.
- Binary fission is the process of cell division in prokaryotes.
- The cell cycle is the repeating of events that make up the life of a cell. The cell cycle consists of cell division and interphase.
- Cell division in eukaryotes includes the division of the nucleus (mitosis) and the division of the cytoplasm (cytokinesis).
- Interphase consists of a phase of growth (G_1), a phase of DNA replication (S), and a phase of preparation for cell division (G_2).
- Mitosis is divided into prophase, metaphase, anaphase, and telophase. Mitosis results in two offspring cells that are genetically identical to the original cell.

Vocabulary

anaphase (150)
binary fission (148)
cell cycle (149)
cell plate (151)
centriole (150)
centrosome (150)
cleavage furrow (151)
cytokinesis (149)
G_0 phase (149)
G_1 phase (149)
G_2 phase (149)
interphase (149)
kinetochore (150)
kinetochore fiber (150)
meiosis (148)
metaphase (150)
mitosis (148)
mitotic spindle (150)
M phase (149)
polar fiber (150)
prophase (150)
S phase (149)
spindle fiber (150)
telophase (151)

8-3

- During meiosis, a cell divides twice.
- Crossing-over during meiosis results in genetic recombination.
- Spermatogenesis is the process by which sperm cells are produced. Oogenesis is the process that produces egg cells (ova).
- Asexual reproduction is the formation of offspring from one parent. Offspring produced by asexual reproduction are genetically identical to the parent.
- Sexual reproduction is the formation of offspring through the union of a sperm and an egg. Offspring produced by sexual reproduction are genetically different from the parents.

Vocabulary

asexual reproduction (156)
crossing-over (153)
gamete (153)
genetic recombination (154)
independent assortment (154)
oogenesis (156)
polar body (156)
sexual reproduction (156)
spermatid (156)
spermatogenesis (156)
synapsis (153)
tetrad (153)

CELL REPRODUCTION 157

Relevant Features That Help You Engage Students

These exciting special feature pages illustrate the process of science, the history of science, and the relevance of science to students. In addition, these features portray the nature of scientific reasoning and the importance of a rational approach to solving problems.

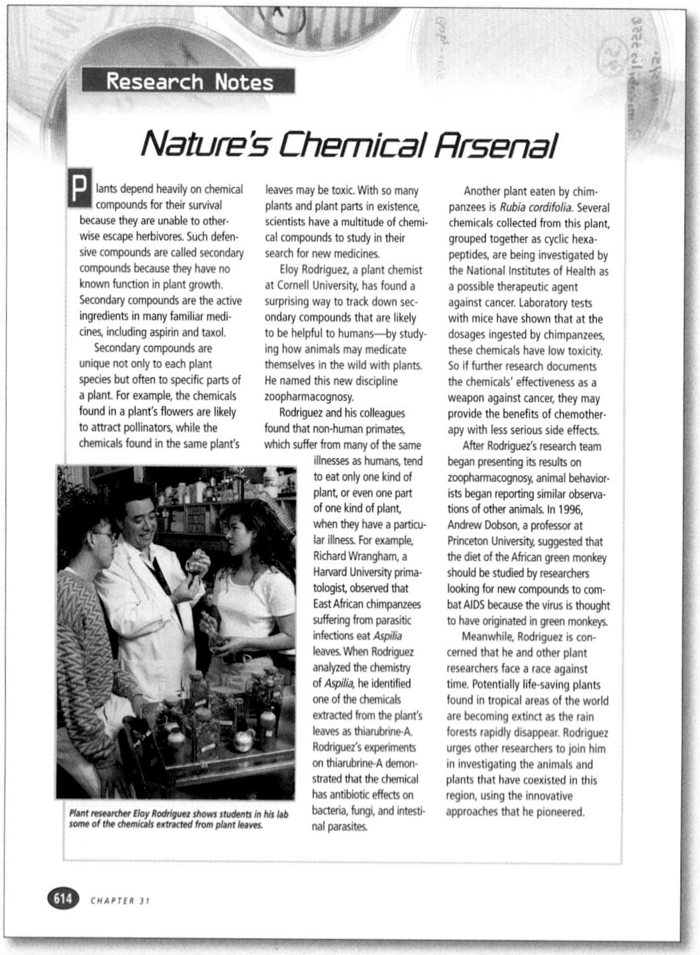

Research Notes

Nature's Chemical Arsenal

Plants depend heavily on chemical compounds for their survival because they are unable to otherwise escape herbivores. Such defensive compounds are called secondary compounds because they have no known function in plant growth. Secondary compounds are the active ingredients in many familiar medicines, including aspirin and taxol.

Secondary compounds are unique not only to each plant species but often to specific parts of a plant. For example, the chemicals found in a plant's flowers are likely to attract pollinators, while the chemicals found in the same plant's leaves may be toxic. With so many plants and plant parts in existence, scientists have a multitude of chemical compounds to study in their search for new medicines.

Eloy Rodriguez, a plant chemist at Cornell University, has found a surprising way to track down secondary compounds that are likely to be helpful to humans—by studying how animals may medicate themselves in the wild with plants. He named this new discipline zoopharmacognosy.

Rodriguez and his colleagues found that non-human primates, which suffer from many of the same illnesses as humans, tend to eat only one kind of plant, or even one part of one kind of plant, when they have a particular illness. For example, Richard Wrangham, a Harvard University primatologist, observed that East African chimpanzees suffering from parasitic infections eat *Aspilia* leaves. When Rodriguez analyzed the chemistry of *Aspilia*, he identified one of the chemicals extracted from the plant's leaves as thiarubrine-A. Rodriguez's experiments on thiarubrine-A demonstrated that the chemical has antibiotic effects on bacteria, fungi, and intestinal parasites.

Another plant eaten by chimpanzees is *Rubia cordifolia*. Several chemicals collected from this plant, grouped together as cyclic hexapeptides, are being investigated by the National Institutes of Health as a possible therapeutic agent against cancer. Laboratory tests with mice have shown that at the dosages ingested by chimpanzees, these chemicals have low toxicity. So if further research documents the chemicals' effectiveness as a weapon against cancer, they may provide the benefits of chemotherapy with less serious side effects.

After Rodriguez's research team began presenting its results on zoopharmacognosy, animal behaviorists began reporting similar observations of other animals. In 1996, Andrew Dobson, a professor at Princeton University, suggested that the diet of the African green monkey should be studied by researchers looking for new compounds to combat AIDS because the virus is thought to have originated in green monkeys.

Meanwhile, Rodriguez is concerned that he and other plant researchers face a race against time. Potentially life-saving plants found in tropical areas of the world are becoming extinct as the rain forests rapidly disappear. Rodriguez urges other researchers to join him in investigating the animals and plants that have coexisted in this region, using the innovative approaches that he pioneered.

Plant researcher Eloy Rodriguez shows students in his lab some of the chemicals extracted from plant leaves.

614 CHAPTER 31

Research Notes

Ten Research Notes features provide information on specific questions in biology. These features also expose students to the tools and techniques that scientists use in developing new products and technologies.

Literature & Life

Dinosaurs Are Extinct—But Why?

The following excerpt is from *The Dinosaur Heresies*, by Robert T. Bakker.

A very well-preserved segment of fossil history . . . permits a computation of how long the average species and genus of dinosaur lasted. And that can be compared with the rates of change computed for warm-blooded animals on the one hand and cold-blooded on the other. Rates of evolutionary change must somehow be linked to the metabolic rate of any given organism. When metabolism is very high, the high physical costs of its living must drive the animal to be an aggressive competitor or predator. And the rapid reproduction typical of warm-blooded animals tends to fill habitats to overcrowded levels much more quickly than the more leisurely breeding schedules characteristic of cold-bloods. An ecological community full of warm-blooded species is therefore a tough environment where the resident species jostle one another for food and water, breeding sites and burrows, all year round. In this sort of environment, the average species will not last long in terms of geological time before it is driven to extinction either by a new species, or by a combination of old species, or by an adverse change of climate.

Clams, by contrast, with their low metabolism, move around very little to accomplish the tasks of their adult lives. It is consequently not surprising that species of clams last much longer than warm-blooded species. Crocodiles and turtles are far more active than clams, but are still sluggish compared to the fully warm-blooded mammals. If dinosaurs resembled cold-blooded reptiles metabolically, their rate of evolutionary change would very nearly match that found in crocodiles or large turtles. But if the dinosaurs' metabolism was heated, then the average life span of one of their species or genera would have to be short, like a mammal's.

One of the best places to study the rate of evolutionary changes in dinosaurs is in the Late Cretaceous deltas of Wyoming, Montana, and Alberta. There the changes through the last ten million years of the dinosaurs' history can be followed. As might be expected, the turtles and crocodiles show very little change in these strata. The genera representing these cold-bloods hold on through formation after formation. But what about the dinosaurs here? Their evolutionary pace stands out as quite different. New species and genera kept appearing and eliminating older ones at quite a brisk tempo, geologically speaking. The average genus of dinosaur lasted for only a fraction of the time of that of the average crocodile. . . .

One part of the orthodox story does appear to be unassailable, an ineradicable fact safe from even the wildest heretic: Dinosaurs are indeed all extinct. The fact of their extinction is the cornerstone underlying the orthodox belief that dinosaurs were maladapted failures.

Dinosaurs are incontrovertibly dead. But that does not prove what orthodoxy believes about them. Paradoxically, the extinction of the dinosaurs is strong evidence that their biology was heated to levels far above those of typical reptiles. The basic principle is simple: The higher the metabolic needs of a group of species, the more vulnerable it is to sudden and catastrophic extinction.

Reading for Meaning
What is Bakker's explanation for the extinction of the dinosaurs?

Read Further
Bakker follows the text excerpted here by describing the best natural design for avoiding extinction. Explain what you think that design might be, and name some existing animals that are similar to that design.

From "Dinosaurs Are Extinct—But Why?" from *The Dinosaur Heresies* by Robert T. Bakker. Copyright © 1986 by Robert T. Bakker. Reprinted by permission of **William Morrow & Company, Inc.**

844 CHAPTER 43

Literature and Life

Literature and Life features spur students' interest in reading about biology outside the classroom. These features also relate issues in biology to many different topics.

Great Discoveries

Each two-page Great Discoveries feature emphasizes the contributions of individual investigators. Each feature includes a personal account of the discovery process and information about a specific contribution to biology.

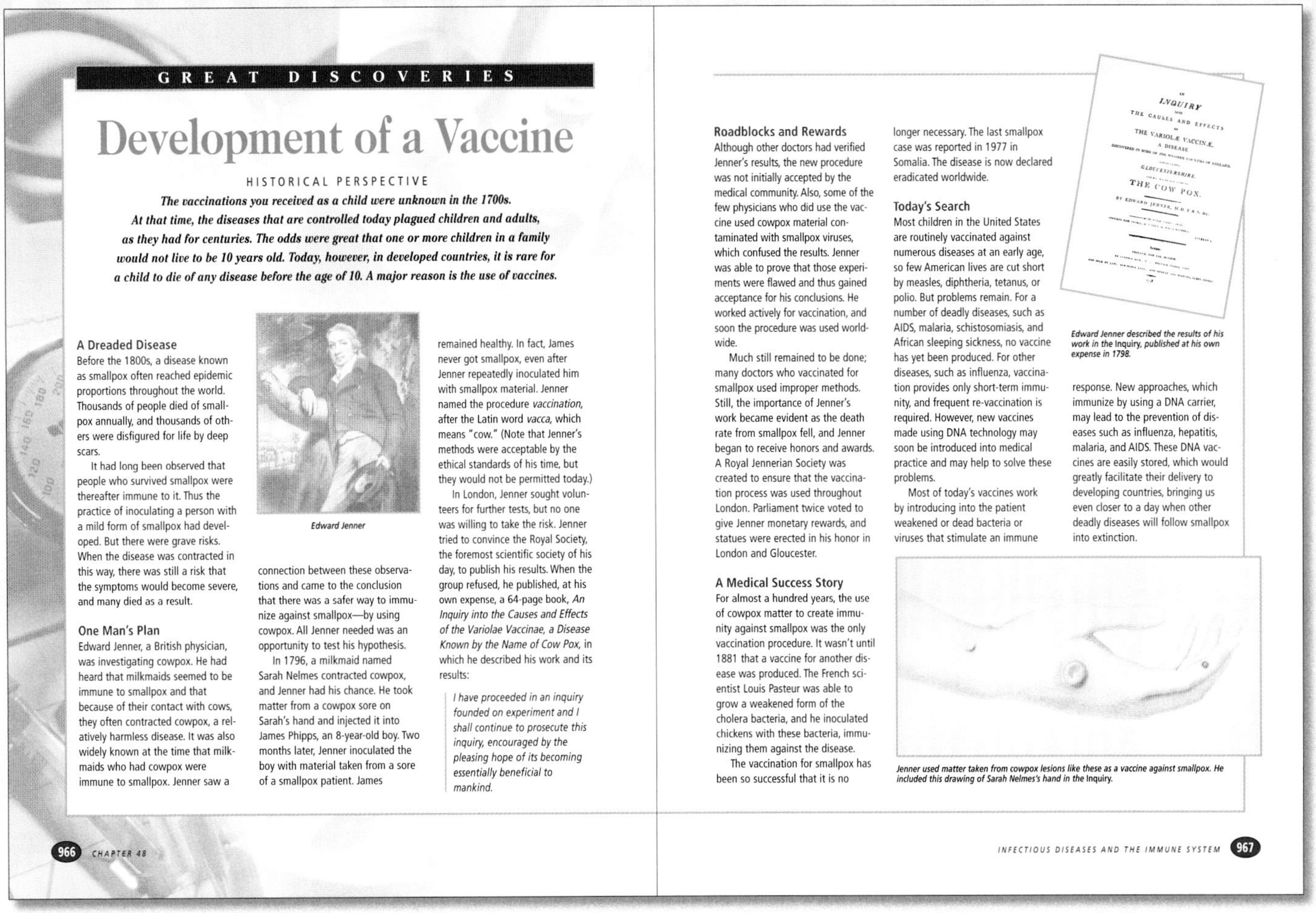
GREAT DISCOVERIES

Development of a Vaccine

HISTORICAL PERSPECTIVE

The vaccinations you received as a child were unknown in the 1700s. At that time, the diseases that are controlled today plagued children and adults, as they had for centuries. The odds were great that one or more children in a family would not live to be 10 years old. Today, however, in developed countries, it is rare for a child to die of any disease before the age of 10. A major reason is the use of vaccines.

A Dreaded Disease

Before the 1800s, a disease known as smallpox often reached epidemic proportions throughout the world. Thousands of people died of smallpox annually, and thousands of others were disfigured for life by deep scars.

It had long been observed that people who survived smallpox were thereafter immune to it. Thus the practice of inoculating a person with a mild form of smallpox had developed. But there were grave risks. When the disease was contracted in this way, there was still a risk that the symptoms would become severe, and many died as a result.

One Man's Plan

Edward Jenner, a British physician, was investigating cowpox. He had heard that milkmaids seemed to be immune to smallpox and that because of their contact with cows, they often contracted cowpox, a relatively harmless disease. It was also widely known at the time that milkmaids who had cowpox were immune to smallpox. Jenner saw a connection between these observations and came to the conclusion that there was a safer way to immunize against smallpox—by using cowpox. All Jenner needed was an opportunity to test his hypothesis.

In 1796, a milkmaid named Sarah Nelmes contracted cowpox, and Jenner had his chance. He took matter from a cowpox sore on Sarah's hand and injected it into James Phipps, an 8-year-old boy. Two months later, Jenner inoculated the boy with material taken from a sore of a smallpox patient. James remained healthy. In fact, James never got smallpox, even after Jenner repeatedly inoculated him with smallpox material. Jenner named the procedure *vaccination,* after the Latin word *vacca,* which means "cow." (Note that Jenner's methods were acceptable by the ethical standards of his time, but they would not be permitted today.)

Edward Jenner

In London, Jenner sought volunteers for further tests, but no one was willing to take the risk. Jenner tried to convince the Royal Society, the foremost scientific society of his day, to publish his results. When the group refused, he published, at his own expense, a 64-page book, *An Inquiry into the Causes and Effects of the Variolae Vaccinae, a Disease Known by the Name of Cow Pox,* in which he described his work and its results:

> *I have proceeded in an inquiry founded on experiment and I shall continue to prosecute this inquiry, encouraged by the pleasing hope of its becoming essentially beneficial to mankind.*

Roadblocks and Rewards

Although other doctors had verified Jenner's results, the new procedure was not initially accepted by the medical community. Also, some of the few physicians who did use the vaccine used cowpox material contaminated with smallpox viruses, which confused the results. Jenner was able to prove that those experiments were flawed and thus gained acceptance for his conclusions. He worked actively for vaccination, and soon the procedure was used worldwide.

Much still remained to be done; many doctors who vaccinated for smallpox used improper methods. Still, the importance of Jenner's work became evident as the death rate from smallpox fell, and Jenner began to receive honors and awards. A Royal Jennerian Society was created to ensure that the vaccination process was used throughout London. Parliament twice voted to give Jenner monetary rewards, and statues were erected in his honor in London and Gloucester.

A Medical Success Story

For almost a hundred years, the use of cowpox matter to create immunity against smallpox was the only vaccination procedure. It wasn't until 1881 that a vaccine for another disease was produced. The French scientist Louis Pasteur was able to grow a weakened form of the cholera bacteria, and he inoculated chickens with these bacteria, immunizing them against the disease.

The vaccination for smallpox has been so successful that it is no longer necessary. The last smallpox case was reported in 1977 in Somalia. The disease is now declared eradicated worldwide.

Today's Search

Most children in the United States are routinely vaccinated against numerous diseases at an early age, so few American lives are cut short by measles, diphtheria, tetanus, or polio. But problems remain. For a number of deadly diseases, such as AIDS, malaria, schistosomiasis, and African sleeping sickness, no vaccine has yet been produced. For other diseases, such as influenza, vaccination provides only short-term immunity, and frequent re-vaccination is required. However, new vaccines made using DNA technology may soon be introduced into medical practice and may help to solve these problems.

Most of today's vaccines work by introducing into the patient weakened or dead bacteria or viruses that stimulate an immune response. New approaches, which immunize by using a DNA carrier, may lead to the prevention of diseases such as influenza, hepatitis, malaria, and AIDS. These DNA vaccines are easily stored, which would greatly facilitate their delivery to developing countries, bringing us even closer to a day when other deadly diseases will follow smallpox into extinction.

AN INQUIRY INTO THE CAUSES AND EFFECTS OF THE VARIOLÆ VACCINÆ, ... THE COW POX. BY EDWARD JENNER

Edward Jenner described the results of his work in the Inquiry, published at his own expense in 1798.

Jenner used matter taken from cowpox lesions like these as a vaccine against smallpox. He included this drawing of Sarah Nelmes's hand in the Inquiry.

966 CHAPTER 48 | INFECTIOUS DISEASES AND THE IMMUNE SYSTEM 967

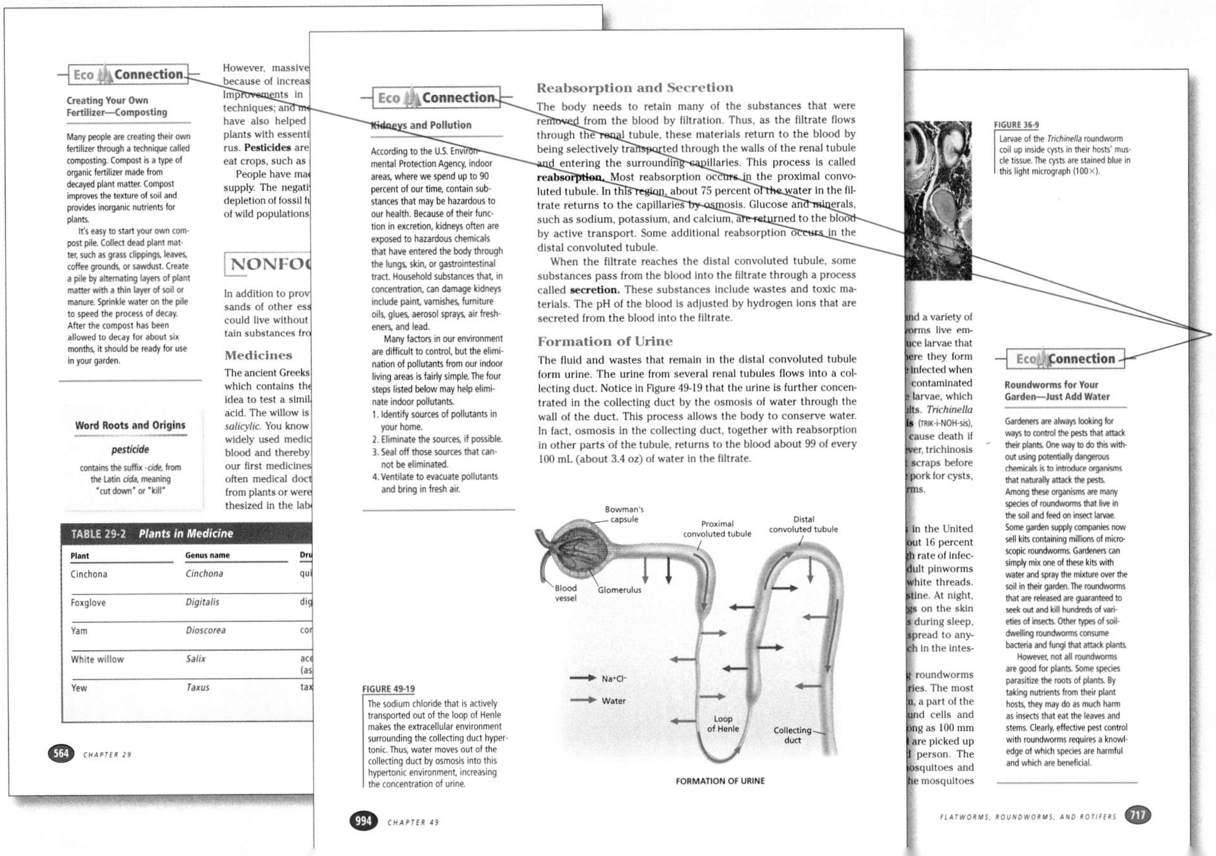
Eco Connection

Creating Your Own Fertilizer—Composting

Many people are creating their own fertilizer through a technique called composting. Compost is a type of organic fertilizer made from decayed plant matter. Compost improves the texture of soil and provides inorganic nutrients for plants.

It's easy to start your own compost pile. Collect dead plant matter, such as grass clippings, leaves, coffee grounds, or sawdust. Create a pile by alternating layers of plant matter with a thin layer of soil or manure. Sprinkle water on the pile to speed the process of decay. After the compost has been allowed to decay for about six months, it should be ready for use in your garden.

Word Roots and Origins

pesticide

contains the suffix *-cide*, from the Latin *cida*, meaning "cut down" or "kill"

TABLE 29-2 Plants in Medicine

564 CHAPTER 29

Eco Connection

Kidneys and Pollution

According to the U.S. Environmental Protection Agency, indoor areas, where we spend up to 90 percent of our time, contain substances that may be hazardous to our health. Because of their function in excretion, kidneys often are exposed to hazardous chemicals that have entered the body through the lungs, skin, or gastrointestinal tract. Household substances that, in concentration, can damage kidneys include paint, varnishes, furniture oils, glues, aerosol sprays, air fresheners, and lead.

Many factors in our environment are difficult to control, but the elimination of pollutants from our indoor living areas is fairly simple. The four steps listed below may help eliminate indoor pollutants.

1. Identify sources of pollutants in your home.
2. Eliminate the sources, if possible.
3. Seal off those sources that cannot be eliminated.
4. Ventilate to evacuate pollutants and bring in fresh air.

Reabsorption and Secretion

The body needs to retain many of the substances that were removed from the blood by filtration. Thus, as the filtrate flows through the renal tubule, these materials return to the blood by being selectively transported through the walls of the renal tubule and entering the surrounding capillaries. This process is called **reabsorption.** Most reabsorption occurs in the proximal convoluted tubule. In this region, about 75 percent of the water in the filtrate returns to the capillaries by osmosis. Glucose and minerals, such as sodium, potassium, and calcium, are returned to the blood by active transport. Some additional reabsorption occurs in the distal convoluted tubule.

When the filtrate reaches the distal convoluted tubule, some substances pass from the blood into the filtrate through a process called **secretion.** These substances include wastes and toxic materials. The pH of the blood is adjusted by hydrogen ions that are secreted from the blood into the filtrate.

Formation of Urine

The fluid and wastes that remain in the distal convoluted tubule form urine. The urine from several renal tubules flows into a collecting duct. Notice in Figure 49-19 that the urine is further concentrated in the collecting duct by the osmosis of water through the wall of the duct. This process allows the body to conserve water. In fact, osmosis in the collecting duct, together with reabsorption in other parts of the tubule, returns to the blood about 99 of every 100 mL (about 3.4 oz) of water in the filtrate.

FIGURE 49-19
The sodium chloride that is actively transported out of the loop of Henle makes the extracellular environment surrounding the collecting duct hypertonic. Thus, water moves out of the collecting duct by osmosis into this hypertonic environment, increasing the concentration of urine.

994 CHAPTER 49

FIGURE 36-9
Larvae of the *Trichinella* roundworm coil up inside cysts in their hosts' muscle tissue. The cysts are stained blue in this light micrograph (100×).

Eco Connection

Roundworms for Your Garden—Just Add Water

Gardeners are always looking for ways to control the pests that attack their plants. One way to do this without using potentially dangerous chemicals is to introduce organisms that naturally attack the pests. Among these organisms are many species of roundworms that live in the soil and feed on insect larvae. Some garden supply companies now sell kits containing millions of microscopic roundworms. Gardeners can simply mix one of these kits with water and spray the mixture over the soil in their garden. The roundworms that are released are guaranteed to seek out and kill hundreds of varieties of insects. Other types of soil-dwelling roundworms consume bacteria and fungi that attack plants.

However, not all roundworms are good for plants. Some species parasitize the roots of plants. By taking nutrients from their plant hosts, they may do as much harm as insects that eat the leaves and stems. Clearly, effective pest control with roundworms requires a knowledge of which species are harmful and which are beneficial.

FLATWORMS, ROUNDWORMS, AND ROTIFERS 717

Eco-Connections

These short features emphasize the interdependence of organisms within the biosphere and provide simple ways for students to positively affect the environment.

A Wide-Ranging In-Text Laboratory Program

The 109 labs in the text provide a variety of options that enable you to give students hands-on experiences. Three different kinds of labs are included in the textbook: Chapter Investigations, Quick Labs, and Interactive Explorations.

Chapter Investigations

Chapter Investigations are traditional laboratory activities that have been bench tested by WARD'S. These labs help students develop and practice basic laboratory techniques. A number of the Chapter Investigations ask students to adapt a technique given in the procedure in order to test their own hypothesis.

Analysis and Conclusions
These questions ask students about their actions and observations during the investigation. These questions also require students to draw conclusions based on their results.

Objectives and Process Skills
The Objectives and Process Skills used in each investigation are highlighted.

Materials
Whenever possible, required materials are kept to a minimum to reduce your costs and preparation time.

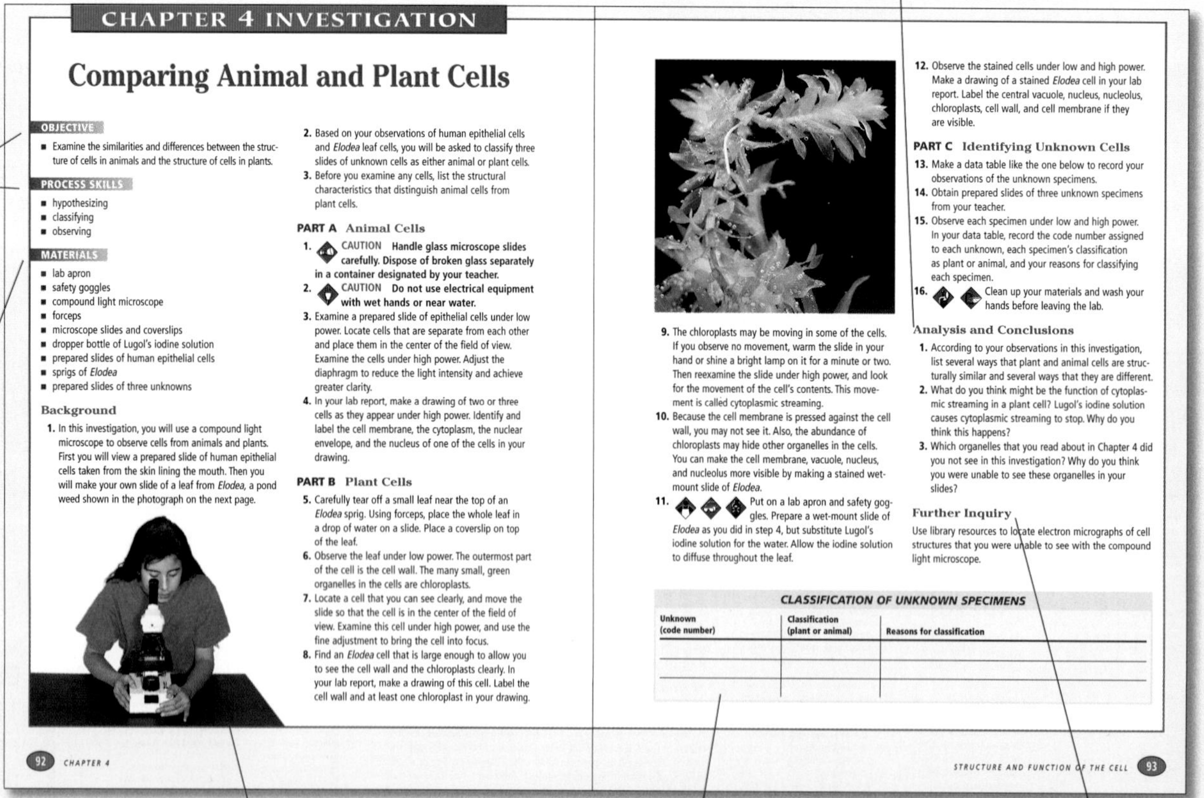

CHAPTER 4 INVESTIGATION

Comparing Animal and Plant Cells

OBJECTIVE

- Examine the similarities and differences between the structure of cells in animals and the structure of cells in plants.

PROCESS SKILLS

- hypothesizing
- classifying
- observing

MATERIALS

- lab apron
- safety goggles
- compound light microscope
- forceps
- microscope slides and coverslips
- dropper bottle of Lugol's iodine solution
- prepared slides of human epithelial cells
- sprigs of *Elodea*
- prepared slides of three unknowns

Background

1. In this investigation, you will use a compound light microscope to observe cells from animals and plants. First you will view a prepared slide of human epithelial cells taken from the skin lining the mouth. Then you will make your own slide of a leaf from *Elodea,* a pond weed shown in the photograph on the next page.
2. Based on your observations of human epithelial cells and *Elodea* leaf cells, you will be asked to classify three slides of unknown cells as either animal or plant cells.
3. Before you examine any cells, list the structural characteristics that distinguish animal cells from plant cells.

PART A Animal Cells

1. CAUTION **Handle glass microscope slides carefully. Dispose of broken glass separately in a container designated by your teacher.**
2. CAUTION **Do not use electrical equipment with wet hands or near water.**
3. Examine a prepared slide of epithelial cells under low power. Locate cells that are separate from each other and place them in the center of the field of view. Examine the cells under high power. Adjust the diaphragm to reduce the light intensity and achieve greater clarity.
4. In your lab report, make a drawing of two or three cells as they appear under high power. Identify and label the cell membrane, the cytoplasm, the nuclear envelope, and the nucleus of one of the cells in your drawing.

PART B Plant Cells

5. Carefully tear off a small leaf near the top of an *Elodea* sprig. Using forceps, place the whole leaf in a drop of water on a slide. Place a coverslip on top of the leaf.
6. Observe the leaf under low power. The outermost part of the cell is the cell wall. The many small, green organelles in the cells are chloroplasts.
7. Locate a cell that you can see clearly, and move the slide so that the cell is in the center of the field of view. Examine this cell under high power, and use the fine adjustment to bring the cell into focus.
8. Find an *Elodea* cell that is large enough to allow you to see the cell wall and the chloroplasts clearly. In your lab report, make a drawing of this cell. Label the cell wall and at least one chloroplast in your drawing.
9. The chloroplasts may be moving in some of the cells. If you observe no movement, warm the slide in your hand or shine a bright lamp on it for a minute or two. Then reexamine the slide under high power, and look for the movement of the cell's contents. This movement is called cytoplasmic streaming.
10. Because the cell membrane is pressed against the cell wall, you may not see it. Also, the abundance of chloroplasts may hide other organelles in the cells. You can make the cell membrane, vacuole, nucleus, and nucleolus more visible by making a stained wet-mount slide of *Elodea.*
11. Put on a lab apron and safety goggles. Prepare a wet-mount slide of *Elodea* as you did in step 4, but substitute Lugol's iodine solution for the water. Allow the iodine solution to diffuse throughout the leaf.
12. Observe the stained cells under low and high power. Make a drawing of a stained *Elodea* cell in your lab report. Label the central vacuole, nucleus, nucleolus, chloroplasts, cell wall, and cell membrane if they are visible.

PART C Identifying Unknown Cells

13. Make a data table like the one below to record your observations of the unknown specimens.
14. Obtain prepared slides of three unknown specimens from your teacher.
15. Observe each specimen under low and high power. In your data table, record the code number assigned to each unknown, each specimen's classification as plant or animal, and your reasons for classifying each specimen.
16. Clean up your materials and wash your hands before leaving the lab.

Analysis and Conclusions

1. According to your observations in this investigation, list several ways that plant and animal cells are structurally similar and several ways that they are different.
2. What do you think might be the function of cytoplasmic streaming in a plant cell? Lugol's iodine solution causes cytoplasmic streaming to stop. Why do you think this happens?
3. Which organelles that you read about in Chapter 4 did you not see in this investigation? Why do you think you were unable to see these organelles in your slides?

Further Inquiry

Use library resources to locate electron micrographs of cell structures that you were unable to see with the compound light microscope.

CLASSIFICATION OF UNKNOWN SPECIMENS

Unknown (code number)	Classification (plant or animal)	Reasons for classification

92 CHAPTER 4

STRUCTURE AND FUNCTION OF THE CELL 93

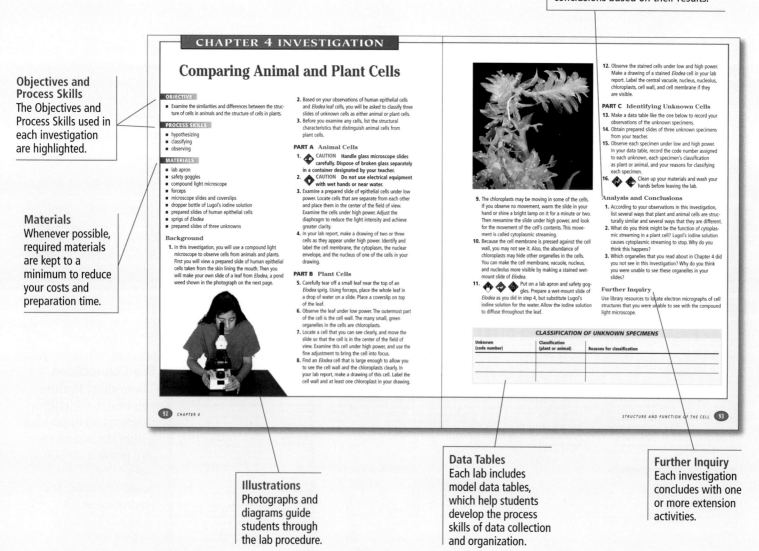

Illustrations
Photographs and diagrams guide students through the lab procedure.

Data Tables
Each lab includes model data tables, which help students develop the process skills of data collection and organization.

Further Inquiry
Each investigation concludes with one or more extension activities.

Quick Labs

Quick Labs help students form concrete models of concepts covered in the adjacent text. These labs provide hands-on experience yet require few materials.

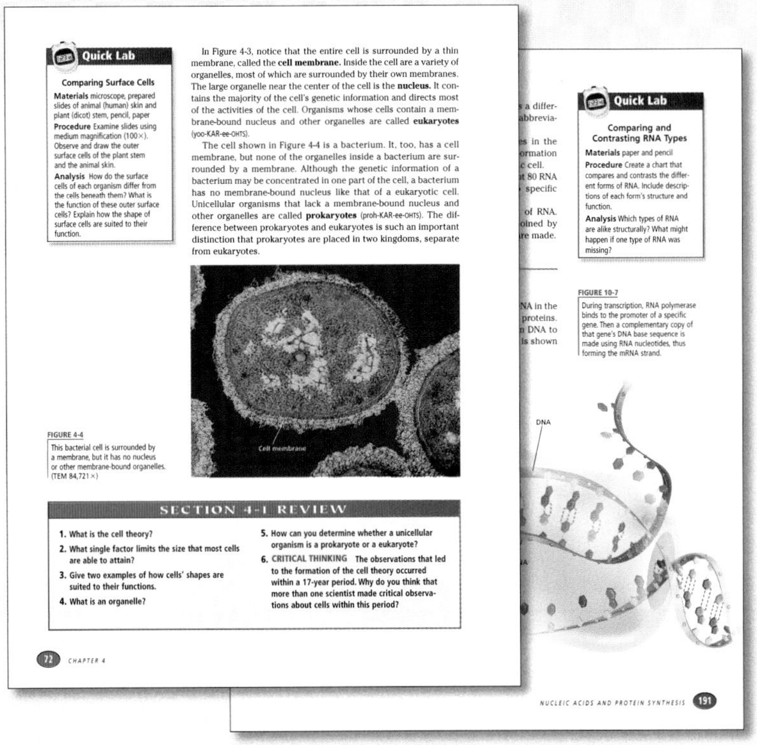

Interactive Explorations

The ***Interactive Explorations in Biology: Human Biology*** and ***Interactive Explorations in Biology: Cell Biology and Genetics*** CD-ROMs allow students to collect and analyze data that would otherwise be difficult or expensive to obtain in a high school laboratory. Students using the Interactive Explorations can work individually, in small groups, or on their own when making up labs missed due to absence. Students can work at their own pace, repeating portions of an activity multiple times if needed.

Background
Each exploration includes an overview that describes the biological concepts presented in the exploration.

Feedback Meters
Students collect data by reading the various feedback meters. Narration is provided in both English and Spanish.

Variables
Students manipulate different variables as they conduct simulated experiments.

Navigation Bar
The navigation bar makes it easy to access other parts of the program.

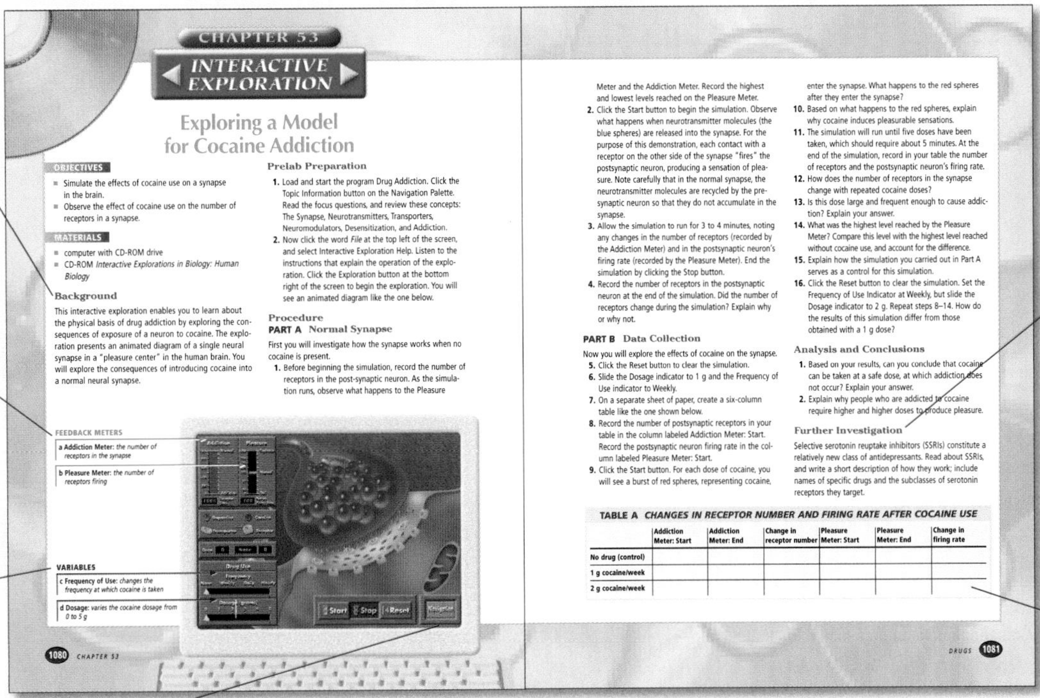

Further Investigation
Each exploration concludes with a suggested relevant reading. All readings are provided on the CD-ROM.

Data Tables
Each exploration includes model data tables, which help students develop the process skills of data collection and organization.

An Additional Five Lab Manuals Bring You Unprecedented Variety

HOLT
BIOSOURCES

The *Holt BioSources* Lab Program brings you an unparalleled collection of 135 labs, extending across a range of student ability levels and equipment requirements. This collection of labs is divided into five lab manuals—ranging from quick, simple labs to labs using techniques in biotechnology. All lab activities in the *Holt BioSources* Lab Program are bench tested by WARD'S technical staff to guarantee successful lab results and experiences.

A Quick Labs

Quick Labs are just that. Inexpensive, easy-to-obtain supplies are used in each of these short lab activities. Use these brief labs to introduce a new topic or to demonstrate a concept.

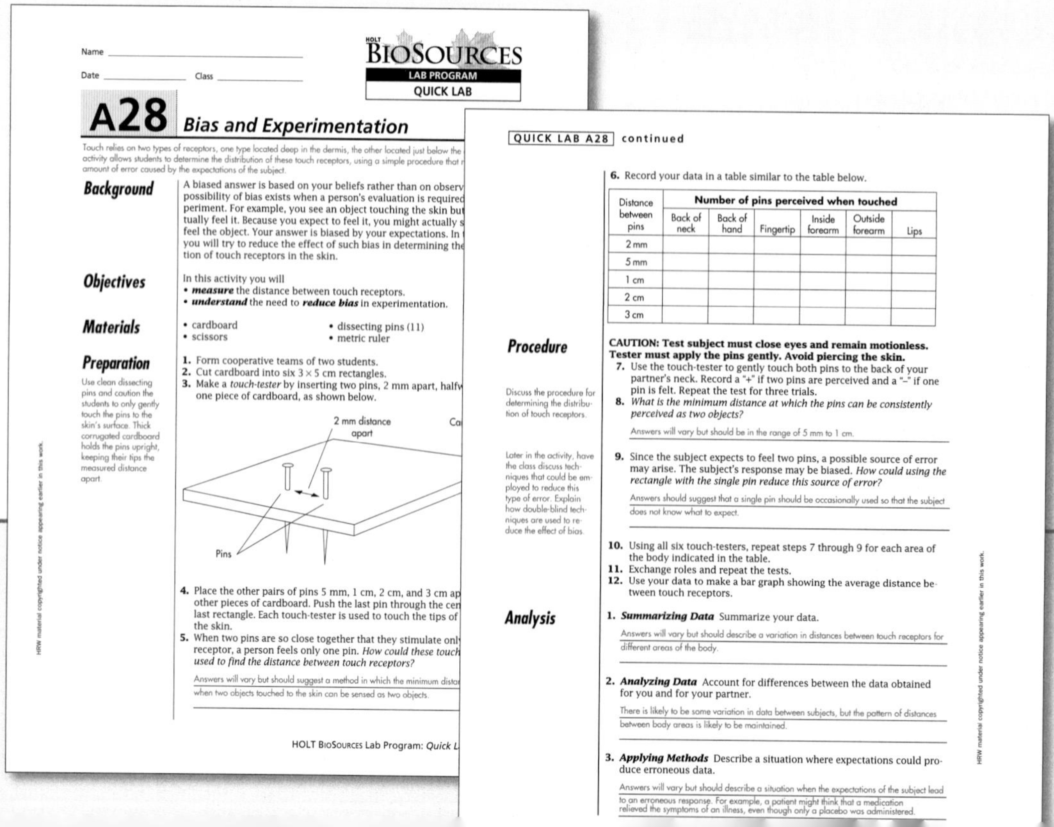
Name ____________

Date ________ Class ________

HOLT BIOSOURCES LAB PROGRAM QUICK LAB

A28 Bias and Experimentation

Touch relies on two types of receptors, one type located deep in the dermis, the other located just below the activity allows students to determine the distribution of these touch receptors, using a simple procedure that amount of error caused by the expectations of the subject.

Background A biased answer is based on your beliefs rather than on observ possibility of bias exists when a person's evaluation is required periment. For example, you see an object touching the skin but tually feel it. Because you expect to feel it, you might actually s feel the object. Your answer is biased by your expectations. In you will try to reduce the effect of such bias in determining the tion of touch receptors in the skin.

Objectives In this activity you will
- ***measure*** the distance between touch receptors.
- ***understand*** the need to ***reduce bias*** in experimentation.

Materials
- cardboard
- scissors
- dissecting pins (11)
- metric ruler

Preparation

Use clean dissecting pins and caution the students to only gently touch the pins to the skin's surface. Thick corrugated cardboard holds the pins upright, keeping their tips the measured distance apart.

1. Form cooperative teams of two students.
2. Cut cardboard into six 3 × 5 cm rectangles.
3. Make a *touch-tester* by inserting two pins, 2 mm apart, halfw one piece of cardboard, as shown below.

2 mm distance apart
Ca
Pins

4. Place the other pairs of pins 5 mm, 1 cm, 2 cm, and 3 cm ap other pieces of cardboard. Push the last pin through the cen last rectangle. Each touch-tester is used to touch the tips of the skin.
5. When two pins are so close together that they stimulate onl receptor, a person feels only one pin. *How could these touch used to find the distance between touch receptors?*

Answers will vary but should suggest a method in which the minimum dista when two objects touched to the skin can be sensed as two objects.

HRW material copyrighted under notice appearing earlier in this work.

HOLT BIOSOURCES Lab Program: *Quick L*

QUICK LAB A28 continued

6. Record your data in a table similar to the table below.

Distance between pins	Number of pins perceived when touched					
	Back of neck	Back of hand	Fingertip	Inside forearm	Outside forearm	Lips
2 mm						
5 mm						
1 cm						
2 cm						
3 cm						

Procedure

CAUTION: Test subject must close eyes and remain motionless. Tester must apply the pins gently. Avoid piercing the skin.

Discuss the procedure for determining the distribution of touch receptors.

7. Use the touch-tester to gently touch both pins to the back of your partner's neck. Record a "+" if two pins are perceived and a "–" if one pin is felt. Repeat the test for three trials.
8. *What is the minimum distance at which the pins can be consistently perceived as two objects?*

Answers will vary but should be in the range of 5 mm to 1 cm.

Later in the activity, have the class discuss techniques that could be employed to reduce this type of error. Explain how double-blind techniques are used to reduce the effect of bias.

9. Since the subject expects to feel two pins, a possible source of error may arise. The subject's response may be biased. *How could using the rectangle with the single pin reduce this source of error?*

Answers should suggest that a single pin should be occasionally used so that the subject does not know what to expect.

10. Using all six touch-testers, repeat steps 7 through 9 for each area of the body indicated in the table.
11. Exchange roles and repeat the tests.
12. Use your data to make a bar graph showing the average distance between touch receptors.

Analysis

1. ***Summarizing Data*** Summarize your data.

Answers will vary but should describe a variation in distances between touch receptors for different areas of the body.

2. ***Analyzing Data*** Account for differences between the data obtained for you and for your partner.

There is likely to be some variation in data between subjects, but the pattern of distances between body areas is likely to be maintained.

3. ***Applying Methods*** Describe a situation where expectations could produce erroneous data.

Answers will vary but should describe a situation when the expectations of the subject lead to an erroneous response. For example, a patient might think that a medication relieved the symptoms of an illness, even though only a placebo was administered.

HRW material copyrighted under notice appearing earlier in this work.

B Inquiry Skills Development

Inquiry Skills Development labs teach students practical skills and techniques they'll need throughout their classroom lab experience—and beyond.

B1 Introduction to Experimental Design and Data Presentation
B2 Cell Structures
B3 Diffusion and Cell Membranes
B4 Plant and Animal Interrelationships
B5 Mitosis
B6 Effect of Environment on Gene Expression
B7 Gene Expression
B8 Fossil Study
B9 Peppered Moth Survey
B10 Ecology Scavenger Hunt
B11 Composting
B12 Classification
B13 Cryobiology
B14 Protists—A Comparison
B15 Flower Structures
B16 Fruits and Seeds
B17 Transpiration and Stem Structure
B18 Mineral Deficiencies in Plants
B19 Gravitropism and Phototropism in Plants
B20 Life in a Pine Cone
B21 Flatworm Behavior
B22 Earthworm Dissection
B23 Live Earthworms
B24 Snails
B25 Crayfish Dissection
B26 Perch Dissection
B27 Schooling Behavior in Fishes
B28 Frog Dissection
B29 Fetal Pig Dissection
B30 Touch Receptors in the Skin
B31 Exploring Vision
B32 Antigen-Antibody Interaction
B33 Transmission of a Communicable Disease
B34 Embryonic Development

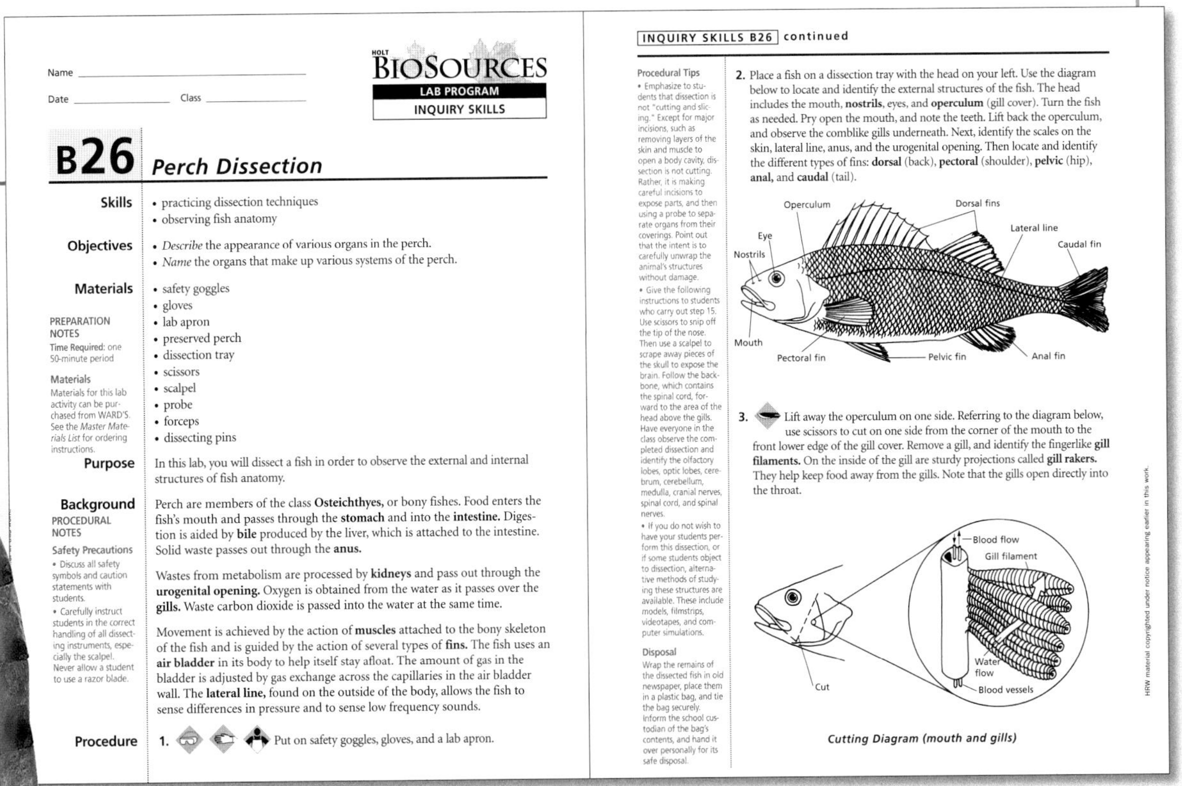

Name ______________________

Date ____________ Class ____________

HOLT BIOSOURCES
LAB PROGRAM
INQUIRY SKILLS

B26 *Perch Dissection*

Skills
- practicing dissection techniques
- observing fish anatomy

Objectives
- *Describe* the appearance of various organs in the perch.
- *Name* the organs that make up various systems of the perch.

Materials
- safety goggles
- gloves
- lab apron
- preserved perch
- dissection tray
- scissors
- scalpel
- probe
- forceps
- dissecting pins

PREPARATION NOTES
Time Required: one 50-minute period

Materials
Materials for this lab activity can be purchased from WARD'S. See the *Master Materials List* for ordering instructions.

Purpose
In this lab, you will dissect a fish in order to observe the external and internal structures of fish anatomy.

Background
Perch are members of the class **Osteichthyes,** or bony fishes. Food enters the fish's mouth and passes through the **stomach** and into the **intestine.** Digestion is aided by **bile** produced by the liver, which is attached to the intestine. Solid waste passes out through the **anus.**

Wastes from metabolism are processed by **kidneys** and pass out through the **urogenital opening.** Oxygen is obtained from the water as it passes over the **gills.** Waste carbon dioxide is passed into the water at the same time.

Movement is achieved by the action of **muscles** attached to the bony skeleton of the fish and is guided by the action of several types of **fins.** The fish uses an **air bladder** in its body to help itself stay afloat. The amount of gas in the bladder is adjusted by gas exchange across the capillaries in the air bladder wall. The **lateral line,** found on the outside of the body, allows the fish to sense differences in pressure and to sense low frequency sounds.

PROCEDURAL NOTES
Safety Precautions
- Discuss all safety symbols and caution statements with students.
- Carefully instruct students in the correct handling of all dissecting instruments, especially the scalpel. Never allow a student to use a razor blade.

Procedure
1. Put on safety goggles, gloves, and a lab apron.

INQUIRY SKILLS B26 continued

2. Place a fish on a dissection tray with the head on your left. Use the diagram below to locate and identify the external structures of the fish. The head includes the mouth, **nostrils**, eyes, and **operculum** (gill cover). Turn the fish as needed. Pry open the mouth, and note the teeth. Lift back the operculum, and observe the comblike gills underneath. Next, identify the scales on the skin, lateral line, anus, and the urogenital opening. Then locate and identify the different types of fins: **dorsal** (back), **pectoral** (shoulder), **pelvic** (hip), **anal,** and **caudal** (tail).

3. Lift away the operculum on one side. Referring to the diagram below, use scissors to cut on one side from the corner of the mouth to the front lower edge of the gill cover. Remove a gill, and identify the fingerlike **gill filaments.** On the inside of the gill are sturdy projections called **gill rakers.** They help keep food away from the gills. Note that the gills open directly into the throat.

Cutting Diagram (mouth and gills)

Procedural Tips
- Emphasize to students that dissection is not "cutting and slicing." Except for major incisions, such as removing layers of the skin and muscle to open a body cavity, dissection is not cutting. Rather, it is making careful incisions to expose parts, and then using a probe to separate organs from their coverings. Point out that the intent is to carefully unwrap the animal's structures without damage.
- Give the following instructions to students who carry out step 15. Use scissors to snip off the tip of the nose. Then use a scalpel to scrape away pieces of the skull to expose the brain. Follow the backbone, which contains the spinal cord, forward to the area of the head above the gills. Have everyone in the class observe the completed dissection and identify the olfactory lobes, optic lobes, cerebrum, cerebellum, medulla, cranial nerves, spinal cord, and spinal nerves.
- If you do not wish to have your students perform this dissection, or if some students object to dissection, alternative methods of studying these structures are available. These include models, filmstrips, videotapes, and computer simulations.

Disposal
Wrap the remains of the dissected fish in old newspaper, place them in a plastic bag, and tie the bag securely. Inform the school custodian of the bag's contents, and hand it over personally for its safe disposal.

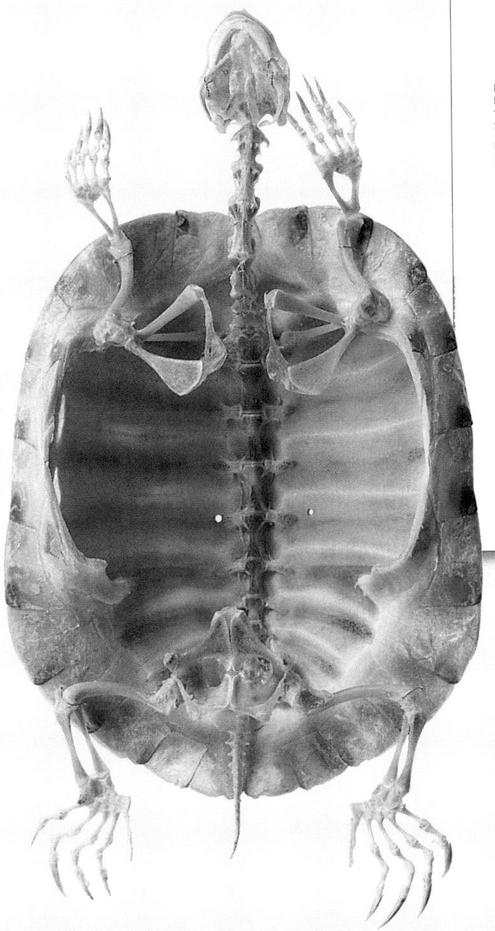

Laboratory Techniques and Experimental Design

Laboratory Techniques and Experimental Design labs require students to apply skills and techniques they've mastered earlier to design their own experiments.

- **C1** Using a Microscope
- **C2** Using a Microscope—Criminal Investigation
- **C3** Using a Microscope—Slowing Protozoans
- **C4** Testing for Vitamin C
- **C5** Observing the Effect of Concentration on Enzyme Activity
- **C6** Observing the Effect of Temperature on Enzyme Activity
- **C7** Measuring the Release of Energy from Sucrose
- **C8** Measuring the Release of Energy—Best Food for Yeast
- **C9** Preparing a Root Tip Squash
- **C10** Preparing a Root Tip Squash—Stopping Meiosis
- **C11** Modeling Mitosis
- **C12** Analyzing Corn Genetics
- **C13** Preparing Tissue for Karyotyping
- **C14** Karyotyping
- **C15** Karyotyping—Genetic Disorders
- **C16** DNA Whodunit
- **C17** Analyzing Blood Serum to Determine Evolutionary Relationships
- **C18** Analyzing Blood Serum—Evolution of Primates
- **C19** Analyzing Amino-Acid Sequences to Determine Evolutionary Relationships
- **C20** Observing Animal Behavior
- **C21** Observing Animal Behavior—Grant Application
- **C22** Examining Owl Pellets
- **C23** Examining Owl Pellets—NW Vs. SE
- **C24** Mapping Biotic Factors in the Environment
- **C25** Assessing Abiotic Factors in the Environment
- **C26** Assessing and Mapping Factors in the Environment
- **C27** Studying an Algal Bloom
- **C28** Studying an Algal Bloom—Phosphate Pollution
- **C29** Classifying Mysterious Organisms
- **C30** Screening for Resistance to Tobacco Mosaic Virus
- **C31** Using Aseptic Technique
- **C32** Gram Staining of Bacteria
- **C33** Gram Staining of Bacteria—Treatment Options
- **C34** Limiting Fungal Growth
- **C35** Staining and Mounting Stem Cross Sections
- **C36** Using Paper Chromatography to Separate Pigments
- **C37** Growing Plants in the Laboratory
- **C38** Growing Plants in the Laboratory—Fertilizer Problem
- **C39** Response in the Fruit Fly
- **C40** Conducting a Bird Survey
- **C41** Evaluating Muscle Exhaustion
- **C42** Blood Typing
- **C43** Blood Typing—Whodunit?
- **C44** Blood Typing—Pregnancy and Hemolytic Disease
- **C45** Screening Sunscreens
- **C46** Identifying Food Nutrients
- **C47** Identifying Food Nutrients—Food Labeling
- **C48** Urinalysis Testing

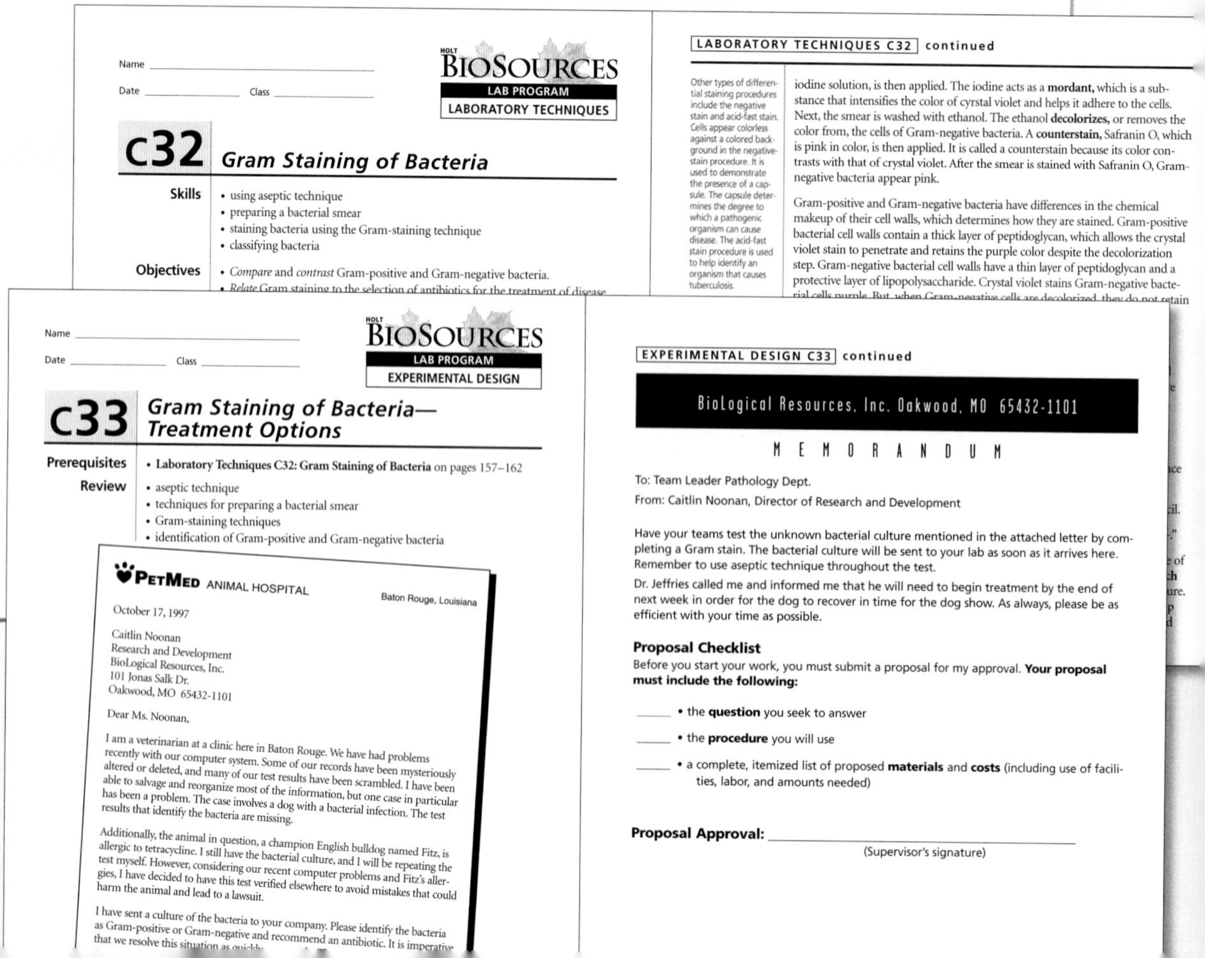

D Biotechnology

Biotechnology labs bring the latest in cutting-edge biotechnology research techniques into your classroom.

- **D1** Laboratory Techniques: Staining DNA and RNA
- **D2** Laboratory Techniques: Extracting DNA
- **D3** Laboratory Techniques: Genetic Transformation of Bacteria
- **D4** Experimental Design: Genetic Transformation—Antibiotic Resistance
- **D5** Laboratory Techniques: Introduction to Agarose Gel Electrophoresis
- **D6** Laboratory Techniques: DNA Fragment Analysis and Restriction Mapping
- **D7** Laboratory Techniques: DNA Ligation
- **D8** Experimental Design: Comparing DNA Samples
- **D9** Laboratory Techniques: Introduction to Fermentation
- **D10** Laboratory Techniques: Ice-Nucleating Bacteria
- **D11** Laboratory Techniques: Oil-Degrading Microbes
- **D12** Experimental Design: Can Oil-Degrading Microbes Save the Bay?

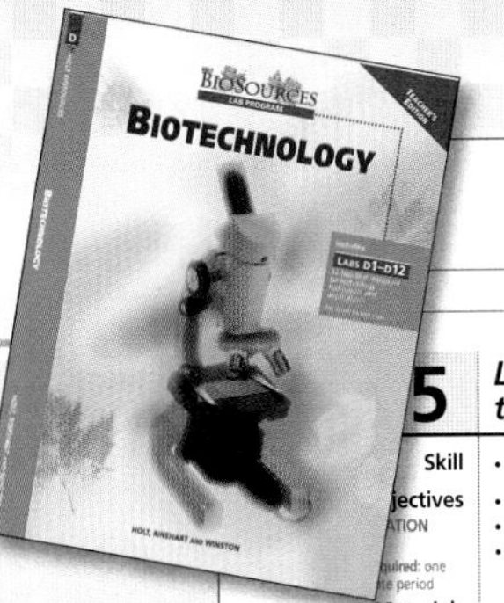

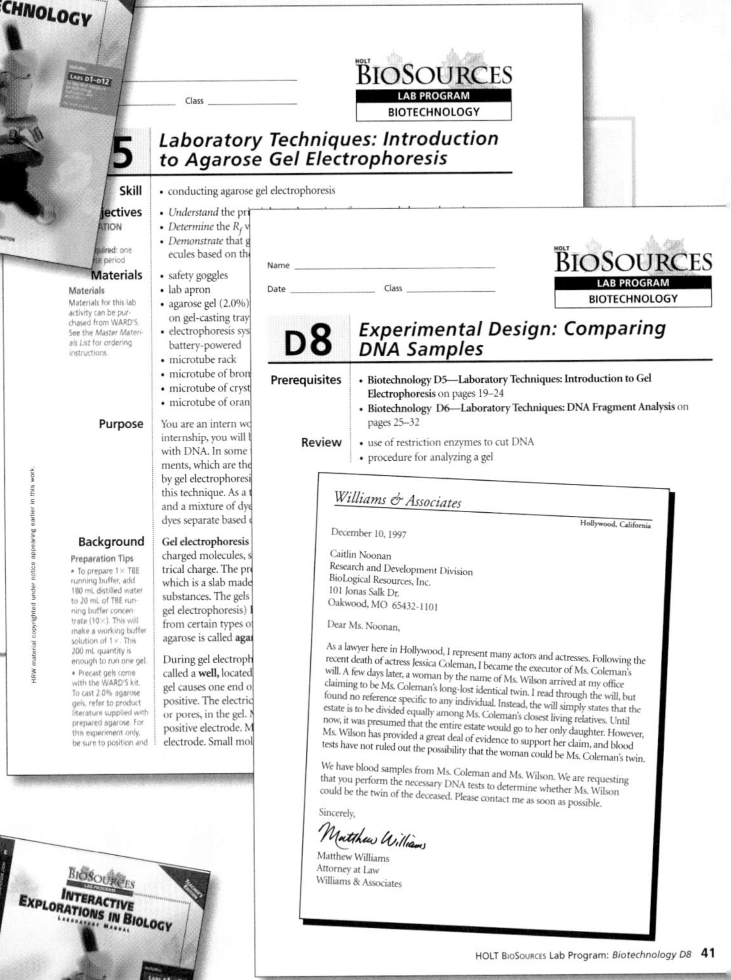

E Interactive Explorations in Biology

Interactive Explorations in Biology enable your students to study phenomena that are difficult or impossible to study in the laboratory. Using CD-ROM technology, students will manipulate variables, predict consequences, and study outcomes in ways never before possible.

- **E1** Cell Size
- **E2** Oxidative Respiration
- **E3** Thermodynamics
- **E4** Gene Regulation
- **E5** Heredity in Families
- **E6** Hemoglobin
- **E7** Diet and Weight Loss

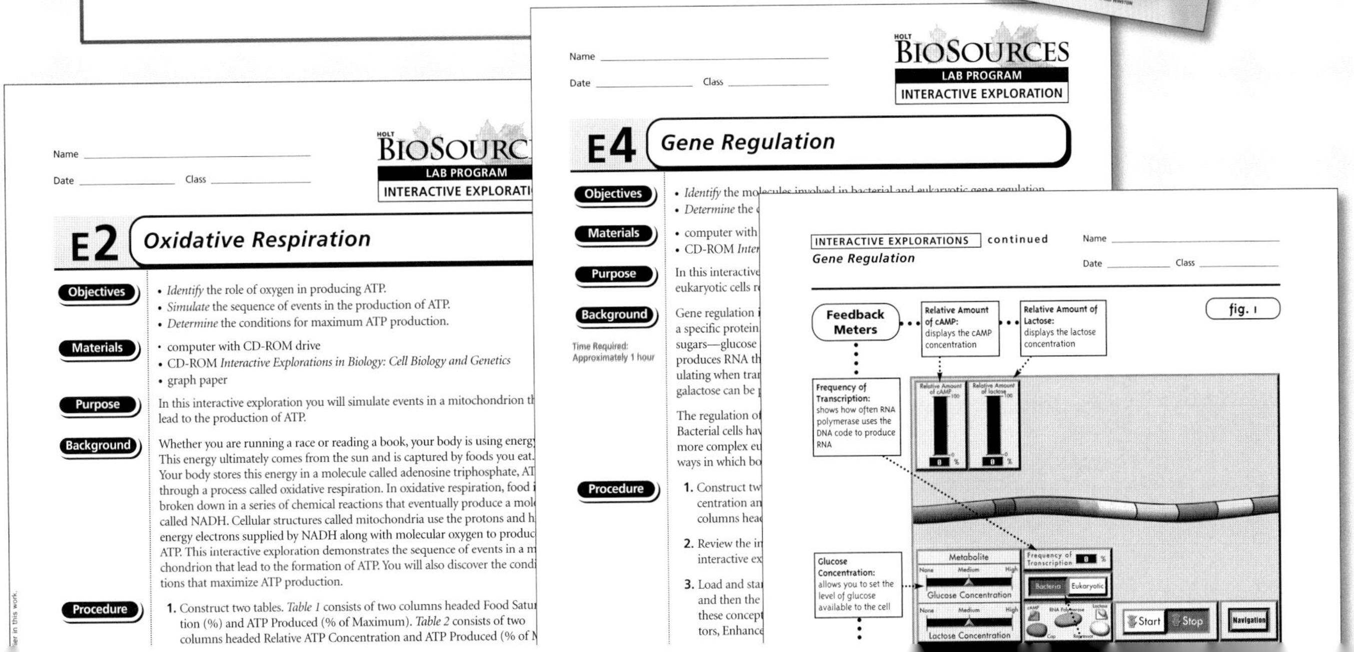

Internet Resources Help You Keep Up to Date

NSTA *sci*LINKS

NSTA *sci*LINKS are found at point of use in the margins of *Modern Biology*. All *sci*LINKS Web site topics are managed and monitored by National Science Teachers Association staff. Sites are selected by teachers for appropriate content and grade level. Many topics include "Good For Teachers," a list of sites that are especially helpful for teachers. Sites are continuously added and deleted from the system so that students don't end up at dead ends.

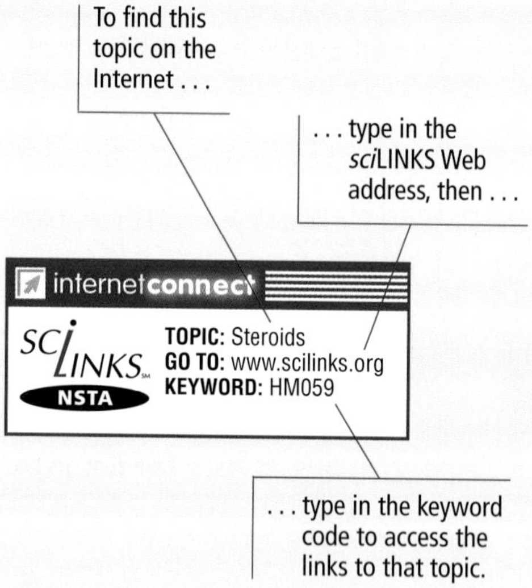

Smithsonian Institution Web resources

The Smithsonian Institution maintains special Web sites for use with *Modern Biology*. Visit **www.si.edu/hrw** for a complete listing of these resources. You can find interactive exhibits, classroom activities, and interviews with scientists, along with an interesting variety of application and extension topics.

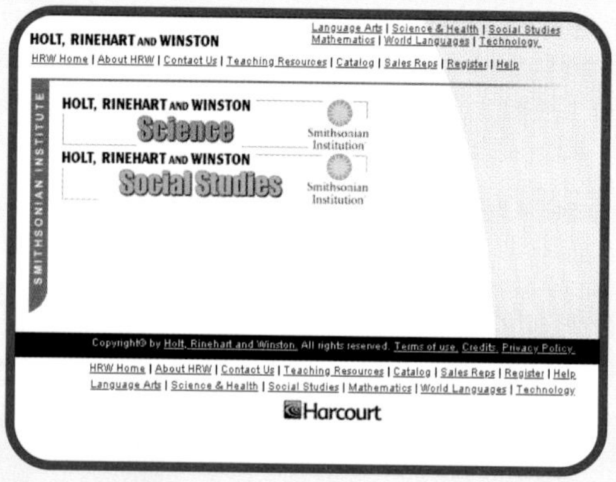

Up-to-the minute news and information from CNNfyi Web resources

Holt, Rinehart and Winston is proud to be a joint sponsor of CNNfyi, today's definitive "real world" site, created specifically for students in grades 6 though 12. Visit **www.cnnfyi.com** for illustrated articles on popular topics, links to related sites, lesson plans, education news, and homework help. Multimedia resources give students access to fascinating video archives of CNN news coverage, 360-degree images from around the globe, and audio programs on popular topics. Includes **CNN Newsroom,** a free instructional program featuring a new video every day on topics ranging from health to the environment—complete with related curriculum guides.

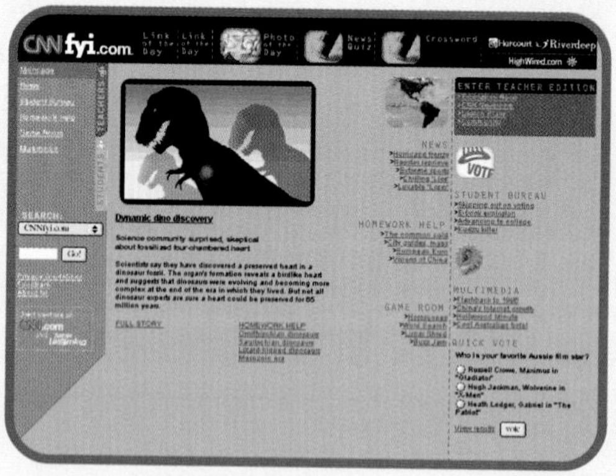

Holt, Rinehart and Winston Web materials for *Modern Biology*

Visit **www.go.hrw.com** for additional resources developed for the *Modern Biology* program. Look for additional articles, activities, and teaching suggestions.

One-Stop Planner CD-ROM Simplifies Planning and Assessment

The *Modern Biology* One-Stop Planner CD-ROM makes lesson planning and assessment simple, quick, and portable. Starting with a suggested lesson plan for each text section, you can see how all resources are hot-linked at appropriate points into that lesson plan. You can print each resource right from the CD-ROM.

For Macintosh®- and Windows®-compatible computers

Lesson Plans

Lesson plans are available as PDF files and as word-processing documents. Lesson formats are available in Microsoft Word, Microsoft Works, WordPerfect (Windows only), and ClarisWorks. Lesson Plans are easily changed to reflect your style and schedule and to match your own objectives. These changes are easily saved as your own files for convenient storage.

Pacing
Pacing suggestions consider block and traditional schedules.

Customization
Make your own custom lesson plan by selecting materials listed in the Other Resource Options section.

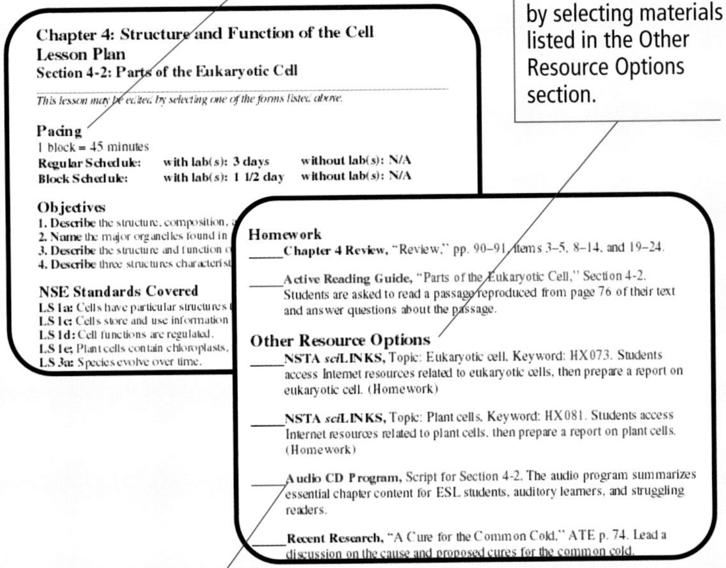

Chapter 4: Structure and Function of the Cell
Lesson Plan
Section 4-2: Parts of the Eukaryotic Cell

This lesson may be edited by selecting one of the forms listed above.

Pacing
1 block = 45 minutes
Regular Schedule: **with lab(s): 3 days** **without lab(s): N/A**
Block Schedule: **with lab(s): 1 1/2 day** **without lab(s): N/A**

Objectives
1. Describe the structure, composition, ...
2. Name the major organelles found in ...
3. Describe the structure and function ...
4. Describe three structures characteri...

NSE Standards Covered
LS 1a: Cells have particular structures ...
LS 1c: Cells store and use information ...
LS 1d: Cell functions are regulated.
LS 1e: Plant cells contain chloroplasts, ...
LS 3a: Species evolve over time.

Homework
_____**Chapter 4 Review,** "Review," pp. 90–91, items 3–5, 8–14, and 19–24.

_____**Active Reading Guide,** "Parts of the Eukaryotic Cell," Section 4-2. Students are asked to read a passage reproduced from page 76 of their text and answer questions about the passage.

Other Resource Options
_____**NSTA *sci*LINKS,** Topic: Eukaryotic cell. Keyword: HX073. Students access Internet resources related to eukaryotic cells, then prepare a report on eukaryotic cell. (Homework)

_____**NSTA *sci*LINKS,** Topic: Plant cells. Keyword: HX081. Students access Internet resources related to plant cells, then prepare a report on plant cells. (Homework)

_____**Audio CD Program,** Script for Section 4-2. The audio program summarizes essential chapter content for ESL students, auditory learners, and struggling readers.

_____**Recent Research,** "A Cure for the Common Cold," ATE p. 74. Lead a discussion on the cause and proposed cures for the common cold.

Hot Links
Titles that appear in blue are hot-linked to the actual resource. Click to view any resource or print any worksheet.

Holt BioSources Teaching Resources

Basic Skills Worksheets are just one of hundreds of additional Holt BioSources Teaching Resources worksheets. Other kinds of worksheets include:

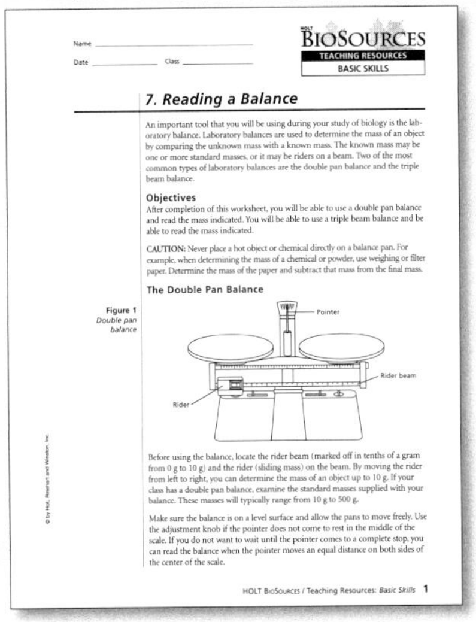

Name __________
Date __________ Class __________

HOLT BIOSOURCES TEACHING RESOURCES BASIC SKILLS

7. Reading a Balance

An important tool that you will be using during your study of biology is the laboratory balance. Laboratory balances are used to determine the mass of an object by comparing the unknown mass with a known mass. The known mass may be one or more standard masses, or it may be riders on a beam. Two of the most common types of laboratory balances are the double pan balance and the triple beam balance.

Objectives
After completion of this worksheet, you will be able to use a double pan balance and read the mass indicated. You will be able to use a triple beam balance and be able to read the mass indicated.

CAUTION: Never place a hot object or chemical directly on a balance pan. For example, when determining the mass of a chemical or powder, use weighing or filter paper. Determine the mass of the paper and subtract that mass from the final mass.

The Double Pan Balance

Figure 1 *Double pan balance*

Before using the balance, locate the rider beam (marked off in tenths of a gram from 0 g to 10 g) and the rider (sliding mass) on the beam. By moving the rider from left to right, you can determine the mass of an object up to 10 g. If your class has a double pan balance, examine the standard masses supplied with your balance. These masses will typically range from 10 g to 500 g.

Make sure the balance is on a level surface and allow the pans to move freely. Use the adjustment knob if the pointer does not come to rest in the middle of the scale. If you do not want to wait until the pointer comes to a complete stop, you can read the balance when the pointer moves an equal distance on both sides of the center of the scale.

© by Holt, Rinehart and Winston, Inc.

HOLT BIOSOURCES / Teaching Resources: *Basic Skills* 1

- Occupational Applications Worksheets
- Problem-Solving Worksheets
- Portfolio Projects
- Reading Strategies Worksheets
- Scoring Rubrics and Classroom Management Checklists
- Safety Contract and Quiz
- Supplemental Reading Guides
- Science Research Paper Worksheets
- Using Gowin's Vee in the Lab
- Answer Keys

Resources Include:

- *Modern Biology* Lesson Plans
- *Modern Biology* Active Reading Guide Worksheets
- *Modern Biology* Chapter Tests
- *Modern Biology* Study Guide
- *Modern Biology* Audio CD Program Scripts (English and Spanish)
- Test Generator Software with Test Item Files
- CNN Biology Connections Critical Thinking Worksheets
- Holt BioSources Lab Program (five laboratory manuals)
- Holt BioSources Teaching Transparencies
- Holt BioSources Teaching Resources

Fully Integrated Technology Options Make Biology Come Alive

For Macintosh®- and Windows®-compatible computers

Holt Biology Interactive Tutor

The power of CD-ROM technology lets you present 42 key topics that are often difficult to teach and are challenging for students to learn. Students can work through the topics at their own pace. Frequent evaluation lets students know if they are on the right track with diagnostic feedback.

UNIT

1 Cell Transport and Homeostasis
- Homeostasis and Diffusion
- Cell Membrane
- Osmosis
- Facilitated Diffusion
- Active Transport
- Active Uptake of Glucose

2 Photosynthesis
- An Overview
- Absorbing Light Energy
- Electron Transport Chain
- Making ATP
- Calvin Cycle
- Regulation

3 Cellular Respiration
- An Overview
- Glycolysis
- Krebs Cycle
- Electron Transport Chain
- Fermentation
- Regulation

4 Cell Reproduction
- Binary Fission
- Cell Cycle
- Chromosomes
- Mitosis
- Meiosis
- Gamete Production

5 Heredity
- Principles of Inheritance
- Determining Inheritance
- Monohybrid Crosses
- Dihybrid Crosses
- Other Patterns of Inheritance
- Pedigrees

6 Gene Expression
- Genes and DNA
- Replication—Copying DNA
- The Genetic Code
- Three Types of RNA
- Transcription—Making RNA
- Translation—Making Proteins

7 Ecosystem Dynamics
- Ecosystems
- Populations
- Succession
- Food Chains
- Food Webs
- Chemical Cycles

Each unit includes six topics.

Each topic contains:

- **BioStory** The unit's tutor, a working professional whose job requires an understanding of the unit's biological content, introduces the topic.
- **Learn** One or more short tutorials deliver the content step-by-step.
- **Try It** and **On Your Own** Interactive activities let students show what they have learned and provide self-assessment that lets them gauge their progress.
- **Review Options** and **Assessment** Review Questions prompt students to make sure they've grasped the content before continuing on to the next topic. Summary Notes list all of the topic's on-screen content summaries.

Also includes:

- **BioWorks**—careers in biology related to each unit
- **Did You Know?**—interesting facts related to each unit
- **Web links**—direct access to go.hrw.com and NSTA *sci*LINKS resources for each unit
- **BioResearch**—areas of ongoing research
- **Glossary**—audio text and definitions for more than 200 terms
- **Student and Teacher Resources**—worksheets and teaching resources in PDF format

Interactive Explorations in Biology

These 19 CD-ROM activities enable your students to observe challenging biological phenomena that are difficult or impossible to study in the laboratory.

For Macintosh®- and Windows®-compatible computers

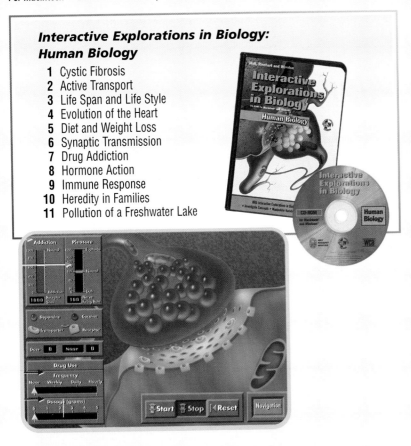

Interactive Explorations in Biology: Human Biology

1 Cystic Fibrosis
2 Active Transport
3 Life Span and Life Style
4 Evolution of the Heart
5 Diet and Weight Loss
6 Synaptic Transmission
7 Drug Addiction
8 Hormone Action
9 Immune Response
10 Heredity in Families
11 Pollution of a Freshwater Lake

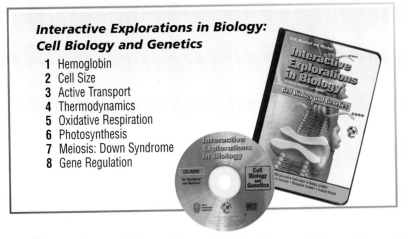

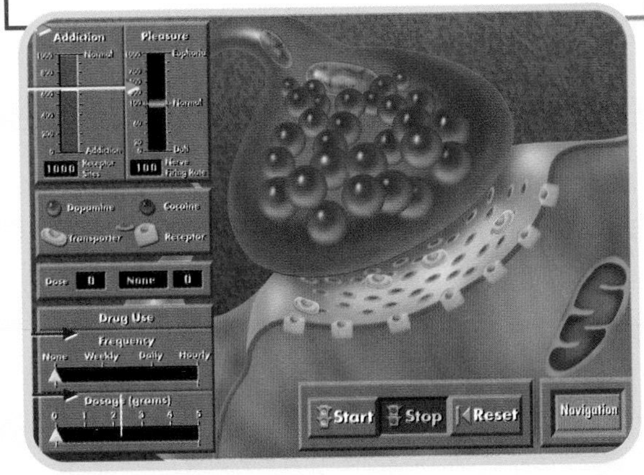

Interactive Explorations in Biology: Cell Biology and Genetics

1 Hemoglobin
2 Cell Size
3 Active Transport
4 Thermodynamics
5 Oxidative Respiration
6 Photosynthesis
7 Meiosis: Down Syndrome
8 Gene Regulation

CNN Videos

CNN Presents Science in the News: Biology Connections includes 35 video clips that show students the newsworthy nature of biology. Includes Teacher's Notes and Critical Thinking Worksheets for each segment

CNN®
Turner Le@rning®
A Time Warner Company

Holt Biology Videodiscs

Holt Biology Videodiscs offers you a comprehensive array of visuals tied directly to lessons found in *Modern Biology*. Includes four double-sided videodiscs containing thousands of images (live-action and animated videos, photographs, illustrations, and glossary terms) that reinforce material covered in the *Modern Biology* text. Also includes **Teacher's Correlation Guide to *Modern Biology*** and an **Image Directory**

Holt BioSources Teaching Transparencies

Teaching Transparencies contains 238 full-color images of key illustrations accompanied by 71 black-line masters. Includes *Transparency Directory with Teacher's Notes*

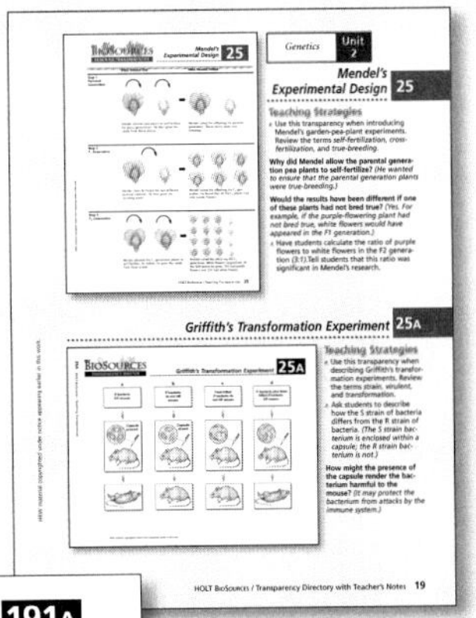

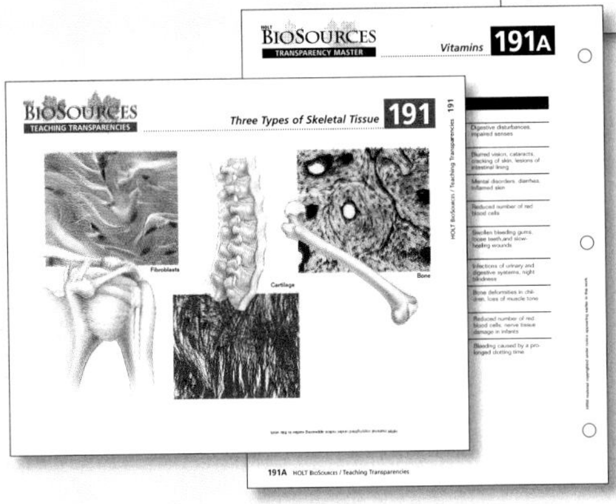

Your *Modern Biology* Teacher's Edition Helps You Plan Efficiently and Effectively

Teacher's Unit Interleaf

Your *Modern Biology* Teacher's Edition is organized to help you plan in advance—on a yearly, weekly, and daily basis. Unit Interleaves correlate each unit with the National Science Education Standards. Each Unit Interleaf also includes an article on a current trend in biology.

Connecting to the Standards correlates each unit with the *National Science Education Standards* for grades 9–12.

Trends in Biology provides you with information on current trends in biology.

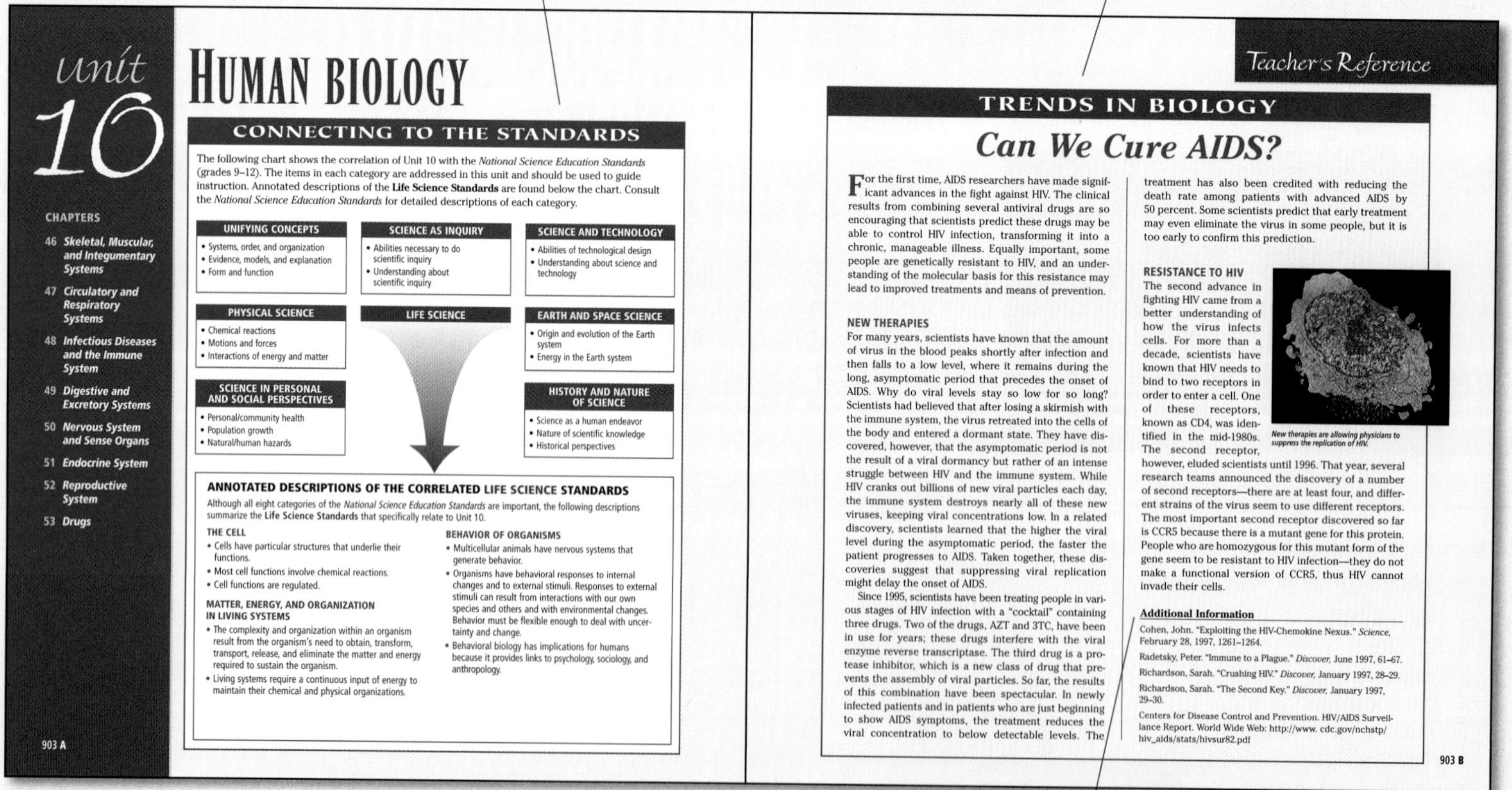

unit 10

CHAPTERS

46 Skeletal, Muscular, and Integumentary Systems
47 Circulatory and Respiratory Systems
48 Infectious Diseases and the Immune System
49 Digestive and Excretory Systems
50 Nervous System and Sense Organs
51 Endocrine System
52 Reproductive System
53 Drugs

903 A

HUMAN BIOLOGY

CONNECTING TO THE STANDARDS

The following chart shows the correlation of Unit 10 with the *National Science Education Standards* (grades 9–12). The items in each category are addressed in this unit and should be used to guide instruction. Annotated descriptions of the **Life Science Standards** are found below the chart. Consult the *National Science Education Standards* for detailed descriptions of each category.

UNIFYING CONCEPTS
- Systems, order, and organization
- Evidence, models, and explanation
- Form and function

SCIENCE AS INQUIRY
- Abilities necessary to do scientific inquiry
- Understanding about scientific inquiry

SCIENCE AND TECHNOLOGY
- Abilities of technological design
- Understanding about science and technology

PHYSICAL SCIENCE
- Chemical reactions
- Motions and forces
- Interactions of energy and matter

LIFE SCIENCE

EARTH AND SPACE SCIENCE
- Origin and evolution of the Earth system
- Energy in the Earth system

SCIENCE IN PERSONAL AND SOCIAL PERSPECTIVES
- Personal/community health
- Population growth
- Natural/human hazards

HISTORY AND NATURE OF SCIENCE
- Science as a human endeavor
- Nature of scientific knowledge
- Historical perspectives

ANNOTATED DESCRIPTIONS OF THE CORRELATED LIFE SCIENCE STANDARDS

Although all eight categories of the *National Science Education Standards* are important, the following descriptions summarize the **Life Science Standards** that specifically relate to Unit 10.

THE CELL
- Cells have particular structures that underlie their functions.
- Most cell functions involve chemical reactions.
- Cell functions are regulated.

MATTER, ENERGY, AND ORGANIZATION IN LIVING SYSTEMS
- The complexity and organization within an organism result from the organism's need to obtain, transform, transport, release, and eliminate the matter and energy required to sustain the organism.
- Living systems require a continuous input of energy to maintain their chemical and physical organizations.

BEHAVIOR OF ORGANISMS
- Multicellular animals have nervous systems that generate behavior.
- Organisms have behavioral responses to internal changes and to external stimuli. Responses to external stimuli can result from interactions with our own species and others and with environmental changes. Behavior must be flexible enough to deal with uncertainty and change.
- Behavioral biology has implications for humans because it provides links to psychology, sociology, and anthropology.

Teacher's Reference

TRENDS IN BIOLOGY

Can We Cure AIDS?

For the first time, AIDS researchers have made significant advances in the fight against HIV. The clinical results from combining several antiviral drugs are so encouraging that scientists predict these drugs may be able to control HIV infection, transforming it into a chronic, manageable illness. Equally important, some people are genetically resistant to HIV, and an understanding of the molecular basis for this resistance may lead to improved treatments and means of prevention.

NEW THERAPIES

For many years, scientists have known that the amount of virus in the blood peaks shortly after infection and then falls to a low level, where it remains during the long, asymptomatic period that precedes the onset of AIDS. Why do viral levels stay so low for so long? Scientists had believed that after losing a skirmish with the immune system, the virus retreated into the cells of the body and entered a dormant state. They have discovered, however, that the asymptomatic period is not the result of a viral dormancy but rather of an intense struggle between HIV and the immune system. While HIV cranks out billions of new viral particles each day, the immune system destroys nearly all of these new viruses, keeping viral concentrations low. In a related discovery, scientists learned that the higher the viral level during the asymptomatic period, the faster the patient progresses to AIDS. Taken together, these discoveries suggest that suppressing viral replication might delay the onset of AIDS.

Since 1995, scientists have been treating people in various stages of HIV infection with a "cocktail" containing three drugs. Two of the drugs, AZT and 3TC, have been in use for years; these drugs interfere with the viral enzyme reverse transcriptase. The third drug is a protease inhibitor, which is a new class of drug that prevents the assembly of viral particles. So far, the results of this combination have been spectacular. In newly infected patients and in patients who are just beginning to show AIDS symptoms, the treatment reduces the viral concentration to below detectable levels. The treatment has also been credited with reducing the death rate among patients with advanced AIDS by 50 percent. Some scientists predict that early treatment may even eliminate the virus in some people, but it is too early to confirm this prediction.

RESISTANCE TO HIV

The second advance in fighting HIV came from a better understanding of how the virus infects cells. For more than a decade, scientists have known that HIV needs to bind to two receptors in order to enter a cell. One of these receptors, known as CD4, was identified in the mid-1980s. The second receptor, however, eluded scientists until 1996. That year, several research teams announced the discovery of a number of second receptors—there are at least four, and different strains of the virus seem to use different receptors. The most important second receptor discovered so far is CCR5 because there is a mutant gene for this protein. People who are homozygous for this mutant form of the gene seem to be resistant to HIV infection—they do not make a functional version of CCR5, thus HIV cannot invade their cells.

New therapies are allowing physicians to suppress the replication of HIV.

Additional Information

Cohen, John. "Exploiting the HIV-Chemokine Nexus." *Science,* February 28, 1997, 1261–1264.

Radetsky, Peter. "Immune to a Plague." *Discover,* June 1997, 61–67.

Richardson, Sarah. "Crushing HIV." *Discover,* January 1997, 28–29.

Richardson, Sarah. "The Second Key." *Discover,* January 1997, 29–30.

Centers for Disease Control and Prevention. HIV/AIDS Surveillance Report. World Wide Web: http://www. cdc.gov/nchstp/hiv_aids/stats/hivsur82.pdf

903 B

Additional Information provides useful sources of information on the topic featured.

Chapter Planning Guides

Each Chapter Interleaf is a master plan for the integration of the many materials available. Whether you teach a traditional schedule or follow a block schedule, you know that an array of teaching strategies, classroom options, and assignment options is essential for sustaining student interest. Chapter Interleaves present options available from *Modern Biology* and *Holt BioSources* for use with each chapter.

Block Scheduling Guides help you keep students engaged throughout the longer class periods required by block scheduling.

Topics provides an outline for each section.

Teaching Resources are options to use when you are teaching.

Lab, Classwork, and Homework includes labs, Study Guide worksheets, and other options for reteaching, reinforcing, and assessing student mastery.

Reading for Content Mastery offers a reading tip followed by strategies and worksheets to help struggling readers.

Technology and Internet Resources gives you options for curriculum-based multimedia tools and for using the Web.

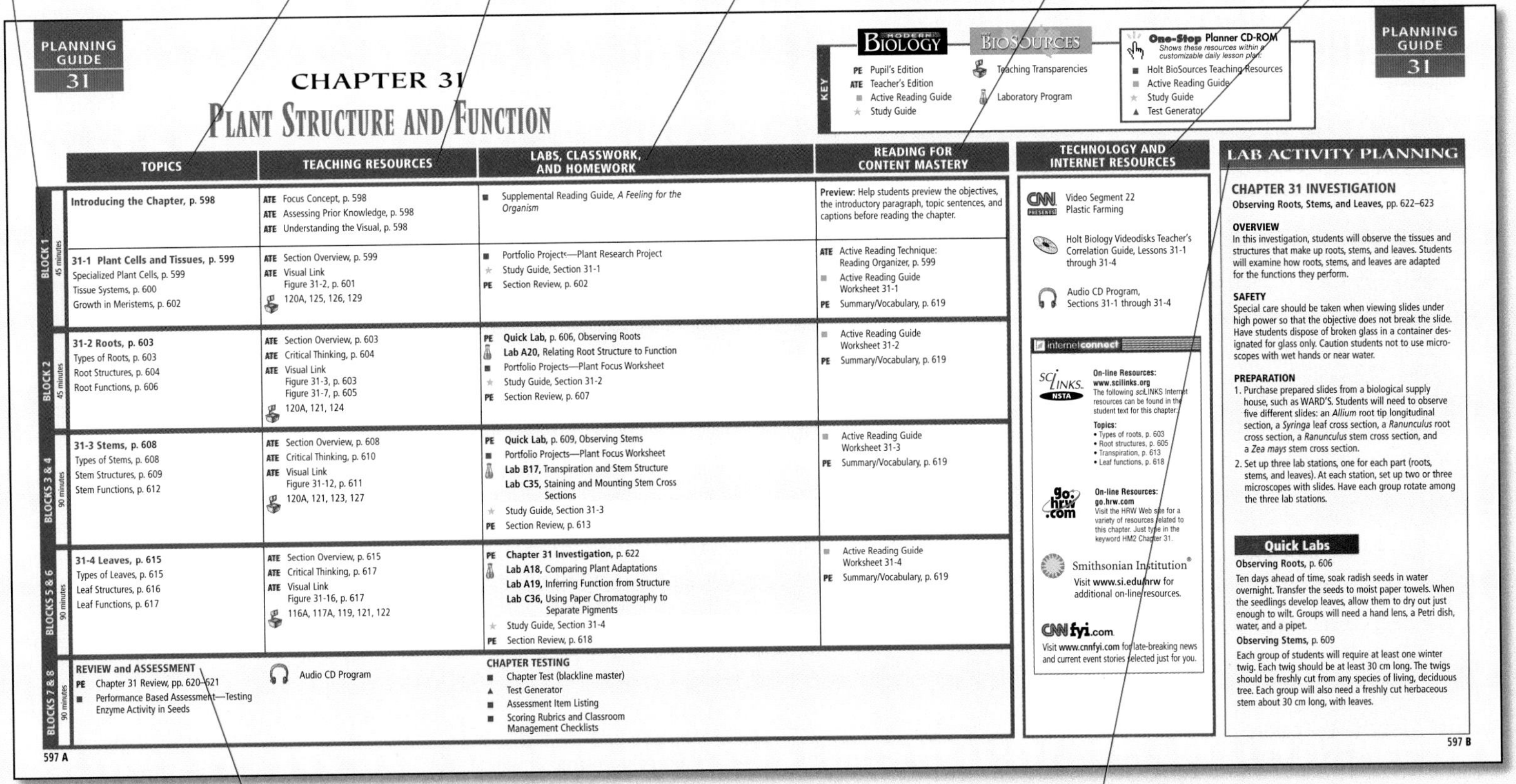

PLANNING GUIDE 31

CHAPTER 31
PLANT STRUCTURE AND FUNCTION

KEY
MODERN BIOLOGY: PE Pupil's Edition; ATE Teacher's Edition; ■ Active Reading Guide; ★ Study Guide
BIOSOURCES: Teaching Transparencies; Laboratory Program
One-Stop Planner CD-ROM — Shows these resources within a customizable daily lesson plan: ■ Holt BioSources Teaching Resources; ■ Active Reading Guide; ★ Study Guide; ▲ Test Generator

Block	Topics	Teaching Resources	Labs, Classwork, and Homework	Reading for Content Mastery
BLOCK 1 (45 minutes)	**Introducing the Chapter, p. 598**	ATE Focus Concept, p. 598 ATE Assessing Prior Knowledge, p. 598 ATE Understanding the Visual, p. 598	■ Supplemental Reading Guide, *A Feeling for the Organism*	**Preview:** Help students preview the objectives, the introductory paragraph, topic sentences, and captions before reading the chapter.
	31-1 Plant Cells and Tissues, p. 599 Specialized Plant Cells, p. 599 Tissue Systems, p. 600 Growth in Meristems, p. 602	ATE Section Overview, p. 599 ATE Visual Link Figure 31-2, p. 601 120A, 125, 126, 129	■ Portfolio Projects—Plant Research Project ★ Study Guide, Section 31-1 PE Section Review, p. 602	ATE Active Reading Technique: Reading Organizer, p. 599 ■ Active Reading Guide Worksheet 31-1 PE Summary/Vocabulary, p. 619
BLOCK 2 (45 minutes)	**31-2 Roots, p. 603** Types of Roots, p. 603 Root Structures, p. 604 Root Functions, p. 606	ATE Section Overview, p. 603 ATE Critical Thinking, p. 604 ATE Visual Link Figure 31-3, p. 603; Figure 31-7, p. 605 120A, 121, 124	PE **Quick Lab,** p. 606, Observing Roots **Lab A20,** Relating Root Structure to Function ■ Portfolio Projects—Plant Focus Worksheet ★ Study Guide, Section 31-2 PE Section Review, p. 607	■ Active Reading Guide Worksheet 31-2 PE Summary/Vocabulary, p. 619
BLOCKS 3 & 4 (90 minutes)	**31-3 Stems, p. 608** Types of Stems, p. 608 Stem Structures, p. 609 Stem Functions, p. 612	ATE Section Overview, p. 608 ATE Critical Thinking, p. 610 ATE Visual Link Figure 31-12, p. 611 120A, 121, 123, 127	PE **Quick Lab,** p. 609, Observing Stems ■ Portfolio Projects—Plant Focus Worksheet **Lab B17,** Transpiration and Stem Structure **Lab C35,** Staining and Mounting Stem Cross Sections ★ Study Guide, Section 31-3 PE Section Review, p. 613	■ Active Reading Guide Worksheet 31-3 PE Summary/Vocabulary, p. 619
BLOCKS 5 & 6 (90 minutes)	**31-4 Leaves, p. 615** Types of Leaves, p. 615 Leaf Structures, p. 616 Leaf Functions, p. 617	ATE Section Overview, p. 615 ATE Critical Thinking, p. 617 ATE Visual Link Figure 31-16, p. 617 116A, 117A, 119, 121, 122	PE **Chapter 31 Investigation,** p. 622 **Lab A18,** Comparing Plant Adaptations **Lab A19,** Inferring Function from Structure **Lab C36,** Using Paper Chromatography to Separate Pigments ★ Study Guide, Section 31-4 PE Section Review, p. 618	■ Active Reading Guide Worksheet 31-4 PE Summary/Vocabulary, p. 619
BLOCKS 7 & 8 (90 minutes)	**REVIEW and ASSESSMENT** PE Chapter 31 Review, pp. 620–621 ■ Performance Based Assessment—Testing Enzyme Activity in Seeds	Audio CD Program	**CHAPTER TESTING** ■ Chapter Test (blackline master) ▲ Test Generator ■ Assessment Item Listing ■ Scoring Rubrics and Classroom Management Checklists	

597 A

TECHNOLOGY AND INTERNET RESOURCES

CNN Presents: Video Segment 22 Plastic Farming

Holt Biology Videodisks Teacher's Correlation Guide, Lessons 31-1 through 31-4

Audio CD Program, Sections 31-1 through 31-4

internetconnect

sciLINKS NSTA — **On-line Resources: www.scilinks.org** The following sciLINKS Internet resources can be found in the student text for this chapter:
Topics:
- Types of roots, p. 603
- Root structures, p. 605
- Transpiration, p. 613
- Leaf functions, p. 618

go.hrw.com — **On-line Resources: go.hrw.com** Visit the HRW Web site for a variety of resources related to this chapter. Just type in the keyword HM2 Chapter 31.

Smithsonian Institution® — Visit **www.si.edu/hrw** for additional on-line resources.

CNNfyi.com — Visit **www.cnnfyi.com** for late-breaking news and current event stories selected just for you.

PLANNING GUIDE 31

LAB ACTIVITY PLANNING

CHAPTER 31 INVESTIGATION
Observing Roots, Stems, and Leaves, pp. 622–623

OVERVIEW
In this investigation, students will observe the tissues and structures that make up roots, stems, and leaves. Students will examine how roots, stems, and leaves are adapted for the functions they perform.

SAFETY
Special care should be taken when viewing slides under high power so that the objective does not break the slide. Have students dispose of broken glass in a container designated for glass only. Caution students not to use microscopes with wet hands or near water.

PREPARATION
1. Purchase prepared slides from a biological supply house, such as WARD'S. Students will need to observe five different slides: an *Allium* root tip longitudinal section, a *Syringa* leaf cross section, a *Ranunculus* root cross section, a *Ranunculus* stem cross section, and a *Zea mays* stem cross section.
2. Set up three lab stations, one for each part (roots, stems, and leaves). At each station, set up two or three microscopes with slides. Have each group rotate among the three lab stations.

Quick Labs

Observing Roots, p. 606
Ten days ahead of time, soak radish seeds in water overnight. Transfer the seeds to moist paper towels. When the seedlings develop leaves, allow them to dry out just enough to wilt. Groups will need a hand lens, a Petri dish, water, and a pipet.

Observing Stems, p. 609
Each group of students will require at least one winter twig. Each twig should be at least 30 cm long. The twigs should be freshly cut from any species of living, deciduous tree. Each group will also need a freshly cut herbaceous stem about 30 cm long, with leaves.

597 B

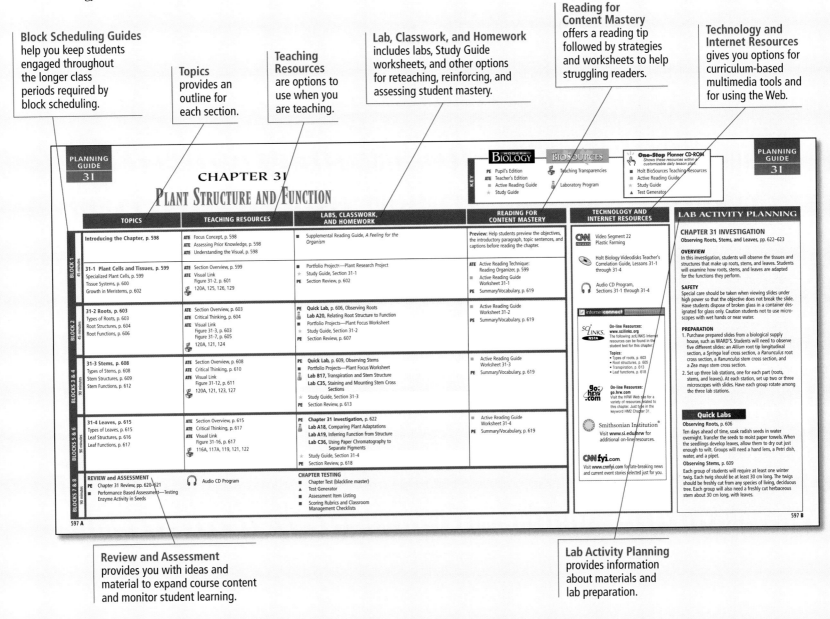

Review and Assessment provides you with ideas and material to expand course content and monitor student learning.

Lab Activity Planning provides information about materials and lab preparation.

Annotated Sidebars Provide Teaching Strategies, Reinforcement, and Extensions

Understanding the Visual provides additional information about the opening photograph that can be used to begin discussing the chapter content.

Focus Concept links the main ideas of the chapter to one of the biological themes that run throughout the text

Section Overview summarizes the main ideas of each section and relates them to the biological themes that recur throughout the textbook.

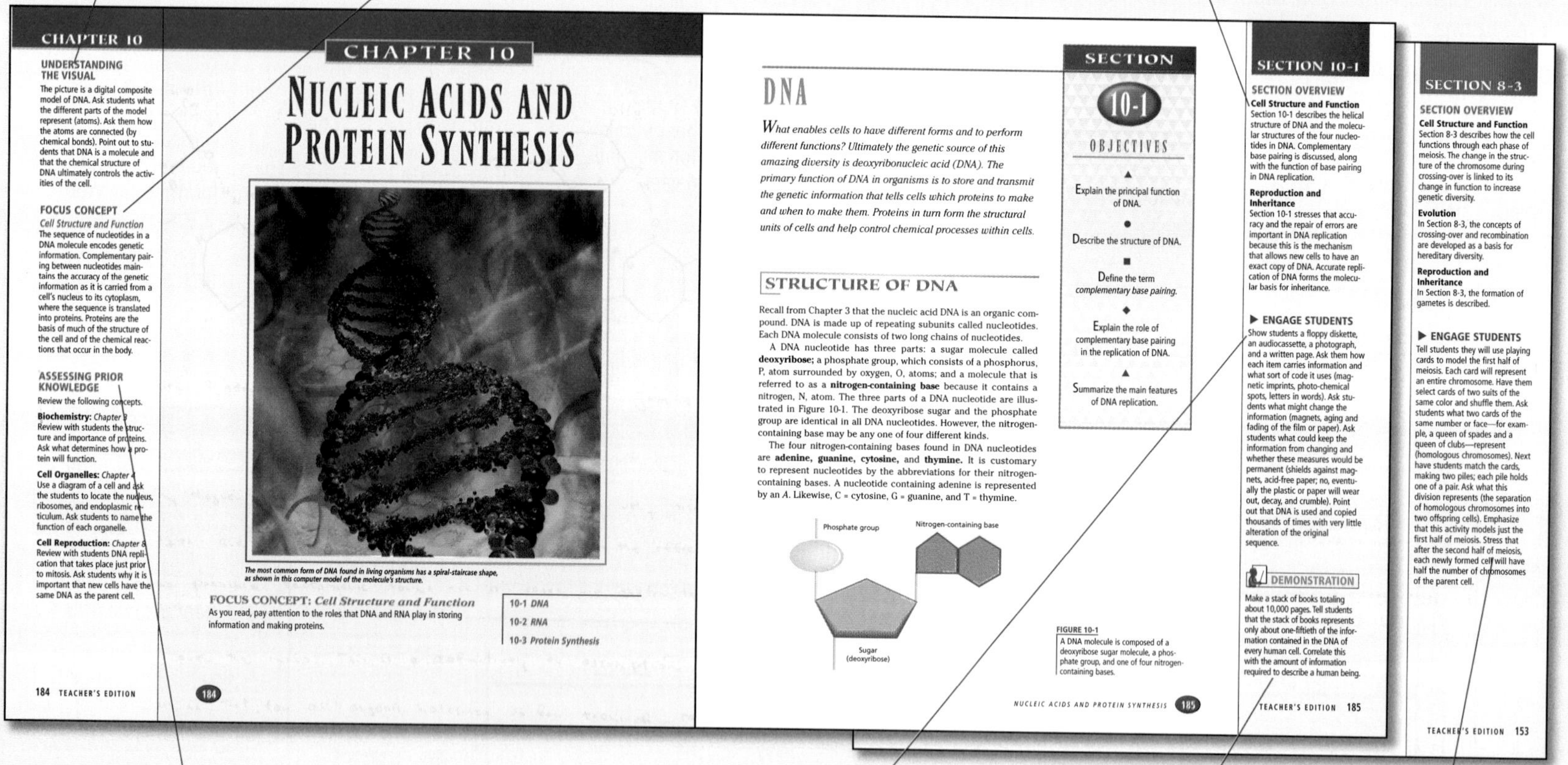

CHAPTER 10

UNDERSTANDING THE VISUAL

The picture is a digital composite model of DNA. Ask students what the different parts of the model represent (atoms). Ask them how the atoms are connected (by chemical bonds). Point out to students that DNA is a molecule and that the chemical structure of DNA ultimately controls the activities of the cell.

FOCUS CONCEPT

Cell Structure and Function

The sequence of nucleotides in a DNA molecule encodes genetic information. Complementary pairing between nucleotides maintains the accuracy of the genetic information as it is carried from a cell's nucleus to its cytoplasm, where the sequence is translated into proteins. Proteins are the basis of much of the structure of the cell and of the chemical reactions that occur in the body.

ASSESSING PRIOR KNOWLEDGE

Review the following concepts.

Biochemistry: *Chapter 3*
Review with students the structure and importance of proteins. Ask what determines how a protein will function.

Cell Organelles: *Chapter 4*
Use a diagram of a cell and ask the students to locate the nucleus, ribosomes, and endoplasmic reticulum. Ask students to name the function of each organelle.

Cell Reproduction: *Chapter 8*
Review with students DNA replication that takes place just prior to mitosis. Ask students why it is important that new cells have the same DNA as the parent cell.

184 TEACHER'S EDITION

CHAPTER 10

NUCLEIC ACIDS AND PROTEIN SYNTHESIS

The most common form of DNA found in living organisms has a spiral-staircase shape, as shown in this computer model of the molecule's structure.

FOCUS CONCEPT: *Cell Structure and Function*

As you read, pay attention to the roles that DNA and RNA play in storing information and making proteins.

10-1 *DNA*

10-2 *RNA*

10-3 *Protein Synthesis*

184

DNA

What enables cells to have different forms and to perform different functions? Ultimately the genetic source of this amazing diversity is deoxyribonucleic acid (DNA). The primary function of DNA in organisms is to store and transmit the genetic information that tells cells which proteins to make and when to make them. Proteins in turn form the structural units of cells and help control chemical processes within cells.

STRUCTURE OF DNA

Recall from Chapter 3 that the nucleic acid DNA is an organic compound. DNA is made up of repeating subunits called nucleotides. Each DNA molecule consists of two long chains of nucleotides.

A DNA nucleotide has three parts: a sugar molecule called **deoxyribose;** a phosphate group, which consists of a phosphorus, P, atom surrounded by oxygen, O, atoms; and a molecule that is referred to as a **nitrogen-containing base** because it contains a nitrogen, N, atom. The three parts of a DNA nucleotide are illustrated in Figure 10-1. The deoxyribose sugar and the phosphate group are identical in all DNA nucleotides. However, the nitrogen-containing base may be any one of four different kinds.

The four nitrogen-containing bases found in DNA nucleotides are **adenine, guanine, cytosine,** and **thymine.** It is customary to represent nucleotides by the abbreviations for their nitrogen-containing bases. A nucleotide containing adenine is represented by an *A*. Likewise, C = cytosine, G = guanine, and T = thymine.

Phosphate group
Nitrogen-containing base
Sugar (deoxyribose)

FIGURE 10-1
A DNA molecule is composed of a deoxyribose sugar molecule, a phosphate group, and one of four nitrogen-containing bases.

SECTION 10-1

OBJECTIVES

▲ Explain the principal function of DNA.

● Describe the structure of DNA.

■ Define the term *complementary base pairing.*

◆ Explain the role of complementary base pairing in the replication of DNA.

▲ Summarize the main features of DNA replication.

NUCLEIC ACIDS AND PROTEIN SYNTHESIS 185

SECTION 10-1

SECTION OVERVIEW

Cell Structure and Function
Section 10-1 describes the helical structure of DNA and the molecular structures of the four nucleotides in DNA. Complementary base pairing is discussed, along with the function of base pairing in DNA replication.

Reproduction and Inheritance
Section 10-1 stresses that accuracy and the repair of errors are important in DNA replication because this is the mechanism that allows new cells to have an exact copy of DNA. Accurate replication of DNA forms the molecular basis for inheritance.

▶ ENGAGE STUDENTS

Show students a floppy diskette, an audiocassette, a photograph, and a written page. Ask them how each item carries information and what sort of code it uses (magnetic imprints, photo-chemical spots, letters in words). Ask students what might change the information (magnets, aging and fading of the film or paper). Ask students what could keep the information from changing and whether these measures would be permanent (shields against magnets, acid-free paper; no, eventually the plastic or paper will wear out, decay, and crumble). Point out that DNA is used and copied thousands of times with very little alteration of the original sequence.

DEMONSTRATION

Make a stack of books totaling about 10,000 pages. Tell students that the stack of books represents only about one-fiftieth of the information contained in the DNA of every human cell. Correlate this with the amount of information required to describe a human being.

TEACHER'S EDITION 185

SECTION 8-3

SECTION OVERVIEW

Cell Structure and Function
Section 8-3 describes how the cell functions through each phase of meiosis. The change in the structure of the chromosome during crossing-over is linked to its change in function to increase genetic diversity.

Evolution
In Section 8-3, the concepts of crossing-over and recombination are developed as a basis for hereditary diversity.

Reproduction and Inheritance
In Section 8-3, the formation of gametes is described.

▶ ENGAGE STUDENTS

Tell students they will use playing cards to model the first half of meiosis. Each card will represent an entire chromosome. Have them select cards of two suits of the same color and shuffle them. Ask students what two cards of the same number or face—for example, a queen of spades and a queen of clubs—represent (homologous chromosomes). Next have students match the cards, making two piles; each pile holds one of a pair. Ask what this division represents (the separation of homologous chromosomes into two offspring cells). Emphasize that this activity models just the first half of meiosis. Stress that after the second half of meiosis, each newly formed cell will have half the number of chromosomes of the parent cell.

TEACHER'S EDITION 153

Assessing Prior Knowledge helps you assess how much your students know—and what misconceptions they may have—before you begin teaching.

Active Reading uses the following strategies to help struggling or unmotivated readers:
- Brainstorming
- K-W-L
- L.I.N.K.
- Paired Reading
- Reading Organizer
- Paired Summarizing
- Discussion
- Anticipation Guide
- Reader Response Logs
- Reading Effectively

Demonstrations present hands-on demonstrations that pertain to content on the pupil's page. Complete instructions, materials, and quantities are provided.

Engage Students suggests activities to start each lesson and help you prepare and motivate students to learn about the subjects covered in the lesson.

Critical Thinking presents high-level questions that prompt students to think critically about the content of the lesson.

Visual Links provide teaching ideas and review questions that directly relate to a figure or a table on the pupil's page.

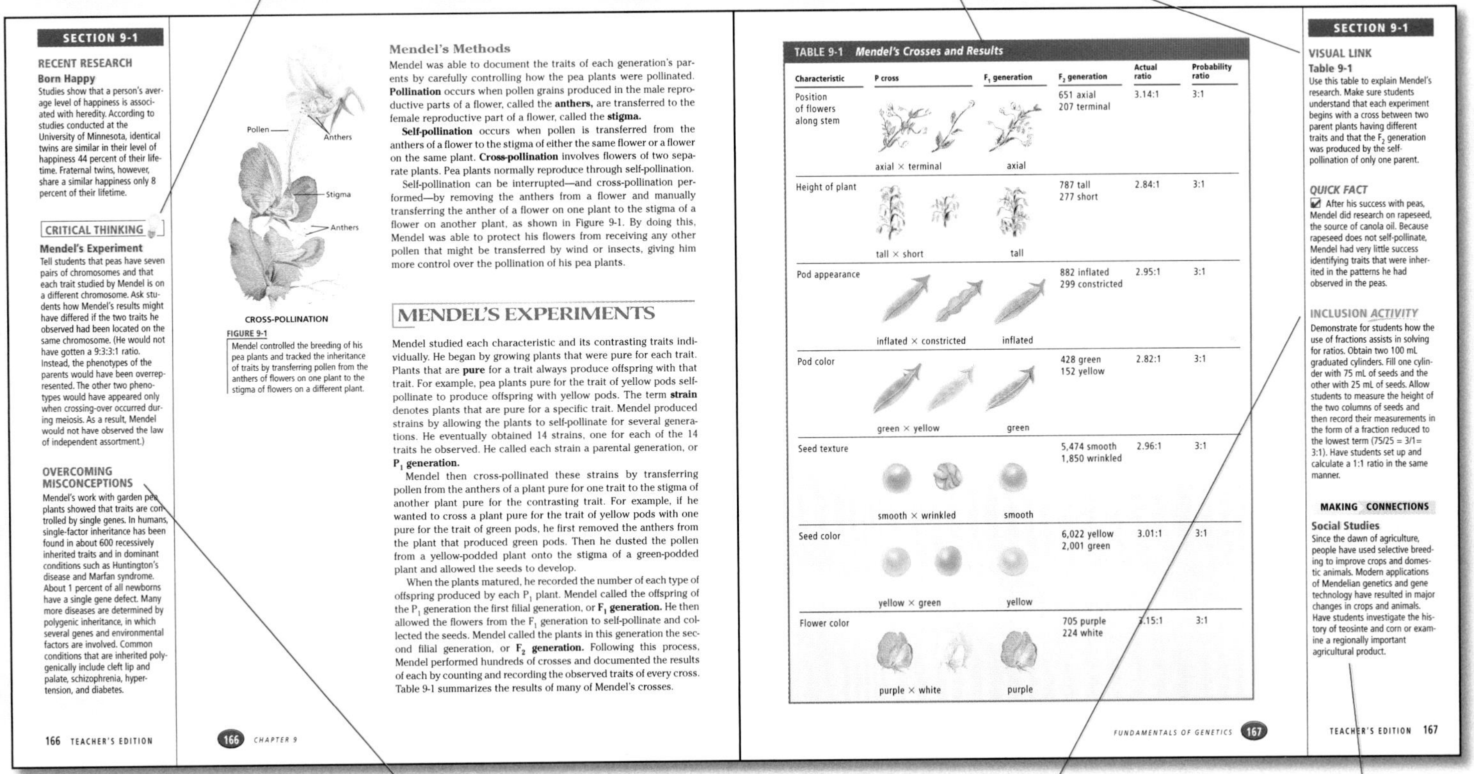

SECTION 9-1

RECENT RESEARCH

Born Happy

Studies show that a person's average level of happiness is associated with heredity. According to studies conducted at the University of Minnesota, identical twins are similar in their level of happiness 44 percent of their lifetime. Fraternal twins, however, share a similar happiness only 8 percent of their lifetime.

CRITICAL THINKING

Mendel's Experiment

Tell students that peas have seven pairs of chromosomes and that each trait studied by Mendel is on a different chromosome. Ask students how Mendel's results might have differed if the two traits he observed had been located on the same chromosome. (He would not have gotten a 9:3:3:1 ratio. Instead, the phenotypes of the parents would have been overrepresented. The other two phenotypes would have appeared only when crossing-over occurred during meiosis. As a result, Mendel would not have observed the law of independent assortment.)

OVERCOMING MISCONCEPTIONS

Mendel's work with garden pea plants showed that traits are controlled by single genes. In humans, single-factor inheritance has been found in about 600 recessively inherited traits and in dominant conditions such as Huntington's disease and Marfan syndrome. About 1 percent of all newborns have a single gene defect. Many more diseases are determined by polygenic inheritance, in which several genes and environmental factors are involved. Common conditions that are inherited polygenically include cleft lip and palate, schizophrenia, hypertension, and diabetes.

166 TEACHER'S EDITION

CROSS-POLLINATION

FIGURE 9-1

Mendel controlled the breeding of his pea plants and tracked the inheritance of traits by transferring pollen from the anthers of flowers on one plant to the stigma of flowers on a different plant.

Mendel's Methods

Mendel was able to document the traits of each generation's parents by carefully controlling how the pea plants were pollinated. **Pollination** occurs when pollen grains produced in the male reproductive parts of a flower, called the **anthers,** are transferred to the female reproductive part of a flower, called the **stigma.**

Self-pollination occurs when pollen is transferred from the anthers of a flower to the stigma of either the same flower or a flower on the same plant. **Cross-pollination** involves flowers of two separate plants. Pea plants normally reproduce through self-pollination.

Self-pollination can be interrupted—and cross-pollination performed—by removing the anthers from a flower and manually transferring the anther of a flower on one plant to the stigma of a flower on another plant, as shown in Figure 9-1. By doing this, Mendel was able to protect his flowers from receiving any other pollen that might be transferred by wind or insects, giving him more control over the pollination of his pea plants.

MENDEL'S EXPERIMENTS

Mendel studied each characteristic and its contrasting traits individually. He began by growing plants that were pure for each trait. Plants that are **pure** for a trait always produce offspring with that trait. For example, pea plants pure for the trait of yellow pods self-pollinate to produce offspring with yellow pods. The term **strain** denotes plants that are pure for a specific trait. Mendel produced strains by allowing the plants to self-pollinate for several generations. He eventually obtained 14 strains, one for each of the 14 traits he observed. He called each strain a parental generation, or **P_1 generation.**

Mendel then cross-pollinated these strains by transferring pollen from the anthers of a plant pure for one trait to the stigma of another plant pure for the contrasting trait. For example, if he wanted to cross a plant pure for the trait of yellow pods with one pure for the trait of green pods, he first removed the anthers from the plant that produced green pods. Then he dusted the pollen from a yellow-podded plant onto the stigma of a green-podded plant and allowed the seeds to develop.

When the plants matured, he recorded the number of each type of offspring produced by each P_1 plant. Mendel called the offspring of the P_1 generation the first filial generation, or **F_1 generation.** He then allowed the flowers from the F_1 generation to self-pollinate and collected the seeds. Mendel called the plants in this generation the second filial generation, or **F_2 generation.** Following this process, Mendel performed hundreds of crosses and documented the results of each by counting and recording the observed traits of every cross. Table 9-1 summarizes the results of many of Mendel's crosses.

166 CHAPTER 9

TABLE 9-1 *Mendel's Crosses and Results*

Characteristic	P cross	F_1 generation	F_2 generation	Actual ratio	Probability ratio
Position of flowers along stem	axial × terminal	axial	651 axial 207 terminal	3.14:1	3:1
Height of plant	tall × short	tall	787 tall 277 short	2.84:1	3:1
Pod appearance	inflated × constricted	inflated	882 inflated 299 constricted	2.95:1	3:1
Pod color	green × yellow	green	428 green 152 yellow	2.82:1	3:1
Seed texture	smooth × wrinkled	smooth	5,474 smooth 1,850 wrinkled	2.96:1	3:1
Seed color	yellow × green	yellow	6,022 yellow 2,001 green	3.01:1	3:1
Flower color	purple × white	purple	705 purple 224 white	3.15:1	3:1

FUNDAMENTALS OF GENETICS 167

SECTION 9-1

VISUAL LINK

Table 9-1

Use this table to explain Mendel's research. Make sure students understand that each experiment begins with a cross between two parent plants having different traits and that the F_2 generation was produced by the self-pollination of only one parent.

QUICK FACT

After his success with peas, Mendel did research on rapeseed, the source of canola oil. Because rapeseed does not self-pollinate, Mendel had very little success identifying traits that were inherited in the patterns he had observed in the peas.

INCLUSION *ACTIVITY*

Demonstrate for students how the use of fractions assists in solving for ratios. Obtain two 100 mL graduated cylinders. Fill one cylinder with 75 mL of seeds and the other with 25 mL of seeds. Allow students to measure the height of the two columns of seeds and then record their measurements in the form of a fraction reduced to the lowest term (75/25 = 3/1= 3:1). Have students set up and calculate a 1:1 ratio in the same manner.

MAKING CONNECTIONS

Social Studies

Since the dawn of agriculture, people have used selective breeding to improve crops and domestic animals. Modern applications of Mendelian genetics and gene technology have resulted in major changes in crops and animals. Have students investigate the history of teosinte and corn or examine a regionally important agricultural product.

TEACHER'S EDITION 167

Overcoming Misconceptions corrects common errors or misconceptions that students may make about biology.

Inclusion Activities offer suggestions for modifying materials and alternative ways to teach the main concepts.

Making Connections connects the content to mathematics, social studies, language arts, fine arts, or technology.

Annotated Sidebars Are Full of Teaching Ideas

Terminology Note provides clarification or additional information on the biological terminology used in *Modern Biology.*

Recent Research summarizes recent biological research that relates to the chapter's topics.

Teaching Strategies are strategies that help you teach the concepts presented in the lesson.

Cultural Connections present relevant and interesting information about people of various cultures who are associated with ideas in the text, or discuss the influence of culture on a biological issue related to the text.

SECTION 19-2

CRITICAL THINKING

Tolerance

Ask students to list ways that people can change their tolerance to physical factors. (Answers may include references to some drugs—increasingly higher doses needed to get the same effect; temperatures—people living in warmer climates tolerate heat more easily than individuals coming from cold climates; elevation—people living at higher altitudes have more lung capacity for obtaining oxygen.) Ask them if this is the same as acclimation. (Yes.)

TERMINOLOGY NOTE

An organism can be both a conformer and a regulator. A fish may be a conformer of environmental temperature but a regulator of internal salt concentration. Do not allow your students to equate *conformer* with *ectotherm* or *regulator* with *endotherm.* These terms are used in different contexts.

RECENT RESEARCH

Acclimation

New studies suggest that there is a downside to an organism's ability to acclimate. Results of these new studies call into question previous laboratory findings that low-level metal contamination makes fish more tolerant of larger doses of metal contamination. David O. Norris, of the University of Colorado, and John D. Woodling, of the Colorado Division of Wildlife, report that brown trout were put in cages for 96 hours in an area so poisoned by mining effluent that no fish lived there. Fish that had resided in water with mining effluent survived only 28 hours, while fish from cleaner waters survived almost 48 hours.

370 TEACHER'S EDITION

Acclimation to Temperature

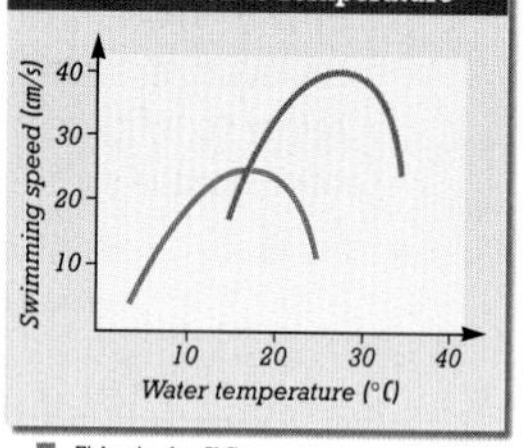

FIGURE 19-9

Fish raised at 25°C are acclimated to higher temperatures ... temperatu

Acclimation

Some organisms can adjust their tolerance to abiotic factors through the process of **acclimation** (AK-luh-MAY-shuhn). For instance, goldfish raised at different temperatures have somewhat different tolerance curves, as shown in Figure 19-9. If you spend a few weeks at a high elevation, you will acclimate to reduced oxygen levels, or "thin air." Over time, the number of red blood cells in your body will increase, thereby increasing the amount of oxygen your blood can carry. Be sure not to confuse acclimation with adaptation. Acclimation occurs within the lifetime of an individual organism. Adaptation is genetic change in a species or population that occurs over many generations.

Control of Internal Conditions

Environments fluctuate in temperature, light, moisture, salinity, and other chemical factors. There are two ways for organ-

FIGURE 19-

An organi... both. The ... conformer ... whereas h... late intern

Body ...

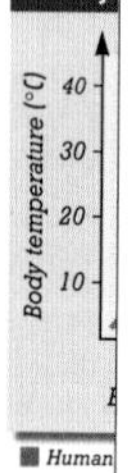

370 CHA...

ORGANS AND ORGAN SYSTEMS

An organ consists of various tissues that work together to carry out a specific function. The stomach, a saclike organ in which food is mixed with digestive enzymes, is composed of the four types of tissues. A single organ, such as the stomach, usually does not function in isolation. Rather, groups of organs interact in an organ system. For example, in the digestive system, the stomach, small intestine, liver, and pancreas all work together to break down food into molecules the body can use for energy. Table 46-1 lists the body's organ systems and names their major structures and functions. As you study the table, think about the ways in which the different organ systems work together to function in an efficient, integrated manner.

TABLE 46-1 *Summary of Organ Systems*

System	Major structures	Functions
Skeletal	bones	provides structure; supports and protects internal organs
Muscular	muscles (skeletal, cardiac, and smooth)	provides structure; supports and moves trunk and limbs; moves substances through body
Integumentary	skin, hair, nails	protects against pathogens; helps regulate body temperature
Circulatory	heart, blood vessels, blood	transports nutrients and wastes to and from all body tissues
Respiratory	air passages, lungs	carries air into and out of lungs, where gases (oxygen and carbon dioxide) are exchanged
Immune	lymph nodes and vessels, white blood cells	provides protection against infection and disease
Digestive	mouth, esophagus, stomach, liver, pancreas, small and large intestines	stores and digests food; absorbs nutrients; eliminates waste
Excretory	kidneys, ureters, bladder, urethra, skin, lungs	eliminates waste; maintains water and chemical balance
Nervous	brain, spinal cord, nerves, sense organs, receptors	controls and coordinates body movements and senses; controls consciousness and creativity; helps monitor and maintain other body systems
Endocrine	glands (such as adrenal, thyroid, and pancreas), hypothalamus	maintains homeostasis; regulates metabolism, water and mineral balance, growth and sexual development, and reproduction
Reproductive	ovaries, uterus, mammary glands (in females), testes (in males)	produces offspring

SKELETAL, MUSCULAR, AND INTEGUMENTARY SYSTEMS 907

SECTION 46-1

VISUAL LINK

Table 46-1

Have students make a table that organizes the tissue types, using Table 46-1 as a model. Have students list organs according to their primary tissue type. (Hint: have students consider the organ's function as they determine the proper tissue category.)

TEACHING STRATEGY

Organ Systems

Have students work in pairs to review the major structures and functions of the 11 organ systems. Suggest that they discuss each system and then quiz each other about the material. For example, one student might ask the other what system is responsible for the regulation of homeostasis (endocrine). Circulate around the room, asking students questions about the major structures and functions of the organ systems.

CULTURAL CONNECTION

Bone China

Bone china is a form of porcelain made from burned animal bones. Bone ash is mixed with kaolin, a white clay found in Jiangxi Province, China, and petuntse, a type of mineral found only in China. The bone ash increases the procelain's translucence. Porcelain was first made in China during the Tang dynasty (618–907). For centuries Chinese porcelain was considered the finest in the world. It was not until the 1700s that Europe began producing its own porcelain. Around 1750, the English found a new way of making porcelain—with bone ash. Today England produces most of the world's bone china.

TEACHER'S EDITION 907

Quick Labs
are simple, quick activities that require few materials.

Content Connection
discusses ways in which a lesson topic relates to material students have studied in a previous chapter.

Reteaching Activities
provide strategies and review questions for reinforcing the main concepts of the lesson.

Gifted Activities
provide opportunities for students to ask questions and extend their study beyond the level presented in the textbook.

Quick Facts
offer interesting tidbits that extend and supplement the text.

SECTION 13-2

Quick Lab

Comparing Unique Characteristics

Time Required 30 minutes

Procedural Tips As an alternative to ink pads, rub the point of a No. 2 pencil on a piece of paper until it makes a solid square of graphite. Have each student roll his or her thumb on the square to coat it with graphite. Press the coated thumb on the sticky side of a piece of transparent tape. Cut the tape, and press it onto the appropriate square of the large sheet. Repeat the procedure to place a print on the small square of paper.

Answers to Analysis All fingerprints have horizontal lines above and below any distinguishing features. Fingerprints are divided into five classes: whorl, right loop, left loop, arch, and tented arch. Students should be able to describe these patterns. Human DNA contains genes common to every person. However, each person also has his or her unique pattern of DNA fragments. A person's DNA fingerprints are as unique as his or her fingerprints.

Quick Lab

Comparing Unique Characteristics

Materials ink pad, paper, pencil, scissors

Procedure

1. Cut four 3 in. squares of plain paper.
2. Draw lines on a full-sized sheet of paper that will divide that sheet into four equal squares.
3. Each team member will press his or her right thumb on the ink pad and then will quickly press the inked thumb in a square on the large sheet of paper and again on one of the small squares of paper.
4. Examine each of the thumbprints, and make a list of characteristics that the prints have in common. Describe how each thumbprint is different from the others. Then shuffle the four individual thumbprints, and try to match each one with its duplicate on the large sheet of paper.

Analysis What characteristics are

double. Every five minutes, the sample of DNA doubles again, resulting in many copies of the sample in a short amount of time, as shown in Figure 13-7.

The new copies of the DNA sample can then be used to make a DNA fingerprint. PCR can use minuscule amounts of a specimen, as few as 50 white blood cells that might be found in a nearly invisible speck of blood, rather than the 5,000 to 50,000 cells needed for RFLP analysis. PCR is important not only for solving violent crimes but also for diagnosing genetic disorders from a few embryonic cells and for studying ancient fragments of DNA found in minute amounts.

HUMAN GENOME PROJECT

Using DNA technology, scientists all over the world are collaborating on one of the most ambitious research efforts in the history of genetics, the **Human Genome Project.** Two of the goals of the Human Genome Project are to determine the nucleotide sequence of the entire human genome—approximately 3 billion nucleotide pairs, or about 100,000 genes—and to map the location of every gene on each chromosome. In addition, the human genome is being compared with the genomes of other organisms in an effort to provide insight into fundamental questions about how genomes are organized, how gene expression is controlled, how cellular growth and differentiation are under genetic control, and how evolution occurs.

This project began in 1990. In 1996, the project had analyzed 1 percent of the 3 billion nucleotide pairs of DNA in the human genome. By knowing even 1 percent of the sequence, it is possible for scien-

246

456

SECTION 23-3

CONTENT CONNECTION

Section 22-4: Estuaries
Remind students that estuaries are areas where rivers flow into the sea. Estuaries are comparable to rain forests in species richness and productivity. The Everglades are breeding grounds for many species with economic value.

RETEACHING *ACTIVITY*
Ask students to imagine they are creating a plan to restore a damaged ecosystem, such as the Everglades. Have them list all the information they would need and the problems they might encounter. (Answers should include an inventory of the current and past conditions and inhabitants of the ecosystem. The plan would also require studies of the ecology of the species in the ecosystem to ensure that the requirements of the species would be met. The soil and water would need to be analyzed. A problem may be that there are no baseline data on what the ecosystem was like before it was altered.)

GIFTED *ACTIVITY*
Have students do group investigations of endangered species that require an estuarine habitat, such as the Everglades. They should make a visual group presentation. For example, whooping cranes have made a comeback, but their refuge in Texas is deteriorating due to dredging and increasing barge traffic through a channel that runs through the refuge. Maintenance of the channel requires that dredge material be piled up in the marsh habitat. These "spoil islands" have become grassland habitats that are not suitable for the cranes.

456 TEACHER'S EDITION

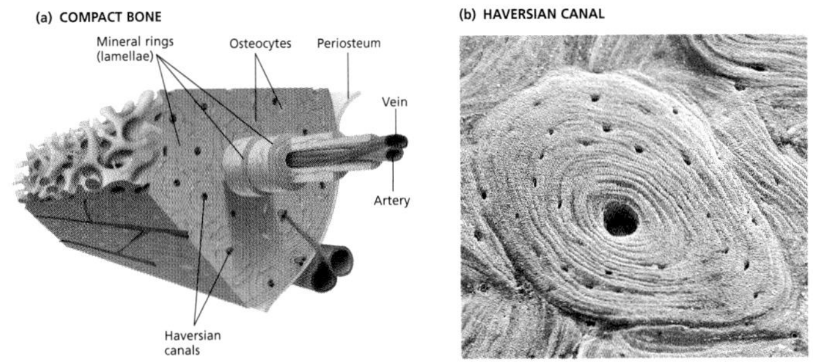

FIGURE 46-5
The cross section in (a) shows the internal structure of compact bone. A micrograph of a Haversian canal (380×) surrounded by lamellae in compact bone is shown in (b).

Under the periosteum is a hard material called **compact bone.** A thick layer of compact bone enables the shaft of the long bone to endure the large amount of stress it receives upon impact with a solid object. In the cross section shown in Figure 46-5a, notice that compact bone is composed of cylinders of mineral crystals and protein fibers called lamellae. In the center of each cylinder is a narrow channel called a **Haversian** (huh-VER-shuhn) **canal,** as shown in Figure 46-5b. Blood vessels run through interconnected Haversian canals, creating a network that carries nourishment to the living bone tissue. Several layers of protein fibers wrap around each Haversian canal. Embedded within the gaps between the protein layers are living bone cells called **osteocytes** (AHS-tee-uh-SIETS).

Inside the compact bone is a network of connective tissue called **spongy bone.** Although its name suggests that it is soft, this tissue is hard and strong. As shown in Figure 46-4, spongy bone has a latticework structure that consists of bony spikes. It is arranged along points of pressure or stress, making bones both light and strong.

Many bones also contain a soft tissue called **bone marrow,** which can be either red or yellow. Red bone marrow—found in spongy bone, the ends of long bones, ribs, vertebrae, the sternum, and the pelvis—produces red blood cells and certain types of white blood cells. Yellow bone marrow fills the shafts of long bones. It consists mostly of fat cells and serves as an energy reserve. It can also be converted to red bone marrow and produce blood cells when severe blood loss occurs.

Injury and Repair

Despite their strength, bones will crack or even break if they are subjected to extreme loads, sudden impacts, or stresses from unusual directions. The crack or break is referred to as a **fracture.** If circulation is maintained and the periosteum survives, healing will occur even if the damage to the bone is severe.

Eco Connection

Bones of Lead

Millions of Americans have been exposed to lead in the environment. Following exposure to lead, the kidneys excrete most of the metal. But 7 to 10 percent of the remaining lead in the body is stored in bone and can stay there for a lifetime. The rapid bone uptake of lead acts as a detoxifying mechanism. But lead is not permanently locked in bone. As people age, bone degeneration may occur, releasing lead into the bloodstream. Even very small concentrations of lead in the bloodstream can cause damage to kidneys, which in turn can cause high blood pressure.

The United States has outlawed the addition of lead to gasoline, water pipes, and paint. As a result, people who are now under age 25 may not accumulate as much lead in their bones as people from earlier generations.

SKELETAL, MUSCULAR, AND INTEGUMENTARY SYSTEMS 911

SECTION 46-2

VISUAL LINK

Figure 46-5
Point out the position of the Haversian canals in the compact bone. Have students describe the function of these canals. (Haversian canals create a network that carries nourishment to living bone tissue.)

RECENT RESEARCH

Bone Repair
"Sea bone" is a synthetic bone material made from coral. The coral is treated with heat to kill the living coral cells and convert its mineral structure (of calcium carbonates) to hydroxyapatite, the salt in human bone. The brittle coral graft is then carved to the desired shape, coated with naturally occurring bone-growth-enhancing proteins, and implanted. Typically 50 percent of sea-bone graft channels are filled with human bone in just three months.

QUICK FACT
☑ Bone tissue is renewed continuously. Every week, we recycle 5 percent to 7 percent of our bone mass, and as much as half a gram of calcium may enter or leave the adult skeleton each day.

GIFTED *ACTIVITY*
Have students research and report on Wolff's law (which states that a bone grows or remodels in response to the forces or stresses placed on it). Students may consider the effects of weightlessness (bone loss), weight lifting (the bones of weight lifters have enormous thickenings at the insertion sites of the most used muscles), or ballet dancing (the foot bones gradually thicken in response to the intense pressure of dancing *en pointe*).

TEACHER'S EDITION 911

Assessment and Practice Opportunities

Assessment

Modern Biology offers teachers a variety of assessment tools to evaluate the progress of students. Assessments range from pencil-and-paper reviews and tests to long-term portfolios and performance-based assessments.

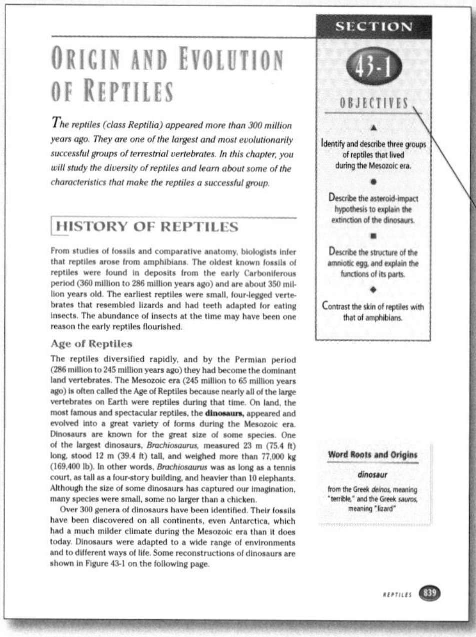

Performance Objectives
Successful assessment requires clearly defined expectations through performance objectives. The objectives found in *Modern Biology* drive the assessment in the Pupil's Edition and the Chapter Tests book.

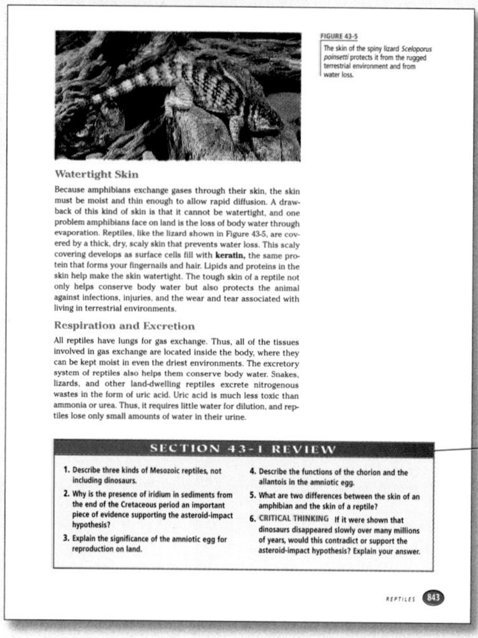

Section Reviews
Six questions at the end of each lesson cover the lesson's learning objectives. A critical thinking question is always included to challenge students.

Chapter Reviews
Each chapter concludes with vocabulary, multiple choice, short answer, and critical thinking questions, which provide an assessment of the entire chapter.

Chapter Summary
Concise summary statements are provided for each chapter. Teachers and students will find this feature invaluable because the statements bring together the important concepts presented in each chapter.

Vocabulary
A list of the vocabulary found throughout the chapter is provided for students. The vocabulary list indicates the page on which each term is defined in context.

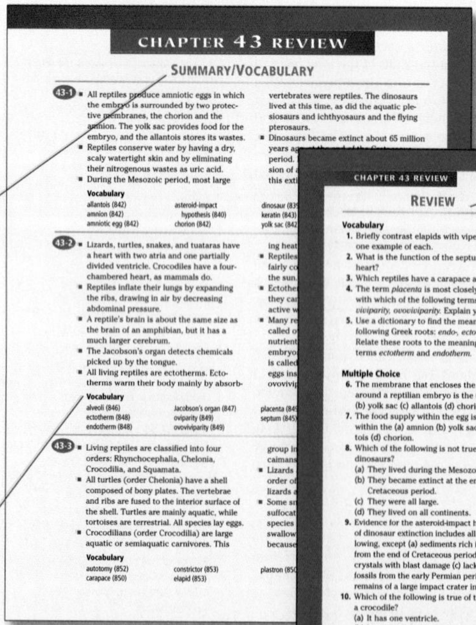

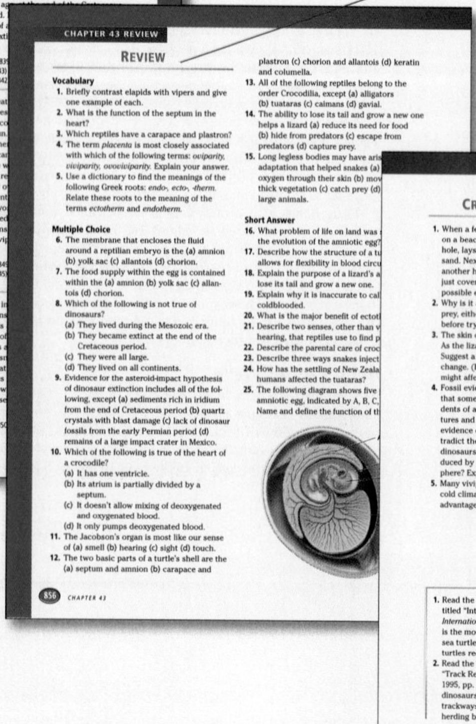

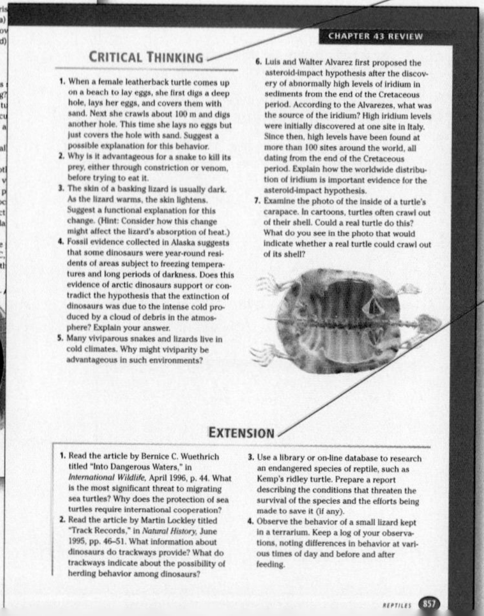

Critical Thinking
Chapter Reviews include several items that require students to extend the boundaries of their thinking. One item is specifically designed to evaluate students' understanding of a picture, graph, or diagram.

Extensions
These items include additional readings, research projects, and hands-on activities. The readings include prereading questions to guide students in reading a scientific article.

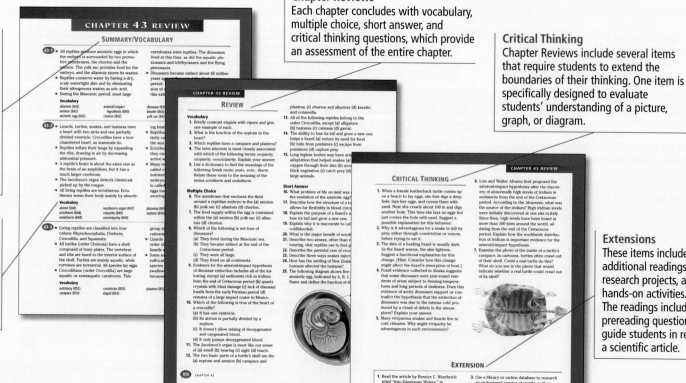

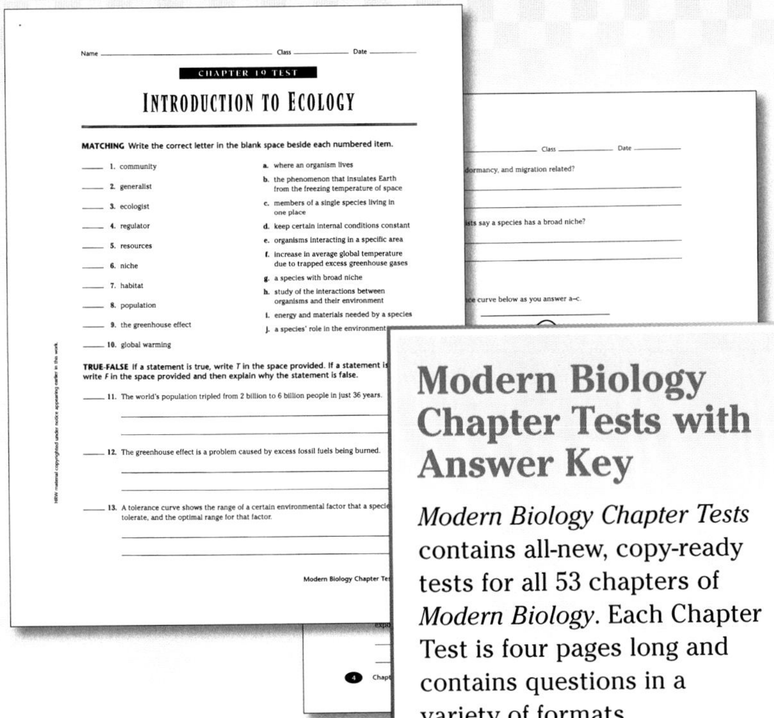

Modern Biology Chapter Tests with Answer Key

Modern Biology Chapter Tests contains all-new, copy-ready tests for all 53 chapters of *Modern Biology*. Each Chapter Test is four pages long and contains questions in a variety of formats.

Additional Content Area Reading Support

Reading science requires some different skills than those required to read a novel or short story. **Holt Science Skills Workshop: Reading in the Content Area** gives your students the strategic help they need to read in science class. Student worksheets develop skills such as comparison context clues, sequence context clues, probability context clues, and the use of cause and effect markers. Teaching strategies and a set of transparencies give you the tools to teach these skills.

Practice

Modern Biology Study Guide and *Modern Biology Active Reading Guide* offer opportunities for practice and reinforcement.

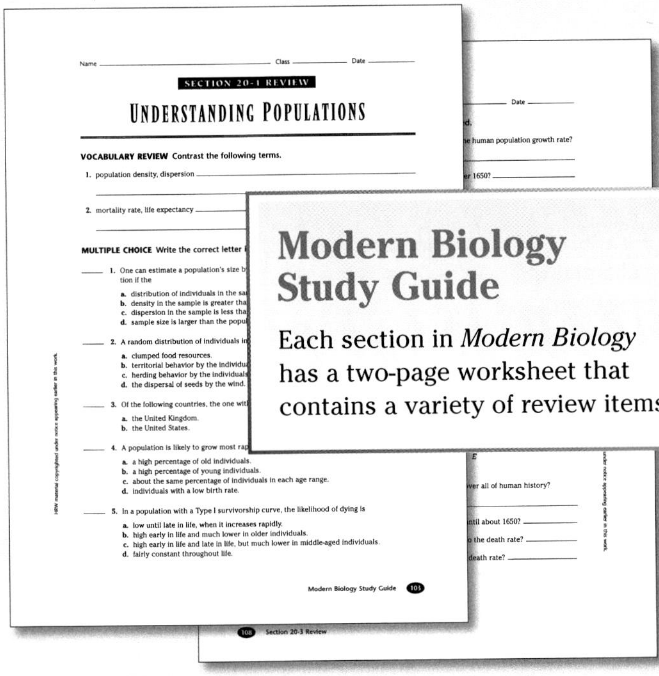

Modern Biology Study Guide

Each section in *Modern Biology* has a two-page worksheet that contains a variety of review items.

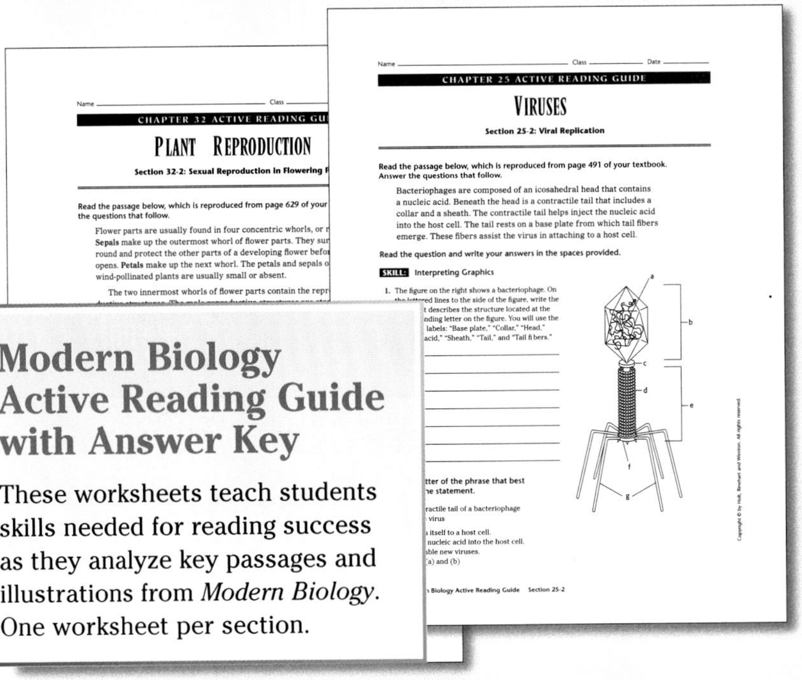

Modern Biology Active Reading Guide with Answer Key

These worksheets teach students skills needed for reading success as they analyze key passages and illustrations from *Modern Biology*. One worksheet per section.

Planning Your Curriculum

Flexibility in Classroom Instruction Time

Today's schools and classrooms are changing. The traditional biology class that meets for a 50-minute period five times a week is no longer the standard in many schools. This revised edition of *Modern Biology* has been designed to maximize the options in scheduling. The Pacing Chart at right has been developed around blocks of time, allowing you to create lessons to fit your schedule.

The Pacing Chart shows you a variety of approaches for teaching *Modern Biology.* You may select one of these approaches or create your own variation. The Pacing Chart offers six curriculum plans, including a basic course and five plans that emphasize a particular strand: cell biology, ecology, zoology, botany, and human biology. Please remember that the suggested time blocks are just that—suggested. For example, you can adjust the Pacing Chart if you do more lab work, spend more time covering a chapter, or add more chapters to your course.

Suggested pacing assumes the school year consists of 180 blocks of time, with each block 45–50 minutes long. More specific information about the blocks suggested for each chapter is found in the Chapter Planning Guides.

Flexibility in Planning for Different Ability Levels

Modern Biology can help you reach a variety of classes, including those classes tracked by ability or those classes that accommodate inclusion students. You can select the topics and content that best meet the needs of individual students and classes. Use the flexible Pacing Chart and the Chapter Planning Guides when making these important decisions.

	Basic Course	Cell Biology Emphasis	Ecology Emphasis	Zoology Emphasis	Botany Emphasis	Human Biology Emphasis
UNIT 1 Biological Principles	12 Blocks	16 Blocks	14 Blocks	13 Blocks	13 Blocks	10 Blocks
Chapter 1	4	6	6	5	5	4
Chapter 2	4	5	4	4	4	3
Chapter 3	4	5	4	4	4	3
UNIT 2 Cells	32 Blocks	42 Blocks	24 Blocks	25 Blocks	28 Blocks	30 Blocks
Chapter 4	8	10	6	7	6	6
Chapter 5	6	8	4	4	4	6
Chapter 6	5	8	4	4	8	5
Chapter 7	5	8	4	4	4	6
Chapter 8	8	8	6	6	6	7
UNIT 3 Genetics	35 Blocks	35 Blocks	26 Blocks	30 Blocks	30 Blocks	27 Blocks
Chapter 9	6	6	6	6	8	5
Chapter 10	8	8	6	8	6	5
Chapter 11	6	6	4	5	4	5
Chapter 12	7	7	4	5	4	7
Chapter 13	8	8	6	6	8	5
UNIT 4 Evolution	25 Blocks	25 Blocks	27 Blocks	23 Blocks	20 Blocks	14 Blocks
Chapter 14	5	6	6	5	6	3
Chapter 15	5	5	5	4	4	3
Chapter 16	5	5	6	4	4	
Chapter 17	5	4	5	4		4
Chapter 18	5	5	5	6	6	4

	Basic Course	Cell Biology Emphasis	Ecology Emphasis	Zoology Emphasis	Botany Emphasis	Human Biology Emphasis
UNIT 5 Ecology	22 Blocks	22 Blocks	52 Blocks	25 Blocks	30 Blocks	10 Blocks
Chapter 19	4	4	10	5	6	4
Chapter 20	5	5	10	5	6	
Chapter 21	4	4	10	5	6	
Chapter 22	5	5	12	5	6	
Chapter 23	4	4	10	5	6	6
UNIT 6 Microorganisms	16 Blocks	28 Blocks	23 Blocks		4 Blocks	20 Blocks
Chapter 24	4	6	5		4	4
Chapter 25	4	6	5			4
Chapter 26	2	5	4			4
Chapter 27	2	5	4			4
Chapter 28	4	6	5			4
UNIT 7 Plants	10 Blocks	9 Blocks	14 Blocks		45 Blocks	4 Blocks
Chapter 29	2	4	4		5	4
Chapter 30	4		5		10	
Chapter 31	4	5	5		10	
Chapter 32					10	
Chapter 33					10	
UNIT 8 Invertebrates	10 Blocks	3 Blocks		36 Blocks	10 Blocks	5 Blocks
Chapter 34	10	3		10	10	5
Chapter 35				4		
Chapter 36				4		

	Basic Course	Cell Biology Emphasis	Ecology Emphasis	Zoology Emphasis	Botany Emphasis	Human Biology Emphasis
UNIT 8 *(Continued)*						
Chapter 37				5		
Chapter 38				5		
Chapter 39				4		
Chapter 40				4		
UNIT 9 Vertebrates	10 Blocks			28 Blocks		
Chapter 41	5			5		
Chapter 42				5		
Chapter 43				6		
Chapter 44				6		
Chapter 45	5			6		
UNIT 10 Human Biology	8 Blocks					60 Blocks
Chapter 46	8					10
Chapter 47						7
Chapter 48						7
Chapter 49						8
Chapter 50						8
Chapter 51						7
Chapter 52						7
Chapter 53						6
TOTAL BLOCKS	180 Blocks	180 Blocks	180 Blocks	180 Blocks	180 Blocks	180 Blocks

Developing Process Skills

The process skills emphasized in each in-text chapter lab are indicated in the matrix at right.

CHAPTERS	PROCESS SKILLS																	
	Observing	Comparing & Contrasting	Identifying	Measuring	Calculating	Collecting Data	Organizing Data	Classifying	Relating	Inferring	Predicting	Modeling	Demonstrating	Analyzing Data	Hypothesizing	Testing Hypotheses	Designing Experiments	Evaluating
1	✓	✓		✓	✓	✓	✓			✓				✓				
2	✓	✓			✓	✓	✓		✓			✓						
3	✓	✓	✓	✓		✓	✓			✓				✓			✓	
4	✓	✓	✓					✓		✓					✓			
5	✓	✓				✓	✓		✓			✓		✓				
6	✓	✓				✓	✓		✓			✓		✓				
7	✓	✓			✓	✓	✓							✓				
8	✓				✓	✓	✓	✓		✓				✓				
9					✓	✓	✓		✓		✓	✓	✓	✓				
10		✓	✓							✓		✓	✓					
11		✓	✓									✓	✓					
12	✓	✓				✓	✓											✓
13	✓	✓	✓	✓	✓	✓	✓	✓		✓				✓				✓
14	✓	✓							✓	✓		✓			✓			
15	✓	✓		✓	✓	✓	✓				✓	✓		✓				
16		✓			✓	✓	✓				✓			✓				
17	✓	✓						✓		✓				✓				
18	✓	✓	✓				✓	✓						✓			✓	✓
19	✓			✓		✓	✓			✓				✓				
20	✓	✓		✓	✓	✓	✓								✓			
21	✓	✓					✓		✓									
22	✓	✓													✓	✓		
23	✓					✓	✓							✓	✓	✓	✓	
24	✓					✓	✓							✓	✓	✓		
25	✓	✓				✓	✓						✓	✓				
26	✓	✓				✓	✓			✓				✓				
27	✓	✓	✓					✓										
28	✓	✓				✓	✓		✓					✓	✓	✓	✓	
29	✓	✓		✓		✓	✓		✓					✓				
30	✓	✓						✓										
31	✓	✓	✓															
32	✓	✓	✓	✓		✓	✓							✓				
33	✓	✓		✓	✓	✓	✓								✓	✓		
34	✓		✓										✓					
35	✓					✓	✓		✓									
36	✓					✓	✓											
37	✓					✓	✓							✓	✓	✓		
38	✓					✓	✓				✓			✓	✓			
39	✓	✓	✓						✓					✓				
40	✓	✓		✓		✓	✓		✓				✓	✓				✓
41	✓					✓	✓		✓									
42	✓	✓				✓	✓							✓				
43	✓	✓				✓	✓							✓	✓	✓	✓	
44	✓	✓				✓	✓		✓									
45	✓	✓				✓	✓			✓								
46	✓	✓		✓	✓	✓	✓		✓					✓				✓
47	✓			✓	✓	✓	✓							✓	✓	✓	✓	
48		✓	✓			✓	✓			✓		✓		✓				✓
49	✓	✓		✓					✓	✓	✓	✓		✓			✓	
50	✓	✓	✓							✓								
51	✓	✓		✓	✓	✓	✓							✓				
52	✓	✓	✓						✓	✓								
53	✓	✓			✓	✓	✓		✓	✓		✓		✓				

Running a Safe and Efficient Laboratory Program

General Safety Guidelines

1. Post laboratory rules in a conspicuous place in the laboratory.
2. Before the class begins an experiment, review safety rules and demonstrate proper procedures.
3. Never permit students to work in your laboratory without your supervision. No unauthorized investigations should ever be conducted, nor should unauthorized materials be in the laboratory.
4. Lock your laboratory (and storage room) when you are not present.
5. Clearly mark locations of the eyewash stations, safety shower, fire extinguishers (ABC tri-class), chemical spill kit, first-aid kit, and fire blanket in the laboratory and storage room. Check this safety equipment prior to conducting each investigation.
6. Post an evacuation diagram and evacuation procedures by every entrance to the laboratory.
7. Provide labeled disposal containers for glass, sharp objects, and waste chemical reagents.
8. Allow no food or beverages in the laboratory. Caution students to keep their hands away from their face and to wash their hands with soap and water before leaving the laboratory.

9. Know the location for the master shut-off for laboratory circuits. Be sure that all outlets have correct polarity and have ground-fault interception. Polarity can be tested with an inexpensive continuity tester, available for about $5 from most electronic hobby shops. All electrical equipment should have three-prong plugs and three-wire cords.
10. Follow prescribed procedures for any safety incident, including full documentation. Remind students that any safety incident, no matter how trivial, must be reported directly to you.

Personal Protective Equipment

Chemical goggles: (Meeting ANSI Standard Z87.1) Chemical goggles should be worn when working with any chemical or chemical solution other than water, when heating substances, when using any mechanical device, and when observing physical processes that could eject an object.

Face shield: (Meeting ANSI Standard Z87.1) Use a face shield in combination with eye goggles when working with corrosives.

Contact lenses: The wearing of contact lenses for cosmetic reasons should be prohibited in the laboratory. If a student must wear contact lenses prescribed by a physician, that student should be instructed to wear eye-cup safety goggles meeting ANSI Standard Z87.1 (similar to swimmer's cup goggles).

Eyewash station: The device must be capable of delivering a copious, gentle flow of water to both eyes for at least 15 minutes. Portable liquid supply devices are not satisfactory and should not be used. A plumbed-in fixture or a perforated spray head on the end of a hose attached to a plumbed-in outlet is suitable if it is designed for use as an eyewash fountain and meets ANSI Standard Z358.1. It must be within a 30-second walking distance from any spot in the room.

Safety shower: (Meeting ANSI Standard Z358.1) A safety shower should be located within a 30-second walking distance from any spot in the room. Students should be instructed in the use of the safety shower in the event of a fire or a chemical splash on their body that cannot simply be washed off.

Gloves: Polyethylene, neoprene rubber, or disposable plastic may be used. Nitrile or butyl rubber gloves are recommended when handling corrosives.

Apron: Rubber-coated cloth or vinyl (nylon coated) halter is recommended.

Emergency Preparedness

What would you do if a student dropped a liter bottle of concentrated sulfuric acid *right now*? Plan how to effectively react *before* you need to.

1. Post the phone numbers of your regional poison control center, fire department, police, ambulance, and hospital on your telephone.
2. Practice fire and evacuation drills. Also have drills on what students MUST do if they are on fire or have an accident involving chemical contact.
3. Assure that all personal and other safety equipment is available and tested frequently.
4. Compile an MSDS file for all chemicals. This reference resource should be readily available in case of a spill or other accident.
5. Provide for spill-control procedures. Handle only those incidents that you feel comfortable handling. Situations of greater severity should be handled by trained hazardous-materials professionals.
6. Students should never fight fires or handle spills.
7. Be trained in first aid and basic life support (CPR) procedures. Have first-aid kits and spill kits readily available.
8. Fully document *any incident* that occurs.

Safety with Animals

It is recommended that teachers follow the "Guidelines for the Use of Live Animals" established by the National Association of Biology Teachers. A copy of this document is printed in the Teacher's Editions of *Holt BioSources Quick Labs* and *Inquiry Skills Development* lab manuals for your convenience.

Safety in Handling Preserved Materials

The following practices are recommended when handling or dissecting any preserved specimen:

1. Wear protective gloves and splash-proof safety goggles at all times when handling preserving fluids or preserved specimens and during dissection.
2. Wear a lab apron. Use of an old shirt or smock is recommended.
3. Conduct dissection activities in a well-ventilated area.
4. Do not allow preservation or body-cavity fluids to contact skin. Fixatives do not distinguish between living and dead tissues.

Biological-supply firms use formalin-based fixatives of varying concentrations to initially fix zoological and botanical specimens. WARD'S Natural Science Establishment provides specimens that are freeze-dried and rehydrated in a 10 percent isopropyl alcohol solution. In these specimens, no other hazardous chemical is present. Many suppliers provide fixed botanical materials in 50 percent glycerin.

Reduction of Free Formaldehyde

Currently, federal regulations mandate a permissible exposure level of 0.75 ppm for formaldehyde. Contact your supplier for a Material Safety Data Sheet that details the amount of formaldehyde present as well as gas-emitting characteristics of individual specimens.

Prewashing specimens in a loosely covered container in running tap water for 1–4 hours will dilute the fixative. Formaldehyde may also be chemically bound (thereby greatly reducing danger) by immersing washed specimens in a 0.5 to 1.0 percent potassium bisulfate solution overnight or by placing them in 1 percent phenoxyethanol holding solutions.

Safety with Microbes

Pathogenic (disease-causing) microorganisms are not appropriate investigational tools in the high school laboratory and should never be used.

Safety with Chemicals

Label student reagent containers with the substance's name and hazard class(es) (flammable, reactive, and so on). Dispose of hazardous waste chemicals according to federal, state, and local regulations. Refer to the Material Safety Data Sheet for recommended disposal procedures.

Remove all sources of flames, sparks, and heat from the laboratory when any flammable material is being used.

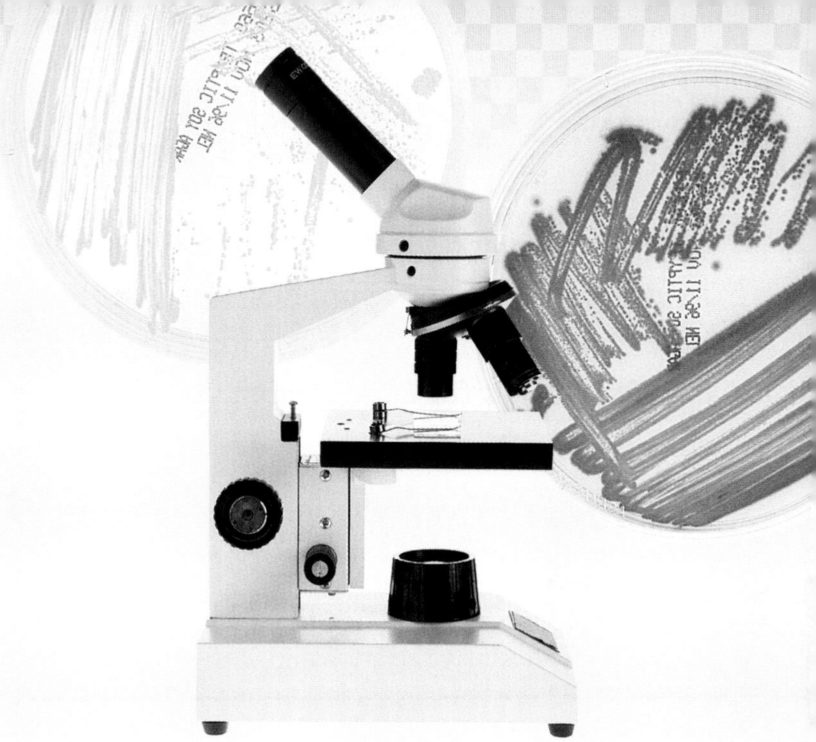

Material Safety Data Sheets

The purpose of a Material Safety Data Sheet (MSDS) is to provide readily accessible information on chemical substances commonly used in the science laboratory or in industry.

The MSDS should be kept on file and referred to *before* handling *any* chemical. The MSDS can also be used to instruct students on chemical hazards, to evaluate spill and disposal procedures, and to warn of incompatibility with other chemicals or mixtures.

Resources

American Chemical Society Health and Safety Service

This service will refer inquiries to appropriate resources for finding answers to questions about health and safety.

American Chemical Society (ACS)
1155 Sixteenth Street, N.W.
Washington, D.C. 20036
1-800-227-5558

Hazardous Materials Information Exchange (HMIX)

Sponsored by the Federal Emergency Management Agency and the U.S. Department of Transportation, HMIX can be accessed through an electronic bulletin board. It provides information regarding instructional material and literature listings, hazardous materials, emergency procedures, and applicable laws and regulations.

Safety Information References

Gerlovich, J. A., et al. *School Science Safety*. Flinn Scientific, Inc., 1984.

Gessner, G. H., ed. *Hawley's Condensed Chemical Dictionary*. 11th Ed. Van Nostrand Reinhold, 1987. (Revised by N. Irving Sax).

A Guide to Information Sources Related to the Safety and Management of Laboratory Wastes from Secondary Schools. New York State Environmental Facilities Corp., 1985.

Lefevre, J.J. *The First Aid Manual for Chemical Accidents*. Dowdwen, 1989. (Revised by Shirley A. Conibeau).

Pipitone, D. ed., *Safe Storage of Laboratory Chemicals*, 2nd Ed. Wiley-Interscience, 1991.

Prudent Practices in the Laboratory: Handling and Disposal of Chemicals. Committee on Prudent Practices for Handling, Storage, and Disposal of Chemicals in Laboratories, National Research Council, National Academy Press, 1995.

Rainis, Kenneth G. *WARD'S MSDS User's Guide*. WARD'S, 1988.

Storing Chemicals

Never store chemicals alphabetically, as this greatly increases the risk of a violent reaction.

Storage Suggestions

1. Always lock the storage room and all its cabinets when not in use.
2. Students should not be allowed in the storage room and preparation area.
3. Avoid storing chemicals on the floor of the storage room.
4. Do not store chemicals above eye level or on the top shelf in the storage room.
5. Be sure shelf assemblies are firmly secured to the walls.
6. Provide anti-roll lips on all shelves.
7. Shelving should be constructed out of wood. Metal cabinets and shelves are easily corroded.
8. Avoid adjustable metal shelf supports and clips. They can corrode, causing shelves to collapse.
9. Acids, flammable chemicals, poisons, and oxidizers should each be stored in their own locking storage cabinet.

Bench-Tested Chapter Investigations

WARD'S, the market leader in the production and distribution of science supplies, is the official materials and equipment supplier for *Modern Biology* and *Holt BioSources Lab Program*.

WARD'S and Holt, Rinehart and Winston Make Teaching Science Easier for You

Great labs are the cornerstone of a superior science textbook. Because of its exclusive relationship with WARD'S, Holt, Rinehart and Winston is proud to bring you the best in textbook lab investigations. This unique partnership guarantees that WARD'S lab materials and supplies are designed to specifically satisfy the requirements of the investigations found in Holt's textbooks and lab manuals.

All Holt textbook and lab manual investigations and activities are bench tested by WARD'S technical staff to guarantee successful lab results and experiences. With exclusive WARD'S products like Simulated Blood™, Simulated Plant Pigments™, and others, WARD'S has eliminated the use of hazardous materials and dangerous chemicals in your classroom.

Our Master Materials List Shows You Everything You Need

The Master Materials List on pp. 53T–57T was prepared by WARD'S, the exclusive science supplier for *Modern Biology.* This list indicates all supplies needed to perform the in-text Investigations and Quick Labs. Items are listed alphabetically under specific categories. Each item is followed by the WARD'S catalog number, the quantity, and the lab number in which the material is used. A number alone indicates an end-of-chapter Laboratory Investigation. **QL** indicates a Quick Lab followed by the page number on which the lab appears. *Local* means that the item should be available locally.

Ordering Materials and Supplies Is Convenient and Easy with WARD'S Software Ordering System.

With WARD'S exclusive software-ordering system specifically designed for use with *Modern Biology,* you can order your materials and supplies quickly and easily.

The software-ordering system lists all required and supplemental materials—per lab group or class size—needed for every lab investigation in *Modern Biology.*

- Click on the products you need, and the software automatically creates your "shopping list," keeping track of the materials ordered and their costs. A Help window explains every function on the program, making it easy to use.
- The software is available for both Macintosh and IBM computers.

Service and Satisfaction

WARD'S is committed to providing you with the highest level of quality and service in the market. WARD'S technical staff is always available to answer your questions.

WARD'S Telephone Directory
Toll-Free Phone: **1-800-962-2660**
Toll-Free Fax: **1-800-635-8439**

WARD'S Natural Science Establishment, Inc.
5100 West Henrietta Road
Rochester, New York 14692-9012

Master Materials List

Biological Supplies

Item	WARD'S No.	Quantity	Lab
bones, fresh	Local	2 bones per team	46
Boston fern with stolons	86 R 5550	1 per class	30
brittle or serpent star, preserved	68 R 7672	1 sea star per team	40
chameleons, American living	87 R 8135	2 anoles per team	43
Chlamydomonas reinhardtii (+)	86 R 0100	1 per team	22, 27
crayfish, living	87 R 6030	1 per team	QL748
crickets, living	87 R 6100	1 per team	42
crickets, living	87 R 6100	6 crickets per team	22
culture, adult brine shrimp	87 R 5105	50 mL per team	19
culture, brown hydra	87 R 2020	3 hydra per team	35
culture, brown *Planaria*	87 R 2500	1 per class	36
cultures, colonial algae set, 3	86 R 3010	1 per class	27
culture, *Daphnia magna*	87 R 5210	1 *Daphnia* per team	35
culture, *Daphnia magna*	87 R 5210	10 *Daphnia* per team	22
culture, duckweed (*Lemna*)	86 R 7650	1 per team	22
culture, *Euglena* sp.	87 R 0100	1 per class	26
culture, *Fontinalis*	87 R 7456	3 strands per team	22
culture, isopods (sowbug)	87 R 5520	6 isopods per team	22, 38
culture, *Paramecium caudatum*	87 R 1310	20 mL per team	23
culture, *Paramecium caudatum*	87 R 1310	1 drop per student	26
culture, plasmodial slime mold	85 R 4750	1 culture per team	QL535
culture, pond water	87 R 9058	1 per team	22
culture, *Spirogyra*	86 R 0650	1 per class	27
earthworms *(Lumbricus)*, living	87 R 4660	1 per team	37
earthworms *(Lumbricus)*, living	87 R 4660	3 per team	22
egg, cooked	Local	1 per team	49
Elodea, living *(Anacharis)*	86 R 7500	3 strands per team	QL118, 22
Elodea, living	86 R 7500	1 per class	4
Equisetum, horsetail, living	86 R 5300	1 per team	22
feather set	69 R 2269	1 set per team	44
fern prothallium—sexual stages (WM)	91 R 4862	1 per team	30
goldfish	87 R 8100	1 per team	QL6
grassfrogs, small, living	87 R 8217	1 frog per team	QL820, 42
grasshopper, lubber, preserved	68 R 4052	1 specimen per team	39
guppies, living	87 R 8110	8 guppies per team	22
isopods, terrestrial, living	87 R 5525	5 pillbugs per team	38
lentils	Local	2 seeds per team	9
lily flower	63 R 1180	1 per class	30
liver, raw	Local	1 piece per team	36
mealworm adults, living	87 R 6255	1 per team	42
mealworm adults, living	87 R 6255	6 worms per team	22
Mnium life history set, pres.	63 R 1005	1 per class	30

Biological Supplies (continued)

Item	WARD'S No.	Quantity	Lab
Norfolk island pine, living	86 R 5515	1 per class	30
peas, fresh or frozen	Local	10 per team	QL37
Planaria, living	87 R 2510	2 per team	QL717
pond snails, living	87 R 4140	2 snails per team	22
pond water	88 R 7010	3000 mL per class	51
potato	Local	1 per team	38
quill feather	Local	1 feather per team	44
ramshorn snails, living	87 R 4160	1 snail per team	22
sea urchin, purple, preserved	68 R 7702	1 sea urchin per team	40
seeds, Kentucky wonder bean—2 oz pack	86 R 8005	2 seeds per team	21, 32, QL653
seeds, kidney bean	86 R 8006	10 seeds per team	33
seeds, oriental mung bean	86 R 8007	10 seeds per team	22
seeds, pea, green arrow—1 lb pkg.	86 R 8251	2 seeds per student	9
seeds, pea, Little Marvel, wrinkled	86 R 8240	7 seeds per team	32
seeds, pea, Little Marvel, wrinkled	86 R 8240	80 seeds per team	7
seeds, ryegrass	86 R 8130	2 pinches per team	22
seeds, sweet corn, untreated	86 R 8080	6 seeds per team	32
seeds, clover	Local	2 pinches per team	22
seeds, rapid radish	86 R 8020	2 seeds per team	21, QL606
seeds, wheat	86 R 8360	12 seeds per team	29
sheep eyes, preserved	69 R 7192	1 eye per team	50
sheep hearts, preserved	69 R 7202	1 per team	34
shell, bivalve	68 R 7142	2–3 per team	QL730
skull, cat	65 R 5220	1 per class	45
skull, human, economy plastic	82 R 3031	1 per class	45
skull, perch	65 R 5121	1 per class	45
skull, small monkey	65 R 7030	1 per class	45
skull, turtle	65 R 5123	1 per class	45
slide, bone-compact (CS&LS) H&E	93 R 3293	1 per team	46
slide, contour-vane feather (WM)	92 R 3805	1 per team	44
slide, dicot stem	92 R 7914	1 per team	QL72
slide, down feather, (WM)	92 R 3806	1 per team	44
slide, generalized animal cell H&E	93 R 2200	1 per class	4
slide, generalized plant cell (SECT)	93 R 2134	1 per class	4
slide, human adult skin—white	93 R 7001	1 per team	QL72, 45
slide, *Mnium* antheridial head (LS)	91 R 4322	1 per team	30
slide, *Mnium* archegonial head (LS)	91 R 4326	1 per team	30
slide, plant mitosis (ls) qs	91 R 7042	1 per student	8
slide, plant mitosis (ls) qs	91 R 7042	1 per team	31
slide, plant mitosis (ls) qs	91 R 7042	1 per class	4
slide, *Ranunculus* (buttercup) stem	91 R 8144	1 per team	31
slide, *Ranunculus* mature root (cs)	91 R 8142	1 per team	31

Master Materials List

Biological Supplies (continued)

Item	WARD'S No.	Quantity	Lab
slide, root nodules (cs) qs	91 R 8192	1 per team	21
slide, squamous epithelium (SM) H&E	93 R 6003	1 per class	4
slide, *Syringa* (lilac) leaf (cs) qs	91 R 8272	1 per team	31
slide, *Zea* (corn) stem (cs) qs	91 R 7448	1 per team	31
Sphagnum, 3 × 4″	86 R 4400	1 per class	30
Sphagnum life history set	63 R 0488	1 per class	30
spinach, strained	Local	1 per team	51
starfish embryology slide set I	95 R 1551	1 set per team	52
tadpoles, living	87 R 8211	9 tadpoles per team	51
tobacco plant, live, 3" pot	86 R 6815	2 plants per team	25
Tradescantia, living	86 R 7300	1 per class	30
yeast, viable	88 R 0929	1 per class	20
Zamia life history set	63 R 0588	1 per class	30

Chemicals and Media

Item	WARD'S No.	Quantity	Lab
agarose, prepared 0.8%	88 R 1207	15 mL per team	13
albumin, egg powder, LABgr	39 R 0197	1 per class	3
amino acid mixture	250 R 3946	3 g per team	14
ammonia solution (household)	37 R 0415	1 per class	37
*Bam*HI restriction enzyme	85 R 1355	1 µL per team	13
beef broth	Local	1 drop per team	35
Benedict's solution, qualitative	37 R 0698	15 mL per team	3, QL995
biuret reagent, urea protein test	37 R 0790	150 drops per team	QL995
bleach, household	37 R 5554	100 mL per team	40
bromothymol blue, aqueous solution	38 R 9100	200 mL per team	QL118, 47
copper sulfate	(see Biuret reagent)	20 drops per team	3
DNA stain for gel electrophoresis	38 R 9014	100 mL per team	13
*Eco*RI restriction enzyme	85 R 1356	1 µL per team	13
ethyl alcohol	39 R 0271	6 mL per team	QL53
gibberellic acid, 0.1% solution	39 R 1451	1 per class	33
glucose solution 15%	37 R 9000	10 drops per team	3, QL995
*Hind*III restriction enzyme	85 R 1358	1 µL per team	13
hydrochloric acid, 1%	37 R 8607	1 per class	49
hydrochloric acid 47%, reagent	37 R 2250	300 mL per team	46
hydroponic nutrient solution	Local	1 per team	29
indophenol solution, 0.1%	37 R 9542	1 mL per student	48
iodine, potassium iodide solution	37 R 2379	1 per class	QL96, 20, QL995
L-Thyroxine, sodium salt	38 R 5330	1 per class	51
lambda phage DNA	85 R 1330	34 µL per team	13
legume inoculant—nitrogen fixation	20 R 6051	0.25 tsp per team	21

Chemicals and Media (continued)

Item	WARD'S No.	Quantity	Lab
loading dye, 6×	38 R 9115	5 mL per team	13
Lugol's iodine solution	39 R 1685	1 mL per student	4
Lugol's iodine solution	39 R 1685	1 drop per team	32
methyl cellulose, 1.5% aq. solution	37 R 7605	1 per class	19, 26
methylene blue chloride, 1% aq. sol	38 R 9522	1 drop per team	35
nutrient agar plates	88 R 0905	1 plate per team	QL18, 24
nutrient agar plates	88 R 0905	2 plates per team	QL18
pepsin, powder, (acidity 1:3000)	39 R 2865	1 per class	49
phenolphthalein solution	39 R 2873	4 drops per team	QL134
phosphate buffer, 0.1 M, pH 6.5	37 R 4848	10 mL per team	25
potato-dextrose media pour pack	88 R 1710	1 per class	28
propionic acid, 99%, BP 140–142C	39 R 2761	1 per class	28
silicon carbide abrasive, 400-grit	28 R 2540	1 g per team	25
silicone culture gum	37 R 9810	1 per team	35
simulated HIV	326 R 0200	1 per class	48
soda lime (4–8 mesh) SCIgr	37 R 5400	3 per team	7
sodium bicarbonate, powder, LABgr	37 R 5464	1 per class	49
sodium chloride, crystals, reagent	37 R 5480	1 per class	14, 26
sodium hydroxide, pellets, LABgr	37 R 5560	1 per class	QL134, 14
sodium hydroxide, solution	(see Biuret reagent)	40 drops per team	3
sodium thiosulfate	37 R 5700	1 per class	QL96
solution, unknown	Local	55 drops per team	3
starch solution	37 R 9001	15 mL per team	QL96
Sudan III (solvent red 23) powder	38 R 8787	1 per class	3
tris-borate-EDTA buffer, 10 × Sol.	37 R 0620	1 per class	13
Urinalysis, Simulated, Testing Kit	36 R 6012	1 per class	QL995
vinegar, white	39 R 0138	1 drop per team	35
water, distilled	88 R 7005	1 mL per team	26
water, distilled	88 R 7005	5 mL per team	24
water, distilled	88 R 7005	10 mL per team	26, QL804, 49
water, distilled	88 R 7005	55 drops per team	3
water, distilled	88 R 7005	1 per class	13

Laboratory Equipment

Item	WARD'S No.	Quantity	Lab
apron, disposable polyethylene	15 R 1050	1 apron per student	QL53, 3, 4, QL96, QL118, QL134, 13, 14, 20, QL472, 25, 28, QL589, 33, 34, 37, 39, 40, QL820, 46, 48, QL995, 49, 50, 51
aquarium/terrarium, small plastic	21 R 2100	2 per team	42
balance, triple beam	15 R 6057	1 per class	QL636, QL804, 46

Master Materials List

Laboratory Equipment (continued)

Item	WARD'S No.	Quantity	Lab
balance, triple beam	15 R 6057	1 per team	1, 14, 49
beaker, low-form 100 mL Griffin	17 R 4020	1 per team	25, 40
beaker, low-form 150 mL Griffin	17 R 4030	1 per team	23
beaker, low-form 150 mL Griffin	17 R 4030	2 per team	32
beaker, low-form 250 mL Griffin	17 R 4040	1 per team	QL804, 46, QL995, QL1019, QL1020
beaker, low-form 400 mL Griffin	17 R 4050	2 per team	23
beaker, low-form 500 mL Griffin	17 R 4060	1 per team	QL421
beaker, low-form 500 mL Griffin	17 R 4060	2 per team	46
beaker, low-form 500 mL Griffin	17 R 4060	3 per team	QL6
beaker, low-form 600 mL Griffin	17 R 4060	1 per team	QL96, 42
beaker, low-form 600 mL Griffin	17 R 4060	6 per team	51
beaker, low-form 1000 mL Griffin	17 R 4080	1 per team	3
calculator, slimline TI-1100+	27 R 3055	1 per team	19
capillary micropipets, pkg/250	15 R 2096	1 per team	13
clamp, Hoffman screw-compressor	15 R 3910	3 per team	19
clamp, Stoddard test-tube	15 R 0841	1 per team	3
corks, size 0	15 R 8360	2 corks per team	19, 23
coverslips, 22 mm plastic	14 R 3555	1 coverslip per team	20
coverslips, 22 mm plastic	14 R 3555	2 per student	4
coverslips, 22 mm plastic	14 R 3555	2 coverslips per team	14, 21, 26
coverslips, 22 mm plastic	14 R 3555	3 coverslips per team	27
deluxe power supply-2 channel	15 R 5340	1 per team	13
dip net, vinyl coated 3"	21 R 3700	1 per team	QL6, 51
dissecting pan, std. polyethylene	18 R 3655	1 per team	39, QL1034
dissection pan set, economy	18 R 3665	1 per team	34, 40, 50
dissection pan set, standard	14 R 7011	1 per team	37
drying tube, U-shaped 150 mm	17 R 4991	1 per team	23
electrophoresis, dual gel chamber	36 R 5160	1 per team	13
Erlenmeyer flask, graduated 125 mL	17 R 2981	2 per team	14, 47
Erlenmeyer flask, 250 mL	17 R 2803	3 per team	QL118
filter paper, medium grade, 9.0 cm	15 R 2815	1 piece per team	35, QL995
filter paper, medium grade, 9.0 cm	15 R 2815	3 pieces per team	38
forceps, dissecting, medium	14 R 1001	1 per student	4
forceps, dissecting, medium	14 R 1001	1 per team	7, 34, 35, 40, 50
forceps, student dissecting, broad	14 R 0512	1 per team	39
freezer	Local	1 per class	13
funnel, 58 mm polypropylene	18 R 1300	1 per team	19
gloves, disposable, medium	15 R 1071	2 pairs per student	QL53, 3, 4, QL96, QL118, QL134, 13, 14, 20, 21, QL421, QL472, 24, 25, 28, QL589, 33, 34, 37, 39, 40, QL820, 46, 48, QL995, 49, QL1019, 50, 51

Laboratory Equipment (continued)

Item	WARD'S No.	Quantity	Lab
gloves, heat defier Kelnit cotton	15 R 1095	1 pair per team	1, 13, 14
graduated cylinder, 10 mL PP	18 R 1705	1 per team	3, 13, 49, 51
graduated cylinder, 50 mL PP	18 R 1720	1 per team	14, 29
graduated cylinder, 100 mL PP	18 R 1730	1 per team	1, QL96, QL118, QL134, 7, QL329, QL362, 19, QL421, QL804, 47
gravity convection laboratory oven	15 R 0061	1 per class	46
hot plate (700 W) single burner	15 R 7999	1 per team	3, 13, 14
lens tissue	15 R 8250	1 sheet per team	46
litmus paper, blue	15 R 3105	5 strips per team	49
litmus paper, red	15 R 3107	5 strips per team	49
magnifier, dual 3× & 6×	24 R 1112	1 per team	23, QL589, QL606, QL609, QL717, 37
meterstick, maple	15 R 4065	1 per team	15, QL399, QL488
microcentrifuge tube rack	18 R 1361	1 per team	13
microfuge tubes, 1.5 mL	18 R 1361	5 per team	13
microscope slide, qual. precleaned	14 R 3500	2 per student	4
microscope slide, qual. precleaned	14 R 3500	1 slide per team	32, 35
microscope slide, qual. precleaned	14 R 3500	2 per team	14, 21, 26
microscope slide, 80 ruled squares	14 R 3120	1 slide per team	20
microscope slide, single concavity	14 R 3510	3 slides per team	27
microscope, the WARD'S scope	24 R 2310	1 per student	8
microscope, the WARD'S scope	24 R 2310	1 per team	QL72, 4, 14, 20, 21, 26, QL535, 27, 30, QL609, 31, 32, 35, 44, 45, 46, 52
mortar, porcelain size 0, 50 mL	15 R 3334	1 per team	25
pestle, porcelain size 0	15 R 3335	1 per team	25
Petri dish, disposable 100 x 15 mm	18 R 7101	1 dish per team	19, QL606, 36, 38, 40, 51
Petri dish, disposable 150 x 15 mm	18 R 7100	1 dish per team	37
Petri dish, disposable 150 x 15 mm	18 R 7100	4 dishes per team	28
Petri dish, disposable 150 x 15 mm	18 R 7100	2 dishes per team	9
pipet tips, Eppendorf disposable	15 R 2089	1 per team	13
pipet, digital, variable volume micro	15 R 1725	1 per team	13
pipet, glass dropping 3"	17 R 0230	1 per team	QL134, 14, 29, QL606, 32, 36, 37, QL995
pipet, glass dropping 3"	17 R 0230	3 per team	26, 27 ,35
pipet, nonsterile 6"	18 R 2971	1 pipet per team	19, 20, 38
pipet, nonsterile 6"	18 R 2971	9 pipets per team	3
pipets, Pasteur 5 3/4"	17 R 1145	1 pipet per team	7
plant tray, nesting 13 × 15 × 3.5"	20 R 3210	2 per team	37
probe & seeker, 5 1/8"	14 R 0950	1 per team	34, 39, 40, 50

Master Materials List

Laboratory Equipment (continued)

Item	WARD'S No.	Quantity	Lab
probe, Huber 6″	14 R 0958	1 per team	36
ruler, 6″ white Vinylite	14 R 0810	1 per team	7, 10, 13, 15, 29, 32, 33
safety goggles, SG34 regular	15 R 3046	1 per student	QL53, 3, 4, QL96, QL118, QL134, 13, 14, 19, 20, QL472, 25, 26, 28, 33, 34, 37, 39, 40, QL820, 46, 47, 48, QL995, 49, 50, QL1034, 51
scalpel, sterile 6 1/2″, #22 blade	14 R 0900	1 per team	21, 32, 34, 49, 50
scissors, student econ. dissecting	14 R 0525	1 per student	10
scissors, student econ. dissecting	14 R 0525	1 per team	15, 34, 38, 39, 40, 50
spirometer	14 R 5070	1 per team	47
stereomicroscope, wide-field	24 R 4601	1 per team	21, 28, 30, 32, 35, 36, 39, 40, 45
stir rods, 6″ glass, 150 × 5 mm	17 R 6005	1 per team	14
stir rods, 6″ glass, 150 × 5 mm	17 R 6005	9 per team	3
stopper, black rubber solid size 0	15 R 8460	4 stoppers per team	19
stopper, black rubber solid size 0	15 R 8460	5 stoppers per team	49
stopwatch, digital	15 R 0512	1 per team	1, 14, QL399, 23, 26, 36, 37, QL932, 47
support ring, 3″ OD nickel plated	15 R 0707	1 per team	7, 14
support, rectangular, 4 × 6″ base	15 R 0719	1 per team	7, 14
swab applicator	14 R 5502	1 per team	QL535, 33
swab applicator	14 R 5502	2 swabs per team	25, 33, 37
swab applicator	14 R 5502	4 swabs per team	24
tag label, for preserved material	480 R 1100	2 per team	46
teasing needle, wood handle	14 R 0650	1 per team	39, 40
test-tube rack, 10 tubes, 2 tiers	18 R 4260	1 per team	3
test-tube rack, 6-well LDPE	18 R 4231	1 per team	QL53, 19, 36, 49
test tube with rim, 13 × 100 mm Pyrex	17 R 0610	1 per student	48
test tube with rim, 13 × 100 mm Pyrex	17 R 0610	1 per team	20, 36
test tube with rim, 13 × 100 mm Pyrex	17 R 0610	4 per team	QL53, 19, QL995
test tube with rim, 13 × 100 mm Pyrex	17 R 0610	5 per team	49
test tube with rim, 13 × 100 mm Pyrex	17 R 0610	9 per team	3
thermometer, lab–20 to 110° C	15 R 1416	1 per team	QL6, 23, 37
thermometer, lab–20 to 110° C	15 R 1416	2 per team	1
tongs, 8″ utility	14 R 0960	1 per team	14, 46
tray, w/cover 4 × 7 × 1 3/4″	18 R 0031	1 per team	13
tubing, dialysis	14 R 4512	1 ft per team	QL96, QL804
tubing, vinyl 3/8 × 1/16″	18 R 5083	44 cm per team	19
volumeter	14 R 8300	3 per team	7

Miscellaneous

Item	WARD'S No.	Quantity	Lab
aluminum foil roll, 12″ wide	15 R 1009	1 12″ piece per team	QL362, 19, 36, 38
bags, resealable zipper, 4 × 6″	18 R 6921	1 per team	13, 46
balloons, red	Local	1 per team	QL842
balloons, yellow	Local	1 per team	QL842
bead, blue wood with hole 7/16″	720 R 4501	6 beads per student	11
beads, 8 mm	36 R 4196	1 cup per team	7
bubble rack, round	18 R 4215	1 per team	29
cardboard (8 1/2 × 11″)	Local	1 per team	7
cheesecloth	15 R 0015	2 10 x 10 cm per team	29
cheesecloth	15 R 0015	1 piece per class	38
clay, blue modeling	15 R 4641	1 per class	11, QL789
clay, red modeling	15 R 4640	1 per class	11, QL789
clay, yellow modeling	15 R 4642	1 per class	11, QL789, QL842, QL853
cloth square, 15 × 15"	15 R 2538	1 piece per team	38
coin	Local	1 coin per team	15
container, cylindrical pint	30 R 0860	3 per team	16
corks, size O	15 R 8360	1 per team	36
corn oil, SClgr.	39 R 2513	25 mL per team	1
cotton, absorbent, nonsterile	15 R 3830	1 per class	7
cup, plastic, 5 oz	18 R 3675	5 cups per team	1
cup, plastic with lid, 9 oz	18 R 3676	2 cups per team	29
cups, paper, 5 oz	15 R 9830	3 cups per team	7
die, 6-sided	Local	1 die per team	15
field guide, mammals	32 R 0400	1 per class	45
flashlight, pen	15 R 4000	1 per team	36, 38
flowerpot, plastic 3″	20 R 2130	2 pots per team	21, 33
food coloring, red	37 R 1597	1 per class	7
gauze	Local	1 per team	46
gravel	45 R 1985	1 per team	22
gravel, rounded aquarium stone	21 R 1805	1 per team	22
heatproof pad	Local	1 per team	46
ice, cubed or crushed	Local	1 per team	QL6, 13, 23, 37
ice bucket	Local	1 per class	13
jar, wide-mouth glass w/cap, 1 gal	18 R 9706	2 per team	43
jar, wide-mouth glass w/cap, 1 gal	18 R 9706	3 per team	22
jellybeans, mixed colors	Local	20 per team	QL173,QL229
lamp, clamp, with reflector	36 R 4168	1 per team	1, 19, 37, 38
light bulb, 150 W, 120 V, Clear	36 R 4173	1 per team	1, 19, 37, 38
marker, black lab	15 R 3083	1 per student	10

Master Materials List

Miscellaneous (continued)

Item	WARD'S No.	Quantity	Lab
marker, black lab	15 R 3083	1 per team	3, 11, 16, 19, 29, 47, 51
marker, black lab	15 R 3083	1 per class	18
marshmallows, mini	Local	1 bag per class	37
mirror, 3 × 4"	15 R 9860	1 per team	45
muslin, unbleached 78 × 76", 46" wide	259 R 0002	1 yd per class	38
paper clips	15 R 9815	54 paper clips per student	10
paper clips, colored	Local	15 of each color per team	QL196
paper towel, 100 sheet 2-ply roll	15 R 9844	1 sheet per team	26, 32
paper towel, 100 sheet 2-ply roll	15 R 9844	5 sheets per team	37
paper, assorted construction	15 R 9841	1 sheet per team	15
paper, black construction	15 R 9825	1 sheet per team	37
paper, brown construction, 6 shades	Local	2 pieces per team	43
paper, graph, 1mm fine division	229 R 3831	1 sheet per team	51
paper, graph, 5 squares/in.	15 R 3835	1 sheet per student	1
paper, graph, 5 squares/in.	15 R 3835	1 sheet per team	22
paper, green construction, 6 shades	Local	2 pieces per team	43
paper, lined	Local	1 sheet per team	49
paper, white	Local	1 sheet per team	7, 11, 32, 36, 44, 52
paper, white	Local	1 sheet per student	11, 17, 18
parafilm M, 4″ wide	15 R 1940	1 per class	46
pen, black wax marker for glass	15 R 1155	1 per student	QL18
pen, black wax marker for glass	15 R 1155	1 per team	QL6, QL53, 23, 24, 25, 28, 32, 33, 36, 43, 46, 49, 51
pencils, colored prismacolor set	15 R 2576	3 pencils per team	QL730, QL820, 51, QL1071
pencils, Ticonderoga #2	15 R 9816	1 per student	QL72, 11, 17, 18, 52
picture, invertebrates	Local	1 per team	QL678
picture, poriferans	Local	1 per team	QL698
picture, vertebrates	Local	1 per team	QL678
pins, "T" nickel-plated	14 R 0201	1 per team	39, 40
pipe cleaners, white 6″	82 R 1115	1 cleaner per student	11
plastic wrap, 12″ wide	15 R 9858	1 piece per team	47
pop-it beads, black 10 cm	36 R 1537	200 beads per team	16

Miscellaneous (continued)

Item	WARD'S No.	Quantity	Lab
pop-it beads, white 10 cm	36 R 1534	200 beads per team	16
pushpins, blue 3/8″	229 R 0506	12 per student	10
pushpins, green 3/8″	229 R 0508	12 per student	10
pushpins, red 3/8″	229 R 0505	12 per student	10
pushpins, white 3/8″	229 R 0509	6 per student	10
pushpins, yellow 3/8″	229 R 0507	12 per student	10
rubber bands, assorted	15 R 9824	2 rubber bands per team	32
sand, black, 32 oz container	45 R 1987	3 oz per team	1
sand, fine white	45 R 1983	3 oz per team	1
screening, fiberglass 36 × 36″	15 R 0002	1 piece per class	19, 38
shoes	Local	10 pair per class	18
soil, garden-potting	20 R 8306	2 cups per team	21, 22, QL653
soil, garden-potting	20 R 8306	50 mL per team	29
soil, garden-potting	20 R 8306	1 per class	33
sticks, Popsicle™	15 R 9893	2 sticks per team	21
straws, plastic-wrapped	15 R 9869	9 straws per student	10
straws, plastic-wrapped	15 R 9869	1 straw per team	QL118, QL134, 15, QL853
straws, plastic-wrapped	15 R 9869	2 straws per team	47
tape, cellophane	15 R 1957	1 per class	QL18, 7, 15, QL362, 24, 38, QL1019
tape, masking 3/4″ × 60 yd roll	15 R 9828	1 per team	28
tape, masking 3/4″ × 60 yd roll	15 R 9828	1 per class	11, 18, 29, QL789
tape, red write-on	15 R 0503	1 per team	3
tape, red write-on	15 R 0503	1 per class	16
tobacco, several brands	Local	1 per team	25
toothpicks, round	15 R 9840	2 toothpicks per team	QL37, QL229, QL399, 28, QL789, QL1019
transparency film, infrared	75 R 9500	3 per team	22
transparency marking pencils, set of 12	26 R 4010	1 per team	22
vegetable oil	37 R 9540	5 drops per team	QL53, 33, QL1020, QL1034
water, hot	Local	1 per team	23
water, tap	Local	2500 mL per team	QL6, 1, QL53, QL96, QL118, QL134, 7
water, warm tap	Local	1 per team	QL6, 33, 37, 38

Unit 1

BIOLOGICAL PRINCIPLES

CHAPTERS

“*Our ideas are only instruments which we use to break into phenomena; we must change them when they have served their purpose, as we change a blunt lancet that we have used long enough.*”

Claude Bernard

Organisms living in this taiga ecosystem are adapted for dry, cold weather and reduced availability of food in winter.

internetconnect

SCILINKS NSTA

National Science Teachers Association *sci*LINKS Internet resources are located throughout this unit.

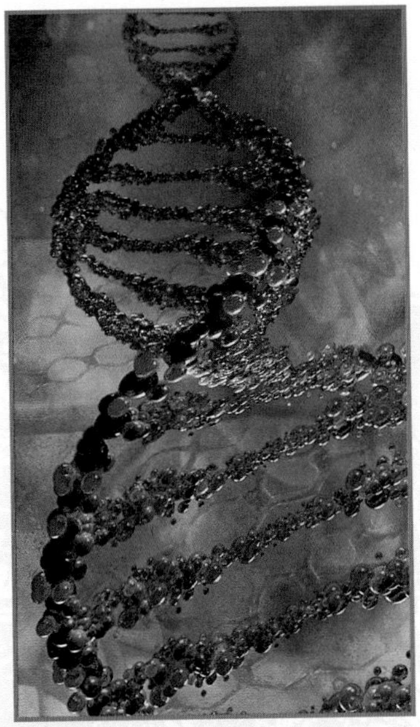

DNA is responsible for transmitting genetic information to offspring.

This giant panda gets its energy by eating bamboo leaves. The Amazon water lily, above left, lives in shallow eutrophic ponds.

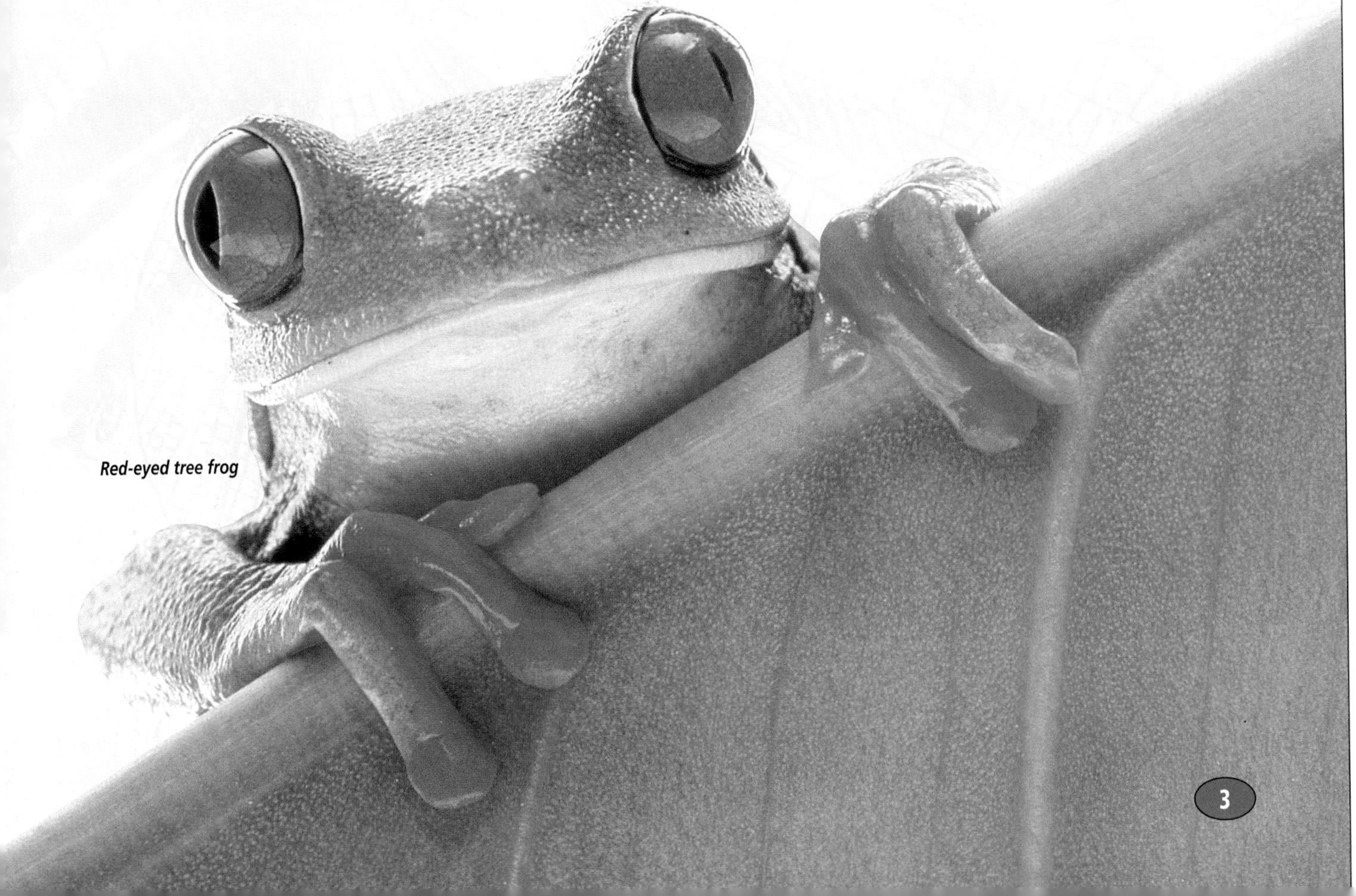

Red-eyed tree frog

unit 1

CHAPTERS

BIOLOGICAL PRINCIPLES

CONNECTING TO THE STANDARDS

The following chart shows the correlation of Unit 1 with the *National Science Education Standards* (grades 9–12). The items in each category are addressed in this unit and should be used to guide instruction. Annotated descriptions of the **Life Science Standards** are found below the chart. Consult the *National Science Education Standards* for detailed descriptions of each category.

UNIFYING CONCEPTS

- Systems, order, and organization
- Change, constancy, and measurement
- Form and function

SCIENCE AS INQUIRY

- Abilities necessary to do scientific inquiry
- Understandings about scientific inquiry

SCIENCE AND TECHNOLOGY

- Abilities of technological design
- Understandings about science and technology

PHYSICAL SCIENCE

- Structure and properties of matter
- Chemical reactions
- Interactions of energy and matter

LIFE SCIENCE

EARTH AND SPACE SCIENCE

- Energy in the Earth system
- Geochemical cycles

SCIENCE IN PERSONAL AND SOCIAL PERSPECTIVES

- Natural resources
- Environmental quality
- Science and technology in local and global challenges

HISTORY AND NATURE OF SCIENCE

- Science as a human endeavor
- Nature of scientific knowledge
- Historical perspectives

ANNOTATED DESCRIPTIONS OF THE CORRELATED LIFE SCIENCE STANDARDS

Although all eight categories of the *National Science Education Standards* are important, the following descriptions summarize the **Life Science Standards** that specifically relate to Unit 1.

THE CELL

- Most cell functions involve chemical reactions.
- Cell functions are regulated.

MOLECULAR BASIS OF HEREDITY

- In all organisms, the instructions for specifying the characteristics of the organism are carried in DNA.

BIOLOGICAL EVOLUTION

- Species evolve over time.
- The millions of different species of organisms that live on Earth today are related by descent from common ancestors.
- Biological classifications are based on how organisms are related.

INTERDEPENDENCE OF ORGANISMS

- The atoms and molecules on the Earth cycle among the living and nonliving components of the biosphere.

MATTER, ENERGY, AND ORGANIZATION IN LIVING SYSTEMS

- The chemical bonds of food molecules contain energy.
- The complexity and organization within an organism result from the organism's need to obtain, transform, transport, release, and eliminate the matter and energy required to sustain the organism.
- Matter and energy flow through different levels of organization of living systems and between living systems.

TRENDS IN BIOLOGY

Life in Unlikely Places

In the 1600s when Anton von Leeuwenhoek looked through his microscope at a droplet of water, he found life where no one had thought it could exist. For the past 30 years or so, Leeuwenhoek's successors, using far more sophisticated technology, have discovered a surprising diversity of organisms inhabiting environments formerly considered too hostile to sustain life—scalding hot springs, polar ice, and solid rock deep underground. Termed *extremophiles,* these organisms are more than just microbiological oddities. Industries from mining to medicine are already exploiting the unique biochemistry of extremophiles, and many more applications are being developed. At a deeper level, scientists hope that extremophiles might shed light on two of biology's most difficult questions: how did life begin, and does life exist elsewhere in the universe?

By human standards, the habitats of extremophiles are forbidding. Some hyperthermophiles (extremophiles that prefer very high temperatures) thrive around deep-sea hydrothermal vents, where temperatures reach 110°C. Other extremophiles, known as halophiles ("salt lovers"), inhabit highly salty lakes, like the Great Salt Lake and the Dead Sea. Still others, called acidophiles, are found in highly acidic areas near volcanoes or in hot springs. Some extremophiles even live trapped in rocks nearly 3 km below the Earth's surface.

Almost all extremophiles are archaebacteria, prokaryotic organisms that scientists have recently recognized as being distinct from traditional bacteria and more closely related to eukaryotes (see **Trends in Biology: Archaebacteria** on page T67B). Extremophiles are not just riding out environmental fluctuations. They have evolved a number of adaptations for coping with their habitats, and they often cannot reproduce under more moderate conditions.

Extremophiles give this hot spring in Yellowstone National Park its characteristic color.

CURRENT AND POTENTIAL APPLICATIONS

Mining companies use acidophilic bacteria to release gold and other metals from low-quality ore. In medicine, genetics, and molecular biology, the polymerase chain reaction (PCR) for copying fragments of DNA relies on DNA polymerase (a DNA-replicating enzyme) from the extremophile *Thermus aquaticus.*

Scientists are just beginning to learn how extremophiles thrive in environments that kill other organisms. For example, how do acidophiles, which sometimes live in habitats with a pH of less than 1, keep their internal pH near neutral? And how do thermophiles prevent their proteins from denaturing and ceasing to function at high temperature? Answering these questions will undoubtedly lead to more practical uses for extremophiles and their enzymes.

Scientists studying the origin of life have turned to extremophiles as models for what the earliest living things might have been like. Perhaps instead of beginning in a "warm little pond," as Darwin once speculated, life got started near a deep-sea vent or deep underground. Moreover, if life can survive in such a range of hostile habitats on Earth, some scientists argue, it may occur elsewhere in the solar system.

Additional Information

Madigan, Michael, and Barry Marrs. "Extremophiles." *Scientific American,* April 1997, 82–87.

Madigan, Michael, John Martinko, and Jack Parker. *Brock Biology of Microorganisms.* 8th ed. Upper Saddle River, New Jersey: Prentice Hall, 1997.

Monastersky, Richard. "Deep Dwellers." *Science News,* March 29, 1997, 192–193.

CHAPTER 1
THE SCIENCE OF LIFE

	TOPICS	TEACHING RESOURCES	LABS, CLASSWORK, AND HOMEWORK
BLOCK 1 45 minutes	**Introducing the Chapter, p. 4**	**ATE** Focus Concept, p. 4 **ATE** Assessing Prior Knowledge, p. 4 **ATE** Understanding the Visual, p. 4	■ Supplemental Reading Guide, *The Andromeda Strain*
	1-1 Themes of Biology, p. 5 The Study of Life, p. 5	**ATE** Section Overview, p. 5 **ATE** Critical Thinking, p. 9 **ATE** Visual Link Figure 1-1, p. 5 Figure 1-2, p. 6	**PE** **Quick Lab,** p. 6, Observing Homeostasis ■ Safety Contract and Quiz ■ Problem-Solving Worksheets—Operations on Small and Large Numbers ★ Study Guide, Section 1-1 **PE** Section Review, p. 10
BLOCK 2 45 minutes	**1-2 The World of Biology, p. 11** Characteristics of Life, p. 11 The Living World, p. 13	**ATE** Section Overview, p. 11 **ATE** Critical Thinking, p. 11 **ATE** Visual Link Figure 1-9, p. 12 Figure 1-11, p. 13	**Lab A2,** Comparing Living and Nonliving Things ■ Occupational Applications Worksheets—Biology Teacher ★ Study Guide, Section 1-2 **PE** Section Review, p. 13
BLOCK 3 45 minutes	**1-3 Scientific Methods, p. 14** Observing, p. 14 Asking a Question, p. 14 Collecting Data, p. 15 Hypothesizing, p. 17 Experimenting, p. 17 Drawing Conclusions, p. 18 Implementing Scientific Methods, p. 20	**ATE** Section Overview, p. 14 **ATE** Critical Thinking, pp. 15–17 **ATE** Visual Link Figure 1-12, p. 15 Figure 1-15, p. 19 1	**PE** **Quick Lab,** p. 18, Predicting Results **Lab A1,** Imagining Solutions: Problem Solving **Lab A13,** Using Random Sampling **Lab A28,** Bias and Experimentation **Lab B1,** Introduction to Experimental Design and Data Presentation ★ Study Guide, Section 1-3 **PE** Section Review, p. 20
BLOCK 4 45 minutes	**1-4 Microscopy and Measurement, p. 21** Microscopes, p. 21 Measurement, p. 23	**ATE** Section Overview, p. 21 **ATE** Critical Thinking, p. 23 **ATE** Visual Link Figure 1-18, p. 22	**PE** **Chapter 1 Investigation,** p. 28 **Lab C1,** Using a Microscope **Lab C2,** Using a Microscope—Criminal Investigation ■ Basic Skills Worksheets—Microscope Magnifications ★ Study Guide, Section 1-4 **PE** Section Review, p. 24
BLOCKS 5 & 6 90 minutes	REVIEW and ASSESSMENT **PE** Chapter 1 Review, pp. 26–27 ■ Performance-Based Assessment—Introduction to Performance-Based Assessment	Audio CD Program	CHAPTER TESTING ■ Chapter Test (blackline master) ▲ Test Generator ■ Assessment Item Listing ■ Scoring Rubrics and Classroom Management Checklists

MODERN BIOLOGY | HOLT BIOSOURCES | One-Stop Planner CD-ROM

KEY

PE Pupil's Edition
ATE Teacher's Edition
■ Active Reading Guide
★ Study Guide

Teaching Transparencies

Laboratory Program

One-Stop Planner CD-ROM
Shows these resources within a customizable daily lesson plan:
■ Holt BioSources Teaching Resources
■ Active Reading Guide
★ Study Guide
▲ Test Generator

READING FOR CONTENT MASTERY

Preview: Have students preview the objectives, the introductory paragraph, topic sentences, and captions before reading the chapter.

ATE Active Reading Technique: Brainstorming, p. 5
■ Active Reading Guide Worksheet 1-1
PE Summary/Vocabulary, p. 25

■ Active Reading Guide Worksheet 1-2
PE Summary/Vocabulary, p. 25

■ Active Reading Guide Worksheet 1-3
PE Summary/Vocabulary, p. 25

■ Active Reading Guide Worksheet 1-4
PE Summary/Vocabulary, p. 25

TECHNOLOGY AND INTERNET RESOURCES

Holt Biology Videodiscs Teacher's Correlation Guide, Lessons 1-1 through 1-4

Audio CD Program, Sections 1-1 through 1-4

internetconnect

sciLINKS NSTA

On-line Resources: www.scilinks.org
The following *sci*LINKS Internet resources can be found in the student text for this chapter:

Topics:
- Homeostasis, p. 8
- Characteristics of life, p. 11
- Scientific method, p. 14
- SI units, p. 23

go.hrw.com

On-line Resources: go.hrw.com
Visit the HRW Web site for a variety of resources related to this chapter. Just type in the keyword HM2 Chapter 1.

Smithsonian Institution®
Visit **www.si.edu/hrw** for additional on-line resources.

CNNfyi.com
Visit **www.cnnfyi.com** for late-breaking news and current event stories selected just for you.

LAB ACTIVITY PLANNING

CHAPTER 1 INVESTIGATION

Using SI Units, pp. 28–29

OVERVIEW
This lab introduces students to the process of taking measurements. Two different experiments allow students to use SI units. This lab also introduces students to basic lab equipment. In Part A, students heat sand and measure its temperature. In Part B, students compare the density of oil and water.

SAFETY
Caution students to wear heat-protective gloves when handling the heat lamp. Do not use the heat lamp near water.

PREPARATION
This is a good lab to use with lab groups. Set up three or four stations for Part A and three or four stations for Part B. Have half of the students complete Part A while the other half of the students complete Part B. Have students switch lab stations after 15–25 minutes.

Quick Labs

Observing Homeostasis, p. 6
Students will need three large beakers (500 mL), a wax pencil, a goldfish, a small fish net for moving the goldfish from beaker to beaker, tap water, a thermometer, very warm water, ice, and a watch or clock with a second hand.

Predicting Results, p. 18
Each group of students will need two Petri dishes that contain sterile nutrient agar. They will also need cellophane tape and a wax pencil to label the Petri dishes.

UNDERSTANDING THE VISUAL

Ask students to study the photograph of the young snowy owl. Point out the owl's gray coloring and explain that adult owls are white. Ask students why this juvenile animal might have different coloring from its parents. (Because snowy owls nest on the ground, their young are vulnerable to predation by a number of animals not able to reach nests in trees. Their gray color is camouflage on the greenish-gray background of the tundra.)

FOCUS CONCEPT

Biological Themes and Processes

Certain themes unify all the biological sciences. Both the scientist who studies a single type of organism and the one who studies the global interactions of millions of organisms will find the same biological themes at work. The themes that will be highlighted in this textbook are cell structure and function; stability and homeostasis; reproduction and inheritance; evolution; interdependence of organisms; and matter, energy, and organization. Living things, in all their great diversity, also share common characteristics. Every living thing, whether an amoeba or an elephant, is composed of one or more cells, has organization, uses energy, maintains a stable internal environment, grows, and reproduces as a species.

ASSESSING PRIOR KNOWLEDGE

Ask students to define the characteristics of living things and to describe how we go about studying living things.

CHAPTER 1

THE SCIENCE OF LIFE

Snowy owls, Nyctea scandiaca, *live on the open tundra and build their nests on the ground. This juvenile owl hatched in a flower-filled alpine meadow.*

FOCUS CONCEPT: ***Biological Themes and Processes***

As you read, take note of the recurring themes of biology and the characteristics that all organisms share.

THEMES OF BIOLOGY

Scientists estimate that 40 million species of organisms inhabit the Earth. But of these 40 million species, only about 2 million have been identified and named, and only a few thousand have been studied in any detail. Thus, much of the world of biology remains to be explored and studied. Fortunately, there are guideposts to help chart a course through this vast and fascinating world. These guideposts are the themes that unify biology. In this chapter, you will read about the unifying themes of biology that allow biologists to study the enormous diversity of life.

SECTION 1-1

OBJECTIVES

▲ List six unifying themes of biology.

● Explain how organisms get the energy they need to survive.

■ Describe the main difference between the structure of a living thing and that of a nonliving thing.

THE STUDY OF LIFE

The first life-form probably arose on Earth more than 3.5 billion years ago. Much evidence suggests that this first **organism**—or living thing—was a single cell, too small to be seen without the aid of a microscope. For millions of years, single-celled organisms floating in the seas were the only life on Earth. No fish swam in the oceans, no birds flew through the air, and no snakes moved across the ground.

Over time, organisms changed. New kinds of organisms arose from older kinds and came to inhabit almost every region of the Earth. Today brilliantly colored birds glide through the trees of tropical rain forests. Odd-looking fish prowl the dark depths of the oceans, emitting a strange red light that helps them locate their prey. Giant sequoia trees tower hundreds of feet above the ferns growing on the forest floors. Nearly blind moles burrow their way underground searching for tiny insects to consume.

These and millions of other types of organisms form the diverse population of Earth. **Biology**—the science of life—is the study of all living things. Biology includes the study of the microscopic structure of single cells. It also includes the study of the global interactions of millions of organisms. It includes the study of the life history of individual organisms as well as the collective history of all organisms.

The science of biology is as varied as the organisms that are its subjects. (Compare the owl shown on the facing page with the algae shown in Figure 1-1.) Biology is unified by certain themes, however, that come into play no matter what organism or what kind of

FIGURE 1-1

This microscopic unicellular alga, *Eremoophaera* sp. (LM, 31×), lives in a water environment. It, together with the owl shown on the facing page, gives an indication of the great diversity found in the living world.

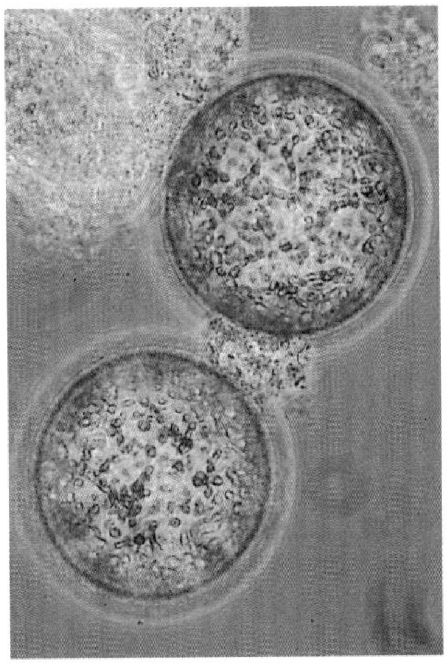

SECTION 1-1

SECTION OVERVIEW

Biological Themes

Section 1-1 explains the six major themes of biology: (1) cell structure and function, (2) stability and homeostasis, (3) reproduction and inheritance, (4) evolution, (5) interdependence of organisms, and (6) matter, energy, and organization.

ACTIVE READING

Technique: Brainstorming

Before students begin Chapter 1, ask them to define the prefix *bio-* (life) and the suffix *-logy* (study of). Next ask the students to generate a list of branches of biology that include the suffix *-logy. (Examples might include microbiology and histology.)* Finally, ask students to describe an area of biology for which they do not know the name. *(An example might include the study of the kidneys.)* Have them use a dictionary to find the specific name *(nephrology).*

VISUAL LINK

Figure 1-1

Refer students to Figure 1-1, and ask them to list the themes of biology listed in this chapter and how they apply to this organism.

Explain that the italicized, Latinized name is the species name of the organism. In this example, the abbreviation sp. indicates that this organism is a member of the genus *Eremoophaera;* the actual species is not indicated. For the owl on the facing page, the first term in the species name is that of the genus and the second term is the species identifier.

VISUAL LINK

Figure 1-2

Refer students to Figure 1-2 and point out that there are different kinds of cells in this photo and that they all perform different but vital functions.

Explain that the abbreviation *LM* refers to light micrograph, which is an image photographed through a light microscope. The notation 240× refers to the degree of enlargement of the image.

Quick Lab

Observing Homeostasis

Time Required 20 minutes

Safety Remind students to handle the fish carefully.

Procedural Tips Allow the tap water to sit overnight before use. Show students how to transport the fish with a dip net. They will have to catch the fish without getting their hands in the water and without bruising the fish. Once the fish is in the net, tilt the handle to close the opening so the fish will not be able to leap out. Place the net with the fish into the beaker and let the fish swim out. Have students prepare the water in their beakers before they move the fish from one beaker to another.

Explain to students that gills are under the hard plates, called opercula, that cover and protect the gills located on each side of the head. See page 808 for an illustration.

Answers to Analysis Gills move faster at higher temperatures. The fish is taking in more water and thus more oxygen. The rate at which the gills move affects the amount of oxygen that enters the blood. In this way, the gills help to maintain the proper balance of gases in the fish's blood.

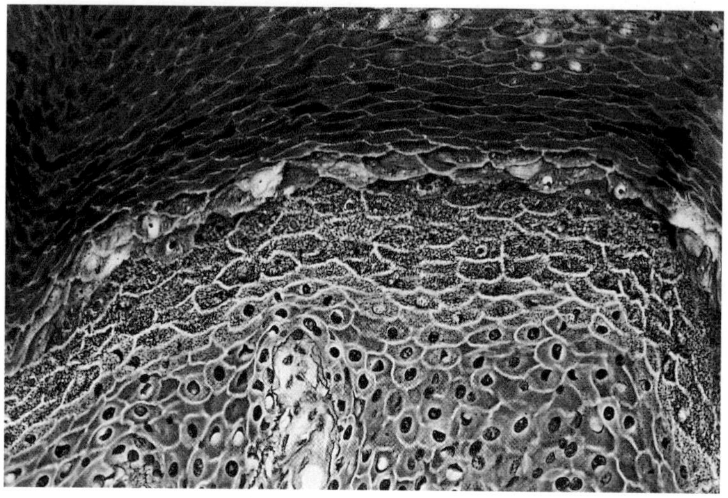

FIGURE 1-2

These tightly packed cells are from a thin section of human skin (LM 240×). There are many different kinds of cells in animal bodies.

interaction is studied. Six major themes recur throughout this book: (1) cell structure and function, (2) stability and homeostasis, (3) reproduction and inheritance, (4) evolution, (5) interdependence of organisms, and (6) matter, energy, and organization. The paragraphs that follow will give you a brief introduction to each theme. You will get a closer look at each theme in later chapters of this book. By reflecting on these themes in your study of biology, you will recognize the patterns that connect all life-forms.

Quick Lab

Observing Homeostasis

Materials 500 mL beakers (3), wax pen, tap water, thermometer, ice, hot water, goldfish, small dip net, watch or clock with a second hand

Procedure

1. Use a wax pen to label three 500 mL beakers as follows: 27°C (80°F), 20°C (68°F), and 10°C (50°F). Put 250 mL of tap water in each beaker. Use hot water or ice to adjust the temperature of the water in each beaker to the temperature indicated on the label.
2. Put the goldfish in the beaker of 27°C water. Record the number of times the gills move in 1 minute.
3. Move the goldfish to the beaker of 20°C water. Repeat observations. Move the goldfish to the beaker of 10°C water. Repeat observations.

Analysis What happens to the rate at which gills move when the temperature changes? Why? How do gills help fish maintain homeostasis?

Cell Structure and Function

The cell is the basic unit of life. All organisms are made of and develop from cells. Some organisms are composed of only one cell. These organisms are called **unicellular** (YOON-uh-SEL-yoo-luhr) **organisms.** Most of the living things that you see around you are **multicellular** (MUHL-ti-SEL-yoo-luhr) **organisms.** That is, they are composed of more than one cell.

Cells are small but highly organized. They contain specialized structures that carry out the cell's life processes. There are many different kinds of cells, but all cells are similar in several ways. They are surrounded by a membrane and contain a set of instructions—in the form of genetic information—necessary for making new cell parts, as well as new cells, from the molecules the cell takes in.

Look again at Figure 1-1, and compare the unicellular organism with the cells shown in Figure 1-2. New cells produced by unicellular organisms are virtually identical to the parent unicellular organisms that produced them. In contrast, mature multicellular organisms, which also begin their lives as one cell, contain not only *many* cells but also many different *kinds* of cells. The cells of multicellular organisms underwent **differentiation.** That is, they became different from each other as they multiplied and followed the various roles supplied for them by their genetic instructions.

Stability and Homeostasis

Living things, from single cells to entire large organisms, maintain very stable internal conditions. You know that the human body operates within a very narrow temperature range, around 37°C (98.6°F). Animal bodies like yours have systems that maintain internal conditions, such as temperature, water content, and even food intake, at very stable levels. This stable level of internal conditions, called **homeostasis** (HOH-mee-oh-STAY-sis), is found in all living things, including single cells.

The seal shown in Figure 1-3 can keep its body temperature very stable in a variety of water temperatures, even in very cold water. It takes in enough food to maintain a thick layer of fat under its skin for insulation. Blood flow through the seal's body, and in its flippers, is adjusted according to the temperature of the water it swims in. Its body can conserve or release heat as needed to remain at a stable temperature.

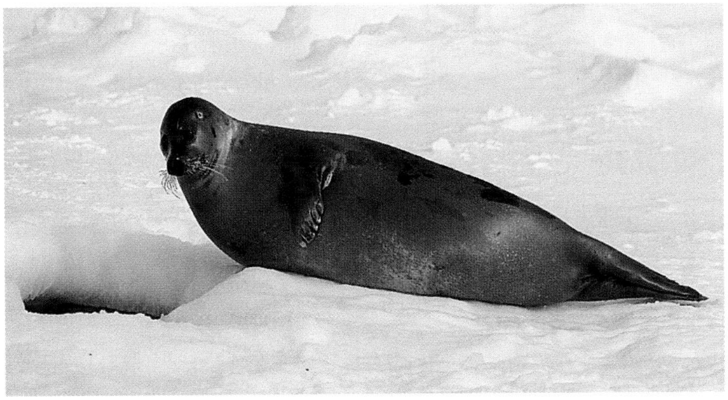

FIGURE 1-3

The shape and composition of the body of this harp seal, *Phoca groenlandica,* together with its efficient, heat-conserving circulatory system, allow the seal to survive in water temperatures that would kill many other animals.

Reproduction and Inheritance

All organisms produce new organisms like themselves. In this process, called **reproduction,** organisms transmit hereditary information to their offspring. This hereditary information is in the form of a large molecule called deoxyribonucleic acid, or **DNA.** In multicellular organisms and in some unicellular organisms, DNA is enclosed within a structure made of membrane. In other unicellular organisms, DNA exists as a loop, like the one shown in Figure 1-4. A short segment of DNA that contains the instructions for the development of a single trait of an organism is called a **gene.**

The DNA of a cell is like a large library. It contains all of the books of instructions—genes—that the cell will ever need. In this way, the cell can make the structures and complex chemicals necessary for life. In multicellular organisms, the DNA of every body cell, even different kinds of body cells, is identical. Each cell uses certain genes from the complete set. For example, cells in your thyroid gland use the set of genes from your DNA that are instructions for making thyroid hormone. Your other body cells ignore these instructions and use only the genes appropriate to their particular functions.

In **sexual reproduction,** hereditary information from two parts of a single organism or from two organisms of the same species is combined. When an egg of a female leopard frog is fertilized by the sperm of a male leopard frog, the egg and sperm join to form a single cell. This joining produces a fertilized egg cell that contains hereditary information from both the female parent and the male parent. This cell begins a series of cell divisions to produce a new leopard frog. With each division, every new cell receives a complete copy of the same hereditary information. Thus, the new organism is composed of cells that contain complete sets of hereditary information from both of its parents.

In **asexual reproduction,** hereditary information from different organisms is not combined. For example, when a bacterium reproduces asexually, it splits in two. As a result, one cell gives rise to two cells. Each of the two cells contains an identical copy of the hereditary information from the original bacterium.

FIGURE 1-4

A loop of DNA (SEM 72,000×) contains the genetic information of the unicellular organism that it is found in. Many genes are linked in the DNA loop.

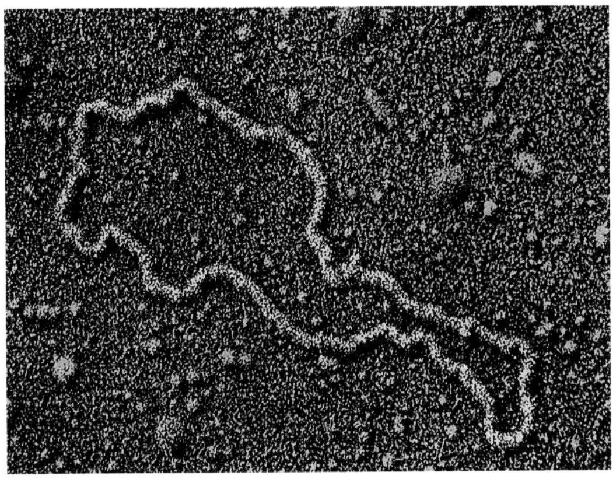

SECTION 1-1

RECENT RESEARCH
Cell Differentiation in the Lab

Developmental biologists have long relied on animal models to help them understand how a fertilized egg cell differentiates into a complete organism. The cells of fruit flies, frogs, and mice have been among those most commonly observed. Recently, researchers have begun observing the cells from zebra fish, a common freshwater aquarium fish. Zebra fish complete their embryonic development in only five days, and their transparent embryos make it easy for scientists to watch tissues as they develop and move about.

RETEACHING *ACTIVITY*

Ask students to clasp their hands together in front of their body. Tell them that the clasped hands represent a cell. Now tell them to pull their hands apart and make two separate fists. Tell them that this represents a cell undergoing asexual reproduction to form two new cells.

Ask students to hold closed fists out in front of their body. Tell them to imagine that one fist is a sperm cell and the other is an egg cell. Now tell them to clasp both hands together to form a single combined cell. This represents an egg cell that has been fertilized by a sperm cell. Now describe how this single fertilized egg will divide many times to form a complete new organism.

QUICK FACT

☑ Tell students that the DNA inside one human cell would extend several feet if stretched out. The human body contains over 50 trillion cells. Ask students to imagine what several hundred trillion feet of DNA might look like.

INCLUSION ACTIVITY

Help students understand how populations of organisms evolve in response to changes in their environment.

Have all but four students get down on their hands and knees. Of the four, have two students stand upright on their knees and the other two stand upright on their feet. The four represent variations within the population. Tell the students that the "fruit" that they must eat to survive is represented by playing cards. Scatter the playing cards on the floor. Each student must pick up a playing card to survive. Point out that all of them were able to pick up a card, although it was easiest for those closest to the ground to do so. These members of the population have a survival advantage.

Now tell the students that 1 million years have passed and the fruit only grows on plants that are of medium height. Scatter the cards on the desktops and tell the students to "feed" once again. This time the students on all fours are not allowed to lift their hands more than 2 in. off the ground. As a result of this limitation, none of the students on all fours should be able to obtain a card. Now the population is made up only of students who are upright to some degree.

Tell the students that another 1 million years have passed and the fruit is found only on the very tallest trees. Tape some playing cards high up on a wall and have the remaining students feed once more. Relate this activity to the adaptations and natural selection that occur in real populations of organisms.

internetconnect

SCILINKS NSTA

TOPIC: Homeostasis
GO TO: www.scilinks.org
KEYWORD: HM008

Evolution

Populations of organisms **evolve,** or change, over generations. The study of **evolution** helps us understand how the many kinds of organisms that have lived on Earth came into existence. It also explains how organisms alive today are related to those that lived in the past. It helps us understand why organisms look and behave the way they do. It provides a basis for exploring the relationships among different groups of living organisms.

Scientists suggest that a process called **natural selection** is the most important driving force in evolution. According to the theory of evolution by natural selection, organisms that have certain favorable traits are better able to successfully reproduce than organisms that lack these traits. For example, white fur is helpful in snowy environments for animals like rabbits and mice, which are prey for many other species. Obviously, an animal that is killed early in life can produce no offspring at all. The animal in Figure 1-5 has a coat that changes color as the day length changes. Thus, it can retain its camouflage during winter and summer.

Evolution by natural selection is driven in part by the competition among individuals for resources necessary for survival, such as food. Not all individuals are able to live and reproduce to the same degree. Differences among individuals of a species can affect their ability to produce offspring. Some individuals have traits that give them advantages for survival and reproduction, and they tend to pass these traits on to their offspring.

The survival of organisms with favorable traits causes a gradual change in populations of organisms over many generations. This "descent with modification" as organisms adapted to different kinds of environments is responsible for the multitude of life-forms we see today.

FIGURE 1-5

Elegant solutions to the year-round problem of predation have evolved in some animals. The very different winter and summer coats of the arctic hare, *Lepus arcticus,* help the hare avoid predators.

Interdependence of Organisms

The study of individual organisms is an important part of biology. To fully understand the biological world, however, scientists study the interactions of organisms with one another and with their environment. This branch of biology is called **ecology.** A scientist may study the ecology of a single species. This would include observing individuals of the species and their interactions with each other and their environment. Scientists may also conduct large ecological studies of environments, studying specific **ecosystems** (EK-oh-SIS-tuhmz), or environmental communities. By studying ecology, biologists have come to recognize the interdependence of organisms with their physical environment. To survive, all organisms need substances such as nutrients, water, and gases from the environment. The stability of the environment in turn depends on the healthy functioning of organisms in that environment.

Scientists now recognize the enormous effect that humans have had on environments all over the world. For thousands of years, tropical rain forests, like the one shown in Figure 1-6, have provided stable—but fragile—environments for their inhabitants. In recent years, humans have cleared vast areas of these forests by cutting and burning vegetation. Roads cross areas that were once havens for tropical organisms. What is the effect of this activity? Plant and animal species native to the forest become extinct. The typically thin topsoil layer becomes depleted of nutrients and hardened by the sun. This makes the land unsuitable for planting or reestablishment of the forest. The atmospheric oxygen that was produced by the dense vegetation is lost. It is replaced by carbon dioxide from the burning of trees. Destruction of these ancient and valuable forests could affect all life on Earth.

FIGURE 1-6

Tropical rain forests, like this one on the eastern slope of the Andes Mountains in Ecuador, support an extraordinary variety and number of plants and animals, typically on a very thin layer of fertile topsoil.

Eco Connection

Saving the Forests

Tropical rain forests, like the one shown in Figure 1-6, exist at equatorial latitudes in Central and South America, Africa, and Asia. They are lush but extremely fragile multistoried ecosystems that support a staggering variety of life-forms. Before humans began cutting trees for timber and clearing land for farming, tropical rain forests covered about 6 million square miles on Earth. In the late 1970s, rain forests were being cleared at a rate of 29,000 square miles per year. By the early 1990s, that figure had climbed to more than 55,000 square miles per year. In less than 100 years, the rain forests may be gone.

Rain-forest products include rubber, coffee, and various kinds of fruits and nuts, as well as wood used for furniture, such as mahogany and teak, and wood used in paper manufacturing. Purchase by consumers of periodically harvested resources such as fruits, nuts, and rubber encourage responsible management of the rain forest renewable resources. Unfortunately, some paper manufacturers are very large-scale consumers of rain-forest wood. By learning about the track record of manufacturers with respect to rain-forest conservation, consumers can make more-responsible decisions about the products they use. Such awareness may help slow damage to these complex and irreplaceable ecosystems.

DEMONSTRATION

Rain-Forest Products

Display various consumer products that come from tropical rain forests, such as Brazil nuts, bananas, coffee, tea, rubber, and chocolate. Explain that humans depend on tropical rain forests not only for products but also for the oxygen that the plants of these lush forests produce. Ask students how tropical rain forests depend on us. (Humans and other animals produce carbon dioxide, which the forest plants need for photosynthesis.)

TEACHING STRATEGY

Soil: The Base of an Ecosystem

Collect a shovelful of soil from an area that contains rich soil. Spread newspaper on the classroom floor and dump the soil onto the paper. Give students hand lenses and make available one or more stereomicroscopes for them to use. Challenge students to find as many different kinds of organisms as possible in the soil. Provide various reference books, such as protozoology, microbiology, and invertebrate zoology textbooks, to help students identify the creatures they discover.

CRITICAL THINKING

Building on Previous Knowledge

Tell students that the branch of biology known as ecology has developed only fairly recently. Then ask them why this might be so? What sort of biological knowledge would be necessary before the study of ecology could begin? (Biologists needed information about specific organisms and their environments in order to study their interrelationships. This information has taken many years to obtain.)

SECTION 1-1

GIFTED ACTIVITY

Green plants are not the only organisms that produce their own energy. Challenge students to research other autotrophs that exist in nature (algae, blue-green bacteria, chemolithotrophic bacteria). Have them prepare one-page reports that compare and contrast the different kinds of autotrophs.

ANSWERS TO SECTION 1-1 REVIEW

1. The six themes discussed in the chapter include cell structure and function; stability and homeostasis; reproduction and inheritance; evolution; interdependence of organisms; and matter, energy, and organization.
2. Organisms transmit hereditary information to their offspring in the form of DNA.
3. (1) Organisms in a population must compete for limited resources. (2) Individuals in a population differ from one another. Some have traits that allow them to reproduce more successfully than other organisms.
4. Rain forests typically support large numbers of plants and animals on a thin layer of topsoil. Removal of vegetation leads to depletion of soil nutrients and hardening of the soil, eliminating the possibility of reforestation.
5. Autotrophs synthesize the nutrients they need from simple substances, in most cases using the energy they trap from the sun. Heterotrophs must rely on other organisms in order to obtain the nutrients they need as their energy source.
6. The traits that are helpful to an organism depend almost entirely on that organism's interaction with the environment.

Word Roots and Origins

photosynthesis

from the Greek *photo,* meaning "light," and *syntithenai,* meaning "to put together"

Matter, Energy, and Organization

Living things have highly organized structures that must be maintained in their orderly state by a constant supply of energy. How organisms obtain, use, and transfer energy is a major topic of study in biology. Almost all the energy for life on Earth comes from the sun. Through the process of **photosynthesis** (FOH-toh-SIN-thuh-sis), plants and some types of unicellular organisms capture the energy from the sun and change it into a form of energy that can be used by living things.

Organisms that obtain their energy by making their own food, like plants, are called **autotrophs** (AWT-oh-TROHFS). Using the energy they trap from the sun, some kinds of autotrophs convert water and carbon dioxide from the environment into energy-rich substances such as sugars and starches. These substances are then used by the organisms for their own energy needs. Other kinds of autotrophs use different chemical processes to get energy. Autotrophs may be consumed and used as an energy source by other living things. **Heterotrophs** (HET-er-oh-TROHFS) are organisms that must take in food to meet their energy needs. Heterotrophs include all animals and fungi as well as many unicellular organisms and a few plant species. Because they cannot make their own food, heterotrophs must consume autotrophs, other heterotrophs, or both for their energy needs.

FIGURE 1-7

Many nonliving things, like this salt crystal (LM 55×), have a very organized structure but do not display the degree of organization and complexity found in living things.

In living things, complex chemicals are broken down. Their parts are then reassembled into chemicals and structures needed by the organism. Living things are set apart by their complexity. Even a simple, small organism, such as the alga shown in Figure 1-1, is far more complex than a nonliving thing, such as the mineral that appears in Figure 1-7. The mineral has an organized molecular and crystalline structure. However, the multitude of different molecular structures found in the alga are much more complex and varied than anything found outside the living world.

SECTION 1-1 REVIEW

1. List six unifying themes of biology.
2. How do organisms produce offspring like themselves?
3. What are two of the driving forces behind the process of natural selection?
4. Why are rain forests considered fragile environments that are vulnerable to permanent destruction?
5. How do autotrophs differ from heterotrophs in obtaining energy?
6. **CRITICAL THINKING** Why do scientists say that the environment "selects" the traits that allow an organism to survive and reproduce?

SECTION

1-2

OBJECTIVES

▲ List six characteristics of life.

● Describe how a living thing is organized.

■ Explain why all living things on Earth are not yet well understood.

THE WORLD OF BIOLOGY

Our world abounds with a great variety of life. Giant tube worms thrive at the bottom of the oceans near bubbling volcanic vents. Red algae spread to cover the surface of arctic glaciers like a carpet. Bacteria grow in every part of the world, even in the pores of your skin.

CHARACTERISTICS OF LIFE

All organisms, no matter how different from each other they may be, share certain features characteristic of all living things.

Cells

All living things are composed of cells. In multicellular organisms, some cells are specialized to play a specific role. The plant stem shown in Figure 1-8 contains many different cells, each separate from the surrounding cells, like many rooms in a large building. Cells are always very small. Large multicellular organisms contain many cells, while smaller organisms contain fewer cells.

Organization

All living things are highly organized at both the molecular and cellular levels. They take in substances from the environment and organize them in complex ways. In cells, specific cell structures carry out particular functions. In most kinds of multicellular organisms, cells and groups of cells are organized by their function. Several different cell types grouped together in the section of the plant stem are easily distinguished in Figure 1-8.

internetconnect

SCILINKS NSTA

TOPIC: Characteristics of life
GO TO: www.scilinks.org
KEYWORD: HM011

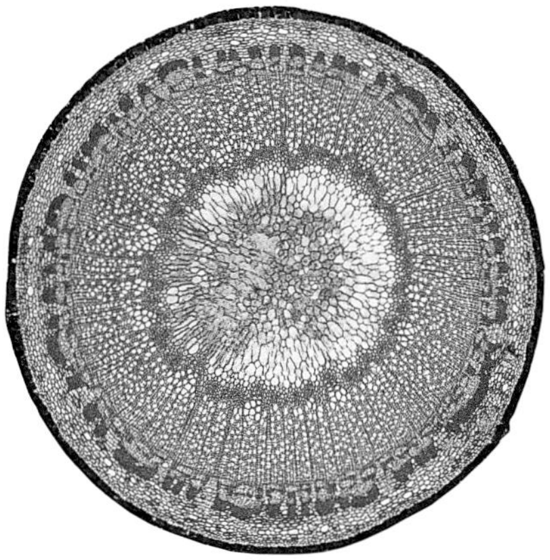

FIGURE 1-8

The different types of cells in this plant stem are organized into structures that perform specific functions. The circular conducting tubes, shown in cross section (LM 19×), transport water and other substances up and down the plant stem.

SECTION 1-2

SECTION OVERVIEW

Cell Structure and Function
Section 1-2 describes the role of cells in living things. Cells are the microscopic, living building blocks of which all organisms are made. Inside cells are even smaller structures that carry out specific functions.

Matter, Energy, and Organization
Section 1-2 describes how living things use energy to maintain their internal organization and to grow and reproduce.

Stability and Homeostasis
Section 1-2 describes how living things, whether single cells or complex multicellular organisms, use various mechanisms to maintain a stable internal environment.

▶ ENGAGE STUDENTS

Have students watch a toy bumper car as it moves about the classroom. Point out that the toy car has organized structure and uses energy. Ask students to explain why the car is not considered a living organism.

CRITICAL THINKING

Levels of Organization in the Human Body
Tell students that the cells in the human body are organized into tissues, organs, and organ systems. Ask them why this kind of specialization works so well. (Cells of one type can function together to accomplish one task more efficiently than if they were separated.)

VISUAL LINK

Figure 1-9

Refer students to Figure 1-9, and ask them what other mechanisms living things may use to maintain stable body temperature (fur, fat, lying in the sun).

DEMONSTRATION

Observing the Product of a Metabolic Pathway

Collect a 500 mL beaker, a 100 mL beaker, a large test tube, and a small test tube that will fit inside the large one. Prepare a fresh lukewarm yeast solution. Add some sugar (glucose) to the solution and stir. Set up a warm-water bath, putting one beaker inside the other. Maintain the temperature of the water in the small beaker at around 38°C. Fill the large and small test tubes to the top with the yeast solution. Invert the small tube and place it inside the large test tube upside down without trapping an air bubble in the bottom of the small tube. You should do this over a sink. Place this "double test tube" setup in the small beaker. Let students observe what happens in the small test tube over the next 10 to 15 minutes. Tell students that you added sugar to the yeast solution. (The sugar is an energy source for the yeast.) Tell students that a bubble forms in the bottom of the small test tube because the yeast cells are using the sugar in a metabolic pathway called cellular respiration. As a result, carbon dioxide gas is produced.

QUICK FACT

✔ The biodiversity of the world has not always been as great as it is today. In fact, there have been periods during Earth's history when the level of biodiversity has sharply decreased. Scientific evidence shows that global mass extinctions have occurred five times throughout the history of life on Earth.

FIGURE 1-9

All living things, even single cells, work to maintain a steady internal environment. Animals may have more than one mechanism that allows them to maintain a stable body temperature. This pine grosbeak, *Pinicola enucleator*, fluffs its feathers in cold weather to stay warm.

Energy Use

All living things use energy in a process called **metabolism** (muh-TAB-uh-LIZ-uhm), which is the sum of all of the chemical processes that occur in the organism. Organisms require energy to maintain their molecular and cellular organization as well as to grow and reproduce. Autotrophic organisms, such as green plants, use the food that they generate from photosynthesis for growth, maintenance, and reproduction.

Homeostasis

All living things maintain stable internal conditions. Even single cells work to keep their internal environment stable. A cell closely controls its water content by taking in or releasing water. A cell that takes in too much water will rupture and die. Multicellular organisms usually have more than one system for maintaining important aspects of their internal environment, like temperature. The feathers of the bird shown in Figure 1-9 become more erect in cold weather. This traps an insulating layer of air next to the bird's body.

Growth

All living things grow, as do many nonliving things. Nonliving things, like crystals or icicles, grow by accumulating more of the material they are made of. Living things grow as a result of cell division and cell enlargement. In unicellular organisms, cell division results in more organisms. **Cell division** is the formation of two cells from an existing cell, as shown in Figure 1-10. Newly divided cells enlarge until they are the size of a mature cell. In multicellular organisms, cell division and cell enlargement together result in growth.

The process by which an adult organism arises is called **development.** Development is produced by repeated cell divisions and cell differentiation. As a result of development, an adult organism is composed of many cells. Your body, for example, is composed of about 50 trillion cells, all of which derived from a fertilized egg.

Word Roots and Origins

metabolism

from the Greek *metabole*, meaning "to change"

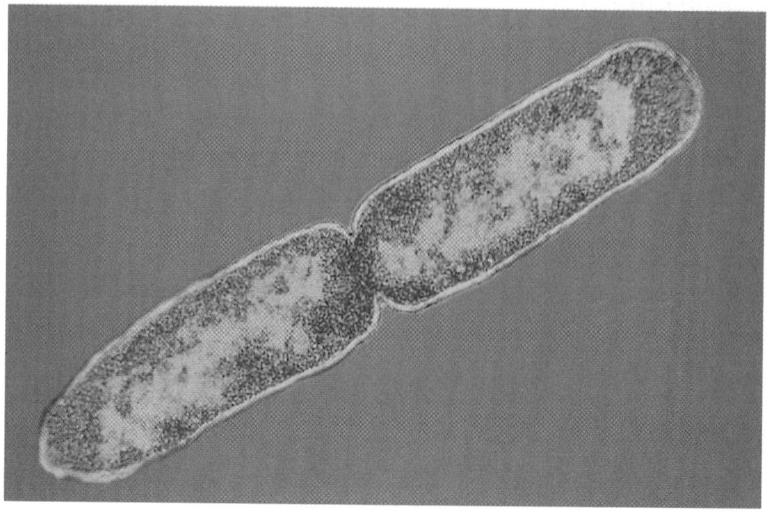

FIGURE 1-10

Cell division in this unicellular organism, *Escherichia coli* (TEM 98,000×), produces two identical offspring cells.

Reproduction

All species of organisms have the ability to reproduce. Reproduction is not essential to the survival of an individual organism. Because no organism lives forever, however, reproduction is essential for the continuation of a species. Glass frogs, such as the one shown in Figure 1-11, will lay many eggs in its lifetime. Of these, only a few offspring will reach adulthood and successfully reproduce.

During reproduction, many species have ways to combine the genetic information of individual members to produce offspring that are not identical to their parents. The frog shown in Figure 1-11 reproduces sexually. The eggs are fertilized by a male glass frog after they are laid. The resulting offspring carry traits of both parents.

FIGURE 1-11

Many species of animals that lay eggs produce large numbers of eggs. Many of the eggs laid by the glass frog, *Centrolenella* sp., will not survive. In contrast, the offspring of animals that give birth to live offspring typically have a high rate of survival.

THE LIVING WORLD

The world of biology is so diverse and so rich that scientists have yet to explore all of it. Many organisms that have not been identified live in regions of the world that are difficult to explore, like dense tropical rain forests and deep ocean waters. Some organisms are so small that they easily escape detection. A single gram of fertile soil, for example, may contain as many as 2.5 billion unicellular organisms. In the dense forests of Vietnam, scientists recently discovered a new species of deer. It weighs between 40 and 50 kg (90 and 110 lb). It is one and a half times as large as any other known Asian deer species.

The wide variety of unstudied organisms shows us how much is still to be learned about life. Wherever you live, the world around you teems with life. This book will help you appreciate the living world and understand how scientists go about exploring and studying it.

SECTION 1-2 REVIEW

1. **Name six characteristics that all living things share.**

2. **What are two levels of organization found in living things?**

3. **How does growth of a nonliving thing differ from growth of a living thing?**

4. **Why is reproduction an important characteristic of life?**

5. **Why are so many organisms yet to be discovered, identified, and described?**

6. **CRITICAL THINKING What would happen if all autotrophic organisms died, leaving only heterotrophic organisms?**

VISUAL LINK

Figure 1-11

Explain to students that though most species of frogs lay their eggs directly in water, some frogs lay their eggs on land in wet, shaded areas, such as in the leaves of a plant. When the eggs hatch, the tadpole may be carried to a body of water by rain or by a parent frog. Students can infer that the mortality of tadpoles would be higher than the mortality of offspring of animals such as those large mammals that give birth to single live young.

ANSWERS TO SECTION 1-2 REVIEW

1. cellular composition, organization, use of energy, homeostasis, growth, and reproduction
2. molecular and cellular organization
3. Some nonliving things, like rock crystals, grow by accumulating more of the materials they are made of. Living things grow by cell enlargement and cell division.
4. Reproduction is not vital to the survival of an individual, but it is vital to the continuation of life.
5. There is a great diversity of life, and many organisms are relatively inaccessible to scientists.
6. The heterotrophic organisms would die too because they depend on autotrophic organisms as a food source.

SECTION 1-3

SECTION OVERVIEW

Cell Structure and Function
Section 1-3 describes the physiological changes that occur in the body of a person infected with Ebola virus. In the advanced stages of this deadly disease, the integrity of the body's tissues breaks down and blood leaks through every body opening, even the pores in the skin.

Interdependence of Organisms
Section 1-3 describes how scientists used scientific methods to find the cause of Ebola hemorrhagic fever. Viruses must exist in living things; thus, a forest animal host probably harbors the virus.

▶ ENGAGE STUDENTS

Go to the library or a store that carries magician's supplies, and get a book on card tricks. Learn a simple trick and perform it for students. After you do the trick, discuss with students why it is important to make careful observations. Ask if anyone can figure out how the trick works. Explain that scientists use keen powers of observation to help them solve scientific problems.

CULTURAL CONNECTION

Unraveling the Mysteries of Life by Touch

Dr. Geerat Vermeij is a Dutch evolutionary biologist who is passionate about shells. This world-renowned scientist has been completely blind since the age of four. Dr. Vermeij makes observations about shells with his hands. By touching a shell, he is able tell what kind it is and where it is from. Feeling the shape and texture of a shell even gives him information about the evolutionary history of the animal that once lived in it.

SECTION 1-3

OBJECTIVES

▲ Define and give examples of *observing, measuring, organizing and analyzing data, inferring,* and *modeling.*

● Explain the relationship between hypothesizing, predicting, and experimenting.

■ Explain why good communication is so important in science.

◆ Describe the methods that scientists use in their work.

internetconnect

SCILINKS NSTA
TOPIC: Scientific method
GO TO: www.scilinks.org
KEYWORD: HM014

SCIENTIFIC METHODS

One of the best ways to begin studying science is to examine how scientists try to solve a problem or answer a question. Whatever they study, all scientists use certain methods to obtain knowledge. Understanding scientific methods will help you formulate strategies to answer questions you will encounter in your own scientific study or in your daily life. Consider how scientists searched for the cause of one of the deadliest infectious diseases in humans—Ebola hemorrhagic fever.

OBSERVING

In 1976, a deadly and contagious human disease appeared in several villages in the central African nation of the Democratic Republic of Congo, then known as Zaire. Local doctors observed a rapidly progressing and consistent sequence of symptoms in disease victims. Most had severe headaches, fever, bloody diarrhea, and vomiting. In the final stages of the disease, victims' internal organs bled uncontrollably and blood leaked through the nose, ears, and even the skin. Death from shock—collapse of the cardiovascular system due to loss of blood—followed rapidly. Teams of scientists from the the Centers for Disease Control and Prevention (CDC), the agency of the United States government responsible for monitoring diseases, and the World Health Organization (WHO) were sent to Zaire. Their job was to investigate the circumstances of this severe outbreak of what seemed to be a new disease.

ASKING A QUESTION

All scientific investigations begin with one or more questions. Think of the many questions that a scientist sent to investigate an outbreak of a disease would have. How is the disease transmitted among people? What caused the disease to appear suddenly in humans? How long after exposure to the agent of infection do the first symptoms of the disease occur? Do all victims die? The most fundamental question to be answered about the outbreak of a new disease is *What is the disease-causing agent?*

COLLECTING DATA

The longest phase of a scientific investigation is usually data collection. **Data** include any and all information that scientists gather in trying to answer their questions. Four important aspects of collecting data are described in the following paragraphs.

Observing

All scientific understanding of the natural world is based on observation. It is the observation of something unusual or unexplained that raises the first questions in a scientific investigation. Observation continues to be essential throughout all stages of scientific investigation. **Observation** typically employs one or more of the five senses to perceive objects or events. A scientist may observe the sound of the call of an animal in the wild. He or she may observe the appearance of an organism under a microscope, as the scientist is doing in Figure 1-12. Most observations in a scientific investigation are direct, although some things, like electricity, must be observed indirectly.

Measuring

Many kinds of observations involve quantitative data—data that can be measured in numbers. Scientists may measure the dimensions of an object, the number of objects in a group, the duration of an event, or other characteristics in precise units.

For example, in Zaire, the scientists recorded several types of quantitative data from their work in the field. These data included the number of people who displayed symptoms of the disease, the number of days that elapsed from the time symptoms first appeared until a victim died, and the number of people who died from the disease after they were infected. These data gave scientists a picture of the outbreak—an indication of how serious the disease was. The numbers they recorded revealed a grim picture. The outbreak had caused nearly 300 deaths. Death usually followed within a week after the first symptoms appeared, and the disease was extremely deadly. From 80 to 90 percent of the people who became infected died from the disease.

Word Roots and Origins

data

the plural of datum; remains unchanged from the Latin *datum,* meaning "a thing given"

FIGURE 1-12
Scientists make many of their observations within the confines of a laboratory.

CRITICAL THINKING

Numbers, Numbers, Numbers!

Tell students that data come in many forms. Sometimes they are descriptive in form, but often they are numerical. Numerical data are called quantitative data. Then ask them what the value of quantifying data is? (Quantitative data can be compared by scientists repeating others' experiments.)

GIFTED *ACTIVITY*

Have students investigate and report on Dr. Edward Jenner's work in the late 1700s. Have them compare and contrast Jenner's use of scientific methods to find a preventor for smallpox with the way modern scientists went about researching Ebola hemorrhagic fever. (Modern scientists tried to prevent the spread of Ebola virus. Jenner intentionally tried to infect a boy after the boy's vaccination.) Encourage them to raise any ethical concerns they might have about the way Dr. Jenner conducted his experiments. (Today Jenner probably would go to prison for trying such an experiment.)

OVERCOMING MISCONCEPTIONS

Many people think the terms *infectious* and *contagious* mean exactly the same thing, but they do not. Infectious diseases comprise a much broader category because they include all diseases transmitted by infected animals or contaminated objects, food, or water. Contagious diseases include only those diseases directly transmitted from person to person.

VISUAL LINK

Figure 1-12

Ask students to examine Figure 1-12 and determine what scientific methods this scientist is using (sampling and observing).

RETEACHING ACTIVITY

Have students use a sampling technique to estimate the student population of their school. First have them determine the total number of students in their classroom. Then ask them to find out how many classrooms are in the school. Multiplying the number of students in their classroom by the number of classrooms in the school will give an estimate of the school's student population. Students can then compare the actual number of students in the school to their estimate to see how accurate their sampling method was. Discuss with students reasons for any discrepancies between the two numbers. (Not all the classrooms may be in use, the number of students per room may vary, or they may have miscounted classrooms.)

CRITICAL THINKING

Why Sample?

Ask students why an accurate representation is not made when a scientist collects only two samples from a population of 30,000. (Statistically, these two samples may not be representative. If more samples are taken, chances increase that the samples collected will be more representative of the population and dilute the error.)

INCLUSION ACTIVITY

Have students practice their measurement, mathematics, and graphing skills. Divide them into pairs. Give each pair a meterstick and let them determine each other's height in centimeters. Have each pair calculate their average height and place their data in a table that you draw on the board. Draw a large *x*-axis and *y*-axis on the board, and let the students display their data by creating a bar graph. Help the students to correctly label the *x*- and *y*-axes and to come up with an accurate descriptive title for their graph.

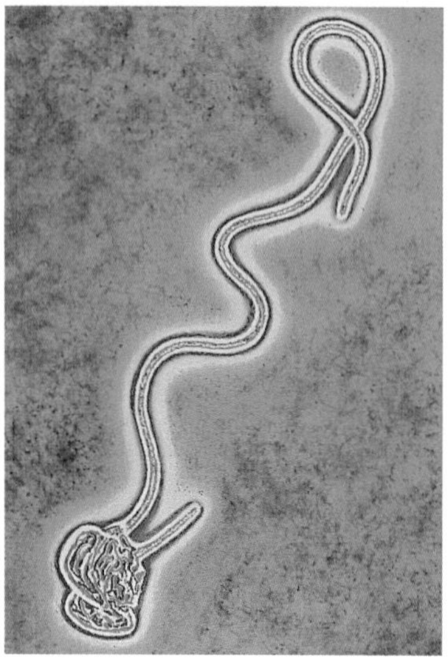

FIGURE 1-13

Because of its threadlike structure, the Ebola virus (TEM 23,000×) is classified as a filovirus. *Filo-* is a prefix meaning "threadlike" and is derived from the Latin word *filum,* which means "thread."

Sampling

Scientific **sampling** is the technique of using a sample, that is, a small part, to represent an entire population. To be useful, samples must be large and random. That is, they should include as many subjects as possible, and scientists must be sure to sample a cross section of the population so that an accurate representation is obtained. The scientists working in Zaire took hundreds of blood samples, both from people stricken with the disease and from people who were apparently healthy.

These blood samples taken from the disease victims provided the first important clue to the nature of the disease. The samples were sent from Zaire to laboratories in Europe and to the CDC in Atlanta, Georgia. Scientists at the CDC and at the University of Anvers, in Belgium, isolated a strange, threadlike virus from the blood samples. Using a powerful electron microscope, the scientists at the CDC photographed the virus, shown in Figure 1-13.

Organizing Data

Data are of little use unless they are organized. Organizing data involves placing observations and measurements in some kind of logical order, such as in a graph, chart, table, or map. The scientists tracking Ebola fever organized many kinds of data. They organized quantitative data into tables and charts. Remember that these quantitative data gave them an initial picture of how deadly the disease was.

Another important goal of the scientists was to trace the identity of the first person who had become infected with the disease. This information would be an important clue for identifying where the virus came from and how it was transmitted to humans. To do this, scientists who track the spread of a disease may organize information on all reported cases of the disease into a flowchart. The scientists can then trace the thread of contagious spread backward in hopes of identifying the first infected person. A flowchart, similar to the one shown in Figure 1-14, serves as a "family history" of the disease.

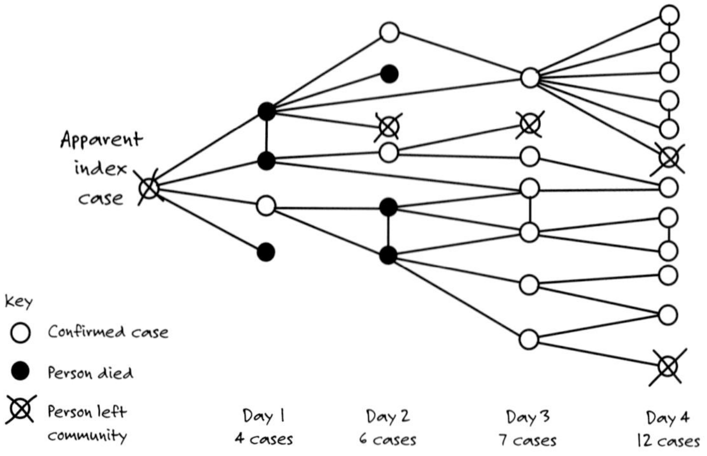

FIGURE 1-14

Scientists who trace the transmission of disease develop flowcharts much like this one so they can track the disease back to its source.

HYPOTHESIZING

When scientists have made many observations and collected sufficient data, they can suggest an explanation for what they have seen and recorded. This explanation, called a **hypothesis,** is a statement that explains their observations *and* can be tested.

Forming a Hypothesis

The central hypothesis in this complicated investigation was that the virus photographed by the CDC scientists caused the hemorrhagic fever. Although it may seem obvious that this is true, cause-and-effect relationships never can be assumed in a scientific investigation. Rather, evidence must be accumulated in every step.

Hypothesizing is a very important step in scientific investigation. A statement is testable if evidence can be collected that either does or does not support it. Recall the scientists' hypothesis about the identity of the infectious agent. It would not be supported if the infection with the virus did not cause symptoms of the disease.

Although a hypothesis may be supported by evidence, it can never be proved true beyond all doubt. At any time, new data might indicate that a previously accepted hypothesis does not hold true in all instances. Scientists often must refine and revise their original hypotheses—or even discard them—as they uncover new evidence.

Predicting

To test a hypothesis, scientists make a prediction that logically follows from the hypothesis. A **prediction** is a statement made in advance that states the results that will be obtained from testing a hypothesis, if the hypothesis is supported. A prediction most often takes the form of an "if-then" statement. In the case of Ebola fever, scientists made a prediction. If the virus were the true disease-causing agent, then introducing the virus into healthy tissue would cause cell death like that found in victims of the disease.

EXPERIMENTING

A hypothesis is often tested by carrying out an **experiment.** Experimenting is the process of testing a hypothesis or prediction by gathering data under controlled conditions.

Conducting a Controlled Experiment

Most experiments in biology are controlled experiments. A **controlled experiment** is based on a comparison of a **control group** with an **experimental group.** The control group and the experimental group are designed to be identical except for one factor. This factor is

CRITICAL THINKING

Hypothesis and Prediction

Tell students that a hypothesis and a prediction are dependent on one another and can never be separated. Ask them to explain this statement. (Because a hypothesis must be supported or proven wrong, the prediction that comes from a hypothesis must be tested. The prediction is dependent on the hypothesis, and the proof is in the experiment.)

GIFTED *ACTIVITY*

Ask students to bring in a recent article from a science magazine or journal that documents a scientific study. As they read their articles, have them circle all passages describing the use of scientific methods. For example, they should circle parts of the article that indicate a hypothesis, a prediction, observations, or experimental designs. Have students make oral presentations to the class in which they summarize the content of their articles as well as point out how various scientific methods were used.

RECENT RESEARCH

Ebola Virus Host

Since the first outbreak of Ebola hemorrhagic fever in Africa in 1976, scientists have tried to identify the animal that harbors the deadly virus in the wild. Knowing the identity of the carrier can help scientists prevent future human outbreaks. After testing numerous wild plants, mammals, insects, and snakes, scientists have found evidence that the virus can live in bats, which are therefore suspected of being a reservoir. Monkeys have been known to become ill and transmit the disease to humans, but they are not suspected of being the reservoir.

Quick Lab

Predicting Results

Time Required 25 minutes

Safety Treat any growth in the dishes as pathogens. Do not allow students to open any of the Petri dishes after incubation.

Procedural Tips Prepare the Petri dishes ahead of time. You may wish to use an incubator to produce a more abundant bacterial growth.

Disposal Discard sealed disposable Petri dishes in a biohazard bag. Glass dishes must be autoclaved. The melted agar can then be discarded as solid waste, and the dishes can be washed and stored as usual.

Answers to Analysis

1. If students' predictions were accurate, students should cite the evidence of visible growth or lack of it on the agar that matches their predictions in detail. (A black, fuzzy growth is not the same as a yellow, shiny growth.)
2. Students might suggest changes to either method or prediction according to their research objective. For example, they might wash their hands before touching the agar or they might narrow or broaden their predictions. Students should be able to justify their changes based on their first result.
3. An unpredicted result is valuable in that it provides information on which to base further predictions and experimentation.

Quick Lab

Predicting Results

Materials 2 Petri dishes with agar, cellophane tape, wax pen

Procedure

1. Open one of the Petri dishes, and streak your finger across the surface of the agar.
2. Replace the lid, and seal it with cellophane tape. Label this Petri dish with your name and the number 1.
3. Seal the second Petri dish without removing the lid. Label this Petri dish with your name and the number 2.
4. Record a description of your actions for each Petri dish. Then write your prediction about what will happen in each dish.
5. Store your dishes as your teacher directs.
6. Record your observations.

Analysis

1. Was your prediction accurate? What evidence can you cite to support your prediction?
2. If you did not obtain the results you predicted, would you change your testing method or your prediction? Explain why and how you would make changes.
3. Evaluate the importance of obtaining a result that does not support your prediction.

called the **independent variable.** During the course of a controlled experiment, a scientist observes or measures another factor in both the control group and the experimental group. This factor is called the **dependent variable**—dependent because it is driven by or results from the independent variable.

A controlled experiment supported the hypothesis that Ebola fever was caused by the previously unknown virus that was isolated by scientists in Belgium and the United States. Scientists at the University of Anvers tried to confirm the link between the virus and the cell death that caused the symptoms of the disease. Because monkeys are genetically similar to humans, kidney cells from a species of African monkey were used in an animal model of human infection. Test tubes containing the monkey kidney cells were divided into an experimental group and a control group. Drops of blood that contained the virus, taken from a victim of Ebola fever in Zaire, were added to each test tube in the experimental group. No blood was added to the test tubes in the control group. In the experiment, the independent variable was the addition of blood to the monkey kidney cells. The dependent variable was the health of the cells.

Analyzing Data

After a scientist has collected and organized data from a field study or an experiment, the data must be analyzed. Analyzing data is the process of determining whether data are reliable and whether they support or do not support a hypothesis or prediction. Scientists analyze data in many ways. For example, they may use statistics to determine relationships between variables, compare the data with those obtained in other studies, and determine possible sources of experimental error.

Eleven days after the virus-containing blood was mixed with the monkey kidney cells, the cells in the experimental test tube were dead. Moreover, addition of liquid from the experimental test tubes to new test tubes containing healthy monkey kidney cells caused the death of those cells also, again after a period of 10 to 11 days.

DRAWING CONCLUSIONS

The goal of scientific investigation is to shed light on something previously not understood. The final step of most investigations is to produce a model.

Modeling

Modeling involves constructing a representation of an object, a system, or a process that helps show relationships among data. A **model** is essentially an *explanation* supported by data. It can be visual, verbal, or mathematical. Scientists sometimes use models to help generate new hypotheses or predictions. For example, scientists have examined the case histories of people who have died from Ebola fever to develop a model of how the virus is transmitted among humans. Close contact with an infected person is necessary. Direct contact with the blood of an infected person seems to be the likeliest way to become infected. Many people who had Ebola fever were infected in crowded hospitals, like the one shown in Figure 1-15, by contaminated needles and instruments. Scientists are not sure whether the virus can penetrate skin or enter through the nose or mouth. Perhaps a break must first be present in the skin or lining of the nose or mouth.

The model of Ebola virus transmission is far from complete. The first human case in the 1976 outbreak in Zaire was never identified. Scientists still do not know the identity of the animal species that harbors the virus in the wild or how the virus is spread to humans. Once scientists have developed a complete model of where the virus originates and how it is transmitted to humans, they can specify the precautions to take to avoid infection.

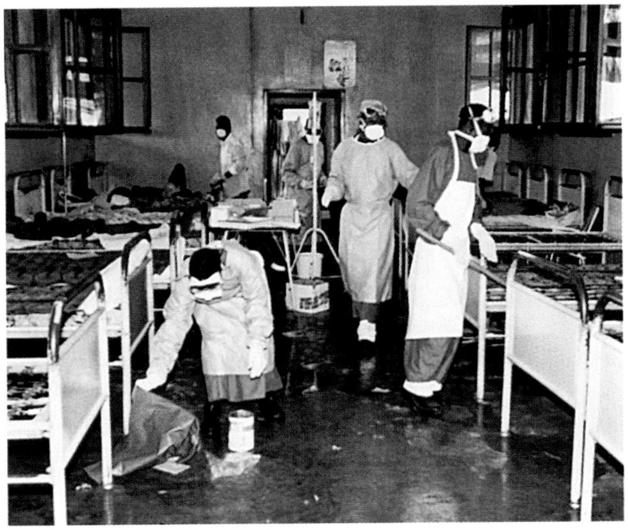

FIGURE 1-15

The Ebola virus spread in hospitals and clinics in Zaire because of overcrowded and unsanitary conditions.

Inferring

An **inference** (IN-fuhr-uhns) is a conclusion made on the basis of facts or premises rather than on direct observations. If you see smoke, you will probably infer that its source is a fire, even if you can't see the fire. In science, inferences are often drawn from data gathered from a field study or experiment, considered together with previous knowledge. Unlike a hypothesis, an inference is not directly testable. The scientists on the trail of the source of the Ebola virus made several important inferences from their work.

For example, based on the observations they made and the data they collected and analyzed, scientists inferred that the Ebola virus is carried by a small, forest-dwelling animal, possibly a bat. The virus may not cause the disease-carrying animals, or hosts, to become ill, as it does to humans. Scientists suspect that the host animals periodically move in search of food into regions inhabited or used by humans. When they do, they come in contact with people, some of whom contract the virus from the animals.

Forming a Theory

In science, a theory may be formed after many related hypotheses have been tested and supported with much experimental evidence. The word *theory* does not mean "wild guess"; it does not even mean "hypothesis." Rather, a **theory** is a broad and comprehensive statement of what is thought to be true. A theory is supported by considerable evidence and may tie together several related hypotheses. Few true theories are produced in science, relative to the volume of work performed and the number of hypotheses tested.

VISUAL LINK

Figure 1-15

Referring to Figure 1-15, ask students to draw conclusions about how Ebola virus was transmitted in crowded hospital conditions such as those depicted in the figure. (Droplets from coughs and sneezes and contact with soiled cloth facilitated the spread of the virus.)

OVERCOMING MISCONCEPTIONS

The word *theory* is frequently used to suggest an idea that is uncertain. For example, a person might say, "It's my theory that wearing red brings me good luck." A theory is a general truth that explains the observations of many scientists. A scientific theory is accepted only after it has been supported by a vast body of experimental data. True scientific theories are rare. Two theories in biology that have been widely accepted by scientists are the cell theory and the theory of evolution by natural selection.

TEACHING STRATEGY

Observation or Inference?

Students often get confused when differentiating between an observation and an inference. Observations are collected firsthand, by using the five senses. Inferences result from past observations or knowledge. Place a green grape in front of each student. Tell students they are not allowed to touch the grape during the activity; they can make only visual observations. Read the five statements below out loud. After each one, ask students to identify the statement as an observation or an inference.

1. The skin is green (observation).
2. The grape is edible (inference).
3. The skin is smooth (observation).
4. The skin is easily broken (inference).
5. There are seeds inside the grape (inference).

SECTION 1-3

RECENT RESEARCH
Mathematical Modeling

Researchers fear that sick cattle in Great Britain might cause an epidemic in humans. People who eat meat from cows that are ill with bovine spongiform encephalopathy (mad cow disease) may contract a similar ailment, which in humans is called Creutzfeldt-Jakob disease (CJD). Although scientists are still not completely sure that the infected cows are the cause of CJD, they have produced mathematical models that predict how many humans will become infected in the next three years if the diseased cows do turn out to be the cause.

ANSWERS TO SECTION 1-3 REVIEW

1. Inferring enables scientists to draw conclusions without making direct observations. Modeling enables scientists to understand the relationships among data they have collected or observations they have made.
2. Hypothesizing is forming a testable statement about observable phenomena. Experimenting is testing a hypothesis or prediction by gathering data under controlled conditions.
3. Communication enables scientists to build on each other's work and findings.
4. Scientists use a variety of methods, depending on the nature of their work.
5. A hypothesis is a testable statement, based on observation. A theory is a broad statement of what is thought to be true, based on the testing of many hypotheses.
6. Scientific theories are tentative because they cannot be proven beyond all doubt.

FIGURE 1-16
Throughout the world, scientists often gather to share what they have discovered. Each gathering is usually devoted to a specific subject or topic.

IMPLEMENTING SCIENTIFIC METHODS

In real-life situations, scientists do not follow one single method of asking questions and seeking answers. Instead, scientists combine some of the processes of the scientific methods in a way that is best suited to answer the questions they ask.

Problem Solving

A field biologist might wish to investigate the animal source of Ebola fever in the rain forest. He or she might use scientific methods that combine observing, hypothesizing, sampling, organizing data, and analyzing data. Consider a laboratory biologist working to understand how the Ebola virus attacks body cells. He or she might use scientific methods that combine observing, measuring, organizing data, hypothesizing, predicting, experimenting, analyzing data, and constructing a model. A public health official would likely be concerned about the spread of Ebola fever in a population. He or she might use scientific methods that combine organizing data, analyzing data, predicting, and inferring.

Communicating

As you have seen in the case of Ebola fever, scientists do not work alone. They share the results of their studies with other scientists. They publish their findings in scientific journals or present them at scientific meetings, like the one shown in Figure 1-16. Sharing information is essential to progress in science. The work of every scientist is subject to examination and verification by other scientists. Communication allows scientists to build on the work of others. Recall that communication among scientists throughout the world enabled them to isolate the Ebola virus. The same kind of communication enabled scientists to recognize the outbreaks of Ebola fever that occurred in Zaire in 1995 and Gabon in 1996. Health-care workers responded much sooner than they had in the earlier outbreaks. Though many people died, a far worse epidemic was avoided thanks to good communication and cooperation among scientists and health-care workers.

SECTION 1-3 REVIEW

1. How are the processes of inferring and modeling useful to scientists?
2. What is the relationship between hypothesizing and experimenting in science?
3. What role does communication play in science?
4. Explain why there is no single scientific method.
5. How does a theory differ from a hypothesis?
6. **CRITICAL THINKING** "Science rarely develops absolute truths that apply to all living things." Based on what you have read about scientific processes, do you think this statement is true?

MICROSCOPY AND MEASUREMENT

To study the living world, biologists need to observe cells and their parts. The development of new tools and new techniques enables biologists to probe even deeper into life's secrets.

SECTION

1-4

OBJECTIVES

▲

Compare light microscopes with electron microscopes in terms of magnification and resolution.

●

Explain the advantage of the Système International d'Unités.

MICROSCOPES

Among the most widely used tools in biology are microscopes. A **microscope** is an instrument that produces an enlarged image of an object. Biologists use microscopes to study organisms, cells, and cell parts that are too small to otherwise be seen. Microscopes both enlarge the image made by an object and show its details. The increase of an object's apparent size is **magnification.** The power to show details clearly refers to **resolution** (REZ-uh-LU-shuhn). Microscopes vary in powers of magnification and resolution.

Light Microscopes

To see small organisms and cells, biologists typically use a **compound light microscope (LM),** as shown in Figure 1-17. To be viewed with a compound light microscope, a specimen is mounted on a glass slide. The specimen must be sliced thin enough to be transparent, or it must be very small. The slide holding the specimen is placed over the opening in the **stage** of the microscope. A light source, a light bulb or mirror in the base, directs light upward. Light passes through the specimen and through the **objective lens,** which is positioned directly above the specimen. The objective lens enlarges the image of the specimen. This magnified image is projected up through the body tube to the **ocular** (AHK-yoo-luhr) **lens** in the eyepiece, where it is magnified further.

FIGURE 1-17

In a compound light microscope, the image of a transparent specimen is enlarged as it passes through the objective and ocular lenses.

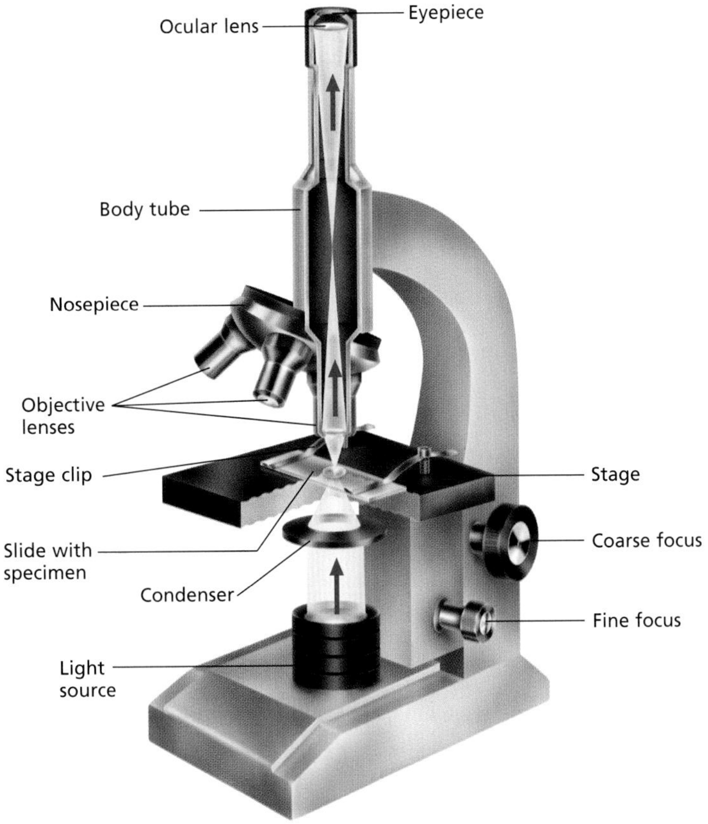

SECTION 1-4

SECTION OVERVIEW

Cell Structure and Function
Section 1-4 describes how light microscopes and electron microscopes are used to view cells and the structures inside them. This section also describes the use of the International System of Measurements, which scientists use to represent scale.

▶ ENGAGE STUDENTS

Hold a sheet of clear plastic wrap over your hand. Then replace the plastic with a sheet of waxed paper. Ask students to identify the material through which it was easier to distinguish individual hand features, such as fingers, hair, and veins (the plastic wrap). The capability to clearly see the individual parts of an object is called resolution.

TEACHING STRATEGY

Making a Wet Mount
Have students practice preparing wet mounts. CAUTION: **Warn students that slides and coverslips break easily and have sharp edges.** Furnish them with glass slides, coverslips, small beakers of water, and eyedroppers. They can start by viewing cut-out letters from newspapers. Challenge every student to make a bubble-free slide. After they accomplish this task, let them prepare wet mounts with drops of pond water. Emphasize that it will be much easier to find and watch organisms if they make their slides bubble free.

INCLUSION ACTIVITY

Have students design a microscope bulletin board in the classroom. Furnish them with a giant picture of a microscope to put up, and have them label its parts. Then have them search through old science magazines, cut out pictures of microscopic organisms and structures, and paste the pictures around the microscope. Use the bulletin board to review the parts of the microscope with students.

GIFTED ACTIVITY

Students who have mastered the basics of using a microscope can learn more advanced microscope techniques. If you have microscopes with oil immersion objective lenses, instruct the students on how to use them. Using oil immersion will enable students to magnify objects almost 1,000 times. Let them practice by viewing prepared slides of bacteria.

VISUAL LINK

Figure 1-18

Refer students to Figure 1-18 and point out that the electron micrograph reveals details that cannot be seen with light microscopes. Point out that *Giardia lamblia* is a unicellular parasite that causes cramps and diarrhea in humans and other animals.

Word Roots and Origins

magnification

from the Latin *magnificus/magnus*, meaning "large" or "great"

Most light microscopes have a set of objective lenses with different powers of magnification. These lenses can be rotated into place by revolving the **nosepiece.** In a typical compound light microscope, the most powerful objective lens produces an image 40 times the actual size of the specimen. This factor of enlargement is called the **power of magnification** of the lens, and it is represented, in this case, as 40×. The standard ocular lens magnifies a specimen 10 times (10×). To compute the power of magnification of a microscope, the power of magnification of the strongest objective lens (in this case 40×) is multiplied by the power of magnification of the ocular lens (10×). This results in a total power of magnification of 400×.

Electron Microscopes

The resolution power of light microscopes is limited by the physical characteristics of light. At powers of magnification beyond about 2,000×, the image of the specimen becomes blurry. To examine specimens even smaller than cells, such as cell parts or viruses, scientists may choose from several types of electron microscopes. In an **electron microscope,** a beam of electrons—rather than a beam of light—produces an enlarged image of the specimen. Electron microscopes are much more powerful than light microscopes. Some electron microscopes can even show the contours of individual atoms in a specimen.

FIGURE 1-18

This unicellular organism, *Giardia lamblia*, looks very different when viewed by (a) transmission electron microscopy and (b) scanning electron microscopy.

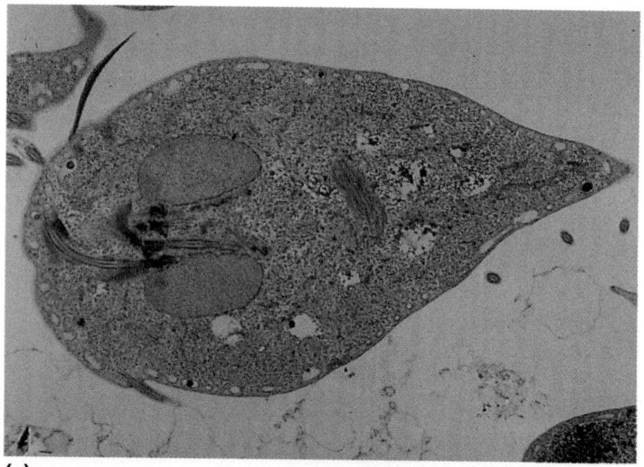

(a)

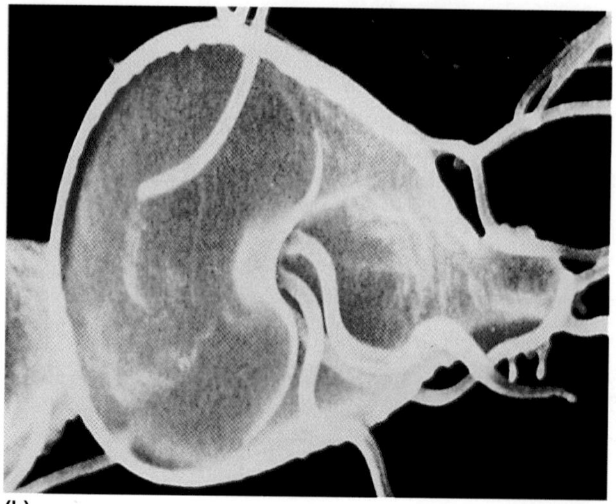

(b)

The **transmission electron microscope,** called a TEM, transmits a beam of electrons through a very thinly sliced specimen. Magnetic lenses enlarge the image and focus it on a screen or photographic plate. This produces an image like the one shown in Figure 1-18a. Transmission electron microscopes can magnify objects up to 200,000 times. But TEMs do have an important limitation—they cannot be used to view living specimens.

The **scanning electron microscope,** called an SEM, provides striking three-dimensional images, as you can see in Figure 1-18b. Specimens are not sliced for viewing. Rather, the surface of the specimen is sprayed with a fine metal coating. A beam of electrons is passed over the specimen's surface. This causes the metal coating to emit a shower of electrons. These electrons are projected onto a fluorescent screen or photographic plate, producing an image of the surface of the object. Scanning electron microscopes can magnify objects up to 100,000 times, and like the TEM, they cannot be used to view living specimens.

MEASUREMENT

Scientists use a single, standard system of measurement. The official name of the measurement system is Système International d'Unités (International System of Measurements), or simply **SI.** You will use the same units when you make measurements in the laboratory.

Base Units

There are seven fundamental **base units** in SI that describe length, mass, time, and other quantities, as shown in Table 1-1. Multiples of a base unit (in powers of 10) are designated by prefixes, as shown in Table 1-2. For example, the base unit for length is the meter. One *kilo*meter is equal in length to 1,000 meters.

TABLE 1-1 ***SI Base Units***

Base quantity	Name	Abbreviation
Length	meter	m
Mass	kilogram	kg
Time	second	s
Electric current	ampere	A
Thermodynamic temperature	kelvin	K
Amount of substance	mole	mol
Luminous intensity	candela	cd

TABLE 1-2 ***Some SI Prefixes***

Prefix	Abbreviation	Factor of base unit
giga	G	1,000,000,000
mega	M	1,000,000
kilo	k	1,000
hecto	h	100
deka	da	10
deci	d	0.1
centi	c	0.01
milli	m	0.001
micro	μ	0.000001
nano	n	0.000000001
pico	p	0.000000000001

internetconnect

SCILINKS NSTA

TOPIC: SI units
GO TO: www.scilinks.org
KEYWORD: HM023

CRITICAL THINKING

Cellular Measurements

Tell students that one micrometer (μm) is equal to 0.000001 m. Then ask them if a cell is 60 μm across, what is its diameter in meters? in centimeters? (0.000060 m, 0.0060 cm)

MAKING CONNECTIONS

Technology

The United States is the only industrialized country that has not adopted SI as its official system of weights and measures. When Great Britain converted to SI in 1963, it became clear that the United States would have to start using SI units in the area of trade and commerce to successfully compete in world economic markets. Since then, SI use has become widespread throughout our economy. Ask students to bring in everyday items that have been manufactured based on SI units. Their collection may include photographic equipment, construction supplies, prescription labels, computer disks, soft-drink containers, and food labels.

RETEACHING *ACTIVITY*

Show students a nickel and a dime. Tell them that the nickel has a mass of about 5 g and the dime is about 1 mm thick. Let the students determine the mass, thickness, and volume of some of their own coins.

MAKING CONNECTIONS

Mathematics

Assign students values of SI units and ask them to convert them into other SI units using Table 1-2. For example, how many centimeters are in 10 millimeters? (1cm)

SECTION 1-4

QUICK FACT

☑ The International Bureau of Weights and Measures in France provides standards of SI measurement. For example, it has a piece of metal that is the standard for the kilogram and a metal bar that is the standard for the meter.

ANSWERS TO SECTION 1-4 REVIEW

1. Some light microscopes magnify up to 2,000×, but as the power of magnification increases, the resolution drops. A TEM microscope magnifies up to 200,000× and can distinguish the internal structure of specimens. An SEM magnifies up to 100,000× and provides three-dimensional images of the surface of a specimen.
2. The power of magnification of a microscope is the product of the power of magnification of the ocular lens and that of the most powerful objective lens.
3. A beam of electrons is scanned across a specimen that has been coated with a thin layer of metal. The electron beam causes the metal to emit a shower of electrons, which produce an image on a fluorescent screen or photographic plate.
4. A beam of electrons is passed through a thinly sectioned specimen. Magnetic lenses enlarge the image and project it onto a screen or photographic plate.
5. Conversions among SI units are simple, allowing scientists to manipulate data much more easily than if they were working with English units.
6. Electron micrographs would provide too much detail for some observations, such as looking at the gross structure of an organism.

TABLE 1-3 *SI Derived Units Often Used in Biology*

Derived quantity	Name	Abbreviation
Area	square meter	m^2
Volume	cubic meter	m^3
Mass density	kilogram per cubic meter	kg/m^3
Specific volume	cubic meter per kilogram	m^3/kg
Celsius temperature	degree Celsius	°C

TABLE 1-4 *Other Units Acceptable for Use with SI*

Name	Abbreviation	Value in SI units
Minute	min	1 min = 60 s
Hour	h	1 h = 60 min = 3,600 s
Day	d	1 d = 24 h = 86,400 s
Liter	L	1 L = 1 dm^3 = 0.001 m^3
Metric ton	t	1 t = 1,000 kg

Derived Units

The base units in Table 1-1 cannot be used to measure surface area or velocity, among other things. Therefore, other important units, derived units, are used. **Derived units** are produced by the mathematical relationship between two base units or between two derived units. Table 1-3 shows some common derived units.

Other Units

Some units of measurement that are not part of the SI are accepted for use with SI units. They are units of time, volume, and mass, as shown in Table 1-4.

SECTION 1-4 REVIEW

1. How do microscopes differ in magnification and resolution?
2. How is the maximum power of magnification computed for a compound light microscope?
3. How does a scanning electron microscope work?
4. How does a transmission electron microscope work?
5. If SI measurement is no more accurate than the English system, why do scientists throughout the world use it?
6. **CRITICAL THINKING** Why might a scientist prefer to use the lower powers of a light microscope to observe aspects of unicellular organisms?

CHAPTER 1 REVIEW

SUMMARY/VOCABULARY

1-1

- Biology is the study of life, ranging from the study of unicellular organisms to the study of the global interactions among millions of organisms.
- Six themes serve to unify the study of biology: (1) cell structure and function; (2) stability and homeostasis; (3) reproduction and inheritance; (4) evolution; (5) interdependence of organisms; and (6) matter, energy, and organization.
- The cell is the basic unit of life. Organisms may be unicellular or multicellular.
- Living things maintain a stable level of internal conditions, called homeostasis.
- Reproduction involves the transmission of hereditary information from organisms to their offspring.
- Populations of organisms evolve over generations primarily by a process called natural selection.
- Organisms interact in important ways with each other and with their environments.
- Living things have highly organized structures that are maintained by a constant input of energy.
- Autotrophs obtain energy by making their own nutrients.
- Heterotrophs obtain energy from the nutrients they obtain from their environment.

Vocabulary

asexual reproduction (7)
autotroph (10)
biology (5)
differentiation (6)
DNA (7)
ecology (9)
ecosystem (9)
evolution (8)
evolve (8)
gene (7)
heterotroph (10)
homeostasis (6)
multicellular organism (6)
natural selection (8)
organism (5)
photosynthesis (10)
reproduction (7)
sexual reproduction (7)
unicellular organism (6)

1-2

- Living things are composed of cells.
- Living things are more highly organized than nonliving structures.
- Living things use energy in a process called metabolism.
- Living things have several mechanisms that help them maintain stable internal conditions.
- When living things grow, their cells enlarge and divide.
- Living things reproduce, producing offspring similar to themselves.
- There is a great deal yet to be learned about the living world.

Vocabulary

cell division (12)
development (12)
metabolism (12)

1-3

- Scientific investigations generally begin with observation.
- Methods scientists use in their work include (1) observing, (2) asking a question, (3) collecting data, (4) hypothesizing, (5) experimenting, and (6) drawing conclusions.
- A hypothesis is a statement that explains observations *and* that can be tested.
- In a controlled experiment, the experimental group is identical to the control group except for one factor called the independent variable.
- Communication is very important in science because scientists build on the work of others.

Vocabulary

control group (17)
controlled experiment (17)
data (15)
dependent variable (18)
experiment (17)
experimental group (17)
hypothesis (17)
independent variable (18)
inference (19)
model (18)
observation (15)
prediction (17)
sampling (16)
theory (19)

CHAPTER 1 REVIEW ANSWERS

REVIEW

1. Sexual reproduction involves the combining of genetic material from different individuals, while asexual reproduction does not.
2. Heterotrophs acquire energy from other organisms, but autotrophs acquire energy by making their own nutrients.
3. Cell division contributes to growth through formation of new cells, and cell enlargement increases cell size.
4. Formulation of a model is based on direct observations that show relationships among data. Inferring is drawing conclusions that are not based on direct observations.
5. A TEM transmits electrons through a thinly sliced specimen and magnifies the image of the specimen up to 200,000×. An SEM scans the surface of a specimen with a beam of electrons and magnifies the image of the surface up to 100,000×.

6. b **7.** d
8. d **9.** b
10. d **11.** c
12. c **13.** b
14. b **15.** b

16. (a) ocular lens, (b) nosepiece, (c) objective lens, (d) stage, (e) light source
17. They have a high level of molecular and cellular organization, showing more complex structure than nonliving things.
18. In a unicellular organism it produces more unicellular organisms. In a multicellular organism it produces a larger organism.
19. Some organisms express traits that enhance their reproduction, and those traits are passed on. Organisms without those traits may not live to reproduce or may not reproduce as successfully.
20. an environmental community
21. Many parts of Earth are remote and difficult to access.
22. A hypothesis is a testable explanation of observable phenomena and data. A prediction is a statement of expected results of testing the hypothesis.
23. One group of test tubes containing monkey kidney cells was inoculated with virus-containing blood. A control group remained uninoculated. The cells in the experimental test tubes died.
24. A model is a representation of an object, system, or process that shows relationships among data.
25. They cannot be used to observe living specimens.

1-4
- Biologists often use a compound light microscope to see small things, like cells.
- Electron microscopes provide higher magnification and better resolution than do light microscopes.
- Système international d'unités (SI) is a standard system of measurement that uses seven fundamental base units.
- All measurements scientists make are done using SI units, derived units, and other acceptable units.

Vocabulary

base unit (23)
compound light microscope (21)
derived unit (24)
electron microscope (22)
magnification (21)
microscope (21)
nosepiece (22)
objective lens (21)
ocular lens (21)
power of magnification (22)
resolution (21)
scanning electron microscope (SEM) (22)
SI (23)
stage (21)
transmission electron microscope (22)

REVIEW

Vocabulary

1. What is the difference between asexual reproduction and sexual reproduction?
2. Compare heterotrophs with autotrophs.
3. How do cell division and enlargement contribute to growth?
4. Explain the difference between formulating a model and inferring.
5. Describe the similarities and differences between a transmission electron microscope and a scanning electron microscope.

Multiple Choice

6. Reproduction involves the transfer of genetic information from (a) autotroph to heterotroph (b) parents to offspring (c) offspring to parents (d) unicellular organism to multicellular organism.
7. The theory of evolution by natural selection helps explain how complex organisms came into existence, why organisms of the past differ from those alive today, and how various groups of living organisms (a) develop (b) reproduce (c) obtain energy (d) are related to each other.
8. The hereditary material in living things is (a) CDC (b) thyroid hormone (c) carbon dioxide (d) DNA.
9. Organisms that obtain energy by taking in food are called (a) autotrophs (b) heterotrophs (c) homeostatic (d) reproducers.
10. Growth occurs by (a) organization and reproduction (b) adaptation and evolution (c) inferring and adapting (d) cell division and cell enlargement.
11. The combination of genetic information from two individuals occurs during (a) homeostasis (b) development (c) reproduction (d) differentiation.
12. The most important driving force in evolution is (a) cellular organization (b) differentiation (c) natural selection (d) development.
13. Data that are quantitative are always (a) described in words (b) represented by numbers (c) recorded on a tape recorder (d) seen through a microscope.
14. A hypothesis is a statement that (a) is identical to a theory (b) can be tested (c) is usually true (d) is always true.
15. The resolution of a microscope refers to (a) its power to increase an object's apparent size (b) its ability to show detail clearly (c) its series of interchangeable objective lenses (d) its power to scan the surface of an object.

Short Answer

16. Name the part of the compound light microscope denoted by each letter.

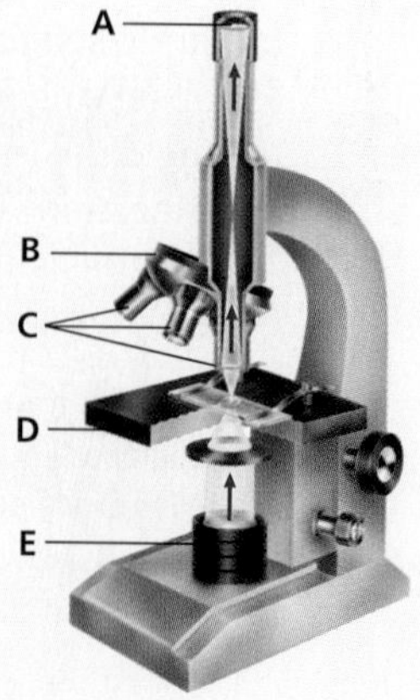

17. Why do biologists say that living things are organized?
18. What are the different results produced by cellular division in a unicellular organism and in a multicellular organism?
19. Explain how the process of natural selection can change traits over time.
20. What is an ecosystem?
21. Why have some animals only recently been discovered?
22. What is the relationship between a hypothesis and a prediction?
23. Describe a controlled experiment that was conducted to confirm the cause of Ebola fever.
24. What is a model?
25. What is an important limitation of SEMs and TEMs in observing organisms?

CRITICAL THINKING

1. One of the first branches of biology to be developed was taxonomy, the naming of organisms. Why is taxonomy important to communication about biology?
2. Rock crystals grow and become larger. How does this process differ from the way living things grow?
3. One of the most important parts of any scientific publication is the part called "Methods and Materials," in which the scientist describes the procedure used in the experiment. Why do you think such details are so important?
4. Scientists know that a disease-causing organism can cause overwhelming illness or death very quickly. Why would it not be adaptive for a disease-causing organism to kill its victims too quickly?
5. Look at the photographs below. The SEM, top, is of *Staphylococcus aureus.* The TEM, bottom, is also of *S. aureus.* Compare and contrast what each electron micrograph tells you about this organism.

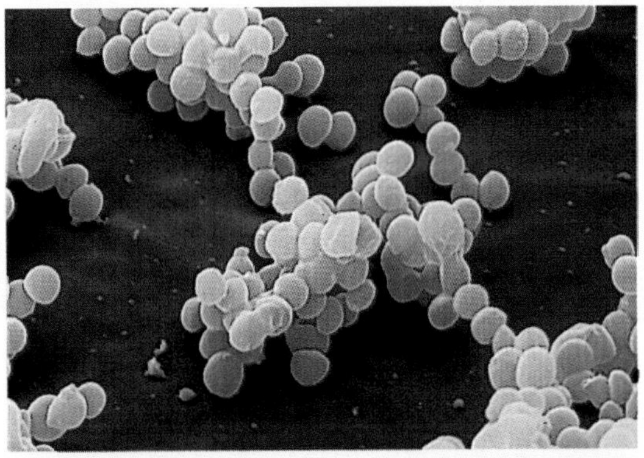

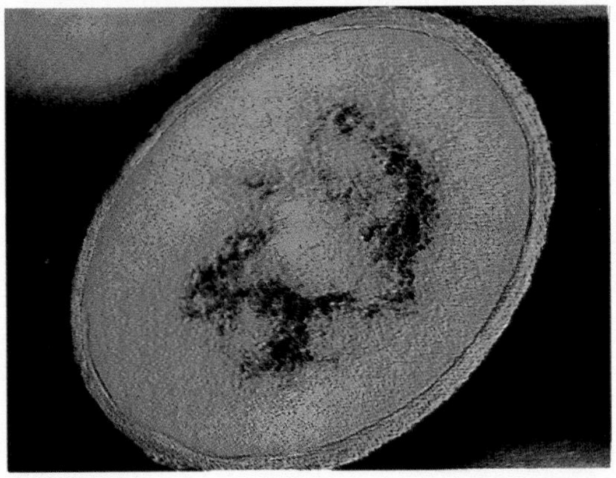

EXTENSION

1. Read the chapter titled "Project Ebola" in Richard Preston's book *The Hot Zone.* The chapter describes the outbreak of the Ebola virus in the Reston, Virginia, primate research facility in 1989. Answer the following questions: Why was the research on Ebola virus done in a Level 4 area? Why was negative air pressure used in the lab? What special precautions did the scientists have to take before entering the Level 4 area?
2. Using a compound light microscope, observe prepared slides of unicellular organisms such as *Euglena* under three different powers of magnification. Draw a picture of each organism as it appears at each power of magnification, and write a brief description of the level of detail that you see.

CRITICAL THINKING

1. Organisms had to be named or described before different scientists studying them independently could accurately compare their results.
2. Crystals accumulate more of the material they are made of. A living thing grows by cell division and enlargement.
3. It allows other scientists to repeat the experiment in an attempt to verify the investigator's conclusions.
4. If it kills its host before it can multiply and be passed to another host, it will die off.
5. Both show that the organism is spherical. The SEM shows that it can be found in clumps, while the TEM reveals that it consists of a single cell with very little internal detail.

EXTENSION

1. Any organism that is potentially lethal to humans and for which there is no known vaccine or cure, such as the Ebola virus, must be contained in a very secure area known as a Level 4 area. Negative air pressure is used so that if a leak occurs, air will flow into the facility and not to the area outside the facility. The scientists donned biological space suits, which are pressurized, heavy duty plastic suits that meet government specifications.
2. Illustrations will vary but should resemble the organism observed and include cell organelles if the organism is a eukaryote.

TIME REQUIRED
One 50-minute class period

SAFETY PRECAUTIONS
Remind students to wear heat-protective gloves for Part A.

QUICK REFERENCES
Lab Preparation Notes are found on page 3B.

Holt BioSources provides a Teaching Resources CD-ROM that contains "Using Gowin's Vee in the Lab" and "Scoring Rubrics."

PROCEDURAL TIPS
1. To save time, have half of the students perform Part A while the other half perform Part B, then switch.
2. Review the use of the balance and graduated cylinders.
3. You may wish to have students record their data in a class data table on the board. Have students calculate the mean of each range of measurement.

ANSWERS TO BACKGROUND
1. Système International d' Unités (International System of Units)
2. meter, kilogram, second, ampere, kelvin, mole, candela

CHAPTER 1 INVESTIGATION

Using SI Units

OBJECTIVES
- Express measurements in SI units.
- Read a thermometer.
- Measure liquid volume using a graduated cylinder.
- Measure mass using a balance.
- Determine the density (mass-to-volume ratio) of two different liquids.

PROCESS SKILLS
- measuring
- calculating

MATERIALS
- 75 mL light-colored sand
- 75 mL dark-colored sand
- 1100 mL graduated cylinder
- Celsius thermometers, alcohol filled (2)
- 5 oz plastic cups (4)
- graph paper
- heat-protective gloves
- light source
- stopwatch or clock
- ring stand or lamp support
- 25 mL corn oil
- 25 mL water
- clear-plastic cup
- balance

Background
1. What does the abbreviation *SI* stand for?
2. List the seven SI base units.

PART A Measuring Temperature
1. In your lab report, prepare a data table similar to Table A, above right.
2. Using a graduated cylinder, measure 75 mL of light-colored sand and pour it into one of the small plastic cups. Repeat this procedure with the dark-colored sand and another plastic cup.
3. Level the sand by placing the cup on your desk and sliding the cup back and forth.

TABLE A *SAND TEMPERATURE*

	Temperature (°C)	
Time (min)	**Dark-colored sand**	**Light-colored sand**
Start		
1		
2		
3		
4		
5		
6		
7		
8		
9		
10		

4. Insert one thermometer into each cup. The zero line on the thermometer should be level with the sand, as shown in the figure below. Re-level the sand if necessary.

5. CAUTION **Wear heat-protective gloves when handling the lamp. The lamp will become very hot and may burn you.** Using a ring stand or lamp support, position the lamp approximately 9 cm from the top of the sand, as shown in the figure below left. Make sure the lamp is evenly positioned between the two containers.
6. Before turning on the lamp, record the initial temperature of each cup of sand in your data table.
7. Note the time or start the stopwatch when you turn on the lamp. The lamp will become hot and warm the sand. Check the temperature of the sand in each container at one-minute intervals for 10 minutes. Record the temperature of the sand after each minute in your data table.

PART B Comparing the Density of Oil and Water

8. In your lab report, prepare a data table similar to Table B below.

TABLE B *DENSITY OF TWO LIQUIDS*

a. Mass of empty oil cup		_______ g
b. Mass of empty water cup		_______ g
c. Mass of cup and oil		_______ g
d. Mass of cup and water		_______ g
e. Volume of oil		25 mL
f. Volume of water		25 mL
Calculating Actual Mass		
Oil	Item c − Item a =	_______ g
Water	Item d − Item b =	_______ g

9. Label one clean plastic cup "oil," and label another "water." Using a balance, measure the mass of each plastic cup, and record the value in your data table.
10. Using a clean graduated cylinder, measure 25 mL of corn oil and pour it into the plastic cup labeled "oil." Using a balance, measure the mass of the plastic cup containing the corn oil, and record the mass in your data table.
11. Using a clean graduated cylinder, measure 25 mL of water and pour it into the plastic cup labeled "water." Using a balance, measure the mass of the plastic cup containing the water, and record the mass in your data table.
12. To find the actual mass of the oil, subtract the mass of the empty cup from the mass of the cup and the oil together.
13. To find the density of the oil, divide the mass of the oil by the volume of the oil, as shown in the equation below:

$$\text{Density of oil} = \frac{\text{mass of oil}}{\text{volume of oil}} = \_\_\_\_\_\_\_ \text{ g/mL}$$

14. To find the mass of the water, subtract the mass of the empty cup from the mass of the cup and the water together.
15. To find the density of the water, divide the mass of the water by the volume of the water, as shown in the equation below:

$$\text{Density of water} = \frac{\text{mass of water}}{\text{volume of water}} = \_\_\_\_\_\_\_ \text{ g/mL}$$

16. Combine the oil and water in the clear cup and record your observations in your lab report.
17. Clean up your materials and wash your hands before leaving the lab.

Analysis and Conclusions

1. Graph the data you collected in Part A. Plot time on the *x*-axis and temperature on the *y*-axis.
2. Based on your data from Part A, what is the relationship between color and heat absorption?
3. How might the color of the clothes you wear affect how warm you are on a sunny 90° day?
4. In Part B, what did you observe when you combined the oil and water in the clear cup? Relate your observation to the densities that you calculated for the two liquids.
5. What could you infer about the value for the density of ice if you observe it to float in water?
6. How would your calculated values for density be affected if you misread the volume measurement on the graduated cylinder?

Further Inquiry

Pumice is a volcanic rock that has a density less than 1.00 g/cm^3. How would you prove this if you did not have a balance to weigh the pumice? (Hint: the density of water is 1.00 g/cm^3.)

ANSWERS TO ANALYSIS AND CONCLUSIONS

1. Students' graphs should show that temperature rises with time.
2. The dark sand absorbs more heat than does the light sand.
3. Wearing light-colored clothing will keep one cooler than wearing dark-colored clothing because light-colored material does not absorb as much heat as does dark-colored material.
4. The oil floated on the water. Therefore, oil is less dense than water.
5. Ice is less dense than water.
6. If the recorded values were too high, then the calculated density would be too low. If the recorded values were too low, then the calculated density would be too high.

CHAPTER 2
CHEMISTRY

	TOPICS	TEACHING RESOURCES	LABS, CLASSWORK, AND HOMEWORK
BLOCK 1 45 minutes	**Introducing the Chapter, p. 30**	**ATE** Focus Concept, p. 30 **ATE** Assessing Prior Knowledge, p. 30 **ATE** Understanding the Visual, p. 30	■ Supplemental Reading Guide, *The Andromeda Strain*
	2-1 Composition of Matter, p. 31 Matter, p. 31 Elements, p. 31 Atoms, p. 32 Compounds, p. 33	**ATE** Section Overview, p. 31 **ATE** Critical Thinking, p. 33 **ATE** Visual Link Figure 2-4, p. 33 2, 4, 8A, 9A	■ Basic Skills Worksheet—Mass and Density ■ Basic Skills Worksheet—Measurement and Scale Drawings ★ Study Guide, Section 2-1 **PE** Section Review, p. 34
BLOCK 2 45 minutes	**2-2 Energy, p. 35** Energy and Matter, p. 35 Energy and Chemical Reactions, p. 36	**ATE** Section Overview, p. 35 **ATE** Visual Link Figure 2-7, p. 36	**PE** **Quick Lab,** p. 37, Modeling Ionic Bonds **Lab A7,** Making Models ■ Occupational Applications Worksheets—Blood Bank Technologist ■ Basic Skills Worksheets—Words and Roots ★ Study Guide, Section 2-2 **PE** Section Review, p. 37
BLOCK 3 45 minutes	**2-3 Solutions, p. 40** Describing Solutions, p. 40 Acids and Bases, p. 41	**ATE** Section Overview, p. 40 1A, 10A	**PE** **Chapter 2 Investigation,** p. 46 ■ Problem-Solving Worksheets—Solution Concentration ■ Problem-Solving Worksheets—Determining Percent Composition ★ Study Guide, Section 2-3 **PE** Section Review, p. 42
BLOCKS 4 & 5 90 minutes	**REVIEW and ASSESSMENT** **PE** Chapter 2 Review, pp. 44–45 ■ Performance-Based Assessment—Introduction to Performance-Based Assessment	Audio CD Program	**CHAPTER TESTING** ■ Chapter Test (blackline master) ▲ Test Generator ■ Assessment Item Listing ■ Scoring Rubrics and Classroom Management Checklists

MODERN BIOLOGY

HOLT BIOSOURCES

KEY

PE Pupil's Edition
ATE Teacher's Edition
■ Active Reading Guide
★ Study Guide

Teaching Transparencies

Laboratory Program

One-Stop Planner CD-ROM
Shows these resources within a customizable daily lesson plan:

■ Holt BioSources Teaching Resources
■ Active Reading Guide
★ Study Guide
▲ Test Generator

READING FOR CONTENT MASTERY

Preview: Have students preview the objectives, the introductory paragraph, topic sentences, and captions before reading the chapter.

ATE Active Reading Technique: Reading Effectively, p. 31
■ Active Reading Guide Worksheet 2-1
PE Summary/Vocabulary, p. 43

■ Active Reading Guide Worksheet 2-2
PE Summary/Vocabulary, p. 43

■ Active Reading Guide Worksheet 2-3
PE Summary/Vocabulary, p. 43

TECHNOLOGY AND INTERNET RESOURCES

Holt Biology Videodiscs Teacher's Correlation Guide, Lessons 2-1 through 2-3

Audio CD Program, Sections 2-1 through 2-3

Interactive Exploration: Cell Biology and Genetics, Thermodynamics

internetconnect

SCILINKS NSTA

On-line Resources: www.scilinks.org
The following *sci*LINKS Internet resources can be found in the student text for this chapter:

Topics:
- Atomic structures, p. 32
- Covalent and ionic bonds, p. 34

go.hrw.com

On-line Resources: go.hrw.com
Visit the HRW Web site for a variety of resources related to this chapter. Just type in the keyword HM2 Chapter 2.

Smithsonian Institution®
Visit **www.si.edu/hrw** for additional on-line resources.

CNNfyi.com
Visit **www.cnnfyi.com** for late-breaking news and current event stories selected just for you.

LAB ACTIVITY PLANNING

CHAPTER 2 INVESTIGATION

Exploring the Activity of Biological Catalysts, pp. 46–47

OVERVIEW

This is the first of six interactive explorations included in the in-text laboratory program. In this interactive exploration, students will use a computer simulation to determine how enzymes respond to changing environmental conditions. This investigation allows students to determine the relationship between temperature and reaction rate for the enzymes carbonic anhydrase and lysozyme, to determine the relationship between pH and reaction rate for the same two enzymes, and to identify the optimal temperature and pH for each enzyme.

The structure of an enzyme is crucial to its function. Each kind of enzyme catalyzes a specific reaction that is determined by its three-dimensional shape. The enzyme's shape is affected by general environmental factors, such as temperature and pH. The efficiency of an enzyme depends on the optimal conditions of its environment.

As an extension activity, the program allows students to investigate additional enzymes, such as penicillinase, ß-galactosidase, and chymotrypsin.

PREPARATION

1. Access to the following materials is needed:
   - computer with CD-ROM drive
   - CD-ROM *Interactive Explorations in Biology: Cell Biology and Genetics.* The title of this program on the CD-ROM is "Thermodynamics."
2. Have students read the Objectives, Background, and Prelab Preparation sections before starting the lab.

Quick Labs

Modeling Ionic Bonds, p. 37
Students will need toothpicks, mini marshmallows, and fresh or frozen peas for this activity. Choose peas that are soft enough that the toothpicks can be pushed into them easily.

UNDERSTANDING THE VISUAL

Have students examine the micrograph of urate crystals. Explain that in some people, certain substances in the urine precipitate, forming crystals. These crystals (of calcium, oxalate, or urate embedded in a protein matrix) form masses called kidney stones. Ask students to name a structure in the body that consists of mineral crystals (bone).

FOCUS CONCEPT

Matter, Energy, and Organization

All organisms are made of matter. In living things, matter is constantly being rearranged through changes called chemical reactions, which involve energy. The chemical properties of matter depend on its organization at various levels: subatomic, atomic, and molecular. Biology is strongly tied to chemistry because chemical structure and reactions are central to all biological processes.

ASSESSING PRIOR KNOWLEDGE

Review the following concepts.

Characteristics of Life: *Chapter 1*
Ask students whether the crystal shown at right is more complex or less complex than a living cell.

Themes of Biology: *Chapter 1*
Have students name and describe the six themes around which this book is organized.

CHAPTER 2

CHEMISTRY

Living things are composed of the same materials and are made by the same processes as nonliving things are. These crystals of a substance called urate (LM 90×) are formed in the human kidney.

FOCUS CONCEPT: ***Matter, Energy, and Organization***

As you read, become aware of how a basic knowledge of chemistry will help you understand and explain biological processes.

SECTION

2-1

COMPOSITION OF MATTER

Earth supports an enormous variety of organisms. You have learned that all organisms share certain characteristics and life processes. The structure and function of all living things are governed by the laws of chemistry. An understanding of the fundamental principles of chemistry will give you a better understanding of living things and how they function.

OBJECTIVES

▲

Define *element, atom, compound,* and *molecule.*

●

Draw a model of the structure of an atom.

■

Explain what determines an atom's stability.

◆

Contrast ionic and covalent bonds.

MATTER

Everything in the universe is made of matter. **Matter** is anything that occupies space and has mass. **Mass** is the quantity of matter an object has. Mass and weight are not the same; the pull of gravity on an object is what gives an object the property of weight. The same mass would have less weight on the moon than it would on Earth because the moon has less gravitational pull.

Chemical changes in matter are essential to all life processes. Biologists study chemistry because all living things are made of the same kinds of matter that make up nonliving things. By learning how changes in matter occur, you will gain an understanding of the life processes of the organisms you will study.

ELEMENTS

Elements are pure substances that cannot be broken down chemically into simpler kinds of matter. More than 100 elements have been identified, though fewer than 30 are important to living things. In fact, more than 90 percent of the mass of all kinds of living things is composed of combinations of just four elements: oxygen, O, carbon, C, hydrogen, H, and nitrogen, N.

Each element has a different chemical symbol. A chemical symbol consists of one or two letters, as shown in Figure 2-1. In most cases, the symbol derives from the first letter or other letters in the name of the element, like C for carbon or Cl for chlorine. The Latin word *natrium* provides the symbol for sodium, Na. The symbol K, for potassium, comes from the Latin word *kalium.*

FIGURE 2-1

All of the elements are arranged on a chart known as the periodic table. Here you see information for three elements from the periodic table. Among the information provided in the periodic table are the atomic number, the chemical symbol, and the atomic mass for each element.

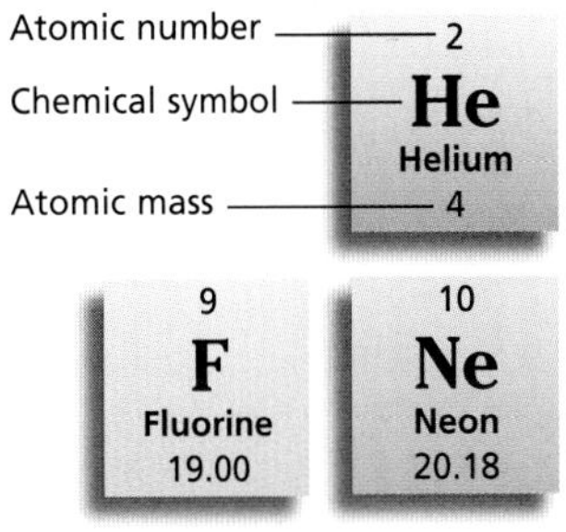

SECTION 2-1

SECTION OVERVIEW

Matter, Energy, and Organization

Section 2-1 introduces basic properties of matter to give students a foundation for understanding biological structures and processes at the cellular level. Students will learn the structure of the atom and the relationships among the subatomic particles—electrons, protons, and neutrons. Section 2-1 also introduces the formation of compounds through ionic and covalent bonding.

ACTIVE READING

Technique: Reading Effectively

Instruct students to take several minutes to study the figures after reading each section. Discuss the importance of reading figure captions. Prepare questions that correspond to each figure. For example, refer to Figure 2-1 on page 31. Ask students to list the three items of information given for each element on the periodic table. Refer to the figures when you talk about the concepts presented in each section.

▶ ENGAGE STUDENTS

Give students a brief introduction to the periodic table. Point out common elements such as iron, Fe; copper, Cu; aluminum, Al; silver, Ag; sodium, Na; gold, Au; nickel, Ni; and chlorine, Cl. Point out that elements in the same column tend to have similar physical and chemical properties. For example, the noble gases (Group 18 or Group 0) are gases at room temperature and are typically nonreactive. Stress that the general trend is toward increased size of atoms from left to right and from top to bottom. The five most important elements for living things are C, H, O, N, and S. Have students locate these elements in the table and note the name of each element.

SECTION 2-1

QUICK FACT

It would take one million hydrogen atoms sitting side by side to equal the thickness of this page.

TEACHING STRATEGY

Probability and the Atom

Inform students that the movement of electrons in an atom is more similar to the movements of a swarm of gnats than to the orbital motion of the planets. Thus, scientists cannot describe the path, position, or momentum of any particular electron, but they can describe the probability of groups of electrons being in particular energy levels. That is, electrons with a certain amount of energy are likely to be found in a certain energy level.

OVERCOMING MISCONCEPTIONS

Students often don't recognize that atoms consist largely of empty space. If the nucleus of an atom were the size of a marble, the first energy level would be about 0.8 km (0.5 mi) away.

GIFTED ACTIVITY

After probing deeper into the structure of the atom, scientists have discovered that protons and neutrons are composed of even smaller particles, called quarks. Have students research these tiny particles, addressing questions like, What are the different types of quarks? How are they arranged within different subatomic particles? How and by whom were they discovered? Have students write reports summarizing what they have learned.

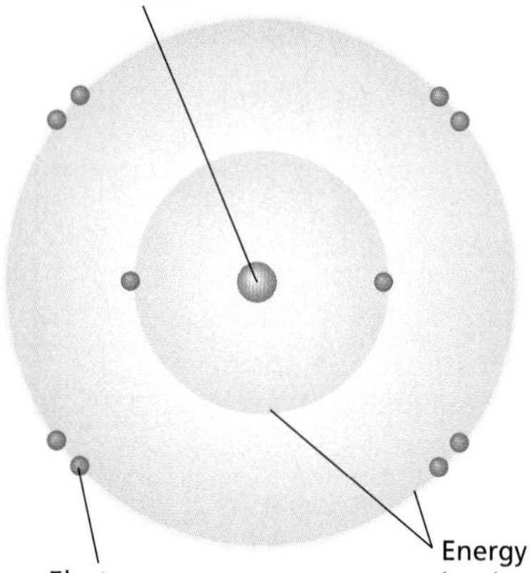

FIGURE 2-2
The electrons in this model of an atom are distributed in two energy levels. The innermost level holds a maximum of two electrons. The second level holds a maximum of eight electrons.

internetconnect
SCILINKS NSTA
TOPIC: Atomic structures
GO TO: www.scilinks.org
KEYWORD: HM032

ATOMS

The simplest particle of an element that retains all of the properties of that element is an **atom.** The properties of different kinds of atoms determine the structure and properties of the matter they compose. Atoms are so small that their true structure cannot be observed. However, scientists have developed models that describe the structure and properties of the atom, as shown in Figure 2-2.

Models of the atom are not meant to show exactly what atoms look like. Rather, they help us understand the structure of atoms and predict how they will act in nature.

The Nucleus

The central core, or **nucleus,** of an atom consists of two kinds of particles. One, the **proton,** has a positive electrical charge. The other, the **neutron,** has no electrical charge. Most of the mass of an atom is concentrated in its nucleus.

All atoms of a given element have the same number of protons. The number of protons in an atom is called the **atomic number** of the element. In the periodic table of elements, the atomic number generally appears directly above the chemical symbol, as shown in Figure 2-1. The atomic number of fluorine is 9, which indicates that each atom of the element fluorine has nine protons.

In an atom, the number of positively charged protons is balanced by an equal number of small, negatively charged particles called **electrons.** The electrical charges of the electrons offset those of the protons, making the net electrical charge of an atom zero.

Electrons

Electrons are high-energy particles with very little mass. They move about the nucleus at very high speeds in one of several different **energy levels,** like those shown in Figure 2-2.

Electrons in outer energy levels have more energy than those in inner energy levels. Each energy level can hold only a certain number of electrons. For example, the first energy level, nearest the nucleus, can hold up to two electrons. This is the outermost energy level for the elements hydrogen and helium. The second energy level can hold up to eight electrons. As shown in Figure 2-3, in most elements, the outer energy level is not filled.

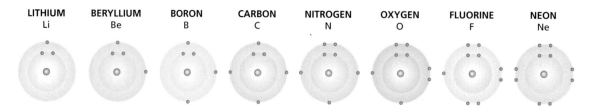

FIGURE 2-3
Below are shown some examples of the number of electrons found in the outermost energy level of elements.

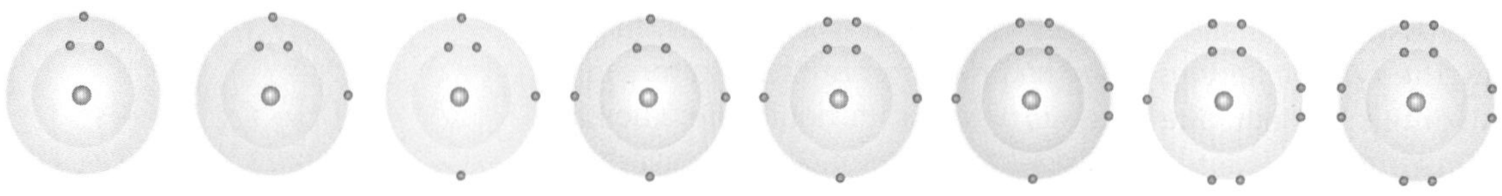

COMPOUNDS

Under natural conditions, most elements do not exist by themselves; most elements readily combine with other elements. A pure substance that is made up of atoms of two or more elements is called a **compound.** In a compound, the proportions of each kind of atom are fixed. A chemical formula shows the kind and proportion of atoms of each element that forms a particular compound. For example, the chemical formula for water, H_2O, indicates that the atoms always combine in a proportion of two hydrogen atoms to one oxygen atom.

The physical and chemical properties of a compound differ from the physical and chemical properties of the individual elements that compose it. In nature, the elements oxygen and hydrogen are usually found as gases with the formulas O_2 and H_2. However, when oxygen gas and hydrogen gas combine to form H_2O, the result is a liquid at room temperature. The tendency of elements to combine and form compounds depends on the number and arrangement of electrons in their atoms. An atom is chemically stable when its outermost energy level is filled. Most atoms are not stable in their natural state. Thus, they tend to react, or combine with other atoms, in ways that make the atoms more stable.

Some elements, such as helium and neon, consist of atoms whose energy levels are filled with electrons. As shown in Figure 2-3, the second energy level of neon is its outermost energy level, and it is filled. Thus, neon tends not to react with other elements. By contrast, carbon, nitrogen, and oxygen consist of atoms with unfilled energy levels. Hence, most elements tend to undergo **chemical reactions,** combining in ways that cause their atoms to become stable. In chemical reactions, chemical bonds are broken, atoms are rearranged, and new chemical **bonds,** or attachments, are formed.

Covalent Bonds

A **covalent bond** forms when two atoms share one or more pairs of electrons. Water is made up of one oxygen atom and two hydrogen atoms held together by covalent bonds. Figure 2-4a shows that an atom of hydrogen needs a second electron to achieve stability, giving it two electrons in its outermost energy level. Oxygen needs two more electrons to give it a stable arrangement of eight electrons. Thus, in the presence of one another, hydrogen atoms and oxygen atoms can achieve stability by sharing pairs of electrons in a ratio of two atoms of hydrogen to one atom of oxygen, as shown in Figure 2-4b. The resulting compound, H_2O (water), is essential to the functioning of all living things.

A **molecule** is the simplest part of a substance that retains all of the properties of the substance and that can exist in a free state. For example, each molecule in hydrogen gas consists of two hydrogen atoms bonded to each other. Figure 2-4c shows a model of a water molecule. Some molecules—particularly many of the molecules that biologists study—are large and complex.

Word Roots and Origins

compound

from the Latin *componere,* meaning "to put together"

FIGURE 2-4

Two atoms of hydrogen and one atom of oxygen share electrons in covalent bonds to become stable. Covalent bonding results in the formation of molecules.

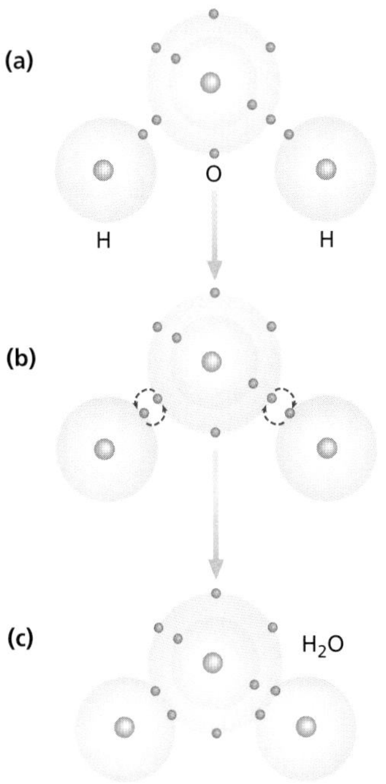

TEACHING STRATEGY

Understanding Compounds

Write on the chalkboard the following chemical formulas: CH_4 (methane), NH_2CH_2COOH (the amino acid glycine), and $(NH_4)_2SO_4$ (ammonium sulfate). Have students determine how many atoms of which elements are in one molecule of each compound. (Methane = 1 C, 4 H; glycine = 1 N, 5 H, 2 C, 2 O; ammonium sulfate = 2 N, 8 H, 1 S, 4 O.)

CRITICAL THINKING

How Many Bonds?

Have students predict the number of covalent bonds that an atom of carbon can form (four). Have students explain how they arrived at their answer. (Carbon has six electrons. Two electrons fill the inner energy level, and the other four occupy the outer energy level, but four more are needed to fill this level. To get these electrons, a carbon atom can form a maximum of four covalent bonds with other atoms.)

VISUAL LINK

Figure 2-4

Using the figure, emphasize that bonding between atoms involves the outermost electrons. Students should note that the inner energy level in the oxygen atom contains two electrons and is therefore filled. The outer energy level contains six electrons and needs two more to be complete. One way this level can be filled is by the formation of covalent bonds with two atoms of hydrogen.

SECTION 2-1

INCLUSION *ACTIVITY*

To demonstrate the attraction between opposite charges, have students gently rub an inflated balloon through their hair. Then have students bring the balloon close to their hair without the two touching. Ask students to describe their observations. Explain that their hair is attracted to the balloon because the negative charges on the balloon are attracted to the positive charges on their hair.

ANSWERS TO SECTION 2-1 REVIEW

1. Elements are substances that cannot be broken down chemically into simpler chemical substances. An atom is the simplest part of an element that retains all of the properties of the element. A compound is a pure substance made up of two or more elements. A molecule is the simplest part of a substance that retains all of the properties of the substance and that can exist in a free state.
2. Protons and neutrons make up the nucleus of the atom. Electrons move around the nucleus.
3. Stable elements have filled outermost energy levels. Elements with unfilled outermost energy levels tend to be reactive.
4. In an ionic bond, ions with opposite electrical charges attract each other. In a covalent bond, the atoms share pairs of electrons.
5. Neon has the maximum number of electrons that its outermost energy levels can hold. Thus, neon is stable and tends not to react with other atoms.
6. Hydrogen is very reactive, while helium is relatively unreactive.

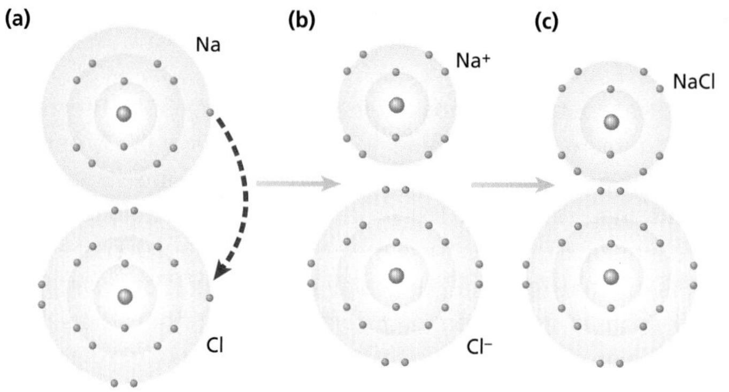

FIGURE 2-5
By losing its outermost electron, a sodium atom becomes an Na^+ ion. By gaining one electron, a chlorine atom becomes a Cl^- ion. Because of their opposite charges, the Na^+ and Cl^- ions are attracted to each other and form an ionic bond.

internet**connect**
SCILINKS NSTA
TOPIC: Covalent and ionic bonds
GO TO: www.scilinks.org
KEYWORD: HM034

Ionic Bonds

As shown in Figure 2-5a, both sodium and chlorine atoms have unfilled outermost energy levels and are therefore reactive. Figure 2-5b shows how both atoms achieve stability in the presence of one another. The one outer electron of a sodium atom is transferred to a chlorine atom. This makes the sodium atom more stable than it was—its new outermost energy level is filled with eight electrons. But it also results in a sodium atom with a net positive electrical charge. The sodium atom has 11 protons (11 positive charges) balanced by only 10 electrons (10 negative charges). An atom or molecule with an electrical charge is called an **ion.** The sodium ion is written as Na^+.

As you can see in Figure 2-5b, by gaining an electron from a sodium atom, a chlorine atom has eight electrons in its outermost energy level, making it more stable. But with this additional electron, chlorine becomes a negatively charged ion called chloride, which is abbreviated as Cl^-.

Because positive and negative electrical charges attract each other, the sodium ion and the chloride ion attract each other. This attraction is called an **ionic bond.** The resulting compound, sodium chloride, NaCl, is an ionic compound and is familiar to you as common table salt.

SECTION 2-1 REVIEW

1. Define *element, atom, compound,* and *molecule.*

2. How are particles arranged in the atom?

3. How can we predict which elements are stable under natural conditions and which elements tend to undergo chemical reactions?

4. How does an ionic bond differ from a covalent bond?

5. Neon seldom, if ever, combines with other elements to form compounds. Why is this so?

6. **CRITICAL THINKING** In the early 1900s, hydrogen gas was used to inflate airships. After one large airship crashed and caught on fire, helium gas began to be used to inflate airships. Why was helium preferred over hydrogen?

ENERGY

One important characteristic of all living things is that they use energy. The amount of energy in the universe remains the same over time, but energy can change in form constantly. It is the flow of energy—from the sun to and through almost every organism on Earth—that biologists seek to understand when they study the chemistry of living things.

SECTION

2-2

OBJECTIVES

▲ List the three states of matter, and explain how matter can change state.

● Describe how energy changes are involved in chemical reactions.

■ Explain how enzymes affect chemical reactions in organisms.

◆ Explain what a redox reaction is.

ENERGY AND MATTER

Scientists define **energy** as the ability to do work or cause change. Energy can occur in various forms, and one form of energy can be converted to another form. In a light bulb's filament, electrical energy is converted to radiant energy (light) and thermal energy (heat).

Energy in Living Things

Some forms of energy important to biological systems include chemical energy, thermal energy, electrical energy, and mechanical energy. Biologists often refer to free energy with respect to living systems. **Free energy** is the energy in a system that is available for work. For example, in a cell, it is the energy that is available to fuel cell processes. As energy flows through a single organism, it may be converted from one form to another. For example, if you ate breakfast this morning, your body is at work now changing the chemical energy found in food into thermal and mechanical energy, among other things.

States of Matter

Although it is not apparent when we observe matter, all the atoms and molecules in any substance are in constant motion. The rate at which atoms or molecules of a substance move determines its **state:** solid, liquid, or gas, as shown in Figure 2-6. Particles of a solid are tightly linked together in a definite shape, where they

FIGURE 2-6
Matter exists as solids, liquids, and gases. You are familiar with all three states of water.

SECTION 2-2

SECTION OVERVIEW

Matter, Energy, and Organization
Section 2-2 describes the different states of matter and explains chemical reactions, including redox reactions. Students will learn how to distinguish exergonic reactions from endergonic reactions and how a catalyst affects the activation energy of a chemical reaction.

▶ ENGAGE STUDENTS
Obtain a cyalume light stick (available from a biological supply house). Fill one 1000 mL beaker with ice water and a second 1000 mL beaker with boiling water. Remove the light stick from its protective wrapper. Explain the construction of the light stick. It contains two chemicals, one of which is kept in a sealed glass vial. Bending the light stick breaks the vial and allows the two chemicals to mix, triggering a chemical reaction that produces light. The color depends on the type of stick—yellow-green, red, blue, and white light sticks are available. Turn off the room lights and place the light stick in the beaker of ice water. Note the dimming of the light. Remove the light stick and place it in the beaker of boiling water. Note the increase in the intensity of light. Point out that temperature is critical to the speed at which a reaction occurs—low temperatures slow reactions, while high temperatures speed up reactions.

CONTENT CONNECTION

Section 1-1: Homeostasis
Remind students that maintenance of internal conditions is one of the characteristics of living things. Ask students why maintaining a relatively constant internal temperature is important.

VISUAL LINK

Figure 2-7

A reversible reaction can proceed in either direction, depending on the concentrations of the reactants and products. Point out that the reversibility of the reaction shown in the figure (carbon dioxide combines with water to yield carbonic acid) helps the body maintain homeostasis. The cells of the body produce carbon dioxide as they break down food. This carbon dioxide diffuses into the blood and drives the formation of carbonic acid. As blood flows through the capillaries of the lungs, carbon dioxide diffuses out of the blood into the lungs. The decline in carbon dioxide concentration favors the dissolution of carbonic acid into water and carbon dioxide. Thus, this reversible reaction helps the body remove carbon dioxide from actively metabolizing tissues and release it into the lungs for elimination.

RECENT RESEARCH

Observing Chemical Bonding

Chemical reactions take place so rapidly that it is usually impossible to observe the interactions between individual molecules. Two chemists from Kansas State University and the University of North Carolina have devised a technique that allows for the detection of the formation of individual bonds. They first placed a solution inside a tiny reaction chamber smaller than a cell, and then ionized the solution with an electric current. Each time two ions of opposite charge met, they emitted a photon that could be detected.

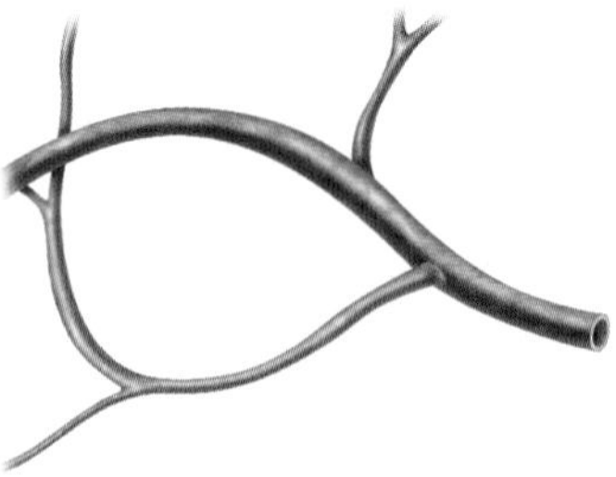

$CO_2 + H_2O \rightleftharpoons H_2CO_3$

FIGURE 2-7

The reaction illustrated in this figure is reversible. Because the products of the reaction remain in the blood, the reaction can proceed either from left to right or from right to left.

vibrate in place. A solid maintains a fixed volume and shape. Particles of a liquid are not as tightly linked as those in a solid. While a liquid maintains a fixed volume, its particles move more freely than those of a solid, giving a liquid its ability to flow and to conform to the shape of any container. Particles of a gas move the most rapidly. Gas particles have little or no attraction to each other, and they fill the volume of the container they occupy. To cause a substance to change from a solid to a liquid and from a liquid to a gas, thermal energy must be added to the substance.

ENERGY AND CHEMICAL REACTIONS

Living things undergo many thousands of chemical reactions as part of their life processes. Many reactions are very complex and are interrelated, involving a multistep sequence. Other reactions are rather simple. The one described in Figure 2-7 takes place in your blood.

The **reactants** are shown on the left side of the equation. In this reaction, the reactants are CO_2 and H_2O. The **products** of the reaction are shown on the right side. In this reaction, the product is H_2CO_3. Notice that the number of each kind of atom must be the same on either side of the arrow. In a chemical reaction, bonds present in the reactants are broken, the elements are rearranged, and new compounds are formed as the products. The two-direction arrow indicates that this chemical reaction can proceed either way. Carbon dioxide and water can combine to form carbonic acid, H_2CO_3, *or* carbonic acid can break down to carbon dioxide and water.

Energy Transfer

Much of the energy your body needs is provided by sugars from foods. Your body continuously undergoes a series of chemical reactions in which sugar and other substances are broken down to carbon dioxide and water. In this process, energy is released for use by your body. Chemical reactions that involve a net release of free energy are called **exergonic** (EKS-uhr-GAHN-ik) **reactions.** Reactions that involve a net absorption of free energy are called **endergonic** (EN-duhr-GAHN-ik) **reactions.**

Activation Energy

For most chemical reactions—both exergonic and endergonic—to begin, energy must be added to the reactants. In many chemical reactions, the amount of energy needed to start the reaction, called **activation energy,** is high. Figure 2-8 shows the activation energy for a hypothetical chemical reaction.

Certain chemical substances, known as **catalysts** (KAT-uh-LISTS), reduce the amount of activation energy that is needed for a reaction,

Word Roots and Origins

catalyst

from the Greek *katalysis,* meaning "dissolution"

Activation Energy With and Without a Catalyst

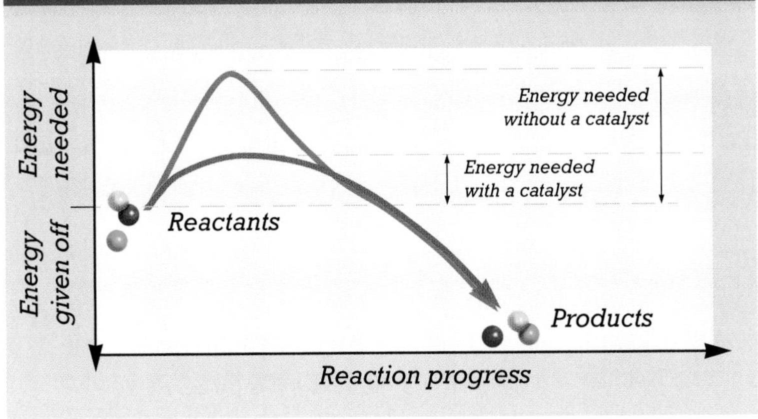

FIGURE 2-8

The blue curve shows the activation energy that must be supplied before this reaction can begin. The activation energy can be reduced, as shown by the pink curve, by adding a catalyst.

as shown in Figure 2-8. A reaction in the presence of the correct catalyst will proceed spontaneously or with the addition of a small amount of energy. **Enzymes** are an important class of catalysts in living things. A single organism may have thousands of different enzymes, each one tailor-made for a different chemical reaction.

Reduction-Oxidation Reactions

You know that there is a constant flow of energy into and throughout living things. Many of the chemical reactions that help transfer *energy* in living things involve the transfer of *electrons.* These reactions in which electrons are transferred between atoms are known as reduction-oxidation reactions, or **redox reactions.** In an **oxidation** (AHKS-uh-DAY-shuhn) **reaction,** a reactant loses one or more electrons, thus becoming more *positive* in charge. For example, remember that a sodium atom loses an electron to achieve stability when it forms an ionic bond, as shown in Figure 2-5. Thus, the sodium atom undergoes oxidation to form an Na^+ ion. In a **reduction reaction,** a reactant gains one or more electrons, thus becoming more *negative* in charge. When a chlorine atom gains an electron to form a Cl^- ion, the atom undergoes reduction. Redox reactions always occur together. An oxidation reaction occurs, and the electron given up by one substance is then accepted by another substance in a reduction reaction.

Quick Lab

Modeling Ionic Bonds

Materials toothpicks, mini marshmallows, peas

Procedure

Use marshmallows to represent chlorine. Use peas to represent sodium. Use toothpicks to create bonds. Make several models of NaCl (sodium chloride).

Analysis Use your models to identify each of the following: a sodium atom, a sodium ion, a chlorine atom, a chloride ion, an ionic bond, and a particle of sodium chloride.

SECTION 2-2 REVIEW

1. What are the three states of matter?
2. How can a substance be changed from a liquid to a gas?
3. State the difference between endergonic and exergonic reactions.
4. Explain how a catalyst affects a reaction.
5. Why does a reduction reaction always accompany an oxidation reaction?
6. CRITICAL THINKING Living things need a constant supply of energy, even though many of the chemical reactions they undergo release energy. Why is this true?

SECTION 2-2

Quick Lab

Modeling Ionic Bonds

Time Required 15 minutes

Safety Remind students not to eat the marshmallows or peas. Toothpicks can puncture the skin. Caution students to handle them with care.

Procedural Tips Choose peas (fresh or frozen) that are soft enough that the toothpicks can be pushed into them easily.

Answers to Analysis An individual marshmallow represents a chlorine atom. An individual pea represents a sodium atom. The toothpick that holds a mini marshmallow and a pea together represents an ionic bond. When the model has been put together, students should be able to identify it as a molecule of sodium chloride. The ions can be identified as subsets of the molecule. For example, a mini marshmallow (attached to a pea) represents a chloride ion and a pea (attached to a mini marshmallow) represents a sodium ion.

ANSWERS TO SECTION 2-2 REVIEW

1. Matter can exist as a solid, liquid, or gas.
2. A substance can be changed from a liquid into a gas by heating.
3. Endergonic reactions absorb free energy. Exergonic reactions release free energy.
4. A catalyst lowers the amount of activation energy that is required for a chemical reaction to begin.
5. If an atom is oxidized (loses electrons), then another atom must be reduced (accept those electrons that have been lost).
6. Even exergonic reactions need a small amount of activation energy to begin.

BACKGROUND

Charles Drew was born on June 3, 1904, in Washington, D.C. He attended Amherst College, McGill University, and Columbia University. While working on a doctorate at Columbia, he researched the properties of blood plasma. He developed effective ways to store and process large amounts of plasma in blood banks. Drew, as the recognized authority on the preservation of human blood for transfusion at the time, directed the blood-bank programs of the United States and Great Britain at the beginning of World War II. Later, he worked as a surgeon and professor of medicine at Freedmen's Hospital, in Washington, D.C., and at Howard University.

It was demonstrated during World War I that stored blood could safely be used in transfusions. However, the discovery of blood types was important in making transfusions successful. It was necessary before the development of blood banking. Before methods of typing blood were available, physicians looked for a match between the patient's blood and that of the patient's relatives and friends by mixing drops of the blood together. If the components of the blood clumped together, the blood did not match. When the proper type was found, the physician bled the donor and transfused the blood to the patient. The ABO blood grouping system discovered by Karl Landsteiner in 1901 is only one of many blood groups. Other blood groups that were discovered later include MNS, Lewis, Kell, Diego, Lutheran, Duffy, and Kidd.

Blood Plasma Meets a Need

HISTORICAL PERSPECTIVE

Throughout history, people have understood the importance of blood, seeing it as the river of life that it is. Prior to the 1900s, severe bleeding often resulted in death. But today blood is stored at blood banks, where people "deposit" blood so that they or others can "withdraw" it when needed. Charles Drew was a pioneer in the work of blood transfusions, especially in the use of plasma and the development of blood banks.

The Need

In the early 1930s, Charles Drew, a medical student at McGill University Medical School in Montreal, Quebec, faced a dilemma. Before him lay a man who needed a blood transfusion so that his leg could be amputated. Compatible blood donors could not be found; even the man's sister had a different blood type. So Drew gave his own blood, a perfect match, and the operation proceeded. After that experience, Drew understood more than ever the importance of finding a way to store blood.

Historical Highlights

Before the twentieth century, successful transfusions like Drew's were almost unknown. Of the few attempts made, some succeeded, but most did not. No one was sure why so many transfusions failed.

The early 1900s marked a breakthrough. The Austrian-born American physiologist Karl Landsteiner determined that there are four blood types—A, B, AB, and O—based on the compatibility of markers on the outer surface of red blood cells. Landsteiner's discovery explained why the outcome of transfusions had been so unpredictable. For his work, Landsteiner received the Nobel Prize in medicine or physiology in 1930.

Charles Drew

Determining a person's blood type—called blood typing—became a vital component of transfusions. Blood types that are not compatible form clumps when mixed together. These clumps can block small blood vessels, causing serious complications and often resulting in the death of the patient.

In 1914, sodium citrate was first added to blood to prevent clotting. This made the storage of blood possible for the first time. Refrigerated blood could be stored for five to seven days. (Today whole blood can be safely stored for 21 to 49 days.) Nevertheless, in the early 1930s, blood banks were still uncommon. Most patients received blood directly from a donor.

In 1937, Bernard Fantus, a physician, collected and distributed blood for transfusions, establishing the first nonprofit blood bank at Cook County Hospital in Chicago. It was Fantus who coined the term *blood bank*. Charles Drew, however, was the one who recognized that a liquid solution in blood called *plasma* could help solve problems associated with storage, making transfusions available on a large scale.

The Composition of Blood

In 1938, Drew and physician John Scudder studied blood chemistry and transfusion, with a focus toward finding a safe way to preserve blood. Blood has two main components: cells and plasma. Three types of cells—red blood cells, white blood cells, and platelets—make up about 45 percent of blood. The other approximately 55 percent of blood is made up of plasma, an amber-colored solution containing more than 100 different solutes, including nutrients, antibodies, hormones, and proteins. Almost 90 percent of plasma is water.

Although plasma contains antibodies that may cause clumping when mixed with incompatible blood types, in most cases transfused plasma dilutes rapidly in the patient's blood, minimizing the risk of clumping. Because red blood cells carry the markers that determine ABO blood types, by removing the red blood cells from blood, the remaining plasma can usually be used safely without type testing. In addition, plasma can be dehydrated and easily stored.

Through his research, Drew concluded that although plasma lacks important components of whole blood, it might be a viable substitute for blood in emergency situations. For example, plasma could save lives on the battlefield.

Response to Wartime Needs

In 1940, Germany's attack on France created a need for a huge blood supply. As experts in transfusion met to decide how to respond, Drew presented his findings on plasma. Although his research was incomplete, the United States began to provide liquid plasma and whole blood for France. However, before any deliveries could be made, France fell to the German army. Soon the "Blood for Britain" program was under way, with Drew as its medical supervisor.

Drew coordinated the American effort. Working with the National Research Council and the American Red Cross, he set up collection centers and was in charge of coordinating the medical aspects of the program, including the establishment of uniform records, standard equipment, and criteria to ensure the safety of the final product. Americans gave blood generously. Drew wanted the blood banks to be well stocked when the United States entered the war—as it soon did.

Charles R. Drew, far left, appears in 1940 with the first mobile blood-collection unit.

The nation's blood supply was ready. Stored blood and plasma has subsequently saved thousands of lives—in both wartime and peacetime. Today blood banks are found in medical facilities worldwide.

Transfusion Update

Early in the 1980s, the practice of blood transfusion had to be reexamined. Some transfused blood was found to carry HIV, causing AIDS in a number of people. Hemophiliacs, whose blood does not clot, were especially vulnerable. It is estimated that half of the hemophiliacs in the United States contracted HIV before adequate testing of donated blood began.

Since 1985, careful screening for HIV, hepatitis, and other diseases has almost entirely removed the risk of receiving contaminated blood. Even so, the level of fear remains high, accounting for a sharp decline in the amount of blood donated. Many people now bank their own blood for later use in surgery. Blood can also be collected during surgery and returned to the patient later.

The AIDS epidemic has triggered a race to create artificial blood. However, blood chemistry is extremely complex, and the process has proven more difficult than expected. Several companies have begun testing potential artificial-blood products. Some of these substitutes make use of chemically treated animal blood and outdated human blood. As a new century begins, the use of artificial blood may be, as one headline put it, "a heartbeat away."

Bags of blood ready for transfusion are kept in cold storage.

Besides being typed to reduce transfusion reaction, blood must be treated with substances such as sodium citrate to prevent clotting. The addition of a small amount of sodium citrate and glucose prevents the coagulation of blood and allows red blood cells to be stored at 4°C for up to 21 days.

DISCUSSION

Guide the discussion by posing the following questions:

1. Before blood-typing methods were developed, what were the dangers involved with blood transfusions? (Blood types that are not compatible form clumps when mixed together. These clumps can block small blood vessels, causing blood flow to be cut off or reduced to parts of the body.)

2. What is the genetic basis of the ABO blood-typing system? (The red blood cells have genetically determined antigens on their cell membranes that "identify" their blood group and antibodies in the plasma that can combine with the antigens on red blood cells of another type.)

3. What are the two main components of blood? (Cells—red blood cells and white blood cells—and the cell fragments—platelets—make up about 45 percent of whole blood. The other 55 percent is plasma.)

FURTHER READING

1. Love, Spencie. *One Blood: The Death and Resurrection of Charles Drew.* Chapel Hill: University of North Carolina Press, 1996.

2. Massaquoi, Hans J. "50 Years of Blacks in Medicine." *Ebony,* July 1995, 120–125.

SECTION OVERVIEW

Stability and Homeostasis

Section 2-3 describes solutions and their components. Solutions are particularly important in biological processes because chemical reactions occur in the aqueous solutions within living things. The section also introduces the concepts of acidity and alkalinity and the importance of pH control in living systems.

▶ ENGAGE STUDENTS

Give each student three small beakers. Have students add just enough water to the first beaker to cover the bottom. They should add the same amount of rubbing alcohol to the second beaker and the same amount of cooking oil to the third beaker. Have students sprinkle a small amount of salt into each beaker and then swirl the beaker gently. Allow the beakers to stand for a minute or two. Have students describe what happens in each beaker. (The salt dissolved only in the beaker containing water). Ask students to name the solute and solvent in each case (solute: salt; solvents: water, rubbing alcohol, cooking oil). Ask them which solvent dissolved the salt (water).

SECTION 2-3

OBJECTIVES

▲ Define *solution, solute, solvent,* and *concentration.*

● Explain the dissociation of water.

■ Contrast properties of acids and bases.

◆ Describe the use of the pH scale.

◆ Explain the action of buffers.

SOLUTIONS

The chemistry of living things involves the study of solutions. A large proportion of the mass of living things is water, and the chemical reactions of life occur for the most part in water solutions. The electricity that courses through your nerves is transmitted through watery pathways of dissolved ions. Chemical messengers that regulate your body's metabolism move through the watery medium of your blood.

DESCRIBING SOLUTIONS

A **solution** is a mixture in which one or more substances are uniformly distributed in another substance. Solutions can be mixtures of liquids, solids, or gases. For example, plasma, the liquid part of blood, is a very complex solution. It is composed of many types of ions and large molecules, as well as gases, that are dissolved in water. The **solute** (SAHL-YOOT) is the substance dissolved in the solution. The particles that compose a solute may be ions, atoms, or molecules. The **solvent** is the substance in which the solute is dissolved. For example, when sugar, a solute, and water, a solvent, are mixed, a solution of sugar water results. Though the sugar dissolves in the water, neither the sugar molecules nor the water molecules are altered chemically. If the water is boiled away, the sugar molecules remain and are unchanged.

Solutions can be composed of various proportions of a given solute in a given solvent. Thus, solutions can vary in concentration. The **concentration** of a solution is the measurement of the amount of solute dissolved in a fixed amount of the solution. For example, a 2 percent saltwater solution contains 2 g of salt dissolved in enough water to make 100 mL of solution. The more solute dissolved, the greater the concentration of the solution. A **saturated solution** is one in which no more solute can dissolve.

Aqueous (AY-kwee-uhs) **solutions**—solutions in which water is the solvent—are universally important to living things. You should be able to think of many different aqueous solutions important to living things. Marine microorganisms spend their lives immersed in the sea, an aqueous solution. Most nutrients that plants need are in aqueous solutions in moist soil. Body cells exist in an aqueous solution of intercellular fluid and are themselves filled with fluid.

Word Roots and Origins

solvent

from the Latin *solvere,* meaning "to loosen"

ACIDS AND BASES

One of the most important aspects of a living system is the degree of its acidity or alkalinity. What do we mean when we say *acid* and *alkaline*?

Dissociation of Water

In water, the force of attraction between molecules is so strong that the oxygen atom of one water molecule can actually remove the hydrogen atom from the other water molecule. This breaking apart of the water molecule into two ions of opposite charge is called **dissociation** and is shown by the chemical equation below.

$$H_2O \rightleftharpoons H^+ + OH^-$$

One water molecule, H_2O, dissociates to form two ions, H^+ and OH^-. The OH^- ion is known as the **hydroxide ion.** The free H^+ ion can react with another water molecule, as shown in the following equation.

$$H^+ + H_2O \rightleftharpoons H_3O^+$$

The H_3O^+ ion is known as the **hydronium ion.** Acidity or alkalinity is a measure of the relative amounts of hydronium ions and hydroxide ions dissolved in a solution. If the number of hydronium ions in a solution equals the number of hydroxide ions, the solution is said to be neutral. Pure water contains equal numbers of hydronium ions and hydroxide ions and is therefore a neutral solution.

Acids

If the number of hydronium ions in a solution is greater than the number of hydroxide ions, the solution is an **acid.** Consider what happens when hydrogen chloride, HCl, a gas, is dissolved in water. Some of its molecules dissociate to form hydrogen ions, H^+, and chloride ions, Cl^-.

$$HCl \rightleftharpoons H^+ + Cl^-$$

These free hydrogen ions combine with water molecules to form hydronium ions, H_3O^+. This aqueous solution contains many more hydronium ions than it does hydroxide ions, making it an acidic solution. Acids tend to have a sour taste. In concentrated forms, they are highly corrosive to some materials, as you can see in Figure 2-9.

FIGURE 2-9

Acids can have a significant impact on our environment. Sulfur dioxide, SO_2, which is produced when fossil fuels are burned, reacts with water in the atmosphere to produce acid precipitation. Acid precipitation can make lakes and rivers too acidic to support life and can even corrode stone, such as this marble carving.

Eco Connection

Acid Precipitation

Acid precipitation, more commonly called *acid rain,* describes rain, snow, sleet, or fog that contains high levels of sulfuric and nitric acids. These acids form when sulfur dioxide gas, SO_2, and nitrogen oxide gas, NO, react with water in the atmosphere to produce sulfuric acid, H_2SO_4, and nitric acid, HNO_3.

Acid precipitation makes soil and bodies of water, such as lakes, more acidic than normal. These high acid levels can harm plant and animal life directly. A high level of acid in a lake may kill mollusks, fish, and amphibians. Even in a lake that does not have a very elevated level of acid, acid precipitation may leach aluminum and magnesium from soils, poisoning water-dwelling species.

Reducing fossil-fuel consumption, such as occurs in gasoline engines and coal-burning power plants, should reduce high acid levels in precipitation. You can help by learning about alternative fuel sources and legislation proposed to encourage or mandate use of non-fossil-fuel sources of energy.

QUICK FACT

✔ Precipitation normally has a pH of about 5.0 to 5.6. In areas of the United States severely affected by acid rain, the pH of precipitation may be as low as 3.

TEACHING STRATEGY

Acid Rain

Acid rain is an environmental problem that can affect areas far from the source of pollution. Prevailing winds can carry pollutants from factories and power plants into undeveloped areas, where they create acid rain, damaging remote forests and bodies of water. Have students research and write a report on the effects of acid rain on forests and lakes in wilderness areas.

DEMONSTRATION

Use pH paper or a pH meter to determine the pH of the following common solutions: lemon juice, household ammonia, vinegar, milk, black coffee, baking soda solution, cola, and milk of magnesia (approximate pH values: lemon juice, 2; ammonia, 11.5; vinegar, 3; milk, 7; black coffee, 5; baking soda solution, 8.5; cola, 3; milk of magnesia, 10.5).

SECTION 2-3

RETEACHING *ACTIVITY*

Have students use Figure 2-10 to find the pH of urine and blood (urine = 6; blood = 7.5). Remind students that a pH change of one unit represents a tenfold change in the concentration of hydronium ions. Ask students how much more concentrated hydronium ions are in stomach acid than in vinegar (10 times) and how much more concentrated hydronium ions are in vinegar than in water (10,000 times).

ANSWERS TO SECTION 2-3 REVIEW

1. A solution is a mixture in which one or more substances are uniformly distributed in another substance.
2. The oxygen atom of one water molecule removes the hydrogen atom of another water molecule. This results in one hydronium ion (H_3O^+) and one hydroxide ion (OH^-).
3. A pH value of 7 is neutral.
4. An acid is a solution that contains more hydronium ions than hydroxide ions. A base contains more hydroxide ions than hydronium ions.
5. A buffer is a chemical substance that neutralizes small amounts of acids or bases added to a solution.
6. The buffer reduces the acidity of the aspirin and thus is less irritating to the stomach.

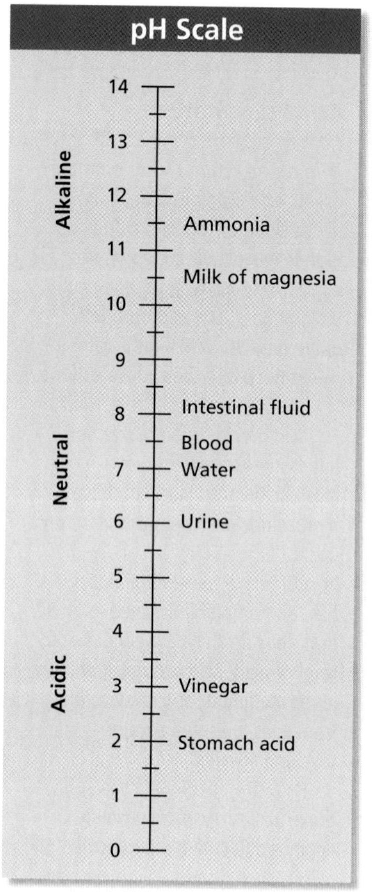

FIGURE 2-10

Some of your body fluids are acidic, while others are alkaline. A solution with a pH above 7 is alkaline, while a solution with a pH below 7 is acidic. Each unit on the pH scale reflects a tenfold change in acidity or alkalinity.

Bases

If sodium hydroxide, NaOH, a solid, is dissolved in water, some of it dissociates to form sodium ions, Na^+, and hydroxide ions, OH^-, as shown in the equation below.

$$NaOH \rightleftharpoons Na^+ + OH^-$$

This solution then contains more hydroxide ions than hydronium ions and is therefore defined as a **base.** The adjective ***alkaline*** refers to bases. Bases have a bitter taste. They tend to feel slippery because the OH^- ions react with the oil on our skin to form a soap. In fact, commercial soap is made by reacting a base with a fat.

pH

Scientists have developed a scale for comparing the relative concentrations of hydronium ions and hydroxide ions in a solution. It is called the **pH scale,** and it ranges from 0 to 14, as shown in Figure 2-10. A solution with a pH of 0 is very acidic, a solution with a pH of 7 is neutral, and a solution with a pH of 14 is very basic. A solution's pH is measured on a logarithmic scale. That is, the change of one pH unit reflects a tenfold change in the acidity or alkalinity. For example, a solution with a pH of 4 has 10 times more H_3O^+ ions than a solution with a pH of 5 and 100 times more H_3O^+ ions than a solution with a pH of 6. The pH of a solution can be measured with litmus paper or with some other chemical indicator that changes color at various pH levels.

Buffers

The control of pH is important for living systems. Enzymes such as those you read about in Section 2-2 can function only within a very narrow pH range. The control of pH in organisms is often accomplished with buffers. **Buffers** are chemical substances that neutralize small amounts of either an acid or a base added to a solution. As Figure 2-10 shows, the composition of your internal environment—in terms of acidity and alkalinity—varies greatly. Some of your body fluids, such as stomach acid and urine, are acidic. Others, such as intestinal fluid and blood, are basic or alkaline. Complex buffering systems maintain the pH values of your body's many fluids at normal and safe levels.

SECTION 2-3 REVIEW

1. What is a solution?

2. Describe the dissociation of water.

3. What pH value is neutral?

4. Define *acid* and *base*.

5. What is a buffer?

6. CRITICAL THINKING The active ingredient in aspirin is acetylsalicylic acid. Why would doctors recommend buffered aspirin for some people, especially those who have a "sensitive" stomach?

CHAPTER 2 REVIEW

SUMMARY/VOCABULARY

2-1

- Elements are substances that cannot be broken down by chemical means into simpler substances.
- Atoms are composed of protons, neutrons, and electrons. Protons and neutrons compose the nucleus of the atom. Electrons travel around the nucleus.
- Compounds consist of atoms of two or more elements that are joined by chemical bonds in a fixed proportion.
- Most elements react to form chemical bonds so that their atoms become stable. Atoms achieve stability when their outermost energy level is filled.
- A chemical reaction is the process of breaking chemical bonds, rearranging the atoms, and forming new bonds.
- A covalent bond is formed when two atoms share electrons.
- A molecule consists of two or more atoms held together by covalent bonds.
- An ionic bond is formed when one atom gives up an electron to another. The positive ion is then attracted to a negative ion to form the ionic bond.

Vocabulary

atom (32)
atomic number (32)
bond (33)
chemical reaction (33)
compound (33)
covalent bond (33)
electron (32)
element (31)
energy level (32)
ion (34)
ionic bond (34)
mass (31)
matter (31)
molecule (33)
neutron (32)
nucleus (32)
proton (32)

2-2

- Addition of thermal energy to a substance can cause its state to change from a solid to a liquid and from a liquid to a gas.
- Chemical reactions that involve a net release of energy are called exergonic reactions. Chemical reactions that involve a net absorption of energy are called endergonic reactions.
- The activation energy is the amount of energy required for a chemical reaction to begin.
- Catalysts lower the amount of activation energy necessary for a reaction to begin.
- A chemical reaction in which electrons are exchanged between atoms is called a reduction-oxidation, or redox, reaction.

Vocabulary

activation energy (36)
catalyst (36)
endergonic reaction (36)
energy (35)
enzyme (37)
exergonic reaction (36)
free energy (35)
oxidation reaction (37)
product (36)
reactant (36)
redox reaction (37)
reduction reaction (37)
state (35)

2-3

- A solution consists of a solute dissolved in a solvent, which is often water. A solution with water as the solvent is known as an aqueous solution.
- An acidic solution contains more hydronium ions than it does hydroxide ions. A basic or alkaline solution contains more hydroxide ions than it does hydronium ions.
- The pH scale indicates the relative concentration of hydronium ions and hydroxide ions in solution. The pH scale ranges from 0 to 14, with 0 being the most acidic, 7 being neutral, and 14 being the most alkaline.
- Buffers are chemicals that neutralize the effects of adding small amounts of either an acid or a base to a solution.

Vocabulary

acid (41)
alkaline (42)
aqueous solution (40)
base (42)
buffer (42)
concentration (40)
dissociation (41)
hydronium ion (41)
hydroxide ion (41)
pH scale (42)
saturated solution (40)
solute (40)
solution (40)
solvent (40)

CHAPTER 2 REVIEW ANSWERS

REVIEW

1. An electron is negatively charged and moves around the nucleus; a neutron has no charge and is part of the nucleus; a proton is positively charged and is part of the nucleus.
2. An element cannot be broken down chemically into simpler substances with the same properties. A compound is composed of the atoms of two or more elements joined by chemical bonds.
3. An acid contains more hydronium ions than hydroxide ions, whereas a base contains more hydroxide ions than hydronium ions.
4. An exergonic reaction involves a net release of free energy.
5. Oxidation involves the giving up of an electron; in reduction, an electron is gained. They are complementary processes.

6. a
7. c
8. c
9. d
10. b
11. d
12. a
13. c
14. b
15. b

16. a. stomach acid
b. intestinal fluid
c. blood
d. stomach acid

17. Each oxygen atom requires two additional electrons to achieve stability. Thus, two oxygen atoms must share four electrons or two pairs in order for each to have eight electrons and become stable.

18. An ion is an atom or a molecule with an electrical charge.

19. In a covalent bond, one or more pairs of electrons are shared by two atoms.

20. Matter exists as a solid, liquid, or gas.

21. A two-direction arrow indicates that the reaction can proceed either way.

22. Most reactions require an infusion of energy to get started.

23. The oxygen atom of one water molecule removes a hydrogen atom from another water molecule.

24. The word *alkaline* means "basic."

25. A buffer neutralizes relatively small changes in the acid/base balance of a solution.

REVIEW

Vocabulary

1. Explain the relationship between electrons, neutrons, and protons.
2. What is the difference between an element and a compound?
3. Distinguish the differences in composition between an acid and a base.
4. Identify the type of reaction that releases free energy.
5. How are the processes of oxidation and reduction related?

Multiple Choice

6. The nucleus of an atom is made up of (a) protons and neutrons (b) protons and electrons (c) elements and compounds (d) negatively charged particles.
7. High-energy particles that circle the nucleus of an atom are (a) ions (b) protons (c) electrons (d) neutrons.
8. The way in which elements bond to form compounds depends on the (a) structural formula of the compound (b) dissociation of the ions in the compound (c) number and arrangement of electrons in the atoms of the elements (d) model of the atom.
9. If an atom is made up of 6 protons, 7 neutrons, and 6 electrons, then its atomic number is (a) 19 (b) 13 (c) 7 (d) 6.
10. Atoms in a solid (a) are fixed in space and show no movement (b) are fixed in space but vibrate in place (c) move rapidly through space (d) repel each other.
11. The amount of energy required for a chemical reaction to begin is called (a) chemical energy (b) mechanical energy (c) electrical energy (d) activation energy.
12. The process in which a chemical reactant loses an electron, becoming more positively charged, is called (a) oxidation (b) reduction (c) metabolism (d) stabilization.
13. In a reduction reaction, an atom gains a(n) (a) proton (b) neutron (c) electron (d) nucleus.
14. An aqueous solution that contains more hydroxide ions than hydronium ions is a(n) (a) acid (b) base (c) gas (d) solid.
15. Acid formed from sulfur dioxide in the atmosphere is present in (a) rocks (b) acid precipitation (c) weak bases (d) pure water.

Short Answer

16. Use the pH scale shown below to answer the following questions:
    a. What is the most acidic body fluid represented?
    b. What is the most alkaline body fluid represented?
    c. What body fluid is closest to being a neutral solution?
    d. Which body fluid is most extremely acidic or basic; that is, which body fluid deviates the most from neutral pH?

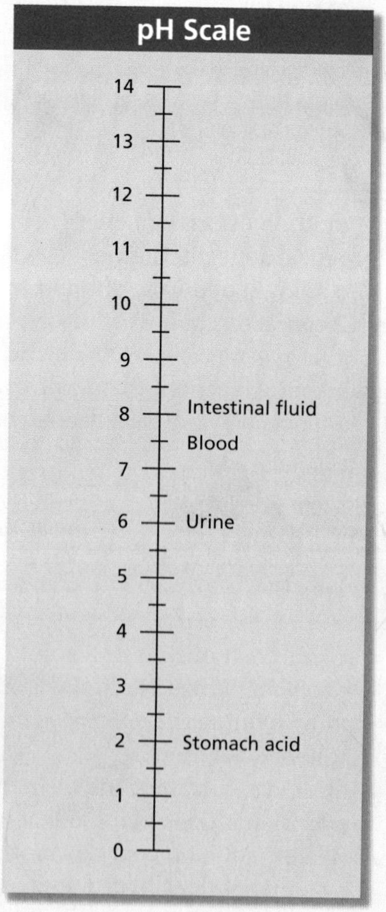

17. An oxygen atom has six electrons in its outermost energy level. Explain why two oxygen atoms must share four electrons when they form a covalent bond.
18. What is an ion?
19. How are electrons distributed in a covalent bond?
20. Name the physical states that matter can exist in.
21. In a chemical equation, what does a two-direction arrow mean?
22. Many reactions in the cell are exergonic. Why, then, do cells need a continuous supply of energy?
23. What happens when water dissociates?
24. What does the word *alkaline* mean?
25. What does a buffer do?

CRITICAL THINKING

1. Hydrogen gas exists as H_2 rather than H. Why is this so?
2. How can a substance be changed from a solid to a liquid? from a liquid to a gas?
3. A magnesium atom has two electrons in its outermost energy level. A sulfur atom has six electrons in its outermost level. How will the two atoms react to form a bond? Explain why this is considered a redox reaction.
4. A dam located on a fast-flowing mountain stream generates electricity. What kind of energy is transformed to create electricity?
5. The table shows melting and boiling points at normal pressure for five different elements or compounds. Above the boiling point, a compound or element exists as a gas. Below the freezing point, a compound or element exists as a solid. Use the table to answer the following questions:
   a. Under normal temperature and pressure conditions, which substances exist as solids? as liquids? as gases?
   b. Which substance exists as a liquid over the broadest range of temperature?
   c. Which substance exists as a liquid over the narrowest range of temperature?
   d. Which one of the substances are you least likely to encounter as a gas?

Melting and Boiling Points at Normal Pressure

Substance	Melting point (°C)	Boiling point (°C)
Aluminum	658	2330
Argon	−190	−186
Chlorine	−104	−34
Mercury	−39	357
Water	0	100

EXTENSION

1. Read "The Femtosecond Camera Shutter" in *Scientific American,* January 2000, on page 15, and answer the following questions: What is the main obstacle scientists encounter in observing chemical reactions step by step? What was the key process Dr. Zewail used to observe chemical reactions? What was the chemical reaction observed using this new technique? What goals may ultimately be achieved using this technique?
2. Dissolve table salt or sugar by stirring it into water at room temperature until it is saturated. (The solution is saturated when excess salt or sugar stays at the bottom of the container.) Remove the clear solution at the top with a spoon, and place the remaining saturated solution in the refrigerator. After two hours, observe the cooled solution. What do your observations tell you about the effect of temperature on molecules in solution?

CRITICAL THINKING

1. A hydrogen atom tends to covalently bond to another hydrogen atom, sharing electrons and becoming stable.
2. If sufficient energy is added to a solid, it will change to a liquid and then to a gas.
3. The magnesium atom will give up its two electrons to the sulfur atom. As a result, both atoms will have eight electrons in their outermost energy levels and become stable. This is a redox reaction because the magnesium atom has lost electrons, and the sulfur atom has gained electrons.
4. Mechanical energy is transformed into electromagnetic energy.
5. a. solids: aluminum; liquids: mercury, water; gases: argon, chlorine
b. aluminum
c. argon
d. aluminum

EXTENSION

1. The difficulty in observing the sequence of chemical reactions is the extreme speed with which they occur. Dr. Zewail developed a laser probe pulse technique to illuminate a chemical reaction in real time. He observed cyanogen iodide split into iodine and cyanide. The long-term goals are to study protein molecules, which are the building blocks of life, and to alter molecules using precisely placed probes.
2. Students will observe more salt or sugar at the bottom of the container after it is left in the refrigerator. Based on this, students may infer that the amount of solute that can dissolve in a solution decreases as temperature decreases. Students may then conclude that the higher the temperature, the more nutrients and gases that dissolve in solution and thus in living cells.

CHAPTER 2
INTERACTIVE EXPLORATION

For an overview of this Interactive Exploration and preparatory information related to this activity, please see the CD User's Guides.

PREPARATION NOTES

Time Required:
One 50-minute class period

PRELAB DISCUSSION

Before students begin the lab, they should be able to answer the following questions:

1. What is an enzyme? (An enzyme is a catalyst that increases the rate of a reaction by reducing the activation energy.)

2. How do enzymes help an organism conserve energy? (By reducing the activation energy, enzymes reduce the amount of energy needed to stimulate chemical reactions within the organism.)

PROCEDURAL TIPS

1. Use the inset of the computer screen to review the functions of each feedback meter before allowing students to begin the lab.

2. A constraint of the program is that the variables must be tested in the following order: enzyme concentration, temperature, pH. At least five values of each variable must be tested before the computer will allow you to move to the next variable. Also remind students that they must click the Plot Point button each time they want to enter a particular value for one of the variables.

3. Remind students that the values for reaction rate are expressed as a percentage of the maximum reaction rate (V_{max}).

INTERACTIVE EXPLORATION

Exploring the Activity of Biological Catalysts

OBJECTIVES

- Simulate the effect of pH, temperature, and enzyme concentration on the activity of two enzymes.
- Determine the optimal conditions for each enzyme.

MATERIALS

- computer with CD-ROM drive
- CD-ROM *Interactive Explorations in Biology: Cell Biology and Genetics*
- graph paper

Background

All organisms rely on enzymes to catalyze chemical reactions. Recall that an enzyme is a biological catalyst that increases the rate of a chemical reaction by lowering the level of activation energy necessary to start the reaction. Without enzymes, many of the chemical reactions that occur within living things would proceed too slowly to be useful. Enzymes speed up these reactions by bringing the reactants into close proximity and facilitating their interaction. Enzyme effectiveness depends on several factors. This interactive investigation allows you to explore how three of these factors—enzyme concentration, temperature, and pH—affect the activity of two enzymes present in the human body, carbonic anhydrase and lysozyme. Carbonic anhydrase helps regulate carbon dioxide levels in red blood cells. Lysozyme breaks large sugar molecules into smaller ones during metabolism.

Prelab Preparation

1. Load and start the program Thermodynamics. You will see an animated diagram like the one below. Click the Navigation button, and then click the Topic Information button with the "key" icon on it. Read the focus questions, and review the following concepts: Reaction Rate, Enzyme Catalysis, Temperature and Reaction Rates, and pH Can Influence Enzyme Shape.
2. Click the word *Help* at the top left of the screen, and select How to Use This Exploration. Listen to the instructions for conducting the exploration. Click the Interactive Exploration button on the Navigation Palette to begin the exploration.

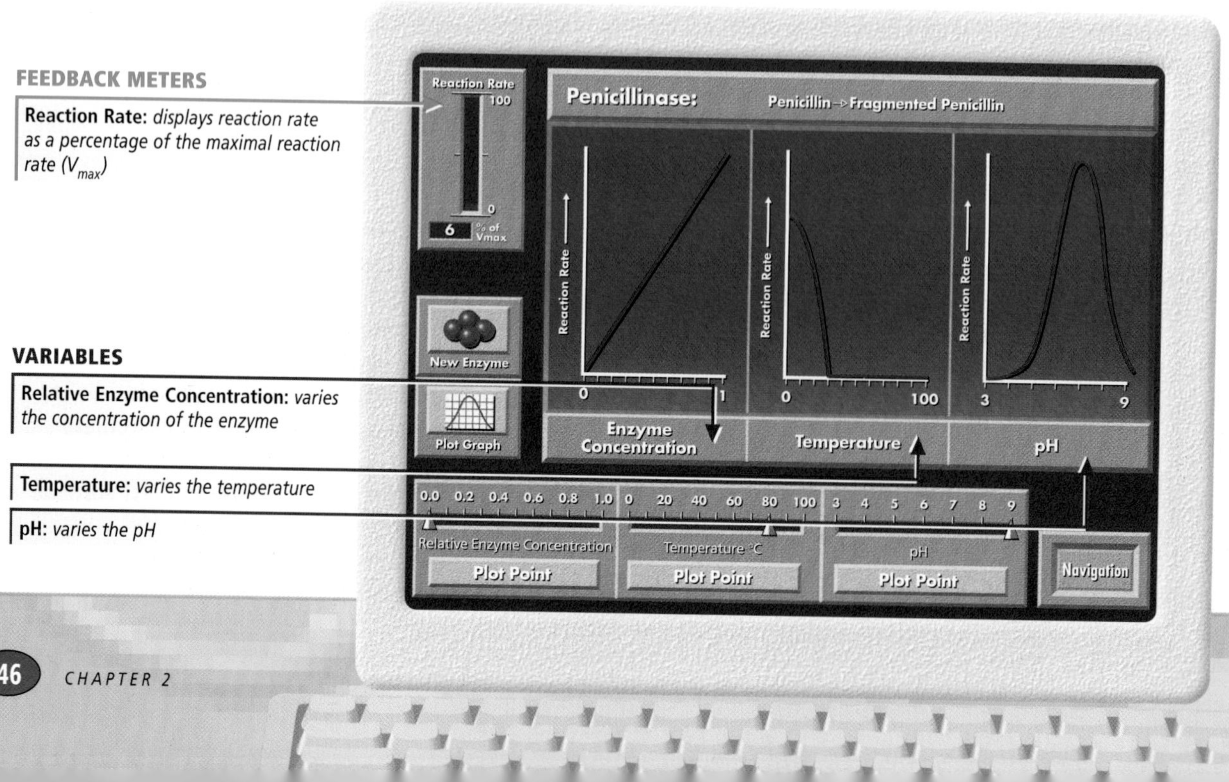

Procedure

PART A Carbonic Anhydrase Activity

1. Click the New Enzyme button until "Carbonic Anhydrase" appears at the top of the screen.
2. Click and drag the indicator so that the Relative Enzyme Concentration is 0.2.
3. Click the Plot Point button. The point that appears on the graph indicates the rate of reaction for this enzyme concentration, expressed as a percentage of V_{max}, the highest possible reaction rate. What is the reaction rate indicated by the Reaction Rate meter?
4. Repeat steps 2 and 3 until you have tested the following relative enzyme concentrations: 0.4, 0.6, 0.8, and 1.0.
5. Click the Plot Graph button. How does increasing the enzyme concentration affect the rate of reaction?
6. Make a table similar to Table A shown below.

TABLE A *EFFECTS OF TEMPERATURE ON REACTION RATE*

Temperature	Carbonic anhydrase % of V_{max}	Lysozyme % of V_{max}
0°C	48%	12%
10°C	90%	23%
20°C	75%	40%
30°C	32%	69%
40°C	0%	98%
50°C	0%	82%

7. Click and drag the indicator so that the Temperature meter indicates 0°C. Click the Plot Point button, and observe the result. In your table, record the reaction rate.
8. Repeat step 7 until you have tested the effects of the following temperatures: 10°C, 20°C, 30°C, 40°C, and 50°C.
9. Use graph paper to make a graph of temperature versus reaction rate. Draw a curve that connects your six points. What can you conclude about the effect of temperature on the activity of carbonic anhydrase?
10. Click the Plot Graph button. Compare your graph with the one shown on the screen. Explain any differences.

TABLE B *EFFECTS OF pH ON REACTION RATE*

pH	Carbonic anhydrase % of V_{max}	Lysozyme % of V_{max}
3	45%	30%
4	91%	84%
5	100%	100%
6	91%	84%
7	45%	30%
8	8%	4%

11. Make a table similar to Table B shown above.
12. Click and drag the indicator so that the pH meter indicates a pH of 3. Click the Plot Point button, and observe the result. In your table, record the reaction rate.
13. Repeat step 12 until you have tested the following pH values: 4, 5, 6, 7, and 8.
14. After testing all the values above, make a graph of pH versus reaction rate. Draw a curve that connects your points. What can you conclude about the effect of pH on the activity of carbonic anhydrase?
15. Click the Plot Graph button. Compare your graph with the graph on the screen. Explain any differences.

PART B Lysozyme Activity

16. Click the New Enzyme button until "Lysozyme" appears at the top of the screen.
17. Repeat steps 2–15 for the enzyme lysozyme, completing Table A and Table B.

Analysis and Conclusions

1. What is the optimal temperature for each enzyme? What is the optimal pH for each enzyme?
2. The optimal pH of pepsin, a stomach enzyme, is about 2, while the optimal pH of trypsin, an enzyme of the small intestine, is about 8. What must occur in your digestive tract between the stomach and the small intestine?

ANSWERS TO PART A

3. The reaction rate is 20 percent.
5. Reaction rate increases linearly with increasing enzyme concentration.
9. Students' graphs should reflect the data shown in the table. Temperature increases reaction rate between 0°C and 10°C. Above 10°C, reaction rate declines with increasing temperature.
10. The graph on the screen is smoother because it includes many more points.
14. Students' graphs should reflect the data in the table. Reaction rate increases with increasing pH and reaches a maximum at a pH of 5. Reaction rate decreases as the pH increases above 5.
15. Students' graphs will not be as smooth because they contain only six points. Students' graphs should reflect these data.

ANSWERS TO PART B

17. For lysozyme, reaction rates rise with temperature, reaching a maximum at 40°C, above which reaction rates begin to decline. Reaction rates rise with increasing pH and reach a maximum at a pH of 5. Above this value, reaction rates decline.

ANSWERS TO ANALYSIS AND CONCLUSIONS

1. The optimal temperature of carbonic anhydrase lies between 10°C and 20°C; for lysozyme, the optimal temperature is 40°C. The optimal pH for carbonic anhydrase and lysozyme is 5.
2. The pH of the contents of the digestive tract must rise.

CHAPTER 3
BIOCHEMISTRY

	TOPICS	TEACHING RESOURCES	LABS, CLASSWORK, AND HOMEWORK
BLOCK 1 (45 minutes)	**Introducing the Chapter,** p. 48	ATE Focus Concept, p. 48 ATE Assessing Prior Knowledge, p. 48 ATE Understanding the Visual, p. 48	■ Supplemental Reading Guide, *The Andromeda Strain* ■ Occupational Applications Worksheets—Forensic Toxicologist
	3-1 Water, p. 49 Polarity, p. 49 Hydrogen Bonding, p. 50	ATE Section Overview, p. 49 ATE Critical Thinking, p. 50 2–3	★ Study Guide, Section 3-1 PE Section Review, p. 51
BLOCK 2 (45 minutes)	**3-2 Carbon Compounds,** p. 52 Carbon Bonding, p. 52 Functional Groups, p. 53 Large Carbon Molecules, p. 53	ATE Section Overview, p. 52 ATE Visual Link Figure 3-7, p. 53 4A	PE **Quick Lab,** p. 53, Demonstrating Polarity ■ Basic Skills Worksheets—Graphing ★ Study Guide, Section 3-2 PE Section Review, p. 54
BLOCK 3 (45 minutes)	**3-3 Molecules of Life,** p. 55 Carbohydrates, p. 55 Proteins, p. 56 Lipids, p. 58 Nucleic Acids, p. 59	ATE Section Overview, p. 55 ATE Visual Link Figure 3-13, p. 57 2A, 3A, 9	PE **Chapter 3 Investigation,** p. 64 **Lab E3,** Thermodynamics ■ Basic Skills Worksheets—Reading a Balance ■ Basic Skills Worksheets—Temperature Conversions ★ Study Guide, Section 3-3 PE Section Review, p. 59
BLOCKS 4 & 5 (90 minutes)	**REVIEW and ASSESSMENT** PE Chapter 3 Review, pp. 62–63 ■ Performance-Based Assessment—Introduction to Performance-Based Assessment	Audio CD Program	**CHAPTER TESTING** ■ Chapter Test (blackline master) ▲ Test Generator ■ Assessment Item Listing ■ Scoring Rubrics and Classroom Management Checklists

MODERN BIOLOGY | HOLT BIOSOURCES | One-Stop Planner CD-ROM

KEY

- PE Pupil's Edition
- ATE Teacher's Edition
- ■ Active Reading Guide
- ★ Study Guide
- Teaching Transparencies
- Laboratory Program

One-Stop Planner CD-ROM *Shows these resources within a customizable daily lesson plan:*

- ■ Holt BioSources Teaching Resources
- ■ Active Reading Guide
- ★ Study Guide
- ▲ Test Generator

READING FOR CONTENT MASTERY

Preview: Have students preview the objectives, the introductory paragraph, topic sentences, and captions before reading the chapter.

- ATE Active Reading Technique: K-W-L, p. 49
- ■ Reading Strategy: K-W-L Worksheet
- ■ Active Reading Guide Worksheet 3-1
- PE Summary/Vocabulary, p. 61

- ■ Active Reading Guide Worksheet 3-2
- PE Summary/Vocabulary, p. 61

- ATE Active Reading Technique: K-W-L , p. 59
- ■ Reading Strategy: K-W-L Worksheet
- ■ Active Reading Guide Worksheet 3-3
- PE Summary/Vocabulary, p. 61

TECHNOLOGY AND INTERNET RESOURCES

Holt Biology Videodiscs Teacher's Correlation Guide, Lessons 3-1 through 3-3

Audio CD Program, Sections 3-1 through 3-3

internetconnect

SCILINKS NSTA

On-line Resources: www.scilinks.org
The following *sci*LINKS Internet resources can be found in the student text for this chapter:

Topics:
- Hydrogen bonding, p. 51
- Carbohydrates, p. 55
- Steroids, p. 59

go.hrw.com

On-line Resources: go.hrw.com
Visit the HRW Web site for a variety of resources related to this chapter. Just type in the keyword HM2 Chapter 3.

Smithsonian Institution®

Visit **www.si.edu/hrw** for additional on-line resources.

CNNfyi.com

Visit **www.cnnfyi.com** for late-breaking news and current event stories selected just for you.

LAB ACTIVITY PLANNING

CHAPTER 3 INVESTIGATION

Identifying Organic Compounds in Food, pp. 64–65

OVERVIEW
In this investigation, students test for carbohydrates, proteins, and lipids in foods.

SAFETY
Instruct students to handle the solutions used in this lab carefully. The solutions can injure skin and eyes or stain skin and clothing. If any chemical gets in a student's eye, flush the eye with water for at least 15 minutes and seek immediate medical attention. Students should use tongs to handle hot test tubes and should not handle the hot plate with wet hands or while near water.

PREPARATION
1. Use low-fat cottage cheese to create the unknown solution. Add 5 g of cottage cheese to 100 mL of distilled water, and stir until mixed thoroughly.
2. Purchase glucose in powder form, and prepare a 10 percent solution (10 g in a final volume of 100 mL of water).
3. Purchase albumin in powder form, and prepare a 10 percent solution (10 g in a final volume of 100 mL of water).
4. Benedict's solution can be purchased from a biological supply house, such as WARD'S.
5. Prepare an 8 percent sodium hydroxide solution (8 g in a final volume of 100 mL of water) and a 1 percent copper(II) sulfate solution (1 g in a final volume of 100 mL of water) separately. Alternatively, purchase biuret reagent from a biological supply house, such as WARD'S.
6. To prepare a 2 percent stock solution of Sudan III, add 2 g of Sudan III and dilute to a final volume of 100 mL with undiluted ethyl alcohol. Just before using, mix the stock solution with an equal amount of 45 percent ethyl alcohol.

Quick Labs

Demonstrating Polarity, p. 53
Students will need disposable gloves, lab aprons, safety goggles, test tubes, test-tube racks, cooking oil, ethanol, and water for this activity.

UNDERSTANDING THE VISUAL

Ask students to name an advantage the jellyfish gains by living in water. Then ask why living things need water. (Water gives the jellyfish structural support. Living things need water to carry out the biochemical reactions vital to life.)

FOCUS CONCEPT

Matter, Energy, and Organization

Water enables temperature regulation in organisms and serves as a reservoir of hydrogen and hydroxide ions needed for important chemical reactions, such as condensation and hydrolysis. Organisms must have energy to move, change, and recombine their carbon compounds. By forming and breaking apart in a continuous process, ATP allows organisms to harvest and use energy. Organization is seen in the wide variety of carbon compounds with specific structures and functions.

ASSESSING PRIOR KNOWLEDGE

Review the following concepts.

Characteristics of Life: *Chapter 1*
Have students discuss how a candle flame's characteristics are similar to those of living things.

Chemistry: *Chapter 2*
Ask students what the charges on protons and electrons are. Have students explain the difference between atoms and ions.

CHAPTER 3

BIOCHEMISTRY

The body of this jellyfish, Pseudorhiza haeckeli, *is almost 99 percent water.*

FOCUS CONCEPT: ***Matter, Energy, and Organization***

As you read this chapter, notice how function depends on structure in each of the compounds you examine.

SECTION

3-1

WATER

Compare the body of the jellyfish shown on the opposite page with your own body. The jellyfish will die if it is removed from its water environment. You can live in the driest parts of Earth. Jellyfish and humans seem utterly unlike each other, yet the bodies of both are made of cells filled with water. The chemical reactions of all living things take place in an aqueous environment. Water has several unique properties that make it one of the most important compounds found in living things.

OBJECTIVES

▲ Describe the structure of a water molecule.

● Explain how water's polar nature affects its ability to dissolve substances.

■ List two of water's properties that result from hydrogen bonding.

POLARITY

Many of water's biological functions stem from its chemical structure. Recall that in the water molecule, H_2O, the hydrogen and oxygen atoms share electrons to form covalent bonds. However, these atoms do not share the electrons equally. An oxygen atom has eight protons in its nucleus and therefore eight positive charges to attract electrons, whereas a hydrogen atom has only one proton and therefore one positive charge. With its greater positive charge, the nucleus of the oxygen atom pulls the shared electrons toward its nucleus and away from the nucleus of the hydrogen atom. As a result, the electrical charge is unevenly distributed, as shown in the models of a water molecule shown in Figure 3-1.

Notice too in Figure 3-1 that the three atoms in a water molecule are not arranged in a straight line as you might expect. Rather, the two hydrogen atoms bond with the single oxygen atom at an angle. Although the total electrical charge on a water molecule is neutral, the region of the molecule where the oxygen atom is located has a

FIGURE 3-1

The oxygen region of the water molecule is weakly negative, and the hydrogen regions are weakly positive. Notice the three very different ways to represent water, H_2O. You are familiar with the electron-energy-level model (a) from Chapter 2. The structural formula (b) is compact and easy to understand. The space-filling model (c) shows the three-dimensional structure of a molecule.

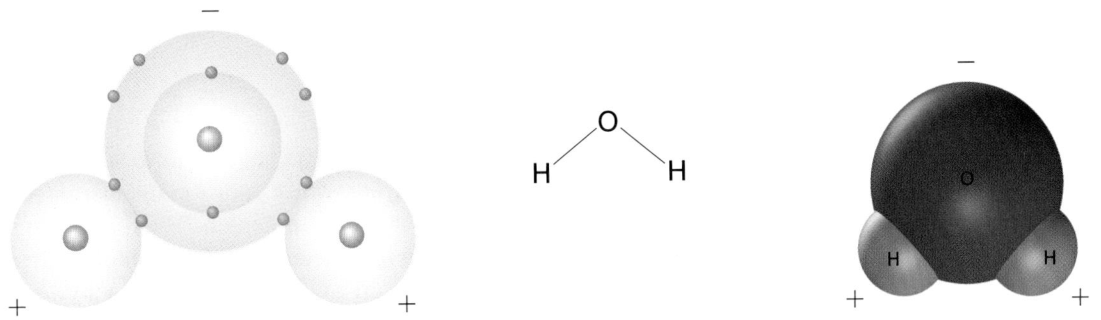

(a) Electron-energy-level model **(b)** Structural formula **(c)** Space-filling model

SECTION 3-1

SECTION OVERVIEW

Matter, Energy, and Organization
Section 3-1 describes the characteristics of water, the most abundant compound in living things. It explains the role of temperature moderation in cells and the organization of the three atoms that make up a water molecule.

ACTIVE READING

Technique: K-W-L
Before students read this chapter, have them make short lists of all the things they already Know (or think they know) about the chemistry of organisms. Ask students to contribute their entries to a group list on a large poster board. Then have students list things they Want to know about chemistry in living things. Have students save their lists as well as the poster boards for later use in the Active Reading activity on page 59.

DEMONSTRATION

Polarity of Water
Run a thin stream of water from a faucet. Rub a glass rod with a silk cloth, and explain that this rubbing gives the rod a positive charge. Then place the rod next to the stream and ask students to explain what happens. (The positive charge on the glass rod attracts the negatively charged oxygen in the water molecules.) Emphasize to students that the + and – symbols in Figure 3-1 indicate weak or partial charges.

SECTION 3-1

TEACHING STRATEGY
Adhesion

CAUTION: **Glass slides can break. Handle glass microscope slides carefully.** Provide students with two microscope slides, string, masking tape, and some small weights, such as washers. Have students tape a 15 cm length of string to the middle of one slide. Then have them dip both slides in water and place the slides together like a sandwich. The string should hang freely from the bottom slide. Students should grasp the edges of the top slide, being careful not to grasp the edges of the bottom slide. Have students tape washers to the hanging string to see how much weight the bottom slide will hold. Ask students what holds the bottom slide to the top slide. (Forces of adhesion between the water and the two slides hold the bottom slide to the top slide.) Have students continue adding weights to the string until they exceed the maximum amount of weight the adhesive forces of water will hold.

QUICK FACT

☑ The human body is composed of about 65 percent water. An elephant and an ear of corn are each made of about 70 percent water, and a tomato is about 95 percent water. Humans will die if they lose more than 20 percent of their normal water content.

CRITICAL THINKING
Spraying Mosquitoes

Tell students that years ago wetlands were sprayed with oil to smother mosquito larvae living just beneath the water's surface. Ask students how oil affects the ability of insects and plants to float on the surface of water. (Oil interferes with hydrogen bonding of the water surface and causes floating organisms to sink.)

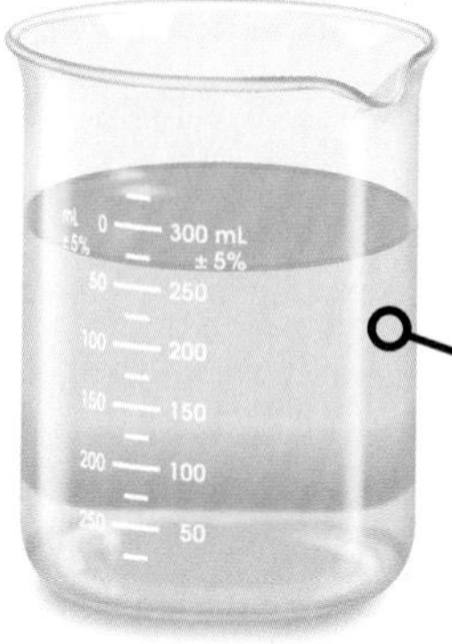

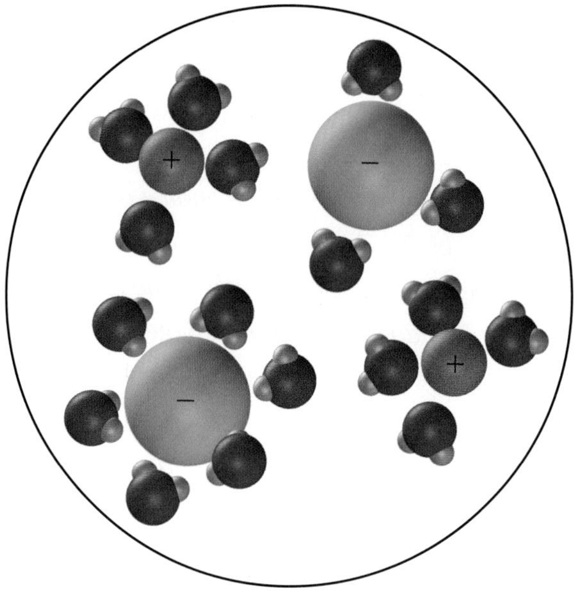

FIGURE 3-2

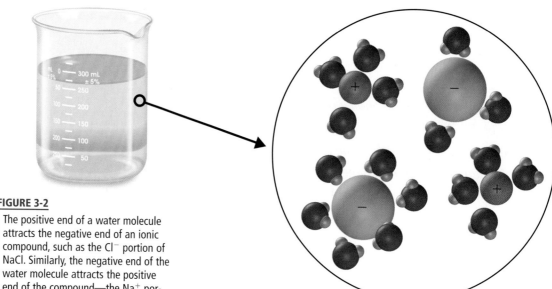

The positive end of a water molecule attracts the negative end of an ionic compound, such as the Cl^- portion of NaCl. Similarly, the negative end of the water molecule attracts the positive end of the compound—the Na^+ portion of NaCl. As a result, NaCl breaks apart, or dissociates, in water.

FIGURE 3-3

The dotted lines in this figure represent hydrogen bonds. A hydrogen bond is a weak force of attraction between a hydrogen atom in one molecule and a negatively charged atom in a second molecule.

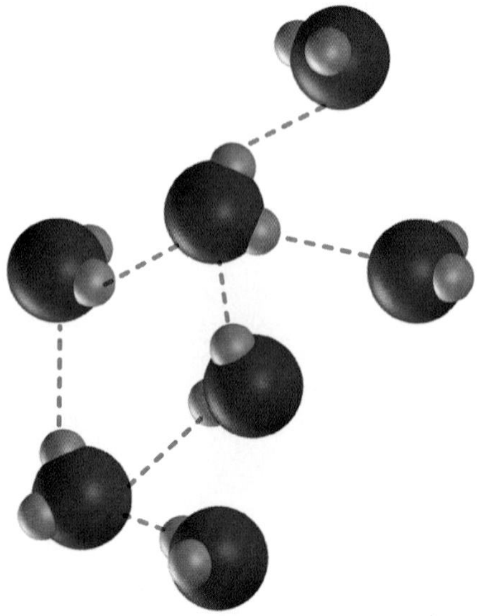

slightly negative charge, while the regions of the molecule where each of the two hydrogen atoms are located have a slightly positive charge. Because of this uneven pattern of charge, water is called a **polar** compound.

It is this polar nature that makes water very effective in dissolving many other substances. Water dissolves other polar substances, including sugars and some proteins, as well as ionic compounds, such as sodium chloride, NaCl. An ionic compound mixed with water tends to dissociate into ions. This is illustrated in Figure 3-2. This breaking up of an ionic compound frees ions to participate in many biological reactions. In your body, both sodium ions and chloride ions are essential to functions like muscle contraction and transmission of impulses in the nervous system. In fact, dissolved, dissociated ions are present in all of the aqueous solutions found in living things. Their concentration is critical to the normal operation of the many systems of your body.

HYDROGEN BONDING

The polar nature of water also causes water molecules to be attracted to one another. The type of attraction that holds two water molecules together is called a **hydrogen bond.** As shown in Figure 3-3, a positive region of one water molecule is attracted to the negative region of another water molecule. Thus, a hydrogen bond tends to form between a hydrogen atom in one molecule and the region of negative charge on another molecule. A hydrogen bond is a weak bond that can be easily broken. Even so, the hydrogen bonds in water exert a significant attractive force, causing water to cling to itself and to other substances.

Cohesion and Adhesion

An attractive force between particles of the same kind is known as **cohesion.** You can see cohesion at work when you observe the surface tension of water. Cohesive forces resulting from water's hydrogen bonding are strong enough to cause water to act as if it has a thin "skin" on its surface. This is why water appears to bulge from the sides of a glass filled to the brim.

Adhesion is the attractive force between unlike sustances. Together, adhesion and cohesion enable water molecules to move upward through narrow tubes against the force of gravity. This property of water is known as **capillarity** (KAP-uh-LER-it-ee). You have seen capillarity at work if you have observed the flow of water into a flower through its stem, such as is shown in Figure 3-4.

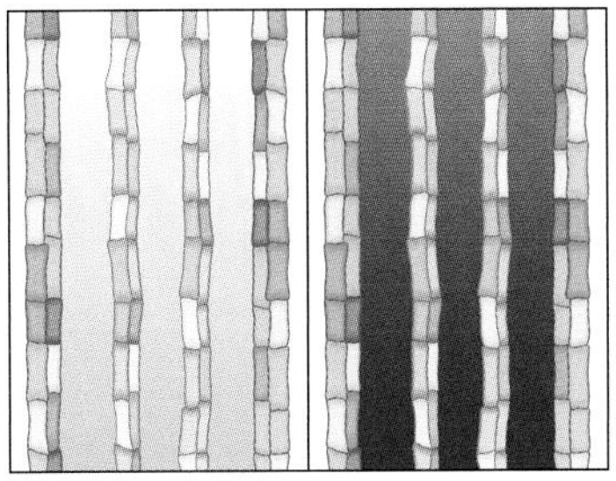

Liquid movement up a stem

FIGURE 3-4

Because of strong cohesive and adhesive forces, water can travel upward from the roots of flowers. In the flower on the right, the water, which has been dyed blue, has moved up through the stem to the flower's petals.

Temperature Moderation

Water must gain or lose a relatively large amount of energy for its temperature to change. When water is heated, most of the thermal energy that the water initially absorbs breaks the hydrogen bonds between the molecules. Only after these bonds have been broken does the thermal energy increase the motion of the molecules and raise the temperature of the water. You read in Chapter 1 that all organisms must maintain homeostasis to live. In organisms, water's ability to absorb large amounts of energy helps keep cells at an even temperature despite temperature changes in the environment.

internet**connect**
SCILINKS NSTA
TOPIC: Hydrogen bonding
GO TO: www.scilinks.org
KEYWORD: HM051

SECTION 3-1 REVIEW

1. Describe the structure of a water molecule.
2. How do molecules of a polar compound differ from those of a nonpolar compound?
3. What happens when ionic compounds are mixed with water?
4. What are two properties of water that result from water's tendency to form hydrogen bonds?
5. What is capillarity?
6. **CRITICAL THINKING** Most automobiles have water-cooled engines. What must be true about a solution that can replace water in the cooling system, such as antifreeze?

GIFTED ACTIVITY

Have students search library or online resources to research how water moves up great distances in plants, such as in giant redwood trees, and have them prepare a short report. (Water is drawn up the plant through evaporation in the leaves. As water moves out of the leaves, cohesion moves the column of water up the trunk of the tree.)

ANSWERS TO SECTION 3-1 REVIEW

1. Water is composed of two hydrogen atoms covalently bonded to one oxygen atom. Water is polar, with the oxygen atom slightly positively charged and the hydrogen atoms slightly negatively charged.
2. Polar molecules have slightly differing charges on different regions, while nonpolar molecules have a uniform charge.
3. Ionic compounds dissociate when they are dissolved in water.
4. Two properties of water that arise from hydrogen bonding are adhesion and cohesion.
5. Capillarity is the ability of water to move up the sides of a narrow tube as a result of water's properties of adhesion and cohesion.
6. Solutions that can replace water as a coolant in automobile engines must have the ability to absorb large quantities of heat.

SECTION OVERVIEW

Matter, Energy, and Organization

Section 3-2 describes how carbon atoms bond to other atoms to form large organic molecules and how organic molecules can also break apart into smaller molecules. This section also describes how functional groups influence the behavior of large molecules and how adenosine triphosphate (ATP) transfers energy to drive chemical reactions.

▶ ENGAGE STUDENTS

Obtain 20 marbles. Glue two pieces of Velcro, one "hook" and one "loop," to opposite sides of each marble. Place the marbles in an empty box. Tell students that the marbles represent small molecules and that shaking the box represents a chemical reaction in which polymers are formed. Ask students what the shaking represents (adding energy) and if shaking the box at different intensities will affect how the "molecules" bond together. (Vigorous shaking gives more energy, which might cause the molecules to bond faster. Gentle shaking might cause the polymers to form more slowly.) Shake the box at two different intensities, and show the results to the class. Ask students to explain how the amount of time the box is shaken affects how the molecules bond. Shake the box for different amounts of time, and reveal the results to the class. (Shorter periods of time might cause the polymers to be smaller; longer periods of time might allow the polymers to grow larger.) If a long chain results, hold it up and show it to the class. Ask students what kind of reaction this represents (condensation).

SECTION 3-2

OBJECTIVES

▲ Define *organic compound* and name three elements often found in organic compounds.

● Explain why carbon forms so many different compounds.

■ Define *functional group* and explain its significance.

◆ Compare a condensation reaction with hydrolysis.

CARBON COMPOUNDS

All of the many compounds discovered can be classified in two broad categories: organic compounds and inorganic compounds. ***Organic compounds*** *contain carbon atoms that are covalently bonded to other carbon atoms and to other elements as well—typically hydrogen, oxygen, and nitrogen. The chemistry of carbon is the chemistry of life.*

CARBON BONDING

A carbon atom has four electrons in its outermost energy level. Remember from Chapter 2 that most atoms become stable when their outermost energy level contains eight electrons. A carbon atom therefore readily forms four covalent bonds with other elements. Unlike other elements, however, carbon also readily bonds with other carbon atoms, forming straight chains, branched chains, or rings, as shown in Figure 3-5. This tendency of carbon to bond with itself results in an enormous variety of organic compounds.

In the symbolic shorthand of chemistry, each line shown in Figure 3-5 represents a covalent bond formed when two atoms share a pair of electrons. A bond formed when two atoms share a pair of electrons is called a single bond. Carbon can also share two or even three pairs of electrons with another atom. Figure 3-6a shows a model for an organic compound in which six carbon atoms have formed a ring. Notice that each carbon atom forms four covalent bonds: a single bond with another carbon atom, a single bond with a hydrogen atom, and a double bond with a second carbon atom. In a double bond—represented by two parallel lines—atoms share two pairs of electrons. A triple bond, the sharing of three pairs of electrons, is shown in Figure 3-6b.

FIGURE 3-5

Carbon can bond in a number of ways to produce molecules of very different shapes, including straight chains, branched chains, and rings. These structures form the backbone of many different kinds of organic molecules.

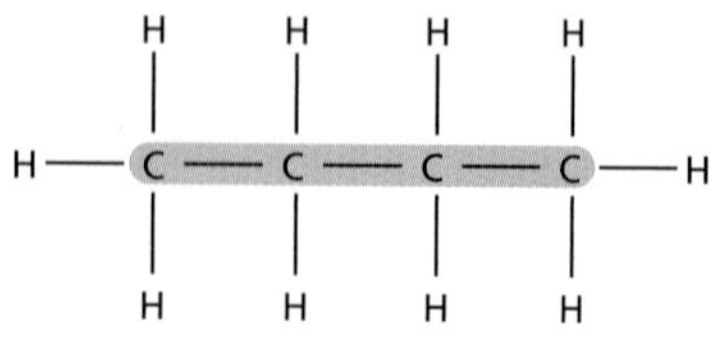

Straight chain

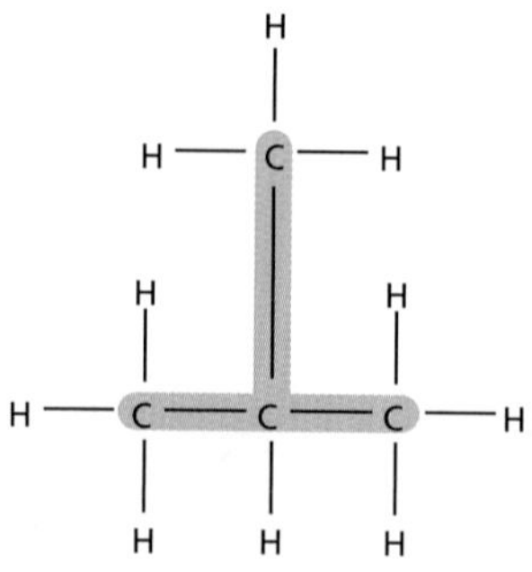

Branched chain

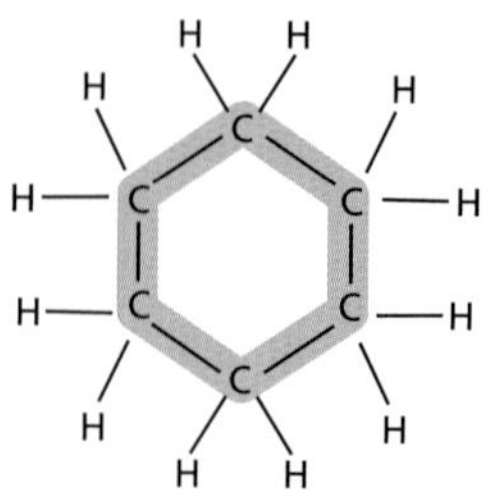

Ring

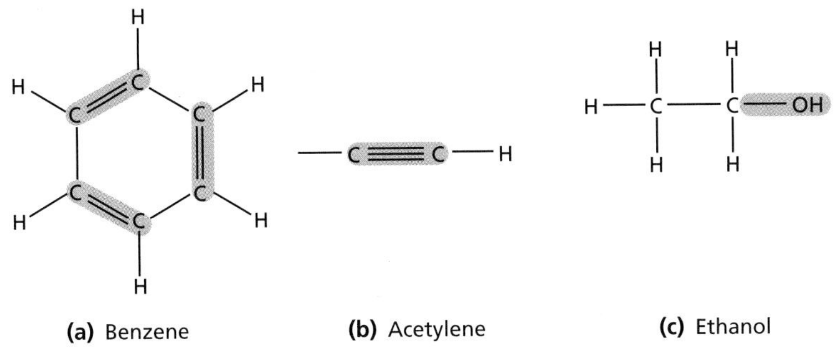

FIGURE 3-6

Carbon can form (a) double or even (b) triple bonds to satisfy its need for eight electrons in its outermost energy level. Organic molecules can have many different shapes and patterns of bonding. Organic molecules can also have many different functional groups, which influence the properties of the molecule they are attached to. Notice the hydroxyl, —OH, group on this model of the alcohol ethanol (c).

FUNCTIONAL GROUPS

In most organic compounds, clusters of atoms, called **functional groups,** influence the properties of the molecules they compose. The functional group is the structural building block that determines the characteristics of the compound. One functional group important to living things, the hydroxyl group, —OH, is shown in Figure 3-6c.

An **alcohol** is an organic compound with a hydroxyl group attached to one of its carbon atoms. Locate the hydroxyl group in the alcohol shown in Figure 3-6c. The hydroxyl group makes an alcohol a polar molecule. Thus, alcohols have some properties similar to water, including the ability to form hydrogen bonds. The alcohol illustrated in Figure 3-6c is ethanol, which is found in alcoholic beverages. Ethanol causes cell death in the liver and brain of humans. The alcohol methanol, also called wood alcohol, can cause blindness or even death when consumed. Some alcohols, however, are needed by organisms to carry out their life processes. Humans, for example, need the alcohol glycerol to assemble certain molecules necessary for life.

LARGE CARBON MOLECULES

In many carbon compounds, the molecules are built up from smaller, simpler molecules known as **monomers,** such as the ones shown in Figure 3-7. As you can also see in Figure 3-7, monomers can bond to one another to form complex molecules known as polymers. A **polymer** consists of repeated, linked units. The units may be identical or structurally related to each other. Large polymers are called **macromolecules.**

Monomers link to form polymers through a chemical reaction called a **condensation reaction.** In the condensation reaction shown

Quick Lab

Demonstrating Polarity

Materials disposable gloves; lab apron; safety goggles; 3 test tubes; test-tube rack; 6 mL each of cooking oil, ethanol, and water

Procedure

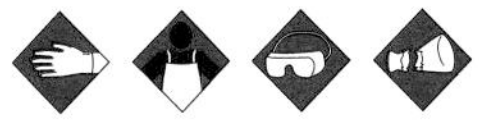

1. Put on your disposable gloves, lab apron, and safety goggles.
2. Label the test tubes "A," "B," and "C."
3. In test tube A, put 3 mL of water and 3 mL of oil.
4. In test tube B, put 3 mL of oil and 3 mL of ethanol.
5. In test tube C, put 3 mL of ethanol and 3 mL of water.
6. With your thumb and middle finger, flick each test tube to mix the contents, and allow it to sit for 10–15 minutes.
7. Record your observations.

Analysis How does this activity demonstrate polarity of molecules that contain the –OH group?

FIGURE 3-7

A polymer is the result of bonding between monomers. The six-sided shape is an organic structural model of a molecule with a central carbon ring. The organic structure of a molecule shows the arrangement of carbon atoms in organic molecules.

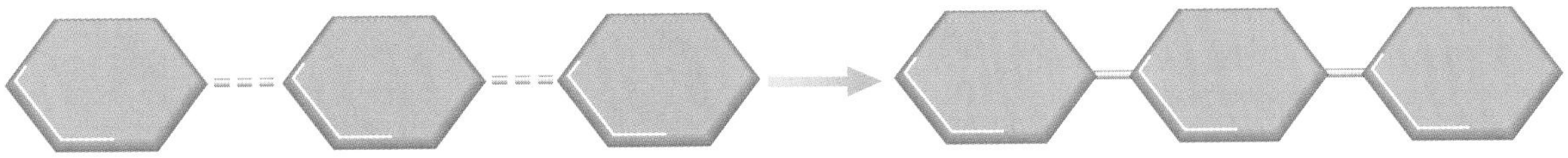

SECTION 3-2

Quick Lab

Demonstrating Polarity

Time Required 20 minutes

Safety Remind students not to drink the solutions and to use care when handling the glass test tubes.

Procedural Tips Food coloring can be added to the water to enhance students' observations.

Disposal All liquids may be poured down the drain followed by several minutes of running water.

Answers to Analysis Water and alcohol are polar molecules. Oil is a nonpolar molecule. Like dissolves like. The alcohol molecules will not remain in suspension with the oil but will quickly settle out.

VISUAL LINK

Figure 3-7

Tell students to compare Figure 3-7 with Figure 3-6. Point out that each hexagon in Figure 3-7 represents a monomer molecule. The models in Figure 3-6 show each atom of the molecule. Ask students why scientists sometimes use geometric shapes to represent monomers in a large organic molecule instead of drawing every atom of the molecule. (The use of geometric shapes or some other symbol to represent monomers simplifies the representation of the molecule, and it allows scientists to more easily differentiate between different types of molecules.)

SECTION 3-2

OVERCOMING MISCONCEPTIONS

Because ATP supplies most of the energy that drives metabolism, it is sometimes called an energy-rich compound and its phosphate bonds are sometimes called high-energy bonds. These terms are misleading because they imply that ATP contains an unusually high amount of energy. In fact, ATP serves as the cell's energy currency because its phosphate bonds are easy to break.

ANSWERS TO SECTION 3-2 REVIEW

1. With few exceptions, organic compounds contain carbon covalently bonded to itself and other atoms.
2. Carbon can form four covalent bonds with any number of atoms, including other carbon atoms. This allows it to form molecules of different shapes.
3. Functional groups are clusters of atoms that influence the properties of the molecules they compose. For example, the hydroxyl group is a functional group.
4. Polymers form through the condensation of monomers.
5. Polymers break down through hydrolysis.
6. Because organic substances contain carbon, this method is more useful for organic substances. Inorganic substances do not necessarily have carbon atoms, and the method therefore does not work for them.

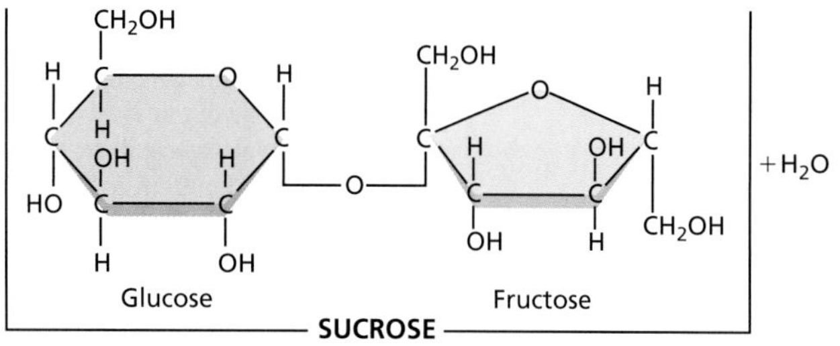

FIGURE 3-8

The condensation reaction of one glucose molecule with one fructose molecule yields sucrose and water. One water molecule is produced each time two monomers form a covalent bond.

in Figure 3-8, two sugar molecules, glucose and fructose, combine to form the sugar sucrose, which is common table sugar. The two sugar monomers become linked by a C–O–C bridge. In the formation of that bridge, the glucose molecule releases a hydroxide ion, OH^-, and the fructose molecule releases a hydrogen ion, H^+. The OH^- and H^+ ions that are released in turn combine to produce a water molecule, H_2O.

The breakdown of some complex molecules, such as polymers, occurs through a process known as **hydrolysis** (HIE-DRAH-luh-sis). Hydrolysis is a reversal of a condensation reaction. The addition of water to some complex molecules, including polymers, under certain conditions can break the bonds that hold them together. As you can see in Figure 3-9, in hydrolysis a large molecule breaks apart.

Energy Currency

Life processes require a constant supply of energy. This energy is available to cells in the form of certain compounds that contain a large amount of energy in their overall structure. One of these compounds is **adenosine** (uh-DEN-uh-SEEN) **triphosphate,** more commonly referred to by its abbreviation, **ATP.**

Figure 3-9 shows the structure of an ATP molecule. Notice the three linked phosphate groups, $—PO_4^-$, that are attached to one another by covalent bonds. The covalent bond that holds the last phosphate group to the rest of the molecule is easily broken. When this bond is broken, much more energy is released than was required to break the bond. This conversion of energy is used by the cell to drive the chemical reactions that enable an organism to function.

FIGURE 3-9

The hydrolysis of ATP yields adenosine diphosphate and inorganic phosphate. In hydrolysis, a hydrogen ion from a water molecule bonds to one of the new molecules, and a hydroxide ion bonds to the other new molecule. Most hydrolysis reactions are exergonic.

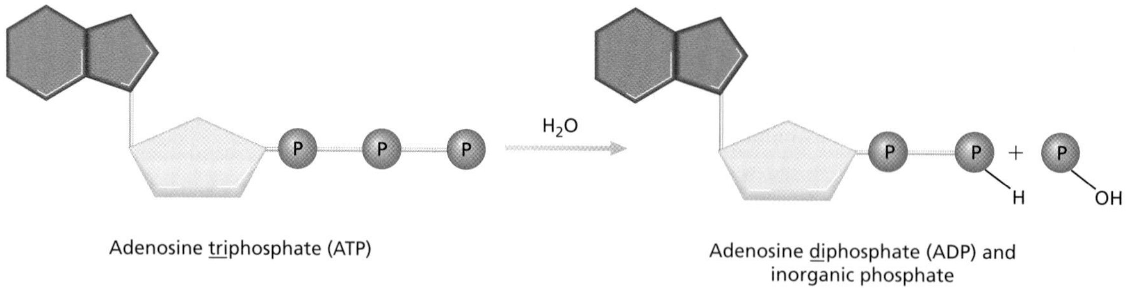

SECTION 3-2 REVIEW

1. What is an organic compound?
2. What property allows carbon compounds to exist in a number of forms?
3. Define *functional group* and give an example.
4. How does a polymer form?
5. How does a polymer break down?
6. CRITICAL THINKING Scientists can determine the age of a substance using a method that compares the amounts of different forms of carbon atoms present in the substance. Is this method more useful for organic substances or inorganic substances?

MOLECULES OF LIFE

Four main classes of organic compounds are essential to the life processes of all living things: carbohydrates, lipids, proteins, and nucleic acids. You will see that although these compounds are built from carbon, hydrogen, and oxygen, the atoms occur in different ratios in each class of compound. Despite their similarities, the different classes of compounds have different properties.

SECTION 3-3

OBJECTIVES

▲ Define *monosaccharide*, *disaccharide*, and *polysaccharide*, and discuss their significance to organisms.

● Relate the sequence of amino acids to the structure of proteins.

■ Relate the structure of lipids to their functions.

◆ List two essential functions of nucleic acids.

CARBOHYDRATES

Carbohydrates are organic compounds composed of carbon, hydrogen, and oxygen in a ratio of about two hydrogen atoms to one oxygen atom. The number of carbon atoms in a carbohydrate varies. Carbohydrates exist as monosaccharides, disaccharides, or polysaccharides.

Monosaccharides

A monomer of a carbohydrate is called a **monosaccharide** (MAHN-oh-SAK-uh-RIED). A monosaccharide—or simple sugar—contains carbon, hydrogen, and oxygen in a ratio of 1:2:1. The general formula for a monosaccharide is written as $(CH_2O)_n$, where *n* is any whole number from 3 to 8. For example, a six-carbon monosaccharide $(CH_2O)_6$ would have the formula $C_6H_{12}O_6$. The most common monosaccharides are glucose, fructose, and galactose, as shown in Figure 3-10. Glucose is a main source of energy for cells. Fructose is found in fruits and is the sweetest of the monosaccharides. Galactose is found in milk and is usually combined with glucose or fructose. Notice in Figure 3-10 that glucose, fructose, and galactose have the same molecular formula, $C_6H_{12}O_6$, but their differing structures determine the slightly different properties of the three compounds. Compounds like these sugars, with a single chemical formula but different forms, are called **isomers** (IE-soh-muhrz).

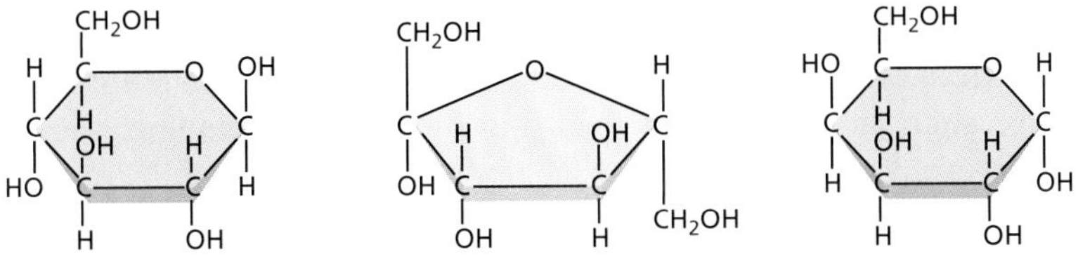

Glucose — Fructose — Galactose

FIGURE 3-10

Although glucose, fructose, and galactose have the same chemical formula, their structural differences result in different properties among the three compounds.

internetconnect

SCILINKS NSTA

TOPIC: Carbohydrates
GO TO: www.scilinks.org
KEYWORD: HM055

SECTION 3-3

SECTION OVERVIEW

Matter, Energy, and Organization

Section 3-3 discusses the structure and function of carbohydrates, proteins, lipids, and nucleic acids. Carbohydrates and lipids store energy in living things. Proteins constitute structural and functional parts of the organism. Lipids form cellular membranes and have important functions in metabolism. Nucleic acids store the information to produce proteins.

▶ ENGAGE STUDENTS

Set up a ring stand to hold a test tube containing 5 mL of water at room temperature. Insert a thermometer in the test tube. Insert the blunt end of a needle into one side of a rubber stopper. Place the rubber stopper under the test tube so that the needle point is directly underneath the bottom of the test tube. Obtain several samples of food that are about the same size, such as cubes of bread, fruit, and meat. Place one sample on the sharp end of the needle so that it is within 1 cm of the bottom of the test tube. Tell students that you are going to assess how much food energy is in each sample by igniting the food with a Bunsen burner and determining how many degrees in temperature the water in the test tube rises. The higher the temperature rises, the more energy there is in the food. Have students record the temperature of the water before the first food is ignited. Ignite the first food, and allow it to burn to ash. Have students record the temperature of the water again. Continue testing the other foods in the same manner. Ask students which foods seemed to have the most energy. (Answers will vary. Foods with high lipid content should raise the water temperature more than other foods.)

SECTION 3-3

OVERCOMING MISCONCEPTIONS

Many people assume that large grazing animals, such as deer, buffalo, and cattle, digest the cellulose in the plants they eat. They do not. Microscopic organisms in their digestive system digest the cellulose and release the nutrients.

TEACHING STRATEGY

Artificial Sweetener

Have students obtain an empty can of a diet soft drink that contained the artificial sweetener aspartame. Explain to students that aspartame is a compound composed of two amino acids, phenylalanine and aspartic acid. Infants born with phenylketonuria (PKU) lack the ability to break down the amino acid phenylalanine. If phenylalanine is not broken down, it can accumulate and retard normal brain development. Ask students why people with PKU cannot consume aspartame. (Their brain functions are impaired by the buildup of phenylalanine.) Have students list other aspartame-containing products, such as sugar-free snacks and artificial sweeteners.

RECENT RESEARCH

Gene for Building Cellulose

Scientists in California and Israel have found three genes that make cellulose synthase, an enzyme responsible for building cellulose from sugar molecules in rice and cotton. Cotton is made up of almost pure cellulose. Textile manufacturers hope that the use of cellulose synthase will help them produce stronger and more-uniform fibers.

Disaccharides and Polysaccharides

In living things, two monosaccharides can combine in a condensation reaction to form a double sugar, or **disaccharide.** As you saw in Figure 3-8, sucrose, which is common table sugar, is composed of fructose and glucose. A **polysaccharide** is a complex molecule composed of three or more monosaccharides.

Animals store glucose in the form of the polysaccharide glycogen. Glycogen consists of hundreds of glucose molecules strung together in a highly branched chain. Much of the glucose that comes from food is ultimately stored in your liver and muscles as glycogen and is ready to be used for quick energy.

In plants, glucose molecules are linked in the polysaccharide starch. Starch molecules have two basic forms—highly branched chains that are similar to glycogen and long, unbranched chains that coil like a telephone cord. The large polysaccharide cellulose is also made by plants. Cellulose, which gives strength and rigidity to plant cells, makes up about 50 percent of wood. In a single cellulose molecule, thousands of glucose monomers are linked in long, straight chains. These chains tend to form hydrogen bonds with each other. The resulting structure is strong and can be broken down by hydrolysis only under certain conditions.

FIGURE 3-11

Amino acids differ only in the type of R group (shown in red) they carry. Glycine (a) has a simpler R group than alanine (b). The R group may be either polar or nonpolar. Amino acids with polar R groups can dissolve in water, while those with nonpolar R groups cannot.

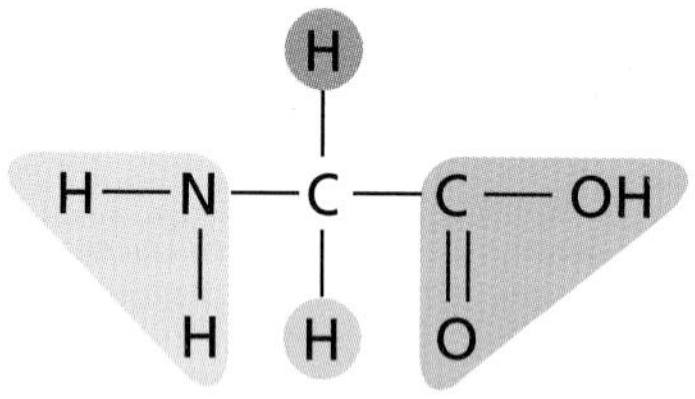

(a) Glycine

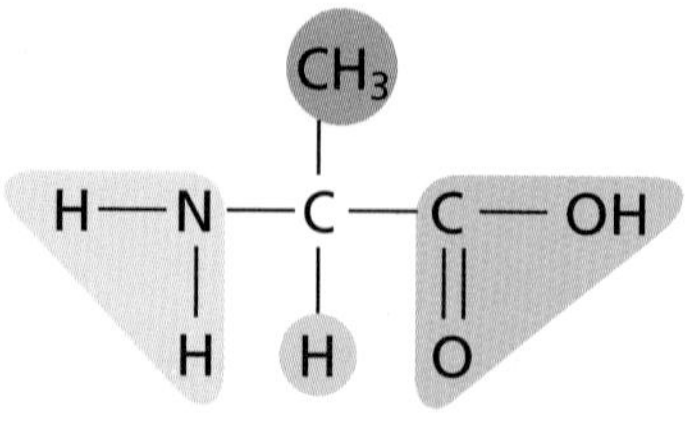

(b) Alanine

PROTEINS

Proteins are organic compounds composed mainly of carbon, hydrogen, oxygen, and nitrogen. Like the other macromolecules, proteins are formed from the linkage of monomers. The skin and muscles of animals are made mostly of proteins, as are many of the catalysts found in both plants and animals.

Amino Acids

The 20 different **amino acids,** the monomer building blocks of proteins, share a basic structure. As Figure 3-11 shows, each amino acid contains a central carbon atom covalently bonded to four other atoms or functional groups. A single hydrogen atom, highlighted in blue on the illustration, bonds at one site. A carboxyl group, —COOH, highlighted in green, bonds at a second site. An amino group, $—NH_2$, highlighted in yellow, bonds at a third site. And a functional group call the R group, highlighted in red, bonds at the fourth site.

The main difference among the different amino acids is found in their R groups. The R group can be as simple as the single hydrogen atom of glycine, shown in Figure 3-11a, or it can be more complex, such as the R group shown in the model of alanine, shown in Figure 3-11b. The differences among the amino acid R groups gives different proteins very different shapes. The different shapes allow proteins to perform many different roles in the chemistry of living things.

Dipeptides and Polypeptides

Figure 3-12 shows how two amino acids bond to form a **dipeptide** (die-PEP-tied). In a condensation reaction, two amino acids form a covalent bond, called a **peptide bond.**

Amino acids can bond to each other one at a time, forming a very long chain called a **polypeptide** (PAH-lee-PEP-tied). Proteins are composed of one or more polypeptides. Some proteins are very large molecules, containing hundreds of amino acids. Often these long proteins are bent and folded upon themselves as a result of interactions—such as hydrogen bonding—among individual amino acids. Protein shape can also be influenced by conditions such as temperature or the type of solvent in which a protein is dissolved. When you cook an egg, heat changes the shape of proteins in the egg white. The firm, opaque result is very different from the clear, runny material you began with.

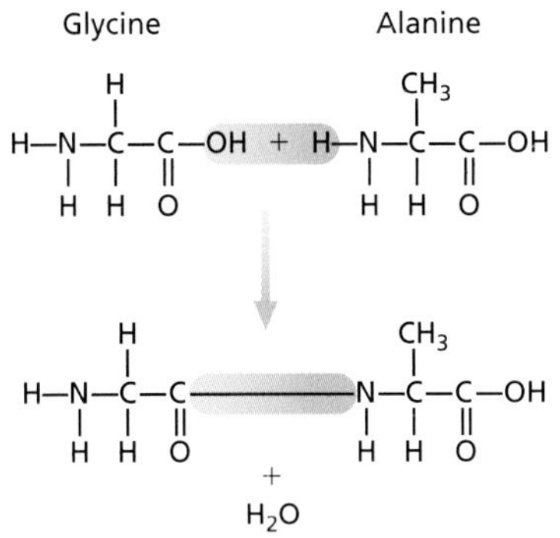

FIGURE 3-12

The peptide bond that binds amino acids together results from a condensation reaction that produces water.

Enzymes

Remember from Chapter 2 that enzymes—organic molecules that act as catalysts—are essential for the functioning of any cell. Most enzymes are proteins.

Figure 3-13 shows a model of enzyme action. Enzyme reactions depend on a physical fit between the enzyme molecule and its **substrate,** the reactant being catalyzed. Notice in Figure 13-3a that the enzyme and substrate have shapes that allow them to fit together like a lock and key. The linkage of the enzyme and substrate causes a slight change in the enzyme's shape, shown in Figure 13-3b. This shape change allows the enzyme to conform to the shape of the substrate and probably weakens some chemical bonds in the substrate, which is one way that enzymes reduce activation energy. After the reaction, the enzyme releases the products, as shown in Figure 13-3c. Like any catalyst, the enzyme itself is unchanged, so it can be used many times.

An enzyme may fail to work if its environment is changed in some way. For example, change in temperature or pH can cause a change in the shape of the enzyme or the substrate. If this happens, the reaction that the enzyme would have catalyzed cannot occur.

FIGURE 3-13

(a) In the lock-and-key model of enzyme action, the enzyme can attach only to a reactant with a specific shape. (b) The enzyme then flexes to conform to the reactant's shape. (c) The enzyme is unchanged by the reaction it participates in and is released to be used again.

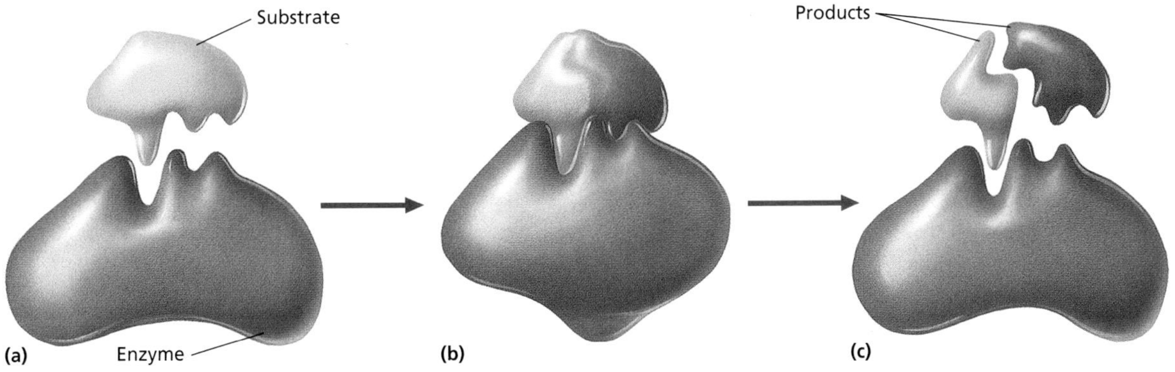

INCLUSION ACTIVITY

Obtain ornament-hanging hooks, or cut paper clips in half and curl the ends into hooks. Mix equal volumes of salt and flour in a container. Add enough water to the mixture to make it the consistency of dough. Have students select several carbon compounds from the text to model, such as alanine, glucose, or a fatty acid. Then have them embed an appropriate number of hooks into balls of dough. They should shape the dough into two models of organic molecules. After the models have dried overnight, they can be linked by the hooks to show how carbon compounds bond.

TEACHING STRATEGY

Enzyme Activity

Prepare a box of gelatin dessert, and pour the hot liquid into small dishes. Have students place small pieces of fresh pineapple in some of the dishes of gelatin and place canned pineapple in the others. Place the gelatin in a refrigerator overnight. The next day, ask students to observe the dishes of gelatin. (The gelatin with the canned pineapple will be solidified. The gelatin with the raw pineapple will be liquid.) Ask students to explain their observations. (Enzymes in the canned pineapple were destroyed by heat and could not digest the gelatin. Enzymes in the raw pineapple were intact and digested the gelatin protein.)

VISUAL LINK

Figure 3-13

Direct students' attention to Figure 3-13, and point out that enzymes are structured to allow only specific substrates to interact with them. Therefore, an enzyme that cleaves protein will not cleave starch or fatty acids.

SECTION 3-3

QUICK FACT

✔ Catalase is an enzyme that catalyzes the breakdown of hydrogen peroxide to oxygen and water. It is found in blood. Catalase breaks down more than 5 million molecules of hydrogen peroxide per minute.

CONTENT CONNECTION

Section 2-1: Matter

Life processes, such as metabolism and reproduction, are able to occur because of the behavior of organic compounds. The behavior of organic compounds is determined by the laws that govern all matter.

TEACHING STRATEGY

Artificial Fat

An artificial fat was approved by the FDA in 1996 for use in certain foods. The benefit of eating foods that contain artificial fat is that they have less than half the calories of the regular food and none of the fat. In some people, the artificial fat causes stomach cramps, diarrhea, and the loss of fat-soluble vitamins and β-carotene. Ask students to weigh the benefits of eating foods with artificial fat against the possible side effects. (Artificial fat would seem to be a healthy alternative to a high-calorie, high-fat food diet, but the associated loss of essential nutrients in growing young people may outweigh these advantages.)

CULTURAL CONNECTION

Fatty Foods in Asia

As a result of adopting a diet high in fat, the Japanese have suffered a higher rate of heart disease than ever before. It is well established that a high-fat diet is linked to higher rates of heart attacks and early death.

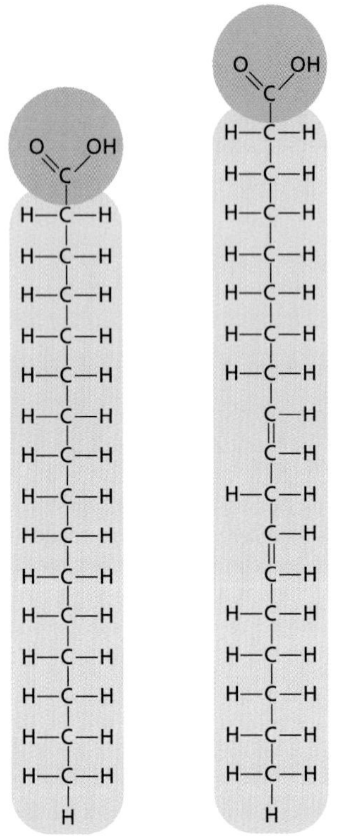

Palmitic acid   Linoleic acid

FIGURE 3-14

Fatty acids have a polar carboxyl head, purple, and a nonpolar hydrocarbon tail, green.

FIGURE 3-15

The lipid bilayer of a cell membrane is made from a double row of phospholipids, arranged with their hydrophobic "tails" facing each other. In this illustration of phospholipid structure, the head represents polar carboxyl heads of two fatty acids.

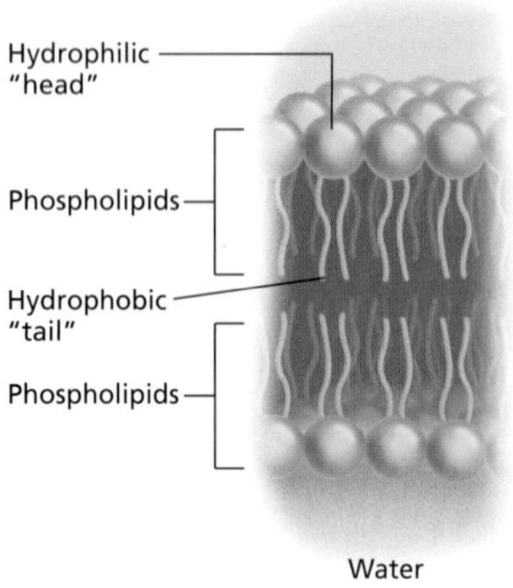

LIPIDS

Lipids are large, nonpolar organic molecules that do not dissolve in water. Lipid molecules have a higher ratio of carbon and hydrogen atoms to oxygen atoms than carbohydrates have. Lipids store energy efficiently. Lipid molecules have large numbers of carbon-hydrogen bonds, which store more energy than the carbon-oxygen bonds common in other organic compounds.

Fatty Acids

Fatty acids are unbranched carbon chains that make up most lipids. The model in Figure 3-14 shows that a fatty acid contains a long, straight carbon chain (from 12 to 28 carbons) with a carboxyl group, —COOH, attached at one end.

The two ends of the fatty-acid molecule have very different properties. The carboxyl end of the fatty-acid molecule is polar and is thus attracted to water molecules. Because of this attraction, the carboxyl end of the fatty-acid molecule is said to be **hydrophilic** (HIE-droh-FIL-ik), which means "water loving." In contrast, the hydrocarbon end of the fatty-acid molecule is nonpolar. This end tends not to interact with water molecules and is said to be **hydrophobic** (HIE-droh-FOH-bik), or "water fearing."

In saturated fatty acids, like palmitic acid, which is pictured in Figure 3-14, each carbon atom is covalently bonded to four atoms. The carbon atoms are in effect full, or "saturated." In contrast, you can see from the structural formula of a molecule of linoleic acid, shown in Figure 3-14, that the carbon atoms are not bonded to the maximum number of atoms that they can bond to. Instead, they have formed double bonds within the carbon chain. This type of fatty acid is said to be unsaturated.

Complex Lipids

Lipids are divided into categories according to their structure. Three classes of lipids important to living things contain fatty acids: triglycerides, phospholipids, and waxes. A **triglyceride** (trie-GLIS-uh-RIED) is composed of three molecules of fatty acid joined to one molecule of the alcohol glycerol. Saturated triglycerides are composed of saturated fatty acids. They typically have high melting points and tend to be solid at room temperature. Common dietary saturated triglycerides include shortening and animal fats. In contrast, unsaturated triglycerides are composed of unsaturated fatty acids and are usually liquid at room temperature. Unsaturated triglycerides are found primarily in plant seeds and fruits, where they serve as an energy and carbon source for sprouting plants.

Phospholipids have two, rather than three, fatty acids joined by a molecule of glycerol. As shown in Figure 3-15, the cell membrane is composed of two layers of phospholipids, which are referred to as the lipid bilayer. The inability of lipids to dissolve in water allows the membrane to form a barrier between the inside and outside of the cell. This bilayer arrangement of molecules produces a stable and effective barrier for a cell.

A **wax** is a type of structural lipid. A wax molecule consists of a long fatty-acid chain joined to a long alcohol chain. Waxes are highly waterproof, and in plants, wax forms a protective coating on the outer surfaces. Wax also forms protective layers in animals. For example, earwax helps prevent microorganisms from entering the middle ear.

Steroids

Unlike most other lipids, which are composed of fatty acids, **steroid** molecules are composed of four fused carbon rings with various functional groups attached to them. Many animal hormones, such as the male hormone testosterone, are steroid compounds. One of the most familiar steroids in humans is cholesterol. Cholesterol is needed by the body for nerve cells and other cells to function normally.

internetconnect
SCILINKS NSTA
TOPIC: Steroids
GO TO: www.scilinks.org
KEYWORD: HM059

NUCLEIC ACIDS

Nucleic acids are very large and complex organic molecules that store important information in the cell. Just as computers use a binary system of zeros and ones to store information, nucleic acids use a system of four compounds to store hereditary information. A sequence of the four compounds arranged in a certain order acts as a code for the genetic instructions of the cell.

Deoxyribonucleic acid, or DNA, contains information that is essential for almost all cell activities, including cell division. **Ribonucleic** (RIE-boh-noo-KLEE-ik) **acid,** or **RNA,** stores and transfers information that is essential for the manufacturing of proteins. Both DNA and RNA are polymers, composed of thousands of linked monomers called **nucleotides** (NOO-klee-uh-TIEDS). As shown in Figure 3-16, each nucleotide is made of three main components: a phosphate group, a five-carbon sugar, and a ring-shaped nitrogen base. You will learn more about these important compounds in Chapter 10.

FIGURE 3-16

A nucleotide consists of a phosphate group, a five-carbon sugar, and a ring-shaped nitrogen base. DNA and RNA are very large molecules formed from thousands of nucleotides strung together in a chain.

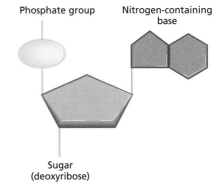

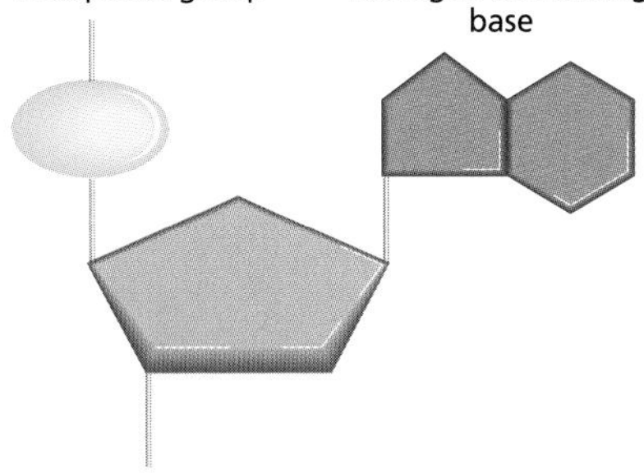

SECTION 3-3 REVIEW

1. Define *monosaccharide, disaccharide,* and *polysaccharide.*
2. Describe the structure of amino acids and proteins.
3. Explain the relationship between an enzyme and its substrate.
4. How do the two ends of a fatty acid differ?
5. Name the two types of nucleic acids, and describe their functions.
6. CRITICAL THINKING High temperatures can weaken bonds between different parts of a protein molecule, thus changing its shape. How might this change alter the effectiveness of an enzyme?

SECTION 3-3

ACTIVE READING
Technique: K-W-L

Tell students to return to the lists of things they Want to know about the role chemistry plays in living organisms, which they made using the Active Reading activity on page 49. Have them place check marks next to the questions they are now able to answer. Find out what students Learned in this chapter that might answer their questions. Lead a class discussion to answer those items without a check mark.

ANSWERS TO SECTION 3-3 REVIEW

1. A monosaccharide is a simple sugar that contains carbon, hydrogen, and oxygen in the ratio of 1:2:1. A disaccharide is composed of two monosaccharides. A polysaccharide is a complex carbohydrate composed of three or more monosaccharides.
2. An amino acid consists of a central carbon atom bonded to a hydrogen atom, an amino group, a carboxyl group, and a variable R group. Proteins are polymers of amino acids.
3. An enzyme attaches physically to the substrate and catalyzes the reaction. The enzyme molecule then releases the resulting products and can be used again.
4. The carboxyl end is polar and therefore hydrophilic. The hydrocarbon end is nonpolar and therefore hydrophobic.
5. DNA stores information essential for almost all cellular activities. RNA stores and transfers information essential for the manufacture of proteins.
6. Changing the shape of an enzyme may affect its ability to function because the enzyme may no longer physically interact appropriately with its substrate.

CHAPTER 3
Research Notes

BACKGROUND

More than 100 amino acids occur in nature, but only 20 are commonly found in proteins. Different types of proteins result from differences in the sequences and number of amino acids in the protein. Although the sequence of amino acids in a protein can be easily determined, its three-dimensional structure is harder to determine. The protein may form a loop, a helix, or irregular coils.

DISCUSSION

Guide the discussion by posing the following questions.

1. Why do scientists want to know the sequence of amino acids in a given protein? (Amino acids carry different charges and therefore influence each other in the folding and arranging of the chains. Knowing the amino acid order helps determine how the chains fold and how the protein works.)

2. Sanger received a Nobel Prize in chemistry rather than in biology. What does this blurring of disciplines mean? (Although Sanger's work had an impact on the biological world, it is fundamentally a chemical-based study. Biology is based in chemistry and thus would be difficult to understand without understanding chemistry too.)

FURTHER READINGS

1. "Science Classics," in *Discover,* April 1997, pp. 32–33. This article discusses the problems involved in determining a protein's three-dimensional structure.

2. "Cow's Milk: New Link to Diabetes," in *Science News,* October 19, 1996, p. 249. β-casein, a protein that makes up 35 percent of protein in cow's milk fed to infants, may stimulate the emergence of diabetes later in life.

Research Notes

The Structure of Insulin

Insulin is a hormone secreted by cells within the pancreas. It is essential in regulating the metabolism of carbohydrates and fats in the body. People with the disorder diabetes mellitus do not produce enough insulin. Some diabetes patients must take insulin injections to maintain normal metabolism.

In 1943, the British biochemist Frederick Sanger set out to analyze the insulin molecule. He was interested in proteins, and he chose insulin as the subject of his research mainly because, as he put it, "It was the only protein you could buy in pure form over the counter." He knew that an understanding of the structure of insulin could have important implications for medical practice. He spent the next 12 years studying the structure of insulin.

Biochemists already knew that proteins consist of combinations of 20 different amino acids linked together in chains. They also knew how to calculate the proportion of each amino acid in a given protein. What they did not know was the order in which the amino acids are linked in a specific protein. They correctly believed that the sequence of amino acids in a protein is crucial to the protein's function. Sanger's goal was to determine the amino acid sequence of insulin, and to do so, he had to develop new laboratory techniques.

Sanger began with a strategy familiar to chemists. He broke the insulin molecule into pieces. After splitting the chains of amino acids into short fragments, Sanger came to understand how they fit together. In his quest to determine the structure of insulin, he devised a new way to label the ends of a protein fragment.

Sanger learned that insulin is made up of two linked chains, one containing 30 amino acids and the other containing 21. He looked for fragments with overlapping sequences, which helped him discover how each chain was put together. By 1952, he had learned the amino acid sequences of both chains, but he still needed to understand how the two chains were linked to make up one insulin molecule. Three years later, Sanger reached his goal of identifying the molecular structure of insulin.

Frederick Sanger was the first scientist to determine the sequence of amino acids in a protein.

Sanger's work with insulin established him as the leader in his field, and in 1958 he received a Nobel Prize in chemistry. By demonstrating that each protein has a unique structure and likewise a unique sequence of amino acids, Sanger paved the way for the development of a technique that enabled the synthesis of insulin in the laboratory. In 1980, he received a second Nobel Prize in chemistry for his work in developing techniques for determining the sequence of nucleotides in molecules of DNA and RNA. Sanger is one of only four people to ever be awarded two Nobel Prizes.

CHAPTER 3 REVIEW

SUMMARY/VOCABULARY

3-1

- Water is a polar molecule in which the electrons are unevenly shared between the hydrogen and oxygen atoms.
- Because of its polar nature, water is effective in dissolving other substances to form solutions.
- Hydrogen bonding is responsible for the cohesion and capillarity that water molecules display.
- Water can absorb a large amount of thermal energy before its own temperature begins to rise.

Vocabulary

adhesion (51)
capillarity (51)
cohesion (51)
hydrogen bond (50)
polar (50)

3-2

- An organic compound contains carbon that is covalently bonded both to other carbon atoms and often to atoms of other elements, including oxygen, hydrogen, and nitrogen.
- A carbon atom forms four covalent bonds with other atoms. Carbon atoms can bond with one another to form straight chains, branched chains, or rings.
- Simple molecules, known as monomers, bond to one another to form complex molecules called polymers. Monomers are joined to form polymers in a condensation reaction. Polymers are broken down into monomers during hydrolysis.
- An alcohol is an organic compound that contains a hydroxyl group, OH^-, attached to one of its carbon atoms.
- Adenosine triphosphate (ATP) makes energy available to a cell.

Vocabulary

adenosine triphosphate (ATP) (54)
alcohol (53)
condensation reaction (53)
functional group (53)
hydrolysis (54)
macromolecule (53)
monomer (53)
organic compound (52)
polymer (53)

3-3

- A carbohydrate is an organic compound composed of carbon, hydrogen, and oxygen atoms in a ratio of about two hydrogen atoms to one oxygen atom. A monomer of a carbohydrate is called a monosaccharide.
- A double sugar is called a disaccharide. A complex sugar made of many monosaccharides is called a polysaccharide.
- A protein is an organic molecule that is formed from amino acids. An amino acid consists of a central carbon atom to which four functional groups are attached.
- Amino acids are joined by peptide bonds. A long chain of amino acids is called a polypeptide.
- Enzymes are catalysts that act in living things. Enzyme action can be explained by the lock-and-key model. Most enzymes are proteins.
- Most lipids contain fatty acids, organic molecules that have a hydrophilic end and a hydrophobic end.
- Unsaturated lipids have one or more pairs of carbon atoms joined by double bonds. Saturated lipids have no double bonds between their carbon atoms.
- Lipids store more energy than the other types of organic molecules.
- Nucleic acids are organic molecules that store genetic information in the cell.

Vocabulary

amino acid (56)
carbohydrate (55)
dipeptide (57)
disaccharide (56)
fatty acid (58)
hydrophilic (58)
hydrophobic (58)
isomer (55)
lipid (58)
monosaccharide (55)
nucleic acid (59)
nucleotide (59)
peptide bond (57)
phospholipid (58)
polypeptide (57)
polysaccharide (56)
protein (56)
ribonucleic acid (RNA) (59)
steroid (59)
substrate (57)
triglyceride (58)
wax (59)

CHAPTER 3 REVIEW ANSWERS

REVIEW

1. Amino acids are joined by peptide bonds to form proteins.
2. Two monosaccharides combine to form a disaccharide; polysaccharides are composed of three or more monosaccharides.
3. Peptide bonds join many amino acids to make a polypeptide.
4. The terms have opposite meanings: hydrophilic substances are attracted to water; hydrophobic substances repel water.
5. The cell membrane is composed of two layers of phospholipids.

6. d **7.** b
8. c **9.** a
10. c **11.** c
12. b **13.** d
14. c **15.** c

16. A—phosphate, B—nitrogen base, C—sugar
17. Isomers are molecules with the same chemical formula but different shapes.
18. Both alcohols and water are polar compounds and thus have some similar properties, such as the ability to form hydrogen bonds.
19. In a condensation reaction, two monomers link to form a polymer, releasing a water molecule; in hydrolysis, a water molecule breaks a polymer down into monomers.
20. Accept diagrams that show steps similar to those shown in Figure 3-13.
21. The carboxyl end is polar and thus hydrophilic; the hydrocarbon end is nonpolar and thus hydrophobic.
22. All three are lipids—they contain carbon, hydrogen, and oxygen in proportions different from those in carbohydrates, and they do not dissolve in water. Triglycerides are composed of three fatty acids joined to one glycerol molecule. Phospholipids consist of two fatty-acid chains joined by a molecule of glycerol. Steroids are composed of four carbon rings.
23. ATP is the cell's main supplier of the energy it needs to carry out its life functions.
24. Waxes are highly waterproof and form protective coatings.
25. Phospholipids are part of the cell membrane.

REVIEW

Vocabulary

1. Explain the relationship between amino acids, peptide bonds, and proteins.
2. How are the structures of monosaccharides, disaccharides, and polysaccharides related to each other?
3. Explain the relationship between a polypeptide and a peptide bond.
4. What is the difference between a hydrophilic substance and a hydrophobic substance?
5. Why is the structure of the cell membrane referred to as a bilayer?

Multiple Choice

6. Water helps keep the temperature of living things (a) high (b) low (c) below the freezing point (d) stable.
7. The distinguishing feature of a molecule of a polar compound is its (a) even distribution of electrical charge (b) uneven distribution of electrical charge (c) even temperature (d) uneven temperature.
8. The element that readily bonds to itself, forming long chains and rings, is (a) hydrogen (b) nitrogen (c) carbon (d) oxygen.
9. Plants store glucose in (a) a polysaccharide called starch (b) long proteins (c) complex lipid molecules called triglycerides (d) simple sugar molecules.
10. A very strong structural molecule in plants that is formed by hydrogen bonding between chains of glucose molecules is (a) starch (b) wax (c) cellulose (d) glycogen.
11. When two amino acids bond, (a) water is taken in by the product (b) hydrolysis occurs (c) a dipeptide is formed through a condensation reaction (d) a triglyceride is formed.
12. Lipids are distinguished from other organic molecules because they (a) contain carbon, hydrogen, and oxygen in a ratio of 1:2:1 (b) do not dissolve in water (c) dissolve easily in water (d) form large protein molecules.
13. Steroids differ from other lipid polymers in that steroids (a) do not occur in varied substances (b) are not hydrophilic (c) are not hydrophobic (d) are not composed of fatty-acid monomers.
14. Most enzymes are (a) lipids (b) phospholipids (c) proteins (d) carbohydrates.
15. A compound that stores hereditary information is (a) ATP (b) alcohol (c) DNA (d) protein.

Short Answer

16. Label the parts of the nucleotide below.

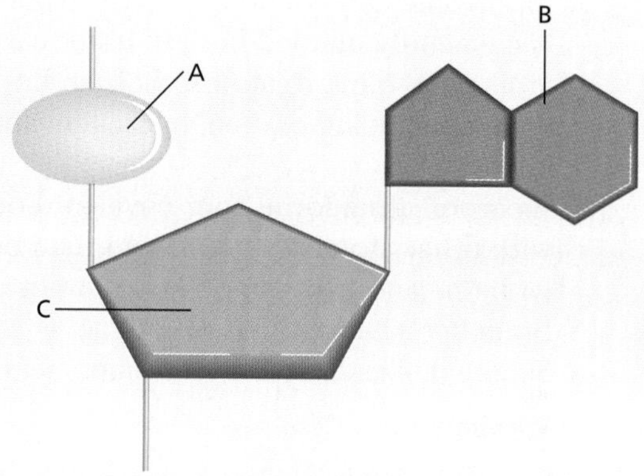

17. What are isomers?
18. What properties do alcohols and water share?
19. Compare and contrast a condensation reaction with a hydrolysis reaction.
20. Use a diagram to show how enzymes work.
21. How does the carboxyl end of the fatty-acid molecule differ from the hydrocarbon end of the molecule?
22. Compare and contrast the structures of triglycerides, phospholipids, and steroids. Which type of lipid is structurally unlike the other two?
23. What role does the compound ATP play in cellular activities?
24. What is an important characteristic of waxes, and why is this valuable to living things?
25. What structural role do phospholipids play in cells?

CRITICAL THINKING

1. Cells contain mostly water. What would happen to the stability of an organism's internal temperature with respect to environmental temperature changes if cells contained mostly oil, which does not have extensive hydrogen bonding?
2. The surface tension of water at room temperature is so great that you can actually "float" a small needle on the surface of water. (The needle doesn't truly float—it is denser than water. It is held in place by the force of hydrogen bonding between water molecules lying below and around the needle.) If the water were heated, what would happen to the needle and why?
3. Starch easily dissolves in water. Cellulose does not. Both substances, however, consist of chains of glucose molecules. What structural difference between starch and cellulose accounts for this different behavior in water?
4. Triglycerides in animals' bodies are usually solid fats, and those in plants are usually oils. However, many animals living in the Arctic and Antarctic have a greater number of triglycerides that are oils than do other animals. What advantage would the storage of body fat as oil instead of solid fat be to animals that live in freezing climates?
5. The specific heat of a substance is the amount of heat that must be added to 1 g of the substance to raise its temperature 1°C. Specific heat is measured in calories. Use the table to answer the following questions.
   a. What substance can absorb the greatest amount of heat before its own temperature rises?
   b. What do the two substances with the lowest specific heat values have in common?
   c. What property of ethanol might account for its relatively high specific heat?
   d. Which would be a better conductor of heat: iron or glass?
   e. What practical use would a substance with a very high specific heat have?

Specific Heats of Common Substances

Substance	Specific heat (cal)
Lead	0.03
Iron	0.10
Glass	0.20
Ethanol	0.60
Water	1.00
Liquid ammonia	1.23

EXTENSION

1. Read "Alien Haven" in *New Scientist*, September 18, 1999, on page 32, and answer the following questions: What are the "Goldilocks criteria" for a planet to be ideal for life? What is the "habitable zone," where a planet that can support life is most likely to be found?
2. Cut fibrous meat into four 1 in. cubes. Sprinkle three of the cubes with equal amounts of meat tenderizer, which contains a protein-splitting enzyme called papain. Place one cube in the refrigerator, leave one at room temperature, and place the other in an incubator at 32°C. For the fourth cube, place the same amount of meat tenderizer and a few tablespoons of water in a container, and boil the mixture for three minutes. (Do not allow the mixture to boil dry. Add water, tablespoon by tablespoon, as needed.) Pour the boiled mixture on the meat. After three hours, observe the texture of all four meat cubes. What do you conclude about the effect of temperature on the enzyme in meat tenderizer? Express your results as a graph showing temperature and the apparent level of activity of the enzyme.

CRITICAL THINKING

1. An organism's internal temperature would likely change more abruptly and drastically in response to environmental temperature changes.
2. Heating the water would cause the water molecules to move about and the hydrogen bonds to break, thus allowing the needle to eventually fall through the surface of the water.
3. Starch occurs in either highly branched chains or long, twisted chains, while cellulose occurs in long, straight chains with extensive hydrogen bonding between them.
4. Oil solidifies at a lower temperature than fat does; storing body fat as oil is helpful in freezing climates because oil is less likely to solidify in these climates than is fat.
5. (a) liquid ammonia (b) They are metals. (c) Its polarity and hydrogen bonding might account for its higher specific heat. (d) Iron would be a better conductor of heat because of its lower specific heat. (e) A substance with high specific heat would make a good insulator.

EXTENSION

1. The "Goldilocks criteria" are that a planet should be not too hot, not too cold, but just right. The "habitable zone" is the region around a star in which a planet could harbor liquid water on its surface.
2. Students should conclude that the rate of enzyme action rises with temperature, within limits. However, extreme heat alters the enzyme, rendering it inactive. Have students use their results to infer the approximate temperature at which papain is most effective in digesting proteins.

CHAPTER 3
INVESTIGATION

TIME REQUIRED

One 50-minute class period

SAFETY PRECAUTIONS

Students should wear a lab apron, protective gloves, and safety goggles at all times during this investigation. Avoid skin and eye contact with Benedict's solution, sodium hydroxide solution, copper sulfate solution, and the Sudan III solution. Do not use Sudan III in the same room with an open flame. Review the use of tongs and safety concerning hot liquids.

QUICK REFERENCES

Lab Preparation Notes are found on page 47B.

Holt BioSources provides a Teaching Resources CD-ROM that contains "Using Gowin's Vee in the Lab" and "Scoring Rubrics."

PROCEDURAL TIPS

1. To save time, you may want to have one-third of the students perform test 1 while the other students perform tests 2 and 3.

CHAPTER 3 INVESTIGATION

Identifying Organic Compounds in Foods

OBJECTIVES

- Determine whether specific nutrients are present in a solution of unknown composition.
- Perform chemical tests using substances called *indicators.*

PROCESS SKILLS

- experimenting
- observing
- measuring

MATERIALS

- lab apron
- safety goggles
- disposable gloves
- 1 L beaker
- hot plate
- 9 test tubes
- labeling tape
- marker
- 10 mL graduated cylinder
- Benedict's solution
- 9 dropping pipets
- glucose solution
- unknown solution
- distilled water
- 9 glass stirring rods
- tongs or test-tube holder
- test-tube rack
- albumin solution
- sodium hydroxide solution
- copper sulfate solution
- vegetable oil
- Sudan III solution

Background

1. Carbohydrates, proteins, and lipids are nutrients that are essential to all living things. Some foods, such as table sugar, contain only one of these nutrients. Most foods, however, contain mixtures of proteins, carbohydrates, and lipids. You can confirm this by reading the information in the "Nutrition Facts" box found on any food label.
2. In this investigation, you will use chemical substances, called indicators, to identify the presence of specific nutrients in an unknown solution. By comparing the color change an indicator produces in the unknown food sample with the change it produces in a sample of known composition, you can determine whether specific organic compounds are present in the unknown sample.
3. Benedict's solution is used to determine the presence of monosaccharides, such as glucose. A mixture of sodium hydroxide and copper sulfate determine the presence of some proteins (this procedure is called the biuret test). Sudan III is used to determine the presence of lipids.

Procedure

CAUTION **Put on a lab apron, safety goggles, and gloves. In this lab, you will be working with chemicals that can harm your skin and eyes or stain your skin and clothing. If you get a chemical on your skin or clothing, wash it off at the sink while calling to your teacher. If you get a chemical in your eyes, immediately flush it out at the eyewash station while calling to your teacher.** As you perform each test, record your data in your lab report, organized in a table like the one on the next page.

Test 1

1. CAUTION **Do not touch the hot plate. Use tongs to move heated objects. Turn off the hot plate when not in use. Do not plug in or unplug the hot plate with wet hands.** Make a water bath by filling a 1 L beaker half full with water. Then put the beaker on a hot plate and bring the water to a boil.
2. While you wait for the water to boil, label one test tube "1-glucose," label the second test tube "1-unknown," and label the third test tube "1-water." Using the graduated cylinder, measure 5 mL of Benedict's solution and add it to the "1-glucose" test tube. Repeat the procedure, adding 5 mL of Benedict's solution each to the "1-unknown" test tube and "1-water" test tube.
3. Using a dropping pipet or eyedropper, add 10 drops of glucose solution to the "1-glucose" test tube. Using a second dropping pipet, add 10 drops of the unknown solution to the "1-unknown" test tube. Using a third dropping pipet, add 10 drops of distilled water to the

TABLE A *IDENTIFICATION OF SPECIFIC NUTRIENTS BY CHEMICAL INDICATORS*

Test	Nutrient in test solution	Nutrient category (protein, lipid, etc.)	Result for known sample	Result for unknown sample	Result for distilled water
1	glucose	carbohydrate	red or orange precipitate	no change	no change
2	albumin	protein	pink-purple color	pink-purple color	no change
3	vegetable oil	lipid	pink color	no change	no change

"1-water" test tube. Mix the contents of each test tube with a clean stirring rod. **(It is important not to contaminate test solutions by using the same dropping pipet or stirring rod in more than one solution. Use a different dropping pipet and stirring rod for each of the test solutions.)**

4. When the water boils, use tongs to place the test tubes in the water bath. Boil the test tubes for 1 to 2 minutes.
5. CAUTION **Do not touch the test tubes with your hands. They will be very hot.** Use tongs to remove the test tubes from the water bath and place them in the test-tube rack. As the test tubes cool, an orange or red precipitate will form if large amounts of glucose are present. If small amounts of glucose are present, a yellow or green precipitate will form. Record your results in your data table.

Test 2

6. Label one clean test tube "2-albumin," label a second test tube "2-unknown," and label a third test tube "2-water." Using a dropping pipet, add 40 drops of albumin solution to the "2-albumin" test tube. Using a second dropping pipet, add 40 drops of unknown solution to the "2-unknown" test tube. Using a third dropping pipet, add 40 drops of water to the "2-water" test tube.
7. Add 40 drops of sodium hydroxide solution to each of the three test tubes. Mix the contents of each test tube with a clean stirring rod.
8. Add a few drops of copper sulfate solution, one drop at a time, to the "2-albumin" test tube. Stir the solution with a clean stirring rod after each drop. Note the number of drops required to cause the color of the solution in the test tube to change. Then add the same number of drops of copper sulfate solution to the "2-unknown" and "2-water" test tubes.
9. Record your results in your data table.

Test 3

10. Label one clean test tube "3-vegetable oil," label a second test tube "3-unknown," and label a third test tube "3-water." Using a dropping pipet, add 5 drops of vegetable oil to the "3-vegetable oil" test tube. Using a second dropping pipet, add 5 drops of the unknown solution to the "3-unknown" test tube. Using a third dropping pipet, add 5 drops of water to the "3-water" test tube.
11. CAUTION **Sudan III solution will stain your skin and clothing. Promptly wash off spills to minimize staining. Do not use Sudan III solution in the same room with an open flame.** Using a clean dropping pipet, add 3 drops of Sudan III solution to each test tube. Mix the contents of each test tube with a clean stirring rod.
12. Record you results in your data table.
13. Clean up your materials and wash your hands before leaving the lab.

Analysis and Conclusions

1. Based on the results you recorded in your data table, identify the nutrient or nutrients in the unknown solution.
2. What are the experimental controls in this investigation?
3. Explain how you were able to use the color changes of different indicators to determine the presence of specific nutrients in the unknown substance.
4. List four potential sources of error in this investigation.

Further Inquiry

Is there a kind of macromolecule that the tests in this lab did not test for? If so, list the kind of macromolecules not tested for, and give one reason why they were not tested for.

ANSWERS TO ANALYSIS AND CONCLUSIONS

1. Students' results should reveal that the main nutrient in the unknown is protein, as evidenced by the purple color in the protein test, test 3. Nonfat cottage cheese also contains a small amount of sugars.
2. The controls are three known foods—glucose, albumin, and vegetable oil.
3. The presence of different nutrients in foods can be detected by color changes in chemical indicators specific for that nutrient.
4. Possible sources of error include inaccurate measurements, impurities in the solutions and samples, unwashed equipment, and, in test 1, inadequately heated or cooled samples.

ANSWERS TO FURTHER INQUIRY

Nucleic acids were not tested for. All foods are made of cells; thus, all foods contain nucleic acids.

Unit 2

CELLS

CHAPTERS

" *The cell is the natural granule of life in the same way as the atom is the natural granule of simple, elemental matter. If we are to take the measure of the transit to life and determine its precise nature, we must try to understand the cell.* "

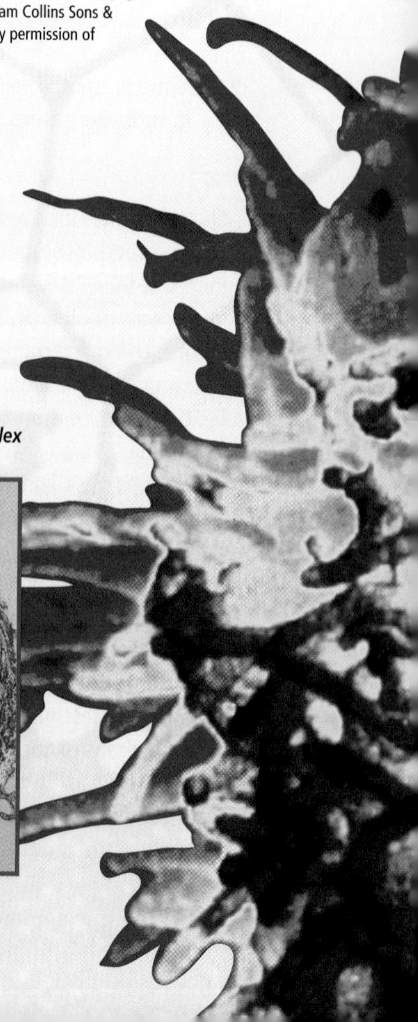

Eukaryotic cells contain a number of complex internal structures.

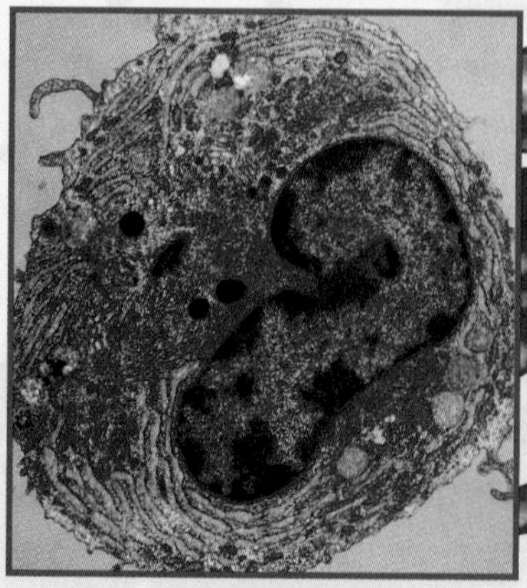

internet**connect**

*sci*LINKS NSTA

National Science Teachers Association *sci*LINKS Internet resources are located throughout this unit.

Most cells are very small, but these frog eggs can be seen with the unaided eye.

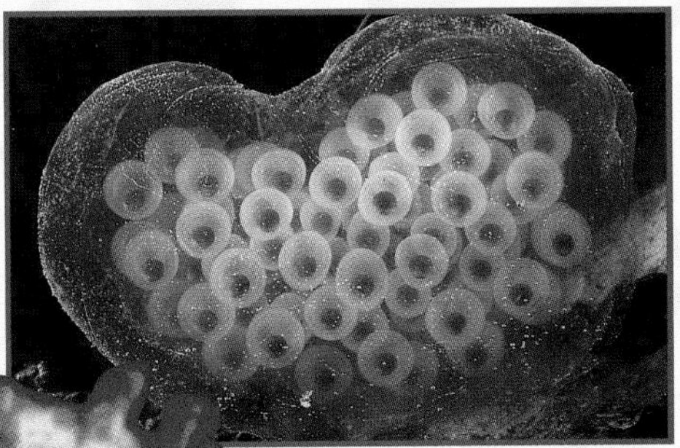

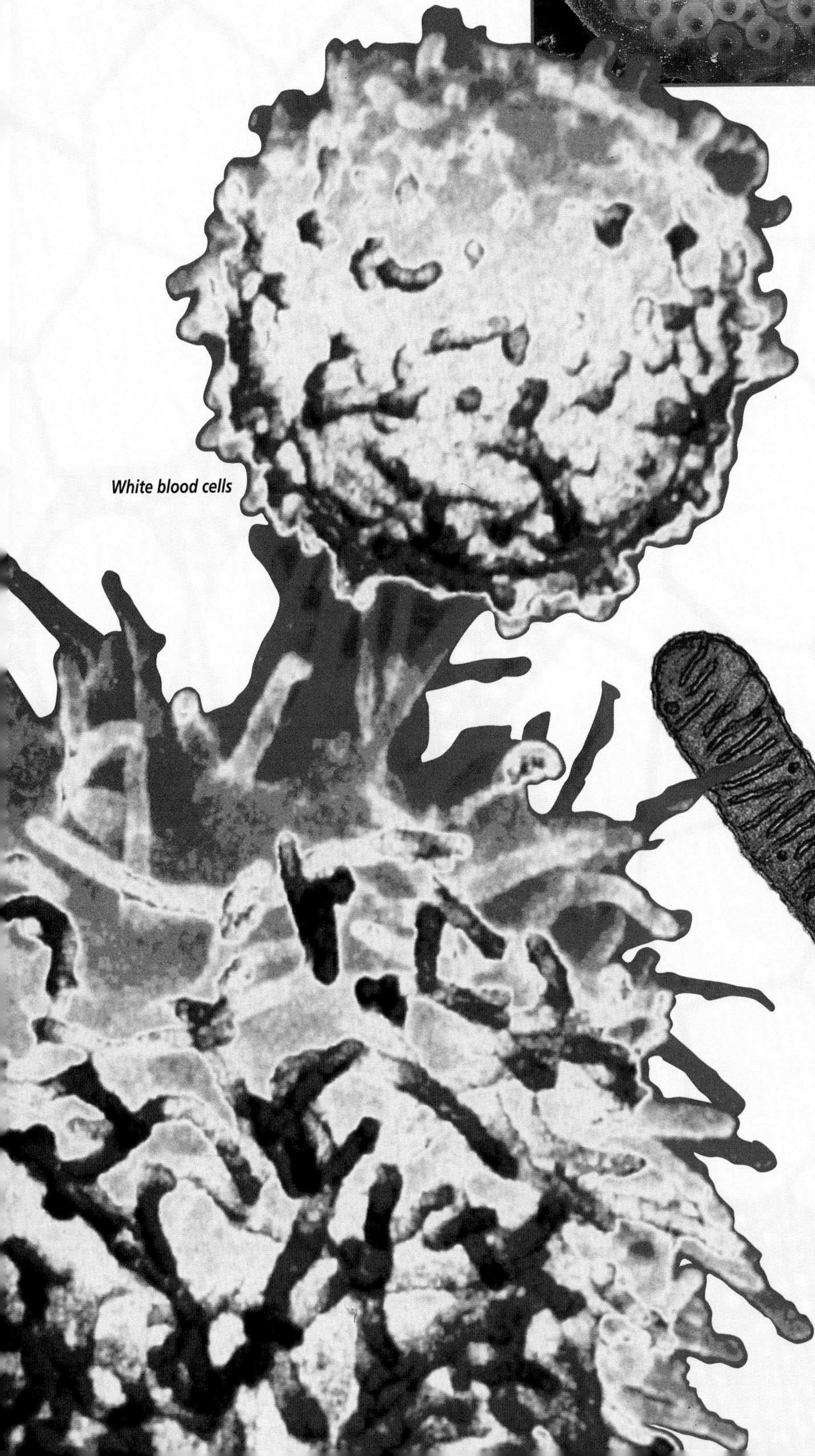

White blood cells

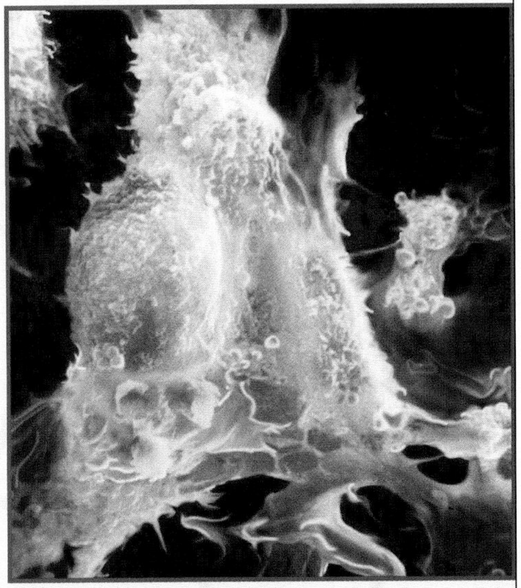

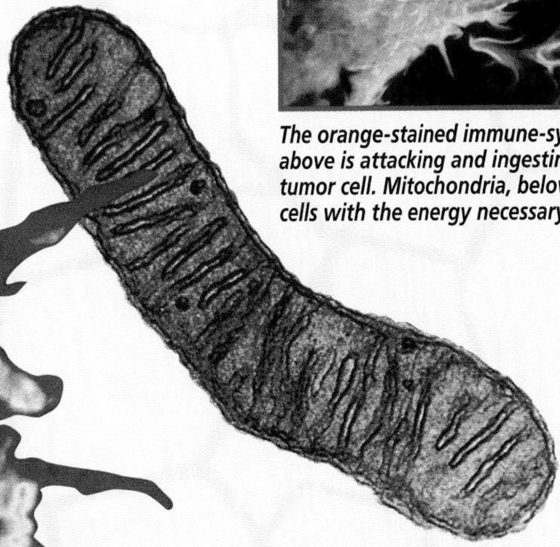

The orange-stained immune-system cell shown above is attacking and ingesting the red-stained tumor cell. Mitochondria, below left, provide cells with the energy necessary for life.

unit 2

CHAPTERS

CELLS

CONNECTING TO THE STANDARDS

The following chart shows the correlation of Unit 2 with the *National Science Education Standards* (grades 9–12). The items in each category are addressed in this unit and should be used to guide instruction. Annotated descriptions of the **Life Science Standards** are found below the chart. Consult the *National Science Education Standards* for detailed descriptions of each category.

UNIFYING CONCEPTS

- Systems, order, and organization
- Change, constancy, and measurement
- Form and function

SCIENCE AS INQUIRY

- Abilities necessary to do scientific inquiry
- Understandings about scientific inquiry

SCIENCE AND TECHNOLOGY

- Abilities of technological design
- Understandings about science and technology

PHYSICAL SCIENCE

- Chemical reactions
- Interactions of energy and matter

LIFE SCIENCE

EARTH AND SPACE SCIENCE

- Energy in the Earth system

SCIENCE IN PERSONAL AND SOCIAL PERSPECTIVES

- Natural/human hazards
- Science and technology in local and global challenges

HISTORY AND NATURE OF SCIENCE

- Science as a human endeavor
- Nature of scientific knowledge
- Historical perspectives

ANNOTATED DESCRIPTIONS OF THE CORRELATED LIFE SCIENCE STANDARDS

Although all eight categories of the *National Science Education Standards* are important, the following descriptions summarize the **Life Science Standards** that specifically relate to Unit 2.

THE CELL

- Cells have particular structures that underlie their functions.
- Most cell functions involve chemical reactions.
- Cells store and use information to guide their functions.
- Cell functions are regulated.
- Plant cells contain chloroplasts, the site of photosynthesis.

MOLECULAR BASIS OF HEREDITY

- In all organisms, the instructions for specifying the characteristics of the organism are carried in DNA.
- Most of the cells in a human contain two copies of each of 22 different chromosomes. One additional pair determines whether the individual will be male or female.

BIOLOGICAL EVOLUTION

- Species evolve over time.
- Biological classifications are based on how organisms are related.

INTERDEPENDENCE OF ORGANISMS

- The atoms and molecules on the Earth cycle among the living and nonliving components of the biosphere.

MATTER, ENERGY, AND ORGANIZATION IN LIVING SYSTEMS

- The energy for life primarily derives from the sun.
- The chemical bonds of food molecules contain energy.
- The complexity and organization within an organism result from the organism's need to obtain, transform, transport, release, and eliminate the matter and energy required to sustain the organism.

TRENDS IN BIOLOGY

Archaebacteria

How many fundamental cell types are there? Until recently, the answer was two—prokaryotes and eukaryotes. However, rapid advances in molecular biology have allowed biologists to detect far more biochemical diversity within the prokaryotes than was previously thought to exist. As a result, most biologists now think this group contains two lineages that are only distantly related to one another. These lineages are the Eubacteria (or bacteria), which includes most of the "traditional" prokaryotes, such as *Escherichia coli* and *Streptococcus,* and the Archaebacteria (or Archaea), which includes mainly inhabitants of extreme environments but also a growing roster of species from milder habitats. The differences between Archaebacteria and Eubacteria are so great that most biologists agree the Archaebacteria represent a third fundamental cell type.

Much of the credit for advancing this new view goes to Carl Woese, a microbiologist at the University of Illinois. In 1977, after determining the ribosomal RNA (rRNA) sequences for a number of eukaryotes and prokaryotes, Woese and his colleague, George Fox, noticed that the sequences fell into three groups rather than two: eukaryotes, traditional bacteria, and a few unusual prokaryotes that Woese and Fox named archaebacteria, which means "ancient bacteria." At the time, little was known about the archaebacteria except that they had some unique metabolic pathways and were typically found in extreme environments, such as hot springs.

ARCHAEBACTERIA ARE UNIQUE

Woese and Fox's proposal that there were three basic kinds of living things drew fire from all quarters, but it also stimulated a closer examination of archaebacteria, which eventually confirmed the distinctiveness of this group. To cite one example, the cell walls of archaebacteria lack peptidoglycan, a component of the cell walls of nearly all forms of bacteria. And when, in 1996, scientists first sequenced the entire genome of an archaebacterium (the deep-sea chemoautotroph *Methanococcus jannaschii*), they found that although some of its genes resemble those of other organisms, more than half of its genes apparently have no counterparts in bacteria or in eukaryotes.

Archaebacteria held two more surprises. First, in many ways, archaebacteria are more similar to eukaryotes than to traditional bacteria. These similarities, which include the mechanisms for transcription, translation, and DNA replication, have convinced most scientists that archaebacteria and eukaryotes are each other's closest relatives.

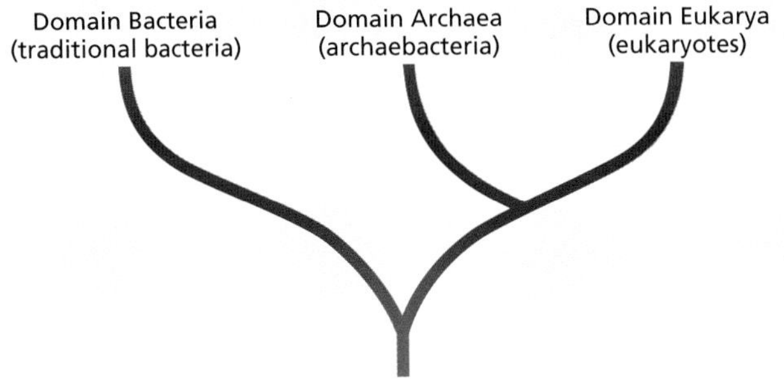

The three-domain classification system proposed by Carl Woese

To reflect this discovery, Carl Woese has proposed creating the domain, a new classification category above the kingdom level in the Linnaean hierarchy. Woese's scheme (shown above) recognizes three domains—Bacteria, Archaea, and Eukarya.

The second surprise is that archaebacteria are neither rare nor confined to harsh environments. Methane-producing archaebacteria live in the human large intestine, and other species seem to be abundant in the soil and oceans.

Additional Information

Cohen, Philip. "Life, But Not As We Know It." *New Scientist,* 31 August 1996, 15.

Madigan, Michael, John Martinko, and Jack Parker. *Brock Biology of Microorganisms.* 8th ed. Upper Saddle River, New Jersey: Prentice Hall, 1997.

University of California Museum of Paleontology. Archaea page. World Wide Web: http://www.ucmp.berkeley.edu.archaea/archae.html

Zimmer, Carl. "Triumph of the Archaea." *Discover,* February 1995, 30–31.

CHAPTER 4
STRUCTURE AND FUNCTION OF THE CELL

	TOPICS	TEACHING RESOURCES	LABS, CLASSWORK, AND HOMEWORK
BLOCKS 1 & 2 (90 minutes)	**Introducing the Chapter, p. 68**	ATE Focus Concept, p. 68 ATE Assessing Prior Knowledge, p. 68 ATE Understanding the Visual, p. 68	■ Supplemental Reading Guide, *The Lives of a Cell: Notes of a Biology Watcher*
	4-1 Introduction to the Cell, p. 69 Discovery of the Cell, p. 69 Cell Diversity, p. 70	ATE Section Overview, p. 69 ATE Critical Thinking, p. 71 5A, 12A	PE **Quick Lab,** p. 72, Comparing Surface Cells **Lab A3,** Modeling Cells—Surface Area to Volume **Lab A32,** Relating Cell Structure to Function ★ Study Guide, Section 4-1 PE Section Review, p. 72
BLOCKS 3–5 (135 minutes)	**4-2 Parts of the Eukaryotic Cell, p. 73** Cell Membrane, p. 73 Organelles, p. 75 Nucleus, p. 81 Plant Cells, p. 81	ATE Section Overview, p. 73 ATE Critical Thinking, pp. 75, 77 ATE Visual Link Figure 4-7, p. 76 7–9	**Lab B2,** Cell Structures **Lab D1,** Staining DNA and RNA ★ Study Guide, Section 4-2 PE Section Review, p. 83
BLOCK 6 (45 minutes)	**4-3 Multicellular Organization, p. 86** Tissues, Organs, and Organ Systems, p. 86 Evolution of Multicellular Organization, p. 87	ATE Section Overview, p. 86 42A	PE **Chapter 4 Investigation,** p. 92 ★ Study Guide, Section 4-3 PE Section Review, p. 88
BLOCKS 7 & 8 (90 minutes)	**REVIEW and ASSESSMENT** PE Chapter 4 Review, pp. 90–91	Audio CD Program BIOLOGY INTERACTIVE TUTOR Unit 1—Cell Transport and Homeostasis	**CHAPTER TESTING** ■ Chapter Test (blackline master) ▲ Test Generator ■ Assessment Item Listing ■ Scoring Rubrics and Classroom Management Checklists

MODERN BIOLOGY · HOLT BIOSOURCES

KEY

PE Pupil's Edition
ATE Teacher's Edition
■ Active Reading Guide
★ Study Guide

Teaching Transparencies

Laboratory Program

One-Stop Planner CD-ROM
Shows these resources within a customizable daily lesson plan:

■ Holt BioSources Teaching Resources
■ Active Reading Guide
★ Study Guide
▲ Test Generator

READING FOR CONTENT MASTERY

Preview: Have students preview the objectives, the introductory paragraph, topic sentences, and captions before reading the chapter.

ATE Active Reading Technique: Paired Reading, p. 69
■ Active Reading Guide Worksheet 4-1
PE Summary/Vocabulary, p. 89

■ Active Reading Guide Worksheet 4-2
PE Summary/Vocabulary, p. 89

■ Active Reading Guide Worksheet 4-3
PE Summary/Vocabulary, p. 89

TECHNOLOGY AND INTERNET RESOURCES

CNN PRESENTS Video Segment 3 Split-Liver Donor

Holt Biology Videodiscs Teacher's Correlation Guide, Lessons 4-1 through 4-3

BIOLOGY INTERACTIVE TUTOR Unit 1—Cell Transport and Homeostasis Topics 1–2

Audio CD Program, Sections 4-1 through 4-3

internetconnect

SCILINKS NSTA

On-line Resources: www.scilinks.org
The following *sci*LINKS Internet resources can be found in the student text for this chapter:

Topics:
- Cell theory, p. 69
- Eukaryotic cells, p. 77
- Plant cells, p. 81

go.hrw.com

On-line Resources: go.hrw.com
Visit the HRW Web site for a variety of resources related to this chapter. Just type in the keyword HM2 Chapter 4.

Smithsonian Institution®
Visit **www.si.edu/hrw** for additional on-line resources.

CNNfyi.com
Visit **www.cnnfyi.com** for late-breaking news and current event stories selected just for you.

LAB ACTIVITY PLANNING

CHAPTER 4 INVESTIGATION

Comparing Animal and Plant Cells, pp. 92–93

OVERVIEW
In this investigation, students will observe plant and animal cells under light microscopes. They will identify cellular organelles and classify unknown cells as either plant or animals cells.

SAFETY
Caution students not to break the slide with the objective when viewing the slide under high power. Have students dispose of broken glass in a container designated for glass only. Students should wear safety goggles and a lab apron for Part B when preparing a wet-mount slide using Lugol's solution.

PREPARATION
1. Obtain prepared slides of epithelial cells for Part A from a biological supply house, such as WARD'S.
2. For Part B, sprigs of *Elodea* can be purchased at a pet store or from a biological supply house. Keep the plants in water under light for 12 hours before the lab to induce cytoplasmic streaming.
3. Each student should examine three or more slides of unknown cells. Prepare the unknowns in advance, covering any labels on the slides with masking tape. Set up several stations for viewing the unknowns.
4. Suggested unknowns include the following prepared slides that can be obtained from WARD'S: human squamous epithelium cells, onion (*Allium*) root-tip cells, and *Psilotum* stem cells.

Quick Labs

Comparing Surface Cells, p. 72
Students will require a microscope, prepared slides of human (animal) skin and a plant stem, a pencil, and paper for this activity.

UNDERSTANDING THE VISUAL

Explain that this photograph shows a cell that has been sliced into a thin section to reveal the interior of the cell. Note that the colors were added by a visual artist to make the various parts clearer. Ask students how many different structures they can find inside the cell. Also ask them to look for structures that appear to be represented more than once in the cell. Explain that each structure is a cell component with a specific function.

FOCUS CONCEPT

Cell Structure and Function

The relationship between structure and function is a unifying theme in biology. In this chapter the relationship is evident at the whole-cell and subcellular levels. Examples at the whole-cell level: nerve cells that carry information between widely separated parts of the body are elongated; epithelial cells that cover body surfaces are flat. Examples at the subcellular level: cells that have a high energy requirement have numerous mitochondria; cells that capture energy in sunlight have chloroplasts; cells that move through a watery medium have cilia or flagella on their surfaces.

ASSESSING PRIOR KNOWLEDGE

Review the following concepts.

Chemistry: *Chapter 2*
Ask students to define the following terms: solution, suspension, colloid.

Biochemistry: *Chapter 3*
Ask students how proteins are related to amino acids, and how polysaccharides are related to monosaccharides. Then ask students what an enzyme is. Finally ask students to describe how lipids react to water.

CHAPTER 4

STRUCTURE AND FUNCTION OF THE CELL

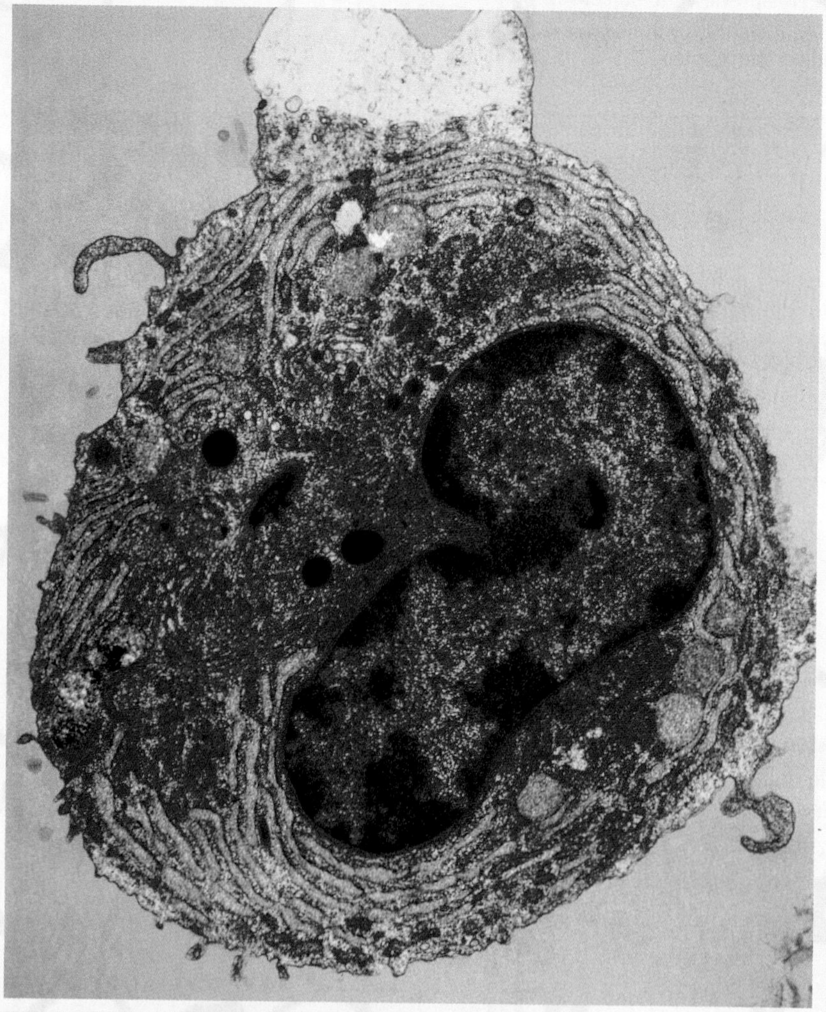

This human bone cell has a complex internal structure. (TEM 17,938×)

FOCUS CONCEPT: ***Cell Structure and Function***

As you read, find examples of how cell structures vary with their functions.

BIOLOGY INTERACTIVE TUTOR

Unit 1—*Cell Transport and Homeostasis*
Topics 1–2

INTRODUCTION TO THE CELL

Both living and nonliving things are composed of molecules made from chemical elements such as carbon, hydrogen, oxygen, and nitrogen. The organization of these molecules into cells is one feature that distinguishes living things from all other matter. A ***cell*** *is the smallest unit of matter that can carry on all of the processes of life.*

SECTION

4-1

OBJECTIVES

▲ Outline the discoveries that led to the development of the cell theory.

● State the cell theory.

■ Identify a limiting factor on the size of cells.

◆ Describe the relationship between cell shape and cell function.

▲ Distinguish between prokaryotes and eukaryotes.

DISCOVERY OF THE CELL

Every living thing—from the tiniest bacterium floating in a drop of water to the largest whale—is made of one or more cells. How did scientists come to this conclusion? The discovery of cells was made possible by the development of the microscope in the early seventeenth century.

In 1665, the English scientist Robert Hooke (1635–1703) used a microscope to examine a thin slice of cork. Hooke wrote, "I could exceedingly plainly perceive it to be all perforated and porous," and he further described it as consisting of "a great many little boxes." When he turned his microscope to the stems of elder trees, carrots, and ferns, Hooke found that each showed a similar formation. These "little boxes" reminded him of the small rooms in which monks lived, so he called them cells.

What Hooke had observed were actually the remains of dead plant cells. The first person to observe living cells was a Dutch microscope maker, Anton van Leeuwenhoek (1632–1723). Although van Leeuwenhoek's microscope was rather simple, in 1673 it was powerful enough to enable him to open up a whole new world—the world of microscopic organisms, which had never before been seen.

The Cell Theory

About 150 years passed before scientists began to organize the observations begun by Hooke and van Leeuwenhoek into a unified theory known as the **cell theory.** This theory has three parts:

- All living things are composed of one or more cells.
- Cells are the basic units of structure and function in an organism.
- Cells come only from the reproduction of existing cells.

internet**connect**

SCiLINKS NSTA
TOPIC: Cell theory
GO TO: www.scilinks.org
KEYWORD: HM069

SECTION OVERVIEW

Cell Structure and Function
The relationship between a cell's function and its size and shape is explained.

▶ ENGAGE STUDENTS

Smear some yogurt on a microscope slide. Mix a drop of water onto the smear and add a coverslip. Examine the slide under high power on a compound microscope to see bacteria of the genus *Lactobacillus.* Compare the size of these cells with that of cells on prepared slides of human and plant tissues.

ACTIVE READING

Technique: Paired Reading
Pair each student with a partner. Have each student identify paragraphs or pages with topics or concepts they find difficult to understand. One student should read the paragraph or section aloud, and then the other student should repeat the information in his or her own words. Both students should agree on the meaning of the topic or the concept they are discussing.

DEMONSTRATION

Demonstrate the relative sizes of prokaryotic and eukaryotic cells using scale models. On a sheet of paper, draw a circle 2 mm in diameter to represent the size of the smallest bacteria (0.2 µm). Using a compass, draw another circle 10 cm in diameter to represent a small eukaryotic cell (10 µm). Use a meterstick to represent a large eukaryotic cell (100 µm). Point out that these models are 10,000 times larger than the cells they represent. Ask students why most cells are microscopic. (Small size allows cells to have a large surface-area-to-volume ratio, which is necessary for cell survival.)

TEACHING STRATEGY

The Effect of Cell Size

Have students cut cubes of different sizes from a potato. Place the cubes in a mixture of water and food dye for 10 minutes. Have students cut the cubes in half and determine how the surface-area-to-volume ratio affects completeness of dye penetration in the cubes. (The dye should penetrate to the same depth in all cubes, but the smaller cubes will be more completely dyed than the larger cubes.) If the cubes were real cells and the dye were a nutrient, the nutrient would take longer to reach the center of the larger cells.

FIGURE 4-1

Although most cells are extremely small, some are large enough to be seen without a microscope. These frog egg cells, for instance, measure about 1.5 mm (0.1 in.) in diameter. Even larger are chicken eggs, which are several centimeters across. The egg cells of both frogs and chickens consist mainly of yolk, which serves as a stored nutrient for the developing embryo.

Early evidence for the cell theory was provided by a trio of German scientists. In 1838, the botanist Matthias Schleiden (1804–1881) concluded that all plants are composed of cells. A year later, the zoologist Theodor Schwann (1810–1882) came to the same conclusion about animals. In 1855, Rudolf Virchow (1821–1902), a physician who had been studying how disease affects living things, reasoned that cells come only from other cells. Over the years, modern scientists have gathered much additional evidence that strongly supports the cell theory.

CELL DIVERSITY

Not all cells are alike. Even cells within the same organism may show enormous diversity in size, shape, and internal organization. Your body, for example, contains at least 200 different cell types.

Size

A few types of cells, such as those shown in Figure 4-1, are large enough to be seen by the unaided eye. The nerve cells that extend down a giraffe's leg, for instance, can be up to 2 m (about $6\frac{1}{2}$ ft) long. However, most plant and animal cells are only about 10 to 50 µm (0.002 in.) in diameter, and some bacterial cells are only 0.2 µm (0.000008 in.) in diameter. Therefore, most cells are visible only with a microscope.

Cells are limited in size by the ratio between their outer surface area and their volume. Table 4-1 shows how growth affects this ratio. For a cuboidal cell, volume increases with the *cube* of the side length, but surface area increases with the *square* of the side length. This means that if a cell keeps the same shape as it grows, its volume will increase more rapidly than its surface area. This trend is important because the nutrients, oxygen, and other materials a cell requires must enter through its surface. Thus, as a cell grows larger, at some point its surface area becomes too small to allow these materials to enter the cell quickly enough to meet the cell's needs.

TABLE 4-1 *Surface Areas and Volumes of Cubes*

Side Length	Surface Area	Volume	Surface Area/Volume Ratio
1 mm	6 mm^2	1 mm^3	6:1
2 mm	24 mm^2	8 mm^3	3:1
3 mm	54 mm^2	27 mm^3	2:1

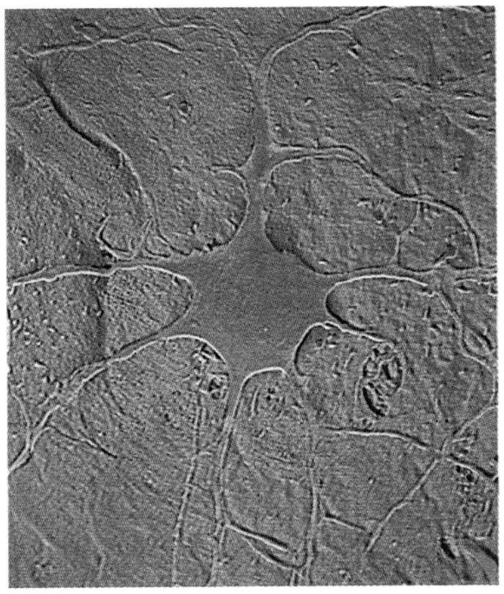

NERVE CELL

SKIN CELLS

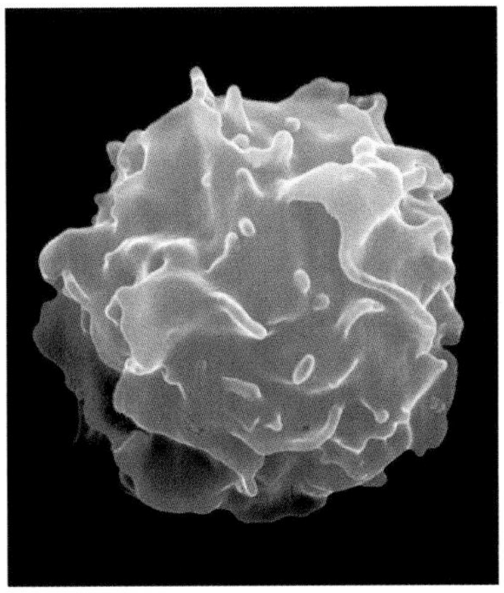

WHITE BLOOD CELL

FIGURE 4-2

These photographs taken with a scanning electron microscope show three different types of cells that are found in the human body. Each cell type has a structure that enables it to perform its function effectively. (left 17,385×, middle 330×, right 23,250×)

Shape

Cells come in a variety of shapes. This diversity of form reflects a diversity of function. Examine the three human cells shown in Figure 4-2. The long extensions that reach out in various directions from the nerve cell enable the cell to receive and transmit nerve impulses. In contrast, the flat shape of dead skin cells is well suited to their function of covering the body surface. Some white blood cells can change shape, leave the blood, and enter the areas surrounding blood vessels. This allows them to move through narrow openings and to isolate, engulf, and destroy bacteria that invade the body.

Internal Organization

The micrographs in Figures 4-3 and 4-4 show that cells contain a variety of internal structures called **organelles.** An organelle is a cell component that performs specific functions for the cell. Just as the organs of a multicellular organism carry out the organism's life functions, the organelles of a cell maintain the life of the cell.

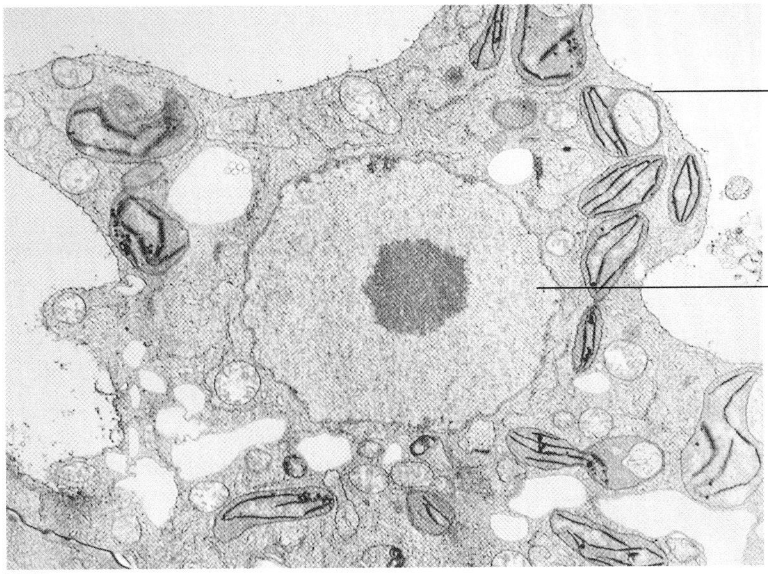

FIGURE 4-3

This plant cell, like other eukaryotic cells, is filled with membrane-bound organelles. The most prominent organelle is the nucleus. The entire cell is surrounded by a membrane. (TEM 105,661×)

CRITICAL THINKING

Overcoming Size Limitations

How might a cell grow large despite the limitations imposed by surface-area-to-volume ratios? (A cell could grow large in one or two dimensions but remain small in the others. For example, parts of nerve cells are very long cylinders with small diameters, and epithelial cells are broad but flat. In both cases, the surface-area-to-volume ratio is quite large.)

INCLUSION *ACTIVITY*

Ask students to sketch a prokaryote and an animal cell, showing the shapes of the organelles present in each. Have the students state both the differences and similarities between these cells.

RETEACHING *ACTIVITY*

Ask students to list major similarities between prokaryotic and eukaryotic cells. (They both have cell membranes, ribosomes, cytoplasm, a similar metabolism, and DNA as their genetic material.) Ask students to list significant differences between prokaryotic and eukaryotic cells. (Eukaryotic cells have a nucleus surrounded by a double membrane, while prokaryotes have a nucleoid, a DNA-containing area in the cytoplasm that is not bound by a nucleus. Eukaryotic cells have other membrane-enclosed organelles, but prokaryotes do not. Eukaryotic cells are approximately 10 to 100 times larger than prokaryotes.)

Quick Lab

Comparing Surface Cells

Time Required 20 minutes

Procedural Tips Slides can be projected on a monitor or screen. This method enables the teacher to control time and ensure that students are examining the correct cell layers.

Answers to Analysis The outer layer of cells is thinner and more flattened than the cells below it. The function is protection. The flat shape of surface cells is suited to its function of covering and protecting the cells below them.

ANSWERS TO SECTION 4-1 REVIEW

1. The cell theory states that all living things are made of one or more cells; cells are the basic units of structure and function; and cells come only from preexisting cells.
2. The ratio between volume and surface area limits cell size.
3. The long extensions of nerve cells allow them to receive and transmit impulses in various directions; the flat shape of dead skin cells enables them to cover the body; the ability of white blood cells to change shape allows them to move through narrow openings.
4. An organelle is a component of a cell that performs specific functions in the cell.
5. If a nucleus and other membrane-bounded organelles are present, then the organism is a eukaryote. If they are absent, it is a prokaryote.
6. Once good-quality microscopes became available, a number of scientists could use them to examine cells.

Quick Lab

Comparing Surface Cells

Materials microscope, prepared slides of animal (human) skin and plant (dicot) stem, pencil, paper

Procedure Examine slides using medium magnification (100×). Observe and draw the outer surface cells of the plant stem and the animal skin.

Analysis How do the surface cells of each organism differ from the cells beneath them? What is the function of these outer surface cells? Explain how the shape of surface cells are suited to their function.

In Figure 4-3, notice that the entire cell is surrounded by a thin membrane, called the **cell membrane.** Inside the cell are a variety of organelles, most of which are surrounded by their own membranes. The large organelle near the center of the cell is the **nucleus.** It contains the majority of the cell's genetic information and directs most of the activities of the cell. Organisms whose cells contain a membrane-bound nucleus and other organelles are called **eukaryotes** (yoo-KAR-ee-OHTS).

The cell shown in Figure 4-4 is a bacterium. It, too, has a cell membrane, but none of the organelles inside a bacterium are surrounded by a membrane. Although the genetic information of a bacterium may be concentrated in one part of the cell, a bacterium has no membrane-bound nucleus like that of a eukaryotic cell. Unicellular organisms that lack a membrane-bound nucleus and other organelles are called **prokaryotes** (proh-KAR-ee-OHTS). The difference between prokaryotes and eukaryotes is such an important distinction that prokaryotes are placed in two kingdoms, separate from eukaryotes.

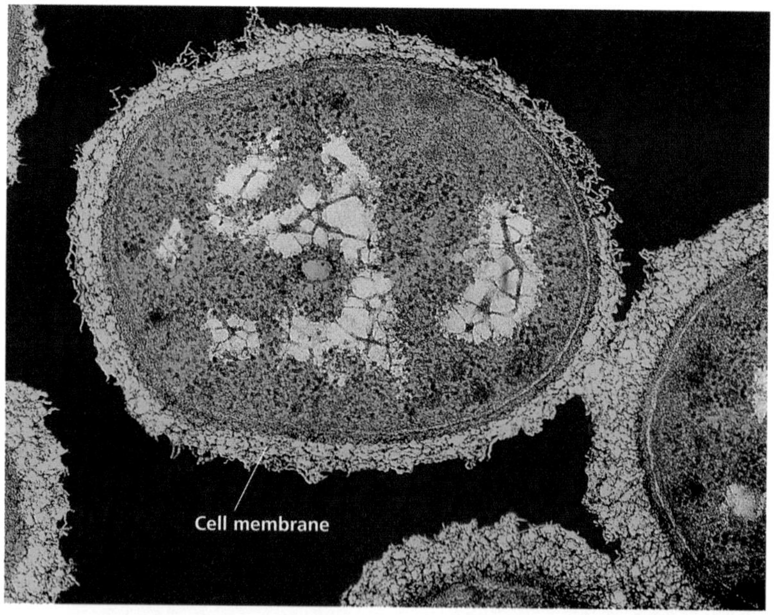

FIGURE 4-4
This bacterial cell is surrounded by a membrane, but it has no nucleus or other membrane-bound organelles. (TEM 84,721×)

SECTION 4-1 REVIEW

1. What is the cell theory?
2. What single factor limits the size that most cells are able to attain?
3. Give two examples of how cells' shapes are suited to their functions.
4. What is an organelle?
5. How can you determine whether a unicellular organism is a prokaryote or a eukaryote?
6. CRITICAL THINKING The observations that led to the formation of the cell theory occurred within a 17-year period. Why do you think that more than one scientist made critical observations about cells within this period?

SECTION

Parts of the Eukaryotic Cell

The structures that make up a eukaryotic cell are determined by the specific functions carried out by the cell. Thus, there is no typical eukaryotic cell. Nevertheless, eukaryotic cells generally have three main components: a cell membrane, a nucleus, and other organelles.

CELL MEMBRANE

A cell cannot survive if it is totally isolated from its environment. All cells must take in nutrients and other materials, and they must also dispose of the wastes they produce. Therefore, both nutrients and wastes must pass through the cell membrane. The cell membrane controls the ease with which substances pass into and out of the cell—some substances easily cross the membrane, while others cannot cross at all. For this reason, the cell membrane is said to be **selectively permeable.**

The structure of the cell membrane depends on the functions the cell performs. In a multicellular organism, for example, some cells secrete materials into their environment for use elsewhere in the organism. Other cells recognize potentially harmful "invaders" and destroy them before they cause any damage. In each case, the cells are surrounded by membranes specialized for that task. No matter what the task, however, all cell membranes are made primarily of lipids and proteins.

Membrane Lipids

One of the major types of lipids in the cell membrane is phospholipid. Recall from Chapter 3 that each phospholipid molecule has a polar "head" and two nonpolar "tails." Because of its hydrophilic nature, the head of a phospholipid will orient itself so that it is as close as possible to water molecules. In contrast, the hydrophobic tails will tend to orient themselves away from water.

Cells are bathed in an aqueous, or watery, environment. Since the inside of a cell is also an aqueous environment, both sides of the cell membrane are surrounded by water molecules. These water molecules cause the phospholipids of the cell membrane to form two layers—a lipid bilayer. As you can see in Figure 4-5,

4-2

OBJECTIVES

▲ Describe the structure, composition, and function of the cell membrane.

● Name the major organelles found in a eukaryotic cell, and describe their functions.

■ Describe the structure and function of the nucleus.

◆ Describe three structures characteristic of plant cells.

SECTION 4-2

SECTION OVERVIEW

Cell Structure and Function
This section explains the relationship between a cell's function and the organelles it contains, and between an organelle's function and its structure.

Matter, Energy, and Organization
The internal organization of eukaryotic cells is described.

▶ ENGAGE STUDENTS

Cytoplasmic streaming, such as students observe in *Elodea,* also can be seen in other cells. Provide students with a culture of live amoebas. Direct the students to place a drop of the culture on a microscope slide and watch the amoebas' organelles flow with the cytoplasm as the amoebas move. Ask students to locate the nucleus. (It should be near the center of the cell.) Which way does the cytoplasm appear to move? (It moves into and out of the cytoplasmic extensions.) Tell the students that most organelles are too small to see.

DEMONSTRATION

To represent the movement of substances through a membrane, pour a few drops of vanilla extract into a balloon. Partially inflate the balloon and tie it off. As the students pass the balloon around, they will be able to smell the vanilla. Ask them how the odor might be able to pass through the balloon. (The vanilla extract has begun to evaporate inside the balloon. Molecules of the extract must either pass through small pores in the balloon's rubber wall or dissolve in the rubber and evaporate from the external surface of the balloon.)

SECTION 4-2

RETEACHING ACTIVITY

Make a model of a cell membrane by cutting off half of one "leg" of several wooden clothespins. Different colors can be used to represent the hydrophobic and hydrophilic ends. Glue the clothespins together. Cardboard tubes can represent channels. Proteins can be represented by buttons or bottle caps. Ask students how this model inaccurately represents a cell membrane. (The button proteins cannot flow between the clothespin phospholipids.)

CONTENT CONNECTION

Section 3-3: Organic Compounds

Three major types of organic compounds are found in cell membranes. Lipids (mostly phospholipids, but also steroids) give membranes their bilayer structure. Proteins are embedded in or located on either side of the bilayer. Carbohydrates are attached to the part of many membrane proteins that is exposed to the outside of the cell.

RECENT RESEARCH

A Cure of the Common Cold

Each year, 26 million school days and 29 million work days are lost in this country due to the common cold. Forty percent of colds are caused by rhinoviruses, which attach to carbohydrates associated with peripheral proteins on cells. Scientists have synthesized the viral molecules that bind to the carbohydrates. Efforts are under way to develop a nasal spray containing the viral molecules. When inhaled, these molecules would bind to the carbohydrates and block attachment by rhinoviruses, preventing infection.

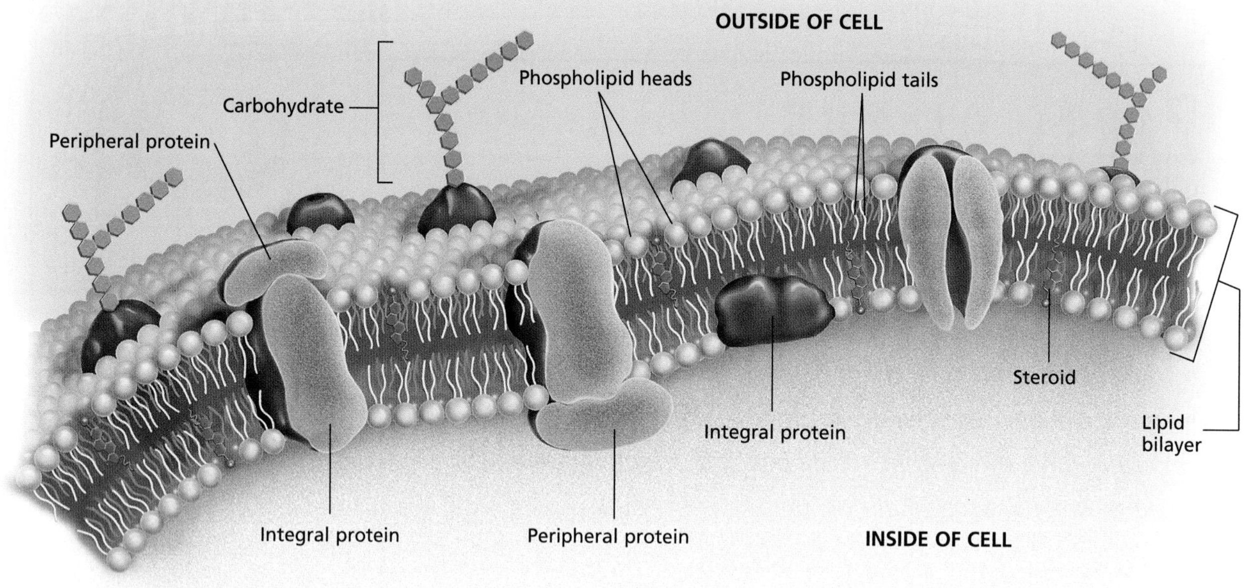

FIGURE 4-5

Cell membranes are composed mostly of a lipid bilayer and two types of proteins. Integral proteins are embedded within the membrane. Peripheral proteins are attached to both surfaces of the membrane.

the phospholipids are arranged so that their heads point outward, while their tails are confined to the interior of the membrane.

Figure 4-5 indicates that eukaryotic cell membranes also contain steroids, another type of lipid you encountered in Chapter 3. In membranes, the steroid molecules fit between the tails of the phospholipids. The major membrane steroid in animal cells is cholesterol. Other steroids are found in the cell membranes of plants.

Membrane Proteins

Some proteins are attached to the surfaces of the cell membrane. As shown in Figure 4-5, these **peripheral proteins** are located on both the interior surface and the exterior surface of the cell membrane. Weak bonds link peripheral proteins to membrane lipids or to other proteins that are embedded in the lipid bilayer. The proteins that are embedded in the bilayer are called **integral proteins.** Figure 4-5 shows that some integral proteins extend across the entire cell membrane and are exposed to both the inside of the cell and the exterior environment. Others extend either only to the inside or only to the exterior surface.

Notice in Figure 4-5 that the integral proteins exposed to the cell's external environment often have carbohydrates attached to them. These carbohydrates may hold adjoining cells together, or they may act as sites where viruses or chemical messengers such as hormones can attach.

Because the cell membrane is selectively permeable, cells must have mechanisms for transporting molecules through the lipid bilayer. Membrane proteins play an important role in this process. For example, some integral proteins form channels or pores through which certain substances can pass. Other proteins bind to a substance on one side of the membrane and carry it to the other side of the membrane. You will learn more about how substances cross the cell membrane in Chapter 5.

Fluid Mosaic Model of Cell Membranes

For many years, scientists thought that the molecular arrangement of lipids and proteins in the cell membrane was relatively static. But with the development of new techniques and instruments, including the scanning electron microscope, scientists have discovered that cell membranes are actually very dynamic. Today scientists use the **fluid mosaic model** to describe the cell membrane. According to this model, the lipid bilayer behaves more like a fluid than a solid. Because of this fluidity, the membrane's lipids and proteins can move laterally within the lipid bilayer, as indicated in Figure 4-6. As a result of such lateral movement, the pattern, or "mosaic," of lipids and proteins in the cell membrane is constantly changing.

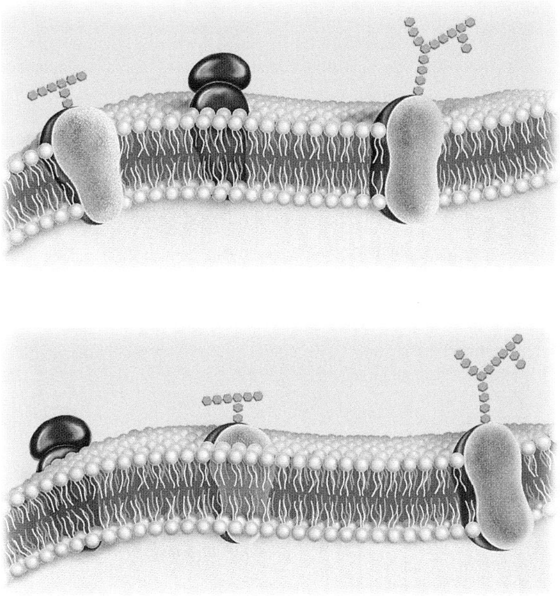

FIGURE 4-6
The cell membrane is a dynamic structure, with both lipids and proteins moving laterally within the lipid bilayer. Scientists therefore describe the structure of the cell membrane in terms of a fluid mosaic model.

ORGANELLES

Between the cell membrane and the nucleus lies the **cytoplasm** (SIET-oh-PLAZ-uhm), which contains the various organelles of the cell. The organelles are bathed in a gelatin-like aqueous fluid called the **cytosol** (SIET-oh-SAWL). Dissolved in the cytosol are salts, minerals, and organic molecules. The major organelles are summarized in Table 4-2, and those that are found in animal cells are illustrated in Figure 4-7.

TABLE 4-2 *Organelles*

Organelle	Function
Mitochondrion	transfers energy from organic compounds to ATP
Ribosome	organizes the synthesis of proteins
Endoplasmic reticulum (ER)	prepares proteins for export (rough ER); synthesizes steroids, regulates calcium levels, breaks down toxic substances (smooth ER)
Golgi apparatus	processes and packages substances produced by the cell
Lysosome	digests molecules, old organelles, and foreign substances
Microfilaments and microtubules	contribute to the support, movement, and division of cells
Cilia and flagella	propel cells through the environment; move materials over the cell surface
Nucleus	stores hereditary information in DNA; synthesizes RNA and ribosomes
Cell wall*	supports and protects the cell
Vacuole*	stores enzymes and waste products
Plastid*	stores food or pigments; one type (chloroplast) transfers energy from light to organic compounds

*Cell walls, large vacuoles, and plastids are found in the cells of plants and some other eukaryotes, but not in the cells of animals.

CRITICAL THINKING

Fluid Mosaic Model

The fluid mosaic model of the cell suggests that the cell membrane is a very dynamic structure. How might a cell benefit from having proteins in its membrane that are able to move laterally? (The proteins can migrate to an area of the membrane where they will be used.)

TEACHING STRATEGY

Observing a Fluid Mosaic Model

Explain to students that phospholipid heads are hydrophilic (attracted to water); in contrast, phospholipid tails attract each other and are hydrophobic (water repellent), forming the interior of the membrane. Have students make a fluid mosaic model by floating a thin layer of vegetable oil on water. Have them sprinkle on top a small amount of an herb, such as oregano, which should float and change position often. Tell students that proteins in a cell membrane appear to move in a similar way. Unlike the herb particles, however, cell membrane proteins are often restricted in their movements by interactions with molecules in the cytoplasm and the external environment of the cell. Also point out that some materials cannot get through the oil layer; like the cell membrane, the oil is semipermeable. You can demonstrate this by carefully placing small drops of colored water on top of the oil layer. The drops should stay on the oil without passing through to the water below.

SECTION 4-2

VISUAL LINK

Figure 4-7

This figure shows most of the organelles that are commonly found in animal cells. Explain that for the sake of clarity, only a few of the microtubules and microfilaments have been drawn. In reality, these structures usually form a dense network called the cytoskeleton inside the cell. Also point out that not all animal cells have flagella. Notice that the nuclear envelope actually consists of two lipid bilayer membranes. At certain points, these two membranes separate, and the outer membrane expands and folds to form the endoplasmic reticulum. Ask students how this cell would look different if it were a plant cell instead of an animal cell. (It would not have a flagellum, but it would have chloroplasts, one or more large vacuoles, and a cell wall outside the cell membrane.)

OVERCOMING MISCONCEPTIONS

Many students think that because plants perform photosynthesis, plant cells do not have mitochondria. It is important for students to understand that plants also carry out cellular respiration, a process that requires mitochondria.

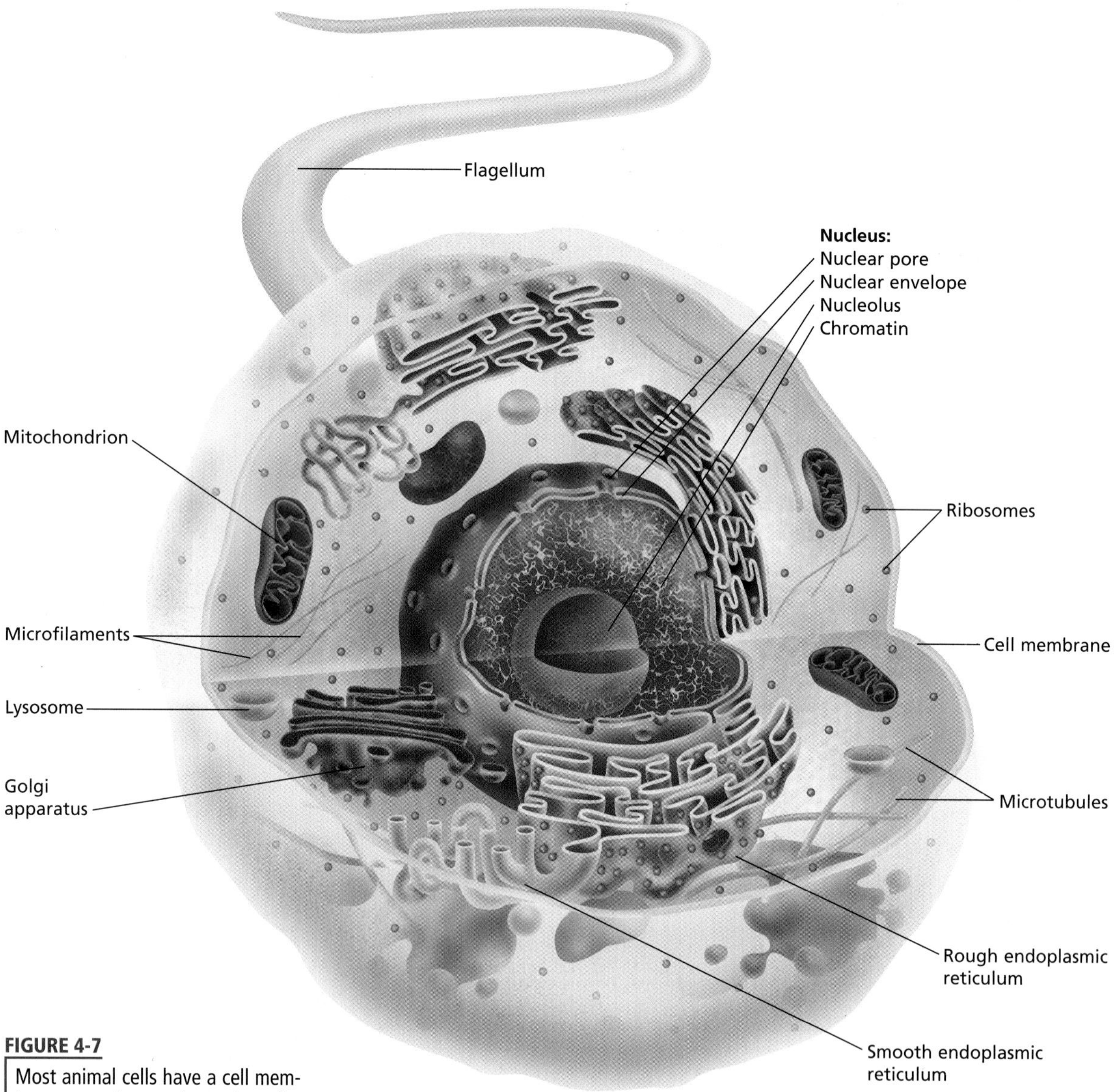

FIGURE 4-7

Most animal cells have a cell membrane, a nucleus, and a variety of other organelles.

Word Roots and Origins

mitochondrion

from the Greek *mitos,* meaning "thread," and *chondrion,* meaning "grain"

Mitochondria

Scattered throughout the cytosol are relatively large organelles called **mitochondria** (MIET-oh-KAHN-dree-uh), illustrated in Figure 4-8. Mitochondria are the sites of chemical reactions that transfer energy from organic compounds to ATP. Remember from Chapter 3 that ATP is the molecule that most cells use as their main energy currency. The energy of ATP ultimately drives most of the chemical reactions that occur in a cell. Therefore, mitochondria are usually more numerous in cells that have a high energy requirement. Liver cells, for instance, carry out a host of biochemical activities, and each cell may contain as many as 2,500 mitochondria. Muscle cells also contain many mitochondria.

If you look closely at Figure 4-8, you will notice that a mitochondrion is surrounded by two membranes. The smooth *outer* membrane serves as a boundary between the mitochondrion and

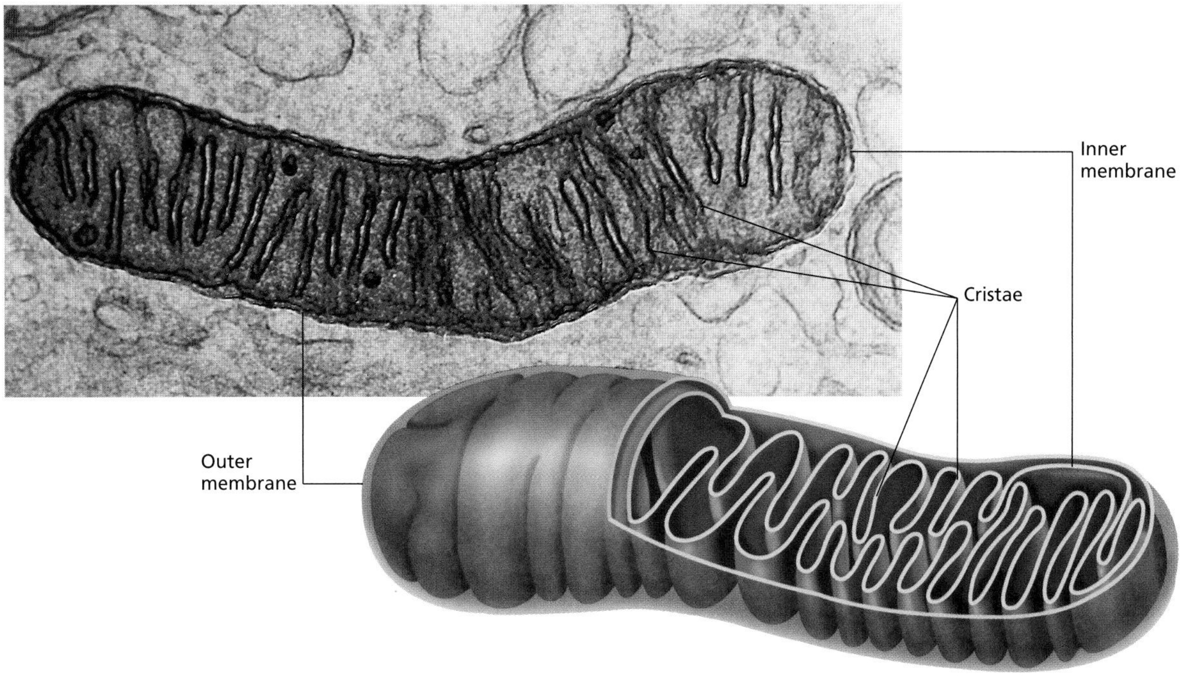

FIGURE 4-8

Mitochondria are surrounded by a double membrane. The inner membrane is composed of many folds called cristae. (TEM 232,000×)

the cytosol. The *inner* membrane has many long folds, known as **cristae** (KRIS-tee). The cristae greatly enlarge the surface area of the inner membrane, providing more space for the chemical reactions that occur in the mitochondrion.

Mitochondria have their own DNA, and new mitochondria arise only when existing ones grow and divide. These observations have led to a theory, discussed further in Chapter 15, that mitochondria developed from prokaryotic cells that lived inside eukaryotic cells. According to this theory, the prokaryotes may have gained protection by living inside the eukaryotes and, in turn, produced energy for the eukaryotes.

Ribosomes

The most numerous organelles in many cells are the **ribosomes** (RIE-buh-SOHMZ). Unlike most other organelles, ribosomes are not surrounded by a membrane. Each ribosome is an assemblage of two organic compounds—proteins and RNA. Inside the cell's nucleus, proteins and RNAs are packaged into ribosomes, which are then transported to the cytosol. Some ribosomes remain free within the cytosol, while others become attached to an organelle called the endoplasmic reticulum. You can see both free and attached ribosomes if you examine Figure 4-9 closely.

Ribosomes play important roles in the synthesis of proteins. Proteins to be used within the cytosol are produced on the ribosomes that are free in the cytosol. Proteins to be inserted into membranes or exported from the cell are produced on the ribosomes that are attached to the endoplasmic reticulum.

internetconnect

SCILINKS NSTA
TOPIC: Eukaryotic cells
GO TO: www.scilinks.org
KEYWORD: HM076

RECENT RESEARCH
Mitochondrial DNA

Researchers are using mitochondrial DNA (mtDNA) to study human evolution. Since mtDNA mutates more quickly than nuclear DNA, it serves as a kind of evolutionary clock. Also, mtDNA is inherited from the mother, eliminating the possibility that changes in this DNA might be the result of sexual recombination. One study found that people of various ethnic groups had a common ancestor 200,000 years ago. Another study found that people living in various regions of the world had common African ancestors about 200,000 years ago and that the original population left Africa about 100,000 years ago.

MAKING CONNECTIONS
Social Studies

In 1992, the Vietnamese government turned over bone fragments of Americans who had died in the Vietnam War. Using mitochondrial DNA, scientists at the Armed Forces Institute of Pathology had been able to identify 37 missing war dead by 1996 and were working to identify another 100 missing people.

CRITICAL THINKING
Mitochondrial Folds

How effective would a mitochondrion be without cristae? (With less surface area to carry out its chemical reactions, it would be much less efficient.) What would happen to an organism if the mitochondria of its cells were suddenly to become half as efficient? (There would be a slowing down of all life processes that require energy.)

RECENT RESEARCH
Packaging ER Proteins

Proteins made in the ER are placed in sacs called vesicles before they are taken to other parts of the cell. Recently researchers discovered how the ER segregates proteins into vesicles and how the vesicles are directed to specific places in the cell. They found that vesicle formation involves a complex addition and deletion of surface proteins and guanosine triphosphate (GTP), an energy-carrying molecule that is much like ATP. The type of proteins on the vesicle determines the vesicle's contents and plays a role in directing the vesicle toward its destination in the cell.

GIFTED *ACTIVITY*

In 1898, the Italian microscopist Camillo Golgi developed a silver stain for use with the light microscope and discovered the organelle that was later named after him. However, the Golgi apparatus was not seriously studied until the development of the electron microscope in the 1950s. Have students use library and online resources to collect electron micrographs and learn about the Golgi apparatus in greater depth.

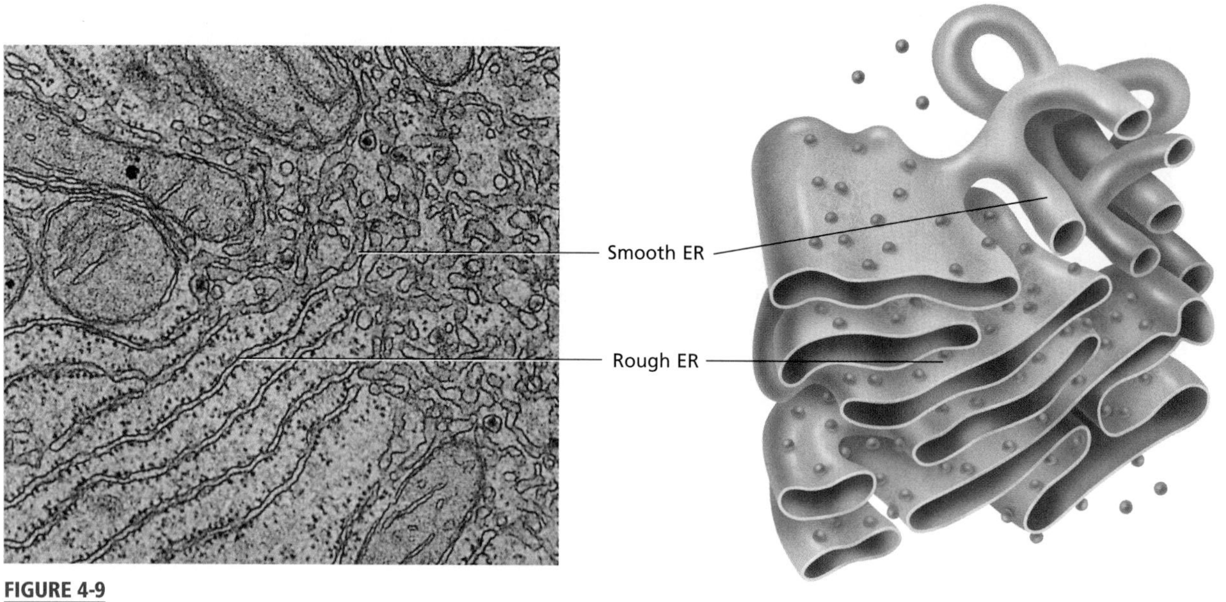

FIGURE 4-9
The small dark dots in this cell are ribosomes. Some are free in the cytosol, while others are attached to the rough ER. The smooth ER lacks attached ribosomes. (TEM 240,000×)

Endoplasmic Reticulum

The **endoplasmic reticulum** (EN-doh-PLAZ-mik ri-TIK-yuh-luhm), abbreviated ER, is a system of membranous tubules and sacs. The dark lines that you see in Figure 4-9 represent the membranes of the ER, while the lighter areas are the channels inside it. The ER functions primarily as an intracellular highway, a path along which molecules move from one part of the cell to another. The amount of ER inside a cell fluctuates, depending on the cell's activity.

A cell usually contains two types of ER, both of which are shown in Figure 4-9. One type appears to be covered with dark dots when viewed with an electron microscope. These dots are ribosomes, and they give the ER a rough appearance. Consequently, this type is known as **rough endoplasmic reticulum,** or rough ER. Rough ER is prominent in cells that make large amounts of proteins to be exported from the cell or inserted into the cell membrane.

The second type of ER is not covered with ribosomes. Because of their absence, this type of ER appears smooth and is therefore called **smooth endoplasmic reticulum,** or smooth ER. Smooth ER is involved in the synthesis of steroids in gland cells, the regulation of calcium levels in muscle cells, and the breakdown of toxic substances by liver cells. As you can see in Figure 4-9, rough ER and smooth ER may be continuous with each other.

Golgi Apparatus

The **Golgi** (GOHL-jee) **apparatus** is the processing, packaging, and secreting organelle of the cell. Like the endoplasmic reticulum, the Golgi apparatus is a system of membranes. Figure 4-10 shows that the Golgi apparatus appears as a series of flattened sacs with a characteristic convex shape in the cytosol. Working in close association with the endoplasmic reticulum, the Golgi apparatus modifies proteins for export by the cell. The role of the Golgi apparatus in protein synthesis will be discussed in Chapter 11.

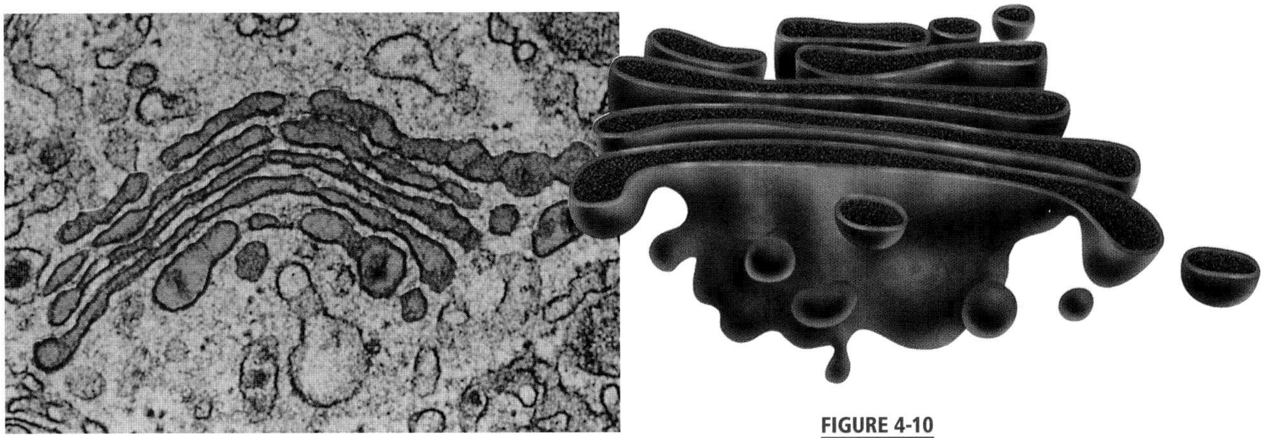

FIGURE 4-10

A collection of smooth, membrane-bound sacs isolated from the endoplasmic reticulum is known as the Golgi apparatus. Although separated from each other, the endoplasmic reticulum and Golgi apparatus work closely together in preparing materials for release by the cell. (TEM 237,250×)

Lysosomes

Lysosomes (LIE-suh-sohmz) are small, spherical organelles that enclose hydrolytic enzymes within single membranes. These enzymes can digest proteins, carbohydrates, lipids, DNA, and RNA. They may also digest old organelles as well as viruses and bacteria that have been ingested by a cell. Lysosomes are common in the cells of animals, fungi, and protists, but they are rare in plant cells. In some multicellular organisms, lysosomes play a role during early development. For example, the human hand begins as a solid structure in the embryo. As the embryo develops, lysosomal enzymes selectively destroy tissue to form the spaces between the fingers.

Cytoskeleton

Just as your body depends on your skeleton to maintain its shape and size, so a cell needs a structure to maintain its shape and size. In many cells, that structure is the **cytoskeleton,** a network of long protein strands located in the cytosol. Like ribosomes, these strands are not surrounded by membranes. In addition to providing support, the cytoskeleton participates in the movement of organelles within the cytosol. Two major components of the cytoskeleton are **microfilaments** and **microtubules.**

Microfilaments are threads made of a protein called **actin.** Each microfilament consists of many actin molecules that are linked together to form a polymer chain. Microfilaments constitute the smallest strands that make up the cytoskeleton. They contribute to cell movement and play a role in the contraction of muscle cells.

The largest strands of the cytoskeleton are hollow tubes known as microtubules. In many cells, microtubules extend outward from a central point near the nucleus to various sites near the cell membrane. When a cell is about to divide, bundles of microtubules come together and extend across the cell. These bundles, known as **spindle fibers,** are thick enough to be visible with a light microscope, as you can see in Figure 4-11. Spindle fibers assist in the movement of chromosomes during cell division. When cell division is complete, the spindle fibers are disassembled, and the microtubules resume their task of providing support to the cell.

FIGURE 4-11

This cell is dividing in two. Its chromosomes, stained orange by a fluorescent dye, are moving to opposite ends of the cell. They are being pulled by the spindle fibers, which are stained green. (LM 3,696×)

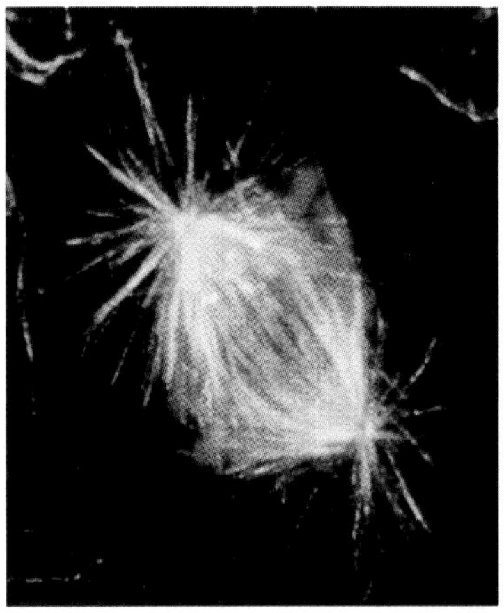

SECTION 4-2

QUICK FACT

☑ Numerous diseases have been attributed to improperly functioning lysosomes. Hurler's disease, characterized by abnormalities of the skeleton and soft tissues, is caused by an excess of hydrolase, an enzyme found in lysosomes. Pompe's disease results when a lysosomal enzyme that breaks down glycogen is absent, causing liver damage.

GIFTED *ACTIVITY*

Have students research the role of the lysosome in diseases such as systemic lupus erythematosus.

QUICK FACT

☑ Tar and hydrogen cyanide (HCN) are among the many thousands of compounds produced when tobacco is burned. While tar increases the production of mucus by the lungs, HCN kills the cilia that would normally push the mucus to the pharynx, where it would be swallowed. The result is a persistent cough, which forces the accumulated mucus out of the trachea.

TEACHING STRATEGY

Observing Cheek Cell Nuclei

Provide students with prepared slides of stained cheek cells from a commercial supplier. Have students observe the cells with a compound microscope. What is the shape of the cheek cells? (The cells are flat and tilelike. Their shape is related to their function, which is to protect underlying cells from friction.) Where is the nucleus located? (It should be close to the center of the cell.) Tell students that chromosomes are too small to see in this kind of preparation.

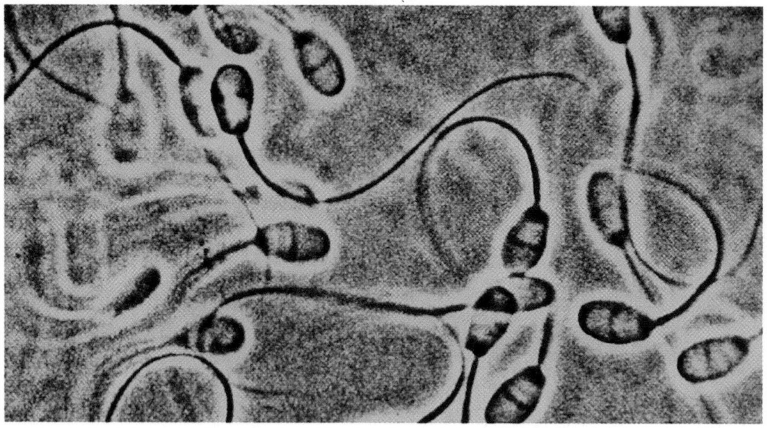

FIGURE 4-12

Sperm cells propel themselves by moving a long flagellum back and forth. (LM 3,350×)

Cilia and Flagella

Cilia (SIL-ee-uh) and **flagella** (fluh-JEL-uh) are hairlike organelles that extend from the surface of the cell, where they assist in movement. Because of the variety of roles they play, cilia and flagella can be found in many eukaryotic cells.

When these organelles are short and present in large numbers on a cell, they are called cilia. The external surfaces of many unicellular organisms are covered with cilia. The movements of the cilia propel these tiny organisms through the water as they search for food or escape from predators. Cilia are also found on the surfaces of cells in multicellular organisms. The cells lining your respiratory tract, for example, bear numerous cilia that trap particles and debris from the air you inhale. As these cilia move, they sweep the trapped materials back up to your throat, where they are removed from your respiratory tract when you swallow.

When the hairlike organelles are long and less numerous on a cell, they are called flagella. On many cells, including the sperm cells illustrated in Figure 4-12, only one flagellum is present. By whipping back and forth, flagella can swiftly propel unicellular organisms or specialized cells in multicellular organisms, such as sperm cells.

Cilia and flagella have a similar internal structure. Notice in Figure 4-13 that both organelles are composed of nine pairs of microtubules arranged around a central pair.

FIGURE 4-13

Microtubules are important components of cilia and flagella. A cross section of a cilium shows that it consists of nine pairs of microtubules that surround a central pair. A flagellum has a similar structure. (TEM 38,000× left, 396,000× right)

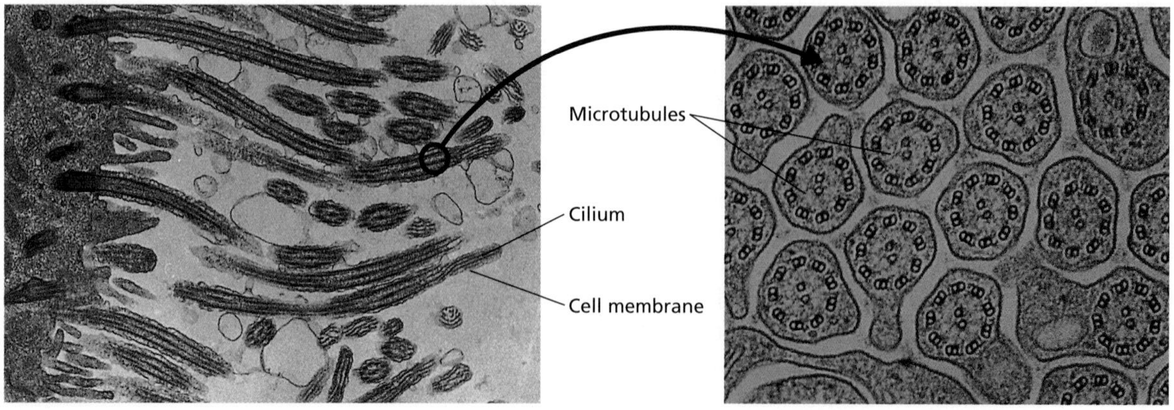

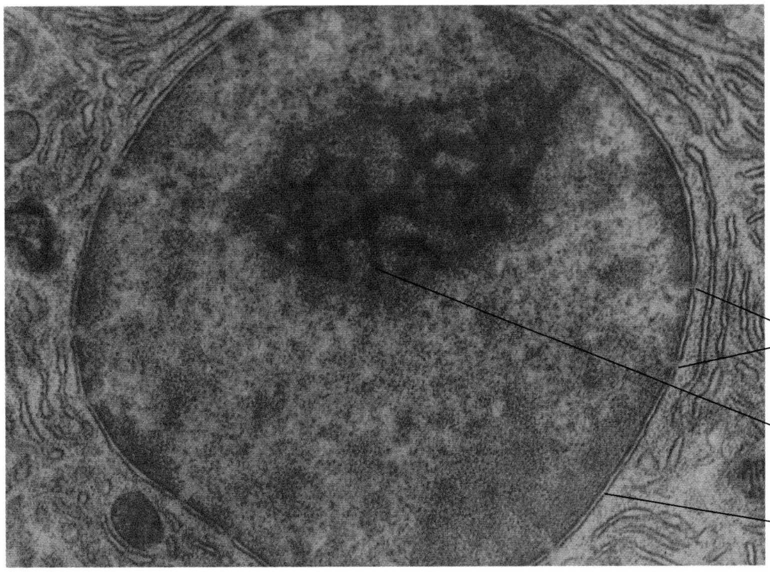

FIGURE 4-14

The most prominent organelle in most eukaryotic cells is the nucleus, which is surrounded by a double membrane perforated with pores. These pores allow materials to be exchanged between the nucleus and the cytosol. Inside the nucleus is the nucleolus, where ribosomes are made before they are transported to the cytosol. (TEM 360,734×)

NUCLEUS

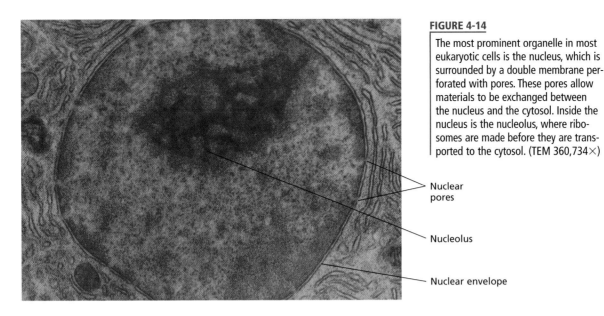

The nucleus is often the most prominent structure within a eukaryotic cell. It maintains its shape with the help of a protein skeleton known as the **nuclear matrix.** As indicated in Figure 4-14, the nucleus is surrounded by a double membrane called the **nuclear envelope.** Inside the nuclear envelope are fine strands of **chromatin,** a combination of DNA and protein. When a cell is about to divide, the chromatin strands coil up and become densely packed, forming **chromosomes.**

The nucleus stores hereditary information in its DNA. The nucleus is also the site where RNA is copied from DNA. In turn, RNA directs the synthesis of proteins, a process that occurs in the cytosol, as you have read. This means that RNA must travel from the nucleus to the cytosol before it can direct protein synthesis. RNA makes this journey by passing through **nuclear pores,** small holes in the nuclear envelope. Most nuclei also contain at least one spherical area called the **nucleolus** (noo-KLEE-uh-luhs). The nucleolus is the site where ribosomes are synthesized and partially assembled before they pass through the nuclear pores to the cytosol. Both the nuclear pores and the nucleolus are visible in Figure 4-14.

Word Roots and Origins

chromosome

from the Greek *chroma,* meaning "color," and *soma,* meaning "body"

internetconnect

SCILINKS NSTA
TOPIC: Plant cells
GO TO: www.scilinks.org
KEYWORD: HM081

PLANT CELLS

Most of the organelles and other parts of the cell just described are common to all eukaryotic cells. However, plant cells may have three additional kinds of structures—cell walls, vacuoles, and plastids—that are extremely important to plant function.

DEMONSTRATION

Role of Cell Wall

To graphically show how a plant cell wall provides strength, inflate a balloon inside a small cardboard box. The balloon represents the cell membrane and the box represents the cell wall. Filling the balloon with air is analogous to filling the vacuole with water.

SECTION 4-2

INCLUSION ACTIVITY

Instruct students to use food to make a model of a eukaryotic cell. Gelatin (cytosol) could be poured into a large, clear cellophane bag (cell membrane). Organelles could be represented by kidney beans (mitochondria), poppy seeds (ribosomes), sections of lasagna noodles (ER and Golgi apparatus), peppercorns (lysosomes), spaghetti noodles (microfilaments and microtubules), and a whole orange (nucleus). To make the model represent a plant cell, peas (chloroplasts) and smaller, sealed cellophane bags full of water (vacuoles) could be added.

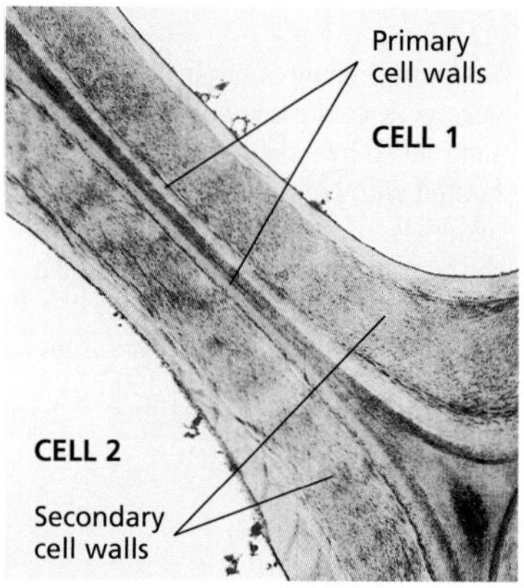

FIGURE 4-15
The two plant cells in this photograph each have their own primary and secondary cell walls. The primary walls are constructed first. Later, the secondary walls are formed inside the primary walls. (TEM 12,750×)

Cell Wall

Plant cells are covered by a rigid **cell wall** that lies outside the cell membrane. The rigidity of cell walls helps support and protect the plant. Cell walls contain long chains of cellulose, one of the complex carbohydrates you read about in Chapter 3. The cellulose is embedded in proteins and other carbohydrates that harden the entire structure. Pores in the cell wall allow ions and molecules to enter and exit the cell.

Figure 4-15 shows that cell walls are of two types—primary and secondary. While a plant cell is being formed, a primary cell wall develops just outside the cell membrane. As the cell expands in length, cellulose and other molecules are added, enlarging the cell wall. When the cell reaches its full size, a secondary cell wall may develop. As you can see in Figure 4-15, the secondary cell wall develops between the primary cell wall and the cell membrane. The secondary cell wall is tough and woody. Therefore, once it is completed, a plant cell can grow no further. When you pick up a piece of wood, you are holding secondary cell walls. The cells inside the walls have died and disintegrated.

Vacuoles

Vacuoles are a second common characteristic of plant cells. These fluid-filled organelles store enzymes and metabolic wastes. Often, as Figure 4-16 shows, they are quite large. In fact, some vacuoles may occupy 90 percent of a plant cell's volume, pushing all of the other organelles up against the cell membrane. Some of the wastes stored by vacuoles are toxic and must be kept away from the rest of the cell. The storage of these materials may be beneficial to a plant in other ways. For instance, the poisons that certain acacia trees have in their vacuoles provide a defense against plant-eating animals.

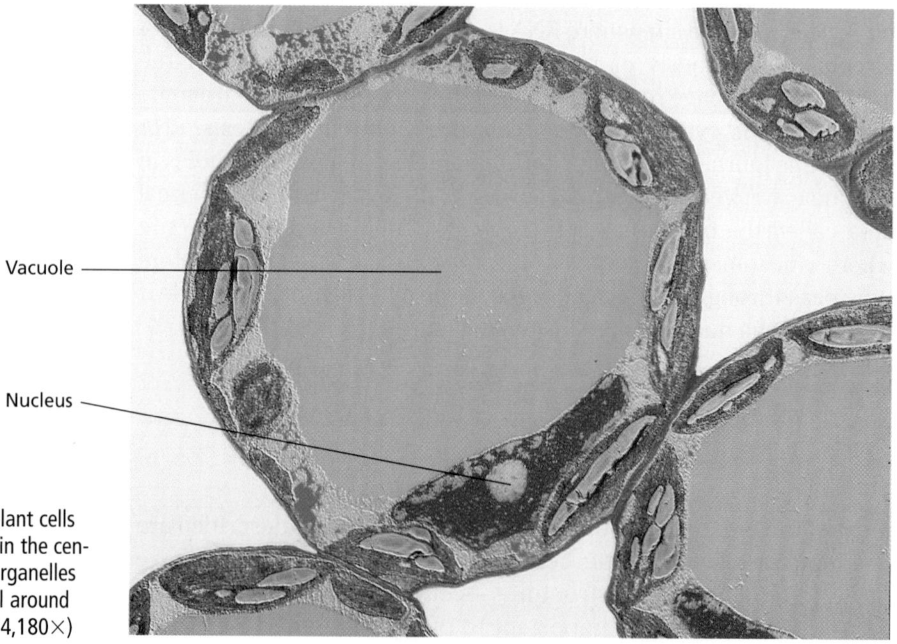

FIGURE 4-16
Much of the volume of these plant cells is occupied by a large vacuole in the center of the cell. The rest of the organelles are confined to a rim of cytosol around the periphery of the cell. (TEM 4,180×)

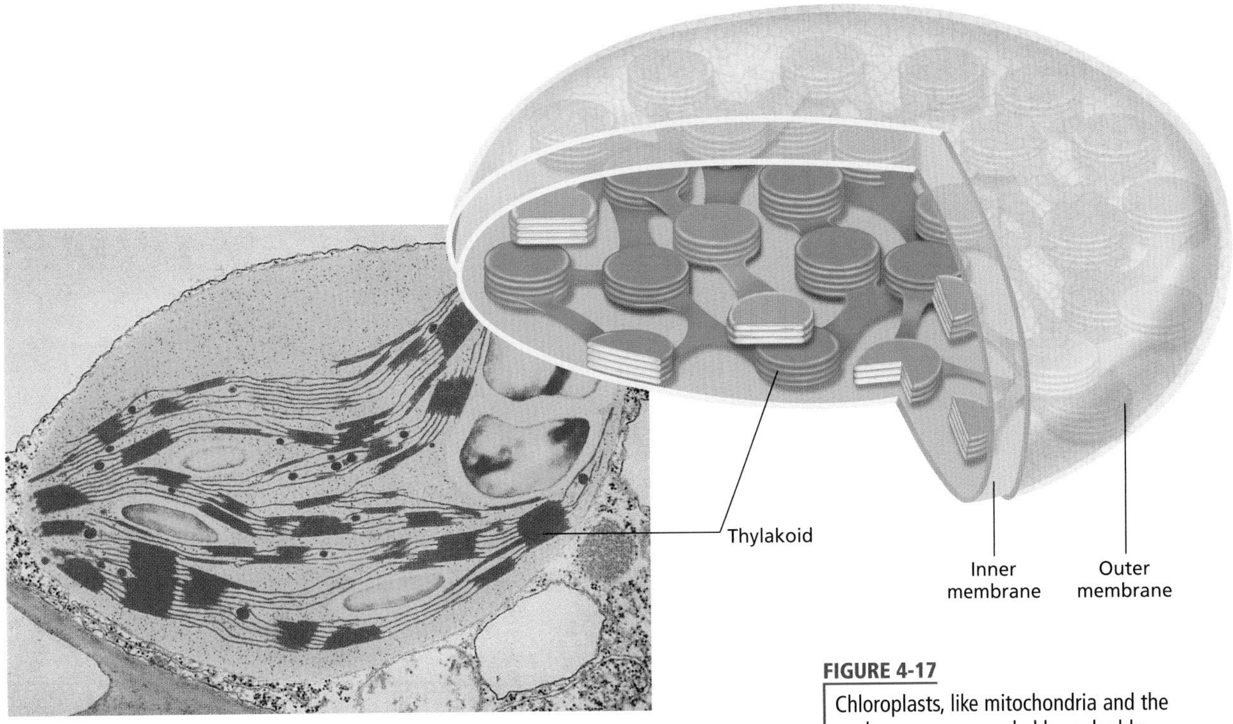

FIGURE 4-17

Chloroplasts, like mitochondria and the nucleus, are surrounded by a double membrane. The thylakoids inside each chloroplast contain pigments involved in photosynthesis. (TEM 14,648×)

Plastids

A third distinguishing feature of plant cells is the presence of **plastids.** Plastids are organelles that, like mitochondria and the nucleus, are surrounded by two membranes and contain DNA. Some plastids store starch or fats, while others contain compounds called pigments, which absorb visible light.

The most familiar type of plastid is the **chloroplast,** illustrated in Figure 4-17. Each chloroplast encloses a system of flattened, membranous sacs called **thylakoids.** Chloroplasts are the organelles in a plant cell in which the energy of sunlight is converted into chemical energy in organic compounds. This conversion occurs in the thylakoids during the process of photosynthesis, which you will study in Chapter 6. Chloroplasts contain large amounts of a green pigment that gives leaves their green color. Different pigments in other types of plastids are responsible for the colors of fruits and flowers.

SECTION 4-2 REVIEW

1. **Name the three main components of a eukaryotic cell.**
2. **Summarize the fluid mosaic model of the cell membrane.**
3. **Distinguish between the nucleus and the nucleolus.**
4. **Which organelles are surrounded by two membranes?**
5. **What structural feature do cilia and flagella have in common?**
6. **CRITICAL THINKING Plant cells have cell walls, but animal cells do not. Why do you think that is so?**

INCLUSION *ACTIVITY*

Ask students to list all the places where membranes are found in cells. (Membranes are found around the entire cell, mitochondria, ER, Golgi apparatus, lysosomes, and nucleus, as well as over cilia and flagella. In plant cells, membranes also surround vacuoles and plastids.) Then ask students to list the parts of a cell that lack membranes. (Ribosomes, microfilaments, microtubules, and cell walls lack membranes.)

ANSWERS TO SECTION 4-2 REVIEW

1. The three main parts of a eukaryotic cell are the cell membrane, nucleus, and other organelles.
2. The protein molecules form patterns, or mosaics, in and on the lipid bilayer of the membrane. Lipids and some proteins move laterally within the bilayer, so the patterns are constantly changing.
3. The nucleolus lies inside the nucleus. The nucleolus is where the ribosomes are synthesized and partially assembled. The nucleus is where DNA directs the synthesis of RNA.
4. Mitochondria, the nucleus, and plastids (in plant cells) are surrounded by two membranes.
5. Both cilia and flagella are composed of nine pairs of microtubules arranged around a central pair.
6. Cell walls may add strength and stability to plants, but they make them immotile. All but a few animals require motility to obtain the nutrients they need to survive.

CHAPTER 4
GREAT DISCOVERIES

BACKGROUND

Anton van Leeuwenhoek was born on October 24, 1632, in Delft, Netherlands. As a young man, van Leeuwenhoek sold cloth for a living, but his real interests lay in making lenses and using them to study tiny objects. He became so engrossed with making the "perfect lens" that he neglected his business, triggering the derision of his family and neighbors.

Scientists of van Leeuwenhoek's time had to overcome what we now call pseudoscience. For example, many people believed diseases were caused by evil spirits that had entered the body. The risks of engaging in scientific thinking and observation were high. Servetus had been burned to death for performing an autopsy, and Galileo had been imprisoned for life for showing that the sun was the center of the solar system. Despite these risks, van Leeuwenhoek continued his observations on a variety of microscopic organisms and parts of animals and plants.

A friend convinced van Leeuwenhoek to write to the British Royal Society in 1673 and describe what he had seen with his microscope. His first communication was titled, "A Specimen of some Observations made by a Microscope contrived by Mr. Leeuwenhoek, concerning Mould upon the Skin, Flesh, etc.; the Sting of a Bee, etc." What van Leeuwenhoek detailed so amazed the members of the Royal Society that they requested more information. He responded by sending more than 300 letters to the Royal Society over the next half-century, contributing to the birth of microscopy, microbiology, and cell biology.

GREAT DISCOVERIES

Discovering a New World

HISTORICAL PERSPECTIVE

The first microscopes, made around 1600, had a single lens and were much like a powerful magnifying glass. Compound microscopes, those with two or more lenses, produced greater enlargement but caused blurred images. This problem, known as chromatic aberration, was not solved until more than a century later. Meanwhile, Anton van Leeuwenhoek, who lived in Holland from 1632 to 1723, made hundreds of simple, high-quality microscopes. He was the first person to view and describe the amazing miniature world of protozoa and bacteria.

Toy or Tool?

About 25 years after van Leeuwenhoek began his work with microscopes, Robert Hooke, a fellow scientist, noted that few scientists besides van Leeuwenhoek considered the microscope an essential tool. "I hear of none that make any other use of that instrument, but for diversion and pastime," Hooke wrote. *Diversion* and *pastime* are not words we usually connect with the microscope, a tool so basic to scientific work today that we can hardly imagine scientists, or students of science, being without one.

Why weren't more scientists using microscopes? The main reason is that early microscopes were of poor quality, and only an unusually diligent and persistent person could find them useful. Van Leeuwenhoek was such a person.

A linendraper by trade, van Leeuwenhoek spent most of his time on his hobby—science—and for him the microscope was a valuable instrument. He ground his own high-quality lenses, more than 400 in his lifetime, and mounted them between thin brass plates. They magnified objects by 50 to 300 times, and van Leeuwenhoek never tired of peering through the lenses and making discoveries about animals, plants, and microorganisms.

Anton van Leeuwenhoek

Over a 50-year period beginning in 1673, van Leeuwenhoek sent letters to the British Royal Society describing his microscopic observations. Most of his letters were published in the society's journal, *Philosophical Transactions.*

Van Leeuwenhoek's first letter described what he called *animalcules,* or "tiny animals," which we now know as protozoa. He explained that he had found them

> *"in rain, which had stood but a few days in a new tub. . . . I saw, after divers observations, that the bodies consisted of 5, 6, 7, or 8 very clear globules."*

He said his animalcules sometimes "stuck out two little horns, which were continually moved, after the fashion of a horse's ears." Later, scientists would name them *Vorticella* and identify their "little horns" as *cilia,* or tiny hairs.

Father of Microbiology

Van Leeuwenhoek examined hundreds of objects, including the lenses of eyes, the striations of muscles, the mouthparts of insects, the detailed structures of plants, and many protozoa and bacteria found in rainwater, pond water, well water, saliva, and other liquids. He was the first person to accurately describe red blood cells.

Van Leeuwenhoek also examined human and animal spermatozoa and made correct guesses about the reproductive process of animals, even though fertilization would not be observed under the microscope until the 1800s. His observations disproved the long-held theory of spontaneous generation—the idea that life could grow out of nonliving substances. Some believed, for example, that fleas grew out of sand or dust, but van Leeuwenhoek refuted that idea in his letter to the Royal Society by describing the flea's development with great precision.

Van Leeuwenhoek's remarkable powers of observation enabled him to make accurate judgments not only of an object's structure but also of its size. For example, he thought 100 human red blood cells in a row would be nearly the width of a grain of coarse sand. That made each red cell about 8.5μm (0.0003 in.) in diameter, close to its actual measurement.

In addition to writing about his observations, van Leeuwenhoek drew sketches of what he saw. In fact, he was the first person to draw an image of bacteria; the drawing was published in *Philosophical Transactions* in 1683. Because of his studies of microorganisms, van Leeuwenhoek is often called the father of microbiology.

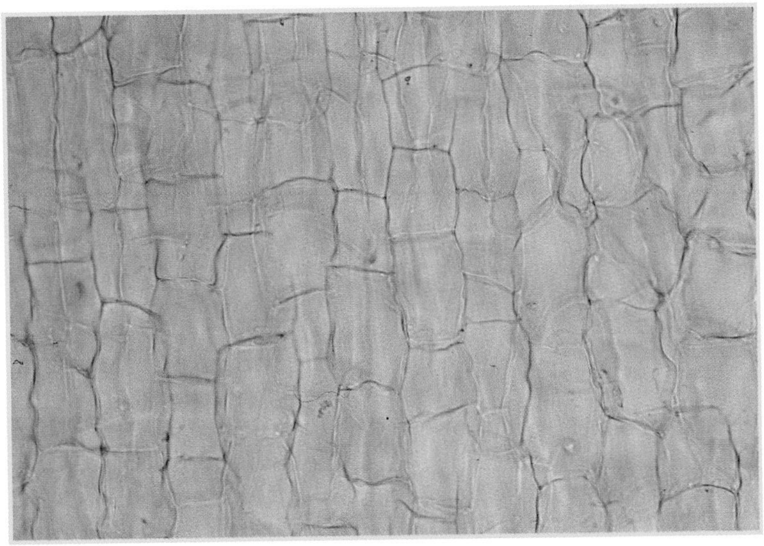

In 1674, van Leeuwenhoek sent a piece of cork, along with other samples, to the British Royal Society. This photograph shows what the piece of cork would have looked like through van Leeuwenhoek's most powerful microscope.

After van Leeuwenhoek

Even though van Leeuwenhoek had published reports about his many discoveries, the scientific community failed to recognize the microscope's significance for many years. In 1733, an amateur optician named Chester Moor Hall found a way to solve the persistent problem of chromatic aberration in compound lenses, and in 1774, the technique was applied to microscopes. By the 1820s, new types of microscopes were available, and the 1800s and 1900s brought frequent improvements. In 1931, the first electron microscope was invented, and in 1981, the scanning tunneling microscope began to reveal objects atom by atom. Today's work in microbiology, protozoology, bacteriology, and other fields depends on advanced microscopes, descendants of those simple lenses through which van Leeuwenhoek first saw his "animalcules."

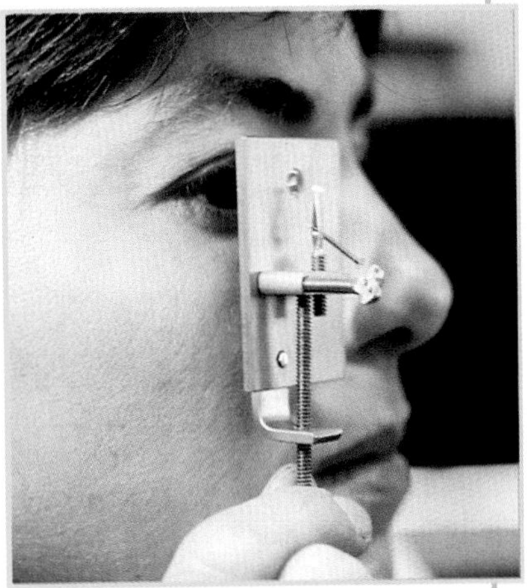

Van Leeuwenhoek's microscope consisted of a single lens. He placed a small drop of liquid on the tip of a fine point and peered through the lens to observe the miniature world hidden in the liquid. By turning a screw, van Leeuwenhoek could move the point closer to the lens, bringing objects in the liquid into focus.

DISCUSSION

Guide the discussion by posing the following questions.

1. Why was the idea of spontaneous generation such a long-held belief? (Before van Leeuwenhoek perfected his microscopes, it was impossible to observe anything as small as the egg of a flea, so it appeared that fleas developed spontaneously from nonliving material.)

2. Why was it rare for scientists of van Leeuwenhoek's time to use microscopes? (Early microscopes were of poor quality, since it was difficult to grind lenses precisely. Van Leeuwenhoek ground his own high-quality lenses, which made his microscopic observations possible.)

3. Van Leeuwenhoek referred to the protozoans he observed as *animalcules,* or "tiny animals." The suffix *-cule* means tiny or small. What other words do you know that have the same suffix, and what do they mean? (Molecule, which means a small mass; minuscule, which means very small; spicule, which means a small, needlelike structure.)

FURTHER READINGS

1. "One Molecule at a Time," in *Discover,* January 1996, p. 72. This article describes a technique of triggering chemical reactions molecule by molecule through the use of an atomic force microscope.

2. "Jolting Crystals into New Nanostructures," in *Science News,* November 2, 1996, p. 276. Researchers at Harvard University have discovered that applying voltage to the tip of a scanning tunneling microscope can alter the surface structure of a nearby target specimen. The electric pulse creates a very small crystalline island with an atomic arrangement different from that of the rest of the material.

SECTION 4-3

SECTION OVERVIEW

Matter, Energy, and Organization
The organization of multicellular organisms into tissues, organs, and organ systems is described.

Evolution
This section also mentions the evolution of eukaryotic cells from prokaryotes and describes how colonial organisms may have evolved from unicellular eukaryotes.

▶ ENGAGE STUDENTS

Collect water from a pond or stream, along with rotting leaves and mud, and store it in a jar. Alternatively, pull some grass out of the ground, complete with roots, and place it in a jar of water for 48 hours prior to this demonstration. Have students use an eyedropper to take a sample of water from the jar and place a drop on a glass slide. Have them observe the drop under a compound microscope. Students are most likely to see protozoa and algae, although bacteria and other organisms will also be present. Are the organisms motile? (Many of them will be.) Direct students to locate a unicellular organism, and ask them if they can see any internal structures. (Internal structures will be hard to see, especially if the organisms are moving. High magnification and proper illumination with high contrast will likely be necessary.)

SECTION 4-3

OBJECTIVES

▲ Distinguish between tissues, organs, and organ systems.

● Describe the features of a colonial organism.

MULTICELLULAR ORGANIZATION

In a unicellular organism, one cell carries out all of the functions of life. In contrast, most cells in a multicellular organism are specialized to perform one or a few functions. Because of cell specialization, the cells of multicellular organisms depend on other cells in the organism for their survival.

TISSUES, ORGANS, AND ORGAN SYSTEMS

In most multicellular organisms, cells are organized into **tissues,** or groups of cells that carry out a specific function. In animals, epithelial tissue consists of sheets of closely packed cells that form surface coverings, such as the outermost living layer of the skin and the inside lining of the nose. The loosely scattered cells of connective tissue serve mainly to support and link together other tissues. Cells that pull against one another by contracting make up muscle tissue. Cells that are specialized for transmitting messages rapidly make up nervous tissue.

Several types of tissues that interact to perform a specific function form an **organ.** The stomach is one example of an organ. In the

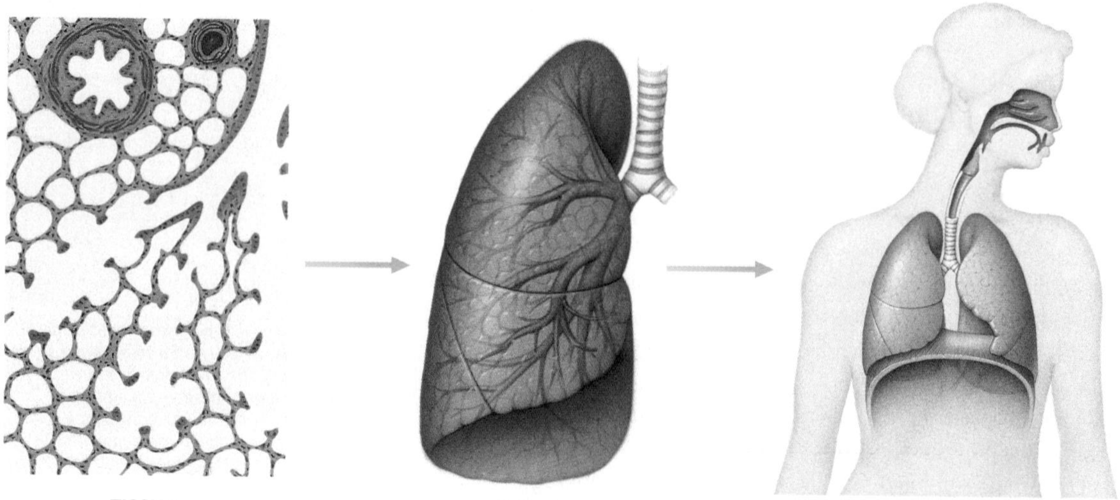

FIGURE 4-18
Spongy tissue makes up the air sacs that in turn make up the lung, which is an organ. The lungs are part of the respiratory system.

stomach, muscle tissue causes movement, epithelial tissue secretes enzymes, connective tissue holds the stomach together, and nervous tissue transmits messages back and forth between the stomach and the brain. All but the simplest animals have organs.

An **organ system** is made up of a group of organs that work together to perform a set of related tasks. For instance, the mouth, esophagus, stomach, intestines, and several other organs make up the digestive system. Each of these organs performs a specific function in the complex process of digesting food. Figure 4-18 illustrates the relationship between tissues and organs in another organ system, the respiratory system.

The different organ systems in a multicellular organism interact to carry out the processes of life. The digestive system, for example, extracts nutrients from food, while the respiratory system obtains oxygen from the environment and eliminates waste carbon dioxide. None of the body's organ systems could survive without the others.

Plants also have tissues and organs, although they are arranged somewhat differently from those in animals. A dermal tissue system forms the outer layer of a plant. A ground tissue system makes up the bulk of roots and stems. A vascular tissue system transports water throughout the plant. The four plant organs are roots, stems, leaves, and flowers.

EVOLUTION OF MULTICELLULAR ORGANIZATION

Fossil evidence suggests that the earliest cells on Earth were simple prokaryotes similar to some present-day bacteria. Like most bacteria, they lacked the internal structures required to synthesize their own nutrients. Instead, they depended on organic nutrients in the environment. As they reproduced and their numbers increased, however, they began to compete for limited environmental resources. Because of the increasing competition, cells with adaptations evolved. Some of these cells were eukaryotic. You will learn more about the evolution of prokaryotes and eukaryotes in Chapter 15.

Colonial Organizations

Eventually, some of the early unicellular eukaryotes may have begun to live in temporary groups, or colonies, with other cells of the same kind. Some of the cells in these colonies may have specialized in performing certain functions, such as converting energy. Biologists refer to such associations of cells as **colonial organisms.** A colonial organism is a collection of genetically identical cells that live together in a closely connected group.

An example of a colonial organism that exists today is the green alga *Volvox,* shown in Figure 4-19. A hollow *Volvox* sphere contains

Eco Connection

The Impact of Single Cells

Most people are familiar with the kinds of effects humans and other multicellular organisms can have on their environment. But single-celled organisms can also produce major environmental changes through their activities. One of the most important examples of such a change occurred about 2.8 billion years ago, when certain types of bacteria began producing oxygen as a byproduct of photosynthesis. As oxygen entered the Earth's atmosphere, the composition of the atmosphere changed dramatically. That set the stage for the evolution of animals and other organisms that require oxygen to live.

Even today, unicellular organisms continue to have a profound ecological impact. For example, about half the organic material produced through photosynthesis is made by single-celled eukaryotes known as algae. These simple but important organisms provide food for countless other organisms in marine and freshwater environments.

QUICK FACTS

☑ Prokaryotes evolved about 3.5 billion years ago, whereas the first eukaryotic cells did not appear until less than 1.5 billion years ago.

☑ In simple animals, some organ systems take on multiple functions that are divided among different organ systems in more-complex animals. In planarians, for example, the multibranched digestive system also circulates materials throughout the body.

INCLUSION *ACTIVITY*

Provide students with terms such as *kidney, epithelial tissue,* and *nervous system,* and have the students show their understanding of hierarchical relationships by stating the level (organ, tissue, organ system) to which each term belongs.

RETEACHING *ACTIVITY*

Draw an analogy between a multicellular organization and a school. Each student represents a cell; all students in one grade level represent a tissue; the school represents an organ; and all of the schools in the district represent an organ system. Have students devise other analogies, such as house, town, state, and country.

GIFTED *ACTIVITY*

Have students use on-line and library resources to research current cell work. Suggest Lynn Margulis's endosymbiosis hypothesis of cell evolution. Ask students to focus on the role of organelles.

CONTENT CONNECTION

Section 2-2: Solutions
Since the cytosol has numerous substances dissolved in it, it qualifies as a solution. However, the cytosol is also an example of a colloid. Like other colloids, the cytosol can act as a fairly firm gel, allowing the cell to maintain its shape. (Microfilaments and microtubules are also important in this function.) The cytosol can also act as a more fluid colloid, giving some cells the ability to assume different shapes.

ANSWERS TO SECTION 4-3 REVIEW

1. A tissue is a group of cells that carry out a specific function.
2. An organ consists of several types of tissue that interact to perform a specialized function.
3. One example would be the digestive system, which consists of the mouth, esophagus, stomach, and intestines.
4. Plant tissue systems include dermal, ground, and vascular. The plant organs are roots, stems, and leaves.
5. Each *Volvox* cell individually maintains its own life, but some cells depend on others for movement and reproduction.
6. Scientists classify *Codium* as multicellular. However, one could argue that it is a unicellular organism with multiple nuclei.

FIGURE 4-19
New *Volvox* colonies are formed in the interior of older ones and are released when the old colony ruptures. (LM 105×)

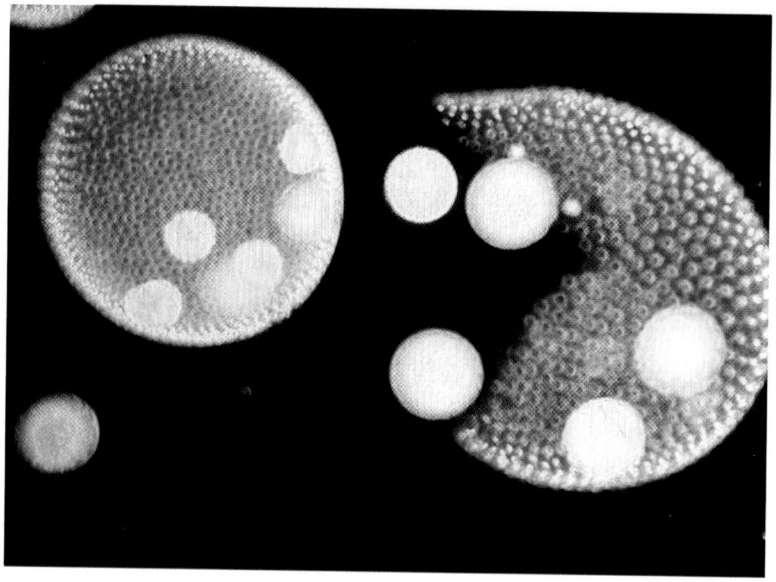

500 to 60,000 cells, each of which maintains its own individual existence. Many of the cells, though, carry out specific functions that benefit the whole colony. The outer cells use their flagella to propel the colony through the water. A few of the other cells are specialized for reproduction. They produce offspring colonies, which you can see in Figure 4-19 as the large green spheres contained within the main sphere.

Colonial organisms, such as *Volvox,* appear to straddle the border between a collection of unicellular organisms and a true multicellular organism. Although they lack tissues and organs, they exhibit the principle of cell specialization, which is found in all multicellular organisms. Biologists may be able to learn how multicellular organisms evolved by studying *Volvox* and other colonial organisms.

Many biologists believe that animals, plants, and fungi probably evolved from different varieties of colonial organisms hundreds of millions of years ago. Initially, all of the cells in each type of colony may have been very similar. As these primitive colonial organisms evolved, however, their cells became more specialized and less capable of living independently.

SECTION 4-3 REVIEW

1. **What is a tissue?**
2. **What is an organ?**
3. **Give an example of an organ system, and name some of the parts that form it.**
4. **Name the plant tissue systems and organs.**
5. **To what extent are the individual cells within a *Volvox* colony independent of one another?**
6. **CRITICAL THINKING Green algae, such as *Codium,* enlarge by dividing the nucleus but do not form cell walls between the parent and offspring cells. Would you call *Codium* a unicellular or a multicellular organism? Explain your answer.**

CHAPTER 4 REVIEW

SUMMARY/VOCABULARY

4-1

- A cell is the smallest unit that can carry on all of the processes of life. The development of the microscope enabled scientists to take their first close look at cells.
- The cell theory states that (1) all living things are composed of one or more cells, (2) cells are the units of structure and function in an organism, and (3) cells come only from preexisting cells.
- The ratio of surface area to volume determines how large a cell can get. Most plant and animal cells are only about 10 to 50 µm in diameter.
- A cell's shape reflects its function.
- Eukaryotic cells contain a nucleus and membrane-bound organelles, but prokaryotic cells have neither.

Vocabulary

cell (69)
cell membrane (72)
cell theory (69)
eukaryote (72)
nucleus (72)
organelle (71)
prokaryote (72)

4-2

- The cell membrane is selectively permeable and consists mostly of lipids and proteins. Both move constantly within the membrane, as described by the fluid mosaic model.
- Mitochondria are organelles in which the energy in organic compounds is transferred to ATP. Mitochondria are surrounded by a double membrane.
- Ribosomes are involved in the synthesis of proteins. Some ribosomes are free in the cytosol. Others are attached to the rough endoplasmic reticulum, which prepares proteins for export from the cell or insertion into the cell membrane. The smooth endoplasmic reticulum lacks ribosomes.
- The Golgi apparatus is the cell's processing, packaging, and secreting organelle.
- Lysosomes contain hydrolytic enzymes that digest organic compounds, old cell parts, and other materials.
- The cytoskeleton includes microfilaments and microtubules, strands of protein that help cells move and maintain their shape.
- Cilia and flagella assist in cell movement. Both are made of nine pairs of microtubules arranged around a central pair.
- The nucleus is surrounded by a double membrane and contains chromatin, a combination of DNA and protein. DNA stores hereditary information and directs the synthesis of RNA. RNA directs the synthesis of proteins in the cytosol.
- Plant cells contain three structures not found in animal cells: cell walls, vacuoles, and plastids.
- A rigid cell wall covers the cell membrane in plant cells and provides support and protection.
- Large fluid-filled vacuoles store enzymes and waste products within plant cells.
- Plastids store starch, fats, and pigments in plant cells. One type of plastid, the chloroplast, is the site where light energy is converted into chemical energy during photosynthesis.

Vocabulary

actin (79)
cell wall (82)
chloroplast (83)
chromatin (81)
chromosome (81)
cilium (80)
crista (77)
cytoplasm (75)
cytoskeleton (79)
cytosol (75)
endoplasmic reticulum (78)
flagellum (80)
fluid mosaic model (75)
Golgi apparatus (78)
integral protein (74)
lysosome (79)
microfilament (79)
microtubule (79)
mitochondrion (76)
nuclear envelope (81)
nuclear matrix (81)
nuclear pore (81)
nucleolus (81)
peripheral protein (74)
plastid (83)
ribosome (77)
rough endoplasmic reticulum (78)
selectively permeable membrane (73)
smooth endoplasmic reticulum (78)
spindle fiber (79)
thylakoid (83)
vacuole (82)

REVIEW

1. Eukaryotes have a true nucleus and prokaryotes do not. The term *prokaryote* suggests that these cells evolved before a nucleus existed.

2. A cell membrane consists of two layers of lipid molecules embedded with proteins. It functions to hold in the cell contents and to regulate the movement of molecules into and out of the cell. The cell wall is located outside the cell membrane of a plant cell and functions in support.

3. The cytoplasm is the space between the cell membrane and the nucleus and is where the other organelles are located. The cytosol is a gelatin-like fluid that bathes the organelles. The cytoskeleton is the internal framework of the cell.

4. Microtubules are hollow protein tubes that form the internal structure of cilia and flagella. Both cilia and flagella make cells mobile, but flagella are longer.

5. Chromatin does not belong in the group, because it is a nuclear material rather than a membrane-bound cytoplasmic organelle like the others.

6. b **7.** a
8. d **9.** c
10. b **11.** c
12. d **13.** c
14. a **15.** b

16. Each advance in microscope technology, from the light microscope to the electron microscope, has enabled scientists to observe more details about the cell and its organelles.

17. The organelle is a mitochondrion, and the structures labeled "A" are cristae. Their long, flattened shape provides a large surface area for the chemical reactions that transfer energy from organic compounds to ATP.

18. Cell size is limited by the ratio of surface area to volume. As a cell becomes larger, this ratio decreases. At some size, a cell's surface area will become too small for materials to enter the cell quickly enough to meet the cell's needs.

19. Phospholipid heads are hydrophilic, but the tails are hydrophobic. This difference causes phospholipids to form lipid bilayers, with the heads on each surface of the bilayer and the tails in the interior.

20. The cell membrane is said to be selectively permeable because it allows some materials to pass through but excludes others.

21. Cells with a high energy requirement should have many mitochondria, since mitochondria are the organelles in which energy in organic compounds is transferred to ATP.

22. Proteins are synthesized on the ribosomes, some of which are attached to the endoplasmic reticulum and others of which are free in the cytosol. The Golgi apparatus modifies the proteins for export by the cell.

23. Answers will vary. Students may focus on various types of marker proteins, enzymes, receptor proteins, and transport proteins.

CRITICAL THINKING

1. A red blood cell is specialized to perform one main function. Without a nucleus or mitochondria, it can carry more hemoglobin and, therefore, more oxygen.

2. The endoplasmic reticulum, the cristae of mitochondria, and the inner folds of chloroplasts have large surface areas similar to those of radiator coils. Such structures provide more area for biochemical reactions.

4-3
- The cells in most multicellular organisms are organized into tissues, organs, and organ systems.
- The earliest cells on Earth were probably single-celled prokaryotes. Unicellular eukaryotes evolved later, followed by multicellular organisms.
- A colonial organism is a group of genetically identical cells that live together in closely connected groups. Some of the cells that make up a colonial organism are specialized to perform certain tasks, like movement or reproduction.

Vocabulary

colonial organism (87) organ (86) organ system (87) tissue (86)

REVIEW

Vocabulary

1. The word part *eu-* means "true," *pro-* means "before," and *kary-* means "nucleus." With this information, explain why the words *prokaryote* and *eukaryote* are good terms for the organisms they describe. What do the terms suggest about the evolution of these organisms?
2. Compare cell membranes with cell walls in terms of their structure and function.
3. Explain the meaning of the terms *cytoplasm, cytosol,* and *cytoskeleton.*
4. Explain the relationship between microtubules, cilia, and flagella.
5. Choose the term that does not belong in the following group, and explain why it does not belong: Golgi apparatus, endoplasmic reticulum, chromatin, and mitochondria.

Multiple Choice

6. A prokaryote has (a) a nucleus (b) a cell membrane (c) membrane-bound organelles (d) all of the above.
7. The growth of cells is limited by the ratio between (a) volume and surface area (b) organelles and surface area (c) organelles and cytoplasm (d) nucleus and cytoplasm.
8. The major components of cell membranes are (a) lipids (b) proteins (c) nucleic acids (d) lipids and proteins.
9. The function of the Golgi apparatus is to (a) synthesize proteins (b) release energy (c) modify proteins for export (d) synthesize lipids.
10. Mitochondria (a) transport materials (b) release energy (c) make proteins (d) control cell division.
11. Ribosomes are (a) surrounded by a double membrane (b) manufactured in the cytosol (c) composed of proteins and RNA (d) attached to the smooth endoplasmic reticulum.
12. Lysosomes function in cells to (a) recycle cell parts (b) destroy viruses and bacteria (c) shape developing body parts (d) all of the above.
13. The nucleolus is (a) the control center of the cell (b) the storehouse of genetic information (c) the site where ribosomes are synthesized (d) none of the above.
14. Plastids (a) store pigments (b) store membranes (c) synthesize proteins (d) secrete proteins.
15. The stomach is an example of (a) a tissue (b) an organ (c) an organ system (d) none of the above.

Short Answer

16. Explain how microscopes have been helpful in the study of cells.
17. Identify the organelle illustrated in the micrograph below, and name the structures labeled *A* inside the organelle. Explain the significance of the shape of these structures.

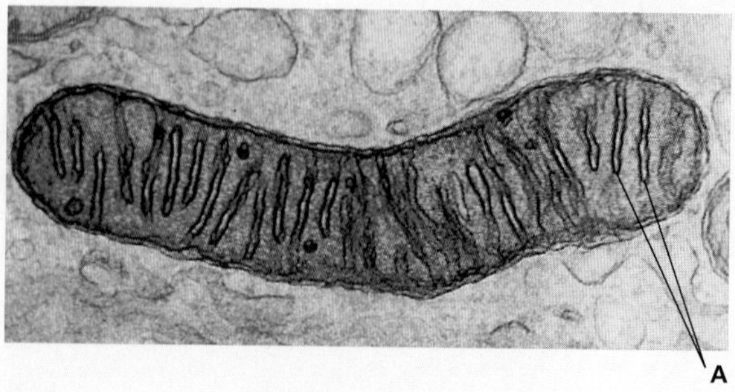

18. What limits the maximum size of cells? Explain your answer.
19. How is the structure of cell membranes influenced by the reaction of phospholipids to water?
20. Why is the cell membrane said to be selectively permeable?
21. If a cell has a high energy requirement, would you expect it to have many or few mitochondria? Explain your answer.
22. Explain how the ribosomes, endoplasmic reticulum, and Golgi apparatus function together in protein synthesis.
23. **Unit 1—Cell Transport and Homeostasis** Write a report summarizing the roles of different types of cell-membrane proteins in the preservation of body organs donated for transplant.

CRITICAL THINKING

1. A mature human red blood cell has no nucleus or mitochondria. It consists primarily of a membrane surrounding hemoglobin, the protein molecule that carries oxygen. Suggest an advantage of the simple organization of human red blood cells.
2. The coils of a radiator provide a large surface area from which heat is radiated into a room. Which cell organelles have a structure similar to that of a radiator? How is their structure related to their function?
3. What characteristic of eukaryotic cells gives them a greater capacity for specialization than prokaryotic cells have? Explain your answer.
4. Livestock in the western United States often die after eating a locoweed, such as *Astragalus toanus.* The chemical that the plant contains is also poisonous to plants. How does locoweed keep from poisoning itself?
5. The graph below illustrates how the ratio between surface area and volume changes as a spherical cell's diameter increases. By what percentage does this ratio change when a cell grows from 1 µm to 2 µm in diameter? What is the maximum diameter the cell could

Area / Volume Ratio Versus Cell Size

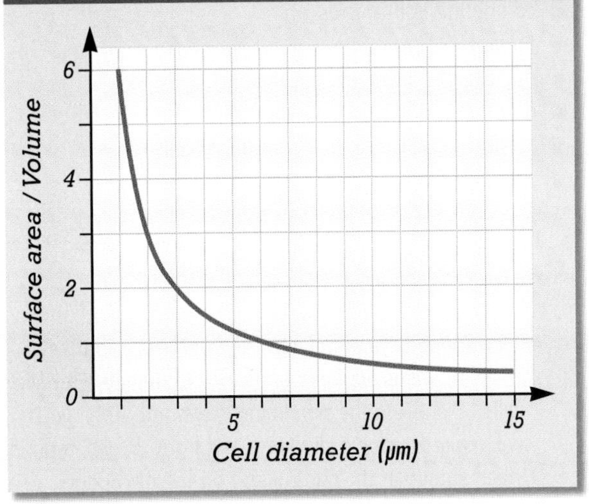

EXTENSION

1. Read "The Nobel Prizes for 1999" in *Scientific American,* January 2000, on page 16. Prepare an oral report on how Günter Blobel, a scientist at Rockefeller University, won the Nobel Prize for Physiology or Medicine.
2. Use the resources in your school or public library to learn more about the work of Schleiden, Schwann, or Virchow. Write a brief report summarizing the processes that the researcher used to arrive at his conclusions about cells.
3. Place a sprig of the aquarium plant *Elodea* in water, and shine a bright light on it for an hour. Then examine a leaf under a light microscope, and make a drawing that indicates the direction of chloroplast movement. How do you think the movement of these organelles in response to light helps the cell to function?

3. The presence of organelles and more membranes in eukaryotic cells makes possible a greater specialization of function.
4. The poisonous chemical is stored in a vacuole, isolating it from the rest of the cell.
5. When a cell grows from 1 mm to 2 mm in diameter, this ratio drops from 6 to 3, a decrease of 50 percent. To keep the ratio greater than or equal to 1, the cell must have a diameter no larger than 6 mm.

EXTENSION

1. Blobel's work in understanding how proteins move across membranes led to the discovery of how proteins find their correct place in cells. This process does not function properly with diseases like cystic fibrosis. Blobel's research may eventually help scientists discover what goes wrong inside the cell with diseases such as cancer and Alzheimer's disease.
2. All three were Germans of the mid-1800s who made numerous microscopic observations of a variety of subjects. Also, the development of organic stains made it possible to study specific cell structures.
3. The movement of chloroplasts, known as cytoplasmic streaming, increases their chances of being exposed to light. Also, movement ensures that reactants and products of photosynthesis are distributed as needed.

CHAPTER 4 INVESTIGATION

TIME REQUIRED

One 50-minute class period

SAFETY PRECAUTIONS

Remind students to handle glass microscope slides carefully and to give them to you for proper disposal if they are broken. If you use lamps for microscope illumination or to warm the slides of *Elodea,* warn students to avoid spilling water on the lamps or their electrical cords. Remind students to wear a lab apron and safety goggles when they handle Lugol's iodine solution.

QUICK REFERENCES

Lab Preparation Notes are found on page 67D.

Holt BioSources provides a Teaching Resources CD-ROM that contains "Using Gowin's Vee in the Lab" and "Scoring Rubrics."

PROCEDURAL TIPS

Obtain prepared slides of human cheek cells from a biological supply house.

Purchase sprigs of *Elodea* at a pet store and keep them in water. To induce cytoplasmic streaming, illuminate the sprigs for 12 hours before class.

For unknowns, make wet-mount slides or purchase prepared slides from a biological supply house. Students should carefully examine the known cells (Parts A and B) before they examine the unknown cells (Part C).

CHAPTER 4 INVESTIGATION

Comparing Animal and Plant Cells

OBJECTIVE

- Examine the similarities and differences between the structure of cells in animals and the structure of cells in plants.

PROCESS SKILLS

- hypothesizing
- classifying
- observing

MATERIALS

- lab apron
- safety goggles
- compound light microscope
- forceps
- microscope slides and coverslips
- dropper bottle of Lugol's iodine solution
- prepared slides of human epithelial cells
- sprigs of *Elodea*
- prepared slides of three unknowns

Background

1. In this investigation, you will use a compound light microscope to observe cells from animals and plants. First you will view a prepared slide of human epithelial cells taken from the skin lining the mouth. Then you will make your own slide of a leaf from *Elodea,* a pond weed shown in the photograph on the next page.

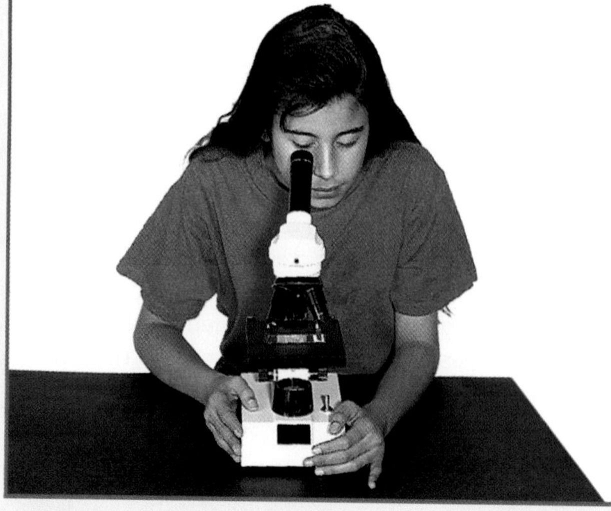

2. Based on your observations of human epithelial cells and *Elodea* leaf cells, you will be asked to classify three slides of unknown cells as either animal or plant cells.
3. Before you examine any cells, list the structural characteristics that distinguish animal cells from plant cells.

PART A Animal Cells

1. CAUTION **Handle glass microscope slides carefully. Dispose of broken glass separately in a container designated by your teacher.**
2. CAUTION **Do not use electrical equipment with wet hands or near water.**
3. Examine a prepared slide of epithelial cells under low power. Locate cells that are separate from each other and place them in the center of the field of view. Examine the cells under high power. Adjust the diaphragm to reduce the light intensity and achieve greater clarity.
4. In your lab report, make a drawing of two or three cells as they appear under high power. Identify and label the cell membrane, the cytoplasm, the nuclear envelope, and the nucleus of one of the cells in your drawing.

PART B Plant Cells

5. Carefully tear off a small leaf near the top of an *Elodea* sprig. Using forceps, place the whole leaf in a drop of water on a slide. Place a coverslip on top of the leaf.
6. Observe the leaf under low power. The outermost part of the cell is the cell wall. The many small, green organelles in the cells are chloroplasts.
7. Locate a cell that you can see clearly, and move the slide so that the cell is in the center of the field of view. Examine this cell under high power, and use the fine adjustment to bring the cell into focus.
8. Find an *Elodea* cell that is large enough to allow you to see the cell wall and the chloroplasts clearly. In your lab report, make a drawing of this cell. Label the cell wall and at least one chloroplast in your drawing.

9. The chloroplasts may be moving in some of the cells. If you observe no movement, warm the slide in your hand or shine a bright lamp on it for a minute or two. Then reexamine the slide under high power, and look for the movement of the cell's contents. This movement is called cytoplasmic streaming.
10. Because the cell membrane is pressed against the cell wall, you may not see it. Also, the abundance of chloroplasts may hide other organelles in the cells. You can make the cell membrane, vacuole, nucleus, and nucleolus more visible by making a stained wet-mount slide of *Elodea.*
11. Put on a lab apron and safety goggles. Prepare a wet-mount slide of *Elodea* as you did in step 4, but substitute Lugol's iodine solution for the water. Allow the iodine solution to diffuse throughout the leaf.
12. Observe the stained cells under low and high power. Make a drawing of a stained *Elodea* cell in your lab report. Label the central vacuole, nucleus, nucleolus, chloroplasts, cell wall, and cell membrane if they are visible.

PART C Identifying Unknown Cells

13. Make a data table like the one below to record your observations of the unknown specimens.
14. Obtain prepared slides of three unknown specimens from your teacher.
15. Observe each specimen under low and high power. In your data table, record the code number assigned to each unknown, each specimen's classification as plant or animal, and your reasons for classifying each specimen.
16. Clean up your materials and wash your hands before leaving the lab.

Analysis and Conclusions

1. According to your observations in this investigation, list several ways that plant and animal cells are structurally similar and several ways that they are different.
2. What do you think might be the function of cytoplasmic streaming in a plant cell? Lugol's iodine solution causes cytoplasmic streaming to stop. Why do you think this happens?
3. Which organelles that you read about in Chapter 4 did you not see in this investigation? Why do you think you were unable to see these organelles in your slides?

Further Inquiry

Use library resources to locate electron micrographs of cell structures that you were unable to see with the compound light microscope.

CLASSIFICATION OF UNKNOWN SPECIMENS

Unknown (code number)	Classification (plant or animal)	Reasons for classification

CHAPTER 4 INVESTIGATION

ANSWERS TO BACKGROUND

3. Plant cells have a cell wall, chloroplasts, and large vacuoles; animal cells lack these structures. Plant cells often have a regular, rectangular shape due to their cell walls, while animal cells may have a variety of shapes.

ANSWERS TO ANALYSIS AND CONCLUSIONS

1. Both animal and plant cells are bound by a cell membrane and have a cytoplasm, nucleus, and other organelles. Plant cells have a cell wall, chloroplasts, and large vacuoles, but animal cells have none of these. Plant cells are more consistent in shape than animal cells.

2. Cytoplasmic streaming speeds the distribution of materials within the cell and the exchange of materials between the cell and its environment. Lugol's iodine solution is a poison that kills the cells and therefore stops cytoplasmic streaming.

3. Students will be unlikely to see any of the organelles except the nuclei in the cheek cells and the chloroplasts, vacuoles, and cell walls in the *Elodea* cells. Some of the unseen organelles are simply too small to be resolved clearly with the light microscope, and others require specific stains to reveal them. The organelles that are visible in the unknown slides will depend on the materials that are selected and how they are stained.

CHAPTER 5
HOMEOSTASIS AND TRANSPORT

	TOPICS	TEACHING RESOURCES	LABS, CLASSWORK, AND HOMEWORK
BLOCKS 1 & 2 90 minutes	**Introducing the Chapter, p. 94**	**ATE** Focus Concept, p. 94 **ATE** Assessing Prior Knowledge, p. 94 **ATE** Understanding the Visual, p. 94	■ Supplemental Reading Guide, *The Lives of a Cell: Notes of a Biology Watcher*
	5-1 Passive Transport, p. 95 Diffusion, p. 95 Osmosis, p. 96 Facilitated Diffusion, p. 99 Diffusion Through Ion Channels, p. 100	**ATE** Section Overview, p. 95 **ATE** Critical Thinking, p. 98 5–6	**PE** **Quick Lab,** p. 96, Observing Diffusion **Lab A4,** Demonstrating Diffusion **Lab A32,** Relating Cell Structure to Function **Lab B3,** Diffusion and Cell Membranes ★ Study Guide, Section 5-1 **PE** Section Review, p. 100
BLOCKS 3 & 4 90 minutes	**5-2 Active Transport, p. 101** Cell Membrane Pumps, p. 101 Endocytosis and Exocytosis, p. 103	**ATE** Section Overview, p. 101 **ATE** Critical Thinking, p. 102 **ATE** Visual Link Figure 5-6, p. 102 6A, 10–11	**PE** **Chapter 5 Investigation,** p. 108 ■ Problem-Solving Worksheets—Solution Concentration ★ Study Guide, Section 5-2 **PE** Section Review, p. 104
BLOCKS 5 & 6 90 minutes	**REVIEW and ASSESSMENT** **PE** Chapter 5 Review, pp. 106–107 ■ Performance-Based Assessment—Diffusion Through a Cell Membrane	Audio CD Program BIOLOGY INTERACTIVE TUTOR Unit 1—Cell Transport and Homeostasis	**CHAPTER TESTING** ■ Chapter Test (blackline master) ▲ Test Generator ■ Assessment Item Listing ■ Scoring Rubrics and Classroom Management Checklists

KEY

MODERN BIOLOGY
- **PE** Pupil's Edition
- **ATE** Teacher's Edition
- ■ Active Reading Guide
- ★ Study Guide

HOLT BIOSOURCES
- Teaching Transparencies
- Laboratory Program

One-Stop Planner CD-ROM
Shows these resources within a customizable daily lesson plan:
- ■ Holt BioSources Teaching Resources
- ■ Active Reading Guide
- ★ Study Guide
- ▲ Test Generator

READING FOR CONTENT MASTERY

Preview: Have students preview the objectives, the introductory paragraph, topic sentences, and captions before reading the chapter.

ATE Active Reading Technique: Anticipation Guide, p. 95
■ Active Reading Guide Worksheet 5-1
PE Summary/Vocabulary, p. 105

ATE Active Reading Technique: Anticipation Guide, p. 104
■ Active Reading Guide Worksheet 5-2
PE Summary/Vocabulary, p. 105

TECHNOLOGY AND INTERNET RESOURCES

CNN PRESENTS — Video Segment 2 Cystic Fibrosis

BIOLOGY INTERACTIVE TUTOR — Unit 1—Cell Transport and Homeostasis Topics 3–6

Audio CD Program, Sections 5-1 through 5-2

Interactive Exploration: Human Biology, Cystic Fibrosis

internetconnect

sciLINKS NSTA
On-line Resources: www.scilinks.org
The following *sci*LINKS Internet resources can be found in the student text for this chapter:

Topics:
- Osmosis, p. 96
- Active transport, p. 101
- Endocytosis, p. 103

go.hrw.com
On-line Resources: go.hrw.com
Visit the HRW Web site for a variety of resources related to this chapter. Just type in the keyword HM2 Chapter 5.

Smithsonian Institution®
Visit **www.si.edu/hrw** for additional on-line resources.

CNNfyi.com
Visit **www.cnnfyi.com** for late-breaking news and current event stories selected just for you.

LAB ACTIVITY PLANNING

CHAPTER 5 INVESTIGATION

Exploring the Role of Osmosis in Cystic Fibrosis, pp. 108–109

OVERVIEW

In this interactive exploration, students will use a computer simulation to study the cellular basis for cystic fibrosis. This activity allows students to relate the function of the cystic fibrosis protein to the cell membrane's permeability to water and chloride ions and allows students to relate the symptoms of cystic fibrosis to the flow of chloride ions and water across the cell membrane.

The cystic fibrosis gene has been discovered and sequenced. The protein encoded by this gene functions in the transportation of chloride ions across the cell membrane. Defects in the protein prevent chloride ions from leaving cells. Cells with abnormally high chloride ion concentrations will retain water, which may account for the thick mucus accumulation in the pancreas, liver, and lungs that causes so many problems for cystic fibrosis patients.

PREPARATION

1. Access to the following materials is needed:
   - computer with CD-ROM drive
   - CD-ROM *Interactive Explorations in Biology: Human Biology.* The title of this program on the CD-ROM is "Cystic Fibrosis."
2. Have students read the Objectives, Background, and Prelab Preparation sections before starting the lab.

Quick Labs

Observing Diffusion, p. 96
Each group of students will need disposable gloves, lab aprons, safety goggles, a 600 mL beaker, dialysis tubing, 10 percent starch solution, IKI solution, water, a graduated cylinder, and string. See page 96 for preparation of solution and disposal instructions.

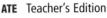

UNDERSTANDING THE VISUAL

The photograph shows a cell of the immune system (orange) ingesting a tumor cell (red) through the process of phagocytosis. Explain that *phago* means "to eat" and *cyto* means "cell." Ask students how this kind of phagocytosis might be useful to an animal. (By ingesting and destroying tumor cells, the immune system cells could stop tumors from growing.)

FOCUS CONCEPT

Stability and Homeostasis

This chapter focuses on transport across cell membranes as a mechanism of maintaining homeostasis at the cellular level. For example, cells regulate the concentration of nutrients and wastes by moving them into or out of the cytoplasm. Paramecia maintain a balanced internal environment by using contractile vacuoles to release excess water. Animal cells use a carrier protein known as the sodium-potassium pump to regulate the concentration of sodium and potassium ions inside the cell.

ASSESSING PRIOR KNOWLEDGE

Review the following concepts.

Energy: *Chapter 2*
Ask students to define energy. (Energy is the ability to do work.) Then ask them to distinguish between potential energy and kinetic energy. (Potential energy is stored energy, or energy of position. Kinetic energy is the energy of motion.)

Membrane Structure: *Chapter 4*
Have students list the two main components of cell membranes (lipids and proteins). Ask students why the cell membrane is said to be selectively permeable. (Some substances easily cross the membrane, while others don't cross at all.)

CHAPTER 5

HOMEOSTASIS AND TRANSPORT

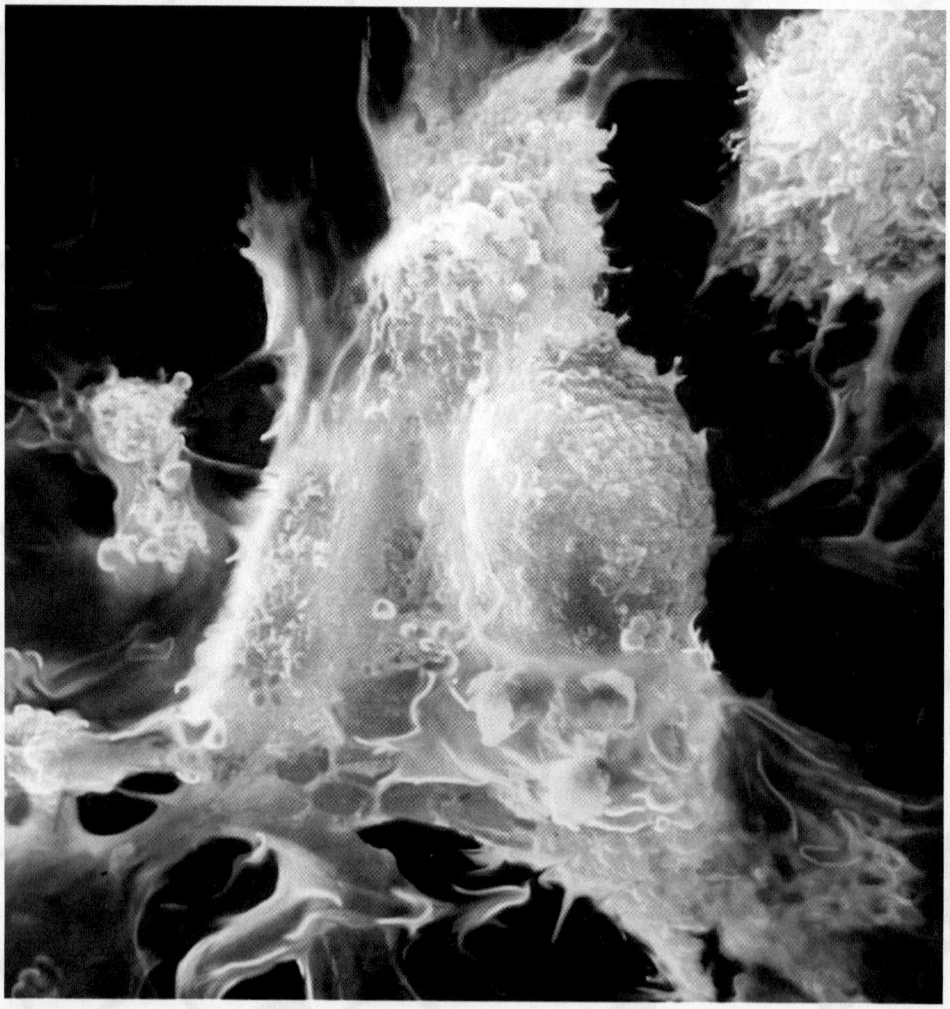

A macrophage engulfs a human tumor cell. (SEM 3,520×)

FOCUS CONCEPT: ***Stability and Homeostasis***

As you read, look for the ways that cells regulate the movement of materials across their membranes and thereby maintain internal balance despite changes in their environment.

INTERACTIVE BIOLOGY TUTOR **Unit 1—*Cell Transport and Homeostasis*** Topics 3–6

PASSIVE TRANSPORT

*Cell membranes help organisms maintain homeostasis by controlling what substances may enter or leave cells. Some substances can cross the cell membrane without any input of energy by the cell. The movement of such substances across the membrane is known as **passive transport.***

SECTION

5-1

OBJECTIVES

▲

Explain how an equilibrium is established as a result of diffusion.

●

Distinguish between diffusion and osmosis.

■

Explain how substances cross the cell membrane through facilitated diffusion.

◆

Explain how ion channels assist the diffusion of ions across the cell membrane.

DIFFUSION

The simplest type of passive transport is diffusion. **Diffusion** is the movement of molecules from an area of higher concentration to an area of lower concentration. This difference in the concentration of molecules across a space is called a **concentration gradient.**

Consider what happens when you add a sugar cube to a beaker of water. As shown in Figure 5-1, the sugar cube sinks to the bottom of the beaker. That makes the concentration of sugar molecules much greater at the bottom of the beaker than at the top. As the cube dissolves, the sugar molecules begin to diffuse slowly through the water, moving from the bottom of the beaker to the top.

Diffusion is driven entirely by the kinetic energy the molecules possess. Because of their kinetic energy, molecules are in constant motion. They move randomly, traveling in a straight line until they hit an object, such as another molecule. When they hit something, they rebound and move off in a new direction, traveling in another straight line. If no object blocks their movement, they continue on their path. Thus, molecules tend to move "down" their concentration gradient, from areas where they are more concentrated to areas where they are less concentrated.

Equilibrium

In the absence of other influences, diffusion will eventually cause the concentration of molecules to be the same throughout the space the molecules occupy. When the concentration of the molecules of a substance is the same throughout a space, a state of **equilibrium** exists. Returning to the example in Figure 5-1, if the beaker of water is left undisturbed, at some point the concentration of sugar molecules will be the same throughout the beaker. The sugar concentration will then be at equilibrium.

FIGURE 5-1

Sugar molecules, initially in a high concentration at the bottom of a beaker, will move about randomly through diffusion. At equilibrium, the sugar concentration will be the same throughout the beaker. Diffusion occurs naturally because of the kinetic energy the molecules possess.

SECTION 5-1

SECTION OVERVIEW

Stability and Homeostasis
Section 5-1 describes the processes of diffusion and osmosis and explains how equilibrium is attained. It also explains how cells respond when placed in different types of solutions.

Matter, Energy, and Organization
This section also explains how substances move into and out of cells without an expenditure of energy by the cells.

ACTIVE READING

Technique: Anticipation Guide
Write the titles of this chapter and its two sections (5-1 and 5-2) on the board. Have students copy these titles in their notebooks, leaving a few lines after each section title. Next ask students to write what they think they will learn in each section. Check their lists for completeness, and tell them to save their lists for later use in the Active Reading activity on page 104.

Quick Lab

Observing Diffusion

Time Required 25 minutes

Safety IKI (iodine potassium iodide) is a poison and irritant. Avoid skin and eye contact; do not ingest. Ask students to use caution with glass beakers.

Procedural Tips Mix a 10 percent starch solution by dissolving 100 g of soluble starch per liter of water. Prepare the starch solution 1 day prior to the lab. Soak the dialysis tubing in water prior to the lab.

Disposal Slowly add 0.1 M sodium thiosulfate to the IKI solution until the solution is fully decolorized. Solutions can then be poured down the drain followed by several minutes of running water.

Answers to Analysis IKI turns dark in the presence of starch. IKI is composed of molecules small enough to pass through the pores in the dialysis tubing and to react with the starch. Starch is composed of large molecules that cannot move out of the dialysis tubing. The solution outside the bag will not change colors.

QUICK FACT

☑ The process of osmosis was first described in 1748 by a French scientist, Abbé Jean Antoine Nollet, who discovered that water spontaneously diffused through a pig bladder membrane into alcohol.

Quick Lab

Observing Diffusion

Materials disposable gloves, lab apron, safety goggles, 600 mL beaker, 25 cm dialysis tubing, 15 mL starch solution (10 percent), 20 drops IKI, 300 mL water, 100 mL graduated cylinder, 20 cm piece of string (2)

Procedure

1. Put on your disposable gloves, lab apron, and safety goggles.
2. Pour 300 mL of water in the 600 mL beaker.
3. Add 20 drops of IKI to the water.
4. Open the dialysis tubing, and tie one end tightly with a piece of string.
5. Using the funnel, pour 15 mL of 10 percent starch solution into the dialysis tubing.
6. Tie the other end of the dialysis tubing tightly with the second piece of string, forming a sealed bag around the starch solution.
7. Place the bag into the solution in the beaker, and observe the setup for a color change.

Analysis What happened to the color in the bag? What happened to the color of the water around the bag? Explain your observations.

It is important to understand that even at equilibrium the random movement of molecules continues. But because there is no concentration gradient, molecules are just as likely to move in one direction as in any other. The random movements of many molecules in many directions balance one another, and equilibrium is maintained.

Diffusion Across Membranes

You learned in Chapter 4 that cell membranes allow some molecules to pass through, but not others. If a molecule can pass through a cell membrane, it will diffuse from an area of higher concentration on one side of the membrane to an area of lower concentration on the other side.

The ability of a molecule to diffuse across a cell membrane depends on the size and type of the molecule and on the chemical nature of the membrane. Remember from Chapter 4 that a membrane is made, in part, of a lipid bilayer and that certain proteins can form pores in the membrane. Molecules that can dissolve in lipids may pass through the membrane by diffusion. For example, because of their nonpolar nature, both carbon dioxide and oxygen dissolve in lipids. Molecules that are very small but not soluble in lipids may diffuse across the membrane by moving through the pores in the membrane.

OSMOSIS

Recall from Chapter 2 that a solution is composed of a solute dissolved in a solvent. In the sugar water described earlier, the solute was sugar and the solvent was water, and the solute molecules diffused through the solvent. It is also possible for solvent molecules to diffuse. In the case of cells, the solutes are organic and inorganic compounds, and the solvent is water. The process by which water molecules diffuse across a cell membrane from an area of higher concentration to an area of lower concentration is called **osmosis** (ahz-MOH-suhs). Because water moves down its concentration gradient, osmosis does not require cells to expend energy. Osmosis, therefore, is a type of passive transport.

internetconnect

SCILINKS NSTA
TOPIC: Osmosis
GO TO: www.scilinks.org
KEYWORD: HM096

Direction of Osmosis

The net direction of osmosis depends on the relative concentration of solutes on the two sides of the membrane. Examine Table 5-1. When the concentration of solute molecules outside the cell is *lower* than the concentration in the cytosol, the solution outside is **hypotonic** to the cytosol. In this situation, water diffuses *into* the cell until equilibrium is established. When the concentration of solute molecules outside the cell is *higher* than the concentration in the cytosol, the solution outside is **hypertonic** to the cytosol. In this situation, water diffuses *out of* the cell until equilibrium is established. When the concentrations of solutes outside and inside

TABLE 5-1 *Direction of Osmosis*

Condition	Net movement of water	
External solution is hypotonic to cytosol	into the cell	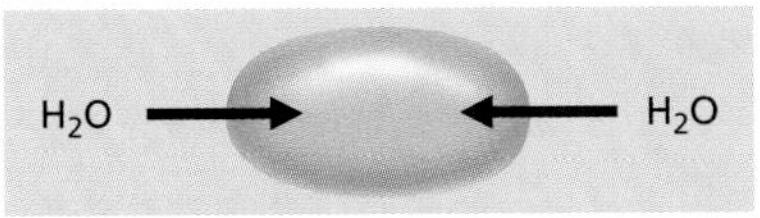
External solution is hypertonic to cytosol	out of the cell	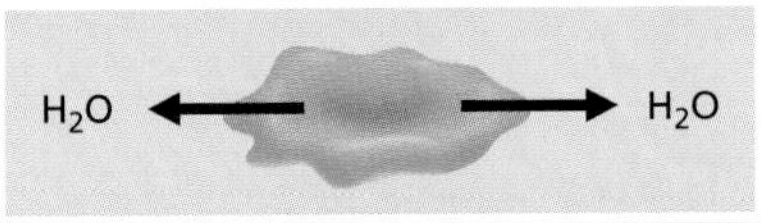
External solution is isotonic to cytosol	none	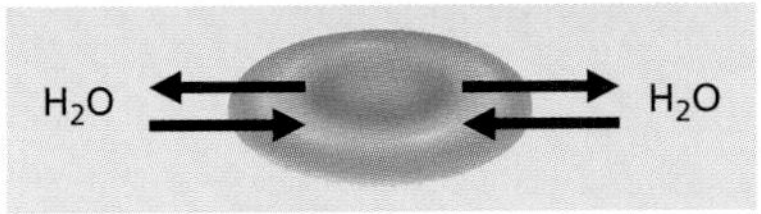

the cell are equal, the outside solution is said to be **isotonic** to the cytosol. Under these conditions, water diffuses into and out of the cell at equal rates, so there is no net movement of water.

Notice that the prefixes *hypo-*, *hyper-*, and *iso-* refer to the relative solute concentrations of two solutions. Thus, if the solution outside the cell is *hypo*tonic to the cytosol, then the cytosol must be *hyper*tonic to that solution. Conversely, if the solution outside is *hyper*tonic to the cytosol, then the cytosol must be *hypo*tonic to the solution. Water tends to diffuse from hypotonic solutions to hypertonic solutions.

How Cells Deal with Osmosis

Cells that are exposed to an isotonic external environment usually have no difficulty keeping the movement of water across the cell membrane in balance. This is the case with the cells of vertebrate animals on land and of most other organisms living in the sea.

In contrast, many cells function in a hypotonic environment. Such is the case for unicellular freshwater organisms. Water constantly diffuses into these organisms. Because they require a relatively lower concentration of water in the cytosol to function normally, unicellular organisms must rid themselves of the excess water that enters by osmosis. Some of them, such as the paramecia shown in Figure 5-2, do this with **contractile vacuoles,** which are organelles that remove water. Contractile vacuoles collect the excess water and then contract, pumping the water out of the cell. Unlike diffusion and osmosis, this pumping action requires the cell to expend energy.

Other cells, including many of those in multicellular organisms, respond to hypotonic environments by pumping solutes out of the cytosol. This

FIGURE 5-2

The paramecia shown below live in fresh water, which is hypotonic to their cytosol. (a) Contractile vacuoles collect excess water that moves by osmosis into the cytosol. (b) The vacuoles then contract, returning the water to the outside of the cell. (LM 315×)

(a)

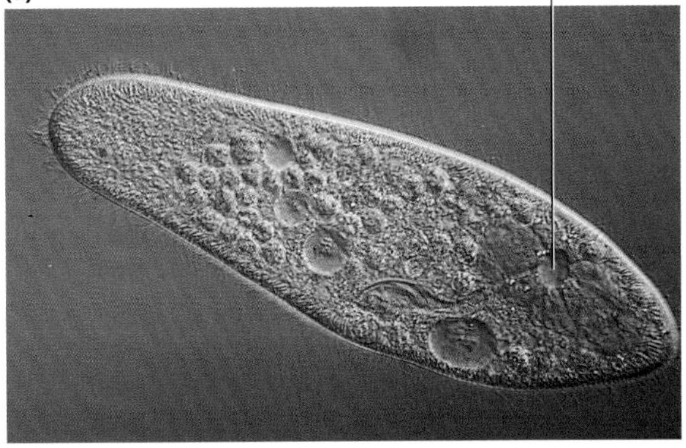

(b)

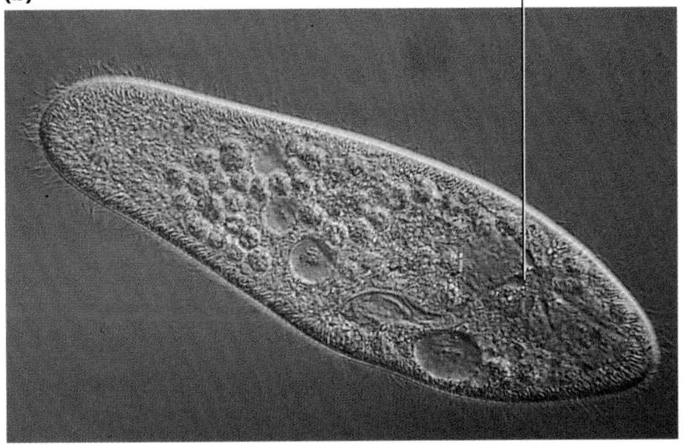

DEMONSTRATION

Place a 10 percent sugar-water solution in a thistle tube that has a semipermeable membrane attached to its wide end. Place a 5 percent sugar-water solution in a beaker. Lower the tube into the beaker, membrane side down, until the tops of the two solutions are level. Ask students to predict whether the top of the solution in the tube will go up or down and to explain why. (Water molecules can diffuse through the membrane, but sugar molecules cannot. The net movement of water will be into the tube because the solution in the tube has a higher solute concentration and therefore a lower water concentration. As water enters the tube, the volume of the solution in the tube increases, so the top of the solution rises.) Ask students to draw the setup and use arrows to show the direction of water movement. (Although the direction of net water movement is into the tube, water molecules actually diffuse in both directions across the membrane, so the arrows should point in both directions.)

GIFTED *ACTIVITY*

Have students use the scientific method to create an experiment to demonstrate hypotonic, isotonic, and hypertonic solutions based on the information and demonstrations presented in this chapter. For example, students could design an experiment from the potato slice activity and supplement their qualitative observations with quantitative measurements.

SECTION 5-1

INCLUSION ACTIVITY

Have students draw pictures representing vocabulary words and key phrases from the chapter. For example, for turgor pressure they could draw a plant standing upright with broad leaves, and for loss of turgor pressure they could draw a wilted plant.

QUICK FACT

✔ Wooden drawers in cabinets and dressers absorb water from the air on humid days due to osmosis. This absorption increases the turgor pressure in the wood, warping the drawers and making them harder to open and close. Water is lost from the wood during dry weather, and the drawers return to their original size.

CRITICAL THINKING

How Cells Deal with Osmosis

A common remedy for a sore throat is to gargle with salt water. Using the concept of osmosis, explain how this remedy might work. (Some of the pain associated with a sore throat is due to swelling of the throat tissues, which is caused by an accumulation of fluid outside cells. When a person gargles with salt water, which has a higher solute concentration than this fluid, water moves by osmosis from the throat tissues into the salt water, decreasing the swelling and relieving some of the pain.)

(a) Hypotonic **(b)** Hypertonic

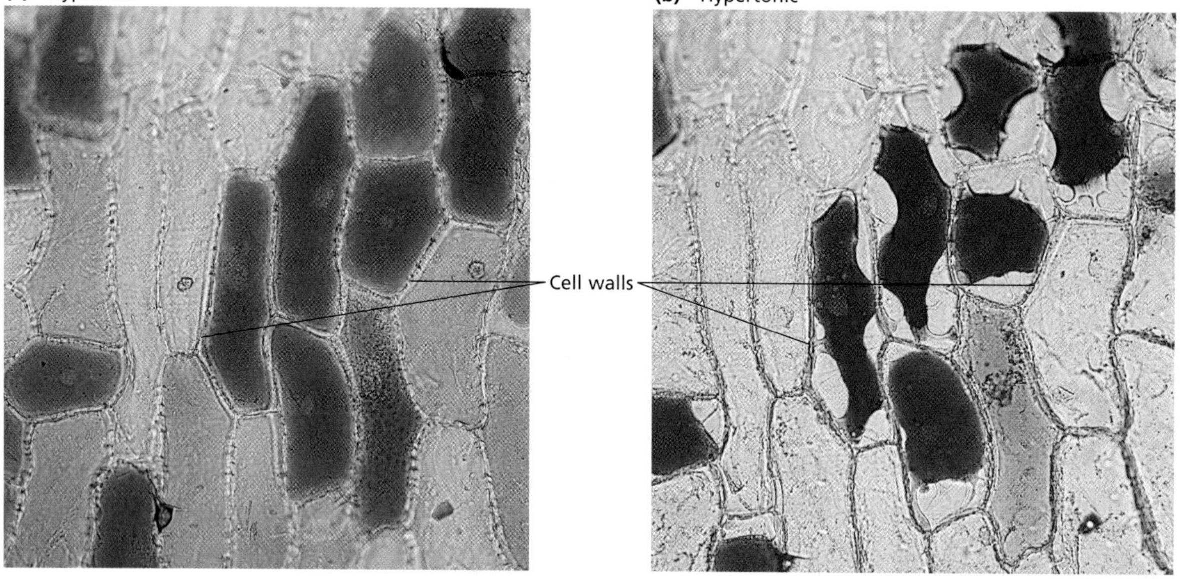

FIGURE 5-3

These two photographs show cells in the skin of a red onion. (a) In a hypotonic environment, the cells are pressed against the cell walls (LM 90×). (b) In a hypertonic environment, the cells contract and pull away from the cell walls (LM 98×).

lowers the solute concentration in the cytosol, bringing it closer to the solute concentration in the environment. As a result, water molecules are less likely to diffuse into the cell.

Most of the time, plant cells also live in a hypotonic environment. In fact, the cells that make up plant roots may be surrounded by water. Water, therefore, moves by osmosis into plant cells, which swell as they fill with water. The swelling stops when the cell membrane is pressed against the inside of the cell wall, as Figure 5-3a shows. The cell wall is strong enough to resist the pressure exerted by the water inside the expanding cell. The pressure that water molecules exert against the cell wall is called **turgor pressure.**

In a hypertonic environment, water leaves the cells through osmosis. As shown in Figure 5-3b, the cells shrink away from the cell walls, and turgor pressure is lost. This condition is called **plasmolysis** (plaz-MAHL-uh-suhs). Plasmolysis is the reason that plants wilt if they don't receive enough water.

Some cells cannot compensate for changes in the solute concentration of their environment. Red blood cells in humans, for instance, lack contractile vacuoles, solute pumps, and cell walls. As you can see in Figure 5-4, these cells lose their normal shape when they are placed in an environment that is not isotonic to their

FIGURE 5-4

(a) In an environment that is isotonic to the cytosol, a human red blood cell keeps its normal shape—round and dimpled (SEM 37,125×). (b) In a hypertonic environment, the cell loses water and becomes shriveled (SEM 39,762×). (c) In a hypotonic environment, the cell gains water and swells (SEM 37,125×).

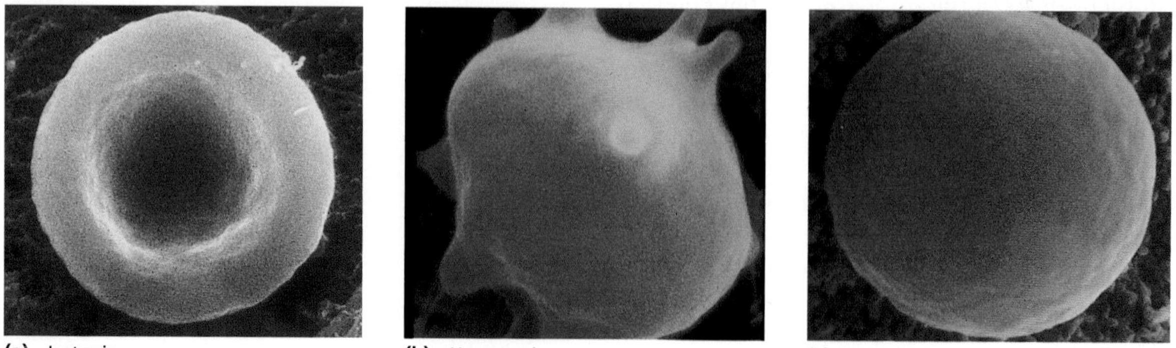

(a) Isotonic **(b)** Hypertonic **(c)** Hypotonic

cytosol. In a hypertonic environment, water leaves the cells, making them shrink and shrivel. In a hypotonic environment, water diffuses into the cells, causing them to swell and eventually burst. The bursting of cells is called **cytolysis** (sie-TAHL-uh-suhs).

Word Roots and Origins

cytolysis

from the Greek *kytos,* meaning "hollow vessel," and *lysis,* meaning "loosening"

FACILITATED DIFFUSION

Another type of passive transport is called **facilitated diffusion.** This process is used for molecules that cannot diffuse rapidly through cell membranes, even when there is a concentration gradient across the membrane. Such molecules may not be soluble in lipids, or they may be too large to pass through the pores in the membrane. In facilitated diffusion, the movement of these kinds of molecules across the cell membrane is assisted by specific proteins in the membrane. These proteins are known as **carrier proteins.**

The carrier proteins that serve in facilitated diffusion transport molecules from an area of higher concentration on one side of the membrane to an area of lower concentration on the other side. Because the molecules are moving down their concentration gradient, facilitated diffusion is passive transport. The cell does not have to supply additional energy to make it happen.

Figure 5-5 shows a model of how facilitated diffusion is thought to work. According to the model, a carrier protein binds to the molecule it transports. As soon as the carrier protein binds to the molecule, the carrier protein changes shape. This altered shape may shield the molecule from the hydrophobic interior of the lipid bilayer. Once shielded, the molecule can be transported across the cell membrane. On the other side of the membrane, the molecule is released from the carrier protein, which then returns to its original shape.

FIGURE 5-5

Facilitated diffusion occurs in three steps. (a) A carrier protein binds to a molecule on one side of the cell membrane. (b) The carrier protein changes shape, shielding the molecule from the interior of the membrane. (c) The molecule is released on the other side of the membrane.

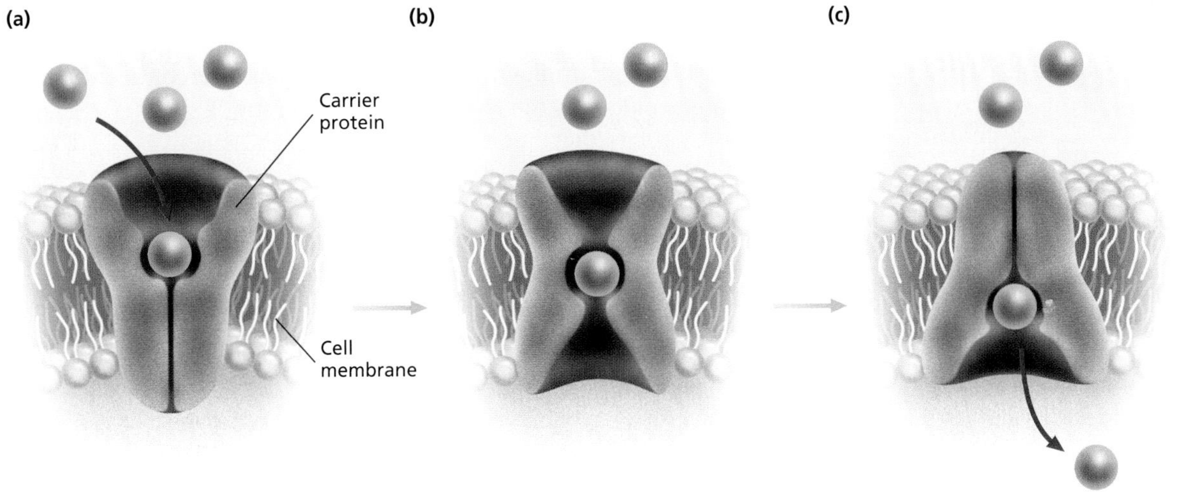

SECTION 5-1

TEACHING STRATEGY

Turgor Pressure in Potato Slices

Have students prepare three potato slices, each approximately 1 cm thick. Place one slice in distilled water, one in tap water, and one in salt water. Ask students to name the solutions using the terms *hypotonic, isotonic,* and *hypertonic.* (Distilled water represents a hypotonic solution, tap water represents an isotonic solution, and salt water represents a hypertonic solution.) Ask them to predict what will happen to each slice. (Turgor pressure increases in hypotonic solutions, so the potato slice in this solution should swell and become very crisp. Turgor pressure stays the same in isotonic solutions, so the potato slice in this solution should be unchanged. Turgor pressure decreases in hypertonic solutions, so the potato slice in this solution should shrink and become limp.)

RETEACHING *ACTIVITY*

To help students understand the various processes in this section, have them develop analogies for specific terms. For example, students might associate diffusion with the experience of detecting odors as they enter a room and osmosis with the image of rainwater oozing through a tent roof. They might associate carrier proteins with the image of trucks carrying important medical shipments across a border.

TERMINOLOGY NOTE

The term *ion channel* replaces the less accurate term *gated channel,* which was used in the previous edition of *Modern Biology.* All ion channels are membrane proteins that allow certain charged molecules to cross from one side of a membrane to the other. A gated ion channel is a channel that can be opened or closed by a chemical or physical stimulus, such as the binding of a messenger molecule to a portion of the channel. Not all ion channels are gated ion channels; some are always open.

ANSWERS TO SECTION 5-1 REVIEW

1. Diffusion eventually leads to equilibrium, which occurs when the concentration of molecules is the same throughout a space.
2. Diffusion is the movement of molecules from an area of higher concentration to an area of lower concentration. Osmosis is the diffusion of water across a membrane.
3. The external solution is hypotonic to the cytosol.
4. Carrier proteins bind to specific substances on one side of the cell membrane, change shape, and then release the substances on the other side of the membrane.
5. Both processes help specific substances move down their concentration gradients from one side of the cell membrane to the other, without an input of energy by cells. Both processes also involve membrane proteins.
6. The solutes in the sea water would make the solution outside the body cells more concentrated. Water would then diffuse out of the cells, causing them to shrink and possibly die.

Eco Connection

Purifying Water with Membranes

The tendency for water molecules to diffuse across membranes can be used to extract pure water from a mixture of water and solutes. If a dilute solution is separated from a more concentrated solution by a selectively permeable membrane, osmosis will occur, as water molecules diffuse from the dilute solution to the concentrated solution. However, if enough external pressure is applied to the concentrated solution, the opposite will happen: water molecules will diffuse from the concentrated solution to the dilute solution. This process, called reverse osmosis, effectively moves most of the water to one side of the membrane and leaves most of the solutes on the other side.

Reverse osmosis was initially developed for desalination plants, which produce fresh water from sea water. It is now also used to purify polluted water from a variety of sources, including manufacturing facilities and sanitary landfills. After the polluted water from these sources is purified through reverse osmosis, it is clean enough to be returned safely to the environment.

A good example of facilitated diffusion is the transport of glucose. As you learned in Chapter 3, many cells depend on glucose for much of their energy needs. But glucose molecules are too large to diffuse quickly across cell membranes. When the level of glucose within a cell is lower than the level of glucose outside the cell, carrier proteins accelerate the movement of glucose into the cell.

The transport of glucose illustrates two important properties of facilitated diffusion. First, facilitated diffusion can help substances move either into or out of a cell, depending on the concentration gradient. Thus, when the level of glucose is higher inside a cell than it is outside the cell, facilitated diffusion speeds the diffusion of glucose out of the cell. Second, the carrier proteins involved in facilitated diffusion are each specific for one type of molecule. For example, the carrier protein that helps with the diffusion of glucose and other simple sugars does not assist with the diffusion of amino acids.

DIFFUSION THROUGH ION CHANNELS

Another type of passive transport involves membrane proteins known as **ion channels.** Ions such as sodium (Na^+), potassium (K^+), calcium (Ca^{2+}), and chloride (Cl^-) are important for a variety of cell functions. Because they are not soluble in lipids, however, ions cannot diffuse across the lipid bilayer without assistance. Ion channels provide small passageways across the cell membrane through which ions can diffuse. Each type of ion channel is usually specific for one type of ion. Thus, most Na^+ ion channels will allow Na^+ ions to pass through them, but they will not accept Ca^{2+} or Cl^- ions.

Some ion channels are always open. Others have "gates" that open to allow ions to pass or close to stop their passage. The gates may open or close in response to three kinds of stimuli: stretching of the cell membrane, electrical signals, or chemicals in the cytosol or external environment. These stimuli therefore control the ability of specific ions to cross the cell membrane.

SECTION 5-1 REVIEW

1. **Toward what condition does diffusion eventually lead, in the absence of other influences?**
2. **How is osmosis related to diffusion?**
3. **If the concentration of solute molecules outside a cell is lower than the concentration in the cytosol, is the external solution hypotonic, hypertonic, or isotonic to the cytosol?**
4. **What role do carrier proteins play in facilitated diffusion?**
5. **How is facilitated diffusion similar to diffusion through ion channels?**
6. **CRITICAL THINKING Sea water has a higher concentration of solutes than do human body cells. Why might drinking large amounts of sea water be dangerous for humans?**

ACTIVE TRANSPORT

In many cases, cells must move materials up their concentration gradient, from an area of lower concentration to an area of higher concentration. Such movement of materials is known as ***active transport.*** *Unlike passive transport, active transport requires a cell to expend energy.*

SECTION

5-2

OBJECTIVES

▲ Distinguish between passive transport and active transport.

● Explain how the sodium-potassium pump operates.

■ Compare and contrast endocytosis and exocytosis.

CELL MEMBRANE PUMPS

Carrier proteins not only assist in passive transport but also help with some types of active transport. The carrier proteins that serve in active transport are often called cell membrane "pumps" because they move substances up their concentration gradients. In other respects, the carrier proteins involved in facilitated diffusion and those involved in active transport are very similar. In both, the protein first binds to a specific kind of molecule on one side of the cell membrane. Once it is bound to the molecule, the protein changes shape, shielding the molecule from the hydrophobic interior of the lipid bilayer. The protein then transports the molecule across the membrane and releases it on the other side.

Sodium-Potassium Pump

One example of active transport in animal cells involves a carrier protein known as the **sodium-potassium pump.** As its name suggests, this protein transports Na^+ ions and K^+ ions up their concentration gradients. To function normally, many types of animal cells must have a higher concentration of Na^+ ions outside the cell and a higher concentration of K^+ ions inside the cell. The sodium-potassium pump works to maintain these concentration differences.

Follow the steps in Figure 5-6 to see how the sodium-potassium pump operates. First, three Na^+ ions bind to the carrier protein on the cytosol side of the membrane, as shown in Figure 5-6a. At the same time, the carrier protein splits a phosphate group from a molecule of ATP. As you can see in Figure 5-6b, the phosphate group also binds to the carrier protein. Figure 5-6c shows how the splitting of ATP supplies the energy needed to change the shape of the carrier protein. With its new shape, the protein carries the three Na^+ ions across the membrane and then releases them outside the cell.

At this point, the carrier protein has the shape it needs to bind two K^+ ions outside the cell, as Figure 5-6d shows. When the K^+ ions bind, the phosphate group is released, as indicated in Figure 5-6e, and the carrier protein changes shape again. This time, the change

internetconnect

SCI*LINKS* NSTA

TOPIC: Active transport
GO TO: www.scilinks.org
KEYWORD: HM101

SECTION 5-2

SECTION OVERVIEW

Matter, Energy, and Organization
Section 5-2 distinguishes between passive and active transport and explains how active transport requires cells to expend energy.

Cell Structure and Function
This section also explains how organelles called vesicles are involved in two types of active transport, endocytosis and exocytosis.

▶ ENGAGE STUDENTS

Create an air current by turning on a window air conditioner or a fan or by opening a window. Ask for a student volunteer to sit where he or she can feel the current. Ask the other students if the volunteer is visibly expending any energy to create the air current (no). Then have the student create an air current by waving a paper fan. Ask if the student is now expending energy to create the air current (yes). Explain that this model illustrates the processes used to transport materials across cell membranes: passive transport occurs without any energy expenditure by the cell, whereas active transport requires the cell to expend energy.

TEACHING STRATEGY

Active Transport
Bring to class an air pump and an inflatable ball, and ask a student to inflate the ball with the pump. Relate this activity to a cell membrane pump. The air pump uses energy supplied by the student to move air against a pressure gradient. Cell membrane pumps use energy supplied by the cell to move substances against a concentration gradient.

SECTION 5-2

VISUAL LINK

Figure 5-6

This figure follows the sodium-potassium pump as it goes through six stages (a–f) in one complete pumping cycle.

Notice that the carrier protein changes shape twice during the cycle. The first shape change occurs in step c, after a phosphate group from ATP binds to the protein. This change exposes the three bound Na^+ ions to the outside of the cell. The sites where the Na^+ ions had been bound are now drawn in a less circular shape, showing that the carrier protein doesn't bind strongly to Na^+ ions in this shape. As a result, the three Na^+ ions leave the protein and diffuse away from the cell. The second shape change occurs in step e, when the phosphate group is released from the carrier protein. This change exposes the two bound K^+ ions to the cytosol. The sites where the K^+ ions had been bound are now drawn smaller, showing that the carrier protein doesn't bind strongly to K^+ ions in this shape. As a result, the two K^+ ions leave the protein and diffuse into the cytosol.

GIFTED *ACTIVITY*

Encourage students to write and perform a play that demonstrates how the sodium-potassium pump works. Major characters in the play should include Na^+ and K^+ ions, the protein pump, ATP, and phosphate groups.

CRITICAL THINKING

Active Transport and Disease

Some cancer cells use active transport to eject medications from their cytoplasm. How would this transport be detrimental to a cancer patient? (The cells would resist chemotherapy and remain cancerous.)

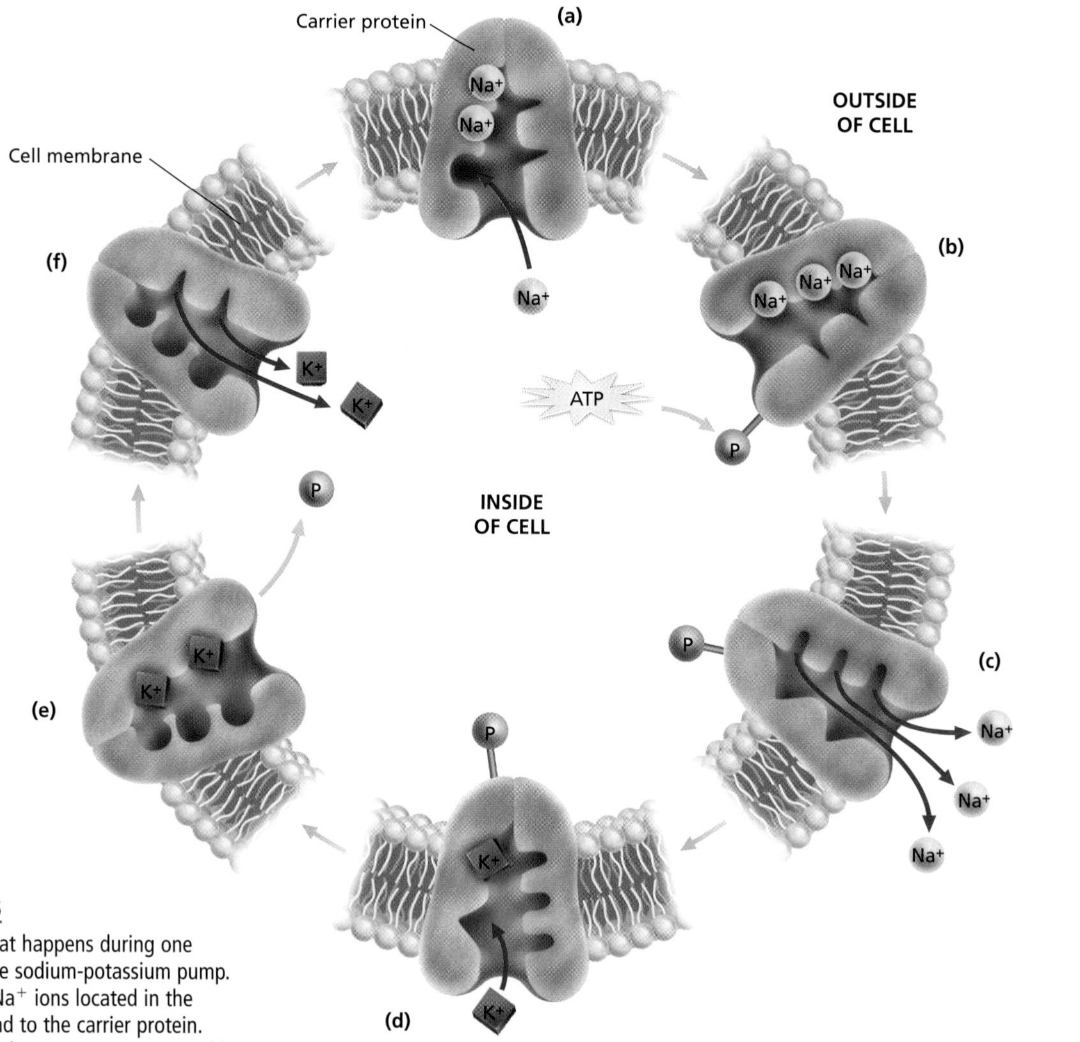

FIGURE 5-6

Follow what happens during one cycle of the sodium-potassium pump. (a) Three Na^+ ions located in the cytosol bind to the carrier protein. (b) A phosphate group, represented by the letter P in the diagram, is removed from ATP and bound to the carrier protein. (c) The binding of the phosphate group changes the shape of the carrier protein, allowing the three Na^+ ions to be released into the cell's environment. (d) Two K^+ ions located outside the cell bind to the carrier protein. (e) The phosphate group is released, changing the shape of the carrier protein again. (f) The two K^+ ions are released into the cytosol, and the cycle is ready to repeat.

in shape causes the carrier protein to release the two K^+ ions inside the cell. As you can see in Figure 5-6f, at this point the carrier protein is ready to begin the process again. Thus, a complete cycle of the sodium-potassium pump transports three Na^+ ions outside the cell and two K^+ ions inside the cell. At top speed, the sodium-potassium pump can transport about 450 Na^+ ions and 300 K^+ ions per second.

The exchange of three Na^+ ions for two K^+ ions creates an electrical gradient across the cell membrane. That is, the outside of the membrane becomes positively charged and the inside of the membrane becomes negatively charged. In this way, the two sides of the cell membrane are like the positive and negative terminals of a battery. This difference in charge is important for the conduction of electrical impulses along nerve cells, as you will learn in Chapter 50. The sodium-potassium pump is only one example of a cell membrane pump. Other pumps work in similar ways to transport important metabolic materials across cell membranes.

ENDOCYTOSIS AND EXOCYTOSIS

Some substances, such as macromolecules and food particles, are too large to pass through the cell membrane by the transport processes you have studied so far. Cells employ two other transport mechanisms—endocytosis and exocytosis—to move such substances across their membranes. Endocytosis and exocytosis are also used to transport large quantities of small molecules into or out of cells at a single time. Both endocytosis and exocytosis require cells to expend energy. Therefore, they are types of active transport.

Endocytosis

Endocytosis (EN-doh-sie-TOH-suhs) is the process by which cells ingest external fluid, macromolecules, and large particles, including other cells. As you can see in Figure 5-7, these external materials are enclosed by a portion of the cell, which folds into itself and forms a pouch. The pouch then pinches off from the cell membrane and becomes a membrane-bound organelle called a **vesicle.** Some of the vesicles fuse with lysosomes, and their contents are digested by lysosomal enzymes. Other vesicles that form during endocytosis fuse with other membrane-bound organelles.

Biologists distinguish two types of endocytosis, based on the kind of material that is taken into the cell: **pinocytosis** (PIN-oh-sie-TOH-suhs) involves the transport of solutes or fluids, while **phagocytosis** (FAG-oh-sie-TOH-suhs) is the movement of large particles or whole cells. Many unicellular organisms feed by phagocytosis. In addition, certain cells in animals use phagocytosis to ingest bacteria and viruses that

internetconnect

SCILINKS NSTA

TOPIC: Endocytosis
GO TO: www.scilinks.org
KEYWORD: HM103

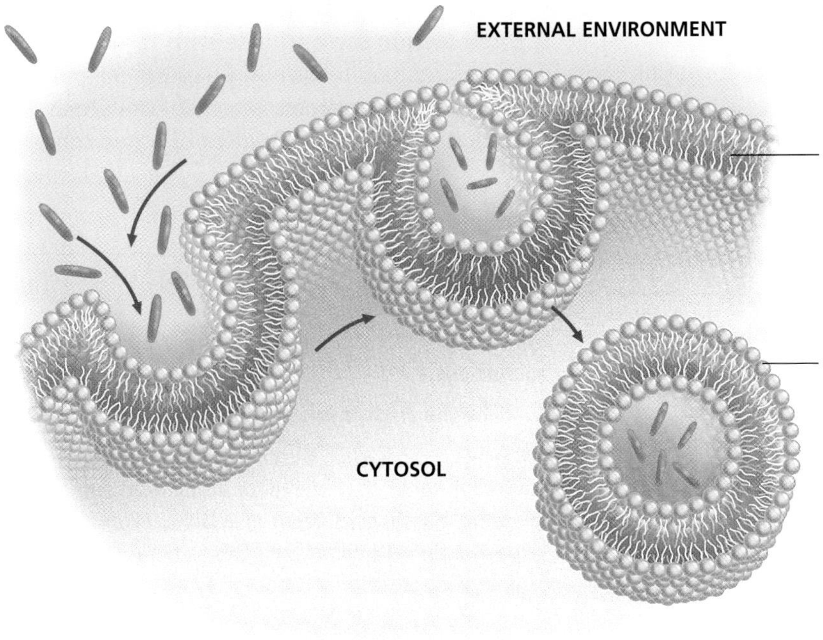

FIGURE 5-7

During endocytosis, the cell membrane folds in and forms a small pouch. The pouch then pinches off from the cell membrane to become a vesicle.

DEMONSTRATION

Endocytosis

Make a transparency of a cell forming a vesicle, and illustrate the stages involved in endocytosis. Cut apart the pieces and demonstrate the stages of endocytosis by manipulating the pieces on an overhead projector.

INCLUSION ACTIVITY

Provide students with laminated pieces of paper that are color-coded to represent cell membranes, vesicles, and lysosomes. Let students manipulate the pieces to demonstrate endocytosis and exocytosis on their desktop.

CULTURAL CONNECTION

Sweat Lodges

Different cultures throughout the world have used water and mineral salts to purify their bodies. Some Native American tribes used sweat lodges or community steam baths for physical and spiritual purification. The dome-shaped sweat lodge was made from a willow tree frame, with burning hot rocks placed in the center of the earthen floor. Small pieces of sage, sweetgrass, or juniper tossed on the rocks burned as incense to purify the air. Bathers drank tea made with medicinal plants and sprinkled it on the rocks. The tea stimulated the sweating process.

RECENT RESEARCH

Proteins and Phagocytosis

Scientists have shown recently that a cytoskeletal protein called coronin accumulates in cellular extensions that protrude from the cell surface during phagocytosis. When phagocytosis is complete, coronin returns to its original location in the cytoplasm.

ACTIVE READING

Technique: Anticipation Guide

Have students return to the lists that they made for the Active Reading activity on page 95. Ask them to check everything they learned that was on their list. Review any unchecked items with the students. Perhaps the material will be covered in a future chapter. If not, spend some time discussing these items.

ANSWERS TO SECTION 5-2 REVIEW

1. Passive transport moves substances down their concentration gradients and, therefore, does not require an input of energy from the cell. Active transport moves substances up their concentration gradients and does require an input of energy from the cell.
2. Carrier proteins bind to specific molecules on one side of the membrane and transport the molecules up their concentration gradients to the other side of the membrane.
3. The removal of a phosphate group from ATP supplies the energy that drives the sodium-potassium pump.
4. Pinocytosis involves the transport of solutes or fluids into cells, whereas phagocytosis is the movement of large particles into cells.
5. During exocytosis, cytoplasmic vesicles fuse with the cell membrane and release their contents outside the cell.
6. The sodium-potassium pump counteracts the accumulation of potassium outside cells by transporting potassium into the cells in exchange for sodium.

FIGURE 5-8
During exocytosis, a vesicle moves to the cell membrane, fuses with it, and then releases its contents to the outside of the cell.

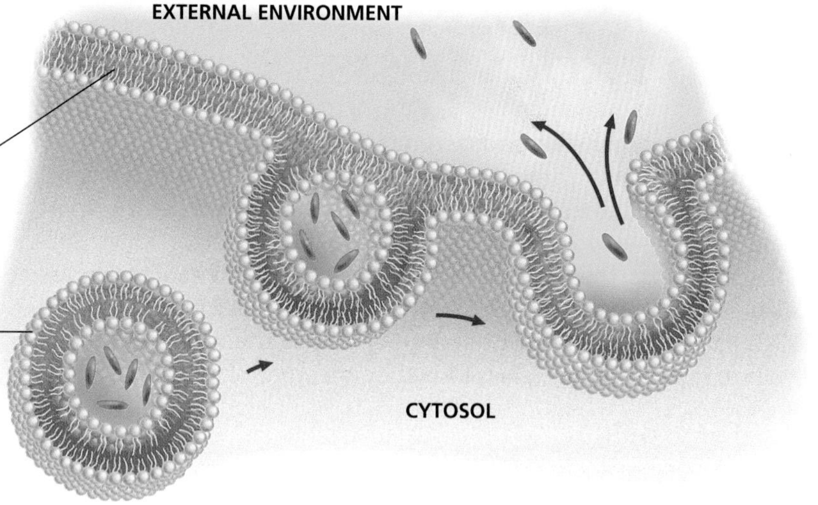

FIGURE 5-9
The vesicle shown here has just fused with the cell membrane, and the vesicle's contents are entering the external environment of the cell. (TEM, 71,250)

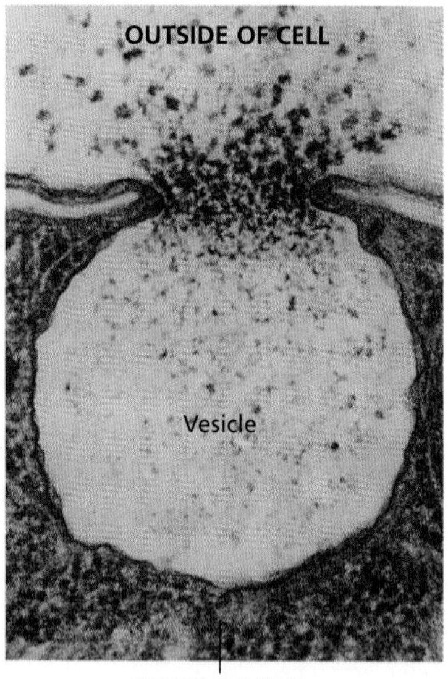

invade the body. These cells, known as **phagocytes,** allow lysosomes to fuse with the vesicles that contain the ingested bacteria and viruses. Lysosomal enzymes then destroy the bacteria and viruses before they can harm the animal.

Exocytosis

Exocytosis (EK-soh-sie-TOH-suhs), illustrated in Figure 5-8, is essentially the reverse of endocytosis. During exocytosis, vesicles in the cytoplasm fuse with the cell membrane, releasing their contents into the cell's external environment. Figure 5-9 shows a vesicle in the process of exocytosis. Cells may use exocytosis to release large molecules such as proteins. Recall that proteins are made on ribosomes and packaged into vesicles by the Golgi apparatus. The vesicles then move to the cell membrane and fuse with it, delivering the proteins outside the cell. As you'll learn in Chapters 50 and 51, cells in the nervous and endocrine systems also use exocytosis to release small molecules that control the activities of other cells.

SECTION 5-2 REVIEW

1. Explain the difference between passive transport and active transport.
2. What function do carrier proteins perform in active transport?
3. What provides the energy that drives the sodium-potassium pump?
4. Explain the difference between pinocytosis and phagocytosis.
5. Describe the steps involved in exocytosis.
6. **CRITICAL THINKING** During intense exercise, potassium tends to accumulate in the fluid surrounding muscle cells. What membrane protein helps muscle cells counteract this tendency? Explain your answer.

CHAPTER 5 REVIEW

SUMMARY/VOCABULARY

5-1

- Passive transport involves the movement of molecules across the cell membrane without an input of energy by the cell.
- Diffusion is the movement of molecules from an area of higher concentration to an area of lower concentration, driven by the molecules' kinetic energy. It eventually leads to equilibrium, a condition in which the concentration of the molecules is the same throughout a space or on both sides of a membrane.
- Molecules can diffuse across a cell membrane by dissolving in the lipid bilayer or by passing through pores in the membrane.
- Osmosis is the diffusion of water across a membrane. The net direction of osmosis is determined by the relative solute concentrations on the two sides of the membrane.
- When the solute concentration outside the cell is lower than that in the cytosol, the solution outside is hypotonic to the cytosol, and water will diffuse into the cell.
- When the solute concentration outside the cell is higher than that in the cytosol, the solution outside is hypertonic to the cytosol, and water will diffuse out of the cell.
- When the solute concentrations outside and inside the cell are equal, the solution outside is isotonic, and there will be no net movement of water.
- To remain alive, cells must compensate for the water that enters the cell in hypotonic environments and leaves the cell in hypertonic environments.
- In facilitated diffusion, a carrier protein binds to a molecule on one side of the cell membrane. The protein then changes its shape and transports the molecule down its concentration gradient to the other side of the membrane.
- Ion channels are proteins that provide small passageways across the cell membrane through which specific ions can diffuse.

Vocabulary

carrier protein (99)
concentration gradient (95)
contractile vacuole (97)
cytolysis (99)
diffusion (95)
equilibrium (95)
facilitated diffusion (99)
hypertonic (96)
hypotonic (96)
ion channel (100)
isotonic (97)
osmosis (96)
passive transport (95)
plasmolysis (98)
turgor pressure (98)

5-2

- Active transport moves molecules across the cell membrane from an area of lower concentration to an area of higher concentration. It requires cells to expend energy.
- Some types of active transport are performed by carrier proteins called cell membrane pumps.
- One example of a cell membrane pump is the sodium-potassium pump. It moves three Na^+ ions into the cell's external environment for every two K^+ ions it moves into the cytosol. ATP supplies the energy that drives the pump.
- Endocytosis and exocytosis are active transport mechanisms in which large substances cross the membrane inside vesicles.
- In endocytosis, the cell membrane folds around something in the external environment and forms a pouch. The pouch then pinches off and becomes a vesicle in the cytoplasm. Endocytosis includes pinocytosis, in which the vesicle contains solutes or fluids, and phagocytosis, in which the vesicle contains large particles or cells.
- In exocytosis, vesicles made by the cell fuse with the cell membrane, releasing their contents into the external environment.

Vocabulary

active transport (101)
endocytosis (103)
exocytosis (104)
phagocyte (104)
phagocytosis (103)
pinocytosis (103)
sodium-potassium pump (101)
vesicle (103)

CHAPTER 5 REVIEW ANSWERS

REVIEW

1. In both processes, molecules move down their concentration gradients. In facilitated diffusion, however, carrier proteins increase the rate of diffusion.
2. They have the same concentration of solutes.
3. As turgor pressure decreases, a plant cell shrinks away from the cell wall. This shrinkage is called plasmolysis.
4. A contractile vacuole is an organelle that collects water from the cytoplasm of a unicellular organism. Contraction of the vacuole expels the water from the cell.
5. In pinocytosis, the cell membrane surrounds liquids or small particles and takes them into the cell. In phagocytosis, the cell membrane surrounds and ingests larger particles.

6. b **7.** d
8. c **9.** a
10. c **11.** b
12. a **13.** b
14. d **15.** b

16. Through diffusion, randomly moving molecules become evenly distributed within a space. This condition is called equilibrium.
17. No, all molecules cannot diffuse through all cell membranes. Some molecules are too large or too insoluble in lipids to diffuse through some membranes.
18. Water would move by osmosis into the organism, which would have no means of removing the excess water. The organism would swell and perhaps burst.
19. The direction of net water movement depends on the relative concentration of solutes on either side of the membrane.
20. The lipid bilayer is hydrophobic, preventing passage of most water-soluble ions and molecules.
21. The cell could use carrier proteins to transport glucose by facilitated diffusion.
22. ATP provides the energy that drives the sodium-potassium

pump, which pumps Na^+ out of the cell and K^+ into the cell.
23. Endocytosis uses vesicles to bring external substances into the cell. Exocytosis uses vesicles to release substances from the cell.
24. Answers will vary. The solutions in which body organs are preserved must have specific concentrations of solutes to keep the organ tissues functioning prior to transplant.

CRITICAL THINKING

1. The air pressure in a balloon resembles turgor pressure in that random molecular motion keeps the outer membrane taut. One difference is that the cell membrane is more permeable to water than the balloon is to air. Also, there is no cell wall surrounding a balloon to limit the balloon's expansion.
2. The organism's selectively permeable membrane can be compared to the porous concrete wall; water crosses both barriers. A contractile vacuole removes water from the organism using energy supplied by the organism, just as the sump pump removes water using electrical energy.
3. It suggests that the cell membrane's inner and outer layers have essentially the same structure and are, therefore, interchangeable.
4. All active transport processes would stop, because they require the cell to expend energy and ATP is the main energy currency of the cell. All passive transport processes would continue until the concentration gradients disappeared.
5. Since hydrogen ions move into the cells along with the sugar molecules, the transport of sugar into the cells would lower the concentration of hydrogen ions in the external environment, raising the pH. Removing hydrogen ions from the external environment would stop the entry of sugar into the cells by this mechanism.

REVIEW

Vocabulary

1. Distinguish between diffusion and facilitated diffusion.
2. What does it mean to say that two solutions are isotonic?
3. How is plasmolysis related to turgor pressure in plant cells?
4. What is a contractile vacuole, and how does it function?
5. The word part *pino-* means "to drink," *phago-* means "to eat," and *cyto-* means "cell." With this information, explain why the words *pinocytosis* and *phagocytosis* are good names for the processes they describe.

Multiple Choice

6. During diffusion, molecules tend to move (a) up their concentration gradient (b) down their concentration gradient (c) in a direction that doesn't depend on the concentration gradient (d) from an area of lower concentration to an area of higher concentration.
7. The part of a cell that functions to maintain homeostasis relative to the cell's environment is the (a) cytosol (b) Golgi apparatus (c) nucleus (d) cell membrane.
8. Ion channels aid the movement of (a) molecules up a concentration gradient (b) carrier proteins within the lipid bilayer (c) ions across a cell membrane (d) water across a cell membrane.
9. Glucose enters a cell most rapidly by (a) facilitated diffusion (b) diffusion (c) osmosis (d) phagocytosis.
10. When the cells in a plant have low turgor pressure, the plant (a) is rigid (b) dies (c) wilts (d) explodes.
11. The sodium-potassium pump transports (a) Na^+ into the cell and K^+ out of the cell (b) Na^+ out of the cell and K^+ into the cell (c) both Na^+ and K^+ into the cell (d) both Na^+ and K^+ out of the cell.
12. A cell must expend energy to transport substances using (a) cell membrane pumps (b) facilitated diffusion (c) ion channels (d) osmosis.
13. Some animal cells engulf, digest, and destroy invading bacteria through the process of (a) exocytosis (b) phagocytosis (c) pinocytosis (d) all of the above.
14. Carrier proteins are important in (a) osmosis (b) endocytosis (c) diffusion (d) facilitated diffusion.
15. The drawing below shows a plant cell after the solute concentration of its environment has been changed. The new external environment is most likely (a) isotonic (b) hypertonic (c) hypotonic (d) none of the above.

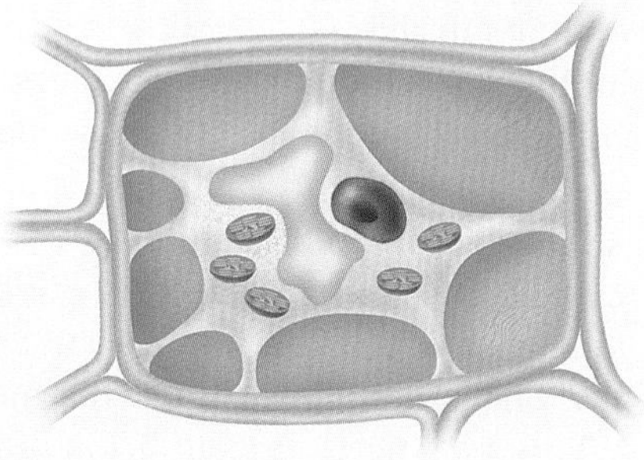

Short Answer

16. What does it mean to say that diffusion eventually results in equilibrium?
17. Can all molecules diffuse through all cell membranes? Explain your answer.
18. Describe three structures characteristic of plant cells. Explain the function of each.
19. What determines the direction of net movement of water across a cell membrane?
20. How does the lipid bilayer of a membrane form a barrier to molecules?
21. How can a cell that consumes glucose speed up its intake of glucose from the environment?
22. How is ATP involved in maintaining the sodium and potassium gradients across a cell membrane?
23. Distinguish between endocytosis and exocytosis.
24. BIOLOGY INTERACTIVE TUTOR Write a report summarizing the roles of osmosis and diffusion in the preservation of body organs donated for transplants. Why must organs be preserved in special solutions prior to a transplant? Find out what kinds of substances these solutions contain.

CRITICAL THINKING

1. There is a higher concentration of air molecules inside an inflated balloon than there is outside the balloon. Because of their constant random motion, the molecules inside press against the balloon and keep it taut. How is the pressure exerted by these air molecules similar to turgor pressure? How is it different?
2. Sometimes water seeps through the concrete wall of a basement after a heavy rain, and the homeowner must remove it with a sump pump. How can this situation be compared to the action of a unicellular organism that lives in a pond?
3. When a cell takes in substances through endocytosis, the cell membrane forms an inside-out vesicle. That is, the outside of the cell membrane becomes the inside of the vesicle. What might this suggest about the structure of the cell membrane?
4. If a cell were exposed to a poison that blocked the cell's ability to manufacture ATP, what effect would that have on the cell membrane's transport processes?
5. Some plant cells have carrier proteins that transport sugar molecules and hydrogen ions (H^+) into the cytosol at the same time. These carrier proteins move sugar molecules up their gradient as hydrogen ions move down their gradient. How would the transport of sugar into these cells affect the pH of the cells' external environment? What would happen to the transport of sugar if hydrogen ions were removed from the external environment?
6. A gelatin block is prepared with a chemical indicator that turns pink in the presence of a base. The block is enclosed in a membrane and placed in a beaker of ammonium hydroxide solution. After half an hour, the block begins to turn pink. Account for the gelatin's pink color.
7. The two curves below show the rate at which glucose is transported across a cell membrane versus the concentration gradient of glucose. One curve represents the diffusion of glucose through the lipid bilayer, and the other curve represents the transport of glucose by facilitated diffusion. Which curve corresponds to facilitated diffusion? Explain your reasoning.

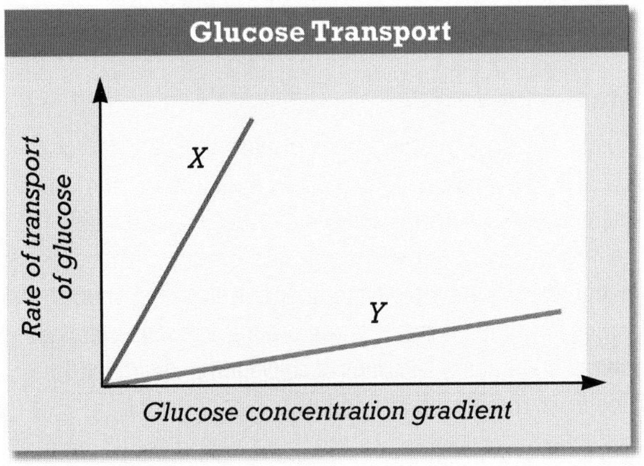

EXTENSION

1. Read "Coming to Grips with the Golgi" in *Science,* December 18, 1999, on page 2169. Describe how the cisternal maturation model of the Golgi system differs from the vesicular transport model.
2. Kidney dialysis is the artificial filtering of blood to remove wastes in patients whose kidneys can no longer function well. Selectively permeable membranes are used in kidney dialysis and in other areas of medicine. Consult your local hospital, and obtain information on how kidney dialysis is performed and on the medical condition of the people who receive dialysis treatment.
3. Examine a number of food items at a grocery store. Find at least four foods for which salt is listed as a preservative. Keeping in mind what you have learned about osmosis, explain why salt is often used to preserve food.

Instead, sugar would leave the cells, because the hydrogen ion gradient would now be reversed.
6. Ammonium hydroxide is small enough to diffuse through the membrane, and because it's a base, it turns the indicator pink.
7. Curve *X* corresponds to facilitated diffusion because it indicates a much higher rate of transport than curve *Y.* Facilitated diffusion increases the rate of transport of substances such as glucose, which are not very soluble in lipids.

EXTENSION

1. There are two views of the role Golgi systems have in transporting newly synthesized proteins. In the vesicular transport model, vesicles bud off of one cisterna and fuse with the next. In the cisternal maturation model, the cisternae move forward while the vesicles move backward to aid in recycling parts of the Golgi and the ER.
2. If their local hospital does not perform or has little information on peritoneal dialysis, students may find more information on the procedure by using an on-line reference source or searching the World Wide Web.
3. Salt can be used to preserve pickles, eggs, and meats. Through osmosis, water will move out of the cells in the food tissues. Because digestion cannot occur in the absence of water, lysosomes and cell enzymes cannot digest the food. Also, salt can kill some bacteria and other microorganisms by causing them to lose too much water.

CHAPTER 5
INTERACTIVE EXPLORATION

For an overview of this Interactive Exploration and preparatory information related to this activity, please see the CD User's Guides.

PREPARATION NOTES

Time required:
One 50-minute class period

PRELAB DISCUSSION

Before students begin the lab, they should be able to answer the following questions:

1. What is the purpose of carrier proteins in the cell membrane? (Carrier proteins allow substances that cannot pass through the cell membrane to enter or leave the cell.)

2. What effect does solute concentration have on the net movement of water across a cell membrane? (There will be a net movement of water in the direction of higher solute concentration.)

PROCEDURAL TIPS

1. Use the inset of the computer screen to review the functions of each feedback meter before allowing students to begin the lab.

2. Remind students that the upper part of the screen (above the cell membrane) represents the outside of the cell and that the lower part of the screen represents the inside of the cell.

ANSWERS TO PART A

2. Water molecules move into and out of the cell through pores in the lipid bilayer of the cell membrane. Chloride ions move into and out of the cell through protein channels. The CF proteins determine whether the chloride ion channels are open or closed.

3. There is no mucus buildup outside the cell membrane because this person has two normal genes. Therefore, the chloride ion channels are working properly.

CHAPTER 5

INTERACTIVE EXPLORATION

Exploring the Role of Osmosis in Cystic Fibrosis

OBJECTIVES

- Simulate the effects of three different genotypes on the movement of water molecules and chloride ions through the cell membrane.
- Relate the symptoms of cystic fibrosis to the ability of water molecules and chloride ions to pass through the cell membrane.

MATERIALS

- computer with CD-ROM drive
- CD-ROM *Interactive Explorations in Biology: Human Biology*

Background

Cystic fibrosis is a fatal genetic disease that results from the failure of chloride ions (Cl^-) to pass through the cell membrane. As a result, chloride ions accumulate inside the cells of the body, drawing water into the cells. When this happens in the lungs, pancreas, and liver, a thick mucus builds up on the lining of these organs. The severe congestion that gradually develops usually causes death by the age of 30 in afflicted individuals.

In this interactive exploration, you will examine the three possible pairs of genes that an individual can have: two normal genes, represented by +/+; one cystic fibrosis gene and one normal gene, represented by cf/+; and two cystic fibrosis genes, represented by cf/cf. Then you will explore how these genetic combinations affect the membrane proteins involved in transporting Cl^- ions through the cell membrane.

Prelab Preparation

1. Load and start the program Cystic Fibrosis. Click the Topic Information button on the Navigation palette. Read the focus questions, and review the following concepts: Transport Channels, Osmosis, Mutation, and Cystic Fibrosis.
2. Click the word *File* at the top left of the screen, and select Interactive Exploration Help. Listen to the instructions for operating the exploration. Click the Exploration button at the bottom right of the screen to begin the exploration. You will see an animated drawing like the one below.

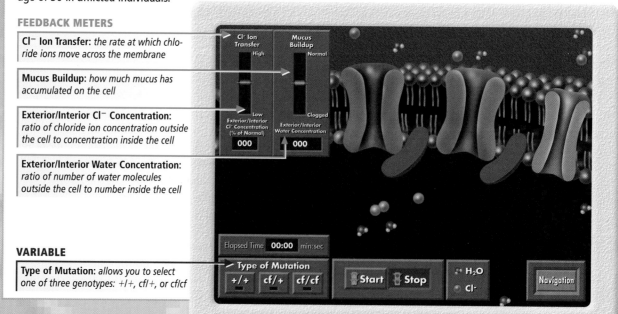

FEEDBACK METERS

Cl^- Ion Transfer: *the rate at which chloride ions move across the membrane*

Mucus Buildup: *how much mucus has accumulated on the cell*

Exterior/Interior Cl^- Concentration: *ratio of chloride ion concentration outside the cell to concentration inside the cell*

Exterior/Interior Water Concentration: *ratio of number of water molecules outside the cell to number inside the cell*

VARIABLE

Type of Mutation: *allows you to select one of three genotypes: +/+, cf/+, or cf/cf*

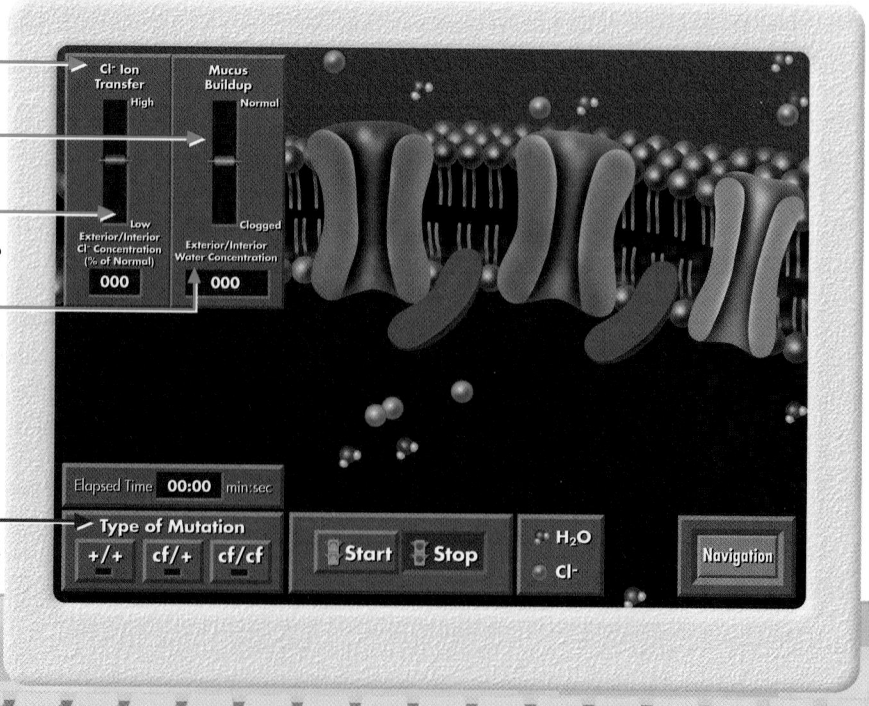

CHLORIDE ION AND WATER CONCENTRATION INSIDE CELLS

Genetic makeup	Exterior/interior Cl^- concentration	Exterior/interior water concentration	Mucus buildup
+/+	100%	normal	normal
cf/+	50%	normal	normal
cf/cf	20%	low	low

Procedure

Create a table like the one above for recording your data.

PART A Two Normal Genes (+/+)

1. Move the pointer to the Type of Mutation box, and click the +/+ button. The transport of Cl^- ions and water molecules in a person with this gene combination is normal. Note that water molecules and Cl^- ions are located both outside the cell (the portion of the screen above the cell membrane) and inside the cell (the portion of the screen below the cell membrane). Also note the position of the two CF proteins in the cell membrane.
2. Click the Start button to begin the simulation. Allow the simulation to run for two to three minutes. Describe how water molecules pass through the cell membrane. How does this differ from the passage of Cl^- ions through the membrane? What appears to be the function of the two CF proteins?
3. Click the Stop button. In your table, record the Exterior/Interior Cl^- Concentration, the Exterior/Interior Water Concentration, and the amount of mucus buildup outside the cell membrane.

PART B One Cystic Fibrosis Gene and One Normal Gene (cf/+)

4. Move the pointer to the Type of Mutation box, and click the cf/+ button. Describe what happens to the CF protein shown on the left and how this affects the Cl^- ion channel.
5. Click the Start button, and allow the simulation to run for two to three minutes. Describe the changes in the way water molecules and Cl^- ions move through the membrane. How will the change in the movement of water molecules and Cl^- ions affect the external surface of the cell?
6. Click the Stop button. In your table, record the same three variables that you recorded in step 3. On which side of the membrane are chloride ions at the higher concentration? On which side is water at the higher concentration? Compare the results with those you obtained in Part A, and explain any differences.

PART C Two Cystic Fibrosis Genes (cf/cf)

7. Move the pointer to the Type of Mutation box, and click the cf/cf button. An individual with this gene combination will be afflicted with cystic fibrosis. Describe what happens to both CF membrane proteins and how this affects the Cl^- ion channels.
8. Click the Start button, and allow the simulation to run for two to three minutes. Describe the changes in the movement of water molecules and Cl^- ions through the membrane. How will the external surface of the cell be affected?
9. Click the Stop button. Describe the difference between the cell's external environment and its internal environment. In your table, record the same variables that you recorded in steps 3 and 6. On which side of the membrane are chloride ions at the higher concentration? Compare the results with those you obtained in Parts A and B, and explain any differences.

Analysis and Conclusions

1. Why are the Cl^- ion channels, but not the water channels, said to have gates?
2. Use the term *hypotonic, hypertonic,* or *isotonic* to describe the cells that line the organs of individuals with cystic fibrosis.
3. How does an accumulation of chloride ions inside a cell result in a thickening of the mucus outside the cell?

ANSWERS TO PART B

4. The CF protein changes shape, closing the channel.
5. With fewer functional channels, fewer chloride ions diffuse through the cell membrane. With more chloride ions remaining inside the cells, more water is drawn into the cells by osmosis. Some buildup of mucus occurs on the outside of the cell membrane.
6. Chloride ions are at a higher concentration inside the cell than outside the cell, and water is drawn into the cell as a result. In Part A, the chloride ions and water molecules were in balance across the cell membrane because all of the chloride ion channels were functioning properly.

ANSWERS TO PART C

7. Both of the CF proteins have changed shape, closing all of the chloride ion channels.
8. No chloride ions can diffuse through the cell membrane. Chloride ions and water will accumulate inside the cell, and mucus will accumulate on the external surface.
9. Chloride ions are at a higher concentration on the inside, and again water is drawn into the cell. Outside the cell, more mucus has built up than in Part B.

ANSWERS TO ANALYSIS AND CONCLUSIONS

1. The chloride ion channels can be closed by changes in the shape of the CF protein. The water channels cannot be closed.
2. The environment inside the cells is hypertonic because of the buildup of chloride ions.
3. The accumulation of chloride ions inside the cell leaches water from the cell's external environment, causing mucus to thicken and clog passageways.

CHAPTER 6

PHOTOSYNTHESIS

	TOPICS	TEACHING RESOURCES	LABS, CLASSWORK, AND HOMEWORK
BLOCKS 1 & 2 90 minutes	**Introducing the Chapter, p. 110**	**ATE** Focus Concept, p. 110 **ATE** Assessing Prior Knowledge, p. 110 **ATE** Understanding the Visual, p. 110	■ Supplemental Reading Guide, *The Lives of a Cell: Notes of a Biology Watcher*
	6-1 Capturing the Energy in Light, p. 111 Energy for Life Processes, p. 111 Light Absorption in Chloroplasts, p. 112 Electron Transport, p. 114 Chemiosmosis, p. 116	**ATE** Section Overview, p. 111 **ATE** Critical Thinking, p. 112 **ATE** Visual Link Figure 6-5, p. 114 11A, 12–13, 15, 18	**Lab B4,** Plant and Animal Interrelationships ★ Study Guide, Section 6-1 **PE** Section Review, p. 116
BLOCKS 3 & 4 90 minutes	**6-2 The Calvin Cycle, p. 117** Carbon Fixation by the Calvin Cycle, p. 117 Alternative Pathways, p. 119 Rate of Photosynthesis, p. 120	**ATE** Section Overview, p. 117 **ATE** Critical Thinking, p. 118 **ATE** Visual Link Figure 6-8, p. 117 19	**PE** **Quick Lab,** p. 118, Analyzing Photosynthesis **PE** **Chapter 6 Investigation,** p. 124 **Lab C36,** Using Paper Chromatography to Separate Pigments ★ Study Guide, Section 6-2 **PE** Section Review, p. 120
BLOCKS 5 & 6 90 minutes	**REVIEW and ASSESSMENT** **PE** Chapter 6 Review, pp. 122–123	Audio CD Program BIOLOGY INTERACTIVE TUTOR Unit 2—Photosynthesis	**CHAPTER TESTING** ■ Chapter Test (blackline master) ▲ Test Generator ■ Assessment Item Listing ■ Scoring Rubrics and Classroom Management Checklists

MODERN BIOLOGY | HOLT BIOSOURCES

KEY

- PE Pupil's Edition
- ATE Teacher's Edition
- ■ Active Reading Guide
- ★ Study Guide
- Teaching Transparencies
- Laboratory Program

One-Stop Planner CD-ROM
Shows these resources within a customizable daily lesson plan:

- ■ Holt BioSources Teaching Resources
- ■ Active Reading Guide
- ★ Study Guide
- ▲ Test Generator

READING FOR CONTENT MASTERY

Preview: Have students preview the objectives, the introductory paragraph, topic sentences, and captions before reading the chapter.

- ATE Active Reading Technique: K-W-L, p. 111
- ■ Reading Strategy: K-W-L Worksheet
- ■ Active Reading Guide Worksheet 6-1
- PE Summary/Vocabulary, p. 121

- ATE Active Reading Technique: K-W-L, p. 120
- ■ Reading Strategy: K-W-L Worksheet
- ■ Active Reading Guide Worksheet 6-2
- PE Summary/Vocabulary, p. 121

TECHNOLOGY AND INTERNET RESOURCES

Holt Biology Videodiscs Teacher's Correlation Guide, Lessons 6-1 through 6-2

BIOLOGY INTERACTIVE TUTOR: Unit 2—Photosynthesis, Topics 1–6

Audio CD Program, Sections 6-1 through 6-2

Interactive Exploration: Cell Biology and Genetics, Photosynthesis

internetconnect

sciLINKS NSTA
On-line Resources: www.scilinks.org
The following *sci*LINKS Internet resources can be found in the student text for this chapter:

Topics:
- Photosynthesis, p. 111
- Calvin cycle, p. 117

go.hrw.com
On-line Resources: go.hrw.com
Visit the HRW Web site for a variety of resources related to this chapter. Just type in the keyword HM2 Chapter 6.

Smithsonian Institution®
Visit **www.si.edu/hrw** for additional on-line resources.

CNNfyi.com
Visit **www.cnnfyi.com** for late-breaking news and current event stories selected just for you.

LAB ACTIVITY PLANNING

CHAPTER 6 INVESTIGATION

Examining the Rate of Photosynthesis, pp. 124–125

OVERVIEW

In this interactive exploration, students will use a computer simulation to study how varying the wavelength and intensity of light affects the rate of photosynthesis. In Part A, students will vary the intensity of light, measured in luxes, while holding the wavelength constant. In Part B, students will vary the wavelength, measured in nanometers, while holding the intensity constant.

A cross section of a thylakoid membrane—the site of light-dependent reactions—within the chloroplast of a cell is presented. Students will manipulate the wavelength and intensity of light that falls on this chloroplast and observe the consequences on the rate of ATP production. The more ATP that is produced—indicated by the percent maximal ATP—the greater the amount of carbohydrate that can be synthesized. Students will graph this information to determine the optimal light intensity and optimal wavelength for photosynthesis.

PREPARATION

1. Access to the following materials is needed:
   - computer with CD-ROM drive
   - CD-ROM *Interactive Explorations in Biology: Cell Biology and Genetics.* The title of this program on the CD-ROM is "Photosynthesis."
2. Have students read the Objectives, Background, and Prelab Preparation sections before starting the lab.

Quick Labs

Analyzing Photosynthesis, p. 118
Students will need disposable gloves, lab aprons, safety goggles, Erlenmeyer flasks, bromothymol blue, and sprigs of *Elodea* for this lab activity.

UNDERSTANDING THE VISUAL

Ask students to study the photograph of the field of corn and to provide two reasons why photosynthetic organisms are vital for the survival of nearly all other living things. (Photosynthetic organisms capture the sun's energy and store it in organic compounds, and they release molecular oxygen, which they and other living things use to harness the energy in those organic compounds.)

FOCUS CONCEPT

Matter, Energy, and Organization

Energy is stored in or released from matter when the chemical bonds holding molecules together are broken and new bonds are formed. In the chemical reactions that constitute photosynthesis, some of the energy in sunlight is captured and stored in organic compounds that are synthesized from carbon dioxide and water. The organization required for these reactions is provided by the structure of chloroplasts.

ASSESSING PRIOR KNOWLEDGE

Review the following concepts.

Chemistry: *Chapter 2*
Ask students to define the following terms: *proton, electron, compound.*

Biochemistry: *Chapter 3*
Ask students to differentiate between inorganic compounds and organic compounds. Ask students to define the term *carbohydrate* and to tell what type of carbohydrate glucose is.

The Cell: *Chapter 4*
Ask students to define the terms *chloroplast* and *thylakoid.* Ask students what chemical process occurs inside chloroplasts.

CHAPTER 6

PHOTOSYNTHESIS

Through photosynthesis, these corn plants obtain energy from the sun and store it in organic compounds.

FOCUS CONCEPT: *Matter, Energy, and Organization*

As you read about photosynthesis, notice the mechanisms in cells that keep the process operating.

BIOLOGY INTERACTIVE TUTOR **Unit 2—*Photosynthesis***
Topics 1–6

CAPTURING THE ENERGY IN LIGHT

All organisms use energy to carry out the functions of life. Some organisms obtain this energy directly from sunlight. They capture part of the energy in light and store it within organic compounds. The process by which this energy transfer takes place is called ***photosynthesis.***

SECTION 6-1

OBJECTIVES

▲ Explain how the structure of the chloroplast relates to its function.

● Describe the role of chlorophylls and other pigments in photosynthesis.

■ Summarize the main events of electron transport.

◆ Describe what happens to a water molecule in photosynthesis.

▲ Explain how ATP is synthesized during the light reactions.

ENERGY FOR LIFE PROCESSES

You learned in Chapter 1 that organisms can be classified according to how they obtain energy. Organisms that manufacture their own food from inorganic substances and energy are autotrophs. Most autotrophs use photosynthesis to convert light energy from the sun into chemical energy, which they then store in various organic compounds, primarily carbohydrates. Plants are the most common example of photosynthetic organisms, but algae and some bacteria can also make their own organic compounds through photosynthesis.

Recall that animals and other organisms that cannot manufacture their own organic compounds from inorganic substances are called heterotrophs. Heterotrophs obtain food by eating autotrophs or by eating other heterotrophs that feed on autotrophs. For example, a caterpillar is a heterotroph that feeds directly on an autotroph, grass. A bird that eats caterpillars is also a heterotroph. The food that fuels the bird originates with autotrophs, but it passes *indirectly* to the bird through the caterpillars. In a similar way, all life ultimately depends on autotrophs.

Photosynthesis involves a complex series of chemical reactions, in which the product of one reaction is consumed in the next reaction. A series of reactions linked in this way is referred to as a **biochemical pathway.**

As you can see in Figure 6-1, autotrophs use the biochemical pathways of photosynthesis to manufacture organic compounds from carbon dioxide (CO_2) and water. During this conversion, molecular oxygen (O_2) is released.

Some of the energy stored in organic compounds is released by cells in another set of biochemical pathways, known as cellular

internetconnect

SCILINKS NSTA

TOPIC: Photosynthesis
GO TO: www.scilinks.org
KEYWORD: HM111

SECTION 6-1

SECTION OVERVIEW

Cell Structure and Function
Section 6-1 describes the structure of chloroplasts and thylakoids and explains how they are involved in the light reactions of photosynthesis.

Matter, Energy, and Organization
This section also explains how the energy in sunlight drives the light reactions.

ACTIVE READING

Technique: K-W-L
Before students read this chapter, have them write short, individual lists of all the things they already Know (or think they know) about photosynthesis. Ask students to contribute their entries to a group list on a large poster board. Then have students list things they Want to know about this process. Have students save their lists as well as the poster board for use later in the Active Reading activity on page 120.

CONTENT CONNECTION

Section 3-2: Organic Compounds
During photosynthesis, CO_2, H_2O, and energy harvested from the sun are used to make organic compounds.

SECTION 6-1

CRITICAL THINKING

Shade-Grown Plants Versus Sun-Grown Plants

Plants grown in the shade produce larger leaves than plants grown in full sun. Propose a hypothesis to explain this. (The larger leaves of shade-grown plants intercept more sunlight, which is advantageous because less light reaches these leaves.)

GIFTED *ACTIVITY*

Have students use the library to research the chemoautotrophic bacteria that live around hot-water vents on the sea floor. The students should explain how these autotrophs manufacture organic compounds without sunlight. (The bacteria obtain energy by oxidizing hydrogen sulfide, H_2S, rather than water, H_2O.)

DEMONSTRATION

Obtain a prism from a scientific supply company. Darken the room and hold the prism in front of a white light source, such as a flashlight or lamp. Position the light source and prism so that the prism projects a spectrum of colors against a white background. Point out that the colors in the spectrum are arranged according to their wavelength. Insert filters of various colors (such as colored cellophane, overhead transparency sheets, or camera filters) between the prism and the background to demonstrate what happens to the spectrum when certain wavelengths are absorbed. (A red filter, for example, should leave red in the spectrum but remove the other colors.)

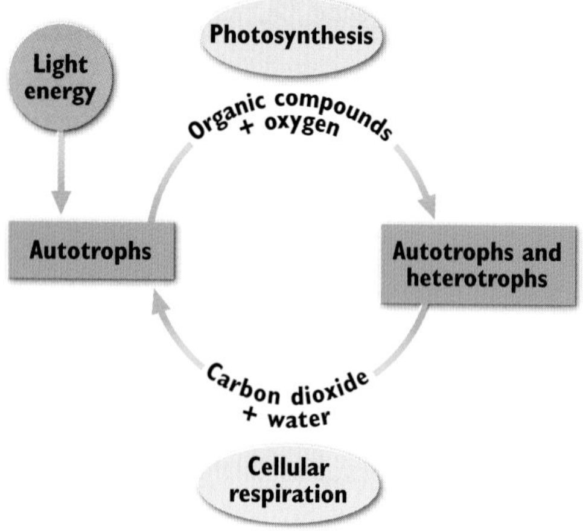

FIGURE 6-1

Many autotrophs produce organic compounds and oxygen through photosynthesis. Both autotrophs and heterotrophs produce carbon dioxide and water through cellular respiration.

respiration. As Figure 6-1 indicates, both autotrophs and heterotrophs perform cellular respiration. During cellular respiration in most organisms, organic compounds are combined with O_2 to produce ATP, yielding CO_2 and water as waste products. Thus, the products of photosynthesis, organic compounds and O_2, are reactants used in cellular respiration. The waste products of cellular respiration, CO_2 and water, are reactants used in photosynthesis. You will learn more about cellular respiration in Chapter 7. The rest of this chapter will focus on photosynthesis in plants.

LIGHT ABSORPTION IN CHLOROPLASTS

In plants, the initial reactions in photosynthesis are known collectively as the **light reactions.** They begin with the absorption of light in chloroplasts. Remember from Chapter 4 that a chloroplast is an organelle found in the cells of plants and in unicellular eukaryotes known as algae. While some algae may contain a single large chloroplast, a cell in the leaf of a plant may have 50 or more chloroplasts.

Most chloroplasts are similar in structure, regardless of the organism in which they are found. As you learned in Chapter 4, each chloroplast is surrounded by a pair of membranes. Inside the inner membrane is another system of membranes, arranged as flattened sacs called thylakoids. Figure 6-2 shows that the thylakoids are interconnected and that some are layered on top of one another to form stacks called **grana** (GRAY-nuh). Surrounding the thylakoids is a solution called the **stroma** (STROH-muh).

FIGURE 6-2

Photosynthesis in eukaryotes occurs inside the chloroplasts. The light reactions of photosynthesis take place in the thylakoids, which are stacked to form grana.

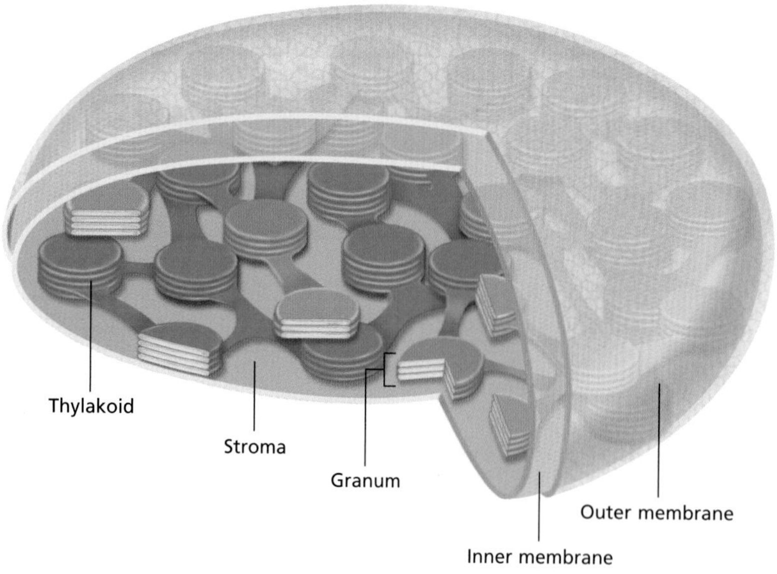

Light and Pigments

To explain how chloroplasts absorb light in photosynthesis, it is important to understand some of the properties of light. Light from the sun appears white, but it is actually composed of a variety of colors. As Figure 6-3 demonstrates, you can separate white light into its component colors by passing the light through a prism. The resulting array of colors, ranging from red at one end to violet at the other, is called the **visible spectrum.**

Light travels through space as waves of energy. These waves are analogous to the waves that travel across a body of water when an object hits the surface. Like water waves, light waves can be measured in terms of their **wavelength,** the distance between crests in a wave. You can see in Figure 6-3 that the different colors in the visible spectrum have different wavelengths.

When white light strikes an object, its component colors can be reflected, transmitted, or absorbed by the object. However, the various colors will react differently if the object contains a **pigment,** which is a compound that absorbs light. Most pigments absorb certain colors more strongly than others. By absorbing certain colors, a pigment subtracts those colors from the visible spectrum. Therefore, the light that is reflected or transmitted by the pigment no longer appears white. For example, the lenses in green-tinted sunglasses contain a pigment that reflects and transmits green light and absorbs the other colors. As a result, the lenses look green.

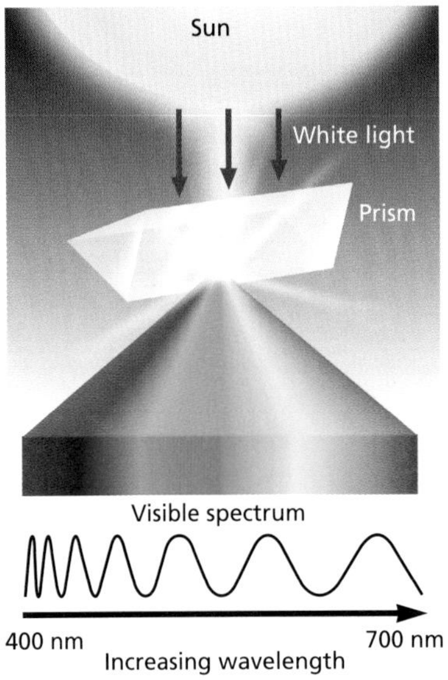

FIGURE 6-3

White light contains a variety of colors called the visible spectrum. Each color has a different wavelength, measured in nanometers.

Chloroplast Pigments

Located in the membrane of the thylakoids are a variety of pigments, the most important of which are called **chlorophylls** (KLOHR-uh-FILZ). There are several different types of chlorophylls. The two most common types are designated chlorophyll *a* and chlorophyll *b*.

A slight difference in molecular structure between chlorophyll *a* and chlorophyll *b* causes the two molecules to absorb different colors of light. As Figure 6-4 shows, chlorophyll *a* absorbs less blue light but more red light than chlorophyll *b* absorbs. Neither chlorophyll *a* nor chlorophyll *b* absorbs much green light. Instead, they allow green light to be reflected or transmitted. That is why young leaves and other plant structures that contain large amounts of chlorophyll look green.

Only chlorophyll *a* is directly involved in the light reactions of photosynthesis. Chlorophyll *b* assists chlorophyll *a* in capturing light energy, and therefore chlorophyll *b* is called an **accessory pigment.** Other compounds found in the thylakoid membrane, including the yellow, orange, and brown **carotenoids** (kuh-RAHT-uhn-OYDZ), also function as accessory pigments. Looking again at Figure 6-4, notice that the pattern of light absorption of one of the carotenoids differs from the pattern of either type of chlorophyll. By absorbing colors that chlorophyll *a* cannot absorb, the accessory pigments enable plants to capture more of the energy in light.

In the leaves of a plant, the chlorophylls are much more abundant and therefore mask the colors of the other pigments. But in the nonphotosynthetic parts of a plant, such as fruits and flowers, the colors of the other pigments may be quite visible. During the fall, many plants lose their chlorophylls, and their leaves take on the rich hues of the carotenoids.

FIGURE 6-4

The three curves on this graph show how three pigments involved in photosynthesis differ in the colors of light they absorb. Where a curve has a peak, much of the light at that wavelength is absorbed. Where a curve has a trough, much of the light at that wavelength is reflected or transmitted.

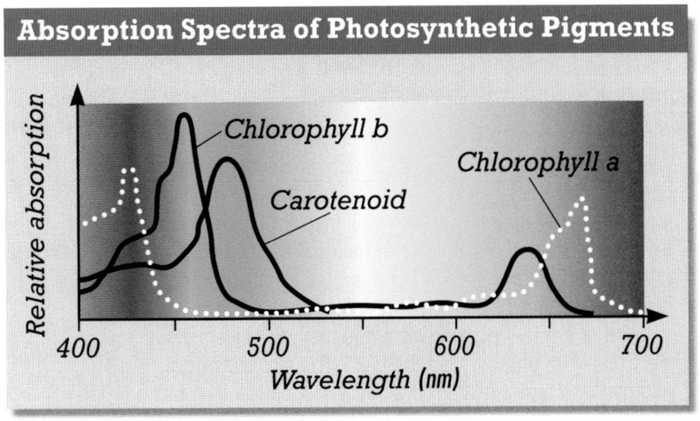

SECTION 6-1

OVERCOMING MISCONCEPTIONS

Many people think that plants are green because they use green light in photosynthesis. In fact, plants absorb and use mainly red and blue light for photosynthesis. Point out that plants appear green because they contain large amounts of chlorophyll, which reflects or transmits most of the green light it intercepts.

CONTENT CONNECTION

Section 2-1: Electron Energy Levels

When light of the right wavelength strikes a chlorophyll molecule, part of the energy in the light is absorbed by an electron in the molecule. That electron moves into a higher energy level, leaving a vacancy in the lower energy level it had occupied. This situation is unstable. In a photosystem, stability returns when chlorophyll donates the excited electron to a primary electron acceptor and fills the vacancy with an electron from another photosystem or from water.

RECENT RESEARCH

Light-Use Efficiency

Scientists studying the dinoflagellate *Amphidinium carterae* have discovered that this alga is nearly 100 percent efficient at transferring light energy that its pigments capture to the electron transport chains. This high efficiency is apparently the result of the alga's chlorophyll and carotenoid pigments being so closely packed. Unlike most plants, *A. carterae* can use blue-green light, the prominent wavelength available in its aquatic environment.

INCLUSION *ACTIVITY*

Have students construct a model of a photosystem, including the thylakoid membrane and the electron transport chain. Ask them to use their imagination to collect ordinary objects for their model. For example, the accessory pigments that absorb light and transfer energy to chlorophyll *a* could be represented by funnels. All of the components of the model should be clearly labeled, and students should include a diagram of a chloroplast showing exactly where each component is located.

RECENT RESEARCH

Photosystems

Recent research on *Chlamydomonas reinhardtii* indicates that a mutant form of this alga carries out the key reactions of photosynthesis—extracting electrons from water and generating O_2, NADPH, and ATP—using only photosystem II. This research suggests that the two-photosystem pathway may be skipped in normal cells under certain conditions.

VISUAL LINK

Figure 6-5

This figure follows the path of electron transport during the light reactions of photosynthesis. It is important that students understand the spatial organization of the structures involved in these reactions. Photosystems I and II, the primary electron acceptors, and the electron transport chains are embedded in the membrane of the thylakoid. Explain that electrons from photosystem II are transferred to photosystem I, not vice versa. At the end of the electron transport path, electrons combine with $NADP^+$ and H^+ to form NADPH. This reaction occurs in the stroma, which lies outside the thylakoids but inside the chloroplast.

ELECTRON TRANSPORT

The chlorophylls and carotenoids are grouped in clusters of a few hundred pigment molecules in the thylakoid membrane. Each cluster of pigment molecules is referred to as a **photosystem.** Two types of photosystems are known: **photosystem I** and **photosystem II.** They are similar in terms of the kinds of pigments they contain, but they have different roles in the light reactions, as you'll soon see.

The light reactions begin when accessory pigment molecules in both photosystems absorb light. By absorbing light, those molecules acquire some of the energy that was carried by the light waves. In each photosystem, the acquired energy is passed quickly to other pigment molecules until it reaches a specific pair of chlorophyll *a* molecules. The events that occur from this point on can be divided into five steps. Refer to Figure 6-5 as you follow these steps.

Step 1. Light energy forces electrons to enter a higher energy level in the two chlorophyll *a* molecules of photosystem II. These energized electrons are said to be "excited."

Step 2. The excited electrons have enough energy to leave the chlorophyll *a* molecules. Because they have lost electrons, the chlorophyll *a* molecules have undergone an oxidation reaction.

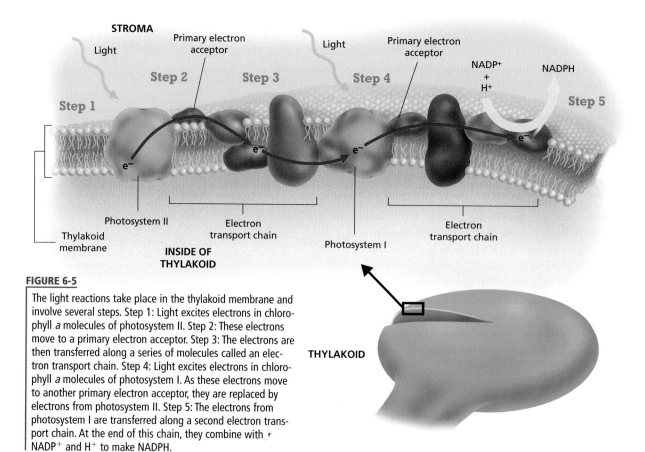

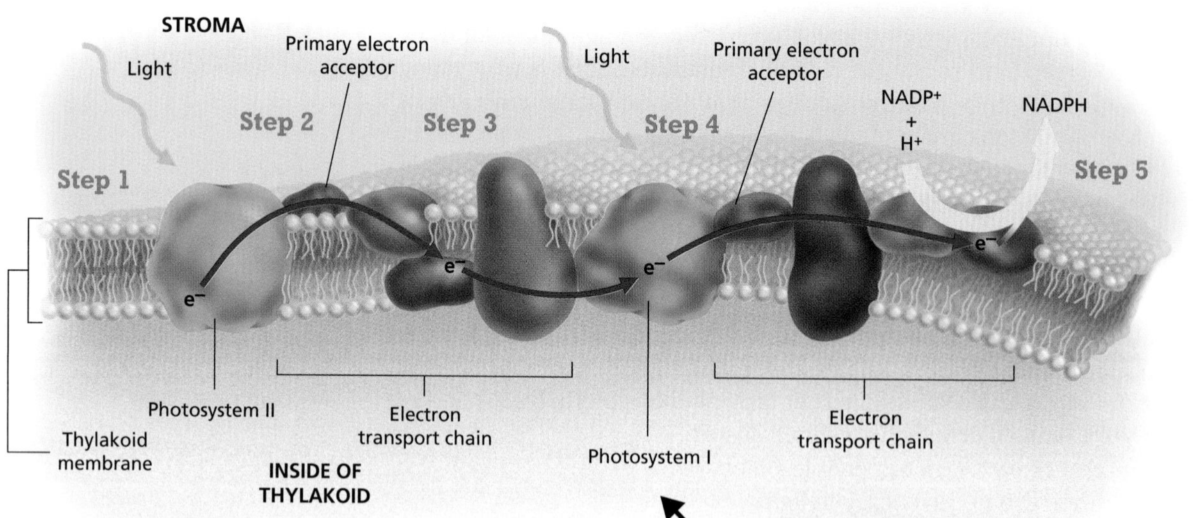

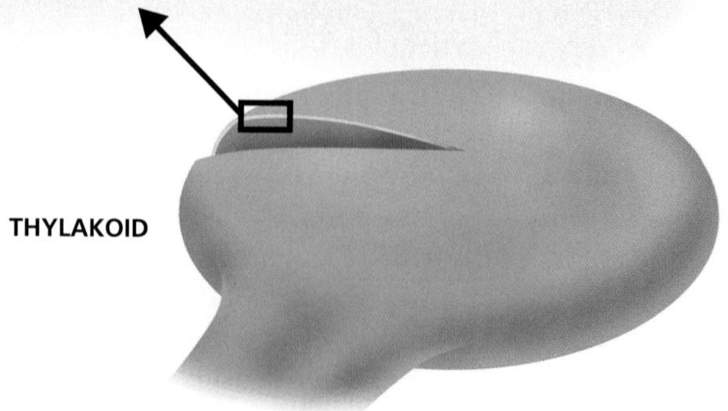

FIGURE 6-5

The light reactions take place in the thylakoid membrane and involve several steps. Step 1: Light excites electrons in chlorophyll *a* molecules of photosystem II. Step 2: These electrons move to a primary electron acceptor. Step 3: The electrons are then transferred along a series of molecules called an electron transport chain. Step 4: Light excites electrons in chlorophyll *a* molecules of photosystem I. As these electrons move to another primary electron acceptor, they are replaced by electrons from photosystem II. Step 5: The electrons from photosystem I are transferred along a second electron transport chain. At the end of this chain, they combine with $NADP^+$ and H^+ to make NADPH.

Remember from Chapter 2 that each oxidation reaction must be accompanied by a reduction reaction. This means that some substance must accept the electrons that the chlorophyll *a* molecules have lost. That substance is a molecule in the thylakoid membrane known as the **primary electron acceptor.**

Step 3. The primary electron acceptor then donates the electrons to the first of a series of molecules located in the thylakoid membrane. This series of molecules is called an **electron transport chain,** because it transfers electrons from one molecule to the next in series. As the electrons pass from molecule to molecule in the electron transport chain, they lose most of the energy that they acquired when they were excited. The energy they lose is harnessed to move protons into the thylakoid.

Step 4. At the same time light is absorbed by photosystem II, light is also absorbed by photosystem I. Electrons move from a pair of chlorophyll *a* molecules in photosystem I to another primary electron acceptor. The electrons that are lost by these chlorophyll *a* molecules are replaced by the electrons that have passed through the electron transport chain from photosystem II.

Step 5. The primary electron acceptor of photosystem I donates electrons to a different electron transport chain. This chain brings the electrons to the side of the thylakoid membrane that faces the stroma. There the electrons combine with a proton and **$NADP^+$**. $NADP^+$ is an organic molecule that accepts electrons during redox reactions. As you can see in Figure 6-5, this reaction causes $NADP^+$ to be reduced to NADPH.

Restoring Photosystem II

You read in Step 4 that electrons from chlorophyll molecules in photosystem II replace the electrons that leave chlorophyll molecules in photosystem I. If the electrons from photosystem II were not replaced, both electron transport chains would stop, and photosynthesis would not occur. The replacement electrons are provided by water molecules. As Figure 6-6 shows, an enzyme inside the thylakoid splits water molecules into protons, electrons, and oxygen. The following equation summarizes the reaction:

$$2H_2O \rightarrow 4H^+ + 4e^- + O_2$$

For every two molecules of water that are split, four electrons become available to replace those lost by chlorophyll molecules in photosystem II. The protons that are produced are left inside the thylakoid, while the oxygen diffuses out of the chloroplast and can then leave the plant. Thus, oxygen can be regarded as a byproduct of the light reactions—it is not needed for photosynthesis to occur. However, as you will learn in Chapter 7, the oxygen that results from photosynthesis is essential for cellular respiration in most organisms, including plants themselves.

Eco Connection

Photosynthesis and the Global Greenhouse

With the beginning of the Industrial Revolution around 1850, the atmospheric concentration of CO_2 started to increase. This increase has resulted largely from the burning of fossil fuels, which releases CO_2 as a byproduct. You might expect plants to benefit from the buildup of CO_2 in the atmosphere. In fact, the rise in CO_2 levels may harm photosynthetic organisms more than it helps them.

CO_2 and other gases in the atmosphere retain some of the Earth's heat, causing the Earth to become warmer. This warming could reduce the amount of worldwide precipitation, creating deserts that would be inhospitable to most plants.

Also, CO_2 in the atmosphere reacts with water to produce acid precipitation, which can kill plants.

FIGURE 6-6

The splitting of water inside the thylakoid releases electrons, which replace the electrons that leave photosystem II when it is illuminated.

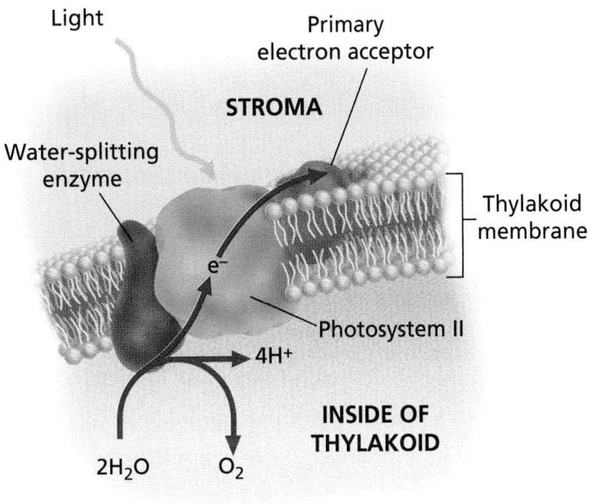

TERMINOLOGY NOTE

Abbreviations are used in the text for compounds whose full chemical names are long and potentially intimidating to students who have not taken a chemistry course. Thus, $NADP^+$ is used for nicotinamide adenine dinucleotide phosphate, RuBP for ribulose bisphosphate, PGA for phosphoglycerate, and PGAL for glyceraldehyde phosphate.

MAKING CONNECTIONS

Technology

Many herbicides work by inhibiting photosynthesis, usually by blocking electron transport in photosystem II. A number of recently developed crop plants are resistant to specific herbicides and can be grown in fields sprayed with that herbicide.

RETEACHING *ACTIVITY*

Help students reinforce the concepts in this section by making flashcards with a structure or process on one side and a description of its role on the other side. Topics should include *pigment, chlorophyll, photosystem, thylakoid, electron transport chain, NADPH, ATP, chemiosmosis,* and *ATP synthase.*

QUICK FACT

☑ In addition to being one of the pigments of photosynthesis, chlorophyll is fluorescent. Fluorescent molecules absorb light of particular wavelengths and emit some of the absorbed energy as light of longer wavelengths. Chlorophyll fluoresces by absorbing ultraviolet light (light with a wavelength shorter than violet light) and emitting red light.

ANSWERS TO SECTION 6-1 REVIEW

1. Thylakoids are flattened membranous sacs inside the chloroplast. They are the structures in which photosynthesis occurs in eukaryotic cells.
2. The accessory pigments absorb colors of light that are not absorbed by chlorophyll *a*. They then transfer the energy in these colors of light to chlorophyll *a* for use in photosynthesis.
3. Electrons from photosystem II are carried along an electron transport chain and replace electrons lost by photosystem I. Electrons from photosystem I are carried along a different electron transport chain and drive the synthesis of NADPH.
4. The breakdown of water produces protons, electrons, and oxygen.
5. ATP synthase makes ATP from ADP and phosphate by using energy provided by the movement of protons from the inside of the thylakoid to the stroma.
6. Without a concentration gradient, there would be no tendency for protons to move from the thylakoid into the stroma. As a result, ATP would not be synthesized by ATP synthase. Also, there would be fewer protons in the stroma to combine with $NADP^+$ and make NADPH.

Word Roots and Origins

chemiosmosis

from the Greek *chemeia,* meaning "alchemy," and *osmosis,* meaning "pushing"

CHEMIOSMOSIS

An important part of the light reactions is the synthesis of ATP through a process called **chemiosmosis** (KEM-ee-ahz-MOH-suhs). Chemiosmosis relies on a concentration gradient of protons across the thylakoid membrane. Recall that some protons are produced from the breakdown of water molecules inside the thylakoid. Other protons are pumped from the stroma to the interior of the thylakoid. The energy required to pump these protons is supplied by the excited electrons as they pass along the electron transport chain of photosystem II. Both of these mechanisms act to build up a concentration gradient of protons. That is, the concentration of protons is higher inside the thylakoid than in the stroma.

The concentration gradient of protons represents potential energy. That energy is harnessed by a protein called **ATP synthase,** which is located in the thylakoid membrane, as Figure 6-7 shows. ATP synthase makes ATP by adding a phosphate group to **adenosine diphosphate,** or **ADP.** The energy that drives this reaction is provided by the movement of protons from the inside of the thylakoid to the stroma. Thus, ATP synthase converts the potential energy of the proton concentration gradient into chemical energy stored in ATP. Remember from Chapter 3 that ATP is the main energy currency of cells.

As you learned earlier, some of the protons in the stroma are used to make NADPH from $NADP^+$. Together, NADPH and ATP provide energy for the second set of reactions in photosynthesis, which are described in the next section.

ATP synthase is a multifunctional protein. By allowing protons to cross the thylakoid membrane, ATP synthase functions as a carrier protein. By catalyzing the synthesis of ATP from ADP, ATP synthase functions as an enzyme. You will encounter other examples of multifunctional proteins in later chapters.

FIGURE 6-7

During chemiosmosis, the movement of protons into the stroma of the chloroplast releases energy, which is used to manufacture ATP.

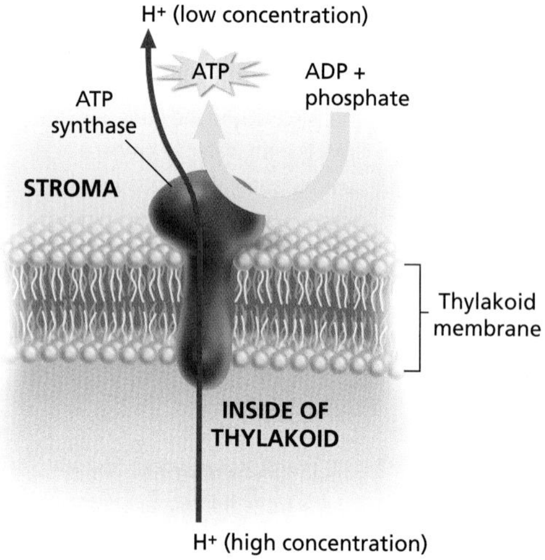

SECTION 6-1 REVIEW

1. **Describe the structure and function of the thylakoids of a chloroplast.**
2. **What role do the accessory pigments play in photosynthesis?**
3. **What happens to the electrons that are lost by photosystem II? What happens to the electrons that are lost by photosystem I?**
4. **Name the three substances that are produced when water molecules are broken down during the light reactions.**
5. **How is ATP made in the light reactions?**
6. **CRITICAL THINKING Explain how the light reactions would be affected if there were no concentration gradient of protons across the thylakoid membrane.**

SECTION

6-2

OBJECTIVES

▲ Summarize the main events of the Calvin cycle.

● Describe what happens to the compounds made in the Calvin cycle.

■ Distinguish between C_3, C_4, and CAM plants.

◆ Explain how environmental factors influence photosynthesis.

THE CALVIN CYCLE

The second set of reactions in photosynthesis involves a biochemical pathway known as the ***Calvin cycle.*** *This pathway produces organic compounds, using the energy stored in ATP and NADPH during the light reactions. The Calvin cycle is named after Melvin Calvin (1911–1997), the American scientist who worked out the details of the pathway.*

CARBON FIXATION BY THE CALVIN CYCLE

In the Calvin cycle, carbon atoms from CO_2 are bonded, or "fixed," into organic compounds. This incorporation of CO_2 into organic compounds is referred to as **carbon fixation.** The Calvin cycle has three major steps, which occur within the stroma of the chloroplast. Refer to Figure 6-8 as you read the following summary of the steps in the Calvin cycle.

Step 1. CO_2 diffuses into the stroma from the surrounding cytosol. An enzyme combines a CO_2 molecule with a five-carbon carbohydrate called **RuBP.** The product is a six-carbon molecule that splits immediately into a pair of three-carbon molecules known as **PGA.**

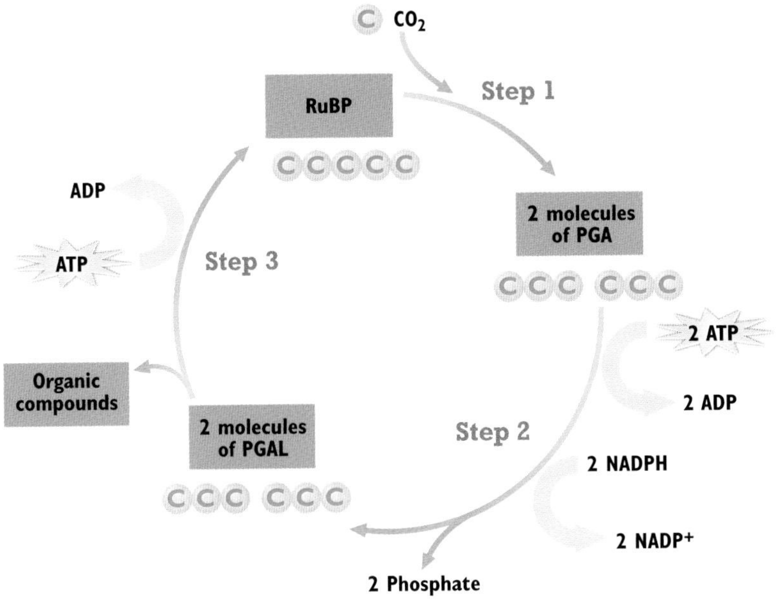

FIGURE 6-8

The Calvin cycle takes place in the stroma of the thylakoid and involves three major steps. Step 1: CO_2 combines with RuBP to form two molecules of PGA. Step 2: Each molecule of PGA is converted into a molecule of PGAL. Step 3: Most of the PGAL is converted back into RuBP, but some PGAL can be used to make a variety of organic compounds.

internetconnect

SCILINKS NSTA

TOPIC: Calvin cycle
GO TO: www.scilinks.org
KEYWORD: HM117

SECTION 6-2

SECTION OVERVIEW

Matter, Energy, and Organization

Section 6-2 explains how the energy that is transferred to NADPH and ATP in the light reactions is used to build organic compounds from CO_2 in the Calvin cycle.

▶ ENGAGE STUDENTS

Have students visualize the reactions of the Calvin cycle using 18 round objects that can be linked together, such as toy beads or plastic-foam balls connected with toothpicks. They should first link three sets of five balls, representing three molecules of RuBP. They should then attach a single ball, representing CO_2, to each 5-ball structure, making three 6-ball structures. Each 6-ball structure should be split immediately into two 3-ball structures (for a total of six structures). These 3-ball structures can represent PGA and, in the next step, PGAL. Have the students set one of the 3-ball structures aside; it represents a PGAL molecule that leaves the Calvin cycle. The other five 3-ball structures should be rearranged to regenerate three 5-ball models of RuBP.

VISUAL LINK

Figure 6-8

Have students count the total number of carbon atoms present at each step in this summary of the Calvin cycle. Explain that CO_2 combines with RuBP, which is regenerated with each turn of the cycle. Emphasize that the main product of the Calvin cycle is *not* glucose but rather the three-carbon compound PGAL.

Quick Lab

Analyzing Photosynthesis

Time Required 25 minutes

Safety Tell students that bromothymol blue is a skin and eye irritant. It can stain skin and clothing.

Disposal Dilute the solution with 20 times as much tap water or until the pH is neutral. Place a beaker containing the diluted solution in the sink and run water in it for 10 minutes.

Answers to Analysis Bromothymol blue is an acid/base indicator. When the student blows into the solutions, CO_2 is added, creating carbonic acid. The acid changes the bromothymol blue to a yellow color. The *Elodea* in the yellow solution will remove CO_2 during the process of photosynthesis. The pH will increase, and the solution's color will change back to blue. The other flasks are controls. The yellow flask without the *Elodea* will not change back to blue because the CO_2 is still in the solution.

CRITICAL THINKING

Calvin Cycle

Why is the Calvin cycle a more efficient biochemical pathway for synthesizing new molecules than a noncyclical pathway would be? (Cyclical pathways regenerate their own starting molecules.)

INCLUSION ACTIVITY

Help students understand the energy transformations that occur in the Calvin cycle by having them point to the primary energy-carrying substances in Figure 6-8. (They should point to NADPH, ATP, and organic compounds.)

Quick Lab

Analyzing Photosynthesis

Materials disposable gloves, lab apron, safety goggles, 250 mL Erlenmeyer flasks (3), bromothymol blue, 5 cm sprigs of *Elodea* (2), water, drinking straw, plastic wrap, 100 mL graduated cylinder

Procedure

1. Put on your disposable gloves, lab apron, and safety goggles.
2. Label the flasks "1," "2," and "3." Add 200 mL of water and 20 drops of bromothymol blue to each flask.
3. Put the drinking straw in flask 1 and blow into the blue solution until the solution turns yellow. Repeat this step with flask 2.
4. Put one *Elodea* sprig in flask 1. Do nothing to flask 2. Put the other *Elodea* sprig in flask 3.
5. Cover all flasks with plastic wrap. Place the flasks in a well-lighted location, and leave them overnight. Record your observations.

Analysis Describe your results. Explain what caused one of the solutions to change color. Why did the other solutions not change color? Which flask is the control in this lab?

Step 2. PGA is converted into another three-carbon molecule, **PGAL,** in a two-part process. First, each PGA molecule receives a phosphate group from a molecule of ATP. The resulting compound then receives a proton from NADPH and releases a phosphate group, producing PGAL. In addition to PGAL, these reactions produce ADP, $NADP^+$, and phosphate. These three products can be used again in the light reactions to synthesize additional molecules of ATP and NADPH.

Step 3. Most of the PGAL is converted back into RuBP in a complicated series of reactions. These reactions require a phosphate group from another molecule of ATP, which is changed into ADP. By regenerating the RuBP that was consumed in Step 1, the reactions of Step 3 allow the Calvin cycle to continue operating. However, some PGAL molecules are not converted into RuBP. Instead, they leave the Calvin cycle and can be used by the plant cell to make other organic compounds, as explained in the next section.

The Balance Sheet for Photosynthesis

How much ATP and NADPH are required to make one molecule of PGAL from CO_2? Each turn of the Calvin cycle fixes one CO_2 molecule. Since PGAL is a three-carbon compound, it takes three turns of the cycle to produce each molecule of PGAL. For each turn of the cycle, two ATP molecules and two NADPH molecules are used in Step 2—one for each molecule of PGAL produced—and one more ATP molecule is used in Step 3. Therefore, three turns of the Calvin cycle use nine molecules of ATP and six molecules of NADPH.

Some of the PGAL and other molecules made in the Calvin cycle are built up into a variety of organic compounds, including amino acids, lipids, and carbohydrates. Among the carbohydrates are the monosaccharides glucose and fructose, the disaccharide sucrose, and the polysaccharides glycogen, starch, and cellulose. Most heterotrophs depend on the chemical energy that is stored in the organic compounds made by plants and other photosynthetic organisms.

Recall that water is split during the light reactions, yielding electrons, protons, and oxygen as a byproduct. Thus, the simplest overall equation for photosynthesis, including both the light reactions and the Calvin cycle, can be written as follows:

$$CO_2 + H_2O + \text{light energy} \rightarrow (CH_2O) + O_2$$

In the equation above, (CH_2O) represents the general formula for a carbohydrate. It is often replaced in this equation by the carbohydrate glucose, $C_6H_{12}O_6$, giving the following equation:

$$6CO_2 + 6H_2O + \text{light energy} \rightarrow C_6H_{12}O_6 + 6O_2$$

Keep in mind, however, that glucose is not actually produced by the pathways of photosynthesis. Glucose is included in the equation mainly to emphasize the relationship between photosynthesis and cellular respiration, which will be discussed in Chapter 7.

ALTERNATIVE PATHWAYS

The Calvin cycle is the most common pathway for carbon fixation. Plant species that fix carbon exclusively through the Calvin cycle are known as **C_3 plants** because of the three-carbon compound, PGA, that is initially formed. Other plant species fix carbon through alternative pathways and then release it to enter the Calvin cycle.

These alternative pathways are generally found in plants that evolved in hot, dry climates. Under such conditions, plants can rapidly lose water to the air. Most of the water loss from a plant occurs through small pores called **stomata** (STOH-muh-tuh), which are usually located on the undersurface of the leaves. As Figure 6-9 shows, plants can partially close their stomata when the air is hot and dry, thereby reducing water loss.

Stomata are also the major passageways through which CO_2 enters and O_2 leaves a plant. Thus, when a plant's stomata are partly closed, the level of CO_2 in the plant falls as CO_2 is consumed in the Calvin cycle. At the same time, the level of O_2 in the plant rises as the light reactions split water and generate O_2. Both of these conditions—a low CO_2 level and a high O_2 level—inhibit carbon fixation by the Calvin cycle. Plants with alternative pathways for carbon fixation have evolved ways of dealing with this problem.

The C_4 Pathway

One alternative pathway enables certain plants to fix CO_2 into four-carbon compounds. This pathway is therefore called the **C_4 pathway,** and plants that use it are known as C_4 plants. During the hottest part of the day, C_4 plants have their stomata partially closed. However, certain cells in C_4 plants have an enzyme that can fix CO_2 into four-carbon compounds even when the CO_2 level is low and the O_2 level is high. These compounds are then transported to other cells, where CO_2 is released and enters the Calvin cycle.

C_4 plants include corn, sugar cane, and crabgrass. Such plants lose only about half as much water as C_3 plants when producing the same amount of carbohydrate.

The CAM Pathway

Cactuses, pineapples, and certain other plants have a different adaptation to hot, dry climates. Such plants fix carbon through a pathway called **CAM.** Plants that use the CAM pathway open their stomata at night and close them during the day—just the opposite of what other plants do. At night, CAM plants take in CO_2 and fix it into a variety of organic compounds. During the day, CO_2 is released from these compounds and enters the Calvin cycle. Because CAM plants have their stomata open at night, when the temperature is lower, they grow fairly slowly. However, they lose less water than either C_3 or C_4 plants.

Word Roots and Origins

stoma

from the Greek *stoma,* meaning "mouth"

FIGURE 6-9

These SEMs show stomata in the leaf of a tobacco plant, *Nicotiana tabacum.* (a) When a stoma is open, water, carbon dioxide, and other gases can pass through it to enter or leave a plant (814×). (b) When a stoma is closed, passage through it is greatly restricted (878×).

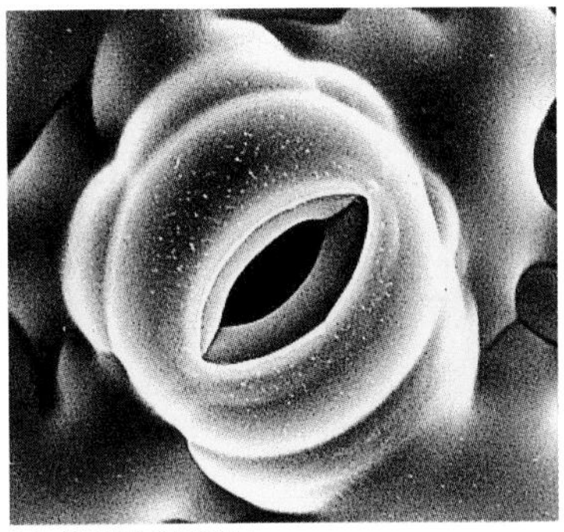

(a) OPEN STOMA

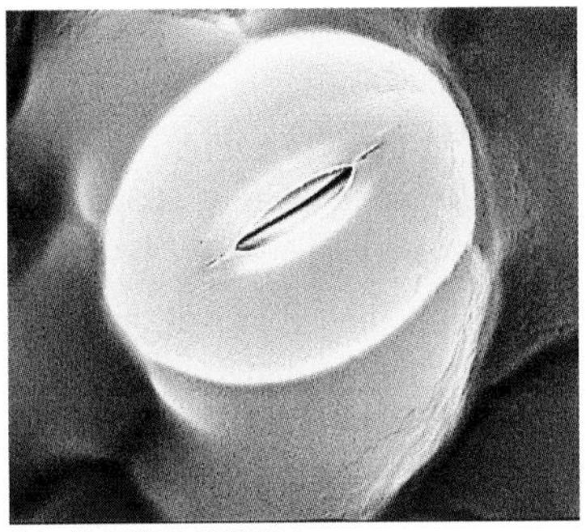

(b) CLOSED STOMA

SECTION 6-2

QUICK FACT

☑ CAM stands for crassulacean acid metabolism and is named for the family Crassulaceae (stonecrops), in which it was discovered.

DEMONSTRATION

Provide students with a variety of C_3 plants (such as bean plants, tomato plants, and petunias) and CAM plants (such as *Kalanchoe* and *Sedum*), along with clear nail polish and forceps. Have them cover part of the surface of a few leaves on each plant with nail polish. Wait about 30 minutes for the polish to dry. Then peel off the nail polish with the forceps and place it on a microscope slide. Add a drop of glycerol and a coverslip. **Caution the students to be careful not to get the glycerol in their eyes or mouth.** Have them examine their specimens under a compound microscope at a magnification of about 40×. They should look for stomata, using the photographs in Figure 6-9 as a guide. Ask students how the stomata on the C_3 plants compare with those on the CAM plants. (If this demonstration is done during the daytime, most of the stomata on the C_3 plants should be open and most of those on the CAM plants should be closed.)

TEACHING STRATEGY

Photosynthetic Products

Use a scalpel to make thin cross sections of a potato. Have students examine the sections under a compound microscope. Ask them to look for starch granules, which should appear as large, translucent structures inside the potato cells. Explain that the organic compounds plants make through photosynthesis are stored mainly as starch.

SECTION 6-2

ACTIVE READING
Technique: K-W-L
Refer to the Active Reading activity on page 111. Have students review their lists and determine if they Learned everything on their lists. If they did not, open a class discussion to answer the questions.

ANSWERS TO SECTION 6-2 REVIEW

1. The Calvin cycle takes place in the stroma.
2. Most of the PGAL is used to make more RuBP, keeping the Calvin cycle operating. Some PGAL is converted into other organic compounds, including amino acids, lipids, and carbohydrates.
3. Producing a molecule of PGAL requires three turns of the Calvin cycle, nine molecules of ATP, and six molecules of NADPH.
4. The stomata control the passage of water and carbon dioxide.
5. A C_4 plant takes in CO_2 and fixes it into four-carbon organic compounds inside certain types of cells. After these compounds are transported to other cells, they release CO_2, which can then enter the Calvin cycle.
6. Increasing the CO_2 concentration makes more CO_2 available to enter the Calvin cycle, thus accelerating photosynthesis. As the CO_2 concentration rises still higher, the rate of photosynthesis begins to be limited by other components in the biochemical pathways of photosynthesis, such as other chemical reactants or enzymes.

RATE OF PHOTOSYNTHESIS

The rate at which a plant can carry out photosynthesis is affected by the plant's environment. One of the most important environmental influences is light intensity. Figure 6-10a shows that as light intensity increases, the rate of photosynthesis initially increases and then levels off to a plateau. This plateau represents the maximum rate of photosynthesis. Higher light intensity causes more electrons in the chlorophyll molecules of both photosystems to become excited. As more electrons are excited, the light reactions occur more rapidly. At some light intensity, however, all of the available electrons are excited, and any further increase in light intensity will not increase the rate of photosynthesis.

CO_2 is another important influence on photosynthesis. Like increasing light intensity, increasing levels of CO_2 around a plant stimulate photosynthesis until the rate of photosynthesis reaches a plateau. Thus, a graph of the rate of photosynthesis versus CO_2 concentration would resemble Figure 6-10a.

A third environmental factor affecting photosynthesis is temperature. Raising the temperature accelerates the various chemical reactions involved in photosynthesis. As a result, the rate of photosynthesis increases as temperature increases, over a certain range. This effect is illustrated by the left half of the curve in Figure 6-10b. The rate of photosynthesis generally peaks at a certain temperature. At that temperature, many of the enzymes that catalyze the reactions in photosynthesis start to become unstable and ineffective. Also, the stomata begin to close, limiting water loss and CO_2 entry into the leaves. These conditions cause the rate of photosynthesis to decrease when the temperature is further increased, as shown by the right half of the curve in Figure 6-10b.

FIGURE 6-10
Environmental factors affect the rate of photosynthesis in plants. (a) As light intensity increases, the rate of photosynthesis increases and then levels off at a maximum. (b) As temperature increases, the rate of photosynthesis increases to a maximum and then decreases with further rises in temperature.

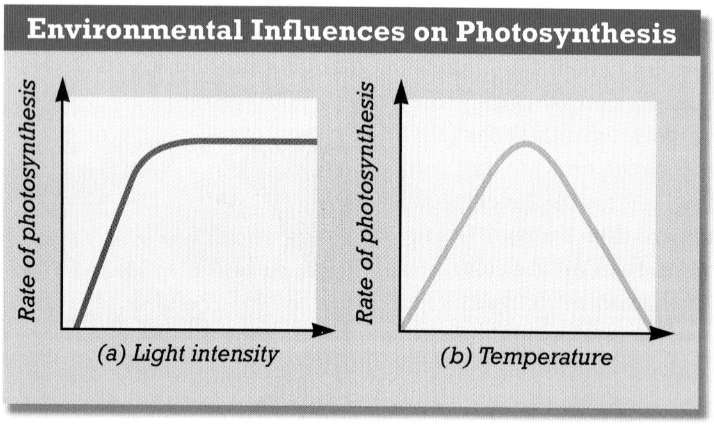

SECTION 6-2 REVIEW

1. In what part of a chloroplast does the Calvin cycle take place?
2. Describe what can happen to PGAL molecules made in the Calvin cycle.
3. How many turns of the Calvin cycle are needed to produce a molecule of PGAL? How many molecules of ATP and NADPH are used in the process?
4. What plant structures control the passage of water out of a plant and carbon dioxide into a plant?
5. What is a C_4 plant?
6. **CRITICAL THINKING** Why does the rate of photosynthesis increase and then reach a plateau as the concentration of CO_2 around a plant increases?

SUMMARY/VOCABULARY

6-1

- Photosynthesis converts light energy into chemical energy through complex series of reactions known as biochemical pathways. Autotrophs use photosynthesis to make organic compounds from carbon dioxide and water.
- In plants and algae, photosynthesis occurs inside the chloroplasts.
- White light from the sun is composed of an array of colors called the visible spectrum. Different colors in the visible spectrum have different wavelengths.
- Pigments absorb certain colors of light and reflect or transmit the other colors.
- The light reactions of photosynthesis begin with the absorption of light by chlorophyll *a* and accessory pigments in the thylakoids.
- Accessory pigments absorb colors of light that aren't absorbed by chlorophyll *a*, and they transfer some of the energy in this light to chlorophyll *a*.
- Excited electrons that leave chlorophyll *a* travel along two electron transport chains, resulting in the production of NADPH. The electrons are replaced when water is split into electrons, protons, and oxygen in the thylakoid. Oxygen is released as a byproduct of photosynthesis.
- As electrons travel along the electron transport chains, a concentration gradient of protons builds up across the thylakoid membrane. The movement of protons down this gradient results in the synthesis of ATP through chemiosmosis.

Vocabulary

accessory pigment (113)
adenosine diphosphate (ADP) (116)
ATP synthase (116)
biochemical pathway (111)
carotenoid (113)
chemiosmosis (116)
chlorophyll (113)
electron transport chain (115)
granum (112)
light reactions (112)
$NADP^+$ (115)
photosynthesis (111)
photosystem (114)
photosystem I (114)
photosystem II (114)
pigment (113)
primary electron acceptor (115)
stroma (112)
visible spectrum (112)
wavelength (113)

6-2

- The ATP and NADPH produced in the light reactions drive the second part of photosynthesis, the Calvin cycle. In the Calvin cycle, CO_2 is incorporated into organic compounds, a process referred to as carbon fixation.
- The Calvin cycle produces a compound called PGAL. Three turns of the Calvin cycle are needed to produce one PGAL molecule.
- Most PGAL molecules are converted into another molecule that keeps the Calvin cycle operating. However, some PGAL molecules are used to make other organic compounds, including amino acids, lipids, and carbohydrates.
- In the overall equation for photosynthesis, CO_2 and water are the reactants, and carbohydrate and O_2 are the products.
- Some plants living in hot, dry climates supplement the Calvin cycle with the C_4 or CAM pathways. These plants carry out carbon fixation and the Calvin cycle either in different cells or at different times.
- The rate of photosynthesis increases and then reaches a plateau as light intensity or CO_2 concentration increases. Below a certain temperature, the rate of photosynthesis increases as temperature increases. Above that temperature, the rate of photosynthesis decreases as temperature increases.

Vocabulary

C_3 plant (119)
C_4 pathway (119)
Calvin cycle (117)
CAM (119)
carbon fixation (117)
PGA (117)
PGAL (118)
RuBP (117)
stoma (119)

REVIEW

1. Chlorophylls absorb relatively little green light, and therefore they look green.
2. The stroma is the part of a chloroplast that lies inside the inner membrane but outside the thylakoids. Stomata are pores that control the passage of water and gases across the surface of plants.
3. Carbon fixation refers to the incorporation of carbon dioxide into organic compounds.
4. A biochemical pathway is a complex series of chemical reactions in which the product of one reaction is consumed in the next reaction.
5. *Calvin cycle* does not belong in the group, because it is not a part of the light reactions.

6. a **7.** c
8. b **9.** d
10. c **11.** d
12. a **13.** c
14. b **15.** b

16. Photosystem I provides electrons that are used to make NADPH from $NADP^+$. Photosystem II provides electrons to replace those that are lost by photosystem I, and it pumps protons into the interior of the thylakoids.
17. In a biochemical pathway, the product of one reaction is consumed in the next reaction. In the Calvin cycle, compounds such as PGA, PGAL, and RuBP fit this description.
18. Most CAM plants are found in hot, dry environments. The CAM pathway enables plants to fix CO_2 at night. Because their stomata are closed in the daytime, CAM plants lose less water.
19. A concentration gradient of protons builds up across the thylakoid membrane. ATP synthase, a protein in the thylakoid membrane, couples the movement of protons down this gradient with the synthesis of ATP.
20. During the summer, plant leaves generally contain a lot of the green pigment, chlorophyll.

During the fall, many plants lose their chlorophyll, and their leaves take on the colors of the carotenoids.

21. *X* is a thylakoid. During photosynthesis, the concentration of protons is higher inside the thylakoids than outside.

22. These plants exchange gases at night when the temperature is cooler. They are able to keep the openings in their leaves closed during the hot part of the day. This allows the plants to reduce water loss.

CRITICAL THINKING

1. Most organisms on Earth depend on photosynthetic autotrophs, which capture light energy using colored pigment molecules.

2. When you eat carrots, you acquire their beta-carotene and convert it into retinal, which can then be used by your eyes to capture light energy.

3. This enzymatic change would inhibit photosynthesis, as CO_2 would no longer be fixed in the Calvin cycle. It would occur whenever the stomata are partially or fully closed, such as during hot, dry weather.

4. Placing these components in the thylakoid membrane allows them to remain relatively close together. This makes it easier for energy and electrons to pass from one component to the next.

5. Because CAM plants open their stomata at night, when temperatures are lower, they grow slowly. In places where water is plentiful, water conservation is less important and faster-growing C_3 plants have an advantage.

REVIEW

Vocabulary

1. The prefix *chloro-* means "green." With this information, explain why the chlorophylls are well named.
2. What is the difference between *stroma* and *stomata*?
3. Explain what is meant by the term *carbon fixation*.
4. What is a biochemical pathway?
5. Choose the term that does not belong in the following group, and explain why it does not belong: electron transport chain, chemiosmosis, Calvin cycle, and photosystem II.

Multiple Choice

6. A product in the overall equation for photosynthesis is (a) O_2 (b) CO_2 (c) H_2O (d) RuBP.
7. A reactant used in the Calvin cycle is (a) H_2O (b) glucose (c) CO_2 (d) O_2.
8. Accessory pigments (a) add color to plants but do not absorb light energy (b) absorb colors of light that chlorophyll *a* cannot absorb (c) receive electrons from the electron transport chain of photosystem I (d) are not involved in photosynthesis.
9. C_4 plants (a) are usually found in cool, moist environments (b) lose more water than C_3 plants during photosynthesis (c) have no stomata (d) fix CO_2 into four-carbon compounds.
10. During photosynthesis, oxygen is produced when (a) PGA is converted into PGAL (b) CO_2 is fixed (c) water is split (d) ATP is converted into ADP.
11. The light reactions take place (a) on the outer membrane of the chloroplast (b) in the stroma (c) in the cytosol (d) on the thylakoid membrane.
12. During chemiosmosis, (a) ATP is synthesized from ADP (b) NADPH is synthesized from $NADP^+$ (c) water is broken down (d) electrons are removed from chlorophyll molecules.
13. Which of the following is NOT part of the light reactions? (a) splitting of water (b) electron transport (c) carbon fixation (d) absorption of light energy
14. Most of the PGAL made in the Calvin cycle is used to (a) synthesize carbohydrates (b) keep the cycle operating (c) convert light energy into chemical energy (d) drive the light reactions.
15. The reactions of the Calvin cycle take place (a) on the outer membrane of the chloroplast (b) in the stroma (c) in the cytosol (d) on the thylakoid membrane.

Short Answer

16. What is the difference between the roles of photosystems I and II in photosynthesis?
17. Explain how the Calvin cycle is an example of a biochemical pathway.
18. In what type of environment are most CAM plants found? How is the CAM pathway advantageous for that type of environment?
19. How is ATP made during photosynthesis?
20. Why do the leaves of some plants look green during the summer and then turn yellow, orange, or brown during the fall?
21. The diagram below shows a portion of a chloroplast. Identify the structure labeled *X* in the diagram. During photosynthesis, is the concentration of protons higher inside this structure or in the space surrounding it?

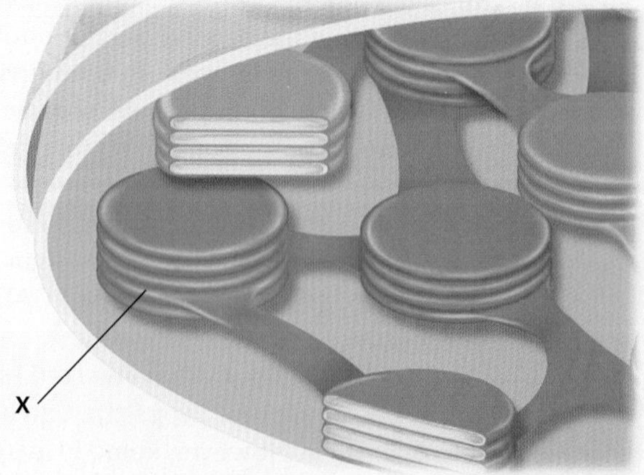

22. **Unit 2—*Photosynthesis***

Many plants have stomata that take in CO_2 at night and release it during the day. Why is this form of photosynthesis an advantage for plants living in a hot, dry climate?

CRITICAL THINKING

1. A famous scientist once said that wherever in the universe life exists, some of those life-forms must be colored. Why would the scientist make such a statement?
2. One of the accessory pigments used in photosynthesis is beta-carotene, a carotenoid found in high concentration in carrots. When one molecule of beta-carotene is split by an enzyme, two molecules of vitamin A are produced. Removal of a hydrogen atom from vitamin A produces retinal, the pigment involved in vision. Explain why eating carrots is important for good vision.
3. When the CO_2 concentration in the cells of a C_3 plant is low compared with the O_2 concentration, an enzyme combines RuBP with O_2 rather than with CO_2. What effect would this enzymatic change have on photosynthesis? Under what environmental conditions would it be most likely to occur?
4. All of the major components of the light reactions, including the pigment molecules clustered in photosystems I and II, are located in the thylakoid membrane. What is the advantage of having these components confined to the same membrane rather than dissolved in the stroma or the cytosol?
5. Cactuses and other CAM plants are very efficient at carrying out photosynthesis while conserving water. Why aren't they more common in environments where water is plentiful?
6. Some bacteria conduct a type of photosynthesis that makes ATP but does not produce NADPH or split water. How might the evolution of cellular respiration have been different if this had been the only type of photosynthesis to evolve?
7. The graph below shows how the percentage of stomata that are open varies over time for two different kinds of plants. One curve represents the stomata of a geranium, and the other curve represents the stomata of a pineapple. Which curve corresponds to the pineapple stomata? Explain your reasoning.

Daily Cycle of Stomatal Opening

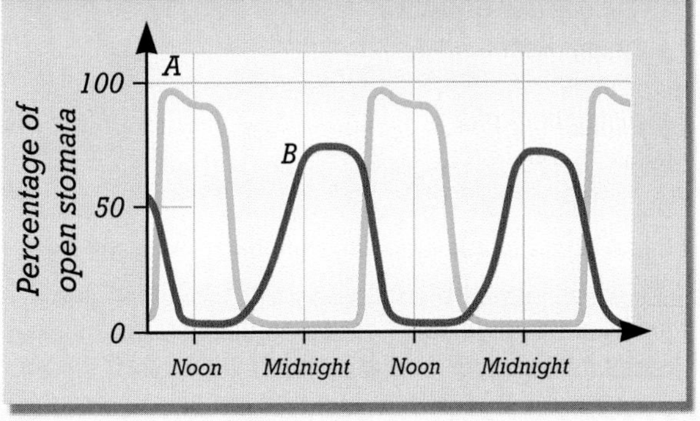

EXTENSION

1. Read "Lake of Dreams" in *New Scientist,* December 4, 1999, on page 35. Write a report describing Lake Vostok and the characteristics that make it so unusual. Explain why some scientists think the lake may be similar to the geothermal vents near the Galápagos Islands.
2. When the sun's rays are blocked by a thick forest, clouds, dust from a volcanic eruption, or smoke from a large fire, what effect do you think this has on photosynthesis? How might it affect the levels of atmospheric carbon dioxide and oxygen? What experiments could scientists conduct in the laboratory to test your predictions?
3. Collect a sample of algae from a local pond. Divide the sample into two covered bowls, and keep one bowl in the dark and the other in the light. Keep both bowls at the same temperature. Every other day for 10 days, observe the algae with a dissecting microscope. Draw and record the organisms and debris that you see. What do your observations tell you about the role of light in pond life?

6. If there were no photosynthetic splitting of water, there would be very little O_2 on Earth, and organisms that use O_2 in cellular respiration probably would not have evolved.
7. Curve *B* corresponds to the pineapple stomata. Pineapples are CAM plants, and CAM plants close their stomata during the day and open them at night, as indicated by curve *B*.

EXTENSION

1. Lake Vostok lies beneath nearly 4 km of ice in the heart of Antarctica. The lake has been isolated from air and sunlight for almost 25 million years. With the discovery of teeming life around the geothermal vents near the Galápagos Islands, scientists learned for the first time that not all life on Earth is dependent on solar energy. Because of this discovery, scientists believe they might discover life in Lake Vostok that has adapted to its cold, airless, and sunless environment.
2. The rate of photosynthesis should decrease, which should result in a rise in atmospheric CO_2 and a fall in atmospheric oxygen. Scientists might study the effect of reduced light intensity using small plants.
3. The number of living organisms, especially autotrophs, in the dark bowl should decrease over time. This suggests that light is necessary to sustain pond life.

CHAPTER 6
INTERACTIVE EXPLORATION

For an overview of this Interactive Exploration and preparatory information related to this activity, please see the CD User's Guides

PREPARATION NOTES

Time required:
One 50-minute class period

PRELAB DISCUSSION

Before students begin the lab, they should be able to answer the following questions:

1. What is chlorophyll? (Chlorophyll is a pigment found in plants and other photosynthetic organisms that transmits or reflects green light and absorbs other colors.)

2. What are chloroplasts, and what function do they perform? (Chloroplasts are organelles in the cells of photosynthetic eukaryotes. Chloroplasts are the site of photosynthesis.)

3. Describe the main steps of photosynthesis. (Light energy is captured by pigments in the chloroplasts. This energy is used to make ATP and NADPH, which are then used to make carbohydrates from CO_2 and water.)

4. Write the simplest overall equation for photosynthesis. ($CO_2 + H_2O +$ light energy $\rightarrow [CH_2O] + O_2$)

PROCEDURAL TIPS

1. The exploration can be run at two speeds. Selecting *Slow Animation* from the *Speed* menu on the toolbar will allow students to follow the events occurring within the chloroplast.

2. Remind students that by measuring the rate of ATP formation instead of the rate of carbohydrate production, they are taking indirect measurements of the rate of photosynthesis. The light reactions are essential for carbohydrate formation, but other factors,

CHAPTER 6

INTERACTIVE EXPLORATION

Examining the Rate of Photosynthesis

OBJECTIVES

- Simulate how the rate of photosynthesis is affected by variations in the intensity and wavelength of light.
- Relate the rate of photosynthesis to the production of ATP in a chloroplast.

MATERIALS

- computer with CD-ROM drive
- CD-ROM *Interactive Explorations in Biology: Cell Biology and Genetics*
- graph paper

Background

Photosynthesis is the process by which plants, algae, and some bacteria use the energy contained in light to make organic compounds, including carbohydrates. In the light reactions of photosynthesis, chlorophyll and other pigments absorb light, and electrons in these molecules are raised to higher energy levels. These excited electrons are passed through electron transport chains, resulting in the production of ATP and NADPH. In the Calvin cycle, ATP and NADPH are used in the synthesis of carbohydrates from carbon dioxide. Because the Calvin cycle cannot occur without ATP, the overall rate of photosynthesis depends on the rate of ATP production in the light reactions. This interactive exploration allows you to see how varying the intensity and wavelength of light affects the rate of ATP production in photosynthesis.

Prelab Preparation

1. Load and start the program Photosynthesis. You will see an animated cross section of a thylakoid membrane like the one below. Click the Navigation button, and then click the Topic Information button. Read the focus questions and review these concepts: Photons, Pigments, and The Action Spectrum of Photosynthesis.
2. Click the word *Help* at the top left of the screen, and select How to Use This Exploration. Listen to the instructions that explain the operation of the exploration. Click the Interactive Exploration button on the Navigation Palette to begin the exploration.

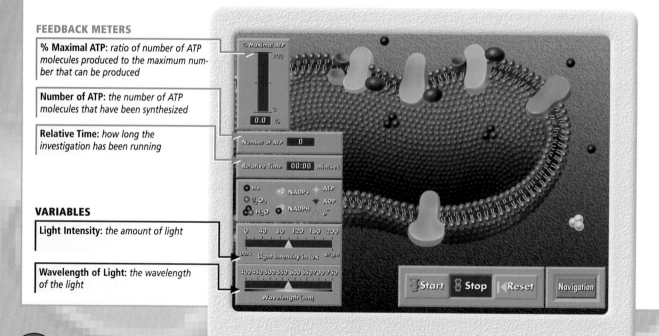

FEEDBACK METERS

% Maximal ATP: *ratio of number of ATP molecules produced to the maximum number that can be produced*

Number of ATP: *the number of ATP molecules that have been synthesized*

Relative Time: *how long the investigation has been running*

VARIABLES

Light Intensity: *the amount of light*

Wavelength of Light: *the wavelength of the light*

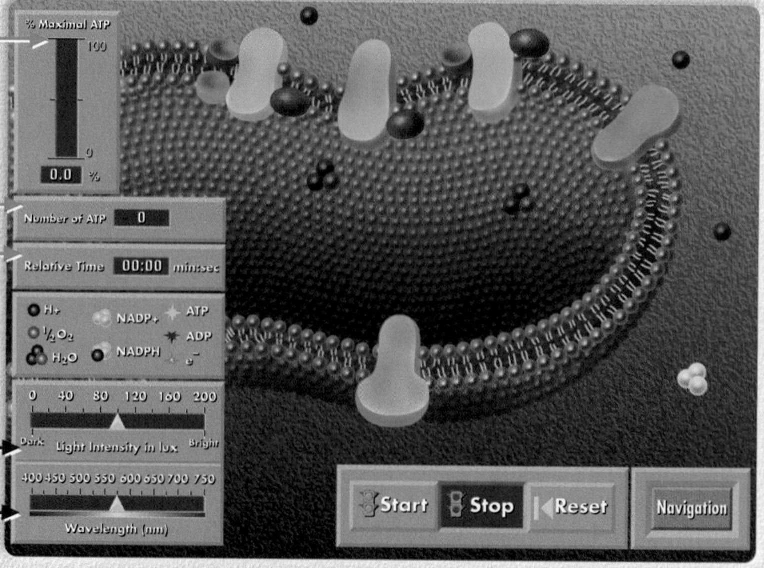

Procedure

PART A Effect of Light Intensity on the Rate of Photosynthesis

1. You will first investigate how varying light intensity affects the rate of photosynthesis. Create a data table like Table A shown below for recording your data.

TABLE A VARYING LIGHT INTENSITY AT A WAVELENGTH OF 650 NM

Light intensity (lux)	% Maximal ATP	Number of ATP
0	0	0
40	17	1
80	34	2
120	51	3
160	68	3
200	85	4

2. Click the word *Speed* at the top of the screen. Select Slow.
3. Click and slide the wavelength indicator to 650. Click and slide the light intensity indicator to 120 lux. A lux is a measure of light intensity.
4. Click the Start button. Observe the diagram of the protein at the bottom of the screen. What molecule is synthesized near this protein?
5. Allow the simulation to run for 30 seconds, as indicated on the relative time meter, and then click the Stop button. In your table, record the % Maximal ATP and the Number of ATP molecules that were produced.
6. Click the Reset button, and then click and slide the light intensity indicator to 160 lux.
7. Click the Start button and allow the simulation to run for 30 seconds, as indicated on the relative time meter.
8. Click the Stop button and record the % Maximal ATP and Number of ATP in your table.
9. Continue testing until you have evaluated all of the light intensities listed on your table.
10. Explain why the wavelength of light is kept constant in this series of steps.

PART B Effect of Wavelength on the Rate of Photosynthesis

11. Click the Reset button. Prepare a data table like Table B shown below.

TABLE B VARYING WAVELENGTH AT A LIGHT INTENSITY OF 200 LUX

Wavelength (nm)	% Maximal ATP	Number of ATP
400	65	3
450	15	1
500	7.5	1
550	15	1
600	20	1
650	85	4
700	5	1
750	3	1

12. Click and slide the light intensity indicator to 200 lux. Click and slide the wavelength of light indicator to 400 nanometers.
13. Click the Start button and allow the simulation to run for 30 seconds, as indicated on the relative time meter.
14. Click the Stop button and record the % Maximal ATP and Number of ATP in your data table.
15. Continue testing until you have evaluated all of the wavelengths of light listed on your table.
16. Explain why you did not vary the intensity of light in this part of the investigation.

Analysis and Conclusions

1. Prepare a bar graph that shows the % Maximal ATP on the *y*-axis and the intensity of light on the *x*-axis. Above each bar, indicate the number of ATP molecules produced.
2. What can you conclude about the effect of light intensity on the rate of photosynthesis?
3. Prepare another bar graph, plotting the % Maximal ATP versus the wavelength of light.
4. Refer to your second graph. What wavelengths of light are most effective for photosynthesis?
5. Which combination of wavelength and intensity of light would produce the highest rate of photosynthesis?

such as availability of water and nutrients, also affect the rate of photosynthesis.

ANSWERS TO PART A

4. ATP is synthesized near the protein.
10. The wavelength is kept constant so that only one variable (light intensity) is evaluated at a time.

ANSWER TO PART B

16. The intensity of light is kept constant so that only one variable (wavelength) is evaluated at a time.

ANSWERS TO ANALYSIS AND CONCLUSIONS

1. The graph should have five bars whose heights are proportional to the last five numbers in the middle column of Table A. The height of the bars should increase incrementally from left to right. For light intensity = zero, there should be a space but no bar, because % maximal ATP = zero.
2. The rate of photosynthesis increases linearly with light intensity.
3. The graph should have eight bars whose heights are proportional to the numbers in the middle column of Table B. The tallest bar should be at 650 nm, and the second-tallest bar should be at 400 nm.
4. The most effective wavelengths are 400 nm and 650 nm.
5. The most effective combination would be 650 nm and 200 lux.

CHAPTER 7
CELLULAR RESPIRATION

Blocks	TOPICS	TEACHING RESOURCES	LABS, CLASSWORK, AND HOMEWORK
BLOCKS 1 & 2 90 minutes	**Introducing the Chapter, p. 126**	ATE Focus Concept, p. 126 ATE Assessing Prior Knowledge, p. 126 ATE Understanding the Visual, p. 126	■ Supplemental Reading Guide, *The Lives of a Cell: Notes of a Biology Watcher*
	7-1 Glycolysis and Fermentation, p. 127 Harvesting Chemical Energy, p. 127 Glycolysis, p. 128 Fermentation, p. 129 Energy Yield, p. 130	ATE Section Overview, p. 127 ATE Critical Thinking, p. 131 ATE Visual Link Figure 7-2, p. 128 11A, 16, 18, 20–21	**Lab C8,** Measuring the Release of Energy—Best Food for Yeast ★ Study Guide, Section 7-1 PE Section Review, p. 131
BLOCKS 3 & 4 90 minutes	**7-2 Aerobic Respiration, p. 133** Overview of Aerobic Respiration, p. 133 The Krebs Cycle, p. 134 Electron Transport Chain, p. 135 Energy Yield, p. 137 Summarizing Cellular Respiration, p. 138	ATE Section Overview, p. 133 ATE Critical Thinking, p. 133 ATE Visual Link Figure 7-7, p. 135 17, 19, 22–23	PE **Chapter 7 Investigation,** p. 142 PE **Quick Lab,** p. 134, Comparing CO_2 Production **Lab E2,** Oxidative Respiration ★ Study Guide, Section 7-2 PE Section Review, p. 138
BLOCKS 5 & 6 90 minutes	**REVIEW and ASSESSMENT** PE Chapter 7 Review, pp. 140–141 ■ Performance-Based Assessment—Fermentation by Yeast	Audio CD Program BIOLOGY INTERACTIVE TUTOR Unit 3—Cellular Respiration	**CHAPTER TESTING** ■ Chapter Test (blackline master) ▲ Test Generator ■ Assessment Item Listing ■ Scoring Rubrics and Classroom Management Checklists

MODERN BIOLOGY

KEY
- PE Pupil's Edition
- ATE Teacher's Edition
- ■ Active Reading Guide
- ★ Study Guide

HOLT BIOSOURCES
- Teaching Transparencies
- Laboratory Program

One-Stop Planner CD-ROM
Shows these resources within a customizable daily lesson plan:
- ■ Holt BioSources Teaching Resources
- ■ Active Reading Guide
- ★ Study Guide
- ▲ Test Generator

READING FOR CONTENT MASTERY

Preview: Have students preview the objectives, the introductory paragraph, topic sentences, and captions before reading the chapter.

- ATE Active Reading Technique: Reading Effectively, p. 127
- ■ Active Reading Guide Worksheet 7-1
- PE Summary/Vocabulary, p. 139

- ■ Active Reading Guide Worksheet 7-2
- PE Summary/Vocabulary, p. 139

TECHNOLOGY AND INTERNET RESOURCES

Holt Biology Videodiscs Teacher's Correlation Guide, Lessons 7-1 through 7-2

BIOLOGY INTERACTIVE TUTOR: Unit 3—Cellular Respiration, Topics 1–6

Audio CD Program, Sections 7-1 through 7-2

internetconnect

sciLINKS NSTA

On-line Resources: www.scilinks.org
The following *sci*LINKS Internet resources can be found in the student text for this chapter:

Topics:
- Fermentation, p. 129
- Krebs cycle, p. 135

go.hrw.com

On-line Resources: go.hrw.com
Visit the HRW Web site for a variety of resources related to this chapter. Just type in the keyword HM2 Chapter 7.

Smithsonian Institution®
Visit **www.si.edu/hrw** for additional on-line resources.

CNNfyi.com
Visit **www.cnnfyi.com** for late-breaking news and current event stories selected just for you.

LAB ACTIVITY PLANNING

CHAPTER 7 INVESTIGATION

Observing Cellular Respiration, pp. 142–143

OVERVIEW
This investigation provides students an opportunity to use a volumeter to determine the rate of cellular respiration of germinating seeds.

SAFETY
Students should wash their hands after handling the seeds. Students should use forceps to pick up the soda-lime packets.

PREPARATION
1. Purchase volumeter kits from a biological supply house.
2. Two days before the lab, begin germinating 50 seeds for each lab setup. Lay a wet paper towel over a sheet of wax paper. Arrange the seeds in five rows across the towel. Lay another wet paper towel over the seeds. Roll the wax paper from left to right, and secure with rubber bands. Stand the cylinder upright in a jar that contains about an inch of water, and store the jar in a dark area. Check the jar twice a day, and add water if necessary.
3. Use food coloring to color the water. One day before the lab, put out several containers of distilled water so that room-temperature water will be available.
4. When preparing the soda-lime packets, work under a fume hood and wear safety goggles and a lab apron. Put about 1/4 teaspoon of soda-lime pellets in the center of a 2 × 2 in. gauze pad. Tie the corners together with string. Each student team will need three packets.

Quick Labs

Comparing CO_2 Production, p. 134
Students will need disposable gloves, lab aprons, safety goggles, a flask, a graduated cylinder, phenolphthalein solution, a pipet, a drinking straw, water, a clock, and sodium hydroxide solution for this lab activity. For preparation tips, see page 134.

UNDERSTANDING THE VISUAL

Have students examine the photograph. Ask them how animals, such as the one shown here, are either directly or indirectly dependent on plants for energy. (Plants capture energy from the sun and store it in carbohydrates. Animals acquire a portion of this energy by eating either the plants themselves or other animals that have eaten the plants.) Tell students that the focus of this chapter is the process by which organisms convert some of the energy in carbohydrates into a form they can use to drive cellular activities.

FOCUS CONCEPT

Matter, Energy, and Organization

Living things are highly organized and require an input of energy to maintain their organization. They obtain the needed energy by breaking down carbohydrates and other macromolecules in chemical pathways that are highly efficient.

ASSESSING PRIOR KNOWLEDGE

Review the following concepts.

Organelles: *Chapter 4*
Ask students to define the terms *mitochondrion* and *cristae.* Ask them to name the energy-carrying compound that is synthesized inside mitochondria.

Photosynthesis: *Chapter 6*
Ask students to write the simplest overall equation for photosynthesis, using CH_2O to represent the general formula for a carbohydrate. Ask them to rewrite the equation after replacing CH_2O with glucose, $C_6H_{12}O_6$.

CHAPTER 7

CELLULAR RESPIRATION

Like other heterotrophs, the giant panda, Ailuropoda melanoleuca, *obtains organic compounds by consuming other organisms. Biochemical pathways within the panda's cells transfer energy from those compounds to ATP.*

FOCUS CONCEPT: ***Matter, Energy, and Organization***

As you read, compare the biochemical pathways described in this chapter with those you studied in the chapter on photosynthesis.

BIOLOGY INTERACTIVE TUTOR **Unit 3—*Cellular Respiration***
Topics 1–6

GLYCOLYSIS AND FERMENTATION

All cells break down complex organic compounds into simpler molecules. Cells use some of the energy that is released in this process to make ATP.

SECTION

7-1

OBJECTIVES

▲ Define *cellular respiration.*

● Describe the major events in glycolysis.

■ Compare and contrast lactic acid fermentation and alcoholic fermentation.

◆ Calculate the efficiency of glycolysis.

HARVESTING CHEMICAL ENERGY

You learned in Chapter 6 that autotrophs, such as plants, use photosynthesis to convert light energy from the sun into chemical energy, which is stored in carbohydrates and other organic compounds. Both autotrophs and heterotrophs depend on these organic compounds for the energy to power cellular activities. By breaking down these compounds into simpler molecules, cells release energy. Some of the energy is used to make ATP from ADP and phosphate. Remember from Chapter 3 that ATP is the main energy currency of cells. The complex process in which cells make ATP by breaking down organic compounds is known as **cellular respiration.**

As you can see in Figure 7-1, cellular respiration begins with a biochemical pathway called **glycolysis** (GLIE-KAHL-uh-suhs), which yields a relatively small amount of ATP. The other products of glycolysis can follow either of two main pathways, depending on whether there is oxygen in the cell. If oxygen is absent, the products of glycolysis may enter fermentation pathways that yield no

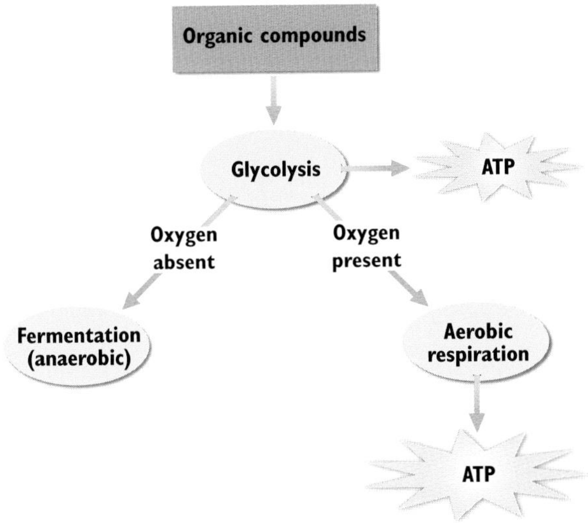

FIGURE 7-1

Cellular respiration harnesses the energy in organic compounds to produce ATP. The initial pathway in cellular respiration, called glycolysis, produces a small amount of ATP. Glycolysis can lead to fermentation if oxygen is absent or to aerobic respiration if oxygen is present. Most of the ATP produced in cellular respiration results from aerobic respiration.

SECTION 7-1

SECTION OVERVIEW

Cell Structure and Function
Section 7-1 describes chemical reactions that take place in the cytosol of cells. These reactions supply the cell with a small amount of ATP, which can be used to power cellular functions.

Matter, Energy, and Organization
Section 7-1 discusses the energy-yielding reactions of glycolysis, which occurs in all cells, and fermentation, which occurs in some cells when oxygen is absent.

ACTIVE READING

Technique: Reading Effectively
Encourage students to take as long as necessary after reading each section to study the figures in the section. The step-by-step charts of the reactions of cellular respiration can help clarify these processes. Prepare questions that correspond to each figure. For example, in reference to Figure 7-2 on page 128, you might ask students to summarize the series of events in their own words. Refer to the figures when you talk about the concepts presented in each section.

CONTENT CONNECTION

Section 2-3: Chemical Energy
Chemical energy can be either absorbed or released when the bonds in molecules are broken and new bonds are formed.

QUICK FACT

☑ Some marine worms, shrimps, and fishes can switch to anaerobic pathways when their environment becomes depleted of oxygen.

RETEACHING *ACTIVITY*

Provide students with pieces of red, yellow, and green construction paper; scissors; and paper clips. Have pairs of students cut the following shapes: six 4 in. squares of red paper; two 1 in. × 2 in. rectangles of yellow paper; two half-inch circles of yellow paper; four 2 in. triangles of green paper; and four half-inch circles of green paper. Have them write *C* on the red squares, NAD^+ on the yellow rectangles, H^+ on the yellow circles, *ADP* on the green triangles, and *P* on the green circles. Have the students lay the red squares in a line and connect each to the next with a paper clip. Tell them that they have just constructed a model of the glucose molecule's carbon skeleton and that they are going to modify it as a cell would through glycolysis. Have them go through the steps of glycolysis, adding phosphate groups (*P*) to the carbon skeleton, then separating the skeleton into two 3-carbon chains, producing ATP from ADP and P, and producing NADH from NAD^+ and H^+.

VISUAL LINK

Figure 7-2

Have students examine Figure 7-2, and ask them to identify the starting material in glycolysis (glucose). Explain that as glucose is broken down in glycolysis, some of the energy contained in glucose is transferred to the products of glycolysis. Ask students to identify those products (pyruvic acid, NADH, and ATP).

FIGURE 7-2

Glycolysis takes place in the cytosol of cells and involves four main steps. Step 1: A molecule of glucose is converted into a new six-carbon compound. Step 2: The new six-carbon compound is split into two molecules of PGAL. Step 3: The two PGAL molecules are oxidized to produce two new three-carbon compounds. Step 4: The new three-carbon compounds are converted into two molecules of pyruvic acid. Note that two ATP molecules are used in Step 1, but four more are produced in Step 4. Thus, glycolysis results in a net production of two ATP molecules.

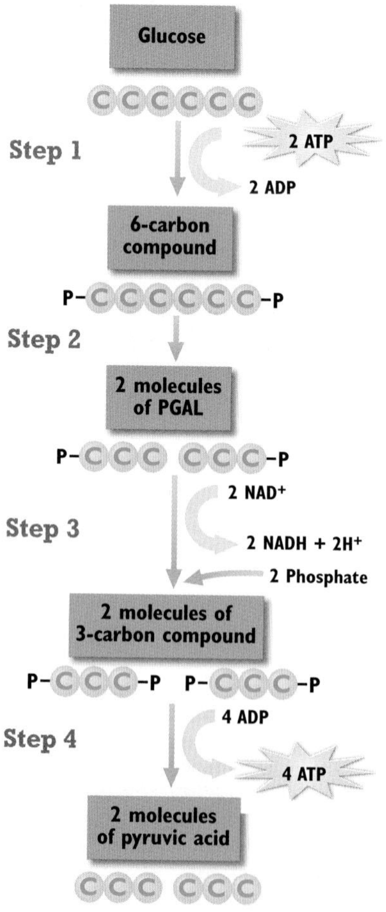

additional ATP. Because they operate in the absence of oxygen, the fermentation pathways are said to be **anaerobic** (AN-uh-ROH-bik) **pathways.** If oxygen is present, the products of glycolysis enter the pathways of aerobic respiration. Aerobic respiration produces a much larger amount of ATP than does glycolysis alone.

Many of the reactions in cellular respiration are redox reactions. Recall from Chapter 2 that in a redox reaction, one reactant is oxidized while another is reduced. Although many kinds of organic compounds can be oxidized in cellular respiration, it is customary to focus on the simple sugar glucose, whose oxidation begins with glycolysis.

GLYCOLYSIS

Glycolysis is a pathway in which one six-carbon molecule of glucose is oxidized to produce two three-carbon molecules of **pyruvic** (pie-ROO-vik) **acid.** Like other biochemical pathways, glycolysis consists of a series of chemical reactions catalyzed by specific enzymes. All of the reactions of glycolysis take place in the cytosol of the cell. These reactions can be condensed into four main steps. Refer to Figure 7-2 as you read about each of these four steps.

Step 1. Two phosphate groups are attached to glucose, forming a new six-carbon compound. The phosphate groups are supplied by two molecules of ATP, which are converted into two molecules of ADP in the process.

Step 2. The six-carbon compound formed in Step 1 is split into two three-carbon molecules of PGAL. Recall from Chapter 6 that PGAL is also produced by the Calvin cycle in photosynthesis.

Step 3. The two PGAL molecules are oxidized, and each receives a phosphate group. The product of this step is two molecules of a new three-carbon compound. As you can see in Figure 7-2, the oxidation of PGAL is accompanied by the reduction of two molecules of $\mathbf{NAD^+}$ to NADH. NAD^+, or nicotinamide adenine dinucleotide, is very similar to $NADP^+$, a compound you encountered in the light reactions of photosynthesis. Like $NADP^+$, NAD^+ is an organic molecule that accepts electrons during redox reactions.

Step 4. The phosphate groups added in Step 1 and Step 3 are removed from the three-carbon compounds formed in Step 3. This reaction produces two molecules of pyruvic acid. Each phosphate group is combined with a molecule of ADP to make a molecule of ATP. Because a total of four phosphate groups were added in Step 1 and Step 3, four molecules of ATP are produced.

Notice that two ATP molecules were used in Step 1, but four were produced in Step 4. Therefore, glycolysis has a net yield of two ATP molecules for every molecule of glucose that is converted into pyruvic acid. What happens to the pyruvic acid depends on the type of cell and on whether oxygen is present.

FERMENTATION

In the absence of oxygen, some cells can convert pyruvic acid into other compounds through additional biochemical pathways that occur in the cytosol. The combination of glycolysis plus these additional pathways is known as **fermentation.** The additional fermentation pathways do not produce ATP. However, they do regenerate NAD^+, which can be used to keep glycolysis going to make more ATP. There are many fermentation pathways, and they differ in terms of the enzymes that are used and the compounds that are made from pyruvic acid. Two common fermentation pathways result in the production of lactic acid and ethyl alcohol.

Lactic Acid Fermentation

In **lactic acid fermentation,** an enzyme converts pyruvic acid into another three-carbon compound, called lactic acid. As Figure 7-3a shows, lactic acid fermentation involves the transfer of two hydrogen atoms from NADH and H^+ to pyruvic acid. In the process, NADH is oxidized to form NAD^+. The resulting NAD^+ is used in glycolysis, where it is again reduced to NADH. Thus, the regeneration of NAD^+ in lactic acid fermentation helps to keep glycolysis operating.

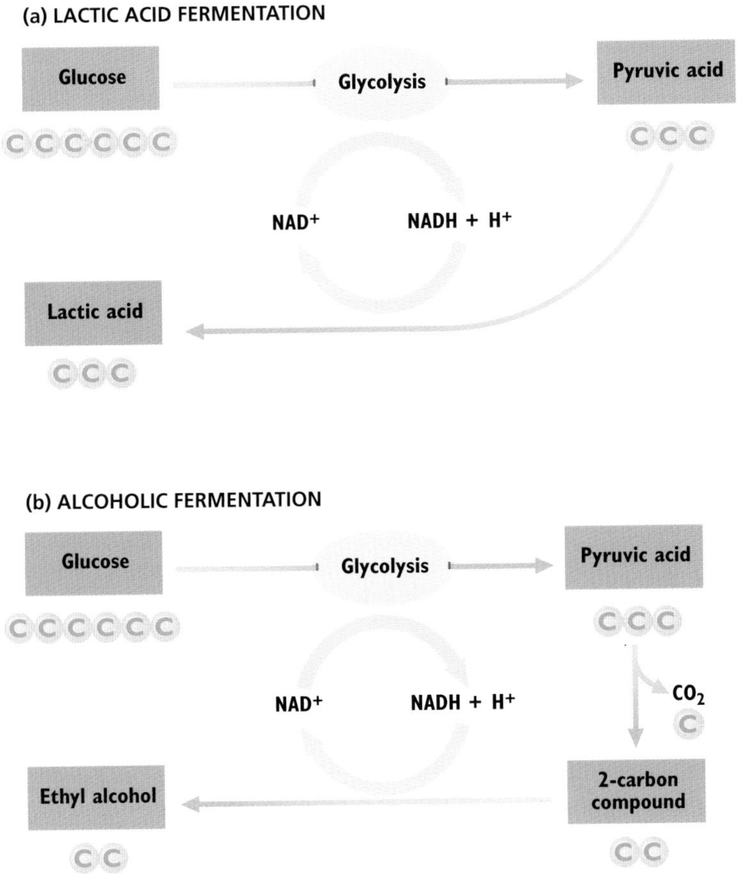

Word Roots and Origins

fermentation

from the Latin *fermentum*, meaning "yeast"

internet**connect**

SCILINKS NSTA

TOPIC: Fermentation
GO TO: www.scilinks.org
KEYWORD: HM129

FIGURE 7-3

(a) Some cells engage in lactic acid fermentation when oxygen is absent. In this process, pyruvic acid is reduced to lactic acid and NADH is oxidized to NAD^+. (b) Other cells engage in alcoholic fermentation, converting pyruvic acid into ethyl alcohol. Again, NADH is oxidized to NAD^+.

TEACHING STRATEGY

Observing Fermentation

Have groups of 3–6 students examine how the rate of fermentation varies with the type of substrate. Each group should add 10 mL of the following solutions to separate 25 mL test tubes: sucrose, fructose, galactose, starch, and gelatin. (Prepare the solutions beforehand by adding the compounds to water at a concentration of 10 percent.) Students should add $\frac{1}{8}$ teaspoon of baker's yeast to each tube, and then seal the tubes with cellophane. Have each group plan a schedule of observation times to record the frequency of bubbling. (Suggest 0.5-, 1-, 4-, and 24-hour intervals.) Ask students: What causes the bubbles? (CO_2, which is a product of fermentation) Which solution bubbles most vigorously? (sucrose) Which solution doesn't bubble at all? (gelatin)

QUICK FACT

✔ Some bacteria carry out fermentation and survive only if oxygen is absent. Other bacteria carry out fermentation, but because oxygen is not toxic to them, they can grow in places where oxygen is present. Many of the latter produce lactic acid, which inhibits the growth of competing organisms that carry out aerobic respiration.

RECENT RESEARCH

Hydrogen from Glucose for Fuel

Using enzymes from bacteria, scientists are developing a process for removing hydrogen atoms from the glucose that is present in old newspapers, grass clippings, and other materials. The hydrogen atoms are transferred to $NADP^+$, forming NADPH, and then removed from NADPH to form molecular hydrogen, H_2. Although H_2 is a clean, safe fuel, until recently there hasn't been an economical way to extract hydrogen from organic compounds.

DEMONSTRATION

Bring some fermentation products or pictures of them to class for students to examine. Note the organism that ferments each product. Examples include ripened cheeses and yogurt (*Streptococcus* and other bacteria), country-cured ham (*Aspergillus,* a fungus), soy sauce (*Aspergillus*), cured olives (*Lactobacillus,* a bacterium), and bread (*Saccharomyces,* yeast, a fungus).

CULTURAL CONNECTION

Consuming Fermentation Products

Various products of fermentation are important components of the diets of many cultures around the world. Examples include yogurt, kefir, and koumiss (eastern Europe); soy sauce, tempeh, and tofu (Asia); and cheeses and alcoholic beverages (many cultures).

INCLUSION *ACTIVITY*

Ask students to draw a eukaryotic cell and label its major organelles. Then have them trace the pathway of a glucose molecule as it enters the cell and is converted into pyruvic acid in glycolysis. (Students should show this pathway occurring in the cytosol.) Ask students to trace the fate of the pyruvic acid in the absence of oxygen. (Students should draw the fermentation pathways, which also occur in the cytosol.)

GIFTED *ACTIVITY*

Have students research and prepare a report on the bacterial genus *Clostridium* and its impact on humans. Three species in this genus are responsible for causing tetanus, gas gangrene, and botulism. All grow only in the absence of oxygen and so carry out only anaerobic pathways.

FIGURE 7-4

In cheese making, fungi or bacteria are added to large vats of milk. The microorganisms carry out lactic acid fermentation, converting some of the sugar in the milk to lactic acid.

Lactic acid fermentation by microorganisms plays an essential role in the manufacture of food products such as yogurt and cheese, as illustrated in Figure 7-4. Lactic acid fermentation also occurs in your muscle cells during very strenuous exercise, such as sprinting at top speed. During this kind of exercise, muscle cells use up oxygen more rapidly than it can be delivered to them. As oxygen becomes depleted, the muscle cells begin to switch from aerobic respiration to lactic acid fermentation. Lactic acid accumulates in the muscle cells, making the cells' cytosol more acidic. The increased acidity may reduce the capacity of the cells to contract, resulting in muscle fatigue, pain, and even cramps. Eventually, the lactic acid diffuses into the blood and is transported to the liver, where it is converted back into pyruvic acid when oxygen becomes available.

Alcoholic Fermentation

Some plant cells and unicellular organisms, such as yeast, use a process called **alcoholic fermentation** to convert pyruvic acid into ethyl alcohol. This pathway requires two steps, which are shown in Figure 7-3b. In the first step, a CO_2 molecule is removed from pyruvic acid, leaving a two-carbon compound. In the second step, two hydrogen atoms are added to the two-carbon compound to form ethyl alcohol. As in lactic acid fermentation, these hydrogen atoms come from NADH and H^+, regenerating NAD^+ for use in glycolysis.

Alcoholic fermentation is the basis of the wine and beer industries. Yeast cells are added to the fermentation mixture to provide the enzymes needed for alcoholic fermentation. As fermentation proceeds, ethyl alcohol accumulates in the mixture until it reaches a concentration that inhibits fermentation. For wine, that concentration is around 12 percent. To make table wines, the CO_2 that is generated in the first step of fermentation is allowed to escape from the mixture through a one-way gas valve. To make sparkling wines, such as champagne, CO_2 is retained within the mixture, "carbonating" the beverage.

Bread making also depends on alcoholic fermentation performed by yeast cells. In this case, the CO_2 that is produced by fermentation makes the bread rise by forming bubbles inside the dough, and the ethyl alcohol evaporates during baking.

Word Roots and Origins

kilocalorie

from the Greek *chilioi,* meaning "thousand," and the Latin *calor,* meaning "heat"

ENERGY YIELD

How efficient are the anaerobic pathways at obtaining energy from glucose and using it to make ATP from ADP? To answer this question, one must compare the amount of energy available in glucose with the amount of energy contained in the ATP that is produced by the anaerobic pathways. In such comparisons, energy is often measured in units of **kilocalories** (kcal). One kilocalorie equals 1,000 calories (cal).

Scientists have calculated that the complete oxidation of a standard amount of glucose releases 686 kcal. Under the conditions that exist inside most cells, the production of a standard amount of ATP from ADP absorbs about 12 kcal. Recall that two ATP molecules are produced from every glucose molecule that is broken down by glycolysis.

$$\text{Efficiency of glycolysis} = \frac{\text{Energy required to make ATP}}{\text{Energy released by oxidation of glucose}}$$

$$= \frac{2 \times 12\ \text{kcal}}{686\ \text{kcal}} \times 100\% = 3.5\%$$

You can see that the two ATP molecules produced during glycolysis receive only a small percentage of the energy that could be released by the complete oxidation of each molecule of glucose. Much of the energy originally contained in glucose is still held in pyruvic acid. Even if pyruvic acid is converted into lactic acid or ethyl alcohol, no additional ATP is synthesized. It's clear that the anaerobic pathways are not very efficient in transferring energy from glucose to ATP.

The anaerobic pathways probably evolved very early in the history of life on Earth. The first organisms were bacteria, and they produced all of their ATP through glycolysis. It took more than a billion years for the first photosynthetic organisms to appear. The oxygen they released as a byproduct of photosynthesis stimulated the evolution of organisms that make most of their ATP through aerobic respiration.

By themselves, the anaerobic pathways provide enough energy for many present-day organisms. However, most of these organisms are unicellular, and those that are multicellular are very small. All of them have limited energy requirements. Larger organisms have much greater energy requirements that cannot be satisfied by the anaerobic pathways alone. These larger organisms meet their energy requirements with the more efficient pathways of aerobic respiration.

SECTION 7-1 REVIEW

1. Define *cellular respiration.*
2. What six-carbon molecule begins glycolysis, and what three-carbon molecules are produced at the end of glycolysis?
3. For each six-carbon molecule that begins glycolysis, how many ATP molecules are used and how many ATP molecules are produced?
4. What condition must exist in a cell for the cell to engage in fermentation?
5. How efficient is glycolysis?
6. CRITICAL THINKING A large amount of ATP in a cell inhibits the enzymes that catalyze the first few steps of glycolysis. How will this inhibition eventually affect the amount of ATP in the cell? Explain your answer.

CRITICAL THINKING

Evolution of Glycolysis and Fermentation

Glycolysis and fermentation take place in the cytosol of cells. What does this suggest about the evolution of these pathways? (They probably evolved in prokaryotes, which do not have organelles.)

ANSWERS TO SECTION 7-1 REVIEW

1. Cellular respiration is a process in which organic compounds are broken down by cells to release energy and make ATP.
2. Glucose begins glycolysis, and two molecules of pyruvic acid are produced at the end of glycolysis.
3. For each molecule of glucose that begins glycolysis, two molecules of ATP are used and four are produced.
4. For a cell to engage in fermentation, oxygen must be either absent or very low in concentration.
5. Glycolysis is not very efficient. Only 3.5 percent of the energy that could be released by the complete oxidation of glucose is transferred to ATP.
6. Inhibiting these enzymes will slow down or even shut off glycolysis, thus reducing the rate at which ATP is made. As ATP continues to be used, the amount of ATP in the cell will fall, until it becomes so low that the enzymes are no longer inhibited and glycolysis resumes.

BACKGROUND

This feature presents the thoughts of Lewis Thomas on the importance of cell organelles, particularly mitochondria and chloroplasts. By using the term *endosymbionts* to refer to these two organelles, Thomas is referring to the endosymbiotic model of eukaryotic evolution, which holds that such organelles evolved from prokaryotes that lived symbiotically within the cytoplasm of larger prokaryotes.

READING FOR MEANING

Thomas was referring to mitochondria and chloroplasts. As he states in the excerpt, "mitochondria and chloroplasts… are… the most important living things on earth. Between them they produce the oxygen and arrange for its use." Chloroplasts produce the oxygen that is needed for aerobic respiration, and mitochondria arrange for its use as an electron acceptor during aerobic respiration.

READ FURTHER

Students' answers will vary. Several diseases are associated with cellular organelles. In cystic fibrosis, for example, a protein needed for the transport of chloride ions out of certain cells remains in the endoplasmic reticulum or the Golgi apparatus and never reaches the cell membrane. As a result, mucus accumulates outside those cells, clogging the airways of the lungs. Another example is Tay-Sachs disease, which results from a deficiency in certain lysosomal enzymes. This deficiency leads to an accumulation of metabolic waste products in the lysosomes of nerve cells. The lysosomes swell and interfere with normal nervous system functioning.

Literature & Life

Organelles as Organisms

This excerpt is from *The Lives of a Cell: Notes of a Biology Watcher,* by Lewis Thomas.

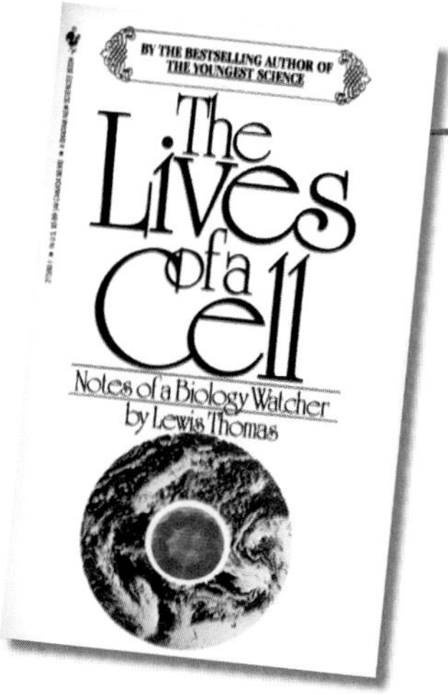

We seem to be living through the biologic revolution, so far anyway, without being upheaved or even much disturbed by it. . . .

It is not too early to begin looking for trouble. I can sense some, for myself anyway, in what is being learned about organelles. I was raised in the belief that these were obscure little engines inside my cells, owned and operated by me or my cellular delegates, private, submicroscopic bits of my intelligent flesh. Now, it appears, some of them, and the most important ones at that, are total strangers.

The evidence is strong, and direct. The membranes lining the inner compartment of mitochondria are unlike other animal cell membranes, and resemble most closely the membranes of bacteria. . . .

The chloroplasts in all plants are, similarly, independent and self-replicating lodgers. . . .

Actually, the suggestion that chloroplasts and mitochondria might be endosymbionts was made as long ago as 1885. . . . There is careful, restrained speculation on how they got there in the first place, with a consensus that they were probably engulfed by larger cells more than a billion years ago and have simply stayed there ever since.

The usual way of looking at them is as enslaved creatures, captured to supply ATP for cells unable to respire on their own, or to provide carbohydrate and oxygen for cells unequipped for photosynthesis. This master-slave arrangement is the common view of full-grown biologists, eukaryotes all. But there is the other side. From their own standpoint, the organelles might be viewed as having learned early how to have the best of possible worlds, with least effort and risk to themselves and their

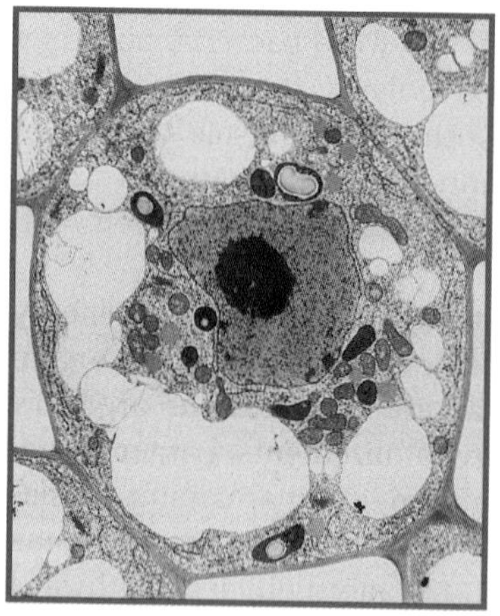

progeny. Instead of evolving as we have done, manufacturing longer and elaborately longer strands of DNA, and running ever-increasing risks of mutating into evolutionary cul-de-sacs, they elected to stay small and stick to one line of work. To accomplish this, and to assure themselves the longest possible run, they got themselves inside all the rest of us.

It is a good thing for the entire enterprise that mitochondria and chloroplasts have remained small, conservative, and stable, since these two organelles are, in a fundamental sense, the most important living things on earth. Between them they produce the oxygen and arrange for its use. In effect, they run the place.

Reading for Meaning

What do you think Thomas means when he says that the organelles "run the place"? What makes them so important?

Read Further

The Lives of a Cell: Notes of a Biology Watcher is a collection of essays by Lewis Thomas. Thomas goes on in this essay to describe things about organelles that trouble him. Some are fairly serious concerns, such as Could my organelles catch a virus? Others are more whimsical, such as Who am I really if I'm mostly organelles? What actual problems might organelles cause?

AEROBIC RESPIRATION

In most cells, the pyruvic acid that is produced in glycolysis does not undergo fermentation. Instead, if oxygen is available, pyruvic acid enters the pathways of ***aerobic*** *(UHR-OH-bik)* ***respiration,*** *or cellular respiration that requires oxygen. Aerobic respiration produces nearly 20 times as much ATP as is produced by glycolysis alone.*

SECTION 7-2

OBJECTIVES

▲ Summarize the events of the Krebs cycle.

● Summarize the events of the electron transport chain.

■ Relate aerobic respiration to the structure of a mitochondrion.

◆ Calculate the efficiency of aerobic respiration.

OVERVIEW OF AEROBIC RESPIRATION

Aerobic respiration has two major stages: the Krebs cycle and the electron transport chain. In the Krebs cycle, the oxidation of glucose that began with glycolysis is completed. As glucose is oxidized, NAD^+ is reduced to NADH. In the electron transport chain, NADH is used to make ATP. Although the Krebs cycle also produces a small amount of ATP, most of the ATP produced during aerobic respiration is made by the electron transport chain. The reactions of the Krebs cycle and the electron transport chain occur only if oxygen is present in the cell.

In prokaryotes, the reactions of the Krebs cycle and the electron transport chain take place in the cytosol of the cell. In eukaryotic cells, however, these reactions take place inside mitochondria rather than in the cytosol. The pyruvic acid that is produced in glycolysis diffuses across the double membrane of a mitochondrion and enters the **mitochondrial matrix.** The mitochondrial matrix is the space inside the inner membrane of a mitochondrion. Figure 7-5 illustrates the relationships between these mitochondrial parts. The mitochondrial matrix contains the enzymes needed to catalyze the reactions of the Krebs cycle.

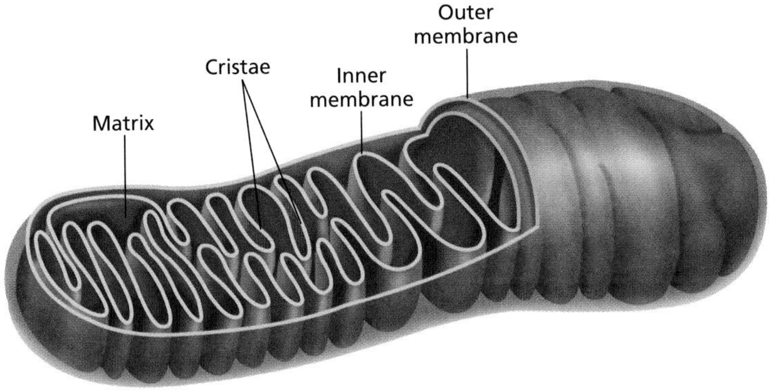

FIGURE 7-5

In eukaryotic cells, the reactions of aerobic respiration occur inside mitochondria. The Krebs cycle takes place in the mitochondrial matrix, and the electron transport chain is located in the inner membrane.

SECTION 7-2

SECTION OVERVIEW

Cell Structure and Function
Section 7-2 discusses the structure of mitochondria and explains how mitochondria are involved in the reactions of the Krebs cycle.

Matter, Energy, and Organization
Section 7-2 describes how the pyruvic acid that results from glycolysis is oxidized in the Krebs cycle and how the energy that is released during this oxidation is harnessed to synthesize ATP.

▶ ENGAGE STUDENTS

Provide students with photographs of a variety of organisms that obtain energy through anaerobic pathways (for example, bacteria such as *Fusobacterium, Salmonella, Enterobacter, Vibrio, Hemophilus,* and *Bacteriodes*) or primarily through aerobic respiration (any eukaryote, including mammals). Ask students to examine the photographs and to describe the major differences they see between the two groups. (They should note that the aerobic organisms are larger, much more complex, and capable of carrying out a wider variety of activities in their bodies. They should conclude that aerobic organisms have much higher energy needs than anaerobic organisms.)

QUICK FACT

☑ Our bodies use ATP at the rate of about one million molecules per cell per second.

CRITICAL THINKING

Mitochondria

Ask students why it is important for glucose to be partially broken down into pyruvic acid before its oxidation can be completed inside mitochondria. (Pyruvic acid is small enough to diffuse across the mitochondrial membranes, but glucose is too large to do so.)

Quick Lab

Comparing CO_2 Production

Time Required 20 minutes

Safety Phenolphthalein solutions are flammable. Ensure that there are no flames or sources of ignition when you are using the phenolphthalein solution. Restrict the maximum amount of phenolphthalein in the room to 100 mL.

Procedural Tips Prepare the solutions ahead of time. To make a 0.4 percent solution of sodium hydroxide (NaOH), add 4 g of NaOH to 1 L of water. NaOH is a strong irritant and is corrosive and toxic. Students should use only eight drops of pre-mixed phenolphthalein. To prepare phenolphthalein solution, add 2 g of phenolphthalein to 100 mL of 95 percent ethanol. Then add 100 mL of water to the ethanol solution. Set out a container for students to dispose of any used phenolphthalein solutions that are left over at the end of the procedure.

Disposal Titrate NaOH with 0.1 M HCl until the pH is between 5 and 9, and pour the solution down the drain. In the disposal container, titrate the phenolphthalein solution with 0.1 M HCl or 0.1 M NaOH until the pH is between 6 and 8, and pour the solution down the drain.

Answers to Analysis
Phenolphthalein is an acid/base indicator. It turns pink when it is exposed to a base. When the student blows into the water, CO_2 is converted to carbonic acid. Adding NaOH will neutralize the acid until the solution becomes basic and the phenolphthalein turns pink. Walking vigorously will require more energy, and the student will produce more CO_2.

When pyruvic acid enters the mitochondrial matrix, it reacts with a molecule called coenzyme A to form **acetyl** (uh-SEET-uhl) **coenzyme A,** abbreviated **acetyl CoA** (uh-SEET-uhl KOH-AY). This reaction is illustrated in Figure 7-6. The acetyl part of acetyl CoA contains two carbon atoms, but as you learned earlier, pyruvic acid is a three-carbon compound. The carbon atom that is lost in the conversion of pyruvic acid to acetyl CoA is released in a molecule of CO_2. Figure 7-6 also indicates that this reaction reduces a molecule of NAD^+ to NADH.

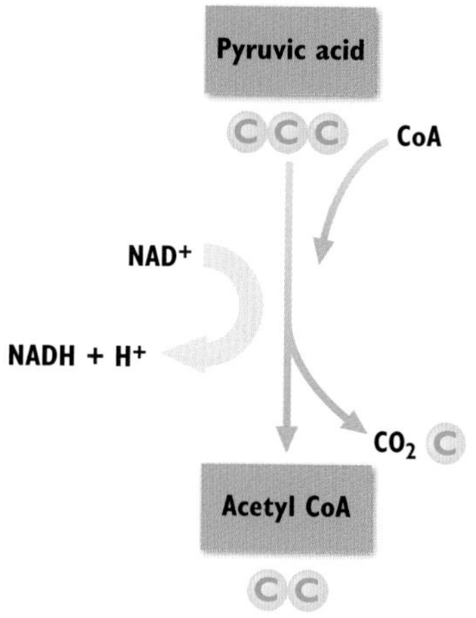

FIGURE 7-6

In aerobic respiration, pyruvic acid combines with coenzyme A to form acetyl CoA. Notice that CO_2, NADH, and H^+ are also produced in this reaction.

Quick Lab

Comparing CO_2 Production

Materials disposable gloves, lab apron, safety goggles, 250 mL flask, 100 mL graduated cylinder, phenolphthalein solution, pipet, drinking straw, water, clock, sodium hydroxide solution

Procedure

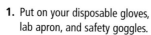

1. Put on your disposable gloves, lab apron, and safety goggles.
2. Add 50 mL of water and four drops of phenolphthalein to the flask.
3. Use the straw to gently blow into the solution for 1 minute. Add the sodium hydroxide one drop at a time, and gently swirl the flask.
4. When the liquid turns pink, stop adding drops. Record the number of drops you used.
5. Empty and rinse your flask as your teacher directs, and repeat step 2. Walk vigorously for 2 minutes, and repeat steps 3 and 4.

Analysis Which trial produced the most carbon dioxide? Which trial used the most energy?

THE KREBS CYCLE

The **Krebs cycle** is a biochemical pathway that breaks down acetyl CoA, producing CO_2, hydrogen atoms, and ATP. The reactions that make up the cycle were identified by Hans Krebs (1900–1981), a German-British biochemist. The Krebs cycle has five main steps. In eukaryotic cells, all five steps occur in the mitochondrial matrix. Examine Figure 7-7 as you read about the steps in the Krebs cycle.

Step 1. A two-carbon molecule of acetyl CoA combines with a four-carbon compound, **oxaloacetic** (AHKS-uh-loh-uh-SEET-ik) **acid,** to produce a six-carbon compound, **citric** (SI-trik) **acid.** Notice that this reaction regenerates coenzyme A.

Step 2. Citric acid releases a CO_2 molecule and a hydrogen atom to form a five-carbon compound. By losing a hydrogen atom with its electron, citric acid is oxidized. The hydrogen atom is transferred to NAD^+, reducing it to NADH.

Step 3. The five-carbon compound formed in Step 2 also releases a CO_2 molecule and a hydrogen atom, forming a four-carbon compound. Again, NAD^+ is reduced to NADH. Notice that in this step a molecule of ATP is also synthesized from ADP.

Step 4. The four-carbon compound formed in Step 3 releases a hydrogen atom to form another four-carbon compound. This time, the hydrogen atom is used to reduce FAD to $FADH_2$. **FAD,** or flavin adenine dinucleotide, is a molecule very similar to NAD^+. Like NAD^+, FAD accepts electrons during redox reactions.

Step 5. The four-carbon compound formed in Step 4 releases a hydrogen atom to regenerate oxaloacetic acid, which keeps the Krebs cycle operating. The hydrogen atom reduces NAD^+ to NADH.

Recall that in glycolysis one glucose molecule produces two pyruvic acid molecules, which can then form two molecules of acetyl CoA. Thus, one glucose molecule causes two turns of the Krebs cycle. These two turns produce six NADH, two $FADH_2$, two ATP, and four CO_2 molecules. The CO_2 is a waste product that diffuses out of the cells and is given off by the organism. The ATP can be used for energy. But note that each glucose molecule yields only two molecules of ATP through the Krebs cycle—the same number as in glycolysis.

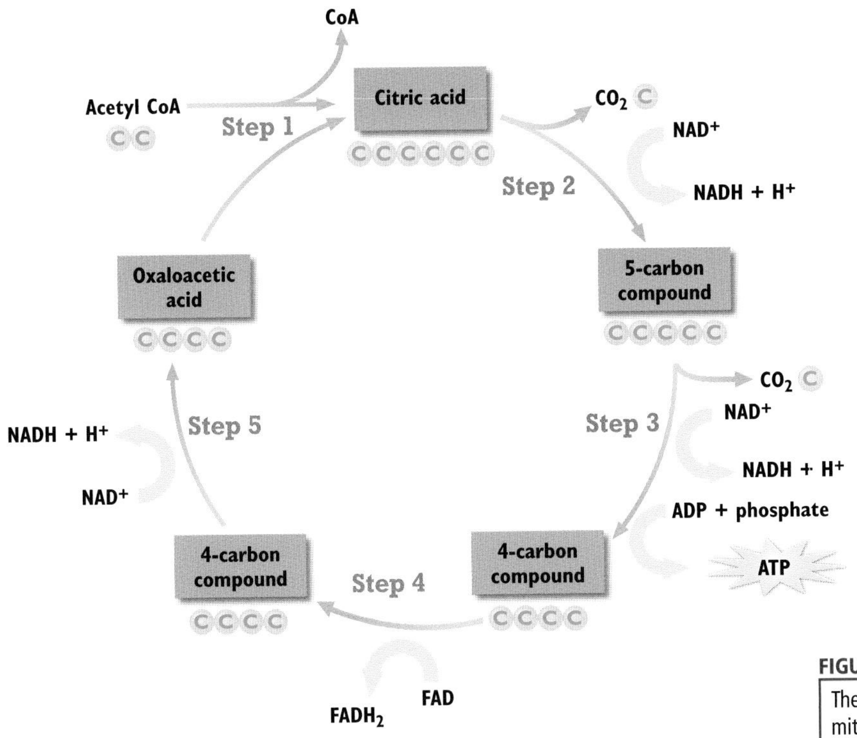

FIGURE 7-7

The Krebs cycle takes place in the mitochondrial matrix and involves five main steps. Step 1: Acetyl CoA combines with oxaloacetic acid to produce citric acid. Step 2: Citric acid releases a CO_2 molecule to form a five-carbon compound. Step 3: The five-carbon compound releases a CO_2 molecule to form a four-carbon compound. Step 4: The four-carbon compound is converted into a new four-carbon compound. Step 5: The new four-carbon compound is converted back into oxaloacetic acid. In addition to CO_2, each turn of the Krebs cycle produces ATP, NADH, and $FADH_2$.

The bulk of the energy released by the oxidation of glucose still has not been transferred to ATP. That transfer requires the NADH and $FADH_2$ made in the pathways you have learned about so far. Recall that glycolysis produces two NADH molecules and that the conversion of pyruvic acid to acetyl CoA produces two more. Adding the six NADH molecules from the Krebs cycle gives a total of 10 NADH molecules for every glucose molecule that is oxidized. These 10 NADH molecules and the two $FADH_2$ molecules from the Krebs cycle drive the next stage of aerobic respiration. That is where most of the energy transfer from glucose to ATP actually occurs.

ELECTRON TRANSPORT CHAIN

The **electron transport chain** constitutes the second stage of aerobic respiration. In eukaryotic cells, the electron transport chain lines the inner membrane of the mitochondrion. Remember from Chapter 4 that the inner membrane has many long folds called cristae. In prokaryotes, the electron transport chain lines the cell membrane. ATP is produced by the electron transport chain when NADH and $FADH_2$ release hydrogen atoms, regenerating NAD^+ and FAD. To understand how ATP is produced, you must follow what happens to the electrons and protons that make up these hydrogen atoms.

internetconnect
SCILINKS NSTA
TOPIC: Krebs cycle
GO TO: www.scilinks.org
KEYWORD: HM135

VISUAL LINK

Figure 7-7

Have students count the total number of carbon atoms present at each step in this summary of the Krebs cycle. Note that both of the carbon atoms that enter the cycle in acetyl CoA leave the cycle in CO_2. Explain that the Krebs cycle completes the breakdown of the carbon skeleton of glucose, which began with glycolysis, and that the starting material (oxaloacetic acid) is regenerated with each turn of the cycle.

RETEACHING *ACTIVITY*

Have five students form a circle, and tell them they will act as the intermediates in the Krebs cycle (refer to Figure 7-7). Give the first student four medium-sized plastic-foam balls held together with toothpicks. Then give this student two more balls held together in the same way. Ask the students to identify these molecules (acetyl CoA and oxaloacetic acid). Have the first student attach the two balls to the four balls. Ask the students to identify the molecule that is formed (citric acid). Continue this process through the entire Krebs cycle, illustrating how carbon atoms are given off as CO_2 and how the starting compound is regenerated.

CONTENT CONNECTION

Section 6-1: Electron Transport

ATP is generated in the light reactions of photosynthesis in nearly the same way it is generated in aerobic respiration. In both processes, electrons are passed along an electron transport chain and picked up by an electron acceptor, while protons are pumped into a membrane-confined area and then passed back through a membrane, driving the synthesis of ATP.

SECTION 7-2

INCLUSION ACTIVITY

Provide students with black permanent markers and about 10 each of red, green, and blue balloons. Have them write *E,* for electron, on the red balloons, *H,* for hydrogen, on the green balloons, and *A,* for ATP, on the blue balloons. Take the students to a stairway with at least 10 or 12 steps. Have several students stand at evenly spaced locations along the steps; tell them they represent carriers in the electron transport chain. Have two students stand near the bottom of the steps; one student, representing ATP synthase, should have several blue balloons, while the other, representing oxygen, should have none. Then have one student stand at the top of the steps with the red and green balloons; he or she will be the hydrogen and electron donor (NADH or $FADH_2$). The rest of the students should stand to one side, at intervals along the steps. Ask the hydrogen donors and electron donors to begin passing red and green balloons, together in pairs, to the first person at the top of the steps. Then have that person pass the balloons to the next "carrier," and so on down the line. At three points along the steps, carriers should toss one of the *H* balloons to the students standing on the sides, who should pass the balloons on to ATP synthase. When ATP synthase receives an *H,* he or she will pass it to oxygen and toss an ATP onto the floor. When oxygen receives two *H* balloons and two *E* balloons (from the last carrier), he or she should announce the formation of water.

OVERCOMING MISCONCEPTIONS

A common misconception is that ATP synthesis means the joining together of all of the various atoms that form ATP. Explain that it actually refers to the addition of a third phosphate group to adenosine diphosphate, ADP.

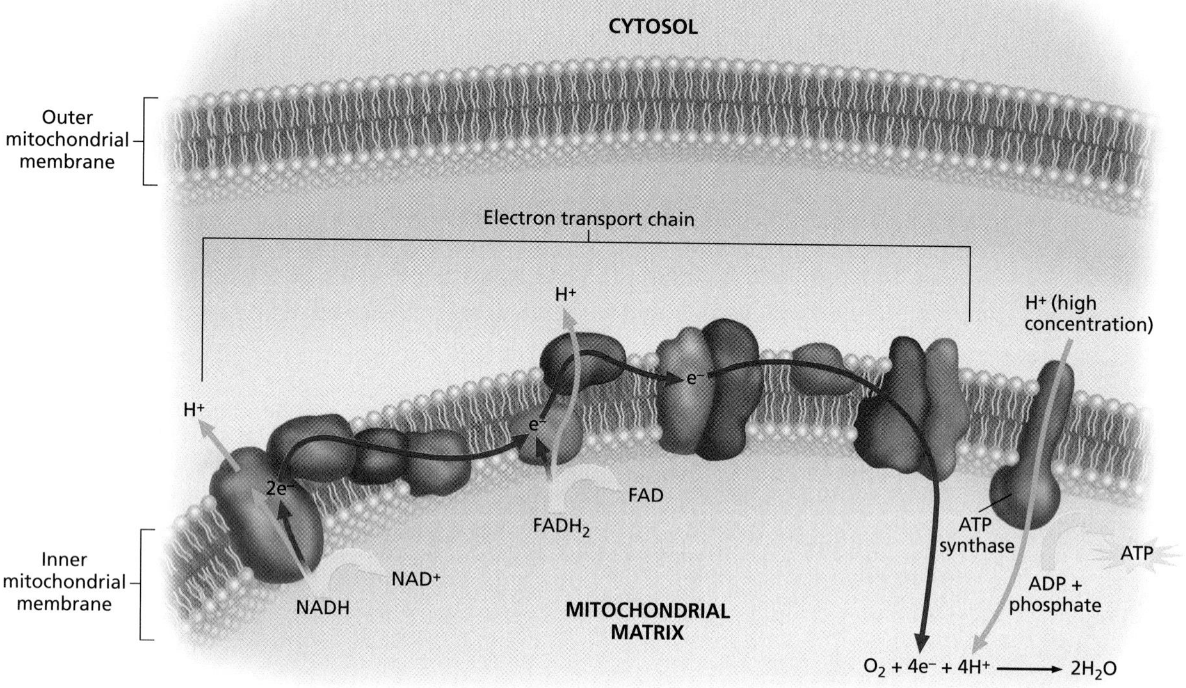

FIGURE 7-8

NADH and $FADH_2$ supply electrons and protons to the electron transport chain. The electrons are passed along the chain from molecule to molecule in a series of redox reactions. The protons are pumped out of the mitochondrial matrix. As the protons return to the mitochondrial matrix through ATP synthase, they release energy, driving the synthesis of ATP. The final acceptor of electrons is oxygen, which also accepts protons in a reaction that forms water.

The electrons in the hydrogen atoms from NADH and $FADH_2$ are at a high energy level. In the electron transport chain, these high-energy electrons are passed along a series of molecules, as shown in Figure 7-8. As they move from molecule to molecule, the electrons lose some of their energy. The energy they lose is used to pump the protons of the hydrogen atoms from the mitochondrial matrix to the other side of the inner mitochondrial membrane. This pumping builds up a high concentration of protons in the space between the inner and outer mitochondrial membranes. In other words, a concentration gradient of protons is created across the inner mitochondrial membrane.

The concentration gradient of protons drives the synthesis of ATP by chemiosmosis, the same process that generates ATP in photosynthesis. As you can see in Figure 7-8, ATP synthase molecules are located in the inner mitochondrial membrane. ATP synthase makes ATP from ADP as protons move down their concentration gradient into the mitochondrial matrix.

The Role of Oxygen

ATP can be synthesized by chemiosmosis only if electrons continue to move from molecule to molecule in the electron transport chain. Obviously, the last molecule in the electron transport chain cannot keep all of the electrons it accepts. If it did, the electron transport chain would come to a halt. Consider what would happen if cars kept entering a dead-end, one-way street. At some point, no more cars could enter the street. Similarly, if the last molecule could not "unload" the electrons it accepts, then no more electrons could enter the electron transport chain and ATP synthesis would stop.

Here is where oxygen comes into play in aerobic respiration. Figure 7-8 shows that oxygen serves as the final acceptor of electrons. By accepting electrons from the last molecule in the electron transport chain, oxygen allows additional electrons to pass along the chain. As a result, ATP can continue to be synthesized by chemiosmosis. Oxygen also accepts the protons that were once part of the hydrogen atoms supplied by NADH and $FADH_2$. By combining with both electrons and protons, oxygen forms water, as shown in the following equation:

$$O_2 + 4e^- + 4H^+ \rightarrow 2H_2O$$

ENERGY YIELD

How many ATP molecules are made in aerobic respiration? Refer to Figure 7-9 as you calculate the total. Recall that glycolysis and the Krebs cycle each produce two ATP molecules for every glucose molecule that is oxidized. Furthermore, each NADH molecule that supplies the electron transport chain can generate three ATP molecules, and each $FADH_2$ molecule can generate two ATP molecules. Thus, the 10 NADH and two $FADH_2$ molecules made through aerobic respiration can produce up to 34 ATP molecules by the electron transport chain. Adding the four ATP molecules from glycolysis and the Krebs cycle gives a maximum yield of 38 ATP molecules per molecule of glucose.

The actual number of ATP molecules generated through aerobic respiration varies from cell to cell. In most eukaryotic cells, the NADH that is made in the cytosol during glycolysis cannot diffuse through the inner membrane of the mitochondrion. Instead, it must be actively transported into the mitochondrial matrix. The active transport of NADH consumes ATP. As a result, most eukaryotic cells produce only about 36 ATP molecules per glucose molecule.

FIGURE 7-9

Follow each pathway to see how one glucose molecule can generate up to 38 ATP molecules in aerobic respiration.

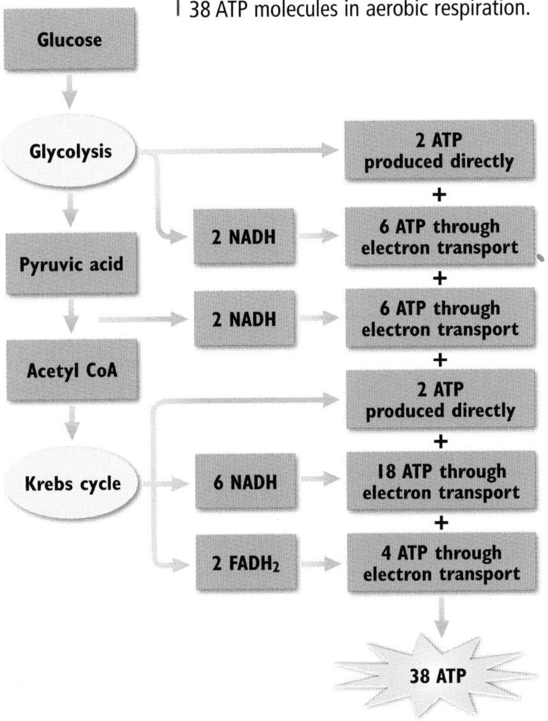

How efficient is aerobic respiration in providing a cell with energy for cellular activities? Consider the efficiency when a cell generates 38 ATP molecules:

$$\text{Efficiency of aerobic respiration} = \frac{\text{Energy required to make ATP}}{\text{Energy released by oxidation of glucose}}$$

$$= \frac{38 \times 12 \text{ kcal}}{686 \text{ kcal}} \times 100\% = 66\%$$

This means that aerobic respiration is nearly 20 times more efficient than glycolysis alone. In fact, the efficiency of aerobic respiration is quite impressive compared with the efficiency of

RECENT RESEARCH

Assisting Oxygen Delivery

Researchers have recently discovered that when hemoglobin, the respiratory pigment carried by red blood cells, picks up oxygen in the lungs, it also picks up nitric oxide, which is produced in the lungs and blood vessels. As the blood circulates through the rest of the body, the nitric oxide dilates blood vessels, speeding oxygen delivery to body tissues for aerobic respiration.

TEACHING STRATEGY

Modeling Mitochondria

Make a model of a mitochondrion using a corrugated cardboard box. Fold corrugated cardboard inside the box to represent the inner mitochondrial membrane and cristae. On separate pieces of paper, write the names of all of the reactants, intermediates, and products of the Krebs cycle and electron transport chain mentioned in the text. Have students place the reactants, intermediates, and products in the proper locations as they trace the steps in the Krebs cycle and electron transport. (For example, pyruvic acid should move from outside the mitochondrion to the mitochondrial matrix, where it combines with CoA to form acetyl CoA; protons should accumulate in the space between the inner and outer mitochondrial membranes.)

GIFTED ACTIVITY

Have students investigate and report on the physiological basis of cirrhosis of the liver, the ninth-leading cause of death in the United States. (High levels of alcohol inhibit the production of NADH by glycolysis and the Krebs cycle. As a result, sugars, amino acids, and fatty acids aren't broken down but are instead converted into fat. Fat buildup in the liver causes cirrhosis.)

SECTION 7-2

MAKING CONNECTIONS

Social Studies

In 1937, Hans Krebs discovered the details of the pathway now known as the Krebs cycle. In 1953, he shared the Nobel Prize in physiology or medicine for his discovery. Have students research and prepare a report on the life of Krebs, who had to interrupt his work and leave Germany prior to World War II because he was Jewish.

ANSWERS TO SECTION 7-2 REVIEW

1. Oxaloacetic acid is regenerated at the end of the Krebs cycle. It combines with acetyl CoA at the start of the Krebs cycle.
2. In both organelles, the passage of electrons along an electron transport chain drives the transport of protons across a membrane. The return of those protons to the other side of the membrane through ATP synthase causes ATP to be made.
3. Oxygen serves at the final electron acceptor, keeping electrons flowing through the electron transport chain so that ATP is continuously made. By combining with electrons and protons, oxygen forms water.
4. The Krebs cycle occurs in the mitochondrial matrix. The electron transport chain is located in the inner mitochondrial membrane.
5. If a cell generates 32 ATP molecules, the efficiency = (energy required to make ATP)/(energy released by oxidation of glucose) = 32 × (12 kcal/686 kcal) × 100% = 56%.
6. The loss of protons would reduce the steepness of the proton concentration gradient across the inner mitochondrial membrane. In turn, this would reduce the rate at which ATP is made through chemiosmosis.

machines that humans have designed and built. An automobile engine, for example, is only about 25 percent efficient in extracting energy from gasoline to move a car. Most of the remaining energy released from gasoline is lost as heat.

SUMMARIZING CELLULAR RESPIRATION

The complete oxidation of glucose in aerobic respiration is summarized by the following equation:

$$C_6H_{12}O_6 + 6O_2 \rightarrow 6CO_2 + 6H_2O + \text{energy}$$

Recall the equations for photosynthesis that you learned in Chapter 6. Notice that the equation above is the opposite of the overall equation for photosynthesis, if glucose is considered to be a product of photosynthesis. That is, the products of photosynthesis are reactants in aerobic respiration, and the products of aerobic respiration are reactants in photosynthesis. However, it is important to remember that aerobic respiration is not the reverse of photosynthesis. As you have seen, these two processes involve different biochemical pathways and occur at different sites inside cells.

Cellular respiration provides the ATP that all cells need to support the activities of life. But providing cells with ATP is not the only important function of cellular respiration. Cells also need specific organic compounds from which to build the macromolecules that compose their own structure. Some of these specific compounds may not be contained in the food a heterotroph consumes. However, the molecules formed at different steps in glycolysis and the Krebs cycle are often used by cells to make the compounds that are missing in food. Thus, another important function of cellular respiration is to provide carbon skeletons that can be built up into larger molecules needed by cells.

SECTION 7-2 REVIEW

1. **What four-carbon compound is regenerated at the end of the Krebs cycle? With what two-carbon compound does it combine at the start of the Krebs cycle?**
2. **How is the synthesis of ATP in the electron transport chain of mitochondria similar to the synthesis of ATP in chloroplasts?**
3. **What role does oxygen play in aerobic respiration? What molecule does oxygen become a part of as a result of aerobic respiration?**
4. **In what part of a mitochondrion does the Krebs cycle occur? In what part of a mitochondrion is the electron transport chain located?**
5. **Calculate the efficiency of aerobic respiration if a cell generates 32 ATP molecules per molecule of glucose.**
6. **CRITICAL THINKING Sometimes protons leak out of a cell or are used for other purposes besides ATP production. How would this affect the production of ATP in aerobic respiration?**

CHAPTER 7 REVIEW

SUMMARY/VOCABULARY

7-1

- Cellular respiration is the process by which cells break down organic compounds to release energy and make ATP. It includes anaerobic pathways, which operate in the absence of oxygen, and aerobic respiration, which occurs when oxygen is present.
- Cellular respiration begins with glycolysis, which takes place in the cytosol of cells. During glycolysis, one glucose molecule is oxidized to form two pyruvic acid molecules. Glycolysis results in a net production of two ATP molecules and four NADH molecules.
- Fermentation is a set of anaerobic pathways in which pyruvic acid is converted into other organic molecules in the cytosol. Fermentation does not produce ATP, but it does regenerate NAD^+, which helps keep glycolysis operating.
- In lactic acid fermentation, an enzyme converts pyruvic acid into lactic acid.
- In alcoholic fermentation, other enzymes convert pyruvic acid into ethyl alcohol and CO_2.
- Through glycolysis, only about 3.5 percent of the energy available from the oxidation of glucose is transferred to ATP.
- The anaerobic pathways probably evolved very early in the history of life on Earth. For more than a billion years, they were the only pathways available for harvesting chemical energy.

Vocabulary

alcoholic fermentation (130)
anaerobic pathway (128)
cellular respiration (127)
fermentation (129)
glycolysis (127)
kilocalorie (130)
lactic acid fermentation (129)
NAD^+ (128)
pyruvic acid (128)

7-2

- In the presence of oxygen, pyruvic acid is converted into acetyl CoA. In eukaryotic cells, this reaction occurs inside the mitochondrial matrix.
- Acetyl CoA enters the Krebs cycle, a biochemical pathway that also takes place in the mitochondrial matrix. Each turn of the Krebs cycle generates three NADH, one $FADH_2$, one ATP, and two CO_2 molecules.
- NADH and $FADH_2$ donate electrons to the electron transport chain, which lines the inner mitochondrial membrane. Electrons are passed from molecule to molecule in the transport chain in a series of redox reactions.
- As electrons pass along the electron transport chain, protons donated by NADH and $FADH_2$ are pumped into the space between the inner and outer mitochondrial membranes. This pumping creates a concentration gradient of protons across the inner mitochondrial membrane. As protons move down their gradient and back into the mitochondrial matrix, ATP synthase uses the energy released by their movement to make ATP.
- During aerobic respiration, oxygen accepts both protons and electrons from the electron transport chain. As a result, oxygen is converted to water.
- Aerobic respiration can produce up to 38 ATP molecules from the oxidation of a single molecule of glucose. This means that up to 66 percent of the energy released by the oxidation of glucose can be transferred to ATP. However, most eukaryotic cells produce only about 36 ATP molecules per molecule of glucose.
- Besides transferring energy to ATP, cellular respiration also provides carbon skeletons that can be built up into larger molecules by cells.

Vocabulary

acetyl coenzyme A (134)
aerobic respiration (133)
citric acid (134)
electron transport chain (135)
FAD (134)
Krebs cycle (134)
mitochondrial matrix (133)
oxaloacetic acid (134)

REVIEW

1. Pyruvic acid is made by glycolysis and used in the final steps of fermentation.
2. NAD^+ is made in the final steps of fermentation and used by glycolysis.
3. NADH and $FADH_2$ are made by the Krebs cycle and are used in the electron transport chain.
4. Oxygen determines which pathway pyruvic acid will enter.
5. The enzymes that are present in the cell determine which fermentation pathway pyruvic acid will enter.

6. c **7.** a
8. a **9.** b
10. d **11.** b
12. b **13.** c
14. a **15.** d

16. Glycolysis yields pyruvic acid, which diffuses from the cytosol into the mitochondrial matrix, where it is converted into acetyl CoA. This reaction produces CO_2 and NADH. Acetyl CoA enters the Krebs cycle and combines with oxaloacetic acid to form citric acid.
17. Most eukaryotic cells must use ATP to actively transport NADH into the mitochondrial matrix. This use of ATP reduces the cells' net production of ATP.
18. The anaerobic pathways occur in the cytosol, whereas the pathways of aerobic respiration occur in the mitochondrial matrix and at the inner mitochondrial membrane.
19. The accumulation of lactic acid, which is produced from pyruvic acid under anaerobic conditions, causes muscle fatigue and sometimes cramps.
20. Photosynthesis supplies both the organic compounds and the oxygen that are used in aerobic respiration.

21. Chemiosmosis is responsible for the production of ATP in the electron transport chain.
22. Oxygen serves as the final electron acceptor, keeping electrons flowing through the electron transport chain. Thus, it allows the continued production of ATP.
23. A has six carbon atoms, B has five, and C, D, and E have four.
24. Answers will vary. The diet of an athlete depends on the energy requirements of the athlete's sport. Some sports, such as weight lifting, involve mainly anaerobic metabolism. Others, such as jogging or swimming, involve aerobic respiration.

CRITICAL THINKING

1. Humans must consume foods that provide an adequate supply of vitamins. Without thiamine, for example, pyruvic acid would not be converted into acetyl CoA, and the Krebs cycle would be halted.
2. The folding of the membrane provides a greater surface area. Thus, the membrane can contain more molecules of ATP synthase and more of the molecules that make up the electron transport chain. In turn, this will allow the mitochondria to make ATP more rapidly.
3. Since fermentation is much less efficient than aerobic respiration, yeast cells must consume much more glucose when oxygen is absent than when oxygen is present.
4. During strenuous exercise, muscle cells switch from aerobic respiration to lactic acid fermentation. The lactic acid is converted back into pyruvic acid when oxygen becomes available. Deeper breathing speeds this conversion. The longer the exercise lasts, the more lactic acid accumulates and the longer it takes for all of the lactic acid to be converted.
5. Aerobic respiration should be more efficient in prokaryotic cells since all of the reactions in aerobic

REVIEW

Vocabulary

1. What molecule made during glycolysis is used in the later steps in fermentation?
2. What molecule made during the later steps in fermentation is used in glycolysis?
3. What molecules does the Krebs cycle make that the electron transport chain uses?
4. What molecule determines whether pyruvic acid will undergo fermentation or be converted for entry into the Krebs cycle?
5. What determines whether pyruvic acid will undergo lactic acid fermentation or alcoholic fermentation?

Multiple Choice

6. Before the Krebs cycle can proceed, pyruvic acid must be converted into (a) citric acid (b) glucose (c) acetyl CoA (d) NADH.
7. The net number of ATP molecules made directly by glycolysis is (a) 2 (b) 6 (c) 32 (d) 38.
8. In lactic acid fermentation, (a) NAD^+ is regenerated for use in glycolysis (b) lactic acid is converted into pyruvic acid (c) oxygen is consumed (d) electrons pass through the electron transport chain.
9. Which of the following is not a product of the Krebs cycle? (a) ATP (b) ethyl alcohol (c) CO_2 (d) $FADH_2$
10. Cellular respiration is similar to photosynthesis in that they both (a) produce ATP (b) involve chemiosmosis (c) make PGAL (d) all of the above.
11. ATP is synthesized in the electron transport chain when which of the following moves across the inner mitochondrial membrane? (a) NADH (b) protons (c) citric acid (d) oxygen
12. By accepting electrons and protons, the oxygen used in aerobic respiration turns into (a) CO_2 (b) H_2O (c) $C_6H_{12}O_6$ (d) ATP.
13. The Krebs cycle occurs in the (a) cytosol (b) outer mitochondrial membrane (c) mitochondrial matrix (d) space between the inner and outer mitochondrial membranes.
14. During each turn of the Krebs cycle, (a) two CO_2 molecules are produced (b) two ATP molecules are consumed (c) pyruvic acid combines with oxaloacetic acid (d) glucose combines with a four-carbon molecule.
15. Most of the ATP synthesized in aerobic respiration is made (a) during glycolysis (b) through fermentation (c) in the cytosol (d) through chemiosmosis.

Short Answer

16. Summarize the events that occur from the end of glycolysis through the first reaction of the Krebs cycle.
17. Why do most eukaryotic cells produce fewer than 38 ATP molecules for every glucose molecule that is oxidized by aerobic respiration?
18. How do the anaerobic pathways differ from the pathways of aerobic respiration, at the sites they occur in eukaryotic cells?
19. What causes your muscles to become fatigued and sometimes develop cramps when you exercise too strenuously?
20. How does aerobic respiration ultimately depend on photosynthesis?
21. What role does chemiosmosis play in aerobic respiration?
22. What role does oxygen play in aerobic respiration?
23. Refer to the diagram of the Krebs cycle shown below. How many carbon atoms are in each of the compounds represented by the letters A–E?

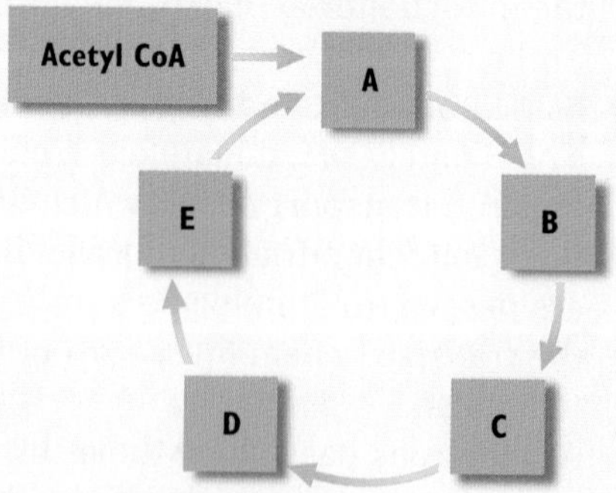

24. **Unit 3—*Cellular Respiration*** INTERACTIVE BIOLOGY TUTOR Write a report summarizing how exercise physiologists regulate the diet and training of athletes. Find out how diet varies according to the needs of each athlete. Research the relationship between exercise and metabolism.

CRITICAL THINKING

1. The enzyme that converts pyruvic acid into acetyl CoA requires vitamin B_1, also called thiamine. Like many other vitamins, thiamine cannot be made in the human body. What can you infer about the nutritional requirements of humans?
2. How does the folding of the inner mitochondrial membrane benefit aerobic respiration?
3. Yeast can produce ATP through either fermentation or aerobic respiration, depending on whether oxygen is present. If oxygen is present, yeast cells consume glucose much more slowly than if oxygen is absent. How can you explain this observation?
4. A person will breathe deeply and rapidly for some time after a period of very strenuous exercise. The longer and more intense the exercise was, the longer the deep breathing will continue after the exercise stops. Using your understanding of cellular respiration, explain why strenuous exercise stimulates deep breathing that continues after the end of exercise.
5. Some eukaryotic cells must use ATP to move NADH into the mitochondrial matrix. Knowing this, would you expect aerobic respiration to be more efficient or less efficient in prokaryotic cells than it is in eukaryotic cells? Explain your answer.
6. The graph below shows the rate of ATP production by a culture of yeast cells over time. At the time indicated by the dashed line, cyanide was added to the culture. Cyanide blocks the flow of electrons to O_2 from the electron transport chain in mitochondria. Explain why adding cyanide has the effect on ATP production that is indicated by the graph.

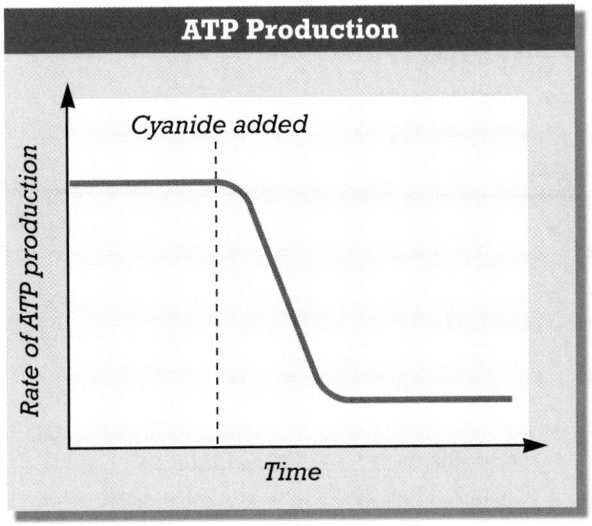

EXTENSION

1. Read "First With a Flower" in *Natural History,* March 2000, on page 12. Explain how some plants are able to obtain energy to grow and bloom under a pack of snow before they are exposed to sunlight and before photosynthesis is able to begin. What adaptation does the skunk cabbage have that enables it to extend its flowering stalk from the snow while the air temperature is still below freezing?
2. Read "Mitochondria Make a Comeback" in *Science,* March 5, 1999, on page 1475. What does the author mean when she says that mitochondria are making a comeback? Describe four more areas of research in which mitochondria play a key role.
3. As you have learned, aerobic respiration is significantly more efficient than glycolysis alone in supplying cells with ATP. Develop a hypothesis to explain why anaerobic organisms have not become extinct but instead have continued to thrive in many regions of the world. Check library and on-line references to locate information about one anaerobic species that relates to your hypothesis.
4. Find a recipe for making leavened, or raised, bread and a recipe for making unleavened bread. What ingredient that is present in the recipe for leavened bread is missing from the recipe for unleavened bread? Explain why that ingredient is omitted from the recipe.

respiration occur in the cytosol and no ATP is used in transporting NADH across a membrane.

6. Blocking the flow of electrons along the electron transport chain will quickly stop the production of ATP by aerobic respiration, as indicated by the sharp drop in the curve. ATP may still be produced at a low rate through glycolysis after the cyanide is added, as indicated by the right portion of the curve.

EXTENSION

1. Some flowers, such as the glacier lily, must use reserve carbohydrates for the energy to grow and to bloom under the snow. Many spring wildflowers have large storage tubers and rhizomes that are full of carbohydrates for energy from the previous growing season. The skunk cabbage has a very high rate of cellular respiration. It actually generates heat, so the plant can be as much as 30°F warmer than its surroundings. The plant is able to melt its way through frozen ground and ice.

2. Mitochondria were first studied to discover the principles of cellular energy metabolism more than 50 years ago. Scientific interest faded after that, but now mitochondria are being studied again in other areas of research. Mitochondria have been shown to play central roles in the regulation of programmed cell death (aging), in evolutionary biology (tracing origins), molecular medicine (enzyme mechanisms), and forensic science (genetic and biochemical inheritance).

3. Anaerobic organisms alive today are adapted to specific environments that generally have little or no oxygen.

4. Unleavened bread is not made with yeast. Since there are no yeast cells to engage in alcoholic fermentation, no CO_2 is produced and the dough does not rise.

CHAPTER 7 INVESTIGATION

TIME REQUIRED

One 50-minute class period

SAFETY PRECAUTIONS

Caution students to wash their hands after handling the seeds and to manipulate the soda lime packets with forceps, not with their hands. Caution students to use care when handling the glassware, especially the capillary tubes, which break easily.

QUICK REFERENCES

Lab Preparation Notes are found on page 125B. *Holt Biosources* provides a Teaching Resources CD-ROM that contains "Using Gowin's Vee in the Lab" and "Scoring Rubrics."

PROCEDURAL TIPS

1. The investigation is written for volumeter kits from WARD'S Natural Science Establishment, Inc. Other kits would also work, although the capillary tubes may be calibrated differently.
2. Before each volumeter is sealed, the plunger in its syringe should be pulled halfway out so that it can be moved in either direction during the experiment.
3. The capillary tubes must be dry; otherwise, the drop of colored water will disperse. If the drop associated with the germinating seeds does not move toward the seeds, check that the tubing and stopper connections are tight. There should be little or no movement associated with the nongerminating seeds, which are dormant.
4. Be sure students understand that each volumeter is a closed system when sealed. As the seeds in the volumeter carry out cellular respiration, they remove O_2 from the system and replace it with CO_2. Since CO_2 is absorbed by the soda lime, the volume of the system decreases. Explain that the rate of O_2 consumption is only one measure of the rate of cellular respiration.

CHAPTER 7 INVESTIGATION

Observing Cellular Respiration

OBJECTIVES

- Measure the rate of cellular respiration in germinating seeds.
- Compare cellular respiration rates in germinating and nongerminating seeds.

PROCESS SKILLS

- experimenting
- collecting data
- analyzing data

MATERIALS

- volumeter jar and screw-on lid
- tap water at room temperature
- ring stand with support ring
- 8.5 in. × 11 in. piece of cardboard
- 8.5 in. × 11 in. sheet of white paper
- cellophane tape
- 40 germinating corn or pea seeds
- 40 nongerminating corn or pea seeds
- 100 mL graduated cylinder
- glass or plastic beads
- 3 plastic or paper cups
- 3 volumeters
- cotton
- 3 soda-lime packets
- forceps
- Pasteur pipet
- colored water
- ruler with millimeter markings

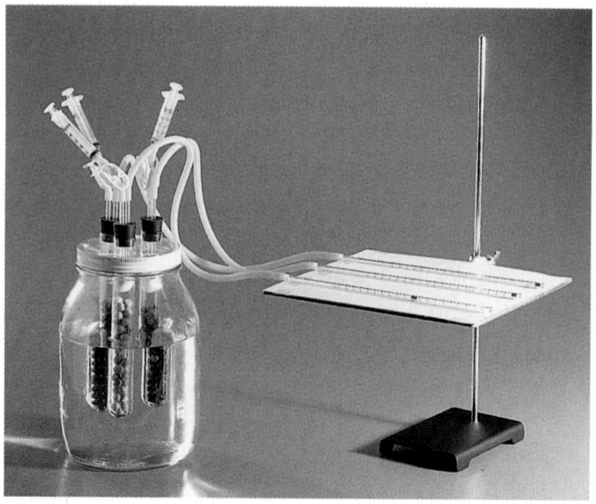

Background

1. When glucose is oxidized in aerobic respiration, what other substance is consumed and what substances are produced?
2. Write the balanced equation for the complete oxidation of glucose in aerobic respiration.

PART A Setting Up the Apparatus

1. Pour room-temperature water into the volumeter jar until the jar is about two-thirds full. Screw on the lid.
2. Place the cardboard on top of the ring stand support ring, and tape a sheet of white paper to the cardboard. Adjust the support ring so that the cardboard is level, as shown in the illustration below.
3. CAUTION **Keep the seeds, which may have been treated with a fungicide, away from your skin.** Place 40 germinating seeds in a graduated cylinder and measure their volume. Do the same for 40 nongerminating seeds.
4. Add beads to the seeds that have the smaller volume until the combined volume is the same as the volume of the other group of seeds. Then transfer both groups (one containing added beads) to separate plastic or paper cups. To a third cup, add a volume of beads equal to the volume of each of the other two cups.
5. Remove the stopper assemblies from three volumeter tubes and transfer the contents of the three cups to separate tubes. Place a 2 cm plug of dry cotton into each tube, leaving a gap of about 1 cm between the cotton and the seeds or beads.
6. CAUTION **Soda lime is corrosive. Do not touch it. If it gets on your skin or clothing, wash it off at the sink. If it gets in your eyes, immediately flush it out at the eyewash station while calling to your teacher.** Using forceps, place a packet of soda lime wrapped in gauze on top of the cotton plug in each tube. Soda lime absorbs the CO_2 that is produced as a result of respiration.
7. Gently but firmly press the stopper assembly into each volumeter tube. Insert the tubes into the volumeter jar through the large holes in the lid.

8. Use a Pasteur pipet to place a small drop of colored water into the three capillary tubes. Tilt two of the tubes slightly until the drops are lined up with the outermost calibration mark. Carefully attach these tubes to the latex tubing on the two volumeter tubes containing seeds. Position the drop in the third capillary tube near the middle of the tube. Attach this tube to the volumeter tube that contains only beads. This volumeter is the control volumeter. Tape all three capillary tubes to the paper on the ring stand.
9. Wait 5 min for the temperature to become uniform throughout the volumeter jar. While you wait, make a data table like the one shown below. Then return the drops in the capillary tubes to their original positions by using the syringes to inject air into or withdraw air from the volumeter tubes, if necessary.

PART B Measuring Respiration Rates

10. On the paper beneath the capillary tubes, mark the position of one end of each drop of colored water. Note the time. Repeat this procedure every 5 min for 20 min. If respiration is rapid, you may have to reposition the drops as you did in Step 9. In which direction would you expect a drop to move if respiration in the volumeter tube were causing it to move?
11. Remove the paper from the ring stand and use a ruler to measure the distance moved by the drops during each time interval. If you repositioned any drops in Step 10, be sure to add this adjustment when you measure the distances. Enter the measurements in the "Uncorrected" columns in your data table.
12. Clean up your materials and wash your hands before leaving the lab.

Analysis and Conclusions

1. No respiration should have occurred in the control volumeter, which contained only beads. Therefore, any movement of the drop in the control volumeter must have been caused by changes in the temperature of the volumeter jar or the air pressure in the classroom. Since these changes would have affected all three volumeters to the same extent, you must subtract the distance you measured for the control volumeter from the distances you measured for the other two volumeters. Do this calculation for each time interval, and enter the results in the "Corrected" columns in your data table.
2. Each capillary tube has a capacity of 0.063 mL between each 1 cm mark on the tube. Use this information to calculate the volume of O_2 consumed by the germinating and nongerminating seeds during each time interval. Enter these results in your data table.
3. Prepare a graph to show the volume of O_2 consumed versus time; use different symbols or colors to distinguish the points for the germinating seeds from those for the nongerminating seeds. Make sure each point represents the cumulative volume of O_2 consumed. For example, the point plotted for the 15–20 min interval should represent the volume consumed during that interval plus the volume consumed during all of the preceding intervals. Draw the best-fit line through the points for each group of seeds. From the slope of this line, calculate the average rate of respiration in milliliters of O_2 per minute for both groups of seeds.
4. Which group of seeds had the higher average rate of respiration? What is the significance of this difference in terms of a seed's ability to survive for long periods?

MEASUREMENTS OF CELLULAR RESPIRATION

Time interval (min)	Distance moved by Drops in Volumeters (mm)					Volume of O_2 consumed (mL)	
	Control	Germinating seeds		Nongerminating seeds			
		Uncorrected	Corrected	Uncorrected	Corrected	Germinating	Nongerminating
0–5							
5–10							
10–15							
15–20							

CHAPTER 7 INVESTIGATION

ANSWERS TO BACKGROUND

1. O_2 is consumed; CO_2 and H_2O are produced.
2. $C_6H_{12}O_6 + 6O_2 \rightarrow 6CO_2 + 6H_2O$ + energy

ANSWER TO PART B

10. The drop should move toward the volumeter tube.

ANSWERS TO ANALYSIS AND CONCLUSIONS

1. If the drop in the control volumeter moves toward the volumeter tube, the distance it moves should be treated as a positive number; if it moves in the opposite direction, the distance it moves should be treated as a negative number. Subtracting these numbers will properly correct the distances measured for the other volumeters.
2. Students should multiply their distance measurements (in mm) by 0.0063 mL/mm.
3. The slope for the nongerminating seeds should be close to zero since these seeds are dormant. The slope for the germinating seeds will vary depending on the type and size of seed, temperature, and other factors. A typical value for peas is about 0.025 mL of O_2 per minute.
4. The germinating seeds should have the higher rate of respiration. When a seed has a low rate of respiration, it uses its stored energy sources slowly, which allows the seed to survive for a long period of time.

CHAPTER 8
CELL REPRODUCTION

	TOPICS	TEACHING RESOURCES	LABS, CLASSWORK, AND HOMEWORK
BLOCKS 1 & 2 90 minutes	**Introducing the Chapter, p. 144**	**ATE** Focus Concept, p. 144 **ATE** Assessing Prior Knowledge, p. 144 **ATE** Understanding the Visual, p. 144	■ Supplemental Reading Guide, *The Lives of a Cell: Notes of a Biology Watcher*
	8-1 Chromosomes, p. 145 Chromosome Structure, p. 145 Chromosome Numbers, p. 146	**ATE** Section Overview, p. 145 14, 29A	**Lab C13,** Preparing Tissue for Karyotyping **Lab C14,** Karyotyping **Lab C15,** Karyotyping—Genetic Disorders ★ Study Guide, Section 8-1 **PE** Section Review, p. 147
BLOCKS 3 & 4 90 minutes	**8-2 Cell Division, p. 148** Cell Division in Prokaryotes, p. 148 Cell Division in Eukaryotes, p. 148	**ATE** Section Overview, p. 148 **ATE** Visual Link Figure 8-5, p. 149 24	**PE** **Quick Lab,** p. 149, Identifying Prefixes and Suffixes **Lab B5,** Mitosis **Lab C9,** Preparing a Root Tip Squash **Lab C10,** Preparing a Root Tip Squash—Stopping Mitosis ★ Study Guide, Section 8-2 **PE** Section Review, p. 151
BLOCKS 5 & 6 90 minutes	**8-3 Meiosis, p. 153** Stages of Meiosis, p. 153	**ATE** Section Overview, p. 153 **ATE** Critical Thinking, p. 155 **ATE** Visual Link Figure 8-9, p. 154 Figure 8-10, p. 154 Figure 8-11, p. 155 27–28	**PE** **Chapter 8 Investigation,** p. 160 **Lab C11,** Modeling Meiosis ★ Study Guide, Section 8-3 **PE** Section Review, p. 156
BLOCKS 7 & 8 90 minutes	**REVIEW and ASSESSMENT** **PE** Chapter 8 Review, pp. 158–159	Audio CD Program BIOLOGY INTERACTIVE TUTOR Unit 4—Cell Reproduction	**CHAPTER TESTING** ■ Chapter Test (blackline master) ▲ Test Generator ■ Assessment Item Listing ■ Scoring Rubrics and Classroom Management Checklists

KEY

MODERN BIOLOGY
- PE Pupil's Edition
- ATE Teacher's Edition
- ■ Active Reading Guide
- ★ Study Guide

HOLT BIOSOURCES
- Teaching Transparencies
- Laboratory Program

One-Stop Planner CD-ROM
Shows these resources within a customizable daily lesson plan:
- ■ Holt BioSources Teaching Resources
- ■ Active Reading Guide
- ★ Study Guide
- ▲ Test Generator

READING FOR CONTENT MASTERY

Preview: Have students preview the objectives, the introductory paragraph, topic sentences, and captions before reading the chapter.

- ATE Active Reading Technique: Anticipation Guide, p. 145
- ■ Active Reading Guide Worksheet 8-1
- PE Summary/Vocabulary, p. 157

- ■ Active Reading Guide Worksheet 8-2
- PE Summary/Vocabulary, p. 157

- ATE Active Reading Technique: Anticipation Guide, p. 156
- ■ Active Reading Guide Worksheet 8-3
- PE Summary/Vocabulary, p. 157

TECHNOLOGY AND INTERNET RESOURCES

CNN PRESENTS Video Segment 4 Organ Cloning

Holt Biology Videodiscs Teacher's Correlation Guide, Lessons 8-1 through 8-3

BIOLOGY INTERACTIVE TUTOR Unit 4—Cell Reproduction, Topics 1–6

Audio CD Program, Sections 8-1 through 8-3

internetconnect

SCILINKS NSTA

On-line Resources: www.scilinks.org
The following *sci*LINKS Internet resources can be found in the student text for this chapter:

Topics:
- Cell cycle, p. 150
- Meiosis, p. 153

go.hrw.com

On-line Resources: go.hrw.com
Visit the HRW Web site for a variety of resources related to this chapter. Just type in the keyword HM2 Chapter 8.

Smithsonian Institution®
Visit **www.si.edu/hrw** for additional on-line resources.

CNNfyi.com
Visit **www.cnnfyi.com** for late-breaking news and current event stories selected just for you.

LAB ACTIVITY PLANNING

CHAPTER 8 INVESTIGATION

Observing Mitosis in Plant Cells, pp. 160–161

OVERVIEW
Students discover that mitosis is a continuous process by examining onion root-tip cells and by identifying the phases of mitosis.

PREPARATION
1. Purchase slides of onion root tips from a biological supply house, or prepare your own slides. Each student group will need one slide.
2. To prepare your own slides, you will need the following materials: 2 M HCl, a glass dish, forceps, aceto-orcein stain, a scalpel, a slide, a coverslip, paper towels, and a cork.
3. Place the root tips of onions in water several days before the lab to initiate root-tip lengthening.
4. Just before the lab, use a sharp scalpel to cut 3 mm root tips. In a fume hood, place the root tips in 2 M HCl for 10 minutes.
5. Use forceps to transfer the soaked root tips to slides. Add a drop of aceto-orcein stain to the slide. Use a scalpel to mince the tips, and then cover the specimens with a coverslip. Put a paper towel and a cork over the coverslip, and press gently but firmly until the tips are flattened. Flattening an improperly prepared root tip can break a glass coverslip. If the tip cannot be flattened, return it to the acid for 10 minutes and try again.

Quick Labs

Identifying Prefixes and Suffixes, p. 149
Students will need a dictionary, eighteen 3 × 5 in. index cards, and a pencil to complete this activity.

UNDERSTANDING THE VISUAL

Ask students what is happening in the micrograph. (A human lymphocyte cell is dividing.) Have the class identify the chromosomes (the large pink structures) and speculate why they see each new cell receiving a copy of the cell's DNA but not copies of its organelles. (Existing organelles in the original cell may be divided between the two new cells, but the new cells' DNA contains the information necessary for each cell to produce new organelles.) Point out that this particular cell was stained to show chromosomes and not cell organelles.

FOCUS CONCEPT

Reproduction and Inheritance
Cell reproduction perpetuates life, allows for the growth and reproduction of organisms, and passes genetic information to future generations.

ASSESSING PRIOR KNOWLEDGE

Review the following concepts.

Biochemistry: *Chapter 3*
Review with students the structure of DNA and RNA. (They are polymers of monomers called nucleotides.) Review the function of nucleic acids in the cell. (Nucleic acids are organic molecules that store genetic information in the cell.)

Cell Organelles: *Chapter 4*
Ask students to review the organelles and their functions and to speculate which one might be directly involved in cell reproduction (nucleus).

CHAPTER 8

CELL REPRODUCTION

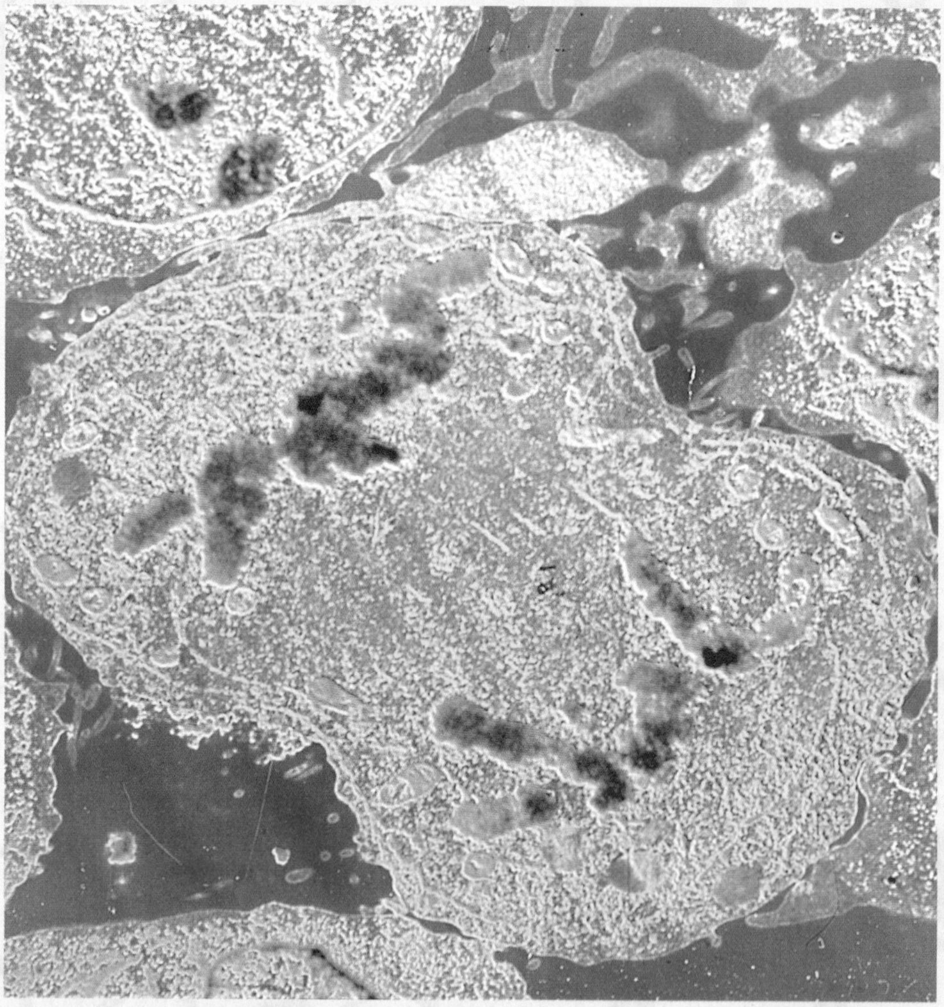

This human lymphocyte cell is dividing into two new cells. (17,687×)

FOCUS CONCEPT: *Reproduction and Inheritance*

In this chapter you will learn how cells reproduce by cell division. Pay attention to the steps of cell division in different kinds of cells.

BIOLOGY INTERACTIVE TUTOR **Unit 4—*Cell Reproduction***
Topics 1–6

CHROMOSOMES

Recall from Chapter 3 that DNA is a long thin molecule that stores genetic information. The DNA in a human cell is estimated to consist of six billion pairs of nucleotides. To visualize the enormity of six billion pairs of nucleotides, imagine increasing a cell nucleus to the size of a basketball. Then imagine taking the DNA out of the basketball-sized nucleus and stretching it into a straight line. That line of DNA would stretch for 64 km (40 mi).

SECTION

8-1

OBJECTIVES

▲ Describe the structure of a chromosome.

● Compare prokaryotic chromosomes with eukaryotic chromosomes.

■ Explain the differences between sex chromosomes and autosomes.

◆ Give examples of diploid and haploid cells.

CHROMOSOME STRUCTURE

During cell division, the DNA in a eukaryotic cell's nucleus is coiled into very compact structures called chromosomes. Chromosomes are rod-shaped structures made of DNA and proteins. In Figure 8-1, you can see the many levels of DNA coiling required to form a chromosome.

The chromosomes of stained eukaryotic cells undergoing cell division are visible as darkened structures inside the nuclear membrane. Each chromosome is a single DNA molecule associated with proteins. The DNA in eukaryotic cells wraps tightly around proteins called **histones.** Histones help maintain the shape of the chromosome and aid in the tight packing of DNA. **Nonhistone** proteins are generally involved in controlling the activity of specific regions of the DNA.

FIGURE 8-1

As a cell prepares to divide, its DNA coils around proteins and twists into rod-shaped chromosomes.

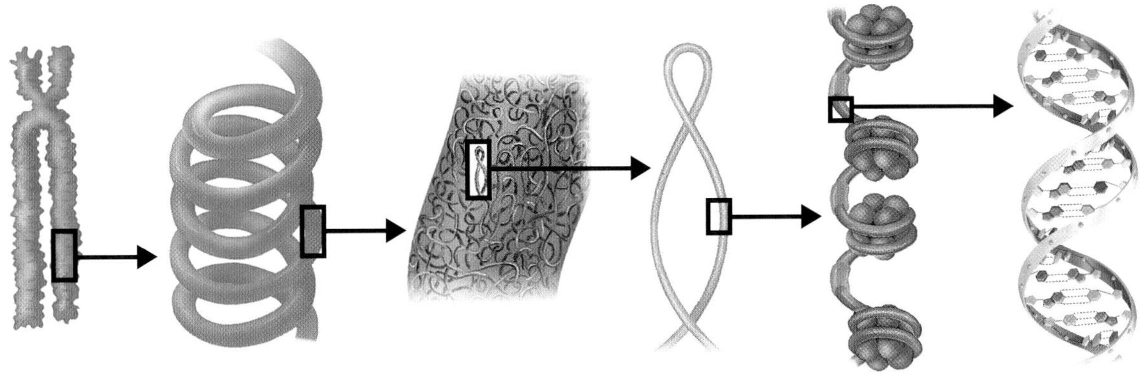

SECTION 8-1

SECTION OVERVIEW

Cell Structure and Function
Section 8-1 describes the structure of chromosomes. This section also compares and contrasts the structure and arrangement of sex chromosomes and autosomes and stresses the difference between diploid and haploid cells.

Matter, Energy, and Organization
Section 8-1 discusses the organization of DNA and its associated proteins. The organization of kinds and numbers of chromosomes is also emphasized.

ACTIVE READING

Technique: Anticipation Guide

Write the following statements on the board:

- Chromosomes from females determine the sex of humans.
- Every human cell contains 46 chromosomes.
- Normal healthy cells cannot become cancerous cells.

Give students time to decide whether they agree or disagree with each statement. Ask volunteers to share their responses. Tell students to save their responses for later use in the Active Reading activity on page 156.

SECTION 8-1

TEACHING STRATEGY
DNA Coiling

Give a group of students a clear plastic bag containing a bundle of four 12 ft strands of yarn, two of one color and two of another color. Tell students that the classroom represents the cell. Tell them that the bag represents the nuclear envelope and the yarn represents the chromatin material for four chromosomes. Ask students why there are two strands of each color. (The two pairs represent homologous strands of DNA.) Have two students twist the ends of one strand in opposite directions. When the coil is tight, have one student take both ends—the yarn will coil up tightly on itself. As the students are repeating these steps with all the yarn, ask them what this represents. (Chromosomes are becoming condensed and visible.) Match up the yarn pairs and join them with a paper clip. Ask students what the clip represents (a centromere). Ask them what each pair represents and why there are two pairs of each color. (Each pair represents a homologous pair. There are two pairs of each color because the chromosomes were replicated.

DEMONSTRATION

Obtain copies of normal human karyotypes from the library or online resources. Enlarge them so that each chromosome is no smaller than 1 in. in length. Paste the copy of the karyotype on poster board. Cut out each chromosome and mix up the pieces. Allow the students to arrange the chromosomes using size, length, position of centromere, and banding patterns. Have them compare the karyotypes with each other and with the karyotype shown in Figure 8-3.

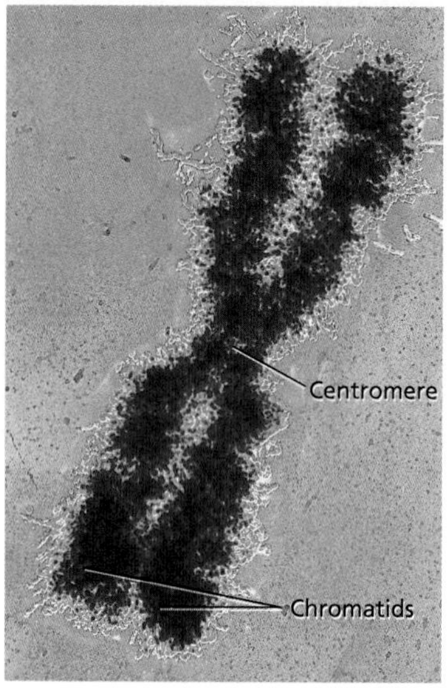

FIGURE 8-2
Chromosomes, like this one isolated from a dividing human cell, consist of two identical chromatids. (TEM 12,542×)

Figure 8-2 shows a chromosome that was isolated from a dividing cell. Notice that the chromosome consists of two identical halves. Each half of the chromosome is called a **chromatid.** Chromatids form as the DNA makes a copy of itself before cell division. When the cell divides, each of the two new cells will receive one chromatid from each chromosome. The constricted area of each chromatid is called a **centromere.** The centromere holds the two chromatids together until they separate during cell division. As you will learn in the next section, centromeres are especially important for the movement of chromosomes during cell division.

Between cell divisions, DNA is not so tightly coiled into chromosomes. Regions of DNA uncoil in between cell divisions so they can be read and so the information can be used to direct the activities of the cell. The less tightly coiled DNA-protein complex is called chromatin.

As you might expect, chromosomes are simpler in prokaryotes than in eukaryotes. The DNA of most prokaryotes comprises only one chromosome, which is attached to the inside of the cell membrane. Prokaryotic chromosomes consist of a circular DNA molecule and associated proteins. As with eukaryotic chromosomes, prokaryotic chromosomes must be very compact to fit into the cell.

CHROMOSOME NUMBERS

Each species has a characteristic number of chromosomes in each cell. Table 8-1 lists the number of chromosomes found in some organisms. Fruit flies, for example, have only eight chromosomes in each cell. Some species of organisms have the same number of chromosomes. For example, potatoes, plums, and chimpanzees all have 48 chromosomes in each cell.

TABLE 8-1 ***Chromosome Numbers of Various Species***

Organism	Number of chromosomes
Adder's tongue fern	1,262
Carrot	18
Cat	32
Chimpanzee	48
Dog	78
Orangutan	48
Earthworm	36
Fruit fly	8
Garden pea	20
Gorilla	48
Horse	64
Human	46
Lettuce	18
Sand dollar	52

Sex Chromosomes and Autosomes

Human and animal chromosomes are categorized as either sex chromosomes or autosomes. **Sex chromosomes** are chromosomes that determine the sex of an organism, and they may also carry genes for other characteristics. In humans, sex chromosomes are either X or Y. Normal females have two X chromosomes, and normal males have an X and a Y chromosome. All of the other chromosomes in an organism are called **autosomes.** Two of the 46 human chromosomes are sex chromosomes, while the remaining 44 chromosomes are autosomes.

Every cell of an organism produced by sexual reproduction has two copies of each autosome. The organism receives one copy of each autosome from each parent. The two copies of each autosome are called **homologous chromosomes,** or homologues. Homologous chromosomes are the same size and shape and carry genes for the same traits. For example, if one chromosome in a pair of homologous chromosomes contains a gene for eye color, so will the other chromosome in the homologous pair. Figure 8-3 shows a **karyotype,** which is a photomicrograph of the chromosomes in a dividing cell found in a normal human. Notice that the 46 human chromosomes exist as 22 homologous pairs of autosomes and two sex chromosomes (XY in males and XX in females). What is the sex of the person whose chromosomes are shown in Figure 8-3?

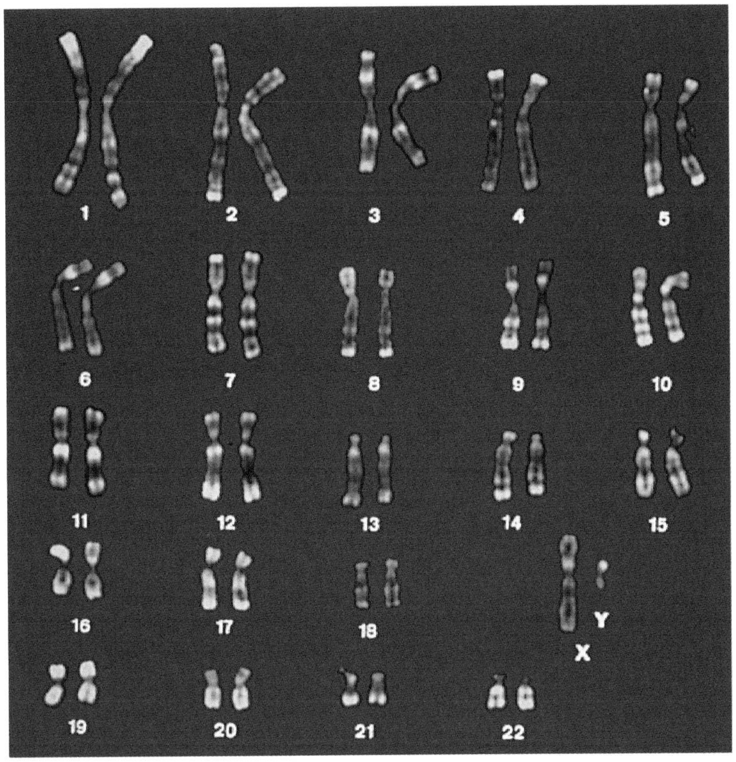

FIGURE 8-3

Karyotypes, like this one, are used to examine an individual's chromosomes. Karyotypes are made from a sample of a person's blood. White blood cells from the sample are treated chemically to stimulate mitosis and to arrest mitosis in metaphase. The chromosomes are then photographed, cut out, and arranged by size and shape into pairs.

Diploid and Haploid Cells

Cells having two sets of chromosomes are said to be **diploid.** Diploid cells have both chromosomes for each homologous pair. Diploid cells also have two sex chromosomes in animals, including humans, and many other organisms that have sex chromosomes. All normal human cells, except reproductive cells (sperm cells and egg cells), are diploid cells. Diploid is commonly abbreviated as $2n$. In humans, the diploid, or $2n$, number of chromosomes is 46–22 pairs of homologous chromosomes and 2 sex chromosomes. If you count the number of chromosomes in the karyotype in Figure 8-3, you should find 46 chromosomes.

Human sperm cells and egg cells are **haploid** cells. These cells contain only one set of chromosomes. Haploid cells have half the number of chromosomes that are present in diploid cells. Thus, human haploid cells have only one chromosome of each homologous pair and only one sex chromosome. Haploid is abbreviated as $1n$. When a sperm cell ($1n$) and an egg cell ($1n$) combine to create the first cell of a new organism, the new cell will be diploid ($2n$). If the reproductive cells were diploid, the new cell would have too many chromosomes and would not be functional.

SECTION 8-1 REVIEW

1. **What are homologous chromosomes?**
2. **Describe the differences between a chromosome and a DNA molecule.**
3. **Compare the structure of prokaryotic chromosomes with that of eukaryotic chromosomes.**
4. **Contrast sex chromosomes with autosomes.**
5. **Using Table 8-1, list the haploid and diploid number of chromosomes for each organism.**
6. **CRITICAL THINKING Is there a correlation between the number of chromosomes and the complexity of an organism? Give support for your answer.**

SECTION 8-1

GIFTED ACTIVITY

Have students do research to determine if the chromosome numbers of species that can interbreed are similar in number. (Yes; for example, a horse has 64 chromosomes and a donkey has 62.)

RETEACHING ACTIVITY

Have students discuss the likely result of gametes in sexually reproducing organisms being $2n$ instead of $1n$. (There would be an increasing number of chromosomes in successive generations.)

ANSWERS TO SECTION 8-1 REVIEW

1. Homologous chromosomes are chromosomes that are similar in shape, size, and genes that they carry.
2. A DNA molecule is a molecule of nucleic acid. A chromosome consists of DNA and proteins.
3. A prokaryotic chromosome consists of a circular DNA molecule and associated proteins and is attached to the inner side of the cell membrane. Eukaryotic chromosomes are rod-shaped, associated with histone and nonhistone proteins, and found within the cell's nucleus.
4. Sex chromosomes are chromosomes that determine the sex of an organism. All of the other chromosomes in an organism are called autosomes.
5. The number of chromosomes given in Table 8-1 is the diploid number. Students should divide that number by 2 to calculate the haploid number of chromosomes.
6. No. By using Table 8-1, students should give examples of simple organisms that have more chromosomes than more-complex organisms.

SECTION 8-2

SECTION OVERVIEW

Cell Structure and Function
Section 8-2 describes the structures involved in prokaryotic and eukaryotic cell division and the function of each structure. Also, this section emphasizes the difference between the structures involved in plant cell mitosis and the structures involved in animal cell mitosis. Each phase of mitosis is discussed, along with its unique structures.

Matter, Energy, and Organization
Section 8-2 discusses the organization of the cell cycle and describes the phases of mitosis as a fluid sequence of events rather than a series of finite stages.

▶ ENGAGE STUDENTS

Ask students what causes a cell to divide. (One reason cells divide is that the ratio of surface area to volume becomes too low as the cell grows.) Lead students in a discussion of what cells have in common with towns and cities (a need for food, water, building materials, leadership, communication, delivery, waste management, etc.). Ask students why taking care of these needs is more complicated in a large city than in a small city (more space to cover, more food and water needed, more waste generated). Remind students that the cell's needs are provided for through osmosis and diffusion through the cell membrane and that when the volume of the cell is too large for the membrane to sufficiently manage, then the cell divides to restore an adequate ratio of surface area to volume.

SECTION

8-2

OBJECTIVES

▲ Describe the events of binary fission.

● Describe each phase of the cell cycle.

■ Summarize the phases of mitosis.

◆ Compare cytokinesis in animal cells with cytokinesis in plant cells.

CELL DIVISION

All cells are derived from the division of preexisting cells. Cell division is the process by which cells produce offspring cells. As you will see, cell division differs in prokaryotes and eukaryotes. In eukaryotes, cell division differs in different stages of an organism's life cycle.

CELL DIVISION IN PROKARYOTES

Binary fission is the division of a prokaryotic cell into two offspring cells. Binary fission consists of three general stages, which are outlined in Figure 8-4.

First the chromosome, which is attached to the inside of the cell membrane, makes a copy of itself, resulting in two identical chromosomes attached to the inside of the prokaryote's inner cell membrane. After the chromosome is copied, the cell continues to grow until it reaches approximately twice the cell's original size. Then a cell wall forms between the two chromosomes and the cell splits into two new cells. Each new cell contains one of the identical chromosomes that resulted from the copying of the original cell's chromosome.

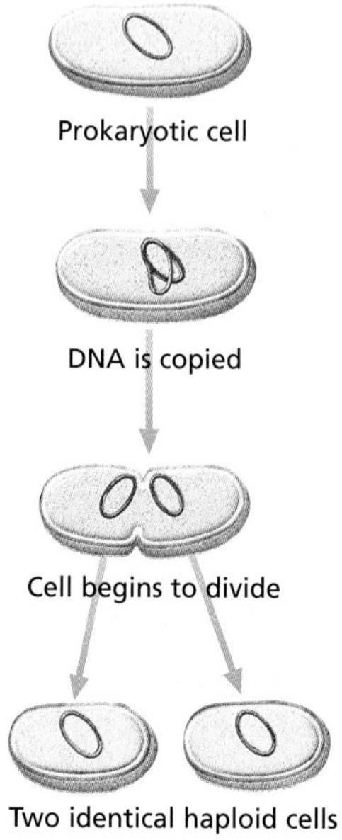

FIGURE 8-4
Most prokaryotes reproduce by binary fission, in which two identical cells are produced from one cell.

CELL DIVISION IN EUKARYOTES

In eukaryotic cell division, both the cytoplasm and the nucleus divide. There are two kinds of cell division in eukaryotes. The first type of cell division that you will learn about is called mitosis. **Mitosis** results in new cells with genetic material that is identical to that of the original cell. Mitosis occurs in the reproduction of unicellular organisms as well as in the addition of cells to a tissue or organ in a multicellular organism.

The type of cell division that you will learn about in Section 8-3 is called meiosis. **Meiosis** reduces the chromosome number by half in new cells. The new cells join together later in the organism's life cycle to produce cells with a complete set of chromosomes.

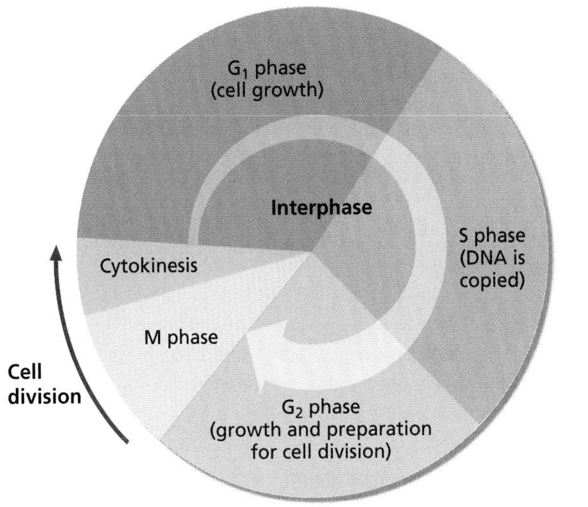

FIGURE 8-5

The life cycle of a cell—called the cell cycle—consists of interphase and cell division. Phases of growth, DNA replication, and preparation for cell division make up interphase. Cell division is divided into mitosis (division of the nucleus) and cytokinesis (division of the cytoplasm).

The Cell Cycle

The **cell cycle** is the repeating set of events that make up the life of a cell. Cell division is one phase of the cell cycle. The time between cell divisions is called **interphase.** Interphase is divided into three phases and cell division is divided into two phases, as illustrated in Figure 8-5.

During cell division, the chromosomes and cytoplasm are equally divided between two offspring cells. Cell division consists of mitosis and cytokinesis. During mitosis, or the **M phase,** the nucleus of a cell divides. **Cytokinesis** is the division of the cytoplasm of the cell.

Interphase

Notice in Figure 8-5 that cells spend most of their lifetime in interphase. Following cell division, offspring cells are approximately half the size of the original cell. During the first stage of interphase—called the **G_1 phase**—offspring cells grow to mature size. G_1 stands for the time gap following cell division and preceding DNA replication. After cells have reached a mature size, they typically proceed into the next phase of interphase, called the **S phase.** During the S phase, the cell's DNA is copied. The **G_2 phase** represents the time gap following DNA synthesis (S phase) and preceding cell division. The G_2 phase is a time during which the cell prepares for cell division.

Cells can also exit the cell cycle (usually from the G_1 phase) and enter into a state called the **G_0 phase.** During the G_0 phase, cells do not copy their DNA and do not prepare for cell division. Many cells in the human body are in the G_0 phase. For example, fully developed cells in the central nervous system stop dividing at maturity and normally never divide again.

Mitosis

Mitosis is the division of the nucleus, which occurs during cell division. Mitosis is a continuous process that is divided into four phases: prophase, metaphase, anaphase, and telophase.

Quick Lab

Identifying Prefixes and Suffixes

Materials dictionary, 3×5 in. index cards (18), pencil

Procedure

1. Write each of the following prefixes and suffixes on separate cards: *pro-, meta-, ana-, telo-, cyto-, oo-, inter-, -kinesis,* and *-genesis.*
2. Use a dictionary to find the definition of each prefix and suffix. Write the definitions on separate cards.
3. Play "Memory" with a partner. Mix the cards and place each one face down on the table. Turn over two cards. If the two cards consist of a prefix or suffix and its definition, pick up the cards and take another turn. If the two cards do not match, turn them face down again and leave them in the same place.
4. Repeat step 3 until no cards remain on the table. The player with the most pairs wins.

Analysis How does knowing the meaning of a prefix or suffix help you to understand the meaning of a word?

VISUAL LINK

Figure 8-5

Use Figure 8-5 to clarify the relationship of cell division and interphase to the cell cycle. Ask students which sections represent interphase (blue) and which represent active cell division (yellow). Stress that cells spend most of their lives in the G_0 stage. Ask students to indicate how and where G_0 might be labeled on the figure. (On the board or overhead, draw the G_0 stage as a detour off G_1 just before the cell enters the S stage. At this point, the cell is mature but not so large that it cannot function, and it can reenter the cell cycle directly into the S stage and begin DNA replication.) Ask students why the cell cycle has a G_0 stage. (Most organisms cannot continue to grow indefinitely. Most cells enter the G_0 stage but remain capable of returning to the cell cycle to replace dead or damaged cells when needed.)

Quick Lab

Identifying Prefixes and Suffixes

Time Required 20 minutes

Procedural Tips If the dictionary provides more than one definition, encourage teams to determine which one is most appropriate to the context of mitosis.

Answers to Analysis The prefix or suffix gives you a clue as to the meaning of half of the word. Then you must put that together with the second half of the word to get a general idea of the meaning. Some prefixes or suffixes, such as *telo-,* which means "end," give clues as to where this phase is in the process of mitosis. *Telo*phase is the last, or end, phase of mitosis. Some prefixes and suffixes, such as *inter-,* which means "between," give clues as to what is happening. *Inter*phase is the phase between cell divisions.

TEACHING STRATEGY
Animating Mitosis

To demonstrate that mitosis is a continual process and not a series of discrete stages, have students create a flip book that animates the phases of mitosis. First have students draw the events of mitosis in no fewer than 25 sketches on sturdy pages about 6 in. square. Stress that each drawing should vary only slightly from the one before it. When the book is flipped through quickly, the process of mitosis should appear to be in motion.

RECENT RESEARCH
Artificial Chromosomes

Researchers report constructing the first wholly synthetic, self-replicating human microchromosome, one-fifth to one-tenth the size of real human chromosomes. Research is being conducted on how these miniature chromosomes might be loaded with genes that are missing or impaired in patients with genetic disorders, such as muscular dystrophy, then introduced into the patient's cells to compensate for the defect.

INCLUSION *ACTIVITY*

Have students act out the process of mitosis. Clear a large area and assign eight students as chromosome pairs 1, 2, 3, and 4. Select eight other students to act as duplicates. Call out "prophase." (Students should mill about in a tight circle, duplicates should step in, chromosomes should locate their duplicates, and the chromosomes should spill out into the cell.) Call out "metaphase." (The pairs of students, should line up across the center of the room.) Call out "anaphase." (The chromosomes and duplicates should separate and move to opposite sides of the cell.) Call out "telophase." (The chromosomes should clump up in each nuclei of the two new cells.)

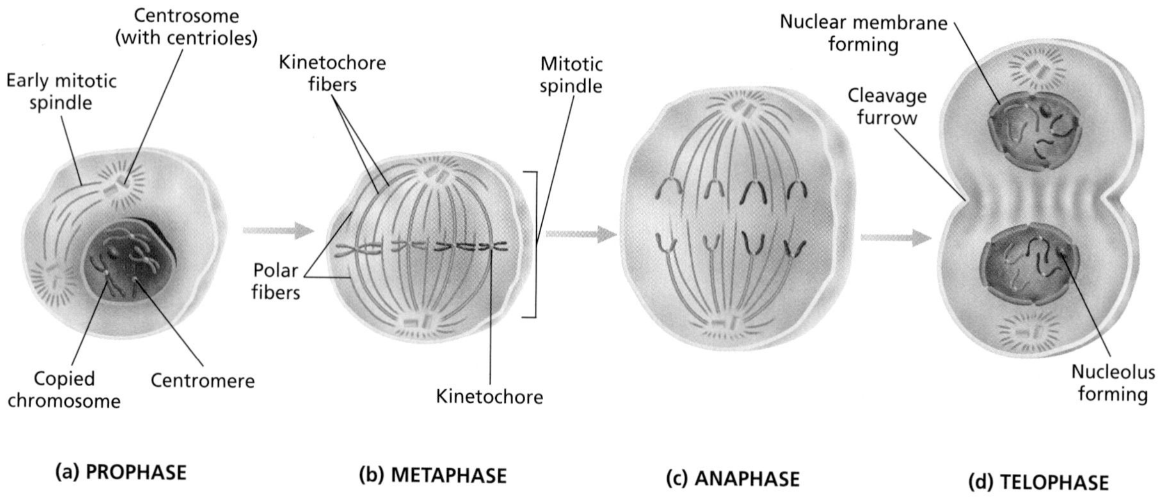

FIGURE 8-6

(a) During prophase, the copied DNA coils into chromosomes. (b) During metaphase, the chromosomes line up along the midline of the dividing cell. (c) During anaphase, the chromatids of each chromosome begin moving toward opposite poles of the cell. (d) During telophase, the chromosomes reach opposite poles of the cell, and the cytoplasm begins to divide.

Prophase is the first phase of mitosis. Prophase, shown in Figure 8-6, begins with the shortening and tight coiling of DNA into rod-shaped chromosomes that can be seen with a light microscope. Recall that during the S phase, each chromosome is copied. The two copies of each chromosome—called chromatids—stay connected to one another by the centromere. At this time, the nucleolus and the nuclear membrane break down and disappear.

Two pairs of dark spots called **centrosomes** appear next to the disappearing nucleus. In animal cells, each centrosome contains a pair of small, cylindrical bodies called **centrioles.** The centrosomes of plant cells lack centrioles. In both animal and plant cells, the centrosomes move toward opposite poles of the cell. As the centrosomes separate, **spindle fibers** made of microtubules radiate from the centrosomes in preparation for mitosis. This array of spindle fibers is called the **mitotic spindle,** which serves to equally divide the chromatids between the two offspring cells during cell division. Two types of spindle fibers make up the mitotic spindle: kinetochore fibers and polar fibers. **Kinetochore fibers** attach to a disk-shaped protein—called a **kinetochore**—that is found in the centromere region of each chromosome. Kinetochore fibers extend from the kinetochore of each chromatid to one of the centrosomes. **Polar fibers** extend across the dividing cell from centrosome to centrosome.

Metaphase is the second phase of mitosis. During metaphase, chromosomes are easier to identify using a microscope than during other phases; thus, karyotypes are typically made from photomicrographs of chromosomes in metaphase. During metaphase, the kinetochore fibers move the chromosomes to the center of the dividing cell. Once in the center of the cell, each chromosome is held in place by the kinetochore fibers.

During **anaphase,** the chromatids of each chromosome separate at the centromere and slowly move, centromere first, toward opposite poles of the dividing cell. When the chromatids separate, they are considered to be individual chromosomes.

internet**connect**

SCILINKS NSTA

TOPIC: Cell cycle
GO TO: www.scilinks.org
KEYWORD: HM150

Telophase is the fourth phase of mitosis. After the chromosomes reach opposite ends of the cell, the spindle fibers disassemble and the chromosomes return to a less tightly coiled chromatin state. A nuclear envelope forms around each set of chromosomes, and a nucleolus forms in each of the newly forming cells.

Cytokinesis

During telophase, the cytoplasm of the cell divides by a process called cytokinesis. In animal cells, cytokinesis begins with a pinching inward of the cell membrane midway between the dividing cell's two poles, as shown in Figure 8-7. The area of the cell membrane that pinches in and eventually separates the dividing cell into two cells is called the **cleavage furrow.** The cleavage furrow pinches the cell into two cells through the action of microfilaments.

Figure 8-8 shows cytokinesis in plant cells. In plant cells, vesicles formed by the Golgi apparatus fuse at the midline of the dividing cell, forming a membrane-bound cell wall called the **cell plate.** When complete, the cell plate separates the cell into two cells. In both animal cells and plant cells, offspring cells are approximately equal in size. Each offspring cell receives an identical copy of the original cell's chromosomes and approximately one-half of the original cell's cytoplasm and organelles.

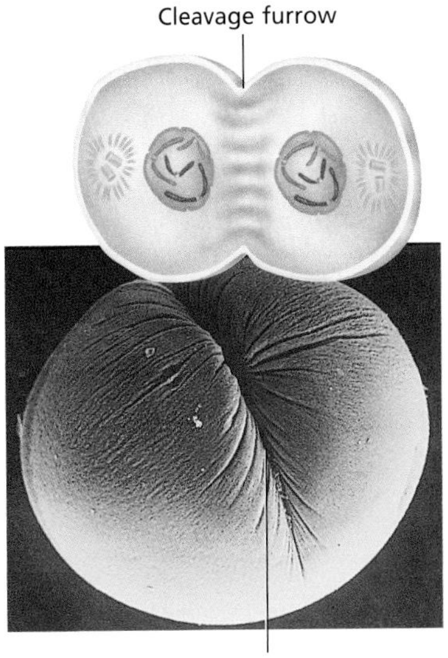

FIGURE 8-7

In animal cells, such as this frog cell, the cell membrane pinches in at the center of the dividing cell, eventually dividing the cell into two offspring cells. (SEM 78×)

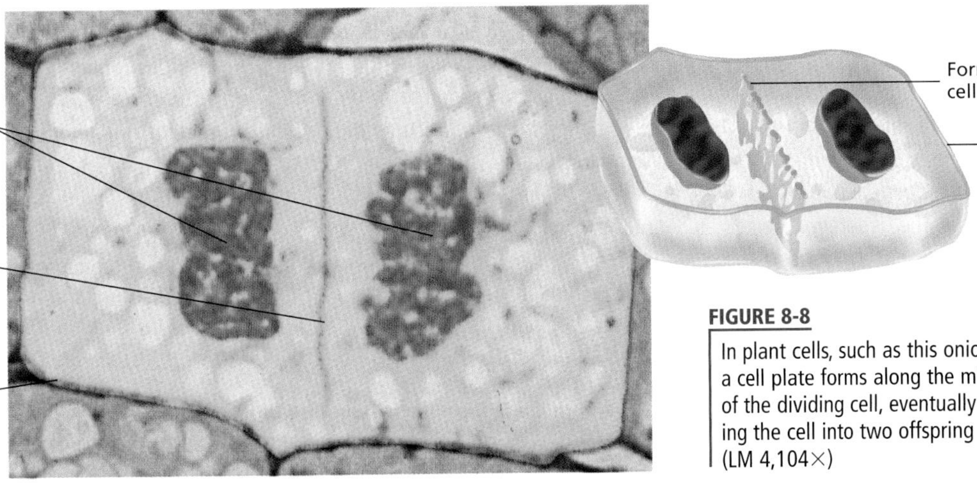

FIGURE 8-8

In plant cells, such as this onion cell, a cell plate forms along the midline of the dividing cell, eventually dividing the cell into two offspring cells. (LM 4,104×)

SECTION 8-2 REVIEW

1. **Describe the events of binary fission.**
2. **During which phase of the cell cycle are chromosomes copied?**
3. **Which phase of the cell cycle could you identify most readily with a light microscope? Explain your answer.**
4. **Describe the structure and function of the mitotic spindle.**
5. **Explain the main differences between cytokinesis in animal cells and cytokinesis in plant cells.**
6. **CRITICAL THINKING What would happen if cytokinesis took place before mitosis?**

SECTION 8-2

RETEACHING ACTIVITY

Have groups of students prepare posters depicting the four stages of the cell cycle. Have them label each stage. For example, the character Gee Suboh could have a bubble over its head saying, "I'll divide again after a while, maybe." Make certain all four phases are defined in some way.

ANSWERS TO SECTION 8-2 REVIEW

1. (1) The chromosome makes a copy of itself. (2) The cell continues to grow until it reaches approximately twice its original size. (3) A cell wall forms between the two chromosomes, and the cell splits into two new cells.
2. During the S phase of interphase, the cell's DNA is copied.
3. Although all phases can be observed with a light microscope, condensed chromosomes can be observed only during the M phase.
4. The mitotic spindle consists of microtubules aligned between centrosomes and between chromosomes and kinetochores. It assists in the movement of chromatids during mitosis.
5. In animal cells, cytokinesis begins with a pinching inward of the cell membrane midway between the dividing cell's two poles. This cleavage furrow eventually separates the dividing cell into two cells. In plant cells, vesicles formed by the Golgi apparatus fuse at the midline of the dividing cell, forming a cell plate. When complete, the cell plate separates the cell into two cells.
6. The nucleus and chromosomes would be in one offspring cell only. The cell with the chromosomes would probably be functional, but the cell lacking chromosomes would not be.

CHAPTER 8
Research Notes

BACKGROUND

Microtubules are tubes formed from molecules of the protein tubulin. They can grow or shrink by the addition or removal of tubulin molecules at the ends of the tubes. Microtubules form the spindle fibers in the cell during cell division. They also help give shape to cells by providing the foundation of the cytoskeleton. They form the major part of cilia and flagella and assist in the movement of molecules from neuron cell bodies to the ends of axons.

DISCUSSION

Guide the discussion by posing the following questions:

1. Griseofulvin, an antibiotic produced by a *Penicillium* mold, is given orally for the treatment of fungal infections of the skin, such as ringworm. It binds to microtubules and disrupts their functions. How might griseofulvin kill the fungus? (By binding to microtubules, griseofulvin disrupts the microtubule structures responsible for forming mitotic spindles thereby preventing cell division.)

2. Microtubules have other functions besides the formation of spindle fibers. What are they? (They help give shape to many cells by providing the foundation of the cytoskeleton. They form the major part of cilia and flagella and also assist in the movement of molecules from neuron cell bodies to the ends of axons.)

FURTHER READINGS

1. "Random Mask Brightens Image," *Nature,* October 31, 1996. This article discusses the history of the confocal scanning laser microscope.

2. "Microscope Yields Sharp 3-D Images," *Science News*, June 10, 1995. This article discusses the development of a computer-assisted microscope.

Research Notes

To View the Invisible

What controls the movement of chromosomes when the cell nucleus divides? This basic question has been studied by biologists for over 100 years. The answer will have great significance for society. Understanding what controls cell division in normal cells will help us understand what happens in cell anomalies, such as cancer.

In 1897, a German anatomist named Walther Flemming began to stain cells with a red dye in order to observe their internal contents during cell division. Because staining kills cells, Flemming had to view mitosis as a series of still images in the various stages of cell division. For many years after Flemming's work, it was not clear whether the spindle fibers, which emerge each time a cell reproduces, were permanent cellular structures. The debate over whether these fibers actively separate the chromosomes during cell division continued among cell biologists for more than 50 years.

During the early 1950s, a Japanese student named Shinya Inoue helped invent the techniques necessary to observe the dynamics of living cells. Inoue worked in the marine biology lab of his professor, Katsuma Dan, who was studying cell division in sea urchins. The electron microscopes they used produced the necessary high-resolution image but required killing and slicing the specimens. Sometimes, part of a cell was altered during the preparation. At the time, there were various light microscopes available that enabled observation of the dynamics of the living system. However, the resolution of those microscopes was not high enough to resolve fine details in the smallest areas of the cells. Dan challenged Inoue to develop a microscope that would allow biologists to study the movement of spindle fibers in dividing cells.

Inoue subsequently developed an improved microscope that enabled him to confirm the existence of spindle fibers in live sea urchin cells and to develop a model for the role the spindle fibers play in cell division. The very instability of the molecules that make up the spindle fibers suggested to Inoue a possible mechanism for chromosome movement. His experiments indicated that these fibers might move the chromosomes that are attached to them by assembling and disassembling. As the subunits of the molecules lengthened or shortened, the chromosomes that were attached to them moved. The bundle of spindle fibers that Inoue observed has since been identified as specialized microtubules. Not until the mid-1970s, however, were these fibers isolated from living cells.

At Woods Hole Oceanographic Institute, Inoue continued to be a pioneer in microscopy techniques for more than 40 years. He developed techniques to bring out the fine structural details of cellular organization, such as showing three-dimensional views of spindle fibers. According to Inoue, this opens up new opportunities for studies of developing embryos and cells undergoing mitosis without destroying the living cells. Such technological advances have brought science one step closer to understanding the complex dynamics of cell division.

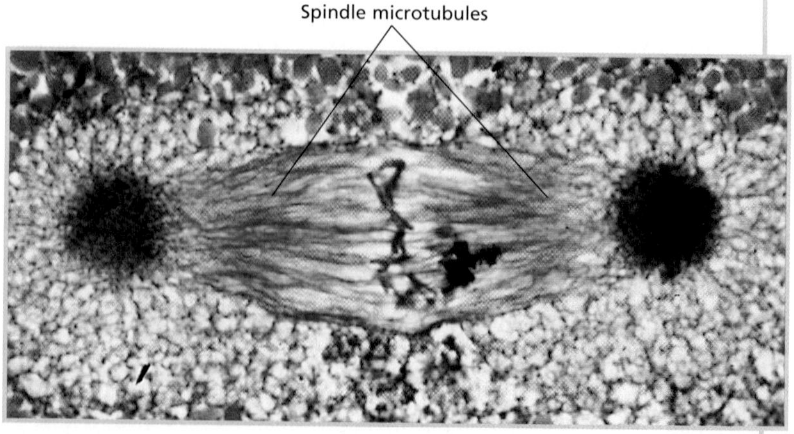

This micrograph (LM 1,080×) of the spindle apparatus during metaphase shows the spindle microtubules studied by Shinya Inoue. The wormlike structures in the center are the chromosomes.

MEIOSIS

Meiosis is a process of nuclear division that reduces the number of chromosomes in new cells to half the number in the original cell. The halving of the chromosome number counteracts a fusion of cells later in the life cycle of the organism. For example, in humans, meiosis produces haploid reproductive cells called **gametes.** *Human gametes are sperm cells and egg cells, each of which contains 23 (1n) chromosomes. The fusion of a sperm and an egg results in a zygote that contains 46 (2n) chromosomes.*

SECTION 8-3

OBJECTIVES

▲ List and describe the phases of meiosis.

● Compare the end products of mitosis with those of meiosis.

■ Explain crossing-over and how it contributes to the production of unique individuals.

◆ Summarize the major characteristics of spermatogenesis and oogenesis.

STAGES OF MEIOSIS

Cells preparing to divide by meiosis undergo the G_1, S, and G_2 phases of interphase. Recall that during interphase, the cell grows to a mature size and copies its DNA. Thus, cells begin meiosis with a duplicate set of chromosomes, just as cells beginning mitosis do. Because cells undergoing meiosis divide twice, diploid (2*n*) cells that divide meiotically result in four haploid (1*n*) cells rather than two diploid (2*n*) cells. The stages of the first cell division are called meiosis I, and the stages of the second cell division are called meiosis II.

Meiosis I

The four phases of meiosis I are illustrated in Figure 8-9 on the next page. Notice how these phases compare with the corresponding phases of mitosis.

During prophase I, DNA coils tightly into chromosomes. As in the prophase of mitosis, spindle fibers appear. Then the nucleus and nucleolus disassemble. Notice how every chromosome lines up next to its homologue. The pairing of homologous chromosomes, which does not occur in mitosis, is called **synapsis.**

Each pair of homologous chromosomes is called a **tetrad.** In each tetrad, chromatids of the homologous chromosomes are aligned lengthwise so that the genes on one chromosome are adjacent to the corresponding genes on the other chromosome. During synapsis, the chromatids within a homologous pair twist around one another, as shown in Figure 8-10. Portions of chromatids may break off and attach to adjacent chromatids on the homologous chromosome—a process called **crossing-over.** This process permits the exchange of genetic material between maternal and paternal

internetconnect

SCILINKS NSTA

TOPIC: Meiosis
GO TO: www.scilinks.org
KEYWORD: HM153

Word Roots and Origins

tetrad

from the Greek *tetras,* meaning "four"

SECTION 8-3

SECTION OVERVIEW

Cell Structure and Function
Section 8-3 describes how the cell functions through each phase of meiosis. The change in the structure of the chromosome during crossing-over is linked to its change in function to increase genetic diversity.

Evolution
In Section 8-3, the concepts of crossing-over and recombination are developed as a basis for hereditary diversity.

Reproduction and Inheritance
In Section 8-3, the formation of gametes is described.

▶ ENGAGE STUDENTS

Tell students they will use playing cards to model the first half of meiosis. Each card will represent an entire chromosome. Have them select cards of two suits of the same color and shuffle them. Ask students what two cards of the same number or face—for example, a queen of spades and a queen of clubs—represent (homologous chromosomes). Next have students match the cards, making two piles; each pile holds one of a pair. Ask what this division represents (the separation of homologous chromosomes into two offspring cells). Emphasize that this activity models just the first half of meiosis. Stress that after the second half of meiosis, each newly formed cell will have half the number of chromosomes of the parent cell.

TEACHING STRATEGY

Chromosome Combinations

Tell students that at metaphase I the orientation of homologous pairs is random: the paternal or maternal homologue can be on either side of the equator. Because of this random orientation, the number of possible chromosome combinations in the gametes is 2^n, where 2 represents the number of homologues in a pair and *n* represents the haploid chromosome number. Using this formula, the number of possible chromosome combinations for 23 chromosomes is 2^{23}. Have students use their calculators to determine how many combinations are possible (8,388,608 combinations).

VISUAL LINK

Figure 8-10

Use this figure to point out how crossing-over produces new genetic combinations. Students should understand that children can never look exactly like their parents because of the random orientation taken by chromosomes at metaphase I and because of crossing-over.

OVERCOMING MISCONCEPTIONS

Be sure students understand the difference between homologous chromosomes and sister chromatids. The two chromosomes of a homologous pair are individual chromosomes that were inherited from different parents. Sister chromatids are a single, replicated chromosome. A good way to count the number of chromosomes is by counting the centromeres. One chromosome has one centromere. Two sister chromatids are joined by one centromere (until Anaphase II when the chromatids separate), and is still counted as one chromosome.

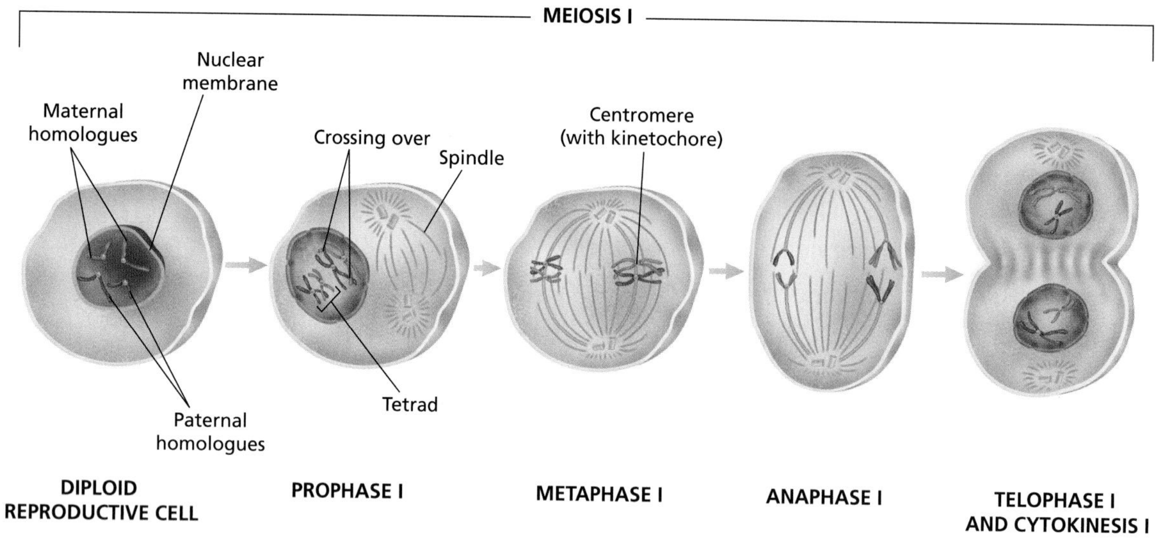

FIGURE 8-9

Meiosis occurs in diploid reproductive cells. Before meiosis begins, the DNA of the diploid reproductive cells is copied. Meiosis I results in two haploid cells.

FIGURE 8-10

Crossing-over occurs when chromosomes that make up a tetrad exchange portions of their chromatids. Crossing-over results in an exchange of genes and in new combinations of genes.

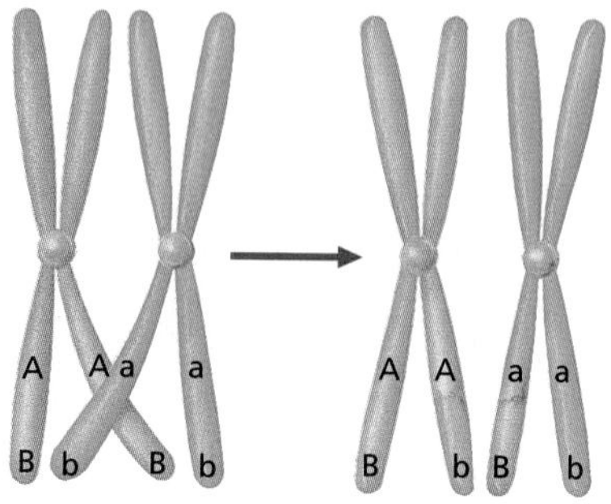

chromosomes. Thus, crossing-over results in **genetic recombination** by producing a new mixture of genetic material.

During metaphase I, the tetrads line up randomly along the midline of the dividing cell. The orientation of the homologous pair of chromosomes is random with respect to the poles of the dividing cell. Spindle fibers from one pole attach to the centromere of one homologous chromosome. Spindle fibers from the opposite pole attach to the other homologous chromosome of the pair.

During anaphase I, each homologous chromosome (consisting of two chromatids attached by a centromere) moves to an opposite pole of the dividing cell, as shown in Figure 8-9. The random separation of the homologous chromosomes is called **independent assortment.** Independent assortment of the chromosomes results in a random separation of the maternal and paternal chromosomes, which results in genetic variation.

Telophase I is the final phase of meiosis I. During telophase I, the chromosomes reach the opposite ends of the cell, and cytokinesis begins. Notice that the new cells contain a haploid number of chromosomes.

During meiosis I, the original cell produces two new cells. Each new cell contains one chromosome from each homologous pair. The new cells contain half the number of chromosomes of the original cell. However, each new cell contains two copies of the chromosome because the original cell copied its DNA before meiosis I.

Meiosis II

Meiosis II occurs in each cell formed during meiosis I and is not preceded by the copying of DNA. The events of meiosis II are shown in Figure 8-11. In some species, meiosis II begins after the nuclear membrane re-forms in the new cells. In other species, meiosis II begins immediately following meiosis I.

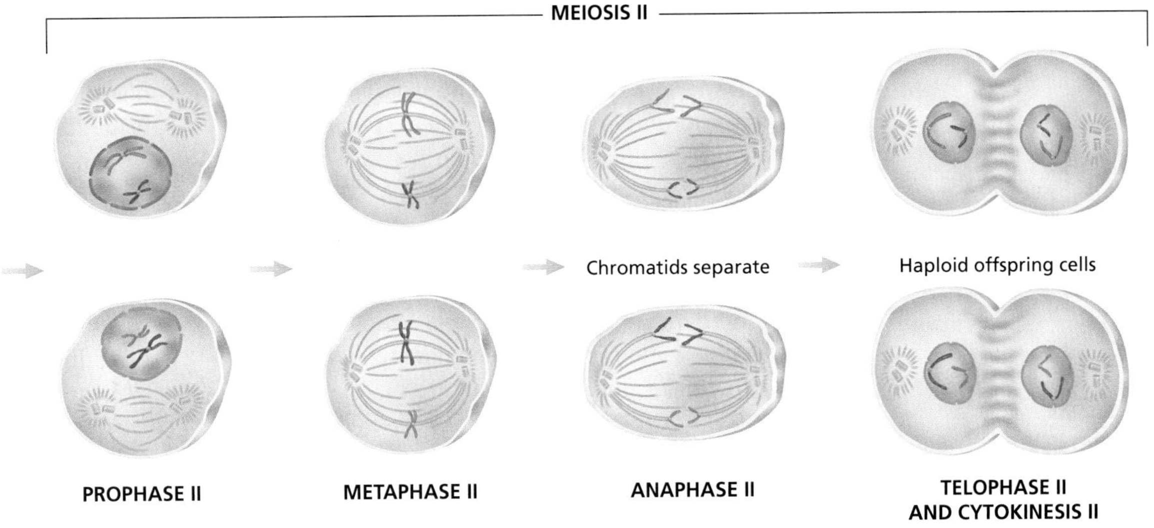

FIGURE 8-11

Meiosis II consists of prophase II, metaphase II, anaphase II, and telophase II. These events closely resemble those of mitosis. Meiosis II results in four haploid offspring cells.

During prophase II, spindle fibers form and begin to move the chromosomes toward the midline of the dividing cell. In metaphase II, the chromosomes move to the midline of the dividing cell, facing opposite poles of the dividing cell. In anaphase II, the chromatids separate and move toward opposite poles of the cell.

In telophase II, a nuclear membrane forms around the chromosomes in each of the four new cells. Cytokinesis II occurs during telophase II, resulting in four new cells, each of which contains half of the original cell's number of chromosomes.

Formation of Gametes

In animals, meiosis produces haploid reproductive cells called gametes, as shown in Figure 8-12. Because only those cells involved in the production of gametes divide by meiosis in animals, meiosis occurs only within their reproductive organs. In humans, meiosis occurs in the testes and in the ovaries.

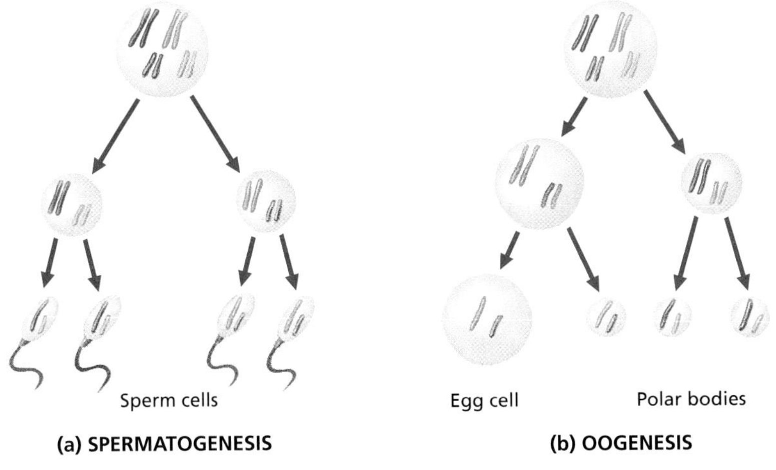

FIGURE 8-12

(a) In the formation of male gametes, the original cell produces four sperm cells by meiosis. (b) In the formation of egg cells, the original cell produces one egg and three polar bodies by meiosis. The egg cell receives most of the original cell's cytoplasm.

VISUAL LINK

Figure 8-9 and Figure 8-11

Ask students to relate Figure 8-9 (meiosis I) to Figure 8-11 (meiosis II). Point out that the two phases sometimes follow each other rapidly; however, in egg development, meiosis II occurs only at the time of ovulation.

CRITICAL THINKING

Surplus Sperm

Ask students what the advantage might be of having one egg and four sperm complete meiosis. (More male gametes increase the probability of fertilization. One egg retains all the nutrients needed for the first cell divisions after fertilization.)

QUICK FACT

☑ Some individuals are born with an extra X or Y chromosome instead of the normal XX or XY. XXX chromosomes are expressed as female, and XYY and XXY chromosomes are usually expressed as male. Also, some humans have only one sex chromosome. An individual with one X (XO) will have some abnormal traits. Having only a Y chromosome (OY) is lethal.

OVERCOMING MISCONCEPTIONS

Throughout history, women have been blamed when no male heir was produced in a family. In fact, the female has no role in the determination of the gender of her child because all of her ova have an X chromosome. Gender is determined by whether the father's gamete contributes an X or a Y chromosome at fertilization.

SECTION 8-3

ACTIVE READING

Technique: Anticipation Guide

Tell students to return to their responses in the Active Reading activity on page 145. Have them change their responses as needed. Discuss the changes.

ANSWERS TO SECTION 8-3 REVIEW

1. Answers include that meiosis results in haploid cells that are not genetically identical to the original cell.
2. telophase I
3. Human gametes normally contain 23 unpaired chromosomes (22 autosomes and 1 sex chromosome).
4. Crossing-over permits the exchange of genetic material between maternal and paternal chromosomes, producing a new mixture of genetic material.
5. Spermatogenesis produces four sperm cells that have approximately equal portions of the original cell's cytoplasm. Oogenesis produces one egg cell that contains most of the original cell's cytoplasm.
6. The chromosomes in the new cells produced by meiosis I have twice the mass of those in the cells produced by meiosis II.

In the testes, meiosis is involved in the production of male gametes known as sperm cells or spermatozoa. In the development of sperm cells, a diploid reproductive cell divides meiotically to form four haploid cells called **spermatids.** Each spermatid then develops into a mature sperm cell. The production of sperm cells is called **spermatogenesis.**

Oogenesis (OH-oh-JEN-uh-sis) is the production of mature egg cells, or ova. During oogenesis, a diploid reproductive cell divides meiotically to produce one mature egg cell (ovum). During cytokinesis I and cytokinesis II of oogenesis, the cytoplasm of the original cell is divided unequally between new cells. As Figure 8-12 shows, one cell, which develops into a mature egg cell, receives most of the cytoplasm of the original cell. As a result, one egg cell is produced by meiosis. The other three products of meiosis, called **polar bodies,** degenerate.

FIGURE 8-13

Many plants, like this kalanchoe, produce offspring plantlets by asexual reproduction. Each plantlet, produced by mitotic cell divisions, is genetically identical to the parent plant.

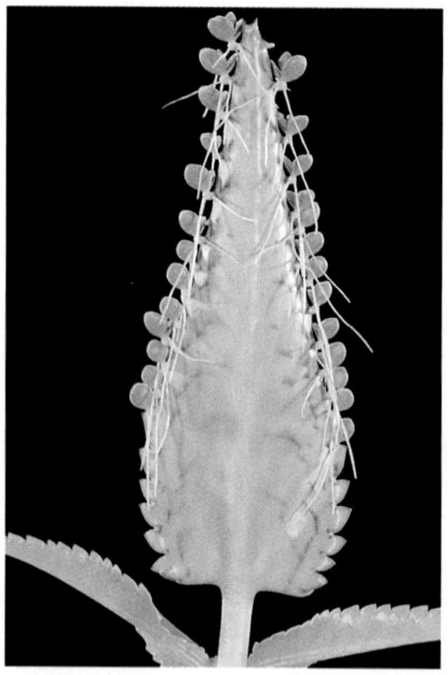

Asexual and Sexual Reproduction

Asexual reproduction is the production of offspring from one parent. Asexual reproduction does not usually involve meiosis or the union of gametes. In unicellular organisms, such as bacteria, new organisms are created by either binary fission or mitosis. Asexual reproduction in multicellular organisms results from the budding off of portions of their bodies, as Figure 8-13 shows. The offspring from asexual reproduction are genetically identical to the parent.

Sexual reproduction is the production of offspring through meiosis and the union of a sperm and an egg. Offspring produced by sexual reproduction are genetically different from the parents because genes are combined in new ways in meiosis. In fact, except in the case of identical twins, sexually produced offspring contain unique combinations of their parents' genes. The evolutionary advantage of sexual reproduction is that it enables species to adapt rapidly to new conditions. For example, if disease strikes a crop of grain, a few plants may have genetic variations that make them resistant to the disease. While many individuals may die, these few resistant plants survive and reproduce.

SECTION 8-3 REVIEW

1. **List two ways that meiosis differs from mitosis.**
2. **During which stage of meiosis is the diploid number of chromosomes reduced to the haploid number of chromosomes?**
3. **How many chromosomes do human gametes normally contain?**
4. **Explain the role of crossing-over in ensuring genetic variation.**
5. **Describe the primary differences between spermatogenesis and oogenesis.**
6. **CRITICAL THINKING Explain why the chromosomes in the haploid cells that are produced by meiosis I look different from those produced by meiosis II.**

CHAPTER 8 REVIEW

SUMMARY/VOCABULARY

8-1

- Chromosomes are tightly coiled DNA molecules and associated proteins.
- In eukaryotes, histone proteins help maintain the compact structure of chromosomes.
- In dividing cells, chromosomes are composed of two identical chromatids constricted together at a centromere.
- Chromosomes are categorized as either sex chromosomes or autosomes.
- Homologous chromosomes consist of one autosomal chromosome from each parent.
- Diploid ($2n$) is the number of chromosomes in cells that have homologous pairs of autosomes and two sex chromosomes.
- Haploid cells ($1n$) have half the number of chromosomes that are present in diploid cells.

Vocabulary

autosome (146)
centromere (146)
chromatid (146)
diploid (147)
haploid (147)
histone (145)
homologous chromosome (146)
karyotype (147)
nonhistone (145)
sex chromosome (146)

8-2

- Cell division is the process by which cells reproduce themselves.
- Binary fission is the process of cell division in prokaryotes.
- The cell cycle is the repeating of events that make up the life of a cell. The cell cycle consists of cell division and interphase.
- Cell division in eukaryotes includes the division of the nucleus (mitosis) and the division of the cytoplasm (cytokinesis).
- Interphase consists of a phase of growth (G_1), a phase of DNA replication (S), and a phase of preparation for cell division (G_2).
- Mitosis is divided into prophase, metaphase, anaphase, and telophase. Mitosis results in two offspring cells that are genetically identical to the original cell.

Vocabulary

anaphase (150)
binary fission (148)
cell cycle (149)
cell plate (151)
centriole (150)
centrosome (150)
cleavage furrow (151)
cytokinesis (149)
G_0 phase (149)
G_1 phase (149)
G_2 phase (149)
interphase (149)
kinetochore (150)
kinetochore fiber (150)
meiosis (148)
metaphase (150)
mitosis (148)
mitotic spindle (150)
M phase (149)
polar fiber (150)
prophase (150)
S phase (149)
spindle fiber (150)
telophase (151)

8-3

- During meiosis, a cell divides twice.
- Crossing-over during meiosis results in genetic recombination.
- Spermatogenesis is the process by which sperm cells are produced. Oogenesis is the process that produces egg cells (ova).
- Asexual reproduction is the formation of offspring from one parent. Offspring produced by asexual reproduction are genetically identical to the parent.
- Sexual reproduction is the formation of offspring through the union of a sperm and an egg. Offspring produced by sexual reproduction are genetically different from the parents.

Vocabulary

asexual reproduction (156)
crossing-over (153)
gamete (153)
genetic recombination (154)
independent assortment (154)
oogenesis (156)
polar body (156)
sexual reproduction (156)
spermatid (156)
spermatogenesis (156)
synapsis (153)
tetrad (153)

CHAPTER 8 REVIEW ANSWERS

REVIEW

1. Chromosomes are rod-shaped structures made of DNA and proteins. A homologous chromosome pair is a pair of chromosomes that are similar in size and shape and carry genes for the same traits.
2. Mitosis is nuclear division that results in genetically identical offspring cells. Meiosis consists of two nuclear divisions, halving the chromosome number in offspring cells. Cytokinesis is the division of cytoplasm.
3. Sex chromosomes determine the sex of an organism. All of the other chromosomes in an organism are autosomes.
4. Kinetochore fibers extend from the kinetochore of each chromatid to one of the centrosomes. Polar fibers extend across the dividing cell from centrosome to centrosome.
5. Cells having two sets of chromosomes are diploid. Cells having one set of unpaired chromosomes are haploid.

6. d **7.** c
8. b **9.** d
10. a **11.** d
12. a **13.** a
14. b **15.** d

16. Binary fission is the division of a prokaryotic cell into two offspring cells. Binary fission occurs in unicellular prokaryotes, such as bacteria.
17. During the G_1 phase—the first stage of interphase—offspring cells grow to mature size. During the G_2 phase, the cell prepares for cell division. Cells that have exited the cell cycle are said to be in the G_0 phase.
18. Haploid cells unite to form a diploid gamete.
19. Meiotic anaphase II separates only half the chromatids that are separated during mitotic anaphase.
20. No. In asexual reproduction, chromosomes are passed from only one parent.

21. Yes. During meiosis II, only half the organism's chromosomes are present. In mitosis, all of the organism's chromosomes are present.
22. Asexual reproduction is the production of offspring from one parent. Asexual reproduction results in genetically identical offspring. Sexual reproduction is the production of offspring through meiosis and the union of gametes. Sexual reproduction results in genetically different offspring.
23. Because gametes are formed, the picture must show meiosis.
24. Answers will vary. Some cancer fighting drugs, such as vincristine and taxol, work by blocking the mitosis spindle microtubules from functioning.

Critical Thinking

1. Yes. The new cell would have two nuclei.
2. Four. The number of chromosomes would double. Each chromosome would consist of two chromatids; thus, the mass would be four times that found in a sperm.
3. Yes. Although the number of chromosomes has been reduced to a haploid number in the metaphase II cell, the chromosomes consist of two chromatids. As a result, the mass of DNA in a meiotic metaphase II cell is the same mass of DNA found in a G_1 cell that contains a diploid number of unreplicated chromosomes.
4. No. A haploid cell contains one chromosome from each homologous pair.
5. As a cell grows, its surface area increases at a slower rate than its volume. Cell division results in a cell returning to a state at which its surface area greatly exceeds its volume.
6. Scientists think that centrioles do not function in mitosis because the plant spindle forms without them and mitosis continues.

Review

Vocabulary

1. Differentiate between a chromosome and a homologous chromosome pair.
2. Distinguish between mitosis, meiosis, and cytokinesis.
3. Distinguish between autosomes and sex chromosomes.
4. Distinguish between kinetochore fibers and polar fibers.
5. Explain the difference in meaning of the terms *haploid* and *diploid.*

Multiple Choice

6. Prokaryotic chromosomes (a) comprise at least two chromosomes (b) are made of DNA wrapped tightly around histone proteins (c) include histone and nonhistone proteins (d) consist of a circular DNA molecule.
7. A chromatid is (a) a dark stain (b) a dense substance within the nuclear membrane of a nondividing cell (c) one of two identical parts that make up a chromosome (d) the point at which each pair of chromatids is joined.
8. Every species has (a) haploid gametes (b) a distinctive number of chromosomes per cell (c) at least eight chromosomes per cell (d) a number of chromosomes that varies with the complexity of the organism.
9. Binary fission is (a) nuclear division of cells (b) eukaryotic cell division (c) sexual reproduction of prokaryotes (d) prokaryotic cell division.
10. Mitosis (a) can increase the number of body cells without changing the information contained in the DNA of those cells (b) is a means of reproducing sexually (c) is never triggered by cell size (d) results in offspring cells that are genetically different from the original cell.
11. Interphase is (a) composed of G_1, G_2, and G_3 (b) the time between meiosis I and meiosis II (c) a small part of the life cycle of a cell (d) a time of cell growth and development.
12. Cytokinesis (a) differs in animal and plant cells (b) does not occur in plant cells (c) immediately precedes mitosis (d) is a process of nuclear division.
13. Spermatogenesis produces (a) four haploid cells (b) four diploid cells (c) one haploid cell and three polar bodies (d) two haploid cells.
14. Oogenesis (a) produces diploid cells (b) requires meiotic cell divisions (c) produces four egg cells (d) produces one diploid cell and three polar bodies.
15. Crossing-over occurs during (a) mitosis (b) interphase (c) meiosis II (d) meiosis I.

Short Answer

16. What does the term *binary fission* refer to? In what type of organism does this type of cell division occur?
17. Discuss the events that occur in G_0, G_1, and G_2 phases.
18. Discuss the role of haploid cells in sexual reproduction.
19. What is the primary functional difference between mitotic anaphase and meiotic anaphase II?
20. Do asexual organisms have homologous chromosomes? Explain your answer.
21. Is there any functional difference between meiosis II and mitosis? Explain your answer.
22. Distinguish between sexual reproduction and asexual reproduction.
23. The photograph below shows cell division in a grasshopper testis. The offspring cells are gametes. Do you think the photograph shows mitosis or meiosis? Explain your answer.

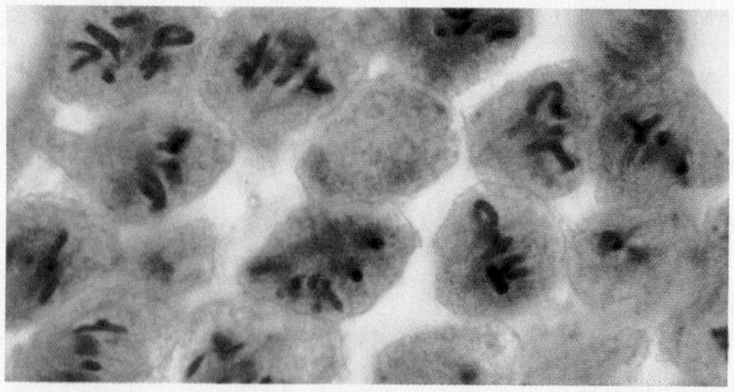

24. **Unit 4—*Cell Reproduction***
INTERACTIVE BIOLOGY TUTOR
Write a report summarizing how different cancer-fighting drugs kill cancer cells by interrupting the life cycle of the cells.

CRITICAL THINKING

1. Can mitosis occur in a cell in the absence of cytokinesis? Support your answer. If your answer is yes, provide a description of how the new cell would appear in the G_1 phase of the cell cycle.
2. If you consider the mass of DNA in a sperm (a haploid cell) to be 1, what would the relative value be for the DNA mass of a cell in the G_2 phase of the cell cycle?
3. Does a cell in metaphase II have the same mass of DNA as a diploid cell in the G_1 phase of the cell cycle? Assume that both cells are from the same animal. Explain your answer.
4. Would a human cell with *any* 23 chromosomes be haploid? Explain your answer.
5. For a cell to function efficiently, the magnitude of its surface area must greatly exceed that of its volume. Explain how cell division functions to maintain this relationship between surface area and volume and in doing so maintains cell homeostasis.
6. The events of mitosis in plants and animals are very similar with the exception of the absence of centrioles in plants. How has the absence of centrioles in plant cells influenced scientists' thinking about the function of centrioles in mitosis?
7. The graph below demonstrates the mass of DNA and chromosome number in each phase of mitosis. Based on the information presented in this graph, at which phase of mitosis are chromatids considered chromosomes? Explain your answer.

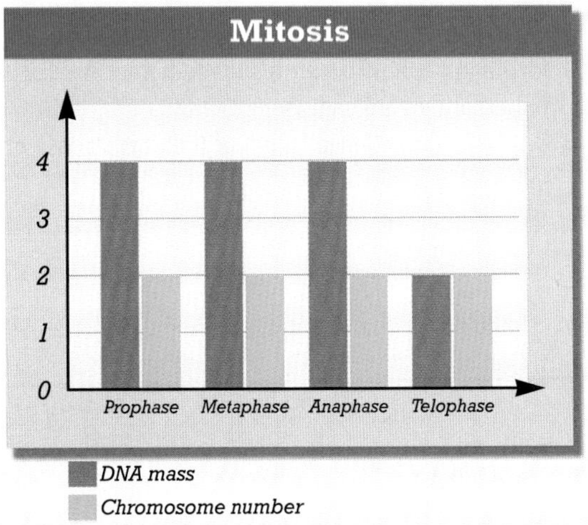

8. Develop graphs to illustrate the mass of DNA and chromosome number in each phase of meiosis I and meiosis II. Your graphs should be similar in format to the one shown for mitosis. For telophase, consider the DNA in only one of the offspring cells at the end of telophase. Let the number 1 represent the mass of DNA that is found in a human ovum. Let the number 1 equal the number of chromosomes found in a human ovum.

EXTENSION

1. Read "Cell Division Gatekeepers Identified" in *Science,* January 23, 2000, on page 477. Identify the cell structures that have been found to regulate the timing of cell division during mitosis. Describe the function of these structures. What conditions in reproductive cells might occur if these structures fail to function during cell division?
2. Read "Dolly's Mixture" in *New Scientist,* September 4, 1999, on page 5. Explain how Dolly might not be the perfect clone she was once thought to be. What problems might this pose for researchers hoping to clone human tissue for transplants?
3. Do library or on-line research to find out how cancer cells differ from normal cells in relation to the cell cycle. Share the results of your research with your classmates.

CHAPTER 8 REVIEW ANSWERS

CONTINUED

7. According to the graph, chromatids are considered chromosomes at telophase because it is at this phase that the mass of chromosomes is no longer double the number of chromosomes.
8. See the following graphs.

EXTENSION

1. Kinetochores have been found

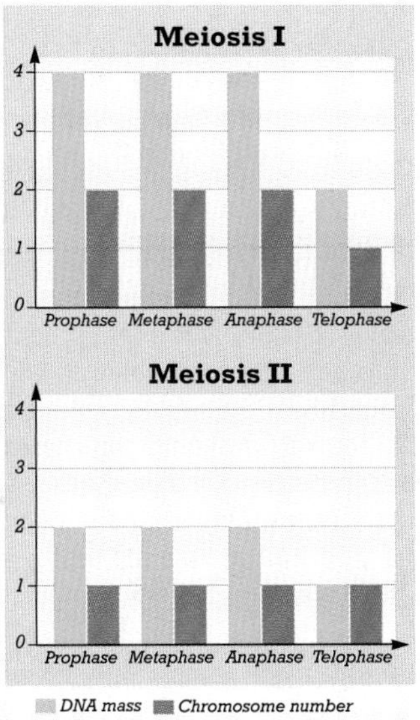

to have a regulatory function as well as a mechanical one in cell division. Kinetochores release a protein that delays anaphase until every chromosome is attached to the spindle. Then another protein is released by the kinetochores that signals anaphase to begin. The result of failure is aneuploidy.
2. Studies show that Dolly has the same nuclear DNA as her donor, but she has the mitochondrial DNA of the egg cell used to host the donated nucleus. Researchers had hoped to use egg cells from cows to clone human tissue for transplants. However, the presence of cow mitochondria could pose unforeseen problems.
3. The cell cycle of cancer cells is unregulated.

TIME REQUIRED

One 50-minute lab period

SAFETY PRECAUTIONS

Have students handle glass slides with care.

QUICK REFERENCES

Lab Preparation Notes are found on page 143B.

Holt BioSources provides a Teaching Resources CD-ROM that contains "Using Gowin's Vee in the Lab" and "Scoring Rubrics."

PROCEDURAL TIPS

1. If you wish to prepare your own slides of *Allium* root tip, see the Lab Preparation Notes on page 143B.
2. Misidentification of the phases of mitosis is a common student error. For instance, the polar view of a cell in anaphase can be confused with metaphase.

ANSWERS TO BACKGROUND

3. Prophase: Chromosomes appear as thick, rod-shaped threads. Metaphase: Chromosomes appear in a line at the center of the dividing cell. Anaphase: The chromatids of each pair separate and begin to move toward opposite poles of the dividing cell. Telophase: A nuclear membrane forms around each set of chromosomes. In animal cells, the cell membrane begins to pinch in. In plant cells, a cell plate forms between the two newly forming nuclear membranes.

ANSWERS TO PART A

5. The number of tally marks beside each stage will vary.

CHAPTER 8 INVESTIGATION

Observing Mitosis in Plant Cells

OBJECTIVES

- Examine the dividing root-tip cells of an onion.
- Identify the phase of mitosis that different cells in an onion root tip are undergoing.
- Determine the relative length of time each phase of mitosis takes in onion root-tip cells.

PROCESS SKILLS

- observing
- classifying
- collecting
- organizing
- analyzing data
- calculating

MATERIALS

- compound light microscope
- prepared microscope slide of a longitudinal section of *Allium* (onion) root tip

Background

1. Mitosis is divided into four phases: prophase, metaphase, anaphase, and telophase.
2. Interphase is not considered a part of mitosis.
3. List the visible characteristics of each phase of mitosis.
4. In many plants, there are growth regions called meristems where mitosis is ongoing. Meristems are found in the tips of plant roots and shoots.

PART A Identifying the Phases of Mitosis

1. Look at the photograph below of a longitudinal section of an onion root tip. Find the meristem on the photograph. As you can see, the meristem is located just behind the root cap.

ONION ROOT TIP

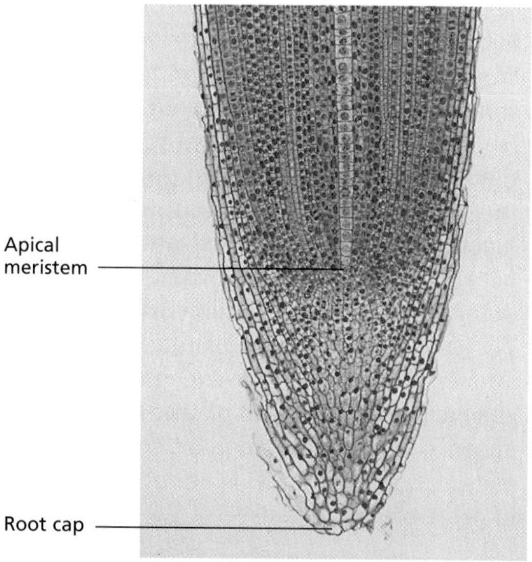

2. CAUTION **Slides break easily. Use caution when handling them.** Using low power on your microscope, bring the meristem region on your slide into focus.

TABLE A *RELATIVE DURATION OF EACH PHASE OF MITOSIS*

Phase of mitosis	Tally marks	Count	Percentage	Time (in minutes)
Prophase	Answers will vary.	Answers will vary.	85	68
Metaphase			8	6
Anaphase			3	2
Telophase			4	3

TABLE B *DATA COLLECTED BY THE ENTIRE CLASS*

Phase of mitosis	Count	Percentage	Time (in minutes)
Prophase	Answers will vary.	85	68
Metaphase		8	6
Anaphase		3	2
Telophase		4	3

3. Examine the meristem carefully. Choose a sample of about 50 cells. Look for a group of cells that appear to have been actively dividing at the time that the slide was made. The cells will appear in rows, so it should be easy to keep track of them. The dark-staining bodies are the nuclei.
4. In your lab report, prepare a data table like Table A.
5. For each of the cells in your sample, identify the stage of mitosis and place a mark in the "Tally marks" column beside the appropriate phase.

PART B Calculating the Relative Length of Each Phase

6. When you have classified each cell in your sample, count the tally marks for each phase and fill in the "Count" column. In which phase of mitosis were the greatest number of cells? In which phase were the fewest number of cells?
7. Calculate the percentage of cells found in each phase. Divide the number of cells in a phase by the total number of cells in your sample, and multiply by 100 percent. Enter the figures under "Percentage."
8. The percentage of cells found in each phase is a measure of how long each phase lasts. For example, if 25 percent of the cells are in prophase, then prophase takes 25 percent of the total time it takes for a cell to undergo mitosis. Mitosis in onion cells takes about 80 minutes. Calculate the actual time for each phase using this information and the percentage you have just determined.

$$\text{Duration of phase (in minutes)} = \frac{\text{percentage}}{100} \times 80 \text{ minutes}$$

9. Record the actual time for each phase in your data table.
10. Make another data table, similar to Table B, shown above. Collect and record the count for each phase of mitosis for the entire class. Fill in the percentage and time information using the data collected by the entire class.
11. Clean up your materials before leaving the lab.

Analysis and Conclusions

1. What color are the chromosomes stained?
2. How can you distinguish between early and late anaphase?
3. According to your data table, which phase of mitosis lasts the longest? Why might this phase require more time than other phases of mitosis?
4. According to your data table, which phase takes the least amount of time?
5. How do your results compare with those of the entire class?
6. In this investigation, you assumed that the percentage of the total time that any given phase takes is equal to the percentage of cells in that phase at any moment. Why might this not be true for very small samples of cells?

Further Inquiry

1. Given the rate of mitosis in a type of animal cells, how could you determine how long each phase of mitosis takes in those cells?
2. Cancerous tissue is composed of cells undergoing uncontrolled, rapid cell division. How could you develop a procedure to identify cancerous tissue by counting the number of cells undergoing mitosis?

ANSWERS TO PART B

6. The greatest number of cells in the sample should be in prophase. The fewest number of cells should be in anaphase.
7. Answers should reflect the data. Roughly 85 percent of the cells should be in prophase, 8 percent in metaphase, 3 percent in anaphase, and 4 percent in telophase.
8. Answers should reflect the data. Roughly 68 min is spent in prophase, 6 min is spent in metaphase, 2 min is spent in anaphase, and 3 min is spent in telophase.
9. Class data and calculations based on class data will vary.

ANSWERS TO ANALYSIS AND CONCLUSIONS

1. Chromosomes are often stained pink or purple, depending on the stain used.
2. In early anaphase, the chromosomes are just beginning to pull away from the equator. In late anaphase, the chromosomes are near the poles.
3. Answers will vary.
4. Answers will vary.
5. Answers will vary.
6. A very small sample may have skewed data. More representative data must come from larger samples.

FURTHER INQUIRY

1. Students should see that they can use the same procedure and calculations they used to determine the relative duration of each mitotic phase in onion root-tip cells.
2. Students would have to find out the percentage of cells undergoing mitosis at any given time in a normal tissue sample and compare it with the cancerous tissue sample.

Unit 3

GENETICS

CHAPTERS

"*. . . [G]enetics has come to occupy an important position at the center of the sciences of life . . . genetics became allied with biochemistry; it revolutionized bacteriology, played a major role in the emergence of the molecular biology of the fifties, resisted the challenge of ecology, took hold of cancer research and is even now reaching out to revolutionize taxonomy and its old rival embryology.*"

These tiger cub siblings look very similar to one another because they have inherited characteristics from their parents.

internetconnect

SciLINKS NSTA

National Science Teachers Association *sci*LINKS Internet resources are located throughout this unit.

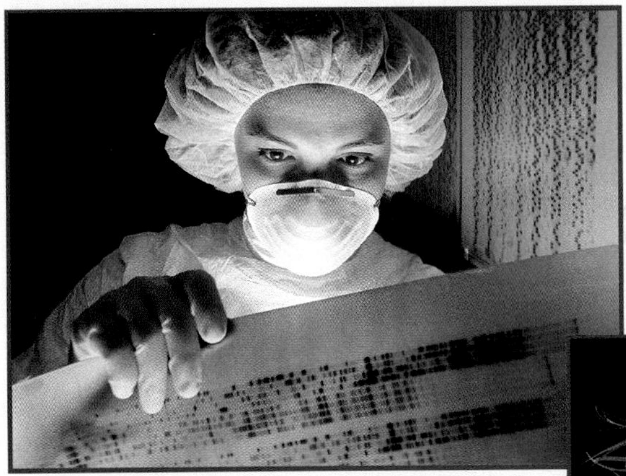

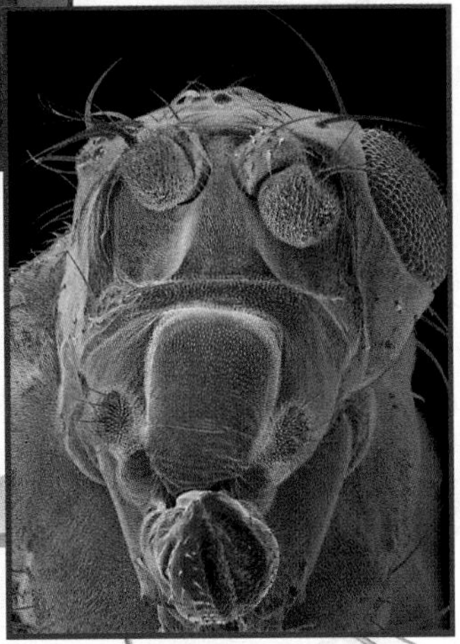

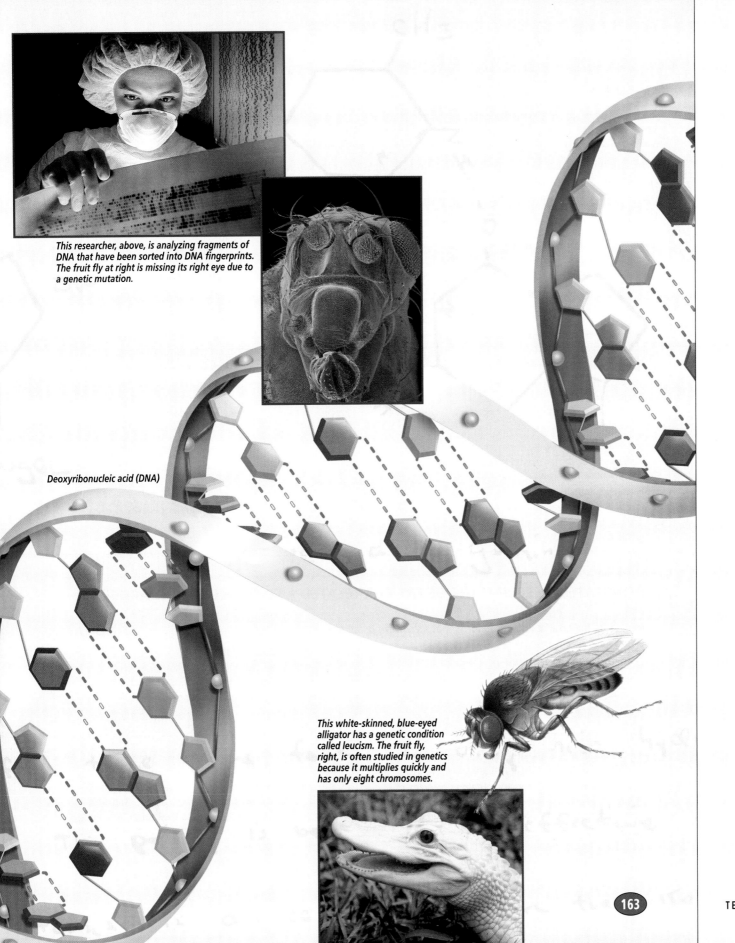

This researcher, above, is analyzing fragments of DNA that have been sorted into DNA fingerprints. The fruit fly at right is missing its right eye due to a genetic mutation.

Deoxyribonucleic acid (DNA)

This white-skinned, blue-eyed alligator has a genetic condition called leucism. The fruit fly, right, is often studied in genetics because it multiplies quickly and has only eight chromosomes.

unit 3

CHAPTERS

GENETICS

CONNECTING TO THE STANDARDS

The following chart shows the correlation of Unit 3 with the *National Science Education Standards* (grades 9–12). The items in each category are addressed in this unit and should be used to guide instruction. Annotated descriptions of the **Life Science Standards** are found below the chart. Consult the *National Science Education Standards* for detailed descriptions of each category.

UNIFYING CONCEPTS

- Systems, order, and organization
- Evidence, models, and explanation
- Change, constancy, and measurement

SCIENCE AS INQUIRY

- Abilities necessary to do scientific inquiry
- Understandings about scientific inquiry

SCIENCE AND TECHNOLOGY

- Abilities of technological design
- Understandings about science and technology

PHYSICAL SCIENCE

- Structure and properties of matter
- Chemical reactions

LIFE SCIENCE

EARTH AND SPACE SCIENCE

- Origin and evolution of the Earth system
- Origin and evolution of the universe

SCIENCE IN PERSONAL AND SOCIAL PERSPECTIVES

- Personal/community health
- Natural/human hazards
- Science and technology in local and global challenges

HISTORY AND NATURE OF SCIENCE

- Science as a human endeavor
- Historical perspectives

ANNOTATED DESCRIPTIONS OF THE CORRELATED LIFE SCIENCE STANDARDS

Although all eight categories of the *National Science Education Standards* are important, the following descriptions summarize the **Life Science Standards** that specifically relate to Unit 3.

THE CELL

- Cell functions are regulated.
- Cells can differentiate. Complex multicellular organisms are formed from highly organized arrangements of differentiated cells.

MOLECULAR BASIS OF HEREDITY

- In all organisms, the instructions for specifying the characteristics of the organism are carried in DNA.
- Most of the cells in a human contain two copies of each of 22 different chromosomes. One additional pair determines whether the individual will be male or female.
- Changes in DNA occur spontaneously at low rates.

BIOLOGICAL EVOLUTION

- Species evolve over time. The millions of different species of organisms that live on Earth today are related by descent from common ancestors.
- Biological classifications are based on how organisms are related.

MATTER, ENERGY, AND ORGANIZATION IN LIVING SYSTEMS

- The complexity and organization within an organism result from the organism's need to obtain, transform, transport, release, and eliminate the matter and energy required to sustain the organism.

TRENDS IN BIOLOGY

Beyond Mendelian Genetics: Genomic Imprinting

Angelman syndrome is a rare developmental disorder characterized by a number of physical and behavioral abnormalities, including a small head, difficulty walking, an enlarged mouth and tongue, frequent seizures, and mental retardation. Although the equally rare Prader-Willi syndrome is also characterized by mental retardation, the other symptoms of this disorder—overeating leading to obesity, sluggishness, abnormally small hands and feet, and underdevelopment of the gonads—do not suggest a connection to Angelman syndrome. So scientists were very surprised to learn that the same defect, a deletion on chromosome 15, caused both syndromes. Even more surprising is the way in which these disorders are inherited. Individuals with Prader-Willi syndrome receive the defective copy of chromosome 15 from their father, while individuals with Angelman syndrome receive the defective chromosome from their mother.

How can this peculiar pattern of transmission be explained? It turns out that Prader-Willi and Angelman syndromes are examples of a newly discovered mechanism called genomic imprinting, in which the effect of a gene depends on which parent it comes from. How does genomic imprinting work? When a gene is imprinted (a process thought to usually occur during gamete formation), it is labeled chemically in a way that prevents its expression in the offspring; the gene, in effect, is silenced. Different genes are imprinted in male and female gametes, meaning that the offspring receives one functional copy and one inactivated copy of each gene.

Prader-Willi and Angelman syndromes are caused by different genes (or clusters of genes) that are closely linked on chromosome 15 and whose products play a key role in controlling development. The diagram at right shows how imprinting and deletion combine to produce Prader-Willi syndrome. The gene involved in Prader-Willi syndrome is normally imprinted in the egg, so the working copy must come from the sperm. If the sperm does not contain a functional copy of this gene—say, because of a deletion—the offspring will not have a version of the gene that can be transcribed and development will not occur normally. The mechanism for Angelman syndrome is similar, but the imprinting is opposite: the responsible gene is normally imprinted in the father.

Only a handful of imprinted genes have been discovered so far, most of them related to rare developmental disorders. However, several kinds of cancer, including Wilms' tumor (kidney), retinoblastoma (eye), and some kinds of lung cancer, have already been shown to contain cells with incorrectly imprinted genes.

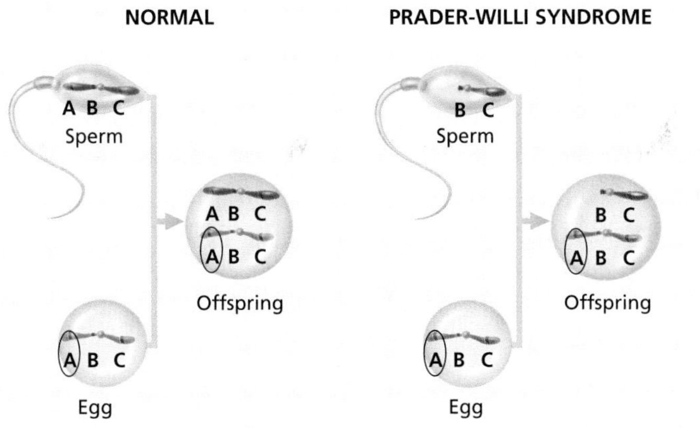

Because the gene involved in Prader-Willi syndrome (shown here as A) is normally imprinted (indicated by circle) in the egg, the sperm must supply a working copy of this gene. If it doesn't happen, the child will have Prader-Willi syndrome.

Additional Information

Vines, Gail. "Where Did You Get Your Brains?" *New Scientist,* May 3, 1997, 34–39.

Lowenstein, Jerold. "Genetic Surprises." *Discover,* December 1992, 82–88.

CHAPTER 9
FUNDAMENTALS OF GENETICS

	TOPICS	TEACHING RESOURCES	LABS, CLASSWORK, AND HOMEWORK
BLOCKS 1 & 2 90 minutes	**Introducing the Chapter, p. 164**	ATE Focus Concept, p. 164 ATE Assessing Prior Knowledge, p. 164 ATE Understanding the Visual, p. 164	■ Supplemental Reading Guide, *The Double Helix*
	9-1 Mendel's Legacy, p. 165 Gregor Mendel, p. 165 Mendel's Experiments, p. 166 Mendel's Results and Conclusions, p. 168 Chromosomes and Genes, p. 169	ATE Section Overview, p. 165 ATE Critical Thinking, p. 166 ATE Visual Link Table 9-1, p. 167 Figure 9-3, p. 169 25–26, 29–32	PE **Quick Lab,** p. 173, Calculating Probability ■ Portolio Projects—Genetics Research Project ■ Portolio Projects Student Worksheets—Genetics Focus Worksheet ★ Study Guide, Section 9-1 PE Section Review, p. 169
BLOCKS 3 & 4 90 minutes	**9-2 Genetic Crosses, p. 172** Genotype and Phenotype, p. 172 Probability, p. 173 Predicting Results of Monohybrid Crosses, p. 174 Predicting Results of Dihybrid Crosses, p. 177	ATE Section Overview, p. 172 ATE Critical Thinking, p. 175 ATE Visual Link Figure 9-9, p. 176 Figure 9-12, p. 178 31–32	PE **Chapter 9 Investigation,** p. 182 PE **Quick Lab,** p. 177, Determining Genotypes **Lab A6,** Interpreting Information on a Pedigree **Lab C12,** Analyzing Corn Genetics ■ Problem-Solving Worksheets—Genetics and Probability ★ Study Guide, Section 9-2 PE Section Review, p. 178
BLOCKS 5 & 6 90 minutes	**REVIEW and ASSESSMENT** PE Chapter 9 Review, pp. 180–181 ■ Performance-Based Assessment—Identifying Dominant and Recessive Traits	Audio CD Program BIOLOGY INTERACTIVE TUTOR Unit 5—Heredity	**CHAPTER TESTING** ■ Chapter Test (blackline master) ▲ Test Generator ■ Assessment Item Listing ■ Scoring Rubrics and Classroom Management Checklists

KEY

MODERN BIOLOGY

- PE Pupil's Edition
- ATE Teacher's Edition
- ■ Active Reading Guide
- ★ Study Guide

HOLT BIOSOURCES

- Teaching Transparencies
- Laboratory Program

One-Stop Planner CD-ROM
Shows these resources within a customizable daily lesson plan:

- ■ Holt BioSources Teaching Resources
- ■ Active Reading Guide
- ★ Study Guide
- ▲ Test Generator

READING FOR CONTENT MASTERY

Preview: Have students preview the objectives, the introductory paragraph, topic sentences, and captions before reading the chapter.

ATE Active Reading Technique: K-W-L, p. 165

■ Reading Strategy: K-W-L Worksheet

■ Active Reading Guide Worksheet 9-1

PE Summary/Vocabulary, p. 179

ATE Active Reading Technique: K-W-L, p. 177

■ Active Reading Guide Worksheet 9-2

PE Summary/Vocabulary, p. 179

TECHNOLOGY AND INTERNET RESOURCES

Holt Biology Videodiscs Teacher's Correlation Guide, Lessons 9-1 through 9-2

BIOLOGY INTERACTIVE TUTOR: Unit 5—Heredity, Topics 1–6

Audio CD Program, Sections 9-1 through 9-2

internetconnect

SCILINKS NSTA

On-line Resources: www.scilinks.org
The following *sci*LINKS Internet resources can be found in the student text for this chapter:

Topics:
- Gregor Mendel, p. 166
- Dominance, p. 176
- Punnett squares, p. 178

go.hrw.com

On-line Resources: go.hrw.com
Visit the HRW Web site for a variety of resources related to this chapter. Just type in the keyword HM2 Chapter 9.

Smithsonian Institution®
Visit **www.si.edu/hrw** for additional on-line resources.

CNNfyi.com
Visit **www.cnnfyi.com** for late-breaking news and current event stories selected just for you.

LAB ACTIVITY PLANNING

CHAPTER 9 INVESTIGATION

Modeling Monohybrid Crosses, pp. 182–183

OVERVIEW

Students will simulate the pairing of alleles as gametes form. Manipulating objects to simulate monohybrid crosses is a good way for students to duplicate Mendelian crosses and to gain an understanding of the mechanisms of genetics. In this investigation, students will simulate the pairing of alleles by combining green peas and lentils. Then they will determine the genotypes, genotypic ratios, phenotypes, and phenotypic ratios of the offspring.

PREPARATION

1. You will need to obtain lentils, green peas, and Petri dishes. Each group of two students will need two lentils, two green peas, and two Petri dishes.
2. You may choose to use pop-it beads, available from WARD'S, instead of lentils and green peas.

Quick Labs

Calculating Probability, p. 173
Each student or group of students will need a small paper sack containing 20 jelly beans (12 white, 6 red, and 2 green) for this activity. Prepare the sacks before class, and instruct students not to look inside their sacks.

Determining Genotypes, p. 177
Students need only pencils and paper for this activity.

UNDERSTANDING THE VISUAL

Tell students that this American alligator is one of 18 white alligators discovered southwest of New Orleans in 1987 by fishermen. Have students point out how the alligator shown differs from other alligators they have seen pictures of. (This alligator is white-skinned and blue-eyed.) Tell students that hide color and eye color are genetically controlled characteristics of alligators. Distinguish *characteristics* from *traits* by pointing out that white and brown are contrasting traits of the characteristic hide color. Examples of characteristics are color, height, and texture. Examples of traits are white/brown, tall/short, smooth/wrinkled.

Emphasize that the white American alligator is not a different species and is not an albino. Albinos have off-white or yellow skin and colorless irises, causing pinkish eyes. The genetic condition of the white alligator shown is called leucism and is more rare than albinism.

FOCUS CONCEPT

Reproduction and Inheritance

Genetics is the study of how individuals inherit characteristics from their parents. Mendel showed that these characteristics are inherited as discrete factors, now called genes, that are not diluted over generations.

ASSESSING PRIOR KNOWLEDGE

Review the following concepts.

Heredity: *Chapter 1*
Ask students to discuss the meaning of the term *gene*.

Crossing-Over: *Chapter 8*
Ask students how crossing-over during meiosis contributes to the physical differences between siblings.

CHAPTER 9

FUNDAMENTALS OF GENETICS

The unique appearance of this white-skinned, blue-eyed alligator is the result of a genetic condition.

FOCUS CONCEPT: ***Reproduction and Inheritance***

As you read, notice how Mendel developed hypotheses to help predict the outcome of different genetic crosses.

BIOLOGY INTERACTIVE TUTOR Unit 5—*Heredity*
Topics 1–6

MENDEL'S LEGACY

***G**enetics* *is the field of biology devoted to understanding how characteristics are transmitted from parents to offspring. Genetics was founded with the work of Gregor Johann Mendel, an Austrian monk who experimented with garden peas. This section describes Mendel's experiments and the principles of genetics that resulted from them.*

SECTION

9-1

OBJECTIVES

▲

Describe the steps involved in Mendel's experiments on garden peas.

●

Distinguish between dominant and recessive traits.

■

State two laws of heredity that were developed from Mendel's work.

◆

Explain the difference between an allele and a gene.

▲

Describe how Mendel's results can be explained by scientific knowledge of genes and chromosomes.

GREGOR MENDEL

In 1842, at the age of 21, Gregor Mendel entered a monastery in Brunn, Austria. His task of tending the garden gave him time to think and to observe the growth of many plants. In 1851 he entered the University of Vienna to study science and mathematics. His mathematics courses included training in the then-new field of statistics. Mendel's knowledge of statistics later proved valuable in his research on **heredity**—the transmission of characteristics from parents to offspring. When Mendel returned to the monastery, he taught in a high school and also kept a garden plot. Although he studied many plants, he is probably remembered most for his experiments with garden peas, *Pisum sativum.*

Mendel's Garden Peas

Mendel observed seven characteristics of pea plants. Each characteristic occurred in two contrasting **traits:** *plant height* (long or short stems), *flower position along stem* (axial or terminal), *pod color* (green or yellow), *pod appearance* (inflated or constricted), *seed texture* (smooth or wrinkled), *seed color* (yellow or green), and *flower color* (purple or white). Mendel used his knowledge of statistics to analyze his observations of these seven characteristics.

Mendel collected seeds from his pea plants, carefully recording the characteristics of the plant from which each seed was collected. The next year he planted the seeds. He observed that purple-flowering plants grew from the seeds obtained from purple-flowering plants, but he noticed that some white-flowering plants also grew from the seeds of purple-flowering plants. And when experimenting with the characteristic of plant height, he observed that while some tall plants grew from seeds obtained from tall plants, some short plants also grew from seeds obtained from tall plants. Mendel wanted to find an explanation for such variations.

Word Roots and Origins

heredity

from the Latin *hereditas,* meaning "heirship"

SECTION 9-1

SECTION OVERVIEW

Reproduction and Inheritance

Section 9-1 discusses Mendel's experiments with pea plants and the laws of heredity that developed from his work. It also relates Mendel's observations to molecular genetics.

ACTIVE READING

Technique: K-W-L

Before students read this chapter, have them write short, individual lists of all the things they already Know (or think they know) about inheritance. Ask them to contribute their entries to a group list on the board or the overhead projector. Then have students list things they Want to know about inheritance. Have students save their lists for later use in the Active Reading activity on page 177.

GIFTED *ACTIVITY*

Have students design a newspaper ad that might have attracted someone like Mendel to purchase *Pisum sativum* for genetic research. Students should include that it is easy to grow, matures quickly, produces a large number of offspring, is able to self-pollinate, and exists in many varieties (differences in flower color, seed color, seed shape, height, position, etc.).

SECTION 9-1

RECENT RESEARCH
Born Happy

Studies show that a person's average level of happiness is associated with heredity. According to studies conducted at the University of Minnesota, identical twins are similar in their level of happiness 44 percent of their lifetime. Fraternal twins, however, share a similar happiness only 8 percent of their lifetime.

CRITICAL THINKING
Mendel's Experiment

Tell students that peas have seven pairs of chromosomes and that each trait studied by Mendel is on a different chromosome. Ask students how Mendel's results might have differed if the two traits he observed had been located on the same chromosome. (He would not have gotten a 9:3:3:1 ratio. Instead, the phenotypes of the parents would have been overrepresented. The other two phenotypes would have appeared only when crossing-over occurred during meiosis. As a result, Mendel would not have observed the law of independent assortment.)

OVERCOMING MISCONCEPTIONS

Mendel's work with garden pea plants showed that traits are controlled by single genes. In humans, single-factor inheritance has been found in about 600 recessively inherited traits and in dominant conditions such as Huntington's disease and Marfan syndrome. About 1 percent of all newborns have a single gene defect. Many more diseases are determined by polygenic inheritance, in which several genes and environmental factors are involved. Common conditions that are inherited polygenically include cleft lip and palate, schizophrenia, hypertension, and diabetes.

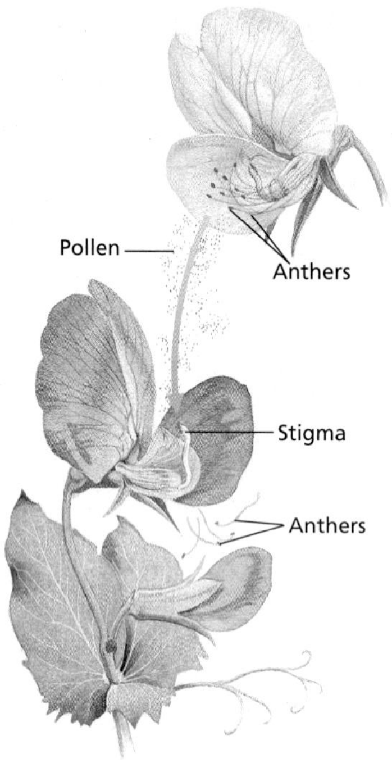

FIGURE 9-1

Mendel controlled the breeding of his pea plants and tracked the inheritance of traits by transferring pollen from the anthers of flowers on one plant to the stigma of flowers on a different plant.

internetconnect

SCLINKS. NSTA
TOPIC: Gregor Mendel
GO TO: www.scilinks.org
KEYWORD: HM166

Mendel's Methods

Mendel was able to document the traits of each generation's parents by carefully controlling how the pea plants were pollinated. **Pollination** occurs when pollen grains produced in the male reproductive parts of a flower, called the **anthers,** are transferred to the female reproductive part of a flower, called the **stigma.**

Self-pollination occurs when pollen is transferred from the anthers of a flower to the stigma of either the same flower or a flower on the same plant. **Cross-pollination** involves flowers of two separate plants. Pea plants normally reproduce through self-pollination.

Self-pollination can be interrupted—and cross-pollination performed—by removing the anthers from a flower and manually transferring the anther of a flower on one plant to the stigma of a flower on another plant, as shown in Figure 9-1. By doing this, Mendel was able to protect his flowers from receiving any other pollen that might be transferred by wind or insects, giving him more control over the pollination of his pea plants.

MENDEL'S EXPERIMENTS

Mendel studied each characteristic and its contrasting traits individually. He began by growing plants that were pure for each trait. Plants that are **pure** for a trait always produce offspring with that trait. For example, pea plants pure for the trait of yellow pods self-pollinate to produce offspring with yellow pods. The term **strain** denotes plants that are pure for a specific trait. Mendel produced strains by allowing the plants to self-pollinate for several generations. He eventually obtained 14 strains, one for each of the 14 traits he observed. He called each strain a parental generation, or **P_1 generation.**

Mendel then cross-pollinated these strains by transferring pollen from the anthers of a plant pure for one trait to the stigma of another plant pure for the contrasting trait. For example, if he wanted to cross a plant pure for the trait of yellow pods with one pure for the trait of green pods, he first removed the anthers from the plant that produced green pods. Then he dusted the pollen from a yellow-podded plant onto the stigma of a green-podded plant and allowed the seeds to develop.

When the plants matured, he recorded the number of each type of offspring produced by each P_1 plant. Mendel called the offspring of the P_1 generation the first filial generation, or **F_1 generation.** He then allowed the flowers from the F_1 generation to self-pollinate and collected the seeds. Mendel called the plants in this generation the second filial generation, or **F_2 generation.** Following this process, Mendel performed hundreds of crosses and documented the results of each by counting and recording the observed traits of every cross. Table 9-1 summarizes the results of many of Mendel's crosses.

TABLE 9-1 *Mendel's Crosses and Results*

Characteristic	P cross	F_1 generation	F_2 generation	Actual ratio	Probability ratio
Position of flowers along stem	 axial × terminal	 axial	651 axial 207 terminal	3.14:1	3:1
Height of plant	 tall × short	 tall	787 tall 277 short	2.84:1	3:1
Pod appearance	 inflated × constricted	 inflated	882 inflated 299 constricted	2.95:1	3:1
Pod color	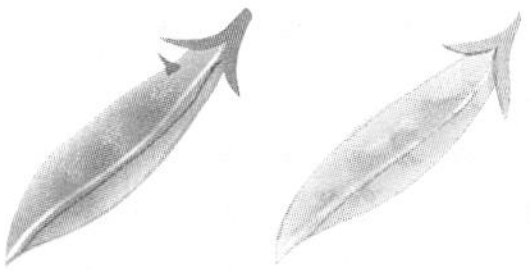 green × yellow	 green	428 green 152 yellow	2.82:1	3:1
Seed texture	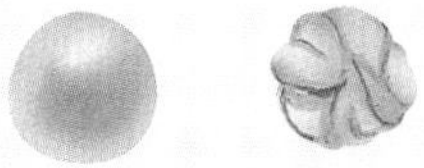 smooth × wrinkled	 smooth	5,474 smooth 1,850 wrinkled	2.96:1	3:1
Seed color	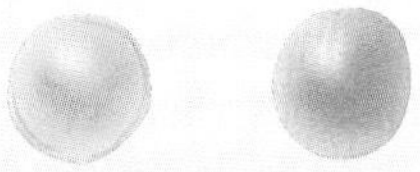 yellow × green	 yellow	6,022 yellow 2,001 green	3.01:1	3:1
Flower color	 purple × white	 purple	705 purple 224 white	3.15:1	3:1

VISUAL LINK
Table 9-1

Use this table to explain Mendel's research. Make sure students understand that each experiment begins with a cross between two parent plants having different traits and that the F_2 generation was produced by the self-pollination of only one parent.

QUICK FACT

☑ After his success with peas, Mendel did research on rapeseed, the source of canola oil. Because rapeseed does not self-pollinate, Mendel had very little success identifying traits that were inherited in the patterns he had observed in the peas.

INCLUSION *ACTIVITY*

Demonstrate for students how the use of fractions assists in solving for ratios. Obtain two 100 mL graduated cylinders. Fill one cylinder with 75 mL of seeds and the other with 25 mL of seeds. Allow students to measure the height of the two columns of seeds and then record their measurements in the form of a fraction reduced to the lowest term (75/25 = 3/1= 3:1). Have students set up and calculate a 1:1 ratio in the same manner.

MAKING CONNECTIONS

Social Studies

Since the dawn of agriculture, people have used selective breeding to improve crops and domestic animals. Modern applications of Mendelian genetics and gene technology have resulted in major changes in crops and animals. Have students investigate the history of teosinte and corn or examine a regionally important agricultural product.

TEACHING STRATEGY

Identifying Genetic Traits

Display large pictures of a few flowering plants or bring in real plants. Ask students to list traits of the plants that could be inherited. Encourage students to think of many different traits, including flower shape, flower color, position of flowers on stems, leaf shape, leaf color, pattern of veins, pattern of stem growth, presence of hairs on stems, and inner structure of flowers.

DEMONSTRATION

Label a container "male gametes" and another container "female gametes." Have students imagine that the containers represent male and female members of a population. Mix 30 white beans and 50 red beans in each container. (Have them imagine that the red represents *R*, the dominant trait for red fur color and that the white represents *r*, the recessive trait for white fur color.) Have a student select one bean from each container, and have them write down the gene pair and resulting fur color (for example, 1 white bean and 1 red bean = *Rr* = red) in a table on the board. Have students repeat the process 10 times. After each selection, students should return the beans to the containers that they were taken from and remix the contents. After students have recorded the results of 10 crosses, have them answer the following questions: Which combinations were the most common? Which were the least common?

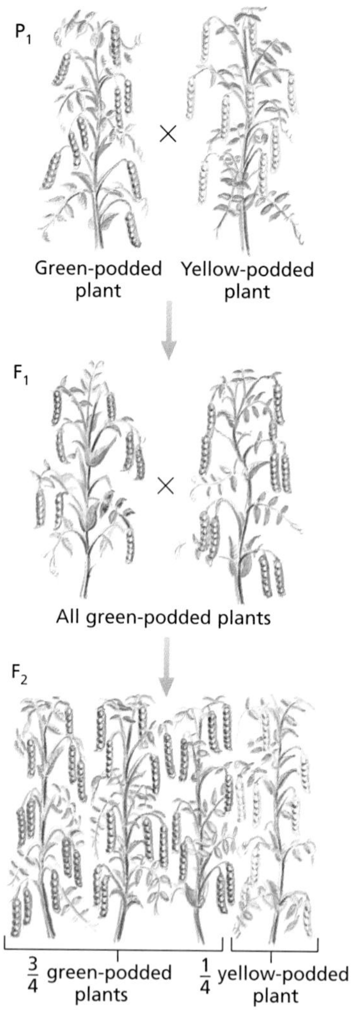

FIGURE 9-2

Pure green-podded pea plants crossed with pure yellow-podded pea plants produce only green-podded plants. Yet when the F_1 generation is permitted to self-pollinate, some yellow-podded plants appear in the F_2 generation.

MENDEL'S RESULTS AND CONCLUSIONS

In one of his experiments, Mendel crossed a plant pure for green pods with one pure for yellow pods, as shown in Figure 9-2. The resulting seeds produced an F_1 generation with only green-podded plants. No yellow pods developed, even though one parent had been pure for yellow pods. Only one of the two traits found in the P_1 generation appeared in the F_1 generation.

Next Mendel allowed the F_1 plants to self-pollinate and planted the resulting seeds. When the F_2 generation plants grew, he observed that about three-fourths of the F_2 plants had green pods and about one-fourth had yellow pods.

Mendel's observations and his careful records led him to hypothesize that something within the pea plants controlled the characteristics he observed. He called these controls *factors*. Mendel hypothesized that each trait was inherited by means of a separate factor. Because the characteristics he studied had two alternative forms, he reasoned that there must be a *pair* of factors controlling each trait.

Recessive and Dominant Traits

Whenever Mendel crossed strains, one of the P_1 traits failed to appear in the F_1 plants. In every case, that trait reappeared in a ratio of about 3:1 in the F_2 generation. This pattern emerged in thousands of crosses and led Mendel to conclude that one factor in a pair may prevent the other from having an effect. Mendel hypothesized that the trait appearing in the F_1 generation was controlled by a **dominant** factor because it masked, or dominated, the other factor for a specific characteristic. The trait that did not appear in the F_1 generation but reappeared in the F_2 generation was thought to be controlled by a **recessive** factor.

Thus, a trait controlled by a recessive factor had no observable effect on an organism's appearance when it was paired with a trait controlled by a dominant factor.

The Law of Segregation

Mendel concluded that the paired factors separate during the formation of reproductive cells. This means that each reproductive cell, or gamete, receives only one factor of each pair. When two gametes combine during fertilization, the offspring have two factors controlling a specific trait. The **law of segregation** states that *a pair of factors is segregated, or separated, during the formation of gametes.*

The Law of Independent Assortment

Mendel also crossed plants that differed in two characteristics, such as in flower color and seed color. The data from these more-complex crosses showed that traits produced by dominant factors

Word Roots and Origins

recessive

from the Latin *recessus*, meaning "to recede"

do not necessarily appear together. A green seed pod produced by a dominant factor could appear in a white-flowering pea plant, as Figure 9-3 shows. Mendel concluded that the factors for different characteristics are not connected. The **law of independent assortment** states that *factors for different characteristics are distributed to gametes independently.*

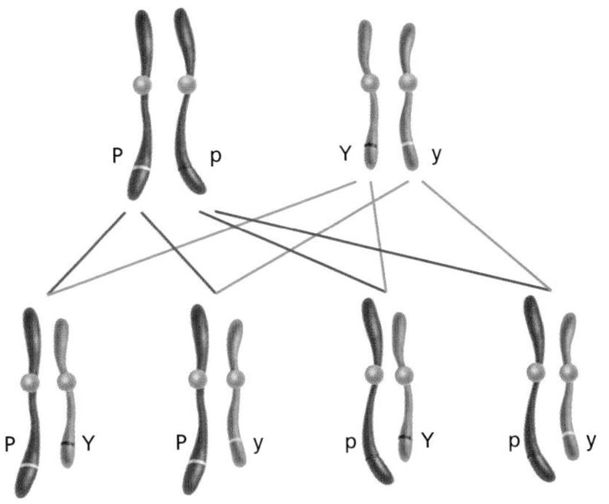

FIGURE 9-3

Independent assortment of these two pairs of homologous chromosomes would give the following allele combinations in gametes. *P* denotes flower color. *Y* denotes seed color.

CHROMOSOMES AND GENES

Most of Mendel's findings agree with what biologists now know about molecular genetics. **Molecular genetics** is the study of the structure and function of chromosomes and genes. Recall from Chapter 8 that a chromosome is a threadlike structure made up of DNA. A gene is the segment of DNA on a chromosome that controls a particular hereditary trait. Because chromosomes occur in pairs, genes also occur in pairs. Each of several alternative forms of a gene is called an **allele.** Mendel's *factors* are now called *alleles.*

Letters are used to represent alleles. Capital letters refer to dominant alleles, and lowercase letters refer to recessive alleles. For example, the dominant allele for the trait of green pod color may be represented by *G,* and the recessive allele for the trait of yellow pod color may be represented by *g.* Whether a letter is capitalized or lowercased is important. The actual letter selected to represent an allele is arbitrary.

Recall from Chapter 8 that during meiosis, gametes receive one chromosome from each homologous pair of chromosomes. This means that when the gametes combine in fertilization the offspring receives one allele for a given trait from each parent.

Mendel's law of independent assortment is supported by the fact that chromosomes segregate independently to gametes during meiosis. Therefore, the law of independent assortment is observed only for genes located on separate chromosomes or located far apart on the same chromosome.

SECTION 9-1 REVIEW

1. **List the steps involved in Mendel's experiments on garden peas.**
2. **Define the terms *dominant* and *recessive.***
3. **Differentiate genes from alleles.**
4. **State in modern terminology the two laws of heredity that resulted from Mendel's work.**
5. **How might Mendel's conclusions have differed if he had studied two traits determined by alleles carried on the same chromosome?**
6. **CRITICAL THINKING What happens during meiosis that would allow genes located on the same chromosome to separate independently of one another?**

VISUAL LINK

Figure 9-3

Ask students to determine from the figure whether a pea plant receiving an allele dominant for one trait will also receive an allele dominant for a second trait. (No, because the alleles sort independently of one another.)

ANSWERS TO SECTION 9-1 REVIEW

1. Mendel first produced pure strains of plants by allowing the plants to self-pollinate over several generations. He then crossed plants that were pure for two contrasting traits. He then allowed the F_1 generation plants to self-pollinate.
2. A trait controlled by a dominant allele masks the effects of other alleles for the same characteristic. A trait controlled by a recessive allele has no observable effect on an organism's appearance when it is paired with a trait controlled by a dominant allele.
3. A gene is a segment of DNA on a chromosome that controls a particular hereditary trait. An allele is an alternative form of a gene pair.
4. The law of segregation states that pairs of alleles are separated during meiosis. The law of independent assortment states that alleles located on separate chromosomes or far apart on the same chromosome separate independently of one another during meiosis.
5. Mendel would not have observed independent assortment occurring for all traits, so he probably would not have formulated his hypothesis of independent assortment.
6. During crossing-over, homologous chromosomes exchange pieces of DNA, enabling alleles to move from one chromosome to a homologous chromosome.

BACKGROUND

Barbara McClintock was an undergraduate and graduate student at Cornell University, where she made significant contributions to cytology and genetics. She continued her research at the University of Missouri and at the Carnegie Institution, where she studied chromosomal mutations. As a result of her achievements, McClintock was elected to the National Academy of Sciences in 1944 and to the Genetics Society of America in 1945. McClintock's work later earned her an unshared Nobel Prize for the discovery of transposable elements 35 years earlier.

Transposable mobile genetic elements are found in both prokaryotes and eukaryotes. These are sequences of nucleotides that can move from one position on a chromosome to another. In doing so, they often carry adjacent nucleotide sequences with them. As a result, transposons can control transcription by increasing or decreasing the production of certain proteins. In McClintock's research, transposons controlled proteins responsible for color development in corn kernels.

Transposase, an enzyme encoded by transposons, is responsible for transposition. Because transposase does not recognize specific nucleotide sequences, the movement of transposons to different sites of the genome is random. Movement can also be a rare event. These elements are responsible for much of the genetic recombination that occurs in organisms and therefore can pose a significant evolutionary impact on organisms.

Jumping Genes

HISTORICAL PERSPECTIVE

The scientific study of heredity began with the work of Gregor Mendel. The rediscovery of Mendel's work in 1900 is said to mark the birth of the science of genetics. Geneticist Barbara McClintock, born in 1902, devoted her life to this new science. Ironically, certain assumptions about genetics, strongly believed but wrong, prevented early acceptance of McClintock's exciting conclusions.

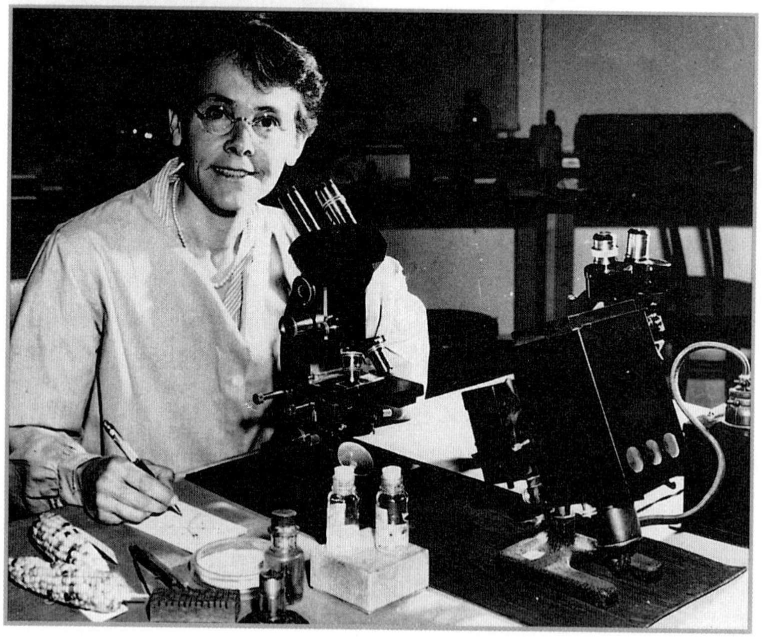

Barbara McClintock

Early Discoveries

Have you ever looked closely at ornamental corn, also called maize? Today it is planted and marketed for decorative use. In the fall, when the corn ripens, its red, brown, and purple ears are harvested and artistically arranged in centerpieces and door hangings.

In the late 1920s and early 1930s, at Cornell University, geneticists harvested the colorful plant for another purpose. They hoped to trace variations among maize plants to specific genes in an effort to discover how chromosomes carry specific traits to new generations.

As a graduate student at Cornell, Barbara McClintock began work on identifying and labeling the 10 chromosomes of maize. She developed new microscopic techniques and was able to visualize the parts of the maize chromosomes. Scientists were thrilled with her discovery because they could identify the chromosome that carried the gene for a particular trait.

After receiving her doctorate in botany in 1927, McClintock stayed at Cornell. Mostly she worked alone, wanting to observe everything for herself. She once said:

> *No two plants are exactly alike. They're all different, and as a consequence, you have to know that difference.... I don't feel I really know the story if I don't watch the plant all the way along.*

Discovering "Jumping Genes"

McClintock taught for several years at Cornell before taking a position as a research scientist at Cold Spring Harbor Laboratory, a scientific community on Long Island about 40 miles east of New York City. She stayed there for the rest of her life, working in maize genetics. There, during the 1940s and 1950s, she did the research and made the discovery for which she won the 1983 Nobel Prize in physiology or medicine.

The prevailing opinion among most geneticists in McClintock's time was that genes were lined up on chromosomes in unchanging places, much like beads on a string. McClintock's observations told her otherwise. The changes she saw in corn kernels and chromosomes led her to conclude that some genes, which she called transposons, are able to move to a new place on a chromosome or to a new chromosome entirely. These "jumping genes," she believed, cause differences to appear in a plant's offspring by either activating or inactivating the genes responsible for kernel color. Unlike with genetic recombination, which occurs during meiosis, transposons change the location of genes on chromosomes in the somatic cells of organisms.

McClintock observed two kinds of transposons: dissociators and activators. The dissociators could jump to a new place when signaled by others called activators. The dissociators would then cause changes in nearby genes on the chromosome and in the color of the kernels in maize. McClintock verified her conclusions through repeated experiments.

Ears of maize similar to these demonstrated to McClintock that certain genetic elements transpose.

She Was Right

In the summer of 1951, at Cold Spring Harbor, McClintock presented her discovery at a meeting of scientists from around the world. Her presentation was not well received. McClintock summarized it best when she stated that her discovery went against "the dogma of the constancy of the genome." Although saddened by the lack of interest in her work, she was not discouraged. She knew that she was right and that the evidence would ultimately convince the world that transposons existed.

Why did most geneticists ignore McClintock's conclusions for so long? One reason was that their attention was directed to the work of the new molecular geneticists. Compared with their work, McClintock's seemed traditional, almost old-fashioned. Her fellow scientists were using electron microscopes capable of seeing molecules. Her methods—an ordinary microscope, crossbreeding, and observation—belonged to an earlier era. In working with maize genetics, she was "paddling upstream against the current of scientific opinion."

In the end, McClintock proved to be not behind her time but ahead of it. Biologist Allan Campbell once said of her discoveries, "The time was not ripe."

By the 1970s, the time was ripe. Molecular biologists, using electron microscopes and other technology, saw bits of DNA "jumping around" in bacteria. Transposons are thought to help bacteria better adapt to new environments. For example, transposons probably play a role in spreading genes for antibiotic resistance among bacteria. Scientists later found transposons in eukaryotes other than maize. Today all organisms, including humans, seem to have transposons.

Like McClintock, many geneticists now expect their understanding of transposons to help solve certain mysteries about evolution: how larger organisms developed from single cells and, in general, where new species come from. Transposons may also have medical applications, such as helping scientists discover how white blood cells make antibodies and what causes cells to sometimes multiply wildly, as in cancer.

DISCUSSION

Guide the discussion by posing the following questions:

1. Why were scientists thrilled with McClintock's new microscopic techniques to visualize the parts of the 10 maize chromosomes? (Scientists were thrilled because they could identify the chromosome that carried the gene for a particular trait.)

2. What are transposons? (Transposons are genes or sequences of nucleotides that are able to move to a new place on a chromosome or to a new chromosome entirely.)

3. What are some of the outcomes that scientists expect of research on transposons? (Many geneticists expect their understanding of transposons to help solve certain mysteries about evolution, such as how larger organisms developed from single cells and where new species come from. Transposons may also have medical applications, such as helping scientists discover why white blood cells make antibodies and why cells sometimes multiply wildly, as with cancer.)

FURTHER READINGS

1. Kittredge, Mary. *Barbara McClintock (American Women of Achievement).* Chelsea House Publishers, 1991. This book includes the biography of Barbara McClintock written by Mary Kittredge.

2. McClintock, Barbara. *The Discovery and Characterization of Transposable Elements: The Collected Papers of Barbara McClintock (Genes, Cells, and Organisms, 17).* Garland Publishers, 1987. This book includes a written collection of Barbara McClintock's research.

SECTION 9-2

SECTION OVERVIEW

Cell Structure and Function
Section 9-2 discusses how alleles are responsible for the genetic makeup, or genotype, of an organism. The genotype is responsible for an organism's physical appearance, which we call the organism's phenotype.

Reproduction and Inheritance
Section 9-2 also discusses how inherited traits passed from the parent cell to the offspring can be predicted with a mathematical model known as the Punnett square.

▶ ENGAGE STUDENTS

Ask students what methods they might use to predict the outcome of a sports event, such as a football game. (Answers might include considering the abilities and experience of the participants, looking at a previous related event, analyzing similar types of events, and using a model such as a coin toss.) Have students discuss some patterns that would help them predict an outcome. (A team might have won two-thirds of its past games. A win might have occurred every time a team played a certain opponent or nearly all the times they played at home.) Ask students if such patterns enable them to be certain of the outcome of future sports events? (No. The prediction of such future outcomes can be based only on probability, not certainty.) Stress that some events, such as sports, are partly affected by chance. Others, such as coin tosses, are totally chance events.

SECTION 9-2

OBJECTIVES

▲ Explain how probability is used to predict the results of genetic crosses.

● Use a Punnett square to predict the results of monohybrid and dihybrid genetic crosses.

■ Explain how a testcross is used to show the genotype of an individual whose phenotype is dominant.

◆ Differentiate a monohybrid cross from a dihybrid cross.

GENETIC CROSSES

Today geneticists rely on Mendel's work to predict the likely outcome of genetic crosses. In this section, you will learn how to predict the probable genetic makeup and appearance of offspring resulting from specified crosses.

GENOTYPE AND PHENOTYPE

The genetic makeup of an organism is its **genotype** (JEN-uh-TIEP). The genotype consists of the alleles that the organism inherits from its parents. For example, the genotype of the white-flowering pea plant in Figure 9-4 consists of two recessive alleles for white flower color, represented as *pp*. The genotype of a purple-flowering pea plant may be either *PP* or *Pp*. Either of these two genotypes would result in a pea plant with purple flowers because the *P* allele is dominant.

The appearance of an organism as a result of its genotype is called its **phenotype** (FEE-noh-TIEP). The phenotype of a *PP* or a *Pp* pea plant is purple flowers, whereas the phenotype of a *pp* pea plant is white flowers. Human phenotypes can appear to be altered by behavior. Hair dye, contact lenses of varying colors, and plastic surgery can all change an individual's appearance, but they do not alter the individual's true phenotype or genotype.

FIGURE 9-4
The genotype of the pea plant on the left is *pp*. Its phenotype is white flowers. The phenotype of the pea plant on the right is purple flowers. Its genotype is either *Pp* or *PP*.

When both alleles of a pair are alike, the organism is said to be **homozygous** (HOH-moh-ZIE-guhs) for that characteristic. An organism may be homozygous dominant or homozygous recessive. For example, a pea plant that is homozygous dominant for flower color would have the genotype *PP*. A pea plant that is homozygous recessive for flower color would have the genotype *pp*. When the two alleles in the pair are different, the organism is **heterozygous** (HET-uhr-OH-ZIE-guhs) for that characteristic. A pea plant that is heterozygous for flower color would have the genotype *Pp*.

PROBABILITY

Probability is the likelihood that a specific event will occur. A probability may be expressed as a decimal, a percentage, or a fraction. Probability is determined by the following equation:

$$\text{Probability} = \frac{\text{number of times an event is expected to happen}}{\text{number of opportunities for an event to happen}}$$

For example, in Mendel's experiments the dominant trait of yellow seed color appeared in the F_2 generation 6,022 times. The recessive trait of green seed color appeared 2,001 times. The total number of individuals was 8,023 (6,022 + 2,001). Using the probability equation above, the probability that the dominant trait will appear in a similar cross is

$$\frac{6{,}022}{8{,}023} = 0.75$$

Expressed as a percentage, the probability is 75 percent. Expressed as a fraction, the probability is 3/4.

The probability that the recessive trait will appear in an F_2 generation is

$$\frac{2{,}001}{8{,}023} = 0.25$$

Expressed as a percentage, the probability is 25 percent. Expressed as a fraction, the probability is 1/4. Fractions can also be expressed as ratios. For example, the ratio 1:4 represents the same probability as 1/4. Probability tells us that there are three chances in four that an offspring of two heterozygous individuals will have the dominant trait and one chance in four that it will have the recessive trait.

The results predicted by probability are more likely to occur when there are many trials. For example, many coin tosses should yield a result of heads 50 percent of the time and tails 50 percent of the time. However, if you toss a coin only a few times, you might not get this result. But *each* time a coin is tossed, the probability of landing tails is 50 percent. Only after many, many tries would you be *likely* to get the percentage of heads predicted on the basis of probability, that is, 50 percent heads and 50 percent tails.

Quick Lab

Calculating Probability

Materials paper sack containing 20 jelly beans in an unknown number of color combinations

Procedure

1. Obtain a sack of 20 jelly beans from your teacher. Do not look into the sack. Do not eat the jelly beans. There are three possible colors of jelly beans that can be pulled from the sack. Pull one jelly bean out, and record the color. Return the jelly bean to the sack, and shake the bag to mix them.
2. Repeat step 1 until you have examined 20 jelly beans.
3. Determine the probability of each color of jelly bean that you pulled from the sack. Compare your results with those of the rest of the class.

Analysis Does anyone have the same probabilities as you? Are there some probabilities that are very close to yours? Are there some probabilities that are very different from yours? Based on these observations, determine the number of jelly beans of each color that are in your sack.

Quick Lab

Calculating Probability

Time Required 20 minutes

Safety Tell students that they should not eat the jelly beans.

Procedural Tips Students should work in pairs. Select three colors (such as white, red, and green) that are easily distinguished. Place the jelly beans (12 white, 6 red, and 2 green) in each sack before class time. Instruct students not to eat the jelly beans and not to look in their sacks.

Answers to Analysis Answers will vary, but the likelihood of similar results increases with class size. There is also the possibility of extreme variations. By averaging all the probabilities for each group, the students should come close to determining the correct number of jelly beans of each color. Below are sample data.

Group 1: 9 white (45%), 6 red (30%), 5 green (25%)

Group 2: 13 white (65%), 5 red (25%), 2 green (10%)

Group 3: 14 white (70%), 5 red (25%), 1 green (5%)

Group 4: 11 white (55%), 9 red (45%), 0 green (0%)

Total number of each color of jelly bean:

White: 9 + 13 + 14 + 11 = 47/4 = 11.75 ~ 12

Red: 6 + 5 + 5 + 9 = 25/4 = 6.25 ~ 6

Green: 5 + 2 + 1 + 0 = 8/4 = 2

GIFTED *ACTIVITY*

Scottish fold cats are heterozygotes whose ears fold down onto their head several days after birth. The homozygote genotype, *rr*, for this trait causes death. Have students construct Punnett squares and determine the genotypic and phenotypic ratios of both the dead and the live kittens (genotypic ratio: 1 *RR* and 2 *Rr;* phenotypic ratio: 1 with unfolded ears and 2 with folded ears). Ask students to determine the most profitable cross (the least amount of death), reminding them that cats with unfolded ears may still be used as pets or breeding stock. (A cross between a homozygous parent and a heterozygous parent would produce live kittens, half with folded ears and half with unfolded ears.)

RETEACHING *ACTIVITY*

Invite three students to participate in this activity. You will need two sets of matching gloves, a red set and a white set; and two T-shirts, a red shirt and a white shirt. The color red will represent the dominant allele, while the color white will represent the recessive allele. Provide each student with a set of gloves, which represents alleles, and a T-shirt of the same color, which represents the phenotype. The students must stand facing the class with their hands behind their back. Each of these students then gives one of the gloves to a third student. The class must then determine what color T-shirt the third student must wear. Continue the activity with crosses between homozygotes and heterozygotes or between two heterozygotes.

QUICK FACT

☑ Anne Boleyn (d. 1536), the second wife of Henry VIII of England, had polydactyly, *pp*. Her daughter, Queen Elizabeth I, did not have this trait. Thus, Anne was probably a heterozygote, *Pp*.

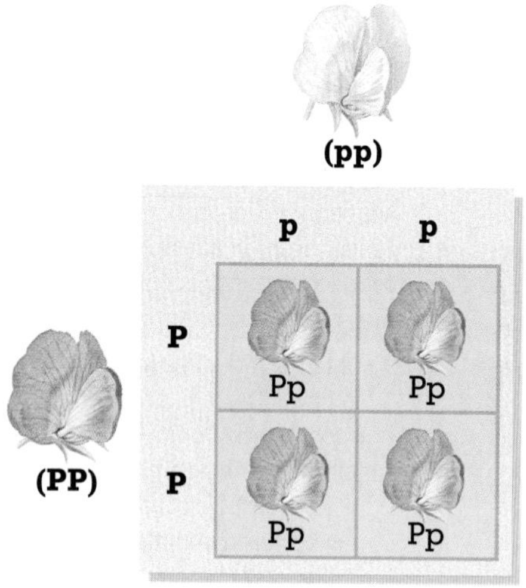

FIGURE 9-5

A pea plant homozygous for purple flowers crossed with a pea plant homozygous for white flowers will produce only purple-flowering offspring. Note that all of the offspring will be heterozygous for flower color.

PREDICTING RESULTS OF MONOHYBRID CROSSES

A cross between individuals that involves one pair of contrasting traits is called a **monohybrid** (MAWN-oh-HIE-brid) **cross.** A cross between a pea plant that is pure for producing purple flowers and one that is pure for producing white flowers is an example of a monohybrid cross. Biologists use a diagram called a **Punnett** (PUHN-uht) **square,** like the one shown in Figure 9-5, to aid them in predicting the probability that certain traits will be inherited by offspring. The following examples show how a Punnett square can be used to predict the outcome of different types of crosses.

Example 1: Homozygous × Homozygous

The cross represented in Figure 9-5 is a monohybrid cross between a pea plant homozygous for purple flower color (*PP*) and a pea plant homozygous for white flower color (*pp*). The alleles contributed by the homozygous dominant parent are represented by *P*s on the left side of the Punnett square. The alleles contributed by the homozygous recessive parent are represented by *p*s across the top of the Punnett square. Each box within the Punnett square is filled in with the letters that are above it and beside it outside the square. The combinations of alleles in the four boxes indicate the possible genotypes that can result from the cross. The predicted genotype is *Pp* in every case. Thus, there is a 100 percent probability that the offspring will have the genotype *Pp* and thus the phenotype purple flower color.

FIGURE 9-6

Crossing a guinea pig homozygous for black coat color with one heterozygous for black coat color produces all black-coated offspring. Note that half of the offspring are predicted to be homozygous for coat color.

Example 2: Homozygous × Heterozygous

Figure 9-6 shows a cross between a guinea pig that is homozygous dominant for the trait of black coat color (*BB*) and a guinea pig that is heterozygous for this trait (*Bb*). The letter *b* stands for the recessive allele. Genotype *bb* results in a brown coat. Notice that there are two possible genotypes that can result from this cross, *BB* or *Bb*. The probability of an offspring with the genotype *BB* is 2/4, or 50 percent. The probability of an offspring with the genotype *Bb* is also 2/4, or 50 percent. In other words, you could expect about 50 percent of the offspring resulting from this cross to be homozygous dominant for the black coat and about 50 percent to be heterozygous dominant for a black coat. The probable phenotype is black coat color in every case; thus, 4/4, or 100 percent, of the offspring are expected to have a black coat. What if the homozygous guinea pig had been homozygous recessive for coat color? In this case, the homozygote would have the genotype *bb*. Crossing a *bb* guinea pig with a *Bb* guinea pig is likely to produce about 50 percent *Bb* offspring and about 50 percent *bb* offspring.

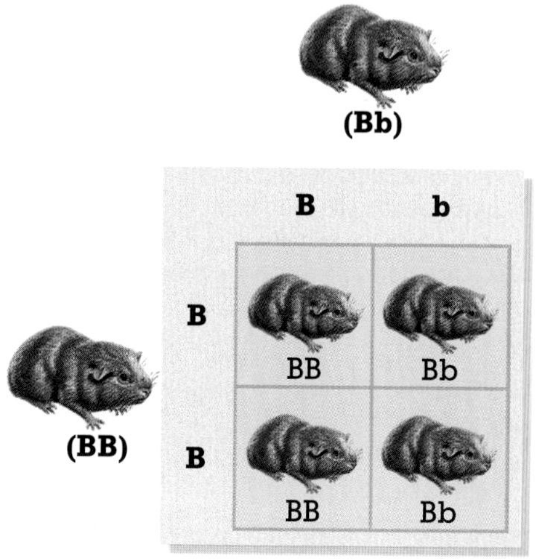

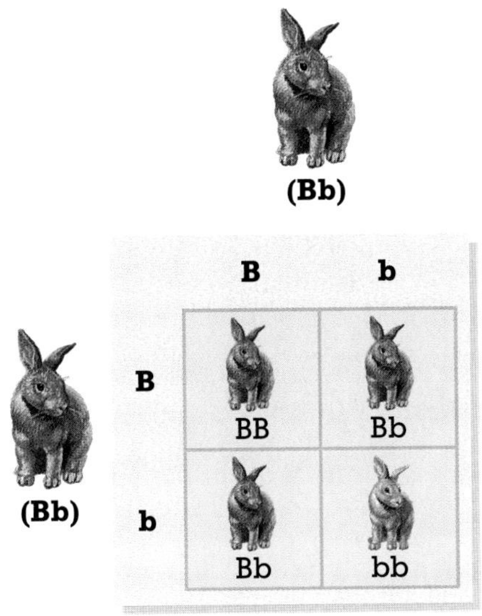

FIGURE 9-7
Crossing two rabbits that are both heterozygous for black coat color tends to produce 50 percent heterozygous black individuals, 25 percent homozygous black individuals, and 25 percent homozygous brown individuals.

Example 3: Heterozygous × Heterozygous

In rabbits, the allele for black coat color (*B*) is dominant over the allele for brown coat color (*b*). The Punnett square in Figure 9-7 shows the predicted results of crossing two rabbits that are both heterozygous (*Bb*) for coat color. As you can see, 1/4 (25 percent) of the offspring are predicted to have the genotype *BB*, 1/2 (50 percent) are predicted to have the genotype *Bb*, and 1/4 (25 percent) are predicted to have the genotype *bb*. Thus, 3/4 (75 percent) of the offspring resulting from this cross are predicted to have a black coat. One-fourth (25 percent) of the offspring are predicted to have a brown coat.

The ratio of the genotypes that appear in offspring is called the **genotypic ratio.** The probable genotypic ratio of the monohybrid cross represented in Figure 9-7 is 1 *BB*:2 *Bb*:1 *bb*. The ratio of the offsprings' phenotypes is called the **phenotypic ratio.** The probable phenotypic ratio of the cross represented in Figure 9-7 is 3 black:1 brown.

Example 4: Testcross

Recall that in guinea pigs, both *BB* and *Bb* result in a black coat. How might you determine whether a black guinea pig is homozygous (*BB*) or heterozygous (*Bb*)? You could perform a **testcross,** in which an individual of unknown genotype is crossed with a homozygous recessive individual. A testcross can determine the genotype of any individual whose phenotype is dominant. You can see from Figure 9-8 that if the unknown genotype is homozygous black, all offspring will be black. If the individual with the unknown genotype is heterozygous black, about half of the offspring will be black. In reality, if the cross produced one brown offspring in a litter of eight, the genotype of the black-coated parent is likely to be heterozygous.

	b	b
B	Bb	Bb
B	Bb	Bb

	b	b
B	Bb	Bb
b	bb	bb

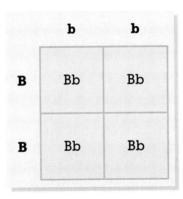

FIGURE 9-8
If a black guinea pig is crossed with a brown guinea pig and even one of the offspring is brown, chances are that the black guinea pig is heterozygous for coat color.

CRITICAL THINKING

Plant Population

Tell students that a variegated plant is crossed with itself; 100 seeds produce green plants, 200 seeds produce variegated plants, and 100 seeds do not germinate. Then ask the students what the genotypes of the F_1 generation are. (Because variegated plants are heterozygous recessive, the F_1 population will include green plants, *GG*, variegated plants, *Gg*, and ungerminated seeds, *gg*.)

DEMONSTRATION

Have students roll their tongue. Most students will be able to do this. Tell students the ability to roll the tongue is caused by a dominant allele. Then ask what the genotype is of a person who cannot roll his or her tongue (*tt*). Have students determine under what circumstances a non-roller parent can produce a tongue-rolling child. (The second parent must have the tongue-rolling phenotype.)

RETEACHING *ACTIVITY*

Give each student a white bead and a purple bead. Ask students to pretend they are peas. Have students begin as the F_1 generation and randomly exchange beads (pollinate) with 10 other people in the classroom by shaking their beads (gametes) behind the back of their partner. Without looking, each set of partners should remove one of the beads from his or her partner's hand. The two partners will then hold out each of their beads to see the results. Each time a pair pollinates, one person from that mating should log the results. After all members of the class have randomly pollinated 10 times, tally the results and have students find the actual and probability ratios.

SECTION 9-2

VISUAL LINK
Figure 9-9

Emphasize that not all phenotypes result from strictly dominant or recessive genes. Ask students what these flowers would look like if the traits for flower color in four o'clocks had complete dominance (red). Ask students to speculate about the color of the flowers if the traits for flower color in four o'clocks were codominant (white and red, not pink).

INCLUSION *ACTIVITY*

Show dominance and recessiveness for a trait while completing a Punnett square on the overhead projector. Write the dominant allele's symbol on a piece of colored cellophane and the recessive allele's symbol on a piece of clear cellophane. Ask the students the following questions. What happens when I stack the pieces of cellophane on top of each other? What color do you see? Why? Then show codominance by writing the two alleles' symbols on different-colored pieces of cellophane. Ask the students to predict the outcome of stacking the two colored pieces of cellophane on top of each other. Demonstrate the results.

TEACHING STRATEGY
The Runt

Give students the following scenario: A female labrador retriever has a litter containing nine black puppies with black noses, three black puppies with pink noses, three white puppies with black noses, and one white puppy that has a pink nose and that dies. Ask students what type of cross this example represents (dihybrid cross). Ask students to suggest a possible explanation for the puppy's death. Lead students to conclude that a puppy with visible recessive genes may have hidden recessive genes that can be lethal.

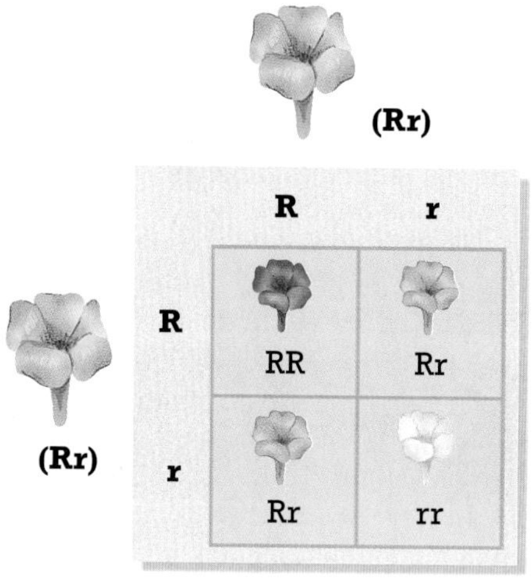

FIGURE 9-9

When red-flowering four o'clocks are crossed with white-flowering four o'clocks, all of the F_1 offspring produce pink flowers, an intermediate between the two phenotypes. When the F_1 generation is interbred, red-flowering, pink-flowering, and white-flowering plants are produced because the trait for red flower color has incomplete dominance over the trait for white flower color.

internetconnect

SCI*LINKS*. NSTA
TOPIC: Dominance
GO TO: www.scilinks.org
KEYWORD: HM176

Example 5: Incomplete Dominance

Recall that in Mendel's pea-plant crosses, one allele was completely dominant over another, a relationship called **complete dominance.** In complete dominance, heterozygous plants and dominant homozygous plants are indistinguishable in phenotype. For example, both pea plants *PP* and *Pp* for flower color have purple flowers.

Sometimes, however, the F_1 offspring will have a phenotype in between that of the parents, a relationship called **incomplete dominance.** Incomplete dominance occurs when two or more alleles influence the phenotype, resulting in a phenotype intermediate between the dominant trait and the recessive trait. In four o'clocks, for example, both the allele for red flowers (*R*) and the allele for white flowers (*r*) influence the phenotype. Neither allele is completely dominant over the other allele. When four o'clocks self-pollinate, red-flowering plants produce only red-flowering offspring, and white-flowering plants produce only white-flowering offspring. However, when red four o'clocks are crossed with white four o'clocks, the F_1 offspring all have *pink* flowers. One hundred percent of the offspring of this cross have the (*Rr*) genotype, which results in a pink phenotype.

What would be the result of crossing two pink-flowering (*Rr*) four o'clocks? As the Punnett square in Figure 9-9 shows, the probable genotypic ratio is 1 *RR*:2 *Rr*:1 *rr*. Given that neither the allele for red flowers (*R*) nor the allele for white flowers (*r*) is completely dominant, the probable phenotypic ratio is 1 red:2 pink:1 white.

FIGURE 9-10

The roan coat of this horse consists of both white hairs and red hairs. Both phenotypes are expressed in individuals heterozygous for coat color when the traits are codominant.

Example 6: Codominance

Codominance occurs when both alleles for a gene are expressed in a heterozygous offspring. In codominance, neither allele is dominant or recessive, nor do the alleles blend in the phenotype. For example, the genes for both white coat color and red coat color are expressed in the horse shown in Figure 9-10. The capital letter *R* is used to indicate red coat color, and *R'* is used to indicate white coat color. Thus, the *RR'* symbol would represent the coat color of the roan horse in Figure 9-10.

PREDICTING RESULTS OF DIHYBRID CROSSES

A **dihybrid** (die-HIE-brid) **cross** is a cross between individuals that involves two pairs of contrasting traits. Predicting the results of a dihybrid cross is more complicated than predicting the results of a monohybrid cross because there are more possible combinations of alleles to work out. For example, to predict the results of a cross involving both seed texture and seed color, you have to consider how the four alleles from each parent can combine.

Homozygous × Homozygous

Suppose that you want to predict the results of a cross between a pea plant that is homozygous for round, yellow seeds and one that is homozygous for wrinkled, green seeds. In pea plants, the allele for round seeds (*R*) is dominant over the allele for wrinkled seeds (*r*), and the allele for yellow seeds (*Y*) is dominant over the allele for green seeds (*y*).

As Figure 9-11 shows, the Punnett square used to predict the results of a cross between a parent of the genotype *RRYY* and a parent of the genotype *rryy* will contain 16 boxes. The independently sorted alleles from one parent—*RY, RY, RY,* and *RY*—are listed along the left side of the Punnett square. The independently sorted alleles from the other parent—*ry, ry, ry,* and *ry*—are listed along the top of the Punnett square. Each box is filled with the letters that are above it and to the left of it outside the square. Notice that the genotype of all the offspring of this cross will be heterozygous for both traits, *RrYy;* therefore, the phenotype of all the offspring will have round, yellow seeds.

rryy

RRYY

	ry	ry	ry	ry
RY	RrYy	RrYy	RrYy	RrYy
RY	RrYy	RrYy	RrYy	RrYy
RY	RrYy	RrYy	RrYy	RrYy
RY	RrYy	RrYy	RrYy	RrYy

FIGURE 9-11

This Punnett square shows a dihybrid cross between a pea plant that is homozygous recessive for wrinkled, green seeds *(rryy)* and a pea plant that is homozygous dominant for smooth, yellow seeds *(RRYY).*

Quick Lab

Determining Genotypes

Materials pencil and paper

Procedure The ability to roll the tongue upward from the sides is a dominant, inherited trait. In one family, both parents and three children are all tongue rollers, while one child is not. Determine the genotype and phenotype of each parent.

Analysis Are the parents homozygous or heterozygous? Are the children homozygous or heterozygous?

Quick Lab

Determining Genotypes

Time Required 10 minutes

Procedural Tips Have students work in pairs. The parents are both tongue rollers with the genotype *Tt.*

Answers to Analysis Both parents are heterozygous. Two of the tongue-roller children are heterozygous *(Tt),* one of the tongue-roller children is homozygous *(TT),* and one child is homozygous *(tt)* and is not a tongue roller.

CULTURAL CONNECTION

Blood Lines

Aristotle associated inheritance with blood. He thought the blood carried information from the body's various structures to the reproductive organs. We know this is untrue, but the idea is ingrained in many languages. For example, "blue blood," "blood stock," and "It is in the blood" (English); "Corre en la sangre" (Spanish); "Bon sang ne peut mentir" and "celle est dans le sang" (French); and "Es liegt im Blute" and "von gutem Blut" (German) all associate inheritance with the blood.

ACTIVE READING

Technique: K-W-L

Tell students to return to their lists of things they **W**ant to know about inheritance, described in the Active Reading activity on page 165. Students should finish by making a list of what they have **L**earned. Conclude by asking students which questions are still unanswered. Ask if they have any new questions.

SECTION 9-2

VISUAL LINK

Figure 9-12

Have students examine the Punnett square in this figure and identify the nine possible genotypes and the four possible phenotypes. Encourage students to compute the ratios for themselves.

ANSWERS TO SECTION 9-2 REVIEW

1. First, cross the purple-flowering pea plant with a white-flowering pea plant. If the offspring produced by the cross include white-flowering pea plants, chances are that the plant in question is heterozygous for flower color.
2. The equation is as follows: number of times an event is expected to happen divided by the number of opportunities for an event to happen. Probability can be expressed in fractions, percentage, decimals, or ratios.
3. 50% pink-flowering plants, 50% white-flowering plants
4. Divide a square into four boxes. Place symbols representing one parent's alleles on the left side and those representing the other parent's alleles across the top. Fill in each box with the appropriate two symbols. The boxes reveal all possible genotypes.
5. A monohybrid cross involves one pair of contrasting traits, such as brown or blue eyes. A dihybrid cross involves two pairs of contrasting traits, such as round and green peas or wrinkled and yellow peas.
6. Each short-tailed parent has two incompletely dominant alleles, one for a long tail and one for no tail.

Heterozygous × Heterozygous

To determine the results of crossing two pea plants heterozygous for round, yellow seeds, the procedure is the same. As Figure 9-12 shows, the offspring of this dihybrid cross are likely to have nine different genotypes. These nine genotypes will result in pea plants with the following four phenotypes:

- 9/16 with round, yellow seeds (genotypes *RRYY, RRYy, RrYY,* and *RrYy*)
- 3/16 with round, green seeds (genotypes *RRyy* and *Rryy*)
- 3/16 with wrinkled, yellow seeds (genotypes *rrYY* and *rrYy*)
- 1/16 with wrinkled, green seeds (genotype *rryy*)

FIGURE 9-12

A dihybrid cross of two individuals heterozygous for both traits is likely to result in nine different genotypes and four different phenotypes.

RrYy

RrYy

	RY	**Ry**	**rY**	**ry**
RY	RRYY	RRYy	RrYY	RrYy
Ry	RRYy	RRyy	RrYy	Rryy
rY	RrYY	RrYy	rrYY	rrYy
ry	RrYy	Rryy	rrYy	rryy

internetconnect

SCILINKS NSTA

TOPIC: Punnett squares
GO TO: www.scilinks.org
KEYWORD: HM178

SECTION 9-2 REVIEW

1. Explain how you might go about determining the genotype of a purple-flowering pea plant.
2. What is the equation used to determine probability? In what ways can probability be expressed?
3. If you were to cross pink-flowering four o'clocks with white-flowering four o'clocks, what results would you expect? Provide a Punnett square to support your answer.
4. Explain how you would use a Punnett square to predict the probable outcome of a monohybrid cross.
5. Explain the difference between a monohybrid cross and a dihybrid cross. Give an example of each.
6. **CRITICAL THINKING** The offspring of two short-tailed cats have a 25 percent chance of having no tail, a 25 percent chance of having a long tail, and a 50 percent chance of having a short tail. Based on this information, what can you hypothesize about the genotypes of the parents?

CHAPTER 9 REVIEW

SUMMARY/VOCABULARY

9-1

- The study of how characteristics are transmitted from parents to offspring is called genetics.
- Self-pollination, in which pollen is transferred from the anthers of a flower to either the stigma of the same flower or the stigma of another flower on the same plant, normally occurs in pea plants. Cross-pollination occurs when pollen is transferred between flowers of two different plants.
- Mendel concluded that inherited characteristics are controlled by factors that occur in pairs. In his experiments on pea plants, one factor in a pair masked the other. The trait that masked the other was called the dominant trait. The trait that was masked was called a recessive trait.
- We now know that the factors that Mendel studied are alleles, or alternative forms of a gene.
- The law of segregation states that a pair of factors is segregated, or separated, during the formation of gametes. The law of independent assortment is observed only for genes that are located on separate chromosomes or far apart on the same chromosome.
- The law of independent assortment states that factors for different characteristics are distributed to gametes independently. You know this to be true from your study of meiosis: during prophase I of meiosis, homologous chromosomes randomly migrate to opposite sides of the dividing cell.

Vocabulary

allele (169)
anther (166)
cross-pollination (166)
dominant (168)
F_1 generation (166)
F_2 generation (166)
genetics (165)
heredity (165)
law of independent assortment (169)
law of segregation (168)
molecular genetics (169)
P_1 generation (166)
pollination (166)
pure (166)
recessive (168)
self-pollination (166)
stigma (166)
strain (166)
trait (165)

9-2

- The genotype is the genetic makeup of an organism. An organism's phenotype is its appearance as a result of its genotype.
- Probability is the likelihood that a specific event will occur. A probability may be expressed as a decimal, a percentage, or a fraction.
- A Punnett square can be used to predict the outcome of genetic crosses.
- A cross between individuals involving one pair of contrasting traits is a monohybrid cross.
- A testcross, in which an individual of unknown genotype is crossed with a homozygous recessive individual, can be used to determine the genotype of an individual whose phenotype is dominant.
- Complete dominance occurs when heterozygous individuals and dominant homozygous individuals are indistinguishable in phenotype.
- Incomplete dominance occurs when two or more alleles influence the phenotype, resulting in a phenotype intermediate between the dominant trait and the recessive trait.
- Codominance occurs when both alleles for a gene are expressed in a heterozygous offspring. Neither allele is dominant or recessive, nor do the alleles blend in the phenotype as they do in incomplete dominance.
- A cross between individuals involving two pairs of contrasting traits is called a dihybrid cross.

Vocabulary

codominance (176)
complete dominance (176)
dihybrid cross (177)
genotype (172)
genotypic ratio (175)
heterozygous (173)
homozygous (173)
incomplete dominance (176)
monohybrid cross (174)
phenotype (172)
phenotypic ratio (175)
probability (173)
Punnett square (174)
testcross (175)

CHAPTER 9 REVIEW ANSWERS

REVIEW

1. The law of segregation states that a pair of factors is segregated during the formation of gametes. The law of independent assortment states that factors for different characteristics are distributed to reproductive cells independently.
2. Self-pollination occurs when pollen grains from an anther alight on the stigma of a flower of the same plant. Cross-pollination occurs when pollen grains from an anther alight on the stigma of a flower on a different plant.
3. The dominant trait is the one that appears in a heterozygous organism. The recessive trait is the one that is hidden in a heterozygous organism.
4. A genotype is the genetic makeup of an organism. A phenotype is the appearance of an organism due to the expression of the alleles of the genotype.
5. In a homozygous organism, alleles for a characteristic are the same. In a heterozygous organism, the alleles for a characteristic are contrasting.

ANSWERS

CONTINUED

6. d **7.** a
8. c **9.** c
10. d **11.** d
12. b **13.** d
14. a **15.** b

16. He wanted to ensure that the plants he was studying were true-breeding for specific traits.

17. (a) dihybrid cross, (b) both parents are *QqTt,* (c) 1/16 *QQTT,* 1/16 *qqtt,* 2/16 *QQTt,* 4/16 *QqTt,* 2/16 *QqTT,* 2/16 *qqTt,* 2/16 *Qqtt,* 1/16 *qqTT,* 1/16 *QQtt*

18. Answers will vary. Animal breeders use genetics to predict how often a trait will appear when two animals are bred. Animals bred for special purposes include dogs, horses, and cattle.

19. P generation refers to the first cross. F_1 generation refers to the offspring of the first cross. F_2 generation refers to the offspring of the second cross.

20. In order for the recessive phenotype to appear, the individual must be homozygous.

21. A cross between individuals that involves one pair of contrasting traits is called a monohybrid cross. A cross between individuals that involves two pairs of contrasting traits is a dihybrid cross.

22. If genes on the same chromosome are close together, the chances of them segregating during crossing-over are small. If the genes are far apart on the chromosome, then the genes will be segregated.

23. The law of segregation states that members of each pair of alleles separate when gametes are formed. During meiosis this occurs physically.

24. Cross the plant with a plant that produces wrinkled seeds. If the plant is homozygous for smooth seeds, the cross will result in all smooth-seed-producing offspring. If the plant is heterozygous about half the offspring will produce wrinkled seeds.

25. genotypic ratio: 100 percent *Bb;* phenotypic ratio: 100 percent black.

REVIEW

Vocabulary

1. State and define the two laws of heredity.
2. Differentiate self-pollination from cross-pollination.
3. What is the difference between a dominant trait and a recessive trait?
4. Differentiate genotype from phenotype.
5. What is the difference between *homozygous* and *heterozygous*?

Multiple Choice

6. A procedure in which an individual of unknown genotype is crossed with a homozygous recessive individual to determine the genotype of the unknown individual is called a (a) monohybrid cross (b) dihybrid cross (c) hybrid cross (d) testcross.
7. A gene is a (a) segment of DNA (b) chromosome (c) segment of RNA (d) protein.
8. An example of a genotype of a heterozygous individual is (a) *pp* (b) *YY* (c) *Zz* (d) none of the above.
9. In a monohybrid cross of two heterozygous parents (*Pp*), one would expect the offspring to be (a) 1 *pp*:3 *PP* (b) 3 *Pp*:1 *pp* (c) 1 *PP*:2 *Pp*:1 *pp* (d) all *Pp*.
10. In a monohybrid cross between a homozygous dominant parent and a homozygous recessive parent, one would predict the offspring to be (a) 3:4 homozygous recessive (b) 2:4 homozygous recessive (c) 1:4 homozygous recessive (d) all heterozygous.
11. In guinea pigs, black fur is dominant. If a black guinea pig is crossed with a white guinea pig and the litter contains a white offspring, the genotype of the black-haired parent is probably (a) homozygous dominant (b) homozygous recessive (c) pure for the trait (d) heterozygous dominant.
12. Segregation of alleles occurs during (a) mitosis (b) meiosis (c) fertilization (d) pollination.
13. In a dihybrid cross between two heterozygous parents, the probability of obtaining an offspring that is homozygous recessive for both traits would be (a) none (b) 9/16 (c) 3/16 (d) 1/16.
14. If two parents with dominant phenotypes produce an offspring with a recessive phenotype, then (a) both parents are heterozygous (b) one parent is heterozygous (c) both parents are homozygous (d) one parent is homozygous.
15. Suppose that you have found a new species of plant. Some of the plants have red flowers and some have yellow flowers. You cross a red-flowering plant with a yellow-flowering plant, and all the offspring have orange flowers. You might assume that the alleles for flower color (a) have complete codominance (b) have incomplete dominance (c) are either dominant or recessive (d) have mutated.

Short Answer

16. Why did Mendel begin his experiments by allowing pea plants to self-pollinate for several generations?
17. Answer the following questions based on the Punnett square shown below:
    a. Does the Punnett square demonstrate a monohybrid cross or a dihybrid cross?
    b. List the genotypes of the parents.
    c. Give the genotypic ratio predicted by the Punnett square for the cross.

	QT	Qt	qT	qt
QT	QQTT	QQTt	QqTT	QqTt
Qt	QQTt	QQtt	QqTt	Qqtt
qT	QqTT	QqTt	qqTT	qqTt
qt	QqTt	Qqtt	qqTt	qqtt

18. **Unit 5—*Heredity***

BIOLOGY INTERACTIVE TUTOR

Write a report summarizing how an understanding of heredity allows animal breeders to develop animals with desirable traits. Find out what kinds of animals are bred for special purposes.

19. Explain the difference between the P generation, F_1 generation, and F_2 generation.
20. When the dominant and recessive traits are known, why is it not necessary to use the term *homozygous* when referring to the genotype of an individual with a recessive phenotype?
21. Explain the difference between a monohybrid cross and a dihybrid cross.
22. How might crossing-over during meiosis affect the segregation of genes on the same chromosome?
23. Relate the events of meiosis to the law of segregation.
24. In pea plants, smooth seed texture is dominant over wrinkled seed texture. A gardener has a pea plant that produces smooth seeds. How can the gardener determine whether the plant is homozygous or heterozygous for the allele that determines seed texture?
25. In rabbits, the allele for black coat color (*B*) is dominant over the allele for brown coat color (*b*). Predict the results of a cross between a rabbit homozygous for black coat color (*BB*) and a rabbit homozygous for brown coat color (*bb*).

CRITICAL THINKING

1. One rule of probability can be expressed as the following: The probability of two independent events occurring simultaneously is the product of the probability of their occurring separately. If, for example, you had a pair of dice and rolled each die one at a time, what would be the probability that you would get two 4s? On the first roll, you would have a 1/6 chance. On the second roll, you would have a 1/6 chance. The probability of obtaining two 4s would be $1/6 \times 1/6 = 1/36$. Suppose you were playing a game with five dice. What is the chance of rolling a 6 on all five dice?
2. A cross between two pea plants with axial flowers and inflated pods gives the following offspring: 20 with axial flowers and inflated pods, 7 with axial flowers and constricted pods, and 5 with terminal flowers and inflated pods. What is the most probable genotype for the two parents? Explain the results and show the Punnett square. How can the explanation be checked?
3. Two black female mice are crossed with the same brown male mouse. Based on the information shown in the table below, answer the following questions:
   a. What are the genotypes of each parent?
   b. Which trait is dominant?
   c. Is the dominance of the trait in (b) completely dominant, incompletely dominant, or codominant? Explain your answer.

Results of Crossing Two Black Female Mice with One Brown Male Mouse

P_1 generation	F_1 generation
female A × male A	9 black, 7 brown
female B × male A	14 black, 0 brown

EXTENSION

1. Read "The Genes of 1998" in *Discover,* January 1999, on page 33. Describe how a mutation on chromosome 8 causes a rare form of baldness. What might be one explanation for why humans no longer have a sharp sense of smell? Explain how a genetic disorder can actually cause a person to display symptoms of psychiatric illness.
2. Toss a penny and a nickel together 32 times, and record the results as they fall: number of times heads comes up for both the penny and the nickel, number of times tails comes up for both the penny and the nickel, number of times the penny is heads and the nickel is tails, and the number of times the penny is tails and the nickel is heads. Explain how this task illustrates independent assortment.

CRITICAL THINKING

1. The chance of rolling a 6 on all five dice would be $1/6 \times 1/6 \times 1/6 \times 1/6 \times 1/6$, or 1/7,776.
2. *Aali;* abnormal ratio probability due to the small sample; do test-crosses using *aaii*
3. (a) Female A is *Bb,* female B is *BB,* and the male is *bb.* (b) Black is dominant over brown. (c) Black is completely dominant over brown.

EXTENSION

1. A mutation on chromosome 8 causes alopecia universalis, a rare type of baldness. People who have two copies of this mutant gene have no body hair, eyelashes, or eyebrows. Humans have up to 1,000 genes for the sense of smell. Scientists have reported that up to 72 percent of these genes are mutated and defective. These mutations have accumulated over millions of years, gradually reducing the human ability to detect odors. People who have a mutation on chromosome 4 suffer from Wolfram syndrome, a rare disease whose symptoms include halucinations and progressive dementia.
2. The nickels and the penny, like chromosomes, are independent of one another. Each combination has a $1/2 \times 1/2 = 1/4$ probability of occurring.

CHAPTER 9 INVESTIGATION

TIME REQUIRED

One 50-minute class period

SAFETY PRECAUTIONS

Remind students to avoid tasting the peas and lentils.

QUICK REFERENCES

Lab Preparation Notes are found on page 163D.

Holt BioSources provides a Teaching Resources CD-ROM that contains "Using Gowin's Vee in the Lab" and "Scoring Rubrics."

PROCEDURAL TIPS

1. You may wish to review the concepts of monohybrid crosses and probability before beginning this investigation.
2. Students can be grouped in pairs for this investigation.
3. In step 1, make sure students place one green pea and one lentil into each Petri dish.
4. In step 3, remind students to return the seeds to their original Petri dishes.

ANSWERS TO BACKGROUND

1. One trait and four alleles are involved in a monohybrid cross.
2. When a dominant allele combines with a recessive allele, the recessive allele will not be expressed. Sometimes the dominant allele is not completely dominant; in such cases, the recessive allele may be partially expressed.
3. When gametes form, the alleles for each trait separate independently of one another.

CHAPTER 9 INVESTIGATION

Modeling Monohybrid Crosses

OBJECTIVES

- Predict the genotypic and phenotypic ratios of offspring resulting from the random pairing of gametes.
- Calculate the genotypic ratio and phenotypic ratio among the offspring of a monohybrid cross.

PROCESS SKILLS

- predicting
- organizing
- analyzing data
- calculating

MATERIALS

- lentils
- green peas
- 2 Petri dishes

Background

1. How many traits are involved in a monohybrid cross? How many alleles are involved?
2. What prevents the expression of a recessive allele?
3. When gametes form, what happens to the alleles for each trait?

PART A Simulating a Monohybrid Cross

1. You will model the random pairing of alleles by choosing lentils and peas from Petri dishes. These dried seeds will represent the alleles for seed color. A green pea will represent *G*, the dominant allele for green seeds, and a lentil will represent *g*, the recessive allele for yellow seeds.
2. Each Petri dish will represent a parent. Label one Petri dish "female gametes" and the other Petri dish "male gametes." Place one green pea and one lentil in the Petri dish labeled "female gametes" and place one green pea and one lentil in the Petri dish labeled "male gametes."
3. Each parent contributes one allele to each offspring. Model a cross between these two parents by choosing a random pairing of the dried seeds from the two containers. Do this by simultaneously picking one seed from each container *without looking.* Place the pair of seeds together on the lab table. The pair of seeds represents the genotype of one offspring.
4. Record the genotype of the first offspring in your lab report in a table like Table A shown below.

TABLE A *GAMETE PAIRINGS*

Trial	Offspring genotype	Offspring phenotype
1		
2		
3		
4		
5		
6		
7		
8		
9		
10		

5. Return the seeds to their original dishes and repeat step 3 nine more times. Record the genotype of each offspring in your data table.
6. Based on each offspring's genotype, determine and record each offspring's phenotype. Assume that the allele for green seeds, *G*, is completely dominant over the allele for yellow seeds, *g*.

PART B Calculating Genotypic and Phenotypic Ratios

7. In your lab report, prepare a data table similar to Table B shown below.
8. Determine the genotypic and phenotypic ratios among the offspring. First count and record the number of homozygous dominant, heterozygous, and homozygous recessive individuals you recorded in Table A. Then record the number of offspring that produce green seeds and the number that produce yellow seeds under "Phenotypes" in your data table.
9. Calculate the genotypic ratio for each genotype using the following equation:

$$\text{Genotypic ratio} = \frac{\text{number of offspring with a given genotype}}{\text{total number of offspring}}$$

10. Calculate the phenotypic ratio for each phenotype using the following equation:

$$\text{Phenotypic ratio} = \frac{\text{number of offspring with a given phenotype}}{\text{total number of offspring}}$$

11. Now pool the data for the whole class, and record the data in your lab report in a table like Table C.
12. Compare your class's sample with your small sample of 10. Calculate the genotypic and phenotypic ratios for the class data, and record them in your data table.
13. Construct a Punnett square showing the parents and their offspring in your lab report.
14. Clean up your materials before leaving the lab.

Analysis and Conclusions

1. What trait is being studied in this investigation?
2. What are the genotypes of the parents? Describe the genotypes of both parents using the terms *homozygous* or *heterozygous,* or both.
3. What does each seed in the Petri dish represent?
4. When the seeds were selected and paired, what did the pairs represent?
5. Did tables B and C reflect a classic monohybrid-cross phenotypic ratio of 3:1?
6. When the class data were tabulated, did a classic monohybrid-cross ratio of a phenotype of 3:1 result?
7. If a genotypic ratio of 1:2:1 is observed, what must the genotypes of both parents be?
8. Show what the genotypes of the parents would be if 50 percent of the offspring were green and 50 percent of the offspring were yellow.
9. Construct a Punnett square for the cross of a heterozygous black guinea pig and an unknown guinea pig whose offspring include a recessive white-furred individual. What are the possible genotypes of the unknown parent?

Further Inquiry

Design a model to demonstrate a dihybrid cross of two parents that are heterozygous for two traits. Construct and complete a Punnett square for this cross.

TABLE B *OFFSPRING RATIOS*

Genotypes	Total	Genotypic ratio
Homozygous dominant (*GG*)		
Heterozygous (*Gg*)		____ : ____ : ____
Homozygous recessive (*gg*)		
Phenotypes		**Phenotypic ratio**
Green seeds		____ : ____ : ____
Yellow seeds		

TABLE C *OFFSPRING RATIOS (Entire Class)*

Genotypes	Total	Genotypic ratio
Homozygous dominant (*GG*)		
Heterozygous (*Gg*)		____ : ____ : ____
Homozygous recessive (*gg*)		
Phenotypes		**Phenotypic ratio**
Green seeds		____ : ____ : ____
Yellow seeds		

ANSWERS TO ANALYSIS AND CONCLUSIONS

1. The trait being investigated is seed, or fruit, color.
2. Both parents are heterozygous green, *Gg*.
3. Each seed represents an allele.
4. The pairs represent the gametes that an offspring will receive.
5. Answers will vary. In a sample size of 10 crosses, no clear ratio may be evident. When combining data from the entire class, the 3:1 ratio should be seen.
6. When combining data from the entire class, the 3:1 ratio should be seen.
7. Both parents must be heterozygous, *Gg*.
8. One parent would be heterozygous, *Gg,* and the other parent would be homozygous recessive, *gg*.
9. Students should draw two Punnett squares. One should show a cross between two heterozygous individuals; the other should show a heterozygous individual crossed with a homozygous recessive individual. If *B* represents the dominant black fur color and *b* stands for white fur color, the unknown parent could be genotypically *Bb* or *bb*.

PLANNING GUIDE 10

CHAPTER 10

NUCLEIC ACIDS AND PROTEIN SYNTHESIS

	TOPICS	TEACHING RESOURCES	LABS, CLASSWORK, AND HOMEWORK
BLOCKS 1 & 2 90 minutes	**Introducing the Chapter, p. 184**	**ATE** Focus Concept, p. 184 **ATE** Assessing Prior Knowledge, p. 184 **ATE** Understanding the Visual, p. 184	■ Supplemental Reading Guide, *The Double Helix*
	10-1 DNA, p. 185 Structure of DNA, p. 185 Replication of DNA, p. 188	**ATE** Section Overview, p. 185 **ATE** Critical Thinking, p. 187 **ATE** Visual Link Figure 10-3, p. 186 Figure 10-5, p. 188 14, 25A, 31A, 33	**Lab D2,** Extracting DNA **Lab D3,** Genetic Transformation of Bacteria ★ Study Guide, Section 10-1 **PE** Section Review, p. 189
BLOCKS 3 & 4 90 minutes	**10-2 RNA, p. 190** Structure of RNA, p. 190 Transcription, p. 191	**ATE** Section Overview, p. 190 **ATE** Critical Thinking, p. 192 **ATE** Visual Link Figure 10-7, p. 191 34	**PE** **Quick Lab,** p. 191, Comparing and Contrasting RNA Types **Lab D1,** Staining DNA and RNA **Lab D2,** Extracting DNA **Lab D3,** Genetic Transformation of Bacteria **Lab D4,** Genetic Transformation—Antibiotic-Resistant Bacteria ★ Study Guide, Section 10-2 **PE** Section Review, p. 192
BLOCKS 5 & 6 90 minutes	**10-3 Protein Synthesis, p. 193** Protein Structure and Composition, p. 193 The Genetic Code, p. 193 Translation, p. 194	**ATE** Section Overview, p. 193 **ATE** Critical Thinking, pp. 193, 194, 196 **ATE** Visual Link Table 10-1, p. 194 Figure 10-9, p. 195 35–37	**PE** **Quick Lab,** p. 196, Modeling Protein Assembly **PE** **Chapter 10 Investigation,** p. 200 **Lab B6,** Effect of Environment on Gene Expression **Lab B7,** Gene Expression **Lab E4,** Gene Regulation ★ Study Guide, Section 10-3 **PE** Section Review, p. 196
BLOCKS 7 & 8 90 minutes	**REVIEW and ASSESSMENT** **PE** Chapter 10 Review, pp. 198–199 ■ Performance-Based Assessment—Identifying Dominant and Recessive Traits	Audio CD Program BIOLOGY INTERACTIVE TUTOR Unit 6—Gene Expression	**CHAPTER TESTING** ■ Chapter Test (blackline master) ▲ Test Generator ■ Assessment Item Listing ■ Scoring Rubrics and Classroom Management Checklists

KEY

MODERN BIOLOGY
- PE Pupil's Edition
- ATE Teacher's Edition
- ■ Active Reading Guide
- ★ Study Guide

HOLT BIOSOURCES
- Teaching Transparencies
- Laboratory Program

One-Stop Planner CD-ROM
Shows these resources within a customizable daily lesson plan:
- ■ Holt BioSources Teaching Resources
- ■ Active Reading Guide
- ★ Study Guide
- ▲ Test Generator

READING FOR CONTENT MASTERY

Preview: Have students preview the objectives, the introductory paragraph, topic sentences, and captions before reading the chapter.

ATE Active Reading Technique: K-W-L, p. 185
■ Reading Strategy: K-W-L Worksheet
■ Active Reading Guide Worksheet 10-1
PE Summary/Vocabulary, p. 197

■ Active Reading Guide Worksheet 10-2
PE Summary/Vocabulary, p. 197

ATE Active Reading Technique: K-W-L, p. 195
■ Reading Strategy: K-W-L Worksheet
■ Active Reading Guide Worksheet 10-3
PE Summary/Vocabulary, p. 197

TECHNOLOGY AND INTERNET RESOURCES

Holt Biology Videodiscs Teacher's Correlation Guide, Lessons 10-1 through 10-3

BIOLOGY INTERACTIVE TUTOR: Unit 6—Gene Expression, Topics 1–6

Audio CD Program, Sections 10-1 through 10-3

internetconnect

SCILINKS NSTA

On-line Resources: www.scilinks.org
The following *sci*LINKS Internet resources can be found in the student text for this chapter:

Topics:
- DNA, p. 185
- Watson and Crick, p. 187
- Genetic code, p. 193

go.hrw.com

On-line Resources: go.hrw.com
Visit the HRW Web site for a variety of resources related to this chapter. Just type in the keyword HM2 Chapter 10.

Smithsonian Institution®
Visit **www.si.edu/hrw** for additional on-line resources.

CNNfyi.com
Visit **www.cnnfyi.com** for late-breaking news and current event stories selected just for you.

LAB ACTIVITY PLANNING

CHAPTER 10 INVESTIGATION

Modeling Replication and Transcription of DNA, pp. 200–201

OVERVIEW
In this investigation, students learn to construct and manipulate models of DNA. Students are guided through the construction of a model that they will analyze and use as a basis for completing a second model. Both models are used to demonstrate DNA replication and transcription. After students have completed the investigation, the models can provide the basis for a detailed discussion of the structure of DNA.

PREPARATION
Each student group will need 3 cm plastic soda straws (54); 54 small, standard-sized paper clips; and 54 colored push-pins (12 red, 12 blue, 12 yellow, 12 green, and 6 white).

Quick Labs

Comparing and Contrasting RNA Types, p. 191
Students will need only paper and pencils for this activity.

Modeling Protein Assembly, p. 196
Students will need paper clips in four colors for this activity. Each color will represent a nitrogenous base.

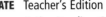

UNDERSTANDING THE VISUAL

The picture is a digital composite model of DNA. Ask students what the different parts of the model represent (atoms). Ask them how the atoms are connected (by chemical bonds). Point out to students that DNA is a molecule and that the chemical structure of DNA ultimately controls the activities of the cell.

FOCUS CONCEPT

Cell Structure and Function

The sequence of nucleotides in a DNA molecule encodes genetic information. Complementary pairing between nucleotides maintains the accuracy of the genetic information as it is carried from a cell's nucleus to its cytoplasm, where the sequence is translated into proteins. Proteins are the basis of much of the structure of the cell and of the chemical reactions that occur in the body.

ASSESSING PRIOR KNOWLEDGE

Review the following concepts.

Biochemistry: *Chapter 3*
Review with students the structure and importance of proteins. Ask what determines how a protein will function.

Cell Organelles: *Chapter 4*
Use a diagram of a cell and ask the students to locate the nucleus, ribosomes, and endoplasmic reticulum. Ask students to name the function of each organelle.

Cell Reproduction: *Chapter 8*
Review with students DNA replication that takes place just prior to mitosis. Ask students why it is important that new cells have the same DNA as the parent cell.

CHAPTER 10

NUCLEIC ACIDS AND PROTEIN SYNTHESIS

The most common form of DNA found in living organisms has a spiral-staircase shape, as shown in this computer model of the molecule's structure.

FOCUS CONCEPT: ***Cell Structure and Function***

As you read, pay attention to the roles that DNA and RNA play in storing information and making proteins.

BIOLOGY INTERACTIVE TUTOR **Unit 6—*Gene Expression***
Topics 1–6

SECTION

10-1

DNA

What enables cells to have different forms and to perform different functions? Ultimately the genetic source of this amazing diversity is deoxyribonucleic acid (DNA). The primary function of DNA in organisms is to store and transmit the genetic information that tells cells which proteins to make and when to make them. Proteins in turn form the structural units of cells and help control chemical processes within cells.

OBJECTIVES

▲ Explain the principal function of DNA.

● Describe the structure of DNA.

■ Define the term *complementary base pairing.*

◆ Explain the role of complementary base pairing in the replication of DNA.

▲ Summarize the main features of DNA replication.

STRUCTURE OF DNA

Recall from Chapter 3 that the nucleic acid DNA is an organic compound. DNA is made up of repeating subunits called nucleotides. Each DNA molecule consists of two long chains of nucleotides.

A DNA nucleotide has three parts: a sugar molecule called **deoxyribose;** a phosphate group, which consists of a phosphorus, P, atom surrounded by oxygen, O, atoms; and a molecule that is referred to as a **nitrogen-containing base** because it contains a nitrogen, N, atom. The three parts of a DNA nucleotide are illustrated in Figure 10-1. The deoxyribose sugar and the phosphate group are identical in all DNA nucleotides. However, the nitrogen-containing base may be any one of four different kinds.

The four nitrogen-containing bases found in DNA nucleotides are **adenine, guanine, cytosine,** and **thymine.** It is customary to represent nucleotides by the abbreviations for their nitrogen-containing bases. A nucleotide containing adenine is represented by an *A*. Likewise, C = cytosine, G = guanine, and T = thymine.

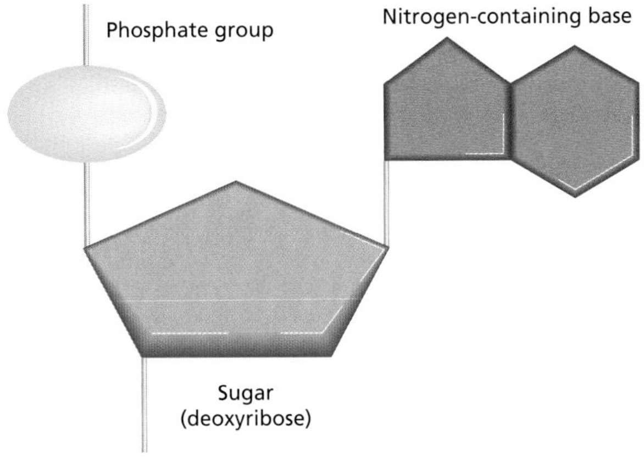

FIGURE 10-1
A DNA molecule is composed of a deoxyribose sugar molecule, a phosphate group, and one of four nitrogen-containing bases.

internetconnect
SCILINKS NSTA
TOPIC: DNA
GO TO: www.scilinks.org
KEYWORD: HM185

SECTION 10-1

SECTION OVERVIEW

Cell Structure and Function
Section 10-1 describes the helical structure of DNA and the molecular structures of the four nucleotides in DNA. Complementary base pairing is discussed, along with the function of base pairing in DNA replication.

Reproduction and Inheritance
Section 10-1 stresses that accuracy and the repair of errors are important in DNA replication because this is the mechanism that allows new cells to have an exact copy of DNA. Accurate replication of DNA forms the molecular basis for inheritance.

ACTIVE READING

Technique: K-W-L
Before students read this chapter, have them write short, individual lists of all the things they already Know (or think they know) about DNA. Ask them to contribute their entries to a group list on the board or overhead. Then have students list things they Want to know about DNA. Have students save their lists for later use in the Active Reading activity on page 195.

DEMONSTRATION

Make a stack of books totaling about 10,000 pages. Tell students that the stack of books represents only about one-fiftieth of the information contained in the DNA of every human cell. Correlate this with the amount of information required to describe a human being.

TEACHING STRATEGY
DNA Extraction

Have students extract DNA from *E. coli* cells. First prepare a meat-extract broth and inoculate it with *E. coli* cells prior to experimentation. Inoculated-broth culture medium is available from biological supply houses, such as WARD's. Have students work in groups of two. Give each group a test tube containing 3 mL of inoculated broth. Add 6 mL of a 50 percent liquid soap solution. Mix well. Put the test tube in a 65°C – 75°C water bath for approximately 15 minutes. Slowly add 3 mL of 95 percent ethanol. Tell students to be very careful not to mix the solution. DNA should appear at the alcohol interface. A glass stirring rod can be used to spool the DNA, which should have the appearance of stringy, cooked egg white.

VISUAL LINK
Figure 10-3

Tell students that the rungs of the spiral staircase of DNA represent nucleotides and that the specific order and number of nucleotides are the code that ensures that a mother mouse will have baby mice and not baby lizards or green beans.

TEACHING STRATEGY
Genome Size

Have students build a model of an *E. coli* genome. The supplies needed per two students are two different-colored threads, a meter-stick, a size 0 gelatin capsule, and scissors. Tell students that 10 m of thread represent the length of the *E. coli* genome enlarged 10,000 times. Have students cut 10 m of two different colors of thread. Students should manipulate the thread until they have it contained within the gelatin capsule. Point out that in an *E. coli* cell, not only is the genome present, but the rest of the cytosol as well.

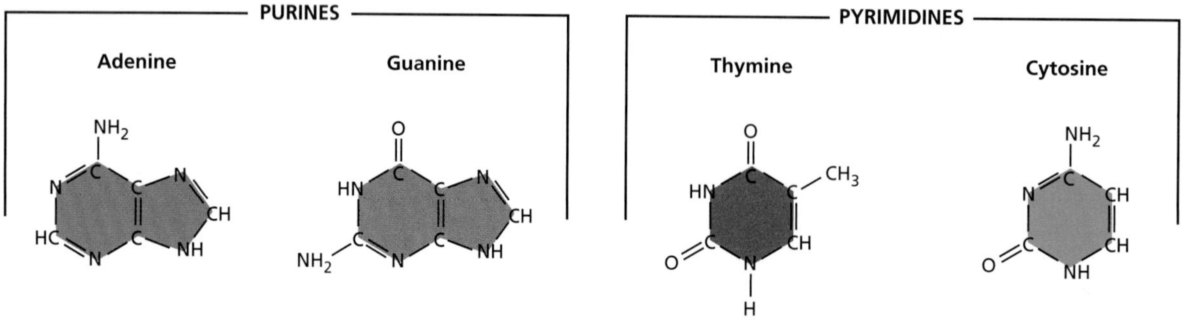

FIGURE 10-2

The four nitrogen-containing bases found in DNA are divided into two groups: purines and pyrimidines. Two-ringed bases are called purines. One-ringed bases are called pyrimidines.

Figure 10-2 shows the structure of these four nucleotides. Notice that adenine and guanine have two rings of carbon, C, and nitrogen, N, atoms. In contrast, cytosine and thymine have only one ring of carbon and nitrogen atoms. Bases that have two rings of carbon and nitrogen atoms, such as adenine and guanine, are called **purines. Pyrimidines** are bases that have one ring of carbon and nitrogen atoms, such as cytosine and thymine.

The Double Helix

In 1953, James Watson and Francis Crick suggested a model for the structure of DNA. The model proposed that DNA is composed of two nucleotide chains that wrap around each other to form a double spiral—similar to a spiral staircase. This shape is called a **double helix.** The double helix structure of DNA is illustrated in Figure 10-3.

FIGURE 10-3

The structure of DNA, which resembles a spiral staircase, is called a double helix.

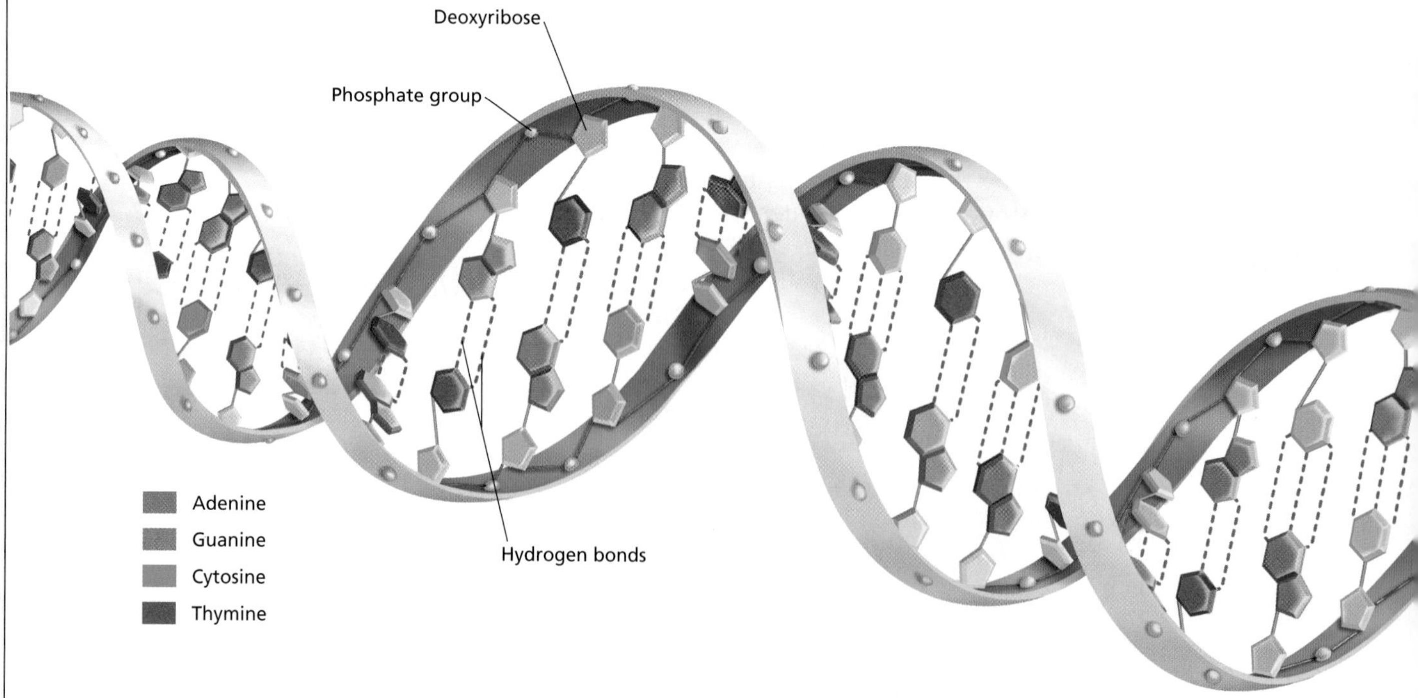

Watson and Crick relied heavily on scientific evidence reported by other scientists to construct the model. The model was inspired in part by X-ray photographs of DNA crystals, like the one shown in Figure 10-4, that had been studied by Rosalind Franklin and Maurice Wilkins. In addition, the model provided an explanation for how copies of DNA could be made and how genetic information might be stored and used within cells. In 1962, Watson, Crick, and Maurice Wilkins received the Nobel Prize in Medicine for their work on DNA. Rosalind Franklin had died in 1958 and thus was not recognized.

Note in Figure 10-3 that individual nucleotides are connected by covalent bonds between the deoxyribose sugar and phosphate molecules. The alternating deoxyribose sugar and phosphate molecules form a "backbone" to which the nitrogen-containing bases attach. Note also that the nitrogen-containing bases face toward the center of the helix and that they are perpendicular to the sugar-phosphate backbone.

By facing toward the center, the bases on one chain of DNA face the bases on the other chain of DNA, with which they form bonds called hydrogen bonds. The locations of the hydrogen bonds are indicated in Figure 10-3 by dotted lines. The hydrogen bonds help hold the two chains together. Hydrogen bonds are different from the covalent bonds and ionic bonds you read about in Chapter 2. A hydrogen bond is a relatively weak bond that usually forms between molecules. Hydrogen bonds form when two atoms share a hydrogen nucleus—one proton. The hydrogen bonds that form between the bases in DNA form between a hydrogen atom and either an oxygen or a nitrogen atom.

Notice that the base pairs are of uniform length because in each case one base is a double-ringed purine and the other is a single-ringed pyrimidine. The form of DNA that is most commonly found in living organisms has a right-hand twist, with each full turn consisting of ten base pairs.

Complementary Base Pairing

As Figure 10-3 shows, cytosine pairs with guanine, and adenine pairs with thymine. The DNA nucleotides normally pair in these combinations. These pairs of bases are called **complementary base pairs.** Two rules—called **base-pairing rules**—describe the pairing behavior of the bases. These rules simply state that cytosine bonds with guanine and adenine bonds with thymine.

Complementary base pairs are connected to each other by hydrogen bonds. Note that cytosine and guanine form three hydrogen bonds and adenine and thymine form two hydrogen bonds. As Figure 10-3 shows, the nucleotide sequence in one nucleotide chain of the DNA molecule is an exact complement of the nucleotide sequence in the other chain.

The complementary nucleotide chains in the DNA model led to suggestions of how DNA might copy itself. The ability of DNA to make exact copies of itself is important because, in most cases, cells that divide must pass exact copies of their DNA to offspring cells.

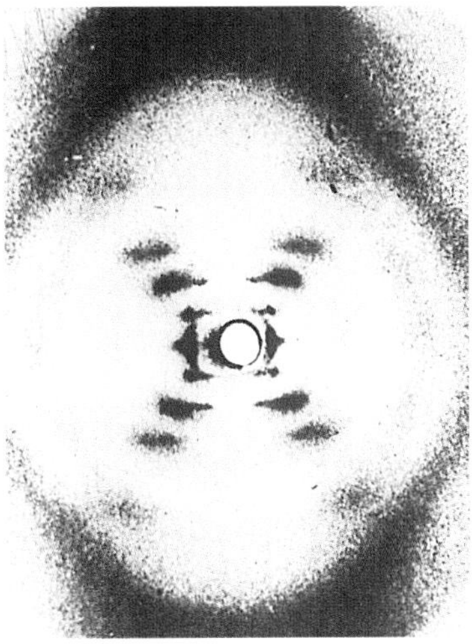

FIGURE 10-4

Rosalind Franklin's X-ray photographs of DNA indicated that DNA is a helix with a sugar-phosphate backbone on the outside and that the helix consists of more than one chain of nucleotides.

internet connect

SciLinks NSTA

TOPIC: Watson and Crick
GO TO: www.scilinks.org
KEYWORD: HM187

TEACHING STRATEGY
Mnemonics

Have students create a mnemonic device to help them identify the four bases. (For example, "**Th**ousands **c**ount **py**ramids" is a reminder that **th**ymine and **cy**tosine are **py**rimidines. "**Ad**s **gua**rantee **pur**ity" may remind students that **ad**enine and **gua**nine are **pur**ines.)

CRITICAL THINKING
Stability of DNA

DNA is considered to be a relatively stable molecule. Ask students what gives it this stability when the hydrogen bonds between the nitrogen bases are easily broken. (The bonds connecting the deoxyribose and phosphate groups, which form the backbone of DNA, are covalent bonds. Covalent bonds are much stronger and keep the strands of nucleotides intact.)

INCLUSION ACTIVITY

Give students various materials and have them work in pairs to create a model of each DNA nucleotide. Then have them use the models to demonstrate complementary base pairing.

GIFTED ACTIVITY

Have students research a genetic disease for which the defective gene has been found through the Human Genome Project, such as cystic fibrosis or Huntington's disease. Students should write or present orally a report that includes a description of the disease, the populations affected, and the prognosis for treatment or cure.

SECTION 10-1

OVERCOMING MISCONCEPTIONS

Students might think that because cells in the same organism appear different and have different functions they must have different DNA. Remind students that every somatic cell in an organism has the same DNA. It is a result of mitosis after fertilization. Other factors in the cell determine which genes can be used for the cell's specific function.

VISUAL LINK

Figure 10-5

Using the overhead, help students identify the replication fork, helicase enzymes, and direction of unwinding and separation of the double strand shown in (a). In (b) have students note the direction of replication on each original strand. Explain that this is because the direction and order of replication is specific and that bidirectional replication is due to the antiparallel nature of DNA. Have students compare (b) with (c) and identify the new strands of DNA in (c).

CULTURAL CONNECTION

True Ancestors of Modern Japanese

The Jamon settled Japan more than 10,000 years ago. It appears that the Yayoi people inhabited Japan at a later date. There is a debate as to which group of people are the true ancestors of the modern Japanese and where these people originated. By studying and comparing the genes in modern Japanese and the people in different parts of Asia, researchers are beginning to discover the lineage of the Japanese people. Data suggest that the Jamon originated in northern Asia and that both the Jamon and Yayoi peoples have contributed genes to the modern Japanese. In order to develop a complete understanding of this history, scientists need to study the DNA of the Jamon and Yayoi as well as archaeological finds.

REPLICATION OF DNA

The process of copying DNA in a cell is called **replication.** During replication, the two nucleotide chains separate by unwinding, and each chain serves as a template for a new nucleotide chain.

DNA replication is illustrated in Figure 10-5. The first step is the separation of the two nucleotide chains. The point at which the two chains separate is called the **replication fork.** The chains are separated by enzymes called **helicases.** As the helicase enzymes move along the DNA molecule, they break hydrogen bonds between the complementary bases, and the chains separate.

As Figure 10-5 shows, enzymes called **DNA polymerases** bind to the separated chains of DNA. As DNA polymerases move along the separated chains, new chains of DNA are assembled using nucleotides in the surrounding medium that are complementary to the existing DNA chains. Nucleotides are joined to the new chains by covalent bonds between deoxyribose sugars and phosphate groups. They are joined to the original nucleotide chain by hydrogen bonds.

FIGURE 10-5

(a) During DNA replication, helicase enzymes separate DNA's two chains of nucleotides. (b) DNA polymerases bind to the separated chains of nucleotides. One nucleotide at a time, the enzyme constructs a new complementary chain of nucleotides. (c) At the end of replication, there are two identical copies of the original DNA molecule. Each DNA molecule is made of one chain of nucleotides from the original DNA molecule and one new chain of nucleotides.

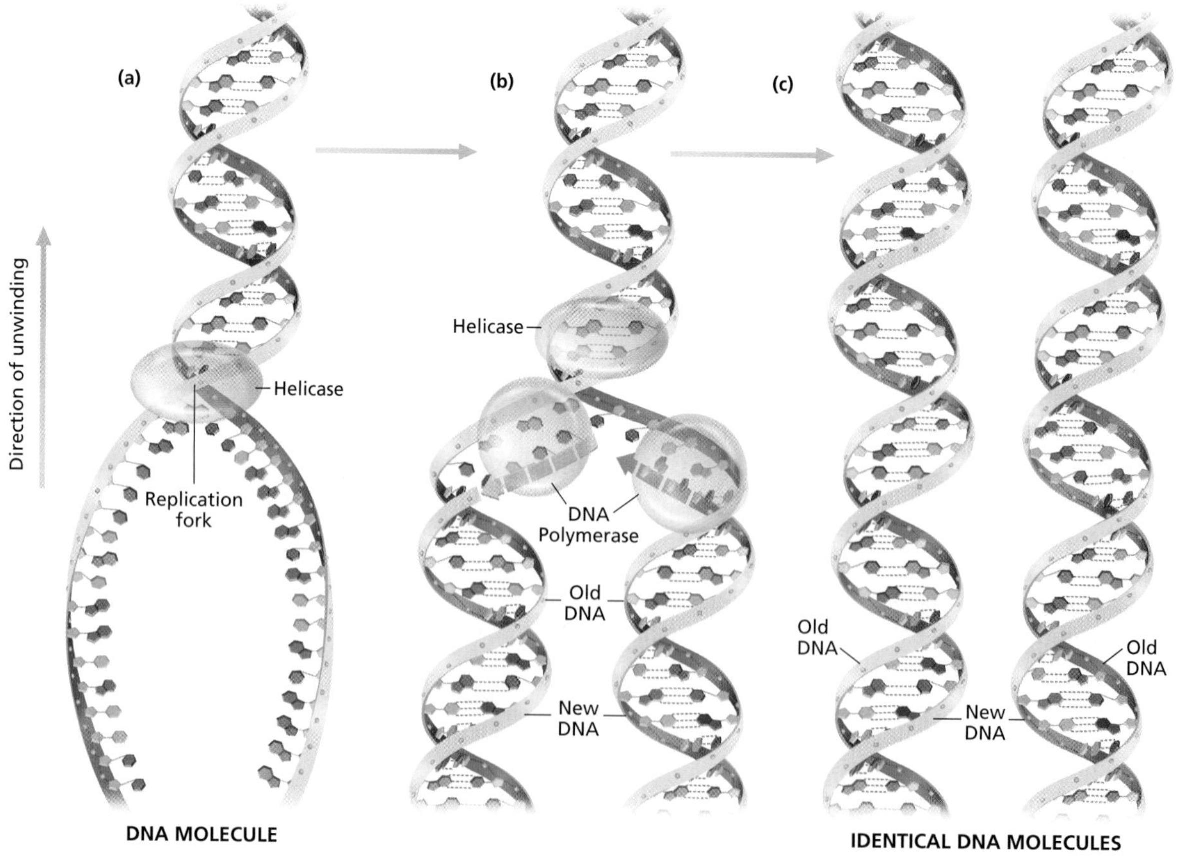

The complementary nature of the two chains of DNA is the foundation for accurate DNA replication. Suppose that the sequence of nucleotides in one chain of the original DNA molecule is A-T-T-C-C-G. DNA polymerases would produce a new nucleotide chain with the sequence of T-A-A-G-G-C.

DNA replication does not begin at one end of the molecule and proceed to the other. Rather, DNA polymerases begin replication simultaneously at many points along the separated nucleotide chains. Replication occurring simultaneously at different sites permits faster DNA replication. For example, replication is initiated simultaneously at about 6,000 sites in fruit fly DNA.

When replication is completed, two new exact copies of the original DNA molecule are produced and the cell is ready to undergo cell division. Each new DNA molecule consists of one new nucleotide chain joined by hydrogen bonds to a nucleotide chain from the original DNA molecule.

Accuracy and Repair

The process of DNA replication occurs with a high degree of accuracy—about one error in every 10,000 paired nucleotides. However, a change in the nucleotide sequence at even one location, called a **mutation,** may have serious effects in new cells. A combination of DNA proofreading and repair processes helps keep the error rate to one error per 1 billion nucleotides.

The number of errors and mutations in DNA replication is reduced as enzymes proofread DNA and repair errors. Most repair enzymes detect errors in the complementary base-paired structure of DNA. If noncomplementary bases are paired, the abnormal DNA structure can be recognized and repaired by specific enzymes.

Although DNA replication and proofreading prevent many replication errors, some errors do occur. In addition, DNA can be damaged by a variety of agents, including chemicals and ultraviolet radiation from the sun. Because cells are continuously proofreading and repairing their DNA, the number of mutations is reduced.

Eco Connection

DNA Repair Enzymes in Frogs

Around the world, frog and toad populations are in decline. Research on two declining species inhabiting the Oregon Cascade Range—Cascade frogs and western toads—suggests that these species' DNA repair enzymes are unable to keep up with mutations caused by increased exposure to ultraviolet radiation. Laboratory experiments showed that the DNA repair enzymes of Pacific tree frogs, whose populations do not appear to be decreasing, were better able to repair mutations than those of Cascade frogs and western toads.

SECTION 10-1 REVIEW

1. **What are the main functions of DNA?**
2. **Identify the types and locations of covalent bonds and hydrogen bonds in a DNA molecule.**
3. **List the base-pairing rules.**
4. **What roles do enzymes play in DNA replication?**
5. **How would the deoxyribose sugar-phosphate backbone of nucleotide chains look if purines paired only with purines and pyrimidines paired only with pyrimidines?**
6. **CRITICAL THINKING A DNA molecule (labeled as A) replicates to produce two new DNA molecules (labeled as B). Both of the B DNA molecules then replicate to form four new DNA molecules (labeled as C). Are any nucleotide chains from A present in the C DNA molecules? Explain your answer. If you believe the answer is yes, how many of the A DNA nucleotide chains are present in the C DNA molecules?**

RETEACHING *ACTIVITY*

Have students work in teams to build their own DNA models. Let students choose their own construction materials to represent the sugars, phosphates, and different nitrogen bases and to connect the components of the molecules. Once the molecules are finished, let students demonstrate how DNA is copied using their own models.

ANSWERS TO SECTION 10-1 REVIEW

1. DNA stores and transmits genetic information.
2. Covalent bonds exist between the deoxyribose sugar and phosphate molecules. Hydrogen bonds join the nitrogen bases.
3. Guanine and cytosine pair. Thymine and adenine pair.
4. Helicase attaches to DNA and breaks hydrogen bonds between the complementary bases and the chains separate. DNA polymerases bind to the separated chains of DNA. DNA polymerase identifies an unpaired base on a template chain, locates a nucleotide that has the complementary base in the surrounding medium, and attaches the free nucleotide to the new nucleotide chain that is being formed. Other enzymes are responsible for identifying and repairing errors in replicated DNA.
5. The backbone would appear uneven, similar to the appearance of a roller coaster.
6. Yes. Each replicated DNA molecule is a hybrid consisting of one new nucleotide chain and one original nucleotide chain. Two of the eight nucleotide chains would have originated from the *A* DNA molecule.

SECTION 10-2

SECTION OVERVIEW

Cell Structure and Function
Section 10-2 describes the three forms of RNA and the role each plays in protein synthesis. The steps in translation are discussed.

Matter, Energy, and Organization
Section 10-2 compares the structure and organization of DNA with the structure and organization of RNA.

▶ ENGAGE STUDENTS

Have students write down what they hear as you play a brief portion of a well-known audiotaped speech. Ask them why the original tape is probably kept by the recording company. (It is kept as a master to preserve the tape's original integrity so that high quality copies of original, important, or historical information can be made from it.) Ask students to list the ways the tape and their transcripts differ. (Their transcripts use a "code" of written words and letters; the tape uses a magnetic "code." Their transcripts could differ if they misheard or misquoted what was on the tape.) Point out to students that DNA and RNA molecules exhibit similarities and differences as much as the tape and their transcripts do—the same information is carried but in different forms. Relate the term *transcription* to the word *transcript.* Stress that RNA is transcribed from DNA.

RECENT RESEARCH

Ribozymes
Research is being conducted using ribozymes to correct genetic errors at the mRNA level instead of at the DNA level. Ribozymes are single-stranded RNA molecules that behave as enzymes in repairing RNA in bacterial cells. Ribozymes have the ability to cut and splice RNA and to add new RNA nucleotides to the original strand.

SECTION 10-2

OBJECTIVES

▲ Explain the primary functions of RNA.

● Compare the structure of RNA with that of DNA.

■ Describe the structure and function of each type of RNA.

◆ Summarize the process of transcription.

RNA

Recall that the nucleotides in DNA molecules are grouped into genes that contain the information needed to make specific proteins. In eukaryotes, the genes directing protein production are in the nucleus, and the enzymes and amino-acid building blocks for protein production are in the cytosol. The nucleic acid called ribonucleic acid (RNA) is responsible for the movement of genetic information from the DNA in the nucleus to the site of protein synthesis in the cytosol.

STRUCTURE OF RNA

Like DNA, RNA is a nucleic acid made up of repeating nucleotides. However, as Figure 10-6 shows, RNA differs from DNA in its structure. The sugar molecule of every RNA nucleotide is **ribose,** whereas DNA nucleotides contain deoxyribose sugar. The name *ribonucleic acid* is derived from the name of its sugar, just as the name *deoxyribonucleic acid* is derived from the name of its sugar.

A second difference between RNA and DNA nucleotides is that thymine is rarely a part of RNA molecules. **Uracil,** a nitrogen-containing pyrimidine base, usually replaces thymine in RNA. As a result, uracil—not thymine—pairs with adenine in RNA.

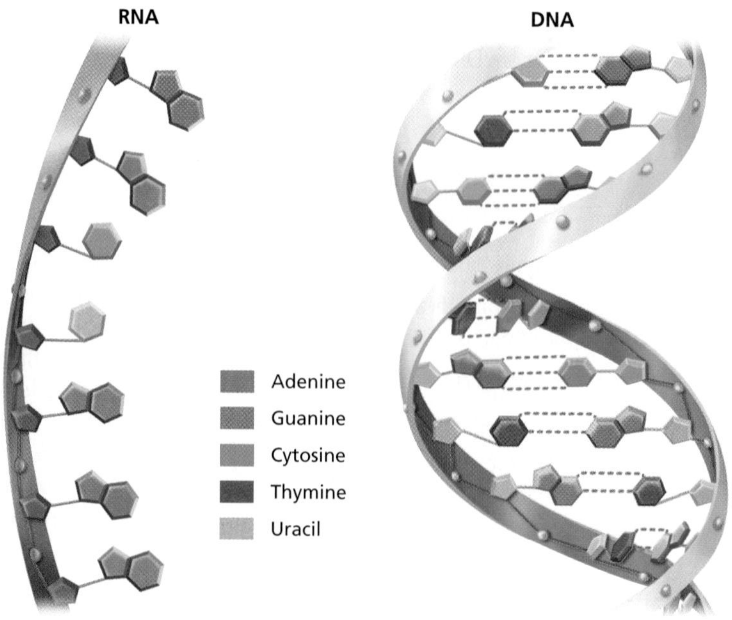

FIGURE 10-6
Like DNA, RNA is a nucleic acid made up of repeating nucleotides. Some forms of RNA involved in protein synthesis are made up of a single chain of nucleotides. Notice also that the nucleotide uracil is found in RNA in place of the DNA nucleotide thymine.

Types of RNA

RNA exists in three different types. Each type of RNA has a different function. The three types of RNA and their common abbreviations are as follows:

Messenger RNA (mRNA) consists of RNA nucleotides in the form of a single uncoiled chain. mRNA carries genetic information from the DNA in the nucleus to the cytosol of a eukaryotic cell.

Transfer RNA (tRNA) consists of a single chain of about 80 RNA nucleotides folded into a hairpin shape that binds to specific amino acids. There are about 45 varieties of tRNA.

Ribosomal RNA (rRNA) is the most abundant form of RNA. rRNA consists of RNA nucleotides in a globular form. Joined by proteins, rRNA makes up the ribosomes where proteins are made.

Quick Lab

Comparing and Contrasting RNA Types

Materials paper and pencil

Procedure Create a chart that compares and contrasts the different forms of RNA. Include descriptions of each form's structure and function.

Analysis Which types of RNA are alike structurally? What might happen if one type of RNA was missing?

TRANSCRIPTION

One function of RNA is to carry genetic information from DNA in the nucleus to the cytosol, where it can be used to produce proteins. The process by which genetic information is copied from DNA to RNA is called **transcription.** The process of transcription is shown in Figure 10-7.

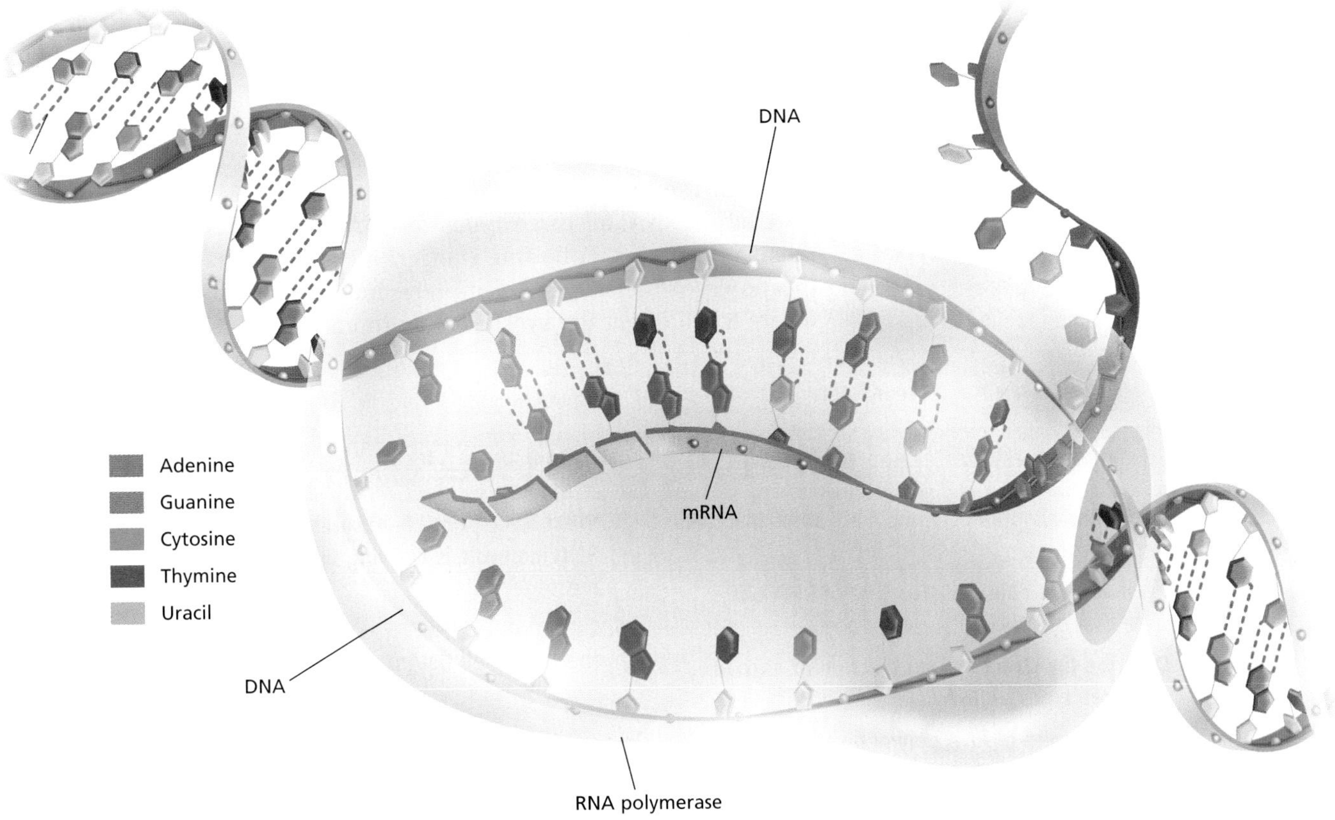

FIGURE 10-7

During transcription, RNA polymerase binds to the promoter of a specific gene. Then a complementary copy of that gene's DNA base sequence is made using RNA nucleotides, thus forming the mRNA strand.

SECTION 10-2

Quick Lab

Comparing and Contrasting RNA Types

Time Required 15 minutes

Procedural Tips Encourage students to create a table that displays their information in a manner that facilitates comparison between the different forms of RNA. Have students work in pairs.

Answers to Analysis Both mRNA and tRNA are alike structurally in that they form single chains. Protein synthesis would cease if a type of RNA was missing.

VISUAL LINK

Figure 10-7

Have students identify the region of the DNA molecule where transcription begins (the promoter region). Ask them what effect RNA polymerase has on the DNA molecule. (It relaxes and unwinds DNA to expose the promoter region and the coding DNA strand.)

CRITICAL THINKING

Termination Codon

Ask students what might happen if the termination signal were absent. (Transcription would continue to the next gene without stopping.)

ANSWERS TO SECTION 10-2 REVIEW

1. Transcription is the process of producing RNA from DNA.
2. RNA usually consists of a single chain of nucleotides, has ribose instead of deoxyribose as its sugar, and contains uracil rather than thymine.
3. mRNA consists of RNA nucleotides in the form of a single uncoiled chain. mRNA carries genetic information from the DNA in the nucleus to the cytosol of a eukaryotic cell. tRNA is a single chain of about 80 nucleotides folded into a hairpin shape that binds to a specific amino acid. rRNA consists of RNA nucleotides in a globular form and makes up the ribosomes where proteins are made.
4. RNA polymerase initiates transcription by binding to promoters. When RNA polymerase binds to a promoter, the DNA molecule in that region separates. RNA polymerase attaches to the first DNA nucleotide of the template chain. Then it begins adding complementary RNA nucleotides to the newly forming RNA molecule. At the termination signal, RNA polymerase releases DNA and RNA.
5. Accuracy of the genetic message is assured by complementary base pairing.
6. Yes. Because templates are complementary, they do not contain identical sequences of nucleotides. A sequence complementary to the template will code for different information.

Word Roots and Origins

transcription

from the Latin *scribere*, meaning "to write," and *trans*, meaning "across"

Steps of Transcription

RNA polymerase, the primary transcription enzyme, synthesizes RNA copies of specific sequences of DNA. RNA polymerase initiates RNA transcription by binding to specific regions of DNA called **promoters.** The promoter marks the beginning of the DNA chain that will be transcribed. In eukaryotes, promoters mark the beginning of a single gene, but in prokaryotes a promoter may mark the beginning of several functionally related genes. When RNA polymerase binds to a promoter, the DNA molecule in that region separates. Only one of the separated DNA chains, called the template, is used for transcription.

RNA polymerase attaches to the first DNA nucleotide of the template chain. Then it begins adding complementary RNA nucleotides to the newly forming RNA molecule. In Figure 10-7, notice that complementary base pairing determines the nucleotide sequence of the RNA chain in transcription, just as it does in DNA replication. The base-pairing rules are identical to those in DNA replication, except that uracil pairs with adenine.

Transcription continues one nucleotide at a time until the RNA polymerase reaches a DNA region called the **termination signal.** The termination signal is a specific sequence of nucleotides that marks the end of a gene in eukaryotes and may mark the end of several functionally related genes in prokaryotes. At the termination signal, RNA polymerase releases both the DNA molecule and the newly formed RNA molecule. All three types of RNA molecules are transcribed in this process.

Products of Transcription

The products of transcription, called transcripts, are the different types of RNA molecules, including mRNA, tRNA, and rRNA. Although the instructions for making a protein are copied from DNA into mRNA, all three types of RNA are involved in the synthesis of proteins. Following transcription, mRNA moves through the pores of the nuclear membrane into the cytosol of the cell, where it will direct the synthesis of proteins.

SECTION 10-2 REVIEW

1. Define transcription. List the main steps involved in this process.
2. In what ways does the structure of RNA differ from that of DNA?
3. Describe the structure and function of each of the three types of RNA.
4. List three roles of RNA polymerase in transcription.
5. What basic principle ensures that the transcribed RNA molecule is carrying the right genetic message?
6. CRITICAL THINKING Does it matter which of the separated DNA chains is used for transcription? Discuss your answer.

SECTION

10-3

OBJECTIVES

▲ Describe the genetic code.

● Distinguish between a codon and an anticodon, and state where each is found.

■ Explain the roles of the start codon and stop codons.

◆ Summarize the process of translation.

PROTEIN SYNTHESIS

Now that you know how RNA is transcribed from DNA, you are ready to learn how the three types of RNA work together to produce proteins. The production of proteins is also called **protein synthesis.** *The amount and kind of proteins that are produced in a cell determine the structure and function of the cell. In this way, proteins carry out the genetic instructions encoded in an organism's DNA.*

PROTEIN STRUCTURE AND COMPOSITION

Like DNA and RNA, proteins are polymers. Proteins are made up of one or more polypeptides, each of which consists of a specific sequence of amino acids linked together by peptide bonds. Recall from Chapter 3 that there are 20 different amino acids that make up proteins. The polypeptides that make up one protein may consist of hundreds or thousands of the 20 different amino acids arranged in a particular sequence. The sequence of the amino acids determines how the polypeptides will twist and fold into the three-dimensional structure of the protein. The function of a protein depends on its ability to bind with other molecules within a cell; that is, the function depends on the protein's three-dimensional structure, which is determined by its amino-acid sequence.

THE GENETIC CODE

During protein synthesis, the sequence of nucleotides in an mRNA transcript is translated into a sequence of amino acids. A correlation between a nucleotide sequence and an amino-acid sequence—called the **genetic code**—is used by most organisms to translate mRNA transcripts into proteins.

The genetic information necessary for making proteins is encoded in series of three mRNA nucleotides. Each combination of three mRNA nucleotides is called a **codon.** Each codon codes for a specific amino acid. Table 10-1 on the next page lists the 64 codons and the amino acids and instructions they code for in almost all organisms. The near-universality of the genetic code supports the idea that all organisms are evolutionarily related.

internet connect

SCILINKS NSTA

TOPIC: Genetic code
GO TO: www.scilinks.org
KEYWORD: HM193

Word Roots and Origins

synthesis

from the Greek *tithenai*, meaning "to put," and *syn*, meaning "together"

SECTION 10-3

SECTION OVERVIEW

Cell Structure and Function
Section 10-3 discusses the structures and functions involved in the steps of translation. This section reviews the structure of proteins and their significance to cell function.

Evolution
Section 10-3 stresses that the near universality of the genetic code supports the idea that all organisms are evolutionarily related.

▶ ENGAGE STUDENTS

Point out that a biochemical pathway, such as respiration, consists of a complex series of reactions and that specific enzymes catalyze specific reactions. Ask students what enzymes are made of and where they come from. (Most enzymes are proteins; a gene in DNA codes for the synthesis of each polypeptide that a cell can make.) Remind students that proteins consist of one or more polypeptide chains.

Show students an unfastened, bilaterally symmetrical necklace made of multicolored beads of graduated sizes. Intertwine the two halves. Ask students how the necklace can be likened to a protein. (Each bead can represent an amino acid. The necklace consists of two symmetrical halves, representing a two-polypeptide protein.) Ask what determines the type of protein formed (the sequence of amino acids). Ask what determines that sequence (DNA).

CRITICAL THINKING

Diversity of Organisms
If DNA is universal, why are all living organisms very different? (The order of the nucleotides in the DNA of each organism is unique. Slight differences in the order can make a very big difference.)

VISUAL LINK

Table 10-1

Have students explain what the table shows and how they can determine which amino acid a specific codon codes for. (A codon consists of three bases. The table gives the possible triplets formed from the four bases and the amino acid that each codon codes for.) Point out that an amino acid may be coded for by more than one codon.

MAKING CONNECTIONS

Mathematics

Have students calculate the largest number of three-letter sequences—codons—that can be made using the four letters of the four different bases ($4^3 = 4 \times 4 \times 4 = 64$). Molecular biologists used this answer as a clue to the size of the codons. With four bases, a sequence of only two letters would give information for 4^2, or 16, amino acids. However, 20 amino acids are known. A sequence of four letters would yield 256 combinations—far too many to code for 20 amino acids. With a sequence of three letters, 64 different combinations are possible—the smallest number possible to code for all 20 amino acids. This information led to the hypothesis that each amino acid could have more than one codon.

CRITICAL THINKING

Polysomes

An mRNA transcript can be translated by many ribosomes at the same time, creating a complex called a polysome. Ask students how this capacity might be an advantage to the cell. (It can quickly produce large amounts of proteins without expending the energy to create new mRNA.)

TABLE 10-1 *Codons in mRNA*

First base	Second base: U	Second base: C	Second base: A	Second base: G	Third base
U	UUU Phenylalanine	UCU Serine	UAU Tyrosine	UGU Cysteine	U
U	UUC Phenylalanine	UCC Serine	UAC Tyrosine	UGC Cysteine	C
U	UUA Leucine	UCA Serine	UAA Stop	UGA Stop	A
U	UUG Leucine	UCG Serine	UAG Stop	UGG Tryptophan	G
C	CUU Leucine	CCU Proline	CAU Histidine	CGU Arginine	U
C	CUC Leucine	CCC Proline	CAC Histidine	CGC Arginine	C
C	CUA Leucine	CCA Proline	CAA Glutamine	CGA Arginine	A
C	CUG Leucine	CCG Proline	CAG Glutamine	CGG Arginine	G
A	AUU Isolecine	ACU Threonine	AAU Asparagine	AGU Serine	U
A	AUC Isolecine	ACC Threonine	AAC Asparagine	AGC Serine	C
A	AUA Isolecine	ACA Threonine	AAA Lysine	AGA Arginine	A
A	AUG Start	ACG Threonine	AAG Lysine	AGG Arginine	G
G	GUU Valine	GCU Alanine	GAU Aspartic acid	GGU Glycine	U
G	GUC Valine	GCC Alanine	GAC Aspartic acid	GGC Glycine	C
G	GUA Valine	GCA Alanine	GAA Glutamic acid	GGA Glycine	A
G	GUG Valine	GCG Alanine	GAG Glutamic acid	GGG Glycine	G

FIGURE 10-8

Each tRNA transports a specific amino acid to the ribosomes during translation.

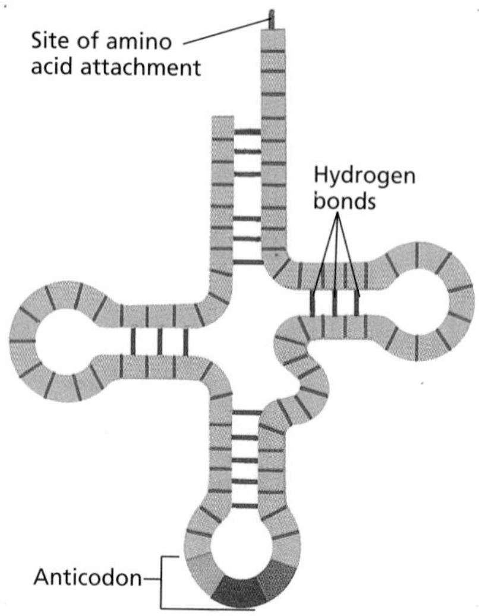

Notice that several codons code for each amino acid listed in Table 10-1. Codons often differ from one another by the nucleotide in the third position. A few codons do not code for amino acids at all. Instead, these codons signal for translation of an mRNA to start or stop. The **start codon** (AUG), which also codes for the amino acid methionine, engages a ribosome to start translating an mRNA molecule. **Stop codons** (UAA, UAG, UGA) cause the ribosome to stop translating an mRNA.

TRANSLATION

The process of assembling polypeptides from information encoded in mRNA is called **translation.** The process of translation begins when mRNA leaves the nucleus through pores in the nuclear membrane. The mRNA then migrates to a ribosome in the cytosol, the site of protein synthesis.

tRNA and Anticodons

Amino acids floating freely in the cytosol are transported to the ribosomes by tRNA molecules. Notice in Figure 10-8 that a tRNA

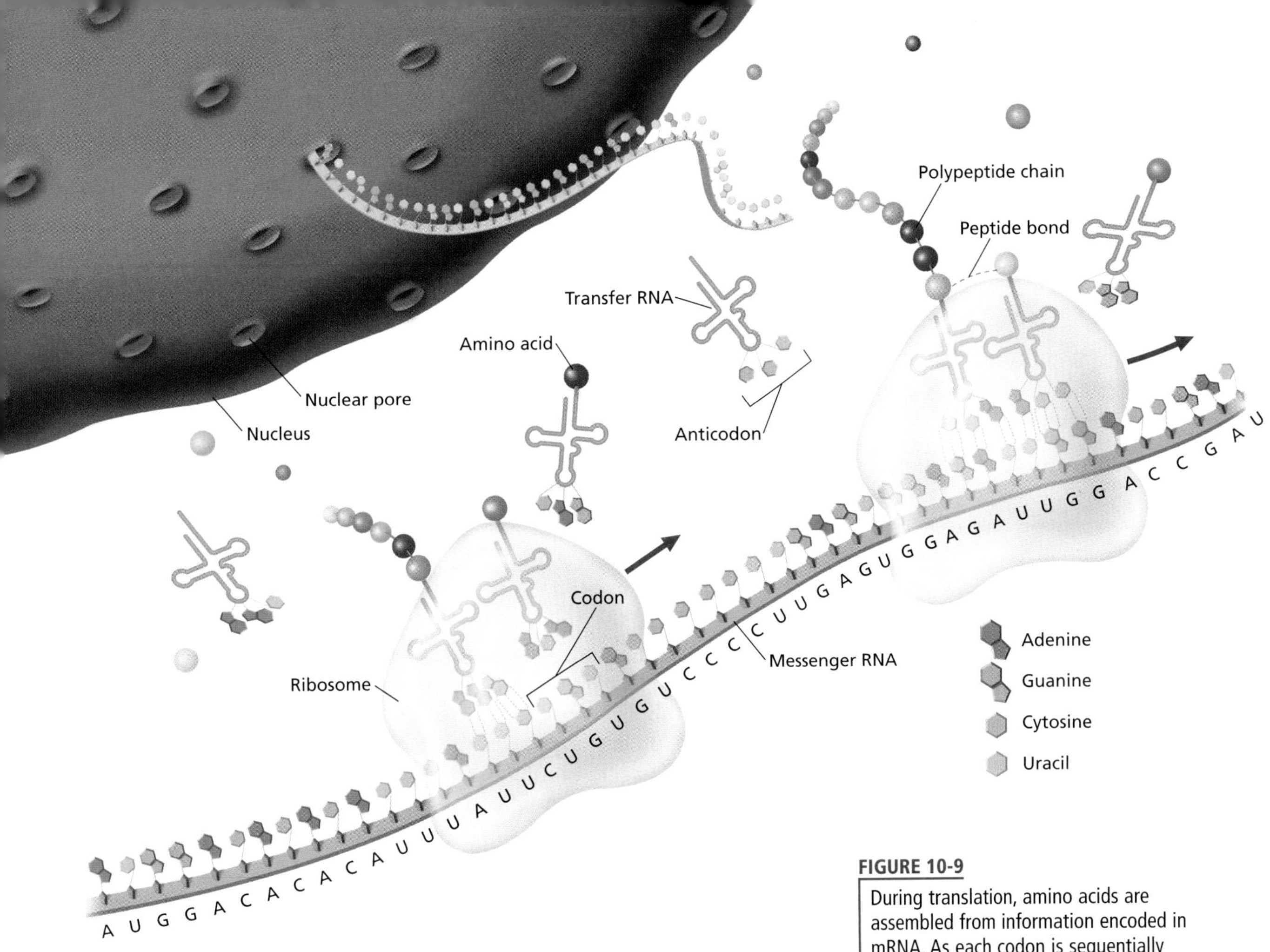

FIGURE 10-9

During translation, amino acids are assembled from information encoded in mRNA. As each codon is sequentially paired with its anticodon, tRNA adds a specific amino acid to the growing polypeptide chain.

molecule has a region that bonds to a specific amino acid. Notice also that the loop opposite the site of amino-acid attachment bears a sequence of three nucleotides called an **anticodon.** The tRNA anticodon is complementary to and pairs with its corresponding mRNA codon.

Notice in Figure 10-9 that the same base-pairing rules followed during transcription are followed during translation. For example, a tRNA with an anticodon sequence of AAA would bind to the mRNA codon sequence of UUU and would be specific for the amino acid phenylalanine. Thus, the pairing of an anticodon with a codon ensures that the amino acids are added to the growing polypeptide in the order prescribed by the mRNA transcript.

Ribosomes

Ribosomes are composed of rRNA and proteins and are usually both free in the cytosol and attached to the endoplasmic reticulum. Ribosomes that are free in the cytosol, as shown in Figure 10-9, produce proteins that will be used within the cell. Membrane proteins and proteins that will be exported for use outside the cell are produced by ribosomes attached to the endoplasmic reticulum.

VISUAL LINK
Figure 10-9

Have students identify all the structures in the figure. Have them make a list of the codons shown, beginning with the start codon AUG on the left end of the mRNA strand. Using Table 10-1, have them identify the four specific amino acids being made by each of the two ribosomes shown in the figure. (From left to right, the amino acids are phenylalanine, cysteine, aspartic acid, and tryptophan.)

RETEACHING *ACTIVITY*

Give students two or three hypothetical DNA sequences, such as AUGAATAGGGTCTCAGCACGCC-CUAA. Have the students encode the mRNA, tRNA, and amino acid sequence. Be sure to include start and stop codons (mRNA—start, UUGAUCCCAGAGUCGUG-CGGG, stop; tRNA—AACUA-GGGUCUCAGC ACGCCC; amino acids—leucine, isoleucine, proline, glutamic acid, serine, cysteine, glycine). Explain to students that the start codon, which codes for the amino acid methionine, is usually removed after translation.

ACTIVE READING
Technique: K-W-L

Have students return to their lists of things they Want to know about DNA, described in the Active Reading activity on page 185. Have students place check marks next to questions that they are able to answer. Students should finish by making a list of what they have Learned. Have students identify unanswered questions as well as new questions they might have.

Quick Lab

Modeling Protein Assembly

Time Required 20 minutes

Procedural Tips Have students work in pairs. Review the base-pairing rules before students begin the activity.

Answers to Analysis The tRNA sequence will be identical to the original DNA except that uracil will replace thymine. Polypeptides are the final product of translation.

ANSWERS TO SECTION 10-3 REVIEW

1. Transcription is the process of producing RNA from DNA. Translation is the process by which polypeptides are assembled from the information encoded in mRNA.
2. A codon is a series of three nucleotides in mRNA that code for a specific amino acid. An anticodon on tRNA is a set of three bases complementary to the codon.
3. The structure of tRNA allows an amino acid at one end to be positioned in a polypeptide when the anticodon at the other end of the tRNA pairs with a codon.
4. serine—glycine—proline—valine
5. The start codon (AUG) engages a ribosome to start translating an mRNA molecule. Stop codons (UAA, UAG, UGA) cause the ribosome to stop translating mRNA. All polypeptides do not begin with the amino acid coded for by the start codon.
6. No protein would be produced because the mRNA begins with a stop codon.

Quick Lab

Modeling Protein Assembly

Materials colored paper clips (blue = adenine, green = guanine, pink = cytosine, yellow = uracil)

Procedure

1. Make the following mRNA sequence by linking the appropriate colored paper clips to represent the following bases: AUG GAC ACA CAU UUA UUC UGA.
2. Make a second chain of paper clips that represents the tRNA anticodons for the mRNA protein sequence shown in step 1. Use the appropriate color of paper clip for the complementary base.

Analysis How is the tRNA base sequence similar to the original DNA? What is the final product of translation?

Ribosomes have three binding sites that are key to translation. One binding site holds an mRNA transcript so that its codons are accessible to rRNA molecules. The other two binding sites hold tRNAs whose anticodons pair with the mRNA codons.

Protein Assembly

The assembly of a polypeptide begins when a ribosome attaches to the start codon (AUG) on an mRNA transcript. Find the start codon on the mRNA transcript shown in Figure 10-9. The start codon pairs with the anticodon UAC on a tRNA. Because the tRNA that bears the UAC anticodon also carries the amino acid methionine, the first amino acid in every polypeptide is initially methionine. However, this first amino acid may be removed later; thus, all polypeptides do not actually begin with methionine.

As a ribosome moves along an mRNA transcript, each mRNA codon is sequentially paired with its tRNA anticodon, as Figure 10-9 shows. The pairing of an anticodon with a codon causes the specified amino acid to attach to the previously translated amino acid with a covalent bond called a peptide bond. In this way, amino acids are joined to a growing polypeptide chain in the order specified by an mRNA transcript. As each amino acid is added to the polypeptide chain, the ribosome moves three nucleotides—one codon—ahead on the mRNA transcript, where the next amino acid will be translated. Eventually, the ribosome reaches a stop codon, bringing translation to an end. At this point the mRNA is released from the ribosome and the polypeptide is complete.

As you can see in Figure 10-9, several ribosomes may simultaneously translate the same mRNA transcript. A new ribosome begins translating an mRNA transcript almost as soon as the preceding ribosome has moved out of the way. These ribosomes are often spaced as close as 80 nucleotides apart on an mRNA.

The polypeptide chain represents the protein's primary structure. As the polypeptide folds and associates with other polypeptides that make up the protein, it assumes the functional structure of the completed protein.

SECTION 10-3 REVIEW

1. Compare transcription with translation.
2. Distinguish a codon from an anticodon, and explain the significance of each.
3. How does the structure of tRNA relate to its function in translation?
4. Using the information in Table 10-1, list the amino acids that are coded for by the codons AGU, GGG, CCU, and GUG.
5. Explain the significance of the start codon and the stop codons. Do all polypeptides begin with the amino acid coded for by the start codon?
6. **CRITICAL THINKING** What would translation of the mRNA transcript UAACAAGGAGCAUCC produce?

CHAPTER 10 REVIEW

SUMMARY/VOCABULARY

10-1

- DNA stores the information that tells cells which proteins to make and when to make them.
- DNA is made up of two chains of nucleotides.
- A DNA nucleotide is composed of a deoxyribose sugar molecule, a phosphate group, and a nitrogen-containing base. The four nitrogen-containing bases found in DNA nucleotides are adenine (A), guanine (G), cytosine (C), and thymine (T).
- Adenine and guanine are called purines. Cytosine and thymine are pyrimidines.
- Cytosine pairs with guanine, and adenine pairs with thymine. Complementary base pairs are connected to each other by hydrogen bonds.
- Before a cell divides, it copies its DNA by a process called replication. Replication results in two exact copies of the cell's DNA.
- Replication begins with the separation of the DNA chains by helicase enzymes. Then as DNA polymerases move along the separated chains, new chains of DNA are assembled using nucleotides in the surrounding medium.

Vocabulary

adenine (185)
base-pairing rule (187)
complementary base pair (187)
cytosine (185)
deoxyribose (185)
DNA polymerase (188)
double helix (186)
guanine (185)
helicase (188)
mutation (189)
nitrogen-containing base (185)
purine (186)
pyrimidine (186)
replication (188)
replication fork (188)
thymine (185)

10-2

- Nucleotides in DNA are grouped into genes, which contain the information required for the production of specific proteins.
- Three forms of RNA are involved in protein synthesis: mRNA, tRNA, and rRNA.
- mRNA carries genetic information from the DNA in the nucleus to the cytosol of a eukaryotic cell. The process of copying genetic information from DNA to mRNA is called transcription.
- tRNA binds to specific amino acids, helping to form polypeptide chains.
- rRNA makes up the ribosomes where proteins are made.

Vocabulary

messenger RNA (mRNA) (191)
promoter (192)
ribose (190)
ribosomal RNA (rRNA) (191)
RNA polymerase (192)
termination signal (192)
transcription (191)
transfer RNA (tRNA) (191)
uracil (190)

10-3

- The production of proteins is also called protein synthesis.
- The genetic code shown in Table 10-1 is used by most organisms to translate mRNA transcripts into proteins.
- The genetic information necessary for making proteins is encoded in mRNA codons. Each codon codes for a specific amino acid.
- The process of assembling polypeptides from information encoded in mRNA is called translation. During translation, tRNA anticodons pair with corresponding mRNA codons, and amino acids are joined together to form a polypeptide.
- As a polypeptide folds and associates with other polypeptides, it assumes the functional structure of a completed protein.

Vocabulary

anticodon (195)
codon (193)
genetic code (193)
protein synthesis (193)
start codon (194)
stop codon (194)
translation (194)

CHAPTER 10 REVIEW ANSWERS

REVIEW

1. Ribosomes are made up of ribosomal RNA and protein.
2. A termination signal occurs on DNA, while a stop codon occurs on mRNA.
3. mRNA consists of RNA nucleotides in the form of a single uncoiled chain. mRNA carries genetic information from the DNA in the nucleus to the cytosol of a eukaryotic cell. tRNA consists of a single chain of RNA nucleotides that are folded into a hairpin shape and that bind to specific amino acids. rRNA consists of RNA nucleotides in a globular form. Joined by proteins, rRNA makes up the ribosomes involved in the production of proteins.
4. Purines, such as adenine and guanine, are bases that have two rings of carbon and nitrogen atoms. Pyrimidines, such as cytosine and thymine, are bases that have one ring of carbon and nitrogen atoms.
5. Protein synthesis is the process of the assembly of amino acids into proteins.

6. a **7.** b
8. c **9.** d
10. c **11.** c
12. d **13.** b
14. c **15.** d

16. The complementary structure of DNA allows it to replicate itself, passing identical copies of itself and the information it encodes to offspring cells. The sequence of four nucleotides allows DNA to encode genetic information.
17. Yes. It decreases the time necessary for DNA replication.
18. Hydrogen bonds are more significant because they are the bonds that hold the nucleotides together. In nucleotide pairs that are not complementary, the hydrogen bonding between the nucleotides does not hold the nucleotides together as tightly as those between complementary nucleotides do.

19. They signal start and stop.
20. The ribosome holds mRNA and is the site where anticodons are matched with codons so that an enzyme can form the peptide bonds between amino acids.
21. The near-universality of the genetic code is evidence that all organisms are evolutionarily related. The genetic code was probably operational in the earliest cells.
22. The DNA is undergoing replication, as evidenced by the replication fork.
23. Reports will vary. Many antibiotics, such as erythromycin and streptomycin, inhibit protein synthesis by combining with ribosomal proteins. Mupirocin and puromycin inhibit protein synthesis at the tRNA level.

Critical Thinking

1. The four codons for alanine differ at the third position. This redundant pattern minimizes errors in translation.
2. methionine (start)—serine—asparagine—lysine—glutamine—alanine—stop
3. mRNA: UUG, AUG, CCA, GAG, UCG, UGA, GGG; tRNA: AAC, UAC, GGU, CUC, AGC, ACU, CCC; Amino acids: leucine, methionine (start), proline, glutamic acid, serine, stop, glycine
4. Possibilities include the following: (a) the mRNA is not translated because the start codon was affected, (b) one amino acid is incorrectly translated, (c) too much of the mRNA is translated if the error occurred in the stop codon, and (d) there is no effect if the nucleotide affected was not between the start codon and the stop codon.

Review

Vocabulary

1. Distinguish between a ribosome and ribosomal RNA.
2. Describe the difference between a termination signal and a stop codon.
3. Distinguish between mRNA, tRNA, and rRNA.
4. Describe the difference between a purine and a pyrimidine.
5. Define the term *protein synthesis.*

Multiple Choice Questions

6. DNA
(a) encodes the information needed for making proteins within the cell
(b) directs RNA to make lipids that are needed by the cell
(c) directs RNA to produce glucose
(d) produces carbohydrates.

7. The physical structure of the genetic code is DNA's sequence of (a) sugars (b) nucleotides (c) phosphates (d) hydrogen bonds.

8. Covalent bonds are found (a) between purines and pyrimidines (b) between nucleotide chains (c) in the sugar-phosphate backbones (d) between RNA and DNA molecules.

9. Complementary base pairing is important for (a) DNA replication (b) RNA transcription (c) RNA translation (d) all of the above.

10. Unlike DNA, RNA
(a) is a polymer made up of nucleotides
(b) contains the nitrogen-containing base thymine
(c) contains the sugar ribose
(d) does not contain the nitrogen-containing base uracil.

11. Codons are (a) sequences of proteins (b) a site on a tRNA (c) three-nucleotide sequences of mRNA (d) sequences of polypeptides.

12. The pattern of base pairing in DNA can be summarized as follows:
(a) purines pair with purines
(b) any purine can pair with any pyrimidine
(c) adenine pairs with guanine
(d) adenine pairs with thymine, and cytosine pairs with guanine.

13. The formation of peptide bonds occurs (a) in the nucleus (b) between amino acids (c) during DNA proofreading (d) during transcription.

14. Transcription occurs (a) at ribosomes in both prokaryotes and eukaryotes (b) in the cytosol of eukaryotes (c) in the nucleus of eukaryotes (d) as the last step of protein synthesis.

15. Transporting amino acids to ribosomes for assembly into needed proteins is the function of (a) DNA (b) mRNA (c) rRNA (d) tRNA.

Short Answer

16. What are some ways that the structure of a DNA molecule is related to its function?
17. Is there an advantage to multiple sites of DNA replication? Explain your answer.
18. Would hydrogen bonds or covalent bonds be more important for the identification of mismatches during DNA replication? Explain your answer.
19. What functions are carried out by those few codons that do not code for amino acids?
20. What is the role of ribosomes in protein synthesis?
21. What is the evolutionary significance of a near-universal genetic code?
22. The photograph below shows the DNA of a mammalian cell. What process is the DNA undergoing? What is the structure shown called? Explain your answer.

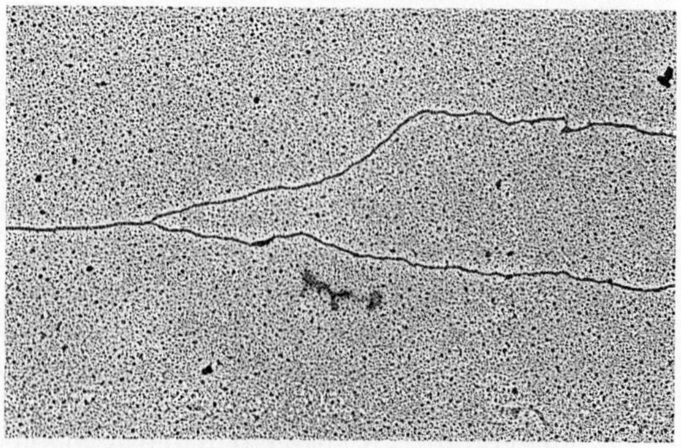

23. **Unit 6—*Gene Expression***
BIOLOGY INTERACTIVE TUTOR
Write a report summarizing how antibiotics inhibit protein synthesis in bacteria. How do some antibiotics interfere with translation?

CRITICAL THINKING

1. Use Table 10-1 to find the codons that code for the amino acid alanine. How many codons code for alanine? How do they differ from one another? What significance might the differences have?
2. Refer again to Table 10-1. Using the table, determine the sequence of amino acids that is specified by the following list of codons: AUG UCU AAC AAA CAG GCU UAA.
3. A segment of DNA has the following sequence: AACTACGGTCTCAGCACTCCC. Write the mRNA transcript of this sequence of DNA. Next, write the tRNA anticodons that would pair with the mRNA transcript. Using Table 10-1, write the names of the amino acids coded for by the mRNA transcript.
4. What would happen to the translation process if one nucleotide were not transcribed correctly? Present a minimum of three possibilities.
5. How is a system composed of three bases per codon better suited to code for 20 amino acids than a system composed of two bases per codon?
6. Genetic engineering involves inserting segments of DNA taken from one organism into the DNA of another organism. What would be the likely result of an experiment in which a scientist replaces a faulty stop codon in the DNA of mouse cells with the codon UAG taken from the DNA of a frog, a pine tree, or a clam? What do the results of this experiment suggest about the evolutionary ancestry of these organisms?
7. The diagram below shows a model of the chemical structure of uracil. Compare this diagram with Figure 10-2. How does the chemical structure of uracil differ from the chemical structure of thymine? How is it possible for both thymine and uracil to bond with adenine? (You may wish to look again at the patterns of hydrogen bonding illustrated in Figure 10-3 to help you answer this question.)

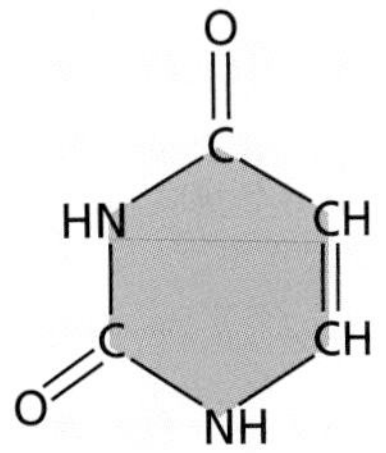

URACIL

EXTENSION

1. Prepare a report on the contributions of one of the following scientists to the discovery of the structure of DNA: Erwin Chargaff, Francis Crick, Rosalind Franklin, Linus Pauling, James Watson, or Maurice Wilkins. The book *The Double Helix,* by James Watson, provides information about Watson and Crick's discovery. *The Eight Days of Creation,* by Horace Freeland, provides information about the work of all of these scientists on DNA.
2. Read "Genetics: Repairing the Genome's Spelling Mistakes" in *Science,* July 16, 1999, on page 316. What new gene-therapy technique is being used to correct genetic disorders in rats? What is the goal of this research if it is found to be safe?
3. Read "Gene Cloned for Stretchiest Spider Silk" in *Science News,* February 21, 1998, on page 119. Describe the protein structure that contributes to the extreme elasticity of spider silk.

5. Three bases per codon provide more than enough units for the 20 amino acids that make up proteins. Two bases would provide only 4^2, or 16, units.
6. The near-universality of the genetic code indicates that the codon UAG would be a stop codon in mouse mRNA, as it is in most organisms, including frogs, trees, and clams. This suggests a common evolutionary ancestor for all organisms.
7. Thymine has a CH_3 (methyl group) attached to one of the carbons in the ring. Uracil has only a hydrogen atom attached at the equivalent site. This difference does not affect bonding. Hydrogen bonds that form with adenine occur on the other side of the molecule.

EXTENSION

1. Student reports should focus on research related to the discovery of the structure of DNA.
2. Researchers are using hybrid molecules of DNA and RNA to trick the cell into replicating good DNA into mutant sites. Researchers hope that someday they can go to the Human Genome Project, read off what the mutated gene should be, and change the "spelling" of the molecule to cure diseases.
3. Spider-silk protein is formed by a chain of thousands of amino acids. At the end of every five amino acids the protein kinks back on itself 180°. The protein forms a spiral that looks like a molecular spring. It is this structure that makes spider silk stretchy and strong.

CHAPTER 10 INVESTIGATION

TIME REQUIRED

One 50-minute lab period

SAFETY PRECAUTIONS

Caution students to be careful of the sharp tip on the pushpins.

QUICK REFERENCES

Lab Preparation Notes are found on page 183B.

Holt BioSources provides a Teaching Resources CD-ROM that contains "Using Gowin's Vee in the Lab" and "Scoring Rubrics."

PROCEDURAL TIPS

1. This lab works well with groups of two or four students. Obtain enough plastic soda straws to make 54 3-cm sections for each lab group.
2. Account for all materials at the end of the lab, and have students store the pushpins by color.

ANSWERS TO BACKGROUND

1. DNA is made of two chains of nucleotides twisted around a central axis. Each nucleotide is made up of the sugar deoxyribose, a phosphate group, and a base.
2. Adenine binds to thymine. Guanine binds to cytosine.
3. The first step is the separation of the two nucleotide chains, forming a replication fork. Then new chains of DNA are assembled using nucleotides in the surrounding medium that are complementary to the existing DNA chains. Nucleotides are joined to the new chains by covalent bonds between deoxyribose sugars and phosphate groups. They are joined to the original nucleotide chain by hydrogen bonds.
4. mRNA carries genetic information from the DNA in the nucleus to the site of protein synthesis in the cytosol of a eukaryotic cell.

CHAPTER 10 INVESTIGATION

Modeling Replication and Transcription of DNA

OBJECTIVES

- Construct and analyze a model of DNA.
- Simulate the process of replication using a model.
- Simulate the process of transcription using a model.

PROCESS SKILLS

- demonstrating
- identifying
- manipulating a model

MATERIALS

- plastic soda straws, cut into 3 cm sections (54)
- metric ruler
- scissors
- permanent marker
- 54 pushpins (12 red, 12 blue, 12 yellow, 12 green, and 6 white)
- 54 paper clips

Background

1. Describe the structure of DNA.
2. State the base-pairing rules.
3. List the steps involved in the copying of DNA before cell division.
4. What is mRNA, and what is its function in protein synthesis?
5. Describe the process of transcription.

PART A Making a Model of DNA

1. CAUTION **Always cut in a direction away from your face and body.** Cut the soda straws into 3 cm pieces to make 54 segments.
2. Insert a pushpin midway along the length of each straw segment as shown in the figure below. Push a paper clip into one end of each straw segment until it touches the pin.

3. Keeping the pins in a straight line, insert the paper clip from a blue-pushpin segment into the open end of a red-pushpin segment. Add additional straw segments to the red-segment end in the following order: green, yellow, blue, yellow, blue, yellow, green, red, red, and green. Use the permanent marker to label the blue-segment end "top." This chain of segments is one-half of your first model.
4. Assign nucleotides to the corresponding pushpin colors as follows: red = adenine, blue = guanine, yellow = cytosine, and green = thymine.
5. Construct the other half of your first model beginning with a yellow segment across from the blue pushpin at the top of your first model. Keep the pins in a straight line. Link segments together according to the base-pairing rules.

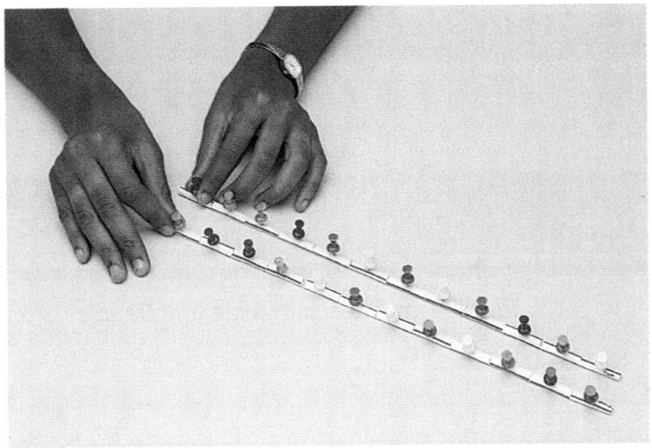

6. When you have completed your model of one DNA segment, make a sketch of the model in your lab report. Use colored pencils or pens to designate the pushpin colors. Include a key that indicates which nucleotide each color represents in your sketch.

PART B Modeling DNA Replication

7. Place the chains parallel to each other on the table with the "top" blue pin of the first chain facing the "top" yellow pin of the second chain.
8. Demonstrate replication by simulating a replication fork at the top pair of pins. Add the remaining straw segments to complete a new DNA model. Be sure to follow the base-pairing rules.
9. Sketch the process of DNA replication in your lab report. Label the replication fork, the segments of original DNA, and the segments of new DNA in your sketch.

PART C Modeling Transcription

10. Place the chains of one of the DNA models parallel to each other on the table. Take the other DNA model apart so that you can use the segments to construct a model of mRNA.
11. Assign the uracil nucleotide to the white pushpins. Using the available pushpins, construct a model of an mRNA transcript of the DNA segment. Begin by separating the two chains of DNA and pairing the mRNA nucleotides with the left strand of DNA as you transcribe from the top of the segment to the bottom of the segment.
12. In your lab report, sketch the mRNA model that you transcribed from the DNA segment.
13. Refer to Table 10-1 on page 194 to determine the sequence of amino acids you transcribed.
14. Clean up your materials before leaving the lab.

Analysis and Conclusions

1. How many nucleotides did the original DNA model contain?
2. Write the base-pair order for the DNA molecule you created, using the following code: red = adenine, blue = guanine, yellow = cytosine, and green = thymine.
3. What is the name given to the point where replication starts on a DNA molecule?
4. How does the replicated model of DNA compare to the original model of DNA?
5. What would the complementary bases be if one side of a DNA molecule had the bases adenine, cytosine, cytosine, thymine, thymine, and adenine?
6. Speculate about what would happen if the nucleotide pairs in the replicated model were not in the same sequence as the original model.
7. Write the mRNA transcription of the DNA sequence presented below.

   CTG TTC ATA ATT

   Next, write the tRNA anticodons that would pair with the mRNA transcription. Finally, write the amino acids coded for by the mRNA transcription using Table 10-1.
8. If you transcribed the "wrong" side of the DNA molecule, what would the result be? How might this affect the proteins that the organism produced?
9. What are the advantages of having DNA remain in the nucleus of eukaryotic cells?

Further Inquiry

Design models for some of the other molecules that are involved in protein synthesis. Use these models along with the DNA model you constructed in this investigation to demonstrate the steps of protein synthesis.

5. RNA polymerase binds to a promoter, separating the DNA chains in that region. RNA polymerase attaches to the first DNA nucleotide of the template chain and begins adding complementary RNA nucleotides. Transcription continues until the RNA polymerase reaches the termination signal.

ANSWERS TO ANALYSIS AND CONCLUSIONS

1. The original DNA model contained 24 nucleotides, with 12 nucleotides on each chain.
2. The base-pair order for the DNA molecule is guanine-cytosine, adenine-thymine, thymine-adenine, cytosine-guanine, guanine-cytosine, cytosine-guanine, guanine-cytosine, cytosine-guanine, thymine-adenine, adenine-thymine, adenine-thymine, thymine-adenine.
3. The point of origin of replication on a DNA molecule is called a replication fork.
4. The replicated model is identical to the original model.
5. The complementary bases would be thymine, guanine, guanine, adenine, adenine, and thymine.
6. If the base pairs were out of sequence, the resulting organism might not be viable or could die as a result of the mutation(s).
7. mRNA: GAC, AAG, UAU, UAA; tRNA: GUC, UUC, AUA, AUU; amino acid: aspartic acid, lysine, tyrosine, stop
8. Transcribing the "wrong," or antisense, strand of the DNA would affect most if not all of the proteins that the organism produced.
9. The nucleus provides an isolated and protected area for storing genetic information. By having DNA in the nucleus, eukaryotes can house more DNA and undergo more-complex cell divisions than prokaryotes can.

CHAPTER 11

GENE EXPRESSION

	TOPICS	TEACHING RESOURCES	LABS, CLASSWORK, AND HOMEWORK
BLOCKS 1 & 2 90 minutes	**Introducing the Chapter, p. 202**	**ATE** Focus Concept, p. 202 **ATE** Assessing Prior Knowledge, p. 202 **ATE** Understanding the Visual, p. 202	■ Supplemental Reading Guide, *The Double Helix*
	11-1 Control of Gene Expression, p. 203 Role of Gene Expression, p. 203 Gene Expression in Prokaryotes, p. 203 Gene Expression in Eukaryotes, p. 206	**ATE** Section Overview, p. 203 **ATE** Critical Thinking, p. 207 27A, 34, 35, 36, 37	**PE** **Quick Lab,** p. 207, Modeling After-Transcription Control **Lab B7,** Gene Expression ★ Study Guide, Section 11-1 **PE** Section Review, p. 208
BLOCKS 3 & 4 90 minutes	**11-2 Gene Expression and Development, p. 209** Cell Differentiation, p. 209 Cancer, p. 211	**ATE** Section Overview, p. 209 **ATE** Critical Thinking, pp. 212, 214 **ATE** Visual Link Figure 11-7, p. 210	**PE** **Chapter 11 Investigation,** pp. 218–219 **Lab D1,** Staining DNA and RNA **Lab D2,** Extracting DNA **Lab D3,** Genetic Transformation of Bacteria **Lab D4,** Genetic Transformation—Antibiotic-Resistant Bacteria **Lab E4,** Heredity in Families ★ Study Guide, Section 11-2 **PE** Section Review, p. 214
BLOCKS 5 & 6 90 minutes	**REVIEW and ASSESSMENT** **PE** Chapter 11 Review, pp. 216–217 ■ Performance-Based Assessment—Identifying Dominant and Recessive Traits	Audio CD Program	**CHAPTER TESTING** ■ Chapter Test (blackline master) ▲ Test Generator ■ Assessment Item Listing ■ Scoring Rubrics and Classroom Management Checklists

KEY

MODERN BIOLOGY
- **PE** Pupil's Edition
- **ATE** Teacher's Edition
- ■ Active Reading Guide
- ★ Study Guide

HOLT BIOSOURCES
- Teaching Transparencies
- Laboratory Program

One-Stop Planner CD-ROM
Shows these resources within a customizable daily lesson plan:
- ■ Holt BioSources Teaching Resources
- ■ Active Reading Guide
- ★ Study Guide
- ▲ Test Generator

READING FOR CONTENT MASTERY

Preview: Have students preview the objectives, the introductory paragraph, topic sentences, and captions before reading the chapter.

- **ATE** Active Reading Technique: Brainstorming, p. 203
- ■ Active Reading Guide Worksheet 11-1
- **PE** Summary/Vocabulary, p. 215

- ■ Active Reading Guide Worksheet 11-2
- **PE** Summary/Vocabulary, p. 215

TECHNOLOGY AND INTERNET RESOURCES

CNN PRESENTS Video Segment 6
Teen Discovery

Holt Biology Videodiscs Teacher's Correlation Guide, Lessons 11-1 through 11-2

Audio CD Program, Sections 11-1 through 11-2

internetconnect

SCILINKS NSTA

On-line Resources: www.scilinks.org
The following *sci*LINKS Internet resources can be found in the student text for this chapter:

Topics:
- Genome, p. 204
- Chromosome, p. 210
- Oncogene, p. 214

go.hrw.com

On-line Resources: go.hrw.com
Visit the HRW Web site for a variety of resources related to this chapter. Just type in the keyword HM2 Chapter 11.

Smithsonian Institution®
Visit **www.si.edu/hrw** for additional on-line resources.

CNNfyi.com
Visit **www.cnnfyi.com** for late-breaking news and current event stories selected just for you.

LAB ACTIVITY PLANNING

CHAPTER 11 INVESTIGATION

Modeling Gene Expression in the *Lac* Operon, pp. 218–219

OVERVIEW
In this investigation, students will model the *lac* operon feedback loop found in *Escherichia coli.* The *lac* operon is a well-known mechanism of gene regulation that enables *E. coli* to build the proteins needed for lactose metabolism. Students will use simple materials to create a model that represents the *lac* operon. The model allows students to examine how various genes regulate the *lac* operon. The terms *operator, repressor,* and *inducer* are explained by this investigation.

PREPARATION
1. Each student group will need six large colored beads and three different colors of modeling clay. Purchase the beads and modeling clay from a local toy store, discount store, or a biological supply house, such as WARD'S. You will need about 1 lb of modeling clay (in each of three colors) for a class of students. Each group will make small round beads out of the modeling clay.
2. Provide each student group with a pipe cleaner, labeling tape, and a marking pen. You can purchase pipe cleaners from a craft store, discount store, or biological supply house, such as WARD'S.

Quick Labs

Modeling After-Transcription Control, p. 207
Students will need felt-tip markers, paper, scissors, and tape for this activity.

UNDERSTANDING THE VISUAL

Have students look at the photograph of the fly. Ask them to point out what is unusual about this fly. (It is missing its right eye.) Explain that the missing eye was caused by a mutation in a gene that scientists call *eyeless*. The gene is so named because it produces no eyes when it is damaged. Genes similar to the fly's *eyeless* have been found in a wide variety of vertebrate and invertebrate organisms. In fact, when a gene similar to *eyeless* that is found in mice was inserted into a fly, eyes formed all over the fly's body. Ask students what this experiment could indicate. (Although mouse eyes and fly eyes are very different, the process of developing eyes seems to be similar in both organisms.)

FOCUS CONCEPT

Cell Structure and Function

In this chapter, students will learn that cells are involved in specialized activities that control an organism's behavior. Cells are composed of genes and gene products that control these activities as well as changes in cell structure.

ASSESSING PRIOR KNOWLEDGE

Review the following concepts.

Cell Structure: *Chapter 4*
Have students describe the structural differences between prokaryotes and eukaryotes.

Protein Synthesis: *Chapter 10*
Ask students to describe the process of protein synthesis in both prokaryotes and eukaryotes.

CHAPTER 11

GENE EXPRESSION

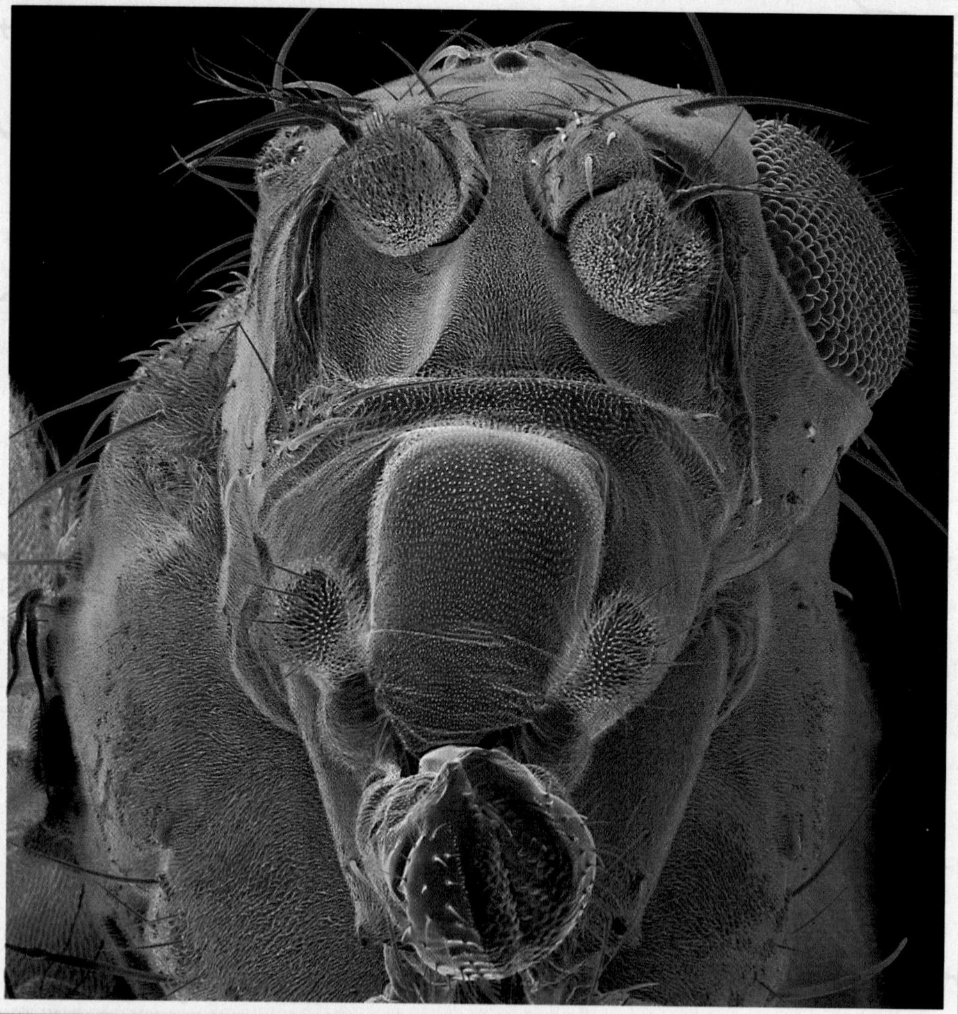

This fruit fly has only one eye due to a mutation in a gene that regulates development.

FOCUS CONCEPT: *Cell Structure and Function*

As you read this chapter, note how gene structure enables prokaryotic and eukaryotic cells to control how and when proteins are produced.

BIOLOGY INTERACTIVE TUTOR **Unit 6—*Gene Expression***
Topics 1–6

CONTROL OF GENE EXPRESSION

Cells use information in genes to build hundreds of different proteins, each with a unique function. But not all proteins are required by the cell at any one time. By regulating gene expression, cells are able to control when each protein is made.

SECTION

11-1

OBJECTIVES

▲ Define the term *gene expression.*

● Describe the regulation of the *lac* operon in prokaryotes.

■ Distinguish between introns and exons.

◆ Describe the role of enhancers in the control of gene expression.

ROLE OF GENE EXPRESSION

Gene expression is the activation of a gene that results in the formation of a protein. A gene is said to be "expressed," or turned "on," when transcription occurs. But cells do not always need to produce all of the proteins for which their genes contain instructions. Recall from Chapter 3 that proteins have many different functions. Some proteins play a structural role. Others are enzymes that act as catalysts in chemical reactions. Mechanisms to control gene expression have evolved to ensure that each protein is produced only when it is needed.

The complete genetic material contained in an individual is called the **genome** (JEE-nohm). By regulating gene expression, cells are able to control which portion of the genome will be expressed and when. Gene expression occurs in two steps, transcription and translation. Gene expression begins when the enzyme RNA polymerase transcribes the DNA nucleotide sequence of a gene into a specific mRNA. During translation, this mRNA then migrates to a ribosome, where it is translated into a specific protein.

Word Roots and Origins

genome

from the words *gene* and *chromosome*

GENE EXPRESSION IN PROKARYOTES

Scientists first studied gene expression in prokaryotes. Much of our initial knowledge of gene expression comes from the work of French scientists Francois Jacob (1920–) and Jacques Monod (1910–1976). Jacob and Monod discovered how genes control the metabolism of the sugar lactose in *Escherichia coli,* a bacterium that lives in the human intestine.

SECTION 11-1

SECTION OVERVIEW

Stability and Homeostasis

Section 11-1 discusses how regulatory genes conserve resources and maintain certain levels of protein production by activating and repressing specific genes.

Matter, Energy, and Organization

In Section 11-1, students will learn how introns and exons influence the production of different gene combinations and therefore different protein structures, depending on the organization of the intron-exon patterns.

ACTIVE READING

Technique: Brainstorming

Pair each student with a partner. Tell students to study pp. 202–205 (Section 11-1) and Figures 11-1 and 11-2 on pp. 204–205. Have student pairs read pp. 202–205 aloud, sharing the reading equally. Then ask the students to answer the following questions:

- What is the subject of this section?
- What is one difference between transcription and translation?
- What is meant by the term *gene expression*?

OVERCOMING MISCONCEPTIONS

Students often assume that all genes are transcribed at the same time in cells. Explain that genes are transcribed only as they are needed so that the cell is able to conserve its energy and resources.

SECTION 11-1

TEACHING STRATEGY

A User-Friendly Operon

Have students develop a strategy for teaching the *lac* operon to other members of the class. Their strategies could include cartoon characters, a story, or a simple flowchart that could make the concept easier to grasp. Have students share their ideas with the class.

CULTURAL CONNECTION

Cultural Dependency on Milk?

From 30 percent to 70 percent of African and Asian adults lack the enzyme lactase and cannot metabolize lactose. Lactose intolerance is less common in Europeans. Biologists believe that this difference may be attributable to the fact that Asian and African diets often do not include the types of dairy products popular in Western diets.

DEMONSTRATION

To demonstrate the turning on and off of genes, plug in a lamp and turn it on and off. Have students compare the lamp to the *lac* operon. Ask the students what parts of the lamp model correspond to the inducer, promoter, operator, and repressor (the lamp switch in the "on" position, electricity, the lamp socket, and the lamp switch in the "off" position, respectively).

INCLUSION *ACTIVITY*

Have students draw and label an operon. Then have them relabel the operon using their own terms (promoter = "start here," operator = "doorway," repressor = "guard," structural genes = "assembly line," etc.).

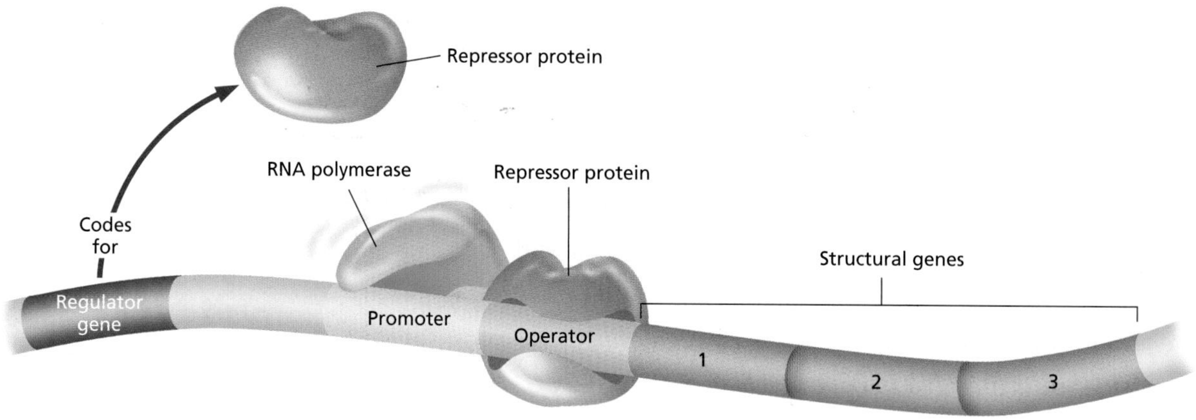

FIGURE 11-1

In the *lac* operon of *E. coli,* three structural genes code for the enzymes needed to utilize lactose. When lactose is absent, a repressor protein attaches to the operator. The presence of the repressor protein prevents RNA polymerase from binding to the structural genes, blocking transcription.

Found naturally in cow's milk, lactose is a disaccharide that is composed of the monosaccharides glucose and galactose. When you drink cow's milk, the presence of lactose stimulates *E. coli* to produce three enzymes. These three enzymes control metabolism of lactose and are adjacent on the chromosome. The production of these enzymes is controlled by three regulatory elements found within the DNA of *E. coli*. These three regulatory elements are as follows:

- **Structural genes** Genes that code for particular polypeptides are called ***structural genes.*** The structural genes studied by Jacob and Monod coded for the enzymes that allow *E. coli* to break down and utilize lactose.
- **Promoter** As you learned in Chapter 10, a promoter is a DNA segment that recognizes the enzyme RNA polymerase and thus promotes transcription.
- **Operator** An **operator** is a DNA segment that serves as a binding site for an inhibitory protein that blocks transcription and prevents protein synthesis from occurring.

The structural genes, the promoter, and the operator collectively form an operon. An **operon** (AHP-uhr-AHN) is a series of genes that code for specific products and the regulatory elements that control these genes. Researchers have found that the clustered arrangement of genes that forms an operon is a pattern that occurs commonly among bacteria. Jacob and Monod named the operon they studied the ***lac* operon** because its structural genes coded for the enzymes that regulate lactose metabolism. The *lac* operon includes the entire segment of DNA required to produce the enzymes involved in lactose metabolism.

In their work with the *lac* operon, Jacob and Monod found that the genes for the enzymes for lactose utilization were expressed *only* when lactose was present. How were the bacteria able to shut off these genes when lactose was absent? Their research showed that gene expression in the *lac* operon exhibits two forms: repression and activation.

internetconnect

SCiLINKS NSTA
TOPIC: Genome
GO TO: www.scilinks.org
KEYWORD: HM204

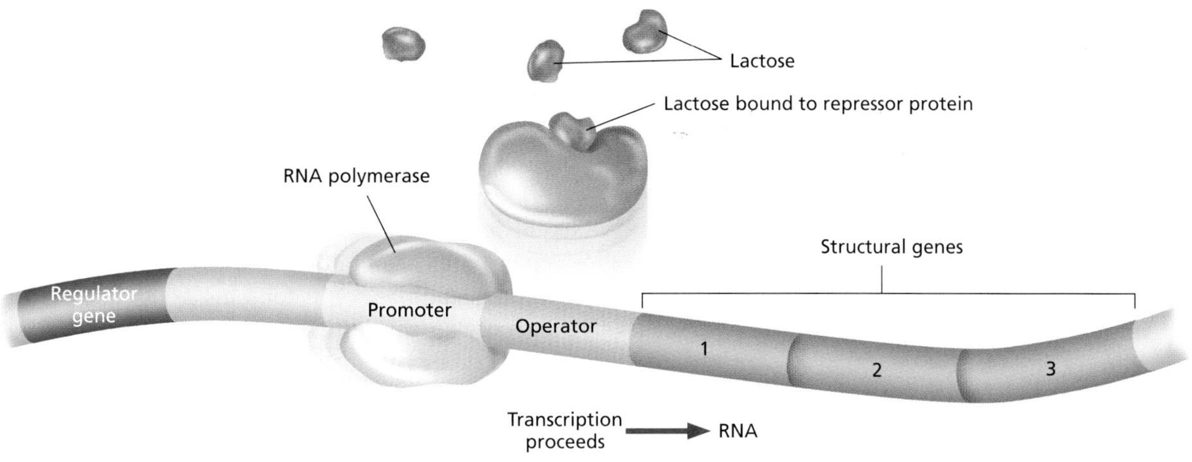

FIGURE 11-2

When lactose is present, it acts as an inducer by binding to the repressor protein and removing it. The removal of the repressor protein allows the transcription of the three structural genes to proceed, producing mRNA.

Repression

In the absence of lactose, a protein called a repressor attaches to the operator. A **repressor protein** is a protein that inhibits a specific gene from being expressed. The attachment of the repressor protein to the operator prohibits RNA polymerase from binding to the structural genes and thus stops transcription from occurring. The blockage of transcription by the action of a repressor protein is called **repression.** Transcription of the structural genes is ultimately controlled by a **regulator gene,** which codes for the production of the repressor protein. The events of repression are summarized in Figure 11-1.

Activation

When lactose is present in the *E. coli* cell, it temporarily binds to the repressor protein on the operator and removes it. The removal of the repressor protein allows RNA polymerase to transcribe the structural genes of the *lac* operon. Since all three structural genes are turned on, all three enzymes required for lactose metabolism are produced. Because it activates, or induces, transcription, lactose acts as an inducer. An **inducer** is a molecule that initiates gene expression. The initiation of transcription by the removal of a repressor protein is called **activation.** Figure 11-2 shows the events that take place when lactose is present in the *E. coli* cell.

The *lac* operon illustrates in simple terms the great advantage of regulating gene expression. Cells of *E. coli* are able to shift between repression and activation, depending on whether lactose is present. Because lactose acts as an inducer, the *lac* operon is transcribed only in the presence of lactose. As a result, lactose induces its own metabolism. When the level of lactose drops, the repressor protein again binds to the operator, shutting off the *lac* operon. The three enzymes used in lactose metabolism are therefore not produced when lactose is not present. By controlling gene expression, *E. coli* conserves resources and produces only those proteins that are needed.

GIFTED ACTIVITY

E. coli can synthesize the amino acid tryptophan. The presence and absence of tryptophan in the cell's growth medium is regulated by the *trp* operon. Like the *lac* operon, the *trp* operon consists of a regulatory gene, a promoter region, an operator region, and structural genes. When the *trp* operon is on, RNA polymerase transcribes structural genes that encode for the enzymatic synthesis of tryptophan. Ask students when RNA polymerase activity would be blocked. Have students describe the steps that take place to block RNA polymerase activity. (RNA polymerase activity is blocked when an abundance of tryptophan is present in the cell's growth medium. The regulatory gene synthesizes a repressor protein that binds to the operator region, thus blocking RNA polymerase's access to the promoter region. This activity turns the operon off so that tryptophan is not produced.)

QUICK FACT

☑ Lactose intolerance is characterized by a person's inability to digest lactose, the principal sugar found in cow's milk and some dairy products. Many people become less tolerant of lactose as they age because the production of lactase, the enzyme that breaks down lactose in their intestines, diminishes. Instead of avoiding dairy products, lactose-intolerant people can treat dairy products with a commercial preparation that breaks down the lactose into disaccharides that can be easily digested.

SECTION 11-1

TEACHING STRATEGY

Gene Expression in Eukaryotes

On a poster board, have students create a sequential diagram that summarizes the steps of gene expression in eukaryotes, beginning with DNA in an active state: (1) inactive DNA is coiled into chromosomes during meiosis; (2) DNA, requiring an active enhancer, uncoils during transcription; (3) introns are removed from pre-mRNA; (4) mRNA moves from the nucleus to the cytoplasm; and (5) active mRNA undergoes translation in the cytoplasm to produce proteins.

RETEACHING *ACTIVITY*

Write a sequence of DNA nucleotides along a strip of adding-machine tape. Then ask students to play the role of the transcriptase and read the base order. Ask students what physical state of the DNA this represents in a eukaryotic cell (uncoiled DNA, known as euchromatin). Then ask students to crumple up a section of the tape. Ask them to read the DNA base order while the tape is still crumpled. (They will be unable to do so.) Ask students how the physical state of the DNA controls gene expression. (Genes will not be expressed because transcription cannot take place.)

CONTENT CONNECTION

Section 10-3: Gene Expression

Transcription is the process in which DNA encodes for RNA. Translation is the process in which RNA encodes for the proteins that carry out cell functions.

GENE EXPRESSION IN EUKARYOTES

Eukaryotes are vastly different from prokaryotes. Their genomes are much larger than those of prokaryotes. In addition, the DNA of eukaryotic cells is located in several individual chromosomes instead of in the single circular chromosome that occurs in prokaryotes. Finally, most eukaryotes are multicellular organisms made of specialized cells. Although each cell type contains a complete set of the organism's genes, only some of these genes are expressed at a given time. Different cell types produce different proteins. Not surprisingly, the control of gene expression in eukaryotes is far more complex than it is in prokaryotes. Although operons are common in prokaryotes, they have not been found in eukaryotes.

Structure of a Eukaryotic Gene

Much of the control of gene expression in eukaryotes occurs at the level of the individual chromosome. In eukaryotes, gene expression is partly related to the coiling and uncoiling of DNA within each chromosome. Recall from Chapter 8 that eukaryotic DNA is organized as fibers of chromatin wrapped around small specialized proteins called histones. Prior to mitosis or meiosis, the DNA and histones coil tightly, forming the structures we recognize as chromosomes. After mitosis or meiosis, certain regions of the DNA coils relax, thus making transcription possible. This uncoiled form, known as **euchromatin** (yoo-KROH-muh-tin), is the site of active transcription of DNA into RNA. However, some portions of the chromatin remain permanently coiled so that their genes can never be transcribed. Thus, the degree to which DNA is uncoiled indicates the degree of gene expression. Figure 11-3 shows transcription as it occurs in a cell.

FIGURE 11-3

Multiple copies of mRNA are made as transcription occurs at several points along this DNA molecule. (11,055×)

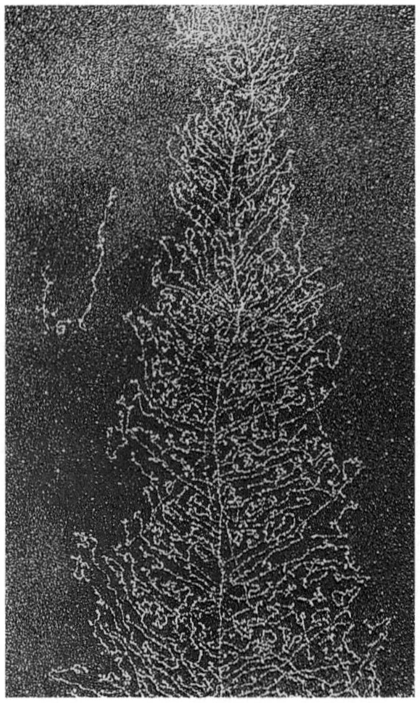

As in the prokaryotes, the promoter is the binding site of RNA polymerase. In the eukaryotic gene, there are two kinds of segments beyond the promoter: introns and exons. **Introns** are the sections of a structural gene that do not code for amino acids and therefore are not translated into proteins. **Exons** are the sections of a structural gene that, when expressed, are translated into proteins.

The benefits of the intron-exon pattern of gene organization are not yet fully understood. However, scientists have noted that the pattern of actively coding portions of DNA interspersed among other noncoding portions of DNA may provide many options for producing different proteins. Perhaps the intron-exon pattern could facilitate the exchange of exons among homologous chromosomes during crossing-over in meiosis. This would result in new combinations of genes, enabling organisms to modify protein structures by replacing one exon with another. If this is the case, the intron-exon pattern of gene organization could serve as an additional source of the genetic diversity that is essential for evolution.

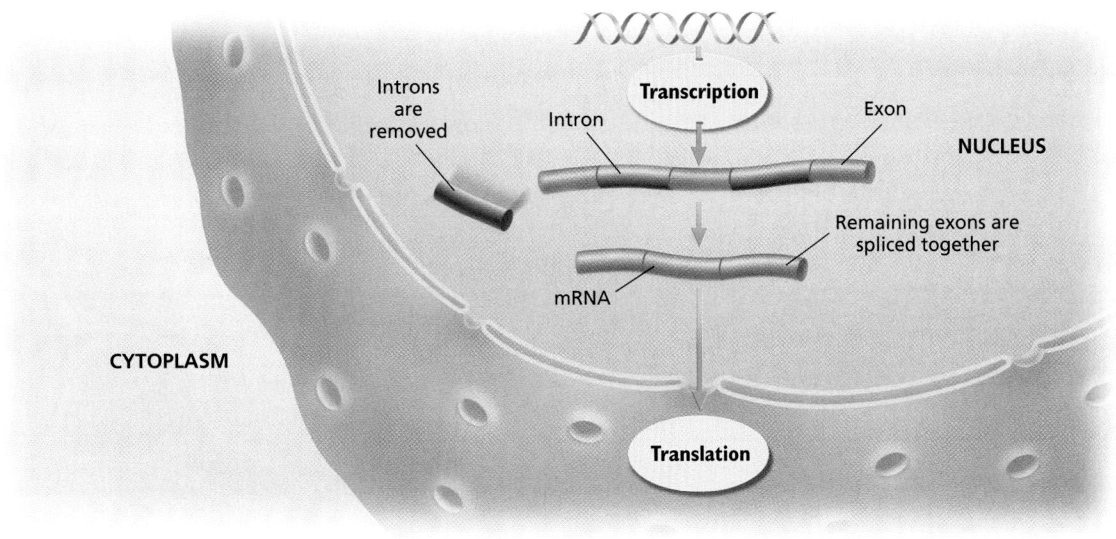

FIGURE 11-4

Both introns and exons are transcribed to form pre-mRNA. Enzymes cut out the introns and join the remaining exons together, forming mRNA.

Control After Transcription

In prokaryotes, transcription and translation occur within the cytoplasm. In eukaryotes, however, transcription occurs in the nucleus, and then mRNA passes through the nuclear envelope and into the cytoplasm, where translation occurs. The physical separation of transcription and translation by the nuclear envelope gives eukaryotes more opportunities to regulate gene expression.

Unlike prokaryotes, eukaryotes can control gene expression by modifying RNA after transcription. When transcription occurs, both introns and exons are transcribed. The result is a large molecule known as pre-mRNA. **Pre-mRNA** is a form of messenger RNA (mRNA) that contains both introns and exons. A molecule of mRNA is formed when introns are removed from pre-mRNA and the remaining exons are spliced (joined) to one another, as shown in Figure 11-4. This occurs when an enzyme splits the pre-mRNA at each end of an intron and joins the exons. The end result is an mRNA containing the exons. The mRNA strand leaves the nucleus and enters the cytoplasm to begin the manufacture of a protein on the ribosomes. When needed, the nucleotides in the removed introns can be used again during the transcription of additional pre-mRNA. Similar RNA splicing occurs following the transcription of transfer RNA and ribosomal RNA.

The removal of introns and splicing of an mRNA molecule has also been found to occur in another way. In the early 1980s, Thomas Cech and his co-workers at the University of Colorado discovered that RNA molecules can act as biological catalysts. The work of Cech and others showed that RNA itself can act as a catalyst to splice introns out of mRNA molecules as they form in the nucleus. Until Cech's discovery, it was thought that all enzymes were proteins.

Quick Lab

Modeling After-Transcription Control

Materials felt-tip markers, paper, scissors, tape

Procedure

1. Write a sentence that contains only three-letter words and makes sense.
2. Hide the words in random places in a long sequence of letters. This sequence should contain random letters and other three-letter words that make no sense in the sentence you are hiding. Print the sequence of letters all the same size, equally spaced, and with no breaks between them.
3. Trade papers with another team. Use scissors to cut out the "introns." Find the message, and reassemble it with tape.

Analysis What represents pre-mRNA in this activity? What represents mRNA?

CRITICAL THINKING

Introns and Exons

Ask students which is more likely to cause a phenotypic change: a mutation in an intron or a mutation in an exon. (exon mutations, because the exons are actually translated into proteins)

Quick Lab

Modeling After-Transcription Control

Time Required 20 minutes

Procedural Tips Have students work in teams. Remind students to print the letters of their sentence and the sequence of letters evenly so that there will be no visual clues to the location of the words they are hiding. For example, if the original sentence is, "The cat ate the rat," students might print it out as follows: ARUNOSRATFTIMCATPQOWTHEJXBOYATEBDGTHEKLAPENUYZ.

The words *boy, run,* and *ape* are nonsense words hidden in the letters above.

As an alternative, word-game tiles or a supply of small, plastic bulletin-board letters can be used. An activity similar to this exercise can be used to demonstrate mutation effects, such as point or frame shift.

Answers to Analysis Pre-mRNA is represented by all the letters of the sequence including the sentence that has been hidden. Pre-mRNA is a form of messenger RNA that contains both introns and exons. The reassembled original sentence represents mRNA after the introns have been removed and the exons have been joined. This is the product that exits the nucleus to be translated.

SECTION 11-1

Language Arts

Jacques Monod was so inspired by the implications of genetics and evolution that he wrote a book, *Chance and Necessity.*

ANSWERS TO SECTION 11-1 REVIEW

1. By controlling gene expression, organisms conserve resources by producing only the proteins they need.
2. The *lac* operon is the segment of DNA on the *E. coli* chromosome that controls the production of enzymes required for lactose utilization.
3. Lactose binds with the repressor molecules, thus freeing the operator regions so that transcription can take place.
4. The loss, addition, or exchange of exons between homologous chromosomes can result in new gene combinations.
5. pre-mRNA contains both introns and exons. mRNA is formed when introns are removed from pre-mRNA and the remaining exons are spliced together.
6. The operator region of a prokaryote is analogous to the enhancer region of a eukaryotic gene. Both operators and enhancers act as a switch that must be turned "on" to activate the expression of a gene.

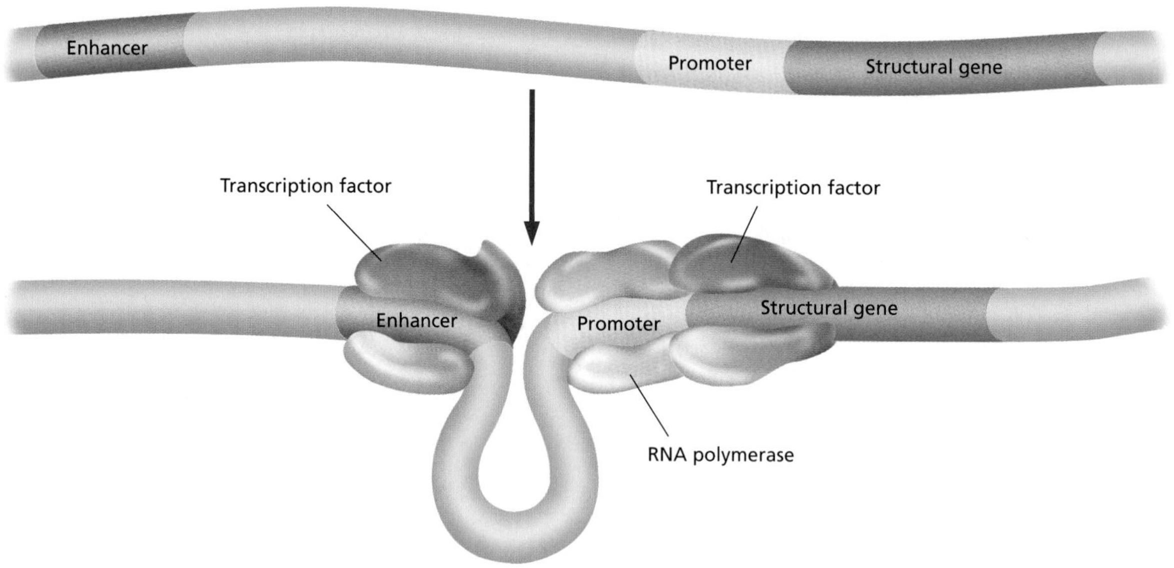

FIGURE 11-5
Many enhancers are located far away from the genes they activate. Transcription factors facilitate transcription by binding to the enhancer and to the RNA polymerase. Bending of the DNA strand brings the enhancer close to the RNA polymerase and the transcription factors associated with it, enabling transcription to begin.

Enhancer Control

Eukaryotic genes on a DNA strand also have noncoding control sequences that facilitate transcription. Such a noncoding control sequence in a eukaryotic gene is called an **enhancer.** An enhancer must be activated for its associated gene to be expressed. Additional proteins, called **transcription factors,** bind to enhancers and RNA polymerase and regulate transcription. Figure 11-5 shows how an enhancer can facilitate transcription, even when it is far away from the gene that it affects.

Activation of enhancers has been studied in the expression of the gene controlling the production of estrogen, a female sex hormone. When estrogen is present in the cell, it binds to a receptor protein in the cytoplasm. The resulting estrogen-receptor protein complex then travels through the pores of the nuclear envelope, entering the nucleus. Once inside the nucleus, the estrogen-receptor protein complex attaches to the enhancer. This activates the enhancer, enabling transcription of specific genes and the synthesis of specific protein molecules to occur.

SECTION 11-1 REVIEW

1. **How is it beneficial for organisms to be able to control gene expression?**
2. **What is the *lac* operon?**
3. **How does lactose affect the functioning of the *lac* operon?**
4. **Describe how the intron-exon pattern of gene organization can serve as a source of genetic diversity.**
5. **Distinguish between mRNA and pre-mRNA.**
6. **CRITICAL THINKING What region of a prokaryotic gene is analogous to the enhancer region of a eukaryotic gene?**

SECTION

11-2

OBJECTIVES

▲ Recognize the relationship between gene expression and morphogenesis.

● Describe the influence of homeotic genes on *Drosophila* development.

■ Summarize the role of the homeobox in eukaryotic development.

◆ List the key characteristics of cancer cells.

▲ Compare and contrast the roles of oncogenes and tumor-suppression genes.

GENE EXPRESSION AND DEVELOPMENT

In Section 11-1 you learned how genes are turned on and off in prokaryotes and eukaryotes. Eukaryotes are far more complex than prokaryotes, and most eukaryotes are multicellular. The control of gene expression plays an important role in the growth of eukaryotes as different cells become specialized to perform different tasks.

CELL DIFFERENTIATION

All multicellular sexually reproducing organisms begin life as a fertilized egg, or zygote. Although every cell in the developing zygote contains the same genes, only a fraction of the genes are expressed. Certain genes are turned on and off as various proteins are needed at different times during the organism's life. For example, as eukaryotes grow, cells become specialized to perform different tasks. Muscle cells specialize in movement, while liver cells specialize in making enzymes that break down fat. The development of cells having specialized function is called **cell differentiation** (DIF-uhr-EN-shee-AY-shun). As organisms grow and develop, organs and tissues develop to produce a characteristic form. This development of form in an organism is called **morphogenesis** (MOR-foh-JEN-uh-sis).

Homeotic Genes

Homeotic (HOH-mee-AHT-ik) **genes** are regulatory genes that determine where certain anatomical structures, such as appendages, will develop in an organism during morphogenesis. Homeotic genes seem to be master genes of development that determine the overall body organization of multicellular organisms. One of the best-known examples of homeotic genes is found in *Drosophila,* the fruit fly. Each homeotic gene of *Drosophila* shares a common DNA sequence of approximately 180 nucleotide pairs. This specific DNA sequence within a homeotic gene regulates patterns of development and is called the **homeobox.** As the *Drosophila* embryo becomes an elongated larva, specific homeoboxes control the morphogenesis of specific regions in the larva. Each of these homeoboxes will also control a specific part of the adult *Drosophila.* As Figure 11-6 shows, a mutation in a homeotic gene can have devastating consequences.

Word Roots and Origins

homeobox

from the Greek *homoio,* meaning "the same"

SECTION 11-2

SECTION OVERVIEW

Cell Structure and Function
In Section 11-2, students will learn that cells are composed of activated genes that influence cells' ability to perform various functions. Morphogenesis is the result of a cell's development into a characteristic form based on its specialized activities.

Stability and Homeostasis
Section 11-2 also discusses how uncontrolled cell proliferation can impose life-threatening outcomes, such as cancer, on an organism.

▶ ENGAGE STUDENTS

Draw or obtain pictures of tadpoles in different stages of development. Remind students that tadpoles begin as fertilized eggs and mature to adult frogs. Ask students to observe the different stages and place them in a logical order of development. Ask why frog development regularly happens in the same pattern. (Genes control the chemical reactions in the cells of the maturing organism.)

SECTION 11-2

TEACHING STRATEGY
Embryo Development

Show students pictures or models of developing embryos from different species. A good source of vertebrate pictures is Kenneth Miller's "One of Us?" in *Life* magazine, November 1996, volume 38. Ask students to identify how the embryos are similar and how they are different. (Characteristics such as having some sort of head are common among the metazoa. This commonality may be due to homeotic genes being highly conserved.)

RECENT RESEARCH
Tumors

Research indicates that cells enclosed in tumors suffer from a loss of oxygen. As cells in tumors grow faster than their blood supply, they begin to lose a significant supply of oxygen, resulting in a condition known as hypoxia. The oxygen-poor environment of tumor cells results in DNA damage. This could explain why multiple-gene mutations and chromosomal abnormalities are associated with tumor cells. This may also explain why tumor cells become more malignant over time.

QUICK FACT

☑ The protein for the homeotic gene that controls antennae development in the fruit fly differs from the protein for one of the homeotic genes of the frog by only one amino acid.

VISUAL LINK
Figure 11-7

Use this figure to point out the genes that control development of anterior and posterior parts of the body in *Drosophila* and mice. Have students identify the genes that are nearly identical in mice and flies.

FIGURE 11-6

(a) The fruit fly shown here is normal. (b) This fruit fly has legs growing out of its head due to a homeotic mutation.

When a homeotic gene is translated, regulatory proteins are formed. It is thought that these proteins regulate development by switching groups of developmental genes on or off. Switching these genes on or off increases or decreases the rates of cell division in various areas of the developing organism. The resultant variation in growth rates in specific areas of the organism produces specific patterns of structural development.

The same or very similar homeobox sequences have been found in homeotic genes of many eukaryotic organisms. It is thought that all organisms may have similar homeoboxes that code for their anatomy. Figure 11-7 shows examples of homologous homeotic genes in *Drosophila* and mice.

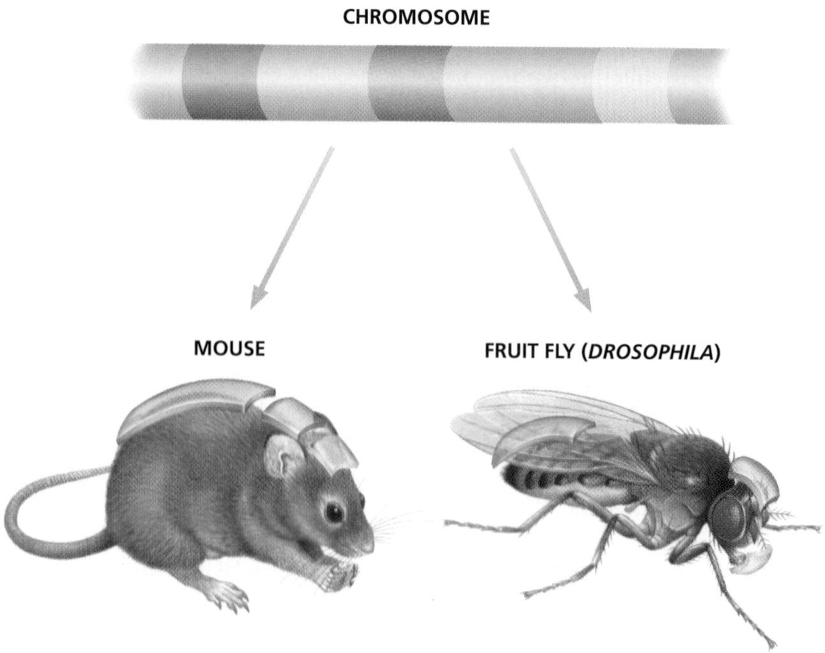

FIGURE 11-7

In many eukaryotes, homeobox sequences govern the development of similar body regions. The gene sequence shown on this chromosome is found in both *Drosophila* and mice. A different color is used to represent each homeobox and the corresponding body region that it controls.

CANCER

A **tumor** is an abnormal proliferation of cells that results from uncontrolled, abnormal cell division. The cells of a **benign** (bi-NIEN) **tumor** remain within a mass. Benign tumors generally pose no threat to life unless they are allowed to grow until they compress vital organs. Examples of benign tumors are the fibroid cysts that occur in a woman's breasts or uterus. Warts are also benign tumors. Most benign tumors can be removed by surgery.

In a **malignant tumor,** the uncontrolled dividing cells invade and destroy healthy tissues elsewhere in the body. Malignant tumors are more commonly known as **cancer. Metastasis** (muh-TAS-tuh-sis) is the spread of cancer cells beyond their original site. When metastasis occurs, the cancer cells break away from the malignant tumor and travel to other parts of the body, where they invade healthy tissue and begin forming new tumors.

Word Roots and Origins

benign

from the Latin *benignus,* meaning "good"

Kinds of Cancer

Malignant tumors can be categorized according to the types of tissues they affect. **Carcinomas** (KAR-si-NOH-muhs) grow in the skin and the tissues that line the organs of the body. Lung cancer, shown in Figure 11-8, and breast cancer are examples of carcinomas. **Sarcomas** (sahr-KOH-muhs) grow in bone and muscle tissue. **Lymphomas** (lim-FOH-muhs) are solid tumors that grow in the tissues that form blood cells. Tumors in blood-forming tissues may cause **leukemia** (loo-KEE-mee-uh), the uncontrolled production of white blood cells. Usually it takes several years for cancer to develop. However, when a vital organ, such as the liver or pancreas, is involved, the symptoms caused by organ dysfunction due to cancer may develop more rapidly.

FIGURE 11-8

(a) Tobacco smoke contains more than 15 known carcinogens. (b) The cancers in this diseased lung are examples of carcinomas. One-third of the cancer deaths in the United States are due to lung cancer. Lung cancer is one of the deadliest forms of cancer; 93 percent of lung cancer patients die within five years after diagnosis. More than 90 percent of lung cancer patients are smokers.

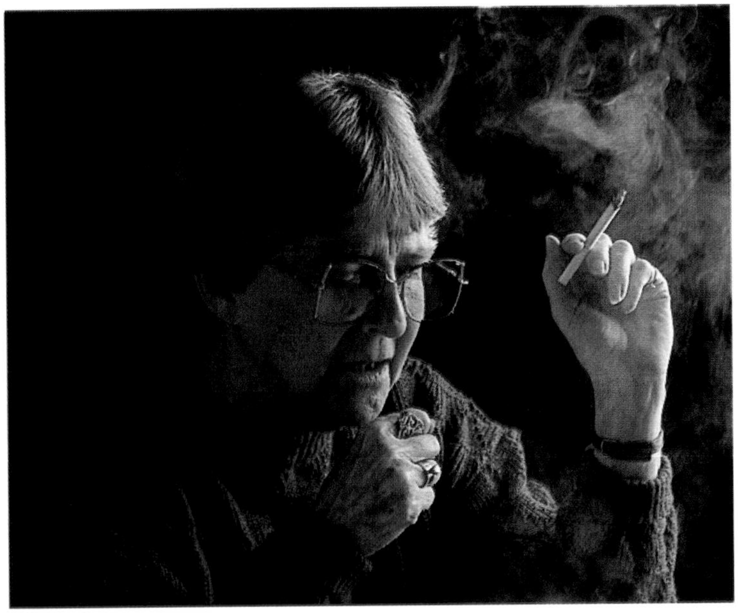

(a)

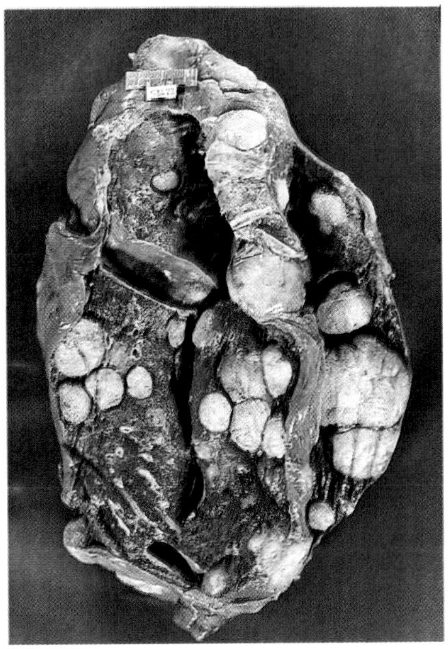

(b)

RECENT RESEARCH

Lung Cancer

Medical research has found that a chemical carcinogen in cigarette smoke, Benzo(a)pyrene, is associated with lung cancer. This chemical has been found to bind to three codon sites, 157, 248, and 273, in the p53 gene. Binding to these sites causes mutations in the p53 gene, thus preventing p53 from stabilizing excessive cell growth. Scientists are currently researching p53 genes to determine whether damaged versions of these genes can be replaced with healthy genes.

QUICK FACTS

- Many of the carcinogens in tobacco are also in marijuana. Marijuana is thought to cause cancers of the mouth and throat.
- Smoking is the leading cause of lung cancer. Smoking is thought to be responsible for 90 percent of lung cancers. It is also linked to cancers of the mouth, pharynx, larynx, esophagus, pancreas, uterine cervix, kidney, and bladder.
- Lung cancer accounts for about 13 percent of all new cancer cases and 29 percent of all deaths from cancer.

DEMONSTRATION

Obtain Materials Safety Data Sheets (MSDS) on carcinogens (e.g., asbestos, benzene, and vinyl chloride) and noncarcinogens (e.g., acetone, isopropyl alcohol, and rosin) from the school's science department, the custodian's office, or the district risk-management office. Have students read both the "Safety Hazard Information" and the "Spill, Leak, and Disposal Procedures." Have students recognize that not all hazardous materials are carcinogenic.

CRITICAL THINKING

Tumor Angiogenesis Factor (TAF)

Cells that produce tumor angiogenesis factor (TAF) form some of the most malignant tumors. TAF affects nearby blood vessels, causing them to grow toward the tumor. Ask students why TAF is important to cancer growth. Based on students' answers, ask them what form of treatment could be used to prevent the growth of this form of cancer. (Cancer cells need adequate nutrition in order to grow and divide. Blood vessels carry nutrition toward, and wastes away from, the cells. Some recent treatments for cancer involve starving the cancer cells of their blood supply.)

MAKING CONNECTIONS

Mathematics

Give the students a chess board and some rice. Have them place one grain of rice on the first square, two on the second square, and so on, doubling the number of grains on each square. After they complete the first eight squares, ask them to predict how much rice will be on the last square ($2^{63} = 9.2 \times 10^{18}$ grains of rice). Have students explain how this activity models cancerous growth. (The cancer cells keep multiplying and eventually overwhelm the organism.)

GIFTED ACTIVITY

Have students contact a local branch of the American Cancer Society (1-800-227-2345) or a cancer-survivors support group in order to research recent developments in cancer treatments. Students may want to consider topics such as early detection and new methods of delivering chemotherapy.

Eco Connection

Secondhand Tobacco Smoke

In 1992, the Environmental Protection Agency (EPA) declared secondhand tobacco smoke, also called environmental tobacco smoke, to be a human carcinogen. According to the EPA standards, secondhand smoke contains more than 4,000 chemical compounds, including four known human carcinogens and several other probable human carcinogens. In fact, the air in an enclosed room of smokers could contain up to six times the air pollution of a busy highway. Thousands of nonsmokers die of lung cancer each year as a result of breathing secondhand tobacco smoke. There is no safe level of tobacco smoke.

As more is learned about the contents and effects of tobacco smoke, regulations to protect people from secondhand smoke are being enacted across the United States. In fact, smoking indoors is now prohibited in many public places.

FIGURE 11-9

The hazardous-chemical workers shown in this photograph are removing asbestos from a refinery. Because it is heat resistant and flexible, asbestos has been used to insulate buildings. When inhaled in large amounts over long periods of time, asbestos particles have been shown to cause lung cancer.

Cancer and the Cell Cycle

In normal cells, the frequency of cell division is governed by several factors. A cell must receive adequate nutrition before it can divide. Also, a cell must be attached to other cells, to a membrane, or to fibers between cells in order to divide. (Not all cells, of course, are destined to divide. Recall that cells of certain tissues, such as nervous tissue, cease dividing once they are formed.) Normal cells will stop dividing when they become too crowded, usually after 20 to 50 cell divisions. Cancer cells, however, continue to divide even when they are very densely packed, seemingly ignoring the normal cellular message to stop dividing. They also continue dividing after they are no longer attached to other cells, a trait that facilitates the spread of cancer cells throughout the body.

Causes of Cancer

What triggers the uncontrolled cell division that characterizes cancer? In normal cells, cell division is governed primarily by genes that regulate cell growth and division. Such genes code for **growth factors,** regulatory proteins that ensure that the events of cell division occur in the proper sequence and at the correct rate. Mutations that alter the expression of genes coding for growth factor proteins can lead to cancer. Such mutations can occur spontaneously but are more likely to occur as a result of the organism's exposure to carcinogens. A **carcinogen** (kahr-SIN-oh-jen) is any substance that increases the risk of cancer. Well-known carcinogens include tobacco, asbestos, and ionizing radiation, such as X rays or ultraviolet light from the sun. For example, cigarette tobacco has been found to be the cause of 90 percent of all lung cancers. Most carcinogens are **mutagens** (MYUT-uh-jens), agents that cause mutations to occur within a cell. Figure 11-9 shows hazardous-chemical workers removing the carcinogen asbestos from a refinery.

Whether a person actually develops cancer seems to depend on many factors. Some families exhibit higher-than-average rates of certain cancers, leading researchers to suspect a genetic predisposition to these types of cancer. With regard to cancers caused by mutagens, the number of exposures to the carcinogen and the amount of carcinogen in each exposure are significant factors. Mutations in gametes (egg or sperm cells) are especially important because these mutations are passed along to offspring.

Usually, more than one mutation is needed to produce a cancer cell. Perhaps this helps to explain why the cancer risk increases with the number of exposures to carcinogens and with the age of the individual. The longer an individual lives, the more mutations he or she will accumulate. Heightened awareness of the causes of cancer, combined with improved detection and treatment of the disease, has resulted in a decline in the number of deaths caused by cancer. In the United States, the cancer death rate fell by about 3 percent between the years 1991 and 1995, according to the National Cancer Institute.

Research Notes

The War on Cancer

Cancer, the leading cause of death in the United States, is a genetic disease caused by mutations in the genes that control cell division. Two kinds of genes that control cell division are proto-oncogenes and tumor-suppressor genes. Proto-oncogenes stimulate cell division, while tumor-suppressor genes inhibit cell division.

Under normal circumstances, these genes work together, enabling the body to replace and repair cells. However, a mutation in either a proto-oncogene or a tumor-suppressor gene can cause a cell to begin dividing uncontrollably. Such a cell may quickly become 2 cells, then 4, then 8, then 16, eventually forming a malignant tumor. The malignant tumor invades healthy tissues, often by spreading into the lymph tissues and blood vessels, which can transport cancer cells to distant sites in the body. A malignant tumor that is not destroyed quickly may become established in vital organs and interfere with their functions.

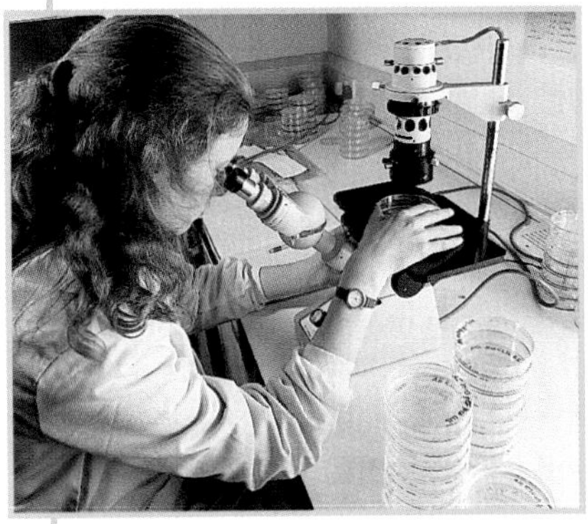

This woman joins other researchers around the world in the search for new cancer treatments.

One strategy for combating cancer is to stop the cell cycle of cancer cells so that they do not continue to divide. Most anticancer drugs interfere with the process of cell division at some step in the cell cycle. Anticancer drugs known as topoisomerase-1 inhibitors prevent the uncoiling of DNA that is necessary for replication. Another class of drugs works by damaging the cell's DNA so that it cannot be copied. Enzymes that monitor DNA replication will activate a series of steps that cause the cell to self-destruct if the DNA is too damaged for enzyme repair. However, these traditional anticancer drugs prevent cell division in all dividing cells, not just in cancer cells; thus, they damage healthy tissues too. A new class of anticancer drugs called farnesyltransferase inhibitors have been shown to target tumor cells but not normal cells.

Treating cancer cells based on their genetic makeup may also be helpful. For example, many cancer cells lack both normal copies of the tumor-suppressor gene p53. Cancer cells that have normal p53 genes tend to be very sensitive to chemotherapy, while those with mutant p53 genes often do not respond to chemotherapy treatment. Using a computer database that contains information about 350,000 potential anticancer compounds, oncologists have identified 112 possible compounds that may be effective against cancers associated with mutant p53 genes.

Simply keeping existing tumors under control is another way that cancer can be treated effectively. Malignant tumors typically stimulate their own growth by constructing a nourishing supply of blood vessels—a process called angiogenesis. Scientists know of more than 20 compounds that prevent angiogenesis. Currently, several of these compounds are being tested on human patients. Perhaps these compounds will be useful in slowing down or even stopping the growth of small clusters of cancer cells that are missed by other cancer treatments.

Combinations of several cancer therapies seem to be the best approach to fighting the war against cancer for now. By combining therapies, physicians may be able to eliminate tiny clusters of cancer cells that could rebound and kill patients. For example, combining chemotherapy and radiation has been shown to improve the condition of patients with advanced cancer of the nose and throat. Three years after treatment, 65 percent of patients who received both therapies were still alive, compared with a 24 percent survival rate among patients who received only radiation. In the coming years, cancer is likely to be considered a serious chronic disease. Like the chronic diseases diabetes and hypertension, perhaps cancers will be controlled with treatment for many years.

CHAPTER 11 Research Notes

BACKGROUND

Although the causes of cancer are diverse and involve the alteration of genes that control cell division, researchers are finally beginning to understand how cancer cells suddenly develop from normal cells. As a matter of fact, a heightened awareness of the causes of cancer, combined with improved detection and treatment of the disease, has resulted in a decline in the number of deaths caused by cancer.

Due to the several forms of detection and treatment that are constantly being implemented in combating this disease, physicians believe that although there may never be a complete cure for cancer, they are getting closer to controlling its gene activity. And as researchers better understand the function of normal cell activity, it becomes more promising that full-blown cancer will be controllable with long-term treatment.

DISCUSSION

1. List the different forms of treatment that are being used to combat the spread of cancer (stopping the cell cycle of cancer, treating cancer cells based on their genetic makeup, stimulating an immune attack against cancer cells, controlling tumors, chemotherapy, and radiation).
2. What may be the best approach to eliminating cancer cells? (a combined treatment of chemotherapy and radiation)

FURTHER READINGS

1. Fackelmann, Kathleen. "Diverse Strategies to Vanquish Cancer." *Science News,* May 1997, pp. 274–275. This article discusses three different approaches that researchers are using in an attempt to control the spread of cancer.

CRITICAL THINKING

Oncogene Activation

Ask students how a virus might activate a host oncogene. (The virus might carry a promoter region that would allow the RNA transcriptase to transcribe the oncogene.)

ANSWERS TO SECTION 11-2 REVIEW

1. The control of gene expression allows certain genes to be expressed, resulting in the formation of particular structures.
2. Homeotic genes determine where certain anatomical structures, such as appendages, will develop in *Drosophila.*
3. The presence of the homeobox suggests that eukaryotes have descended from a common ancestor.
4. Cancer cells divide when they are densely packed. Cancer cells continue to divide even when they are no longer attached to other cells. Cancer cells do not stop dividing after 20 to 50 cell divisions, but continue dividing indefinitely.
5. Normally, proto-oncogenes code for proteins that regulate the rate of the cell cycle, and tumor-suppressor genes code for proteins that prevent the uncontrolled rate of the cell cycle. A mutation could cause a proto-oncogene or a tumor-suppresser gene to mutate into an oncogene.
6. The ovaries and testes contain rapidly dividing cells that will become egg and sperm cells, respectively. A mutation due to X-ray exposure could thus be passed on to offspring.

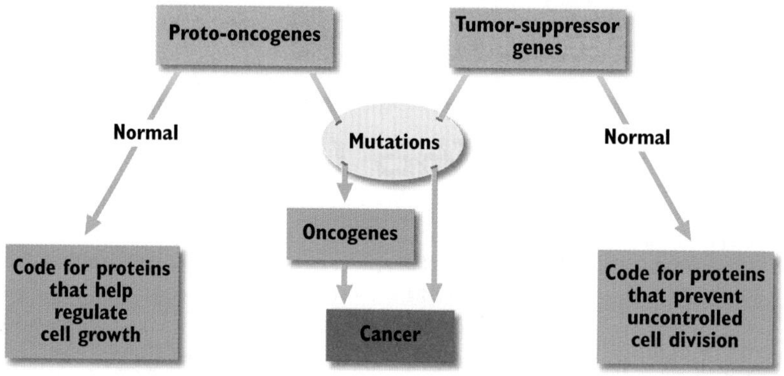

FIGURE 11-10
Mutations in proto-oncogenes or tumor-suppressor genes can destroy normal gene functioning, possibly resulting in cancer. A mutation in a proto-oncogene can cause it to become an oncogene, a gene that triggers cancer.

Oncogenes

An **oncogene** (AHN-koh-jeen) is a gene that causes cancer or other uncontrolled cell proliferation. Oncogenes begin as normal genes called **proto-oncogenes** that control a cell's growth and differentiation. Normally, proto-oncogenes code for proteins that regulate the rate of the cell cycle by controlling cell growth, cell division, and the ability of cells to adhere to one another. A mutation in a proto-oncogene may cause it to produce more protein or a protein that seems to be unusually active in triggering cell division. The rate of the cell cycle then increases, and cancer occurs as a result.

Other human genes suppress tumor formation. **Tumor-suppressor genes** code for proteins that prevent the uncontrolled rate of cell division. When tumor-suppressor genes mutate, the proteins for which they code are either defectively expressed or not expressed at all, causing a predisposition to cancer. Figure 11-10 summarizes the actions of oncogenes and tumor-suppressor genes.

internet**connect**
SCILINKS NSTA
TOPIC: Oncogene
GO TO: www.scilinks.org
KEYWORD: HM214

Viruses and Cancer

Certain viruses can cause cancer in plants and animals. Many viral genes are actually oncogenes. Viruses can also stimulate uncontrolled growth in host cells by causing mutations in proto-oncogenes or tumor-suppressor genes, thus accelerating the rate of cell division in the host cell. Or they may activate the cell's own oncogenes. Viruses have been found to cause various kinds of leukemia.

SECTION 11-2 REVIEW

1. How can morphogenesis be affected by the control of gene expression?
2. What is the role of homeotic genes in *Drosophila*?
3. What does the presence of the homeobox among eukaryotes suggest about evolution?
4. List three ways in which cancer cells differ from normal cells.
5. Describe how mutations in proto-oncogenes or tumor-suppressor genes could cause cancer.
6. CRITICAL THINKING Why might X rays be more dangerous to an ovary or a testis than to muscle tissue?

CHAPTER 11 REVIEW

SUMMARY/VOCABULARY

11-1

- Gene expression is the activation of a gene that results in the formation of a protein. Only a fraction of any cell's genes are expressed at any one time.
- Structural genes code for a particular product. Regulatory elements (a promoter and an operator) regulate the transcription of structural genes. In prokaryotes, the structural genes, the promoter, and the operator collectively form an operon.
- A promoter is the segment of DNA that recognizes the enzyme RNA polymerase and thus promotes transcription. An operator is the segment of DNA that can block transcription and thus prevent protein synthesis from occurring.
- In the bacterium *E. coli,* the *lac* operon codes for the three enzymes required for lactose metabolism.
- A repressor protein can inhibit a specific gene from being expressed. The blockage of transcription by the action of a repressor protein is called repression. A regulator gene controls the expression of a particular gene by coding for a repressor protein.
- An inducer is a molecule that initiates gene expression. In *E. coli,* lactose serves as an inducer.
- Eukaryotes do not have operons. The genomes of eukaryotes are larger and more complex than those of prokaryotes.
- Eukaryotic genes are organized into noncoding sections called introns and coding sections called exons.
- An enhancer must be activated for a eukaryotic gene to be expressed. Transcription factors initiate transcription by binding to enhancers and RNA polymerases.

Vocabulary

activation (205)
enhancer (208)
euchromatin (206)
exon (206)
gene expression (203)
genome (203)
inducer (205)
intron (206)
lac operon (204)
operator (204)
operon (204)
pre-mRNA (207)
regulator gene (205)
repression (205)
repressor protein (205)
structural gene (204)
transcription factor (208)

11-2

- The development of specialized cells is called cell differentiation. The development of form in an organism is called morphogenesis. Both cell differentiation and morphogenesis are governed by gene expression.
- Homeotic genes are regulatory genes that determine where anatomical structures will be placed during development. Within each homeotic gene, a specific DNA sequence known as the homeobox regulates patterns of development. The homeoboxes of many eukaryotic organisms appear to be very similar.
- Cancer is the uncontrolled growth of abnormal cells. Cancerous cells may form a tumor. Tumors may be benign or malignant.
- A carcinogen is any substance that increases the risk of cancer. Mutagens are substances that cause mutations.
- Mutations of proto-oncogenes or tumor-suppressor genes may lead to cancer.
- Some viruses can cause cancer.

Vocabulary

benign tumor (211)
cancer (211)
carcinogen (212)
carcinoma (211)
cell differentiation (209)
growth factor (212)
homeobox (209)
homeotic gene (209)
leukemia (211)
lymphoma (211)
malignant tumor (211)
metastasis (211)
morphogenesis (209)
mutagen (212)
oncogene (214)
proto-oncogene (214)
sarcoma (211)
tumor (211)
tumor-suppressor gene (214)

CHAPTER 11 REVIEW ANSWERS

REVIEW

1. An operator is a DNA segment that serves as a binding site for an inhibitory protein that blocks transcription. A promoter is a DNA segment that recognizes the enzyme RNA polymerase and thus promotes transcription.
2. A gene is a short segment of DNA that codes for a polypeptide or protein. A genome is the complete genetic material in an individual.
3. In eukaryotes, an exon is a section of a structural gene that, when expressed, codes for a protein. An intron is a section of a structural gene that does not code for a protein. After being transcribed into mRNA, the introns must drop out before the mRNA leaves the nucleus and protein synthesis begins.
4. A gene is a short segment of DNA that codes for a polypeptide or protein. A homeotic gene is a regulatory gene that determines where certain anatomical structures, such as appendages, will develop in an organism.
5. A carcinogen is any substance that increases the risk of cancer. A mutagen is a substance that causes mutations to occur within a cell. (Most carcinogens are mutagens.) An oncogene is a gene that induces cancer or other uncontrolled cell proliferation.

6. b
7. d
8. a
9. c
10. a
11. c
12. b
13. d
14. a
15. c

16. The *lac* operon was studied in *E. coli.* This bacterium lives in the human intestines.
17. The *lac* operon shuts off when lactose is not present in the cell and the repressor molecule attaches to the operator.

ANSWERS

CONTINUED

18. Because protein synthesis requires materials and energy, it is more efficient to make the enzymes needed for lactose utilization only when lactose is in the cell.
19. After estrogen binds to a receptor protein in the cell cytoplasm, the estrogen-receptor protein complex enters the nucleus; there it attaches to an enhancer, which initiates transcription.
20. Cell differentiation is the development of a specialized function in a cell. Morphogenesis is the development of form in an organism.
21. When a malignant tumor metastasizes, cancerous cells break away from the original tumor and form new tumors in other parts of the body.
22. tobacco, asbestos, ultraviolet light
23. In some families, certain forms of cancer appear at higher-than-normal frequencies, suggesting that some family members have inherited an unusually high risk for developing a particular form of cancer.
24. The presence of identical or very similar homeoboxes among many different eukaryote species suggests a common evolutionary origin for eukaryotes.
25. a. The promoter recognizes RNA polymerase and promotes transcription. The operator serves as a binding site for the repressor protein. The structural genes encode for particular polypeptides. **b.** When genes are able to be transcribed by RNA polymerase, they are considered to be turned "on." **c.** Yes.

CRITICAL THINKING

1. The results are probably correct because different genes were expressed in the brain and liver tissues, resulting in the production of two types of mRNA.
2. The genes that control the enzymes are switched off or have

CHAPTER 11 REVIEW

REVIEW

Vocabulary

Explain the difference between the terms in each of the following sets:

1. operator, promoter
2. gene, genome
3. intron, exon
4. gene, homeotic gene
5. carcinogen, mutagen, oncogene

Multiple Choice

6. A repressor protein is coded for by a(n) (a) structural gene (b) regulator gene (c) promoter (d) enhancer.
7. Active DNA transcription in eukaryotes occurs within (a) an operon (b) an operator (c) an intron (d) euchromatin.
8. When present in an *E. coli* cell, lactose (a) binds with a repressor (b) stops transcription from occurring (c) binds with mRNA (d) acts as a regulator.
9. In eukaryotes, introns and exons are interspersed regions of a(n) (a) protein (b) DNA molecule (c) RNA molecule (d) transcription factor.
10. Enhancers (a) initiate transcription (b) join amino acids (c) unwind DNA (d) transport proteins into the nucleus.
11. The control of gene expression enables organisms to (a) reproduce more quickly (b) remove mutations from their DNA (c) produce proteins only when needed (d) form new combinations of genes.
12. Unlike eukaryotes, prokaryotes (a) have introns and exons (b) lack nuclei (c) have chromosomes of RNA (d) have enhancer sequences.
13. In eukaryotic cells, pre-mRNA contains (a) introns only (b) exons only (c) neither introns nor exons (d) both introns and exons.
14. A malignant tumor found in bone or muscle tissue is an example of (a) a sarcoma (b) leukemia (c) a carcinoma (d) a lymphoma.
15. The transformation of a normal gene into an oncogene may result in (a) a loss of cell division (b) the release of new viruses (c) cancer (d) the transcription of introns.

Short Answer

16. In which organism did Jacob and Monod study the *lac* operon? Where is this organism commonly found?
17. What causes the *lac* operon to shut off? Why can this mechanism be considered a feedback mechanism?
18. How does *E. coli* benefit by making enzymes to utilize lactose only when lactose is in the cellular environment?
19. Describe the events that take place after estrogen binds to a receptor protein in the cytoplasm.
20. How does cell differentiation differ from morphogenesis?
21. What happens when a malignant tumor undergoes metastasis?
22. Name three commonly known carcinogens.
23. What does it mean for a person to be "genetically predisposed" to cancer?
24. What does the study of homeoboxes suggest regarding the evolutionary relationships among eukaryotes?
25. Study the diagram of the *lac* operon shown below.
    a. Describe the role of the following elements shown in the diagram: promoter, operator, structural genes.
    b. What does it mean to say that a gene is turned "on"?
    c. Are the structural genes turned "on" in the diagram of the *lac* operon shown below?

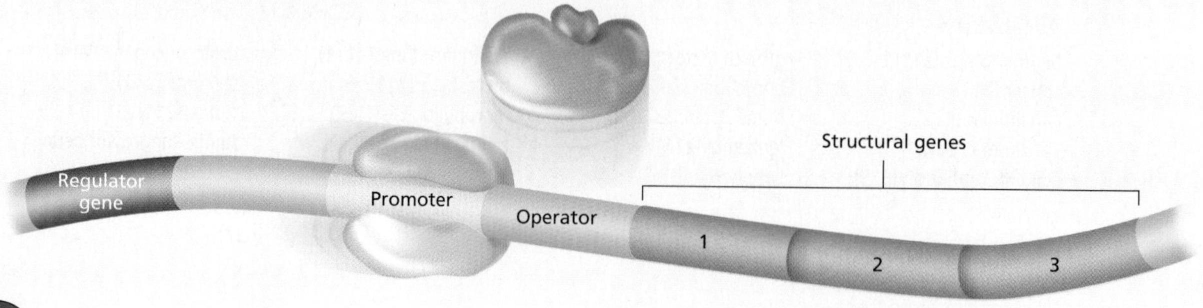

CRITICAL THINKING

1. A molecular biologist isolates mRNA from the brain and liver of a mouse and finds that the two types of mRNA are different. Can these results be correct, or has the biologist made an error? Explain your answer.
2. Kwashiorkor is a disease in children caused by a diet high in carbohydrates but lacking in complete protein. When children with kwashiorkor are suddenly put on a diet rich in protein, they may become very ill with ammonia poisoning, and some even die. The high level of ammonia in their blood is due to the inadequate metabolism of protein. What does this tell you about the enzymes that metabolize protein?
3. Mutations may occur in gametes or in body cells. In which cell type is a mutation likely to be a source of genetic variation for evolution? Why?
4. *Drosophila* feed on fermenting fruit, which often contains a large amount of alcohol. If *Drosophila* are fed a diet that has a high alcohol content, there is an increase in the amount of dehydrogenase, an enzyme that metabolizes alcohol, in the digestive tract. What does this increase tell you about the enzyme?
5. The graphs at right show the number of cigarettes smoked per capita per year since 1900 and the annual incidence of lung cancer among men and women. What is the relationship between the number of cigarettes smoked and the incidence of lung cancer? Why do you think the incidence of lung cancer among women has increased sharply in the last 30 years while the incidence of lung cancer among men has remained relatively stable during this period?

Cigarette Smoking and Lung Cancer

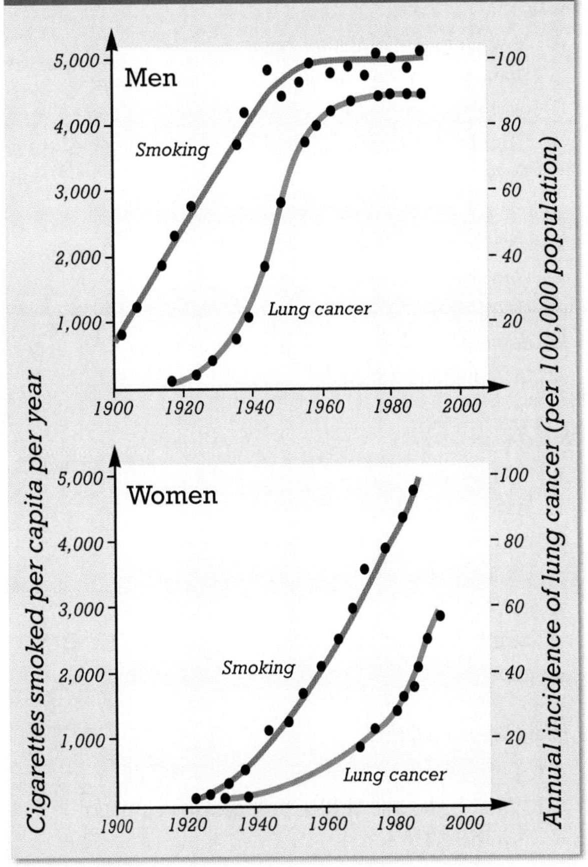

EXTENSION

1. Using your library, research forms of human cancer associated with viruses, such as leukemia (human T lymphotrophic virus), liver cancer (hepatitis B virus), Burkitt's lymphoma (Epstein-Barr virus), and cervical cancer (human Papillomavirus). Write a report on the symptoms, possible treatments, and prognoses of patients with these forms of cancer.
2. Construct a chart in which you compare and contrast the qualities in bacteria and fruit flies that make them ideal experimental tools for the scientist. Make sure to include similarities as well as differences between the two species.
3. Read "Got Cancer Killers?" in *Discover,* June 1999, on page 68. Write a short report about the potential cancer cure being researched at Lund University, in Sweden. Explain how the cancer-killing protein alpha-lactalbumin protects nursing infants. Describe the next step toward making a treatment available to cancer victims.

ANSWERS

CONTINUED

very low activity when the diet lacks adequate protein. The genes are switched on by a high-protein diet, but the earlier lack of protein temporarily results in inadequate protein metabolism.

3. A mutation in gametes is likely to be a source of genetic variation because gametes pass on the mutation to the next generation when forming a zygote.

4. The expression of the gene that codes for dehydrogenase is controlled by the amount of alcohol present.

5. The incidence of lung cancer increased as the number of cigarettes increased. The incidence of lung cancer in women has increased sharply in the last 30 years because the number of women who smoke cigarettes has increased sharply during this period. The incidence of lung cancer in men has remained relatively stable during this period because the rate of smoking among men has not increased.

EXTENSION

1. Answers will vary, but students should accurately compare and contrast the four cancers listed.

2. Charts should examine factors such as generation time, ease of culturing, lab equipment required, ease of manipulation, cost, etc.

3. Human mothers' milk has been found to contain a protein (alpha-lactalbumin or alpha-lac) that kills cancer cells that are exposed to it. Infancy is marked by extremely rapid growth. Some of an infant's rapidly dividing cells may grow out of control. Alpha-lac targets cancer cells plus all kinds of rapidly growing cells, leaving stable cells alone. Alpha-lac also helps an infant's immune system to develop. The next step in the research plan is to test the protein on human subjects.

CHAPTER 11
INVESTIGATION

TIME REQUIRED

One 50-minute class period

SAFETY PRECAUTIONS

Remind students to take precautionary measures to avoid pricking their finger or poking their eyes with the pipe cleaner.

QUICK REFERENCES

Lab Preparation Notes are found on page 201B.

Holt BioSources provides a Teaching Resources CD-ROM that contains "Using Gowin's Vee in the Lab" and "Scoring Rubrics."

PROCEDURAL TIP

Students may want to create a key to keep track of what the color-coded pieces of clay represent in their models.

ANSWERS TO BACKGROUND

1. A gene is a segment of DNA that codes for a particular protein.
2. RNA polymerase is the enzyme that manufactures mRNA.
3. Protein synthesis occurs in the cell cytoplasm. The nucleic acid mRNA transports the genetic code from the DNA in the nucleus to the cell cytoplasm.
4. Ribosomes assemble amino acids in the correct sequence.
5. The operator serves as a binding site for an inhibitory protein that blocks transcription. The promoter is a segment of DNA that recognizes DNA polymerase and enables transcription to begin. The structural genes within the *lac* operon code for the enzymes that *E. coli* needs to utilize lactose.
6. Lactose binds to the repressor protein and removes it, causing transcription of the structural genes to proceed. When lactose is absent, the repressor protein remains bound to the operator and transcription cannot occur.

CHAPTER 11 INVESTIGATION

Modeling Gene Expression in the *Lac* Operon

OBJECTIVES

- Make a model of the *lac* operon.
- Demonstrate the mechanisms that regulate gene expression in the *lac* operon of *Escherichia coli.*
- Simulate the transcription of the structural genes in the *lac* operon.

PROCESS SKILLS

- comparing and contrasting
- identifying
- demonstrating
- manipulating a model

MATERIALS

- pipe cleaner
- large colored beads, 6
- colored modeling clay in three colors
- labeling tape
- marking pen
- pencil
- paper

Background

1. Define *gene.*
2. What is the role of RNA polymerase in protein synthesis?
3. Where does protein synthesis occur? What is the function of mRNA in protein synthesis?
4. What is the role of ribosomes during protein synthesis?
5. What are the roles of the operator, promoter, and structural genes within the *lac* operon?
6. How does the presence or absence of lactose affect the *lac* operon?
7. What is a regulator gene?

PART A Making a Model of the *Lac* Operon

1. In this investigation, you will use the materials provided to make a model of a *lac* operon. Allow the pipe cleaner to represent the portion of DNA that constitutes the *lac* operon.
2. Thread the pipe cleaner through three beads of similar size, shape, and color. These three beads represent the structural genes of the *lac* operon.
3. Add one bead to represent the operator portion of the *lac* operon. Also add beads to represent the promoter and the regulatory gene, respectively.
4. Using labeling tape and a marking pen, label each of the beads you have placed on the pipe cleaner. This represents a model of the *lac* operon.
5. Compare the sequence of the labeled beads on the pipe cleaner with the sequence of segments in the diagram of the *lac* operon in Figure 11-1. When your model of the *lac* operon correctly reflects the parts of the *lac* operon in the figure, proceed to Part B.

PART B The *Lac* Operon When It Is Turned Off

6. Choose one color of modeling clay to represent the enzyme RNA polymerase and choose another color to represent the repressor molecule. Use the modeling clay to mold an RNA polymerase molecule and a repressor molecule.
7. Using the molecules you made out of clay in step 6, modify your model of the *lac* operon so that it shows the *lac* operon when it is turned off.
8. In your lab report, draw your model of the *lac* operon when it is turned off. Label all parts of your drawing. How does the presence of the repressor molecule prevent transcription of the structural genes?

PART C The *Lac* Operon When It Is Turned On

9. Choose a third color of modeling clay to represent the inducer molecule. Use the modeling clay to form an inducer molecule.
10. Using the inducer molecule you made out of clay, modify your model of the *lac* operon so that it shows the *lac* operon when it is turned on.
11. Simulate the activation of the *lac* operon and the transcription of the structural genes.
12. In your lab report, prepare a diagram of your model that shows the expression of the structural genes in the *lac* operon. Include ribosomes and mRNA in your diagram. Label all parts of your diagram.
13. The graphic organizer below shows the sequence of steps that occurs after lactose enters *E. coli* cells. Copy this graphic organizer in your lab report. Complete the graphic organizer by describing what takes place during step 2 and step 3. Explain how the end product affects the events shown in the graphic organizer.
14. Clean up your materials before leaving the lab.

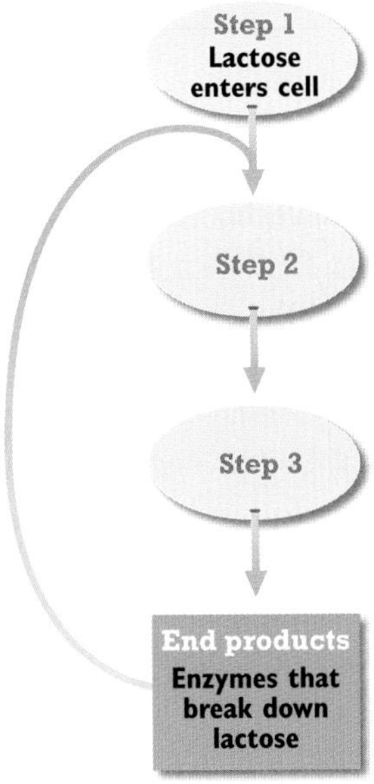

Analysis and Conclusions

1. Compare the processes of repression and activation.
2. What substance serves as an inducer in the *lac* operon?
3. How might a mutation in the regulator gene affect the *lac* operon?
4. Look at the diagram you made in step 12. Refer to your diagram, and predict what will happen when the inducer is no longer present.
5. How would the loss of the promoter site from the operon affect the production of the enzymes needed to utilize lactose?
6. In homes and apartments, a consistent temperature is maintained by means of a thermostat, which regulates when heating (or air conditioning) is turned on or off. In what way does the *lac* operon function like a thermostat?
7. Biological processes often take place in a series of sequential steps called a biochemical pathway. Many biochemical pathways are controlled by feedback inhibition. In feedback inhibition, a pathway's end product affects an earlier step in the pathway and causes the pathway to stop. Explain why the *lac* operon in *E. coli* is considered an example of feedback inhibition.

Further Inquiry

1. Use classroom or library references to find additional examples of feedback inhibition in biology. Describe why models of feedback inhibition are sometimes called feedback loops.
2. The products of the *lac* operon are produced when lactose is present. In this way, the presence of a specific molecule stimulates transcription of the structural genes. In contrast, some operons are repressed when a specific molecule is present. Use classroom or library references to find out how the *trp* operon functions in *E. coli.* Then compare activation and repression in the *trp* operon with activation and repression in the *lac* operon.

7. A regulator gene is a gene that codes for a repressor protein.

ANSWERS TO ANALYSIS AND CONCLUSIONS

1. Repression is the blockage of transcription by the action of a repressor protein. Activation is the initiation of transcription by the removal of a repressor protein.
2. lactose
3. A mutation in a regulator gene could result in a defective repressor protein, which would be unable to bind to the operator. This would cause the *lac* operon to remain on.
4. When the inducer is no longer present, the repressor protein will remain bound to the operator. Transcription cannot proceed, and the enzymes that break down lactose will not be produced.
5. If the promoter were lost, RNA polymerase could not bind to the DNA. Transcription of the structural genes could not take place and the enzymes needed to utilize lactose would not be produced.
6. The *lac* operon responds to the end product and turns off, much as a thermostat recognizes when a room is sufficiently warm (or, in the case of air conditioning, sufficiently cool).
7. The end product stops the pathway.

FURTHER INQUIRY

1. Examples could include the control of blood sugar by insulin release or the control of body temperature by the release of thyroid-stimulating hormone from the pituitary gland. Models of feedback inhibition are feedback loops because the end product loops back to affect an earlier step in the pathway.
2. The presence of tryptophan inhibits its own production by activating the repressor. The repressor is inactivated when no tryptophan is present.

CHAPTER 12
Inheritance Patterns and Human Genetics

	TOPICS	TEACHING RESOURCES	LABS, CLASSWORK, AND HOMEWORK
BLOCKS 1 & 2 90 minutes	**Introducing the Chapter, p. 220**	**ATE** Focus Concept, p. 220 **ATE** Assessing Prior Knowledge, p. 220 **ATE** Understanding the Visual, p. 220	■ Supplemental Reading Guide, *The Double Helix*
	12-1 Chromosomes and Inheritance, p. 221 Sex Determination, p. 221 Linkage Groups, p. 222 Chromosome Mapping, p. 224 Mutation, p. 224	**ATE** Section Overview, p. 221 **ATE** Critical Thinking, p. 224 **ATE** Visual Link Figure 12-2, p. 223 Figure 12-4, p. 223 28A	★ Study Guide, Section 12-1 **PE** Section Review, p. 226
BLOCKS 3–5 135 minutes	**12-2 Human Genetics, p. 227** Studying Human Inheritance, p. 227 Genetic Traits and Disorders, p. 228 Detecting Human Genetic Disorders, p. 231	**ATE** Section Overview, p. 227 **ATE** Critical Thinking, pp. 228, 231 **ATE** Visual Link Figure 12-9, p. 227 Figure 12-11, p. 232 30A, 38	**PE** **Chapter 12 Investigation,** pp. 236–237 **PE** **Quick Lab,** p. 229, Modeling Linkage **Lab C13,** Analyzing Corn Genetics **Lab C14,** Karyotyping **Lab C15,** Karyotyping—Genetic Disorders **Lab D1,** Staining DNA and RNA **Lab D2,** Extracting DNA **Lab D3,** Genetic Transformation of Bacteria **Lab D4,** Genetic Transformation—Antibiotic-Resistant Bacteria **Lab E5,** Heredity in Families ★ Study Guide, Section 12-2 **PE** Section Review, p. 232
BLOCKS 6 & 7 90 minutes	**REVIEW and ASSESSMENT** **PE** Chapter 12 Review, pp. 234–235 ■ Performance-Based Assessment—Identifying Dominant and Recessive Traits	Audio CD Program BIOLOGY INTERACTIVE TUTOR Unit 5—Heredity	**CHAPTER TESTING** ■ Chapter Test (blackline master) ▲ Test Generator ■ Assessment Item Listing ■ Scoring Rubrics and Classroom Management Checklists

KEY

MODERN BIOLOGY
- **PE** Pupil's Edition
- **ATE** Teacher's Edition
- ■ Active Reading Guide
- ★ Study Guide

HOLT BIOSOURCES
- Teaching Transparencies
- Laboratory Program

One-Stop Planner CD-ROM
Shows these resources within a customizable daily lesson plan:
- ■ Holt BioSources Teaching Resources
- ■ Active Reading Guide
- ★ Study Guide
- ▲ Test Generator

READING FOR CONTENT MASTERY

Preview: Have students preview the objectives, the introductory paragraph, topic sentences, and captions before reading the chapter.

- **ATE** Active Reading Technique: Anticipation Guide, p. 221
- ■ Active Reading Guide Worksheet 12-1
- **PE** Summary/Vocabulary, p. 233

- ■ Active Reading Guide Worksheet 12-2
- **PE** Summary/Vocabulary, p. 233

TECHNOLOGY AND INTERNET RESOURCES

CNN PRESENTS Video Segment 8 Alzheimer's Mutation

Holt Biology Videodiscs Teacher's Correlation Guide, Lessons 12-1 through 12-2

BIOLOGY INTERACTIVE TUTOR Unit 5—Heredity, Topics 5–6

Audio CD Program, Sections 12-1 through 12-2

Interactive Exploration: Cell Biology and Genetics, "Meiosis: Down Syndrome"

internetconnect

SCILINKS NSTA
On-line Resources: www.scilinks.org
The following *sci*LINKS Internet resources can be found in the student text for this chapter:

Topics:
- X-linked traits, p. 222

go.hrw.com
On-line Resources: go.hrw.com
Visit the HRW Web site for a variety of resources related to this chapter. Just type in the keyword HM2 Chapter 12.

Smithsonian Institution®
Visit **www.si.edu/hrw** for additional on-line resources.

CNNfyi.com
Visit **www.cnnfyi.com** for late-breaking news and current event stories selected just for you.

LAB ACTIVITY PLANNING

CHAPTER 12 INVESTIGATION

Meiosis—Down Syndrome, pp. 236–237

OVERVIEW

In this interactive exploration, students will use a computer simulation to determine whether parental age affects the probability of having a child with Down Syndrome. This investigation allows students to explore the cause of Down Syndrome. This condition is a consequence of the failure of chromosomes to separate normally during meiosis. Meiosis and nondisjuction are explained in this program by using animation.

Students will generate graphs that show the probability of having a child with Down Syndrome based on the age of each parent. Students will analyze the graphs by creating a data table for each parent that shows the incidence of Down Syndrome per 1,000 live births.

PREPARATION

1. Access to the following materials is needed:
   - computer with CD-ROM drive
   - CD-ROM *Interactive Explorations in Biology: Cell Biology and Genetics.* The title of this program on the CD-ROM is "Meiosis: Down Syndrome."
2. Have students read the Objectives, Background, and Prelab Preparation sections before starting the lab.
3. Each student will need one sheet of graph paper (5 squares per inch).

Quick Labs

Modeling Linkage, p. 229

Students will need one kind of candy, such as gumdrops, in two colors (20 pieces of each color). They will need another kind of candy, such as jelly beans, in two other colors (20 pieces of each color). Students will also need toothpicks, pencils, and paper.

UNDERSTANDING THE VISUAL

Students should recognize the rod-shaped structures shown in the photograph as chromosomes. Have students identify the parts of one of the chromosomes shown. (They should be able to point out the chromatids and the centromere region.) Ask students how many chromosomes are found in the normal human genome (46). Point out that each chromosome contains many genes, which control the development and function of the body. Then ask students if they can see the individual genes on the chromosomes (no).

FOCUS CONCEPT

Reproduction and Inheritance
In humans, the inheritance of an X or Y chromosome from the male gamete determines the sex of a child. The genetic information on chromosomes also determines various patterns of inheritance, ranging from genetic disorders to sex-influenced traits.

ASSESSING PRIOR KNOWLEDGE

Review the following concepts.

Genetic Traits: *Chapter 9*
Ask students to distinguish between dominance and recessiveness. Have them give examples of dominant and recessive inheritance.

Gene Expression: *Chapter 10*
Ask students how transcription and translation contribute to an individual's inheritance of various traits.

CHAPTER 12

INHERITANCE PATTERNS AND HUMAN GENETICS

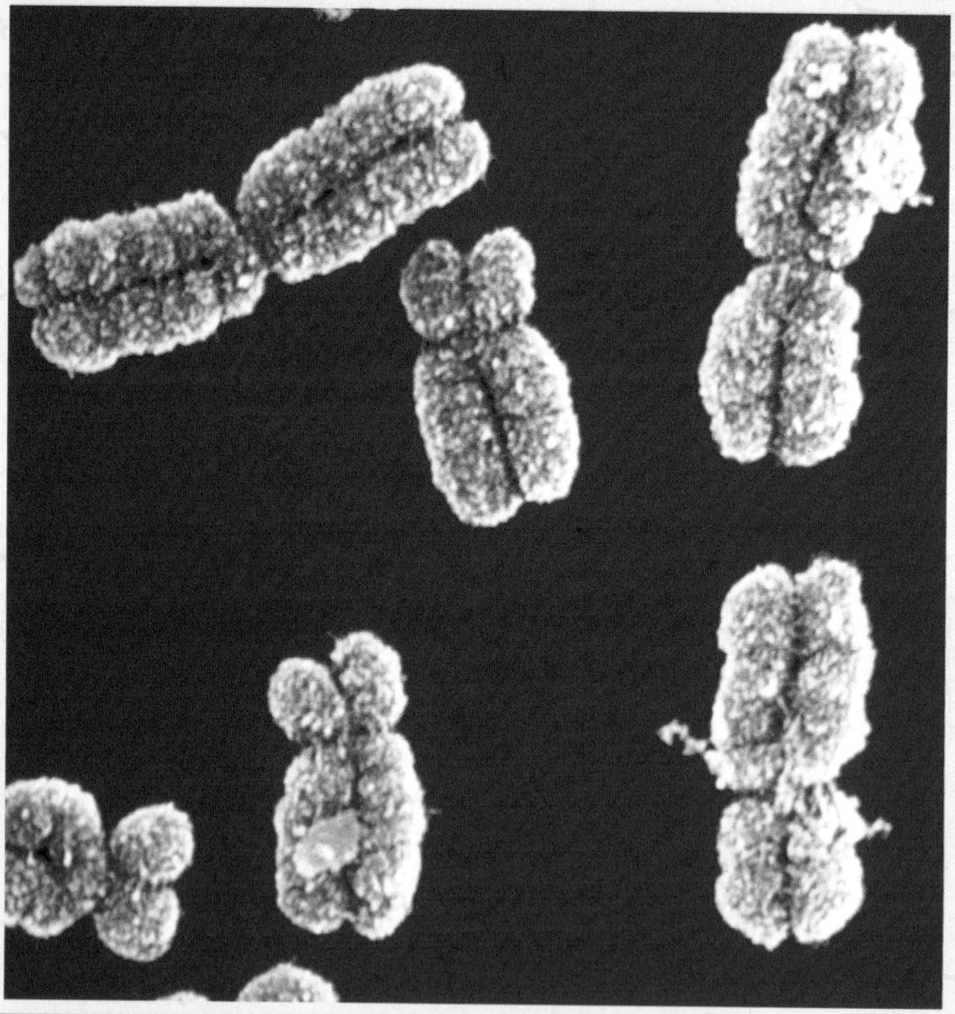

These rod-shaped structures are human chromosomes as seen through a scanning electron microscope. (SEM, 5,148×)

FOCUS CONCEPT: ***Reproduction and Inheritance***
As you read this chapter, note how Mendel's theories of inheritance (Chapter 9) and our knowledge of chromosome structure and protein synthesis (Chapters 10 and 11) have helped shape the study of genetics.

BIOLOGY INTERACTIVE TUTOR
Unit 6—*Gene Expression*
Topics 1–6

CHROMOSOMES AND INHERITANCE

You have learned how DNA in chromosomes contains instructions for protein synthesis and how chromosomes are transmitted from one generation to the next. In this chapter you will learn how biologists use their knowledge of DNA and chromosome behavior to study how traits are inherited and expressed.

SECTION

12-1

OBJECTIVES

▲

Explain the role of sex chromosomes in sex determination.

●

Describe how sex linkage affects the inheritance of traits.

■

Explain the effect of crossing-over on the inheritance of genes in linkage groups.

◆

Summarize the procedure involved in constructing a chromosome map.

▲

Distinguish between chromosome mutations and gene mutations.

SEX DETERMINATION

In the early 1900s, geneticist Thomas Hunt Morgan, of Columbia University, began breeding experiments with *Drosophila,* the fruit fly. *Drosophila* has four pairs of homologous chromosomes. Morgan observed that one chromosome pair was different in males than it was in females. In females, the two chromosomes in the pair were identical. In males, however, while one chromosome looked like those of the corresponding female pair, the other chromosome was shorter and hook-shaped. Morgan called the chromosome that appeared to be the same in males and females the X chromosome. He called the shorter, hook-shaped chromosome the Y chromosome. Morgan correctly hypothesized that the X and Y chromosomes are sex chromosomes. Recall from Chapter 8 that sex chromosomes determine an individual's sex. All of the other chromosomes, those not involved in sex determination, are called autosomes.

Like other chromosomes, the sex chromosomes form pairs and segregate into separate cells during meiosis I. As a result, the gametes that form during meiosis II each have *either* an X chromosome or a Y chromosome. In mammals and most insects, males have one X chromosome and one Y chromosome, which are symbolized XY, and females have two X chromosomes, which are symbolized XX. Gametes produced by males can contain either an X chromosome or a Y chromosome, whereas gametes produced by females contain only an X chromosome.

In humans and fruit flies, an egg that is fertilized by a sperm with an X chromosome will be a female zygote (XX), and an egg that is fertilized by a sperm with a Y chromosome will be a male zygote (XY). As shown in Figure 12-1, this system of sex determination means that 50 percent of the offspring of any mating will be male and 50 percent will be female.

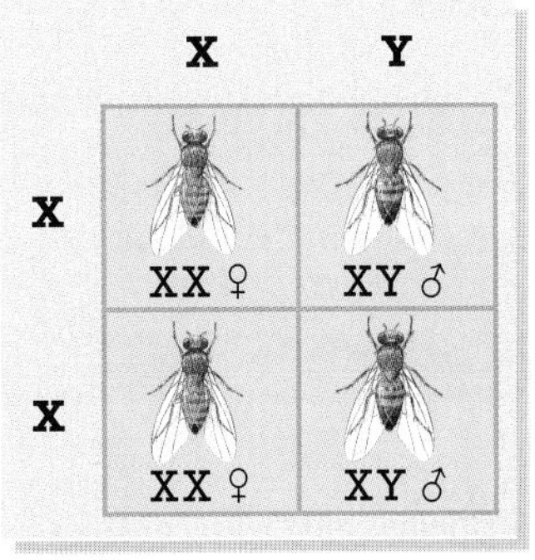

FIGURE 12-1

As this Punnett square shows, approximately half of all *Drosophila* offspring are likely to be female (♀), and half are likely to be male (♂).

SECTION 12-1

SECTION OVERVIEW

Reproduction and Inheritance

Section 12-1 discusses how sex chromosomes determine the sex of organisms and how the presence of sex-linked genes on sex chromosomes determines inheritance patterns unrelated to sex.

Stability and Homeostasis

Section 12-1 also discusses the different types of genetic mutations, where these genetic mutations occur, and the mechanisms that are responsible for their occurrence.

ACTIVE READING

Technique: Anticipation Guide

Write the titles of this chapter and Sections 12-1 and 12-2 on the board. Have students copy these titles in their notebooks, leaving a few lines after each section title. Next ask students to write what they think they will learn in each section. Check their lists for completeness, and tell them to save their lists for use later in the Active Reading activity on page 231.

TEACHING STRATEGY

Sex Determination

Have students set up a Punnett square using the sex chromosomes to demonstrate a cross between a male and female. Ask students what the chances are of the offspring being male and female, respectively. (Fifty percent of the offspring are likely to be male, and fifty percent are likely to be female.) Then ask the students which parent determines the sex of the offspring (the father).

OVERCOMING MISCONCEPTIONS

Students may believe that many sex-linked traits are on the Y chromosome since sex-linked traits are commonly expressed by males. However, sex-linked traits are often associated with recessive, X-linked phenotypes. These phenotypes appear in males more frequently because men are hemizygous. (Men have only one X chromosome and therefore only one locus for expressing the gene. Women have two X chromosomes and therefore have to be homozygous in order to express a recessive sex-linked phenotype.)

RETEACHING *ACTIVITY*

Have students work in small groups to compare the following crosses involving eye color of *Drosophila:* $X^RX^r \times X^rY$, $X^rX^r \times X^RY$, and $X^RX^R \times X^rY$. Ask students to specify the genotypes and phenotypes expected for female and male offspring in each cross. (The offspring expected in the first cross would be X^RX^r, X^rX^r, X^rY, and X^RY—both males and females would be red-eyed and white-eyed.)

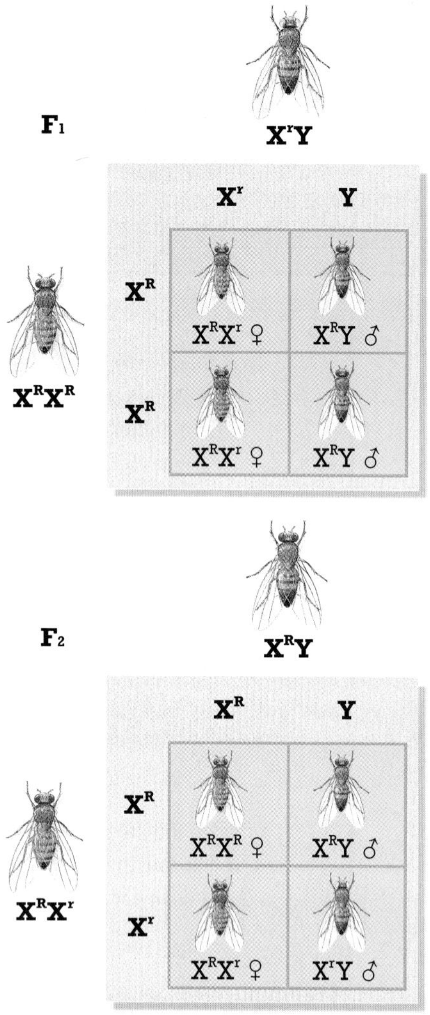

FIGURE 12-2

Eye color is an X-linked trait in fruit flies. Alleles for eye color are carried on the X chromosome—*R* for red and *r* for white.

internet**connect**

SCILINKS NSTA

TOPIC: X-linked traits
GO TO: www.scilinks.org
KEYWORD: HM222

Sex Linkage

The discovery of the different sizes and shapes of the X and Y chromosomes led Morgan to hypothesize that more genes could be carried by the X chromosome than by the smaller Y chromosome. Genes found on the X chromosome are said to be **X-linked genes.** Genes found on the Y chromosome are **Y-linked genes.** The presence of a gene on a sex chromosome is called **sex linkage.**

Morgan's *Drosophila* experiments confirmed the existence of X-linked traits. Although most fruit flies have red eyes, a few males have white eyes. When Morgan crossed a white-eyed male with a red-eyed female, the results of the cross followed Mendel's predictions: the F_1 generation all had red eyes. Morgan next crossed members of the F_1 generation. The F_2 generation that resulted exhibited the expected ratio of three red-eyed flies to one white-eyed fly. However, all of the white-eyed flies were male.

Why were there no white-eyed females? Morgan hypothesized that the gene for eye color is carried on the X chromosome. Figure 12-2 shows his reasoning. Assume that the X chromosome carries the gene for eye color, either X^R (red eyes) or X^r (white eyes). If an X^RX^R female (red-eyed) is crossed with an X^rY male (white-eyed), all of the F_1 females will be X^RX^r (red-eyed) and all of the F_1 males will be X^RY (red-eyed). Note that the Y chromosome does not carry a gene for eye color. In the F_2 generation, half the females will be X^RX^R, and the other half will be X^RX^r; therefore, all of the females will be red-eyed. Half of the F_2 males will be X^RY (red-eyed), but the other half will be X^rY (white-eyed). Morgan correctly concluded that eye color in *Drosophila* is an X-linked trait.

LINKAGE GROUPS

Because there are thousands more genes than there are chromosomes, each chromosome carries many genes. The genes located on one chromosome form a **linkage group.** Two or more genes that are found on the same chromosome are thus said to be linked. Because they are on the same chromosome, linked genes tend to be inherited together.

Morgan demonstrated the existence of linkage groups in his work with *Drosophila.* In *Drosophila* the allele *G* for gray body is dominant to the allele *g* for black body. The allele *L* for long wings is dominant to the allele *l* for short wings. Morgan crossed homozygous gray, long-winged (*GGLL*) flies with homozygous black, short-winged (*ggll*) flies to produce an F_1 generation of heterozygous gray, long-winged (*GgLl*) flies. Then he crossed members of this F_1 generation with one another (*GgLl* × *GgLl*) to produce an F_2 generation. Morgan knew that if the alleles for body color and wing length were on

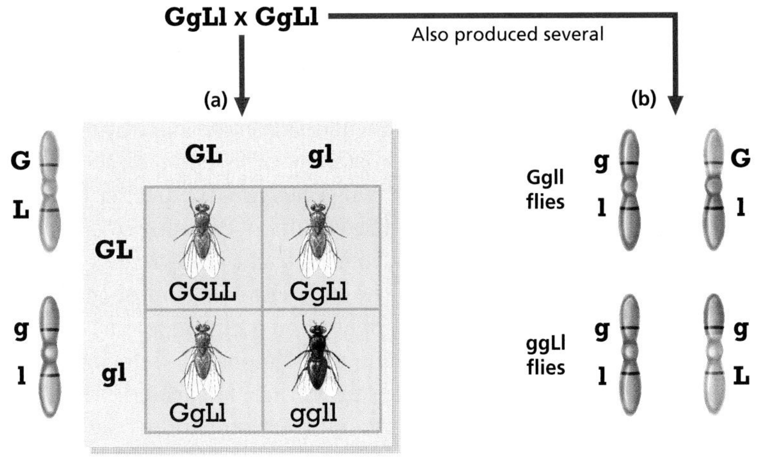

FIGURE 12-3

(a) Because the genes for body color and wing length are linked, the cross *GgLl* × *GgLl* produces a 3:1 phenotypic ratio. (b) Crossing-over produces gametes with new combinations of alleles, leading to unexpected phenotypes among the offspring.

different chromosomes, they should assort independently and produce an F_2 generation with a phenotypic ratio of 9:3:3:1. Morgan predicted that if the alleles were on the same chromosome, however, an F_2 generation with a phenotypic ratio of 3 gray, long-winged flies: 1 black, short-winged fly would result. The results of the cross closely approximated this 3:1 ratio. Morgan hypothesized that the genes for body color and wing length are linked, as shown in Figure 12-3a.

Unexpectedly, the cross also produced several gray, short-winged (*Ggll*) flies and several black, long-winged flies (*ggLl*). If the genes for body color and wing length were indeed on the same chromosome, how could they be separated? Morgan realized that these results could have occurred only if the alleles had somehow changed or become rearranged. Mutations could have caused the changes, but mutations usually only occur in one individual out of tens of thousands. The alleles must therefore have become rearranged. Morgan inferred that this rearrangement had taken place through crossing-over. Recall from Chapter 8 that crossing-over is the exchange of pieces of DNA between homologous chromosomes. Figure 12-3b shows how crossing-over accounts for the unexpected phenotypes in the F_2 generation of Morgan's experiment.

Crossing-over does not create new genes or delete old ones. Instead, it changes the locations of genes among the chromosomes that carry them, thus producing new gene combinations, as shown in Figure 12-4.

FIGURE 12-4

When traits do not appear according to the expected ratio in offspring, crossing-over may have occurred. Parental gametes contain alleles on the same chromosome that have been inherited together. Crossover gametes contain new combinations of alleles that are not found in the parental gametes.

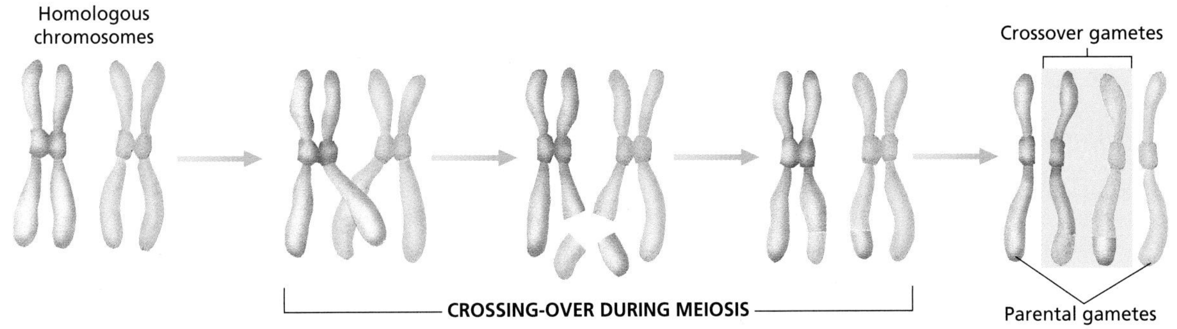

SECTION 12-1

RETEACHING *ACTIVITY*

Give students the following problem: Linked genes on homologous chromosomes are known to be *ABC* and *abc*, respectively, with genes *a*, *b*, and *c* in the sequence a-b-c. In crosses of heterozygous parents, some of the offspring receive chromosomes with *ABc* or *abC* and other offspring receive *Abc* or *aBC*, with *ABc/abC* occurring twice as often as *Abc/aBC*. Ask students to use this crossing-over data to compare how far apart the genes are. (Because crossing-over occurs twice as frequently between *b* and *c* as it does between *a* and *b*, genes *b* and *c* are twice as far apart as genes *a* and *b*.)

VISUAL LINK

Figures 12-2 and 12-4

Take students through the sex linkage of eye color in *Drosophila* (Figure 12-2) and the crossing-over of homologous chromosomes (Figure 12-4). Ask students to compare the role of X and Y chromosomes in transmitting sex-linked traits.

OVERCOMING MISCONCEPTIONS

Students sometimes have the misconception that all traits associated with sex are controlled by genes on the sex chromosomes, while genes controlling traits that have nothing to do with sex are located on the autosomes. Point out that the genes for hemophilia and red-green colorblindness are located on the sex chromosomes, while those controlling the production of sex hormones are found on the autosomes.

SECTION 12-1

INCLUSION ACTIVITY

Have students make homologous chromosomes out of strips of paper and mark them with symbols for genes. Show students how crossing-over can result in different genetic combinations in offspring receiving an altered chromosome. For example, if crossing-over occurs between *AB/ab* homologues, the resulting chromosomes will be *Ab* and *aB.* Therefore, in a cross of *AaBb* with *aabb,* some offspring will show both the *A* and *b* phenotypes, and others will show both the *a* and *B* phenotypes.

MAKING CONNECTIONS

Mathematics

Provide students with this chromosome map: (genes: black body = 48.5, purple eyes = 54.5, cinnabar eyes = 57.7, safranin eyes = 60.0, pink wings = 64.0, vestigial wings = 67.0). Have students calculate how many units apart the genes are for black body and pink wings (15.5 units), for cinnabar eyes and safranin eyes (2.3 units), and for purple eyes and vestigial wings (12.5 units).

CRITICAL THINKING

Map Units

In *Drosophila* the alleles for red eyes *(R)* and brown body *(B)* are dominant to the alleles for white eyes *(r)* and yellow body *(b).* Suppose Morgan crossed a heterozygous red-eyed, brown-bodied female *(RrBb)* with a white-eyed, yellow-bodied male *(rrbb).* Of 500 offspring, 7 have red eyes and yellow bodies, and 8 have white eyes and brown bodies. Ask the students how many map units apart the two alleles are. (Since there was a 3 percent crossover—15 out of 500—the two alleles are 3 map units apart.)

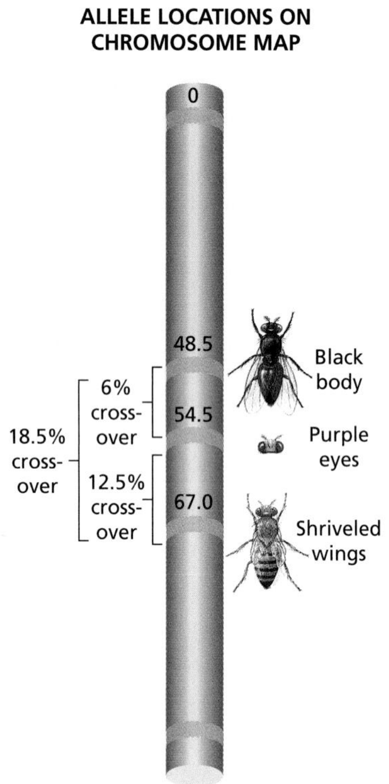

FIGURE 12-5

The genes for black body and purple eyes are separated by crossing-over 6 percent of the time, while the genes for purple eyes and shriveled wings are separated by crossing-over 12.5 percent of the time. The genes for black body and shriveled wings are separated by crossing-over 18.5 percent of the time.

CHROMOSOME MAPPING

The likelihood that crossing-over will result in the separation of two genes depends on the genes' distance from each other on the chromosome. The farther apart two genes are, the likelier they are to be separated by crossovers. The results of crossing-over appear in the offspring as new combinations of traits. The greater the percentage of offspring that show the new combination of traits, the farther apart the two genes are located on a chromosome.

Scientists can conduct breeding experiments to determine how frequently genes for particular traits are separated from one another among the offspring. This information can be used to prepare chromosome maps. A **chromosome map** is a diagram that shows the linear sequence of genes on a chromosome. Alfred H. Sturtevant, one of Morgan's students, used crossing-over data to construct a chromosome map of *Drosophila.*

To construct his chromosome map, Sturtevant compared the frequency of crossing-over for several genes. The percentage of crossing-over between the genes for two traits is proportional to the distance between them on a chromosome. For example, two genes that are separated by crossing-over 1 percent of the time are considered to be one **map unit** apart. Figure 12-5 illustrates how crossover percentages can be used to determine the positions of the genes for black body, purple eyes, and shriveled wings in *Drosophila.*

In recent years, new techniques of gene mapping have been developed. These techniques have helped to map the chromosomes of many species, even those with many chromosomes and tens of thousands of genes.

MUTATION

As you learned in Chapter 10, a change in the DNA of an organism is called a mutation. Mutations can involve an entire chromosome or a single DNA nucleotide, and they may take place in any cell. **Germ-cell mutations** occur in an organism's germ cells (gametes). Germ-cell mutations do not affect the organism itself, but they may be passed on to offspring if the affected gamete is fertilized. **Somatic** (soh-MAT-ik) **mutations** take place in an organism's body cells and can therefore affect the organism. Certain types of human skin cancer and leukemia result from somatic mutations. Somatic mutations are not passed on to offspring.

Lethal mutations cause death, often before birth. However, some mutations result in phenotypes that are beneficial. Organisms with beneficial mutations have a better chance of reproducing and therefore have an evolutionary advantage. Such mutations provide the variation upon which natural selection acts.

Word Roots and Origins

somatic

from the Greek *sōmatikos,* meaning "body"

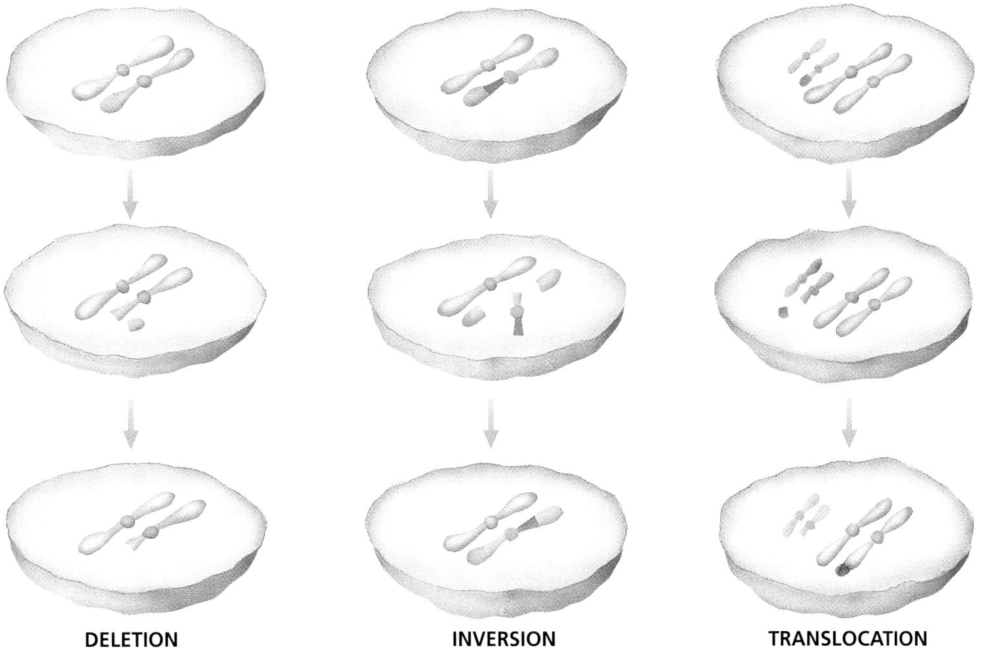

FIGURE 12-6
In many chromosome mutations, a piece of chromosome breaks off. The piece may be lost or may reattach in a reversed position. In translocations, a broken piece attaches to a nonhomologous chromosome.

Chromosome Mutations

Chromosome mutations are either changes in the structure of a chromosome or the loss of an entire chromosome. A **deletion** is the loss of a piece of a chromosome due to chromosomal breakage. As a consequence, all of the information carried by the missing piece may be lost. An **inversion** is a chromosome mutation in which a chromosomal segment breaks off and then reattaches in reverse orientation to the same chromosome. A chromosome mutation in which a chromosome piece breaks off and reattaches to another, nonhomologous chromosome is a **translocation.** Figure 12-6 shows these mutations.

Some types of chromosomal mutations alter the number of chromosomes found in the cell. **Nondisjunction** (NON-dis-JUHNK-shuhn) is the failure of a chromosome to separate from its homologue during meiosis. When nondisjunction occurs, one gamete receives an extra copy of a chromosome and the other gamete lacks the chromosome entirely. Human disorders resulting from nondisjunction are discussed in Section 12-2.

FIGURE 12-7
In a substitution, one nucleotide replaces another, forming a new codon that may signal the insertion of the wrong amino acid.

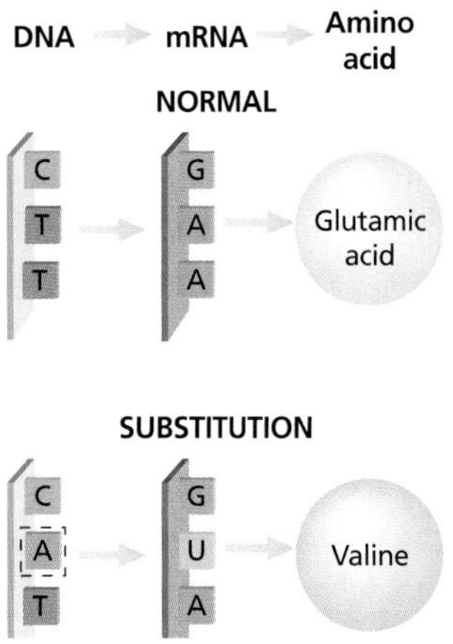

Gene Mutations

Gene mutation may involve large segments of DNA or a single nucleotide within a codon. As you will recall from Chapter 10, a codon consists of three nucleotides that cause a specific amino acid to be inserted into a polypeptide during protein synthesis. The substitution, addition, or removal of a single nucleotide is called a **point mutation.**

In point mutations called **substitutions,** one nucleotide in a codon is replaced with a different nucleotide, resulting in a new codon. An example of a substitution mutation is shown in Figure 12-7. If the new codon codes for the same amino acid as the original codon, the

DEMONSTRATION

Demonstrate point mutations using the following sentence: The cat ate the big rat. Show students point mutations that have resulted from substitutions (the bat ate the big rat, the cat ate the big rag) and additions or removals that have resulted in frameshifts (the aca tat eth ebi grat, the ata tet hib igr at).

TEACHING STRATEGY
Mutations

Give each student four pieces of colored ribbon (red, white, green, and yellow), a pair of scissors, and tape. Have students make a mark in the middle of each ribbon to represent a centromere. Then have them model deletions by cutting off a piece of the red ribbon. Have students throw away the piece of red ribbon that doesn't have a centromere. Next, have students model inversion by cutting the white ribbon to the right of the centromere. Have students turn the piece of white ribbon marked with the centromere 180° and then tape the two pieces of white ribbon together. Finally, have them model translocation by cutting off a piece of the green ribbon and taping the piece without a centromere to the yellow ribbon.

SECTION 12-1

QUICK FACT

☑ Heterozygotes for sickle cell anemia seldom experience problems with their red blood cells. More important, these carriers for sickle cell anemia are resistant to malaria, unlike homozygous dominant individuals, who are susceptible to malaria.

ANSWERS TO SECTION 12-1 REVIEW

1. The male sex chromosomes are X and Y. One-half of male gametes will contain a Y chromosome and one-half will contain an X chromosome; the number of male offspring and the number of female offspring are approximately equal as a result.
2. The mutant allele for white eyes is X-linked, so a female *Drosophila* must have two copies of the mutant allele in order to be white-eyed. A male who has only one mutant allele for white eyes will be white-eyed because he has only one X chromosome.
3. Traits known to be controlled by genes located on the same chromosome are normally inherited together, but they can be inherited individually if crossing-over has separated them.
4. The farther apart two alleles are on a chromosome, the more often they will be separated by crossing-over and thus the more often they will appear in new combinations in the offspring.
5. A point mutation is the substitution, addition, or removal of a single nucleotide within a segment of DNA.
6. Just before nuclear division, chromatin condenses to form chromosomes, which move throughout the cell during the different stages of nuclear division and thus are at higher risk of breakage.

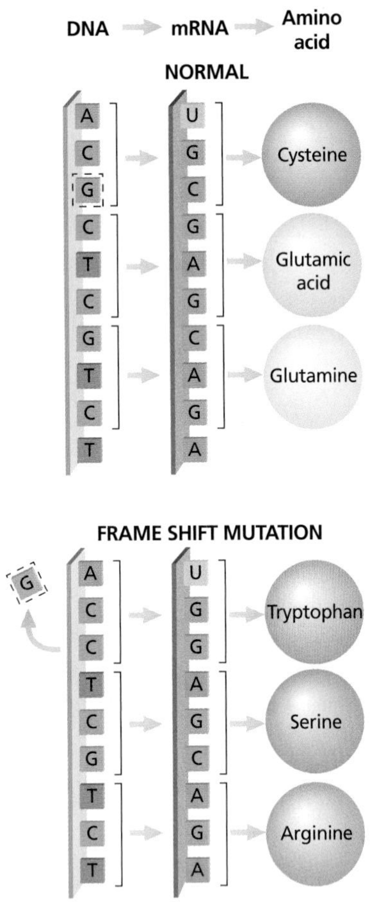

FIGURE 12-8

Deleting a nucleotide causes all subsequent codons to be incorrectly read, resulting in a frame shift mutation. Adding a nucleotide will have the same effect.

substitution will have no effect. However, if the new codon is a "stop" codon or if it codes for a different amino acid, the resulting protein will be affected.

The genetic disorder **sickle cell anemia** (also called sickle cell disease) is caused by a point mutation that substitutes adenine for thymine in a single DNA codon, as shown in Figure 12-7. This substitution results in a defective form of the protein hemoglobin. Hemoglobin is found within red blood cells, where it binds with oxygen and transports it throughout the body. The red blood cells of people with two copies of the mutant allele for sickle cell anemia have a distorted, sickle shape, causing anemia (loss of red blood cells) and circulatory problems. Because of this, children with sickle cell anemia may suffer damage to the brain, heart, lungs, and many other organs and tissues. The sickle cell allele is especially widespread among African Americans. In the United States, approximately 1 out every 500 African Americans has sickle cell anemia.

Individuals who are heterozygous for the sickle cell allele make both normal and altered forms of hemoglobin and are generally healthy. In the United States, approximately 1 out of 10 African Americans is heterozygous for the sickle-cell allele and is therefore a carrier for sickle cell anemia. A simple blood test can reveal the presence of the allele.

Two forms of point mutations are nucleotide insertions and nucleotide deletions. In nucleotide deletion mutations, one or more nucleotides in a gene are lost. In insertion mutations, one or more nucleotides are added to a gene. Deletion and insertion mutations tend to have more serious effects than substitution mutations. Because each codon consists of a group of three sequential nucleotides, the addition or deletion of a single nucleotide causes the remaining codons to be incorrectly grouped, resulting in a **frame shift mutation,** as shown in Figure 12-8. In fact, frame shift mutations occur anytime the number of nucleotides inserted or deleted is not a multiple of three. Figure 12-8 shows a frame shift mutation that results from the deletion of one nucleotide. Thus, a frame shift mutation occurring near the beginning of a gene will prevent a number of codons from coding for the proper amino acids.

SECTION 12-1 REVIEW

1. **How does the inheritance of sex chromosomes result in approximately equal numbers of males and females among the offspring of fruit flies?**
2. **Offer an explanation for why Morgan did not find white-eyed female *Drosophila* in the F_2 generation when he crossed white-eyed males with red-eyed females.**
3. **How does crossing-over show that genes are found on chromosomes?**
4. **How can crossing-over between two alleles be used to map their locations on chromosomes?**
5. **What are point mutations?**
6. **CRITICAL THINKING Biologists have observed that chromosome mutations often occur during nuclear division. Why do you think this is so? Explain your answer.**

HUMAN GENETICS

Biologists have studied many different organisms in their search to discover the fundamental principles of genetics. These fundamental principles also apply to humans. Much of what we have discovered about the inheritance of traits has been learned in the study of human genetics.

STUDYING HUMAN INHERITANCE

Patterns of inheritance are significantly more complicated to study among humans than among *Drosophila.* Humans have up to 20 times as many genes as *Drosophila,* and our 23 pairs of chromosomes are made up of about 100,000 genes. Geneticists have often focused on studying disease-causing genes because such genes are easily traced from one generation to the next and because they are of great concern to the human population.

Pedigree Analysis

Biologists discover how traits are inherited by studying phenotypes among members of the same species from one generation to the next. In particular, they often study members of the same family. In such studies, geneticists often prepare a **pedigree,** a family record that shows how a trait is inherited over several generations. Figure 12-9 shows a pedigree illustrating a trait that is inherited as

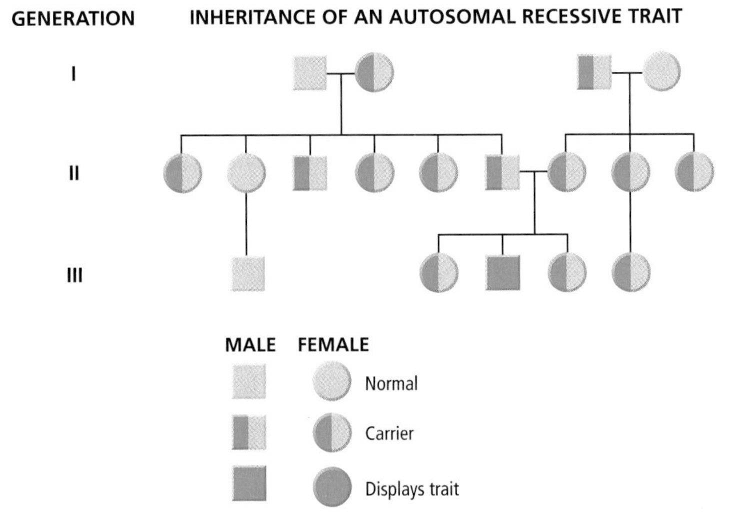

FIGURE 12-9

This pedigree shows the inheritance of an autosomal recessive trait. Pedigrees are particularly important if the trait is a genetic disorder and family members wish to know if they are carriers or if their children will have the disorder.

SECTION

12-2

OBJECTIVES

▲ Show how pedigree analysis can be used to illustrate the inheritance of traits.

● Explain the inheritance of ABO blood groups.

■ Give examples of traits or disorders transmitted by autosomal dominant, autosomal recessive, polygenic, and X-linked recessive inheritance.

◆ Compare sex-linked traits with sex-influenced traits.

▲ Explain how nondisjunction can cause human genetic disorders.

SECTION 12-2

SECTION OVERVIEW

Evolution

Section 12-2 focuses on the inheritance patterns of humans. This section shows students how inherited traits can be traced from generation to generation through pedigree analysis.

Reproduction and Inheritance

Section 12-2 also discusses how genetic disorders, which can be traced through patterns of inheritance, can be detected.

▶ ENGAGE STUDENTS

Provide a list of traits that students recognize and that can be designated with a pair of alleles (free ear lobe, dimples, hair on knuckles, curved pinkie finger, etc.). Have students use the list of characteristics to describe a potential mate. Assume that each characteristic is governed by a single gene. Students can design a person or find a celebrity to represent that person. Students should then assign their potential mate a genotype for each trait. They should also make a genotype sheet for themselves. Collect and shuffle the genotypic and phenotypic descriptions of the potential mates, and redistribute them among the students. Have students use a coin toss to determine which alleles the offspring will receive. After the genotype is known, have students describe the phenotype of the "child."

VISUAL LINK

Figure 12-9

Ask students how they can determine if the trait is recessive. (Offspring will demonstrate the trait when their parents do not.) Ask students how can they determine if the trait is autosomal. (Daughters of unaffected males will be able to demonstrate the trait if it is autosomal.)

SECTION 12-2

TEACHING STRATEGY

Design a Pedigree

Have students develop stories from which a pedigree can be drawn. Encourage them to be creative in thinking of the characters, and their traits, that they choose to follow through several generations. To illustrate their pedigree stories, students can add "family portraits."

QUICK FACT

☑ Huntington's disease can be traced back to two brothers who immigrated to North America from England in the 1600s because of accusations of witchcraft in their family. The family members of these brothers were persecuted because of their brothers' strange behaviors, which are now understood to be the symptoms of Huntington's disease.

CRITICAL THINKING

Blood Type

Ask students to list the possible genotypes of parents who produce offspring with type O blood (*ii*, I^Ai, and I^Bi) and with type AB blood (all except *ii*).

RECENT RESEARCH

Cystic Fibrosis

The latest medical research indicates that the bacterium *Pseudomonas aeruginosa* enhances the disease cystic fibrosis by activating a chain of enzymatic reactions. The activated enzymes cause increased production of mucin in the lungs, which becomes a breeding ground for *Pseudomonas.* In addition, the lungs' antibiotic, defensin, is inactivated due to a genetic defect that prevents salt entry into the lung cells. This dual activity causes the lungs to turn into bloated cysts.

a recessive autosomal trait. When analyzing pedigrees, biologists find that certain phenotypes are usually repeated in predictable patterns from one generation to the next. These patterns are called **patterns of inheritance.** Individuals who have one copy of a recessive autosomal allele are called **carriers.** Carriers usually do not express the recessive allele, but they can pass it along to their offspring.

GENETIC TRAITS AND DISORDERS

Genes controlling human traits exhibit several patterns of inheritance. Some of these genes have been subjects of intense scientific interest because they cause genetic disorders. **Genetic disorders** are diseases or debilitating conditions that have a genetic basis. Table 12-1 lists several patterns of inheritance and examples of traits or disorders for each.

TABLE 12-1
Patterns of Inheritance for Several Human Traits

Single Allele (Dominant)
Huntington's disease (HD)
Achondroplasia (dwarfism)
Cataracts
Polydactyly (extra fingers or toes)
Single Allele (Recessive)
Albinism
Cystic fibrosis
Phenylketonuria (PKU)
Hereditary deafness
X-Linked
Colorblindness
Hemophilia
Muscular dystrophy
Icthyosis simplex (scaly skin)
Polygenic
Skin, hair, and eye color
Foot size
Nose length
Height
Multiple Alleles
ABO blood groups

Traits Controlled by a Single Allele

Single-allele traits are controlled by a single allele of a gene. Geneticists have discovered that more than 200 human traits are governed by a single dominant allele. **Huntington's disease (HD)** is caused by a dominant allele located on an autosome and is therefore said to show an autosomal-dominant pattern of inheritance. The first symptoms of HD—mild forgetfulness and irritability—appear in victims in their thirties or forties. In time, HD causes loss of muscle control, uncontrollable physical spasms, severe mental illness, and eventually death. Unfortunately, most people who have the HD allele do not know they have the disease until after they have had children. Thus, the disease is unknowingly passed on from one generation to the next.

Recently, however, scientists have discovered a genetic marker for the HD allele. A **genetic marker** is a short section of DNA that is known to have a close association with a particular gene located nearby. Thus, the presence of a genetic marker in a DNA sample is a strong indicator that a particular allele (in this case, the allele for HD) is also present. People who have the genetic marker for the HD allele have a 96 percent chance of developing HD. By conducting a test on sample cells, geneticists are now able to inform a person of the presence of the marker before he or she becomes a parent.

More than 250 other single-allele traits are controlled by homozygous recessive alleles. Cystic fibrosis (CF) and sickle cell anemia are examples of single-allele recessive traits. Single-allele recessive traits are fully expressed only when the individual has two copies of the recessive allele and is thus homozygous recessive. Because cystic fibrosis and sickle cell anemia are caused by recessive alleles located on autosomes, they are said to show an autosomal-recessive pattern of inheritance.

Traits Controlled by Multiple Alleles

Multiple-allele traits are controlled by three or more alleles of the same gene that code for a single trait. In humans, the ABO blood groups are controlled by the three alleles I^A, I^B, and *i*. Each individual's ABO blood group genotype consists of two of these alleles, which determine his or her ABO blood type. The alleles I^A and I^B are codominant (both are expressed when together), and both are dominant to the *i* allele. Table 12-2 shows the possible ABO blood group genotypes and the ABO blood types they produce.

TABLE 12-2
ABO Blood Types

Genotype	Blood type
I^AI^A	A
I^Ai	A
I^BI^B	B
I^Bi	B
I^AI^B	AB
ii	O

Polygenic Traits

Although some human traits are governed by a single gene, most human characteristics are controlled by several genes. A trait that is controlled by two or more genes is called a **polygenic** (PAHL-ee-JEHN-ik) **trait.** Skin color, for example, is influenced by the additive effects of three to six genes. Each gene results in a certain amount of a brownish black pigment called melanin. The more melanin produced, the darker the skin; the less melanin produced, the lighter the skin. Eye color is also a polygenic trait. Light-blue eyes have very little melanin. Very dark brown eyes have a great deal of melanin. Polygenic traits show many degrees of variation.

It is important to note that the expression of many traits, particularly those that are governed by many genes, is influenced by the environment. For example, human height is a polygenic trait controlled by an unknown number of genes that play a role in determining the growth of the skeleton. However, height is also influenced by environmental factors, such as nutrition and disease.

X-Linked Traits

Colorblindness is a recessive X-linked disorder in which an individual cannot distinguish between certain colors. Recall that genes for X-linked traits are found only on the X chromosome. Although many forms of colorblindness exist, the most common is the inability to distinguish red from green. About 8 percent of males are colorblind.

Hemophilia (HEE-moh-FIL-ee-uh) is another recessive X-linked disease that occurs almost exclusively in males. This disorder impairs the ability of the blood to clot following a cut, bruise, or other injury. Another recessive X-linked trait in humans is **Duchenne muscular dystrophy** (DIS-troh-fee), a form of muscular dystrophy that weakens and progressively destroys muscle tissue. These and several other important genetic disorders are described in Table 12-3 on the following page.

Not all or even most X-linked traits are diseases. There are hundreds of genes on the X chromosome, but only a few are associated with diseases. The others code for proteins that perform many normally needed functions in the body.

Sex-Influenced Traits

The presence of male or female sex hormones influences the expression of certain human traits, called **sex-influenced traits.** In

Quick Lab

Modeling Linkage

Materials two kinds of candy, toothpicks, pencil, paper

Procedure Use two kinds of candy to represent genes for two traits. Long noses are dominant over short noses. Large ears are dominant over small ears. One color of candy will represent the dominant allele, and a different color candy will represent the recessive allele. Use these materials to determine the outcome of a cross between two individuals, each heterozygous for both traits. Your teacher will tell you if the genes are linked or not linked.

1. Draw a large Punnett square on your paper. Use the appropriate alleles to make gametes for each individual. Then place the allele combinations in each square representing the possible zygotes from that cross.
2. If your team is working with genes that are linked, you must first use toothpicks to link the genes together before you arrange the gametes on your Punnett square.

Analysis What is the offspring's phenotypic ratio when the genes are not linked? What is the phenotypic ratio when the genes are linked? Explain the difference.

RETEACHING *ACTIVITY*

A man who carries an allele for cystic fibrosis marries a woman who is not a carrier. Ask students what the chances are of the parents having a child with cystic fibrosis. (None. But half of their children would be expected to be carriers, like the father.) Ask students what the chances are if the mother is a carrier of cystic fibrosis and the father is not. Have them explain their answers (no chance, because cystic fibrosis is an autosomal trait).

Quick Lab

Modeling Linkage

Time Required 20 minutes

Procedural Tips Have students work in small teams. You will need one kind of candy, such as gumdrops, in two colors, such as red and black. You will need another kind of candy, such as jelly beans, in two other colors, such as yellow and green. Each team will need 20 candies of each color. If the cross is worked correctly, several pieces of each color will not be used. Tell some teams that they are working with genes that are linked. For example, short ears are linked to small noses, and large ears are linked to long noses. Tell the other teams that they are working with genes that are not linked. Demonstrate how to attach the candies together with a toothpick by skewering them together. Have teams compare their results.

Answers to Analysis The phenotypic ratio when the genes are not linked is nine long nose/large ears, three long nose/small ears, three short nose/large ears, and one short nose/small ears. The phenotypic ratio when the genes are linked is three long nose/large ears and one short nose/small ears. When the genes are not linked, independent assortment is possible, so four phenotypes are seen. When the genes are linked, independent assortment is not possible. Only two phenotypes are seen.

SECTION 12-2

RETEACHING ACTIVITY

Have students review sex-linked traits and sex-influenced traits. Ask students to compare sex-linked traits with sex-influenced traits and to provide examples of each of these traits. Provide students with the following crosses, and have them identify the crosses as either sex-linked or sex-influenced: $X^CX^C \times X^cY$ (sex-linked; the alleles are expressed on the sex chromosome); $X^CX^c \times X^CY$ (sex-linked; the alleles are expressed on the sex chromosome); and $H^BH^B \times H^BH^N$ (sex-influenced; the alleles are expressed on the autosomes, not on a sex chromosome).

TEACHING STRATEGY

Genetic Disorders

Divide the class into groups of two to five students. Designate each student as an "expert" on one of the following genetic disorders: Huntington's disease, cystic fibrosis, sickle cell anemia, Tay-Sachs disease, phenylketonuria, hemophilia, and muscular dystrophy. Ask students to research the disorders and share their information with other group members. Then ask students to develop quizzes to test other students' understanding of the genetic disorders.

CONTENT CONNECTION

Section 8-1: Gametes

Human sperm cells and egg cells are haploid, which means that they contain only one set of chromosomes.

TABLE 12-3 *Some Important Genetic Disorders*

Disorder	Symptom	Defect	Pattern of inheritance	Frequency among human births
Huntington's disease	gradual deterioration of brain tissue in middle age; shortened life expectancy	production of an inhibitor of brain cell metabolism	autosomal dominant	1/10,000
Cystic fibrosis	mucus clogs lungs, liver, and pancreas; victims usually don't survive to adulthood	failure of chloride ion transport mechanism	autosomal recessive	1/2,080 (whites)
Sickle cell anemia	impaired blood circulation, organ damage	abnormal hemoglobin molecules	autosomal recessive	1/500 (African Americans)
Tay-Sachs disease	deterioration of central nervous system in infancy; death occurs in early childhood	defective form of enzyme hexosaminidase A	autosomal recessive	1/1,600 (Jews of European descent)
Phenylketonuria	failure of brain to develop in infancy; if untreated, causes death in childhood	defective form of enzyme phenylalanine hydroxylase	autosomal recessive	1/18,000
Hemophilia	failure of blood to clot	defective form of blood-clotting factor	X-linked recessive	1/7,000
Muscular dystrophy	wasting away of muscles; shortened life expectancy	muscle fibers degenerate	X-linked recessive	1/10,000

a sex-influenced trait, males and females have different phenotypes, even when they share the same genotype. For example, pattern baldness is controlled by the allele *B*, which is dominant in males but recessive in females. Both men and women who are homozygous (*BB*) will eventually lose their hair. The allele *B´* codes for the normal, nonbald phenotype. A heterozygous (*BB´*) woman will not lose her hair, but a heterozygous (*BB´*) man will. These differences in gene expression are due to higher levels of the male sex hormone testosterone in men, which interacts with the genotype *BB´* to produce baldness in men. The alleles that code for most sex-influenced traits are located on the autosomes. Therefore, males and females can have the same genotype, but the sex-influenced trait will be expressed in only one sex.

Disorders Due to Nondisjunction

Nondisjunction occurring during meiosis can cause gametes to lack a chromosome or to have an extra chromosome, as shown in Figure 12-10. If nondisjunction occurs during egg formation in humans, one egg will have 22 chromosomes instead of the normal 23 ($1n$) chromosomes. Another egg will have 24 chromosomes. If one of these eggs combines with a normal sperm, the resulting zygote will have either 45 or 47 chromosomes instead of the normal 46 ($2n$) chromosomes. A zygote with 45 chromosomes has only one copy of a particular chromosome, a condition called **monosomy** (MAHN-uh-SOH-mee). A zygote with 47 chromosomes has three copies of a particular chromosome, a condition called **trisomy** (trie-SOH-mee).

Such abnormalities in chromosomal number are often lethal. However, some instances of monosomy or trisomy allow development to proceed. An extra chromosome 21 results in **Down syndrome** (also called **trisomy-21**), which includes mild to severe mental retardation, characteristic facial features, muscle weakness, heart defects, and short stature.

Nondisjunction can also affect the sex chromosomes. Males with an extra X chromosome have Klinefelter's syndrome (XXY). These persons have some feminine characteristics and are sometimes mentally retarded and infertile. Individuals who have a single X chromosome instead of a pair of sex chromosomes have Turner's syndrome (XO, the O meaning that there is only one sex chromosome). People with Turner's syndrome have a female appearance, but they do not mature sexually and they remain infertile. Zygotes that receive only a single Y chromosome do not survive because the X chromosome contains information that is essential for development.

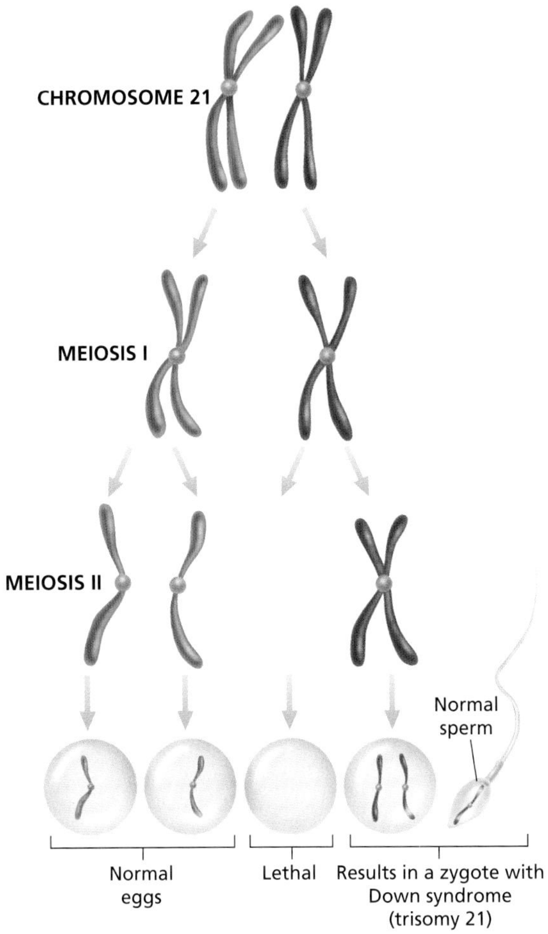

FIGURE 12-10

Nondisjunction can cause abnormalities in chromosome number. Normal gametes have one copy of each chromosome. Down syndrome can result when a gamete (an egg, in the example shown here) containing two copies of chromosome 21 is fertilized by a normal sperm.

DETECTING HUMAN GENETIC DISORDERS

A person with a family history of genetic disorders may wish to undergo **genetic screening** before becoming a parent. Genetic screening is an examination of a person's genetic makeup. It may involve constructing a karyotype, a picture of an individual's chromosomes grouped in pairs and arranged in sequence. Other techniques test an individual's blood for the presence or absence of certain proteins. These procedures can detect potential genetic disorders that might be passed on to children. Couples at risk may wish to undergo **genetic counseling,** a form of medical guidance that informs them about problems that could affect their offspring.

Word Roots and Origins

monosomy

from the Greek *mono,* meaning "one," and *soma,* meaning "body" or "chromosome"

INCLUSION *ACTIVITY*

To give students a concrete representation of nondisjunction, hold up a meterstick in each hand. Place the two metersticks next to each other, and slowly move them apart to demonstrate what happens normally during disjunction. Then show students what happens during nondisjunction by slowly moving both metersticks in the same direction.

DEMONSTRATION

Review nondisjunction with students by using pipe cleaners to represent the chromosomes and paper clips to represent the centromeres. Have students model meiosis I and II for both normal and nondisjunction gametes. Then have them share their gametes with a neighbor to produce zygotes. Have students note how the abnormal gametes produce aneuploid zygotes.

ACTIVE READING

Technique: Anticipation Guide

Have students return to the lists that they made for the Active Reading activity on page 221. Ask them to place check marks next to everything they learned that was on their list. Review any unchecked items with the students. Let students know if the material will be covered in a future chapter. If not, spend some time discussing these items.

SECTION 12-2

VISUAL LINK

Figure 12-11

Have students note the procedural differences in amniocentesis and chorionic villi sampling. Tell students that amniocentesis is usually performed during the 14th through 16th weeks of pregnancy, while chorionic villi sampling can be performed at approximately the 8th or 10th week of pregnancy. Hence, possible abnormalities can be detected earlier.

ANSWERS TO SECTION 12-2 REVIEW

1. Pedigrees should show that the parents have a normal dominant allele and the recessive CF allele and thus are carriers for CF. Their son is homozygous recessive; their daughter has at least one copy of the normal dominant allele.
2. Their children could have the ABO blood types A or B.
3. Huntington's disease (HD) is inherited as a dominant autosomal allele. Anyone who receives the allele for HD will have the disease. Sickle cell anemia is inherited as a recessive autosomal allele. A person heterozygous for sickle cell anemia is a carrier of the disease.
4. Pattern baldness is a sex-influenced trait because it is controlled by an autosomic allele, but it is expressed differently in males and females.
5. Nondisjunction can produce gametes with either one too few or one too many chromosomes. If such a gamete is fertilized, a monosomic or trisomic individual can result.
6. Because colorblindness is an X-linked recessive trait, a female would have to receive two copies of the allele for colorblindness, one from each parent, in order to exhibit colorblindness.

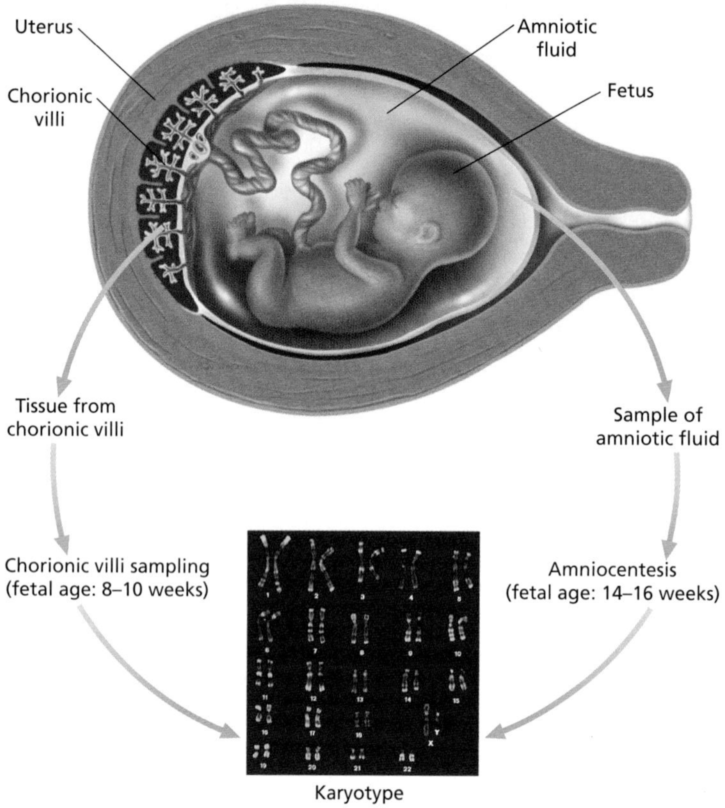

FIGURE 12-11

Fetal cells obtained by amniocentesis or chorionic villi sampling can be used to prepare fetal karyotypes, enabling physicians to diagnose chromosomal abnormalities before a child's birth.

Physicians can also diagnose more than 200 genetic disorders in the fetus using a variety of tools and techniques. Figure 12-11 shows amniocentesis and chorionic villi sampling, two forms of prenatal (before birth) testing that can reveal fetal abnormalities. During **amniocentesis** (AM-nee-OH-sen-TEE-sis) the physician removes a small amount of amniotic fluid from the amnion, the sac that surrounds the fetus, between the fourteenth and sixteenth weeks of pregnancy. Fetal cells and proteins from the fluid then can be analyzed and a karyotype can also be prepared. During **chorionic** (KOR-ee-ON-ik) **villi** (VIL-IE) **sampling** the physician obtains a sample of the chorionic villi, a tissue that grows between the mother's uterus and the placenta, between the eighth and tenth week of pregnancy. The villi have the same genetic makeup as the fetus because they were both coded for by fetal DNA. Tissue samples from the villi can be used to produce a karyotype.

In some cases, genetic screening is performed immediately after birth. In the United States, approximately 1 out of every 10,000 babies is afflicted with **phenylketonuria** (FEN-uhl-KEET-oh-NUHR-ee-uh), or **PKU,** a genetic disorder in which the body cannot metabolize the amino acid phenylalanine. The accumulation of excess phenylalanine causes severe brain damage. PKU can be detected by means of a blood test administered to infants during the first few days of life. The dangers of PKU can be eliminated by placing these infants on a special diet that lacks phenylalanine.

SECTION 12-2 REVIEW

1. **A husband and wife are heterozygous for cystic fibrosis. Their son has cystic fibrosis. Their second child, a daughter, does not. Prepare a pedigree for this family.**
2. **A husband and wife have the ABO blood group genotypes I^AI^B and *ii*. What ABO blood types can their children have?**
3. **Compare the inheritance of Huntington's disease with the inheritance of sickle cell anemia.**
4. **Is pattern baldness a sex-linked trait or a sex-influenced trait? Explain your answer.**
5. **How can nondisjunction change chromosome number?**
6. **CRITICAL THINKING Colorblindness is rare among females. Why? Explain your answer.**

CHAPTER 12 REVIEW

SUMMARY/VOCABULARY

12-1

- Sex chromosomes determine the sex of an organism.
- A sex-linked gene is found on one sex chromosome but not on the other. Genes found on the X chromosome are X-linked genes. Genes found on the Y chromosome are Y-linked genes.
- The genes found on one chromosome make up a linkage group. The genes found on a chromosome are said to be linked.
- Linked genes can be separated by crossing-over. The percentage of crossing-over between two genes is directly proportional to the distance between them.
- The frequency of crossing-over between genes can be used to construct a chromosome map.
- Crossing-over creates new combinations of genes and therefore serves as a source of genetic variation.
- Germ-cell mutations occur in gametes and can be passed on to offspring. Somatic mutations occur in body cells and affect only the individual organism.
- Chromosome mutations are changes in the structure of a chromosome or the loss of an entire chromosome. Gene mutations are changes in one or more of the nucleotides in a gene.

Vocabulary

chromosome map (224)
deletion (225)
frame shift mutation (226)
germ-cell mutation (224)
inversion (225)
lethal mutation (224)
linkage group (222)
map unit (224)
nondisjunction (225)
point mutation (225)
sex linkage (222)
sickle cell anemia (226)
somatic mutation (224)
substitution (225)
translocation (225)
X-linked gene (222)
Y-linked gene (222)

12-2

- A pedigree is a family record that shows how a trait is inherited over several generations.
- Single-allele traits are controlled by a single allele of a gene. Multiple-allele traits are controlled by three or more alleles of a gene.
- Polygenic traits are controlled by two or more different genes.
- The genes for X-linked human traits, such as colorblindness, Duchenne muscular dystrophy, and hemophilia, are found on the X chromosome.
- A sex-influenced trait is expressed differently in men than it is in women, even when individuals have the same genotype.
- Nondisjunction during meiosis can cause gametes to have one too few or one too many chromosomes. If such a gamete is fertilized, a genetic disorder such as monosomy ($2n$=45) or trisomy ($2n$=47) can occur.
- Genetic screening examines a person's genetic makeup and can identify people at risk of passing on genetic disorders to their children. Genetic counseling informs these individuals about problems that could affect their prospective offspring.
- A karyotype can be used to reveal chromosomal abnormalities.
- Amniocentesis and chorionic villi sampling enable physicians to test a fetus for the presence of genetic disorders.

Vocabulary

amniocentesis (232)
carrier (228)
chorionic villi sampling (232)
colorblindness (229)
Down syndrome (231)
Duchenne muscular dystrophy (229)
genetic counseling (231)
genetic disorder (228)
genetic marker (228)
genetic screening (231)
hemophilia (229)
Huntington's disease (HD) (228)
monosomy (231)
multiple-allele trait (229)
pattern of inheritance (228)
pedigree (227)
phenylketonuria (PKU) (232)
polygenic trait (229)
sex-influenced trait (229)
single-allele trait (228)
trisomy (231)
trisomy-21 (231)

CHAPTER 12 REVIEW ANSWERS

REVIEW

1. Germ-cell mutations occur in the gametes. Somatic mutations take place in body cells.
2. Nondisjunction is the failure of a chromosome to separate from its homologue during meiosis.
3. Multiple-allele traits are controlled by three or more alleles of the same gene that code for the same trait. Polygenic traits are controlled by two or more different genes.
4. A sex-linked trait is controlled by a gene that is found only on one sex chromosome. A sex-influenced trait is expressed differently in men and women who have the same genotype for the trait.
5. In amniocentesis, a small amount of amniotic fluid is removed at some time during the 14th through 16th weeks of pregnancy; fetal cells and proteins in the fluid can then be analyzed. Fetal cells from chorionic villi sampling are also obtained and analyzed. A karyotype of the fetus can be prepared using fetal cells obtained by either method.

6. a **7.** b
8. c **9.** b
10. d **11.** c
12. b **13.** a
14. b **15.** a

16. Morgan crossed a female fruit fly, heterozygous for eye color, with a red-eyed male. He noted that all of the female offspring were red-eyed and half of the male offspring were white-eyed. Morgan's observations could be explained only if the allele for eye color was carried on the X chromosome.
17. Hemophilia is an X-linked recessive disorder. Therefore, females would have to inherit two copies of the allele in order to express the disease. Males are affected by inheriting just one copy of the gene since the Y chromosome does not carry this gene.

18. A linkage group consists of all the genes on a pair of homologous chromosomes. Humans have 23 linkage groups.
19. A chromosome mutation is either a change in the structure of a chromosome or the loss of an entire chromosome. A point mutation involves the substitution, addition, or deletion of a single nucleotide.
20. autosomal dominant
21. A genetic marker is a short section of DNA that is known to have a close association with a particular gene of interest located nearby.
22. I^AI^A and I^Ai
23. No; heterozygous
24. The karyotype of a normal human male would show chromosomes in pairs of two, with each chromosome paired with its homologue. However, the karyotype of a human male with Down syndrome would show three copies of chromosome 21.
25. The karyotype of an XXY individual would show three sex chromosomes—two Xs and one Y—instead of two sex chromosomes. The karyotype of an XO individual would show only one sex chromosome, an X.

CRITICAL THINKING

1. phenotypic ratio—3 gray-bodied long-winged : 1 black-bodied short-winged; genotypic ratio—1 *GGLL* : 2 *GgLl* : 1 *ggll*
2. Individuals who are heterozygous for sickle cell anemia have normal and sickled red blood cells. Since sickled red blood cells cannot transport hemoglobin effectively, individuals can experience circulatory difficulties under low-oxygen conditions.
3. A broken portion of an extra copy of chromosome 21 has become attached to the end of another, nonhomologous chromosome, giving $2n = 46$. The parent of the offspring is phenotypically

REVIEW

Vocabulary

1. Distinguish between germ-cell mutations and somatic mutations.
2. Define the term *nondisjunction.*
3. Distinguish between multiple-allele traits and polygenic traits.
4. Describe the difference between a sex-linked trait and a sex-influenced trait.
5. Distinguish between amniocentesis and chorionic villi sampling.

Multiple Choice

6. If a heterozygous red-eyed female *Drosophila* is crossed with a red-eyed male *Drosophila,* (a) half the male offspring will be red-eyed and half will be white-eyed (b) three-fourths of the male offspring will be red-eyed and one-fourth will be white-eyed (c) all the offspring will be red-eyed (d) one-fourth of the male offspring will be red-eyed and three-fourths will be white-eyed.
7. A chromosomal map shows (a) whether a gene is autosomal or recessive (b) the positions of genes along a chromosome (c) the presence of mutant alleles (d) the sex of the individual.
8. The deletion of a single nucleotide results in (a) nondisjunction (b) monosomy (c) a frame shift mutation (d) a translocation.
9. Geneticists can use pedigrees to determine (a) environmental effects on trait expression (b) whether someone could be a carrier for a genetic disorder (c) the frequency of a gene in a population (d) the position of a gene on a chromosome.
10. A geneticist working with *Drosophila* discovers a mutant phenotype that appears only in males who are offspring of males showing the same phenotype. This information suggests that the mutant phenotype is (a) autosomal dominant (b) autosomal recessive (c) X-linked (d) Y-linked.
11. A man and a woman have the same genotype for a particular trait, yet only one of them expresses that trait. This evidence suggests that the trait is (a) sex-linked (b) polygenic (c) sex-influenced (d) multiple-allele.
12. Klinefelter's syndrome is an example of (a) monosomy (b) trisomy (c) a point mutation (d) translocation.
13. A karyotype can reveal (a) chromosomal abnormalities (b) blood type (c) point mutations (d) carriers for recessive traits.
14. Testing for phenylketonuria (PKU) enables physicians to (a) recognize abnormalities in chromosome number (b) prescribe a special diet (c) correct fetal defects before birth (d) determine the sex of a fetus.
15. The greatest amount of phenotypic variation is seen in (a) polygenic traits (b) single-allele traits (c) multiple-allele traits (d) sex-linked traits.

Short Answer

16. What evidence led Morgan to hypothesize that the gene for eye color in *Drosophila* is carried on the X chromosome?
17. Why is hemophilia carried by females but expressed in males and rarely in females?
18. What is a linkage group? How many linkage groups do humans have?
19. How does a chromosome mutation differ from a point mutation?
20. What is the pattern of inheritance of Huntington's disease?
21. What is a genetic marker?
22. List the possible genotypes for a person whose ABO blood group is type A.
23. The photograph below shows red blood cells taken from an individual. Notice that one of the red blood cells is bent, while the other one is normal. Does this person have sickle-cell anemia? What is this person's genotype with respect to sickle cell anemia?

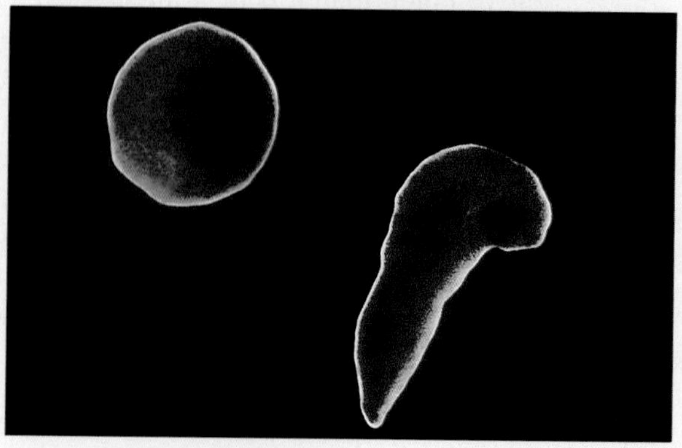

24. Explain how you would distinguish the karyotype of a normal human male from that of a human male with Down syndrome.
25. Describe how the karyotype of an XXY human would differ from that of an XO human.

CRITICAL THINKING

1. In *Drosophila* the genes for body color and wing length are on the same chromosome. Gray body (*G*) is dominant to black body (*g*), and long wings (*L*) are dominant to short wings (*l*). Assume that both dominant alleles are on the same chromosome. Draw a Punnett square representing the cross *GgLl* × *GgLl*. Write the phenotypic and genotypic ratios that would be expected among the offspring, assuming that crossing-over does not occur.
2. Individuals who are heterozygous for sickle cell anemia generally have no symptoms of the disease. However, they should avoid extreme conditions that severely reduce the amount of oxygen available to the body, such as playing vigorous sports at high elevations. Explain why this would be advisable.
3. The karyotypes of some Down syndrome individuals show that they have 46 chromosomes. However, close examination of these karyotypes reveals that three copies of chromosome 21 are indeed present in these individuals. Explain how a translocation can account for this rare form of Down syndrome.
4. A 20-year-old man diagnosed with muscular dystrophy has a sister who is soon to be married. If you were the man, what would you tell your sister?
5. The individual shown in green in the pedigree below is afflicted with a genetic disorder. State the pattern of inheritance for the disorder and whether the disorder is autosomal or sex-linked. Explain your answer.

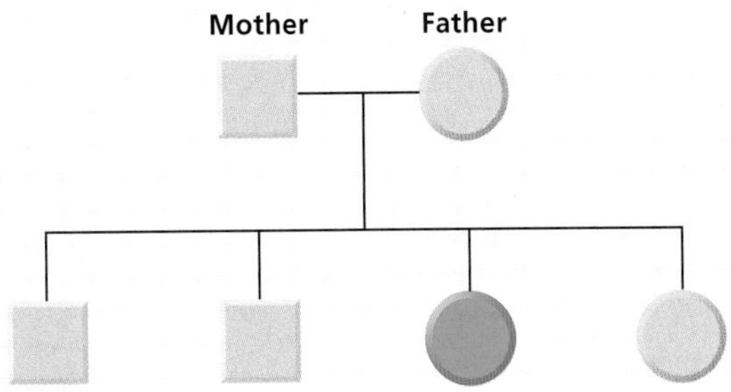

6. What advice might a genetic counselor give to the unaffected brother and sisters shown in the pedigree in question 5?

EXTENSION

1. Read "Chromosomes Show Plants' Secret Complexity" in *Science News,* December 18, 1999, on page 389. Explain why the mustard plant *Arabidopsis thaliana* is a favorite laboratory plant of geneticists.
2. Read "High Risk Defenses" in *Natural History,* February 1999, on page 40. Explain what Pythagoras meant when he said, "Do not eat broad beans!" Describe how the mutant gene for cystic fibrosis is a protection against typhoid fever.
3. Biologists have long been interested in studying the inheritance of traits in twins, particularly identical twins who have been raised apart. Do library or on-line research, and prepare a report on the study of twins in genetics.
4. Do library or on-line research, and prepare a report on the history of the use of *Drosophila* in genetic studies. Conclude your report with a list of the advantages of using *Drosophila*.

normal but is a heterozygous carrier for the translocation.
4. Muscular dystrophy is a sex-linked recessive disorder. The man's sister may be a carrier of muscular dystrophy and may wish to consult a genetics counselor before having children.
5. The disorder is inherited as an autosomal recessive-linked pattern.
6. The siblings may wish to undergo genetic screening for the recessive allele for the disorder. If any sibling is a carrier for this allele, that individual's spouse should also undergo genetic screening. If both the husband and the wife are carriers, there is a 25 percent chance that any future child will have the disorder.

EXTENSION

1. *Arabidopsis thaliana* is easy to study. The plants have only five small chromosomes that contain relatively little junk DNA. The chromosomes are also packed with genes—over 8,000 genes are found in chromosomes 2 and 4.
2. In Greece, people who have two copies of a specific mutant gene can develop a severe, sometimes fatal anemia from eating broad beans. People who have only one copy of the mutant gene for cystic fibrosis produce an intestinal enzyme that bars entry to the organism that causes typhoid fever, yet these people do not have cystic fibrosis, the condition that results from having two copies of the mutant gene.
3. Student reports should note the difficulty of separating environmental influences from the effects of heredity.
4. Thomas Hunt Morgan was the first to use *Drosophila* in genetic studies. *Drosophila* display easily distinguishable traits, a short life cycle, a large number of offspring from each mating, and only four pairs of chromosomes. In addition, they are inexpensive to maintain.

CHAPTER 12
INVESTIGATION

TIME REQUIRED

One 50-minute class period

SAFETY PRECAUTIONS

Remind students to avoid handling any of the electrical cords or power strips used in operating the computers.

QUICK REFERENCES

Lab Preparation Notes are found on page 219B.

Holt BioSources provides a Teaching Resources CD-ROM that contains "Using Gowin's Vee in the Lab" and "Scoring Rubrics."

PROCEDURAL TIPS

1. If enough computers are available, students can work individually; otherwise, they should work in teams of two to four.
2. Be sure that students are familiar with the instructions for loading and running an interactive investigation before they begin the lab.
3. Remind students that the animations of normal and abnormal meiosis are greatly simplified to facilitate understanding. Only two pairs of chromosomes are shown.
4. Students may find the computer's graph of parental age versus incidence of Down syndrome confusing. Inform them that the program plots two points for each combination of parents. The blue point represents the incidence of Down syndrome for the father's age, and the red point represents the incidence for the mother's age. Inform students that the meter labeled "Incidence of Down Syndrome per 1,000 Live Births," located at the top left of the screen, provides one value for the incidence of Down syndrome, when given the ages of both parents. Students should use this value when recording their data.

CHAPTER 12

INTERACTIVE EXPLORATION

Meiosis: Down Syndrome

OBJECTIVES

- Compare the events of a normal meiotic division with those resulting in Down syndrome.
- Determine the relationship between parental age and the likelihood that an offspring will have Down syndrome.

MATERIALS

- computer with CD-ROM drive
- CD-ROM *Interactive Explorations in Biology: Cell Biology and Genetics*
- graph paper

Background

This interactive exploration allows you to learn about the cause of Down syndrome, an inherited condition that results in mental retardation, short stature, stubby hands and feet, and a characteristic heavy eyefold. This condition is a consequence of nondisjunction, or the failure of chromosomes to separate normally during meiosis. In this exploration, you will have the opportunity to examine how the age of each parent affects the events of meiosis.

Prelab Preparation

1. Load and start the program "Meiosis: Down Syndrome." You will see an animated diagram like the one below. Click the Navigation button, and then click the Topic Information button. Read the focus questions, and review the concepts of ploidy, meiosis, aneuploidy, and nondisjunction.
2. Now click the word *Help* at the top left of the screen, and select How to Use This Exploration. Listen to the instructions for operating the exploration. Click the Interactive Exploration button on the Navigation Palette to begin the exploration.

FEEDBACK METERS

Incidence of Down Syndrome: *displays the frequency of Down syndrome births for the combination of parents chosen*

VARIABLES

Age of Mother: *allows you to set the mother's age*

Age of Father: *allows you to set the father's age*

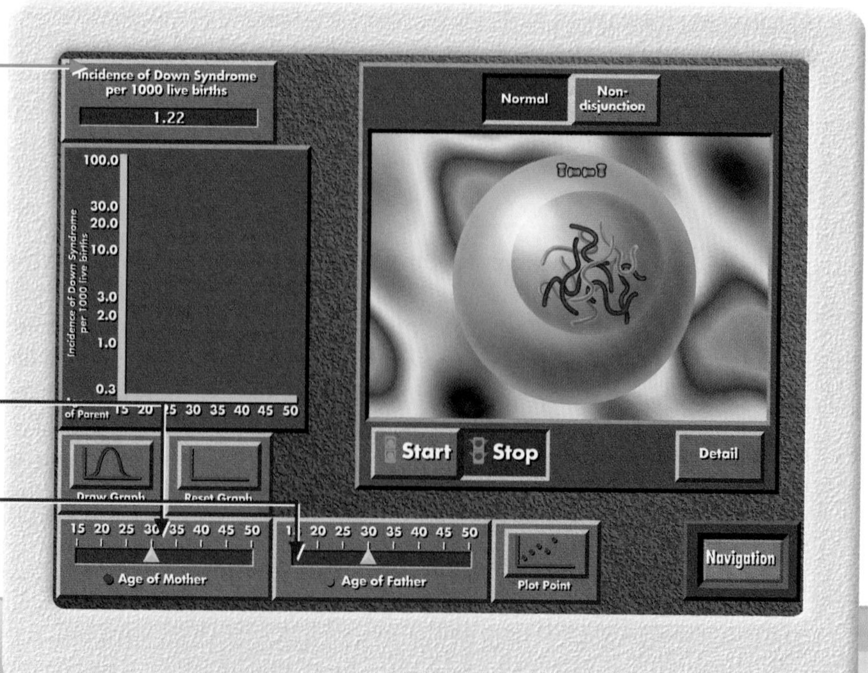

Procedure

PART A Normal Meiosis

1. Click the Normal button to observe a simulation of normal meiosis. For simplicity, this animation shows only two pairs of chromosomes.
2. Click the Detail button. Then click the Forward button (the one with the right-facing arrow) to progress through the eight steps of a normal meiotic division. As you view each step, read the caption and study the illustration.
3. Click the Close button.
4. Click the Start button to observe the same meiotic events in continuous motion. How many gametes formed after meiosis was completed?

PART B Nondisjunction

5. Click the Nondisjunction button.
6. Click the Detail button. Then click the Forward button to observe nondisjunction. As you view each step, read the caption and study the illustration. What differences between normal meiosis and nondisjunction do you observe in frame 4?
7. Click the Close button.
8. Click the Start button to observe nondisjunction in continuous motion.

PART C Effect of the Father's Age

Make a table like Table A shown below.

TABLE A *EFFECT OF THE FATHER'S AGE*

Age of father	Incidence of Down syndrome
25	0.70
30	0.70
35	0.70
40	0.70
45	0.70
50	0.70

9. Click and slide the left-hand indicator so that the age of the mother is 25.
10. Click and slide the other indicator so that the age of the father is 25.
11. Click the Plot Point button. In your table, record the probability that these parents will have a child with Down syndrome, shown by the meter labeled "Incidence of Down Syndrome per 1000 Live Births."
12. Repeat steps 10–11 five times, each time increasing the age of the father by five years. Make a graph of the age of the father versus the incidence of Down syndrome. What can you conclude about the effect of the father's age on the likelihood of having a child with Down syndrome?

PART D Effect of the Mother's Age

Make a table like Table B shown below.

TABLE B *EFFECT OF THE MOTHER'S AGE*

Age of mother	Incidence of Down syndrome
25	0.70
30	1.22
35	3.38
40	12.3
45	56.0
50	145.0

13. Click the Reset Graph button.
14. Click and slide the indicator so that the age of the father is 25.
15. Click and slide the indicator so that the age of the mother is 25.
16. Click the Plot Point button. In your table, record the incidence of Down syndrome per 1,000 live births.
17. Repeat steps 15–16 five times, each time increasing the age of the mother by five years. Be sure to record the incidence of Down syndrome for each age. Make a graph of the mother's age versus the incidence of Down syndrome. What can you conclude about the effect of the mother's age on the likelihood of having a child with Down syndrome?

Analysis and Conclusions

1. In step 12, why did you hold the mother's age constant while varying the father's age?
2. In the United States, the age at which women have their first child has been increasing. How might this affect the incidence of Down syndrome?

ANSWERS TO PROCEDURE

4. Four gametes have formed.
12. The father's age apparently has no effect on the incidence of Down syndrome because the incidence does not vary with the age of the father.
17. The likelihood of having a child with Down syndrome rises with the mother's age. Student's graphs should reflect the data shown above.

ANSWERS TO ANALYSIS AND CONCLUSIONS

1. The mother's age was held constant so that the influence of the father's age could be evaluated.
2. This may lead to an increase in the number of children with Down syndrome.

CHAPTER 13
DNA TECHNOLOGY

	TOPICS	TEACHING RESOURCES	LABS, CLASSWORK, AND HOMEWORK
BLOCKS 1 & 2 90 minutes	**Introducing the Chapter, p. 238**	**ATE** Focus Concept, p. 238 **ATE** Assessing Prior Knowledge, p. 238 **ATE** Understanding the Visual, p. 238	■ Supplemental Reading Guide, *The Double Helix*
	13-1 The New Genetics, p. 239 Manipulating Genes, p. 239 Transplanting Genes, p. 241 Expression of Cloned Genes, p. 242	**ATE** Section Overview, p. 239 **ATE** Visual Link Figure 13-1, p. 239 Figure 13-2, p. 240 39, 40	**Lab C16,** DNA Whodunit **Lab D4,** Antibiotic Resistance Study Guide, Section 13-1 **PE** Section Review, p. 242
BLOCKS 3 & 4 90 minutes	**13-2 DNA Technology Techniques, p. 243** DNA Fingerprints, p. 243 Human Genome Project, p. 246	**ATE** Section Overview, p. 243 **ATE** Critical Thinking, p. 246	**PE** **Quick Lab,** p. 246, Comparing Unique Characteristics **Lab D5,** Introduction to Agarose Gel Electrophoresis **Lab D6,** DNA Fragment Analysis **Lab D7,** DNA Ligation **Lab D8,** Comparing DNA Samples Study Guide, Section 13-2 **PE** Section Review, p. 248
BLOCKS 5 & 6 90 minutes	**13-3 Practical Uses of DNA Technology, p. 249** Producing Pharmaceutical Products, p. 249 Genetically Engineered Vaccines, p. 250 Increasing Agricultural Yields, p. 251 Safety and Environmental Issues, p. 251	**ATE** Section Overview, p. 249 **ATE** Critical Thinking, p. 251 **ATE** Visual Link Figure 13-10, p. 251 32A, 33A, 41	**PE** **Chapter 13 Investigation,** pp. 256–257 **Lab A8,** Making a Genetic Engineering Model **Lab E5,** Heredity in Families Study Guide, Section 13-3 **PE** Section Review, p. 252
BLOCKS 7 & 8 90 minutes	**REVIEW and ASSESSMENT** **PE** Chapter 13 Review, pp. 254–255 ■ Performance-Based Assessment—Identifying Dominant and Recessive Traits	Audio CD Program	**CHAPTER TESTING** ■ Chapter Test (blackline master) ▲ Test Generator ■ Assessment Item Listing ■ Scoring Rubrics and Classroom Management Checklists

MODERN BIOLOGY

HOLT BIOSOURCES

One-Stop Planner CD-ROM
Shows these resources within a customizable daily lesson plan:

KEY

- PE Pupil's Edition
- ATE Teacher's Edition
- ■ Active Reading Guide
- ★ Study Guide
- Teaching Transparencies
- Laboratory Program
- ■ Holt BioSources Teaching Resources
- ■ Active Reading Guide
- ★ Study Guide
- ▲ Test Generator

READING FOR CONTENT MASTERY

Preview: Have students preview the objectives, the introductory paragraph, topic sentences, and captions before reading the chapter.

- **ATE** Active Reading Technique: K-W-L, p. 239
- ■ Reading Strategy: K-W-L Worksheet
- ■ Active Reading Guide Worksheet 13-1
- **PE** Summary/Vocabulary, p. 253

- ■ Active Reading Guide Worksheet 13-2
- **PE** Summary/Vocabulary, p. 253

- **ATE** Active Reading Technique: K-W-L, p. 252
- ■ Active Reading Guide Worksheet 13-3
- **PE** Summary/Vocabulary, p. 253

TECHNOLOGY AND INTERNET RESOURCES

CNN PRESENTS
Video Segment 5—Cloning Mice
Video Segment 7—Gene Progress

Holt Biology Videodiscs Teacher's Correlation Guide, Lessons 13-1 through 13-2

Audio CD Program, Sections 13-1 through 13-2

internetconnect

sciLINKS NSTA

On-line Resources: www.scilinks.org
The following *sci*LINKS Internet resources can be found in the student text for this chapter:

Topics:
- DNA fingerprinting, p. 245
- Herbicides, p. 251

go.hrw.com

On-line Resources: go.hrw.com
Visit the HRW Web site for a variety of resources related to this chapter. Just type in the keyword HM2 Chapter 13.

Smithsonian Institution®
Visit **www.si.edu/hrw** for additional on-line resources.

CNNfyi.com
Visit **www.cnnfyi.com** for late-breaking news and current event stories selected just for you.

LAB ACTIVITY PLANNING

CHAPTER 13 INVESTIGATION

Gel Electrophoresis, pp. 256–257

OVERVIEW
Students will prepare and analyze a gel to identify an unknown restriction enzyme used to cut a DNA sample.

SAFETY
Students will wear disposable gloves, lab aprons, and safety goggles at all times. If a chemical gets in a student's eye, flush the eye with water for 15 minutes and seek immediate medical attention. Follow safety instructions for use of the electrophoresis chamber. Provide a container for broken glass. Consult Material Safety and Data Sheets (MSDS) for the disposal of solutions and gels.

PREPARATION
1. Purchase 0.125 μg/μL Lambda DNA. Each student group will need six microtubules: one labeled "A," one labeled "B" (with 1 μL *Hin*dIII), one labeled "C" (with 1 μL *Bam*HI), one labeled "D" (with 1 μL *Eco*RI), one labeled "E" (with 1 μL *Hin*dIII), and one labeled "Lambda DNA" (with 34 μL 0.125 μg/μL Lambda DNA).
2. To make the 1 × Tris-Borate-EDTA (TBE) buffer, dilute 250 mL of 10 × TBE buffer to 2.5 L with distilled water.
3. Provide each group with 15 mL of 0.8 percent agarose, and add 1.6 g of agarose to 200 mL of TBE buffer. Stir to suspend the agarose, then heat on a hot plate or in a microwave oven until dissolved. Cover and keep in a 65°C water bath until ready for use.
4. To prepare loading dye, dissolve 0.25 g bromophenol blue, 0.25 g xylene cyanol, 50.0 g sucrose, and 1.0 mL 1 M Tris (pH 8) in 60 mL distilled water. Dilute with distilled water to make a total volume of 100 mL.

Quick Labs

Comparing Unique Characteristics, p. 246
Students will need an ink pad, paper, pencil, and scissors for this activity.

UNDERSTANDING THE VISUAL

Have students examine the tomatoes in the picture. Ask them how these tomatoes differ in appearance from tomatoes they have seen in grocery stores or at home. (These tomatoes are not visibly different from any other tomato.) Tell them that the tomatoes in the picture contain genetic information that keeps them from spoiling as quickly as other tomatoes do. Tell students that these tomatoes are the result of genetic engineering. Ask students to name some advantages to growing food crops that do not ripen and spoil quickly. (More food may be received by consumers in better condition and without loss of flavor or nutrients.)

FOCUS CONCEPT

Reproduction and Inheritance
Methods have been developed to analyze DNA and isolate genes. Some of these genes can be manipulated to change the biochemical activities that normally occur in a cell. These changes can be passed on to the next generation of cells. Genetic technology such as this has led to the production of pharmaceutical products, improvement in agricultural crops, and gene therapy.

ASSESSING PRIOR KNOWLEDGE

Review the following concepts.

Cell Reproduction: *Chapter 8*
Ask students to identify the phases of the cell cycle. Emphasize the S phase, and remind them that this is the phase during which DNA is replicated and that it is therefore important in genetic engineering.

Gene Expression: *Chapter 10*
Review with students the roles of mRNA and tRNA in the making of a protein. Emphasize that a gene is expressed by the making of a protein.

CHAPTER 13

DNA TECHNOLOGY

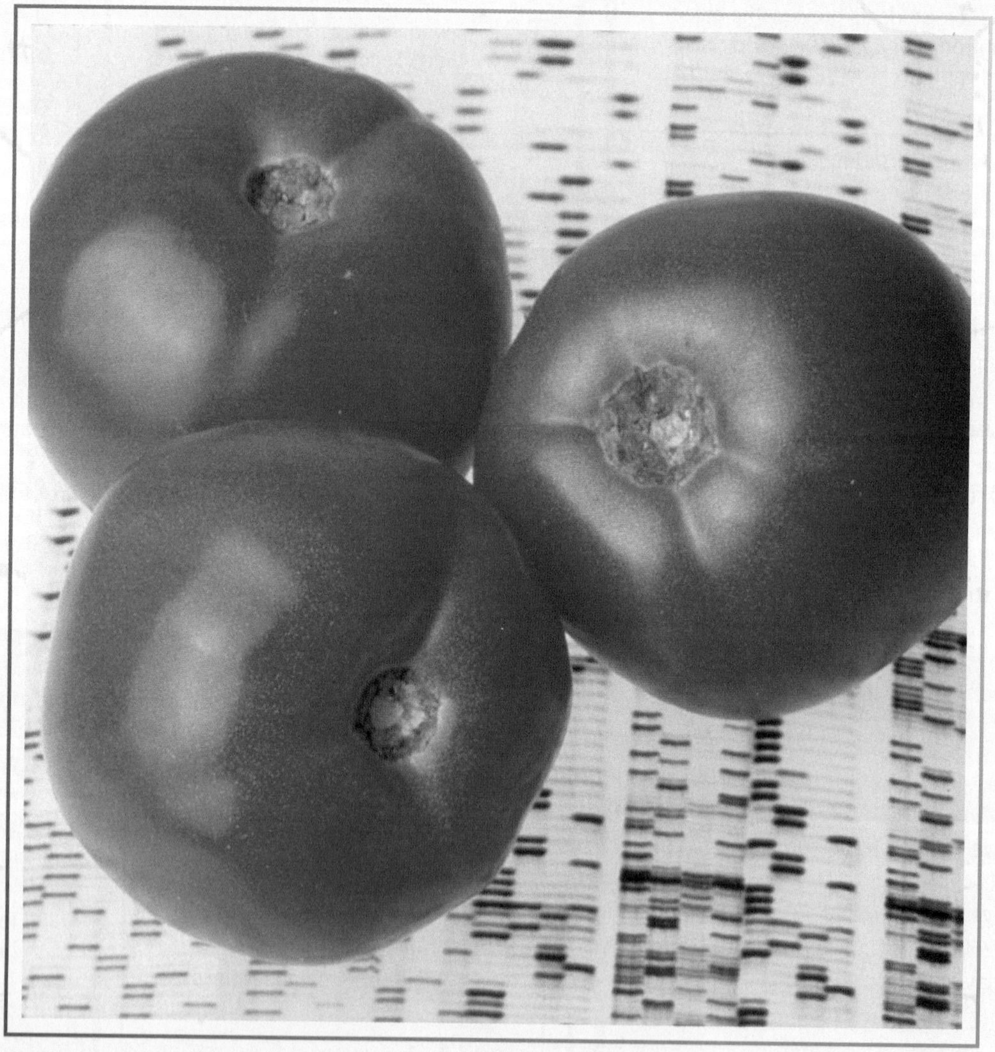

The alteration of a gene that controls ripening gives these genetically engineered tomatoes a better taste and a longer shelf life than other commercially grown tomatoes.

FOCUS CONCEPT: ***Reproduction and Inheritance***
As you read, notice the ways that scientists are endowing organisms with traits that they did not inherit naturally by changing the organisms' genes.

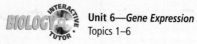

BIOLOGY INTERACTIVE TUTOR **Unit 6—*Gene Expression***
Topics 1–6

THE NEW GENETICS

An understanding of the molecular basis of inheritance has led to a new form of applied genetics called genetic engineering. ***Genetic engineering*** *is the application of molecular genetics for practical purposes. Genetic engineering can be used to identify genes for specific traits or to transfer genes for a specific trait from one organism to another organism.*

SECTION

13-1

OBJECTIVES

▲ Define *genetic engineering.*

● Explain how restriction enzymes can be used to make recombinant DNA.

■ Explain how cloning vectors can be used to clone and transfer genes.

◆ List the steps in a gene-transfer experiment.

MANIPULATING GENES

The technology involved in genetic engineering is called **DNA technology.** DNA technology can be used to cure diseases, to treat genetic disorders, to improve food crops, and to do many other things that may improve the lives of humans. Before you can understand DNA technology, you should understand restriction enzymes and cloning vectors, which genetic engineers use to manipulate genes.

Restriction Enzymes

Recall that a DNA molecule is a long sequence of nucleotides. Genetic engineers use certain bacterial enzymes—called **restriction enzymes**—to cut DNA molecules into more manageable pieces. As Figure 13-1 shows, restriction enzymes recognize

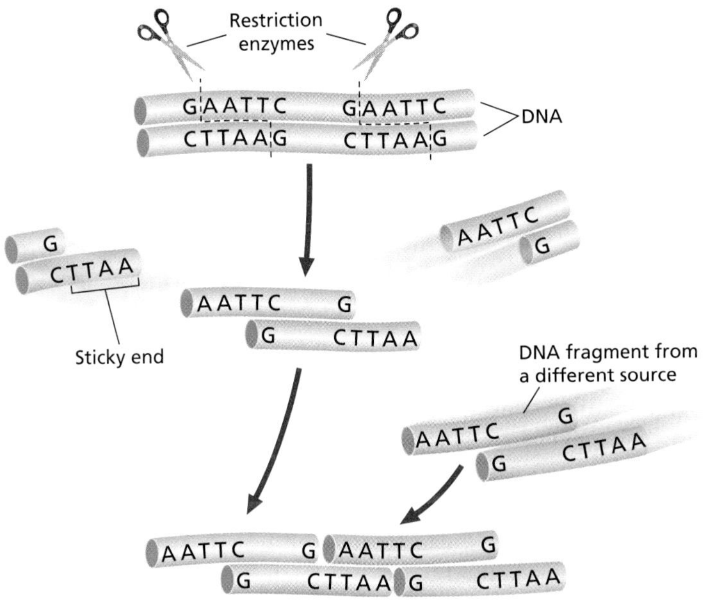

FIGURE 13-1

The restriction enzyme represented by the scissors in this figure recognizes the sequence CTTAAG on each chain of DNA. Then it cuts each chain between the A nucleotide and the G nucleotide, producing DNA fragments with sticky ends.

SECTION 13-1

SECTION OVERVIEW

Cell Structure and Function
Section 13-1 describes the plasmid of a bacterial cell.

Reproduction and Inheritance
Section 13-1 describes how donor DNA can be spliced, inserted into the DNA of the host cell, and ultimately expressed in all cells produced by cell division.

ACTIVE READING

Technique: K-W-L
Before students read this chapter, have them write short lists of all the things they already Know (or think they know) about gene technology. Ask them to contribute their entries to a group list on the board. Then have students list things they Want to know about gene technology. Have students save their lists for use later in the Active Reading activity on page 252.

VISUAL LINK

Figure 13-1
A restriction enzyme recognizes a specific sequence and cuts between specific nucleotides. For example, the enzyme in this illustration, *Eco*RI, recognizes the sequence CTTAAG and its complement GAATTC and cuts between the A and G nucleotides. Other restriction enzymes recognize other DNA sequences. Another restriction enzyme is *Hae*III, which recognizes the sequence CCGG and cuts between the C and G nucleotides.

SECTION 13-1

VISUAL LINK

Figure 13-2

Use this diagram to reinforce the function of restriction enzymes, the importance of sticky ends, and the process for inserting donor DNA into host or plasmid DNA. Emphasize that the plasmid is the cloning vector. Once the gene is inserted and the plasmid is returned to the bacterial cell, the plasmid will divide every time the bacterium divides, thus making copies of the donor gene.

QUICK FACT

☑ Restriction enzymes apparently evolved in bacteria as a defense against DNA invaders. The enzymes destroyed the foreign DNA before it could interfere with cell functions.

TERMINOLOGY NOTE

Restriction enzyme nomenclature follows specific rules that assist the reader in identifying the source of the enzyme. The first letter is capitalized and italicized and represents the abbreviation of the bacterial genus from which the enzyme was isolated. The letters following the genus abbreviation, which are also italicized, represent the species of bacterium. Following these may be a letter that is not italicized; this represents the strain of the particular species from which the enzyme was isolated. Roman numerals designate the order in which an enzyme was isolated from the particular species and strain of bacterium. For example, *Eco*RI was isolated from *Escherichia (E) coli (co),* strain R, and it was the first restriction endonuclease isolated from this strain.

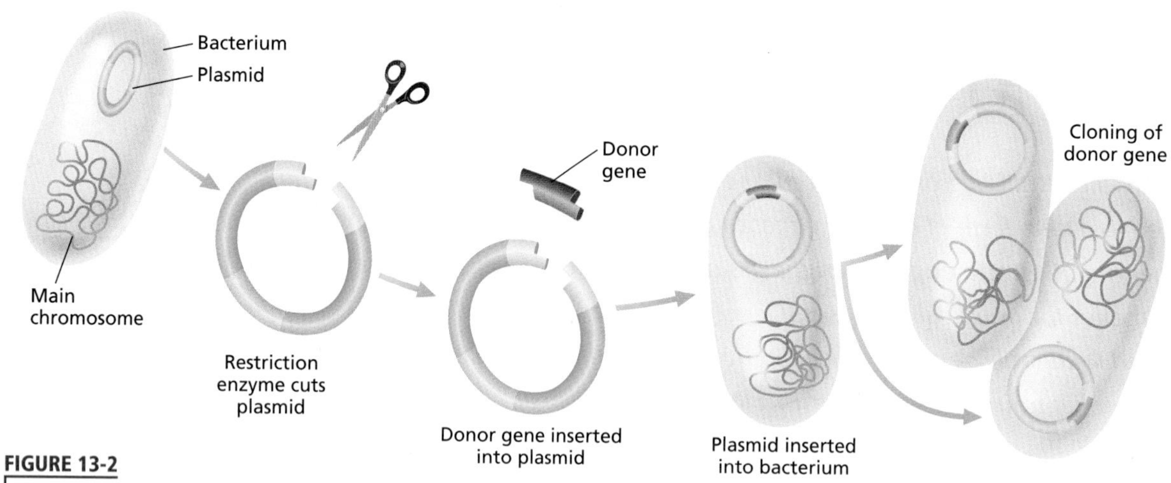

FIGURE 13-2

A donor gene from another organism can be spliced into a plasmid. The plasmid containing the donor gene is then placed inside a bacterium. As the bacterium divides, clones of the donor gene are made.

specific sequences of nucleotides. Then they cut the DNA at a specific site within the sequence.

The restriction enzyme shown in Figure 13-1 recognizes the nucleotide sequence CTTAAG on both chains of DNA. In one chain, the sequence runs from left to right; on the complementary chain, the sequence runs from right to left. The restriction enzyme cuts each chain separately between the G nucleotides and the A nucleotides. In this way, single-chain "tails" of DNA, called **sticky ends,** are created on each DNA segment cut by the restriction enzyme. Sticky ends readily bind to complementary chains of DNA. Thus, pieces of DNA that have been cut with the same restriction enzyme can bind together to form a new sequence of nucleotides.

FIGURE 13-3

The tumor-causing gene in this plasmid can be replaced with a donor gene. When the harmless plasmid containing the donor gene is placed in bacteria, it can be used to transfer new genes into plants.

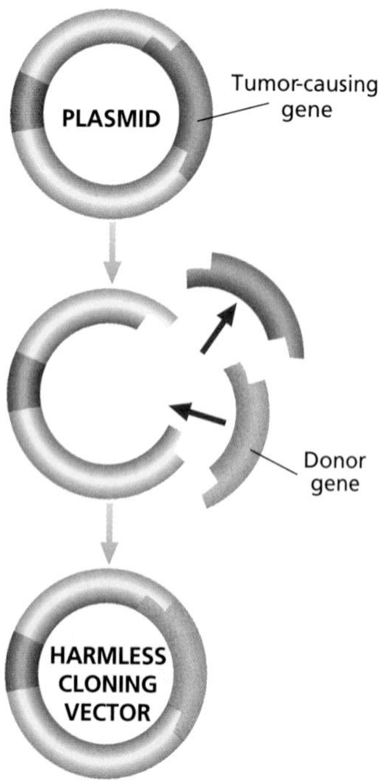

Cloning Vectors

Restriction enzymes can be used to isolate a specific gene. Once a gene has been isolated, it can be transferred by a cloning vector to an organism. A **cloning vector** is a carrier that is used to clone a gene and transfer it from one organism to another. Many bacteria contain a cloning vector called a plasmid. A **plasmid** is a ring of DNA found in a bacterium in addition to its main chromosome.

To be used as a cloning vector in gene-transfer experiments, a plasmid is isolated from a bacterium, as shown in Figure 13-2. Using restriction enzymes, the plasmid is then cut and a **donor gene**—a specific gene isolated from another organism—is spliced into it. Then the plasmid is returned to the bacterium, where it is replicated as the bacterium divides, making copies of the donor gene. Once many copies of the donor gene have been made, plasmids with the donor gene can be isolated from bacteria. Each plasmid now contains a **gene clone,** an exact copy of a gene. One example of a cloning vector, shown in Figure 13-3, is a plasmid that carries a gene that causes tumors in some plants. This plasmid can be modified to no longer cause tumors by replacing the tumor-causing gene with a donor gene. The modified plasmid can be used to transfer the donor gene into plants by infecting the plants with bacteria that carry the plasmid.

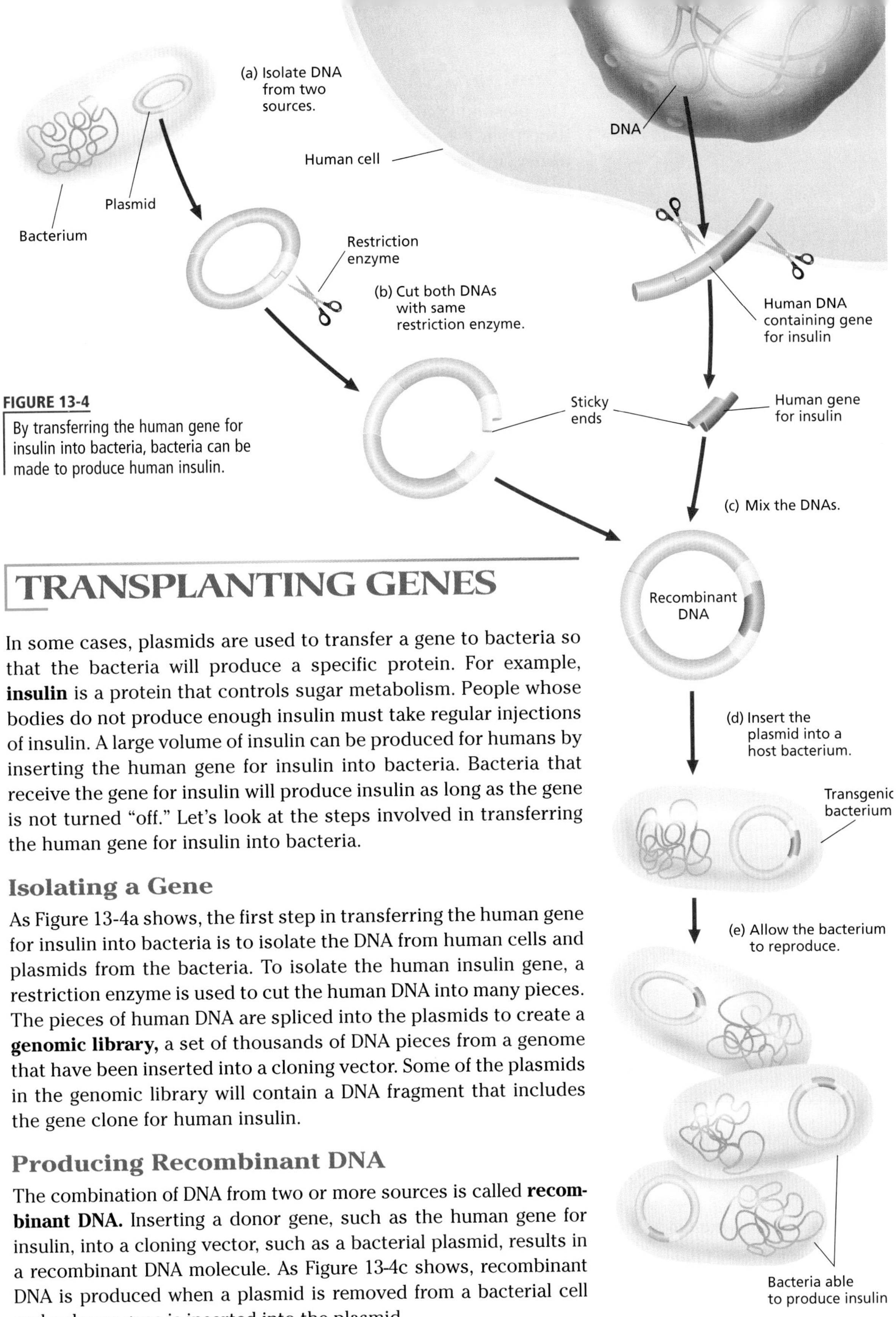

FIGURE 13-4

By transferring the human gene for insulin into bacteria, bacteria can be made to produce human insulin.

TRANSPLANTING GENES

In some cases, plasmids are used to transfer a gene to bacteria so that the bacteria will produce a specific protein. For example, **insulin** is a protein that controls sugar metabolism. People whose bodies do not produce enough insulin must take regular injections of insulin. A large volume of insulin can be produced for humans by inserting the human gene for insulin into bacteria. Bacteria that receive the gene for insulin will produce insulin as long as the gene is not turned "off." Let's look at the steps involved in transferring the human gene for insulin into bacteria.

Isolating a Gene

As Figure 13-4a shows, the first step in transferring the human gene for insulin into bacteria is to isolate the DNA from human cells and plasmids from the bacteria. To isolate the human insulin gene, a restriction enzyme is used to cut the human DNA into many pieces. The pieces of human DNA are spliced into the plasmids to create a **genomic library,** a set of thousands of DNA pieces from a genome that have been inserted into a cloning vector. Some of the plasmids in the genomic library will contain a DNA fragment that includes the gene clone for human insulin.

Producing Recombinant DNA

The combination of DNA from two or more sources is called **recombinant DNA.** Inserting a donor gene, such as the human gene for insulin, into a cloning vector, such as a bacterial plasmid, results in a recombinant DNA molecule. As Figure 13-4c shows, recombinant DNA is produced when a plasmid is removed from a bacterial cell and a donor gene is inserted into the plasmid.

TEACHING STRATEGY
Sticky Ends

Provide students with copies of the model sequences below. Use different colors for each sequence. Have them cut out each strand and position them on their desktop so that the nucleotides are paired correctly. Explain to students that these sequences represent the template strand on top and its complementary strand on the bottom. The arrows indicate the direction students must follow to locate the correct sequence. Explain to students that they will use the restriction enzyme *Eco*RI, which recognizes only the sequence GAATTC (and its complement CTTAAG) and cuts between the G and A nucleotides, to make appropriate cuts in the sequences.

1. Animal DNA containing a specified gene:
CCTGAATTCCCTTGAAGAATTCTT
GGACTTAAGGGAACTTCTTAAGAA
2. A portion of circular bacterial plasmid DNA:
→GCTGCCCGAATTCCCC
CGACGGGCTTAAGGGG←

Have students locate the correct sequence and make the appropriate cuts with scissors. They should note a four-base overhang in each sequence. Let them splice the animal DNA into the bacterial DNA to produce a recombinant molecule. (<u>AATT</u>CCCTTGAAG
GGGAACTTC<u>TTAA</u>
The sticky ends are underlined. The cuts in the plasmid are as follows:
GCTGCCCG<u>AATT</u>CCCC
CGACGGGC<u>TTAA</u>GGGG)

Cloning DNA

The plasmid containing recombinant DNA is inserted into a host bacterium. A host organism receiving recombinant DNA is called a **transgenic organism.** In this case, the transgenic organism is a bacterium containing both bacterial DNA and human DNA.

The transgenic bacterium is placed in a nutrient medium where it can grow and reproduce, as illustrated in Figure 13-4e. Within each bacterium, the plasmid is copied many times, making clones of the gene for insulin. Thousands of bacteria are produced very quickly through cell division, resulting in thousands of bacteria that carry the gene for insulin. The transgenic bacteria can then be used to produce large amounts of insulin.

EXPRESSION OF CLONED GENES

Once a donor gene, such as the human gene for insulin, is transferred to a host cell, it is transcribed and translated as though it were in its own cell. However, not all of the genes in a cell's genome are expressed. Genes are often turned off until the proteins they code for are needed. As you can imagine, it can be difficult to induce host cells to express foreign genes. One way to induce a prokaryotic host cell to express a foreign gene is to also transfer the sequences (promoters) that turn on the foreign gene. In another method, genetic engineers insert a donor gene into prokaryotes beside a gene that is normally produced in large quantities within a cell. That way, the donor gene is expressed along with the host cell's frequently expressed gene.

An obstacle in the expression of eukaryotic genes in prokaryotic hosts is the difference in gene processing. As you know, eukaryotic DNA contains noncoding sequences that are not found in prokaryotic DNA. In order for a prokaryotic cell to receive a functional donor gene, only the coding sequences should be inserted into the host prokaryotic cell. Otherwise, the prokaryotic cell will not remove the noncoding sequences of the gene and the protein will not be produced. Thus, successful gene-transfer experiments require extensive knowledge of the processes required to express genes in different types of cells.

SECTION 13-1 REVIEW

1. Define *genetic engineering.*
2. What role do restriction enzymes play in genetic engineering?
3. How do sticky ends function?
4. Explain the role of cloning vectors in genetic engineering.
5. What steps are used to produce insulin using recombinant DNA and bacteria?
6. **CRITICAL THINKING** List three ways that genetic engineering could be used to improve the lives of humans.

SECTION 13-1

ANSWERS TO SECTION 13-1 REVIEW

1. Genetic engineering is the application of molecular genetics for practical purposes.
2. Restriction enzymes are bacterial enzymes used to cut DNA molecules into more-manageable pieces. Each restriction enzyme cleaves DNA at a specific sequence, producing DNA fragments with sticky ends. A known restriction enzyme may be used to isolate a specific gene for further research.
3. When a DNA fragment is formed by a restriction enzyme, free base pairs are left on each end of the fragment. These single chains, called sticky ends, readily bind to complementary chains of DNA. If the same restriction enzyme is used on a plasmid, the unpaired bases will be on the ends of the plasmid. The sticky ends of the donor gene will then pair up with the DNA in the plasmid, and a bond will form.
4. A cloning vector—a plasmid or a virus—is a carrier used to clone a gene and transfer it from one organism to another.
5. (1) The gene coding for insulin protein is isolated. (2) Recombinant DNA is formed with the plasmid and the human gene coding for insulin. (3) The plasmid is inserted into a bacterium. (4) As the bacterium divides, the gene for insulin is cloned. Insulin is produced by new generations of bacteria and collected.
6. Answers will vary and may include improvements in agriculture, medicine, and forensics.

DNA TECHNOLOGY TECHNIQUES

In Section 13-1, you learned about some of the tools and procedures genetic engineers use to manipulate genes and transplant them into other organisms. In this section, you will learn about some of the techniques used to analyze the nucleotide sequence of an individual's DNA.

SECTION 13-2

OBJECTIVES

▲ Explain what a DNA fingerprint is and how it is prepared.

● Distinguish between the following laboratory techniques: RFLP analysis, gel electrophoresis, and polymerase chain reaction.

■ Describe the purpose of the Human Genome Project and the potential uses of the information collected in the project.

◆ Explain how gene therapy may be used in humans.

DNA FINGERPRINTS

A **DNA fingerprint** is a pattern of bands made up of specific fragments from an individual's DNA. The banding patterns of DNA fragments from two different individuals may be compared to establish whether they are related. The DNA fingerprints of members of two different species can also be compared to determine how closely the species are related. Using DNA fingerprints to compare samples of blood or tissue found at a crime scene with a suspect's blood sample may even help solve a crime. The fragments of DNA in a DNA fingerprint appear as fuzzy bands of stain arranged in columns, like those shown in Figure 13-5.

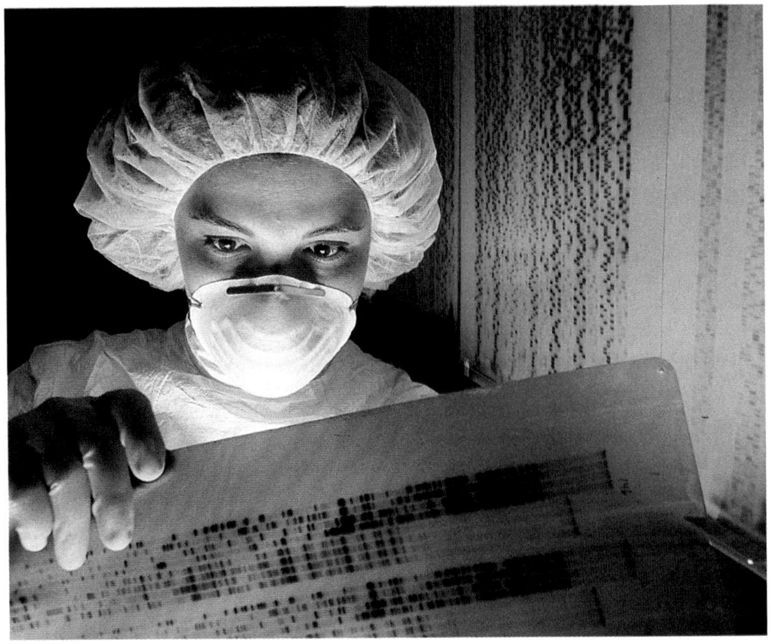

FIGURE 13-5

If you look carefully at the DNA fingerprints that this researcher is examining, you will see many columns of bands. The pattern of bands in each column can be analyzed to establish whether two individuals are related.

SECTION 13-2

SECTION OVERVIEW

Reproduction and Inheritance
Section 13-2 discusses DNA fingerprinting and explains how it can be used in the identification of inherited relationships and diseases.

Cell Structure and Function
Section 13-2 describes the goals and progress of the Human Genome Project. This section also emphasizes the strengths of the Human Genome Project and discusses some of the ethical issues surrounding the project.

▶ ENGAGE STUDENTS

Create a crime-scene scenario. For example, have the class imagine that a glass cabinet has been broken into and a microscope has been taken. Take the fingerprints of four student volunteers by using an ink pad and unlined paper. Some of the fingerprints should be single prints, and others should be overlapping prints of more than one of the "suspects." Tape the papers containing the prints around the area where the microscope disappeared. Inform the class that a microscope has disappeared and that four students have been named as suspects. Tell the students that the area has been dusted for fingerprints and that the fingerprints are represented by the pieces of paper containing the fingerprints. Have two students take the fingerprints of each of the suspects by using an ink pad and paper. Have the class attempt to identify the culprit. The students should not be able to come up with a positive identification because the fingerprints of all the suspects were found at the scene of the crime. Afterward, discuss the importance of fingerprinting in solving crimes, and explain that fingerprints do not always lead to the identification of a criminal.

DEMONSTRATION

Fill a graduated cylinder at least 1 L in size with large beads that are equal in size and color. Be certain that spaces exist between the beads. This represents the gel in an electrophoresis apparatus. Now prepare a mixture of smaller beads. Choose two types, one larger than the other but both smaller than the gel beads. These beads represent the DNA fragments. Pour the bead mixture through the "gel." Students should observe that the smaller beads flow through faster than the larger beads. This is analogous to the way smaller DNA fragments flow through faster than larger ones in gel electrophoresis.

CULTURAL CONNECTION

The Dead Sea Scrolls

The Dead Sea Scrolls, which are over 2,000 years old, have survived in fragments that are excellently preserved. Scott Woodward, a microbiologist from Brigham Young University, has shown that DNA can still be extracted from these scrolls, which are on parchment made of animal skin. Analysis of this DNA will help determine not only what species of animal the parchment came from but also what herd the animal came from. So far, most scrolls have been shown to come from goat skin. If the DNA comes from different herds, then they were probably written in different places. This will determine if the scrolls were written by the Essenes, a Jewish fundamentalist sect, or if they were a collection of important Jewish works from various places that were hidden together to protect them from invaders.

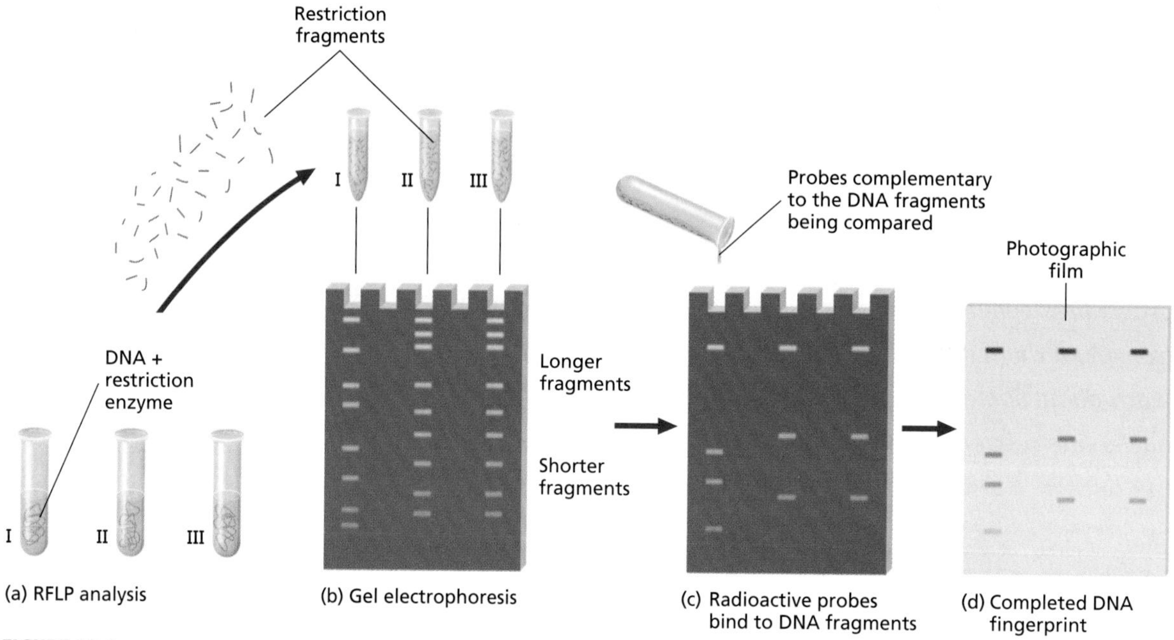

FIGURE 13-6

(a) To make a DNA fingerprint, the DNA sample is first cut into many fragments by restriction enzymes. (b) The DNA fragments are then separated according to their size by gel electrophoresis. (c) Radioactive probes then bind to DNA fragments that have been selected for comparison. (d) Photographic film allows visualization of the radioactive probes that are bound to the DNA fragments, producing a DNA fingerprint.

Making a DNA Fingerprint

The method for preparing a DNA fingerprint is called restriction fragment length polymorphism (RFLP) analysis. **RFLP analysis** involves extracting DNA from a specimen of blood or other tissue and cutting it into fragments using restriction enzymes, as shown in 13-6a. The number of fragments and the length of each fragment vary from person to person.

The fragments of DNA are then separated using a technique called **gel electrophoresis.** Gel electrophoresis separates nucleic acids or proteins, primarily according to their size and charge. To make a DNA fingerprint, samples of the DNA being compared are placed in wells made on the gel, as shown in Figure 13-6b. An electric current is then run through the gel for a given period of time. The DNA fragments, which are negatively charged, migrate toward the positively charged end of the gel, but not all at the same rate. The pores in the gel allow smaller DNA fragments to migrate faster—and thus farther across the gel—than longer fragments, separating the fragments by size.

The final step in preparing a DNA fingerprint is making visible only the bands that are being compared. The DNA fragments that have been separated on the gel are split into single chains and blotted onto filter paper. Then **probes**—radioactive segments of DNA that are complementary to the segments being compared—are added to the filter paper. The probes bind to complementary fragments of DNA in the samples, forming visible bands when they are exposed to photographic film, as shown in Figure 13-6c. The exposed film is developed to reveal a DNA fingerprint. The bands can then be analyzed visually or by a computer.

Word Roots and Origins

electrophoresis

from the Latin *electrocus,* meaning "electricity," and the Greek *phoresis,* meaning "to carry"

Accuracy of DNA Fingerprints

The accuracy of DNA fingerprints depends on how unique the prints are. The complete nucleotide sequence of each individual is certainly unique for each person, except in the case of identical twins, who share identical DNA. However, complete sets of DNA are not compared in DNA fingerprints. The DNA technology in use today deals only with a small portion of a person's DNA. Nevertheless, DNA fingerprints are very accurate because they compare segments of DNA that tend to vary the most from person to person. The segments of DNA that vary the most are noncoding segments where the DNA repeats over and over. These repeated patterns are found throughout the genome.

A repeating pattern might consist of the bases GGAT on one chain, matched by the complementary sequence CCTA on the other chain. DNA fingerprinting typically compares the repeat patterns at five different sites. It is highly unlikely—less than one chance in a million—that all five sites compared in DNA fingerprints will match exactly between two people who are not identical twins. The odds against a match occurring through sheer chance multiply as more sites are compared.

Polymerase Chain Reaction

To make a DNA fingerprint, a certain amount of DNA is needed. If only a very tiny amount of DNA is available, the **polymerase chain reaction (PCR)** can be used to quickly make many copies of selected segments of the available DNA.

Notice in Figure 13-7 that PCR requires a DNA molecule or a fragment of DNA, a supply of the four DNA nucleotides, DNA polymerase (the enzyme involved in DNA replication), and primers. A **primer** is an artificially made single-stranded sequence of DNA required for the initiation of replication. When these ingredients are combined and incubated, the selected regions of DNA quickly

FIGURE 13-7

The polymerase chain reaction (PCR) can be used to multiply selected regions of DNA in a sample. Notice that the number of copies of the selected regions of DNA doubles with each cycle of the reaction.

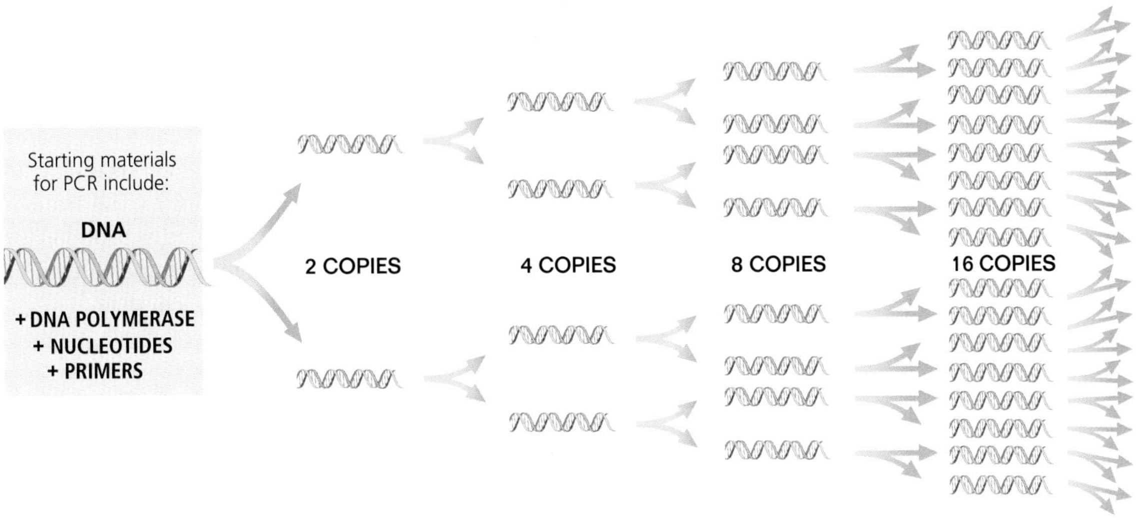

INCLUSION ACTIVITY

Have students make ink impressions of their fingerprints that can be posted in the room. Have students compare their fingerprints with those of their classmates. Lead students to conclude that each person has a different pattern of lines that make up his or her fingerprint. Then compare the fingerprints made by a person's fingers with those made by a sample of DNA. Emphasize that in both kinds of fingerprints the pattern of lines is unique for each person.

MAKING CONNECTIONS

Social Studies

Czar Nicholas II, emperor of Russia from 1894 to 1917, his wife, and their three children were shot to death by the Bolsheviks in 1918 in the aftermath of the Russian Revolution. Their bodies were not found until 1991, when Russian anthropologists discovered their bones. To be certain of the identity of the remains, DNA samples were taken from the corpses and from living members of the family. The DNA fingerprints of the discovered remains matched those of the surviving family members. The remains were determined to be those of Czar Nicholas II and his family.

TEACHING STRATEGY

DNA Fingerprinting

This is an extension of the Engage Students on page 243. Tell students that a few drops of blood were found in the area where the microscope disappeared. Give a prepared simulated DNA fingerprint to each of the suspects. Make copies and distribute them to the students in class. Place one of the samples at the scene of the crime. Have students compare the DNA fingerprints to determine the culprit. Briefly compare and contrast the methods of regular fingerprinting with those of DNA fingerprinting.

Quick Lab

Comparing Unique Characteristics

Time Required 30 minutes

Procedural Tips As an alternative to ink pads, rub the point of a No. 2 pencil on a piece of paper until it makes a solid square of graphite. Have each student roll his or her thumb on the square to coat it with graphite. Press the coated thumb on the sticky side of a piece of transparent tape. Cut the tape, and press it onto the appropriate square of the large sheet. Repeat the procedure to place a print on the small square of paper.

Answers to Analysis All fingerprints have horizontal lines above and below any distinguishing features. Fingerprints are divided into five classes: whorl, right loop, left loop, arch, and tented arch. Students should be able to describe these patterns. Human DNA contains genes common to every person. However, each person also has his or her unique pattern of DNA fragments. A person's DNA fingerprints are as unique as his or her fingerprints.

CRITICAL THINKING

Human Genome Project

The genetic information contained in the human genome may not be immediately understood. Only 5 percent of the genome codes for protein. Scientists are attempting to locate and isolate the genes responsible for the production of those proteins, the nucleotide sequence of the genes, and the function of the proteins. Ask students why the human genome should be deciphered. (Knowing the function of the proteins may help cure or treat genetic disorders; DNA probes can be designed to detect genetic disorders; the history of human evolution and the mechanism of human development may be studied.)

Quick Lab

Comparing Unique Characteristics

Materials ink pad, paper, pencil, scissors

Procedure

1. Cut four 3 in. squares of plain paper.
2. Draw lines on a full-sized sheet of paper that will divide that sheet into four equal squares.
3. Each team member will press his or her right thumb on the ink pad and then will quickly press the inked thumb in a square on the large sheet of paper and again on one of the small squares of paper.
4. Examine each of the thumbprints, and make a list of characteristics that the prints have in common. Describe how each thumbprint is different from the others. Then shuffle the four individual thumbprints, and try to match each one with its duplicate on the large sheet of paper.

Analysis What characteristics are common to all thumbprints? What characteristics made each thumbprint unique? How do a person's fingerprints relate to his or her DNA fingerprints?

double. Every five minutes, the sample of DNA doubles again, resulting in many copies of the sample in a short amount of time, as shown in Figure 13-7.

The new copies of the DNA sample can then be used to make a DNA fingerprint. PCR can use minuscule amounts of a specimen, as few as 50 white blood cells that might be found in a nearly invisible speck of blood, rather than the 5,000 to 50,000 cells needed for RFLP analysis. PCR is important not only for solving violent crimes but also for diagnosing genetic disorders from a few embryonic cells and for studying ancient fragments of DNA found in minute amounts.

HUMAN GENOME PROJECT

Using DNA technology, scientists all over the world are collaborating on one of the most ambitious research efforts in the history of genetics, the **Human Genome Project.** Two of the goals of the Human Genome Project are to determine the nucleotide sequence of the entire human genome—approximately 3 billion nucleotide pairs, or about 100,000 genes—and to map the location of every gene on each chromosome. In addition, the human genome is being compared with the genomes of other organisms in an effort to provide insight into fundamental questions about how genomes are organized, how gene expression is controlled, how cellular growth and differentiation are under genetic control, and how evolution occurs.

This project began in 1990. In 1996, the project had analyzed 1 percent of the 3 billion nucleotide pairs of DNA in the human genome. By knowing even 1 percent of the sequence, it is possible for scientists to establish the identity of and determine the function of about 16,000 genes. Genome research continues internationally at universities, laboratories, scientific institutions, and private companies.

It is hoped that the knowledge gained from the Human Genome Project will improve diagnoses, treatments, and even cures for the approximately 4,000 human genetic disorders. Scientists have already discovered specific genes responsible for several genetic disorders, including cystic fibrosis, Duchenne muscular dystrophy, and colon cancer. As many as 5 percent of the babies born in the United States each year are afflicted with a genetic disorder that could one day be treated or prevented using the information provided by the Human Genome Project. Identifying these genes and the defective proteins for which they code may make it possible to design therapies aimed at correcting the gene defects responsible for genetic disorders.

Gene Therapy

Treating a genetic disorder by introducing a gene into a cell or by correcting a gene defect in a cell's genome is called **gene therapy.** In 1990, doctors first began to develop and test gene therapies on

Literature & Life

Recombinant-DNA Technology

This excerpt is from *Genethics: The Clash Between the New Genetics and Human Values,* by David Suzuki and Peter Knudtson.

Already recombinant DNA technology, along with rapid DNA sequencing and other related laboratory techniques, has begun to offer profound new insights into the workings of genes. First, it has helped to establish the virtual universality of both the genetic code and the graceful molecular dances of genes in nature. Second, it has revolutionized the study of genes of multicellular organisms, including humans, that have long been inaccessible. In so doing, it has led to dramatic discoveries concerning the structure and function of genes in complex plants and animals. One such discovery is that their coded instructions for assembling amino acids into proteins—called *exons*—are often interspersed with long intervening segments of DNA—called *introns*—of largely unknown function. Introns are neatly removed from nuclear RNA molecules prior to protein synthesis. A second discovery is that the comparison of DNA sequences in different species can often serve as a faint chronological record of otherwise hidden evolutionary relationships. Third, recombinant DNA technology has already led to the complete chemical characterization of a number of relatively small viral genomes and has initiated that process in more complex species, including our own. Finally, by facilitating the exchange of genetic components between cells of different species, it has exploded the boundaries of comparative genetics. We are now just beginning to grasp the evolutionary ingenuity with which the genes of different species—however interchangeable their codes—are variously packaged, punctuated and controlled.

But the rewards from recombinant DNA technology will be practical as well as intellectual. Genetic engineering holds great promise in agriculture—for modifying the productivity, growth requirements and nutrition of food crops. In medicine, it is opening up new vistas in the design of vaccines, in the diagnosis of disease and in genetic counseling, and even the possibility of new genetic therapies for ancient hereditary diseases. In the biotechnology industry, it already has established its value in the manufacture of antibiotics, hormones and an assortment of other biologically active substances.

For the first time, we find that we can command individual genes to do our bidding. In the future, we will forever be faced with the temptation to harness our knowledge of hereditary processes long before we have resolved the possible long-term consequences of our applications. If history is any guide, we can expect no shortage of ingenious schemes to reap rewards from our genetic engineering skills. But each of us has a responsibility to ensure that equal imagination is devoted to the search for ways to apply this knowledge wisely and to share its fruits with all humankind.

Reading for Meaning

Suzuki and Knudtson point out some general applications for genetic engineering. List possible uses that you think would benefit humankind. What might be advantages and disadvantages of each item on your list?

Read Further

Suzuki and Knudtson's book, *Genethics,* raises concerns about modern technology's amazing techniques for manipulating genes and how those techniques will be used. When and for what purposes do you think we should perform such genetic engineering?

BACKGROUND

This feature presents some recent advances in recombinant DNA technology. The term *recombinant* DNA means the joining, or recombining, of two pieces of DNA from two different species. The technology has brought about major advances in the understanding of the molecular nature of genes and their functions. It has helped establish the universality of the genetic code and has expanded the field of comparative genetics. It has led to dramatic discoveries concerning the structure and function of genes in complex organisms and the complete chemical analysis of a number of viral genomes.

READING FOR MEANING

Students' answers will vary as to the advantages and disadvantages of genetic engineering. For example, the advantage of greater productivity in agriculture is that there will be more food produced. A disadvantage to genetic engineering in agriculture is that a crop plant may pass its genetically engineered resistance to disease and herbicides to wild plants and thus make the wild plants "superweeds."

READ FURTHER

Students' answers will vary but may include such purposes for genetic engineering as making advances in medicine and curing diseases. These purposes involve manipulating genes within the body cells. A further reading of *Genethics* will show that the authors feel that the manipulation of genes within the gametic cells and gamete-producing cells should be prohibited. They believe that the effects of the manipulation of genes in humans should be confined to a single generation and not passed on to future generations.

SECTION 13-2

RETEACHING *ACTIVITY*

Under conditions of PCR amplification, DNA replicates about every five minutes. Ask students to determine how many copies of a DNA fragment will result from 85 minutes of PCR. (In 85 minutes, DNA will replicate approximately 17 times, doubling the number of copies each time. In 85 minutes there will theoretically be 131,072 DNA fragments replicated.)

ANSWERS TO SECTION 13-2 REVIEW

1. A DNA fingerprint is a pattern of bands made up of specific fragments from an individual's DNA.
2. DNA is extracted from a specimen and cut into fragments using restriction enzymes. The fragments of DNA are then separated according to their size and charge using gel electrophoresis. The fragments are then split into single chains of DNA, identified by radioactive complementary chains, and visualized.
3. Both PCR and DNA replication require the same ingredients to make copies of DNA.
4. The Human Genome Project is an international collaborative effort whose goal is to determine the nucleotide sequence of the entire human genome and to map the location of every gene on each chromosome.
5. Gene therapy is the treatment of a genetic disorder by introducing a gene into a cell or correcting a gene defect in a cell's genome. Gene therapy differs from other treatments in that it aims to treat or cure disorders at a molecular level.
6. Answers will vary. Possible responses include cancer, allergies, mental illness, and any disease caused by a single defective gene.

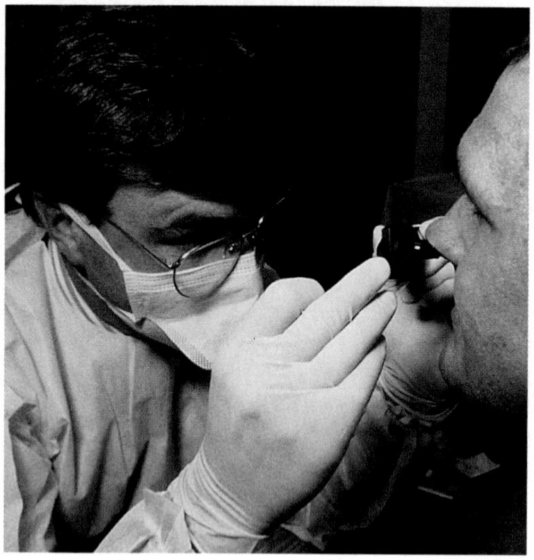

FIGURE 13-8

This patient is receiving gene therapy for cystic fibrosis. Dr. Terrence Flotte (left), of the University of Florida College of Medicine, administers healthy copies of the gene responsible for cystic fibrosis through the patient's nose.

humans. Gene therapy is well suited for treating genetic disorders that result from a deficiency of a single enzyme or protein. As you learned in Chapter 11, the genetic disorder cystic fibrosis is caused by one defective gene, resulting in the malfunction of a single protein. Some success has been reported with nasal sprays that carry a normal cystic fibrosis gene to the cells in the nose and lungs—cells that are particularly affected by a defective cystic fibrosis protein. Figure 13-8 shows the gene-therapy procedure that has been used with some cystic fibrosis patients. The treatment has to be repeated periodically because the cystic fibrosis gene is not inserted on a chromosome but rather in the nucleus of cells; thus, it is not passed on to future generations. Gene-therapy possibilities for hemophilia B, lung cancer, AIDS, ovarian cancer, brain tumors, malignant melanoma, and several other terminal diseases are also under way. Until the obstacles regarding how and where to insert genes safely and directly into eukaryotic chromosomes are overcome, gene therapy may offer only limited and temporary success as a treatment. But researchers hope to overcome these obstacles and to one day provide permanent cures for genetic disorders using gene therapy.

Ethical Issues

The Human Genome Project is producing much information about human susceptibility to disease. However, a genetic susceptibility to a disease does not ensure that the person will get the disease. Nor does such a gene indicate the severity of the disease, should it occur, or how the person would respond to treatment. Many people worry about how personal genetic information will be used. Some think this type of information could be used by insurance companies and employers to discriminate against individuals who are genetically susceptible to a disease. For example, an insurance company may decide to deny insurance to people who are predisposed to diseases that are expensive to treat. The information provided by the Human Genome Project will undoubtedly involve ethical decisions about how society should use the information.

SECTION 13-2 REVIEW

1. **Define *DNA fingerprint*, and discuss the accuracy of using DNA fingerprints to establish relatedness.**

2. **Describe the major steps and techniques involved in preparing a DNA fingerprint.**

3. **Compare the polymerase chain reaction and DNA replication.**

4. **Describe the purpose of the Human Genome Project, and name the potential uses of the information collected in the project.**

5. **What is gene therapy, and how does it differ from traditional treatments for diseases?**

6. **CRITICAL THINKING What genetic disorders not discussed in this chapter might be treated using DNA technology?**

PRACTICAL USES OF DNA TECHNOLOGY

The production of insulin by inserting the human gene for insulin into bacteria is one commercial use of DNA technology. DNA technology can also be used to produce prescription drugs and vaccines and to improve food crops.

SECTION 13-3

OBJECTIVES

▲ Explain how DNA technology can be used to produce medical products.

● Describe some ways that DNA technology can be used to improve crop yields and the food supply.

■ Discuss some environmental and ethical issues in genetic engineering.

PRODUCING PHARMACEUTICAL PRODUCTS

Many medicines are proteins. In the past, these proteins were derived from animal tissues or plants. Today, some proteins can be produced far more inexpensively using DNA technology. Insulin is one of the many products now produced through DNA technology. Table 13-1 lists several other pharmaceutical products produced in this way. As in the production of insulin, these pharmaceutical products can be produced in bulk by bacteria. Just as yeast are fermented in large vats in the production of beer, bacteria can be grown in large vats to produce specific proteins.

TABLE 13-1 ***Medicines Produced by DNA Technology***

Product	Medical uses
Colony-stimulating factors	used to treat immune-system deficiencies by stimulating the production of white blood cells
Erythropoietin	used to treat anemia by stimulating the production of red blood cells
Growth factors	used to promote the healing of wounds by stimulating growth and differentiation of different types of cells
Human growth hormone	used as a treatment for dwarfism
Interferons	used to treat viral infections and some cancers by preventing the replication of viruses
Interleukins	used to treat a variety of conditions, including HIV infections, cancer, and immune deficiencies, by activating and stimulating different kinds of white blood cells
Tissue plasminogen activator	used to prevent heart attacks and strokes by dissolving blood clots
Atrial peptides	used to treat high blood pressure and kidney failure

SECTION 13-3

SECTION OVERVIEW

Stability and Homeostasis
Section 13-3 discusses the use of DNA technology to genetically engineer medicines and vaccines that help maintain stable internal conditions in the human body.

Interdependence of Organisms
Section 13-3 describes how genetic engineering can increase agricultural yields worldwide. Section 13-3 also emphasizes some of the environmental and ethical issues in DNA technology and some of the safeguards that have been devised regarding the stewardship of this new science.

▶ ENGAGE STUDENTS

Select scenic and human-interest pictures from magazines. Distribute these pictures to the class and have the students examine them. Have them list one or two items that appear in the picture that someday might be affected by DNA technology. Students must state whether DNA technology will help or harm the item and whether the change will be beneficial or harmful to society.

RECENT RESEARCH
Polynucleotide Immunization

Polynucleotide immunization involves the splicing of a gene that codes for an immunity factor into plasmid DNA and then injecting the plasmid DNA into an organism. No vector is involved, only naked DNA. Researchers say this method, which has shown some success in animal studies, is promising. Advantages include the reduction or elimination of booster immunizations and an immunity product that is more precise when produced *in vivo.* There are several major safety issues, however, including the possibility that the plasmid DNA may incorporate itself into the host DNA, causing mutations.

RECENT RESEARCH
The Banana Vaccine

Researchers at the Boyce Thompson Institute for Plant Research at Cornell University are creating a banana that can vaccinate children against viral diseases, such as hepatitis B. Scientists inject an altered form of the virus into banana saplings. The genetic material of the virus becomes a permanent part of the plant cells. As the plant grows, its cells produce the virus protein, not the virus. When a human eats the banana, the body makes antibodies to the foreign viral proteins, resulting in immunization against that disease.

TEACHING STRATEGY
Selective Breeding

Discuss the advantages of DNA technology over the advantages of selective breeding. (One example is that new varieties can be developed in a shorter period of time using DNA technology.)

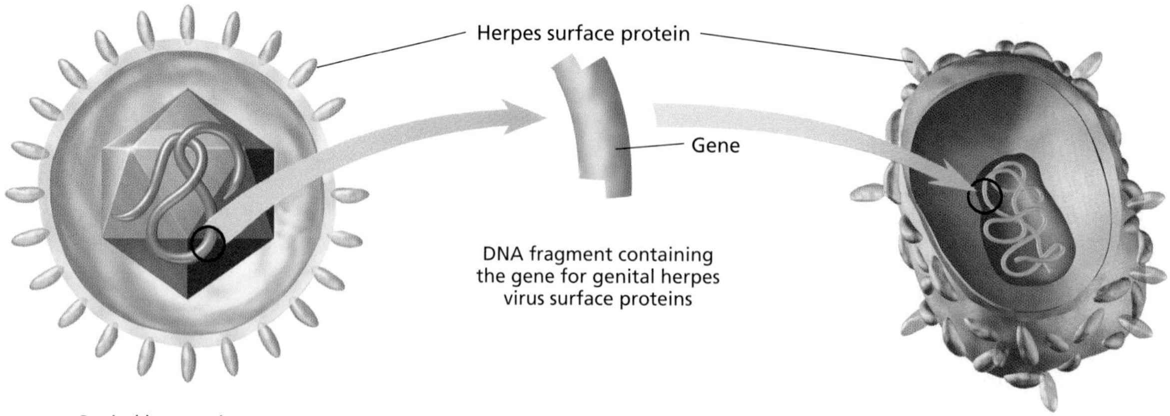

FIGURE 13-9

The gene that codes for the surface proteins of the genital herpes virus can be transferred to a cowpox virus—a virus that is harmless to humans—to produce a vaccine against genital herpes. The cowpox virus then produces the surface proteins characteristic of the genital herpes virus. A person receiving the vaccine will be protected against genital herpes virus infections in the future.

GENETICALLY ENGINEERED VACCINES

Many viral diseases, such as AIDS, smallpox, and polio, cannot be treated effectively by existing drugs. Instead, many viral diseases are combated by prevention, using vaccines. A **vaccine** is a solution that contains a harmless version of a virus or a bacterium. Traditionally, vaccines have been made of disease-causing agents—also called **pathogens**—that have been treated (chemically or physically) so that they can no longer cause disease. Vaccines can also be produced using active pathogens that carry surface proteins that are the same as or very similar to a more-harmful virus. When a person receives a vaccine, his or her body recognizes the pathogen's surface proteins and mobilizes against the pathogen. In the future, if the same pathogen enters the body, the body is prepared to combat it quickly and to prevent or weaken the pathogen's effects.

Usually, a person who receives a vaccine does not become ill. However, on rare occasions a vaccine may cause the disease it is intended to protect people against. DNA technology can be used to produce effective vaccines, which may be safer than some traditionally prepared vaccines. As Figure 13-9 shows, the genes for a disease-causing virus's surface proteins can be inserted into a harmless virus. The transplanted genes cause the harmless virus to produce the surface proteins that alert the body to the presence of the disease-causing virus. DNA technology can also be used to alter the genome of a pathogen so that it no longer causes a disease. The altered pathogens can then be used as a vaccine against unaltered forms of the pathogen.

Word Roots and Origins

pathogen

from the Greek *pathos,* meaning "suffering"

INCREASING AGRICULTURAL YIELDS

DNA technology has been used to develop new strains of plants, which in turn can be used to improve food crop yields. By transferring genes for enzymes that are harmful to hornworms into tomato plants, scientists can make tomato plants toxic to hornworms and effectively protect the plants from these pests, which otherwise could seriously damage them. Figure 13-10 shows a hornworm on a tomato plant that is not toxic to hornworms.

The crop cassava, used to make tapioca pudding and other food products, is highly susceptible to pests and diseases, which often cost farmers up to 80 percent of a cassava crop. Using DNA technology, scientists have made cassava plants resistant to some diseases. In addition, strains of wheat, cotton, and soybeans that are resistant to weed-controlling chemicals—called **herbicides**—have been developed. Such herbicide-resistant crops can be protected from weeds more easily and less expensively than crops that are susceptible to the herbicides.

FIGURE 13-10

Tomato hornworms, like the member of *Manduca quinque maculata* shown here, feed on the leaves and fruits of tomatoes and other related plants. A large hornworm is capable of stripping a tomato plant of its foilage.

Crops That Do Not Need Fertilizer

To make proteins and nucleic acids, all plants require the element nitrogen, N. Although nitrogen is the most abundant element found in the atmosphere, it is in a form that plants are unable to use. Some plants, such as soybeans and peanuts, have bacteria living in their roots that convert, or "fix," nitrogen in the atmosphere into forms of nitrogen that plants can use. Other types of plants take up nitrogen directly from the soil.

To help crops grow, farmers apply expensive nitrogen fertilizers to the soil. Currently, researchers are working to isolate and clone genes from nitrogen-fixing bacteria and transplant the genes into plants. It is hoped that transgenic food crops containing genes for nitrogen fixation will be able to grow in nitrogen-poor soil, where crops often are grown only after the soil has been treated with nitrogen fertilizers.

internet**connect**

SciLINKS NSTA

TOPIC: Herbicides
GO TO: www.scilinks.org
KEYWORD: HM251

SAFETY AND ENVIRONMENTAL ISSUES

Many people are concerned about procedures of genetic engineering and the safety of genetically engineered products. In the United States, genetic engineering is regulated by the Food and Drug Administration (FDA), the National Institutes of Health Recombinant DNA Advisory Committee, the Department of Agriculture (USDA), and the Environmental Protection Agency (EPA). These organizations set standards for safety procedures and products and require special permits and labels for the sale and use of certain products.

Word Roots and Origins

herbicide

from the Latin *herba*, meaning "plant," and *cida*, meaning "to kill"

SECTION 13-3

VISUAL LINK

Figure 13-10

Inform students that as much as 50 percent of the world food supply is lost to pests. Pesticide use is effective but does have many disadvantages. Target pests can develop genetic resistance, thus stronger pesticides may be needed; nontarget insects and animals may be harmed or killed; there may be an upset in the balance of the ecosystem; and the pesticide may be harmful to humans. Genetically engineered crops present an alternative to the use of pesticides.

OVERCOMING MISCONCEPTIONS

Some people assume that a genetically altered fruit would contain genes from another fruit or a bacterium. This assumption is not necessarily correct. Experiments are under way in which animal genes are spliced with plant genes. Scientists hope to extend the shelf life or increase the nutrient or antioxidant level of the food product.

CRITICAL THINKING

Patenting Life

In 1980, the Supreme Court ruled that new life-forms could be patented. A utility company received a patent for a bacterium that it discovered could clean up oil spills. Genes can be patented if their usefulness has been determined, but gene fragments cannot. Have students debate how the patenting of genes or genetically engineered organisms affects society.

SECTION 13-3

ACTIVE READING
Technique: K-W-L
Have students return to their lists of things they Want to know about gene technology, which they made at the beginning of their study of this chapter. Have students place check marks next to each question they are now able to answer. They should finish by making a list of what they have Learned. Have students identify any new questions they might have.

ANSWERS TO SECTION 13-3 REVIEW

1. drugs, vaccines
2. A vaccine is a solution that contains a harmless version of a pathogen, such as a virus. Vaccines help mobilize the body against specific pathogens without causing disease. In some cases, vaccines produced using DNA technology are less expensive and safer than vaccines treated by attenuating pathogens.
3. Answers may include resistance to pathogens, insects, and herbicides.
4. Plants that contain a nitrogen-fixing gene transplanted from bacteria may be able to grow in very poor soils, where other kinds of plants are unable to grow without ample amounts of nitrogen fertilizers.
5. Answers may include allergies to or toxins in foods and superweeds that can wipe out native species.
6. Answers must be rationally supported. Some students may feel that as long as the FDA is confident that the products are identical to traditionally produced products, they are safe and do not require labels. Other students may cite concerns about unsuspected allergens or toxins in the food products.

Genetically Engineered Foods

Some people are concerned that foods produced by genetic engineering could contain toxic proteins or substances that cause allergies in people who consume them. One precaution required by the FDA requires manufacturers of genetically engineered foods to provide scientific evidence that allergy-inducing properties have not been transferred to new foods when the new foods contain a gene transplanted from a food known to cause allergic reactions, such as peanuts. In general, however, foods produced by transgenic crops can be sold without special permits or labels if the product is identical to products produced by nontransgenic crops. However, if a genetically engineered food contains a new protein, carbohydrate, or fat, the FDA generally requires additional approval before the product can be introduced to the food supply.

Notice that the genetically-engineered tomatoes in Figure 13-11 look just like other tomatoes. Scientists isolated and cloned the gene that codes for an enzyme necessary for ripening in tomatoes. By changing this one gene in the genetically-engineered tomatoes, they were able to make a tomato that ripens without becoming soft. Typically, supermarket tomatoes are picked before they are ripe because ripe tomatoes are soft and bruise easily.

FIGURE 13-11
Plant physiologist Athanasios Theologis, shown here, genetically engineered tomatoes to ripen without becoming soft. His genetically engineered tomatoes are comparable to greenhouse-grown tomatoes, which taste better than commercially grown tomatoes that are picked while they are still green.

Genetically Engineered Crops

Some people are concerned that genetically engineered crops could spread into the wild and wipe out native plant species. In addition, they worry that transgenic crop plants could transmit their new genes to other species in neighboring areas. Rice and lawn grasses, for example, exchange genes in their pollen with native plants that are related to them. Exchanging these types of genes with wild plant species could cause powerful new "superweeds" that could take over large areas of land because they would have advantages over native plant species. Regulatory agencies sometimes require labels on the packages of transgenic plants to indicate their proper use and the risks of using transgenic plants that may pose hazards to the environment.

SECTION 13-3 REVIEW

1. **List two types of medical products that can be produced using DNA technology.**
2. **What is a vaccine, and what are the benefits of producing vaccines using DNA technology?**
3. **List three traits for which genes are being transplanted into crop plants.**
4. **What benefit are plants that contain a nitrogen-fixing gene transplanted from bacteria?**
5. **Describe two potential safety and environmental problems that could result from genetic engineering.**
6. **CRITICAL THINKING The FDA does not require special labels for genetically engineered food products that are identical to similar products produced by traditional breeding techniques. Do you think that genetically engineered food products should be labeled as such? Why or why not?**

CHAPTER 13 REVIEW

SUMMARY/VOCABULARY

13-1

- Genetic engineering brings DNA technology and molecular genetics together for practical purposes.
- Restriction enzymes, which are used to isolate and transfer genes, are specific in the nucleotide sequences they recognize and cut. When a restriction enzyme cuts a piece of DNA, it creates single chains—called sticky ends—on the ends of each piece of DNA.
- Sticky ends bind to complementary sticky ends, forming recombinant DNA molecules out of DNA fragments from two or more organisms.
- Cloning vectors, such as plasmids and viruses, are used to clone and transplant genes from one organism to another one.
- Transgenic organisms contain a donor gene and express it just as they do their own DNA as long as the donor gene is turned on.

Vocabulary

cloning vector (240)
DNA technology (239)
donor gene (240)
gene clone (240)
genetic engineering (239)
genomic library (241)
insulin (241)
plasmid (240)
recombinant DNA (241)
restriction enzyme (239)
sticky end (240)
transgenic organism (242)

13-2

- A DNA fingerprint is a pattern of bands that represent certain fragments from an individual's DNA. These fragments are typically noncoding repeating sequences that vary from person to person.
- DNA fingerprints of individuals can be compared to establish relatedness. Chances are less than one in one million that five sites of DNA fingerprints from two people will match, unless the two people are identical twins. As the number of sites that are compared in a DNA fingerprint increases, so does validity of the DNA fingerprint.
- Gel electrophoresis is a technique used to separate DNA or protein fragments by their length and charge.
- Polymerase chain reaction (PCR) can be used to quickly make additional copies of segments of DNA when the original sample is small. The results of PCR can then be used to make a DNA fingerprint.
- The goal of the Human Genome Project is to determine the nucleotide sequence of the entire human genome and to map the location of all genes. This information might then be used to diagnose, treat, cure, and prevent human genetic disorders.
- Gene therapy refers to treating genetic disorders by correcting a defect in a gene or by providing a normal form of a gene. It is hoped that gene therapy can be used to cure genetic disorders in the future.

Vocabulary

DNA fingerprint (243)
gel electrophoresis (244)
gene therapy (246)
Human Genome Project (246)
polymerase chain reaction (PCR) (245)
primer (245)
probe (244)
RFLP analysis (244)

13-3

- DNA technology is used to produce medical products that are often safer and less expensive than those produced by conventional means.
- DNA technology is used to produce disease-resistant, pest-resistant, and herbicide-resistant crops. Perhaps these crops will improve the quality and quantity of the human food supply.
- Current research involves efforts to isolate genes from nitrogen-fixing bacteria and to transplant them into plants. Such transgenic plants may be able to grow in nitrogen-poor soil without the addition of fertilizers.
- Many safety, environmental, and ethical issues involved in the new DNA technology have not been resolved.

Vocabulary

herbicide (251)
pathogen (250)
vaccine (250)

CHAPTER 13 REVIEW ANSWERS

REVIEW

1. It is a molecule of DNA that originated from two sources.
2. Restriction enzymes are used to make a genomic library. More than one genomic library can exist for a genome because different restriction enzymes cut the DNA in different places.
3. A host organism receives a donor gene to become a transgenic organism.
4. PCR amplifies the number of copies of selected segments of DNA in a sample. RFLP analysis is used to extract a sample of DNA and cut it into pieces to make a DNA fingerprint.
5. Primers are needed to start DNA replication and are used in PCR. Probes bind to single chains of complementary DNA and are used to visualize specific bands in a DNA fingerprint.

6. d **7.** a
8. b **9.** c
10. d **11.** b
12. d **13.** a
14. d **15.** a

16. It cuts the DNA molecule at specific nucleotide sequences.
17. (1) Extract DNA from a specimen and cut it into fragments using restriction enzymes; (2) separate the fragments of DNA using gel electrophoresis; (3) make visible only the bands that are being compared using probes and photographic film.
18. PCR can be used to make DNA fingerprints from small samples found at a crime scene, to diagnose genetic disorders from a few embryonic cells, and to study ancient fragments of DNA found in minute amounts.
19. (1) Samples of DNA are placed in lanes on the gel; (2) an electric current is turned on for a given period of time; and (3) the DNA fragments migrate across the gel. The pores in the gel allow smaller DNA fragments to migrate faster than longer fragments, separating the fragments by size.

20. It is a carrier used to clone a gene and transfer it from one organism to another.
21. Suspect 2. Although the blood sample does not prove that suspect 2 was involved in the crime, it is very likely that the blood belongs to him or her.
22. It would allow us to map genes that cause genetic diseases and help in diagnosis, treatment, and prevention.
23. Nasal sprays have been used to carry a normal cystic fibrosis gene to the cells in the nose and lungs of cystic fibrosis patients. The treatment has to be repeated periodically because the cystic fibrosis gene is inserted not on a chromosome but in the nucleus of cells and is not passed on to future generations of cells.
24. The PCR is a technique that allows a small sample of DNA to be copied many times. The DNA is treated with DNA polymerase and small segments of nucleic acid that are required for the initiation of replication. When the DNA is incubated it copies itself repeatedly resulting in billions of copies. Analysis of the patterns (DNA fingerprinting) can then be performed as well as the search for mutations on specific genes

CRITICAL THINKING

1. advantage: it selects for a total phenotype instead of considering only one gene; thus, the plant will be better-adapted; disadvantage: selection may take longer than genetic engineering
2. concerns that genetically engineered organisms might cause disease or have harmful effects on the environment
3. Genetic engineering alters the course of natural selection. The fittest members of a naturally evolving population might not be able to compete with individuals containing new genes.

REVIEW

Vocabulary

1. Define *recombinant DNA.*
2. Relate restriction enzymes to genomic libraries.
3. Relate a host organism, a donor gene, and a transgenic organism.
4. Distinguish between PCR and RFLP.
5. Distinguish between a primer and a probe.

Multiple Choice

6. A plasmid is (a) a cell (b) a nucleus (c) a bacterium (d) an extra ring of DNA.
7. A molecule containing DNA from two sources is called (a) recombinant DNA (b) RFLP DNA (c) a DNA clone (d) a plasmid.
8. Enzymes that cut DNA molecules in specific locations are called (a) sticky ends (b) restriction enzymes (c) cloning vectors (d) cloning enzymes.
9. Carriers used for transferring DNA from one organism to another are called (a) sticky ends (b) restriction enzymes (c) cloning vectors (d) cloning enzymes.
10. The host organism into which a cloning vector is placed is called a (a) plasmid (b) virus (c) donor (d) transgenic organism.
11. A eukaryotic donor gene is likely to be expressed in a prokaryotic host organism if (a) the donor gene is a promoter gene (b) the donor gene is transferred along with the genes that turn it on (c) the donor gene is placed beside a gene that is not expressed (d) all of these.
12. DNA technology has been used to (a) produce pharmaceutical products (b) produce herbicide-resistant crops (c) treat genetic diseases (d) all of these.
13. Gel electrophoresis is used to (a) separate DNA fragments (b) produce DNA fragments with sticky ends (c) cut DNA into fragments (d) quickly make copies of DNA.
14. PCR is used to (a) separate DNA fragments (b) produce DNA fragments with sticky ends (c) cut DNA into fragments (d) quickly make copies of DNA.
15. Genetic engineering has been used to make plants (a) toxic to insects (b) immune to vaccines (c) able to produce insulin (d) all of these.

Short Answer

16. Explain what happens when a restriction enzyme is used on DNA.
17. How is a DNA fingerprint prepared?
18. List three practical uses of PCR.
19. Explain how gel electrophoresis can be used to separate fragments of DNA.
20. What is a cloning vector?
21. The photograph below shows 8 columns on a gel. Several of these columns contain DNA fingerprints of samples taken from a crime scene, a victim, and four suspects. Indicate which suspect's DNA fingerprint matches the blood found at the crime scene. How likely is it that blood found at the crime scene belongs to the suspect?

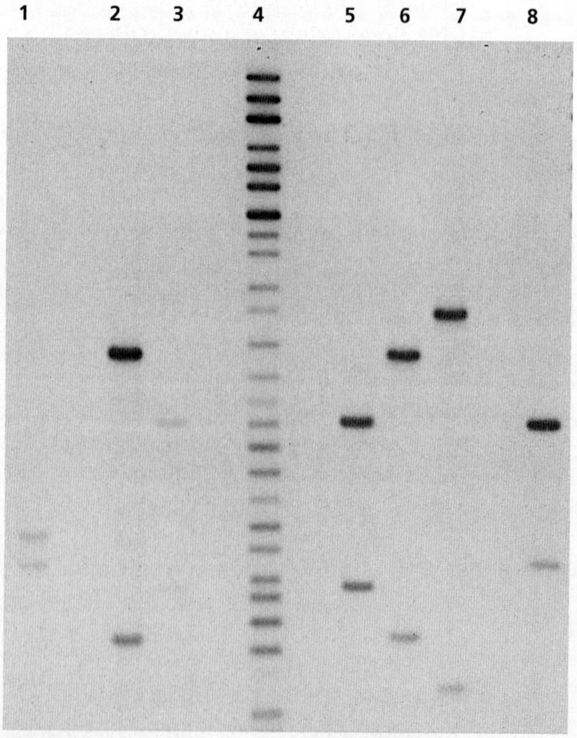

KEY
1 Control
2 Blood
3 Victim
4 Ladder
5 Suspect 1
6 Suspect 2
7 Suspect 3
8 Suspect 4

22. Discuss the accuracy of DNA fingerprints.
23. State two possible benefits of the work being completed by the Human Genome Project.
24. Explain how gene therapy has been used to treat cystic fibrosis. Explain why gene therapy has not been used to cure cystic fibrosis.
25. Describe how genetic engineering can be used to produce vaccines.
26. List two ways that genetic engineering may increase crop yields.

CRITICAL THINKING

1. In the past, breeders developed new plant and animal varieties by selecting an organism with desirable traits, breeding it with another organism with similar traits, and then breeding the offspring so that the desirable traits would be passed on to the strain. Name an advantage and a disadvantage of genetic engineering techniques over traditional breeding techniques.
2. The United States government has stringent regulations requiring researchers to confine genetically engineered organisms to the laboratory. What concerns do you think might have led to the enactment of these regulations?
3. Natural selection is a mechanism of evolution whereby the members of a population who are best adapted to their environment survive and produce offspring. How is natural selection affected by genetic engineering?
4. How might an insect-resistant crop that is the product of genetic engineering affect a population of insects that depends on the crop for food?
5. Why do you think the United States authorized limited gene therapy trials?
6. Examine the restriction map of the bacterial plasmid pBR 322, shown below. This is a commonly used plasmid composed of 4,363 base pairs. The map shows sites at which certain restriction enzymes cut the DNA of the plasmid. For example, the restriction enzyme called *Sph*I cuts the plasmid at base pair 566. Suppose you want to isolate from the plasmid the gene that codes for resistance to the antibiotic tetracycline, which is indicated as Tc^R. What restriction enzymes would you use? How many base pairs make up the Tc^R gene? How might you check to be certain that your procedure was successful?

Restriction Map of pBR 322 DNA

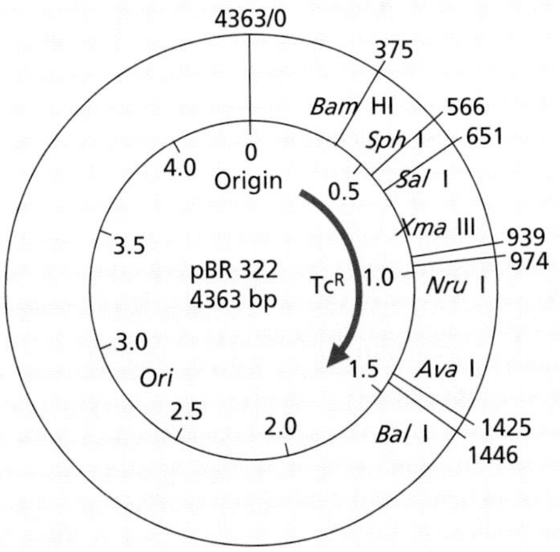

EXTENSION

1. Read "DNA Detectives" in *Popular Science,* August 1999, on page 48. What is an advantage of providing hand-held gene readers to police investigators? What is a disadvantage?
2. Read "Genetic-Code Breaker" in *Discover,* March 2000, on page 22. According to Francis Collins, one of the Human Genome Project's chief researchers, what is the main reason that we should know the sequences of our 46 chromosomes?
3. Read "Who Owns Our Genes?" in *Time,* January 11, 1999, on page 51. Briefly explain the controversy in economics and ethics over the patenting of human genes as they are sequenced by laboratories.

4. The insect population will lose a food source. The insects may then become extinct or they may adapt by finding other habitats or food sources.
5. It allows individuals suffering from terminal diseases to weigh the promise of treatment offered by recombinant DNA research against the risks.
6. Use the restriction enzymes *Bam* HI and *Bal* I. The gene would be 1,071 base pairs long. To determine whether the procedure was successful, you would reinsert the plasmid into the bacterial cell, clone the bacterium, and expose the new bacterial colony to tetracycline. If the bacteria are not resistant to tetracycline, the procedure has been successful.

EXTENSION

1. An advantage to providing gene readers to police investigators is that DNA left at a crime scene can be analyzed within minutes. Perpetrators of violent crimes may be apprehended before they commit more crimes. A disadvantage is the cost of product development and the training of police officers in the proper use of gene readers.
2. The human genome is the instruction book for human biology. Knowing the genomic sequence will help us understand everything about our growth and development from conception to adulthood, the diseases we suffer from, and the ways we fight off diseases.
3. Laboratories working to sequence the human genome hope to make some profit from medical advances derived from the results of their work. Ethicists question whether anyone should claim a patent to a part of the human body.

TIME REQUIRED

3 to 4 50-minute lab periods

SAFETY PRECAUTIONS

Have students wear safety goggles, disposable gloves, and a lab apron. Remind students to use chemicals only as directed and to carefully read the labels on containers of chemicals to be sure they have chosen the proper chemical. When using the WARD'S battery-powered gel chamber, use no more than five 9-volt batteries connected in series. Do not alter the electrical lead jacks to accommodate any power supply plugs. Do not use with any other power source. If another type of power supply is used, follow the safety instructions for the operation of that equipment.

QUICK REFERENCES

Lab Preparation Notes are found on page 237B.

Holt BioSources provides a Teaching Resources CD-ROM that contains "Using Gowin's Vee in the Lab" and "Scoring Rubrics."

PROCEDURAL TIPS

1. Prior to the lab, discuss with students the concept of gel electrophoresis—how it works and why.
2. Tell students that an agarose gel contains agarose, which is a polysaccharide obtained from seaweed.
3. Discuss with students how to analyze a gel. You may do this by placing a gel on an overhead projector and showing students how to recognize the bands and calculate R_f values.
4. DNA banding patterns can be drawn by students. Photos of gels can be taken after staining and can then be attached to a piece of paper.

Gel Electrophoresis

OBJECTIVES

- Use restriction enzymes to cut DNA.
- Separate DNA fragments of different sizes.

PROCESS SKILLS

- conducting agarose gel electrophoresis

MATERIALS

- protective clothing
- crushed ice
- ice bucket
- microtube rack
- microtube A—Uncut DNA
- microtube B—*Hind*III
- microtube C—*Bam*HI
- microtube D—*Eco*RI
- microtube E—Unknown
- permanent marker
- 0.5–10 µL micropipetter
- micropipetter tips
- 10× restriction buffer for each restriction enzyme
- 34 µL Lambda virus DNA
- 37°C water bath
- gel-casting tray
- 6-well gel comb
- 65°C hot-water bath
- 0.8% agarose
- hot mitt
- zipper-lock plastic bag
- 10 mL graduated cylinder
- 1× TBE buffer
- freezer
- 5 µL loading dye
- gel chamber and power supply
- WARD'S DNA stain
- staining trays
- distilled water
- metric ruler

Background

1. DNA has a negative charge and flows toward the positive end of a gel during electrophoresis. Small DNA fragments move faster than larger fragments.
2. The distance that each DNA fragment moves is used to calculate the R_f, or relative mobility, of a fragment. The R_f is used to calculate the number of base pairs in the fragment.

PART A Cutting DNA

1. Wear safety goggles, gloves, and a lab apron at all times.
2. Fill an ice bucket with ice. Obtain one each of the following microtubes: A—Uncut DNA, B—*Hind*III, C—*Bam*HI, D—*Eco*RI, and E—Unknown. Microtubes B–D contain 1 µL of the indicated restriction enzyme. Place all microtubes in the ice. *Restriction enzymes MUST be kept on ice until step 6.*
3. With a permanent marker, write the initials for everyone in your group *on the top* of microtubes A–E.
4. CAUTION **If you get a chemical on your skin or clothing, wash it off at the sink while calling to your teacher.** Set a micropipetter to 1 µL, and put a tip on the end of the micropipetter, as shown in (a) below. Using a new tip for each microtube, add 1 µL of the corresponding 10× restriction buffer to each of microtubes B–D. Place the buffer on the side of the tube. *Do not touch the micropipetter tips to the solutions in the microtubes.*

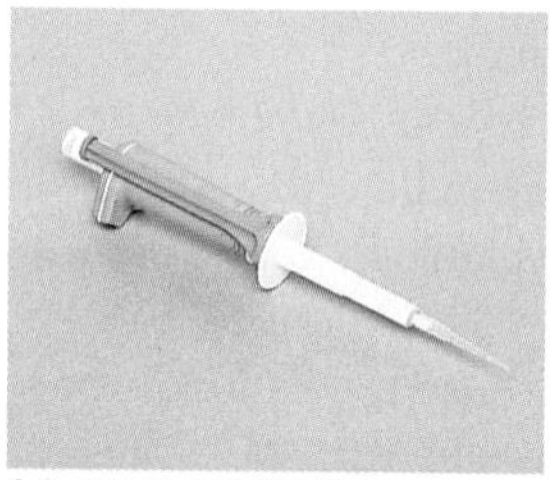

(a) MICROPIPETTER

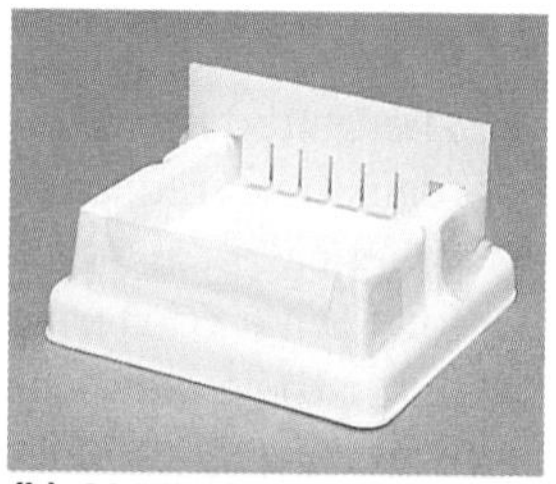

(b) CASTING TRAY

5. Reset the micropipetter to 8 µL. Using the micropipetter and a new tip for each microtube, add 8 µL of the Lambda virus DNA to the side of each of microtubes B–D. Gently tap each microtube on your lab table until the solutions are thoroughly mixed. *Do not shake the microtubes!* Reset the micropipetter to 10 µL, and add 10 µL of Lambda DNA to microtube A.
6. Place all of the microtubes into a 37°C water bath. After 50–60 minutes, remove the microtubes from the water bath, and immediately put them into a freezer. If the class period ends before 50 minutes has passed, your teacher will give you further directions. While the restriction enzymes are working, go to Part B.

PART B Preparing an Agarose Gel

7. Set up a gel-casting tray, as shown in (b). Place a gel comb in the grooves of the gel-casting tray. Make sure

that the comb does not touch the bottom of the tray. If it does, get another comb from your teacher.

8. Write the names of the members of your group on a paper towel. Carry your tray to the table with the melted agarose, and place your tray on the paper towel.
9. Using a hot mitt, pour melted 0.8% agarose into your gel-casting tray until the agarose reaches a depth of 3 mm. Make sure that the agarose spreads evenly throughout the tray. *Do not move your gel tray before the agarose solidifies.*
10. Let the gel cool (about 20–30 minutes) until the agarose solidifies.
11. While the gel is cooling, write your name, the date, and your class period on a zipper-lock plastic bag. Pour 5 mL of 1× TBE buffer into the bag.
12. When the gel has solidified, carefully remove the gel comb by pulling it straight up. If the comb does not come up easily, pour a little 1× TBE buffer on the comb area. After removing the gel comb, open the plastic bag and carefully slide the gel tray into the bag. *Do not remove the gel from the gel-casting tray.* Store the gel according to your teacher's instructions.

PART C Running a Gel

13. Retrieve your microtubes (A–E) and your gel. If the materials in the microtubes are frozen, hold each tube in your hand until the solutions thaw.
14. Set a micropipetter to 1 µL, and place a tip on the end. Add 1µL of loading dye to each microtube. *Use a new tip for each microtube.* Gently tap each microtube on your lab table to thoroughly mix the solutions. *Do not shake the microtubes.*
15. Remove your gel (still in the gel-casting tray) from the plastic bag, and place it in a gel chamber. Orient the gel so that the wells are closest to the black wire, or anode.
16. Set a micropipetter to 10 µL, and place a new tip on the end. Open microtube A, and remove 10 µL of solution. Carefully place the solution into the well in lane 1, the left-most lane. To do this, place both elbows on the lab table, lean over the gel, and slowly lower the micropipetter tip into the opening of the well before depressing the plunger. *Do not jab the micropipetter tip through the bottom of the well.*
17. *Using a new micropipetter tip for each tube,* repeat step 16 for each of the remaining microtubes. Use lane 2 for microtube B, lane 3 for microtube C, lane 4 for microtube D, and lane 5 for microtube E.
18. Very slowly fill the gel chamber with 1× TBE buffer until the level of the buffer is approximately 1–2 mm above the surface of the gel.
19. CAUTION **Follow all of the manufacturer's precautions regarding the use of this equipment.** Close the gel chamber and connect it to a power supply according to your teacher's instructions.
20. Allow an electric current to flow through the gel. You will see a blue line moving away from the wells. When the blue line is approximately 5 mm from the end of the gel, disconnect the power supply and remove the gel. Store the gel overnight in the plastic bag.

PART D Analyzing a Gel

21. To stain a gel, carefully place the gel (wells up) into a staining tray. Pour WARD'S DNA stain into the staining tray until the gel is completely covered. Cover the staining tray, and label it with your initials. Allow the stain to sit for at least 2 hours. Next, carefully pour the stain into the sink drain, and flush it down the drain with water. *Do not let the stained gel slip out of the staining tray.*
22. To destain a gel, cover the gel with distilled water by pouring water to one side of the gel. Let the gel sit overnight (or at least 8–12 hours). The bands of DNA will appear as purple lines against a light background.
23. Calculate the R_f for each fragment using the following equation:

$$R_f = \frac{\text{distance in mm that DNA fragment migrated}}{\text{distance in mm from well to the dye}}$$

24. Dispose of your materials according to your teacher's directions, and wash your hands before leaving the lab.

Analysis and Conclusions

1. Which two samples appear to have the same pattern of DNA bands?
2. Which restriction enzyme cut the DNA in the unknown sample? Justify your answer.
3. What are some measures that you took to prevent contamination of your DNA samples during this lab?

5. Gels can be dried on filter paper.
6. A variety of equipment can be used to perform gel electrophoresis. The equipment discussed in this lab is the micropipetter, the single-tray gel box, and a battery-operated power supply. Be sure that you discuss the equipment used in your classroom and that you instruct students to follow the safety instructions specifically for that equipment. **Model the proper use of all equipment for the students prior to the lab.**

ANSWERS TO ANALYSIS AND CONCLUSIONS

1. The two samples that appear to have the same pattern of DNA bands are the *Hind*III (Sample B) and the unknown (Sample E).
2. *Hind*III cut the DNA in the unknown sample because the fragments of both samples produced the same pattern during separation by gel electrophoresis.
3. Answers will vary but might include that they were careful to use a new tip for each microtube, to not jab the micropipetter tip through the bottom of the well, or to not touch the micropipetter tips to solutions in the microtubes or to the gels.

Unit 4

EVOLUTION

CHAPTERS

"The past, the finite greatness of the past! For what is the present, after all, but a growth out of the past."

Walt Whitman

Charles Darwin studied the giant tortoises, above, on the Galápagos Islands. Observations Darwin made during his voyage on the H.M.S. Beagle, right, led him to look for a mechanism by which evolution occurs.

internet**connect**

sciLINKS NSTA

National Science Teachers Association *sci*LINKS Internet resources are located throughout this unit.

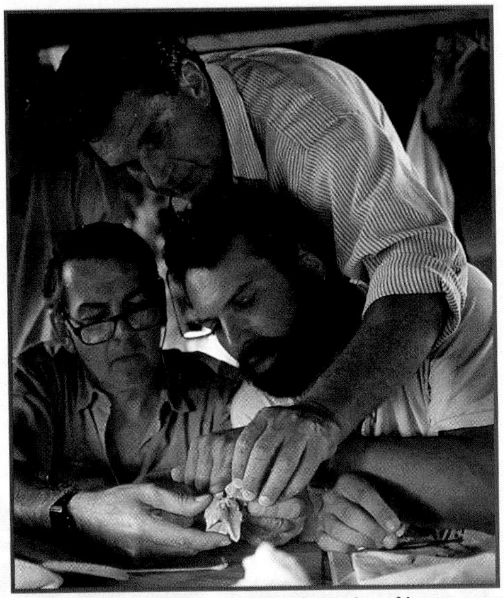

Archaeologists scrutinize fossilized remains of human ancestors for clues to human evolution. Darwin observed a number of finches, right, on the Galápagos Islands and found that each species was adapted to a different food source.

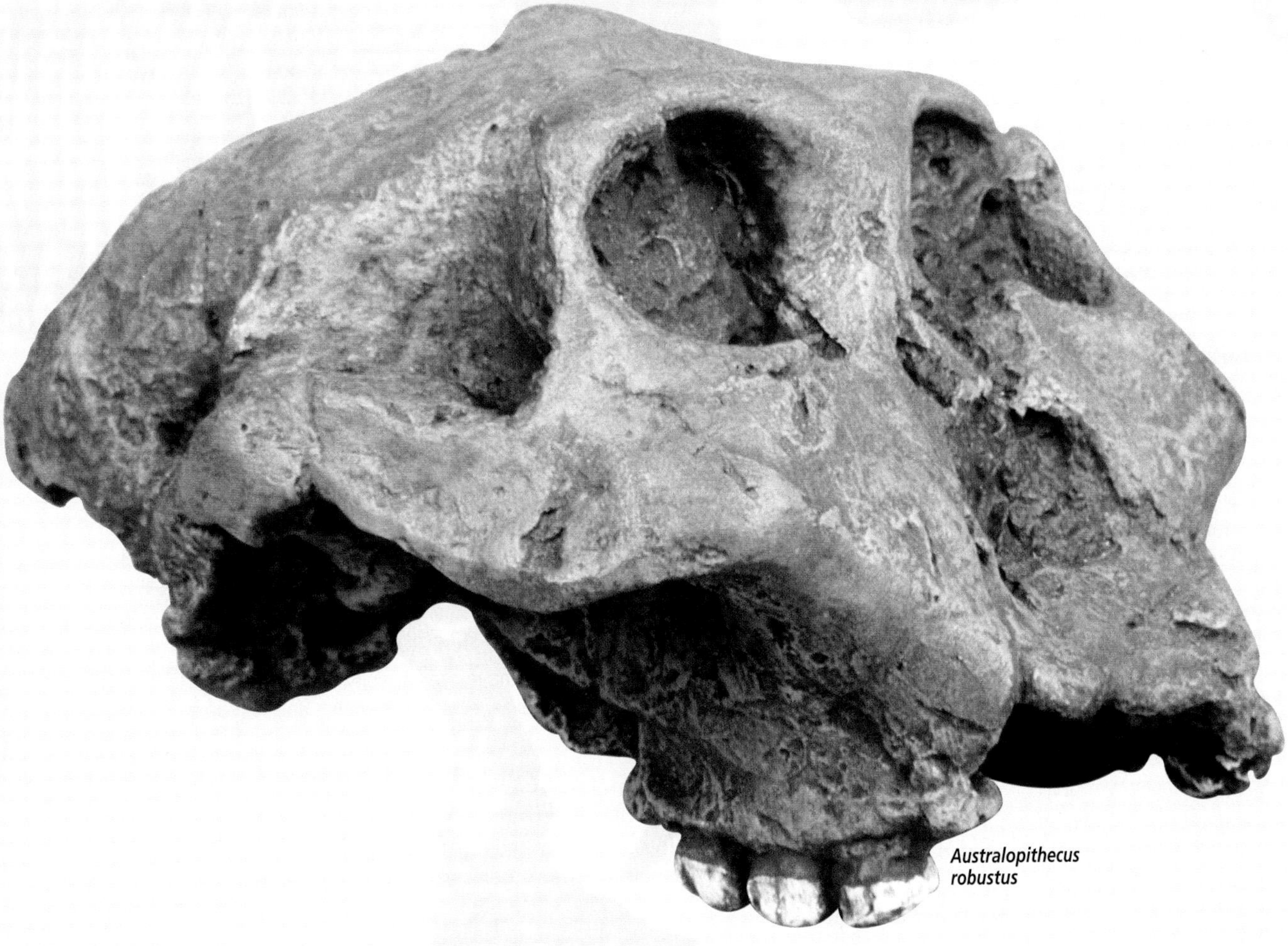

Australopithecus robustus

unit 4

CHAPTERS

14 ***Origin of Life***

15 ***Evolution: Evidence and Theory***

16 ***The Evolution of Populations and Speciation***

17 ***Human Evolution***

18 ***Classification***

EVOLUTION

CONNECTING TO THE STANDARDS

The following chart shows the correlation of Unit 4 with the *National Science Education Standards* (grades 9–12). The items in each category are addressed in this unit and should be used to guide instruction. Annotated descriptions of the **Life Science Standards** are found below the chart. Consult the *National Science Education Standards* for detailed descriptions of each category.

UNIFYING CONCEPTS

- Systems, order, and organization
- Evidence, models, and explanation
- Change, constancy, and measurement
- Evolution and equilibrium

SCIENCE AS INQUIRY

- Abilities necessary to do scientific inquiry
- Understandings about scientific inquiry

SCIENCE AND TECHNOLOGY

- Abilities of technological design
- Understanding about science and technology

PHYSICAL SCIENCE

- Structure and properties of matter

LIFE SCIENCE

EARTH AND SPACE SCIENCE

- Origin and evolution of the Earth system
- Origin and evolution of the universe

SCIENCE IN PERSONAL AND SOCIAL PERSPECTIVES

- Science and technology in local, national, and global challenges

HISTORY AND NATURE OF SCIENCE

- Nature of scientific knowledge
- Historical perspectives

ANNOTATED DESCRIPTIONS OF THE CORRELATED LIFE SCIENCE STANDARDS

Although all eight categories of the *National Science Education Standards* are important, the following descriptions summarize the **Life Science Standards** that specifically relate to Unit 4.

THE CELL

- Cells can differentiate. Complex multicellular organisms are formed from highly organized arrangements of differentiated cells.

MOLECULAR BASIS OF HEREDITY

- In all organisms, the instructions for specifying the characteristics of the organism are carried in DNA.
- Changes in DNA occur spontaneously at low rates.

BIOLOGICAL EVOLUTION

- Species evolve over time.
- The great diversity of organisms is the result of more than 3.5 billion years of evolution that has filled every available niche with life-forms.
- Natural selection provides a scientific explanation for the fossil record of ancient life-forms.
- The millions of different species of plants, animals, and microorganisms that live on Earth today are related by descent from common ancestors.
- Biological classifications are based on how organisms are related.

INTERDEPENDENCE OF ORGANISMS

- Living organisms have the capacity to produce populations of infinite size, but environments and resources are finite.

MATTER, ENERGY, AND ORGANIZATION IN LIVING SYSTEMS

- The complexity and organization within an organism result from the organism's need to obtain, transform, release, and eliminate the matter and energy required to sustain the organism.

TRENDS IN BIOLOGY

Observing Evolution

Although Darwin was renowned as a keen observer of nature, he never tried to demonstrate natural selection or measure evolution in the wild. Why? Evolution, to his mind, occurred through the slow accumulation of imperceptible changes and was driven by an "insensibly working" mechanism. "We see nothing of these slow changes in progress," he wrote of evolution, "until the hand of time has marked the lapse of ages." Over the last 30 years, more than 2,000 studies of organisms ranging from bacteria to birds have shown that evolution does happen fast enough to be measured—in some cases, a detectable difference appears within a generation.

DARWIN'S FINCHES REVISITED

For more than 25 years, Peter and Rosemary Grant, of Princeton University, and their students have been studying the finches of the small Galápagos island of Daphne Major which Darwin never visited. They have kept up a rotating watch on the island, collecting as much data on the morphology and ecology of the finches as possible.

Two episodes of unusual weather triggered bouts of rapid evolution among the birds (shown on the diagram at left). The first was a severe drought in 1977. The vegetation on the island withered, and the birds' normal food supply of small seeds was quickly exhausted. The finches were forced to turn to other food sources, such as larger, tougher seeds that were more difficult to open. Many finches couldn't open them at all—about 85 percent of the finches died before the drought was over.

The Grants' data showed that a subtle change had taken place during the drought. The surviving finches were about 6 percent larger than the finches that died—the population had evolved. Statistical analysis revealed that one measure was the most important in determining whether a finch survived: beak depth (the height of the beak). On average, the beak depth of the surviving finches was 0.5 mm greater than that of the dead birds. The birds with larger beaks had been able to open large seeds, and thus they got more food than the birds with smaller beaks.

Five years after the drought, the island was drenched during the El Niño of 1982–1983. The finch population boomed, quintupling within a few months. But when conditions returned to normal, the island was not able to support the large finch population, and birds began to die. This time, the smaller birds were more likely to survive. They not only required less food than the larger birds, but also were better at opening the small seeds that were abundant at the time. The population evolved again, but this time toward smaller body and beak size.

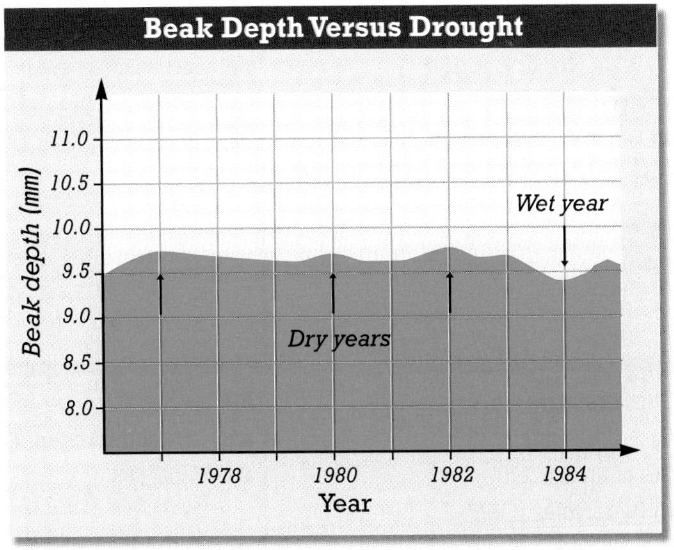

The finches of Daphne Major evolved rapidly during the drought of 1977 and after El Niño of 1982–1983.

Additional Information

Grant, Peter. "Natural Selection and Darwin's Finches." *Scientific American,* October 1991, 82–87.

Weiner, Jonathan. *The Beak of the Finch.* New York: Knopf, 1994.

Morrell, Virginia. "Predator-Free Guppies Take an Evolutionary Leap Forward." *Science*, March 28, 1997, 1880.

Ridley, Mark. *Evolution.* 2nd ed. Oxford, U.K.: Blackwell, 1966.

CHAPTER 14
ORIGIN OF LIFE

	TOPICS	TEACHING RESOURCES	LABS, CLASSWORK, AND HOMEWORK
BLOCK 1 45 minutes	**Introducing the Chapter, p. 260**	ATE Focus Concept, p. 260 ATE Assessing Prior Knowledge, p. 260 ATE Understanding the Visual, p. 260	■ Supplemental Reading Guide, *Origin of Species*
	14-1 Biogenesis, p. 261 Redi's Experiment, p. 261 Spallanzani's Experiment, p. 262 Pasteur's Experiment, p. 263	ATE Section Overview, p. 261 ATE Critical Thinking, p. 262 ATE Visual Link Figure 14-3, p. 263 42, 42A, 43, 46	★ Study Guide, Section 14-1 PE Section Review, p. 263
BLOCKS 2 & 3 90 minutes	**14-2 Earth's History, p. 264** The Formation of Earth, p. 264 The First Organic Compounds, p. 266 From Molecules to Cell-Like Structures, p. 268	ATE Section Overview, p. 264 45	PE **Quick Lab,** p. 267, Inferring Probability ★ Study Guide, Section 14-2 PE Section Review, p. 268
BLOCK 4 45 minutes	**14-3 The First Life-Forms, p. 269** The Origin of Heredity, p. 269 The Roles of RNA, p. 270 The First Prokaryotes, p. 270 The First Eukaryotes, p. 272	ATE Section Overview, p. 269 ATE Critical Thinking, p. 270 ATE Visual Link Figure 14-11, p. 272 47	PE **Chapter 14 Investigation,** pp. 276–277 ★ Study Guide, Section 14-3 PE Section Review, p. 272
BLOCKS 5 & 6 90 minutes	**REVIEW and ASSESSMENT** PE Chapter 14 Review, pp. 274–275	Audio CD Program	**CHAPTER TESTING** ■ Chapter Test (blackline master) ▲ Test Generator ■ Assessment Item Listing ■ Scoring Rubrics and Classroom Management Checklists

MODERN BIOLOGY — HOLT BIOSOURCES

KEY

- **PE** Pupil's Edition
- **ATE** Teacher's Edition
- ■ Active Reading Guide
- ★ Study Guide
- Teaching Transparencies
- Laboratory Program

One-Stop Planner CD-ROM
Shows these resources within a customizable daily lesson plan:

- ■ Holt BioSources Teaching Resources
- ■ Active Reading Guide
- ★ Study Guide
- ▲ Test Generator

READING FOR CONTENT MASTERY

Preview: Have students preview the objectives, the introductory paragraph, topic sentences, and captions before reading the chapter.

ATE Active Reading Technique: K-W-L, p. 261
■ Reading Strategy: K-W-L Worksheet
■ Active Reading Guide Worksheet 14-1
PE Summary/Vocabulary, p. 273

■ Active Reading Guide Worksheet 14-2
PE Summary/Vocabulary, p. 273

ATE Active Reading Technique: K-W-L, p. 271
■ Active Reading Guide Worksheet 14-3
PE Summary/Vocabulary, p. 273

TECHNOLOGY AND INTERNET RESOURCES

Holt Biology Videodiscs Teacher's Correlation Guide, Lessons 14-1 through 14-3

Audio CD Program, Sections 14-1 through 14-3

internetconnect

sciLINKS NSTA

On-line Resources: www.scilinks.org
The following *sci*LINKS Internet resources can be found in the student text for this chapter:

Topics:
- Radioactive dating, p. 266

go.hrw.com

On-line Resources: go.hrw.com
Visit the HRW Web site for a variety of resources related to this chapter. Just type in the keyword HM2 Chapter 14.

Smithsonian Institution®
Visit **www.si.edu/hrw** for additional on-line resources.

CNNfyi.com
Visit **www.cnnfyi.com** for late-breaking news and current event stories selected just for you.

LAB ACTIVITY PLANNING

CHAPTER 14 INVESTIGATION

Making Microspheres, pp. 276–277

OVERVIEW
In this investigation, students will produce microspheres, observe them, and compare them with living cells.

SAFETY
Students will wear disposable gloves, lab aprons, and safety goggles during the lab. If a chemical gets in a student's eye, flush the eye with water for 15 minutes and seek immediate medical attention. Instruct students to handle hot objects with tongs or hot pads. Keep ice on hand for scalds or burns. Dispose of broken glass in a container designated for that purpose.

PREPARATION
Each student group will need 3 g each of six different amino acids. You can purchase an amino acid mixture from a biological supply house.

Prepare a 1 percent NaCl solution by adding 1 g of sodium chloride to a 500 mL graduated cylinder and diluting with distilled water to a final volume of 100 mL. Prepare a 1 percent NaOH solution by adding 1 g of sodium hydroxide to a 500 mL graduated cylinder and diluting with distilled water to a final volume of 100 mL.

Each group will need a 500 mL beaker, two 125 mL Erlenmeyer flasks, a ring stand with clamp, a 50 mL graduated cylinder, a glass stirring rod, a hot plate, tongs, a dropper, and a microscope slide and coverslip. Groups will need access to a clock, a balance, and a microscope.

Quick Labs

Inferring Probability, p. 267
Each pair of students will need a set of 12 cards. Label 3 × 5 cards as follows: "CH_3– "(3 cards), "–CH_2– "(3 cards), "NH_2– "(3 cards), and "–COOH" (3 cards).

UNDERSTANDING THE VISUAL

Point out that volcanic eruptions such as the one shown in the opening photograph were very common when Earth was young. Remind students that chemical reactions need input of energy to start, and ask students how volcanic eruptions may have been important in the origin of life on Earth. (The mixture of hot gases and other matter may have helped provide the raw materials and energy for the formation of organic compounds—precursors of life on Earth.)

FOCUS CONCEPT

Matter, Energy, and Organization

Life on Earth arose from nonliving matter in a process that required the input of energy. New life-forms evolved from the earliest life-forms, requiring energy input and resulting in highly complex organisms.

ASSESSING PRIOR KNOWLEDGE

Review the following concepts.

Organic Compounds: *Chapter 4*

Have students name the four types of organic compounds found in living things and state a major role of each. Have students name the elements common to all organic compounds.

DNA: *Chapter 9*

Have students describe the role of RNA in inheritance. Remind students that transfer RNA, for example, is an elaborately folded *t*-shape. Ask how the shape of RNA is important to its function.

CHAPTER 14

ORIGIN OF LIFE

Molten rock breaks through Earth's crust in the form of a volcano. Conditions in a volcano are similar to those thought to have been present on early Earth.

FOCUS CONCEPT: ***Matter, Energy, and Organization***

As you read, look for examples of the many scientific processes that help answer questions about how life began.

BIOGENESIS

The principle of ***biogenesis*** *(BIE-oh-JEN-uh-sis), which states that all living things come from other living things, seems very reasonable to us today. Before the seventeenth century, however, it was widely thought that living things could also arise from nonliving things in a process called* ***spontaneous generation.*** *This seemed to explain why maggots appeared on rotting meat and why fish appeared in ponds that had been dry the previous season—people thought mud might have given rise to the fish. In attempting to learn more about the process of spontaneous generation, scientists performed controlled experiments. As you read about these experiments, refer to the figures that show the experimental design.*

SECTION

14-1

OBJECTIVES

▲ Define *spontaneous generation,* and list some of the observations that led people to think that life could arise from nonliving things.

● Summarize the results of experiments by Redi and by Spallanzani that tested the hypothesis of spontaneous generation.

■ Describe how Pasteur's experiment disproved the hypothesis of spontaneous generation.

REDI'S EXPERIMENT

Flies have often been viewed as pesky creatures. Most people are too busy trying to get rid of them to even think about studying them. In the middle of the seventeenth century, however, the Italian scientist Francesco Redi (1626–1697) noticed and described the different developmental forms of flies. Redi observed that tiny wormlike maggots turned into sturdy oval cases, from which flies eventually emerge. He also observed that maggots seemed to appear where adult flies had previously landed. These observations led him to question the commonly held belief that flies were generated spontaneously from rotting meat.

Figure 14-1 shows an experiment that Redi conducted in 1668 to test his hypothesis that meat kept away from adult flies would

FIGURE 14-1

In Redi's experiment, maggots were found only in the control jars because that was the only place where adult flies could reach the meat to lay eggs.

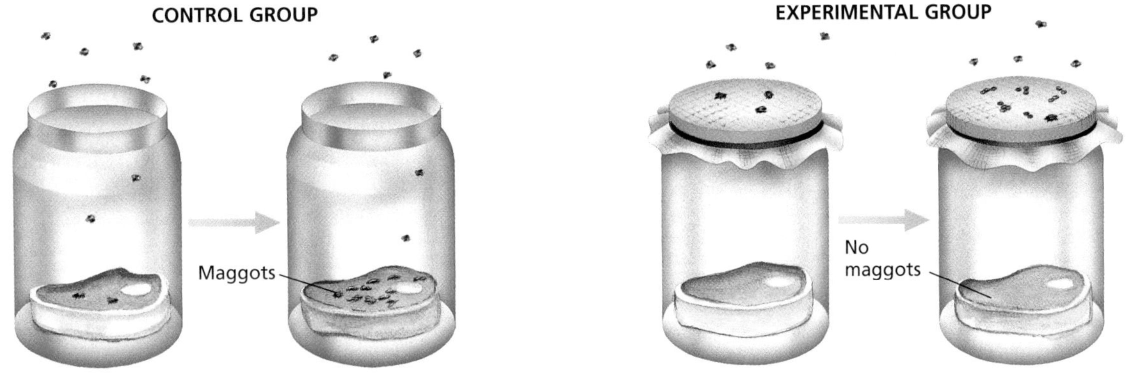

SECTION 14-1

SECTION OVERVIEW

Reproduction and Inheritance

Section 14-1 recounts important experiments performed by early scientists to ascertain whether life arises spontaneously or must come from existing life. Students will see how relatively simple experimental designs yielded a great deal of information and confirmed that living things arise only from other living things.

ACTIVE READING

Technique: K-W-L

Before students read this chapter, have them write short lists of all the things they already **K**now (or think they know) about how life began and how it developed into what exists today. Ask them to contribute their entries to a group list placed on the board or overhead. Then have students list things they **W**ant to know about the development of life on this planet. Have students save their lists for use later in the Active Reading activity on page 271.

CRITICAL THINKING

Experimental Variables

Ask students to identify the independent and dependent variables in Redi's experiment. (The independent variable was the presence or absence of a cover on the jar; the dependent variable was the presence or absence of maggots on the meat.)

CRITICAL THINKING

Pasteurization

Ask students to explain how Pasteur's experiments led to the development of the pasteurization process. (He demonstrated that heating could kill bacteria such as those that cause food spoilage.)

QUICK FACT

Some of Pasteur's original vessels from his experiments are on display at the Pasteur Institute in Paris. After more than 100 years, they still show no sign of contamination.

INCLUSION *ACTIVITY*

Have students redraw Figure 14-3 and add symbols representing bacteria in places where bacteria should be found in any of the flasks. (There should be no bacteria in the first flask; there should be bacteria in the curves of the second and third flasks and bacteria entering and inside the fourth and fifth flasks.)

remain free of maggots. The experimental group consisted of netting-covered jars that contained meat. The control group consisted of uncovered jars that also contained meat. The netting allowed air to enter and prevented flies from landing on the meat. After a few days, maggots swarmed over the meat in the open jars, while the net-covered jars remained free of maggots. Redi's experiment showed convincingly that flies come only from eggs laid by other flies. Redi's hypothesis was confirmed, and a major blow was struck against the hypothesis of spontaneous generation.

SPALLANZANI'S EXPERIMENT

FIGURE 14-2

In Spallanzani's experiment, he boiled meat broth in open flasks. Then he sealed the flasks of the experimental group by melting the glass necks of the flasks closed. The broth inside remained uncontaminated by microorganisms.

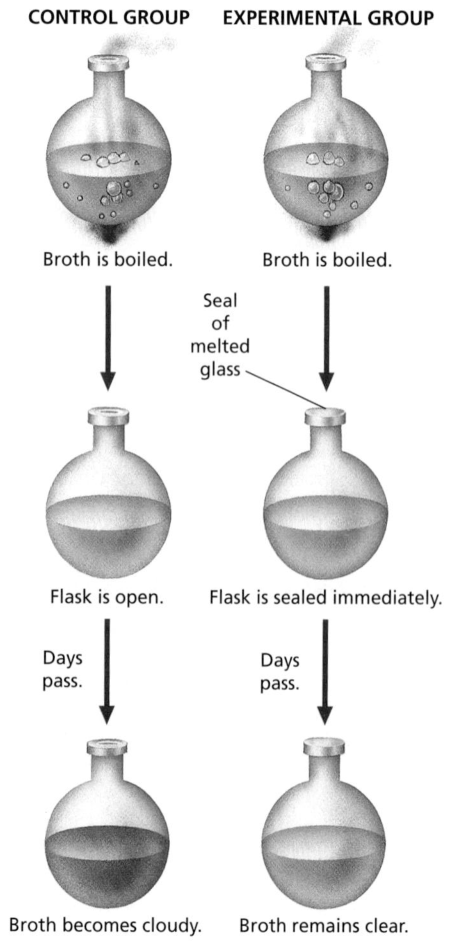

At about the same time that Redi carried out his experiment, other scientists began using a new tool—the microscope. Their observations with the microscope revealed that the world is teeming with tiny creatures. They discovered that microorganisms are simple in structure and amazingly numerous and widespread. Many investigators at the time thus concluded that microorganisms arise spontaneously from a "vital force" in the air.

In the 1700s, another Italian scientist, Lazzaro Spallanzani (1729–1799), designed an experiment to test the hypothesis of spontaneous generation of microorganisms, as shown in Figure 14-2. Spallanzani hypothesized that microorganisms formed not from air but from other microorganisms. He knew that microorganisms grew easily in food. Therefore, he decided to test their growth in meat broth. Spallanzani reasoned that boiling broth in a flask would kill all the microorganisms in the broth, on the inside of the glass, and in the air in the flask. For his experimental group, Spallanzani boiled clear, fresh broth until the flasks filled with steam. He then sealed the flasks by melting their glass necks closed while the broth was hot. The control-group flasks of broth were left open. The broth in the sealed flasks remained clear and free of microorganisms, while that in the open flasks became cloudy due to contamination with microorganisms.

Spallanzani concluded that the boiled broth became contaminated only when microorganisms from the air entered the flask. Spallanzani's opponents, however, objected to his method and disagreed with his conclusions. They claimed that Spallanzani had heated the experimental flasks too long, destroying the "vital force" in the air inside them. Air lacking this "vital force," they claimed, could not generate life. Thus, those who believed in spontaneous generation of microorganisms kept the idea alive for another century.

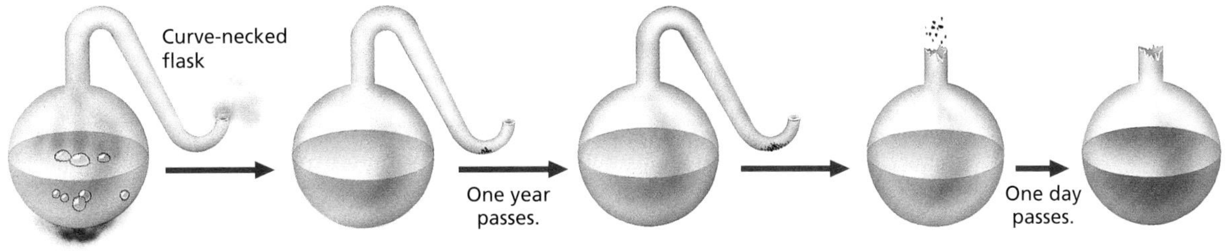

FIGURE 14-3

In Pasteur's experiment, a flask with a curved but open neck prevented microorganisms from entering. Broth boiled in the flasks became contaminated by microorganisms only when the curved necks were removed from the flasks.

PASTEUR'S EXPERIMENT

By the mid-1800s the controversy over spontaneous generation had grown fierce. The Paris Academy of Science offered a prize to anyone who could clear up the issue once and for all. The winner of the prize was the French scientist Louis Pasteur (1822–1895).

Figure 14-3 shows how Pasteur set up his prize-winning experiment. To answer objections to Spallanzani's experiment, Pasteur made a curve-necked flask that allowed the air inside the flask to mix with air outside the flask. The curve in the neck of the flask prevented solid particles, such as microorganisms, from entering the body of the flask. Broth boiled inside the experimental curve-necked flasks remained clear for up to a year. But when Pasteur broke off the curved necks, the broth became cloudy and contaminated with microorganisms within a day. Pasteur reasoned that the contamination was due to microorganisms in the air.

Those who had believed in the spontaneous generation of microorganisms gave up their fight. With Pasteur's experiment, the principle of biogenesis became a cornerstone of biology.

Word Roots and Origins

biogenesis

from the Greek *bioun,* meaning "to live," and *gignesthai,* meaning "to be born"

SECTION 14-1 REVIEW

1. **What does the term *spontaneous generation* mean?**
2. **Explain how Redi's experiment disproved the hypothesis of spontaneous generation in flies.**
3. **What caused people to think there was a "vital force" in the air that produced living organisms?**
4. **Did Spallanzani's experiment disprove the hypothesis that microorganisms could arise spontaneously from a "vital force" contained in the air? Explain why Spallanzani's procedure did or did not disprove such a possibility.**
5. **In conducting his experiment, Spallanzani demonstrated a technique that would become universally used in the preservation of food. What was this technique?**
6. **CRITICAL THINKING What would have happened if Pasteur had tipped one of his flasks so that the broth in the flask came into contact with the curve of the neck? Explain how this result would or would not have supported his conclusion.**

VISUAL LINK

Figure 14-3

Ask students to explain what the control group and the experimental group were in Pasteur's experiment, as shown in Figure 14-3. (The intact flasks were the control group, and they became the experimental group when their curved necks were removed.)

ANSWERS TO SECTION 14-1 REVIEW

1. In spontaneous generation, living things arise from nonliving things.
2. In Redi's experiment, meat kept away from adult flies did not produce maggots, proving that maggots arise from existing flies rather than from the meat itself.
3. Early investigators discovered microorganisms so numerous and widespread that the scientists concluded they arose from a "vital force" in the air.
4. Spallanzani's experiment did not disprove the hypothesis that microorganisms could arise from a "vital force," assuming that the "vital force" could be damaged by heat, as Spallanzani's critics claimed.
5. Spallanzani demonstrated sealing of food using sterile technique, such as is used in canning.
6. If Pasteur had tipped his flask so that the broth inside had come in contact with microorganisms trapped in the curve of the neck, the broth would have been contaminated and microorganisms would have grown in the flask. This would have supported Pasteur's conclusion that microorganisms came not from the air but from other microorganisms.

SECTION 14-2

SECTION OVERVIEW

Matter, Energy, and Organization
Section 14-2 explains the accretion model of the formation of Earth. The principles of radioactive dating are explained with respect to the age of Earth. Oparin's hypothesis regarding the origin of organic compounds on early Earth and Miller and Urey's experiments to test Oparin's hypothesis are introduced.

Reproduction and Inheritance
Section 14-2 introduces the terms *coacervate* and *microsphere* and explains that while these are nonliving structures because they do not have inheritance, they do share several properties with living cells.

▶ ENGAGE STUDENTS

Provide students with a long, narrow piece of paper and have them mark time intervals from 5 billion years ago to the present. Ask them to write in at the appropriate points on their chart the events they read about in this section (5 billion years ago—formation of the solar system; 4.6 billion years ago—formation of Earth). Have students hang the timeline up on a wall and complete it after reading Section 14-3.

SECTION 14-2

OBJECTIVES

▲ Outline the modern scientific understanding of the formation of Earth.

● Summarize the concept of *half-life.*

■ Describe the production of organic compounds in the Miller-Urey apparatus.

◆ Summarize the possible importance of cell-like structures produced in the laboratory.

EARTH'S HISTORY

The phylogenetic tree of all living things is a history of cell-based life. Cells may evolve and change, but new cells always arise from existing cells. How, then, did the first cells originate? Using models thought to approximate the conditions found on early Earth, scientists attempt to reconstruct the processes that gave rise to the first cellular life.

THE FORMATION OF EARTH

Evidence from computer models of the sun suggests that about 5 billion years ago, our solar system was a swirling mass of gas and dust, as shown in Figure 14-4. Over time, most of this material collapsed inward, forming the sun. The remaining gas, dust, and debris circled the young sun. The planets are thought to have been formed from repeated violent collisions of this space debris. During its 400-million-year-long period of formation, Earth grew increasingly large as it was bombarded by debris. These collisions not only added to Earth's mass but also released a great deal of thermal energy. Each collision between the young and growing Earth and a large piece of space debris, for example, would have released enough energy to melt the entire surface of Earth.

Earth's Age

The estimated age of Earth, more than 4 billion years, is about 700,000 times as long as the period of recorded history. It is about 50 million times as long as an average human life span. How can we determine what happened so long ago? Scientists have drilled deep into Earth and examined its many layers to establish a fairly complete picture of its geologic history. Early estimates of Earth's age were made from studying layers of sediment in Earth's crust. The age of Earth could not be estimated accurately, however, until the middle of the twentieth century, when modern methods of establishing the age of materials were developed.

Radioactive Dating

Methods of establishing the age of materials include the techniques known as **radioactive dating.** Recall from Chapter 2 that the atomic number of an element is the number of protons in the nucleus. All atoms of an element have the same atomic number, but their number of neutrons can vary. Atoms of the same element that differ in the number of neutrons they contain are called **isotopes**

FIGURE 14-4
It took about one-half of a billion years for modern Earth to form from a swirling mass of gases.

(IE-suh-TOHPS). Most elements have several isotopes. The **mass number** of an isotope is the total number of protons and neutrons in the nucleus. The mass number of the most common carbon isotope is 12. If you recall that the atomic number of carbon is 6, you can calculate that this carbon isotope has six protons and six neutrons. Isotopes are designated by their chemical name followed by their mass number; for example, carbon exists as both *carbon-12* and *carbon-14.*

Some isotopes have unstable nuclei, which tend to undergo **radioactive decay;** that is, their nuclei tend to release particles or radiant energy, or both. Such isotopes are called **radioactive isotopes.** Rates of decay of radioactive isotopes have been determined for many isotopes. The length of time it takes for one-half of any size sample of an isotope to decay is called its **half-life.** Depending on the isotope, half-lives vary from a fraction of a second to billions of years.

The age of a material can be determined by measuring the amount of a particular radioactive isotope it contains. This quantity is compared with the amount of some other substance in the fossil that remains constant over time. For example, relatively young fossils can be dated by measuring the ratio of the amount of carbon-14, a radioactive isotope, to the amount of a stable isotope, carbon-12. Living things take carbon into their bodies constantly. Most of the carbon is in the form of carbon-12. A very small proportion of it, however, is in the form of carbon-14, which undergoes decay. This ratio of carbon-14 to carbon-12 is a known quantity for

Word Roots and Origins

isotope

from the Greek *iso,* meaning "equal," and *topos,* meaning "place"

RECENT RESEARCH

New Planets Discovered

From mid-1995 through 1996, scientists discovered more than a dozen planets. This suggests that there are many undiscovered planets, some of which may have some form of life on them.

QUICK FACT

☑ Earth's atmosphere is quite different from the rest of the cosmos in its relative abundance of elements. It has much lower amounts of hydrogen and noble gases than the cosmos does. Earth's early atmosphere likely formed from the release of volatile substances trapped by the crust and mantle.

MAKING CONNECTIONS

Technology

Have students research the use of radioisotopes in medicine and biological research and report on their research. (Radioisotopes are used to produce X rays, which are used as a treatment for cancer and to diagnose injury and disease. Radioisotopes are used in research to trace the biochemical activities of various atoms and molecules.)

MAKING CONNECTIONS

Mathematics

Calculate the age of two specimens tested for their radioisotope contents: a human body buried in ice in which 87.5 percent of the carbon-14 has decayed, and a meteorite in which 75 percent of the thorium-230 has decayed. (Computations can be made by half-lives: 87.5 percent decay is three half-lives; 75 percent is two half-lives. The results are: human—17,145 years old; meteorite—150,000 years old.)

INCLUSION *ACTIVITY*

Provide students with 64 chips or markers that are different colors on the two sides to reinforce the concept of radioactive decay. Tell students they will be making calculations of half-lives, as scientists do with radioisotopes. They will use a hypothetical half-life of one minute. Place all the chips on a table with one color up. Have students mark the start time and one minute later turn 32 chips over. Tell students these chips have now decayed and are at a stable energy level. After another minute, students should turn 16 more chips over. They should continue the exercise until they need to have halves of their chips (six half-lives or six minutes).

internetconnect

SCILINKS NSTA
TOPIC: Radioactive dating
GO TO: www.scilinks.org
KEYWORD: HM266

TABLE 14-1 *Some Isotopes Used in Radioactive Dating*

Isotope	Half-life
Carbon-14	5,730 y
Thorium-230	75,000 y
Potassium-40	1,300,000,000 y
Uranium-238	4,500,000,000 y

living organisms. When an organism dies, its uptake of carbon stops, and decay of the existing carbon-14 continues. Thus, over time, the amount of carbon-14 declines with respect to the amount of the stable carbon-12. After 5,730 years, half of the carbon-14 in a sample will have decayed. After another 5,730 years, half of the remaining carbon-14 in the sample likewise will have decayed. Use of carbon-14 dating is limited to organic remains less than about 60,000 years old, like the leather quiver and wooden bow and arrows shown in Figure 14-5. Isotopes with longer half-lives are used to date older fossils and rocks. Some of the isotopes commonly used in radioactive dating procedures appear in Table 14-1.

Scientists have estimated Earth's age by using a dating method that is based on the decay of uranium and thorium isotopes in rock crystals. Collisions between Earth and large pieces of space debris probably caused the surface of Earth to melt many times as the planet was formed. Therefore, the age of the oldest unmelted surface rock should tell us when these collisions stopped and the cooling of Earth's surface began. Scientists have found zircon crystals that are 4.2 billion years old. We can infer that organic molecules could have survived and begun to accumulate sometime after this.

FIGURE 14-5

This deerskin quiver, with a wooden bow and arrows, is about 3,000 years old. Carbon-14 dating methods can be used for organic materials less than 60,000 years old.

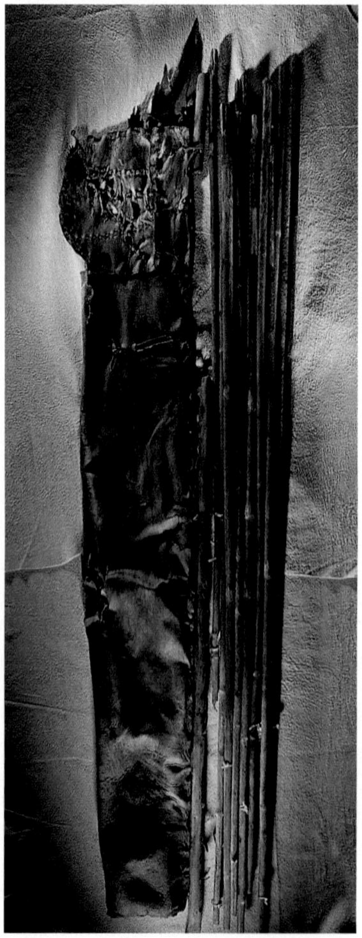

THE FIRST ORGANIC COMPOUNDS

All of the elements found in organic compounds are thought to have existed on Earth and in the rest of the solar system when the Earth formed. But how and where were these elements assembled into the organic compounds found in life? One of the most popular hypotheses proposed to solve this puzzle was developed by the Soviet scientist Alexander I. Oparin in 1923. Oparin (1894–1980) suggested that the atmosphere of the primitive Earth was very different from that of today. Oparin thought the early atmosphere contained ammonia, NH_3; hydrogen gas, H_2; water vapor, H_2O; and compounds made of hydrogen and carbon, such as methane, CH_4. At temperatures well above the boiling point of water, these gases might have formed simple organic compounds, such as amino acids. According to Oparin, when Earth cooled and water vapor condensed to form lakes and seas, these simple organic compounds would have collected in the water. Over time these compounds could have entered complex chemical reactions, fueled by energy from lightning and ultraviolet radiation. These reactions, Oparin reasoned, ultimately would have resulted in the macromolecules essential to life, such as proteins.

The Experimental Synthesis of Organic Compounds

Oparin carefully developed his hypotheses, but he did not perform experiments to test them. So in 1953, an American graduate student, Stanley L. Miller (1930–), and his professor, Harold C. Urey (1893–1981), set up an experiment using Oparin's hypotheses as a starting point. Their apparatus, illustrated in Figure 14-6, included a chamber containing the gases Oparin assumed were present in the young Earth's atmosphere. As the gases circulated in the chamber, electric sparks, substituting for lightning, supplied energy to drive chemical reactions. The Miller-Urey experiment, and other variations that have followed, produced a variety of organic compounds, including amino acids.

Since the 1950s, scientists have continued to explore the origin of simple organic compounds. Their experiments have produced a variety of compounds, including various amino acids, ATP, and the nucleotides in DNA. Such results suggest many ways that vital organic compounds might have formed on the young Earth.

In recent years, new hypotheses regarding early Earth's atmosphere have been proposed by investigators who study planet formation. In contrast to Oparin's hypotheses, it has been suggested that the atmosphere of early Earth was composed largely of carbon dioxide, CO_2; nitrogen, N_2; and water vapor, H_2O. Laboratory simulations of these atmospheric conditions have shown that both carbon dioxide and oxygen gas interfere with the production of organic compounds. Therefore, it is thought that conditions in areas protected from the atmosphere, such as those that exist in undersea hot springs, might have favored the production of organic compounds.

Organic Compounds from Beyond Earth

Recently, a broad mixture of organic compounds was found in a newly fallen meteorite that was recovered before it was contaminated with organic compounds from Earth. These compounds, which had not been destroyed by heat as the meteoroid entered Earth's atmosphere, must have formed in space. Some scientists hypothesize that after the period of Earth's formation, some organic compounds may have accumulated on the surface of Earth in this way, carried by space debris rather than originating here.

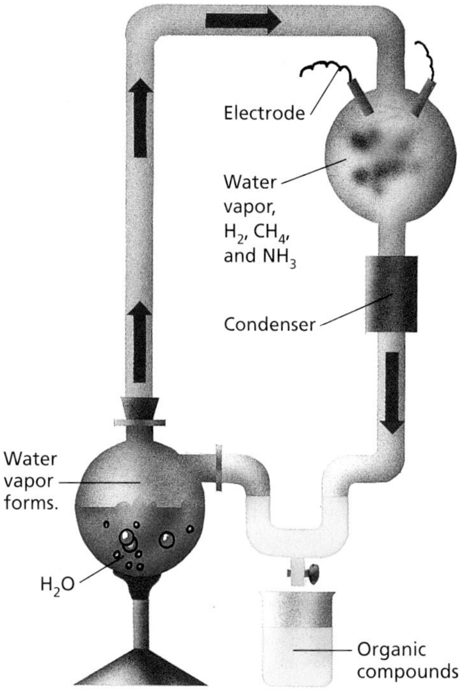

FIGURE 14-6

Miller and Urey's apparatus was a model for the atmospheric and temperature conditions of early Earth.

Quick Lab

Inferring Probability

Materials 3 × 5 in. cards (12) labeled with organic compounds

Procedure

1. Deal three cards, and try to make one of the following combinations: $NH_2–CH_2–COOH$, $CH_3–COOH$, or $CH_3–CH_2–COOH$. Each of these combinations represents an organic molecule.
2. Record your results. Replace the dealt cards in the set and shuffle the cards. Repeat the procedure 19 times.
3. Count the number of molecules you were able to form. Then calculate the probability of forming a molecule with each deal.

Analysis How can you compare a simple game of chance to the synthesis of organic compounds?

TEACHING STRATEGY

Constructing Organic Molecules

Provide students with a chemistry modeling set or make one using colored construction paper. (Cut circles of different-colored pieces of paper to represent carbon, oxygen, hydrogen, and nitrogen atoms. Provide tape for attaching atoms to one another.) Have the students put different-colored circles together to form molecules of NH_3, H_2, H_2O, and CH_4. Next have them take apart their molecules and reuse their atoms to form simple sugars, lipids, amino acids, and nucleotides.

Quick Lab

Inferring Probability

Time Required 30 minutes

Procedural Tips Have students work in small groups. Each group needs a set of 12 cards. To make a set of 12 cards, label 3 cards with each of the following compounds: CH_3–, –CH_2–, NH_2–, and –COOH. The dashes represent where the functional groups can be bonded together. Write the following equation on the board or overhead: Probability = number of organic compounds made ÷ 20 deals. The probability is 9/20 that students will be able to form an organic molecule with each deal of three cards.

Answers to Analysis The cards represent compounds similar to those Oparin proposed to be present in the primitive environment. Oparin and Miller based their hypotheses on the probability that certain organic compounds existed on the early Earth. Conditions in the atmosphere had to be just right (by chance) for organic synthesis to occur.

SECTION 14-2

QUICK FACT

☑ Although Oparin didn't conduct the experiments that he supposed would produce organic substances from inorganic ones, he did produce coacervate droplets. Some of these contained chlorophyll and conducted photosynthesis.

ANSWERS TO SECTION 14-2 REVIEW

1. Earth probably formed and cooled to a temperature below the melting point of rock between 4.6 and 4.2 billion years ago.
2. 1 billion years: 50 percent; 2 billion years: 25 percent; 3 billion years: 12.5 percent; 4 billion years: 6.25 percent
3. Simple organic compounds may have formed from gases in the high-temperature atmosphere of early Earth. Alternatively, organic compounds may have formed in space and been deposited on Earth by meteorites.
4. Oparin hypothesized that simple organic compounds formed in the atmosphere of early Earth, which, he proposed, was composed of ammonia, hydrogen gas, water, and hydrocarbon compounds. Miller and Urey tested Oparin's hypothesis using an apparatus that simulated the atmospheric conditions of early Earth.
5. Microspheres are membrane-bound structures. Microspheres can bud to form smaller microspheres, and coacervates can grow.
6. Radioactive isotopes with very short half-lives would not be useful for dating fossils because after a relatively short period, the amount of the undecayed isotope would be too small to give an accurate indication of age.

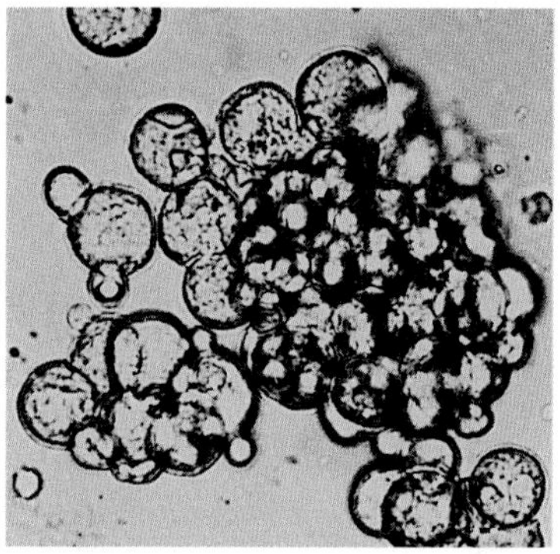

FIGURE 14-7

Membrane-bound structures, such as these, have been formed in the laboratory under conditions that may have existed on early Earth. Structures such as these may have enclosed replicating molecules of RNA and may have been the forerunners of the first cells.

FROM MOLECULES TO CELL-LIKE STRUCTURES

Sidney Fox (1912–) and others have done extensive research on the physical structures that may have given rise to the first cells. These cell-like structures, like the ones shown in Figure 14-7, form spontaneously in the laboratory from solutions of simple organic chemicals. The structures include **microspheres,** which are spherical in shape and are composed of many protein molecules that are organized as a membrane, and **coacervates** (coh-AS-uhr-vayts), which are collections of droplets that are composed of molecules of different types, including linked amino acids and sugars.

For many years, it had been assumed that all cell structures and the chemical reactions of life required enzymes that were specified by the genetic information of the cell. Both coacervates and microspheres, however, can form spontaneously under certain conditions. For example, the polymers that form microspheres can arise when solutions of simple organic chemicals are dripped onto the surface of hot clay. The heat vaporizes the water, encouraging polymerization. Coacervates and microspheres have a number of life-like properties, including the ability to take up certain substances from their surroundings. Coacervates can grow, and microspheres can bud to form smaller microspheres. These properties of coacervates and microspheres show that some important aspects of cellular life can arise without direction from genes. Thus, these studies suggest that the gap between the nonliving chemical compounds and cellular life may not be quite as wide as previously thought.

When considering the evolution of cells from simpler structures, it is important to remember that microspheres and coacervates could not have responded to natural selection. Recall from Chapter 1 that natural selection is an important driving force of evolution—which is descent with modification, or change over generations. The laboratory-produced cell-like structures do not have hereditary characteristics. Thus, although these cell-like structures have some of the properties of life, they are not alive because they do not have heredity.

SECTION 14-2 REVIEW

1. **The oldest rocks on Earth date from about 4.2 billion years ago. What does this suggest about the interval between 4.6 billion years ago, when the Earth started to form, and 4.2 billion years ago?**
2. **If a radioactive isotope had a half-life of 1 billion years, how much of it would be left after each of the following intervals of time: 1 billion years, 2 billion years, 3 billion years, and 4 billion years.**
3. **What are two possible sources of simple organic compounds on the early Earth?**
4. **What was Oparin's hypothesis, and how was it tested?**
5. **What properties do microspheres and coacervates share with cells?**
6. **CRITICAL THINKING Some radioactive isotopes that are used in medicine as tracers in the bloodstream have very short half-lives, often only a few years or less, rather than thousands of years. Would these isotopes also be useful in dating fossils? Why or why not?**

THE FIRST LIFE-FORMS

A remote and desolate corner of Australia was nicknamed the North Pole by disappointed gold prospectors in the 1800s. But this region has been a "gold mine" after all for twentieth-century scientists. It was there that the oldest known cellular fossils—3.5-billion-year-old traces of early unicellular organisms—were found.

SECTION

14-3

OBJECTIVES

▲

Explain the importance of the chemistry of RNA in relation to the origin of life.

●

List three inferred characteristics that describe the first forms of cellular life on Earth.

■

Name two types of autotrophy and explain the difference between them.

◆

Explain how photosynthesis and aerobic respiration are thought to be related.

▲

Define *endosymbiosis,* and explain why it is important in the history of eukaryotes.

THE ORIGIN OF HEREDITY

Chapter 10 provides a detailed explanation of how hereditary information affects the phenotype of cells. Recall that the hereditary information contained in a DNA molecule is first transcribed into an RNA message, and then the RNA message is translated into a protein, as shown in Figure 14-8. Thus, DNA serves as the template for RNA, which in turn serves as the template for specific proteins.

In recent years, scientists have taken a closer look at the DNA-RNA-protein sequence. Why is RNA necessary for this process? Why doesn't DNA, which is a template itself, carry out protein synthesis directly? The clues to a more complete understanding of RNA function may be found in its shape. Unlike DNA, RNA molecules can take on a great variety of shapes, for example, the t shape of transfer RNA, shown in Figure 14-8. These shapes are dictated by hydrogen bonds between particular nucleotides in an RNA molecule, much as the shapes of proteins depend on hydrogen bonds between particular amino acids. These questions and observations led to the speculation that some RNA molecules might actually behave like proteins and catalyze chemical reactions.

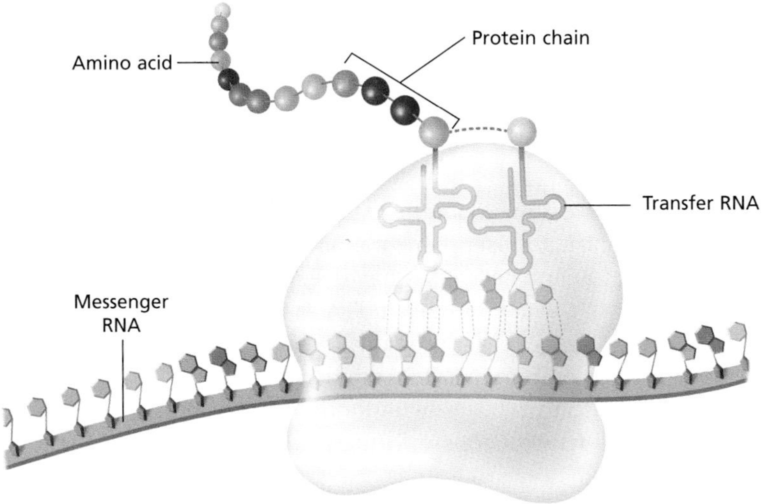

FIGURE 14-8

Messenger RNA is transcribed from a DNA template. Transfer RNA translates the three-base codons in the mRNA, assembling a protein from the specified amino acids.

SECTION 14-3

SECTION OVERVIEW

Reproduction and Inheritance

Section 14-3 explains currently favored hypotheses that the first hereditary molecule might have been RNA. The catalytic properties of some types of RNA are introduced, and the possibility that early RNA might have catalyzed its own replication is raised.

Interdependence of Organisms

Section 14-3 explains the stepwise evolutionary progression thought to have occurred: from heterotrophic prokaryotes to autotrophic prokaryotes and finally to modern eukaryotes. The hypothesis of the evolution of mitochondria and chloroplasts by endosymbiosis is explained.

▶ ENGAGE STUDENTS

Provide students with pictures of a variety of unicellular organisms. Ask them to study the pictures and identify features that the cells share. Then have students discuss the role of each of the features in life processes. (The cells have some kind of outer barrier that contains the contents of the cell. Some of the cells [the eukaryotes] have a nucleus and a variety of organelles.) Ask the students to recall from Chapter 4 that the cell theory states that all cells arise from preexisting cells. Ask them how an existing cell would carry out the processes that must occur to construct a new cell. (It would use building-block molecules and enzymes to make new cell membranes, new DNA, etc.) Ask them to postulate how the first cells on Earth might have arisen. (Answers will vary, depending on students' creativity and prior knowledge.)

CULTURAL CONNECTION

The theme that Earth gave rise to life appears in the legends of many cultures. For example, according to a Hindu proverb, "The Earth is our Mother, and we are all her children." A Hopi proverb states, "Earth gives life and seeks the man who walks gently upon it."

RETEACHING *ACTIVITY*

Ask students to draw a schematic diagram of DNA function. (They may need to refer to Chapters 9 and 10. The diagram should show DNA directing its own replication and the synthesis of RNA, which in turn directs the synthesis of proteins.) Then have the students draw another diagram including only RNA and proteins. (The diagram should show RNA directing its own replication and the synthesis of proteins.)

CRITICAL THINKING

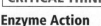

Enzyme Action

What kind of bonds form between regions of a folded protein and a folded nucleic acid? (hydrogen bonds) How would the relative strength of these bonds enable these molecules to function as enzymes? (They can break and re-form easily, allowing the molecule to change shape as a substrate is acted on.)

RECENT RESEARCH

RNA-DNA Link

Scientists have analyzed an enzyme from a bacterium that converts the building blocks of RNA into the building blocks of DNA. The function of this enzyme suggests that early life-forms could have changed from being RNA-based to being DNA-based.

Eco Connection

Archaebacteria

Some species of archaebacteria, such as *Methanosarcina barkeri* pictured in Figure 14-9, are referred to as methanogens. Within these bacteria, hydrogen gas reacts with carbon dioxide to produce methane, a simple carbon compound. Methanogens are poisoned by oxygen, but they can live in watery environments where other bacteria have consumed all free oxygen, such as in swamps and even the intestines of animals.

Methanogens may prove useful to humans in two significant ways: they are currently used in the cleanup of organic waste, such as sewage, and they may eventually be harnessed for large-scale production of methane for use as a fuel source.

Other species of archaebacteria are being used in the cleanup of petroleum spills into soil, such as occur when underground gasoline tanks develop leaks. This technique, called bioremediation, often relies on bacteria already present in the soil. These bacteria are activated by application of nutrient-rich solutions formulated to their taste. As the bacteria multiply, they metabolize petroleum, releasing harmless byproducts.

THE ROLES OF RNA

In the early 1980s, researcher Thomas Cech (1947–) found that a type of RNA found in some unicellular eukaryotes is able to act as an enzyme. Cech used the term **ribozyme** (RIE-boh-ziem) for an RNA molecule that can act as an enzyme and promote a specific chemical reaction. Hypothetically, a ribozyme could act as an enzyme *and* have the ability to replicate itself.

Recent studies based on Cech's discovery have indicated that life may have started with self-replicating molecules of RNA. RNA molecules would have heredity and would be able to respond to natural selection and thus evolve. How could a single molecule respond to natural selection? Replication—or reproduction of the RNA molecule—might involve competing with other similar, but not identical, RNA molecules for a fixed number of available nucleotides. An RNA molecule that is more successful in getting nucleotides from its environment has an advantage over other RNA molecules. This advantage would then be passed on to the "offspring" of the RNA molecules, the new RNA molecules created by replication.

Since Cech's discovery, other ribozymal activities have been discovered, and it is clear that RNA plays a vital role in DNA replication, protein synthesis, RNA processing, and other basic biochemistry. Perhaps most or all of the chemistry and genetics of early cells were based on RNA.

As exciting as these discoveries have been, there are several questions left unanswered. For one thing, investigators still have not made or found a ribozyme capable of producing other ribozymes. Moreover, it is unclear how such RNA molecules could have evolved into cellular life. Perhaps self-replicating molecules of RNA started to evolve inside cell-like structures similar to microspheres or coacervates. If these RNA molecules were able to alter the phenotype of the cell-like structure that carried them, cellular life could have begun. The self-replicating RNA would have provided the hereditary information that the cell-like structures lack.

Word Roots and Origins

archaebacteria

from the Greek *arche,* meaning "the beginning," and *bactron,* meaning "a staff"

THE FIRST PROKARYOTES

What clues do we have about the nature of the first cellular life? When the first organisms arose, there was little or no oxygen gas in existence. Thus, the first cells must have been anaerobic. The small size of the oldest of the microfossils indicates that these early cells were prokaryotes. These cells probably were heterotrophs taking in organic molecules from their environment.

We can reason that a growing population of heterotrophs that depended on spontaneously formed organic molecules for food eventually would have removed most of these molecules from the environment. At this point, there would have been strong environmental pressure for autotrophs to evolve. These first autotrophs, however, probably did not depend on photosynthesis the way that most autotrophs do today.

Chemosynthesis

If we look for living organisms that may be similar to these early organisms, we find the **archaebacteria** (AR-kuh-bak-TIR-ee-uh). The archaebacteria constitute a kingdom of unicellular organisms, many of which thrive under extremely harsh environmental conditions. *Methanosarcina barkeri,* the archaebacterium shown in Figure 14-9, lives in anaerobic marine sediments. Many species of archaebacteria are autotrophs that obtain energy by chemosynthesis (KEE-moh-SIN-thuh-sis) instead of photosynthesis. In the process of **chemosynthesis,** CO_2 serves as a carbon source for the assembly of organic molecules. Energy is obtained from the oxidation of various inorganic substances, such as sulfur.

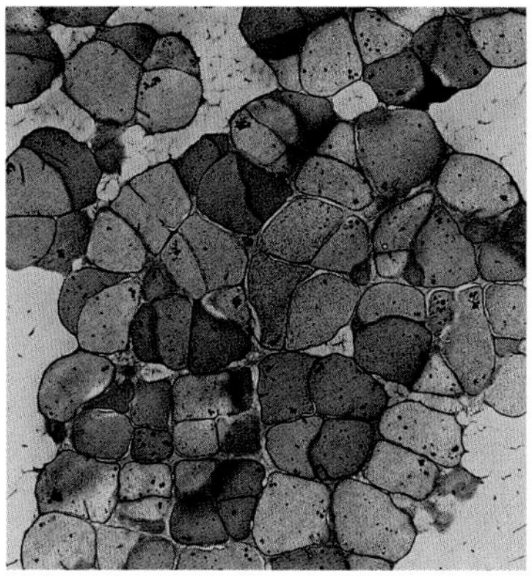

FIGURE 14-9

This archaebacterium species, *Methanosarcina barkeri,* produces methane during metabolism. Archaebacteria are thought to be similar to the types of cellular life that first populated Earth about 4 billion years ago.

Photosynthesis and Aerobic Respiration

Oxygen, a byproduct of photosynthesis, was damaging to many early unicellular organisms. Oxygen could destroy some coenzymes essential to cell function. Within some organisms, however, oxygen bonded to other compounds, thereby preventing the oxygen from doing damage. This bonding was one of the first steps in aerobic respiration. Thus, an early function of aerobic respiration may have been to prevent the destruction of essential organic compounds by oxygen.

Some forms of life had become photosynthetic by 3 billion years ago. In the mid-1990s, scientists discovered traces of carbon in geological formations off the coast of Greenland. These traces were likely left by some type of photosynthetic organism more than 3.8 billion years ago. The 3.5-billion-year-old microfossils found in Australia, shown in figure 14-10, are probably of photosynthetic unicellular organisms that are related to modern **cyanobacteria** (SIE-uh-no-bak-TIR-ee-uh), a group of photosynthetic unicellular prokaryotes.

It took a billion years or more for oxygen gas levels to reach today's levels. The oxygen, O_2, eventually reached the upper part of the atmosphere, where it was bombarded with sunlight. Some wavelengths of sunlight can split O_2 to form highly reactive single oxygen atoms, O. These react with O_2 and form **ozone,** O_3. Ozone is poisonous to both plant and animal life, but in the upper atmosphere, a layer of ozone absorbs intense ultraviolet radiation from the sun. Ultraviolet radiation damages DNA, and without the protection of the ozone layer, life could not have come to exist on land.

FIGURE 14-10

These filament-shaped microfossils from the Fig Tree Chert formation in northern Australia represent the oldest-known cellular life. These organisms lived 3.5 billion years ago.

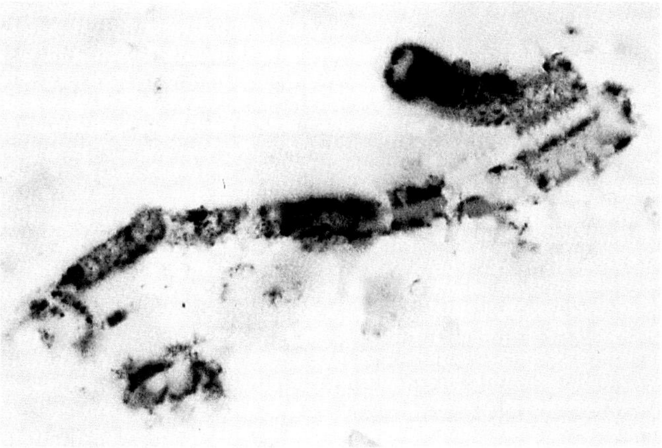

CONTENT CONNECTION

Chapters 6 and 7—Photosynthesis and Respiration

Both photosynthesis and respiration are series of many chemical reactions. The overall reactions are essentially opposite to each other: photosynthesis produces glucose and O_2 from H_2O, CO_2, water, and light energy; respiration produces H_2O, CO_2, water, and energy as ATP from glucose and O_2. Nearly all organisms on Earth depend on the complementary relationship between these two major processes.

RECENT RESEARCH

Eukaryotes from Archaebacteria

The complete genome of an archaebacterium was sequenced and compared with the genome of yeast and two eubacterial genomes that have been sequenced. Because of a number of gene similarities, scientists propose that eukaryotes may have evolved from an archaebacterium rather than from a eubacterium.

ACTIVE READING

Technique: K-W-L

Tell students to return to their lists of things they **W**ant to know about how life began, which they made at the beginning of the study of this chapter. Have them place check marks next to the questions that they are now able to answer. Students should finish by making a list of what they have **L**earned. Ask students which questions are still unanswered. Ask them if they have any new questions.

SECTION 14-3

VISUAL LINK

Figure 14-11

Trace the steps depicted in the process of endosymbiosis shown in Figure 14-11. Ask the students what process accounts for the uptake of large structures from the environment (endocytosis). Point out that eukaryotic cells do not contain bacteria but rather that engulfed bacteria evolved over time and became mitochondria and chloroplasts.

ANSWERS TO SECTION 14-3 REVIEW

1. The different shapes of RNA molecules depend on hydrogen bonding between different nucleotides.
2. The first cells arose when there was little or no oxygen in the atmosphere.
3. Aerobic respiration causes the bonding of oxygen to other compounds. This may have prevented it from damaging early unicellular organisms.
4. Mitochondria replicate independently of the replication cycle of the rest of the cell, and mitochondria have some of their own genes, which are different from those of the cell.
5. The first eukaryotic cells may have evolved when small aerobic prokaryotes invaded them and began to live and reproduce inside the larger cells, eventually evolving in mitochondria. Likewise, chloroplasts may have originated from photosynthetic prokaryotic invaders.
6. Damage to the ozone would allow more ultraviolet radiation to reach Earth. Ultraviolet radiation damages living things.

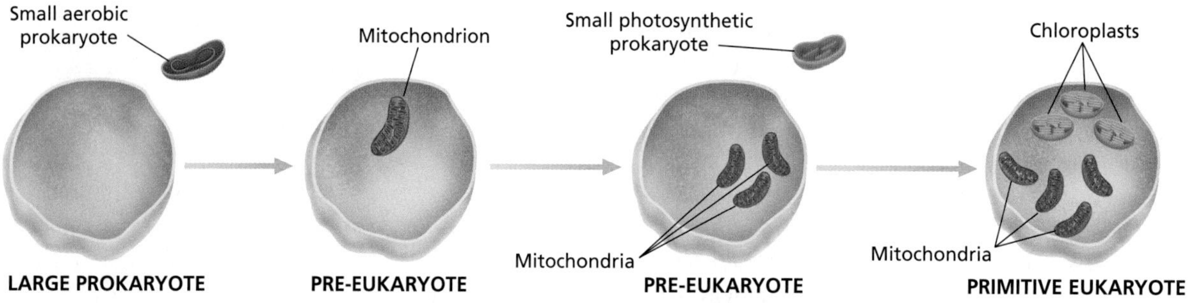

FIGURE 14-11

According to the hypothesis of endosymbiosis, large prokaryotic, unicellular organisms were invaded by smaller prokaryotic, unicellular organisms. These smaller organisms eventually gave rise to modern mitochondria and chloroplasts, which carry their own DNA and replicate independently from the rest of the cell that contains them.

THE FIRST EUKARYOTES

Recall from Chapter 4 that eukaryotic cells differ from prokaryotic cells in several ways. Eukaryotic cells are larger, their DNA is organized into chromosomes in a cell nucleus, and they contain membrane-bound organelles. How did such a complex type of cell evolve from the simple prokaryotes? A large body of evidence now suggests that between about 2.0 and 1.5 billion years ago, a type of small aerobic prokaryote entered and began to live and reproduce *inside* larger, anaerobic prokaryotes. Researcher Lynn Margulis (1938–) has proposed that what may have started as an invasion became a successful, mutually beneficial relationship—called **endosymbiosis** (EN-doh-sim-bee-OH-sis), shown in Figure 14-11. It is thought that the aerobic prokaryote eventually gave rise to modern mitochondria, which are the site of aerobic respiration in eukaryotic cells.

Sometime later, there was a second successful invasion of pre-eukaryotic cells. This time, the invader was a relative of modern photosynthetic cyanobacteria. These invaders eventually gave rise to chloroplasts, the sites of photosynthesis. There is compelling evidence to support this hypothesis of eukaryotic evolution. Both chloroplasts and mitochondria replicate independently from the replication cycle of the cell that contains them. Moreover, chloroplasts and mitochondria contain some of their own genes, which are different from those of the rest of the cell. These genes are found in the organelles themselves, in a circular piece of DNA, an arrangement characteristic of prokaryotic, though not eukaryotic, DNA.

SECTION 14-3 REVIEW

1. **Why do different molecules of RNA assume many different shapes?**
2. **Why do scientists think the first forms of life on Earth were anaerobic?**
3. **How might aerobic respiration have protected early cells from damage?**
4. **What evidence supports the hypothesis that mitochondria were once free-living prokaryotic cells?**
5. **Explain how the first eukaryotes may have evolved.**
6. **CRITICAL THINKING Some forms of air pollution damage Earth's ozone layer. How might such damage affect life?**

CHAPTER 14 REVIEW

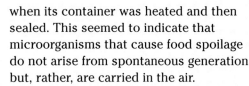

SUMMARY/VOCABULARY

14-1

- Before the 1600s, it was generally thought that organisms could arise from nonliving material by spontaneous generation.
- Redi showed in 1668 that rotting meat kept away from flies would not produce new flies. Maggots appeared only on meat that had been exposed to flies.
- Spallanzani showed in the 1700s that microorganisms would not grow in broth when its container was heated and then sealed. This seemed to indicate that microorganisms that cause food spoilage do not arise from spontaneous generation but, rather, are carried in the air.
- Pasteur used a variation of Spallanzani's design to prove that microorganisms are carried in the air and do not arise by spontaneous generation.

Vocabulary

biogenesis (261) spontaneous generation (261)

14-2

- Earth is thought to have been formed by repeated collisions of debris moving through space.
- The mass number of an element is the total number of protons and neutrons in the nucleus. The number of neutrons in atoms of an element can vary. Atoms with varying numbers of neutrons are called isotopes.
- An isotope's half-life is the time it takes for one-half of a sample of the isotope to decay.
- The ages of objects like fossils and rocks can be determined by measuring the amount of radioactive decay that has occurred in radioactive isotopes found in the sample.
- The first simple organic compounds may have been formed at high temperatures on early Earth in an atmosphere of ammonia, hydrogen gas, and water vapor.
- Macromolecules important to life may have been assembled from simple organic compounds. Lightning may have supplied the energy for these chemical reactions.
- Some organic compounds may have been deposited on Earth by meteorites.
- Cell-like structures, including microspheres and coacervates, form spontaneously in certain kinds of solutions. These structures do not show heredity.

Vocabulary

coacervate (268) isotope (264) microsphere (268) radioactive decay (265)
half-life (265) mass number (265) radioactive dating (264) radioactive isotope (265)

14-3

- In addition to its role as a template for protein assembly, some RNA molecules can act as enzymes.
- The first molecule that held hereditary information may have been RNA rather than DNA.
- RNA can assume different shapes, much as proteins do. These shapes depend on areas of attraction between the RNA nucleotides.
- The first cells that formed on Earth were probably heterotrophic prokaryotes.
- The first autotrophic cells probably used chemosynthesis to make food.
- An important initial function of aerobic respiration may have been to bind oxygen and prevent it from doing damage to early organisms.
- Eukaryotic cells may have evolved from large prokaryotic cells that were invaded by smaller prokaryotic cells. These small prokaryotic invaders may have been the ancestors of organelles, including mitochondria and chloroplasts.

Vocabulary

archaebacteria (271) cyanobacteria (271) ozone (271) ribozyme (270)
chemosynthesis (271) endosymbiosis (272)

CHAPTER 14 REVIEW ANSWERS

REVIEW

1. In the process of biogenesis, life arises from existing life. In the process of spontaneous generation, life arises from nonliving matter.
2. The half-life of a radioactive isotope is the period of time it takes for one-half a sample of the isotope to break down by radioactive decay.
3. coacervates and microspheres
4. Ribozymes can catalyze certain chemical reactions, as enzymes do.
5. Photosynthesis is the process of making organic molecules from inorganic molecules and energy from sunlight. In chemosynthesis, organic molecules are made from inorganic molecules and the oxidation of inorganic compounds.

6. a **7.** c
8. b **9.** b
10. b **11.** c
12. a **13.** c
14. a **15.** a

16. The gases CH_4, H_2, and NH_3 simulate the composition of the early atmosphere, according to Oparin's hypothesis. The electric spark simulates lightning, which may have supplied the energy needed to drive the chemical reactions that produced the first organic macromolecules.
17. The theory that life arises only from other life does not explain how the first life arose.
18. Archaebacteria are thought to be the modern organisms most similar to the first life-forms on Earth.
19. Removal by heterotrophs of most of the existing, spontaneously formed organic molecules from the environment would provide environmental pressure for the evolution of autotrophs.
20. The atmosphere of Earth at the time of the synthesis of the first organic molecules probably contained very little oxygen gas.
21. Unbound oxygen can damage living things; therefore, organisms that could bind oxygen (as part of cellular respiration) could avoid damage.
22. The ozone layer shields Earth from much of the damaging ultraviolet radiation from the sun.
23. Some nonphotosynthetic autotrophs use chemosynthesis.
24. RNA may have existed before DNA, RNA can act as a template, and RNA can catalyze some chemical reactions.
25. Investigators have not found a ribozyme capable of producing other ribozymes.

CRITICAL THINKING

1. Students' answers will vary but should be logical. For example, remove a large sample of mud from a dry lake that regularly has fish during the wet season. Place the mud in an aquarium with water and see if fish develop. If they do, repeat the experiment, but divide a second sample of mud

REVIEW

Vocabulary

1. Explain the difference between biogenesis and spontaneous generation.
2. What is the relationship between radioactive decay and the half-life of an isotope?
3. Name two nonliving, cell-like structures that can form in certain solutions.
4. How is a ribozyme like an enzyme?
5. What is photosynthesis. What is chemosynthesis?

Multiple Choice

6. In the seventeenth and eighteenth centuries, the hypothesis of spontaneous generation was used to explain (a) how new life started (b) how simple organic compounds formed (c) how coacervates and microspheres formed (d) how eukaryotes evolved.
7. Redi's experiment was important because it showed that (a) maggots give rise to microorganisms (b) flies swarm on rotting meat (c) flies do not form from rotting meat (d) air contains a "vital force."
8. People objected to Spallanzani's experiment because he (a) used an open jar of meat as a control (b) boiled his flasks of broth for a long time (c) drew the necks of his flasks into a curved shape (d) did not have a control group.
9. The neck of Pasteur's flasks (a) allowed both air and particles to enter the flask (b) allowed air to enter the flask but kept particles out (c) allowed particles to enter but kept air out (d) kept both air and particles out.
10. During the first half-billion years of its existence, Earth and the other planets of the solar system grew by a process involving (a) the synthesis of organic molecules (b) collisions with space debris (c) flames from the sun (d) tidal forces generated on the moon.
11. The oldest fossil of cellular life found on Earth is (a) about 4.6 billion years old (b) about 4.2 billion years old (c) about 3.5 billion years old (d) less than 2 billion years old.
12. Coacervates and microspheres cannot evolve because they have no (a) genetic information (b) cell membrane (c) complex organic molecules (d) fluid in their interior.
13. Miller and Urey's experiment (a) proved Oparin's hypothesis about the origin of life (b) disproved Oparin's hypothesis about the origin of organic molecules (c) provided support for Oparin's hypothesis about the origin of organic molecules (d) is irrelevant to any hypothesis regarding the origin of life.
14. Coacervates and microspheres are (a) collections of organic molecules enclosed within a boundary (b) identical to the first forms of life (c) the oldest microfossils (d) new forms of bacteria.
15. The generation of organisms from nonliving material does not occur today mostly because (a) the presence of oxygen in the atmosphere prevents organic compounds from forming spontaneously (b) coacervates and microspheres take up all the extra nutrients on Earth (c) coacervates and microspheres cannot form today (d) there is not enough energy to drive the chemical reactions needed to form the complex organic compounds necessary for life.

Short Answer

16. Explain the role of the gases CH_4, H_2, and NH_3 and the role of the electric spark in the apparatus shown.

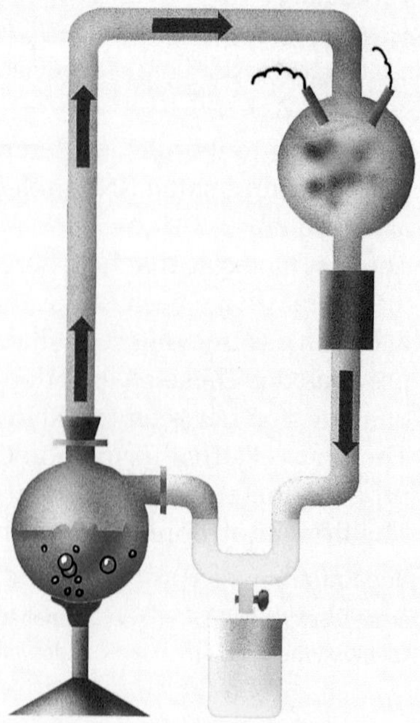

17. Why did the theory of biogenesis pose a dilemma regarding the origin of life?

18. What modern organisms are thought to be most like the first life-forms on Earth?

19. What environmental factors probably favored the evolution of autotrophs?

20. Organic compounds will not form in the Miller-Urey apparatus if O_2 is present. Why is this not a serious problem for scientists who study the origin of life?

21. How might the bonding of oxygen gas have served evolving organisms?

22. How did the formation of the ozone layer permit organisms to colonize land?

23. What energy sources do some nonphotosynthetic autotrophs use?

24. Why have many scientists who investigate the origin of life focused on RNA chemistry?

25. What problems remain with the hypothesis that ribozymes or similar molecules may have been the first self-replicating structures?

CRITICAL THINKING

1. People once believed fish could form from the mud in a pond that sometimes dried up. How could you demonstrate that this conclusion is false?

2. According to a recent hypothesis, lightning may not have existed on early Earth. How could you modify the Miller-Urey experiment to reflect this new idea? What sources of energy could you use to replace lightning?

3. RNA is copied the same way that DNA is copied—by base pairing. RNA, unlike double-stranded DNA, exists as a single strand. If a ribozyme is copied, would the copy be another identical ribozyme or something else? If the copy is not identical to the original, how could the original be replicated?

4. The term *chemical evolution* is sometimes applied to the series of events that might have resulted in the spontaneous origin of life. How do the events that are thought to have produced living organic matter from nonliving inorganic matter represent a process of evolution?

5. The graph below represents radioactive decay of an isotope. If the half-life of carbon-14 is 5,730 years, how many years would it take for ⅞ of the original amount of carbon-14 in a sample to decay?

Amount Remaining of Radioactive Sample

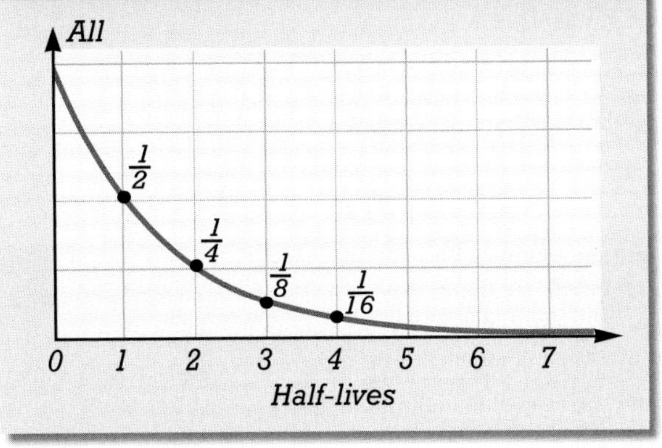

EXTENSION

1. Read "To Hell and Back" in *Discover,* July 1999, on page 76, and answer the following questions: What were some of the extreme conditions in the environment where the microbes lived? What substances besides oxygen do some deep-Earth-dwelling bacteria use for respiration? What precautions were taken to protect the organisms in the rock samples from being poisoned by the surface atmosphere?

2. Investigators studying the atmosphere have observed a large, thin area in the ozone layer above the continent of Antarctica. Use current science magazines to find out about this phenomenon. Present a report to your class summarizing different hypotheses about the causes of this thinning of the ozone layer.

into two parts. Put one part in a water-filled aquarium and cook the other part in an oven before placing it into a separate water-filled aquarium. The cooked mud cannot give rise to fish, while the uncooked mud should.

2. Other sources of energy, such as an ultraviolet light (to simulate ultraviolet radiation) or a heat lamp (to simulate geothermal energy), could be used with the Miller-Urey apparatus.

3. The copy would be the complement of the original; it would not be identical. The original could be replicated by using the copy as a template.

4. In order for RNA molecules to become self-replicating, they would have to compete with other RNA molecules for free nucleotides in their environment. RNA molecules that were better than other RNA molecules at getting nucleotides would have a selective advantage, and that type of RNA would tend to be replicated more, producing evolution in the population of RNA molecules.

5. 17,145 years

EXTENSION

1. The microbes were found living more than 2 mi (3.2 km) beneath the surface of the Earth at temperatures of 120°F (48.8°C) or more, in the absence of oxygen. Some bacteria use iron, nitrate, manganese, sulfur, chromium, or cobalt for respiration. The microbe samples were placed in an airtight tent filled with 98 percent nitrogen and 2 percent hydrogen.

2. Students' reports should mention the industrial production of chlorofluorocarbons (CFCs) as a culprit in the breakdown of the ozone layer.

CHAPTER 14
INVESTIGATION

TIME REQUIRED

One 50-minute class period

SAFETY PRECAUTIONS

Students should wear a lab apron, protective gloves, and safety goggles at all times during this investigation. Review the use of tongs and safety procedures when handling hot liquids. Avoid skin, eye, and mouth contact with amino acids, and have students wash their hands after handling the amino acid mixture. Review safety procedures when using an electric hot plate. Caution students to make sure their hands and their work area are dry when using electrical equipment. Electrical cords must not be frayed or kinked and must not dangle from the work stations. Remind students that the surface of the hot plate is very hot. Students should turn off and unplug hot plates before leaving the lab.

QUICK REFERENCES

Lab Preparation Notes are found on page 259D.

Holt BioSources provides a Teaching Resources CD-ROM that contains "Using Gowin's Vee in the Lab" and "Scoring Rubrics."

PROCEDURAL TIPS

You may want to review with students the structure of amino acids and how they combine to form proteins.

ANSWERS TO BACKGROUND

2. Microspheres differ from living cells in that they do not have a cell membrane that consists of phospholipids and proteins, and they lack heredity.
3. Microspheres are similar to living cells in that their contents

CHAPTER 14 INVESTIGATION

Making Microspheres

OBJECTIVES

- Make microspheres from amino acids by simulating the conditions found on early Earth.
- Compare the structure of microspheres with the structure of living cells.

PROCESS SKILLS

- observing
- comparing and contrasting
- modeling
- relating

MATERIALS

- safety goggles
- lab apron
- heat-protective gloves
- 500 mL beaker
- hot plate
- 125 mL Erlenmeyer flasks, 2
- ring stand with clamp
- balance
- amino acid mixture (of at least six different amino acids)
- glass stirring rod
- tongs
- clock or timer
- 1% sodium chloride (NaCl) solution
- 50 mL graduated cylinder
- dropper
- microscope slide
- coverslip
- compound light microscope
- 1% sodium hydroxide (NaOH) solution

Background

1. Microspheres are very small, spherical vessels that are bounded by a membranelike layer of amino acids. Microspheres can be created in the laboratory under controlled conditions.
2. How do microspheres differ from living cells?
3. How do microspheres resemble living cells?
4. What role might have been played by microspheres or similar structures before life began on Earth?

Procedure

1. Put on safety goggles, a lab apron, and heat-protective gloves before beginning this investigation.
2. CAUTION **Do not plug in or unplug the hot plate with wet hands. Use care to avoid burns when working with the hot plate. Do not touch the hot plate. Use tongs to move heated objects. Turn off the hot plate when not in use.** Fill a 500 mL beaker half full with water, and heat it on a hot plate. You will use the beaker as a hot-water bath. Leave space on the hot plate for a 125 mL Erlenmeyer flask, to be added later.
3. While waiting for the water to boil, clamp a 125 mL Erlenmeyer flask to a ring stand. Add 6 g of the amino acid mixture to the flask.
4. When the water in the beaker begins to boil, move the ring stand carefully so that the flask of amino acids sits in the hot-water bath.
5. When the amino acids have heated for 20 minutes, measure 10 mL of NaCl solution in a graduated cylinder and pour the solution into a second Erlenmeyer flask. Place the second flask on the hot plate beside the hot-water bath.

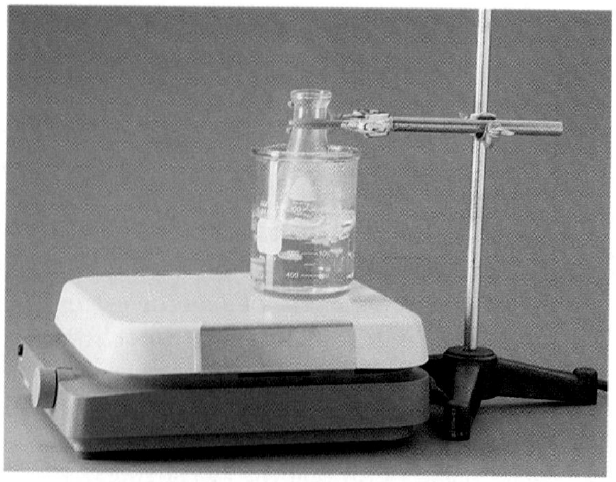

6. When the NaCl solution begins to boil, use tongs to remove the flask containing the NaCl solution from the hot plate. Then, while holding the flask with tongs, slowly add the NaCl solution to the hot amino acids while stirring.
7. Let this NaCl–amino acid solution boil for 30 seconds.
8. Remove the solution from the water bath, and allow it to cool for 10 minutes.
9. CAUTION **Slides break easily. Use caution when handling them.** Use a dropper to place a drop of the solution on a microscope slide, and cover the drop with a coverslip.
10. Place the slide on the microscope stage. Examine the slide under low power for tiny spherical structures. Then examine the structures under high power. These tiny sphere-shaped objects are microspheres.
11. CAUTION **If you get the sodium hydroxide (NaOH) solution on your skin or clothing, wash it off at the sink while calling to your teacher. If you get the sodium hydroxide solution in your eyes, immediately flush your eyes at the eyewash station while calling to your teacher.** Place a drop of 1% NaOH solution at the edge of the coverslip to raise the pH as you observe the microspheres. What happens?
12. In your lab report, make a table similar to the one shown below. Based on your observations of microspheres and cells, complete your table. Consider the appearance of microspheres and cells, their method of reproduction, their interaction with their environment, and any other characteristics that you observe.
13. Clean up your lab materials and wash your hands before leaving the lab.

Analysis and Conclusions

1. Suggest how the microspheres formed.
2. What did you observe when the pH was raised in step 11?
3. What does this suggest about the relationship of pH to microsphere formation?
4. Compare and contrast microspheres with living cells.
5. What characteristics would microspheres have to exhibit in order to be considered living?
6. How might the conditions you created in the lab be similar to those that are thought to have existed when life first evolved on Earth?
7. Predict what would happen to microspheres if they were placed in hypotonic and hypertonic solutions.

Further Inquiry

1. What do you think would happen if you added too much or too little heat? What happens to proteins at high temperatures? How can you test for the right amount of heat to use?
2. Do you think your microsphere experiment would have worked if you had substituted other amino acids? How can you test your hypothesis?

TABLE A ***COMPARING MICROSPHERES WITH CELLS***

Cell-like characteristics	Characteristics that are not cell-like
Answers will vary. Answers may include membrane-like envelope, replication by fission, shape and size similar to living cells, and selectively permeable.	**Answers will vary. Answers may include lacks organelles and lacks genetic material.**

differ from their surroundings and have a boundary.
4. Microspheres or similar structures might have enclosed replicating RNA, and may have been the precursors to living cells.

ANSWERS TO ANALYSIS AND CONCLUSIONS

1. Answers will vary. Encourage speculation. Students might suggest that the amino acids were attracted to each other and formed clumps that later became spheres.
2. The microspheres split in two.
3. Microspheres may be larger in acidic solutions and smaller in alkaline solutions.
4. Like living cells, microspheres have a membrane-like boundary and undergo fission. Unlike cells, microspheres have no organelles or nucleus.
5. Before they could be considered living, microspheres would have to have the ability to replicate themselves.
6. Answers will vary, but students should mention that the flask contained a mixture of amino acids, like the early oceans, and that the temperatures were extremely hot, as they were on early Earth.
7. Students should predict that microspheres will swell in hypotonic solutions and shrink in hypertonic solutions.

CHAPTER 15

EVOLUTION: EVIDENCE AND THEORY

	TOPICS	TEACHING RESOURCES	LABS, CLASSWORK, AND HOMEWORK
BLOCK 1 45 minutes	**Introducing the Chapter, p. 278**	ATE Focus Concept, p. 278 ATE Assessing Prior Knowledge, p. 278 ATE Understanding the Visual, p. 278	■ Supplemental Reading Guide, *Origin of Species*
	15-1 The Fossil Record, p. 279 Nature of Fossils, p. 279 Distribution of Fossils, p. 280	ATE Section Overview, p. 279 ATE Visual Link Table 15-1, p. 280	**Lab B8,** Fossil Study ★ Study Guide, Section 15-1 PE Section Review, p. 281
BLOCKS 2 & 3 90 minutes	**15-2 Theories of Evolution, p. 283** Lamarck's Explanation, p. 283 The Beginning of Modern Evolutionary Thought, p. 284 Darwin's Theories, p. 286	ATE Section Overview, p. 283 48, 49, 52	PE **Quick Lab,** p. 284, Analyzing Relationships ★ Study Guide, Section 15-2 PE Section Review, p. 288
BLOCK 4 45 minutes	**15-3 Evolution in Process, p. 289** Evidence of Evolution, p. 289 Patterns of Evolution, p. 291	ATE Section Overview, p. 289 ATE Critical Thinking, p. 291 ATE Visual Link Figure 15-9, p. 291 50	PE **Chapter 15 Investigation,** pp. 296–297 **Lab C19,** Analyzing Amino Acid Sequences to Determine Evolutionary Relationships ★ Study Guide, Section 15-3 PE Section Review, p. 292
BLOCKS 5 & 6 90 minutes	PE Chapter 15 Review, pp. 294–295	Audio CD Program	**CHAPTER TESTING** ■ Chapter Test (blackline master) ▲ Test Generator ■ Assessment Item Listing ■ Scoring Rubrics and Classroom Management Checklists

MODERN BIOLOGY

- **PE** Pupil's Edition
- **ATE** Teacher's Edition
- ■ Active Reading Guide
- ★ Study Guide

HOLT BIOSOURCES

- Teaching Transparencies
- Laboratory Program

One-Stop Planner CD-ROM
Shows these resources within a customizable daily lesson plan:

- ■ Holt BioSources Teaching Resources
- ■ Active Reading Guide
- ★ Study Guide
- ▲ Test Generator

KEY

READING FOR CONTENT MASTERY

Preview: Have students preview the objectives, the introductory paragraph, topic sentences, and captions before reading the chapter.

ATE Active Reading Technique: Reader Response Logs, p. 279
■ Active Reading Guide Worksheet 15-1
PE Summary/Vocabulary, p. 293

■ Active Reading Guide Worksheet 15-2
PE Summary/Vocabulary, p. 293

■ Active Reading Guide Worksheet 15-3
PE Summary/Vocabulary, p. 293

TECHNOLOGY AND INTERNET RESOURCES

CNN PRESENTS
Video Segment 9
Dino Egg Discovery

Holt Biology Videodiscs Teacher's Correlation Guide, Lessons 15-1 through 15-3

Audio CD Program, Sections 15-1 through 15-3

internetconnect

sciLINKS NSTA

On-line Resources: www.scilinks.org
The following *sci*LINKS Internet resources can be found in the student text for this chapter:

Topics:
- Extinction, p. 281
- Natural selection, p. 288

go.hrw.com

On-line Resources: go.hrw.com
Visit the HRW Web site for a variety of resources related to this chapter. Just type in the keyword HM2 Chapter 15.

Smithsonian Institution®
Visit **www.si.edu/hrw** for additional on-line resources.

CNNfyi.com
Visit **www.cnnfyi.com** for late-breaking news and current event stories selected just for you.

LAB ACTIVITY PLANNING

CHAPTER 15 INVESTIGATION

Modeling Selection, pp. 296–297

OVERVIEW
In this investigation, students will model selection by using a hypothetical animal, the Egyptian Origami Bird. Students will simulate the breeding of several generations of the birds and observe the effect of various phenotypes on the evolutionary success of these animals. The random nature of mutations is demonstrated by randomly changing the anterior and posterior wing position and wing circumference of the birds.

PREPARATION
Each student group will need one sheet of construction paper, a meter stick, a ruler, a pair of scissors, a straw, cellophane tape, a coin, and a die.

Quick Labs

Analyzing Relationships, p. 284
Each group of students will need pencil, paper and a set of 20 hardware items for this activity. Put together hardware items such as spiral-thread decking nails, wood and sheet metal screws, round-headed screws, flat-headed screws, slotted screws, Phillips-headed screws, large and small nails, threaded nails, a variety of washers, and a variety of nuts.

UNDERSTANDING THE VISUAL

Have students examine the photograph of the Galápagos tortoises. Point out that Darwin discovered that these tortoises were different from tortoises on the mainland of South America and also that tortoises on the various Galápagos islands were different from each other. His observations of these and other organisms of the Galápagos provided evidence and inspiration for his theories of evolution by natural selection.

FOCUS CONCEPT

Evolution

The theory of evolution states that species change over time. The primary mechanism for this change is natural selection. Point out that fossils provide one line of evidence for evolution. Through similarities in morphology; development; and DNA, RNA, and protein sequences, modern organisms also provide evidence for evolution.

ASSESSING PRIOR KNOWLEDGE

Review the following concepts.

Protein Structure and Function: *Chapter 3*
Have students name the subunit of proteins. Ask how the type and order of these subunits determine a protein's structure and function.

Translation: *Chapter 10*
Have students explain how the DNA in an organism determines the structure and function of the proteins that the organism produces. Ask how a change in an organism's DNA might result in a change in a protein it produces.

CHAPTER 15

EVOLUTION: EVIDENCE AND THEORY

Giant tortoises, Geochelone elephantopus, *crowd a puddle on the Galápagos Islands.*

FOCUS CONCEPT: ***Evolution***

As you read, note the many lines of evidence that support evolutionary theory.

THE FOSSIL RECORD

Fossil evidence shows a long history of life on Earth. The fossil record shows that forms of organisms appeared, lasted for long periods of time, and then disappeared, only to be followed by newer forms of life that also eventually disappeared. The history of life is one of constant change and a tremendous diversity of life-forms.

SECTION

15-1

OBJECTIVES

▲ Define *fossil,* and tell how the examination of fossils led to the development of evolutionary theories.

● Explain the law of superposition and its significance to evolutionary theory.

■ Describe how early scientists inferred a succession of life-forms from the fossil record.

◆ Tell how biogeographic observations suggest descent with modification.

NATURE OF FOSSILS

A **fossil** is a trace of a long-dead organism. Fossils are often found in layers of sedimentary rock, which is formed when **sediment,** such as dust, sand, or mud, is deposited by wind or water. Sedimentary fossils usually develop from the hard body parts of an organism, such as the shell, bones, teeth, or, in the case of plants, the woody stem. Over long periods of time, hard minerals replace the tissue of the organism, leaving rocklike structures.

A type of fossil called a **mold** is essentially an imprint in rock in the shape of an organism. Limestone owes its spongelike texture to the many molds scattered throughout its structure. Some molds eventually are filled with hard minerals, forming a **cast,** a rocklike model of the organism.

As you can see in Figure 15-1, many different kinds of fossils have been found. Some fossils are in the form of lacy carbon tracings of a fern, captured for all time on a flat plate of slate. Others, such as the tracks of animals or the fossilized marks on the bones of some human ancestors, are evidence of behavior.

How did people in the past regard these natural curiosities? Perhaps the most widely held view was that fossils were simply a naturally occurring part of rocks. Beginning in the late seventeenth century, however, a flurry of scientific investigation occurred. In

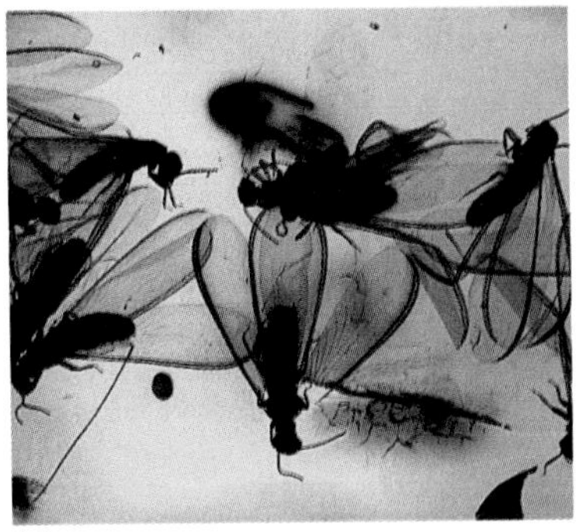

(a)

(b)

FIGURE 15-1

There are several different types of fossils. (a) Amber, the fossilized sap of trees, holds well-preserved insects. (b) This fossilized shell is that of a trilobite, *Modocia typicalus,* a dominant life-form in early seas.

SECTION 15-1

SECTION OVERVIEW

Matter, Energy, and Organization

Section 15-1 describes how fossils are formed. Steno's law of superposition is introduced, along with the concept of biogeography. The inference of Earth's age from examination of geological strata is discussed, and the geological history of Earth is presented in a table.

ACTIVE READING

Technique: Reader Response Logs

Students may already hold opinions about the origin of living things. So that students can make personal responses to the concepts presented in this chapter, have them make a Reader Response Log. Ask students to divide a sheet of paper in half lengthwise. On the left side of the paper, have them copy a word, phrase, or passage from the text. On the right side, have them write their reactions, thoughts, or questions about the entries on the left side of the paper.

RECENT RESEARCH

Oldest Multicellular Fossils

Scientists have discovered fossilized multicellular organisms that resemble brown algae in 1.7-billion-year-old rocks. These are 800 million years older than other known fossils of multicellular organisms. The first multicellular organisms were probably colonial protists that developed cell specialization. Some modern algae show a degree of cell specialization—reproductive cells differ from other cell types.

TEACHING STRATEGY

Making "Fossils"

Have students cover leaves or shells with petroleum jelly and set them in plaster of Paris (mixed to the consistency of a thick milkshake). After the plaster sets, they can remove the objects and reveal the molds.

VISUAL LINK

Table 15-1

Have students examine Table 15-1. Ask them to note the major groups of organisms, when they arose, and whether they are represented by similar, modern organisms. Have students note the mass extinctions, and ask them what might have caused these. (Volcanic eruptions and meteorites may have raised thick clouds of dust, cooling Earth and halting photosynthesis in large areas.) Ask students what happened after the mass extinctions. (New forms of organisms evolved from those that survived the extinctions.)

INCLUSION *ACTIVITY*

Using old magazines and news papers, have students collect photographs and drawings of organisms that lived during each of the periods in Table 15-1. Have them attach these illustrations to the wall on a long strip of paper and re-create the text's table on a horizontal scale. Have them discuss increases in diversity that have occurred over Earth's history.

QUICK FACT

☑ The mass extinctions that occurred at the end of the Permian period eliminated 95 percent of all species on Earth. The extinctions that wiped out the dinosaurs at the end of the Cretaceous period eliminated only 50 percent of Earth's species.

1668, Robert Hooke (1635–1703) published his conclusion that fossils are the remains of plants and animals. Hooke was one of the first scientists to study fossils, principally petrified wood, with the aid of a microscope. Hooke thought the detail he saw with the microscope was too fine and precise to have been formed by the rock itself. He hypothesized that living organisms had somehow been turned to rock.

DISTRIBUTION OF FOSSILS

Hooke's view was shared by another scientist of his time, Nicolaus Steno (1638–1686). Steno made an important contribution toward a modern understanding of Earth's geological and biological history. In 1669, he proposed the **law of superposition** (SOO-puhr-puh- ZISH-uhn), which states that successive layers of rock or soil were deposited on top of one another by wind or water. The lowest **stratum,** or layer, in a cross section of Earth is the oldest, while the top stratum

TABLE 15-1 *Geological History of Earth*

Millions of years ago	Era	Period	Epoch	Organisms
	Cenozoic	Quaternary	Recent	modern humans arise
0.01			Pleistocene	humans arise
1.8		Tertiary	Pliocene	large carnivores arise
5.3			Miocene	mammals diversify
23.8			Oligocene	diverse grazing animals arise
33.7			Eocene	early horses arise
54.8			Paleocene	more modern mammals arise
65	Mesozoic	Cretaceous		dinosaurs go extinct; mass extinction
144		Jurassic		dinosaurs diversify; birds arise
208		Triassic		primitive mammals arise; mass extinction; dinosaurs arise
245	Paleozoic	Permian		seed plants arise; reptiles diversify; mass extinction
290		Carboniferous		reptiles arise
354		Devonian		amphibians arise; mass extinction
417		Silurian		land plants arise
443		Ordovician		fishes arise; mass extinction
490		Cambrian		marine invertebrates arise
540	Precambrian			prokaryotes, then eukaryotes arise
4,600	Formation of the Earth			

is the most recent. Thus, fossils within a single stratum are of the same approximate age. Using Steno's law, observers could establish the **relative age** of a fossil; that is, they could say that a given fossil was younger or older than another fossil. The fossil's **absolute age** (its age in years) could be estimated from radiological evidence.

Eventually the application of these geologic principles, coupled with modern technological methods, told a compelling story. The history of Earth, as shown in Table 15-1, is more than 100,000 times longer than recorded human history.

internet**connect**

SCILINKS NSTA
TOPIC: Extinction
GO TO: www.scilinks.org
KEYWORD: HM281

Succession of Forms

Fossil-bearing strata show that species of organisms appeared, existed for a while, and then disappeared, or became **extinct.** In turn, newer species continued to arise. In Table 15-1, notice the order in which different types of organisms arose, beginning with prokaryotes in the Precambrian era. The fossil record indicates that there were several **mass extinctions,** brief periods during which large numbers of species disappeared. Some of these life-forms were unlike any organisms alive today. Look back at Figure 15-1 to see the fossil of a trilobite, which lived during the Paleozoic era. Most trilobite species disappeared during the Permian extinction, 245 million years ago. Mass extinctions probably resulted from drastic changes in the environment, perhaps following periods of volcanic activity or collisions with asteroids. Ash and dust in the atmosphere may have blocked sunlight for long periods of time and caused a decrease in temperatures around the world.

Biogeography

The study of the geographical distribution of fossils and of living organisms is called **biogeography** (BIE-oh-jee-AH-gruh-fee). A comparison of recently formed fossil types with types of living organisms in the same geographic area shows that new organisms arise in areas where similar forms already lived. Thus, armadillos appeared in North and South America, where glyptodonts, shown in Figure 15-2, lived in the past. Modern kangaroos appeared only in Australia, where the now-extinct giant kangaroo had lived.

FIGURE 15-2

The modern armadillo (top) resembles the extinct glyptodont (bottom), although the glyptodont was much larger. Life-forms arise in the same areas where similar, older forms once existed.

SECTION 15-1 REVIEW

1. Define the term *fossil,* and name three different kinds of fossils.
2. What was Robert Hooke's contribution to the understanding of fossils?
3. How does the law of superposition allow paleontologists to assign relative ages to fossils?
4. In what geological era did the first organisms arise?
5. How do biogeographic observations suggest that a modification process caused new species to arise?
6. CRITICAL THINKING What can you conclude if you find identical examples of a fossilized organism in two adjacent geologic strata?

RECENT RESEARCH

Bizarre Fossils Linked to Modern Organisms

The 600-million-year-old Ediacaran fossils have puzzled scientists for over a century. They constitute a menagerie of fronds, pouches, blobs, and disks that are not readily classified as either plant or animal. Thought by many to represent an evolutionary dead end, new evidence extending their existence into the Cambrian period suggests that these organisms may be ancestral to modern organisms.

ANSWERS TO SECTION 15-1 REVIEW

1. A fossil is a trace of a long-dead organism. Examples of fossils include molds and casts of the bodies of organisms, molds and casts of footprints, amber, and stone tools.
2. After Hooke studied the fine detail of many fossils with the aid of a microscope, he concluded that they are the remains of organisms.
3. Using the law of superposition, paleontologists can infer that fossils found in deep geologic strata are older than fossils found in shallower strata.
4. The first organisms arose in the Precambrian era.
5. Biogeographic observations show that organisms tend to arise in areas where similar forms lived in the past, suggesting that the new organisms are modified versions of the older organisms.
6. You can conclude that the organism lived, relatively unchanged, during the period during which the two strata were deposited.

BACKGROUND

The Burgess Shale and other deposits like it are important keys to understanding the magnitude of animal diversity that appeared during the Cambrian period. The Burgess Shale's yield of the fossils of soft-bodied organisms, in addition to fossils of organisms with hard body parts (which make up typical fossil deposits), shows that hard-bodied animals constituted only about 40 percent of the animals alive at the time. Thus, the ecology of Cambrian seas was much more complex than originally thought.

DISCUSSION

Guide the discussion by posing the following questions:

1. What did the Burgess Shale fossils reveal about the earliest animals? (The fossils show that the earliest animals included a much broader range of body plans than expected—some so different that they cannot be classified as part of any group of modern animals.

2. What other evidence besides the Burgess Shale supports the hypothesis that an "explosion" of life-forms took place during the Cambrian period? (Scientists have unearthed similarly diverse fossils from sites in China, Australia, and Greenland.)

FURTHER READINGS

1. "When Life Was Odd," in *Discover,* March 1997, pages 52–61. This article discusses some of the organisms of the Cambrian period that were once thought to be evolutionary dead ends but may have been ancestral to modern forms.

2. "The Origin of Animal Body Plans," in *American Scientist,* March–April 1997, pages 126–137. This article discusses recent fossil finds and the new insight they provide into animal development.

Research Notes

Solving the Burgess Puzzle

Were the animals that lived in the seas 550 million years ago, during the Cambrian period, more diverse than those that inhabit modern seas? A small limestone quarry tucked away 800 m above sea level in the Canadian Rocky Mountains is slowly yielding answers to this and other evolutionary puzzles. These rocks, which contain the fossil remains of organisms from a Cambrian sea buried by an ancient mudslide, are known collectively as the Burgess Shale (named for the Burgess Pass, where the quarry was found).

Discovered in 1909 by Charles Doolittle Walcott, head of the Smithsonian Institution at the time, the Burgess fossils remained in museum drawers until 1971, when Cambridge University professor Harry Whittington and two graduate students reexamined them.

Attaching a device known as a *camera lucida* to a microscope, the scientists used mirrors to project enlarged images of the fossils onto a flat surface. Then they carefully made illustrations of the ancient animal forms. In the process, they discovered that they could learn more from the large-scale drawings than they could from the actual tiny specimens.

The very fine-grained silt that surrounded the organisms helped preserve soft tissues in addition to bony structures. Thus, the remains of organisms that might have otherwise remained unknown were preserved. The variety of organisms found in the Burgess Shale is evidence of the explosion of life-forms that took place during the Cambrian period.

The Burgess organisms lived much as organisms do today. They lived in an ecosystem that included a wide range of animals, each with specialized organs for movement and feeding. Many of the organisms were ancestors of common sea creatures such as mollusks, segmented worms, and crustaceans. Whittington's team also discovered fossils of animals that were completely different from any inhabiting modern oceans.

What does this mean? The shale fossils reveal that the earliest multicellular animals included a much broader range of body plans than expected—some so different that they cannot be classified into any group of modern animals.

These conclusions about Cambrian life-forms are supported by recent discoveries in China, Australia, and Greenland. Scientists have unearthed similarly diverse and complex fossils dating back to the beginning of the Cambrian period, 550 million years ago. Thus, it appears that a high level of diversity and complexity developed in animals in a relatively short span of geologic time.

One of the most exciting Burgess finds was the discovery of the 5 cm (2 in.) long Pikaia gracilens. *It is a very early representative of phylum Chordata. Organizationally, it resembles the simple living chordate* Amphioxus.

THEORIES OF EVOLUTION

The word evolution *refers to an orderly succession of changes. Biological evolution is the change of populations of organisms over generations. Early scientists noticed that new life-forms appeared to be modifications of fossil forms found in the same geographical area. This strongly implied that a natural modification process was at work.*

SECTION

15-2

OBJECTIVES

▲ Define *evolution.*

● Explain Lamarck's theory of evolution, and describe how it was flawed.

■ List some of the evidence that led Darwin to his idea of how species might change over time.

◆ Explain Darwin's two major theories.

LAMARCK'S EXPLANATION

The French scientist Jean Baptiste de Lamarck (1744–1829) was one of the first to propose a unifying hypothesis of species modification. Lamarck proposed that similar species descended from a common (the same) ancestor—thus, living species were descended from similar extinct species evident in the fossil record. Lamarck cataloged an extensive collection of invertebrates (animals without a backbone), and he related fossil forms to living animals based on their similar appearance.

To explain how species change, he hypothesized that acquired traits were passed on to offspring. An **acquired trait** is one that is not determined by genes. Instead, it arises during an organism's lifetime as a result of the organism's experience or behavior.

Lamarck made a study of the habits and physiology of shore birds, and he believed that the webbed foot of water birds resulted from repeated stretching of the membrane between the toes. Over time, Lamarck said, this produced a broad webbed foot, a trait that would be preserved by reproduction. In other words, offspring of parents who had acquired webbed feet would have webbed feet as well. In the same way, an organism that did not use some part of its anatomy, such as a tail, would produce offspring with a smaller version of that body part.

Lamarck's hypotheses were fiercely attacked, primarily by scientists who rejected the idea of evolution itself. Although Lamarck's hypothesis of the passage of acquired traits was easily disproved, his work was an important forerunner of modern evolutionary theory. Lamarck was the first to clearly state that types of organisms change over time and that new types of organisms are modified descendants of older types. This idea was presented more convincingly more than 50 years later by Charles Darwin.

SECTION 15-2

SECTION OVERVIEW

Evolution

Section 15-2 introduces the work of Lamarck, Darwin, and Wallace. Darwin's voyage to the Galápagos and his observations there are described. Darwin's theories of descent with modification (evolution) and natural selection are explained.

▶ ENGAGE STUDENTS

Using an overhead projector or reference books, display for the class illustrations of several kinds of dinosaurs (examples should include sauropods, therapods, and orinithopods). Have students identify specialized adaptations of various kinds (such as teeth adapted for tearing in carnivores and grinding in herbivores, body armor and tusks for protection, large hind limb relative to front limb for bipedal locomotion, and so forth). Have the class consider why there were so many different kinds of dinosaurs (many different habitats, many different food sources, etc.). Ask students to discuss what type of organism dinosaurs arose from, what happened to the dinosaurs, and why. (They arose from earlier reptiles. They became extinct. Their environments changed, but they could not adapt quickly enough.)

SECTION 15-2

Quick Lab

Analyzing Evolutionary Relationships

Time Required 20 minutes

Procedural Tips Have students form teams of three or four. Prepare a group of 20 hardware objects including nails, screws, washers, and nuts, for each team ahead of class time. A spiral-thread decking nail has characteristics of both nails and screws. The assortment of screws can include wood and sheet metal screws, round-headed screws, flat-headed screws, slotted screws, and Phillips-headed screws. Nails can include large and small sizes, large or small heads, threaded nails, and smooth nails. Washers can vary in diameter, hole size, or material, such as plastic or metal. Glue a nut to a washer for an object with common characteristics. Nuts can be octagonal, hexagonal, different thicknesses, flat nuts, and capping nuts. The number and variety of objects may be changed as you desire.

Answers to Analysis The decking nail is hammered like a nail, and it has threads like a screw. The washer and nut combination also has characteristics of two groups. Criteria depend on the variety of objects used in the activity—size, number of threads on the screws, pointed or flat ends of screws, types of heads on the nails, size and shape of the washers, or the material from which the objects are made. Most students will form a group of nails and a group of screws. The objects demonstrate an increase in complexity as the tree and branches form.

FIGURE 15-3

Charles Darwin began the work that would lead to his theory of evolution when he was a young man.

Quick Lab

Analyzing Relationships

Materials set of hardware items, paper, pencil

Procedure

1. Find one object that has characteristics common to all the objects, and place it at the middle of the paper near the bottom. This is the "common ancestor" of all the hardware.
2. Separate the remaining objects into two groups that have a single, basic difference.
3. Form two branches from the common ancestor by arranging the members of each group on the paper. Each of the two branches may then form smaller branches.
4. Draw lines connecting each object to the one above and below it, forming a tree.

Analysis What characteristic did the common ancestor have that made it similar to both groups? What criteria did you use to separate the two groups? What was the trend in the evolution of each group?

THE BEGINNING OF MODERN EVOLUTIONARY THOUGHT

In the mid-1800s, both Charles Darwin (1809–1882) and Alfred Wallace (1823–1913) independently proposed the hypothesis that species were modified by natural selection. In the process of **natural selection,** organisms best suited to their environment reproduce more successfully than other organisms. Thus, over generations, the proportion of organisms with favorable traits increases in a **population,** or interbreeding single-species group. While Darwin and Wallace announced their hypotheses at the same time, Darwin's name became more associated with evolutionary theory after his book *The Origin of Species* was published in 1859.

Charles Darwin

Darwin, shown as a young man in Figure 15-3, was born in 1809, the son of wealthy British physician. Darwin attended medical school at the University of Edinburgh and then enrolled at Cambridge University to study for the clergy. He had little enthusiasm for either course of study. At Cambridge University, a friendship with a professor of botany, John Henslow, awakened Darwin's interest in natural history. In 1831, Darwin sailed on the ship H.M.S. *Beagle.* The *Beagle* was chartered for a five-year mapping and collecting expedition to South America and the South Pacific. Darwin assumed the post of ship naturalist, which required that he collect specimens and keep careful records of his observations.

Voyage of the Beagle

Soon after leaving England, Darwin read a geology book that had been given to him by Henslow. This book, *Principles of Geology,* by Charles Lyell, emphasized the great age of Earth and the principles of **uniformitarianism** (YOON-uh-FORM-uh-TER-ee-uhn-IZ-uhm). The principles of uniformitarianism hold that the geological structure of Earth resulted from cycles of observable processes and that these same processes operate continuously through time. For example, silt is deposited by modern rivers in the same way it was deposited by ancient rivers. Modern volcanoes spew forth lava and ash in the same way those on early Earth did. Lyell's book spurred Darwin's interest in the study of geology, and it allowed him to consider the possibility that the modification of environments might be a very slow process requiring long periods of time. In Chile, Darwin observed the results of an earthquake. The land around a harbor had been lifted more than a meter (3.3 ft). In the nearby Andes Mountains, Darwin observed fossil shells of marine organisms in rock beds about 4,300 m (14,100 ft) above sea level. These observations convinced Darwin that Lyell was correct in saying that geologic changes, such as the elevation of the Andes, required many millions of years. Darwin reasoned that the formation of mountain ranges would slowly change habitats, requiring organisms that

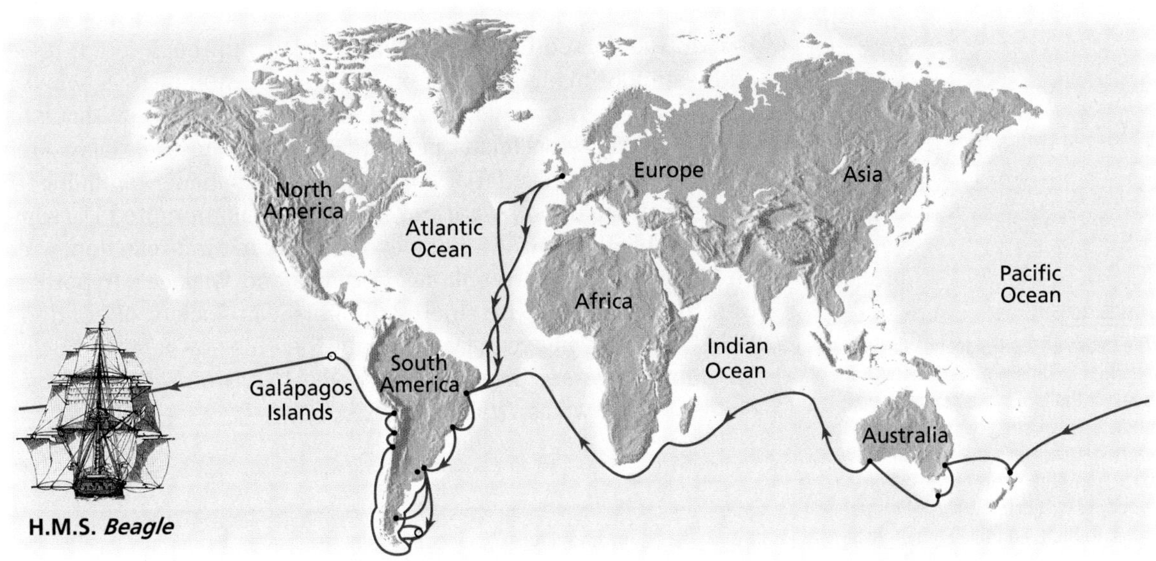

FIGURE 15-4

Darwin spent five years as the ship naturalist on the H.M.S. *Beagle* as the ship sailed around the world.

lived there to adapt to these changes. Furthermore, he reasoned that the pace of those adaptations would be so slow with respect to a human life span that they would be difficult to detect.

During the *Beagle*'s five-year trip, as shown in Figure 15-4, Darwin often left the ship at one port and was picked up months later at another port. During his time on land, Darwin trekked hundreds of kilometers through unmapped regions. He collected many different types of fossils and observed thousands of species of organisms.

Analysis of Darwin's Data

When Darwin returned to England, in October 1836, his collections from the voyage were praised by experts from the scientific community. A bird specialist who studied Darwin's collection of finches from the Galápagos Islands reported that Darwin had collected 13 similar but separate species of finches. Each finch species has a distinctive bill that is specialized for a particular food source. Despite the bill differences, the overwhelming similarities of the Galápagos finches implied that the finches shared a recent common ancestor, meaning they descended from a single species. Over a period of years after returning to England, Darwin analyzed his data. Darwin considered the possibility that all of the islands' finches had descended from a few birds or even a single female that had been blown off course from South America, 1,000 km (620 mi) to the east. Because the Galápagos are geologically young islands (they are about 5 million years old), Darwin assumed that the offspring of the original finches had been adapting to different environments and food sources for a relatively short time. Darwin reasoned, therefore, that over many millions of years, many large differences could accumulate between species.

RECENT RESEARCH
Deflating the Biological Big Bang

Biologists since Darwin have been struggling to explain the sudden appearance of a wide variety of animal fossils in the early Cambrian period. New evidence gathered from the analysis of DNA sequences indicates that the appearance of new animal phyla began earlier and occurred over a longer period of time than was previously thought. The lack of fossil evidence may be because many were soft-bodied animals, which do not readily fossilize.

DEMONSTRATION

Provide students with a phylogenetic tree depicting all 13 of Darwin's finches. Point out the three major branches of the tree, and ask students to describe the characteristics of the birds on each branch. Ask them to deduce from each bird's beak size and shape what kind of food it eats. Ask them to describe the characteristics they would expect the birds' common ancestor to have had. Point out that these finches include both ground and tree dwellers, suggesting that the finches filled available habitats as they evolved.

QUICK FACT

☑ Are humans the only animals that use tools? This was considered by many biologists to be a trait unique to humans. However, some birds, including one of Darwin's finch species, use twigs held in their beaks to dig insects out of trees. While they don't make the tools to fit the task, as humans do, this is a use of tools.

GIFTED ACTIVITY

Have students research and report on some of the other organisms and topics that Darwin studied. His extensive investigations included studies of barnacles, beetles, island geology, and plant movements.

RETEACHING ACTIVITY

Provide students with pictures of a modern horse and its ancestors, and provide the dates when each ancestor lived. The pictures should include *Equus, Merychippus, Mesohippus,* and *Hyracotherium (Eohippus).* Ask students how the adaptations of horse evolution might have been advantageous, increasing the animals' odds of survival. (Longer legs and single hooves enabled them to run faster over hard ground.)

OVERCOMING MISCONCEPTIONS

The word *theory,* as used here, refers to an accepted explanation for numerous observations that has stood the tests of time and experimentation. It does not imply that evolution (descent with modification) itself is hypothetical any more than the atomic theory implies that atoms are hypothetical or the theory of gravity implies that gravity is hypothetical. Evidence that organisms evolve is overwhelming. Hypotheses about the mechanisms of change that result in evolution are continually being refined and updated, but the general theory stands firm.

Publication of *The Origin of Species*

Darwin was forced to abandon his leisurely approach to the review and testing of his hypothesis on natural selection in 1858, when he was approached by a young naturalist, Alfred Wallace. Wallace had been collecting specimens in South America and the Malay Archipelago. Wallace asked Darwin to review his paper outlining the process of evolution by natural selection. This prompted Darwin to finally publish his own work on evolution by natural selection, which he had spent 21 years refining. Darwin's and Wallace's hypotheses were presented side by side to the Linnaean Society of London in 1858. The following year, Darwin published his now-famous book on the subject, *On the Origin of Species by Means of Natural Selection,* commonly known as *The Origin of Species.*

FIGURE 15-5

The beaks of the Galápagos finches are adapted to different food sources. The beak of the large ground finch, *Geospiza magnirostris,* center, is suited to cracking the seeds that compose the bird's diet. The narrower beaks of the woodpecker finch, *Camarynchus pallidus,* top left, and the warbler finch, *Certhidea olivacea,* bottom right, are specialized for capturing insects.

DARWIN'S THEORIES

Darwin's ideas about evolution and natural selection are summed up in two theories. In his first theory, Darwin stated that evolution, which is defined as descent with modification, occurs. In his second theory, Darwin proposed that a process he called natural selection causes evolution.

Descent with Modification

Darwin's first theory—*descent with modification*—states that the newer forms appearing in the fossil record are actually the modified descendants of older species. Recall that Lamarck had made the same argument. But Darwin's theory goes further. Darwin inferred that *all* species had descended from one or a few original types of life. Darwin thought that the different species of finches on the Galápagos Islands, shown in Figure 15-5, had descended from a recent common ancestor. He then reasoned that all living things probably had

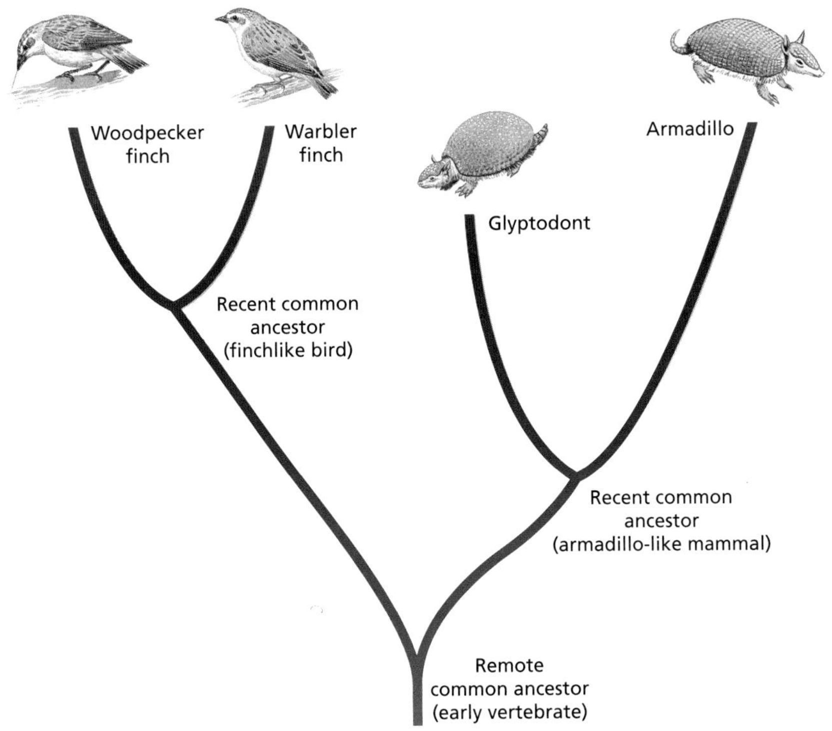

FIGURE 15-6

Darwin reasoned that if the Galápagos finches were similar to each other because of recent common ancestry, then organisms that are more dissimilar, such as finches and armadillos, share a more remote ancestor.

likewise descended from one, or perhaps a few, remote common ancestors that lived a very long time ago. For example, all vertebrates, such as the birds and mammals shown in Figure 15-6, probably descended from a vertebrate that lived in the distant past. Bear in mind that the terms *recent ancestor* and *remote ancestor* are used in the context of geologic history—the span of time represented in Table 15-1—rather than in the context of a single human lifetime.

Darwin's first theory accounted for the fact that similar organisms arise in the same geographic location. Darwin's theory of evolution explained the observation that organisms give rise to others similar to themselves. Thus, it would be natural for modern kangaroos to evolve from a now-extinct ancestor very much like them.

Modification by Natural Selection

Darwin's second theory—*modification by natural selection*—states how evolution occurs. Darwin was heavily influenced by the English clergyman Thomas Malthus (1766–1834), who had published a thesis pointing out that populations (Malthus was referring to humans) have the potential of doubling and redoubling their numbers. Malthus proposed that the growth of human populations was limited by adverse conditions, such as war, disease, or a limited supply of resources.

Agreeing with Malthus's views, Darwin noted that although populations of all organisms have the potential to grow unchecked, most do not. He reasoned that the environment limits the growth of populations by increasing the rate of death or decreasing the rate of reproduction, or both.

Eco Connection

Galápagos Islands

The exotic and fragile ecosystem of the Galápagos that fascinated Charles Darwin in the mid-1830s has been in danger since the discovery of the Galápagos in 1535. For 400 years, the islands were a favorite stopping place for pirates and whalers. Herman Melville, the author of *Moby Dick,* saw and wrote of the giants of the islands, the Galápagos tortoises, while on a whaling expedition in 1841. The tortoises, pictured in the opening photograph of this chapter, were a frequent target of sailors. The animals were valued as a meat source because they could live for long periods aboard ship with little or no food or water. Over the years, the tortoise population has been reduced from about 250,000 animals to about 15,000 animals.

Over the past 40 years, the government of Ecuador has taken steps to preserve and restore the Galápagos and their wildlife. The Charles Darwin Research Station conducts breeding programs for endangered animals and has bred and released several thousand tortoises, representing most of the 14 tortoise species found on the islands.

INCLUSION *ACTIVITY*

Ask students to examine Figure 15-6. Remind them that the only living organisms depicted on the tree are those at the tips of the branches. Ask students how scientists inferred the relationships between these organisms in order to formulate this kind of tree. (They used fossils and comparative anatomy to determine the common ancestors of the organisms.)

CONTENT CONNECTION

Section 8-3: Meiosis

Meiosis generates genetic variability in sexually reproducing organisms, and variability is necessary in order for evolution to occur. Variability is a result of genetic recombination caused by crossing-over between homologous chromosomes and random segregation of homologous pairs of chromosomes.

QUICK FACT

✔ Darwin's theories of evolution by natural selection were based on the supposition that hereditary variability existed in species and could be acted on by natural selection. The ultimate source of this variability, mutations, wasn't known until 1901. Darwin died in 1882.

OVERCOMING MISCONCEPTIONS

Because natural selection tends to eliminate characteristics that are detrimental to a species, it is often assumed that all of the characteristics in a successful organism confer some advantage. In fact, many characteristics are neutral and confer no benefit.

SECTION 15-2

CONTENT CONNECTION

Section 10-1: Mutation

Mutations are permanent changes in an organism's DNA. While most mutations are harmful, the process generates genetic variability, which is a prerequisite for evolution.

QUICK FACT

☑ Scientists sometimes irradiate plant seeds to induce mutations and thus produce new varieties of peppers, soy beans, cotton, sugar cane, sunflowers, roses, irises, chrysanthemums, azaleas and many other plants.

ANSWERS TO SECTION 15-2 REVIEW

1. An acquired characteristic is one that arises during the lifetime of an organism. It does not change the organism's genotype.
2. Upon reading *Principles of Geology,* Darwin understood that geologic processes, such as the formation of a mountain range, occurred over long periods of time and could change the environment significantly, requiring local organisms to adapt.
3. Answers will vary. *The Origin of Species* was published in 1859.
4. Darwin's theories do not contradict the principles of biogeography. Darwin hypothesized that modern organisms arose from similar, older forms.
5. Darwin thought the environment limits the growth of populations.
6. A favorable trait that does not enhance reproductive success does not contribute to evolution.

internet connect
SCILINKS NSTA
TOPIC: Natural selection
GO TO: www.scilinks.org
KEYWORD: HM288

Darwin proposed that the environment may affect individual organisms in a population in different ways because individuals of a species are not identical. Some organisms have traits that make them better able to cope with their environment. Organisms that have a greater number of these favorable traits tend to leave more offspring than organisms with fewer beneficial traits. Darwin called the different degrees of successful reproduction among organisms in a population natural selection.

If a trait both increases the reproductive success of an organism *and* is inherited, then that trait will tend to be passed on to many offspring. A population of organisms **adapt** to their environment as their proportion of genes for favorable traits increases. The resulting change in the genetic makeup of a population is evolution. In an evolving population, a single organism's genetic contribution to the next generation is termed **fitness.** Thus, an individual with high fitness is well adapted to its environment and reproduces more successfully than an individual with low fitness.

Bear in mind that natural selection is not an active process. Organisms do not purposefully acquire traits that they need, although it may seem that this is true. The environment "selects" the traits that will increase in a population. The kinds of traits that are favorable depend on the demands of the environment. An organism may be able to run fast, or it may be strong or have coloring that acts as camouflage from predators. Traits that are favorable for some organisms in some environments are not necessarily favorable for all organisms or all environments. For example, the large body size of large mammals such as the elephant would not be beneficial to a species of flying birds if size prevented flight. A favorable trait is said to give the organism that has it an **adaptive advantage.**

Selection conditions change as the demands of the environment change. For example, a significant change in climate or available food can cause rapid evolutionary change as populations adapt to the change. If the environmental change is too extreme, however, populations cannot adapt quickly enough and they become extinct.

SECTION 15-2 REVIEW

1. What is an acquired characteristic? Do acquired characteristics change the genotype of an organism?
2. How did Lyell's book *Principles of Geology* help Darwin see that natural selection over many generations could explain species modification?
3. How many years have passed since the publication of *The Origin of Species*? Why is its publication still considered so important?
4. Do Darwin's theories of evolution by natural selection contradict the principles of biogeography? Why or why not?
5. What did Darwin think limits the growth of populations?
6. **CRITICAL THINKING** If a favorable trait increases the life span of an organism without affecting reproductive success, does it contribute to evolution?

EVOLUTION IN PROCESS

Evolution is a continuous process. By examining genotypic and phenotypic evidence in modern organisms, we can see evidence that evolution has occurred. By considering species in relation to one another, we can also detect definite patterns of evolution.

SECTION 15-3

OBJECTIVES

▲ Describe the difference between homologous, analogous, and vestigial structures.

● Tell how similarities in macromolecules and embryos of different species suggest a relationship between them.

■ Explain the difference between coevolution, and divergent and convergent evolution.

EVIDENCE OF EVOLUTION

If organisms change through a process of gradual modifications, we should be able to see evidence of this process. Living things, in fact, display many different clues to their evolutionary history.

Homologous and Analogous Structures

It seems obvious that the beaks of the finches shown in Figure 15-5 derive from the same embryonic structure of each different species. Moreover, they are modifications of a feature found in an ancestor common to all birds. Similar features that originated in a shared ancestor are described as **homologous** (hoh-MAHL-uh-guhs) features.

Some examples of homologous features are not as obvious as the beaks of birds. Compare the forelimbs shown in Figure 15-7. Although the limbs look strikingly different and vary greatly in function, they are very similar in skeletal structure, and they derive from the same structures in the embryo. Homologous features can result from modifications that change an original feature

FIGURE 15-7

The forelimbs of the penguin, alligator, bat, and human all derive from the same embryological structures.

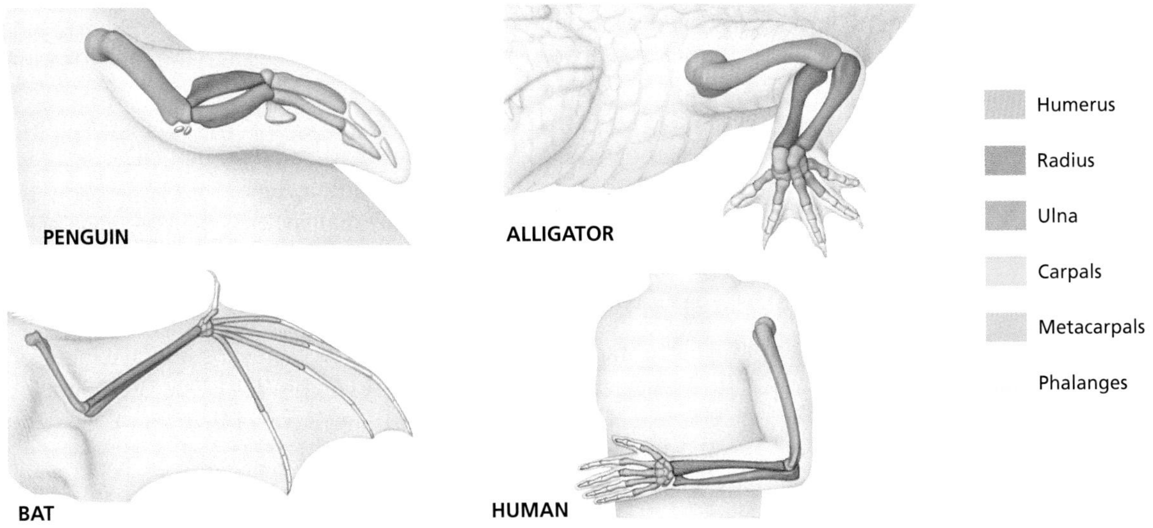

SECTION 15-3

SECTION OVERVIEW

Evolution

Section 15-3 presents major lines of evidence supporting evolution, including the presence of homologous, analogous, and vestigial structures in different species; similarities in the early embryos of different species; and similarities of certain macromolecules (including DNA, RNA, and proteins) in different species.

Interdependence of Organisms

Section 15-3 introduces patterns of evolution, including coevolution, convergent evolution, and divergent evolution. All three patterns are driven by the interaction of the evolving organism with its environment, including the other organisms of its environment.

▶ ENGAGE STUDENTS

Provide students with pictures of a shark, a porpoise, and a fossil ichthyosaur; a flying insect, a bat, and a bird; and a cactus and a thorned euphorbia (such as a spurge). Ask students to explain how the organisms in each of these groups are similar in appearance even though they are not closely related. (Students should conclude that all show adaptation to a similar environment: an aquatic environment, a terrestrial environment, and a dry desert.)

RETEACHING ACTIVITY

Have students examine the two animals shown in Figure 15-8. Ask them how a scientist might classify these animals incorrectly if the wings were assumed to be homologous structures. (The scientist would assume that they had a recent common ancestor.) Ask what other kind of evidence the scientist might collect to help determine how closely related these animals are (characteristics of their development patterns, biochemical evidence, and similarities of body structures).

QUICK FACT

☑ Between 1989 and 1994, three fossils were found of animals that have many characteristics in common with modern whales. All three have functional hind legs. These fossils provide further evidence that today's whales evolved from a land-dwelling, four-legged ancestor.

RECENT RESEARCH

Behavioral Evidence for Evolution

Paleontologists have discovered a fossil of a dinosaur sitting on a nest of eggs. This finding of brooding behavior, characteristic of birds, supports the hypothesis that birds evolved from dinosaurs.

FIGURE 15-8

The wings of the hummingbird, *Selasphorus rufus*, top, and the humming moth, *Hemaris thysbe*, bottom, serve similar functions, but they differ in structure and derive from different embryological structures.

to two extremely different types, such as a wing and an arm. The presence of homologous features in different species indicates that the species shared a fairly recent common ancestor.

Analogous (uh-NAL-uh-guhs) features serve identical functions, and they look somewhat alike. They have very different embryological development, however, and may be very different in internal anatomy. Figure 15-8 shows a hummingbird and a humming moth. Both organisms can hover to feed on the sugar-rich nectar from flowers. There is no anatomical or embryological similarity, however, between their wings. Birds and insects differ greatly in anatomy and in embryological development. Although birds and insects do share a very remote ancestor, we can infer that their wings evolved independently and differently in more-recent ancestors of each animal.

Vestigial Structures

Many organisms have features that seem to serve no useful function. For example, humans have a tailbone at the end of the spine that is of no apparent use. The human appendix, a small, fingerlike projection from the intestine, also has no known function. Some snakes have tiny pelvic bones and limb bones. Whales also have pelvic bones, along with a four-chambered stomach like that of a cow.

These apparently useless features are said to be vestigial. **Vestigial** (ves-TIJ-ee-uhl) features were useful to an ancestor, but they are not useful to the modern organism that has them. The vestigial tailbone in humans is homologous to the functional tails of other vertebrate species. A vestigial feature in a modern organism is evidence that the structure *was* functional in some ancestor of the modern organism. Moreover, an organism with a vestigial feature probably shares common ancestry with an organism that has a functional version of the same feature.

So what sort of evolutionary clues can vestigial features provide? Consider that normal sperm whales, like all whales, have small pelvic bones but no hind legs. A very small percentage of sperm whales, however, have vestigial leg bones, and some sperm whales even have bone-supported bumps protruding from their body.

Whales probably are descended from an ancestor that lived on land. In the whales' genome, many of the genes needed to make hind legs have been **conserved,** or have remained unchanged. In normal whales, the genes for hind legs are turned off. In rare cases, however, the genes are partially turned on, and vestigial hind legs form. Thus, whales and other living things may display their evolutionary history in the usually unexpressed genes they carry.

Similarities in Embryology

The early stages of different vertebrate embryos are strikingly similar to each other, as you can see in Figure 15-9. The German zoologist Ernst Haeckel (1834–1919), who was also struck by these similarities, declared that "ontogeny recapitulates phylogeny." This statement can be translated to "embryological development repeats evolutionary history." We now know that this is a bit of an exaggeration. For example, during no stage of development does a gorilla look like an adult fish. In the early stages of development, all vertebrate embryos are similar, but those similarities fade as development proceeds. Nevertheless, the similarities in early embryonic stages of vertebrates can be taken as yet another indication that vertebrates share a common ancestry.

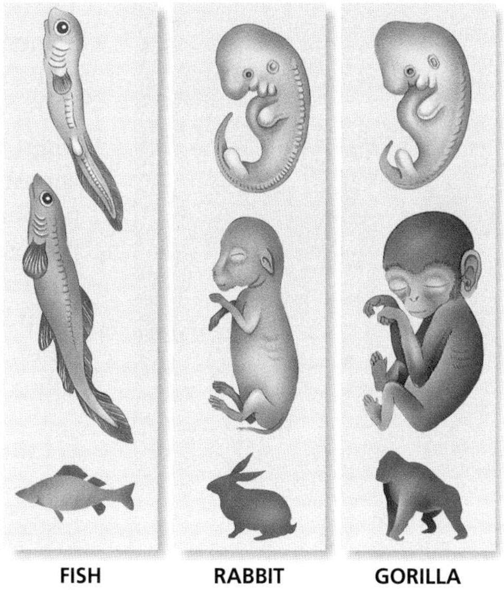

FIGURE 15-9

This is a modernized version of Haeckel's drawings of embryological stages in different species. Although modern embryologists have discovered that Haeckel exaggerated some features in his drawings, it is true that early embryos of many different vertebrate species look remarkably similar.

Similarities in Macromolecules

Darwin hypothesized that more-similar forms of organisms have a more recent common ancestor than do less-similar forms. He arrived at this hypothesis by observing anatomical features only. He could not have known how true this rule would prove at the molecular level—for homologous proteins, as well as RNA and DNA molecules. For example, many species have the red-blood-cell protein, hemoglobin. The amino acid sequences in the hemoglobin molecules of different species are similar, but not identical. The amino acid sequences in human hemoglobin and gorilla hemoglobin differ by one amino acid, while the hemoglobin molecules of humans and frogs differ by 67 amino acids. The number of amino acid differences in homologous proteins of two species is proportional to the length of time that has passed since the two species shared a common ancestor. Thus, the more-similar the homologous proteins are in different species, the more closely related the species are thought to be. Information provided by molecular biology can confirm the evolutionary histories suggested by fossils and anatomy.

PATTERNS OF EVOLUTION

There are several ways that species can change to adapt to their habitats. The pattern and speed of evolutionary change result from the changing requirements of the environment.

Coevolution

The change of two or more species in close association with each other is called **coevolution.** Predators and their prey sometimes coevolve, parasites and their hosts often coevolve, and plant-eating animals and the plants they feed on also coevolve. One example of coevolution is plants and the animals that pollinate them.

In tropical regions, some species of bats feed on the nectar of flowers, as shown in Figure 15-10. These bats have a slender muzzle and a long tongue with a brushlike tip, which aid them in feeding. The fur on the bat's face and neck picks up pollen, which the bat takes to the next flower. Flowers that have coevolved with these bats are light in color, enabling the bats, which are active at night, to easily locate them. The flowers also have a fruity odor that is attractive to bats.

FIGURE 15-10

Some species of bats, such as this long-nosed fruit bat, have coevolved with the flowers they feed on.

VISUAL LINK
Figure 15-9

Ask students to study the embryos shown in Figure 15-9. Ask them what characteristics the earliest stages of the embryos shown have in common (tail, skull, eyes, etc.). Ask them to infer what the common ancestor of the vertebrates looked like. (It probably had a skull and other bones, a tail, and prominent eyes, etc.)

CRITICAL THINKING
Cytochrome *c* Similarities

Human and chimpanzee cytochrome *c* proteins are identical in all 104 of their amino acids. How can you explain the lack of difference between these two animals' cytochrome *c*? (This protein is only one of thousands of proteins in each of these animals. Two animals are different from each other because of the sum of all their biochemical differences.)

TEACHING STRATEGY
Coevolutionary Relationships

Obtain a variety of pictures of flowering plants and the insect pollinators with which they have coevolved. First show the students the pictures of the flowers, and ask them to study them and predict what kind of insect might pollinate each. Then show students the pictures of the insect pollinators, and have them try to match up the flowers with their pollinators. (Some examples to include are bee orchids and their bee pollinators, wasp orchids and their wasp pollinators, carrion flowers and their fly pollinators, and figs and their fig-wasp pollinators. Botany and entomology textbooks would be good sources of these pictures.)

SECTION 15-3

CULTURAL CONNECTION

Artificial Selection of Animals

Humans have been breeding and using artificial selection to create new types of animals for centuries. The Arabian horse was bred in the Middle East at least 2,000 years ago. It is delicate-looking but was bred for stamina for desert riding. Large-bodied draft horses, such as the Clydesdale, from Scotland, were bred in Europe and Great Britain from the Middle Ages onward.

ANSWERS TO SECTION 15-3 REVIEW

1. The mouthparts of the two species are homologous. Because they are feeding structures on the heads of similar organisms, they probably originated in a shared ancestor but were modified by evolution for different functions.
2. The wings of birds and bees are analogous features. The embryos of vertebrates (such as the bird) are very different from those of arthropods (such as the bee). The wings of each species evolved independently and are structurally dissimilar as well.
3. It suggests that a relatively short period of time has passed since humans and gorillas shared a common ancestor.
4. It suggests that frogs and humans are more closely related than either species is to the fruit fly.
5. Any aspect of phenotype can be acted on by natural selection, including vestigial structures.
6. coevolution

FIGURE 15-11
The rate of divergent evolution among dogs has been increased by artificial selection by humans.

Convergent Evolution

Sometimes organisms that appear to be very similar, such as a shark and a porpoise, are not closely related at all. This kind of similarity is the result of **convergent evolution.** Convergent evolution occurs when the environment selects similar phenotypes, even though the ancestral types were quite different from each other. Sharks and porpoises have very different origins. Sharks are fishes, and porpoises are mammals. Many features of these animals are similar, however, and have been selected by the environment they share. Their large, streamlined bodies and even their fins resemble each other. Analogous structures, such as similar fins in very different animals, are associated with convergent evolution.

Divergent Evolution

In **divergent evolution,** two or more related populations or species become more and more dissimilar. Divergence is nearly always a response to differing habitats, and it can ultimately result in new species.

One important type of divergent evolution is adaptive radiation. In **adaptive radiation,** many related species evolve from a single ancestral species. The Galápagos finches are an example of adaptive radiation. They diverged in response to the availability of different types of food in their different habitats.

Sometimes the process of divergence can be sped up artificially, through **artificial selection.** All domestic dogs are the same species, *Canis familiaris.* Dogs have been bred by humans for certain phenotypic characteristics, resulting in different breeds with different traits, as you can see in Figure 15-11. Thus, the process of divergent evolution in this species has sped up many times beyond what could have occurred in nature. Divergent evolution operating over very long periods of time has produced the seemingly endless variety of species alive today.

SECTION 15-3 REVIEW

1. The mouthparts of an adult horsefly are modified for biting. The mouthparts of a mosquito are modified for piercing skin and sucking blood. Are the mouthparts of the two species homologous or analogous? Explain your answer.
2. Birds and bees have wings. Are their wings homologous features or analogous features? Explain your answer.
3. The hemoglobin of humans is nearly identical to that of a gorilla. What does this suggest about the length of time that has passed since the last common ancestor of humans and gorillas lived?
4. Fruit fly embryos and frog embryos differ from each other more than frog embryos and human embryos do. What does this tell us about how the three species are related?
5. Are vestigial structures acted on by natural selection?
6. CRITICAL THINKING Some monarch butterflies contain chemicals that are toxic to birds. Another species of butterfly, the viceroy, has some protection from predation because it closely resembles the monarch. What pattern of evolution is illustrated by this example?

CHAPTER 15 REVIEW

SUMMARY/VOCABULARY

15-1

- A fossil is a trace of a long-dead organism.
- The law of superposition states that new geologic strata are deposited on top of older strata.
- The history of Earth and its life-forms can be inferred by examining the fossil record.
- The fossil record shows that new life-forms have arisen continually during the history of life on Earth.
- The study of biogeography shows that organisms arise in areas where similar, now-extinct organisms once lived.

Vocabulary

absolute age (281)
biogeography (281)
cast (279)
extinct (281)
fossil (279)
law of superposition (280)
mass extinction (281)
mold (279)
relative age (281)
sediment (279)
stratum (280)

15-2

- Lamarck proposed that species evolve over time. He incorrectly hypothesized that species modification is the result of acquired characteristics and that these characteristics can be passed on to offspring.
- Charles Darwin began his work on evolution when he was employed as a naturalist for a voyage of the H.M.S. *Beagle*.
- Darwin was influenced by Charles Lyell, who proposed the principles of uniformitarianism, which hold that the structure of Earth results from cycles of observable processes.
- Darwin found evidence of species modification in both modern and extinct species.
- Darwin hypothesized that related species, such as the Galápagos finches, descended from a common ancestor.
- Darwin wrote *The Origin of Species,* in which he proposed that natural selection is the principal driving force behind evolution.
- A population of organisms adapt to their environment as their proportion of genes for favorable traits increases.
- Evolution is the change in the genetic makeup of a population over generations.

Vocabulary

acquired trait (283)
adapt (288)
adaptive advantage (288)
fitness (288)
natural selection (284)
population (284)
uniformitarianism (284)

15-3

- Evidence supporting evolution is found in the body structures of living organisms. Homologous structures have a common evolutionary origin. Analogous structures are similar in function but have different evolutionary origins.
- A species with a vestigial structure probably shares evolutionary origins with a species that has a functional form of the structure.
- Similar embryological development among species indicates a common evolutionary history.
- Similarity in macromolecules such as RNA, DNA, and proteins indicates a common evolutionary history.
- In coevolution, two or more closely associated species, such as a predator and its prey, change in response to each other.
- In convergent evolution, organisms that are not closely related resemble each other because they have responded to similar environments.
- In divergent evolution, related populations become less similar as they respond to different environments. Adaptive radiation is a type of divergent evolution.

Vocabulary

adaptive radiation (292)
analogous (290)
artificial selection (292)
coevolution (291)
conserve (290)
convergent evolution (292)
divergent evolution (292)
homologous (289)
vestigial (290)

CHAPTER 15 REVIEW ANSWERS

REVIEW

1. A mold is a hollow fossil formed by sediment encasing a dead organism. A cast is a rocklike model of the organism.
2. The age of deeper geologic strata is greater relative to shallower strata.
3. An acquired trait is not determined by genes and arises during an organism's lifetime. A genetic trait is determined by genes.
4. Homologous structures originated in a shared ancestor and derive from the same structures in the embryo. Analogous structures do not have the same embryological origins and may be very different structurally.
5. Adaptive radiation is a kind of divergent evolution in which many related species evolve from a single ancestral species.

6. a
7. a
8. b
9. c
10. c
11. a
12. b
13. c
14. c
15. b

16. Hooke said the degree of fine detail found in fossils could not have been formed in rock, but must have arisen in living things.
17. Malthus stated that populations have the potential for unlimited growth.
18. Darwin stated that species change because individuals differ and that those with more favorable traits reproduce more, increasing those traits in a population. Lamarck argued that a single organism could change in response to its environment and pass that change on to its offspring.
19. Very similar embryological forms in different species suggests that those species shared a relatively recent ancestor.
20. The traits that are favorable for a particular species are influenced by the environment. A white coat may be helpful in a snowy environment, but might not be favorable in a tropical rain forest.
21. The Galápagos finches are an example of divergent evolution or adaptive radiation.
22. There are several vestigial structures in humans, including the tailbone and the appendix.
23. Species having more-similar proteins are probably more closely related, or share a more recent common ancestor, than species with less-similar proteins.
24. Nectar-feeding bats and the flowers they use as food are an example of coevolution.
25. Yes, a non-genetic characteristic could be selected by the environment, but it would not contribute to evolution.

CRITICAL THINKING

1. Organisms with a short generation time could be modified more rapidly than those with a long generation time. For example, a bacterium with a generation time of one hour would have almost 9 million generations in 1,000 years, whereas humans

REVIEW

Vocabulary

1. What is the difference between a mold and a cast?
2. Explain the relationship between geologic strata and relative age.
3. What is the difference between an acquired trait and a genetic trait?
4. Distinguish between homologous structures and analogous structures.
5. How are divergent evolution and adaptive radiation related?

Multiple Choice

6. Lamarck's explanation for the modification of species depended on (a) inheritance of acquired characteristics (b) convergent evolution (c) the law of superposition (d) natural selection.
7. The idea that processes occurring now on Earth are much the same as those that occurred long ago is called (a) uniformitarianism (b) relativism (c) evolutionism (d) convergent evolution.
8. The observation that organisms arise in locations where similar, extinct organisms lived is referred to as (a) superposition (b) biogeography (c) uniformitarianism (d) evolution.
9. The similarities in the Galápagos finches implied (a) coevolution (b) convergent evolution (c) adaptive radiation (d) descent from different remote ancestors.
10. Difference in reproductive success is (a) an acquired trait (b) adaptive radiation (c) natural selection (d) coevolution.
11. Great similarity between species implies (a) recent common ancestry (b) remote common ancestry (c) successful reproduction (d) extinction.
12. Features that were useful in ancestors but are no longer useful are called (a) analogous features (b) vestigial features (c) homologous features (d) favorable traits.
13. Similar features in different species that originated in a shared ancestor are called (a) vestigial features (b) analogous features (c) homologous features (d) unexpressed genes.
14. A hummingbird and a humming moth have a number of superficial features in common with each other. This is an example of (a) divergent evolution (b) coevolution (c) convergent evolution (d) superposition.
15. The phylogenetic tree below implies that modern finches and armadillos (a) are unrelated (b) share a remote common ancestor (c) share a recent common ancestor (d) did not evolve from older forms of life.

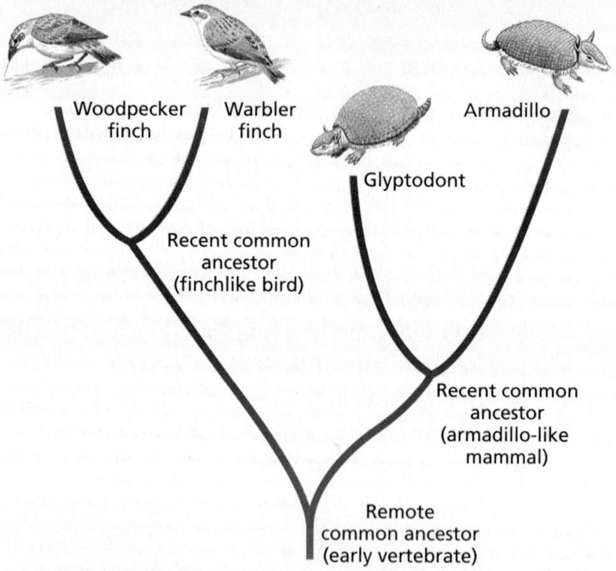

Short Answer

16. What led Hooke to argue that fossils had once been living organisms?
17. What did Malthus observe about the potential for growth in populations?
18. How was Darwin's theory of evolution different from Lamarck's theory?
19. What does the existence of very similar embryological forms among species imply?
20. Why are some traits favorable for some species and not for others?
21. What pattern of evolution is demonstrated by the Galápagos finches?
22. What is an example of a vestigial structure in humans?
23. How does the structure of macromolecules, such as proteins, act as an index of relatedness between species?
24. What kind of evolution is demonstrated by nectar-feeding bats and the flowers they feed on?

25. Could a characteristic that is not controlled by genetics be selected by the environment? Would this characteristic contribute to the evolution of the organism that has it?

CRITICAL THINKING

1. In recent years, paleontologists have claimed that in some cases the evolution of a new species occurs quite suddenly—in less than a thousand years. Darwin stated that evolution was a gradual process. What effect does generation time have on evolution rate?
2. The process of natural selection throughout the history of life on Earth has resulted in the success of some species and the extinction of other species. Why has natural selection not resulted in the existence of a single best-adapted species?
3. Many vestigial traits, such as the human tailbone, seem to be largely neutral, that is, neither beneficial nor harmful. The appendix is an example of a vestigial structure in humans. How might having an appendix be harmful to humans?
4. The graph below shows the diversity of different groups of animals over time. The width of the colored-in areas is proportional to the number of different types of bivalves, mammals, roundworms, and trilobites that were alive during different eras. Use the graph to answer the following questions:
   a. Which group was the last to evolve?
   b. Which group is or was the most diverse? The least diverse?
   c. Which group diversified rapidly soon after evolving?
   d. Which group(s) became extinct?
   e. Which groups did not live at the same time?

Animal Diversity

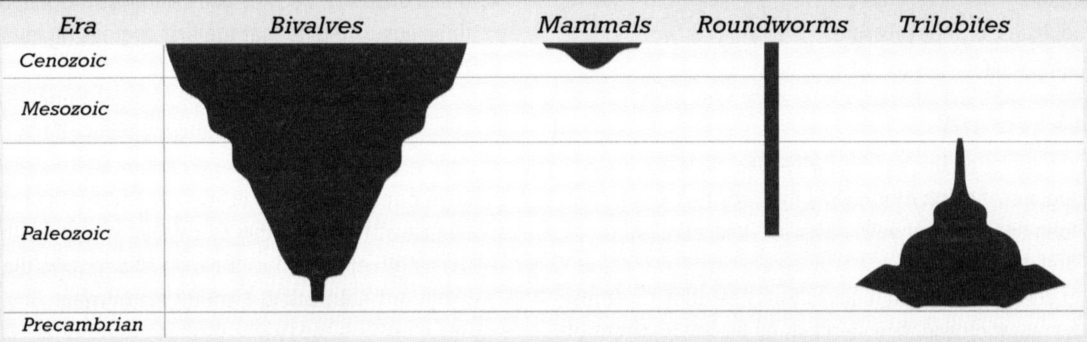

EXTENSION

1. Read "A Dinosaur Named Sue" in *National Geographic,* June 1999, on page 47, and answer the following questions: What technique was used to examine the inside of the skull of the 67-million-year-old *Tyrannosaurus rex* fossil? How many *T. rex* fossils have ever been discovered, and what makes Sue unique? When and where was Sue discovered, and what was the principal reason Sue's ownership was in question?
2. Though he never became as famous as Charles Darwin, Alfred Wallace made many important contributions to the field of biology in the late nineteenth and early twentieth centuries. Use biographies, autobiographies, and an encyclopedia to find information on the life of Alfred Wallace. Write a short report highlighting Wallace's contributions to biology.

would have about 40 generations in 1,000 years.
2. There are many different environments and many different demands on species. A species adapted to one environment might fare poorly in a different environment. Moreover, different types of organisms are interrelated in complex ways. For example, animals depend on plants for their existence, and many plants depend on the animals that feed on them for pollination.
3. The appendix can become infected, and if left untreated, this condition is life-threatening.
4. a. Mammals evolved last.
b. Bivalves are the most diverse; roundworms are the least diverse.
c. Trilobites diversified rapidly soon after evolving.
d. Trilobites became extinct.
e. Trilobites and mammals did not live at the same time.

EXTENSION

1. The skull was examined by CT scan, which enabled scientists to see details of its internal structure without using a dissecting saw. Of all the 22 *T. rex* ever discovered, Sue is the largest, estimated to have weighed 6 to 8 tons. Sue was discovered on the Cheyenne River Indian Reservation in South Dakota in August 1990. The United States government claimed ownership after the fossil was sold to its discoverer by the landowner, a Sioux citizen.
2. Student reports should reflect that Wallace began his work as a naturalist collecting plants and insects. He explored the Amazon in the late 1840s and traveled to Borneo in the late 1850s. Based on his observations over these years, Wallace formulated theories of evolution and natural selection that were nearly identical to Darwin's, and both men's work was presented to the scientific community simultaneously.

CHAPTER 15 INVESTIGATION

TIME REQUIRED
One 50-minute class period

SAFETY PRECAUTIONS
Remind students to be careful with scissors and to always cut in a direction away from their face and body.

QUICK REFERENCES
Lab preparation notes are found on page 277B.

Holt Biosources provides a Teaching Resources CD-ROM that contains "Using Gowin's Vee in the Lab" and "Scoring Rubrics."

PROCEDURAL TIPS
1. Have students work in pairs or small groups.
2. To save time, have students create their data tables before beginning the lab.
3. You may wish to review and clarify the principles of selection. This investigation reinforces the concept of natural selection as differential *reproduction* rather than merely differential survival.

CHAPTER 15 INVESTIGATION

Modeling Selection

OBJECTIVES
- Simulate the generation of variation.
- Model the selection of favorable traits in new generations.

PROCESS SKILLS
- observing
- testing
- measuring

MATERIALS
- construction paper
- cellophane tape
- soda straws
- penny, or other coin
- six-sided die
- scissors
- meterstick or tape measure
- metric ruler

Background
1. The Egyptian Origami Bird *(Avis papyrus)* lives in arid regions of North Africa. Only the birds that can fly the long distances between oases live long enough to breed successfully.
2. Successful evolution requires the generation of variety by mutation and then selection by the environment of the most-fit individuals.

PART A Parental Generation
1. First cut two strips of paper, 2 cm × 20 cm each. Make a loop with one strip of paper, overlapping the paper 1 cm, and tape the loop closed. Repeat for the other strip. Tape one loop 3 cm from each end of the straw, as shown in the figure at the top of the next column. Mark the front end of the bird with a felt-tip marker. This bird will represent the parental generation.
2. In your lab report, prepare a data table like the one shown on the facing page.
3. Test how far your bird can fly by releasing it with a gentle overhand pitch. Test the bird twice. Then record the bird's average flight distance in your data table.

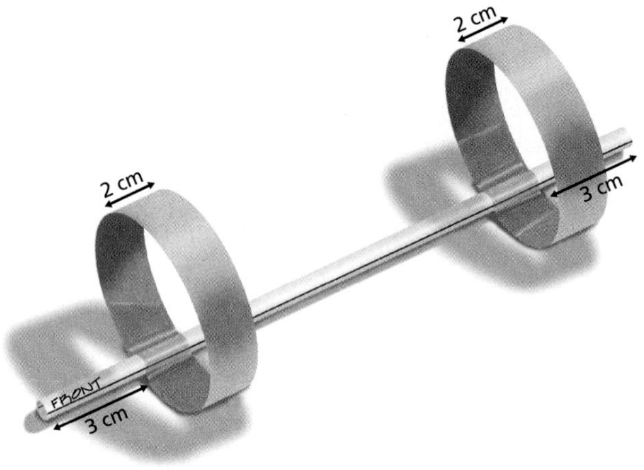

PART B F_1 Generation
4. Breed offspring. Each Origami Bird lays a clutch of three eggs. Assume that the first egg has no mutations. It is a clone of the parent.
5. Assume that the other two chicks have mutations. Follow the steps below to determine the effects of each mutation. Record the mutations and the dimensions of each offspring in your data table. The circumference of the wings can be calculated by measuring the length of the strips of paper used to form the wings and subtracting 1 cm for the overlap.

Step A A coin flip determines which part of the bird is affected by the mutation.
Heads = anterior (front)
Tails = posterior (back)

Step B A die throw determines how the mutation affects the wing.

 (1) = The wing position changes 1 cm toward the end of the straw.

 (2) = The wing position changes 1 cm toward the middle of the straw.

 (3) = The circumference of the wing increases 2 cm.

 (4) = The circumference of the wing decreases 2 cm.

 (5) = The width of the wing increases 1 cm.

 (6) = The width of the wing decreases 1 cm.

Step C A mutation that results in a wing falling off or a wing with a circumference smaller than that of the straw is lethal. If you get a lethal mutation, disregard it and breed another chick.

6. Test the birds. Release each bird with a gentle overhand pitch. It is important to release the birds as uniformly as possible. Test each bird at least twice.
7. The most successful bird is the one that flies the farthest. Record the flight distance of each offspring bird in your data table.

PART C F_2 Generation

8. Assume that the most successful bird in the F_1 generation is the sole parent of the next (F_2) generation. Continue to breed, test, and record data for 10 generations.
9.  Clean up your materials before leaving the lab.

Analysis and Conclusions

1. Did your selection process result in birds that fly better?
2. Describe two aspects of this investigation that model evolution of biological organisms.
3. Your most successful bird has a different lineage from the most successful bird of your neighboring groups. Compare your winning bird with those of your neighbors. How does it differ?
4. What might happen to your last bird if the environmental conditions change?
5. How might this lab help explain the observations Darwin made about finches on the Galápagos Islands?

Further Inquiry

A flock of Origami Birds is blown off the mainland and onto a very small island. These birds face little danger on the ground, but they experience significant risk when flying because they can be blown off the island. Birds that cannot fly at all are most likely to survive and reproduce. Continue the experiment for several generations, selecting birds that can't fly.

TABLE A *MUTATIONS AMONG OFFSPRING*

Generation	Coin flip (heads or tails)	Die throw (1–6)	Measurements of most successful offspring		Distance flown
Parental	not applicable	not applicable	anterior wing	width 2 cm	_____ m
				circumference 19 cm	
			posterior wing	width 2 cm	
				circumference 19 cm	
			anterior wing position	from front 3 cm	
			posterior wing position	from back 3 cm	
F_1	heads or tails	1–6	anterior wing	width _____ cm	_____ m
				circumference _____ cm	
			posterior wing	width _____ cm	
				circumference _____ cm	
			anterior wing position	from front _____ cm	
			posterior wing position	from back _____ cm	
F_2	heads or tails	1–6	anterior wing	width _____ cm	_____ m
				circumference _____ cm	
			posterior wing	width _____ cm	
				circumference _____ cm	
			anterior wing position	from front _____ cm	
			posterior wing position	from back _____ cm	

ANSWERS TO ANALYSIS AND CONCLUSIONS

1. Most students should answer "yes" because the best-flying birds are selected as the sole parents of the next generation.
2. Mutation and selection are modeled by this investigation.
3. Answers will vary.
4. The last bird might become more successful or less successful, depending on the type of environmental change.
5. This lab demonstrates that from the starting point of a single phenotype (the ancestral bird of the finches or of the Origami Birds), significant changes can accrue over generations.

CHAPTER 16
The Evolution of Populations and Speciation

	TOPICS	TEACHING RESOURCES	LABS, CLASSWORK, AND HOMEWORK
BLOCK 1 45 minutes	**Introducing the Chapter, p. 298**	**ATE** Focus Concept, p. 298 **ATE** Assessing Prior Knowledge, p. 298 **ATE** Understanding the Visual, p. 298	■ Supplemental Reading Guide, *Origin of Species*
	16-1 Genetic Equilibrium, p. 299 Variation of Traits in a Population, p. 299 Allele Frequencies and the Gene Pool, p. 300 Hardy-Weinberg Genetic Equilibrium, p. 302	**ATE** Section Overview, p. 299 **ATE** Visual Link Figure 16-3, p. 301	■ Problem-Solving Worksheets—Genetics and Probability ■ Problem-Solving Worksheets—Population Size ★ Study Guide, Section 16-1 **PE** Section Review, p. 302
BLOCK 2 45 minutes	**16-2 Disruption of Genetic Equilibrium, p. 304** Mutation, p. 304 Migration, p. 304 Genetic Drift, p. 305 Nonrandom Mating, p. 306 Natural Selection, p. 306	**ATE** Section Overview, p. 304 **ATE** Critical Thinking, p. 304 **ATE** Visual Link Figure 16-5, p. 305 63, 64	**PE** **Quick Lab,** p. 306, Evaluating Selection ★ Study Guide, Section 16-2 **PE** Section Review, p. 308
BLOCK 3 & 4 90 minutes	**16-3 Formation of Species, p. 309** The Concept of Species, p. 309 Isolating Mechanisms, p. 310 Rates of Speciation, p. 312	**ATE** Section Overview, p. 309 **ATE** Critical Thinking, p. 311 51	**PE** **Chapter 16 Investigation,** pp. 316–317 **Lab B9,** Peppered Moth Survey ★ Study Guide, Section 16-3 **PE** Section Review, p. 312
BLOCKS 5 & 6 90 minutes	**REVIEW and ASSESSMENT** **PE** Chapter 16 Review, pp. 314–315	Audio CD Program	**CHAPTER TESTING** ■ Chapter Test (blackline master) ▲ Test Generator ■ Assessment Item Listing ■ Scoring Rubrics and Classroom Management Checklists

MODERN BIOLOGY — HOLT BIOSOURCES — One-Stop Planner CD-ROM

KEY

- **PE** Pupil's Edition
- **ATE** Teacher's Edition
- ■ Active Reading Guide
- ★ Study Guide
- Teaching Transparencies
- Laboratory Program

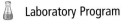

One-Stop Planner CD-ROM
Shows these resources within a customizable daily lesson plan:

- ■ Holt BioSources Teaching Resources
- ■ Active Reading Guide
- ★ Study Guide
- ▲ Test Generator

READING FOR CONTENT MASTERY

Preview: Have students preview the objectives, the introductory paragraph, topic sentences, and captions before reading the chapter.

ATE Active Reading Technique: Anticipation Guide, p. 299
■ Active Reading Guide Worksheet 16-1
PE Summary/Vocabulary, p. 313

■ Active Reading Guide Worksheet 16-2
PE Summary/Vocabulary, p. 313

■ Active Reading Guide Worksheet 16-3
PE Summary/Vocabulary, p. 313

TECHNOLOGY AND INTERNET RESOURCES

CNN PRESENTS Video Segment 10 Galápagos

Holt Biology Videodiscs Teacher's Correlation Guide, Lessons 16-1 through 16-3

Audio CD Program, Sections 16-1 through 16-3

internetconnect

sciLINKS NSTA
On-line Resources: www.scilinks.org
The following *sci*LINKS Internet resources can be found in the student text for this chapter:

Topics:
- Population genetics, p. 299
- Species formation, p. 312

go.hrw.com
On-line Resources: go.hrw.com
Visit the HRW Web site for a variety of resources related to this chapter. Just type in the keyword HM2 Chapter 16.

Smithsonian Institution®
Visit **www.si.edu/hrw** for additional on-line resources.

CNNfyi.com
Visit **www.cnnfyi.com** for late-breaking news and current event stories selected just for you.

LAB ACTIVITY PLANNING

CHAPTER 16 INVESTIGATION

Predicting Allele Frequency, pp. 316–317

OVERVIEW
In this investigation, students will simulate a monohybrid cross to determine genotype and phenotype in a population of hypothetical animals. The monohybrid cross involves coat color (the allele for black coat color is dominant and the allele for white coat color is recessive). Students will compare the frequencies of alleles in a random mating simulation and a nonrandom mating simulation in which selection against white coat color occurs.

Students should understand that natural selection is a process by which organisms with favorable variations, such as a black coat color, survive and reproduce at a higher rate. Natural selection causes allele frequencies to change from one generation to the next, resulting in changes in the gene pool of a population.

PREPARATION
Each student group will need three 1 pt containers, 200 black beads, 200 white beads, a marker, and tape. You can purchase the beads at a local craft store, discount store, or biological supply house, such as WARD'S.

Quick Labs

Evaluating Selection, p. 306

Each student group will need a sheet of plain, unlined paper, colored markers, and 25 pieces of colored candy. The candy can be any kind, and each piece should be about the size of a dime.

UNDERSTANDING THE VISUAL

Have students examine the photograph of the male mandrill baboon. Tell them that the females of this species do not have this vivid facial coloration. Ask them if they can think of other examples of marked differences between males and females. (Examples include the tails of peacocks, facial hair in men, bright coloration of many male birds, larger body size of many male mammals, and antlers on male deer.) Ask students how these differences could affect the reproductive success—and therefore the fitness—of individuals. (One sex may select a mate based on coloration, size, or other characteristics that differ between the sexes.)

FOCUS CONCEPT

Reproduction and Inheritance
When organisms reproduce, they pass their genetic information on to their offspring. In sexually reproducing organisms, offspring receive genetic information from both parents. The distribution of genes in successive generations is determined by the characteristics of the genes and by the interaction between genes and environment.

ASSESSING PRIOR KNOWLEDGE

Review the following concepts.

Genetics: *Chapter 9*
Tell students that in tigers, a recessive allele causes an absence of fur pigmentation, producing an albino, or white, tiger. Ask students what the genotype of an orange, heterozygous tiger is.

Natural Selection: *Chapter 14*
Ask students whether the white phenotype in tigers would be favored by natural selection. Have students explain their reasoning.

CHAPTER 16

The Evolution of Populations and Speciation

Sexual selection, which is one variation of natural selection, influences the development of extreme phenotypic traits, particularly in males. The vibrant red stripe on the blue muzzle of this male mandrill baboon, Mandrillus sphinx, *does not appear in females.*

FOCUS CONCEPT: ***Reproduction and Inheritance***

As you read, pay attention to the steps that lead to the formation of new species.

GENETIC EQUILIBRIUM

By the time of Darwin's death, in 1882, the idea of evolution by natural selection had gained wide acceptance among scientists. But with the birth of the field of genetics in the early 1900s, spurred by the rediscovery of Mendel's work on the mechanics of inheritance, many questions about evolution and natural selection resurfaced.

SECTION

16-1

OBJECTIVES

▲ Explain the importance of the bell curve to population genetics.

● Describe two causes of genotypic variation in a population.

■ Explain how to compute allele frequency and phenotype frequency.

◆ Explain Hardy-Weinberg genetic equilibrium.

VARIATION OF TRAITS IN A POPULATION

Population genetics is the study of evolution from a genetic point of view. Evolution is a gradual change in the genetic material of a population. Recall from Chapter 15 that a population consists of a collection of individuals of the same species that routinely interbreed. Populations are important to the study of evolution because a population is the smallest unit in which evolution occurs.

Within a population, individuals may vary in observable traits. For example, fish of a single species in a pond may vary in size. Biologists often study variation in a trait by measuring that trait in a large sample. Figure 16-1 shows a common result of such measurements. It is a graph of the frequency of lengths in a population of mature fish. Because the shape of the curve looks like a bell, it is called a **bell curve.** The bell curve shows that while a few fish in this population are extremely short and a few are extremely long, most are of average length. In nature, many quantitative traits in a population—such as height and weight—tend to show variation that follows a bell curve pattern.

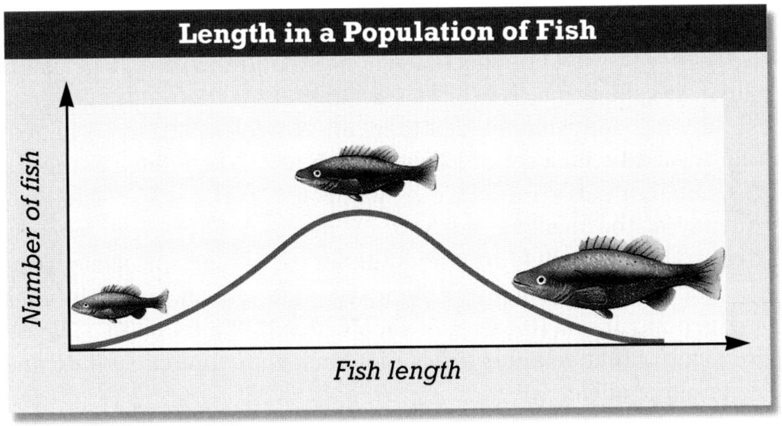

FIGURE 16-1
A bell curve illustrates that most members of a population have similar values for a given, measurable trait. Only a few individuals display extreme variations of the trait.

internetconnect
SCILINKS NSTA
TOPIC: Population genetics
GO TO: www.scilinks.org
KEYWORD: HM299

SECTION 16-1

SECTION OVERVIEW

Reproduction and Inheritance
Section 16-1 describes the most common pattern of variation in a population and explains how to calculate the frequencies of alleles and phenotypes in a population.

Evolution
Section 16-1 also explains the Hardy-Weinberg genetic equilibrium, which describes situations in which allele frequencies remain constant because no evolutionary forces are acting on the population.

ACTIVE READING

Technique: Anticipation Guide
Before students begin this chapter, write the following statements on the board or overhead. Have students write "true" or "false" on a sheet of paper for each statement. Students should then read the chapter to see if their predictions are correct.

1. Very small populations are more likely to become extinct than larger populations. *(true)*
2. Natural selection always takes millions of years to produce a different phenotype. *(false)*

TEACHING STRATEGY

Graphing Human Height
Have students make a histogram of the following data on human height: 11 individuals are 160 cm tall; 61 are 165 cm tall; 149 are 170 cm tall; 199 are 175 cm tall; 153 are 180 cm tall; 58 are 185 cm tall; and 9 are 190 cm tall. Ask students to describe the shape of this distribution and to explain what this shape suggests about the inheritance of human height (the curve is bell-shaped). Ask them to calculate the mean height for this sample of humans (175 cm).

SECTION 16-1

DEMONSTRATION

A few human traits are controlled by a single gene with only two alleles. One such trait is the form of the earlobe. Attached earlobes, which do not hang free, are recessive, while unattached earlobes are dominant. Using student volunteers, show each type of earlobe to the class. Determine the number of individuals in the class with each type of earlobe. Have students calculate the class's phenotype frequencies for these two traits.

QUICK FACT

On average, between two and three crossover events occur on each pair of human chromosomes during meiosis.

INCLUSION *ACTIVITY*

Remind students how to convert between fractions, decimals, and percentages. Have students practice their skills by converting the following fractions to decimals and percentages: 1/8, 1/16, 3/4, 1/9, 3/8 (0.125, 12.5 percent; 0.0625, 6.25 percent; 0.75, 75 percent; 0.111; 11.1 percent; 0.375, 37.5 percent).

RECENT RESEARCH

Genetic Variability

A wasp species was discovered to commonly harbor a bacterium that kills male progeny but not females. It is passed down from one female to another. This increases the fitness of the infected females by releasing resources to females that would have gone to males and by increasing genetic variability by reducing the number of males and thus the likelihood of inbreeding.

FIGURE 16-2

Many varied but similar phenotypes occur within families because members of a family share some alleles but not others.

Causes of Variation

What causes variation in traits? Some variations are influenced by environmental factors, such as the amount or quality of food available to an organism. Variation is often influenced by heredity. Usually, both factors play a role.

To consider variability, think about phenotypes within a single human family. Two parents, each with a distinct genotype, may produce several children. In the picture of the family in Figure 16-2, the two young-adult brothers are not identical to each other, even though their genotypes are combinations of the genotypes of the same two parents. Both young men resemble their father, though in different traits. The baby resembles his young father, his grandfather, and his uncle. Thus, these males representing three generations look similar but not identical. What causes the genes to vary? The answer lies in the way gametes are produced and in the way gametes fuse with each other.

Variations in genotype arise in three main ways. (1) *Mutation* results from flawed copies of individual genes. (2) *Recombination* is the reassociation of genes in a diploid individual. Recombination occurs during meiosis by the independent assortment of genes on nonhomologous, or different, chromosomes and by crossing-over between genes located on homologous chromosomes. (3) The *random fusion of gametes* is essentially a game of chance played by individual gametes. Often there are hundreds of millions of sperm involved in a mating. The one that actually fertilizes an egg is largely a matter of chance. These processes ensure that offspring are not carbon copies of their parents.

ALLELE FREQUENCIES AND THE GENE POOL

Population geneticists use the term **gene pool** to describe the total genetic information available in a population. It is easy to imagine genes for the next generation as existing in an imaginary pool. If you could inventory this pool and know the alleles that are present, then you could apply a simple set of rules based on probability theory to predict expected genotypes and their frequencies for the next generation.

Suppose, for example, that there are two forms of a hypothetical allele, A and a, in a set of 10 gametes. If half the gametes in the set (5 gametes) carry the allele A, we would say that the allele frequency of the A allele is 0.5, or 50 percent. **Allele frequency** is determined by dividing the number of a certain allele (five instances of the A allele) by the total number of alleles of all types in the population (10 gametes, each with either an A or an a allele). Remember that gametes are haploid and therefore carry only one form of the allele.

Predicting Phenotype

The population of four o'clock flowers, shown in Figure 16-3, illustrates how phenotype can change from generation to generation. Homozygous *RR* flowers are red. Homozygous *rr* flowers are white. Heterozygous *Rr* flowers are pink rather than red, as you might expect. These flowers show incomplete dominance for color, meaning heterozygotes show a trait that falls between the dominant trait and the recessive trait. Thus, homozygotes and heterozygotes can be easily identified by observing the phenotype.

Compare the parent generation with the offspring generation of the four o'clock flowers shown in Figure 16-3. There are equal numbers of plants with the *RR* genotype and the *Rr* genotype in the first generation. You can compute the phenotype frequencies from the figure. A **phenotype frequency** is equal to the number of individuals with a particular phenotype divided by the total number of individuals in the population. Phenotype frequencies in the first generation are 0.5 pink (4 pink plants out of a total of 8 plants), 0.5 red (4 red plants out of a total of 8 plants), and 0.0 white. Recall that allele frequencies are computed using the same principle: the allele frequencies in the first generation plants are 0.75 *R* (12 *R* alleles out of a total of 16 alleles) and 0.25 *r* (4 *r* alleles out of a total of 16 alleles).

We now can predict the genotypes and phenotypes of the second generation. If a male gamete encounters a female gamete, they will produce a new four o'clock plant whose genotype is the combination of both parental gametes. Thus, an *R* male gamete combined with an *R* female gamete will produce a plant with the *RR* genotype, which has red flowers. According to the laws of probability, the chance of an *R* gamete (a single allele) meeting with another *R* gamete is the arithmetic product of their allele frequencies in the gene pool:

$$\text{frequency of } R \times \text{frequency of } R = \text{frequency of } RR \text{ pair}$$
$$0.75 \times 0.75 = 0.5625$$

The expected frequency of the *rr* genotype is then

$$\text{frequency of } r \times \text{frequency of } r = \text{frequency of } rr \text{ pair}$$
$$0.25 \times 0.25 = 0.0625$$

FIGURE 16-3

Although the four o'clock flowers differ phenotypically from generation to generation, the allele frequencies remain the same.

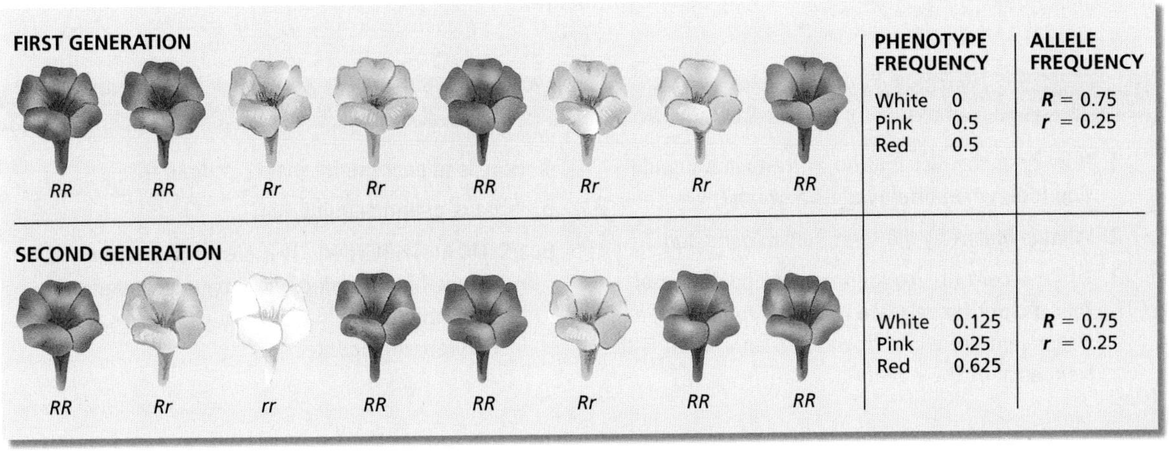

SECTION 16-1

TEACHING STRATEGY
Flower Color

Divide the class into groups of three or four. Provide each group with 12 red and 4 white squares of paper. Tell them these squares will represent white, *r*, and red, *R*, alleles, respectively, in four o'clocks. Ask students what the allele frequencies are for this population ($r = 0.25$, $R = 0.75$). Have the groups place their squares in a pile on the desk. One student from each group should close his or her eyes and take eight pairs of squares, representing eight individuals, from the pile. Have each group write down the genotypes of the eight individuals and determine the phenotypes (1 *R* allele and 1 *r* allele make a pink phenotype) and phenotype frequencies. Tell each group to compare their results with those of different groups and those in the text.

VISUAL LINK
Figure 16-3

Ask students to use this figure to explain how genotype frequency is determined. (The first step is to count the alleles of each type. In the first generation, 4 out of 16, or 25 percent, are *r*, and 12 out of 16, or 75 percent, are *R*. In the second generation, there are still only 4 out of 16 *r* alleles; they are merely distributed differently.) Ask students how a white flower could appear in the second generation when there were no white flowers in the first generation. (The white flower is the offspring of two pink flowers, each of which contributed an *r* allele.)

SECTION 16-1

GIFTED ACTIVITY

The Hardy-Weinberg genetic equilibrium is often expressed as an equation: $p^2 + 2pq + q^2 = 1$. Have students use this equation to calculate the expected number of heterozygotes and homozygous dominant individuals in a population of 100,000 tigers, 10 of which are white ($q^2 = 10/100{,}000$, or 0.0001; $q = 0.01$; $p = 1-q$, or 0.99; the frequency of homozygous dominant individuals is p^2, or 0.9801, or 98,010 individuals out of 100,000; the frequency of heterozygotes is 2pq, or 0.0198, or 1,980 individuals out of 100,000).

RETEACHING ACTIVITY

Determine the phenotype frequency of the class for attached and unattached earlobes. For the purposes of this exercise, assume that 50 percent of the people with unattached earlobes are heterozygous. Ask students to calculate the allele frequencies within the gene pool.

ANSWERS TO SECTION 16-1 REVIEW

1. It forms a bell-shaped curve.
2. The human gene pool contains all the alleles for all traits found in the human population.
3. The frequency is 0.5.
4. by determining the number of individuals with that phenotype and then dividing by the number of individuals in the population
5. Genetic equilibrium is a state in which the allele frequencies of a population do not change from generation to generation.
6. Incomplete dominance is easier because heterozygotes have a different phenotype from homozygous dominant individuals.

FIGURE 16-4
This flock of mallards, *Anas platyrhynchos,* likely violates some or all of the conditions necessary for Hardy-Weinberg genetic equilibrium.

The frequencies of all types expected in the second generation must add up to 1.0, just as fractions of a whole must add up to 1. Having established the probabilities of getting an *RR* and an *rr* plant, we can compute the expected frequency of the *Rr* plants. All those plants that are neither *RR* nor *rr* will be *Rr,* so

$$1.0 - \text{frequency of } RR - \text{frequency of } rr = \text{frequency of } Rr$$
$$1.0 - 0.5625 - 0.0625 = 0.375$$

HARDY-WEINBERG GENETIC EQUILIBRIUM

It is clear from the example of the four o'clock flowers that phenotype frequencies can change dramatically from generation to generation. But what happens to allele frequencies over generations? A German physician, Wilhelm Weinberg (1862–1937), and a British mathematician, Godfrey Hardy (1877–1947), independently showed that allele frequencies in a population tend to remain the same from generation to generation unless acted on by outside influences. This is referred to as **Hardy-Weinberg genetic equilibrium,** and it is based on a set of assumptions about an ideal hypothetical population that is not evolving:

1. No net mutations occur; that is, allele frequencies do not change overall because of mutation.
2. Individuals neither enter nor leave the population.
3. The population is large (ideally, infinitely large).
4. Individuals mate randomly.
5. Selection does not occur.

Bear in mind that true genetic equilibrium is a theoretical state. Real populations, such as the flock of mallards in Figure 16-4, may violate conditions necessary for genetic equilibrium. By providing a model of how genetic equilibrium is maintained, Hardy-Weinberg genetic equilibrium allows us to consider what forces disrupt equilibrium.

Word Roots and Origins

equilibrium

from the Latin *aequilibris,* meaning "evenly balanced"

SECTION 16-1 REVIEW

1. **How does the distribution of traits in a population look when displayed as a graph?**
2. **What is meant by the term *human gene pool*?**
3. **Fifty percent of an experimental population of four o'clock flowers are red-flowered plants, and 50 percent are white-flowered plants. What is the frequency of the *r* allele?**
4. **How is phenotype frequency computed?**
5. **What is genetic equilibrium?**
6. **CRITICAL THINKING Is it easier to analyze genotype by observing phenotype in organisms with complete dominance or in organisms with incomplete dominance?**

Literature & Life

Opus 100

This excerpt is from the 100th in a regular series of columns by Stephen Jay Gould published in *Natural History* magazine. It was reprinted in a collection of Gould's columns called *The Flamingo's Smile.*

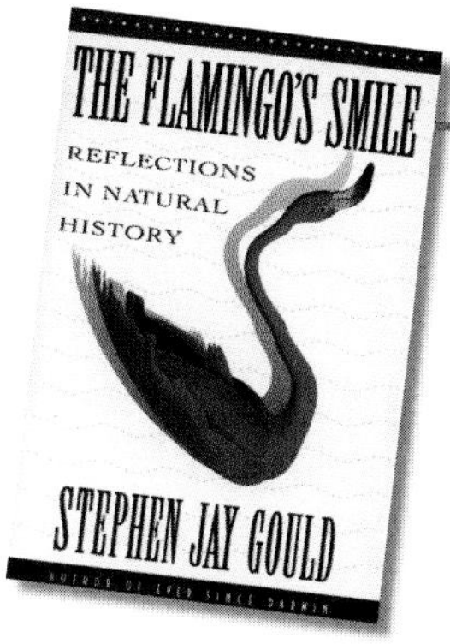

This passage describes some of the discoveries of Gould and fellow biologist David Woodruff as they began to study various species of Cerion, *a land snail of the West Indian islands.*

About fifteen names had been proposed for the *Cerions* of Grand Bahama and neighboring Abaco Island. After a week, Woodruff and I recognized that only two distinct populations inhabited these islands, each restricted to a definite and different environment. . . .

As Woodruff and I moved from island to island on Great Bahama Bank, we found the same pattern of two different populations, always in the same distinctive environments. On Little Bahama Bank, a dozen invalid names had fallen into this pattern. On Great Bahama Bank, they collapsed, literally by the hundred. About one-third of all *Cerion* "species" (close to 200 in all) turned out to be invalid names based on minor variants within this single pattern. We had reduced a chaos of improper names to a single, ecologically based order. . . .

Bahamian islands have two different kinds of coastlines. Major islands lie at the edge of their banks. The banks themselves are very shallow across their tops but plunge precipitously into deep ocean at their edges. Thus, bank-edge coasts abut the open ocean and tend to be raw and windy. Dunes build along windy coasts and solidify eventually into rock (often mistakenly called "coral" by tourists). Bank-edge coasts are, therefore, usually rocky as well. By contrast, coastlines that border the interior parts of banks—I will call them bank-interior coasts—are surrounded by calm, shallow waters that extend for miles and do not promote the building of dunes. Bank-interior coasts, therefore, tend to be vegetated, low, and calm.

Woodruff and I found that bank-edge coasts in the northern Bahamas are invariably inhabited by thick-shelled, strongly ribbed, uniformly colored (white to darkish brown), relatively wide, and parallel-sided *Cerions.* To avoid writing most of the rest of this column in Latin, I will skip the formal names and refer to these forms as the "ribby populations." Bank-interior coasts are the home of thin-shelled, ribless or weakly ribbed, variegated (usually with alternating blotches of white and brown), narrow, and barrel-shaped *Cerions*—the "mottled populations." (Mottled *Cerions* also live away from coasts in the centers of islands, while ribby *Cerions* are confined exclusively to bank-edge coasts.)

This pattern is so consistent and invariable that we can "map" hybrid zones even before we visit an island, simply by looking at a chart of bathymetry. Hybrid zones occur where bank-edge coasts meet bank-interior coasts. . . .

The distinction of mottled and ribby resolved nearly all the two hundred names previously given to *Cerions* from the northern Bahamas. . . .

Reading for Meaning

Make a chart comparing snails of the two snail populations Gould and Woodruff identified.

Read Further

Think about what you have learned about adaptations. What factors might have caused the differences between the two species of *Cerion* that Gould wrote about in this passage?

CHAPTER 16
Literature & Life

BACKGROUND

This feature describes some of the discoveries made by Gould and Woodruff as they investigated separated populations of a West Indian land snail. Land snails typically live on or near the ground. Arboreal forms are usually brightly colored; terrestrial forms are generally drab.

READING FOR MEANING

Students' answers will vary.

Ribby population shells	Mottled population shells
thick	thin
strongly ribbed	weakly ribbed or ribless
uniformly colored	variegated
wide	narrow
parallel-sided	barrel-shaped

READ FURTHER

Students' answers will vary.
The environment in which the two populations live is reflected in their anatomy. The ribby populations live on the bank-edge coasts. Their thick, ribbed shells allow them to survive in their "raw and windy" environment. In contrast, the mottled populations live in the calm, shallow waters of the bank-interior coasts. They are not subjected to the harsh conditions of the bank-edge coasts.

SECTION 16-2

SECTION OVERVIEW

Evolution
Section 16-2 describes the five forces that can disrupt genetic equilibrium and cause evolution: mutation, migration, genetic drift, nonrandom mating, and natural selection. It also describes four major types of natural selection: stabilizing selection, directional selection, disruptive selection, and sexual selection.

▶ ENGAGE STUDENTS

Describe the symptoms of sickle cell anemia, a recessively inherited disease that is usually fatal before adulthood, to the class. (A college biology text is a good source of information.) In areas of the world where malaria is endemic, the heterozygous condition protects against malaria, conferring a survival advantage over the homozygous dominants. Draw a Punnett square of a cross between two heterozygotes. Ask students to assume that all homozygous recessive offspring die before reproducing and that all homozygous dominant offspring could die from malaria. Ask them which offspring would be most likely to reproduce (the heterozygotes). Then ask them the percentage of heterozygotes in the next generation (50 percent). Tell them that in parts of Africa where malaria is common, the percentage of heterozygotes is close to 40 percent.

CRITICAL THINKING

Source of Dwarfism Trait
Achondroplastic dwarfism is a dominantly inherited disorder in humans. Nearly all homozygous dominant individuals die before birth. Only about 20 percent of dwarfs have a family history of dwarfism. Ask students where the other 80 percent of dwarfs come from (new mutations).

SECTION

16-2

OBJECTIVES

▲ List five conditions that can cause evolution to take place.

● Give an example of how migration can affect evolution.

■ Define *genetic drift,* and tell how it affects endangered species.

◆ Contrast the effects of stabilizing, directional, and disruptive selection on variations in a trait over time.

▲ Give an example of sexual selection.

DISRUPTION OF GENETIC EQUILIBRIUM

Evolution is the change in a population's genetic material over generations, that is, a change of the population's allele frequencies or genotype frequencies. Any violation of the five conditions necessary for Hardy-Weinberg equilibrium can result in evolution.

MUTATION

The first requirement of genetic equilibrium is that allele frequencies not change overall because of mutations. Spontaneous mutations occur constantly, at a very low rate and under normal conditions. But if an organism is exposed to mutagens—mutation-causing agents such as radiation and certain chemicals—mutation rates can increase significantly. Mutations can affect genetic equilibrium by producing totally new alleles for a trait. Many, if not most, mutations are harmful. Because natural selection operates only on genes that are expressed, it is very slow to eliminate harmful recessive mutations. In the long run, however, beneficial mutations are a vital part of evolution.

MIGRATION

The second requirement of genetic equilibrium is that the population remain constant. **Immigration,** the movement of individuals into a population, and **emigration,** the movement of individuals out of a population, can change gene frequencies.

The behavioral ecology of some animal species encourages immigration and emigration. Common baboons live on the savannas of central Africa in social and breeding groups called troops. A troop is dominated by a few adult males, and it may have from 10 to 200 members. Females tend to remain with the troop they are born into, however, younger or less dominant males leave their birth troop, eventually joining another troop. This constant movement of male animals ensures gene flow. **Gene flow** is the process of genes moving from one population to another.

Word Roots and Origins

immigration

from the Latin *immigrare,* meaning "to go into"

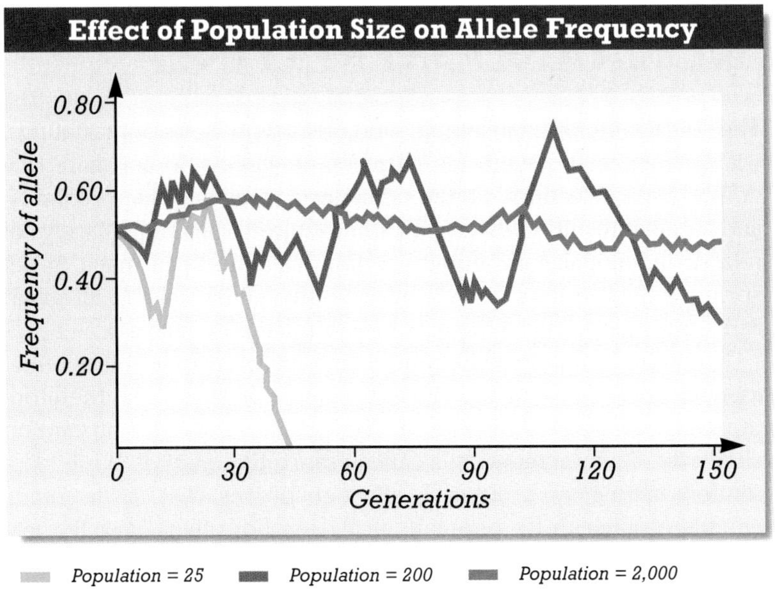

FIGURE 16-5

Genetic drift is significant only in small and medium-sized populations. In a small population, a particular allele may disappear completely over a few generations. In a larger population, a particular allele may vary widely in frequency due to chance but still be present in enough individuals to be maintained in the population. In a much larger population, the frequency of a particular allele may vary slightly due to chance but remain relatively stable over generations.

GENETIC DRIFT

The third requirement of genetic equilibrium is the presence of a large population. The Hardy-Weinberg principle is based on the laws of probability, which do not necessarily hold for small and medium-sized populations. **Genetic drift** is the phenomenon by which allele frequencies in a population change as a result of random events, or chance. In small populations, the failure of even a single organism to reproduce can significantly disrupt the allele frequency of the population, as can greater-than-normal reproduction by an individual, resulting in genetic drift.

Figure 16-5 shows a graph of genetic drift in populations of three differing sizes. Small populations can undergo abrupt changes in allele frequencies, exhibiting a large degree of genetic drift, while large populations retain fairly stable allele frequencies, maintaining a small degree of genetic drift. In the smallest population shown in Figure 16-5, the frequency of the example allele reaches zero at about the 45th generation. If we assume that we started with two alleles for a trait, then only one allele is left and every individual is homozygous for the remaining allele. Once this happens, the population is in danger of becoming extinct because there is no variation for natural selection to act on. For example, a new disease could wipe out the entire population. This is why endangered species, like the northern elephant seal, *Mirounga angustirostris,* shown in Figure 16-6, remain in peril of extinction even if their numbers increase significantly from near-extinction.

FIGURE 16-6

Individuals of the once nearly extinct northern elephant seal, *Mirounga angustirostris,* have lost genetic variability—they are homozygous for all of their genes that have been tested. This result of genetic drift could cause extinction because it limits the species' ability to further evolve.

CONTENT CONNECTION

Chapter 12-2: Mutation

Mutations are changes in DNA. Different types of mutations include gene mutations, in which nucleotides of DNA are altered or deleted, and chromosome mutations, in which pieces of chromosomes are rearranged, lost, or moved to other chromosomes.

OVERCOMING MISCONCEPTIONS

A common misconception is that individuals evolve, changing to match changes in the environment. The unit of evolution is the population. Evolution is defined as a change in gene frequency in a population.

VISUAL LINK

Figure 16-5

Have students study the figure and answer the following questions. What is the graph comparing? (the frequency of an allele in three populations of different size over time) Which line on the graph represents a population in which the frequency of the allele changes very little over many generations? (the blue line) What is significant about the green line's intersecting the *x*-axis? (It means that the allele frequency becomes zero for the smallest breeding population after roughly 45 generations.)

CULTURAL CONNECTION

The Old-Order Amish

Have students research the old-order Amish. This group provides a good example of genetic drift in a small population that has split off from a larger group. The Amish isolate themselves from other Americans geographically and culturally. The incidence of some genetic diseases in the Amish is higher than in other populations of European descent. Have students report their findings to the class.

Quick Lab

Evaluating Selection

Time Required 20 minutes

Safety Instruct students not to eat the candy.

Procedural Tips The candy can be any type, and each piece should be about the size of a dime. As an alternative to using all the colors found in the bag, the candy colors can be limited to one or two colors. You can include at least one white candy in a group's "population." This activity works best if the candies are brightly colored and the pencils are of similar colors.

Answers to Analysis Answers will vary, but there should be more color matches on the colored background. The populations are more successfully camouflaged on the colored background. Students will recognize that the greater the diversity of a population is, the greater the number of survivors in diverse environments. Stabilizing selection favoring the color successfully camouflaged most often will tend to increase a population's fitness for that environment.

TEACHING STRATEGY

Mimicking Natural Selection

Obtain two red and two green sheets of poster board. Hang one of each on the wall. Have students cut out 40 moth shapes of each color from the other pieces of poster board. Tape 20 red and 20 green "moths" to each sheet on the wall. Have students come to the front of the room one at a time and quickly remove one moth from each sheet of poster board. When all the students have finished, add up the number of each color of moth that was removed from each board. Ask students to explain the results.

Quick Lab

Evaluating Selection

Materials unlined paper, colored pencils, 25 colored candies

Procedure

1. Fold a sheet of unlined paper in half, top over bottom. Using colored pencils, decorate half the paper with different colored patterns. Make each colored pattern about the size of a quarter.
2. Scatter your "population" of candies over the undecorated half of the sheet of paper. Count and record how many candies match the background color.
3. Now scatter the candies over the decorated half of the sheet of paper. Count and record how many candies match the background color.
4. Candies that match the background color are camouflaged. Calculate the ratio of camouflaged candies to uncamouflaged candies in steps 2 and 3.
5. Repeat steps 2–4 two times, and average your results.
6. Exchange paper with another group, and repeat steps 2–5.

Analysis Was your population more successfully camouflaged on the white background or on the colored background? How did color diversity affect your population's success on the colored background? Based on your results, predict which type of selection might increase your population's fitness for a multicolored environment.

NONRANDOM MATING

The fourth requirement of genetic equilibrium is random matings, without regard to genetic makeup. Many species do not mate randomly. Mate selection is often influenced by geographic proximity, and this can result in mates with some degree of kinship. Matings of related individuals can amplify certain traits and can result in offspring with disorders caused by recessive genes, which, although rare, may be present in the genomes of related individuals.

In another example of nonrandom mating, individuals often select a mate that has similar physical characteristics and therefore probably has some similar genes. The selection of a mate based on similarity of characteristics is called **assortative mating.** While nonrandom mating can profoundly affect genotypes, that is, the *combinations* of alleles of a population, it does not affect overall allele frequencies.

NATURAL SELECTION

The fifth requirement of genetic equilibrium is the absence of natural selection. Natural selection is an ongoing process in nature, and it is the single most significant factor that disrupts genetic equilibrium. As you learned in Chapter 15, as a result of natural selection, some members of a population are more likely to contribute their genes to the next generation than are other members. Any of several broad types of natural selection—including stabilizing, directional, disruptive, and sexual—can cause evolution.

Stabilizing Selection

In **stabilizing selection,** individuals with the average form of a trait have the highest fitness. The average represents the optimum for most traits; extreme forms of most traits confer lower fitness on the individuals that have them. Consider a hypothetical species of lizard in which larger-than-average individuals might be more easily spotted, captured, and eaten by predators. On the other hand, lizards that are smaller than average might not be able to run fast enough to escape.

Figure 16-7a shows the effect of stabilizing selection on body size in these lizards. The red curve shows the initial variation in lizard size as a standard bell curve. The blue curve represents the variation in body size several generations after a new predator was introduced. This predator easily captured the large, visible lizards and the small, slower lizards. Thus, selection against these extreme body types reduced the size range of the lizards. Stabilizing selection is the most common kind of selection. It operates on most traits and results in very similar morphology between most members of a species.

(a)

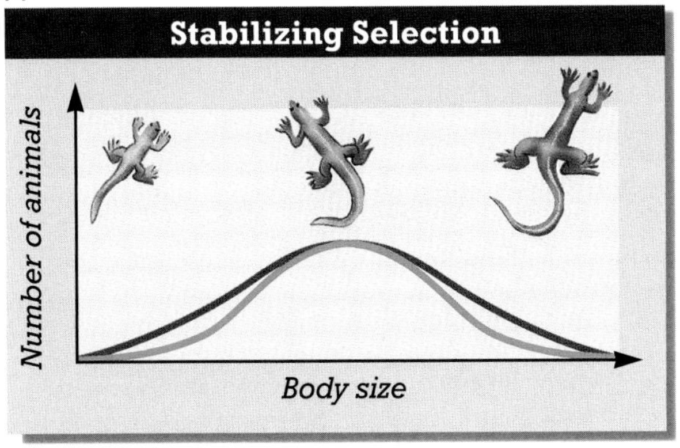

(b)

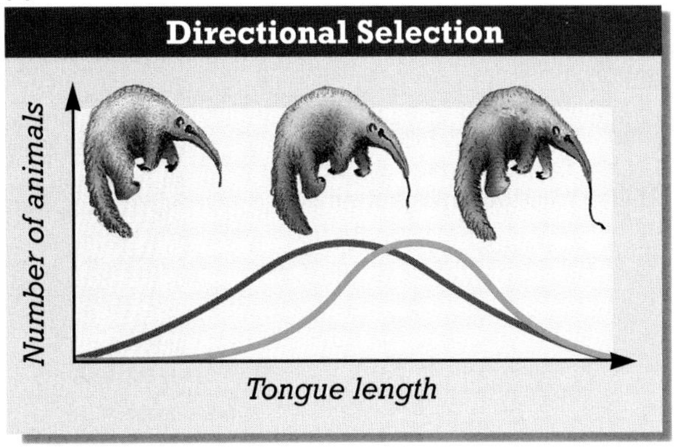

(c)

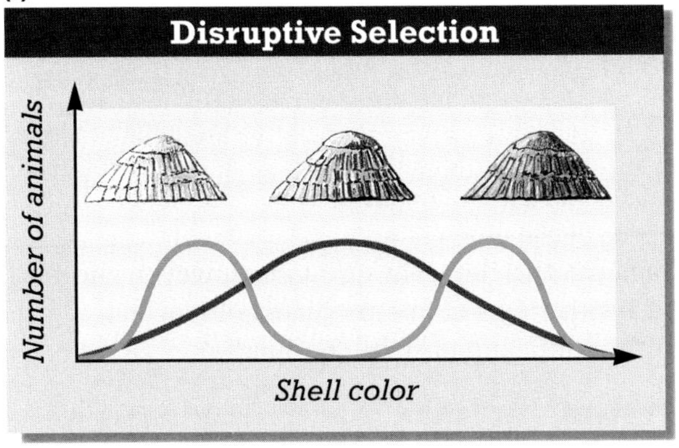

FIGURE 16-7

Stabilizing (a), directional (b), and disruptive (c) selection can all be illustrated as changes (shown in blue) of the normal bell curve (shown in red).

Directional Selection

In **directional selection,** individuals that display a more extreme form of a trait have greater fitness than individuals with an average form of the trait. Figure 16-7b shows the effects of directional selection on tongue length in anteaters. Anteaters feed by breaking open termite nests, extending their sticky tongue into the nest, and lapping up termites. Suppose that an area was invaded by a new species of termite that built very deep nests. Anteaters with long tongues could more effectively prey on these termites than could anteaters with short or average tongues. Thus, directional selection would act to direct the trait of tongue length away from the average and toward one extreme.

Disruptive Selection

In **disruptive selection,** individuals with either extreme variation of a trait have greater fitness than individuals with the average form of the trait. Figure 16-7c shows the effect of disruptive selection on shell color in marine animals called limpets. The shell color of limpets varies from pure white to dark tan. White-shelled limpets that are on rocks covered with goose barnacles, which are also white, are at an advantage. Birds that prey on limpets have a hard time distinguishing the white-shelled limpets from the goose barnacles. On bare, dark-colored rocks, dark-shelled limpets are at an advantage. Again, the limpet-eating birds have a hard time locating

GIFTED *ACTIVITY*

Some scientists hypothesize that because we humans are capable of drastically altering our environment, we will eventually become immune to the pressures of natural selection. Have students write an essay arguing for or against this hypothesis.

RETEACHING *ACTIVITY*

Ask students to graph, illustrate, and identify the following selection patterns in populations of snails whose shell colors vary from tan to dark brown: selection against snails that are unusually light or dark, selection against snails with lighter-colored shells, and selection against snails of intermediate coloration. (Graphs should depict stabilizing, directional, and disruptive selection, respectively.)

SECTION 16-2

RECENT RESEARCH
Odor Preferences

Scientists have found that human females show a distinct preference for particular male body odors. They found that the major histocompatibility complex (MHC) genes confer specific odors and that females have a preference for males whose MHC genes are most unlike their own. Matings between humans with dissimilar MHC genes may help ensure that offspring have very diverse and therefore very effective immune systems.

ANSWERS TO SECTION 16-2 REVIEW

1. Genetic drift is a change in the gene frequency of a population due to chance. Genetic drift is most important in small and medium-sized populations.
2. Mutation can disrupt genetic equilibrium by creating new alleles in a population. Immigration can disrupt genetic equilibrium because the immigrants might have different alleles or different allele frequencies. Adding the immigrants' alleles to the gene pool of the population changes the relative abundances of alleles.
3. Stabilizing selection favors intermediate phenotypes. Directional selection favors one extreme phenotype. Disruptive selection favors both extreme phenotypes.
4. Sexual selection is a form of natural selection in which traits that increase mating success are favored.
5. mutation, migration, genetic drift, nonrandom mating, natural selection
6. This is an example of stabilizing selection because the intermediate phenotype is favored.

FIGURE 16-8
Males sometimes display extreme traits, like the large tail of this peacock, *Pavo cristatus.* This trait is favorable if it attracts females and increases the reproductive fitness of the male.

the dark shells against the dark background. However, the birds easily spot limpets with shells of intermediate color, which are visible against both the white and dark backgrounds.

Sexual Selection

In many species of birds, the males are brightly colored and often heavily plumed, like the peacock shown in Figure 16-8. These elaborately decorated males are easy for predators to see. Why would natural selection work in favor of an organism being conspicuous to a predator? Females tend to choose the males they mate with based on certain traits. This is referred to as **sexual selection.** In order to leave offspring, a male must be selected by the female, and the peacock's gaudy plumage increases his chances of being selected. Extreme traits, such as heavy, brightly colored plumage, may give the female an indication of the quality of the male's genes. While survival to reproductive maturity is necessary, survival alone is not enough to further evolution. The genes of successful *reproducers,* rather than those of merely successful *survivors,* are amplified through natural selection.

SECTION 16-2 REVIEW

1. What is genetic drift?

2. Explain how mutation and immigration disrupt genetic equilibrium.

3. Compare and contrast stabilizing, directional, and disruptive selection.

4. What is sexual selection?

5. Name the five violations of the conditions necessary for Hardy-Weinberg genetic equilibrium that can cause evolution to occur.

6. CRITICAL THINKING Human newborns with either a very high or a very low birth weight are more likely to die in infancy. What type of selection does this seem to be?

FORMATION OF SPECIES

How many species of organisms exist on Earth today? Undiscovered species are so numerous that we have no accurate answer. For example, even small areas of tropical rain forests can contain thousands of species of plants, animals, and microorganisms. New species are discovered and others become extinct at an incredible rate. In this section, you will learn how one species can become two through a process called speciation.

SECTION

16-3

OBJECTIVES

▲ Explain the difference between the morphological concept of species and the biological species concept.

● Define *geographic isolation,* and explain how it can lead to speciation.

■ Name three kinds of reproductive isolation.

◆ Summarize the punctuated equilibrium hypothesis, and contrast it with the hypothesis of gradual change.

THE CONCEPT OF SPECIES

You have learned that existing species are essentially changed versions of older species. The process of species formation, **speciation** (SPEE-shee-AY-shun), results in many related populations of organisms. Some are very similar to their shared ancestral species, while other populations become quite different.

Morphological Concept of Species

For many years, scientists used the internal and external structure and appearance of an organism—its **morphology** (mor-FAHL-uh-jee)—as the chief criterion for classifying it as a species. Using the morphological concept of species, a species is defined primarily according to its structure and appearance. Because morphological characteristics are easy to observe, making species designations based on morphology proved convenient.

The morphological concept of species has limitations, however. There can be phenotypic differences among individuals in a single population. Notice, for example, the variation between the two red-tailed monkeys shown in Figure 16-9. To further complicate the matter, some organisms that appear different enough to belong to different species interbreed in the wild and produce fertile offspring. In response to the capacity of dissimilar organisms to reproduce, the biological species concept arose.

FIGURE 16-9
Individual red-tailed monkeys, *Cercopithecus ascanius,* can have different facial features.

SECTION 16-3

SECTION OVERVIEW

Evolution
Section 16-3 explains the mechanisms of species formation and describes two competing hypotheses about the rate at which evolution occurs—punctuated equilibrium and gradualism.

▶ ENGAGE STUDENTS

Show students pictures of different breeds of dogs. Ask students to select several breeds they are familiar with and list as many differences between these breeds as possible. Tell students to consider behavioral as well as physical traits. Ask students if breeds of dogs should be considered separate species. Describe how the different breeds of dogs were produced—through artificial selection—and explain how this process is similar to natural selection.

TEACHING STRATEGY

Morphological Concept of Species
Obtain a variety of pictures of organisms that illustrate different sexes and different stages of development. (Some to include are deer or elk, frogs, metamorphic insects, a tree and a tree seedling, and mice.) Ask students to classify these organisms into different species. Ask them if a buck with antlers can be called a different species from a doe with no antlers (no). Ask them if a young rat with no fur can be called a different species from an adult rat with fur (no). Point out to students that these examples demonstrate some of the difficulties of using the morphological concept of species.

SECTION 16-3

TEACHING STRATEGY

Classifying by Morphology

Provide students with a variety of flower types or vegetable types (ones that are actually fruits, such as beans, tomatoes, squashes, etc.), including similar varieties of a species. Ask students to classify the flowers into different species based on their morphological characteristics. Ask them to identify some of the difficulties in classifying these organisms in this manner. (You cannot tell by looking at two similar flowers of different color whether they can interbreed and should be considered members of the same species.)

RECENT RESEARCH

Isolation and Speciation Rate

Scientists developed a model to predict how the number of geographically isolated populations in a species affects the rate of speciation. The model predicts that speciation occurs more rapidly when natural selection acts on two large subpopulations than when it acts on many small subpopulations.

GIFTED ACTIVITY

F_1 hybrids are very vigorous and desirable as crop plants. However, they must be produced each year from the parental types. The commercial production of F_1 hybrid corn seed used to require removal of the pollen-bearing tassels so that plants could not self-pollinate. Have students investigate the discovery and use of hybrid corn seed made with cytoplasmic male sterility factors, which block the formation of male reproductive structures and thus promote genetic diversity.

(a)

(b)

FIGURE 16-10

Two types of pupfish that have limited ranges in the western United States are (a) *Cyprinodon macularius* and (b) *Cyprinodon nevadensis.*

The Biological Species Concept

According to the **biological species concept,** as proposed by German-born, American biologist Ernst Mayr (1904–), a species is a population of organisms that can successfully interbreed but cannot breed with other groups. While this definition is useful for living animals, the biological species concept does not provide a satisfactory definition for species of extinct organisms, whose reproductive compatibility cannot be tested. Nor is it useful for organisms that do not reproduce sexually. Thus, our modern definition of species includes components of both the morphological and biological species concepts. A species is a single type of organism. Members of a species are morphologically similar and can interbreed to produce fully fertile offspring. The many species alive today diverged from a smaller number of earlier species.

ISOLATING MECHANISMS

How do species give rise to other, different species? Speciation begins with isolation. In isolation, two parts of a formerly interbreeding population stop interbreeding. Two important types of isolation frequently drive speciation.

Geographic Isolation

Geographic isolation is the physical separation of members of a population. Populations may be physically separated when their original habitat becomes divided. A deep canyon could develop, a river could change course, or a drying climate in a valley could force surviving fragments of an original population into separate mountain ranges. Once the subpopulations become isolated, gene flow between them stops. Natural selection and genetic drift cause the two subpopulations to diverge, eventually making them incompatible for mating.

In pupfish, small freshwater fish shown in Figure 16-10, speciation following geographic isolation apparently took place in parts of the western United States, including the desert of Death Valley. Death Valley has a number of isolated ponds formed by springs. Each pond contains a species of fish that lives only in that one pond, but the fish species of various ponds in the area are quite similar.

How did these different populations of fish become isolated in Death Valley? Geologic evidence indicates that most of Death Valley was covered by a lake during the last ice age. When the ice age ended, the region became dry, and only small, spring-fed ponds remained. Members of a fish species that previously formed a single population in the lake may have become isolated in different ponds. The environments of the isolated ponds differ enough that the separate populations of fish diverged. Eventually, the fishes in the different ponds diverged enough to be considered separate species.

Reproductive Isolation

Sometimes groups of organisms within a population become genetically isolated without being geographically isolated. **Reproductive isolation** results from barriers to successful breeding between population groups in the same area. Reproductive isolation and the species formation that follows it may sometimes arise through disruptive selection. Remember that in disruptive selection, the two extremes of a trait in a given population are selected for and the organisms begin to diverge. Once successful mating is prevented between members of the two subpopulations, the effect is the same as what would have occurred if the two subpopulations had been geographically isolated. There are two broad types of reproductive isolation: **prezygotic** (pree-zie-GAHT-ik) **isolation,** which occurs *before* fertilization, and **postzygotic isolation,** which occurs *after* fertilization.

If two potentially interbreeding species mate and fertilization occurs, success is measured by the production of healthy, fully fertile offspring. But this may be prevented by one of several types of postzygotic isolation. The offspring of interbreeding species may not develop completely and may die early, or, if healthy, they may not be fertile. From an evolutionary standpoint, if death or sterility of offspring occurs, the parent organisms have wasted their gametes producing offspring that cannot, in turn, reproduce.

This situation favors prezygotic mechanisms, such as incompatible behavior, that reduce the chance of hybrid formation. For example, a mating call that is not recognized as such by a potential mate can contribute to isolation. Differences in mating times are another type of prezygotic isolation. Both mechanisms are in effect for the frogs shown in Figure 16-11. Their mating calls and peak mating times, as shown in the graph, differ, reducing the chance of interbreeding. As a result, the wood frog and the leopard frog are reproductively isolated. Though these two frogs interbreed in captivity, they do not interbreed where their ranges overlap in the wild. As you can see in Figure 16-11, the wood frog usually breeds in late March and the leopard frog usually breeds in mid-April.

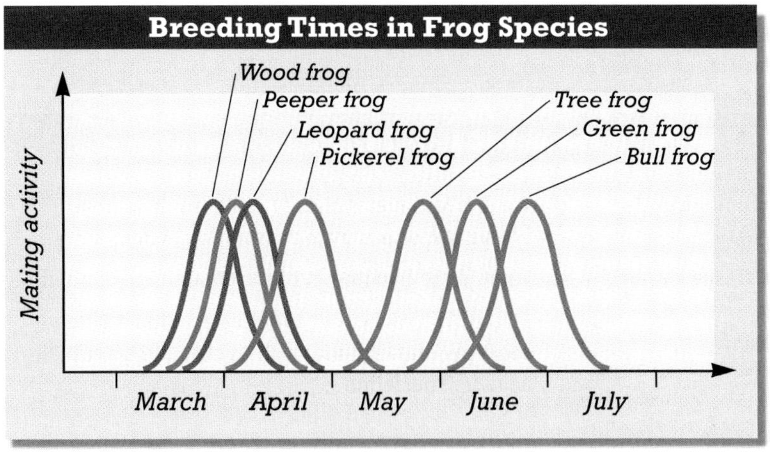

Word Roots and Origins

prezygotic

from the Latin *prae,* meaning "before," and the Greek *zygotos,* meaning "yoked"

FIGURE 16-11

As the graph shows, peak mating activity in frog species can vary widely. Such variance, coupled with different calls, has led to reproductive isolation in the wood frog, *Rana sylvatica,* top, and the leopard frog, *Rana pipiens,* bottom.

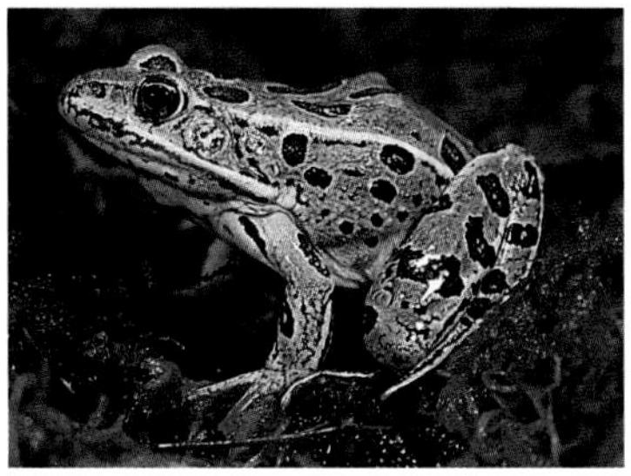

CRITICAL THINKING

Coevolution

Many plant species have coevolved with insect species, often to the point that one plant species is pollinated by only one insect species and one insect species pollinates only one plant species. Ask students what kind of species isolating mechanism this is, and ask them to explain their answers. (It is prezygotic because a plant cannot be pollinated with pollen from any species other than its own.)

RETEACHING ACTIVITY

Ask students to draw a diagram illustrating different isolating mechanisms. Have them first write "gametes" and draw pictures of an egg and sperm. Then have them draw a long arrow from the gametes to a "zygote," also illustrated, and a long arrow from the zygote to an "adult," also illustrated (e.g., a flower or an insect). Ask students to write names of prezygotic isolating mechanisms they have studied in this chapter beneath the first long arrow (different mating seasons, incompatible behavior). Ask them to write names of postzygotic isolating mechanisms they have studied beneath the second long arrow (failure of zygote to develop, infertility).

QUICK FACT

☑ Hybrid sterility is often caused by "complementary genes." A mutation in one gene of a complementary pair is not harmful. However, when two organisms that have each picked up a mutation in a different one of these genes mate, the combination is lethal. This is a type of postzygotic isolating mechanism that results in the formation of two new species.

RECENT RESEARCH

Big Changes from a Few Mutations

Scientists studying plants known as monkeyflowers, *Mimmulus* spp., found that only a few genes were responsible for determining large differences in flower color, size, and nectar quantity between two species. The flowers are so different that one is pollinated by a hummingbird and the other by a bumblebee.

ANSWERS TO SECTION 16-3 REVIEW

1. The distinguishing feature of a species, reproductive isolation from other species, cannot be tested for extinct species and does not apply to species that do not reproduce sexually.
2. When segments of a population become physically isolated, each segment adapts to its own environment. Eventually, the segments diverge so much genetically that their members cannot interbreed. Natural selection is the principal cause of speciation.
3. The offspring of two individuals of two different species may not develop normally or may die early. If the offspring reaches adulthood, it may be sterile.
4. Prezygotic isolation is less costly because an organism does not "waste" gametes producing offspring that fail to develop, die early, or are sterile.
5. Punctuated equilibrium is the hypothesis that evolution proceeds at an irregular rate, with short periods of rapid evolution during speciation followed by long periods during which no evolution occurs.
6. A very short generation time speeds up the rate of evolution and therefore can increase the rate of speciation.

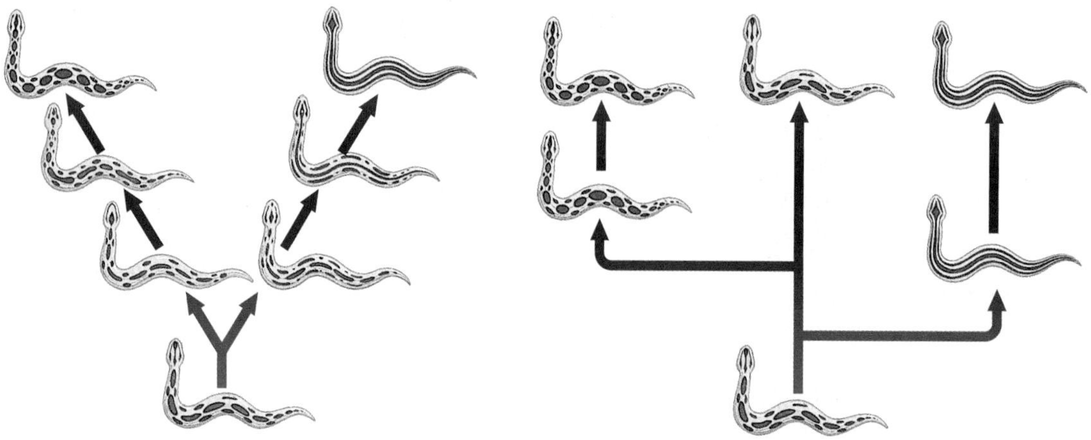

FIGURE 16-12
In the model of speciation presented on the left, species evolve gradually, at a stable rate. In the punctuated equilibrium model of speciation, illustrated on the right, species arise abruptly and are quite different from the root species. These species then change little over time.

internet**connect**
SCILINKS NSTA
TOPIC: Species formation
GO TO: www.scilinks.org
KEYWORD: HM312

RATES OF SPECIATION

Speciation sometimes requires millions of years. But apparently some species can form more rapidly. For example, Polynesians introduced banana trees to the Hawaiian Islands about a thousand years ago. Today, there are several species of moths that are unique to the Hawaiian Islands and that feed only on bananas. Because these species are closely related to other plant-eating moths in Hawaii, it seems likely that they have descended from ancestral moths during the past thousand years, since bananas were introduced to Hawaii.

Divergence of organisms and thus speciation may not occur smoothly and gradually. Indeed, the fossil record suggests that rapid speciation may be the norm rather than the exception. The fossil record seems to indicate that many species existed without change for long periods of time. The periods of stability were separated by an "instant" change in terms of geologic time. That is, a change occurred in a few thousand, rather than a few million, years. Scientists call this pattern of species formation **punctuated equilibrium.** The *punctuated* part of this term refers to the sudden shift in form that is often seen in the fossil record. Figure 16-12 shows two contrasting models, punctuated and gradual, of the evolution of two hypothetical species of snakes.

SECTION 16-3 REVIEW

1. What are two shortcomings of the biological species concept?
2. How can geographic isolation lead to speciation? What is the principal cause of such speciation?
3. Give two examples of postzygotic isolation mechanisms.
4. What is less metabolically costly to an animal, prezygotic or postzygotic isolation? Why?
5. What is the hypothesis of punctuated equilibrium?
6. CRITICAL THINKING What effect would a very short generation time, such as that of bacteria, have on speciation?

CHAPTER 16 REVIEW

SUMMARY/VOCABULARY

16-1

- Biologists study many different traits in populations, such as size and color.
- Traits vary and can be mapped along a bell curve, which shows that most individuals have average traits, while a few individuals have extreme traits.
- Variations in genotype arise by mutation, recombination, and the random fusion of gametes.
- The total genetic formation available in a population is called the gene pool.
- Allele frequencies in the gene pool do not change unless acted upon by certain forces.
- The Hardy-Weinberg genetic equilibrium, a theoretical model of a population in which no evolution occurs, tends to maintain the population as it is.

Vocabulary

allele frequency (300)
bell curve (299)
gene pool (300)
Hardy-Weinberg genetic equilibrium (302)
phenotype frequency (301)
population genetics (299)

16-2

- Evolution can take place if the genetic equilibrium of a population is disrupted.
- Immigration can bring new genes into a population, causing evolution.
- Nonrandom mating can alter the genotypes of a population, but it does not affect allele frequencies.
- Genetic drift operates in small populations; the contribution or lack of contribution of the genes of one or a few organisms can change the population's gene pool significantly.
- Stabilizing selection encourages the formation of average traits.
- Directional selection encourages the formation of more-extreme traits, such as a very long tongue in anteaters.
- Disruptive selection selects for extreme traits rather than average traits.
- In sexual selection, the development of traits that may seem harmful can actually enhance reproductive fitness if they encourage mating.

Vocabulary

assortative mating (306)
directional selection (307)
disruptive selection (307)
emigration (304)
gene flow (304)
genetic drift (305)
immigration (304)
sexual selection (308)
stabilizing selection (306)

16-3

- According to the biological species concept, a species is a population of organisms that can successfully interbreed and cannot breed with other groups.
- *Speciation* means species formation, and it always begins with a population that has become isolated.
- Geographic isolation results from the division of an original population.
- Reproductive isolation results from barriers to successful breeding. Prezygotic isolation occurs before fertilization. Postzygotic isolation occurs after fertilization and results in wasted gametes.
- Some scientists think that enormous phenotypic changes in species occur in sharp (punctuated) steps, rather than along a gradual curve, as Darwin proposed.

Vocabulary

biological species concept (310)
geographic isolation (310)
morphology (309)
postzygotic isolation (311)
prezygotic isolation (311)
punctuated equilibrium (312)
reproductive isolation (311)
speciation (309)

CHAPTER 16 REVIEW ANSWERS

REVIEW

1. Plotting the proportion of the population with each value of the trait gives a bell-shaped curve.
2. Allele frequency is the proportion of alleles in a population that are a particular type. Phenotype frequency is the fraction of the population with a particular phenotype.
3. Directional selection causes the distribution of traits in a population to move toward one extreme. Disruptive selection causes the distribution to move toward both extremes, reducing the abundance of intermediate phenotypes.
4. prezygotic isolation, postzygotic isolation
5. Punctuated equilibrium is an evolutionary pattern in which a species forms rapidly but then changes little, if any, for the remainder of its existence. Gradual evolution is the pattern in which species change slowly and constantly throughout their existence.

6. b
7. a
8. c
9. b
10. d
11. b
12. b
13. c
14. d
15. c

16. Variations can be influenced by the environment, by genes, or by both.
17. The necessary conditions for Hardy-Weinberg equilibrium are large population size (ideally infinitely large), no migration, random mating, no selection, and no net accumulation of mutations.
18. Gene flow is the introduction or removal of genes from a population by migration.
19. Immigrants may add alleles to a population, thus increasing the frequency of these alleles.
20. Nonrandom mating does not affect allele frequencies, but it does affect the frequencies of the different phenotypes—for example, it sometimes increases the frequency of individuals with rare diseases caused by recessive alleles.
21. This is stabilizing selection.
22. Disruptive selection can result in speciation.
23. Sexual selection is a form of natural selection whose driving force is success at attracting mates.
24. Prezygotic isolating mechanisms keep an organism from expending time, energy, and resources on offspring that will not be capable of reproduction.
25. Natural selection is the primary driving force for evolution.

REVIEW

Vocabulary

1. Explain why the term *bell curve* is appropriate for a graph of a normal distribution of traits.
2. Explain the relationship between allele frequency and phenotype frequency.
3. Distinguish between the terms *directional selection* and *disruptive selection.*
4. Name the two broad types of reproductive isolation.
5. Explain the difference between punctuated equilibrium and gradual evolution.

Multiple Choice

6. Phenotypic traits often vary between two extremes, with most individuals having an average version of the trait. This can be graphed as a (a) Punnet square (b) bell curve (c) straight line (d) genotype frequency table.
7. Variations in genotype arise by random fusion of gametes, mutation, and (a) recombination (b) translation (c) transcription (d) sorting by phenotype.
8. The total genetic information in a population is called the (a) allele frequency (b) phenotype frequency (c) gene pool (d) distribution of traits.
9. Saint Bernards and Chihuahuas (two breeds of domestic dogs) cannot mate normally owing to great differences in size. Thus, they are reproductively isolated to some extent. What type of isolating mechanism is operating here? (a) developmental (b) prezygotic (c) postzygotic (d) geographic.
10. If a population is in genetic equilibrium,
    (a) evolution is occurring
    (b) speciation is occurring
    (c) allele frequencies change from one generation to the next
    (d) allele frequencies remain the same from one generation to the next.
11. Mutations affect genetic equilibrium by (a) maintaining it (b) introducing new alleles (c) causing immigration (d) causing emigration.
12. Directional selection, disruptive selection, and stabilizing selection are all examples of (a) genetic equilibrium (b) natural selection (c) mutation (d) speciation.
13. The most common way for new species to form is through (a) mutation (b) stabilizing selection (c) geographic and reproductive isolation (d) genetic equilibrium.
14. The tendency for males to develop extreme versions of traits that appeal to females is a result of (a) random mating (b) speciation (c) reproductive isolation (d) sexual selection.
15. In the population of four o'clock flowers shown below, what is the allele frequency of the *R* allele? (a) 33% (b) 25% (c) 50% (d) 67%

Short Answer

16. What causes variations in the traits of organisms?
17. What conditions are necessary for Hardy-Weinberg equilibrium?
18. What is gene flow?
19. How can immigration alter allele frequencies in a population?
20. What results can be expected from nonrandom mating?
21. What type of selection is shown when the bell curve narrows over time?
22. What kind of selection can result in speciation?
23. What is the relationship between natural selection and sexual selection?
24. Why do prezygotic isolating mechanisms have an advantage over postzygotic isolating mechanisms?
25. What is the relationship between evolution and natural selection?

CRITICAL THINKING

1. Where populations of two related species of frogs overlap geographically, their mating calls are different. Where the species don't overlap, their calls are identical. What type of isolating mechanism is in operation?
2. Freeways may provide an effective geographic isolating mechanism for some slow-moving animals. Why are such artificial barriers not likely to result in complete speciation?
3. The most common definition of *species* states that a species is a group of organisms that can interbreed and produce fertile offspring in nature. A mule is a sterile offspring of a horse and a donkey. By the definition above, do a horse and a donkey belong to the same species? Explain your answer.
4. In the late nineteenth century, hunting reduced the population of the northern elephant seal to about 20 individuals. How might such a reduction in population have disrupted genetic equilibrium?
5. The graph below shows change in phenotype of two hypothetical species, A and B, over time. Use the graph to answer the following:
   a. What kind of evolution, punctuated or gradual, does the curve for species A represent?
   b. What kind of evolution does the curve for species B represent?
   c. Are the overall rates of change different for species A and B?
   d. What might have caused the vertical parts of the curve for species B?
   e. What do the horizontal parts of the curve for species B represent?

Phenotypic Change over Time in Two Species

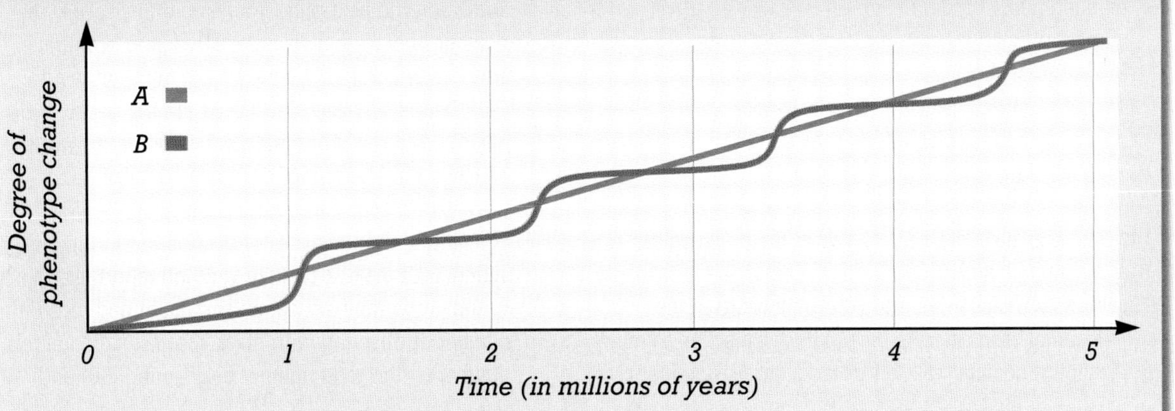

EXTENSION

1. Read "Evolving Backward" in *Discover,* September 1998, on page 64. Describe the two hypotheses that Dr. Diamond has developed to explain why humans have lost evolutionary traits, such as tails and body hair. Describe the eyes of a blind mole rat, and explain how the mole rat uses them.
2. Visit an area where plants or animals are bred. Possible places include farms, zoos, arboretums, seed companies, and nurseries. Find out how the breeders manipulate the genetic makeup of the plants or animals, and prepare an oral report on how this manipulation speeds up or slows down evolution. If you cannot visit one of the suggested locations, look in the *Readers' Guide to Periodical Literature* or use a CD-ROM-based index of periodic literature for articles on plant and animal breeding.

ANSWERS

CRITICAL THINKING

1. This is a prezygotic isolating mechanism—it prevents mating between the two species.
2. Such artificial barriers are unlikely to be in place long enough for speciation to occur.
3. A donkey and a horse belong to different species because their mating does not produce fertile offspring.
4. Some of the alleles present in the original elephant seal population may have been lost during this catastrophic die-off. Thus, the genetic diversity of the small population is almost certainly much lower than that of the original population.
5. a. gradual
b. punctuated
c. no
d. periods of rapid environmental change
e. periods of little phenotypic change

EXTENSION

1. Unrepaired mutations may occur in unused structures. These mutations have a neutral effect on the organism, but they can lead to the eventual loss of the unused structure. The loss of an unused structure, such as a tail, may free up space and energy that could be used for a useful new structure, such as a bigger brain. A blind mole rat has eyes that are buried under hair and skin and that are the size of a pinhead. The eyes have no irises or lenses, only a greatly reduced retina. The function of these remnants of eyes is to detect the daily photoperiod and the seasonal changes in day length.
2. Selective breeding generally speeds up the process of change within a species.

CHAPTER 16 INVESTIGATION

TIME REQUIRED

One 50-minute class period

QUICK REFERENCES

Lab Preparation notes are found on page 297B.

Holt BioSources provides a Teaching Resources CD-ROM that contains "Using Gowin's Vee in the Lab" and "Scoring Rubrics."

PROCEDURAL TIPS

1. Have students work in pairs, with one student drawing the beads while the other records the results. Students should switch roles after completing Part A.
2. To save time, have students create their tables before beginning the lab.
3. Remind students about the conditions under which the Hardy-Weinberg genetic equilibrium holds, and describe the forces that can disrupt this equilibrium.

ANSWERS TO BACKGROUND

1. Natural selection is a process by which organisms with favorable variations survive and reproduce at a higher rate.
2. Natural selection causes allele frequencies to change from one generation to the next, resulting in changes in the gene pool of a population.

ANSWERS TO PART A

3. Selecting beads without looking simulates random mating.

CHAPTER 16 INVESTIGATION

Predicting Allele Frequency

OBJECTIVES

- Demonstrate the effect of natural selection on genotype frequencies.

PROCESS SKILLS

- modeling
- predicting
- calculating
- analyzing

MATERIALS

- 200 black beads
- 200 white beads
- 3 containers
- labeling tape
- marking pen

Background

1. What is natural selection?
2. What is the result of natural selection?

PART A Random Mating

1. Obtain three containers, and label them "Parental," "Offspring," and "Dead."
2. Place 200 black beads and 200 white beads in the "Parental" container. Assume that each black bead represents a dominant allele for black coat *(B)* and that each white bead represents a recessive allele for white coat *(b)* in a hypothetical animal. Assume that the container holds gametes from a population of 200 of these hypothetical animals: 50 *BB,* 100 *Bb,* and 50 *bb.*
3. Without looking, remove two beads from the "Parental" container. What does this simulate?
4. In your lab report, make a data table like Table A below. Record the genotype and phenotype of the resulting offspring in your data table. Then put the alleles into the "Offspring" container.
5. Repeat steps 3 and 4 forty-nine times. Record the genotype and phenotype of each offspring in your data table.
6. Calculate the frequencies of alleles in the offspring. First make a table in your lab report like Table B shown on the next page. Then count and record the number of black beads in the "Offspring" container. This number divided by the total number of beads (100) and multiplied by 100% is the frequency of *B* alleles. Then count and record the number of white beads in the "Offspring" container. Determine the frequency of *b* alleles as you did with the *B* alleles.
7. Calculate the frequencies of phenotypes in the offspring. First make a table in your lab report like Table C, shown on the next page. Then count and record the number of offspring with black coat color. Divide this number by the total number of offspring (50) and

TABLE A *MATING*

Trial	Random mating		Nonrandom mating	
	Offspring genotype	**Offspring phenotype**	**Offspring genotype**	**Offspring phenotype**
1	Answers will vary.	Answers will depend on genotype.	Answers will vary.	Answers will depend on genotype.
2				
3				
4				
5				

TABLE B *ALLELE FREQUENCIES*

Generation	Number of *B* alleles	*B* alleles/ total alleles	*B* allele frequency	Number of *b* alleles	*b* alleles/ total alleles	*b* allele frequency
Parental	200	200/400	50%	200	200/400	50%
Offspring	Answers will vary.	Answers will vary.	Answers should approximate 50%.	Answers will vary.	Answers will vary.	Answers should approximate 50%.

TABLE C *PHENOTYPE FREQUENCIES*

Generation	Number of black animals	Black animals/ total animals	Frequency of black animals	Number of white animals	White animals/ total animals	Frequency of white animals
Parental	100	150/200	75%	50	50/200	25%
Offspring	Answers will vary.	Answers will vary.	Answers will vary.	Answers will vary.	Answers will vary.	Answers will vary.

multiply by 100% to determine the frequency of black phenotype. Repeat this calculation to determine the frequency of white coat color among the offspring.

PART B Nonrandom Mating

8. Return the beads in the "Offspring" container to the container labeled "Parental."
9. Assume that animals with white-coat phenotype are incapable of reproducing. What is the genotype of animals with a white coat? To simulate this situation, remove 100 white beads from the container labeled "Parental," and set them aside.
10. Start by removing two beads from the container labeled "Parental," and record the results in your Table A. If the offspring has a white-coat phenotype, put its alleles in the container labeled "Dead." If the offspring has a black-coat phenotype, put its alleles in the container labeled "Offspring."
11. If animals with a white-coat phenotype cannot reproduce, predict what would happen to allele frequency if step 10 were repeated until the parental gene pool was empty. Write your prediction in your lab report.
12. Repeat step 10 until the parental gene pool is empty. Record the results of each pairing in your lab report in Table A. Compare your results with your prediction.
13. Transfer the beads from the "Offspring" container to the "Parental" container. Leave the beads that you have placed in the "Dead" container in that container. Do not return those beads to the parental container.
14. Repeat step 10 again until the parental pool is empty. Record your results in your data table.
15. Repeat steps 13 and 14 two more times.
16. Calculate the frequencies of the final genotypes produced, as you did in Part A. Compare the results with your prediction from step 11.
17.  Clean up your materials before leaving the lab.

Analysis and Conclusions

1. Compare the frequency of recessive alleles produced in Part A with that produced in Part B. Did you correctly predict the frequencies?
2. Did the frequency of the *b* allele change uniformly through all generations? If not, what happened?
3. Why did you remove 100 white beads from the "Parental" container in Step 9?
4. How did this change the phenotype frequency of white animals in the parental generation from the original ratio of 50/200?

Further Inquiry

If you continued Part B, would you eventually eliminate the *b* allele? Form a hypothesis and test it.

CHAPTER 16 INVESTIGATION

ANSWERS TO PART B

9. The genotype of white-coated animals is *bb*.

ANSWERS TO ANALYSIS AND CONCLUSIONS

1. The recessive allele declined in frequency in Part B but not in Part A. Students should state whether their data accorded with their predictions.
2. No, the decrease in the frequency of the *b* allele began to slow in later generations.
3. Animals with white coats (genotype *bb*) cannot reproduce and are not included in the parental generation.
4. It reduced the number of white parental animals to zero.

CHAPTER 17
HUMAN EVOLUTION

Block	TOPICS	TEACHING RESOURCES	LABS, CLASSWORK, AND HOMEWORK
BLOCK 1 45 minutes	**Introducing the Chapter, p. 318**	**ATE** Focus Concept, p. 318 **ATE** Assessing Prior Knowledge, p. 318 **ATE** Understanding the Visual, p. 318	■ Supplemental Reading Guide, *Origin of Species* ■ Supplemental Reading Guide,—*Through a Window: My Thirty Years with the Chimpanzees of Gombe*
	17-1 The Study of Human Origins, p. 319 The Hominid Fossil Record, p. 319 Primate Characteristics, p. 320	**ATE** Section Overview, p. 319 53, 54	**Lab A11,** Comparing Primate Features **Lab C17,** Analyzing Blood Serum to Determine Evolutionary Relationships ★ Study Guide, Section 17-1 **PE** Section Review, p. 321
BLOCKS 2 & 3 90 minutes	**17-2 Fossil Evidence of Hominid Evolution, p. 324** The First Hominids, p. 324 *Australopithecus,* p. 324	**ATE** Section Overview, p. 324 **ATE** Critical Thinking, p. 324 54, 55, 57	**Lab C18,** Analyzing Blood Serum—Evolution of Primates ★ Study Guide, Section 17-2 **PE** Section Review, p. 326
BLOCK 4 45 minutes	**17-3 Hypotheses of Hominid Evolution, p. 327** Early Members of *Homo*, the Human Genus, p. 327 *Homo sapiens,* p. 328 Spread of Modern Humans, p. 329	**ATE** Section Overview, p. 327 **ATE** Critical Thinking, p. 329 62, 64	**PE** **Chapter 17 Investigation,** pp. 334–335 **PE** **Quick Lab,** p. 329, Comparing Cranial Capacities ★ Study Guide, Section 17-3 **PE** Section Review, p. 330
BLOCKS 5 & 6 90 minutes	**REVIEW and ASSESSMENT** **PE** Chapter 17 Review, pp. 332–333	Audio CD Program	**CHAPTER TESTING** ■ Chapter Test (blackline master) ▲ Test Generator ■ Assessment Item Listing ■ Scoring Rubrics and Classroom Management Checklists

MODERN BIOLOGY | HOLT BIOSOURCES

KEY

- PE Pupil's Edition
- ATE Teacher's Edition
- ■ Active Reading Guide
- ★ Study Guide
- Teaching Transparencies
- Laboratory Program

One-Stop Planner CD-ROM
Shows these resources within a customizable daily lesson plan:

- ■ Holt BioSources Teaching Resources
- ■ Active Reading Guide
- ★ Study Guide
- ▲ Test Generator

READING FOR CONTENT MASTERY

Preview: Have students preview the objectives, the introductory paragraph, topic sentences, and captions before reading the chapter.

ATE Active Reading Technique: Brainstorming, p. 319
■ Active Reading Guide Worksheet 17-1
PE Summary/Vocabulary, p. 331

■ Active Reading Guide Worksheet 17-2
PE Summary/Vocabulary, p. 331

■ Active Reading Guide Worksheet 17-3
PE Summary/Vocabulary, p. 331

TECHNOLOGY AND INTERNET RESOURCES

Holt Biology Videodiscs Teacher's Correlation Guide, Lessons 17-1 through 17-2

Audio CD Program, Sections 17-1 through 17-3

internet**connect**

sciLINKS NSTA

On-line Resources: www.scilinks.org
The following *sci*LINKS Internet resources can be found in the student text for this chapter:

Topics:
- Fossil evidence of hominid evolution, p. 324
- *Homo erectus*, p. 327

go.hrw.com

On-line Resources: go.hrw.com
Visit the HRW Web site for a variety of resources related to this chapter. Just type in the keyword HM2 Chapter 17.

Smithsonian Institution®
Visit **www.si.edu/hrw** for additional on-line resources.

CNNfyi.com
Visit **www.cnnfyi.com** for late-breaking news and current event stories selected just for you.

LAB ACTIVITY PLANNING

CHAPTER 17 INVESTIGATION

Relating Amino Acid Sequences to Evolutionary Relationships, pp. 334–335

OVERVIEW
In this investigation, students will observe the amino acid sequences of two proteins found in vertebrates, cytochrome c and hemoglobin. Students will then use the degree of similarity between amino acid sequences in these homologous proteins to infer evolutionary relationships between species. In Part A, students will examine the amino acid sequence of cytochrome *c* in nine vertebrate species to determine the degree of similarity with human cytochrome *c*. In Part B, students will examine the amino acid sequence for hemoglobin in four vertebrate species to determine the degree of similarity with human hemoglobin.

PREPARATION
Each student will need one sheet of paper and a pencil.

Quick Labs

Comparing Cranial Capacities, p. 329
Each student or student group will need a calculator, four 2 L plastic soda bottles, a graduated cylinder, water, paper, pencil, and a marking pen.

UNDERSTANDING THE VISUAL

Ask students to examine the picture of the field laboratory work table. Ask them what they recognize among the items shown, and then ask them to speculate about what the scientists were doing. (There are several jawbones [these are casts of actual fossils], which the scientists used for comparison with a new fossil find.) Ask students why the teeth of fossil organisms are often well preserved, while the bones and other body parts may not be. (Teeth are covered by enamel, which is a very hard, durable material.)

FOCUS CONCEPT

Evolution

The organisms alive on Earth today, including humans, evolved from other organisms. During their evolution, they acquired characteristics that distinguish them from their ancestors and from other organisms alive today. Scientists rely heavily on careful interpretation of fossil evidence in making hypotheses about human evolution.

ASSESSING PRIOR KNOWLEDGE

Review the following concepts.

Interpreting Fossils: *Chapter 14*
Ask students how a scientist would study a fossil to assess the phylogenetic relationships of the fossilized organism to other organisms, both extinct and modern.

Speciation: *Chapter 16*
Ask students how new species form. Ask how scientists determine that a new species has been formed.

CHAPTER 17

HUMAN EVOLUTION

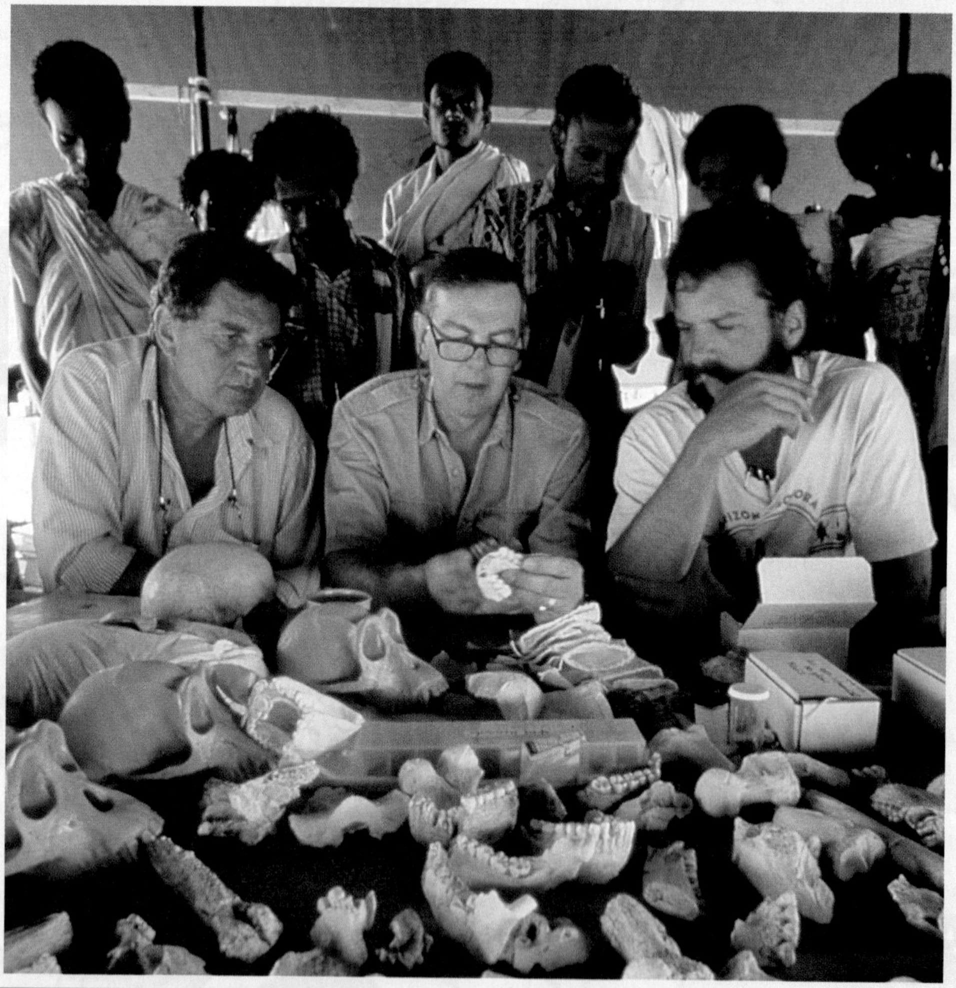

Scientists Yoel Rak (seated at left), Donald Johanson (center), and William Kimbel (right), together with a group of Afar tribesmen, examine an early hominid jawbone at a field laboratory at Hadar, Ethiopia.

FOCUS CONCEPT: *Evolution*

As you read, pay attention to how scientific methods are used to formulate conclusions about human origins.

THE STUDY OF HUMAN ORIGINS

To understand the story of human evolution, we must understand both our ancestry and our relationship to our closest living kin. Humans are members of the ancient mammalian order Primates. Primates have grasping hands, acute vision, and large brains. Primate parents provide extended periods of intense care for their young, and many primate species live in complex social groups. As you will see, many of our behaviors and characteristics are similar to those of other primates, and some are uniquely human.

SECTION 17-1

OBJECTIVES

▲ Describe how paleoanthropologists gather evidence of human ancestry.

● List some traits shared by all primates.

■ Name two distinguishing characteristics of anthropoids.

◆ Give examples of traits unique to humans.

THE HOMINID FOSSIL RECORD

Scientists who study fossil evidence of human evolution are called **paleoanthropologists** (PAY-lee-oh-AN-thro-PAHL-uh-jists) or biological anthropologists. Just as detectives try to solve mysteries, paleoanthropologists piece together an assortment of clues to construct models of how and when different stages of human evolution occurred. Much of the information available about human evolution comes from the fossilized bones of early **hominids** (HAHM-uh-nidz), a group that comprises humans and their immediate ancestors. Fossilized hominid remains are seldom complete skeletons—often only fragments of fossilized bone are found. Scientists pay close attention to subtle clues in these fossils. For example, the curvature of the spine, the position at which the spine attaches to the skull, and the shape of the **pelvis,** or hipbones, can indicate whether an organism walked upright. Similarly, a skull fragment can be used to estimate brain size, and wear on a fossil tooth can give some indication of an organism's diet.

Often the immediate surroundings of a fossil give important clues to how the species lived. Sometimes stone tools and the bones of prey are found with the fossil, and the geologic stratum in which the fossil is found can give the approximate age of the fossil. Other information, such as climate, forestation, and food sources prevalent at the time the fossil species lived, can sometimes be determined from traces of plant remains and pollen grains.

Word Roots and Origins

hominid

from the Latin *homo,* meaning "human being," and the Greek *-ides,* meaning "a thing belonging to"

SECTION 17-1

SECTION OVERVIEW

Evolution
Section 17-1 describes the importance of the fossil record in determining human origins.

Reproduction and Inheritance
Section 17-1 also describes the characteristics of primates, focusing on anthropoid primates. Features of modern humans are compared with features of modern apes and the concept of a common ancestry is introduced.

ACTIVE READING

Technique: Brainstorming
Before students begin reading, explain that the focus of this chapter is the evolution of humans. Work with the class to develop a list of characteristics that humans share with other living primates and characteristics that are unique to humans. Incorporate amendments that students suggest as they read the chapter.

INCLUSION ACTIVITY

To help students realize the advantages of an opposable thumb, have them restrict the use of their thumbs. Provide masking tape to secure their thumbs against their forefingers. Provide various objects for them to try to hold. Have them think of how they would perform everyday activities, such as eating an apple. Have them try to write something and fold a piece of paper. Ask them if they could get dressed in the morning or button their shirts.

RECENT RESEARCH

Anthropoid Locomotion

D. M. Doran, a scientist investigating locomotion in gorillas and chimpanzees, found that both species frequently walk bipedally, climb, and hang from their arms when they are juveniles. As the animals mature, these behaviors decrease and quadrupedal behavior increases. Doran also noted that the locomotive behavior of gorillas and chimpanzees is very similar despite their very different sizes.

QUICK FACT

☑ All New World monkeys (those that live in Central and South America) are arboreal. They have tails adapted for grasping, and the end of the tail resembles the palm of a hand. Old World (African and Asian) monkeys have tails that are not prehensile and some, such as various species of baboons and the Patas monkey of the central African plains, are primarily terrestrial.

FIGURE 17-1

The grasping fingers and toes and front-facing eyes of the tarsier, *Tarsius* sp., are primate characteristics that serve well for a life in the trees. The tarsier, like many prosimian primates, is nocturnal.

FIGURE 17-2

Mobile arm-and-shoulder anatomy allows anthropoids such as this gibbon, *Hylobates lar*, to swing by their arms through trees.

PRIMATE CHARACTERISTICS

Hominids belong to the order of mammals known as **primates.** Two large divisions of modern primates are recognized. The **anthropoid** (AN-thruh-POID) **primates** include marmosets, monkeys, apes, and humans. **Prosimian** (proh-SIM-ee-uhn) **primates,** many of which resemble very early primate forms, include lemurs, lorises, and tarsiers, like the one shown in Figure 17-1. Fossils of extinct primates reveal that the majority of them lived in trees, as do most modern species. Many of the characteristics that primates share apparently evolved as adaptations to life in trees.

Primates have movable fingers and toes, and most have flattened nails rather than claws. The hands and, in some species, the feet are **prehensile** (pree-HEN-sil), or grasping. Unlike most mammals, primates have color vision. This may have arisen when primates became more active during the day than at night, a change that occurred about 60 million years ago. The front-facing eyes found in primates result in broadly overlapping fields of vision. This allows primates to perceive depth—a useful trait for an animal that moves by swinging or jumping from branch to branch in trees.

Characteristics of Anthropoids

Anthropoid primates, such as the gibbon shown in Figure 17-2, have a well-developed collarbone, rotating shoulder joints, and partially rotating elbow joints. Anthropoids also have an **opposable thumb**—a thumb that can be positioned opposite the other fingers. This arrangement of fingers results in increased precision in the use of the hands. Additionally, nonhuman anthropoids have an opposable big toe, as seen on the chimpanzee in Figure 17-3, and this prehensile foot is an important aid to climbing.

All anthropoids have a similar **dental formula,** or number and arrangement of teeth. In humans, apes, and African and Asian monkeys, each half of the upper and lower dental arches includes two incisors, one canine, two premolars, and three molars, as shown in Figure 17-3.

Compared with other primates, anthropoids have a large brain relative to their body size. The fossil record shows that as primates evolved, brain size increased. Humans and the **great apes** (gibbons, orangutans, gorillas, and chimpanzees) have a larger cranial capacity relative to body size and a more complex brain structure than other primates have.

Of the anthropoid species, the chimpanzees may be the most closely related to humans. Comparisons of chimpanzee and human DNA have shown a very high degree of similarity. This similarity suggests that humans and chimpanzees may have shared an ancestor less than 6 million years ago. It is important to understand, however, that humans are not descended from chimpanzees or from any other modern ape. Rather, modern apes and humans are probably descended from a more primitive apelike ancestor.

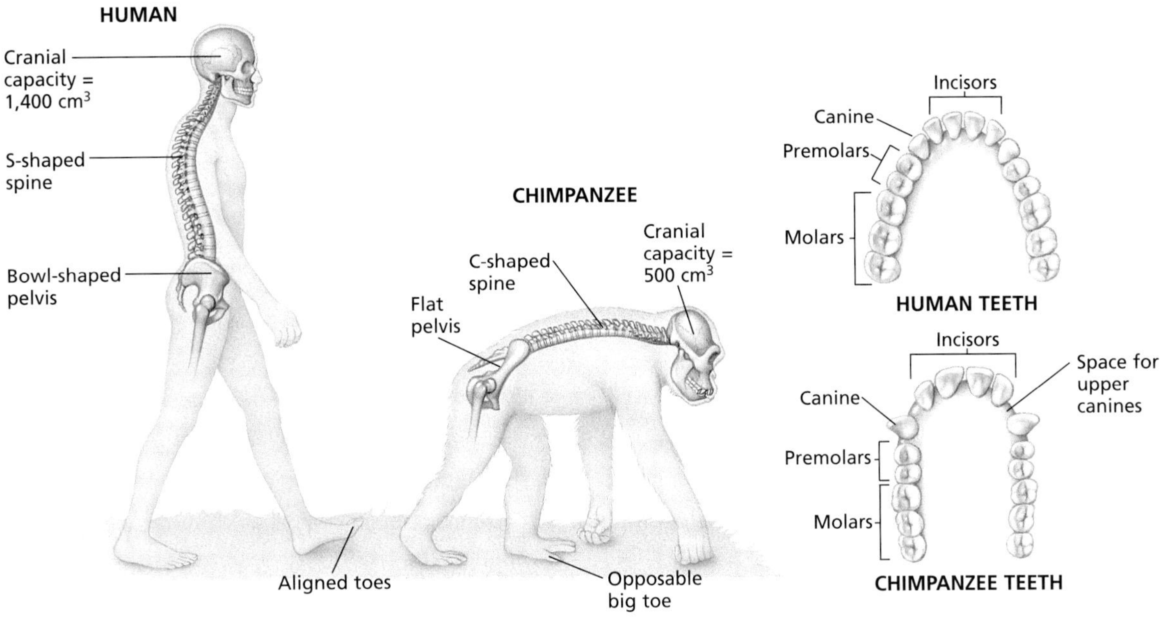

FIGURE 17-3

Some human physical characteristics differ markedly from those of the chimpanzee, a modern ape. The human jaw is rounder than the U-shaped jaw of the ape. The human pelvis is cup-shaped, compared with the flatter ape pelvis. The human spine is S-shaped, compared with the single curve of the ape spine. The human foot has short, aligned toes, compared with the longer, grasping toes—and the opposable big toe—of the ape.

Characteristics of Humans

Bipedalism (bie-PEED-uhl-iz-uhm), the ability to walk primarily on two legs, is a uniquely human trait among mammals. Figure 17-3 shows that the cup-shaped human pelvis supports the internal organs during upright walking. The human spine has two curves, resulting in an S shape that allows for upright posture.

In the human foot, the toes are much shorter than those of apes and are aligned with each other. Because humans are the only primates that have this foot structure, we can infer that the shape of the human foot is a specific adaptation for bipedalism.

The enlargement of the brain in humans has resulted in a more vertical face than that found in apes. Among other differences, the larger human brain has extensive areas devoted to the production and understanding of speech. Apes have homologous areas in their brains that are important in the production of sounds used in communication, and apes can also be taught to mimic certain forms of sign language. However, apes living in the wild have not developed any complex, flexible set of signals that can compare to those that make up the languages of humans.

SECTION 17-1 REVIEW

1. What do paleoanthropologists study, and how do they gather their information?
2. What might a paleoanthropologist infer from the surroundings of a fossilized hominid?
3. Name two characteristics of all primates, and explain how these characteristics appear to be adaptations to life in the trees.
4. What features distinguish anthropoids from the other primates?
5. What are two specifically human traits?
6. **CRITICAL THINKING** How might the acquisition of language account for the very fast cultural and intellectual development that has occurred in the evolution of humans?

SECTION 17-1

GIFTED *ACTIVITY*

Have students conduct research and prepare a report on the ability of chimpanzees and gorillas to learn and use sign language. Ask them to include such information as the number of words that these apes have been taught, whether the animals can formulate new words by combining others, and the limitations of language use in these animals.

ANSWERS TO SECTION 17-1 REVIEW

1. Paleoanthropologists study evidence of human evolution by studying the fossils of early hominids.
2. A paleoanthropologist might infer the age of the fossil from the geologic stratum it is found in. The presence of tools or bones of prey indicate how the fossil species lived; plant remains can indicate climate, forestation, and food sources.
3. Primates have prehensile hands and (except for humans) prehensile feet. Primates have color vision and front-facing eyes, which give them depth perception.
4. Anthropoids have mobile shoulder-and-arm anatomy, an opposable thumb, and a large brain. Most have a dental formula that includes two incisors, one canine, two premolars, and three molars.
5. Bipedalism and the use of language are traits unique to humans.
6. In social groups with language, information can be passed on and accumulated, which significantly speeds up cultural and intellectual development.

BACKGROUND

This feature profiles the Leakeys, a family of prominent paleoanthropologists who have collectively made many important discoveries of fossil evidence of early hominids. Louis S. B. Leakey was born in 1903 in Kabete, Kenya, and grew up in Kenya. He was educated at the University of Cambridge, where he earned a doctorate in anthropology. He began his research in East Africa and Olduvai Gorge in 1931. In 1936, he met and married English-born archaeologist and paleoanthropologist Mary Douglas Nicol.

Over the next half-century, the Leakeys and their sons supervised the excavation of various archaeological sites in Africa. Their discoveries and research have formed the basis for subsequent research into the origins of human life. The first of these discoveries was made by Mary in 1947. She discovered the skull of an apelike animal that may have been ancestral to both apes and early humans, *Proconsul africanus.* In 1959, she discovered the skull of an early hominid, *Zinjanthropus* (since reclassified as *Australopithecus*). From 1960 to 1963, the Leakey family and others found hominid-fossil remains at Olduvai Gorge sites that were between 2.5 and 1.5 million years old. Hand-bone discoveries suggest that these hominids, called *Homo habilis,* were capable of precisely manipulating objects. *H. habilis* is generally accepted as the earliest member of the genus *Homo.*

Later fossil finds by Richard Leakey include an entire skeleton of *Homo erectus,* a species that lived about 1.8 million to 50,000 years ago. *H. erectus* appears to have dispersed widely from Africa.

Currently, a major question in the study of human evolution is how and where *Homo erectus* evolved into early *Homo sapiens.*

Africa: Cradle of Humanity

HISTORICAL PERSPECTIVE

In 1871, Charles Darwin predicted that the remains of human ancestors would be found in Africa because our closest living relatives—the great apes—are found there today. But Darwin may have been in the minority. Before 1925, many scientists believed that the first humans had evolved in Europe or Asia no more than 200,000 years ago. Kenyan-born British archaeologist Louis Leakey was firmly convinced that he would find evidence to support Darwin's prediction. Louis Leakey, working together with his wife, Mary Leakey, compiled substantial evidence that Africa, not Europe or Asia, was the birthplace of humans. Moreover, the Leakeys' many fossil discoveries suggested that the first hominids were far older than previously believed. The Leakeys were drawn to Olduvai Gorge, where they carried out much of their lifework, by the abundance of primitive stone tools that had been found in the vicinity.

A Gorgeous Site

Olduvai Gorge is a steep-sided ravine in northern Tanzania. It lies in the East African Rift Valley, an area extending from the Red Sea to Mozambique. Two million years ago, the site of the gorge was occupied by a lake, which later filled with layers of sediment and volcanic ash that were ideal conditions for fossil preservation. Layering of sediment continued for more than 2 million years, resulting in seven major layers, referred to as "beds." Cross sections of these seven beds, exposed by the erosion of the river that cut the gorge, have yielded an extraordinarily rich record of the animal and plant life that existed in the area from 2.1 million to 15,000 years ago.

Mary Leakey vividly described the landscape of Olduvai in her 1984 autobiography, *Disclosing the Past:*

Louis and Mary Leakey

> *As one comes over the shoulder of the volcanic highlands to start the steep descent...suddenly one sees the Serengeti, the plains stretching away to the horizon like the sea....Away to the right are Precambrian outcrops and an almost moon-like landscape. To the left, the great slopes of the extinct volcano Lemagrut dominate.*

Together the Leakeys excavated Olduvai Gorge and found that it yielded more information about human ancestors than any other site that had been found.

Uncovering Africa's Secrets

In July of 1959, after years of searching at Olduvai Gorge, Mary Leakey discovered a hominid skull, leg bone, jaw, and tooth. The skull was in fragments but was otherwise well preserved. After the Leakeys spent many hours patiently fitting the fragments together, the skull took shape. The teeth were large, and a massive bony ridge dominated the top of the skull, indicating that the hominid had very strong neck and jaw muscles. Most important, the hominid bones, later called *Australopithecus boisei,* were accompanied by stone artifacts and animal bones. The task that lay ahead was to determine their age.

To date the skull, the Leakeys used a relatively new technique that

Louis and Mary Leakey were first drawn to Tanzania's Olduvai Gorge by the abundance of primitive stone tools that had been found in the vicinity.

measures the ratio of potassium to argon in volcanic rocks. This method determines age by measuring the radioactive decay of isotopes of potassium with sophisticated equipment. With this technique, the skull revealed that the origin of humans dated back at least 1.75 million years.

In 1960, the Leakeys discovered fossils of a hominid species that dated as far back as 2 million years. Details of the hominid's anatomy, such as tooth characteristics and size and shape of the skull, jaw, and leg bones, were quite different from those of the earlier fossils—they were much more like those of modern humans. The species was relatively large-brained and was probably a tool user. The species was therefore named *Homo habilis,* or "handy human."

Because some of the *Homo habilis* fossils came from the same rock layers as earlier finds, it appeared that this new species may have been a contemporary of *Australopithecus.* This was an important discovery because it implies that the course of human evolution was not necessarily an orderly progression, with one species giving rise to the next. Rather, the family tree of modern humans probably has many branches. Several species may have lived at once. Most early species probably died out without leaving descendants. Modern humans probably arose from a single ancestral line that successfully adapted and evolved while its cousins perished. *Homo habilis* was accepted as an early ancestor of modern humans.

In 1978, Mary Leakey made the dramatic discovery of several sets of bipedal footprints in volcanic ash at Laetoli, a site located south of Olduvai Gorge. When the ash was analyzed, the footprints were dated at 3.6 million years old, clearly demonstrating that human ancestors became bipedal very early.

The Research Continues

The Leakeys' early work formed the basis for much of the research into human origins that continues today. In addition to paleoanthropology, Louis Leakey had a lively interest in modern primates. He directly influenced Jane Goodall's research on chimpanzee behavior, Birute Galdikas Brindamour's research on orangutan behavior, and Dian Fossey's research on mountain gorilla behavior.

Louis and Mary Leakey's son Richard Leakey and his wife, Meave Leakey, are still actively excavating sites in eastern Africa. In 1984, at Lake Turkana, near Kenya's border with Ethiopia, Richard unearthed a skull and later an entire skeleton that at the time were the earliest known specimens of *Homo erectus,* the species that directly preceded *Homo sapiens.* In 1996, Meave Leakey and her team announced the discovery of the oldest known hominid species—*Australopithecus anamensis.* These discoveries underscore the elder Leakeys' contention that Africa is the ancestral home of the human family, the place where humans first became tool users, and ultimately is the cradle of humanity.

Years of painstaking work by the Leakeys produced a great variety of hominid fossils from several different periods.

CHAPTER 17
GREAT DISCOVERIES

The Further Readings selections discuss two models of the evolution of modern humans from *H. erectus.*

DISCUSSION

Guide the discussion by posing the following questions.

1. What makes Olduvai Gorge unique as an archaeological site? (The seven beds of sediment within the gorge were laid down over the course of 2 million years. The cross section of the beds, exposed by the erosion of the river that cut the gorge, are rich with animal fossils.)

2. What are the major differences between *Australopithecus* and *Homo habilis*? (The teeth, skull, jaw, and leg bones of *H. habilis* were more humanlike than those of *Australopithecus.* Moreover, *H. habilis* was relatively large-brained and was probably a tool user.

3. In his later years, Louis Leakey became very interested in nonhuman primate behavior. What work in modern primatology did he influence? (He directly influenced Jane Goodall's research on chimpanzee behavior, Birute Galdikas Brindamour's research on orangutan behavior, and Dian Fossey's research on mountain gorilla behavior.)

FURTHER READINGS

1. "Unusual Unity," *Natural History,* April 1997, pp. 20–23, 69–71. In this essay, Stephen Jay Gould discusses five discoveries and arguments that support one of the two proposed models of human evolution.

2. "The African Emergence and Early Asian Dispersal of the Genus *Homo,*" *American Scientist,* November–December 1996, pp. 538–551. This article discusses the driving force behind the dispersal from Africa of early hominids almost 1 million years ago.

SECTION 17-2

SECTION OVERVIEW

Evolution

Section 17-2 enumerates major hominid fossil finds and describes the traits of the earliest hominid genus, *Australopithecus.* It also discusses the possibility that a recent find, *Ardipithecus,* may have been bipedal and thus a root species of the hominid line.

▶ ENGAGE STUDENTS

Hominid brain sizes are mentioned throughout the last two sections of this chapter, and increasing brain size is an important trend in human evolution. To help students appreciate how much hominid brain size has increased, label six 2,000 mL beakers with the names of the species discussed in this chapter, and add water to each beaker as follows: *A. afarensis* (Lucy)—490 mL, *A. robustus/A. boisei*—525 mL, *H. habilis*—700 mL, *H. erectus*—975 mL, *H. sapiens* (modern)—1400 mL, and *H. sapiens* (Neanderthal)—1450 mL. (Add a drop of food coloring to each beaker to make the water level easier to see.) Arrange the beakers in the order that these hominids evolved, and tell students that the water levels represent the different brain sizes of the hominids. Point out the steady increase of brain size (with the puzzling exception of the Neanderthals).

CRITICAL THINKING

Bipedalism

What are some advantages of bipedalism over quadrupedalism? (The forelimbs are freed for a variety of activities, such as carrying and manipulating objects, carrying offspring who cannot yet walk, and defending oneself.)

SECTION

17-2

OBJECTIVES

▲ Explain how the discovery of Lucy changed hypotheses about the evolution of bipedalism.

● Explain the significance of finding fossils of hominids that are not ancestral to modern humans.

■ List the fossil finds of 1995, and discuss their significance regarding the evolution of bipedalism in hominids.

FOSSIL EVIDENCE OF HOMINID EVOLUTION

Scientists who study the fossil remains of early hominids have inferred the evolutionary trends toward a larger brain and bipedalism. The fossils of hominids, unlike those of apes and their ancestors, show a whole spectrum of unique adaptations for upright walking.

THE FIRST HOMINIDS

Bipedalism is the principal trait that defines the hominid line. Modern nonhuman anthropoid primates are **quadrupedal** (kwah-DROO-pi-duhl); that is, they walk on four limbs. The apelike ancestors of the first hominids likely were quadrupedal as well. How long ago did the first bipedal primate, that is, the first hominid, evolve? This question has not yet been answered conclusively.

In 1974, a 3.2-million-year-old fossil of a primate was found in the Afar Valley region of eastern Africa by Donald Johanson, shown in the photograph on the first page of this chapter, and his colleagues. The fossil, shown in Figure 17-4, is unusually well preserved. The cranial capacity, which is used as an approximation of brain size, is about equal to that of a chimpanzee (475 cm^3), or about one-third that of a modern human (1,400 cm^3). The fossil primate's height ranged from 1 to 1.5 m, probably varying according to gender. The pelvis and leg bones, however, clearly indicated that the fossil organism was an upright-walking hominid.

This find changed many ideas about the evolution of humans. It was generally thought that the hallmarks of hominids—bipedalism and a large brain with areas dedicated to higher reasoning and the production of speech—had all evolved at the same time. But the new fossil showed that upright walking had apparently come before many of these other adaptations that make the hominids unique among the anthropoid primates.

Word Roots and Origins

quadrupedal

from the Latin *quattour,* meaning "four," and *pes,* meaning "foot"

internet**connect**

SCILINKS NSTA

TOPIC: Fossil evidence of hominid evolution
GO TO: www.scilinks.org
KEYWORD: HM324

AUSTRALOPITHECUS

The new fossil find was given the species name *Australopithecus afarensis* (abbreviated *A. afarensis*), which means "southern ape of the Afar Valley." Unofficially this female fossil is called **Lucy.** Since 1974, other specimens of the same species have been discovered. They date from about 3 million to 3.9 million years ago.

Other Australopithecines

A number of other fossils that are similar to *A. afarensis* (Lucy) have been discovered from time to time. All of these finds have now been designated **australopithecines** (aw-STRAY-loh-PITH-uh-seenz)—organisms from the genus *Australopithecus. Australopithecus africanus,* which dates from about 2.3 to 3 million years ago, probably descended from *A. afarensis. A. africanus* was taller and heavier than Lucy and had a slightly larger cranial capacity, between 430 and 550 cm^3.

Two more-recent species, *Australopithecus robustus,* shown in Figure 17-5, and *Australopithecus boisei,* date from about 1 to 2.6 million years ago. These species had heavier skulls and larger teeth than *A. afarensis.* The cranial capacity in *A. robustus* and *A. boisei* ranged from 450 to 600 cm^3. The general appearance of these other australopithecines suggests that while they may have descended from *A. afarensis* (Lucy's species), they were probably not ancestral to modern humans.

In the past two decades, fossil finds of hominids have increased dramatically due to increased research on human origins. In 1995, Meave Leakey and her colleagues at the National Museums of Kenya announced the discovery of a new fossil representing a species distinct from and older than *A. afarensis.* This species, named *Australopithecus anamensis,* is 300,000 years older than any hominid fossil previously found. While the head and neck of *A. anamensis* share several features with modern chimpanzees, a shinbone found at the site indicates that *A. anamensis* was bipedal. Fossil evidence of very early bipedal primates is not limited to fossilized bones. Fossilized footprints dating back 3.6 million years further confirm that bipedalism occurred very early, before—and not as a result of—the rapid enlargement of the hominid brain.

FIGURE 17-4

The original fossil find of *Australopithecus afarensis* consisted of a partial skeleton. The fossil was given the nickname Lucy by the team of investigators who discovered it.

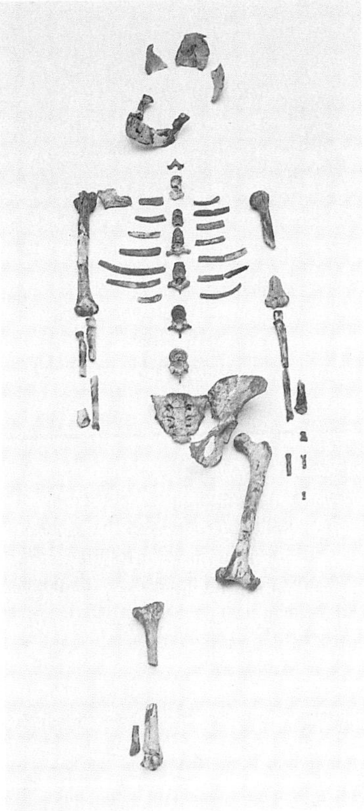

FIGURE 17-5

This well-preserved *Australopithecus robustus* skull shows the heavy bone structure and large teeth characteristic of the species.

RECENT RESEARCH

Savanna Hypothesis

An old hypothesis that bipedalism arose in response to an ecological shift to grassland dominance has not been supported by recent evidence. Scientists studying areas in Kenya that are rich in ape and hominid remains have analyzed the carbon isotope signatures of area soils. Their findings indicate no evidence of the sudden shift to grasslands that had been assumed for many years. *A. afarensis* (Lucy) and related fossils dating from 3.4 million years ago have curved fingers similar to those of arboreal apes and have been found in ancient grasslands and woodland areas. *Ardipithecus ramidus,* dated at 4.4 million years old, lived in a densely wooded habitat. If it turns out that this species was bipedal, then the savanna hypothesis will have to be discarded.

CONTENT CONNECTION

Section 14-2: Radioactive Dating

The most accurate method for dating fossils is measuring the amount of the decay product of an element's radioactive isotope and comparing this to the amount of undecayed isotope. The radioactive isotope chosen for analysis is determined by its having a half-life appropriate for the fossil being studied.

SECTION 17-2

INCLUSION ACTIVITY

Have students study the human phylogenetic tree in Figure 17-6. Ask them to name the oldest hominid genus shown (*Ardipithecus* or *Australopithecus*). Ask them which species of this genus is or are shown as ancestral to modern humans (*Ard. ramidus* or *A. anamensis* and *A. afarensis*). Ask them to name the genus that includes modern humans (*Homo*). Then ask them which species of this genus is or are shown as a direct ancestor of modern humans (*H. habilis* and *H. erectus*).

ANSWERS TO SECTION 17-2 REVIEW

1. The discovery of Lucy changed the assumption that bipedalism and a large brain evolved at the same time.
2. *A. robustus* and *A. boisei* are thought to have died out without leaving descendant lines.
3. The existence of hominid species not ancestral to modern humans implies that the hominid phylogenetic tree is quite bushy, with many species that were evolutionary dead ends.
4. The fossils finds of 1995 are older than previous hominid finds. *A. anamensis* is the oldest australopithecine known, and *Ard. ramidus* may be the oldest hominid, if analysis shows that it was bipedal.
5. Bipedalism distinguishes hominids from prehominids.
6. The location of the foramen magnum indicates the angle at which the spinal cord exits the skull, a clear indicator of posture.

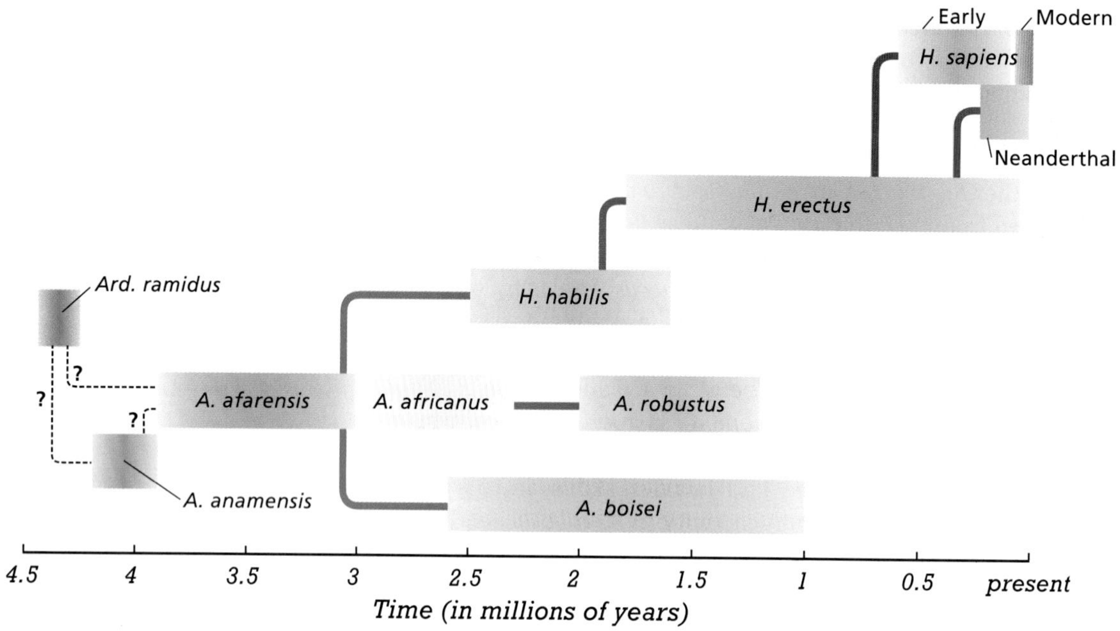

FIGURE 17-6

Physical anthropologists sometimes offer conflicting models of the progression of forms in human evolution. This phylogenetic tree represents one popular view.

An Older Hominid?

In 1995, Tim White (1950–), of the University of California–Berkeley, and his colleagues announced the discovery of fossils representing a new genus that predates the earliest known australopithecines by 200,000 years. Examination of these fossils may ultimately indicate whether the newly discovered primate, *Ardipithecus ramidus,* was bipedal and thus a hominid. It remains to be determined whether this species was ancestral to the australopithecines or whether it was one of several that died out during the course of human evolution.

It is important to understand that human evolution did not occur in a single, uninterrupted parade of increasingly humanlike forms. Rather, there is fossil evidence of several hominid forms that arose and died out, leaving no descendants. Moreover, as Figure 17-6 shows, it is clear that different species of hominids lived at the same time and, in some cases, in the same area. Thus, the human phylogenetic tree has many branches.

SECTION 17-2 REVIEW

1. Lucy was a small-brained hominid. What assumption did her discovery change?
2. What were the probable fates of *Australopithecus robustus* and *A. boisei*?
3. What does the existence of hominid species that were not ancestral to modern humans imply?
4. How do the ages of the fossil finds of 1995 compare with the ages of previous hominid finds?
5. *Ardipithecus* may be a hominid or a prehominid. What characteristic distinguishes hominids?
6. CRITICAL THINKING When analyzing fossils, scientists examine the foramen magnum, the opening for the spinal cord in the base of the skull. What does location of the foramen magnum on a skull tell about the posture of an animal?

HYPOTHESES OF HOMINID EVOLUTION

Starting in the early part of the twentieth century, scientists have worked continually to establish a robust fossil record of human evolution. This work has been rewarded by the discovery of many hominid life-forms, some of which are clearly identifiable as a known type, while others appear to be transitions between known types. As scientists fill in the puzzle of human evolution, they have been surprised by some of their discoveries: dead-end branches of the family tree, as well as evidence that two or more quite different hominid forms may have coexisted.

SECTION 17-3

OBJECTIVES

▲ Name two behavioral advances made by *Homo* species.

● Describe where Neanderthals are placed on the hominid phylogenetic tree.

■ Contrast the multiregional hypothesis of the evolution of modern humans with the hypothesis of recent African origin.

EARLY MEMBERS OF *HOMO*, THE HUMAN GENUS

In the early 1960s, paleoanthropologists working in East Africa found a hominid skull with a much larger brain case than that of the australopithecines. This fossil was the first evidence of an early hominid species that had a cranial capacity as large as 600 to 800 cm^3 but was only slightly taller than *A. afarensis.* The human-like morphology apparent in these fossils resulted in the grouping of this species in the human genus, *Homo.*

Homo habilis

Unlike australopithecine fossils, these newly discovered remains were found along with stone tools. This finding led to the naming of the new species as *Homo habilis,* meaning "handy human." Additional remains of *H. habilis* were later found in southern and eastern Africa, where the first fossils were found. These fossils are between 1.6 million and 2.5 million years old. Some studies of *H. habilis* skulls indicate that a region of the brain essential to speech may have existed in this species. Tool marks on animal bones found near the hominid fossils suggest that *H. habilis* ate meat.

Homo erectus

Fossils of *Homo erectus* (meaning "upright human") were found as early as 1891. They were originally found on the Pacific island of Java and have since been found in China, Europe, and Africa. Thus,

internetconnect

SCILINKS NSTA

TOPIC: *Homo erectus*
GO TO: www.scilinks.org
KEYWORD: HM327

SECTION 17-3

SECTION OVERVIEW

Evolution
Section 17-3 introduces the human genus, *Homo. H. habilis* and *H. erectus* are discussed. Early *H. sapiens,* including Neanderthals and Cro-Magnons, are described, and speculation is given about the fate of the Neanderthals.

Reproduction and Inheritance
Section 17-3 also introduces the two major hypotheses for the spread of modern humans across Earth: the multiregional hypothesis and the hypothesis of recent African origin. The genetic evidence in favor of the second hypothesis is explained.

▶ ENGAGE STUDENTS

Provide the students with replicas or pictures of tools used by early humans. Try to include nonstone tools in addition to stone tools. Ask the students to examine the tools and try to determine what they were used for, possibly by simulating an activity while holding a tool replica. Prod them by asking what kinds of things these humans would have needed tools for. Then tell the students what the investigators who discovered the tools inferred they were used for.

CULTURAL CONNECTION

Stone Tools and Weapons

As recently as the last century, Native Americans used stone to make tools and weapons such as ax heads and arrow points. Metal, such as iron, bronze, and steel, had replaced stone in Europe for these functions many years before that. Arrowheads, which are small, were finely worked, usually in a type of stone that cleaved easily, such as flint.

SECTION 17-3

RECENT RESEARCH

Earliest *Homo* and Tools

Archaeologists recently discovered some *Homo* remains along with tools. The fossils were found to be 2.33 million years old. This is the oldest known association of hominid remains with stone tools.

GIFTED ACTIVITY

Have students research and report on the Piltdown hoax, including the discovery of this fake "missing link" in human ancestry and the analyses done to expose the hoax.

TEACHING STRATEGY

Mapping Human Fossil Sites

Have students conduct library research to find the locations and species names of major sites where human fossils have been discovered. (They may find articles in *Nature, Discover,* and *National Geographic.* The information should include the age or age range of fossils found at each site. Ask them to include fossils dating until about 30,000 years ago. Have the students make a map of the locations of these fossil sites. Ask the students to form a hypothesis that explains their results. (Humans originated in Africa and later emigrated to other locales.)

QUICK FACT

☑ Some scientists think that the Neanderthals had long noses to correspond with the large nasal cavities seen in their skulls. Remind students that we do not know what the soft tissue of extinct hominids looked like. These large noses would serve to warm cold air and may have been an adaptation that enabled Neanderthals to live in the cool northern climates of Europe.

FIGURE 17-7

This reconstruction of a *Homo erectus* skull (a) shows a prominent brow, low forehead, and large, protruding teeth. Contrast these traits with those of the skull of a modern *Homo sapiens* (b).

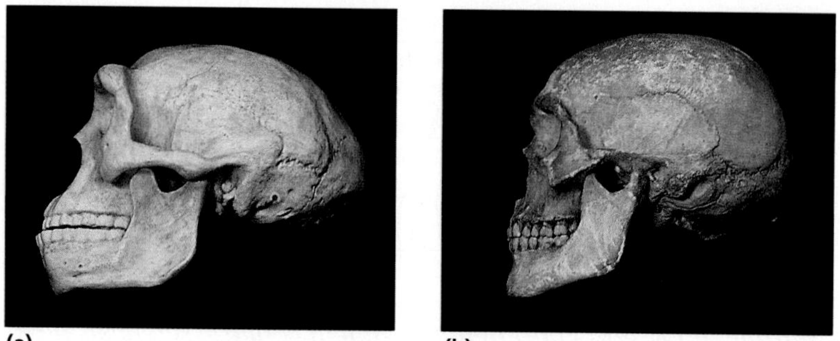

(a) (b)

H. erectus was apparently the first hominid to travel out of Africa. *H. erectus* fossils range from 1.8 million to less than 50,000 years old.

Compared with modern humans, *H. erectus* had a thick skull, large brow ridges, a low forehead, and large, protruding teeth, as shown in Figure 17-7. Their cranial capacity ranged from 700 to 1,250 cm^3, so the average brain size was about two-thirds that of modern humans. The fossil of an almost complete *H. erectus* skeleton, called the Turkana boy, is the remains of a 12-year-old male who stood about 1.7 m (5 ft 7 in.) tall. This means that *H. erectus* adults could easily have been as tall as modern humans.

Traces of charred bones indicate that *H. erectus* were hunters who used fires for cooking and probably for warmth. To survive in the colder climates of Europe and northern Asia, many *H. erectus* groups lived in caves.

FIGURE 17-8

Neanderthals had larger brains than modern humans, but they were shorter and stockier. They lived and hunted in groups. They disappeared about 35,000 years ago.

HOMO SAPIENS

An early, now-extinct form of our species, *Homo sapiens,* probably arose from *Homo erectus* about 800,000 years ago. Over the years, evidence of hominid forms that were transitional between *H. erectus* and *H. sapiens* has been found. Some skulls have the large brow ridges of *H. erectus* and the large cranial capacity of *H. sapiens.* When *H. sapiens* arose, they did not completely replace *H. erectus* right away. Recent finds of *H. erectus* fossils indicate that this species existed until as recently as 50,000 to 35,000 years ago. Thus, *H. sapiens* and *H. erectus* may have coexisted for more than 700,000 years.

Neanderthals

Many examples of a distinctive type of hominid fossil skeleton dating from 230,000 to 30,000 years ago have been found in Europe and Asia. They belong to a group of early *H. sapiens* called Neanderthals (nee-AND-uhr-TAHLZ). Neanderthals, shown in Figure 17-8, had heavy bones, thick brow ridges, and protruding teeth. However, the cranial capacity of Neanderthals averaged 1,450 cm^3. This is slightly larger than the cranial capacity of modern humans. Neanderthals stood about 1.5 m (5 ft) tall but were heavily built. They lived in caves and stone shelters during the last ice age. Their carefully shaped stone tools, probably used to scrape animal hides, have led scientists to speculate that they wore clothing made of skins. Many paleoanthropologists think that Neanderthals were not ancestral to modern humans. In Europe, Neanderthals disappeared at approximately the same time that modern *H. sapiens* arrived in large numbers, leading some scientists to hypothesize that the Neanderthals were killed off violently or by disease and replaced by modern humans.

Modern *Homo sapiens*

The first fossil skeletons bearing a distinct resemblance to modern humans were discovered in caves in southwestern France. The fossils are about 35,000 years old. Older fossils of the same type have been found elsewhere in Europe, as well as in Africa, Asia, and even Australia. These early modern humans are referred to as Cro-Magnons, named for the cave where they were originally discovered. Cro-Magnons had a cranial capacity equal to that of modern humans, 1400 cm^3. They are distinguished from Neanderthals by their high forehead and lack of protruding brow ridge and teeth. Taller than Neanderthals, they stood about 1.8 m (6 ft) tall.

The oldest truly modern *H. sapiens* fossils yet found are about 100,000 years old and were found in Africa. Modern *H. sapiens* probably coexisted with the Neanderthals for about 70,000 years and with *H. erectus* for more than 50,000 years. As the Neanderthals declined, modern humans became more advanced and prevalent. About 50,000 years ago, they became more efficient hunters and home builders, and their tools became distinctly more sophisticated.

SPREAD OF MODERN HUMANS

All modern humans belong to a single species, *Homo sapiens*. There are phenotypic differences, such as skin color, associated with people living in different regions. However, all *H. sapiens* are genetically similar enough to produce offspring together. How did these different phenotypes arise?

Multiregional Evolution

Some anthropologists propose that modern humans evolved in parallel all over Earth from different populations of *Homo erectus*. For this process to result in a single species of modern human, as actually exists, constant gene flow between the different populations would be necessary. Without exchanges of genes during the transition from *H. erectus* to *H. sapiens*, the different populations would tend to speciate into separate groups, in response to local environmental pressures.

If the **multiregional hypothesis,** outlined in Figure 17-9a, is correct, it would suggest that regional differences in phenotype have been developing for well over a million years. Some investigators

Quick Lab

Comparing Cranial Capacities

Materials calculator, 2 L plastic soda bottles (4), graduated cylinder, water, paper, pencil, wax marking pencil

Procedure

1. Using the data provided on pp. 327–329, calculate the average cranial capacity for *Homo habilis, Homo erectus,* and both groups of *Homo sapiens* (Neanderthals and modern humans).
2. Convert your averages to milliliters. (1 cm^3 = 1 mL)
3. Label one of the plastic soda bottles *"Homo habilis."* Label another bottle *"Homo erectus."* Label the third bottle *"Homo sapiens* (Neanderthal)" and the fourth bottle "Modern *Homo sapiens."*
4. Fill a graduated cylinder with an amount of water equal to the average cranial capacity of *Homo habilis*. Pour the water in the appropriately labeled bottle.
5. Repeat step 4 for each of the remaining three species, and record your observations.
6. Calculate in cubic centimeters the change in average cranial capacity between each species over time.

Analysis Based on your calculations and observations of cranial capacities, which species had the smallest brain? How small was it? Which species had the largest brain? How large was it? Between which two species did you find the greatest change in cranial capacity? What trend can you observe in the change in cranial capacity over time? What might explain such a trend?

Quick Lab

Comparing Cranial Capacities

Time Required 25 minutes

Procedural Tips Ask students to bring empty 2 L plastic bottles to class. The bottles should be clear, if possible. A few drops of food coloring may be placed in the water to help students see the differences in cranial capacity more clearly. Students should work in small groups, especially if bottles and calculators are in short supply. If time permits, have groups put their calculations on the board.

Answers to Analysis By observing the water in the bottles, students should be able to visualize a trend of increasing cranial capacity and, by inference, brain size over time. *Homo habilis* had the smallest brain, with an average cranial capacity of 700 cm^3. Neanderthals had the largest brain, with an average cranial capacity of 1,450 cm^3. The greatest change was between *Homo erectus* and Neanderthals, an increase in cranial capacity of 475 cm^3. The average cranial capacity of *Homo erectus* was 975 cm^3. Possible explanations for increasing brain size are that the genus *Homo* was becoming more sophisticated intellectually and was becoming larger in body size. Neanderthals were the exception to the trend.

DEMONSTRATION

Early Human Art

Collect pictures of early human art. Cave paintings and other forms of art can be found in magazines such as *National Geographic* and *Discover*. Include information about the age of the art and the type of human who made it. Ask students to try to interpret what these early humans were trying to express about their world.

SECTION 17-3

OVERCOMING MISCONCEPTIONS

Many people think that the proposal of a "mitochondrial Eve" means that the human evolutionary tree has a single female ancestor. Rather, the data suggest that the ancestral population was sufficiently small that only one set of mitochondrial genes was passed to later generations.

ANSWERS TO SECTION 17-3 REVIEW

1. Some clues considered by paleoanthropologists are brain size, evidence of tool use and toolmaking, knowledge of fire, the wearing of clothing, hunting, and the building of shelter.
2. *H. erectus* had fire and was the first hominid to leave Africa.
3. Neanderthals were shorter and stockier than modern humans, and they had larger brains.
4. The fact that all human mitochondria have very similar genes indicates that they arose in a population only 100,000 to 200,000 years old.
5. The multiregional hypothesis of human evolution accounts for the origin of regional phenotypic differences.
6. Fossil evidence of Neanderthals who met a violent death in areas where Cro-Magnons later flourished would constitute evidence that the Neanderthals were killed off by Cro-Magnons. Fossil evidence of a body type representing a blend of Neanderthal and Cro-Magnon features would support the hypothesis that the two groups interbred.

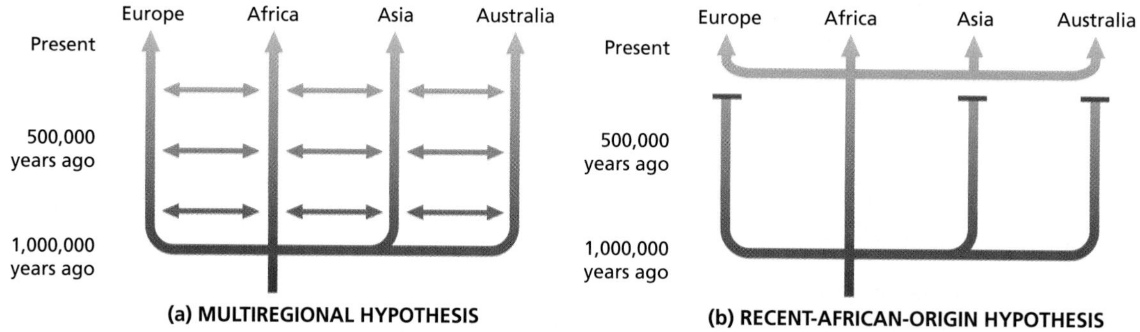

FIGURE 17-9

According to the multiregional hypothesis of human evolution (a), parallel populations of *H. sapiens* evolved from different *H. erectus* populations around the world. The hypothesis of recent African origin (b) states that modern *H. sapiens* evolved in Africa and spread throughout the world, replacing populations of *H. erectus* and early *H. sapiens.*

claim that Asian fossils of *Homo erectus* show the high cheek bones seen in modern *Homo sapiens* living in Asia.

Out of Africa

The more widely supported hypothesis, the **recent-African-origin hypothesis,** states that modern *Homo sapiens* originated in Africa only about 100,000 to 200,000 years ago and then, like *Homo erectus* before them, left Africa. They colonized the world, displacing and causing the extinction of *Homo erectus* and early *Homo sapiens,* such as the Neanderthals.

The recent-African-origin hypothesis gets much of its support from studies of the genes found in mitochondria. Because mitochondria reproduce asexually, their genes are not subject to the mixing caused by gene flow and meiosis. If humans were one large population dating back to over a million years ago, we should find human mitochondria that show a million years of accumulated mutational differences. Instead, most human mitochondria have very similar genes. The period of time needed for mitochondria to accumulate the differences actually seen is only 100,000 to 200,000 years—far short of a million years. Because all human mitochondria are so similar, supporters of this hypothesis infer that all modern humans came from one small group in Africa a fairly short time ago—100,000 or 200,000 years ago.

SECTION 17-3 REVIEW

1. **What clues do paleoanthropologists look for when they try to determine the habits and capabilities of early hominids?**
2. **What are two behavioral advances that distinguished *Homo erectus* from *H. habilis*?**
3. **How did the body types of Neanderthals and modern humans differ?**
4. **What observations of mitochondrial DNA are used to support the hypothesis of recent African origins of modern *Homo sapiens*?**
5. **Which hypothesis of human evolution attempts to explain the origin of regional phenotypic differences?**
6. **CRITICAL THINKING According to the existing fossil evidence, Neanderthals died out about 30,000 years ago. Some anthropologists hypothesize that the Neanderthals were killed off by Cro-Magnons. Others hypothesize that the two groups interbred. What evidence would you look for to evaluate these two hypotheses?**

CHAPTER 17 REVIEW

SUMMARY/VOCABULARY

17-1

- Paleoanthropologists gather data from the fossilized remains of early hominids and their ancestors. The shapes of the bones may indicate whether the organism was bipedal. Other clues may indicate diet or habits.
- Humans belong to the order of mammals known as primates. There are two divisions of primates: anthropoid primates and prosimian primates.
- Most primates have nails, instead of claws, and prehensile hands and feet. Primates have color vision and depth perception.
- Anthropoid primates include marmosets, monkeys, apes, and humans. Prosimian primates include lemurs, lorises, and tarsiers.
- Anthropoid primates have large brains relative to their body size.
- Bipedalism and language are two important traits found only in humans.

Vocabulary

anthropoid primate (320)
bipedalism (321)
dental formula (320)
great ape (320)
hominid (319)
opposable thumb (320)
paleoanthropologist (319)
pelvis (319)
prehensile (320)
primate (320)
prosimian primate (320)

17-2

- Bipedalism is the defining characteristic for the hominids—human ancestors.
- The oldest known genus of hominids is *Australopithecus.* Its members are called australopithecines.
- The earliest known hominid, *Australopithecus anamensis,* lived more than 4 million years ago.
- At least two australopithecine lines, *A. africanus* and *A. boisei,* probably were not ancestral to modern humans. They became extinct more than 1 million years ago.
- The discovery of Lucy, a nearly half-complete fossil of an early hominid, *Australopithecus afarensis,* implies that hominids became bipedal before their brains began to dramatically enlarge.
- *Ardipithecus ramidus* is a recent discovery. It is not clear whether it was bipedal. *Ard. ramidus* is 4.4 million years old.
- The existence of hominid species not ancestral to modern humans implies that the hominid phylogenetic tree is bushy in appearance, with many branches representing species that died out, leaving no descendants.

Vocabulary

australopithecine (325)
Lucy (325)
quadrupedal (324)

17-3

- Early members of the genus *Homo, H. habilis* and *H. erectus,* probably were ancestral to modern humans. They had larger brains than the australopithecines and may have had speech.
- Our species, *Homo sapiens,* probably evolved about 800,000 years ago.
- The brains of members of the genus *Homo* were much larger than those of the australopithecines.
- Neanderthals were early *Homo sapiens.* They may be ancestral to modern humans, or they may have died out and been replaced by modern humans.
- Some anthropologists think that *H. sapiens* evolved in parallel from populations of *H. erectus* all over the world.
- Some anthropologists propose that *H. sapiens* descended from *H. erectus* in Africa and then dispersed across Earth.

Vocabulary

multiregional hypothesis (329)
recent-African-origin hypothesis (330)

CHAPTER 17 REVIEW ANSWERS

REVIEW

1. A bipedal animal walks on two feet; a quadrupedal animal walks on four feet.
2. The dental formula refers to the type and number of teeth found in an animal.
3. The primate's opposable thumb can be moved to lie opposite the fingers, making a grasping hand.
4. Cranial capacity is used as a rough estimate of brain size.
5. *Australopithecus* means "southern ape."

6. b **7.** b
8. c **9.** b
10. b **11.** d
12. a **13.** c
14. c **15.** d

16. The ape skull is larger; the ape jaw is more massive, with very large canines; the ape face is less vertical; the bottom of the skull (where the spinal cord exits) faces backward in the ape and downward in the human.
17. It would be difficult to tell where tree branches are without depth perception.
18. Upright walking allows humans to see long distances and frees the hands to manipulate objects.
19. Behavioral traits unique to humans include language, tool making, and wearing clothing.
20. The human jaw is rounder than the U-shaped jaw of apes.
21. Paleoanthropologists use skulls or skull fragments to estimate the brain size of hominids.
22. Most hominid fossil remains have been found in Africa.
23. Neanderthals lived in caves and stone shelters and probably wore clothing.
24. Lucy's brain was about one-third the size of that of a modern human's, or about the size of a chimpanzee's brain.
25. The human spine has an S-shape; the paired curves allow humans to stand upright. The ape spine has a single curve; apes stand curled forward, with their weight on their forelimbs.

Critical Thinking

1. It is not inevitable that descendant species of *H. sapiens* will have larger brains than modern humans. Stabilizing selection tends to keep traits in an average range. The size of the human female pelvis must enlarge along with any enlargement of the head.
2. The ape's anatomy would restrict its ability to use spoken language because speech production requires a great degree of control of the voice box and facial muscles.

Review

Vocabulary

1. What is the difference between a bipedal animal and a quadrupedal animal?
2. What does the term *dental formula* mean?
3. Why is a primate thumb described as "opposable"?
4. What is the relationship between cranial capacity and brain size?
5. What does the genus name *Australopithecus* mean?

Multiple Choice

6. The earliest primates probably lived in (a) grasslands (b) forests (c) deserts (d) tundra.
7. An important difference between anthropoids and prosimians is that anthropoids have (a) larger eyes (b) a larger brain (c) a longer tail (d) no depth perception.
8. The dental formula found in humans is found (a) in all animals (b) only in hominids (c) in apes and African and Asian monkeys. (d) only in marmosets.
9. Humans are better adapted to upright walking than chimpanzees are because humans have (a) a large brain (b) a cup-shaped pelvis (c) a big toe that is opposable to the other toes (d) the ability to use language.
10. The similarity of human and chimpanzee DNA suggests that the two species (a) are identical (b) shared a relatively recent ancestor (c) are not related (d) have stopped evolving.
11. Paleoanthropologists determine the manner of walking and the physical features of human ancestors on the basis of (a) fingerprint patterns (b) comparisons of living species (c) written descriptions in ancient sources (d) shapes of fossilized bones.
12. The earliest known bipedal human ancestor was (a) *A. anamensis* (b) *A. afarensis* (c) *A. africanus* (d) *A. boisei.*
13. The species *A. afarensis* probably (a) died out (b) gave rise to the genus *Ardipithecus* (c) was ancestral to later australopithicenes (d) lived after *A. africanus.*
14. *H. habilis* was so named because evidence indicates that this species (a) was bipedal (b) used language (c) made tools (d) ate meat.
15. Cro-Magnons are regarded as (a) a type of *Homo erectus* (b) a transitional species between *Ardipithecus* and the australopithecines (c) ancestors of Neanderthals (d) modern humans.

Short Answer

16. Study the figure below of the ape skull, top, and the human skull, bottom, and name three major differences between the two skulls.

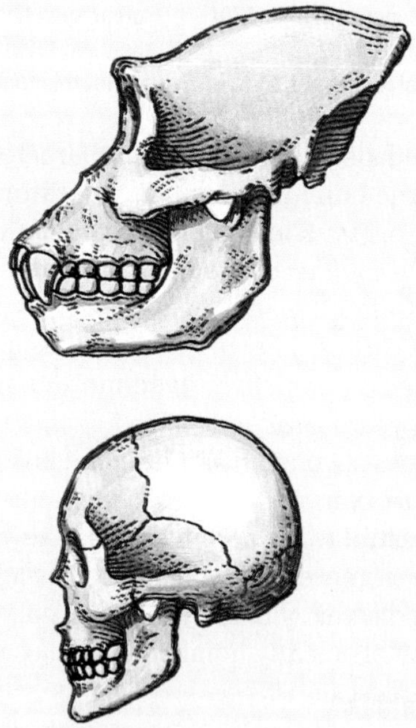

17. Why is depth perception handy for climbing in trees?
18. How is upright walking a beneficial trait for humans?
19. What are two behavioral traits unique to humans?
20. How does the jaw of humans differ from that of apes?
21. How do paleoanthropologists use fossil evidence to draw conclusions about the brain size of hominids?
22. On what continent have most fossil remains of hominids been found?
23. How did Neanderthals accommodate the cold climate of northern Europe?

24. How does the brain size of *A. afarensis* (Lucy) compare with that of chimpanzees and that of modern humans?
25. How is the human spine adapted for standing erect, compared with the spine of a chimpanzee?

CRITICAL THINKING

1. From *H. habilis* to modern *H. sapiens,* there has been a trend for brain size to increase. Is it inevitable that any future descendant species of *H. sapiens* will have larger brains than we have? Explain your answer in terms of some form of natural selection, such as directional selection.
2. Compared with humans, apes do not have a well-developed voice box or well-developed facial muscles. How might the ape's anatomy affect its ability to use spoken language?
3. According to fossil evidence, *H. habilis* made stone tools. They traveled long distances to collect specific types of rock and minerals, and then they shaped the rocks by chipping them at the edges. What does this reveal about *H. habilis*'s ability to use foresight?
4. Cro-Magnon remains have been found with reindeer bones in certain areas of southern Europe. What does this fact suggest about the diet of Cro-Magnons and the environment in which they lived?
5. The two phylogenetic trees shown below express two different views of the relationship of Neanderthals to modern humans. If Neanderthals were very genetically different from modern humans, which tree would more likely be the correct one and why?

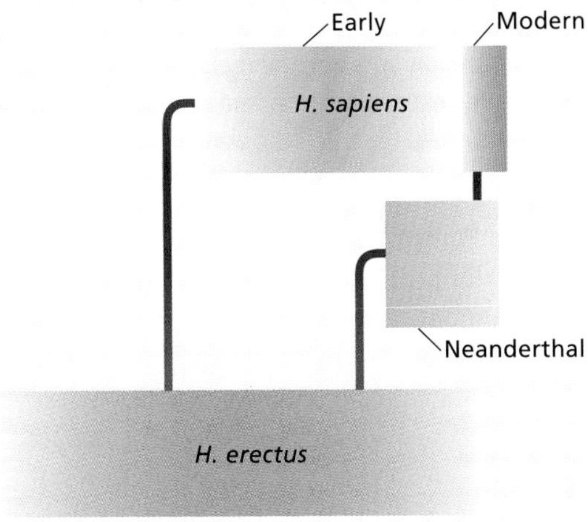

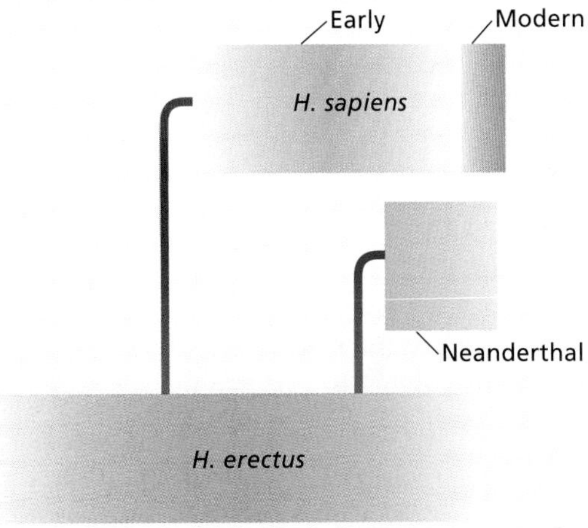

EXTENSION

1. Read "No, After You, Afarensis" in *Discover,* January 1999, on page 81. What physical characteristics of *Australopithecus africanus* cause Berger and McHenry to disagree about the human ancestry of *A. africanus*? What reasons are given to defend the current theory that *A. afarensis* is ancestral to *A. africanus*? How long ago did each of the three species of *Australopithecus* live?
2. Visit a local zoo to observe the behavior of monkeys, apes, or other primates. Pay close attention to the facial expressions of the animals. Notice the ways in which the animals interact. Take notes on what you observe. What similarities and differences do you see between the behavior of the primates you observed and that of humans? If it is not possible to visit a zoo, study pictures or videotapes of primates. What inferences can you make about primate behavior from the pictures?

3. *H. habilis* must have used foresight to plan trips and to make tools before they were needed.
4. Cro-Magnons probably ate reindeer meat, and because reindeer are adapted to cold climates, the environment in which Cro-Magnons and reindeer coexisted was probably cold.
5. The tree showing Neanderthals as not interbreeding with modern humans would probably be more correct than the other.

EXTENSION

1. *Australopithecus africanus* had more apelike proportions than *A. afarensis,* which existed long before *A. africanus. A. africanus* was bipedal, but it spent more time in the trees than *A. afarensis. A. afarensis* had a smaller brain, a more primitive skull, and more primitive teeth than *A. africanus,* indicating that *A. afarensis* was ancestral to *A. africanus. A. anamensis* lived from 4.1 million to 3.5 million years ago, *A. afarensis* lived from 3.6 million to 3 million years ago, and *A. africanus* lived from 3 million to 2.6 million years ago.
2. Students may notice that the animals groom each other, feed in the order of dominance in the primate group (dominant animals feed first), display aggression by showing their teeth and feinting attacks on other animals, and are protective of juveniles. Humans behave differently from other primates in some respects. Humans do not do extensive social grooming and they do not have rigid feeding behavior. A human smile is a display of teeth that would signal aggression, not friendship, to a nonhuman primate. Human behavior toward juveniles, however, is very similar to that of other primates.

CHAPTER 17 INVESTIGATION

TIME REQUIRED
One 50-minute class period

QUICK REFERENCES
Lab Preparation Notes are found on page 317B.

Holt Biosources provides a Teaching Resources CD-ROM that contains "Using Gowin's Vee in the Lab" and "Scoring Rubrics."

PROCEDURAL TIPS
1. To save time, have students create their data tables before beginning the lab.

2. Remind students that proteins are three-dimensional and that their function depends on their shape. Thus, it is conceivable that some amino acids in the sequence are less important for function than others. (Some amino acids are attracted to other parts of the molecule, causing bending of the amino acid chain.) If different amino acids make the same contribution to shape, or if a slight change in shape does not interfere with the function of the protein, different amino acids may appear at that location as a result of random genetic change, such as mutation.

ANSWERS TO BACKGROUND
4. This method analyzes organisms on a molecular level rather than on the macroscopic levels of morphology and embryology. Moreover, this method is based on a dynamic model of molecular change over time, and certain assumptions about the rate of change of different proteins and of homologous proteins in different organisms are made.

CHAPTER 17 INVESTIGATION

Relating Amino Acid Sequences to Evolutionary Relationships

OBJECTIVES
- Observe the amino acid sequence of hemoglobin and cytochrome *c* in several species.
- Compare the amino acid sequences of the same protein in different species.
- Deduce evolutionary relationships among species.

PROCESS SKILLS
- comparing and contrasting
- classifying
- analyzing
- inferring

MATERIALS
- pencil
- paper

Background
1. Hemoglobin and cytochrome *c* are two proteins commonly studied by scientists attempting to deduce evolutionary relationships from differences in amino acid sequences.
2. Researchers believe that the greater the similarity that exists between the amino acid sequences of two species, the more closely related the two species are evolutionarily.
3. The greater the differences that exist in the amino acid sequences of two species, the more distantly related the two species are.
4. The longer two species have been diverging from a common ancestor, the greater the difference that can be expected in their amino acid sequences. This principle is based on the assumption that the rate of

Cytochrome *c* Amino Acid Sequences

	Horse	Chicken	Tuna	Frog	Human	Shark	Turtle	Monkey	Rabbit
42	Gln	Gln	Gln	Gln	Gln	Gln	Gln	Gln	Gln
43	Ala	Ala	Ala	Ala	Ala	Ala	Ala	Ala	Ala
44	Pro	Glu	Glu	Ala	Pro	Gln	Glu	Pro	Tyr
46	Phe	Phe	Tyr	Phe	Tyr	Phe	Phe	Tyr	Pro
47	Thr	Ser	Ser	Ser	Ser	Ser	Ser	Ser	Ser
49	Thr	Thr	Thr	Thr	Thr	Thr	Thr	Thr	Thr
50	Asp	Asp	Asp	Asp	Ala	Asp	Asp	Ala	Asp
53	Lys	Lys	Lys	Lys	Lys	Lys	Lys	Lys	Lys
54	Asn	Asn	Ser	Asn	Asn	Ser	Asn	Asn	Asn
55	Lys	Lys	Lys	Lys	Lys	Lys	Lys	Lys	Lys
56	Gly	Gly	Gly	Gly	Gly	Gly	Gly	Gly	Gly
57	Ile	Ile	Ile	Ile	Ile	Ile	Ile	Ile	Ile
58	Thr	Thr	Val	Thr	Ile	Thr	Thr	Ile	Thr
60	Lys	Gly	Asn	Gly	Gly	Gln	Gly	Gly	Gly
61	Glu	Glu	Asn	Glu	Glu	Gln	Glu	Glu	Glu
62	Glu	Asp	Asp	Asp	Asp	Glu	Glu	Asp	Asp
63	Thr	Thr	Thr	Thr	Thr	Thr	Thr	Thr	Thr
64	Leu	Leu	Leu	Leu	Leu	Leu	Leu	Leu	Leu
65	Met	Met	Met	Met	Met	Arg	Met	Met	Met
66	Glu	Glu	Glu	Glu	Glu	Ile	Glu	Glu	Glu
100	Lys	Asp	Ser	Ser	Lys	Lys	Asp	Lys	Lys
101	Ala	Ala	Ala	Ala	Ala	Thr	Ala	Ala	Ala
102	Thr	Thr	Thr	Gly	Thr	Ala	Thr	Ala	Thr
103	Asn	Ser	Ser	Ser	Asn	Ala	Ser	Asn	Asn
104	Glu	Lys	—	Lys	Glu	Ser	Lys	Glu	Glu

change of a specific amino acid sequence is the same in all species. Think of other methods used to determine evolutionary relationships. How is this method different?

PART A Cytochrome *c*

1. Cytochrome *c*, a protein found in the mitochondria of many species, consists of a chain of 104 amino acids. The figure on page 334 shows the corresponding parts of noncontinuous parts of the cytochrome *c* amino acid sequences of nine vertebrate species. The numbers along the left side of the figure refer to the position of these sequences in the chain. The letters identify the specific amino acids in the chain.
2. Make a table to record your data in your lab report. Label the columns of your data table "Species" and "Number of Differences from Human Cytochrome *c*."

Hemoglobin Amino Acid Sequences

	Human	Chimpanzee	Gorilla	Monkey	Horse
87	Thr	Thr	Thr	Gln	Thr
88	Leu	Leu	Leu	Leu	Leu
89	Ser	Ser	Ser	Ser	Ser
90	Glu	Glu	Glu	Glu	Glu
91	Leu	Leu	Leu	Leu	Leu
92	His	His	His	His	His
93	Cys	Cys	Cys	Cys	Cys
94	Asp	Asp	Asp	Asp	Asp
95	Lys	Lys	Lys	Lys	Lys
96	Leu	Leu	Leu	Leu	Leu
97	His	His	His	His	His
98	Val	Val	Val	Val	Val
99	Asp	Asp	Asp	Asp	Asp
100	Pro	Pro	Pro	Pro	Pro
101	Glu	Glu	Glu	Glu	Glu
102	Asn	Asn	Asn	Asn	Asn
103	Phe	Phe	Phe	Phe	Phe
104	Arg	Arg	Lys	Lys	Arg
105	Leu	Leu	Leu	Leu	Leu
106	Leu	Leu	Leu	Leu	Leu
107	Gly	Gly	Gly	Gly	Gly
108	Asn	Asn	Asn	Asn	Asn
109	Val	Val	Val	Val	Val
110	Leu	Leu	Leu	Leu	Leu
111	Val	Val	Val	Val	Ala
112	Csy	Cys	Cys	Cys	Leu
113	Val	Val	Val	Val	Val
114	Leu	Leu	Leu	Leu	Val
115	Ala	Ala	Ala	Ala	Ala
116	His	His	His	His	Arg

For each vertebrate species, count the amino acids in the sequence that differ from the human sequence. List these in your data table.

3. In your lab report, list the eight vertebrate sequences in descending order according to the degree of similarity of their cytochrome *c* with that of humans. According to your analysis of the amino acid sequences, which species listed is most closely related to humans? Which species is least closely related to humans?

PART B Hemoglobin

4. Look at the hemoglobin sequences for the five species shown in the figure at left. Hemoglobin is the oxygen-carrying molecule of red blood cells. Only the portion of the chain between amino acid numbers 87 and 116 is shown in the figure.
5. In your lab report, make a table to record your data. Label the columns of your data table "Species" and "Number of Differences from Human Hemoglobin." In each species' sequence, count the number of amino acids that differ from the human sequence and list them in your data table, as you did in Part A.
6. In your lab report, list the four vertebrate sequences in descending order according to the degree of similarity of their hemoglobin with that of humans. According to your analysis of the amino acid sequences, which species listed is most closely related to humans? Which species is least closely related to humans?

Analysis and Conclusions

1. Why can it be said that proteins behave like molecular clocks?
2. There is a difference of only one amino acid in one portion of hemoglobin of gorillas and humans. What could have been responsible for this change?
3. If the amino acid sequences are similar in gorillas and humans, why would you expect the DNA of these two organisms to also be similar?

Further Inquiry

Would you expect to find the same number of differences in the cytochrome *c* and hemoglobin amino acid chains when comparing organisms? Or might one of these proteins have changed at a faster rate than the other? Why might the rates of change differ among proteins?

ANSWERS TO ANALYSIS AND CONCLUSIONS

1. Proteins are said to behave like molecular clocks because homologous proteins in different species accumulate differences over time caused by the mutation of DNA. The more time that has passed since two species diverged from a common ancestor, the more differences that could have accumulated between homologous proteins.
2. A mutation could have been responsible for the difference.
3. DNA is the template for amino acid sequences in proteins, thus similar amino acid sequences probably have similar DNA templates.

ANSWER TO FURTHER INQUIRY

Cytochrome *c* and hemoglobin would not necessarily change at the same rate. The proteins have very different functions in the body, and there might be different degrees of evolutionary pressure on the two proteins, producing more-rapid evolution in one.

CHAPTER 18
CLASSIFICATION

	TOPICS	TEACHING RESOURCES	LABS, CLASSWORK, AND HOMEWORK
BLOCK 1 45 minutes	**Introducing the Chapter, p. 336**	ATE Focus Concept, p. 336 ATE Assessing Prior Knowledge, p. 336 ATE Understanding the Visual, p. 336	■ Supplemental Reading Guide, *A Journey to the Ants*
	18-1 History of Taxonomy, p. 337 Early Systems of Classification, p. 337 Linnaeus's System, p. 337	ATE Section Overview, p. 337 56, 57, 58	PE **Quick Lab,** p. 338, Practicing Classification **Lab A15,** Grouping Things You Use Daily **Lab C29,** Classifying Mysterious Organisms ★ Study Guide, Section 18-1 PE Section Review, p. 339
BLOCK 2 45 minutes	**18-2 Modern Phylogenetic Taxonomy, p. 342** Systematics, p. 342 Cladistics, p. 345	ATE Section Overview, p. 342 ATE Critical Thinking, p. 343 44A, 59	**Lab A11,** Comparing Primate Features ★ Study Guide, Section 18-2 PE Section Review, p. 346
BLOCK 3 45 minutes	**18-3 Two Modern Systems of Classification, p. 347** Six-Kingdom System, p. 347 Three-Domain System, p. 350	ATE Section Overview, p. 347 ATE Critical Thinking, p. 348 ATE Visual Link Figure 18-11, p. 350 45A, 57, 88	PE **Chapter 18 Investigation,** pp. 354–355 ■ Occupational Applications Worksheets—Botanist ★ Study Guide, Section 18-3 PE Section Review, p. 350
BLOCKS 4 & 5 90 minutes	**REVIEW and ASSESSMENT** PE Chapter 18 Review, pp. 352–353 ■ Performance-Based Assessment—Exploring Invertebrates	Audio CD Program	**CHAPTER TESTING** ■ Chapter Test (blackline master) ▲ Test Generator ■ Assessment Item Listing ■ Scoring Rubrics and Classroom Management Checklists

KEY

MODERN BIOLOGY

PE Pupil's Edition
ATE Teacher's Edition
■ Active Reading Guide
★ Study Guide

HOLT BIOSOURCES

Teaching Transparencies

Laboratory Program

One-Stop Planner CD-ROM
Shows these resources within a customizable daily lesson plan:

■ Holt BioSources Teaching Resources
■ Active Reading Guide
★ Study Guide
▲ Test Generator

READING FOR CONTENT MASTERY

Preview: Have students preview the objectives, the introductory paragraph, topic sentences, and captions before reading the chapter.

ATE Active Reading Technique: Brainstorming, p. 337
■ Active Reading Guide Worksheet 18-1
PE Summary/Vocabulary, p. 351

■ Active Reading Guide Worksheet 18-2
PE Summary/Vocabulary, p. 351

■ Active Reading Guide Worksheet 18-3
PE Summary/Vocabulary, p. 351

TECHNOLOGY AND INTERNET RESOURCES

CNN PRESENTS Video Segment 11
Something Worth Saving

Holt Biology Videodiscs Teacher's Correlation Guide, Lessons 18-1 through 18-3

Audio CD Program, Sections 18-1 through 18-3

internetconnect

sciLINKS NSTA

On-line Resources: www.scilinks.org
The following *sci*LINKS Internet resources can be found in the student text for this chapter:

Topics:
- Taxonomy, p. 339
- Phylogenetic tree, p. 343

go.hrw.com

On-line Resources: go.hrw.com
Visit the HRW Web site for a variety of resources related to this chapter. Just type in the keyword HM2 Chapter 18.

Smithsonian Institution®
Visit **www.si.edu/hrw** for additional on-line resources.

CNN fyi.com
Visit **www.cnnfyi.com** for late-breaking news and current event stories selected just for you.

LAB ACTIVITY PLANNING

CHAPTER 18 INVESTIGATION

Using and Formulating Dichotomous Keys, pp. 354–355

OVERVIEW
In this investigation, students will use a dichotomous key to identify plant leaves. Students will also learn how to make their own dichotomous key to identify the shoes of their classmates. This investigation emphasizes that a dichotomous key uses pairs of contrasting descriptive statements that lead to the identification of an organism or object.

PREPARATION
1. In Part B, 10 students will need to volunteer one shoe each.
2. You may want to bring in a few extra pairs of different shoes to help ensure that the dichotomous key uses a variety of descriptive statements.
3. Provide the class with a marker and masking tape. Each student will need one sheet of paper and a pencil.

Quick Labs

Practicing Classification, p. 338
Students will need only paper and pencil for this activity.

UNDERSTANDING THE VISUAL

Ask students to study the photograph of the pangolin and note any characteristics it shares with the armadillo pictured in Figure 15-2 of Chapter 15. At one time, pangolins were classified in the same orders as aardvarks, sloths, and armadillos. This classification was changed when scientists concluded that the pangolin's similarity to these other anteating mammals was a result of convergent evolution rather than recent, shared ancestry. Explain to students that classification systems change as new information arises about organisms. Ask students to recall from Chapter 15 what information is considered when scientists determine evolutionary relationships between animals (morphology, including the presence of homologous or vestigial structures; embryological development; similarity of DNA, amino acid sequences, and proteins; and the fossil record).

FOCUS CONCEPT

Evolution

All organisms on Earth are thought to have descended from an ancient common ancestor. Many scientists now recognize six major evolutionary pathways and classify organisms into six kingdoms.

ASSESSING PRIOR KNOWLEDGE

Review the following concepts.

DNA: *Chapter 9*

Ask students what molecule in organisms is passed from one generation to the next. Also ask why similarities in this molecule between different organisms would be a good basis for classification.

Early History of Life: *Chapter 14*

Ask students to identify the first kind of organisms that appeared on Earth and to explain how these organisms obtained organic molecules for food.

CHAPTER 18

CLASSIFICATION

The pangolin, Manis temmincki, *is a species of scaly anteater found in eastern and southern Africa. Pangolins move slowly and, when threatened, curl into a ball, as this one has done.*

FOCUS CONCEPT: ***Interdependence of Organisms***

As you read, consider how the classification of a species reflects its relationships with many related species.

HISTORY OF TAXONOMY

Every year, thousands of new species are discovered. Biologists use the characteristics of each newly discovered species to classify it with organisms having similar characteristics. The ways we group organisms continue to change, and today these methods reflect the evolutionary history of organisms.

EARLY SYSTEMS OF CLASSIFICATION

Taxonomy (taks-AHN-uh-mee) is the branch of biology that names and groups organisms according to their characteristics and evolutionary history. Organisms were first classified more than 2,000 years ago by the Greek philosopher Aristotle. Aristotle classified living things as either plants or animals. He grouped animals into land dwellers, water dwellers, and air dwellers. He also grouped plants into three categories, based on differences in their stems.

As modern science developed in the fifteenth and sixteenth centuries, Aristotle's system at first seemed adequate. Then, in a period of rapid scientific exploration, many new organisms were discovered. Biologists realized that Aristotle's categories were not adequate. They also found that using a common name, such as *robin* or *fir tree,* for an organism presented its own problems; common names varied from one locale to the next, just as they do today. Moreover, common names may not describe species accurately. For example, a jellyfish is not a fish at all. Some early scientists devised scientific names that consisted of long descriptions in Latin, but these names were difficult to remember and suggested nothing about how organisms were related to other organisms.

LINNAEUS'S SYSTEM

In response to the need for organization, the Swedish naturalist Carolus Linnaeus (1707–1778) devised a system of grouping organisms into hierarchical categories. For the most part, Linnaeus used an organism's morphology, that is, its form and structure, to categorize it.

SECTION

18-1

OBJECTIVES

▲ Describe Aristotle's classification system, and explain why it was replaced.

● Explain Linnaeus's system of classification, and identify the main criterion he used to classify organisms.

■ List Linnaeus's levels of classification from the most general to the most specific.

◆ Name the primary criterion that modern taxonomists consider when they classify an organism.

Word Roots and Origins

morphology

from the Greek *morphe,* meaning "form," and *logos,* meaning "word"

SECTION 18-1

SECTION OVERVIEW

Evolution

Section 18-1 describes the Linnaean system of classification. Linnaeus classified organisms based on their morphological characteristics, but biologists now attempt to classify organisms based on their presumed evolutionary relationships. This involves obtaining information about embryological development and various chemicals, such as DNA and proteins.

ACTIVE READING

Technique: Brainstorming

Point out that we often organize things—such as products in a store, books in a library, and subjects in a school curriculum—into fundamental classes or groups. Work with the class as a whole to develop a list of other things that are organized into classes or groups. After students have made a list, tell them that biologists have organized all life-forms into six major groups, or kingdoms.

TEACHING STRATEGY

Sorting Species

Classifying organisms in a hierarchical system is analogous to sorting mail by country, ZIP code, street, and house number. Ask students which of these categories is analogous to the species (house number).

Quick Lab

Practicing Classification

Time Required 15 minutes

Procedural Tips Have students work in pairs. If time permits, ask for volunteers who would like to share their work.

Answers to Analysis Accept all reasonable answers for classifications that include all seven levels of organization. The least specific name is at the kingdom level. The most specific name is at the species level. (Remind students that genus and species must always be used together to identify an organism. Refer to the Demonstration on page 339.) Aristotle grouped organisms into only two groups, plants and animals. His system did not use the seven levels of Linnaeus's system.

RETEACHING ACTIVITY

Help students understand the hierarchical nature of the Linnaean system of classification by having them look up information on organisms related to those shown in Table 18-1. Have them choose one of these organisms and find the names of one or more other phyla in the same kingdom, one or more classes in the same phylum, and so forth. A college biology textbook's appendix would be a good source for this project.

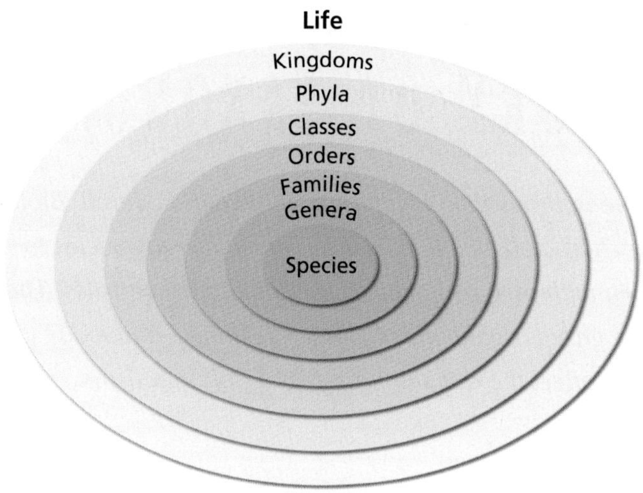

FIGURE 18-1
Linnaeus's categorization scheme used a nested hierarchy. Seven levels of organization, each more specific than the last, allowed organisms to be grouped with similar organisms.

Quick Lab

Practicing Classification

Materials paper, pencil

Procedure Using Table 18-1 as a model, classify a fruit or vegetable you would find in a grocery store. Use all seven levels of classification.

Analysis At which level did you assign the least specific name? At which level did you assign the most specific name? Would Aristotle have classified your item differently? Explain your answer.

Levels of Classification

Linnaeus devised a nested hierarchy of seven different levels of organization, as is shown in Figure 18-1. Linnaeus's largest category is called a **kingdom.** There are two kingdoms, plant and animal, which are the same as Aristotle's main categories. Each subset within a kingdom is known as a **phylum** (FIE-luhm), in the animal kingdom, or a **division,** in the plant kingdom. Within a phylum or division, each subset is called a **class,** and each subset within a class is called an **order.** Still smaller groupings are the **family** and then **genus** (JEE-nuhs). The smallest grouping of all, which contains only a single organism type, is known as the **species** (SPEE-sheez). Table 18-1 shows an example of how two similar organisms and one very different one fit into this classification system.

TABLE 18-1 ***Classification Hierarchy of Organisms***

	Bobcat	Lion	Shaggy mane mushroom
Kingdom	Animalia	Animalia	Fungi
Phylum/division	Chordata	Chordata	Basidiomycota
Class	Mammalia	Mammalia	Homobasidiomycetae
Order	Carnivora	Carnivora	Agaricales
Family	Felidae	Felidae	Copricaceae
Genus	*Lynx*	*Panthera*	*Coprinus*
Species	*Lynx rufus*	*Panthera leo*	*Coprinus comatus*

Binomial Nomenclature

In Linnaeus's system, the **species name** (also called the scientific name) of an organism has two parts. The first part of the name is the genus, and the second part is the **species identifier,** usually a descriptive word. Thus, we humans are known by our genus, *Homo,* and by our species identifier, *sapiens,* which means "wise." This system of two-part names is known as **binomial nomenclature** (bi-NOH-mee-uhl NOH-muhn-KLAY-chuhr). By custom, the genus name is capitalized and both names are underlined or written in italics. Linnaeus classified thousands of organisms, and a version of Linnaeus's system of classification and binomial nomenclature is still used today. Because species names are Latinized, they are the same in every language. This enables scientists around the world to identify organisms by the same name.

A species name may describe the organism. The microscopic amoeba *Chaos chaos,* shown in Figure 18-2, might never look the same way twice. Sometimes a scientific name is chosen to honor a person, or it may suggest the geographic range of the organism. *Linnaea borealis,* a species of flower that grows in northern regions, was Linnaeus's favorite. *Borealis* means "northern."

Linnaeus's choice of seven levels of classification was arbitrary. Significant variation in some species has led taxonomists to establish additional levels of organization. Botanists sometimes split species into subsets known as **varieties.** Peaches and nectarines are fruits of two slightly different varieties of the peach tree, *Prunus persica.* Zoologists refer to variations of a species that occur in different geographic areas as **subspecies.** The variety or subspecies name follows the species identifier. *Terrapene carolina triungui* is a subspecies of the common eastern box turtle, *Terrapene carolina,* and gets its name from having three, rather than four, toes on its hind feet.

To classify organisms, modern taxonomists consider the **phylogeny** (fie-LAHJ-uh-nee), or evolutionary history, of the organism. Much of Linnaeus's work in classification is relevant today, even in this phylogenetic context. By concentrating on morphology, Linnaeus focused on features that are largely influenced by genes and that are clues of common ancestry.

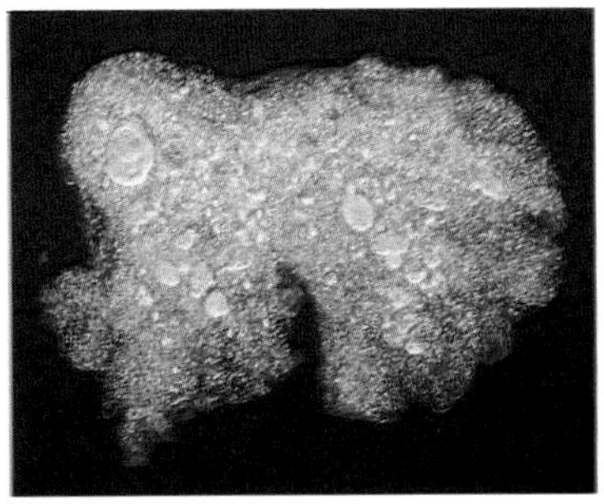

FIGURE 18-2

Names selected for some organisms reflect traits of the organism. The amoeba *Chaos chaos* (LM 56×) changes its shape constantly.

internetconnect

SCILINKS NSTA

TOPIC: Taxonomy
GO TO: www.scilinks.org
KEYWORD: HM339

SECTION 18-1 REVIEW

1. How did Aristotle classify organisms, and why did his method prove inadequate?
2. What criterion did Linnaeus use to classify organisms?
3. What are the seven levels of organization that Linnaeus used to categorize organisms?
4. What are two reasons that species names are more precise than common names?
5. What criterion do modern taxonomists use to classify organisms?
6. **CRITICAL THINKING** Linnaeus's work was done many years before that of Darwin and Mendel. Explain why many of Linnaeus's categories are still relevant in light of genetic and evolutionary relationships among organisms.

DEMONSTRATION

Write the following names on the board: *Thamnophis melanogaster* (a garter snake) and *Drosophila melanogaster* (a fruit fly); *Canis lupus* (the wolf) and *Anarrhichas lupus* (the wolf fish); and *Phaseolus vulgaris* (the common bean) and *Triticum vulgaris* (wheat). Ask students to explain why only the binomial designation for a species specifically identifies it. (The generic name often applies to more than one species, and the specific identifier may be used for more than one genus.)

ANSWERS TO SECTION 18-1 REVIEW

1. Aristotle classified organisms as either plants or animals. After scientists had discovered many more organisms, some of which could not be placed into any of Aristotle's groups, they recognized that this system was inadequate.
2. Linnaeus used the morphology of organisms as the basis for classification.
3. Linnaeus grouped organisms in a hierarchical system that included kingdom, phylum, class, order, family, genus, and species.
4. Species names, or scientific names, are the same in all languages. Species names place the organism in the hierarchy devised by Linnaeus and indicate the organism's relationship to other organisms.
5. Modern taxonomists use an organism's phylogeny to classify it.
6. Linnaeus based his system on morphological traits of an organism. Morphological traits are heavily influenced by genes, which are clues to common ancestry.

BACKGROUND

A system of nomenclature is necessary for biologists to communicate with each other about organisms. Before Linnaeus, most taxonomists classified organisms by dividing them into large groups and then subdividing these into progressively smaller groups. Linnaeus, however, began with the species, which he organized into genera. Then he arranged genera to form families and related families to form orders and classes. Today the Linnaean system is the internationally accepted taxonomic nomenclature. The international codes of nomenclature for botany were established in 1901. They were established for zoology in 1906 and for microbiology in 1948. There is also a code, established in 1953, for naming cultivated plants, many of which have no wild type. The codes include the following elements: the naming of species by two words using Latin, and a law of priority that states that the first valid binomial name published for a given taxon will prevail.

Carolus Linnaeus was the son of a clergyman. He developed a love of flowers at an early age. At the age of eight he was nicknamed "the little botanist." He studied at the University of Lund and later received a degree in medicine at Uppsala University. The work he published after his exploration of Lapland, *Flora Lapponica,* along with the publication of *Systema Naturae* firmly established his reputation as a botanist. *Systema Naturae* was published in 12 editions during his lifetime. In 1738 Linnaeus settled in Stockholm, Sweden, as a practicing physician. He was married in 1739, and in 1740 was appointed to the Uppsala University chair of medicine. In 1741 he was selected for the position he

Creating Order Out of Chaos

HISTORICAL PERSPECTIVE

During the fifteenth and sixteenth centuries, the exploration of new lands brought large numbers of unknown animals and plants to the attention of naturalists. European explorers returned from other parts of the world with so many unidentified organisms that it became difficult to keep track of them all. Before the introduction of Carolus Linnaeus's binomial system, there was no accepted method for naming and classifying animals and plants. Linnaeus provided a system for grouping organisms in a manner that reflected the relationships between the organisms.

A Rose by Any Other Name . . .

People all over the world use familiar or common names for plants. Sometimes they use different names for the same plant or similar names for different plants. Imagine how difficult it must have been for people from different countries to share their knowledge about the natural world before there was a standard naming system for plants and other organisms.

These early naturalists needed a system for naming living things and placing them into groups of related organisms. The names needed to be short and descriptive—and they needed to be written in a language that was widely understood and accepted. Although there had been some earlier attempts at devising classification systems, Carolus Linnaeus was the first to develop a system that was widely used and accepted.

Carolus Linnaeus

Seeds of Change

Born in Sweden in 1707, Carolus Linnaeus had a great love of plants and nature, instilled in him by a father who educated him about the natural world and taught him the names of many plants. In 1732, while a lecturer at the University of Uppsala, Linnaeus undertook a trip to the then largely unexplored region of northern Scandinavia known as Lapland.

The Lapland journey helped to focus Linnaeus's attention on the need for a standard system of classification, a task that became his lifework. Using earlier research by the German botanist Rudolph Camerius, Linnaeus divided all flowering plants into 23 classes. These classes were based on the number, length, and arrangement of the stamens and pistils. His 24th class included the nonflowering plants, such as mosses.

Taking Root

The prominent Dutch botanist Jan Fredrick Gronovius was greatly impressed with Linnaeus's early botanical work. Gronovius paid for the publication of Linnaeus's *Systema Naturae* (1735), which contained the beginnings of Linnaeus's system for classifying animals and plants.

Linnaeus's greatest contribution to biology was the introduction of the binomial system, in which the

An illustration made to accompany Linnaeus's Systema Naturae *shows the division of flowering plants into 23 classes, based on the number, relative length, and arrangement of the stamens and pistil.*

species name of an organism consists of two parts based on Latin word roots. The concept of the genus name came from the French naturalist Joseph Tournefort. The concept of the second word, the species identifier, came from Linnaeus. Prior to the introduction of this two-part naming system, each type of plant was characterized by a 12-word description. Linnaeus was the first to apply the uniform use of binomials to all organisms. The advantage of this method is that it provides a standardized label for each kind of organism in place of the common name.

Linnaeus's classification system did not escape criticism by his contemporaries, however. He was denounced for imposing an artificial system on nature. Critics claimed his method of classification was based on a single, and perhaps the least important, characteristic of flowers—the arrangement of flower parts. Linnaeus acknowledged that he had sacrificed natural principles to some extent in order to devise a useful sorting principle. But his organizational system had brought together related groups.

Linnaeus first introduced his binomial system in the work *Species Planterum* (1753). This two-volume work contained every plant that Linnaeus was familiar with, and it demonstrated the utility of his system. After publication of *Species Planterum,* the binomial system became the most widely used system in botanical works. As Linnaeus commented in a letter to a friend:

> *Now the whole world is obsessed with writing in the field of botany, now they can go ahead without difficulty, thanks to my method.*

Fruits of His Labor

Since Linnaeus first devised the binomial classification, the most dramatic change has been the systematic effort to make modern taxonomic schemes reflect evolutionary relationships. As Linnaeus himself did, modern scientists have proposed revisions of the traditional classification system. In light of recent research, they have proposed creating new kingdom designations for unicellular organisms. The names of these organisms, however, are still based on the Latin binomial system devised by Linnaeus.

Mertensia virginica, *known in the United States as the bluebell, illustrates the problem with using common names. In Europe and Asia, plants of the genus* Endymion *are called bluebells. Elsewhere, certain species of the genera* Campanula, Clematis, *and* Polemonium *are also commonly known as bluebells.*

most desired, the chair of botany. In *Species Plantarum,* which was published in 1753, Linnaeus gave descriptions of the 6,000 species known at the time. He classified not only organisms but also minerals. He even began a classification of the diseases known in his day.

DISCUSSION

Guide the discussion by posing the following questions:

1. What was Linnaeus's greatest contribution to biology? (Linnaeus's most significant contribution to biology was the introduction of the binomial, or two-name, system, in which the scientific name of an organism consists of two parts—the genus name and its specific identifier.

2. What method did Linnaeus use to classify the flowering plants? (Linnaeus divided flowering plants into 23 classes, based on the number, length, and arrangement of the stamens and pistil.)

3. What changes are presently being proposed in taxonomy? (One proposal is the systematic effort to make modern taxonomic schemes reflect evolutionary relationships. New kingdoms have been proposed for unicellular organisms.)

FURTHER READINGS

1. "How Taxonomy Helps Us Make Sense Out of the Natural World," in *Smithsonian,* May 1996, pages 140–152. This article describes the collectors and collections of the Smithsonian's National Museum of Natural History and the overwhelming task curators face in classifying the specimens in their collections.

2. "A Backbone Runs Through It," in *Natural History,* June 1996, page 3. This article describes the American Museum of Natural History's fossil halls.

SECTION 18-2

SECTION OVERVIEW

Evolution

Section 18-2 describes the evidence from morphology, the fossil record, embryology, and molecular biology that is used to construct phylogenies of organisms. It also describes the new method of cladistics, in which biologists infer phylogenies from patterns of shared, derived characters.

▶ ENGAGE STUDENTS

Show students a picture or specimen of an organism that is difficult to classify, such as a coral or a sponge. Ask if the specimen is a plant or an animal, and ask students to defend their classification. Ask students to consider what observations they would need to make before they could make a firm classification. Point out that advances in technology now enable us to observe distinctions between organisms that Aristotle and Linnaeus could not have observed.

GIFTED *ACTIVITY*

Classification of Regional Plants

Have students make a collection of plant parts from a variety of plants native to your area. Provide them with plant keys, and ask them to identify as many of the plants as possible. Have them present their results and discuss any difficulties they had in making identifications. (They should find that it is extremely difficult to identify plants solely by leaf characteristics and that flowers or fruits are necessary for accurate identifications. Students should find that information about size, presence or absence of wood, time of flowering, and other characteristics are also helpful.)

SECTION 18-2

OBJECTIVES

▲ Define *phylogenetic tree,* and explain what information a phylogenetic tree shows.

● List four types of evidence used to organize organisms in systematic taxonomy.

■ Name two differences found in the embryos of vertebrates and arthropods that suggest a very different phylogenetic history.

◆ Explain cladistic taxonomy, and identify one conclusion that is in conflict with classical, systematic taxonomy.

MODERN PHYLOGENETIC TAXONOMY

More than 200 years ago, Linnaeus based his classification system on the most evident characteristics of organisms—their morphology. Today, the young field of molecular biology can provide a wealth of information about an organism's molecular nature. When placing an organism into a taxonomic category, modern taxonomists may consider its morphology, chromosomal characteristics, nucleotide and amino acid sequences, and embryological development. These features are almost entirely inherited. Thus, consideration of all of them, together with information from the fossil record, is likely to yield reliable information about the phylogeny of an organism.

SYSTEMATICS

Most modern taxonomists agree that the classification of organisms should reflect their phylogeny. This phylogenetic approach is a cornerstone of a branch of biology called systematic taxonomy, or, more commonly, systematics. **Systematics** organizes the tremendous diversity of living things in the context of evolution. Systematic taxonomists use several lines of evidence to construct a phylogenetic tree.

A **phylogenetic tree** is a family tree that shows the evolutionary relationships thought to exist among groups of organisms. A phylogenetic tree represents a hypothesis, and it is generally based on several lines of evidence. Systematic taxonomists may evaluate an organism's morphology with respect to the morphology of similar and possibly ancestral organisms in the fossil record. Likewise, they may compare its morphology with that of living organisms. Patterns of embryological development, along with the degree of similarity of an organism's chromosomes and certain macromolecules to those of other organisms, provide further clues to phylogenetic relationships. A phylogenetic tree is subject to change, as is any hypothesis, as new information arises. A phylogenetic tree that shows possible relationships among phyla in the kingdom Animalia appears in Figure 18-3.

Word Roots and Origins

phylogenetic

from the Greek *phylon,* meaning "tribe," and *gignesthai,* meaning "to be born"

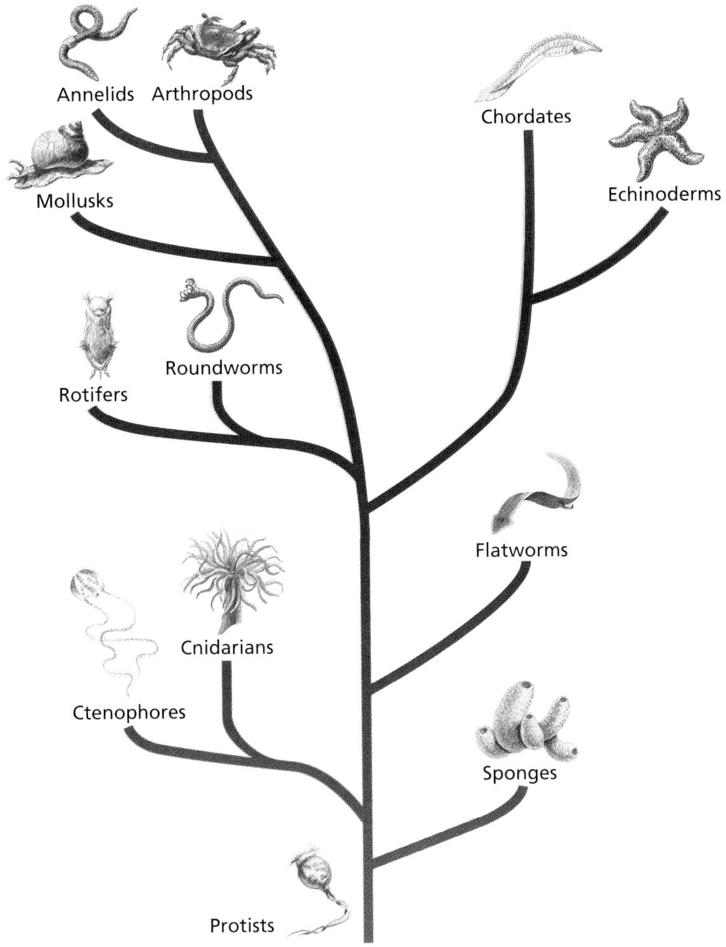

FIGURE 18-3

The branches of this phylogenetic tree show that chordates and echinoderms shared a common ancestor more recently than did echinoderms and other animals, including mollusks and arthropods. Phylogenetic trees are generally derived from several lines of evidence, including morphological, embryological, and macromolecular similarities among organisms.

The Fossil Record

The fossil record often provides clues to evolutionary relationships, but it is important to understand that the fossil record cannot be read like a history book. Some organisms, such as some ocean-living invertebrates, have fairly complete fossil records. Other organisms have incomplete fossil records; there may be series of strata in which no fossils of the organism appear. The fossil record may provide the framework of a phylogenetic tree, but a systematic taxonomist would seek to confirm the information it provided with other lines of evidence.

Morphology

Taxonomists study an organism's morphology and compare it with the morphology of other living organisms. Recall from Chapter 15 that homologous features suggest descent from a common ancestor. Naturally, it is essential to separate those features that are *truly* homologous from those that *seem* homologous but are actually analogous. For example, insects, which are arthropods, and mammals, which are vertebrates, both have legs. But it is clear from the fossil record that legs evolved independently in the two groups. The greater the number of homologous morphological features two organisms share, the more closely related they are thought to be.

internetconnect

SCILINKS NSTA

TOPIC: Phylogenetic tree
GO TO: www.scilinks.org
KEYWORD: HM343

INCLUSION ACTIVITY

Use Figure 18-3 to show students how to read a phylogenetic tree. Point out to students that the organism at the base of the tree is the common ancestor of all the others in the tree. Tell them that branch points indicate the evolution of some characteristic that splits a group of organisms into two groups. Also point out that the groups shown at the tips of the branches include organisms that have evolved most recently.

RETEACHING ACTIVITY

Provide students with skeletons or skulls (or pictures of these) of a variety of vertebrates. Ask them to arrange the items in an order that indicates their degree of relatedness, based on similarities they observe. They should consider the number, shape, and placement of the bones and teeth.

CRITICAL THINKING

Determining Relatedness

Ask students to answer the following question: If you were trying to determine how closely two organisms were related, what would provide the most reliable evidence—morphological characteristics, patterns of embryological development, or amino acid sequences of proteins? Have students defend their answers.

TEACHING STRATEGY

Homology Versus Analogy

Ask students whether each of the following pairs contains homologous or analogous characters: fly's wing and bat's wing (analogous), bat's wing and bird's wing (homologous). Ask students to explain how they made their decisions. (Homologous characters have a common basic structure, while analogous characters usually differ greatly in structure.

DEMONSTRATION

Obtain an illustration comparing embryonic development in several animals, such as a sea urchin, a frog, a chick, and a human. (A college-level developmental biology text would be a good source.) Ask students to identify the first stage at which they notice a major difference between these animals (cleavage—the chick cells do not separate from each other). Ask them how the embryos differ at the blastula stage (the sea urchin and frog embryos are hollow balls, whereas the chick and human embryos are disks of cells on a yolk or attached to a uterine wall). Ask them how the animals differ at the stage when the organs are first formed. (The sea urchin looks entirely different from the other animals; the frog is like the chick and human in having a head, backbone, limbs, etc.; and the chick and human have very few differences at this stage.)

QUICK FACT

☑ Cytochrome *c* is a protein found in all aerobic organisms. The amino acid sequences of cytochrome *c* have been determined for a wide variety of organisms. The sequences for humans and chimpanzees match exactly for all 104 amino acids. Human cytochrome *c* differs from that of a dog by 13 amino acids.

CONTENT CONNECTION

Section 10-1: DNA Structure

DNA is the genetic material of all living things, and it contains instructions for making the proteins that determine the characteristics of an organism. Therefore, the most accurate way to assess evolutionary relationships between organisms is to compare their DNA or the DNA's immediate products, proteins.

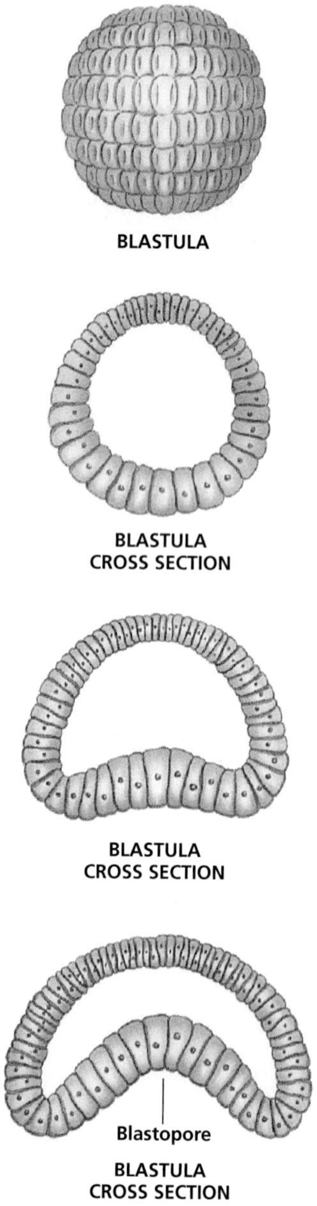

FIGURE 18-4

The blastopore, an indentation of the blastula, forms in early embryological development. The blastopore becomes the posterior end of the digestive system in chordates (which include the vertebrates) and echinoderms. In other animal phyla, the blastopore becomes the anterior end of the digestive system.

Embryological Patterns of Development

Early patterns in embryological development provide evidence of phylogenetic relationships. They also provide a means of testing hypotheses about relationships that have been developed from other lines of evidence. Refer to Figure 15-12 in Chapter 15, which shows the great similarity among embryos of three species of vertebrates.

Differences among animal phyla may appear very early in embryological development. As development begins, the zygote starts to divide by mitosis. Within a matter of hours, a ball of cells called a **blastula** (BLAS-tyoo-luh), shown in Figure 18-4, forms. Soon after that, a small indentation, the **blastopore** (BLAS-toh-POR), develops on the outside of the blastula. Eventually this indentation will develop into the digestive system. In most animal phyla, the blastopore becomes the anterior end of the digestive system (the mouth). But in **echinoderms** (e-KIE-noh-DURHMZ), such as starfish and sand dollars, the blastopore becomes the posterior end of the digestive system, as it does in chordates, which include vertebrates. This pattern of development suggests that echinoderms are more closely related to vertebrates than they are to other invertebrates, such as mollusks, as shown in Figure 18-3.

Echinoderms and vertebrates share other important characteristics as well. In both phyla, each cell of the early embryo is potentially capable of forming the entire organism. For example, identical twins are formed when the early embryo splits in two. Each half becomes a complete individual, and the two individuals share the same genetic information. In contrast, when early embryos of fruit flies (phylum Arthropoda) are experimentally split, each part is already committed to becoming a certain part of the organism, such as the head. Thus, splitting an arthropod embryo will cause the two halves to die.

Chromosomes and Macromolecules

Taxonomists use comparisons of macromolecules such as DNA, RNA, and proteins as a kind of "molecular clock." Recall from Section 15-3 that scientists compare amino acid sequences for homologous protein molecules of different species. The number of amino acid differences is a clue to how long ago two species diverged from a shared evolutionary ancestor. This molecular-clock model is not a perfect one. It assumes that all changes in amino acid sequence are random and are not affected by natural selection. This is probably not true. Moreover, sequences of amino acids can change at different rates in different organisms. But the molecular-clock model is used, together with other kinds of data, to estimate degrees of relatedness between different species.

In a similar kind of analysis, biologists compare the karyotypes, or patterns of chromosomes, of two related species. Regions of chromosomes that have the same pattern of banding are clues to the degree of relatedness of organisms. The chromosomes of humans and chimpanzees show a surprising degree of similarity, as

shown in Figure 18-5. One of the chromosomes in humans is homologous to two smaller chimpanzee chromosomes. This fused human chromosome and six inverted chromosome segments are the only observed chromosomal differences between the two species. In fruit flies, nearly that much chromosomal variation can be found within one species.

The comparison of chimpanzee and human chromosomes prompted several biologists to reevaluate the accepted estimate of how long ago chimpanzees and humans last shared an ancestor. Before chromosomal analysis, it was widely thought that the ancestors of humans and chimpanzees diverged from a common ancestor about 25 million years ago. After the comparisons of karyotypes and amino acid sequences in proteins, molecular biologists decreased the estimate from 25 million years to as little as 5 million years.

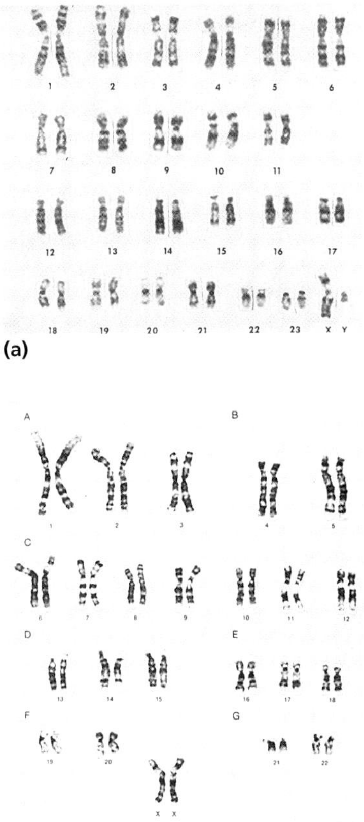

(a)

(b)

FIGURE 18-5

Chimpanzees are genetically very similar to humans. (a) A karyotype of a chimpanzee's 24 chromosome pairs is remarkably similar to (b) a karyotype of a human's 23 chromosome pairs.

CLADISTICS

One relatively new system of phylogenetic classification is called **cladistics** (kluh-DIS-tiks). Cladistics uses certain features of organisms, called shared derived characters, to establish evolutionary relationships. A **derived character** is a feature that apparently evolved only within the group under consideration. For example, if the group being considered is birds, one example of a derived character is feathers. Most animals do not have feathers; birds are the only animals that do. Therefore, it is safe to assume that feathers evolved within the bird group and were not inherited from some distant ancestor of the birds.

Cladistic taxonomists agree that organisms that share a derived character—like feathers—probably share it because they inherited

MAKING CONNECTIONS

Technology

Have students research DNA mapping. This process involves the use of restriction enzymes to cleave DNA into specific sequences of nucleotides, separation of DNA fragments by electrophoresis, and comparisons of the results to those obtained for other species. The separation patterns of the fragments will be most similar for those species that are most closely related. Have students write a paper or give an oral presentation on their research.

TEACHING STRATEGY

Cladograms

Have students use biology books to collect information about humans, turtles, gorillas, lampreys, lungfish, trout, and cats. They should determine whether each animal has hair, jaws, limbs, lungs, and a tail. Have them construct a cladogram using this information. Tell them that the loss of a character can be a derived character. (The first branch on the cladogram should be labeled lamprey, followed by trout, lungfish, turtle, cat, gorilla, and human.)

SECTION 18-2

ANSWERS TO SECTION 18-2 REVIEW

1. A phylogenetic tree is a diagram showing the evolutionary relationships among a group of organisms.
2. Systematic taxonomy is a branch of biology whose purpose is to organize biological diversity within an evolutionary framework. Taxonomists use morphology, fossil evidence, embryology, chromosomal similarities, and biochemical evidence to construct phylogenies.
3. Embryological evidence might reveal similarities between organisms that are not apparent in adult organisms or fossils. For instance, the developmental fate of the blastopore suggests a close relationship between vertebrates and echinoderms.
4. The molecular clock model assumes that all changes of nucleotide and amino acid sequences are random and not affected by natural selection and that homologous proteins in all species change at the same rate.
5. A shared, derived character is found only within a certain group, presumably because it evolved in the common ancestor of the group.
6. Members of the class Reptilia appear to have evolved from several different ancestral lines. "Reptiles" consist of several groups that arose in parallel.

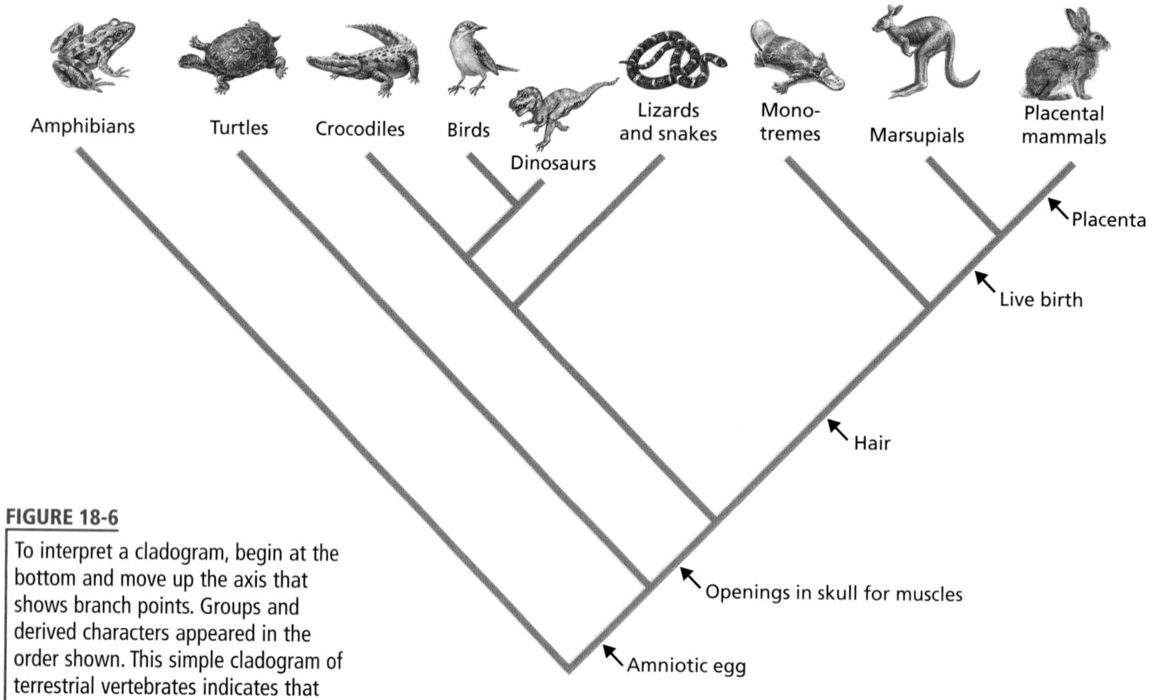

FIGURE 18-6

To interpret a cladogram, begin at the bottom and move up the axis that shows branch points. Groups and derived characters appeared in the order shown. This simple cladogram of terrestrial vertebrates indicates that amphibians arose first, turtles arose next, and placental mammals arose last. Only groups branching *above* a listing of a derived character share that character. Thus, amphibians do not have amniotic eggs, but all other groups do. Likewise, monotremes, marsupials, and placental mammals have hair, though no other groups do.

it from a common ancestor. So shared derived characters, particularly a *group* of several shared derived characters, are strong evidence of common ancestry between organisms that share them. Ancestry diagrams made by means of cladistic analysis, such as the one shown in Figure 18-6, are called **cladograms** (KLAD-uh-GRAMZ).

The application of cladistic taxonomy leads to a number of nontraditional conclusions. One of the most notable is that birds, crocodiles, and dinosaurs are more closely related to each other than any one of them is to a snake or lizard. In the more-classical systematic scheme used in this textbook, snakes, lizards, and crocodiles are classified in the reptile class, while birds are in a class by themselves. A related cladistic conclusion, which differs from that of classical taxonomy, is that the reptiles did not all spring from one common ancestor. Rather, reptiles are a composite of several branches that have occurred during the evolution of the vertebrates, as you can see in the cladogram of vertebrates shown in Figure 18-6.

SECTION 18-2 REVIEW

1. **Define the term *phylogenetic tree.***
2. **What is systematic taxonomy, and what kinds of data are used by a systematic taxonomist?**
3. **How can embryological evidence be used to show phylogenetic relationships that are not evident from either the study of morphology or the study of the fossil record?**
4. **What are two flaws of the molecular clock model in determining relatedness between species?**
5. **What is a shared derived character?**
6. **CRITICAL THINKING Why does the cladistic approach to classification suggest that the class Reptilia (reptiles) is not a phylogenetic classification?**

TWO MODERN SYSTEMS OF CLASSIFICATION

Aristotle classified organisms as either plants or animals, but today we recognize that many forms of life are neither. In this section, you will read about two alternative classification systems that are in current use. But remember, organizational systems are imposed by humans and therefore may be flawed. As is true of everything in science, they are subject to change as new information arises.

SIX-KINGDOM SYSTEM

A classification system based on five kingdoms of organisms was preferred by taxonomists for many years. But further studies of bacteria have shown that there are two important subtypes with very different morphologies and properties. Recognition of these two broad types of bacteria has driven the acceptance of a newer, six-kingdom system, illustrated in Table 18-2, which is used in this textbook.

TABLE 18-2 *Six Kingdoms of Life*

Kingdom	Cell type	Number of cells	Nutrition
Archaebacteria	prokaryotic	unicellular	autotrophy and heterotrophy
Eubacteria	prokaryotic	unicellular	autotrophy and heterotrophy
Protista	eukaryotic	unicellular and multicellular	autotrophy and heterotrophy
Fungi	eukaryotic	unicellular and multicellular	heterotrophy
Plantae	eukaryotic	multicellular	autotrophy and (rarely) heterotrophy
Animalia	eukaryotic	multicellular	heterotrophy

SECTION 18-3

OBJECTIVES

▲ Describe the six-kingdom system of classification.

● List the characteristics that distinguish archaebacteria from eubacteria.

■ Explain why the protists are grouped together in the six-kingdom system in spite of having differences that are greater than those between plants and animals.

◆ Describe the evidence that prompted the creation of the three-domain system of classification.

▲ Explain the principal difference between the six-kingdom system and the three-domain system of classification.

SECTION 18-3

SECTION OVERVIEW

Evolution

Section 18-3 describes the characteristics of the six kingdoms of organisms—archaebacteria, eubacteria, protists, plants, fungi, and animals. It also describes an alternative scheme for classifying life into three broader groups, called domains.

▶ ENGAGE STUDENTS

Provide students with a variety of preserved or living specimens from each of the six kingdoms of living things. Provide microscopes, hand lenses, and dissecting tools for them to use in studying the specimens. Ask them to classify each organism in one of the kingdoms and to support their choice by identifying at least three characteristics it possesses that are characteristic of the kingdom.

MAKING CONNECTIONS

Social Studies

Ask students to choose one kingdom. Using encyclopedias and other reference books as sources, students should investigate the importance of several organisms in this kingdom to humans. They should consider benefits we receive from the organisms, such as food, oxygen, fibers, structural materials, clothing, and pharmaceuticals. They should also consider detrimental effects of the organisms, such as diseases, food spoilage, property damage, and destruction of crops. Have students make an oral presentation on what they learn.

CRITICAL THINKING

Kingdom Characteristics

Have students imagine that they have been asked to classify an organism in one of the six kingdoms. The organism is multicellular and heterotrophic. Ask them what other piece of information they would need to know in order to classify it in the correct kingdom (whether it obtains food by ingestion, and is thus an animal, or by absorption, and is thus a fungus).

OVERCOMING MISCONCEPTIONS

Many people say that they have a "bug" of some kind when they are ill with an infectious disease. There are insects that are classified as true bugs (order Hemiptera). However, infectious diseases are usually caused by viruses, bacteria, and protists.

CONTENT CONNECTION

Section 14-3: The First Eukaryotes

The hypothesis of endosymbiosis suggests that certain prokaryotes were engulfed by precursors of eukaryotic cells and eventually became organelles within today's eukaryotic cells. The hypothesis proposes that mitochondria came from aerobic prokaryotes and chloroplasts came from photosynthetic prokaryotes. The presence of DNA in these two organelles supports this hypothesis.

QUICK FACT

☑ Most protists are harmless, and many are very beneficial to humans. Algae generate more than half of the world's oxygen. However, protists in the genus *Plasmodium,* which cause malaria, are responsible for the death of more than 1 million people each year.

FIGURE 18-7

Archaebacteria often live in very hostile environments that cannot support other forms of life, such as this hot spring, Morning Glory Pool, in Yellowstone National Park.

Kingdom Archaebacteria

The members of the **kingdom Archaebacteria** (AHR-kee-bak-TIR-ee-uh) are unicellular prokaryotes with distinctive cell membranes as well as biochemical and genetic properties that differ from all other kinds of life. Some species of archaebacteria are autotrophic, producing food by chemosynthesis. Their waste products may include flammable gases, such as methane. Many archaebacteria live in harsh environments such as sulfurous hot springs, as shown in Figure 18-7, and very salty lakes, and in anaerobic environments, such as in the intestines of mammals.

The prefix *archae-* comes from the Greek word for "ancient." Modern archaebacteria may be directly descended from and very similar to the first organisms on Earth, which flourished before the evolution of photosynthesis. These early archaebacterial ancestors evolved before the release of large amounts of oxygen gas into the environment.

FIGURE 18-8

Some protists are multicellular. This ocean-living giant kelp resembles a plant, but it lacks the organization of tissues found in true plants.

Kingdom Eubacteria

The *eu* part of **eubacteria** (YOO-bak-TEER-ee-uh) means "true." Eubacteria are unicellular prokaryotes. Most of the bacteria that affect your life—those that cause tooth decay, turn milk into yogurt, and cause food poisoning—are members of the **kingdom Eubacteria.** Most species of eubacteria use oxygen, but a few species cannot live in the presence of oxygen.

At first glance, both eubacteria and archaebacteria may seem unimportant. But remember that they include the greatest number of living things on Earth. Moreover, their ancestors, which may have been quite similar to modern bacteria, were probably the first living things on Earth.

Both archaebacteria and eubacteria reproduce by binary fission, but they do have some ways to recombine genes, allowing evolution to occur. The very short generation times of bacteria (as little as 30 minutes) allow rapid evolutionary response to environmental change. If you have ever had an antibiotic-resistant bacterial infection, you have had experience with the remarkably fast evolution of bacteria.

Kingdom Protista

The **kingdom Protista** (proh-TIS-tuh) is made up of a variety of eukaryotic, mostly single-celled organisms. Some species of **protists** (PROH-tists) exist as multicellular organisms, like the giant kelp shown in Figure 18-8. Although they look much like plants, multicellular protists lack specialized tissues. Being eukaryotes, they have a membrane-bound true nucleus with linear chromosomes, and they have membrane-bound organelles.

It is difficult to make generalizations about the protists because many protist species are more distantly related to each other than plants are to animals. The kingdom Protista contains all eukaryotes that are not plants, animals, or fungi, more than 50,000 species in all. The sexual cycles of many protists are unknown, but most are thought to have some process of genetic recombination.

Euglena and the amoebas are common types of unicellular protists. *Euglena,* shown in Figure 18-9, can feed on other organisms in the manner of an animal, but it also has chloroplasts and can perform photosynthesis if light is available. Amoebas, such as the one shown in Figure 18-2, feed on other organisms and respond to touch and light. Yet *Euglena* is not a plant, and amoebas are not animals—both are protists.

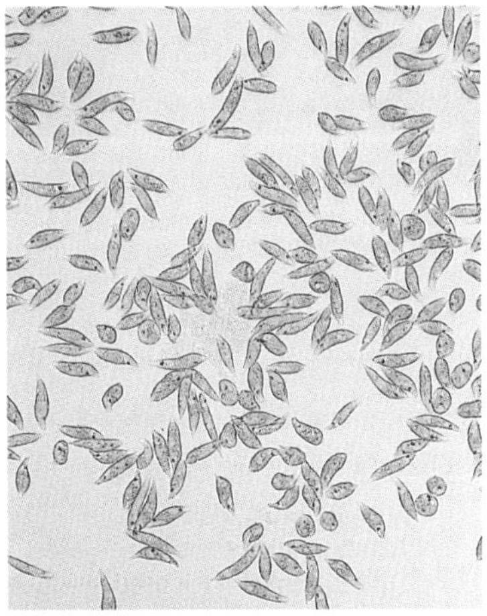

FIGURE 18-9

Euglena gracilis (LM 580×), like all species of *Euglena,* is a unicellular protist that can be autotrophic or heterotrophic, depending on its environmental conditions.

Kingdom Fungi

The **kingdom Fungi** (FUHN-jee) is made up of heterotrophic unicellular and multicellular eukaryotic organisms. Fungi absorb nutrients rather than ingesting them the way some protists, such as amoebas, do. While sexual cycles are not known for many fungi, it is likely that all species have some way of promoting gene recombination. The well-studied mold *Neurospora* has a standard sexual cycle and has been used extensively in the study of meiosis and in the genetic control of physiological functions. There are over 100,000 species of fungi, including mushrooms, puffballs, rusts, smuts, mildews, and molds.

Kingdom Plantae

As you might have expected, the **kingdom Plantae** (PLAN-tee) consists of multicellular plants. All except for a few parasitic forms are autotrophic and use photosynthesis as a source of energy. Most plants live on land, and most have a sexual cycle based on meiosis. More than 350,000 species of plants have been identified. They include mosses, ferns, conifers, and flowering plants, like the orchid shown in Figure 18-10.

FIGURE 18-10

Orchids, like this *Paphiapedilum rothschildionum,* are found in tropical climates. Flowers contain the plant's gametes, or reproductive cells.

Kingdom Animalia

The **kingdom Animalia** (AN-uh-MAH-lee-uh) is made up of eukaryotic, multicellular heterotrophic organisms. Most animals have symmetrical body organization and move about their environment. Almost all animals have a standard sexual cycle that employs meiosis for the recombination of genes.

DEMONSTRATION

Provide the students with pictures and descriptions of a variety of protists. Be sure to include at least one plantlike protist (for example, *Ulva*), one animal-like protist (for example, *Paramecium*), and one fungus-like protist (for example, *Phytopthora infestans*). Ask them to identify the characteristics each organism shares with members of another kingdom. (Plantlike protists are autotrophs and have cell walls; animal-like protists are ingestive heterotrophs and lack cell walls; and fungus-like protists are absorptive heterotrophs and have cell walls.)

RECENT RESEARCH

Fungi May Be Closely Related to Animals

At one time, fungi were classified in the plant kingdom. Scientists analyzing ribosomal RNA from a wide variety of organisms, however, found fungi to be more similar to animals than to plants.

INCLUSION ACTIVITY

Have students collect pictures of members of the six kingdoms of life, using photographs or drawings cut from old newspapers and magazines. Have them glue all the pictures for one kingdom on one page and write the major characteristics of that kingdom on the paper.

RECENT RESEARCH

A New Phylum of Animals

A tiny invertebrate has been identified that has such unique characteristics that it cannot be classified in any of the animal kingdom's 35 or so phyla. This animal attaches itself to the mouthparts of lobsters and uses cilia around its mouth to capture food intended for the lobster. This creature has been assigned to a new phylum, called Cycliophora.

SECTION 18-3

VISUAL LINK
Figure 18-11

Have students examine Figure 18-11 as you summarize the characteristics of the members of each category in the two systems of classification. Ask students why they think many biologists find the six-kingdom system more useful. (It contains more categories, so it is more useful in distinguishing different organisms.)

ANSWERS TO SECTION 18-3 REVIEW

1. The six kingdoms of the six-kingdom system are Archaebacteria, Eubacteria, Protista, Fungi, Plantae, and Animalia.
2. Archaebacteria grow in hostile environments that are difficult to replicate in the laboratory. Some produce explosive or noxious gases, such as methane.
3. Kingdom Protista is the most morphologically heterogeneous.
4. The conclusion that organisms fall into three broad categories, or domains, is based on the sequence of nucleotides in ribosomal RNA molecules.
5. These organisms are all eukaryotes and share the characteristics of true nuclei, linear chromosomes, and membrane-bound organelles.
6. Answers will vary, but students might suggest that because bacteria are small and fairly simple, they have few characters on which to base taxonomic decisions. Alternatively, they might suggest that differences between eubacteria and archaebacteria are mainly molecular and the technology for identifying such differences has only recently been developed.

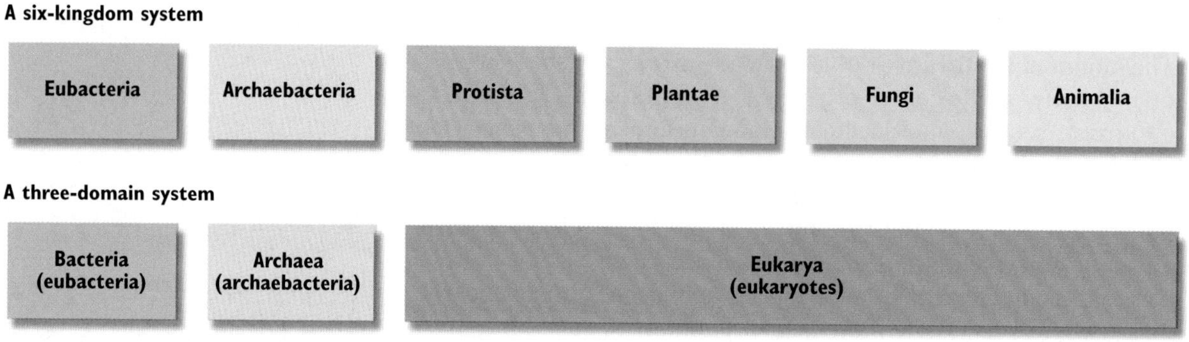

FIGURE 18-11

The three-domain system highlights the importance of archaebacteria as a life-form. This system is often used by molecular biologists. Notice that the domain Eukarya includes members of the kingdoms Protista, Plantae, Fungi, and Animalia, which are all made up of eukaryotic organisms.

Word Roots and Origins

domain

from the Latin *dominium,* meaning "right of ownership"

THREE-DOMAIN SYSTEM

The young science of molecular biology has led to an alternative to the six-kingdom system. By comparing sequences of ribosomal RNA in different organisms, molecular biologist Carl Woese (1928–), of the University of Illinois, has estimated how long ago pairs of different organisms shared a common ancestor. Because all organisms, even prokaryotes, have ribosomes, the ribosomal RNA molecule can be used to study the degree of relationship between any two living things. The phylogenetic tree drawn from these data shows that living things seem to fall naturally into three broad groups, or **domains.** Plants, animals, and fungi are one small twig of a large branch—a domain—that includes all the eukaryotes. These domains are shown in comparison with the six kingdoms in Figure 18-11.

Domain Archaea (ahr-KEE-uh) is known in the six-kingdom system as kingdom Archaebacteria. **Domain Bacteria** is known in the six-kingdom system as the kingdom Eubacteria.

Domain Eukarya (yoo-KAR-ee-uh) consists of the protists, the fungi, and the plants and animals. All eukaryotes have true nuclei with linear chromosomes and membrane-bound organelles. Most of the variation in this domain is among the protists. Surprisingly, when considered from the perspective of the complete diversity of life on Earth, the fungi, plants, and animals are quite similar to each other.

SECTION 18-3 REVIEW

1. What are the six kingdoms in the six-kingdom system of classification?
2. What are two things that make archaebacteria difficult to study?
3. What is the most heterogeneous kingdom in terms of morphology?
4. What kind of evidence indicates that organisms fall naturally into three broad domains?
5. Why do protists, fungi, plants, and animals share a domain in the six-kingdom system?
6. **CRITICAL THINKING** In the five-kingdom system, which is still used by some scientists, all species of bacteria are grouped in the kingdom Monera. Why might there have been only one bacteria kingdom recognized in the past?

CHAPTER 18 REVIEW

SUMMARY/VOCABULARY

18-1

- Taxonomy is the science of grouping organisms according to their morphology and evolutionary history.
- Carolus Linnaeus originated a seven-level hierarchy system for classifying organisms according to their morphology. Moving from the most general to the most specific, the levels are called kingdom, phylum, class, order, family, genus, and species. A version of this system is still in use.
- A species name consists of the genus name together with a species identifier. A species denotes a single organism type.

Vocabulary

binomial nomenclature (339)
class (338)
division (338)
family (338)
genus (338)
kingdom (338)
order (338)
phylogeny (339)
phylum (338)
species (338)
species identifier (339)
species name (339)
subspecies (339)
taxonomy (337)
variety (339)

18-2

- A modern approach to taxonomy is systematics, which analyzes the diversity of organisms in the context of their evolutionary history.
- Scientists consider several lines of evidence when classifying organisms according to their evolutionary history.
- An organism's relationship to organisms in the fossil record as well as to living organisms is taken into account in the formulation of a phylogenetic tree.
- Similarities in patterns of embryological development provide clues to the degree of relatedness of different organisms.
- Molecular similarities, such as those found in homologous proteins of different organisms, also indicate how closely organisms are related.
- Shared derived characters, those traits that developed within a certain group, are clues to the degree of relatedness among organisms. The system that uses shared derived characters to deduce evolutionary history is called *cladistics.*

Vocabulary

blastopore (344)
blastula (344)
cladistics (345)
cladogram (346)
derived character (345)
echinoderm (344)
phylogenetic tree (342)
systematics (342)

18-3

- Many modern taxonomists use the six-kingdom system of classification, which recognizes the unique nature of the archaebacteria.
- Archaebacteria, some of which live in extremely harsh environments, have been largely ignored until recently. Scientists now think archaebacteria closely resemble the first kinds of organisms to live on Earth.
- An alternative classification system that employs three broad domains groups all eukaryotic organisms under the domain Eukarya. Eubacteria (domain Bacteria) and archaebacteria (domain Archaea) form each of the other two domains.

Vocabulary

domain (350)
domain Archaea (350)
domain Bacteria (350)
domain Eukarya (350)
eubacteria (348)
kingdom Animalia (349)
kingdom Archaebacteria (348)
kingdom Eubacteria (348)
kingdom Fungi (349)
kingdom Plantae (349)
kingdom Protista (348)
protist (348)

CHAPTER 18 REVIEW ANSWERS

REVIEW

1. Common names can vary by locale and language. Species names are the same in all languages, and they place an organism in an organizational scheme based on evolutionary history.
2. The word *division* is used by those who study plants. *Phylum* is preferred for other organisms.
3. *Phylon* means "tribe," and *geny* means "to make" (as in "generate"). *Phylogeny,* which literally means "tribe-making," is a good word to describe the science of grouping organisms.
4. A variety is a subdivision of species in plants. Likewise, a subspecies is a subdivision of species in organisms other than plants.
5. *Echino* comes from a Greek word that means "hedgehog"—an animal with protective spines. Many species of echinoderms, such as sea urchins, have spines.

6. d
7. a
8. c
9. c
10. b
11. c
12. c
13. c
14. a
15. b

16. (a) kingdom, (b) phylum or division, (c) class, (d) order, (e) family, (f) genus, (g) species
17. Both grouped organisms according to observable characteristics.
18. Communication about organisms would be impossible without a standard naming system. The species name also tells what genus the organism is grouped into.
19. Plants are usually autotrophic and photosynthetic. Fungi are heterotrophic.
20. Differences in amino acid sequences for homologous proteins in two different species are a clue to how long ago the species shared an ancestor.
21. cladistics
22. who studied and named it, where it lives, and what it looks or acts like
23. Similarities in embryological development generally indicate that two species shared an ancestor relatively recently.
24. by chemosynthesis
25. Plants, fungi, and animals are all eukaryotes and share a common cellular organization, with a nucleus, linear chromosomes, and membrane-bound organelles.

Critical Thinking

1. Scientists use different methods and different lines of reasoning to attempt to understand what has occurred over evolutionary time, beginning billions of years ago. There are probably some flaws in all systems of determining phylogenetic histories of organisms, and methods are constantly improved. Different methodologies sometimes produce different results, and new methods and lines of reasoning, such as those used in cladistic taxonomy, are often not accepted immediately.
2. The results of macromolecular comparisons should resemble the results of cladistic analysis, since both are designed to determine

Review

Vocabulary

1. Distinguish between a species name and a common name.
2. What is the difference between a phylum and a division?
3. Using a dictionary, find the meaning of the word parts *phylo-* and *-geny* in *phylogeny.* Explain why this term is appropriate for discussing common ancestry of organisms.
4. What is the difference between a subspecies and a variety?
5. Look up the roots of the word *echinoderm.* How does what you found relate to the properties of a group that includes starfish, sand dollars, and sea urchins?

Multiple Choice

6. A species name includes information about (a) species and phylum (b) division and genus (c) genus and order (d) genus and species.
7. Aristotle classified plants on the basis of differences in their (a) stems (b) flowers (c) leaves (d) roots.
8. Linnaeus classified organisms based on similarities of their (a) genes (b) homologous proteins (c) morphology (d) embryology.
9. A group of related classes of organisms make up a (a) genus (b) order (c) phylum (d) kingdom.
10. The kingdom Animalia is divided into phyla. At the same level of organization, the kingdom Plantae is divided into (a) classes (b) divisions (c) species (d) genera.
11. Some animal species are divided into (a) identical species (b) varieties (c) subspecies (d) twin species.
12. Classifying organisms according to their presumed evolutionary history is called a (a) six-kingdom approach (b) morphological approach (c) phylogenetic approach (d) three-domain approach.
13. The kingdom Protista includes (a) bacteria (b) plants (c) algae (d) mushrooms.
14. Some protists are similar to plants in that they (a) carry on photosynthesis (b) have plantlike organization of tissues (c) ingest nutrients (d) are unicellular.
15. Taxonomists can use data from RNA-sequencing techniques to (a) predict future changes in species (b) estimate when two species diverged from a common ancestor (c) determine species name (d) explain the origin of life.

Short Answer

16. List the seven levels of Linnaeus's classification hierarchy from most general, *A*, to most specific, *G*.

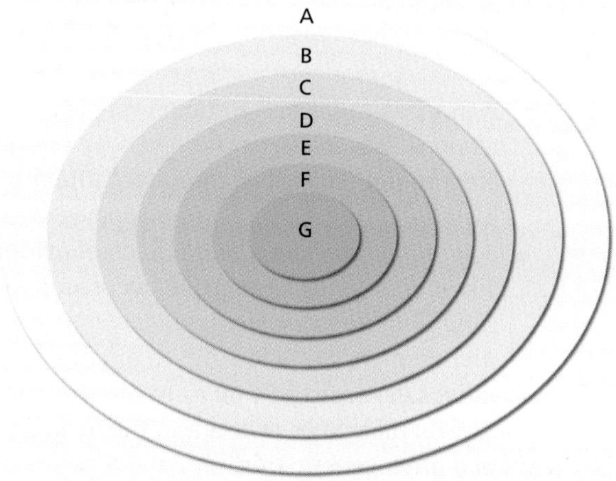

17. How were Aristotle's and Linnaeus's classification systems for organisms similar?
18. Why are species names important in scientific work?
19. What are the differences between plants and fungi?
20. How do amino acid sequences function as a "biological clock"?
21. What do we call the system of classification that is based on an analysis of shared derived characters?
22. Name three things you might learn about an organism by investigating the meaning of its scientific name.
23. How might a taxonomist use embryological evidence in classifying an organism?
24. How do some archaebacteria produce food?
25. What do plants and fungi have in common with animals?

CRITICAL THINKING

1. Scientists agree that evolution has occurred and continues to occur, and that all of the organisms on Earth are related to each other to varying degrees. It is also obvious that the course of evolution proceeded in one way only, yet scientists often disagree about phylogenetic histories of organisms. Cladistic taxonomists regard reptiles in a different light than do more classical taxonomists. Why might scientists disagree with each other about the course of evolution?
2. The evolutionary history of reptiles can be studied using comparisons of their sequences of macromolecules. The degree of difference can be related to the time that has passed since any two species descended from a common ancestor. Would the phylogenetic tree derived from macromolecular comparisons probably more closely resemble the results of cladistic analysis or the standard classification of all reptiles as a single class of vertebrates?
3. Biologists think that there are probably millions of undescribed and unclassified species on Earth. Why might so many species still be undescribed or unclassified today?
4. Legs are an example of a shared derived character in vertebrates. Arthropods, such as lobsters and crickets, also have legs, but they are not accepted as a character shared with vertebrates. Why?
5. A number of years ago scientists found a living fish, called a coelacanth and pictured below, that was thought to have become extinct about 65 million years ago. The earliest fossils of coelacanths are about 350 million years old. Thus, the appearance of the coelacanth has remained unchanged for 350 million years. Although it is impossible to compare macromolecules such as proteins of a 350-million-year-old fossil coelacanth with that of a freshly caught coelacanth, what would you expect to find if you could?

EXTENSION

1. Read "Robust About Face" in *Science News,* April 24, 1999, on page 267. Describe the features of the head of *Australopithecus robustus*. Where do some researchers place *A. robustus* in the human lineage? According to Melanie McCollum, why is this cladistic reasoning misleading?
2. Visit the zoo, and list the scientific names of all the animals you see, or use your library to research 10 organisms. Record the scientific and common names of these organisms. For each animal, list a trait that led taxonomists to classify the organism in its particular genus or family.
3. Collect half a cup of water from a shallow pond. Using a microscope, study several samples from the water. Draw the organisms you find, and classify them as best you can into kingdom and phylum.

which organisms descended from a common ancestor.
3. Many of the unclassified species may be few in number or located in remote places.
4. Arthropod legs are structurally very different from the legs of vertebrates. Arthropod legs are covered by an exoskeleton and do not have an internal skeleton of bone.
5. Since the modern coelacanth is very similar morphologically to an ancient coelacanth, it is likely that it would be similar molecularly as well.

EXTENSION

1. *Australopithecus robustus* had massive jaws, large peglike teeth with smaller ones in front, a small brain case, and a bony crest on top of the skull. Some researchers view *A. robustus* as a deadend in human evolution, and they place *A. robustus* in its own genus, *Paranthropus.* Cladistics can compare only those traits that have evolved independently of one another. McCollum proposes that the unusual head traits of *A. robustus* have evolved as a result of its unusual set of teeth.
2. Answers will vary. Students will probably list obvious traits of external morphology. You may wish to photograph the animals that the students identify to provide them with a visual aid when they do their library research.
3. Answers will vary. Students may see various protists, such as *Paramecium, Amoeba,* and *Stentor.* They likely will not be able to observe bacteria without culturing and staining them, but point out that there are probably many bacteria present in the drop of water on their slide.

TIME REQUIRED

One 50-minute class period

QUICK REFERENCES

Lab Preparation Notes are found on page 335B.

Holt BioSources provides a CD-ROM that contains "Using Gowin's Vee in the Lab" and "Scoring Rubrics."

PROCEDURAL TIPS

1. You may want to have students complete Parts A and B on two different days. Part A will take about 10 minutes to complete, and Part B will require about 40 minutes.
2. Emphasize that a dichotomous key includes pairs of descriptions that lead to the identification of an object. Having students make dichotomous keys allows them to practice observing, recording, and organizing data. Lead students through a discussion of using the process of elimination when working with a dichotomous key.
3. For Part B, encourage students to select characteristics that are useful in distinguishing one shoe from another. Characteristics might include type of shoe, style, color, size, or function.

ANSWERS TO BACKGROUND

1. Carolus Linnaeus developed the classification system used by scientists.
3. Classification enables scientists to identify a specific organism and to communicate accurate information efficiently.

Using and Formulating Dichotomous Keys

OBJECTIVES

- Use a dichotomous key to identify leaves.
- Construct a dichotomous identification key.

PROCESS SKILLS

- identifying
- classifying
- designing
- interpreting
- organizing data
- comparing and contrasting

MATERIALS

- pencil
- paper
- shoes
- masking tape
- marker

Background

1. Who developed the classification system that scientists use today to classify and group organisms according to their inherited traits?
2. Taxonomy is the science of naming and classifying organisms according to their characteristics and evolutionary history.
3. Why is classification essential to biology?
4. A dichotomous key uses pairs of contrasting, descriptive statements to lead to the identification of an organism (or other object).
5. The principle behind dichotomous keys—the forced choice—is used in many different situations to narrow the path toward an answer. If you have ever had your eyes examined for corrective lenses, you are familiar with the series of forced choices that end with the choice of the correct lenses for your eyes.

PART A Using a Dichotomous Key

1. Field guides often use dichotomous keys to identify organisms. Use the dichotomous key shown here to identify the tree leaves below. Begin with the paired descriptions 1a and 1b, and follow the directions. Proceed through the list of paired descriptions until you identify the leaf in question. In your lab report, write the names of the leaves as you identify them.

Dichotomous Key for Identifying Common Leaves

1a. If the edge of the leaf has no teeth, or lobes, go to 2 in the key.
1b. If the edge of the leaf has teeth, or lobes, go to 3 in the key.

2a. If the leaf has slightly wavy edges, it is a shingle oak.
2b. If the leaf has smooth edges, go to 4 in the key.

3a. If the leaf edge is toothed, it is a Lombardy poplar.
3b. If the leaf edge has lobes, go to 5 in the key.

4a. If the leaf is heart-shaped with veins branching from the base, it is a redbud.
4b. If the leaf is not heart-shaped, it is a live oak.

5a. If the leaf edge has a few large lobes, it is an English oak.
5b. If the leaf edge has many small lobes, it is a chestnut oak.

PART B Making a Dichotomous Key

2. Gather 10 different single shoes, and use masking tape and a marker to label the soles of the shoes with the owner's name. The labeled shoes should then be placed on a single table in the classroom.
3. Form small groups. Discuss the appearance of the shoes. In your lab report, make a table like the one below that lists some general characteristics of the shoes, such as the type and size. Also list the names of the students who own the shoes. Complete the chart by describing the characteristics of each person's shoe.
4. Use the information in your table to make a dichotomous key that can be used to identify the owner of each shoe. Remember that a dichotomous key includes pairs of opposing descriptions. At the end of each description, the key should either identify an object or give directions to go to another specific pair of descriptions. Write your dichotomous key in your lab report.
5. After all groups have completed their key, exchange keys with a member of another group. Use the key to identify the owner of each shoe, and then verify the accuracy of your identification by reading the label on the shoe. If the key has led you to an inaccurate identification, return the key so that corrections can be made.
6. Clean up your materials before leaving the lab.

Analysis and Conclusions

1. What other characteristics might be used to identify leaves with a dichotomous key?
2. Were you able to identify the shoes using another group's key? If not, describe the problems you encountered.
3. How was it helpful to list the characteristics of the shoes before making the key?
4. Does a dichotomous key begin with general descriptions and then proceed to more specific descriptions, or vice versa? Explain your answer, giving an example from the key you made.
5. Are dichotomous keys based on a phylogenetic or morphological approach to classification? Explain your answer.

Further Inquiry

List characteristics that might be used to identify birds or other animals using a dichotomous key. Compare your list of characteristics with those used in a dichotomous key in a field guide for identifying birds or other animals.

TABLE A *DISTINGUISHING FEATURES OF A SAMPLE OF SHOES*

	Left/right	Men's/women's	Laced/slip-on	Color	Size	Owner
1						
2						
3						
4						
5						

ANSWERS TO ANALYSIS AND CONCLUSIONS

1. Other leaf characteristics might include whether the leaf is compound or simple, whether or not the leaf is needlelike, and the arrangement of leaves on the branches.
2. Answers will vary. Students might have discovered that some of the opposing statements were not contrasting, that some of the shoes were not correctly described, or that the key did not include a sufficient number of descriptions to distinguish all the shoes from one another.
3. Listing characteristics encourages organization of the information that will be elicited on the dichotomous key.
4. A dichotomous key begins with general descriptions and proceeds to specific descriptions as possible identities for the object being identified are discarded. Examples will vary.
5. Dichotomous keys are based on a morphological approach to classification because they depend only on observable characteristics of the object being identified.

Unit 5

ECOLOGY

CHAPTERS

internetconnect

SciLINKS NSTA

National Science Teachers Association *sci*LINKS Internet resources are located throughout this unit.

"*We and our fellow vertebrates are largely along for the ride on this planet. If we want to perpetuate the dream that we are in charge of our destiny and that of our planet, it can only be by maintaining biological diversity—not by destroying it. In the end, we impoverish ourselves if we impoverish the biota.*"

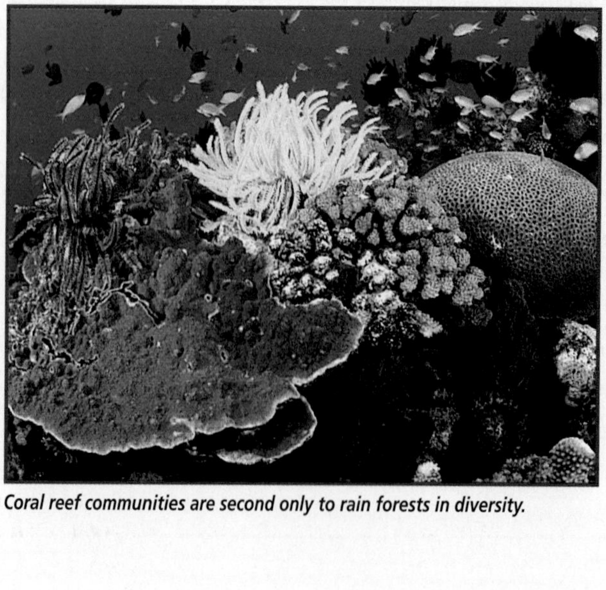

Coral reef communities are second only to rain forests in diversity.

Bears are among the largest terrestrial predators.

The biosphere

Mimicry helps this mantid hide from both predators and potential prey.

Tropical rain forests are richer in species than other areas of Earth.

unit 5

CHAPTERS

ECOLOGY

CONNECTING TO THE STANDARDS

The following chart shows the correlation of Unit 5 with the *National Science Education Standards* (grades 9–12). The items in each category are addressed in this unit and should be used to guide instruction. Annotated descriptions of the **Life Science Standards** are found below the chart. Consult the *National Science Education Standards* for detailed descriptions of each category.

UNIFYING CONCEPTS
- Systems, order, and organization
- Evidence, models, and explanation
- Evolution and equilibrium

SCIENCE AS INQUIRY
- Abilities necessary to do scientific inquiry
- Understandings about scientific inquiry

SCIENCE AND TECHNOLOGY
- Abilities of technological design
- Understandings about scientific inquiry

PHYSICAL SCIENCE
- Conservation of energy and increase in disorder
- Interactions of energy and matter

LIFE SCIENCE

EARTH AND SPACE SCIENCE
- Energy in the Earth system
- Geothermic cycles
- Origin and evolution of the Earth system

SCIENCE IN PERSONAL AND SOCIAL PERSPECTIVES
- Population growth
- Natural resources
- Environmental quality

HISTORY AND NATURE OF SCIENCE
- Science as a human endeavor
- Nature of scientific knowledge

ANNOTATED DESCRIPTIONS OF THE CORRELATED LIFE SCIENCE STANDARDS

Although all eight categories of the *National Science Education Standards* are important, the following descriptions summarize the **Life Science Standards** that specifically relate to Unit 5.

BIOLOGICAL EVOLUTION
- Species evolve over time.
- The millions of different species that live on Earth today are related by descent from common ancestors.

INTERDEPENDENCE OF ORGANISMS
- The atoms and molecules on the Earth cycle among the living and nonliving components of the biosphere.
- Energy flows through ecosystems in one direction, from photosynthetic organisms to herbivores to carnivores and decomposers.
- Organisms both cooperate and compete in ecosystems.
- Living organisms have the capacity to produce populations of infinite size, but environments and resources are finite.
- Human beings live within the world's ecosystems.

MATTER, ENERGY, AND ORGANIZATION IN LIVING SYSTEMS
- The energy of life primarily derives from the sun.
- The chemical bonds of food molecules contain energy.
- The complexity and organization within an organism result from the organism's need to obtain, transform, release, and eliminate the matter and energy required to sustain the organism.
- The distribution and abundance of organisms and populations in ecosystems are limited by the availability of matter and energy and the ability of the ecosystem to recycle materials.
- Matter and energy flow through different levels of organization of living systems and between living systems.

TRENDS IN BIOLOGY

Protecting the Ozone Layer

Chemist Sherwood Rowland remembers coming home one day from the lab and telling his wife, "The work is going very well, but it looks like the end of the world." The grim possibility that Rowland and his colleague Mario Molina foresaw and described in a 1974 paper was the destruction of the ozone layer by CFCs. Today, even though measurements of the ozone layer have substantiated Rowland and Molina's warning and revealed much greater ozone loss than previously imagined, many ozone experts predict that the ozone layer will fully recover within 50 years.

Once evidence of ozone depletion was discovered, the international community acted swiftly to control CFCs and other ozone-destroying chemicals. This effort is unusual not only for its speed but also for the unprecedented cooperation among scientists, governments, and industry.

PHASING OUT CFCs

The regulation of ozone-destroying chemicals began in the late 1970s when the United States, Canada, Sweden, and Norway banned the use of CFCs in aerosol cans. Little else was done at the time, however, because most scientists and governments thought that ozone depletion was not an urgent problem. In 1985, British scientists destroyed this complacency when they discovered an enormous region of reduced ozone levels—the famous ozone "hole" over Antarctica.

One result of the discovery of the ozone hole was the Montreal Protocol, an international treaty signed in 1987 that restricted the manufacture of CFCs and other ozone-depleting chemicals. The treaty has been amended several times and now calls for the elimination of several kinds of ozone-destroying chemicals, including CFCs, methyl bromide (a pesticide), carbon tetrachloride (used in dry cleaning), and, eventually, hydrochlorofluorocarbons, which are interim replacements for CFCs. The current version of the treaty mandated that industrialized countries stop producing CFCs by January 1, 1996, but the developing countries were given until 2010 to meet this requirement.

Some positive effects of the Montreal Protocol can already be seen. World production of CFCs has fallen by more than 75 percent (see graph below). Moreover, the amount of CFCs in the lower atmosphere, which increased every year between 1975 (the year monitoring began) and 1994, decreased by 1.5 percent in 1995, suggesting that CFC levels may have peaked.

POTENTIAL PROBLEMS

Still, scientists are cautiously optimistic because a number of events can derail the recovery of the ozone layer. Perhaps the most likely problem is a widespread failure to comply with the Montreal Protocol. Another problem is that CFCs are produced legally in developing countries and smuggled into industrialized countries, where they are sold on the black market.

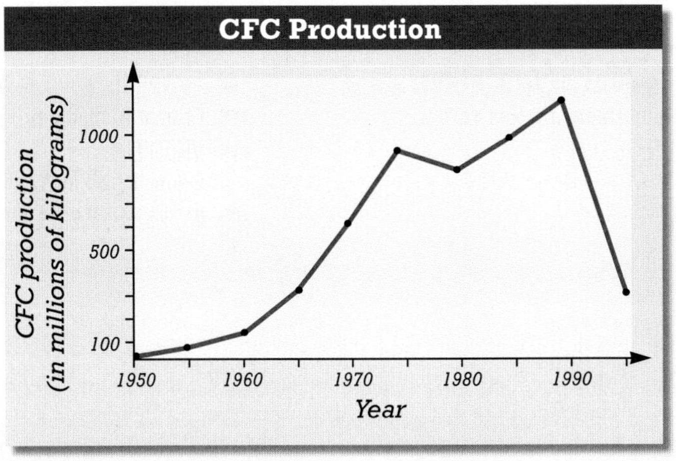

CFC production has fallen drastically as a result of the Montreal Protocol of 1987.

Additional Information

French, Hilary. "Learning from the Ozone Experience." *State of the World 1997.* Washington, D.C.: Worldwatch Institute. 151–171.

Svitil, Kathy. "A Slow-Healing Wound." *Discover,* January 1997. 63.

Gibbs, W. Wayt. "The Treaty That Worked—Almost." *Scientific American,* September 1995, 20–21.

CHAPTER 19
INTRODUCTION TO ECOLOGY

BLOCKS	TOPICS	TEACHING RESOURCES	LABS, CLASSWORK, AND HOMEWORK
BLOCKS 1 & 2 90 minutes	**Introducing the Chapter, p. 358**	**ATE** Focus Concept, p. 358 **ATE** Assessing Prior Knowledge, p. 358 **ATE** Understanding the Visual, p. 358	■ Supplemental Reading Guide, *Silent Spring*
	19-1 Ecology, p. 359 Today's Environment, p. 359 Levels of Organization, p. 362 A Key Theme in Ecology, p. 364 Ecological Models, p. 365	**ATE** Section Overview, p. 359 **ATE** Critical Thinking, pp. 362, 364 **ATE** Visual Link Figure 19-2, p. 360 Figure 19-6, p. 364 65, 66, 82	**PE** **Quick Lab,** p. 362, Modeling the Greenhouse Effect **Lab B10,** Ecology Scavenger Hunt ★ Study Guide, Section 19-1 **PE** Section Review, p. 365
BLOCKS 3 & 4 90 minutes	**19-2 Ecology of Organisms, p. 368** Biotic and Abiotic Factors, p. 368 The Niche, p. 371	**ATE** Section Overview, p. 368 **ATE** Critical Thinking, pp. 369, 370 **ATE** Visual Link Figure 19-13, p. 372 71, 72, 73, 74, 86, 87	**PE** **Chapter 19 Investigation,** pp. 376–377 **Lab C22,** Examining Owl Pellets **Lab C24,** Mapping Biotic Factors in the Environment **Lab C25,** Assessing Abiotic Factors in the Environment ★ Study Guide, Section 19-2 **PE** Section Review, p. 372
BLOCKS 5 & 6 90 minutes	**REVIEW and ASSESSMENT** **PE** Chapter 19 Review, pp. 374–375 ■ Performance-Based Assessment—Testing Pond Water for Carbon Dioxide	Audio CD Program BIOLOGY INTERACTIVE TUTOR Unit 7—Ecosystem Dynamics	**CHAPTER TESTING** ■ Chapter Test (blackline master) ▲ Test Generator ■ Assessment Item Listing ■ Scoring Rubrics and Classroom Management Checklists

MODERN BIOLOGY — HOLT BIOSOURCES

KEY

- **PE** Pupil's Edition
- **ATE** Teacher's Edition
- ■ Active Reading Guide
- ★ Study Guide
- Teaching Transparencies
- Laboratory Program

One-Stop Planner CD-ROM
Shows these resources within a customizable daily lesson plan:

- ■ Holt BioSources Teaching Resources
- ■ Active Reading Guide
- ★ Study Guide
- ▲ Test Generator

READING FOR CONTENT MASTERY

Preview: Have students preview the objectives, the introductory paragraph, topic sentences, and captions before reading the chapter.

ATE Active Reading Technique: Anticipation Guide, p. 359

■ Active Reading Guide Worksheet 19-1

PE Summary/Vocabulary, p. 373

■ Active Reading Guide Worksheet 19-2

PE Summary/Vocabulary, p. 373

TECHNOLOGY AND INTERNET RESOURCES

CNN PRESENTS Video Segment 28 What's Slithering in Guam?

Holt Biology Videodiscs Teacher's Correlation Guide, Lessons 19-1 through 19-2

BIOLOGY INTERACTIVE TUTOR Unit 7—Ecosystem Dynamics, Topic 1

Audio CD Program, Sections 19-1 through 19-2

internetconnect

SCILINKS NSTA

On-line Resources: www.scilinks.org
The following *sci*LINKS Internet resources can be found in the student text for this chapter:

Topics:
- Global warming, p. 363
- Threats to ecosystems, p. 365
- Niches, p. 372

go.hrw.com

On-line Resources: go.hrw.com
Visit the HRW Web site for a variety of resources related to this chapter. Just type in the keyword HM2 Chapter 19.

Smithsonian Institution®
Visit **www.si.edu/hrw** for additional on-line resources.

CNNfyi.com
Visit **www.cnnfyi.com** for late-breaking news and current event stories selected just for you.

LAB ACTIVITY PLANNING

CHAPTER 19 INVESTIGATION

Observing Habitat Selection, pp. 376–377

OVERVIEW
In this investigation, students will observe how changes in fluorescent light affect brine shrimp. This investigation demonstrates how organisms select habitats in which conditions such as temperature, light levels, salinity, and pH are within their tolerance limits.

SAFETY
Students will wear safety goggles for Part B. If methyl cellulose solution gets in a student's eye, flush the eye with water for 15 minutes. Remind students to handle living animals with care.

PREPARATION
1. If you choose to use brine shrimp eggs, place them in salt solution and allow 2 days for the shrimp to hatch. Each lab group will need 50 mL of brine shrimp culture.
2. Plastic tubing (3/8 × 1/16 in.) can be purchased from a hardware store or a biological supply house. You will need 44 cm of tubing for each lab group.
3. Each lab group will need a roll of aluminum foil, fourteen 25 × 25 cm pieces of screening, a calculator, three screw clamps, a small funnel, a 100 mL graduated cylinder, a Petri dish, a pipet, six stoppers (size 0), four test tubes, a test-tube rack, a fluorescent lamp, and a marker.
4. Methyl cellulose solution can be purchased from a biological supply house, such as WARD'S. You will need a 60 mL bottle for each class of students.

Quick Labs

Modeling the Greenhouse Effect, p. 362
Each student group will need two 1 qt jars, a graduated cylinder, aluminum foil, transparent tape, a thermometer, and water.

UNDERSTANDING THE VISUAL

Have students examine the photograph of Earth from space. Ask students to infer the relationship between the planet and ecology. The photograph represents the biosphere, where all living organisms interact.

FOCUS CONCEPT

Interdependence of Organisms

Emphasize for students that a set of crucial interactions takes place between living organisms and the living and nonliving environment. Have students brainstorm for examples of interdependence.

ASSESSING PRIOR KNOWLEDGE

Review the following concepts.

Evolution: *Chapter 14, 16, and 17*
Ask students how a constantly changing environment might affect natural selection. Ask them if people have an impact on the environment.

Matter, Energy, and Organization: *Chapter 5*
Ask students to trace backward the source of the food they had for their last meal. Tell them that plants are in the first level of a one-way series of energy transfers that fuel an ecosystem.

CHAPTER 19

INTRODUCTION TO ECOLOGY

This is a view of the biosphere as seen from space.

FOCUS CONCEPT: ***Interdependence of Organisms***

As you read, notice that all organisms affect and are affected by the living and nonliving components of their environment.

ECOLOGY

***Ecology** is the study of the interactions between organisms and the living and nonliving components of their environment. The Earth includes a tremendous variety of living things. Each organism depends in some way on other living and nonliving things in its environment. Ecology involves collecting information about organisms and their environments, looking for patterns, and seeking to explain these patterns.*

SECTION

19-1

OBJECTIVES

▲ Define the term *ecology,* and explain why ecology is important.

● List and describe three human-caused environmental problems.

■ Identify the five different levels of organization in ecology.

◆ Explain the theme of interconnectedness.

▲ Identify the importance of models to ecology.

TODAY'S ENVIRONMENT

Although the field of ecology was not named until 1866, ecological information and understanding have always been crucial to humans. Before the development of agriculture, about 10,000–12,000 years ago, our ancestors obtained all of their food by hunting animals and gathering grains, seeds, berries, nuts, and plants. For them, survival depended on practical knowledge about the environment. While our understanding of the environment has grown much more sophisticated, our need for information has grown more urgent. Over the past few decades, humans have changed the environment on a greater scale than ever before in history. Learning how to improve our effect on the environment is critical to the survival of our species.

The Exploding Human Population

The most significant environmental change is probably the rapid increase in the number of people on Earth. The world's human population has tripled from 2 billion in 1930 to 6 billion people in 1999. This increase in population is unprecedented in the history of our species. According to recent projections by the United Nations, in the year 2050 the world's population will be between 7.8 billion and 12.5 billion. Figure 19-1 on the following page shows that our present numbers are already causing severe crowding in some areas. An increasing population requires increasing amounts of energy, food, and space for the disposal of waste. Providing for the needs of this growing population will take an increasingly greater share of Earth's resources.

The Sixth Mass Extinction

As the human population has increased, many other species have declined in number or become extinct. For example, on the

Word Roots and Origins

ecology

from the Greek words *oikos,* meaning "house," and *logos,* meaning "study of"

SECTION 19-1

SECTION OVERVIEW

The Interdependence of Organisms

Section 19-1 addresses some key issues in ecology, such as population growth, the ozone layer, and climate changes. The levels of organization in ecology and the use of models in science are discussed. This section emphasizes that no organism is isolated.

Evolution

Section 19-1 connects the evolution of a species and the challenges the species faces in its natural environment.

Stability and Homeostasis

Section 19-1 discusses the changing environment and how disturbances in ecosystems can cause instability in unexpected ways.

ACTIVE READING

Technique: Anticipation Guide

Before students read this chapter, write the following on the board or overhead:

1. All living things interact with other organisms in their surroundings and with the nonliving portion of their environment.
2. Scientists have found evidence in the fossil record of five major mass extinctions. The sixth mass extinction is happening now.

Let the students think a minute about the statements and decide if they agree with them. Have students read the chapter to see if their opinions are confirmed.

MAKING CONNECTIONS

Social Studies

Environmental changes have occurred throughout the world due to human impact. Ask students to research such changes and show where they have occurred on maps of the world. Some examples are the gradual desertification in Africa's Sahel over the last century and the extensive destruction of rain forests in Central or South America during the same period.

RECENT RESEARCH

The Greenhouse Effect and Carbon Dioxide

Climatologists have calculated that the mean global temperature has increased about 1°C since 1900. The U.S. National Research Council recently estimated that by the year 2035 the average surface air temperature will increase between 1.5°C and 4.5°C. It was also suggested that the increase might be much greater because of the number of trace gases, such as methane, nitrous oxide, ozone, and chlorofluorocarbons, which are increasing rapidly and have warming or greenhouse effects similar to carbon dioxide.

VISUAL LINK

Figure 19-2

Tell students that no actual photographs are available of the birds depicted in Figure 19-2. They are looking at an artist's rendering from an ornithologist's description or sketch.

FIGURE 19-1

Most of our planet is already very crowded, as this photograph of people in New York City shows. Scientists are concerned about how our quality of life will be affected if our population doubles within the next 50 years. If our population continues to grow at its present rate, the environment that supports us may collapse.

Hawaiian Islands, where isolation fostered the evolution of a diverse and unique set of species, 60 of the 100 species of native birds have disappeared since the first human colonists arrived. They were eliminated by habitat destruction, overhunting, and introduced diseases and predators. Two of these extinct birds are shown in Figure 19-2. Worldwide, about 20 percent of the species of birds have become extinct in the last 2,000 years. Today extinction is occurring most rapidly in the tropics.

Paleontologists have found evidence in the fossil record of five major mass extinctions, episodes in which a large percentage of Earth's species became extinct in a relatively short time. The sixth mass extinction is happening now. Currently, species are disappearing faster than at any time since the last mass extinction—the disappearance of the dinosaurs 65 million years ago. Scientists estimate that about one-fifth of the species in the world may disappear in just the next century.

FIGURE 19-2

The impact of humans has destroyed the habitats of many of the bird species native to the Hawaiian Islands. This destruction has caused the extinction of about 60 species of birds, including the birds shown below.

(a) Nukupu'u, *Heterorhynchus lucidus*

(b) Greater Koa, *Telespiza palmeri*

The Thinning Ozone Layer

The ozone layer in the upper atmosphere protects the living organisms on Earth by absorbing ultraviolet (UV) radiation from the sun. Industrial chemicals called chlorofluorocarbons (CFCs) and some related chemicals react with ozone and are destroying the ozone layer. In 1985, British scientists discovered an ozone "hole," a region of an abnormally low ozone level, over Antarctica, as shown in Figure 19-3. This discovery led to a treaty in 1992 to ban CFCs and other ozone-destroying chemicals.

Although ozone depletion is occurring to some degree over most of the planet, the reduction in the ozone level over Antarctica by 1996 was about 50 percent of its maximum density. About 1 percent of the UV radiation from the sun gets through the ozone shield to reach Earth's surface. That 1 percent is responsible for all sunburns and for more than a half-million skin cancers each year.

Climatic Changes

Although carbon dioxide and water vapor are transparent to visible light, they intercept much of the reflected heat and direct it back toward Earth. This natural phenomenon, known as the **greenhouse effect,** is the mechanism that insulates Earth from the deep freeze of space. However, human activities are changing the composition of the atmosphere and thereby influencing climate. One prediction of possible future changes in climate is shown in Figure 19-4. When fossil fuels, such as coal, oil, and natural gas, are burned to provide energy, carbon dioxide is released. The burning of large amounts of fossil fuels has caused the concentration of carbon dioxide in the atmosphere to increase by about 25 percent in the last 100 years. A

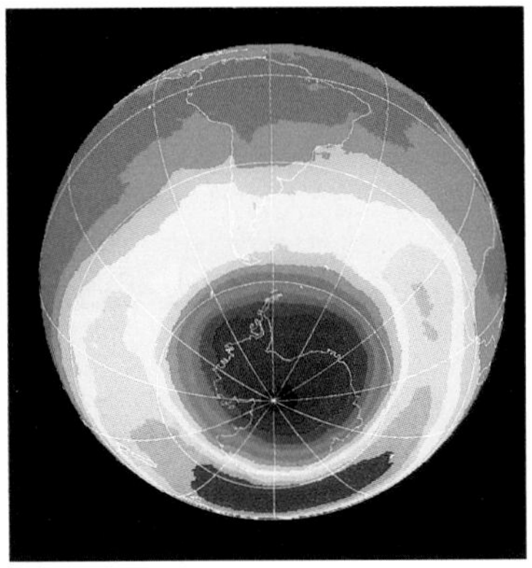

FIGURE 19-3

The ozone shield over Antarctica fluctuates in density seasonally, sometimes to a low of half the original density. The ozone shield is diminishing all over the planet as well.

FIGURE 19-4

The regions of the world that are likely to become wetter or drier as a result of global warming are shown below. Global warming and changes in rainfall are likely to be variable and unpredictable, so predictions cannot be entirely accurate. Predictions for the darker colored areas have a higher probability of being correct.

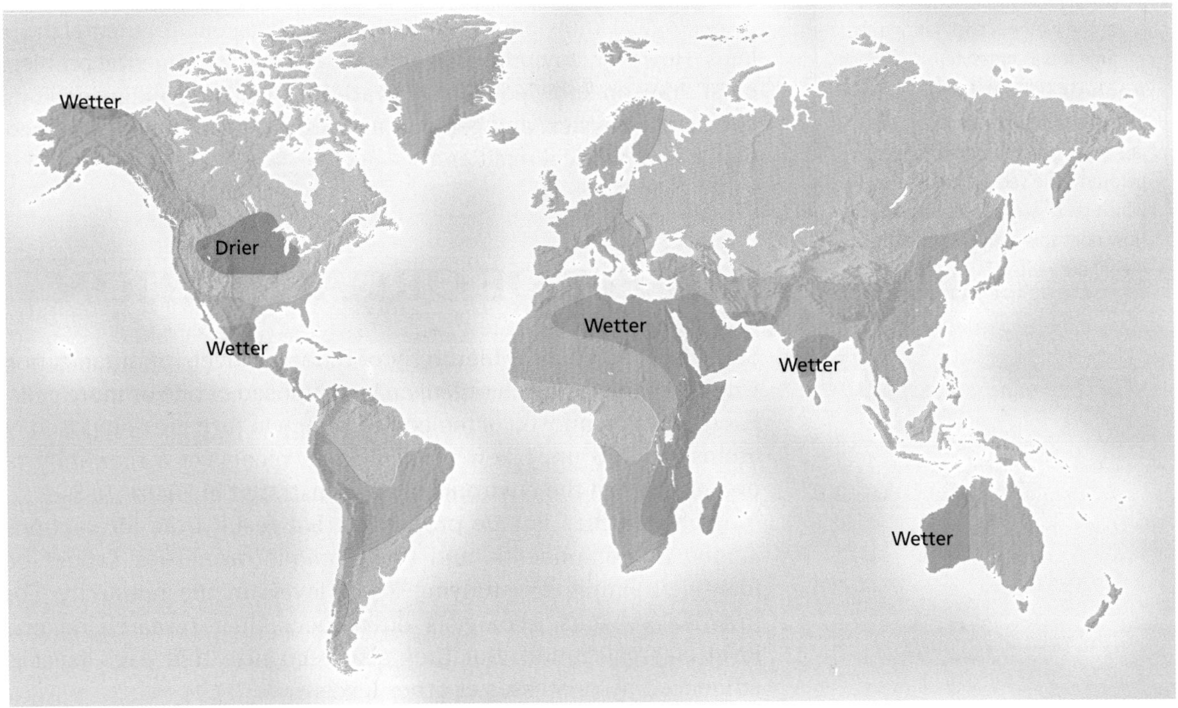

QUICK FACT

☑ Ultraviolet radiation is a serious human health concern. It is estimated that every 1 percent drop in the atmospheric ozone content leads to a 6 percent increase in the incidence of skin cancers.

CULTURAL CONNECTION

Chile

The southernmost city in the world is Punta Arenas, Chile. It is located underneath the Antarctic ozone hole, and its people and animals have had disturbing health problems. Many people in Punta Arenas wear sunglasses because they have noticed an increase in blindness among those who do not. The farmers raise many sheep, and the sheep now have a high incidence of blindness. The incidence of skin cancer is also very high in Punta Arenas, and hats are very popular in this land of sheep, potatoes, and penguins.

MAKING CONNECTIONS

Technology

Arrange for the class to visit a house where solar panels are used to produce hot water. Students may be surprised to learn how simple this technology is. A flat-plate solar panel is little more than an insulated black box covered with one or more layers of clear glass or plastic. Inside the box is a plate of black metal or plastic. The plate absorbs sunlight and converts it to heat, which is trapped under the glass. Water flows through tubes in the plate and absorbs heat from the plate. Water then may be used at once to heat a house or stored in a holding tank.

Quick Lab

Modeling the Greenhouse Effect

Time Required 25 minutes

Safety Instruct students to use caution with glassware.

Procedural Tips If the classroom lacks natural light, provide plant lights, not heat lamps.

Answers to Analysis Students will find that the water in the covered jar will be warmer than the water in the uncovered jar. Carbon dioxide, methane, and other gases in the atmosphere are transparent to incoming light, but they trap reflected heat. The glass of the jars is transparent to incoming light, but it is not transparent to reflected heat. Heat is trapped and absorbed by the water in the jars.

CRITICAL THINKING

The Greenhouse Effect and Methane

The atmospheric level of methane has more than doubled since 1951. Ask students what has caused this increase in atmospheric methane. (Methane is a product of the bacterial decomposition of organic matter in the absence of oxygen. This decomposing action has occurred mainly in rice paddies and in the digestive tracts of termites, which are now present in greater numbers because of all the forests that are being destroyed. Students may be amused to learn that belching by cattle and sheep is another source of greenhouse gases.) Help students understand the interconnectedness of changes in our environment.

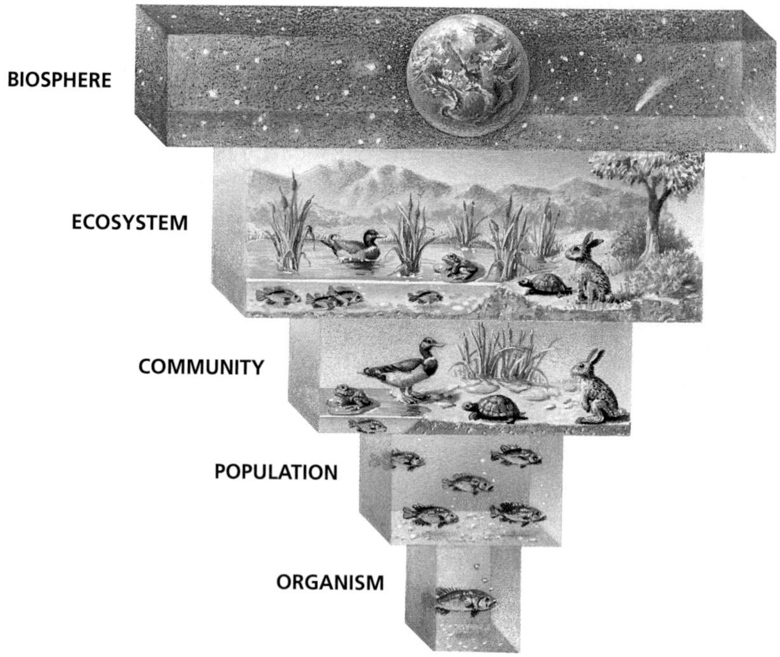

FIGURE 19-5
Ecology has been organized into several levels because of the complexity of the science.

Quick Lab

Modeling the Greenhouse Effect

Materials 1 qt jars (2), graduated cylinder, aluminum foil, transparent tape, thermometer, water

Procedure

1. Put 750 mL of cold tap water in each jar.
2. Measure and record the initial water temperature in each jar, and record the time it was measured.
3. Cover one jar with a single layer of aluminum foil, and secure the foil with tape. Leave the top of the other jar uncovered.
4. Place both jars in the same sunny window or under a plant light.
5. Before the end of the class period, record the temperature of the water in each jar and the time it was measured.

Analysis Did you discover a temperature difference between the jars when you measured the temperatures the second time? If so, which jar of water was warmer? How does this activity model the greenhouse effect?

global warming of the atmosphere is now occurring because of heat trapped by excess "greenhouse" gases, such as carbon dioxide.

Since 1860, the average global temperature has risen about 0.6°C (1°F), and most scientists agree that this increase was caused by higher levels of carbon dioxide in the atmosphere. Furthermore, scientists project that the average global temperature will increase 1.5°C to 4.5°C (3°F to 8°F) by the year 2100. This increase in average temperature may change global weather patterns and cause rising sea levels due to melting polar icecaps.

Ecological knowledge is essential for solving environmental problems. However, any proposed solution to an environmental problem must draw on knowledge from a variety of fields, including ecology, chemistry, physics, and geology, and it must take into account economic and political realities.

LEVELS OF ORGANIZATION

Scientists recognize a hierarchy of different levels of organization within organisms. Each organism is composed of one or more cells. Each cell is composed of molecules, which in turn are composed of atoms, and so on. Likewise, ecologists recognize a hierarchy of organization in the environment, as illustrated in Figure 19-5.

Each level has unique properties that result from interactions among its components, and these unique properties cannot be identified simply by studying lower levels in the hierarchy. For practical reasons, ecologists often focus their research on one level of organization. But they also recognize that each level is influenced by processes at other levels.

The Biosphere

The broadest, most inclusive level of organization is the **biosphere** (BIE-oh-SFEER), the thin volume of Earth and its atmosphere that supports life. All organisms are found within the biosphere. It is about 20 km (13 mi) thick and extends from about 8 to 10 km (5 to 6 mi) above the Earth's surface to the deepest parts of the oceans. In comparison, the Earth's diameter is about 12,700 km (7,900 mi), or more than 600 times the thickness of the biosphere. If Earth were the size of an apple, the biosphere would only be as thick as the apple's skin. Ecologists often describe the biosphere as a thin film of life covering an otherwise lifeless planet. Living things are not distributed evenly throughout the biosphere. Most organisms are found within a few meters of the surface of the land or oceans.

Ecosystems

The biosphere is composed of smaller units called ecosystems. An **ecosystem** (EK-oh-SIS-tuhm) includes all of the organisms and the nonliving environment found in a particular place. Consider a pond ecosystem. It contains a variety of living things, such as fish, turtles, aquatic plants, algae, insects, and bacteria. These organisms interact in ways that affect their survival. For instance, insects and fish eat aquatic plants, and turtles eat fish. The pond ecosystem also includes all the nonliving (physical and chemical) aspects of the pond that influence its inhabitants. The chemical composition of the pond—how much dissolved oxygen and carbon dioxide it contains, its supply of nitrogen, its pH—helps to determine what kinds of organisms live in the pond and how abundant they are. A very important physical factor is the amount of sunlight the pond receives, because sunlight is the ultimate source of energy for the pond's inhabitants.

Communities, Populations, and Organisms

While an ecosystem contains both the living and nonliving components, a community includes only organisms. A **community** is all the interacting organisms living in an area. For instance, all the fish, turtles, plants, algae, and bacteria in the pond described above make up a community. Although it is less inclusive than an ecosystem, a community is still very complex, and it may contain thousands of species. Ecologists studying a community often focus on how species interact and how these interactions influence the nature of the community. Remember that the word *community* has a specific meaning in biology that differs from its meaning in everyday speech.

Below the community level of organization is the population level, where the focus is on the members of a single species. A **population** includes all the members of a species that live in one place at one time.

The simplest level of organization in ecology is that of the organism. Research at this level concentrates on the adaptations that allow organisms to overcome the challenges of their environment.

Eco Connection

Global Warming and Disease

Ecologists are working with climatologists to investigate the relationship between climatic changes caused by global warming and the outbreak of disease.

In 1993, a virus began killing young people in the southwestern United States. An unusually mild winter and a wet spring sent piñon trees blooming, providing virus-carrying mice with a plentiful supply of pine nuts. The mice population increased tenfold. The abundant mice found their way into people's homes and spread the virus to humans. Half the people infected with the virus died.

Several of the deadliest diseases known, such as malaria, yellow fever, and encephalitis, are carried by mosquitoes, whose population levels are affected by even small changes in temperature and rainfall. Researchers are carefully monitoring mosquitoes, mice, and other disease carriers and charting climatic changes in an effort to predict if, when, and where the next outbreak of disease might occur.

internet**connect**

SCILINKS NSTA
TOPIC: Global warming
GO TO: www.scilinks.org
KEYWORD: HM363

TEACHING STRATEGY

The Effects of Global Temperature Changes

Ask students to determine whether organisms change in response to temperature changes. Divide the class into groups of three and have them do the following: Place a paramecium culture in a 12 cm long V-shaped glass tubing that can be corked at both ends. Gently place one end of the glass tubing in a beaker of very hot tap water (60°C) and the other end in ice water. Have the students determine the temperature range using a thermometer. Have the students wait 5–10 min and then observe with a dissecting scope or magnifying glass any differences in the response of the paramecia to their environments. Have students also note the number of paramecia that congregate in the hot and cold ends of the tubing as opposed to the middle of the tubing. Ask students what temperature they think paramecia prefer. (They prefer the temperature in the middle of the tube.) Ask students where paramecia live in nature. (They live in every area on Earth except the polar regions and thermal hot spots.)

INCLUSION *ACTIVITY*

Have students name the types of communities of organisms that might be found in your local area. List the different communities on the board. Then have students name all the organisms they can think of that belong to each community, and you list the organisms on the board. Have students copy the flow chart you have created in their notebooks.

CRITICAL THINKING

Mosquitoes

In the late 1960s, tiger mosquito larvae were accidentally brought to the United States in a shipment of tires. These mosquitoes are carriers of tropical diseases, such as encephalitis, dengue fever, and yellow fever. The tiger mosquito can be controlled, but not eradicated, by spraying pesticides. Ask students what they think of the health implications of the mosquitoes' spread or the efforts to control it. (Students must choose between being susceptible to epidemic tropical diseases from which the United States has been relatively free in the past and the threat of human exposure to the harmful effects of pesticides.)

RETEACHING *ACTIVITY*

Have students make a graphic organizer of the components of the biosphere. Have them begin by drawing a small circle on a sheet of paper. Ask them to write their name inside the circle along with the level of organization that the individual belongs to (organism). Then have them draw another circle outside the first one to represent population. Have students label this circle with a population to which he or she belongs. Continue this activity, using increasingly larger circles, until the entire diagram is encircled by the level of biosphere.

VISUAL LINK

Figure 19-6

Ask students how the caterpillar might affect the incidence of Lyme disease. (During seasons when there is an abundance of the caterpillar, there may also be a decrease in acorns. When there are fewer acorns, there are fewer deer and mice, and consequently, fewer ticks.)

A KEY THEME IN ECOLOGY

An important fact to keep in mind as you study ecology is that no organism is isolated. The theme of the interconnectedness of all organisms is central to the study of ecology. All organisms interact with other organisms in their surroundings and with the nonliving portion of their environment. Their survival depends on these interactions. Thus, each ecosystem is a network in which organisms are linked to other organisms and to the nonliving environment. Ecologists refer to this quality as interconnectedness or interdependence. For example, you could not survive without the plants and other photosynthetic organisms that produce oxygen. Your cells need oxygen to release the energy in food, and cells will die if deprived of oxygen for even a few minutes. Conversely, photosynthetic organisms depend on the release of carbon dioxide gas by the cellular respiration of other organisms, such as humans, and geochemical processes, such as volcanic eruptions. Carbon dioxide gas is an essential raw material for making carbohydrates.

Disturbances in Ecosystems

An important consequence of interconnectedness is that any disturbance or change in an ecosystem can spread through the network of interactions and affect the ecosystem in widespread and often unexpected ways. To see one example, look at Figure 19-6, which shows some of the interrelationships among species in an oak forest. The number of people who get sick with Lyme disease,

FIGURE 19-6

All of the very different species shown are ecologically connected in the forest. An unusually plentiful crop of acorns helps support a large population of deer and mice, which help support a large population of ticks. Ticks carry the bacterium that causes Lyme disease. They pass on the disease to humans who visit the forest.

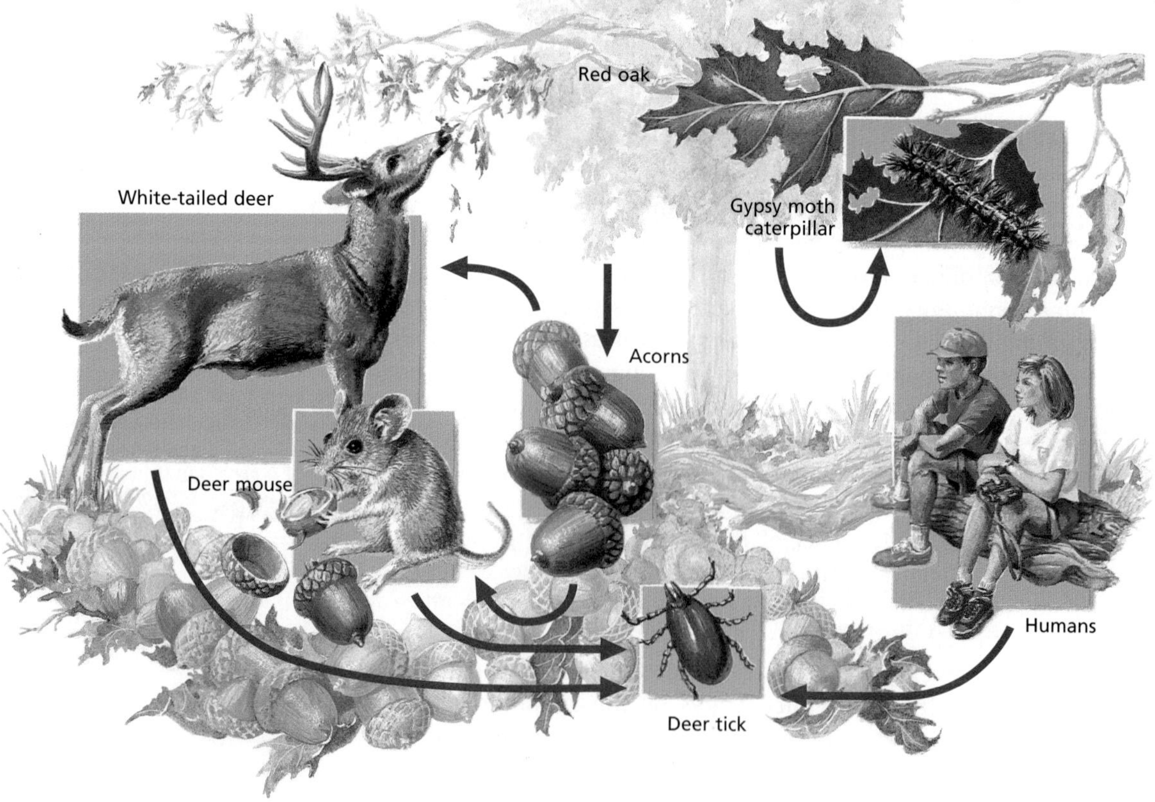

a bacterial infection that can damage the nervous system, is related to acorn production in forests in the eastern United States.

Oak trees typically produce few acorns (or none at all) in most years. Every few years, however, they produce a huge crop of acorns, setting off a chain of events within the ecosystem. The abundance of acorns enables deer and mice, which feed on acorns, to have more offspring that survive, and their populations grow. More deer and mice can support more ticks, so the tick population also increases. Lyme disease is spread by the bite of the deer tick. The number of people bitten by ticks (and potentially exposed to Lyme disease) depends on the number of ticks and the number of people in the forest. In general, as the population of ticks increases, the number of Lyme disease cases increases.

internetconnect

SciLINKS NSTA

TOPIC: Threats to ecosystems
GO TO: www.scilinks.org
KEYWORD: HM365

ECOLOGICAL MODELS

Ecological knowledge is vital, but ecosystems are extremely complex and difficult to study. One way that ecologists deal with this complexity is to use models. A model may be visual, verbal, or mathematical. An ecologist might use a graphical model to show how sunlight, rainfall, and temperature affect the growth of plants. Ecologists construct models to help them understand the environment and to make predictions about how the environment might change. Figure 19-4 is a model based on scientific predictions about how climates may change. In ecology, models are typically expressed as graphs, diagrams, or mathematical equations. Figure 19-5 is a model in the form of a diagram.

A model can be used to test a hypothesis about an ecosystem. Scientists use models to make predictions about the future behavior of the ecosystem, and these predictions can be tested by comparing them with observations from the natural world. Models are widely used to help plan and evaluate solutions to environmental problems. It is important to keep in mind, however, that models are simplified systems designed only to mimic the behavior of the natural world. A model is limited in its application, therefore, because it cannot account for the influence of every variable in a real environment.

SECTION 19-1 REVIEW

1. Why was knowledge about the environment important to the earliest members of our species? Why is such knowledge important now?
2. What is causing the thinning of the ozone layer?
3. How does a population differ from a community?
4. Describe one example of interdependence.
5. Why are models so useful in ecology?
6. CRITICAL THINKING Why is it important to understand challenges in a species' environment in order to understand the species' evolution? Relate your answer to what Darwin referred to as "the struggle for survival."

SECTION 19-1

QUICK FACT

☑ Models of cell lines, tissue types, cell abnormalities, and diseases are kept by the American Type Culture Collection in Bethesda, Maryland, to help scientists to identify unusual organisms or cell types.

ANSWERS TO SECTION 19-1 REVIEW

1. Our ancestors needed to know how and where to find food and how to avoid dangerous animals and poisonous plants. Today ecology is important in understanding the likely effects of human alterations to the environment.
2. The main cause of ozone depletion is a class of industrial chemicals called CFCs.
3. A population contains individuals of only one species. A community contains populations of several species.
4. One possible answer is the following. A farmer sprays insecticide on a field. This causes reduced fertility in a species of bird that preys on a species of snake that preys on field mice. The farmer's home becomes infested with mice.
5. Models enable scientists to simplify and focus on one or a few processes within the ecosystem.
6. Darwin believed that natural selection occurred only in the presence of a struggle for survival or in the presence of challenges and changes in a species' environment. If there were no struggle, adaptation would be unnecessary and a species would not evolve. Chance mutations would still occur, but they would not contribute to a species being better fit in a theoretical unchanging or unchallenging environment.

BACKGROUND

Edward O. Wilson earned his doctorate at Harvard University in 1955 with a taxonomic analysis of the ant genus *Lasius.* He is recognized today as the world's leading authority on ants.

The late Robert H. MacArthur, a professor of biology at Princeton University, was the ecologist who developed the categories of "*r*" or "*K*" to describe the life and reproductive strategies of species and who first described the niche strategy of resource partitioning.

Biogeography is the study of the geographic distribution of plants and animals in an area. In formulating their theory of island biogeography, Wilson and MacArthur realized that islands offer a controlled area for studying the factors that affect species' distribution. Their theory predicts the number of species that can inhabit an island. The theory also predicts the species diversity of "islands" on land, such as mountaintops, lakes, and patches of forest left after logging.

Ask students if the graphical models associated with this theory are practical.

DISCUSSION

Guide the discussion by posing the following questions:

1. What two observations form the basis of MacArthur and Wilson's theory of island biogeography? (1. Large islands can have more species; small islands have fewer. 2. Remote islands—those far from the mainland or a large island—have fewer species than less-remote ones.)

2. MacArthur and Wilson used the theory of island biogeography to predict the number of bird species on the island of Krakatau. How close was their prediction? (They predicted that the birds at the point of equilibrium would number about 30 species. When

GREAT DISCOVERIES

The Rise and Fall of Island Species

HISTORICAL PERSPECTIVE

In the mid-1800s, Charles Darwin observed and collected organisms in the Galápagos Islands. His fieldwork led to the theory of natural selection as a mechanism for evolution. About a century later, mathematician and ecologist Robert H. MacArthur also studied species found on islands. With taxonomist and zoogeographer Edward O. Wilson, MacArthur developed the theory that provided the title to their book, **The Theory of Island Biogeography.** ***Biogeography is the study of the geographical distribution of plants and animals. Today this field has increasing relevance for ecologists and for all who care about the survival of species.***

A Project in Biogeography

By the mid-1950s, Edward O. Wilson had studied insect species he had found on a number of islands. His specialty was myrmecology, the biology of ants. After many years of study and fieldwork, Wilson wanted to go beyond the simple collection and description of organisms, a desire that meshed well with the thinking of Robert H. MacArthur.

MacArthur began his career as a mathematician, but he later moved to mathematical ecology. He was interested in patterns and ideas and in a bigger picture of nature. He wanted to understand the fundamental principles in the science of ecology.

MacArthur and Wilson each brought specific strengths to the field of biogeography. Soon after the two men met, in 1959, they began the project that led to their book. In their book, MacArthur and Wilson explain why they chose to concentrate on the study of species on islands: "In the science of biogeography, the island is the first unit that the mind can pick out and begin to comprehend."

Robert H. MacArthur and Edward O. Wilson

Seeing the Patterns

Wilson noticed that the number of ant species on an island tends to correlate with the size of the island. He also noticed that when a new ant species arrives on an island, one of the species already on the island becomes extinct, which, in this context, means that they cease to exist on the island. However, the total number of ant species remains constant. Wilson and MacArthur called that constant number of species an equilibrium.

When they examined the data closely, MacArthur and Wilson found that the same pattern exists among bird species in the Philippines, Indonesia, and New Guinea. In each case, when one species of animal disappeared from an island, a new one moved in, but the total number of species remained constant.

MacArthur and Wilson graphed their data, as shown on the facing page. The downward slope of the immigration (colonization) line shows that the number of species

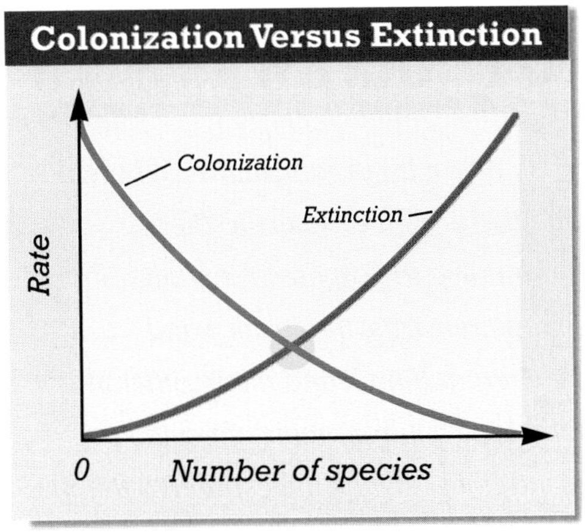

This simple model represents the equilibrium of numbers of island species. Note that the rate of colonization is highest when the number of species is lowest, as shown on the model at the y-axis. The rate of extinction is highest and the rate of colonization is lowest when the number of species is at a maximum. At what point are the rates of colonization and extinction equal?

that moved in decreased as the island became crowded. That is, the rate of colonization slowed with an increase in the number of species. The upward slope of the extinction line shows that as the island grew more crowded, more species became extinct. In other words, the rate of extinction increased with an increase in the number of species. The crossing point is an equilibrium, the constant number of species for that island.

Constructing a Model

MacArthur and Wilson developed a mathematical model to explain their observations. The mathematics of the theory is complex, but the broad outlines center on two observable patterns:

1. Large islands have more species than small islands have.
2. Remote islands—those located far from the mainland or from a larger island—have fewer species than less-remote ones.

Earlier biogeographers had explained these phenomena from a historical perspective. They reasoned, for example, that a remote island took eons to fill up with species, so an island with few species probably has a relatively short history.

In MacArthur and Wilson's model, size and isolation—not the age of the island—regulate the number of immigrations and extinctions.

Testing and Predicting

Could their model be used to predict the number of species on an island? MacArthur and Wilson decided to test their model on Krakatau, an island in Indonesia on which a volcano had erupted in 1883, killing every living thing on the island. The eruption left Krakatau much like a newborn island. Equally important, the return of plant and animal life to the island had been carefully recorded since Krakatau was first revisited, in 1886.

Using their model, MacArthur and Wilson predicted that at the point of equilibrium, the number of bird species would be about 30. After examining the records of bird life at Krakatau, they learned that their prediction had come close. The number of species had climbed to 27 before leveling off. Five species were new to the island, but an equal number had become extinct, maintaining the equilibrium number of 27.

New Ways of Thinking

The response from other scientists to *The Theory of Island Biogeography,* published in 1967, was mixed. At the very least, the theory introduced in the book has stimulated further research in the field of biogeography. This in turn has led to increased knowledge about how an environment's geography affects its ecology. MacArthur and Wilson looked beyond isolated facts and events to investigate similarities, patterns, and ongoing processes. By reflecting on the larger picture, perhaps biogeographers can help increase awareness of the importance of biodiversity.

This model shows effects that the size and remoteness of an island have on species equilibrium. Note that far (remote), small islands reach an equilibrium (shown at 1) with fewer species than near, large islands (shown at 4).

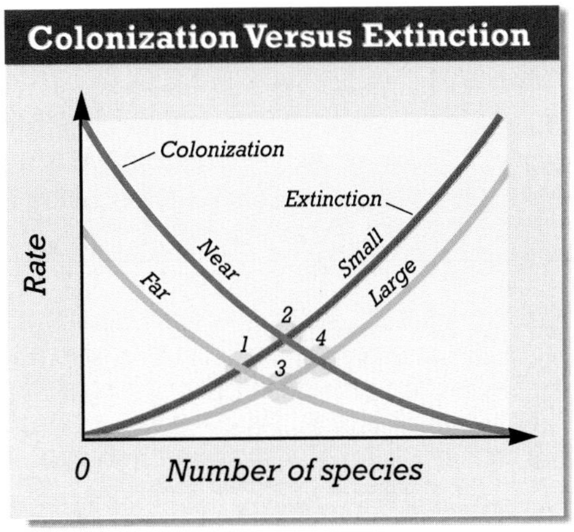

they examined the records of bird life at Krakatau, they found that the number of species was 27.)

3. Why did MacArthur and Wilson use the term *equilibrium* to refer to the constant number of species found on an island? (They used the term to refer to the stable or balanced state of the number of species.)

FURTHER READINGS

1. Read the article "Why Care? Here's Why," in *Outdoor Life,* June 1996, Vol. 197, Issue 6, page 14 (2 pages). This article expands Wilson and MacArthur's theory on biogeography and includes not only islands in the traditional sense, but also islands created on land by fragmentation of ecosystems.

2. Read the article "Life in Equilibrium," in *Discover,* March 1996, Vol. 17, Issue 3, page 66 (10 pages). This article is an interview with Edward O. Wilson by David Quammen, author of *Song of the Dodo.* The interview reveals the profound respect that existed between Wilson and MacArthur and the circumstances leading up to their development of the theory of island biogeography.

SECTION 19-2

SECTION OVERVIEW

Interdependence of Organisms
Section 19-2 describes how organisms are always interacting with the biotic and abiotic factors of their environment and explains that the need to meet biological needs underlies all interactions.

Stability and Homeostasis
Section 19-2 also discusses how changes are always occurring in the environment and explains that organisms respond to changes through adaptations that help provide biological stability.

▶ ENGAGE STUDENTS

Have students name as many abiotic factors as they can. Then have them explain how biotic factors interact with the named abiotic factors and with each other. (Sample answers: In the rainforest canopy there is plenty of sun and rain, and trees don't lose their leaves. In the tundra, where only the top of the ground thaws in the spring, the plants grow shallow roots and grow low to the ground.)

SECTION

19-2

OBJECTIVES

▲ Contrast abiotic factors with biotic factors, and list two examples of each.

● Explain the importance of tolerance curves.

■ Describe some adaptations that allow organisms to avoid unfavorable conditions.

◆ Explain the concept of the niche.

▲ Contrast the fundamental niche with the realized niche.

ECOLOGY OF ORGANISMS

Some basic questions to ask about an organism are, Where does it live? and Why does it live there? The answers to these questions are complex and involve the organism's evolutionary history, its tolerances and requirements, the history and conditions of its ***habitat*** *(where it lives), and many other factors. In this section you will study how environments affect the distribution of organisms and how organisms respond to their environments.*

BIOTIC AND ABIOTIC FACTORS

Ecologists separate the environmental factors that influence an organism into two classes. The living components of the environment are called **biotic** (bie-AHT-ik) **factors.** Biotic factors include all of the living things that affect the organism. The nonliving factors, called **abiotic** (AY-bie-AHT-ik) **factors,** are the physical and chemical characteristics of the environment. Important abiotic factors include temperature, humidity, pH, salinity, oxygen concentration, amount of sunlight, availability of nitrogen, and precipitation. The importance of each factor varies from environment to environment.

Abiotic and biotic factors are not independent; organisms change their environment and are influenced by those changes. For example, the availability of nitrogen in the soil affects how fast plants can grow, and plants affect nitrogen availability by absorbing nitrogen compounds from the soil.

The Changing Environment

Abiotic factors are not constant. They vary from place to place and over time, as shown in Figure 19-7. Consider temperature, which is probably the most important abiotic factor. Temperature changes on several different scales. It varies from hour to hour, from day to day, from season to season, and from year to year. A glance at a weather map in the newspaper will confirm that temperature also varies from place to place in a predictable way. Phoenix is usually warmer than Los Angeles and nearly always warmer than Seattle. Also important to organisms are the small differences in temperature within a habitat, such as the difference between an area in the shade of a tree and an area exposed to direct sunlight.

FIGURE 19-7

These pictures show the same area of forest at different times of the year. On the left, the forest displays spring foliage. On the right, the same area is covered with snow in winter.

Responses to a Changing Environment

Organisms are able to survive within a wide range of environmental conditions. Some heat-tolerant bacteria live far below Earth's surface at temperatures as high as 110°C (230°F), while some plants can withstand temperatures as low as −70°C (−94°F). However, no single organism can survive both of these extremes. Rather, organisms are adapted to function within a specific range of temperatures. It is possible to determine this range for an organism by measuring how efficiently it performs at different temperatures. A graph of performance versus values of an environmental variable, such as temperature, is called a **tolerance curve.** Figure 19-8 shows a tolerance curve for a species of fish. Here, performance was measured by the fish's maximum possible swimming speed. Notice that the fish's swimming speed is highest at intermediate temperatures, within what is called its optimal range. The fish can survive and function at temperatures outside its optimal range, but its performance is greatly reduced. The fish cannot survive outside its tolerance limits.

An organism cannot live in areas where it is exposed to conditions that fall outside its tolerance limits. In some cases, an organism's range may be determined by its tolerance to just one factor, such as temperature. In most cases, however, the levels of several factors, such as pH, temperature, and salinity, must fall within the organism's tolerance range.

FIGURE 19-8

The tolerance curve below shows that fish are capable of swimming fastest when the temperature of the water is within their optimal range. When the water is too warm or too cool, the fish are stressed and may not survive.

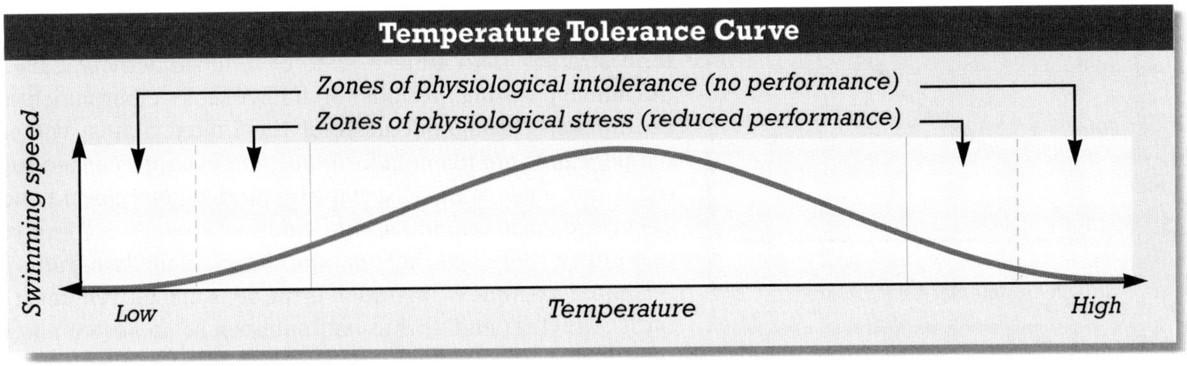

CRITICAL THINKING

Rainfall Variations

Ask students what happens to the vegetation when the rainfall of an area changes from season to season. (During times of little rainfall, plants are smaller and fewer and may die, but some come back when it rains again.) Ask students what happens to the vegetation when rainfall amounts gradually decrease over centuries. (Plants gradually become permanently smaller and fewer and may die out forever. The ecosystem may change from forest to grassland to desert over time.)

CRITICAL THINKING

Migration

Ask students how they think migrations and hibernations are different solutions to the same problem. (Both are ways of coping with adverse climatic conditions.) Ask students how birds benefit by laying their eggs in the spring. (The young will be grown and ready to migrate by winter.)

TEACHING STRATEGY

Changes in the Environment

Have students make a list of the ways humans have recently changed the environment through pollution. (Students may list thermal pollution, by heating lakes and streams; water pollution, by disposing of detergents and fertilizer; air pollution, as evidenced by acid rain and smog; or the ozone hole, from release of CFCs.) Ask students to list organisms that have been affected by these changes or that potentially could be affected. (This list could be extensive.) Ask students to discuss any changes that have occurred in their neighborhood or town or county.

CRITICAL THINKING

Tolerance

Ask students to list ways that people can change their tolerance to physical factors. (Answers may include references to some drugs—increasingly higher doses needed to get the same effect; temperatures—people living in warmer climates tolerate heat more easily than individuals coming from cold climates; elevation—people living at higher altitudes have more lung capacity for obtaining oxygen.) Ask them if this is the same as acclimation. (Yes.)

TERMINOLOGY NOTE

An organism can be both a conformer and a regulator. A fish may be a conformer of environmental temperature but a regulator of internal salt concentration. Do not allow your students to equate *conformer* with *ectotherm* or *regulator* with *endotherm.* These terms are used in different contexts.

RECENT RESEARCH

Acclimation

New studies suggest that there is a downside to an organism's ability to acclimate. Results of these new studies call into question previous laboratory findings that low-level metal contamination makes fish more tolerant of larger doses of metal contamination. David O. Norris, of the University of Colorado, and John D. Woodling, of the Colorado Division of Wildlife, report that brown trout were put in cages for 96 hours in an area so poisoned by mining effluent that no fish lived there. Fish that had resided in water with mining effluent survived only 28 hours, while fish from cleaner waters survived almost 48 hours.

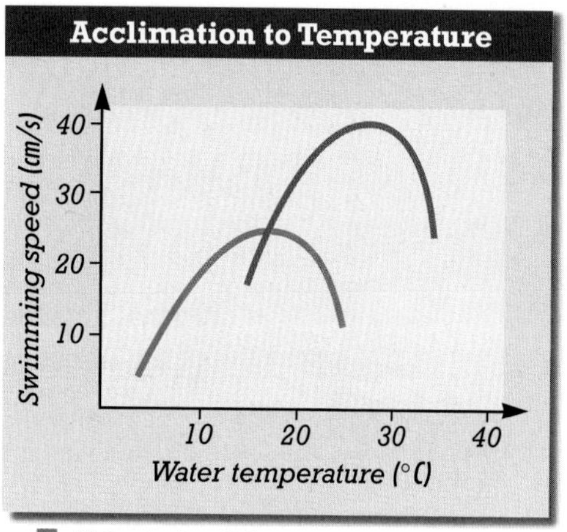

FIGURE 19-9

Fish raised at 25°C are acclimated to higher temperatures and are able to tolerate much warmer temperatures than the fish raised at 5°C.

FIGURE 19-10

An organism can be a regulator or a conformer, or both. The lizard, represented by the blue line, is a conformer with regard to body temperature, whereas humans, represented by the red line, regulate internal temperature.

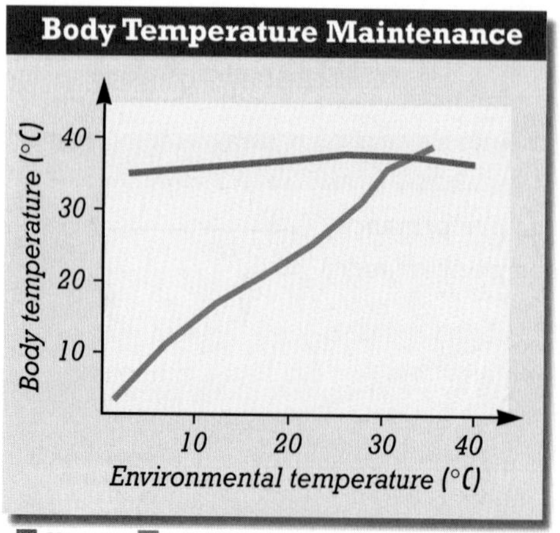

Acclimation

Some organisms can adjust their tolerance to abiotic factors through the process of **acclimation** (AK-luh-MAY-shuhn). For instance, goldfish raised at different temperatures have somewhat different tolerance curves, as shown in Figure 19-9. If you spend a few weeks at a high elevation, you will acclimate to reduced oxygen levels, or "thin air." Over time, the number of red blood cells in your body will increase, thereby increasing the amount of oxygen your blood can carry. Be sure not to confuse acclimation with adaptation. Acclimation occurs within the lifetime of an individual organism. Adaptation is genetic change in a species or population that occurs over many generations.

Control of Internal Conditions

Environments fluctuate in temperature, light, moisture, salinity, and other chemical factors. There are two ways for organisms to deal with some of these changes in their environment. **Conformers** are organisms that do not regulate their internal conditions; they change as their external environment changes. For example, the body temperature of desert lizards rises and falls with the temperature of the lizards' environment, as shown in Figure 19-10. The internal conditions of a conformer remain within the optimal range only as long as environmental conditions remain within that range.

In contrast, **regulators** are organisms that use energy to control some of their internal conditions. Regulators can keep an internal condition within the optimal range over a wide variety of environmental conditions. Your body temperature, for instance, remains within a few degrees of 37°C (99°F) throughout the day. Pacific salmon, which spend part of their lives in salt water and part in fresh water, are conformers to environmental temperatures but are regulators of their internal salt concentration.

Escape from Unsuitable Conditions

Some species can survive unfavorable environmental conditions by escaping from them temporarily. For example, desert animals usually hide underground or in the shade during the hottest part of the day. Many desert species are active at night, when temperatures are much lower. A longer-term strategy is to enter a state of reduced activity, called **dormancy,** during periods of unfavorable environmental conditions. Throughout most of the United States, winter temperatures are too cold for reptiles and amphibians to tolerate, but these animals survive by hiding underground and becoming dormant until spring. Another strategy is to move to another, more favorable habitat. This is called **migration.** A familiar example of migration is the seasonal movements of birds, which spend spring and summer in cooler climates

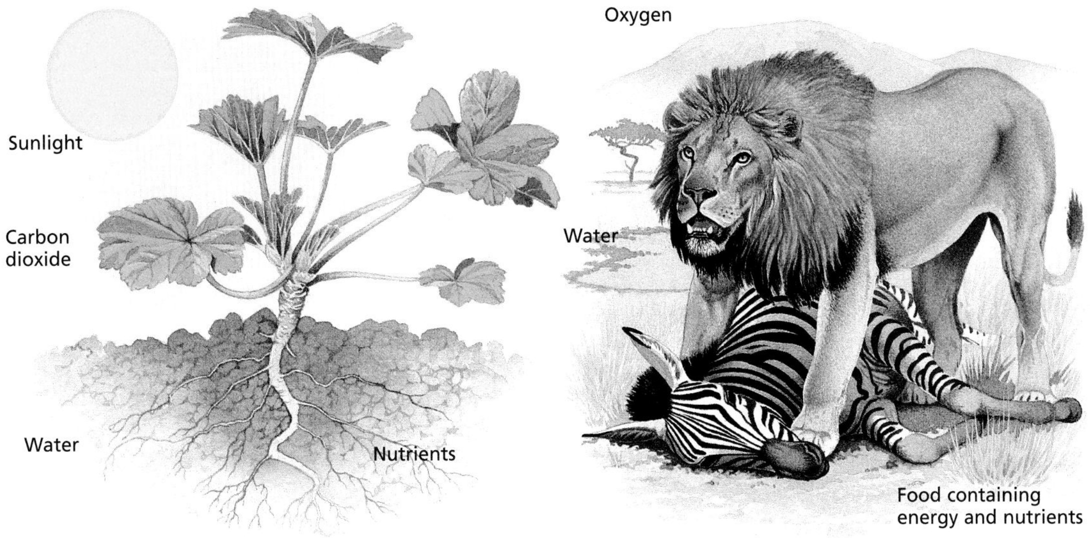

FIGURE 19-11

Plants and animals are able to share the same spaces because they each have different requirements for survival.

and then migrate to warmer climates in the fall. They remain in the warmer climate until spring, when they return to the cooler climate. Birds thus avoid the low temperatures and food scarcity of winter.

Resources

Whether a species can survive in a particular habitat depends on the suitability of environmental conditions and also on the availability of **resources,** the energy and materials the species needs. Food, energy, nesting sites, water, and sunlight are examples of resources that are often required by animals or plants. The resources essential for survival differ from species to species. As you can see in Figure 19-11, plants require a very different set of resources than animals do.

THE NICHE

A species' **niche** (NICH) is its way of life, or the role the species plays in its environment. The niche includes the range of conditions that the species can tolerate, the methods by which it obtains needed resources, the number of offspring it has, its time of reproduction, and all of its other interactions with its environment. When studying the niche of a species, scientists usually concentrate on a few features that can be readily measured, such as where the species lives, what time of day it is active, and what it eats. Figure 19-12 shows an aspect of the feeding behavior of a common species of songbird.

The **fundamental niche** is the range of conditions that a species can potentially tolerate and the range of resources it can potentially use. In reality,

FIGURE 19-12

The graph below illustrates the blue-gray gnatcatcher's feeding behavior. The darkest shade in the center of the contour lines indicates that most prey are captured between 3 m and 5 m above ground. The prey most frequently captured averaged about 4 mm in length.

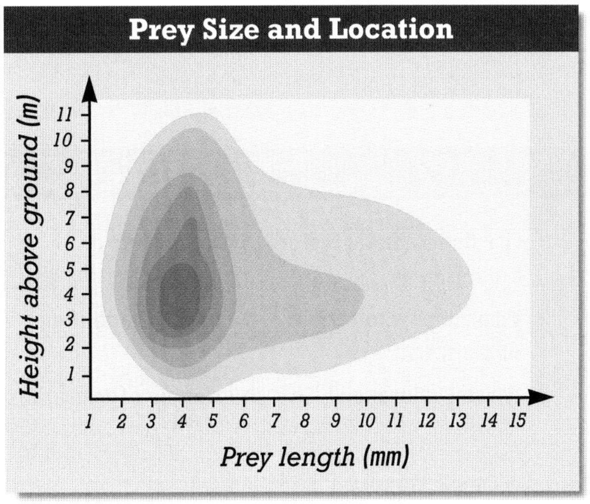

TEACHING STRATEGY
Plants

Show students pictures of deciduous trees in the summer and in the winter. Ask students what might be the selective advantage of not having leaves in the winter. Ask students how shedding leaves is an adaptation to avoid unsuitable conditions. Tell them that precipitation in winter in a deciduous forest is mostly frozen and that little water is available to the tree until spring.

INCLUSION *ACTIVITY*

On a map of the world or of the Western Hemisphere, show students that many of the songbirds we think of as North American migrate to Central and South America in the winter. Ask students if they have seen migrating flocks of birds. Have students describe how these flocks appear. Ask students to suggest ways that migrating birds might navigate. (They may follow shorelines, magnetic fields, stars or the sun, or prevailing wind direction.)

GIFTED *ACTIVITY*

Divide the class into four groups. Have one group study circadian rhythms, such as human sleep cycles; have another study annual rhythms, such as estrus (reproductive) cycles in deer. Another group can study tidal rhythms and how they relate to sea turtles that come ashore to lay eggs, and the fourth group can study migratory rhythms, such as the movement of the whooping crane or salmon. Ask each group to read about the organism in the library, and write a report describing the rhythm and the role it plays in the organism's life. Have each group report about their rhythm to the rest of the class. Have each group discuss how each type of biological rhythm is an adaptation to a rhythmic change in the environment.

SECTION 19-2

VISUAL LINK

Figure 19-13

Ask students to think of an organism and then draw a diagram similar to Figure 19-13 for the organism they chose. Have them label niche parameters and indicate any other organisms that could exist in the same fundamental niche.

ANSWERS TO SECTION 19-2 REVIEW

1. Possible factors include weather, pH, temperature, sunlight, nitrogen, carbon dioxide, and oxygen.
2. A tolerance curve shows the range of conditions within which an organism can function.
3. Migration is when animals move to a more favorable habitat, such as the migration of North American songbirds to the tropics in the fall and the annual migration of monarch butterflies to Mexico.
4. Habitat is where an organism lives; niche is how an organism lives—how it obtains resources and what conditions it can tolerate.
5. An organism might restrict its use of resources because of competition or because of the threat of predation.
6. Species have unique niches in order to reduce competition for resources. Two different species cannot occupy the same niche. One species will be better at competition, and eventually the other species will be eliminated.

FIGURE 19-13

Each of these circles represents the fundamental and realized niche of a species. The center of each circle represents optimal conditions. Farther from the center, conditions are less ideal. Note that the fundamental niches can overlap but the realized niches cannot. Ecologically similar species do not occupy the same realized niche.

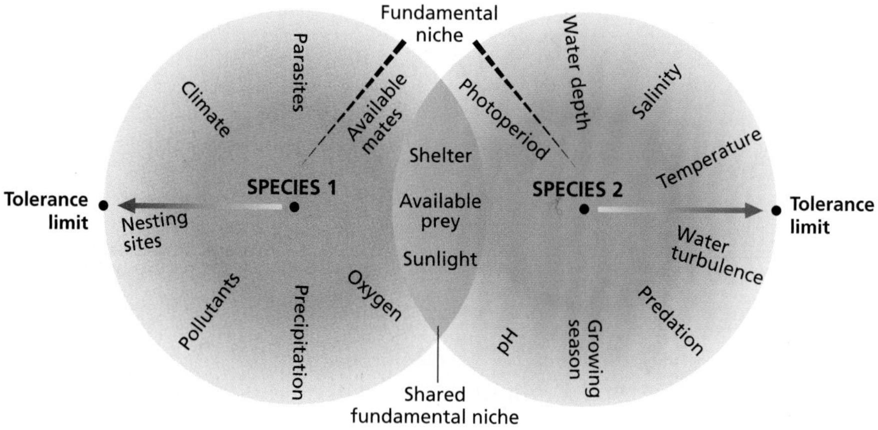

internet connect

SCI LINKS NSTA
TOPIC: Niches
GO TO: www.scilinks.org
KEYWORD: HM372

Word Roots and Origins

niche

from the Old French *nichier,* meaning "to nest"

a species may have to restrict its activity to avoid predators, or competition with other species may prevent it from using a resource. The **realized niche** of a species is the range of resources it actually uses. The fundamental niche will usually include a much broader set of conditions than the realized niche. The darkest blue area in Figure 19-12 indicates a realized niche. The progressively lighter shades indicate increasing intolerance to environmental conditions as the birds' niche changes from realized to fundamental.

Niche Differences

A species' niche can change within a single generation. For example, caterpillars eat the leaves of plants. After feeding for some time, they transform into butterflies, which feed on nectar.

Generalists are species with broad niches; they can tolerate a range of conditions and use a variety of resources. An example of an extreme generalist is the Virginia opossum, which is found across much of the United States. The opossum feeds on almost anything, from eggs and dead animals to fruits and plants. In contrast to the opossum is the koala of Australia, which feeds only on leaves from a few species of eucalyptus trees. Species that have narrow niches, such as the koala, are called **specialists.**

SECTION 19-2 REVIEW

1. List three abiotic factors that can affect an organism.

2. What does a tolerance curve indicate about an organism?

3. What is migration? Give an example of migration.

4. How is an organism's niche different from its habitat?

5. List two factors that might cause an organism to restrict its use of a resource.

6. **CRITICAL THINKING** Explain why two different species do not occupy exactly the same niche.

CHAPTER 19 REVIEW

SUMMARY/VOCABULARY

19-1

- Ecology is the study of the relationships between organisms and their environment, including both the living and nonliving components.
- Ecological knowledge was crucial to the survival of our early human ancestors. Ecology is essential today because humans are changing the world rapidly and on a global scale.
- The human population is growing extremely rapidly. It has tripled in size in about 70 years.
- Rapid growth of the human population has caused the extinction of many other species of organisms. The rate of extinction today is higher than at any time since the extinction of the dinosaurs.
- Industrial chemicals released into the atmosphere are destroying the ozone layer, the protective shield around Earth that absorbs harmful ultraviolet radiation from the sun.
- Burning of fossil fuels has increased atmospheric levels of carbon dioxide. Most scientists think this is causing global warming, or a rise in global temperatures.
- The science of ecology is usually organized into five levels, each of which has unique properties: organism, population, community, ecosystem, and biosphere.
- Species in ecosystems interact with other species and with their nonliving environment. As a result, a disturbance that affects one species can spread to other species in the ecosystem.
- Because ecosystems are so complex, ecologists rely on models, simplified systems that mimic the behavior of the natural world.

Vocabulary

biosphere (363)
community (363)
ecology (359)
ecosystem (363)
greenhouse effect (361)
population (363)

19-2

- An organism's habitat is where it lives.
- Two kinds of factors influence organisms: biotic factors, which are living things, and abiotic factors, which are nonliving things or processes, such as climate, sunlight, and pH.
- The environment changes over time and from place to place.
- Each organism can tolerate a range of environmental conditions. A graph showing this range is called a tolerance curve.
- Over a short period of time, organisms can adjust their tolerance curves through the process of acclimation.
- Species follow two strategies in dealing with environmental change. Regulators control their internal conditions, while conformers' internal conditions vary with the environment.
- Species often escape unfavorable environmental conditions by migrating to a new habitat or by becoming dormant and waiting out the unfavorable conditions.
- A species' niche is its way of life, or its role in an ecosystem.
- A species' fundamental niche is the range of conditions that it can potentially tolerate and the resources that it can potentially use. In the face of competition or predation, a species may use a smaller range of resources or live in a reduced range of conditions. These actual resources used and conditions best tolerated make up a species' realized niche.
- Species with broad niches are called generalists, while species with narrow niches are called specialists.

Vocabulary

abiotic factor (368)
acclimation (370)
biotic factor (368)
conformer (370)
dormancy (370)
fundamental niche (371)
generalist (372)
habitat (368)
migration (370)
niche (371)
realized niche (372)
regulator (370)
resource (371)
specialist (372)
tolerance curve (369)

CHAPTER 19 REVIEW ANSWERS

REVIEW

1. Ecology is the study of the interactions between the living and nonliving environment.
2. An ecosystem includes both living things and the nonliving things in an area. A community contains only the organisms found in a particular place.
3. Answers will vary, but might include sunlight, climate, and oxygen.
4. Migration is when animals escape from unfavorable conditions by moving to another area. Dormancy is when they escape from unfavorable conditions by becoming inactive and waiting them out.
5. A regulator can control an internal condition, while a conformer allows an internal condition to vary with the environment.

6. d **7.** a
8. c **9.** b
10. d **11.** c
12. b **13.** d
14. c **15.** c

16. In the years after oaks produce a large number of acorns, the number of ticks increases. The health official should increase the amount of money budgeted for Lyme disease treatment during years following large acorn crops.
17. Models help ecologists reduce the complexity of ecosystems and focus on a few important factors. Models are only as good as the information on which they were based. All models are simplified and do not account for all variables in a real environment.
18. Answers will vary, but students might mention that animals withdraw oxygen, and plants release oxygen. Also, plants absorb carbon dioxide from the atmosphere, and animals release carbon dioxide.
19. A tolerance curve shows an organism's performance across a range of environmental conditions. Tolerance curves are important for ecology because an

organism's range is often determined by its tolerance of environmental conditions.

20. Answers will vary, but students will probably list extremely cold temperatures and reduced availability of food.

21. Within the optimum range, the species functions at its best because all the species' requirements are met. Within the stress zone, at least one biotic or abiotic factor is not optimal. The species can survive, but it does not perform at its best level. Within the intolerance zone, the species experiences stressful conditions in at least one biotic or abiotic factor. The species' survival is threatened. Outside the tolerance limit, the species cannot survive.

22. Acclimation is a change in an organism's tolerance in response to environmental conditions. Adaptation is a long-term change in the genetic makeup of a species or population. Acclimation within the life of the organism and involves no genetic change.

23. Organisms may migrate, or move to a more favorable habitat. They may sit out periods of unfavorable conditions by entering dormancy, or becoming inactive.

24. Humans are generalists because they feed on many kinds of foods, and they live in almost every habitat on Earth.

25. Some abiotic factors are length of growing season, precipitation, sunlight, and soil nutrients.

CRITICAL THINKING

1. By migrating south, songbirds escape the extreme cold and food shortages of the winter, but this behavior costs a large amount of energy in flying long distances.

2. An outbreak of gypsy moth caterpillars would damage the oak trees and probably reduce the amount of acorns they produce. As a result, populations of deer mice and deer could fall, causing a reduction in the population of

REVIEW

Vocabulary

1. Define the term *ecology*.
2. How is a community different from an ecosystem?
3. List two abiotic factors that influence you.
4. In what way is migration similar to dormancy?
5. Distinguish between regulators and conformers.

Multiple Choice

6. The number of species on Earth (a) is exactly 1.5 million (b) has increased over the last 2,000 years (c) is smaller than the number that occurred during the time of the dinosaurs (d) is decreasing rapidly.
7. The thinning of the ozone layer is caused by (a) CFCs (b) carbon dioxide (c) oxygen gas (d) carbon monoxide.
8. Which of the following is not one of the five main levels of organization within ecology? (a) biosphere (b) ecosystem (c) atom (d) community (e) population
9. An ecosystem includes (a) all the members of one species (b) all the living and nonliving factors in an environment (c) all parts of Earth where life exists (d) all members of a species in the same area.
10. Which of the following is not true of models? (a) They are simplified systems that mimic the natural world. (b) They can be tested by comparison with real situations. (c) They help scientists make predictions. (d) They exactly duplicate reality.
11. Abiotic factors in an ecosystem can include (a) plants (b) animals (c) sunlight (d) microorganisms.
12. An animal could migrate to avoid (a) a hurricane (b) cold temperatures in winter (c) an extreme cold snap (d) a volcanic eruption.
13. A species' fundamental niche (a) cannot be larger than its realized niche (b) changes in response to predators or competitors (c) can be larger or smaller than its realized niche (d) is always as large as or larger than its realized niche.
14. Which of the following is not a resource needed by animals? (a) water (b) food (c) carbon dioxide (d) oxygen
15. Which of the following is not true of an organism's tolerance for an environmental variable?
(a) Its performance is usually best at intermediate values.
(b) Its performance can be illustrated with a tolerance curve.
(c) Its tolerance levels cannot vary over the organism's lifetime.
(d) Tolerance levels can change through acclimation.

Short Answer

16. Explain how an understanding of interconnectedness in ecosystems might help a health official determine how much money to allocate for the treatment of people with Lyme disease.
17. Why are models so valuable to ecologists? What are some limitations of models?
18. The biotic and abiotic factors in an ecosystem can interact. Give two examples of such interactions.
19. What is a tolerance curve? What application does it have to ecology?
20. Give two examples of environmental conditions that bears avoid by hibernating in winter.
21. Examine the diagram below of a tolerance curve. Briefly describe the conditions in each zone of tolerance and the reactions a species may have to them.
22. What is acclimation? How is acclimation different from adaptation by natural selection?

Salinity Tolerance Curve

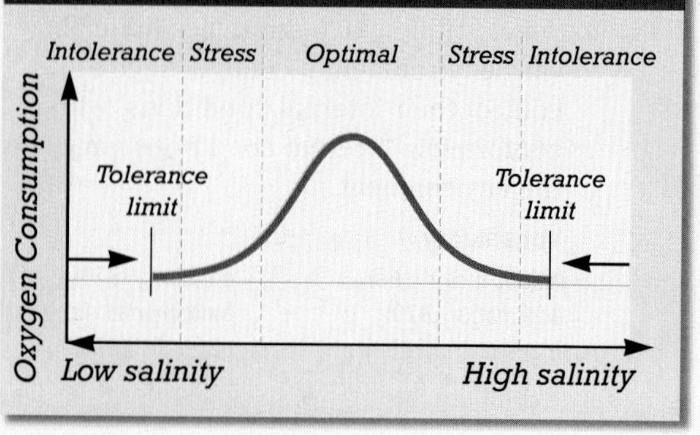

23. Describe two methods, not including acclimation, that organisms use to respond to unfavorable conditions in their environment.
24. Ecologically speaking, humans are considered generalists rather than specialists. Explain why.
25. What abiotic factors might account for differences in net primary productivity in ecosystems?

CRITICAL THINKING

1. In the fall, many kinds of songbirds migrate from the United States to Central America or South America. Explain the benefits of migration for songbirds. What are some consequences of this behavior?
2. The gypsy moth is a destructive insect pest. In the ecosystem illustrated in Figure 19-6, the caterpillars of the gypsy moth negatively affect oak trees by consuming their leaves. The number of caterpillars in the forest fluctuates, and every few years the caterpillar population increases dramatically. Trace the effects that a large increase in the caterpillar population might have on the other members of this ecosystem. How might an increase in caterpillars affect the incidence of Lyme disease? Explain your answer.
3. Ecologists have identified several characteristics that increase the likelihood that a species will become extinct. Specialization is one such characteristic. Explain why a very specialized species is likely to be more vulnerable to extinction.
4. Evaluate the following statement: Though the field of ecology had not been defined in Charles Darwin's day, Darwin was an ecologist.
5. Explain why scientists believe that ecology and evolution are tightly linked.
6. Study the diagram of the niche of a species of songbird below. Where does the bird capture most of its food? Which sizes of insects does it prefer?

Prey Size and Location

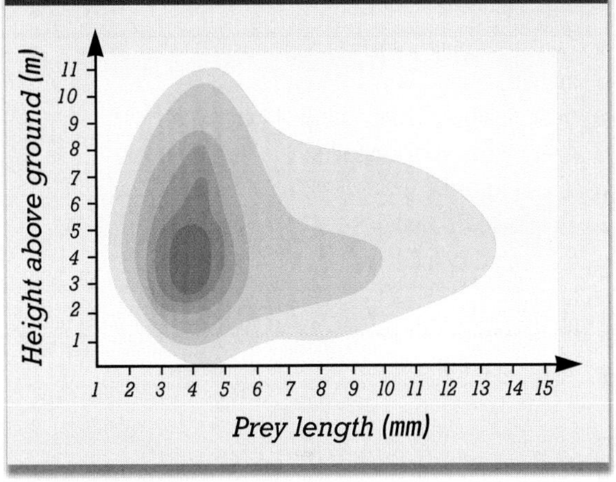

EXTENSION

1. Read "The Final Countdown" in *Audubon*, November–December 1999, on page 65. According to E. O. Wilson, how does biodiversity measure the value of an ecosystem? How will the loss of biodiversity affect human life?
2. Pick a small outdoor area—a lawn, or window box. With a string, mark off a 1,000 cm^2 area and carefully count the plants and animals found there. Note the number of species found, and identify as many as you can. Compare your findings with those of your classmates.
3. Using library resources or an on-line database, research the Biosphere 2 project. What is Biosphere 2, and where is it located? What were the goals and outcomes of the first two-year experiment in Biosphere 2? What scientific criticisms were leveled at the project?

deer ticks as well. Such an outbreak also would likely reduce the incidence of Lyme disease.
3. A specialized species requires a narrow range of environmental conditions and resources. Such a species is vulnerable to environmental fluctuations that shift conditions outside the species' tolerance range.
4. Darwin collected information about the ranges, habits, and interrelationships of organisms during his voyage. Furthermore, the central theme of natural selection is the interaction between the organism and its environment.
5. Changing environmental conditions set the stage for adaptation through the evolutionary mechanism of natural selection.
6. The bird captures prey between 3 mm and 5 mm in length, and between 3 m and 5 m above the ground.

EXTENSION

1. The more species of plants, animals, and other organisms there are in an ecosystem, the more resistant the ecosystem is to destruction and the better it can perform its roles, such as cleaning water, enriching the soil, stabilizing the climate, and generating oxygen. Wilson expects that future loss of biodiversity will have severe global effects on the quality of life-support systems, such as watersheds, air quality, and rainfall.
2. Answers will vary, but students should identify as many species as possible.
3. Biosphere 2 is a sealed artificial ecosystem located near Oracle, Arizona. In 1991, eight Biospehereans were sealed inside to show that it could function as a self-sustaining system. During its initial two year run, Biosphere 2 failed to achieve their goal. Scientists criticized the project for concentrating on what was called "New Age theatrics" while producing little of scientific value.

CHAPTER 19 INVESTIGATION

TIME REQUIRED

Two 50-minute class periods

SAFETY PRECAUTIONS

Remind students to wear safety goggles for Part B. Avoid eye contact with methyl cellulose. Remind students that they are working with live animals and to treat them gently.

QUICK REFERENCES

Lab Preparation Notes are found on page 357D.

Holt BioSources provides a Teaching Resources CD-ROM that contains "Using Gowin's Vee in the Lab" and "Scoring Rubrics."

PROCEDURAL TIPS

1. Order brine shrimp in advance from a biological supply house. Allow extra time to grow large cultures.
2. Use 0.5 in. internal diameter plastic tubing cut into 44 cm lengths. When tubing is divided as directed, it allows for space taken up by stoppers and forms equal quarters.
3. Detain™ microlife slowing agent works well to slow the movement of brine shrimp without killing them. You may substitute methyl cellulose, which does not work as well.
4. For the reaction to light, soft white fluorescent bulbs are recommended. These will give sufficient light without significantly changing the temperature. Grow lights will also work well, and even natural lighting is acceptable. Avoid incandescent bulbs, which pose the hazard of burns.
5. Fiberglass screening is available at hardware stores or may be ordered from a biological supply house.
6. Divide the class into teams of six, with two students in each

CHAPTER 19 INVESTIGATION

Observing Habitat Selection

OBJECTIVES

- Assess the effect of light on habitat selection by brine shrimp.

PROCESS SKILLS

- observing
- measuring
- collecting data
- organizing data
- analyzing data

MATERIALS

- safety goggles
- marking pen
- clear, flexible plastic tubing (44 cm long)
- 4 test tubes with stoppers
- test-tube rack
- 2 corks to fit tubing
- graduated cylinder
- funnel
- brine shrimp culture
- aluminum foil
- 3 screw clamps
- 1 pipet
- Petri dish
- methyl cellulose
- fluorescent lamp or grow light
- 14 pieces of screening
- calculator

Background

1. Recall that a species' habitat is a specific area where it lives.
2. A species habitat selection depends on how well the location fits within the species' tolerance range. The more optimal all limiting factors are within a portion of an organism's range, the more likely the organism is to select that area for its habitat.
3. What limiting factors might be involved in habitat selection?
4. What is a niche?

PART A Setting Up

1. Mark the plastic tubing at 12 cm, 22 cm, and 32 cm from one end so that you will have the tube divided into four sections. Starting at one end, label the sections 1, 2, 3, and 4. Label four test tubes 1, 2, 3, and 4.
2. Place a cork in one end of the tubing. Use a graduated cylinder and a funnel to transfer about 50 mL of brine shrimp into the tubing. Cork the open end, and lay the tubing on the desktop.
3. You and your partner will complete either Part B or Part C and then share your results with the other students on your team. CAUTION **You will be working with live animals. Be sure to treat them gently and to follow directions carefully.**

PART B Control Group

1. Cover the tubing with aluminum foil, and let it remain undisturbed for 30 minutes. While you are waiting, create a data table like Table A, below, in your lab report to record the numbers of shrimp in each section of the tubing.

TABLE A CONTROL GROUP

Test tube number	Count 1	Count 2	Count 3	Count 4	Count 5	Average number of shrimp in test tube
1						
2						
3						
4						

2. After 30 minutes have passed, attach screw clamps to each spot that you marked on the tubing. While your partner holds the corks firmly in place, tighten the middle clamp first, and then tighten the outer clamps.
3. Immediately pour the contents of each section of tubing into the test tube labeled with the corresponding number.

4. CAUTION **Put on safety goggles before handling methyl cellulose. If you get methyl cellulose in your eyes, immediately flush it out at the eyewash station while calling to your teacher.** Stopper test tube 1 and invert it gently to distribute the shrimp. Use a pipet to draw a 1 mL sample of shrimp culture and transfer the culture to a Petri dish. Add a few drops of methyl cellulose to the Petri dish to slow down the shrimp. Count the live shrimp, and record the count in your lab report.
5. Dispose of the shrimp as your teacher directs. Repeat step 4 four more times for a total of five counts from test tube 1.
6. Calculate the average number of shrimp in test tube 1, and record the result in the data table you made in your lab report.
7. Repeat steps 4–6 for the contents of each of the remaining test tubes.
8. Clean up your materials and wash your hands before leaving the lab.
9. In your lab report, make a histogram showing the total number of shrimp you counted in each section of tubing.

PART C Experimental Group

1. Set a fluorescent lamp 20 cm away from the tubing.
2. Cover section 1 of the tubing with eight layers of screening. Place four layers of screening on section 2 and two layers of screening on section 3. Leave section 4 uncovered. Leave this setup in place for 30 minutes. While you are waiting, create a data table in your lab report like Table B, below, to record the numbers of shrimp in each section of the tubing.

TABLE B EFFECTS OF LIGHT ON BRINE SHRIMP

Test tube number	Count 1	Count 2	Count 3	Count 4	Count 5	Average number of shrimp in test tube
1						
2						
3						
4						

3. After 30 minutes have passed, attach screw clamps to each spot that you marked on the tubing. While your partner holds the corks firmly in place, tighten the middle clamp first, and then tighten the outer clamps.
4. Immediately pour the contents of each section of tubing into the test tube labeled with the corresponding number.
5. CAUTION **Put on safety goggles before handling methyl cellulose. If you get methyl cellulose in your eyes, immediately flush it out at the eyewash station while calling to your teacher.** Stopper test tube 1 and invert it gently to distribute the shrimp. Use a pipet to draw a 1 mL sample of shrimp culture and transfer the culture to a Petri dish. Add a few drops of methyl cellulose to the Petri dish to slow down the shrimp. Count the live shrimp, and record the count in your lab report.
6. Dispose of the shrimp as your teacher directs. Repeat Step 5 four more times for a total of five counts from test tube 1.
7. Calculate the average number of shrimp in test tube 1 and record the result in the data table you made in your lab report.
8. Repeat steps 5–7 for the contents of each of the remaining test tubes.
9. Clean up your materials and wash your hands before leaving the lab.
10. In your lab report, make a histogram showing the number of shrimp in each section of tubing. Identify each section with the amount of screening.

Analysis and Conclusions

1. How did the brine shrimp react to the changes in light?
2. Why was a control (Part B) necessary?
3. How do the histograms made by different groups compare?

Further Inquiry

Design an experiment to test the reaction of brine shrimp to a gradient of heat.

team working on each part (A, B, or C).
7. Students may count shrimp by viewing them in the petri dish or by holding the pipet up to the light.

ANSWERS TO BACKGROUND

3. Organisms select habitats in which conditions such as temperature, light levels, salinity, and pH are within their tolerance limits. Brine shrimp, for example, select for temperature, light, and pH.
4. The niche is all of an organism's interactions with its living and nonliving environment, including its habitat preferences.

ANSWERS TO ANALYSIS AND CONCLUSIONS

1. Answers will vary, depending on the species of *Artemia* used.
2. The control was necessary to show that the brine shrimp did not inherently prefer one part of the tube—the ends of the middle, for example.
3. Answers will vary, depending on the species of *Artemia* used.

FURTHER INQUIRY

Designs will vary but should include subjecting the brine shrimp to temperature extremes, and should include a method of eliminating all other variables, such as placing all test animals under the same light level.

CHAPTER 20
POPULATIONS

	TOPICS	TEACHING RESOURCES	LABS, CLASSWORK, AND HOMEWORK
BLOCKS 1 & 2 90 minutes	**Introducing the Chapter, p. 378**	**ATE** Focus Concept, p. 378 **ATE** Assessing Prior Knowledge, p. 378 **ATE** Understanding the Visual, p. 378	■ Supplemental Reading Guide, *Silent Spring*
	20-1 Understanding Populations, p. 379 Properties of Populations, p. 379 Population Dynamics, p. 381	**ATE** Section Overview, p. 379 **ATE** Critical Thinking, p. 381 **ATE** Visual Link Figure 20-2, p. 381 Figure 20-3, p. 382 63A	**Lab A13,** Using Random Sampling ★ Study Guide, Section 20-1 **PE** Section Review, p. 382
BLOCKS 3 & 4 90 minutes	**20-2 Measuring Populations, p. 383** Population Growth Rate, p. 383 The Exponential Model, p. 384 The Logistic Model, p. 385 Population Regulation, p. 386	**ATE** Section Overview, p. 383 **ATE** Visual Link Figure 20-5, p. 384 64A, 75, 76	■ Problem-Solving Worksheets—Population Size ★ Study Guide, Section 20-2 **PE** Section Review, p. 387
BLOCKS 5 & 6 90 minutes	**20-3 Human Population Growth, p. 388** History of Human Population Growth, p. 388	**ATE** Section Overview, p. 388 **ATE** Visual Link Figure 20-11, p. 389 75, 76	**PE** **Chapter 20 Investigation,** pp. 394–395 **PE** **Quick Lab,** p. 389, Demonstrating Population Doubling ■ Occupational Applications Worksheets—Forestry Technician ★ Study Guide, Section 20-3 **PE** Section Review, p. 390
BLOCKS 7 & 8 90 minutes	**REVIEW and ASSESSMENT** **PE** Chapter 20 Review, pp. 392–393 ■ Performance-Based Assessment—Testing Pond Water for Carbon Dioxide	Audio CD Program BIOLOGY Interactive Tutor Unit 7—Ecosystem Dynamics	**CHAPTER TESTING** ■ Chapter Test (blackline master) ▲ Test Generator ■ Assessment Item Listing ■ Scoring Rubrics and Classroom Management Checklists

MODERN BIOLOGY | HOLT BIOSOURCES | One-Stop Planner CD-ROM

KEY

PE Pupil's Edition
ATE Teacher's Edition
■ Active Reading Guide
★ Study Guide

Teaching Transparencies

Laboratory Program

One-Stop Planner CD-ROM
Shows these resources within a customizable daily lesson plan:
■ Holt BioSources Teaching Resources
■ Active Reading Guide
★ Study Guide
▲ Test Generator

READING FOR CONTENT MASTERY

Preview: Have students preview the objectives, the introductory paragraph, topic sentences, and captions before reading the chapter.

ATE Active Reading Technique: Reading Effectively, p. 379
■ Active Reading Guide Worksheet 20-1
PE Summary/Vocabulary, p. 391

■ Active Reading Guide Worksheet 20-2
PE Summary/Vocabulary, p. 391

■ Active Reading Guide Worksheet 20-3
PE Summary/Vocabulary, p. 391

TECHNOLOGY AND INTERNET RESOURCES

CNN PRESENTS Video Segment 12 Britain's Living Plan

Holt Biology Videodiscs Teacher's Correlation Guide, Lessons 20-1 through 20-3

BIOLOGY INTERACTIVE TUTOR Unit 7—Ecosystem Dynamics, Topic 2

Audio CD Program, Sections 20-1 through 20-3

internetconnect

sciLINKS NSTA
On-line Resources: www.scilinks.org
The following *sci*LINKS Internet resources can be found in the student text for this chapter:

Topics:
- Factors affecting population growth, p. 383
- History of population growth, p. 388

go.hrw.com
On-line Resources: go.hrw.com
Visit the HRW Web site for a variety of resources related to this chapter. Just type in the keyword HM2 Chapter 20.

Smithsonian Institution®
Visit **www.si.edu/hrw** for additional on-line resources.

CNNfyi.com
Visit **www.cnnfyi.com** for late-breaking news and current event stories selected just for you.

LAB ACTIVITY PLANNING

CHAPTER 20 INVESTIGATION

Studying a Yeast Population, pp. 394–395

OVERVIEW
In this investigation, students will observe the growth and decline in a population of yeast cells and apply the underlying principles in order to understand changes in human populations. Students will use a simple sampling technique to count the members of the yeast population.

SAFETY
Students will wear disposable gloves, lab aprons, and safety goggles during the lab. If iodine gets in a student's eye, flush the eye with water for at least 15 minutes and seek immediate medical attention.

PREPARATION
1. Up to 4 hours before lab time, dissolve a fresh cake of yeast in 80 mL of warm water. Add 20 g of sucrose, glucose, or fructose, and stir. Fresh or dry yeast is available in grocery stores, or you may obtain freeze-dried cultures from WARD'S. Maintain the culture in a warm, dark area for the duration of this investigation.
2. Each student group will need one ruled microscope slide.
3. Each student group will also need a 1 mL pipet, a test tube, iodine in a dropper bottle, coverslips, and access to a compound microscope.

Quick Labs

Demonstrating Population Doubling, p. 389
Each student will need a pencil, paper, and a sheet of newspaper (two pages across).

UNDERSTANDING THE VISUAL

Ask students to count the number of individual dolphins in this picture. Ask them if they think this picture represents all of this population of dolphins. (Students might say that some dolphins may be out of the picture.) Ask students if they think it would be easy to take a census of a population. (It is often hard to find or see all the members of a population.) Ask students if they think everyone was counted in the last population census of the United States. (Some may say the census was not accurate because people move around frequently or may not want to be counted.) Ask students why counting a population of dolphins or citizens an important. (Population data tells about birth and death rates, migration, and the overall prosperity of a population, which could be helpful in planning policies for the future or to protect a species.)

FOCUS CONCEPT

Interdependence of Organisms

Tell students that for populations to grow and survive they must compete with each other and with other populations for food and space. When some populations increase dramatically, they may endanger other populations. Carrying-capacity fluctuations may be due to species interactions, such as predation.

ASSESSING PRIOR KNOWLEDGE

Review the following concepts.

Evolution: *Chapter 16*
A gene pool is all the genes present in a population. Ask students to predict what might happen to a population of animals if it is very small and a serious disease strikes.

CHAPTER 20

POPULATIONS

This is part of a population of bottlenose dolphins, Tursiops truncatus.

FOCUS CONCEPT: *Interdependence of Organisms*

As you read the chapter, notice how even very simple models can help you understand population processes.

UNDERSTANDING POPULATIONS

The human population of the world was about 6 billion in 1996, three times its size in 1900. During this period of rapid human population growth, populations of many other species have decreased dramatically. Will the human population continue to grow? Will populations of other species continue to get smaller? Will other species continue to become extinct? An understanding of populations is crucial to answering these questions.

SECTION 20-1

OBJECTIVES

▲ Explain the differences between population size, density, and dispersion.

● Describe the three main patterns of population dispersion.

■ Explain the importance of a population's age structure.

◆ Contrast the three main types of survivorship curves.

PROPERTIES OF POPULATIONS

A population is a group of organisms that belong to the same species and live in a particular place at the same time. All of the bass living in a pond during a certain period of time constitute a population because they are isolated in the pond and do not interact with bass living in other ponds. The boundaries of a population may be imposed by a feature of the environment, such as a lake shore, or they can be arbitrarily chosen to simplify a study of the population.

The properties of populations differ from those of individuals. An individual may be born, it may reproduce, or it may die. A population study focuses on a population as a whole—how many individuals are born, how many die, and so on.

Population Size

A population's size is the number of individuals it contains. Size is a fundamental and important population property, but it can be difficult to measure directly. If a population is small and composed of immobile organisms, such as plants, its size can be determined simply by counting individuals. More often, though, individuals are too abundant, too widespread, or too mobile to be easily counted, and scientists must estimate the number of individuals in the population. Suppose that a scientist wants to know how many oak trees live in a 10 km^2 patch of forest. Instead of searching the entire patch and counting all the oak trees, the scientist could count the trees in a smaller section of the forest, such as a 1 km^2 area, and use this

SECTION 20-1

SECTION OVERVIEW

Matter, Energy, and Organization
Section 20-1 describes how species are organized into populations within communities of an ecosystem. These populations have common properties, including size, density, dispersion, and population dynamics (such as mortality, birth rate, and age structure), that are driven by the energy input in the population.

Reproduction and Inheritance
Section 20-1 also discusses how populations are continually changing in size, density, birth and death rate, and dispersion. Population change can be measured with growth curves.

ACTIVE READING

Technique: Reading Effectively
Instruct students to take a break after reading each section to study the figures and tables in the section. Discuss the importance of reading figure captions. Prepare questions that correspond to each figure. For example, refer to Figure 20-4 on page 382, and ask students to list their own examples of each type of survivorship. Refer to the figures when you talk about the concepts presented in each section.

TEACHING STRATEGY
Dispersion Patterns

Have students select and make a "buttongram" of one of the three types of dispersion: random, clumped, and even. Provide groups of students a plain piece of paper, glue or tape, and a bag of buttons. Each bag of buttons should include buttons of a single color. (As an alternative to buttons, small squares cut out of colored poster board or fabric scraps can be used to represent individuals.) Tell students the piece of paper represents a habitat and each button represents an individual member of a population. Have students arrange and fasten the buttons on the paper to illustrate one of the dispersion types. Display each group's buttongram, and allow the class to identify the type of dispersion each buttongram represents.

TEACHING STRATEGY
Population Density

At each lab table, set out a clear 2 L bottle containing rocks, soil, and leaf litter. Add live pill bugs, crickets, or other small invertebrates to each bottle. Each bottle can have a different number of organisms. Tell students the bottle represents a country or area of the world. Ask students to calculate the density of the population. (Students should count the number of pill bugs per liter.) Ask students if the population was difficult to count and, if so, why. Ask if it is easy to get accurate population density figures for any species. (It was probably difficult to count the population density of each bottle because the invertebrates were hiding or moving around.)

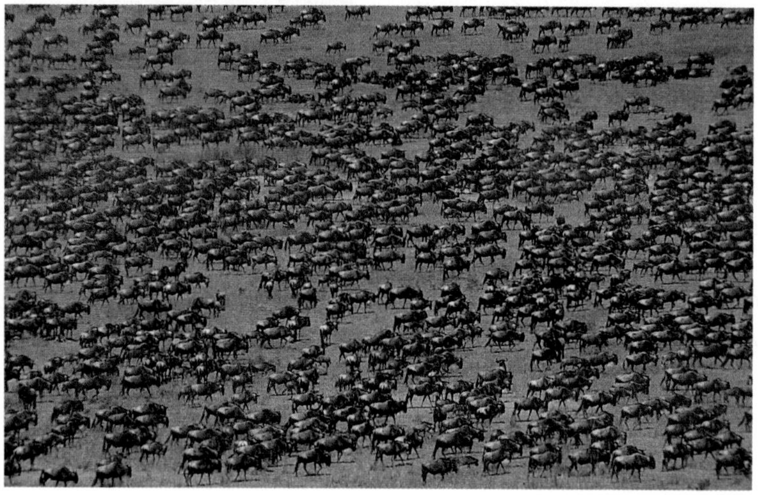

FIGURE 20-1

These migrating wildebeests in East Africa are too numerous and mobile to be counted. Scientists must use sampling methods at several locations to monitor changes in the population size of the animals.

value to estimate the population of the larger area. If the small patch contains 25 oaks, an area 10 times larger likely would contain 10 times as many oak trees, so a reasonable estimate for the population's size is 250 oak trees. A similar kind of sampling technique must be used to estimate the size of the population shown in Figure 20-1.

This kind of estimate assumes that the distribution of trees in the forest is the same as that of the sampled patch. If this assumption is not accurate, the estimate will be inaccurate. Estimates of population size are based on certain key assumptions, so they all have the potential for error.

Population Density

Population density measures how crowded a population is. Population density is always expressed as the number of individuals per unit of area or volume. For example, the population density of humans in the United States is about 30 people per square kilometer. Table 20-1 shows the population densities for several countries. These estimates are calculated for the total land area. Some areas of a country may be sparsely populated, while other areas are very densely populated.

TABLE 20-1
Population Densities

Country	Population density (individuals/km^2)
Japan	330
United Kingdom	240
Kenya	50
Mexico	50
United States	30
Russia	10

Dispersion

A third population property is dispersion. **Dispersion** (di-SPUHR-zhuhn) is the spatial distribution of individuals within the population. In a clumped distribution, individuals are clustered together. In an even distribution, individuals are separated by a fairly consistent distance. In a random distribution, each individual's location is independent of the locations of other individuals in the population. The three possible patterns of dispersion are illustrated in Figure 20-2a. Clumped distributions often occur when resources such as food or living space are clumped. Clumped distributions may also occur because of a species' social behavior, such as when zebras gather into herds or birds form flocks. Even distributions usually result from social interactions, but the interactions result in individuals getting as far away from each other as possible. For example, each gannet, which are the birds shown in Figure 20-2c, stakes out a small area on the coast and defends it from other gannets. Each gannet tries to maximize its distance from all of its neighbors, resulting in an even distribution of individuals. A random distribution usually results from seed dispersal by the wind or by birds, as in the third illustration in Figure 20-2a. Forests or a field of wildflowers result from random seed dispersal.

FIGURE 20-2

Illustrated in (a) are the three dispersion patterns—clumped, even, and random. Turtles commonly clump together to bask in the sun. Birds often are observed in even dispersions as a result of social interactions. A forest is an example of random dispersion. Close up, the fishes in (b) may appear to be in an even distribution, but further away, they can be seen to be clumped. The birds in (c) are evenly distributed, but at a great distance they appear to be clumped.

The dispersion pattern of a population sometimes depends on the scale at which the population is observed. The gannets shown in Figure 20-2c are evenly distributed on a scale of a few meters. If the scale of observation is the entire island on which the gannets live, however, the distribution appears clumped because the birds live only near the shore.

POPULATION DYNAMICS

All populations are dynamic—they change in size and composition over time. To understand these changes, more knowledge is needed about the population than its size, density, and dispersion. One important measure is the **birth rate,** the number of births occurring in a period of time. In the United States, for example, there are about 4 million births per year. A second important measure is the **death rate,** or **mortality rate,** which is the number of deaths in a period of time. The death rate for the United States is about 2.4 million deaths per year. Another important statistic is **life expectancy,** or how long on average an individual is expected to live. In the United States in 1996, the life expectancy for a man was 72 years, and for a woman it was 79 years.

VISUAL LINK

Figure 20-2

Point out to students that, as shown in Figure 20-2b, fish distribution may be clumped as fishes swim about in a school. Ask students why the spatial distribution of this fish population is important. (If offers the fishes protection from predators and more opportunity for getting food.)

CRITICAL THINKING

Distinguishing Populations

Ask students if it is always possible to tell which individuals belong to a population by visual characteristics. Ask students how they could distinguish populations that seem to be alike. (In plants and animals, some species are so similar that they appear to be the same, but on closer inspection, their nutrient requirements, behavior patterns, and habitat may differ. To be certain that two similar populations are different, DNA comparisons could be done).

MAKING CONNECTIONS

Social Studies

Point out to students that developed, industrialized countries have aging populations. Life expectancy is at record highs, and fertility is at or below replacement level. There is concern about the declining number of younger workers and the rising number of elderly people on pensions or social security. Have students research the political issue of social security or pension taxes and how they relate to changes in populations. Have students present their findings and opinions in a forum on population trends and the funding of pension plans.

VISUAL LINK

Figure 20-3

Explain to students that gender statistics are seldom symmetrical. Ask them what it might imply if the right side of the graph for Country B never fell below 2.5 through the oldest age class. (Women live longer in Country B.)

ANSWERS TO SECTION 20-1 REVIEW

1. A population's density depends on its number and the amount of space it occupies. Two populations of the same size can have different densities if they occupy different amounts of space.
2. Even distributions result when organisms attempt to get as far away from each other as possible.
3. There are more elderly individuals in Country B.
4. Country A has a very bottom-heavy age structure, indicating that many individuals have been added to the youngest (and most reproductive) age classes in recent years.
5. A large fraction of the individuals die while young, so each surviving individual must have many offspring to overcome this catastrophic mortality.
6. Some populations of migratory birds are difficult to count because the flocks are too numerous, too widespread, or too mobile. Students may have some interesting ideas for determining the population size of migratory birds. Answers should include a method of isolating and counting a few and then using a proportion to obtain an estimate of the full population.

FIGURE 20-3

These two diagrams show the age structure by gender of two countries. A comparison indicates that Country A has a higher percentage of young people and a lower percentage of elderly people than Country B does.

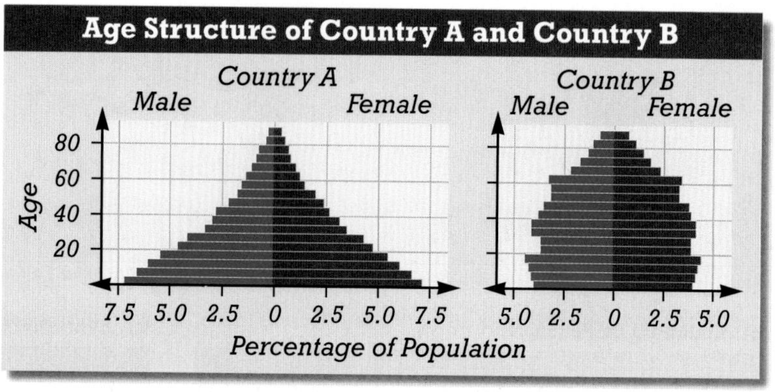

Age Structure

The distribution of individuals among different ages in a population is called **age structure.** Age structures are often presented in graphs, as in Figure 20-3. Many important population processes vary with age. In many species, including humans, very old individuals do not reproduce. Populations with a high percentage of young individuals have a greater potential for rapid growth.

FIGURE 20-4

Humans have a Type I survivorship curve. Some species of birds have a Type II survivorship curve. Some species of fish are examples of a Type III survivorship.

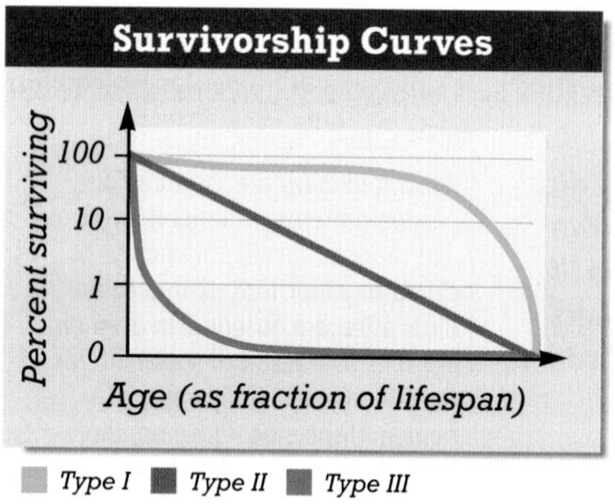

Patterns of Mortality

The mortality rate data of different species tend to conform to one of three curves on a graph, as shown in Figure 20-4. These curves are called **survivorship curves** because they show the likelihood of survival at different ages throughout the lifetime of the organism. In humans or elephants, for instance, the likelihood of dying is small until late in life, when mortality increases rapidly. This pattern of mortality produces the Type I survivorship curve. For other organisms, such as some species of birds, the probability of dying does not change throughout life, giving a linear, or Type II, survivorship curve. Many organisms are very likely to die when young. If an individual survives this early period, however, it has a good chance of surviving to old age. This type of survivorship curve, called Type III, is characteristic of animals such as oysters, salmon, and many insects.

SECTION 20-1 REVIEW

1. **Explain how two populations can be the same size but have different densities.**

2. **Explain why even distributions usually result from social interactions between individuals.**

3. **Explain how it is possible to conclude from Figure 20-3 that the life expectancy of individuals in Country B is greater than that of individuals in Country A.**

4. **How does Figure 20-3 indicate which country's population has the greatest potential for rapid growth?**

5. **Explain why natural selection might favor a high reproduction rate in organisms with Type III survivorship curves.**

6. **CRITICAL THINKING** **Explain two difficulties an ecologist might have in counting a population of migratory birds. Develop and explain a method for estimating the size of such a population.**

SECTION

20-2

MEASURING POPULATIONS

Charles Darwin calculated that a single pair of elephants could increase to a population of 19 million individuals within 750 years. The fact that the world is not overrun with elephants is evidence that some factor or factors restrain the population growth of elephants. In this section you will study how populations grow and what factors limit their growth.

POPULATION GROWTH RATE

Demographers, scientists who study population dynamics, define the **growth rate** of a population as the amount by which a population's size changes in a given time.

Whether a population grows, shrinks, or remains the same size depends on four processes: birth, death, emigration, and immigration. **Immigration** (im-uh-GRAY-shuhn) is the movement of individuals into a population, and **emigration** (em-i-GRAY-shuhn) is the movement of individuals out of the population. Two of these processes—birth and immigration—add individuals to a population, while the other two processes—death and emigration—subtract individuals from the population. For simplicity's sake, demographers usually assume that immigration and emigration are zero when calculating a population's growth rate. By making this simple assumption, a population's growth rate can be described mathematically and can be graphed.

It is customary for demographers to divide large populations into groups of 1,000 and to present data per capita, meaning per individual. Birth rates, death rates, and growth rates for a large population are usually expressed per capita. For example, if there are 52 births and 14 deaths per 1,000 individuals in a large population in one year, the per capita birth rate would be $^{52}/_{1,000}$, or 0.052 births per individual per year. The per capita death rate would be $^{14}/_{1,000}$, or 0.014 deaths per individual per year. The per capita growth rate can be found by the following simple equation:

$$\text{birth rate} - \text{death rate} = \text{growth rate}$$

OBJECTIVES

▲ Describe the exponential model of population growth.

● Compare the similarities and differences between the logistic model and the exponential model.

■ Distinguish between density-dependent and density-independent regulatory factors.

◆ List three reasons why small populations are more vulnerable to extinction.

internet**connect**

SCILINKS NSTA

TOPIC: Factors affecting population growth
GO TO: www.scilinks.org
KEYWORD: HM383

SECTION 20-2

SECTION OVERVIEW

Interdependence of Organisms
Section 20-2 stresses that density-independent factors of a population usually involve more than one kind of population within a community. Predation and competition are density-dependent factors and cause fluctuations around carrying capacity.

Stability and Homeostasis
Section 20-2 describes stability and equilibrium of a population as dependent on growth-rate factors, such as birth rate, death rate, immigration, emigration, and carrying capacity. There are two models of growth, the exponential model and the logistic model.

▶ ENGAGE STUDENTS

Have students look out the window for a few minutes and write down as many different plant or animal populations as they can see. Ask them to describe how these populations might change over time and why. (You could have students watch a short video on the rain forest if your classroom doesn't have a window or if the scene from the window does not overlook nature. An example of window observations might be cardinals and blue jays.)

INCLUSION *ACTIVITY*

Put a picture of an animal on the overhead. Write the words *birth, death, immigration,* and *emigration,* one in each corner. Decide whether an increase or decrease in population results from the process represented by each word. Draw an arrow from the word to the picture if the population increases. Draw an arrow from the animal to the word if the population decreases. (Birth and immigration increase a population; death and emigration decrease it.)

VISUAL LINK

Figure 20-5

Have students look at Figure 20-5. Ask them if a population with a J-shaped graph will most likely continue its rapid exponential rise. Ask them to predict what shape the line might have in the future. (The population will not be able to rise indefinitely because food, space, or some other limiting factor will eventually run out. At that point, the population will either crash or reach equilibrium, and the curve will change dramatically.)

TEACHING STRATEGY

Observing an Animal Population

Set up lab stations with small aquariums, bottles, or cages with a population of some kind of small animal, such as mice, fruit flies, fish, or snails. Ask students to watch the behavior of the animals for 5 min. Then ask them to write down what things might keep the population from growing, based on their observations. Have students share their observations with the class and discuss them. Ask students why these are called limiting factors and ask whether they are density dependent or density independent. (Limiting factors are things that, when scarce, keep a population from growing. Food, water, or nesting material are density-dependent factors. A sudden shift to an extreme temperature can decimate an entire population. This is a density-independent factor).

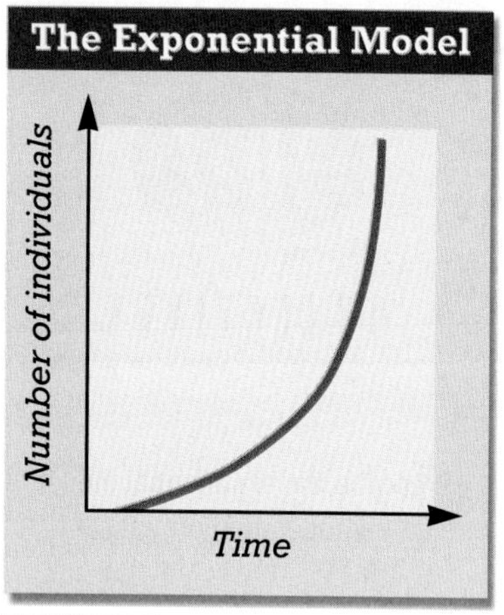

FIGURE 20-5

The graph of exponential population growth has a characteristic J shape. The exponential model indicates infinite, constantly increasing population growth.

FIGURE 20-6

The population increase of bacteria in the laboratory produces a characteristic graph of exponential growth. With this kind of graph, the size of the population of bacteria at any future time can be predicted if the culture is provided with unlimited resources, such as food.

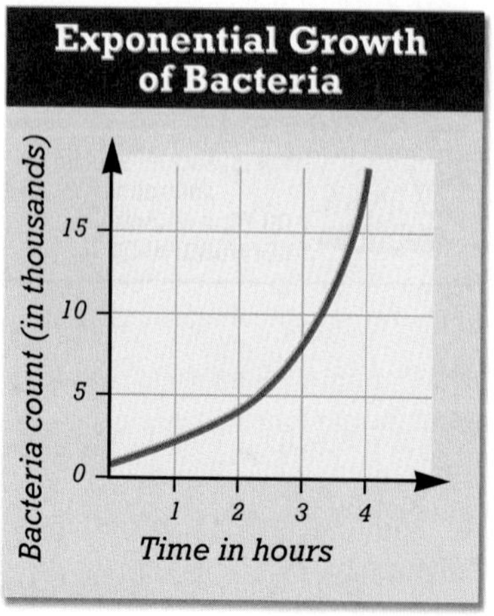

Using the same example, we can calculate the per capita growth rate as follows:

$$0.052 \text{ (births per capita)} - 0.014 \text{ (deaths per capita)} = 0.038 \text{ (growth per capita)}$$

To find the number of new individuals that will be added to the population in a year, simply multiply the per capita growth rate by the number of individuals in the population. If the population in our example numbers 50,000, the population will increase by 1,900 individuals in one year.

$$0.038 \times 50{,}000 = 1{,}900$$

If the growth rate is a positive number, the population is increasing. If it is a negative number, the population is shrinking.

THE EXPONENTIAL MODEL

The **exponential** (EKS-poh-NEN-shuhl) **model** of population growth describes a population that increases rapidly after only a few generations; the larger the population gets, the faster it grows. This is called **exponential growth.** In constructing an exponential model of population growth, it is assumed that birth rates and death rates remain constant, however large the population becomes.

Predictions Based on the Exponential Model

One way to understand predictions based on the exponential model is to look at a graph of population size over time. Population growth, according to the exponential model, follows the characteristic J-shaped curve shown in Figure 20-5. As you can see on the graph, the population grows slowly when it is small, but its growth speeds up as more and more individuals are added to the population. We can predict that the population will grow indefinitely and at an increasingly rapid rate based on the exponential model. Figure 20-6 shows the exponential growth of bacteria in a laboratory culture.

Limitations of the Exponential Model

The ultimate test of an exponential model is how well it matches the growth pattern of a real population. Do populations grow exponentially? The answer is yes, but only under rare conditions and for short periods of time. For example, populations of bacteria and other microorganisms can grow exponentially in the laboratory

Data Table

Time	0	1	2	3	4
Bacteria count	1,000	2,000	4,000	8,000	16,000

if they are provided with an abundance of food and space and if their waste is removed.

In reality, populations cannot grow indefinitely because the resources they depend on become scarce and wastes accumulate. All populations are limited by their environment. A factor that restrains the growth of a population is called a **limiting factor.**

As a population grows, competition among individuals for the shrinking supply of resources intensifies, and each individual, on average, obtains a smaller share. This reduces each individual's ability to fight off disease, to grow, and to reproduce. The results for the population are a declining birth rate and an increasing mortality rate. Figure 20-7 shows how increasing population size affected the growth of a species of plant.

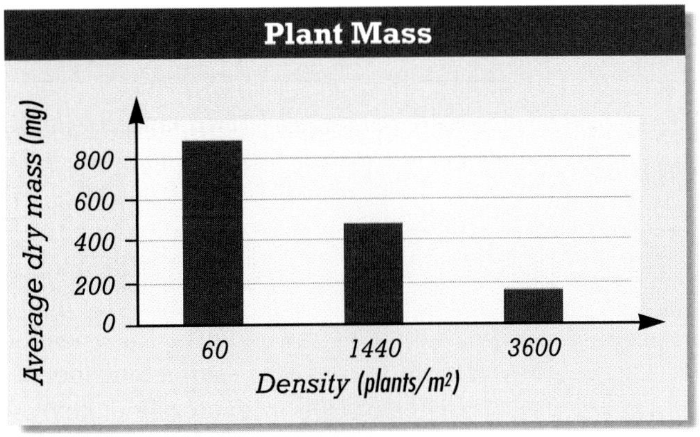

FIGURE 20-7
The graph displays the results of an experiment that tests how the growth of a plant is affected under three conditions of crowding. Under the condition of least crowding, the plants were observed to grow larger.

THE LOGISTIC MODEL

The **logistic** (loh-JIS-tik) **model** of population growth builds on the exponential model but accounts for the influence of limiting factors. Birth rates and death rates are not constant, but vary with population size: birth rates decline and death rates rise as the population grows. The logistic model includes a new term, **carrying capacity** (symbolized by K), the number of individuals the environment can support over a long period of time.

A graph of logistic growth looks like a stretched-out letter *S*. Examine Figure 20-8. When the population size is small, birth rates are high and death rates are low, and the population grows at very near the exponential rate, as shown in region *A*. As the population size approaches the carrying capacity, however, the population growth rate falls due to the falling birth rate and the increasing death rate, as shown in region *B*. When a population size is at its carrying capacity, the birth rate equals the death rate and growth stops, as shown in region *C*. This type of growth is known as **logistic growth.**

The logistic model, like the exponential model, contains some assumptions. One such assumption is that the carrying capacity is constant and does not fluctuate with environmental changes. In reality, carrying capacity does fluctuate. It is greater when prey is abundant, for instance, and smaller when prey is scarce. The logistic and exponential models are not accurate representations of real populations, but they are an important tool that scientists use to study population growth and regulation.

FIGURE 20-8
The graph of logistic population growth describes a population stabilizing around *K*, the carrying capacity. The logistic graph is similar to the exponential graph in region A. In region B, the graph begins to indicate a slowing rate of increase. In region C, the graph indicates that the population has become stable, neither growing nor getting smaller.

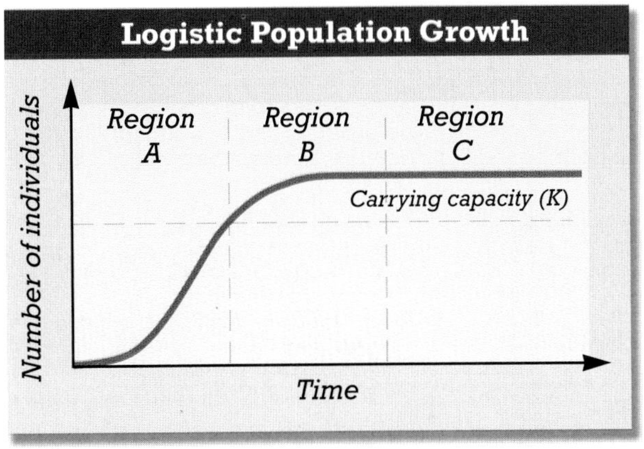

DEMONSTRATION

Place 1 g of filamentous algae from pond water in a flask or jar with 250 mL of distilled water. Do the same with another flask. Add 0.5 g of commercial plant food to the first flask and swirl. Put both flasks on the overhead and observe. Have students note differences and similarities. After one week, observe the two flasks on the overhead again. Ask students to compare them. Then filter the flasks' contents and determine the mass of the filtrate (algae). Ask students what conclusions can be drawn from this demonstration. (On the first day, the only observable difference may be a slight color difference due to the addition of the fertilizer. After a week, the fertilizer solution should appear cloudier or green and should produce a greater mass. Populations of similar or identical species may vary in size due to the availability of a limiting factor, such as the nutrients in the fertilizer in the first flask.)

RECENT RESEARCH
Dolphin Reproduction

In many terrestrial mammals of higher latitudes, births are timed to occur during the season with the greatest food availability. Mammals of lower latitudes show less of this tendency. In a study of wild bottlenose dolphin populations, the seasonal reproduction of dolphins was found to be related mostly to local environmental conditions, not to the latitude or the season the population was in. In this study, bottlenose dolphins from all over the world were studied. Populations of bottlenose dolphins in the Gulf of Mexico were found to have different birth peaks but were located in latitudes similar to those of the bottlenose dolphin populations in the Pacific Ocean.

TEACHING STRATEGY

Posters

Have students make posters that illustrate density-independent factors on the top half and density-dependent factors on the bottom half. Pictures can be drawn or cut from magazines. Have them label the two factors that regulate populations. Students should present the posters to the class with an explanation. (Floods, tornadoes, fires, and snowstorms are examples of density-independent factors. Food, vitamins, and housing might be examples of density-dependent factors.)

DEMONSTRATION

Ask students to bring small stuffed toy bunnies for this demonstration. At least a dozen are needed. Several stuffed lions, tigers, or wolves are also needed. Use the stuffed animals to demonstrate how predator and prey populations are cyclic and the predator population lags behind the prey population. Ask students if predation is the only factor that may control a prey population size. (No, because other limiting factors, such as food and disease, also may limit or decrease the prey population.)

RECENT RESEARCH

Endangered Habitat

A study of snow geese has shown that snow goose populations have dramatically increased. One population numbered around 2,000 pairs in the 1960s, and in the 1990s, a similar population numbered over 22,000 pairs. The birds are destroying sensitive tundra by trampling and digging out tundra food plants faster than the plants can regrow. Wildlife managers have now proposed decreasing the snow goose population by 5 percent to protect the many other species of tundra animals that are suffering from an abundance of snow geese.

POPULATION REGULATION

Two kinds of limiting factors, which control population size, have been identified. **Density-independent factors,** such as weather, floods, and fires, reduce the population by the same proportion, regardless of the population's size. For example, if a forest fire destroys a population of chipmunks, it does not matter if the population of chipmunks is 1 or 100. An unseasonable cold snap is a density-independent factor because its severity and duration are completely independent of population size. **Density-dependent factors** include resource limitations, such as shortages of food or nesting sites, and are triggered by increasing population density.

Population Fluctuations

All populations fluctuate in size. Some population fluctuations are clearly linked to environmental changes. For example, a drought may reduce a population of deer living in a forest. Some population fluctuations are not obviously connected to environmental fluctuations, and explaining their occurrence is much more difficult. The famous population cycles of the snowshoe hare, first described by Charles Elton, one of the pioneers of ecology, is shown in Figure 20-9. Elton obtained more than 70 years of records showing the number of snowshoe hare pelts the Hudson's Bay Company of Canada purchased from trappers. He assumed that the number of pelts purchased in a year indicated the size of the snowshoe hare population. The records showed that the hare population underwent a very regular cycle, with about 10 years between peaks in population size. When Elton examined the records for the number of lynx pelts purchased, he found that the lynx, a medium-sized species of cat that preys on snowshoe hares, also followed a population cycle. The peaks in the lynx population usually occurred one or two years after peaks in the hare population.

Elton thought that each species was the cause of the other's cycle. Thus, when the population of snowshoe hares increased, providing more food for the lynxes, the lynx population also increased. The increased lynx population then ate more hares, so the hare population decreased. With less food, more lynxes starved and the lynx population declined, allowing the hare population to increase and start the cycle over again. However, the observation that the same cycles occur in snowshoe hare populations living on islands without lynxes indicates that this explanation is inaccurate. Another possible explanation is that the lynx cycle is dependent on the hare population but the hare cycle is dependent on some other factor.

FIGURE 20-9

The hare and the lynx (a) were observed by Elton to have parallel changes in their population cycles. The graph below (b) shows the data recorded by Elton supporting his idea that each animal controlled the other animal's cycle. You can see that the cycles fluctuate together. Because hares show the same population cycles when there are no lynxes present, it is now known that lynxes are not controlling factors in the hares' cycles.

(a)

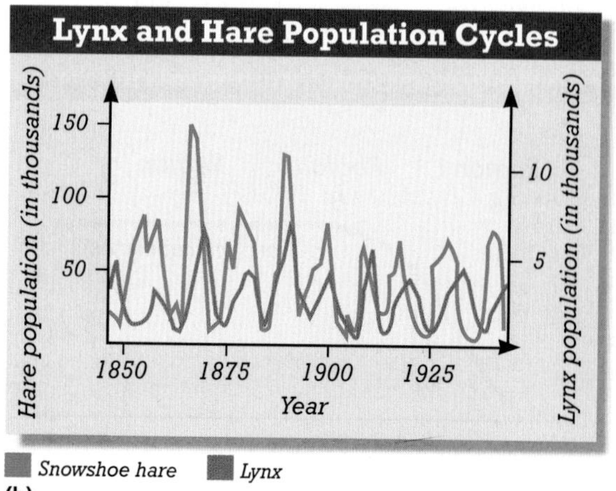

(b)

Perils of Small Populations

The rapidly growing human population has caused extreme reductions in the populations of some other species and subspecies. For instance, fewer than 200 Siberian tigers remain in the wild due to overhunting and habitat destruction. Even greater reductions have been experienced by the California condor, which was once found throughout the southwestern United States. By the 1980s, the condor's wild population was down to nine individuals.

Small populations, such as the cheetahs shown in Figure 20-10, are particularly vulnerable to extinction. Environmental disturbances, such as storms, fires, floods, or disease outbreaks, can kill off the entire population or leave too few individuals to maintain the population. Also, the members of a small population may be descended from only a few individuals, increasing the likelihood of **inbreeding,** or mating with relatives. Offspring of related parents often have fewer offspring, are more susceptible to diseases, and have a shorter life span. Inbreeding in small populations often leads to decreased genetic variability, and, over evolutionary time, may reduce the population's ability to adapt to changing environmental conditions.

FIGURE 20-10

Biologists estimate that there are fewer than 15,000 cheetahs remaining in the wild. Inbreeding in small populations, such as the one shown, leads to a loss of genetic diversity or variability. Cheetahs are bred in captivity in an effort to maintain their remaining genetic variability. Some of these cheetahs may eventually be released into the wild.

SECTION 20-2 REVIEW

1. Explain what is described by the exponential model.
2. According to the exponential model, how do birth and death rates change with population size?
3. Describe two differences between the exponential model and the logistic model.
4. List two density-independent factors.
5. Explain how inbreeding can threaten the survival of a small population.
6. CRITICAL THINKING Write a brief paragraph supporting the assumption that immigration and emigration are insignificant factors when studying global human population growth.

RETEACHING ACTIVITY

Tell students they are going on a journey to an unknown planet to study its populations. Ask them to imagine and describe the humanoid inhabitants and to identify the most likely age and cause of death among the inhabitants. Then show the three survivorship curves and ask students to identify the curve that fits their story. (An answer might describe humanoids whose most likely cause of death is being eaten by sand dragons. Death can come at any time for these humanoids. This would be a Type II survivorship curve.)

ANSWERS TO SECTION 20-2 REVIEW

1. The exponential model describes a population experiencing rapid, unlimited growth.
2. In the exponential model, birth and death rates are constant.
3. The logistic model allows birth and death rates to vary with population size. Also, it includes a limit on population size, the carrying capacity.
4. Density-independent factors include climate and natural disasters, such as hurricanes, volcanic eruptions, and floods.
5. Inbreeding can result in offspring that are unhealthy, have reduced fertility, and little genetic variability. The population may not be able to produce enough individuals to maintain itself.
6. As far as anyone has been able to prove, immigration and emigration to and from Earth has not yet occurred. It has not yet been proven that even microorganisms exist elsewhere in the universe, although some scientists believe there is evidence of ancient bacterial life in meteorite fossils from Mars.

SECTION 20-3

SECTION OVERVIEW

Reproduction and Inheritance
Section 20-3 discusses the principles of human population growth, birth rates, the population explosion, and projected human population growth. Some populations are so small that there may not be enough genetic variability for survival.

Interdependence of Organisms
Section 20-3 discusses how humans are challenged to use resources wisely so that Earth's ecosystems are not further compromised.

▶ ENGAGE STUDENTS

Designate each corner of the room as a developed country, such as Japan, the United States, Great Britain, and Canada. Tell students to assume the role of a poor farmer's son or daughter living in Central America. Then ask students who want to immigrate to go to the corner of the room representing the country they want to immigrate to. When the student immigration is complete, ask students to observe how the population of the Central American country has changed. Discuss the choices students made and why. (Students will probably observe that many more people immigrated than stayed in the Central American country. Point out to students that the actual percentage of people who leave their country is small. Many people choose to stay with their families or they have no way to leave.)

SECTION 20-3

OBJECTIVES

▲ Explain how the development of agriculture changed the pattern of human population growth.

● Describe the change in human population growth that began around 1650.

■ Describe how growth rates have changed since World War II.

◆ Compare the general standard of living in developed countries with that in developing countries.

internetconnect
SCILINKS NSTA
TOPIC: History of population growth
GO TO: www.scilinks.org
KEYWORD: HM388

HUMAN POPULATION GROWTH

In the time it takes you to read this chapter, the human population will grow by about 10,000 people. The rapid growth of the human population over the last several centuries is unprecedented in history. What caused this rapid growth? How long can it continue? This section examines these and other questions about the human population explosion.

HISTORY OF HUMAN POPULATION GROWTH

From the origin of *Homo sapiens,* more than 500,000 years ago, until about 10,000–12,000 years ago, the human population grew very slowly. During this time, humans lived in small nomadic groups and obtained food by hunting animals and gathering roots, berries, nuts, shellfish, and fruits. This way of life is called the **hunter-gatherer lifestyle.** By studying the few hunter-gatherer societies that exist today, scientists have learned that the low rate of population growth results from small populations and high mortality rates, especially among infants and young children, who never reach reproductive maturity.

The Development of Agriculture

The hunter-gatherer lifestyle began to change about 10,000 to 12,000 years ago, when humans discovered how to domesticate animals and cultivate certain plants for food. This dramatic change in lifestyle is called the **agricultural revolution,** and it led to profound changes in every aspect of life. Most important, agriculture greatly stabilized and increased the available food supply. As a result, the human population began to grow faster. About 10,000 years ago, there were between 2 million and 20 million people on Earth. By about 2,000 years ago, the population had increased to between 170 million and 330 million.

The Population Explosion

As you can see in Figure 20-11, human population growth continued through the Middle Ages, despite some short-term reversals. The outbreak of bubonic plague in 1347–1352 is thought to have killed about 25 percent of the population of Europe. Human population growth began to accelerate after 1650, primarily because of a sharp decline in death rates. There are many reasons for the decline in death rates, including better sanitation and hygiene, control of disease, increased availability of food, and improved economic conditions. While death rates fell, birth rates remained high, resulting in rapid population growth. The human population was about 500 million in 1650 and had risen to about 1 billion by 1800 and 2 billion by 1930.

Mortality rates fell sharply again in the decades immediately following World War II because of improvements in health and hygiene in the world's poorer countries. Birth rates in these countries remained high, pushing the per capita growth rate to its highest values. It took most of human history for the human population to reach 1 billion, but the population grew from 3 billion to 5 billion in just the 27 years between 1960 and 1987.

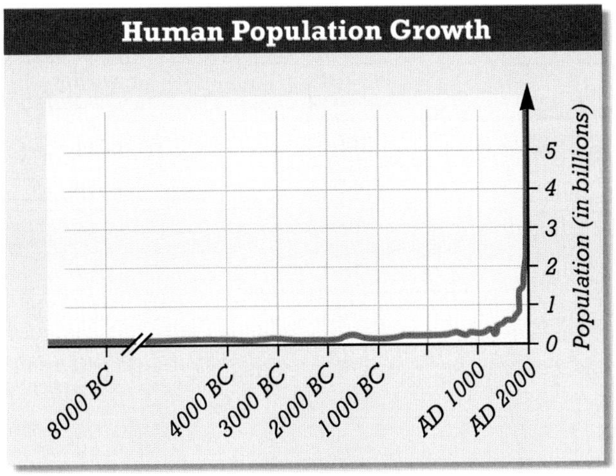

FIGURE 20-11

The J shape of the graph is characteristic of exponential growth. Many ecologists agree that the current human population growth rate is not sustainable.

Population Growth Today

The global growth rate peaked in the late 1960s at about 0.021 per capita. Because birth rates have decreased in many countries, the growth rate has gradually declined slowly to its current level of about 0.014 per capita. This decline has led some people to mistakenly conclude that the population is not increasing. In fact, the number of people that will be added to the world population this year is larger than it was when the growth rate was at its peak. This is simply a function of today's greater population size. For example, in 1970 there were about 3.7 billion people, and the growth rate was about 0.0196. In 1970, therefore, about $3{,}700{,}000{,}000 \times 0.0196$, or about 73 million people, were added to the world's population. In 1999 there were 6 billion people and the growth rate was 0.014 per capita, so the number of people added to the population was $6{,}000{,}000{,}000 \times 0.014$, or 84 million.

Today about 20 percent of the world's population live in **developed countries.** This category includes all of the world's modern, industrialized countries, such as the United States, Japan, Germany, France, the United Kingdom, Australia, Canada, and Russia. On average, people in developed countries are better educated, healthier, and live longer than the rest of the world's population. Population growth rates in developed countries are very low —less than 0.01 per capita. The populations of some of these countries, such as Russia, Germany, and Hungary, are shrinking because death rates exceed birth rates.

Most people (about 80 percent of the world's population) live in **developing countries,** a category that includes most countries in Asia and all of the countries in Central America, South America, and Africa. In general, these countries are poorer than the more-

Quick Lab

Demonstrating Population Doubling

Materials pencil, paper, sheet of newspaper

Procedure

1. Make a data table. Label the columns "Fold number," "Number of layers," and "Power of 2." Write the numbers 1-10 in the first column.
2. Fold a sheet of newspaper repeatedly in half, as your teacher demonstrates. Fill in your data table after each fold.

Analysis What can you conclude about the speed at which populations grow by doubling?

VISUAL LINK

Figure 20-11

Have students examine the graph showing human population growth. Ask them to predict what the graph may look like after 20 years. Ask what factors might cause a reduction in population (food shortages and famines, or new deadly viral diseases).

Quick Lab

Demonstrating Population Doubling

Time Required 20 minutes

Procedural Tips Demonstrate the folding procedure to students. Fold a full spread (two pages across) of newspaper in half. One layer has become two (2^1) layers. Fold the newspaper in half again. Two layers are now four (2^2) layers. The third fold produces eight (2^3) layers. Continue folding the newspaper in half until you cannot fold it anymore (≈7 folds). Students will need to complete the table by calculation. (Fold 8 = 256 layers [2^8], Fold 9 = 512 layers [2^9]) Tell students that in just 10 doublings they would be trying to fold the equivalent of their textbook—2^{10} = 1,024 pages. (Have students graph their data.)

Answers to Analysis Students will see that anything that increases by doubling grows exponentially.

SECTION 20-3

RETEACHING ACTIVITY

Have students make illustrated time lines of human population growth and present them to the class. (The first illustration might include tribes of primitive people hunting large animals and gathering nuts and berries. The second illustration might show primitive people sowing seed and gathering a crop. The next illustration might be of people using primitive farming implements, such as a mule-drawn plow. The last illustration could be of people at work in industries or with modern farming implements, such as tractors.)

ANSWERS TO SECTION 20-3 REVIEW

1. The agricultural revolution stabilized the amount of food available, and the rate of population growth increased.
2. The mortality rate fell around 1650 because of improved hygiene, diet, and economic conditions.
3. After World War II, improvements in medicine and sanitation were transferred to the world's poorer countries, causing a decline in death rates.
4. Population growth rates in the developing countries are much higher than in the developed countries.
5. Generally, living standards are much higher in developed countries. On average, people are healthier and better educated.
6. Growth rate is the change in size of a population over a period of time. Fertility rate is the average number of children the females in a population have; it does not consider deaths, as growth rate does. A lower fertility rate results in a lower birth rate. A lower birth rate will lead to a lower growth rate if the death rate remains constant.

TABLE 20-2 Population Statistics for Selected Countries

Country	Population (millions)	Growth rate (per capita)	Projected population in 2025 (millions)	Estimated doubling time (years)
China	1,200	0.011	1,500	64
India	950	0.019	1,400	37
United States	265	0.006	335	117
Brazil	161	0.017	202	41
Russia	148	−0.005	153*	N/A
Japan	126	0.002	126	350
Nigeria	104	0.031	246	23
Mexico	95	0.022	142	32
Germany	82	−0.001	79*	N/A
United Kingdom	59	0.002	63	350
France	58	0.003	64	233
South Africa	45	0.023	70	30
Kenya	28.2	0.027	49	26
Australia	18	0.008	23	88
Haiti	7.3	0.023	11	30

* projection assumes per capita growth rate will increase to a positive value
N/A = Not applicable

developed countries, and their populations are growing much faster—at a growth rate of more than 0.02 per capita. Table 20-2 shows some population statistics for several countries.

The human population explosion will eventually stop. The only questions still to be answered are How large will the population be? and Will the planet be able to support the population over the long term? The answers to both questions depend on whether we use our resources wisely so that the future productivity of Earth's ecosystems is not compromised.

SECTION 20-3 REVIEW

1. **What effect did the agricultural revolution have on the growth of the human population?**

2. **Explain why mortality rates began to decline rapidly around 1650.**

3. **Why did population growth rates increase rapidly after World War II?**

4. **How do population growth rates in the developing countries compare with those of the developed countries?**

5. **How do living standards in the developing countries compare with those in the developed countries?**

6. **CRITICAL THINKING** **In five or six sentences, differentiate between growth rate and birth rate, and evaluate the statement that a decreasing birth rate may lead to a decreasing growth rate.**

CHAPTER 20 REVIEW

SUMMARY/VOCABULARY

20-1

- A population is a group of individuals of the same species living in the same place in the same time period.
- A population's size is the number of individuals it contains. Population density is a measure of how crowded the population is. The dispersion pattern—random, even, or clumped—indicates the distribution of individuals within the population.
- A population's age structure indicates the percentage of individuals at each age.
- Populations show three patterns of mortality: Type I (low mortality until late in life), Type II (constant mortality throughout life), and Type III (devastating mortality early in life followed by low mortality for the remainder of the life span).

Vocabulary

age structure (382)
birth rate (381)
death/mortality rate (381)
dispersion (380)
life expectancy (381)
population density (380)
survivorship curve (382)

20-2

- Four processes determine whether a population will shrink or grow: birth, death, immigration, and emigration.
- The exponential model describes perpetual growth at an increasing rate in a population. The model assumes constant birth and death rates and no immigration or emigration.
- In the logistic model, birth rates fall and death rates climb as the population grows.
- The carrying capacity is the number of individuals the environment can support for an indefinite period of time. At the carrying capacity, birth and death rates are equal, and the population size is stable.
- Density-independent factors kill the same percentage of a population regardless of its size. Density-dependent factors kill more individuals in large populations than in small ones.
- Populations fluctuate over time due to environmental changes.
- Small populations are less able to rebound from environmental changes, they are more likely to experience inbreeding, and their genetic diversity is often low.

Vocabulary

carrying capacity (385)
density-dependent factor (386)
density-independent factor (386)
emigration (383)
exponential growth (384)
exponential model (384)
growth rate (383)
immigration (383)
inbreeding (387)
limiting factor (385)
logistic growth (385)
logistic model (385)

20-3

- About 10,000–12,000 years ago, human population growth was slow.
- The development of agriculture increased the growth rate of the population. Improvements in hygiene, diet, and economic conditions around 1650 further accelerated population growth.
- The population grew at its fastest rate in the decades immediately after World War II, largely because of better sanitation and medical care in the poorer countries.
- Today, population growth is faster in the less-developed countries and slower in the more-developed countries.

Vocabulary

agricultural revolution (388)
developed countries (389)
developing countries (389)
hunter-gatherer lifestyle (388)

CHAPTER 20 REVIEW
ANSWERS

REVIEW

1. A density-dependent factor varies with population size. A density-independent factor always has the same effect, regardless of size.
2. Density is a measure of the number of individuals per unit of volume or area. Dispersion indicates how the organisms are distributed.
3. Exponential growth is continuously accelerating, and birth and death rates are constant. In logistic growth, birth and death rates vary with population size and growth stops at the carrying capacity.
4. Carrying capacity is the number of individuals the population can support. Density is the number of individuals per unit area.
5. Developed countries are industrialized and economically wealthy; their citizens are generally healthier and better educated, and population growth rates are slower than in developing countries.

6. b **7.** b
8. c **9.** b
10. d **11.** a
12. b **13.** c
14. a **15.** c

16. The turtles might have gathered to exploit some resource, such as food, or for social interactions.
17. In an even distribution, the distance between individuals is fairly constant. In a random distribution, each individual's position is independent of the locations of other individuals.
18. A survivorship curve shows the likelihood of surviving to each age within the life span. The Type I survivorship curve is typical of humans; mortality is low until late in life, when it increases dramatically. Some birds have a Type II curve, in which mortality is fairly constant throughout life. In Type III survivorship curves, found in oysters, mortality is very high for

young individuals but then drops to low levels and remains fairly constant throughout the remainder of life.

19. The exponential model assumes that there is no immigration or emigration and that birth and death rates are constant and describes a population that will continue to grow at an ever-accelerating rate.

20. Chance events can wipe out the population, inbreeding may reduce the number of healthy offspring, and low genetic diversity may prevent the population from adapting to long-term environmental changes.

21. The logistic model assumes that resources are limited and that birth and death rates change with population size. Exponential growth is indicated in region A. As the population grows, the birth rate falls and the death rate increases, as indicated in region B. At the carrying capacity, shown in region C, the birth and death rates are equal and population size is stable.

22. Elton used data on the number of pelts purchased by the Hudson's Bay Company. Snowshoe hares living on islands that have no lynxes show similar population cycles.

23. The agricultural revolution, which occurred when humans learned to grow crops and raise animals for food, increased population growth. Before then, people hunted and gathered fruits, berries, nuts, and other foods.

24. Three factors that caused a decline in death rates were better nutrition, improved sanitation and hygiene, and improved economic conditions.

25. Developing countries are economically poorer. Individuals are not as healthy, receive less education, and die younger than individuals in developed countries.

REVIEW

Vocabulary

Explain the difference between the terms in each of the following pairs:

1. density-dependent factor, density-independent factor
2. density, dispersion
3. exponential growth, logistic growth
4. carrying capacity, density
5. developing countries, developed countries

Multiple Choice

6. A clumped distribution (a) occurs when individuals are evenly spaced (b) may occur when resources are concentrated (c) occurs only because of social interactions between individuals (d) occurs only among plants.
7. Life expectancy (a) refers to the maximum life span of an individual (b) is the average life span (c) depends only on birth rates (d) is the same for all species.
8. Population growth (a) is a group of individuals of the same species living in the same place (b) is the spatial distribution of organisms in a population (c) is the change in the number of individuals in a population over a period of time (d) occurs only if all individuals in the population survive and reproduce.
9. In the exponential model, the growth rate (a) is the same as the birth rate (b) is the change in population size after birth and death rates have been accounted for (c) changes with population size (d) is zero at the carrying capacity.
10. According to the exponential model, (a) population growth stops at the carrying capacity (b) population growth increases and then decreases (c) the immigration rate falls with increasing population size (d) population growth continues indefinitely.
11. Which of the following is not true of the carrying capacity in the logistic model? (a) varies with population size (b) remains constant (c) represents the maximum sustainable population (d) is the population size at which the birth rate equals the death rate.
12. Which of the following is not a density-independent factor for a population of deer in a forest? (a) a period of freezing weather (b) the number of cougars in the forest (c) a drought (d) a landslide
13. Inbreeding can be harmful to a population because it (a) increases the genetic variability of the population (b) increases the rate of evolution in the population (c) can increase mortality rate of offspring (d) decreases the carrying capacity.
14. During the hunter-gatherer period of human history, (a) death rates were high (b) crops were cultivated to provide food (c) the population grew to 1 billion individuals (d) population growth rates were high.
15. The cause of the decline in death rates following World War II was (a) introduction of genetically engineered crops (b) a decrease in life expectancy (c) improved hygiene and medical care (d) declining birth rates.

Short Answer

16. A scientist observes that the population of turtles in a pond shows a clumped distribution. Explain two reasons why turtles might show this kind of distribution.
17. Describe how an even distribution differs from a random distribution. Draw an example of each type of distribution.
18. What is a survivorship curve? What are the three types of survivorship curves? Name an organism that shows each kind of curve.
19. Explain one key assumption of the exponential model of population growth. What does the model help scientists predict about changes in population size over time?
20. Explain three reasons why small populations are particularly vulnerable to extinction.
21. Examine the graph below. Explain how the logistic model describes population changes in region A, region B, and region C.

Logistic Population Growth

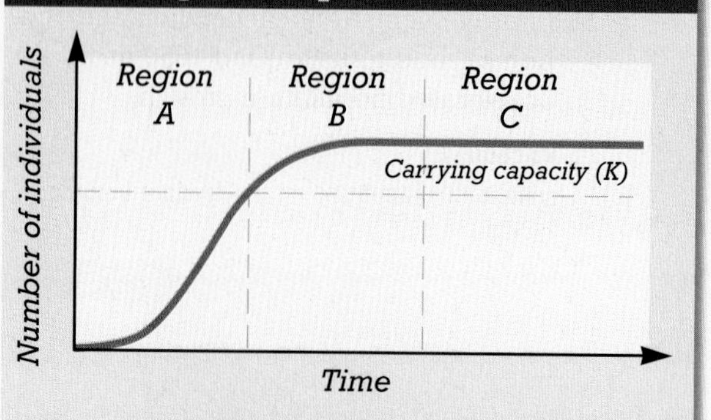

22. What was the agricultural revolution? What effect did it have on population growth? Before the agricultural revolution occurred, how did people obtain food?
23. The human population began to grow very rapidly about 1650. Describe three factors that caused this change in the rate of population growth.
24. **Unit 7–Ecosystem Dynamics** Write a report summarizing how the availability of resources on a ranch determine the carrying capacity of the ranch. What are the risks of maintaining a population of livestock that exceeds the carrying capacity of the ranch?

CRITICAL THINKING

1. Because we humans have more power to alter our environment than other animals do, we can affect the carrying capacity of our environment. How do we increase or decrease the carrying capacity of our local area?
2. The cause of the population cycle of the snowshoe hare is still a mystery. Suggest two possible explanations for why this cycle occurs. Describe how you would test each possibility.
3. Explain how disease could be a density-dependent factor in a population.
4. Describe an imaginary population, and name at least two density-dependent factors that affect the population when it has reached carrying capacity. Draw a graph of what you think logistic population growth would look like for the population. Include the effects that a density-independent event might have on the graph.
5. The population of country X is projected to grow rapidly in the next few decades, while slow growth is projected for country Y. Using only the age structures in the figure below, explain why these projections are plausible.

Age Structure of Country X and Country Y

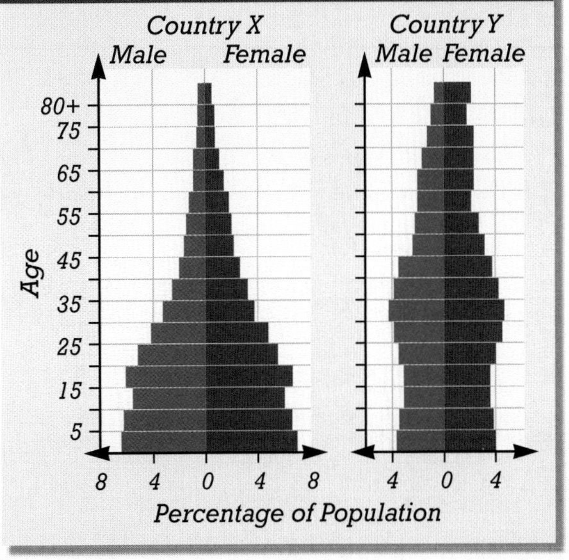

EXTENSION

1. Read "Japan's Harsh Reality Check" in *Newsweek,* January 10, 2000, on page 57. According to the article, how does Japan's aging society and low birthrate influence how Japan will rank as a world power in the twenty-first century? Why is the slowed birthrate in Japan unlikely to change?
2. Using library resources or an on-line database, obtain information about how the United States population census is conducted. Examine the results of the 1990 census, and answer these questions. What was the population of the United States in 1990? How much larger was the 1990 population than the 1980 population? Which state grew by the greatest percentage? Which states showed reductions in population?
3. Obtain population records for your town, city, or county. Try to get data that cover as long a period of time as possible. Make a graph that shows changes in population size. Describe any patterns you see in the data.

CHAPTER 20 REVIEW

ANSWERS

CRITICAL THINKING

1. Humans have increased carrying capacity by improving crop yields, health care, and sanitation. Pollution and overexploitation of resources have reduced carrying capacities in some areas.
2. Crashes in the snowshoe hare population may occur at the hares' carrying capacity. Food could be supplied to the hares when the population is large.
3. Crowding promotes contagion. Accumulated waste promotes lack of hygiene and sanitation.
4. Food shortages, disease, and space are density-dependent factors. Graphs may be typical logistic curves, but they should show some fluctuation around *K*.
5. Country X has far more people in the youngest and most reproductive age groups than in the oldest groups. In Country Y there are more old people and fewer young people. The number of deaths should be closer to the number of births, and the population should grow slowly.

EXTENSION

1. Japan's aging society and low birthrate threaten Japan's future through progressive depopulation. In Japan, women are having an average of 1.5 children. The present population is expected to decrease by 25 million people in the next 45–50 years, and a large portion of the population will be too elderly to have children.
2. The 1990 census resulted in a United States population count of 248,709,873. The population grew by 9.8 percent between 1980 and 1990. The fastest-growing state was Nevada. Four states decreased in population: Iowa, North Dakota, West Virginia, and Wyoming.
3. Answers should offer explanations for any changes in population size shown on their graphs.

CHAPTER 20 INVESTIGATION

TIME REQUIRED

15–20 minutes each day for 5 consecutive days

SAFETY PRECAUTIONS

Students should wear a lab apron, safety goggles, and disposable gloves while working with iodine. Caution students to avoid skin and eye contact with iodine.

QUICK REFERENCES

Lab preparation notes are found on page 377B. *Holt BioSources* provides a Teaching Resources CD-ROM that contains "Using Gowin's Vee in the Lab" and "Scoring Rubrics."

PROCEDURAL TIPS

1. Instructions for the preparation of the yeast culture is in the Chapter 20 Interleaf, p. 377B.
2. Have students work in pairs.

ANSWERS TO BACKGROUND

1. A population is a group of individuals of the same species living in the same place at the same time.
2. Properties of populations include size, density, and dispersion.
3. Carrying capacity is the maximum number of individuals the environment can support for an indefinite period of time.
4. Common limiting factors may include food, water, space, presence of predators, and extremes of temperature.
5. To arrive at a reasonably accurate count, a population can be divided into several smaller units. Several of these units are randomly chosen, and then counted and averaged. The average is multiplied by the number of units to give an estimate for the entire population.

CHAPTER 20 INVESTIGATION

Studying a Yeast Population

OBJECTIVES

- Observe the growth and decline in a population of yeast cells.
- Apply the underlying principles to changes in human populations.

PROCESS SKILLS

- calculating
- measuring
- collecting data
- organizing data
- analyzing data

MATERIALS

- yeast culture
- 1 mL pipet
- test tube
- iodine in dropper bottle
- microscope slide ruled in 2 mm squares
- safety goggles
- coverslip
- compound microscope

Background

1. What is a population?
2. List three properties of populations.
3. What is carrying capacity?
4. What are some common limiting factors that prevent populations from exceeding their carrying capacity?
5. How can populations be sampled to achieve an accurate count?

PART A Day 1

1. CAUTION **Put on a lab apron, goggles, and disposable gloves.**
2. Your teacher will transfer approximately 1 mL of the yeast culture to a test tube and then add 2 drops of iodine to the test tube. CAUTION **Work with care—iodine is a poison and an eye irritant. If you get iodine on your skin or clothing, wash it off at the sink while calling to your teacher. If you get iodine in your eyes, immediately flush your eyes with water at the eyewash station while calling to your teacher.**
3. Use the 1 mL pipet to transfer 0.1 mL (one drop) from the test tube to a ruled microscope slide. Carefully lower a coverslip over the drop. CAUTION **Slides break easily. Use caution when handling them.**
4. Using the compound light microscope, view your slide under low power. Perform the following steps to estimate the total number of yeast cells in 0.1 mL. Note the appearance of your yeast culture. Your culture should look similar to the one shown in the photograph labeled (a) below.

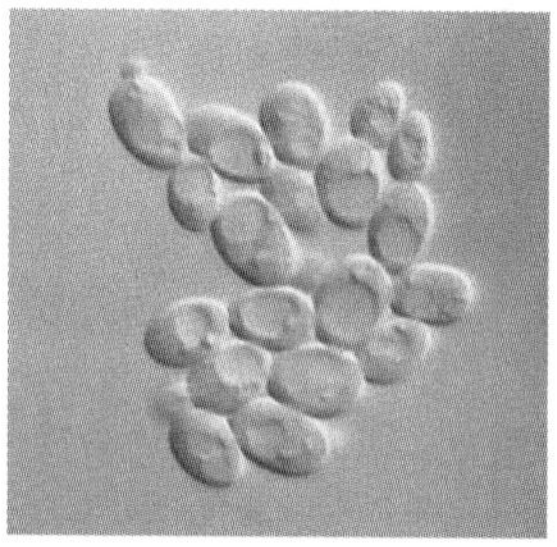
(a)

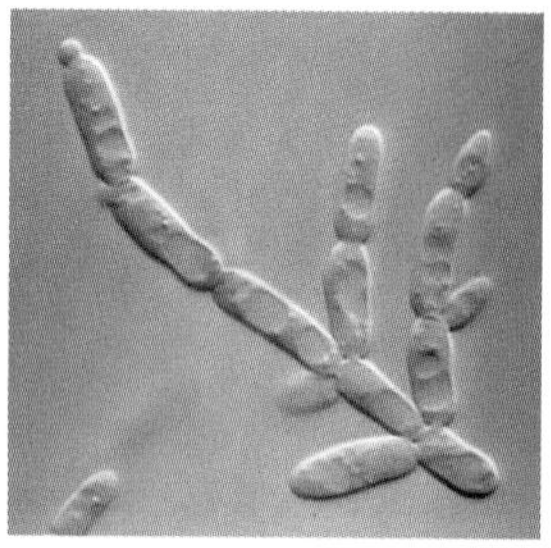
(b)

5. Each slide is ruled in 2 mm squares. The figure below shows only 3 of these 2 mm squares. You will be counting the cells in six 2 mm squares. Focus on the yeast cells and notice the black grid lines on the slide. Switch to the 400× objective, and align the slide so that you can just see the top left corner of one 2 mm square, area 1, shown in the figure below.

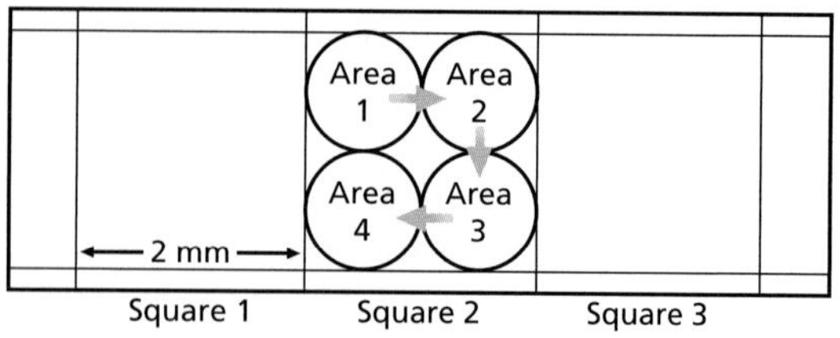

6. Count all the yeast cells in area 1, and record the number in a data table similar to the one on the facing page.

ESTIMATED NUMBER OF YEAST CELLS IN 0.1 mL OF CULTURE

Time (hours)	Number of cells per 2 mm square						Average	Population size (cells/0.1 mL)
	1	2	3	4	5	6		
0								
24								
48								
72								
96								

7. Move the slide to area 2. Continue counting cells and recording data in your lab report until you have counted the cells in each of the four areas that make up one 2 mm square. Add the total number of cells in the square, and record this number in your lab report.
8. Switch the microscope to low power, and move the slide one square to the left. Under high power, count the cells and record the number in your lab report.
9. Repeat steps 5–8 until you have counted the cells in a total of six squares. Add the total number of cells you counted in the six squares. Calculate the average number of yeast cells in a 2 mm square by dividing the total by 6. Record this number in your data table under "Average."
10. Clean up your materials and wash your hands before leaving the lab.

PART B Days 2–5

11. Repeat steps 1–10 each day for four more days. Record your data for 24, 48, 72, and 96 hours. Note any changes in the appearance of your yeast culture in your lab report. If possible, repeat steps 1–10 again after the weekend. The yeast culture may have changed in appearance to resemble the photograph labeled (b) on the facing page.
12. To find the total population of yeast cells in 1 mL of yeast culture (the amount in the test tube), multiply the average number of cells counted in a 2 mm square by 2,500.
13. Graph your data using the values you calculated for the entire test-tube population over the five-day period.
14. Clean up your materials and wash your hands before leaving the lab.

Analysis and Conclusions

1. What effect does iodine have on yeast cells?
2. Why were several areas counted and then averaged each day?
3. When was the most rapid growth? the slowest growth?
4. When was the growth peak reached?
5. How did the yeast population change over time?
6. What limiting factors probably caused the yeast population to decline?
7. If you kept counting for several more days, what do you think would happen to the yeast population?
8. What limiting factors that affect yeast also affect human populations?
9. Did the growth rate of the yeast population more closely follow the predictions of the exponential model or the logistic model? Defend your answer.
10. In what ways do population growth and decline in a yeast population resemble growth and decline in a human population? In what ways do they differ?
11. To find the total number of cells in 1 mL, you multiplied the number of cells you counted in a 2 mm square by 2,500. To see why this works, calculate the volume of fluid in one of the small squares. The fluid forms a layer about 0.1 mm thick on the microscope slide. So the volume is $2 \text{ mm} \times 2 \text{ mm} \times 0.1 \text{ mm} = 0.4 \text{ mm}^3$. Now convert to milliliters, remembering that $1 \text{ mL} = 1 \text{ cm}^3$ and $1 \text{ cm}^3 = 1{,}000 \text{ mm}^3$. Finally, divide your answer into 1 mL to find what multiplier you need to use to go from the number of cells per square to the number of cells per milliliter.

Further Inquiry

Is it possible to set up a population of yeast that continues to grow for a week without declining? Would it be possible to keep the population size growing indefinitely?

CHAPTER 20 INVESTIGATION

ANSWERS TO ANALYSIS AND CONCLUSIONS

1. Iodine kills the cells. Because dead cells no longer move, they are easier to count. Iodine also stains the cells, making them easier to see.
2. An average was taken to allow for variation within the population.
3. The most rapid growth was between 24 and 48 hours. Growth was slowest from 0 to 24 hours.
4. The growth peak was reached at 72 hours.
5. The yeast population grew rapidly and then declined.
6. The yeast cells were limited by a lack of food and space. They also could have been poisoned by their own waste products.
7. The yeast population would die out when the food supply was exhausted.
8. The limiting factors of yeast—lack of food and space and excess waste—are some of the same limiting factors that affect human populations.
9. The yeast population increased rapidly until it reached carrying capacity and then declined. This pattern more closely resembles the logistic model.
10. Human population growth is exponential and rapid and resembles the early, expanding phase of the yeast population. So far, human population has not crashed like the yeast population did.
11.

$$\frac{\text{number of cells}}{0.4 \text{ mm}^3} \times \frac{1{,}000 \text{ mm}^3}{1 \text{ cm}^3} \times \frac{1 \text{ cm}^3}{1 \text{ mL}} = \text{number of cells} \times \frac{2{,}500}{1 \text{ mL}}$$

CHAPTER 21
COMMUNITY ECOLOGY

	TOPICS	TEACHING RESOURCES	LABS, CLASSWORK, AND HOMEWORK
BLOCKS 1 & 2 (90 minutes)	**Introducing the Chapter, p. 396**	ATE Focus Concept, p. 396 ATE Assessing Prior Knowledge, p. 396 ATE Understanding the Visual, p. 396	■ Supplemental Reading Guide, *Silent Spring*
	21-1 Species Interactions, p. 397 Predation, p. 397 Parasitism, p. 399 Competition, p. 400 Mutualism and Commensalism, p. 402	ATE Section Overview, p. 397 ATE Critical Thinking, pp. 398, 400 78, 82, 87	PE **Quick Lab,** p. 399, Analyzing Predation **Lab C26,** Assessing and Mapping Factors in the Environment ★ Study Guide, Section 21-1 PE Section Review, p. 402
BLOCKS 3 & 4 (90 minutes)	**21-2 Properties of Communities, p. 403** Species Richness and Diversity, p. 403	ATE Section Overview, p. 403 ATE Critical Thinking, p. 404 83	**Lab B4,** Plant and Animal Interrelationships ★ Study Guide, Section 21-2 PE Section Review, p. 405
BLOCKS 5 & 6 (90 minutes)	**21-3 Succession, p. 406** Successional Changes in Communities, p. 406 The Complexity of Succession, p. 408	ATE Section Overview, p. 406 ATE Critical Thinking, pp. 406, 408 80	PE **Chapter 21 Investigation,** pp. 412–413 ★ Study Guide, Section 21-3 PE Section Review, p. 408
BLOCKS 7 & 8 (90 minutes)	**REVIEW and ASSESSMENT** PE Chapter 21 Review, pp. 410–411 ■ Performance-Based Assessment—Testing Pond Water for Carbon Dioxide	Audio CD Program BIOLOGY INTERACTIVE TUTOR Unit 7—Ecosystem Dynamics	**CHAPTER TESTING** ■ Chapter Test (blackline master) ▲ Test Generator ■ Assessment Item Listing ■ Scoring Rubrics and Classroom Management Checklists

KEY

MODERN BIOLOGY
- **PE** Pupil's Edition
- **ATE** Teacher's Edition
- ■ Active Reading Guide
- ★ Study Guide

HOLT BIOSOURCES
- Teaching Transparencies
- Laboratory Program

One-Stop Planner CD-ROM
Shows these resources within a customizable daily lesson plan:
- ■ Holt BioSources Teaching Resources
- ■ Active Reading Guide
- ★ Study Guide
- ▲ Test Generator

READING FOR CONTENT MASTERY

Preview: Have students preview the objectives, the introductory paragraph, topic sentences, and captions before reading the chapter.

ATE Active Reading Technique: Anticipation Guide, p. 397
■ Active Reading Guide Worksheet 21-1
PE Summary/Vocabulary, p. 409

■ Active Reading Guide Worksheet 21-2
PE Summary/Vocabulary, p. 409

■ Active Reading Guide Worksheet 21-3
PE Summary/Vocabulary, p. 409

TECHNOLOGY AND INTERNET RESOURCES

CNN PRESENTS Video Segment 13 Fat Wolves

Holt Biology Videodiscs Teacher's Correlation Guide, Lessons 21-1 through 21-3

BIOLOGY INTERACTIVE TUTOR Unit 7—Ecosystem Dynamics, Topic 3

Audio CD Program, Sections 21-1 through 21-3

internetconnect

sciLINKS NSTA
On-line Resources: www.scilinks.org
The following *sci*LINKS Internet resources can be found in the student text for this chapter:

Topics:
- Symbiosis, p. 398
- Competition, p. 401
- Species interaction and richness, p. 405
- Succession, p. 406

go.hrw.com
On-line Resources: go.hrw.com
Visit the HRW Web site for a variety of resources related to this chapter. Just type in the keyword HM2 Chapter 21.

Smithsonian Institution®
Visit **www.si.edu/hrw** for additional on-line resources.

CNNfyi.com
Visit **www.cnnfyi.com** for late-breaking news and current event stories selected just for you.

LAB ACTIVITY PLANNING

CHAPTER 21 INVESTIGATION

Nitrogen Fixation in Root Nodules, pp. 412–413

OVERVIEW
In this investigation, students will observe the roots of legume plants (beans) and nonlegume plants (radishes) to determine if nitrogen-fixing bacteria *(Rhizobium)* are present. Students will explore the concept of mutualism.

SAFETY
Instruct students to keep their hands away from their faces so that no sap or plant juices will get in their eyes. Students will wear disposable gloves, lab aprons, and safety goggles at all times in the lab. Remind students to use care with the scalpel and glass. Have students dispose of broken glass in a container designated for that purpose.

PREPARATION
1. To save time, you may want to plant the seeds and grow the plants 6 to 8 weeks ahead of the scheduled investigation.
2. Bean seeds can be obtained at a local garden store or from a biological supply house such as WARD'S. Each group will need two seeds of each type, potting soil, and 3 in. pots.
3. Order cultures and slides of root nodules of *Rhizobium leguminosarum* from a biological supply house.
4. Provide students access to microscopes and stereoscopes for this investigation.

Quick Labs

Analyzing Predation, p. 399
Each pair of students will need 4.1 m of white string or twine, 4 stakes, a stopwatch or kitchen timer, a meterstick, and 40 toothpicks (10 each of the following colors: red, blue, green, and yellow).

UNDERSTANDING THE VISUAL

Ask students to offer suggestions about how animals in the picture are using the coral reef (as a home, as a place to find food, for protection from predators, or for surface area to grow on). Then ask why the coral reef community is so rich in number and kinds of organisms. (They may say that it provides many opportunities for food and safety or that it brings many species together.)

FOCUS CONCEPT

Interdependence of Organisms

The adaptations that evolve to enable species to occupy all niches seem to be directly related to selective pressures brought about by interactions among species and their nonliving environment. Emphasize to students that no organism exists alone. Each organism affects and is affected by other species and the environment.

ASSESSING PRIOR KNOWLEDGE

Review the following concepts.

Natural Selection: *Chapter 16*
Ask students how an owl and a hawk are alike. Review the process of natural selection and relate it to the evolution of the predatory features of hawks and owls.

Population Dynamics: *Chapter 20*
Density-dependent factors, such as the presence or absence of food and predators, limit a population's growth. Ask students to imagine that their neighborhood is being overrun by deer or mice. Ask them what factors could have caused the population explosion and whether they are density-dependent or independent factors.

CHAPTER 21

COMMUNITY ECOLOGY

This coral-reef community rivals a tropical rain forest in number of species.

FOCUS CONCEPT: *Interdependence of Organisms*

As you read, note how evolution shapes the interactions among species in a community.

BIOLOGY INTERACTIVE TUTOR Unit 7—*Ecosystem Dynamics*
Units 1–6

SECTION

21-1

OBJECTIVES

▲ Distinguish predation from parasitism.

● Evaluate the importance of mimicry as a defense mechanism.

■ Describe two ways plants defend themselves against herbivores.

◆ Explain how competition can affect community structure.

▲ Contrast mutualism with commensalism, and give one example of each type of relationship.

SPECIES INTERACTIONS

Just as populations contain interacting members of a single species, communities contain interacting populations of many species. This chapter introduces the five major types of close interactions, or ***symbioses,*** *among species—predation, parasitism, competition, mutualism, and commensalism. These symbiotic relationships help determine the nature of communities.*

PREDATION

Predation is a powerful force in a community. In predation, one individual, the **predator,** captures, kills, and consumes another individual, the **prey.** Predation influences where and how species live by determining the relationships in the food web. Predation is also an effective regulator of population size.

Predators, Prey, and Natural Selection

Natural selection, the major mechanism of evolution, favors adaptations that improve the efficiency of predators at finding, capturing, and consuming prey. For example, rattlesnakes have adaptations for locating their prey with an acute sense of smell and with specialized heat-sensitive pits located below each nostril, as shown in Figure 21-1. These pits enable a rattlesnake to aim its strike at warm-bodied prey with great accuracy, even in the dark. Other predator adaptations include the webs of spiders; the sharp, flesh-cutting teeth of wolves and coyotes; and the striped pattern of a tiger's coat, which provides camouflage in the tiger's grassland habitat.

FIGURE 21-1

A rattlesnake is able to detect a variation in temperature of as little as 0.1°C, which helps it locate prey. To disable its prey, the snake injects a strong, fast-acting poison through sharp, hollow fangs.

SECTION 21-1

SECTION OVERVIEW

Matter, Energy, and Organization

Section 21-1 discusses the organization of species interactions within a community. This section stresses five species interactions: predation, competition, mutualism, parasitism, and commensalism.

Evolution

Section 21-1 describes evolutionary adaptations through natural selection that enable species to survive and function within the five kinds of species interactions.

ACTIVE READING

Technique: Anticipation Guide

Before students read this chapter, write the following on the board or overhead:

1. The stability of a community—its resistance to change—is directly related to its species richness. Communities with more species are less susceptible to disturbance.

2. After a disturbance in a community, such as a forest fire, the community usually proceeds through a predictable series of stages until it reaches a stable end point.

Let students think a minute about the statements and decide if they agree with them. Have students read the chapter to see if their opinions are confirmed.

TEACHING STRATEGY

Predation Defense

Have students look at live freshwater or land snails. Explain that these snails eat plant material. Ask students what adaptations these slow-moving snails have for protection from predators. (Student responses might include that they are slimy, they might taste bad, and they are covered with a shell to hide in.)

SECTION 21-1

TEACHING STRATEGY

Observing Predation

Have students observe a live predator, such as a lizard, toad, or nonpoisonous snake. **CAUTION: Do not allow students to handle the animals.** Provide food for the animal and have students observe how these predators catch their prey. Students should list the adaptations these predators have for catching prey. (These might include a long, sticky tongue, a hinged mouth, small pointed teeth for grasping, or eyesight that is sensitive to movement.) Ask how these predators affect their prey population. (They often control the population size of prey species in the community.)

OVERCOMING MISCONCEPTIONS

In most animal stories, weasels are notorious fiends that randomly kill all the gentle creatures of their community. People who are dishonest are often called weasels. The fact is that weasels are predators of rats and mice. Weasels are very important in preventing population explosions of these disease-carrying animals.

CRITICAL THINKING

Fish Predation

Ask students why a school of fish clump together rather than break up in all directions when threatened by a predator. (Most predators will hesitate to attack a large school that behaves as one. The clumped school is probably perceived as a single large organism.)

(a)

(b)

FIGURE 21-2

Favorable coloration helps many organisms escape predation. The mantis in (a), *Chaenadodus rhombicollis,* cannot readily be detected among the leaves. The frog in (b), *Dendrobates matrimaculatus,* is brightly colored to warn other organisms that it is extremely poisonous.

A predator's survival depends on its ability to capture food, but a prey's survival depends on its ability to avoid being captured. Therefore, natural selection has favored ways for the prey to avoid, escape, or otherwise ward off predators. Some organisms flee when a predator approaches. Others escape detection by hiding or by resembling an inedible object. Can you find the mantis shown in Figure 21-2a, for example? Some organisms, like the frog shown in Figure 21-2b, are poisonous and use bright colors to warn other organisms of their toxicity.

Mimicry

Deception is important in antipredator defenses. In a defense called **mimicry,** a harmless species resembles a poisonous or distasteful species. The harmless mimic is protected because it is often mistaken to be its dangerous look-alike. The king snake shown in Figure 21-3b is a mimic of the poisonous coral snake shown in Figure 21-3a.

Another form of mimicry is when two or more dangerous or distasteful species look similar. Many kinds of bees and wasps have similar patterns of alternating yellow and black stripes, for example. This kind of mimicry benefits each species involved because a predator that encounters an individual of one species will avoid similar individuals.

Word Roots and Origins

parasite

from the Latin word *parasitus,* meaning "one who eats at the table of another"

Plant-Herbivore Interactions

Animals that eat plants are known as **herbivores.** Ecologists usually classify the relationship between plants and herbivores as a form of predation.

Through natural selection, plants have evolved adaptations that protect them from being eaten. Physical defenses, such as sharp thorns, spines, sticky hairs, and tough leaves, can make the plant more difficult to eat. Plants have also evolved a range of chemical defenses. They synthesize chemicals from products of their metabolism, called **secondary compounds,** that are poisonous, irritating,

internetconnect

SCILINKS NSTA

TOPIC: Symbiosis
GO TO: www.scilinks.org
KEYWORD: HM398

(a) Eastern coral snake, *Micrurus fulvius fulvius* **(b)** Scarlet king snake, *Lampropeltis triangulum elapsoides*

FIGURE 21-3

The scarlet king snake is easy to differentiate from the coral snake. In the coral snake, the red and yellow rings are together. In the king snake, they are separated by a black ring. Also, the coral snake has a black snout, while the king snake has a red snout.

or bad-tasting. Some examples of secondary compounds that provide a defensive function include strychnine (found in the leaves of plants of the genus *Strychnos*) and nicotine (which is toxic to insects and is found in tobacco leaves). Poison ivy and poison oak produce an irritating chemical that causes a rash on most people. Although secondary compounds are usually toxic, many also have medicinal uses. A large number of drugs, including morphine, atropine, codeine, taxol, and quinine, are derived from the secondary compounds of plants.

PARASITISM

Parasitism is a species interaction that resembles predation in that one individual is harmed while the other individual benefits. In parasitism, one individual, known as the **parasite,** feeds on another individual, known as the **host.** But while most forms of predation immediately remove an individual of the prey species from the population, parasitism usually does not result in the immediate death of the host. Often, the parasite feeds on the host for a long time instead of killing it.

Parasites can be grouped into two general categories, based on how they interact with their host. **Ectoparasites** (EK-toh-PER-uh-siets) are external parasites; they live on their host but do not enter the host's body. Examples of ectoparasites are ticks, fleas, lice, leeches, lampreys, and mosquitoes. **Endoparasites** (EN-doh-PER-uh-siets) are internal parasites, and they live inside the host's body. Familiar endoparasites are disease-causing bacteria, protists such as malaria parasites, and tapeworms.

Evolution of Parasites and Their Hosts

Parasites can have a strong negative impact on their host, affecting both the health and reproduction of the host. Consequently, parasitism has stimulated the evolution of a variety of defenses in hosts. Skin is an important defense that prevents most parasites from entering the body. Openings through which parasites could pass,

Quick Lab

Analyzing Predation

Materials 4.1 m white string, 4 stakes, 40 colored toothpicks, stopwatch or kitchen timer, meterstick

Procedure

1. Mark off a 1 m square in a grassy area using the stakes and string.
2. One partner will scatter the toothpicks randomly throughout the square. The other partner will have 1 minute to pick up as many toothpicks as possible, one at a time. Repeat this procedure until each team has performed five trials.
3. Record your team's results in a data table.

Analysis Which colors of toothpicks were picked up most often? Which were picked up least often? How do you account for this difference?

DEMONSTRATION

Simmer one-fourth of a cup of tea leaves in one cup of water for about five minutes in a covered cooking pot. Then cool the liquid and drain it into a clean glass. Have students observe the clarity of the tea. It should be very dark and opaque. Ask a volunteer to taste the tea. It will be very strong and bitter. Explain to students that the bitter taste in the tea comes from tannin, a secondary compound produced by many plants to prevent herbivory.

Quick Lab

Analyzing Predation

Time Required 30 minutes

Safety Remind students not to place any toothpicks in their mouths.

Procedural Tips Have students work in pairs. Each team should have a set of 40 toothpicks, 10 in each of the following colors: red, blue, green, and yellow. Select a grassy area on the school grounds ahead of time. If the activity is to be done indoors or on a hard surface, strips of overhead transparency sheets, cut to approximately 1 × 4 cm and colored with permanent markers, work better than toothpicks. Some of the strips should be left uncolored.

Answers to Analysis In a grassy area, red and blue toothpicks are usually picked up in greater numbers. Green and yellow are picked up least often. Green and yellow provide favorable coloration, which represents a camouflage adaptation that helps organisms escape predation. Indoors, answers will vary according to the background colors. In general, the strips that most closely match the background will be picked up the least.

SECTION 21-1

CULTURAL CONNECTION

Ancient Community Ecology

In Thailand, a huge mound stands out in a flat river floodplain. It is the site of a prehistoric village with a culture that exploited many natural resources. Scientists have been able to reconstruct the plant communities of this culture and determine how they interacted with the environment by studying pollen grains in the layers of sediment in the mound. Researchers found that the area was originally a mangrove swamp, with periods of grassland and rice farming. Sediments yielded evidence that this ancient culture practiced burning to clear areas to grow rice and grasses for food. There was also an abundance of fish and shellfish. Even in ancient cultures humans played a dominant role in reshaping the environment. The study of pollen grains can be used to understand previously unknown ancient cultures and their effect on the succession of their environment.

CRITICAL THINKING

Butterfly Defenses

One species of butterfly develops a large eyespot pattern on its wings if it develops in a warm, wet season. The same species of butterfly will develop small eyespots if it develops in a dry, cool season. Ask students what could cause two very different wing patterns in the same species. (The crysalis temperature regulates the gene for eyespot expression. Natural selection favors this adaptation as a defense against predation. Predators, such as lizards, prey on butterflies with large eyespots during the dry season. The butterfly with small eyespots has a selective advantage during the dry season.)

(a)

(b)

FIGURE 21-4

The ticks in (a) are ectoparasites that can sicken and occasionally kill their host by infection or disease. A tapeworm, the endoparasite shown in (b), can grow to 20 m or more in length and can cause illness and death by intestinal blockage and by robbing its host of nutrition.

such as the eyes, mouth, and nose, are defended chemically by tears, saliva, and mucus. Parasites that get past these defenses still may be attacked by the cells of the immune system. Figure 21-4 shows two examples of parasites.

Adaptations of Parasites

Natural selection favors adaptations that allow a parasite to efficiently exploit its host. Parasites are usually specialized anatomically and physiologically. Tapeworms are so specialized for a parasitic lifestyle that they do not even have a digestive system. They live in the small intestine of their host and absorb nutrients directly through their skin.

COMPETITION

Competition results from fundamental niche overlap—the use of the same limited resource by two or more species. Some species of plants release toxins into the soil that prevent individuals of other species from growing nearby, restricting the living space of the other species. More often, one organism will be able to use a resource more efficiently, leaving less of the resource available to the other species.

Research Studies on Competition

Soviet ecologist G. F. Gause was one of the first to study competition in the laboratory. In test tubes stocked with a food supply of bacteria, Gause raised some species of paramecia, separately and in various combinations. *Paramecium caudatum* and *Paramecium aurelia* thrived when grown separately. As you can see in Figure 21-5, when the two species were combined, *Paramecium caudatum* always died out because *Paramecium aurelia* was a more efficient predator of bacteria. Ecologists use the principle of **competitive exclusion** to describe situations in which one species is eliminated from a community because of competition for the same limited resource. In competitive exclusion, one species uses the resource

FIGURE 21-5

The graph shows the negative effect of competition on populations. When *Paramecium aurelia* and *Paramecium caudatum* were grown separately, they each grew to twice the density that they reached when they were grown together, as shown. When grown together, *Paramecium aurelia* proved to be the better competitor for the available food resource. Eventually, *Paramecium caudatum* died out because it could not compete for food.

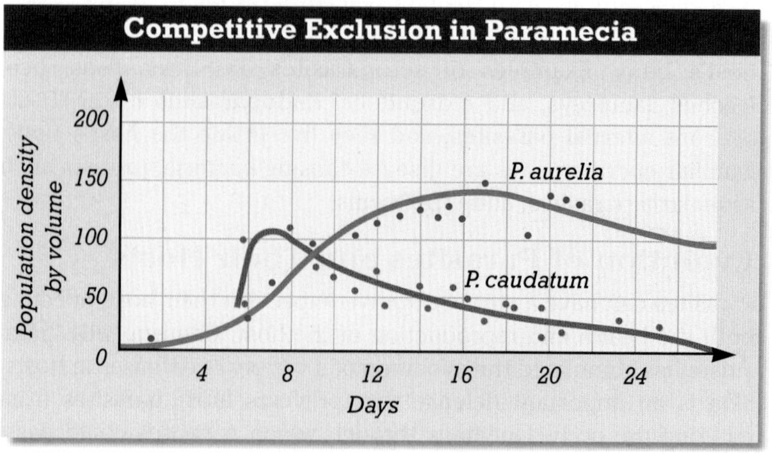

more efficiently and has a reproductive advantage that eventually eliminates the other species.

Joseph Connell's study of barnacles along the Scottish coast in the 1960s demonstrates competition in the wild. Connell studied two species of barnacles, *Semibalanus balanoides* and *Chthamalus stellatus,* that live in the intertidal zone, the portion of the shore that is exposed during low tide. Each species formed a distinct band within the intertidal zone, as shown in Figure 21-6, with *Chthamalus* living higher on the rocks than *Semibalanus* did. Connell demonstrated that this difference was partly due to competition. When a rock covered with *Chthamalus* was transplanted to the lower zone, *Chthamalus* was clearly able to tolerate the conditions in the lower zone. However, *Semibalanus* settled on the rock and eventually crowded out *Chthamalus.* Connell concluded that competition restricted the range of *Chthamalus.* Although it could survive lower on the rocks, competition from *Semibalanus* prevented it from doing so. Intolerance to long periods of drying when the tide was out probably restricted the range of *Semibalanus* to the lower zone.

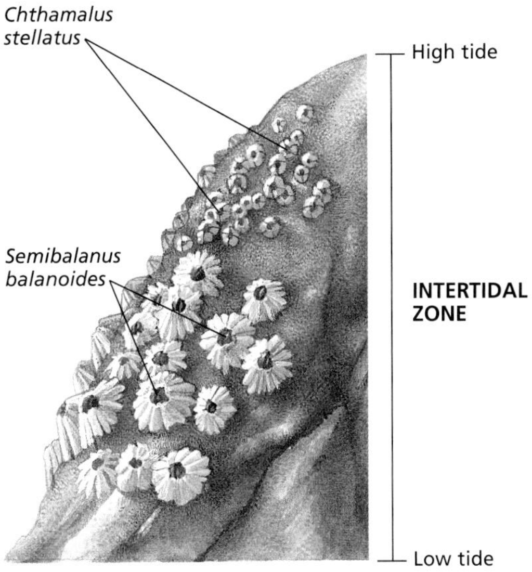

FIGURE 21-6

Although *Chthamalus* is capable of surviving in the lower intertidal zones, it is crowded out by *Semibalanus. Semibalanus* is unable to survive in the upper zone, which is frequently above the tidal level for extended periods. *Semibalanus* is the most efficient competitor in the areas that are usually under water.

Competition and Community Structure

Competition has the potential to be an important influence on the nature of a community. The composition of a community may change through competitive exclusion. Competitors may also evolve niche differences or anatomical differences that lessen the intensity of competition. Natural selection favors differences between potential competitors. These differences are often greatest where the ranges of potential competitors overlap. This phenomenon is called **character displacement.**

An example of character displacement is provided by Darwin's finches. These finches feed on seeds. Birds with larger beaks can crack open and eat larger seeds, so beak size is a good indicator of diet. Two species of Darwin's finches are very similar to each other. On islands where only one of two similar species is found, birds of the two species have the same size beak. But on an island where both birds are found, their beak sizes are different. They have evolved a character displacement that enables them to feed on different-sized seeds and reduce competition.

As Darwin noted, competition is likely to be most intense between closely related species that require the same resources. When similar species coexist, each species uses only part of the available resources. This pattern of resource use is called **resource partitioning.** For example, consider three species of warblers that live in spruce and fir trees and feed on insects. The late Princeton ecologist Robert MacArthur discovered that the warblers differ in where they forage. Each kind of warbler hunts for insects only in a particular section of the tree. As a result, competition among the species is reduced.

internetconnect

SCILINKS NSTA
TOPIC: Competition
GO TO: www.scilinks.org
KEYWORD: HM401

SECTION 21-1

TEACHING STRATEGY
Observing Flowers

Have students look at flower specimens or pictures of flowers. Ask them what adaptations these flowers have to attract insects to transfer their pollen. (Attractants include pollinator guides, such as bold color markings or lines that lead to the nectar source; bright colors, such as blue and yellow, or, for night-flying moths, pale colors; and landing sites for insects.) Have students select one specimen to draw and color, and have them label the parts that attract pollinators. Ask students what type of species interaction exists between a flower and a pollinator (mutualism).

GIFTED ACTIVITY

Have each student investigate one of the different types of species interactions. Students should write a report that contains at least five detailed examples of adaptation for the kind of species interaction selected.

INCLUSION ACTIVITY

Take students on a short walk to a natural area on the campus. Have them point out any plant adaptations they observe that might prevent a plant from being eaten. Ask them what kind of species interaction this is. (Students may point out plants with thorns, spines, or a strong smell.) Point out any other species interactions observed, such as a bird catching an insect, a dragonfly searching for prey, or lichens growing on a tree. Ask students to name each type of interaction. When students return to class, have them discuss the interactions they observed.

SECTION 21-1

RETEACHING ACTIVITY

Have students work in groups. Have each group pick a specific community and make posters that illustrate each of the five types of interactions within that community. The pictures could be drawn or cut from magazines. Have students label the interactions and draw arrows between the interacting species.

ANSWERS TO SECTION 21-1 REVIEW

1. Predators always kill their prey. Parasites often do not kill their prey immediately. Examples of predators include sharks, tigers, and eagles. Examples of parasites include tapeworms, ticks, and leeches.
2. This mechanism is called mimicry. The flies are harmless, but they resemble a dangerous species. Flies are avoided by predators in the same way that the predators avoid the dangerous species.
3. Plants produce poisons, such as strychnine and nicotine. Poison ivy plants produce an irritating substance that causes a severe rash.
4. Character displacement occurs when two species use the same resources. In the zone where the ranges of the two species overlap, natural selection favors differences in resource use, which can lead to anatomical differences in the species.
5. No, it would be a case of mutualism.
6. The birds' behavior may be an example of resource partitioning. One species may feed on insects, and the other may feed on seeds. They may nest on different levels of the tree. At least one factor in the birds' niches must differ to allow them to exist closely together.

MUTUALISM AND COMMENSALISM

Mutualism is a cooperative relationship in which both species derive some benefit. Some mutualistic relationships are so close that neither species can survive without the other. An example of such mutualism involves ants and a shrub called bull's horn acacia. The ants nest inside the acacia's large thorns and receive food from the plant. The ants protect the acacia from predation by herbivores and trim back vegetation that shades the shrub.

Pollination is one of the most important mutualistic relationships on Earth. Animals such as bees, butterflies, flies, beetles, bats, and birds pollinate many flowering plants. Animals that carry pollen are called **pollinators.** A flower is a lure for pollinators, which are attracted by the flower's color, pattern, shape, or scent. The plant usually provides food for its pollinators. As the animal feeds in the flower, it picks up a load of pollen, which it will carry to the next flower of the same species it visits, as shown in Figure 21-7.

FIGURE 21-7

Some bats are active at night and locate their food by sound or smell instead of by sight. Flowers pollinated by nectar-feeding bats do not need to be brightly colored, but they usually do have a strong fragrance. As the bat feeds on the flower's nectar, it becomes smeared with pollen, as shown. The bat then carries the pollen to the next flower it feeds on.

Commensalism is an interaction in which one species benefits and the other is not affected. Some cases of commensalism may be mutualisms in which the benefit to the second organism hasn't yet been identified. One example of commensalism is the relationship between cattle egrets and Cape buffalo in Tanzania. The birds feed on small animals such as insects and lizards that are forced out of their hiding places by the buffalo's movement through the grass.

SECTION 21-1 REVIEW

1. Explain how predators differ from parasites. Give an example of each kind of organism.
2. Some harmless flies resemble bees and wasps. What is this mechanism called? Evaluate its importance as a defense mechanism.
3. Describe two chemical defenses of plants.
4. Explain the advantage of character displacement and give an example.
5. If cattle egrets removed ticks from Cape buffalo, would their relationship still be considered commensalism? Explain your answer.
6. CRITICAL THINKING Explain how two similar species of birds are able to inhabit the same area and even nest in the same tree without occupying the same niche.

PROPERTIES OF COMMUNITIES

The investigation of community properties and interactions is an active area of ecology. What properties are most significant in structuring a community? What determines species richness and abundance? These questions are central to a study of communities.

SECTION

21-2

OBJECTIVES

▲ Explain the difference between species richness and species diversity.

● Describe how species richness varies with latitude, and explain a hypothesis for this pattern.

■ Explain the cause and consequences of the species-area effect.

◆ Explain the two main views of the relationship between species richness and stability.

SPECIES RICHNESS AND DIVERSITY

One characteristic of a community is its **species richness,** the number of species it contains. A related measure is **species diversity,** which relates the number of species in the community to the relative abundance of each species. These two measures provide slightly different information. Species richness is a simple count of the species in the community. Each species contributes one count to the total, regardless of whether the species' population size is one or 1 million. In contrast, species diversity suggests each species' importance because it takes into account how common each species is in the community. To calculate the species diversity of a community, an ecologist must measure or estimate the population size of all the species in the community.

Patterns of Species Richness

Species richness varies with latitude (distance from the equator). As a general rule, the closer a community is to the equator, the more species it will contain. Species richness is greatest in the tropical rain forests. For example, entomologists E. O. Wilson and Terry Erwin identified nearly as many species of ants in a single tree in Peru as can be found in the entire British Isles.

Why are there more species in the tropics than there are in the temperate zones? One hypothesis is that temperate habitats are younger, having formed since the last ice age. Therefore, tropical habitats were not disturbed by the ice ages, while habitats at latitudes farther north were. Also, the climate is more stable in the tropics. This stability allowed species to specialize to a greater degree than they could in temperate regions, where the climate is more variable. Another hypothesis suggests that because plants

SECTION 21-2

SECTION OVERVIEW

Matter, Energy, and Organization
Section 21-2 discusses the organization of species within communities and stresses the concepts of richness and diversity. This section also describes how the species-area effect influences distribution of species.

Stability and Homeostasis
Section 21-2 describes how communities return to stability after major disturbances. This section stresses that a high species richness improves a community's stability.

▶ ENGAGE STUDENTS

Have students use microscopes to look at a drop of pond water that is rich in algae. Have them sketch as many species as they can find on the slide. Ask what the species richness of this freshwater community is (count the number of different species).

TEACHING STRATEGY

Species Diversity

Describe the following two communities to the class:
Community A has a total of 1,000 individuals and contains 25 different species. Each species has 40 individuals. Community B also has a total of 1,000 individuals and also contains 25 different species. Ten species have 15 members each. Twelve species have 10 members each. One species has 30 members, one species has 100 members, and one species has 600 members. Ask students which community has more species diversity. (Community A is more diverse because all species are equally important, whereas community B is dominated by a single species.)

SECTION 21-2

RECENT RESEARCH

Shade-Plantation Diversity

Research on coffee production in Latin American countries has centered on old-style coffee plantations that require a canopy of trees over the coffee plants. These shade plantations are providing islands of biodiversity in a landscape of destroyed rain forests. Studies suggest these old-style plantations should be continued as a haven for species diversity, as opposed to the new-style coffee farms which are not shaded and grow only coffee plants.

CRITICAL THINKING

Native-Species Diversity

Ask students if the loss of native species diversity can cause changes in the environment, such as changes affecting the water cycle. (Yes, a decrease in species means a drastic change in water-cycling patterns in a community, and more water may be used or lost. In South Africa, an economic analysis was done on the loss of native shrubs to exotic Australian trees. The findings showed that there was a 30 percent reduction in water supply because the trees used more water than the native shrubs.)

CRITICAL THINKING

New Species

Ask students where they would go, and why, if they wanted to discover a new species of bird. (The tropics have the greatest species diversity in the world, and new species are frequently being discovered. For example, in 1996, an ornithologist discovered a previously unknown species of bird, the barbet, high in the clouds in the forests of Peru.)

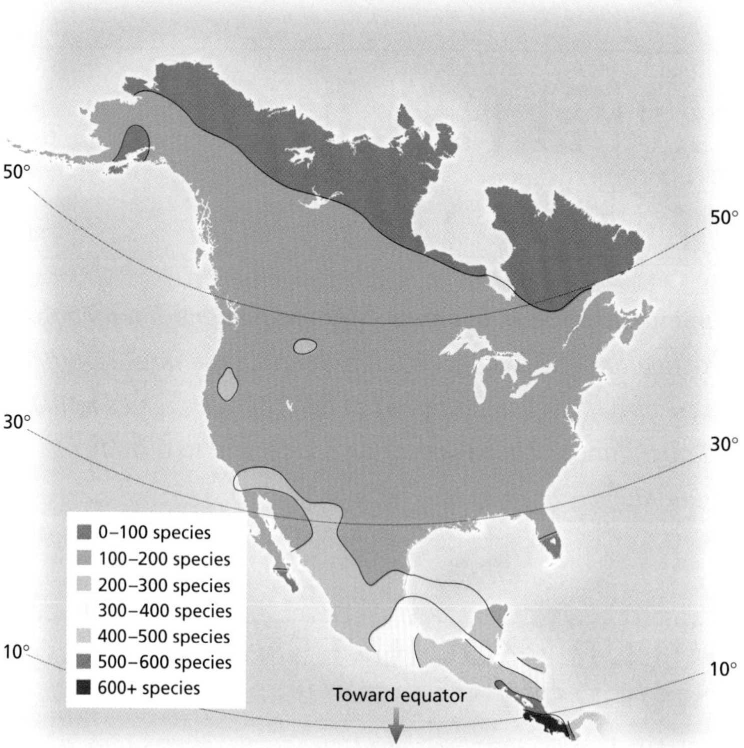

FIGURE 21-8

This species-richness map of North American and Central American birds shows that fewer than 100 species of birds inhabit arctic regions, whereas more than 600 species occupy some tropical regions. This is evidence that species richness increases nearer the equator. Equatorial rain forests are biologically the richest habitats on Earth.

can photosynthesize year-round in the tropics, there is more energy available to support more organisms. It is likely that the high diversity of species in the tropics, as indicated in Figure 21-8, is the result of several factors acting together.

The Species-Area Effect

Another pattern of species richness is that larger areas usually contain more species than smaller areas do. This relationship is called the **species-area effect.** The species-area effect is most often applied to islands, where area is clearly limited by geography. In the Caribbean, for example, more species of reptiles and amphibians live on large islands, such as Cuba, than on small islands, such as Redonda, as shown in Figure 21-9. Because all of these islands are close together, differences in species richness cannot be due to differences in latitude. Why does species richness increase with increasing area? Larger areas usually contain a greater diversity of habitats and thus can support more species.

The species-area effect has one very important practical consequence—reducing the size of a habitat reduces the number of species it can support. Today, natural habitats are shrinking rapidly under pressure from the ever-growing human population. About 2 percent of the world's tropical rain forests is destroyed each year, for example. The inevitable result of the destruction of habitats is the extinction of species.

FIGURE 21-9

As shown in the graph, the large islands of Cuba and Hispaniola have about 100 species (of reptiles and amphibians). The small island of Redonda has only about five species of reptiles and amphibians. In general, species diversity increases as available habitat area increases. This principle is true whether the boundaries of the area are created by shoreline, as on an island, or by encroaching human population, as in the construction of housing in a natural area.

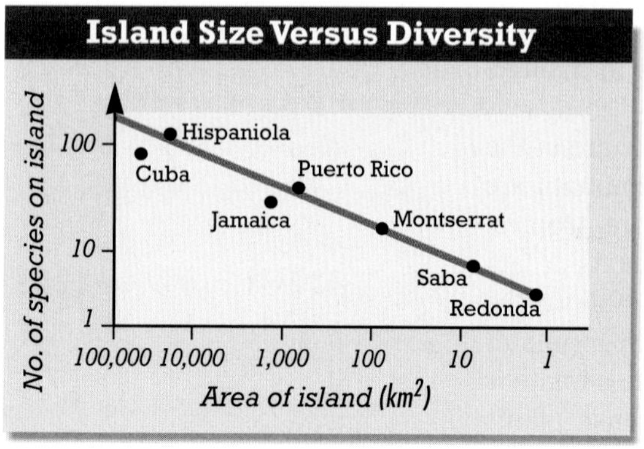

Species Interactions and Species Richness

Interactions among species sometimes promote species richness. Several studies have demonstrated that predators can prevent competitive exclusion from occurring among their prey. In the 1960s, Robert Paine of the University of Washington showed the importance of the sea star *Pisaster,* shown in Figure 21-10, in maintaining the species richness of communities on the Washington coast. Paine removed all *Pisaster* individuals from one site and for several years prevented any new *Pisaster* individuals from settling there. This change caused a dramatic shift in the community. The mussel *Mytilus,* which had previously coexisted with several other species, became much more abundant and spread over the habitat, crowding out other species. The species richness of the community fell from 15 to 8 during the course of the study. Evidently, *Mytilus* was the superior competitor for space on the rocks, but its population was normally held in check by predation from *Pisaster.*

FIGURE 21-10

When the sea star *Pisaster* was removed from an area where the sea stars had preyed on the mussel *Mytilus,* the mussels crowded out many of the other competing species in the area. Predation by the sea star on the mussel promoted diversity by controlling the superior competitor—the mussel.

Community Stability

One of the most important characteristics of a community is how it responds to disturbance. The **stability** of a community indicates its resistance to change. For many years, most ecologists agreed that stability was directly related to species richness. Communities with more species, they presumed, would contain more links between species. These links would, in a sense, disperse the effects of the disturbance and prevent disruption of the community. One line of evidence cited in support of this view was the vulnerability of agricultural fields, which usually contain one species of crop plant, to outbreaks of insect pests.

David Tilman of the University of Minnesota and John Dowling of the University of Montreal investigated how the response to drought in a small plot of grasses was affected by species richness. They grew varying numbers of species in many small plots of land and subjected each to drought. They found that plots with more species present lost a smaller percentage of plant mass than plots with fewer species. The plots with more species also took less time to recover from the drought. They concluded that species richness improves a community's stability.

internet**connect**

SCILINKS NSTA

TOPIC: Species interaction and richness
GO TO: www.scilinks.org
KEYWORD: HM405

SECTION 21-2 REVIEW

1. What is the difference between species richness and species diversity?
2. Explain the relationship between species richness and latitude.
3. Why is the species-area effect important in efforts to conserve species?
4. Describe how predation can affect species diversity.
5. Explain how species richness contributes to community stability.
6. CRITICAL THINKING Explain how the example of agricultural fields supports the idea that species richness promotes stability.

RETEACHING *ACTIVITY*

Show students some photographic slides of different biomes and each biome's inhabitants. Then have them list as many different species as they can for each biome. Discuss why more available energy in the tropics leads to greater species richness.

ANSWERS TO SECTION 21-2 REVIEW

1. Species richness is the number of species in a community. Species diversity indicates how abundant each species is.
2. The closer to the equator a community is, the greater its species richness. Tropical regions might be older and have had more time to accumulate species. Their greater productivity, resulting from year-round photosynthesis, might allow more species to coexist on the greater abundance of resources.
3. Smaller areas usually support fewer species. Destruction and alteration of natural landscapes are rapidly shrinking the area of available habitat throughout the world.
4. Some predators feed on species that are superior competitors. Predation keeps their population smaller and prevents them from crowding out other species.
5. Communities with more species contain more links between species. These links disperse the effects of disturbances.
6. Agricultural fields contain a single species of crop. They can be prone to plagues of destructive insects. If an insect that feeds on wheat grain invades a wheat field, the entire field may be destroyed. If the field contains more plant species than just wheat, the plants not used by the insects would survive.

SECTION 21-3

SECTION OVERVIEW

Stability and Homeostasis
Section 21-3 discusses the stages of successional changes in a community and points out that the climax community is a stable end point of succession.

Matter, Energy, and Organization
Section 21-3 describes succession as an organized, sequential replacement of species.

▶ ENGAGE STUDENTS

Bring to class a variety of fast-growing weeds for students to examine. Ask them what characteristics these weeds have in common. (Possible answers are that they have a single, large taproot; they are fast-growing; they have green, soft stems; they have a small size; and they have leafy upper parts.) Ask students what role these kinds of plants might play in succession. (Fast-growing weeds are primary or early-succession plants that follow the pioneer plants because they grow fast even in poor soil.)

CRITICAL THINKING

Exotic Weeds
Ask students how exotic weeds can destroy the natural succession of communities. (Weeds are some of the first invaders of an area following a natural disaster. They are aggressive, fast-growing plants that require minimal nutrients. If foreign weeds are introduced to an area where they have no natural checks, they quickly take over and prohibit the natural order of succession.)

SECTION 21-3

OBJECTIVES

▲ Distinguish between primary and secondary succession.

● Identify some of the characteristics of pioneer species.

■ Describe the sequence of changes occurring at Glacier Bay.

◆ Explain the successional changes that can occur when an existing community is disrupted.

internet**connect**

SCILINKS NSTA
TOPIC: Succession
GO TO: www.scilinks.org
KEYWORD: HM406

SUCCESSION

Disturbances such as fires, landslides, hurricanes, and floods trigger a sequence of changes in the composition of a community. Certain species flourish immediately after the disturbance, then are replaced by other species, which are replaced by still others. Over time, the composition of the community changes.

SUCCESSIONAL CHANGES IN COMMUNITIES

During the summer and early fall of 1988, fires burned large areas of Yellowstone National Park, affecting nearly 300,000 hectares (720,000 acres). If you visit Yellowstone today, you will find that regrowth is well under way in the burned areas. In time, if there are no further disturbances, the burned areas of Yellowstone National Park will undergo a series of regrowth stages. The gradual, sequential regrowth of species in an area is called **succession** (SUHK-SE-shuhn). You can see early stages of succession in abandoned fields, in vacant lots, along roads, and even in sidewalks or parking lots where weeds are pushing up through cracks in the concrete.

Ecologists recognize two types of succession. **Primary succession** is the development of a community in an area that has not supported life previously, such as bare rock, a sand dune, or an island formed by a volcanic eruption. **Secondary succession** is the sequential replacement of species that follows disruption of an existing community. The disruption may stem from a natural disaster, such as a forest fire or a strong storm, or from human activity, such as farming, logging, or mining.

Any new habitat, whether it is a pond left by heavy rain, a freshly plowed field, or newly exposed bedrock, is an invitation to many species that are adapted to be good pioneers. The species that predominate early in succession—called the **pioneer species**—tend to be small, fast-growing, and fast-reproducing. Pioneer species are well suited for invading and occupying a disturbed habitat. For example, they are often very good at dispersing their seeds, which enables them to quickly reach disrupted areas. Some of the pioneers that might be found in a vacant lot or abandoned field in the eastern United States include horseweed, crabgrass, and ragweed.

(a) (b) (c)

FIGURE 21-11

Ecologists study the process of primary succession by examining a variety of areas at different successional stages. These photos were taken at different locations at Glacier Bay, Alaska; the changes they illustrate take about 200 years. Shown in (a) is lifeless glacial "till" (pulverized bare rocks) left in the wake of the retreating glacier. Shown in (b) is an early stage of succession in which small plants and shrubs are growing on the site. A mature forest is shown in (c), an end stage of succession.

Primary Succession

Primary succession often proceeds very slowly because the minerals necessary for plant growth are unavailable. For example, when glaciers last retreated from eastern Canada, about 12,000 years ago, they left a huge stretch of barren bedrock from which all the soil had been scraped. This geologic formation, called the Canadian Shield, was a place where plants and most animals could not live. Repeated freezing and thawing broke this rock into smaller pieces. In time, lichens—mutualistic associations between fungi and either algae or cyanobacteria—colonized the barren rock. Acids in the lichens and mildly acidic rain washed nutrient minerals from the rock. Eventually, the dead organic matter from decayed lichens along with minerals from the rock began to form a thin layer of soil in which a few grasslike plants could grow. These plants then died, and their decomposition added more organic material to the soil. Soon shrubs began to grow, and then trees appeared. Today much of the Canadian Shield is densely populated with pine, balsam, and spruce trees, whose roots cling to soil that in some areas is still only a few centimeters deep. A similar series of changes has been documented at Glacier Bay, Alaska, shown in Figure 21-11.

INCLUSION *ACTIVITY*

Have students paste pictures from magazines, advertisements, newspapers, brochures, and other sources on note cards showing different stages of succession for a specific community and species representative of those stages. Have students use their textbook to arrange the cards consecutively according to the appropriate stages of succession. Then have them close their books, shuffle their cards, repeat the process, and check their accuracy.

OVERCOMING MISCONCEPTIONS

A common misconception is that all wildfires are bad. Wildlife managers have learned that fire is necessary for the succession of many communities. For example, most pine species of a mature forest in a climax community cannot germinate without the high temperatures that a fire produces. After a fire, the succession of the forest begins again. Grasslands are also maintained by fires. Without fires the grassland would revert to woody species.

QUICK FACT

✔ Some pioneer plants are used as medicines. The purple cornflower, a native of the plains, was once used in herbal medicine. It contains a compound that stimulates the immune system and kills pain.

SECTION 21-3

CRITICAL THINKING

Rain-Forest Succession

Ask students if a rain forest will return to its original state through stages of succession if it is cut or burned to the ground. (The natural succession to a mature rain forest may not be possible once it is destroyed because the nutrients of the system will be lost and may not become available for natural succession.)

ANSWERS TO SECTION 21-3 REVIEW

1. Primary succession is the formation of a new community on newly formed habitat. Secondary succession is the recovery of an existing community after a disturbance.
2. Weeds grow rapidly, reproduce quickly, produce many offspring, and disperse well.
3. The initial colonists at Glacier Bay include lichens and small plants, such as mosses.
4. Pioneer species become established first in a community, then succession proceeds through a series of regrowth stages to a stable end point called a climax community.
5. Frequent disturbances, such as fires, can interrupt succession and return the community to an earlier stage.
6. Soil formation begins with weathering and with the breaking of rock into smaller pieces. Lichens eventually colonize the rock. Repeated generations of growth and decay of lichens and leaching of minerals from the rock produce an accumulation of organic material from which soil is formed. Lichens are pioneers in primary succession. After soil formation, herbaceous plants can grow. Also after soil formation, further disturbance of the area would result in secondary succession.

(a)

(b)

(c)

FIGURE 21-12

In (a), a recently abandoned agricultural field is being pioneered by weeds. Eventually, taller plants and shrubs will shade out the pioneers, as shown in (b). Next, a forest of pine or cottonwood may be succeeded by a hardwood forest. The whole process, if there are no further disturbances, takes about 100 years.

Secondary Succession

Secondary succession occurs where an existing community has been cleared by a disturbance, such as agriculture, but the soil has been left intact. In secondary succession, it commonly takes about 100 years for the original ecosystem to return through a series of well-defined stages. In eastern temperate regions, secondary succession typically begins with annual grasses, mustards, and dandelions. Succession proceeds with perennial grasses and shrubs, continues with trees like dogwoods, and often continues to a deciduous forest, as shown in Figure 21-12.

THE COMPLEXITY OF SUCCESSION

The traditional description of succession is that the community proceeds through a predictable series of stages until it reaches a stable end point, called the **climax community.** The organisms in each stage alter the physical environment in ways that make it less favorable for their own survival but more favorable for the organisms that eventually succeed them. In a sense, each stage paves the way for the next, leading ultimately to the climax community, which remains constant for a long period of time.

When ecologists began to study and document many instances of succession, they found a complex picture. Some so-called climax communities, for example, are not stable and continue to change. Instead of proceeding inevitably toward the climax community, succession may be regularly "reset" by disturbances. For example, many grasslands give way to forests, but periodic fires prevent the forests from developing. There may be many possible successional pathways in a particular area. The actual path followed may depend on the identities of the species present, the order in which they arrive, the climate, and many other factors. Ecologists agree that the idea of a single successional pathway ending in a stable climax community is too simple to describe what actually occurs in nature.

SECTION 21-3 REVIEW

1. What is the difference between primary succession and secondary succession?
2. How are plants we think of as weeds, such as ragweed, well adapted to be pioneer species?
3. Identify one of the initial colonists in succession at Glacier Bay.
4. Describe secondary succession, and distinguish between pioneer species and a climax community.
5. How do frequent fires alter succession in a community?
6. CRITICAL THINKING Describe the geological process of soil formation. Explain its importance to succession.

CHAPTER 21 REVIEW

SUMMARY/VOCABULARY

21-1

- Ecologists recognize five major kinds of species interactions in communities: predation, parasitism, competition, mutualism, and commensalism.
- The interaction in which one organism kills and eats another is predation. Predators have evolved many ways to efficiently find and capture prey. Prey have evolved many ways to defend themselves against predators.
- Mimicry is a defense mechanism in which a harmless species gains protection by its resemblance to a poisonous or distasteful species, or in which two or more poisonous or distasteful species resemble each other.
- Parasitism involves one organism feeding on, but not always killing, another. Parasites are grouped into two general categories—external parasites (ectoparasites) and internal parasites (endoparasites).
- Competition occurs when two or more species use the same scarce resource.
- Competition may cause competitive exclusion, the extinction of one competitor from the community. It may also be a strong selective force in the evolution of niche differences among competitors.
- In mutualism, both interacting species benefit. The relationship between flowering plants and their pollinators is an example of mutualism.
- In commensalism, one species benefits and the other is not affected.

Vocabulary

character displacement (401)
commensalism (402)
competition (400)
competitive exclusion (400)
ectoparasite (399)
endoparasite (399)
herbivore (398)
host (399)
mimicry (398)
mutualism (402)
parasite (399)
parasitism (399)
pollinator (402)
predator (397)
prey (397)
resource partitioning (401)
secondary compound (398)
symbiosis (397)

21-2

- Species richness is the number of species in a community.
- Species diversity incorporates both species richness and the abundance of each species.
- As a general rule, species richness is greatest near the equator.
- Larger areas generally support more species. This is called the species-area effect.
- Species interactions such as predation can promote species richness.
- Species richness improves a community's stability.

Vocabulary

species-area effect (404)
species diversity (403)
species richness (403)
stability (405)

21-3

- Succession is a change in the species composition of a community. Primary succession is the assembly of a community on newly created habitat. Secondary succession is the change in an existing community following a disturbance.
- Pioneer species are the initial colonists of a disturbed area. They are usually small organisms that grow fast, reproduce quickly, and disperse their seeds well.
- Succession at Glacier Bay proceeded from bare debris left behind by a retreating glacier to small plants such as mosses, to alders, to spruce and hemlocks.
- Secondary succession occurs in areas where the original ecosystem has been cleared by a disturbance, such as clearing for agriculture. It proceeds from plants commonly thought of as weeds to a stable end point called a climax community.

Vocabulary

climax community (408)
pioneer species (406)
primary succession (406)
secondary succession (406)
succession (406)

CHAPTER 21 REVIEW ANSWERS

REVIEW

1. Mutualism is an interaction in which both participants benefit. In commensalism, one species benefits and the other is not affected.
2. Resource partitioning is an adaptation that enables species to use different resources when they coexist in an area. Competition occurs when species use the same limited resource. Resource partitioning reduces competition.
3. Ectoparasites live on the outer surface of their host. Endoparasites live within their host.
4. Competitive exclusion is a change in the composition of species in a community in which inferior competitors can become extinct. Character displacement is an evolutionary change in two competing species. They evolve physical differences that reduce their use of common resources.
5. Species richness is the number of species in a community. The species-area effect is a global pattern in species richness: larger areas usually contain more species.

6. c
7. b
8. d
9. b
10. c
11. a
12. c
13. a
14. b
15. c

16. Answers will vary, but students may describe the heat-sensitive pits of rattlesnakes, a spider's web, the flesh-cutting teeth of wolves and coyotes, or the striped pattern of a tiger's coat.
17. Secondary compounds are defensive chemicals synthesized by plants. Strychnine and nicotine are two examples.
18. The skin keeps out parasites. Tears, saliva, and sweat contain an antibacterial enzyme. Cells within the body detect and attack invaders.
19. Gause grew some species of paramecia alone and in various combinations. He found that one pair of species—*Paramecium*

caudatum and *P. aurelia*—could not coexist. *P. caudatum* could coexist with *P. bursaria,* but both species' populations were well below their carrying capacity.

20. *Chthamalus* could live higher on the rocks, in a zone that was exposed to longer periods of drying, which *Semibalanus* couldn't tolerate.

21. Ants receive places to build their nests, nectar, and protein-rich food. The acacia gains protection from herbivores, and the ants trim back branches that shade the acacia.

22. Island B is larger and closer to the equator. If all other factors are equal between the two islands, Island B will have a larger number of species.

23. Answers will vary. Artificial wastewater-treatment ecosystems tend to undergo eutrophication, just as natural wetlands do. However, the high-nutrient levels in wastewater promote rapid algae growth. If the systems are not managed, they will eventually fill with algae and decaying organic matter, form a marsh and eventually a meadow.

CRITICAL THINKING

1. This conclusion is not valid. Past competition may have caused some species to go extinct in the community and may have stimulated character displacement and evolution of niche differences. All of these mechanisms would reduce competition among the members of a community, possibly to the point where competition is no longer detectable.

2. The plant's pollen is more likely to be transported to a member of the same species because the pollinator does not visit any other species. If the pollinator becomes extinct, however, the plant may be doomed.

3. To calculate species diversity, an estimate is needed for the population size of every species in the

REVIEW

Vocabulary

1. Explain the difference between mutualism and commensalism.
2. Describe resource partitioning and competition, and explain how they are related.
3. What is the difference between ectoparasites and endoparasites?
4. How do competitive exclusion and character displacement differ?
5. Distinguish between species richness and the species-area effect.

Multiple Choice

6. In which of the following interactions do both species benefit? (a) competition (b) predation (c) mutualism (d) commensalism
7. An example of a parasite is (a) a lion (b) a tick (c) a deer (d) a snake.
8. Which of the following is not a plant defense against herbivores? (a) thorns (b) tough leaves (c) secondary compounds (d) lysozyme
9. The evolution of anatomical differences in response to competition is called (a) competitive exclusion (b) character displacement (c) niche release (d) convergence.
10. Which of the following is true of a mimic? (a) It is poisonous or distasteful. (b) It is identical to a poisonous species. (c) It closely resembles a poisonous species. (d) An example is the coral snake.
11. Species richness (a) is greater on large islands than it is on small islands (b) is lowest in the tropics (c) is increasing worldwide (d) is greater on small islands than on large islands.
12. In the experiment with the sea star *Pisaster* and the mussel *Mytilus,* (a) elimination of the mussel *Mytilus* increased the number of sea stars (b) species richness of the community increased over time (c) removal of the sea star led to a decrease in species richness (d) removal of the sea star led to an increase in species diversity.
13. Primary succession occurs (a) on bare rock (b) in a disrupted habitat (c) after a forest fire (d) only on dry land.
14. The feeding behavior of warblers is an example of (a) niche overlap (b) resource partitioning (c) mutualism (d) character displacement.
15. Which is not true of secondary compounds? (a) They are part of a defense mechanism. (b) They can be poisonous and taste bad. (c) They are metabolized as food by plants. (d) They compose the irritant produced by poison ivy.

Short Answer

16. Describe two evolutionary adaptations that enable organisms to be efficient predators.
17. What are secondary compounds, and what is their function? List two examples of secondary compounds.
18. Explain some of the ways your body defends itself against parasites.
19. Describe the experiments G. F. Gause conducted on competition in *Paramecium.* What were the results of these experiments?
20. In the study of competition between two species of barnacles, *Semibalanus* was clearly the superior competitor, yet *Chthamalus* was not excluded from the community. Explain why.
21. What benefits do ants derive from their relationship with the bull's horn acacia? What benefits does the acacia receive from the ants?
22. Examine the island map below. Which island will probably have more species of plants and animals, Island A or Island B? Explain your answer.

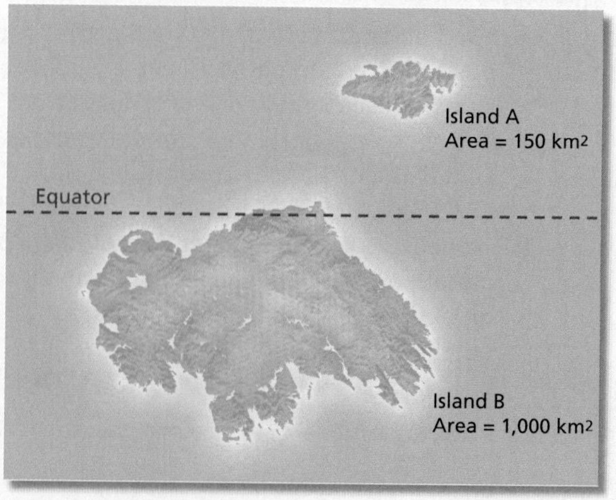

23. Unit 7—*Ecosystem Dynamics*

Write a report summarizing how artificial ecosystems, used in the management and treatment of wastewater and pollutants, can demonstrate succession.

CRITICAL THINKING

1. A scientist studies a community and finds no evidence of competition. The scientist concludes that competition has never had an effect on the structure of the community. Is this a valid conclusion? Explain your answer.
2. Some plants are pollinated by only one pollinator. Explain why this might be advantageous to the plant. Explain why this specificity could also lead to the extinction of the plant species.
3. Explain why it is usually harder to measure the species diversity of a community than it is to measure its species richness.
4. Examine the figure of the warblers on the right.
   a. Each bird is shown in the part of the tree where it usually nests. Explain how three closely related species of warbler are able to coexist in the same tree. What phenomenon is demonstrated by their coexistence?
   b. What might happen if only a single species of insect inhabited the spruce tree? Explain the phenomenon that might be demonstrated in this case.
   c. Suppose the birds are seed-eating finches and the tree produces large and small seeds. Each of the birds has either a small beak or a large beak. Explain the evolutionary phenomenon demonstrated in this case. Would a bird with a medium-sized beak be able to coexist in the tree? Explain why or why not.

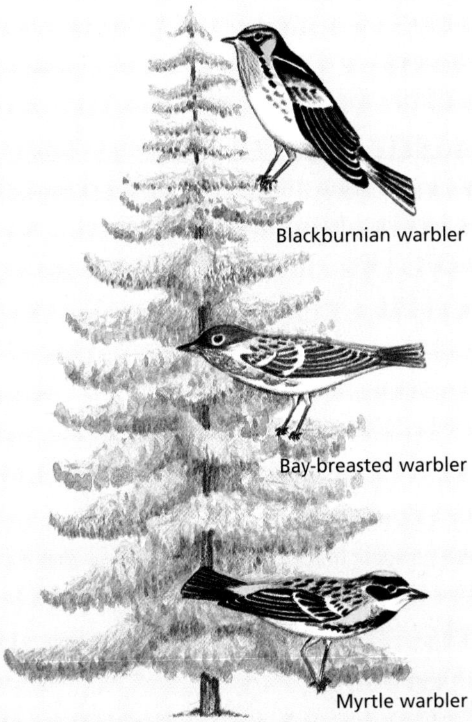

EXTENSION

1. Ants have mutualistic relationships with a variety of different organisms. Use an on-line database or conduct library research to identify two mutualisms (besides those mentioned in the text) between ants and other species. Write a report that summarizes what you have learned. Be sure to explain the nature of each interaction and the benefits each species receives.
2. Read "Hidden Existence" in *Audubon,* March–April 1999, on page 56. Describe two ways some animals are able to hide in plain view. Explain how an animal might combine camouflage coloration with specialized behavior to avoid predators. Camouflage is important to prey animals, but how does it help predators?
3. Choose two flowering plants that live in your area. Find out what pollinators visit the flowers. Make drawings that show the flowers and their pollinators.

community. To calculate species richness, you only need to know how many species are found there.

4. a. The three species of birds can coexist because they forage in different areas of the tree. This is called resource partitioning. b. If all species had only one source of food, they would compete for that limited resource. Eventually, one of the species would drive the others out. This is called competitive exclusion. c. The birds are demonstrating character displacement. They have developed anatomical differences (beak size) that allow them to take advantage of different resources (the small and large seeds). A bird with a medium-sized beak probably would not survive. It would not be able to eat the large seeds and would be out-competed for the small seeds.

EXTENSION

1. Ants are involved in many mutualistic relationships. Students may mention the relationship between ants and fungi or the relationship between ants and aphids.
2. Some animals are able to hide using camouflage to match their surroundings. Some animals press themselves flat to avoid casting a shadow, or they might also freeze in place. With camouflage, these animals are almost invisible. Predators use camouflage in order to ambush prey.
3. Answers will vary but may include bees, bats, or hummingbirds as pollinators. Some students may want to display their art.

CHAPTER 21 INVESTIGATION

TIME REQUIRED

Part A—one 50-minute lab period
Part B—one 50-minute lab period
Parts C and D—one 50-minute lab period

SAFETY PRECAUTIONS

Remind students to handle scalpels with care. Students with plant allergies should be cautioned to avoid contact with plant juices.

QUICK REFERENCES

Lab Preparation Notes are found on page 395B.

Holt BioSources provides a Teaching Resources CD-ROM that contains "Using Gowin's Vee in the Lab" and "Scoring Rubrics."

PROCEDURAL TIPS

1. To save time, you may want to plant the seeds and grow the plants six to eight weeks ahead of the scheduled lab.
2. *Rhizobium* and legume seeds are available from a biological supply house. Noninfected plants, used as a control, must be grown from sterile seeds in sterile soil. Label the plants *infected* and *non-infected* so that students can select one of each.
3. If students plant the seeds in class, be sure to have them monitor the plants each day for moisture.
4. Point out to students that once a *Rhizobium* bacterium is committed to fixing nitrogen, nearly all its other genes are turned off, so it can no longer reproduce.
5. Point out that a single family of bacteria (which includes *Rhizobium*) contains the only organisms on Earth able to utilize the nitrogen in the atmosphere. Without them, no nitrogen would be available to other organisms, and there would be no life on Earth.

Nitrogen Fixation in Root Nodules

OBJECTIVES

- Examine root nodules in legumes.
- Investigate the differences between legume (bean) and a nonlegume (radish).
- View active cultures of *Rhizobium.*

PROCESS SKILLS

- recognizing relationships
- hypothesizing
- comparing and contrasting

MATERIALS

- protective gloves
- 2 three-inch flowerpots
- 2 cups of soil
- 2 mixing sticks
- 1.2 mL (1/4 tsp) of *Rhizobium* bacteria per pot
- 2 bean seeds
- 2 radish seeds
- 2 microscope slides
- 2 coverslips
- 1 prepared reference slide of a legume root-nodule cross-section infected with *Rhizobium*
- compound light microscope
- stereoscope or magnifying glass
- scalpel

Background

1. Define *symbiosis,* and give an example of three types of symbiotic associations.
2. Nitrogen-fixing bacteria and leguminous plants have a symbiotic relationship.
3. Root nodules are swellings in the roots of leguminous plants that are infected with nitrogen-fixing bacteria.
4. *Rhizobium* is a genus of nitrogen-fixing bacteria. *Rhizobium* exists in soil and infects the root nodules of leguminous plants.
5. Green root nodules indicate actively reproducing bacteria that are not fixing nitrogen. Pink nodules indicate bacteria that are actively fixing nitrogen but not reproducing.

PART A Growing the Test Plants

1. Fill two flowerpots with soil. Using a mixing stick, stir approximately 1/4 teaspoon of the *Rhizobium* mixture into each pot.
2. Plant two bean seeds in one pot and label it "bean," and two radish seeds in the other and label it "radish." Water each pot so that the soil is moist but not saturated. Label the two pots.
3. Place the plants where they will receive direct sunlight. Water the soil when necessary to keep the soil moist but not saturated. Do not fertilize these plants.
4. After approximately one week, check to see if both seeds germinated in each pot. If they did, remove the smaller seedling. The plants will be ready to be examined after six to eight weeks. Monitor the plants each day. Give them enough water to keep the soil slightly moist, and keep them in direct sunlight. Note: The radish seeds will germinate faster than the bean seeds.
5. Clean up your materials and wash your hands before leaving the lab.

PART B Observing the Roots of Beans

6. Prepare a data table similar to the one below. As you work, record your observations in your data table.
7. **CAUTION Wear disposable gloves while handling plants. Do not rub any plant part or plant juice on your eyes or skin.** Remove the bean plant from the pot by grasping the bottom of the stem and gently pulling the plant out. Be careful not to injure the plant. Carefully remove all dirt from the roots.
8. View the roots of the bean plant under a stereoscope. Compare the appearance of the bean root system

OBSERVATIONS OF RHIZOBIUM

Appearance of nodule	
Color of nodule	All answers will vary.
Number of nodules	
Number of pink nodules	
Number of green nodules	

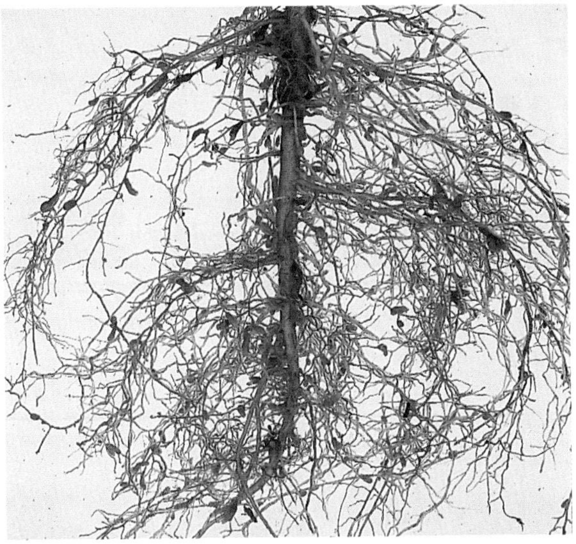

with the photograph above. Note the formation of any nodules on your bean plant's roots. Draw a root with a nodule in your lab report, and label each structure.

9. CAUTION **Use the scalpel with care. A scalpel is a very sharp instrument. Always make cuts with the blade facing away from your body. If you cut yourself, quickly apply direct pressure to the wound and call for your teacher.** Remove a large nodule from the bean root and carefully cut it in half with a scalpel. The pink nodules contain active nitrogen-fixing bacteria. The green nodules contain bacteria but cannot fix nitrogen because they are actively reproducing. *Rhizobium* will begin fixing nitrogen only after it stops reproducing. View the cross section under a stereoscope. Note the arrangement of bacteria within the cell. Draw a cell infected with bacteria in your lab report. Label the nodule, cell, and bacteria.

PART C Observing the Roots of Radishes

10. CAUTION **Wear disposable gloves while handling plants. Do not rub any plant part or plant juice on your eyes or skin.** Remove the radish plant from the pot by grasping the bottom of the stem and gently pulling the plant out. Be careful not to injure the plant. Carefully remove all dirt from the roots.
11. Examine the roots of the radish plant under a stereoscope. Compare the radish roots with the roots of the bean plant that you have already examined. Are there any nodules on the roots of the radish plant? Draw the root of the radish plant in your lab report. Label the drawing "Radish."

PART D Preparing a Wet Mount of *Rhizobium*

12. CAUTION **Handle the slide and coverslip carefully. Glass slides break easily and the sharp edges can cut you.** Prepare a wet mount by placing part of a nodule from the root of a bean plant on a microscope slide, adding a drop of water, and covering with a coverslip.
13. Place the slide on a flat surface. Gently press down on the slide with your thumb; use enough pressure to squash the nodule. Make sure the coverslip does not slide.
14. Examine the slide under a microscope. Draw and label a cell and the arrangement of bacteria in the cell in your lab report. Note what power of magnification you used.
15. Compare your wet mount preparation with the prepared reference slide of *Rhizobium* and the photograph on the right. A cell infected with *Rhizobium* should have a similar appearance to the photograph at right.

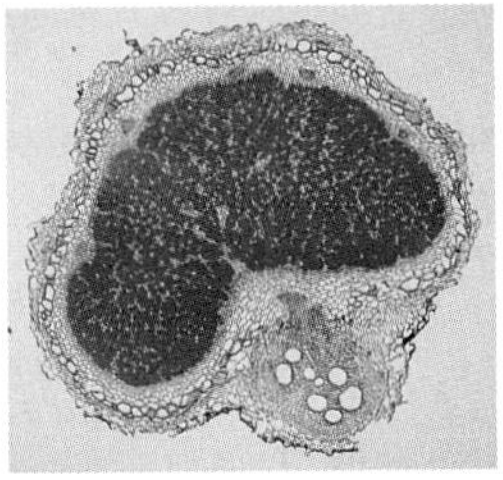

16. Clean up your materials and wash your hands before leaving the lab.

Analysis and Conclusions

1. Which plant had the most nodules?
2. How many nodules were found on the radish plants?
3. How do legumes become infected with bacteria in nature?
4. What kind of relationship exists between the legume plant and *Rhizobium*? How does this relationship benefit the legume plant? How does this relationship benefit the bacteria?
5. If you were to grow legumes without root nodules to use as experimental controls, why should you plant the seeds in sterile soil?

Further Inquiry

Perform the experiment using beans with and without *Rhizobium*. Count the number of leaves on each plant, and measure the mass of each whole plant as well as the masses of roots, stems, and leaves separately. Predict which part of the plant will have the greatest difference in mass.

ANSWERS TO BACKGROUND

1. Symbiosis is the relationship that exists between two organisms that live in a close association with each other. Symbiotic relationships can be parasitic, commensalistic, or mutualistic. The bull's horn acacia has a mutualistic relationship with ants. Ticks, fleas, and tapeworms are examples of parasites. Cape buffalo and cattle egrets have a commensalistic relationship. Parasitism is a symbiosis in which one organism benefits from the relationship and the other organism is harmed. Commensalism is a symbiosis in which one organism benefits and the other is neither harmed nor benefited by the relationship. Mutalism is a symbiosis in which both organisms benefit from the relationship.

ANSWERS TO ANALYSIS AND CONCLUSIONS

1. Only the beans have nodules because they are legumes and have the ability to fix nitrogen.
2. There are none. Radishes do not have the ability to form nodules and cannot shelter nitrogen-fixing bacteria.
3. Nitrogen-fixing bacteria live in the soil.
4. *Rhizobium* and legume plants have a mutualistic relationship. In the legume-*Rhizobium* relationship, the legume benefits from the nitrogen compounds produced by the bacteria, and the bacteria benefit by receiving food in the carbohydrates from the plant's photosynthesis.
5. Nonsterile soil might contain *Rhizobium* that could infect the young plants.

CHAPTER 22
ECOSYSTEMS AND THE BIOSPHERE

	TOPICS	TEACHING RESOURCES	LABS, CLASSWORK, AND HOMEWORK
BLOCK 1 45 minutes	**Introducing the Chapter, p. 414**	**ATE** Focus Concept, p. 414 **ATE** Assessing Prior Knowledge, p. 414 **ATE** Understanding the Visual, p. 414	■ Supplemental Reading Guide, *Silent Spring*
	22-1 Energy Transfer, p. 415 Producers, p. 415 Consumers, p. 416 Energy Flow, p. 417	**ATE** Section Overview, p. 415 **ATE** Critical Thinking, p. 418 **ATE** Visual Link Figure 22-1, p. 416 67, 68, 69, 82, 84, 85	**Lab A12,** Making a Food Web ★ Study Guide, Section 22-1 **PE** Section Review, p. 419
BLOCK 2 45 minutes	**22-2 Ecosystem Recycling, p. 420** The Water Cycle, p. 420 The Carbon Cycle, p. 421 The Nitrogen Cycle, p. 422	**ATE** Section Overview, p. 420 **ATE** Critical Thinking, p. 422 **ATE** Visual Link Figure 22-6, p. 421 65A, 71, 72, 73, 74, 81	**PE** **Quick Lab,** p. 421, Modeling Ground Water **Lab B10,** Ecology Scavenger Hunt ★ Study Guide, Section 22-2 **PE** Section Review, p. 423
BLOCK 3 45 minutes	**22-3 Terrestrial Ecosystems, p. 424** The Seven Major Biomes, p. 424 Tundra, Taiga, Temperate Deciduous Forests, Temperate Grasslands, Deserts, Savannas, and Tropical Rain Forests, pp. 425–429	**ATE** Section Overview, p. 424 **ATE** Critical Thinking, pp. 425, 428 **ATE** Visual Link Table 22-1, p. 425 Figure 22-11, p. 427 65	**Lab C22,** Examining Owl Pellets ★ Study Guide, Section 22-3 **PE** Section Review, p. 429
BLOCKS 4 & 5 90 minutes	**22-4 Aquatic Ecosystems, p. 431** Ocean Zones, p. 431 Freshwater Zones, p. 434	**ATE** Section Overview, p. 431 **ATE** Critical Thinking, p. 432 **ATE** Visual Link Figure 22-16, p. 431 Figure 22-17, p. 432 Figure 22-18, p. 432 65, 77, 79	**PE** **Chapter 22 Investigation,** p. 438 **Lab C23,** Examining Owl Pellets—NW vs. SE ★ Study Guide, Section 22-4 **PE** Section Review, p. 434
BLOCKS 6 & 7 90 minutes	**REVIEW and ASSESSMENT** **PE** Chapter 22 Review, pp. 436–437 ■ Performance-Based Assessment—Testing Pond Water for Carbon Dioxide	Audio CD Program BIOLOGY INTERACTIVE TUTOR Unit 7—Ecosystem Dynamics	**CHAPTER TESTING** ■ Chapter Test (blackline master) ▲ Test Generator ■ Assessment Item Listing ■ Scoring Rubrics and Classroom Management Checklists

MODERN BIOLOGY — HOLT BIOSOURCES — One-Stop Planner CD-ROM

KEY

- PE Pupil's Edition
- ATE Teacher's Edition
- ■ Active Reading Guide
- ★ Study Guide
- Teaching Transparencies
- Laboratory Program

One-Stop Planner CD-ROM
Shows these resources within a customizable daily lesson plan:

- ■ Holt BioSources Teaching Resources
- ■ Active Reading Guide
- ★ Study Guide
- ▲ Test Generator

READING FOR CONTENT MASTERY

Preview: Have students preview the objectives, the introductory paragraph, topic sentences, and captions before reading the chapter.

- ATE Active Reading Technique: Brainstorming, p. 415
- ■ Active Reading Guide Worksheet 22-1
- PE Summary/Vocabulary, p. 435

- ■ Active Reading Guide Worksheet 22-2
- PE Summary/Vocabulary, p. 435

- ■ Active Reading Guide Worksheet 22-3
- PE Summary/Vocabulary, p. 435

- ■ Active Reading Guide Worksheet 22-4
- PE Summary/Vocabulary, p. 435

TECHNOLOGY AND INTERNET RESOURCES

CNN PRESENTS Video Segment 14
Tropical Reforestation

Holt Biology Videodiscs Teacher's Correlation Guide, Lessons 22-1 through 22-4

BIOLOGY INTERACTIVE TUTOR Unit 7—Ecosystem Dynamics, Topics 1, 3–6

Audio CD Program, Sections 22-1 through 22-4

internetconnect

sciLINKS NSTA
On-line Resources: www.scilinks.org
The following *sci*LINKS Internet resources can be found in the student text for this chapter:

Topics:
- Producers, and consumers p. 415
- Temperate forests, p. 426
- Aquatic ecosystems, p. 432

go.hrw.com
On-line Resources: go.hrw.com
Visit the HRW Web site for a variety of resources related to this chapter. Just type in the keyword HM2 Chapter 22.

Smithsonian Institution®
Visit **www.si.edu/hrw** for additional on-line resources.

CNNfyi.com
Visit **www.cnnfyi.com** for late-breaking news and current event stories selected just for you.

LAB ACTIVITY PLANNING

CHAPTER 22 INVESTIGATION

Constructing and Comparing Ecosystems, pp. 438–439

OVERVIEW
In this investigation, students will observe closed ecosystems and will observe the interaction of organisms in them.

SAFETY
Instruct students to use care in handling living organisms.

PREPARATION
1. This investigation will take about 4 weeks to complete, after the initial 50-minute class period during which the ecosystems are prepared (Parts A, B, and C). Allow 1 week before beginning Part D to let the plants and algae grow in their ecosystems. For Part D, students will need a few minutes each day for a period of 3 or 4 weeks to make observations.
2. Before collecting materials, decide what you would like to have your class study. Rather than having the class observe several different habitats, you may wish to have the entire class observe a single ecosystem with one variable, or you may wish to study variation within similar populations.
3. If you use tap water, allow it to stand in an open jar for at least 24 hours so that the chlorine can escape. For containers, gallon-size glass jars work well. Your school cafeteria or a local deli are good sources for glass jars. Students could also use 5 gal water bottles or small aquariums.

Quick Labs

Modeling Ground Water, p. 421
Each student group will need disposable gloves, lab aprons, a 3 L plastic bottle (cut in half), small pebbles, a chunk of dry sod with grass, water, a graduated cylinder, and a 500 mL beaker.

UNDERSTANDING THE VISUAL

Have students examine the trees in the picture of a tropical rain forest. Ask if the leaves on the lower trees have the same kind of adaptations as the leaves at the top, which receive more sunlight. (The leaves at lower levels must be able to conduct photosynthesis in reduced light much as house-plants do.) Have students observe any animals in the visual. Ask them what adaptations they all have. (Adaptations for living in trees is common to most rain-forest organisms. They might have specialized limbs for perching and climbing or mouth parts for eating fruits and leaves in the canopy.)

FOCUS CONCEPT

Interdependence of Organisms

Energy relationships within a food web are intricate. Each consumer organism is dependent on another consumer or producer organism for energy. Nutrient cycling in ecosystems often involves unique symbiotic associations.

ASSESSING PRIOR KNOWLEDGE

Review the following concepts.

Respiration: *Chapter 7*
Ask students how respiration relates to the carbon cycle. Review with them that the product of respiration is carbon dioxide, which is released in the atmosphere as a gas. Remind students that carbon is the main element in all compounds that make up living organisms.

Symbiosis: *Chapter 21*
Ask students to describe how atmospheric nitrogen is converted to a useable form for living things. Ask them what kind of relationship between organisms is involved in this conversion.

CHAPTER 22

ECOSYSTEMS AND THE BIOSPHERE

Tropical rain forests are the most biologically diverse of all the biomes on Earth. Shown is the scarlet macaw, Ara macao.

FOCUS CONCEPT: *Interdependence of Organisms*

As you read, note how organisms interact with each other and with their environment to survive in the different kinds of ecosystems on Earth.

BIOLOGY INTERACTIVE TUTOR **Unit 7—*Ecosytem Dynamics***
Topics 1, 3–6

ENERGY TRANSFER

All organisms need energy to carry out essential functions, such as growth, movement, maintenance and repair, and reproduction. In an ecosystem, energy flows from the sun to autotrophs, then to organisms that eat the autotrophs, then to organisms that feed on other organisms. The amount of energy an ecosystem receives and the amount that is transferred from organism to organism have an important effect on the ecosystem's structure.

PRODUCERS

Autotrophs, which include plants and some kinds of protists and bacteria, manufacture their own food. Because autotrophs capture energy and use it to make organic molecules, they are called **producers.** Most producers are photosynthetic, so they use solar energy to power the production of food. However, some autotrophic bacteria do not use sunlight as an energy source. These bacteria carry out **chemosynthesis** (KEE-mo-SIN-thuh-sis), which means they produce carbohydrates by using energy from inorganic molecules. In terrestrial ecosystems, plants are usually the major producers. In aquatic ecosystems, photosynthetic protists and bacteria are usually the major producers.

Measuring Productivity

Gross primary productivity is the rate at which producers in an ecosystem capture energy. Photosynthetic producers use the energy they capture to make sugar. Some of the sugar is used for cellular respiration, some for maintenance and repair, and some to make new organic material through either growth or reproduction. Ecologists refer to the organic material in an ecosystem as **biomass.** Producers add biomass to an ecosystem by making organic molecules.

Only energy stored as biomass is available to other organisms in the ecosystem. Ecologists often measure the rate at which biomass accumulates, and this rate is called **net primary productivity.** Net primary productivity is typically expressed in units of energy per unit area per year (kcal/m^2/y) or in units of mass per unit area per year (g/m^2/y). Net primary productivity equals gross primary productivity minus the rate of respiration in producers.

Figure 22-1 shows that net primary productivity can vary greatly from one ecosystem to another. For example, the average rate of net

SECTION

22-1

OBJECTIVES

▲ Contrast producers with consumers.

● Explain the important role of decomposers in an ecosystem.

■ Contrast a food web with a food chain.

◆ Explain why ecosystems usually contain only a few trophic levels.

internet**connect**

SCILINKS NSTA

TOPIC: Producers and consumers
GO TO: www.scilinks.org
KEYWORD: HM415

SECTION 22-1

SECTION OVERVIEW

Reproduction and Inheritance
Section 22-1 describes how energy in an ecosystem is acquired by organisms for growth and reproduction.

Stability and Homeostasis
Section 22-1 emphasizes that additional loads of carbon dioxide in the atmosphere may alter the stability of the nitrogen/carbon balance in an ecosystem.

Matter, Energy, and Organization
Section 22-1 describes how energy in an ecosystem is organized into trophic levels with corresponding producers and consumers. The acquisition of energy, the storing of biomass, and the process of making energy available at another trophic level are discussed in this section.

ACTIVE READING

Technique: Brainstorming
Before students begin reading, work with the class to create a list of all the ways that two species in an ecosystem can interact. Then develop a list of all the different types of communities/ecosystems that students can recall. Can any of the ecosystems be grouped into larger systems (biomes)? Have students read the chapter to see how scientists have defined interaction and biomes.

VISUAL LINK

Figure 22-1

Ask students why an estuary or wetland is more productive than a freshwater lake. Explain that an estuary receives nutrients drained from larger landmasses and supports organisms that are marine, freshwater, and brackish.

INCLUSION ACTIVITY

Have students draw or cut out pictures to make a poster of a specific group of animals that are herbivores, carnivores, or omnivores, showing the animals in their environment. For example, if herbivores are the topic, the poster could include caribou from the tundra or antelopes from the grasslands. Have students label each animal on the poster and title it by the group of organisms it represents.

TEACHING STRATEGY

Freshwater Producers

Obtain a sample of pond or stream water that has a heavy growth of algae. Divide the sample into Petri dishes, and provide slides, droppers, and coverslips so that students can observe the water under the microscope. Have students count the number of green things they see in the square of the coverslip. They should also count the number of animal-like organisms. Ask students which organism is greatest in number, the algae or the animals, and why. Ask which species is more numerous. (Algae are producers, and animals are consumers. Producers have more energy available to them than consumers, so they are usually in greater number.)

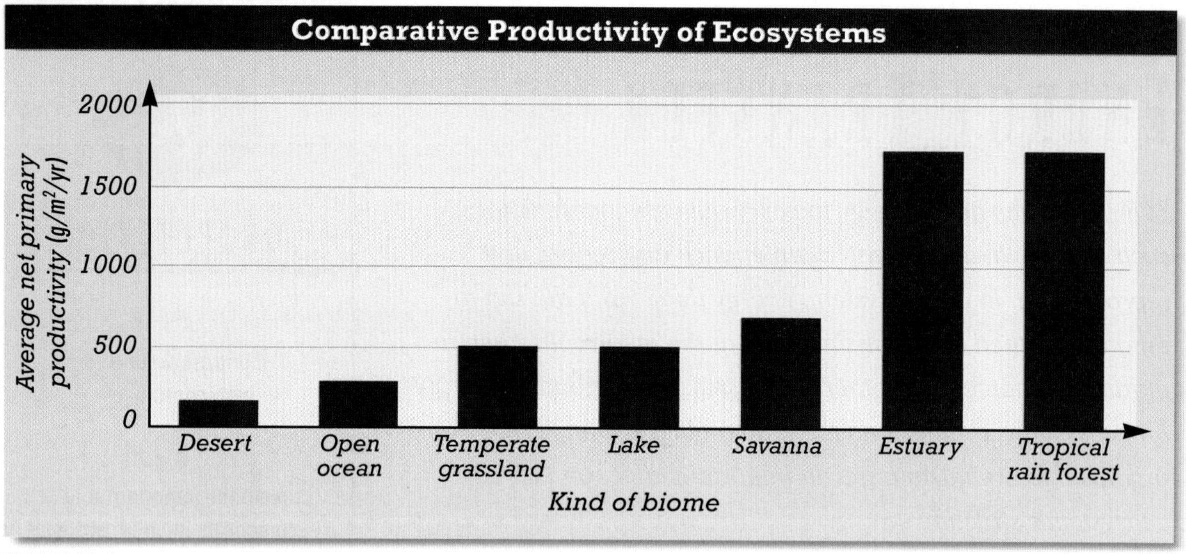

FIGURE 22-1

As the histogram shows, the net primary productivity in a tropical rain forest is very similar to the net primary productivity in an estuary. Temperate grasslands and freshwater lakes are also very similar in productivity.

primary productivity in a tropical rain forest is 25 times greater than the rate in a desert of the same size. Although rain forests occupy only 5 percent of Earth's surface, they account for almost 30 percent of the world's net primary productivity. Variations in three factors—light, temperature, and precipitation—account for most of the variation in productivity among terrestrial ecosystems. An increase in any of these three variables usually leads to an increase in productivity. In aquatic ecosystems, productivity is usually determined by only two factors: light and the availability of nutrients.

CONSUMERS

All animals, most protists, all fungi, and many bacteria are heterotrophs. Unlike autotrophs, heterotrophs cannot manufacture their own food. Instead they get energy by eating other organisms or organic wastes. Ecologically speaking, heterotrophs are **consumers.** They obtain energy by consuming organic molecules made by other organisms. Consumers can be grouped according to the type of food they eat. **Herbivores** eat producers. An antelope that eats grass is a herbivore. So are the minute zooplankton that feed on phytoplankton floating in oceans and lakes. **Carnivores** eat other consumers. Lions, bald eagles, cobras, and praying mantises are examples of carnivores. **Omnivores** eat both producers and consumers. The grizzly bear, whose diet ranges from berries to salmon, is an omnivore.

Word Roots and Origins

omnivore

from the Latin *omnis,* meaning "all," and *-vore,* meaning "one who eats"

Detritivores (dee-TRIE-ti-vorz) are consumers that feed on the "garbage" of an ecosystem, such as organisms that have recently died, fallen leaves and branches, and animal wastes. A vulture is an example of a detritivore. Bacteria and fungi belong to a class of detritivores called decomposers. **Decomposers** cause decay by breaking down the complex molecules in dead tissues and wastes into simpler

TROPHIC LEVELS

FIGURE 22-2

In an ecosystem, all organisms that feed on the same kind of food are in the same trophic level. In this figure, the autotrophs (grass and tree) are on the first trophic level, herbivores (zebra and giraffe) are on the second trophic level, and carnivores (lion and leopard) are on the third trophic level.

molecules. Some of the molecules released during decay are absorbed by the decomposers, and some of them are returned to the soil or water. The action of the decomposers makes the nutrients contained in the dead bodies and wastes of organisms available to autotrophs. Thus, the process of decomposition recycles chemical nutrients.

ENERGY FLOW

Whenever one organism eats another, molecules are metabolized and energy is transferred. As a result, energy flows through an ecosystem, moving from producers to consumers. One way to follow the pattern of energy flow is to group organisms in an ecosystem based on how they obtain energy. An organism's **trophic** (TROH-fik) **level** indicates the organism's position in the sequence of energy transfers, as shown in Figure 22-2. For example, all producers belong to the first trophic level. Herbivores belong to the second trophic level, and the predators of herbivores belong to the third level. Most ecosystems contain only three or four trophic levels.

Food Chains and Food Webs

A **food chain** is a single pathway of feeding relationships among organisms in an ecosystem that results in energy transfer. A food chain may begin with grass, which is a primary producer. The chain may continue with a consumer of grass seeds—a meadow mouse. Next on the chain may be a carnivorous snake, which kills and eats the mouse. A hawk then may eat the snake.

TEACHING STRATEGY
Observing Plant Detritivores

Collect some leaf litter that has been moist for some time. Place the litter in dishes for observation under dissecting microscopes. Have students observe the leaf litter for signs of mold. Ask them how they can identify a mold and what role a mold plays in a forest ecosystem. Ask them what kingdom a mold belongs to. (Mold usually has a fuzzy appearance. Mold is a detritivore in a forest ecosystem, breaking down leaves and other plant debris to organic molecules. Molds belong to the kingdom Fungi.)

TEACHING STRATEGY
Airborne Detritivores

CAUTION: **Treat all growths on Petri dishes as pathogens. Do not allow students to open Petri dishes once they are sealed.** Prepare Petri dishes of nutrient agar. A little added starch may enrich the media. After plates are poured, place open Petri dishes of agar around the room. Leave them in position for a day and then seal them with tape and place them in an incubator for 24 hours. Have students observe the Petri dishes for growth of mold and bacteria. Ask them how the mold can be distinguished from the bacteria and if mold and bacteria can both be detritivores. (The mold will look fuzzy under a dissecting scope and the bacteria will look shiny or crusty. Both mold and bacteria are detritivores called decomposers.)

CRITICAL THINKING

Chemical Energy

Chemosynthetic bacteria often live in habitats so extreme that most other organisms cannot exist there. Photosynthetic producers are often rare in these places. Ask students where they might find these types of bacteria (hot springs, thermal pools, salt flats, or vents of superheated water on the bottom of the ocean floor).

VISUAL LINK

Figure 22-3

Figure 22-3 shows a simplified food web. Have students describe and create a similar diagram for a food web that might exist in their schoolyard or in their own yard. (Grass, trees, shrubs, weeds, and flowers are the primary producers. Songbirds eat seeds, but some also eat insects. The insects are also eaten by lizards, which in turn are caught by cats. Some insects eat other insects, and some feed on plants. Birds may be eaten by cats. The more organisms students think of, the more complex the web will become.)

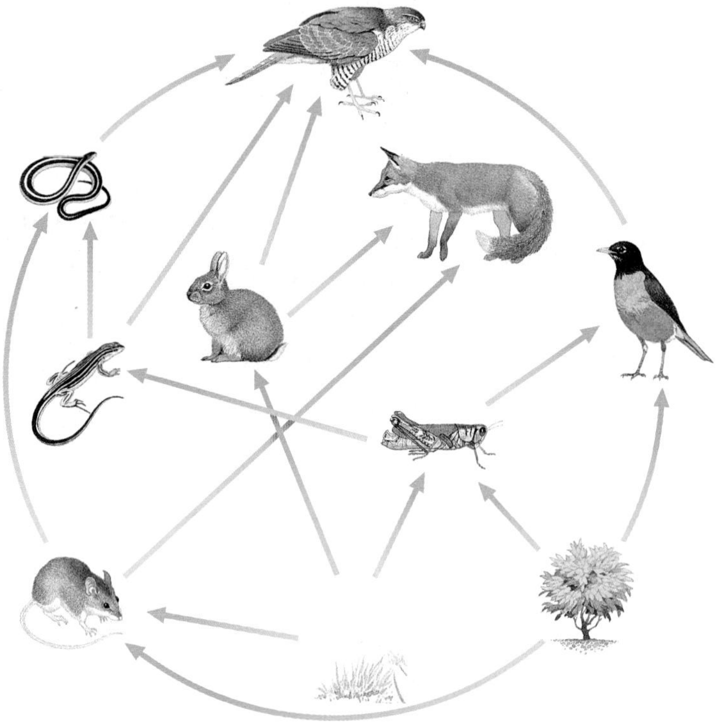

FIGURE 22-3

This food web shows how some organisms in an ecosystem might relate to each other. Because a large carnivore may be at the top of several food chains, it is often more helpful for ecologists to diagram as many feeding relationships as possible in an ecosystem. You can imagine how complicated this food web would be if it were possible to catalog every species present in an ecosystem.

The feeding relationships in an ecosystem are usually too complex to be represented by a single food chain. Many consumers eat more than one type of food. In addition, more than one species of consumer may feed on the same organism. Many food chains interlink, and a diagram of the feeding relationships among all the organisms in an ecosystem would resemble a web. For this reason, the interrelated food chains in an ecosystem are called a **food web.** Figure 22-3 shows a simplified food web.

Quantity of Energy Transfers

Roughly 10 percent of the total energy consumed in one trophic level is incorporated into the organisms in the next level. The ability to maintain a constant body temperature, the ability to move, and a high reproductive rate are functions that require a lot of energy. The kinds of organisms that have those characteristics will transfer less energy to the next trophic level than organisms that do not. For example, grass transfers more energy to a moose than the moose transfers to a wolf. The measured values for energy transfer from one trophic level to the next can range from 20 percent to less than 1 percent. In a recent study of wolves and moose on Isle Royale, Michigan, ecologists found that only 1.3 percent of the total energy consumed by the moose on the island was transferred to the wolves through the wolves' predation on the moose. Figure 22-4 represents the rate at which each trophic level in an ecosystem stores energy as

FIGURE 22-4

This diagram represents energy transfer through four trophic levels. The amount of energy transferred from one level to another can vary, so the structure shown can vary. What is always true, however, is that the top level is much smaller than the lowest level. Hence, energy-transfer diagrams are always roughly pyramidal in shape.

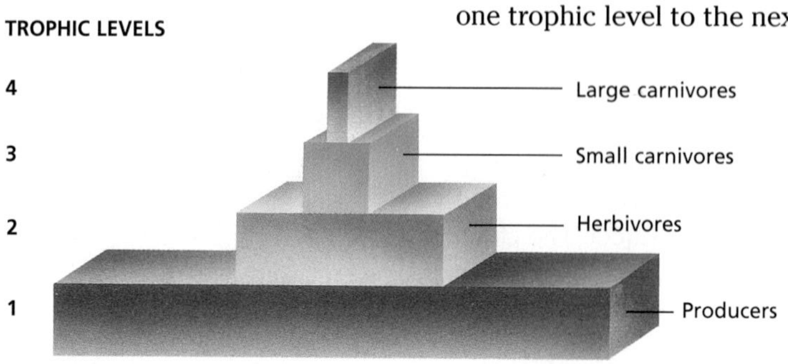

organic material. The pyramid shape of the diagram represents the low percentage of energy transfer from one trophic level to another.

Why is the percentage of energy transfer so low? One reason is that some of the organisms in a trophic level escape being eaten. They eventually die and become food for decomposers, but the energy contained in their bodies does not pass to a higher trophic level. Even when an organism is eaten, some of the molecules in its body will be in a form that the consumer cannot break down and use. For example, a cougar cannot extract energy from the antlers, hooves, and hair of a deer. Also, the energy used by prey for cellular respiration cannot be used by predators to synthesize new biomass. Finally, no transformation or transfer of energy is 100 percent efficient. Every time energy is transformed, such as during the reactions of metabolism, some energy is lost as heat.

To understand this idea better, consider what happens when a deer eats 1,000 kcal of leaves. About 350 kcal are eliminated as urine, dung, and other wastes. Another 480 kcal are lost as heat. Only about 170 kcal are actually stored as organic matter, mostly as fat. Producers use and transfer energy in a similar way. A plant stores only about 1–5 percent of the solar energy that it converts to sugar as organic material. The rest is reflected off the plant, used in its life processes, or lost in the form of heat.

Short Food Chains

The low rate of energy transfer between trophic levels explains why ecosystems rarely contain more than a few trophic levels. Because only about 10 percent of the energy available at one trophic level is transferred to the next trophic level, there is not enough energy in all the organisms at the highest trophic level to support additional levels.

Organisms belonging to the lowest trophic level are usually much more abundant than organisms belonging to the highest level. If you go on a safari in Kenya or Tanzania, for example, you will see about 1,000 zebras, gazelles, wildebeest, and other herbivores for every lion or leopard you see, and there are far more grasses, trees, and shrubs than there are herbivores. Higher trophic levels contain less energy, and, as a consequence, they can support fewer individuals.

SECTION 22-1 REVIEW

1. **Why are autotrophs essential components of an ecosystem?**
2. **What role do decomposers play in an ecosystem? Why is this role important?**
3. **How does a food chain differ from a food web?**
4. **Give two reasons for the low rate of energy transfer within ecosystems.**
5. **Explain why the same area of land can support more herbivores than carnivores.**
6. **CRITICAL THINKING Suppose you remove the rabbits, grasshoppers, birds, and mice (the herbivores) from a food web that also includes grass, mushrooms, lizards, and hawks. Which organisms would be affected and how?**

RETEACHING ACTIVITY

Give students a picture of a mouse, or place a picture on the overhead. Have them construct a food web with as many parts as possible that connect to the mouse. Have students explain their food web to other students.

ANSWERS TO SECTION 22-1 REVIEW

1. Autotrophs are primary producers that capture energy from the sun (and occasionally from inorganic molecules) and store it in complex molecules. They make energy available to all other organisms in an ecosystem.
2. Decomposers cause decay, which releases the nutrients contained in the organism. They return these nutrients to the ecosystem.
3. A food chain shows a single pathway of energy flow in an ecosystem. A food web shows several pathways of energy flow in the ecosystem.
4. Energy transfer is low because many organisms avoid being eaten, contain indigestible molecules, and use energy for their own life processes; and because no energy transfers can be 100 percent efficient.
5. Because the transfer of energy from one trophic level to the next is small, more herbivores than carnivores are required in an area to support the energy requirements of the carnivores.
6. All of the organisms in the web would be affected. The flow of energy would be interrupted by the removal of the herbivores. The lizards and hawks would have to search for food elsewhere. The grass would grow unaffected by predation from herbivores.

SECTION 22-2

SECTION OVERVIEW

Stability and Homeostasis
Section 22-2 describes the predictable patterns of cycling that water, carbon, and nitrogen take through an ecosystem. This section also emphasizes how the stability of biogeochemical cycles may be disrupted by human influence.

Interdependence of Organisms
Section 22-2 discusses the mutualism between bacteria and legumes as an essential process to all living things.

▶ ENGAGE STUDENTS

Obtaining enough water for survival is a challenge for desert animals. Ask students how an animal can get water if there is no running water or ponds and if rain falls only once or twice in several years. (Desert plants have adaptations to keep them from losing water, such as thick waxy leaves and stomata that open at night. Some desert animals can get all their water from eating these plants and do not need to drink at all.)

SECTION

22-2

OBJECTIVES

▲ Define *biogeochemical cycle.*

● Trace the steps of the water cycle.

■ Summarize the major steps in the nitrogen cycle.

◆ Describe the steps of the carbon cycle.

ECOSYSTEM RECYCLING

While energy flows through an ecosystem, water and minerals, such as carbon, nitrogen, calcium, and phosphorus, are recycled and reused. Each substance travels through a ***biogeochemical*** *(BIE-oh-GEE-oh-KEM-i-kuhl)* ***cycle,*** *moving from the abiotic portion of the environment, such as the atmosphere, into living things, and back again.*

THE WATER CYCLE

Water is crucial to life. Cells contain 70 percent to 90 percent water, and water provides the aqueous environment in which most of life's chemical reactions occur. The availability of water is one of the key factors that regulate the productivity of terrestrial ecosystems. However, very little of the available water on Earth is trapped within living things at any given time. Bodies of water such as lakes, rivers, streams, and the oceans contain a substantial percentage of the Earth's water. The atmosphere also contains water—in the form of water vapor. In addition, some water is found below ground. Water in the soil or in underground formations of porous rock is known as

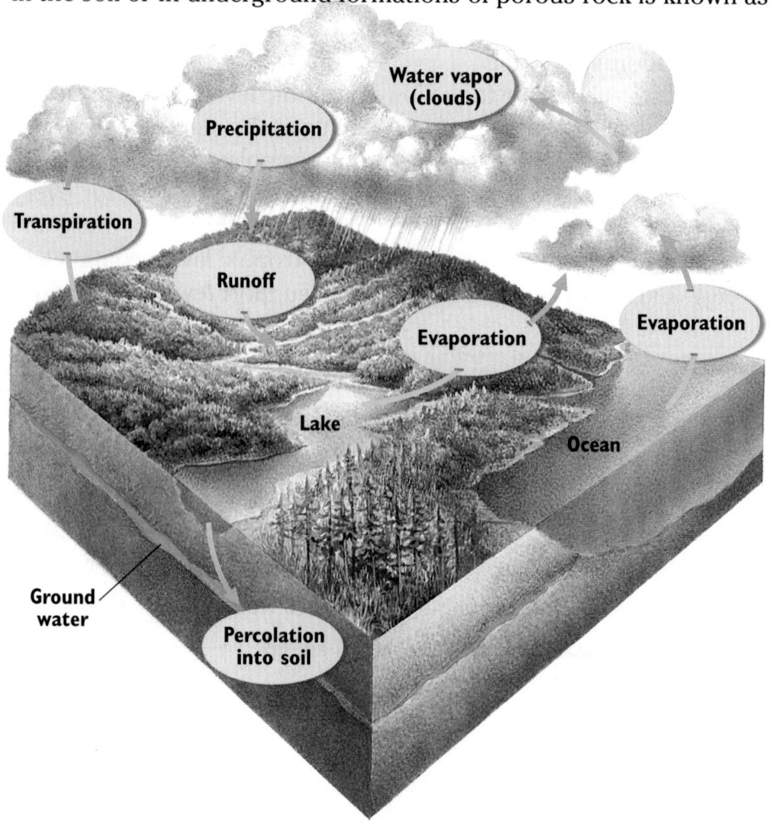

FIGURE 22-5

In the water cycle, water falls to Earth's surface as precipitation. Some water reenters the atmosphere by evaporation and transpiration. Some water runs into streams, lakes, rivers, and oceans. Other water seeps through the soil and into the ground water. Follow the pathways of the water cycle in the figure.

ground water. The movement of water between these various reservoirs, known as the **water cycle,** is illustrated in Figure 22-5.

Three important processes in the water cycle are evaporation, transpiration, and precipitation. Evaporation adds water as vapor to the atmosphere. Heat causes water to evaporate from the oceans and other bodies of water, from the soil, and from the bodies of living things. At least 90 percent of the water that evaporates from terrestrial ecosystems passes through plants in a process called **transpiration.** In transpiration, plants take in water through their roots, and they release water and take in carbon dioxide through the stomata in their leaves. Animals also participate in the water cycle, but their impact is less significant than that of plants. Animals drink water or obtain it from their food. They release this water when they breathe, sweat, or excrete.

Water leaves the atmosphere through precipitation. The amount of water the atmosphere can hold depends on abiotic factors, such as temperature and air pressure. Once the atmosphere becomes saturated with water vapor, precipitation occurs in the form of rain, snow, sleet, hail, or fog.

THE CARBON CYCLE

Together, photosynthesis and cellular respiration form the basis of the **carbon cycle,** which is illustrated in Figure 22-6. During photosynthesis, plants and other autotrophs use carbon dioxide (CO_2), along with water and solar energy, to make carbohydrates. Both autotrophs and heterotrophs use oxygen to break down carbohydrates during

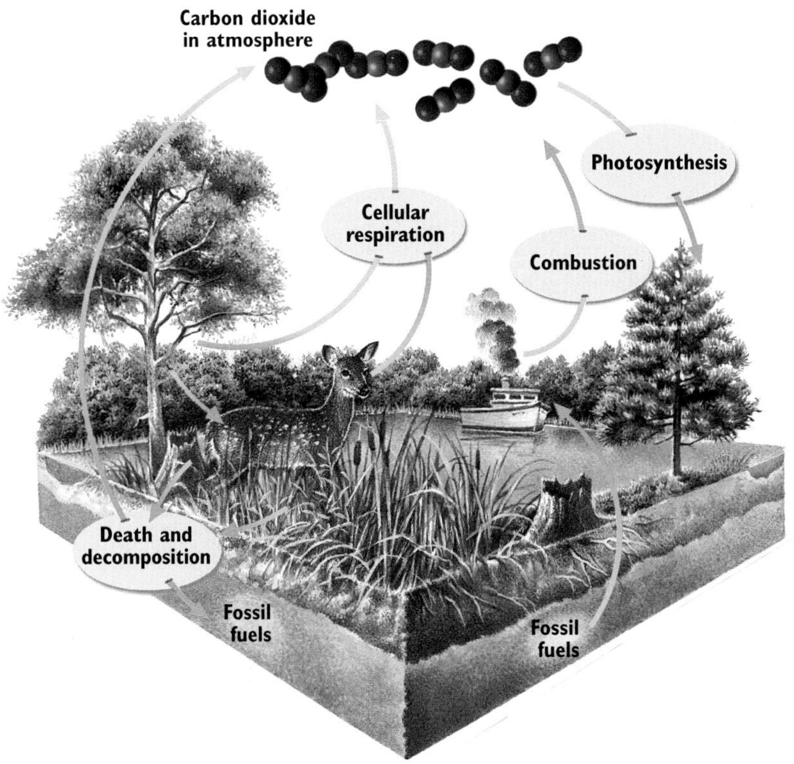

FIGURE 22-6

Carbon exists in the atmosphere in the form of carbon dioxide. Cellular respiration, combustion, and decomposition of organic matter are the three major sources of carbon dioxide. By burning large amounts of fossil fuels, humans are increasing the amount of carbon dioxide in the atmosphere.

Quick Lab

Modeling Ground Water

Materials disposable gloves, lab apron, 3 L plastic bottle (cut in half), small stones (250 mL), dry sod with grass, water, graduated cylinder, 500 mL beaker

Procedure

1. Put on your lab apron and disposable gloves.
2. Invert the top half of the plastic bottle and place it inside the bottom half of the bottle to form a column.
3. Place the stones in the bottom of the inverted top half of the bottle. Place a chunk of dry sod with grass on top of the stones.
4. Pour 250 mL of water over the sod, and observe how the water penetrates the soil and moves through the column.
5. When the water is no longer draining, remove the top half of the column and pour the water from the bottom of the column into a beaker. Measure the volume of liquid in the beaker.

Analysis What is the volume of the water that drained through the sod? How much of the water remained in the soil? Where does the water go when applied to a real lawn or crop? What might the fate of fertilizer or pesticides be that are applied to a lawn or crop?

Quick Lab

Modeling Ground Water

Time Required 25 minutes

Preparation Tips Have students work in teams of three or four. Each team will need a plastic soda bottle prepared ahead of time. Punch several small holes around the neck of the bottle and in the bottle cap. Use a small awl or a large nail and a hammer to punch the holes. Cut the bottle in half. Obtain squares of sod from a nursery, or dig grassy soil from a local area. The sod must be dry.

Answers to Analysis Answers will vary according to the type of soil used. The water will take longer to drain through clay soil, but it will drain quickly through sandy soil. Clay will hold the most water, and sand will hold the least amount of water. Loamy garden soil will hold approximately 100 mL of water. In nature, the water will drain through the soil down to the ground water where dissolved fertilizer or pesticides might contaminate the ground water.

VISUAL LINK

Figure 22-6

Ask students to examine the carbon cycle illustrated in Figure 22-6. Ask them what kind of compounds plants produce that require carbon. (Plants produce carbohydrates in the form of monosaccharides, such as the 6-carbon glucose.)

CRITICAL THINKING

Marine Links to the Carbon Cycle

Have students discuss the role of marine organisms in the carbon cycle. (Coral organisms build reefs of calcium carbonate that contains carbon. The carbon comes from food produced by photosynthetic algae that live inside the coral. Other organisms, such as many mollusks, produce a calcium carbonate shell. When these animals die, their shells become part of the bottom sediment that may eventually become layers of rock. The calcium carbonate eventually decomposes to carbon dioxide and calcium oxide. The carbon dioxide is freed into the atmosphere.)

INCLUSION *ACTIVITY*

Take students on a walk around the campus to point out various examples of the carbon, nitrogen, and water cycle. You can point out that cars emit carbon dioxide and nitrogen oxide into the atmosphere. If the school grounds have grass, point out that fertilizers may be used to make it grow, and show students where the fertilizers might run off. If a stream, pond, or gutter is nearby, have students speculate on where the water goes and how it relates to the water cycle. Have students discuss their observations in class.

VISUAL LINK

Figure 22-7

Point out that the nitrogen cycle depends on bacteria for nitrogen conversion. Ask students how human activities could disrupt this cycle. (The addition of toxic chemicals to soil or the use of pesticides and herbicides may alter the bacterial composition of soil. Construction and deforestation also cause soil erosion that would eliminate soil bacteria.)

cellular respiration. The byproducts of cellular respiration are carbon dioxide and water. Decomposers release carbon dioxide into the atmosphere when they break down organic compounds.

Human Influence on the Carbon Cycle

In the last 150 years, the concentration of carbon dioxide in the atmosphere has risen nearly 30 percent. Almost half of this increase has occurred in the last 40 years. Human activities are responsible for the increase. Our industrial society depends on the energy released by the burning of fossil fuels—coal, oil, and natural gas. Fossil fuels are the remains of organisms that have been transformed by decay, heat, and pressure into energy-rich organic molecules. Burning releases the energy in these molecules, but it also releases carbon dioxide. Carbon dioxide is also added to the atmosphere by the burning of vegetation. Today large areas of tropical rain forest are being burned to create farmland and pasture for cattle. The burning of vegetation adds carbon dioxide to the atmosphere, and the destruction of vegetation removes plants that could have absorbed carbon dioxide from the atmosphere through photosynthesis.

THE NITROGEN CYCLE

All organisms need nitrogen to make proteins and nucleic acids. The complex pathway that nitrogen follows within an ecosystem is called the **nitrogen cycle.** Consider how nitrogen cycles within the terrestrial ecosystem shown in Figure 22-7. Nitrogen gas, N_2, makes up about 78 percent of the atmosphere, so it might seem that nitrogen would be readily available for living things. However, shortages

FIGURE 22-7

This figure shows the cycling of nitrogen within an ecosystem. Bacteria are responsible for many of the steps in the nitrogen cycle, including the conversion of atmospheric nitrogen into ammonia. Nitrogen-fixing bacteria live in the soil or in the roots of plants. Plants take up the ammonia produced by the bacteria. Animals get nitrogen by eating plants or other animals.

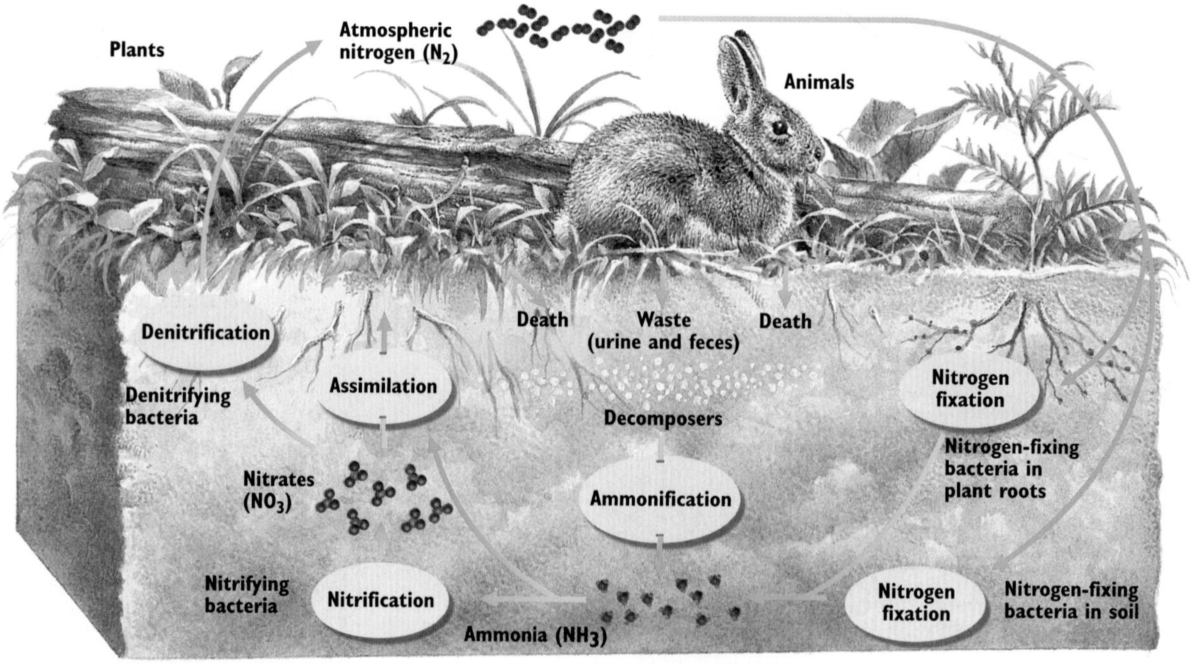

of nitrogen often limit the productivity of plants—and therefore the productivity of ecosystems. Most plants can use nitrogen only in the form of nitrate. The process of converting nitrogen gas to nitrate is called **nitrogen fixation.** Organisms rely on the actions of bacteria that are able to transform nitrogen gas into a usable form. Separate groups of **nitrogen-fixing bacteria** convert nitrogen gas into ammonia, then nitrite, and then nitrate, which plants can use.

Nitrogen-fixing bacteria live in the soil and in the roots of some kinds of plants, such as beans, peas, clover, and alfalfa. These plants have evolved a complex mutualistic relationship with nitrogen-fixing bacteria. The plant provides the bacteria with a home—airtight swellings on its roots—and supplies them with carbohydrates. In exchange, the bacteria produce usable nitrogen for the plant. Excess nitrogen produced by the bacteria is released into the soil.

Recycling Nitrogen

The bodies of dead organisms contain nitrogen, mainly in proteins and nucleic acids. Urine and dung also contain nitrogen. Decomposers break down the corpses and wastes of organisms and release the nitrogen they contain as ammonia. This process is known as **ammonification** (ah-MAHN-i-fi-KAY-shuhn). Through ammonification, nitrogen that would otherwise be lost is reintroduced into the ecosystem.

Bacteria in the soil take up ammonia and oxidize it into nitrites, NO_2^-, and nitrates, NO_3^-. This process, called **nitrification** (NIE-tri-fi-KAY-shuhn), is carried out by bacteria. The erosion of nitrate-rich rocks also releases nitrates into an ecosystem. Plants use nitrates to form amino acids. Nitrogen is returned to the atmosphere through **denitrification.** Denitrification occurs when anaerobic bacteria break down nitrates and release nitrogen gas back into the atmosphere.

Plants can absorb nitrates from the soil, but animals cannot. Animals obtain nitrogen in the same way they obtain energy—by eating plants and other organisms and then digesting the proteins and nucleic acids.

SECTION 22-2 REVIEW

1. Describe the biogeochemical cycle.
2. Where do nitrogen-fixing bacteria live? What crucial function do they perform?
3. Describe the role of decomposers in the nitrogen cycle.
4. How has the burning of fossil fuels affected the carbon cycle?
5. Through what process does most water vapor enter the atmosphere? Explain the process.
6. CRITICAL THINKING Explain two ways that the burning of vegetation affects carbon dioxide levels in the atmosphere. How do you think the removal of vegetation affects oxygen levels in the atmosphere?

RECENT RESEARCH

Interdependence of Cycles

A recent study of a grassland linked the carbon, nitrogen, and water cycles together. It was found that an increased amount of carbon dioxide stimulated the availability of soil nitrogen and its uptake in grasses. Plant transpiration decreased with elevated CO_2 levels, and soil moisture increased. Changing one cycle may alter or influence other cycles.

ANSWERS TO SECTION 22-2 REVIEW

1. A biogeochemical cycle is the movement of water, carbon, nitrogen, or other mineral from the abiotic portion of the environment into living things and back again.
2. Nitrogen-fixing bacteria live in the soil and in plant roots. They transform unusable nitrogen gas into ammonia, which plants can use.
3. Decomposers break down the nitrogen-containing molecules in dead organisms and wastes and release nitrogen, making it available to other organisms.
4. The burning of fossil fuels releases carbon dioxide and thereby increases the amount of CO_2 in the atmosphere.
5. About 90 percent of the water that reenters the atmosphere is from the transpiration of plants. Plants take up water through their roots and release it to the atmosphere through their stomata.
6. The burning of vegetation contributes carbon dioxide to the atmosphere through the process of combustion. Also, the removal of vegetation by burning (or by any other method) eliminates the plants that absorb carbon dioxide and produce oxygen during photosynthesis.

SECTION 22-3

SECTION OVERVIEW

Matter, Energy, and Organization
Section 22-3 describes the organization of each terrestrial biome and stresses that the boundaries of biomes are designated on the basis of the similarity of their components, such as climate and typical organisms.

Evolution
Section 22-3 discusses selective pressures in each biome that have resulted in evolutionary adaptations among the typical organisms in the biome.

▶ ENGAGE STUDENTS

Have students examine Figure 22-8 and read the caption. Ask them to speculate about why Antarctica has no terrestrial biomes. Show a picture of seals on the snow and ice of Antarctica. Ask students if these seals belong to a terrestrial biome. (No, they belong to an aquatic ecosystem, as do most life-forms on Antarctica.) Discuss how the biomes change in relationship to the equator. Have students locate and identify the biome where they live.

SECTION

22-3

OBJECTIVES

▲ Describe the differences between tundra and taiga biomes.

● Contrast temperate grassland with savanna.

■ Describe three water-conservation adaptations of desert organisms.

◆ Compare tropical rain forests with temperate deciduous forests.

TERRESTRIAL ECOSYSTEMS

Biomes *(BIE-ohmz) are very large terrestrial ecosystems that contain a number of smaller but related ecosystems within them. A certain biome may exist in more than one location on Earth, but similar biomes have similar climates and tend to have inhabitants with similar adaptations.*

THE SEVEN MAJOR BIOMES

Biomes are distinguished by the presence of characteristic plants and animals, but they are commonly identified by their dominant plant life. For example, hardwood trees, such as beeches and maples, are the dominant form of plant life in the deciduous forest biome. Most ecologists recognize seven major biomes, shown on the map in Figure 22-8, and several minor biomes. In this section, you will learn about the characteristics of the seven major biomes: tundra, taiga, temperate deciduous forest, temperate grassland, desert, savanna, and tropical rain forest.

Because abiotic factors change gradually over a landscape, biomes seldom have distinct boundaries. As climate varies over the Earth's surface, for example, deserts tend to gradually change into grasslands, tundra into taiga, and so on. Figure 22-8 shows how the

FIGURE 22-8
The seven biomes cover most of the Earth's land surface. Antarctica is not shown because it has no biomes.

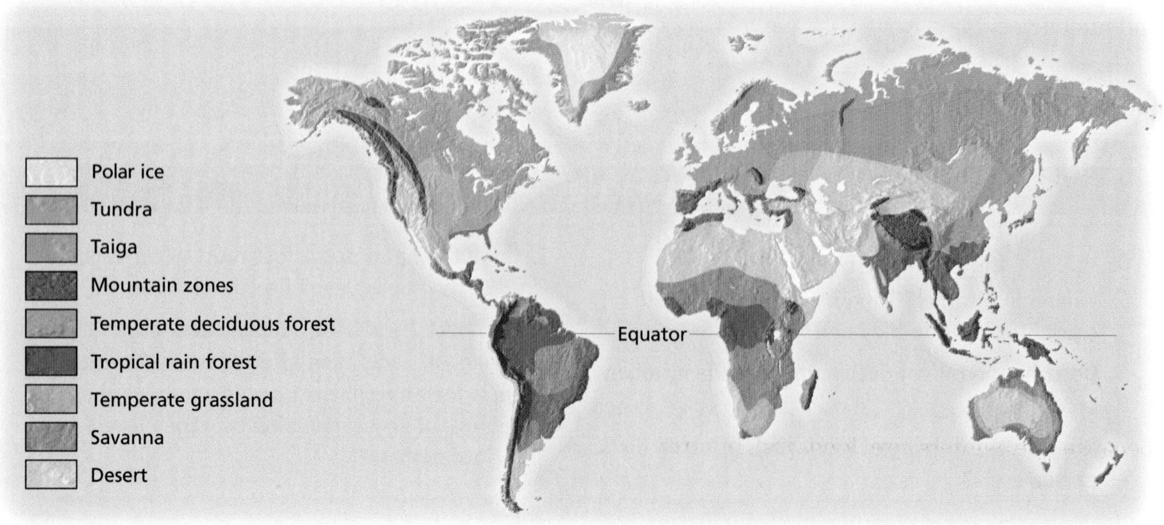

TABLE 22-1 Characteristics of the Major Biomes

Biome	Average yearly temperature range	Average yearly precipitation	Soil	Vegetation
Tundra	−26°C to 12°C	<25 cm	moist, thin topsoil over permafrost; nutrient-poor; slightly acidic	mosses, lichens, dwarf woody plants
Taiga	−10°C to 14°C	35–75 cm	low in nutrients; highly acidic	needle-leaved evergreen trees
Temperate deciduous forest	6°C to 28°C	75–125 cm	moist; moderate nutrient levels	broad-leaved trees and shrubs
Temperate grassland	0°C to 25°C	25–75 cm	deep layer of topsoil; very rich in nutrients	dense, tall grasses in moist areas; short clumped grasses in drier areas
Desert	7°C to 38°C	<25 cm	dry, often sandy; nutrient-poor	succulent plants and scattered grasses
Savanna	16°C to 34°C	75–150 cm	dry, thin topsoil; porous, low in nutrients	tall grasses, scattered trees
Tropical rain forest	20°C to 34°C	200–400 cm	moist, thin topsoil; low in nutrients	broad-leaved evergreen trees and shrubs

seven major biomes are distributed over the Earth. The figure also shows the areas that cannot be classified into one of the seven major biomes. Because climate varies with elevation, mountains contain a variety of communities and do not belong to any one biome. Table 22-1 describes the major biomes and lists their average annual temperature and rainfall.

TUNDRA

The **tundra** (TUHN-druh) is a cold and largely treeless biome that forms a continuous belt across northern North America, Europe, and Asia. It is the largest and northernmost biome, covering about one-fifth of the world's land surface. **Permafrost,** a permanently frozen layer of soil under the surface, characterizes the tundra. Even the surface soil above the permafrost remains frozen for all but about eight weeks of the year. Figure 22-9 shows some of the plants of the tundra.

Trees do not usually grow in the tundra because the winters are long and bitterly cold and because permafrost prevents their roots from penetrating far into the soil. The

FIGURE 22-9
This photograph shows that the tundra is usually bleak and uniform in appearance, although there may be bright patches of color in the summer season.

SECTION 22-3

VISUAL LINK
Table 22-1
Have students examine Table 22-1. Ask them which biome has the greatest temperature range. (The tundra has the greatest temperature range because of the extremely low temperatures that sometimes occur and the occasionally balmy days in summer.) Ask which biome is similar to the tundra in amount of precipitation. (The desert biome is similar to the tundra in amount of precipitation. Plants and animals of both biomes must have adaptations to conserve water.)

CRITICAL THINKING
Biomass
Variations in light, temperature, and precipitation can affect the amount of biomass produced. Ask students why the biomass produced in an arctic region is much less than in a temperate or tropic region, even though the amount of precipitation may be high. (Permafrost prevents the development of deep roots, precipitation is usually frozen, temperatures are lower, and winter seasons are long and dark.)

RECENT RESEARCH
Adaptations
The desert adaptations of brown hares (*Lepus capenses*) of the Negev desert in Israel have been compared to the adaptations of European hares of the same species. The European hares were from a temperate forest region in France. It was found that the desert hares required only 75 percent of the food consumed by the temperate hares, and their metabolic rate was 61 percent that of the temperate hares. The desert hares used half the amount of water per day that the temperate hares used. The desert hares could drink a salt solution of 6 percent, but the temperate hares could tolerate only a 2.5 percent salt solution. The urine of desert hares was twice as concentrated as that of the temperate hares. The desert hare is an example of how animals of different biomes meet their nutritional requirements through physiological adaptations that efficiently exploit their resources.

TEACHING STRATEGY
Observing Tundra Plants
Obtain living specimens of mosses, grasses, sedges, and lichens. Have students observe these plants with a hand lens. Explain that these are typical of plants that survive in the tundra. Ask why most tundra plants are small. (They must survive very harsh conditions of cold and wind. So much energy is expended in surviving that little is left to produce biomass.) Ask students if these plants have anything in common with desert plants. (Like desert plants, tundra plants must prevent water loss, because most precipitation is frozen.)

FIGURE 22-10
Organisms of the taiga are adapted for dry, cold conditions and a reduced availability of food in the winter season. The conifers have needles, which are leaf adaptations to conserve water.

tundra receives little precipitation and has a very short growing season of about two months. Cold temperatures retard decay, so the soil is often low in nutrients. For these reasons, tundra plants are usually small and grow slowly. Grasses, sedges, and mosses are common.

Animals that inhabit the tundra include caribou, musk oxen, snowy owls, arctic foxes, lemmings, and snowshoe hares. In the brief summer, the frozen soil thaws, creating a patchwork of ponds and bogs. Swarms of mosquitoes and black flies appear, and ducks, geese, and predatory birds arrive in great numbers to feed.

TAIGA

South of the tundra is the **taiga** (TIE-guh), a forested biome dominated by cone-bearing evergreen trees, such as pines, firs, hemlock, and spruce. Taiga stretches across large areas of northern Europe, Asia, and North America. During the long winter, snow covers and insulates the ground, protecting tree roots against freezing.

Plants living in the taiga are adapted for long and cold winters, short summers, and nutrient-poor soil. On the cone-bearing evergreen trees, called conifers, the waxy needles remain all winter long. The shape of the needle is a leaf adaptation that reduces water loss. The stomata of conifers are partially enclosed in the needle, which helps the tree conserve water. Typical mammals of this biome include moose, bears, wolves, and lynx. Some animals stay in the forest year-round, while others migrate to warmer climates in the fall and return in the spring. Many species hibernate six to eight months of the year. Figure 22-10 shows a representative area of taiga.

Word Roots and Origins

taiga

from the Russian word *taiga*, meaning "a transitional plant community"

internet**connect**
SCILINKS NSTA
TOPIC: Temperate forests
GO TO: www.scilinks.org
KEYWORD: HM426

TEMPERATE DECIDUOUS FORESTS

Temperate deciduous forests are characterized by trees that lose all of their leaves in the fall. Deciduous forests stretch across eastern North America, much of Europe, and parts of Asia and the Southern Hemisphere. These regions have pronounced seasons, with precipitation evenly distributed throughout the year. Compared with taiga, temperate deciduous forests have warmer winters and longer summers, and they receive more precipitation. Deciduous trees have broad, thin leaves with a large surface area that permits maximum

FIGURE 22-11

A temperate deciduous forest is characterized by trees that lose their leaves in the winter. This adaptation to conserve water also creates an energy loss for the tree. The tree will never recover the energy that went into its leaf formation, but the decaying leaves on the forest floor will add nutrients to the soil. Temperate deciduous forests are inhabited by a variety of animals, such as cardinals and white-tailed deer.

light absorption. Familiar deciduous trees include the birch, beech, maple, oak, hickory, sycamore, elm, willow, and cottonwood. White-tailed deer, foxes, raccoons, and squirrels are typical mammals of the temperate deciduous forests. Large areas of temperate deciduous forest in the United States, Europe, and Asia have been cut for timber or cleared to make way for farms, towns, and cities. Figure 22-11 shows a stand of trees in a temperate deciduous forest.

TEMPERATE GRASSLANDS

Temperate grasslands are, as the name suggests, dominated by grasses. Temperate grasslands usually form in the interior of continents, at about the same latitude as temperate deciduous forests. However, rainfall patterns make these areas too dry to support trees. This biome once covered large areas of North America, Asia, Europe, Australia, and South America. Grasslands are known by different names in different parts of the world: *prairie* in North America, *steppes* in Asia, *pampas* in South America, and *veldt* in southern Africa.

Temperate grasslands have rich, fertile soil. In areas that have remained relatively undisturbed by humans, grasslands support large herds of grazing mammals, such as the bison shown in Figure 22-12. Grass can survive continuous grazing by animals and occasional fires that sweep across the area because the actively growing part of the plant is at or below the ground, rather than at the tip of the stem. Because grasslands have such rich soil, much of the world's temperate grassland has been transformed into farmland for growing crops such as wheat and corn. Only fragments of undisturbed prairie remain in the United States.

FIGURE 22-12

Temperate grasslands once covered a large portion of the United States and supported huge herds of herbivores, such as the bison shown below. Today, several conservation organizations are actively attempting to preserve this valuable, endangered biome.

VISUAL LINK
Figure 22-11

Ask students to name the biome where an oak tree might be found (temperate forest). Ask students to name other organisms that might be at home in the photograph in Figure 22-11. Have them describe and discuss any adaptations these organisms have that help them survive in the temperate forest.

GIFTED *ACTIVITY*

Have students create a video documentary about a biome that they live in or near. The video should be 8 to 10 minutes long and narrated. It should show typical plants and animals of the biome with descriptions of their adaptations. The video must also include descriptions of environmental problems associated with the biome, such as the loss of habitat in the biome from clearing of land to build a housing subdivision. Videos should be shown to the class.

RETEACHING *ACTIVITY*

Ask groups of students to construct a biome-bag display. Have each group obtain a large, resealable plastic bag. They should fill the bag with as many different plants, objects, or pictures that illustrate a specific biome as possible. For example, a bag representing a deciduous forest might contain leaves, twigs, logs, ferns, and mushrooms or other fungi.

CRITICAL THINKING

Savannas

The savannas of Africa have many large herds of herbivores. Ask students what herds of herbivores can do to obtain food when the dry season comes and the vegetation has been eaten. (Migration is an adaptation for survival in a semi-arid region. Many African animals, such as wildebeests, can migrate to a place where food and water are more abundant.)

TEACHING STRATEGY

Observing Desert Plants

Provide students with several potted cactus plants. Have them wear heavy gloves as a protection from the cactus spines. Ask students what the spines on a cactus do and what helps them to prevent water loss. Provide students with slices of a cactus to observe under a dissecting scope. Ask students to describe what they see and to speculate on the purposes of each structure. (Cactuses are thick-bodied plants. They have green bodies that are actually stems. They are covered with spines, which are leaf adaptations. Spines can protect by shading, preventing plants from being eaten, and preventing drying from wind exposure. A thick outer covering helps cactuses prevent water loss. The inside of the cactus looks fibrous or spongy and is very moist because it must keep its water inside.)

FIGURE 22-13

A desert biome may appear lifeless at first glance. But on closer inspection, the desert reveals many living things. All of the organisms of the desert biome are adapted to dry, often hot conditions and are adapted for conserving energy. The saguaro cactus shown stores water from the infrequent rains of the desert.

DESERTS

Deserts are areas that receive an average of less than 25 cm (9.9 in.) of rainfall per year. Large parts of North Africa, central Australia, southwestern North America, and eastern Asia are desert. Contrary to popular belief, deserts are not hot all the time. So-called cold deserts, such as the Great Basin in the western United States and the Gobi in eastern Asia, are hot in summer but cold in winter. Even in hot deserts, temperatures may fall by as much as 30°C (54°F) at night because the dry air is a poor insulator, allowing the heat that builds up during the day to escape.

Desert vegetation is often sparse and consists mainly of plants that have adapted to the dry climate. The leaves of the creosote bush of the United States and Mexico, for example, have a waxy coating that reduces evaporation. To limit water loss through transpiration, some desert plants open their stomata only at night. The saguaro cactus, which is found in southern Arizona and southeastern California, as well as in northern Mexico, has an expandable body, an adaptation that allows it to store water. A single saguaro can hold about 1,000 kg (2,200 lb) of water. The saguaro, as seen in Figure 22-13, has evolved sharp protective spines, a leaf adaptation that protects the plant from thirsty herbivores.

Desert animals must conserve water just as the plants do. Many animals avoid the heat of the day by hiding in small spots of shade or by burrowing into the ground. Others, such as kit foxes and some kinds of lizards and snakes, are active only at night, when loss of water to evaporation is low.

FIGURE 22-14

The savanna biome is an area rich in wildlife, where great herds of large herbivores exist. Because of the large number of herbivores, this area also supports many large carnivores. The climate of a savanna experiences only two seasons—wet and dry.

SAVANNAS

Savannas (suh-VAN-uhz) are tropical or subtropical grasslands with scattered trees and shrubs. The savannas of Africa are the best known, but this biome also occurs in South America and Australia. Savannas receive more rainfall than deserts, but less than tropical rain forests. Alternating wet and dry seasons characterize savannas. Like temperate grasslands, savannas support large numbers of herbivores, such as zebras, wildebeest, giraffes, and gazelles, as shown in Figure 22-14. Large carnivores, such as lions, leopards, and cheetahs, feed on these herbivores.

Because most of the rain falls during the wet season, the plants and animals of the savanna must be able to deal with prolonged periods of drought. Some trees of the savanna shed their leaves during the dry season to conserve water, and the above-ground parts of grasses often die during the dry season and regenerate after a period of rain.

TROPICAL RAIN FORESTS

Tropical rain forests are characterized by tall trees like those shown in Figure 22-15. Tropical rain forests are found near the equator in Asia, Africa, South America, and Central America. The stable, year-round growing season and abundant rainfall combine to make the tropical rain forest the most productive biome.

Competition for light is intense in the tropical rain forest. Most of the plants are trees, and some have evolved to grow as tall as 50 to 60 m (164 to 197 ft). The treetops form a continuous layer, called the **canopy,** that shades the forest floor. Though you may think of the tropical rain forest as an impenetrable jungle, much of the forest floor is relatively free of vegetation because so little sunlight reaches the ground. The very dense growth known as jungle is found along riverbanks and in disrupted areas where sunlight can reach the forest floor. Small plants called **epiphytes,** as shown in Figure 22-15a, which would not get enough light on the forest floor, often live on the branches of tall trees. These plants, which include mosses and orchids, use other organisms as support, but they are not parasitic because they make their own food.

Tropical rain forests have the highest species richness of all the biomes. One hectare of tropical rain forest (about the size of two football fields) may contain as many as 300 species of trees. An area of temperate deciduous forest of the same size, by contrast, would probably contain fewer than 10 species of trees. Animal life is also very diverse in the tropical rain forest. The sloth, pictured in Figure 22-15b, is a rain forest mammal. Colorful birds, such as parrots and toucans, many kinds of monkeys, and a variety of snakes and lizards are among the vertebrate inhabitants of this biome. The variety of insects is particularly diverse in tropical rain forests. There may be more than 8 million species of tree-dwelling beetles in the tropical rain forest biome alone. Overall, tropical rain forests probably contain about one-fifth of the world's known species.

FIGURE 22-15

Animal life is as diverse as plant life in the rain forest, as reflected by the epiphyte, shown in (a), and the three-toed sloth, a tree dweller of the rain forest, shown in (b).

(a)

(b)

SECTION 22-3 REVIEW

1. Why would it be unusual to find a tree in the tundra?
2. What two similarities are shared by temperate grassland and savanna?
3. Describe two adaptations of the saguaro cactus for water conservation.
4. List at least one animal that lives in each of these biomes: desert, temperate grassland, and tundra.
5. Why do many animals in the tropical rain forest live in trees?
6. **CRITICAL THINKING** Explain the benefits deciduous trees gain from shedding their leaves in the fall. Describe some possible disadvantages of shedding leaves.

SECTION 22-3

RETEACHING *ACTIVITY*

Prepare sets of 4 to 6 pictures of each terrestrial biome. Prepare an answer key that names the biome represented by each set. Have students form groups and provide a different set of pictures to each group. Have them name the biome that each set illustrates. Let them check their accuracy using the key, and then have them trade pictures with another group.

ANSWERS TO SECTION 22-3 REVIEW

1. Permanent ice below the surface of the soil prevents roots of trees from penetrating into the soil. The climate is dry and cold, and the soil is deficient in nutrients.
2. Both biomes are dominated by grasses and support large herds of grazing mammals.
3. A saguaro cactus has an expandable body that can store large amounts of water. Its spines protect its moist tissues from herbivores.
4. Animals of the desert include lizards, snakes, and kangaroo rats. An animal of the temperate grassland is the bison. Tundra animals include caribou, snowy owls, arctic foxes, snowshoe hares, mosquitoes, and geese.
5. The canopy formed by the vegetation of the tall trees shades the forest floor. Most of the production occurs in the treetops, and animals live there to be near the food supply.
6. Transpiration cannot occur if the leaves are absent. Deciduous trees conserve water by shedding their leaves. The energy and materials that went into growing and maintaining the leaves are lost to the tree when they are shed.

BACKGROUND

Rain forests display a high degree of vertical stratification in plant and animal life. The tree canopy extends to heights of 30 to 50 meters. Many of the larger animals rarely come down to ground level. They have developed numerous means of moving about to escape predators and obtain food, including climbing, gliding, leaping, and swinging. The middle layer of the rain forest is filled with small trees and epiphytic plants—such as ferns, orchids, and bromeliads—and the animals that feed on them and each other. The rain-forest floor is virtually bare, except for a thin layer of fallen leaves and rotting vegetation. Underneath the soil surface can be found burrowing animals and the microorganisms that decompose the organic matter that accumulates.

READING FOR MEANING

Bates reminds us that plankton consists of both plant and animal components. The leaves of the trees correspond to the phytoplankton of the sea, and the insects correspond to the zooplankton.

READ FURTHER

Comparing the seemingly dissimilar organisms in their different biomes illustrates one of the unifying themes in biology. That is, organisms interact with their environment. Those interactions form the web of life that is the ecosystem. The two biomes have similarities. Both environments are relatively stable. They both have vertical gradients in light, temperature, and air or water movement. Therefore, understanding one environment might help in understanding the other.

The Forest and the Sea

The following account is from *The Forest and the Sea,* by Marston Bates.

When a botanist visited the South American forest station where Marston Bates was working, Bates and the visitor climbed up onto a platform high in the forest canopy. Mosquitoes began buzzing about them.

In the course of our mosquito studies we had found that each different species had its characteristic flight habits. Some kinds were found only near the ground, others only high in the trees; some that were most common high in the trees in the morning or afternoon would come down near the ground during the midday hours, showing a sort of daily vertical migration.

While I was explaining this to my friend, it struck me, that this is just the way animals act in the sea. Most life is near the top, because that is where the sunlight strikes and everything below depends on this surface. Life in both the forest and the sea is distributed in horizontal layers.

The analogy, once thought of, was easily developed. The vocabulary for life in the sea could be transferred to the forest. In the treetops we were in what marine students call the pelagic zone—the zone of active photosynthesis, where sunlight provides the energy to keep the whole complicated biological community going. Below, we had been in the benthos, the bottom zone, where organisms live entirely on second-hand materials that drift down from above—on fallen leaves, on fallen fruits, on roots and logs. Only a few special kinds of green plants were able to grow in the rather dim light that reached the forest floor.

My mosquitoes acted in some ways like the microscopic floating life of the sea, the plankton. Each species among the plankton organisms has a characteristic vertical distribution: some living only near the surface, others only at considerable depths, and so forth. The plankton organisms in general show a daily vertical migration, coming to the surface at night and sinking during the day: a migration to which my mosquitoes were only a feeble counterpart. But insects on land are only partially analogous with the plankton of the sea. A major portion of the plankton consists of microscopic plants, busy using the energy of the sun and the dissolved carbon dioxide of the water to build up starch and thus provide the basis for all the rest of the life of the sea. These microscopic plants would correspond not to the insects of the forest, but to the leaves of the trees.

The forest insects would correspond only to the animal component of the plankton: to the copepods and tiny shrimp and larval fish which live directly on the plants or on each other at the very beginning of the endless chain of who eats whom in the biological community.

Reading for Meaning

Bates compared living forms of the forest with those of the sea. Why does he say that mosquitoes and plankton are "only partly analogous"?

Read Further

In *The Forest and the Sea,* Bates presents observations based on his field studies in two very different biomes. What advantages might be found in comparing seemingly dissimilar organisms?

Aquatic Ecosystems

As terrestrial organisms, we tend to focus on the other land-dwelling organisms we see around us. But life may have arisen in the sea and only recently (in geological terms) colonized the land. Water covers about three-fourths of Earth and is home to a variety of organisms. In this section we will look at some of the inhabitants of aquatic ecosystems.

SECTION

22-4

OBJECTIVES

▲

Contrast the aphotic and photic zones in the ocean.

●

Describe the differences between the neritic zone and the oceanic zone.

■

Explain how organisms near deep-sea vents obtain energy.

◆

Contrast eutrophic lakes with oligotrophic lakes.

OCEAN ZONES

The ocean covers about 70 percent of Earth's surface and has an average depth of 3.7 km (2.3 mi). The deepest parts of the ocean are about 11 km (6.8 mi) deep. The water contains about 3 percent salt, mostly sodium chloride, a factor that profoundly affects the biology of the organisms that live there. Another important variable affecting marine organisms is the availability of light. Because water absorbs light, sunlight penetrates only the upper few hundred meters of the ocean. The **photic** (FOH-tik) **zone** is the part of the ocean that receives sunlight. The rest of the ocean falls within the **aphotic** (AY-FOH-tik) **zone,** the cold and dark depths where sunlight cannot penetrate. Photosynthesis cannot occur in the aphotic zone because of the lack of sunlight.

Ecologists recognize three zones extending out from land, as illustrated in Figure 22-16. Along ocean shores, the tides produce a rhythmic rise and fall of the water level in an area called the **intertidal zone.** Farther out is the **neritic** (nee-RI-tik) **zone,** which extends

FIGURE 22-16

This diagram shows the various zones of the ocean. The neritic zone generally extends from the intertidal zone to the point where water depth is about 180 m (590 ft). The photic zone varies in depth, depending on how far light penetrates into the water.

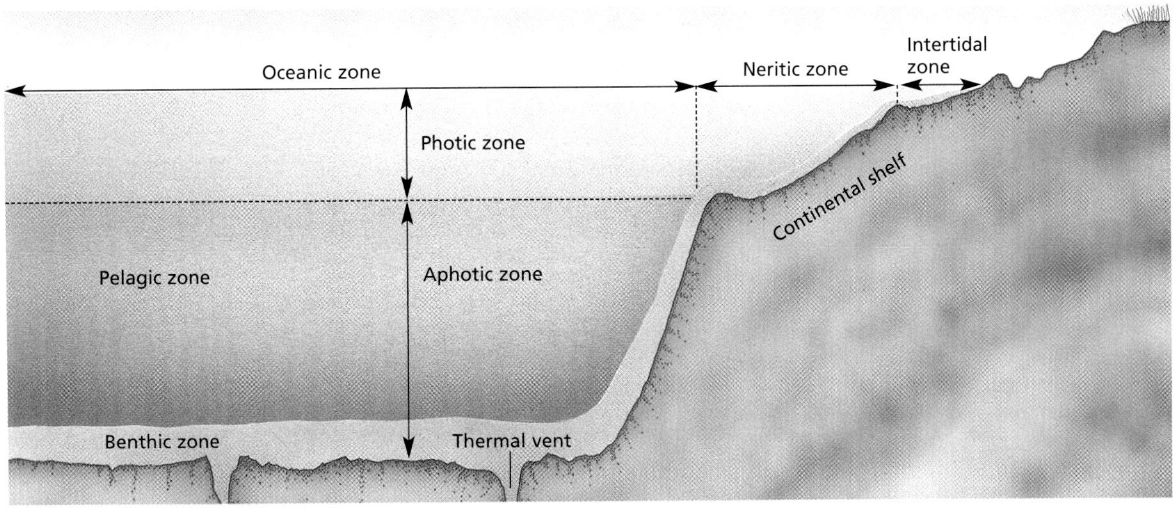

SECTION 22-4

SECTION OVERVIEW

Matter, Energy, and Organization

Section 22-4 describes the organization of aquatic biomes and the further zonation of the oceanic biome. This section also discusses how energy is acquired by the chemosynthetic organisms of deep-sea vents.

Evolution

Section 22-4 emphasizes evolutionary adaptations that enable organisms to survive in their environment.

▶ ENGAGE STUDENTS

Collect as many specimens or photographs of marine organisms, such as coral, shells, and starfish, as possible. Show them one at a time to the class, and ask students what part or zone of the ocean each organism is adapted to. (For example, a piece of coral would be found in the photic zone and neritic zone. A starfish would be found in the intertidal zone.)

VISUAL LINK

Figure 22-16

Have students look at the diagram in Figure 22-16. Ask them why most organisms live in the neritic zone. (It is not very deep and is in the photic zone. It also receives more nutrients from runoff from land.)

VISUAL LINK

Figure 22-17

Ask students to speculate about what advantages marine organisms might have in the harsh environment of a rocky shoreline. Explain to them that the intertidal zone provides sheltering crevices and surfaces for organisms to attach to. The intertidal zone is usually rocky and provides many sessile organisms a solid foothold. The waves also bring nutrients to the intertidal organisms.

INCLUSION ACTIVITY

Have students draw an oceanic profile that illustrates the different zones of the ocean. Each zone should be labeled and should include a picture of an animal that is representative of the zone.

CRITICAL THINKING

Productivity of the Ocean

The oceans do not have flowering or many large plants for photosynthesis, yet they account for half the photosynthetic productivity on Earth. Have students explain this. (The oceans are very large, providing a huge surface area for phytoplankton to grow in the photic zone.)

VISUAL LINK

Figure 22-18

Have students examine Figure 22-18. Ask them to name the adaptations coral reefs have evolved for living in the photic zone. (Coral and photosynthetic algae have adapted to form a symbiotic relationship. The algae require the light of the photic zone to make food for the coral animal. The coral adds its skeletal structure to the reef, which keeps the algae close to the surface of the water.)

over the continental shelf. The water in the neritic zone is relatively shallow (no more than a few hundred feet deep). Beyond the continental shelf is the **oceanic zone,** which is the deep water of the open sea. The neritic and oceanic zones are further divided. The open ocean is known as the **pelagic** (pi-LA-jik) **zone,** while the ocean bottom is known as the **benthic zone.**

FIGURE 22-17

The intertidal zone shown here is similar to intertidal zones all over the planet. Some are artificially created by the construction of rocky protective bulkheads along former stretches of beach. An open beach is part of the intertidal zone, but it is usually distinguished from the rocky intertidal zones and not as rich in species as rocky intertidal zones.

The Intertidal Zone

Organisms in this zone are adapted to periodic exposure to air during low tide. Crabs avoid dehydration by burrowing into the sand or mud. Clams, mussels, and oysters retreat into their shells at low tide. Organisms living in the intertidal zone must be able to withstand the force of crashing waves. Sea anemones cling to rocks with a muscular disk, and sea stars use tube feet to adhere to surfaces. Figure 22-17 shows some of the organisms living in the intertidal zone of the California coast.

The Neritic Zone

The neritic zone is the most productive zone in the ocean, supporting more species and numbers of organisms than any other zone. The water throughout most of the neritic zone is shallow enough for photosynthesis to occur. Strong currents called upwellings carry nutrients from the ocean bottom and mix them with nutrients contained in runoff from land. These waters are rich in **plankton,** communities of small organisms that drift with the ocean currents. Plankton is consumed by many larger organisms. Numerous fishes, squid, sea turtles, and other animals also live in these waters.

Coral reefs form in the neritic zone of tropical areas. Like tropical rain forests, coral reefs are very productive and rich in species. A coral reef is built over a long time by coral animals. They construct external skeletons of a hard chemical compound called calcium carbonate. As the animals grow and die, the skeletons accumulate over time to form a reef, as shown in Figure 22-18, that is home to many species of fishes, crustaceans, mollusks, and other animals. Some species of coral have a mutualistic relationship with photosynthetic protists, from which they receive food.

internet**connect**

SCILINKS NSTA

TOPIC: Aquatic ecosystems
GO TO: www.scilinks.org
KEYWORD: HM432

FIGURE 22-18

Coral reefs, such as the one pictured here, are among the most diverse ecosystems on Earth. Coral reefs are built from the skeletal remains of tiny marine animals and are continually growing just below the surface of warm sunlit seas.

The Oceanic Zone

The oceanic zone contains fewer species than the neritic zone. Even in photic areas, nutrient levels are too low to support as much life. Although the productivity per square meter of open ocean is very low because the ocean covers such a vast area, the total productivity of the oceanic zone is high. About half of the photosynthesis that occurs on Earth takes place in the oceanic zone. The producers of the upper parts of the oceanic zone are protists and bacteria in the plankton. Animals living in the oceanic zone include fishes, mammals such as whales, and many invertebrates.

In the aphotic zone, animals feed primarily on sinking plankton and dead organisms. Organisms living deep in the ocean must cope with near-freezing temperature and crushing pressure. Deep-sea organisms, like the squid shown in Figure 22-19, have slow metabolic rates and reduced skeletal systems. Fishes in these depths have large jaws and teeth and expandable stomachs that can accommodate the rare prey that they can catch.

In the 1970s, scientists found diverse communities living near volcanic vents 2,500 m (8,200 ft) below the surface. These vents release water that is rich in minerals and often exceeds 750°C. Chemosynthetic bacteria that use energy contained in hydrogen sulfide (H_2S) are the producers for this ecosystem. An assortment of unique clams, crabs, and worms feed on these bacteria. Clamworms living near thermal vents have lost their digestive system over evolutionary time and receive all of their food directly from chemosynthetic bacteria living in their bodies.

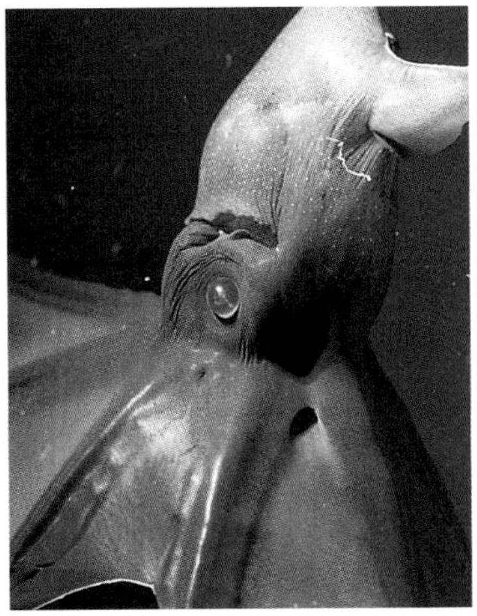

FIGURE 22-19

Organisms in the deep sea have many adaptations to their environment. The squid *Vampyroteuthis* sp., in the photograph is adapted to eating a large quantity of food at once because prey is hard to find.

Estuaries

An **estuary** (ES-tyoo-eree) occurs where freshwater rivers and streams flow into the sea. Examples of estuary communities include bays, mud flats, and salt marshes. The shallow water ensures plenty of light, and rivers deposit large amounts of mineral nutrients. However, the interaction between fresh water and salt water causes great variation in temperature and salinity. In addition, like the intertidal zone, much of the ground surface of an estuary is exposed during low tide. Inhabitants of estuaries are adapted for frequent change. For example, some kinds of mangrove trees have special glands on their leaves that eliminate excess salt water taken up by the roots. Soft-shell clams lie buried in mud with only their long siphons protruding above the surface. The siphon filters plankton from the salt water at high tide and detects predators at low tide, contracting whenever it senses danger. Figure 22-20 is an example of an estuary.

FIGURE 22-20

Estuaries are almost as species rich as tropical rain forests. The plant life is often uniform, but the bird and aquatic diversity is enormous. Estuaries are the ocean's nurseries. Many marine organisms hatch and spend their juvenile life stages in the estuary. The dense vegetation protects them from pounding wave action and gives them cover from predators. Estuaries are vital to marine animals that are used as food by people all over the world. Examples of food animals that begin life in estuaries are shrimp, mullet, redfish, and anchovies.

VISUAL LINK

Figure 22-19

This rare picture is from a videotape that was recorded by a deep-sea research vessel at a depth of 600 m (2000 ft). Tell students that this squid, *Vampyroteuthis infernalis,* is adapted for the deep, dark sea with large eyes and sensitive vision. It has barbed arms equipped with photophores (bioluminescent organs) and webbing between its arms that prevents prey from escaping. It has been observed at depths of 15,840 ft (5,000 m).

TEACHING STRATEGY

Watersheds and the Water Cycle

Obtain a U.S.G.S. map of a local area. Divide the map into parts that are one page in size, and photocopy it. Working in groups, have students color the waterways (creeks, lakes, rivers, streams, and reservoirs). Have students use a ruler and the scale on the map to calculate the total length of all the drainage shown on their portion of the map. Have them highlight the main body of water (river or stream) that all the small tributaries empty into. Explain to students that a watershed is an area that drains into one major tributary or river. Ask them where all the water comes from that goes into the watershed. Ask students where the water will eventually go once it reaches a major waterway or tributary. Ask them to describe how the water will return to the watershed. (The water comes from precipitation and runoff. It will eventually go to the ocean and will evaporate, cool, and fall as rain again.) Have groups put the map parts together again and share findings with the rest of the class.

SECTION 22-4

QUICK FACTS

☑ Nearly all the fish and shellfish Americans eat begin their lives in estuarine spawning grounds.

☑ Zooplankton and fish that live in the oceanic zone are either transparent or blue in color, making them invisible in their aquatic biome.

ANSWERS TO SECTION 22-4 REVIEW

1. The photic zone is the upper few hundred meters of the ocean that receive sunlight. The aphotic zone, below the photic zone, receives no sunlight.
2. The neritic zone receives nutrients from the ocean floor and from the land.
3. Chemosynthetic bacteria are producers for these ecosystems. They capture the energy contained within hydrogen sulfide gas.
4. Oligotrophic lakes are clear and low in organic matter. Eutrophic lakes are murky but rich in organic material.
5. Organisms living in fast-moving water tend to have adaptations to keep them from being swept away, but they probably won't be adapted to the lakelike conditions created by a dam. Thus, they probably would disappear.
6. Nutrient levels are much higher in estuaries than they are in the open ocean. Estuaries also contain many predators that feed on the plentiful juvenile organisms.

Word Roots and Origins

oligotrophic

from the Greek *oligos,* meaning "small," and the Greek *trophikos,* meaning "food"

FRESHWATER ZONES

Low levels of dissolved salts characterize freshwater ecosystems. The salt content of fresh water is about 0.005 percent. Examples of freshwater ecosystems include lakes, ponds, clear mountain streams, and slow sediment-rich rivers.

Lakes and Ponds

Ecologists divide lakes and ponds into two categories. **Eutrophic** (yoo-TRAH-fik) lakes are rich in organic matter and vegetation, making the waters relatively murky. The giant water lily pictured in Figure 22-21 is growing in a shallow eutrophic pond. **Oligotrophic** (AH-li-goh-TRAH-fik) **lakes** contain little organic matter. The water is much clearer, and the bottom is usually sandy or rocky. Fishes inhabit both eutrophic and oligotrophic lakes. Freshwater lakes and ponds also support mammals, such as the otter and muskrat, and birds, such as ducks and loons.

FIGURE 22-21

The Amazon water lily, *Victoria amazonica,* is adapted to life in shallow eutrophic ponds. As organic matter accumulates, the lake or pond will eventually fill in and disappear.

Rivers and Streams

A river is a body of water that flows down a gradient, or slope, toward its mouth. Water flows swiftly down steep gradients, and organisms are adapted to withstand powerful currents. For example, the larvae of caddis flies and the nymphs of mayflies cling to the rocky bottom, while brook trout and other fishes have evolved the strength to face upstream while feeding on drifting invertebrates. Slow-moving rivers and their backwaters are richer in nutrients and therefore support a greater diversity of life. Rooted plants and the fishes that feed on them are adapted to the weaker currents of slow-moving rivers.

SECTION 22-4 REVIEW

1. Distinguish between the photic and aphotic zones.
2. What are two sources of nutrients in the neritic zone?
3. What role do chemosynthetic bacteria play in deep-sea-vent ecosystems?
4. What is the main difference between oligotrophic and eutrophic lakes?
5. What might happen to organisms living in a fast-moving river if a dam is built on the river?
6. **CRITICAL THINKING** Estuaries serve as breeding sites and nurseries for thousands of species of marine animals. What characteristics of estuaries make them advantageous places for marine organisms to reproduce? What are some possible disadvantages?

CHAPTER 22 REVIEW

SUMMARY/VOCABULARY

22-1

- Autotrophic organisms are the primary producers. They manufacture carbohydrates using energy from the sun. Consumers obtain energy by eating other organisms.
- Gross primary productivity is the rate at which autotrophs capture energy. Net primary productivity is the rate at which primary producers make new biomass.
- Decomposers feed on dead bodies and wastes and release the nutrients they contain.
- A single pathway of energy transfer is called a food chain. A network showing all paths of energy transfer is called a food web.
- Food chains are short because a large amount of energy is used at each trophic level. There are fewer individuals and less biomass at higher trophic levels.

Vocabulary

biomass (415)
carnivore (416)
chemosynthesis (415)
consumer (416)
decomposer (416)
detritivore (416)
food chain (417)
food web (418)
gross primary productivity (415)
herbivore (416)
net primary productivity (415)
omnivore (416)
producer (415)
trophic level (417)

22-2

- Materials such as carbon, nitrogen, and water are cycled within ecosystems.
- The three key processes in the water cycle are evaporation, transpiration, and precipitation.
- Few organisms can use nitrogen directly from the environment. Nitrogen-fixing bacteria in the soil and in plant roots transform nitrogen gas into ammonia, which plants can use.
- Photosynthesis and cellular respiration are the two main processes of the carbon cycle. Cellular respiration adds carbon dioxide to the atmosphere, while photosynthesis removes it.
- By burning large amounts of fossil fuels and vegetation, humans are disrupting the carbon cycle. Many scientists think the rising level of carbon dioxide in the atmosphere will lead to global warming.

Vocabulary

ammonification (423)
biogeochemical cycle (420)
carbon cycle (421)
denitrification (423)
ground water (421)
nitrification (423)
nitrogen cycle (422)
nitrogen fixation (423)
nitrogen-fixing bacteria (423)
transpiration (421)
water cycle (421)

22-3

- On land, there are seven major types of ecosystems, known as biomes.
- Tundra is a cold biome characterized by permafrost under the surface.
- Taiga is warmer than tundra and receives more precipitation. It is dominated by conifer forests.
- The trees in temperate deciduous forests shed all of their leaves in the fall.
- Temperate grasslands occur in areas with cold winters and hot summers. They are dominated by grasses and herds of grazing animals.
- Deserts receive less than 25 cm (9.9 in.) of precipitation per year. Their inhabitants have adaptations for water conservation.
- Savannas are tropical grasslands with alternating wet and dry seasons. They are dominated by herds of grazing animals.
- Tropical rain forests receive abundant rainfall and have a year-round growing season. They contain more species than any other biome.

Vocabulary

biome (424)
canopy (429)
desert (428)
epiphyte (429)
permafrost (425)
savanna (428)
taiga (426)
temperate deciduous forest (426)
temperate grassland (427)
tropical rain forest (429)
tundra (425)

CHAPTER 22 REVIEW ANSWERS

REVIEW

1. A producer captures solar or chemical energy to make carbohydrates. A decomposer feeds on the complex molecules contained in dead organisms and wastes.
2. Gross primary productivity is the rate at which producers capture energy. Net primary productivity is the rate at which producers make biomass.
3. Tundra is a cold, treeless biome where the soil is permanently frozen below the surface. Taiga is a slightly warmer and moister biome dominated by conifers.
4. Nitrogen fixation is the production of ammonia from atmospheric nitrogen. Ammonification is the release of ammonia due to the decomposition of dead organisms or organic wastes.
5. Biomass is the amount of organic material in an ecosystem. A biome is a widespread general type of terrestrial ecosystem.

6. a **7.** b
8. b **9.** a
10. d **11.** c
12. c **13.** a
14. c **15.** b

16. A herbivore eats producers. Examples include deer, mice, grasshoppers, and zooplankton. A carnivore feeds on herbivores or other carnivores. Examples include cougars, snakes, sharks, and hawks. Omnivores feed on producers and on other consumers. Humans and raccoons are omnivores.
17. Although all organisms would be affected by the removal of any one of the organisms in the food web, removal of the primary producer would destroy the food web. The green plant is an autotroph and the foundation of the food web.
18. A food chain shows only one sequence of energy transfer. Most organisms feed on more than one type of food, so the actual pathway of energy flow branches and resembles a web.

19. Detritivores include all organisms that feed on wastes and dead organisms. Decomposers are detritivores that cause decay.
20. Energy flows through an ecosystem, from producers to consumers, and some energy is continually being lost from the ecosystem. Nutrients cycle within an ecosystem and can be used again.
21. Nitrogen-fixing bacteria receive a place to live and sugars from their host plant. The plant receives usable nitrogen compounds from the bacteria.
22. Plants keep their stomata open to absorb carbon dioxide, which they need for photosynthesis.
23. Tropical rain forests are often burned, and that adds carbon dioxide to the atmosphere. The removal of large amounts of vegetation reduces the absorption of carbon dioxide from the air.
24. Answers will vary. People create artificial wetland ecosystems to treat wastewater. Wetland plants take up nitrogen and phosphorous, and both plants and soils absorb heavy metals. Cities use such artificial wetlands to treat municipal wastewater.

Critical Thinking

1. These plants contain nitrogen-fixing bacteria in their roots. The bacteria release any excess nitrogen they fix into the soil.
2. At each trophic level, energy is dissipated as heat, a form of energy organisms cannot use. Thus, energy is continually lost to the ecosystem.
3. Photosynthetic plankton in the ocean account for about 50 percent of the photosynthesis on Earth. If their population is reduced, carbon dioxide levels will likely rise, intensifying the greenhouse effect.
4. The selective pressures at great depths include the absence of light, scarcity of prey, slippery prey

22-4

- The photic zone in the ocean receives light, while the aphotic zone does not. There are three major zones in the ocean—the intertidal zone, the neritic zone, and the oceanic zone.
- In the intertidal zone, organisms must be able to tolerate drying and pounding by waves.
- The neritic zone receives nutrients from the bottom of the ocean and from land. It is the ocean's richest zone in terms of the number of species and individuals.
- Production in the oceanic zone is limited by a shortage of nutrients.
- Estuaries are very productive areas where rivers and streams flow into the sea.
- Oligotrophic lakes are clear and lacking in nutrients. Eutrophic lakes are rich in nutrients and are often murky.
- Rivers and streams are characterized by a gradient in elevation.

Vocabulary

aphotic zone (431)
benthic zone (432)
estuary (433)
eutrophic (434)
intertidal zone (431)
neritic zone (431)
oceanic zone (432)
oligotrophic lake (434)
pelagic zone (432)
photic zone (431)
plankton (432)

Review

Vocabulary

Explain the difference between the terms in each of the following pairs.

1. producer, decomposer
2. gross primary productivity, net primary productivity
3. tundra, taiga
4. nitrogen fixation, ammonification
5. biomass, biome

Multiple Choice

6. The major producers found in aquatic ecosystems are (a) photosynthetic protists (b) chemosynthetic bacteria (c) aquatic plants (d) heterotrophic protists.
7. Decomposers benefit an ecosystem by (a) manufacturing energy (b) returning nutrients to the soil (c) controlling the population (d) removing toxic substances.
8. Which of the following organisms from the African savanna would you expect to be the least abundant? (a) grass (b) lion (c) zebra (d) grasshopper
9. Which of the following is not true of the nitrogen cycle?
   (a) Plants absorb nitrogen gas directly from the atmosphere.
   (b) Bacteria convert nitrogen gas into ammonia.
   (c) Plants absorb ammonia from the soil.
   (d) Animals obtain nitrogen by eating other organisms.
10. The combustion of fossil fuels has increased atmospheric levels of (a) ammonia (b) nitrogen (c) CFCs (d) carbon dioxide.
11. One reason trees are unusual in the tundra is that (a) large herbivores eat them (b) there is not enough rainfall to support them (c) permafrost prevents root growth (d) grass and shrubs crowd them out.
12. Which of the following is not true of tropical rain forests?
   (a) They are found near the equator.
   (b) They have the highest species richness of any biome.
   (c) They show wide seasonal changes in temperature.
   (d) They are rapidly disappearing.
13. The neritic zone (a) receives nutrients from land (b) receives little sunlight (c) is exposed to the air by low tide (d) supports very few species.
14. Estuaries are among the most productive areas on Earth because (a) they are home to vast coniferous forests (b) they support great herds of herbivores (c) they have shallow, nutrient-laden water (d) they support many large predators.
15. Eutrophic lakes are (a) clear (b) murky (c) swift (d) small.

Short Answer

16. Explain the difference between a herbivore, a carnivore, and an omnivore. Give an example of each type of organism.

17. Examine the diagram of the food web below. How do you think the food web would be affected if the grass and the shrub were both eliminated? Explain your answer.

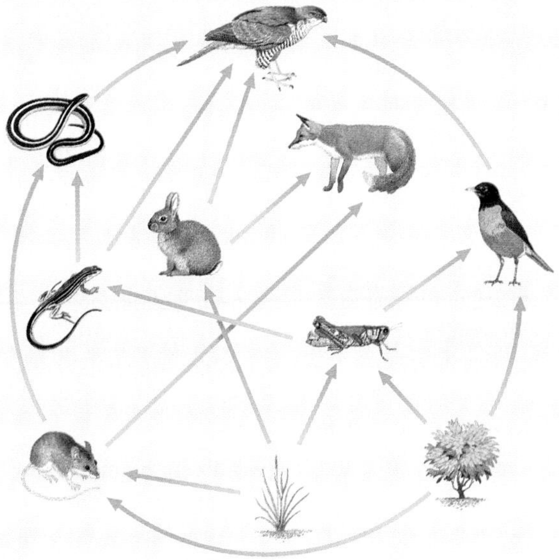

18. Why is a food web a more complete picture of the feeding relationships in an ecosystem than a food chain is?

19. Explain the difference between the terms *decomposer* and *detritivore.*

20. How does the transfer of energy in an ecosystem differ from the transfer of nutrients?

21. Describe the benefits nitrogen-fixing bacteria gain from the plants they inhabit. What does the plant receive from the bacteria?

22. Transpiration accounts for most of the water loss in plants. Explain why plants cannot shut their stomata for long periods of time.

23. Give two reasons why the destruction of tropical rain forests contributes to an increase in carbon dioxide levels in the atmosphere.

24. Unit 7—*Ecosystem Dynamics*

Write a report summarizing how artifical ecosystems are used to eliminate pollutants from wastewater. Find out what kinds of factories or plants use artificial ecosystems.

Critical Thinking

1. Explain why farmers often grow alfalfa, clover, or beans in a field after they have grown corn.

2. Nitrogen, water, and carbon are recycled and reused within an ecosystem, but energy is not. Explain why energy cannot be recycled.

3. Thinning of the ozone layer may lead to reduced populations of photosynthetic plankton in the ocean. Explain how this would affect the carbon cycle.

4. This rare species of squid has several adaptations for living in very deep water. Explain some of the selective pressures that exist at great depths.

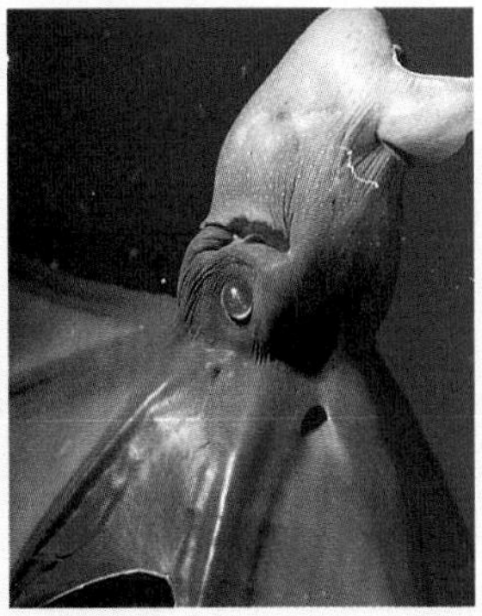

Extension

1. Study an area near your home or school. Make a list of all the consumers, producers, and decomposers you observe. Draw a food web using the organisms you have observed.

2. Read "When Nature Goes Nuts" in *National Wildlife,* October–November 1999, on page 48. Choose three different animals mentioned in the article, and explain how each animal depends on the oak tree. What does the U.S. Forest Service blame for the decline of the oak?

3. Use library resources or an on-line database to research eutrophication of lakes. What is eutrophication? What causes eutrophication, and how does it affect the lake's inhabitants?

that is hard to see, and extremely high pressure.

Extension

1. Because the area is small, most students will find that grass or weeds are the producers. Herbivores might be caterpillars, snails, or other leaf-eating insects. Ants should be identified as detritivores. Some students will assume there are bacteria as decomposers. Some carnivores may be small lizards, spiders, or millipedes.

2. Gray squirrels store acorns to ensure their supply of food over the winter. Acorn woodpeckers store acorns in trees to supplement their diets when insects are in short supply. White-tailed deer eat oak foliage as well as the acorns. (At least 150 species of animals depend on acorns as food.) The U.S. Forest Service points to logging, deer browsing, wildfire suppression, gypsy moth defoliation, and drought as reasons for the decline and mortality of mature oaks and for regrowth failures.

3. Eutrophication is essentially the overfertilization of a lake. Untreated sewage and runoff from farmland carry high concentrations of nutrients into the lake. In response, the population of algae booms. The algae use up the nutrients and die. Bacteria that decompose the algae use up large amounts of oxygen, and many of the lake's other inhabitants suffocate.

CHAPTER 22 INVESTIGATION

TIME REQUIRED

Parts A, B, and C—one 50-minute lab period
Part D—5 minutes daily for several weeks and one 50-minute lab period

SAFETY PRECAUTIONS

Caution students to handle glass jars carefully to avoid cuts. Also remind students that they must handle living organisms gently.

QUICK REFERENCES

Lab Preparation Notes are found on page 413B.

Holt BioSources provides a Teaching Resources CD-ROM that contains "Using Gowin's Vee in the Lab" and "Scoring Rubrics."

PROCEDURAL TIPS

1. After completion of Parts A, B, and C, allow one week for plants and algae to grow before beginning Part D. For Part D, students will need a few minutes each day for a period of a few weeks to make observations.
2. Encourage students to think about the ecological roles that organisms have, such as primary producers, herbivores, and carnivores.
3. Recording the number of organisms may be tricky in some cases, so estimates may be required. Discuss the sources of error in estimates versus direct counts.
4. The ecosystems may be kept for the rest of the school year and then given to students to take home over the vacation.

ANSWERS TO BACKGROUND

1. Organisms are limited by abiotic factors, such as the type of soil, rock, or water that they live on or in, the extremes of temperature, and the amount and frequency of precipitation. Organisms can also

CHAPTER 22 INVESTIGATION

Constructing and Comparing Ecosystems

OBJECTIVES

- Observe the interaction of organisms in a closed ecosystem.
- Compare this ecosystem with others observed in nature.

PROCESS SKILLS

- observing
- recognizing relationships
- hypothesizing

MATERIALS

- large glass jar with lid for each ecosystem
- pond water or dechlorinated tap water
- gravel, rocks, and soil
- graph paper
- $8\frac{1}{2} \times 11$ in. acetate sheets
- several colored pens for overhead transparencies

Ecosystem 1

- pinch of grass seeds
- pinch of clover seeds
- 10 mung-bean seeds
- 3 earthworms
- 4–6 isopods
- 6 mealworms
- 6 crickets

Ecosystem 2

- strands of *Anacharis, Fontinallis,* and foxtail
- duckweed
- *Chlamydomonas* culture
- black ram's horn snail
- 4 guppies or platys

Ecosystem 3

- small, clear glass container for water environment
- pinch of grass seeds
- pinch of clover seeds
- pond snails
- 2 *Anacharis* strands
- 10 *Daphnia*
- 3 *Fontinallis* strands
- *Chlamydomonas*
- 4 guppies or platys

Background

1. How are living things affected by nonliving things in the environment?
2. Relate the theme of interdependence among organisms to ecosystems.
3. How do different types of organisms interact?

Experiment Setup

1. In this investigation, you will observe organisms in one of three different ecosystems. Look at the organisms listed in the materials list and choose the ecosystem you would like to observe. Which organisms might be the most numerous? Which organisms might decrease in number? Hypothesize how the organisms will interact.
2. Form a group with classmates who have all chosen the same ecosystem. As a group, prepare the environment in a jar with a chosen substrate. Go to Part A, Part B, or Part C for set-up instructions for your ecosystem.
3. Put lids on the containers, and let the ecosystem remain undisturbed in indirect sunlight for a week. Go to Part D after one week.

PART A Setting Up Ecosystem 1

1. Place the soil in the bottom of the jar about 4 or 5 in. deep. Clean soil off the inside of the jar.
2. Place rocks in and on top of the soil in a natural-looking arrangement.
3. Moisten the soil carefully. Do not saturate the soil.
4. Plant the seeds according to the package instructions.
5. Clean up your materials and wash your hands before leaving the lab.

PART B **Setting Up Ecosystem 2**

6. Place about 2 in. of clean gravel in the bottom of the jar.
7. Gently fill the jar with pond water or dechlorinated tap water until the jar is about three-fourths full.
8. Add the plants and algae to your ecosystem.
9. Clean up your materials and wash your hands before leaving the lab.

PART C **Setting Up Ecosystem 3**

10. Place a small, clear glass container, such as a mayonnaise jar, inside the larger jar and against one side.
11. Carefully place soil in the larger jar, filling around the smaller jar so that you can see through both jars on one side. Do not place any soil on the side where the two jars touch. Clean any soil from the sides of the jar that may obstruct your view.
12. Place some rocks on the soil, and moisten the soil carefully.
13. Place 1 in. of gravel inside the smaller jar and gently fill with pond water or dechlorinated tap water to the level of the soil.
14. Place the algae and the aquatic plants in the water. Plant the grass seeds and clover seeds in the soil in separate areas of the jar.
15. Clean up your materials and wash your hands before leaving the lab.

PART D **Observing Your Ecosystem**

16. Place the chosen animals in the jar and loosely replace the lid.
17. Observe the jar for a few minutes daily.
18. Make a chart in your lab report to record the original number of each species in your ecosystem. Record daily any changes you observe.
19. Make a graph for each species in your chart, and plot the number of organisms as a function of time. Place a clear acetate sheet over each graph. Using a pen used for overhead transparencies, trace each graph onto the acetate sheet. Use a different colored pen and a different acetate sheet for each organism.
20. Compare the acetate sheets of two organisms that you hypothesized would interact—a predator and its prey, for example. Hold one sheet on top of the other and analyze both graphs. Record the results in your lab report.
21. Clean up your materials and wash your hands before leaving the lab.

Analysis and Conclusions

1. What happened to the organisms in your ecosystem? How did their population sizes change?
2. What are some possible causes of the change in the populations you observed?
3. Construct a food chain for the ecosystem you observed. Which organisms were producers? Which organisms were consumers?
4. What could you learn if you set up more than one jar in an identical manner?
5. How does your ecosystem resemble a natural ecosystem? How does it differ?
6. How did your observation compare with your hypothesis? If the results differed from what you expected, explain what might have caused the difference.
7. Look at your graphs, and determine what kind of relationship exists between predator and prey populations.
8. How would you modify the ecosystem if you were to repeat this investigation?

Further Inquiry

Develop an experiment to study the effects of certain abiotic factors—including temperature, light, and moisture—on the organisms in the ecosystem you constructed.

CHAPTER 22 INVESTIGATION

be killed by extreme weather conditions, such as storms, or by major changes in climate.
2. Interdependence among organisms implies feeding relationships and allotment of other resources. In the laboratory ecosystem, resources for each organism may be limited. Predator/prey interactions are readily observable. The same kinds of relationships and interdependence exists in a real ecosystem, only without the boundaries of the laboratory. In a real ecosystem, there are usually more organisms, and the food web is usually far more complex.
3. Predation, competition, mutualism, and commensalism are the four main ways in which organisms interact.

ANSWERS TO ANALYSIS AND CONCLUSIONS

1. Answers will vary.
2. Answers will vary but may include that plant and algae populations decreased after animals started eating them.
3. Answers will vary.
4. It would help to determine whether the observed changes were random or were the result of the interactions among the ecosystem's inhabitants.
5. Real ecosystems and laboratory ecosystems contain a diversity of organisms from several trophic levels, have both living and nonliving components, and depend on the sun for energy. The laboratory ecosystem is less diverse than a real ecosystem.
6. Answers will vary.
7. Students should note that as the number of predators increases, the number of prey decreases; the trend reverses as the number of prey reaches a minimum.
8. Answers will vary. Students may choose to set up a desert ecosystem or use different organisms.

CHAPTER 23
ENVIRONMENTAL SCIENCE

	TOPICS	TEACHING RESOURCES	LABS, CLASSWORK, AND HOMEWORK
BLOCK 1 45 minutes	**Introducing the Chapter, p. 440**	**ATE** Focus Concept, p. 440 **ATE** Assessing Prior Knowledge, p. 440 **ATE** Understanding the Visual, p. 440	■ Supplemental Reading Guide, *Silent Spring*
	23-1 Humans and the Environment, p. 441 A Global Connection, p. 441 Human Influences on Global Systems, p. 443 Future Population Growth, p. 446	**ATE** Section Overview, p. 441 **ATE** Critical Thinking, p. 445 **ATE** Visual Link Figure 23-1, p. 441 Figure 23-2, p. 442 74, 75, 76	**PE** **Quick Lab,** p. 444, Calculating CO_2 Production ■ Occupational Applications—Wildlife Biologist ★ Study Guide, Section 23-1 **PE** Section Review, p. 446
BLOCK 2 45 minutes	**23-2 The Biodiversity Crisis, p. 448** Biodiversity, p. 448 Measuring Earth's Biodiversity, p. 449 The Importance of Biodiversity, p. 451	**ATE** Section Overview, p. 448 **ATE** Critical Thinking, p. 449	**Lab C27,** Studying an Algal Bloom **Lab C28,** Studying an Algal Bloom—Phosphate Pollution ★ Study Guide, Section 23-2 **PE** Section Review, p. 451
BLOCKS 3 & 4 90 minutes	**23-3 Taking Action, p. 452** Conservation and Restoration Biology, p. 452 Conserving Migratory Birds, p. 453 Reintroduction of the Wolf, p. 455 Restoring the Everglades Ecosystem, p. 456 Getting Involved, p. 458	**ATE** Section Overview, p. 452	**PE** **Chapter 23 Investigation,** p. 462 **Lab A14,** Determining the Amount of Refuse **Lab D11,** Oil-Degrading Microbes **Lab D12,** Can Oil-Degrading Microbes Save the Bay? ★ Study Guide, Section 23-3 **PE** Section Review, p. 458
BLOCKS 5 & 6 90 minutes	**REVIEW and ASSESSMENT** **PE** Chapter 23 Review, pp. 460–461 ■ Performance-Based Assessment—Testing Pond Water for Carbon Dioxide	Audio CD Program BIOLOGY INTERACTIVE TUTOR Unit 7—Ecosystem Dynamics	**CHAPTER TESTING** ■ Chapter Test (blackline master) ▲ Test Generator ■ Assessment Item Listing ■ Scoring Rubrics and Classroom Management Checklists

KEY

MODERN BIOLOGY
- PE Pupil's Edition
- ATE Teacher's Edition
- ■ Active Reading Guide
- ★ Study Guide

HOLT BIOSOURCES
- Teaching Transparencies
- Laboratory Program

One-Stop Planner CD-ROM
Shows these resources within a customizable daily lesson plan:
- ■ Holt BioSources Teaching Resources
- ■ Active Reading Guide
- ★ Study Guide
- ▲ Test Generator

READING FOR CONTENT MASTERY

Preview: Have students preview the objectives, the introductory paragraph, topic sentences, and captions before reading the chapter.

- ATE Active Reading Technique: Brainstorming, p. 441
- ■ Active Reading Guide Worksheet 23-1
- PE Summary/Vocabulary, p. 459

- ■ Active Reading Guide Worksheet 23-2
- PE Summary/Vocabulary, p. 459

- ■ Active Reading Guide Worksheet 23-3
- PE Summary/Vocabulary, p. 459

TECHNOLOGY AND INTERNET RESOURCES

CNN PRESENTS Video Segments 13 and 15
Fat Wolves
Greening Sudbury

Holt Biology Videodiscs Teacher's Correlation Guide, Lessons 23-1 through 23-3

BIOLOGY INTERACTIVE TUTOR Unit 7—Ecosystem Dynamics, Topics 1–6

Audio CD Program, Sections 23-1 through 23-3

internetconnect

SCILINKS NSTA

On-line Resources: www.scilinks.org
The following *sci*LINKS Internet resources can be found in the student text for this chapter:

Topics:
- El Niño, p. 442
- Global warming, p. 445
- Conservation, p. 452
- Wolf, p. 455

go.hrw.com

On-line Resources: go.hrw.com
Visit the HRW Web site for a variety of resources related to this chapter. Just type in the keyword HM2 Chapter 23.

Smithsonian Institution®
Visit **www.si.edu/hrw** for additional on-line resources.

CNNfyi.com
Visit **www.cnnfyi.com** for late-breaking news and current event stories selected just for you.

LAB ACTIVITY PLANNING

CHAPTER 23 INVESTIGATION

Testing the Effects of Thermal Pollution, pp. 462–463

OVERVIEW
In this investigation, students will model the effects of thermal pollution on living organisms.

SAFETY
Demonstrate the careful and safe handling of a hot beaker with tongs or a hot pad. For burns or scalds, have ice on hand for immediate application if necessary. Make sure that hot plates have an on-off switch and an indicator light. Instruct students not to handle the hot plate.

PREPARATION
1. *Paramecium* cultures may be mixed with bottled spring water or tap water that has been left in an open container for at least 24 hours to eliminate the chlorine.
2. Each group will need a 400 mL beaker filled with ice water, a 400 mL beaker filled with 60°C tap water, an alcohol-filled thermometer, a 30 cm long U-shaped glass tube, corks to fit both ends of the glass tube, a hand lens, and a glass-marking pen or wax pencil.

Quick Labs

Calculating CO_2 Production, p. 444
Students will need only a pencil and paper for this activity.

UNDERSTANDING THE VISUAL

The photograph shows a gorilla in the rain forest of Africa. Remind students that gorillas are one of our closest relatives (only bonobos and chimpanzees are closer). They live in small groups dominated by one large male that can weigh over 200 kg. Gorillas live in moist forests, and their range has shrunk dramatically since they were first brought to the attention of scientists in the mid-1800s. Today, one of the largest and best-studied populations lives in eastern Democratic Republic of Congo (formerly Zaire) and Rwanda. Civil war, poverty, and rapid population growth in these countries threaten the survival of the gorillas there.

FOCUS CONCEPT

Interdependence of Organisms

Through technology, humans now have the power to change the global systems on which all living things depend. Humans may be changing the global climate by altering the composition of the atmosphere. These climatic changes may influence habitats throughout the world.

ASSESSING PRIOR KNOWLEDGE

Review the following concepts.

Population growth: *Chapter 20*
Ask students the following questions: Where is human population growth currently fastest? Where is it slowest?

Carbon cycle: *Chapter 22*
Ask students the following questions: What processes add carbon dioxide to the atmosphere? What processes remove it? How is the burning of fossil fuels affecting the concentration of carbon dioxide in the atmosphere?

CHAPTER 23

ENVIRONMENTAL SCIENCE

Gorillas are one species whose survival is threatened by the growing human population.

FOCUS CONCEPT: ***Interdependence of Organisms***

As you read, notice how a knowledge of biology helps us understand larger environmental processes.

BIOLOGY INTERACTIVE TUTOR **Unit 7—*Ecosytem Dynamics***
Topics 1–6

HUMANS AND THE ENVIRONMENT

*In Chapter 22, you learned how environmental factors influence organisms within particular ecosystems. Large-scale environmental forces also have significant effects on human populations. A new field of study called **environmental science** uses biological principles to look at the relationships between humans and the Earth. Environmental science is becoming increasingly important because humans are rapidly changing the global environment.*

SECTION

23-1

OBJECTIVES

▲ Give an example of how global systems are linked together.

● Identify several effects of El Niño on human populations.

■ Describe two ways that humans have modified the composition of the atmosphere, and identify the possible consequences of these changes.

◆ Explain how future human population growth could affect the environment.

A GLOBAL CONNECTION

You might not guess that wind patterns off the coast of Australia would have anything to do with fish harvests in South America or winter weather in the United States, but they do. What connects these events is a complex interaction between air and water currents that is first set in motion by solar energy. When the sun's rays strike air near the surface of the Earth, the molecules are heated and the air, which becomes less dense, starts to rise. As the air rises, it gets closer to outer space and eventually starts to cool. Cooling air becomes more dense and sinks. We can draw this pattern of rising and falling air using a loop called a **convection** (kuhn-VEK-shuhn) **cell.** Groups of convection cells form the system of global air circulation that helps determine climate, as shown in Figure 23-1.

In the southern Pacific Ocean, the usual pattern of convection cells creates a wind that blows from east to west and pushes warm surface water

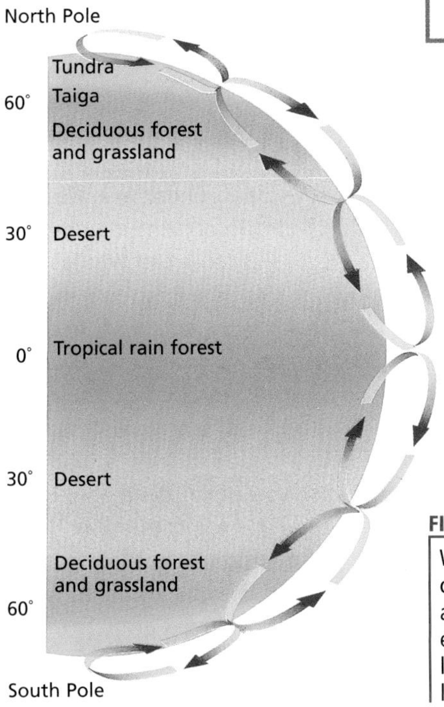

FIGURE 23-1

Warming and cooling air form loops called convection cells in Earth's atmosphere. These air currents influence where different ecosystems are located and also create oceanic circulation patterns.

SECTION 23-1

SECTION OVERVIEW

Interdependence of Organisms
Section 23-1 discusses how weather patterns are organized, how they affect the plankton populations that provide the energy for food webs, and how the disruption of the normal weather pattern affects organisms, including humans.

Stability and Homeostasis
Section 23-1 also explores the causes and possible effects of global warming and ozone depletion.

ACTIVE READING

Technique: Brainstorming
Most students will be familiar with or will have participated in efforts to preserve the environment. Working with the class, create a student-generated list of issues and efforts. Tell students that they will read about global environmental problems in this chapter and will learn how they can help solve these problems.

VISUAL LINK

Figure 23-1
Draw students' attention to the convection cells at the equator. Point out that as equatorial air rises, it loses most of its moisture as rain. The result is a zone of heavy rainfall that stretches around the globe. All of the world's tropical rain forests lie within a few degrees of the equator, within this zone of heavy rainfall.

VISUAL LINK

Figure 23-2

Have students use an atlas or world map to locate the South Pacific Ocean, South America, Peru, and Australia.

GIFTED *ACTIVITY*

Ask students to do a research project on major weather disasters in the United States in the last 20 years. Have them divide this time period into five-year segments for comparison. They should use magazine and newspaper articles, the magazine *Weatherwise,* almanacs, and books as sources for data. Have students make graphs for each five-year period showing the storm type, location, severity (lives and money lost), and size of areas affected. Students should discuss their findings as they relate to issues of global warming and climate change.

RETEACHING *ACTIVITY*

Have students make a flowchart that shows the sequence of events during a normal year and a separate flowchart that illustrates the events of an El Niño year.

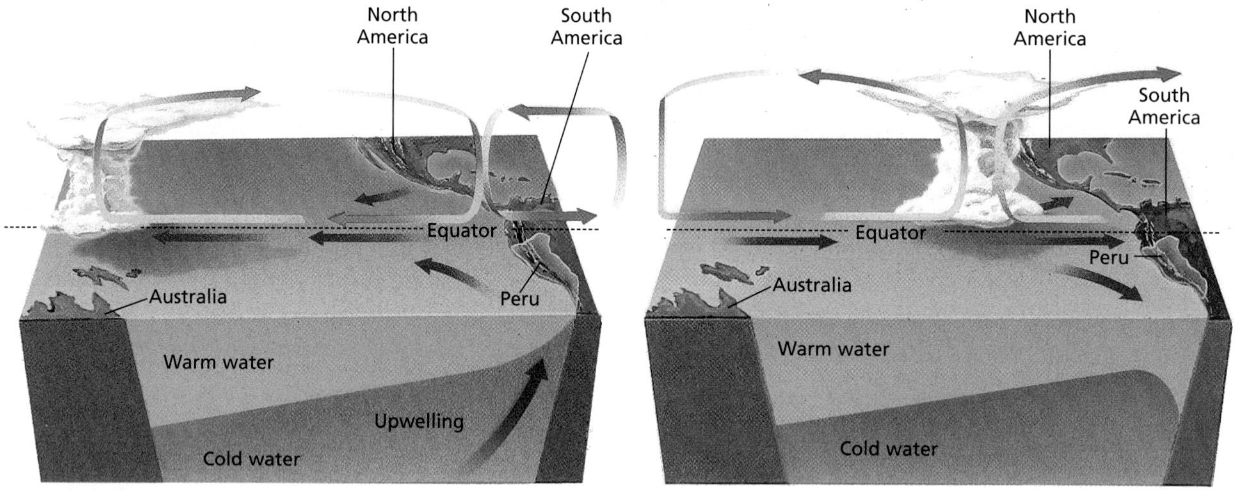

FIGURE 23-2

These two drawings show a comparison of water and atmospheric circulation in normal (left) and El Niño (right) years. The red arrows indicate the movement of warm oceanic currents, and the yellow arrows indicate the movement of air. In an El Niño year, the cold upwelling is blocked by warm water moving toward the coast of Peru.

from South America toward Australia, as shown in Figure 23-2. Along the South American coast, cold water rises from deeper in the ocean and replaces the warm water. This rising current is called an **upwelling,** and it brings with it organic material and nutrients that support an abundance of plankton. The plankton, in turn, support large populations of fish.

One kind of fish that is abundant in these conditions is the anchovy, a main export of the South American country of Peru and a popular pizza topping. Anchovies provide a livelihood not only to fishermen but also to packers, shippers, netmakers, boat builders, and other workers. Diving sea birds eat anchovies too. When these birds come back to shore to roost and digest their food, they excrete a white phosphorus-rich substance called guano. Some guano deposits are dug up and sold as high-quality fertilizer, another major Peruvian export.

El Niño

The Peruvian economy, along with sea birds, depends on normal atmospheric conditions. But sometimes, usually in December, the normal east-to-west winds do not form over the Pacific Ocean. Instead, winds push warm water eastward toward the coast of South America, as illustrated in Figure 23-2. When these conditions occur, the warm surface water cuts off the upwelling of nutrients. This event is called **El Niño** (meaning "the child") because it usually happens near Christmas.

With fewer nutrients, the fish populations decline and Peruvian anchovy exports decrease. Fewer anchovies mean fewer birds and reduced guano production. Because all convection cells are linked in the atmosphere, the effects of El Niño extend beyond Peru. Under a strong El Niño, northeastern Australia can suffer summer drought, leading to reduced grain production there. The southeastern United States gets higher rainfall in El Niño years, boosting agriculture while also decreasing forest fires.

internet**connect**

SCILINKS NSTA
TOPIC: El Niño
GO TO: www.scilinks.org
KEYWORD: HM442

HUMAN INFLUENCES ON GLOBAL SYSTEMS

Global systems, such as the interconnected system of convection cells, link people and economies from across the world. Examine the following examples of how humans have unintentionally changed the global systems on which we depend.

Declining Ozone

As you learned in Chapter 14, ozone, O_3, is a naturally occurring gas that is vital to life on Earth. Ozone in the upper atmosphere screens out most of the ultraviolet radiation from the sun that causes mutations. Humans, if exposed to ultraviolet radiation, suffer increased rates of skin cancer and cataracts (clouding of the lens of the eye). People are advised to wear sunscreen and sunglasses to protect against the ultraviolet radiation that is not blocked by ozone.

Several kinds of human-made chemicals are diminishing the ozone shield, allowing more ultraviolet radiation to reach the Earth's surface. The most important of these ozone-destroying chemicals are known as **chlorofluorocarbons,** or CFCs. Originally thought to be harmless, CFCs have been used as coolants in refrigerators and air conditioners and as the propellant in aerosol spray cans. They have also been used to make plastic-foam products and to clean electronic equipment. In the upper atmosphere, CFCs act as catalysts that break down ozone much faster than it is formed by natural processes. Scientists have estimated that a single CFC molecule can help destroy up to 100,000 ozone molecules.

Beginning in the 1980s, atmospheric measurements have indicated some alarming declines in ozone levels. Ozone destruction is most severe over the Earth's polar regions, and for a few weeks every year, an ozone "hole," a zone of very low ozone concentration, forms over Antarctica, as shown in Figure 23-3. In 1991 an international study estimated that even a 10 percent worldwide decrease in ozone levels would cause 300,000 new cases of skin cancer in humans. Other organisms, including plants and photosynthetic algae, are also harmed by high levels of ultraviolet radiation, so ozone depletion could modify entire ecosystems over time.

The evidence of widespread and worsening damage to the ozone layer led to international agreements to stop production of CFCs by the end of 1995. Because the substitutes for CFCs are initially more expensive, the developed countries are contributing to a fund to help the developing countries make the changeover. A global environmental success story, these agreements have already cut CFC production by more than 75 percent. Environmental scientists estimate that if the terms of the agreements continue to be followed, the ozone layer will start to build up again, perhaps recovering completely in 50 to 100 years. The effort to protect the ozone layer is a good example of how scientists and policy makers can work together to solve an environmental problem.

FIGURE 23-3
These computer-generated images are based on satellite measurements of ozone levels over Antarctica.

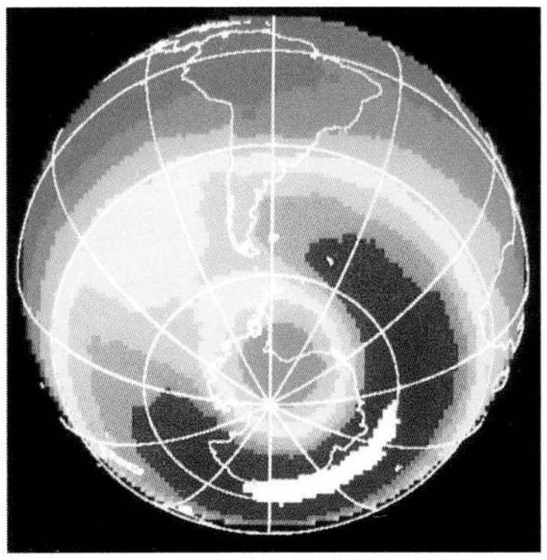

OCTOBER 1979

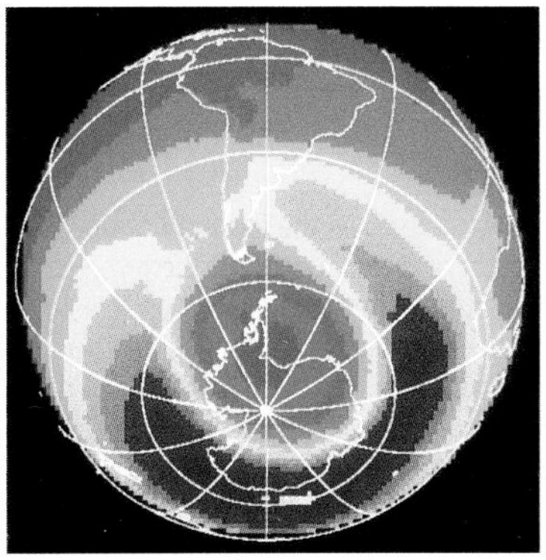

OCTOBER 1988

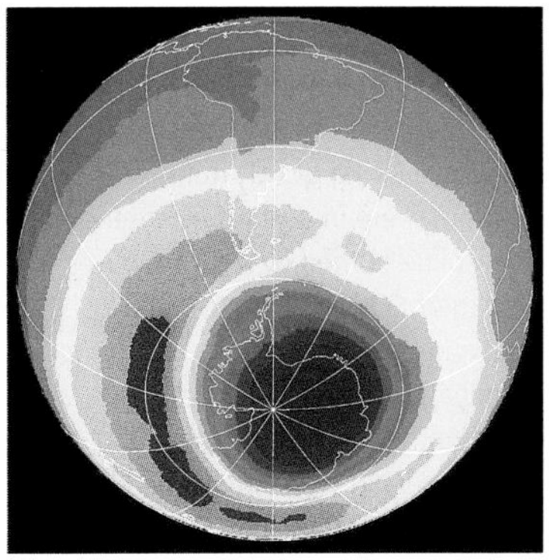

OCTOBER 1996

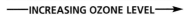

INCREASING OZONE LEVEL →

TEACHING STRATEGY
"Good Ozone" and "Bad Ozone"

Tell students that the ozone layer lies between 17 and 26 km (11–16 mi) above the Earth's surface. The ozone layer absorbs about 99 percent of the ultraviolet radiation that reaches Earth. Ozone also forms in the lower atmosphere, typically when emissions from automobiles or factories react with sunlight. Ozone is a corrosive, reactive substance that is a serious pollutant in the lower atmosphere.

TEACHING STRATEGY
Comparing Sunscreens

Bring a variety of sunscreens to class and pass them around the room. On the board, make a list of ingredients for each sunscreen and circle the ones that are common to all of them. (Ingredients common to all are usually oxybenzone, 2-ethylhexyl salicylate, and ethylhexyl p-methoxycinnamate.) Ask students what type of electromagnetic radiation is blocked by sunscreens. (They block UVA and UVB rays, which are two types of ultraviolet radiation.) Point out that the SPF rating of the sunscreen indicates how much protection it provides: higher values give greater protection.

DEMONSTRATION

Show students photos of the three main types of skin cancer: squamous cell carcinoma, basal cell carcinoma, and malignant melanoma. A health book is a possible source for photos, or you may be able to obtain an illustrated pamphlet from a dermatologist. Emphasize to students that malignant melanoma is one of the most dangerous cancers. It spreads rapidly and is usually fatal unless diagnosed and removed early.

TEACHING STRATEGY

Greenhouse Gases

Inform students that carbon dioxide, which constitutes only about 0.035 percent of the atmosphere, is one of several naturally occurring greenhouse gases. Others include water vapor, methane, and nitrous oxide. Greenhouse gases help keep the average temperature of the Earth at 17°C (63°F). Without these greenhouse gases, scientists estimate the Earth's average temperature would be about –25°C (–13°F), about the same as on Mars.

Quick Lab

Calculating CO_2

Time Required 20 minutes

Procedural Tips Students can work separately or in pairs. Have students check their work with a partner or another group.

Answers to Analysis You will need 19.2 L of gasoline for a 250 km trip in a car that gets 13 km/L. (250 km ÷ 13 km/L = 19.2 L) During your trip, 57.6 kg of CO_2 are produced. (3 kg/L × 19.2 L = 57.6 kg) More than five young trees are needed to remove the CO_2 produced by a single trip. (57.6 kg ÷ 11 kg/tree = 5.24 trees) In 1 year 524 young trees would be required to remove the CO_2 produced by 100 cars taking a 250 km trip. Answers to the discussion questions will vary. The car is fairly efficient in terms of kilometers per liter. Students might mention that the number of people in the car would be a factor. Public transportation will probably be the suggested alternative.

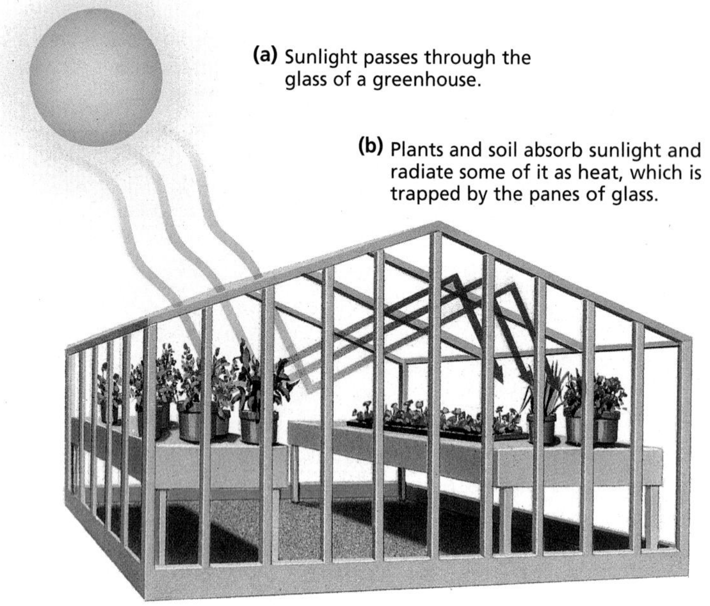
(a) Sunlight passes through the glass of a greenhouse.
(b) Plants and soil absorb sunlight and radiate some of it as heat, which is trapped by the panes of glass.

FIGURE 23-4

Solar energy in the form of light penetrates a greenhouse, but the reradiated heat does not immediately escape. In the atmosphere, gases such as carbon dioxide trap heat and warm the surface.

Increasing Carbon Dioxide

Carbon dioxide, CO_2, is a naturally occurring gas that is a raw material of photosynthesis and a byproduct of cellular respiration. It is also released when fossil fuels, such as natural gas, coal, and petroleum, are burned in homes, power plants, and motor vehicles. Around the middle of the nineteenth century, before humans began to substantially increase their use of fossil fuels, carbon dioxide levels in the atmosphere were fairly stable, at about 280 parts per million (ppm). Since 1850, carbon dioxide levels have risen nearly 30 percent, to 360 ppm, and they seem likely to increase as the world's use of fossil fuels grows. According to some projections, by the year 2100 the concentration of CO_2 in the atmosphere might be twice what it was in 1850.

The concentration of carbon dioxide in the atmosphere influences how much heat from the sun is trapped by the atmosphere. The atmosphere's ability to retain heat is called the greenhouse effect. As you can see in Figure 23-4, the glass in a greenhouse allows the sun's rays in but prevents heat from escaping. The same process happens in cars left in the sun with the windows rolled up.

Quick Lab

Calculating CO_2 Production

Materials pencil, paper

Procedure A young tree can remove 11 kg of CO_2 from the atmosphere in a year. Each liter of gasoline burned by a car produces 3 kg of CO_2. Suppose you are taking a trip by car. Your round-trip mileage is 250 km, and you get 13 km/L of gas. Calculate how many young trees are needed in 1 year to remove the CO_2 that was produced from your single trip.

Analysis In 1 year, how many trees are required to remove the CO_2 produced by 100 cars taking the same trip and getting the same number of kilometers per liter? Discuss with your classmates whether your trip was ecologically efficient. What alternative transportation methods might be more ecologically efficient?

Effects of Rising Carbon Dioxide Levels

If you look at climatic records derived from ice-core samples, shown in Figure 23-5, it is apparent that temperature changes over the last 160,000 years correspond to changes in atmospheric carbon dioxide concentration. Scientists call this type of matching relationship a **correlation.** Because temperature changes correlate with changes in carbon dioxide level, the long-term pattern suggests that a **cause-and-effect relationship** might exist between these two variables. In a cause-and-effect relationship, a change in one variable directly leads to a change in the other variable. Even a strong correlation does not prove the existence of a cause-and-effect relationship, but experiments enable scientists to establish a connection between two variables.

It is not possible to carry out a global experiment that would conclusively demonstrate a cause-and-effect relationship between carbon dioxide levels and rising temperatures. Instead, environmental scientists rely on computer models that simulate the Earth's climate. Because so many factors besides carbon dioxide levels have to be taken into account (wind patterns, ocean temperatures, and effects of ecosystems on CO_2 levels, to name a few), these models are very complicated. It is like predicting the weather hundreds of years in advance.

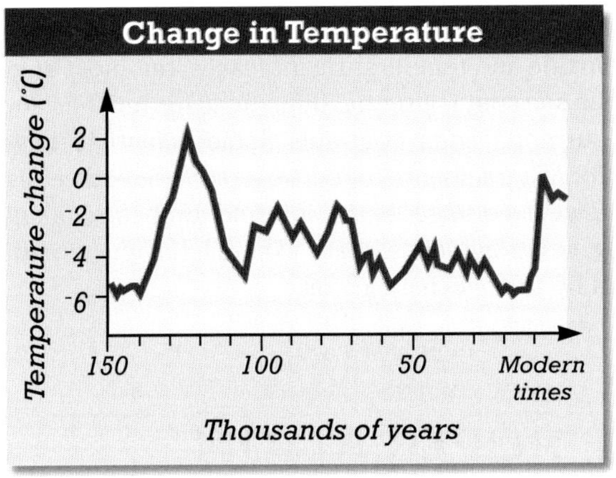

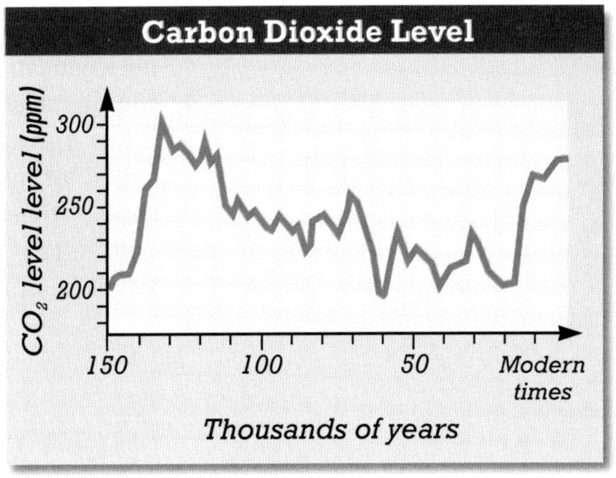

FIGURE 23-5

These graphs show data representing 150,000 years of Earth's climatic history. Although the correlation is not perfect, in general, high levels of carbon dioxide correlate with temperature increases and low levels correlate with temperature decreases. As of 1995, carbon dioxide levels had risen to 360 ppm, higher than any part of this record.

By varying carbon dioxide levels in these climate models, scientists can simulate the possible effects of CO_2 on temperature. The results of these simulations are consistent: doubling carbon dioxide concentration leads to higher average global temperatures—in the range of 1.0°C to 4.5°C (2°F to 8°F). The exact temperature change depends on the particular assumptions built into the model, but nearly all scientists studying this problem have concluded that increased carbon dioxide levels cause temperature increases in the atmosphere.

Measurements of temperatures from around the world support this conclusion. The recent rapid increase in atmospheric carbon dioxide has been accompanied by higher global temperatures, as shown in Figure 23-6. Today the average global temperature is about 0.6°C (1°F) higher than it was in 1860. An international panel of scientists notes that temperatures are expected to rise an additional 2°C (4°F) within the next century. Although this may not seem significant, this increase can have global effects on rainfall patterns, soil moisture, and sea level. These factors might shift agricultural regions of the world and disrupt natural ecosystems. Specific effects on a particular country or state cannot be predicted yet, but most models project that the effects will be felt most strongly in temperate and polar regions.

internetconnect

SCILINKS NSTA

TOPIC: Global warming
GO TO: www.scilinks.org
KEYWORD: HM445

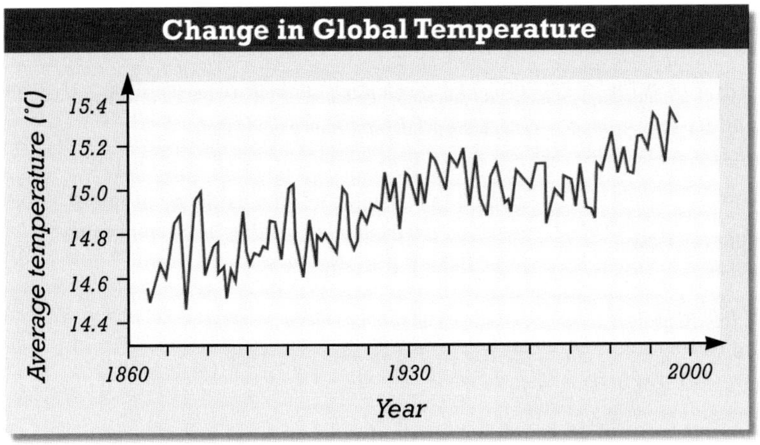

FIGURE 23-6

Although temperatures fluctuate from year to year, a general warming trend is evident over the last 140 years. During this same period, carbon dioxide levels have risen 30 percent, mainly due to the increased use of fossil fuels.

CRITICAL THINKING

Economics of Global Warming

One predicted effect of global warming is that storms may become more severe and more frequent. Ask students what economic impact this might have. (One answer is that insurance companies might charge higher rates to compensate for greater storm losses; this is already happening with some insurance companies because of the unusually high number of disastrous floods and storms in recent years. Also, areas that are prone to floods or severe weather might require the construction of expensive dikes or drainage systems to prevent storm damage. Farmers and ranchers might lose more crops and livestock because of severe storms.)

CULTURAL CONNECTION

Ancient Pollution

Widespread pollution is usually associated with industrialization, which began with the Industrial Revolution in the late 1700s. However, some recent studies have extended the history of large-scale pollution back several thousand years. Sediment samples from the beds of Swedish lakes show that lead concentrations increased well above natural levels beginning about 2,600 years ago. Ice cores removed from Greenland's ice cap also show higher-than-normal lead levels beginning around the same time. What caused this pollution? It is most likely fallout from lead-smelting conducted by the ancient Greek and Roman civilizations.

SECTION 23-1

ANSWERS TO SECTION 23-1 REVIEW

1. Winds blowing across the Pacific push warm water westward. Cold water rises to take its place, carrying nutrients that nourish large plankton populations, which are eaten by anchovies that, in turn, support a fishing industry. The excrement of birds that feed on the anchovies is sold as a fertilizer.
2. El Niño stops the upwelling of nutrients, causing declines in anchovy populations, reductions in the anchovy catch, and declines in the production of guano. Northern Australia usually experiences drought, while the southeastern United States gets more precipitation than normal.
3. CFCs are destroying the ozone layer that protects the Earth from ultraviolet light, which causes skin cancer and cataracts.
4. Computer models enable scientists to experiment on the atmosphere by altering the levels of carbon dioxide and then determining the effects on climate.
5. Continued growth of the human population will almost certainly mean greater use of fossil fuels and a growing demand for land, water, roads, schools, landfill space, and other necessities. This will mean reductions in the amount of habitat available for other species.
6. It is possible that rising carbon dioxide levels might increase the rate of photosynthesis and speed up plant growth. Also, rising temperatures may expand the growing season.

We do not fully understand all aspects of the relationship between carbon dioxide and temperature increases, but most environmental scientists see some potential difficulties for humans and other species if current trends continue. Some scientists have called for international agreements to reduce carbon dioxide emissions. Others think we need to get better information before taking any action, noting some evidence that "excess" carbon dioxide might be absorbed by the oceans and soil.

FUTURE POPULATION GROWTH

You learned in Chapter 20 that the Earth's human population is currently 6 billion and growing at the rate of about 90 million people per year. The United Nations estimates that by the year 2050 the world's population could more than double, to 12.5 billion. How would a doubling of the Earth's population affect ozone levels, carbon dioxide concentrations, and other environmental conditions, such as the availability of clean water, open space, and wildlife habitats?

Although ozone levels may increase and stabilize, doubling the number of people will likely mean more fossil-fuel use and more clearing and burning of forests, which in turn will increase carbon dioxide levels and hasten global warming. Fresh water constitutes less than 3 percent of the water on Earth, and two-thirds of the world's population already lacks a reliable source of drinking water. Doubling the population could also require twice as many homes, schools, hospitals, landfills, and roads, all of which decrease the amount of undeveloped land. Humans already are using 40 percent of the net primary productivity of the Earth; if we double our use, many other organisms will be unable to survive.

SECTION 23-1 REVIEW

1. **What causes the upwelling off the coast of South America? How does this upwelling affect the economy of Peru?**
2. **Describe the consequences of El Niño.**
3. **How have CFCs affected the atmosphere? How might this change affect humans?**
4. **How have scientists used computer models to help understand the effects of rising carbon dioxide levels?**
5. **Describe some possible effects of a continued increase in the human population.**
6. **CRITICAL THINKING Some scientists have argued that rising carbon dioxide levels might lead to increased food production. Explain the logic behind this argument.**

Research Notes

Environmental Eyes in the Skies

Governments, educational institutions, and even private corporations from around the world have sent satellites into orbit to monitor global environmental changes. The satellites use various devices to collect and measure information about Earth's surface and relay it back to Earth, often in "real time." That is, observers on Earth can view the images at the same time the satellite's instruments are observing them.

One system of polar-orbiting satellites, which circle Earth in a north-to-south orbit each day, sends information to the National Oceanic and Atmospheric Administration. These satellites carry equipment that measures clouds, provides temperature and moisture data from Earth's surface through the atmosphere, measures the energy emitted by the sun, and measures radiation striking and leaving the Earth. The satellites also receive emergency signals from people in distress. ARGOS, a French-provided satellite system, collects information from sensors placed in various locations, such as on ships, buoys, weather balloons, and even birds and other animals. The ARGOS satellites provide both short-term information, such as storm warnings, and ongoing information, such as global changes in pollution levels and vegetation cover.

Hans Tømmervik conducts research projects in Tromsø, Norway, a busy seaport located north of the Arctic Circle. Tømmervik's primary concern has been the effect of air pollution on the natural environment of the Norwegian-Russian border area, located hundreds of kilometers east of Tromsø. In one project, Tømmervik and fellow scientists used information from remote-sensing satellites to map changes in the area's vegetation cover from 1973 to 1988. They compared vegetation-cover maps with sulfur dioxide emission levels for the same years. A time sequence of satellite images showed that the area with lichen-dominated vegetation had decreased by 85 percent over 15 years, while desert-like conditions had increased by more than five times. During the same time, emission levels of sulfur dioxide had increased dramatically. The scientists developed a pollution-impact map on the basis of the vegetation maps. The map clearly shows how the impact of pollution varies with distance from the pollution source.

Satellite monitoring of the environment has made large gains since the end of the Cold War between the United States and the former Soviet Union. Satellites that these nations previously used for spying on each other are now being used to monitor environmental conditions. These satellites have much better resolution than satellites currently used for environmental-science information gathering. Some are capable of viewing objects that are less than 15 cm (6 in.) across. The scientific community stands to benefit not only from better equipment but also from international sharing of expertise in designing sensors and interpreting data.

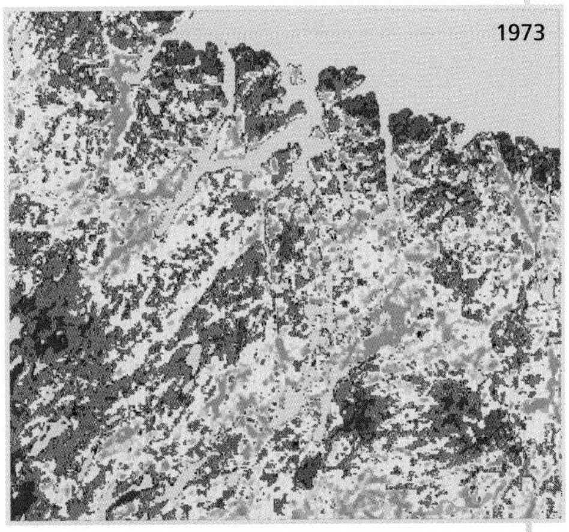

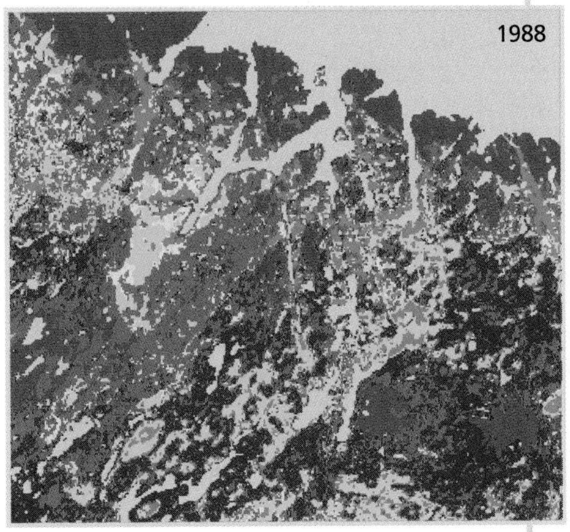

Between 1973 and 1988, a satellite monitored the effects of pollution on a 10,000 km² area along the border between Russia and Norway. One of the main discoveries was that lichen-covered areas (light green on maps) decreased by 85 percent, while there was a fivefold increase in bare, eroded, and damaged areas (violet on maps).

CHAPTER 23
Research Notes

BACKGROUND

Between 1972 and 1992, the United States launched six Landsat scientific satellites. The original satellites carried cameras and infrared sensors that could photograph an area of the Earth's surface as large as 184 km^2 (115 mi^2). The later models contained sensors that could map areas as small as 30 m (100 ft) on one side. The satellites were designed to gather information about the Earth's natural resources.

DISCUSSION

Guide the discussion by posing the following questions.

1. Some satellites send information back to the Earth in "real time." What does this mean? ("Real time" means that observers on Earth can view the images at the same time the satellite's instruments are collecting and measuring information about the Earth's surface.)

2. What kind of data are collected by the satellites described in the feature? (Some satellites measure clouds, provide temperature and moisture data from the Earth's surface, or measure the energy emitted by the sun and the radiation striking and leaving the Earth. Other satellites receive emergency signals from people in distress and signals from sensors placed in various locations, such as on ships, on buoys, in weather balloons, and even on birds and other animals.)

FURTHER READINGS

1. "Rich Pickings from Russia's Polluted Soils," in *New Scientist,* September 28, 1996, page 5.

2. "Ozone Hole Starts Strong, Fades Quickly," in *Science News,* October 19, 1996, page 246.

SECTION 23-2

SECTION OVERVIEW

Interdependence of Organisms
Section 23-2 describes what biodiversity is and how to measure it. It also explains the human impact on biodiversity and several ways that scientists and others are trying to preserve biodiversity.

▶ ENGAGE STUDENTS

Ask students to bring bags of gummy bears or worms to class. Divide students into groups and have all members of each group combine their gummy bears or worms into one population on a piece of paper. Have each different-colored bear or worm represent a different species. Ask each group to determine the species richness of their bear or worm population (three to five is usual). Record the values on the board, and ask students which group has the highest species richness. Then have students count the number of individuals of each species to determine the evenness of the population. (Answers will vary.) Also record these results on the board, and ask students which group has the greatest evenness. Point out to students that biologists use both of these measurements—species richness and evenness—as measures of biodiversity.

CONTENT CONNECTION

Section 21-2: Measuring Diversity
Remind students that species richness and species diversity are two ways to measure the diversity of a community. Species richness is a count of the total number of species and species diversity is a mathematical expression that takes into account both the total number of species and the relative abundance of each species.

SECTION 23-2

OBJECTIVES

▲ Define *biodiversity,* and explain three ways to measure it.

● Describe global patterns of biodiversity.

■ Identify two strategies for conserving biodiversity in developing countries.

◆ Distinguish between utilitarian and nonutilitarian reasons for conserving biodiversity.

THE BIODIVERSITY CRISIS

Although some extinctions of species are natural events that have been going on since life began, environmental scientists have noted that humans are now causing species to become extinct much faster than in the past. Because extinction is irreversible and stops the evolution of future species, biologists are urgently trying to learn more about how we can conserve species.

BIODIVERSITY

Biodiversity refers to the variety of organisms in a given area. Biodiversity can be measured in several ways. Looking at Figure 23-7, it seems easy to say that Site A has more biodiversity than Site B and less than Sites C and D. Recall from Chapter 21 that the number of species in an area is called species richness. In this example, species richness for Site A is 3, for Site B is 1, for Site C is 4, and for Site D is 4. For quick comparisons between sites, biologists often find that species richness is a very useful estimate of biodiversity.

Now compare Site C with Site D. Each site has four moth species, but the moth communities are not the same. Site C has three individuals of each species of moth, while Site D has one individual of

FIGURE 23-7
This chart shows the number of individuals of four moth species captured at four sites.

each of three species and nine individuals of the fourth species. Even though the species richness (4) and the total number of individuals (12) are the same, biologists would expect these two communities to behave differently. Thus, biologists often determine how many individual organisms belong to each species, a measure called **evenness.** In our example, Site C has greater evenness than Site D. For detailed comparisons between communities, biodiversity is sometimes expressed as a quantity called species diversity (a concept introduced in Chapter 21), which combines species richness and evenness.

Because evolution depends on the presence of genetic variation within a population, as you learned in Chapter 15, some biologists would want to know the genetic makeup of each moth. With this information, biologists can calculate the **genetic diversity,** or amount of genetic variation, for each site. In the long run, genetic diversity might be the most important measure of biodiversity, but it is the one we know the least about.

MEASURING EARTH'S BIODIVERSITY

If we use the species-richness measure of biodiversity, how much biodiversity is there on Earth? Estimates vary, but most biologists are confident that there are at least 10 million species on Earth—and possibly as many as 30 million. These are staggering numbers, especially because, in about 200 years of cataloging, scientists have named and described fewer than 3 million species.

When most people think of nature, they tend to focus on mammals. Humans seem to have a basic attraction to large mammals, especially to those that have big heads, big eyes, and a cuddly appearance like pandas. Yet mammals are a very small fraction of biodiversity. Insects and plants are far more representative of life on Earth, as illustrated in Figure 23-8.

FIGURE 23-8

This diagram shows the known species richness of the Earth, with representatives drawn to a size that is proportional to their abundance. Although people tend to think mainly of mammals, there are many more species of insects and plants. Data from tropical forests, which are still largely unexplored, suggest that insects may represent an even greater fraction of the Earth's biodiversity than is shown here.

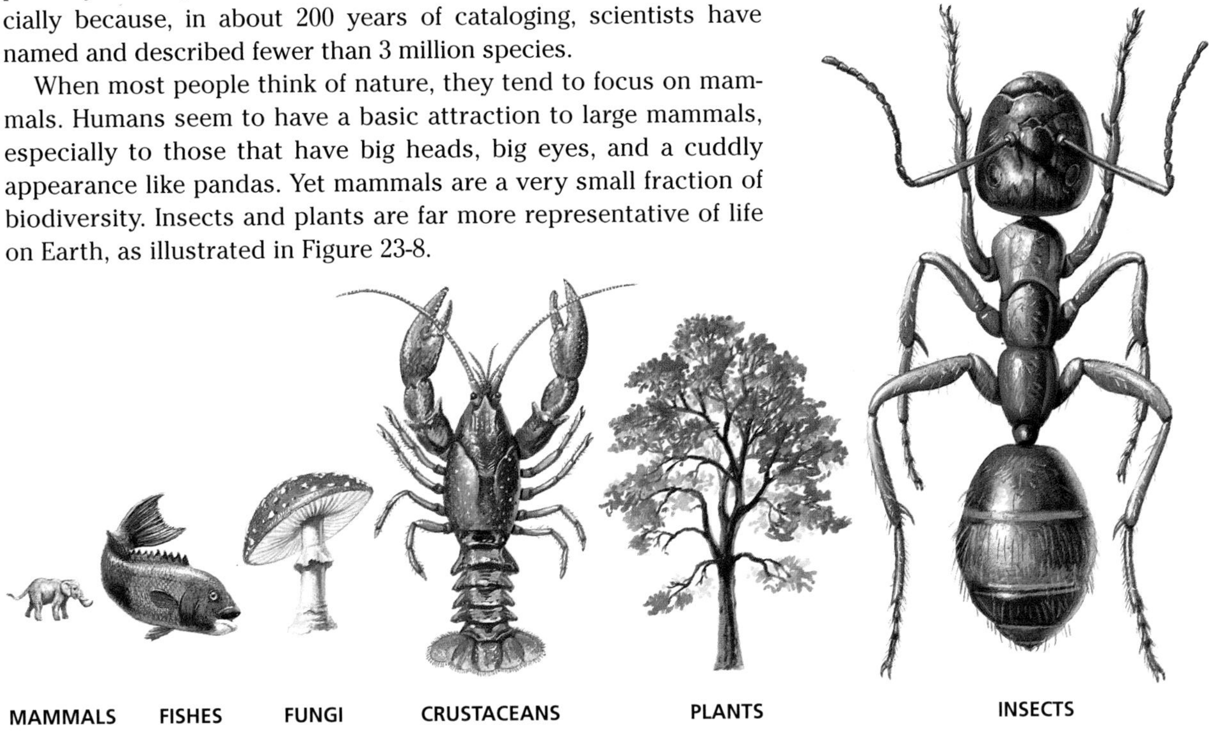

DEMONSTRATION

To demonstrate genetic diversity within a species, show students photographs or samples of corn cobs of different sizes, colors, and patterns. Point out that the differences between varieties of corn are the result of genetic variation. By selectively crossing corn plants that have desirable traits, humans have manipulated the genetic diversity within this species to produce the varieties we see today.

RECENT RESEARCH

Genetic Diversity and Disease Resistance

Genetic diversity within our species is also a valuable resource. For example, a gene discovered in 1996 may lead to a vaccine against HIV. The normal version of this gene codes for a cell-surface protein that HIV must bind to before it can enter a cell. A mutant allele encodes a variant protein that blocks HIV from binding, thus preventing infection. Homozygotes for this allele apparently are resistant to infection by HIV. By studying how the mutant allele repels HIV, scientists hope to find a way to prevent HIV infection.

CRITICAL THINKING

Seed Banks

Ask students to think of reasons for having a seed bank of native plants and for cultivating native plants. (Ideas should include the need for genetic diversity to ensure vigorous plants; the possibility of transferring genes from native plants to crop plants; and the potential for replacing current crop species with other, hardier native species.)

SECTION 23-2

CONTENT CONNECTION

Section 15-1: Mass Extinctions

Remind students that mass extinctions have altered Earth's biodiversity at least five times in the past. The mass extinction at the end of the Cretaceous period is the most famous because it involved the dinosaurs, but the largest occurred at the end of the Permian period. Up to 96 percent of the world's species disappeared at this time. Have students identify this mass extinction in Figure 23-9.

TEACHING STRATEGY

Balancing Development and Conservation

Inform students that another way to reconcile the needs of rural populations with species conservation is through extractive reserves, areas in which the habitat is preserved but the sustainable harvesting of natural products, such as fruits, seeds, and rubber, is permitted. Extractive reserves benefit local populations by providing a long-lasting source of income. Also, because continued production depends on keeping the natural habitat intact, species are protected. Ecologists Charles Peters, Robert Mendelsohn, and the late Alwyn Gentry calculated that one hectare of Peruvian forest managed as extractive reserve could yield products worth about $400 per year. If the same area were logged, the value of the timber would be higher—about $1,000—but the area might not produce any additional income for at least several decades (if ever), until the forest regenerated.

FIGURE 23-9
This graph illustrates the changes in biodiversity, measured as the number of families of marine organisms, over time. At present, biodiversity is at an all-time high. Arrows point to the five known mass extinctions. A sixth extinction is under way due to human activities.

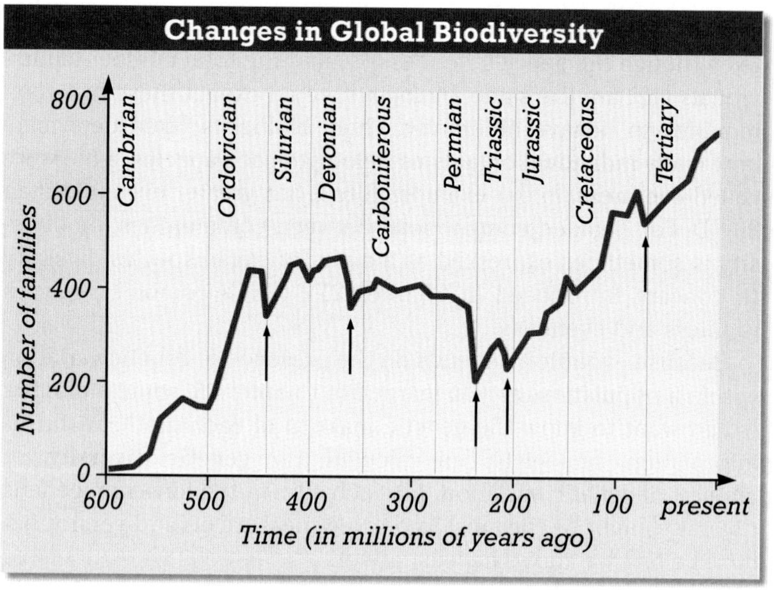

Reducing Biodiversity

You can see in Figure 23-9 that we are living in a time of very high biodiversity. But we are also living in a time of very rapid extinction. Biologists estimate that up to 20 percent of existing species may become extinct by the year 2030. As you learned in Chapter 15, five mass extinctions, large and relatively rapid declines in biodiversity, have occurred in the history of life. The mass extinction currently under way is different because humans are the cause. We do not know what the consequences of eliminating millions of species will be.

The greatest threat to biodiversity is the rapid destruction of natural habitats to provide for the needs and wants of the growing human population. In general, humans convert complex, self-sustaining natural ecosystems into simplified systems, such as farmland and urban areas, that cannot sustain as many species. For example, since the discovery of agriculture 10,000 years ago, more than half the world's tropical rain forests have been destroyed, and half of what remains is likely to be gone by the year 2020. Because tropical rain forests have the highest species richness of any biome, containing up to one-fifth of all species on Earth, their destruction is especially damaging to biodiversity.

Ways to Save Biodiversity

The United States became prosperous partly by converting forests and native prairies into farms. The tropics and other regions of high biodiversity often include some of the economically poorest countries on Earth. These countries are trying to use their natural resources to build their economies and raise the standard of living for their citizens, just as the United States did. Several conservation strategies offer ways for developing countries to benefit economically from preserving their biodiversity. For example, in a **debt-for-nature swap,** richer countries or private conservation organizations pay off some of the debts of a developing country. In exchange, the developing country agrees to take steps to protect its biodiversity, such as setting up a preserve or launching an education program for its citizens. Another idea to help local people make money from an intact ecosystem is to set up a national park to attract tourists. People who want to see the ecosystem and its unique organisms will pay money for nature guides, food, and lodging. This idea is called **ecotourism.**

Word Roots and Origins

utilitarian

from the Latin *utilitas,* meaning "useful"

THE IMPORTANCE OF BIODIVERSITY

One way to weigh the importance of biodiversity is called **utilitarian value;** it involves thinking of the economic benefits biodiversity provides to humans. For example, different plants and animals can be harvested for food, and trees can be cut to build homes and provide fuel. Some species are valuable as sources of medicines. Given that most of the world's species have not been named or described, it is reasonable to expect that some undiscovered species will also have medical benefits. Ecosystem functions crucial to our survival, such as the water cycle and nitrogen cycle, depend on living organisms. Harvard University biologist E. O. Wilson summed up the importance of biodiversity when he said, "Biological diversity is the key to the maintenance of the world as we know it."

Another way to weigh the importance of biodiversity is called **nonutilitarian value.** Basically, some people believe that life-forms have value simply because they exist, apart from any human uses of them. Intrinsic value is often associated with moral or religious beliefs that are beyond the scope of biology. Many people attach both utilitarian and nonutilitarian value to biodiversity.

Everyone is familiar with the Declaration of Independence, which led to the separation of the United States from England. Some biologists have called for a "Declaration of Interdependence," an acknowledgment of the connections between organisms, including humans, and their environments. Recognizing interdependence is what environmental science is all about.

SECTION 23-2 REVIEW

1. Explain how species richness and evenness differ.

2. What is genetic diversity?

3. How many species of organisms are there?

4. What is a debt-for-nature swap?

5. List three utilitarian uses of biodiversity.

6. CRITICAL THINKING Explain why the conservation of genetic diversity is necessary for the long-term conservation of biodiversity.

RECENT RESEARCH
Birds and Ecotourism

A study by the University of Arizona Department of Agriculture and Economics found that, in one year, visits to local nature preserves by tourists generated $3 million in total revenue, 56 new jobs, and $890,000 in wages for a nearby small town.

ANSWERS TO SECTION 23-2 REVIEW

1. Species richness is the number of species in an area. Evenness takes into account the number of individuals of each species.
2. Genetic diversity is the amount of genetic variation in a population or species.
3. No one is certain how many species there are. Fewer than 3 million have been named and described, but the total number may be between 10 million and 30 million.
4. In a debt-for-nature swap, a developing country takes some measures to conserve its biodiversity—setting up a preserve, educating its citizens, or the like. In return, some of its debts are paid off by a developed country or a conservation organization.
5. Species provide food, timber, medicine, recreation, scientific knowledge, and ecosystem "services," such as water purification and oxygen renewal.
6. Because genetic variation is necessary for evolution, populations or species with low genetic diversity may not be able to adapt to changing environmental conditions, and they may become extinct as a result. Over the longer term, speciation also depends on genetic diversity, so a species with little genetic diversity may not give rise to any other species.

SECTION OVERVIEW

Interdependence of Organisms

Section 23-3 focuses on three efforts to restore or maintain biodiversity: migratory bird conservation, the reintroduction of wolves to Yellowstone National Park, and the protection and expansion of the Everglades.

▶ ENGAGE STUDENTS

Ask students to list as many adjectives as possible that describe wolves. On the board, make a list of all student responses, dividing them into positive and negative attributes. Ask students to describe personal experiences or media images that have shaped their views of wolves. Then have students research the ecology and behavior of wolves. After students have completed their research, ask them if what they learned has changed their views of wolves and, if so, what information caused them to change their mind.

INCLUSION ACTIVITY

Have students design and make their own bird feeders or houses. They should be simple enough to be constructed in class. Have students bring materials for the feeders. Hang the bird feeders and houses for display and then let students take them home or hang them on trees on campus.

SECTION

23-3

OBJECTIVES

▲ Contrast conservation biology with restoration biology.

● Describe current efforts to conserve migratory birds.

■ Discuss the biological principles and social issues related to wolf reintroduction.

◆ Explain the plan to restore the Everglades.

TAKING ACTION

Although biologists have just begun to learn how nature works, they are now being called upon to help conserve threatened wildlife and restore ecosystems. Science, public involvement, and new partnerships have led to several environmental success stories. You too can contribute. Becoming aware of your local environment can link you to community action and global issues.

CONSERVATION AND RESTORATION BIOLOGY

As the human population has increased, the influence of humans on natural ecosystems has also increased. In the United States during the last 200 years, over 99 percent of native prairies have been replaced with farmland or urban development, and most of the old-growth forests have been cut. Loss of so much of these vegetation types has meant losses of biodiversity.

Biologists are being asked to develop plans to protect and manage the remaining areas that still have much of their biodiversity. A new discipline, called **conservation biology,** seeks to identify and maintain natural areas. In areas where human influence is greater—such as agricultural areas, former strip mines, and drained wetlands—biologists may have to reverse major changes and replace missing ecosystem components. For example, returning a strip-mined area to a grassland may involve contouring the land surface, introducing bacteria to the soil, planting grass and shrub seedlings, and even using periodic fires to manage the growth of vegetation. Dealing with a more extreme case like this is called **restoration biology.**

At present, even the best scientific efforts may not be enough to completely restore an area to its original condition. But by using their understanding of ecological principles such as energy flow, species interactions, and biogeochemical cycling, biologists can often make improvements. Let's look at three examples of conservation and restoration biology: the conservation of migratory birds, the reintroduction of the gray wolf to Yellowstone National Park, and new plans to restore the Everglades.

internetconnect

SCILINKS NSTA

TOPIC: Conservation
GO TO: www.scilinks.org
KEYWORD: HM452

CONSERVING MIGRATORY BIRDS

Imagine weighing as little as 20 g (just a bit more than a floppy disk) and flying nonstop over open ocean for 100 hours before reaching land to rest and feed. This is exactly what blackpoll warblers like the one shown in Figure 23-10 do. The blackpoll warbler is one of the 200 species of **migratory birds** that travel twice each year between North America and Latin America. Migratory birds take advantage of long days and abundant prey in northern tundra and forest ecosystems, where they breed and raise their young. Then, as autumn approaches and the food supply decreases, they fly south to warmer ecosystems that can sustain them during nonbreeding months. Examples of migratory birds you may recognize include the Canada goose, sandhill crane, barn swallow, and scarlet tanager.

Most migratory birds tend to follow generally north-south routes along rivers, mountains, and coastlines. These routes are called **flyways.** Figure 23-11 shows the four major flyways in North America. As many as 5 billion individual birds depend on suitable habitat being available at each end of their migratory journey. Some birds fly along coastlines or over land and make several stopovers for food or rest along the way. If food or habitat is lacking along their way or at their destination, they may not breed and may even die.

FIGURE 23-10
Each fall, blackpoll warblers migrate more than 2,500 km (1,600 mi) over the ocean to reach their wintering grounds in South America.

FIGURE 23-11
This map shows the four flyways commonly used by North American migratory birds. Birds tend to follow landscape features such as rivers, mountains, and coastlines.

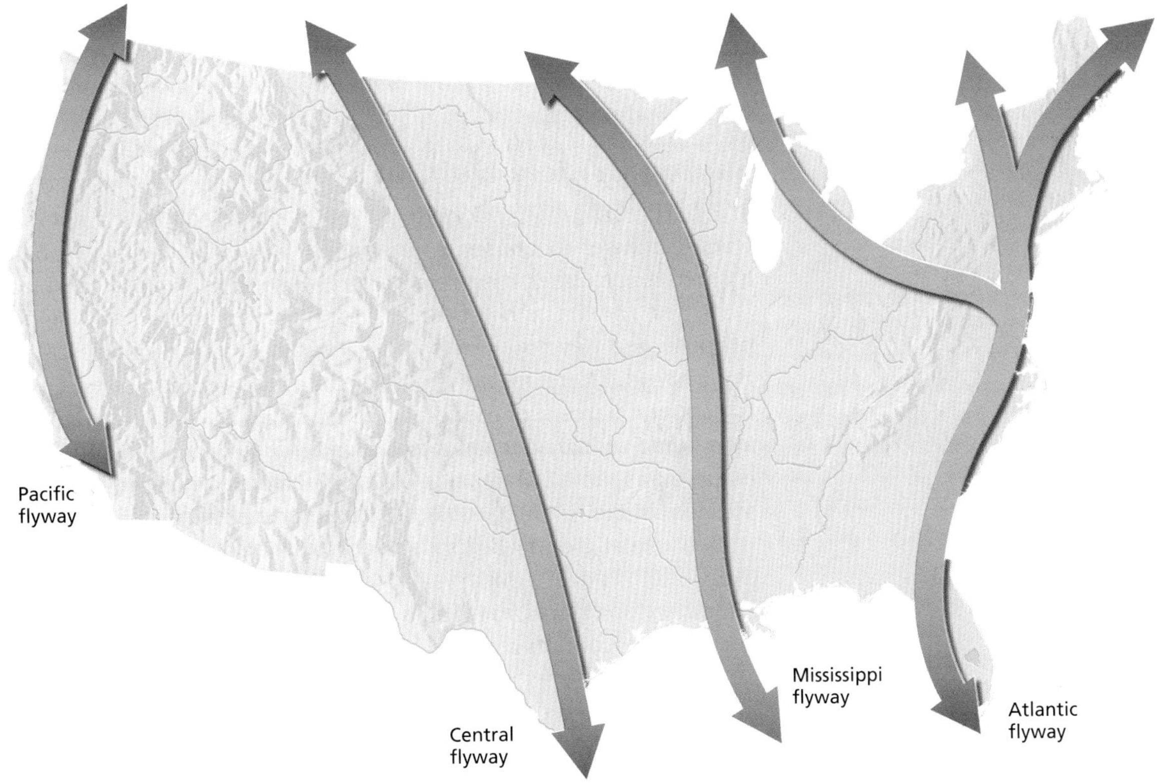

TEACHING STRATEGY
Bird Watching
Explain the utilitarian and non-utilitarian value of birds and bird-watching. Tell students that in many states bird-watching is a more popular sport than hunting. According to some estimates, more than 80 million people participate in bird-watching each year. Bird-watching brings in millions of tourist dollars. The interest in bird-watching is helping to protect bird habitats. Encourage interested students to participate in this activity. Have them observe the birds at their homes or at school, and make a list of the species they see.

TEACHING STRATEGY
Songbirds
Have students listen to a tape or compact disc of bird songs. Select several species (preferably species found in your area), play their songs several times, and then test the students to see if they can identify each species by its song. Point out to students that songs serve an important function for birds. Male songbirds sing to attract females and to deter other males from entering their territory.

MAKING CONNECTIONS
Fine Arts
Ask students to draw or paint a picture of a species of bird from their area. Their work should be labeled with the bird's scientific name, common name, range, and conservation status. Display the paintings and drawings in class.

SECTION 23-3

GIFTED ACTIVITY

Have students participate in a local Christmas Bird Count sponsored by the Audubon Society. Amateurs are welcome to participate in a learning capacity. During this annual event, bird-watchers across the country count all the numbers and species of birds they see within an assigned 15-mile radius on a selected winter day. The data they collect are used by conservation biologists to determine population trends and to pinpoint environmental problems. For more information, contact a local Audubon chapter.

TEACHING STRATEGY

The Species-Area Effect

Remind students that the number of species that an area can support is proportional to its size. The species-area effect, discussed in Chapter 21, has important consequences for the conservation of biodiversity. As a general rule, the smaller a natural area is, the fewer species it can support. When designing preserves, scientists must take the species-area effect into consideration to ensure that the preserve is large enough to sustain populations of the species it is intended to protect.

QUICK FACT

☑ Endangered whale species may have a better chance of survival since the Antarctic Marine Sanctuary was established by the International Whaling Commission in 1994.

FIGURE 23-12

Nationwide surveys of breeding populations of ducks indicate long-term declines in some species.

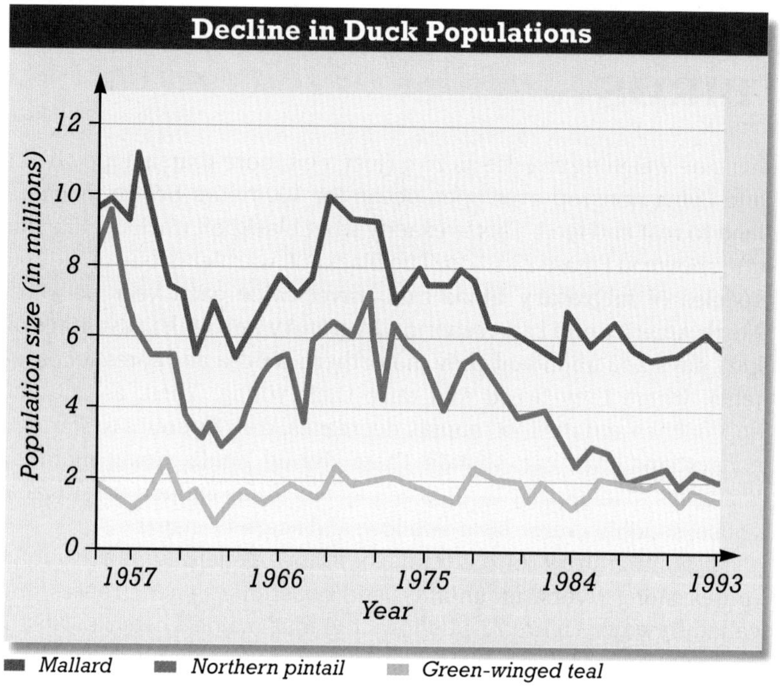

In recent years, scientists, with the help of outdoor recreationists such as hunters and birdwatchers, have documented significant declines in the populations of some migratory birds, including ducks, shorebirds, and songbirds. The number of ducks recorded in the winter of 1993 was the lowest since the surveys began, in 1955, as shown in Figure 23-12. This period was marked by the loss of 60 to 90 percent of prairie wetlands and over 50 percent of all United States wetlands.

Saving Critical Habitat

In 1903, President Theodore Roosevelt established the first national wildlife refuge, to conserve wading birds in Florida. As we have learned more about the preferred travel routes of migratory birds and their habitat and food requirements, biologists have helped propose and develop new wildlife refuges at critical places along the flyways. There are now 500 refuges in the United States, covering about 4 percent of the total area of our country. The refuges are also home to 220 species of mammals, 250 species of reptiles and amphibians, and 200 species of fish, including one-third of all the species on the threatened and endangered species lists.

Because migratory birds also depend on winter habitats outside the United States, conservation efforts have to be international. Migratory songbirds are the focus of a new U.S. Fish and Wildlife Service program, called Partners in Flight–Aves de las Americas, covering the United States and Mexico. The Western Hemisphere Shorebird Reserve Network operates in Central America and South America. Deforestation and coastal development projects remain major threats to migratory birds in these areas.

REINTRODUCTION OF THE WOLF

Gray wolves, shown in Figure 23-13, formerly ranged over most of the United States. For nearly a century, wolves were shot, trapped, and poisoned by people who feared for their own safety or who wanted to protect their livestock. Today the gray wolf is an endangered species, protected by law. In the contiguous 48 states, it is found primarily in Montana and Minnesota. The 1995 total population of wolves in these states was about 2,500 animals. Alaska and Canada have approximately 62,000 wolves.

The current status of wolves in the United States has much to do with human attitudes. Many people fear wolves because of childhood stories. Although sick wolves and wolf-dog hybrids have attacked people, there is no documented case of a healthy wild wolf ever killing a person in North America. Some ranchers regard wolves as threats to their livelihood. As humans have reduced natural-prey populations, some wolves have occasionally attacked livestock to survive. However, only a tiny percentage of livestock in the United States is ever lost to wolf predation. Other people associate positive qualities with wolves. Some view the wolf as a symbol of wilderness, even though wolves are adaptable and do not require wilderness areas to survive.

To a biologist, the wolf is a top carnivore that is dynamically involved with prey species, such as elk and moose. Restoration ecologists became interested in reintroducing wolves to Yellowstone National Park, shown in Figure 23-14, because until about 60 years ago wolves were the top predators of deer, moose, and elk within the park. With the eradication of the wolf—and because no hunting is allowed in national parks—elk populations had grown so large that they may have exceeded the carrying capacity of the park. A proposal was made to reintroduce the wolf to help control elk numbers, to restore a well-known species to the park, and to increase enjoyment of the park for its 3 million annual visitors.

FIGURE 23-13

The gray wolf is an endangered species that is the ancestor of the domestic dog. Adult wolves weigh about 45 kg (100 lb) and are about 2 m (6 ft) long from the nose to the tip of the tail. Wolves are social and commonly form packs with two to eight members.

internetconnect

SCILINKS NSTA
TOPIC: Wolf
GO TO: www.scilinks.org
KEYWORD: HM455

FIGURE 23-14

Yellowstone National Park is the oldest national park in the United States. Wolves lived there until about 60 years ago, when they were exterminated by hunting.

TEACHING STRATEGY

Carnivore Teeth

Show students the skull of a dog. Allow them to examine and touch the teeth. Tell them that the modern dog is a descendant of the wolf and that both species have similar skulls and teeth. Ask students to explain the function of the canine teeth. (The long, sharp canines are used for biting and holding onto prey.) Draw students' attention to the shape of the premolars and molars. Show students the shearing action of these teeth as the jaw closes. The teeth act like a pair of scissors to slice flesh off the bone. Also point out that a wolf or dog can crush bones between its molars.

TEACHING STRATEGY

Other Wolf Reintroduction Projects

Inform students that the Yellowstone wolf project is one of several intended to return wolves to at least part of their historical range. Red wolves have been reintroduced to preserves in North Carolina. As in Yellowstone, a few wolves were shot as soon as they were released; others were lost to disease or injury. But the population is increasing. Now scientists plan to release Mexican wolves, a subspecies of the gray wolf, into areas in southeastern Arizona and southwestern New Mexico. All the released wolves will come from captive populations. Encouraged by the success of the Yellowstone project, several other areas in the United States are now considering wolf reintroductions, including Colorado, Maine, and the Adirondack region of New York.

SECTION 23-3

CONTENT CONNECTION

Section 22-4: Estuaries

Remind students that estuaries are areas where rivers flow into the sea. Estuaries are comparable to rain forests in species richness and productivity. The Everglades are breeding grounds for many species with economic value.

RETEACHING *ACTIVITY*

Ask students to imagine they are creating a plan to restore a damaged ecosystem, such as the Everglades. Have them list all the information they would need and the problems they might encounter. (Answers should include an inventory of the current and past conditions and inhabitants of the ecosystem. The plan would also require studies of the ecology of the species in the ecosystem to ensure that the requirements of the species would be met. The soil and water would need to be analyzed. A problem may be that there are no baseline data on what the ecosystem was like before it was altered.)

GIFTED *ACTIVITY*

Have students do group investigations of endangered species that require an estuarine habitat, such as the Everglades. They should make a visual group presentation. For example, whooping cranes have made a comeback, but their refuge in Texas is deteriorating due to dredging and increasing barge traffic through a channel that runs through the refuge. Maintenance of the channel requires that dredge material be piled up in the marsh habitat. These "spoil islands" have become grassland habitats that are not suitable for the cranes.

FIGURE 23-15

Before being released in Yellowstone Park, the wolves were kept in pens for three months to allow them to become accustomed to their new surroundings. This strategy is called soft release.

The Wolf Reintroduction Plan

After many years of public hearings involving people with a wide range of opinions, the National Park Service agreed to reintroduce wolves to Yellowstone. Special precautions were taken to protect the interests of ranchers opposed to the effort. First, ranchers are permitted to kill wolves that are seen attacking their livestock. Second, individual wolves that become a nuisance to humans will be relocated. Third, a private conservation organization, Defenders of Wildlife, set up a $100,000 fund to reimburse ranchers for any economic losses caused by wolves. This group also offers $5,000 to private landowners who agree to let wolves breed on their property. These financial arrangements were key factors in the approval of the reintroduction program.

Before reintroduction, biological information about wolves and Yellowstone was gathered to give the wolves their best chance of survival. In 1995, 14 wolves from Canada were placed in pens at release sites within the park, as shown in Figure 23-15. Fifteen more were released in central Idaho. In 1996, 17 more wolves were released in Yellowstone, and though 11 wolves were killed (some illegally), about 20 pups from both years survived. As of January 1997, the total number of wolves in and around Yellowstone was 40, and their breeding success continues to be monitored.

RESTORING THE EVERGLADES ECOSYSTEM

An enormous amount of biological information must be gathered in order to reintroduce a single species into an ecosystem and monitor it over time. But imagine what is involved in restoring an entire ecosystem that has been severely damaged. A new federal, state, and local partnership has been formed to restore one of the largest and most species-rich national parks in the United States—the Everglades, shown in Figure 23-16.

Early in this century, land developers were attracted to the large beaches and semitropical climates of southern Florida. They found extensive wetlands that were hard to build on and mosquito populations that discouraged new residents. Over time, a series of drainage canals was dug to intercept water moving south from Lake Okeechobee and divert it to the ocean, drying out the land. The non-native melaleuca tree was also planted to take large amounts of water out of the soil.

At that time, few people could foresee the ecological consequences of these actions, but Marjorie Stoneman Douglas did. She helped acquire land to create Everglades National Park and in 1947 wrote a book, *The Everglades: River of Grass,* that explained how the Everglades functioned. It is not stagnant water, but rather a unique, slow-moving river 80 km (50 mi) wide and 15 cm (6 in.) deep. The Everglades is nearly flat and is dominated by saw grass, yet it is home to over 100 species of water birds. Douglas led a lifetime crusade for the Everglades, which in just 50 years experienced a 50 percent reduction in the amount of wetlands, a doubling of salinity in Florida Bay that killed sea grass and shrimp nurseries, and a 90 percent reduction in populations of wading birds. The diversion of water also prevented the ground water from being replenished, creating water shortages for farmers and residents of Miami. Rain washed fertilizers from agricultural fields and polluted the water that did make it to the park. Heavy metals, such as mercury, poisoned park life.

The newly approved 20-year plan for the Everglades ecosystem includes eliminating some of the drainage canals, restoring the Kissimmee River to its original channel, cutting back stands of melaleuca trees, and purchasing more than 40,000 hectares (100,000 acres) for park protection. It is the most ambitious ecosystem-restoration project attempted in the United States.

FIGURE 23-16
Everglades National Park contains only about 20 percent of the Everglades ecosystem. The map below shows some of the key features related to the decline and restoration of Everglades National Park.

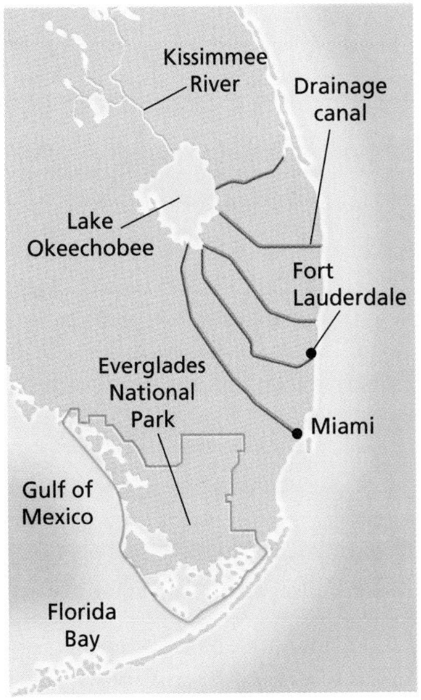

SECTION 23-3

TEACHING STRATEGY

Population Growth in Florida

Inform students that the problems occurring in the Everglades are partly the result of very rapid population growth in Florida. Between 1960 and 1990, Florida's population increased from just under 5 million to nearly 13 million. Population growth has been particularly rapid in the Miami–Fort Lauderdale metropolitan area, which borders the Everglades. The population of Broward and Dade counties grew by nearly 8 percent just between 1990 and 1995, and the greater Miami area now has a population of about 3.5 million, making it the 11th largest metropolitan area in the country.

RECENT RESEARCH

Where Have the Plankton Gone?

South Florida is not the only coastal area that has experienced reduced productivity. A study by the Scripps Institution of Oceanography found that zooplankton populations have dropped by as much as 80 percent along the California coast. The data were collected by 222 ocean cruisers called the California Cooperative Fisheries Investigation. They also recorded a decrease in nutrient upwelling from cold, deep water and more stratification of water, which leads to less mixing and fewer nutrients for plankton. It is not clear whether this population decline is a result of global warming or other factors.

SECTION 23-3

ANSWERS TO SECTION 23-3 REVIEW

1. Conservation biology is concerned with protecting areas that retain most of their biodiversity, while restoration biology is concerned with rebuilding badly damaged ecosystems. Both disciplines use biological knowledge to help conserve biodiversity.
2. Migratory birds are declining because their habitats—including their breeding grounds, overwintering sites, and stopover sites during migration—are being altered or destroyed.
3. Wolves are being restored to Yellowstone National Park to help control herbivore populations, to provide enjoyment for park visitors, and to return a part of the park's former biodiversity.
4. Ranchers will be compensated for any losses of livestock. Also, wolves that become a nuisance will be relocated. Though endangered and protected by law, wolves can be shot if they are seen attacking livestock.
5. The diversion of water from the Everglades reduced the recharge of ground water, leading to water shortages in Miami and surrounding areas. Florida Bay, south of the Everglades, became excessively salty because it no longer received the fresh water that had passed through the Everglades.
6. More tourists might visit the park and spend money in stores in the surrounding areas. Also, farmers and ranchers might benefit from the reduction in elk populations caused by the wolves. Elk would be less likely to damage crops, and domestic animals would face less competition for forage.

GETTING INVOLVED

It is important for individuals to get involved in conservation, and the best place to start is at home. The first step is to learn about your local environment. For example, apply these questions to where you live.

1. *Name five native plants, and determine their seasons. Can they be used for landscaping homes or businesses?*
2. *Name five resident birds and five migratory birds. Are there any special laws that protect them?*
3. *Name two major agricultural crops. How do farmers or ranchers obtain water for crops or livestock?*
4. *Trace the path of water that you use from when it falls as precipitation to when it flows from your faucet. Where does the water go after you have used it, and how is it treated?*
5. *Name three endangered species in your area. If any species have become extinct, what was the cause?*
6. *Trace the path of garbage after it is collected. Does your sanitation department support recycling?*
7. *Describe the primary geological processes that helped form the land where you live. If the land was shaped by water, wind, glaciers, or volcanoes, are any of these still contributing to geologic change?*
8. *Find the names and addresses of two nongovernment conservation associations that are active in your region. Will they allow you to act as a volunteer?*

Exploring these questions may lead you to your own ideas about what you can do to maintain biodiversity or improve the ecological integrity in your area. A new environmental field, called **urban ecology,** involves people who are interested in the challenge of increasing biodiversity in the most heavily developed areas.

SECTION 23-3 REVIEW

1. **How are conservation biology and restoration biology different? How are they similar?**

2. **Why are some populations of migratory birds declining?**

3. **Give three reasons why wolves are being restored to Yellowstone National Park.**

4. **What rules have been made to protect the interests of ranchers concerned about wolves?**

5. **How did the diversion of water from the Everglades lead to environmental problems?**

6. **CRITICAL THINKING What benefits might people living near Yellowstone National Park receive from the reintroduction of wolves?**

CHAPTER 23 REVIEW

SUMMARY/VOCABULARY

23-1

- Currents of air and water are linked into a global system that is responsible for climate.
- El Niño events occur when the normal east-to-west winds across the southern Pacific ocean reverse, causing a variety of effects on organisms worldwide, including humans.
- Over a short time period, humans have affected global systems, including altering the composition of the atmosphere by decreasing ozone levels and increasing carbon dioxide levels.
- Industrial chemicals called CFCs are destroying the ozone layer. A treaty to ban CFC production has been signed.
- From the results of computer models of the atmosphere, a large majority of scientists have concluded that increased carbon dioxide levels have resulted in warmer surface temperatures on the Earth. Scientists expect temperatures to continue to rise as fossil fuel use and carbon dioxide levels increase.

Vocabulary

cause-and-effect relationship (444)
chlorofluorocarbons (443)
convection cell (441)
correlation (444)
El Niño (442)
environmental science (441)
upwelling (442)

23-2

- *Biodiversity* refers to the variety of life found in a given area and can be measured in different ways, including by species richness, evenness, and genetic diversity.
- Scientists estimate that there are at least 10 million species on Earth and may be as many as 30 million. Scientists have described fewer than 3 million species so far.
- Insects and plants make up the majority of species on Earth, especially in tropical rain forests, which are rapidly being destroyed. Two new ideas for conserving tropical biodiversity are debt-for-nature swaps and ecotourism.
- People value biodiversity for utilitarian reasons, which emphasize economic benefits from species. Some of these benefits include medicines, foods, and other useful products, as well as ecosystem services. Nonutilitarian reasons for conserving biodiversity draw on the assertion that living things have intrinsic value. This assertion often derives from moral or religious beliefs.

Vocabulary

biodiversity (448)
debt-for-nature swap (450)
ecotourism (451)
evenness (449)
genetic diversity (449)
nonutilitarian value (451)
utilitarian value (451)

23-3

- Conservation biology and restoration biology are two new disciplines. Conservation biologists are concerned with identifying and maintaining areas that are still relatively undisturbed, whereas restoration biologists are usually involved with repairing badly damaged ecosystems.
- Populations of some migratory birds appear to be in decline due to habitat destruction by humans, but they are being helped by new refuges and international partnerships.
- After a 60-year absence, the gray wolf has been successfully reintroduced in small numbers to Yellowstone National Park to help control elk populations and to increase public enjoyment.
- A new 20-year plan to restore the Everglades ecosystem has recently been approved.

Vocabulary

conservation biology (452)
flyway (453)
migratory bird (453)
restoration biology (452)
urban ecology (458)

CHAPTER 23 REVIEW ANSWERS

REVIEW

1. Environmental science looks at the large-scale relationships between humans and the Earth.
2. Flyways are the routes followed by migratory birds.
3. Chlorofluorocarbons contain chlorine (chloro-), fluorine (-fluoro-), and carbon.
4. Ecotourism involves preserving an area to attract tourists, who in turn spend money to support the local economy.
5. An upwelling is a rising current that brings nutrients to the surface. El Niño shuts off the upwelling off the coast of Peru.

6. a **7.** b
8. a **9.** c
10. b **11.** a
12. d **13.** d
14. a **15.** d

16. Moist air heated at the equator rises and releases large amounts of water, creating moist conditions that favor the development of tropical rain forests. To the north and south of the equator, falling air in convection cells is drier and associated with deserts.
17. The usual pattern of water circulation in the Pacific Ocean is reversed, cutting off the upwelling of nutrients. El Niño causes drought in Australia and brings more rain to the southeastern United States.
18. Interdependence means that each species in an ecosystem depends on other species for its survival. Anchovies depend on plankton nourished by the upwelling of nutrients. Humans and birds depend on anchovies.
19. A correlation is a mathematical relationship between two variables in which high and low values correspond. In a cause-and-effect relationship, a change in one variable leads directly to a change in the other variable.
20. It could lead to an increased demand for water, food, and agricultural lands, with decreased resources for other organisms.

It could increase pollution and perhaps worsen global warming due to greater fossil fuel use.

21. In the past, most people feared wolves or saw them as competitors, leading to the near eradication of the species from the contiguous United States. Now many people have more positive views of wolves, associating them with wilderness, and this has led to the reintroduction of wolves into Yellowstone Park.

22. Community A has the most species richness. Community B has the highest evenness.

23. Answers will vary. Humans can conserve ground water by planting crop, yard, or garden plants suited to the local climate, among other things. Using and recycling ground water efficiently would help preserve ecosystems that depend on springs for water.

CRITICAL THINKING

1. With fewer nutrients available, competition may increase among anchovies, and the populations of anchovy predators will likely decline. Some of these predators may switch to other prey species.

2. Photosynthetic organisms were largely responsible for the addition of oxygen to the early atmosphere, and they therefore made the formation of the ozone layer possible.

3. Estimate how much water each person needs for drinking, cooking, bathing, and other personal uses. Then calculate the amount of water needed, per person, for agriculture and the manufacture of critical goods. Adding these two numbers gives the total amount of water needed per person. Divide the total amount of fresh water available by this number to estimate carrying capacity. New technology could increase this figure by reducing the amount of water needed for agriculture, industry, or personal use.

REVIEW

Vocabulary

1. Define the term *environmental science.*
2. What are flyways?
3. What chemical elements would you expect to find in chlorofluorocarbons?
4. Describe ecotourism.
5. What is an upwelling, and what connection does it have to El Niño?

Multiple Choice

6. Which of the following is a consequence of El Niño? (a) decreased anchovy production (b) greater number of fires in the southern United States (c) increased guano production (d) ozone depletion
7. Ozone protects organisms from (a) meteor impacts (b) harmful radiation (c) salt depletion (d) cold temperatures.
8. Which of the following is not a use for CFCs? (a) fuel (b) coolant in refrigerators (c) propellant in aerosol cans (d) cleaning electronics.
9. Since 1850, carbon dioxide levels have (a) decreased by 30 percent (b) remained about the same (c) increased by 30 percent (d) not been measured accurately.
10. Which of the following is *not* true of biodiversity?
    (a) It is decreasing.
    (b) It consists mostly of mammals and reptiles.
    (c) It has declined sharply at least five times in the past.
    (d) It is higher in tropical rain forests than in any other biome.
11. Biologists estimate that most species on Earth are (a) insects (b) plants (c) mammals (d) fungi.
12. Benefits from biodiversity include (a) medicines (b) useful products (c) water purification (d) all of the above.
13. Stopovers are used by migratory birds to (a) breed (b) seek new habitat (c) avoid predators (d) feed and rest.
14. Wolves (a) are carnivores (b) need wilderness areas (c) are usually solitary (d) usually attack humans.
15. Which of the following is true about the Everglades ecosystem?
    (a) It is a stagnant swamp.
    (b) It has doubled in size over the last 50 years.
    (c) It supports very few species.
    (d) It has been damaged by pollution.

Short Answer

16. Explain how convection cells are associated with biomes.
17. What happens during El Niño? Describe two ways that El Niño affects the economy of countries other than Peru.
18. What do environmental scientists mean by *interdependence*? Give an example of interdependence from this chapter.
19. Explain the difference between a correlation and a cause-and-effect relationship. How can scientists distinguish between the two?
20. How could future human population growth affect the environment?
21. Describe some of the different attitudes toward wolves. How have these attitudes affected the status of wolves?
22. Which of the communities in the table below has the highest species richness? the greatest evenness? Explain your answers.

Species Richness and Evenness

Community	*Number of individuals of each species* 1	2	3	4
A	7	1	1	1
B	3	3	4	0
C	0	9	0	1

23. **Unit 7—*Ecosystem Dynamics***
    BIOLOGY INTERACTIVE TUTOR Write a report summarizing ways that humans can work to reduce the depletion and pollution of ground water. How would more efficient use of ground water benefit ecosystems?

CRITICAL THINKING

1. You are aware of some of the connections between plankton, anchovies, humans, and birds. Suggest other species interactions that are likely to be influenced by El Niño.
2. The formation of the ozone layer depended on the presence of oxygen in the atmosphere. Drawing on what you have learned about the history of life, explain how organisms affected and were affected by the forming ozone layer.
3. There are no substitutes for clean, fresh water, which is in short supply worldwide. Environmental scientists think fresh water may become a limiting factor on human population growth. Explain how you could estimate Earth's carrying capacity for humans based just on the availability of fresh water. What information do you need in order to make this estimate? How might technological advances change your estimate?
4. What question would you add to the list shown on page 458? Explain your choice. Share your question with your classmates.
5. As part of the global treaty to eliminate CFCs, developed countries are contributing to a fund to help the developing countries buy CFC substitutes. What benefits do the developed countries receive from this investment?
6. It is a widely held belief that the birth rate declines when countries make progress toward industrialization. However, birth rate has declined in Sri Lanka and Costa Rica, countries with minimal industrial development. How might these data be explained?
7. Look at the graph below. Notice that although the long-term trend is toward higher carbon dioxide levels, the carbon dioxide concentration fluctuates during each year, falling in the spring and summer and rising in the fall. What do you think causes this fluctuation?

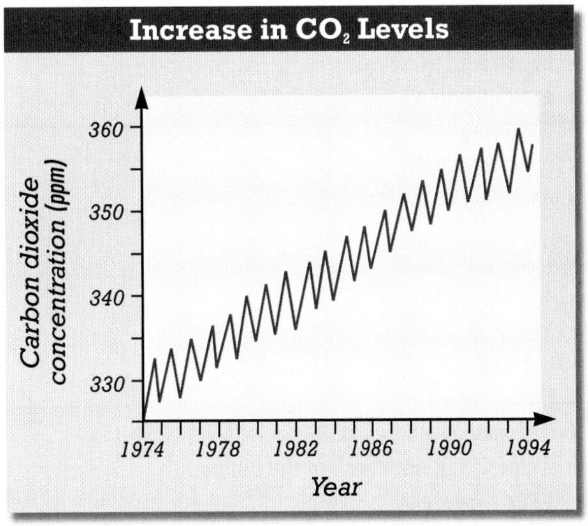

EXTENSION

1. In 1995, three scientists—Paul Crutzen, Sherwood Rowland, and Mario Molina—were awarded the Nobel Prize in chemistry for their work on the ozone layer. Use library resources or an on-line database to research each scientist's contribution to our understanding of the ozone layer. Write a short report that summarizes what you have learned.
2. Form a cooperative team with another member of your class. Work with your partner to answer the questions listed on page 458. You may need to consult several sources of information, including the library, an on-line database, local government agencies, and a nearby university, zoo, or botanical garden. When you have answered all of the questions, prepare a poster that displays what you have learned.
3. Read "In the Wake of the Spill" in *National Geographic,* March 1999, on page 96. Describe the ecological disaster that occurred in Alaska's Prince William Sound just after midnight on March 24, 1989. Which was more effective in the recovery of the ecosystem, nature or cleaning crews? Explain. What other unrelated ecological changes are happening in Alaska?

4. Answers will vary. The question "What will your local environment look like in 50 years?" is a possible answer.
5. The developed countries hasten the recovery of the ozone layer and reduce the number of skin-cancer cases and the agricultural damage caused by ozone depletion.
6. Student responses will vary but should center on factors that may be more closely linked to birth rate than to per-capita income. Some possibilities include adequate nutrition, basic health care, proper sanitation, and education of women.
7. Carbon dioxide levels rise and fall with the seasons because the rate of photosynthesis varies seasonally.

EXTENSION

1. In 1970, Crutzen was the first researcher to show that chemicals produced at the Earth's surface could reach the upper atmosphere and destroy ozone. In 1974, Rowland and his student Molina sounded the first alarm about CFCs and the ozone layer.
2. Answers will vary.
3. On March 24, 1989, a fully loaded oil tanker ran up hard on the shoals of Bligh Reef deep inside Prince William Sound. Alaskan crude oil (11 million gallons) began to flow from the tanker's ruptured hull. Many thousands of birds and marine mammals were poisoned, and 500 mi of Alaskan beaches were spoiled. Over the years the natural flushing action of waves and weather have been more effective at restoration than all the cleanup crews. Alaska has warmed 5°F in the past 30 years. Glaciers are retreating, and the food web is changing.

TIME REQUIRED

One 50-minute class period

SAFETY PRECAUTIONS

Be sure that students wear safety goggles during the lab. Also remind them to be very careful when handling the hot water, as it can scald.

QUICK REFERENCES

Lab Preparation notes are found on page 439B.

Holt BioSources provides a Teaching Resources CD-ROM that contains "Using Gowin's Vee in the Lab" and "Scoring Rubrics."

PROCEDURAL TIPS

1. Students can work in groups of two to four.
2. The U-shaped tubes must be completely filled to distribute the paramecia. If large air bubbles form in the tubing, the bubbles will move to the bend in the tube when it is inverted into the beakers, and the paramecia will not be able to move freely.
3. Check that glass tubing has no cracks or fractures before students begin the investigation.

ANSWERS TO BACKGROUND

2. A pollutant is anything introduced into the environment that can be harmful.
3. If heat harms living things, it is considered a pollutant.
4. Power plants can cool the water before returning it to the environment.

CHAPTER 23 INVESTIGATION

Testing the Effects of Thermal Pollution

OBJECTIVES

- Model the effects of thermal pollution on living organisms.
- Apply the underlying scientific principles to environmental issues.

PROCESS SKILLS

- hypothesizing
- experimenting
- observing
- organizing data
- analyzing data

MATERIALS

- 400 mL beakers, 2
- ice
- hot water
- thermometer
- U-shaped glass tubing, 30 cm long
- 2 corks to fit both ends of the tubing
- 125 mL beaker
- water
- *Paramecium* culture
- hand lens
- stopwatch or clock
- glass-marking pen or wax pencil

Background

1. Some power plants use water from rivers as a coolant. After the water is used, it is much hotter than the water in the river.
2. What is a pollutant?
3. How can heat be considered a pollutant?
4. How can power plants release nonharmful water?

Procedure

1. Discuss the objectives of this investigation with your partners. Develop a hypothesis concerning the effect of temperature on *Paramecium.*
2. Design an experiment using the given materials to test your hypothesis. In your experiment, *Paramecium* will be contained in the U-shaped tube. One large beaker will be filled with ice, and the other large beaker will be filled with hot water. Other materials that you can use in your experiment are listed in your materials list.
3. In designing your experiment, decide which factor will be an independent variable. Plan how you will vary your independent variable. In your lab report, list your independent variable and your method of varying it.
4. Decide which factor will be the dependent variable in your experiment. Plan how you will measure your dependent variable. In your lab report, list your dependent variable and your method of measuring it.
5. In most experiments, a control is necessary. Plan your control, and describe it in your lab report.
6. Discuss your planned experiment with your teacher. Proceed with your experiment only after you have received your teacher's permission to do so.
7. CAUTION **Water hotter than 60°C can scald. Be careful handling hot water, and alert your teacher if you burn yourself.** Fill a 400 mL beaker with ice and water. Make sure that ice remains in the beaker for the entire experiment. Fill another 400 mL beaker with 60°C tap water.
8. In a 125 mL beaker, gently swirl 20 mL of water and 20 mL of *Paramecium* culture. Your teacher will provide aged tap water or spring water for you to use during this step. (Chlorinated water would kill *Paramecium.*)
9. While your partner holds the test tube steady, carefully pour the *Paramecium*-and-water mixture into the U-shaped tube. Fill the tube completely, leaving just enough room for a cork at each end of the tube. Make sure there are no large air bubbles in the tube. Place a cork at each end of the tube.
10. Create a table similar to the one at right to record your data. For example, the table below is designed to

record the number of *Paramecia* in three parts of the U-shaped tube over time. Design your data table to fit your own experiment. Remember to allow plenty of space for recording your data.

11. Proceed with your experiment, using the tube, ice water, hot water, and hand lens to observe any response of *Paramecium* to the environment.
12. As you conduct your experiment, record your results, including the number of *Paramecium* and the time involved, in your data table. Organize your data so that others reading your lab report will be able to understand the results of your experiment.
13. Clean up your materials and wash your hands before leaving the lab.

Analysis and Conclusions

1. Did the results of your experiment support your hypothesis? Explain your answer.
2. What effect did heat and cold have on *Paramecium* in your experiment?
3. What evidence do you have that *Paramecium* preferred one temperature range to another?
4. What are some possible sources of error in your experiment?
5. How might a pollutant cause an increase in the number of organisms? Explain.
6. Judging from your experiment, how do you think other organisms might react to a change in water temperature?
7. How could a power plant change the type of organisms that live in the water where it releases its cooling water?

Further Inquiry

Develop a hypothesis about the effects of acid rain on *Paramecium,* and design an experiment to test your hypothesis.

OBSERVATIONS OF PARAMECIA

		Number of *Paramecia*		
		Cold end	**Hot end**	**Middle**
	0			
Elapsed	**15**			
time	**30**			
(seconds)	**45**			
	60			
	75			

ANSWERS TO ANALYSIS AND CONCLUSIONS

1. Answers will vary depending on students' hypotheses.
2. The paramecia avoided both heat and cold, moving to the center of the U-shaped tube.
3. They move away from sections of the tube with extreme temperatures.
4. Answers will vary. Sources of error might include jarring the tube, unhealthy paramecia, use of temperatures that do not affect the paramecia, inaccurate counting, or giving the paramecia insufficient time to respond.
5. A pollutant might be harmful to some species but beneficial to others. For instance, those organisms suited to higher temperatures will increase when thermal pollution raises water temperatures.
6. Organisms might move out of the area, die off, or increase in number, depending on the type of organism and the exact temperature.
7. Only organisms that can tolerate the higher temperatures are able to live in thermally polluted water.

Unit

6

CHAPTERS

MICROORGANISMS

"*In the leaves of every forest, in the flowers of every garden, in the waters of every rivulet, there are worlds teeming with life...*"

From *Thoughts on Animalcules*, or *A Glimpse of the Invisible World Revealed by the Microscope*, by G. A. Mantell, as quoted in Primo Levi's book *Other People's Trades.*

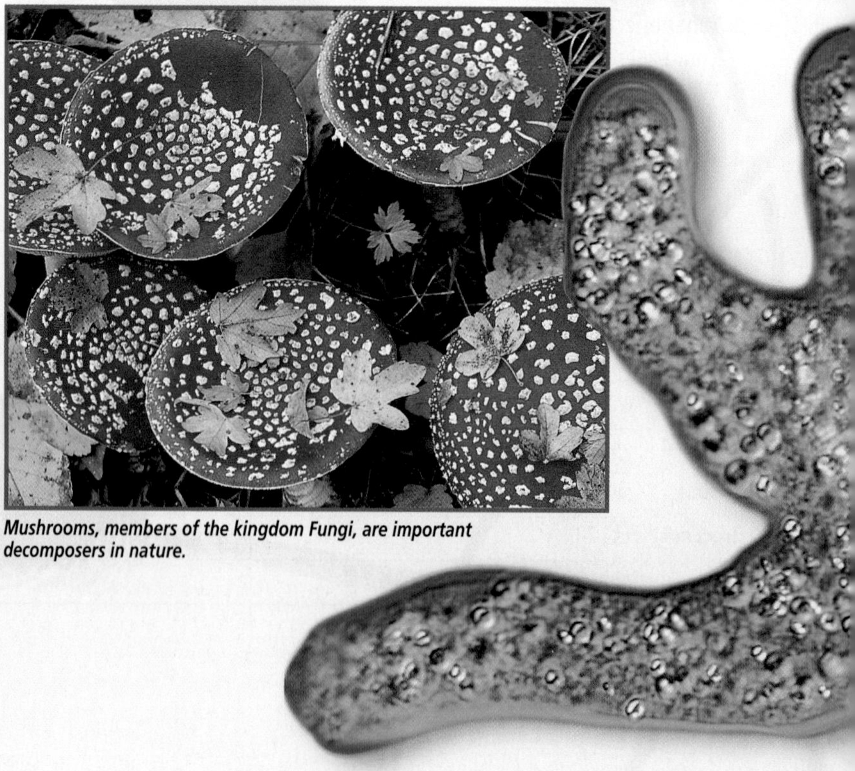

Mushrooms, members of the kingdom Fungi, are important decomposers in nature.

internet**connect**

SCiLINKS NSTA

National Science Teachers Association *sci*LINKS Internet resources are located throughout this unit.

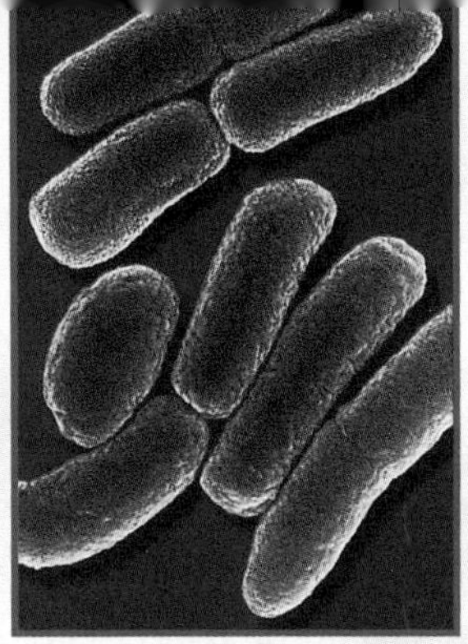

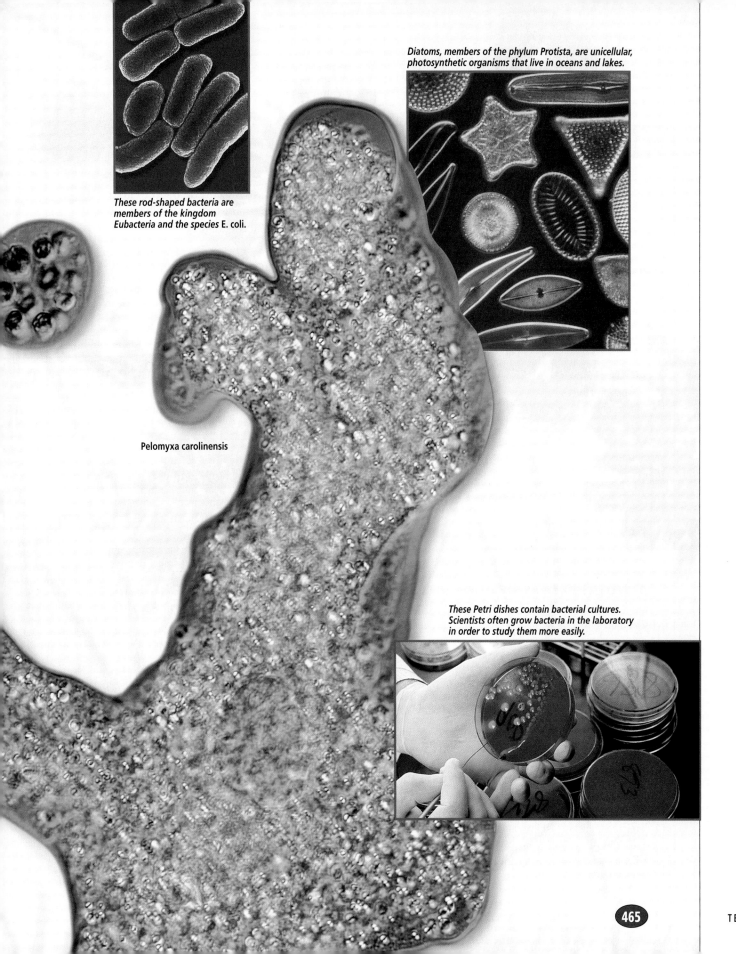

These rod-shaped bacteria are members of the kingdom Eubacteria and the species E. coli.

Diatoms, members of the phylum Protista, are unicellular, photosynthetic organisms that live in oceans and lakes.

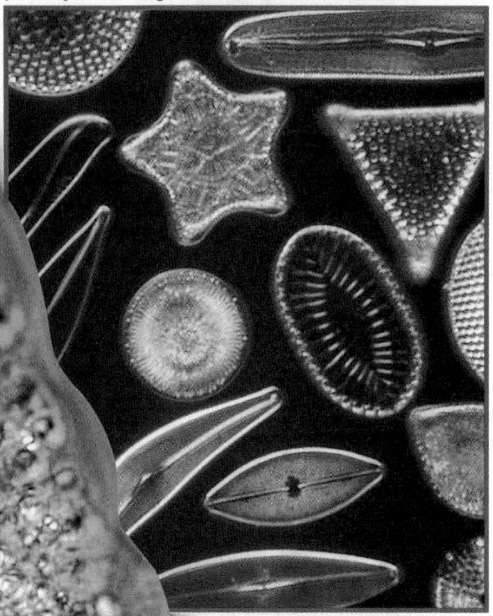

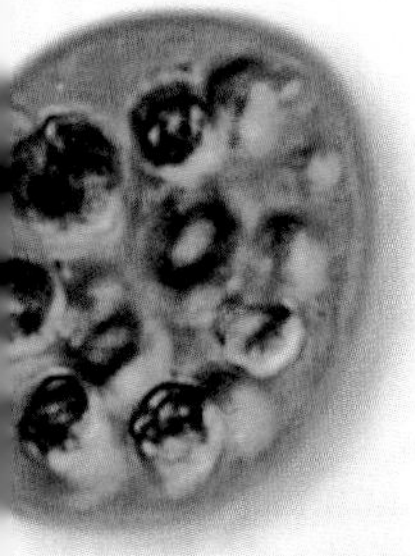

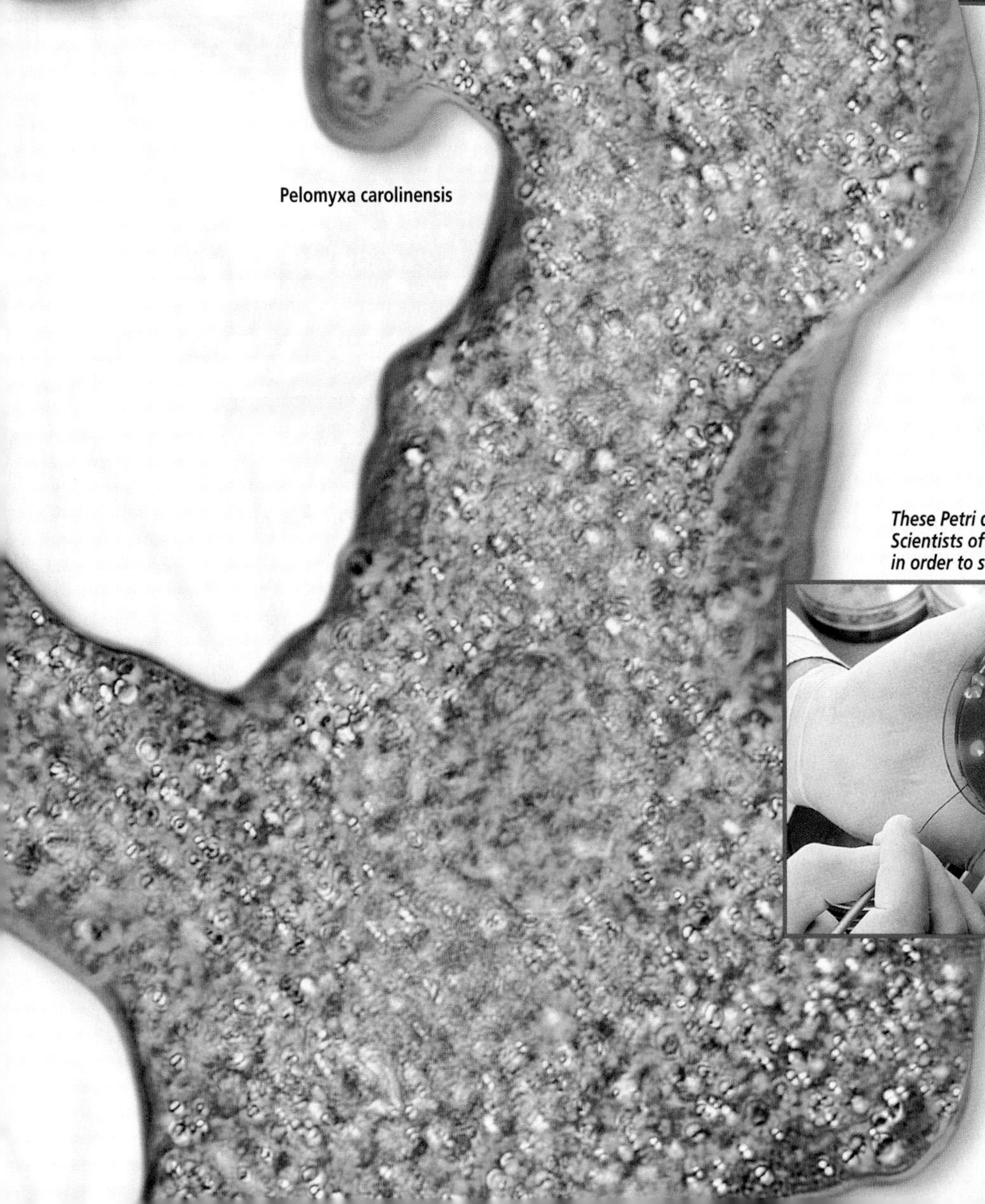

Pelomyxa carolinensis

These Petri dishes contain bacterial cultures. Scientists often grow bacteria in the laboratory in order to study them more easily.

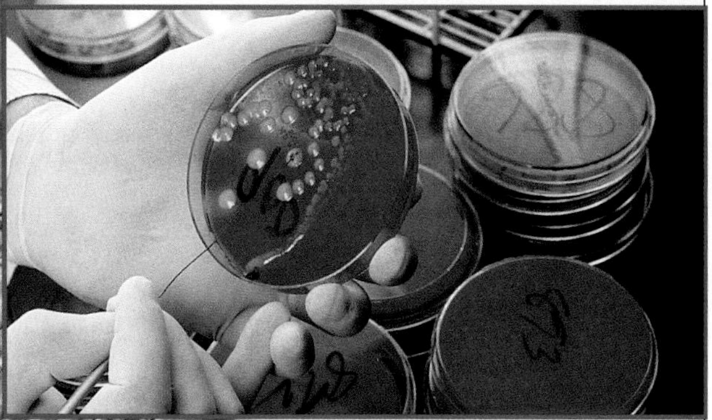

unit 6

MICROORGANISMS

CHAPTERS

CONNECTING TO THE STANDARDS

The following chart shows the correlation of Unit 6 with the *National Science Education Standards* (grades 9–12). The items in each category are addressed in this unit and should be used to guide instruction. Annotated descriptions of the **Life Science Standards** are found below the chart. Consult the *National Science Education Standards* for detailed descriptions of each category.

UNIFYING CONCEPTS
- Systems, order, and organization
- Evolution and equilibrium
- Form and function

SCIENCE AS INQUIRY
- Abilities necessary to do scientific inquiry
- Understandings about scientific inquiry

SCIENCE AND TECHNOLOGY
- Abilities of technological design
- Understanding about science and technology

PHYSICAL SCIENCE
- Motions and forces
- Interactions of energy and matter

LIFE SCIENCE

EARTH AND SPACE SCIENCE
- Energy in the Earth system
- Geochemical cycles

SCIENCE IN PERSONAL AND SOCIAL PERSPECTIVES
- Personal/community health
- Science and technology in local and global challenges

HISTORY AND NATURE OF SCIENCE
- Science as a human endeavor
- Historical perspectives

ANNOTATED DESCRIPTIONS OF THE CORRELATED LIFE SCIENCE STANDARDS

Although all eight categories of the *National Science Education Standards* are important, the following descriptions summarize the **Life Science Standards** that specifically relate to Unit 6.

THE CELL
- Cells have particular structures that underlie their functions.
- Cells can differentiate. Complex multicellular organisms are formed from highly organized arrangements of differentiated cells.

BIOLOGICAL EVOLUTION
- The great diversity of organisms is the result of more than 3.5 billion years of evolution that has filled every available niche with life-forms.
- The millions of different species of plants, animals, and microorganisms that live on Earth today are related by descent from common ancestors.
- Biological classifications are based on how organisms are related.

INTERDEPENDENCE OF ORGANISMS
- Living organisms have the capacity to produce populations of infinite size, but environments and resources are finite.

MATTER, ENERGY, AND ORGANIZATION IN LIVING SYSTEMS
- The energy of life primarily derives from the sun.
- The distribution and abundance of organisms and populations in ecosystems are limited by the availability of matter and energy and by the ability to recycle materials.
- Matter and energy flow through different levels of organization of living systems and between living systems.

TRENDS IN BIOLOGY

Emerging Diseases

Every few years, it seems, a new deadly disease captures front-page headlines and triggers widespread fear in the public. In 1996, for instance, the British government set off a near panic among its citizens by announcing a possible connection between bovine spongiform encephalopathy (mad cow disease) and 10 cases of Creutzfeldt-Jakob disease, a fatal degeneration of the brain. In 1993, the United States experienced its own disease outbreak, when 24 people in the Southwest became ill and 11 died from a previously unrecognized lung disease, hantavirus pulmonary syndrome.

Scientists refer to newly discovered diseases like these as emerging diseases. The appearance of so many emerging diseases (see table below) in recent years raises a number of important questions: Where do such diseases come from? What environmental and social conditions favor or inhibit their spread? How much of a threat do they pose to public health? Can we anticipate and prevent outbreaks? The field of emerging diseases is young, and scientists are just beginning to gather the data necessary to answer these questions.

Some Recently Discovered Diseases

Disease	Pathogen	Transmitted by	First identified
AIDS	virus	contact with blood or body fluids	United States, 1981
Ebola fever	virus	contact with blood or body fluids	Zaire, 1976
Hantavirus pulmonary syndrome	virus	rodents	Southwestern U.S.,1993
Lassa fever	virus	rodents	Nigeria, 1969
Legionnaires' disease	bacterium	air and water borne	Pennsylvania, 1976
Lyme disease	bacterium	ticks	Connecticut, 1976

ORIGIN OF EMERGING DISEASES

An emerging disease may be a truly new disease (one caused by a pathogen that has not previously made people sick), or it may just be new to science. A disease can escape detection for a long time if it is confined to a remote area with inadequate medical facilities, if there are few cases, or if the tools necessary for diagnosis have only recently been developed.

A case in point is hantavirus pulmonary syndrome (HPS). After the cause of the 1993 outbreak was pinned down (a virus now called Sin Nombre), epidemiologists searched for evidence of the virus in preserved tissues and blood samples from people who had died from undiagnosed respiratory diseases. They identified 50 cases of HPS from as far back as 1959. Each year between 50,000 and 150,000 people die from unexplained respiratory disease. It is not surprising that 50 cases of HPS went unnoticed.

For most people, probably the most important question about emerging diseases is Should we be worried? Absolutely, say several recent popular books, which breathlessly prophesy that a worldwide epidemic in which millions (or even billions) could die is just around the corner. Most scientists conclude that while the predictions of our species' imminent demise are greatly exaggerated, there is still cause for concern. Even without a global plague, emerging diseases will cause a great deal of sickness and death in the future. And the appearance and rapid spread of AIDS (a disease recognized only in 1981) shows how swiftly an emerging disease can become a serious health threat.

Additional Information

Garrett, Laurie. *The Coming Plague.* New York: Penguin, 1995.

Lanchester, John. "A New Kind of Contagion." *New Yorker,* December 2, 1996, 70–81.

Le Guenno, Bernard. "Emerging Viruses." *Scientific American,* October 1995, 56–64.

National Center for Infectious Diseases. *Emerging Infectious Diseases:* On-line journal. World Wide Web: http://www.cdc.gov/ncidod/EID/eid.html

CHAPTER 24
BACTERIA

	TOPICS	TEACHING RESOURCES	LABS, CLASSWORK, AND HOMEWORK
BLOCKS 1 & 2 90 minutes	**Introducing the Chapter, p. 466**	ATE Focus Concept, p. 466 ATE Assessing Prior Knowledge, p. 466 ATE Understanding the Visual, p. 466	■ Supplemental Reading Guide, *Microbe Hunters*
	24-1 Bacterial Evolution and Classification, p. 467 Evolution, p. 467 Kingdom Archaebacteria, p. 468 Kingdom Eubacteria, p. 470	ATE Section Overview, p. 467 ATE Critical Thinking, p. 468 ATE Visual Link Figure 24-4, p. 471 88, 88A, 89, 90, 91	PE **Quick Lab**, p. 472, Predicting the Spread of Disease **Lab C32,** Gram Staining of Bacteria **Lab C33,** Gram Staining of Bacteria—Treatment Options ★ Study Guide, Section 24-1 PE Section Review, p. 473
BLOCK 3 45 minutes	**24-2 Biology of Bacteria, p. 474** Structure, p. 474 Nutrition and Growth, p. 476 Genetic Recombination, p. 477	ATE Section Overview, p. 474 ATE Critical Thinking, p. 475 91	**Lab B11,** Composting **Lab C31,** Using Aseptic Technique **Lab D10,** Ice-Nucleating Bacteria ★ Study Guide, Section 24-2 PE Section Review, p. 477
BLOCKS 4 & 5 90 minutes	**24-3 Bacteria and Humans, p. 478** Bacteria and Disease, p. 478 Useful Bacteria, p. 480	ATE Section Overview, p. 478 ATE Critical Thinking, pp. 478, 479 ATE Visual Link Figure 24-6, p. 480 92	PE **Chapter 24 Investigation,** p. 484 **Lab A16,** Using Bacteria to Make Food **Lab D11,** Oil-Degrading Microbes **Lab D12,** Can Oil-Degrading Microbes Save the Bay? ■ Occupational Applications Worksheets—Sanitarian ★ Study Guide, Section 24-3 PE Section Review, p. 480
BLOCKS 6 & 7 90 minutes	**REVIEW and ASSESSMENT** PE Chapter 24 Review, pp. 482–483	Audio CD Program	**CHAPTER TESTING** ■ Chapter Test (blackline master) ▲ Test Generator ■ Assessment Item Listing ■ Scoring Rubrics and Classroom Management Checklists

MODERN BIOLOGY | **HOLT BIOSOURCES**

KEY

- PE Pupil's Edition
- ATE Teacher's Edition
- ■ Active Reading Guide
- ★ Study Guide
- Teaching Transparencies
- Laboratory Program

One-Stop Planner CD-ROM
Shows these resources within a customizable daily lesson plan:
- ■ Holt BioSources Teaching Resources
- ■ Active Reading Guide
- ★ Study Guide
- ▲ Test Generator

READING FOR CONTENT MASTERY

Preview: Have students preview the objectives, the introductory paragraph, topic sentences, and captions before reading the chapter.

- ATE Active Reading Technique: Reading Organizer, p. 467
- ■ Active Reading Guide Worksheet 24-1
- PE Summary/Vocabulary, p. 481

- ■ Active Reading Guide Worksheet 24-2
- PE Summary/Vocabulary, p. 481

- ■ Active Reading Guide Worksheet 24-3
- PE Summary/Vocabulary, p. 481

TECHNOLOGY AND INTERNET RESOURCES

CNN PRESENTS Video Segment 33 Salmonella Outbreak

Holt Biology Videodiscs Teacher's Correlation Guide, Lessons 24-1 through 24-3

Audio CD Program, Sections 24-1 through 24-3

internetconnect

SCILINKS NSTA

On-line Resources: www.scilinks.org
The following *sci*LINKS Internet resources can be found in the student text for this chapter:

Topics:
- Eubacteria, p. 471
- Proteobacteria, p. 473

go.hrw.com

On-line Resources: go.hrw.com
Visit the HRW Web site for a variety of resources related to this chapter. Just type in the keyword HM2 Chapter 24.

Smithsonian Institution®
Visit **www.si.edu/hrw** for additional on-line resources.

CNNfyi.com
Visit **www.cnnfyi.com** for late-breaking news and current event stories selected just for you.

LAB ACTIVITY PLANNING

CHAPTER 24 INVESTIGATION

Culturing Bacteria, pp. 484–485

OVERVIEW
In this investigation, students will test common surfaces for the presence of bacteria.

SAFETY
Treat all growth in the Petri dishes as pathogenic. Once Petri dishes have been sealed, do not allow students to open them again. Students should dispose of the cotton swabs in a biohazard bag.

PREPARATION
1. Each group of students will need one Petri dish containing nutrient agar. To prepare the Petri dishes, melt the nutrient agar in a microwave or hot-water bath and pour the contents into the bottom lid of each Petri dish. Sterilize the Petri dishes at 15 psi for 20 minutes. Allow the Petri dishes to cool for at least 30 minutes. Refrigerate dishes until lab time. After refrigeration, let the Petri dishes stand at room temperature for 30 minutes to 1 hour before beginning the lab.
2. Each student group will also need four sterile cotton swabs, distilled water, a glass-marking pen, and tape.

Quick Labs

Predicting the Spread of Disease, p. 472
Students will need disposable gloves, lab aprons, safety goggles, clear plastic cups, and water for this activity. Additionally, one student will need a 50 percent solution of household ammonia (half household ammonia and half water). You will need 3–4 drops of phenolphthalein per student.

UNDERSTANDING THE VISUAL

Have students look at the micrograph of *Clostridium perfringens.* Ask what structural information they can infer about bacteria from the photograph (small, no internal organelles present). Ask students to describe the shape of the bacteria (rod-shaped chains).

FOCUS CONCEPT

Structure and Function
In this chapter the relationships between bacterial structures and their functions are explored. Examples include the cell wall, which protects bacteria, and flagella, which enable movement.

ASSESSING PRIOR KNOWLEDGE

Review the following concepts.

Cell Structure: *Chapter 4*
Ask students to distinguish between prokaryotic cells and eukaryotic cells.

Respiration: *Chapter 7*
Ask students to list two metabolic pathways that cells use to harvest energy from nutrients.

Classification: *Chapter 18*
Ask students to name the kingdoms that bacteria compose.

CHAPTER 24

BACTERIA

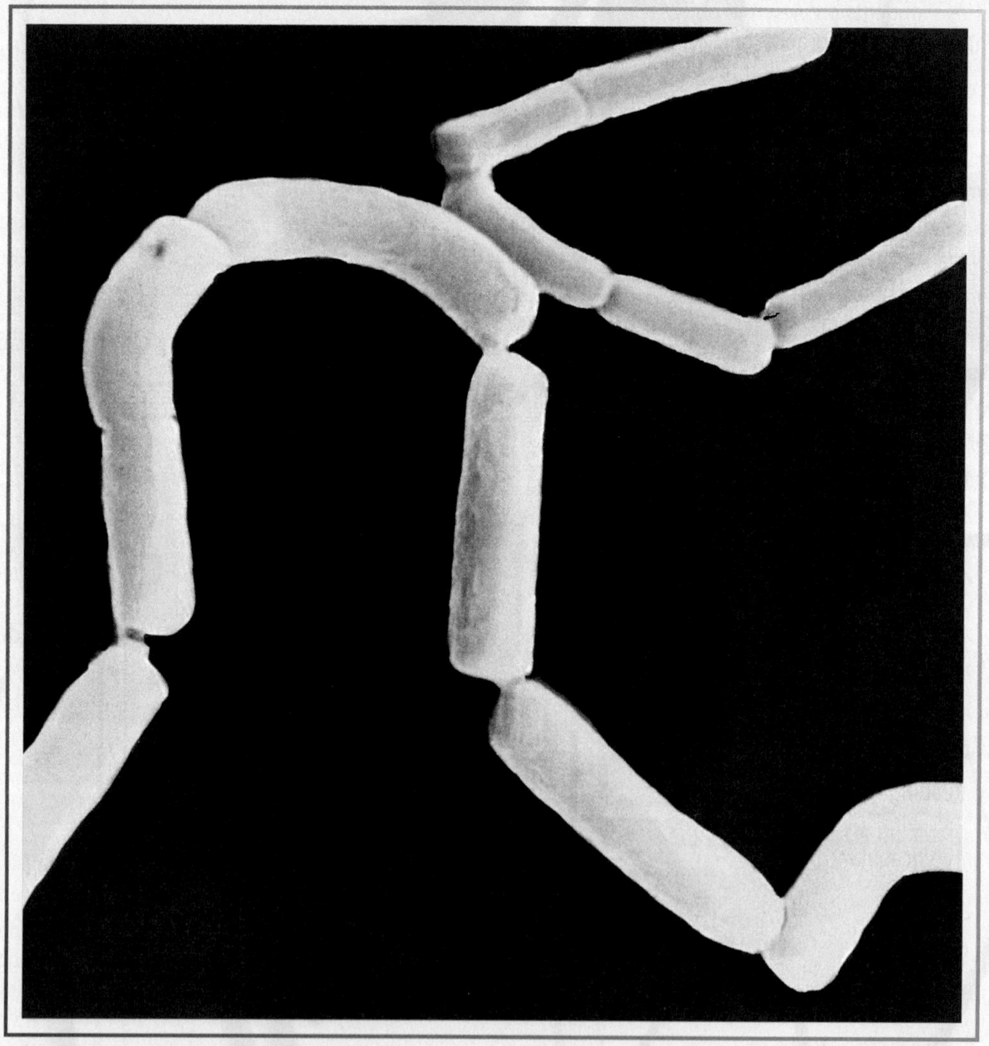

These bacteria are called Clostridium perfringens. *They are commonly found in soil and are the cause of the disease gas gangrene. (SEM 17,100×)*

FOCUS CONCEPT: *Cell Structure and Function*

As you read, notice that bacteria have complex structural, nutritional, and genetic properties even though they are single-celled microorganisms.

BACTERIAL EVOLUTION AND CLASSIFICATION

Bacteria are the most numerous organisms on Earth as well as the most ancient—they were probably the first forms of life. It is likely that all other organisms evolved from bacteria. The earliest fossils show that bacteria existed long before other forms of life evolved.

SECTION

24-1

OBJECTIVES

▲ Define *bacteria, eubacteria,* and *archaebacteria,* and note the relationships between them.

● Describe the methods used to classify bacteria.

■ Name and describe three known types of archaebacteria.

◆ Distinguish Gram-positive bacteria from Gram-negative bacteria.

▲ Describe the significance of cyanobacteria in the formation of the Earth's present atmosphere.

EVOLUTION

Bacteria are microscopic prokaryotes. Rock deposits found in Australia contain fossils of bacteria that existed on Earth about 3.5 billion years ago. Evidence in the fossil record indicates that eukaryotes are about 2.5 billion years old and modern humans arose about 100,000 years ago.

Bacteria have evolved into many different forms, and they are now a part of nearly every environment on Earth. They have even been found at the bottom of oceanic trenches 9.6 km (6 mi) below the water's surface and in Arctic and Antarctic regions. Evolution has yielded hundreds of thousands of species of bacteria that are adapted to places where no other organisms can live.

Classification

Unlike most other organisms, bacteria have few morphological differences that can be used to classify them. For example, bacteria do not vary in size and shape to the extent that other types of organisms do. Traditionally, bacteria have been grouped based on their structure, physiology, molecular composition, and reaction to specific types of stains, rather than on their evolutionary relationships. By comparing ribosomal RNA sequences, scientists have found that there are two vastly different types of bacteria. The bacteria that we generally refer to as germs are classified in the kingdom Eubacteria. In this text, members of the kingdom Eubacteria are sometimes referred to as **eubacteria** (YOO-bak-TEER-ee-uh), but more frequently, members of this kingdom are simply called bacteria. The other type of bacteria are called **archaebacteria** (AHR-kee-bak-TEER-ee-uh). These bacteria, which are more ancient than the eubacteria, are classified in the kingdom Archaebacteria.

SECTION OVERVIEW

Evolution
Section 24-1 describes the evolution and classification of bacteria.

Structure and Function
Section 24-1 also describes the Gram-staining technique and how it is used to identify bacteria whose cell walls differ in composition.

ACTIVE READING

Technique: Reading Organizer
Explain that in situations with an overwhelming amount of information, a useful classification system might include both broad and specific categories. Biology provides a good example. Ask students to recall the seven levels of taxonomic classification they learned about in Chapter 18.
As students read, ask them to make a diagram, flowchart, or concept map of the classifications of bacteria. Have them start with the kingdoms (Eubacteria and Archaebacteria). Have them assign the phyla appropriately and list the characteristics of each phylum.

TERMINOLOGY NOTE

Analyses of ribosomal RNA and genome sequences indicate that prokaryotes constitute two distinct groups—the archaebacteria and the eubacteria. The classification of all prokaryotes into a single kingdom, Monera, is based on phenotypic similarities. Because molecular data are thought to more accurately reflect evolutionary relationships than phenotypic characteristics, the molecular studies conducted in the 1980s and 1990s justified splitting the kingdom Monera into two kingdoms. Although some scientists still refer to "kingdom Monera," the majority now accept the six-kingdom organization. The classification of bacteria into two separate kingdoms—Eubacteria and Archaebacteria—is discussed in this chapter.

TEACHING STRATEGY

Comparing Kingdoms

Make a two-column table on the board titled "Kingdom Characteristics." Label the columns "Eubacteria" and "Archaebacteria." Ask volunteers to list the characteristics of each kingdom (eubacteria—peptidoglycan in cell walls, no introns; archaebacteria—cell walls without peptidoglycan, found in extreme environments, unusual lipids in cell membranes, genes that resemble eukaryotic genes, and that contain introns).

CRITICAL THINKING

Exobiology on Earth

Tell students that scientists consider archaebacteria to be a good model of life-forms that may be on other planets. Ask students why scientists think so. (Because many archaebacteria live in harsh environmental conditions similar to those believed to exist on other planets, scientists think life-forms similar to archaebacteria may exist there.)

RECENT RESEARCH

Ubiquitous Archaea

Using molecular DNA techniques, scientists in several countries have found evidence that archaebacteria are in soil and water. The polymerase chain reaction has enabled scientists to detect archaebacterial genes in soil and water samples. Traditionally, bacteria are identified by growing them and testing their biochemical reactions. But because scientists have not been able to grow all archaebacteria in the lab, they have been unable to identify and count most species of archaebacteria. One study estimates that standard culture methods have detected less than 0.1 percent of observed microorganisms.

KINGDOM ARCHAEBACTERIA

Scientists treat archaebacteria as a separate kingdom because these organisms are so different from other bacteria. For example, archaebacteria have unusual lipids in their cell membranes and have introns in their DNA. Their cell walls are also characterized by the absence of **peptidoglycan** (PEP-tuh-doh-GLIE-KAHN), a protein-carbohydrate compound found in the cell walls of eubacteria. While archaebacteria have some genes that resemble eubacterial genes, they also have genes that closely resemble those found in eukaryotes. This suggests that archaebacteria probably evolved from an ancestral organism that gave rise to other forms of life on Earth.

Archaebacteria were first discovered in extreme environments, such as swamps, salt lakes, and hot springs. Until recently, scientists believed that archaebacteria lived only in these extreme environments. However, by testing samples of surface water in the North Pacific and Antarctic Oceans for the presence of archaebacterial genomic sequences, scientists have discovered that archaebacteria may be more common than once thought.

Methanogens (meth-AN-uh-jenz), a broad phylogenetic group of archaebacteria, are named for their unique method of harvesting energy by converting H_2 and CO_2 into methane gas. Because oxygen is a poison to them, methanogens can live only in anaerobic conditions, such as at the bottom of a swamp and in sewage, where they are the source of marsh gas, as shown in Figure 24-1. They can also be found thriving in the intestinal tracts of humans and other animals, such as cows.

Extreme halophiles (HAL-uh-FIELZ), which are salt-loving archaebacteria, live in environments with very high salt concentrations, such as the Great Salt Lake and the Dead Sea. High salt concentrations would kill most bacteria, but this high concentration is beneficial to the growth of extreme halophiles, and these organisms use salt to generate ATP.

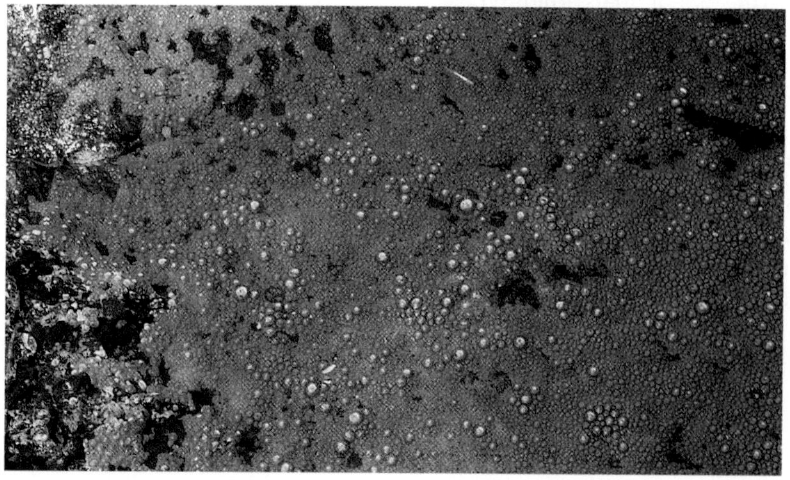

FIGURE 24-1

Some archaebacteria, such as the methanogens that are found in the muddy bottom of this swamp, live in anaerobic conditions. Methanogens produce methane, which you can see bubbling up through the water.

Thermoacidophiles (THUHR-moh-uh-SID-oh-fielz), a third group of archaebacteria, live in extremely acidic environments that have extremely high temperatures, such as hot springs. Some thermoacidophiles thrive at temperatures up to 110°C (230°F) and at a pH of less than 2. Thermoacidophiles live near volcanic vents on land or near hydrothermal vents, cracks in the ocean floor miles below the surface that leak scalding acidic water.

Research Notes

A Geologic Hot Spot

In 1977, the submarine *Alvin* drifted slowly through the cold, dark waters off the coast of Ecuador, in the Galápagos Rift of the South Pacific. On board was a team of marine geologists from the Woods Hole Oceanographic Institution led by Robert D. Ballard. They were seeking hydrothermal vents—fissures in the Earth's crust that release heat and minerals into the surrounding water. At an ocean depth of 2550 m (8,366 ft), they found hydrothermal vents and underwater hot springs teeming with marine life.

At the time, scientists believed that no living organism could survive the harsh combination of extremely high temperatures, high pressure, and total darkness. To their surprise, when the geologists developed their film, they found ghostly images of huge clamshells and previously unknown giant tube worms thriving among the bare lava. At those depths, no sunlight penetrates to support photosynthesis. What were the vent organisms using for food?

When chemist John Edmond, who was aboard the *Alvin,* analyzed the water samples, he found that they contained a large amount of dissolved hydrogen sulfide. Later, researchers following up on Ballard's deep-sea find confirmed that bacteria from vent waters, when cultured under pressures and temperatures found in a deep-sea environment, metabolize hydrogen sulfide as an energy source in a process known as chemosynthesis. To test the idea that the larger organisms used these bacteria for food, marine biologists continued to observe the organisms living in vent communities.

These biologists discovered that the dominant vent animals—clams, mussels, and giant tube worms—have a symbiotic relationship with the vent bacteria. Using microscopes, they found that the tube worms, for example, have colonies of bacteria living in their tissues. Tube worms are red because they are filled with hemoglobin. In humans, hemoglobin transports oxygen to cells. In tube worms, hemoglobin binds with hydrogen sulfide and carries it to the bacteria. The bacteria then oxidize the hydrogen sulfide, producing carbon compounds that in turn nourish the worms.

Similar communities of organisms have since been found at hundreds of geologic hot spots around the world. In December 1993, an *Alvin* expedition to the East Pacific Rise, an underwater mountain range southwest of Acapulco, Mexico, was conducted to study a recently formed hydrothermal vent.

At the site, thickets of 1.2 m (4 ft) long tube worms were found planted on the ocean floor. Measurements indicated that these tube worms grew at a rate of 84 cm (33 in.) per year, making them the fastest-growing marine organisms known.

The 1993 expedition was only one in a series of dives to the East Pacific Rise. The first visit had taken place in 1989, when scientists first discovered the vent community. Then, in April 1991, the team of scientists was surprised to find another hydrothermal vent at the site. A recent volcanic eruption at the site produced the hottest hydrothermal vent ever recorded. Scientists named the site Tube Worm Barbecue after retrieving tube worms with charred flesh. One of the scientists reported that bacteria in the vent communities were so plentiful that there appeared to be a snow blizzard. After the eruption, the only living organisms that remained on the ocean floor were centimeters of bacterial mats. However, by March 1992 the bacterial mats had been replaced by new life.

With the discovery of hydrothermal vents, there are many questions to be asked concerning their significance. As research continues, scientists will study the light source that these vents give off and its effect on the organisms that reside there.

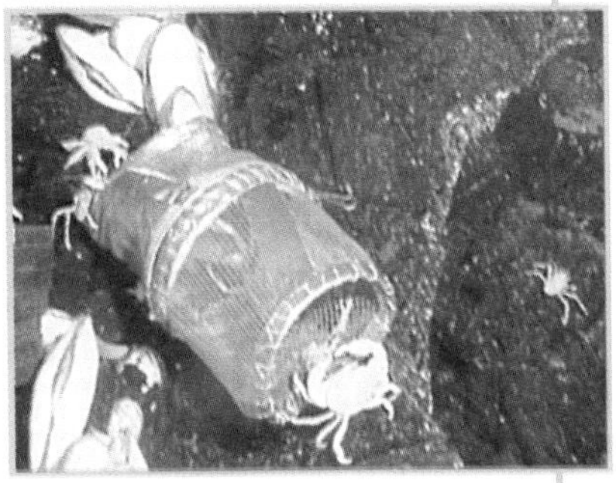

***A vent-dwelling crab is spotted by the remote-operated sub* Alvin.**

CHAPTER 24
Research Notes

BACKGROUND

The *Alvin* is a deep-submergence research vehicle. It can carry a pilot and two observers to a depth of 4,000 m. Hydrothermal vents form when cold sea water seeps down into fissures in the oceanic crust. When water contacts the hot planetary crust, the heated water carries dissolved minerals up to the ocean floor. The water that wells up from the hydrothermal vents around the Galápagos Rift may be 20°C hotter than the surrounding water. The emerging hot water carries sulfides of zinc, iron, and copper.

DISCUSSION

Guide the discussion by posing the following questions:

1. What are hydrothermal vents? (underwater fissures in Earth's crust that release heat and minerals into the surrounding water)

2. What is unique about the bacteria in water surrounding the hydrothermal vents? (The bacteria metabolize hydrogen sulfide as an energy source through the process of chemosynthesis.)

3. What is the nature of the symbiotic relationship between the vent bacteria and the giant tube worms? (Tube worms have hemoglobin, which binds hydrogen sulfide and carries it to the bacteria. The bacteria oxidize the hydrogen sulfide and produce carbon compounds that nourish the worms.)

FURTHER READINGS

1. "Archaea Tells All," in *Discover,* January 1997, p. 37. This article discusses Archaea.

2. "Live Long and Prosper," in *Science News,* September 28, 1996, p. 201. This article discusses the unusual life-forms found at ocean seeps.

SECTION 24-1

TEACHING STRATEGY
Eubacteria Shapes

To teach prefixes and shape names, make spheres, rods, and spirals from paper, or use models such as beads, straws, and springs. Place the shapes on the overhead and ask students to name them (rods—bacilli; spirals—spirilla; spheres—cocci). Ask students what several spherical bacteria in a chain are called (streptococci). Tell students that the prefix *diplo-* means "two." Then place two spheres on the overhead and ask students to name them (diplococci). Place clusters of spheres on the overhead and ask students to name them (staphylococci).

MAKING CONNECTIONS

Section 18-1 Binomial Nomenclature

Tell students that binomial nomenclature is used to name bacteria. Then remind them that both the genus and species names must be used to avoid confusion. For example, *Escherichia coli* is a gram-negative bacterium, but *Entamoeba coli* is a protist.

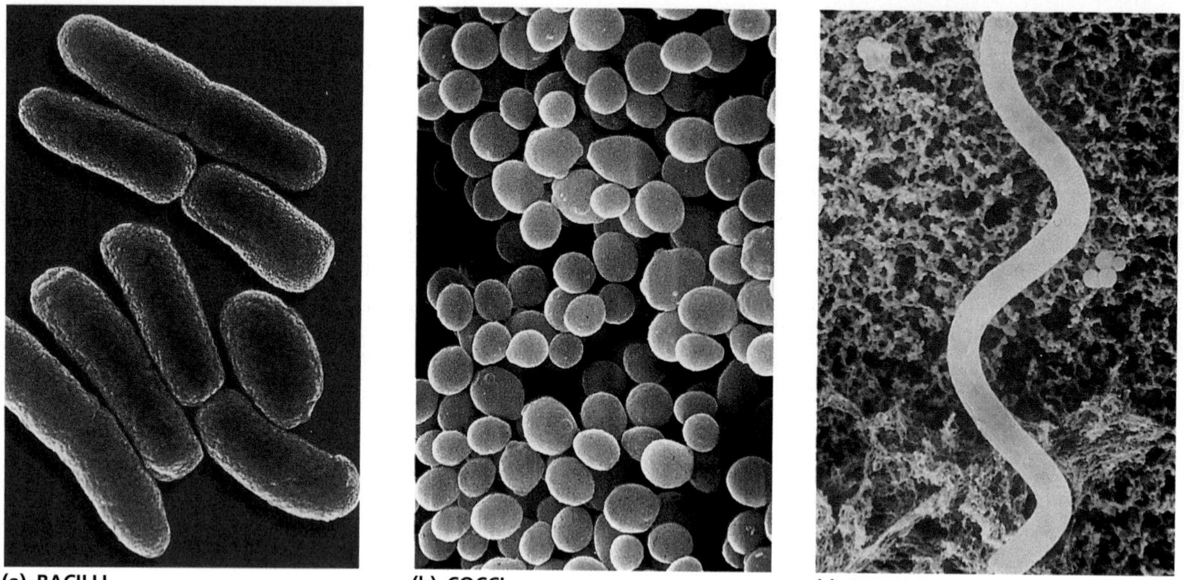

(a) BACILLI (b) COCCI (c) SPIRILLA

FIGURE 24-2

The most common shapes among bacteria are represented here by (a) *Escherichia coli,* bacilli; (b) *Micrococcus luteus,* cocci (SEM 117,300×); and (c) *Spirillum volutans,* spirilla. (SEM 19,900×)

KINGDOM EUBACTERIA

Eubacteria account for most bacteria; they occur in many shapes and sizes and have distinct biochemical and genetic characteristics. Most eubacteria have one of three basic shapes, as shown in Figure 24-2. Eubacteria that are rod-shaped are called **bacilli** (buh-SIL-ie). Sphere-shaped eubacteria are called **cocci** (KAHK-sie), and spiral-shaped eubacteria are called **spirilla** (spie-RIL-uh). When cocci occur in chains, they are called **streptococci** (STREP-tuh-KAHK-sie); grapelike clusters of cocci are called **staphylococci** (STAF-uh-loh-KAHK-sie).

Eubacteria can be divided into as many as 12 different phyla according to their evolutionary relationships. Because bacteria diversified so long ago, scientists disagree on how they should be classified phylogenetically. Table 24-1 lists several generally recognized phyla of bacteria and their properties.

TABLE 24-1 *Some Phyla of Bacteria and Their Properties*

Phylum	Shape	Motility	Metabolism	Gram reaction
Cyanobacteria	bacilli, cocci	gliding; some nonmotile	aerobic, photosynthetic autotrophic	Gram-negative
Spirochetes	spirals	corkscrew motion	aerobic and anaerobic; heterotrophic	Gram-negative
Gram-positive bacteria	bacilli, cocci	flagella; some nonmotile	aerobic and anaerobic; heterotrophic, photosynthetic	mostly Gram-positive
Proteobacteria	bacilli, cocci, spiral	flagella; some nonmotile	aerobic and anaerobic; heterotrophic; unusual metabolism; and photosynthetic autotrophic	Gram-negative

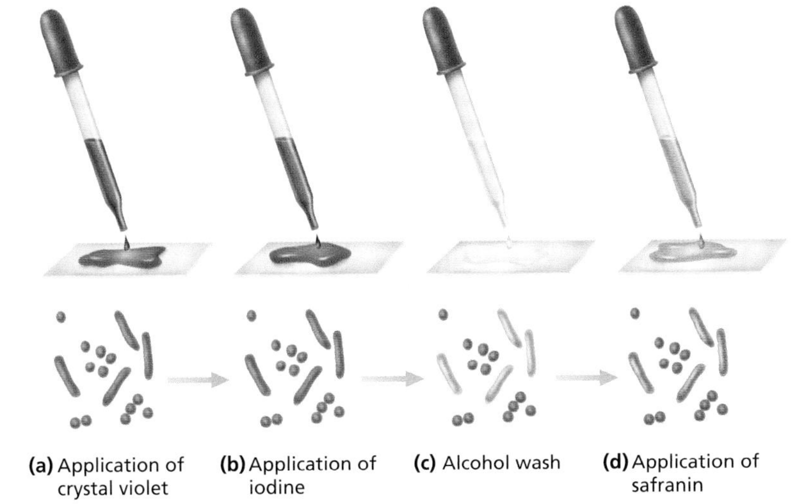

FIGURE 24-3

In the Gram-stain procedure, bacteria that have been placed on a slide are stained with a purple dye solution called crystal violet (a). The purple dye is washed off with water, and then a solution of iodine is added to the slide (b). The bacteria are rinsed with alcohol (c) and then restained with a pink dye solution called safranin (d). Gram-positive bacteria will retain the purple dye and appear purple, while Gram-negative bacteria will appear pink from the pink dye solution.

Gram Stain

Most species of eubacteria can also be grouped into two categories based on their response to a laboratory technique called the **Gram stain,** which is shown in Figure 24-3. **Gram-positive** bacteria retain the Gram stain and appear purple under the microscope. **Gram-negative** bacteria do not retain the purple stain and take up a second pink stain instead. Because Gram-positive bacteria have a thicker layer of peptidoglycan in their cell wall than Gram-negative bacteria do, they are able to retain the Gram stain. Figure 24-4 compares the cell walls of these two groups of bacteria. Gram-positive and Gram-negative bacteria also differ in several other ways. For example, they have different susceptibilities to antibacterial drugs, they produce different toxic materials, and they react differently to disinfectants. For these reasons, the Gram stain is useful for identifying and grouping eubacteria.

internetconnect

SCILINKS NSTA
TOPIC: Eubacteria
GO TO: www.scilinks.org
KEYWORD: HM471

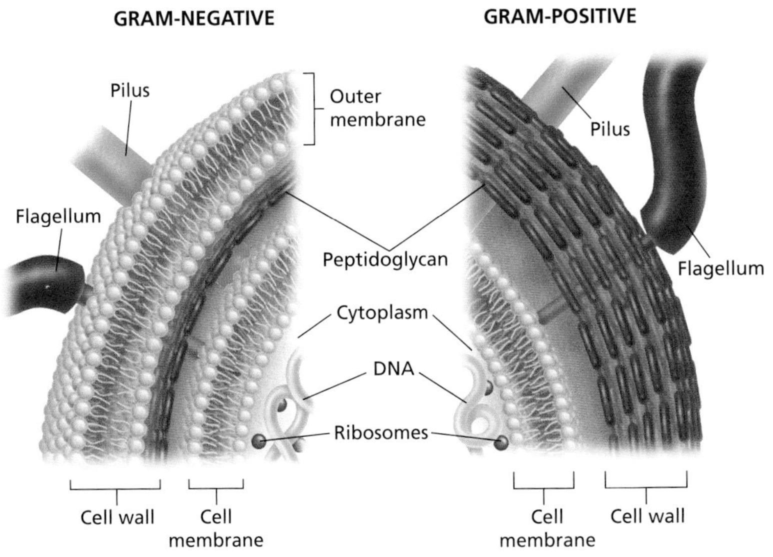

FIGURE 24-4

The drawing demonstrates the similarities and differences between Gram-negative bacteria and Gram-positive bacteria.

SECTION 24-1

INCLUSION ACTIVITY

Provide students with scissors, tape, and magazines that can be cut up (nature and wildlife magazines, such as *National Geographic, Discover,* or *Science News*). Have students prepare a poster with pictures of various environments that bacteria inhabit (lake, soil, volcano, cow, swamp).

QUICK FACT

☑ Danish physician Hans Christian Gram, the developer of the Gram stain, originally developed his staining technique to differentiate bacteria from mammalian nuclei in the slides of patients' tissues.

VISUAL LINK

Figure 24-4

Point out to students that Gram-negative bacteria differ from Gram-positive bacteria in some important aspects. Ask students to point out what structural differences differentiate Gram-positive bacteria from Gram-negative bacteria. (The layer of peptidoglycan in the Gram-positive cell wall is thicker than it is in the Gram-negative cell wall. Gram-negative bacteria have an outer membrane but Gram-positive bacteria do not.)

GIFTED ACTIVITY

Have students research the topic of eutrophication. Instruct students to find out the locations, dates, and effects of population blooms. Students should display their results in the form of a newspaper front page.

Quick Lab

Predicting the Spread of Disease

Time Required 20 minutes

Safety Remind students not to drink or sniff the solution.

Procedural Tips Prepare the cups ahead of time. Fill each cup about one-third full of water. Select one student who will not mind being the "infected" student. Secretly give that student a cup that contains 50 percent household ammonia and 50 percent water. You will need an eyedropper and three or four drops of phenolphthalein per student to test students' solutions. After the activity, lead a discussion about how easily a contagious disease can turn into an epidemic. Relate the ease of contagion to tuberculosis, meningitis, and other bacterial diseases familiar to students.

Disposal Collect the students' solutions in a large container, and tell students to recycle the cups. Test the pH of the solution with litmus or another indicator. If necessary, adjust the pH to neutral using small amounts of 1.0 M acid. Run water in the container for several minutes, flushing the drain.

Answers to Analysis After students have exchanged solutions, test the originally infected student's solution with three or four drops of phenolphthalein. The water will turn pink. Test the other students' cups. Have students count how many cups changed color. (Almost every cup will have changed color.) Many students will underestimate the possibility that they had become infected. Because the students did not touch hands and the disease was still spread, direct contact was not needed.

Word Roots and Origins

heterocyst

from the Greek *hetero,* meaning "other or different," and *kystis,* meaning "sac or bladder"

Quick Lab

Predicting the Spread of Disease

Materials disposable gloves, lab apron, and safety goggles; clear plastic cup with liquid

Procedure

1. Put on your disposable gloves, lab apron, and safety goggles.
2. Obtain a cup of liquid from your teacher.
3. Pour the contents of your cup into a classmate's cup. Have the classmate pour half the liquid back into your cup. Circulate around the room and repeat this step with two other classmates at random. Choose classmates you do not know well, and do not touch hands with anyone.
4. If one of your classmates had a highly contagious bacterial disease, predict how likely it is that you have "caught" the disease. Then your teacher will test the liquid in each student's cup to see who else has caught the disease. Count the number of students who are "infected."

Analysis How did the actual results compare with your prediction? How many students became infected? Was direct contact needed?

Phylum Cyanobacteria

The cyanobacteria are photosynthetic; that is, they use photosynthetic chemicals to capture sunlight, harvest the energy to produce carbohydrates, and give off oxygen as a waste product. Recall that the atmosphere of early Earth was filled with the oxygen produced by cyanobacteria, which allowed aerobic organisms to develop.

Cyanobacteria, once classified as blue-green algae, are now considered eubacteria because they lack a membrane-bound nucleus and chloroplasts. Unlike other eubacteria, however, the cyanobacteria are encased in a jellylike substance and often cling together in colonies.

Certain cyanobacteria grow in chains. Some of these cells form specialized cells called **heterocysts.** Heterocysts contain enzymes for fixing atmospheric nitrogen. Cyanobacteria that form heterocysts make nitrogen available to plants in a form that plants can use.

Cyanobacteria, such as *Anabaena,* thrive on phosphates and nitrates that accumulate in a body of water. The sudden increase in the number of cyanobacteria due to a high availability of nutrients is called **eutrophication** (YOO-troh-fuh-KAY-shuhn), or **population bloom.** Following eutrophication, many of the cyanobacteria die and are decomposed by heterotrophic bacteria. The increasing population of heterotrophic bacteria consume available oxygen in the water, causing other organisms in the water, such as fish, to die from lack of oxygen.

Phylum Spirochetes

Spirochetes (SPIE-roh-KEETS) are Gram-negative, spiral-shaped heterotrophic bacteria. Some spirochetes are aerobic, and some are anaerobic. They all move by means of a corkscrew-like rotation. Spirochetes live freely, symbiotically, or parasitically. One well-known spirochete is *Treponema pallidum,* which causes the sexually transmitted disease syphilis.

Phylum Gram-Positive Bacteria

Despite its name, not all members of this phylum are Gram-positive. A few species of Gram-negative bacteria are also grouped in this phylum because they share molecular similarities with Gram-positive bacteria. Members of this phylum include the species of streptococci that causes strep throat.

Milk becomes yogurt when certain Gram-positive bacilli grow in milk and produce lactic acid. Gram-positive bacilli are also found in the oral cavity and in the intestinal tract, where they retard the growth of disease-causing bacteria. Lactobacilli, Gram-positive bacilli found on the teeth, are known to cause tooth decay through the release of acid.

Actinomycetes (AK-tuh-noh-MIE-seets) are Gram-positive bacteria that form branching filaments. They grow in the soil and produce many **antibiotics,** chemicals that inhibit the growth of or kill other microscopic organisms.

Phylum Proteobacteria

The proteobacteria make up one of the largest and most diverse phylum among bacteria. This group is divided into several subdivisions, including enteric bacteria, chemoautotrophic bacteria, and nitrogen-fixing bacteria.

Enteric bacteria are Gram-negative heterotrophic bacteria that inhabit animal intestinal tracts and can live in either aerobic or anaerobic conditions. This group includes the well-known organism *Escherichia coli* (abbreviated *E. coli*). *E. coli* lives in the human intestine where it produces vitamin K and assists enzymes in the breakdown of foods. Other genera of enteric bacteria are responsible for many diseases. For example, bacteria in the genus *Salmonella* are responsible for many cases of food poisoning.

Chemoautotrophs (KEEM-oh-AW-toh-TROHFS) are Gram-negative bacteria that can extract energy from minerals by oxidizing the chemicals in these minerals. For example, iron-oxidizing bacteria, live in freshwater ponds that contain a high concentration of iron salts. The iron bacteria oxidize the iron in the salts to obtain energy.

As you learned in Chapter 22, nitrogen-fixing bacteria, which include some Gram-negative rods, live both freely and symbiotically with plants. *Rhizobium* is one example of nitrogen-fixing bacteria that live symbiotically with plants.

Nitrogen-fixing bacteria are vital to the success of many ecosystems. Although almost 80 percent of Earth's atmosphere is composed of nitrogen gas, N_2, plants and animals are unable to use it. *Rhizobium* is able to convert nitrogen gas to a form of nitrogen that plants can most easily use. Nitrogen-fixing bacteria colonize many plants, such as beans, soybeans, peas, alfalfa, and clover. By inducing plants to form nodules on their roots, nitrogen-fixing bacteria receive organic compounds that they need while providing the plants with nitrogen in a form they can use. *Rhizobium* is considered a major microbial source of soil nitrogen. *Rhizobium* can produce about 250 kg (114 lb) of fixed nitrogen per hectare (2.5 acres) of alfalfa in a single year, whereas the free-living nitrogen-fixing *Azotobacter* genus can produce only 2.5 kg (1.1 lb) of fixed nitrogen per hectare of alfalfa in a year.

internet**connect**

SCI LINKS NSTA
TOPIC: Proteobacteria
GO TO: www.scilinks.org
KEYWORD: HM473

SECTION 24-1 REVIEW

1. **Explain how the terms *bacteria, eubacteria,* and *archaebacteria* relate to one another.**
2. **List the characteristics that are used to classify bacteria.**
3. **List the habitats of three types of archaebacteria.**
4. **Distinguish between Gram-positive bacteria and Gram-negative bacteria.**
5. **Describe the significance of cyanobacteria in the formation of the Earth's atmosphere.**
6. **CRITICAL THINKING Why are methanogens and cyanobacteria unable to live in the same environment?**

DEMONSTRATION

Show students the roots of a pea plant, airplane plant, or asparagus fern. Point out the nodules on the roots. Explain to students that the nodule is the place where nitrogen-fixing bacteria convert nitrogen gas to ammonia, the form of nitrogen that plants can use.

ANSWERS TO SECTION 24-1 REVIEW

1. Bacteria are microscopic prokaryotes. Eubacteria are prokaryotes that have a peptidoglycan cell wall and lack introns. Archaebacteria are prokaryotes that lack peptidoglycan and have introns. They can be found in extreme environments.
2. Bacteria are classified according to their structure, physiology, molecular composition, and staining reactions.
3. Archaebacteria inhabit anaerobic environments, such as the bottoms of swamps and sewage; high salt environments, such as the Great Salt Lake and Dead Sea; and very hot environments, such as hot springs and volcanic vents.
4. Gram-positive and Gram-negative bacteria differ with respect to the chemical components of their cell wall, their susceptibility to antibiotics, and the toxins they produce.
5. Cyanobacteria produced oxygen as a waste product of photosynthesis. This allowed for the development of oxygen-breathing organisms.
6. Methanogens are obligate anaerobes; thus, if cyanobacteria were present, the oxygen generated by cyanobacteria would be toxic to the methanogens.

SECTION 24-2

SECTION OVERVIEW

Structure and Function
Section 24-2 describes structures specific to bacterial cells and their function.

Matter, Energy, and Organization
Section 24-2 compares autotrophic and heterotrophic modes of nutrition in bacteria.

Reproduction and Inheritance
Section 24-2 also describes genetic recombination in bacteria.

▶ ENGAGE STUDENTS

Show students a video of bacteria, such as *Salmonella, Proteus,* or *Escherichia,* moving in an aqueous environment. Videos may be obtained from a biological supply company. After students watch the video, ask them how the bacteria are able to move. (Bacteria have flagella.) Then show students a transmission electron micrograph of a bacterium. Ask students if they can see any organelles. Point out that the functions that chloroplasts and mitochondria carry out in eukaryotes are maintained in the internal foldings of the bacterial cell membrane.

TEACHING STRATEGY

Bacterial Membrane Function

Tell students that in addition to maintaining homeostasis, the bacterial cell membrane also serves to create a proton gradient that helps produce ATP. Ask students how eukaryotes create a proton gradient. (Eukaryotes use mitochondrial membranes to make the proton gradient that helps make ATP.)

SECTION

24-2

OBJECTIVES

▲ Describe the structure of a bacterial cell.

● Describe three ways that bacteria move.

■ Compare the heterotrophic modes of nutrition in bacteria with the autotrophic modes.

◆ Discuss the various types of environments that bacteria occupy.

▲ List three types of genetic recombination used by bacteria.

BIOLOGY OF BACTERIA

Viewed through a light microscope, bacteria appear to be relatively simple rods, spheres, and other forms. However, the electron microscope reveals a great amount of detailed structure within each form. These detailed structures are responsible for the activities carried out by bacteria.

STRUCTURE

Bacteria are typically composed of a cell wall, a cell membrane, and cytoplasm. Some bacteria have distinctive structures, such as endospores, capsules, and outer membranes. The variety of structures among bacteria is due to adaptations to individual niches. A summary of bacterial structures is presented in Table 24-2.

Cell Wall

With a few exceptions, both eubacteria and archaebacteria have a cell wall. Unlike plant cell walls, eubacterial cell walls are made of peptidoglycan. Peptidoglycan is composed of short chains of amino acids, or peptides, and carbohydrates. Archaebacterial cell walls are composed of a different compound. In Gram-negative eubacteria, the cell wall includes an outer membrane that is composed of a layer of lipids and sugars. The outer membrane protects these bacteria against some kinds of antibiotics by preventing their entry into the cell. Thus, many antibiotics have no effect on Gram-negative bacteria.

Cell Membrane and Cytoplasm

The bacterial cell membrane, which is composed of a lipid bilayer, is similar to the eukaryotic cell membrane. However, in bacteria, the cell membrane contains the enzymes that catalyze the reactions of cellular respiration. Because bacteria do not have mitochondria, they use their cell membranes to create proton gradients and to carry out cellular respiration.

The cell membranes of photosynthetic bacteria have internal foldings called thylakoids. These structures are homologous to the thylakoids found in plant chloroplasts. Like those in plant chloroplasts, bacterial thylakoids contain photosynthetic pigments and carry out the function of harvesting light energy.

Unlike eukaryotic cells, bacterial cells do not contain membrane-bound organelles. The cytoplasm of bacterial cells is made of a viscous solution of ribosomes and DNA. The bacterial DNA is arranged

TABLE 24-2 Structural Characteristics of a Bacterial Cell

Structure	Function
Cell wall	protects the cell and gives it shape
Outer membrane	protects the cell against some antibiotics (only present in Gram-negative cells)
Cell membrane	regulates movement of materials into and out of the cell; contains enzymes important to cellular respiration
Cytoplasm	contains DNA, ribosomes, and organic compounds required to carry out life processes
Chromosome	carries genetic information inherited from past generations
Plasmid	contains some genes obtained through genetic recombination
Capsule and slime layer	protect the cell and assist in attaching the cell to other surfaces
Endospore	protects the cell against harsh environmental conditions, such as heat and drought
Pilus	assists the cell in attaching to other surfaces, which is important for genetic recombination
Flagellum	moves the cell

in a single, closed loop. In addition to the main chromosome, some species of bacteria also have plasmids, self-replicating loops of DNA, in their cytoplasm.

Capsules and Pili

Many bacterial species produce an outer covering called a **capsule.** The capsule is made of polysaccharides that cling to the surface of the cell and protect it against drying or harsh chemicals. It also protects an invading bacterium from the host body's white blood cells, which could otherwise engulf it. When a capsule consists of a fuzzy coat of sticky sugars, it is called a **glycocalyx** (GLIE-koh-KAY-liks). The glycocalyx enables bacteria to attach to the surface of host cells and tissues.

Pili (PIL-ee) are short, hairlike protein structures found on the surface of some species of bacteria. Pili help bacteria adhere to host cells. Pili are also used to transfer genetic material from one bacterium to another.

Endospores

A bacterial **endospore** is a dormant structure that is produced by some Gram-positive bacterial species that are exposed to harsh environmental conditions. Endospores consist of a thick outer covering that surrounds the cell's DNA. Although the original cell may be destroyed by harsh conditions, its endospore will survive. Endospores are not reproductive cells. Instead, they help bacteria resist high temperatures, harsh chemicals, radiation, drying, and other environmental extremes. When conditions become favorable, the endospore will open, allowing the living bacterium to emerge and to begin multiplying. Endospores can be formed by species of the genera *Bacillus* and *Clostridium.*

Word Roots and Origins

pili

from the Latin *pilus,* meaning "hair"

SECTION 24-2

INCLUSION ACTIVITY

Give each pair of students construction paper, various colors of yarn, poster board, and glue. Ask each pair to make a model of a typical bacterium using the structures listed in Table 24-2. Students can use different colors of paper and yarn for the various structures, which should be labeled. Call on each group to describe their model and explain how it differs from the others.

CRITICAL THINKING

Endospore Resistance

Tell students that disease-causing bacterial cells usually die in boiling water within a few minutes. Ask them why hospitals sterilize surgical instruments by exposing them to steam and pressure for at least 15 minutes. (While vegetative bacterial cells are killed quickly, the endospores can survive much longer and therefore require more heat to be killed.)

DEMONSTRATION

Point out to students that flagella are shaped crookedly, which helps them move. To graphically illustrate the principle of bacterial flagellar movement, obtain a pipe cleaner and bend it in several places. Poke one end through a small paper bag and hold the interior part with your hand. Tell students that the bag is a model of a bacterial cell and the pipe cleaner is a flagellum. Then, with your hand inside the bag, twist the pipe cleaner to demonstrate bacterial flagellar movement.

TEACHING STRATEGY

Comparing Terms

Tell students that obligate aerobes are organisms that require oxygen. Then tell them that facultative anaerobes can exist with or without oxygen. Ask students to define an obligate heterotroph (an organism that can obtain nutrition only by eating other organisms). Then ask them to define facultative photoautotroph (an organism that may obtain nutrition through autotrophy, but does not require it).

DEMONSTRATION

CAUTION **Do not open Petri dishes once they have been sealed.** The day before class, obtain three sterile Petri dishes and inoculate each with *Bacillus subtilis,* which can be obtained from a biological supply company. Seal each Petri dish with tape. Label each Petri dish as a biohazard. Place one in the refrigerator, one at room temperature, and one in an incubator set at 100°C. The next day, show students the three Petri dishes. (Students should see little or no growth in the refrigerated and heated Petri dishes. There should be some growth in the dish stored at room temperature.)

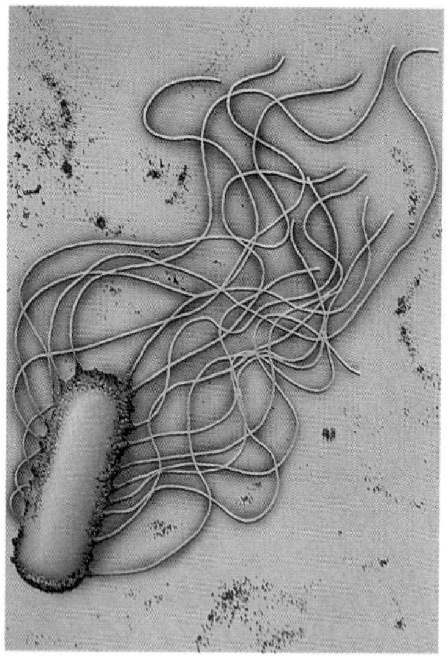

FIGURE 24-5
Some bacteria, such as this Gram-negative *Salmonella,* move by means of flagella.

Movement Structures

Many bacteria use flagella to move. Flagella, which are made of protein, turn and propel the bacterium in an erratic, "run-and-tumble" motion. Bacteria can have a single flagellum or a tuft of flagella. Some species of bacteria have flagella at both ends of the cell, and other species of bacteria are completely surrounded by flagella, as is the bacterium shown in Figure 24-5.

The bacteria that lack flagella have other methods of movement. For example, myxobacteria produce a layer of slime and then glide through it. Wavelike contractions of the outer membrane propel the organism through the slime. Some spiral-shaped bacteria move by a corkscrew-like rotation. These organisms have flexible cell walls and filaments within the cell walls that, when contracted, cause the bacterium to turn and move ahead.

NUTRITION AND GROWTH

Depending on the species, bacteria may be heterotrophic or autotrophic. Heterotrophic bacteria use organic matter as a source of nutrition. Heterotrophic bacteria that feed on dead and decaying material are called **saprophytes** (SAP-ruh-FIETS). Autotrophic bacteria obtain their energy from sunlight or minerals.

Bacteria that use sunlight as an energy source, such as cyanobacteria, are called **photoautotrophs** (FOH-toh-AW-toh-trohfs). These autotrophs use light-trapping compounds similar to those used by plants to obtain energy. As you learned in the previous section, chemoautotrophs, another group of autotrophic organisms, oxidize inorganic compounds to obtain energy. For example, members of the genus *Nitrosomonas* oxidize ammonia, NH_3, to form nitrite, NO_2, and harvest the resulting energy.

Many bacteria are **obligate anaerobes** (AHB-luh-git AN-uhr-OHBZ), which means they cannot survive in the presence of oxygen. The obligate anaerobe *Clostridium tetani* causes tetanus. **Facultative anaerobes** (FAK-uhl-TAY-tiv AN-uh-ROHBZ) can live with or without oxygen. *Escherichia coli,* which is common in the human digestive tract, is a facultative anaerobe. Bacteria that cannot survive without oxygen are called **obligate aerobes.** The obligate aerobe *Mycobacterium tuberculosis* lives in the lungs and causes tuberculosis.

Bacteria have varying temperature requirements for growth. Some bacteria grow best in cold temperatures of 0–20°C (32–68°F). Other bacteria grow best in temperatures that range from 20°C (68°F) to 40°C (104°F). **Thermophilic** (THUHR-moh-FIL-ik) bacteria grow best in temperatures between 40°C (104°F) and 110°C (230°F). Most bacterial species grow best at a pH of 6.5 to 7.5 (7.0 is neutral). Certain species, such as those used to produce yogurt and sour cream, prefer acidic environments of a pH of 6.0 or lower.

TABLE 24-3 A Comparison of Three Genetic Recombination Methods

Characteristic	Transformation	Conjugation	Transduction
Method of DNA transfer	across cell wall and cell membrane of recipient	through a conjugation bridge between two cells	by a virus
Plasmid transfer	yes	yes	not likely
Chromosome transfer	no	sometimes	no
Antibiotic resistance acquired	yes	yes	sometimes

GENETIC RECOMBINATION

Table 24-3 summarizes three nonreproductive ways that bacteria can acquire and express new combinations of genetic material. **Transformation** occurs when a bacterial cell takes in DNA from its external environment. The new DNA is then substituted for a similar DNA fragment in the chromosome of the bacterial cell.

Bacterial **conjugation** is the process by which two living bacteria bind together and one bacterium transfers genetic information to the other. In order for conjugation to take place, the genetic donor must have a specialized plasmid and pilus. The specialized pilus binds to a recipient bacterium that does not have the specialized plasmid and forms a **conjugation bridge,** a passageway for the transfer of the genetic information. The plasmid replicates in the donor bacterium, and one copy of the plasmid passes through the conjugation bridge to the recipient bacterium. After the DNA transfer, the cells detach.

In **transduction,** a virus obtains a fragment of DNA from a host bacterium. As the viruses replicate inside the bacterium, they produce new copies of the bacterial DNA fragment. After the newly formed viruses have been released, they carry the bacterial gene to a new bacterium, where it can be expressed by the recipient bacterium.

SECTION 24-2 REVIEW

1. List the various structures of the bacterial cell, and describe their function.

2. Describe three types of movement among bacteria.

3. What specific terms are used to describe the oxygen requirements of bacteria?

4. In what key way do photoautotrophs and chemoautotrophs differ?

5. List and summarize three methods of genetic recombination in bacteria.

6. **CRITICAL THINKING** What eukaryotic structures are functionally similar to the bacterial cell membrane and its infoldings?

SECTION 24-2

ANSWERS TO SECTION 24-2 REVIEW

1. cell wall—protects bacteria; outer membrane—protects cell from some antibiotics; cell membrane—regulates movement into and out of cell, contains respiratory enzymes; chromosome—carries genetic information; plasmid—contains genetic information from genetic recombination; capsule/slime layer—protects the cell and assists in attaching cell to surfaces; endospores—protects the cell from harsh environments; pili—assists in attaching to surfaces and used in genetic recombination; flagella—movement.

2. Motility found among bacteria include the propulsion provided by flagella, the gliding movement of myxobacteria, and the corkscrew-like rotation of spirochetes.

3. Bacteria that must have oxygen are obligate aerobes; obligate anaerobes cannot tolerate oxygen; and those that can tolerate both situations are facultative anaerobes.

4. A photoautotroph is an autotroph that synthesizes nutrients using light energy from the sun; a chemoautotroph synthesizes nutrients using the energy of inorganic substances.

5. In transformation, bacteria acquire DNA fragments from the environment and incorporate them into their genome. In conjugation, a fragment of DNA passes from the donor bacterium to the recipient bacterium through a conjugation bridge. In transduction, a virus picks up a fragment of DNA from a donor and transports it into the cytoplasm of a recipient bacterium.

6. Mitochondria and chloroplasts are homologous with the bacterial cell membrane and bacterial thylakoids.

SECTION 24-3

SECTION OVERVIEW

Cell Structure and Function
Section 24-3 describes how bacteria cause disease and are used to produce foods, drugs, and industrial chemicals.

Stability and Homeostasis
Section 24-3 also describes how antibiotics interfere with the function of bacteria and kill them.

▶ ENGAGE STUDENTS

Begin by explaining to students that disease-causing bacteria are not malevolent. The bacteria simply need nutrition, and humans, animals, and plants often fit the nutritional needs of bacteria. Then ask students how bacteria are beneficial. (They are recyclers in ecosystems and are used in genetic engineering.)

GIFTED *ACTIVITY*

Have students contact their local health agency or the Centers for Disease Control and Prevention, in Atlanta, to find out how bacterial diseases are tracked and if there is a pattern to disease outbreaks in the United States. Have them prepare a report or poster to communicate their findings.

CRITICAL THINKING

Food Poisoning
Ask students if it is safe to eat potato salad that was made with homemade mayonnaise and left unrefrigerated. Have them explain their answer. (Mayonnaise is made with raw eggs, which are often contaminated with *Salmonella* bacteria. Warm potato salad provides a favorable environment for *Salmonella,* which grow and produce a toxin that can cause severe intestinal illness.)

SECTION

24-3

OBJECTIVES

▲ Describe the ways that bacteria can cause disease in humans.

● Specify how antibiotic resistance has come about, and describe ways that bacteria resist antibiotics.

■ List three ways that bacteria are helpful to humans.

BACTERIA AND HUMANS

Much of our knowledge about bacteria is a result of the study of the diseases they cause in humans. In addition to what we have learned about pathogenic bacteria and how they cause disease, we have also learned how bacteria benefit us. Bacteria are used in food preparation and in environmental, chemical, and mining processes.

BACTERIA AND DISEASE

The scientific study of disease is called **pathology** (path-AHL-uh-jee). Bacteria that cause disease are called pathogens. Some diseases caused by bacteria are listed in Table 24-4.

Some bacteria cause disease by producing poisons called **toxins** (TAHK-sins). **Exotoxins** (EKS-oh-TAHK-sins) are toxins that are made of protein. Exotoxins are produced by Gram-positive bacteria and are secreted into the surrounding environment. For example, tetanus is a disease caused by an exotoxin.

TABLE 24-4 ***A Summary of Bacterial Diseases***

Disease	Pathogen	Areas affected	Mode of transmission
Botulism	*Clostridium botulinum*	nerves	improperly preserved foods
Cholera	*Vibrio cholerae*	intestine	contaminated water
Dental caries (tooth decay)	*Streptococcus mutans, sanguis,* and *salivarius*	teeth	bacteria enter the mouth from the environment
Gonorrhea	*Neisseria gonorrhoeae*	urethra, fallopian tubes, epididymis	person-to-person by sexual contact
Lyme disease	*Borrelia burgdorferi*	skin, joints, heart	tick bite
Rocky Mountain spotted fever	*Rickettsia ricketsii*	blood, skin	tick bite
Salmonella food poisoning	*Salmonella*	intestine	contaminated water and food
Strep throat	*Streptococcus pyogenes*	upper respiratory tract, blood, skin	person-to-person by sneezes, coughs, or direct contact
Tetanus	*Clostridium tetani*	nerves at synapses	contaminated wounds
Tuberculosis	*Mycobacterium tuberculosis*	lung, bones, other organs	person-to-person by coughs

Endotoxins, toxins made of lipids and carbohydrates, are associated with the outer membrane of Gram-negative bacteria, such as *E. coli.* While exotoxins are steadily released by living Gram-positive bacteria, endotoxins are not released by Gram-negative bacteria until the bacteria die. Once released, endotoxins cause fever, body aches, and weakness, and they damage the vessels of the circulatory system.

Bacteria also cause disease by destroying body tissues. As bacteria adhere to cells, they secrete digestive enzymes. These enzymes allow further tissue invasion. For example, some *Streptococci* bacteria produce a blood-clot-dissolving enzyme that allows infectious bacteria to spread to other tissues.

Word Roots and Origins

endotoxin

from the Greek *endon,* meaning "within," and *toxicon,* meaning "poison"

Antibiotics

Antibiotics are drugs that combat bacteria by interfering with various cellular functions. For example, **penicillin** interferes with cell-wall synthesis. **Tetracycline** (TE-truh-SIE-klin) interferes with bacterial protein synthesis. Many antibiotics are derived from chemicals that bacteria and fungi produce. Antibiotics protect bacteria and fungi from other microscopic invaders. Other antibiotics, such as **sulfa drugs,** are synthesized in laboratories. Many antibiotics are able to affect a wide variety of organisms; they are called **broad-spectrum antibiotics.** Some antibiotics, their mechanisms of action, and target bacteria are listed in Table 24-5.

Antibiotic Resistance

When a population of bacteria is exposed to an antibiotic, the bacteria that are most susceptible to the antibiotic die first. However, a few mutant bacteria that are resistant to the antibiotic may continue to grow. A resistant population then grows from these mutant bacteria through reproduction and genetic recombination. In this

TABLE 24-5 *A Summary of Common Antibiotics*

Antibiotic or synthetic drug	Mechanism of action	Target bacteria
Penicillin	inhibits cell-wall synthesis	Gram-positive bacteria
Ampicillin	inhibits cell-wall synthesis	broad spectrum
Bacitracin	inhibits cell-wall synthesis	Gram-positive bacteria; used as a skin-ointment
Cephalosporin	inhibits cell-wall synthesis	Gram-positive bacteria
Tetracycline	inhibits protein synthesis	broad spectrum
Streptomycin	inhibits protein synthesis	Gram-negative bacteria, tuberculosis
Sulfa drug	inhibits cell metabolism	bacterial meningitis, urinary-tract infections
Rifampin	inhibits RNA synthesis	Gram-positive bacteria and some Gram-negative bacteria
Quinolines	inhibit DNA synthesis	urinary-tract infections

TEACHING STRATEGY
Toxins

Have students make a two-column table to compare exotoxins and endotoxins. Have them title the rows "Composition," "Gram-Stain Reaction," "When Released," and "Example." As this section is discussed, have students complete their table (Exotoxin—protein, Gram-positive, secreted into surrounding environment, tetanus; Endotoxin—lipids and carbohydrates, Gram-negative, when bacteria die, *Escherichia coli*).

CRITICAL THINKING
Overdosing Toxins

Tell students that when a physician treats a patient for a bacterial infection, antibiotics are prescribed for several days. Ask students why a physician would do this. You may remind students that bacterial endotoxins are released when a bacterium dies. (If the bacteria make toxins and the bacteria are quickly killed, an overwhelming amount of toxins could suddenly be released, causing a toxic shock that could kill the patient.)

QUICK FACT

☑ Food poisoning affects millions of people each year. Seafood, poultry, and eggs are responsible for many cases of food poisoning, so it is important to thaw these foods in a refrigerator rather than at room temperature and to cook these types of foods thoroughly.

TEACHING STRATEGY
Bacterial Diseases

Assign students a bacterial disease from Table 24-4. Have them research the disease's history, characteristics, cause, frequency of occurrence, and methods of prevention.

SECTION 24-3

VISUAL LINK

Figure 24-6

Tell students that the clear zones indicate no bacterial growth. The larger the zone, the more sensitive the bacteria are to that antibiotic. In each Petri dish, the right disk contains penicillin, the lowest disk contains erythromycin, and the upper-left disk contains streptomycin. Ask students which culture is most sensitive to penicillin (lower left) and what its Gram reaction is (positive, because of its penicillin susceptibility).

ANSWERS TO SECTION 24-3 REVIEW

1. Bacteria cause disease by producing toxins and digestive enzymes that destroy tissue.
2. Antibiotics are thought to function in nature as a mechanism of bacteria and fungi to protect themselves from microscopic invaders.
3. Answers may vary. See Table 24-5 for complete details.
4. Bacteria transform sewage into simpler organic compounds, produce and process food, produce industrial chemicals, and recycle nutrients in the environment. Bacteria can also be used to leach minerals from ore, as insecticides, and to clean chemical and oil spills.
5. Antibiotic resistance in bacteria occurs when mutant bacteria survive an antibiotic treatment and give rise to a resistant population. Bacteria resist antibiotics by producing antibiotic-destroying enzymes or by having a cell wall that prevents entry of antibiotics into the bacterial cell.
6. Bacteria make pickles from cucumbers by fermentation, which is an anaerobic process. Oxygen would allow the bacteria to utilize aerobic respiration, which would not yield fermentation products.

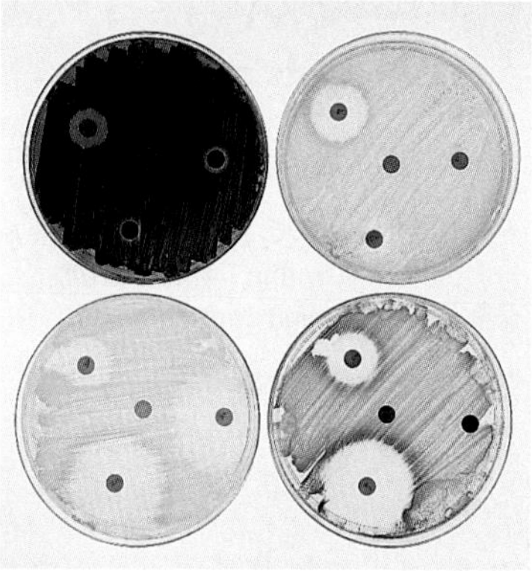

FIGURE 24-6

Bacteria can be tested for their sensitivity to antibiotics by growing them in a Petri dish with paper disks containing different antibiotics. As the antibiotics diffuse into the agar, the bacteria's growth will be inhibited by the antibiotic if the bacteria are sensitive to that antibiotic.

way, antibiotics provide a selective advantage to antibiotic-resistant bacteria. Because antibiotics have been overused, many diseases that were once easy to treat are becoming more difficult to treat.

The mechanisms of antibiotic resistance vary. In some bacteria, cell walls prevent passage of the antibiotic. Still other bacteria secrete enzymes that destroy or alter the antibiotic, as penicillin-resistant bacteria do. Figure 24-6 shows a method of testing bacterial resistance and sensitivity toward antibiotics.

USEFUL BACTERIA

Bacteria affect our lives in many positive ways. In sewage treatment, for example, bacteria break down the remains of organic matter in dead plant and animal waste, recycling carbon and nitrogen. Bacteria also turn sewage into simpler organic compounds. Bacteria, along with other microorganisms, recycle compounds from dead organisms, making them available to other organisms through the process of decay. Many bacteria are also able to fix carbon dioxide and create organic compounds.

Bacteria are also useful in producing and processing food. For example, bacteria ferment the lactose in milk to produce sour-milk products such as buttermilk, sour cream, and yogurt. Other species of bacteria digest the protein in milk and produce unripened cheeses such as ricotta and cottage cheese. Fermented foods are also produced by bacteria. Sauerkraut is produced when bacteria digest the carbohydrates in cabbage, and pickles result from the fermentation of cucumbers.

Bacteria are also used in industrial chemical production. They produce organic chemicals and fuels. Some are useful in mining for minerals and in petroleum recovery, while other bacteria and their products are used as insecticides. Bacteria are also used to help clean up environmental disasters caused by humans, such as chemical and oil spills.

SECTION 24-3 REVIEW

1. **Describe ways that bacteria cause human disease.**
2. **Describe the function of antibiotics in nature.**
3. **List three antibiotics used to treat disease and the mechanism of action of each.**
4. **List some of the positive uses of bacteria.**
5. **Explain how antibiotic resistance arises in bacteria and how bacteria resist antibiotics.**
6. **CRITICAL THINKING Why would a pickle processor carry out the preparation of pickles in anaerobic conditions?**

CHAPTER 24 REVIEW

SUMMARY/VOCABULARY

24-1
- Bacteria are prokaryotic single-celled organisms. They occur in several variations of three basic shapes: rods, spheres, and spirals.
- Bacteria are the oldest and most populous organisms. They are believed to have existed on Earth for about 3.5 billion years.
- Bacteria are classified in two kingdoms: Archaebacteria, which includes ancient forms of life, and Eubacteria, which includes most bacteria.
- The archaebacteria include the methanogens, which produce methane gas; the extreme halophiles, which live in very salty environments; and the thermoacidophiles, which live in extremely acidic environments at extremely high temperatures.
- The Gram stain is used to group bacteria into two groups; Gram-positive and Gram-negative bacteria.
- The cyanobacteria are photosynthetic bacteria that probably produced much of the oxygen in the Earth's atmosphere.

Vocabulary

actinomycete (472)
antibiotic (472)
archaebacterium (467)
bacillus (470)
chemoautotroph (473)
coccus (470)
enteric bacterium (473)
eubacterium (467)
eutrophication (472)
extreme halophile (468)
Gram-negative (471)
Gram-positive (471)
Gram stain (471)
heterocyst (472)
methanogen (468)
peptidoglycan (468)
population bloom (472)
spirillum (470)
spirochete (472)
staphylococcus (470)
streptococcus (470)
thermoacidophile (468)

24-2
- The major structures of the bacterial cell include a cell wall, a cell membrane, cytoplasm, a capsule, pili, endospores, ribosomes, and movement structures.
- Aerobic and anaerobic bacteria differ in whether they need an oxygen-rich environment or an oxygen-free environment. Different species live under different temperature conditions, ranging from 0°C to 110°C. Most bacterial species grow best at a neutral pH.
- Genetic recombination in bacteria can occur through transformation, conjugation, and transduction.

Vocabulary

capsule (475)
conjugation (477)
conjugation bridge (477)
endospore (475)
facultative anaerobe (476)
glycocalyx (475)
obligate aerobe (476)
obligate anaerobe (476)
photoautotroph (476)
pilus (475)
saprophyte (476)
thermophilic (476)
transduction (477)
transformation (477)

24-3
- Many bacteria are pathogens. Diseases may result from toxins produced by bacteria, from the destruction of body tissues, or from bacterial enzymes interfering with normal body processes.
- Antibiotics inhibit the growth of bacteria. Antibiotic-resistant bacteria destroy antibiotics, or prevent entry of the antibiotic into the cytoplasm.
- Helpful bacteria are used to convert sewage into simpler organic compounds, to produce and process food, to produce industrial chemicals, to mine for minerals, to produce insecticides, and to clean up chemical and oil spills.

Vocabulary

broad-spectrum antibiotic (479)
endotoxin (479)
exotoxin (478)
pathology (478)
penicillin (479)
sulfa drug (479)
tetracycline (479)
toxin (478)

CHAPTER 24 REVIEW ANSWERS

REVIEW

1. *Chemoautotroph* does not belong because the other terms describe bacteria that use food produced by other organisms.
2. *Methanogen* is an archaebacterium. The other terms refer to eubacteria.
3. *Archaebacterium* refers to a group of bacteria, while the other terms are related to disease.
4. *Conjugation* is a process, and the other terms are structures. A pilus is also used in starting conjugation—a process that does not involve an endospore.
5. *Anaerobe* refers to an organism living under a certain environmental condition. The other terms refer to groups of bacteria.

6. c **7.** b
8. c **9.** d
10. d **11.** a
12. b **13.** b
14. a **15.** a

16. spirochetes
17. The methanogens make methane as a byproduct, the halophiles require a very salty environment, and the thermoacidophiles require a high temperature and a low pH.

18. Cyanobacteria are prokaryotic and lack other organelles.
19. The capsule is a protective layer of polysaccharides that surrounds the bacterial cell wall. It protects the cell from harsh environments.
20. By decomposing dead organic matter, saprophytic bacteria release carbon and other elements for reuse by other organisms.
21. Chemoautotrophs harvest energy from minerals by oxidizing the chemicals.
22. In conjugation, a portion of DNA from one bacterium passes across a conjugation bridge into another bacterium. In transduction, a virus transfers genetic material from one bacterium to another. In transformation, a bacterial cell picks up DNA from its environment.
23. Bacteria use fermentation, which is an anaerobic pathway.
24. Answers will vary but should resemble the information in Table 24-4.
25. *A:* cell wall, *B:* DNA, *C:* cell membrane, and *D:* ribosome

CRITICAL THINKING

1. Fossil bacteria are very small, and their cells do not contain bones or other hard materials that would be easily preserved.
2. A deep puncture wound is more likely to be contaminated by tetanus bacilli because oxygen is less likely to be present in the deep skin layer.
3. Gram-positive bacteria have more peptidoglycan in their cell wall, so the effect on cell-wall synthesis is more pronounced. Gram-negative bacteria also have an outer layer that retards the movement of antibiotics into the cell.
4. They consume the excess food in the intestinal tract, and by doing so, they may prevent the establishment of pathogenic bacteria.

REVIEW

Vocabulary

In each of the following sets, choose the term that does not belong, and explain why it does not belong.

1. heterotroph, saprophyte, chemoautotroph
2. methanogen, spirochetes, enteric bacteria
3. archaebacterium, exotoxin, pathogen
4. pilus, conjugation, endospore
5. cyanobacterium, anaerobe, enteric bacteria

Multiple Choice

6. Bacteria produce yogurt from milk by (a) conjugation (b) aerobic respiration (c) fermentation (d) fixing nitrogen.
7. Rod-shaped bacteria are called (a) cocci (b) bacilli (c) halophiles (d) spirilla.
8. Thermoacidophiles are (a) eubacteria (b) cyanobacteria (c) archaebacteria (d) spirochetes.
9. Gram-positive bacteria stain (a) blue (b) pink (c) red (d) purple.
10. Eutrophication is the result of (a) antibiotics (b) pathogens (c) bacterial conjugation (d) population explosions.
11. Bacterial DNA is (a) a closed loop (b) encased in a capsule (c) linear (d) found in the nucleus.
12. The glycocalyx helps bacteria (a) survive unfavorable environmental conditions (b) stick to surfaces (c) metabolize gaseous nitrogen (d) ingest food.
13. During nitrogen fixation, gaseous nitrogen is converted to (a) carbon (b) ammonia (c) nitrate (d) methane.
14. An organism that must have oxygen to survive is (a) an obligate aerobe (b) a facultative anaerobe (c) a facultative aerobe (d) an obligate anaerobe.
15. Genetic recombination in bacteria can occur during the process of (a) conjugation (b) heterocyst formation (c) binary fission (d) endospore production.

Short Answer

16. Which bacteria move about by rotation?
17. List one distinguishing characteristic of each of the three main groups of archaebacteria.
18. Why are cyanobacteria no longer classified as algae?
19. Describe the capsule of a bacterium and its function.
20. Explain how saprophytic bacteria contribute to the recycling of nutrients in the environment.
21. How do chemoautotrophs harvest energy from their environment?
22. Describe one way bacteria can exchange genetic information.
23. Identify the metabolic process that bacteria use to make food products such as pickles and sauerkraut.
24. List some diseases caused by bacteria, and list the organs they affect.
25. Label the parts of the bacterium below.

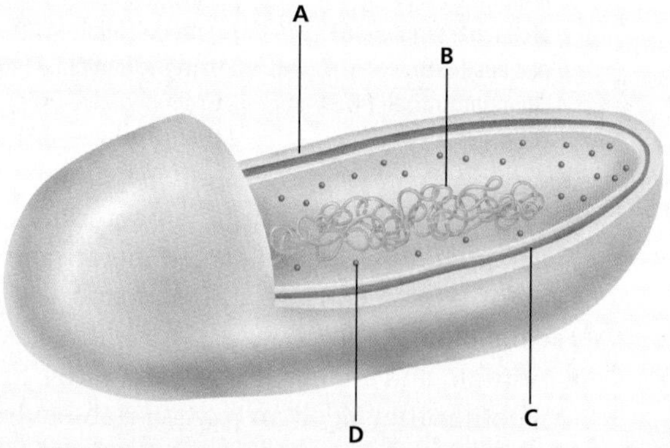

CRITICAL THINKING

1. Scientists have only recently discovered fossilized bacteria. Explain why this discovery may have taken so long.
2. *Clostridium tetani,* the bacterium that causes tetanus, is an obligate anaerobe. From this information, would you infer that a deep puncture wound or a surface cut would be more likely to become infected by tetanus bacteria? Explain the reason for your inference.
3. Penicillin works by interfering with the ability of bacteria to polymerize the peptidoglycan cell wall. Given this fact, explain why Gram-positive bacteria are more susceptible to the effects of penicillin than are Gram-negative bacteria.

4. Some of the bacteria that are normally found in the human intestinal tract are beneficial. For example, *E. coli* produces vitamin K. However, *E. coli* can also cause diarrhea under exceptional circumstances, and it can cause serious infections if it invades other parts of the body. Other bacteria found in the digestive tract do not produce substances the body can use, but they do not produce substances that are harmful either. What positive role might they play?
5. Thermal-vent communities, which are extremely hot regions of ocean in areas where the Earth's crust is open, have some of the densest and most productive populations known to exist. What might explain this?
6. Over the last 20 years, the number of antibiotic-resistant bacterial pathogens has steadily increased. This is thought to be a result of antibiotic abuse by patients and doctors. Doctors tend to overprescribe antibiotics for patients who demand a quick fix for their illness. Just recently, the World Health Organization has established a global computer database so that doctors can report outbreaks of antibiotic resistance. What are the potential benefits of this database?
7. The antibiotic-sensitivity report (above right) lists several antibiotics that a bacterium from a patient is susceptible to. Based on these data, what do you infer is the Gram reaction of the pathogen? A "−" sign indicates that the bacterium is not sensitive to the antibiotic, and a "+" sign means that it is sensitive to the antibiotic. A "+++" sign means that it is very sensitive to the antibiotic.

Antibiotic-Sensitivity Report

Antibiotic	Sensitivity
Penicillin	−
Cephalosporin	−
Streptomycin	+++
Tetracycline	++
Bacitracin	−
Rifampin	−
Ampicillin	+++

8. Examine the photograph below of bacteria that have been treated with a Gram stain. Would you hypothesize that these bacteria produce endotoxins? Explain your answer.

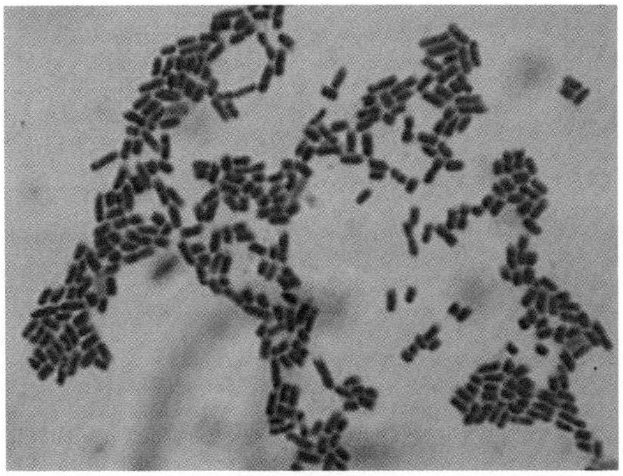

EXTENSION

1. Read "The Cholera Lesson" in *Discover*, February 1999, on page 71. According to microbiologist Rita Colwell, where does the cholera-causing organism *Vibrio cholerae* live between outbreaks? Describe the methods she is designing to predict and prevent widespread disease.
2. Research and write a report on the use of bacteria in either the processing of foods or the treatment of sewage.
3. Visit or telephone your local public-health department, and ask for information on bacterial diseases that have been reported in your area in recent weeks. Report your findings to the class.

5. In vent communities, predators that might limit bacterial growth are practically nonexistent. Thus, the bacterial communities can grow to the maximum that the food supply will allow.
6. Doctors will have almost immediate access to information about bacterial pathogens that are resistant to various antibiotics. Over time, the database will help reduce the number of unnecessary prescriptions written for antibiotics. This, in turn, should help preserve the effectiveness of antibiotics.
7. Refer to Table 24-5, and note that the organism in question is sensitive to antibiotics that are effective against Gram-negative bacteria and is resistant to antibiotics that affect Gram-positive bacteria. Therefore, it is a Gram-negative bacterium.
8. The bacteria are stained pink and must be Gram negative. Since most Gram-negative bacteria produce endotoxins, it is likely that these do too.

EXTENSION

1. *Vibrio cholerae* goes into a dormant, sporelike state when conditions are not right for replication and it lives all over the world in association with a copepod, a form of zooplankton. When the plankton bloom in the world's oceans, so does cholera. These blooms are more frequent with global warming. Colwell has devised a water-filtering system that isolated people can use. She is researching the possibility of using satellites to warn of plankton blooms.
2. Have students make presentations to the class about these beneficial bacteria. To follow up, invite a food scientist or specialist to address the class on industrial microbiology.
3. To follow up, invite a public-health specialist to address the class on epidemic bacterial diseases.

CHAPTER 24
INVESTIGATION

TIME REQUIRED

One 50-minute class period and 10 minutes for seven classes

SAFETY PRECAUTIONS

Students should not open the Petri dishes after they have been sealed. Used Petri dishes and used cotton swabs should be disposed of after autoclaving or by incineration. A local hospital or university laboratory may be willing to dispose of the used Petri dishes for you.

QUICK REFERENCES

Lab Preparation Notes are found on page 465D.

Holt BioSources provides a Teaching Resources CD-ROM that contains "Using Gowin's Vee in the Lab" and "Scoring Rubrics."

PROCEDURAL TIPS

1. Have students complete the background in advance, so that they arrive with ideas about places to test.
2. You may want to have students get your approval on their test locations before allowing them to proceed to step 3.
3. Students probably will not be able to check their Petri dishes for seven consecutive days. Encourage students to maintain an accurate record of their observations regardless of how often they make them.

ANSWERS TO BACKGROUND

2. Favorable conditions include a fairly high moisture content, temperatures between 20°C and 40°C, pH near neutral, and plenty of nutrients, including carbohydrates, lipids, vitamins, and minerals.
3. Answers will vary, but may include locations that include

CHAPTER 24 INVESTIGATION

Culturing Bacteria

OBJECTIVES

- Test common surfaces for the presence of bacteria.
- Learn a simple procedure for culturing bacteria.

PROCESS SKILLS

- observing
- hypothesizing
- experimenting
- collecting data
- analyzing data

MATERIALS

- Petri dish with nutrient agar
- glass-marking pen
- 4 sterile cotton swabs
- distilled water
- tape
- protective gloves

Background

1. Culturing bacteria involves growing microorganisms in a nutrient medium that is favorable for growth.
2. What are favorable conditions for the growth of most bacteria?
3. List some places in your school where bacteria are likely growing.
4. If bacteria reproduce once an hour and you start with one bacterium, how many bacteria do you have after 24 hours?

PART A Setting Up the Experiment

1. Discuss with your laboratory partner(s) places in school that would be likely to contain bacteria or that have conditions under which bacteria grow best. Make a list of three places where you will test for the presence of bacteria. Do not choose places outside your laboratory without your teacher's permission.
2. Obtain a Petri dish with nutrient agar from your teacher. Divide your Petri dish into four equal quadrants by writing on the outer bottom surface of the dish with the marking pen. Label the quadrants 1, 2, 3, and 4, as shown in the photograph at left. Also label the bottom of the dish with your group name or number. Do not mark on the top of the Petri dish because it will rotate with respect to the agar.
3. Use quadrant 1 as a control. Take a sterile cotton swab, and moisten it with distilled water. Be careful not to touch the cotton swab to any other surfaces, including your fingers. Remove the cover to the Petri dish, and rub the cotton swab across the nutrient agar in quadrant 1. Be careful to not tear the surface of the agar. Replace the cover immediately.
4. Take another sterile cotton swab, and moisten it with distilled water. Swipe the moistened swab across a surface that you have decided to test for the presence of bacteria. Be careful not to touch the cotton swab to any other surface. Touching the swab to other surfaces could contaminate the swab with bacteria other than that from your selected surface.

5. Remove the cover to the Petri dish, and rub the cotton swab across the nutrient agar in quadrant 2. This process transfers bacteria from the surface you have sampled to the nutrient agar in the Petri dish. Replace the cover immediately.
6. Repeat steps 4–6 for quadrants 3 and 4, using a clean cotton swab each time. Be sure to note in your lab report the areas that you swipe for each quadrant.
7. After you have swiped a sample in each quadrant of the Petri dish, seal the Petri dish with tape. CAUTION **Do not open the Petri dish again. Treat the contents of the Petri dish as you would any other pathogens.**
8. Place the Petri dish in a warm place with the cover side down for 24 hours.
9. Dispose of the cotton swabs according to your teacher's directions.
10. Clean up your lab materials and wash your hands before leaving the lab.

PART B Collecting Data

11. In your lab report, create a data table similar to the model shown below. Allow plenty of space to record your observations for each quadrant swabbed.
12. Check the Petri dish daily for bacterial growth until you no longer find new bacterial colonies (about seven days). Check your Petri dish and record your observations in your data table. What is happening that enables you to see bacteria? What does each colony represent?
13. Discard your Petri dish as directed by your teacher.

Analysis and Conclusions

1. On which surfaces did you find the most bacteria? the fewest bacteria? Did your results conform to your expectations? Explain.
2. Compare the colonies of bacteria that grew in each quadrant. What can you tell about the bacteria from the kind of colonies they produced?
3. What are some possible sources of error in the procedure you followed?
4. Combine the data obtained by the entire class. Which surfaces yielded the most bacteria?
5. Which of the surfaces that you sampled would you prefer to use as a food-preparation area? Explain your choice.
6. Would the amount of surface area that you sampled with your swab affect the number of colonies that grew on your dish? Explain your answer.
7. Can you tell by looking at the colonies on the dish if they would cause disease? Why or why not?
8. What test described in this chapter could you use to partially identify your bacteria? What characteristics of the bacteria could you learn from this test?

Further Inquiry

Design an experiment to test the effects of one variable, such as temperature or the presence of anitbacterial soap, on the growth of bacteria. What would you use as a control in your experiment?

PRESENCE OF BACTERIA ON COMMON SURFACES

Surface swabbed	Day 1	Day 2	Day 3	Day 4	Day 5	Day 6	Day 7

CHAPTER 24 INVESTIGATION

moisture, favorable temperature, and some source of nutrition.
4. Assuming no bacteria die, then there will be 16,777,216 bacteria after 24 hours. (2^{24})

ANSWER TO PART B

12. Bacteria are increasing in population and are therefore becoming visible. Each colony represents an individual bacterial cell that multiplied enough times to become visible.

ANSWERS TO ANALYSIS AND CONCLUSIONS

1. Answers will vary. Students should take into account frequency of cleaning of a tested surface, temperature, availability of nutrients and moisture, and sufficiency of the nutrient agar to support growth.
2. If the colonies are spread, they are probably motile organisms. Some bacteria may produce pigments, such as *Staphylococcus aureus* (yellow).
3. Error could result if the plates or swab was contaminated, if the incubator was not set at the correct temperature, if the swabbed surface had been recently treated with an antibacterial detergent, if the agar contained insufficient nutrient, or if the atmosphere was insufficient to sustain growth.
4. Answers will vary.
5. Answers will vary.
6. Yes, because a larger sample area would yield more bacteria.
7. No, because complicated biochemical tests would be needed in order to identify the bacteria.
8. The Gram stain would identify its Gram reaction and shape.

CHAPTER 25
VIRUSES

	TOPICS	TEACHING RESOURCES	LABS, CLASSWORK, AND HOMEWORK
BLOCK 1 (45 minutes)	**Introducing the Chapter, p. 486**	**ATE** Focus Concept, p. 486 **ATE** Assessing Prior Knowledge, p. 486 **ATE** Understanding the Visual, p. 486	■ Supplemental Reading Guide, *Microbe Hunters*
	25-1, Structure, p. 487 Advent of Virology, p. 487 Characteristics of Viruses, p. 488 Grouping Viruses, p. 489	**ATE** Section Overview, p. 487 **ATE** Visual Link Figure 25-1, p. 488 Figure 25-3, p. 490 94, 95	**PE** **Quick Lab,** p. 488, Calculating Nanometers ★ Study Guide, Section 25-1 **PE** Section Review, p. 490
BLOCK 2 (45 minutes)	**25-2 Viral Replication, p. 491** The Bacteriophage, p. 491 The Lytic Cycle, p. 492 The Lysogenic Cycle, p. 493 Evolution, p. 495	**ATE** Section Overview, p. 491 **ATE** Visual Link Figure 25-7, p. 494 93, 96	★ Study Guide, Section 25-2 **PE** Section Review, p. 495
BLOCKS 3 & 4 (90 minutes)	**25-3 Viruses and Human Disease, p. 496** Infectious Diseases, p. 496 Prevention and Treatment, p. 497 Emerging Viruses, p. 498 Viruses and Cancer, p. 500	**ATE** Section Overview, p. 496 **ATE** Critical Thinking, p. 496 **ATE** Visual Link Figure 25-12, p. 498 94, 97	**PE** **Chapter 25 Investigation,** p. 504 **Lab C30,** Screening for Resistance to Tobacco Mosaic Virus ■ Occupational Applications Worksheet—Forensic Toxicologist ★ Study Guide, Section 25-3 **PE** Section Review, p. 500
BLOCKS 5 & 6 (90 minutes)	**REVIEW and ASSESSMENT** **PE** Chapter 25 Review, pp. 502–503	Audio CD Program	**CHAPTER TESTING** ■ Chapter Test (blackline master) ▲ Test Generator ■ Assessment Item Listing ■ Scoring Rubrics and Classroom Management Checklists

MODERN BIOLOGY

HOLT BIOSOURCES

KEY

PE Pupil's Edition
ATE Teacher's Edition
■ Active Reading Guide
★ Study Guide

Teaching Transparencies

Laboratory Program

One-Stop Planner CD-ROM
Shows these resources within a customizable daily lesson plan:
■ Holt BioSources Teaching Resources
■ Active Reading Guide
★ Study Guide
▲ Test Generator

READING FOR CONTENT MASTERY

Preview: Have students preview the objectives, the introductory paragraph, topic sentences, and captions before reading the chapter.

ATE Active Reading Technique: K-W-L, p. 487
■ Reading Strategy: K-W-L Worksheet
■ Active Reading Guide Worksheet 25-1
PE Summary/Vocabulary, p. 501

■ Active Reading Guide Worksheet 25-2
PE Summary/Vocabulary, p. 501

ATE Active Reading Technique: K-W-L, p. 500
■ Active Reading Guide Worksheet 25-3
PE Summary/Vocabulary, p. 501

TECHNOLOGY AND INTERNET RESOURCES

CNN PRESENTS Video Segment 16
Depression Virus

Holt Biology Videodiscs Teacher's Correlation Guide, Lessons 25-1 through 25-3

Audio CD Program, Sections 25-1 through 25-3

internetconnect

sciLINKS NSTA

On-line Resources: www.scilinks.org
The following *sci*LINKS Internet resources can be found in the student text for this chapter:

Topics:
- Lytic cycle, p. 492
- AIDS virus, p. 498

go.hrw.com

On-line Resources: go.hrw.com
Visit the HRW Web site for a variety of resources related to this chapter. Just type in the keyword HM2 Chapter 25.

Smithsonian Institution®
Visit **www.si.edu/hrw** for additional on-line resources.

CNNfyi.com
Visit **www.cnnfyi.com** for late-breaking news and current event stories selected just for you.

LAB ACTIVITY PLANNING

CHAPTER 25 INVESTIGATION

Tobacco Mosaic Virus, pp. 504–505

OVERVIEW
In this investigation, students will test whether tobacco from cigarettes can infect tobacco plants with tobacco mosaic virus (TMV). Each group will set up a controlled experiment that uses two tobacco plants.

SAFETY
Students need to thoroughly wash their hands and all laboratory equipment that has come into contact with tobacco. Students should conduct this investigation away from gardens and house plants because TMV spreads from plant to plant. Students should wear disposable gloves, lab aprons, and safety goggles during the lab. If dibasic potassium phosphate solution and carborundum powder get in a student's eye, immediately flush the eye out with water and seek medical attention.

PREPARATION
1. Plant tobacco seeds 2 weeks prior to the investigation. Tomato plants may be substituted for tobacco plants.
2. To prepare 0.1 M dibasic potassium phosphate solution, add 6.8 g KH_2PO_4 and dilute to 500 mL with distilled water.
3. Extract tobacco from cigarettes of various brands before the lab. Place each brand in a labeled, separate container. Students will create a mixture of tobacco from the different brands of cigarettes. You may want to prepare the mixture of tobacco in advance.

Quick Labs

Calculating Nanometers, p. 488
Students will need a meterstick with millimeter marks, paper, scissors, tape, and a pencil for this activity.

UNDERSTANDING THE VISUAL

Have students study the photograph of virus-cell interaction. Tell them that the pink structures are HIV and the brownish mass is a T-lymphocyte. Ask students to describe what they think is happening in the photograph. (The T-lymphocyte has produced HIV, and the viruses are leaving the cell.) Ask students how the viruses, pictured leaving the T-lymphocyte, entered the cell. (All of the viruses pictured did not enter the T-lymphocyte. The genetic material from a single HIV particle that entered the T-lymphocyte was responsible for the replication of all the viruses pictured.)

FOCUS CONCEPT

Cell Structure and Function

In this chapter, students learn that viruses depend on cells to replicate and that this viral dependence on cells affects cell structure and function. For example, during the process of viral replication, viruses take over the host cell's metabolic machinery and may later completely destroy the host cell as the new viruses are released.

ASSESSING PRIOR KNOWLEDGE

Review the following concepts.

Characteristics of Life: *Chapter 1*
Have students list the characteristics of living organisms.

Protein Synthesis: *Chapter 10*
Have students summarize the steps involved in protein synthesis. Ask students to identify the location of each step within a eukaryotic cell.

CHAPTER 25

VIRUSES

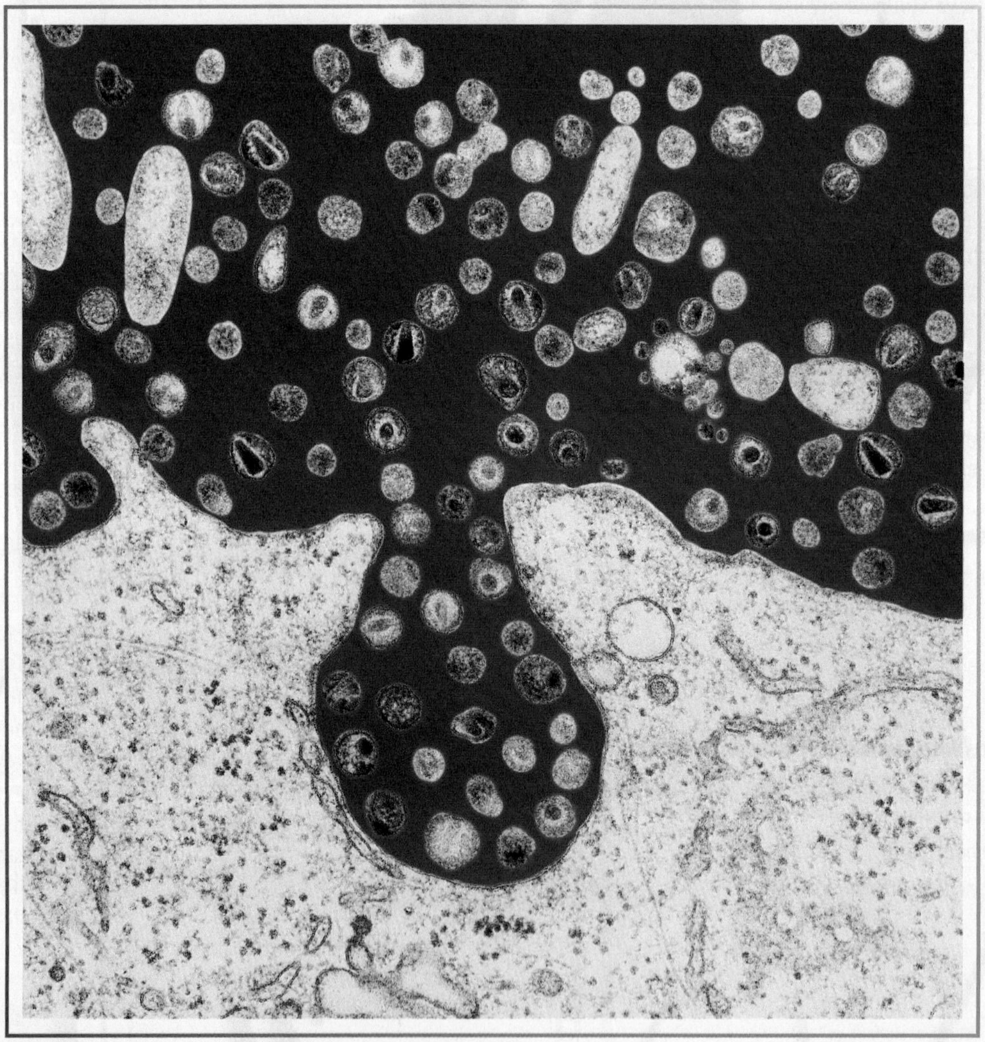

The human immunodeficiency virus (HIV), shown here stained pink, is the cause of acquired immune deficiency syndrome (AIDS). (TEM 29,640×)

FOCUS CONCEPT: ***Interdependence of Organisms***

As you read, note how the structure and replication cycle of viruses distinguish them from living organisms.

STRUCTURE

SECTION 25-1

*A **virus** is a nonliving particle composed of a nucleic acid and a protein coat. Although viruses are not living organisms, they are of interest to biologists because they cause many diseases in living organisms. Viruses are also quite useful in genetic research because they can change how a cell functions. The study of viruses is called **virology** (vie-RAH-luh-jee).*

OBJECTIVES

▲ Describe the structure and classification of viruses.

● Describe the achievement of Wendell Stanley in the development of virology.

■ Identify the range of sizes and shapes among viruses.

◆ List the characteristics used to group viruses.

▲ Compare and contrast viroids and prions with viruses.

ADVENT OF VIROLOGY

By the late 1800s, scientists knew that some factor smaller than bacteria could transmit disease. But they lacked the technology to see these structures and to study them in depth. Scientists gained more awareness of the nature of viruses in 1935, when Wendell Stanley (1904–1971) crystallized the tobacco mosaic virus, an agent responsible for the mosaic mottling and withering of tobacco leaves. His work suggested that viruses might be chemicals rather than tiny cells. Until Stanley's crystallization of a virus, viruses were thought to be primitive cells that were perhaps ancestors of bacteria. Table 25-1 compares viruses with cells.

Virology now provides clues to the biochemistry of living organisms, including mutation, the combination of genetic material from different sources, and other essential processes of genetics. Also, pharmaceutical companies use viruses to develop new antiviral medications, and researchers continue to investigate the mechanisms of viral replication.

TABLE 25-1 *A Comparison of Viruses and Cells*

Characteristics of life	Virus	Cell
Growth	no	yes
Homeostasis	no	yes
Metabolism	no	yes
Mutation	yes	yes
Nucleic acid	DNA or RNA	DNA
Reproduction	only within host cell	independently by cell division
Structure	nucleic acid core, protein covering, and, in some cases, an envelope	cytoplasm, cell membrane, cytoskeleton, and, in the eukaryotic cell, organelles

SECTION 25-1

SECTION OVERVIEW

Cell Structure and Function
Section 25-1 describes the structure of viruses. This section also discusses the importance of nucleic acids in the production of viral proteins as well as in the grouping of viruses.

Reproduction and Inheritance
Section 25-1 explains that reverse transcriptase plays an important role in the replication of retroviruses because this enzyme is necessary to make DNA from RNA before proteins can be produced and used in the assembly of retroviruses.

ACTIVE READING

Technique: K-W-L
Before students read this chapter, have them write short individual lists of all the things they already Know (or have heard) about viruses. When students have finished their lists, ask them to contribute their entries to a class list on the board or overhead. Then have students list things they Want to know about viruses. Have students save their lists to determine what they have Learned at the end of the chapter.

QUICK FACT

☑ Charles Chamberland, a collaborator of Pasteur and inventor of the autoclave, developed a porcelain bacterial filter in 1884 that allowed viruses to be isolated.

VISUAL LINK

Figure 25-1

Refer students to the figure, and ask them what they think reverse transcriptase is and what it does. (It is an enzyme that catalyzes the production of a viral DNA chain from an RNA template.) Explain to students that it is called reverse transcriptase because it catalyzes the transcription of DNA from an RNA template. Tell students that viruses that have reverse transcriptase are called retroviruses. Examples of retroviruses include HIV, murine leukemia virus, and feline immunodeficiency virus.

Quick Lab

Calculating Nanometers

Time Required 20 minutes

Procedural Tips Have students work in pairs. If metersticks with millimeter marks are not available, provide students with metersticks and centimeter rulers.

Answers to Analysis

1. 1 m = 1,000,000,000 nm
2. 2 m = 2,000,000,000 nm
3. 10,000,000 nm/cm; 20,000,000 nm/2 cm; 200,000,000 nm/20 cm; 2 cm = 20,000,000 nm; 20 cm = 200,000,000 nm
4. 1,000 mm/m; 1,000,000 nm/mm; 2,000,000 nm/2 mm; 2 mm = 2,000,000 nm

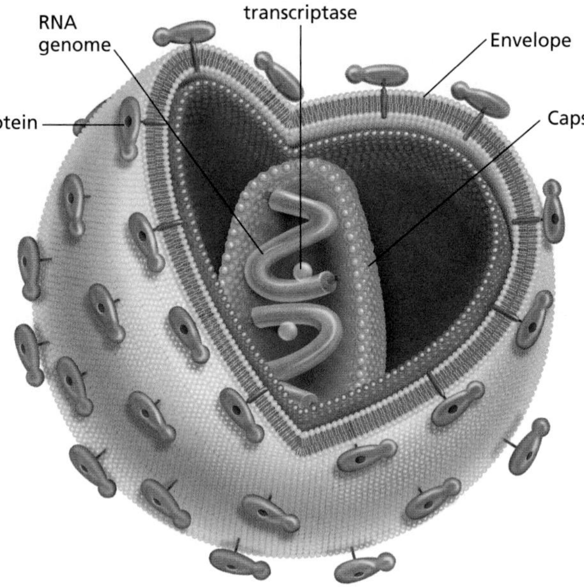

FIGURE 25-1

This view of the human immunodeficiency virus (HIV) displays some of the virus's structural features.

Word Roots and Origins

virus

from the Greek *ios*, meaning "poison"

Quick Lab

Calculating Nanometers

Materials meterstick with millimeter marks, paper, scissors, tape, pencil

Procedure Cut the paper into strips, and tape the strips together to form one strip that is 2 m long. On the paper, mark and label lines at 1 m, 20 cm, 2 cm, and 2 mm.

Analysis

1. Write an equation beside the 1 m line to show that 1 m contains 1 billion nanometers.
2. Write an equation at the end of the paper strip that shows the relationship between the length of the paper in meters and the length of the paper in nanometers.
3. How many nanometers are in 1 cm? How many nanometers are in 2 cm? in 20 cm? Write equations beside the 2 cm and the 20 cm marks to show the relationship of centimeters to nanometers.
4. How many millimeters are in 1 m? How many nanometers are in 1 mm? How many nanometers are in 2 mm? Write an equation by the 2 mm mark to show its relationship to nanometers.

CHARACTERISTICS OF VIRUSES

Viruses are among the smallest biological particles capable of causing disease in living organisms. Viruses range in size from the extremely small poliovirus, which is 20 nm in diameter, to the large smallpox virus, which is about 250 nm in diameter—about the same size as the smallest bacterium. One nanometer is equal to 0.001 µm (0.00000004 in.).

Viruses are constructed of compounds usually associated with cells, but they are not considered living organisms. They have some, but not all, of the characteristics of life listed in Chapter 1. They have no nucleus, cytoplasm, organelles, or cell membrane, and they are not capable of carrying out cellular functions. Moreover, viruses are able to replicate only by infecting cells and using the organelles and enzymes within cells.

Viral Structure

All viruses have two essential features—a nucleic acid and a protein coat surrounding it. Viral nucleic acid may be either DNA or RNA, but not both. The shape of the nucleic acid may be helical, a closed loop, or a long strand, depending on the virus. The protein coat surrounding the nucleic acid is called a **capsid** (KAP-sid).

Some viruses have a membrane-like structure outside the capsid called an **envelope.** The envelope, which is made mostly of lipids, is taken from a host cell membrane during replication. The envelope allows new viruses to infect host cells during the first stage of viral replication. Enveloped viruses include influenza, chickenpox, herpes simplex, and **HIV** (human immunodeficiency virus). Refer to the model of HIV in Figure 25-1 to identify each of the viral structures.

On the surface of the envelope are projections made of glycoprotein. These **glycoprotein** (GLIE-koh-PROH-teen) projections are protein-containing sugar chains that the virus uses to attach to a host cell.

Viral Shape

A virus's shape may be determined by its capsid or its nucleic acid. Figure 25-2 shows two examples of virus shapes. Some viruses have the shape of an **icosahedron** (ie-KOH-suh-HEE-druhn), which is a geometric shape with 20 triangular faces. The viral capsid forms this shape. Icosahedral viruses include those that cause herpes simplex, chickenpox, and polio.

Other viruses are shaped like a **helix** (HEE-liks). A helix resembles a coiled spring. The viral nucleic acid is responsible for this shape. The rabies, measles, and tobacco mosaic viruses are helical viruses.

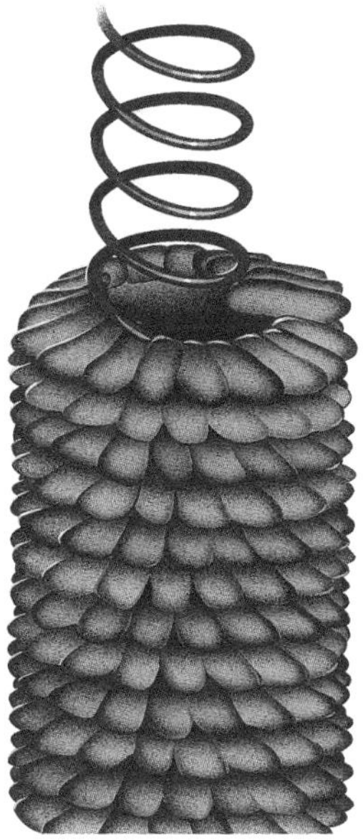

TOBACCO MOSAIC VIRUS
(helical)

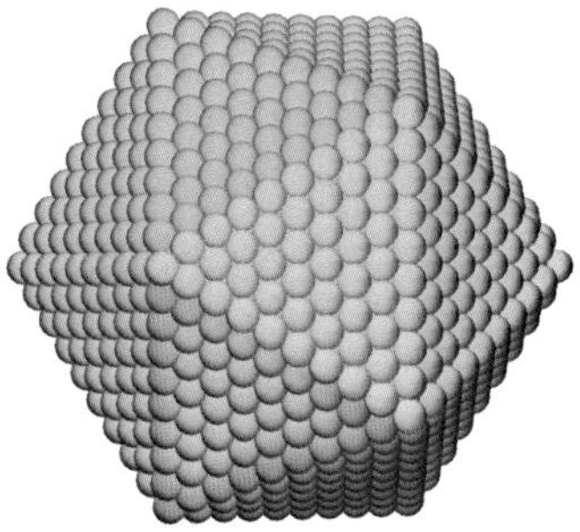

POLIO VIRUS
(icosahedral)

FIGURE 25-2

Viruses occur in different shapes. Helical and icosahedral are two examples of virus shapes.

GROUPING VIRUSES

The grouping of viruses is based on the presence of a capsid structure and an envelope. Viruses are also grouped according to whether they contain RNA or DNA and whether the nucleic acid is single-stranded or double-stranded. Therefore, viruses are grouped based on their shape and structure. Some common viral groups are presented in Table 25-2.

Virus Types

DNA and RNA viruses differ in the way they use the host cell's machinery to produce new viruses. For example, upon entering the host cell, a DNA virus may act in one of two ways: the virus may directly produce RNA that then makes more viral proteins or it may join with the host cell's DNA to direct the synthesis of new viruses.

TABLE 25-2 ***Some Common Viruses of Humans***

Viral group	Nucleic acid	Shape and structure	Examples of diseases they cause
Papovaviruses	DNA	icosahedral, non-enveloped	warts, cancer
Adenoviruses	DNA	icosahedral, non-enveloped	respiratory and intestinal infections
Herpesviruses	DNA	icosahedral, enveloped	herpes simplex, chickenpox, shingles, infectious mononucleosis
Poxviruses	DNA	complex brick-shaped, enveloped	smallpox, cowpox
Picornaviruses	RNA	icosahedral, non-enveloped	poliomyelitis, hepatitis, cancer
Myxoviruses	RNA	helical, enveloped	influenza A, B, and C
Rhabdoviruses	RNA	helical, enveloped	rabies
Retroviruses	RNA	icosahedral, enveloped	AIDS, cancer

TEACHING STRATEGY

Virus Structure

Place students in groups, and have them design a virus model using the following items: plastic-foam ball, stick pins, pipe cleaner, colored markers, clear plastic wrap, round beads, tape, and scissors. Have the groups present their models to the class and discuss the functions of all the structures in the model.

RETEACHING *ACTIVITY*

Show students a picture of a plant cell and a picture of a virus. Have students compare the components of a plant cell with the components of a virus. Review with students the structures that may not be visible, such as DNA or RNA.

RECENT RESEARCH

Accidental Vaccine

A report indicates that a goat virus known as caprine arthritis-encephalitis, a relative of HIV, might provide protection from HIV. This virus has been found to produce antibodies that react with HIV. If further studies indicate that these antibodies provide protection from HIV infection, the goat virus may be an accidental vaccine for HIV.

GIFTED *ACTIVITY*

Explain to students that viruses can be cultured only in living tissues, whereas bacteria can be grown on nutrient agar. Ask students why viruses cannot be grown on nutrient agar. (Viruses cannot carry out the cellular functions necessary for independent growth; they can reproduce only within cells of living tissues.)

VISUAL LINK

Figure 25-3

Ask students to describe what they see in the TEM of the viroid. Have them compare the structures associated with the viroid with the structures associated with the HIV in Figure 25-1. Ask the students what is missing from the viroid. (It has neither a capsid nor an envelope.)

ANSWERS TO SECTION 25-1 REVIEW

1. The two essential components of a virus are a protein coat and nucleic acid.
2. Wendell Stanley crystallized the tobacco mosaic disease agent. His work indicated that viruses might be chemicals rather than tiny cells.
3. Viruses are classified according to their envelope, capsid structure, and nucleic acid content.
4. Nucleic acids in viruses help scientists group different viruses together. This grouping is based on whether viruses contain DNA or RNA and whether the nucleic acid is single- or double-stranded.
5. Viroids, prions, and viruses are all nonliving particles that are capable of replicating and causing disease. Viruses and viroids are made of nucleic acid. Viroids have no protein. Prions are made of protein and have no nucleic acids.
6. Viruses generally are not considered to be living organisms because they do not have all of the characteristics of life given in Chapter 1. Students should justify their answer by listing which characteristics of life viruses lack. For example, viruses have no growth, homeostasis, metabolism, or organelles.

FIGURE 25-3

Viroids, which cause certain plant diseases, are infective RNA strands without capsids.

RNA viruses replicate differently from DNA viruses. Upon entering the host cell, viral RNA is released into the host cell's cytoplasm. There, it uses the host cell's ribosomes to produce new viral proteins.

Some RNA viruses, known as **retroviruses** (RE-troh-VIE-ruhs-uhz), contain an enzyme called reverse transcriptase in addition to RNA. **Reverse transcriptase** (tran-SKRIP-tays) uses RNA as a template to make DNA. The viral DNA is integrated into the host genome. The DNA then makes an RNA transcript of itself. This RNA is then translated into proteins that become part of new viruses. Reverse transcriptase is so named because it reverses the normal process of transcription, in which DNA serves as a template for producing RNA.

Viroids and Prions

Even simpler than viruses are the disease-causing agents called viroids and prions. **Viroids** (VIE-roidz) are the smallest known particles that are able to replicate. A viroid consists of a short, single strand of RNA and has no capsid, as shown in Figure 25-3. These simple RNA molecules are able to disrupt plant cell metabolism and damage entire crops. Economically important plants that have been affected by viroids include potatoes, cucumbers, avocados, and oranges.

Prions (PREE-ahnz) are abnormal forms of proteins that clump together inside a cell. This clumping activity eventually kills the cell, perhaps by blocking the cell's molecular traffic. Found on the surface of mammalian cells and in the brain of hosts, prions are composed of about 250 amino acids and have no associated nucleic acid.

Prions have been linked to certain diseases of the brain in humans and animals, such as scrapie. Scrapie is a disease in sheep that is characterized by slow degeneration of the nervous system. As the nervous system decays, the animals develop tremors and scrape their bodies against fence posts and tree trunks.

Bovine spongioform encephalopathy (BSE), or "mad cow disease," is a fatal brain disease of cattle that may be linked to prions. The prion that is thought to cause mad cow disease may be similar to one implicated in a human brain disease called Creutzfeld-Jakob (KROITZ-felt-YAK-ohb) disease (CJD).

SECTION 25-1 REVIEW

1. What are the two essential components of a virus?
2. What was the accomplishment of Wendell Stanley?
3. Explain how viruses are grouped.
4. What role do nucleic acids play in the grouping of viruses?
5. How do viroids and prions compare with viruses?
6. **CRITICAL THINKING** Are viruses considered to be living organisms? Justify your answer by referring to the characteristics of life presented in Chapter 1.

VIRAL REPLICATION

SECTION

25-2

OBJECTIVES

▲ Describe a bacteriophage.

● Summarize the five phases of the lytic cycle.

■ Compare the lytic and lysogenic cycles of viral replication.

◆ Differentiate between a prophage and a provirus.

▲ Summarize how viruses may have evolved.

Because viruses are not cells, they can replicate only by invading a host cell and using the enzymes and organelles of the host cell to make more viruses. Because they depend on host cells for replication, viruses are called ***obligate intracellular parasites.*** *Outside the host cell, a virus is a lifeless particle with no control over its movements. It is spread by the wind, in water, in food, or via blood or other body secretions.*

THE BACTERIOPHAGE

In the 1950s, scientists gained a better understanding of viral replication through their work with **bacteriophages** (bak-TEER-ee-uh-fay-juz), which are viruses that infect bacteria. Bacteriophage replication cycles have been found to be similar to those of the viruses that cause colds, measles, and acquired immune deficiency syndrome. The most commonly studied bacteriophages, T phages, are known to infect a bacterium found in the human digestive tract, *Escherichia coli.*

Examine the structure of the bacteriophage particle in Figure 25-4. Bacteriophages are composed of an icosahedral head that contains a nucleic acid. Beneath the head is a contractile tail that includes a collar and a sheath. The contractile tail helps inject the nucleic acid into the host cell. The tail rests on a base plate from which tail fibers emerge. These fibers assist the virus in attaching to a host cell. As you read about the replication cycles of bacteriophages, notice how the structure of the bacteriophage suits its function.

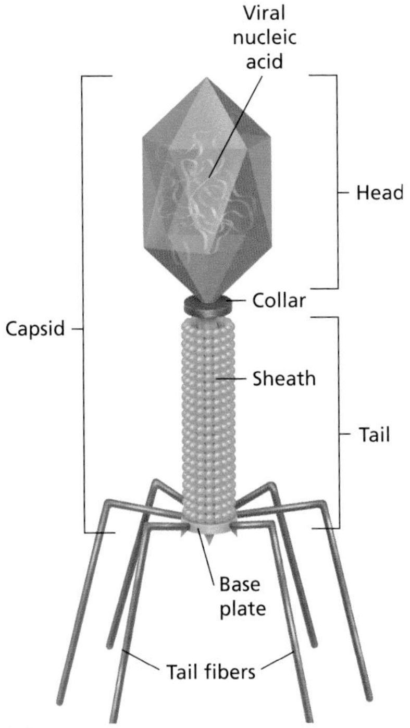

(a)

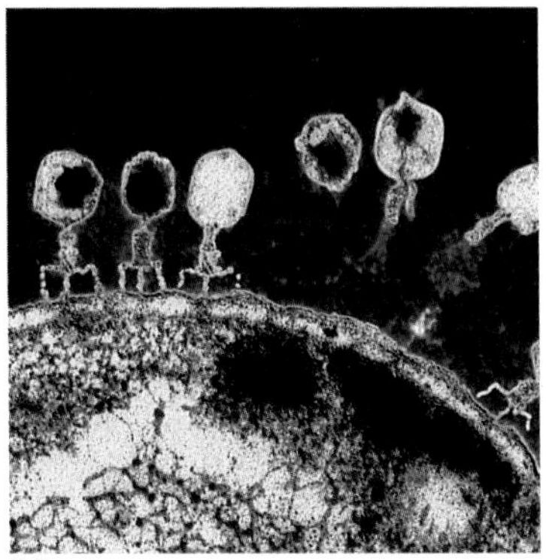

(b)

FIGURE 25-4

(a) This diagram shows the structural complexity of a bacteriophage. (b) This transmission electron micrograph shows a cross section of an *E. coli* cell being attacked by several bacteriophages. Some of the bacteriophages can be seen developing within the cell's cytoplasm, and some can be seen outside the cell. (TEM 138,600×)

SECTION 25-2

SECTION OVERVIEW

Reproduction and Inheritance

Section 25-2 describes how viruses use the host cell to replicate through two different cycles: the lytic cycle and the lysogenic cycle. The text explains that depending on the replication pathway taken by the virus, the virus replicates immediately or lies dormant in the cell until stimulated to begin replication.

Evolution

Section 25-2 also discusses how viruses are constantly evolving due to their ability to replicate quickly and evade the immune system.

▶ ENGAGE STUDENTS

Ask students to imagine that they work for a rock music company; perhaps they are recording artists, marketers, or bookkeepers. One day a new person takes over the company and signs a billion-dollar contract to produce classical music. This person insists that everyone be transferred to the production department to begin processing and packaging classical music CDs. After reaching their billion-dollar quota, the company runs out of raw materials and goes bankrupt. Ask students how this event is similar to a viral infection. (The new person is analogous to a virus; he or she takes over the company to make lots of money at the expense of losing the company. In the same way, a virus takes over a host cell to make many copies of itself at the expense of destroying the host cell.)

SECTION 25-2

TEACHING STRATEGY
Stages of Replication

Guide students through the stages of the lytic cycle. Have students list the five stages of the lytic cycle and describe each stage in their own words. Then have them draw pictures that symbolize the meaning of their definitions (examples: a key and lock = attachment; parking a car in the garage = entry; a photocopy of paper = replication; putting a puzzle together = assembly; kids running out of the classroom for recess = release).

DEMONSTRATION

Simulate the activity of a bacteriophage's contractile tail during replication. Inflate a white balloon. (Do not over-inflate.) Obtain an oversized hypodermic needle and syringe. (This can be obtained from a cooking specialty shop.) Mix food coloring (blue or another dark color) in a specific volume of water that can fill the entire syringe. Before filling the syringe with the colored water, lubricate the inside of the barrel with petroleum jelly to reduce friction. Next, fill the syringe with the colored water and insert the needle into the end of the balloon where the rubber is thick. Slowly inject the contents of the syringe into the balloon and carefully withdraw the needle. Tell students that when comparing this model to a real virus, the nucleic acid is a more discrete unit, unlike the water used in the simulation.

MAKING CONNECTIONS
Mathematics

A single virus can infect a bacterial cell and produce about 100 virus particles in about 20 minutes. Have students calculate the number of viruses that would be produced four hours after a single virus cell infected a single bacterial cell (10^{18}).

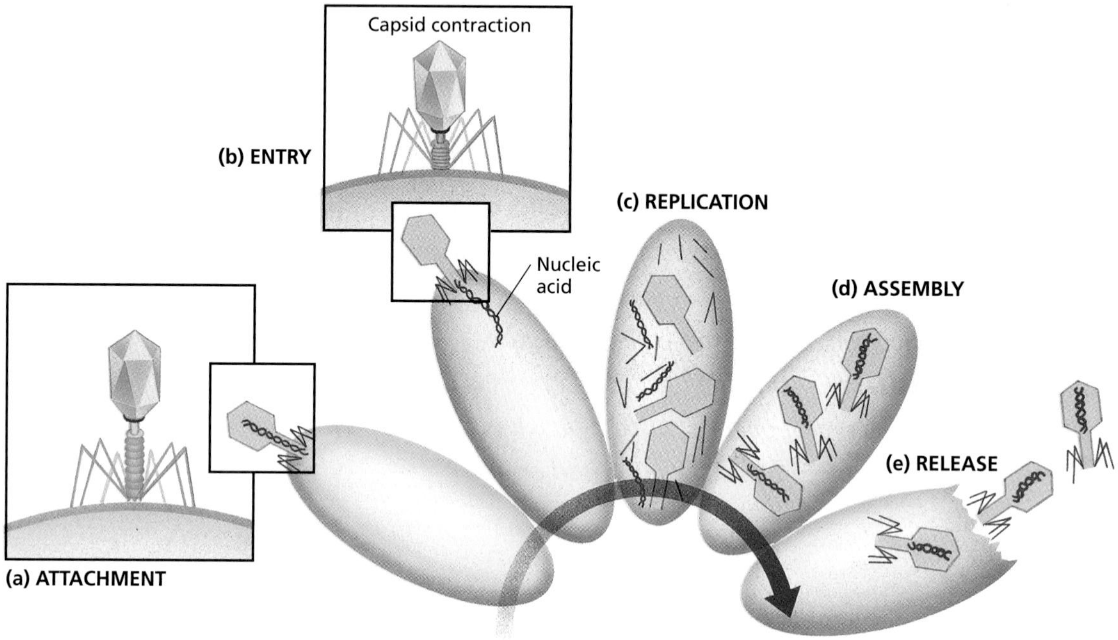

FIGURE 25-5

The lytic cycle of a virulent bacteriophage includes five steps: (a) attachment of the virus to a susceptible host cell, (b) entry of the viral DNA into the host cell, (c) replication of viral DNA, (d) assembly of new viruses, and (e) release of new viruses from a lysed host cell.

internetconnect

SCILINKS NSTA
TOPIC: Lytic cycle
GO TO: www.scilinks.org
KEYWORD: HM492

THE LYTIC CYCLE

During the **lytic cycle,** a virus invades a host cell, produces new viruses, destroys the host cell, and releases newly formed viruses. Viruses that undergo the lytic cycle are called **virulent** because they cause disease. The lytic cycle consists of five phases, as shown in Figure 25-5.

As Figure 25-5a shows, the bacteriophage first attaches to a susceptible bacterium by attaching its tail fibers to a receptor site. **Receptor sites** are specific sites that viruses recognize and attach to on the host cell's surface. If the bacteriophage does not find a receptor site, it cannot infect the cell. This specificity is true for many other viruses. For example, hepatitis virus infects only liver cells, and the glycoproteins on the surface of HIV and influenza enable them to attach only to specific types of cells.

Next the bacteriophage releases an enzyme that weakens a spot in the cell wall of the host, as shown in Figure 25-5b. Then the phage presses its sheath against the cell and injects its DNA into the host cell through the weak spot in the cell wall. The bacteriophage leaves its capsid outside the host cell.

As Figure 25-5c shows, the virus then takes control of the host's protein-synthesizing mechanisms, transcribing mRNA from the viral DNA. The resulting bacteriophage mRNA is translated by ribosomes and enzymes in the host cell into viral proteins and enzymes that form bacteriophage capsids. The viral DNA in the host bacteria is also replicated during this phase.

The replicated viral genes are enclosed in the newly created virus capsids, as shown in Figure 25-5d. The assembly of new virus particles usually occurs in the cytoplasm, but it also may take place in a eukaryotic host cell's nucleus.

During the last phase of the lytic cycle, one of the enzymes that is produced by the bacteriophage genome causes the host cell to disintegrate, releasing the new bacteriophages. Cell disintegration, which is shown in Figure 25-5e, is called **lysis** (LIE-sis). In enveloped viruses, the newly formed viruses move to the cell surface and force their way through the cell membrane. As a result, the virus leaves the cell with a piece of the host cell membrane attached to the capsid. This "borrowed" cell membrane fragment becomes the viral envelope.

Word Roots and Origins

lysis

from the Greek *lysis,* meaning "loosening" or "dissolving"

THE LYSOGENIC CYCLE

Some viruses can infect a cell without causing its immediate destruction. Viruses that stay in their host cell for an extended period of time—days, months, or years—are in a **lysogenic** (lie-soh-JEN-ik) **cycle.** A virus that replicates through the lysogenic cycle and does not kill the host cell immediately is called a **temperate** virus. Examples of a DNA and an RNA virus replicating through the lysogenic cycle are discussed below.

Lysogeny in Bacteriophages

Temperate bacteriophages enter bacteria in the same way that a virulent bacteriophage does, as shown in Figures 25-6a and 25-6b. The tail fibers of the temperate bacteriophage attach to a specific receptor site on the bacterial cell wall. Then the bacteriophage injects its DNA into the host cell. Instead of immediately creating new RNA and viral proteins, however, the bacteriophage DNA integrates itself into the host cell's DNA, as shown in Figure 25-6c. The bacteriophage DNA molecule that integrates itself into a specific

FIGURE 25-6

The lysogenic cycle of a temperate bacteriophage involves (a) the attachment of the virus to the host cell, (b) injection of viral DNA, (c) integration of the viral DNA into the host genome, and (d) multiplication of the host cell with the viral DNA.

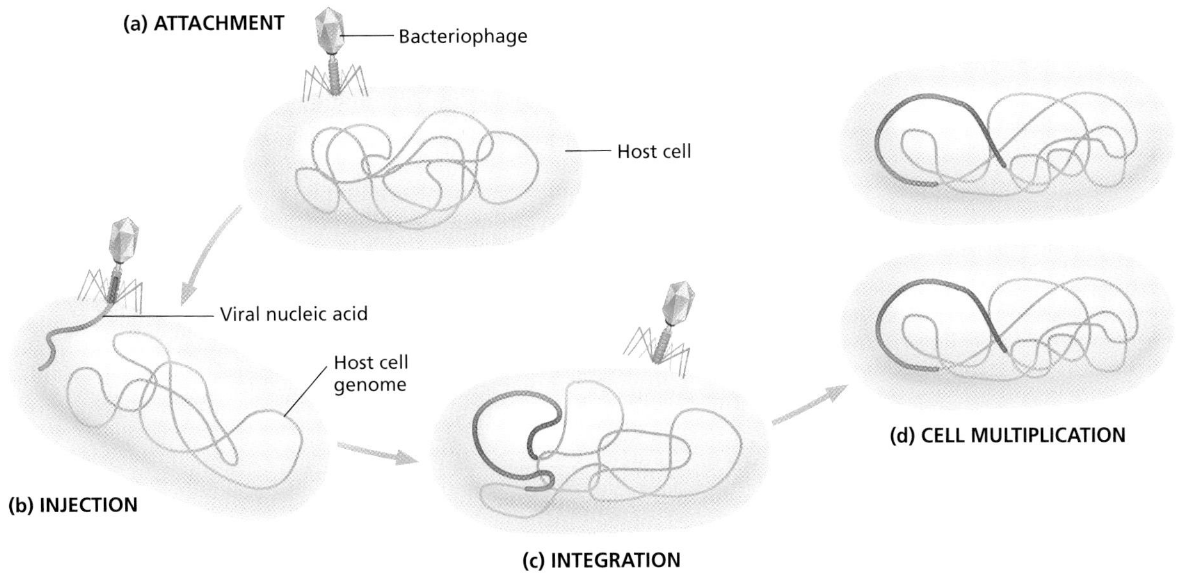

RETEACHING *ACTIVITY*

Have students research the following viral diseases: common cold, AIDS, rubella, cold sores, mumps. Ask students to categorize the diseases as being caused by virulent (lytic cycle) or temperate (lysogenic cycle) viruses. (Virulent: common cold, rubella, and mumps; temperate: AIDS and cold sores.) Ask students how the onset of disease is affected by temperate viruses compared with how it is affected by virulent viruses. (Virus reproduction and associated disease manifestations are delayed as a result of the virus being in a temperate stage of infection.)

INCLUSION *ACTIVITY*

Provide students with several pieces of construction paper shaped into 2.5 cm hexagons (head), 30 cm of 0.5 cm wide ribbon (nucleic acid), one sheet of construction paper (in a contrasting color), several pieces of 1 cm string (tail fibers), and glue. Have students make a virus by gluing a 2 cm length of ribbon and pieces of string to the hexagon. Then have them attach the virus to the construction paper (host cell) with the ribbon extending from the virus onto the construction paper. Have students glue pieces of string, hexagons, and ribbon to the construction paper in a group. Finally, have students construct two or three complete viruses leaving the paper. Once the model is constructed, have students identify the different steps of replication and the structures represented by the materials provided.

QUICK FACT

☑ The science of molecular genetics was launched by virus research. And studies of virus reproduction have helped scientists understand how genes function and how genetic information is translated from DNA into proteins.

VISUAL LINK
Figure 25-7

Have students compare Figure 25-7 with Figure 25-8. (Both replication cycles end in lysing of the host cell.) Ask students why people infected with HIV appear healthy for many years even though Figure 25-8 shows cell lysis, which means that disease symptoms would be detected in the HIV patient. (The virus has entered the lysogenic cycle.)

GIFTED *ACTIVITY*

Have students design a drug strategy that would inhibit viral infection. Tell students to consider the different steps involved in viral replication and to create a drug that could inhibit one of these steps. (The drug may be targeted at preventing the virus from binding to the host cell, degrading the reverse transcriptase if the virus is a retrovirus, or interrupting virus assembly.) Ask students to provide a diagram of virus–host cell interaction and to identify the step at which the drug begins working to prevent infection.

RECENT RESEARCH
Interaction Between Viral Pathogens

Studies propose that HIV may gain entry into host cells through the use of chemokine receptors. Chemokines are immune-system signaling molecules. Evidence also suggests that cytomegalovirus (CMV) infection of a host cell stimulates production of the chemokine receptor US28, which may act as a binding site for HIV. Statistics now indicate that 80 percent of the general population is infected with CMV. These results suggest that researchers may now have to consider HIV infection as a result of a cooperative interaction with other viral pathogens.

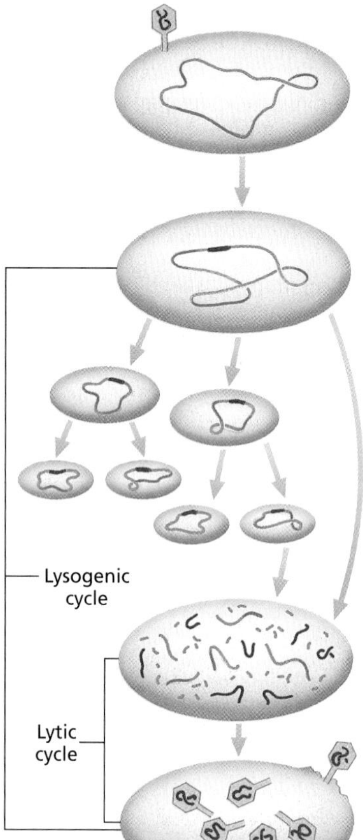

FIGURE 25-7

In the lysogenic cycle, a bacteriophage enters a cell and remains inactive in the host's genome until an external stimulus causes the virus to enter the lytic cycle.

site of the host cell's genome is called a **prophage** (PROH-fahj). The prophage replicates whenever the host bacterium reproduces, as shown in Figure 25-6d. As the host DNA replicates, so does the viral DNA, and each bacterial offspring is infected with a prophage. During the lysogenic cycle, the prophage does not harm the host cell. However, radiation or certain chemicals can cause a prophage to become virulent. Figure 25-7 shows that when the prophage becomes virulent, it enters the lytic cycle, proceeding with replication and destroying the host cell.

Lysogeny in HIV

The normal lysogenic cycle usually involves DNA viruses, but it can also involve RNA viruses, such as HIV, the virus that causes AIDS. When HIV infects a susceptible white blood cell, it attaches to receptor sites on the host cell's surface and enters the cell by fusing with the cell membrane. Next, viral RNA and reverse transcriptase are released into the cell's cytoplasm. Reverse transcriptase then transcribes the viral RNA into DNA. Recall from Section 25-1 that reverse transcriptase uses RNA as a template to make DNA.

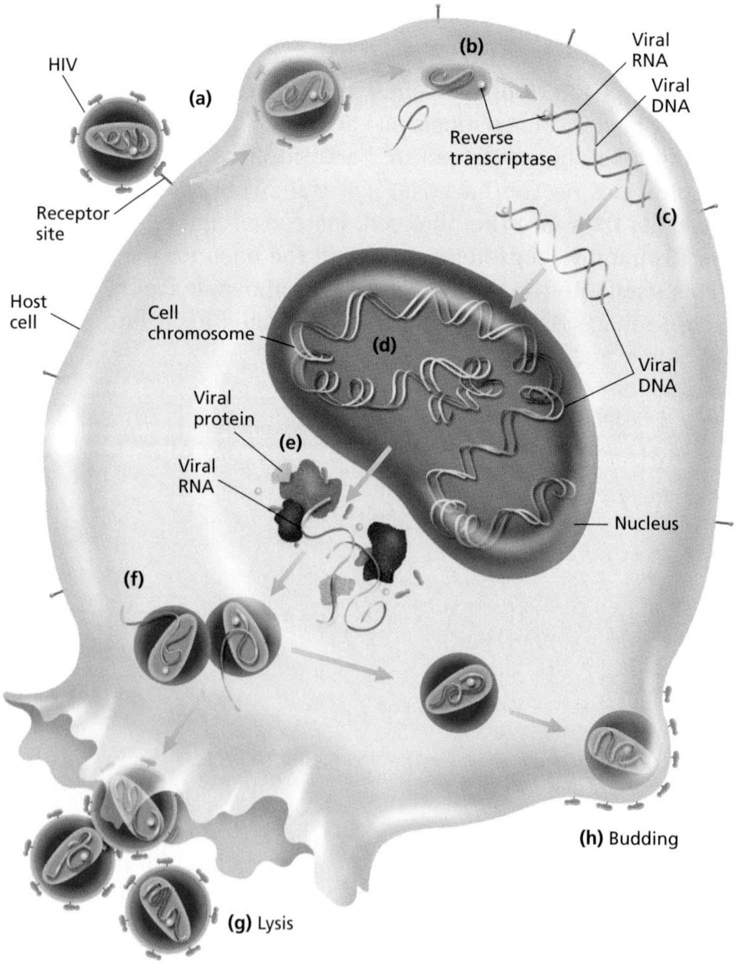

FIGURE 25-8

This figure traces the steps of HIV replication in a host cell. (a) HIV attaches to the receptor site. (b) The viral RNA is released from the capsid, and (c) reverse transcription takes place. (d) The viral DNA is integrated into the host genome. (e) Viral RNA and viral proteins are produced, and (f) viral assembly takes place. The assembled viral particles leave the cell through lysis (g) or budding (h).

As shown in Figure 25-8, the viral DNA molecule produced by reverse transcriptase inserts itself into the host cell's genome for an unspecified period of time. This viral DNA molecule is called a **provirus** (proh-VIE-rus). The lysogenic cycle ends when the provirus is transcribed into RNA and translated into viral proteins. These proteins are used in the assembly of new HIV particles containing capsids and enzymes. The new HIV particles are released after they lyse the cell or bud from the cell after taking a part of the cell's membrane for their viral envelope.

EVOLUTION

Because viruses depend on cells to replicate, most scientists reason that viruses evolved from early cells. The first viruses were probably naked pieces of nucleic acid that were able to travel from one cell to another, perhaps through damaged cell membranes. Over time, genes that allowed viruses to invade healthy cells evolved.

For example, when influenza viruses invade the human body, the immune system will destroy most of the viruses. However, a few of the viruses will escape destruction, enter cells, and begin producing thousands of copies of themselves within a few days.

How were a few viruses able to avoid destruction and begin replicating? These viruses are usually the result of mutations in the viral nucleic acid. Just as mutations in bacterial cells enable bacteria to become resistant to antibiotics, mutations in viruses can enable viruses to evade the immune system. Viruses that mutate quickly, such as the influenza virus and HIV, make it difficult for the immune system to immediately recognize and destroy them. The immune system will respond to the new strain of viruses eventually, but not until many new viruses are formed. Thus, it is difficult to develop vaccines that prevent these viral infections over long periods of time. To avoid this problem, the flu vaccine targets a different strain of influenza virus each year.

SECTION 25-2 REVIEW

1. Discuss the activities of a virus during the lytic cycle.

2. Compare and contrast the lytic and lysogenic cycles.

3. Describe the structure of a bacteriophage.

4. Explain the difference between a prophage and a provirus.

5. Discuss how the earliest viruses may have originated.

6. CRITICAL THINKING During the lytic cycle, the assembly of new viral particles can sometimes take place in the host cell's nucleus. This assembly in the host cell's nucleus does not occur with bacteriophage particles. Why?

SECTION 25-2
ANSWERS TO SECTION 25-2 REVIEW

1. During the lytic cycle, the virus is involved in a series of activities: attachment of the virus to the host cell, entry of the viral genome into the host cytoplasm, replication of the viral genome that encodes for new viruses, assembly of new viral parts to form new viruses, and release of new viruses as the host cell disintegrates.

2. The lytic and lysogenic cycles both have attachment and entry steps, but the lysogenic virus integrates its DNA into the host genome and does not produce viral particles immediately, whereas the lytic virus produces virus particles immediately and lyses the host cell.

3. The bacteriophage structure consists of a head region, contractile tail, base plate, and tail fibers. The head region contains the viral nucleic acid, the contractile tail region assists in injecting the viral nucleic acid into the host cell, the base plate provides a platform for the contractile tail and tail fibers, and the tail fibers assist the virus in attaching to a host cell.

4. A prophage is a bacteriophage genome that has inserted itself into a specific site on the bacterial chromosome. A provirus is a viral DNA molecule that has inserted itself into the eukaryotic host genome.

5. The earliest viruses may have originated from the nucleic acid fragments of early cells. These nucleic acid fragments may have entered host cells through a damaged cell membrane.

6. Bacteriophages attack only bacteria, which are prokaryotic cells. Prokaryotic cells do not have a nucleus.

SECTION 25-3

SECTION OVERVIEW

Structure and Function
Section 25-3 discusses how cell function is interrupted by viral disease processes.

Evolution
Section 25-3 discusses how viruses living in an isolated habitat can adapt to infect new hosts, such as humans, when the environment changes.

▶ ENGAGE STUDENTS

Obtain a vaccination card to show to your students. (If you do not have one, your physician, a school nurse, or the county health department can probably provide you with one.) Many of these cards are multilingual and carry the imprint of the World Health Organization. Ask students why many countries want visitors to be vaccinated. (They do not want a disease to be introduced into their country. Since people can now travel quickly by air, a person showing no symptoms of disease could easily carry a virus to another location. A vaccinated person, however, would be less likely to be a carrier.)

CRITICAL THINKING

Shingles
Explain that shingles is a disease that may develop years after a person has been infected by the chickenpox virus. Ask students to use their knowledge of viral reproduction to explain the appearance of shingles. (The virus enters a lysogenic stage after the initial infection until it is reactivated to cause symptoms.)

SECTION

25-3

OBJECTIVES

▲ Name four viral diseases that result in serious illness in humans.

● Compare the two types of viral vaccines, and discuss other forms of viral-disease prevention.

■ Discuss the relationship between viruses and cancer.

◆ Outline the onset of a virus outbreak.

VIRUSES AND HUMAN DISEASE

Viral diseases are among the most widespread illnesses in humans. These illnesses range from mild fevers to some forms of cancer and include several other severe and fatal diseases. Transmission of these illnesses varies; some are transmitted by human contact, while others are transmitted through water or an insect bite.

INFECTIOUS DISEASES

Many significant viral diseases are caused by viruses that use humans as their natural hosts. Some of the most common human viral diseases include the common cold, chickenpox, measles, mumps, polio, rabies, and hepatitis. Viral infections can affect various organs of the human body, including the brain, liver, heart, lungs, and the skin.

Rabies is transmitted from the bite of an infected animal, which carries the virus in its saliva. When a person is infected, the virus travels from the wound to the central nervous system. Symptoms of rabies include fever, headache, throat spasms, paralysis, and coma. Rabies is so lethal that few people have survived its effects.

Chickenpox is a highly contagious viral disease. The virus multiplies in the lungs and uses the network of blood vessels to reach the skin. Symptoms include fever and a skin rash. Transmission occurs from direct contact with the skin rash, which is the source of infectious virus particles, and through the air. Fortunately, the disease is usually mild, and recovery is usually followed by a lifelong resistance to reinfection. However, if all of the chickenpox virus is not destroyed, it can persist in the nerve cells as a provirus and cause a disease called shingles later during adulthood. Shingles results in a more severe case of chickenpox; the fever is higher, the immune system weakens, and pneumonia can occur. Like chickenpox, shingles is also defined by a skin rash. However, this painful rash is limited to an area of skin serviced by a particular neural pathway. For instance, the rash has been known to occur on only one side of the chest. The shingles rash can shed chickenpox viruses and infect susceptible children and adults. Note the characteristic skin rash of shingles in Figure 25-9.

FIGURE 25-9
The shingles rash, shown below, usually affects only one part of the body.

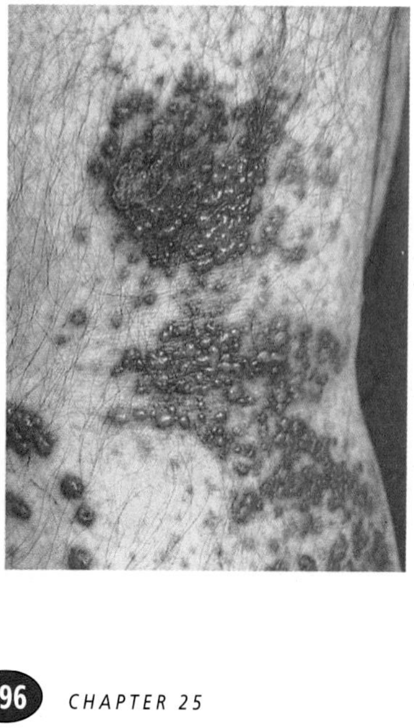

PREVENTION AND TREATMENT

The control of viral diseases is accomplished in two ways: vaccination to prevent disease and administration of **antiviral drugs**—drugs that interfere with viral nucleic acid synthesis—to infected patients. Unfortunately, there are few antiviral drugs compared with drugs used to treat bacterial, fungal, and parasitic infections. The most successful approach to controlling viral diseases has been prevention through vaccination. The Centers for Disease Control and Prevention in Atlanta, Georgia, is committed to the control and prevention of disease through research, as shown in Figure 25-10, and education.

Types of Virus Vaccines

As you will recall from Chapter 13, a vaccine is a preparation of pathogens or other materials that stimulates the body's immune system to provide protection against that pathogen. Some vaccines consist of inactivated or attenuated viruses. **Inactivated** viruses do not replicate in a host system. **Attenuated** viruses are viruses that have been genetically altered so that they are incapable of causing disease under normal circumstances. In general, vaccines made from attenuated viruses are preferred over those made from inactivated viruses because protection is greater and lasts longer. Additional doses of some vaccines, called booster shots, can extend a person's protection against some viruses.

By the 1960s, vaccines for measles, mumps, and rubella had been developed. The hepatitis B vaccine became available in the 1980s, and the chickenpox and hepatitis A vaccines were developed in the 1990s. Scientists continue to work on the development of an AIDS vaccine, but the genetic diversity and mutability of the virus create a problem for vaccine development. Educating people about HIV transmission is currently the best approach to slowing the spread of AIDS.

Smallpox Eradication Program

Smallpox once killed 40 percent of the people it infected, leaving the other 60 percent scarred and often blind. Smallpox is a DNA virus that is transmitted by nasal droplets emitted during sneezing and coughing. Symptoms include fever, headache, backache, and the development of a skin rash. This virus is hardy enough to be spread through infected blankets and clothing. Figure 25-11 shows a person infected with smallpox.

Vaccination played an important role in the eradication of smallpox. The World Health Organization began the smallpox eradication program in 1967. It ended in 1980, with the official declaration that smallpox had been eradicated. The program included vaccination and the quarantine of infected people. The last naturally acquired smallpox case occurred in Somalia in 1977.

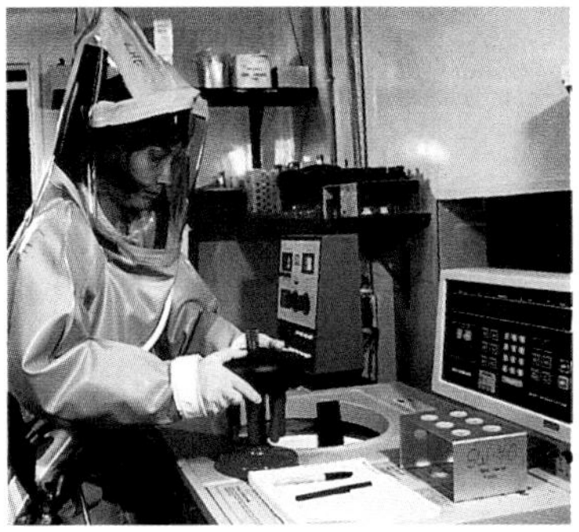

FIGURE 25-10

This technician is working in a level-4 laboratory at the Centers for Disease Control and Prevention (CDC). Workers in a level-4 laboratory use the highest level of security procedures to protect themselves from the world's deadliest viruses, such as Lassa fever virus, Ebola virus, and smallpox virus.

Word Roots and Origins

vaccine

from the Latin *vaccinus*, which means "pertaining to cows"

FIGURE 25-11

Smallpox is characterized by lesions covering the face, shoulders, chest, and in a later stage, arms and legs.

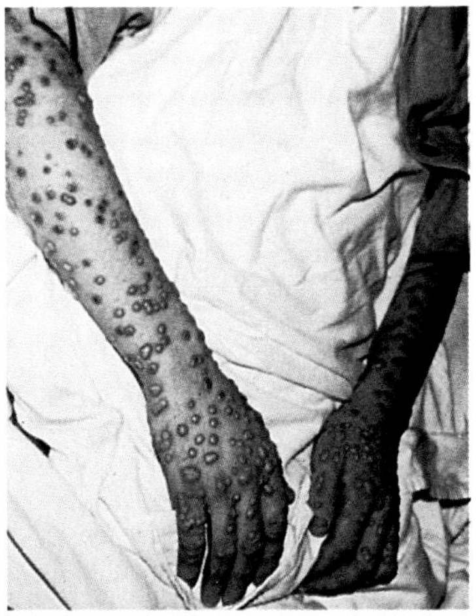

TEACHING STRATEGY
Spread of Infection

Have students place all of the red checkers on a checkerboard in both the black and white squares. The red checkers represent healthy people. Remind students to make sure that some of the red checkers have neighbors (red checkers in squares whose sides or corners touch the sides or corners of squares containing other red checkers), while some of the red checkers do not have neighbors. Have students add a black checker to the board to represent an infected person. Tell students to crown any of the neighboring checkers of the black checker with black and then crown the neighboring checkers of the newly "infected" checkers with black.

DEMONSTRATION

Ask students to bring newspaper and magazine articles about specific bacterial and viral infections to class. Make a list on the chalkboard of the diseases discussed in each article, and have students indicate whether the pathogen responsible for the disease is a bacterium or virus. Have students discuss the means of contracting each disease, the people commonly affected by it, and potential treatments.

QUICK FACT

☑ Edward Jenner, an English physician, noticed that milkmaids who contracted cowpox did not catch smallpox. To test this hypothesis, he inoculated a young boy with cowpox and later inoculated him with smallpox. The child did not contract smallpox.

VISUAL LINK
Figure 25-12

The patient in the picture is being treated for AIDS. Ask students if the mask the health professional is wearing is a necessary health precaution. (No, because HIV is not thought to be spread through the air.) Ask students what precautions the health professional should take to prevent the spread of AIDS (avoid contact with blood and other body fluids by wearing gloves).

RETEACHING *ACTIVITY*

Advertisements for cold medicines may imply that the medicines cure the cold. However, the medicines do not disable the virus; they merely relieve the cold symptoms. Ask students how a real cure for the common cold could work. (A cure needs to disrupt the replication cycle. For example, it could prevent the virus from entering the cell, replicating, or translating capsid proteins.)

TEACHING STRATEGY
New Viruses

The principal method of fighting viruses is by vaccination. Unfortunately, mutations in existing viruses are a major source of new virus development. Therefore, new viruses pose greater health risks, since new vaccines must be developed to combat them. New viruses are discovered every day. For example, epidemiologists have identified 50 new viruses in the Brazilian rain forest. Researchers estimate that hundreds more viruses remain undetected in the forest. Ask students how viruses in the Brazilian rain forest could affect them. (Answers might discuss how the mobility of the world's populations can spread the new viruses among different populations or how studying new viruses can provide scientists with insight into fighting viral diseases.)

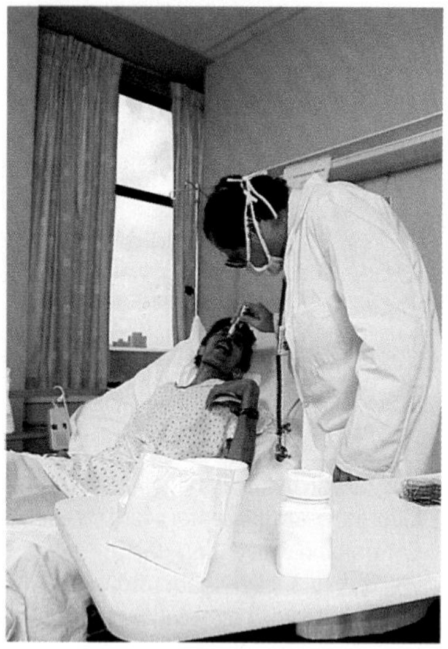

FIGURE 25-12

This physician is administering two different antiviral drugs—DDI (didexoyinosine) and AZT (azidothymidine)—to this AIDS patient. Because of HIV's ability to rapidly adapt to antiviral drugs, it is often necessary to administer more than one drug at a time in order to reduce the amount of virus in a patient.

Other Antiviral Approaches

One important part of viral disease prevention is the control of animals that spread viral disease. Yellow fever was completely eradicated in the United States through mosquito-control programs. Annual rabies vaccinations keep pets free of infection and protect humans too. Wildlife officials in the western part of the United States set out meat that contains an oral rabies vaccine to control the spread of rabies in coyotes and wolves.

Treatment of virally infected patients includes the use of antiviral agents. **Acyclovir** (ay-SIE-kloh-VEER) is used against herpes simplex and chickenpox, and **azidothymidine** (ay-ZIED-oh-THIE-mi-deen) **(AZT)** inhibits the reverse transcriptase of retroviruses, such as HIV. These drugs interfere with the synthesis of viral nucleic acids. Another class of drugs, called **protease inhibitors,** interferes with the synthesis of viral capsids during viral replication. Combinations of protease inhibitors and AZT have been shown to be helpful in slowing the progression from HIV infection (lysogenic phase) to AIDS (lytic phase). Figure 25-12 shows an AIDS patient undergoing treatment.

Antibiotics are not effective in the treatment of viral diseases. Antibiotics are used specifically to attack bacterial cells' metabolic machinery. Because viruses use only the host cell's machinery, antibiotics are of no use in destroying viruses.

EMERGING VIRUSES

As medical researchers work to find cures for existing viral diseases, newly discovered viruses are emerging in different parts of the world. Emerging viruses are viruses that exist in isolated habitats, but infect humans when these habitats are developed. For example, Figure 25-13 shows a tropical forest in the Democratic Republic of Congo (formerly Zaire), where the emerging Ebola virus has been known to exist. When these forests are cleared, humans may be exposed to virus-infected animals. If these viruses can infect humans, they may spread with deadly consequences.

internet**connect**

SCiLINKS NSTA
TOPIC: AIDS virus
GO TO: www.scilinks.org
KEYWORD: HM498

FIGURE 25-13

The remote tropical forests of the Democratic Republic of Congo (formerly Zaire) are the hiding places of the deadly Ebola virus.

Literature & Life

The AIDS Virus

The following is the introduction to Robert C. Gallo's article "The AIDS Virus," from the January 1987 issue of *Scientific American.*

It is a modern plague: the first great pandemic of the second half of the 20th century. The flat, clinical-sounding name given to the disease by epidemiologists—acquired immune deficiency syndrome—has been shortened to the chilling acronym AIDS. First described in 1981, AIDS is probably the result of a new infection of human beings that began in central Africa, perhaps as recently as the 1950's. From there it probably spread to the Caribbean and then to the U.S. and Europe. By now as many as two million people in the U.S. may be infected. In the endemic areas of Africa and the Caribbean the situation is much worse. Indeed, in some areas it may be too late to prevent a disturbingly high number of people from dying.

In sharp contrast to the bleak epidemiological picture of AIDS, the accumulation of knowledge about its cause has been remarkably quick. Only three years after the disease was described its cause was conclusively shown to be the third human retrovirus: human *T*-lymphotropic virus III (HTLV-III), which is also called human immunodeficiency virus (HIV). Like other retroviruses, HTLV-III has RNA as its genetic material. When the virus enters its host cell, a viral enzyme called reverse transcriptase exploits the viral RNA as a template to assemble a corresponding molecule of DNA. The DNA travels to the cell nucleus and inserts itself among the host's chromosomes, where it provides the basis for viral replication.

In the case of HTLV-III the host cell is often a *T4* lymphocyte, a white blood cell that has a central role in regulating the immune system. Once it is inside a *T4* cell, the

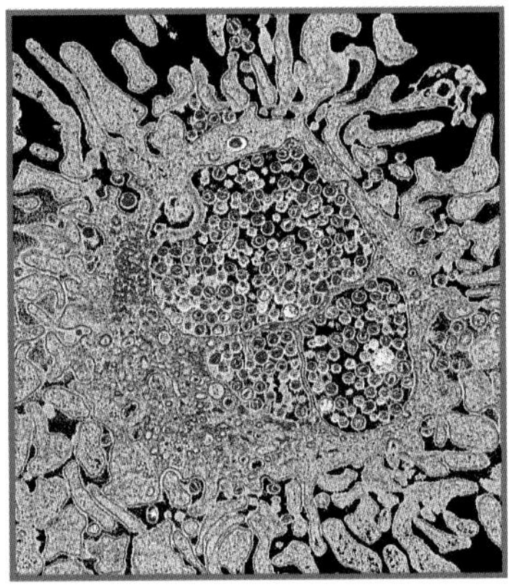

virus may remain latent until the lymphocyte is immunologically stimulated by a secondary infection. Then the virus bursts into action, reproducing itself so furiously that the new virus particles escaping from the cell riddle the cellular membrane with holes and the lymphocyte dies. The resulting depletion of *T4* cells—the hallmark of AIDS—leaves the patient vulnerable to "opportunistic" infections by agents that would not harm a healthy person.

How HTLV-III manages to replicate in a single burst after lying low, sometimes for years, is one of the most fundamental questions confronting AIDS researchers. Another important question is the full spectrum of diseases with which the virus is associated. Although most of the attention given to the virus has gone to AIDS, HTLV-III is also associated with brain disease and several types of cancer. In spite of such lingering questions, more is known about the AIDS virus than is known about any other retrovirus. The rapidity of that scientific advance was made possible partly by the discovery in 1978 of the first human retrovirus, HTLV-I, which causes leukemia. In its turn the new knowledge is making possible the measures that are desperately needed to treat AIDS and prevent its spread.

Reading for Meaning

What does Gallo mean by "opportunistic" infections? Use Gallo's description and the context of the article to figure it out.

Read Further

Gallo's article tells what was known about the AIDS virus in 1987. What new information might be included in a follow-up article about the AIDS virus today?

CHAPTER 25
Literature & Life

BACKGROUND

This feature presents Robert C. Gallo's ideas concerning acquired immune deficiency syndrome (AIDS). In a few descriptive words, he depicts the distribution and prevalence of AIDS as becoming a modern plague. The word *pandemic* describes how the disease has spread across the world. *Endemic* describes how the disease has become established in the Caribbean and in Africa. Finally, he uses the word *epidemiologist* to describe individuals who are studying the AIDS outbreak.

In Robert Gallo's discussion of this disease, it is pointed out that medical researchers have accumulated more knowledge and information on the human immunodeficiency virus (HIV), the retrovirus responsible for AIDS, than on any other retrovirus.

Gallo calls HIV human T-lymphotropic virus because the virus was known to infect T-cells and lymph tissue preferentially. After the discovery of HTLV-III, other HTLVs that cause AIDS were found. These viruses were designated HIV. Several HIV types (HIV-1, HIV-2) have been identified.

READING FOR MEANING

Students' answers will vary. An "opportunistic" infection is an infection by an agent (bacteria, fungus, protist, etc.) that would not usually harm a healthy person with a normal immune system.

READ FURTHER

Students' answers will vary. The information and knowledge about HIV seem to change daily. New therapies that involve a closely regulated and monitored regimen of several drugs have met with success. Better tests of exposure to and infection with HIV are also currently available.

ACTIVE READING
Technique: K-W-L
Tell students to return to the list that they made at the beginning of this chapter. Have them check off the questions that they know the answers to. Students should then make a list of what they have learned. Ask students if they have any questions that are still unanswered or if they have any new questions.

ANSWERS TO SECTION 25-3 REVIEW

1. chickenpox and rabies
2. Attenuated viruses have been genetically altered so that they are incapable of causing disease under ordinary conditions. Inactivated viruses are viruses incapable of replication. Attenuated viruses are found to provide longer-lasting protection in vaccines compared with inactivated viruses.
3. Antibiotics are used to target bacterial cells' metabolic machinery. Viruses use only the host cell's metabolic machinery. The host cell's metabolic machinery is not a part of the virus structure or genetic makeup; therefore, antibiotics do not work in destroying viruses.
4. Humans contribute to the development of emerging viruses by entering the environment of isolated pathogenic viruses, thereby enabling the viruses to spread.
5. the lysogenic cycle
6. Emerging viruses are isolated existing viruses that have spread to human hosts.

TABLE 25-3 ***Viruses Linked to Human Cancers***

Type of virus	Mode of transmission	Type of cancer
Human T lymphotropic	mother-to-child transmission, body fluids, sexual contact	leukemia
Hepatitis B	body fluids, sexual contact, mother-to-child transmission	liver cancer
Epstein-Barr	physical contact	Burkitt's lymphoma
Human papillomavirus	sexual contact	cervical cancer

Examples of emerging viruses include hantavirus, which caused an outbreak of pneumonia in the southwestern United States in 1993; Machupo virus, which exists in South America; HIV, which originated in Africa; Ebola virus, which is found in Africa; and Lassa fever virus, which exists in West Africa. There are four known strains of the Ebola virus, all of which probably evolved from a common ancestor. One of these strains sickens only monkeys, but the other three strains are deadly to humans. Ebola's sudden emergence within the population of the Democratic Republic of Congo has left medical researchers puzzled over the identity of the animal host that carried the original Ebola virus strain.

VIRUSES AND CANCER

Recall that cancer is a condition that results from the uncontrolled reproduction of cells, which invade surrounding tissue. Scientists believe that cancers may be traced to genes within normal cells. When these genes are mutated by an outside agent, such as cigarette smoke, asbestos, sunlight, chemicals, or radiation, they may stimulate the cells to multiply uncontrollably. Cancer genes may also be triggered by certain lysogenic viruses. Some viruses associated with cancer are summarized in Table 25-3.

SECTION 25-3 REVIEW

1. **List two diseases that are the result of a viral attack on the human nervous system.**
2. **Discuss the differences between attenuated and inactivated viruses.**
3. **Why are antibiotics not effective in the treatment of viral diseases?**
4. **Explain how human actions have contributed to the increase of emerging viral diseases.**
5. **Which replication cycle are viruses found in when they stimulate the activation of cancer genes?**
6. **CRITICAL THINKING Consider how emerging viruses develop. Would you consider emerging viruses to be new viruses? Why or why not?**

CHAPTER 25 REVIEW

SUMMARY/VOCABULARY

25-1

- Viruses are biological particles composed of nucleic acid and a protein coat. Enveloped viruses also have a membrane enclosing them.
- Viruses are not usually considered living organisms because they lack most of the characteristics of living things.
- Wendell Stanley was the first scientist to report the crystallization of tobacco mosaic virus in 1935. This suggested that viruses might be chemicals rather than primitive cells.
- Viruses range in size from about 20 nm to about 250 nm in diameter.
- Many viruses have the shape of an icosahedron, which is a geometric figure containing 20 triangular faces. Other viruses take the shape of a helix, which resembles a coiled spring.
- Some viruses include a membrane-like envelope, with glycoprotein projections extending from the envelope.
- Viruses are grouped into families based on their nucleic acid type, their capsid structure, and the presence or absence of an envelope.
- Viruses probably originated from fragments of host-cell nucleic acid material.
- Viroids are viruslike particles composed of RNA only. Prions are pathogenic particles composed of protein only.

Vocabulary

capsid (488)	helix (489)	prion (490)	viroid (490)
envelope (488)	HIV (488)	retrovirus (490)	virology (487)
glycoprotein (489)	icosahedron (489)	reverse transcriptase (490)	virus (487)

25-2

- Bacteriophages are viruses that infect bacteria. Their discovery has increased scientists' understanding of virus replication.
- Replication by viruses occurs by either the lytic cycle or the lysogenic cycle.
- During the lytic cycle, the viral genome is released into the host cell, and replication of the virus follows immediately. Cellular components are used to make new viruses. A viral enzyme then causes host cell lysis and death.
- HIV infects specific white blood cells and remains in them as proviruses. As the immune system begins to fail, opportunistic infections occur; this condition is called AIDS.
- In the lysogenic cycle, the nucleic acid of the virus becomes part of the host cell's chromosome and remains with the cell in this form for many generations. HIV follows this pattern.

Vocabulary

bacteriophage (491)	lytic cycle (492)	prophage (494)	temperate (493)
lysis (493)	obligate intracellular parasite (491)	provirus (495)	virulent (492)
lysogenic cycle (493)		receptor site (492)	

25-3

- Vaccination and antiviral drug therapy are two major approaches to controlling and preventing the spread of viral diseases.
- Emerging viruses do not usually infect humans, but they can when environmental conditions favor their contact with and infection of human populations.
- Several viruses are implicated in the development of cancers such as leukemia, Burkitt's lymphoma, and liver cancer.

Vocabulary

acyclovir (498)	attenuated (497)	inactivated (497)
antiviral drug (497)	azidothymidine (AZT) (498)	protease inhibitor (498)

CHAPTER 25 REVIEW ANSWERS

REVIEW

1. The word *obligate* means "restricted to a single condition of life"; *intracellular* refers to something occurring within a cell; and a *parasite* is an organism that lives at the expense of another organism inside of or on the host. A virus is an infectious particle that multiplies only within a living cell.
2. Receptor sites are specific sites that viruses recognize and attach to on the host cell's surface. If the virus does not find a receptor site, then it cannot infect a host cell.
3. A *virus* is an infectious particle outside its host cell, while a *provirus* is the viral DNA integrated in the host cell's chromosome.
4. The term *lysis* refers to "a process of disintegration." The lytic cycle ends with disintegration of the host cell. In the term *lysogenic cycle,* the stem *-genic* means "capable of," and the virus is capable of causing cell disintegration.
5. Reverse transcriptase is an enzyme that catalyzes the production of DNA from RNA, the opposite of transcription, hence the term *reverse transcriptase.*

6. c **7.** c
8. b **9.** c
10. a **11.** b
12. b **13.** d
14. d **15.** c

16. Reverse transcriptase is an enzyme that uses RNA as a template for DNA synthesis. This process is the opposite of transcription, which is catalyzed by the enzyme transcriptase.
17. HIV has an RNA genome, a protein capsid, an envelope with glycoprotein, and reverse transcriptase.

18. DNA viruses may produce RNA that is used to make more viral proteins, or its DNA may integrate with the cell's DNA and become a provirus. Viral RNA is released into the host cell's cytoplasm, where it uses the cell's ribosomes to produce new viral proteins. Viral RNA may also use reverse transcription to make DNA, which becomes a provirus.
19. Both scrapie and mad cow disease are degenerative brain diseases associated with prion activity. Scrapie occurs in sheep, and mad cow disease occurs in cattle.
20. (a) attachment, (b) injection, (c) integration, (d) cell multiplication, lysogenic.
21. Viruses provide clues to the biochemistry of living organisms, and they can be used in the development of antiviral medications.
22. A virulent virus is a virus that replicates via the lytic cycle. A temperate virus is a virus that can replicate via the lysogenic cycle.
23. External stimuli such as radiation or certain chemicals can be responsible.
24. Vaccines are used to prevent viral infections, and disease transmission is prevented by controlling animals that carry viruses. Certain viral diseases can be treated with antiviral drugs.
25. lysogenic viruses, cigarette smoke, asbestos, sunlight, chemicals, and radiation

CRITICAL THINKING

1. AIDS is caused by a retrovirus that uses the enzyme reverse transcriptase in the replication process. AZT blocks the activity of reverse transcriptase and prevents viral replication.
2. Although researchers are not absolutely certain, the virus probably remains in the liver cells as a temperate virus and becomes activated by toxic substances such as aflatoxin.
3. Infection by a second phage could kill the cell if the second

REVIEW

Vocabulary

1. In a dictionary, look up the words *obligate, intracellular,* and *parasite,* and relate virus activity to the meaning of these words.
2. Viral infection is usually very specific. Identify the structure of the cell surface that permits only specific viruses to infect it.
3. Distinguish a virus from a provirus.
4. Use what you know about the meaning of the word *lysis* to explain the meanings of *lytic* and *lysogenic* cycles. If necessary, use your dictionary to help you.
5. Explain the meaning of the term *reverse transcriptase.*

Multiple Choice

6. A virus is a biologically active particle composed of (a) enzymes and fats (b) mitochondria and lysosomes (c) protein and nucleic acid (d) carbohydrates and ATP.
7. Which of the following was a key event in the development of virology? (a) the discovery of ribosomes (b) the discovery of nutrient agar (c) the crystallization of a virus (d) the discovery of protein structure
8. The term *icosahedron* refers to the structure of the viral (a) nucleic acid (b) capsid (c) lipid layer (d) receptor site.
9. Certain viruses have reverse transcriptase, an enzyme that (a) synthesizes RNA using DNA as a template (b) unites viral DNA with host DNA (c) synthesizes DNA using RNA as a template (d) assists the release of viruses from the host cell.
10. An essential aspect of viral replication is the (a) release of the viral nucleic acid in the cytoplasm of the host cell (b) entry of the viral envelope to the cytoplasm of the host cell (c) union of the viral envelope with the nuclear membrane of the cell (d) release of the viral capsid into the host cell nucleus.
11. Viroids differ from viruses in their (a) larger size (b) absence of a capsid (c) absence of nucleic acids (d) ability to cause disease in plants.
12. The head region of a bacteriophage is used to (a) attach the phage to the bacterial wall (b) enclose the nucleic acid (c) transmit the nucleic acid to the bacterium (d) take over the genetic machinery of the cell.
13. Lysis refers to the disintegration of (a) the bacteriophage capsid on entering the cell (b) the DNA of the host cell (c) a bacteriophage enzyme (d) the host cell.
14. Temperate viruses are those that (a) take over the genetic machinery of the host cell (b) are composed solely of protein (c) attach themselves to the ribosomes of the host cell (d) integrate their viral genes into the DNA of the host cell.
15. One virus that participates in the lysogenic cycle is (a) the measles virus (b) the rabies virus (c) HIV (d) the polio virus.

Short Answer

16. Explain the activity of reverse transcriptase.
17. Describe the structure of HIV.
18. DNA and RNA viruses differ in the way that they utilize the host cell's machinery to produce new viruses. Explain these differences.
19. Name the animal diseases that result from prion activity.
20. Identify the lettered steps in the figure below and identify the cycle.

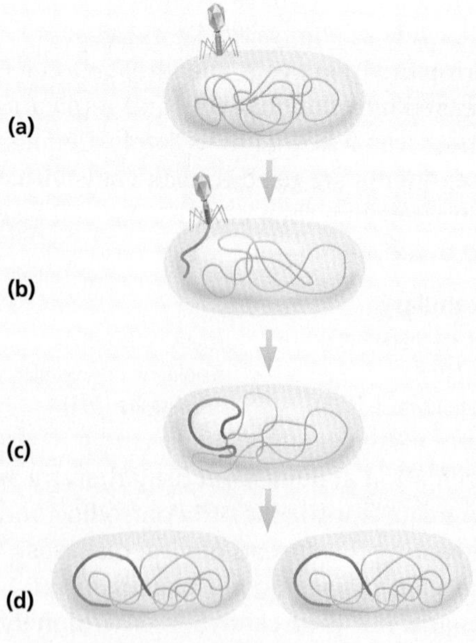

_______________ CYCLE

21. What are some of the useful attributes of viruses?
22. Distinguish between a virulent virus and a temperate virus.
23. What causes a temperate bacteriophage to become virulent?
24. Explain the methods humans use to control viral diseases.
25. What agents can trigger cells to multiply uncontrollably?

CRITICAL THINKING

1. Some people suggest that the drug AZT (azidothymidine) can help patients infected with HIV. This drug blocks the enzyme reverse transcriptase. Explain how AZT might help HIV-infected patients.
2. People who have had a hepatitis B viral infection have a greater chance of developing liver cancer later in life, especially if they are exposed to aflatoxin. Aflatoxin is a mold toxin found in some foods, such as contaminated peanuts. What does this relationship suggest to you about the role of the hepatitis virus in causing cancer?
3. Lambda is a bacteriophage that attacks *E. coli* and may enter a lysogenic cycle. Once it enters the lysogenic cycle, the phage inhibits the entry of more phages. What is the evolutionary advantage to the phage of "immunizing" its host against further lambda infections?
4. Shingles is a disease caused by the same herpes virus that causes chickenpox. How do you account for the fact that shingles often appears years after the initial chickenpox attack?
5. Based on your knowledge of HIV structure and replication, describe one way to interrupt the replication of HIV.
6. Based on your understanding of virus replication, how might scientists cultivate or artificially reproduce viruses in the laboratory?
7. Look at the graph below. Discuss how the sharp jump in the number of viruses outside the cell corresponds to the phases of the lytic cycle.

Growth Curve of Viruses

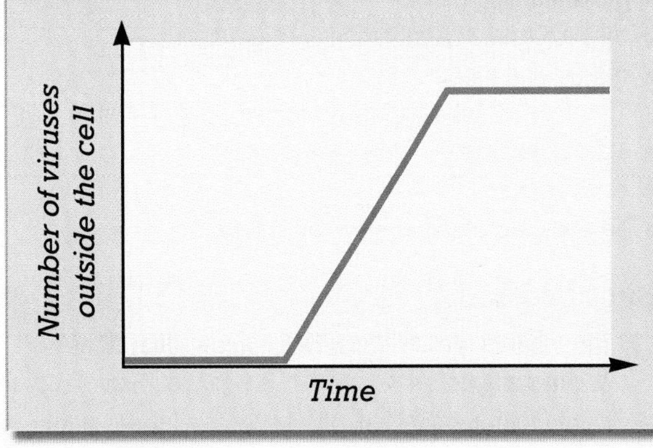

EXTENSION

1. Read "What New Things Are Going to Kill Me?" in *Time,* November 8, 1999, on page 86. Explain why Richard Preston thinks that new microbes are the ticking time bomb in the new century.
2. Call your local hospital or family doctor, and ask how viral diseases are diagnosed. If possible, visit the clinical laboratory of a hospital to see how the tests are done.
3. Research and write a report on a preventable viral disease, such as polio or smallpox. In your report, discuss the process scientists followed in identifying the cause of the disease, isolating the virus, formulating a vaccine, and testing the vaccine.
4. Read "Down but Not Out" in *New Scientist,* February 5, 2000, on page 20. The World Health Organization plans to eradicate polio by 2005 and stop vaccinating people. Explain why some polio experts believe that stopping vaccinations is unwise, even though the disease has been eradicated.

phage is virulent. A virulent phage would start a lytic cycle and cause the death of the host bacterium.
4. Shingles is caused by the activation of a temperate chickenpox virus to become lytic after it has been dormant in the cells for many years.
5. Answers might include preventing the virus from binding to white blood cells, degrading the retrovirus's reverse transcriptase, or interrupting the replication cycle.
6. by infecting an animal such as a mouse or by infecting cell culture systems that are artificially grown in a test tube
7. The sharp rise in the graph corresponds to the release phase of the lytic cycle, when there is a dramatic increase of viruses emerging from the host cell.

EXTENSION

1. Preston has observed that in nature, when populations soar and become densely packed, viral diseases tend to break out. Then the population drops. This is one of nature's population-control mechanisms. As headway is made against an old disease, a new disease might come along to take its place.
2. You may wish to visit a medical laboratory as a class field trip. Have students prepare questions for the laboratory personnel in advance.
3. Invite students to present their reports in class, and have the class compare the processes that scientists follow for the various diseases discussed.
4. The most effective and widely used vaccine for polio is the oral polio vaccine that uses a living but weakened form of the virus. Polio experts are concerned that the virus could persist after the disease-causing virus is gone and might occasionally revert to the dangerous type. If that happens, unvaccinated children will be unprotected.

CHAPTER 25 INVESTIGATION

TIME REQUIRED

One 50-minute class period and 10 minutes from seven consecutive days

SAFETY PRECAUTIONS

Remind students to keep their hands away from their eyes and mouth.

Ask students not to smell or inhale the pinches of tobacco from the cigarettes.

QUICK REFERENCES

Lab Preparation Notes are found on page 485B.

Holt BioSources provides a Teaching Resources CD-ROM that contains "Using Gowin's Vee in the Lab" and "Scoring Rubrics."

PROCEDURAL TIPS

1. It would be a good idea to have students label the control and experimental plants at the time the plants are distributed.
2. Have students pour a small amount of the dibasic potassium phosphate solution into a small container for their personal use so that they do not risk contaminating the original bottle of dibasic potassium phosphate with TMV.
3. Have students prepare their data tables ahead of time.

ANSWERS TO BACKGROUND

3. Outside of host cells, viruses are lifeless particles with no control of their movements.
8. Viruses are spread via wind, insects, water, food, blood, and other bodily secretions.

ANSWERS TO PART A

6. The powder abrades the leaf and facilitates infection.

CHAPTER 25 INVESTIGATION

Tobacco Mosaic Virus

OBJECTIVES

- Study the effect of cigarette tobacco on leaves of tobacco plants.

PROCESS SKILLS

- observing
- comparing and contrasting
- experimenting

MATERIALS

- lab apron
- protective gloves
- safety goggles
- 2 tobacco or tomato plants
- glass-marking pencil
- tobacco from several brands of cigarettes
- mortar and pestle
- 10 mL 0.1 M dibasic potassium phosphate solution
- 100 mL beaker
- cotton swabs
- 400 grit carborundum powder

Background

1. The tobacco mosaic virus, TMV, infects tobacco as well as other plants.
2. Plants that are infected with TMV have lesions and yellow patches on their leaves.
3. In what form do viruses exist outside host cells?
4. The tobacco mosaic virus is an RNA virus with rod-shaped capsids and proteins arranged in a spiral.
5. Plants damaged by wind, low temperatures, injury, or insects are more susceptible to plant viruses than healthy plants are.
6. Plant viruses are transmitted by insects, gardening tools, inheritance from the parent, and sexual reproduction.
7. Most viral diseases in plants have no known cures, but scientists are breeding various genetic forms of plants that are resistant to certain viruses.
8. Describe how viruses spread among organisms.
9. In this investigation, you will test whether tobacco from cigarettes can infect tobacco plants with TMV.

PART A Setting Up the Experiment

1. Put on a lab apron, gloves, and goggles before beginning this investigation.
2. Obtain two tobacco plants that have not been infected with TMV. Label one of the plants "control plant." Label the other plant "experimental plant."
3. CAUTION **Use poisonous chemicals with extreme caution. Keep your hands away from your face when handling plants or chemical mixtures.** Place pinches of tobacco from different brands of cigarettes into a mortar. Add 5 mL of dibasic potassium phosphate solution, and grind the mixture with a pestle as shown in the figure at left.
4. Pour the mixture into a labeled beaker. This mixture can be used to test whether cigarette tobacco can infect plants with TMV.
5. Wash your hands and all laboratory equipment used in this step with soap and water to avoid the accidental spread of the virus.
6. Moisten a sterile cotton swab with the mixture, and sprinkle a small amount of carborundum powder onto the moistened swab. Apply the mixture to two leaves on the "experimental" plant by swabbing the surface of the leaves several times. Why do you think swabbing the leaves with carborundum powder might facilitate infection?

7. Moisten a clean swab with dibasic potassium phosphate solution, and sprinkle a small amount of carborundum powder onto the moistened swab. This swab should *not* come into contact with the mixture of cigarette tobacco. Swab over the surface of two leaves on the control plant several times.
8. Do not allow the control plants to touch the experimental plants. Keep both plants away from other plants that may be in your investigation area, such as houseplants or garden plants. Wash your hands after handling each plant to avoid the accidental spread of TMV.
9. Treat both plants in precisely the same manner. The only difference between the two plants should be the experimental factor—exposure to cigarette tobacco. Both plants should receive the same amount of light and water.
10. Clean up your materials according to your teacher's instructions and wash your hands before leaving the lab.

PART B Collecting Data

11. In your lab report, create a data table similar to the model shown below. Allow plenty of space to record your observations of each plant.
12. Check the control and experimental plants each day for one week. Record your observations of each plant in your lab report. Wash your hands after handling each plant to prevent contaminating your results.

Analysis and Conclusions

1. What differences, if any, did you detect in the two plants after one week?
2. Did the plants exposed to cigarette tobacco become infected with the tobacco mosaic virus?
3. Why do you think it was necessary to use tobacco from different brands of cigarettes?
4. What are some of the possible sources of error in the experiment?
5. Greenhouse operators generally do not allow smoking in their greenhouses. Aside from health and safety issues, how might your results support this practice?

Further Inquiry

The tobacco mosaic virus is capable of infecting different species of plants. Design an experiment to determine which of several types of plants are susceptible to the virus.

OBSERVATIONS OF TOBACCO PLANTS

Day	Control plants	Experimental plants
1		
2		
3		
4		
5		
6		
7		

CHAPTER 25 INVESTIGATION

ANSWERS TO ANALYSIS AND CONCLUSIONS

1. Answers should be supported by the data.
2. Answers should be supported by the data.
3. It increases the chance of obtaining TMV.
4. Errors could include insufficient grinding of the leaves to obtain virus, inappropriate amount of carborundum, uneven amounts or uneven application of virus mixture, insufficient sampling of each leaf type, and contamination due to carelessness.
5. TMV in cigarettes or cigars may infect plants.

FURTHER INQUIRY

Students may suggest that the virus mixture be applied to various plants and use an identical group of plants as controls. The plants could then be observed for several days for signs of infection.

CHAPTER 26
PROTOZOA

	TOPICS	TEACHING RESOURCES	LABS, CLASSWORK, AND HOMEWORK
BLOCK 1 (45 minutes)	**Introducing the Chapter, p. 506**	ATE Focus Concept, p. 506 ATE Assessing Prior Knowledge, p. 506 ATE Understanding the Visual, p. 506	Supplemental Reading Guide, *Microbe Hunters*
	26-1 Overview of Protozoa, p. 507 Characteristics, p. 507	ATE Section Overview, p. 507 ATE Critical Thinking, p. 507 ATE Visual Link Figure 26-2, p. 508 89A, 90A, 98	**Lab A17,** Observing Protists Occupational Applications Worksheets—Sanitarian Study Guide, Section 26-1 PE Section Review, p. 509
BLOCKS 2 & 3 (90 minutes)	**26-2 Protozoan Diversity, p. 512** Phylum Sarcondina, p. 512 Phylum Ciliophora, p. 514 Phylum Zoomastigina, p. 516 Phylum Sporozoa, p. 517	ATE Section Overview, p. 512 ATE Critical Thinking, pp. 517, 518 ATE Visual Link Figure 26-5, p. 513 Figure 26-7, p. 514 98, 99, 100, 101, 102	PE **Chapter 26 Investigation,** p. 522 **Lab B13,** Cryobiology **Lab B14,** Protists—A Comparison **Lab C29,** Classifying Mysterious Organisms Study Guide, Section 26-2 PE Section Review, p. 518
BLOCKS 4 & 5 (90 minutes)	**REVIEW and ASSESSMENT** PE Chapter 26 Review, pp. 520–521	Audio CD Program	**CHAPTER TESTING** Chapter Test (blackline master) Test Generator Assessment Item Listing Scoring Rubrics and Classroom Management Checklists

MODERN BIOLOGY

HOLT BIOSOURCES

KEY

PE Pupil's Edition
ATE Teacher's Edition
■ Active Reading Guide
★ Study Guide

Teaching Transparencies

Laboratory Program

One-Stop Planner CD-ROM
Shows these resources within a customizable daily lesson plan:
■ Holt BioSources Teaching Resources
■ Active Reading Guide
★ Study Guide
▲ Test Generator

READING FOR CONTENT MASTERY

Preview: Have students preview the objectives, the introductory paragraph, topic sentences, and captions before reading the chapter.

ATE Active Reading Technique: L.I.N.K., p. 507
■ Reading Strategy: LINK Teacher Notes
■ Active Reading Guide Worksheet 26-1
PE Summary/Vocabulary, p. 519

■ Active Reading Guide Worksheet 26-2
PE Summary/Vocabulary, p. 519

TECHNOLOGY AND INTERNET RESOURCES

Holt Biology Videodiscs Teacher's Correlation Guide, Lessons 26-1 through 26-2

Audio CD Program, Sections 26-1 through 26-2

internetconnect

sciLINKS NSTA

On-line Resources: www.scilinks.org
The following *sci*LINKS Internet resources can be found in the student text for this chapter:

Topics:
- Ciliophora, p. 514
- Zoomastigina, p. 516

go.hrw.com

On-line Resources: go.hrw.com
Visit the HRW Web site for a variety of resources related to this chapter. Just type in the keyword HM2 Chapter 26.

Smithsonian Institution®
Visit **www.si.edu/hrw** for additional on-line resources.

CNNfyi.com
Visit **www.cnnfyi.com** for late-breaking news and current event stories selected just for you.

LAB ACTIVITY PLANNING

CHAPTER 26 INVESTIGATION

Observing *Paramecium*, pp. 522–523

OVERVIEW
In this investigation, students will compare structural differences and modes of locomotion between *Paramecium* and *Euglena*. In addition, the effects of a hypotonic solution and a hypertonic solution on the contractile vacuoles of the *Paramecium* will be observed.

SAFETY
Have students dispose of broken glass in a container designated for that purpose. Instruct students to wear safety goggles during the lab. If methyl cellulose gets in a student's eye, immediately flush the eye out with water.

PREPARATION
1. Aerate the cultures of *Paramecium* and *Euglena* as soon as they arrive from the biological supply house. Squirt air into the water with a medicine dropper. Keep the cultures capped loosely. Add a rice grain to the cultures and they should keep for several weeks.
2. Feed *Euglena* to the *Paramecium* culture at least 20 to 30 minutes before the start of each class to ensure that students will observe full food vacuoles.
3. Prepare a 10 percent methyl cellulose solution by adding 10 g of methyl cellulose into a 500 mL graduated cylinder, and dilute with distilled water until the final volume is 100 mL. Prepare a 5 percent NaCl solution by adding 5 g of sodium chloride into a 500 mL graduated cylinder, and dilute with distilled water until the final volume is 100 mL.
4. Each group will need access to a stopwatch or clock to monitor the pulses of the contractile vacuoles.

UNDERSTANDING THE VISUAL

Have students closely examine the photograph of *Stentor.* Ask students to name all the structures that they see, based on their previous knowledge of organelles (cilia, nuclei, food vacuole, and cell membrane). Then explain that protozoa are unicellular organisms with a variety of adaptations that enable them to live in many diverse environments.

FOCUS CONCEPT

Cell Structure and Function

This chapter shows that unicellular organisms have structures that perform specific functions. For example, cilia, flagella, and pseudopodia function to help protozoa feed or propel themselves.

ASSESSING PRIOR KNOWLEDGE

Review the following concepts.

Cellular Organelles: *Chapter 4*
Ask students to name five cellular organelles and their functions.

Diffusion: *Chapter 5*
Ask students to predict what will happen if a protozoa-like amoeba is placed in distilled water.

Parasitism: *Chapter 21*
Ask students to define *parasite* and *host.*

CHAPTER 26

PROTOZOA

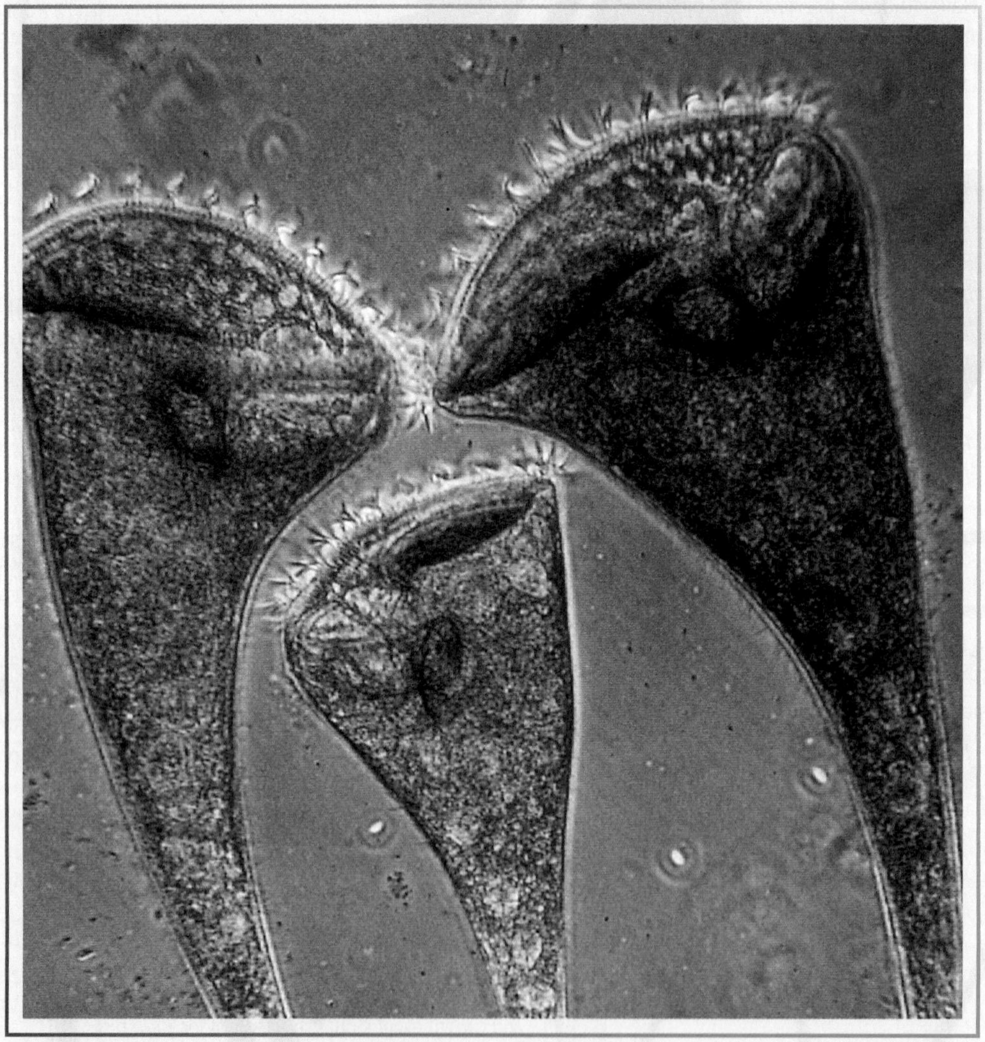

Stentor *displays two of the characteristics of protozoa, unicellularity and lack of tissue differentiation.*

FOCUS CONCEPT: ***Structure and Function***

As you read this chapter, note the wide variations in the shape, size, structure, and adaptations of protozoa as well as the features they have in common.

SECTION

26-1

OVERVIEW OF PROTOZOA

The kingdom Protista contains a diverse collection of eukaryotic organisms—protozoa, algae, slime molds, and water molds. Collectively, these organisms are called ***protists.*** *Protists are sometimes described as animal-like, plantlike, and funguslike. However, all protists are eukaryotic and lack tissue differentiation.*

OBJECTIVES

▲ Describe the characteristics of protozoa.

● Explain the role some protozoa play in aquatic ecosystems.

■ Discuss a classification scheme used to identify protozoa.

◆ Name an adaptation that enables some protozoa to survive harsh environmental conditions.

▲ Briefly explain the evolution of protozoa.

CHARACTERISTICS

Protozoa are single-celled microscopic organisms that are noted for their ability to move independently. Biologists have identified about 65,000 species of protozoa, almost half of which are extinct species identified from fossils. Protozoa live in many different environments; they can drift in the ocean, creep across vegetation in freshwater rivers and ponds, crawl in deep soil, and even reproduce in the bodies of other organisms. The protozoan shown in Figure 26-1 lives in the gut of termites.

Most protozoans are heterotrophic, obtaining their nutrients by ingesting small molecules or cells. These particles are usually broken down in **food vacuoles,** membrane-bound chambers that contain digestive enzymes.

Many species of protozoa are free-living, while others are parasitic. Free-living protozoa live in any habitat where water is available at some time during the year. Many species make up **zooplankton,** a population of organisms that constitutes one of the primary sources of energy in aquatic ecosystems. Other free-living protozoa live in the soil. Parasitic protozoa usually have complex life cycles that take place in the cells, tissues, and bloodstream of their hosts. Several species cause a variety of serious human diseases, including malaria, amebic dysentery, and giardiasis.

Reproduction

All protozoa are capable of asexual reproduction, usually by binary fission. During binary fission, a protozoan divides into two essentially identical individuals. Some species reproduce by **multiple fission,** a form of cell division that results in a number of identical individuals.

While all protozoa can reproduce asexually, a few species also reproduce sexually, through **conjugation.** During conjugation in protozoa, individuals from opposite mating strains pair and exchange genetic material. Conjugation in protozoa is a more complex process than conjugation in bacteria.

FIGURE 26-1

The phylum Zoomastigina includes *Trichonympha,* a protozoan that inhabits the gut of termites and helps the termite digest the cellulose in its diet.

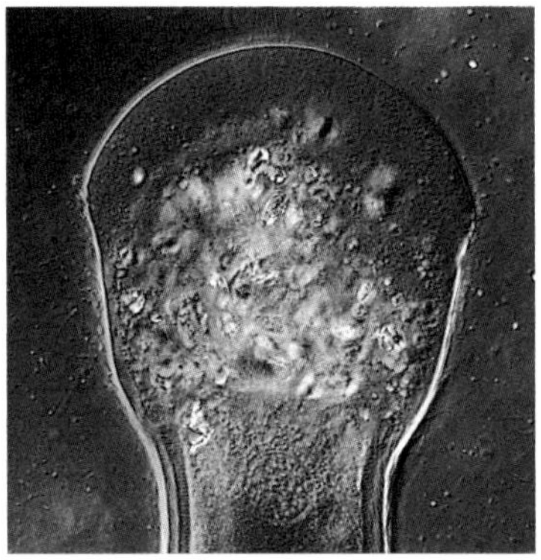

SECTION 26-1

SECTION OVERVIEW

Cell Structure and Function

Section 26-1 describes the wide variety of protozoa, the classification based on type of movement, and the structure and function of eyespots and cysts.

Evolution

Section 26-1 describes the adaptations that protozoa have evolved in response to their environment, such as the ability to form a cyst in harsh environmental conditions.

ACTIVE READING

Technique: L.I.N.K.

Write the word *protist* on the board, and have students List on their own paper all the words, phrases, and ideas that they associate with the word. After several minutes, ask the students to share one or two of their ideas as you record them. Facilitate a discussion in which students Inquire of each other or the teacher to clarify each of the ideas listed. After all discussion is completed, have students write Notes on paper about what they remember. Allow about 1 minute. Have them look over their notes to see what they Know about protists based on experience and discussion.

CRITICAL THINKING

Endosymbionts

Termites are pests that cause great destruction. Ask students why an entomologist assigned the task of controlling termite infestation might think it is a good idea to find a way to control *Trichonympha*. (If *Trichonympha* could be controlled or eradicated, then termites, which are symbiotic with *Trichonympha* and depend on them to digest the cellulose in wood, could not digest their food. The termites would then die.)

VISUAL LINK

Figure 26-2

Have students examine the four organisms in the figure. Ask them to name the structure each organism uses for movement and, based on its type of locomotion, the phylum that each belongs to. (Answers should be as follows: (a) cilia, Ciliophora; (b) nonmotile, Sporozoa; (c) flagella, Zoomastigina; (d) pseudopodia, Sarcodina.)

OVERCOMING MISCONCEPTIONS

A common misconception is that protozoa live only in aquatic environments. Point out that protozoa live in many environments, including in soil and in the digestive tracts of many animals.

DEMONSTRATION

To show that protozoa live in many environments, add a handful each of soil and dead grass to a jar. Then fill the jar with distilled water, cover it, and place it in indirect light. A day or two later, prepare wet-mount slides of water from the jar. Let students examine the samples under a microscope. Students should see protozoa in the samples. Ask students how the protozoa got in the water. (Students should infer that the protozoa were in the soil and on the dead grass, possibly in cysts, and that when conditions became favorable, the protozoa became active in the water.)

GIFTED *ACTIVITY*

Have students research library and on-line resources for the ecological significance of zooplankton and present their findings to the class. Students should use visual aids to illustrate the variety of zooplankton. Also have students include organisms that depend on zooplankton for food, such as filter-feeding invertebrates.

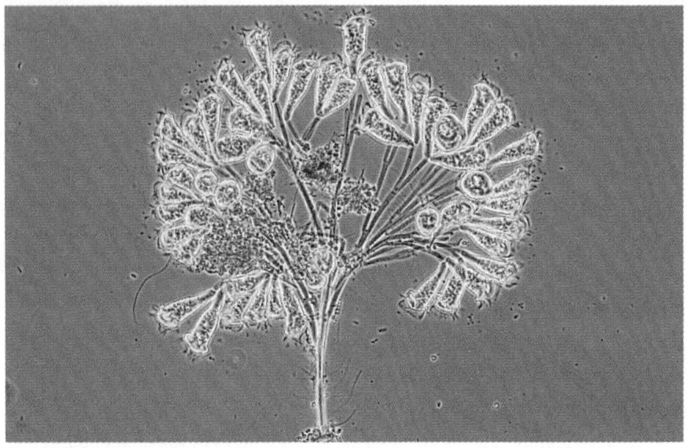

(a) *Zoothamnium,* a ciliate

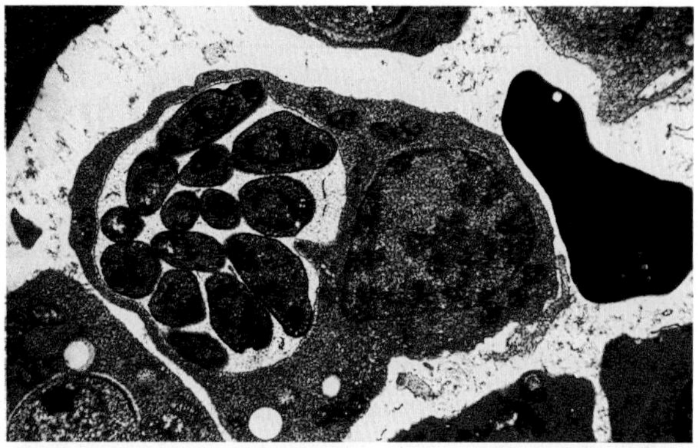

(b) *Plasmodium,* a sporozoan

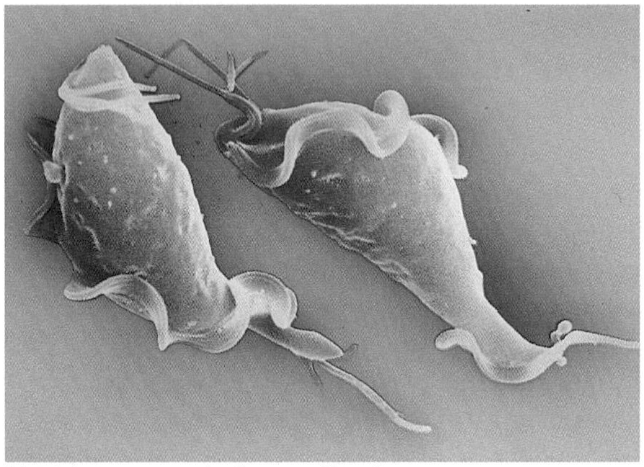

(c) *Trichomonas vaginalis,* a zooflagellate

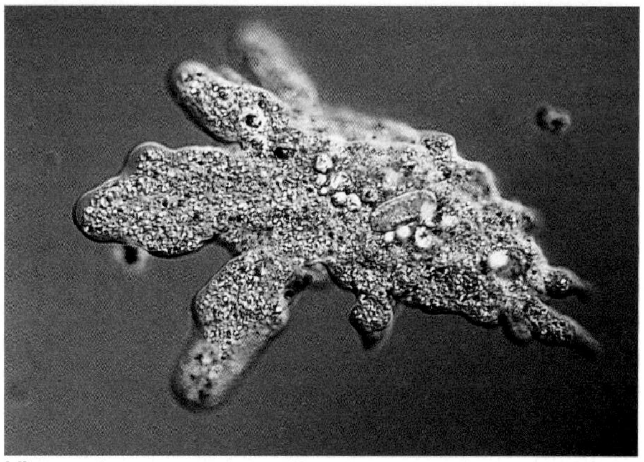

(d) *Amoeba proteus,* a sarcodine

FIGURE 26-2

These organisms are representatives of the four phyla of protozoa.

Classification

Protozoa are members of the kingdom Protista, along with algae, slime molds, and water molds. Research into the evolutionary relationships among these organisms is constantly yielding new information and altering protistan classification. For this reason, biologists are always reevaluating the classification scheme. Protozoa are sometimes classified into four phyla: Sarcodina (SAHR-kuh-DIEN-uh), Ciliophora (sil-ee-AHF-uhr-uh), Zoomastigina (ZOH-uh-mas-tuh-JIEN-uh), and Sporozoa (spor-uh-ZOH-uh). A representative of each group is shown in Figure 26-2. The characteristics of these four phyla, along with representative genera, are summarized in Table 26-1.

Adaptations

Many species of protozoa have physiological mechanisms for monitoring conditions in their environment. For example, many free-living species have a localized region of pigment called an **eyespot.** Eyespots detect changes in the quantity and quality of light. Certain protozoan species also sense physical and chemical changes or obstacles in their environment. For instance, some protozoa will back up and bypass a noxious chemical.

Most protozoa are separated from their environment only by their delicate external cell membrane. These fragile organisms can survive in environments where extreme conditions often exist.

TABLE 26-1 A Summary of Protozoa

Phylum	Common name	Locomotion	Nutrition type	Representative genera
Sarcodina	sarcodines	pseudopodia	heterotrophic; some parasitic	*Amoeba* *Radiolaria* *Naegleria*
Ciliophora	ciliates	cilia	heterotrophic; some parasitic	*Paramecium* *Tetrahymena* *Balantidium*
Zoomastigina	zooflagellates	flagella	heterotrophic; some parasitic	*Trypanosoma* *Leishmania* *Giardia* *Trichonympha*
Sporozoa	sporozoans	(none in adult)	heterotrophic; some parasitic	*Plasmodium* *Toxoplasma*

Their hardiness is in part due to their ability to form cysts. A **cyst** is a dormant form characterized by a hardened external covering in which metabolic activity has ceased. Many species of protozoa form cysts in response to changes in the environment, such as nutrient deficiency, drought, decreased oxygen concentration, or pH or temperature changes. Cyst formation is extremely important to many protozoa that must survive such conditions between hosts. When favorable environmental conditions return, a protozoan emerges from the cyst and resumes metabolic activities.

Evolution

As you learned in Chapter 14, the first prokaryotes evolved more than 3.5 billion years ago. These organisms were the only life-forms on Earth for almost 2 billion years. About 1.5 billion years ago, the first eukaryotic organisms evolved. Protozoa are the descendants of these early eukaryotes. The first eukaryotes probably evolved through endosymbiosis, a process in which one prokaryote lives inside another and gradually both host and guest become dependent on one another.

Word Roots and Origins

cyst

from the Greek *kystis*, meaning "sac"

SECTION 26-1 REVIEW

1. **What kind of organisms are found in the kingdom Protista? What characteristics do they share?**
2. **What are protozoa? How do they reproduce? What kinds of environments do they inhabit?**
3. **What role do some protozoa play in aquatic environments?**
4. **What is a cyst? Under what conditions might certain protozoa form cysts?**
5. **Why are protozoa considered some of the oldest existing life-forms?**
6. **CRITICAL THINKING On what basis can protozoa be classified? Does this classification scheme reflect evolutionary relationships among protozoa?**

INCLUSION ACTIVITY

Provide each student with a blank "bingo card." Use card stock to make cards with five rows and five columns. The spaces should be large enough to write vocabulary words. Instruct students to fill in the card with key words from the chapter, including phylum names, organelles, locomotion types, and the names of organisms. Have students circle the terms on their cards as you read one- or two-word definitions aloud. For example, if you say, "flagella," then students can circle "zoomastigina" on the card.

ANSWERS TO SECTION 26-1 REVIEW

1. Protozoa, algae, slime molds, and water molds are found in the kingdom Protista. They are eukaryotic and lack tissue differentiation. Most are single-celled.
2. Protozoa are single-celled eukaryotes that can move at some stage of life. They reproduce asexually by binary fission and multiple fission and sexually by conjugation. They live in any environment that occasionally contains water.
3. Some protozoa, such as zooplankton, are a primary food source.
4. A cyst is a dormant stage that some protozoa can produce in response to harsh environmental conditions.
5. Protozoa are considered to be among the oldest living life-forms because they are thought to have descended from the first eukaryotes.
6. Protozoa can be classified based on their type of locomotion. No, because a similar type of locomotion in two species may have developed through convergent evolution.

BACKGROUND

Margulis was born March 5, 1938, in Chicago. She received a bachelor's degree from the University of Chicago in 1957 and a master's degree from the University of Wisconsin in 1960. She received a Ph.D. from the University of California–Berkeley in 1965. She is a professor of biology at the University of Massachusetts, in Boston.

Kwang W. Jeon (1934–) received his B.S. and M.S. from Seoul National University, in Korea, and his Ph.D. from the University of London. He is now a biology professor at the University of Tennessee.

Endosymbiosis theory states that primitive eukaryotic cells evolved from a symbiotic relationship between two prokaryotic cells. It is thought that one bacterium parasitized another bacterium. The interior bacterium formed a symbiotic relationship with the larger bacterium. The smaller cell received nutrients from the larger cell, which in turn received energy from the interior cell. Aerobic eukaryotic cells probably appeared about 1.5 billion years ago, after these prokaryotic symbioses were first established. Chloroplasts are thought to have developed from primitive photosynthetic endosymbiotic cyanobacteria.

The endosymbiosis hypothesis was first proposed in the early twentieth century and was rejected because of lack of evidence.

Evidence established in the 1960s and 1970s that supports the endosymbiosis theory includes the following: mitochondria and chloroplasts are self-dividing, are approximately the same size as prokaryotes, and are able to make proteins and lipids; the DNA within mitochondria and chloroplasts is similar to bacterial rather than eukaryotic DNA; the

GREAT DISCOVERIES

The Origin of Eukaryotic Cells

HISTORICAL PERSPECTIVE

In the late nineteenth and early twentieth centuries, French, Russian, and American scientists speculated that mitochondria contained hereditary information similar to that found in the nucleus. They hypothesized that mitochondria evolved as a result of a symbiotic relationship between different types of bacteria. Attempts to culture mitochondria were unsuccessful, and given the popular view of bacteria as agents of disease, scientists generally disregarded the hypothesis that mitochondria evolved from symbiotic bacteria.

A "Dead" Hypothesis Reappears

In the 1950s, as an undergraduate at the University of Chicago, Lynn Margulis became interested in genetics and inheritance through her readings of the works of noted scientists. Although she had started college when she was only 15 years old, Margulis was not intimidated by the complexity of genetics. As a graduate student in genetics at the Universities of Wisconsin and California, she began to question well-established ideas about heredity and evolution. While at the University of Wisconsin, Margulis became aware of patterns of inheritance in components of the cytoplasm in plants and algae that did not seem to result from genes in the nucleus. After discussing her observations with her advisor, cell biologist Hans Ris, she concluded that the cytoplasm contains genes that function separately from those in the nucleus.

Using an electron microscope, Margulis and Ris were able to study chloroplasts and mitochondria in detail. They observed that both of these organelles have properties in common with modern bacteria. Both chloroplasts and mitochondria contain ribosomes and circular DNA, and they reproduce by binary fission.

These observations, along with the papers of biologists who had earlier suggested that hereditary information could be found in the cytoplasm, led Margulis to infer that the genes in chloroplasts and mitochondria may have had a different evolutionary origin than the genes in the nucleus. Margulis developed a hypothesis referred to as serial endosymbiosis. In this hypothesis, she proposed that eukaryotes evolved from symbiotic relationships established between anaerobic bacteria and intracellular aerobic bacteria.

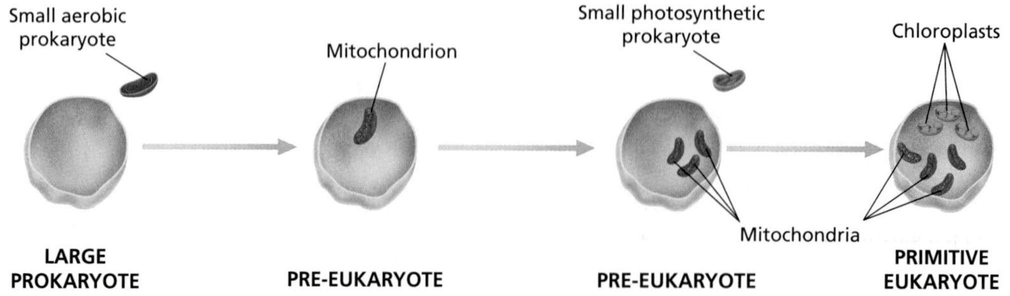

The Evolution of Mitochondria and Chloroplasts

To explain the selective pressure for endosymbiosis, Margulis hypothesized that there was an "oxygen holocaust" on Earth about two billion years ago. The amount of oxygen in the atmosphere increased sharply, probably as a result of bacterial photosynthesis. The rise in oxygen levels killed most bacteria—but not all bacteria. Some small, oxygen-utilizing bacteria invaded larger, anaerobic or nonoxygen-utilizing bacteria. Although many of the anaerobic bacteria were unable to adapt to their new "guests" and died, a few survived and became a living food source for the invaders. In effect, the larger bacteria became hosts to parasitic aerobic bacteria. In turn, the invading, aerobic bacteria supplied some of the energy derived from their aerobic respiration to the host cell. Margulis proposed that these aerobic invaders may have been the evolutionary forerunners of modern mitochondria.

Margulis further hypothesized that chloroplasts evolved from an ancient invasion—this time by small, photosynthetic bacteria similar to modern blue-green bacteria. Thus, in addition to energy-producing components (the primitive mitochondria), the large host cells now had components that could produce food using simple molecules and the energy from sunlight. These increasingly complex host cells were primitive eukaryotes, the ancestors of modern algae and plants. What began as a parasitic relationship evolved to the mutual benefit of both host and parasite, resulting in the emergence of eukaryotic cells.

An Unpopular Idea

Margulis's article presenting her hypothesis was rejected by 15 journals before it was published in 1966 in the *Journal of Theoretical Biology.* In 1970, when she published an expanded account of her hypothesis in a book called *Origin of Eukaryotic Cells,* most biologists still rejected her ideas because of the absence of fossil evidence to support it.

In defending her hypothesis, Margulis cited experiments performed by Kwang Jeon, of the University of Tennessee. Jeon had infected a species of amoeba with a bacterial parasite. By culturing those amoebas that survived the infection, Jeon eventually produced a line of amoebas that were actually dependent on the bacteria for survival. Essentially, symbiosis and the selection that followed produced a new kind of amoeba in five years. Other studies in the late 1960s supported Margulis's endosymbiosis hypothesis.

By 1981, when Margulis published a revised version of her book, titled *Symbiosis in Cell Evolution,* her hypothesis was supported by a wealth of evidence and had become widely accepted. Since then, many more reports of endosymbiotic relationships have been published.

In one recent study, scientists at the University of Pennsylvania found evidence of endosymbiosis between protozoa and green algae. DNA from a previously unknown organelle in protozoa was identified as being similar to DNA found in algae plastids.

A Fast-Changing Science

In her preface to the 1992 edition of *Symbiosis in Cell Evolution,* Margulis refers to significant changes in biology. She calls the difference between prokaryotes and eukaryotes the most fundamental division in the organization of the living world. This certainly is a big change from the time when biologists considered the division of plants and animals to be the most basic division. As Margulis and others continue to explore microorganisms, there will undoubtedly be other changes in our understanding of the evolution of life.

Kwang Jeon provided experimental evidence that supported Margulis's hypothesis of evolution through endosymbiosis.

ribosomes in mitochondria and chloroplasts are more similar to bacteria than to those in eukaryotes in size, are sensitive to antibiotics like bacteria, and synthesize protein using metabolic steps that bacteria use; and chloroplasts produce energy in the same way that some bacteria produce energy.

DISCUSSION

Guide the discussion by posing the following questions:

1. According to the serial endosymbiotic model, what might have been the stimulus for the development of eukaryotic cells? (About two billion years ago, the amount of oxygen increased sharply, killing off most bacteria—but not all. Some small, aerobic bacteria invaded larger, anaerobic bacteria. Many of the anaerobic bacteria died, but a few survived and became a food source for the invaders. In this way, the previously anaerobic bacteria were able to survive in an oxygen-rich environment.)

2. What phenomenon led Margulis to investigate the origin of cellular organelles? (Margulis became interested in heredity that was unrelated to nuclear inheritance patterns.)

3. Why was it difficult for many scientists to accept Lynn Margulis's hypothesis? (the absence of fossils, the lack of evidence, unsuccessful attempts to culture mitochondria, and the history of rejection of the hypothesis by early twentieth-century biologists)

4. How would one determine if Jeon's amoebas were dependent on the endosymbiotic bacteria for survival? (If an antibiotic were added to the culture, the intracellular bacteria would die. Their deaths would be followed by death of the amoebas.)

SECTION 26-2

SECTION OVERVIEW

Evolution
Section 26-2 describes how some members of phylum Sarcodina have contributed to Earth's geological features.

Cell Structure and Function
Section 26-2 describes how specialized structures allow some protozoa to reproduce sexually.

Interdependence of Organisms
Section 26-2 describes the life cycle of some parasitic protozoa and how they are transmitted through insect carriers.

▶ ENGAGE STUDENTS

Tell students that one championship college basketball team employed what they called the amoeba defense. Tell students that the amoeba defense is a zone defense. Rather than staying in fixed positions, team members move about within certain regions to match up with opposing players.

Use a microprojector or camera attached to a television set to show students how these protists project portions of the cell during movement. Then ask five volunteers to demonstrate how they might copy amoeba behavior when playing defense in basketball.

SECTION 26-2

OBJECTIVES

▲ Explain how sarcodines have contributed to Earth's geological features.

● Describe a type of sexual reproduction that occurs in ciliates.

■ Identify four human diseases caused by zooflagellates.

◆ Discuss the life cycle of *Plasmodium.*

PROTOZOAN DIVERSITY

The broad diversity of protozoa is evident in the four major phyla identified by biologists. The phyla are distinguished by their form of locomotion, and each group contains a number of parasites that cause serious human diseases. The complexity of protozoa sets them apart from the relatively simple structures of bacteria and viruses.

PHYLUM SARCODINA

Biologists have classified 40,000 species of protozoa in the phylum Sarcodina (SAHR-kuh-DIE-nuh). Sarcodines include hundreds of species of amoebas, which inhabit fresh water, salt water, and soil. One of the most unusual sarcodines is *Pelomyxa carolinensis,* shown in Figure 26-3. These protozoa are found living on mud, rocks, and other surfaces in shallow, slow-moving streams and ponds.

Most sarcodines have flexible cell membranes and are constantly thrusting out **pseudopodia** (SOO-duh-POH-dee-uh). Pseudopodia are large, rounded cytoplasmic extensions that function in movement. A pseudopodium forms when the **endoplasm,** the inner portion of the cytoplasm, pushes the **ectoplasm,** the outer layer, forward to create a blunt, armlike extension. Simultaneously, other pseudopodia retract, and the cytoplasm flows in the direction of the new pseudopodium. This form of movement, referred to as **ameboid movement,** is illustrated in Figure 26-4. Ameboid movement is a form of **cytoplasmic streaming,** the internal flowing of a cell's cytoplasm.

Sarcodines also use pseudopodia for feeding. Sarcodines live on other protists, which they engulf by phagocytosis. When a sarcodine feeds, it surrounds the food with its pseudopodia. A portion of the cell membrane then pinches together and surrounds the food in a food vacuole, in a process called endocytosis. Enzymes from the cytoplasm then enter the vacuole and digest the food. Undigested food leaves the cell in a reverse process called exocytosis.

Most freshwater sarcodines have an internal structure that is similar to that of the amoeba shown in Figure 26-4. Notice the **contractile vacuole,** an organelle that expels fluid from the cell. Freshwater organisms are usually hypertonic relative to their environment, so water diffuses into them. To maintain homeostasis, many freshwater protozoa have contractile vacuoles that rid the cell of excess water.

FIGURE 26-3
Sarcodines, such as this *Pelomyxa,* are characterized by cytoplasmic extensions called pseudopodia. By thrusting pseudopodia outward, a sarcodine can move across a surface or capture prey. (240×)

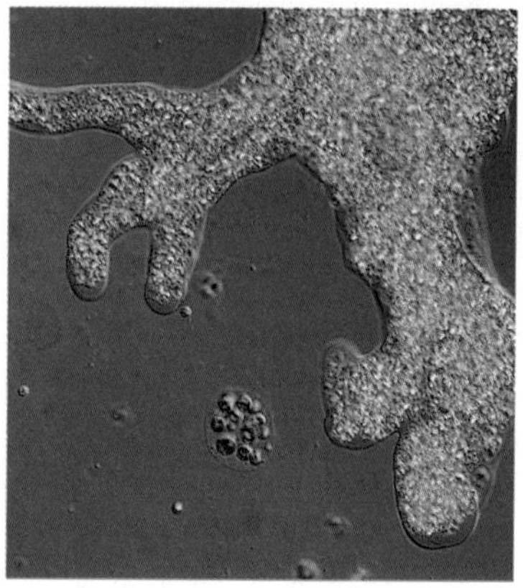

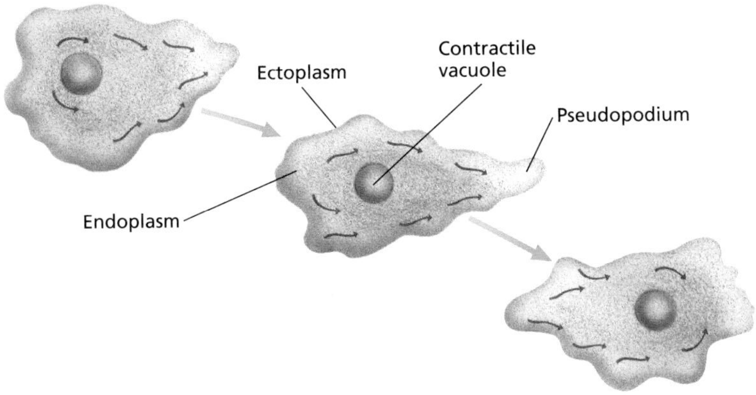

FIGURE 26-4

Ameboid movement is powered by cytoplasmic streaming (indicated here with arrows). As the cytoplasm surges forward to form a new tubelike pseudopodium, other pseudopodia retract.

Ecological Role

Many sarcodines are "naked," that is, their cell membranes are exposed directly to the environment. Some have their delicate cell membranes covered with a protective **test,** or shell. For example, **foraminifera** (FOHR-a-MIN-i-fohr-uh) are an ancient group of shelled sarcodines found primarily in oceans. Their tests, shown in Figure 26-5, have many chambers and are made of calcium carbonate. Slender pseudopodia extend through tiny openings in the test. The group of sarcodines called **radiolarians** (RAY-dee-OH-LER-ee-uhnz) is among the oldest known. Most radiolarians live in shallow, open water. Their tests contain silicon dioxide and usually have a radial arrangement of spines that extend through the shell. One example of their intricate tests is shown in Figure 26-6.

Foraminifera and radiolarians have existed since Precambrian times (more than 600 million years ago) and have left excellent fossil records. For millions of years, the tests of dead foraminifera have been sinking to the bottom of the ocean, where they have built up a calcium-rich layer of sediment. These sedimentary layers can be seen as limestone and chalk deposits that formed in the sea and later emerged as dry land. The chalk deposits in many areas of England, including the White Cliffs of Dover, were formed in this way. The Great Pyramids of Egypt were built with stones quarried from limestone beds that are made of a type of large foraminiferan that flourished during the early Tertiary period (between 7 and 70 million years ago). The tests of dead radiolarians have contributed to the formation of a type of rock called chert.

FIGURE 26-5

Foraminifera once inhabited these hard shells called tests.

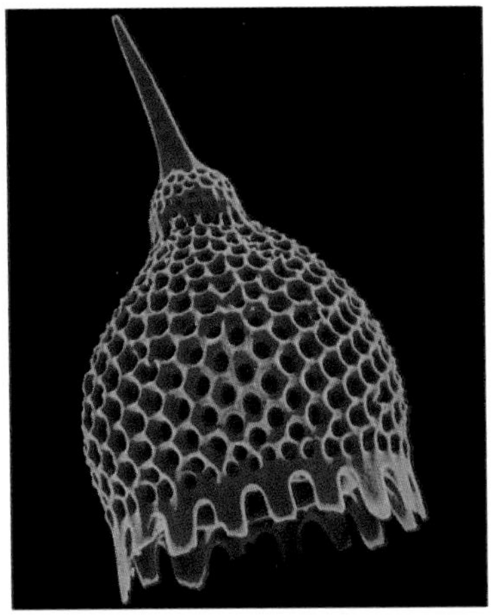

FIGURE 26-6

The phylum Sarcodina includes radiolarians, such as the one shown above, which are also covered by protective tests.

Human Diseases

Although most amoebas are free-living, some species live in the intestines of humans or other animals. One such amoeba, *Entamoeba histolytica,* can cause serious illness in humans. This amoeba enters the body via contaminated food and water. It lives in the large intestine, where it secretes enzymes that attack the intestinal lining and cause deep ulcers. If this occurs, a sometimes fatal disease called **amebic dysentery** may result. Affected individuals feel intense pain, and complications arise when the amoebas are carried by the blood to the liver and other organs.

TEACHING STRATEGY

Living Protozoa

Ask students to collect samples of pond, stream, puddle, or aquarium water and bring it to class. Have students make wet-mount slides of their samples and use microscopes to look for protozoa. Provide visual guides to help students identify the protozoa.

VISUAL LINK

Figure 26-5

Draw students' attention to Figure 26-5, and ask them to compare this figure with Figure 26-6. Ask students to identify their similarities (both have tests) and differences (radiolarian tests are made of silicon dioxide, and foraminiferan tests are made of calcium carbonate).

MAKING CONNECTIONS

Social Studies Activity

Have students research library and on-line resources for information about the fossil remains of foraminifera as related to the construction of the pyramids in Egypt. Also assign students to research the White Cliffs of Dover, and ask them to compare the fossil evidence of both places for similarities and differences. (Limestone, which is found in the White Cliffs of Dover and in the pyramids, is made of fossilized foraminifera. Limestones differ according to the types of clay and other organisms that were sedimented with the fossilized foraminifera.)

VISUAL LINK
Figure 26-7

Have students study Figures 26-7 and 26-8. Ask them what primary structures distinguish a paramecium from an amoeba (the pellicle and cilia). Ask students what two functions cilia serve. (Cilia provide movement for the organism, and they move food into the oral groove.)

TEACHING STRATEGY
Summarizing Information

Have students make a table similar to Table 26-2 and summarize the information in Section 26-2. Tell students to add other headings to their columns, such as "Specialized structures" and "Diseases caused."

QUICK FACT
☑ *Ichthyophthirius multifiliis,* a parasitic ciliate large enough to be seen without a microscope (0.1–1.0 mm), is responsible for white spot disease in freshwater fish. The free-swimming ciliate attaches to a fish's skin and feeds on the fish. After the ciliate matures, it leaves the fish, sinks to the bottom of the lake, and forms a gelatinous capsule in which it multiplies by mitosis. The resulting protozoa leave the capsule in search of new fish to parasitize. As a result of the wound left by the parasite, fish sometimes acquire lethal bacterial and fungal infections.

INCLUSION ACTIVITY
Provide students with index cards, glue, and pictures of a typical protozoan from each of the four phyla. Have students cut out and glue a picture of a protozoan on one side of a card. On the other side, have them write the names and phonetic pronunciations of the phylum, genus, and locomotive structures of the protozoan. Students can use the cards as study aids.

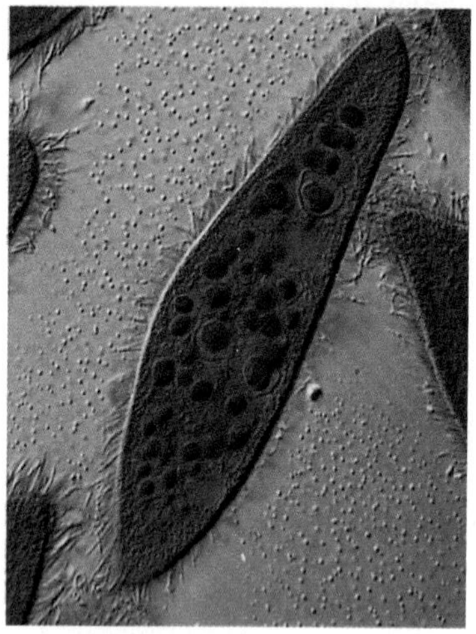

FIGURE 26-7
Like all ciliates, paramecia such as *Paramecium* above move by using short, hairlike projections called cilia.

internetconnect
SCILINKS NSTA
TOPIC: Ciliophora
GO TO: www.scilinks.org
KEYWORD: HM514

PHYLUM CILIOPHORA

The 8,000 species that make up the phylum Ciliophora (SIL-EE-AWF-uh-ruh) swim by means of cilia, which are short, hairlike cytoplasmic projections that line the cell membrane. Members of the genus *Paramecium* are among the most thoroughly studied ciliates. Like all paramecia, *Paramecium,* shown in Figure 26-7, is abundant in ponds and slow-moving streams that contain plants and decaying organic matter. As you can see, a paramecium has cilia arranged in rows across its cell membrane. The cilia beat in synchronized strokes that pass in waves across the cell, causing the protozoan to rotate on its axis. Paramecia and other ciliates feed on bacteria, algae, and other small organisms that are found in all types of marine and freshwater habitats.

Internal Structure

Ciliates have the most elaborate organelles of any protozoa. Figure 26-8 illustrates the basic structure of a paramecium. A clear, elastic layer of protein, called a **pellicle,** surrounds the cell membrane. The pellicle has a funnel-like depression called an **oral groove,** which is lined with cilia. These beating cilia create water currents that sweep food down the groove to the **mouth pore.** The mouth pore opens into a **gullet,** which forms food vacuoles that circulate throughout the cytoplasm. Organic molecules in the vacuole are digested and absorbed. Molecules that are not digested move to the **anal pore,** where they are expelled.

Ciliates are multinucleate; that is, they have at least one macronucleus and one micronucleus. The large **macronucleus** contains

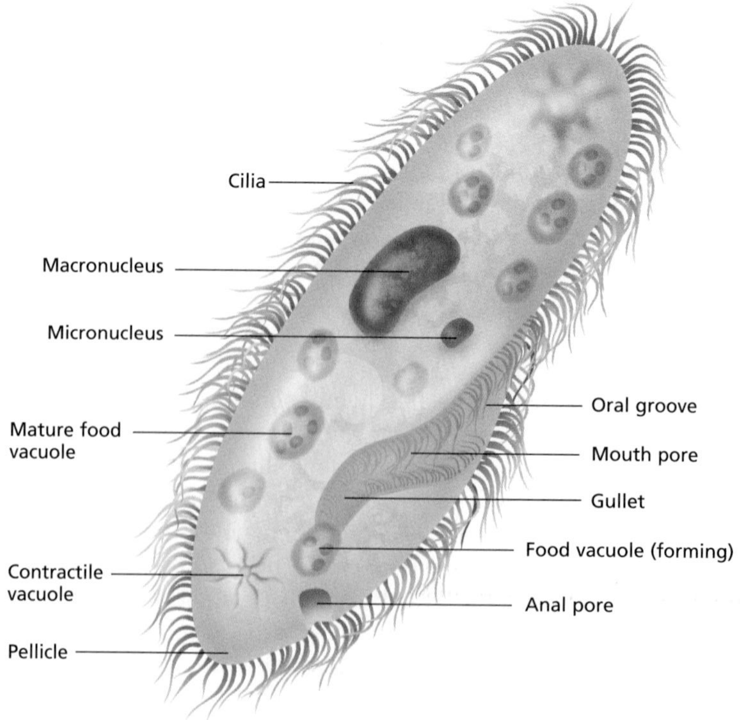

FIGURE 26-8
Paramecia have two types of nuclei: a large macronucleus and one or more small micronuclei. Paramecia have an oral groove, mouth pore, and gullet, into which food particles are drawn by currents produced by the beating cilia. They also have an anal pore, through which undigested waste is expelled from food vacuoles.

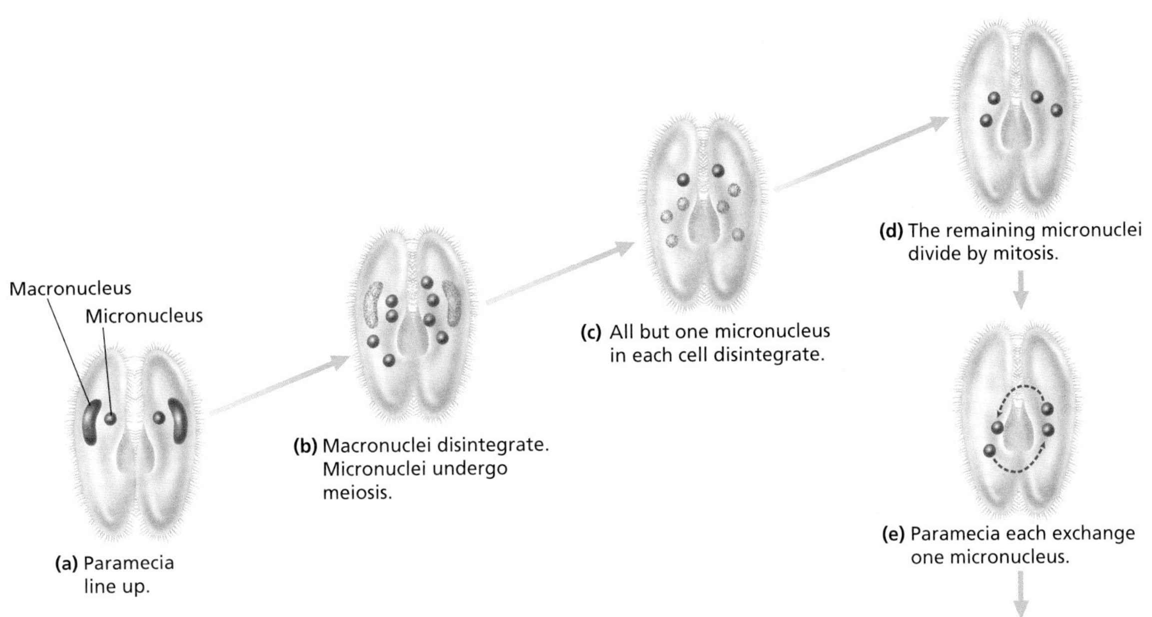

multiple copies of DNA and is responsible for metabolic and developmental functions. The macronucleus is also necessary for asexual reproduction. The smaller **micronucleus** participates in the exchange of genetic material between individuals during conjugation.

Reproduction

Asexual reproduction in ciliates occurs by binary fission. In this process, only the micronucleus divides by mitosis. The macronucleus, which has up to 500 times more DNA than the micronucleus, simply elongates and splits in half. One of the two halves goes to each new cell.

Sexual reproduction occurs in many ciliates by the process of conjugation. Conjugation in paramecia is illustrated in Figure 26-9. Conjugation begins when two organisms from opposite mating strains line up and join together. The macronucleus in each paramecium disintegrates. Each diploid micronucleus then undergoes meiosis, producing four haploid micronuclei. In each paramecium, all but one micronucleus disintegrate. The remaining micronucleus divides by mitosis, producing two identical haploid micronuclei. The two paramecia then exchange one micronucleus. The two micronuclei in each paramecium then fuse to form one diploid micronucleus. The two paramecia separate, and a macronucleus forms in each paramecium from products of mitotic divisions of the micronucleus. Although genetic material is exchanged during conjugation, no new cells are produced.

Following conjugation, each paramecium divides, producing a total of four genetically identical paramecia. Because genetic material is exchanged between the two original paramecia, the four offspring paramecia are genetically different from either original paramecium.

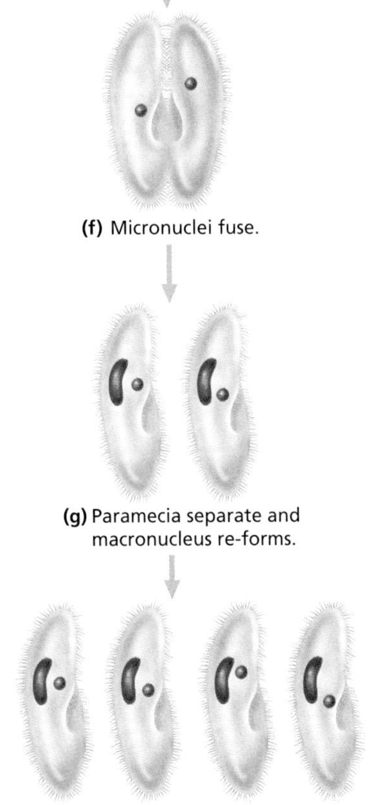

FIGURE 26-9
Paramecium reproduces by conjugation, a form of sexual reproduction in which genetic material is exchanged by two mating organisms.

TEACHING STRATEGY

Conjugation in Paramecia

Separate students into groups of three to five. Assign each group the task of teaching another group about conjugation in paramecia using Figure 26-9 as a guide. Provide construction paper, markers, and scissors. Tell students they have 10 minutes to use the supplies provided to visually demonstrate conjugation. At the end of 10 minutes, pair groups together for an exchange of information. After each group has presented its visuals, ask the students to display the paramecia offspring that were produced. Then point out that during conjugation, genetic material is exchanged. Using their models as an example, have students show that the offspring are genetically different from the original paramecia pair.

RECENT RESEARCH

Possible Key to Apoptosis

Scientists at the University of Rochester, in New York, have identified the proteins in ciliates that dissolve the macronucleus during conjugation. *Tetrahymena* species produce a protein that apparently dissolves the parental genes, leaving only the recombined genes of the micronucleus. Scientists think that understanding this mechanism of selective genetic destruction in protozoa may lead to an understanding of programmed cell death (apoptosis) in animals.

CONTENT CONNECTION

Section 24-2: Conjugation

Conjugation in paramecia differs from that in prokaryotes. In prokaryotes, only a portion of the DNA transfers from one cell to the receiving cell, and the transfer occurs through a temporary bridge between the two organisms.

RECENT RESEARCH
Curing Vectors

Scientists at the Yale University School of Medicine have genetically engineered a symbiotic bacteria that normally lives in the gut of kissing bugs to produce a protein that kills *Trypanosoma cruzi.* The bacteria, *Rhodococcus rhodnii,* are spread when young bugs eat dried fecal pellets left by older insects. As the genetically engineered bacteria spread, the number of *T. cruzi* infected bugs will decrease.

QUICK FACT
☑ The East African form of African sleeping sickness is an acute disease caused by *Trypanosoma brucei rhodesiense.* Illness and death occur within a few weeks of infection. The West African form is a chronic disease caused by *Trypanosoma brucei gambiense.* Illness may not occur until months or years after infection.

TEACHING STRATEGY
Disease Carriers

Show students pictures of sand flies, tsetse flies, kissing bugs, and *Anopheles* mosquitoes. Have students name the disease associated with each insect (sand flies—leishmaniasis; tsetse flies–African sleeping sickness; kissing bugs—Chagas' disease; anopheles mosquitoes—malaria).

RECENT RESEARCH
New Vaccines

Scientists have developed experimental vaccines, called DNA vaccines, against several protozoan parasites (*Toxoplasma, Leishmania,* and *Plasmodium*). These new vaccines use parasite DNA instead of a protein or carbohydrate. It is thought that parasite DNA is taken up by host cells and expressed. This stimulates a protective host response.

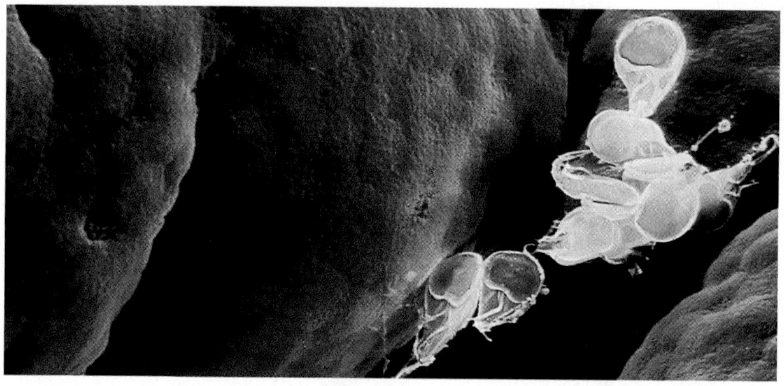

FIGURE 26-10

These zooflagellates are called *Giardia lamblia.* The protozoa pictured in this scanning electron micrograph are shown in their natural habitat—the human intestine.

PHYLUM ZOOMASTIGINA

The 2,500 species that make up the phylum Zoomastigina (ZOO-uh-MAST-uh-JIE-nuh) are characterized by the presence of one or more flagella, long hairlike structures that are made up of microtubules and are used for moving. The rapid whipping motion of the flagella pushes or pulls the protozoan through water. Many zooflagellates are free-living species that move through lakes and ponds, where they feed on small organisms. Some of the most primitive protozoa are found in this phylum. *Giardia lamblia* is shown in Figure 26-10.

internetconnect
SCILINKS NSTA
TOPIC: Zoomastigina
GO TO: www.scilinks.org
KEYWORD: HM516

Human Disease

Some of the most important protozoan parasites are zooflagellates. Many of them belong to the genus *Trypanosoma.* They live in the blood of fish, amphibians, reptiles, birds, and mammals and are carried from host to host by bloodsucking insects, such as flies. Some species are nonpathogenic, but others produce severe diseases in humans and animals. For example, two species of *Trypanosoma* can cause African **trypanosomiasis** (TRIP-uh-NOH-soh-MIE-uh-sis), or sleeping sickness. Trypanosomiasis is transmitted by the tsetse fly, which lives only in Africa. The disease is characterized by increasing fever, lethargy, mental deterioration, and coma. Another species, called *Trypanosoma cruzi,* causes **Chagas' disease.** Chagas' disease is transmitted by an insect called the "kissing bug." Patients with Chagas' disease suffer from fever and severe heart damage.

A zooflagellate called *Leishmania donovani* is transmitted by sand flies. It causes leishmaniasis, a blood disease that afflicts millions of people in Africa, Asia, and Latin America. The disease is characterized by disfiguring skin sores, and it can be fatal. A leishmanial sore is shown in Figure 26-11.

FIGURE 26-11

This person has leishmaniasis, which is caused by several species of *Leishmania. Leishmania* infects the skin, and some species infect major internal organs.

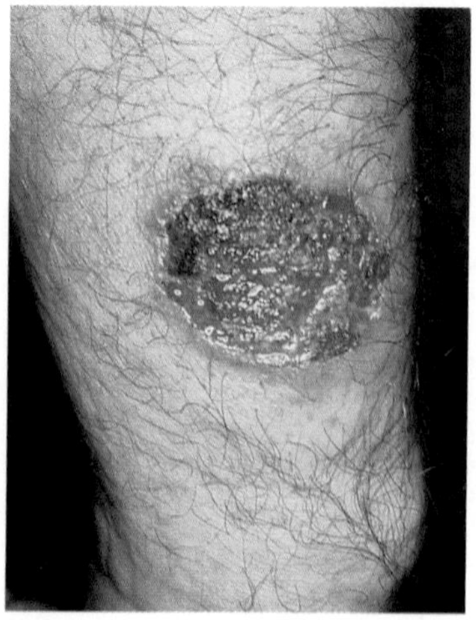

Giardia lamblia causes **giardiasis** (JEE-ahr-DIE-uh-sis), an illness characterized by severe diarrhea and intestinal cramps. Several kinds of animals carry the parasite and contaminate water with their feces. Hikers and others who are likely to drink contaminated water are susceptible to giardiasis, and thousands of cases occur annually in the United States. The disease is usually not fatal, and drugs hasten recovery.

PHYLUM SPOROZOA

All 6,000 species in the phylum Sporozoa (SPOHR-uh-ZOH-uh) have adult forms with no means of locomotion. Most species are parasitic and have complex life cycles in which they develop a spore, an infective form protected by a resistant coat.

Sporozoans are carried in the blood and other tissues of their hosts, where they absorb nutrients and destroy host cell tissues. The sporozoan shown in Figure 26-12 is *Toxoplasma gondii,* a parasite found in birds, rodents, and domestic cats. In humans, it can cause **toxoplasmosis** (TAHKS-oh-plaz-MOH-sis), a disease that causes few or no symptoms in adults with healthy immune systems but that can be dangerous to a developing fetus or newborn. Some adults can become seriously ill with flu-like symptoms. One species of sporozoan closely related to *Toxoplasma* causes coccidiosis, a deadly disease that affects birds and young cattle.

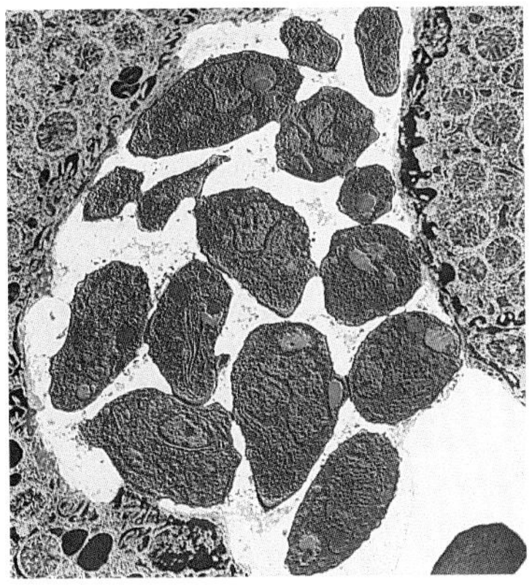

FIGURE 26-12

The sporozoan *Toxoplasma gondii* is pictured above in a section of human liver.

Plasmodium

The best-known sporozoan is *Plasmodium,* the protozoan that causes malaria. **Malaria** is a very serious disease characterized by severe chills, fever, sweating, fatigue, and great thirst. Victims die of anemia, kidney failure, or brain damage. The genus *Plasmodium* has caused more human deaths than any other genus in history. According to the World Health Organization, malaria afflicts 500 million people annually, killing as many as 2.7 million.

Four species of *Plasmodium* infect humans, and all have life cycles that involve the female *Anopheles* mosquito. When an infected mosquito bites a person, *Plasmodium* **sporozoites** enter the bloodstream and travel to liver cells, where they divide repeatedly. New spores called **merozoites** emerge and infect red blood cells, where they reproduce asexually. At regular intervals, the merozoites burst out of the red blood cells and release toxins into the blood. The destruction of red blood cells and the release of toxin in the blood cause the fever, anemia, and other symptoms of malaria. The merozoites infect other red blood cells and again reproduce asexually. This asexual reproduction can happen many times over a long period of time. Merozoites of some species remain in the liver and do not come out for months or years. Thus, an infected person could take antimalarial drugs and cure the infection in the blood, only to become ill again when the merozoites leave the liver cells. There are, however, antimalarial drugs that prevent the reoccurrence of malaria by killing the liver-stage parasites.

Some of the merozoites in the blood develop into specialized cells called **gametocytes.** When a female *Anopheles* bites the infected person, it ingests these gametocytes. In the mosquito's digestive system, the sperm and eggs combine to form a zygote. The nucleus of the zygote divides repeatedly to form more sporozoites. When the zygote bursts, the sporozoites migrate to the

Eco Connection

Protozoan Biocontrol

Scientists from the United States Department of Agriculture are conducting field studies with the protozoan *Edhazardia aedis* to test its effectiveness as a control agent of disease-carrying mosquitoes. The chemical DDT was used previously, but it was banned in the 1970s after it was discovered to be harmful to bird populations. *E. aedis,* which was discovered in Argentina, infects mosquito larvae in the water and kills them. Scientists think that this protozoan will prevent disease while causing no environmental damage.

CRITICAL THINKING

Toxoplasma Safety

Tell students to imagine that they live in a neighborhood that has many cats and that they also have a newborn living in their household. Their neighbor, who is a parasitologist, tells them that it would be a good idea to keep the baby away from the backyard sandbox. Ask students to explain what the parasitologist is concerned about. (Cats can harbor *Toxoplasma gondii,* a parasite. In humans, the parasite causes toxoplasmosis, a disease dangerous to newborns. If a cat is a carrier of the parasite, it can be a risk to the health of a newborn.)

MAKING CONNECTIONS

Language Arts

The ancient Romans gave us the name *malaria,* which means "bad air." The Romans had noticed that whenever the wind blew from the swamps, people became ill. The air blew in swamp gas, which smells bad, and it also blew in malaria-laden mosquitoes.

CULTURAL CONNECTION

Cinchona Bark

The first effective treatment for malaria was quinine, a drug derived from the bark of the cinchona tree. Native Americans first discovered the medical use of cinchona to treat malaria and introduced it to European colonists. Today, modern antimalarial drugs, such as chloroquine and primaquine, are derivatives of quinine.

GIFTED ACTIVITY

Have students prepare a report on early malaria research by searching library or on-line resources for information on Sir Ronald Ross, who first determined the life cycle of malaria.

CRITICAL THINKING

Successful Parasitism

First tell students that the most successful parasites harm their hosts the least. Then ask students to explain how this could be an advantage to the parasite. (The adaptive advantage for the parasite is that it preserves its host—a favorable environment. This strategy also permits the parasite to produce more offspring.)

ANSWERS TO SECTION 26-2 REVIEW

1. Pseudopodia are cytoplasmic extensions of a cell. They are used to move and feed the cell.
2. The tests of foraminifera and radiolarians have contributed to the layers of sediment covering the ocean floor, to formation of a rock called chert, and to the deposits of limestone that later emerged as dry land.
3. Conjugation is sexual reproduction that occurs in ciliates. Conjugation permits genetic recombination, thus increasing the genetic diversity of the species.
4. Zooflagellates can cause trypanosomiasis, Chagas' disease, leishmaniasis, and giardiasis.
5. See Figure 26-13. Like all sporozoans, *Plasmodium* forms spores.
6. Students may suggest ways of controlling the organisms that carry the protists; they may suggest that the water supply could be improved by isolating sewage from drinking supplies; or they may suggest that people at risk of contracting these diseases be vaccinated against the protists, if vaccines are developed in the future.

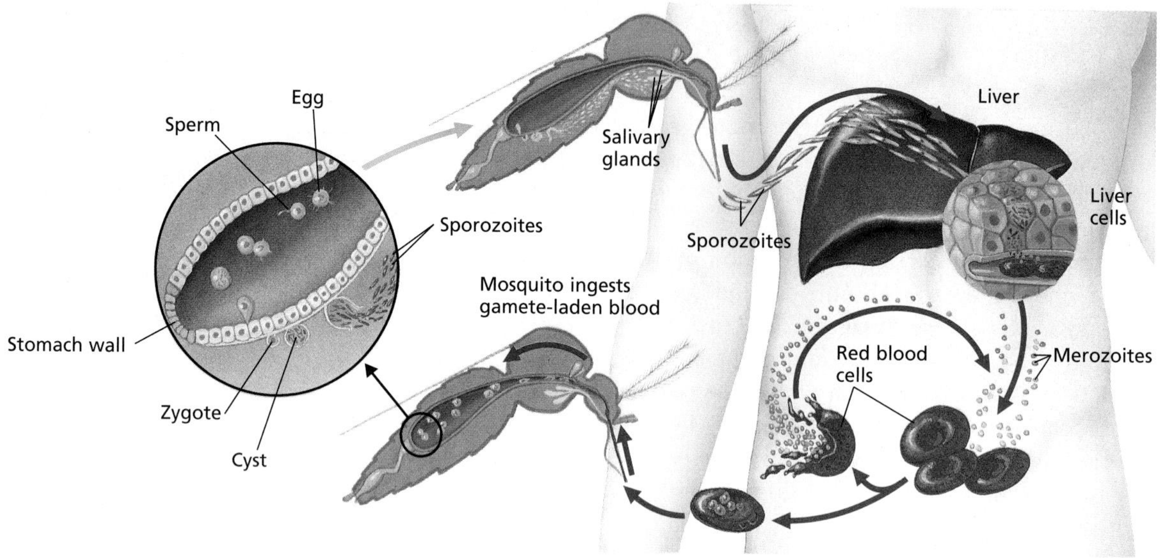

FIGURE 26-13

Malaria is caused by a sporozoan of the genus *Plasmodium. Plasmodium* is transmitted from host to host by female *Anopheles* mosquitoes. The host experiences attacks of chills and fever each time infected red blood cells burst and release the parasites that have multiplied within them. Some parasites develop into spores, which are picked up by uninfected *Anopheles* mosquitoes that bite the host.

body cavity and salivary glands of the mosquito. When the insect bites another person, the life cycle begins again. One effective way to reduce human deaths from malaria is to control mosquito populations. Without the mosquito hosts, the *Plasmodium* protozoans cannot complete their life cycle. The malaria life cycle is illustrated in Figure 26-13.

Malaria is usually cured with a drug derived from the cinchona tree, which is native to the Americas. This drug, quinine, has been used to treat malaria for over 500 years. As you learned in Chapter 24, bacteria can evolve resistance to antibiotics; similarly, sporozoans can evolve resistance to antimalarial drugs. Resistance to antimalarial drugs is a significant worldwide problem. Many more people will die from malaria despite the availability of drugs. Pharmaceutical companies have developed new drugs related to quinine, such as chloroquine and primaquine, but *Plasmodium* has evolved resistance to these newer quinine-related drugs. Although experimental malaria vaccines have not been successful, molecular techniques may offer new hope that a successful vaccine will be developed.

SECTION 26-2 REVIEW

1. **What are pseudopodia? What function do they serve in sarcodines?**
2. **How have sarcodines built geological features of the environment?**
3. **What is conjugation? How is this process advantageous for ciliates, such as *Paramecium*?**
4. **What kinds of disease can zooflagellates cause in humans?**
5. **Describe the life cycle of *Plasmodium,* the sporozoan that causes malaria. What features typical of sporozoans does this life cycle exhibit?**
6. **CRITICAL THINKING How might health workers attempt to control diseases caused by protozoa?**

CHAPTER 26 REVIEW

SUMMARY/VOCABULARY

26-1
- Protozoa are unicellular eukaryotic organisms classified in the kingdom Protista. They are found in moist habitats, and they include free-living and parasitic forms. Most species of protozoa are heterotrophic organisms that obtain nutrients by the process of phagocytosis.
- Many scientists believe that protozoa evolved from prokaryotes about 1.5 billion years ago.
- Many species of protozoa have adaptations for responding to changes in the environment. Such adaptations include eyespots and cyst formation.
- Protozoa are placed into four groups, according to the type of locomotion they display. The sarcodines move by means of pseudopodia, the ciliates move by means of cilia, the zooflagellates move by means of flagella, and the sporozoans are unable to move in the adult form.

Vocabulary

conjugation (507)	eyespot (508)	multiple fission (507)	protozoa (507)
cyst (509)	food vacuole (507)	protist (507)	zooplankton (507)

26-2
- The phylum Sarcodina consists of protozoa that move by means of pseudopodia. Sarcodines include amoebas, foraminifera, and radiolarians.
- Sarcodines move by means of cytoplasmic streaming. In this process, the endoplasm pushes the ectoplasm outward to form a pseudopodium. Pseudopodia are also used for phagocytosis.
- Most sarcodines are free-living, but the sarcodine *Entamoeba histolytica* is parasitic and causes the human disease amebic dysentery.
- The phylum Ciliophora consists of protozoa that move by means of cilia. Ciliates include the well-studied *Paramecium.* Paramecia have a complex array of organelles, including a macronucleus, a micronucleus, an oral groove, and an anal pore.
- Ciliates reproduce by binary fission as well as by sexual reproduction. Ciliates exchange genetic material by a process called conjugation.
- The phylum Zoomastigina consists of protozoa that move by means of flagella. Zooflagellates include *Trypanosoma,* a species that causes African sleeping sickness.
- Other important human diseases caused by zooflagellates are Chagas' disease, leishmaniasis, and giardiasis.
- The phylum Sporozoa is made up of protozoa that have complex life cycles in which they develop a spore. Virtually all species of sporozoans are parasites in humans and other animals.
- The sporozoan *Plasmodium* causes the disease malaria. Plasmodium has a complex life cycle. The *Anopheles* mosquito transmits the parasite, which causes extensive damage to the red blood cells in a victim.

Vocabulary

amebic dysentery (513)	endoplasm (512)	merozoite (517)	radiolarian (513)
ameboid movement (512)	foraminiferan (513)	micronucleus (515)	sporozoite (517)
anal pore (514)	gametocyte (517)	mouth pore (514)	test (513)
Chagas' disease (516)	giardiasis (516)	oral groove (514)	toxoplasmosis (517)
contractile vacuole (512)	gullet (514)	pellicle (514)	trypanosomiasis (516)
cytoplasmic streaming (512)	macronucleus (514)	pseudopodium (512)	
ectoplasm (512)	malaria (517)		

CHAPTER 26 REVIEW
ANSWERS

REVIEW

1. Protists include protozoa, algae, slime molds, and water molds. Protozoa are one group of protists; *protozoan* is the singular form of *protozoa.*
2. *Pseudopodium* means "false foot," which appropriately defines these temporary extensions used for locomotion.
3. Ectoplasm is the outer layer of cytoplasm; endoplasm is its inner region.
4. *Zoo-* is a prefix that means "animal." *Planktos* is a Greek word that means "drifting." Thus, *zooplankton* means "drifting animal-like" organisms.
5. *Ciliate, zooflagellate,* and *sporozoan* are common names of members of the phyla Ciliophora, Zoomastigina, and Sporozoa, respectively.

6. c	**7.** b
8. b	**9.** b
10. c	**11.** c
12. a	**13.** a
14. c	**15.** b

16. In ameboid movement the endoplasm pushes the ectoplasm outward, forming a new pseudopodium. As the new pseudopodium stretches outward, old pseudopodia retract. As a result, the amoeba moves in one direction. Pseudopodia surround a food particle so that the amoeba can obtain the nutrients.
17. The contractile vacuole removes excess water from the cell, enabling a sarcodine to maintain homeostasis in an aquatic environment.
18. Amebic dysentery is caused by *Entamoeba histolytica.* It is characterized by painful ulcers in the intestines and, in serious cases, liver and other organ infections.
19. Two cells join at the oral groove. The macronucleus in each cell disintegrates; the micronucleus in each cell divides both

meiotically and mitotically, eventually producing a micronucleus. Micronuclei are exchanged and fuse. In bacterial conjugation there is a transfer of genetic material, not a mutual exchange of genes.

20. Ciliates have cilia that sweep food particles down the oral groove into the mouth pore. The mouth pore leads to a gullet, which forms a food vacuole that circulates throughout the cytoplasm.

21. Ciliates have the most complex organelles of any protozoa.

22. Zooflagellates infect their hosts via a blood-sucking insect, such as a tsetse fly, kissing bug, or sand fly, which transmit African sleeping sickness, Chagas' disease, or leishmaniasis, respectively.

23. Sporozoans are not motile during their asexual phase.

24. Drawings should resemble Figure 26-13.

25. These protozoa are foraminifera with tests rich in calcium. The tests of dead foraminifera are a major component of the ocean floor sediment and of limestone and chalk deposits.

Critical Thinking

1. Paramecia have more organelles than do amoebas; they have an oral groove, gullet, anal pore, and cilia. Paramecia can also reproduce sexually through conjugation.

2. The proximity of food would be a stimulus that might cause a pseudopodium to form. An irritant may also cause pseudopodia to form.

3. The advantage of genetic variation is that members of the species will vary enough to enable the species to adapt to changing selective pressures. Although the cost in energy is high, the adaptive advantage is worth the cost.

4. If the vaccine is successful, then gametocytes will be destroyed by a host immune

Review

Vocabulary

1. Compare the following terms: protozoa, protozoan, and protist.
2. Research the meaning of the word *pseudopodium.* Why is this an appropriate term for the structure it defines?
3. Distinguish between the two regions of cytoplasm known as ectoplasm and endoplasm.
4. Research the meaning of the word *zooplankton.* Explain the relationship between the word's meaning and its roots.
5. Define the words *ciliate, zooflagellate,* and *sporozoan.*

Multiple Choice

6. Most scientists believe that protists evolved from (a) free-living worms (b) trypanosomes (c) prokaryotes (d) euglenoids.
7. Protozoan habitats are characterized by the presence of (a) algae (b) moisture (c) blood (d) soil.
8. Some protozoa monitor light quality with (a) pseudopodia (b) eyespots (c) cilia (d) contractile vacuoles.
9. Pseudopodia are extensions of a sarcodine's (a) pellicle (b) cytoplasm (c) cilia (d) test.
10. A pellicle is a characteristic of the (a) zooflagellates (b) sarcodines (c) ciliates (d) sporozoans.
11. Ciliates, such as *Paramecium,* have (a) hard outer tests (b) parasitic life cycles that involve *Anopheles* mosquitoes (c) a complex array of organelles (d) silicon dioxide in their cell membranes.
12. Flagella are characteristic of members of the phylum (a) Zoomastigina (b) Sarcodina (c) Sporozoa (d) Ciliophora.
13. African sleeping sickness is transmitted by (a) tsetse flies (b) *Anopheles* mosquitoes (c) kissing bugs (d) muskrats.
14. Members of the phylum Sporozoa do not cause (a) toxoplasmosis (b) malaria (c) giardiasis (d) coccidiosis.
15. Malaria is transmitted by (a) tsetse flies (b) *Anopheles* mosquitoes (c) kissing bugs (d) *Trypanosoma cruzi.*

Short Answer

16. Describe the process of ameboid movement and how it helps with the amoeba's nutrition.
17. What adaptive significance does the contractile vacuole have in the freshwater sarcodine?
18. Describe a potentially fatal human disease caused by an amoeba.
19. Explain the process of conjugation in *Paramecium.* How does it differ from the process of conjugation in bacteria?
20. How does a ciliate, such as *Paramecium,* capture and digest food?
21. Why are ciliates considered the most complex group of protozoa?
22. Explain how parasitic zooflagellates infect their hosts. Give two examples.
23. What characteristic of sporozoans distinguishes them from the three other protozoan groups?
24. Draw a simplified life cycle of *Plasmodium,* showing the protozoan activity in both the mosquito and the human hosts.
25. Identify the protozoa that once inhabited the shells shown in the figure below. How have they contributed to the geological formation of Earth?

CRITICAL THINKING

1. Many scientists suggest that paramecia are more complex than amoebas. What adaptations in paramecia would justify this claim?
2. Many protozoa feed by means of pseudopodia. What stimulus might cause these cytoplasmic extensions to form?
3. The process of conjugation is complex. It requires an expenditure of energy and other resources. Relate the high biological cost of conjugation to the adaptive advantage of exchanging genetic material.
4. Scientists are trying to develop a vaccine against malaria, but because malaria has several life stages, scientists must decide which life stage to develop a vaccine against. Some scientists are trying to develop a vaccine against gametocytes as a way of controlling malaria. If they are successful in developing this vaccine, how will it help people living in areas where malaria regularly occurs? Explain your answer.
5. Many parasitic protozoa, such as *Entamoeba histolytica,* are able to form cysts whenever they pass out of a host. Why is cyst formation an advantage to a parasitic protozoan?
6. Some protozoan parasites are very difficult to grow in test tubes because they lose the ability to produce certain key enzymes or growth factors when outside their host. How is this host-parasite relationship similar to that described in the hypothesis of endosymbiosis? How is it different?
7. Study the micrograph below showing the sporozoan *Plasmodium*. What evolutionary advantage does *Plasmodium* gain by forming gametocytes?

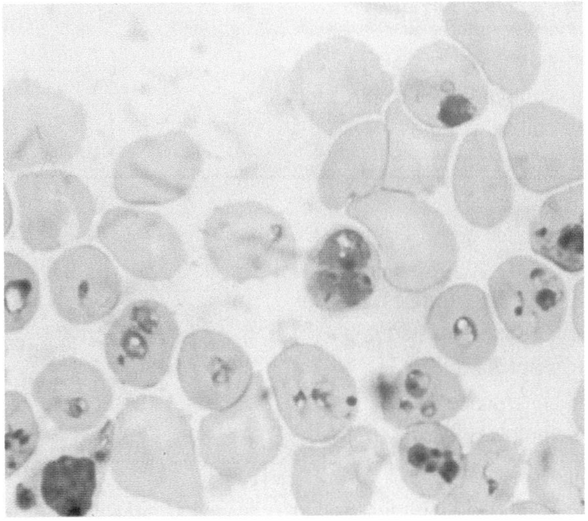

EXTENSION

1. Read "Intruder in the Heart" in *Discover,* December 1998, on page 38. Why is Chagas' disease so difficult to diagnose? Describe how humans become infected with Chagas' disease. What is the main reason heart transplantation is seldom an effective treatment for Chagas' disease?
2. Write a report on the types of protozoa found in zooplankton in the ocean. In your report, include information about the species of fish that depend on these protozoa as a food source.
3. Collect water from at least three sources—ponds, lakes, taps, or ditches—and examine the samples under a microscope. Count the different kinds of protozoa in each sample, and draw sketches of all the types you see.
4. Read "New Combination Vaccine May Fight Malaria" in *Science News,* February 20, 1999, on page 117. How many people die of malaria each year? Describe how the new vaccine works against malaria.

response. This will keep the infection from spreading to other people.

5. The environment outside the host may be hostile, and cyst formation enables a parasite to survive until a host is encountered.

6. This is similar to endosymbiosis because the endosymbiotic prokaryote provides its host with necessary enzymes that the host does not produce. It is different because a parasite does not necessarily provide an advantage to its host and is often harmful to its host.

7. The evolutionary advantage is that gametocytes allow *Plasmodium* to reproduce sexually. Sexual reproduction produces genetic variability and therefore the ability to adapt to changing environments.

EXTENSION

1. Chagas' disease often presents the same symptoms as coronary heart disease does. Insects called reduviids often nest in poor, rural dwellings. These insects feed on the blood of sleeping humans and then defecate, leaving parasite-laden feces next to the wound they made to feed from. Heart transplants require drugs to prevent organ rejection. These drugs allow the parasite to spread out of control inside the patient.

2. Have students identify other organisms, besides protozoa, that make up zooplankton (larvae, jelly fish). Challenge students to create a food web that includes zooplankton.

3. Make sure students label the drawings with the location where the protozoa were found. Provide students with college texts and lab manuals to help them identify the protozoa.

4. Three million people die of malaria each year. The vaccine produces antibodies that combat the parasite at each stage of its life cycle.

CHAPTER 26 INVESTIGATION

TIME REQUIRED

One 50-minute class period

SAFETY PRECAUTIONS

Have students wear safety goggles during this investigation. They should wash their hands thoroughly after handling the cultures. Remind them to handle coverslips with care because they can break easily.

QUICK REFERENCES

Lab Preparation Notes are found on page 505B.

Holt BioSources provides a Teaching Resources CD-ROM that contains "Using Gowin's Vee in the Lab" and "Scoring Rubrics."

PROCEDURAL TIP

To save time, have half the students perform Part A while the other half perform Part B, then have them switch.

ANSWERS TO BACKGROUND

1. Protists are eukaryotic, mostly unicellular organisms that lack tissue differentiation. Most protists reproduce asexually, although some reproduce sexually.
2. Protists are eukaryotic organisms; examples include protozoa, algae, slime molds, and water molds. Protozoa are single-celled microscopic organisms that can move independently.
3. Ciliophora have cilia for movement and very complex organelles, such as a pellicle, oral groove, mouth pore, anal pore, macronucleus, micronucleus, and gullet.
4. Ciliates are heterotrophs that eat bacteria and some protozoa. They consume food by using their cilia to sweep it into their oral groove, to the mouth pore, and into a gullet. The food is taken

CHAPTER 26 INVESTIGATION

Observing *Paramecium*

OBJECTIVES

- Observe protozoa under a compound light microscope.
- Compare *Paramecium* with *Euglena.*
- Test the effects of different solutions on the movement of a paramecium's contractile vacuoles.

PROCESS SKILLS

- observing
- comparing and contrasting
- collecting data
- analyzing data

MATERIALS

- safety goggles
- medicine droppers
- *Paramecium* culture
- *Euglena* culture
- microscope slides
- methyl cellulose
- coverslips
- compound light microscope
- stopwatch or clock with second hand
- distilled water
- paper towels
- salt water

Background

1. List some of the general characteristics of the kingdom Protista.
2. Distinguish between protists and protozoa.
3. Distinguish members of the phylum Ciliophora from other phyla of protozoa.
4. What foods do ciliates eat? How do they take in and digest food?
5. What function does the contractile vacuole serve in the *Paramecium*?

PART A Comparing *Paramecium* with *Euglena*

1. CAUTION **Put on safety goggles. If you get methyl cellulose in your eyes, immediately flush it out at the eyewash station while calling to your teacher. Slides break easily. Use caution when handling them.** Using a medicine dropper, place one drop of methyl cellulose on a microscope slide.
2. Place one drop of *Paramecium* culture on the methyl cellulose to slow the paramecia down, and cover them with a coverslip.
3. Use the low-power setting of your microscope to locate the paramecia and observe their movement. Then change to the high-power setting.
4. Examine one paramecium under the high-power setting of your microscope. Check for the structural details of the paramecium by changing the fine adjustment and altering the light conditions. If the organism moves out of view, move the slide to keep up with it. If you are unsuccessful, return to low power to scan a larger field, and begin again.
5. Make a sketch of a paramecium in your lab report. Label the cilia, oral groove, mouth pore, and gullet in your sketch. Note the color of the food vacuoles.
6. Remove the coverslip. Add one drop of *Euglena* culture to the slide. Cover and examine under high power. *Euglena* is a protist that you will study in Chapter 27. Its green color is due to chloroplasts.
7. As you view these two protists, note the relative sizes of the two kinds of organisms. Compare their means of locomotion.
8. Observe the food vacuoles in the paramecia. Did they change color after you added the *Euglena* culture to the slide?

PART B Observing the Contractile Vacuoles of *Paramecium*

9. In your lab report, prepare a data table similar to the one shown below.
10. Find a paramecium that is easy to observe, and identify its contractile vacuoles—the two roundish forms at each end of the organism. Count how many times the contractile vacuoles fill and empty in one minute. Record your data in your data table in the row labeled "Culture medium."
11. Repeat step 10, observing two different paramecia.
12. Use a clean medicine dropper to put some distilled water at one edge of the coverslip. Draw the distilled water to the other side of the slide by placing a paper towel at the opposite edge of the coverslip.
13. Count how many times the contractile vacuoles fill and empty in one minute when the paramecium is in distilled water. Record the number of pulses per minute in your data table in the row labeled "Distilled water." Repeat the process using two different paramecia.
14. Using a clean medicine dropper, put some salt water at one edge of the coverslip. Draw the salt water to the other side of the slide by placing a paper towel at the opposite edge of the coverslip as you did in step 12. Then count the number of times the contractile vacuoles fill and empty in one minute in the saltwater environment. Record the number of pulses per minute in your data table in the row labeled "Salt water." Repeat the process using two different paramecia.
15. Average the number of pulses per minute by adding the counts you obtained for each environment and dividing by 3. Record the average pulses per minute in the last column of your data table.
16. Clean up your lab materials and wash your hands before leaving the lab.

Analysis and Conclusions

1. Based on your observations in Part A, is *Paramecium* autotrophic or heterotrophic? Explain your answer.
2. Explain any changes in the color of the food vacuoles in Part A.
3. What effect did the distilled water have on the paramecium's contractile vacuoles in Part B?
4. What effect did the salt water have on the paramecium's contractile vacuoles?
5. Did you use a control in Part B? Explain your answer.
6. Return to the data table you completed in Part B. Using the terms *isotonic, hypotonic,* and *hypertonic,* describe the three different environments you observed the paramecium in.
7. Explain how the external environment affected the number of pulses of the contractile vacuoles. (Hint: Compare the effects of a hypotonic solution with those of a hypertonic solution.)
8. Explain how contractile vacuoles are important to the survival of paramecia living in freshwater environments.
9. *Paramecium* is among the protists called protozoa, which means "first animals." In what way is this description accurate or misleading?

Further Inquiry

Design an experiment to test the responses of protozoa to a range of temperatures or a range of pH.

CONTRACTILE VACUOLE PULSES

Environment	Number of pulses per minute			Average pulses per minute
	paramecium 1	paramecium 2	paramecium 3	
Culture medium				
Distilled water				
Salt water				

into a food vacuole, where it is digested by enzymes.
5. The contractile vacuole pumps out water so that the *Paramecium* maintains homeostasis with its environment.

ANALYSIS AND CONCLUSIONS

1. *Paramecium* is heterotrophic.
2. The food vacuoles change color as they are filled with different-colored food.
3. The contractile vacuoles operated more frequently than before.
4. The contractile vacuoles operated less frequently than before.
5. Yes, the first observation with the paramecium in its culture medium was the control.
6. The culture medium was isotonic, distilled water was hypotonic, and the salt water was hypertonic.
7. The hypertonic solution caused water to diffuse out of the paramecium, and therefore the contractile vacuoles did not have to pump frequently. The hypotonic solution caused water to diffuse into the paramecium, and the contractile vacuoles therefore had to pump more frequently in order to maintain homeostasis. The isotonic solution did not create a net gain or loss of water for the cell; therefore, the contractile vacuoles did not have to pump often.
8. In freshwater environments, the isotonic environment will pump water into the cell. Therefore, the contractile vacuole must pump out the excess water, or the cell will burst.
9. Among the protozoa, paramecia are very advanced; therefore calling them "first animals" may be misleading.

CHAPTER 27
ALGAE AND FUNGUSLIKE PROTISTS

	TOPICS	TEACHING RESOURCES	LABS, CLASSWORK, AND HOMEWORK
BLOCK 1 45 minutes	**Introducing the Chapter, p. 524**	**ATE** Focus Concept, p. 524 **ATE** Assessing Prior Knowledge, p. 524 **ATE** Understanding the Visual, p. 524	■ Supplemental Reading Guide, *Microbe Hunters*
	27-1 Overview of Algae, p. 525 Characteristics, p. 525 Structure, p. 525 Classification, p. 526 Reproduction, p. 527	**ATE** Section Overview, p. 525 **ATE** Critical Thinking, p. 528 **ATE** Visual Link Figure 27-1, p. 526 Table 27-1, p. 527 Figure 27-2, p. 528 Figure 27-4, p. 529 89A, 90A, 103, 104, 105	**Lab A17,** Observing Protists **Lab C27,** Studying an Algal Bloom **Lab C28,** Studying an Algal Bloom—Phosphate Pollution ★ Study Guide, Section 27-1 **PE** Section Review, p. 529
BLOCK 2 45 minutes	**27-2 Algal Diversity, p. 530** Phylum Chlorophyta, p. 530 Phylum Phaeophyta, p. 531 Phylum Rhodophyta, p. 531 Phylum Bacillariophyta, p. 532 Phylum Dinoflagellata, p. 532 Phylum Chrysophyta, p. 533 Phylum Euglenophyta, p. 533	**ATE** Section Overview, p. 530 **ATE** Critical Thinking, p. 532 **ATE** Visual Link Figure 27-7, p. 531 99	★ Study Guide, Section 27-2 **PE** Section Review, p. 533
BLOCK 3 45 minutes	**27-3 Funguslike Protists, p. 534** Slime Molds, p. 534 Water Molds, p. 535	**ATE** Section Overview, p. 534 **ATE** Critical Thinking, pp. 535, 536 **ATE** Visual Link Figure 27-12, p. 534	**PE** **Chapter 27 Investigation,** p. 540 **PE** **Quick Lab,** p. 535, Observing Slime Mold **Lab B14,** Protists—A Comparison ★ Study Guide, Section 27-3 **PE** Section Review, p. 536
BLOCKS 4 & 5 90 minutes	**REVIEW and ASSESSMENT** **PE** Chapter 27 Review, pp. 538–539	Audio CD Program	**CHAPTER TESTING** ■ Chapter Test (blackline master) ▲ Test Generator ■ Assessment Item Listing ■ Scoring Rubrics and Classroom Management Checklists

MODERN BIOLOGY

HOLT BIOSOURCES

One-Stop Planner CD-ROM
Shows these resources within a customizable daily lesson plan:

KEY

- **PE** Pupil's Edition
- **ATE** Teacher's Edition
- ■ Active Reading Guide
- ★ Study Guide
- Teaching Transparencies
- Laboratory Program

One-Stop Planner CD-ROM:

- ■ Holt BioSources Teaching Resources
- ■ Active Reading Guide
- ★ Study Guide
- ▲ Test Generator

READING FOR CONTENT MASTERY

Preview: Have students preview the objectives, the introductory paragraph, topic sentences, and captions before reading the chapter.

ATE Active Reading Technique: Reading Effectively, p. 525
■ Active Reading Guide Worksheet 27-1
PE Summary/Vocabulary, p. 537

■ Active Reading Guide Worksheet 27-2
PE Summary/Vocabulary, p. 537

■ Active Reading Guide Worksheet 27-3
PE Summary/Vocabulary, p. 537

TECHNOLOGY AND INTERNET RESOURCES

CNN PRESENTS Video Segment 17
Killer Algae

Holt Biology Videodiscs Teacher's Correlation Guide, Lessons 27-1 through 27-3

Audio CD Program, Sections 27-1 through 27-3

internetconnect

SCILINKS NSTA
On-line Resources: www.scilinks.org
The following *sci*LINKS Internet resources can be found in the student text for this chapter:

Topics:
- Algae, p. 527
- Bioluminescence, p. 533

go.hrw.com
On-line Resources: go.hrw.com
Visit the HRW Web site for a variety of resources related to this chapter. Just type in the keyword HM2 Chapter 27.

Smithsonian Institution®
Visit **www.si.edu/hrw** for additional on-line resources.

CNNfyi.com
Visit **www.cnnfyi.com** for late-breaking news and current event stories selected just for you.

LAB ACTIVITY PLANNING

CHAPTER 27 INVESTIGATION

Classifying Green Algae, pp. 540–541

OVERVIEW
In this investigation, students will observe several live specimens of green algae. Students will also compare unicellular, colonial, thalloid, and filamentous green algae. Students will use a dichotomous key to classify the colonial green algae according to their multicellularity and appearance.

SAFETY
Special care should be taken when viewing slides under high power so that the objective does not break the slide. Have students dispose of broken glass in a container designated for glass only. Do not allow students to use microscopes with wet hands or near water.

PREPARATION
1. You will need to purchase *Chlamydomonas, Spirogyra, Eudorina, Gonium, Pandorina, Volvox,* and *Hydrodictyon.* Aerate the cultures as soon as they arrive by squirting air into the water with a medicine dropper. Keep the cultures loosely capped.
2. Create a mixed culture of colonial green algae. Combine equal amounts of *Eudorina, Gonium, Pandorina, Volvox,* and *Hydrodictyon* in a specimen dish. Aerate the culture periodically.
3. Each group will also need three depression slides with coverslips, three medicine droppers, and access to a compound light microscope.

Quick Labs

Observing Slime Mold, p. 535
Each student team will need a plasmodial slime mold culture. Plasmodial slime mold cultures are available from a biological supply house. Each team will also need oatmeal, vinegar, a cotton swab, and a dissecting microscope.

UNDERSTANDING THE VISUAL

Have students examine the kelp in the photograph. Explain that these organisms are multicellular algae. Identify the kelp's spherical air bladders, and ask students to infer these structures' function. (Air bladders help leaflike portions of the algae float near the water's surface, where they are exposed to sunlight.)

FOCUS CONCEPT

Structure and Function

In this chapter, the relationship between structure and function is evident in protist adaptations. Examples in algae are zoospores, zygospores, and gametangia that function in reproduction. Examples in funguslike protists include fruiting bodies and pseudoplasmodia that function in reproduction.

ASSESSING PRIOR KNOWLEDGE

Review the following concepts.

Cellular Organelles: *Chapter 4*
Ask students to explain the functions of the following organelles: nucleus, chromosome, chloroplast, mitochondrion, and contractile vacuole.

Homeostasis: *Chapter 5*
Have students define the following terms: *diffusion, equilibrium,* and *transportation across membranes.* Ask students to name the structures that ensure homeostasis in cells.

Cell Reproduction: *Chapter 8*
Have students review the processes of mitosis and meiosis. Ask them why sexual reproduction is advantageous for unicellular algae.

Protozoa: *Chapter 26*
Have students review the characteristics of animal-like protists such as *Amoeba* and *Paramecium.*

CHAPTER 27

ALGAE AND FUNGUSLIKE PROTISTS

Algae, such as these kelp, Macrocystis pyrifera, *off the coast of California, are giants among protists. They grow in massive groves that support large numbers of aquatic organisms.*

FOCUS CONCEPT: ***Structure and Function***

As you read this chapter, note that algae and funguslike protists are extremely diverse groups of organisms that vary in structure, reproduction, and biochemistry.

Overview of Algae

SECTION **27-1**

Algae are plantlike organisms that belong to the kingdom Protista. Although most species of algae are unicellular, some, such as the Macrocystis pyrifera *shown on the opposite page, are large, multicellular organisms. Algae differ from protozoa, which are also classified in the kingdom Protista, in that they manufacture their food through the process of photosynthesis. This section explores the basic characteristics of algae.*

OBJECTIVES

▲ Compare algae with other protists.

● Explain how algae differ from plants.

■ Describe the various body structures of algae.

◆ Identify the characteristics used to classify algae into seven phyla.

▲ Summarize the events of asexual and sexual reproduction in representative genera of algae.

CHARACTERISTICS

Algae are a diverse group of protists. They range in size from microscopic single-celled organisms to large seaweeds that may be hundreds of feet long. Unlike protozoa, which are heterotrophic, **algae** are autotrophic protists—they have chloroplasts and produce their own carbohydrates by photosynthesis. In the past, some classification systems placed algae in the plant kingdom. However, algae lack tissue differentiation and thus have no true roots, stems, or leaves. The reproductive structures of algae also differ from those of plants; they form gametes in single-celled **gametangia** (GAM-uh-TAN-jee-uh), or gamete chambers. Plants, by contrast, form gametes in multicellular gametangia. For these reasons, algae are classified as protists.

Despite their diversity, different kinds of algae have several features in common. For example, most algae are aquatic and have flagella at some point in their life cycle. In addition, algal cells often contain **pyrenoids** (pie-REE-NOYDZ), organelles that synthesize and store starch.

STRUCTURE

The body portion of an alga is called a **thallus** (THAL-uhs). The thallus of an alga is usually haploid. A variety of thallus formats characterize algae. In some species, the thallus consists of a single cell. In other species, it is made up of many cells in varying arrangements. Four types of algae are recognized, based on the following body structures: unicellular, colonial, filamentous, and multicellular.

Unicellular algae have a structure that consists of a single cell. Most unicellular algae are aquatic organisms that compose the **phytoplankton,** a population of photosynthetic organisms that

Word Roots and Origins

algae

from the Latin *alga,* meaning "seaweed"

SECTION 27-1

SECTION OVERVIEW

Structure and Function
Section 27-1 describes the characteristics and structures of algae and the basis for their classification. Many types of protists have adaptations for moving through their environments, surviving extreme conditions, and reproducing asexually and sexually.

Reproduction and Inheritance
Section 27-1 also describes asexual and sexual reproduction in several species of algae. Asexual reproduction produces high numbers of genetically identical offspring; sexual reproduction provides for genetic diversity.

ACTIVE READING

Technique: Reading Effectively
Before students begin to read this chapter, have them write down all the vocabulary words for each of the three sections of the chapter. Ask them to write a definition next to each word that they have heard of. As the students read the chapter, have them write definitions next to the words that they did not know previously and modify as needed any definitions of words they knew.

QUICK FACT

☑ Phytoplankton (*phyton,* meaning "plant," and *plankton,* meaning "wanderer") float near the surface of oceans, lakes, and ponds so that they can be exposed to sunlight. They have different strategies for staying afloat. Some phytoplankton swim by spinning flagella or waving cilia; some thrust out pseudopodia, which increase their surface areas and keep them from sinking. Other types of phytoplankton have oil reserves, which act as a flotation device. The oil also serves as a food supply.

SECTION 27-1

VISUAL LINK

Figure 27-1

Instruct students to examine the organisms in Figure 27-1. Ask them to name the characteristics the organisms have in common. (They are all classified as algae and contain chlorophyll in chloroplasts.) Ask students to identify the thallus types in each of the photos (a—unicellular, b—colonial, c—filamentous, d—multicellular).

DEMONSTRATION

Use a microprojector to show slides of the four types of thallus structure–unicellular, colonial, filamentous, and multicellular. Be sure students can identify each of the four types. Then, using prepared slides, set up microscope stations where students can observe samples of each type of thallus. Have students sketch their observations and label major algal structures. (Examples may include *Macrocystis*—multicellular, dinoflagellates—unicellular, *Spirogyra*—filamentous, and *Hydrodictyon*—colonial.)

INCLUSION *ACTIVITY*

Divide the class into small groups, and provide each group with photographs of algae. Be sure the four types of thallus structures are represented. Have each group place their pictures into categories based on thallus structure. (Examples may include *Plumaria*—multicellular, *Euglena*—unicellular, *Pandorina*—colonial, and *Oedogonium*—filamentous.)

(a)

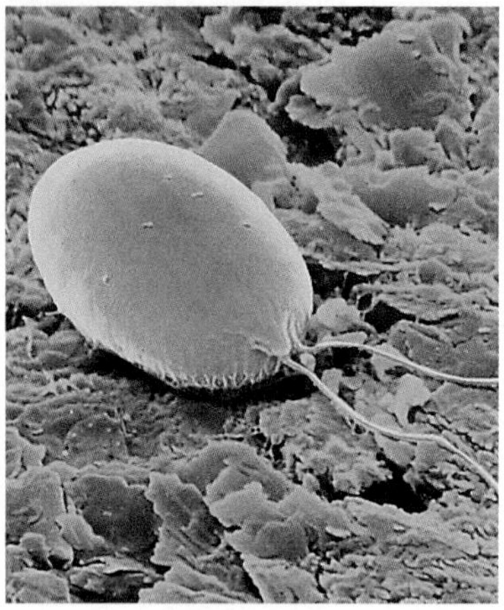

(b)

(c)

(d)

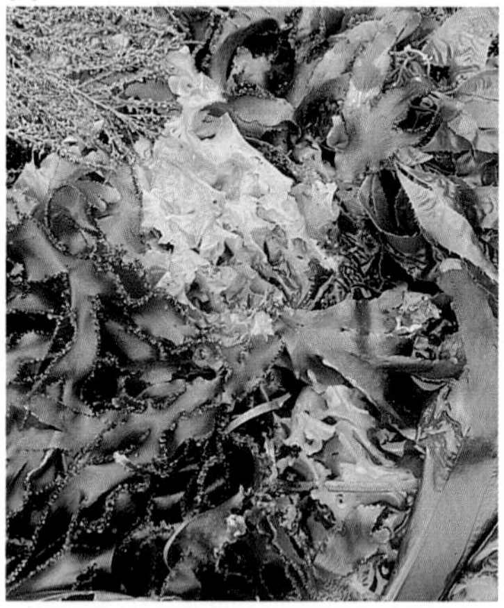

FIGURE 27-1

Algae are characterized by a variety of body structures. (a) *Chlamydomonas* is a flagellated unicellular organism (2,905×). (b) An example of a colonial green alga is *Volvox* (400×). (c) *Spirogyra* is a green alga with a filamentous body form (291×). (d) Some multicellular algae are sheetlike; *Ulva* has a thin, leaflike form.

forms the foundation of aquatic food chains. Through photosynthesis, phytoplankton produces almost half the world's carbohydrates, thereby providing important nutrients for numerous aquatic organisms. Such unicellular algae are also among the major producers of oxygen in the atmosphere. The *Chlamydomonas* in Figure 27-1a is an example of a unicellular alga.

Colonial algae, such as *Volvox,* shown in Figure 27-1b, have a structure that consists of groups of cells acting in a coordinated manner. Some of these cells become specialized. This division of labor allows colonial algae to move, feed, and reproduce efficiently.

Filamentous algae, such as *Spirogyra,* shown in Figure 27-1c, have a slender, rod-shaped thallus composed of rows of cells joined end to end. Other species of filamentous algae have specialized structures that anchor the thallus to the ocean bottom. This adaptation secures the alga in one place as it grows toward the sunlight at the water's surface.

Multicellular algae often have a large, complex thallus. For instance, *Ulva,* shown in Figure 27-1d, has a leaflike thallus that may be several centimeters wide but only two cells thick. The giant kelp, *Macrocystis,* is among the largest of the multicellular algae. It has rubbery leaflike portions, stemlike regions, and enlarged air bladders.

CLASSIFICATION

Algae are classified into seven phyla, based on their color, type of chlorophyll, form of food-storage substance, and cell wall composition. All known phyla contain the light-absorbing photosynthetic pigment chlorophyll *a.* However, different types of algae also contain other forms of chlorophylls—such as chlorophylls *b, c,* or *d*—that absorb slightly different wavelengths of light. Some phyla also have accessory pigments that give them their characteristic color. The seven phyla of algae are summarized in Table 27-1.

TABLE 27-1 Seven Phyla of Algae

Phylum	Thallus format	Photosynthetic pigments	Form of food storage	Cell wall composition
Chlorophyta (green algae, 7,000 species)	unicellular, colonial, filamentous, and multicellular	chlorophylls *a* and *b*, carotenoids	starch	polysaccharides, primarily cellulose
Phaeophyta (brown algae, 1,500 species)	multicellular	chlorophylls *a* and *c*, carotenoids, fucoxanthin	laminarin (an oily carbohydrate)	cellulose with alginic acid
Rhodophyta (red algae, 4,000 species)	multicellular	chlorophyll *a*, phycobilins, carotenoids	starch	cellulose or pectin, many with calcium carbonate
Bacillariophyta (diatoms, 11,500 species)	mostly unicellular; some colonial	chlorophylls *a* and *c*, carotenoids, xanthophyll	leucosin (an oily carbohydrate)	pectin, many with silicon dioxide
Dinoflagellata (dinoflagellates, 1,100 species)	unicellular	chlorophylls *a* and *c*, carotenoids	starch	cellulose
Chrysophyta (golden algae, 850 species)	mostly unicellular; some colonial	chlorophylls *a* and *c*, xanthophyll, carotenoids	laminarin (an oily carbohydrate)	cellulose
Euglenophyta (euglenoids, 1,000 species)	unicellular	chlorophylls *a* and *b*, carotenoids, xanthophyll	paramylon (a starch)	no cell wall, protein-rich pellicle

REPRODUCTION

Many species of algae reproduce both asexually and sexually. Sexual reproduction in algae is often triggered by environmental stress. Some species reproduce only asexually.

Both asexual and sexual reproduction have been studied extensively in the unicellular green alga *Chlamydomonas.* As a mature organism, *Chlamydomonas* exists as a flagellated haploid cell. During asexual reproduction, the alga first absorbs its flagellum. Then the haploid cell divides mitotically up to three times, and from two to eight haploid flagellated cells called **zoospores** (ZOH-oh-SPOHRZ) develop within the parent cell. These motile asexual reproductive cells break out of the parent cell, disperse, and eventually grow to full size.

Sexual reproduction in *Chlamydomonas* also begins by haploid cells dividing mitotically to produce either "plus" or "minus" gametes. (The *plus* and *minus* terminology describes gametes that look similar but differ in chemical composition.) A plus gamete and a minus gamete come into contact with one another and shed their cell walls. They fuse and form a diploid zygote, which develops a thick protective wall. A zygote in such a resting state is called a **zygospore** (ZIE-go-SPOHR). A zygospore can withstand unfavorable environmental conditions. When favorable conditions exist, the thick wall opens and

internet**connect**

SCiLINKS NSTA

TOPIC: Algae
GO TO: www.scilinks.org
KEYWORD: HM527

VISUAL LINK
Table 27-1

Use Table 27-1 to highlight the major characteristics of the seven algal phyla. Ask students the following questions. Which phyla contain only algae with a unicellular thallus format? (Dinoflagellata and Euglenophyta) Which phylum has algae that contain the pigment fucoxanthin? (Phaeophyta) the pigment phycobilin? (Rhodophyta) Which phylum has algae with cells that lack cell walls? (Euglenophyta)

QUICK FACT

☑ The names of four algal phyla have prefixes that describe the color of the phyla members. The prefix *chloro-* means "green," *phaeos-* means "brown," *rhodo-* means "red," and *chryso-* means "gold."

RETEACHING ACTIVITY

On the board, draw a table titled "Forms of Food Storage." Add columns for the following carbohydrates: starch, laminarin, leucosin, and paramylon. Have students complete the table by writing the phyla names under each column (starch—Chlorophyta, Rhodophyta, and Dinoflagellata; laminarin—Phaeophyta, Chrysophyta; leucosin—Baccillariophyta; paramylon—Euglenophyta).

CONTENT CONNECTION

Section 6-1: Chlorophyll and Photosynthesis

Autotrophs use the sun's energy to power the process of photosynthesis. Algae are autotrophic, plantlike protists that contain chlorophyll *a* and other pigments that absorb various wavelengths of light.

SECTION 27-1

VISUAL LINK

Figure 27-2

Explain that in asexual reproduction, mitosis forms haploid cells. During sexual reproduction, a "plus" gamete fuses with a "minus" gamete to form a zygospore. When favorable conditions exist, the zygospore undergoes meiosis, forming haploid cells.

CRITICAL THINKING

Algal Adaptations

Ask students why it is advantageous for algae to undergo sexual reproduction during times of environmental stress. (Students might speculate that spores produced during sexual reproduction can withstand extreme conditions, increasing the likelihood of survival. Students might also include the advantages of the genetic diversity that results from gamete fusion.)

RECENT RESEARCH

Algal Populations

Researchers have developed a technique for analyzing aquatic ecosystems using algal accessory pigments. Algal pigments have unique spectral lines, so the composition of an algal population can be detected using remote sensing equipment in satellites or aircraft. This new technique, combined with accessory pigment analysis, will enable scientists to study large areas of aquatic ecosystems. Such research will enable scientists to assess algal identity and determine their numbers without the time-consuming cell counts that were previously performed.

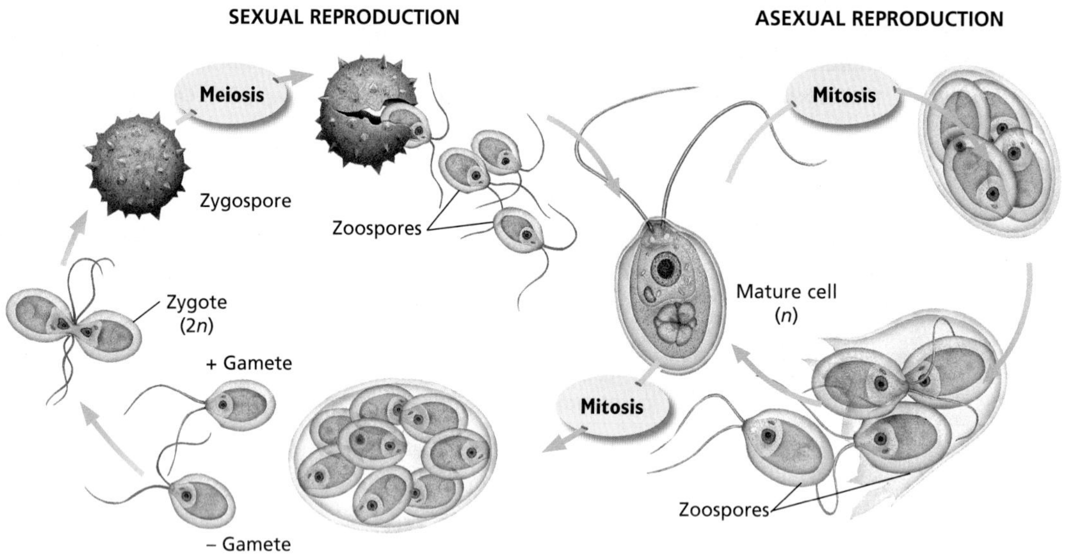

FIGURE 27-2

The unicellular green alga *Chlamydomonas* reproduces asexually by mitosis. It also reproduces sexually when gametes of opposite mating types fuse.

the living zoospore emerges. It then undergoes meiosis, forming numerous haploid *Chlamydomonas* cells that grow into mature organisms. The life cycle of *Chlamydomonas* is illustrated in Figure 27-2.

Reproduction in Multicellular Algae

Reproduction of multicellular algae varies widely among the phyla. Reproduction in red and brown algae is particularly complex, with that of the red algae often involving three states in a sexual life cycle. Examination of the following algal life cycles illustrates some of the reproductive variation in algae.

Oedogonium is a filamentous green alga. As shown in Figure 27-3, *Oedogonium* has cells specialized for producing gametes. Recall that the modified cells that produce and hold the gametes are called unicellular gametangia. The male unicellular gametangium, called an **antheridium** (AN-thuh-RID-ee-um), produces sperm. The female unicellular gametangium, called an **oogonium** (OH-uh-GOH-nee-um), produces

FIGURE 27-3

Oedogonium reproduces sexually by producing male and female gametes. The sperm, released into the surrounding water, swim to the egg.

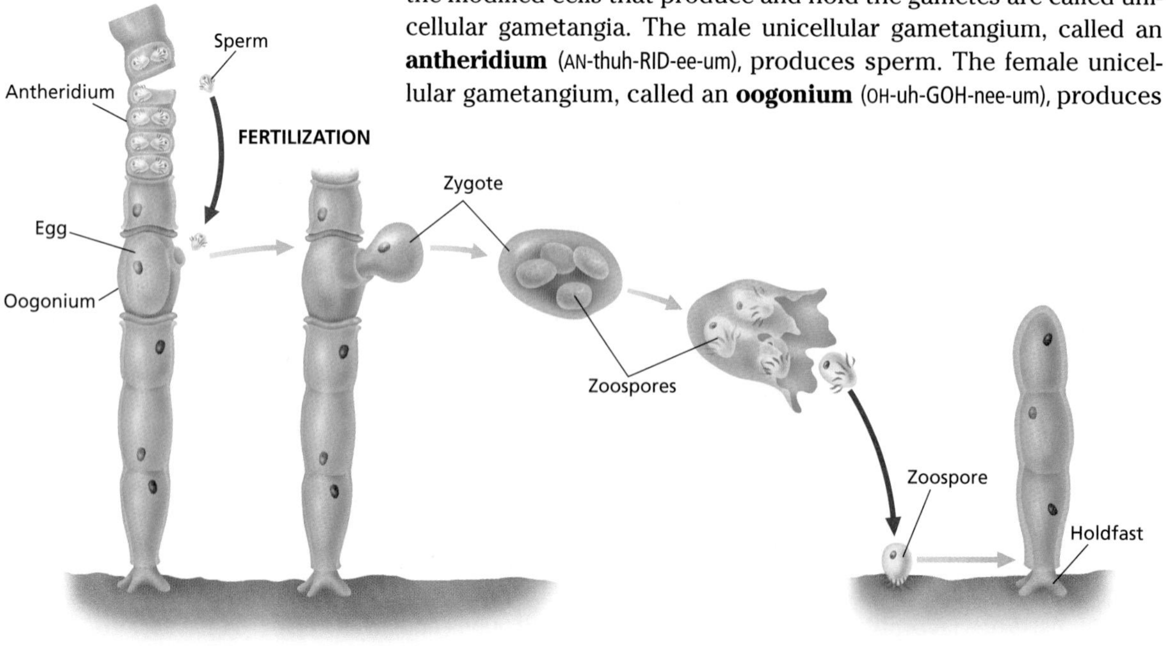

an egg. The antheridium releases flagellated sperm into the surrounding water, where they swim to an oogonium and enter through small pores. After fertilization, the resulting zygote is released from the oogonium and forms a thick-walled, resting spore. The diploid spore then undergoes meiosis, forming four haploid zoospores that are released into the water. Each zoospore settles and divides. One of the new cells becomes a rootlike **holdfast,** and the other divides and forms a new filament.

The filamentous green alga *Spirogyra* reproduces sexually by a process called conjugation. During conjugation, two filaments align side by side. The walls of adjacent cells then dissolve and a conjugation tube forms between the cells. One cell contains a plus gamete. Its contents move through the conjugation tube, enter the adjacent cell, and fuse with the minus gamete. After fertilization, the resulting zygote develops a thick wall, falls from the parent filament, and becomes a resting spore. It later produces a new *Spirogyra* filament.

The leaflike alga *Ulva* has a sexual reproductive cycle that is characterized by a pattern called **alternation of generations.** As illustrated in Figure 27-4, a life cycle that exhibits alternation of generations has two distinct multicellular phases—a haploid, gamete-producing phase called a **gametophyte** (guh-MEE-tuh-FIET) and a diploid, spore-producing phase called a **sporophyte** (SPOHR-uh-FIET). The adult sporophyte has reproductive cells called **sporangia** (spoh-RAN-jee-uh), which produce haploid zoospores by meiosis. The zoospores divide mitotically and form motile spores, which settle on rocks and grow into multicellular, haploid gametophytes. Note that the gametophyte looks exactly like the sporophyte. The gametophyte produces gametangia and then produces plus and minus gametes that unite and form zygotes. As shown in Figure 27-4, the diploid zygote completes the cycle by dividing mitotically into a new diploid sporophyte.

The phenomenon of alternation of generations in green algae is important because it also occurs in more-complex land plants. However, in plants the gametophyte and sporophyte generations do not resemble each other as they do in *Ulva*. Also, the male and female gametes (sperm and egg) in plants are developed in multicellular reproductive structures, rather than in the unicellular gametangia seen in *Ulva*.

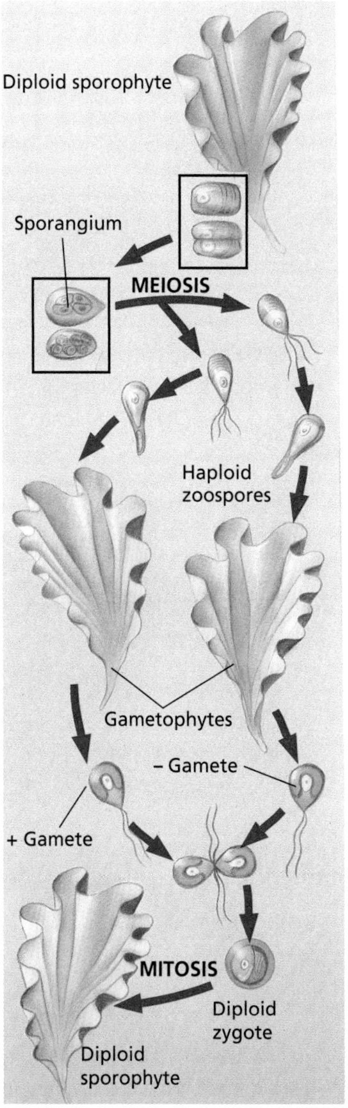

FIGURE 27-4

The multicellular green alga *Ulva* has a life cycle characterized by alternation of generations. Its haploid and diploid forms are equal in appearance and duration.

SECTION 27-1 REVIEW

1. In what ways do algae differ from protozoa?
2. How are algae similar to plants? How are they different?
3. How is body structure used to group algae?
4. What characteristics are used to classify algae into seven phyla?
5. Describe the general process of reproduction in the unicellular alga *Chlamydomonas.*
6. CRITICAL THINKING What adaptive advantage do you think different types of chlorophylls confer on algae?

SECTION 27-1

VISUAL LINK

Figure 27-4

Ask students to identify the haploid structures in *Ulva* (zoospores, the gametophyte, and the "plus" and "minus" gametes). Then ask them which structures are diploid (the zygote, the sporophyte, and the sporangium).

ANSWERS TO SECTION 27-1 REVIEW

1. Algae are autotrophic protists, and most have cell walls.
2. Like plants, algae have chloroplasts and make their own carbohydrates through photosynthesis. Unlike plants, algae lack tissue differentiation and have unicellular gametangia.
3. Algae can be grouped according to their thallus formats–unicellular, colonial, filamentous, or multicellular.
4. Algae are classified into seven phyla based on their color, type of chlorophyll, form of food-storage substance, and cell-wall composition.
5. During asexual reproduction, a haploid cell divides by mitosis to produce genetically identical zoospores. During sexual reproduction, two gametes of opposite mating types fuse to form a diploid zygote, which undergoes meiosis and gives rise to haploid cells.
6. Each chlorophyll type absorbs light at a different wavelength, and different wavelengths of light penetrate varying depths of water. So different types of chlorophylls enable algae to live at different depths in the ocean.

SECTION 27-2

SECTION OVERVIEW

Structure and Function
Section 27-2 discusses the structural characteristics of the different phyla of algae and how these characteristics contribute to their function.

Evolution
Section 27-2 also discusses the evidence that green algae gave rise to land plants.

▶ ENGAGE STUDENTS

Set up microscope stations that show slides of members of each of the seven algal phyla. Have students draw and label each organism they observe. Examples may include Chlorophyta—*Chlamydomonas;* Phaeophyta—*Fucus;* Rhodophyta—*Porphyra;* Bacillariophyta—*Synedra;* Dinoflagellata—*Peridinium;* Chrysophyta—*Synura;* and Euglenophyta—*Euglena.*

TEACHING STRATEGY

Evolutionary Evidence
Write "Phylum Chlorophyta" and "Land plants" on the board. Ask students to explain the evidence that suggests green algae gave rise to land plants. (Students may explain that both green algae and land plants have chloroplasts that contain chlorophylls *a* and *b,* have many carotenoids, store their food as starch, and have cell walls made of cellulose.)

DEMONSTRATION

Algae Products
Set up a display of the following products: salad dressing, seasoned salt, agar, sausage casing, toothpaste, and ice cream. Ask students to explain what the samples have in common. (They all contain carageenan, an ingredient derived from algae.)

SECTION 27-2

OBJECTIVES

▲ Explain why the phylum Chlorophyta is considered the most diverse phylum of algae.

● Describe the characteristics of the members of the phyla Phaeophyta and Rhodophyta.

■ Describe the essential characteristics of diatoms, and explain their industrial importance.

▲ List the important structural characteristics associated with dinoflagellates.

● Discuss why *Euglena* is considered both a protozoan and an alga.

ALGAL DIVERSITY

The seven phyla of algae are a showcase of diversity. Microscopic forms, such as diatoms, dinoflagellates, and euglenoids, differ enough to be placed in separate phyla. Pigmentation distinguishes the green algae, red algae, brown algae, and golden algae, which are also placed in separate phyla.

PHYLUM CHLOROPHYTA

The phylum Chlorophyta (KLOH-rah-FIED-uh) contains more than 7,000 identified species of organisms called green algae. Members of this phylum have an amazing number of forms and reproductive methods. Their body structures range from single cells and colonial forms to multicellular filaments and sheets. Most species, such as the *Caulerpa racemosa* shown in Figure 27-5a, are aquatic. However, some species, such as the *Protococcus* shown in Figure 27-5b, inhabit moist terrestrial environments, such as the soil, rock surfaces, and tree trunks. Some species live as symbiotic partners in the cells of invertebrates or as part of organisms called lichens.

Biologists reason that green algae gave rise to land plants. Evidence supporting this idea includes the fact that both groups of organisms have chloroplasts that contain chlorophylls *a* and *b*. Both also have many carotenoids and store their food as starch. In addition, both green algae and plants have cell walls made of cellulose.

(a)

(b)

FIGURE 27-5
The bunch-of-grapes (a) is an example of a marine green algae. These Chlorophyta are found worldwide. Other Chlorophyta, such as the green *Protococcus* growing on this tree (b), are found in moist environments on land.

PHYLUM PHAEOPHYTA

The phylum Phaeophyta contains approximately 1,500 species of organisms called brown algae. Brown algae are mostly marine, and they include the plantlike organisms called seaweeds and kelps. They are most common along rocky coasts where ocean water is cool. A few species, such as *Sargassum,* are found far offshore, where they form dense floating mats.

The brown algae contain chlorophylls *a* and *c* and a large amount of pigment called **fucoxanthin** (FYOO-koh-ZAN-thin), which gives them their characteristic brown color. The food they produce is stored as **laminarin,** a carbohydrate with glucose units that are linked differently than those in starch.

All brown algae are multicellular, and most are large, often reaching lengths of more than 45 m (147 ft). Some of the largest algae known are classified in the phylum Phaeophyta. The large brown alga shown in Figure 27-6 is *Macrocystis,* a genus that thrives in intertidal zones. Individual alga can grow to a length of 100 m (328 ft). The thallus is anchored to the ocean bottom by a rootlike holdfast. The stemlike portion of the alga is called the **stipe.** And the leaflike region, modified to capture sunlight for photosynthesis, is called the **blade.** The cell walls of *Macrocystis* contain alginic acid, a source of a commercially important substance called **alginate.** Alginate is used in cosmetics and various drugs, as food, and as a stabilizer in most ice creams.

FIGURE 27-6

Macrocystis is an example of brown algae. Also known as giant kelp, this species is a member of the phylum Phaeophyta.

Word Roots and Origins

stipe

from the Latin *stipes,* meaning "log" or "trunk of a tree"

PHYLUM RHODOPHYTA

The phylum Rhodophyta contains 4,000 species of organisms called red algae. A few species of red algae live in fresh water or on land, but most are marine seaweeds. Red algae are usually smaller than brown algae and are found as deep as 200 m (657 ft). The red alga *Corallina* is shown in Figure 27-7.

Red algae contain chlorophyll *a* and pigments called **phycobilins.** Phycobilins play an important role in absorbing light for photosynthesis. These pigments can absorb the wavelengths of light that penetrate deep into the water. Thus, they may make it possible for red algae to live at depths where algae lacking these pigments cannot survive. Despite their common name, not all red algae are red. The depth at which they live in the ocean determines the amount of pigment they have. Because Rhodophyta live at varying depths, their color also varies.

Certain species of red algae have cell walls that are coated with a sticky substance called **carageenan** (KAR-uh-GEEN-uhn). Carageenan is a polysaccharide used in the production of cosmetics, gelatin capsules, and some types of cheese. **Agar,** which is used as a gel-forming base for culturing microbes, is also extracted from the cell walls of red algae.

FIGURE 27-7

Although not as large as the Phaeophyta, the Rhodophyta, such as this *Corallina,* are often referred to as seaweeds.

OVERCOMING MISCONCEPTIONS

A common misconception is that seaweeds are aquatic plants. Seaweeds, such as the kelp discussed in this section, lack organized tissue and are therefore classified as algae.

CULTURAL CONNECTION

Seaweed as Food

In many cultures, seaweed and other algae are important dietary components. For example, in Japan, a red alga called *Porphyra* is dried and wrapped around portions of rice to make sushi.

GIFTED *ACTIVITY*

Have students research the topic of algal pigments and find information about each pigment's absorption spectrum (see Section 6-1). Instruct students to visually display the results of their research, making sure to include phylum names.

RECENT RESEARCH

Radioactive Algae

By measuring the amount of radioactive uranium in fossil algae, scientists at Texas A&M University have determined the amount of time that elapsed between ice ages. As algae grow and incorporate calcium into their cell walls, they also incorporate uranium from the sea water. The uranium slowly decays.

VISUAL LINK

Figure 27-7

Have students examine *Corallina,* the red alga in Figure 27-7. Ask students to explain the advantage of the alga's filamentous shape. (The filaments increase the thallus's surface area, providing more area for light absorption.)

RECENT RESEARCH

Diatoms and Water Quality

Researchers at the Philadelphia Academy of Natural Sciences are developing ways to use diatoms as environmental indicators. By studying which diatom species are present in various water samples, scientists can develop guidelines for determining water pH, salinity, and nutrient pollution. For example, the diatom *Eunotia exigua* prefers acidic conditions. Therefore, if this organism is present, the water must be acidic, which could have resulted from acid rain or another form of pollution. By using diatoms as water-quality indicators, officials can more effectively monitor environmental conditions.

CRITICAL THINKING

Algal Blooms

Tell students that during an algal bloom, algae multiply rapidly and then die off, depleting the water's oxygen supply. This event can be caused by the presence of large amounts of nutrients from treated sewage or from runoff from fertilized soil. Have students determine how algal blooms affect fish and other aquatic life-forms. (Students may reason that without adequate supplies of oxygen, fish and other aquatic organisms suffocate and die.)

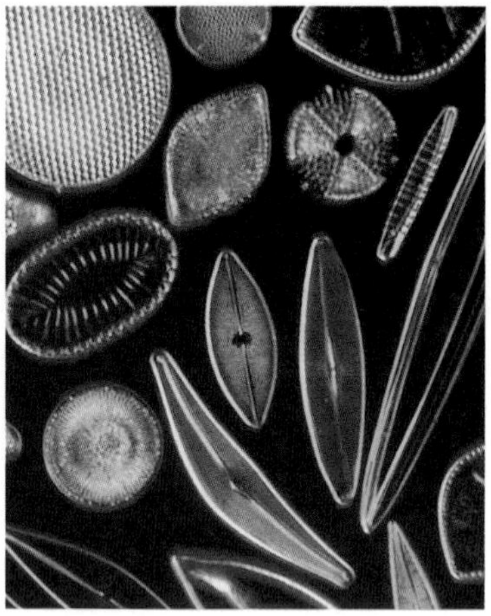

FIGURE 27-8

Diatoms, such as these, usually reproduce asexually. Sexual reproduction among diatoms is rare. (320×)

PHYLUM BACILLARIOPHYTA

The phylum Bacillariophyta contains 11,500 species of organisms called diatoms. **Diatoms** are abundant in both freshwater and marine environments. Their cell walls, commonly called shells, consist of two pieces that fit together like a box with a lid. Each half is called a **valve.** The shells contain silicon dioxide. Figure 27-8 shows two types of diatoms. **Centric diatoms** have circular or triangular shells and are most abundant in marine environments. **Pennate diatoms** have rectangular shells and are most abundant in freshwater ponds and lakes. Some pennate diatoms move by secreting threads that attach to the surface of the water. When these threads contract, they pull the diatom forward.

Diatoms are an abundant component of phytoplankton and are important producers in freshwater and marine food webs. They are an essential source of nutrients for microscopic heterotrophs. In addition, they release an abundance of oxygen.

When diatoms die, their shells sink and accumulate in large numbers, forming a layer of material called **diatomaceous** (DIE-uh-tuh-MAY-shuhs) **earth.** Diatomaceous earth is slightly abrasive and is a major component of many commercial products, such as detergents, paint removers, fertilizers, insulators, and some types of toothpaste.

PHYLUM DINOFLAGELLATA

The phylum Dinoflagellata contains 1,100 species of organisms called dinoflagellates. Dinoflagellates are small, usually unicellular organisms. Most are photosynthetic, but a few species are colorless and heterotrophic. Along with diatoms, dinoflagellates are one of the major producers of organic matter in marine environments.

Photosynthetic dinoflagellates usually have a yellowish green to brown color due to a large amount of pigments called carotenoids as well as other pigments and chlorophylls *a* and *c*. Most dinoflagellates have two flagella of unequal length, as shown in Figure 27-9. Each flagellum fits into one of two grooves that run perpendicular to each other. Movement of the flagella causes the dinoflagellate to spin like a top through the water. Dinoflagellates' cell walls are made of cellulose plates that look like armor when seen under a microscope.

Some species of dinoflagellates, such as *Noctiluca,* can produce **bioluminescence,** a display of sparkling light often seen in ocean water at night. Other species produce toxins and red pigments. When their populations explode, they turn the water brownish red, resulting in a phenomenon known as **red tide.** Red tides are fairly common in the Gulf of Mexico off the coast of Florida. When shellfish, including oysters, feed on the dinoflagellates, they also consume the toxins, which are dangerous to humans who eat the shellfish.

FIGURE 27-9

Dinoflagellates, such as these, often harbor endosymbiotic cyanobacteria. Dinoflagellates can also be endosymbionts in sponges, jellyfish, corals, and some types of protozoa. (450×)

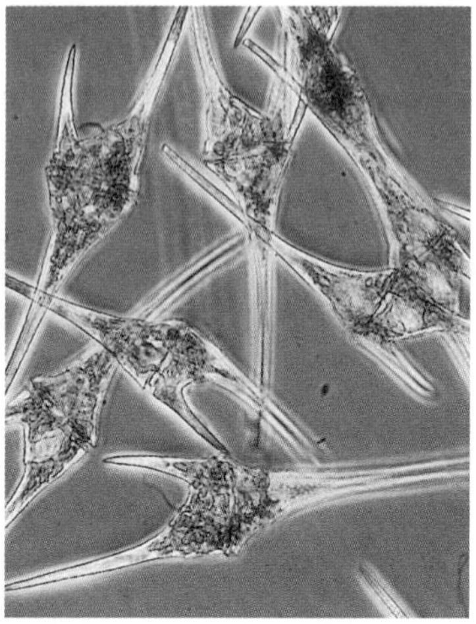

PHYLUM CHRYSOPHYTA

The phylum Chrysophyta contains about 850 species of organisms called golden algae. Most of the golden algae live in fresh water, but a few are found in marine environments. The cells form highly resistant cysts that enable them to survive beneath frozen surfaces of lakes in winter and dry lake beds during summer. Two flagella of unequal length are located at one end of each cell.

Most of the species placed in this phylum are some shade of yellow or brown due to the presence of large amounts of carotenoids. They also have chlorophylls *a* and *c*. Golden algae store much of their surplus energy as oil and are important in the formation of petroleum deposits.

PHYLUM EUGLENOPHYTA

The phylum Euglenophyta contains approximately 1,000 species of flagellated unicellular algae called **euglenoids.** Euglenoids show both plantlike and animal-like characteristics. They are plantlike in that many have chlorophyll and are photosynthetic; they are animal-like in that they lack a cell wall and are highly motile. Euglenoids contain chlorophylls *a* and *b* and the pigments known as carotenoids. Most species live in fresh water, but a few are found in moist environments, such as in soil or in the digestive tracts of certain animals.

A familiar genus of euglenoids is *Euglena,* shown in Figure 27-10. *Euglena* is abundant in fresh water, especially water polluted by excess nutrients. This protist has an elastic, transparent pellicle made of protein just beneath its cell membrane. It also has a contractile vacuole to rid the cell of excess water. Because *Euglena* lacks a cell wall, it is fairly flexible and can change its shape as it swims about. Although usually photosynthetic, if *Euglena* is raised in a dark environment, it will not form chloroplasts and will become heterotrophic.

internetconnect

SCILINKS NSTA
TOPIC: Bioluminescence
GO TO: www.scilinks.org
KEYWORD: HM533

FIGURE 27-10

Euglena gracilis, shown below, is a familiar type of euglenoid. It is propelled by a long flagellum. An eyespot guides it toward light.

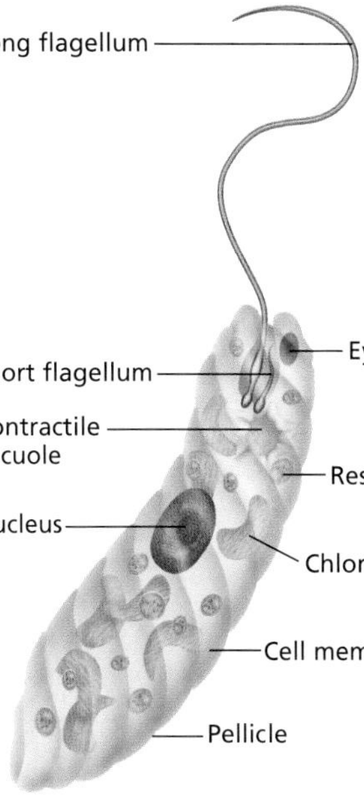

SECTION 27-2 REVIEW

1. Why are green algae considered a diverse group of algae?
2. Which phylum contains the largest multicellular forms of algae? In what way are these algae commercially important?
3. What characteristics enable red algae to exist in deep marine environments?
4. What is a diatom? What useful commercial products do the shells of these algae yield?
5. Which phylum is composed of unicellular organisms that usually have two flagella and a cell wall made of cellulose plates?
6. **CRITICAL THINKING** If a *Euglena* that is raised in the dark loses its chloroplasts, will it grow new ones if it is returned to the light?

SECTION 27-2

RETEACHING *ACTIVITY*

Divide the class into groups, and have each group make a poster displaying the characteristics of the seven phyla of algae, important facts about each phylum, and a photograph or illustration representative of each phylum.

ANSWERS TO SECTION 27-2 REVIEW

1. The phylum Chlorophyta contains algae that have an amazing number of forms and reproductive methods and live in a variety of environments.
2. The largest multicellular algae are brown algae in the phylum Phaeophyta. The brown alga *Macrocystis* has cell walls that contain alginic acid, a source of alginate. Alginate is used in cosmetics, foods, and drugs.
3. In addition to chlorophyll *a,* red algae contain pigments called phycobilins, which can absorb the light that penetrates deep water.
4. Diatoms are single-celled algae whose shells contain silicon dioxide. These shells are the main ingredient of diatomaceous earth, which is used in detergents, paint removers, fertilizers, insulators, and some types of toothpaste.
5. phylum Dinoflagellata
6. Answers may vary. Students may suggest that the euglena will not make new chloroplasts. Like mitochondria, chloroplasts divide autonomously within the cell, and the cell's nuclear genome cannot code for a new chloroplast.

SECTION 27-3

SECTION OVERVIEW

Structure and Function
Section 27-3 describes the characteristics of slime molds and water molds. The unique life cycles of these funguslike protists exemplify kingdom Protista's diverse nature.

Reproduction and Inheritance
Section 27-3 also describes the reproduction of slime molds. Both groups produce a spore-bearing structure called a fruiting body.

▶ ENGAGE STUDENTS

Have students study Figures 27-11 and 27-12. Ask them to list the characteristics of slime molds and speculate about why these organisms are called funguslike protists. (Students should note that these organisms have an amoeba-like appearance and stalks called fruiting bodies. The amoeba-like appearance groups them with protists, while their fruiting body resembles a fungal reproductive structure.)

DEMONSTRATION

Amoeboid Movement
Obtain a *Physarum* culture from a biological supply house. After the culture is established, use a microprojector to show cytoplasmic streaming in the *Physarum* plasmodium. Ask students to name another organism that displays cytoplasmic streaming (amoeba).

VISUAL LINK

Figure 27-12
Have students explain what needs to occur for the life cycle to be complete. (The fruiting bodies must break open, the wind must disperse the spores, and the spores must develop into individual amoeboid cells.)

SECTION 27-3

OBJECTIVES

▲ Describe the two forms that characterize the life cycle of the slime mold.

● Describe the environment in which slime molds live.

■ Outline the basic life cycles of the two groups of slime molds.

◆ Point out the unique characteristics of water molds.

FUNGUSLIKE PROTISTS

The kingdom Protista contains a number of funguslike protists in addition to the algae. Among these are the slime molds and water molds. These organisms have unique life cycles that set them apart from the protozoa, algae, and fungi. However, like all of these groups, they are eukaryotic. They are multicellular or large multinucleate heterotrophic organisms with very little tissue specialization.

SLIME MOLDS

Slime molds are a curious group of organisms. They spend part of their lives in a mobile, amoeba-like feeding form, engulfing organic matter and bacteria much as protozoa do. However, they also produce funguslike reproductive structures, which is why they were once classified as fungi. Slime molds are typically found growing on damp soil, rotting logs, decaying leaves, or other decomposing organic matter in moist areas. These organisms appear as glistening, viscous masses of slime: some are white, but most are yellow or red.

Biologists recognize two groups of slime molds—the cellular slime molds (phylum Acrasiomycota) and the plasmodial slime molds (phylum Myxomycota). These groups are not closely related, but they share certain characteristics. For example, both types have life cycles with two phases: a mobile feeding stage and a stationary reproductive stage. During reproduction, slime molds produce a spore-bearing structure called a **fruiting body.**

Phylum Acrasiomycota

The phylum Acrasiomycota (ah-KRAYZH-ee-oh-mie-KOH-tuh) comprises about 65 species of cellular slime molds. **Cellular slime molds** live as individual haploid cells that move about like amoebas. Each cell moves as an independent organism, creeping over rotting logs and soil or swimming in fresh water, ingesting bacteria and other food.

When food or water becomes scarce, the cells release a chemical that attracts nearby cells and causes them to gather by the hundreds or thousands to form a dense structure called a pseudoplasmodium, as shown in Figure 27-11. A **pseudoplasmodium** is a coordinated colony of individual cells that resembles a slug, and it leaves a slimy trail as it crawls over decaying logs, leaves, and twigs. During this stage, the cells move as one unit, even though each cell retains its cell membrane and identity. Eventually, the

FIGURE 27-11
The well-known cellular slime mold *Dictyostelium discoideum* can be easily grown in the laboratory. Approximately 65 other species of cellular slime molds are known to exist. (80×)

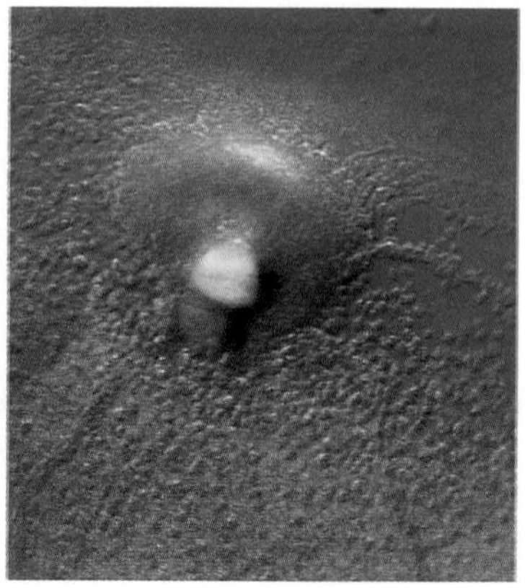

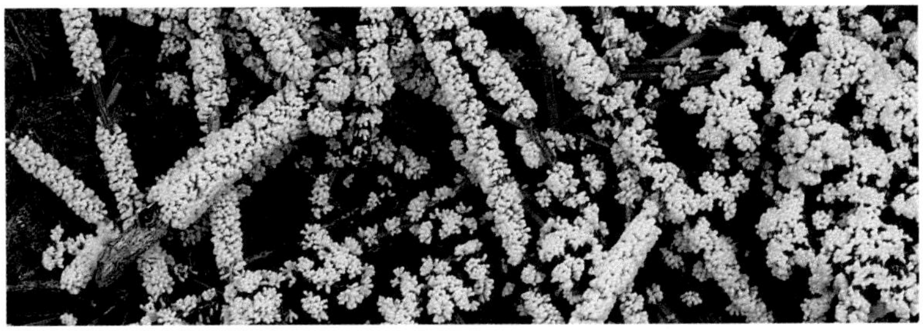

pseudoplasmodium settles and forms a fruiting body in which haploid spores develop. When the fruiting body breaks open, the wind disperses the spores to new locations. Each spore may grow into an individual amoeboid cell, thus completing the life cycle.

Phylum Myxomycota

Approximately 450 species of **plasmodial slime molds** compose the phylum Myxomycota (MIKS-oh-mie-KOH-tuh). During the feeding stage of its life cycle, a plasmodial slime mold is a mass of cytoplasm called a **plasmodium,** and it may be as large as several square meters. Each plasmodium is **multinucleate,** meaning it contains thousands of nuclei. As the plasmodium moves along the forest floor, it engulfs decaying leaves and other debris by the process of phagocytosis.

When food or water is scarce, the plasmodium crawls to an exposed surface and begins to reproduce, as shown in Figure 27-12. It forms stalked fruiting bodies in which haploid spores form by meiosis. The spores are very resistant to adverse conditions. Under favorable conditions, they crack open and give rise to haploid reproductive cells. Two such cells fuse, and their nuclei combine to form a diploid nucleus. Repeated divisions by mitosis follow, but the cells do not undergo cytokinesis, so the result is the multinucleated cytoplasm characteristic of the plasmodium.

FIGURE 27-12

The plasmodial slime mold *Physarum* has a bright yellow plasmodium. This naked mass of cytoplasm feeds on bacteria and other microorganisms. The reproductive structures are stalked fruiting bodies.

Quick Lab

Observing Slime Mold

Materials slime mold culture, oatmeal, vinegar, cotton swab, dissecting microscope

Procedure

1. Observe the movement of the slime mold. Identify the plasmodium, the nuclei, and the food vacuoles. Draw and label the structures you have observed.
2. Drop a small bit of oatmeal near an outside edge of the slime mold.
3. Dip the swab in vinegar and touch it to the slime mold.

Analysis How does a slime mold move? How does it react to oatmeal? To vinegar?

WATER MOLDS

A **water mold** is a funguslike organism composed of branching filaments of cells. Most water molds are aquatic and are commonly found in bodies of fresh water. However, some live in the soil, and some are parasites. For example, water molds are familiar as the white fuzz on diseased aquarium fish or on organic matter floating on water.

Phylum Oomycota

The phylum Oomycota includes a number of organisms that are pathogenic to plants. For example, the water mold *Phytophthora infestans* causes late blight in potatoes and was responsible for the Irish potato famine in the mid-1800s. **Blight** is a disease of plants characterized by quickly developing decay and discoloring of leaves, stems, and flowers. Late blight is demonstrated in Figure 27-13.

FIGURE 27-13

Potato blight is caused by the water mold *Phytophthora infestans.* Between 1845 and 1849, the late blight of potatoes devastated farms in Ireland, forcing the mass emigration of Irish people to countries such as Australia and the United States.

Quick Lab

Observing Slime Mold

Time Required 20 minutes

Procedural Tips Have students work in small groups. Obtain plasmodial slime mold cultures from a biological supply house. Each group will need a few drops of vinegar and a pinch of oatmeal. Instruct students to wash their hands after working with the slime mold.

Disposal Culture dishes containing slime molds should be placed in a 5 percent solution of household bleach for 24 hours. They can then be placed in a solid-waste container.

Answers to Analysis The slime mold's cytoplasm streams outward in one direction and the rest of the organism follows. The slime mold should engulf the oatmeal by phagocytosis, but the mold will move away from vinegar.

GIFTED ACTIVITY

Have students find out more about several types of water molds, such as those that cause diseases in fish and blight in plants. Have them display their results visually, using a cause-and-effect format.

CRITICAL THINKING

Rapid Reproduction in Water Molds

Water molds are able to reproduce rapidly. Ask students why it is difficult to eliminate water molds, such as the oomycete that caused the Irish potato famine. (The damp weather in Ireland created perfect growing conditions for *Phytophthora infestans,* making it very difficult to control its life cycle.)

QUICK FACT

☑ *Plasmopara viticola,* a downy mildew, is native to North America and was accidentally introduced in Europe in the late 1870s. French winemakers, in an attempt to control aphid infestation, introduced aphid-resistant vine strains from the United States to graft to their vines. The introduced strain was infected with downy mildew and almost wiped out the entire French winemaking industry.

CRITICAL THINKING

Regulatory Laws for Vegetation Transport

Many countries and states have strict laws on the vegetation that can be brought into countries or across state lines. Ask students to discuss why such laws exist. (Students might conclude that the laws are an attempt to prevent the spread of plant diseases, such as potato blight and downy mildew.)

ANSWERS TO SECTION 27-3 REVIEW

1. Slime molds have a mobile, amoeba-like feeding form and a reproductive spore-bearing form.
2. phylum Acrasiomycota and phylum Myxomycota
3. A pseudoplasmodium is a sluglike colony of cells that is characteristic of cellular slime molds (phylum Acrasiomycota).
4. Plasmodial slime molds form fruiting bodies when food or water is scarce.
5. Water molds are composed of branching filaments of cells. They are found in water and soil.
6. They produce spores, which can be carried to locations with more-abundant resources.

FIGURE 27-14

Downy mildew on grapes is caused by *Plasmopara viticola,* which grows on the leaves and fruit of grapes. This protist can cause significant economic harm if it is not carefully controlled.

FIGURE 27-15

Chytrids, such as this *Allomyces,* are considered by many scientists to be more like fungi than like protists, but the presence of motile cells has caused them to be classified as protists. Although most are aquatic, some live in moist soil. (70×)

Another example of a plant pathogenic oomycete is *Plasmopara viticola.* This organism infects grape plants, attacking the leaves and the fruit. *Plasmopara* infection causes the grape leaves to wilt and die. Young grape plants are also attacked and may also wilt and die from *Plasmopara* infection. *Plasmopara* may also infect vegetables and other fruits. Figure 27-14 shows a grape plant infected with *Plasmopara viticola.*

Water molds reproduce asexually and sexually. During asexual reproduction, they produce motile, flagellated reproductive zoospores. Zoospores germinate into threadlike cells, which accumulate to form a matlike mass. Some zoospores form a reproductive structure called a sporangium, in which new zoospores are produced.

During sexual reproduction, the cells of the water mold develop egg-containing and sperm-containing structures. Fertilization tubes grow between the two types of structures, enabling haploid sperm cells to fertilize haploid egg cells and form diploid zygotes. A zygote develops into a new mass of filaments, from which asexual sporangia as well as sexual oogonia and antheridia form.

Phylum Chytridiomycota

Biologists place approximately 750 protist species in the phylum Chytridiomycota (kie-TRID-ee-oh-mie-KOH-tuh). The **chytrids** (KIE-tridz) are primarily aquatic protists characterized by gametes and zoospores with a single, posterior flagellum. Most chytrids are unicellular. Some chytrids have long filamentous bodies that anchor the organism. Many chytrids are parasites on algae, plants, and insects, while others are saprophytes.

Until recently, biologists classified this phylum with the protists, but many biologists now think that chytrids should be classified as fungi. Chytrids and fungi have similar characteristics; they have similar methods for absorbing nutrients through the cell wall, cell walls that are made of the same type of material, and long filamentous bodies. Fungi also share similar types of enzymes and biochemical pathways with chytrids. Because of all of these similarities, biologists think that chytrids are a link between protists and fungi. An example of a chytrid is shown in Figure 27-15.

SECTION 27-3 REVIEW

1. **What two phases are found in the life cycle of slime molds?**
2. **In which phyla are the slime molds classified?**
3. **What are the distinctive features of the pseudoplasmodium, and which organisms form this structure?**
4. **When do the plasmodial slime molds form their fruiting body?**
5. **What are the characteristic features of water molds?**
6. **CRITICAL THINKING What evolutionary advantage do pseudoplasmodia gain by producing fruiting bodies?**

CHAPTER 27 REVIEW

SUMMARY/VOCABULARY

27-1

- The kingdom Protista includes algae, which are mostly aquatic organisms that contain chlorophyll. Algae include microscopic single cells and giant marine kelps.
- Algae produce large amounts of organic matter, which serves as nutrients for other organisms. Algae also add an enormous amount of oxygen to the atmosphere.
- The body portion of an alga is the thallus. It may consist of a single cell, a colony of cells, a filament, or a complex multicellular arrangement.
- Algae can be classified into seven phyla, based on color, type of chlorophyll, and form of food-storage substances.
- Algae reproduce asexually by mitosis and sexually by more-complex methods.
- In the leaflike alga *Ulva,* an alternation of generations takes place, with a gametophyte and sporophyte phase.

Vocabulary

algae (525)
alternation of generations (529)
antheridium (528)
colonial alga (526)
filamentous alga (526)
gametangium (525)
gametophyte (529)
holdfast (529)
multicellular alga (526)
oogonium (528)
phytoplankton (525)
pyrenoid (525)
sporangium (529)
sporophyte (529)
thallus (525)
unicellular alga (525)
zoospore (527)
zygospore (527)

27-2

- The phylum Chlorophyta consists of green algae. Green algae contain chlorophylls *a* and *b,* carotenoids, and cell walls made of cellulose.
- The phylum Phaeophyta is made up of brown algae. Brown algae contain chlorophylls *a* and *c* and carotenoids including fucoxanthin.
- The phylum Rhodophyta consists of red algae. Red algae contain chlorophyll *a* and phycobilins.
- The phylum Bacillariophyta is made up of diatoms. These unicellular algae have shells that contain silicon dioxide.
- The phylum Dinoflagellata includes the dinoflagellates. Dinoflagellates contain chlorophylls *a* and *c* and carotenoids. Most species have two flagella.
- The phylum Chrysophyta consists of the golden algae. Golden algae contain chlorophylls *a* and *c* and large amounts of carotenoids.
- The phylum Euglenophyta is made up of euglenoids, such as *Euglena.* Euglenoids show both plantlike and animal-like characteristics.

Vocabulary

agar (531)
alginate (531)
bioluminescence (532)
blade (531)
carageenan (531)
centric diatom (532)
diatom (532)
diatomaceous earth (532)
euglenoid (533)
fucoxanthin (531)
laminarin (531)
pennate diatom (532)
phycobilin (531)
red tide (532)
stipe (531)
valve (532)

27-3

- Slime molds are eukaryotes. Their life cycles include a creeping amoeba-like form and a reproductive spore-bearing form.
- Slime molds are classified into phylum Acrasiomycota and phylum Myxomycota.
- The phylum Oomycota contains funguslike organisms called water molds. Water molds are composed of branching filaments.

Vocabulary

blight (535)
cellular slime mold (534)
chytrid (536)
fruiting body (534)
multinucleate (535)
plasmodial slime mold (535)
plasmodium (535)
pseudoplasmodium (534)
water mold (535)

CHAPTER 27 REVIEW ANSWERS

REVIEW

1. fruiting body because gametophytes and sporophytes are found in algae, whereas fruiting bodies form in funguslike protists
2. phytoplankton because they are organisms, whereas chlorophyll *a* and carotenoids are pigments
3. plasmodium because plasmodium is a stage in the slime mold life cycle, whereas a blade and a holdfast are parts of an algal thallus
4. kelp because kelp are algae, whereas water molds and slime molds are funguslike protists
5. diatomaceous earth because diatomaceous earth is formed from diatoms, whereas red tide and bioluminescence are associated with dinoflagellates

6. c
7. a
8. d
9. a
10. d
11. b
12. c
13. d
14. c
15. b

16. Like protozoa, algae are eukaryotic and mostly unicellular. Unlike protozoa, algae can manufacture their own carbohydrates through photosynthesis. Like plants, algae have chloroplasts and are autotrophic. Unlike plants, algae do not have tissue differentiation, so they have no true stems, roots, or leaves and form gametes in single-celled gametangia.
17. Algae are placed into seven phyla based on their color, type of chlorophyll, form of food-storage substance, and cell-wall composition. All algae contain chlorophyll *a.*
18. The types of food-storage substance include starch, laminarin, leucosin, and paramylon. Starch is the most common form of food-storage substance.
19. The types of pigments include chlorophylls *a, b,* and *c;* carotenoids; fucoxanthin; phycobilins; and xanthophyll. Chlorophyll *a* is the most common pigment.
20. A thallus is the body portion of an alga. Thallus formats can be unicellular, colonial, filamentous, or multicellular.
21. See Figures 27-2, 27-3, and 27-4 for details. Sexual reproduction occurs in unicellular and multicellular algae.
22. Euglenoids are plantlike in that many have chlorophyll and are photosynthetic; they are animal-like in that they lack a cell wall and are highly motile. When *Euglena* is raised in a dark environment, it will not form chloroplasts and will become heterotrophic.
23. A fruiting body is a spore-bearing structure that forms during slime mold reproduction.
24. Multinucleate describes the plasmodium of plasmodium slime molds, which contain thousands of nuclei.
25. *A*—flagellum, *B*—eyespot, *C*—nucleus, *D*—chloroplast; this is a euglenoid, or *Euglena gracilis*

REVIEW

Vocabulary

In the following groups of terms, choose the term that does not belong, and explain why it does not belong.

1. gametophyte, sporophyte, fruiting body
2. chlorophyll *a,* carotenoids, phytoplankton
3. plasmodium, blade, holdfast
4. water mold, cellular slime mold, kelp
5. diatomaceous earth, red tide, bioluminescence

Multiple Choice

6. The green alga *Chlamydomonas* reproduces asexually by forming (a) a zygote (b) a gametophyte (c) zoospores (d) an opposite mating type.
7. The green alga *Ulva* forms a sporophyte that has structures called (a) sporangia (b) antheridia (c) gametangia (d) carageenan.
8. The green alga *Oedogonium* produces a unicellular female gametangium called (a) an antheridium (b) a zygospore (c) a conjugation tube (d) an oogonium.
9. Red algae contain an accessory pigment called (a) phycobilin (b) fucoxanthin (c) carotene (d) xanthophyll.
10. The phenomenon know as red tide is caused by a population explosion of (a) diatoms (b) red algae (c) water molds (d) dinoflagellates.
11. Many types of algae have flagellated reproductive cells called (a) sporangia (b) zoospores (c) zygospores (d) sporophytes.
12. A pyrenoid is an organelle that (a) lends golden algae their yellow color (b) anchors seaweeds to the ocean floor (c) makes and stores starch (d) enables some types of algae to produce light.
13. Biologists believe that the most likely ancestors of land plants are members of the phylum (a) Chrysophyta (b) Euglenophyta (c) Phaeophyta (d) Chlorophyta.
14. The feeding stage of a plasmodial slime mold is called (a) an amoeba (b) a fruiting body (c) a plasmodium (d) a zygospore.
15. In addition to protozoa and algae, the kingdom Protista contains organisms called (a) euglenoids (b) slime molds (c) dinoflagellates (d) diatoms.

Short Answer

16. How are algae similar to protozoa? How are they different? What characteristics do algae share with plants? What characteristics differ?
17. List the characteristics used to classify algae into seven phyla.
18. Compare the types of food-storage molecules found in the seven types of algae. What is the most common food-storage molecule used by algae?
19. List the types of pigments found in the seven phyla of algae. What is the most common photosynthetic pigment found in algae?
20. What is a thallus? What kinds of thallus formats are found in algae?
21. Summarize three types of sexual reproduction that occur in algae. Does sexual reproduction in algal protists occur in only multicellular algae?
22. Why are euglenoids described as both plant-like and animal-like organisms? Explain how euglenoids can be both heterotrophic and autotrophic.
23. What is a fruiting body? At what point of their life cycle do slime molds form fruiting bodies?
24. Define the term *multinucleate.* What kind of protists are characterized by this form of cell structure?
25. Identify the structures in the figure below. What is the name of this organism?

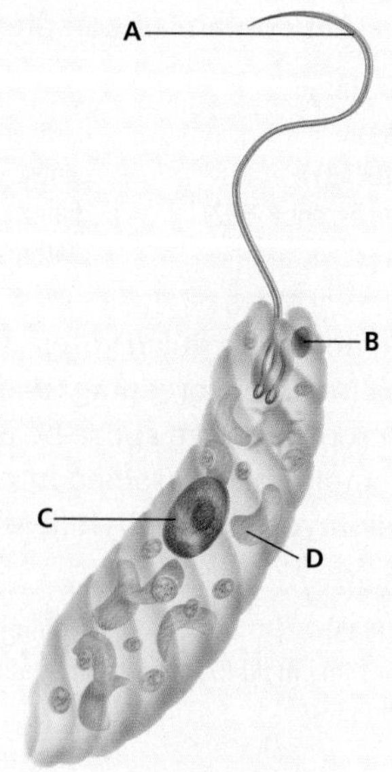

CRITICAL THINKING

1. Several years ago, many botanists classified algae as plants. They also classified flagellated *Euglena* as photosynthetic protozoa. Algae are now classified as protists, and *Euglena* is now considered an alga. What do these changes tell you about the classification systems of living things?
2. In the Pacific Northwest of the United States, people believe it is unsafe to collect and eat shellfish during the months that lack an *r* in their name. Name these months, and give a possible scientific explanation for the belief. Keep in mind that shellfish obtain their food by filtering water.
3. The horsetail plant, *Equisetum,* has silicon dioxide in the outer cells of its stem. Diatoms also contain silicon dioxide in their shells. Explain whether the presence of this compound in both organisms indicates a close relationship. (Hint: Consider the types of evolution described in Chapter 15.)
4. Scientists often use a pigment called phycoerythrin to label parts of cells so that they can be seen with a special type of microscope that uses ultraviolet light. Phycoerythrin fluoresces under ultraviolet light. Because ultraviolet light is a higher frequency of light, it can penetrate deeper into water than can visible light. Based on this information and what you read in this chapter, which phylum of algae do you think produces phycoerythrin? Justify your answer.
5. Explain how the absence of a cell wall in *Euglena* makes the function of the contractile vacuole critical.
6. Examine the drawing of lateral conjugation in *Spirogyra* in which adjacent cells of the same filament have conjugated. Explain whether this type of conjugation offers more or less genetic recombination than does sclariform conjugation, in which conjugation is between the cells of two different filaments.

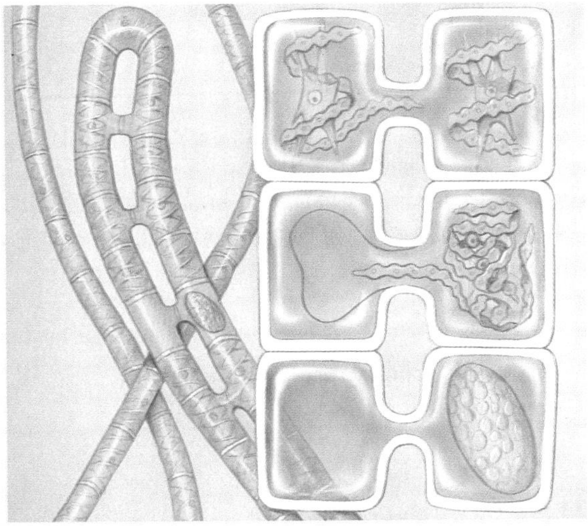

EXTENSION

1. Read "The Lurking Perils of *Pfiesteria*" in *Scientific American,* August 1999, on page 42. Prepare a report that discusses the role *Pfiesteria piscicida* has in fish kills in several major estuaries. Include in your report a description of the toxic and nontoxic forms *Pfiesteria* can assume. Describe the toxin's indirect effects on the survival of fish. Explain how *Pfiesteria* can be dangerous to people.
2. Visit your local grocery store or supermarket, and find at least three items that contain carageenan, a polysaccharide derived from red algae.
3. Gather samples of soil, dead leaves, old grass, and fresh leaves. Place each sample in a jar or test tube filled with clean water, and leave the samples near a light source. Examine the samples under a microscope once a week, and make drawings of the types of organisms you find. What do your observations reveal about the different habitats of algae?

CRITICAL THINKING

1. The classification of living things changes as scientists learn more about different species.
2. May, June, July, and August because during these warm months, the number of *Gonyaulax* increases. Shellfish feed on these algae and accumulate their toxins.
3. This is an example of convergent evolution. The two genera are not closely related.
4. Because phycoerythrin is excited by light that is able to penetrate deep water, it follows that phycoerythrin is produced by a deep water algae. Deep water algae are the red algae.
5. Because *Euglena* lacks a cell wall, it has no mechanical support to help prevent the cell membrane from bursting when excess water enters the cell. The contractile vacuole removes excess water by pumping it out.
6. Less. The cells of one filament have the same genetic makeup. The cells of two different filaments may have come from different zygospores and thus have different genetic compositions.

EXTENSION

1. *Pfiesteria* is a dinoflagellate that produces some of the most potent poisons found in nature and is responsible for killing billions of fish in estuaries in North Carolina and Maryland. According to their available food supply, *Pfiesteria* can assume a nontoxic amoeboid form or a hibernating cyst. *Pfiesteria* can undermine the ability of fish to reproduce or resist disease. *Pfiesteria* can be toxic to humans who breathe the air over fish kills.
2. Possible items are chocolate, jams, jellies, instant coffee, wine, and ice cream.
3. Algae grow in all these samples, indicating that algae grow in a variety of terrestrial habitats.

CHAPTER 27 INVESTIGATION

TIME REQUIRED

One 50-minute class period

SAFETY PRECAUTIONS

Remind students that microscope slides may break and that they should handle the slides with care.

QUICK REFERENCES

Lab Preparation Notes are found on page 523B.

Holt BioSources provides a Teaching Resources CD-ROM that contains "Using Gowin's Vee in the Lab" and "Scoring Rubrics."

PROCEDURAL TIPS

1. Refer students to pictures of *Chlamydomonas, Spirogyra,* and *Volvox* in the textbook so that they can identify their algae.
2. Have students prepare their data collection tables ahead of time.

ANSWERS TO BACKGROUND

1. *Protozoa* refers to organisms in the kingdom Protista that are heterotrophic. *Algae* refers to protists that are autotrophic.
3. Because both groups have chlorophylls *a* and *b,* biologists think green algae gave rise to plants.
4. Green algae occur as unicellular, filamentous, or multicellular forms. They have carotenoids and chlorophylls *a* and *b.* Most species are aquatic, store food as starch, and have cell walls composed of polysaccharides. They differ from plants in that they lack tissue differentiation.
5. Green algae differ primarily in that they have chlorophylls *a* and *b.* The euglenophyta also have these chlorophylls, but they have no cell wall, whereas green algae do.

CHAPTER 27 INVESTIGATION

Classifying Green Algae

OBJECTIVES

- Observe live specimens of green algae.
- Compare unicellular, colonial, thalloid, and filamentous green algae.
- Classify genera of colonial green algae.

PROCESS SKILLS

- observing
- comparing and contrasting
- classifying
- inferring

MATERIALS

- culture of *Chlamydomonas*
- colonial green algae culture *(Chlamydomonas, Eudorina, Gonium, Pandorina, Volvox,* and *Hydrodictyon)*
- culture of *Spirogyra*
- 3 depression slides
- 3 coverslips
- 3 medicine droppers
- compound light microscope

Background

1. Distinguish between the terms *protozoa* and *algae.*
2. Green algae are either unicellular, colonial, filamentous, or thalloid.
3. Explain why green plants are thought to have evolved from green algae.
4. List characteristics of green algae. Include ways that algae differ from plants.
5. How do green algae differ from other algae?

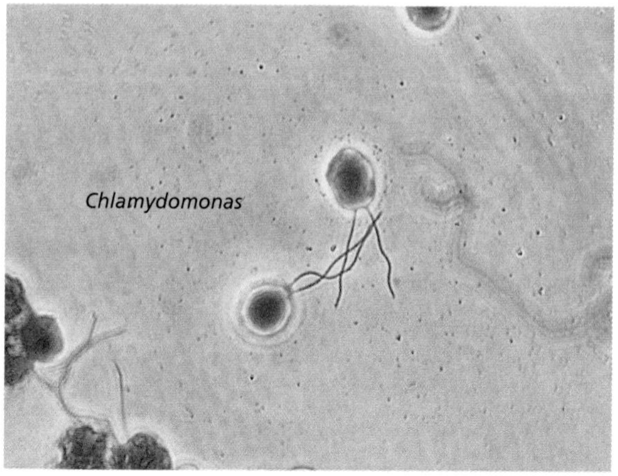

PART A Observing Unicellular Green Algae

1. Make a table similar to the one below in your lab report. Allow substantial space in your data table for labeled sketches of the different kinds of green algae you will view in this investigation. Use your data table to record your observations of each kind of green algae that you view.
2. CAUTION **Slides break easily. Use caution when handling them.** Prepare a wet mount of the *Chlamydomonas* culture by placing a drop of the culture on a microscope slide with a medicine dropper and placing a coverslip on top of the specimen.

OBSERVATIONS OF GREEN ALGAE

Genus	Sketch of organism	Type of green algae (unicellular, colonial, filamentous, or thalloid)

3. Examine the slide of *Chlamydomonas,* first under low power and then under high power.
4. In your lab report, make a sketch of *Chlamydomonas.* Label the cell wall, flagella, nucleus, and chloroplasts, if they are visible.

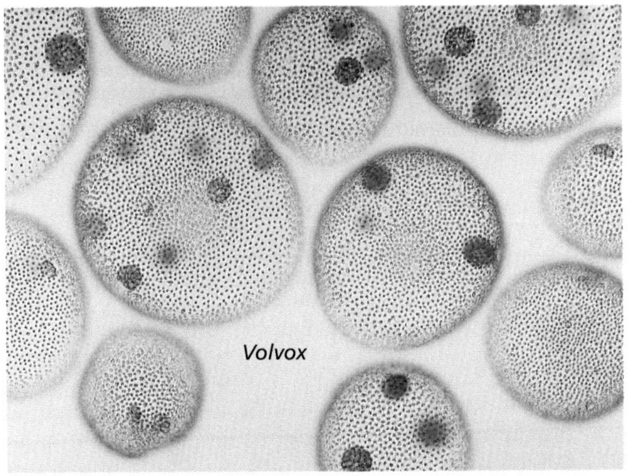

PART B Observing and Identifying Colonial Green Algae

5. CAUTION **Slides break easily. Use caution when handling them.** Prepare a wet mount of the colonial green algae culture, using a clean medicine dropper, a clean microscope slide, and a clean coverslip.
6. Examine the slide of mixed colonial green algae under low power, switching to the high-power setting as needed for clarity. Your slide should resemble the photograph of *Volvox* above. How many different kinds of colonial green algae can you find?
7. In your data table, draw each type of colonial green algae you observe. How are these algae different in appearance from *Chlamydomonas*?
8. Use the dichotomous key, above right, to identify each type of colonial green algae. Choose an alga, and start at 1. Decide which choice describes the alga. If the alga is not identified, go to the next number and make the next decision. Continue until you identify the alga. Then label your drawing of the alga according to the identification you made.

PART C Observing Filamentous Green Algae

9. CAUTION **Slides break easily. Use caution when handling them.** Prepare a wet mount of the *Spirogyra* culture, using a clean medicine dropper, a clean microscope slide, and a clean coverslip.

DICHOTOMOUS KEY FOR COLONIAL GREEN ALGAE

1a	single cells	*Chlamydomonas*
1b	colony of cells	go to 2
2a	colony flattened or netlike	go to 3
2b	colony round	go to 4
3a	netlike colony	*Hydrodictyon*
3b	flattened colony	*Gonium*
4a	more than 100 cells in colony	*Volvox*
4b	fewer than 100 cells in colony	go to 5
5a	cells close together	*Pandorina*
5b	cells apart from each other	*Eudorina*

10. Examine *Spirogyra* under low power. Switch to the high-power setting if you need more magnification to see the specimen clearly. How does *Spirogyra* differ in appearance from the other kinds of algae you have observed in this investigation?
11. In your data table, make a sketch of *Spirogyra.* Label the filaments and the individual cells in your drawing. Also label the spiral-shaped chloroplasts if they are visible.
12. Clean up your lab materials and wash your hands before leaving the lab.

Analysis and Conclusions

1. What characteristics did all of the algae you viewed have in common?
2. Describe examples of specialization in the different kinds of algae you viewed. In which types of algae do some cells depend on others?
3. What differences did you observe between small and large colonies of algae?
4. Which specializations in algae are characteristic of green plants?

Further Inquiry

How might you search for evidence of evolutionary relationships among algae and between algae and green plants?

CHAPTER 27 INVESTIGATION

ANSWERS TO ANALYSIS AND CONCLUSIONS

1. Answers will vary according to the type of cultures the students observe.
2. Answers will vary according to the type of cultures the students observe.
3. The small colonial algae are simple, but the large colonial algae are complex and show signs of organization.
4. A specialization in algae that is characteristic of green plants is chloroplasts.

FURTHER INQUIRY

Evolutionary relationships among algae and between algae and plants can be established by determining the homology in the DNA from their chloroplasts and by determining the homology in certain types of enzymes that commonly occur in algae and plants.

CHAPTER 28
FUNGI

	TOPICS	TEACHING RESOURCES	LABS, CLASSWORK, AND HOMEWORK
BLOCK 1 45 minutes	**Introducing the Chapter, p. 542**	**ATE** Focus Concept, p. 542 **ATE** Assessing Prior Knowledge, p. 542 **ATE** Understanding the Visual, p. 542	■ Supplemental Reading Guide, *Microbe Hunters*
	28-1 Overview of Fungi, p. 543 Characteristics, p. 543	**ATE** Section Overview, p. 543 **ATE** Visual Link Figure 28-3, p. 544 106, 107, 108, 109, 110	★ Study Guide, Section 28-1 **PE** Section Review, p. 545
BLOCKS 2 & 3 90 minutes	**28-2 Classification, p. 546** Phylum Zygomycota, p. 546 Phylum Basidiomycota, p. 547 Phylum Ascomycota, p. 548 Mycorrhizae and Lichens, p. 549	**ATE** Section Overview, p. 546 **ATE** Visual Link Figure 28-6, p. 548 106, 107, 108, 109, 110, 111	**Lab C34,** Limiting Fungal Growth ★ Study Guide, Section 28-2 **PE** Section Review, p. 549
BLOCK 4 45 minutes	**28-3 Fungi and Humans, p. 550** Fungi and Human Disease, p. 550 Fungi in Industry, p. 551	**ATE** Section Overview, p. 550 **ATE** Visual Link Figure 28-10, p. 550	**PE** **Chapter 28 Investigation,** p. 556 ■ Occupational Applications Worksheets—Sanitarian ★ Study Guide, Section 28-3 **PE** Section Review, p. 552
BLOCKS 5 & 6 90 minutes	**REVIEW and ASSESSMENT** **PE** Chapter 28 Review, pp. 554–555 ■ Performance-Based Assessment—Fermentation by Yeast	Audio CD Program	**CHAPTER TESTING** ■ Chapter Test (blackline master) ▲ Test Generator ■ Assessment Item Listing ■ Scoring Rubrics and Classroom Management Checklists

MODERN BIOLOGY

KEY

PE Pupil's Edition
ATE Teacher's Edition
■ Active Reading Guide
★ Study Guide

HOLT BIOSOURCES

Teaching Transparencies

Laboratory Program

One-Stop Planner CD-ROM
Shows these resources within a customizable daily lesson plan:

■ Holt BioSources Teaching Resources
■ Active Reading Guide
★ Study Guide
▲ Test Generator

READING FOR CONTENT MASTERY

Preview: Have students preview the objectives, the introductory paragraph, topic sentences, and captions before reading the chapter.

ATE Active Reading Technique: Anticipation Guide, p. 543
■ Active Reading Guide Worksheet 28-1
PE Summary/Vocabulary, p. 553

■ Active Reading Guide Worksheet 28-2
PE Summary/Vocabulary, p. 553

■ Active Reading Guide Worksheet 28-3
PE Summary/Vocabulary, p. 553

TECHNOLOGY AND INTERNET RESOURCES

CNN PRESENTS
Video Segment 18
Lethal Mushrooms

Holt Biology Videodiscs Teacher's Correlation Guide, Lessons 28-1 through 28-3

Audio CD Program, Sections 28-1 through 28-3

internetconnect

SCILINKS NSTA

On-line Resources: www.scilinks.org
The following *sci*LINKS Internet resources can be found in the student text for this chapter:

Topics:
- Fungi, p. 545

go.hrw.com

On-line Resources: go.hrw.com
Visit the HRW Web site for a variety of resources related to this chapter. Just type in the keyword HM2 Chapter 28.

Smithsonian Institution®
Visit **www.si.edu/hrw** for additional on-line resources.

CNNfyi.com
Visit **www.cnnfyi.com** for late-breaking news and current event stories selected just for you.

LAB ACTIVITY PLANNING

CHAPTER 28 INVESTIGATION

Observing Fungi on Food, pp. 556–557

OVERVIEW
In this investigation, students will observe the effect of propionic acid on the growth of fungi.

SAFETY
Students will wear disposable gloves, lab aprons, and safety goggles throughout the lab. Have students dispose of the toothpicks and Petri dishes in a biohazard bag. Treat all growth in the Petri dishes as pathogenic. Once Petri dishes have been sealed, do not allow students to open them again. Remind students to wash their hands before and after completing the lab.

PREPARATION

1. Each lab group will require two Petri dishes with propionic acid and two Petri dishes without propionic acid. Remind students that they should not touch the inside of a Petri dish or let it be exposed to the air.
2. Fungal samples may be grown on food items, such as unpreserved bread. Place a piece of damp filter paper under the food item in a bowl. Store the bowl in an incubator at room temperature for a week. Keep the filter paper moist. To prevent contamination when putting the cultures out in the lab, cover each bowl with plastic wrap.
3. Purchase a 500 g bottle of potato-dextrose agar and a small bottle of 0.3 percent propionic acid mold inhibitor from a biological supply house. Melt the agar in a microwave or hot-water bath and divide the contents evenly between two flasks. Add 0.75 mL of the propionic acid to one labeled flask. Sterilize both flasks and Petri dishes at 15 psi for 20 minutes (or buy sterile, disposable dishes). Label the Petri dishes. Pour the contents of each flask into the dishes. Make sure you flame the lip of the flask before you pour the media. Refrigerate dishes until lab time.

UNDERSTANDING THE VISUAL

Ask students to examine the photograph and describe what role the mushrooms have in an ecosystem. Tell students that the mushrooms recycle organic matter on the forest floor through cells that penetrate the Earth and dead matter on the ground.

FOCUS CONCEPT

Cell Structure and Function

In this chapter the relationship between fungal structure and function is illustrated at the macroscopic and microscopic level. During most of the life cycle of a fungus, its cells are similar and able to meet their own needs for survival. When fungi reproduce, they produce specialized structures that aid in the fertilization and dispersal of offspring.

ASSESSING PRIOR KNOWLEDGE

Review the following concepts.

Cell Structure: *Chapter 4*
Ask students what organelles are found in eukaryotic cells. Then ask them if all eukaryotic cells have a cell wall. Finally, ask them to explain why cells from different organisms vary in structure.

Cellular Reproduction: *Chapter 8*
Ask students what the chromosome number is in cells produced by mitosis and by meiosis. Then ask students what the chromosome number would be in the offspring cells if a cell produced by meiosis were to undergo mitosis.

CHAPTER 28

FUNGI

Fungi, such as these mushrooms, are important decomposers in nature.

FOCUS CONCEPT: *Cell Structure and Function*

As you read, note how the distinctive traits of fungi, such as their structure and physiology, enable them to affect their environment and thus human health.

OVERVIEW OF FUNGI

SECTION

28-1

OBJECTIVES

▲ Describe the origin and evolution of fungi.

● Compare fungi with other eukaryotic organisms.

■ Describe how fungi obtain nutrients.

◆ Distinguish between a hypha and a mycelium.

In the six-kingdom classification of organisms, fungi are in their own kingdom. They differ from other organisms in several ways, including in structure, in method of reproduction, and in methods of obtaining nutrients.

CHARACTERISTICS

Fungi are eukaryotic, nonphotosynthetic organisms, and most are multicellular heterotrophs. Most fungi are microscopic molds or yeasts. **Molds,** such as the fungus that grows on bread and oranges, are tangled masses of filaments of cells. **Yeasts** are unicellular organisms whose colonies resemble those of bacteria. Yeasts are best known as the microorganisms that make bread rise.

Filaments of fungi are called **hyphae** (HIE-fee). The cell walls of hyphae contain **chitin** (KIE-tin), a complex polysaccharide not found in bacteria, protists, or other microorganisms but found in insects. The presence of chitin distinguishes cell walls of fungi from those of plants, which have cellulose but no chitin. Fungi range in size from the microscopic yeast to the largest organism in the world—the fungus *Armillaria,* which lives underground and occupies a space of up to eight hectares (861,000 ft^2). The study of fungi is called **mycology** (mie-KAHL-uh-jee).

While animals and many microorganisms ingest their nutrients before digesting them, fungi secrete enzymes and then absorb the digested nutrients through their cell wall. Like animals, fungi store energy in the form of glycogen. Most fungi are saprophytic, that is, they live on organic compounds that they absorb from dead organisms in the environment. This characteristic makes fungi a very important recycler of organic material in nature.

Structure of Fungi

A mat of hyphae visible to the unaided eye is a **mycelium** (mie-SEE-lee-uhm). The hyphae of fungi that commonly grow on old bread and fruit form mycelia. In some species, the cells that make up hyphae are divided by cross sections called **septa** (SEP-tuh). Hyphae whose cells are divided by septa are called septate hyphae. The hyphae of species that do not have septa are called **coenocytic** (SEE-noh-SIT-ik). The general structures of septate and nonseptate hyphae are shown in Figure 28-1. Hyphae increase in length by cellular growth and division at the tip. As the hyphae grow, the size of the mycelium increases. When hyphae encounter organic matter, such

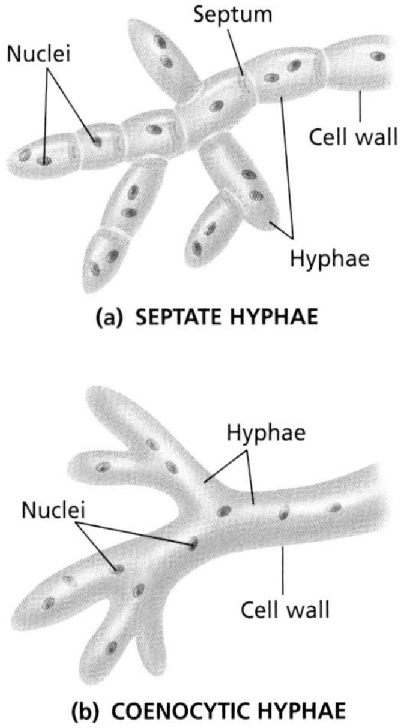

FIGURE 28-1

The hyphae of some fungi have separating walls called septa (a). The hyphae of other fungi do not have septa and are called coenocytic (b).

SECTION 28-1

SECTION OVERVIEW

Cell Structure and Function
Section 28-1 describes the structures of fungi and how these structures enable fungi to grow, reproduce, and take in nutrients.

Reproduction and Inheritance
Section 28-1 describes the ways fungi reproduce, spread their spores, and increase their genetic variability.

ACTIVE READING

Technique: Anticipation Guide

Before students begin this chapter, write the following statements on the board or overhead:

1. Fungi are closely related to plants.

2. Fungi are economically valuable.

3. Many fungi are beneficial to other organisms.

Give students a minute or two to consider the statements and to decide whether they agree or disagree with each. Ask volunteers to share their opinions with the class. Have students read the chapter to see if their opinions are confirmed or changed.

DEMONSTRATION

Prepare a display of pictures of different fungi, including saprobes (mushrooms growing on dead wood), parasites (rust on wheat plants; smut on corn plants; and ringworm, thrush, or athlete's foot on humans), and predators (*Arturobotrys conoides,* a predator of certain nematodes). Ask students to recall the definitions of *saprophyte, parasite,* and *predator.* (A saprophyte breaks down dead organic matter, a parasite feeds on a living host, and a predator captures prey for food.)

SECTION 28-1

RECENT RESEARCH
Evolving Wood Rot

Scientists studying wood-rotting fungi have found that the decay known as soft rot was probably the first type to evolve. It forms longitudinal cavities in wood and has only a limited ability to break down lignin, the strengthening compound in cell walls. The decay known as white rot evolved later and has the ability to degrade lignin. Brown rot is the most recently evolved type of decay, and it avoids the problem of lignin breakdown by oxidizing carbohydrates without enzymes; this type of digestion causes extensive decay in trees.

TEACHING STRATEGY
Observing Fungi

Provide students with prepared slides of a number of fungi that can be obtained from a biological supply company. Examples include *Rhizopus nigricans* (zygomycetes), *Coprinus cinereus* (basidiomycetes), *Penicillium expansum* and *Aspergillus flavus* (ascomycetes), and *Fusarium oxysporum* (deuteromycetes). Have students use dissecting microscopes to examine the fungi. Ask them to make drawings of mycelia and any reproductive structures they see. (They will not likely observe any sexual reproductive structures.)

VISUAL LINK
Figure 28-3

Have students examine Figures 28-3a, b, and c. Ask them what all of the asexual structures depicted have in common. (They all produce single cells that can grow into new fungi.) Then ask how these structures differ from each other. (A yeast cell buds off a new cell that grows, sporangiospores are produced within sacs, and conidia are produced on conidiophores.)

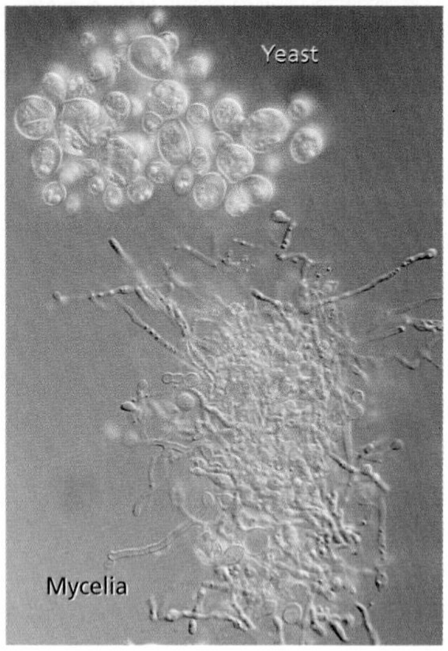

FIGURE 28-2

This light micrograph shows the dimorphic fungus *Paracoccidioides brasiliensis* as it changes from its unicellular yeast form to a mycelial form. Note how the hyphae grow out from the yeast.

FIGURE 28-3

Fungi reproduce asexually in many different ways. Yeasts (a) pinch off a piece of their cell to produce new cells. The common bread mold *Rhizopus stolonifer* (b) produces a hyphal stalk to disperse its spores. *Penicillium* produces unprotected conidia (c).

as a tree trunk or dead animal, they secrete digestive enzymes and then absorb the digested nutrients.

Several species of fungi are able to change their form in response to changes in their environment. For example, *Histoplasma capsulatum,* which causes a severe disease in humans that can resemble tuberculosis, normally grows as a mycelium on the ground. However, when it invades a human, the increased temperature and available nutrients cause the fungus to grow unicellularly as a yeast. This ability to change, demonstrated in Figure 28-2, is called **dimorphism** (die-MOR-FIZ-uhm).

Asexual Reproduction

Most fungi reproduce both sexually and asexually. Asexually, fungi produce thousands of genetically identical haploid spores, usually on modified cells of the hyphae. When these spores are placed in favorable environmental conditions, they germinate and grow new hyphae, which will form a mycelium and can produce thousands of new asexual spores.

A variety of asexual spores are formed by different fungi. For example, **sporangiophores** (spoh-RAN-jee-oh-FOHRZ) are specialized hyphae that look like upright stalks. On top of a sporangiophore is an enclosing sac called a **sporangium** (spoh-RAN-jee-UHM). Inside each sporangium, spores called **sporangiospores** (spoh-RAN-jee-oh-SPOHRZ) are made. *Rhizopus,* the black mycelial fungus that is commonly found growing on bread, is an example of a sporangiospore-forming fungus.

Other fungi form spores called **conidia** (koh-NID-ee-uh), which are formed without the protection of an enclosing sac. Conidia are formed on top of a stalklike structure called a **conidiophore** (koh-NID-ee-uh-FOHR). *Penicillium,* which produces penicillin and cheese, is a fungus that reproduces asexually by means of conidia.

Asexual reproduction may also occur by **fragmentation.** In this process, a septate hypha dries and shatters, releasing individual cells that act as spores. The fungus that causes athlete's foot reproduces this way.

Yeast reproduce by a process called budding. **Budding** is an asexual process in which part of a yeast cell pinches itself off to produce a small offspring cell. Budding may be repeated many times. Three types of asexual reproduction are shown in Figure 28-3.

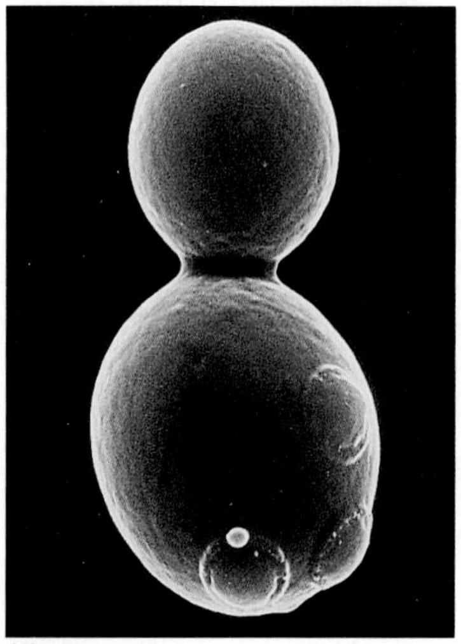

(a)

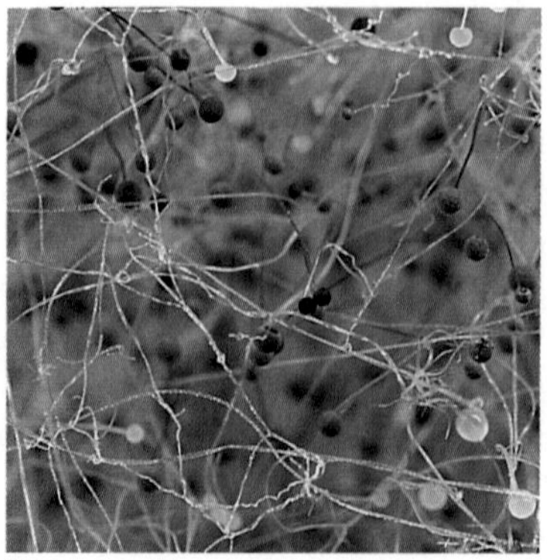

(b)

(c)

TABLE 28-1 The Three Phyla of Fungi

Phylum and number of species	Structure	Asexual reproduction	Sexual reproduction (where identified)	Examples
Zygomycota 600 species	coenocytic hyphae	spores from sporangia	conjugation results in zygospores	*Mucor, Rhizopus, Penicillium* species
Basidiomycota 25,000 species	septate hyphae	rare	basidia produce basidiospores	*Puccinia, Ustilago* (mushrooms)
Ascomycota 60,000 species	septate or unicellular hyphae	conidia, budding	asci produce ascospores	bread yeast, morel

Sexual Reproduction

Many but not all species of fungi are also able to reproduce sexually. Fungi are neither male nor female. They occur in mating types that are sometimes called "minus" and "plus." When two different mating types of the same species encounter one another, the hyphae of one mating type fuse with the hyphae of the opposite mating type. These fused hyphae give rise to a specialized structure, which produces and scatters genetically diverse spores. Unlike most eukaryotes, most fungi are haploid throughout most of their life cycle.

The ability of some fungi to reproduce both sexually and asexually provides an adaptive advantage. When the environment is favorable, rapid asexual reproduction ensures an increased spread of the species. During environmental stress, sexual reproduction ensures genetic recombination, increasing the likelihood that offspring will be better adapted to the new environmental conditions.

Evolution

The first fungi were probably unicellular organisms that might have clung together after mitosis to form a long filament of cells. Biologists think fungi colonized dry land at about the same time that early plants did. They reason that fungi, like other eukaryotes, arose from prokaryotes, possibly by endosymbiosis. According to the fossil record, all modern phyla of fungi had evolved by about 300 million years ago. These phyla are listed in Table 28-1.

Eco Connection

Fungi in the Food Chain

The well-known northern spotted owl depends indirectly on a forest fungus for its survival. The owls prey on northern flying squirrels, and the squirrels depend on truffles for the bulk of their diet. Truffles are the sexual reproductive structure of some fungal species.

internet connect

SCILINKS NSTA
TOPIC: Fungi
GO TO: www.scilinks.org
KEYWORD: HM545

SECTION 28-1 REVIEW

1. **How do fungi obtain nutrients?**
2. **Why does a fungus that reproduces both sexually and asexually have an evolutionary advantage over an organism that reproduces only asexually?**
3. **Describe dimorphism.**
4. **What characteristic makes fungi an important resource recycler?**
5. **Compare how fungi obtain nutrients with how protists obtain nutrients.**
6. **CRITICAL THINKING Based on fossil evidence, scientists think that fungi adapted to land dwelling at about the same time plants did. Explain why fungi might have adapted to land dwelling before plants did.**

SECTION 28-1

RETEACHING ACTIVITY

Provide students with microscopes and prepared slides of *Rhizopus* and mushroom hyphae. Have students draw pictures of the hyphae of both types of fungi and describe how the hyphae differ. (The hyphae in the mushroom are made up of cells divided by septa; the hyphae in the bread mold do not have septa.)

ANSWERS TO SECTION 28-1 REVIEW

1. Fungi are heterotrophic saprophytes. They secrete extracellular enzymes and absorb nutrients from digested dead material in their environment.
2. The advantage is that sexual reproduction encourages genetic variability; thus, the species is more diverse and is therefore able to more effectively react to changing environments than is a species that reproduces only asexually.
3. Dimorphism is the ability of fungi to switch from growing unicellularly (yeast form) to growing multicellularly (mycelia), and vice versa.
4. Fungi are saprophytic and have the ability to turn organic nutrients from dead organisms into resources that other organisms can use.
5. Protists ingest and then digest their food; fungi digest their food extracellularly, then absorb the digested nutrients.
6. It is possible that bacteria and algae may have colonized dry land before fungi or plants. These organisms would have provided a food base for primitive saprophytic fungi prior to the development of mycorrhizal fungi.

SECTION 28-2

SECTION OVERVIEW

Structure and Function
Section 28-2 describes the specialized structures characteristic of different phyla of fungi.

Interdependence of Organisms
Section 28-2 describes two types of endosymbiotic relationships that involve fungi: mycorrhizae and lichens.

▶ ENGAGE STUDENTS

Provide students with a variety of pictures of sexual and asexual reproductive structures of fungi. (College biology textbooks are good sources.) For zygomycetes and deuteromycetes, provide micrographs of reproductive structures. For basidiomycetes and ascomycetes, provide photographs of the sexual reproductive structures, such as ascogonium or basidia, and asexual reproductive structures, such as sporangiophores and conidia. Have students examine the pictures and observe the tremendous diversity of fungal types. Point out that classification of fungi into the phyla of this kingdom traditionally has been based on the type of sexual reproductive structures produced or their absence.

CONTENT CONNECTION

Section 3-3: Enzyme Action

Enzymes are biological catalysts that speed up the rates of biological reactions. They are involved in breakdown reactions, such as the digestion of food into nutrient particles that can be absorbed by an organism's cells.

SECTION 28-2

OBJECTIVES

▲ List the characteristics that distinguish the three phyla of fungi.

● Identify the common sexual reproductive traits of the three phyla of fungi.

■ Define *mycorrhiza* and *lichens*, and distinguish between them.

◆ Explain the importance of mycorrhizae and lichens to the environment.

CLASSIFICATION

The approximately 100,000 species of fungi are classified in three phyla. Traditionally, fungi have been classified according to their structure and form of sexual reproduction. While these are no longer the sole bases of classification, these characteristics are still useful in identifying fungi.

PHYLUM ZYGOMYCOTA

Most species in the phylum Zygomycota (ZIE-goh-MIE-koh-tuh) are terrestrial organisms found primarily in soil that is rich in organic matter. The hyphae of zygomycetes are coenocytic.

Rhizopus stolonifer, the common bread fungus, which is illustrated in Figure 28-4, is in phylum Zygomycota. The hyphae that anchor the mold to the surface of the bread and that penetrate the bread's surface are called **rhizoids** (RIE-zoydz). Digestive enzymes produced by rhizoids break down organic compounds to release nutrients in the bread. Other hyphae, called **stolons** (STOH-lahnz), grow across the surface of the bread.

Sexual reproduction in zygomycetes is called conjugation. Conjugation in fungi occurs when two compatible mating types meet and hyphae from each mating type line up next to each other. Short branches form on the hyphae of both strains and grow outward until they touch each other. A septum forms near the tip of each branch, and a cell called a gametangium develops. A **gametangium** (GAM-ee-TAN-jee-uhm) is a sexual reproductive structure that contains a nucleus of a mating type. The nucleus within each parent's gametangium divides several times. When the gametangia

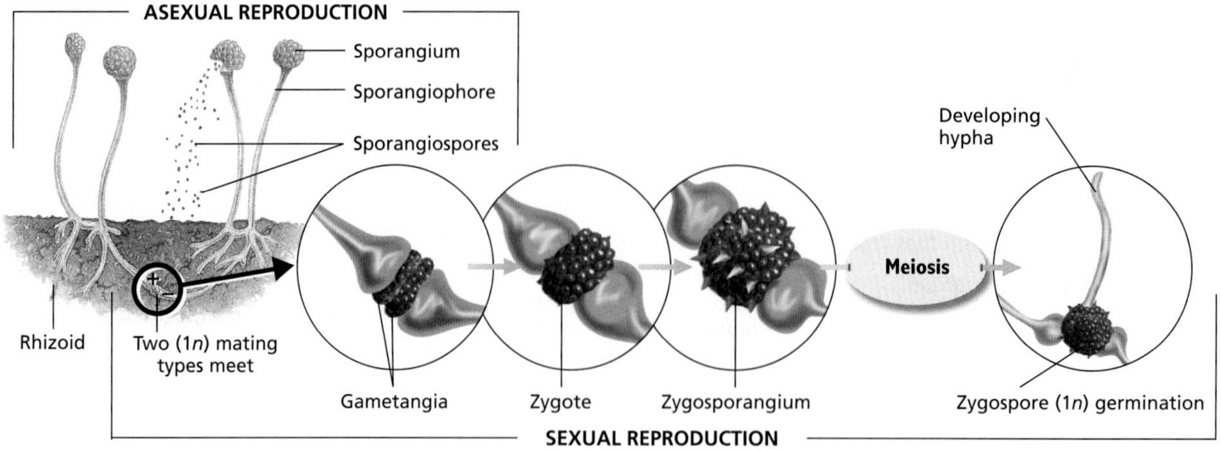

FIGURE 28-4
Zygomycetes can reproduce both sexually and asexually. Sexually, when two compatible mating types meet, they produce specialized cells called gametangia. After the gametangia fuse, the nuclei from both types fuse and eventually produce genetically diverse zygospores.

fuse, the nuclei mix and fuse in pairs, each pair containing a nucleus from each mating type. The fused gametangia, called a **zygosporangium** (ZIE-goh-spohr-AN-jee-uhm), forms a thick wall and becomes dormant. Germination depends on environmental conditions. A sporangiophore grows from the diploid zygosporangium and produces a sporangium, which ruptures and releases haploid spores.

Word Roots and Origins

rhizoid

from the Greek *rhiza,* meaning "root"

PHYLUM BASIDIOMYCOTA

Basidiomycetes (bah-SID-ee-oh-MIE-seets) are often called club fungi because they produce small clublike reproductive structures called **basidia** (bah-SID-ee-uh) during sexual reproduction. Sexual reproduction in basidiomycetes is outlined in Figure 28-5.

The spore-bearing structure of basidiomycetes is an aboveground structure called the **basidiocarp** (bah-SID-ee-oh-KARP). Mushrooms are basidiocarps. The basidiocarp consists of a stem called a stalk and a flattened structure known as a cap. On the underside of the cap are rows of gills that radiate out from the center. Each gill is lined with thousands of dikaryotic basidia. Cells containing two nuclei are called **dikaryotic** (die-KAR-ee-OH-tik). In each basidium, two nuclei fuse to form a zygote ($2n$). The zygote undergoes meiosis to form four haploid nuclei. These develop into four **basidiospores** (bah-SID-ee-oh-SPOHRZ), which are then released into the air. Under favorable environmental conditions, basidiospores germinate to produce haploid mycelia that grow underground. When compatible mating types encounter one another, their hyphae fuse and form a basidiocarp, which emerges aboveground.

FIGURE 28-5

On the gills of the basidiocarp, thousands of dikaryotic basidia form (a). The haploid nuclei inside each basidium fuse to form a diploid nucleus (b). The diploid nucleus undergoes meiosis, producing basidiospores. The basidiocarp releases the basidiospores (c), which fall to the ground and germinate. Underground, the hyphae of compatible mating types fuse and form a mycelium (d). Secondary mycelia intertwine and grow to form a basidiocarp (e).

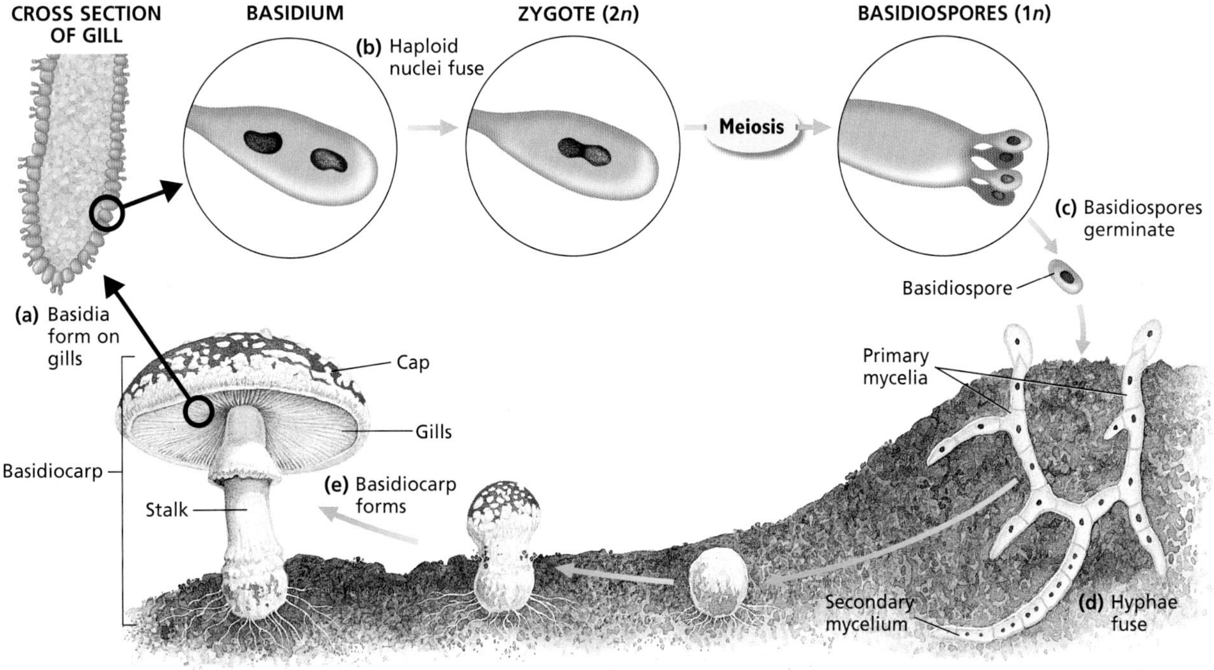

TEACHING STRATEGY

Inducing Sexual Reproduction

CAUTION **Students should wear lab aprons, disposable gloves, and goggles.** Obtain "+" and "–" strains of a zygomycete, either a *Rhizopus stolonifer* or a *Phycomyces blakesleeanus,* from a biological supply company, and prepare enough potato dextrose agar Petri dishes to provide one plate for each pair of students. Provide students with sterile swabs. Have students divide the bottom of their plates with a permanent marker and write + on one side and – on the other side. Have them remove a small sample of the + strain of the fungus and place it on the + half with the swab. Have them do the same with the – strain using another swab. Have them label their plates, seal them with tape, and set them in a warm place to grow. After one week have them examine their plates, without opening them, and look for zygosporangia. CAUTION **Students should not open petri dishes once they are sealed.** Tell students to consult Figure 28-4 for an example of a zygosporangium. Dispose of the plates and swabs by incineration or autoclaving.

QUICK FACT

☑ The spores of fungi are different from the endospores of bacteria. A bacterium forms only one endospore, which can survive adverse conditions and grow into one bacterial cell. Endospore formation is not reproduction because it does not increase the number of bacteria. A fungus produces many reproductive spores. These spores can survive adverse conditions and grow.

TEACHING STRATEGY

Reproductive Structures

CAUTION Remind students that scalpels are sharp and should be handled with care. Provide students with a mushroom, shelf fungus (basidiomycetes), morel, cup fungus (ascomycetes), scalpel, hand lens, slides, and compound microscopes. Review microscopic technique with the class. Have students cut the specimens and examine the internal structures with a hand lens. Help them identify the basidiocarps and the ascocarps. Ask students what the functions of these structures are (fertilization, formation and dispersal of spores). Have students make wet mounts of thin sections of the mushroom and morel. Ask them to examine the slides, looking in particular for basidiospores and ascospores, and make drawings of what they see.

VISUAL LINK

Figure 28-6

Have students examine Figure 28-6. Ask students to explain the stages of the ascomycete life cycle, noting where fertilization (fusion) of haploid nuclei occurs and where meiosis occurs, and the ploidy level of all structures shown. (Fusion and meiosis occur in (c) and the ploidy level is 1*n* until they fuse and become 2*n*. Following meiosis, they are 1*n*.) Then ask them which structures are visible with the unaided eye (the ascocarp or cuplike structure).

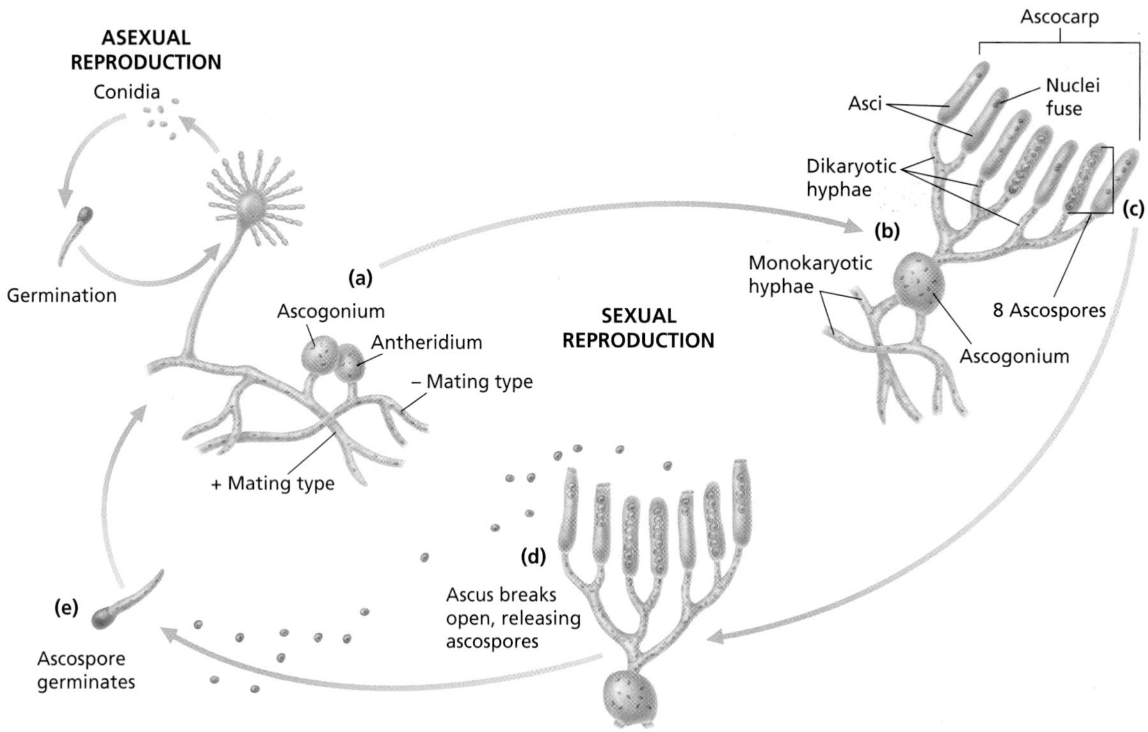

FIGURE 28-6

Compatible mating types form special structures (a) that fuse to form an ascogonium (b). From the ascogonium, dikaryotic hyphae grow and intertwine with monokaryotic hyphae to form an ascocarp. The tips of the dikaryotic hyphae form asci inside, and ascospores form (c). When the ascus ruptures (d), it releases the ascospores, which then germinate (e) to form new monokaryotic hyphae.

FIGURE 28-7

The ascocarp of this ascomycete, *Sarcoscypha coccinea,* is the sexual reproductive structure from which ascospores are released.

PHYLUM ASCOMYCOTA

Ascomycetes (ASK-oh-MIE-seets) are distinguished by the presence of saclike compartments where sexually produced spores form. Ascomycetes, also called sac fungi, live parasitically and in various habitats, including salt water, fresh water, and land.

Sexual reproduction in the ascomycetes begins when the hyphae of two compatible mating types form male and female haploid gametangia. The female gametangium is called an **ascogonium** (AS-koh-GOH-nee-uhm), and the male gametangium is called an **antheridium** (AN-thuhr-ID-ee-uhm), as shown in Figure 28-6. As the ascogonium and antheridium approach one another, a tube forms between them and the nuclei from the antheridium cross the tube and enter the ascogonium. Dikaryotic hyphae grow out of the ascogonium and intertwine with the monokaryotic hyphae of the original fungi (parents) to form a visible cuplike structure called the **ascocarp** (AS-koh-KAHRP). Cells that contain one nucleus are called **monokaryotic** (mah-noh-KAR-ee-OH-tik). An example of an ascocarp is shown in Figure 28-7.

Within the ascocarp, sacs called **asci** (AS-kee) develop at the tips of the dikaryotic hyphae. Within the asci, the haploid nuclei fuse. The zygotes undergo meiosis once and divide again by mitosis to form eight haploid nuclei. The eight nuclei form walls and become **ascospores** (AS-koh-SPOHRZ), which are released. When an ascospore germinates, a new haploid hypha emerges.

The traditional brewer's and baker's yeasts *(Saccharomyces cerevisiae)* are ascomycetes. *S. cerevisiae* makes bread rise and ferments grapes to make wine and grain to make beer.

Deuteromycota

Fungi that do not have a sexual stage are placed in a group called **fungi imperfecti,** or deuteromycota. Recent phylogenetic analyses have led some mycologists to place these fungi in the other three established phyla. Most species of fungi that were formerly classified as fungi imperfecti can now be classified in the phylum Ascomycota. However, some biologists disagree with this reclassification.

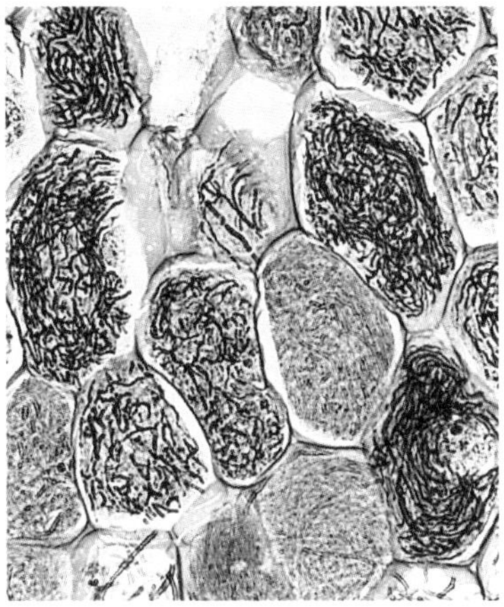

FIGURE 28-8

In this light micrograph, the stained hyphae of the mycorrhizae have infiltrated the plant root cells. (300×)

MYCORRHIZAE AND LICHENS

A **mycorrhiza** (MIE-koh-RIE-zah) is a symbiotic association between a fungus and plant roots, as shown in Figure 28-8. Over 90 percent of plants contain such fungi on their roots. The fungus absorbs and concentrates phosphate and other ions for delivery to the plant root and provides a secondary root system. In turn, the fungi receive sugars synthesized by the plant during photosynthesis. Many zygomycetes and basidiomycetes form mycorrhizae. These mycorrhizal relationships coevolved with plants.

Lichens (LIE-kenz) represent symbiotic relationships between a fungus and a photosynthetic partner (usually a cyanobacterium or green alga). Most fungi in lichens are ascomycetes. The photosynthesizer synthesizes sugars for the fungus, while the fungus provides moisture, shelter, and anchorage for the photosynthesizer. The fungus produces acids that decompose rocks making minerals available to the lichen. The chemical decomposition of rocks by lichens contributes to the production of soil.

Lichens are identified according to their distribution and structure. **Crustose** lichens grow as a layer on the surface of rocks and trees. **Fruticose** lichens are shrublike, and some grow up to 1.5 m (5 ft) in length. **Foliose** lichens live on flat surfaces, where they form matlike growths with tangled bodies. One example of a lichen is shown in Figure 28-9.

FIGURE 28-9

Lichens are grouped according to their type of body. For example, this red lichen is a crustose lichen that grows on rocks.

SECTION 28-2 REVIEW

1. Explain why some fungi are called club fungi, and identify their phylum.
2. Compare and contrast the basidiocarp with the ascocarp.
3. Explain why lichens are important to the environment.
4. Describe the Ascomycota life cycle.
5. Explain the benefits plants and fungi derive from a mycorrhizal relationship.
6. **CRITICAL THINKING** Morels are ascomycetes that resemble mushrooms. Even though they resemble Basidiomycota they are classified as Ascomycota. What evolutionary mechanism causes fungi of one phylum to resemble fungi of another phylum?

MAKING CONNECTIONS

Technology

Tell students that two of the earliest processes of civilization are bread making and wine making and that both depend on yeast. Then explain that yeast cells produce enzymes that break down sugar into alcohol and CO_2. Have students work in pairs to research and prepare a report on the commercial uses of yeast today (production of protein dietary supplements, vitamins, beverages and industrial alcohol, various enzymes, and genetic engineering).

ANSWERS TO SECTION 28-2 REVIEW

1. The basidiomycetes produce a club-shaped cell called a basidium.
2. The basidiocarp is produced from dikaryotic mycelia, and each basidium produces only four basidiospores. The ascocarp is produced from dikaryotic and monokaryotic mycelia, and each ascus produces eight ascospores.
3. Lichens help break down the minerals in rocks and contribute to the production of soil.
4. Compatible mating types produce an ascogonium and an antheridium. Nuclei from the antheridium pass to the ascogonium through a connecting tube. The ascogonium produces a dikaryotic mycelium. The dikaryotic mycelium forms an ascocarp, in which asci form. Ascospores develop within the asci and are dispersed. Asexually, the hyphae can develop conidia, which are capable of dispersion and germination.
5. Plants obtain minerals and moisture from fungi, and fungi obtain carbohydrates from plants.
6. convergent evolution

SECTION 28-3

SECTION OVERVIEW

Interdependence of Organisms

Section 28-3 describes how fungi can affect other organisms through disease in humans and production of useful products, such as breads, antibiotics, and cheeses.

▶ ENGAGE STUDENTS

Provide students with pictures of a variety of harmful and beneficial fungi. Include as many of those discussed in this section as possible. Examples include *Alternaria alternata* (a common allergen), *Pneumocystis carinii* (which causes pneumonia in AIDS patients), *Ceratocystis ulmi* (which causes Dutch elm disease in American elm trees), *Fusarium* species (which cause wilt diseases in many plants), *Botrytis cinerea* (which causes rot on many ornamental and fruit crops), *Gymnosporangium juniperi-virginianae* (which causes apple rust), and *Penicillium* species (which are a source of many antibiotics). Photographs of many of these fungi can be found in mycology and microbiology textbooks. Describe each organism and the effect it has. Point out that fungi affect humans in many different ways.

VISUAL LINK

Figure 28-10

Ask students if they think the symptoms shown in the figure resemble those that could be produced by insect bites or contact with an allergy-inducing plant. Point out that skin irritations are very difficult to diagnose and that, unlike insect bites and allergic reactions, fungal infections can be spread from person to person.

SECTION

28-3

OBJECTIVES

▲ Describe three ways that fungi cause disease in humans.

● Describe the types of food that fungi provide.

■ Provide examples of fungi's industrial importance.

FUNGI AND HUMANS

Fungi are important to humans. Some fungi cause devastating human and plant diseases, while others serve as important food sources for humans. Fungi are also used to produce chemicals, fuels, and pharmaceutical compounds.

FUNGI AND HUMAN DISEASE

Fungi can sometimes attack the tissues of living plants and animals and cause disease. Fungal disease is a major concern for humans because fungi attack not only us but also our food sources, making fungi competitors with humans for nutrients.

Mold spores can cause mild to serious allergies in some people. Billions of mold spores can become airborne and may then be inhaled, triggering an allergic reaction. Sniffling, sneezing, and respiratory distress are symptoms of an allergic reaction. Fungi can also infect and poison humans. Table 28-2 lists some infectious human fungal diseases.

Fungal Skin Infections

Fungi may infect the skin, hair, nails, and tissues of the body. For example, fungi on the skin can cause athlete's foot or ringworm, as shown in Figure 28-10. Ringworm, so named because people once thought they saw living worms in the rings on the skin, can occur almost anywhere on the skin. Athlete's foot occurs on the foot and between the toes.

Another fungal pathogen is *Candida albicans.* This yeast is commonly found in the mouth, intestine, and, in women, in the vaginal tract. It exists in balance with other microorganisms, such as bacteria. However, when conditions change, such as when some antibiotics are used or when pregnancy or illness occurs, then *Candida albicans* can flourish.

FIGURE 28-10

Ringworm is a fungal infection of the skin. The dried skin that falls off a lesion is contaminated with fungal spores. These spores can infect other people and spread the infection.

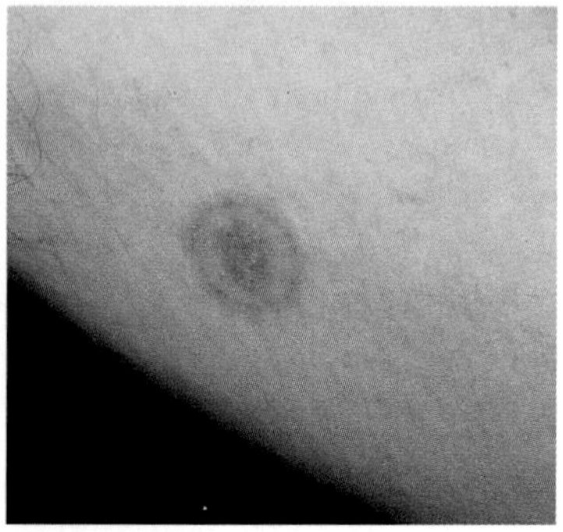

Other Fungal Illnesses

Serious fungal diseases that involve the internal organs are often caused by dimorphic fungi such as *Histoplasma capsulatum, Paracoccidiodes brasiliensis, Coccidioides immitis,* and *Blastomyces dermatitidis.* If their spores are inhaled, they can cause severe respiratory illness and spread to many other organs, sometimes resulting in death. They grow as a mold on the ground, but when they infect a human, they become unicellular. *H. capsulatum,* which often

TABLE 28-2 Summary of Human Fungal Diseases

Disease	Symptoms	Fungus	Route of transmission
Athlete's foot	fluid-filled blisters, scaly skin, itching	*Trichophyton* species (Ascomycete) or *Epidermophyton* species	contact with skin lesions or contaminated floors
Ringworm	ring-shaped skin lesions	*Microsporum, Trichophyton* (Ascomycetes)	contact with skin lesions, contaminated floors, or contaminated objects
Vaginal yeast infection	burning sensation, itching, discharge	*Candida*	contact with fecal material, diabetes; antibiotic treatments increase susceptibility
Tinea cruris (jock itch)	intense itching, ring-shaped lesions	*Microsporum, Trichophyton* (Ascomycetes)	contact with skin lesions, contaminated floors, or contaminated objects
Histoplasmosis	fever, chills, headache, body aches, chest pains, nonproductive cough	*Histoplasma capsulatum* (Ascomycete)	inhalation of airborne conidia

grows as a mold in the feces of birds, can become airborne when dried feces are disturbed and can then be inhaled.

Sometimes humans accidentally eat poisonous mushrooms. For example, *Amanita* species contain extremely dangerous toxins that can destroy a person's liver within one week. Figure 28-11 shows an example of an *Amanita* mushroom. The danger of *Amanita* mushrooms is reflected in their nicknames—"death angel" and "destroying angel."

Other fungal poisons include the **aflatoxins** (AF-luh-TAHKS-ins), poisons produced by some species of *Aspergillus.* Aflatoxins cause liver cancer. Fungi that make aflatoxin may be found as contaminants in peanuts and in grains such as corn and grain sorghum.

FIGURE 28-11

Poisonous mushrooms, such as this *Amanita virosa,* harm people when they are mistaken for edible mushrooms. Intense abdominal pain, vomiting, and diarrhea occur, followed by a short recovery period. Damage occurs in the liver, kidneys, and muscles. The symptoms persist for about six to eight days, and death occurs in about 50 to 90 percent of the cases.

FUNGI IN INDUSTRY

Fungi produce many products used in nonfood industries. For example, *Penicillium* species produce penicillin, and *Cephalosporium* species produce cephalosporin antibiotics. *Rhizopus* causes chemical transformations of specific chemicals to make cortisone and similar drugs. Cortisone is used to reduce joint swelling.

The yeast *Saccharomyces cerevisiae* is an important tool in genetic engineering. For example, the vaccine for hepatitis B was developed by inserting hepatitis B genes into yeast plasmids. The yeast uses the inserted viral genes to produce viral proteins that are used as vaccines. Yeast is also used to produce ethanol, a main ingredient in the automobile fuel gasohol.

RECENT RESEARCH

Genetic Susceptibility

Scientists studying infections of the fungus *Candida albicans* in mice have found that nine strains of mice showed mild tissue damage and two strains showed severe tissue damage. Their results indicate that two mouse genes can confer resistance to severe infection. The disease caused by this infection in mice resembles the human disease; the brain and the kidney are major targets, and the lesions are similar. Candidiasis has increased over 500 percent from 1986 to 1996.

CULTURAL CONNECTION

Witches and Ergotism

Ergot is the common name of the fungus *Claviceps purpurea,* which infects the seeds of rye, wheat, and other grains. Alkaloids produced by the fungus are derivatives of lysergic acid and, when eaten, may cause gangrene, convulsions, mental disturbances, and perceptual distortions. The disease symptoms are referred to as ergotism, or St. Anthony's fire. Scientists think that the women accused of being witches during the 1692 Salem witch trials in Massachusetts may have been suffering from ergotism.

QUICK FACT

☑ Of the more than 100,000 species of fungi known, only about 100 are pathogenic to humans and other animals. However, thousands of fungi are pathogenic to plants. Almost every economically important plant is attacked by one or more fungal species.

SECTION 28-3

RETEACHING ACTIVITY

Have students prepare poster boards for each phylum of fungi, describing and illustrating examples that either benefit or harm humans. Zygomycetes should include pathogens, antibiotics, and allergens; basidiomycetes should include edible and poisonous mushrooms; and ascomycetes should include many pathogens and yeast products. A college microbiology textbook would be a good resource.

ANSWERS TO SECTION 28-3 REVIEW

1. Fungal spores may trigger an allergic reaction, fungi may infect a person's body and cause disease, and a person may eat poisonous mushrooms, which can cause severe illness or death.
2. Athlete's foot is caused by *Trichophyton,* and vaginal yeast infection is caused by *Candida albicans.*
3. Food from fungi may include cheeses, alcoholic beverages, soy products, nutritional yeast, breads, and mushrooms.
4. Nonfood products of fungi include antibiotics, vaccines, cortisone, and ethanol for gasohol.
5. Fungi compete with humans for food resources, such as fruits, vegetables, grains, and the nutrients in our own bodies (infection is a direct competition for the nutrients in the body).
6. Bacteria keep the yeast numbers in check through competition for available nutrients and by the chemicals that they secrete. So, in the presence of an ingested antibiotic that decreases the numbers of bacteria, the resident yeast population is no longer held in check and begins to flourish, sometimes causing disease.

FIGURE 28-12

Morels and truffles are prized by gourmets for their delicate flavor. Truffles (top) grow in association with oak trees. Truffle hunters often use specially trained pigs or dogs to help find truffles. Morels (bottom) grow wild in the Americas and are usually found in the spring.

TABLE 28-3 *Food Products and Fungi*

Type of food	Fungus
Cheeses: blue, brie, Camembert, Gorgonzola, Limburger, Roquefort	*Penicillium* species
Beer, wine	*Saccharomyces carlsbergensis, Saccharomyces cerevisiae*
Soy products: miso (Japanese), soy sauce, tempeh (Indonesian), tofu (Japanese)	*Aspergillus oryzae, Rhizopus* species, *Mucor* species
Nutritional yeast	*Saccharomyces* species
Breads	*Saccharomyces cerevisiae*

Fungi and Food Industries

Many fungi are valuable food sources for humans. Yeast, such as *Saccharomyces,* is an important nutritional supplement because it contains vitamins, minerals, and other nutrients. Mushrooms are also an important food. *Agaricus* (white button), shiitake, and portabella mushrooms are often found in grocery stores in the United States. In other parts of the world, people prize the taste of other fungi, such as truffles and morels, which are pictured in Figure 28-12. Truffles and morels are ascocarps found near the roots of trees. Table 28-3 summarizes some of the uses of fungi in food.

Fungi not only can add value to food but also can take value away. Many fungi are important plant pathogens that attack grain or fruit. For example, wheat rust is a basidiomycete that attacks wheat grains. Other fungi can attack food crops such as corn, beans, onions, squashes, and tomatoes.

Fungi also produce several chemical compounds that are important to the food-processing industry, such as citric and gluconic acids. Citric acid is used in soft drinks and candies. Gluconic acid is fed to chickens to enhance the hardness of eggshells. *Ashbya gossypii* is a producer of vitamin B_2, an important nutritional supplement.

SECTION 28-3 REVIEW

1. Explain how fungi cause disease in humans.

2. Which fungi cause athlete's foot and vaginal yeast infection?

3. List the types of foods that are derived from fungi.

4. Name three nonfood products produced by fungi.

5. Explain how fungi compete with humans for nutrients.

6. CRITICAL THINKING Why would upsetting the balance of microorganisms in the body, such as occurs during antibiotic treatments, cause a yeast infection?

CHAPTER 28 REVIEW

SUMMARY/VOCABULARY

28-1

- Fungi are eukaryotic, nonphotosynthetic organisms that can be unicellular or multicellular.
- Fungi are among the most important decomposers of organic matter in the soil. Fungi secrete extracellular enzymes that digest material and absorb simple organic molecules from the environment.
- All modern phyla of fungi had evolved by 300 million years ago. Fungi probably evolved from the prokaryotes and then adapted to various terrestrial environments.
- Hyphae are tangled masses of fungal filaments. Some species have partitions called septa in their hyphae that separate the individual cells.
- Most fungi reproduce asexually and sexually.

Vocabulary

budding (544)	conidiophore (544)	mold (543)	sporangiophore (544)
chitin (543)	dimorphism (544)	mycology (543)	sporangiospore (544)
coenocytic (543)	fragmentation (544)	mycelium (543)	sporangium (544)
conidium (544)	hypha (543)	septa (543)	yeast (543)

28-2

- The phylum Zygomycota is coenocytic. Asexual sporangiospores form within sacs called sporangia. Sexual reproduction results in zygospores.
- The phylum Basidiomycota includes mushrooms. Mushrooms, or basidiocarps, are sexual reproductive structures. Basidiocarps produce basidia, which produce basidiospores.
- Most fungi are found in the phylum Ascomycota, also called the sac fungi. The intertwining of the hyphae produces a visible ascocarp, which resembles a cup. Mating produces ascospores.
- Yeast are unicellular Ascomycota. They reproduce asexually by budding. Yeast are used for brewing, baking, and genetic engineering.
- Mycorrhizae are symbiotic relationships between plant roots and fungi. Fungi provide nutrients to the plant and derive nutrients from the plant.
- Lichens represent symbiotic relationships between fungi and cyanobacteria, or green algae. Fungi dissolve nutrients from rock. Algae and cyanobacteria provide fungi with carbohydrates. Lichens are highly sensitive to environmental changes.

Vocabulary

antheridium (548)	basidium (547)	foliose (549)	monokaryotic (548)
ascocarp (548)	basidiocarp (547)	fruticose (549)	mycorrhiza (549)
ascogonium (548)	basidiospore (547)	fungi imperfecti (549)	rhizoid (546)
ascospore (548)	crustose (549)	gametangium (546)	stolon (546)
ascus (548)	dikaryotic (547)	lichen (549)	zygosporangium (547)

28-3

- *Candida albicans* is an opportunistic pathogen that causes disease in the oral, intestinal, and vaginal tissues of humans.
- Fungi cause diseases such as athlete's foot, ringworm, and jock itch, and these diseases are easily spread.
- Pathogenic fungi that cause serious disease include *Histoplasma capsulatum, Blastomyces dermatitidis,* and *Coccidioides immitis. H. capsulatum* is associated with bird droppings.
- Some fungi, such as portabella mushrooms and yeasts, are edible. *Amanita* mushrooms, among others, are poisonous.
- In industry, fungi produce antibiotics, fuels, and foods. Yeasts are also valuable genetic-engineering research tools.

Vocabulary

aflatoxin (551)

CHAPTER 28 REVIEW ANSWERS

REVIEW

1. A mycorrhiza is a symbiotic relationship between a fungus and a plant. A rhizoid is a rootlike hypha of a fungus.
2. A sporangiophore is a stalk of hypha on which a sporangium is produced. The sporangium holds sporangiospores inside.
3. An ascocarp is a cuplike sexual reproductive structure formed by ascomycetes; an ascogonium is a female gametangium of ascomycetes; and an ascus is a spore-containing sac found on an ascocarp.
4. *Monokaryotic* refers to a cell that contains only one nucleus; *dikaryotic* refers to a cell that contains two nuclei.
5. Basidia are spore-forming structures found in sexually reproducing basidiomycetes; a basidiocarp is the aboveground spore-bearing structure of basidiomycetes; and a basidiospore is the sexually produced spore produced by basidia.

6. a **7.** a
8. b **9.** a
10. c **11.** d
12. d **13.** a
14. c **15.** d

16. The three phyla produce gametangia, which fuse and produce hyphae, which in turn produce spore-bearing structures. In zygomycetes, the fused structures are called zygosporangia; in basidiomycetes, they are called basidiocarps; in ascomycetes, they are called ascocarps.
17. because fossil evidence indicates that they both occupied the land at about the same time
18. In coenocytice hyphae, the cells are not separated by walls; in septate hyphae, they are. *Mucor* and *Rhizopus* are coenocytic, and *Trichophyton* and *Agaricus* are septated.
19. Saprophytic fungi digest organic materials, solubilize nutrient-containing minerals, and return nutrient molecules to the

ecosystem to be used by other organisms.
20. Mutualism is a type of symbiotic relationship in which both parties benefit. Mycorrhizae are mutualistic relationships between a fungus and plant roots; lichens are mutualistic relationships between a fungus and an alga or a cyanobacterium.
21. Rust is a plant disease caused by parasitic Basidiomycota. Because it infects food crops, it competes with humans for nutrients.
22. They are both symbiotic relationships between a fungus and another organism.
23. Fungi can cause an allergic response, infect human tissue, or cause poisoning in humans (who eat contaminated food or a poisonous mushroom). Fungi can be eaten or used to process food, produce antibiotics, or produce fuels, and it can be used in genetic engineering.
24. When dried feces is disturbed, the dust of feces and fungal spores, such as *Histoplasma capsulatum,* become airborne and a person can breathe them in.
25. (a) rhizoid (b) sporangiospores (c) sporangium (d) sporangiophore (e) zygosporangium Zygomycota

CRITICAL THINKING

1. No, because this is an instance of convergent evolution.
2. The advantage is that mammals will not eat poisonous fungi, allowing the fungus to complete its sexual reproduction.
3. Moldy bread sometimes contains antibiotic-producing molds. The antibiotic would have controlled infections.
4. It helps fungi compete for food by reducing the number of competing organisms in a given area, thus freeing more resources for the fungus.
5. Because 37°C is human body temperature, it would grow in a

REVIEW

Vocabulary

1. Distinguish between a mycorrhiza and a rhizoid.
2. Differentiate between a sporangiophore, a sprangiospore, and a sporangium.
3. Differentiate between an ascocarp, an ascogonium, an ascospore, and an ascus.
4. Distinguish between *monokaryotic* and *dikaryotic.*
5. Differentiate between basidia, a basidiocarp, and a basidiospore.

Multiple Choice

6. Hyphae of the phylum Zygomycota are considered (a) coenocytic (b) monokaryotic (c) nonnuclear (d) plasmodial.
7. The common edible mushroom found in your grocery store is classified in the phylum (a) Basidiomycota (b) Ascomycota (c) Oomycota (d) Zygomycota.
8. The female gametangium in sac fungi is called an (a) oogonium (b) ascogonium (c) antheridium (d) oospore.
9. Fungi that feed on decaying organic matter are said to be (a) saprophytic (b) parasitic (c) mutualistic (d) symbiotic.
10. The fungus *Rhizopus* usually can be found growing on (a) soil (b) fruit (c) bread (d) decaying logs.
11. The walls that separate cells in fungal hyphae are known as (a) rhizoids (b) gills (c) asci (d) septa.
12. Lichens represent symbiotic associations of fungi and (a) roots (b) roundworms (c) water molds (d) green algae.
13. The gills of a mushroom are the sites of spore-bearing structures known as (a) basidia (b) conidiophores (c) basidiocarps (d) ascocarps.
14. In a mycorrhiza, a fungus lives in a symbiotic relationship with a (a) virus (b) slime mold (c) plant (d) bacterium.
15. All of the following human diseases are caused by a species of fungus, except (a) ringworm (b) athlete's foot (c) histoplasmosis (d) influenza.

Short Answer

16. How do the sexual reproductive structures in the three phyla of fungi resemble one another? How do they differ?
17. Explain why biologists think fungi colonized dry land at the same time as plants did.
18. How does a coenocytic hypha differ from a septate hypha? Name two examples of fungi that are coenocytic and two more that are septate.
19. Explain how fungi contribute to nutrient cycling in the environment.
20. Describe mutualism, and give an example of such a relationship involving fungi.
21. Explain what a rust is and how it affects humans.
22. What similarities are there between a mycorrhiza and a lichen?
23. Describe three ways that fungi cause disease in humans and three ways that fungi are useful or beneficial to humans.
24. Describe how a fungus growing in feces can infect a person's lungs. Name an example of a fungus that can do this.
25. Identify the lettered structures in the figure below. What phylum does this organism represent?

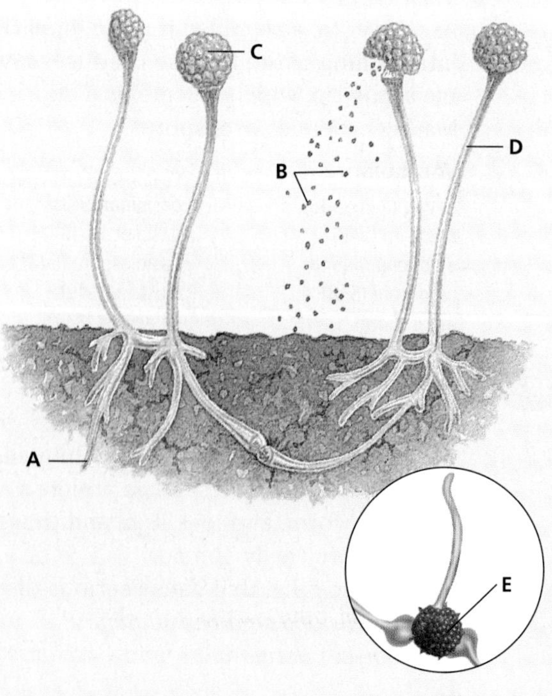

CRITICAL THINKING

1. The cell walls of fungi and the exoskeletons of insects and crustaceans contain chitin. Tell whether this is of phylogenetic significance, and explain why.
2. Many fungi are fatally poisonous to mammals. What adaptive advantage do toxins give to a fungus?
3. Long before antibiotics were discovered, it was common practice to place a piece of moldy bread on wounds. Explain why this practice might have helped the wounds heal.
4. Fungi such as *Penicillium* engage in a kind of chemical warfare against other microorganisms by producing chemicals that diffuse outward and kill nearby organisms. Explain how producing antibiotics is an adaptive advantage for fungi.
5. Most fungi grow best at temperatures of about 15°C to 21°C. *Aspergillus fumigatus,* however, can grow well at 37°C. Knowing this, where would you expect *A. fumigatus* to grow?
6. *Drosophila,* the common fruit fly, ingests juices from ripe fruit and yeast. How might the fruit fly help speed the natural decay of fruit?
7. When transplanting a wild plant to garden soil, it is very important to include some of the soil from the original habitat. Give a possible explanation, based on the information from this chapter.
8. A winemaker experimenting with three different strains of yeast charts the number of viable yeast as the grapes ferment. The three strains were grown in separate containers, at the same temperature, and with the same type of grape. Why does strain C survive longer than strains A or B?

Growth of Three Strains of Yeast

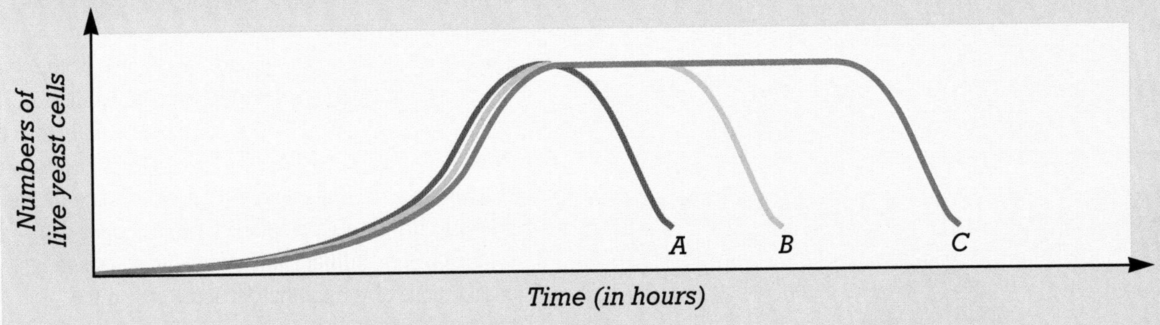

EXTENSION

1. Read "Joyous Mushrooms" in *Natural History,* September 1999, on page 48. Describe the work of natural historian, amateur mycologist, and artist Mary Elizabeth Banning.
2. Use the library to research the ways that medical science has been influenced by the study of fungi. Investigate the discovery of penicillin and other drugs derived from fungi. Investigate the role played by fungi in various diseases.
3. Prepare a yeast culture by adding a pinch of baker's yeast to a mixture of nine parts water and one part molasses. Allow the culture to ferment, and examine it under a microscope. Make a drawing of several yeast cells, and identify any visible cell parts.
4. Read "The Lichen Connection" in *Popular Science,* July 1998, on page 22. Describe how lichens are valuable to earthquake researchers.

human as a pathogen or as a commensal organism.

6. The yeasts stick to its body; as that fruit fly flies from fruit to fruit, it spreads the yeasts, which ferment the sugars in the fruit.

7. The original soil serves as a source of fungi needed to form mycorrhizae.

8. Some strains of yeast produce more alcohol than others. Alcohol is fatal to yeast at a certain concentration, and the alcohol that yeast produce eventually kills them. Therefore, strain C produces less alcohol, or strain C may be more tolerant of higher alcohol concentrations than strains A or B.

EXTENSION

1. Banning devoted the years from 1860 to 1890 to making field notes and painting watercolors of the fungi of Maryland. She described 175 species, 23 of which were new to science. Most of her descriptions can still be used for field identification.

2. Some drugs that students might research include the cephalosporin drugs produced by fungi. Some fungal diseases that students can find out about include those mentioned in this chapter, as well as plant diseases, such as the ergot disease of rye plants.

3. Students may observe many offspring cells budding and then separating from parent cells. The odor is typical of beer or wine, and carbon dioxide bubbles will form in the liquid.

4. Lichens begin to colonize freshly exposed rock after an earthquake. Because lichens have a slow, predictable growth rate, researchers can pinpoint when an earthquake occurred in an area within the last 1,000 years.

CHAPTER 28 INVESTIGATION

TIME REQUIRED

One 50-minute class period and 20 minutes from one class period seven days later

SAFETY PRECAUTIONS

Have students wear safety goggles, disposable gloves, and a lab apron during this investigation. They should wash their hands thoroughly after handling food samples or any fungal samples. Be sure that all Petri dishes are sealed with clear tape after fungal samples have been added and are not opened again. All fungal samples should be destroyed by incineration or in an autoclave.

QUICK REFERENCES

Lab Preparation Notes are found on page 541B.

Holt BioSources provides a Teaching Resources CD-ROM that contains "Using Gowin's Vee in the Lab" and "Scoring Rubrics."

PROCEDURAL TIPS

1. Have students prepare their data collection tables ahead of time.
2. You may want to homogenize food samples in a blender ahead of time in order to make it easier for students to obtain a small sample for their Petri dishes.

ANSWERS TO BACKGROUND

1. Multicellular fungi obtain nutrients by secreting digestive enzymes and absorbing the nutrients that they break down in the environment.
2. Multicellular fungi reproduce asexually by producing genetically identical spores and sexually by fusing with a compatible mating type and making structures that produce spores.

CHAPTER 28 INVESTIGATION

Observing Fungi on Food

OBJECTIVES

- Recognize fungal growth on food.
- Identify environmental conditions that favor the growth of fungi on food and those that inhibit it.

PROCESS SKILLS

- designing an experiment
- collecting data
- organizing data
- analyzing results

MATERIALS

- safety goggles
- lab apron
- disposable gloves
- 2 sterile Petri dishes with nutrient medium (such as potato dextrose agar)
- 2 sterile Petri dishes with nutrient medium and propionic acid
- fungal samples
- stereomicroscope
- toothpicks
- wax pencil
- masking tape

Background

1. How do multicellular fungi, such as molds, obtain nutrients?
2. How do multicellular fungi reproduce and grow?
3. What ecological role do fungi fulfill?
4. How might fungi growing on food harm someone who eats it?

PART A Experimental Setup

1. CAUTION **Put on safety goggles, a lab apron, and disposable gloves.** Obtain four sterile Petri dishes, two with nutrient medium and two with nutrient medium plus propionic acid. Label the dishes for the presence or absence of propionic acid.
2. Examine the fungal samples through a dissecting microscope. Select a dense growth of a fungus from which you will take samples.
3. Use a toothpick to scoop up a small sample of the fungus you have selected. In the two Petri dishes without propionic acid, gently touch the sample to the medium in four places. Raise the lids of the dishes as little as possible while doing this. Do the same for the two Petri dishes with propionic acid, using a clean toothpick and another small sample of the same fungus. Dispose of the toothpicks according to the teacher's instructions.
4. Place a piece of masking tape on opposite sides of each dish to hold the lid and bottom together. Label each Petri dish with your name.
5. Design an experiment to determine which of two opposite environmental conditions is better for fungal growth. Some possible combinations are warm/cold, light/dark, and moist/dry. Label one of the two environmental conditions you selected on a dish with propionic acid. Label the other environmental condition on the other two Petri dishes (one with propionic acid and one without). Then incubate all four dishes under the appropriate conditions.

FUNGAL GROWTH IN DIFFERENT ENVIRONMENTS

Dish	Environmental condition	Propionic acid?	Source of fungus	Growth
1				
2				
3				
4				

6. Create a data table similar to the model above to record your experimental observations for your lab report. For example, the table above is designed to record any growth that may occur under two different environmental conditions (warm/cold, sunlight/dark, moist/dry), source of the fungus, and if propionic acid is present. Design your data table to fit your own experiment. Remember to allow plenty of space to record your observations. Be sure to label your Petri dishes with the appropriate environmental conditions and with the date on which the experiment was begun.

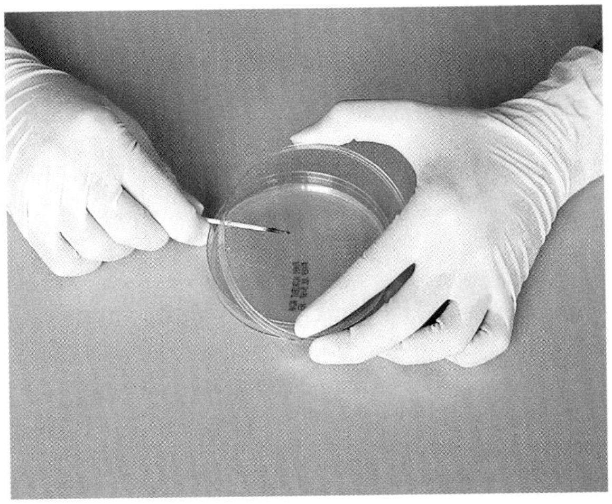

PART B Comparing Amounts of Growth

7. After one week, examine each dish under the stereomicroscope without opening the dish.
8. Record your observations in your data table in your lab report.
9. Examine dishes belonging to other groups, especially those grown under different environmental conditions.
10. Add your observations of those dishes to your lab report.
11. Clean up your materials and wash your hands before leaving the lab. Dispose of all materials according to instructions provided by your teacher.

Analysis and Conclusions

1. Besides the environmental conditions you chose, what additional factor was tested in your experiment?
2. What steps were taken in your experiment to avoid contamination of the plates?
3. How would contamination of the plates affect the results of your experiment? What are the possible sources of contamination of your experiment?
4. What does extensive fungal growth on a plate indicate? What does lack of fungal growth on a plate indicate?
5. What effect does propionic acid have on fungal growth? How do you know?
6. Why do you think propionic acid is added to foods?
7. Which environmental conditions favor fungal growth? Which inhibit it?
8. Compare the results for the different kinds of fungi grown. Did fungi from certain food sources grow more rapidly than others?
9. Based on your conclusions, under what conditions would you keep a nonsterile food product if you wanted to prevent it from becoming moldy?
10. Has this lab changed your attitude toward food storage? If so, how?

Further Inquiry

Design an experiment that determines the kinds of food that best support fungal growth and whether the presence of chemicals inhibits fungal growth.

3. Fungi are recyclers, breaking down the organic material of dead organisms and making them available to other organisms.
4. If the fungus produces a toxin, then it will harm the organism that consumes it.

ANALYSIS AND CONCLUSIONS

1. whether fungi could grow in the presence of propionic acid
2. opening the Petri dishes only slightly and using a clean toothpick for each food sample
3. Contamination of the plate would give a false indication about the types of fungi found on that food. It could also give a false indication about the ability of fungi to grow in those particular conditions. Possible sources of contamination could include dirty toothpicks, air currents, contaminated agar, or bad technique.
4. The environmental conditions and the nutrients in the agar are favorable to fungal growth. It also indicates that the food sample was probably heavily contaminated with fungal spores. Lack of growth indicates that the conditions were not favorable to fungal growth or that something in the agar was preventing growth from occurring.
5. Propionic acid retards the growth of fungi. The plates with propionic acid should have little or no fungal growth.
6. Propionic acid is added to foods to prevent spoilage.
7. Answers will vary according to the results that students obtain.
8. Answers will vary according to the types of food that students use.
9. Cold storage and preservatives best prevent fungal growth.
10. Students should understand the importance of cold storage and the use of preservatives in preventing spoilage.

Unit 7

PLANTS

CHAPTERS

internetconnect

SCiLINKS NSTA

National Science Teachers Association *sci*LINKS Internet resources are located throughout this unit.

“*The day passed delightfully. Delight itself, however, is a weak term to express the feelings of a naturalist who, for the first time, has wandered by himself in a Brazilian forest. The elegance of the grasses, the novelty of the parasitical plants, the beauty of the flowers, the glossy green of the foliage, but above all the general luxuriance of the vegetation, filled me with admiration.*”

Charles Darwin, *Voyage of the Beagle*

These autumn leaves show bright colors because they have lost most of their chlorophyll.

Rafflesia arnoldii has the largest known flowers. Although they are large and beautiful, these flowers emit the smell of rotting meat, which attracts flies as pollinators.

Strawberry plants have aboveground stems that are able to form new plants.

Orchid cactus

This cotton boll developed from the ovary of a flower of a cotton plant.

unit 7

CHAPTERS

PLANTS

CONNECTING TO THE STANDARDS

The following chart shows the correlation of Unit 7 with the *National Science Education Standards* (grades 9–12). The items in each category are addressed in this unit and should be used to guide instruction. Annotated descriptions of the **Life Science Standards** are found below the chart. Consult the *National Science Education Standards* for detailed descriptions of each category.

UNIFYING CONCEPTS
- Systems, order, and organization
- Evolution and equilibrium
- Form and function

SCIENCE AS INQUIRY
- Abilities necessary to do scientific inquiry
- Understanding about scientific inquiry

SCIENCE AND TECHNOLOGY
- Abilities of technological design
- Understanding about science and technology

PHYSICAL SCIENCE
- Structure and properties of matter
- Chemical reactions
- Motions and forces

LIFE SCIENCE

EARTH AND SPACE SCIENCE
- Energy in the Earth system
- Geochemical cycles
- Origin and evolution of the Earth system

SCIENCE IN PERSONAL AND SOCIAL PERSPECTIVES
- Natural Resources
- Environmental quality

HISTORY AND NATURE OF SCIENCE
- Nature of scientific knowledge
- Historical perspectives

ANNOTATED DESCRIPTIONS OF THE CORRELATED LIFE SCIENCE STANDARDS

Although all eight categories of the *National Science Education Standards* are important, the following descriptions summarize the **Life Science Standards** that specifically relate to Unit 7.

THE CELL
- Cells have particular structures that underlie their functions.
- Plant cells contain chloroplasts, the site of photosynthesis.
- Cells can differentiate. Complex multicellular organisms are formed from highly organized arrangements of differentiated cells.

BIOLOGICAL EVOLUTION
- The great diversity of organisms is the result of more than 3.5 billion years of evolution.
- Biological classifications are based on how organisms are related.

INTERDEPENDENCE OF ORGANISMS
- Organisms both cooperate and compete in ecosystems.

MATTER, ENERGY, AND ORGANIZATION IN LIVING SYSTEMS
- The chemical bonds of food molecules contain energy.
- The complexity and organization within an organism result from the need to obtain, transform, transport, release, and eliminate the matter and energy used to sustain the organism.
- The distribution and abundance of organisms and populations in ecosystems are limited by the availability of matter and energy and by the ability of the ecosystem to recycle materials.

TRENDS IN BIOLOGY

Tropical Deforestation Continues

It has been more than a decade since conservationists first sounded the alarm about the deforestation of the tropics. The area covered by tropical forests, which includes rain forests, dry forests, and deciduous forests, had already been reduced by 50 percent, and more than 15 million hectares—almost 1 percent of the remaining forested area—were being destroyed every year. Although public attention (and condemnation) focused on Brazil's clearing of the Amazonian rain forest, a greater percentage of Asian and Central American forests were being cleared. During the 1980s, for instance, the Philippines and Thailand allowed their forests to be cut at a rate of more than 3 percent per year. Continued deforestation, conservationists warned, would be an ecological and economic disaster for tropical countries, and it would have worldwide repercussions by accelerating global warming and by causing a mass extinction of species.

What is the status of the tropical forests today? The sad truth is that they are in worse shape than they were a decade ago. Although measuring the rate of deforestation has proved difficult, according to the latest estimates it is still occurring at about the same rate as it did during the peak years of the late 1980s. Since cutting and burning have continued unabated, there is much less forested area remaining today than there was a decade ago, and more disappears every day. In Amazonia, the largest tract of rain forest in the world, the rate of deforestation has been increasing in recent years after decreasing in the late 1980s and early 1990s.

The news about tropical deforestation is not all bad. Scientists and conservationists cite the establishment of a number of extractive reserves, areas in which sustainable harvesting of fruits, nuts, rubber, and other products is permitted, allowing local people to benefit economically without cutting the forest. They also point to the increasing amount of forest protected in parks and reserves and the small but growing trend toward sustainable forestry, a method of forest management in which a limited amount of timber is cut and the forest is then allowed to regenerate. Most significant is the growing awareness of the urgency of the problem among politicians, government officials, and ordinary people in countries around the world. The challenge is to translate this awareness into action.

Rapid destruction of forest continues throughout the tropics.

THE CAUSES OF DEFORESTATION

Clearing of land for agriculture accounts for more than 50 percent of the deforestation in the tropics. Logging, mining, cutting of wood for fuel, and creation or expansion of settlements account for the rest. Discussions of deforestation often focus on these causes, which are well known. But to bring tropical deforestation under control, the underlying causes—the economic, social, and political forces that drive people to exploit the forest—will have to be addressed. Although these causes vary from country to country—and even between regions within a country—the two most important are poverty and rapid population growth.

Additional Information

Myers, Norman. *The Ultimate Source: Tropical Forests and Our Future.* 2nd ed. New York: W.W. Norton, 1992.

Rice, Richard, Raymond Gullison, and John Reid. "Can Sustainable Management Save Tropical Forests?" *Scientific American,* April 1997, 44–49.

Sponsel, Leslie, Thomas Headland, and Robert Bailey, eds. *Tropical Deforestation: The Human Dimension.* New York: Columbia University Press, 1996.

CHAPTER 29
THE IMPORTANCE OF PLANTS

Block	TOPICS	TEACHING RESOURCES	LABS, CLASSWORK, AND HOMEWORK
BLOCK 1 45 minutes	**Introducing the Chapter, p. 560**	**ATE** Focus Concept, p. 560 **ATE** Assessing Prior Knowledge, p. 560 **ATE** Understanding the Visual, p. 560	■ Supplemental Reading Guide, *A Feeling for the Organism*
	29-1 Plants and People, p. 561 Plants as Food, p. 561 Food Crops, p. 562 Nonfood Uses of Plants, p. 564	**ATE** Section Overview, p. 561 112, 112A, 113, 114	**Lab C36,** Using Paper Chromatography to Separate Pigments ■ Portfolio Projects—Plant Research Project ■ Portfolio Projects—Plant Focus Worksheet ★ Study Guide, Section 29-1 **PE** Section Review, p. 567
BLOCK 2 45 minutes	**29-2 Plants and the Environment, p. 570** Plant Ecology, p. 570 Harmful Plants, p. 572	**ATE** Section Overview, p. 570 **ATE** Critical Thinking, p. 571	**PE** **Chapter 29 Investigation,** p. 576 **Lab A18,** Comparing Plant Adaptations **Lab C37,** Growing Plants in the Laboratory **Lab C38,** Growing Plants in the Laboratory—Fertilizer Problem ■ Occupational Applications Worksheets—Forestry Technician ★ Study Guide, Section 29-2 **PE** Section Review, p. 572
BLOCKS 3 & 4 90 minutes	**REVIEW and ASSESSMENT** **PE** Chapter 29 Review, pp. 574–575 ■ Performance-Based Assessment—Testing Enzyme Activity in Seeds	Audio CD Program	**CHAPTER TESTING** ■ Chapter Test (blackline master) ▲ Test Generator ■ Assessment Item Listing ■ Scoring Rubrics and Classroom Management Checklists

KEY

MODERN BIOLOGY
- **PE** Pupil's Edition
- **ATE** Teacher's Edition
- ■ Active Reading Guide
- ★ Study Guide

HOLT BIOSOURCES
- Teaching Transparencies
- Laboratory Program

One-Stop Planner CD-ROM
Shows these resources within a customizable daily lesson plan:
- ■ Holt BioSources Teaching Resources
- ■ Active Reading Guide
- ★ Study Guide
- ▲ Test Generator

READING FOR CONTENT MASTERY

Preview: Have students preview the objectives, the introductory paragraph, topic sentences, and captions before reading the chapter.

ATE Active Reading Technique: Reading Organizer, p. 561
■ Active Reading Guide Worksheet 29-1
PE Summary/Vocabulary, p. 573

■ Active Reading Guide Worksheet 29-2
PE Summary/Vocabulary, p. 573

TECHNOLOGY AND INTERNET RESOURCES

CNN PRESENTS Video Segment 21 Maple Syrup

Holt Biology Videodiscs Teacher's Correlation Guide, Lessons 29-1 through 29-2

Audio CD Program, Sections 29-1 through 29-2

internetconnect

SCILINKS NSTA
On-line Resources: www.scilinks.org

go.hrw.com
On-line Resources: go.hrw.com
Visit the HRW Web site for a variety of resources related to this chapter. Just type in the keyword HM2 Chapter 29.

Smithsonian Institution®
Visit **www.si.edu/hrw** for additional on-line resources.

CNNfyi.com
Visit **www.cnnfyi.com** for late-breaking news and current event stories selected just for you.

LAB ACTIVITY PLANNING

CHAPTER 29 INVESTIGATION

Comparing Soil-Grown Plants with Hydroponic Plants, pp. 576–577

OVERVIEW
In this investigation, students will compare the growth of plants that have been cultivated hydroponically with the growth of plants that have been cultivated in soil. Students will observe the germination of wheat seeds over a 2-week period.

SAFETY
Students should wear safety goggles during the lab. Students should keep their hands away from their faces when handling plants. Caution students not to get any sap or plant juice in their eyes, skin, or mucous membranes. Remind students to wash their hands thoroughly after completing the lab.

PREPARATION
1. This investigation will take 14 days to complete.
2. Order wheat seeds in advance from a biological supply house, such as WARD'S, or get them from a local nursery or garden center. Each group of students will need 12 wheat seeds.
3. Plastic foam floaters can be purchased from WARD'S in a hydroponic kit. Plastic foam floaters can also be constructed from materials acquired from a hardware store or a craft supply store. Cut a piece of flat plastic foam that is 1/8–1/4 in. thick into 2 in. squares or circles. Punch six holes through the plastic foam with a pencil. Each hole can then be used to cradle the seed in nutrient solution.
4. You will also need to provide potting soil and nutrient solution for the investigation. Nutrient solution can be purchased from garden shops, nurseries, or a biological supply house, such as WARD'S.

UNDERSTANDING THE VISUAL

Have students look at the photograph and ask them to notice how the rice paddies are compartmentalized. Focus students' attention on the standing water. Help students recognize that rice needs abundant water to grow. Farmers can utilize various techniques to conserve water in areas that do not receive adequate rainfall. Also help students recognize that rice is an important crop worldwide, especially because of its high nutritional value.

FOCUS CONCEPT

Interdependence of Organisms

Crucial interactions take place between plants and people. Plants provide people with food, medicines, commercial materials, and oxygen. Although people cultivate plants, they also destroy vegetation, which has a critical effect on the biosphere.

ASSESSING PRIOR KNOWLEDGE

Review the following concepts.

Artificial Selection: *Chapter 14*
Ask students to list some plant characteristics that would be advantageous to select for when planting a garden.

Natural Resources: *Chapter 19*
Ask students to explain how people use fertilizer, irrigation, and greenhouses to affect plants they wish to select.

Populations and Ecosystems: *Chapter 20*
Ask students to explain how the expansion of communities across the United States was tied to the development of agriculture.

CHAPTER 29

THE IMPORTANCE OF PLANTS

Rice fields, often called paddies, thrive in regions with abundant rainfall. Rice is a prime food source for more than 60 percent of the world's population. About 90 percent of the world's rice is grown in Asia, primarily in China and India.

FOCUS CONCEPT: *Interdependence of Organisms*

As you read, pay attention to the relationship between plants and humans. Recognize how plants have changed over time. Describe how plants have been used to improve human existence.

PLANTS AND PEOPLE

Plants are essential to our survival because they produce virtually all our food. We eat plants either directly, in the form of fruits, vegetables, and grains, or indirectly, by eating animals that consume plants. Plants also provide medicines, clothing, paper, cosmetics, and many other products. Plants play a major role in the continuous cycling of the Earth's water, oxygen, carbon dioxide, and mineral nutrients. The study of plants is called ***botany*** *(BAHT-nee).*

SECTION

29-1

OBJECTIVES

▲ Describe ways that people use plants.

● Distinguish between cereals, root crops, legumes, fruits, and vegetables.

■ Explain how humans have increased food production in the world.

◆ List three plants that are widely used as medicines.

PLANTS AS FOOD

Of the more than 350,000 plant species, people use at least 10,000 species for food. Incredibly, fewer than 20 plant species provide more than 90 percent of our food supply. The cultivation of plants for food probably began about 11,000 years ago in the Middle East. Wheat, barley, lentils, and peas were the first domesticated food crops. Growing plants and raising animals for human use is called **agriculture** (AG-ri-KUHL-chuhr). People propagated, or reproduced, individual plants that had valuable characteristics, such as plants that produced the largest or tastiest fruits.

In the 11,000 years that humans have been cultivating plants, we have changed many of the plants so much that they could not grow and survive without us. For example, the wild wheat stalk, as shown in Figure 29-1, breaks easily in the wind, an adaptation that increases the dispersal of its seeds. But early farmers used seeds from plants with stalks that did not break easily for replanting. When these plants were grown, the seeds could be harvested before they fell from the plant. This form of selection—with people acting as selecting agents—has resulted in high-quality food plants.

You have probably eaten Thompson Seedless grapes, McIntosh apples, or Valencia oranges. They are just three examples of the several hundred thousand different cultivars. The word *cultivar* is a contraction of the terms *cultivated* and *variety.* **Cultivars** (KUHL-ti-VAHRZ) are selected by people, and they have at least one distinguishing characteristic that sets them apart from other members of their species. The famous Japanese flowering cherry trees in Washington, D.C., Yoshino cherries, are another example of a cultivar.

FIGURE 29-1

Wheat is one of the world's most important food crops. It is used to make breads, crackers, macaroni, and spaghetti.

SECTION 29-1

SECTION OVERVIEW

Interdependence of Organisms
Section 29-1 describes how humans depend on plants for food, medicines, oxygen, building materials, and many other uses.

Evolution
Section 29-1 also discusses how people have affected plant evolution since early humans began growing crops for food.

ACTIVE READING

Technique: Reading Organizer
Before students begin this chapter, explain that it provides an overview of the types of plants and how plants are used by humans. Have students survey each section before they read it and identify any subtitles, headings, and captions that signal the topic of discussion. As students read, ask them to locate other words in the body of the text that signal the sequential pattern.

QUICK FACT

✔ Of the thousands of species grown for food today, three—wheat, rice, and corn—account for more than half the human food intake from plants. If sugar from cane and beets is added, five plants account for two-thirds of the plant-derived food consumed by humans.

MAKING CONNECTIONS

Social Studies

Ask students to research famines of the past 25 years and to identify their causes and effects. (Examples are drought in the Sudan and Biafra due to the lack of water for plant growth; floods in Bangladesh that destroyed plants and land; war in Ethiopia and Iraq that took farmers from their land and destroyed crops; chemical-farming techniques and slash-and-burn agriculture in West Africa that depleted the soils; and insect plagues in North Africa in which plants were devoured before they could be harvested.) Students should consider the impact famine has on the history, sociology, and politics of each affected country

DEMONSTRATION

Bring a wall map to class and trace the development of wheat and wheat cultivars in the Middle East. Point out the area known as the Fertile Crescent, and note the countries that are there today, including Turkey, Syria, Iraq, Iran, Saudi Arabia, Oman, and Yemen.

TEACHING STRATEGY

Types of Flour

Bring to class two or three different varieties of flours, including all-purpose white, bread flour, cake flour, and whole-wheat flour. Allow students to touch the different types of flour. Ask students to look at the labels and compare the nutritional values of each type of flour. Ask students to explain the differences in protein content. (For example, whole-wheat flour contains more protein because it is made from the entire wheat kernel including the outer covering and the endosperm, while white flour is made only from the endosperm. The protein content of each flour will range from 8 to 11 percent.

Word Roots and Origins

agriculture

from the Latin *ager,* meaning "field," and *cultura,* meaning "cultivation"

FOOD CROPS

Food crops are usually classified partly by use and partly by family. The classification system in Table 29-1 is not like the taxonomic classification used by scientists because most categories contain species that are not closely related. Also, many crops fit into more than one category. For example, corn is a cereal, but it can also be classified as an oil crop, a sweetener, a vegetable, and a beverage.

Cereals

Cereals are grasses that contain grains. Grains are the edible, dry fruits of a cereal. Over half of the world's cultivated land is devoted to cereal crops, such as rice, wheat, corn, oats, sorghum, rye, and millet. Worldwide, cereals provide about 50 percent of the calories in the average human diet. In addition, much of the harvested grain is used for animal feed, so it is indirectly consumed by people as meat, poultry, eggs, and dairy products.

Wheat and corn are produced in the largest amounts. Wheat grows well in moderate to cold climates, including parts of the United States, Russia, and Canada. The United States is the leading producer of corn, also called maize. Rice is different from other cereals because it grows best in shallow water, as shown on page 560. Rice thrives in areas with warm temperatures.

TABLE 29-1 *Food Crops*

Category	Example plants
Cereals	rice, wheat, corn, oats, sorghum, rye, barley, millet
Root crops	potato, cassava, sweet potato, yam, taro
Legumes	soybean, peanut, bean, pea, alfalfa, lentils
Fruits	apple, peach, banana, grape, orange, blueberry, pineapple, cherry, mango, pear
Nuts	peanut, walnut, cashew, pecan, coconut, almond, macadamia, filbert, pistachio
Vegetables	spinach, cabbage, sweet corn, pea, turnip, asparagus, tomato, artichoke, zucchini
Forages	cereals, legumes, grasses
Oils	cottonseed, rapeseed, palm, sesame, soybean, corn, safflower, sunflower
Beverages	coffee, tea, cola, cacao, fruit juice, grape (wine), corn (whiskey), barley and hops (beer)
Sweeteners	sugar cane, sugar beet, sugar maple, corn
Spices	pepper, cinnamon, vanilla, paprika, cloves, saffron, nutmeg, ginger, allspice
Herbs	rosemary, thyme, sage, dill, basil, oregano, mint
Flavorings	cacao (chocolate), coconut, carob, licorice, quinine
Colorings	red beet, anatto, turmeric, saffron, carrot
Additives	guar, locust bean, citrus (pectin), gum arabic, chicle tree
Garnishes	sesame, caraway, and poppy seeds; parsley; pimento
Snacks	popcorn, sunflower seeds, pumpkin seeds

Root Crops

Root crops are roots or underground stems that are rich in carbohydrates. In many parts of the world, root crops substitute for cereals in providing the major part of the diet. However, diets of root crops or cereals alone are usually low in some important amino acids. To correct this deficiency, other foods, such as legumes or animal protein, must be eaten.

Root crops include potatoes, beets, carrots, radishes, rutabagas, and turnips. You may have eaten tapioca pudding, which comes from a plant grown in the tropics called cassava, shown in Figure 29-2.

Legumes

Legumes are members of the pea family and bear protein-rich seeds in pods. Soybean, shown in Figure 29-3, is the most important legume crop because it is produced in the largest amount and has many important uses. Soybean is used to make vegetable oil, soy milk, soy sauce, tofu, and margarine. Alfalfa and clover are legumes used mainly as feed for livestock. Legumes are important in agriculture because they improve the nitrogen content of soil. Recall from Chapter 24 that the bacteria *Rhizobium* form a symbiotic relationship with many legumes.

Fruits, Vegetables, and Nuts

Many "vegetables" we know, such as tomatoes, green beans, and squash, are actually taxonomically classified as fruits. A **fruit** is the part of a flowering plant that usually contains seeds. Foods derived from the leaves, stems, seeds, and roots of nonwoody plants are often called **vegetables.** Vegetables are excellent sources of many important vitamins and minerals, making them an essential part of a healthy diet. Most **nuts** have a hard outer layer and contain a dry, one-seed fruit. Nuts include almonds, walnuts, pecans, and hazelnuts. Peanuts are commonly considered to be nuts but are taxonomically classified as legumes.

Spices, Herbs, and Flavorings

Other food crops add variety and pleasure to our diet by flavoring our water, beverages, and food. More than half the population get daily stimulation from caffeine in coffee, tea, and cola drinks. Both **spices** and **herbs** are used to add taste to food. In general, spices come from plant parts other than the leaf and are tropical. Herbs usually come from leaves and usually can be grown in a home garden. Flavorings, such as chocolate and coconut, are not usually considered spices or herbs and are therefore placed in a separate category. Another flavor, quinine, is used to make tonic water. **Quinine** comes from the bark of the cinchona tree and is used to treat malaria.

Food Production

For decades, experts have been predicting widespread food shortages due to the continuing increase in the world population.

FIGURE 29-2

An important root crop in South America is cassava, which has thick roots that are eaten like potatoes. The starch-filled roots of cassava can be 30–120 cm (1–4 ft) long.

FIGURE 29-3

Soybean is an important legume crop grown in the midwestern and southern parts of the United States. The soybean plant is covered with short, fine hairs and is usually 60–120 cm (2–4 ft) tall. It is an inexpensive and useful source of protein.

QUICK FACT

☑ Modern agricultural practices reduce the genetic diversity of crops, making them more susceptible to pests and disease. In the United States, only 9 percent of the corn varieties that were available 100 years ago are being cultivated.

TEACHING STRATEGY

Storing Starch

CAUTION **Iodine-potassium iodide solution is a poison that is harmful if inhaled or swallowed. This corrosive solution causes irritation to skin, eyes, and mucous membranes.** Have students wear safety goggles, protective gloves, and a lab apron during this activity. Have students add a drop of iodine-potassium iodide solution to carrot, potato, and turnip slices to observe where starch is stored. Have students observe starch granules through a dissecting microscope. The iodine-potassium iodide solution will stain the starch granules in the vegetables a bluish black color.

CONTENT CONNECTION

Section 3-3: Proteins

Review the molecular structure of protein. Ask students what units make up a protein. Remind students that a protein is an organic molecule that is formed from amino acids. Amino acids are joined by peptide bonds.

INCLUSION ACTIVITY

On a sheet of paper, have students make two columns, one labeled "Fruits" and the other labeled "Vegetables." Have students list plant foods in each column according to the definitions of the terms. Then have students compare their grouping with the common meanings of these terms.

SECTION 29-1

CULTURAL CONNECTION

The cuisines of various cultures include combinations of foods that provide a complete and protein-balanced diet. Many foods lack adequate amounts of one or more of the essential amino acids. But when certain foods that contain incomplete proteins are eaten together, they provide a complete set of amino acids. Examples of these protein-balanced combinations include beans and corn in the Americas; wheat and yogurt, or wheat and beans (or lentils) in the Middle East; and rice and soy in Asia.

TEACHING STRATEGY
The Green Revolution

Between 1950 and 1970, researchers concentrated on improving the productivity of land already under cultivation. Their work led to a great increase in worldwide food production, known as the Green Revolution. Ask students to organize a table showing the advantages (increased food production, less starvation) and disadvantages (expensive, difficult to improve soil) of the Green Revolution.

TEACHING STRATEGY
Medicines

Have each student choose a medicine derived from a plant and create a chart that shows the medicine's discovery and uses. You could consult with a pharmacist for suggestions and literature. (Examples include aspirin, from willow and spirea; digitalis, from foxglove; and belladonna, from deadly nightshade.) Ask students why plants are good chemical resources. (Animals have the ability to flee from bad situations, while plants must develop chemical weapons against herbivores.)

Eco Connection

Making Your Own Fertilizer—Composting

Many people are making their own fertilizer through a technique called composting. Compost is a type of organic fertilizer made from decayed plant matter. Compost improves the texture of soil and provides inorganic nutrients for plants.

It's easy to start your own compost pile. Collect dead plant matter, such as grass clippings, leaves, coffee grounds, or sawdust. Make a pile by alternating layers of plant matter with a thin layer of soil or manure. Sprinkle water on the pile to speed the process of decay. After the compost has been allowed to decay for about six months, it should be ready for use in your garden.

Word Roots and Origins

pesticide

contains the suffix *-cide*, from the Latin *cida*, meaning "cut down" or "kill"

However, massive food shortages have not occurred mainly because of increased use of irrigation, fertilizers, and pesticides. Improvements in cultivars; farm machinery; food preservation techniques; and methods of controlling diseases, weeds, and pests have also helped improve food production. **Fertilizers** supply plants with essential mineral nutrients like nitrogen and phosphorus. **Pesticides** are chemicals that kill undesirable organisms that eat crops, such as some insects.

People have made many trade-offs to support an adequate food supply. The negative consequences include massive soil erosion, depletion of fossil fuel and water supplies, pollution, and destruction of wild populations of plants and animals as more land is cultivated.

NONFOOD USES OF PLANTS

In addition to providing us with food, plants provide us with thousands of other essential products. It is hard to imagine how we could live without plants, given the variety of products that contain substances from plants.

Medicines

The ancient Greeks treated headaches with the bark of white willow, which contains the chemical salicin. This use gave scientists the idea to test a similar chemical, acetylsalicylic (uh-SEET-uhl-SAL-uh-SIL-ik) acid. The willow is in the genus *Salix,* hence the names *salicin* and *salicylic.* You know acetylsalicylic acid as **aspirin,** the world's most widely used medicine. Besides pain relief, aspirin is used to thin blood and thereby prevent heart attacks and strokes. Plants were our first medicines, and early plant biologists, like Linnaeus, were often medical doctors. Many modern medicines either still come from plants or were originally obtained from plants and are now synthesized in the laboratory. Table 29-2 lists examples of plants that

TABLE 29-2 *Plants in Medicine*

Plant	Genus name	Drug	Use
Cinchona	*Cinchona*	quinine	treat malaria and certain disorders of heart rhythm
Foxglove	*Digitalis*	digitalis	treat heart disease, help regulate heart rate
Yam	*Dioscorea*	cortisone	treat inflammation and allergies
White willow	*Salix*	acetylsalicylic acid (aspirin)	relieve pain, prevent heart attacks and strokes
Yew	*Taxus*	taxol	treat ovarian cancer, breast cancer, and some types of lung cancer

(a)

(b)

FIGURE 29-4

(a) Taxol, originally derived from the bark of the Pacific yew, is a recently discovered cancer drug. This evergreen tree or shrub produces seeds that look like berries. (b) Foxglove, the source of digitalis, is used in the treatment of heart disease. The beautiful flowers grow in a cluster.

are used in medicine. Two of these plants, yew and foxglove, are shown in Figure 29-4. Scientists are currently evaluating thousands of plant species that may have medicinal properties. Scientists are very concerned about the destruction of rain forests because many rain-forest plant species have yet to be researched. In addition to medicines, plants provide many other products, which are summarized in Table 29-3 on page 566.

Your local health-food store carries a wide range of plant products that claim to prevent disease or improve health. These substances are not regulated by the Food and Drug Administration (FDA). For example, it is reported that garlic and *Echinacea* (purple coneflower) may improve immune response. Consumers should remember that the effectiveness and safety of herbal remedies have not been confirmed by the rigorous scientific testing that new medicines must undergo before receiving FDA approval. The FDA, pharmaceutical companies, and health-care providers are working together to investigate the claims of those who market these remedies.

Clothing and Fabric Dyes

Figure 29-5 shows cotton, which is used to make most of our clothing. The original jeans of Levi Strauss were made with sailcloth woven from hemp. Strauss later switched to cotton. Some expensive clothing is woven with linen, which is made from the flax plant. Artificial fabrics, like rayon, arnel, and cellulose acetate, are made from processed wood fibers. Leather is made from animal hides, but it is usually treated with tannin, a chemical obtained from many tree species. Tanning makes leather stronger and prevents it from rotting.

Prior to the mid-1800s, fabrics were dyed with natural plant dyes. Today most clothing is colored with dyes manufactured from coal, which consists of the remains of ancient plants.

FIGURE 29-5

Cotton, the world's most widely used source of clothing, consists of hairs attached to the seed.

RECENT RESEARCH
Fabric Futures

Tencel™ is the first new fiber to be developed in over 30 years. It is a type of rayon made entirely from cellulose found in the wood pulp of trees grown on managed tree farms. Because Tencel is 100 percent cellulose, it is completely biodegradable. It is very strong and does not shrink much in water. It can be blended with other fibers and turned into knits.

RECENT RESEARCH
New Antibiotic Fibers

Micro Safe™ is a new acetate fiber for the activewear clothing market. This antibiotic fiber controls organisms such as bacteria, yeast, fungi, and mildew, which are found in the humid environment of health clubs or in clothing after excessive perspiration. The target uses for this fiber include undergarments, lingerie, hospital gowns, bedding, gloves, airline pillows, mattress fabrics, and sleeping bags.

SECTION 29-1

QUICK FACT

☑ Most of the steroids used in cortisone and birth-control pills were originally produced by a plant. Although we know how to synthesize steroids, plants do it much more efficiently. Cortisol, which is marketed as cortisone and is used to treat inflammation, is found in wild yam roots *(Dioscorea)* obtained mostly from Mexico. Yams are also commercially grown in India, Japan, and China.

TEACHING STRATEGY

Textiles from Plants

Divide the class into groups of three. Have one student in each group research the techniques used to transform plant fibers into thread. Have a second student research dyeing, weaving, and knitting techniques that are used to create textiles. Have the third student investigate how textiles are made into clothing. Ask the groups to assemble their findings and present a report to the class. Each group can determine what they will present, such as a poster illustrating yarn and fabric dyeing with plant-based dyes; the history of making linen from flax; or the story of how cotton is transformed from a boll to a T-shirt or blue jeans.

TABLE 29-3 *Nonfood Uses of Plants*

Use	Example plants
Brooms/brushes	broomcorn, palms, coconut
Building materials	trees, bamboo, reeds, palms, grasses
Carpets/mats	jute, coconut (coir), cotton, trees
Clothing	cotton, flax (linen), ramie, pineapple, trees (rayon and arnel)
Cosmetics	corn, avocado, carrot, almond, cacao, soybean, macadamia, aloe
Fabric dyes	indigo (blue), madder (red), onion (yellow), black walnut (brown), peach (green), maple (pink)
Fuels	trees, bamboo, water hyacinth, grain alcohol, vegetable oils, gopher plant
Furniture	redwood, oak, rattan, teak, willow (wicker), rushes
Hair dyes	henna, rhubarb, chamomile, black walnut
Incense	frankincense, myrrh, cinnamon
Inks	soybean, flax (linseed oil), tung-oil tree
Leather	black wattle, quebracho, Spanish chestnut (tannin)
Lipstick	jojoba, castor bean, carnauba palm, soybean, coconut
Medicines	foxglove (digitalis), cinchona (quinine), yew (taxol), opium poppy (morphine, codeine), yam (cortisone), aloe, ipecac, ginseng, ginkgo, guarana, purple coneflower, kudzu, saw palmetto
Miscellaneous	cork oak (cork), incense cedar (pencil shafts), trees (disposable baby diapers and cellulose acetate plastic), kapok (life preserver stuffing), rosary pea (bead necklaces), water hyacinth (water purification), lignum vitae (submarine engine bearings)
Musical instruments	ebony (black piano keys), maple (violins), reed (woodwind reeds), African blackwood (woodwinds)
Ornamentals	shade trees, shrubs, lawns, cut flowers, Christmas trees, houseplants
Paints	flax (linseed oil), tung-oil tree, soybean, pine (turpentine)
Paper/cardboard	trees, cotton, flax, hemp, bamboo, papyrus
Perfumes	rose, orange, lavender, orchids, sandalwood, lilac, jasmine, lily of the valley, pine
Pesticides/repellents	tobacco (nicotine sulfate), derris (rotenone), chrysanthemum (pyrethrum), citronella grass, garlic, citrus
Rope	hemp, agave (sisal)
Rubber	rubber tree, guayule
Shampoo	palm oil, coconut, jojoba, aloe, trees, herbs, fruits
Soaps	coconut, palm oil, cacao, lavender, herbs, fruits
Sports equipment	balata (golf balls), persimmon (golf club heads), ash (baseball bats), ebony and ash (pool cues)
Toothpaste	mint, wheat, palm oil, coconut
Tourist attractions	redwoods, giant sequoias, saguaro cactuses, fall foliage, Holland flower bulbs
Waxes	carnauba palm, cauassu, candelilla, bayberry

Fuels

Most of the energy we use for heat, electricity, and machine fuel comes from fossil fuels—coal, oil, and natural gas. Fossil fuels are composed of stored photosynthetic energy from millions of years ago. In developing nations, much of the fuel comes from wood or other plant materials. For example, grains can be fermented into alcohol and mixed with gasoline to make **gasohol.** Gasohol, which is made of about 10 percent alcohol, is an alternative fuel for automobiles.

FIGURE 29-6

The redwood trees that grow along the West Coast of the United States are a majestic sight. Redwoods usually grow 60–84 m (200–275 ft) high. The bark is very thick, making the trees resistant to fires.

Other Uses of Plants

Ornamental trees, shrubs, and other plants outside our homes do much more than provide beauty. Besides their decorative function, they improve the environment by preventing soil erosion, reducing noise, providing habitats for wild animals, acting as windbreaks, providing shade, and moderating temperatures, which, in turn, reduces home heating and cooling costs. Scientists have also found that ornamental plants improve our mental well-being. Gardening has long been a popular hobby in the United States, and it is an important form of exercise for millions of people.

Many plants have become major tourist attractions, such as the California redwoods shown in Figure 29-6. Every fall, people visit the forests of the northeastern United States to view the spectacular leaf colors.

Plants are essential to our survival because they produce virtually all of our food, and they enhance our lives in many ways. Growing cut flowers is now a multibillion-dollar-a-year industry, and it is only a small part of the huge business of growing and using plants. Plants can also provide the inspiration to develop innovative products. The cocklebur plant provided the idea for hook and loop fasteners when the hooked fruit was caught in the inventor's clothing. Plants have made our lives better in numerous ways, and they undoubtedly will continue to do so in the future.

SECTION 29-1 REVIEW

1. **Describe ways that people use plants.**
2. **Distinguish between cereals, root crops, legumes, fruits, and vegetables.**
3. **List three plants that are widely used as medicines.**
4. **List several personal grooming products that use ingredients from plants.**
5. **Explain three environmental benefits of having ornamental trees around a house.**
6. **CRITICAL THINKING How might transferring specific genes from legumes into rice plants help reduce malnutrition?**

SECTION 29-1

ANSWERS TO SECTION 29-1 REVIEW

1. People use plants for many purposes, including food, medicines, cosmetics, clothing, and gardening.
2. Cereals, such as wheat and corn, provide the bulk of the human food supply. Legumes, such as soybeans and beans, are an important protein source for much of the world population. Root crops, such as potatoes and sweet potatoes, store energy in underground roots or stems. Fruits, such as bananas and oranges, are the parts of a flowering plant that usually contain seeds. Vegetables, such as celery and lettuce, are foods derived from soft plant parts, such as leaves, stems, and roots.
3. Quinine, from cinchona trees, is used to treat malaria. Digitalis, from foxglove, is used to treat heart disease. Taxol, from yew, is used to treat cancer.
4. Plants provide ingredients for many personal grooming products, including soap, perfume, shampoo, toothpaste, cosmetics, and hair dye.
5. Ornamental trees provide shade, reduce noise, prevent soil erosion, and modify temperatures, which reduces home heating and air-conditioning costs.
6. Transferring genes from legumes into rice plants could result in rice that contains a complete protein. This could help reduce malnutrition by supplying more of the nutrients needed for a healthy diet, even when other types of foods are unavailable.

BACKGROUND

George Washington Carver was born around 1861 into a slave family, near Diamond Grove, Missouri. As a teenager, he supported himself as a household worker, hotel cook, and farm laborer. In his late twenties, Carver completed his high school education in Kansas while working as a farmhand. He studied piano and art at Simpson College in Iowa. Later he transferred to Iowa State Agricultural College in Ames, Iowa, where he earned a bachelor's degree in agricultural science in 1894 and a master of science degree in 1896. In the fall of 1896, Carver left Iowa for Alabama to direct the new Department of Agriculture at Tuskegee Institute. He spent most of his career teaching and conducting research at Tuskegee Institute.

At Tuskegee Institute, Carver investigated the commercial uses of the peanut and sweet potato. He developed more than 300 peanut products, including milk substitute, face powder, printer's ink, and soap. From sweet potatoes, he developed 118 products.

In 1921, Carver gained national attention when he testified before a Congressional tariff committee. He spoke about the merits of peanut farming. His personality and presentation charmed the audience. Eventually a high peanut tariff was passed, which helped protect U.S. peanut farmers.

In 1923, Carver received the Spingarn Medal for distinguished service in agricultural chemistry from the National Association for the Advancement of Colored People (NAACP).

George Washington Carver: Healer of the Soil

HISTORICAL PERSPECTIVE

Throughout the 1800s, cotton was the main cash crop and was called king of the South. As the century drew to a close, few people saw any reason for change. One man who did, however, was George Washington Carver. In 1896, Carver came to Alabama to teach scientific agriculture and direct the research department at Tuskegee Institute. But he had greater goals—to free African Americans from the ignorance and poverty left by slavery and to free the South of its dependence on cotton.

Scientific Agriculture

After the Civil War, the South remained an agricultural economy, with its recovery dependent on its farmers. Former slaves and masters devoted themselves to raising cotton to sell to eager markets. But George Washington Carver spent his life teaching another, better way.

All his life, Carver was intrigued by plants. He collected them, carried them around, identified and studied them, and even drew them. Like his teachers at Iowa State College of Agriculture (now Iowa State University), Carver believed American agriculture could be transformed by applying scientific principles.

After Carver earned a master's degree in agriculture and botany in 1896, Booker T. Washington, the founder of Tuskegee Institute (now Tuskegee University), invited him to head the school's agriculture department. Carver knew he had found a place where he could carry on his research.

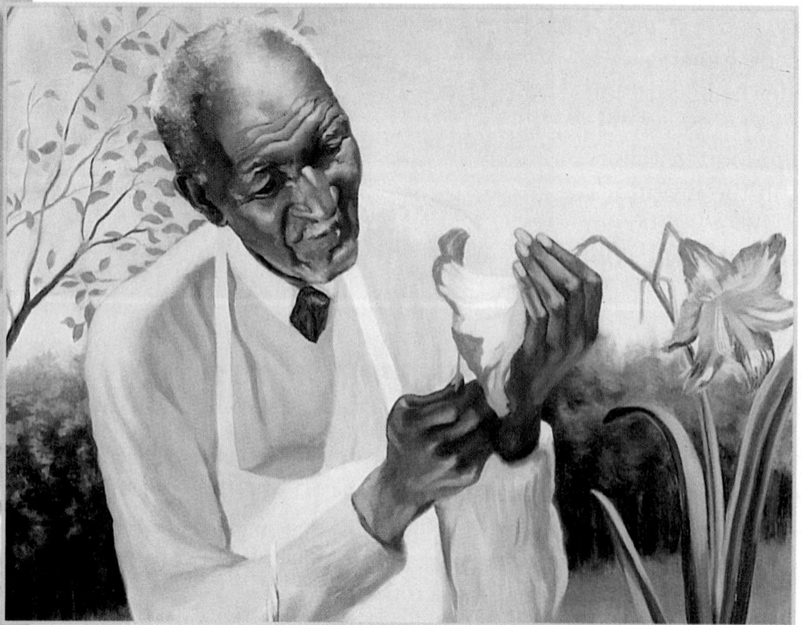

George Washington Carver

Better Soil, Bigger Crops

When Carver saw the desolate fields around Tuskegee, he understood the problem. Planting the same crop in the same soil year after year had taken its toll. Great forests had been felled to make room for cotton, and without tree roots to hold the soil, millions of tons had washed away, along with valuable plant nutrients.

Carver and his students started a 20-acre school "farm." Later Carver said,

> *They told me it was the worst soil in Alabama, and I believed them. But it was the only soil I had. I could either sit down and cry over it or I could improve it.*

The first year, the soil yielded five bales of cotton, 120 bushels of sweet potatoes, and a net loss of $16.50. Each year thereafter, the soil and the harvest improved as Carver and his students made compost, spread fertilizer, planted various crops, and rotated the crops from year to year. After a few growing seasons, some cotton bushes bore as many as 275 giant white bolls.

The farm demonstrated what Carver had told his students: each plant needs certain things, the soil carries certain things, and it's up to the farmer to make any necessary changes to the soil. The depleted soil lacked nitrogen because cotton is a voracious nitrogen consumer, but legumes, such as cowpeas, peanuts, and soybeans, extract nitrogen from air and return it to the soil. The roots of legumes make nitrogen in the air available in a form that plants can use, and thereby enrich the soil with nitrogen.

From Shampoo to Axle Grease

Carver's contributions to southern farming were many, but he is best known for the many new uses he found for the peanut. Peanuts, he said, restore the soil's nitrogen and are easy to plant, grow, and harvest. Carver traveled throughout the Southern states, trying to promote the growing of peanuts.

At one point, Carver wondered if he had made a mistake. Farmers were growing a surplus of peanuts, but was there a market? Farmers would continue to plant peanuts only if they could sell them. Carver returned to his laboratory, where he created more than 300 peanut products. By 1950, the soil-enriching peanut was the second largest crop in the South. Carver experimented with other plants, including soybeans and sweet potatoes. Over 100 products were developed from the sweet potato plant by Carver. In fact, the Center for Sweet Potato Research is located on the Tuskegee University campus.

Working in his laboratory, George Washington Carver developed more than 300 products using peanuts. He also developed flour, molasses, rubber, and postage stamp glue from sweet potatoes.

Presidential Praises

Carver answered hundreds of letters, wrote a series of farm bulletins, displayed his products at fairs, and met with agriculture secretaries of the United States—and even the president. President Theodore Roosevelt once told him, "There is no more-important work than what you are doing." In 1930, the Roosevelt Medal was given to Carver for his contributions to science. Carver spent his life leading African Americans forward. In the process, he paved the way for a better life for an entire region.

Another reason to appreciate Carver's work is that he taught conservation long before that word became popular. He showed that certain "weeds" could make useful products, reminding us that seemingly useless plants may someday serve humanity as vital medicines and food products. Carver once said, "We can learn to synthesize materials for every human need from the things that grow."

In 1951, a national monument was established to honor George Washington Carver. This tribute to his extraordinary life in science is located on a Missouri farm—where he was born.

DISCUSSION

Guide the discussion by posing the following questions:

1. What were Carver's main agricultural concerns when he arrived at Tuskegee? (soil erosion, depletion of the nutrients in the soil, and the reliance on cotton as the cash crop)

2. Calculate the size of Carver's 20-acre school farm at Tuskegee Institute in square meters. Use the following equivalencies to solve the problem: 1 acre $= 43,560\ ft^2$ $= 4,047\ m^2$ (20 acres $\times$ $4,047\ m^2$/acre $= 80,940\ m^2$).

3. What were some of the agricultural methods that Carver encouraged southern farmers to employ? (Carver encouraged southern farmers to obtain agricultural training, to diversify and rotate crops, and to conserve the soil.)

FURTHER READINGS

1. Read the book *Fruits of Creation: A Look at Global Sustainability as Seen Through the Eyes of George Washington Carver,* John S. Ferrell, Macalester Park Publishing, 1995.

2. Read the book *George Washington Carver: Scientists and Symbol,* Linda O. McMurray, Oxford Press, 1981.

3. Read the book *George Washington Carver in His Own Words,* George Washington Carver, University of Missouri Press, 1981.

SECTION OVERVIEW

Interdependence of Organisms

Section 29-2 describes how plants absorb carbon dioxide from the air and release oxygen into the air. This section also emphasizes the beneficial associations that plants have with a variety of organisms.

▶ ENGAGE STUDENTS

Have students observe a classroom terrarium. Discuss the role of plants in the biosphere by asking what an animal needs to live and where the animal gets what it needs (oxygen from the air or water; water from rain or bodies of water; food from plants or animals). Emphasize the systems of interaction between plants and animals. Ask students where plants get the inorganic nutrients (from soil or fertilizer) and CO_2 (from air) they need to live.

GIFTED ACTIVITY

Genetic erosion refers to the decline in the number of plant varieties grown by traditional farmers. This process usually coincides with the introduction of new and highly uniform commercial cultivars of plants. Have students explore the concept of genetic erosion. Ask students which characteristics of plants are viewed as more desirable to farmers and to consumers? (more disease- and insect-resistant, more hardy, more tolerant of cold weather, better tasting, and less susceptible to bruising) Ask students to report on other causes of genetic erosion, such as deforestation, overexploitation, population pressure, and overgrazing.

SECTION

29-2

OBJECTIVES

▲ Define *plant ecology.*

● Give two examples of how plants recycle elements in the environment, and explain why this recycling is essential to humans.

■ Explain how plants benefit from interactions with animals.

◆ Explain the beneficial interactions between plants, fungi, and bacteria.

▲ Describe how people have damaged wild plant populations.

PLANTS AND THE ENVIRONMENT

Based solely on weight, algae and photosynthetic bacteria are dominant organisms in the oceans, and plants are dominant on land. Photosynthetic plants are called producers because they make food for other living things. Organisms that eat other organisms, like animals, are called consumers and depend on plants for a source of organic compounds.

PLANT ECOLOGY

The study of the interactions between plants and the environment is called **plant ecology.** The most important interaction involves the ability of plants to capture solar energy through photosynthesis. In photosynthesis, plants absorb carbon dioxide from the air, produce sugar and starch, and break apart water, releasing oxygen into the air. Consumers use sugar and oxygen in aerobic respiration and produce carbon dioxide and water. Organic compounds from plants provide consumers with energy, building blocks, and essential molecules like vitamins and fiber.

Plants also provide organisms with inorganic nutrients. Plant roots are very efficient at mining the soil for inorganic nutrients, such as nitrogen, phosphorus, potassium, iron, and magnesium. Plants use these inorganic nutrients in the organic compounds they make. Consumers ingest these organic compounds and incorporate the inorganic nutrients into their own bodies. Eventually, these same

FIGURE 29-7

About half of the world's species of plants and animals live in tropical rain forests. Scientists are very concerned about the destruction of rain forests because many plant species have yet to be researched, and some may have medicinal properties.

inorganic nutrients are returned to the soil when the consumer's waste material or dead body is decomposed by bacteria and fungi. Plants thus play a major role in the continuous cycling of the Earth's water, oxygen, carbon dioxide, and inorganic nutrients.

Plants are also responsible for the formation and maintenance of soil. Roots bind soil particles together, leaves reduce the soil-eroding impact of wind and rain, and dead plant parts add organic matter to the soil.

Plant-Animal Interactions

Plants interact with animals in many fascinating ways. Many flowering plants attract pollinators, animals that carry pollen from one plant to another. Usually the pollinator gets a reward for its efforts in the form of food from nectar. The size, shape, color, and odor of many flowers make them attractive to their pollinators. For example, Figure 29-8 shows that in some orchid species, the flowers have evolved to look and smell like the female of their wasp or bee pollinators. A male wasp or bee lands on a flower believing he has located a mate. The pollen he touches sticks to his body and is transferred to the next orchid he visits. In this case, the flower lures the pollinator with the promise of a mate, but fools the insect into picking up pollen without receiving a reward.

FIGURE 29-8

Some orchid species have evolved to resemble their wasp or bee pollinators.

Plant-Microbe Interactions

Two important aspects of plant ecology are plant interactions with fungi and with bacteria. Plant-microbe interactions may be harmful to plants, as in the case of fungal and bacterial diseases. Diseases often cause major crop losses. However, bacteria and fungi also form important beneficial relationships with plants.

The majority of plant species form mycorrhizae, which are symbiotic relationships between fungi and the roots of a plant. A mycorrhizal fungus infects a root, often changing the root structure. However, the fungus does not harm the root. Instead, it greatly increases the root's ability to absorb water and other inorganic nutrients, such as phosphorus and potassium. In return, the root supplies the fungus with energy.

The roots of many plant species also form beneficial associations with bacteria. Some bacteria can take nitrogen gas from the air and "fix" it, or convert it to a form that plants can use. Plants of the legume family, such as peas, beans, and peanuts commonly host bacteria that fix nitrogen.

FIGURE 29-9

The water hyacinth has become a weed that clogs waterways in the southeastern United States.

Protecting Native Plants

We protect and care for many plants that provide us with food, clothing, shelter, and other products. However, humans have drastically changed natural plant populations by introducing foreign plant species, diseases, and animals. Introduced plants, like the water hyacinth shown in Figure 29-9, kudzu, crabgrass, and dandelion have become widespread weeds. **Weeds** are undesirable plants that often crowd out crop plants or native plant species. For

TEACHING STRATEGY
Organisms in the Soil

Have groups of students set up a Baermann funnel, a large funnel attached to a ring stand by rubber tubing that has been clamped off on the end of the funnel. Have students bring in some topsoil collected near their homes. Put a little of the topsoil in a piece of cheesecloth, and put a rubber band around the cheesecloth to create a pouch. Put the pouch in the funnel and add water until the soil in the pouch is submerged. Leave this setup in the classroom for 24 to 72 hours. Then have students decant a little water from the rubber tubing at the end of the funnel into a petri dish, and have them observe the water under a dissecting scope. Have students observe, list, and draw all the different kinds of organisms they discover living in the soil.

TEACHING STRATEGY
Soils

Have students collect different types of soils. Have them observe the soil types they collect and classify soil types by particle size (coarse sand, fine sand, silt, and clay—soils are mixtures of these sizes). Have them check the pH of each soil with pH paper or a pH meter. Discuss what impact acidity and size of soil particles have on plant growth (water retention and absorption, and the availability of inorganic nutrients for plant growth).

CRITICAL THINKING
Relationships

Have students classify the following relationships between animals and plants: a butterfly drinking nectar and pollinating a flower (mutualism); cutworms eating plant stems (predation); and mistletoe on a plant (parasitism).

SECTION 29-2

ANSWERS TO SECTION 29-2 REVIEW

1. Plant ecology is the study of the interactions between plants and their environment.
2. Plants absorb carbon dioxide from the air and water and inorganic nutrients from the soil. In photosynthesis, the water is split, producing oxygen gas, which is released to the air. Then carbon dioxide and hydrogen from water are combined to form organic compounds. Animals breathe in the oxygen produced by plants and eat the plants as food. In cellular respiration, animals use the oxygen and most of the organic compounds provided by plants to produce energy, carbon dioxide, and water. When the animal's waste material or its dead body is decayed by bacteria and fungi, the inorganic nutrients are returned to the soil.
3. Plants often recruit animals to carry or protect their pollen or seeds.
4. Plants interact with mycorrhizal fungi by producing mycorrhizae, which greatly increase the roots' ability to absorb inorganic nutrients and water. Roots of most legumes interact with bacteria to form root nodules, which convert nitrogen gas from the air into a form that the plants can use.
5. People can be harmed by eating or touching poisonous parts of certain plants. The pollen of some flowering plants causes allergies, including hay fever, that affect people.
6. People have damaged wild plant populations by introducing foreign plants that became weeds and crowded out native plants. Human introduction of certain foreign diseases and animals has also devastated wild plant populations.

example, water hyacinths float on lakes and rivers, growing so fast and dense that they impede boats and shade underwater plants. The introduction of a fungal disease, chestnut blight, in 1904 virtually wiped out the American chestnut as a dominant forest tree in the eastern United States. Government inspectors now carefully screen plant materials entering the country to prevent the introduction of new plant pests.

FIGURE 29-10

Giant ragweed, which can grow to more than 4 m (13 ft.) tall, produces massive amounts of pollen that is a major cause of hay fever. The small, dull flowers indicate that it is wind-pollinated.

HARMFUL PLANTS

Despite the many benefits plants provide, some plants can also cause harm. Many deaths are caused by addictive plant products, such as tobacco, cocaine, opium, and alcohol. Some plant species are poisonous when eaten or touched. Poison ivy and poison oak give an itchy rash to millions of Americans each year. Children are often poisoned, though usually not fatally, when they eat the leaves or colorful berries of house or garden plants. Despite widespread reports to the contrary, the popular Christmas plant poinsettia is not deadly, but its sap may cause skin irritation. However, holly berries and all parts of American mistletoe are poisonous.

Tens of millions of people suffer from pollen allergies, one cause of hay fever. **Hay fever** is an allergic reaction that results in sneezing, a runny nose, and watering eyes. Pollen allergies occur at three seasons. In early spring, deciduous trees, such as oak, ash, birch, and sycamore release pollen. In late spring or early summer, it is mainly wild and pasture grasses that cause allergy problems. Of the cereal crops, only rye pollen seems to be an important cause of allergies. In late summer and fall, the highly allergenic pollen of ragweeds, shown in Figure 29-10, affects people. Contrary to popular belief, large, colorful flowers do not cause hay fever. Pollen that causes allergies comes from small, drab flowers that are wind-pollinated.

SECTION 29-2 REVIEW

1. **What is plant ecology?**
2. **How do plants continuously recycle the Earth's inorganic nutrients?**
3. **How do plants benefit from their interactions with animals?**
4. **How do plants benefit from their interactions with fungi and bacteria?**
5. **How do plants harm people?**
6. **CRITICAL THINKING Explain how people have damaged wild plant populations by introducing foreign organisms.**

CHAPTER 29 REVIEW

SUMMARY/VOCABULARY

29-1

- The branch of biology that studies plants is called botany. The practical applications of botany are evident in agriculture, which is the raising of crops and livestock for food or other uses.
- Humans have cultivated plants for at least 11,000 years and have changed many plant species so much, via selection, that these plants can no longer survive in the wild.
- There are several hundred thousand plant cultivars, or cultivated varieties, that are given names, such as McIntosh apple.
- Food crops can be classified in many ways, including by their usage and by their taxonomic classification.
- The major part of the human diet is provided by a few cereal crops in the grass family, especially corn, wheat, and rice.
- The identification of fruits and vegetables is sometimes confusing because everyday definitions are different from botanical definitions. Many of our common vegetables, such as green beans, tomatoes, squash, and pumpkins, are actually fruits. Botanically speaking, a fruit is the part of a flowering plant that usually contains seeds.
- Plants provide many important medicines, such as digitalis, quinine, morphine, and cancer drugs.
- Several factors have increased food production, including the use of fertilizers and pesticides. As land is cultivated to produce an adequate food supply, the health of the environment is compromised by soil erosion, depleted water supplies, and pollution.
- Plants provide thousands of nonfood products, including clothing, fabric dye, lumber, paper, cosmetics, fuel, cork, rubber, turpentine, and pesticides.
- Ornamental plants improve the human environment in many important ways: they provide shade, minimize soil erosion, reduce noise, and lower home energy costs.

Vocabulary

agriculture (561)	cultivar (561)	herb (563)	quinine (563)
aspirin (564)	fertilizer (564)	legume (563)	root crop (563)
botany (561)	fruit (563)	nut (563)	spice (563)
cereal (562)	gasohol (567)	pesticide (564)	vegetable (563)

29-2

- Based on weight, plants are the dominant organisms on land.
- Photosynthetic plants are producers, and animals are consumers.
- Plants play a major role in recycling the Earth's water, oxygen, carbon dioxide, and inorganic nutrients.
- Plants provide animals with inorganic nutrients as well as organic nutrients.
- Most plant roots are infected with beneficial mycorrhizal fungi, which greatly increase the roots' ability to absorb inorganic nutrients.
- Most nitrogen in living organisms must first be fixed by bacteria, which may live in association with plant roots, especially the roots of legumes.
- Plants associate with animals in many mutually beneficial ways, such as providing food to animals that protect them or carry their pollen.
- People have negatively affected wild plant populations by introducing foreign species of plants, animals, and disease organisms.
- Hay fever, or pollen allergies, is caused by the small flowers of certain wind-pollinated plants, not by ornamental flowers.

Vocabulary

hay fever (572)	plant ecology (570)	weed (571)

CHAPTER 29 REVIEW ANSWERS

REVIEW

1. Botany is the study of plants. Agriculture is growing plants and raising animals for food.
2. A cultivar is a plant that has at least one distinguishing characteristic that sets it apart from other members of its species. Examples include Thompson Seedless grapes and Valencia oranges.
3. Answers may include three of the following: rice, wheat, corn, oats, sorghum, rye, and barley.
4. Answers may include three of the following: foxglove (digitalis), cinchona (quinine), yew (taxol), opium poppy (morphine, codeine), yam (cortisone), and aloe.
5. Mycorrhizae result from the symbiotic relationships between fungi and the roots of a plant. Mycorrhizae greatly increase the roots' ability to absorb water and inorganic nutrients, such as phosphorus and potassium.

6. d	**7.** d
8. b	**9.** a
10. c	**11.** a
12. c	**13.** d
14. c	**15.** c

16. A fruit is the part of a flowering plant that usually contains seeds. A vegetable is derived from the leaves, stems, seeds, and roots of plants.
17. Examples include cotton in clothing, wood for paper, and palm oil for soap.
18. A diet of only cereals lacks adequate protein, vitamins, and inorganic nutrients. A complete diet would also require legumes or animal products for protein, and fruits and vegetables for vitamins, inorganic nutrients, and fiber.
19. Because herbal remedies are not regulated by the FDA, their effectiveness and safety may not be supported by scientific experiments.
20. Fungi and plant roots form mycorrhizae, which increase plant absorption of inorganic nutrients and water. Legume roots and bac-

teria form nodules in which the bacteria fix nitrogen, which the plant uses. In both types of interactions, the microbes get energy from the plant.
21. People propagate and protect the most desirable plants, such as those that produce the most food or the tastiest food. The selected plants will survive, while the unselected plants may not.
22. Examples include soap, perfume, shampoo, toothpaste, cosmetics, lipstick, and hair dye.
23. Colorful flowers attract pollinators and increase the chance of sexual reproduction. People may also be more likely to cultivate them.
24. Tobacco, cocaine, opium, and alcohol are addictive plant products.
25. Ornamental trees act as windbreaks, provide shade that can reduce the incidence of skin cancer, reduce noise, and reduce home heating and air-conditioning costs by moderating temperatures.

Critical Thinking

1. Pole beans benefit by using the corn plants as support. Beans are legumes and have nitrogen-fixing root nodules, which will provide some nitrogen fertilizer for the corn and squash. The squash will cover the ground and act as a living mulch to reduce soil erosion, smother weeds, and conserve soil moisture. Corn, a cereal, provides adequate amounts of carbohydrate but has inadequate protein. The beans make up for the lack of protein in the corn. Squash provides vitamins and mineral nutrients.
2. People could grow plants under electric lights to produce food and oxygen. Nuclear, wind, geothermal, or hydroelectric power would need to be used.
3. On the positive side, widespread deforestation by humans would end and plants would no longer be eaten or trampled by animals. On the negative side,

Review

Vocabulary

1. Distinguish between botany and agriculture.
2. Define *cultivar,* and give two examples of popular cultivars.
3. List three cereals.
4. List three medicinal plants.
5. What are mycorrhizae, and how do they benefit a plant?

Multiple Choice

6. Plants provide us with (a) energy (b) oxygen (c) inorganic elements (d) all of the above.
7. Some plants attract pollinators by producing (a) spines (b) grains (c) toxic chemicals (d) modified flowers.
8. Organisms that use plants as food are referred to as (a) predators (b) consumers (c) producers (d) pollinators.
9. The cultivation of plants by humans probably began in (a) the Middle East (b) South America (c) North America (d) Northern Europe.
10. The foxglove plant aids in the treatment of heart disease because it is the source of a medication called (a) quinine (b) morphine (c) digitalis (d) taxol.
11. Edible underground structures that store nutrients are called (a) root crops (b) legumes (c) forage crops (d) grains.
12. Legumes can produce seeds that are protein-rich even in unfertilized soil because of their associations with bacteria that (a) break down carbohydrates (b) absorb oxygen (c) fix nitrogen (d) produce carbon.
13. Plants that are the major source of food for the world today are called (a) spices (b) root crops (c) legumes (d) cereals.
14. Pollen allergies most likely would be caused by pollen from (a) a bee-pollinated flower (b) a large colorful flower (c) a wind-pollinated flower (d) a fragrant flower.
15. During photosynthesis, plants produce (a) carbon dioxide (b) quinine (c) oxygen (d) water.

Short Answer

16. Distinguish between a fruit and a vegetable.
17. List three nonfood, nonmedicinal uses of plants.
18. Why can't people survive solely on a diet of cereals?
19. Why are pharmaceutical companies and health care providers sometimes skeptical of herbal remedies?
20. Explain how plants and microbes interact in mutually beneficial ways.
21. Describe how people act as selecting agents in the evolution of food plants.
22. List several grooming products that use ingredients from plants.
23. What adaptive advantage might colorful flowers have for plants?
24. Name three addictive plant products.
25. Describe how ornamental trees improve the environment.

Critical Thinking

1. Native American farmers often grow corn, pole beans, and squash together. The drawing below shows how the pole beans climbed the corn while the squash carpeted the ground around the corn. What advantage might each plant gain from this arrangement? What advantage is this to the human diet?

2. One explanation for the sudden extinction of dinosaurs is that they starved after an asteroid hit the Earth. The impact created huge dust clouds that blocked out the sunlight and prevented photosynthesis. How would people obtain food and oxygen if the same thing were to happen today?
3. If all animals disappeared from the Earth, what effects, positive and negative, would this have on plants?
4. During the rainy season in the Brazilian rain forest, the rivers flood the land. Many fish from these rivers then swim among the land plants and eat their fruit. How might this intermingling of fish and plants help the plants?
5. Suppose a friend asks you why corn, which he or she considers to be a vegetable, is listed as a cereal crop in the encyclopedia. To answer this question, write a paragraph that explains why corn is a cereal crop, agriculturally, and why it is also a fruit, botanically. Include other examples of foods that are classified as vegetables or grains, but are also fruits.
6. How did artificial selection by humans play a role in the origin of agricultural crops? How is artificial selection similar to natural selection?
7. Many athletes consume carbohydrates before a competition to increase their endurance. The table below shows a variety of legumes and grains that can be purchased at a grocery store. Compare their prices and nutritional backgrounds. Determine which food is the least expensive source of carbohydrates.

Common Legumes and Grains

				Per serving			
Food	Package size	Price	Serving (g)	Calories	Total fat (g)	Total carbohydrate (g)	Total protein (g)
Navy beans	454 g	.69	45	80	0	23	8
Rice	907 g	.79	45	150	0	35	3
Barley	454 g	.49	45	100	0	24	3
Soybeans	454 g	.89	45	170	8	14	15
Spaghetti	907 g	.79	45	200	1	34	6

EXTENSION

1. Examine the labels of some processed foods, such as ice cream, yogurt, crackers, and margarine. Determine the types of plants that were used in their preparation.
2. Read "The Bromeliads of the Atlantic Forest" in *Scientific American,* March 2000, on page 86. Describe at least one interaction between bromeliads and an animal of the forest. How are these plants endangered?
3. Read "Redwoods in the Mist" in *National Wildlife,* April/May 1999, on page 8. Discuss how Todd Dawson of Cornell University explains why tall redwood trees might not have to transport water all the way from their roots to their tops. What role does fog play in the process?
4. Using field guides from a library, determine which plants in your school grounds or in a nearby park are native and which are foreign. For foreign species, list the continent of origin.

ANSWERS

CONTINUED

plants would no longer have animals to carry their pollen or seeds. Animals would no longer protect, fertilize, and propagate plants.

4. The fish may help to disperse the seeds of the fruits.

5. Agriculturally, corn is a cereal crop because it is a grass that produces seeds in grains; botanically, it is a fruit because each grain on an ear of corn is an individual fruit with one seed. All other cereals, such as rice and wheat, as well as vegetables, such as tomatoes and peppers, are botanically fruits.

6. Agricultural crops were derived by selecting certain plants with desirable characteristics to become the parents of the next generation. In both artificial selection and natural selection, offspring of certain individuals survive to become the parents of the next generation. Thus, over time, a population's characteristics become more similar to the characteristics of the individuals selected.

7. At about two cents per gram, barley is the least expensive source of carbohydrates shown.

EXTENSION

1. Answers may include cereals, legumes, nuts, oil, sweeteners, spices, and flavorings.

2. Many bromeliads provide life support for over 900 species. A frog, *Hyla venulosa*, hibernates through the dry season inside the bromeliad *Billbergia zebrina*. Because of farming and logging, only 8 percent of the bromeliad habitat remains.

3. Fog condenses in large quantities on the needles of giant redwoods, and the trees might be able to use the water directly by taking it in through their leaves. The fog may also slow the evaporation rate in the leaves.

4. Help students use the field guides. Encourage students to find examples of the major groups of plants.

CHAPTER 29 INVESTIGATION

TIME REQUIRED

One 50-minute class period and 10 minutes a day for about two weeks

SAFETY PRECAUTIONS

Remind students to wear safety goggles. Avoid eye contact with the nutrient solution.

QUICK REFERENCES

Lab preparation notes are found on page 559D.

Holt BioSources provides a Teaching Resources CD-ROM that contains "Using Gowin's Vee in the Lab" and "Scoring Rubrics."

PROCEDURAL TIPS

1. Hydroponic activity kits are available from WARD'S Natural Science Establishment, Inc.
2. If wheat seeds are difficult to obtain, try using lettuce, tomato, or marigold seeds. These seeds can be purchased in packets at garden shops or nurseries. Other good choices are dried beans from the grocery store, Wisconsin Fast Plants, or chia from ChiaPets™.
3. The cheesecloth may not provide adequate moisture for seed germination, especially if placed in a dry location. It may be necessary to cover the top of the hydroponics cup with plastic wrap to raise the humidity level.
4. Plastic-foam floaters can be easily constructed. You can purchase plastic foam from craft supply stores. Cut a piece of flat plastic foam that is 1/8 in.–1/4 in. in thickness into 2 in. squares or circles. Punch six holes through the plastic foam with a pencil. Each hole can then be used to cradle a seed in nutrient solution.
5. Complete nutrient solution can be purchased from garden shops or nurseries. You can also purchase it from science supply companies.

CHAPTER 29 INVESTIGATION

Comparing Soil-Grown Plants with Hydroponic Plants

OBJECTIVES

- Compare hydroponic plant-cultivation techniques with conventional plant-cultivation techniques.
- Observe the germination of wheat seeds over a two-week period.

PROCESS SKILLS

- analyzing data
- measuring
- comparing and contrasting

MATERIALS

- 2 clear plastic cups
- plastic-foam floater with 6 holes in it
- 50 mL of potting soil
- cheesecloth (must be large enough to cover the plastic-foam floater)
- 12 wheat seeds
- 50 mL of complete nutrient solution
- plastic dropper
- labeling tape
- marking pen
- 50 mL graduated cylinder
- metric ruler

Background

1. Hydroponic cultivation is a technique for growing plants in a solution that contains all of the inorganic nutrients the plant needs. Plants that are grown hydroponically do not require soil.
2. The beginning of growth in a seed is called *germination.*

PART A Day 1

1. Using the marking pen and the labeling tape, label one clear plastic cup "Soil Cultivated," and label the other plastic cup "Hydroponically Cultivated."
2. Fill the cup labeled "Soil Cultivated" halfway with moist potting soil. Place six wheat seeds on the surface of the soil; use the distance between the holes in the foam floater as a guide to determine the spacing of the wheat seeds. (Do not place the floater on the soil.)
3. Press the seeds into the soil until they are approximately 0.5 cm below the surface. Cover the seeds with soil, and press down firmly.
4. Water the seeds with 10 mL of distilled water.
5. Add 50 mL of complete nutrient solution to the cup labeled "Hydroponically Cultivated," and place the plastic-foam floater on the surface of the solution, as shown in the figure below.
6. Place the cheesecloth on top of the floater, as illustrated in the figure. Press lightly at the location of the holes in the floater to moisten the cheesecloth.

7. Place the remaining six wheat seeds on top of the cheesecloth in the cup labeled "Hydroponically Cultivated." Position the seeds so that each one lies in an indentation formed by the cheesecloth in a hole in the floater. Press each seed lightly into the hole until the seed coat is moistened.
8. Place both cups in a warm, dry location. Water the soil-cultivated seeds as needed, monitoring the amount of water added. Aerate the roots of the hydroponic plants every day by using a clean plastic dropper to blow air into the nutrient solution.
9. In your lab report, prepare data tables similar to Table A and Table B. Write your observations of the seeds in your data tables.
10. Clean up your lab materials and wash your hands before leaving the lab.

TABLE A OBSERVATIONS OF SOIL-GROWN PLANTS

Day	Appearance of seedlings	Average height (mm)
1		
2		
3		
4		
5		
6		
7		
8		
9		
10		
11		
12		
13		
14		

TABLE B OBSERVATIONS OF HYDROPONICALLY GROWN SEEDS

Day	Appearance of seedlings	Average height (mm)
1		
2		
3		
4		
5		
6		
7		
8		
9		
10		
11		
12		
13		
14		

PART B Days 2–14

11. Compare the contents of each cup every day for two weeks, and record the appearance of the wheat seedlings in your data tables. If you are unable to observe your seedlings over the weekend, be sure to note in your data table that no observations were made on those days.
12. Each time that you observe the seedlings after they have begun to grow, measure their height and record the average height of the seedlings in each cup in your data tables. To find the average height for one cup, add the heights of each seedling in the cup together and divide by the number of seedlings (6).
13. After the seeds in the cup containing nutrient solution have germinated and formed roots, allow an air pocket to form between the floater and the surface of the nutrient solution. A portion of the roots should still be submerged in the nutrient solution. The air pocket allows the roots of the seeds to absorb the oxygen necessary for metabolic processes while continuing to absorb nutrients from the nutrient solution. Continue to observe and record the progress of the seedlings in each cup on a daily basis.
14. Clean up your lab materials and wash your hands before leaving the lab.

Analysis and Conclusions

1. Based on the data you recorded, which seeds germinated more quickly? Which seeds attained the greatest height?
2. Compare your results with those of your classmates. Were the results the same for each group of students?
3. You planted six seeds in each cup. Why do you think the lab had you do this instead of having you plant a single seed in each cup? Why is the use of more than one sample important?
4. How could hydroponic growing techniques be useful to countries that have either a growing season that is too short to grow a variety of crops or soil that does not support most agricultural crops?

Further Inquiry

The nutrient solution you used in this investigation should have provided all of the inorganic nutrients that the wheat seeds needed for proper growth. How could you determine exactly which inorganic nutrients a plant requires by using hydroponic cultivation?

ANSWERS TO ANALYSIS AND CONCLUSIONS

1. In most cases, seeds grown hydroponically will germinate quicker and attain a greater height than soil-grown seeds.
2. Class results will vary. Most student groups will find that the hydroponically grown plants germinate faster and grow taller than the soil-grown plants.
3. If you used only one seed for each treatment, the results could be unreliable. Not all seeds grown hydroponically or with soil will germinate. It is better to collect data from a large sample group to lessen the probability that a particular effect occurred by chance.
4. Hydroponic growing techniques are not dependent on weather, climate, lighting, or soil conditions. A country that has unfavorable soil-growing conditions can use hydroponic growing techniques to support agricultural crops. Besides growing plants in containers filled with water, growers can use coarse sand, gravel, peat, perlite, or vermiculite to which nutrients have been added.

FURTHER INQUIRY

Different nutrient solutions could be made by altering the quantity of a specific nutrient. By varying the amounts of nutrients, scientists can find the optimal ratio of nutrients for successful growth.

CHAPTER 30
Plant Evolution and Classification

	TOPICS	TEACHING RESOURCES	LABS, CLASSWORK, AND HOMEWORK
BLOCKS 1 & 2 90 minutes	**Introducing the Chapter, p. 578**	**ATE** Focus Concept, p. 578 **ATE** Assessing Prior Knowledge, p. 578 **ATE** Understanding the Visual, p. 578	■ Supplemental Reading Guide, *A Feeling for the Organism*
	30-1 Overview of Plants, p. 579 Adapting to Land, p. 579 Classifying Plants, p. 580	**ATE** Section Overview, p. 579 **ATE** Visual Link Figure 30-2, p. 581 113A, 114A, 115A	**Lab A18,** Comparing Plant Adaptations ■ Portfolio Projects—Occupational Applications Worksheets—Wildlife Biologist ★ Study Guide, Section 30-1 **PE** Section Review, p. 582
BLOCK 3 45 minutes	**30-2 Nonvascular Plants, p. 583** Classifying Bryophytes, p. 583	**ATE** Section Overview, p. 583 **ATE** Critical Thinking, p. 583 **ATE** Visual Link Figure 30-4, p. 583 113A, 114A, 115A	■ Supplemental Reading Guide—*Silent Spring* ■ Portfolio Projects—Occupational Applications Worksheets—Botanist ★ Study Guide, Section 30-2 **PE** Section Review, p. 584
BLOCKS 4 & 5 90 minutes	**30-3 Vascular Plants, p. 586** Seedless Vascular Plants, p. 586 Vascular Seed Plants, p. 588	**ATE** Section Overview, p. 586 **ATE** Critical Thinking, pp. 587, 589 **ATE** Visual Link Figure 30-11 through 30-15, p. 589 113A, 114A, 115A	**PE** **Chapter 30 Investigation,** p. 596 **PE** **Quick Lab,** p. 589, Examining Ferns ■ Portfolio Projects—Plant Research Project ■ Portfolio Projects—Plant Focus Worksheet ★ Study Guide, Section 30-3 **PE** Section Review, p. 592
BLOCKS 6 & 7 90 minutes	**REVIEW and ASSESSMENT** **PE** Chapter 30 Review, pp. 594–595 ■ Performance-Based Assessment—Testing Enzyme Activity in Seeds	Audio CD Program	**CHAPTER TESTING** ■ Chapter Test (blackline master) ▲ Test Generator ■ Assessment Item Listing ■ Scoring Rubrics and Classroom Management Checklists

MODERN BIOLOGY

HOLT BIOSOURCES

One-Stop Planner CD-ROM

Shows these resources within a customizable daily lesson plan:

KEY

- **PE** Pupil's Edition
- **ATE** Teacher's Edition
- ■ Active Reading Guide
- ★ Study Guide
- Teaching Transparencies
- Laboratory Program

One-Stop Planner CD-ROM:

- ■ Holt BioSources Teaching Resources
- ■ Active Reading Guide
- ★ Study Guide
- ▲ Test Generator

READING FOR CONTENT MASTERY

Preview: Have students preview the objectives, the introductory paragraph, topic sentences, and captions before reading the chapter.

ATE Active Reading Technique: Reading Organizer, p. 579

■ Active Reading Guide Worksheet 30-1

PE Summary/Vocabulary, p. 593

■ Active Reading Guide Worksheet 30-2

PE Summary/Vocabulary, p. 593

■ Active Reading Guide Worksheet 30-3

PE Summary/Vocabulary, p. 593

TECHNOLOGY AND INTERNET RESOURCES

CNN PRESENTS Video Segment 19 Magnolia DNA

Holt Biology Videodiscs Teacher's Correlation Guide, Lessons 30-1 through 30-3

Audio CD Program, Sections 30-1 through 30-3

internetconnect

SCILINKS NSTA

On-line Resources: www.scilinks.org

The following *sci*LINKS Internet resources can be found in the student text for this chapter:

Topics:

- Vascular plants, p. 587
- Monocots/dicots, p. 592

go.hrw.com

On-line Resources: go.hrw.com

Visit the HRW Web site for a variety of resources related to this chapter. Just type in the keyword HM2 Chapter 30.

Smithsonian Institution®

Visit **www.si.edu/hrw** for additional on-line resources.

CNNfyi.com

Visit **www.cnnfyi.com** for late-breaking news and current event stories selected just for you.

LAB ACTIVITY PLANNING

CHAPTER 30 INVESTIGATION

Observing Plant Diversity, pp. 596–597

OVERVIEW

In this investigation, students will observe the structural adaptations of mosses, ferns, conifers, and angiosperms. Students will compare the characteristics and life cycle of vascular plants with the characteristics and life cycle of nonvascular plants.

SAFETY

Caution students not to get any sap or plant juice in their eyes, skin, or mucous membranes. Students should wear disposable gloves when handling any wild plant. Remind students to wash their hands thoroughly after completing the lab.

PREPARATION

1. Set up eight display stations—two stations each for mosses, ferns, conifers, and angiosperms. Initially, direct students to spread out among the stations. They do not need to begin the investigation at Station 1.
2. Use a variety of plants and plant parts that are available in your local area. Specimens that can be ordered from WARD'S include mosses (*Polytichum, Sphagnum*); ferns (Boston fern, woodland fern); conifers (Norfolk Island pine, *Podocarpus*); and angiosperms (*Tradescentia, Rex begonia*).

Quick Labs

Examining Ferns, p. 589

Each student group will need disposable gloves, lab aprons, safety goggles, a potted fern, a hand lens, and water for this activity.

UNDERSTANDING THE VISUAL

The tropical flower shown in the photograph is one of about 90 species of heliconia, or lobster claw, plants that belong to the genus *Heliconia.* Because both heliconias and banana plants grow clumped together and have similar leaves, heliconias were once classified in the banana family. However, heliconias do not produce banana-like fruit, and they were later reclassified into their own family, Heliconiaceae.

Almost all of the species of *Heliconia* are native to Central America and South America. These exotic plants are commonly used in flower arrangements in many parts of the world. On Caribbean islands and in Mexico, heliconia leaves are used for roof thatching and for wrapping food. The roots and seeds of heliconias have also been used for medicinal purposes.

FOCUS CONCEPT

Evolution

All organisms in the kingdom Plantae are multicellular. The majority are photosynthetic and live on land. The first true plant is thought to have been similar to a green alga, with adaptations that enabled it to survive the dry conditions on land. From that algal ancestor, a wide variety of plants have evolved.

ASSESSING PRIOR KNOWLEDGE

Review the following concepts.

Natural Selection: *Chapter 15*
Ask students to explain how successive generations become adapted to different conditions and evolve into different species.

Classification: *Chapter 18*
Ask students to explain how organisms are classified. Have students discuss what types of information are included in a phylogenetic tree.

CHAPTER 30

PLANT EVOLUTION AND CLASSIFICATION

Red-eyed tree frogs are climbing on this colorful Heliconia flower. These organisms are from the tropical rain forest of Belize, a country in Central America. Worldwide, tropical rain forests are home to almost half the world's species of plants and animals.

FOCUS CONCEPT: ***Evolution***

As you read, look for the characteristics that reveal how plants have adapted to conditions on land.

SECTION

30-1

OVERVIEW OF PLANTS

Plants dominate the land and many bodies of water. Plants exhibit tremendous diversity. Some plants are less than 1 mm (0.04 in.) in width, and some plants grow to more than 100 m (328 ft) in height. The 12 phyla, or divisions, of kingdom Plantae include more than 270,000 species. Some plants complete their life cycle in a few weeks, while others may live nearly 5,000 years.

OBJECTIVES

▲ Compare and contrast green algae and plants.

● Name three adaptations plants have made to life on land.

■ Compare vascular plants with nonvascular plants.

◆ Define and describe alternation of generations.

ADAPTING TO LAND

Although life had flourished in the oceans for more than 3 billion years, no organisms lived on land until about 430 million years ago, when a layer of ozone formed. The ozone protected organisms from the sun's ultraviolet radiation. Eventually, small club-shaped plants began to grow in the mud at the water's edge.

Preventing Water Loss

The move from water to land offered some organisms distinct advantages, including more exposure to sunlight for photosynthesis, increased carbon dioxide levels, and a greater supply of inorganic nutrients. However, the land environment also presented challenges. Plants on land are susceptible to drying out through evaporation.

One early adaptation to life on land was the **cuticle** (KYOOT-ih-kuhl), a waxy protective covering on plant surfaces that prevents water loss. Although the cuticle protects a plant by keeping water in the plant, it also keeps out carbon dioxide. Plants that had small openings in their surfaces, called stomata, were able to survive. Stomata allow the exchange of carbon dioxide and oxygen.

Reproducing by Spores and Seeds

Successful land plants also developed structures, such as spores and seeds, that helped protect reproductive cells from drying out. A **spore** is a haploid reproductive cell surrounded by a hard outer wall. Spores allowed the widespread dispersal of plant species. Eventually, most plants developed seeds. A **seed** is an embryo surrounded by a protective coat. Seeds also contain **endosperm,** a tissue that provides nourishment for the developing plant. Figure 30-1 shows the unusual seed adaptation of the sugar maple tree; the winged seeds get caught in the wind and twirl to the ground. Seeds are more effective at dispersal than spores are.

FIGURE 30-1

The leaves of a sugar maple tree are covered with a waxy cuticle that prevents water loss. The seeds are found inside a winged fruit. The wings help the seeds disperse away from the parent tree, usually twirling down to the ground.

SECTION 30-1

SECTION OVERVIEW

Evolution

Section 30-1 describes plant adaptations to life on land. The diversity of plants ranges from small, simple plants with inconspicuous reproductive structures to tall, woody plants with showy reproductive structures.

Cell Structure and Function

Section 30-1 also explains that the 12 phyla of plants are classified by many characteristics, including the presence of specialized vascular tissue and the type of reproductive cells.

ACTIVE READING

Technique: Reading Organizer

Have students survey the chapter, noting the headings and subheadings. Also have students note the boldface words and the use of illustrations. Discuss with students how the chapter is structured.

TEACHING STRATEGY

Observing Adaptations to Land

Have students work in groups to create a poster of pictures that show adaptations found in some land plants. Pictures of microscopic guard cells and stomata, vascular tissues, and tissue sections showing the cuticle can be found in college biology textbooks. Have them include pictures of leaves with a waxy coating, herbaceous and woody stems, rhizoids of nonvascular plants, and seeds and fruits. Have them label all the examples.

TEACHING STRATEGY

Simulating Cuticle and Stomata

Provide groups of students with three filter papers and four wax papers (all 4 in. in diameter), tape, three clothespins, a ring stand with one ring, and a balance. Have students saturate the filters with water and weigh them, adding water as needed so that all three filter papers weigh the same. Have students make 30 holes (1/8 in. in diameter) in two pieces of wax paper, place the pieces on either side of one filter, and tape the circles together at the edges. Have them place two unaltered pieces of wax paper on either side of another filter and tape the edges together. Ask them to suspend the three filters by paper clips from the ring stand. When the uncovered filter paper is dry, have students take down all of the filters, remove the wax papers, and reweigh the filters. Ask students to compare the water loss from the different filter papers and discuss the adaptations of land plants this simulates. (The wax paper represents the cuticle, which helps reduce water loss due to evaporation. The holes in the wax paper represent stomata; they allow air to enter the space occupied by the filter paper, which represents the cells.) Ask students how various adaptations would affect water loss. (A thicker cuticle would allow less water loss, and more stomata per square inch would allow more water loss.)

MAKING CONNECTIONS

Fine Arts

Have students conduct research on the kinds of plants that inhabited Earth during the time of the dinosaurs (about 213 to 144 million years ago). Have students paint a large poster illustrating the typical flora of this period. They may also want to add pictures of dinosaurs to their poster.

Word Roots and Origins

vascular

from the Latin *vasculum*, meaning "small vessel"

Transporting Materials Throughout the Plant

Certain species of plants evolved **vascular** (VAS-kyu-luhr) **tissue,** a type of tissue that transports water and dissolved substances from one part of the plant to another. Two types of specialized tissue make up vascular tissue. **Xylem** (ZIE-luhm) carries water and inorganic nutrients in one direction, from the roots to the stems and leaves. **Phloem** (FLOH-uhm) carries organic compounds, such as carbohydrates, and some inorganic nutrients in any direction, depending on the plant's needs. Vascular tissue also helps support the plant, which is an important function for land plants. Aquatic plants are mainly supported by the water around them.

Some plants developed woody tissue and grew to great heights, giving them an advantage in gathering light. **Woody tissue** is formed from several layers of xylem, usually concentrated in the center of the stem. Woody stems are usually brown and rigid. Nonwoody plants are usually called **herbaceous** because they have soft, usually green stems. Because the vascular tissue is not surrounded by rigid sclerenchyma cells, the stems of herbaceous plants are flexible.

CLASSIFYING PLANTS

Study the classification of plants in Table 30-1. The 12 phyla of plants, formerly referred to as *divisions,* can be divided into two groups based on the presence of vascular tissue. The three phyla of

TABLE 30-1 ***The 12 Phyla of the Plant Kingdom***

Type of plant	Phylum	Common name	Approximate number of species
Nonvascular	Bryophyta	mosses	10,000
	Hepatophyta	liverworts	6,500
	Anthocerophyta	hornworts	100
Vascular, seedless	Psilotophyta	whisk ferns	10–13
	Lycophyta	club mosses	1,000
	Sphenophyta	horsetails	15
	Pterophyta	ferns	12,000
Vascular, seed			
Gymnosperms	Cycadophyta	cycads	100
	Ginkgophyta	ginkgoes	1
	Coniferophyta	conifers	550
	Gnetophyta	gnetophytes	70
Angiosperms	Anthophyta	flowering plants	240,000
	class Monocotyledones	monocots	70,000
	class Dicotyledones	dicots	170,000

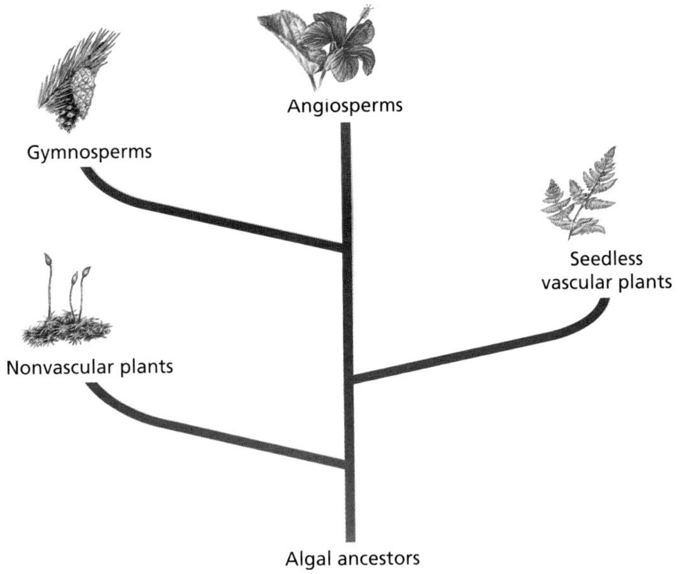

FIGURE 30-2

This phylogenetic tree shows how plants evolved from green algae. The earliest plants were nonvascular and radiated into more-complex forms of vascular plants.

nonvascular plants have neither true vascular tissue nor true roots, stems, or leaves. Most members of the nine phyla of **vascular plants** have vascular tissue and true roots, stems, and leaves.

Notice in Table 30-1 that vascular plants can be further divided into two groups, seedless plants and seed plants. Seedless plants include the phylum of ferns and three phyla made up of plants closely associated with ferns. **Seed plants**—plants that produce seeds for reproduction—include four phyla of gymnosperms and one phylum of angiosperms. **Gymnosperms** (JIM-nuh-SPUHRMZ), which include pine trees, produce seeds that are not enclosed in fruits. **Angiosperms** (AN-jee-uh-SPUHRMZ), also known as flowering plants, produce seeds within a protective fruit. Examples are apple and orange trees.

The Fossil Record of Plants

Figure 30-2 shows the origins of major plant groups. Much of what we know about plant phylogeny has come from studying the fossil record. The fossil record is incomplete, but scientists think that plants evolved from an ancestor of green alga. The strongest evidence supporting this hypothesis lies in the similarities between modern green algae and plants. Both have the same photosynthetic pigments—chlorophylls *a* and *b*, both store energy as starch, and both have cell walls made of cellulose.

Alternating Life Cycles

All plants have a life cycle that involves two phases. The first phase consists of a diploid ($2n$) **sporophyte** (SPOR-uh-FIET) plant that produces spores. The second phase consists of a haploid ($1n$) **gametophyte** (guh-MEET-uh-FIET) plant that produces eggs and sperm. The two plant phases are named for the type of reproductive cells they produce. This type of life cycle, which alternates between the gametophyte phase and sporophyte phase, is called **alternation of generations.**

Eco Connection

Reforestation Efforts

The process of replacing trees that have died or been cut down is called reforestation. Natural reforestation occurs when seeds grow into new seedlings. Throughout the world, many governments and private landowners do not replace trees after land has been cleared to produce timber, build roads, and construct buildings.

Deforestation is occurring at a rate of several hundred thousand square miles per year. Although this rate seems overwhelming and daunting, people can help reforestation efforts by planting seeds or seedlings in their own community. Trees can be planted in yards, in pots on patios or balconies, and along streets. It's best to choose trees that are well adapted to your area.

Many neighborhoods and local organizations sponsor tree-planting programs. Besides providing beautification, these programs can help educate people about the importance of trees. For more information, call your local or state parks department, the U.S. Forest Service, a county extension agent, or a nearby college forestry department.

VISUAL LINK

Figure 30-2

Have students examine Figure 30-2. Ask students the following questions. Which plants are thought to be the first to appear on Earth? (nonvascular) In what order did land plants appear? (nonvascular plants, seedless vascular plants, gymnosperms, and finally angiosperms)

QUICK FACT

☑ Spores are formed by prokaryotes, fungi, protists, and plants.

RECENT RESEARCH

Ancient Internal Plant Parasite

A paleobiologist at the National Museum of Natural History, in Washington, D.C., has found a 300-million-year-old fossil of a fern that contains the feces of insect grubs. The feces were found in an area where the fern had curled up to enclose the parasitic grubs. No fossil grubs were found, apparently because the grubs chewed their way out before the fern fossilized. This finding indicates that ferns adapted very early in their evolutionary history to insect invaders.

RETEACHING *ACTIVITY*

Have students select one nonvascular plant and one vascular plant from Table 30-1 and find pictures from Section 30-3 or other textbooks of the sporophytes and gametophytes of these plants. Ask students to draw a diagram similar to those in Figure 30-3, depicting the life cycle of each plant.

SECTION 30-1

MAKING CONNECTIONS

Language Arts

Linnaeus's system of classification, developed in the mid-1700s, classified plants according to their reproductive parts. He chose this system because reproductive parts vary less from plant to plant than do nonreproductive parts. A recent reexamination of Linnaeus's writings showed that he made detailed comparisons between plant and human reproductive structures. He described "courtships" and "nuptials" of plants, he wrote of shrubs donning "wedding gowns," and he wrote of flowers becoming "husbands" and "wives."

ANSWERS TO SECTION 30-1 REVIEW

1. Plants and green algae have chlorophylls *a* and *b*, store energy as starch, and have cellulose cell walls.
2. The cuticle, a waterproof covering, reduced the plants' water loss. It allowed early land plants to survive in drier environments.
3. Spores are water-resistant reproductive structures that can survive and be dispersed in a dry land environment.
4. Vascular plants have vascular tissues called xylem and phloem, which are specialized for transporting water, inorganic nutrients, and organic compounds. Nonvascular plants lack xylem and phloem.
5. See Figure 30-3. In nonvascular plants, the small sporophyte is dependent on the larger gametophyte. In vascular plants, a small gametophyte and a large, independent sporophyte are present.
6. They have the most efficient system for gathering and distributing water and mineral nutrients.

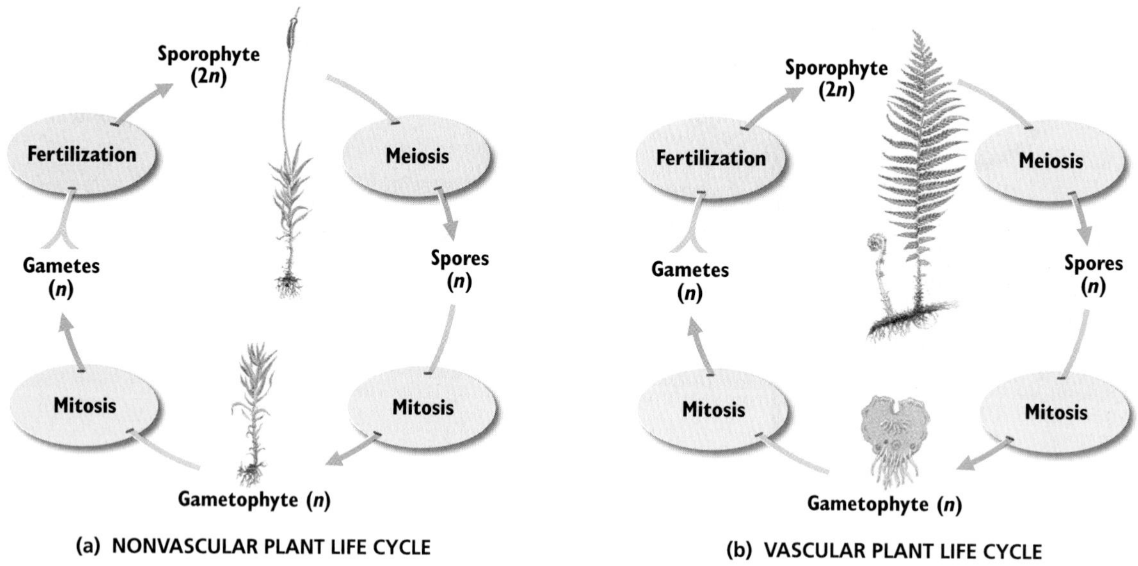

FIGURE 30-3

In the life cycle of a plant, there is an alternation of the haploid gametophyte generation and the diploid sporophyte generation. (a) The life cycle of a nonvascular plant, such as a moss, is characterized by a gametophyte that is larger than the sporophyte. (b) The life cycle of a vascular plant, such as a fern, is characterized by a large sporophyte and a very small gametophyte.

Figure 30-3 shows the life cycles of a nonvascular plant and a vascular plant. In alternation of generations, the gametophyte ($1n$) produces structures that form gametes—eggs and sperm—by mitosis. Once an egg is fertilized by a sperm and produces a zygote, the plant begins the diploid phase of its life cycle. The zygote divides by mitosis to form a sporophyte plant. The sporophyte ($2n$) produces structures that undergo meiosis to form haploid spores. These spores are released by most seedless plants but are retained by seed plants. The life cycle begins again when spores divide by mitosis to form new gametophytes.

In nonvascular plants, the gametophyte is the dominant phase. In contrast, the sporophyte is the dominant phase of vascular plants. Oak trees are large sporophytes that dominate some parts of our landscape. In seedless vascular plants, the gametophyte is usually a separate small organism quite different from the sporophyte. In seed plants, the gametophyte is a very small parasite of the sporophyte. For example, gametophytes of flowering plants are microscopic parts of their flowers.

SECTION 30-1 REVIEW

1. **What are the similarities between today's plants and green algae?**
2. **How does the cuticle represent an adaptive advantage for early land plants?**
3. **How has the evolution of spores contributed to the chances of success for land plants?**
4. **What is the main difference between vascular plants and nonvascular plants?**
5. **Describe alternation of generations, and explain how it differs in vascular plants and nonvascular plants.**
6. **CRITICAL THINKING Why are vascular plants the most successful land plants?**

NONVASCULAR PLANTS

The three phyla of nonvascular plants are collectively called **bryophytes.** *Botanists have identified 16,600 species of bryophytes. They lack vascular tissue and do not form true roots, stems, and leaves. These plants usually grow on land near streams and rivers.*

CLASSIFYING BRYOPHYTES

Bryophytes are the most primitive type of plants. Overall, their characteristics are more closely related to plants than to algae. Bryophytes are mostly terrestrial and have an alternation-of-generations life cycle. Bryophytes are seedless, and they produce spores. Because they do not have vascular tissue, they are very small, usually 1–2 cm (less than 1 in.) in height.

Bryophytes need water to reproduce sexually because the sperm must swim through water to an egg. In dry areas, bryophytes can reproduce sexually only when adequate moisture is available. The asexual production of haploid spores does not require water.

Phylum Bryophyta

Almost every land environment is home to at least one species of moss in the phylum Bryophyta (brie-AHF-uh-tuh). The thick green carpets of moss you see on shady forest floors actually consist of thousands of tiny moss gametophytes. Each gametophyte is attached to the soil by rootlike structures called **rhizoids** (RIE-zoidz). Unlike roots, rhizoids do not have vascular tissue. But rhizoids do function like roots by anchoring the moss and by absorbing water and inorganic nutrients.

Moss gametophytes are usually less than 3 cm (1 in.) tall, as shown in Figure 30-4. The moss sporophyte is attached to and dependent on the larger gametophyte. Gametophytes may be male, female, or contain both male and female reproductive parts.

Mosses are called *pioneer plants* because they are often the first species to inhabit a barren area. This is an important environmental function because mosses gradually accumulate inorganic and organic matter on the surface of rocks, creating a layer of soil in which other plants can grow. In areas devastated by fire, volcanic action, or human activity, pioneering mosses can help trigger the development of new biological communities. They also help prevent soil erosion by covering the soil surface and absorbing water.

SECTION 30-2

OBJECTIVES

▲ Name three types of plants that make up the bryophytes.

● List distinguishing characteristics shared by nonvascular plants.

■ Compare sporophytes in bryophytes with gametophytes in bryophytes.

◆ Describe the environmental importance of bryophytes.

▲ Name the main ways people use *Sphagnum* moss.

FIGURE 30-4

The leafy carpet of moss gametophytes is topped by sporophytes. Mature sporophytes release their spores, which can then grow into a new generation of gametophytes.

SECTION 30-2

SECTION OVERVIEW

Evolution

Section 30-2 surveys the three phyla of bryophytes and describes the characteristics that bryophytes share with algae.

▶ ENGAGE STUDENTS

Demonstrate the remarkable water-absorbing qualities of bryophytes. Grab a handful of dry sphagnum moss (available from florists or garden shops), and place it in a beaker of water. After 10–15 minutes, remove the moss. It can absorb many times its mass in water. Ask students how a bryophyte's ability to absorb water is related to its usefulness in the garden. (Its water-holding ability allows for less-frequent watering. This makes it a good additive to gardening soil.)

VISUAL LINK

Figure 30-4

Have students examine Figure 30-4. Ask them which structure in the photograph is diploid (sporophyte). Ask students which structures are haploid (spores and gametophytes). Ask students how the sporophytes and gametophytes of vascular plants are different from those of nonvascular plants. (Sporophytes are larger than gametophytes in vascular plants and smaller than gametophytes in nonvascular plants.)

CRITICAL THINKING

Acidic Pioneer Plants

Ask students to explain how the acid secreted by mosses helps them to be good pioneer plants. (The acid helps erode rock into soil, making the ground suitable for growth of other plants.)

SECTION 30-2

RECENT RESEARCH

Hornworts

Scientists have constructed a phylogenetic tree of the bryophytes and their relationship to vascular plants based on rRNA sequences. The tree places the hornworts at the base. The common ancestor to the liverworts and mosses branches off next, and then these two groups split. Finally, the vascular plants appear. This phylogenetic tree differs from trees based on morphological characteristics, which place the liverworts, rather than hornworts, at the base of the tree.

ANSWERS TO SECTION 30-2 REVIEW

1. Having separate male and female gametophytes ensures that a gametophyte will not fertilize its own eggs. This increases the chances that the offspring will have new combinations of genes.
2. Bryophytes do not have xylem and phloem; they lack true roots, stems, and leaves; they are seedless; they require water for sexual reproduction; and they have a dominant gametophyte phase.
3. Mosses help soil to form on bare rocks, enabling other plant species to take root.
4. Gardeners use sphagnum peat moss in potting soils to retain moisture.
5. Bryophytes are mostly terrestrial and have an alternation-of-generations life cycle. Tissues are specialized into rootlike rhizoids, stemlike parts, and leaflike parts.
6. Sphagnum moss lacks vascular tissue, which would allow it to grow taller.

FIGURE 30-5
(a) A leafy liverwort has two rows of leaves growing on a stem. (b) A thalloid liverwort has very flat, thin leaves.

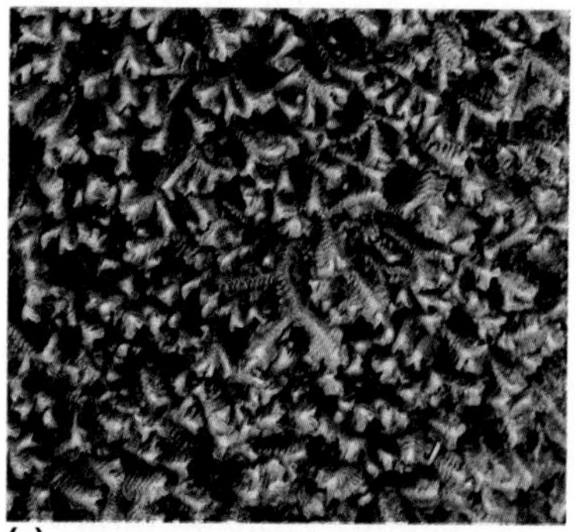
(a)

(b)

Sphagnum (SFAG-nuhm), or peat moss, is a genus of moss that is a major component of peat bogs in northern parts of the world. Peat moss consists of partially decomposed plant matter. In many northern European and Asian countries, peat moss is mined and dried for use as fuel. Sphagnum produces an acid that slows down decomposition in the swamplike bogs. Sphagnum is widely used to enhance the water-retaining ability of potting and gardening soils. Sphagnum is also used by florists to pack bulbs and flowers for shipping.

FIGURE 30-6
Hornworts grow in warm, moist, shaded habitats. Look for hornworts growing along roads or near streams.

Phyla Hepatophyta and Anthocerophyta

Phylum Hepatophyta (HEP-uh-TAHF-uh-tuh) includes the liverworts, unusual-looking plants that grow in moist, shady areas. Most liverworts have thin, transparent leaflike structures arranged along a stemlike axis, as shown in Figure 30-5a. Some liverworts have a **thalloid** (THAL-oid) form—that is, a flat body with distinguishable upper and lower surfaces, as shown in Figure 30-5b. All liverworts lie close to the ground. This is an adaptation that allows them to absorb water readily. In some species, the gametophyte is topped by an umbrella-shaped structure that holds the reproductive cells.

Phylum Anthocerophyta (AN-thoh-suh-RAHF-uh-tuh) includes the hornworts, which resemble liverworts, as shown in Figure 30-6. They grow in moist, shaded areas. They share an unusual characteristic with algae: each cell usually has a single large chloroplast rather than numerous small ones.

SECTION 30-2 REVIEW

1. **What advantage do bryophyte gametophytes that are either male or female have over bryophyte gametophytes that have both male and female structures?**
2. **List two characteristics shared by all bryophytes.**
3. **What role do mosses play in the early development of biological communities?**
4. **What are the main uses of sphagnum moss?**
5. **Why are bryophytes classified with plants instead of with algae?**
6. **CRITICAL THINKING Why can't sphagnum moss grow as large as maple or oak trees?**

Literature & Life

Earth's Green Mantle

This excerpt is from Rachel Carson's book *Silent Spring.*

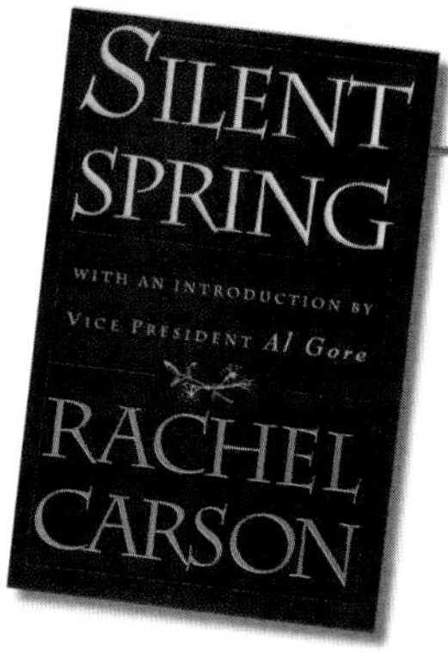

The earth's vegetation is part of a web of life in which there are intimate and essential relations between plants and the earth, between plants and other plants, between plants and animals. Sometimes we have no choice but to disturb these relationships, but we should do so thoughtfully, with full awareness that what we do may have consequences remote in time and place. But no such humility marks the booming "weed killer" business of the present day, in which soaring sales and expanding uses mark the production of plant-killing chemicals.

One of the most tragic examples of our unthinking bludgeoning of the landscape is to be seen in the sagebrush lands of the West, where a vast campaign is on to destroy the sage and to substitute grasslands. If ever an enterprise needed to be illuminated with a sense of the history and meaning of the landscape, it is this. For here the natural landscape is eloquent of the interplay of forces that have created it. It is spread before us like the pages of an open book in which we can read why the land is what it is, and why we should preserve its integrity. But the pages lie unread.

The land of the sage is the land of the high western plains and the lower slopes of the mountains that rise above them, a land born of the great uplift of the Rocky Mountain system many millions of years ago. It is a place of harsh extremes of climate: of long winters when blizzards drive down from the mountains and snow lies deep on the plains, of summers whose heat is relieved by only scanty rains, with drought biting deep into the soil, and drying winds stealing moisture from leaf and stem.

As the landscape evolved, there must have been a long period of trial and error in which plants attempted the colonization of this high and windswept land. One after another must have failed. At last one group of plants evolved which combined all qualities needed to survive. The sage—low-growing and shrubby—could hold its place on the mountain slopes and on the plains, and within its small gray leaves it could hold moisture enough to defy the thieving winds. It was no accident, but rather the result of long ages of experimentation by nature, that the great plains of the West became the land of the sage. . . .

So in a land which nature found suited to grass growing mixed with and under the shelter of sage, it is now proposed to eliminate the sage and create unbroken grassland. Few seem to have asked whether grasslands are a stable and desirable goal in this region. Certainly nature's own answer was otherwise. The annual precipitation in this land where the rains seldom fall is not enough to support good sod-forming grass; it favors rather the perennial bunchgrass that grows in the shelter of the sage.

Reading for Meaning

In this passage, Carson asserts that altering the natural plant life of a region may be a change for the worse. What facts does she use to support her argument?

Read Further

Carson's book *Silent Spring* describes how human activities, including the use of pesticides, endanger our environment. What are some questions we should ask before we disturb the relationships between ourselves and the plants and animals with whom we share Earth?

CHAPTER 30
Literature & Life

BACKGROUND

The popular best-selling book *Silent Spring,* by Rachel Carson (1907–1964), was first published in 1962 and was reissued in 1982. The book warned people about the dangers inherent in the indiscriminate use of pesticides.

This feature describes Rachel Carson's concerns about the use of chemicals to replace native sage with grassland for cattle grazing. In her book, she questions whether grasslands are a stable and desirable goal in that particular region of the country.

READING FOR MEANING

Students' answers will vary. Carson states that nature gives us the answer to why the area is not already a grassland: "The annual precipitation in this land where the rains seldom fall is not enough to support good sod-forming grass; it favors rather the perennial bunchgrass that grows in the shelter of the sage." Altering the native plant life would destabilize the landscape—a landscape that took millions of years to evolve.

READ FURTHER

Students' answers will vary. The effects on the ecosystem's web of life must be considered. How would other species of plants and animals be affected by the elimination of a plant or animal species? What would be the long-term effects of disturbing an ecosystem? Too many times in the past, only the short-term gains have been considered, with little thought given to the long-term effects.

SECTION 30-3

SECTION OVERVIEW

Evolution

This section surveys the vascular plants and describes how various adaptations have allowed this group of plants to flourish.

▶ ENGAGE STUDENTS

Bring to class a living specimen of *Psilotum,* a common whisk fern in the tropics and subtropics, and pictures of fossils of *Rhynia,* an early, extinct vascular plant. Have students examine the specimen and pictures. Have them identify the similarities between them. (Both produce underground stems; lack roots and leaves; branch; and produce spore-shedding sporangia at the growing tips.)

GIFTED ACTIVITY

Have students research the dominant plant forms of the Carboniferous period, from 360 to 286 million years ago. The dominant plants at that time were lycopod trees and giant horsetails, which disappeared during the Permian period. Have students create a picture of a Carboniferous swamp forest. Have them identify the plants they depict. (A college botany textbook would be a good source.)

RECENT RESEARCH

Vascular Plants

Scientists used analyses of the complete subunit rRNA sequences of all living seedless vascular plants and of many seed plants to develop a phylogenetic tree of vascular plants. Their results indicate that lycopods were the earliest diverging group of vascular plants and that angiosperms were the latest. Their phylogenetic tree agrees with the evolutionary relationships of land plants as interpreted from the fossil record.

SECTION

30-3

OBJECTIVES

▲ List two main characteristics of vascular plants.

● Distinguish between seedless plants and seed plants.

■ Distinguish between gymnosperms and angiosperms.

◆ Summarize the adaptive advantages of seeds.

▲ Distinguish between monocots and dicots.

VASCULAR PLANTS

Vascular plants contain specialized conducting tissues (xylem and phloem) that transport water and dissolved substances from one part of the plant to another. Vascular plants can grow larger and live in more environments than nonvascular plants. The strong stems of vascular plants allow the plants to grow tall, enabling them to rise above other plants and receive more sunlight than shorter plants do.

SEEDLESS VASCULAR PLANTS

Seedless vascular plants dominated the Earth until about 200 million years ago. Characteristics of the four phyla of seedless vascular plants are summarized in Table 30-2. The first three phyla are called fern allies, while members of the last phylum are ferns. Spores are the mobile sexual reproductive parts of all seedless plants.

Phylum Psilotophyta

The phylum Psilotophyta (sie-lah-TAHF-uh-tuh) is represented by whisk ferns, illustrated in Figure 30-7. Despite their name, whisk ferns are not ferns at all. They have no roots or leaves and produce spores on the ends of short branches. These features suggest that whisk ferns resemble early land plants. Whisk ferns are epiphytes, which means they grow on other plants, but they are not considered parasites.

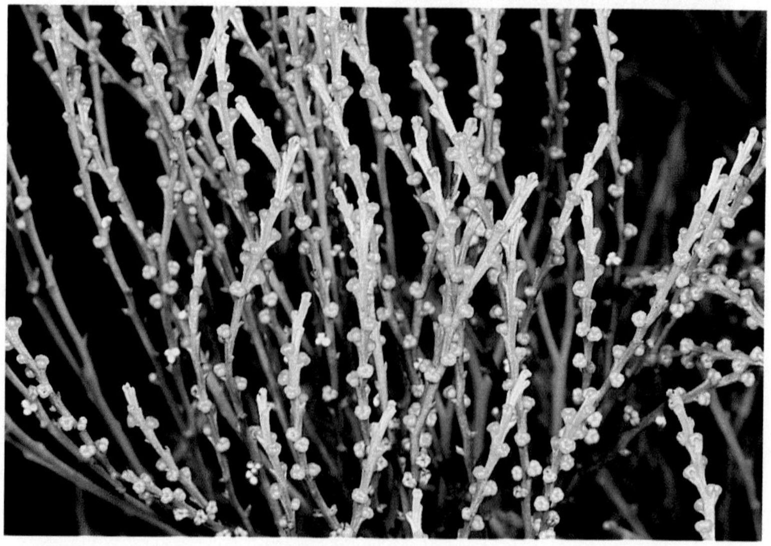

FIGURE 30-7

The whisk fern has branched stem tips. These rare plants are found in tropical and subtropical regions.

TABLE 30-2 Seedless Vascular Plants

Example plant	Phylum	Features	Size	Location
Whisk ferns	Psilotophyta	• produce reproductive structures on the ends of forked branches • no roots or leaves	• about 30 cm (1 ft) tall	• tropical and temperate regions, as far north as South Carolina
Club mosses	Lycophyta	• evergreens that produce spores in cones • have roots	• about 5 cm (2 in.) tall	• tropical and temperate regions, on forest floors, in swamps, or as epiphytes
Horsetails	Sphenophyta	• jointed stems • outer cells of stems contain silica, the major component of sand	• about 60–90 cm (2–3 ft) tall	• tropical and temperate regions, usually in moist soil
Ferns	Pterophyta	• leaves • most have an underground stem • most produce spores on the underside of their leaves	• range from less than 1 cm (0.4 in.) to 25 m (82 ft) tall	• all climates, on forest floors, as epiphytes, some in full sun, some aquatic

Phylum Lycophyta

The phylum Lycophyta (lie-KAHF-uh-tuh) contains the club mosses, shown in Figure 30-8. Because they look like miniature pine trees, club mosses are also called ground pines. They produce a **strobilus** (STROH-bi-luhs), or cone, which is a cluster of sporangia-bearing modified leaves. Club mosses were once widely collected as Christmas decorations.

Another member of phylum Lycophyta is a spike moss called *Selaginella lepidophylla,* native to the American Southwest. *Selaginella* turns brown and curls up in a ball during drought. However, when moistened, the plant uncurls and turns green again after a few hours.

internet**connect**

SCILINKS NSTA
TOPIC: Vascular plants
GO TO: www.scilinks.org
KEYWORD: HM587

FIGURE 30-8

The club moss, sometimes known as a ground pine, is a member of the phylum Lycophyta. The tips of the aerial stems contain conelike structures.

CONTENT CONNECTION

Section 15-2: Natural Selection

Heritable variation that occurs naturally because of mutation and genetic recombination is acted on by environmental factors, which select for individuals that are best adapted to the environment.

CRITICAL THINKING

Carboniferous Plants

Seedless vascular plants, such as tree ferns, giant horsetails, and lycopod trees, dominated the plant life on Earth at the same time that amphibians dominated the animal life (during the Carboniferous period, from 360 to 286 million years ago). Ask students to discuss the characteristics these organisms have in common that help explain their dominance during this period. (The Carboniferous period was a very wet time, and swamps were common. Seedless vascular plants and amphibians require moist environments for reproduction because their sperm cells must swim to egg cells in water.)

MAKING CONNECTIONS

Language Arts

William Shakespeare made many references to plants in his literary works. He referred to ornamental plants, herbals, and poisonous plants. In a number of botanical gardens throughout the world, there are special "Shakespeare gardens," where some of these plants are grown, along with a quotation from Shakespeare referring to the plants. Have students find two or three plant references in the works of Shakespeare, write them down, and find pictures of the plants that are referenced.

CRITICAL THINKING

Using Native Plants

Insects cause significant crop losses by consuming plants and infecting plants with microbial diseases. Insecticides are effective at reducing losses caused by insects, but they are usually harmful to pollinators and other animals. The use of native plants for landscaping reduces plant losses from insects because these plants are usually less susceptible to damage by native insects. But many native plants are not considered desirable as ornamentals. Have students discuss whether they would choose native plants or nonnative plants if they were planting a home landscape. (Answers will vary, depending on individual opinions and the region considered.)

DEMONSTRATION

Bring to class a variety of cones as well as pictures of the gymnosperms that produce the cones and the names of the gymnosperms. Have students organize the cones and pictures into the four phyla of gymnosperms, noting the basis for their classification. Point out that cones contain gametophytes, that pollen grains are usually wind-dispersed, and that seeds mature within female cones and are then wind-dispersed.

RECENT RESEARCH

Anticancer Drug

Taxol, a chemical found in the bark of the Pacific yew tree, has been found to have anticancer properties. Scarcity of the plant and difficulties in extracting the chemical have caused scientists to look into creating synthetic derivatives of taxol. One of these derivatives has been found to be more potent than the parent chemical, and it is cheaper to obtain and easier to use.

FIGURE 30-9

The horsetail, *Equisetum,* has hollow, jointed stems that contain silica. About 300 million years ago, *Equisetum* was a large tree, growing with large club moss trees and ferns in steaming swamps. Over millions of years, the trees and other plant life in the swamps died, became buried, and turned into coal.

FIGURE 30-10

Tree ferns, such as the *Dicksonia antarctica* shown here, look like palm trees but are actually the largest living ferns. Sometimes epiphytic ferns grow on the trunk of tree ferns. Orchids are often cultivated on sections of trunk cut from tree ferns. Tree ferns live in the tropics and subtropics.

Phylum Sphenophyta

The phylum Sphenophyta (sfee-NAHF-uh-tuh) includes horsetails, or *Equisetum.* Horsetails have jointed photosynthetic stems that contain silica, with scalelike leaves at each joint. American pioneers used horsetails to scrub pots and pans; hence, they are frequently called scouring rushes. As you can see in Figure 30-9, the shoots are often highly branched and remind some people of a horsetail.

Phylum Pterophyta

Ferns probably originated over 350 million years ago. Ferns belong to the phylum Pterophyta (tuhr-AHF-uh-tuh) and represent a diverse group. Some are floating plants that are less than 1 cm (0.4 in.) across. Ferns also grow above the Arctic Circle and in desert regions. The largest living ferns are tree ferns, shown in Figure 30-10. These ferns can reach 25 m (82 ft) in height, and some have leaves 5 m (16 ft) long.

Most ferns have an underground stem called a **rhizome** (RIE-zohm). The fibrous rhizomes of some ferns are commonly used as a growing medium for orchids. The tightly coiled new leaves of ferns are called **fiddleheads.** The young fiddleheads of some species are eaten by humans as a vegetable. Fiddleheads uncoil and develop into mature leaves called **fronds.**

VASCULAR SEED PLANTS

The mobile sexual reproductive part of seed plants is the multicellular seed. Seeds are an evolutionary success story. Plants with seeds have a greater chance of reproductive success than seedless plants. Inside the tough, protective outer coat of a seed is an embryo and a nutrient supply. When conditions are too hot or too cold, or too wet or too dry, the seed remains inactive. When conditions favor growth, the seed sprouts, or **germinates**—that is, the embryo begins to grow into a young plant, called a **seedling.**

There are two main groups of seed-bearing vascular plants, gymnosperms and angiosperms. The four phyla of gymnosperms produce naked seeds, which means the seeds are not enclosed and protected in fruits. Most gymnosperms are evergreen and bear their seeds in cones. A **cone** is a reproductive structure composed of hard scales. The seeds of a conifer lie open on the surface of the scales. The one phylum of angiosperms produces seeds that are enclosed and protected in fruits. Angiosperms are commonly referred to as flowering plants. Cones serve some of the same functions for gymnosperms that flowers serve for angiosperms.

Phylum Cycadophyta

Cycads (SIE-KADZ) are gymnosperms of the phylum Cycadophyta. Although cycads flourished during the age of the dinosaurs, only about 100 species survive today. Most are native to the tropics and

FIGURE 30-11

The cycad is a gymnosperm that looks like a palm or fern. Cycads can sometimes grow to 18 m (60 ft) in height. Some cycads live for almost a thousand years.

grow slowly. Many are endangered because of habitat loss, overcollection, and their slow growth. Most cycads have fernlike, leathery leaves at the top of a short, thick trunk. Cycad plants are either male or female, and they bear large cones like those shown in Figure 30-11. Cycads are mostly used as ornamental plants.

Phylum Ginkgophyta

Like cycads, ginkgoes (GINK-ohz) flourished during the time of the dinosaurs. The only species existing today is *Ginkgo biloba,* which is native to China. It is called a living fossil because it closely resembles fossil ginkgoes that are 125 million years old. The ginkgo tree has fan-shaped leaves that fall from the tree at the end of each growing season—an unusual characteristic for a gymnosperm. Trees that lose their leaves at the end of the growing season, like the ginkgo, are called **deciduous.** Most gymnosperms are evergreens and retain their leaves year-round.

Ginkgoes are tolerant of air pollution, making them good plants for urban settings. Ginkgo seeds are considered a delicacy in China and Japan. Notice the plum-shaped, fleshy seeds on the ginkgo shown in Figure 30-12. They are often mistakenly called berries or fruits.

Phylum Coniferophyta

The conifers (KOHN-uh-fuhrz), which are gymnosperms of the phylum Coniferophyta, include pine, cedar, redwood, fir, spruce, juniper, cypress, and bald cypress trees. They are very important sources of wood, paper, turpentine, resin, ornamental plants, and Christmas trees. Gin is flavored with juniper seeds. Amber is yellow or brownish yellow fossilized resin that once flowed from ancient conifers. Prehistoric insects are often preserved in amber, a fact that figured prominently in the plot of the novel *Jurassic Park,* by Michael Crichton.

Quick Lab

Examining Ferns

Materials disposable gloves, lab apron, potted fern, hand lens, water

Procedure

1. Put on your disposable gloves and lab apron.
2. Choose a frond of the fern, and examine its underside for the structures that contain spores.
3. Wash the soil from the underground structures. Examine the fern's horizontal stems and roots.

Analysis How do ferns differ from nonvascular plants? What enables ferns to surpass nonvascular plants in height and size? From what part of the fern do the fronds grow?

FIGURE 30-12

The ginkgo, *Ginkgo biloba,* has unusual fan-shaped leaves and large seeds. This gymnosperm tree can reach heights of 24 m (80 ft).

Quick Lab

Examining Ferns

Time Required 20 minutes

Preparation Tips Purchase fern plants from a florist, a greenhouse, or a grocery store. Have students work in small groups.

Disposal The plants, pots, and soil can be discarded in a solid-waste container. However, the ferns can be repotted if you wish to keep them in the classroom. (Ferns will be covered in detail in Chapter 32.)

Answers to Analysis Ferns have phloem and xylem running through their roots, fronds, and stems. Nonvascular plants have no roots, stems, or true leaves. Ferns can grow taller than nonvascular plants because ferns have vascular tissues. New fronds grow from underground stems. Students should see roots growing out of the underground stems.

QUICK FACT

☑ Ginkgo trees are widely planted in the United States for their value as ornamental trees. However, only the male trees are planted. The seeds from female trees fall on sidewalks and decompose, producing butyric acid, which has a noxious odor. This chemical is also found in rancid butter.

VISUAL LINK
Figures 30-11 through 30-15

Have students study Figures 30-11 through 30-15. Ask them to identify the common characteristics of the four phyla of gymnosperms. (Although they range from shrubs to trees, they all have a dominant sporophyte. The cycads and conifers have needle-like leaves, while the ginkgoes and gnetophytes have broad leaves or no leaves. Most bear cones, but those of the ginkgoes are covered with a fleshy material.)

SECTION 30-3

RETEACHING ACTIVITY

Have students examine Figures 30-7 through 30-17 and Table 30-1. Ask them to draw and label pictures showing the distinguishing characteristics of each phylum of vascular plants. Have them include reproductive structures as well as nonreproductive structures.

TEACHING STRATEGY

Reproductive Structures

Bring examples of a variety of gymnosperms and angiosperms to class, and point out their reproductive structures. Also bring in illustrated diagrams of their life cycles. (These can be found in Chapter 32 and in college biology textbooks.) Have students examine the specimens and identify the sporophytes, male and female gametophytes, seeds, cones or fruits, and function of each structure (for example, male gametophyte produces sperm). Have students determine whether each structure is haploid or diploid. (Sporophytes are diploid; gametophytes, spores, and gametes are haploid.)

RECENT RESEARCH

New Flowering Plant

Scientists have identified an unknown flowering plant by using comparisons of DNA sequences. They have classified the new plant to the family level, and its vegetative characteristics are typical of the identified family. The plant has not yet flowered, so it has not been identified by traditional methods. Scientists suggest that molecular plant identification may become commonplace in the future.

(a) Fir needles and cones

(b) Pine needles and cones

(c) Yew needles and seeds

FIGURE 30-13

The needles and cones of conifers come in many shapes and sizes. (a) The fir tree displays its female cones. Its needle-shaped leaves grow evenly all around the branch. (b) The pine tree shows its small male and larger female cones. Some pines reach heights of 60 m (200 ft). (c) The seed of the yew tree is surrounded by a red covering that looks like a berry. Its leaves are flat, pointed needles that are dark green on top and pale green underneath.

Conifers are woody plants, and most have needle or scalelike leaves, as shown in Figure 30-13. A conifer usually bears both male and female cones. Small male cones typically grow in clusters. Male cones release clouds of dustlike pollen, and then the cones fall from the branches. The pollen falls or blows into the larger female cones, where the egg cells are attached to the scales of the cone. After pollination, the female cone closes up tightly. This protects the developing seeds, which mature after one or two years. The mature seeds are released when the female cone opens.

Redwoods and giant sequoia trees provide a majestic forest setting along the West Coast of the United States. These conifers are the Earth's tallest and most massive living organisms. The tallest living coastal redwood, *Sequoia sempervirens,* is about 110 m (360 ft) tall, the height of a 30-story building. The most massive tree is a giant sequoia, *Sequoiadendron giganteum,* estimated to weigh 5,600 megagrams (6,200 tons).

Phylum Gnetophyta

An odd group of cone-bearing gymnosperms called gnetophytes (NEE-tuh-FIETS) have vascular systems that more closely resemble those of angiosperms. As Figure 30-14 shows, *Ephedra* (ih-FED-ruh) is a genus of desert shrubs with jointed stems that look like horsetails. It is the source of the drug **ephedrine,** which is used as a decongestant.

FIGURE 30-14

Ephedra viridis, called Mormon tea, grows on the rim of the Grand Canyon. This highly branched shrub has small, scale-like leaves. It is the source of the medicine ephedrine and is brewed to make a tea.

Figure 30-15 shows the unique *Welwitschia mirabilis* plant. The plant's stem is only a few centimeters tall but can grow to 1 m (3.3 ft) in diameter. Two leaves grow from the stem. The leaves elongate from their base and then become tattered and split lengthwise by the wind. A mature leaf may be nearly 1 m (3.3 ft) wide and 3 m (10 ft) long. *Welwitschia* grows in the Namib Desert of southwestern Africa. The Namib Desert lies near the Atlantic Ocean, so a thick night fog often rolls in over the desert. *Welwitschia* apparently gets most of its water from the dew that condenses from the fog.

FIGURE 30-15

Welwitschia mirabilis has a short, wide stem and twisting leaves. The female plants of this unusual gymnosperm bear large seed cones that are bright scarlet in color.

Phylum Anthophyta

The largest phylum of plants, Anthophyta (an-THAHF-uh-tuh), includes over 240,000 species of flowering plants. Angiosperms, or the flowering plants, are seed plants characterized by the presence of a flower and fruit. You may recall from Chapter 29 that botanists define a fruit as a ripened ovary that surrounds the seeds of angiosperms. The **ovary** is the female part of the flower that encloses the egg(s).

Angiosperms grow in many forms and occupy diverse habitats. Some are herbaceous plants with showy flowers, such as violets and impatiens. Others, such as rose bushes, are shrubs. Some angiosperms are vines, like grape and ivy plants. Oak, aspen, and birch trees are all flowering plants that have woody stems, although you may never have noticed their small flowers. Grasses are also angiosperms, but you must look closely to see their small, highly modified flowers. The world's largest flower, which can grow to 1 m (3.3 ft) in diameter, is produced by *Rafflesia,* shown in Figure 30-16.

FIGURE 30-16

The stinking-corpse lily, *Rafflesia arnoldii,* has the world's largest flowers but no leaves or stems. The flowers can be male or female, and they are pollinated by flies. The plant lacks chlorophyll and is parasitic on a woody vine native to Southeast Asia.

The Evolution of Angiosperms

Angiosperms first appeared in the fossil record about 135 million years ago. By about 90 million years ago, angiosperms had probably begun to outnumber gymnosperms. What led to the success of this new kind of plant? Several factors were probably involved. In many angiosperms, seeds germinate and produce mature plants, which in turn produce new seeds, all in one growing season. This is a tremendous advantage over gymnosperms, which often take 10 or more years to reach maturity and produce seeds. Also, the fruits of flowering plants protect seeds and aid in their dispersal. Angiosperms also have a more efficient vascular system and are more likely to be associated with mycorrhizae than gymnosperms are. Angiosperms also may gain an advantage by using animal pollination rather than the less-efficient wind pollination method used by gymnosperms. However, wind pollination is used by many successful angiosperms, such as the grasses and many deciduous trees. Finally, angiosperms are more diverse than gymnospems, so they occupy more niches, such as in aquatic, epiphytic, and parasitic environments.

OVERCOMING MISCONCEPTIONS

It is a common misconception that trees don't produce flowers. It is true that gymnosperm trees and tree ferns do not have flowers, but angiosperm trees do produce flowers. In many of these plants, the flowers are very small and pollen is carried by the wind. Other angiosperm trees, such as magnolias, tulip trees, and catalpas, produce very large, showy flowers.

CULTURAL CONNECTION

Peyote

The peyote plant, *Lophophora williamsii,* is a cactus plant native to the area extending from southern Texas to central Mexico. One of the earliest North American hallucinogens, it was already solidly established in religious rites before the Spanish conquest. It was used by the Aztecs, and it is still used by their descendants in Mexico and by members of the Native American Church in the United States. Usually, dried brown pieces of the aboveground part of the cactus, known as mescal buttons, are eaten. Peyote induces chills, vomiting, terror, anxiety, and visual distortion, followed by intense hallucinogenic activity.

TEACHING STRATEGY

Dicots and Monocots

Bring to class a variety of dicot and monocot plants. If possible, most of the plants should be in the flowering stage. Have students use hand lenses and Table 30-3 to classify each plant as either a monocot or a dicot. Have them prepare a chart that includes the name of each plant, the distinguishing characteristics they observed (number of flower parts, leaf venation pattern, number of cotyledons), and the classification of the plant as either a monocot or a dicot.

SECTION 30-3

ANSWERS TO SECTION 30-3 REVIEW

1. Vascular plants contain xylem and phloem. Xylem primarily transports water and inorganic nutrients. The phloem primarily transports organic compounds, like sugars, and some inorganic nutrients.
2. Gymnosperms lack the flowers and fruits found in angiosperms. They also usually have more-primitive vascular tissues than angiosperms have.
3. Male cones produce pollen and are short-lived. Female cones contain eggs, which develop into seeds after fertilization.
4. Monocots usually have one cotyledon, while most dicots have two. Monocot leaves usually have parallel veins, while dicot leaves usually have net venation. Monocot flowers usually have parts in threes, while dicot flowers have parts in fours or fives.
5. Seed plants have multicellular seeds, which usually contain an embryo and a supply of nutrients in a protective coating. Seeds are further protected by fruits or cones and are aided in their dispersal by wind, water, or animals. A seed can rapidly sprout into a sporophyte. In contrast, seedless plants have no dispersible embryos or fruits. Seedless plants are adapted for wind dispersal by single-celled spores that have no nutrient supply. In seedless vascular plants, the spores go through several stages before the sporophyte is produced.
6. Seedless vascular plants are mainly used as ornamentals. Seed plants provide virtually all of our food, clothing, and fuel, and they provide much of our shelter, medicines, paints, inks, and dyes.

TABLE 30-3 ***Comparing Monocots and Dicots***

Plant type	Embryos	Leaves	Stems	Flower parts	Examples
Monocots	One cotyledon 	Parallel venation 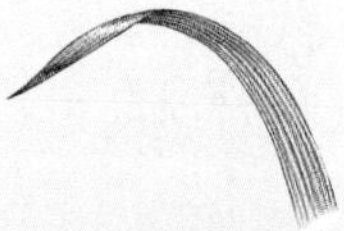	Scattered vascular bundles 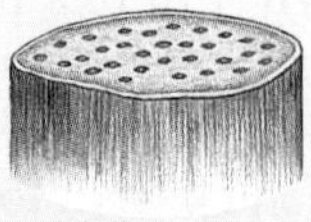	Usually occur in threes 	lilies, irises, orchids, palms, tulips, bananas, pineapples, onions, bamboo, coconut, wheat, corn, rice, oats, barley, sugarcane
Dicots	Two cotyledons 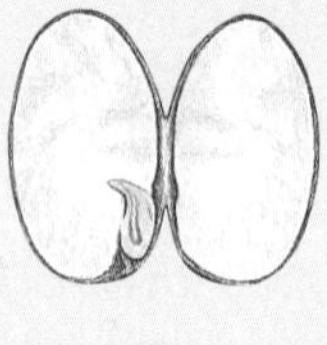	Net venation 	Radially arranged vascular bundles 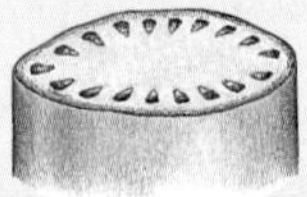	Usually occur in fours or fives 	beans, lettuce, oaks, maples, elms, roses, carnations, cactuses, most broad-leaved forest trees

Monocots and Dicots

The flowering plants classified under phylum Anthophyta are divided into two classes—Monocotyledones (monocots) and Dicotyledones (dicots). The primary feature that distinguishes these two groups is the number of **cotyledons** (KAHT-e-LEE-duhnz), or seed leaves, in a plant embryo. **Monocots** (MAHN-uh-KAHTS) usually have one cotyledon in their embryo, while **dicots** (DIE-KAHTS) typically have two. By comparison, gymnosperms usually have two or more cotyledons.

internet**connect**
SCI*LINKS* NSTA
TOPIC: Monocots/dicots
GO TO: www.scilinks.org
KEYWORD: HM592

Characteristics used to identify monocots and dicots are shown in Table 30-3. Most mature monocot leaves have several main **veins,** or bundles of vascular tissue, running roughly parallel to each other. This is called **parallel venation.** Most dicots have one or more nonparallel main veins that branch repeatedly, forming an interconnected network. This is called **net venation.** When viewed in cross section under a microscope, most monocot stems have scattered vascular bundles, while the vascular bundles of dicot stems are arranged in a circle. It is best to use more than one characteristic to determine whether a species is a monocot or a dicot.

SECTION 30-3 REVIEW

1. What are the two main tissues of all vascular plants?

2. What are the main differences between gymnosperms and angiosperms?

3. How do male cones differ from female cones?

4. What are three characteristics that distinguish monocots from dicots?

5. What reproductive advantages do seed plants have over seedless vascular plants and nonvascular plants?

6. **CRITICAL THINKING** How do humans use seedless vascular plants and seed plants?

CHAPTER 30 REVIEW

SUMMARY/VOCABULARY

30-1

- All plants are multicellular, most live on land, and almost all are photosynthetic.
- There are 12 phyla and over 270,000 species of living plants. Three phyla are nonvascular plants, and the rest are vascular plants.
- Plants originated about 430 million years ago, when a green algae adapted to land.
- The evolution of spores and seeds helped plants colonize land.
- The main function of xylem is to carry water and inorganic nutrients. Phloem carries organic compounds and some inorganic nutrients.
- All plants have a life cycle known as alternation of generations. In alternation of generations, a haploid gametophyte produces gametes, which unite and give rise to a diploid sporophyte. Through meiosis, the sporophyte produces haploid spores, which develop into gametophytes.
- The gametophyte is the dominant phase in nonvascular plants, while the sporophyte is dominant in vascular plants.

Vocabulary

alternation of generations (581)
angiosperm (581)
cuticle (579)
endosperm (579)
gametophyte (581)
gymnosperm (581)
herbaceous (580)
nonvascular plant (581)
phloem (580)
seed (579)
seed plant (581)
spore (579)
sporophyte (581)
vascular plant (581)
vascular tissue (580)
woody tissue (580)
xylem (580)

30-2

- The three phyla of nonvascular plants are called bryophytes. These plants do not have true roots, stems, or leaves. They are very small and often live in moist areas.
- Sphagnum moss, found in peat bogs, is an important bryophyte because of its acidity and moisture-holding capacity.

Vocabulary

bryophyte (583)
rhizoid (583)
sphagnum (584)
thalloid (584)

30-3

- All nine phyla of vascular plants contain vascular tissue, xylem, and phloem. The sporophyte generation is dominant in the life cycle of all vascular plants.
- Whisk ferns have no true roots or leaves.
- Ferns are the dominant phylum of seedless plants, and they have sporangia on their leaves.
- Seed plants are either gymnosperms, which are characterized by naked seeds and no flowers, or angiosperms, which have flowers and seeds enclosed by a fruit.
- Conifers, the dominant living gymnosperms, form vast forests in the Northern Hemisphere.
- Various species of flowering plants have numerous adaptations that give them an advantage over gymnosperms.
- Angiosperms, or flowering plants, are the dominant phylum today, with over 240,000 species.
- Dicots are distinguished from monocots on the basis of several characteristics: cotyledon number, leaf venation, arrangement of stem vascular tissue, and number of flower parts.

Vocabulary

cone (588)
cotyledon (592)
deciduous (589)
dicot (592)
ephedrine (590)
fiddlehead (588)
frond (588)
germinate (588)
monocot (592)
net venation (592)
ovary (591)
parallel venation (592)
rhizome (588)
seedling (588)
strobilus (587)
vein (592)

CHAPTER 30 REVIEW ANSWERS

REVIEW

1. A gametophyte is haploid and produces eggs and sperm. A sporophyte is diploid and produces spores.
2. Angiosperms are flowering plants that produce seeds within a protective fruit. Gymnosperms produce seeds that are not enclosed in a fruit.
3. Monocots have one cotyledon in their embryo, and dicots have two cotyledons in their embryo. Monocots usually have parallel venation, and dicots usually have net venation. Monocots have scattered vascular bundles in their stems, and dicots have radially arranged vascular bundles in their stems. Monocot flowers usually have parts in threes, while dicot flowers have parts in fours or fives.
4. Xylem tissue transports water and inorganic nutrients upward in the plant, from the roots to the stems and leaves. Phloem tissue transports organic compounds and some mineral nutrients throughout the plant.
5. A spore is a haploid reproductive cell surrounded by a hard outer wall. A seed is an embryo surrounded by a protective coat.

6. b **7.** b
8. b **9.** c
10. b **11.** c
12. d **13.** d
14. a **15.** c

16. The first land plants probably developed about 430 million years ago.
17. Life on land was not possible until the atmospheric oxygen concentration became high enough to produce a protective ozone layer that blocks fatal ultraviolet rays.
18. Green algae of phylum Chlorophyta are believed to represent the ancestors of land plants because they share many characteristics with land plants. Shared characteristics include the presence of chlorophylls *a* and *b*, the ability to store energy as starch, and cellulose cell walls.

19. The gametophyte is haploid, and the sporophyte is diploid.
20. The gametophyte is the dominant organism in nonvascular plants, while the sporophyte is dominant in vascular plants.
21. Nonvascular plants are limited in height because they lack vascular tissue specialized for internal transport and support.
22. Nonvascular plants usually live in moist environments because they lack true roots. As a consequence, they are inefficient at absorbing water from deep in the soil. Also, they require water on their surfaces for sexual reproduction.
23. The principal distinguishing feature of vascular plants is the vascular tissue—xylem and phloem—which is specialized for internal transport.
24. Vascular plants have xylem, phloem, and true stems. Most vascular plants also have true roots and leaves.
25. Ferns are seedless plants, and flowering plants are seed plants.

Critical Thinking

1. Any plant phyla except Anthophyta would be appropriate. For trees, use ginkgoes, conifers, and tree ferns. For shrubs, cycads would be best. For ground cover, use ferns, mosses, horsetails, club mosses, and whisk ferns.
2. Tree fossils might be easier to find simply because they are larger than fossils of small plants. Also, because tree tissue is more resistant to decay, it might have a better chance to form a fossil than small plants, which decay quickly.
3. The cuticle should be thicker on the upper surface because the purpose of the cuticle is to prevent evaporation. The lower leaf surface is in contact with water, so the cuticle would not need to be as thick.
4. Luminous moss grows in low light. The lens-shaped cells help

Review

Vocabulary

1. Explain the difference between a gametophyte and a sporophyte.
2. Explain the difference between angiosperms and gymnosperms.
3. List two ways that monocots and dicots differ.
4. Describe the functions of xylem and phloem.
5. Compare and contrast spores and seeds.

Multiple Choice

6. The phylum Bryophyta includes (a) mosses and club mosses (b) hornworts and liverworts (c) liverworts and cycads (d) conifers and ginkgoes.
7. Mosses help start new biological communities by (a) detecting air pollution (b) forming new soil (c) producing spores (d) slowing decomposition.
8. Look at the diagram of the plant life cycle below. The process that occurs at *X* is (a) fertilization (b) meiosis (c) mitosis (d) meiosis and mitosis.

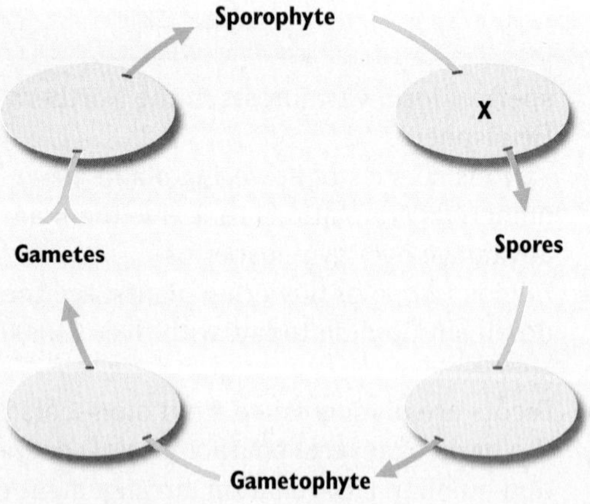

9. The presence of silica in the outer cells of the plant is a characteristic of (a) ferns (b) mosses (c) horsetails (d) club mosses.
10. Flowering plants are in the phylum (a) Psilotophyta (b) Anthophyta (c) Gnetophyta (d) Sphenophyta.
11. Seedless vascular plants include all of the following *except* (a) ferns (b) horsetails (c) cycads (d) club mosses.
12. Gymnosperms include all of the following *except* (a) cycads (b) gnetophytes (c) ginkgoes (d) horsetails.
13. The great success of angiosperms is due in part to (a) a highly efficient vascular system (b) seeds protected by fruits (c) animal dispersion of pollen, fruits, and seeds (d) all of the above.
14. A monocot has (a) parallel leaf venation (b) two seed leaves (c) four-part flowers (d) five-part flowers.
15. The distinguishing feature of all plants is the presence of (a) vascular tissue (b) pollen grains (c) a cuticle (d) rhizoids.

Short Answer

16. When did the first land plants probably develop?
17. Why did it take about 3 billion years after life developed in the oceans for plants to first appear on land?
18. Which phylum of organisms is believed to represent the ancestors of land plants? Why?
19. What is the genetic difference between the gametophyte and the sporophyte?
20. How does alternation of generations in nonvascular plants differ from that in vascular plants?
21. What factor limits the size of nonvascular plants?
22. Why do nonvascular plants usually live in moist environments?
23. What is the principal distinguishing feature of vascular plants?
24. What are the structural features found in vascular plants that are lacking in nonvascular plants?
25. Why are ferns and flowering plants placed in different phyla?

CRITICAL THINKING

1. Steven Spielberg has hired you to landscape his new $110 million Jurassic Park theme ride with living plants similar to those that grew during the Jurassic period. What plant phyla could you use? Give examples of specific plants that you would use as shade trees, shrubs, and low-growing ground cover.
2. Fossil trees are easier to find than fossils of small plants. Give two possible explanations.
3. The water lily, shown below, is an aquatic vascular plant. Do you think its cuticle would be thicker on the upper surface of the leaf or on the lower surface? Explain.

4. Luminous moss, *Schistostega pennata,* has cells shaped like lenses, with chloroplasts spread out behind the curving cell membrane. In what sort of environment would you expect to find this moss?
5. Cactuses are vascular plants that are adapted to dry environments. What specific adaptations might have been selected for during evolution?
6. In the chart below, indicate which groups of plants have vascular tissue. Also indicate which groups of plants have a dominant sporophyte as part of their life cycle.

Plant Evolution

Phylum	Has vascular tissue	Has dominant sporophyte
Anthophyta		
Bryophyta		
Coniferophyta		
Hepatophyta		
Pterophyta		

EXTENSION

1. Read "Tasty Brazil Nuts Stun Harvesters and Scientists" in *Smithsonian,* April 1999, on page 38. Describe a Brazil nut pod, and compare it to the familiar Brazil nuts we find in the grocery store. How are the nuts harvested? Explain the role of the agouti in the survival of the Brazil nut.
2. Go to your local garden center and get a handful of sphagnum moss. Let it air-dry. Determine the mass of the moss to the nearest 0.1 g. Then submerge the moss in water, and let it sit overnight. Remove the moss, let the excess water drain off, and determine the mass of the moss again. Calculate the percentage of increase in mass.
3. Collect one or more fern fronds with mature but unopened sporangia, and place them in an envelope to dry. Soak a clean 7.5 cm diameter clay flowerpot in distilled water or rainwater overnight. Fill the pot with wet sphagnum moss, and place it upside down in a pan of distilled water. Sprinkle the fern spores from the envelope on the flowerpot's upturned surface, and cover the pot with a bell jar or clear 2 L plastic soda bottle with its base cut off. Place the setup in indirect light, and keep it watered. Watch for the development of fern gametophytes in a week or two.

ANSWERS

CONTINUED

to concentrate light on the chloroplasts.

5. Cactus adaptations to dry environments include (1) thick water-storing stems; (2) a thick cuticle; (3) the reduction of leaves to spines, which minimizes surface area for evaporation and protects the plant's stored water from animal predators; (4) stomata that open only at night, when a lower temperature reduces the amount of water that will be lost to evaporation; and (5) shallow and wide-spreading root systems that can take advantage of small rain showers.

6. Bryophyta, Pterophyta, Coniferophyta, and Anthophyta have vascular tissue; Pterophyta, Coniferophyta, and Anthophyta have a dominant sporophyte.

EXTENSION

1. A Brazil nut pod weighs 4–6 lb (1.8–2.7 kg). Each pod holds from 10 to 25 seeds—the familiar Brazil nuts—that are arranged inside the pod like sections of an orange. After Brazil nuts fall to the ground, they are gathered by castañeros who make their living harvesting these nuts. The agouti gathers Brazil nuts and buries them for later use. Many of these buried nuts germinate and grow into new trees.

2. Answers will vary, depending on the initial mass of sphagnum moss. The moss will increase in mass by the absorption of water.

3. Fern gametophytes should develop in one or two weeks.

CHAPTER 30 INVESTIGATION

TIME REQUIRED

One 50-minute class period

QUICK REFERENCES

Lab Preparation notes are found on page 577B.

ANSWERS TO BACKGROUND

1. Most common plants are flowering plants, which are more complex than green algae but which have the same types of chlorophyll and cell walls.
2. Vascular plants contain xylem tissue and phloem tissue. Vascular tissue supports the plants allowing them to grow taller.
3. Alternation of generations is the alternating of a haploid gametophyte with a diploid sporophyte in the life cycle of all plants.
4. All plants produce spores and seeds. Seeds have an advantage over spores—they contain a nutrient supply for the developing embryo.
5. Students should suggest that brightly colored flowers attracted animal pollinators.

ANSWERS TO STATIONS

2. The leafy green part is the haploid moss gametophyte. The bare stalk with a spore capsule is the diploid moss sporophyte.
3. Eggs are produced by the female gametophyte. Sperm are produced by the male gametophyte.
4. No. Mosses obtain water and nutrients from their surroundings through osmosis and diffusion.
5. (a) Water travels through ferns in vascular tissues. Ferns have veins in their leaves and roots. (b) The leafy green frond is diploid. (c) spores
6. (a) Male gametophytes produce sperm, while female gametophytes produce eggs. (b) haploid
7. Ferns are like bryophytes in that their sporophytes grow directly from their gametophytes. They are

Observing Plant Diversity

OBJECTIVES

- Compare similarities and differences among phyla of living plants.
- Relate structural adaptations to the evolution of plants.

PROCESS SKILLS

- observing
- comparing and contrasting
- classifying
- relating structure and function

MATERIALS

- live and preserved specimens representing four plant phyla
- stereomicroscope or hand lens
- compound light microscope
- prepared slides of male and female gametophytes of mosses and ferns

Background

1. How do plants you commonly see compare with their ancestors, the green algae?
2. What are the differences between nonvascular plants and vascular plants? How do these differences relate to the size of a plant?
3. What is alternation of generations? Is it found in all plants?
4. Do all plants produce spores? Do all plants produce seeds? What are the advantages of producing seeds?
5. What do you think was the evolutionary pressure that resulted in colorful flowers?

Procedure

1. CAUTION **Keep your hands away from your face while handling plants.** You will travel to four stations to observe plants that are representatives of four phyla of plants. Record the answers to the questions in your lab report.

STATION 1 Mosses

2. Use a stereomicroscope or a hand lens to examine the samples of mosses, which are bryophytes. Which part of the moss is the gametophyte? Which part of the moss is the sporophyte? Make a sketch of your observations in your lab report. In your drawing, label the gametophyte and sporophyte portions of the moss plant and indicate whether each is haploid or diploid.
3. Use a compound microscope to look at the prepared slides of male and female gametophytes. What kinds of reproductive cells are produced in each of these structures? Draw the cells in your lab report.
4. Do mosses have roots? How do mosses obtain water and nutrients from the soil?

STATION 2 Ferns

5. Look at the examples of ferns at this station. The fern leaf is called a frond. Use the hand lens to examine the fronds.
   a. How does water travel throughout a fern? List observations supporting your answer.
   b. Make a drawing of the fern plant in your lab report.

Indicate whether the leafy green frond in your drawing is haploid or diploid.

c. Search the underside of the fern fronds for evidence of reproductive structures. Make a drawing of your findings in your lab report. What kind of reproductive cells are produced by these structures?

6. Examine the examples of fern gametophytes.

a. Locate and identify the reproductive organs found on the gametophytes. In your lab report, sketch and label these organs and identify the reproductive cells produced by each.

b. Are the gametophytes haploid or diploid?

7. In what ways are ferns like bryophytes? In what ways are they different?

STATION 3 Conifers

8. The gymnosperms most familiar to us are conifers. Look at the samples of conifers at this station.

a. When you look at the limb of a pine tree, which portion (gametophyte or sporophyte) of the plant life cycle are you seeing?

b. In what part of the conifer would you find reproductive structures?

9. Name an evolutionary advancement found in gymnosperms but lacking in ferns.

STATION 4 Angiosperms

10. Draw one of the representative angiosperms at this station in your lab report. Label the representative angiosperm as a monocot or a dicot, and list at least two characteristics you used to identify it.

11. Name an evolutionary development that is present in both gymnosperms and angiosperms but absent in bryophytes and ferns.

12. How do the seeds of angiosperms differ from those of gymnosperms?

13. Examine the fruits found at this station. How have fruits benefited angiosperms?

14. Clean up your materials and wash your hands before leaving the lab.

Analysis and Conclusions

1. In bryophytes, how do the sperm travel from the male gametophyte to the female gametophyte?

2. In angiosperms, how do the sperm get to the part of the flower containing the egg?

3. Which portion of the plant life cycle is dominant in bryophytes? Which portion is dominant in ferns, gymnosperms, and angiosperms?

4. What is a seed? Why is the seed a helpful adaptation for terrestrial plants?

5. Why are gymnosperms referred to as *naked* seed plants?

6. Which group of plants is the most successful and diverse today? What are some adaptations found among members of this group?

Further Inquiry

1. Find out how the geographic distribution of the phyla of living plants relates to their structures.

2. Research the deforestation of tropical rain forests. How are the different groups of plants affected by deforestation?

different in that the fern sporophyte is the predominant generation, while the gametophyte is the predominant bryophyte generation.
8. (a) sporophyte (b) cones
9. Answers may include seeds and pollen.
10. Monocots have parallel venation, scattered vascular bundles, and flower parts occurring in threes. Dicots have net venation, radially arranged vascular bundles, and flower parts occurring in fours or fives.
11. Answers may include seeds and pollen.
12. The seeds of angiosperms develop within fruits.
13. Fruits benefit angiosperms by protecting seeds and promoting seed dispersal.

ANSWERS TO ANALYSIS AND CONCLUSIONS

1. Sperm swim through water from the male gametophyte to the female gametophyte.
2. After pollination occurs, a pollen tube grows from each pollen grain to the ovary, and the sperm nuclei pass through the pollen tube to the ovule.
3. The gametophyte is the dominant portion of the bryophyte life cycle. The sporophyte is the dominant portion of the life cycle of ferns, gymnosperms, and angiosperms.
4. A seed is a sporophyte embryo and nutrient supply surrounded by a protective coat. Inside the seed, the embryo is protected from drying out and it can develop into a new plant when conditions are favorable for growth.
5. The seeds of a gymnosperm are not enclosed by a fruit.
6. Flowering plants are the most successful and diverse. Flowers are adaptations for pollination and fertilization without water. Fruits are adaptations for efficient seed dispersal. Seeds with stored nutrients are adaptations for nourishing young plants as they grow.

CHAPTER 31

PLANT STRUCTURE AND FUNCTION

	TOPICS	TEACHING RESOURCES	LABS, CLASSWORK, AND HOMEWORK
BLOCK 1 45 minutes	**Introducing the Chapter, p. 598**	ATE Focus Concept, p. 598 ATE Assessing Prior Knowledge, p. 598 ATE Understanding the Visual, p. 598	■ Supplemental Reading Guide, *A Feeling for the Organism*
	31-1 Plant Cells and Tissues, p. 599 Specialized Plant Cells, p. 599 Tissue Systems, p. 600 Growth in Meristems, p. 602	ATE Section Overview, p. 599 ATE Visual Link Figure 31-2, p. 601 120A, 125, 126, 129	■ Portfolio Projects—Plant Research Project ★ Study Guide, Section 31-1 PE Section Review, p. 602
BLOCK 2 45 minutes	**31-2 Roots, p. 603** Types of Roots, p. 603 Root Structures, p. 604 Root Functions, p. 606	ATE Section Overview, p. 603 ATE Critical Thinking, p. 604 ATE Visual Link Figure 31-3, p. 603 Figure 31-7, p. 605 120A, 121, 124	PE **Quick Lab,** p. 606, Observing Roots **Lab A20,** Relating Root Structure to Function ■ Portfolio Projects—Plant Focus Worksheet ★ Study Guide, Section 31-2 PE Section Review, p. 607
BLOCKS 3 & 4 90 minutes	**31-3 Stems, p. 608** Types of Stems, p. 608 Stem Structures, p. 609 Stem Functions, p. 612	ATE Section Overview, p. 608 ATE Critical Thinking, p. 610 ATE Visual Link Figure 31-12, p. 611 120A, 121, 123, 127	PE **Quick Lab,** p. 609, Observing Stems ■ Portfolio Projects—Plant Focus Worksheet **Lab B17,** Transpiration and Stem Structure **Lab C35,** Staining and Mounting Stem Cross Sections ★ Study Guide, Section 31-3 PE Section Review, p. 613
BLOCKS 5 & 6 90 minutes	**31-4 Leaves, p. 615** Types of Leaves, p. 615 Leaf Structures, p. 616 Leaf Functions, p. 617	ATE Section Overview, p. 615 ATE Critical Thinking, p. 617 ATE Visual Link Figure 31-16, p. 617 116A, 117A, 119, 121, 122	PE **Chapter 31 Investigation,** p. 622 **Lab A18,** Comparing Plant Adaptations **Lab A19,** Inferring Function from Structure **Lab C36,** Using Paper Chromatography to Separate Pigments ★ Study Guide, Section 31-4 PE Section Review, p. 618
BLOCKS 7 & 8 90 minutes	**REVIEW and ASSESSMENT** PE Chapter 31 Review, pp. 620–621 ■ Performance-Based Assessment—Testing Enzyme Activity in Seeds	Audio CD Program	**CHAPTER TESTING** ■ Chapter Test (blackline master) ▲ Test Generator ■ Assessment Item Listing ■ Scoring Rubrics and Classroom Management Checklists

MODERN BIOLOGY

HOLT BIOSOURCES

KEY

PE Pupil's Edition
ATE Teacher's Edition
■ Active Reading Guide
★ Study Guide

Teaching Transparencies

Laboratory Program

One-Stop Planner CD-ROM
Shows these resources within a customizable daily lesson plan:

■ Holt BioSources Teaching Resources
■ Active Reading Guide
★ Study Guide
▲ Test Generator

READING FOR CONTENT MASTERY

Preview: Help students preview the objectives, the introductory paragraph, topic sentences, and captions before reading the chapter.

ATE Active Reading Technique: Reading Organizer, p. 599
■ Active Reading Guide Worksheet 31-1
PE Summary/Vocabulary, p. 619

■ Active Reading Guide Worksheet 31-2
PE Summary/Vocabulary, p. 619

■ Active Reading Guide Worksheet 31-3
PE Summary/Vocabulary, p. 619

■ Active Reading Guide Worksheet 31-4
PE Summary/Vocabulary, p. 619

TECHNOLOGY AND INTERNET RESOURCES

CNN PRESENTS Video Segment 22
Plastic Farming

Holt Biology Videodisks Teacher's Correlation Guide, Lessons 31-1 through 31-4

Audio CD Program, Sections 31-1 through 31-4

internet**connect**

SCILINKS NSTA

On-line Resources: www.scilinks.org
The following *sci*LINKS Internet resources can be found in the student text for this chapter:

Topics:
- Types of roots, p. 603
- Root structures, p. 605
- Transpiration, p. 613
- Leaf functions, p. 618

go.hrw.com

On-line Resources: go.hrw.com
Visit the HRW Web site for a variety of resources related to this chapter. Just type in the keyword HM2 Chapter 31.

Smithsonian Institution®
Visit **www.si.edu/hrw** for additional on-line resources.

CNN fyi.com
Visit **www.cnnfyi.com** for late-breaking news and current event stories selected just for you.

LAB ACTIVITY PLANNING

CHAPTER 31 INVESTIGATION

Observing Roots, Stems, and Leaves, pp. 622–623

OVERVIEW
In this investigation, students will observe the tissues and structures that make up roots, stems, and leaves. Students will examine how roots, stems, and leaves are adapted for the functions they perform.

SAFETY
Special care should be taken when viewing slides under high power so that the objective does not break the slide. Have students dispose of broken glass in a container designated for glass only. Caution students not to use microscopes with wet hands or near water.

PREPARATION
1. Purchase prepared slides from a biological supply house, such as WARD'S. Students will need to observe five different slides: an *Allium* root tip longitudinal section, a *Syringa* leaf cross section, a *Ranunculus* root cross section, a *Ranunculus* stem cross section, and a *Zea mays* stem cross section.
2. Set up three lab stations, one for each part (roots, stems, and leaves). At each station, set up two or three microscopes with slides. Have each group rotate among the three lab stations.

Quick Labs

Observing Roots, p. 606
Ten days ahead of time, soak radish seeds in water overnight. Transfer the seeds to moist paper towels. When the seedlings develop leaves, allow them to dry out just enough to wilt. Groups will need a hand lens, a Petri dish, water, and a pipet.

Observing Stems, p. 609
Each group of students will require at least one winter twig. Each twig should be at least 30 cm long. The twigs should be freshly cut from any species of living, deciduous tree. Each group will also need a freshly cut herbaceous stem about 30 cm long, with leaves.

UNDERSTANDING THE VISUAL

Tell students that the photograph shows a cluster of individual flowers found in the *Dahlia* "flower." *Dahlia* is a popular garden plant that is a member of a family of plants called the composites, which also includes sunflowers and daisies. Tell students that the flower parts of a plant are actually modified stems and leaves. Point out to students that the function of each of the plant organs (the roots, stems, and leaves) is dependent on the structures found within each organ.

FOCUS CONCEPT

Cell Structure and Function

The function of each plant cell is linked to its structure. Plant cells vary in specialization. They include unspecialized cells, specialized cells for photosynthesis, and highly specialized cells that regulate gas exchange. Plant cells are organized into tissues, which in turn are organized into organs. All cells function in a coordinated manner.

ASSESSING PRIOR KNOWLEDGE

Review the following concepts.

Plant Cells: *Chapter 4*
Have students discuss how plant cells compare with animal cells. Encourage students to suggest the functions of the different organelles. For example, chloroplasts are found only in algae and plant cells and serve as the site for making carbohydrates through photosynthesis.

Cell Division: *Chapter 8*
Ask students to define the terms *mitosis* and *meiosis.* Lead students to review the two processes and their respective end products.

CHAPTER 31

PLANT STRUCTURE AND FUNCTION

This Dahlia "flower" is really a cluster of small, individual flowers. Dahlia is a member of the family of plants called the composites, which also includes sunflowers and daisies.

FOCUS CONCEPT: ***Cell Structure and Function***

As you read, notice how the roots, stems, and leaves perform different functions for a plant. Plant tissues are organized for specialized functions.

PLANT CELLS AND TISSUES

Plants have adapted to a range of environments over the course of their evolution. As plants grow, their cells become specialized for particular functions. The patterns of specialized tissues vary in each of the plant organs—the root, the stem, and the leaf. They also vary depending on the plant's stage of growth and taxonomic group. This chapter examines the structure and function of roots, stems, and leaves.

SECTION

31-1

OBJECTIVES

▲ Describe the three kinds of plant cells.

● Explain the differences between the three plant tissue systems.

■ Describe the main types of meristems.

◆ Differentiate between monocot and dicot meristems.

▲ Differentiate between primary and secondary growth.

SPECIALIZED PLANT CELLS

All organisms are composed of cells. Recall from Chapter 4 that plant cells have unique structures, including a central vacuole, plastids, and a thick cell wall that surrounds the cell membrane. These common features are found in the three types of specialized plant cells—parenchyma, collenchyma, and sclerenchyma—which are shown in Figure 31-1. Small changes in the structure of these plant cells help make different functions possible. The three types of plant cells are arranged differently in roots, stems, and leaves.

Parenchyma (puh-REN-kuh-muh) cells are usually loosely packed cube-shaped or elongated cells that contain a large central vacuole and have thin, flexible cell walls. Parenchyma cells are involved in many metabolic functions, including photosynthesis, storage of water and nutrients, and healing. These cells usually form the bulk of nonwoody plants. For example, the fleshy part of an apple is made up mostly of parenchyma cells.

The cell walls of **collenchyma** (koh-LEN-kuh-muh) cells are thicker than those of parenchyma cells. Collenchyma cell walls are also irregular in shape. The thicker cell walls provide more support for

FIGURE 31-1

Plants are composed of three types of specialized cells. (a) Parenchyma cells are usually cube-shaped with thin walls. (b) Collenchyma cells are elongated with irregularly thickened cell walls. (c) Sclerenchyma cells are cube-shaped or elongated and have thick, rigid cell walls.

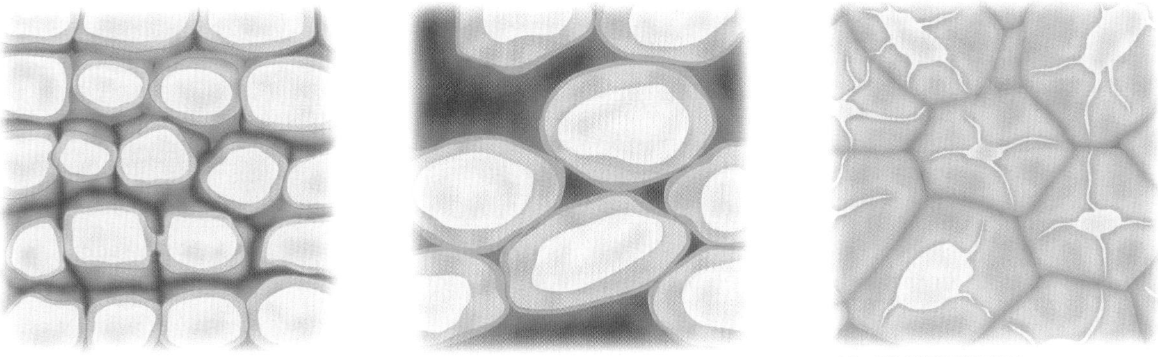

(a) PARENCHYMA (b) COLLENCHYMA (c) SCLERENCHYMA

SECTION 31-1

SECTION OVERVIEW

Cell Structure and Function
Section 31-1 describes how the nonreproductive parts of plants are composed of specialized cells that are organized into three types of tissues. These tissues either provide protection, function in photosynthesis or storage, or are involved in the transport of materials.

Matter, Energy, and Organization
Section 31-1 also provides a description of the three organs found in plants: roots, stems, and leaves. Meristematic cells continue to divide and provide new specialized cells for growing plant organs.

ACTIVE READING

Technique: Reading Organizer
Before students read this chapter, have them survey the subtitles, headings, captions, and words in boldface type. Ask students to identify the purpose of this chapter. *(The purpose of this chapter is to describe the three major parts of the plant body—roots, stems, and leaves—and to describe their functions.)* As students read this chapter, have them work individually or in small groups to develop a descriptive reading organizer.

SECTION 31-1

TEACHING STRATEGY

Specialized Plant Cells

Provide students with a collection of plant parts containing the three types of specialized plant cells. CAUTION **Students may be allergic to some plants. Have students wash their hands thoroughly after handling a plant part.** To show parenchyma cells, provide an apple (the flesh), a potato (the flesh), or a rose or other flower (the petals are mostly parenchyma). To show collenchyma cells, provide celery stalks (the outer ribs are collenchyma). To show sclerenchyma cells, provide a piece of woody stem, a piece of herbaceous stem, a peach pit, or a walnut shell. Have students examine the plant parts and discuss their functions in relation to their structure. Students should note that plant parts with parenchyma cells generally function in storage, photosynthesis, and protection; those parts with collenchyma cells function in support; and those parts with sclerenchyma cells are usually functional after they die, providing support or a conduit for the transport of materials.

QUICK FACT

☑ Plants differ fundamentally from animals in that they continue to grow in size throughout their life. Animals grow only until they reach maturity. Plants also differ from animals in that they do not have reproductive structures during their entire mature lifetime. Plants typically produce reproductive structures at annual intervals or in response to some other environmental cue, or simply when they reach a certain size. Animals typically have reproductive structures by the time they reach maturity and then retain the structures for the rest of their lives.

the plant. Collenchyma cells are usually grouped in strands. They are specialized for supporting regions of the plant that are still lengthening. Celery stalks contain a great amount of collenchyma cells.

Sclerenchyma (skluh-REN-kuh-muh) cells have thick, even, rigid cell walls. They support and strengthen the plant in areas where growth is no longer occurring. This type of cell usually dies at maturity, providing a frame to support the plant. The gritty texture of a pear fruit is due to the presence of sclerenchyma cells.

TISSUE SYSTEMS

Cells that work together to perform a specific function form a tissue. Tissues are arranged into systems in plants, including the dermal system, ground system, and vascular system, which are summarized in Table 31-1. These systems are further organized into the three major plant organs—the roots, stems, and leaves. The organization of each organ reflects adaptations to the environment.

Dermal Tissue System

The **dermal tissue system** forms the outside covering of plants. In young plants, it consists of the **epidermis** (EP-uh-DUHR-muhs), the outer layer made of parenchyma cells. In some species, the epidermis is more than one cell layer thick. The outer epidermal wall is often covered by a waxy layer called the **cuticle,** which prevents water loss. Some epidermal cells of the roots develop hairlike extensions that increase water absorption. Openings in the leaf and stem epidermis are called stomata. Stomata regulate the passage of gases and moisture into and out of the plant. In woody stems and roots, the epidermis is replaced by dead cork cells.

TABLE 31-1 ***Characteristics of Plant Tissue Systems***

Tissue system	Type of cells	Location	Function in roots	Function in stems	Function in leaves
Dermal tissue system	flat, living parenchyma (epidermal cells) in nonwoody parts; flat, dead parenchyma (cork cells) in woody parts	outermost layer(s) of cells	absorption, protection	gas exchange, protection	gas exchange, protection
Ground tissue system	mostly parenchyma, usually with some collenchyma and fewer sclerenchyma	between dermal and vascular in nonwoody plant parts	support, storage	support, storage	photosynthesis
Vascular tissue system	elongated cells—dead xylem and living phloem, also parenchyma and sclerenchyma (fibers)	tubes throughout plant	transport, support	transport, support	transport, support

Ground Tissue System

Dermal tissue surrounds the **ground tissue system,** which consists of all three types of plant cells. Ground tissue functions in storage, metabolism, and support. Parenchyma cells are the most common type of cell found in ground tissue. Nonwoody roots, stems, and leaves are made up primarily of ground tissue. Cactus stems have large amounts of parenchyma cells for storing water in dry environments. Plants that grow in waterlogged soil often have parenchyma with large air spaces that allow air to reach the roots. Nonwoody plants that must be flexible to withstand wind have large amounts of collenchyma cells. Sclerenchyma cells are found where hardness is an advantage, such as in the seed coats of hard seeds and in the spines of cactuses.

Vascular Tissue System

Ground tissue surrounds the **vascular tissue system,** which functions in transport and support. Recall from Chapter 30 that the term *vascular tissue* refers to both xylem and phloem. Xylem tissue conducts water and mineral nutrients primarily from roots upward in the plant. Xylem tissue also provides structural support for the plant. Phloem tissue conducts organic compounds and some mineral nutrients throughout the plant. Unlike xylem, phloem is alive at maturity.

In angiosperms, xylem has two major components—tracheids and vessel elements. Both are dead cells at maturity. Look at Figure 31-2a. A **tracheid** (TRAY-kee-id) is a long, thick-walled sclerenchyma cell with tapering ends. Water moves from one tracheid to another through **pits,** which are thin, porous areas of the cell wall. A **vessel element**, shown in Figure 31-2b, is a sclerenchyma cell that has either large holes in the top and bottom walls or no end walls at all. Vessel elements are stacked to form long tubes called **vessels.** Water moves more easily in vessels than in tracheids. The xylem of most seedless vascular plants and most gymnosperms contains only tracheids, which are considered a primitive type of xylem cell. The vessel elements in angiosperms probably evolved from tracheids. Xylem also contains parenchyma cells and sclerenchyma cells.

The conducting parenchyma cell of angiosperm phloem is called a **sieve tube member.** Look at Figure 31-2c. Sieve tube members are stacked to form long **sieve tubes.** Compounds move from cell to cell through end walls called **sieve plates.** Each sieve tube member lies next to a specialized parenchyma cell, the **companion cell,** which assists in transport. Phloem also usually contains sclerenchyma cells called fibers. Commercially important hemp, flax, and jute fibers are phloem fibers.

Vascular tissue systems are also modified for environmental reasons. For example, xylem forms the wood of trees, providing the plants with strength while conducting water and mineral nutrients. In aquatic plants, such as duckweeds, xylem is not needed for support or water transport and may be nearly absent from the mature plant.

FIGURE 31-2

(a) Tracheids are long and thin, and they contain pits in their cell walls. (b) Vessel elements are shorter and wider than tracheids. Both tracheids and vessel elements transport water. (c) Sieve tube members are long and tubular, and they contain pores in their cell walls. Sugar is transported through sieve tube members and companion cells.

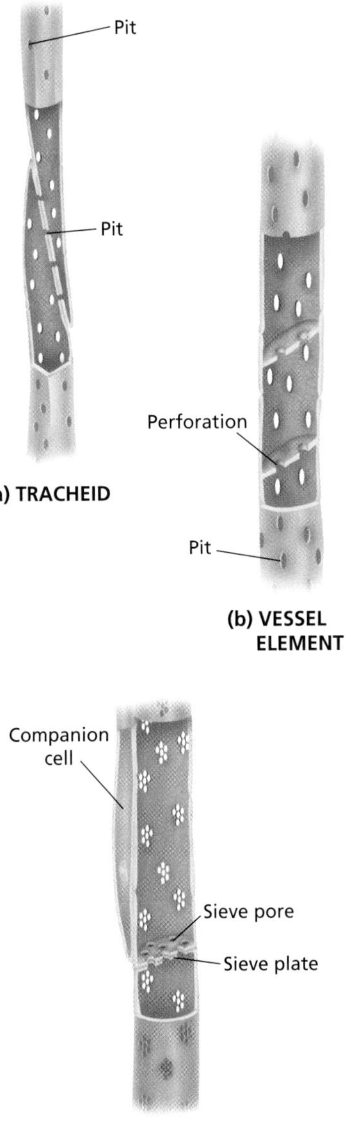

RECENT RESEARCH
Programmed Cell Death

Scientists have recently discovered that apoptosis, or programmed cell death, occurs in plant cells. Apoptosis is caused by the degeneration of nuclear DNA and has been known to occur in animal cells. Scientists hope that by learning more about the genes that regulate apoptosis, they can engineer plant cells to avoid accelerated apoptosis caused by microorganisms.

RETEACHING *ACTIVITY*

Ask students to write down the following terms: *dermal tissue, vascular tissue, tracheid.* Ask students to compare the functions of the similarly named plant and animal structures. (Dermal tissue in animals has the same name and function as it does in plants—protection. Vascular tissue in animals has the same name and function as it does in plants—transport of materials. Tracheids, which transport water in plants, have a name similar to the trachea of animals, an organ that transports air to the lungs in many animals.)

VISUAL LINK
Figure 31-2

Have students examine Figure 31-2. Ask students how the tracheids and vessel elements of xylem are specialized to transport water. (The cells are dead, leaving the cell wall to form a conduit for water movement; the cells are connected end to end, and the end walls are almost completely open to allow water to move freely from cell to cell.) Ask students how the sieve tube members of phloem are specialized for the transport of sugars. (The cells are long and thin and connected end to end with pores in the end walls; this allows movement of sugars from cell to cell.)

SECTION 31-1

INCLUSION ACTIVITY

Have students draw a picture of a young tree. Ask them to color in red any part of their tree that would contain meristematic tissue. (These include the shoot apical meristem, root apical meristem, vascular cambium in the trunk, and cork cambium in the bark.)

ANSWERS TO SECTION 31-1 REVIEW

1. Tracheids are long, narrow cells with pointed ends. Vessel elements are usually shorter and wider, with holes in the end walls or no end walls. Water travels from tracheid to tracheid through pits. Water travels from vessel element to vessel element through the holes in the end walls. Most gymnosperms have only tracheids. Angiosperms have both tracheids and vessel elements.
2. Both have apical meristems at the tips of stems and roots. Monocots often have intercalary meristems. Dicots usually have a vascular cambium and cork cambium.
3. No. Sclerenchyma cells are dead, so they cannot elongate as cells near meristems must.
4. No. Sclerenchyma and collenchyma cells primarily function in support. Most roots are supported by soil and do not need as much supporting tissue.
5. An increase in a tree's height is called primary growth. An increase in a tree's diameter is called secondary growth.
6. The transfer of water among cells is more efficient when xylem tissue is aligned. The pits of a tracheid must be aligned to a neighboring tracheid. Also, the holes of one vessel element must be joined with the holes of other vessel elements.

TABLE 31-2 *Types of Meristems*

Type	Location	Function
Apical meristem	tips of stems and roots	growth; increase length at tips
Intercalary meristem	between the tip and base of stems and leaves	growth; increase length between nodes
Lateral meristem	sides of stems and roots	growth; increase diameter

GROWTH IN MERISTEMS

Plant growth originates mainly in **meristems** (MER-i-STEMZ), regions where cells continuously divide. Look at Table 31-2. Most plants grow in length through **apical** (AP-i-kuhl) **meristems** located at the tips of stems and roots. Some monocots have **intercalary** (in-TUHR-kah-ler-ee) **meristems** located above the bases of leaves and stems. Intercalary meristems allow grass leaves to quickly regrow after being grazed or mowed.

Gymnosperms and most dicots also have **lateral meristems,** which allow stems and roots to increase in diameter. Lateral meristems are located near the outside of stems and roots. There are two types of lateral meristems, the vascular cambium and the cork cambium. The **vascular cambium,** located between the xylem and phloem, produces additional vascular tissues. The **cork cambium,** located outside the phloem, produces cork. Cork cells replace the epidermis in woody stems and roots, protecting the plant. **Cork** cells are dead cells that provide protection and prevent water loss.

You have probably noticed how trees grow taller and wider over time. Growth in length is called **primary growth** and is produced by apical and intercalary meristems. Growth in diameter is called **secondary growth** and is produced by the lateral meristems—that is, by the vascular cambium and cork cambium.

SECTION 31-1 REVIEW

1. **What are the differences between tracheids and vessel elements? How does water travel through each structure?**
2. **What kinds of meristems do monocots and dicots have in common? What kinds do they not share?**
3. **Would you expect to find sclerenchyma cells near meristems? Why or why not?**
4. **Would you expect to find sclerenchyma and collenchyma cells in roots?**
5. **Distinguish between the primary growth and secondary growth in a tree.**
6. **CRITICAL THINKING Describe the factors that influence the transport of water through xylem tissue.**

ROOTS

SECTION

31-2

OBJECTIVES

▲ List the three major functions of roots.

● Explain the difference between a taproot system and a fibrous root system.

■ Distinguish between primary growth and secondary growth in roots.

◆ Describe primary root tissues.

Plants have three kinds of organs—roots, stems, and leaves. Roots are the structures that typically grow underground. Roots are important because they anchor the plant in the soil. They also absorb and transport water and mineral nutrients. The storage of water and organic compounds is provided by roots.

TYPES OF ROOTS

When a seed sprouts, it produces a primary root. If this first root becomes the largest root, it is called a **taproot,** as illustrated in Figure 31-3a. Many plants, like carrots and certain trees, have taproots. Contrary to what you might think, even taproots rarely penetrate the ground more than a meter or two. A few species, such as cottonwoods, do have some roots that grow 50 m (164 ft) deep to tap into underground water supplies. However, tall trees rarely have roots as deep as their height aboveground.

In some plants, the primary root does not become large. Instead, numerous small roots develop and branch to produce a **fibrous root system,** like that shown in Figure 31-3b. Many monocots, such as grasses, have fibrous root systems. Fibrous roots of monocots often develop from the base of the stem rather than from other roots.

(a) TAPROOT SYSTEM

(b) FIBROUS ROOT SYSTEM

FIGURE 31-3
Plants can have either a taproot system or a fibrous root system. (a) Many dicots, including the radish, have a large central taproot with small lateral roots. (b) Most monocots, including grasses, have a highly branched fibrous root system.

internet**connect**

SCiLINKS NSTA
TOPIC: Types of roots
GO TO: www.scilinks.org
KEYWORD: HM603

SECTION 31-2

SECTION OVERVIEW

Cell Structure and Function
This section describes the types of roots, the relationship between root structures and their functions, and the growth of roots.

▶ ENGAGE STUDENTS

Bring in several plants or pictures of plants that show different kinds of root systems. Have students observe the different roots and discuss the structural adaptations for carrying out particular functions. Select from storage roots (carrots or parsnips), taproots (carrots), fibrous roots (those of grasses), adventitious roots (corn prop roots), and aerial roots (those of bromeliads or orchids). Wait until just before class to remove any roots from soil, and keep them moist.

VISUAL LINK

Figure 31-3
Have students examine Figure 31-3 and ask them which type of root is well adapted for soils that frequently dry out (taproot). Ask students which type of root is likely to be found on a grass plant (fibrous root). Ask students which type of root is likely to be adapted for shallow soils (fibrous root).

OVERCOMING MISCONCEPTIONS

It is commonly thought that because plants consume carbon dioxide and produce oxygen during photosynthesis, they do not require oxygen for survival. In fact, plants require oxygen for aerobic respiration in mitochondria, which converts the sugars that the plants produce into ATP. Chloroplasts generate enough oxygen for aerobic respiration. In roots, however, oxygen must be supplied from the soil. This is why plants die if overwatered or if the soil becomes too compact.

SECTION 31-2

QUICK FACT

☑ Many trees will form root grafts, which occur when two roots grow together and their cambia and vascular tissues become united. Natural root grafting commonly occurs between roots of different individuals of the same species. Certain diseases, such as Dutch elm disease and oak wilt (both caused by fungi), readily spread through root grafts of closely planted trees.

CRITICAL THINKING

Nitrogen Deficiency Versus Iron Deficiency

Both nitrogen deficiency and iron deficiency cause yellowing, or chlorosis, of leaves. Nitrogen can be moved from one part of a plant to another, but iron cannot. Ask students how the symptoms of nitrogen deficiency can be distinguished from those of iron deficiency. (Nitrogen deficiency expresses itself as chlorosis of older leaves; nitrogen can be moved into newer leaves. Iron deficiency expresses itself as chlorosis of newer leaves because iron cannot be moved from older leaves.)

INCLUSION *ACTIVITY*

An understanding of cross and longitudinal sections is essential to understanding tissue relationships in three dimensions. Bring in a sharp knife and several different types of vegetables and fruits, such as squash, a cucumber, an apple, a kiwi, and beans. Cut them in both cross sections and longitudinal sections so that students will know what is meant by these terms. Have students draw scientific illustrations of both types of specimens. Then ask them to draw cross sections and longitudinal sections of familiar objects.

(a)

(b)

FIGURE 31-4
Some plants grow adventitious roots from aboveground parts, including stems and leaves. (a) Corn plants grow prop roots at their base to provide additional stability. (b) This orchid displays another type of adventitious root, called an air root. These plants live on tree branches and absorb water and mineral nutrients from the surface of the tree and from the air.

Specialized roots that grow from stems and leaves are called **adventitious roots.** Figure 31-4a shows the prop roots of corn, which help keep the plant's stems upright. The air roots of an epiphytic orchid, shown in Figure 31-4b, obtain water and mineral nutrients from the air. Air roots on the stems of ivy and other vines enable them to climb walls and trees.

ROOT STRUCTURES

Study Figure 31-5. Notice that the root tip is covered by a protective **root cap,** which covers the apical meristem. The root cap produces a slimy substance that functions like lubricating oil, allowing the root to move more easily through the soil as it grows. Cells that are crushed or knocked off the root cap as the root moves through the soil are replaced by new cells produced in the apical meristem, where cells are continuously dividing. Look at Figure 31-6 on the next page. **Root hairs,** which are extensions of epidermal cells, increase the surface area of the root and thus increase the plant's ability to absorb water and mineral nutrients.

Root structures are adapted for several functions. Root hairs on the epidermis greatly increase the surface area available for absorption. Most roots also form partnerships with mycorrhizal fungi, whose threadlike hyphae also increase the surface area for absorption. The spreading, usually highly branched root system increases the amount of soil that the plant can "mine" for water and mineral nutrients and aids in anchoring the plant. The large amount of root parenchyma usually functions in storage and general metabolism. Roots are dependent on the shoots for their energy, so they must store starch to use as an energy source during periods of little or no photosynthesis, such as winter.

FIGURE 31-5
The root cap protects the apical meristem of a root tip. The primary growth of a root is caused by cell division occurring in the apical meristem.

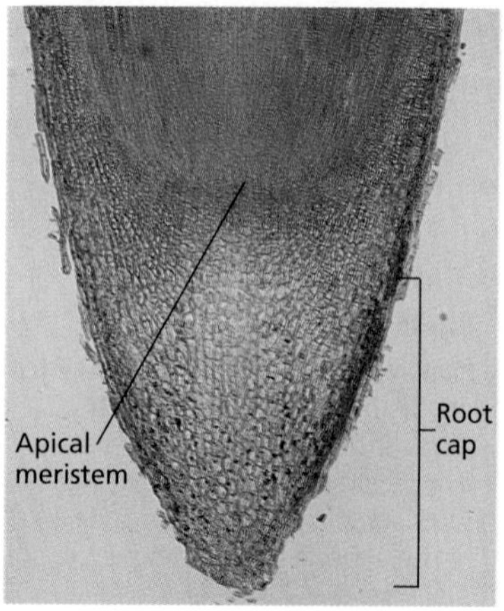

FIGURE 31-6

The root tip of this seedling has many root hairs, which help the plant absorb water and mineral nutrients from the soil. Root hairs grow from the outer layer of cells, called the epidermis.

Primary Growth in Roots

Roots increase in length through cell division, elongation, and maturation in the root tip. Dermal tissue matures to form the epidermis, the outermost cylinder of the root. Ground tissue in roots matures into two specialized regions: the cortex and the endodermis. The **cortex** is located just inside the epidermis, as you can see in Figure 31-7. This largest region of the primary root is made of loosely packed parenchyma cells.

The innermost cylinder of the cortex is the **endodermis** (EN-doh-DUHR-mis), also shown in Figure 31-7. Endodermal cell walls contain a narrow band of a waterproof substance that stops further movement of water through the cell walls. To enter farther into the root than the endodermis, water and dissolved substances must pass through the selectively permeable membrane of a root cell. Once inside the cell membrane, dissolved substances can move from cell to cell via small channels in cell walls that interconnect the cytoplasm of adjoining cells.

Vascular tissue in roots matures to form the innermost cylinder of the root. In most dicots and gymnosperms, xylem makes up the central core of the root, as shown in Figure 31-7b. Dicot root xylem

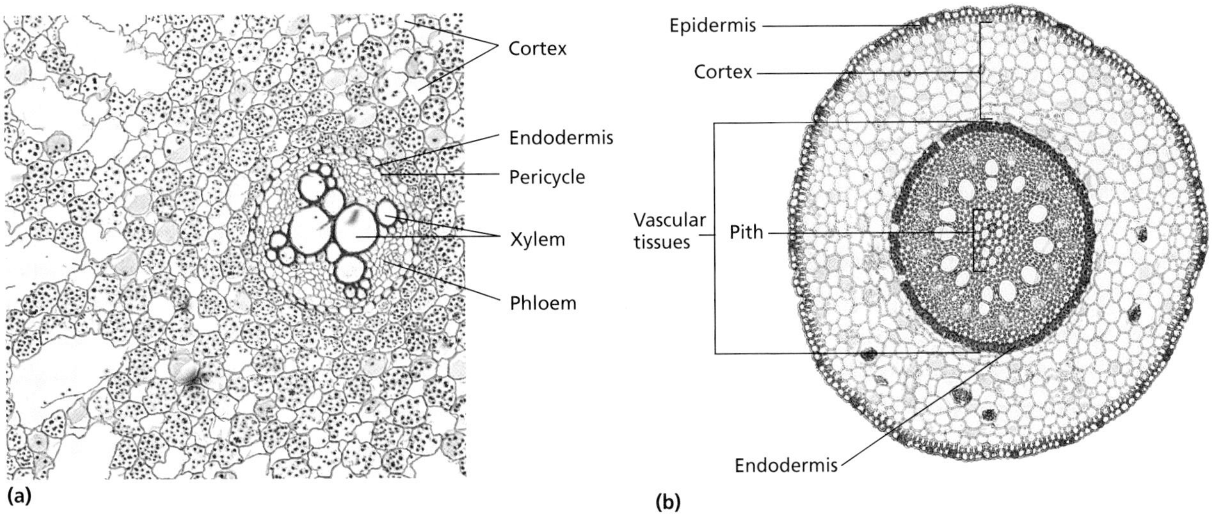

FIGURE 31-7

(a) This cross section of a dicot root shows the arrangement of vascular tissue and ground tissue. Note how the xylem tissue forms an X in the center. The cortex and the endodermis, which are composed of ground tissue, surround the vascular tissue. (b) This cross section of a monocot root shows a prominent endodermis, the innermost layer of the cortex. The center of the root, called the pith, is made up of parenchyma cells.

RECENT RESEARCH
Genetically Engineered Roots

Scientists at the Salk Institute have found a group of genes that regulate the rate of root growth. The genes, called cyclins, were studied in *Arabidopsis thaliana* but are also known in other plants. The scientists spliced the cyclin genes onto a region of the plant's DNA that would activate the gene throughout the cell cycle, not just before cell division. They found that the genetically engineered roots grew 40 percent faster than the unaltered roots. Scientists hope to use this knowledge to manipulate genes to speed up root development.

CONTENT CONNECTION
Section 8-2: Mitosis

New cells for growth are produced by mitosis. The nuclear DNA is first replicated and then separated evenly into two new cells that are identical to each other and to the original cell.

VISUAL LINK
Figure 31-7

Have students examine Figure 31-7. Ask students the following questions: Which part of the root transports water? (xylem) Which part transports sugars? (phloem) Which part regulates movement of water and minerals into the vascular tissue? (endodermis) Which part stores starch? (cortex)

SECTION 31-2

RETEACHING ACTIVITY

Have students draw pictures of a cross section and longitudinal section of a nonwoody root. Ask them to label the following structures in each drawing: epidermis, xylem, phloem, endodermis, cortex, root apical meristem, root cap, and pericycle. Refer students to Figures 31-5 and 31-7. Ask students to describe the function of each of these structures (epidermis—protection from drying out; xylem—water transport; phloem —sugar transport; endodermis—regulation of materials entering vascular tissues; cortex—storage; root apical meristem—production of new root cells; root cap—protection of root tip cells; and pericycle—production of lateral roots).

Quick Lab

Observing Roots

Time Required 20 minutes

Procedural Tips Each student or group will need three or four seedlings. Ten days ahead of time, soak radish seeds in water overnight. Transfer the seeds to moist paper towels. Keep the paper towels damp. When the seedlings develop leaves, allow the seedlings to dry out just enough to wilt.

Disposal Radish seedlings can be put in a compost heap or thrown in the trash.

Answers to Analysis The wilted seedlings become firm after the roots are covered with water. The seedlings absorb water through the roots. Roots take in water and minerals and hold the plant in the soil.

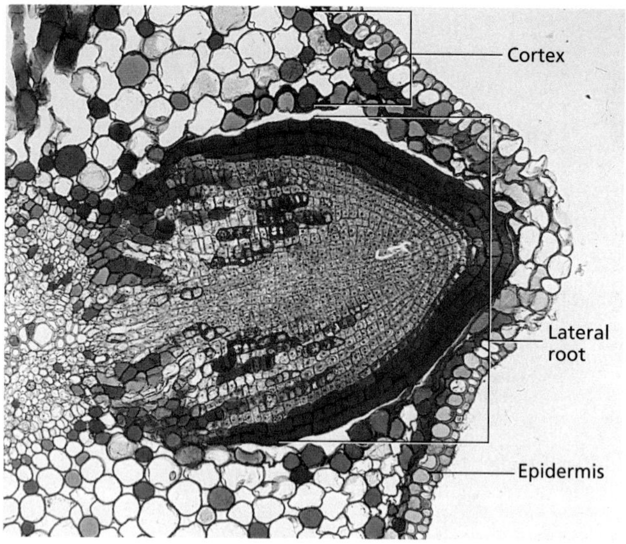

FIGURE 31-8

The vascular core of a primary root is surrounded by the pericycle, a tissue that produces lateral roots.

usually has a fluted structure with pockets of phloem between the xylem lobes. In monocots, the center of the root usually contains a pith of parenchyma cells, as Figure 31-7b shows. Monocot root xylem occurs in many patches that circle the pith. Small areas of phloem occur between the xylem patches.

The outermost layer or layers of the central vascular tissues is termed the **pericycle** (PER-i-SIE-kuhl). Lateral roots are formed by the division of pericycle cells. The developing lateral root connects its vascular tissues and endodermis to those of the parent root. Figure 31-8 shows how a lateral root grows out through the parent root's endodermis and cortex, finally emerging from the epidermis.

Secondary Growth in Roots

Dicot and gymnosperm roots often experience secondary growth. Secondary growth begins when a vascular cambium forms between primary xylem and primary phloem. Pericycle cells form the vascular cambium at the ends of the xylem lobes, where no phloem is located. The vascular cambium produces secondary xylem toward the inside of the root and secondary phloem toward the outside. The expansion of the vascular tissues in the center of the root crushes all the tissues external to the phloem, including the endodermis, cortex, and epidermis. A cork cambium develops in the pericycle to replace the crushed cells with cork.

Quick Lab

Observing Roots

Materials wilted radish seedlings, hand lens, Petri dish, water, pipet

Procedure

1. Place the wilted radish seedlings in a Petri dish. Observe them with the hand lens. Record your observations.
2. Using a pipet, cover only the roots with water. Observe the seedlings with the hand lens every 5 minutes for 15 minutes. Record each of your observations.
3. Use the hand lens to observe the roots. Draw and label what you see.

Analysis What happened to the wilted seedlings when you put them in water? How can you explain what happened? Describe two functions of a root.

ROOT FUNCTIONS

Besides anchoring a plant in the soil, roots serve two other primary functions. They absorb water and a variety of minerals or mineral nutrients that are dissolved in the soil. Roots are selective about which minerals they absorb. Roots absorb some minerals and exclude others. Table 31-2 lists the 13 minerals that are essential for all plants. They are absorbed mainly as ions. Carbon, hydrogen, and oxygen are not listed because they are absorbed as water and carbon dioxide.

Plant cells use some minerals, such as nitrogen and potassium, in large amounts. These elements are called **macronutrients** and are required in relatively large amounts (usually more than 1,000 mg/kg of dry matter). Plant cells use other minerals, such as manganese, in smaller amounts. These are called **micronutrients** and are required in relatively small amounts (usually less than 100 mg/kg of dry matter).

Adequate amounts of all 13 mineral nutrients in Table 31-3 are required for normal growth. Plants with deficiencies show

TABLE 31-3 Essential Mineral Nutrients in Plants

Macronutrients

Element	Absorbed as	Use in plants
Nitrogen	NO_3^-, NH_4^+	part of proteins, nucleic acids, chlorophyll, ATP
Phosphorus	$H_2PO_4^-$	part of nucleic acids, ATP, phospholipids, coenzymes
Potassium	K^+	required for stomatal opening and closing, enzyme cofactor
Calcium	Ca^{2+}	part of cell walls and cell membranes
Magnesium	Mg^{2+}	part of chlorophyll
Sulfur	SO_4^{2-}	part of proteins

Micronutrients

Element	Absorbed as	Use in plants
Iron	Fe^{2+}	part of cytochromes in electron transport
Manganese	Mn^{2+}	required by many enzymes
Boron	$B(OH)_3$	thought to be involved in carbohydrate transport
Chlorine	Cl^-	required to split water in photosynthesis
Zinc	Zn^{2+}	essential part of many enzymes
Copper	Cu^{2+}	essential part of many enzymes
Molybdenum	MoO_4^{2-}	required for nitrogen metabolism

characteristic symptoms and reduced growth. Severe mineral deficiencies can kill a plant. Excess amounts of some essential mineral nutrients also can be toxic to a plant.

Roots are often adapted to store carbohydrates or water. Phloem tissue carries carbohydrates made in leaves to roots. Carbohydrates that roots do not immediately use for energy or building blocks are stored. In roots, these excess carbohydrates are usually converted to starch and stored in parenchyma cells. You are probably familiar with the storage roots of carrots, turnips, and sweet potatoes. The roots of some species in the pumpkin family store large amounts of water, which helps the plants survive during dry periods.

SECTION 31-2 REVIEW

1. **What are the major functions of the root system?**

2. **What are the differences between a taproot system and a fibrous root system?**

3. **What is the difference between primary growth and secondary growth? What types of tissues are involved in each type of growth?**

4. **Explain how root hairs increase the ability of a plant to absorb water from the soil.**

5. **Name two areas of the root where you would probably find parenchyma cells.**

6. **CRITICAL THINKING Why might a taproot system be an advantage to some plants, while a fibrous root system might be an advantage to others?**

MAKING CONNECTIONS

Mathematics

Bring bags of various fertilizers to class. Tell students that the labels contain an analysis of the contents of the bag and that the analysis of a mixed fertilizer is the percentage of nitrogen, phosphorus (as P_2O_5), and potassium (as K_2O) in the bag. A fertilizer containing 30 percent nitrogen, 10 percent phosphorus, and 10 percent potassium is labeled "30-10-10." Ask students to convert the percentages given on the labels as compounds into percentages of pure elements. Instruct students that to do this, they should multiply the percentage of phosphorus as P_2O_5 by 0.44 (percentage of P in P_2O_5) and multiply the percentage of potassium as K_2O by 0.83 (percentage of K in K_2O). Nitrogen is already expressed in its elemental form.

ANSWERS TO SECTION 31-2 REVIEW

1. Roots anchor the plant, absorb water and mineral nutrients, and can store carbohydrates or water.

2. A taproot system has one large central taproot with small lateral roots. A fibrous root system has many small branched roots.

3. Primary growth is the increase in root length and is caused by apical meristems. Secondary growth is the increase in root diameter and is caused by lateral meristems.

4. Root hairs greatly increase the root surface area in contact with the soil.

5. root cap, cortex, and vascular tissue

6. Taproots offer large plants support and enable plants to reach water deep in the ground. Fibrous roots offer support in shallow soils and maximize water collection while they hold soil and prevent erosion.

SECTION 31-3

SECTION OVERVIEW

Cell Structure and Function
This section describes the types of stems, how their structures are related to their functions, and the growth of stems.

▶ ENGAGE STUDENTS

Provide students with a variety of stems displaying various modifications. Have students examine the specimens and discuss the adaptations they possess and their advantages. Include modifications such as the following: for storage—potato tubers and the fleshy rhizomes of irises and ginger; for photosynthesis—asparagus; for propagation—strawberry stolons, spider plants, and the rhizomes of many grasses; and for defense—the thorns of barberry. Ask students how they can tell that all of these structures are stems and not roots. (Stems have nodes, which roots do not have.)

RECENT RESEARCH

Insects Keeping Ahead
If an insect pest chews on a plant for a few days, the plant will produce inhibitors to the insect's digestive enzymes. Scientists in the Netherlands have recently genetically engineered tobacco plants to produce the inhibitors all the time in an effort to protect from crop loss. However, the scientists soon discovered that the insects began producing large amounts of a backup digestive enzyme and that this enzyme was immune to the plants' inhibitors. These results suggest that genetically engineering plants for insect resistance may prove more difficult than originally thought.

SECTION

31-3

OBJECTIVES

▲ Describe the differences between monocot stems and dicot stems.

● List five differences and five similarities between the structure of roots and the structure of stems.

■ Explain how annual rings are formed.

◆ Describe the pressure-flow model for organic-compound movement in the phloem.

▲ Describe the cohesion-tension theory for water movement in the xylem.

STEMS

In contrast to roots, which are mainly adapted for absorption and anchoring, stems are usually adapted to support leaves. Whatever their sizes and shapes, stems also function in transporting materials and providing storage.

TYPES OF STEMS

The various differences in stem shape and growth represent adaptations to the environment. Several types of stems are shown in Figure 31-9. Strawberry runners grow along the soil surface and produce new plants at their nodes. Stems such as the edible white potato tuber are modified for storing energy as starch. Cactuses have green fleshy stems that both store water and carry on all the plant's photosynthesis. Stems of the black locust and the honeylocust develop sharp thorns to protect the plant from animals. These stem modifications are related to environmental adaptations.

(a) (b) (c)

FIGURE 31-9
Stems provide a supporting framework for leaves. Some plants produce stems that are modified for other functions. (a) A strawberry plant has stolons, which are horizontal, aboveground stems that form new plants. (b) The potato plant has tubers, which are enlarged, short, underground stems used for storing starch. (c) A cactus is called a succulent because of its fleshy, water-storing stems.

STEM STRUCTURES

Stems have a more complex structure than roots, yet they are similar in many ways. Did you know that a sign nailed 2 m (7 ft) high on a tree will remain 2 m high, even though the tree may grow much taller? That is because most stems, like roots, grow in length only at their tips, where apical meristems produce new primary tissues. Stems, like roots, grow in circumference through lateral meristems.

The surfaces of stems have several features that roots lack, as Figure 31-10 shows. Stems are divided into segments called **internodes.** At each end of an internode is a **node.** Initially, one or more leaves are attached at each node. At the point of attachment of each leaf, the stem bears a lateral bud. A **bud** is capable of developing into a new shoot. A bud contains an apical meristem and is enclosed by specialized leaves called **bud scales.** The tip of each stem usually has a terminal bud. When growth resumes in the spring, the terminal bud opens, and the bud scales fall off. The bud

Quick Lab

Observing Stems

Materials winter twig, fresh stem with leaves, hand lens or dissecting microscope

Procedure

1. Observe the twig with the hand lens. Locate several buds, and identify the bud scales. Locate leaf scars. Identify a node and an internode. Draw and label the twig.
2. Observe the fresh stem with the hand lens. Locate and identify the stem, nodes, internodes, buds, and a leaf. Draw and label the fresh stem.

Analysis How are a node and an internode related? How are a bud and a node related? How are the two stems alike? How are they different?

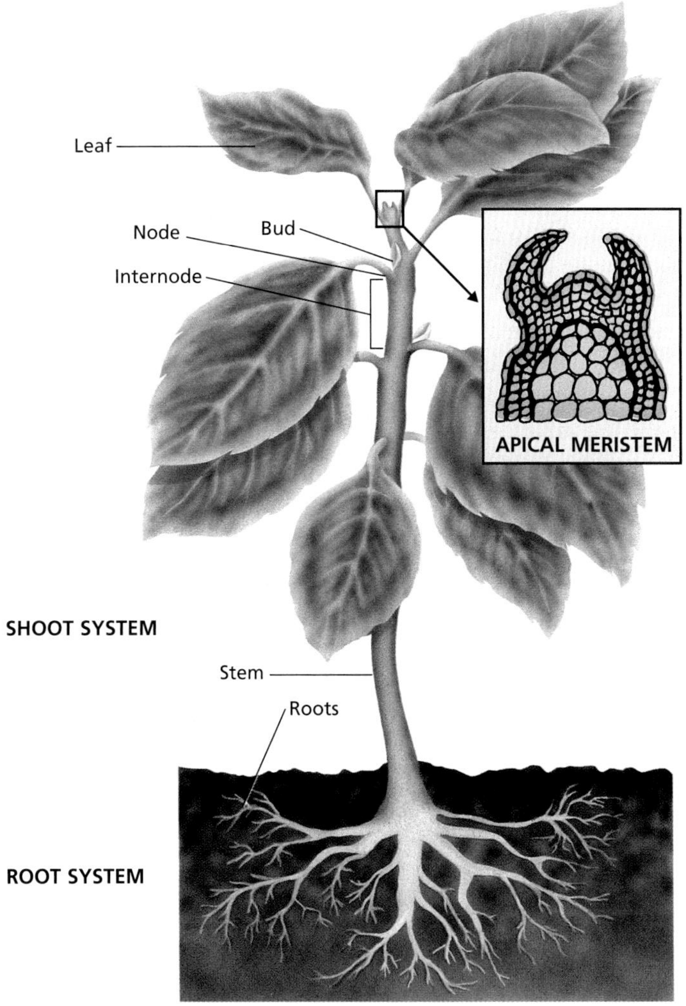

FIGURE 31-10

Apical meristems, responsible for the primary elongation of the plant body, are found in the shoot tips and root tips. Each leaf is attached to the stem at a node. The space between nodes, called the internode, is much larger in the older part of the stem. When the shoot tip is pinched off, the lateral buds begin to grow, resulting in a bushier plant.

Quick Lab

Observing Stems

Time Required 20 minutes

Procedural Tips Have students work in small groups. Each group will need a winter twig (dormant with no leaves) that is at least 30 cm long. Each group will need a freshly cut, herbaceous stem with leaves that is also about 30 cm long.

Disposal The stems and twigs may be put in a compost heap or thrown in the trash.

Answers to Analysis An internode is the space between two nodes. A lateral bud is found at each node. Both stems have buds, nodes, and internodes. The twig has leaf scars. The fresh stem has leaves.

GIFTED *ACTIVITY*

Have students research and report on scientists' use of aphids to discover how sugars are transported in plants. These tiny insects insert stylets into the cells of the phloem and collect sap from the plant. Many aphids exude some of the sap from their anal openings. Scientists noticed this phenomenon and used it to determine that phloem is the sugar-transporting tissue in plants and that sugars are transported under pressure.

SECTION 31-3

RETEACHING *ACTIVITY*

Have students examine Figure 31-11. Ask students to note the placement of the vascular bundles in the monocot stem (they are scattered throughout the center) and the dicot stem (they form a ring around the perimeter). Ask students to explain the significance of this difference between monocots and dicots. (Dicots contain vascular cambium in the vascular bundles. It divides and produces new cells, which encircle the entire stem in a ring of meristematic cells. These cells can then divide to produce secondary growth—new xylem toward the inside of the stem and new phloem toward the outside of the stem. Monocots lack vascular cambium and therefore most do not produce secondary growth in stems.)

CRITICAL THINKING

Lateral Water Transport

Ask students the following question: How do water and minerals get from the xylem in a tree trunk to the living cells in the phloem? (They move laterally in a woody trunk through pits in the side walls of tracheids and vessel elements.)

QUICK FACT

☑ The chemical quinine is made from the bark of the Cinchona tree (*Cinchona* spp.). Discovered in the 1600s, it has been used to treat malaria, which has killed more humans than any other disease except typhus and even today kills about 2 million people a year. The tree is killed when its bark is stripped for quinine. The plant almost disappeared from many of its native areas before people began growing it in plantations. Today, synthetic derivatives of quinine are also used to treat malaria.

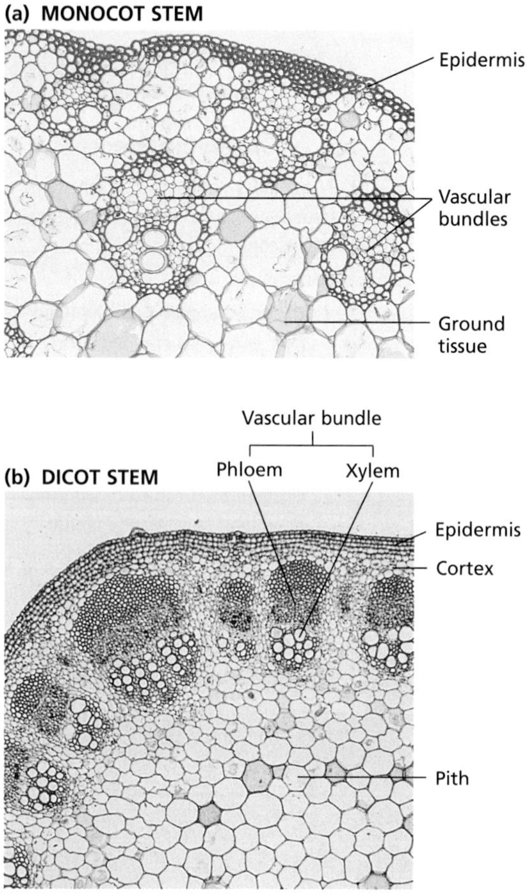

FIGURE 31-11

Compare the two basic types of stems. (a) The cross section of a herbaceous stem of corn, a monocot, shows the vascular bundles scattered throughout the ground tissue. (b) In the sunflower, a dicot, the vascular tissues appear as a single ring of bundles between the cortex and the pith.

scales leave scars on the stem surface. Trees and shrubs often are identified in winter by these twig characteristics.

Root tips have a permanent protective layer, the root cap. The stem apical meristem is protected by bud scales only when the stem is not growing. A surface bud forms very close to the stem tip with one or more buds at each node. In contrast, lateral roots originate farther back from the root tip and form deep inside the root at no particular location along the root axis.

Primary Growth in Stems

As in roots, apical meristems of stems give rise to the dermal, ground, and vascular tissues. Locate each of these tissues in Figure 31-11. As you can see, the dermal tissue is represented by the epidermis, or the outer layer of the stem. Its main functions are to protect the plant and to reduce the loss of water to the atmosphere while still allowing gas exchange through stomata.

In gymnosperm and dicot stems, ground tissue usually forms a cortex and a pith. The cortex lies just inside the epidermis, as it does in the root. The cortex frequently contains flexible collenchyma cells. The **pith** is located in the center of the stem. The ground tissue of monocot stems is usually not clearly separated into pith and cortex.

Vascular tissue formed near the apical meristem occurs in bundles—long strands that are embedded in the cortex. Look at the vascular bundles in Figure 31-11. Each bundle contains xylem tissue and phloem tissue. Xylem is usually located toward the inside of the stem, while phloem is usually located toward the outside.

Compare the arrangement of vascular bundles in monocots and dicots. Monocot stem vascular bundles are usually scattered throughout the ground tissue. Stem vascular bundles of dicots and gymnosperms usually occur in a single ring. Most monocots have no secondary growth, and they retain the primary growth pattern their entire lives. However, in many dicots and gymnosperms, the primary tissues are eventually replaced by secondary tissues.

Secondary Growth in Stems

Stems increase in thickness due to the division of cells in the vascular cambium. The vascular cambium in dicot and gymnosperm stems first arises between the xylem and phloem in a vascular bundle. Eventually the vascular cambium forms a cylinder. The vascular cambium produces secondary xylem to the inside and secondary phloem to the outside. It usually produces much more secondary xylem than it does secondary phloem. Secondary xylem is called **wood.** Occasionally, the vascular cambium produces new cambium cells, which increase its diameter.

Older portions of the xylem eventually stop transporting water. They often become darker than the newer xylem due to the accumulation of resins and other organic compounds produced by the few live cells remaining in the xylem. This darker wood in the center of a tree is called **heartwood,** as shown in Figure 31-12a. The functional, often lighter-colored wood nearer the outside of the trunk is **sapwood.** In a large-diameter tree, the heartwood keeps getting wider while the sapwood remains about the same thickness.

The phloem produced near the outside of the stem is part of the bark. **Bark** is the protective outside covering of woody plants. It consists of cork, cork cambium, and phloem. The cork cambium produces cork near the outside. However, cork cells are dead at maturity and cannot elongate, so the cork ruptures as the stem continues to expand in diameter. This results in the bark of some trees, such as oaks and maples, appearing rough or irregular in texture.

During spring, when water is plentiful, the vascular cambium forms new xylem tissue with cells that are wide and thin walled. This wood is called **springwood.** In summer, when water is more limited, the vascular cambium produces **summerwood,** which has smaller cells with thicker walls. In a stem cross section, the abrupt change between small summerwood cells and the following year's large springwood cells produces an **annual ring.** The circles that look like rings on a target in Figure 31-12 are the annual rings of the stem. Because one ring is usually formed each year, you can estimate the age of the stem by counting its annual rings. Annual rings also form in dicot and gymnosperm roots, but they are often less pronounced because the root environment is more uniform. Annual rings often do not occur in tropical trees because of their uniform year-round environment.

FIGURE 31-12

(a) This cross section of a mature woody stem shows secondary growth, which results in a thicker stem. Wood consists of secondary xylem. The dark-colored wood is called heartwood and the light-colored wood is called sapwood. (b) You can see the annual rings produced by alternating springwood cells and summerwood cells. The abundant water of the spring season helps expand the cells walls of springwood cells.

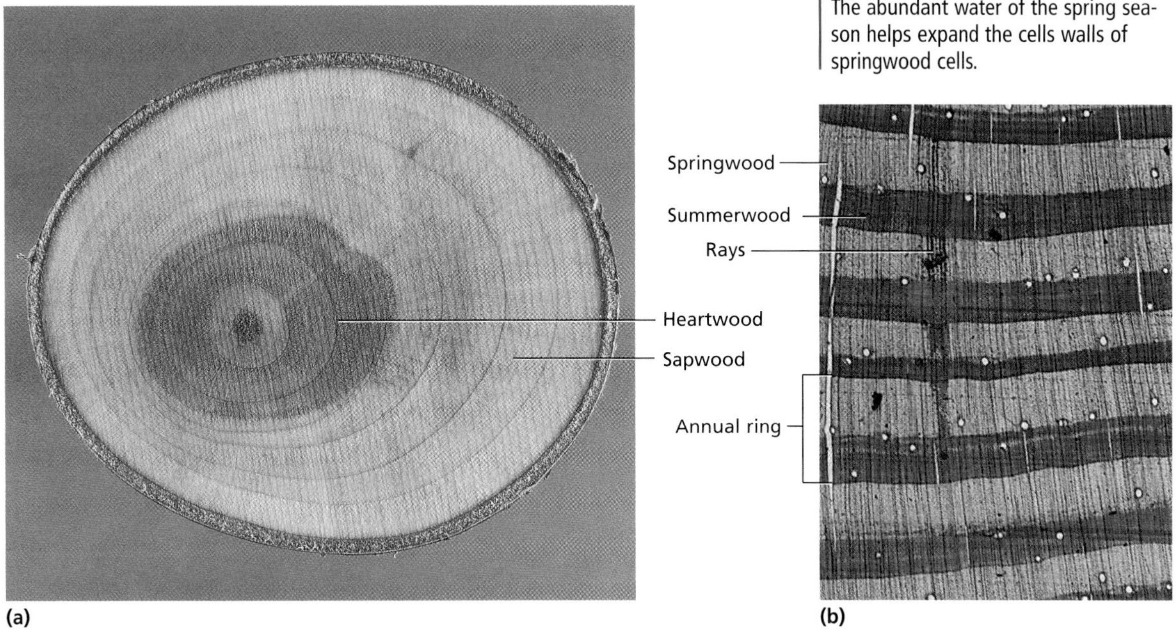

SECTION 31-3

CULTURAL CONNECTION

Uses of Bamboo Stems

Bamboo plants, of which there are more than 700 known species, are native to tropical and some temperate areas of Asia. These plants are giant grasses; some attain a height of 120 ft and a stem diameter of 1 ft. Products made from the stems of bamboo plants are essential for people of many Asian cultures. These include houses, furniture, sandals, floormats, bed covers, animal cages, fences, rafts, sails, tow rope, paper, fishing poles, ornamental screens, cooking utensils, tools, baskets, and food (young sprouts).

TEACHING STRATEGY

Examining Annual Rings

Present cross sections of a tree to the class. (Obtain cross sections of a tree from a landowner, a tree-cutting service, or the parks department, or use pictures of a cross section.) Ask students how old the tree was when it was cut down and how the age of a tree can be determined from the cross section. (Each ring represents one year's growth.) Point out that tree rings are actually composed of secondary xylem produced by vascular cambium.

VISUAL LINK

Figure 31-12

Have students examine Figure 31-12. Tell students that girdling of a tree involves removing a ring of bark down to the wood. Ask them why this would kill a tree. (Sugars would not be transported from the leaves to the roots and the roots would die without an energy source.)

TEACHING STRATEGY
Comparing Water Loss

Provide students with 10 identical plants grown in 4 in. pots. (Beans grown from seed for about two weeks are suitable.) Have students water the plants until water flows out the drain holes and wait about 10 minutes for the excess water to drain off. Have students place the plants in saucers and set five of them in a sunny window and five of them in an area of the classroom that receives no direct sunlight. Have students observe the plants for several days to a week, depending on the temperature and humidity of the room air, and determine when the plants in each group of pots begin to wilt and when they completely wilt. Do not water the plants. Ask students to discuss any differences they observe between the plants set in the sunny site and those placed in low light, and have them explain these differences. (The plants in the sunny site should wilt first. There is a greater pull on the water in these plants from the sun, causing more evaporation from the leaves than from the leaves of those placed in low light.)

INCLUSION *ACTIVITY*

Demonstrate water transport in a stem by splitting the stem of a white flower, such as a carnation, in half lengthwise, up to the base of the flower. Place one half in plain water and the other in water with food coloring. Within a few hours, half the flower will turn the color of the food coloring.

QUICK FACT

☑ More than 90 percent of the water taken up by the roots of most plants is lost in transpiration and is never involved in metabolic activities of the plant.

STEM FUNCTIONS

Stems function in the transportation and storage of nutrients and water, and they support the leaves. Carbohydrates, some plant hormones, and other organic compounds are transported in the phloem. The movement of carbohydrates occurs from where the carbohydrates are made or have been stored, called a **source,** to where they are stored or used, called a **sink.** Botanists use the term **translocation** to refer to the movement of carbohydrates through the plant. For example, in most plants, carbohydrates move from the leaves to the roots, to the shoot apical meristems, and to the developing flowers or fruits. Carbohydrates may be newly made in photosynthetic cells or may have been stored as starch in roots or stems.

Movement in the phloem is explained by the **pressure-flow hypothesis,** which states that carbohydrates are actively transported into sieve tubes. Look at Figure 31-13. As carbohydrates enter the sieve tubes, water is also transported in by osmosis. Thus, a positive pressure builds up at the source end of the sieve tube. This is the pressure part of the hypothesis.

At the sink end of the sieve tube, this process is reversed. Carbohydrates are actively transported out, water leaves the sieve tube by osmosis, and pressure is reduced at the sink. The difference

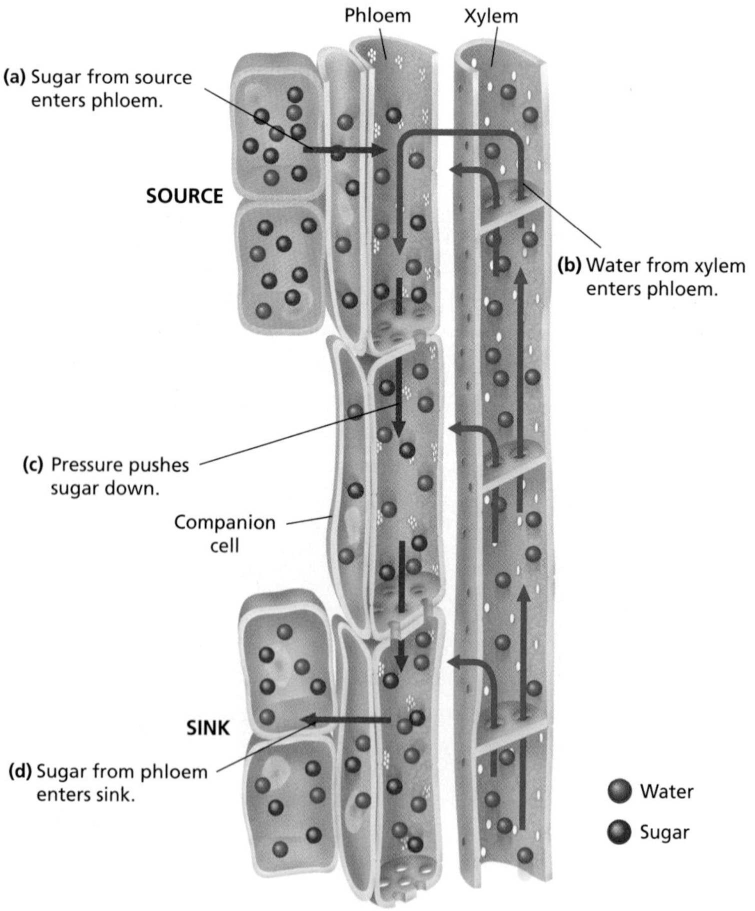

FIGURE 31-13
In the pressure-flow model, carbohydrates are pushed through the phloem by the pressure that results from the movement of water into the phloem by osmosis. The blue arrows indicate water movement, and the red arrows indicate carbohydrate movement.

in pressure causes water to flow from source to sink—carrying dissolved substances with it. Transport in the phloem can occur in different directions at different times.

Transport of Water

The transport of water and mineral nutrients occurs in the xylem of all plant organs, but it occurs over the greatest distances in the stems of tall trees. During the day, water is constantly evaporating from the plant, mainly through leaf stomata. This water loss is called **transpiration** (TRAN-spuh-RAY-shuhn). The large amount of water lost from the plant is a result of the plant's need to obtain carbon dioxide from the air. Exactly how do huge trees, like redwoods, move water and mineral nutrients up their 100 m tall trunks?

According to the **cohesion-tension theory,** water is pulled up the stem xylem by the strong attraction of water molecules to each other, a property of water called cohesion. The movement also depends on the rigid xylem walls and the strong attraction of water molecules to the xylem wall, which is called adhesion. The thin, continuous columns of water extend from the leaves through the stems and into the roots. As water evaporates from the leaf, the water column is subject to great tension. However, the water column does not break because of cohesion and it does not pull away from the xylem walls because of adhesion. The only other possibility is for it to be pulled upward. The pull at the top of the tree extends all the way to the bottom of the column. As water is pulled up the xylem, more water enters the roots from the soil to replace the lost water.

Storing Water and Nutrients

With abundant parenchyma cells in the cortex, plant stems are adapted for storage in most species. In some species, storage is a major function. Cactus stems are specialized for storing water. The roots of a cactus are found close to the soil surface, enabling them to absorb rainwater quickly and transport it to the cactus's fleshy stem. Sugar-cane stems store large amounts of sucrose. The edible white potato is a stem that is specialized for storing starch. The "eyes" of white potatoes are buds that have the ability to develop into new shoots.

Word Roots and Origins

transpiration

from the Latin *trans,* meaning "across," and the Latin *spirare,* meaning "to breathe"

internet**connect**

SCiLINKS. NSTA

TOPIC: Transpiration
GO TO: www.scilinks.org
KEYWORD: HM613

SECTION 31-3 REVIEW

1. How does the arrangement of vascular tissues in a stem differ between monocots and dicots?
2. Describe five differences and five similarities between the structure of roots and the structure of stems.
3. Explain the difference between summerwood and springwood and tell how they form annual rings.
4. Describe how water moves through xylem tissue.
5. Describe the movement of carbohydrates within phloem tissue.
6. CRITICAL THINKING Some squirrels damage trees by stripping off portions of the bark. Why would a squirrel eat bark?

QUICK FACT

☑ The process of making syrup from the sap of sugar-maple trees takes advantage of the fact that the trees store sugars produced by photosynthesis in the form of starch in the roots. In the spring, the starch is converted back into sugar and moves upward in the vascular tissue to the buds on the stems. The sap provides the buds with energy to grow until they become photosynthetic. Maple-sugar producers tap these trees, collect the sap, and process it into syrup. Maple trees yield an average of 70 L (18 gal) of sap.

ANSWERS TO SECTION 31-3 REVIEW

1. Monocot stems usually have vascular bundles scattered in the ground tissue. Dicot stems usually have a ring of vascular bundles separating the internal pith from the external cortex.
2. Both roots and stems have a cylindrical shape, possess an epidermis, contain vascular tissue, provide storage, and produce lateral growth. Only roots contain an endodermis, a pericycle, and a root cap. Only stems are photosynthetic and contain vascular bundles, nodes, internodes, and buds.
3. Springwood consists of large-diameter, thin-walled cells. Summerwood consists of smaller-diameter, thicker-walled cells. An annual ring is the abrupt change between the summerwood and the following year's springwood.
4. Water enters the plant from the roots and is pulled up the xylem by cohesion.
5. Carbohydrates move from leaves (or storage structures) to roots, shoot apical meristems, and developing flowers or fruits.
6. The bark contains phloem, which provides food for the squirrel.

CHAPTER 31
Research Notes

BACKGROUND

Eloy Rodriguez has combined his scientific studies in phytochemistry with a knowledge of ethnobotany in his search for plants of medicinal importance.

DISCUSSION

Guide the discussion by posing the following questions to students.

1. Eloy Rodriguez coined the term *zoopharmacognosy.* What does it mean? (Zoopharmacognosy is the science of how wild animals use plants as medicine.)

2. Dobson suggested that the diet of the African green monkey should be studied by researchers. Why? (HIV is thought to have originated in green monkeys, so studying their diet may lead to new compounds to combat AIDS.)

3. Rodriguez feels that he is in a race against time to find possible life-saving plants in tropical areas. Why? (Possible life-saving plants in tropical areas of the world are becoming extinct as the rain forests rapidly disappear.)

FURTHER READINGS

1. "Clean Sweep for Leaf-eating Chimps." *New Scientist,* August 24, 1996. 17. This article presents another theory for why East African chimpanzees eat *Aspilia* leaves: a physical trapping of the intestinal parasites by the leaf surfaces.

2. "Monkeys with Zest." *Discover,* July 1996. 30. This article describes how capuchin monkeys in Costa Rica rub themselves with lemons, limes, and oranges, apparently to repel insects.

Research Notes

Nature's Chemical Arsenal

Plants depend heavily on chemical compounds for their survival because they are unable to otherwise escape herbivores. Such defensive compounds are called secondary compounds because they have no known function in plant growth. Secondary compounds are the active ingredients in many familiar medicines, including aspirin and taxol.

Secondary compounds are unique not only to each plant species but often to specific parts of a plant. For example, the chemicals found in a plant's flowers are likely to attract pollinators, while the chemicals found in the same plant's leaves may be toxic. With so many plants and plant parts in existence, scientists have a multitude of chemical compounds to study in their search for new medicines.

Eloy Rodriguez, a plant chemist at Cornell University, has found a surprising way to track down secondary compounds that are likely to be helpful to humans—by studying how animals may medicate themselves in the wild with plants. He named this new discipline zoopharmacognosy.

Rodriguez and his colleagues found that non-human primates, which suffer from many of the same illnesses as humans, tend to eat only one kind of plant, or even one part of one kind of plant, when they have a particular illness. For example, Richard Wrangham, a Harvard University primatologist, observed that East African chimpanzees suffering from parasitic infections eat *Aspilia* leaves. When Rodriguez analyzed the chemistry of *Aspilia,* he identified one of the chemicals extracted from the plant's leaves as thiarubrine-A. Rodriguez's experiments on thiarubrine-A demonstrated that the chemical has antibiotic effects on bacteria, fungi, and intestinal parasites.

Plant researcher Eloy Rodriguez shows students in his lab some of the chemicals extracted from plant leaves.

Another plant eaten by chimpanzees is *Rubia cordifolia.* Several chemicals collected from this plant, grouped together as cyclic hexapeptides, are being investigated by the National Institutes of Health as a possible therapeutic agent against cancer. Laboratory tests with mice have shown that at the dosages ingested by chimpanzees, these chemicals have low toxicity. So if further research documents the chemicals' effectiveness as a weapon against cancer, they may provide the benefits of chemotherapy with less serious side effects.

After Rodriguez's research team began presenting its results on zoopharmacognosy, animal behaviorists began reporting similar observations of other animals. In 1996, Andrew Dobson, a professor at Princeton University, suggested that the diet of the African green monkey should be studied by researchers looking for new compounds to combat AIDS because the virus is thought to have originated in green monkeys.

Meanwhile, Rodriguez is concerned that he and other plant researchers face a race against time. Potentially life-saving plants found in tropical areas of the world are becoming extinct as the rain forests rapidly disappear. Rodriguez urges other researchers to join him in investigating the animals and plants that have coexisted in this region, using the innovative approaches that he pioneered.

LEAVES

SECTION 31-4

OBJECTIVES

▲ Identify the difference between a simple leaf and a compound leaf.

● Describe the tissues that make up the internal structure of a leaf.

■ Describe adaptations of leaves for special purposes.

◆ Explain the importance of stomata.

Most leaves are thin and flat, an adaptation that helps them capture sunlight for photosynthesis. Although this structure may be typical, it is certainly not universal. Like roots and stems, leaves are extremely variable. This variability represents adaptations to environmental conditions.

TYPES OF LEAVES

Look at the leaves in Figure 31-14a. The coiled structure is a **tendril,** a specialized leaf found in many vines, such as peas and pumpkins. It wraps around objects to support the climbing vine. In some species, like grape, tendrils are specialized stems.

An unusual leaf modification occurs in carnivorous plants such as the pitcher plant, shown in Figure 31-14b. In carnivorous plants, leaves function as food traps. These plants grow in soil that is poor in several mineral nutrients, especially nitrogen. The plant receives substantial amounts of mineral nutrients when it traps and digests insects and other small animals.

Leaves, or parts of leaves, are often modified into spines that protect the plant from being eaten by animals, as shown in Figure 31-14c. Because spines are small and nonphotosynthetic, they greatly reduce transpiration in desert species such as cactuses.

(a)

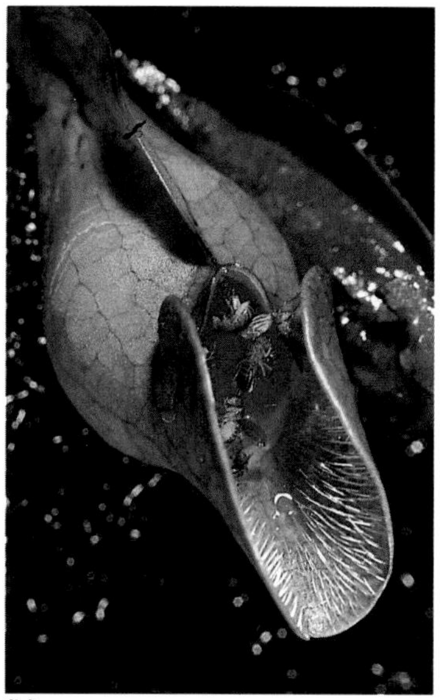
(b)

(c)

FIGURE 31-14
Many plants have developed leaf adaptations. (a) The pea plant has tendrils that climb. (b) The pitcher plant has tubular leaves that trap insects. (c) The barberry has spines that protect against herbivores.

SECTION 31-4

SECTION OVERVIEW

Cell Structure and Function
This section describes the types of leaves, how leaf structures are related to their functions, and the growth of leaves.

▶ ENGAGE STUDENTS

Bring a variety of leaves to class for students to examine. Perhaps include samples of maple, pine, corn, oak, spider plant, iris, cactus with spines, African violet, palm, lettuce, rhubarb, geranium, Venus' flytrap, onion bulb, lawn grass, elephant ears, coleus, and clover. Ask students to identify the external features of the leaves. On the board, list the differences that students notice from their examination of the leaves. Point out to students that plants show such variety in their leaves because they have adapted to a wide range of habitats.

TEACHING STRATEGY

Insect Capture
Divide the class into groups of six. Give each group one of the following plants: Venus' flytrap, pitcher plant, sundew plant (available from supply houses, florists, or nurseries). Have each group experiment to find out how each kind of leaf works to trap insects. (Venus' flytrap: one of the trigger hairs must be touched two or three times in succession; pitcher plant: insect falls in, and downward-pointing hairs prevent its crawling out; sundew: insect gets stuck on the sticky tendrils, and the sundew curls over it.) After completing the experiment, each group should report their findings to the class and share their observations.

TEACHING STRATEGY

Tree Identification

Take students to a nearby natural area or park. Provide each pair of students with a dichotomous key for identifying trees in your area by their leaves. Ask students to identify 10 trees by their leaves. CAUTION **Students may be allergic to some plants. Have students wash their hands thoroughly after handling leaves.** Give each pair a magnifying lens for observing leaf hairs and other microscopic structures. Check students' identifications with the key when they are finished.

QUICK FACT

☑ Coumarin is a chemical compound found in freshly cut grass and hay that gives it a distinctive smell. A fungus that can grow on cut hay synthesizes dicoumerol (two coumarin molecules bonded together) from coumarin. Dicoumerol prevents blood clotting, which can cause hemorrhaging in livestock and is useful in treating heart disease and preventing strokes.

OVERCOMING MISCONCEPTIONS

Most conifers and some other tree species retain their leaves over winter and are thus known as evergreen trees. These trees don't retain their leaves for their entire lives, however. Typically, new leaves are produced every year and each leaf stays on the plant for 2 to 3 years.

DEMONSTRATION

Use a projecting microscope to display several prepared slides of leaves (cross sections). Point out to students the following structures: upper and lower epidermis, palisade parenchyma, spongy mesophyll, cuticle, vascular bundle, guard cell, and stoma.

(a) SIMPLE LEAF

(b) COMPOUND LEAF

(c) DOUBLY COMPOUND LEAF

FIGURE 31-15

(a) A sugar-maple leaf is called a simple leaf because it has only one blade. (b) A white clover leaf is called a compound leaf because the leaf blade is divided into distinct leaflets. (c) The honeylocust has a doubly compound leaf because each leaflet is subdivided into smaller leaflets.

LEAF STRUCTURES

Leaves come in a wide variety of shapes and sizes and are an important feature used for plant identification. Leaves can be round, straplike, needlelike, or heart-shaped. The broad, flat portion of a leaf, called the **blade,** is the site of most photosynthesis. The blade is usually attached to the stem by a stalklike **petiole.** The maple leaf shown in Figure 31-15a is a **simple leaf;** it has a single blade. In **compound leaves,** such as the white clover in Figure 31-15b, the blade is divided into **leaflets.** In some species, the leaflets themselves are divided. The result is a doubly compound leaf, such as that of the honeylocust shown in Figure 31-15c.

Leaves consist of three tissue systems. The dermal tissue system is represented by the epidermis. In most leaves the epidermis is a single layer of cells coated with a nearly impermeable cuticle. Water, oxygen, and carbon dioxide enter and exit the leaf through stomata in the epidermis. Epidermal hairs are often present and usually function to protect the leaf from insects and intense light.

The number of stomata per unit area of leaf varies by species. For example, submerged leaves of aquatic plants have few or no stomata. Corn leaves have up to 10,000 stomata per square centimeter on both upper and lower surfaces. Scarlet oak has over 100,000 stomata per square centimeter on the lower leaf surface and none on the upper surface. Regardless of their exact distribution, stomata are needed to regulate gas exchange.

In most plants, photosynthesis occurs in the leaf **mesophyll** (MEZ-oh-FIL), a ground tissue composed of chloroplast-rich parenchyma cells. In most plants, the mesophyll is organized into two layers, which are shown in Figure 31-16. The **palisade mesophyll** layer occurs directly beneath the upper epidermis and is the site of most photosynthesis. Palisade cells are columnar and appear to be packed tightly together in one or two layers. However, there are air spaces between the long side walls of palisade cells. Beneath the palisade layer is the **spongy mesophyll.** It consists of irregularly shaped cells surrounded by large air spaces, which allow oxygen, carbon dioxide, and water to diffuse into and out of the leaf.

The vascular tissue system of leaves consists of vascular bundles called **veins.** Veins are continuous with the vascular tissue of the stem and the petiole, and they lie embedded in the mesophyll. Veins branch repeatedly so that each cell is usually less than 1 mm (0.04 in.) from a vein.

Venation is the arrangement of veins in a leaf. Leaves of most monocots, such as grasses, have **parallel venation,** meaning that several main veins are roughly parallel to each other. The main veins are connected by small, inconspicuous veins. Leaves of most dicots, such as sycamores, have **net venation,** meaning that the main vein or veins repeatedly branch to form a conspicuous network of smaller veins.

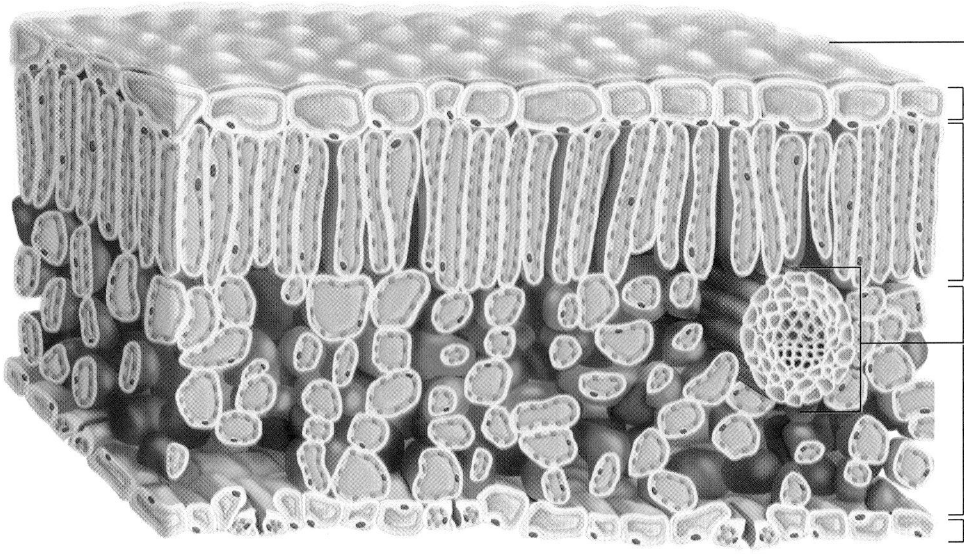

FIGURE 31-16

Cells from all three tissue systems are represented in the internal structure of a leaf. The epidermis is part of the dermal system, and the vascular bundle is part of the vascular system. The mesophyll is ground tissue made of parenchyma cells, and it generally contains chloroplasts. Note that the palisade mesophyll layer is more dense than the spongy mesophyll layer.

LEAF FUNCTIONS

Leaves are the primary site of photosynthesis in most plants. Mesophyll cells in leaves use light energy, carbon dioxide, and water to make carbohydrates. Light energy also is used by mesophyll cells to synthesize amino acids, fats, and a variety of other organic molecules. Carbohydrates made in a leaf can be used by the leaf as an energy source or as building blocks. They also may be transported to other parts of the plant, where they are either stored or used for energy or building blocks.

A major limitation to plant photosynthesis is insufficient water due to transpiration. For example, about 98 percent of the water that is absorbed by a corn plant's roots is lost through transpiration. However, transpiration may benefit the plant by cooling it and by speeding the transport of mineral nutrients through the xylem.

Modifications for Capturing Light

Leaves absorb light, which, in turn, provides the energy for photosynthesis. Leaves often adapt to their environment to maximize light interception. On the same tree, leaves that develop in full sun are thicker, have a smaller area per leaf, and have more chloroplasts per unit area. Shade-leaf chloroplasts are arranged so that shading of one chloroplast by another is minimized, while sun-leaf chloroplasts are not.

In dry environments, plants often receive more light than they can use. These plants often have structures that reduce the amount of light absorbed. For example, many desert plants have evolved dense coatings of hairs that reduce light absorption. The window plant shown in Figure 31-17 protects itself from its dry environment by growing underground. Only its transparent leaf tips protrude above the soil to gather light for photosynthesis.

FIGURE 31-17

The unusual transparent leaf tips of this window plant channel light to the plant's buried leaves.

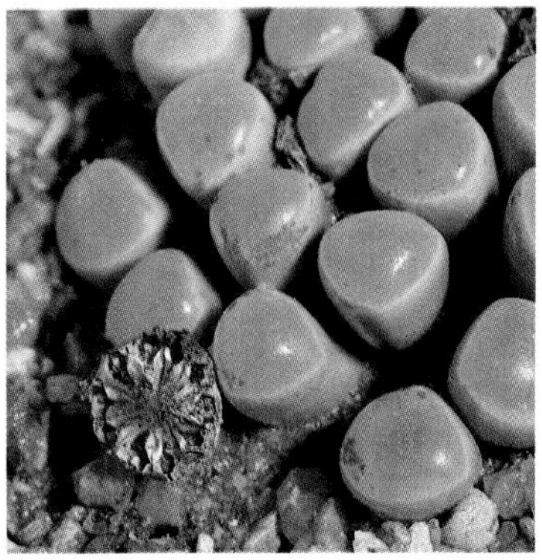

SECTION 31-4

VISUAL LINK

Figure 31-16

Have students notice the shape and arrangement of the palisade mesophyll. This arrangement increases the exposure of chloroplasts to light.

CRITICAL THINKING

Stomatal Density

After discussing gas exchange, ask students: Would leaves of rain-forest plants have more or fewer stomata per square centimeter of surface than leaves of temperate or cold-climate plants? (Humidity is the determining factor. Tropical rain-forest plants have leaves with many stomata. Leaves of plants in colder, drier regions have fewer stomata.)

CONTENT CONNECTION

Section 6-1: Photosynthesis

In photosynthesis, light energy excites electrons in chlorophyll. The electrons are then passed along a carrier chain, making ATP and NADPH in the process. In the Calvin cycle, carbon dioxide molecules are linked to form sugars using ATP and NADPH from the energy-capturing reactions.

RECENT RESEARCH

Unusual Association

Scientists at Stanford University have found that an epiphytic plant (one that grows on but does not harm another plant) in Malaysia, *Dischidia major,* has an unusual association with ants that live in its sacklike leaves. In exchange for shelter, ants provide the plants with 39 percent of the carbon from carbon dioxide that the ants exhale. They also provide 30 percent of the plants' nitrogen in the ants' debris, which includes feces, dead ants, and scavenged insect parts.

SECTION 31-4

ANSWERS TO SECTION 31-4 REVIEW

1. A simple leaf has a single blade. A compound leaf has the blade divided into two or more leaflets.
2. The leaf epidermis allows gas exchange for photosynthesis while minimizing water loss. The ground-tissue system functions in photosynthesis. The vascular system transports water and mineral nutrients to the leaf and transports organic compounds made in photosynthesis out of the leaf.
3. Leaves of carnivorous plants are tubular and trap animals. Leaves of cactuses are spines that protect the plant from animals. Leaves of some vines are tendrils that climb.
4. When guard cells gain water by osmosis, they swell and bow out, causing the stoma to open. When guard cells lose water by osmosis, they shrink and the stoma closes. The fact that water loss causes the stoma to close helps prevent water loss when the plant experiences a water shortage.
5. Plants in greenhouses in winter rarely grow as fast as the same type of plant grown outside in the summer because the light intensity is much lower in the winter than in the summer. Also, because a greenhouse is often closed tightly in winter to conserve heat, plant photosynthesis can quickly use up all the carbon dioxide in the greenhouse.
6. Stomata on the lower leaf surface are not as easily clogged by dust or rain.

(a)

(b)

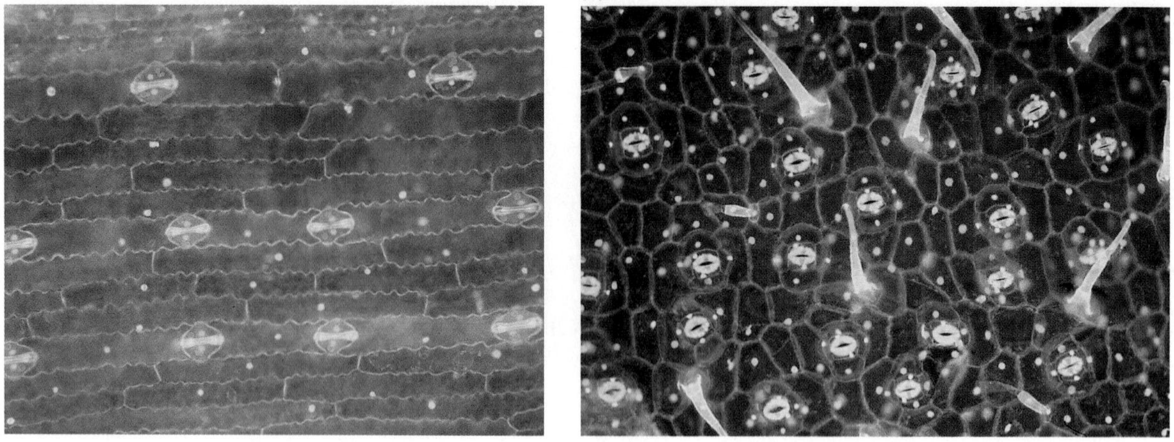

FIGURE 31-18

Scanning electron micrographs of monocot and dicot leaves show how the arrangement of stomata in each is different. (a) In the corn leaf, the stomata are in a parallel arrangement, which is typical of monocots. (b) In the potato leaf, the stomata are in a random arrangement, which is typical of dicots. For both monocots and dicots, the movement of water in the guard cells of each stoma is regulated by potassium ions (K^+).

Gas Exchange

Plants obtain carbon dioxide for photosynthesis from the air. Plants must balance their need to open their stomata to receive carbon dioxide and release oxygen with their need to close their stomata to prevent water loss through transpiration. A stoma is bordered by two kidney-shaped guard cells. **Guard cells** are modified cells found on the leaf epidermis that regulate gas and water exchange. Figure 31-18 shows how the stomata are arranged differently in monocots and dicots.

The stomata of most plants open during the day and close at night. The opening and closing of a stoma is regulated by the amount of water in its guard cells. When epidermal cells of leaves pump potassium ions (K^+) into guard cells, water moves into the guard cells by osmosis. This influx of water makes the guard cells swell, which causes them to bow apart to form a pore. During darkness, potassium ions are pumped out of the guard cells. Water then leaves the guard cells by osmosis. This causes the guard cells to shrink slightly and the pore to close.

Stomata also close if the plant experiences a shortage of water. The closing of stomata greatly reduces further water loss and may help the plant survive until the next rain. However, stomata closure virtually shuts down photosynthesis by cutting off the supply of carbon dioxide.

internetconnect
SCILINKS NSTA
TOPIC: Leaf functions
GO TO: www.scilinks.org
KEYWORD: HM618

SECTION 31-4 REVIEW

1. What is the difference between a simple leaf and a compound leaf?
2. Describe the basic function of each of the three leaf tissues.
3. What are three types of specialized leaves?
4. Explain the function of the guard cells in regulating stomatal opening and closing.
5. Why do plants grown in greenhouses in winter rarely grow as fast as the same type of plant grown outside in the summer even if the temperatures are the same in both locations?
6. **CRITICAL THINKING** Why is it an advantage for a plant to have most of its stomata on the underside of a horizontal leaf?

CHAPTER 31 REVIEW

SUMMARY/VOCABULARY

31-1

- Plants consist of three types of cells—parenchyma, sclerenchyma, and collenchyma.
- The dermal system consists of the epidermis, or the outermost layer of cells; it functions in absorption and protection in the roots and in gas exchange and protection in stems and leaves.
- The bulk of leaves, nonwoody stems, and nonwoody roots is ground tissue, which functions in storage, metabolism, and support.
- Vascular tissue consists of xylem, which carries water and mineral nutrients, and phloem, which transports organic compounds and some mineral nutrients.
- An increase in length, called primary growth, occurs mainly at the tips of stems and roots in the apical meristems.
- In secondary growth, the stems and roots increase in diameter in the lateral meristems.

Vocabulary

apical meristem (602)
collenchyma (599)
companion cell (601)
cork (602)
cork cambium (602)
cuticle (600)
dermal tissue system (600)
epidermis (600)
ground tissue system (601)
intercalary meristem (602)
lateral meristem (602)
meristem (602)
parenchyma (599)
pit (601)
primary growth (602)
sclerenchyma (600)
secondary growth (602)
sieve plate (601)
sieve tube (601)
sieve tube member (601)
tracheid (601)
vascular cambium (602)
vascular tissue system (601)
vessel (601)
vessel element (601)

31-2

- Roots anchor the plant and store and absorb water and mineral nutrients from the soil.
- A taproot system has a large primary root, and a fibrous root system has many small branching roots.
- Young roots produce root hairs, which are extensions of epidermal cells that increase the surface area for absorption.
- The root endodermis prevents substances from entering or leaving the root vascular tissue without passing through a cell membrane.

Vocabulary

adventitious root (604)
cortex (605)
endodermis (605)
fibrous root system (603)
macronutrients (606)
micronutrients (606)
pericycle (606)
root cap (604)
root hair (604)
taproot (603)

31-3

- Stems function in the transportation and storage of nutrients and water.
- The stems of monocots usually have scattered vascular bundles and usually lack secondary growth.
- Dicot stems have vascular bundles arranged in a ring and often produce abundant secondary growth.
- Secondary growth consists primarily of secondary xylem, called wood.
- In nontropical areas, secondary xylem in stems forms one annual ring each year.
- The outer bark of trees consists of cork produced by the cork cambium.
- The cohesion-tension theory describes the process of how water is pulled up through the xylem tissue.
- The pressure-flow hypothesis describes the process of how organic compounds are transported in the phloem tissue.

Vocabulary

annual ring (611)
bark (611)
bud (609)
bud scale (609)
cohesion-tension theory (612)
heartwood (611)
internode (609)
node (609)
pith (610)
pressure-flow hypothesis (613)
sapwood (611)
sink (613)
source (613)
springwood (611)
summerwood (611)
translocation (612)
transpiration (613)
wood (611)

CHAPTER 31 REVIEW ANSWERS

REVIEW

1. epidermis; it is not a secondary growth tissue.
2. mesophyll; it is not vascular tissue.
3. vascular cambium; it is not related to primary growth.
4. collenchyma; it is not related to water absorption and transport.
5. cork; it is not found in leaves.

6. c
7. c
8. d
9. a
10. c
11. c
12. b
13. b
14. d
15. c

16. Plant stems and roots usually grow in diameter due mainly to the action of the vascular cambium, which produces phloem toward the epidermis and xylem toward the pith. A cork cambium also helps to increase the diameter of stems and roots by producing cork cells.
17. Heartwood is the usually darker inner xylem that no longer functions in transport. Sapwood is the usually lighter-colored outer xylem that still functions in transport.
18. Monocot stems usually have vascular bundles scattered throughout the ground tissue, which is not separated into cortex and pith. Primary growth in dicot stems usually has a ring of vascular bundles separating the outer cortex from the inner pith. Monocot leaves usually have main veins that are roughly parallel. The main veins are connected by inconspicuous minor veins. Dicot leaves usually have net venation, in which one to several main veins branch repeatedly to form a conspicuous network of veins.
19. The phloem mainly transports carbohydrates, other organic molecules, and several mineral nutrients.
20. Woody plants, such as oak and pine, have secondary growth that contains a large amount of xylem or wood. Nonwoody plants, such as coleus and grass, have

ANSWERS

CONTINUED

either no secondary growth or secondary growth that contains little hard xylem.

21. Primary growth is an increase in the length of roots, stems, and leaves. Secondary growth is an increase in the diameter of roots and stems.

22. The cohesion-tension theory states that water in the xylem is pulled up the plant as water evaporates (transpires) from leaves. The pull occurs because water molecules are strongly attracted to each other and to the rigid xylem walls.

23. Succulent leaves are adapted for storage. Leaves of carnivorous plants are adapted for trapping animals. Spines of cacti are adapted for protection from animals. Vine tendrils are adapted for climbing.

24. A sweet potato is a modified root while a white potato is a modified stem. Both are underground structures specialized for storage.

25. When surrounding cells pump potassium ions into a pair of guard cells, the guard cells swell as water moves into them by osmosis and the stoma opens. When potassium ions are pumped out of guard cells, the guard cells shrink as water moves out of them by osmosis and the stoma closes.

CRITICAL THINKING

1. a. stomata; b. xylem and phloem or vascular tissues; c. usually roots, leaves in some species like Spanish moss and carnivorous plants; d. usually leaves, stems in some species like cactuses; e. epidermis or cork; f. secondary xylem (wood) and collenchyma; g. parenchyma

2. Roots and root hairs function primarily in water and mineral-nutrient absorption. Root hairs and young roots are easily damaged if soil is removed. Plant leaves require huge amounts of

CHAPTER 31 REVIEW

31-4

- The identification of plants is based on the shape, size, and arrangement of leaf blades. Plants with simple leaves or compound leaves are common.
- Photosynthesis occurs mostly in the palisade mesophyll, which consists of rows of closely packed cells, and the loosely packed spongy mesophyll.
- Gas exchange in leaves is controlled by stomata, or small openings in the leaf.
- Most of the water absorbed by a plant evaporates from the stomata during transpiration.
- Two guard cells surround each stoma. When the guard cells gain water, the stoma opens. When the guard cells lose water, the stoma closes.

Vocabulary

blade (616)	mesophyll (616)	petiole (616)	tendril (615)
compound leaf (616)	net venation (616)	simple leaf (616)	vein (616)
guard cell (618)	palisade mesophyll (616)	spongy mesophyll (616)	venation (616)
leaflet (616)	parallel venation (616)		

REVIEW

Vocabulary

In each of the following sets, choose the term that does not belong and explain why.

1. wood, bark, epidermis, cork
2. sieve tube member, mesophyll, tracheid, vessel element
3. apical meristem, primary xylem, vascular cambium, endodermis
4. root hair, transpiration, vessel element, collenchyma
5. blade, cork, petiole, vein

Multiple Choice

6. Nonfunctional xylem in the center of a tree trunk is (a) summerwood (b) sapwood (c) heartwood (d) springwood.
7. Mesophyll is the site of (a) water absorption (b) storage (c) photosynthesis (d) secondary growth.
8. Water moves between tracheids through (a) end walls (b) stomata (c) vessel elements (d) pits.
9. Stomata open and close due to water pressure changes in the (a) guard cells (b) sieve tube member (c) root hairs (d) cortex.
10. The root apical meristem is protected by the (a) bud scales (b) cuticle (c) root cap (d) cortex.
11. Collenchyma and sclerenchyma function mainly in (a) storage (b) photosynthesis (c) support (d) transport.
12. Most monocots do not have (a) primary growth (b) secondary growth (c) xylem (d) phloem.
13. A waterproof substance occurs in cell walls of (a) cortex (b) endodermal cells (c) palisade mesophyll cells (d) vessel elements.
14. The main site of photosynthesis in cactuses is the (a) root (b) leaf (c) flower (d) stem.
15. A structure found in stems but not in roots is (a) epidermis (b) vascular tissue (c) node (d) cortex.

Short Answer

16. What causes a plant stem or root to grow in diameter?
17. Explain the difference between heartwood and sapwood.
18. Describe the different arrangements of vascular bundles in the stems and leaves of monocots and dicots.
19. What substances are transported through the tissue shown in the photograph below?
20.

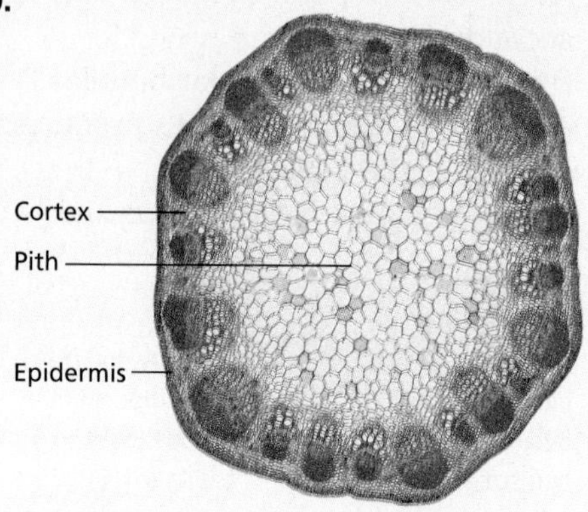

Distinguish between woody plants and nonwoody plants, and give an example of each.

21. Explain the difference between primary growth and secondary growth.

22. Briefly describe the cohesion-tension theory of water movement in the xylem.

23. List three modified types of leaves.

24. How are a sweet potato and a white potato different? How are they similar?

25. Explain how guard cells regulate the opening and closing of stomata.

CRITICAL THINKING

1. Copy the chart shown below on a sheet of paper. Fill in the plant structures corresponding to the human structures listed in the chart.

Common Structural Functions

Function	Human structure	Plant structure
Gas exchange	lungs	a.
Circulation	blood vessels	b.
Water intake	mouth	c.
Energy intake	mouth	d.
Protective covering	skin	e.
Internal support	skeleton	f.
Energy storage	fat cells	g.

2. When transplanting a plant, it is important that you not remove any more soil than necessary from around the roots. From your knowledge of the function of roots and root hairs, why do you think is this so important?

3. Suppose you examine a tree stump and notice that the annual rings are thinner and closer together 50 rings in from the edge. What would you conclude about the climate in the area 50 years ago?

4. Girdling is the removal of a narrow strip of bark all the way around a tree trunk. What effects might girdling have on a plant's shoots and roots?

5. What causes a knot in a board, like that shown in the photograph below?

EXTENSION

1. Read "Why Tulips Can't Dance" in *Science News,* February 5, 2000, on page 95. Describe how the stem structures of tulips and daffodils compare. How does the stem structure of daffodils contribute to the plants' ability to bend in gusty winds?

2. Measure transpiration in a common nonwoody plant. Buy two small flowering plants or tomato plants from a local nursery. Get plants as identical in size as possible. Make sure the plants are well watered, and then cover the pots with a plastic bag. Tie the bag shut around the base of the plant's stem without injuring the stem. Weigh each pot to the nearest gram and record the mass. Place one plant in the dark and one plant in bright light. Weigh the pots again after two hours, and calculate the amount of transpiration in grams as follows: transpiration = initial mass − final mass. If there is no difference in transpiration, extend the length of the experiment. Which treatment resulted in the most transpiration? Why?

water. Thus, root damage can slow photosynthesis because stomata close if there is a shortage of water. Severe root damage during transplanting could even result in plant death due to drying out of the leaves or stems.

3. Thinner annual rings 50 years ago indicate that growth was reduced by some environmental factor. Usually lack of rain is the cause; however, it could be anything that reduces photosynthesis, such as animals eating leaves, abnormally cold weather, or root injury due to construction or disease.

4. Girdling disrupts movement in the phloem, so carbohydrates cannot move from leaves to roots. In a girdled plant, carbohydrates accumulate above the girdle. Unless or until the phloem regenerates, the roots must rely on stored energy. Roots usually have enough stored energy to survive at least one growing season. Since carbohydrates cannot be transported to roots, girdled plants often produce more or larger fruits. Thus, growers often intentionally girdle some woody plants, such as grapes.

5. A knot represents the base of a branch or lateral stem enclosed by the wood of the trunk or the main stem.

EXTENSION

1. A daffodil's stem is elliptical in a cross section. A tulip's stem is round. The elliptical shape of the daffodil's stem enables the daffodil to twist to face downward in a breeze, reducing the flower's wind drag by 30 percent. A tulip's stem is not very bendable.

2. The plant that received bright light transpired more heavily because of greater evaporation.

CHAPTER 31
INVESTIGATION

TIME REQUIRED

One 50-minute class period

SAFETY PRECAUTIONS

Remind students to be careful when handling a microscope and glass slides. Special care should be taken when viewing slides under high power so that the objective does not break the slide. Have students dispose of broken glass in a container designated only for glass. Caution students not to use microscopes with wet hands or near water.

QUICK REFERENCES

Lab Preparation notes are found on page 597B.

Holt BioSources provides a Teaching Resources CD-ROM that contains "Using Gowin's Vee in the Lab" and "Scoring Rubrics."

PROCEDURAL TIP

Set up three lab stations, one for each part (roots, stems, and leaves). Divide the class into groups, and have each group rotate among the three lab stations.

ANSWERS TO BACKGROUND

1. Root hairs absorb water and mineral nutrients.
2. Translocation within phloem tissue transports sugar throughout a plant.
3. Herbaceous stems are soft and usually green. Woody stems are rigid and not green. Both woody stems and herbaceous stems are cylindrical and have dermal, ground, and vascular tissues.
4. The vascular tissues, xylem and phloem, are continuous in the root, stem, and leaf.
5. The closing of stomata by guard cells and the cuticle covering the epidermis are two ways that a leaf conserves water.

CHAPTER 31 INVESTIGATION

Observing Roots, Stems, and Leaves

OBJECTIVES

- Observe the tissues and structures that make up roots, stems, and leaves.
- Explain how roots, stems, and leaves are adapted for the functions they perform.

PROCESS SKILLS

- observing
- identifying
- relating structure and function

MATERIALS

- prepared slides of the following tissues:
  - *Allium* root tip longitudinal section
  - *Syringa* leaf cross section
  - *Ranunculus* root cross section
  - *Ranunculus* stem cross section
  - *Zea mays* stem cross section
- compound light microscope

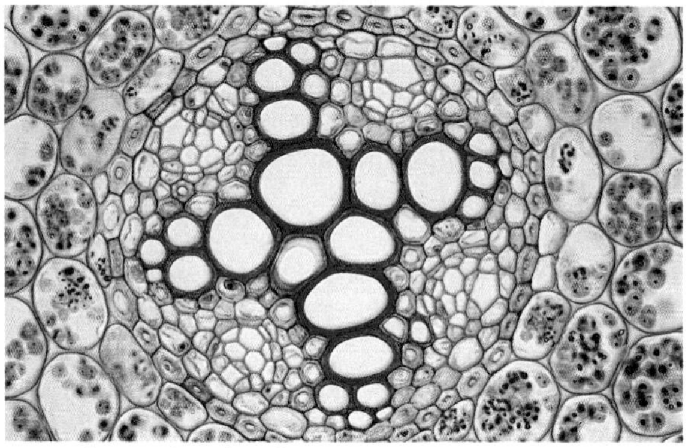

Ranunculus root (275×)—a dicot

Background

1. Which plant tissues are responsible for the absorption of water and mineral nutrients?
2. How is sugar, which is produced in the leaf, moved to other parts of the plant?
3. How are woody and herbaceous stems different, and how are they alike?
4. What tissues are continuous in the root, stem, and leaf?
5. How does the leaf conserve water?

PART A Roots

1. **CAUTION Glass slides may break and cut you.** Observe the prepared slide of the *Allium* root tip under low power. Locate the root cap and the root tip meristematic cells. You may look at the photograph on page 604 for help. Note the long root hairs in the area above the root tip.
2. In your lab report, draw the root tip that you observe, and label the root cap, meristem, and root hairs in your drawings.
3. Change slides to the root cross section of the dicot *Ranunculus*. This slide should look similar to the photograph shown above. The inner core is the vascular tissue, which is surrounded by the endodermis. This area of the tissue is involved in the transport of water, mineral nutrients, and organic compounds. Look for the star-shaped xylem and the smaller phloem cells surrounding the xylem.
4. In your lab report, draw what you see, and identify the xylem tissue and the phloem tissue in your drawing.
5. Locate the cortex, where starch is stored, which surrounds the vascular cylinder. Outside the cortex, find the epidermal cells and their root hairs. Draw a one-quarter section of the root tissues. Label all the tissues in your lab report.

PART B Stems

6. Observe a prepared slide of the stem of *Zea mays,* a monocot, and find the epidermis and the photosynthetic layer. It should look like the photograph on the next page. In the center, look for the vascular bundles made up of xylem and phloem. Draw a diagram showing the location of the vascular bundles and the epidermis layer as they appear when viewed under low power.
7. Switch to high power, and observe a vascular bundle. Draw the vascular bundle, and label the tissues.
8. Observe a cross section of *Ranunculus,* a herbaceous dicot stem. Compare your slide with the photograph of

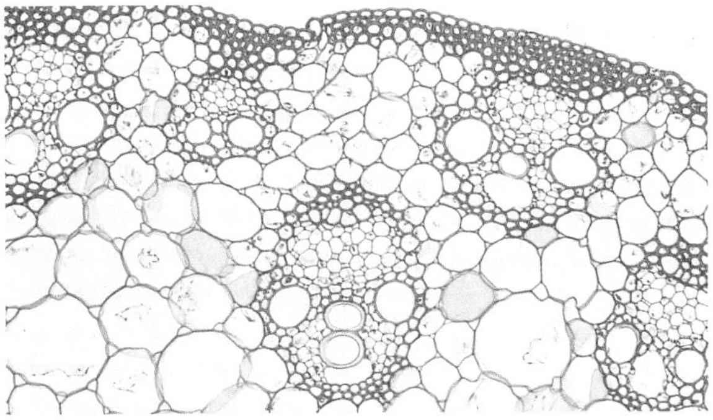

Zea mays stem (130×)—a monocot

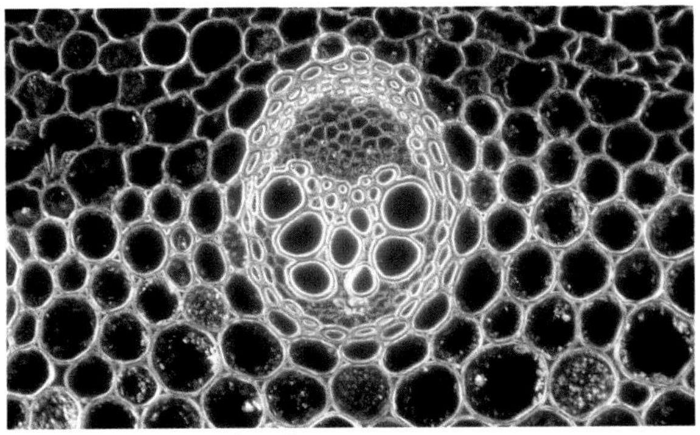

Ranunculus stem (151×)—a dicot

Ranunculus shown above. Look for the epidermis and cortex layers. Notice that in a dicot, the stem is more complex than the root. Note the arrangement of the vascular bundles. In your lab report, draw what you observe. Label the epidermis, cortex, and individual vascular bundles.

9. Focus on a vascular bundle under high power. Draw and label a diagram of a vascular bundle in your lab report.

PART C Leaves

10. Observe a prepared slide of a lilac leaf, *Syringa,* shown below, under low power, and find the lower epidermis.

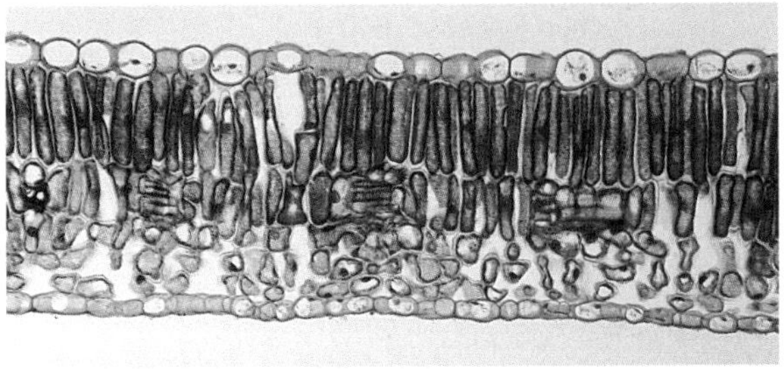

Syringa leaf (530×)

11. Identify the stomata on the lower epidermis of the lilac leaf. Find the guard cells that open and close a particular stoma. Locate an open stoma and a closed stoma. Draw and label diagrams of the stomata and the guard cells in your lab report.

12. Look at the center part of the cross section. Note the spongy texture of the mesophyll layer. Locate a vein containing xylem and phloem. Now identify the palisade layer, the upper epidermis, and finally the clear, continuous, noncellular layer on top. This layer is called the cuticle. Draw and label a diagram of your observations in your lab report.

13. Clean up your materials before leaving the lab.

Analysis and Conclusions

1. In the dicot root that you observed, where are phloem and xylem located?
2. Where are the xylem and phloem found in the herbaceous stem that you observed?
3. How are the vascular bundles different in the monocot and dicot stems that you observed?
4. How are the root cap cells different from the root tip meristematic cells?
5. What is the function of the root hairs?
6. How do the arrangements of xylem and phloem differ in roots, stems, and leaves?
7. What is the function of a stoma?
8. What is the function of the air space in the mesophyll of the leaf?
9. Which leaf structures help to conserve water?
10. Which tissues of the leaf are continuous with the stem and root tissues? How is this functional?
11. Look at the various tissues found in your drawings of roots, stems, and leaves. Classify each tissue as either dermal tissue, ground tissue, or vascular tissue.

Further Inquiry

The parts of a flower are actually modified stems and leaves. Design—but do not carry out—a procedure for dissecting a flower. Include a diagram of the parts of the flower to be viewed. Use references from the library to determine which kinds of flowers are best for dissection.

ANSWERS TO ANALYSIS AND CONCLUSIONS

1. Xylem and phloem are located in a cylinder at the center of the root.
2. In herbaceous stems, xylem and phloem are found in vascular bundles, with phloem toward the outside of the stem and xylem toward the center of the stem.
3. In monocots, vascular bundles are scattered throughout the stem. In dicots, vascular bundles are arranged in an orderly ring near the outside of the stem.
4. Root cap cells are large and loosely packed. Meristematic cells are very small and tightly packed.
5. Root hairs absorb water and mineral nutrients from the soil.
6. Xylem and phloem are found in the center of roots, in vascular bundles either scattered throughout or arranged in rings near the outside of stems, and in the veins of leaves.
7. A stoma allows water vapor, carbon dioxide, and oxygen to pass into and out of a leaf.
8. Air spaces enable gases to diffuse between the outside of a leaf and the cells inside the leaf.
9. The guard cells and the cuticle help conserve water.
10. Xylem and phloem are continuous from root to stem to veins of leaves, an arrangement that facilitates the movement of water from the roots to the rest of the plant and the movement of sugars from where they are made or stored to where they are needed.
11. Dermal tissue is found in the outermost layer of cells in roots, stems, and leaves. The bulk of leaves, nonwoody roots, and nonwoody stems consists of ground tissue. Vascular tissue consists of xylem and phloem, which is found in roots, stems, and leaves.

CHAPTER 32
PLANT REPRODUCTION

BLOCKS	TOPICS	TEACHING RESOURCES	LABS, CLASSWORK, AND HOMEWORK
BLOCKS 1 & 2 90 minutes	**Introducing the Chapter, p. 624**	ATE Focus Concept, p. 624 ATE Assessing Prior Knowledge, p. 624 ATE Understanding the Visual, p. 624	■ Supplemental Reading Guide, *A Feeling for the Organism*
	32-1 Plant Life Cycles, p. 625 The Life Cycle of Mosses, p. 625 The Life Cycle of Ferns, p. 626 The Life Cycle of Conifers, p. 627	ATE Section Overview, p. 625 ATE Visual Link Figure 32-1, p. 625 115A, 117, 118, 133	■ Portfolio Projects—Plant Research Project ★ Study Guide, Section 32-1 PE Section Review, p. 628
BLOCKS 3 & 4 90 minutes	**32-2 Sexual Reproduction in Flowering Plants, p. 629** Parts of a Flower, p. 629 Ovule Formation, p. 630 Pollen Grain Formation, p. 631 Pollination, p. 631 Fertilization, p. 632	ATE Section Overview, p. 629 ATE Critical Thinking, p. 632 ATE Visual Link Figure 32-6, p. 630 Figure 32-7, p. 631 116, 128, 131, 132	**Lab B15,** Flower Structures **Lab B16,** Fruits and Seeds **Lab C37,** Growing Plants in the Laboratory ★ Study Guide, Section 32-2 PE Section Review, p. 633
BLOCK 5 45 minutes	**32-3 Dispersal and Propagation, p. 634** Dispersal of Fruits and Seeds, p. 634 Types of Fruits, p. 635 Structure of Seeds, p. 636 Seed Germination, p. 637 Asexual Reproduction, p. 639 Vegetative Propagation, p. 639	ATE Section Overview, p. 634 ATE Critical Thinking, pp. 634, 637 ATE Visual Link Figure 32-12, p. 636 115, 128	PE **Chapter 32 Investigation,** p. 644 PE **Quick Lab,** p. 636, Predicting Seed Dispersal **Lab C38,** Growing Plants in the Laboratory—Fertilizer Problem ■ Portfolio Projects—Plant Focus Worksheet ★ Study Guide, Section 32-3 PE Section Review, p. 640
BLOCKS 6 & 7 90 minutes	**REVIEW and ASSESSMENT** PE Chapter 32 Review, pp. 642–643 ■ Performance-Based Assessment—Testing Enzyme Activity in Seeds	Audio CD Program	**CHAPTER TESTING** ■ Chapter Test (blackline master) ▲ Test Generator ■ Assessment Item Listing ■ Scoring Rubrics and Classroom Management Checklists

KEY

MODERN BIOLOGY
- PE Pupil's Edition
- ATE Teacher's Edition
- ■ Active Reading Guide
- ★ Study Guide

HOLT BIOSOURCES
- Teaching Transparencies
- Laboratory Program

One-Stop Planner CD-ROM
Shows these resources within a customizable daily lesson plan:
- ■ Holt BioSources Teaching Resources
- ■ Active Reading Guide
- ★ Study Guide
- ▲ Test Generator

READING FOR CONTENT MASTERY

Preview: Have students preview the objectives, the introductory paragraph, topic sentences, and captions before reading the chapter.

- ATE Active Reading Technique: K-W-L, p. 625
- ■ Reading Strategy: K-W-L Worksheet
- ■ Active Reading Guide Worksheet 32-1
- PE Summary/Vocabulary, p. 641

- ■ Active Reading Guide Worksheet 32-2
- PE Summary/Vocabulary, p. 641

- ATE Active Reading Technique: K-W-L, p. 639
- ■ Reading Strategy: K-W-L Worksheet
- ■ Active Reading Guide Worksheet 32-3
- PE Summary/Vocabulary, p. 641

TECHNOLOGY AND INTERNET RESOURCES

CNN PRESENTS Video Segment 20 Migrant Bees

Holt Biology Videodiscs Teacher's Correlation Guide, Lessons 32-1 through 32-3

Audio CD Program, Sections 32-1 through 32-3

internetconnect

SCILINKS NSTA

On-line Resources: www.scilinks.org
The following *sci*LINKS Internet resources can be found in the student text for this chapter:

Topics:
- Pollination, p. 631
- Seeds, p. 637
- Seed germination, p. 638

go.hrw.com

On-line Resources: go.hrw.com
Visit the HRW Web site for a variety of resources related to this chapter. Just type in the keyword HM2 Chapter 32.

Smithsonian Institution®

Visit **www.si.edu/hrw** for additional on-line resources.

CNNfyi.com

Visit **www.cnnfyi.com** for late-breaking news and current event stories selected just for you.

LAB ACTIVITY PLANNING

CHAPTER 32 INVESTIGATION

Comparing Seed Structure and Seedling Development, pp. 644–645

OVERVIEW
In this investigation, students will observe the structures of dicot and monocot seeds and will compare the development of their embryos.

SAFETY
Lugol's solution is an irritant. If Lugol's iodine solution gets in a student's eye, flush the eye with water for 15 minutes and get medical attention. If Lugol's solution is swallowed and the person is conscious, give the person one or two glasses of water to drink, induce vomiting, and call a physician immediately. Instruct students to use scalpels carefully to avoid injuring themselves or others.

PREPARATION
1. Each lab group will need one pea seed, six bean seeds, and six corn kernels.
2. To make the seed coats easy to remove, soak the seeds in water overnight prior to the investigation.
3. Provide each lab group with a dropper bottle containing Lugol's iodine solution.
4. Each lab group will need a scalpel, two rubber bands, two 150 mL beakers, a medicine dropper, a glass-marking pen, a metric ruler, and a microscope slide. Each lab group will need access to a compound light microscope and a stereomicroscope.

Quick Labs

Predicting Seed Dispersal, p. 636
For this activity, you will need to provide at least 10 different fruits from which students can choose. Students will also need a balance or scale.

UNDERSTANDING THE VISUAL

Have students examine the picture of the two pollen grains from goose grass, *Galium aparine*. Goose grass is commonly known as bedstraw. Its hairy leaves often cling to other plants or attach to the skin or clothing of passersby. The pollen grains in this SEM are approximately 100 μm wide. (About 25,000 μm equals 1 in.)

Tell students that pollen grains contain two haploid cells produced by meiosis. Ask students to describe the shape, size, and surface features of the pollen grains shown in the photograph. Tell students that these types of characteristics are unique to each plant species. Ask students to hypothesize about how the pollen grains shown in this photograph are dispersed (wind, insects, people).

FOCUS CONCEPT

Evolution

Plants have evolved elaborate reproductive adaptations over millions of years. Because plants are not mobile, they must rely on wind, water, or animals to disperse their sperm cells. Plants also rely on these same mechanisms to disperse their offspring.

ASSESSING PRIOR KNOWLEDGE

Review the following concepts.

Cell Reproduction: *Chapter 8*
Begin a review of mitosis and meiosis by asking students the following questions. How do organisms make new cells that are exact copies of existing cells? How do organisms make new cells with half the genetic material of the parent cells? What results when two haploid cells combine?

CHAPTER 32

PLANT REPRODUCTION

Two pollen grains are on the stigma of a flower of goose grass, Galium aparine. *A pollen tube, seen growing out of one of these pollen grains, will travel toward the ovary.*

FOCUS CONCEPT: *Evolution*

As you read, note the many plant adaptations that help ensure the successful reproduction, protection, and dispersal of offspring.

PLANT LIFE CYCLES

A life cycle includes all of the stages of an organism's growth and development. Recall from Chapter 30 that a plant's life cycle involves two alternating multicellular stages—a diploid (2n) sporophyte stage and a haploid (1n) gametophyte stage. This type of life cycle is called alternation of generations. Also recall that the size of gametophytes and sporophytes varies among the plant groups.

SECTION

32-1

OBJECTIVES

▲ Describe the life cycle of a moss.

● Describe the life cycle of a typical fern.

■ Describe the life cycle of a gymnosperm.

◆ Compare and contrast homospory and heterospory.

THE LIFE CYCLE OF MOSSES

The dominant form of a moss is a clump of leafy green gametophytes. Look at the moss life cycle illustrated in Figure 32-1. Moss gametophytes produce gametes in two types of reproductive structures—antheridia and archegonia. An **antheridium** (AN-thuhr-ID-ee-uhm) is a male reproductive structure that produces hundreds of flagellated sperm by mitosis. An **archegonium** (AWR-kuh-GOH-nee-uhm) is a female reproductive structure that produces a single egg by mitosis. During moist periods, sperm break out of the antheridia and swim to the archegonia. One sperm fertilizes the egg at the base of an archegonium, forming a diploid zygote. Through repeated mitotic divisions, the zygote forms an embryo and develops into a sporophyte.

FIGURE 32-1

The life cycle of a moss includes a relatively large, leafy green gametophyte, which produces gametes, and a smaller sporophyte, which grows from the tip of the gametophyte and produces only one type of spore. The flagellated sperm must swim to the eggs.

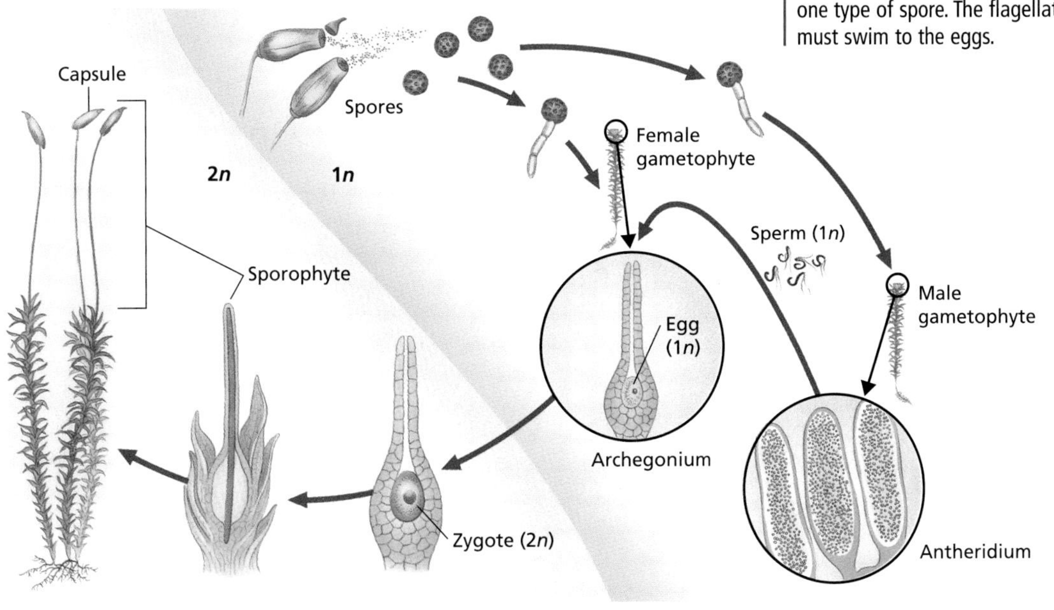

SECTION 32-1

SECTION OVERVIEW

Reproduction and Inheritance

Section 32-1 presents an overview of plant life cycles. All plants have a life cycle in which the sporophyte phase alternates with the gametophyte phase. All plants have diploid sporophytes that produce spores, which eventually develop into gametophytes. Usually the gametophytes are either male or female and produce sperm or eggs, respectively.

ACTIVE READING

Technique: K-W-L

Before students read this chapter, have them write short individual lists of all the things they already **K**now (or think they know) about plant life cycles and reproduction. Ask them to contribute their entries to a group list on the board or overhead projector. Then have students list things they **W**ant to know about plant life cycles and reproduction. Have students save their lists for use later in the Active Reading exercise on page 639.

VISUAL LINK

Figure 32-1

Have students identify ways that mosses differ from flowering plants. (Mosses are smaller than most flowering plants. The structure we recognize as the plant is the gametophyte in mosses and the sporophyte in flowering plants. Mosses release spores into the air, and flowering plants release pollen.)

TEACHING STRATEGY
Germinating Moss Spores

Obtain some fresh, mature moss sporophyte capsules (available from a biological supply house). Provide the students with dissecting needles, and instruct them to gently open the capsules and collect spores. Place the spores on a microscope slide. Have students place a round piece of filter paper on the base of a Petri dish and saturate the paper with distilled water. Then have students add the spores to the dish, place the cover on the dish, label the dish, and set the dish in a lighted location. Tell students that spore germination requires light. Have students observe their dishes under a dissecting microscope daily. Add water as necessary to keep the filter paper moist. Germination should occur within 24 hours, and the protonemas should be visible within 96 hours. Have students make daily drawings and record their observations.

GIFTED *ACTIVITY*

Have students conduct research and report on the now-extinct seed ferns that were common in the Carboniferous period. These plants had fernlike foliage and gymnosperm-like seeds attached to the leaf margins. (A paleobotany textbook would be a good source.)

QUICK FACT

☑ Ferns in the genus *Davallia* are characterized by creeping stems (rhizomes) covered with coarse tan or brown hairs that resemble animal feet. Cultivated species include *D. trihomanioides* (squirrel's foot fern), *D. canariensis* (deer's foot fern), and *D. fejeensis* (rabbit's foot fern).

A moss sporophyte begins as a thin stalk that grows from the tip of a gametophyte. The sporophyte remains attached to the gametophyte and depends on it for nourishment. Soon, cells at the tip of a stalk form a sporangium, called the capsule. Cells in the capsule undergo meiosis to form haploid spores, which are all the same. The production of one type of spore is called **homospory** (hoh-MAHS-puh-ree). Therefore, the life cycle of mosses is called *homosporous* (hoh-MAHS-puh-ruhs) *alternation of generations.* When the spores are mature, the capsule splits open, and the spores are carried away by the wind. Spores that land in favorable environments grow into new gametophytes.

THE LIFE CYCLE OF FERNS

The life cycle of a fern, shown in Figure 32-2, is similar to the moss life cycle. Like mosses, most ferns are homosporous. And as in mosses, the fern sporophyte grows from the gametophyte. But in the fern life cycle, the sporophyte, not the gametophyte, is the dominant generation. Fern gametophytes are tiny (about 10 mm, or 0.5 in., in diameter), flat plants that are anchored to the soil by rhizoids. Both antheridia and archegonia form on the lower surface of a fern gametophyte. When water is present, sperm released by antheridia swim to archegonia. One sperm fuses with the egg in an archegonium, forming a zygote, which is the first cell of a new sporophyte.

The zygote grows into an embryo and then into a young fern sporophyte through mitotic cell division. The sporophyte crushes the gametophyte as it grows. A mature fern sporophyte usually has compound leaves called fronds, which grow from an underground

FIGURE 32-2

The life cycle of most ferns includes a large sporophyte, which produces only one type of spore, and a small gametophyte, which produces gametes. Both the eggs and sperm are produced on the same gametophyte. The flagellated sperm must swim to the eggs.

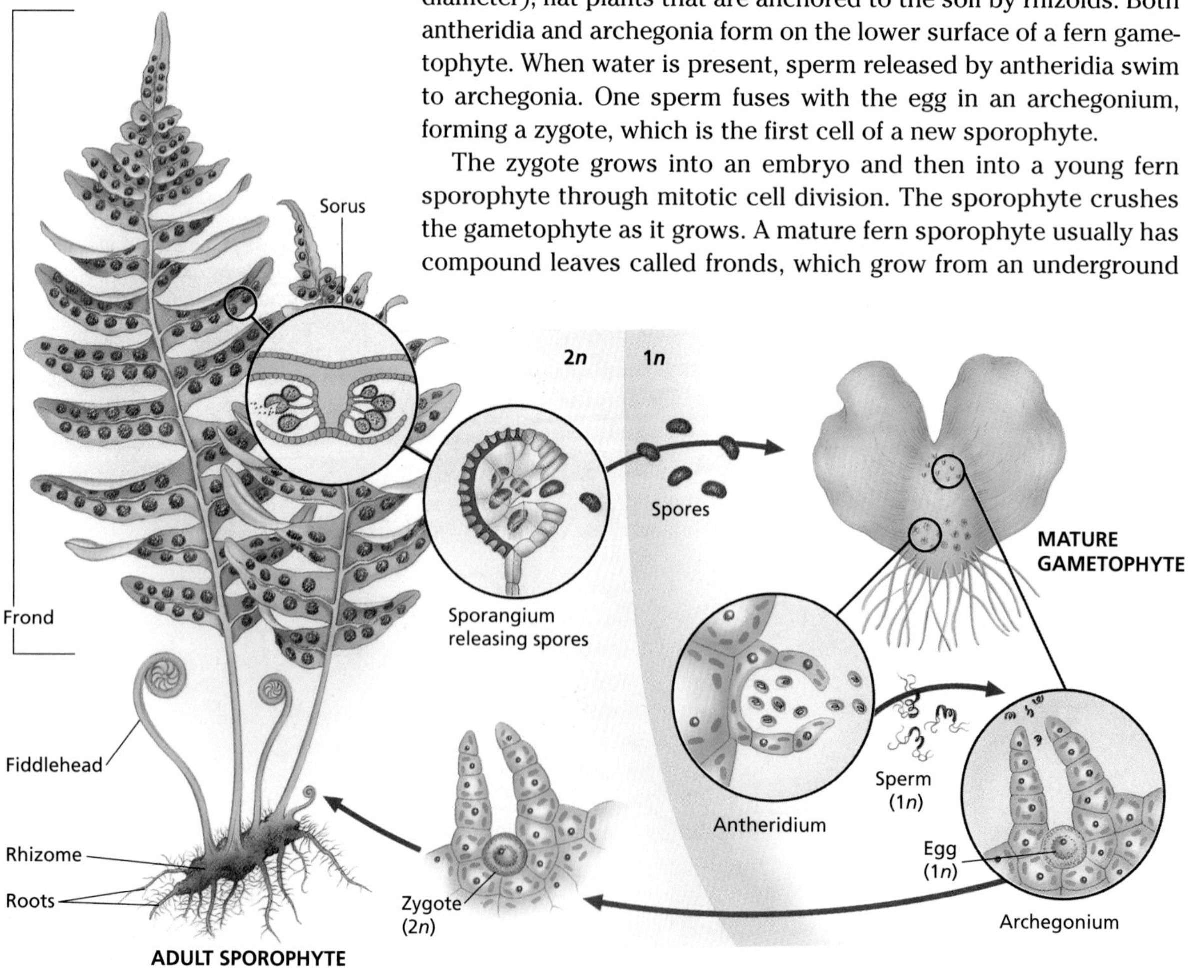

stem, or rhizome. In most ferns, certain cells on the underside of the fronds develop into sporangia. In many ferns, the sporangia are grouped together in clusters called **sori** (SOH-ree). Cells inside a sporangium undergo meiosis, forming haploid spores. At maturity, the sporangium opens and the spores are catapulted 1 cm (0.4 in.) or more and are then carried away by air currents. When the spores land, and if conditions are right, they grow into new gametophytes.

THE LIFE CYCLE OF CONIFERS

Unlike mosses and most ferns, gymnosperms produce two types of spores—male **microspores** and female **megaspores.** Microspores grow into male gametophytes, while megaspores grow into female gametophytes. The production of different types of spores is called **heterospory** (HET-uhr-AHS-puh-ree). Thus, the gymnosperm life cycle is called *heterosporous* (HET-uhr-AHS-puh-ruhs) *alternation of generations.* All seed plants, spike mosses, quillworts, and a few fern species also have heterospory, which ensures that a sperm will fertilize an egg from a different gametophyte and increases the chance that new combinations of genes will occur among the offspring.

Figure 32-3 shows the life cycle of a pine, which is a conifer, the most common kind of gymnosperm. The conifer life cycle illustrates how seed plants reproduce without water for sperm to swim through. The familiar pine tree is a sporophyte. Like humans, pines cannot sexually reproduce until they are mature. Depending on the species, pines take from 3 to more than 30 years to reach adulthood.

In pines, sexual reproduction takes more than two years. During the first summer, a mature pine tree produces separate male and

FIGURE 32-3

The life cycle of a gymnosperm includes a large sporophyte, which produces two types of spores, and microscopic gametophytes, which produce gametes. Female gametophytes produce eggs, and male gametophytes produce sperm. The nonflagellated sperm reach the eggs through a pollen tube.

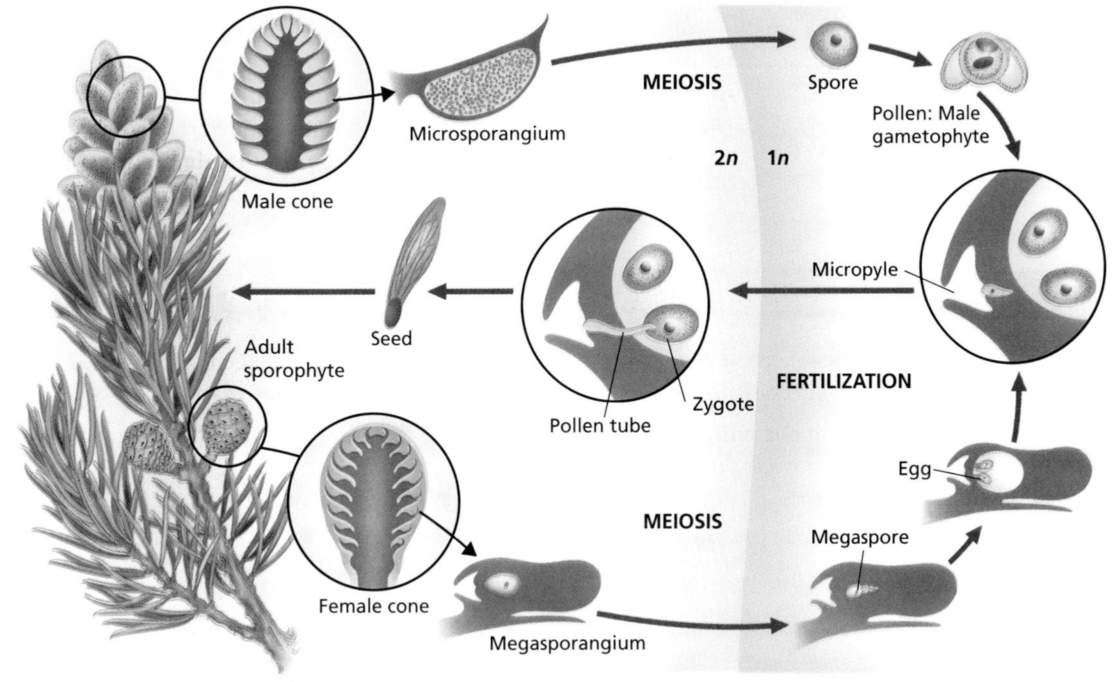

SECTION 32-1

RECENT RESEARCH

Fern Spore Germination

Germination of fern spores typically requires light. Scientists in Japan recently discovered that the first cell division in germination requires red light. This red-light induction of mitosis involves the pigment phytochrome, which is also involved in the regulation of many plant responses. As in other phytochrome responses, the red-light induction can be reversed by exposure to blue light. Subsequent cell divisions also require light, but not of a particular wavelength.

RETEACHING ACTIVITY

Provide students with living specimens or pictures of a moss, a fern, and a conifer. Show the various stages of each plant's life cycle. Hold up each plant specimen and ask students whether they think the structure is haploid or diploid. Help students arrange the specimens according to their stage in the plant life cycle. Glue the specimens or pictures onto a piece of poster board in their proper order. Have students draw arrows to indicate the cycle and draw a line through the cycle to separate the haploid and diploid phases. Point out to students the major differences between the life cycles. (The gametophyte is larger in mosses; the sporophyte is larger in ferns and conifers. Mosses and ferns do not produce seeds; conifers do. Mosses and ferns have swimming sperm; conifers do not.)

SECTION 32-1

ANSWERS TO SECTION 32-1 REVIEW

1. An antheridium is the male reproductive structure that produces sperm. An archegonium is the female reproductive structure that produces an egg.
2. See Figure 32-3 for a generalized life cycle diagram and Figure 32-1 for a diagram of the moss life cycle.
3. Most ferns are homosporous, and their spores are released from the sporophyte. The fern gametophyte is a small photosynthetic organism, and its sperm are flagellated. A pine is heterosporous, and its spores are not released from the sporophyte. The female gametophyte is microscopic and lives within the sporophyte, and the sperm lack flagella.
4. Seed plants produce two types of spores that remain inside the sporophyte. Bryophytes and most ferns produce one type of spore, and the spores are released from the sporophyte.
5. Enclosure of the male gametophyte produced pollen, which greatly increased sperm mobility. Thus, sperm could be easily transported long distances, making cross-pollination more likely. Free water was no longer required for fertilization, allowing sexual reproduction to occur in dry environments.
6. Some areas in the desert can provide the moisture necessary for mosses and ferns to live. Standing water can be found in cracks of rocks and in shady areas.

female cones. The male cones produce microsporangia, while the female cones produce megasporangia. The following spring, cells in all sporangia undergo meiosis and divide to produce haploid spores. These spores never leave the parent to develop independently. **Megasporangia** produce megaspores, which develop into **megagametophytes,** or female gametophytes. A thick layer of cells called an **integument** (in-TEG-yoo-muhnt) surrounds each megasporangium. The integument has a small opening called the **micropyle** (MIE-kroh-PIEL). Together, a megasporangium and its integument form a structure called an **ovule** (AHV-yool). Two ovules develop on each scale of a female cone. **Microsporangia** produce microspores, which develop into **microgametophytes,** or male gametophytes. A **pollen grain** is a microgametophyte of a seed plant.

The male cones of a pine release huge numbers of pollen grains, as seen in Figure 32-4. Pine pollen travels on the wind, and only a few grains may land on a female cone. The pollen grains drift between the cone scales until they reach the ovules. The arrival of a pollen grain at the micropyle of a pine ovule is called pollination. A drop of fluid at the micropyle captures the pollen grain. As the fluid dries, the pollen grain is drawn into the micropyle. After pollination, the female gametophyte within the ovule produces archegonia and eggs.

After pollination, the pollen grain begins to grow a **pollen tube,** a slender extension of the pollen grain that enables sperm to reach an egg. Unlike the sperm of seedless plants, pine sperm do not have flagella and they do not swim to an egg. The pollen tube takes about a year to reach an egg only a few millimeters away. During this time, two sperm develop in the pollen tube. When the pollen tube reaches an archegonium, one sperm unites with an egg to form a zygote. The other sperm and the pollen tube die. Over the next few months, the zygote develops into an embryo as the ovule matures into a seed.

Seed plants benefit from having a life cycle with microscopic gametophytes. The microscopic gametophytes are enclosed in structures produced by the much larger sporophyte. These structures provide a way for the sperm to fertilize the eggs without water. Sexual reproduction in seed plants can therefore take place independent of seasonal rains or other periods of moisture.

FIGURE 32-4

The male cones (pollen cones) of pines produce millions of pollen grains. The male cones then die. Wind-pollinated species typically produce large amounts of pollen. Large numbers of pollen grains increase the odds that female cones (seed cones) will be pollinated.

SECTION 32-1 REVIEW

1. Distinguish between an antheridium and an archegonium, and state the function of each.
2. Draw a generalized diagram of alternation of generations showing the haploid and diploid phases of the moss life cycle.
3. List three differences between the life cycle of a fern and that of a pine.
4. How are the spores of seed plants different from the spores of bryophytes and most ferns?
5. In what way did the enclosure of the male gametophyte contribute to the great evolutionary success of seed plants?
6. **CRITICAL THINKING** Many mosses and ferns live in the desert. Why do you think this is so?

SECTION

32-2

SEXUAL REPRODUCTION IN FLOWERING PLANTS

You have probably admired flowers for their bright colors, attractive shapes, and pleasing aromas. These characteristics are adaptations that help ensure successful sexual reproduction by attracting animal pollinators. But some flowers are not colorful, large, or fragrant. Such flowers rely on wind or water for pollination.

OBJECTIVES

▲ Identify the four main flower parts, and state the function of each.

● Describe ovule formation and pollen formation in angiosperms.

■ Relate flower structure to methods of pollination.

◆ Describe fertilization in flowering plants.

▲ Compare and contrast the gymnosperm and angiosperm life cycles.

PARTS OF A FLOWER

Recall from Chapter 30 that early land plants lacked leaves and roots and consisted only of stems. Leaves evolved from branches of stems. Botanists consider flowers to be highly specialized branches and the parts of a flower to be specialized leaves. All of these specialized leaves form on the swollen tip of a floral "branch," which is called the **receptacle.**

Flower parts are usually found in four concentric whorls, or rings. Figure 32-5 shows a classic flower with all of the flower parts. **Sepals** (SEE-puhlz) make up the outermost whorl of flower parts. They surround and protect the other parts of a developing flower before it opens. **Petals** make up the next whorl. Most animal-pollinated flowers have brightly colored petals. The petals and sepals of wind-pollinated plants are usually small or absent.

The two innermost whorls of flower parts contain the reproductive structures. The male reproductive structures are **stamens** (STAY-muhnz), each of which consists of an anther and a filament. An **anther** contains microsporangia, which produce microspores that develop into pollen grains. A stalklike **filament** supports an anther. The innermost whorl contains the female reproductive structures, which are called **carpels** (KAHR-puhlz). One or more carpels fused together make up the structure called a **pistil.** The enlarged base of a pistil is called the **ovary.** A **style,** which is usually stalklike, rises from the ovary. The tip of the style is called the **stigma.** Generally, a stigma is sticky or has hairs, enabling it to trap pollen grains. Most species of flowering plants have flowers with both stamens and pistils. However, some species have flowers with only stamens (male flowers) or pistils (female flowers).

FIGURE 32-5

This diagram shows a flower with all four whorls of flower parts—sepals, petals, stamens, and carpels. Many flowers lack one or more of these whorls.

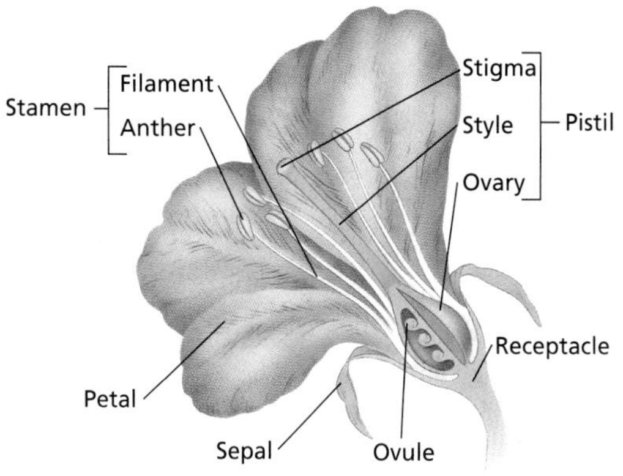

SECTION OVERVIEW

Reproduction and Inheritance
Section 32-2 illustrates how sexual reproduction occurs in flowering plants.

Cell Structure and Function
Section 32-2 describes the structures of a flower and relates the structures to the various types of pollination.

▶ ENGAGE STUDENTS

Provide students with a variety of flowers or pictures of flowers. Include large, colorful flowers, scented flowers, inconspicuous or small flowers, and flowers with nectaries. Make sure that some of the flowers have visible stamens and pistils. Day-old flowers are often available from florists, garden centers, and grocery stores. Have students examine the flowers and discuss the diversity they observe. Ask students to describe the adaptive advantage of bright coloration, large flowers, fragrance, or nectar (attracts pollinators). Ask students why wind-pollinated flowers often have small petals. (Large petals block air currents and waste energy.)

TEACHING STRATEGY

Floral Structures

Provide students different kinds of flowers. Wildflowers are preferable because cultivated flowers often lack floral parts. Avoid using daisies, chrysanthemums, and other composites because the individual flowers of these plants are very tiny. Have students examine the flowers with a hand lens and a dissecting microscope. Have students remove and examine the outermost structures of the flower and then progressively remove and identify the inner structures. Ask students to make labeled drawings of the structures they see.

VISUAL LINK

Figure 32-6

Take students through ovule formation step by step, as shown in Figure 32-6. Stress the roles of meiosis and mitosis in alternation of generations. Ask students if the megaspore mother cell is haploid or diploid (diploid). Ask students how many cells meiosis produces and if they are haploid or diploid (four, haploid). Ask students what kinds of cells are produced by mitosis (haploid if from a haploid cell, diploid if from a diploid cell).

RECENT RESEARCH

"Warm-Blooded" Flowers

Scientists studying the sacred lotus, *Nelumbo nucifera,* have discovered that this plant is able to warm up its flowers. Scientists found that as the flower buds open, the flower temperature increases to an average of 32°C (90°F), even if the surrounding air temperature is as cool as 5°C (41°F). The high temperature is held steady throughout flowering. The flowers are pollinated by beetles that must warm up their bodies to about 30°C (86°F) in order to fly and carry pollen to another flower. They do this by shivering. Scientists think that this plant's warm flower temperature allows the beetles to spend more time pollinating.

CONTENT CONNECTION

Section 30-1: Ploidy

In the life cycle of a plant, there is an alternation between diploid and haploid stages. A diploid zygote grows into a multicellular diploid sporophyte, which produces haploid spores by meiosis. The haploid spores develop into multicellular haploid gametophytes, which produce haploid gametes by mitosis. The union of two haploid gametes produces a new diploid zygote.

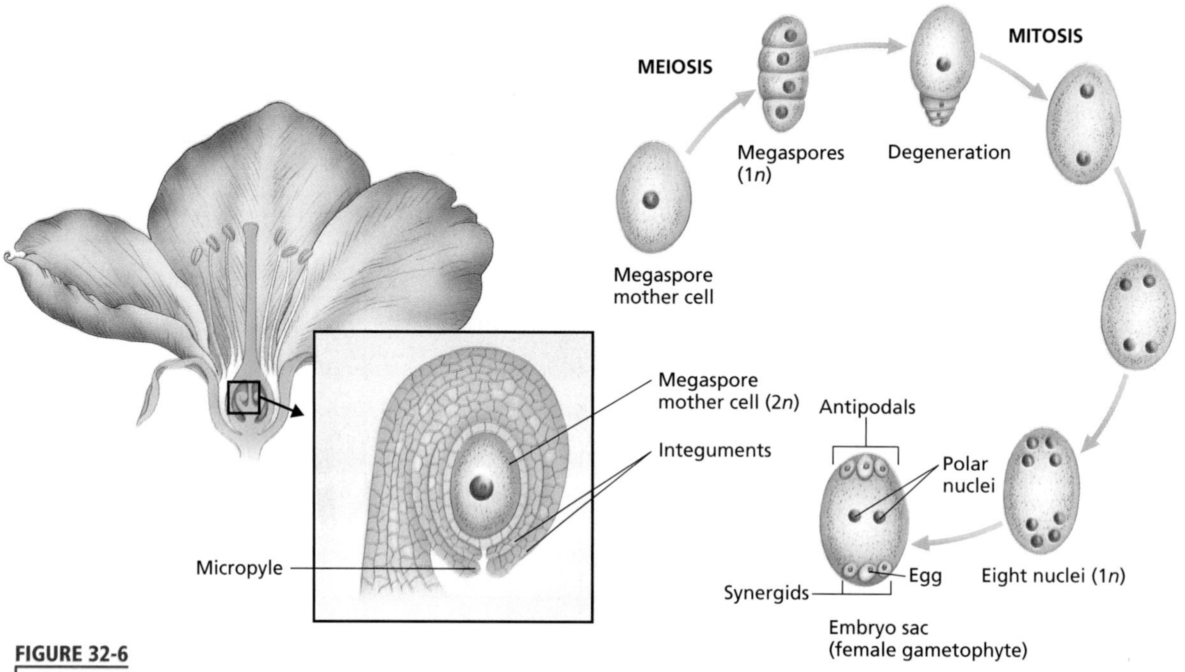

FIGURE 32-6

A cross section of an immature ovule from a flower reveals a single large cell, a megaspore mother cell. This cell undergoes meiosis and produces four megaspores. One of the megaspores undergoes a series of divisions that result in the formation of an embryo sac (a female gametophyte).

OVULE FORMATION

In flowering plants, ovules form in the ovary of a pistil. As Figure 32-6 shows, an angiosperm ovule consists of a megasporangium surrounded by two integuments. Like the integument of a pine ovule, these two integuments do not completely enclose the megasporangium. At one end of the ovule is the micropyle, through which a pollen tube can enter.

Initially, an ovule contains a large diploid cell called a megaspore mother cell. A **megaspore mother cell** undergoes meiosis and produces four haploid megaspores. Only one of the megaspores enlarges. The other three die and disappear. In most species, the remaining megaspore undergoes three consecutive mitotic divisions to produce eight haploid nuclei. These nuclei migrate to certain locations within the ovule. As you can see in Figure 32-6, the nuclei are initially arranged in two groups of four, with one group at each end of the cell. Here is what happens next:

- One nucleus from each group migrates to the center of the cell. These two nuclei are called **polar nuclei** because they came from the ends, or poles, of the cell.
- Cell walls form around each of the remaining six nuclei.
- One of the three cells nearest the micropyle enlarges and becomes the egg. The remaining five cells die after fertilization.

The resulting structure, which is microscopic and usually contains seven cells and eight nuclei, is called an **embryo sac.** The embryo sac is a mature female gametophyte, or megagametophyte. The embryo sac is another feature seen in angiosperms but not in gymnosperms.

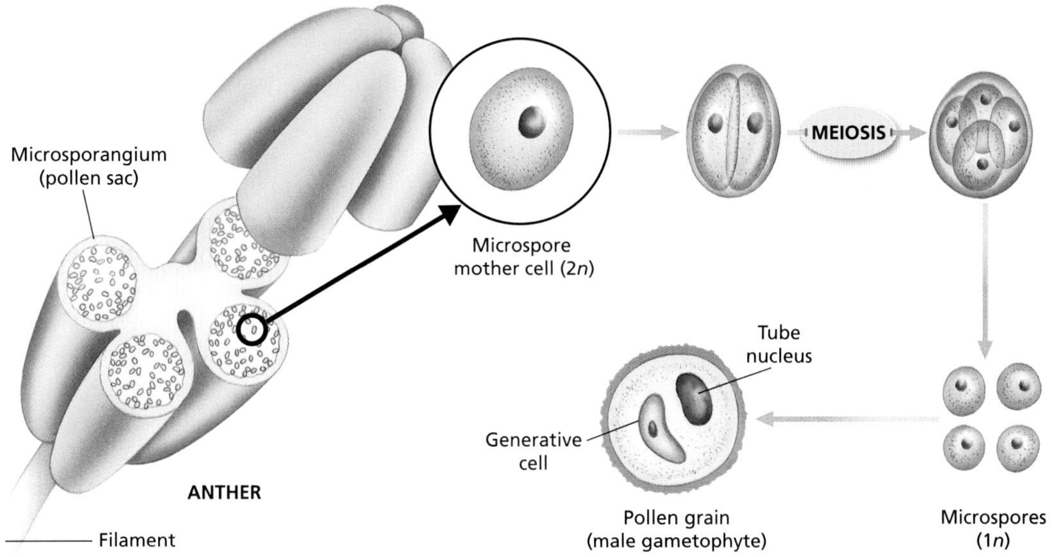

FIGURE 32-7

A cross section of the anther of a flower reveals four pollen sacs (microsporangia). Microspore mother cells within the microsporangia undergo meiosis and produce microspores. Each microspore then develops into a two-celled pollen grain (a male gametophyte).

POLLEN GRAIN FORMATION

In flowering plants, pollen grains form in the anthers of stamens. Figure 32-7 shows how pollen forms in anthers. An anther contains four microsporangia, or pollen sacs. Initially, the pollen sacs contain many diploid cells called **microspore mother cells.** Each of these cells undergoes meiosis and produces four haploid microspores. A microspore undergoes mitosis and divides into two haploid cells. A thick wall then develops around the microspore. The resulting two-celled structure is a pollen grain, which is the male gametophyte, or microgametophyte. The larger of the two cells is the **tube cell,** which contains the tube nucleus. When a pollen grain germinates, the tube nucleus causes the tube cell to grow through the style, forming a pollen tube. The smaller of the two cells is the **generative cell,** which divides by mitosis to form two sperm.

internetconnect

SCiLINKS NSTA

TOPIC: Pollination
GO TO: www.scilinks.org
KEYWORD: HM631

POLLINATION

Before a sperm can fertilize the egg contained in an embryo sac, pollen must be transferred from an anther to a stigma. This process is called pollination. Recall from Chapter 9 that pollination involving the same flower, flowers on the same plant, or flowers from two genetically identical plants is called self-pollination. In contrast, pollination involving two genetically different plants is called cross-pollination. Plants produced by cross-pollination are called hybrids.

VISUAL LINK

Figure 32-7

Take students through pollen formation step by step, as shown in Figure 32-7. Ask students if the microspore mother cell is haploid or diploid (diploid). Ask students how many cells meiosis produces and if they are haploid or diploid (four, haploid). Ask students what happens to these cells. (Each cell divides by mitosis to produce two haploid cells, which then become surrounded by a thick wall. These two-celled structures are pollen grains.)

DEMONSTRATION

Provide students with a variety of prepared slides and electron micrographs of pollen. (Prepared slides are available from biological supply houses, and micrographs can be found in college biology or botany textbooks.) Have students examine the slides under microscopes and study the micrographs. Ask students to speculate on the adaptive advantages of the particular pollen shapes and surface characteristics. (Pollen with spikes could be easily transported by furry animals; smooth pollen could be easily transported by air currents; pollen arranged in packets could be easily transported on the backs of insects.)

RECENT RESEARCH

Fruit-Setting

Scientists at the University of Texas found that the number of fruits that set in *Phlox drummondii* depends on the number of pollen grains that germinate on the stigma of the flowers. They found that flowers that produced fruit had almost five times as many pollen tubes per stigma as flowers that did not produce fruit.

SECTION 32-2

RECENT RESEARCH

Rewards for Pollinators

Scientists working with nine different species of insect-pollinated flowering plants found that all of the species underwent a flower-color change over time—from yellow to orange or orange-red—that was correlated with a loss of food rewards for the pollinators. Scientists concluded that flower-color change is an adaptation that attracts first-time visitors and provides them with cues for associating one color with a reward.

QUICK FACT

It has been estimated that about 10,000 species of orchids produce an odor that mimics the smell of a food substance, such as sweet nectar. This common mechanism lures insect pollinators, such as bees, to their flowers. Also, more than 100 species of orchids produce flowers that mimic female insects. The male insects try to mate with the "females," spreading pollen in the process.

GIFTED ACTIVITY

Have students conduct research and report on a coevolutionary relationship between a flowering plant and its pollinator. Examples include the carrion flower and the fly; the American sausage tree and bats; yuccas and the yucca moth; fig trees and fig wasps; and wasp or fly orchids and a wasp or fly.

CRITICAL THINKING

Color of Hybrid Fruit

If a flower of a Red Delicious apple plant is pollinated by a Yellow Delicious apple, what would the fruit look like? (It would look like a typical Red Delicious apple. Because a fruit is a ripened ovary, which is maternal tissue, it would have the characteristics of the female parent.)

FIGURE 32-8

These long, tubular flowers are adapted for pollination by hummingbirds, which have long beaks adapted for reaching the nectar located deep in the flowers.

The type of pollination is of great concern to growers of many fruit crops. Some cultivars of fruit crops, such as apples and peaches, will produce fruit when they are self-pollinated. Others must be cross-pollinated in order to produce fruit. Date palms and kiwis produce male and female flowers on separate plants. Therefore, a few male plants must be grown to pollinate the mostly female crop.

The pollination of flowering plants occurs in several ways. Flower structure promotes self-pollination in some plants, such as peas and beans, which have flowers with petals that completely enclose both the male and female flower parts. Aquatic plants, such as sea grasses, often have pollen that is dispersed by water. Many plants, such as oaks and grasses, release their pollen into the air. The flowers of wind-pollinated angiosperms are small and lack showy petals and sepals. Successful wind pollination depends on four conditions: the release of large amounts of pollen, the ample circulation of air to carry pollen, the relative proximity of individuals to one another, and dry weather to ensure that pollen is not washed from the air by rain.

Most plants with colorful or fragrant flowers are pollinated by animals. Bright petals and distinctive odors attract animals that feed on pollen and **nectar,** which is a nourishing solution of sugars. When animals gather nectar, the pollen sticks to their bodies. The animals deposit some of the pollen on other flowers as they collect more nectar. Pollinators include bats, bees, beetles, moths, butterflies, hummingbirds, monkeys, mosquitoes, and people. Of course, people pollinate to obtain desirable seeds or fruits, not to obtain nectar. The long beak of a hummingbird, such as the one shown in Figure 32-8, is adapted to collect nectar from flowers. As the hummingbird collects the nectar, pollen from the flowers is deposited on the hummingbird's body.

FERTILIZATION

Fertilization, which is the union of gametes, follows pollination. In order for fertilization to occur, a pollen tube must grow to an egg, and sperm must form. Unlike a conifer's pollen tubes, which take about a year to reach an egg, an angiosperm's pollen tubes usually reach an egg a day or two after pollination. The sequence of events leading to angiosperm fertilization is shown on the next page in Figure 32-9.

When a pollen grain germinates, the nucleus of its tube cell forms a pollen tube that grows through the stigma and style toward the ovary. As the pollen tube grows, its generative cell divides mitotically to form two haploid sperm. The pollen tube grows to an ovule within the ovary and enters it through the micropyle. After the pollen tube penetrates the ovule's embryo sac, the sperm can reach the egg through the passageway that has been formed.

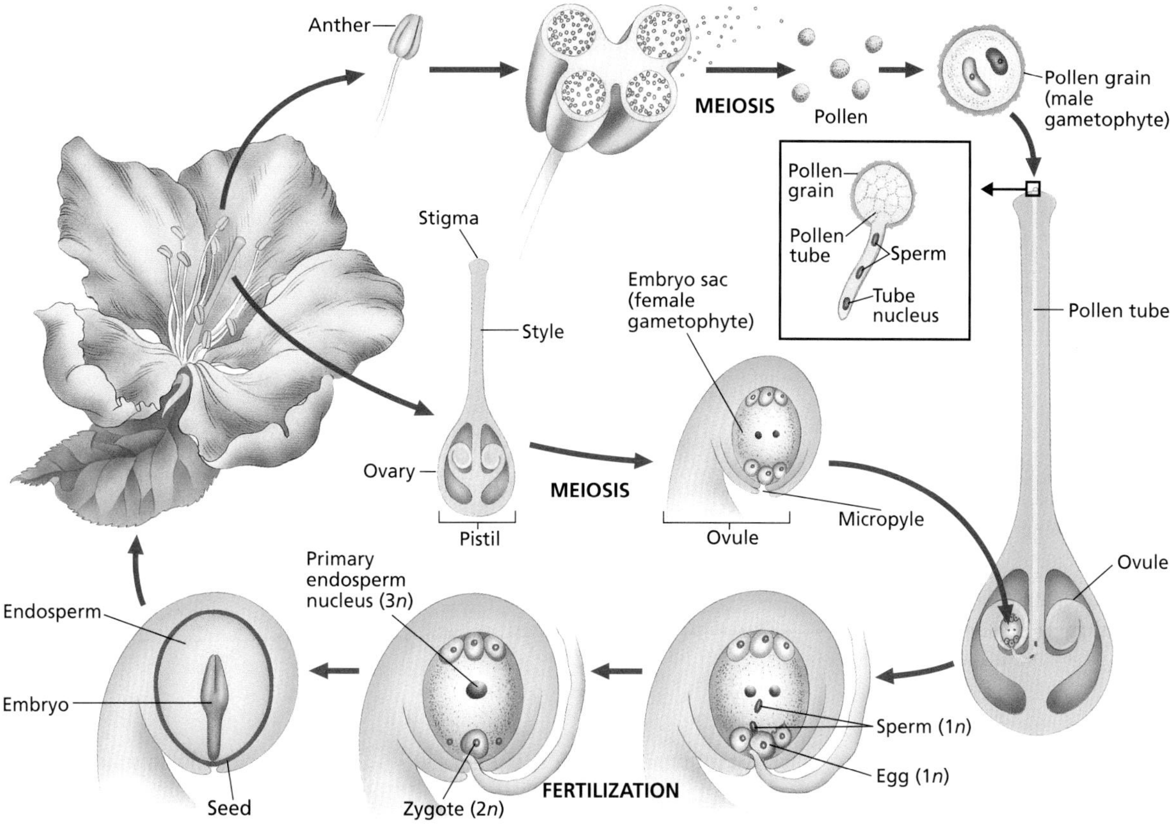

FIGURE 32-9

Following pollination, a pollen grain germinates and forms a pollen tube, which grows through the style and enters an ovule through its micropyle. Two sperm travel down the pollen tube. One fertilizes the egg in the ovule's embryo sac, forming a zygote. The other sperm fertilizes the two polar nuclei, forming the endosperm. This process is called double fertilization.

One of the two sperm fuses with the egg, forming a diploid zygote. The zygote eventually develops into an embryo. The second sperm fuses with the two polar nuclei, producing a triploid ($3n$) nucleus. This nucleus then develops into a tissue called endosperm. The endosperm provides nourishment for the embryo. Kernels of corn are mostly endosperm. In many plants, however, the endosperm is absorbed by the embryos as the seeds mature. As you can see, two types of cell fusion occur in the embryo sac: one that produces the zygote and one that produces the endosperm. This process, which is called **double fertilization,** is unique to angiosperms.

SECTION 32-2 REVIEW

1. **Draw a generalized flower and show the four types of flower parts in relation to each other. Be sure to label each structure.**
2. **Which process in angiosperms, ovule formation or pollen formation, is more similar to the process as it occurs in gymnosperms? Justify your answer.**
3. **How do the flowers of wind-pollinated plants differ from the flowers of animal-pollinated plants?**
4. **Why is fertilization in flowering plants called double fertilization?**
5. **How are the conifer and angiosperm life cycles similar? How are they different?**
6. **CRITICAL THINKING Why is cross-pollination considered to be an adaptive advantage over self-pollination?**

SECTION 32-2

ANSWERS TO SECTION 32-2 REVIEW

1. See Figure 32-5 for a complete flower drawing.
2. Pollen formation is more similar than ovule formation. In both angiosperms and gymnosperms, diploid cells in the microsporangia undergo meiosis to form microspores. The nucleus of a microspore divides by mitosis to produce two cells, a generative cell and a tube cell. In angiosperms, ovules form enclosed within an ovary. The megaspore in an ovule undergoes a series of mitotic divisions to form an embryo sac. In gymnosperms, ovules form on the scales of cones, and no embryo sac forms.
3. The flowers of wind-pollinated plants usually lack scents, nectar, and large petals and sepals, while the flowers of animal-pollinated flowers usually have scents, nectar, and large, colorful petals or sepals that attract animals.
4. Two sperm fuse with cells in the embryo sac. One sperm fertilizes the egg and the other sperm fuses with the two polar nuclei.
5. Both produce pollen and seeds. In conifers, archegonia are produced, the seeds are naked, and the tissue of the haploid female gametophyte nourishes the embryo. In angiosperms, no archegonia are produced, the seeds are enclosed by a fruit, and a triploid endosperm nourishes the embryo. Angiosperms have flowers, which conifers lack.
6. Cross-pollination usually produces greater genetic diversity in the offspring than does self-pollination. Genetic variability increases the chances that at least some of the offspring will have characteristics that will make them better able to survive and produce offspring.

SECTION 32-3

SECTION OVERVIEW

Interdependence of Organisms

Section 32-3 describes how plants depend on animals, wind, water, and gravity to disperse their fruits and seeds.

▶ ENGAGE STUDENTS

Provide students with a variety of seeds and fruits. Include tiny seeds, such as those of carrots or lettuce; large seeds, such as beans; tiny fruits, such as those of grasses; and large fruits, such as pineapple and watermelons. Also try to include monocot seeds, such as those of grasses and corn, and dicot seeds, such as beans, peanuts, and squash seeds. Include animal-dispersed burred fruits; animal-dispersed brightly colored fruits, such as apples and red peppers; wind-dispersed dandelion or maple fruits; and a water-dispersed coconut. Ask students to arrange the fruits and seeds according to their likely method of dispersal (wind, water, or animal), and have them infer what the specific agent of dispersal is. Ask students to speculate on the advantages of the different adaptations. (For example, tiny seeds may be an adaptation that fits more seeds inside the fruit and facilitates wind dispersal.)

CRITICAL THINKING

Fruit Color Advantages

Ask students to explain the adaptive advantage of fruits remaining green until the seeds are mature and then turning colors, such as red, orange, or yellow. (Green fruits are camouflaged with the foliage around them and are thus less likely to be eaten by animals. Red, orange, or yellow fruits contrast with green foliage, inviting animals to eat the fruit containing the ripe seeds.)

SECTION 32-3

OBJECTIVES

▲ Name different types of fruits.

● Describe several adaptations for fruit and seed dispersal.

■ Compare and contrast the structure and germination of different types of seeds.

◆ Recognize the advantages and disadvantages of asexual reproduction.

▲ Describe methods of vegetative propagation.

DISPERSAL AND PROPAGATION

Fruits and seeds normally result from sexual reproduction in flowering plants. Fruits are adapted for dispersing seeds, while seeds function in the dispersal and propagation of plants. Many plants also propagate (produce new individuals) through asexual reproduction.

DISPERSAL OF FRUITS AND SEEDS

Recall from Chapter 20 that one property of populations is dispersion, the spacial distribution of the individuals. If individual plants are too close together, they must compete with each other for available water, nutrients, and sunlight. One reason for the success of the seed plants is the development of structures that are adapted for dispersing offspring—fruits and seeds.

Fruits and seeds are dispersed by animals, wind, water, forcible discharge, and gravity. You may have walked through a field and unwittingly collected burrs, or stickers, on your shoes or socks. These burrs are fruits, and you helped disperse them. The smell, bright color, or flavor of many fruits attract animals. When animals eat such fruits, the seeds often pass unharmed through their digestive system.

Fruits and seeds dispersed by wind or water are adapted to those methods of dispersal. Orchids have tiny, dustlike seeds that can easily be carried by a slight breeze. Figure 32-10 shows an example of seeds adapted for wind dispersal. Many plants that grow near water produce fruits and seeds that contain air chambers, which allow them to float. Coconuts, for example, may float thousands of kilometers on ocean currents.

FIGURE 32-10
Milkweed seeds have "parachutes" that help them drift with the wind.

The most dramatic method of seed dispersal occurs in plants that forcibly discharge their seeds from their fruits. The tropical sandbox tree, which has fruits that hurl seeds up to 100 m (328 ft), seems to hold the distance record.

Although gymnosperms do not produce fruits, their cones may help protect seeds and aid in seed dispersal. Pine seeds are often dispersed when gravity causes cones to drop and roll away from the parent tree. Pine seeds have wings that aid in wind dispersal.

TYPES OF FRUITS

Botanists define a fruit as a matured ovary. Many different types of fruits have evolved among the flowering plants. Figure 32-11 shows examples of some of these fruit types. Fertilization usually initiates the development of fruits. Fruits protect seeds, aid in their dispersal, and often delay their sprouting. Fruits are classified mainly on the basis of how many pistils or flowers form the fruit and whether it is dry or fleshy. Table 32-1 presents a classification system for fruits. Notice that fruits with common names that include "nut" or "berry" may not be nuts or berries. You may have heard the fleshy seeds of ginkgo, juniper, and yew referred to as berries. These names are misleading because ginkgo, juniper, and yew are gymnosperms, which do not form fruits.

FIGURE 32-11
A pea pod is a simple fruit. A raspberry is an aggregate fruit. A pineapple is a multiple fruit.

TABLE 32-1 *Fruit Classification*

Major categories and types of fruits	Examples
I. **Simple fruit**—formed from one pistil of a single flower	
A. Dry at maturity	
1. Usually splits open	
a. **Legume**—splits along two sides to form two halves	pea, peanut, black locust
b. **Follicle**—splits along one side	milkweed, columbine
c. **Capsule**—splits in a variety of other ways	poppy, tulip
2. Usually does not split open	
a. **Grain**—thin ovary wall fused to seed coat	corn, wheat
b. **Nut**—thick, woody ovary wall not fused to single seed	oak, chestnut
c. **Achene**—thin ovary wall not fused to single seed	sunflower, dandelion
d. **Samara**—like an achene but with a thin, flat wing	ash, elm, maple
B. Fleshy at maturity and usually not opening	
1. Usually contains only one seed	
a. **Drupe**—stony inner layer around the seed	cherry, coconut, pecan
2. Usually contains many seeds	
a. **Pome**—core with seeds surrounded by papery ovary walls; outer part formed from sepals	apple, pear
b. **Typical berry**—thin skin	grape, tomato, banana
c. **Pepo**—berry with a thick, hard rind	watermelon, cucumber
d. **Hesperidium**—berry with leathery, easily removed skin	orange, grapefruit, lemon
II. **Aggregate fruit**—formed from several pistils of a single flower	
A. Dry at maturity	tulip tree, magnolia
B. Fleshy at maturity	raspberry, strawberry
III. **Multiple fruit**—formed from several flowers growing together	
A. Dry at maturity	sweetgum, sycamore
B. Fleshy at maturity	pineapple, fig

TEACHING STRATEGY
Classifying Fruit Types

Provide students with a variety of fruit types, and have them classify the fruit by using the key in Table 32-1. The following fruits can also be used: honeylocust, bean (legumes); delphinium (follicle); any grain (caryopsis); pecan, walnut (nuts); zinnia (achenes); okra, azalea (capsules); peach, plum, olive (drupes); quince (pome); blueberry, cranberry (berries); lime (hesperidium); squash, muskmelon (pepos); blackberry (aggregate fruit). Ask students to think of other examples of fruits that fit into each category.

QUICK FACTS

☑ Not all plants sprout from a seed with the characteristics of the adult plants. Some plants, such as English ivy and philodendrons, have both juvenile and adult forms. The leaves of both forms are distinctly different.

☑ The fruits of orchids are pods that contain millions of tiny seeds. The mature seeds are at a very early stage of development and usually are composed of only about 64 cells. Most orchid seeds cannot germinate without the aid of mycorrhizal fungal associations.

CULTURAL CONNECTION

Kola Nuts

Kola nuts are cultivated in the West Indies, West Africa, and South Africa. They are the fruits of several types of evergreen trees. People who live in African countries call the nuts *guru* or *goora* nuts and chew them like gum. Kola nuts are also used to make cola soft drinks and medicines. The nuts contain caffeine, a stimulant, and theobromine, a toxic alkaloid.

SECTION 32-3

VISUAL LINK

Figure 32-12

Ask students what structures monocot and dicot seeds have in common (plumule, hypocotyl, radicle, cotyledon). Ask students to compare the corn kernel with the bean seed with respect to nutrient storage (corn kernel—endosperm, bean seed—cotyledons).

Quick Lab

Predicting Seed Dispersal

Time Required 30 minutes

Safety Instruct them not to eat the fruit.

Procedural Tips Provide at least 10 fruits from which students can choose. Include some fruits that are commonly considered vegetables, such as tomatoes or whole green beans. Cut open hard fruits for students.

Disposal Fruit and seed remains may be added to a compost heap or thrown in the trash.

Answers to Analysis Students' hypotheses will vary. Each hypothesis should be reasonable and testable. For example, one hypothesis might be that apples (or other fleshy fruits) are eaten by an animal and the seeds are dispersed after traveling through the animal's digestive system. Hard seeds might be buried by animals or carried in the wind. Students should be able to describe a test for their hypothesis and predict the outcome. Accept reasonable answers.

RECENT RESEARCH

Photosynthesis in Space

Scientists at NASA grew wheat plants, *Triticum aestivum,* from seeds on the space shuttle *Discovery.* After 10 days in space, the photosynthetic rates of wheat seedlings were 25 percent lower than those of the control plants. The scientists concluded that it is possible to produce food in space but yields may be reduced because of the microgravity environment.

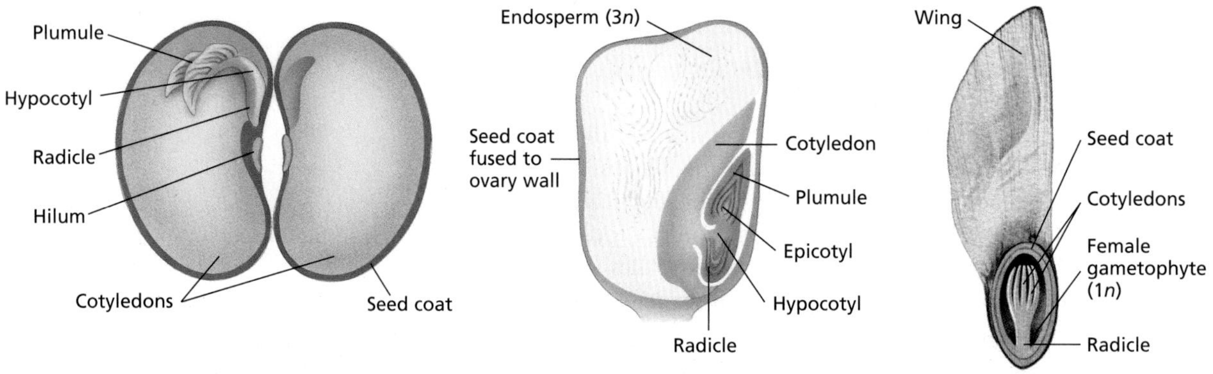

(a) BEAN SEED **(b) CORN GRAIN** **(c) PINE SEED**

FIGURE 32-12

(a) A bean seed has two cotyledons and no endosperm. (b) A corn grain contains a single seed, which has one cotyledon and endosperm. (c) A pine seed has eight cotyledons and tissue from the female gametophyte.

STRUCTURE OF SEEDS

As you learned in Chapter 30, a seed is a plant embryo surrounded by a protective coat called the **seed coat.** The structure of seeds differs among the major groups of seed plants—angiosperms, which include monocots and dicots, and gymnosperms. To understand some of the differences, examine the seeds shown in Figure 32-12.

Look at the bean seed, which has been opened to reveal the structures inside. Most of the interior of a bean seed is filled by two large, fleshy cotyledons (seed leaves), which are part of the embryo. Recall that angiosperms are classified as either monocots or dicots, based on the number of cotyledons in their embryos. Therefore, beans are dicots. A mature bean seed has no endosperm. The endosperm was absorbed by the fleshy cotyledons.

Between the two cotyledons of a bean seed are the three parts that make up the rest of the embryo: the **radicle,** or embryonic root; the **hypocotyl** (HIE-poh-KAHT-uhl), which is the stem between the cotyledons and the radicle; and the **epicotyl** (EP-i-KAHT-uhl), which is the stem above the cotyledons. The epicotyl, along with any embryonic leaves, is called the **plumule** (PLOO-MYOOL). Along the concave edge of the seed is the **hilum** (HIE-luhm), which is a scar that marks where the seed was attached to the ovary wall.

Now examine the corn grain in Figure 32-12b. Technically, a corn grain is a fruit, but the seed occupies almost the entire grain. The wall of the fruit is very thin and is fused to the seed coat. A single umbrella-shaped cotyledon is pressed close to the endosperm. The cotyledon of a monocot seed does not store nutrients, as bean cotyledons do. Instead, it absorbs nutrients from the endosperm and transfers them to the embryo.

Finally, look at the pine seed in Figure 32-12c. A pine seed contains an embryo with an average of eight needle-like cotyledons. The embryo is surrounded by tissue of the female gametophyte. Like the triploid endosperm of angiosperm seeds, the haploid tissue of the female gametophyte functions as a source of nourishment for the embryo.

Quick Lab

Predicting Seed Dispersal

Materials five different fruits, balance or scale

Procedure

1. Create a data table that has at least five rows. Your table should have six columns with the following headings: "Fruit name," "Fruit type" (from Table 32-1, p. 635), "Dry/fleshy," "Seed mass in grams," "Whole fruit mass in grams," and "Dispersal method."
2. Examine your fruits, and fill in your data table. Discuss with your group how characteristics of fruits and seeds might relate to dispersal methods.

Analysis Form a hypothesis about a dispersal method for one of the fruits you have examined. Describe how you might test your hypothesis.

SEED GERMINATION

Many plants are easily grown from seeds. Although its embryo is alive, a seed will not germinate, or sprout, until it is exposed to certain environmental conditions. Delaying germination often assures the survival of a plant. For example, if seeds that mature in the fall were to sprout immediately, the young plants could be killed by cold weather. Similarly, if a plant's seeds were to sprout all at once and all of the new plants died before producing seeds, the species could become extinct. Many seeds will not germinate even when exposed to conditions ideal for germination. Such seeds exhibit **dormancy,** which is a state of reduced metabolism. The longevity of dormant seeds is often remarkable. Recently, a botanist germinated lotus seeds that were almost 1,000 years old.

Conditions Needed for Germination

Environmental factors, such as water, oxygen, and temperature, trigger seed germination. Most mature seeds are very dry and must absorb water to germinate. Water softens the seed coat and activates enzymes that convert starch in the cotyledons or endosperm into simple sugars, which provide energy for the embryo to grow. As the embryo begins to grow, the softened seed coat cracks open. This enables the oxygen needed for cellular respiration to reach the embryo. In addition, seeds germinate only if exposed to temperatures within a certain range. Many small seeds need light for germination. This adaptation prevents the seeds from sprouting if they are buried too deeply in the soil.

Some seeds germinate only after being exposed to extreme conditions, as illustrated in Figure 32-13. Some seeds, such as black locust seeds, germinate after passing through the digestive systems of animals. Acids in the digestive system wear away the hard seed coat,

FIGURE 32-13

Many animals, including this raccoon, eat apples or other fruits. The seeds found in a fruit are swallowed by an animal and are exposed to acids as they pass through the animal's digestive system. The acids wear away the seed coat, allowing water and oxygen to enter and enabling the growing embryo to break out.

Word Roots and Origins

dormancy

from the Latin *dormire,* meaning "to sleep"

internetconnect

SCILINKS NSTA

TOPIC: Seeds
GO TO: www.scilinks.org
KEYWORD: HM637

CRITICAL THINKING

Dormancy Hormone

Many seeds that exhibit dormancy contain a hormone in the seed coat that inhibits growth of the embryo. Ask students what must happen in these seeds before they can germinate. (The hormone must be washed away or broken down.)

RECENT RESEARCH

Starch Storage

Scientists in France have been studying the buildup of nutrient reserves in pea seeds, *Pisum sativum.* Scientists found that the enzyme sucrose synthase catalyzes the first step in the conversion of sucrose into starch. Sucrose moves into the seed from photosynthetic parts of the plant. The scientists found that the activity of sucrose synthase was strongly correlated with starch concentration.

QUICK FACT

☑ Seeds can remain viable for years if they are kept dry and cold. The record for seed viability is held by a species of *Lupinus.* Some *Lupinus* seeds found buried in arctic ice and dated at 11,000 years old have germinated when supplied with water and warm temperatures.

GIFTED ACTIVITY

Have students investigate the rapidly expanding area of plant patents. United States law allows a person who develops a new plant type (not just discovers a naturally occurring new type) to apply for a patent on that plant. A patented plant cannot be propagated and sold without the permission of the patent holder and without payment of royalties to this person. There are a number of horticultural plants that have patents on them such as Erica Sunset heather.

SECTION 32-3

TEACHING STRATEGY
Germinating Bluebonnets

Obtain seeds of Texas bluebonnets (*Lupinus texensis*). Have students examine the seeds, noting their hard seed coat. Have students divide the seeds into two groups (100 seeds per group), and give each group of students a cookie sheet. Tell students that they are going to test the effectiveness of some treatments in promoting germination of the seeds. Have the students place their seeds on the cookie sheet. (**CAUTION** Have students wear protective gloves when handling the hot container, and be careful when pouring the boiling water so that the water does not spill.) Heat 1 L of water on a hot plate or in a microwave oven. Then have one group of students (the treatment group) pour boiling water over the seeds to cover them. Have the other group (the control group) pour room-temperature water over their seeds. Have students place the treatment seed tray in a freezer and place the control tray on a counter in the classroom. Have them remove the tray from the freezer the next day, allow the water to thaw, and pour off the water. They should also pour the water off the control seeds. Have them repeat all of these procedures two more times. Then have the students plant the seeds in pots or outdoors, labeling them appropriately. It will take several weeks for seeds to germinate, depending on the weather. When a high percentage of the treatment seeds have germinated, have students collect data for both groups of seeds. (They should find less than 20 percent germination in the control group and around 80 percent germination in the treatment group.) Ask students to discuss their results. Ask them what they have simulated with the boiling water and freezing treatments (rainfall, freezing, and thawing).

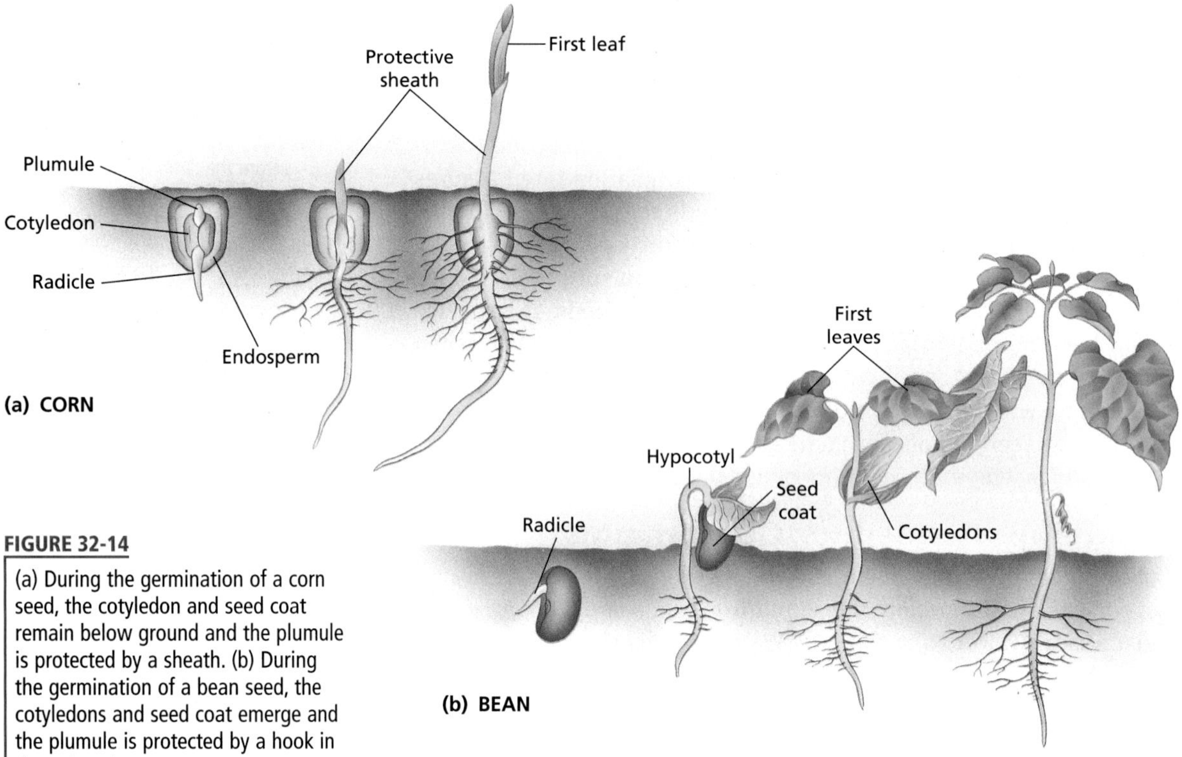

FIGURE 32-14

(a) During the germination of a corn seed, the cotyledon and seed coat remain below ground and the plumule is protected by a sheath. (b) During the germination of a bean seed, the cotyledons and seed coat emerge and the plumule is protected by a hook in the epicotyl.

and as an added bonus, the seed is deposited with a bit of natural fertilizer. Other seeds, such as apple seeds, must be exposed to near-freezing temperatures for several weeks before they sprout. This temperature requirement prevents the seeds from germinating in the fall and thus ensures that the seedlings will not be killed by the cold temperatures of winter. The cold temperatures cause chemical changes within the seed. These changes enable the embryo to grow.

Process of Germination

Figure 32-14 compares seed germination in corn and beans. The first visible sign of seed germination is the emergence of the radicle. In beans, the entire root system develops from the radicle. In corn, most of the root system develops from the lower part of the stem. Soon after the radicle breaks the seed coat, the shoot begins to grow.

In some seeds, such as bean seeds, the hypocotyl curves and becomes hook-shaped. Once the hook breaks through the soil, the hypocotyl straightens. This straightening pulls the cotyledons and the plumule into the air. The plumule's embryonic leaves unfold, synthesize chlorophyll, and begin photosynthesis. After their stored nutrients are used up, the shrunken bean cotyledons fall off.

In contrast, the cotyledon of the corn seed remains underground and transfers nutrients from the endosperm to the growing embryo. Unlike the bean shoot, the corn hypocotyl does not hook or elongate, and the cotyledon remains below ground. Instead, the corn plumule is protected by a sheath as it pushes through the soil. When the shoot breaks through the soil surface, the leaves of the plumule unfold.

internetconnect

SCILINKS NSTA
TOPIC: Seed germination
GO TO: www.scilinks.org
KEYWORD: HM638

ASEXUAL REPRODUCTION

Recall that asexual reproduction is the production of an individual without the union of gametes. Elaborate technology has recently yielded some success in the asexual reproduction, or cloning, of mammals. Even so, movies and books that depict people who create **clones,** or exact duplicates of themselves, are still science fiction. However, asexual reproduction is common in the plant kingdom. Asexual reproduction can be an advantage to individuals that are well-adapted to their environment. Many new individuals can be produced in a short period of time, enabling the clones to spread rapidly and fill the available space. A disadvantage of asexual reproduction is the lack of genetic variation among the offspring. All of the offspring are genetically identical to the parent and to each other. Thus, the clones have the same tolerance to an environment and are attacked by the same diseases and pests. Many of our cultivars are clones.

In nature, plants reproduce asexually in a variety of ways. One type of asexual reproduction is shown in Figure 32-15. Reproduction with usually nonreproductive parts, such as leaves, stems, or roots, is termed **vegetative reproduction.** Many different types of structures specialized for vegetative reproduction have evolved among the plants. Table 32-2 lists some of these structures.

FIGURE 32-15

New plants are produced from the runners (or stolons) of an airplane plant. This plant is sometimes called a spider plant. Each new plant found on the runner can be put into a glass of water to grow roots. This is an example of vegetative propagation.

VEGETATIVE PROPAGATION

People often use vegetative structures to propagate plants. The use of vegetative structures to produce new plants is called **vegetative propagation.** Many species of plants are vegetatively propagated from specialized structures such as runners, rhizomes, bulbs, and tubers. People have also developed several methods of propagating plants from other vegetative parts. These methods include layering, grafting, and using cuttings and tissue cultures.

TABLE 32-2 ***Plant Structures Adapted for Vegetative Reproduction***

Name	Description	Examples
Runner (stolon)	horizontal, aboveground stem that produces leaves and roots at its nodes; a new plant can grow from each node	strawberry, spider plant, Boston fern
Rhizome	horizontal, belowground stem that produces leaves and roots at its nodes; a new plant can grow from each node	ferns, horsetails, iris, ginger, sugar cane
Bulb	very short, underground monocot stem with thick, fleshy leaves adapted for storage; bulbs divide naturally to produce new plants	tulip, daffodil, onion, garlic, hyacinth
Tuber	underground, swollen, fleshy stem specialized for storage; the buds on a tuber can grow into new plants	potato, caladium, Jerusalem artichoke

DEMONSTRATION

Provide students with examples of plants that propagate vegetatively from specialized structures. Any of the examples given in Table 32-2 would be suitable. Ask students what plant organs could be produced from these specialized structures (all). Ask students how they can tell they are all stems (they have nodes). Ask students what kind of tissue is contained in the stems that enables them to produce new roots, stems, and leaves (meristematic). Ask students how plants benefit from vegetative propagation (it requires less energy than producing flowers and fruits).

GIFTED *ACTIVITY*

Methods to use for the commercial propagation of crops depend on the characteristics of the crop plant and the grower's resources. Have students select a crop plant, either a food plant or an ornamental, and investigate the requirements for the production of this crop. Ask students to develop a production plan for the crop and include in it their decision about whether it would be best to propagate it by seed or vegetatively.

ACTIVE READING
Technique: K-W-L

Have students return to their lists of things they **W**ant to know about plant life cycles and reproduction, which they made for the Active Reading activity on page 625. Have them place check marks next to questions that they are able to answer. Students should finish by making a list of what they have **L**earned. Have students identify unanswered questions as well as new questions they may have.

SECTION 32-3

ANSWERS TO SECTION 32-3 REVIEW

1. See Table 32-1 for answers.
2. The wind disperses some fruits and seeds, such as the maple and the orchid. Coconuts are dispersed by water. Animals disperse seeds and fruits that become caught in their fur. Animals also eat fruits and distribute the seeds in their droppings. Some fruits catapult their seeds.
3. A bean seed has two cotyledons, which store nutrients and emerge when the seed germinates. The epicotyl forms a hook, which helps to protect the plumule as it passes through the soil. A corn seed has one cotyledon that remains below ground, endosperm, and a sheath that protects the plumule as it passes through the soil.
4. Asexual propagation quickly produces plants that are identical to the parent plant. Sexual reproduction is often slow because flowering usually occurs once a year and seed dormancy must be overcome. Asexual reproduction requires only a single plant and does not require animals, wind, or water for pollination. Sexual reproduction often requires two plants and depends on unreliable forces such as wind, water, and animals for pollination.
5. Structures for vegetative propagation include bulbs, runners, tubers, and rhizomes. Methods of vegetative propagation include cuttings, layering, grafting, and tissue cultures.
6. Seed dormancy prevents seeds from germinating while still on the plant, while still in the fruit, or while in the stomach of an animal. It also prevents seeds from germinating at the wrong time of year.

Cuttings

In some plants, roots will form on a cut piece of a stem, or shoots will form on a piece of a root. Pieces of stems and roots that are cut from a plant and used to grow new plants are called **cuttings.** Plants such as African violets can be grown from leaf cuttings, which will form both roots and shoots. Cuttings are widely used to propagate houseplants, ornamental trees and shrubs, and fruit crops such as grapes, figs, and olives.

Layering

In some species, such as raspberries, roots form on stems where they make contact with the soil. People often stake branch tips to the soil or cover the bases of stems with soil to propagate such plants. The process of causing roots to form on a stem is called **layering.** Air layering, which involves wounding a stem and placing moist sphagnum moss around the wound, is another common form of layering.

Grafting

Grafting is the joining of two or more plant parts to form a single plant. In grafting, a bud or small stem of one plant is attached to the roots or stems of a second plant. The vascular cambium of both parts must be aligned for a graft to be successful. Grafting enables the desirable characteristics of two cultivars to be combined. For example, an apple cultivar with excellent fruit can be grafted onto one with disease-resistant roots. Grafting is used to propagate virtually all commercial cultivars of fruit and nut trees and many ornamental trees and shrubs.

FIGURE 32-16

Tissue culture can be used to grow round-leaved sundews, *Drosera rotundifolia.* The jelly in the Petri dish is a sterile soil substitute that contains needed nutrients.

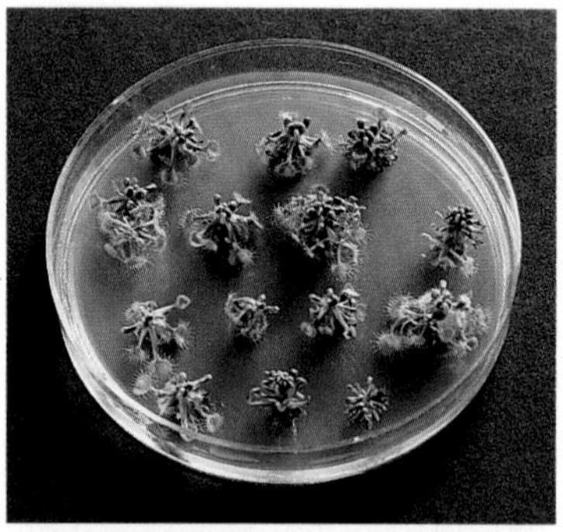

Tissue Culture

Figure 32-16 shows plants grown by **tissue culture,** the production of new plants from pieces of tissue placed on a sterile nutrient medium. Unlike most animal cells, plant cells contain functional copies of all the genes needed to produce a new plant. Thus, it is possible for a whole plant to regrow from a single cell. Millions of identical plants can be grown from a small amount of tissue. Tissue culture is used in the commercial production of orchids, houseplants, cut flowers, fruit plants, and ornamental trees, shrubs, and nonwoody plants.

SECTION 32-3 REVIEW

1. **List 10 different types of fruits, and give an example of each type.**
2. **Name three common methods of fruit and seed dispersal, and give an example of each method.**
3. **How do the structure and germination of a bean seed differ from the structure and germination of a corn seed?**
4. **Compare asexual reproduction with sexual reproduction.**
5. **Make a table that compares structures and methods used for the vegetative propagation of plants.**
6. **CRITICAL THINKING Why is seed dormancy an evolutionary advantage?**

CHAPTER 32 REVIEW

SUMMARY/VOCABULARY

32-1

- Plants have a life cycle called alternation of generations, in which a multicellular haploid gametophyte stage alternates with a multicellular diploid sporophyte stage.
- Mosses and most ferns are homosporous (produce only one type of spore).
- In the moss life cycle, a spore develops into a leafy green gametophyte that produces eggs in archegonia and swimming sperm in antheridia. A moss sporophyte grows from a gametophyte and is dependent on it for nourishment.
- In the fern life cycle, a spore develops into a small flat gametophyte that produces eggs in archegonia and swimming sperm in antheridia. A sporophyte grows from a gametophyte but later crushes it and is not dependent on it for nourishment.
- All seed plants, spike mosses, and quillworts as well as a few fern species are heterosporous (produce two types of spores, male microspores and female megaspores).
- The conifer life cycle features a reduced male gametophyte (pollen grain) and sperm without flagella. A sperm reaches an egg through a pollen tube that grows into a female gametophyte.

Vocabulary

antheridium (625)	integument (628)	microgametophyte (628)	ovule (628)
archegonium (625)	megagametophyte (628)	micropyle (628)	pollen grain (628)
heterospory (627)	megasporangium (628)	microsporangium (628)	pollen tube (628)
homospory (626)	megaspore (627)	microspore (627)	sorus (627)

32-2

- Flowers are reproductive structures of angiosperms. Most familiar flowers consist of four whorls of parts—protective sepals, colorful petals, pollen-producing stamens, and egg-containing carpels.
- Many flowering plants have flowers adapted for animal pollination or for wind pollination.
- In angiosperms, the female gametophyte is a microscopic embryo sac that usually has eight nuclei and is found within an ovule.
- Double fertilization is a unique feature of angiosperms. Two sperm reach the embryo sac through a pollen tube. One sperm combines with the egg to form a zygote. A second sperm combines with two polar nuclei to form a triploid nutritive tissue, the endosperm.

Vocabulary

anther (629)	generative cell (631)	petal (629)	stamen (629)
carpel (629)	megaspore mother cell (630)	pistil (629)	stigma (629)
double fertilization (633)	microspore mother cell (631)	polar nuclei (630)	style (629)
embryo sac (630)	nectar (632)	receptacle (629)	tube cell (631)
filament (629)	ovary (629)	sepal (629)	

32-3

- Angiosperm seeds are enclosed by fruits, which protect seeds and aid in dispersion.
- Seeds need water, oxygen, suitable temperatures, and sometimes light to germinate.
- Asexual reproduction enables plants to spread rapidly in a favorable environment.
- Plants naturally reproduce vegetatively with specialized structures such as bulbs, rhizomes, runners, and tubers.
- People propagate plants asexually by using cuttings, layering, grafting, and tissue culture.

Vocabulary

clone (639)	grafting (640)	plumule (636)	vegetative propagation (639)
cutting (640)	hilum (636)	radicle (636)	vegetative reproduction (639)
dormancy (637)	hypocotyl (636)	seed coat (636)	
epicotyl (636)	layering (640)	tissue culture (640)	

CHAPTER 32 REVIEW ANSWERS

REVIEW

1. A megaspore, which is a reproductive cell, is produced in a megasporangium.
2. A haploid gametophyte produces gametes, and a diploid sporophyte produces spores.
3. Both ovules and ovaries are found within flowers, are part of a pistil, and may develop into edible structures—seeds and fruits.
4. Pomes, berries, and drupes are all fleshy simple fruits.
5. The prefix *epi-* means "over." The epicotyl is the part of the stem above the attachment of the cotyledons. The prefix *hypo-* means "below." The hypocotyl is the part of the stem below the attachment of the cotyledons.

6. a **7.** b
8. b **9.** c
10. b **11.** d
12. d **13.** c
14. a **15.** a

16. The life cycle of a typical fern is homosporous; gametophytes are separate, photosynthetic plants; water is required for fertilization; and no seeds are produced. The life cycle of a seed plant is heterosporous; gametophytes are microscopic; water is not needed for fertilization; and seeds are produced.
17. Heterospory enables more genetic variation among the offspring.
18. production of large amounts of pollen, strong wind, a lack of rain, and plants that grow close together
19. See Figure 32-9.
20. Fertilization in flowering plants requires pollination, growth of a pollen tube, and fusion of two sperm with nuclei in the embryo sac. One sperm fuses with the egg, and the other sperm fuses with two polar nuclei.
21. Fertilization in conifers occurs more than a year after pollination. One sperm fuses with the egg in an ovule while the other sperm dies. Fertilization in flowering

plants occurs within a few days of pollination. Two sperm fuse with cells in the embryo sac, one sperm fuses with the egg and the other sperm fuses with the two polar nuclei.

22. Dustlike orchid seeds are dispersed by wind. Coconut seeds are dispersed by water in floating fruits. Seeds of fleshy fruits, such as mulberry, are dispersed by birds.

23. The hook protects the plumule by pushing aside the soil and pulling the plumule through the soil so that the cotyledons stay folded around it or so that the plumule's leaves stay folded together.

24. Asexual reproduction has the advantages of being more rapid and producing exact copies of the single parent. Its disadvantages are that there is no variation in the offspring and the offspring are often not as easily dispersed as seeds. Sexual reproduction has the advantage of producing variable offspring, which are more likely to be adapted to a range of environments. Another advantage is that the mobile, long-lived seed can survive until it senses an ideal environment for germination. Sexual reproduction is a slower, more-complex process than asexual propagation.

25. Layering involves the formation of roots on intact stems. Propagation by cuttings involves the formation of roots on a piece of stem, shoots on a piece of root, or shoots and roots on a detached leaf. Grafting involves joining pieces of two genetically distinct plants. Tissue culture involves the growth of small bits of plant tissue in a lab. Some plants reproduce by forming bulbs, runners, rhizomes, tubers, or plantlets on their leaves.

REVIEW

Vocabulary

1. How is a megaspore related to a megasporangium?
2. Explain the differences between a gametophyte and a sporophyte.
3. List three similarities between an ovule and an ovary.
4. What do pomes, berries, and drupes have in common?
5. Using a dictionary, find the meaning of the word roots in *epicotyl* and *hypocotyl.* Explain why these terms are appropriate for the structures they describe.

Multiple Choice

6. Moss sperm are produced in (a) antheridia (b) anthers (c) archegonia (d) sori.
7. In pollen, the generative cell forms (a) polar nuclei (b) sperm (c) the epicotyl (d) endosperm.
8. The integuments of an ovule are interrupted by (a) a microspore (b) a micropyle (c) a hilum (d) a stamen.
9. The pistil of a flower is made up of (a) styles (b) ovaries (c) carpels (d) anthers.
10. The microspores in an anther become (a) megaspores (b) pollen grains (c) eggs (d) sperm.
11. Pollination occurs when (a) a sperm fuses with an egg (b) insects ingest nectar (c) a spore leaves a sporangium (d) pollen lands on a stigma.
12. Which of the following terms applies to angiosperms but not to gymnosperms? (a) integument (b) pollen (c) seed (d) double fertilization
13. The epicotyl in a dicot seed is part of the (a) hilum (b) hypocotyl (c) plumule (d) cotyledon.
14. Asexual reproduction in plants occurs in all but which of the following ways? (a) Sperm unite with eggs. (b) New plants form on runners. (c) Seed plants form bulbs. (d) New plants grow in tissue culture.
15. The process shown in the photograph below is termed (a) grafting (b) cutting (c) tissue culture (d) layering.

Short Answer

16. What are two important differences between the life cycle of a typical fern and that of a seed plant?
17. How is heterospory an advantage over homospory?
18. List four factors that promote successful wind pollination.
19. Draw a diagram showing the events and plant structures involved in pollination.
20. How does the process of fertilization occur in flowering plants?
21. How does the process of fertilization in conifers differ from the process of fertilization in flowering plants?
22. List three types of seed dispersal, and give an example of each.
23. In many seeds, the epicotyl forms a hook as the embryo emerges from the seed. What is the advantage of this hook?
24. Compare asexual reproduction with sexual reproduction in terms of their advantages and disadvantages.
25. Describe several methods of vegetative reproduction.

CRITICAL THINKING

1. How is the growth habit of mosses an advantage for their method of sexual reproduction?
2. Suppose that a type of fern normally has a $2n$ chromosome number of 12. Also suppose that meiosis does not occur in the formation of spores, which grow into new gametophytes. What will be the chromosome number of the gametes produced by these gametophytes? What will be the chromosome number of a zygote that results from the fusion of two of these gametes?
3. Why do you think there are more heterosporous plants than there are homosporous plants?
4. In many flowers with both stamens and pistils, the stigma is located well above the anthers. What is the value of such an arrangement?
5. Following the self-pollination of some plants, the pollen tubes die before reaching ovules. What is the significance of this event?
6. Why are fruits and seeds a nutritious source of food for humans and other animals? How is this an advantage for plants?
7. Why do plants with fleshy fruits usually have seeds with hard seed coats?
8. The acorns (nuts) of many oak trees often germinate on the surface of ground that is hard or covered with grass. The new root may grow several centimeters before it reaches a crack or soft spot where it can enter the ground. How is the germinating seed able to survive and grow in such conditions?
9. If you were to discover a new type of rose, would you use seeds or vegetative parts to propagate the rose and produce large numbers of identical plants? Justify your answer.
10. In the photograph below, the houseplant on the left shows a *Sansevieria trifasciata,* which has yellow-edged leaves. A section of the leaf, or a leaf cutting, can be used to produce new plants. However, as shown on the right side of the photograph, only shoots with all-green leaves and no yellow edges will form. How do you explain this?

EXTENSION

1. Buy a 5 cm (2 in.) or longer piece of ginger root at a supermarket. Examine it closely and determine if it is really a root. Plant it about 1 cm (0.4 in.) below the surface in a container of potting soil, keep it watered, and place it in a lighted area. Describe what happens over a period of two months.
2. Visit a commercial fruit-tree orchard and find out how the trees are propagated and pollinated.
3. Read "Life in Bloom" in *Natural History,* May 1999, on page 66. What are three factors that contribute to the life span of a flower?
4. Read "She Loves Me, She Loves Me Not" in *National Wildlife*, January 1999, on page 10. Describe the mystery that plant geneticists have discovered about the instant and strong attraction flowers have for their own species' pollen.
5. Try air-layering an overly tall houseplant, such as a dieffenbachia or a dracena.

CRITICAL THINKING

1. Mosses grow together in tightly packed clumps or mats. Sperm are able to reach eggs before they reach the edge of a water droplet or before the water evaporates.
2. 12; 24
3. Heterosporous plants are more likely to be cross-fertilized, which promotes genetic variation.
4. In flowers in which the stigma is elevated above the stamens, cross-pollination is more likely to occur, which promotes genetic variation in offspring.
5. The death of pollen tubes prevents fertilization by keeping sperm from reaching the egg.
6. Seeds contain stored nutrients. Animals disperse the seeds.
7. A hard seed coat protects the embryo as a seed passes through an animal's digestive system.
8. The germinating seed will be able to survive by using the nutrients stored in the large fleshy cotyledons of the nut.
9. The new rose should be propagated vegetatively to produce large numbers of identical plants.
10. A shoot developing from the green part of the leaf will produce an all-green shoot. To retain the yellow leaf edge during propagation, cut off a piece of the rhizome containing at least one leaf.

EXTENSION

1. Students should describe the formation of the root and shoot.
2. Orchards use a variety of techniques.
3. Three factors that contribute to the life span of a blossom are the rate of photosynthesis, rate of water loss, and the abundance of pollinators.
4. No visible structure provides clues to the strong attachment that forms between a flower's stigma and its pollen grains.
5. Growth should occur within weeks.

CHAPTER 32 INVESTIGATION

TIME REQUIRED

Three 30-minute class periods over five days

SAFETY PRECAUTIONS

Remind students to wear safety goggles and a lab apron at all times during the lab.

Caution students about Lugol's iodine solution, which can cause eye, nose, and throat irritation. If Lugol's iodine solution gets on skin or clothing, rinse it off with water. If Lugol's iodine solution gets in the eyes, flush it out immediately with water for 15 minutes and get medical attention. If it is swallowed (and the person is conscious), give the person one or two glasses of water to drink, induce vomiting, and call a physician immediately.

Caution students to use scalpels carefully to avoid injuring themselves and others.

QUICK REFERENCES

Lab Preparation notes are found on page 623B.

Holt BioSources provides a Teaching Resources CD-ROM that contains "Using Gowin's Vee in the Lab" and "Scoring Rubrics."

PROCEDURAL TIP

Soak seeds in water prior to the lab. This will allow the seed coats to be removed easily.

ANSWERS TO BACKGROUND

1. A seed is made up of an embryo, stored nutrients, and a seed coat.
2. Like the parent plant, the embryo is a sporophyte.
3. Monocots have one cotyledon, flowers with parts in multiples of three, leaves with parallel veins, and stems with scattered vascular

Comparing Seed Structure and Seedling Development

OBJECTIVES

- Observe the structures of dicot and monocot seeds.
- Compare the structure of dicot and monocot embryos.
- Compare the development of dicot and monocot seedlings.

PROCESS SKILLS

- relating structure to function
- comparing and contrasting
- identifying
- collecting data

MATERIALS

- 1 pea seed soaked overnight
- 6 bean seeds soaked overnight
- 6 corn kernels soaked overnight
- stereomicroscope
- scalpel
- Lugol's iodine solution in dropper bottle
- paper towels
- 2 rubber bands
- 2 150 mL beakers
- glass-marking pen
- metric ruler
- microscope slide
- medicine dropper
- compound light microscope

Background

1. What are the parts of a seed?
2. In what ways are seeds like their parent plant?
3. How do monocotyledons and dicotyledons differ?
4. What changes occur as a seed germinates?

PART A Seed Structure

1. Obtain one each of the following seeds—pea, bean, and corn.
2. Remove the seed coats of the pea and bean seeds. Open the seeds, and locate the two cotyledons in each seed.
3. Using the stereomicroscope, examine the embryos of the pea seed and the bean seed.
4. In your lab report, draw the pea and bean embryos and label all of the parts that you can identify.
5. CAUTION **Use the scalpel carefully to avoid injury.** Examine a corn kernel, and find a small, oval, light-colored area that shows through the seed coat. Use the scalpel to cut the seed in half along the length of this area. Place a drop of iodine solution on the cut surface.
6. Use the stereomicroscope to examine the corn embryo. In your lab report, sketch the embryo and label all the parts that you can identify.

PART B Seedling Development

7. Set five corn kernels on a folded paper towel. Roll up the paper towel, and put a rubber band around the roll. Stand the roll in a beaker with 1 cm of water in the bottom. The paper towel will soak up water and moisten the corn. Keep water at the bottom of the beaker, but do not allow the corn kernels to be covered by the water.

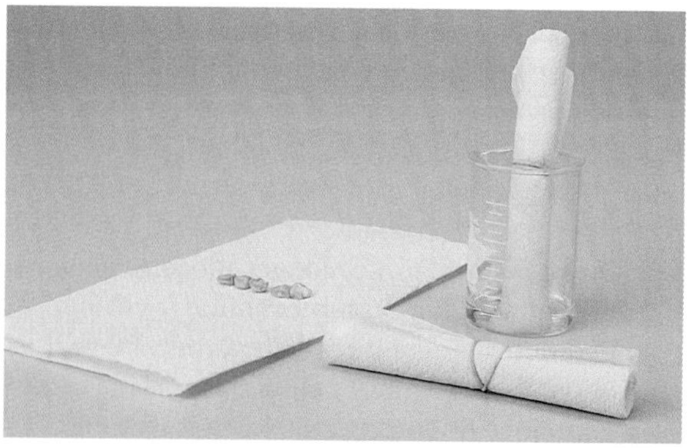

8. Repeat step 7 with five bean seeds.
9. After three days, unroll the paper towels and examine the corn and bean seedlings. Use a glass-marking pen to mark the roots and shoots of the developing

seedlings. Starting at the seed, make a mark every 0.5 cm along the root of each seedling. And again starting at the seed, make a mark every 0.5 cm along the stem of each seedling. Measure the distance from the last mark on the root to the root tip of each seedling. Also measure the distance from the last mark on the stem to the shoot tip of each seedling. Record these data in your lab report.

10. Draw a corn seedling and a bean seedling in your lab report. Using a fresh paper towel, roll up the seeds, place the rolls in the beakers, and add fresh water to the beakers.
11. Make a data table similar to the one shown below in your lab notebook. Expand your table by adding columns under each "Roots" and "Stems" head to account for every section of the roots and stems of your seedlings.
12. After two days, reexamine the seedlings. Measure the distance between the marks, and record the data in your data table.
13. CAUTION **Use the scalpel carefully to avoid injury.** Using the scalpel, make a cut about 2 cm from the tip of the root of a bean seedling. Place the root tip on a microscope slide and add a drop of water. Using a compound light microscope on low power, observe the root tip. In your lab report, draw the root tip.
14. Clean up your materials and wash your hands before leaving the lab.

Analysis and Conclusions

1. What protects the tips of corn shoots as they push through the soil? What protects the bean shoots?
2. What types of leaves first appear on the bean seedling?
3. What substance does the black color in the corn kernel indicate? Why might you expect to find this substance in the seed?
4. Examine the data you recorded for steps 9 and 12. Has the distance between the marks changed? If so, where has it changed?
5. What parts of the embryo were observed in all seeds on the third day?
6. How does the structure and development of the corn kernel differ from the structure and development of the pea and bean seeds?
7. What was the source of nutrients for each seed embryo? What is your evidence?
8. Describe the growth in the seedlings you observed.
9. Corn and beans are often cited as representative examples of monocots and dicots, respectively. Relate the seed structure of each to the terms *monocotyledon* and *dicotyledon.*

Further Inquiry

Design an experiment to find out how monocots and dicots compare in general plant growth and in the structure of their leaves and flowers.

CORN AND BEAN SEEDLING GROWTH AFTER TWO DAYS

	Corn seedlings				Bean seedlings			
	Roots		Stems		Roots		Stems	
Seedling	Seed to first mark	Root tip to last mark	Seed to first mark	Shoot tip to last mark	Seed to first mark	Root tip to last mark	Seed to first mark	Shoot tip to last mark
1								
2								
3								
4								
5								

bundles. Dicots have two cotyledons, flowers with parts in multiples of four or five, leaves with netted veins, and stems with vascular bundles arranged in rings.
4. The first visible evidence of germination is the emergence of the embryo's root. The shoot tip then emerges from the soil.

ANSWERS TO ANALYSIS AND CONCLUSIONS

1. Corn shoots are protected by a sheath as they push through the soil. Bean shoots are protected by a hook in the embryonic stem that pushes through the soil before the cotyledons.
2. Cotyledons are the first leaves to appear on a seedling.
3. The black color indicates that starch is contained in the corn kernel. You might expect to find starch in a corn kernel because much of the kernel is a nutrient for the embryo, and starch is the main storage form of carbohydrates.
4. Yes. The distance between the marks has changed at the tips of the stems and the tips of the roots.
5. Embryonic leaves and roots were observed in all seeds on the third day.
6. Peas and beans have two cotyledons, while corn has one. Corn seeds have endosperm, while mature bean and pea seeds do not. The shoots of peas and beans hook as they germinate, while a corn shoot grows straight up.
7. Nutrients are stored in the endosperm of the corn kernel and in the cotyledons of the beans and peas. The iodine test for starch confirms this.
8. The seedlings grow at the tips of their roots and stems.
9. Corn seeds have only one cotyledon and thus are monocotyledons, while bean seeds have two cotyledons and thus are dicotyledons.

CHAPTER 33
PLANT RESPONSES

	TOPICS	TEACHING RESOURCES	LABS, CLASSWORK, AND HOMEWORK
BLOCK 1 45 minutes	**Introducing the Chapter,** p. 646	ATE Focus Concept, p. 646 ATE Assessing Prior Knowledge, p. 646 ATE Understanding the Visual, p. 646	■ Supplemental Reading Guide, *A Feeling for the Organism*
	33-1 Plant Hormones, p. 647 Groups of Hormones, p. 647	ATE Section Overview, p. 647 ATE Critical Thinking, p. 649 ATE Visual Link Figure 33-1, p. 647 119A	**Lab B19,** Gravitropism and Phototropism in Plants ★ Study Guide, Section 33-1 PE Section Review, p. 651
BLOCK 2 45 minutes	**33-2 Plant Movements,** p. 652 Tropisms, p. 652 Nastic Movements, p. 654	ATE Section Overview, p. 652 119A	PE **Quick Lab,** p. 653, Visualizing Phototropism **Lab B19,** Gravitropism and Phototropism in Plants ★ Study Guide, Section 33-2 PE Section Review, p. 655
BLOCK 3 45 minutes	**33-3 Seasonal Responses,** p. 656 Photoperiodism, p. 656 Vernalization, p. 658	ATE Section Overview, p. 656 118A, 130	PE **Chapter 33 Investigation,** p. 662 ★ Study Guide, Section 33-3 PE Section Review, p. 658
BLOCKS 4 & 5 90 minutes	**REVIEW and ASSESSMENT** PE Chapter 33 Review, pp. 660–661 ■ Performance-Based Assessment—Testing Enzyme Activity in Seeds	Audio CD Program	**CHAPTER TESTING** ■ Chapter Test (blackline master) ▲ Test Generator ■ Assessment Item Listing ■ Scoring Rubrics and Classroom Management Checklists

MODERN BIOLOGY — HOLT BIOSOURCES

KEY

- PE Pupil's Edition
- ATE Teacher's Edition
- ■ Active Reading Guide
- ★ Study Guide
- Teaching Transparencies
- Laboratory Program

One-Stop Planner CD-ROM
Shows these resources within a customizable daily lesson plan:

- ■ Holt BioSources Teaching Resources
- ■ Active Reading Guide
- ★ Study Guide
- ▲ Test Generator

READING FOR CONTENT MASTERY

Preview: Have students preview the objectives, the introductory paragraph, topic sentences, and captions before reading the chapter.

- **ATE** Active Reading Technique: Anticipation Guide, p. 647
- ■ Active Reading Guide Worksheet 33-1
- **PE** Summary/Vocabulary, p. 659

- ■ Active Reading Guide Worksheet 33-2
- **PE** Summary/Vocabulary, p. 659

- ■ Active Reading Guide Worksheet 33-3
- **PE** Summary/Vocabulary, p. 659

TECHNOLOGY AND INTERNET RESOURCES

Holt Biology Videodiscs Teacher's Correlation Guide, Lessons 33-1 through 33-3

Audio CD Program, Sections 33-1 through 33-3

internetconnect

SCILINKS NSTA

On-line Resources: www.scilinks.org
The following *sci*LINKS Internet resources can be found in the student text for this chapter:

Topics:
- Plant growth regulators, p. 651
- Nastic movements, p. 655
- Photoperiodism, p. 657

go.hrw.com

On-line Resources: go.hrw.com
Visit the HRW Web site for a variety of resources related to this chapter. Just type in the keyword HM2 Chapter 33.

Smithsonian Institution®
Visit **www.si.edu/hrw** for additional on-line resources.

CNNfyi.com
Visit **www.cnnfyi.com** for late-breaking news and current event stories selected just for you.

LAB ACTIVITY PLANNING

CHAPTER 33 INVESTIGATION

Testing the Effect of Gibberellin on Plant Growth, pp. 662–663

OVERVIEW
In this investigation, students will set up a controlled experiment, collect data, and formulate a hypothesis about the effects of the plant growth hormone gibberellin on bean seedlings.

SAFETY
Gibberellic acid is a flammable irritant. Store in a cool, dry place away from strong oxidizing materials and fire hazards. If gibberellic acid solution gets in a student's eye, flush the eye with water for 15 minutes and get medical attention.

PREPARATION
1. Purchase a 0.1percent gibberellic acid solution from a biological supply house, such as WARD'S. Each lab group will need about 20 mL of gibberellic acid solution in a labeled dropper bottle.
2. Each lab group will need 10 bean seedlings that are at least 3 cm tall.
3. Soak the bean seeds for 6–8 hours. To prevent the growth of bacteria and mold, soak the seeds for 15 minutes in a solution of 5 mL chlorine bleach and 95 mL distilled water. Always grow more seedlings than you need so that plants of equal size and vigor may be selected. Allow 2–3 days before students begin for the seedlings to grow at least 3 cm tall.

Quick Labs

Visualizing Phototropism, p. 653
Each student group will need two 2 in. pots containing potting soil, four bean seeds, and a cardboard box for this activity. Small, paper drinking cups can substitute for 2 in. pots.

UNDERSTANDING THE VISUAL

Have students examine the photograph of Rio Cuchujaqui. Tell them that the trees in this area are adapted to both very dry and very wet conditions. Some trees have surface roots to take advantage of surface water from floods, and some trees have taproots that reach very deep into the aquifer to help them survive the dry seasons. A bridge once spanned the stream near where this picture was taken. The bridge was washed out by a flood, but the trees were undisturbed because of their amazing surface-root system.

FOCUS CONCEPT

Evolution

Plants have evolved a number of very complex adaptations to their environment that help them to survive and reproduce. Plants cannot move to new environments like animals can, so they must be able to tolerate unfavorable conditions. Plants respond to a wide variety of signals in their environment.

ASSESSING PRIOR KNOWLEDGE

Review the following concepts.

Plant Structure: *Chapter 31*
Have students discuss stem and leaf structure. Ask students to identify leaf and stem adaptations that may be involved in plant movements.

Plant Growth: *Chapter 31*
Have students draw a plant on a piece of paper. Ask students to circle the sites of primary growth. Ask students to explain what happens when a plant experiences secondary growth.

CHAPTER 33

PLANT RESPONSES

Montezuma baldcypress trees, Taxodium mucronatum, *with their tangled roots, line the banks of the Rio Cuchujaqui in the southern region of the Mexican state of Sonora.*

FOCUS CONCEPT: *Evolution*

As you read, notice how plants respond to their environment and to hormones. Plant responses to environmental conditions are adaptive advantages.

Plant Hormones

The growth and development of a plant are influenced by genetic factors, external environmental factors, and chemicals inside the plant. Plants respond to many environmental factors, such as light, gravity, water, inorganic nutrients, and temperature.

SECTION

33-1

OBJECTIVES

▲

List the five major types of plant hormones, and give some effects of each.

●

Give examples of the adaptive advantages of hormonal responses.

■

Describe three agricultural or gardening applications for each of three classes of plant hormones.

GROUPS OF HORMONES

Plant **hormones** are chemical messengers that affect a plant's ability to respond to its environment. Hormones are organic compounds that are effective at very low concentrations; they are usually synthesized in one part of the plant and transported to another location. They interact with specific target tissues to cause physiological responses, such as growth or fruit ripening. Each response is often the result of two or more hormones acting together.

Because hormones stimulate or inhibit plant growth, many botanists also refer to them as plant **growth regulators.** Many hormones can be synthesized in the laboratory, increasing the quantity of hormones available for commercial applications. Botanists recognize five major groups of hormones: auxins, gibberellins, ethylene, cytokinins, and abscisic acid. These groups of hormones are examined in Table 33-1 on the next page.

Auxins

Auxins (AWK-suhnz) are hormones involved in plant-cell elongation, apical dominance, and rooting. A well known natural auxin is **indoleacetic** (IN-DOHL-uh-SEET-ik) **acid,** or **IAA.** Developing seeds produce IAA, which stimulates the development of a fleshy fruit. Figure 33-1 shows how the removal of seeds from a strawberry prevents the fruit from enlarging. The application of IAA after removing the seeds causes the fruit to enlarge normally.

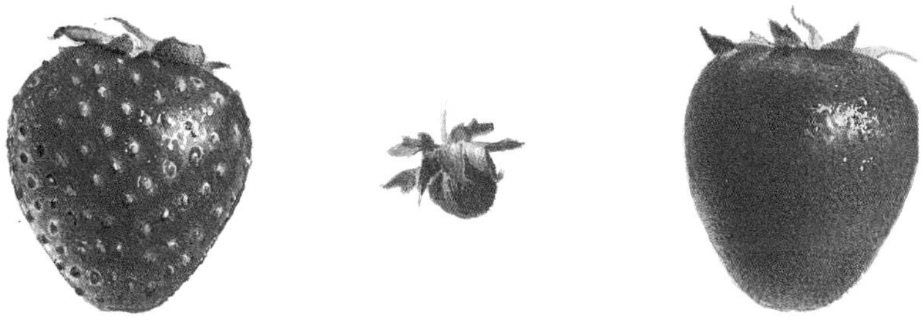

(a) (b) (c)

FIGURE 33-1

(a) Strawberry plants produce a fleshy, heart-shaped fruit that is covered with yellow seeds. (b) When the seeds are removed from the fruit, the growth of the fruit is retarded. (c) When the seeds are removed and the fruit is treated with an auxin, growth is normal.

SECTION 33-1

SECTION OVERVIEW

Stability and Homeostasis

Section 33-1 discusses how plant growth, development, and reproduction are regulated by five different types of hormones. Each type of hormone can produce a variety of effects, depending on the tissue it is acting on. A hormone can also interact with another hormone, producing different plant responses.

ACTIVE READING

Technique: Anticipation Guide

Before students read this chapter, write the following on the board or overhead:

1. Plants, like animals, produce hormones that regulate some biological functions.

2. Plants are adapted to respond to conditions in their environment.

Let students think for a minute about these statements and decide if they agree with the statements. Have students read the chapter to see if their opinions are confirmed.

VISUAL LINK

Figure 33-1

Have students examine Figure 33-1, and ask them what causes strawberry fruits to develop (an auxin produced by the developing seeds). Ask students why the fruit has not developed in Figure 33-1b. (The fruit has not been exposed to an auxin because the source of the auxin, the seeds, has been removed.)

SECTION 33-1

RETEACHING ACTIVITY

Have students examine Table 33-1, and ask them to make a list on the board of all of the growth-promoting effects of each type of hormone. (All have growth-promoting effects except abscisic acid.) Then ask students to list all of the growth-inhibiting effects of each hormone. (All have growth-inhibiting effects except gibberellin and cytokinin.) Ask students to pair growth-promoting effects with contrasting growth-inhibiting effects. (Auxins inhibit lateral bud growth and cytokinins promote bud growth. Abscisic acid inhibits seed germination and gibberellins promote seed germination.) Ask students what determines a plant's response when it is regulated by paired hormones (the relative amounts of the two hormones).

RECENT RESEARCH

Insects Stimulate Plant Growth

Scientists at the University of Georgia isolated six proteins from the digestive system of grasshoppers and exposed new shoots of sorghum plants to the proteins. The scientists found that one of the proteins more than quadrupled plant growth. The protein appears to make auxins more effective in stimulating epidermal growth.

GIFTED ACTIVITY

Have each student pick a plant hormone and research and report on the commercial uses of that hormone. Have students list the crops that commonly use the hormone. (A horticulture textbook would be a good source.) Some examples include the use of gibberellins to increase the size of seedless grapes and the use of auxins to inhibit sprouting.

TABLE 33-1 *Five Groups of Plant Hormones*

Plant hormone	Function	Features	Examples
Auxins	• promote cell growth • promote root formation on stem and leaf cuttings • promote apical dominance • increase number of fruits • prevent dropping of fruit • prevent sprouting of stored potatoes and onions	• produced in growing regions of plant (shoot tips, young leaves, developing fruit) • important role in tropisms	• indoleacetic acid, IAA (natural) • naphthalene acetic acid, NAA (synthetic) • herbicides 2,4-D and Agent Orange (synthetic)
Gibberellins (GA)	• promote elongation growth • promote germination, and seedling growth • increase size of fruit • overcome bud dormancy • substitute for long-day or vernalization requirements for flowering	• produced in all parts of plant, especially in immature seeds • more than 80 types	• GA_3 (natural)
Ethylene	• promotes ripening of fruit • promotes flowering in mangoes and pineapples • promotes abscission	• produced in fruits, flowers, leaves, and roots • colorless gas	• ethephon (synthetic) breaks down and releases ethylene (natural)
Cytokinins	• promote cell division • promote lateral bud growth in dicots	• produced in developing roots, fruits, and seeds • auxin-to-cytokinin ratio is important	• zeatin (natural) • kinetin (synthetic) • benzyladenine (synthetic)
Abscisic acid (ABA)	• promotes stomatal closure • promotes dormancy • inhibits other hormones • blocks growth	• produced in leaves • expensive to synthesize	• ABA (natural or synthetic)

IAA is produced in actively growing shoot tips and developing fruit, and it is involved in elongation. Before a cell can elongate, the cell wall must become less rigid so that it can expand. IAA triggers an increase in the plasticity, or stretchability, of cell walls, allowing elongation to occur.

Synthetic Auxins

Chemists have synthesized several inexpensive compounds similar in structure to IAA. Synthetic auxins, like **naphthalene** (NAF-thuh-LEEN) **acetic** (uh-SEET-ik) **acid,** or **NAA,** are used extensively to promote root formation on stem and leaf cuttings, as shown in Figure 33-2. Gardeners often spray auxins on tomato plants to increase the number of fruits on each plant.

(a) (b)

FIGURE 33-2

(a) This coleus plant, *Coleus hybridus,* is growing roots in pure water. (b) This coleus stem cutting has been treated with the auxin NAA. Roots form more rapidly when the stem is treated with NAA. Both cuttings were taken at the same time.

When NAA is sprayed on young fruits of apple and olive trees, some of the fruits drop off so that the remaining fruits grow larger. When NAA is sprayed directly on fruits—such as apples, pears, and citrus fruits—several weeks before they are ready to be picked, NAA prevents the fruits from dropping off the trees before they are mature. The fact that auxins can have opposite effects—causing fruit to drop or preventing fruit from dropping—illustrates an important point. The effects of a hormone on a plant often depend on the stage of the plant's development.

NAA is used to prevent the undesirable sprouting of stems from the base of ornamental trees. As discussed in Chapter 32, stems contain a lateral bud at the base of each leaf. In many stems, these buds fail to sprout as long as the plant's shoot tip is intact. The inhibition of lateral buds by the presence of a shoot tip is called **apical dominance.** If the shoot tip of a plant is removed, the lateral buds begin to grow. If IAA or NAA is applied to the cut tip of the stem, the lateral buds remain dormant. This adaptation is manipulated to cultivate beautiful ornamental trees. NAA is used commercially to prevent buds from sprouting on potato tubers during storage.

Another important synthetic auxin is **2,4-D,** which is a herbicide, or weedkiller. It selectively kills dicots, such as dandelions and pigweed, without injuring monocots, such as lawn grasses and cereal crops. Given our major dependence on cereals for food, 2,4-D has been of great value to agriculture. A mixture of 2,4-D and another auxin, called **Agent Orange,** was used to defoliate jungles in the Vietnam War. A nonauxin contaminant in Agent Orange has caused severe health problems in many people who were exposed to it.

Gibberellins

In the 1920s, scientists in Japan discovered that a substance produced by the fungus *Gibberella* caused fungus-infected plants to grow

CONTENT CONNECTION

Section 31-3: Types of Buds

Review the location of the terminal bud and lateral buds on a stem. Tell students that in many plants, lateral buds are dormant as long as the terminal bud is actively growing.

CRITICAL THINKING

Auxin-Cytokinin Interactions

Ask students if they have noticed that trees usually produce more side branches on the lower half of the tree than on the upper half of the tree. Ask students to explain this growth pattern. Lead students in a discussion of auxins and cytokinins, including the site of production and the function of each hormone. (Auxin is produced at the shoot tip and diffuses downward, while cytokinin is produced at the root tip and diffuses upward. Toward the top of a tree, there is more auxin than cytokinin, so no lateral buds will develop into branches. On the lower part of the tree, there is less auxin than cytokinin, so lateral buds will develop into branches.)

RECENT RESEARCH

Stomatal Closure

Scientists in Japan have found that abscisic acid induces stomatal closure in the bean *Vicia faba* by inhibiting the pumping of H^+ out of the guard cells, which in turn prevents the pumping of K^+ into the guard cells. Stomatal opening occurs following the pumping of K^+ into guard cells, which causes them to take up water, swell, and open the stomatal pores.

DEMONSTRATION

Provide students with pictures of tissue-cultured plants grown in different combinations of auxins and cytokinins. (These are common in college biology, botany, and horticulture textbooks.) The cultures will exhibit root formation when only an auxin is present, shoot formation when only a cytokinin is present, and undifferentiated growth when both hormones are present in high amounts. Ask students how a plant grower would treat cultured plant tissues with hormones in order to induce the formation of normal, whole plants. (He or she would supply the cultures initially with auxins to stimulate root formation and then with cytokinins to stimulate shoot formation.) Point out to students that plant growth occurs as a result of interactions between different hormones present in the tissues.

MAKING CONNECTIONS

Technology

A bacterium causes the formation of tumors in a plant disease known as crown gall disease by inducing a high level of an auxin and a cytokinin in the infected plant, which stimulates massive cell proliferation. It does this by inserting genes for auxin and cytokinin biosynthesis into host cells, effectively engineering them to form tumors. Scientists have been conducting research on using this bacterium to insert beneficial genes into crop plants.

FIGURE 33-3

The bird's ... nidus, show... age.

(a) (b)

abnormally tall. The substance, named **gibberellin** (JIB-uh-REL-uhn), was later found to be produced in small quantities by plants themselves. It has many effects on a plant, but it primarily stimulates elongation growth. Spraying a plant with gibberellins will usually cause the plant to grow to a larger than normal height, as shown in Figure 33-3.

Like auxins, gibberellins are a class of hormones that have important commercial applications. Almost all seedless grapes are sprayed with gibberellins to increase the size of the fruit, as Figure 33-4 shows. Beer makers use gibberellins to increase the alcohol content of beer by increasing the amount of sugar produced in the malting process. Gibberellins are also used to treat seeds of some food crops because they break seed dormancy and promote uniform germination.

FIGURE 33-4

Almost all of the raisins produced in California are made from Thompson Seedless grapes. Normal Thompson Seedless grapes are smaller than those treated with a gibberellin hormone. The grapes on the left were treated with a gibberellin. As you can see, the addition of a gibberellin to grapes stimulates the grapes to grow larger.

Ethylene

The hormone **ethylene** (ETH-uh-LEEN) is responsible for the ripening of fruits. Unlike the other four classes of plant hormones, ethylene is a gas at room temperature. Ethylene gas diffuses easily through the air from one plant to another. The saying "One bad apple spoils the barrel" has its basis in ethylene gas. One rotting apple will produce ethylene gas, which stimulates nearby apples to ripen and eventually spoil.

Ethylene is usually applied in a solution of **ethephon** (ETH-uh-fohn), a synthetic chemical that breaks down to release ethylene gas. It is used to ripen bananas, honeydew melons, and tomatoes. Oranges, lemons, and grapefruits often remain green when they are ripe. Although the green fruit tastes good, consumers will not usually buy them. The application of ethylene to green citrus fruits causes the development of desirable citrus colors, such as yellow and orange.

In some plant species, ethylene promotes **abscission,** which is the detachment of leaves, flowers, or fruits from a plant. Cherries and walnuts are harvested with mechanical tree shakers. Ethylene treatment increases the number of fruits that fall to the ground when the trees are shaken. Leaf abscission is also an adaptive advantage for the plant. Dead, damaged, or infected leaves drop to the ground rather than shading healthy leaves or spreading disease. The plant can minimize water loss in the winter, when the water in a plant is often frozen.

Cytokinins

Cytokinins (sie-toh-KIE-nuhnz) promote cell division in plants. Produced in the developing shoots, roots, fruits, and seeds of a plant, cytokinins are very important in the culturing of plant tissues in the laboratory. A high ratio of auxins to cytokinins in a tissue-culture medium stimulates root formation. A low ratio promotes shoot formation. Cytokinins are also used to promote lateral bud growth of flower crops.

Abscisic Acid

Abscisic (ab-SIS-ik) **acid,** or **ABA,** generally inhibits other hormones, such as IAA. It was originally thought to promote abscission, hence its name. Botanists now know that ethylene is the main abscission hormone. ABA helps to bring about dormancy in a plant's buds and maintains dormancy in its seeds. ABA causes the closure of a plant's stomata in response to drought. Water-stressed leaves produce large amounts of ABA, which triggers potassium ions to be transported out of the guard cells. This causes the stomata to close. It is too costly to synthesize ABA for agricultural use.

Other Growth Regulators

Many growth regulators are widely used on ornamental plants. These substances do not fit into any of the five classes of hormones. For example, utility companies in California often apply **growth retardants,** chemicals that prevent plant growth, to trees to prevent them from interfering with overhead utility lines. It is less expensive to apply these chemicals than to prune the trees. Also, azalea growers sometimes apply a chemical to kill an azalea's terminal buds rather than hand-pruning them. Scientists are still searching for a hormone to slow the growth of lawn grass so that it does not have to be mowed so often.

Word Roots and Origins

abscisic

from the Latin *abscisus,* meaning "to cut off"

internetconnect

SCILINKS NSTA

TOPIC: Plant growth regulators
GO TO: www.scilinks.org
KEYWORD: HM651

SECTION 33-1 REVIEW

1. What are important commercial uses of auxins, gibberellins, and ethylene?
2. Explain one way ethylene differs from the other plant hormones.
3. How are a plant's shoot tip and lateral buds affected by apical dominance?
4. Why are growth retardants often sprayed on potted flowers?
5. Explain how auxins and cytokinins interact.
6. **CRITICAL THINKING** What factors influence the effect of a hormone on target cells?

SECTION 33-1

ANSWERS TO SECTION 33-1 REVIEW

1. Auxins are used to kill weeds, stimulate root growth of cuttings, prevent sprouting of potato tubers, increase the size or number of tomato fruit, reduce the number of young apples or olives, and prevent fruit from falling prematurely. Gibberellins are used to increase the size of seedless grapes and stimulate seed germination. Ethylene is used to ripen bananas and tomatoes, color ripe citrus fruits, promote the dropping of mechanically harvested fruits, and promote the flowering of pineapples.
2. Ethylene is a gas, while the other hormones are dissolved solids.
3. Apical dominance keeps lateral buds in reserve. If the terminal bud is injured, the dormant lateral buds start growing. By keeping the lateral buds dormant, the stem can use more energy to grow taller. This allows it to get more light.
4. Flower crops often grow very tall. Growth retardants keep them more compact. Compact plants have stronger stems, which help prevent the plants from being damaged as easily, and are smaller, which reduces shipping costs.
5. In tissue cultures, a high auxin-to-cytokinin ratio promotes root formation while a low ratio promotes shoot formation. In a whole plant, a high auxin-to-cytokinin ratio inhibits lateral bud growth, while a low ratio promotes lateral bud growth.
6. The effect of a hormone on a target cell is influenced by the hormone's concentration and its interaction with other hormones.

SECTION 33-2

SECTION OVERVIEW

Stability and Homeostasis

Section 33-2 describes how plants move in response to stimuli. Plants maintain homeostasis by responding to various environmental influences.

▶ ENGAGE STUDENTS

Display diagrams of the famous experiments that Fritz Went conducted in 1926 on phototropism. (These can be found in college biology and botany textbooks.) Tell students that Went hypothesized that a chemical messenger produced in the plant's shoot tip diffused downward and affected the elongation of cells behind the shoot tip. Discuss Went's experimental design and how it enabled him to test his hypothesis. Went removed the shoot tip of a plant to see if phototropism continued to occur. He collected chemicals from the severed shoot tip and placed the chemicals on the tip of the cut stem. Went found that the chemical alone could induce growth or phototropism. Draw diagrams on the board to illustrate the experiments. Discuss Went's results and conclusions with the class.

RECENT RESEARCH

Light Receptor in Phototropism

Scientists in Germany have studied corn shoot tips and found that a sensor in these tips detects light in the blue region of the electromagnetic spectrum. This blue-light photoreceptor is located in the cell membrane of the cells at the shoot tip. The photoreceptor gets a phosphate group added to it by ATP when exposed to blue light. This triggers a series of events that leads to auxin transport to the side of the shoot away from the light source.

SECTION

33-2

OBJECTIVES

▲ Explain the difference between tropisms and nastic movements.

● List the environmental stimuli to which plants respond and the tropism for each stimulus.

■ Explain the current hypotheses regarding auxins and their function in phototropism and gravitropism.

◆ Describe common nastic movements and explain how they help a plant survive.

PLANT MOVEMENTS

Plants appear immobile because they are rooted in place. However, time-lapse photography reveals that parts of plants frequently move. Most plants move too slowly for us to notice. Plants move in response to several environmental stimuli, such as light, gravity, and mechanical disturbances. These movements fall into two groups, tropisms and nastic movements.

TROPISMS

A **tropism** (TROH-piz-uhm) is a plant movement that is determined by the direction of an environmental stimulus. Movement toward an environmental stimulus is called a *positive* tropism, and movement away from a stimulus is called a *negative* tropism. Each kind of tropism is named for its stimulus. For example, a plant movement in response to light coming from one direction is called **phototropism.** Thus, in Figure 33-5, the shoot tips of a plant that grow toward the light source are positively phototropic. The various types of tropisms are summarized in Table 33-2.

Phototropism

Phototropism is illustrated by the movement of the sprouts in Figure 33-5. Light causes the hormone auxin to move to the shaded side of the shoot. The auxin causes the cells on the shaded side to elongate more than the cells on the lighted side. As a result, the shoot bends toward the light and exhibits positive phototropism. In some plant stems, phototropism is not caused by auxin movement.

FIGURE 33-5

The way these new sprouts of the flowering shamrock, *Oxalis rubra,* grow toward a light is an example of positive phototropism. The auxin hormone stimulates the cells on the shaded side of the plant stem to elongate.

TABLE 33-2 *Various Plant Tropisms*

Tropism	Stimulus	Positive example
Phototropism	light	plant leans toward light
Thigmotropism	contact with object	vine twines around a tree
Gravitropism	gravity	roots grow downward
Chemotropism	chemical	pollen tube grows toward ovule

In these instances, light causes the production of a growth inhibitor on the lighted side. Negative phototropism is sometimes seen in vines that climb on flat walls where coiling tendrils have nothing to coil around. These vines have stem tips that grow away from light, toward the wall. This brings adventitious roots or adhesive discs in contact with the wall on which they can cling.

Solar tracking is the motion of leaves or flowers as they follow the sun's movement across the sky, as shown in Figure 33-6. By continuously facing toward a moving light source, the plant maximizes the light available for photosynthesis.

Thigmotropism

Thigmotropism is a plant's growth response to touching a solid object. Tendrils and stems of vines, such as morning glories, coil when they touch an object. Thigmotropism allows some vines to climb other plants or objects, increasing its chance of intercepting light for photosynthesis. It is thought that an auxin and ethylene are involved in this response.

FIGURE 33-6

These sunflower plants face the sun as it sets in the evening. In the morning, they face the sun as it rises, then they follow it as it moves across the sky. This form of phototropism is called solar tracking.

Quick Lab

Visualizing Phototropism

Materials 2 in. pots (2) containing potting soil, 4 bean seeds, cardboard box

Procedure

1. Plant two bean seeds in each pot. Label each pot for your group. Place one pot in a window or under a plant light.
2. Cut a rectangular window in the box, and place the box over the second pot so that the window faces the light. Place the box in a different location from the open pot. Keep both pots moist for several days.
3. Remove the box 2–3 days after the seedlings have emerged from the soil. Compare the seedlings grown in the light with the ones grown in the box. Sketch your observations.

Analysis How are the seedlings different? How do you account for the difference? Describe or draw what you think the cells inside the curved part of the stem look like.

Word Roots and Origins

thigmotropism

from the Greek *thiga,* meaning "touch," and *tropos,* meaning "turning"

Quick Lab

Visualizing Phototropism

Time Required 30 minutes

Preparation Tips Paper drinking cups (6 oz) are a good substitute for 2 in. flower pots. The cups or pots should be two-thirds filled with moist potting soil. You might want to start the seeds so that they have just germinated before the start of this lesson. Shoe boxes are good for this activity and are easy to cut. You might want to prepare the boxes ahead of time. Students will observe phototropism in 2–3 days after the seedlings in the box emerge from the soil.

Answers to Analysis The seedlings grown in the box will be bent toward the source of light. The seedlings grown in the light will be more upright. The difference is due to phototropism. Students should describe or draw longer, curved cells on the side of the stem that is opposite the light source and shorter cells on the side of the stem that faces the light.

DEMONSTRATION

Provide students with a plant that is capable of solar tracking, such as cotton, beans, or alfalfa. Place the plant outside early on a sunny morning. Have students observe the position of the leaves relative to the sun's position in the sky every hour over the course of the day. Ask them to check the plant early the next morning and record their observations. (The leaves should be oriented in the direction of the morning sun; about an hour before sunrise these plants orient their leaves toward the point on the horizon where the sun will rise.)

SECTION 33-2

RECENT RESEARCH

Shoot Gravitropism

Scientists in Israel have found that both auxins and ethylene are involved in shoot gravitropism in snapdragons. They found that cell-membrane-bound calcium ions act as messengers, signaling the redistribution of auxins in response to gravity. Auxin production stimulates ethylene production, which, along with auxins, promotes shoot elongation.

TEACHING STRATEGY

Inducing Tropisms

Have students plant water-soaked corn seeds in various positions, including upright, upside down, and sideways, against the side of a glass jar lined with wet paper towels. After five days, have students record the growth and direction of the roots and stems and identify the stimuli and tropisms involved in this activity (gravitropism: roots grow down because of gravity; negative gravitropism: stems grow up because of gravity; phototropism: stems grow toward light because light is coming from only one direction).

GIFTED ACTIVITY

Have students research and report on the acid growth hypothesis of an auxin's role in cell elongation. (According to this hypothesis, auxins stimulate proton pumps in cells that pump protons from the inside of the cell into the cell wall. This lowers the pH of the wall, which weakens cellulose cross-links and makes the wall more plastic. This is followed by the uptake of water by osmosis, which results in elongation of the cell.)

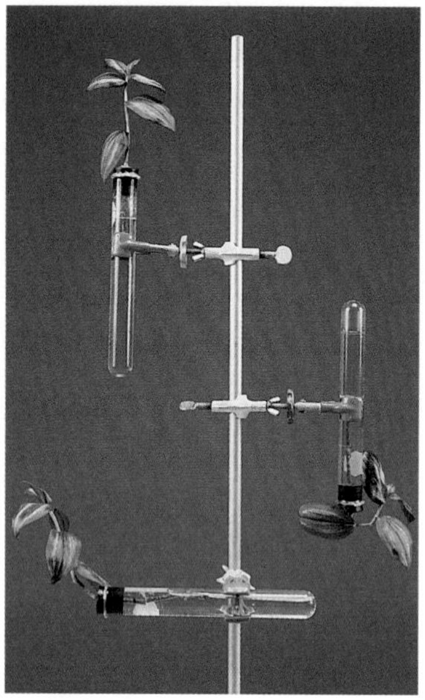

FIGURE 33-7

This photograph was taken seven days after three cuttings of the inch plant, *Zebrina pendula,* were placed in growing tubes in different orientations. Notice that the two lower cuttings have started to grow upward, while the cutting on top, which was placed upright to begin with, has not changed its direction of growth. The growth movement of the lower cuttings is caused by the plants' response to gravity and is called negative gravitropism. When the roots turn and begin to grow in a downward direction, the movement is called positive gravitropism.

Gravitropism

Gravitropism is a plant's growth response to gravity. A root usually grows downward and a stem usually grows upward—that is, roots are positively gravitropic and stems are negatively gravitropic.

Like phototropism, gravitropism appears to be regulated by auxins. One hypothesis proposes that when a seedling is placed horizontally, auxins accumulate along the lower sides of both the root and the stem. This concentration of auxins stimulates cell elongation along the lower side of the stem, and the stem grows upward. A similar concentration of auxins inhibits cell elongation in the lower side of the root, and the root grows downward, as shown in Figure 33-7.

Chemotropism

A plant's growth in response to a chemical is called **chemotropism.** After a flower is pollinated, a pollen tube grows down through the stigma and style and enters the ovule through the micropyle. The growth of the pollen tube in response to chemicals produced by the ovule is an example of a chemotropism.

NASTIC MOVEMENTS

Plant movements that occur in response to environmental stimuli but that are independent of the direction of stimuli are called **nastic movements.** These movements are regulated by changes in the water pressure of certain plant cells.

Thigmonastic Movements

Thigmonastic (THIG-mah-NAS-tik) **movements** are a type of nastic movement that occurs in response to touching or shaking a plant. Many thigmonasties involve rapid plant movements, such as the closing of the leaf trap of a Venus' flytrap or the folding of a plant's leaves in response to being touched. Figure 33-8 shows how the leaves of a sensitive plant fold within a few seconds after being touched. This movement is caused by the rapid loss of turgor pressure in certain cells, a process similar to that which occurs in guard cells. Physical stimulation of the plant leaf causes potassium

(a) (b)

FIGURE 33-8

(a) *Mimosa pudica,* also called the sensitive plant, is a small shrub that has leaflets. (b) When a leaflet is touched, it folds together. This rapid movement is a thigmonastic movement.

(a)

(b)

FIGURE 33-9

A common houseplant is the prayer plant, *Maranta leuconeura.* (a) During the day, the leaf blades of the prayer plant are oriented horizontally in response to light. (b) During the night, the leaf blades are oriented vertically. This movement is called nyctinastic movement.

ions to be pumped out of the cells at the base of leaflets and petioles. Water then moves out of the cells by osmosis. As the cells shrink, the plant's leaves move. It is believed that the folding of a plant's leaves in response to touch discourages insect feeding.

In addition, thigmonastic movements may help prevent water loss in plants. When the wind blows on a plant, the rate of transpiration is increased. So if the leaves of a plant fold in response to the "touch" of the wind, water loss is reduced. This could be an important adaptive advantage to a plant.

Nyctinastic Movements

Nyctinastic (NIK-tuh-NAS-tik) **movements** are plant movements in response to the daily cycle of light and dark. Nyctinastic movements involve the same type of osmotic mechanism as thigmonastic movements, but the changes in turgor pressure are more gradual. Nyctinastic movements occur in many plants, including bean plants, honeylocust trees, and silk trees. The prayer plant, shown in Figure 33-9, gets its name from the fact that its leaf blades are vertical at night, resembling praying hands. During the day, the leaf blades of the prayer plant are horizontal. The botanist Linnaeus planted many different species of plants with nyctinastic movements in a big circle to make a "flower clock." The nyctinastic movements of each species occurred at a specific time of day.

SECTION 33-2 REVIEW

1. **How do a tropism and a nastic movement differ?**
2. **Define the three major plant tropisms, and give an example of each.**
3. **How does auxin cause plant movements in response to light and gravity?**
4. **What adaptive advantage might nastic movements provide a plant?**
5. **How do nastic movements occur?**
6. **CRITICAL THINKING Why is it an advantage for plant growth and development to be regulated by stimuli received from the environment?**

SECTION 33-2

ANSWERS TO SECTION 33-2 REVIEW

1. A tropism is determined by the direction of the environmental stimulus. A nastic movement is not determined by the direction of the environmental stimulus.
2. Phototropism is a response to unidirectional light. Leaves and stems are usually positively phototropic. Gravitropism is a response to gravity. Roots are usually positively gravitropic. Thigmotropism is a response to touch. Twining vine stems are usually positively thigmotropic.
3. Auxin moves into the cells of the shoot tip that are not directed toward the light. Auxin causes the cells to elongate, which causes the shoot to lean toward the light.
4. Nastic movements allow carnivorous plants, such as Venus' flytraps, to trap insects. They also enable sensitive plant leaves to discourage insect predators.
5. Nastic movements can occur quickly. Potassium ions move out of certain cells, causing osmotic movement of water, which makes the cells shrink or swell. Shrinking or swelling of cells causes leaf movements.
6. It is an advantage for plant growth to be regulated by environmental stimuli because plants are unable to move to a new environment when conditions become unfavorable.

SECTION 33-3

SECTION OVERVIEW

Stability and Homeostasis
Section 33-3 describes how the changing seasons affect plant growth and development. Seasons are distinguished by changes in the lengths of day and night. These changes influence a variety of plant responses.

▶ ENGAGE STUDENTS

Ask students to describe the four seasons. Show students photographs of the seasonal changes that plants exhibit. Some examples include spring-flowering magnolias and tulips, summer-flowering evening primrose and spinach, fall-flowering poinsettia and Christmas cactus, fall foliage in woody plants and deciduous plants dropping their leaves. Ask students what factors might cause such plant responses. (Length of day and night, rainfall, and temperature affect flowering, leaf growth and death, dormancy, storage of carbohydrates, and seed germination.)

GIFTED ACTIVITY

Have students investigate the hypothetical flowering hormone that has been named *florigen.* (Although the flowering hormone has not been isolated, a message is apparently transported from the leaves to the buds. Most plant physiologists think that the messenger is a hormone. For example, if a plant that has been induced to flower by the appropriate photoperiod is grafted to another plant not exposed to the appropriate photoperiod, the result is that both plants will flower. Some scientists think that flowering may be due to a mixture of hormones.)

SECTION 33-3

OBJECTIVES

▲ Define *critical night length,* and explain its role in flowering.

● Describe vernalization, and give an example of a human application of this phenomenon.

■ Describe what causes the spectacular fall leaf colors.

◆ What is the role of phytochrome in plant responses?

SEASONAL RESPONSES

In nontropical areas, plant responses are strongly influenced by seasonal changes. For example, many trees shed their leaves in the fall, and most plants flower only at certain times of the year. How do plants sense seasonal changes? Although temperature is involved in some cases, plants mark the seasons primarily by sensing changes in night length.

PHOTOPERIODISM

A plant's response to changes in the length of days and nights is called **photoperiodism.** Photoperiodism affects many plant processes, including the formation of storage organs and bud dormancy. However, the most-studied photoperiodic process is flowering. Some plants require a particular night length to flower. In other species, a particular night length merely makes the plant flower sooner than it otherwise would.

Critical Night Length

Researchers have found that the important factor in flowering is the amount of darkness, or night length, that a plant receives. Each plant species has its own specific requirement for darkness, called the **critical night length.** Although scientists now know that night length, and not day length, regulates flowering, the terms short-day plant, and long-day plant are still used. A **short-day plant** flowers when the days are short and the nights are longer than a certain length. Conversely, a **long-day plant** flowers when the days are long and the nights are shorter than a certain length.

Responding to Day Length and Night Length

Plants can be divided into three groups, depending on their response to the photoperiod, which acts as a season indicator. One group, called **day-neutral plants** (DNPs), are not affected by day length. DNPs for flowering include tomatoes, dandelions, roses,

TABLE 33-3 *Flowering Photoperiodism*

Type of plant	Conditions needed for flowering	Seasons of flowering
Day-neutral plant (DNP)	not affected by day-night cycle	spring to fall
Short-day plant (SDP)	short days (long nights)	spring, fall
Long-day plant (LDP)	long days (short nights)	summer

corn, cotton, and beans. DNPs and the other two groups of plants are summarized in Table 33-3.

Short-day plants (SDPs) flower in the spring or fall, when the day length is short. For example, ragweed flowers when the days are shorter than 14 hours, and poinsettias flower when the days are shorter than 12 hours. Chrysanthemums, goldenrods, and soybeans are SDPs for flowering.

Long-day plants (LDPs) flower when days are long, usually in summer. For example, wheat flowers only when the days are longer than about 10 hours. Radishes, asters, petunias, and beets are LDPs for flowering.

Adjusting the Flowering Cycles of Plants

Examine Figure 33-10, which compares the flowering of an SDP with that of an LDP. With an 8-hour night, the LDP flowers but the SDP does not. With a 16-hour night, the SDP flowers and the LDP does not. However, if a 16-hour night is interrupted in the middle by one hour of light, the LDP flowers and the SDP does not. This response shows that the length of uninterrupted darkness is the important factor. Even though there is a daily total of 15 hours of darkness, the SDP does not flower because of that one hour of light. Flower growers who want to obtain winter flowering of LDPs simply expose them to a low level of incandescent light in the middle of the night. Summer flowering of SDPs is obtained by covering the plants in the late afternoon with an opaque cloth so that the SDPs receive enough darkness.

Regulation by Phytochrome

Plants monitor changes in day length with a bluish, light-sensitive pigment called **phytochrome** (FIET-uh-KROHM). Phytochrome exists in two forms, based on the wavelength of the light that it absorbs. The form that absorbs red rays is called P_r, and the form that absorbs far red (infrared) rays is called P_{fr}. Daylight converts P_r to P_{fr}. In the dark, P_{fr} is converted to P_r. Besides photoperiodism, phytochrome is involved in bud dormancy and seed germination.

internetconnect

SCILINKS NSTA
TOPIC: Photoperiodism
GO TO: www.scilinks.org
KEYWORD: HM657

FIGURE 33-10

The figure compares a short-day plant (SDP) with a long-day plant (LDP) in three variations of night length. The SDP has a critical night length of 14 hours. The LDP has a critical night length of 10 hours.

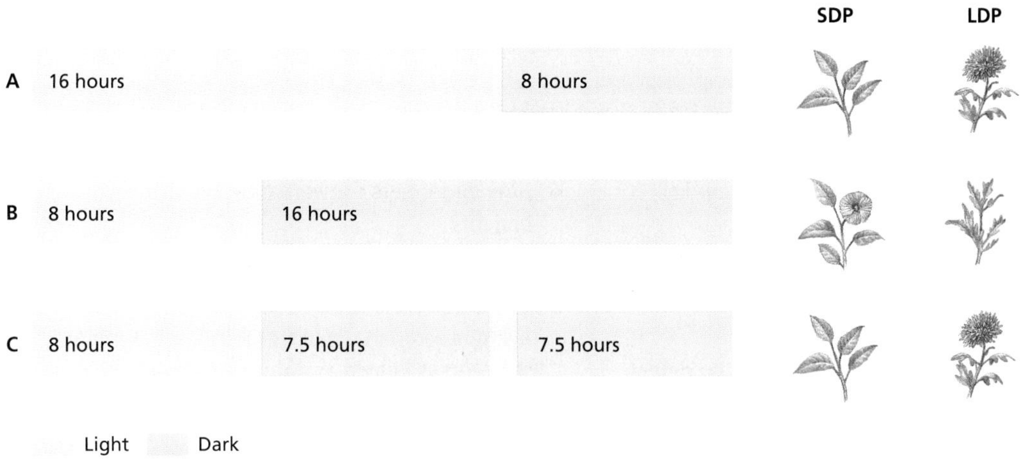

SECTION 33-3

TEACHING STRATEGY

Photoperiod Requirements

Provide students with the following information, and ask them to describe the growing conditions that each plant would need in order to flower at the desired time. Tell students that most plants require two weeks or more to develop flowers after they have been induced. Flowers desired include chrysanthemums, SDPs with critical night length of 16 hours, for Mother's Day; and poinsettias, SDPs with critical night length of 12 hours, for Christmas. (Chrysanthemums: grow during the winter with light flashes in the middle of the night until several weeks before Mother's Day. This will delay blooming until the appropriate time. Poinsettias: grow during the fall, and shade the plants in the late afternoon beginning several weeks before Christmas.) Have students draw diagrams similar to those shown in Figure 33-10.

TERMINOLOGY NOTE

Have students examine Figure 33-10. Remind students that a plant's photoperiod is traditionally referred to in terms of day length. But in fact, it is night length that is critical. A mere several minutes of light can prevent flowering as shown in C, Figure 33-10. Explain that the terminology makes little difference because when days are long in summer, nights are short. Likewise, when days are short in winter, nights are long. Therefore, a long-day plant is by default a short-night plant.

QUICK FACT

☑ Many short-day plants are extremely sensitive to light exposure during the night. In some plants, even the headlights from a vehicle or light from a flashlight can prevent flowering.

SECTION 33-3

ANSWERS TO SECTION 33-3 REVIEW

1. Long-day plants flower when they are exposed to short nights. Short-day plants flower when they are exposed to long nights.
2. Growers cover the plants each afternoon with opaque cloth so that the plant receives its critical night length. The cloth is removed each morning.
3. Phytochrome exists in two interconvertible forms. In light, it is converted to one form. In the dark, it converts to the other form. This adaptation allows the plant to respond to the length of darkness.
4. Vernalization is a low-temperature stimulation of flowering. Farmers plant winter wheat in the fall so that it gets vernalized by the cold weather. The wheat flowers in the spring about two months earlier than if it were sown in the spring. This allows farmers to avoid common summer droughts.
5. As nights get longer in the fall, leaves stops synthesizing chlorophyll and the chlorophyll degrades. The loss of chlorophyll reveals yellow or orange carotenoids and xanthophylls that had been masked by the chlorophyll. In some species, cool, sunny weather also stimulates synthesis of red leaf pigments called anthocyanins.
6. *Nyctoperiodism* is a more accurate term than *photoperiodism. Nycto-* means night. *Photo-* means light. Plants respond to the length of the daily dark period, not the daily light period.

VERNALIZATION

Vernalization is a low-temperature stimulation of flowering. Vernalization is important for fall-sown grain crops, such as winter wheat, barley, and rye. For example, wheat seeds are sown in the fall and survive the winter as small seedlings. Exposure to cold winter temperatures causes the plants to flower in early spring, and an early crop is produced. If the same wheat is sown in the spring, it will take about two months longer to produce a crop. Thus, cold temperatures are not absolutely required for most cultivars, but they do quicken flowering. Farmers often use vernalization to grow and harvest their crops before a summer drought sets in.

A **biennial** plant is a plant that lives for two years, usually producing flowers and seeds during the second year. Biennial plants, such as carrots, beets, celery, and foxglove, survive their first winter as short plants. In the spring, their flowering stem elongates rapidly, a process called **bolting.** Most biennials must receive vernalization before they flower during the second year. They then die after flowering. One botanist kept a biennial beet alive for over three years by protecting it from exposure to cold temperatures. Treating a biennial with gibberellin is sometimes a substitute for cold temperatures and can stimulate the plant to flower.

FIGURE 33-11
The colors of the carotenoids are visible in these autumn leaves, which have lost most of their chlorophyll.

Fall Colors

Some tree leaves are noted for their spectacular fall color. The changing **fall colors** are caused mainly by a photoperiodic response but also by a temperature response. As nights become longer in the fall, leaves stop producing chlorophyll. As the chlorophyll degrades, it is not replaced. Other leaf pigments, the carotenoids (kuh-RAHT-uhn-OYDZ), become visible as the green chlorophyll degrades. Carotenoids include the orange carotenes and the yellow xanthophylls (ZAN-thuh-filz). The carotenoids were always in the leaf; they were just hidden by the more abundant chlorophyll. Another group of pigments found in leaves, the anthocyanins (AN-thoh-SIE-uh-ninz), are produced in cool, sunny weather. Anthocyanins produce beautiful red and purplish-red colors.

SECTION 33-3 REVIEW

1. **Distinguish between short-day and long-day plants.**
2. **How do flower growers get short-day plants, such as chrysanthemums, to flower at any time of year?**
3. **How does phytochrome enable plants to detect changes in seasons?**
4. **How do farmers use vernalization to their benefit?**
5. **Explain why tree leaves change color in the fall.**
6. **CRITICAL THINKING Why would the term *nyctoperiodism* be a better term for describing the process known as photoperiodism?**

CHAPTER 33 REVIEW

SUMMARY/VOCABULARY

33-1

- Small concentrations of plant hormones are formed in many plant parts and transported to other locations in the plant, where they regulate almost all aspects of growth and development.
- Hormones are natural chemicals. Many hormones can be synthesized in the laboratory.
- There are five major groups of plant hormones: auxins, gibberellins, ethylene, cytokinins, and abscisic acid.
- Synthetic auxins are used for many purposes, including to promote rooting of cuttings, to kill weeds, to prevent bud sprouting, and to stimulate or prevent fruit drop.
- Apical dominance is the inhibition of lateral bud growth by auxin produced by the terminal bud.
- Gibberellins are used to increase the size of seedless grapes, to stimulate seed germination, and to brew beer.
- Growth retardants interfere with gibberellins and are widely used to reduce plant height.
- Ethylene, the only gaseous hormone, promotes abscission, fruit ripening, and pineapple flowering.
- Cytokinins stimulate cell division and growth of lateral buds.
- Abscisic acid promotes dormancy and stomatal closing in response to water shortage.

Vocabulary

2,4-D (649)
abscisic acid (ABA) (651)
abscission (650)
Agent Orange (649)
apical dominance (649)
auxin (647)
cytokinin (651)
ethephon (650)
ethylene(650)
gibberellin (GA) (650)
growth regulator (647)
growth retardant (651)
hormone (647)
indoleacetic acid (IAA) (647)
naphthaleneacetic acid (NAA) (648)

33-2

- Tropisms are plant-growth movements in which the direction of growth is determined by the direction of the environmental stimulus.
- Most stems and leaves are positively phototropic, but some vine stems are positively thigmotropic or negatively phototropic, which allows them to climb walls.
- Roots are usually positively gravitropic, and shoots are usually negatively gravitropic.
- Nastic plant movements result from an environmental stimulus and are independent of the direction of the stimuli.

Vocabulary

chemotropism (654)
gravitropism (654)
nastic movement (654)
nyctinastic movement (655)
phototropism (652)
solar tracking (653)
thigmonastic movement (654)
thigmotropism (653)
tropism (652)

33-3

- Photoperiodism is the triggering of a plant response, such as flowering or dormancy, by relative length of light and darkness.
- Plants fit in one of three photoperiodic classes for flowering: day-neutral, short-day, and long-day.
- Plants sense night and day lengths using the pigment phytochrome.
- Vernalization is the promotion of flowering by cold temperatures.
- Fall colors in tree leaves are caused by chlorophyll degradation, which reveals yellow pigments that have always been present, and by the synthesis of red pigments.

Vocabulary

biennial (658)
bolting (658)
critical night length (656)
day-neutral plant (DNP) (656)
fall color (658)
long-day plant (LDP) (657)
photoperiodism (656)
phytochrome (657)
short-day plant (SDP) (657)
vernalization (658)

CHAPTER 33 REVIEW
ANSWERS

REVIEW

1. Plant movements are caused by various stimuli, including touch (thigmotropism), gravity (gravitropism), and chemicals (chemotropism).
2. Hormones are natural chemical messengers. Growth regulators include hormones and other synthetic compounds. Both hormones and growth regulators interact with target tissues, causing a plant response.
3. Vernalization and photoperiodism are seasonal responses that can lead to flowering or to abscission, which is the detachment of leaves, flowers, or fruits from a tree.
4. Positive phototropism and negative phototropism are plant movements that are dependent on the direction of a stimulus. Nastic movements are also plant movements, but they are independent of the direction of a stimulus.
5. Short-day plants and long-day plants have different responses to a critical night length. Short-day plants flower when the night length is long. Long-day plants flower when the night length is short.

6. b **7.** a
8. d **9.** b
10. d **11.** b
12. b **13.** a
14. b **15.** c

16. Auxins produced by the terminal bud move down the stem and inhibit the growth of lateral buds behind the shoot tip.
17. A biennial plant is a plant that lives for two years, usually producing flowers and seeds during the second year.
18. The critical night length is the daily uninterrupted period of darkness that causes a plant to flower. The critical night length depends on the particular species or cultivar.
19. Ethylene can be used to promote flowering in mangoes and pineapples, to ripen bananas and

honeydew melons, to cause mechanically harvested cherries to drop, and to color citrus fruits.

20. Auxins are often used as herbicides to kill broadleaf (dicot) weeds in lawns, to promote root formation on leaf or stem cuttings, to increase the size or number of tomatoes, to cause young fruits or seeds to drop from ornamental trees before they become messy, and to prevent the sprouting of stems from the bases of trees.

21. Leaf abscission allows a plant to easily discard a dead or damaged leaf. This prevents dead leaves from shading functional leaves. It may also slow the spread of diseases. Fall leaf abscission allows the plant to minimize water absorption during the winter, when water often cannot be absorbed or transported in the plant because it is frozen. Fall shedding of leaves reduces the surface area that can accumulate ice or snow. Accumulations of ice and snow often break tree limbs.

22. Hormone effects may depend on the hormone concentration, the concentration of other hormones, and the stage of plant growth.

23. Corn is a monocot, and many weeds in cornfields are dicots. Concentrations of 2,4-D that kill dicots do not damage monocots.

24. Increasing night length triggers leaves to stop making chlorophyll. As the chlorophyll degrades, yellow or orange carotenoids are revealed. Cool, sunny weather promotes the synthesis of red leaf pigments, also called the anthocyanins.

25. Negative phototropism of stem tips can allow the vine to climb higher than neighboring plants to obtain more light.

Critical Thinking

1. The bag containing the banana and pear will ripen sooner. Ripe

Review

Vocabulary

In the following sets of related words, explain the differences between the terms.

1. thigmotropism, gravitropism, chemotropism
2. hormone, plant growth regulator
3. vernalization, abscission, photoperiodism
4. positive phototropism, negative phototropism, nastic movement
5. critical night length, short-day plant, long-day plant

Multiple Choice

6. Exposure of some plants to cold promotes flowering, a process called (a) photoperiodism (b) vernalization (c) dormancy (d) thermotropism.
7. The growth of roots above the soil surface, where oxygen would be more available, is called (a) negative gravitropism (b) negative chemotropism (c) positive phototropism (d) positive chemotropism.
8. Nastic movements occur (a) without a stimulus (b) toward a stimulus (c) away from a stimulus (d) independent of the direction of the stimulus.
9. Cytokinins promote cell (a) aging (b) division (c) storage (d) transport.
10. Fall color changes in tree leaves are caused by the synthesis of (a) chlorophyll (b) carotenoids (c) xanthophylls (d) anthocyanins.
11. Seedless grapes are usually treated with (a) cytokinins (b) gibberellins (c) ethylene (d) abscisic acid.
12. A ripe avocado will cause nearby avocados to ripen through the release of (a) auxin (b) ethylene (c) abscisic acid (d) gibberellin.
13. Abscission layers are areas where (a) leaves fall off stems (b) plants sense night length (c) abscisic acid is produced (d) cuttings are produced.
14. The tropism that is a response to touch is (a) phototropism (b) thigmotropism (c) chemotropism (d) gravitropism.
15. A hormone that is a gas is (a) auxin (b) gibberellin (c) ethylene (d) abscisic acid.

Short Answer

16. Explain how shoot tips and lateral buds are influenced by apical dominance.
17. Define a biennial plant, and explain why vernalization is important in its life cycle.
18. Define critical night length, and explain how it applies to photoperiodism.
19. How might a fruit grower use ethylene on crops?
20. Describe three ways a home gardener could use auxin.
21. What is an adaptive advantage of leaf abscission?
22. Name two factors that affect the influence a hormone will have.
23. Why is 2,4-D effective as a weedkiller in cornfields?
24. Explain why tree leaves change color in autumn.
25. Look at the photograph below. Explain how negative phototropism could result in greater absorption of sunlight for photosynthesis.

CRITICAL THINKING

1. Suppose you placed a green banana in each of several plastic bags, placed a ripe pear in half of the bags, and then sealed all of the bags. Which group of bananas (with or without pears) do you think would ripen sooner? Justify your answer.
2. Suppose that a friend who lives in North Dakota gives you some seeds from a plant that you admired when you saw it growing in your friend's yard. You plant the seeds at your home in Georgia, but they fail to germinate. Based on your knowledge of seed germination, what might be preventing the germination of the seeds?
3. If abscisic acid were inexpensive, what would be some of its possible agricultural or gardening uses?
4. If a whole potato tuber is planted, only one or two buds at one end will sprout. However, if the potato is cut into pieces that each have a bud, all the buds will sprout. Explain why.
5. Potted poinsettias purchased for Christmas will often survive in people's homes for many years but will rarely bloom again. Explain why this happens.
6. The growth of most deciduous trees in the northern United States and Canada, where winters are severe, is regulated strictly by photoperiodism. That is, temperature plays no part in the regulation of their yearly growing cycle. Explain why this is significant.
7. The seasonal loss of leaves by trees and shrubs serves the adaptive advantage of conserving nutrients. What other adaptive advantages might loss of leaves provide?
8. Suppose you notice that your neighbor has a perfect lawn and that some of the trees in your yard have distorted leaves, like those in the photo shown below. What is the probable cause of the leaves' unusual appearance?

EXTENSION

1. Read the book *The Private Life of Plants: A Natural History of Plant Behaviour,* by David Attenborough (Princeton University Press). This book accompanies a six-hour television series that is available on videotape. Attenborough spent three years taking incredible time-lapse photography to show how plants respond to various stimuli.
2. Try taking time-lapse photos to capture plant movements using either a regular camera or a video camera with an automatic time-lapse feature.
3. Select several plant species, and plant six seeds of each species in separate pots of soil. Allow the seeds to sprout and grow until the plants are about 10 cm tall. Lay the pots on their sides and observe how long it takes the plants to point upward again. Do the species differ in their rate of bending?
4. Read about Charles Darwin's famous experiments on Venus' flytraps in his book *Insectivorous Plants.* If you can obtain a potted Venus' flytrap, repeat Darwin's experiments to determine how many touches are needed to make the leaf trap close.

fruits give off ethylene, which causes ripening in nearby fruit.
2. Possible answers include the following: the seeds did not receive the required cold period, they did not receive the right amount of light, they received too much water, and they got too hot.
3. Abscisic acid could be used as a herbicide on lawns because it inhibits seed germination. Because it causes dormancy, it might be used to delay fruit-tree growth in the spring to prevent frost damage. Because it closes stomates, it could be used to slow plant growth or conserve water during drought.
4. If each piece has only one bud, there will be no apical dominance and all buds will sprout.
5. Poinsettias have a critical night length of about 12 hours. Most people use electric lights at night and sleep only 8 hours. The night length is less than 12 hours, so the poinsettias do not flower.
6. If a tree only used temperature to determine when spring began, it might start growing in the middle of winter. Its sprouting buds would then be killed when the temperatures dropped below freezing again.
7. The loss of leaves conserves water, which is difficult to absorb or transport in winter because it is often frozen.
8. The trees were probably damaged by an auxin that was used on the neighbor's lawn as a herbicide.

EXTENSION

1. The book and videotape offer a review of plants through time-lapse and slow-motion photography.
2. Consult a local camera shop for tips on how to take time-lapse photographs.
3. Phototropism rates differ among plant species.
4. Students should find that two touches are needed to make the flytrap leaf close.

CHAPTER 33 INVESTIGATION

TIME REQUIRED

One 50-minute class period

SAFETY PRECAUTIONS

Remind students to wear safety goggles, a lab apron, and disposable gloves at all times during the lab. Caution students to keep the seeds away from their skin because the seeds may have been treated with a fungicide, which can cause skin irritation.

Caution students about gibberellic acid solution, which can cause eye, nose, and throat irritation. Ingestion of 100 mL of gibberellic acid solution can be fatal because it is made with isopropyl alcohol. Gibberellic acid solution is flammable and should be stored in a cool, dry place away from strong oxidizing materials and fire hazards.

If gibberellic acid solution gets on skin or clothing, rinse it off with water. If gibberellic acid solution gets in the eyes, immediately flush it out with water for 15 minutes and get medical attention. If a person swallows gibberellic acid and remains conscious, give him or her one or two glasses of water to drink, induce vomiting, and call a physician.

QUICK REFERENCES

Lab Preparation notes are found on page 645B.

Holt BioSources provides a Teaching Resources CD-ROM that contains "Using Gowin's Vee in the Lab" and "Scoring Rubrics."

PROCEDURAL TIPS

1. Each pot should be labeled as either the experimental setup or the control setup. Students should write their name on the label.
2. Make sure that students use different swabs for the two solutions.

CHAPTER 33 INVESTIGATION

Testing the Effect of Gibberellin on Plant Growth

OBJECTIVE

- Study the effect of gibberellin on the growth of bean seedlings.

PROCESS SKILLS

- observing
- hypothesizing
- measuring
- comparing and contrasting
- collecting data
- organizing data

MATERIALS

- safety goggles
- lab apron
- disposable gloves
- 2 flowerpots, 20 cm or larger
- marking pen
- labeling tape
- potting-soil mixture
- 10 bean seedlings, at least 3 cm tall
- centimeter ruler
- cotton swabs
- gibberellic acid solution (contains gibberellin)
- water

Background

1. Gibberellins are found in all parts of plants, but are most concentrated in immature seeds. Gibberellins are only one group of many hormones that affect plant functions.
2. What is the major effect of gibberellins on plant tissues?

PART A Setting Up

1. Discuss the objective of this investigation with your lab partners. You will apply gibberellic acid solution to the leaves and shoot tips of one group of bean seedlings, and you will apply only water to an identical group of bean seedlings. Develop a hypothesis about the effect of gibberellic acid solution on bean seedlings. Record your hypothesis in your lab report.
2. CAUTION **Wear safety goggles, a lab apron, and disposable gloves at all times during this investigation.**
3. Obtain two flowerpots, and label them A and B. Also write your initials on the flowerpots.
4. Fill each flowerpot with potting-soil mixture. The pots should be filled to the same level.
5. CAUTION **Keep the seeds, which may have been treated with a fungicide, away from your skin.** Plant five bean seedlings in each flowerpot, spacing them evenly around the edge. Each seedling's roots should be at least 1 cm below the soil surface.
6. Measure the height of each plant in pot A to the nearest millimeter. Measure from the soil surface to the shoot tip. In your lab report, record your measurements in a data table similar to the table on the facing page.
7. Average the height of the five seedlings. In your lab report, record your average measurement under "Initial height" in a data table like the one shown.
8. Repeat steps 6 and 7 using the plants in pot B.
9. Wet a cotton swab with water. Rub the water over the leaves and shoot tips of each seedling in pot A.
10. CAUTION **Gibberellic acid is an irritant. If you get gibberellic acid solution on your skin or clothing, rinse it off at the sink while calling to your teacher. If you get gibberellic acid solution in your eyes, flush it out immediately at an eyewash station while calling to your teacher.** Wet a clean cotton swab with gibberellic acid solution. Apply the solution to the seedlings in pot B by rubbing the cotton swab on the leaves and shoot tips.
11. Place both flowerpots in a window or other area where they are likely to get light. The seedlings should receive the same amount of water, sunlight, darkness, and air circulation.
12. Clean up your materials and wash your hands before leaving the lab.

PART B Observing the Results

13. Each day over the next 10 days, measure the height of each seedling. Average the height of the seedlings in each flowerpot, and record the average height in your data table.

14. Calculate the average change in height for each day. For example, the average change in height during the second day is equal to the average height on day 2 minus the average height on day 1.

15. If you are unable to make measurements over the weekend, simply record in your data table the number of days that elapsed since the last measurement was taken. To calculate the average change in height for the days when measurements were not taken, you will need to divide the amount of change by the number of days that have elapsed since your last measurement.

16. At the end of 10 days, clean up your materials according to your teacher's instructions.

Analysis and Conclusions

1. Use the data you have recorded to calculate the percentage increase—between your first measurement and your last measurement—in the average height of the two groups of plants.

2. Use your data to make a line graph of height as a function of time. Label the *x*-axis of your graph *Time,* and measure time in days. Label the *y*-axis *Height,* and measure height in millimeters.

3. Do your data support your hypothesis? Explain why or why not.

4. What differences, other than height, did you detect in the bean seedlings?

5. What were some of the possible sources of error in this experiment?

6. How might a knowledge of plant regulatory chemicals be of use to farmers and other plant growers?

Further Inquiry

Design an experiment to determine whether gibberellic acid accelerates the germination of bean seeds.

FLOWERPOT A: *EFFECTS OF WATER ON BEAN SEEDLINGS*

Day	Height (mm)	Total change in height (mm)
Initial		
1		
2		
3		
4		
5		
6		
7		
8		
9		
10		

FLOWERPOT B: *EFFECTS OF GIBBERELLIC ACID ON BEAN SEEDLINGS*

Day	Height (mm)	Total change in height (mm)
Initial		
1		
2		
3		
4		
5		
6		
7		
8		
9		
10		

ANSWER TO BACKGROUND

2. Gibberellins promote plant growth.

ANSWERS TO ANALYSIS AND CONCLUSIONS

1. Students should compare their two averages.
2. The line graphs should have a slow-rising slope.
3. Students should determine whether their data support their hypotheses.
4. Students' answers may include differences in the numbers of leaves or buds, differences in the size and thickness of the leaves and stems, and differences in color.
5. Sources of error include individual genetic differences among seeds; differences in the environments of the two pots; disease or insect predation; accidents or damage to individual plants; errors in measuring, labeling, or calculating; and incorrect application of the solutions.
6. Farmers and plant growers can use hormones to stimulate or inhibit the growth and development of plants. Hormones can help with the timing of harvesting commercial crops.

Unit 8

INVERTEBRATES

CHAPTERS

“It has taken biologists some 230 years to identify and describe three quarters of a million insects; if there are indeed at least thirty million, as Erwin (Terry Erwin, the Smithsonian Institute) estimates, then, working as they have in the past, insect taxonomists have ten thousand years of employment ahead of them.”

internet connect

sciLINKS NSTA

National Science Teachers Association *sci*LINKS Internet resources are located throughout this unit.

This Sally lightfoot crab lives on bare volcanic rock on the Galápagos Islands.

This sea star clearly demonstrates pentaradial symmetry.

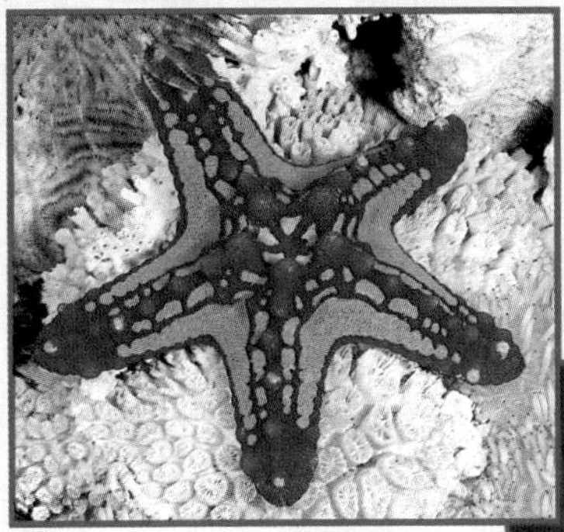

The leaf-footed bug, Diactor billineatus, is a colorful member of the diverse world of insects.

The blood fluke, Schistosoma mansoni, infects 200 million people worldwide. To feed, the fluke attaches its suckers to a host's blood vessels.

Leaf-cutter ants

unit 8

INVERTEBRATES

CHAPTERS

CONNECTING TO THE STANDARDS

The following chart shows the correlation of Unit 8 with the *National Science Education Standards* (grades 9–12). The items in each category are addressed in this unit and should be used to guide instruction. Annotated descriptions of the **Life Science Standards** are found below the chart. Consult the *National Science Education Standards* for detailed descriptions of each category.

UNIFYING CONCEPTS

- Systems, order, and organization
- Evolution and equilibrium
- Form and function

SCIENCE AS INQUIRY

- Abilities necessary to do scientific inquiry
- Understanding about scientific inquiry

SCIENCE AND TECHNOLOGY

- Abilities of technological design
- Understanding about science and technology

PHYSICAL SCIENCE

- Motions and forces
- Conservation of energy and increase in disorder

LIFE SCIENCE

EARTH AND SPACE SCIENCE

- Energy in the Earth system
- Origin and evolution of the Earth system

SCIENCE IN PERSONAL AND SOCIAL PERSPECTIVES

- Natural resources
- Natural/human-induced hazards

HISTORY AND NATURE OF SCIENCE

- Nature of scientific knowledge
- Historical perspectives

ANNOTATED DESCRIPTIONS OF THE CORRELATED LIFE SCIENCE STANDARDS

Although all eight categories of the *National Science Education Standards* are important, the following descriptions summarize the **Life Science Standards** that specifically relate to Unit 8.

THE CELL

- Cell functions are regulated.
- Cells have particular structures that underlie their functions.
- Cells can differentiate. Complex multicellular organisms are formed from highly organized arrangements of differentiated cells.

BIOLOGICAL EVOLUTION

- The great diversity of organisms is the result of more than 3.5 billion years of evolution.
- Biological classifications are based on how organisms are related.

INTERDEPENDENCE OF ORGANISMS

- Organisms both cooperate and compete in ecosystems.

MATTER, ENERGY, AND ORGANIZATION IN LIVING SYSTEMS

- The complexity and organization within an organism result from the organism's need to obtain, transform, transport, release, and eliminate the matter and energy required to sustain the organism.
- The distribution and abundance of organisms and populations in ecosystems are limited by the availability of matter and energy and by the ability of the ecosystem to recycle materials.

THE BEHAVIOR OF ORGANISMS

- Multicellular animals have nervous systems that generate behavior.
- Organisms have behavioral responses to internal changes and external stimuli.

TRENDS IN BIOLOGY

The Origin of Animals

The Cambrian explosion, during which nearly all phyla of animals burst onto the evolutionary scene within just a few million years, has troubled biologists since Darwin's day because of its suddenness. Darwin recognized that the apparent lack of ancestral forms posed a serious problem for his theory, but he could only speculate about what happened before the Cambrian explosion. Today we know much more about life prior to the Cambrian period than Darwin did, including the fact that animals appeared before the Cambrian explosion. However, many fundamental questions about the origin of animals remain unanswered. For instance, how old is the animal kingdom, and what role, if any, did the unusual creatures known as Ediacarans play in the evolution of animals?

Animals first appeared before the beginning of the Cambrian period. Late Precambrian rocks contain ample evidence of animals—although what sort of animals left these fossils is still not known. Fossilized tracks and burrows show that bilaterally symmetrical animals were crawling over and digging into the ocean floor more than 550 million years ago. Since these tracks suggest that the animals went a specific direction rather than wandering aimlessly, their makers probably had sense organs to register their position and a nervous system to guide their movements. But no fossils of the animals themselves have been found. Also present are what scientists call "small, shelly fossils," an assortment of tubes and other hard parts that cannot be attributed to any known animal.

WHERE DO THE EDIACARANS FIT IN?

The Precambrian fossil record also contains the impressions of a group of strange animals called Ediacarans. Most Ediacarans bear little resemblance to any other modern or extinct animal (see photograph at right). Their thin, flat bodies show no evidence of a gut, a mouth, an anus, sense organs, or even an anterior or posterior end. Some resemble ferns, while others have a quilted appearance. A few paleontologists have even questioned whether the Ediacarans were animals at all. They propose that they may have been lichens or members of a previously unknown kingdom of life. The majority of paleontologists won't go that far, but they acknowledge that most Ediacarans don't have obvious connections to modern groups of animals. Only the modern cnidarians known as sea pens seem to be descended from Ediacarans.

The Ediacaran Dickinsonia *grew up to a meter in length.*

Adding the Ediacarans to the animal lineage takes their origins back to a little over 600 million years ago. What happened before that? The fossil record is silent on this issue, so scientists have turned to molecular data, which have proven valuable in determining times of divergence in other cases when fossil evidence is absent. In a recent study, a protein molecular clock was used to estimate that animals originated about 1 billion years ago. Although all molecular clocks are subject to some uncertainty, this date is about the same as estimates based on the fossil record.

Additional Information

Levinton, Jeffrey. "The Big Bang of Animal Evolution." *Scientific American,* November 1992, 84–91.

Monastersky, Richard. "The Ediacaran Enigma." *Science News,* July 8, 1995, 28–30.

Raff, Rudolf. *The Shape of Life.* Chicago: University of Chicago Press, 1996.

Wright, Karen. "When Life Was Odd." *Discover,* March 1997, 52–61.

CHAPTER 34
INTRODUCTION TO ANIMALS

	TOPICS	TEACHING RESOURCES	LABS, CLASSWORK, AND HOMEWORK
BLOCK 1 45 minutes	**Introducing the Chapter, p. 666**	ATE Focus Concept, p. 666 ATE Assessing Prior Knowledge, p. 666 ATE Understanding the Visual, p. 666	■ Supplemental Reading Guide, *Journey to the Ants: A Story of Scientific Exploration*
	34-1 The Nature of Animals, p. 667 Characteristics, p. 667 Origin and Classification, p. 669	ATE Section Overview, p. 667 ATE Critical Thinking, p. 667 134, 137A	■ Occupational Applications Worksheets—Veterinary Technician ★ Study Guide, Section 34-1 PE Section Review, p. 669
BLOCK 2 45 minutes	**34-2 Animal Bodies, p. 670** Body Structure, p. 670 Animal Diversity, p. 672	ATE Section Overview, p. 670 ATE Visual Link Figure 34-4, p. 671 Figure 34-5, p. 672 134–135, 146	**Lab A21,** Recognizing Patterns of Symmetry ★ Study Guide, Section 34-2 PE Section Review, p. 673
BLOCKS 3 & 4 90 minutes	**34-3 Comparison of Invertebrates and Vertebrates, p. 674** Invertebrate Characteristics, p. 674 Vertebrate Characteristics, p. 677	ATE Section Overview, p. 674 ATE Critical Thinking, pp. 676, 678 ATE Visual Link Figure 34-9, p. 676 Figure 34-11, p. 677 134, 137A, 167	PE **Quick Lab,** p. 678, Identifying Animal Characteristics **Lab A23,** Observing Some Major Animal Groups ★ Study Guide, Section 34-3 PE Section Review, p. 679
BLOCKS 5 & 6 90 minutes	**34-4 Fertilization and Development, p. 682** Fertilization and Early Development, p. 682 Gastrulation, p. 683 Patterns of Development, p. 684	ATE Section Overview, p. 682 ATE Critical Thinking, p. 684 ATE Visual Link Figure 34-16, p. 685 143	PE **Chapter 34 Investigation,** p. 690 **Lab A22,** Comparing Animal Eggs ★ Study Guide, Section 34-4 PE Section Review, p. 686
BLOCKS 7 & 8 90 minutes	**REVIEW and ASSESSMENT** PE Chapter 34 Review, pp. 688–689	Audio CD Program	**CHAPTER TESTING** ■ Chapter Test (blackline master) ▲ Test Generator ■ Assessment Item Listing ■ Scoring Rubrics and Classroom Management Checklists

MODERN BIOLOGY

HOLT BIOSOURCES

One-Stop Planner CD-ROM
Shows these resources within a customizable daily lesson plan:

- ■ Holt BioSources Teaching Resources
- ■ Active Reading Guide
- ★ Study Guide
- ▲ Test Generator

KEY

- **PE** Pupil's Edition
- **ATE** Teacher's Edition
- ■ Active Reading Guide
- ★ Study Guide
- Teaching Transparencies
- Laboratory Program

READING FOR CONTENT MASTERY

Preview: Have students preview the objectives, the introductory paragraph, topic sentences, and captions before reading the chapter.

ATE Active Reading Technique: Reading Organizer, p. 667
■ Active Reading Guide Worksheet 34-1
PE Summary/Vocabulary, p. 687

■ Active Reading Guide Worksheet 34-2
PE Summary/Vocabulary, p. 687

■ Active Reading Guide Worksheet 34-3
PE Summary/Vocabulary, p. 687

■ Active Reading Guide Worksheet 34-4
PE Summary/Vocabulary, p. 687

TECHNOLOGY AND INTERNET RESOURCES

CNN PRESENTS Video Segment 29 Animal Communication

Holt Biology Videodiscs Teacher's Correlation Guide, Lessons 34-1 through 34-4

Audio CD Program, Sections 34-1 through 34-4

internetconnect

SciLINKS NSTA
On-line Resources: www.scilinks.org
The following *sci*LINKS Internet resources can be found in the student text for this chapter:

Topics:
- Multicellular organisms, p. 667
- Vertebrates, p. 673
- Invertebrates, p. 675

go.hrw.com
On-line Resources: go.hrw.com
Visit the HRW Web site for a variety of resources related to this chapter. Just type in the keyword HM2 Chapter 34.

Smithsonian Institution®
Visit **www.si.edu/hrw** for additional on-line resources.

CNNfyi.com
Visit **www.cnnfyi.com** for late-breaking news and current event stories selected just for you.

LAB ACTIVITY PLANNING

CHAPTER 34 INVESTIGATION

Sheep's Heart Dissection, pp. 690–691

OVERVIEW
In this investigation, students will observe the external structures of a sheep's heart, which is similar to a human heart. Students will dissect the heart to study its internal structures.

SAFETY
Students will wear disposable, latex gloves, lab aprons, and safety goggles throughout the lab. Instruct students in the correct handling of all dissecting instruments, such as the scalpel. Remind students to wash their hands after completing the lab.

PREPARATION
1. Each lab group will need one heart, a dissection tray, a blunt probe, fine dissection scissors, a scalpel, and tweezers.
2. To dispose of the sheeps' hearts, wrap them in old newspaper and close them securely in a plastic bag. Inform the school custodian of the bag's contents, and hand it over personally for its safe disposal.
3. If dissection is not practical for your classroom, there are alternative ways of studying heart structures, including models, filmstrips, videotapes, and computer simulations.

Quick Labs

Identifying Animal Characteristics, p. 678
Students will need 20 note cards (3 × 5 in.), 5 pictures of vertebrates, and 5 pictures of invertebrates for this activity.

UNDERSTANDING THE VISUAL

Have students look closely at the photograph of the Galápagos crab. Have students name some other crabs they are familiar with. Ask them how the crab shown in the photo differs from these crabs. Ask students to list some of the anatomical differences between crabs and humans that are obvious from the photograph.

FOCUS CONCEPT

Matter, Energy, and Organization

All animals are multicellular and have cells specialized to perform particular functions needed for survival and reproduction. In most animals, cells are organized into larger units—tissues, organs, organ systems—and their functions are integrated by means of chemical and electrical signals.

ASSESSING PRIOR KNOWLEDGE

Review the following concepts.

Cellular Respiration: *Chapter 6*
Ask students to name the reactants and products of cellular respiration. Ask them why cellular respiration is necessary for animals.

Phylogenetic Trees: *Chapter 18*
What is a phylogenetic tree, and what information does it contain?

CHAPTER 34

INTRODUCTION TO ANIMALS

The diversity of animal life is staggering. Animals have adapted to Earth's lushest environments and to its harshest environments. This Sally Lightfoot crab, Grapsus grapsus, *lives on the bare volcanic rock of the geologically young Galápagos Islands.*

FOCUS CONCEPT: ***Matter, Energy, and Organization***

As you read about the characteristics of different animals, notice the relationship between structure and function in animals' bodies.

THE NATURE OF ANIMALS

If you are asked to name an animal, you might respond with the name of a familiar large-bodied ***vertebrate****—an animal with a backbone—such as a horse, a shark, or an eagle. But the kingdom Animalia is extraordinarily diverse, and most of its members are not vertebrates and do not even live on land. In Unit 8, you will read about* ***invertebrates,*** *animals without a backbone, which account for more than 95 percent of all animal species alive today. In Unit 9, you will read about the vertebrates, which are less numerous but more familiar to us; humans are vertebrates.*

SECTION 34-1

OBJECTIVES

▲ Define the terms *invertebrate* and *vertebrate.*

● Identify four important characteristics of animals.

■ List two kinds of tissues found only in animals.

◆ Explain how the first animals might have evolved from unicellular organisms.

CHARACTERISTICS

Animals are multicellular heterotrophic organisms that lack cell walls. Most members of the animal kingdom share other important characteristics, including sexual reproduction and movement.

Multicellular Organization

The bodies of animals are multicellular. Some animals contain large numbers of cells. For example, the body of an adult human contains about 50 trillion cells. Unlike the cells of unicellular organisms, the cells of multicellular organisms do not lead independent lives. Each cell depends on the presence and functioning of other cells.

In all but the simplest animal phyla, there is a division of labor among cells. **Specialization** is the adaptation of a cell for a particular function. Just as a general contractor makes use of carpenters, electricians, and plumbers to build a house, a multicellular organism makes use of specialized cells to perform particular functions, such as digesting food or reproducing. Recall from Chapter 4 that a tissue is a group of similar cells specialized for a specific task. Most animal bodies are composed of combinations of different kinds of tissues. The formation of tissue from many individual cells is made possible by **cell junctions,** connections between cells that hold the cells together as a unit. The members of most animal phyla have organs, body structures that are composed of more than one type of tissue and that are specialized for a certain function.

Without multicellularity, the enormous variety found in the animal kingdom would not exist. The size of unicellular organisms is limited. Moreover, all of their functions, such as reproduction and

internet**connect**

SCILINKS NSTA

TOPIC: Multicellular organisms
GO TO: www.scilinks.org
KEYWORD: HM667

SECTION 34-1

SECTION OVERVIEW

Matter, Energy, and Organization

Section 34-1 relates how the main activities of animals, such as movement, reproduction, and ingestion, are made possible by multicellularity.

Evolution

This section also discusses the hypothesis that the first animals may have originated from colonial flagellates.

ACTIVE READING

Technique: Reading Organizer

Have students survey the chapter noting the green headings at the beginnings of Sections 34-1, 34-2, 34-3, and 34-4. Then have students note the blue subheadings. Also have students note the boldface words and the use of illustrations. Discuss with students how the chapter is structured.

CRITICAL THINKING

Unicellular to Multicellular

Have students explain why unicellular organisms cannot be as large as multicellular organisms. (As a cell becomes larger, its surface area increases more slowly than its volume. If the cell continues to grow, eventually it reaches a point at which its surface area is not large enough to absorb enough nutrients and oxygen for all of its cytoplasm.)

INCLUSION ACTIVITY

Have students make a sketch of a number of cells in a tissue by drawing 20 to 30 circles on a piece of paper. Instruct students to draw an arrow from one cell to two other cells. Then have them draw arrows to other cells from those cells. Have students continue until all cells have arrows between them. This diagram shows how cells use chemical signals to communicate with other cells in a tissue.

RETEACHING ACTIVITY

Show students a photo of a red blood cell and a photo of a neuron. Have them point out as many differences between these two cell types as possible. Inform students that the red blood cells of mammals have no nucleus and typically live for only a few weeks, while neurons often live for the lifetime of the organism. Emphasize that both cells are descended from the same fertilized egg and have become distinct through the process of differentiation.

OVERCOMING MISCONCEPTIONS

Inform students that the process of differentiation probably does not (except in some immune system cells) involve the loss or gain of any genes. A neuron and a liver cell contain the same set of genes, but different genes are activated in the two types of cells.

digestion, must be handled within a single cell. Multicellularity and cell specialization have enabled organisms to evolve and adapt to many environments.

Heterotrophy

Plants and some unicellular organisms are autotrophic. They make food using simple molecules from their environment and an energy source, such as the sun. Animals, on the other hand, are heterotrophic. They must obtain complex organic molecules from other sources. Most animals accomplish this by ingestion. During **ingestion,** an animal takes in organic material, usually in the form of other living things. Digestion then occurs within the animal's body, and carbohydrates, lipids, amino acids, and other organic molecules are extracted from the material or cells the animal has ingested.

Sexual Reproduction and Development

Most animals can reproduce sexually, and some can also reproduce asexually. Recall from Chapter 8 that in sexual reproduction, two haploid gametes fuse. This diploid **zygote,** the first cell of a new individual, then undergoes repeated mitotic divisions. Mitotic division of a cell produces two identical offspring cells. How does an adult animal, with its many different organs, tissues, and cell types, arise from a single cell? In the process called development, the enlarging mass of dividing cells undergoes differentiation. During **differentiation** (dif-uhr-EN-shee-AY-shuhn), cells become different from each other. For example, some cells may become blood cells, while others may become bone cells. The process of differentiation is the path to cell specialization.

Movement

Although some animals, such as barnacles, spend most of their lives attached to a surface, most animals move about in their environment. The ability to move results from the interrelationship of two types of tissue found only in animals: nervous tissue and muscle tissue. Nervous tissue allows an animal to detect stimuli in its environment and within its own body. Cells of nervous tissue, called **neurons,** conduct electrical signals throughout an animal's body. Multiple neurons work together in circuits to take in information, transmit and process it, and initiate an appropriate response. Often this response involves muscle tissue, which can contract and exert a force to move specific parts of the animal's body. The bat shown in Figure 34-1 continuously processes information about its position in space and the position of its prey. It can adjust its muscular responses so rapidly that it can intercept insects in flight.

FIGURE 34-1

Capturing fast-moving prey requires exquisitely timed coordination between the nervous tissue and muscle tissue in the body of this heart-nosed bat, *Cardioderma cor.*

ORIGIN AND CLASSIFICATION

The first animals probably arose in the sea. The structural characteristics of invertebrates suggest that they were the first multicellular animals and that they evolved from protists. Because protists are both heterotrophic and eukaryotic, scientists have inferred that multicellular invertebrates may have developed from colonies of loosely connected, flagellated protists, like the one shown in Figure 34-2.

What path did cell specialization take in these early organisms? Colonial protists may have lost their flagella over the course of evolution as individual cells in the colony grew more specialized. They may have been similar to modern colonial protists that do show some degree of cell specialization, such as some species of algae. In these species, the gametes are distinct from nonreproductive cells. A similar division of labor in early colonial protists may have been the first step toward multicellularity.

Taxonomists have grouped animals into several phyla, based on their evolutionary history, which is inferred from morphology and other factors. Recall from Chapter 18 that taxonomy is an ever-changing branch of science. Therefore, it should not be surprising that the actual number and names of animal phyla continue to be debated. Many taxonomists recognize 30 or more different animal phyla, though some phyla contain a very small number of species. Eleven animal phyla will be discussed in detail in this unit and in Unit 9. Ten of these phyla include only invertebrates. The eleventh phylum, Chordata, includes all vertebrate species as well as a small number of invertebrate marine chordates. Although vertebrates are particularly conspicuous to us, they make up a small segment of kingdom Animalia.

FIGURE 34-2

The first animals may have evolved from colonial protists similar to the one shown in this drawing. Colonial organisms like these may have exhibited basic cell specialization early in evolutionary history.

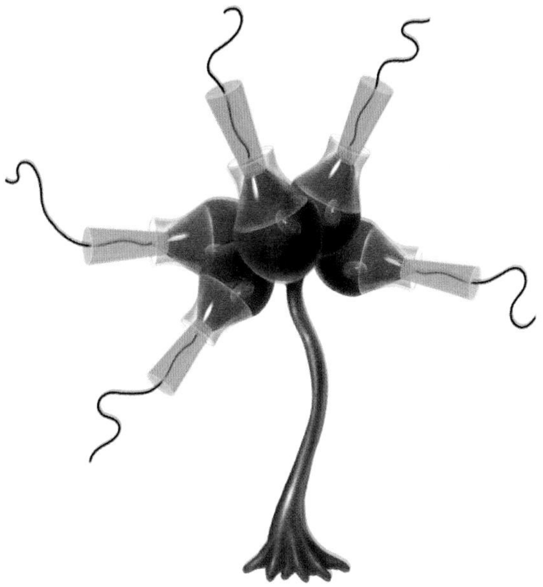

SECTION 34-1 REVIEW

1. **Define the words *invertebrate* and *vertebrate*.**
2. **What are four important characteristics common to most animals?**
3. **How is cell specialization related to multicellularity?**
4. **How are nervous tissue and muscle tissue interrelated, and why are they important to animals?**
5. **What unicellular organisms are thought to have been the immediate ancestors of the first animals?**
6. **CRITICAL THINKING Why is colonialism, the grouping together of like organisms, thought to be one of the first steps in the evolution of multicellularity?**

SECTION 34-1

ANSWERS TO SECTION 34-1 REVIEW

1. An invertebrate is an animal without a backbone; a vertebrate is an animal with a backbone.
2. Animals are multicellular, and heterotrophic, and they lack cell walls. Animals have nervous tissue and muscle tissue. In addition, most animals can reproduce sexually and are able to move.
3. Cell specialization allows different cells of a multicellular organism to assume different tasks, such as reproducing or digesting food. Multicellular organisms would not be able to grow beyond a few cells without the division of labor produced by cell specialization.
4. Nervous tissue allows an animal to detect stimuli in its environment and muscle tissue allows it to respond to those stimuli.
5. Protists may have been the immediate ancestors of the first animals.
6. Colonialism brought similar unicellular organisms together as a unit. This may have encouraged specialization of different cells for different tasks. Thus, an aggregation of specialized cells working as a unit might have been the precursor to the first multicellular organisms.

SECTION 34-2

SECTION OVERVIEW

Matter, Energy, and Organization
Section 34-2 discusses the fundamental organization of the animal body, introducing the concepts of symmetry, germ layers, and body cavities.

Evolution
Section 34-2 explains how animals can be classified based on similarities in their embryological development and morphology.

▶ ENGAGE STUDENTS

Show students a picture of a radially symmetrical animal, such as a jellyfish or sea anemone. Ask students to examine the photo and point out what they think is the head of the animal. Then explain to students that radially symmetrical animals, unlike most bilaterally symmetrical animals, do not have a definite head end. Contrast radial symmetry with bilateral symmetry by showing students a photo of a bilaterally symmetrical animal, such as a fish. Emphasize that in bilaterally symmetrical animals, sensory organs tend to be concentrated in the head.

TEACHING STRATEGY

Internal Bilateral Symmetry
Point out to students that bilateral symmetry often is evident not only in external structure but also in the arrangement of the internal organs of the body. In humans, for instance, one kidney, one lung, and one gonad are located on each side of the body. Also point out that not all internal organs show this arrangement. The liver, pancreas, and heart, for instance, are not divided evenly between the two sides of the body.

SECTION 34-2

OBJECTIVES

▲ Define the terms *dorsal, ventral, anterior,* and *posterior.*

● Describe two types of symmetry found in animals.

■ Name the trait that is strongly associated with bilateral symmetry.

◆ List two functions of the body cavity in animals.

▲ List three structural features that taxonomists use to classify animals.

● List four features found only in chordates.

ANIMAL BODIES

Taxonomic organization is based on phylogenetic relationships. Today, systematic taxonomists classify animals according to similarities in morphology and other criteria, including the similarity of embryological development and the similarity of certain macromolecules. Recall from Chapter 18 that taxonomists since Linnaeus's time have used an organism's morphology to classify it with similar organisms. Morphology, however, is not confined to external appearance. A survey of an animal's morphology also assesses the internal structure of the body and organization at the level of fundamental tissue types.

BODY STRUCTURE

The term **symmetry** refers to a consistent overall pattern of structure. The simplest animals, sponges, display no symmetry, as is shown in Figure 34-3a. Moreover, although sponges are multicellular, their cells are not organized into tissues. Animal bodies range from those that lack true tissues and an organized body shape, such as that of a sponge, to those that have very organized tissues and a consistent body shape, as is found in most other animal phyla.

Patterns of Symmetry

Some animals have a top side and a bottom side, but no front and back end or right and left end. These animals are said to display radial symmetry. In **radial symmetry,** similar parts branch out in all directions from a central line. Cnidarians, such as sea anemones, jellyfishes, and the hydra pictured in Figure 34-3b, are radially symmetrical.

Most animals have a **dorsal** (top) and **ventral** (bottom) side, an **anterior** (head) and **posterior** (tail) end, and a right and left side, as shown in the moth pictured in Figure 34-3c. Such animals have two similar halves on either side of a central plane and are said to display **bilateral symmetry.** Bilaterally symmetrical animals tend to exhibit cephalization. **Cephalization** (SEF-uh-li-ZAY-shun) is the concentration of sensory and brain structures in the anterior end of the animal; a cephalized animal has a head. As a cephalized animal swims, burrows, walks, or flies through its environment, the head precedes the rest of the body, sensing danger, prey, or a potential mate.

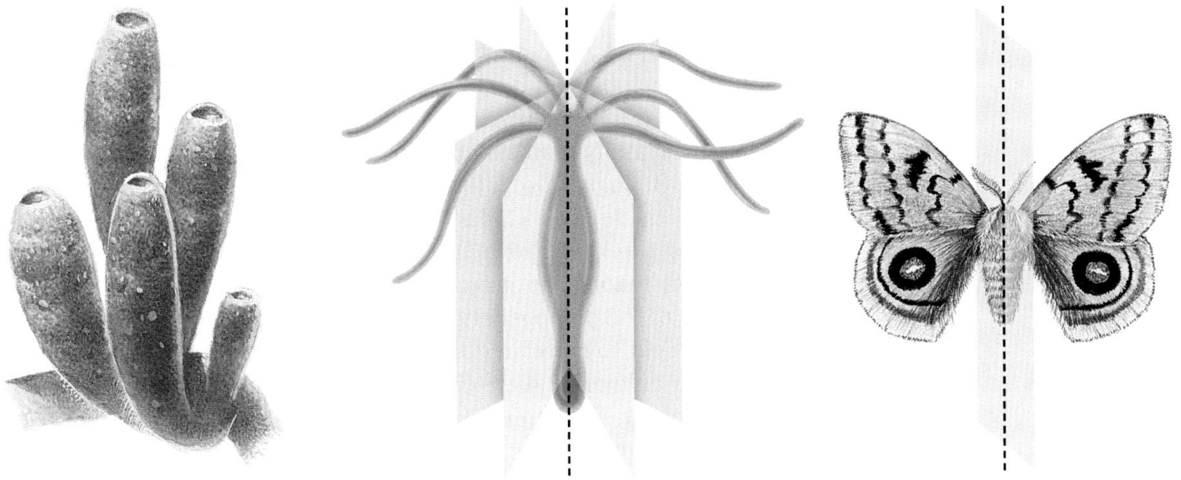

(a) NO SYMMETRY

(b) RADIAL SYMMETRY

(c) BILATERAL SYMMETRY

FIGURE 34-3

(a) The sponge lacks a consistent pattern of structure. (b) The hydra, an aquatic animal, displays radial symmetry. (c) The moth displays bilateral symmetry and cephalization.

Germ Layers

Germ layers are fundamental tissue types found in the embryos of all animals except sponges, which have no true tissues. The embryos of cnidarians and ctenophores have only two germ layers, but all other animals form three distinct germ layers very early in their development. Every body feature, organ, and tissue—from teeth to toenails—arises from one of these germ layers.

Body Cavities

Most animals have some type of body cavity, a fluid-filled space that forms between the digestive tract and the outer wall of the body during development. Some animals, such as flatworms, have three germ layers but have a solid body. These animals lack a body cavity.

In the roundworm shown in Figure 34-4, the body cavity aids in movement by providing a firm structure against which muscles can contract. The body cavity also allows some degree of movement of the exterior of the body with respect to the internal organs, resulting in more freedom of movement for the animal. Finally, the fluid in the body cavity acts as a reservoir and medium of transport for nutrients and wastes, which diffuse into and out of the animal's body cells.

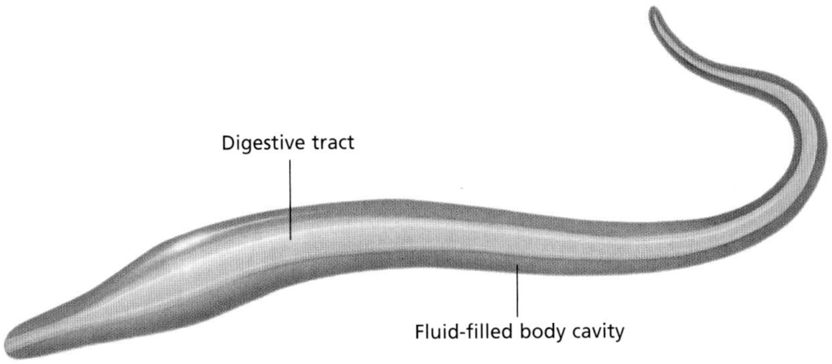

FIGURE 34-4

The body of a roundworm is held erect by its fluid-filled body cavity, which is firm but flexible, like a balloon filled with water.

DEMONSTRATION

Germ-Layer Clay Models

Have students make clay models of the animal germ layers. The models should consist of different colors of clay to show the germ-layer development characteristic of embryos of different phyla. For example, the students can make a model of a sponge using only one color of clay, but they will need two layers for cnidarians and three layers for all other phyla.

VISUAL LINK

Figure 34-4

Fill a long balloon with water and tie off the end. Use the balloon to demonstrate the function of the coelom in an animal's body. By bending and squeezing the balloon, show students how the fluid-filled body cavity resists compression and thus stiffens the animal's body, forming a hydrostatic skeleton against which the muscles can work.

MAKING CONNECTIONS

Fine Arts

Give students construction paper and ink to create ink-blot animals. Have students fold the paper in half, open the paper and apply drops of ink to the crease, then refold and press the paper together. Upon opening the paper, students should be able to observe bilaterally symmetrical "animals."

RETEACHING *ACTIVITY*

Use a mirror to show students how a mirror image illustrates bilateral symmetry. Place the mirror's edge against a student's hand or other object so both images can be seen. Point out to students that the two images are like the right and left sides of a bilaterally symmetrical animal.

VISUAL LINK

Figure 34-5

Have students examine the phylogenetic tree and then answer the following questions: Which phylum is most closely related to the chordates? (echinoderms) Which phyla are bilaterally symmetrical? (all except cnidarians, sponges, and echinoderms) Which phylum is most closely related to the first animals? (sponges)

GIFTED ACTIVITY

Have students research an invertebrate that was not discussed in the text and prepare a short, illustrated report describing its ecology and evolution.

RECENT RESEARCH

Timing the Origin of Animals

Fossil evidence for most living animal phyla first appears in rocks from the Cambrian period, but the fossil record reveals little about the evolution of animals prior to this time. Recently, a team of scientists from the University of California at San Diego has attempted to fill in this gap by studying molecular evidence from living organisms. They constructed a molecular clock, based on differences in amino acid sequences in 57 proteins common to all living things, that allows them to estimate the times at which major groups of organisms diverged. This molecular clock suggests that animals originated about 1 billion years ago and that the protostome-deuterostome split occurred about 670 million years ago. Both dates agree with extrapolations based on measurements of the rate of evolution in the fossil record.

ANIMAL DIVERSITY

Similarities in body plans and patterns of development allow biologists to classify animals and hypothesize about their evolutionary history. Biologists use information from modern species and extinct species to develop cladograms and classical phylogenetic trees, such as the one shown in Figure 34-5, that show relationships between taxonomic groups. Animal phyla shown on the same branch of the phylogenetic tree, such as roundworms and rotifers, are thought to be related to each other more closely than they are to other animals and are characterized by important similarities in morphology. Conversely, animals shown in different parts of the tree are thought to be more distantly related.

- *Multicellularity and a limited degree of cell specialization* characterize the sponges. Sponges have no organized body shape and no true tissues.
- *True tissues in two layers* are found in the cnidarians and the ctenophores.
- *True tissues in three layers and bilateral symmetry* characterize all of the other animal phyla. Among these, phylogenetic categories are based on the absence or presence and type of body cavity and on fundamental patterns of development.

FIGURE 34-5

This phylogenetic tree of animals shows the evolutionary relationships thought to exist among 11 major animal phyla.

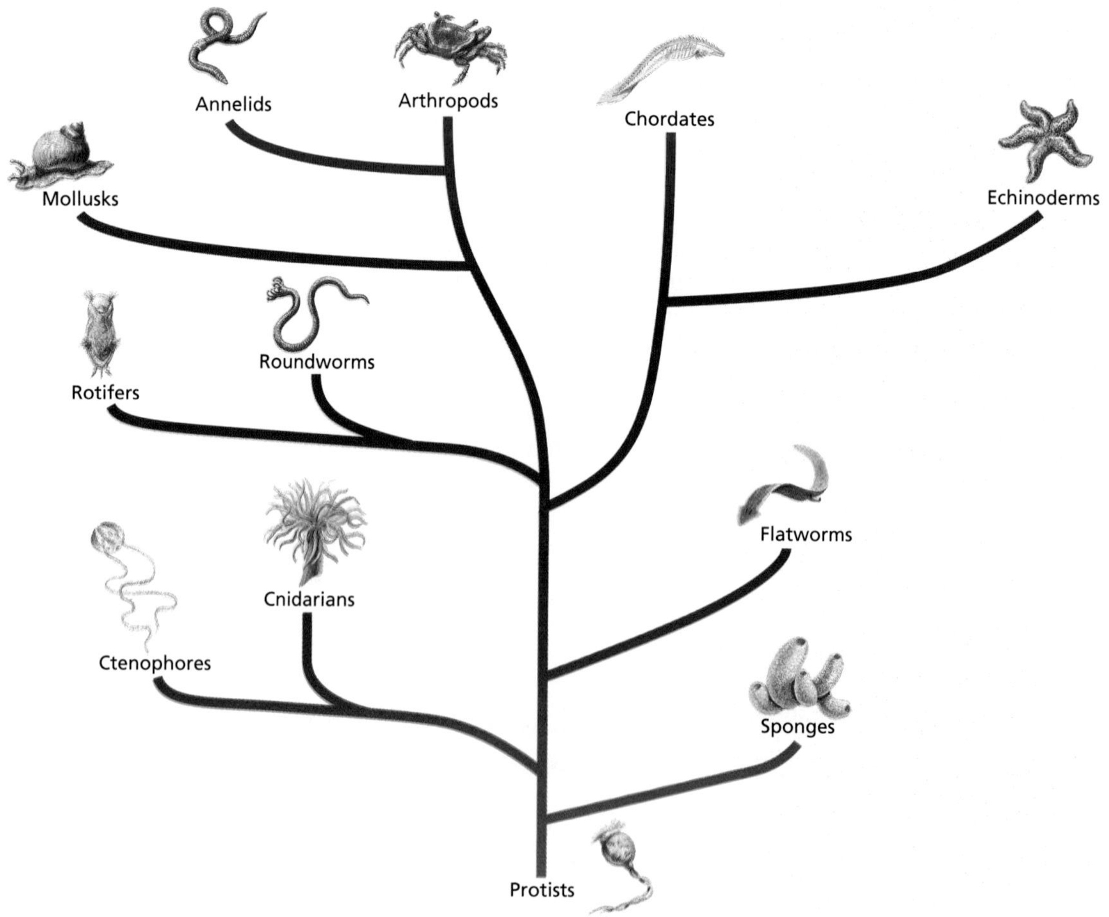

Invertebrates

The 10 invertebrate phyla pictured in Figure 34-5 are a remarkably heterogeneous group. Their body plans range from the complete absence of body symmetry and true tissues, as is found in sponges, to the bilateral symmetry and specialized body parts found in arthropods, such as the spider shown in Figure 34-6. In fact, the primary trait that links all invertebrates is the absence of a backbone. Invertebrates constitute the greatest number of animal species as well as most of the individual animals alive today.

Chordates

The eleventh phylum shown in Figure 34-5 is Chordata. The name **chordate** (KOHR-dayt) refers to the **notochord,** a firm, flexible rod of tissue located in the dorsal part of the body. At some stage of their development, all chordates have a notochord, as well as a dorsal nerve cord, pharyngeal pouches, and a postanal tail. The **dorsal nerve cord** is a hollow tube lying just above, or dorsal to, the notochord. **Pharyngeal** (fuh-RIN-jee-uhl) **pouches** are small outpockets of the anterior part of the digestive tract. The **postanal tail** consists of muscle tissue and lies behind the posterior opening of the digestive tract.

A few chordate species retain their early chordate characteristics all their lives. In most vertebrates, a subphylum of the chordates, the dorsal nerve cord develops into the brain and the spinal cord, which runs within the hollow backbone. In aquatic vertebrates, such as fishes and amphibians, the pharyngeal pouches have evolved into gills, which are used for breathing. In most vertebrates, the notochord appears only in the embryonic stage. It is replaced by the backbone early in development.

Vertebrates

Although vertebrates constitute only one subphylum of the phylogenetic tree of all animals, they merit extensive discussion from a human perspective. Humans are vertebrates, and the ecology of humans includes extensive interaction with other vertebrate species. For example, fishes, birds, and many species of mammals are primary food sources for humans.

FIGURE 34-6

The palm spider, *Nephila* sp., is an arthropod, with a segmented body and body parts specialized for trapping, killing, and eating its prey.

SECTION 34-2 REVIEW

1. **Define the terms *anterior* and *posterior.* What type of body symmetry includes both an anterior end and a posterior end?**
2. **What common feature is not found in radially symmetric animals?**
3. **What are two functions of the body cavity?**
4. **In determining an animal's phylogenetic history, what are three body features that taxonomists consider?**
5. **What happens to the notochord and dorsal nerve cord as most vertebrates develop?**
6. **CRITICAL THINKING How does having a head help an animal?**

TEACHING STRATEGY
Chordate Characteristics in Humans

Emphasize to students that while all chordates have a dorsal nerve cord, a notochord, pharyngeal pouches, and a postanal tail at some time during their life, all four features may not be present in the adult animal. In humans, pharyngeal pouches, the notochord, and the postanal tail almost completely disappear before birth.

ANSWERS TO SECTION 34-2 REVIEW

1. *Anterior* refers to the head end of an animal. *Posterior* refers to the tail end. Bilaterally symmetric animals have an anterior end and a posterior end.
2. Radially symmetric animals do not have a head.
3. The body cavity provides a firm base against which muscles can contract; the body cavity allows the interior of an animal to move somewhat independently of its exterior; the body cavity acts as a reservoir and transport medium for substances in the body.
4. presence of tissue, number of germ layers, presence and type of body cavity, type of development, and presence of other features, such as a backbone.
5. In most vertebrates, the notochord is replaced by the backbone, and the dorsal nerve cord develops into the brain and the spinal cord.
6. The head is the site of many sensory receptors and the brain structures that process sensory information. Thus, a cephalized animal moving head first can perceive its environment before its body enters, giving the animal a chance to retreat if necessary.

SECTION 34-3

SECTION OVERVIEW

Matter, Energy, and Organization

Section 34-3 describes invertebrate and vertebrate strategies for some of the fundamental problems of survival, such as supporting the body, acquiring and transporting oxygen, digesting food, eliminating metabolic wastes, and obtaining and evaluating information about the environment.

▶ ENGAGE STUDENTS

Obtain an earthworm or other invertebrate, and have students compare it with a vertebrate, such as a fish. Have students make comparisons between the body systems of the two animals. (The earthworm is supported by a coelom, while the fish has vertebrae and an internal skeleton. Gas exchange occurs in both animals, but earthworms absorb oxygen through their skin, while vertebrates have lungs or gills.) Have students make comparisons between other systems, such as the digestive and reproductive systems. Explain to students that although humans tend to focus on vertebrates, there are far more invertebrates—in number of species and number of individual organisms.

QUICK FACT

☑ Although radially symmetrical animals are common in the ocean, all land animals are bilaterally symmetrical.

SECTION

34-3

OBJECTIVES

▲

Compare the body plans and development of invertebrates and vertebrates.

●

Define the term *segmentation,* and name a phylum of segmented animals.

■

Explain the difference between an open circulatory system and a closed circulatory system.

◆

Describe the digestive system found in most invertebrate phyla, and compare it with vertebrate digestive systems.

FIGURE 34-7

The sea hare, *Aplysia californica,* is a shell-less mollusk with a simple nervous system.

COMPARISON OF INVERTEBRATES AND VERTEBRATES

Comparative anatomy, the study of the structure of animal bodies, is one of the oldest disciplines in biology. Some modern scientists work to establish the relationships between different animals, while others try to establish the relationships between the form and function of morphological features of animals and the role of these features in animal ecology.

INVERTEBRATE CHARACTERISTICS

While it may be difficult for us to see many similarities between a clam and an octopus, they are classified in the same phylum. Adult invertebrates show a tremendous amount of morphological diversity.

Symmetry

Invertebrates display radial or bilateral symmetry. The radial symmetry of a jellyfish, which drifts rather than swims, allows the animal to receive stimuli from all directions. Most invertebrates have bilateral symmetry, which is an adaptation to a more motile lifestyle. Bilateral symmetry allows for cephalization, which is present in varying degrees in different animals. Some bilaterally symmetric invertebrates, like the sea hare shown in Figure 34-7, are not highly cephalized. *Aplysia* does not have a true, centralized brain and is capable of only simple responses to its environment. Other invertebrates, such as squids and octopuses, are highly cephalized, with a distinct head and a nervous system dominated by a well-organized brain.

Segmentation

Animals in some invertebrate phyla are segmented. **Segmentation** in animals refers to a body composed of a series of repeating similar units. Segmentation is seen in its simplest form in the earthworm, an annelid in which each unit of the body is very similar to the next one. Within the phylum Arthropoda, however, segments may look different and have different functions. In many arthropod species, two or more segments are fused into larger functional units. In the arthropod shown in Figure 34-8, fusion of the anterior segments has resulted in a large structure that includes the animal's head and chest regions.

Support of the Body

Invertebrate bodies have diverse means of support. Sponges have a simple skeleton that supports their soft tissue; the dried, brown, irregularly shaped "natural sponge" found in stores is this skeleton. The bodies of some other invertebrates, such as roundworms, are supported by the pressure of their fluid-filled body cavity.

An **exoskeleton** is a rigid outer covering that protects the soft tissues of many animals, including arthropods, such as lobsters, and mollusks, such as clams. An exoskeleton, however, limits the size and impedes the movement of the organism. Moreover, an exoskeleton does not grow, but must be shed and replaced as the animal grows.

FIGURE 34-8

In animals such as this crayfish, *Procambarus* sp., segments are fused, producing larger structures. The head and chest structure in this animal results from the fusion of several segments. Segments may also give rise to other structures, such as limbs.

Respiratory and Circulatory Systems

Animals produce carbon dioxide, CO_2, as a byproduct of metabolism. Therefore, carbon dioxide in the blood must be exchanged with oxygen, O_2, from the environment. This process is called **gas exchange,** and it occurs most efficiently across a moist membrane. In the simplest aquatic invertebrates, gas exchange occurs directly through the covering of the body. Aquatic arthropods and mollusks, however, have **gills,** organs specialized for gas exchange in water.

In most animals, the **circulatory system** moves blood or a similar fluid through the body to transport oxygen and nutrients to cells. At the same time, carbon dioxide and wastes are transported away from the cells. Sponges and cnidarians have no circulatory system, so nutrients and gases are exchanged directly with the environment by diffusion across cell membranes. Arthropods and some mollusks have an **open circulatory system,** in which blood-like circulatory fluid is pumped from vessels in the body into the body cavity, and then is returned to the vessels. Annelids and other mollusks have a closed circulatory system. In a **closed circulatory system,** blood circulates through the body in tubular vessels. The exchange of gases, nutrients, and wastes occurs between body cells and very small blood vessels that lie near each cell.

Digestive and Excretory Systems

In sponges, digestion occurs within individual cells. In cnidarians, a central chamber with one opening serves as the digestive system.

internet**connect**

SCILINKS NSTA

TOPIC: Invertebrates
GO TO: www.scilinks.org
KEYWORD: HM675

DEMONSTRATION

Invertebrate and Vertebrate Segmentation

Obtain a live earthworm and a cat skeleton or other mounted vertebrate skeleton. Ask students to compare their patterns of segmentation. (The segments are similar in that they are repeating units. In the cat, the segments are exclusively internal, and only part of the skeleton is segmented. The earthworm's entire body, including its exterior surface, is segmented.)

TEACHING STRATEGY

Natural Versus Artificial Sponges

Obtain a natural sponge and an artificial sponge. Pass them around the room, and have students compare their structure and texture. Inform students that although a few sponges purchased in the store may be derived from sponge animals, most are artificial copies made from plastic or cellulose. Artificial sponges resemble natural sponges in structure because they have been designed to match the absorbency of natural sponges.

CONTENT CONNECTION

Section 31-3: Transportation of Materials in Plants

Remind students how nutrients are transported in plants. Ask students to contrast the movement of nutrients in plants and animals. (In plants, mineral nutrients are carried in water that rises from the roots to the leaves, drawn by evaporation from the stomata. Sugars are transported in the phloem by a mechanism that is not yet understood. In most animals, sugars and other nutrients are transported in a circulatory system in which blood is propelled by the contraction of one or more hearts.)

CRITICAL THINKING

Advantages of the Larval Stage

In many animals, the larva is morphologically and ecologically distinct from the adult. The larva may live in a different habitat from the adult, feed on different foods, or be active at different times of the day or year. For example, caterpillars feed on vegetation, while butterflies feed on nectar. Have students propose a hypothesis to explain the adaptive advantage of such ecological differences. (These differences were probably favored by natural selection because they reduce competition between adults and larvae.)

QUICK FACT

☑ Sea hares were named for the long tentacles that protrude from their head and resemble a rabbit's ears.

VISUAL LINK

Figure 34-9

Insects provide the most-familiar examples of direct and indirect development. Inform students that the life cycle of a beetle illustrates indirect development. The egg hatches into the larval form, which eats and grows for some time. Eventually, it enters a quiescent state, the pupa stage, during which it transforms into the adult form. Insects with direct development include grasshoppers and cockroaches. A newly hatched grasshopper is a miniature of the adult and undergoes no radical anatomical change during maturation.

Most other invertebrates, however, have a digestive tract, or **gut,** running through their body. In these animals, food is broken down and nutrients are absorbed by specialized cells that line the gut.

In simple aquatic invertebrates, wastes are excreted as dissolved ammonia, NH_3. In terrestrial invertebrates, specialized excretory structures filter ammonia and other wastes from the body cavity. The ammonia is then converted to less toxic substances and water is reabsorbed by the animal before the waste is excreted.

Nervous System

The extraordinary degree of diversity among invertebrates is reflected in their nervous systems. Sponges have no neurons, although individual cells can react to environmental stimuli in much the same way protozoa can. Neurons evolved in cnidarians, which have a very simple, loosely connected nervous system. Within a single invertebrate phylum, Mollusca, we can trace a stepwise progression of cephalization and the evolution of the brain. The mollusks have very diverse nervous systems. Recall the sea hare, shown in Figure 34-7. Although its head is not well defined, and its nervous system can perform only simple information processing, the sea hare can learn to contract a part of its body in response to certain stimuli. Contrast this simple behavior with that of a highly cephalized mollusk, such as the octopus. The octopus shows very complex decision-making behavior, and it can build a shelter from debris it finds on the ocean floor.

Reproduction and Development

Invertebrates are capable of some form of sexual reproduction, and many can also reproduce asexually. Some invertebrates, such as earthworms, are hermaphrodites. A **hermaphrodite** (huhr-MAF-roh-DIET) is an organism that produces both male and female gametes, allowing a single individual to function as both a male and a female.

Invertebrates may undergo indirect or direct development. Animals that undergo **indirect development** have an intermediate larval stage, as is shown in Figure 34-9. A **larva** is a free-living, immature form of an organism. Many insects, which constitute a class of arthropods, have indirect development.

FIGURE 34-9

Animals with indirect development, such as this beetle, have an intermediate, larval stage. A larva is an immature form that exhibits physical traits different from the adult form.

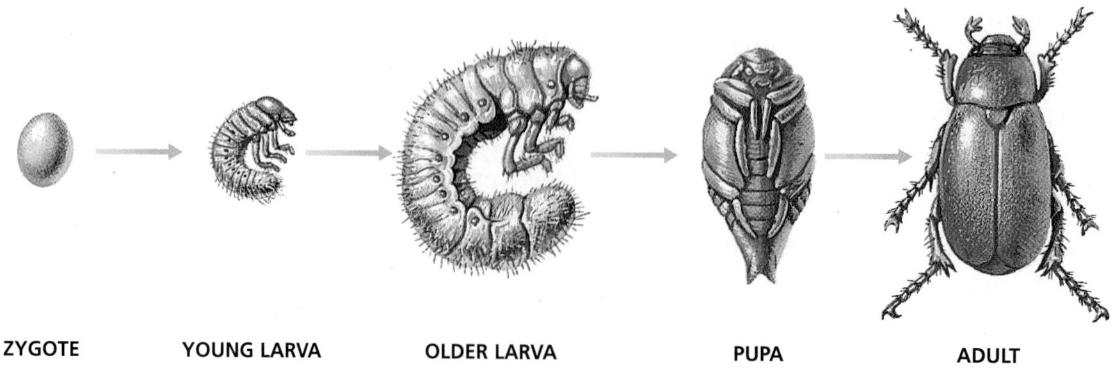

In contrast, in **direct development,** the young animal is born or hatched with the same appearance and way of life it will have as an adult; no larval stage occurs. Although most invertebrates undergo indirect development, a few, such as grasshoppers, undergo direct development.

VERTEBRATE CHARACTERISTICS

Classes of vertebrates include fishes, amphibians, reptiles, birds, and mammals. All vertebrate classes except fishes spend part or all of their life on land. Many characteristics of terrestrial vertebrates are adaptations to life on land and fall into two broad categories: support of the body and conservation of water.

Support of the Body

In addition to a backbone, vertebrates have an **endoskeleton,** an internal skeleton that can support a large, heavy body. The endoskeleton grows as the animal grows.

Although it is not immediately apparent, vertebrates are segmented animals. Segmentation is evident in the ribs and the **vertebrae** (VUHR-tuh-BREE), the repeating bony units of the backbone, as shown in Figure 34-10. As terrestrial vertebrates evolved from aquatic vertebrates, their limbs and associated muscles evolved to give the animals better support and greater mobility. For example, the legs of amphibians, the first land vertebrates to evolve, are positioned away from the body, as shown in Figure 34-11a. However, the legs of mammals, such as the deer shown in Figure 34-11b, are positioned directly beneath the body, allowing the animal to move faster and with a longer stride. Humans show an extreme version of this trait: we are bipedal, and our head is positioned directly over our body.

FIGURE 34-10

Vertebrates, such as birds, are segmented animals. Segmentation is evident in the repeating structures of the animal's backbone and ribs.

FIGURE 34-11

(a) The legs of amphibians, such as this frog, *Agalychnis saltator,* are sharply bent and positioned away from the body. (b) The legs of terrestrial mammals, such as this deer, *Odocoileus virginianus,* are straighter than those of amphibians, providing greater mobility and speed.

(a) (b)

TEACHING STRATEGY
Vertebrate Segmentation

Ask students to point out examples of segmentation in the external structure of the human body. Students will find this difficult, so tell them that although vertebrates are considered segmented animals, vertebrate segmentation is strictly internal. The backbone and rib cage are composed of repeating units, and the muscles are made up of multiple blocks of tissue. Segmentation is also evident in the development of the brain.

VISUAL LINK
Figure 34-11

Have students stand up and compare the position of their legs (directly beneath the body) with the position of the frog's legs (angled out from the body) and the position of the deer's legs (also directly beneath the body). Point out that the mammalian limb orientation provides more support for body weight and enables mammals to run faster. Also point out that frogs are specialized for jumping, and ask students to identify some adaptations for jumping that are visible in the photo. (The frog has long, powerful hind limbs that propel it forward.)

RETEACHING *ACTIVITY*

Divide the class into two groups. Tell one group to make a crossword puzzle using invertebrate characteristics. Tell the other group to make a crossword puzzle using vertebrate characteristics. When the puzzles are finished, have the groups exchange puzzles and fill them in.

SECTION 34-3

Quick Lab

Identifying Animal Characteristics

Time Required 30 minutes

Procedural Tips Students should use characteristics given in the text and illustrations. Some characteristics may be common to both vertebrates and invertebrates. The textbook will also serve as the reference in case of dispute. Use pictures of animals described in the textbook that have well-defined characteristics. Prepare the pictures ahead of time. Each pair of students will need a set of pictures.

Answers to Analysis Animals are identified by characteristics that are easily recognized, such as size, outer covering, number of limbs, and body plan. Unrelated animals that look alike or related species that appear different may be misidentified if identification is based solely on morphology.

TEACHING STRATEGY

Adaptations of the Digestive System

Point out to students that the length of the digestive tract in vertebrates varies with diet. Carnivores typically have a shorter digestive system (relative to body size) than do herbivores or omnivores. Two factors contribute to this difference. First, vegetation is harder to digest than flesh because it is composed partly of cellulose, which vertebrates cannot break down (unless aided by microorganisms). Second, vegetation typically contains fewer nutrients per gram than flesh does. The longer the digestive tract is, the more time that is available for digestion and the greater is the surface area for absorbing nutrients.

Quick Lab

Identifying Animal Characteristics

Materials 3×5 in. note cards (20), 5 pictures of vertebrates, 5 pictures of invertebrates

Procedure

1. Working in pairs, one partner will write one different vertebrate characteristic on each of 10 note cards. The other partner will write one different invertebrate characteristic on each of 10 note cards.
2. Designate one partner as the dealer. Place the animal pictures upside down in a stack. The dealer will shuffle and deal all the cards. Turn over one animal picture.
3. The nondealer plays first by laying down as many cards as possible that describe characteristics of the pictured animal. If no card matches, the play is passed to the other player. When neither partner can play, another picture is turned up and play continues.
4. Play ceases when neither student can play or when no pictures are left. The player who is holding the fewest number of cards wins.

Analysis Why are morphological characteristics used to identify organisms? What are the disadvantages of using only morphological characteristics to identify an organism?

Body Coverings

The outer covering of an animal is called the **integument** (in-TEG-yoo-muhnt). While the integuments of fishes and most amphibians are adapted only to moist environments, the integuments of most terrestrial vertebrates are adapted to hold water inside the body. All animal bodies are composed of water-filled cells, and if the water content of the cells is reduced appreciably, the animal will die. The outer covering of terrestrial vertebrates, such as reptiles, birds, and mammals, is largely watertight. Integuments also serve other purposes. The moist skin of an amphibian functions as a respiratory organ for the exchange of gases. The scales of a reptile help protect it from predators. The feathers of birds and the fur of mammals efficiently insulate the body.

Respiratory and Circulatory Systems

Gas exchange occurs in the gills of aquatic vertebrates, including fishes and larval amphibians, but these gills do not function out of water. **Lungs** are organs for gas exchange composed of moist, membranous surfaces deep inside the animal's body. Lungs evolved in terrestrial vertebrates.

Vertebrates have a closed circulatory system with a multichambered, pumping heart. In some vertebrates, the multichambered heart separates oxygenated and deoxygenated blood, improving the efficiency of the circulatory system over that found in other vertebrates and many invertebrates.

Digestive and Excretory Systems

Digestion in vertebrates occurs in the gut, which runs from the mouth, at the anterior end, to the anus, at the posterior end. In many vertebrates, the gut is very long with respect to the length of the body, increasing the surface area over which nutrients are absorbed. The human digestive tract is nearly 7 m (23 ft) long. It is folded to fit into a body one-fourth its length.

Vertebrates have the same waste-disposal problems as invertebrates. They must deal with the very toxic ammonia their bodies produce, and most vertebrates must expel wastes while conserving water. Like invertebrates, most vertebrates convert ammonia to less toxic substances. In most vertebrates, organs called **kidneys** filter wastes from the blood while regulating water levels in the body.

Nervous System

Vertebrates have highly organized brains and the control of specific functions occurs in specific centers in the brain. The structure and function of the nervous system varies among vertebrate orders. For example, within the brain of a fish, much of the tissue processes sensory information. In Figure 34-12, the elongated structure that projects in front of the fish's eye processes only information about smell. Fishes have limited neural circuitry devoted to decision making. A fish's responses to stimuli in its environment are rigid, that is, they vary little from situation to situation and from fish to fish.

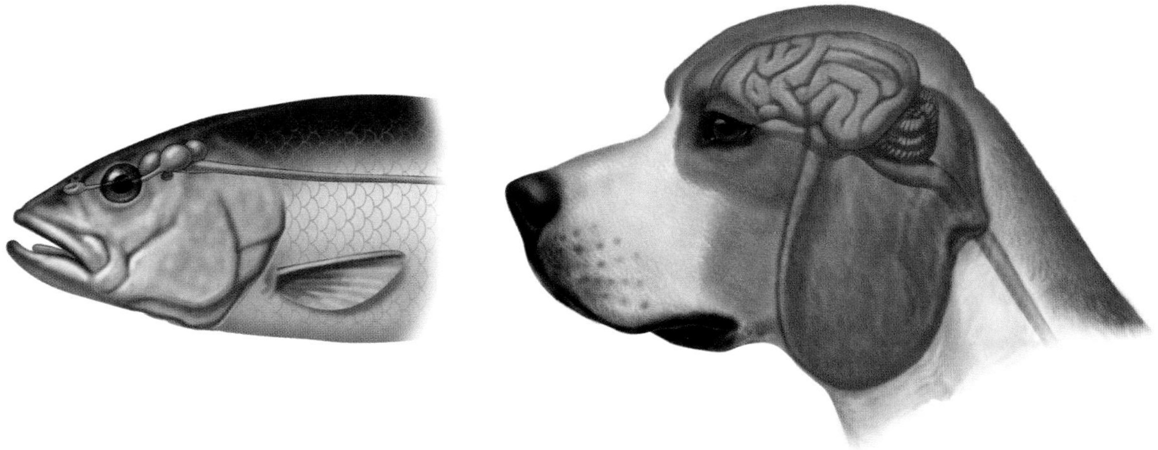

FIGURE 34-12

A comparison of the brain of a fish with the brain of a mammal, such as a dog, shows that the size of the brain with respect to the body is larger in the dog. In the dog, the areas devoted to complex processing and decision making, such as the heavy, wrinkled structure that dominates its brain, are also larger.

Other animals, such as dogs, display complex and flexible behavior. Much of the tissue in the dog's brain is given over to neural circuitry that is involved in decision making, and its brain is large with respect to body size, as Figure 34-12 shows.

Reproduction and Development

In most fish and amphibian species, eggs and sperm are released directly into the water, where fertilization takes place. In reptiles, birds, and mammals, the egg and sperm unite within the body of the female, increasing the likelihood that the egg will be fertilized.

The fertilized egg—the zygote—of many fishes, amphibians, reptiles, and birds develops outside the body. The developing embryo is nourished by the yolk of the egg and protected by jellylike layers or by a shell. The zygotes of some species of fishes, amphibians, and reptiles remain inside the body of the female, nourished by the yolk, until they hatch. In contrast, most mammals give birth to live offspring. Embryos of placental mammals develop in the female's body, nourished by the mother's blood supply, until the young are born.

With the exception of amphibians and some fishes, vertebrates undergo direct development. Over the course of direct development, the young and the adults can share the same resources—an advantage if those resources are plentiful.

SECTION 34-3 REVIEW

1. **What is segmentation, and what trend is apparent in the segmented bodies of arthropods?**
2. **How does a closed circulatory system function, and how does it differ from an open circulatory system?**
3. **How are vertebrates related to chordates?**
4. **How is the digestive chamber of a cnidarian different from that of most other invertebrates and all vertebrates?**
5. **What is a primary function of the body covering of terrestrial animals?**
6. **CRITICAL THINKING How is the degree of an animal's cephalization related to its behavior?**

SECTION 34-3

ANSWERS TO SECTION 34-3 REVIEW

1. *Segmentation* refers to a body composed of repeating similar units. In arthropods, some segments are fused into larger units, and some segments give rise to specialized structures, such as legs.
2. In a closed circulatory system, blood circulates in tubular vessels. In an open circulatory system, bloodlike circulatory fluid is pumped from the body vessels to the body cavity, and then is returned to the body vessels.
3. Vertebrates are a subset of chordates.
4. A cnidarian's digestive chamber has only one opening, unlike most other invertebrates and all vertebrates.
5. The body covering of most terrestrial animals is adapted to hold water inside the body.
6. Animals that are more strongly cephalized generally exhibit more flexible, complex behavior than have animals that are only slightly cephalized.

BACKGROUND

Ernest Everett Just was born in 1883 in Charleston, South Carolina. He attended Kimball Union Academy, in Vermont, and graduated from Dartmouth in 1907 as the only *magna cum laude* in his class. In 1916, he earned a Ph.D. in zoology at the University of Chicago. He went on to teach at Howard University until 1929. From 1909 to 1929, he spent his summers working at the Marine Biological Laboratory at Woods Hole, Massachusetts. During this time, he was elected to the American Association for the Advancement of Science, the American Society of Zoologists, and the American Society of Naturalists. In the 1930s he conducted research at numerous European laboratories. Just published over 60 papers and two books. He died of pancreatic cancer in 1941.

Just's research was in fertilization and the importance of the cell membrane in the process. During fertilization, the fusion of a sperm and an egg is initiated by the acrosome of the sperm. The acrosome releases hyaluronidase, an enzyme that dissolves the egg's protective coverings and permits the sperm to reach the egg. The membrane of the acrosome lengthens into a tube that extends to the surface of the egg and fuses with the egg's membrane. The tube provides a tunnel through which the sperm nucleus can reach the interior of the egg. This union of the two membranes stimulates the cortical reaction. The cortical reaction effectively forms a barrier to polyspermy, that is, it blocks the entrance of more sperm. The sperm nucleus and the female pronucleus meet in the center of the egg. Here the chromatin from the two nuclei organize into chromosomes, thus creating a zygote. Cleavage follows.

GREAT DISCOVERIES

The Cell Surface: Embryonic Development and Beyond

HISTORICAL PERSPECTIVE

The mysteries of embryonic growth and development have engaged scientists since ancient times. Until the last century, it was largely unknown how a large, multicellular animal was produced from a source too small to see. Early scientists debated whether the embryo is pre-formed as a miniature individual (a homunculus) or is entirely undifferentiated in form, becoming more specialized only as it grows and develops. The field of developmental biology, which advanced rapidly in the late 1800s and throughout the 1900s, owes much to the pioneering efforts of biologist Ernest Everett Just. Just demonstrated that the differentiation of the embryonic cells during early development is the result of the interaction between the cytoplasm and the nucleus of the egg. Just's later work helped establish that the cell membrane is more than a boundary layer. Rather it plays a critical role as a gatekeeper and conduit for information vital to the cell.

First Steps

In 1899, the German-born American scientist Jacques Loeb, who worked at the University of Chicago and at Woods Hole Marine Biological Laboratory in Cape Cod, Massachusetts, began a study of the early development of eggs. Loeb made the surprising discovery that certain environmental disturbances could initiate development of unfertilized sea urchin eggs. When Loeb pricked the eggs with a needle or changed the salt concentration of the solution surrounding them, the eggs began to divide as if they had been fertilized by sperm. This phenomenon is known as *parthenogenesis,* and it is now known that it occurs in nature in every major group of invertebrates. To explain his findings of forced parthenogenesis in the laboratory, Loeb initially hypothesized that the physical and chemical agents he used in his experiments mimicked sperm function and that agents from outside the egg—such as sperm—are necessary for development to begin.

Ernest Everett Just

Challenging a Well-Established Model

A decade later, a young biologist named Ernest Just began conducting his own research in cell biology. In 1911, while working at the Woods Hole laboratories, Just made an important discovery about cell cleavage, the successive cell divisions leading to the formation of the embryo. Just conducted experiments in chemical-induced parthenogenesis in sea urchins, marine worms, and sand dollars, and he also observed normal fertilization of eggs by sperm cells. In contrast to Loeb's original hypothesis—that sperm or something like it is necessary to trigger development—Just proposed that the egg contains the necessary mechanism for development. Just's subsequent work confirmed the role of the egg cytoplasm in the initiation of cleavage. This new understanding of the importance of the cytoplasm to development prompted Just to take a closer look at the individual components of the egg cell. He studied the relationship between the nucleus and the cytoplasm as well as the arrangement of different cytoplasmic components throughout the egg. In his later work, he focused on the role of the cell surface.

More Than a Boundary

Just maintained that there is a membrane around animal cells and that this membrane is a fundamental part of the living system of the cell.

Initially, some scientists doubted that the cell membrane existed because its cellular structure could not be easily distinguished using the microscopes of the day. Just, however, was correct about the existence of the cell membrane and its extraordinary functions as a gateway for passage into and out of the cell and as an active participant in cell-to-cell communication.

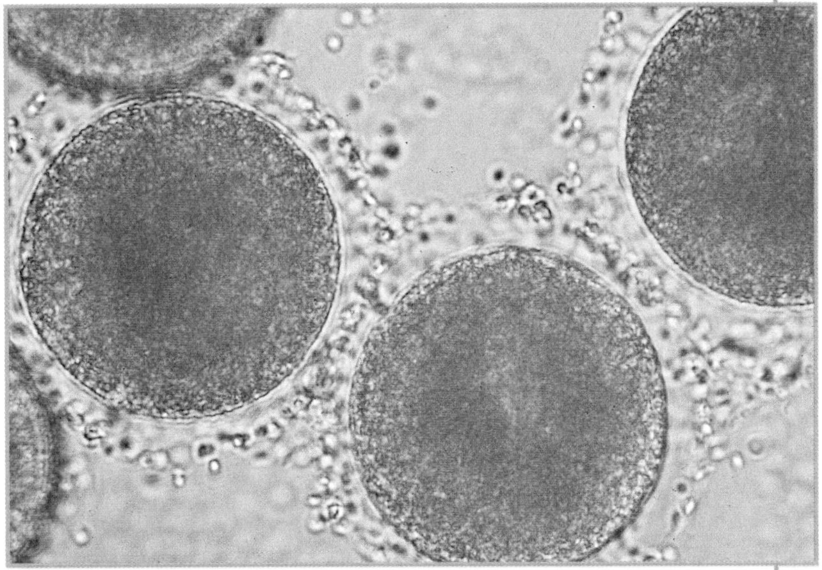

Some animal eggs, such as these sea urchin eggs, can be induced to begin development without being fertilized by sperm.

In his 1939 book, *The Biology of the Cell Surface,* Just wrote:

> *It [the cell surface] is keyed to the outside world as no other part of the cell. It stands guard over the peculiar form of the living substance, and is a buffer against the attacks of the surroundings and the means of communication with it.*

Meeting Modern Challenges

Just's experiments, which highlighted the relationship between the nucleus and the cytoplasm and the significance of the cell membrane, have proved helpful in understanding how a cell carries out its basic functions. Today, we understand the role of the cytoplasm in tissue differentiation as that of controlling the transcription of genetic information from the nucleus. Modern researchers using electron microscopes have resolved the fine structure of the cell membrane, revealing it to be an active cellular component. Despite the limited technology available to him, Just's pioneering research in cell biology has proved sound, and his techniques and discoveries are relevant to research in cell biology today.

In 1915, when he was 32 years old, Just was awarded the first Spingarn Medal by the National Association for the Advancement of Colored People for his work in cell biology. This medal, the organization's highest honor, continues to be presented in recognition of outstanding achievements by African Americans. From 1920 to 1931, Just held a National Research Rosenwald fellowship in biology. He served as vice-president of the American Zoological Society and as associate editor of the *Journal of Morphology,* of *The Biological Bulletin,* and of *Physiological Zoology.* Ernest Everett Just died in 1941.

DISCUSSION

Guide the discussion by posing the following questions.

1. What were some of the early techniques used to cause cleavage in eggs, and what is egg cleavage? (Sea urchin eggs and frog eggs could be forced to develop by pricking them with a needle or changing the salt concentration of the culture medium. Cleavage is the successive cell divisions that lead to the formation of the embryo.)

2. What is the importance of Just's experiments? (Just's experiments demonstrated the important relationship between the nucleus and the cytoplasm and the significance of the cell membrane.)

FURTHER READINGS

1. "How to Coax New Life," in *Time,* Fall 1996 Special Issue, pages 36–40. This article discusses some of the advances in artificial reproductive techniques. The article also discusses the cost and success rate of each technique.

2. "Brave New Egg," in *Discover,* April 1997, pages 76–80. This article describes how biologists are able to raise mouse eggs in Petri dishes.

SECTION 34-4

SECTION OVERVIEW

Matter, Energy, and Organization

Section 34-4 describes fertilization and the process of development, which occurs in a series of specific, sequential stages. Differences in development are to a large degree responsible for the anatomical differences among the phyla of animals.

▶ ENGAGE STUDENTS

Obtain a beach ball or soccer ball that can be easily deflated. Tell students that after fertilization and the process of cleavage, the zygote turns into a hollow ball of cells, called a blastula, that is very similar in appearance to an inflated soccer or beach ball. Explain to students that the central cavity is called the blastocoel. Draw a small circle on the ball and indicate that this region is called the blastopore. Tell them that the next stage in development is formation of the gastrula, or gastrulation, in which cells on the outside of the blastula begin migrating into the blastocoel. At this time, you might partially deflate the ball to represent the infolding process of gastrulation. Continue the discussion of these processes by referring to photographs or diagrams in the textbook.

SECTION

34-4

OBJECTIVES

▲ List the steps of fertilization and development through gastrulation.

● Identify the three primary germ layers, and list two body parts formed from each germ layer.

■ Define *protostome* and *deuterostome.*

◆ Contrast spiral cleavage with radial cleavage, and name the category of organisms that undergo each type of cleavage.

▲ Contrast schizocoely with enterocoely.

FERTILIZATION AND DEVELOPMENT

Development of a multicellular animal from an egg cell is a truly remarkable process. Each cell has the same set of genes that are used to build an animal, yet animals have many different kinds of cells. From the fertilized egg come large numbers of cells—many millions in humans—that consistently give rise to structural features of the animal body.

FERTILIZATION AND EARLY DEVELOPMENT

In most animals, fertilization is the union of female and male reproductive cells, the eggs and sperm. Fertilization results in the combination of haploid sets of chromosomes from two individuals into a single diploid cell, the zygote.

Gametes

In most animal species, the sperm cell, shown in Figure 34-13, is specialized for movement—it is very streamlined and small. The head of the sperm contains chromosomes, while the tail of the sperm is composed of a large flagellum.

The egg is typically large, owing to its large store of cytoplasm and yolk. The size of a species egg seems to depend on how long the food supply in the yolk must last. In aquatic animals, in which the embryo begins to feed itself early, eggs are small and there is little yolk. In sharp contrast, the embryos of birds must live on the yolk until they hatch. In these eggs, the yolk volume is very large.

Fertilization

At the start of fertilization, the sperm's cell membrane fuses with the egg's cell membrane, and the nucleus of the sperm enters the cytoplasm of the egg. The fusion of the cell membranes of the egg and sperm causes an electrical change that blocks entry to the egg by other sperm cells. The sperm nucleus merges with the egg nucleus to form the diploid nucleus of the zygote. Once a zygote is formed, replication of DNA begins, and the first cell division soon follows.

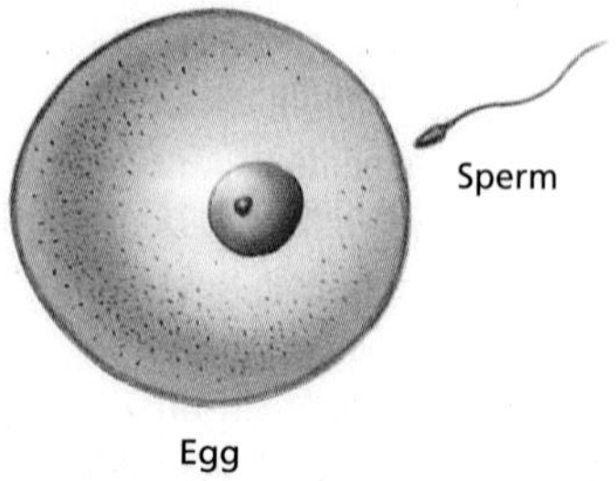

FIGURE 34-13

The small, flagellated sperm is adapted for motility and speed. It must seek out and fertilize the much larger, yolk-filled egg.

Cleavage and Blastula Formation

The divisions of the zygote immediately following fertilization are termed **cleavage.** As Figures 34-14a–c show, as cleavage progresses, the number of cells increases, from 2 to 4, then to 8, and so on. During cleavage, mitotic divisions rapidly increase the number of cells, but the cells do not grow in size. Thus, the cell divisions of cleavage yield smaller and smaller individual cells. Cleavage increases the surface-area-to-volume ratio of each cell, which enhances gas exchange and other environmental interactions.

In most species, cleavage produces a raspberry-shaped mass of 16 to 64 cells. As the number of dividing cells further increases, the mass becomes a hollow ball of cells called a **blastula,** as shown in Figure 34-14d. The central cavity of a blastula is called the **blastocoel** (BLAS-toe-SEEL), shown in Figure 34-14e.

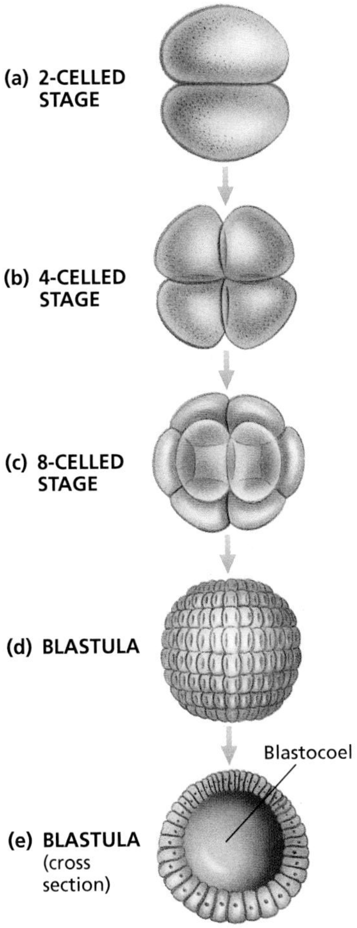

FIGURE 34-14
During cleavage, the zygote divides repeatedly (a–c) without undergoing cell growth, producing a many-celled hollow blastula (d and e).

GASTRULATION

At the start of the next stage of development, shown in Figure 34-15a, an area of the blastula begins to collapse inward, much the way a partially inflated soccer ball would collapse if a fist were pressed into it. This infolded region of the blastula, called the **blastopore,** is shown in Figure 34-15b. A fundamental reorganization of the cells of the hollow blastula begins with the formation of the blastopore. This process, called **gastrulation,** transforms the blastula into a multilayered embryo, called the **gastrula.** Gastrulation is marked by changes in the shape of cells, together with changes in their adhesion to other cells.

As the inward folding, shown in Figure 34-15c, continues, the now cup-shaped embryo enlarges and a deep cavity, called the **archenteron,** develops. This cavity will function as the gut. Forming the outer layer of the gastrula is the outer germ layer, the **ectoderm,** shown in blue in Figure 34-15c. The inner germ layer, the **endoderm,** is shown in yellow. In most phyla, the gastrula does not remain a two-layer structure. As development progresses, a third layer, the **mesoderm,** forms between the endoderm and the ectoderm.

Recall that each of the three germ layers formed during gastrulation is a forerunner to certain parts of the fully formed body. The archenteron, which is surrounded by the endoderm, forms the throat passage, including gills or lungs, and the gut and its

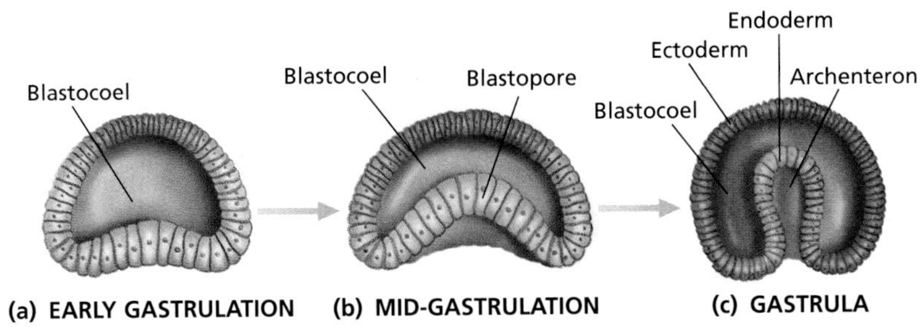

FIGURE 34-15
Echinoderms undergo the gastrulation process shown here. The blastula is reorganized (a and b), forming the cup-shaped gastrula (c). Other phyla have somewhat different patterns of gastrulation.

DEMONSTRATION

Sea Urchin Embryo Development

Show video clips, slides, or laserdisc images of sea urchin embryo development. Discuss stages such as fertilization, cleavage, formation of the blastula, gastrulation, and neural tube formation. As you proceed through the discussion, stop periodically and ask students to predict what the next image will look like.

TEACHING STRATEGY

The Three Germ Layers

Help students remember the names of the three germ layers by explaining the roots of each term. All three share the root *-derm,* which is derived from the Greek *derma,* meaning "skin." *Ectoderm* comes from the Greek *ecto,* meaning "outside," while *endoderm* comes from the Greek *endon,* which means "within." *Mesoderm* is derived from the Greek *mesos,* which means "middle."

MAKING CONNECTIONS

Mathematics

Have students calculate how many cells will be present in the blastula after 3, 5, and 10 divisions (8, 32, 1024). Have students write a general formula for calculating the number of cells at any time during cleavage (number of cells = 2^n, where n = the number of divisions). Ask students what assumption must be made in order for this formula to be valid. (Every cell in the blastula must divide during each division. Actually, asynchronous division occurs during the early stages of many species, so this model holds for only early divisions in some species.)

TEACHING STRATEGY

Developmental Flip Books

Give students 10 to 20 note cards that have been cut in half. Instruct them to trace around a coin, such as a quarter, in the center of each card. Give them reference books that show stages of vertebrate embryonic development. On each card, they should sketch in the circle one stage of development, beginning with a fertilized egg on the first card. When finished with their sketches, they should stack the cards in order. They should then offset each card in the stack by about 1 mm from its predecessor so they can flip through the sketches. Staple the cards together. When the cards are flipped, they appear to show embryonic development in motion.

CRITICAL THINKING

Developmental Biology

Ask students to imagine that they are developmental biologists who have been given the task of determining which cells give rise to particular structures in the adult mouse. Ask them what experiments they might perform on a blastula to manipulate an effect at the end of development. (Individual cells of the blastula from specific locations could be removed or transplanted to other areas. Changes in the adult organism could be traced back to changes made in the blastula.)

associated organs, such as the pancreas and liver. The ectoderm forms the outer layer of the skin, the hair, nails, and the nervous system.

The versatile mesoderm forms a multitude of body parts, including the skeleton, muscles, inner layer of the skin, the circulatory system, and the lining of the body cavity.

PATTERNS OF DEVELOPMENT

A body cavity completely lined by mesoderm is called a **coelom.** Most phyla have a coelom, and, like patterns of symmetry and number of germ layers, the coelom is a feature which taxonomists use to classify animals with similar phylogenetic origins. The distinct patterns of cleavage and coelom formation found in different animal phyla are additional clues to their phylogenetic history.

Blastopore Fate and Cleavage

Recall from Figure 34-5 that echinoderms and chordates share a branch of the phylogenetic tree of animals, and mollusks, annelids, and arthropods share another branch. There are two distinct patterns of development in animals that have a coelom. In the embryos of mollusks, arthropods, and annelids, the blastopore develops into a mouth, and another opening eventually arises and develops into an anus. These organisms are called **protostomes** (PROHT-oh-STOHMZ), which means "first mouth." Many protostomes undergo **spiral cleavage,** in which the cells divide in a spiral arrangement. Each cell in the blastula nestles between two cells of the adjacent row, as is shown in the top left frame of Figure 34-16.

In contrast, in the embryos of echinoderms and chordates, the blastopore develops into an anus, and a second opening in the embryo becomes the mouth. These organisms are called **deuterostomes** (DOOT-uhr-oh-STOHMZ), which means "second mouth." Most deuterostomes undergo **radial cleavage,** in which the cell divisions are parallel to or at right angles to the axis from one pole of the blastula to the other, as shown in the top right frame of Figure 34-16.

Protostomes and deuterostomes also differ in how early the cells of the embryo specialize. If the cells of some protostome embryos are separated at the four-cell stage of development, each cell will develop into only one-fourth of a complete embryo and the developing organism will die. Thus, the path of each cell is determined early in the development of the protostome in a pattern called **determinate cleavage.** In contrast, if the cells of most four-celled deuterostome embryos are separated, each cell will embark on its own path to become a separate organism. This type of development is called **indeterminate cleavage.** Indeterminate cleavage very early in embryo development in humans can result in identical twins.

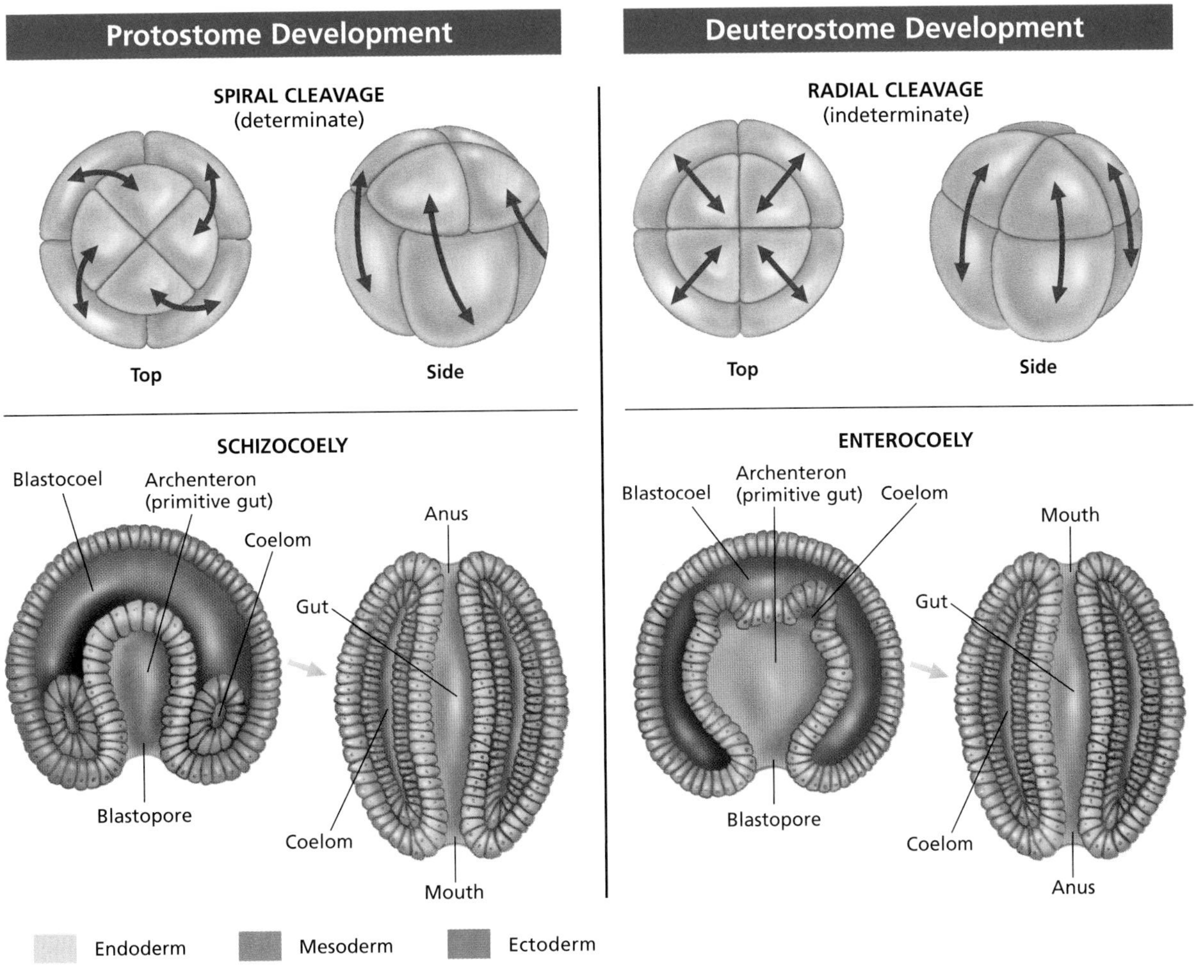

FIGURE 34-16

Many protostomes undergo spiral cleavage during early development, while many deuterostomes undergo radial cleavage. In protostomes, the coelom arises by schizocoely, and the blastopore becomes the mouth. In deuterostomes, the coelom arises by enterocoely, and the blastopore becomes the anus.

Coelom Formation

The way in which the coelom forms in many protostomes differs from the way it forms in many deuterostomes. The lower left frame of Figure 34-16 shows coelom formation in protostomes. Cells located at the junction of the endoderm and ectoderm (at the rim of the cup-shaped embryo) split away toward the interior of the gastrula. Rapid division of these cells (shown in pink) in the blastocoel forms the mesoderm. This process of mesoderm formation is called **schizocoely** (SKIZ-oh-SEEL-ee), or "split body cavity."

The lower right frame of Figure 34-16 shows coelom formation in deuterostomes. The mesoderm forms when the cells lining the dorsal, or top, part of the archenteron begin dividing rapidly. These rapidly dividing cells (shown in pink) roll outward into the blastocoel, forming the mesoderm. This process of mesoderm formation is called **enterocoely** (EN-tuhr-oh-SEEL-ee), meaning "gut body cavity." During both enterocoely and schizocoely, mesodermal cells spread out to completely line the coelom, and the blastocoel disappears. Thus, in both protostomes and deuterostomes, mesoderm lines the interior of the outer body wall and surrounds the gut.

TEACHING STRATEGY
How Twins Form

Inform students that there are two kinds of twins. Monozygotic twins result when the zygote splits early in cleavage and gives rise to two genetically identical embryos. Dizygotic twins result when two eggs are ovulated and fertilized. Because dizygotic twins develop from different eggs and different sperm, they are not genetically identical—in fact, they are no more similar, genetically speaking, than are two siblings who are born years apart.

VISUAL LINK
Figure 34-16

Ask students what two layers the mesoderm layer splits in schizocoely. (The mesoderm splits the endoderm and ectoderm.) Ask where the mesoderm came from in enterocoely. (It came from the dorsal part of the archenteron.)

TEACHING STRATEGY
Models of Development

Have students use three colors of clay to make models of the steps of cleavage in protostomes and deuterostomes. Students should base their models on the drawings in Figure 34-16. Students should label each stage with a tag that describes the events of that stage.

SECTION 34-4

ANSWERS TO SECTION 34-4 REVIEW

1. fertilization, cleavage, and gastrulation (which includes formation of the blastopore, differentiation of the endoderm and the ectoderm, and formation of the mesoderm)
2. endoderm: throat, gut and associated organs; mesoderm: skeleton, muscles, circulatory system; ectoderm: outer layer of skin, hair, nails, nervous system
3. A protostome is an animal in which the blastopore develops into the mouth; a deuterostome is an animal in which the blastopore develops into the anus. Protostomes undergo spiral cleavage, while deuterostomes undergo radial cleavage.
4. Indeterminate cleavage can result in the formation of identical offspring.
5. In schizocoely, the mesoderm forms by rapid division of cells that lie at the boundary of the endoderm and ectoderm. In enterocoely, the mesoderm forms by rapid division of cells that line the dorsal part of the blastopore.
6. If an early embryo of an organism that undergoes indeterminate cleavage is split, it can produce two or more normal organisms. In contrast, splitting the early embryo of an organism with determinate cleavage results in the death of the embryo.

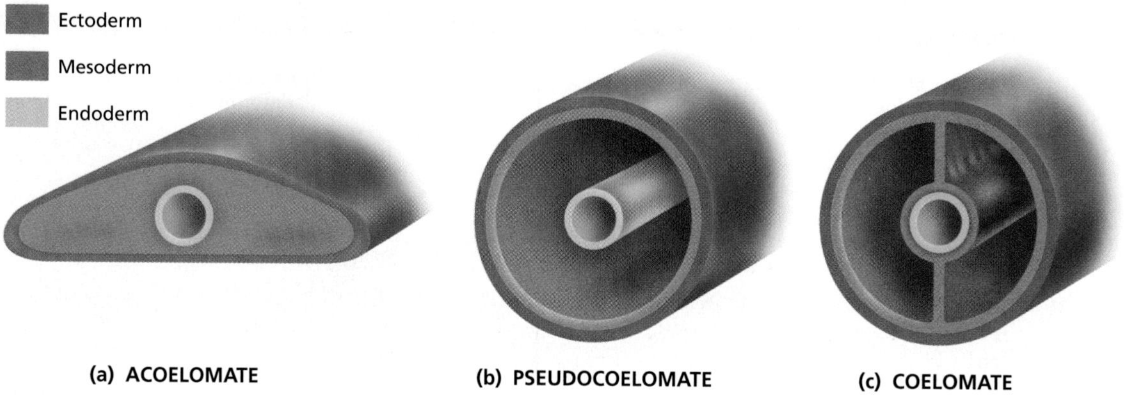

FIGURE 34-17
In three-layered acoelomates (a), the endodermic gut is surrounded by a solid layer of mesoderm. In pseudocoelomates (b), the endodermic gut is suspended in a fluid-filled cavity which is surrounded by mesoderm. In coelomates (c), the endodermic gut is surrounded and suspended by mesoderm, which also surrounds the coelom.

Types of Body Cavities

Compare the cross sections of an acoelomate, a pseudocoelomate, and a coelomate shown in Figure 34-17. In **acoelomates** (UH-SEE-luh-mayts), such as flatworms, the body cavity is absent. The interior of the animal is solid, as shown in Figure 34-17a. The endodermic gut, shown in yellow, and the outer covering of the animal, shown in blue, are connected by the solid tissue of the mesoderm.

In some phyla, including rotifers and roundworms, the mesoderm lines the interior of the coelom but does not surround the exterior of the endodermic gut. This type of body cavity, shown in Figure 34-17b, is called a **pseudocoelom** (SOO-doh-SEE-luhm), which means "false body cavity." In **pseudocoelomates,** such as the one shown in Figure 34-17b, mesoderm lines the fluid-filled coelom, and the endodermic gut is suspended in the fluid of the coelom.

In **coelomates** (SEE-luh-MAYTS), animals with a true coelom, such as the one shown in Figure 34-17c, mesoderm lines the body cavity and surrounds and supports the endodermic gut. The mesoderm also forms the tissues of attachment for the organs located in the coelom, such as the liver and the lungs. Mollusks, annelids, arthropods, chordates, and echinoderms are coelomates.

Look back at the phylogenetic tree in Figure 34-5 and locate the acoelomate, pseudocoelomate, and coelomate phyla. How does their placement on the phylogenetic tree reflect their body type?

SECTION 34-4 REVIEW

1. **Beginning with fertilization, list the steps of development through mesoderm formation.**
2. **What are the three germ layers formed in all embryos except those of sponges, cnidarians, and ctenophores? Name two body parts that arise from each germ layer.**
3. **What is a protostome? What is a deuterostome? How does cleavage in these groups differ?**
4. **Humans and other vertebrates sometimes produce two or more identical offspring. What type of cleavage can result in the formation of identical offspring?**
5. **How is the mesoderm formed in schizocoely, and how does this process differ from enterocoely?**
6. **CRITICAL THINKING What adaptive advantage is associated with indeterminate cleavage?**

SUMMARY/VOCABULARY

34-1

- Animals are multicellular and heterotrophic, and their cells lack walls. Most animals reproduce sexually and can move.
- Animals have cells that are specialized for different functions.
- Most animals ingest their food and digest it within their bodies.
- Most animals reproduce sexually, though some also reproduce asexually.
- Movement and response to the environment are governed by an animal's nervous tissue and muscle tissue.
- The first animals may have evolved from colonial protists.

Vocabulary

cell junction (667)
differentiation (668)
ingestion (668)
invertebrate (667)
neuron (668)
specialization (667)
vertebrate (667)
zygote (668)

34-2

- Sponges, the simplest animals, have no true tissue and no body symmetry. All other animals have tissue.
- In the bodies of animals with radial symmetry, similar parts branch out in all directions from a central line. Animals with bilateral symmetry have similar halves. Bilateral symmetry is associated with cephalization, that is, having a head.
- Most animals have three germ layers. All body features arise from one of the germ layers.
- At some stage of their lives, all members of phylum Chordata have a notochord, a dorsal nerve cord, a postanal tail, and pharyngeal pouches.

Vocabulary

anterior (670)
bilateral symmetry (670)
cephalization (670)
chordate (673)
dorsal (670)
dorsal nerve cord (673)
germ layer (671)
notochord (673)
pharyngeal pouch (673)
postanal tail (673)
posterior (670)
radial symmetry (670)
symmetry (670)
ventral (670)

34-3

- Invertebrates have no body symmetry or are radially or bilaterally symmetrical; vertebrates are bilaterally symmetrical.
- A segmented body is composed of repeating similar units. Some invertebrates and all vertebrates are segmented.
- Some invertebrates have an exoskeleton. All vertebrates have an endoskeleton.
- The simplest invertebrates have no circulatory system. Arthropods and some mollusks have an open circulatory system. Other mollusks, annelids, and vertebrates have a closed circulatory system.
- Sponges digest food within individual cells. Cnidarians digest food in a central chamber. Other invertebrates and all vertebrates have a gut.
- Some invertebrates have loosely connected circuits of neurons. Some invertebrates and all vertebrates have a well-defined brain.
- Most invertebrates and vertebrates are capable of some form of sexual reproduction, and some invertebrates can also reproduce asexually.

Vocabulary

circulatory system (675)
closed circulatory system (675)
direct development (677)
endoskeleton (677)
exoskeleton (675)
gas exchange (675)
gill (675)
gut (676)
hermaphrodite (676)
indirect development (676)
integument (678)
kidney (678)
larva (676)
lung (678)
open circulatory system (675)
segmentation (674)
vertebra (677)

REVIEW

1. Cell specialization is the assumption of different tasks by different cells in a multicellular organism. Differentiation is the developmental process in which cells become different from each other according to their genetic instructions as they divide.
2. Cell junctions allow the formation of tissue from separate cells.
3. In radial symmetry, similar structures are arranged about a central line. In bilateral symmetry, similar structures are arranged on either side of a central plane.
4. A coelomate is an animal that has a coelom, a body cavity lined with mesoderm. An acoelomate does not have a body cavity.
5. In determinate cleavage, the fate of cells of the early embryo are set, and division of the embryo will result in the death of the cells. In indeterminate cleavage, division of the early embryo can result in the production of multiple identical organisms.

6. c
7. a
8. d
9. b
10. d
11. d
12. a
13. b
14. d
15. d

16. Neural tissue allows an animal to perceive a stimulus, and muscle tissue allows the animal to respond to the stimulus.
17. Early colonial flagellates probably lost their flagella, and different cells became specialized for different tasks, such as reproduction.
18. Two phyla represented on the same branch of a phylogenetic tree are thought to be more closely related to each other than to phyla found on other branches.
19. At some stage of their lives, all chordates have a notochord, a dorsal nerve cord, pharyngeal pouches, and a postanal tail. In the adult human, the notocord has developed into a backbone, and the dorsal nerve cord has developed into the spinal cord and the brain. The pharyngeal pouches and tail disappear.
20. In animals with direct development, the immature animal looks like a small version of the adult. Animals with indirect development have an intermediate larval stage that looks quite different from the adult. Except for amphibians, vertebrates undergo direct development.
21. As vertebrates adapted to land, the bodies of later-evolving animals, such as mammals, were situated directly over their legs.
22. In some aquatic species, body waste diffuses directly through the integument. In terrestrial animals, body waste must be concentrated to conserve water in the body, then this concentrated waste must be expelled from the body.
23. In a pseudocoelomate, the body cavity is lined with mesoderm only on the interior body wall; the endodermic gut is not covered by mesoderm. In coelomates, the entire body cavity is lined with mesoderm.
24. endoderm: throat, gills, lungs, gut and associated organs; mesoderm: skeleton, muscles, circula-

34-4
- During the first cell divisions in the zygote, called cleavage, cells divide repeatedly.
- The mass of cells produced by cleavage continues to divide, producing the blastula.
- During gastrula formation, the germ layers —the ectoderm, the endoderm, and in most phyla, the mesoderm—are defined.
- The endoderm-lined central cavity of the embryo forms the throat passage, gills or lungs, and the gut and its accessory organs.
- In most protostomes, each cell of the blastula nestles between two cells of adjacent rows; the blastopore develops into the mouth. Separation of cells of the early embryo cause it to die. The mesoderm forms from the division of cells at the junction of the endoderm and the ectoderm in a process called schizocoely.
- In most deuterostomes, each cell in the blastula rests directly over the cell of the next row, and the blastopore develops into the anus. Separation of cells of the early embryo results in the development of multiple embryos. The mesoderm forms from cell divisions at the top of the archenteron in a process called enterocoely.
- Acoelomates have no body cavity. Pseudocoelomates have a body cavity partially lined with mesoderm, and coelomates have a coelom.

Vocabulary

acoelomate (686)
archenteron (683)
blastocoel (683)
blastopore (683)
blastula (683)
cleavage (683)
coelom (684)
coelomate (686)
determinate cleavage (684)
deuterostome (684)
ectoderm (683)
endoderm (683)
enterocoely (685)
gastrula (683)
gastrulation (683)
indeterminate cleavage (684)
mesoderm (683)
protostome (684)
pseudocoelom (686)
pseudocoelomate (686)
radial cleavage (684)
schizocoely (685)
spiral cleavage (684)

REVIEW

Vocabulary

1. Distinguish between cell specialization and differentiation.
2. Identify the structures that allow the formation of tissue from many separate cells.
3. Distinguish between radial symmetry and bilateral symmetry.
4. What is the difference between a coelomate and an acoelomate?
5. Compare determinate cleavage with indeterminate cleavage.

Multiple Choice

6. The process that occurs during development in multicellular organisms and leads to cell specialization is (a) asexual reproduction (b) evolution (c) differentiation (d) fertilization
7. Animals must eat because they are (a) not autotrophic (b) not heterotrophic (c) neither autotrophic nor heterotrophic (d) both autotrophic and heterotrophic.
8. Animals that have no true tissues are the (a) chordates (b) ctenophores (c) cnidarians (d) sponges.
9. Fundamental tissue types in the embryo are called (a) notochords (b) germ layers (c) pharyngeal pouches (d) coeloms.
10. Gas exchange in many aquatic phyla takes place in the (a) lungs (b) gut (c) kidneys (d) gills.
11. The repeating units of the backbone are called the (a) integument (b) exoskeleton (c) notochord (d) vertebrae.
12. A feature in vertebrates that is an adaptation to life on land is (a) lungs (b) gills (c) kidneys (d) a gut.
13. The process that takes place as the zygote begins to divide immediately after fertilization is (a) gastrulation (b) cleavage (c) meiosis (d) organ formation.
14. Organisms in which the blastopore develops into the anus are called (a) protostomes (b) gastrulae (c) pseudocoelomates (d) deuterostomes.

15. The process of mesoderm formation by division of the cells at the top of the archenteron is called (a) indeterminate cleavage (b) determinate cleavage (c) schizocoely (d) enterocoely.

Short Answer

16. Explain how neural tissue and muscle tissue work together in an animal's body to allow the animal to respond to its environment.
17. What probable changes did early colonial flagellates undergo as they evolved into the first animals?
18. What can you infer if two phyla are represented on the same branch of a phylogenetic tree?
19. What are four features common to all chordates at some time in their life, and what has happened to two of these features in an adult human?
20. How does indirect development differ from direct development, and which type is found in vertebrates?
21. What happened to the position of the body with respect to the legs as vertebrates adapted to life on land?
22. How do the problems of disposal of body waste differ in terrestrial and aquatic animals?
23. How does the body cavity of a pseudocoelomate differ from that of a coelomate?
24. Name two body parts formed by each of the following: endoderm, mesoderm, and ectoderm.
25. What structure does the archenteron become in a developing animal?

CRITICAL THINKING

1. From the perspective of a single cell, what is one advantage of cell specialization and one disadvantage of cell specialization?
2. Considering that an endoskeleton can support more weight than an exoskeleton, would a large-bodied animal with an exoskeleton be more likely to live in the water or on land? Why?
3. On mammals and birds, the head is positioned higher with respect to the body than it is on amphibians and reptiles. Why might it be helpful to have a head positioned over the body?
4. Observe the body of the animal pictured below, and answer the following questions.
   a. What kind of symmetry does the animal display?
   b. Is it cephalized?
   c. How many germ layers does it have?
   d. How many openings does its digestive system have?
   e. Does it have neurons?

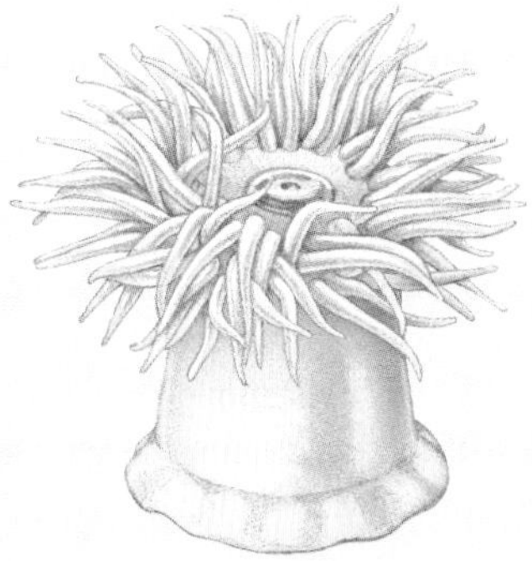

EXTENSION

1. Read "Early Learning" in *New Scientist*, January 29, 2000, on page 6, and answer the following questions: What is the focus of the study by Gilly and Preuss at Hopkins Marine Station in California? What reasons did scientists give for choosing to study squids to find out more about adult humans? How does the behavior of an adult squid differ from the behavior of a juvenile squid in reaction to a mild shock?
2. Some human disorders result from problems in very early stages of embryo development. Research two disorders related to embryonic development, and write a brief report outlining the probable cause of each disorder, the embryonic stage at which the problem occurs, and the physical manifestations of the disorder in the child or adult.

tory system; ectoderm: outer layer of skin, hair, nails, nervous system
25. the gut

CRITICAL THINKING

1. Answers will vary. For example, cell specialization implies that other specialized cells are available to carry out activities that this cell is not tailored for, which is an advantage to a single cell. A disadvantage is that the cell cannot exist alone.
2. A very large-bodied animal with an exoskeleton would be more likely to live in the water than on land because water would help support its body.
3. A head positioned over the body allows an animal to see relatively long distances, which would help it to evade predators and find food.
4. a. radial, b. no, c. two, d. one, e. yes

EXTENSION

1. Gilly and Preuss hope to find out more about how early learning affects brain chemistry, which in turn affects adult behavior and abilities. Squids have a simple nervous system, and they exhibit behavior similar to that of vertebrates. When shocked, juveniles fire their giant axons and jet away instantly. Adults have learned to suppress this behavior in favor of a more subdued response that is more favorable for hunting.
2. You may wish to suggest that students read about the relatively common disorders related to malformation of the neural crest, such as cleft palate, scoliosis, and spina bifida. Accept any topic that pertains to a disorder of embryonic development, including those deriving from problems in the uterine environment (such as exposure to teratogens) and those arising from genetic origins.

CHAPTER 34 INVESTIGATION

TIME REQUIRED

one 50-minute class period

SAFETY PRECAUTIONS

Tell students to handle the sharp and pointed instruments—including the scalpel, tweezers, scissors, and the probe—carefully. At the end of the investigation, instruct students to place the sheep's hearts in a plastic biohazard bag. Secure the bag, and hand it over personally to the school custodian for safe disposal. Also, place the used scalpel blades in a biohazard sharps container for proper disposal.

QUICK REFERENCES

Lab Preparation notes are found on page 665D.

Holt BioSources provides a Teaching Resources CD-ROM that contains "Using Gowin's Vee in the Lab" and "Scoring Rubrics."

PROCEDURAL TIP

During the dissection, have students compare the sheep's heart to a diagram of the human heart shown in Chapter 47.

ANSWERS TO BACKGROUND

1. Multi-chambered hearts efficiently separate oxygenated and deoxygenated blood.
2. Gas exchange occurs in the lungs; carbon dioxide diffuses out of the blood, and oxygen diffuses into the blood.

CHAPTER 34 INVESTIGATION

Sheep's Heart Dissection

OBJECTIVES

- Describe the appearance of the external and internal structures of a sheep's heart.
- Name the structures and functions of a sheep's heart.

PROCESS SKILLS

- observing structures
- identifying
- demonstrating

MATERIALS

- sheep's heart
- dissecting tray
- blunt metal probe
- scissors
- scalpel
- tweezers

Background

1. The heart has a left and a right side. It has two upper chambers, the left and right atria, and two lower chambers, the left and right ventricles. Why do multiple chambers result in a more efficient heart?
2. Blood enters the heart from the body through the superior or inferior vena cava. The blood then enters the right atrium and flows through valves into the right ventricle. Blood flows from the right ventricle through the pulmonary artery to the lungs. What process occurs in the lungs?
3. Oxygenated blood flows from the lungs through the pulmonary veins to the left atrium. Then it flows through valves into the thick-walled left ventricle. Blood flows from the left ventricle through the large aorta to the rest of the body.

Procedure

1. In this lab, you will observe the external structure of a sheep's four-chambered heart and dissect the heart to study its internal structure.
2. Put on safety goggles, gloves, and a lab apron.
3. Place a sheep heart in a dissecting tray. Turn the heart so that the ventral surface is facing you, as shown in the diagram below. Use the diagram of the ventral view to locate the left and right atria, the left and right ventricles, the aorta, the superior and inferior vena cava, and the pulmonary arteries. Turn the heart over. Use the diagram of the dorsal view to locate once again the structures just named, as well as the pulmonary veins.
4. Use a blunt metal probe to explore the blood vessels that lead into and out of the chambers of the heart.
5. Locate a diagonal deposit of fat along the lower two-thirds of the heart. This serves as a guideline to mark the wall between the two ventricles. Use this fatty deposit to guide your incision into the heart.

(a) DORSAL VIEW

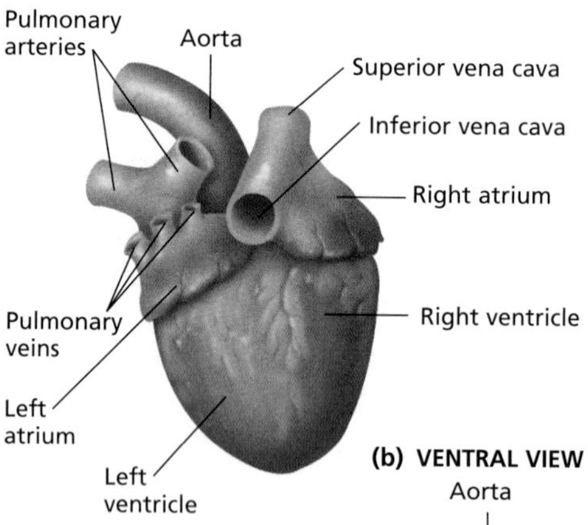

(b) VENTRAL VIEW

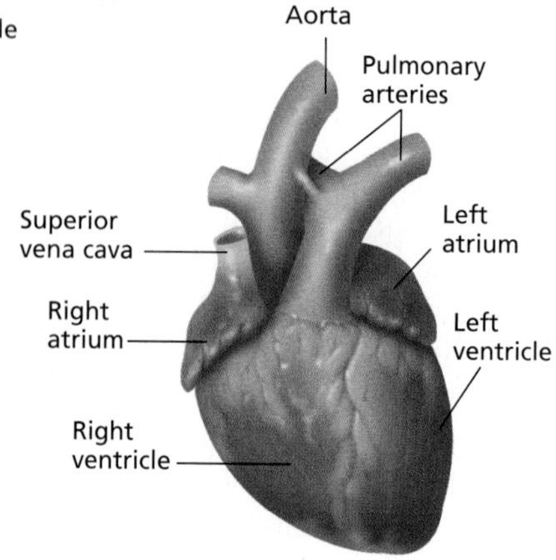

6. Follow the cutting diagram below very carefully to study the anatomy of the *right* side of the heart.

7. Again turn the heart with the ventral surface facing you and the apex pointing downward. Use scissors to cut along line 1. **CAUTION Always cut in a direction away from your face and body.** Cut just deep enough to go through the atrial wall. Continue the cut into the right ventricle. With a probe, push open the heart at the cut, and examine the internal structure.

(a) VENTRAL VIEW **(b) VENTRAL VIEW**

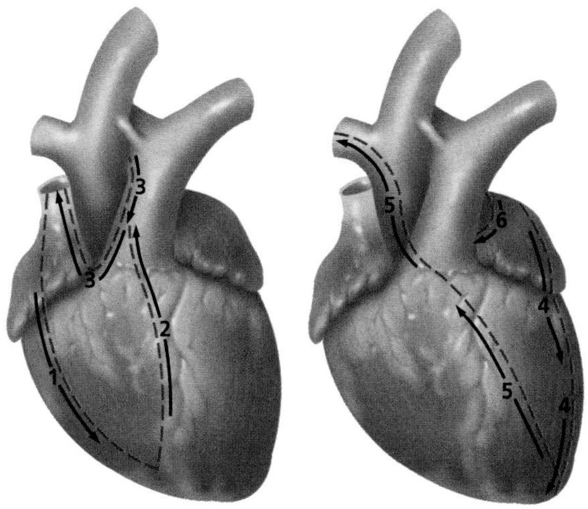

8. Cut along line 2, and extend the cut upward toward the pulmonary artery. Cut just deep enough to go through the ventricle wall. Complete the cut on line 3. Cut downward along the pulmonary artery, around through the wall of the right atrium, and upward along the right superior vena cava.

9. With tweezers, carefully lift the resulting flap to expose the structures underneath.

10. Follow the cutting diagram above very carefully to study the anatomy of the *left* side of the heart.

11. Start to cut on line 4 at the top of the left atrium, and continue into the left ventricle. Cut just deep enough to go through the ventricle wall.

12. Cut on line 5 across the middle of the left ventricle into the aorta. Leave a small margin between this cut and the cut previously made for line 2. Begin to cut on line 6 on the left atrium where cut 4 began. Extend this cut around and through the pulmonary artery upward on the aorta to the right of cut 5.

13. With tweezers, carefully lift up the resulting flap to expose the structure underneath.

14. Observe the thick septum dividing the left and right ventricles. Also note the greater thickness of the walls of the left ventricle.

15. Locate the tricuspid valve between the right atrium and ventricle. Locate the mitral valve between the left atrium and ventricles. Observe that the valves are connected by fibers to the inner surface of the ventricle. Use a probe to explore the openings in the valves.

16. With a scalpel, cut across a section of the aorta and a section of the vena cava. Compare the thickness of their walls.

17. Dispose of your materials according to the directions from your teacher.

18. Clean up your work area and wash your hands before leaving the lab.

Analysis and Conclusions

1. Trace the path of blood from the right atrium to the aorta.
2. Pulmonary circulation carries blood between the heart and the lungs. Systemic circulation carries blood to the rest of the body. In what chambers of the heart does pulmonary circulation begin and end? In what chamber does systemic circulation begin and end?
3. What is the function of the septum separating the left and right ventricles?
4. What is the function of the mitral and tricuspid valves?
5. Why are the walls of the left ventricle thicker than the walls of the right ventricle?

Further Inquiry

The heartbeat originates in a small bundle of tissue in the right atrium. This bundle is the sinoatrial, or S-A, node. Read about the S-A node. What does it do? Why is the S-A node known as the pacemaker?

ANSWERS TO ANALYSIS AND CONCLUSION

1. right atrium, right ventricle, pulmonary arteries, lungs, pulmonary veins, left atrium, left ventricle, aorta
2. Pulmonary circulation begins in the right ventricle and ends in the left atrium. Systemic circulation begins in the left ventricle and ends in the right atrium.
3. The septum keeps the deoxygenated blood in the right ventricle separate from oxygenated blood in the left ventricle.
4. The mitral valve separates the left atrium from the left ventricle; the tricuspid valve separates the right atrium from the right ventricle.
5. The muscular left ventricle must supply sufficient pressure to pump blood out of the heart and through the body.

FURTHER INQUIRY

The S-A node regulates heartbeat by coordinating activation of muscle fibers in the heart. It is called the "pacemaker" because of its timing and coordination function.

CHAPTER 35

SPONGES, CNIDARIANS, AND CTENOPHORES

	TOPICS	TEACHING RESOURCES	LABS, CLASSWORK, AND HOMEWORK
BLOCK 1 45 minutes	**Introducing the Chapter, p. 692**	**ATE** Focus Concept, p. 692 **ATE** Assessing Prior Knowledge, p. 692 **ATE** Understanding the Visual, p. 692	■ Supplemental Reading Guide, *Journey to the Ants: A Story of Scientific Exploration*
	35-1 Porifera, p. 693 Structure and Function of Sponges, p. 693	**ATE** Section Overview, p. 693 **ATE** Visual Link Figure 35-2, p. 695 134–135, 144–145	**Lab A23,** Observing Some Major Animal Groups ★ Study Guide, Section 35-1 **PE** Section Review, p. 695
BLOCK 2 45 minutes	**35-2 Cnidaria and Ctenophora, p. 696** Structure and Function of Cnidarians, p. 696 Class Hydrozoa, p. 698 Class Scyphozoa, p. 700 Class Anthozoa, p. 701 Phylum Ctenophora, p. 702	**ATE** Section Overview, p. 696 **ATE** Critical Thinking, p. 697 **ATE** Visual Link Figure 35-4, p. 697 134–135, 143, 147–148, 153–155	**PE** **Chapter 35 Investigation,** p. 706 **PE** **Quick Lab,** p. 698, Identifying Poriferans, Ctenophorans, and Cnidarians **Lab A23,** Observing Some Major Animal Groups ★ Study Guide, Section 35-2 **PE** Section Review, p. 702
BLOCKS 3 & 4 90 minutes	**REVIEW and ASSESSMENT** **PE** Chapter 35 Review, pp. 704–705 ■ Performance-Based Assessment—Exploring Invertebrates	Audio CD Program	**CHAPTER TESTING** ■ Chapter Test (blackline master) ▲ Test Generator ■ Assessment Item Listing ■ Scoring Rubrics and Classroom Management Checklists

MODERN BIOLOGY — HOLT BIOSOURCES — One-Stop Planner CD-ROM

KEY

- **PE** Pupil's Edition
- **ATE** Teacher's Edition
- ■ Active Reading Guide
- ★ Study Guide
- Teaching Transparencies
- Laboratory Program

One-Stop Planner CD-ROM
Shows these resources within a customizable daily lesson plan:

- ■ Holt BioSources Teaching Resources
- ■ Active Reading Guide
- ★ Study Guide
- ▲ Test Generator

READING FOR CONTENT MASTERY

Preview: Have students preview the objectives, the introductory paragraph, topic sentences, and captions before reading the chapter.

- **ATE** Active Reading Technique: Reading Effectively, p. 693
- ■ Active Reading Guide Worksheet 35-1
- **PE** Summary/Vocabulary, p. 703

- ■ Active Reading Guide Worksheet 35-2
- **PE** Summary/Vocabulary, p. 703

TECHNOLOGY AND INTERNET RESOURCES

CNN PRESENTS Video Segment 24 Year of the Reef

Holt Biology Videodiscs Teacher's Correlation Guide, Lessons 35-1 through 35-2

Audio CD Program, Sections 35-1 through 35-2

internetconnect

SCILINKS NSTA

On-line Resources: www.scilinks.org
The following *sci*LINKS Internet resources can be found in the student text for this chapter:

Topics:
- Sponges, p. 693
- Hydra, p. 699

go.hrw.com

On-line Resources: go.hrw.com
Visit the HRW Web site for a variety of resources related to this chapter. Just type in the keyword HM2 Chapter 35.

Smithsonian Institution®

Visit **www.si.edu/hrw** for additional on-line resources.

CNNfyi.com

Visit **www.cnnfyi.com** for late-breaking news and current event stories selected just for you.

LAB ACTIVITY PLANNING

CHAPTER 35 INVESTIGATION

Observing Hydra Behavior, pp. 706–707

OVERVIEW
In this investigation, students will observe live hydras, their structures, feeding behaviors, and responses to stimuli such as touch.

SAFETY
Students will wear a lab apron and safety goggles during the lab. Instruct students to keep their hands away from their eyes while handling the hydra and *Daphnia* cultures. Methylene blue and vinegar are irritants. If either chemical gets in a student's eye, flush the eye with water for 15 minutes and get medical attention.

PREPARATION
1. Hydras can be obtained from a pond or a biological supply house and can be maintained in a freshwater aquarium. Feed them *Daphnia* daily. Do not feed the hydras for 1–2 days before the lab. Each group will need three hydras.
2. Each lab group will need one *Daphnia*.
3. Each lab group will need a microscope slide, two medicine droppers, forceps, one drop of concentrated beef broth, one drop of vinegar, and filter paper.
4. Purchase reusable silicone culture gum to create a microaquarium on a glass microscope slide.

Quick Labs

Identifying Poriferans, Ctenophorans, and Cnidarians, p. 698
Students will need pencils, paper, and pictures of poriferans, ctenophorans, and cnidarians for this activity.

UNDERSTANDING THE VISUAL

Ask students if they think the coral shown in the photograph is a plant or an animal. Explain that corals are animals, even though they are stationary and some, like the one in the photograph, are branched. Ask students what process plants can carry out that animals cannot (photosynthesis). Point out that since stationary animals cannot make food through photosynthesis or move about in search of food, they must either collect food that passes by them or live symbiotically with photosynthetic organisms.

FOCUS CONCEPT

Stability and Homeostasis

The invertebrates in this chapter achieve stability and homeostasis several ways. The collar cells of sponges regulate the flow of water and nutrients into the sponge, while the osculum directs the outflow of water and wastes. The cnidocytes of cnidarians and colloblasts of ctenophores capture food, providing the nutrients these organisms need to maintain their structure and grow.

ASSESSING PRIOR KNOWLEDGE

Review the following concepts.

Cilia and Flagella: *Chapter 4*
Ask students to explain the difference between cilia and flagella. Ask them to describe the types of functions that cilia and flagella perform.

Reproduction: *Chapter 8*
Ask students to explain the difference between sexual and asexual reproduction. Ask them which process results in offspring that are genetically identical to the parents.

CHAPTER 35

SPONGES, CNIDARIANS, AND CTENOPHORES

Delicate polyps of Monet's tube coral,* Dendrophyllia gracilis, *extend from a hard skeleton.

FOCUS CONCEPT: ***Stability and Homeostasis***

As you read, notice how specialized functions related to feeding, reproduction, and other activities are performed in animals with simple body plans and few or no organs.

PORIFERA

Invertebrates are animals that do not have a backbone. Rather than being classified according to shared characteristics, invertebrates are an arbitrary classification of extremely diverse animals that share the absence of a characteristic. Invertebrates comprise more than a dozen phyla and more than a million species. About 97 percent of all animal species known are invertebrates, among the simplest of which are sponges.

SECTION

35-1

OBJECTIVES

▲ Define *invertebrates,* and explain why they are such a diverse group.

● Describe the basic body plan of a sponge.

■ Describe the process of filter feeding in sponges.

◆ Contrast the processes of sexual and asexual reproduction in sponges.

STRUCTURE AND FUNCTION OF SPONGES

Sponges are aquatic animals that make up the phylum Porifera (pohr-IF-uhr-uh). These simple organisms clearly represent the transition from unicellular to multicellular life. Sponges have no gastrula stage, exhibit less cell specialization than most other animals, and have no true tissues or organs. There are about 10,000 species of sponges. About 150 species live in fresh water, while the rest are marine.

Early biologists thought sponges were plants, and most sponges do resemble plants in some ways. Adult sponges are **sessile,** which means they attach themselves firmly to a surface and do not move. Sponges grow in many shapes, sizes, and colors, and they often look like mossy mats, cactuses, or blobs of fungus. They can be as small as 1 cm (0.4 in.) in length or as large as 2 m (6.6 ft) in diameter.

Body Plan

The basic body plan of a sponge, as shown in Figure 35-1, suggests many relationships between structure and function. The body wall consists of two layers of cells separated by a jellylike substance. In simple sponges, the body wall surrounds a hollow cylinder that is closed at the bottom and open at the top. The interior of the cylinder is lined with **collar cells.** By beating their flagella, collar cells draw water into the sponge through numerous pores that penetrate the body wall. In fact, the name *Porifera* comes from a Latin word meaning "pore-bearer." The water that is pumped into the interior of the sponge leaves through the **osculum** (AHS-kyoo-luhm), the opening at the top of the sponge that you can see in Figure 35-1.

A sponge would collapse without some type of supporting structure. In some sponges, support is provided by a simple skeleton made of a network of protein fibers called **spongin** (SPUHN-jin). Other sponges have skeletons consisting of **spicules,** tiny, hard particles

internet**connect**

SCILINKS NSTA

TOPIC: Sponges
GO TO: www.scilinks.org
KEYWORD: HM693

SECTION 35-1

SECTION OVERVIEW

Cell Structure and Function
Section 35-1 discusses the functions of two types of cells that make up sponges: collar cells and amebocytes.

Matter, Energy, and Organization
Section 35-1 describes how the sponge body is organized into two layers of cells and is supported by a skeleton of protein fibers, mineral crystals, or both.

ACTIVE READING

Technique: Reading Effectively

Before students begin to read this chapter, have them write down all of the key words for Sections 35-1 and 35-2. The key words can include the vocabulary words and the main topic words of an introductory paragraph. Ask students to write a definition next to each word that they have heard of. As the students read the chapter, have them write definitions next to the words that they did not previously know and modify as needed any definitions of words they knew.

DEMONSTRATION

Fill a small aquarium with water, and use an aquarium pump to circulate the water through a filter. Add enough pond sediment or other particulate debris to make the water cloudy. Show students how the pump and the filter work together to clear the water. Ask students what structures in sponges are analogous to the pump and the filter. (The flagella on collar cells serve as a pump. The pores and collar cells serve as a filter.) Explain that both the aquarium pump and the sponge must expend energy to move water. The pump uses electrical energy, while the sponge uses chemical energy in the form of ATP.

SECTION 35-1

QUICK FACT

☑ A sponge that is 10 cm tall must filter more than 100 L of water a day to stay alive.

CULTURAL CONNECTION

Using Sponges

For centuries, people around the world have used natural sponges with spongin skeletons for cleaning and bathing, taking advantage of these sponges' soft, flexible, and highly porous bodies. The ancient Greeks also used sponges as padding inside their helmets and as a part of their first-aid supplies for wounds received in battle.

RECENT RESEARCH

Freshwater Sponges and Zebra Mussels

Freshwater sponges are becoming very abundant in the Great Lakes. Recent studies indicate that this upsurge in the sponge population is due to the zebra mussel, a non-native species of mollusk that was introduced to the area. Zebra mussels stir up water currents around them, thereby diverting more food particles to the sponges, which increases the sponges' reproductive success.

RETEACHING *ACTIVITY*

Show students a diagram of a sponge, and provide them with self-sticking labels bearing the names of sponge cells and structures. Have students describe the path of water as it flows through a sponge, the transport of nutrients within a sponge, and the process of sexual reproduction in sponges. As students name structures in their descriptions, have them attach the corresponding labels to the structures on the diagram.

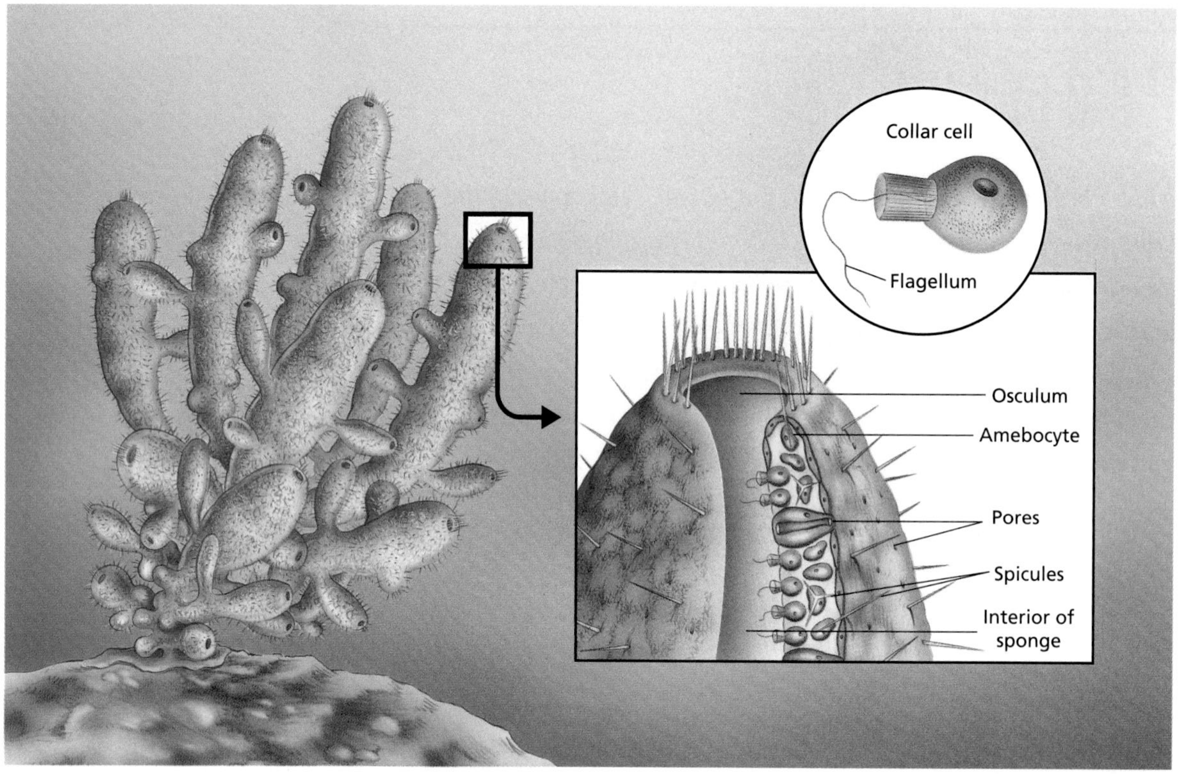

FIGURE 35-1

The body of a sponge is a hollow cylinder. Water is drawn into the cylinder through many small pores and exits through the opening at the top of the sponge, called the osculum.

of calcium carbonate or silicon dioxide that are often shaped like spikes. Calcium carbonate is one of the compounds that give bones and teeth their hardness, and silicon dioxide is the major component of glass and quartz. Still other sponges have a combination of spongin and spicules. Differences in the composition of the skeleton have enabled biologists to divide sponges into three classes.

Feeding and Digestion

Because they are sessile, sponges cannot pursue their food. Instead, most sponges feed by screening food out of the water that the collar cells pump through their body. This feeding method is called **filter feeding.** The food of such sponges includes bacteria, protozoans, unicellular algae, and bits of organic matter. However, scientists have discovered one species of sponge that uses movable filaments covered with hooked spicules to snare small shrimp. The shrimp are then absorbed into the sponge's body.

The food that a sponge collects is engulfed and digested by collar cells. Nutrients then pass from the collar cells to other cells that crawl about within the body wall, bringing the nutrients to the rest of the body. Scientists call these crawling cells **amebocytes** (uh-MEE-buh-siets) because they resemble amoebas. Locate the amebocytes in Figure 35-1. This procedure for distributing food shows how sponges are related to their protozoan ancestors. Carbon dioxide and other wastes produced by the sponge's cells diffuse into the water passing through the sponge. The water carries these wastes away as it flows out through the osculum.

Word Roots and Origins

amebocyte

from the Greek *amoibe,* meaning "change," and *kytos,* meaning "hollow vessel"

Reproduction

Sponges can reproduce asexually by forming small buds that break off and live separately. The sponge illustrated in Figure 35-1 has many buds that are still attached. During droughts or cold weather, some freshwater sponges produce internal buds called **gemmules** (JEM-yoolz). Each gemmule is a food-filled ball of amebocytes surrounded by a protective coat made of organic material and spicules. Gemmules can survive harsh conditions that may kill the adult sponge that formed them. When conditions improve, the sponge cells emerge from the gemmules and grow into new sponges.

Sponges also have remarkable powers of **regeneration,** the ability to regrow missing parts. In fact, a small piece of a sponge can regenerate a complete new sponge. In some species, even particles small enough to pass through a cloth strainer can regenerate.

Sponges can also reproduce sexually. As you can see in Figure 35-2, sperm released into the water from one sponge enter the pores of another sponge. Collar cells in the second sponge engulf the sperm and transfer them to amebocytes, which carry the sperm to an egg. After the egg is fertilized, it develops into a larva. A larva is an immature stage of an animal that is usually very different in form from the adult. Flagella on the larva's surface enable the larva to leave the parent sponge and swim about as it is carried along by the current. Eventually, the larva settles and attaches to an object. Its cells then reorganize to form an adult sponge.

Some species of sponges have separate sexes, but in most species each individual produces both eggs and sperm. Any individual organism that produces both eggs and sperm is called a hermaphrodite (her-MAF-roh-diet). Self-fertilization rarely, if ever, occurs in hermaphroditic species. Instead, the sperm of one individual usually fertilize the eggs of another individual. Since all hermaphrodites produce eggs, the chances of successful fertilization are greater than they would be if only half of the population (females) produced eggs. Hermaphroditism is common in many invertebrates that are sessile, that move slowly, or that live in low-density populations.

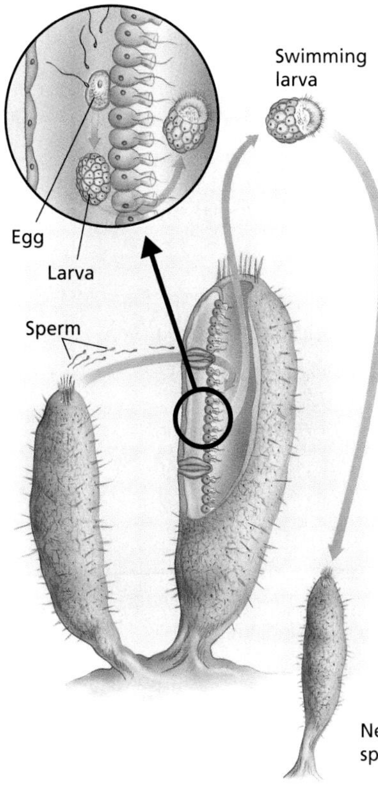

FIGURE 35-2

The union of a sponge egg and sperm ultimately results in a swimming larva that escapes from the parent sponge and grows into a new individual.

SECTION 35-1 REVIEW

1. Define *invertebrate.* Why is it not surprising that a group defined in this way includes a wide diversity of species?
2. What is the difference between spongin and spicules? What function do they both perform?
3. Describe how most sponges feed.
4. How do gemmules help some freshwater sponges survive unfavorable conditions?
5. What role do amebocytes play in the sexual reproduction of sponges?
6. CRITICAL THINKING Since sponges are sessile animals, how is it possible for a population of sponges to spread out into a larger area?

SECTION 35-1

VISUAL LINK
Figure 35-2

Have students examine Figure 35-2 and identify three types of cells or structures in the figure that bear flagella (sperm, collar cells, and swimming larvae). Ask them to explain the functions served by the flagella in each case. (Flagella propel sperm and larvae through the water. Flagella on collar cells circulate water through the sponge.) Point out that eggs, which lack cilia and flagella, are nonmotile. They remain in the sponge that produces them, where, if fertilized, they develop into larvae.

ANSWERS TO SECTION 35-1 REVIEW

1. *Invertebrate* means "lacking a backbone." Because this group is defined by what its members lack, the characteristics that its members possess can be extremely varied.
2. Spongin is a protein that forms a network in some sponges. Spicules are tiny, hard particles of calcium carbonate or silicon dioxide. Both spongin and spicules provide support for sponges.
3. Collar cells screen small organisms and food particles out of the water that they pump through the sponge's body. The collar cells engulf and digest the food, releasing nutrients for amebocytes to carry throughout the body.
4. Gemmules can survive harsh conditions that may kill adult sponges. When more-favorable conditions return, the gemmules can develop into new sponges.
5. Amebocytes carry sperm to an egg inside the sponge.
6. Sponge larvae have flagella that enable them to escape from their parents and swim. Water currents may carry the larvae far from the parents before the larvae attach to an object.

SECTION OVERVIEW

Cell Structure and Function
Section 35-2 describes the functions of specialized cells: cnidocytes, which are used for feeding and defense by cnidarians; and colloblasts, which are used for feeding by ctenophores.

Stability and Homeostasis
Section 35-2 describes the organization of the nerve net, a system of nerve cells that allows cnidarians to perform coordinated movements during feeding, locomotion, and defense.

▶ ENGAGE STUDENTS

Bring several pieces of hard coral to class, and allow students to observe and feel them. Ask the students how these coral pieces might differ from living corals in the ocean. (Living corals have small, soft polyps that protrude from their calcium carbonate skeletons.) Explain that the skeletons of corals, like the bones and shells of other animals, provide protection and support for the soft tissues of these organisms.

QUICK FACT

✔ The venom of the box jellyfish of Australia can kill a human in 4 min or less. No other poisonous animal can kill a human this quickly.

SECTION

35-2

OBJECTIVES

▲ Name and describe the two body forms of cnidarians.

● Describe the common characteristics of cnidarians.

■ Identify the three classes of cnidarians, and give an example of each.

◆ Describe the common characteristics of ctenophores.

CNIDARIA AND CTENOPHORA

Cnidaria (nie-DER-ee-uh) and Ctenophora (tee-NAHF-uhr-uh) are two phyla of radially symmetrical invertebrates. The animals in these phyla are somewhat more complex than the sponges. Their cells are organized into tissues, and they have a few simple organs. All members of the phyla Cnidaria and Ctenophora are aquatic, and most live in the ocean.

STRUCTURE AND FUNCTION OF CNIDARIANS

Tiny freshwater hydra, stinging jellyfish, and flowerlike coral all belong to the phylum Cnidaria. Animals in this phylum are called cnidarians. As you can see in Figure 35-3, the body of a cnidarian may be either vase-shaped or bell-shaped. The vase-shaped form, called a **polyp** (PAHL-ip), is specialized for a sessile existence. In contrast, the bell-shaped **medusa** (me-DOO-suh) is specialized for swimming.

Figure 35-3 also shows that all cnidarians have bodies constructed of two cell layers—an outer **epidermis** and an inner **gastrodermis.** Between these layers is a jellylike material known as **mesoglea** (mez-uh-GLEE-uh). In the center of the body is a hollow gut called the **gastrovascular cavity,** which has a single opening, or mouth. Surrounding the mouth are numerous flexible extensions called **tentacles.**

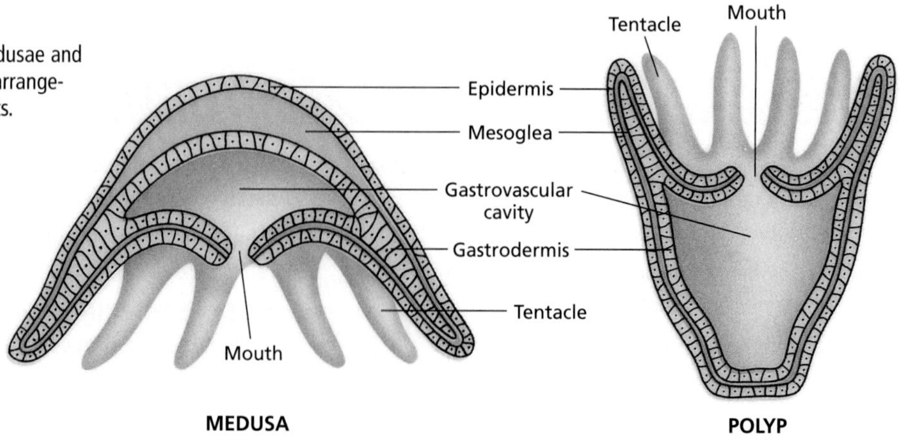

FIGURE 35-3
The contrasting forms of medusae and polyps result from different arrangements of the same body parts.

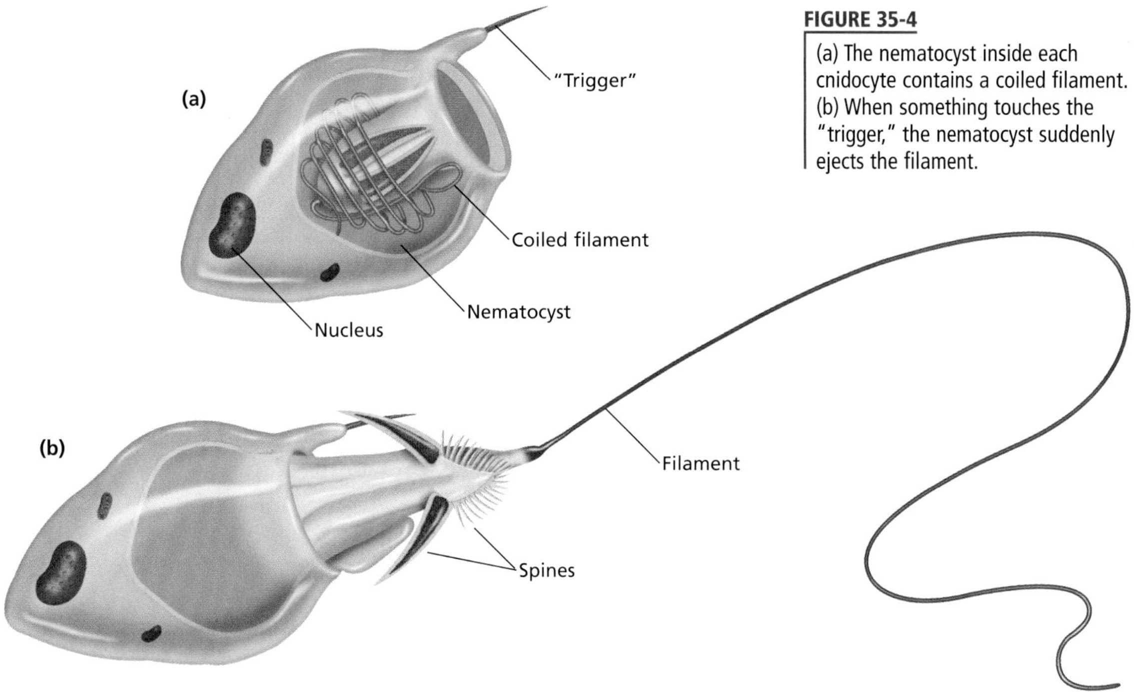

FIGURE 35-4

(a) The nematocyst inside each cnidocyte contains a coiled filament. (b) When something touches the "trigger," the nematocyst suddenly ejects the filament.

Feeding and Defense

One of the distinguishing features of cnidarians is the presence of **cnidocytes** (NIE-duh-siets), which give the phylum its name. Cnidocytes are specialized cells used for defense and capturing prey. Figure 35-4 reveals that each cnidocyte contains an organelle called a **nematocyst** (nuh-MAT-uh-sist), which has a long filament coiled up inside it. In some cnidarians, the cnidocytes are concentrated in the epidermis, especially on the tentacles. When an object brushes against the "trigger" on a cnidocyte, the nematocyst inside it suddenly pushes the filament out of the cell with great force. Some nematocysts have filaments with sharp tips and spines that puncture the object and inject poison. Others have filaments that adhere to the object by wrapping around it.

The relationship between structure and function is clearly seen in the way cnidarians feed. The tentacles capture small animals with their nematocysts and paralyze them with the poison they inject. The tentacles then push the prey into the gastrovascular cavity through the mouth. After enzymes inside the gastrovascular cavity break up the prey, cells lining the cavity absorb the nutrients. Undigested food and waste are expelled through the mouth.

FIGURE 35-5

The interconnected nerve cells in the nerve net of this cnidarian coordinate the animal's responses to its environment.

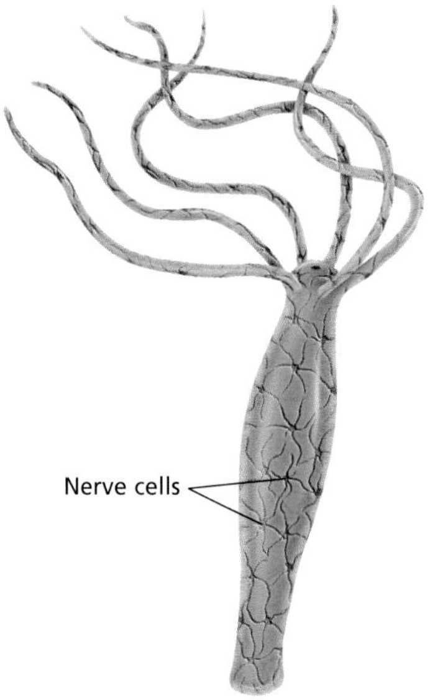

Nervous System

The nervous system in cnidarians is composed of a diffuse web of interconnected nerve cells called a **nerve net.** In many cnidarians, like the polyp shown in Figure 35-5, the nerve net is distributed uniformly throughout the entire body. There is no brain or similar structure that controls the rest of the nerve net. In the medusa form of some cnidarians, however, some of the nerve cells are clustered in rings around the edge of the bell-shaped body.

SECTION 35-2

VISUAL LINK

Figure 35-4

This figure shows a single cnidocyte before (a) and after (b) it has discharged its nematocyst. Be sure students understand that a cnidocyte is a cell that contains an organelle called a nematocyst. In response to a specific stimulus, the cnidocyte suddenly turns its nematocyst inside out, causing the filament within the nematocyst to be ejected like a harpoon. As the figure illustrates, many nematocysts have spines that ensnare prey or inject poison.

CONTENT CONNECTION

Section 5-2: Endocytosis and Exocytosis

Although digestion begins in the gastrovascular cavity of a cnidarian, it is completed inside the gastrodermal cells that line the cavity. These cells take in particles of food through endocytosis. Undigested particles are released back into the cavity through exocytosis.

CRITICAL THINKING

The Sessile Lifestyle

All sessile animals, including sponges and many varieties of cnidarians, live in aquatic environments. Why are there no sessile animals on land? (Sessile animals must obtain their nourishment from the food that passes by them. Food that swims or is suspended in water is more plentiful than food that flies or is light enough to be suspended in air.)

Quick Lab

Identifying Poriferans, Ctenophorans, and Cnidarians

Time Required 30 minutes

Procedural Tips Have students work in groups of four. The information students use for making the paired statements should be taken from the textbook. The source of the pictures of the different phyla and classes may come from the textbook or other sources available. Prepare the pictures ahead of time. Each group should have one picture.

Answers to Analysis
Poriferans and hydrozoans are colonial organisms. Cnidarians have two body types—medusae and polyps—and they capture prey by using their nematocysts. Ctenophorans move by cilia and have colloblasts that secrete a sticky substance to capture their prey. Hydrozoans are colonial polyps or both medusae and polyps. Scyphozoans are medusae. Anthozoans are polyps.

DEMONSTRATION

Make models of nematocysts using lengths of string. Nematocysts that wrap around prey, adhere to prey, or impale prey with spines can be made by attaching washers, adhesive tape, or darts to the end of the string. Explain that these models represent different kinds of nematocysts used by cnidarians. Make models of small invertebrates out of plastic foam, foam rubber, or crumpled newspaper, and show how these different types of nematocysts can be used to capture prey.

Quick Lab

Identifying Poriferans, Ctenophorans, and Cnidarians

Materials pencil, paper, a picture of a poriferan, a ctenophoran, or a cnidarian

Procedure

1. Prepare a dichotomous key to differentiate between poriferans, ctenophorans, and cnidarians. (Refer to pp. 354 and 355.)
2. Write paired statements for the phyla Porifera, Ctenophora, and Cnidaria and for the classes Hydrozoa, Scyphozoa, and Anthozoa.
3. Statement (1a) describes cnidarians and leads to the paired statements (3). Statement (1b) will lead to the paired statements (2).
4. Statement (2a) describes the phylum Porifera and (2b) describes the phylum Ctenophora.
5. Statement (3a) describes Hydrozoa and (3b) leads to paired statements (4).
6. Statement (4a) describes Scyphozoa and (4b) describes Anthozoa.
7. Get a picture from your teacher. Identify the animal's phylum and class using your dichotomous key. Exchange keys with another group, and use the new key to identify your animal.

Analysis What do sponges and hydrozoans have in common? How do comb jellies differ from cnidarians? What is the dominant body form in each of the three classes of cnidarians?

The nerve net enables cnidarians to respond to specific stimuli in their environment. For example, when cells in the epidermis are touched, they relay a signal to nerve cells. The nerve cells, in turn, transmit a signal via the nerve net to contractile cells, which can cause the animal to withdraw from the stimulus. In cnidarians with the simplest nerve nets, a stimulus anywhere on the body causes signals to be sent through the nerve net in all directions. These signals bring about a contraction of the entire body.

The nerve net also coordinates the complex activities of the body that are necessary for feeding and traveling through the environment. The movements by which the tentacles bring prey to the mouth and push it into the gastrovascular cavity are controlled by the nerve net, as are the rhythmic contractions of the body that propel swimming medusae through the water.

Classification

Scientists divide cnidarians into three classes—Hydrozoa, Scyphozoa, and Anthozoa—and the members of these classes are known as hydrozoans, scyphozoans, and anthozoans, respectively. Some species of hydrozoans live only as polyps, some live only as medusae, some alternate between these two forms, and some live as mixed colonies of polyps and medusae. Scyphozoans spend most of their lives as medusae, while anthozoans live only as polyps.

CLASS HYDROZOA

The class Hydrozoa includes about 3,700 species, most of which live as colonial organisms in the oceans. One example of a colonial hydrozoan is *Obelia.* As Figure 35-6a illustrates, *Obelia* has many polyps attached to branched stalks. Some of the polyps function in gathering food, while others are responsible for reproduction.

Perhaps the most remarkable hydrozoan is the Portuguese man-of-war, *Physalia,* shown in Figure 35-6b. This hydrozoan exists as a colony of medusae and polyps. Its gas-filled float, which can

(a)

(b)

FIGURE 35-6

Obelia (a) and *Physalia,* the Portuguese man-of-war (b), are two examples of colonial hydrozoans. *Obelia* consists solely of polyps, while *Physalia* is made of both polyps and medusae. (*Obelia:* LM 13×)

measure as much as 30 cm (1 ft) across, keeps the colony at the surface of the ocean. The polyps in the colony are specialized for feeding, digestion, or sexual reproduction. Tentacles up to 20 m (65 ft) long dangle from the feeding polyps and carry large numbers of cnidocytes. The Portuguese man-of-war preys mostly on small fish, but its cnidocytes contain a poison that can be painful and even fatal to humans.

Hydras

One hydrozoan that has been extensively studied is the hydra. Hydras are not typical hydrozoans because they exist only as polyps, they are not colonial, and they live in fresh water. Hydras range from 1 to 4 cm (0.4 to 1.6 in.) in length. Most hydras are white or brown, but some, like the one shown in Figure 35-7, appear green because of the algae that live symbiotically inside cells of their gastrodermis. Hydras can be found in quiet ponds, lakes, and streams. They attach themselves to rocks or water plants by means of a sticky secretion produced by cells at the hydra's base.

A hydra can leave one place of attachment and move to another. This can happen when the base secretes bubbles of gas, which cause the hydra to float upside down on the surface of the water. Hydras can also move by somersaulting. This peculiar movement occurs when the tentacles and the mouth end bend over and touch the bottom while the base pulls free.

During warm weather, hydras generally reproduce asexually. Small buds, like the one you can see in Figure 35-8, develop on the outside of the hydra's body. These buds grow their own tentacles and then separate from the body and begin living independently.

Sexual reproduction usually occurs in the fall, when low temperatures trigger the development of eggs and sperm. The eggs are produced by meiosis along the body wall in swellings called

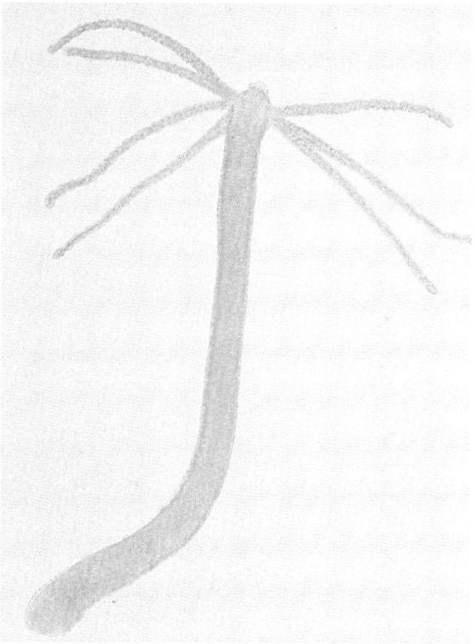

FIGURE 35-7

The green color of this hydra, *Chlorohydra viridissima,* comes from the algae that live inside the hydra's cells. (LM 30×)

internetconnect

SCILINKS NSTA

TOPIC: Hydra
GO TO: www.scilinks.org
KEYWORD: HM699

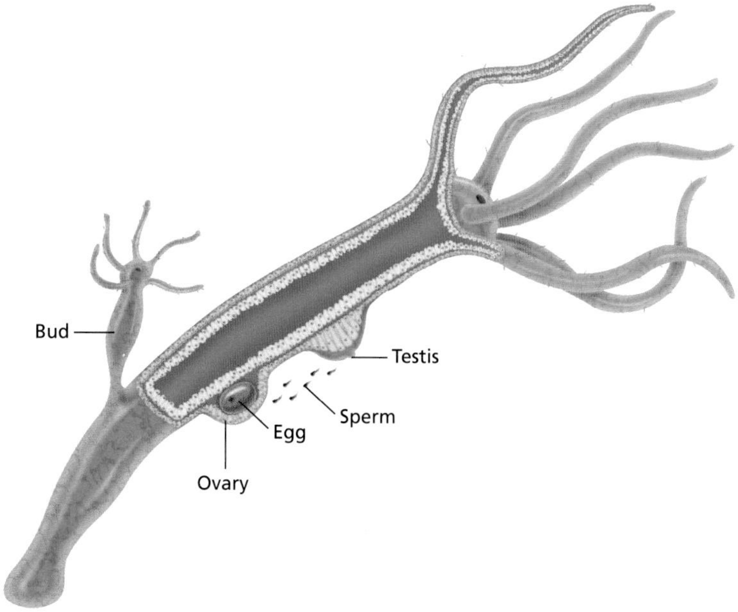

FIGURE 35-8

Hydras can reproduce either asexually, by forming buds, or sexually, by producing sperm that fertilize eggs.

RECENT RESEARCH
Hydras to Humans

At least a half-dozen genes that had been associated only with vertebrates have recently been identified in hydra. This discovery provides evidence of a genetic link between hydras and vertebrates.

INCLUSION *ACTIVITY*

Divide students into pairs. Have one student in each pair name a cnidarian, and have the other student make a hand sign to indicate whether the cnidarian exists as a polyp, a medusa, or both. The hand sign for a polyp is a cupped hand, palm-side up, and the hand sign for a medusa is a cupped hand, palm-side down. Have students name as many cnidarians as they can, and have them switch roles during the activity.

TEACHING STRATEGY
Aurelia Scramble

Have students make detailed sketches of all of the stages and key events in the life cycle of the jellyfish *Aurelia* on separate note cards. On the back of each card, have students write the name of each stage or event. Tell students to scramble the cards and then place them in the proper sequence, first using only the sketches and then using only the written information on the back of the cards.

MAKING CONNECTIONS

Mathematics

Scientists drill holes in coral reefs to study their formation, composition, and other characteristics. In 1951, a hole was drilled 1,389 m to the bottom of the reef around the Eniwetok Atoll. From fossils in the area, the reef was estimated to be 60 million years old. On average, how many meters were added to the height of the reef each million years? (about 23 m)

ovaries. Motile sperm are formed by meiosis in similar swellings called **testes.** In some species, eggs and sperm are produced in the same hermaphroditic individual, as indicated in Figure 35-8. In other species, the individuals are either male or female. In either case, sperm are released into the water, and those that reach ovaries can fertilize egg cells. Each fertilized egg then divides and grows into an embryo. A hard covering protects the embryo through the winter, and in the spring the embryo hatches out and develops into a new hydra.

CLASS SCYPHOZOA

The name *Scyphozoa* (sie-foh-ZOH-uh) means "cup animals," which describes the medusa, the dominant form of the life cycle of this class. There are more than 200 species of scyphozoans, known commonly as jellyfish. The cups of the medusae range from 2 cm (0.8 in.) to 4 m (13 ft) across, and some species have tentacles that are several meters long. Pulsating motions of the cup propel the jellyfish through the water. Like the Portuguese man-of-war, some jellyfish carry poisonous nematocysts that can cause severe pain and even death in humans.

The common jellyfish *Aurelia* is a scyphozoan whose life cycle includes both medusa and polyp forms. As you can see in Figure 35-9, adult medusae release sperm and eggs into the water, where fertilization occurs. The resulting zygote divides many times to form a blastula. The blastula then develops into a ciliated larva called a **planula** (PLAN-yuh-luh), which attaches to the ocean bottom. The planula becomes a polyp by developing a mouth and tentacles at the unattached end. As the polyp grows, it forms a stack of medusae that detach and develop into free-swimming jellyfish.

FIGURE 35-9

The common jellyfish *Aurelia* reproduces when sperm from an adult male medusa fertilize eggs from an adult female medusa. Each fertilized egg produces a blastula, which develops into a larva known as a planula. The planula forms a polyp, which produces more medusae.

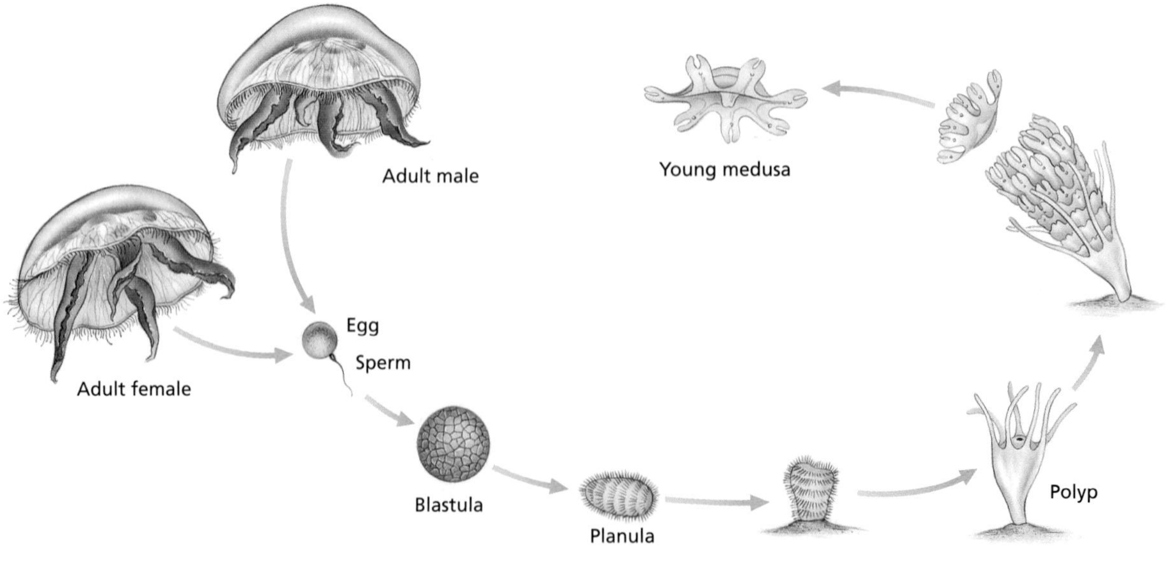

CLASS ANTHOZOA

The name *Anthozoa* means "flower animals," which is a fitting description for the approximately 6,100 marine species in this class. Two examples of anthozoans are sea anemones and corals, which are shown in Figure 35-10.

Sea anemones are polyps commonly found in coastal areas, where they attach themselves to rocks and other submerged objects. Anemones feed on fishes and other animals that swim within reach of their tentacles. However, some anemones in the Pacific Ocean have a symbiotic relationship with the clownfish, as Figure 35-11 demonstrates. The two animals share food and protect each other from predators. The movements of the clownfish also help prevent sediments from burying the anemone. For reasons that are not entirely clear, the anemone does not fire its nematocysts when the clownfish touches its tentacles.

Corals are small polyps that usually live in colonies. Each polyp cements its calcium carbonate skeleton to the skeletons of adjoining polyps in the colony. When the polyps die, their hardened skeletons remain, serving as the foundation for new polyps. Over thousands of years, these polyps build up large, rocklike formations known as **coral reefs.** Only the top layer of the reef contains the living polyps. Coral reefs provide food and shelter for an enormous and colorful variety of fishes and invertebrates.

Coral reefs are restricted to a band of ocean within 30 degrees north or south of the equator. They form only at shallow depths in warm, clear waters. These conditions are necessary in order for photosynthesis to be carried out by algae that live symbiotically inside the coral cells. The corals depend on the algae to provide oxygen and to speed up the accumulation of calcium from the sea water. The algae in turn depend on the corals to supply vital nutrients.

(a)

(b)

FIGURE 35-10

Anthozoans, including this crimson anemone, *Cribrinopsis fernaldi* (a), and this golden cup coral, *Tubastraea* (b), live as polyps along ocean coasts.

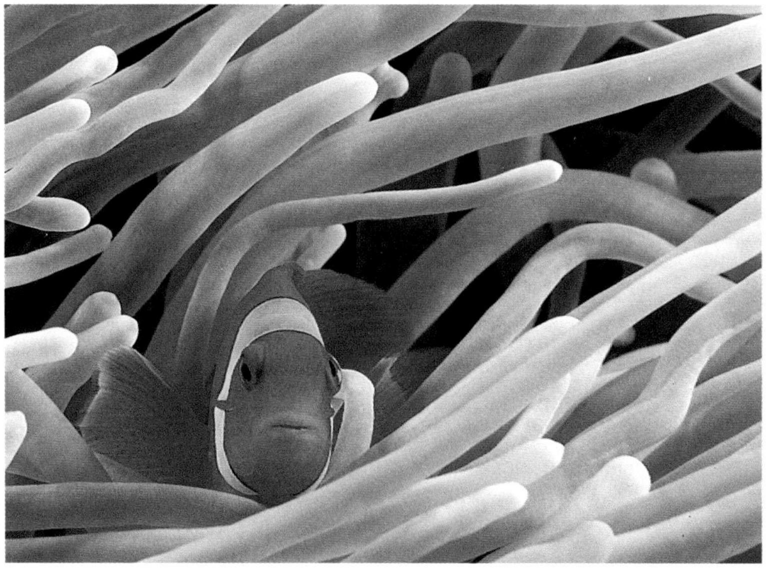

FIGURE 35-11

The clownfish, *Amphiprion ocellaris,* lives symbiotically among the tentacles of sea anemones. The anemone's stinging tentacles protect the clownfish from predators. The clownfish, in turn, drives away other fish that try to feed on the anemone.

RECENT RESEARCH

Anemone Territoriality

Scientists have found that neighboring colonies of sea anemones may actually fight each other for territory. These colonies have specific stinger tentacles for fighting, use guerrilla tactics to compete for territory on the ocean floor, and overproduce individuals to compensate for losses resulting from competition.

GIFTED ACTIVITY

Have students collect information and write a report on the symbiotic relationship between some species of sea anemones and certain types of fish, such as the clownfish and the damsel fish. Students should investigate the possible mechanisms by which these fish avoid discharging the nematocysts of the anemones.

RETEACHING ACTIVITY

Have students make a table to list similarities and differences between four cnidarians: hydra, jellyfish, sea anemone, and coral. (Similarities include radial symmetry, an aquatic existence, and the presence of cnidocytes. Differences include polyp or medusa form and solitary or colonial lifestyle.) One portion of the table should be used for a small sketch of each organism.

SECTION 35-2

ANSWERS TO SECTION 35-2 REVIEW

1. A cnidarian may exist as a vase-shaped polyp or a bell-shaped medusa.
2. Common characteristics of cnidarians include a body made of two cell layers (epidermis and gastrodermis), a gastrovascular cavity with a single opening, cnidocytes for defense and capturing prey, and a nerve net.
3. The three classes of cnidarians are Hydrozoa (including *Obelia,* Portuguese man-of-war, and hydras), Scyphozoa (jellyfishes), and Anthozoa (including sea anemones and coral).
4. Symbiosis involving cnidarians includes algae that live inside the epidermal cells of some hydras, algae that live inside the cells of corals, and clownfish that live among the tentacles of sea anemones.
5. Ctenophores are propelled through the water by eight rows of beating cilia, capture prey by means of sticky cells called colloblasts, sense their orientation in the water with an apical organ, and are bioluminescent.
6. Since medusae can swim, they can move to other areas more quickly than the mostly sessile polyps. These other areas may be more suitable for reproduction and growth.

Word Roots and Origins

colloblast

from the Greek *kollo,* meaning "glue," and *blastos,* meaning "bud"

PHYLUM CTENOPHORA

The phylum Ctenophora includes about 100 species of marine animals known as ctenophores. A typical ctenophore is shown in Figure 35-12. The name *Ctenophora* means "comb holder" and refers to the eight comblike rows of cilia that run along the outside of the animal. Because ctenophores resemble jellyfish, they are often called comb jellies.

Ctenophores differ from jellyfish and other cnidarians in several ways. Rather than pulsating like jellyfish, they move through the water by beating their cilia. They are the largest organisms to move in this fashion. Also, most ctenophores do not have cnidocytes. Instead, they have cells called **colloblasts,** which secrete a sticky substance that binds to their prey. Colloblasts are usually located on two tentacles. Ctenophores also have a sensory structure called an **apical organ** at one end of their body. This organ enables a ctenophore to sense its orientation in the water. Nerves running from the apical organ coordinate the beating of the cilia. Most ctenophores are hermaphroditic.

One of the most striking features of ctenophores is their **bioluminescence** (BIE-oh-LOO-muh-NES-ens), or production of light. Bioluminescent ctenophores often occur in large swarms near the surface of the ocean, creating a spectacular display at night.

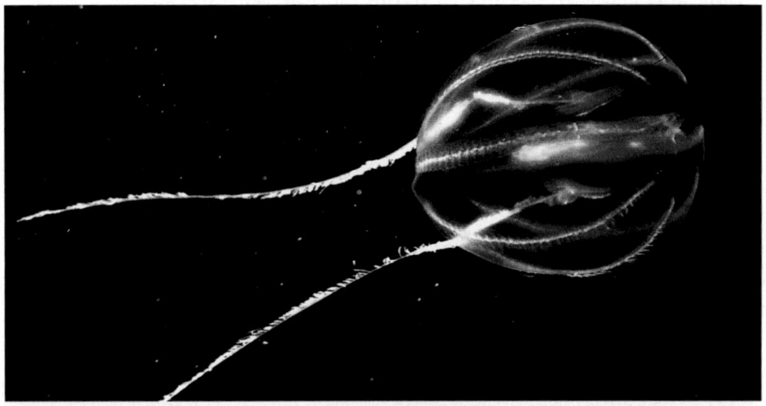

FIGURE 35-12

A ctenophore, *Pleurobrachia pileus,* moves through the water by beating its cilia, trailing its two long tentacles behind it.

SECTION 35-2 REVIEW

1. What are the two basic body forms a cnidarian may have?
2. What are three characteristics that all cnidarians have in common?
3. Name the three classes of cnidarians, and give an example of each.
4. Describe two examples of symbiosis found among cnidarians.
5. Describe three characteristics of ctenophores.
6. CRITICAL THINKING For a cnidarian, what might be the advantage of having a life cycle consisting of both medusae and polyps, instead of just polyps?

CHAPTER 35 REVIEW

SUMMARY/VOCABULARY

35-1

- Invertebrates, animals without a backbone, are a diverse group of over a million species.
- The phylum Porifera is made up of sponges, sessile invertebrates that have no true tissues or organs. The simplest sponges are hollow cylinders.
- The body wall of a sponge is composed of two layers of cells. The body is supported by a skeleton made of spongin, spicules, or both.
- Collar cells lining the inside of a sponge beat their flagella, drawing a current of water into the sponge through pores in the body wall. Water leaves through the osculum, an opening at the top of the sponge.
- Sponges feed by filtering small organisms and organic matter out of the water that passes through their body. Nutrients are distributed through the body by amebocytes, which crawl about within the body wall.
- Sponges can reproduce asexually, through budding or regeneration, as well as sexually. Most sponges are hermaphroditic, meaning that a single animal can produce both eggs and sperm.

Vocabulary

amebocyte (694)	gemmule (695)	sessile (693)	spongin (693)
collar cell (693)	osculum (693)	spicule (693)	
filter feeding (694)	regeneration (695)	sponge (693)	

35-2

- Animals in the phylum Cnidaria can be either sessile polyps or swimming medusae. Some cnidarians alternate between polyp and medusa stages during their life cycles.
- The body of a cnidarian consists of two cell layers—an outer epidermis and an inner gastrodermis—separated by a jellylike mesoglea.
- Cnidarians have cells called cnidocytes, which contain organelles known as nematocysts. When a cnidocyte is stimulated, its nematocyst ejects a filament that can paralyze or ensnare prey.
- Cnidarians feed by capturing small animals with their nematocysts and pushing the animals into their gastrovascular cavity with their tentacles.
- The cnidarian nervous system is a diffuse web of interconnected nerve cells called a nerve net.
- The three classes of cnidarians are Hydrozoa (which includes animals such as the hydra), Scyphozoa (jellyfish), and Anthozoa (which includes sea anemones and corals).
- Hydrozoans may live as polyps, medusae, or mixed colonies of polyps and medusae. Scyphozoans spend most of their lives as medusae. Anthozoans live only as polyps.
- Animals in the phylum Ctenophora move through the water by beating the cilia that occur in eight rows on the outside of their body.
- Ctenophores capture prey with a sticky substance secreted by cells called colloblasts, which are usually located on a pair of tentacles.
- An apical organ at one end of the body enables ctenophores to sense their orientation in the water. Most ctenophores are hermaphroditic and bioluminescent.

Vocabulary

apical organ (702)	epidermis (696)	mesoglea (696)	planula (700)
bioluminescence (702)	gastrodermis (696)	nematocyst (697)	polyp (696)
cnidocyte (697)	gastrovascular cavity (696)	nerve net (697)	tentacle (696)
colloblast (702)	medusa (696)	ovary (700)	testis (700)
coral reef (701)			

CHAPTER 35 REVIEW
ANSWERS

Review

1. *Porifera* means "pore bearer." The body wall of sponges has numerous pores.
2. Most sponges strain small organisms and organic matter from the water that passes through their bodies. This method is called filter feeding.
3. The vase-shaped form is called a polyp; examples include *Obelia,* hydras, sea anemones, and corals. The bell-shaped form is called a medusa; the most prominent examples are jellyfish.
4. A cnidocyte is a type of cell found in cnidarians (and in one species of ctenophore). A nematocyst is an organelle found inside a cnidocyte.
5. The apical organ enables a ctenophore to sense its orientation in the water. Nerves running from the apical organ coordinate the beating of the cilia.

6. c	**7.** c
8. a	**9.** b
10. d	**11.** c
12. b	**13.** a
14. c	**15.** b

16. Sponges have no gastrula stage, exhibit less cell specialization than most other animals, and have no true tissues or organs.
17. Nematocysts eject their filaments, which entangle and paralyze the prey. The tentacles then move the prey into the gastrovascular cavity through the mouth.
18. Driven by the beating of flagella on the collar cells, water enters through pores in the body wall and exits via the osculum at the top of the sponge.
19. Most sponges obtain food by filter feeding, which they can do even as they remain in one place.
20. The Portuguese man-of-war is a colony of medusae and polyps. The polyps are specialized for feeding, digestion, or sexual reproduction.

21. Sperm produced by one sponge are shed into the water and enter the pores of another sponge. The sperm are then engulfed by collar cells and carried to an egg by amebocytes. A fertilized egg develops into a larva, which escapes from the sponge, attaches to some object, and develops into an adult sponge.
22. The clownfish shares some of its food with the anemone, protects the anemone from predators, and prevents sediments from burying the anemone. The anemone also shares some of its food with the clownfish, and its stinging tentacles protect the clownfish from predators.
23. Corals depend on the algae that live inside them. Beyond a certain depth, there is not enough sunlight to support the algae.
24. Unlike cnidarians, ctenophores move through the water by beating their cilia. Most ctenophores capture prey with colloblasts rather than with cnidocytes. Ctenophores have an apical organ, which cnidarians lack.
25. *A*, gastrodermis; *B*, epidermis; *C*, mesoglea; *D*, cnidocyte

CRITICAL THINKING

1. Radial symmetry is fine for animals that filter food out of the water, since that food may come from any direction. Animals that live on land generally must move about to find food, and a bilateral body plan with cephalization is better for that lifestyle.
2. The production of eggs and sperm at different times prevents self-fertilization, which could increase the frequency of recessive traits in sponge populations.
3. The life cycles are similar only in that both include two distinct body forms. The major difference is that in moss a diploid form alternates with a haploid form, while in *Aurelia* both the medusa and the polyp are diploid.

REVIEW

Vocabulary

1. Explain why *Porifera* is a good name for the phylum of sponges.
2. Explain how sponges feed. What term describes this method of feeding?
3. Describe and name the two body forms of cnidarians. For each body form, give an example of a cnidarian that has that form.
4. What is the difference between a cnidocyte and a nematocyst?
5. What is the function of a ctenophore's apical organ?

Multiple Choice

6. All invertebrates (a) live in water (b) have tentacles (c) have no backbone (d) are hermaphrodites.
7. A hydra shows (a) asymmetry (b) spherical symmetry (c) radial symmetry (d) bilateral symmetry.
8. The structures that propel ctenophores through the water are (a) rows of cilia (b) colloblasts (c) apical organs (d) polyps.
9. Spongin and spicules are important to a sponge because they (a) trap food (b) provide support (c) remove wastes (d) produce offspring.
10. Both feeding and sexual reproduction in sponges involve (a) gemmules and collar cells (b) spicules and gemmules (c) amebocytes and spongin (d) collar cells and amebocytes.
11. The primary basis for classifying sponges is (a) their shape (b) their color (c) the composition of their skeleton (d) the method by which they reproduce.
12. Characteristics of cnidarians include all of the following *except* (a) nematocysts (b) collar cells (c) gastrovascular cavity (d) tentacles.
13. The cnidarian body form that is specialized for swimming is called a (a) medusa (b) polyp (c) planula (d) blastula.
14. The movements of cnidarians are coordinated by their (a) mesoglea (b) nematocysts (c) nerve net (d) gastrodermis.
15. Colloblasts are (a) cells that draw water through sponges (b) cells that secrete a sticky substance in ctenophores (c) stages in the life cycle of a jellyfish (d) medusae that live in colonies.

Short Answer

16. Describe three ways that sponges represent the transition from unicellular to multicellular life.
17. How does a cnidarian capture and ingest its prey?
18. Describe the path of water through a sponge, and explain what causes water to flow through a sponge.
19. How is a sponge's usual method of feeding suited to its sessile lifestyle?
20. Describe the specialization that is found among the individuals that make up a Portuguese man-of-war.
21. Describe the process of sexual reproduction in sponges.
22. What benefit does a sea anemone obtain from its symbiotic relationship with a clownfish? What benefit does the clownfish obtain from this relationship?
23. Why do coral reefs form only at shallow depths?
24. Describe two ways that ctenophores differ from cnidarians.
25. In the diagram of a hydra shown below, identify the cell layers labeled *A* and *B*, the material between them labeled *C*, and the cell labeled *D*.

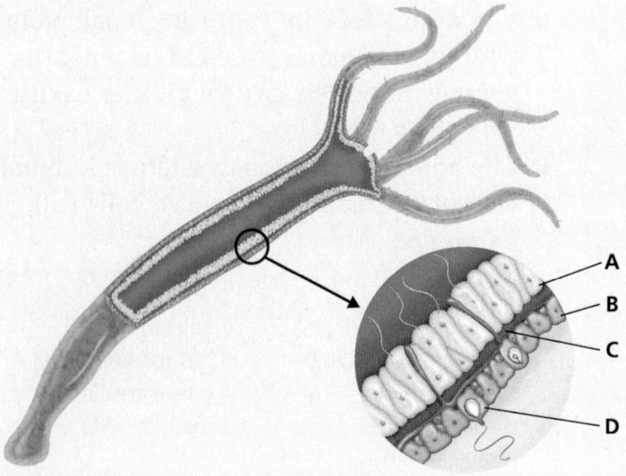

CRITICAL THINKING

1. Radial symmetry is found only among animals that live in water. Why is radial symmetry better suited to animals that live in water than to animals that live on land?
2. A hermaphroditic sponge generally produces eggs and sperm at different times. Based on your knowledge of sexual reproduction, why do you think it is evolutionarily advantageous for a sponge not to produce eggs and sperm at the same time?
3. Compare and contrast the life cycle of *Aurelia* with alternation of generations in mosses, which you studied in Chapter 32. Be sure to list similarities and differences.
4. Hydras generally reproduce asexually during warm weather and sexually in the fall. Based on what you have learned about the hydra embryo, what do you think is the advantage to the hydra of reproducing sexually when the weather turns cool?
5. Sponge larvae have flagella on the outside of their body, whereas adult sponges have flagella in their internal cavity. How is this structural difference related to functional differences between the larval and adult stages in the life cycle of a sponge?
6. A single species of sponge may assume different appearances due to differences in substrate, availability of space, and the velocity and temperature of water currents. How might these factors make the classification of sponges confusing? What features besides outward appearance can biologists use to classify sponges and eliminate some of this confusion?
7. What would happen to a coral reef if pollution or sediment caused the water around the reef to become less clear? Explain your answer.
8. The pie chart below shows the relative numbers of species in each class of the phylum Cnidaria. Which segment of the chart represents scyphozoans? How do you know?

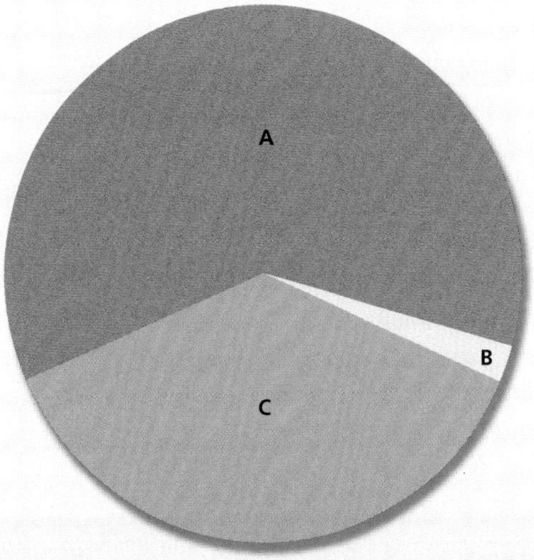

EXTENSION

1. Read "Transparent Animals" in *Scientific American,* February 2000, on pp. 80–89. Describe some of the survival benefits of having a body made of gelatinous material that is nearly transparent. What is the major drawback to having a transparent and gelatinous body? Why is scooping these organisms up from the depths of the ocean using large nets almost impossible?
2. Create a chart that summarizes information about all of the phyla you will study in this unit. Use headings such as Symmetry, Feeding and Digestion, Reproduction, and Habitats. Begin your chart by filling in the information for the phyla Porifera, Cnidaria, and Ctenophora. Continue your chart as you work through the unit.
3. Natural sponges are preferred to synthetic ones for some uses, such as in surgery. Find out what methods are used to gather and prepare sponges for use by surgeons and others. Write a report based on your findings, and present it to your class.
4. Research the most common ways to neutralize a jellyfish sting. Investigate why these treatments are effective, and present the information to the class.

4. Sexual reproduction results in an embryo whose covering can protect it during the winter.
5. Sponge larvae use flagella to swim through the water, whereas adult sponges use flagella to pump water through the body and filter food from water.
6. A single species may appear in many different forms, and different species may assume the same form in similar environments. Biologists can use internal features, such as the composition of the skeleton, to eliminate some of the confusion.
7. If the water around a reef were less clear, less light would reach the algae living symbiotically inside the coral cells. The algae would therefore grow more slowly, and because the coral cells depend on the algae, the reef would also grow more slowly.
8. Segment B represents scyphozoans because there are far fewer scyphozoans than hydrozoans or anthozoans.

EXTENSION

1. A gelatinous body is made almost entirely of incompressible water, which protects the animal from the crushing pressures of the ocean. Transparency is a camouflage advantage at depths where there is little light. The drawback is that they are delicate and are easily torn apart by collectors' nets.
2. This activity is self-explanatory if the students complete the chart as they go through the chapter.
3. The report should include the fact that natural sponges are gathered from the ocean, crushed to break up the rigid skeleton, washed, and dried. Sponges to be used in surgery are then sterilized.
4. The most common way to neutralize a jellyfish sting is to douse the wound with vinegar. The mild acidity of vinegar changes the shape of the venom protein, which causes it to stop stinging. Acid also disables the cnidocytes.

CHAPTER 35 INVESTIGATION

TIME REQUIRED

One 50-minute class period

QUICK REFERENCES

Lab Preparation Notes are found on page 691B.

Holt BioSources provides a Teaching Resources CD-ROM that contains "Using Gowin's Vee in the Lab" and "Scoring Rubrics."

PROCEDURAL TIPS

1. Instruct students to handle the hydras gently and to watch patiently—it may take time for certain behaviors to occur.
2. Review with students the organization of the cnidarian nervous system, and relate this information to the behavior of hydras.

ANSWERS TO BACKGROUND

1. Animals may move toward or away from the stimulus, release chemicals, or exhibit other behaviors.
2. A hydra is a freshwater cnidarian in the class Hydrozoa.
3. Shared characteristics include tentacles, nematocysts, a gastrovascular cavity, two cell layers with a jellylike substance between them, and a nerve net.
4. Sessile animals filter food from the water or actively capture food as it passes by them.
5. A nematocyst is an organelle inside a cnidocyte. Each nematocyst contains a long, coiled filament.

ANSWER TO PART A

5. The bumps should disappear. Vinegar is not a normal part of a hydra's diet. Its acidity makes the nematocysts discharge.

CHAPTER 35 INVESTIGATION

Observing Hydra Behavior

OBJECTIVES

- Observe live specimens of hydra.
- Determine how hydras respond to different stimuli.
- Determine how hydras capture and feed on prey.

PROCESS SKILLS

- observing
- relating structure to function

MATERIALS

- silicone culture gum
- microscope slide
- hydra culture
- 2 medicine droppers
- compound microscope
- methylene blue solution
- vinegar
- stereomicroscope
- filter paper cut into pennant shapes
- forceps
- concentrated beef broth
- *Daphnia* culture

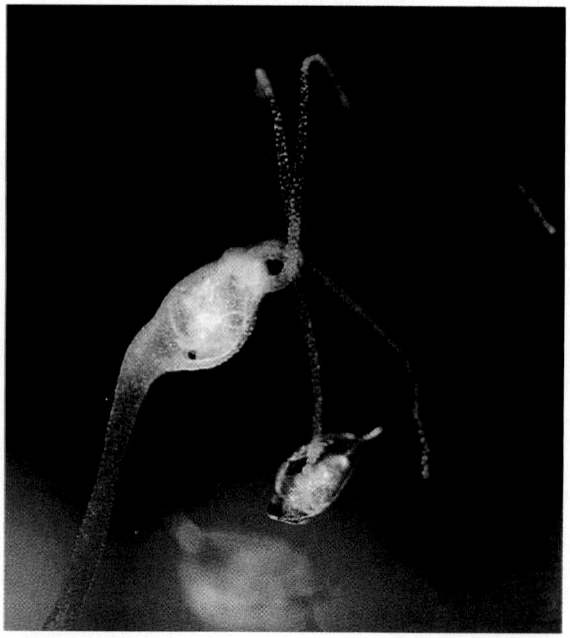

Background

1. How do animals respond to stimuli in their environment?
2. What is a hydra?
3. What characteristics do hydras share with other cnidarians?
4. How does a sessile animal, such as a hydra, obtain food?
5. What is a nematocyst?

PART A Close-up Examination of a Hydra

1. CAUTION **Slides break easily. Use care when handling them.** Using a long piece of silicone culture gum, make a circular "well" on a microscope slide, as shown in the illustration below.

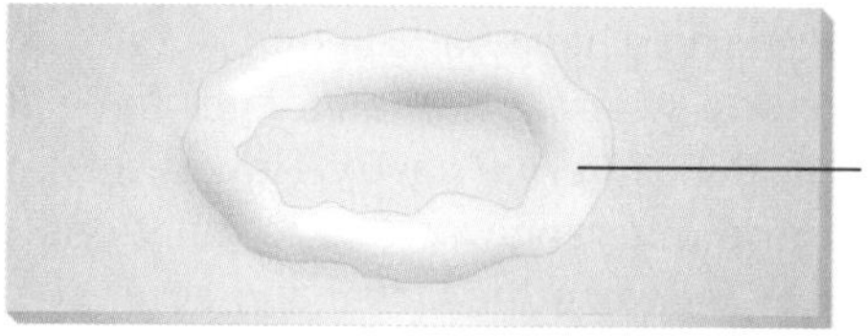

2. CAUTION **You will be working with a live animal. Be sure to treat it gently and to follow directions carefully.** With a medicine dropper, gently transfer a hydra from the culture dish to the well on the slide, making sure the hydra is in water. The hydra should be transferred quickly; otherwise, it may attach itself to the medicine dropper. Allow the hydra to settle. As you go through the following steps, add water to the slide periodically to replace water that has evaporated, and keep the hydra wet.
3. Examine the hydra under the low-power setting of a compound microscope. Add a drop of methylene blue solution to the well containing the hydra to make the tentacles more visible. Identify and draw the hydra's body stalk, mouth, and tentacles in your lab report.
4. In your lab report, make a data table like the one shown on the next page. As you complete the following steps, record your observations in your data table.
5. As you continue to observe the hydra at low power, add a drop of vinegar to the well. What happens to

the bumps on the tentacles? These bumps are cnidocytes. Is vinegar a normal part of a hydra's diet? Why do you think vinegar is used in this step?

6. Transfer the hydra to the culture dish labeled "Used hydras." Rinse the well on your microscope slide with water to remove all traces of methylene blue and vinegar.

PART B Feeding Behavior

7. Hydras eat small invertebrates, such as *Daphnia.* With a medicine dropper, gently transfer another hydra to the well on your slide. Then transfer live *Daphnia* to the well in the same manner.
8. Observe the hydra carefully with the high-power setting of the stereomicroscope. Watch for threadlike nematocysts shooting out from the hydra. Some nematocysts release a poison that paralyzes prey. If the hydra does not respond after a few minutes, obtain another hydra from the culture dish and repeat this procedure. Do you think the hydra's response is triggered by water vibrations, by chemicals that *Daphnia* releases, or by *Daphnia* touching the hydra?
9. Observe the way the hydra captures and ingests *Daphnia,* and record your observations in your data table. How long does it take for a hydra to ingest a *Daphnia*?
10. Transfer the hydra to the culture dish labeled "Used hydras." Rinse the well on your microscope slide with water to remove the *Daphnia.*

PART C Response to Stimuli

11. Transfer another hydra to the well on your slide, and examine it with the high-power setting of the stereomicroscope. Using forceps, move the long tip of a pennant-shaped piece of filter paper near the hydra's tentacles. Be careful not to touch the hydra with the filter paper. Observe the hydra's response to the filter paper, and record your observations in your data table.
12. Now observe how the hydra responds to a chemical stimulus. Dip the same piece of filter paper in beef broth, and repeat the procedure in step 11. Again, be careful not to touch the hydra. Record the hydra's response to the beef broth in your data table.
13. Finally, investigate how the hydra responds to touch. Using the long tip of a clean pennant-shaped piece of filter paper, touch the hydra's tentacles, mouth, and body stalk. CAUTION **Touch the hydra gently.** Record your observations in your data table.
14. Transfer the hydra to the culture dish labeled "Used hydras." Clean up your materials and wash your hands before leaving the lab.

Analysis and Conclusions

1. Based on your observations, how do you think a hydra behaves when it is threatened in its natural habitat?
2. Describe a hydra's feeding behavior.
3. What happens to food that has not been digested by a hydra?
4. What was the purpose of using the clean filter paper in step 11?
5. Did the hydra show a feeding response or a defensive response to the beef broth? Explain.
6. How is a hydra adapted to a sessile lifestyle?
7. How is the feeding method of a hydra different from that of a sponge?

Further Inquiry

Design an experiment to determine how hydras respond to other stimuli, such as light.

OBSERVATIONS OF HYDRA BEHAVIOR

Behavior	Observations
Response to vinegar	
Feeding behavior	
Response to filter paper	
Response to beef broth on filter paper	
Response to touch with filter paper	

ANSWERS TO PART B

8. The experiments in Part C are designed to answer this question.
9. Typical data table entry: tentacles wrap around *Daphnia* and bring it toward the hydra's mouth. It may take a half-hour or longer for a hydra to ingest a *Daphnia.*

ANSWERS TO PART C

11. Typical data table entry: no response
12. Typical data table entry: hydra bends toward the filter paper
13. Typical data table entry: hydra contracts by shortening tentacles and body stalk

ANSWERS TO ANALYSIS AND CONCLUSIONS

1. A hydra contracts its body when it is threatened.
2. A hydra feeds by capturing prey with its tentacles and nematocysts, which sting the prey. The tentacles then push the food into the hydra's expanded mouth.
3. Undigested food is released from the hydra's mouth.
4. The clean filter paper served as a control to show that the hydra was responding to the beef broth rather than to the filter paper.
5. The hydra should show a feeding response to the beef broth, which could include expansion of its mouth, movement of its tentacles, elongation of its body, and the release of nematocysts.
6. Tentacles and nematocysts enable a hydra to capture prey that drift past it in slow-moving water. Its ability to contract when touched protects the hydra from predators.
7. A hydra's feeding behavior is more active than that of a sponge, which passively filter feeds. Sponges have no specialized organs for capturing food. Hydras have tentacles with nematocysts for capturing food.

CHAPTER 36

FLATWORMS, ROUNDWORMS, AND ROTIFERS

	TOPICS	TEACHING RESOURCES	LABS, CLASSWORK, AND HOMEWORK
BLOCK 1 45 minutes	**Introducing the Chapter, p. 708**	**ATE** Focus Concept, p. 708 **ATE** Assessing Prior Knowledge, p. 708 **ATE** Understanding the Visual, p. 708	■ Supplemental Reading Guide, *Journey to the Ants: A Story of Scientific Exploration*
	36-1 Platyhelminthes, p. 709 Structure and Function of Flatworms, p. 709 Class Turbellaria, p. 710 Class Trematoda, p. 711 Class Cestoda, p. 713	**ATE** Section Overview, p. 709 **ATE** Critical Thinking, pp. 712, 713 **ATE** Visual Link Figure 36-2, p. 710 134–135, 149–150, 156–157	**Lab B21,** Flatworm Behavior ★ Study Guide, Section 36-1 **PE** Section Review, p. 714
BLOCK 2 45 minutes	**36-2 Nematoda and Rotifera, p. 715** Phylum Nematoda, p. 715 Phylum Rotifera, p. 718	**ATE** Section Overview, p. 715 **ATE** Visual Link Figure 36-8, p. 716 134–135, 158	**PE** **Chapter 36 Investigation,** p. 722 **PE** **Quick Lab,** p. 717, Comparing Flatworms and Roundworms ★ Study Guide, Section 36-2 **PE** Section Review, p. 718
BLOCKS 3 & 4 90 minutes	**REVIEW and ASSESSMENT** **PE** Chapter 36 Review, pp. 720–721 ■ Performance-Based Assessment—Exploring Invertebrates	Audio CD Program	**CHAPTER TESTING** ■ Chapter Test (blackline master) ▲ Test Generator ■ Assessment Item Listing ■ Scoring Rubrics and Classroom Management Checklists

MODERN BIOLOGY | HOLT BIOSOURCES | One-Stop Planner CD-ROM

KEY

PE Pupil's Edition
ATE Teacher's Edition
■ Active Reading Guide
★ Study Guide

Teaching Transparencies

Laboratory Program

One-Stop Planner CD-ROM
Shows these resources within a customizable daily lesson plan:

■ Holt BioSources Teaching Resources
■ Active Reading Guide
★ Study Guide
▲ Test Generator

READING FOR CONTENT MASTERY

Preview: Have students preview the objectives, the introductory paragraph, topic sentences, and captions before reading the chapter.

ATE Active Reading Technique: K-W-L, p. 709
■ Reading Strategy: K-W-L Worksheet
■ Active Reading Guide Worksheet 36-1
PE Summary/Vocabulary, p. 719

ATE Active Reading Technique: K-W-L, p. 718
■ Reading Strategy: K-W-L Worksheet
■ Active Reading Guide Worksheet 36-2
PE Summary/Vocabulary, p. 719

TECHNOLOGY AND INTERNET RESOURCES

Holt Biology Videodiscs Teacher's Correlation Guide, Lessons 36-1 through 36-2

Audio CD Program, Sections 36-1 through 36-2

internetconnect

sciLINKS NSTA

On-line Resources: www.scilinks.org
The following *sci*LINKS Internet resources can be found in the student text for this chapter:

Topics:
- Tapeworms, p. 713
- Hookworms, p. 716

go.hrw.com

On-line Resources: go.hrw.com
Visit the HRW Web site for a variety of resources related to this chapter. Just type in the keyword HM2 Chapter 36.

Smithsonian Institution®
Visit **www.si.edu/hrw** for additional on-line resources.

CNNfyi.com
Visit **www.cnnfyi.com** for late-breaking news and current event stories selected just for you.

LAB ACTIVITY PLANNING

CHAPTER 36 INVESTIGATION

Observing Flatworm Responses to Stimuli, pp. 722–723

OVERVIEW
In this investigation, students will observe live specimens of a flatworm, *Dugesia* (brown planaria), noting its symmetry and external features. Students will also observe the planarian's feeding behavior and response to various stimuli, such as touch, gravity, and light.

SAFETY
Instruct students to wear lab aprons and safety goggles during the lab and to keep their hands away from their eyes while handling the *Dugesia* cultures. Remind students to wash their hands after completing the lab.

PREPARATION
1. Purchase *Dugesia* from a biological supply house. Each lab group will need one *Dugesia*.
2. Each lab group will need a small piece of raw liver.
3. Each lab group will need a culture dish, a medicine dropper, a blunt probe, a test tube, a test-tube rack, a 4 × 4 in. piece of aluminum foil, a wax pencil, a cork stopper, and one sheet of white paper. Each lab group will need access to a small flashlight, stopwatch, or clock with a second hand, and a stereomicroscope.

Quick Labs

Comparing Flatworms and Roundworms, p. 717
Students will need living planarians, preserved specimens of tapeworms, male and female *Ascaris*, and a hand lens for this activity.

UNDERSTANDING THE VISUAL

Have students observe the rotifer shown in the photograph. Ask them how they think rotifers got their name. (The crown of cilia surrounding a rotifer's mouth looks like a pair of rotating wheels.) Ask students how they think rotifers differ from sponges and cnidarians based on what they can see in the photograph. (Rotifers are structurally more complex and have an anterior and a posterior end.)

FOCUS CONCEPT

Matter, Energy, and Organization

The body organization of the animals discussed in this chapter is more complex than that of sponges and cnidarians. Flatworms, roundworms, and rotifers are bilaterally symmetrical and cephalized, and they have tissues and organs derived from three germ layers during embryonic development. Roundworms and rotifers have a pseudocoelom, a fluid-filled body cavity that provides a space for internal organs.

ASSESSING PRIOR KNOWLEDGE

Review the following concepts.

Multicellular Organization: *Chapter 4*
Ask students to define *tissue, organ,* and *organ system.* Ask them to give an example of each.

Body Symmetry: *Chapter 34*
Ask students to state which body areas correspond to the terms *anterior, posterior, dorsal,* and *ventral.* Ask them to define *cephalization.*

CHAPTER 36

FLATWORMS, ROUNDWORMS, AND ROTIFERS

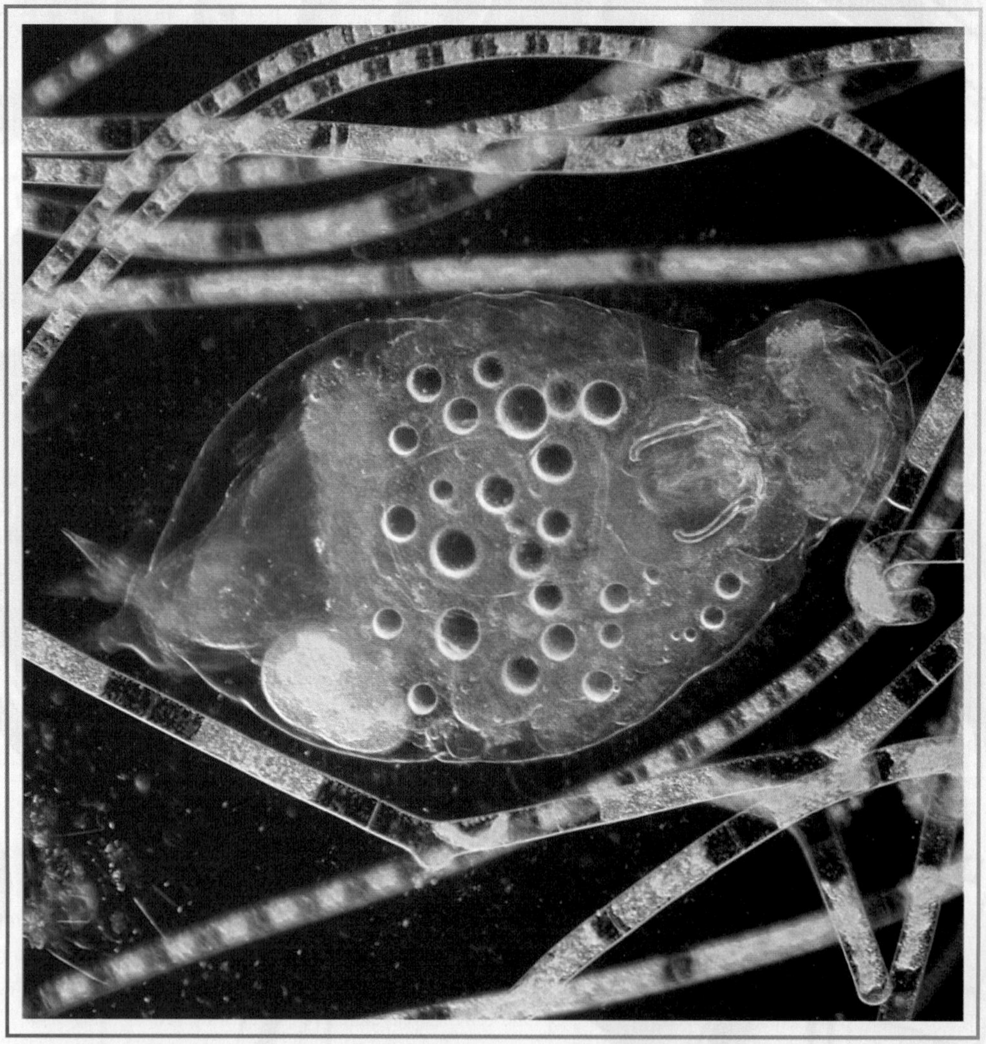

Many of the internal organs of this rotifer are visible through its transparent skin. (LM 675×)

FOCUS CONCEPT: *Matter, Energy, and Organization*

As you read, note the increased complexity of body plans in these phyla compared with those of sponges, cnidarians, and ctenophores. Also notice how organisms are modified for a free-living or parasitic way of life.

PLATYHELMINTHES

Members of the phylum Platyhelminthes (PLAT-ee-hel-MINTH-eez) are called ***flatworms.*** *Their bodies develop from three germ layers and are more complex than those of sponges, cnidarians, and ctenophores. Flatworms have bilaterally symmetrical bodies, with dorsal and ventral surfaces, right and left sides, and anterior and posterior ends.*

SECTION

36-1

OBJECTIVES

▲ State the distinguishing characteristics of flatworms.

● Describe the anatomy of a planarian.

■ Compare and contrast free-living and parasitic flatworms.

◆ Diagram the life cycle of a fluke.

▲ Describe the life cycle of a tapeworm.

STRUCTURE AND FUNCTION OF FLATWORMS

Flatworms are the simplest animals with bilateral symmetry. You learned in Chapter 34 that the tissues in bilaterally symmetrical animals develop from three germ layers: ectoderm, mesoderm, and endoderm. In flatworms, the three germ layers are pressed against one another to form a solid body. Since flatworms do not have a hollow cavity between the endoderm and the mesoderm, they belong to a group of animals known as acoelomates, a name that means "without a hollow."

The acoelomate body plan gives flatworms the thin, flat bodies for which they are named. This body shape ensures that no cell in a flatworm is very far from the animal's external environment. Thus, the cells can exchange oxygen and carbon dioxide directly with the environment through diffusion, allowing flatworms to survive without a circulatory system or respiratory system. Like cnidarians, most flatworms have a gastrovascular cavity, a gut with a single opening. Food is taken in and digested in the gastrovascular cavity, and any undigested material is eliminated through the same opening. Most of the sensory organs and nerve cells of flatworms, such as the marine species shown in Figure 36-1, are located at the anterior end of the body. As you learned in Chapter 34, this characteristic is known as cephalization.

The more than 18,000 species of flatworms are divided into three classes: Turbellaria, Trematoda, and Cestoda. Trematodes and cestodes (SES-tohdz) live as parasites on or inside other animals, whereas almost all turbellarians are nonparasitic, or free-living. Parasites probably evolved from free-living organisms. As they evolved, some organs that were advantageous to free living became modified for parasitism, while others were lost entirely. Thus, parasitic flatworms are structurally simpler than their free-living relatives.

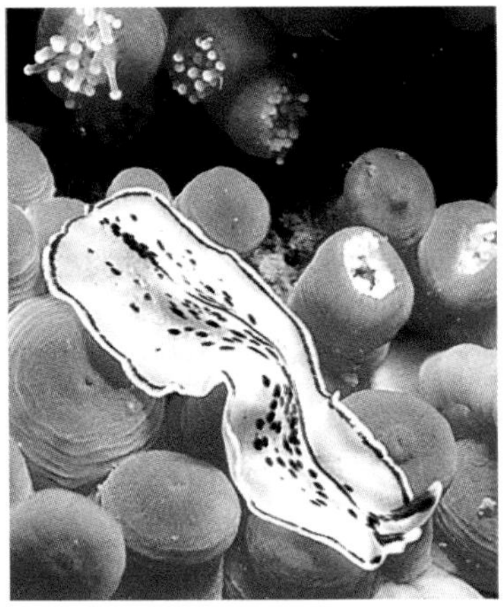

FIGURE 36-1

Many of the sensory organs in this marine flatworm, *Euryletta,* are concentrated in the two tentacles at the anterior end of its body. This characteristic is an example of cephalization.

SECTION 36-1

SECTION OVERVIEW

Reproduction and Inheritance
Section 36-1 describes the process of reproduction in the free-living flatworm, in the planarian, and in two parasitic flatworms, the blood fluke and the beef tapeworm.

Matter, Energy, and Organization
Section 36-1 explains how the body organization of a flatworm is closely related to its lifestyle.

ACTIVE READING

Technique: K-W-L

Before students read the chapter, display a list of the following phyla: Platyhelminthes, Nematoda, and Rotifera. Have students make a list of everything they **K**now (or think they know) about the members of each phylum. Have them work together to form a class list. Then have students list things they **W**ant to know about the organisms in these phyla. Have students save their lists for use later in the Active Reading activity on page 718.

GIFTED *ACTIVITY*

Have students research the hypothesis that parasitic flatworms originated as free-living organisms. Then have them develop a script for a television nature program illustrated with a sequence of sketches in which they discuss the evidence for and against the hypothesis.

SECTION 36-1

QUICK FACT

☑ Approximately 40 percent of the world's human population is infected with parasitic worms.

MAKING CONNECTIONS

Social Studies

The health problems associated with infections by parasitic worms are often compounded by social problems such as poverty, improper sanitation, environmental disturbance, and lack of governmental services to help infected people. Have students conduct research on some of the countries that are most affected by these parasites and report on medical or social breakthroughs in those countries. Students could devise plans to alleviate the health and social problems.

CONTENT CONNECTION

Section 34-2: Patterns of Symmetry

The bilateral symmetry of flatworms is apparent externally in paired structures, such as eyespots, and internally in the arrangement of the digestive, nervous, excretory, and reproductive systems.

VISUAL LINK

Figure 36-2

This figure indicates that an animal as small and simple as a flatworm can have a body whose tissues are arranged at the organ system level. Have students compare Figure 36-2 with Figures 35-1 and 35-3, which show the much simpler body organization of sponges and cnidarians, respectively. Contrast the planarian's body structure with that of the tapeworm, shown in Figure 36-5, and explain that the demands of a free-living lifestyle require specific body structures that are not required in an internal parasite.

Word Roots and Origins

pharynx

from the Greek *pharynx,* meaning "throat"

CLASS TURBELLARIA

The majority of the 4,500 species in the class Turbellaria live in the ocean. However, the most familiar turbellarian is the freshwater planarian *Dugesia,* which is shown in Figure 36-2. Planarians have a spade-shaped anterior end and a tapered posterior end. They move through the water by swimming with a wavelike motion of their body. Over solid surfaces, planarians glide on a layer of mucus that they secrete, propelled by the cilia that cover their body.

Digestion and Excretion in Planarians

Planarians feed by scavenging for bits of decaying plant or animal matter. They also prey on smaller organisms, such as protozoa. The food is ingested through a muscular tube, the **pharynx** (FAR-eenks), which the planarian extends from the middle of its body. As Figure 36-2a indicates, the pharynx leads to the highly branched gastrovascular cavity. Cells lining the cavity secrete digestive enzymes and absorb nutrients and small pieces of food. The nutrients then diffuse to other body cells.

Organisms that live in fresh water must deal with the water that constantly enters their bodies by osmosis. Planarians eliminate excess water through a network of excretory tubules that run the length of the body. Figure 36-2b shows that each tubule is connected to several **flame cells,** which are so named because they enclose tufts of beating cilia that resemble flickering candle flames. The flame cells collect the excess water, which is then transported through the tubules and excreted from numerous pores scattered over the body surface.

FIGURE 36-2

The organ systems of a planarian carry out the functions that allow it to maintain its free-living existence. (a) The digestive system consists of the pharynx and gastrovascular cavity, which has many branches. (b) In the excretory system, flame cells collect excess water, which travels through excretory tubules to pores on the surface of the body. (c) The nervous system is a ladderlike arrangement of nerves with two cerebral ganglia at the anterior end. (d) Since planarians are hermaphrodites, their reproductive system includes both testes and ovaries.

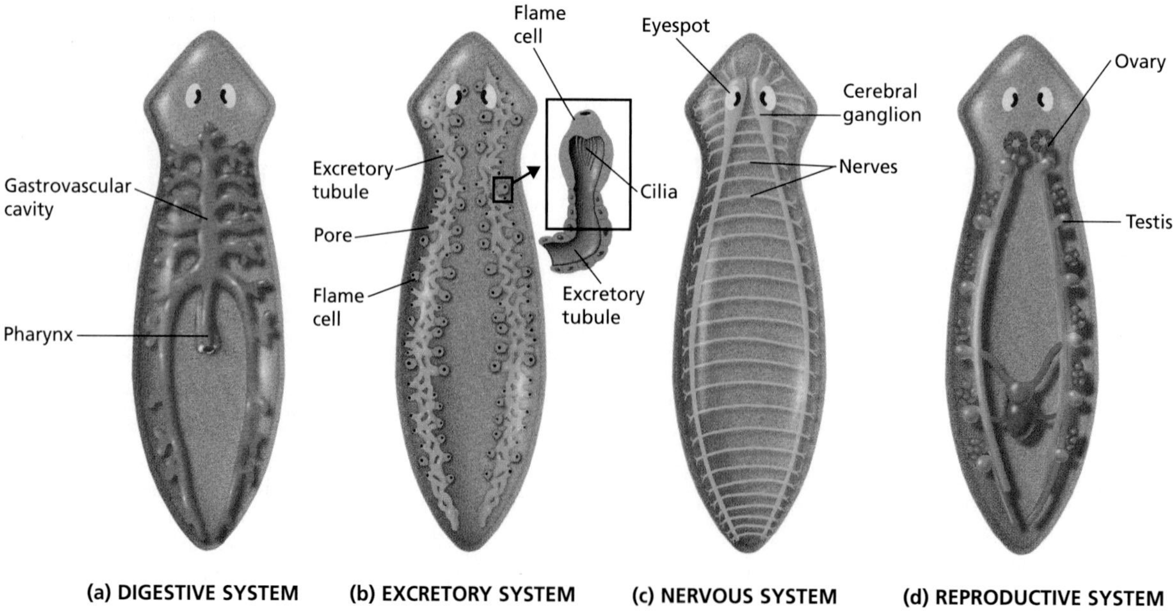

Neural Control in Planarians

The planarian nervous system is more complex than the nerve net of cnidarians, as Figure 36-2c illustrates. Two clusters of nerve cells at the anterior end, the **cerebral ganglia** (suh-REE-bruhl GAN-glee-uh), serve as a simple brain. They receive information from sensory cells and transmit signals to the muscles along a ladderlike arrangement of nerves. A planarian's nervous system gives it the ability to learn. For example, a planarian normally moves away from light, but it can be trained to remain still when illuminated.

Planarians sense the intensity and direction of light with two cup-shaped **eyespots** located near the cerebral ganglia. You can see the eyespots in Figure 36-2c. Other sensory cells respond to touch, water currents, and chemicals in the environment. These cells are distributed over the body, but most are concentrated at the anterior end.

Reproduction in Planarians

Because planarians are free-living and motile, they can encounter and mate with other individuals of the same species. As Figure 36-2d shows, planarians are hermaphrodites—they have both male sex organs (testes) and female sex organs (ovaries). When two planarians reproduce sexually, they simultaneously fertilize each other. Their eggs are laid in protective capsules that stick to rocks or debris and hatch in two to three weeks.

Planarians also reproduce asexually, generally during the summer. During asexual reproduction, the body constricts just behind the pharynx. While the posterior part of the worm is attached to a solid surface, the anterior part moves forward until the worm splits in two. This type of asexual reproduction is known as **fission.** The two halves then regenerate their missing parts to produce two complete planaria. During regeneration, each part of the planarian retains information about its original orientation in the body. If a piece is cut from the middle of a planarian, the anterior end will always regenerate a head and the posterior end will regenerate a tail.

CLASS TREMATODA

The class Trematoda consists of about 9,000 species of **flukes,** leaf-shaped flatworms that parasitize many kinds of animals, including humans. Some flukes live inside their host, in the blood, intestines, lungs, liver, or other organs. Others live on the external surface of aquatic hosts such as fish and frogs.

Structure of Flukes

A fluke clings to the tissues of its host by means of an anterior sucker and a ventral sucker, which are shown in Figure 36-3. The anterior sucker surrounds the fluke's mouth, which draws the host's body fluids into the gastrovascular cavity. A fluke's nervous

FIGURE 36-3

Suckers on this blood fluke, *Schistosoma mansoni,* attach the fluke to the blood vessels of its host. (SEM 550×)

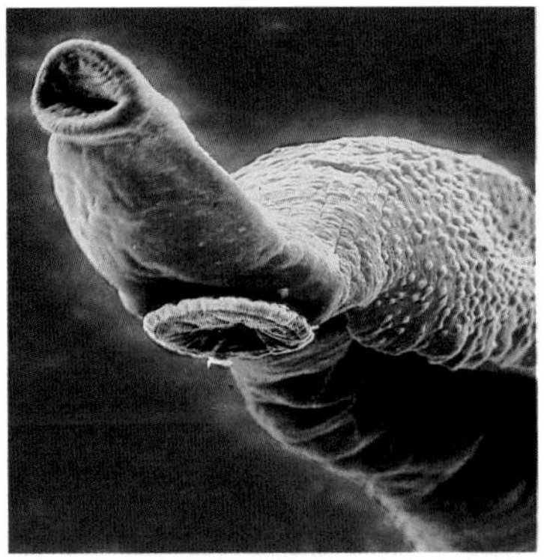

SECTION 36-1

CONTENT CONNECTION

Section 4-3: Multicellular Organization

The organization of the flatworm body into organ systems is accompanied by a greater cell specialization than is present in sponges, cnidarians, and ctenophores. Individual cells in each organ system carry out a fairly restricted range of functions, but the variety of different cell types results in greater efficiency for the organism as a whole.

DEMONSTRATION

Obtain preserved specimens of tapeworms and male and female *Ascaris* worms from a biological supply company, and circulate them around the class inside their sealed containers. Explain that both organisms live as parasites in the intestines of vertebrates, including humans. Ask students to contrast the long, ribbonlike body of the tapeworm (a flatworm) with the shorter, thicker body of *Ascaris* (a roundworm). Have students locate the knob-shaped scolex at the anterior end and the numerous reproductive body sections (proglottids) at the posterior end of the tapeworm. Ask students to distinguish between male and female *Ascaris* specimens. (The male is much smaller than the female and has a hooked posterior end.)

CRITICAL THINKING

Blood Flukes

At what points in the life cycle of a blood fluke do you think it would be easiest to interrupt the cycle and thereby prevent the spread of schistosomiasis? (These points would include when eggs enter the water and when ciliated larvae enter snails.) What could be done to interrupt the cycle at each point? (Prevent human feces and urine from contaminating waterways, and reduce snail populations in areas where humans enter the water.)

INCLUSION ACTIVITY

Separate students into groups, and give each group a sheet of white wrapping paper 6–7 ft long. Ask one student in each group to lie down on the paper so that the other members of the group may trace the outline of that student's body on the paper. Have students use marking pens to draw the body structures where the flatworm and roundworm parasites discussed in the chapter reside and to use colored construction paper to make models of these parasites. As students describe the life cycle of each parasite, have them move the model of that parasite to the appropriate body structure.

TEACHING STRATEGY

Worm Identification

Divide students into groups, and assign each group a different flatworm or roundworm to investigate. Ask the students in each group to write several questions pertaining to their assigned organism. After the groups have finished their investigations, have each group give an oral presentation describing the characteristics and life cycle of their organism without mentioning its identity. The rest of the class should identify the organism and answer the questions.

system is similar to a planarian's, but flukes have no eyespots, and their other sensory structures are very simple.

The entire external surface of a fluke is covered by a continuous sheet of fused cells called the **tegument.** The outer zone of the tegument consists of a layer of proteins and carbohydrates that make the fluke resistant to the defenses of the host's immune system. The tegument also protects the fluke against the enzymes secreted by the host's digestive tract.

Reproduction and Life Cycle of Flukes

Most flukes have highly developed reproductive systems and are hermaphroditic. The eggs are stored in a long, coiled tube called the **uterus** until they are ready to be released. Each fluke may produce tens of thousands of eggs at a time.

Flukes have complicated life cycles. A good example is provided by the blood flukes of the genus *Schistosoma.* Follow the steps in Figure 36-4 to see how the schistosome life cycle operates. Adult schistosomes live inside human blood vessels. Therefore, a human is the schistosome's **primary host,** the host from which the adult parasite derives its nourishment and in which sexual reproduction occurs. Unlike most flukes, schistosomes have separate sexes. Eggs produced by the female are fertilized by the male. Some of the fertilized eggs make their way to the host's intestine or bladder and are excreted with the feces or urine. Those that enter fresh water develop into ciliated larvae that swim about. If the larvae encounter a snail of a particular species within a few hours, they burrow into the snail's tissues and begin to reproduce asexually. The snail

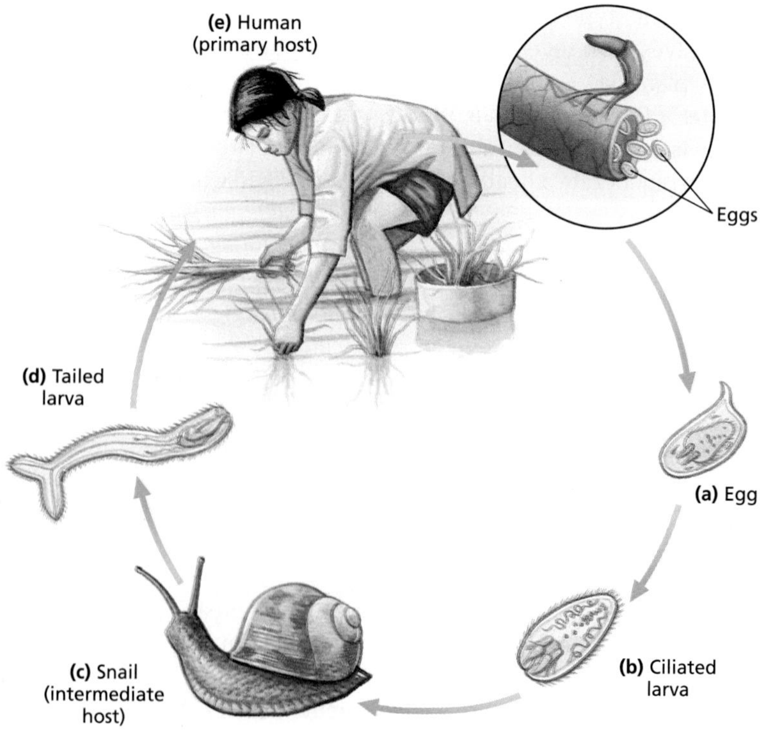

FIGURE 36-4

Adult blood flukes of the genus *Schistosoma* live in the blood vessels of humans, their primary hosts. As the inset shows, fertilized *Schistosoma* eggs are released into the blood vessels. (a) The eggs pass out of the primary host in feces or urine. (b) In water, the eggs develop into ciliated larvae. (c) The larvae burrow into snails, which serve as intermediate hosts. (d) The larvae develop tails, escape from the snail, and swim about. (e) The tailed larvae bore through the exposed skin of a person and settle in the blood vessels. There they develop into adults, and the cycle repeats.

serves as the schistosome's **intermediate host,** the host from which the larvae derive their nourishment. Eventually, the larvae develop tails and escape from the snail. These tailed larvae swim through the water, and if they find the bare skin of a human, they penetrate the skin, enter a blood vessel, and develop into adults. At this point, the cycle begins again.

Not all schistosome eggs leave the human body, however. Many are carried by the blood to the lungs, intestines, bladder, and liver, where they may block blood vessels and cause irritation, bleeding, and tissue decay. The resulting disease, called **schistosomiasis** (SHIS-tuh-soh-MIE-uh-suhs), can be fatal. It affects 200 to 300 million people, mostly in Asia, Africa, and South America.

Other kinds of flukes cause less-serious diseases in humans. For example, a small brown fluke common in freshwater lakes of North America is responsible for **swimmer's itch,** a condition characterized by minor skin irritation and swelling.

internetconnect

SCILINKS NSTA

TOPIC: Tapeworms
GO TO: www.scilinks.org
KEYWORD: HM713

CLASS CESTODA

About 5,000 species of **tapeworms** make up the class Cestoda. Tapeworms live in the intestines of almost all vertebrates. Humans may harbor any of seven different species. Tapeworms enter their host when the host eats raw or undercooked food containing eggs or larvae. A tapeworm infection may cause digestive problems, weight loss, lack of energy, and anemia, which is a decrease in the number of red cells in the blood.

Structure of Tapeworms

Like flukes, tapeworms are surrounded by a tegument that protects them from their host's defenses. As Figure 36-5 shows, at the anterior end of a tapeworm is a knob-shaped organ, the **scolex** (SKOH-leks), which is adorned with hooks and suckers that enable the worm to attach to its host. A short neck connects the scolex with a long series of body sections called **proglottids** (proh-GLAHT-idz). As a tapeworm grows, it adds proglottids just behind the neck, pushing the older proglottids toward the rear. A single tapeworm may have 2,000 proglottids and exceed 10 m (33 ft) in length.

The excretory system and nervous system of a tapeworm are similar to those of other flatworms. However, tapeworms lack eyespots and other light-sensitive structures, and they have no mouth, gastrovascular cavity, or other digestive organs. They absorb nutrients directly from the host's digestive tract through their tegument. The tegument is highly folded, which increases the surface area available for absorption.

FIGURE 36-5

A tapeworm grows by adding proglottids behind its scolex. Each proglottid contains both male and female reproductive organs.

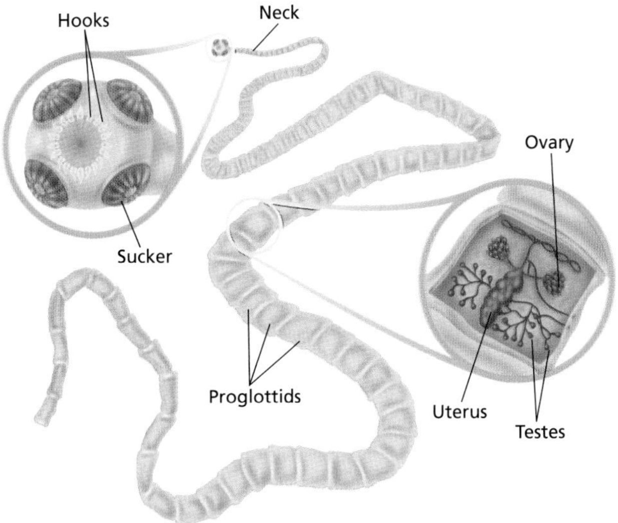

SECTION 36-1

RECENT RESEARCH

Schistosomiasis Vaccine

Scientists have recently made great advances in the development of vaccines and vaccine-enhancing substances to treat schistosomiasis. One vaccine that gives only partial resistance to this infection has now been found to reduce egg production by female flukes.

OVERCOMING MISCONCEPTIONS

Students may assume that the relatively simple body plan of the tapeworm suggests that more-complex, free-living flatworms, such as planarians, evolved from parasitic flatworms. In fact, parasitic flatworms probably evolved from free-living forms and gradually lost structures that were not advantageous to a parasitic lifestyle.

CRITICAL THINKING

Tapeworms

How is a tapeworm like an intestine turned inside out? (A tapeworm absorbs nutrients from its host directly through its body wall.)

CULTURAL CONNECTION

The Ayuveda

Parasitic worms have plagued human civilization for centuries, and treatments for worm infections have a long history. The Ayuveda, or "knowledge of life," philosophy of India began thousands of years ago. It includes the use of medicines extracted from plants, such as the bitter gourd, flame-of-the-forest, and golden champa, which are administered to rid the body of intestinal worms.

SECTION 36-1

RETEACHING ACTIVITY

Have students compare and contrast the structure and function of cnidarians with the structure and function of both free-living and parasitic flatworms. Ask students to consider body symmetry and complexity, feeding, digestion, sensing of the environment, neural control, and reproduction.

QUICK FACT

☑ A human tapeworm can release more than a million eggs per day.

ANSWERS TO SECTION 36-1 REVIEW

1. Flatworms do not have a hollow cavity between the endoderm and the mesoderm. Therefore, they are called acoelomates, which means "without a hollow."
2. A flame cell is a ciliated cell that collects excess water in a planarian. It is a part of the planarian excretory system.
3. Planarians have cerebral ganglia and eyespots located at the anterior end of the body. Other sensory cells are concentrated at the anterior end as well.
4. Blood flukes block blood vessels in the lungs, intestines, bladder, and liver, causing irritation, bleeding, and tissue decay.
5. Tapeworms absorb nutrients directly from the host's digestive tract through their tegument.
6. A parasite depends on its host for nutrition and for protection against adverse conditions and predators. If the parasite killed its host, the parasite would also die.

Reproduction and Life Cycle of Tapeworms

Nearly all tapeworms are hermaphrodites. You can see in Figure 36-5 that each proglottid contains both male and female reproductive organs, but little else. As the proglottids move to the rear of the tapeworm, they grow, mature, and begin producing eggs. The oldest proglottids are almost completely filled with 100,000 or more eggs. Eggs in one proglottid are usually fertilized by sperm from a different proglottid, either in the same individual or a different individual if the host has more than one tapeworm.

The life cycle of the beef tapeworm, *Taeniarhynchus,* is illustrated in Figure 36-6. Like the blood fluke, the beef tapeworm has two hosts. The primary host is a human. In the human intestine, mature proglottids break off from the adult and are eliminated with the host's feces. If the feces are deposited on the ground, the proglottids crawl out of the feces and onto nearby vegetation. The eggs they release may remain alive for several months before the vegetation is eaten by a cow, the intermediate host. Inside the cow, the eggs develop into larvae that burrow through the cow's intestine and enter the bloodstream. The larvae then make their way to muscle tissue and form **cysts,** or dormant larvae surrounded by protective coverings. Humans become infected when they eat beef that has not been cooked sufficiently to kill the worms inside the cysts. Once a cyst enters the human intestine, the cyst wall dissolves and releases the worm. The worm then attaches to the intestinal wall and develops into an adult, beginning the cycle again.

Another tapeworm that infects humans is the pork tapeworm, *Taenia solium.* Its life cycle is similar to that of the beef tapeworm, except that a pig serves as the intermediate host.

FIGURE 36-6

Adult beef tapeworms live in the intestines of humans, their primary hosts. (a) Proglottids pass out of the primary host in feces, crawl onto vegetation, and release their eggs. (b) A cow, the intermediate host, ingests the eggs when it eats the vegetation. The eggs hatch into larvae that form cysts in the cow's muscles. (c) When a person eats undercooked beef, the larvae develop into adult tapeworms in the person's intestine, and the cycle repeats.

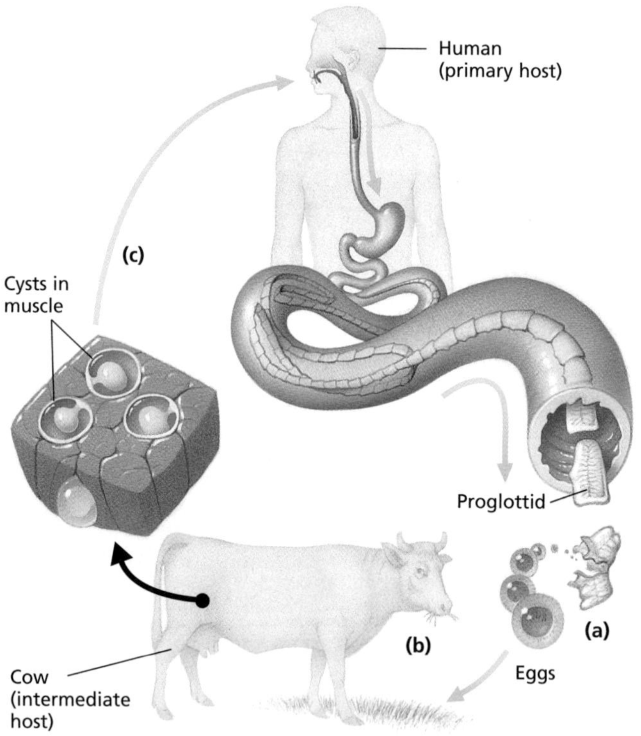

SECTION 36-1 REVIEW

1. **Why are flatworms called acoelomates?**
2. **What is a flame cell?**
3. **Describe two ways in which a planarian shows cephalization.**
4. **How does the blood fluke *Schistosoma* affect the human body?**
5. **How do tapeworms obtain nutrients without a mouth or a digestive system?**
6. **CRITICAL THINKING Why is it an adaptive advantage that a parasite not kill its host?**

NEMATODA AND ROTIFERA

Members of the phyla Nematoda (NEE-muh-TOHD-uh) and Rotifera (roh-TIF-uhr-uh) have bilaterally symmetrical bodies that contain a fluid-filled space. This space holds the internal organs and serves as a storage area for eggs and sperm. It also supports the body and provides a structure against which the muscles can contract.

SECTION

36-2

OBJECTIVES

▲ Describe the body plan of a pseudocoelomate.

● Explain the relationship between humans and three types of parasitic roundworms.

■ Describe the anatomy of a rotifer.

PHYLUM NEMATODA

The phylum Nematoda is made up of **roundworms,** worms with long, slender bodies that taper at both ends. Roundworms are among several phyla of animals known as pseudocoelomates. As you learned in Chapter 34, pseudocoelomates are so named because they have a pseudocoelom, a hollow, fluid-filled cavity that is lined by mesoderm on the outside and endoderm on the inside.

Roundworms range in length from less than 1 mm to 120 cm (4 ft). In contrast to cnidarians, ctenophores, and flatworms, which have a gastrovascular cavity with a single opening, roundworms have a digestive tract with two openings. Food enters the digestive tract through the mouth at the anterior end, and undigested material is eliminated from the **anus** (AY-nuhs) at the posterior end. A digestive tract represents a significant advancement over a gastrovascular cavity because food moves through the tract in only one direction. This allows different parts of the tract to be specialized for carrying out different functions, such as enzymatic digestion and absorption of nutrients. Most roundworms have separate sexes and are covered by a protective, noncellular layer called the **cuticle** (KYOO-ti-kuhl).

About 80,000 species of roundworms are known, but biologists estimate that there may be 500,000 or more species. The vast majority of roundworm species are free-living on land, in salt water, and in fresh water. One free-living roundworm, *Caenorhabditis elegans,* is a favorite organism of scientists studying developmental biology. However, about 150 species of roundworms are parasites of plants and animals. Humans are host to about 50 roundworm species. As you read about these roundworms, notice their adaptations for parasitism.

Eco Connection

Roundworms for Your Garden—Just Add Water

Gardeners are always looking for ways to control the pests that attack their plants. One way to do this without using potentially dangerous chemicals is to introduce organisms that naturally attack the pests. Some garden supply companies now sell kits containing millions of microscopic roundworms. The roundworms that are released are guaranteed to seek out and kill hundreds of varieties of insects. Other types of soil-dwelling roundworms consume bacteria and fungi that attack plants.

However, not all roundworms are good for plants. Some species parasitize the roots of plants. Effective pest control with roundworms requires a knowledge of which species are harmful and which are beneficial.

SECTION 36-2

SECTION OVERVIEW

Reproduction and Inheritance
Section 36-2 discusses the reproduction and life cycles of several species of parasitic roundworms.

Matter, Energy, and Organization
Section 36-2 introduces nematodes and rotifers, the simplest animals that have a fluid-filled body cavity. The cavity in these animals, lined by mesoderm on the outside and endoderm in the inside, is called a pseudocoelom to distinguish it from a "true" coelom, which is completely surrounded by mesoderm.

▶ ENGAGE STUDENTS

Obtain a preserved dog heart with heartworms from a veterinarian or a biological supply company. (A photograph from an invertebrate zoology textbook may be substituted.) Tell students that dogs are often host to parasitic roundworms and can be given pills during mosquito season to prevent heartworm infection. Ask students to speculate on the role mosquitoes play in such infections. (Mosquitoes pick up the parasite when they bite one host and transfer the parasite to another host when they take another blood meal.) Explain that many other animals, including humans, can be infected by roundworms and that some roundworm infections can cause very serious diseases.

TERMINOLOGY NOTE

Although the text uses the term *roundworm* to refer to members of the phylum Nematoda, another commonly used term for these animals is *nematodes.*

GIFTED ACTIVITY

Although parasitic roundworms are discussed more often in textbooks, most roundworms are free-living inhabitants of the soil. Have students research a free-living roundworm species, such as *Tubatrix* (the vinegar eel), and write a report on that species.

RECENT RESEARCH

Hookworm Clotting Agent

Scientists have recently isolated and purified a new clotting agent that prevents the formation of blood clots in the arteries of cardiac disease patients. The agent comes from the hookworm *Ancyclostoma caninum,* which infects dogs and humans.

RECENT RESEARCH

Roundworm Parasites and Allergy

Current research suggests that the allergic reaction many people have to pollen and dust mites may have evolved from an immune response to parasitic worms. This idea is supported by the finding that patients who have been treated successfully for roundworm infections have an increased sensitivity to allergens.

VISUAL LINK

Figure 36-8

This figure shows a section of the anterior end of a hookworm that is attached to a dog intestine. The hookworm's cutting plates have pinched off a portion of the intestinal wall, anchoring the parasite and rupturing capillaries in the wall. Explain that hookworms feed on the blood that is released by these ruptured capillaries, not on the food that passes through the intestine. Contrast this method of obtaining nutrients with the method used by tapeworms, described in Section 36-1.

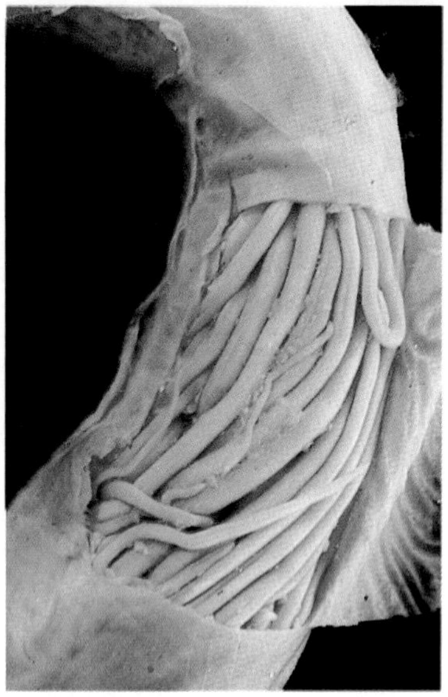

FIGURE 36-7

This pig intestine is completely blocked by *Ascaris* roundworms.

Ascaris

Ascaris (AS-kuh-ris) is a genus of roundworm parasites that live in the intestines of pigs, horses, and humans. These roundworms feed on the food that passes through the intestines of their host. As Figure 36-7 shows, they can become so numerous that they completely block the host's intestines if left untreated. The adult female can reach lengths of up to 30 cm (1 ft). The much smaller male has a hooked posterior end that holds the female during mating.

One *Ascaris* female can produce up to 200,000 eggs every day. The fertilized eggs escape from the host's body in the feces. If they are not exposed to direct sunlight or high temperatures, they can remain alive in the soil for years. *Ascaris* eggs enter the body of another host when the host ingests contaminated food or water. The eggs develop into larvae in the intestines, and the larvae bore their way into the bloodstream and are carried to the lungs and throat. There they are coughed up, swallowed, and returned to the intestines, where they mature and mate, completing the life cycle. If the infection is severe, the larvae in the lungs can block air passages and cause bleeding from small blood vessels.

Hookworms

Hookworms are another group of intestinal parasites. As you can see in Figure 36-8, a hookworm's mouth has cutting plates that clamp onto the intestinal wall. Hookworms feed on their host's blood, and because they remove much more blood than they need for food, a heavy hookworm infection can cause anemia. Hookworm infections in children can result in slowed mental and physical development.

Like *Ascaris,* hookworms release their eggs in the host's feces. The eggs produce larvae in warm, damp soil, and the larvae enter new hosts by boring through the host's feet. They then travel through the blood to the lungs and throat. Swallowing takes them to the intestines, where they develop into adults. Hookworms infect more than 400 million people worldwide. Approximately 90 percent of all infections occur in tropical and semitropical regions.

internetconnect
SCILINKS NSTA
TOPIC: Hookworms
GO TO: www.scilinks.org
KEYWORD: HM716

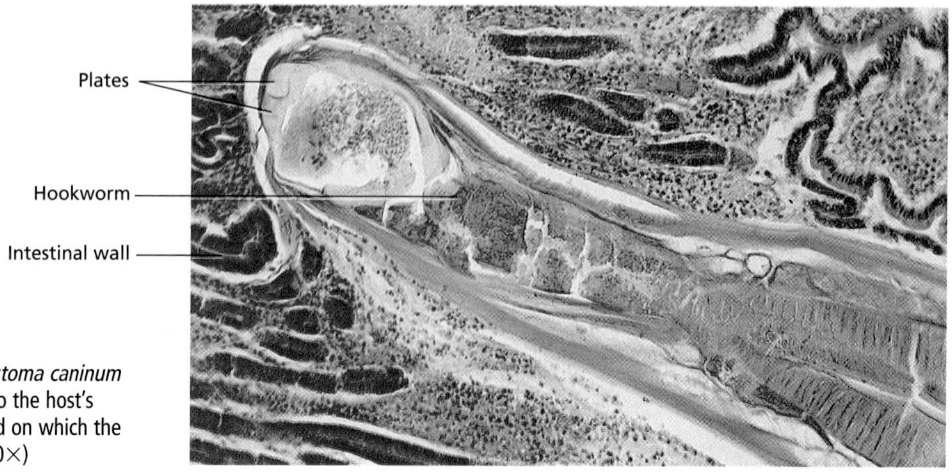

FIGURE 36-8

The hookworm *Ancyclostoma caninum* uses its plates to cut into the host's intestine, releasing blood on which the hookworm feeds. (LM 40×)

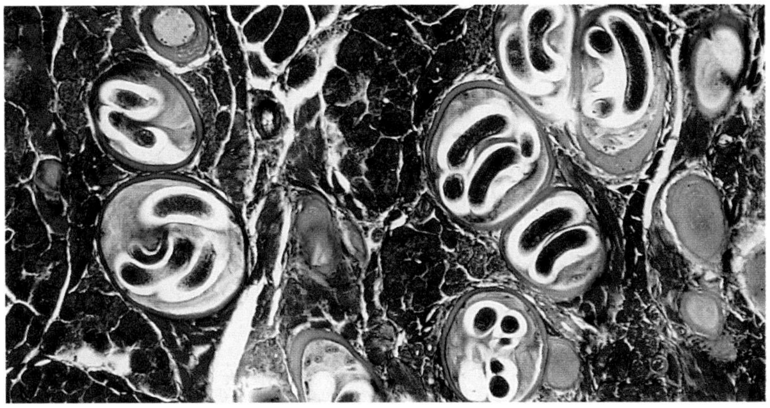

FIGURE 36-9

Larvae of the *Trichinella* roundworm coil up inside cysts in their hosts' muscle tissue. The cysts are stained blue in this light micrograph (100×).

Trichinella

Roundworms of the genus *Trichinella* infect humans and a variety of other mammals, including pigs. Adult *Trichinella* worms live embedded in the walls of the host's intestine. They produce larvae that travel through the bloodstream to the muscles, where they form cysts. Figure 36-9 shows one such cyst. People become infected when they eat undercooked meat—usually pork—that is contaminated with cysts. After they are eaten, the cysts release the larvae, which burrow into the intestinal wall and mature into adults. *Trichinella* infections are responsible for the disease **trichinosis** (TRIK-i-NOH-sis), which causes muscle pain and stiffness. It can even cause death if large numbers of cysts form in the heart muscle. However, trichinosis is now rare in the United States. Farmers cook meat scraps before feeding them to hogs, government inspectors examine pork for cysts, and meatpackers generally freeze pork, killing the worms.

Other Parasitic Roundworms

The most common roundworm parasite of humans in the United States is the **pinworm,** *Enterobius,* which infects about 16 percent of adults and 30 percent of children. Despite their high rate of infection, pinworms do not cause any serious disease. Adult pinworms are 5–10 mm (0.2–0.4 in.) in length and resemble white threads. They live and mate in the lower portion of the intestine. At night, the females migrate out of the intestine and lay eggs on the skin around the anus. When an infected person scratches during sleep, the eggs are picked up by the person's hands and spread to anything the person touches. Eggs that are ingested hatch in the intestine, where the worms develop into adults.

Filarial (fuh-LAR-ee-uhl) **worms** are disease-causing roundworms that infect over 250 million people in tropical countries. The most dangerous filarial worms live in the lymphatic system, a part of the circulatory system that collects excess fluid around cells and returns it to the blood. The adult worms can be as long as 100 mm (4 in.). The larvae they produce enter the blood and are picked up by mosquitoes that draw blood from an infected person. The larvae develop into an infective stage inside the mosquitoes and are injected into the blood of another person when the mosquitoes

Quick Lab

Comparing Flatworms and Roundworms

Materials living planarian, preserved specimens of tapeworms, male and female *Ascaris*, hand lens

Procedure Examine specimens of tapeworms, *Ascaris*, and the living planarian. Try to locate the following structures on each worm: anterior end, posterior end, mouth, eyespot, hooks, suckers, auricles, and anus. Draw each worm and label each of the features you located.

Analysis

1. List the features found on the posterior and anterior end of each worm.
2. Which worm has a separate mouth and an anus?
3. Which worm "absorbs" digested nutrients?
4. Which worm has a digestive tract instead of a gastrovascular cavity?

Quick Lab

Comparing Flatworms and Roundworms

Time Required 25 minutes

Safety Remind students to handle living planarians with care. Tell them not to open containers that contain preserved specimens.

Procedural Tips Preserved specimens of tapeworms and *Ascaris* as well as living cultures of planarians are available from biological supply companies.

Answers to Analysis

1. *Ascaris* has a mouth with three "lips" at the anterior end. The tapeworm has hooks and suckers, and the planarian has auricles and eyespots but no mouth. The posterior of *Ascaris* has an anus. The posterior of the male is sharply curved and contains two hooklike structures.
2. *Ascaris*
3. Tapeworm
4. *Ascaris*

TEACHING STRATEGY

Map of Parasitic Diseases

Have students collect information about human diseases caused by flatworms and roundworms and about the geographic areas where these diseases are prevalent. Give students world maps with the continents outlined and have them use colored pencils or pens to mark areas where there is a high incidence of these diseases.

SECTION 36-2

ACTIVE READING

Technique: K-W-L

Tell students to return to their lists of things they **W**ant to know about organisms in the phyla Platyhelminthes, Nematoda, and Rotifera, which they made using the Active Reading exercise on page 709. Have them place check marks next to the questions that they are now able to answer. Students should finish by making a list of what they have **L**earned. Conclude by asking students which questions are still unanswered. Ask if they have any new questions.

ANSWERS TO SECTION 36-2 REVIEW

1. Roundworms and rotifers have a hollow, fluid-filled cavity between the endoderm and the mesoderm called a pseudocoelom and a digestive tract with two openings: the mouth and the anus.
2. *Ascaris* eggs escape from the first host's body in the feces. They enter the second host's body when that host ingests food or water that contains the eggs.
3. Hookworm eggs escape from the first host's body in the feces. They enter the second host's body by boring through that host's feet.
4. Trichinosis is a disease caused by the *Trichinella* roundworm, which usually enters the body when undercooked pork is consumed.
5. The cilia move food into the rotifer's mouth.
6. A roundworm's body is long and slender and has a pointed tip at either end. This shape enables a parasitic roundworm to move easily through the tissues of its host.

feed again. Inside the new host, the larvae complete their development and settle in the lymphatic system. When they are present in large numbers, filarial worms can block the lymphatic vessels, causing fluid to accumulate in the limbs. In severe cases, the limbs become extremely swollen and filled with connective tissue, a condition known as **elephantiasis.** Another type of filarial worm infects dogs. It lives in the heart and large arteries of the lungs and is responsible for **heartworm disease.**

PHYLUM ROTIFERA

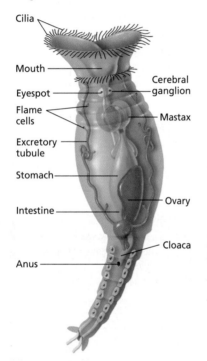

FIGURE 36-10
Cilia surrounding the mouth of a rotifer sweep food into the animal's digestive tract.

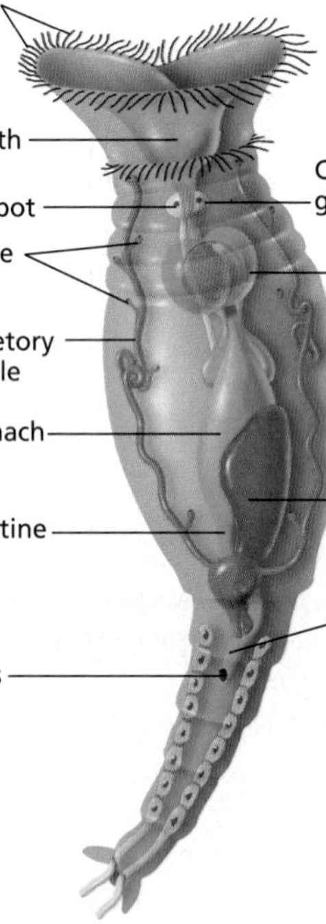

Another group of pseudocoelomates are the 1,750 species in the phylum Rotifera. Members of this phylum are called **rotifers.** Most rotifers are transparent, free-living animals that exist in fresh water. They typically range in length from 100 to 500 µm, with males being much smaller than females. Many rotifers can survive without water for long periods, during which they dry up and look like grains of sand. When wet conditions return, they absorb water and resume their activities.

As Figure 36-10 illustrates, rotifers have a crown of cilia surrounding their mouth. Under a microscope, the crown with its beating cilia looks like a pair of rotating wheels. Rotifers use their cilia to sweep food—algae, bacteria, and protozoans—into their digestive tract. You can follow the path of food through the tract by referring to Figure 36-10. Food moves from the mouth to the **mastax,** a muscular organ that breaks the food into smaller particles. The food is further digested in the stomach, and the nutrients are absorbed in the intestine. Indigestible material passes from the intestine to the **cloaca** (kloh-AY-kuh), a common chamber into which the digestive, reproductive, and excretory systems empty. Like planarians, rotifers use flame cells and excretory tubules to collect excess water in the body. The excess water, along with wastes from the intestine and eggs from the ovaries of females, leaves the cloaca through the anus. Rotifers exhibit cephalization, with a pair of cerebral ganglia and, in some species, two eyespots at the anterior end of the body.

SECTION 36-2 REVIEW

1. **What are two features of roundworms and rotifers that flatworms do not have?**
2. **How are *Ascaris* roundworms transmitted from one host to another?**
3. **How are hookworms transmitted from one host to another?**
4. **What is *trichinosis*?**
5. **What is the function of the cilia on the anterior end of a rotifer?**
6. **CRITICAL THINKING How is the body shape of a parasitic roundworm adapted to the worm's way of life?**

CHAPTER 36 REVIEW

SUMMARY/VOCABULARY

36-1

- The phylum Platyhelminthes is made up of flatworms, the simplest animals with bilateral symmetry. Flatworms are acoelomates, animals that develop from three germ layers and that lack a hollow cavity between the endoderm and the mesoderm.
- Most flatworms have a gastrovascular cavity, a gut with a single opening. These cephalized animals also have excretory, nervous, and reproductive systems.
- The class Turbellaria consists mostly of nonparasitic flatworms, including the freshwater planarian. Planarians are sexually reproducing hermaphrodites that can also reproduce asexually, by splitting in two and regenerating the missing parts.
- The class Trematoda consists of parasitic flukes. Flukes have complex life cycles in which they alternate between two types of hosts: a primary host, from which the adults derive their nourishment and in which sexual reproduction occurs, and an intermediate host, from which the larvae derive their nourishment. Some flukes live in the internal organs of their host, while others live on the external surface of their host.
- The class Cestoda consists of parasitic tapeworms. Adult tapeworms live inside the intestines of vertebrates. Lacking a digestive system, tapeworms absorb nutrients through their body surface. Tapeworms have a long series of body sections called proglottids, each of which contains reproductive structures. Tapeworm life cycles also involve primary and intermediate hosts.

Vocabulary

cerebral ganglion (711)	flatworm (709)	proglottid (713)	tapeworm (713)
cyst (714)	fluke (711)	schistosomiasis (713)	tegument (712)
eyespot (711)	intermediate host (713)	scolex (713)	uterus (712)
fission (711)	pharynx (710)	swimmer's itch (713)	
flame cell (710)	primary host (712)		

36-2

- Members of the phyla Nematoda and Rotifera are pseudocoelomates. They have a hollow, fluid-filled cavity called a pseudocoelom between the mesoderm and the endoderm. They also have a digestive tract with a mouth at the anterior end and an anus at the posterior end.
- The phylum Nematoda consists of roundworms. Most roundworms are free-living, but some are parasites of plants and animals.
- *Ascaris* infects people who consume food or water containing *Ascaris* eggs. The eggs develop into larvae that migrate through the body and mature in the intestines.
- Hookworm larvae in the soil burrow through a person's feet. The larvae migrate through the body and mature in the intestines.
- *Trichinella* infects people who consume undercooked meat containing *Trichinella* cysts. The cysts release larvae, which burrow into the intestinal wall and develop into adults.
- Pinworms live in the lower intestine and lay eggs on the skin around the anus. After being transmitted by the hands to other objects, the eggs may be ingested. They then hatch in the intestine, where the worms mature.
- Filarial worms include species that live in the human lymphatic system. Their larvae are transmitted between hosts by mosquitoes.
- Rotifers are small animals, and most of them live as nonparasites in fresh water. The cilia surrounding their mouth sweep food into their digestive tract.

Vocabulary

anus (715)	elephantiasis (718)	hookworm (716)	rotifer (718)
cloaca (718)	filarial worm (717)	mastax (718)	roundworm (715)
cuticle (715)	heartworm disease (718)	pinworm (717)	trichinosis (717)

REVIEW

1. An acoelomate is an animal that develops from three germ layers but doesn't have an internal body cavity. A pseudocoelomate has a hollow, fluid-filled cavity between the mesoderm and the endoderm.
2. The term *proglottid* does not belong in the group. The first three terms refer to structures found in planarians, but proglottids are found only in tapeworms.
3. A tegument is an external layer found in flukes and tapeworms that protects these parasitic flatworms from the immune system and digestive enzymes of their host.
4. The mastax is a muscular organ that breaks food into smaller particles.
5. Rotifera is a good name for the animals in this phylum because the animals' crown with its beating cilia looks like a pair of rotating wheels.

6. d	**7.** c
8. d	**9.** a
10. c	**11.** b
12. b	**13.** a
14. d	**15.** a

16. Because a flatworm is so flat, none of its cells are very far from the external environment. Thus, the cells can exchange carbon dioxide for oxygen in the environment through diffusion.
17. The cerebral ganglia at the anterior end of a planarian serve as a simple brain. They receive information from sensory cells and transmit signals to the muscles along a ladderlike arrangement of nerves.
18. Planarians are hermaphrodites that reproduce sexually by simultaneously fertilizing each other. They also reproduce asexually by dividing in two and regenerating the missing parts from the two halves.
19. The correct order of steps is g, b, e, d, a, c, f.

20. One adaptation of flukes is the tegument. Others are suckers with which they attach to the primary host; a pharynx with which they feed on the primary host's body fluids; and the existence of ciliated and tailed larval stages, during which they can move through the water to increase their chances of encountering a host.

21. Tapeworms have a protective tegument as well as adhesive suckers and hooks. Their tegument is highly folded, which increases the surface area across which they can absorb nutrients from the host's digestive tract.

22. Roundworms and rotifers both have bilaterally symmetrical bodies with pseudocoeloms, digestive tracts with two openings, and some degree of cephalization.

23. A primary host is an animal from which an adult parasite derives its nourishment. An intermediate host is an animal from which a larval parasite derives its nourishment.

24. *Ascaris* adults live in the intestines of pigs, horses, and humans. They release fertilized eggs that leave the body with the feces of the host. If the eggs are ingested by another host in contaminated food or water, they will develop into larvae in that host's intestine. These larvae will travel through the blood to the lungs. They will then be coughed up, swallowed, and returned to the intestine, where they will mature.

25. People can avoid beef tapeworm infections by ensuring that any beef they eat has been thoroughly cooked. If the beef contains tapeworm cysts, thorough cooking will kill the larvae inside the cysts.

CRITICAL THINKING

1. Asexual reproduction would be advantageous whenever the flatworm has little likelihood of find-

REVIEW

Vocabulary

1. Distinguish between the terms *acoelomate* and *pseudocoelomate.*
2. Choose the term that does not belong in the following group, and explain why it does not belong: eyespot, flame cell, gastrovascular cavity, proglottid.
3. What is a tegument, and where is it found?
4. What is the function of the mastax in a rotifer?
5. The word part *roti-* means "wheel," and the suffix *-fera* means "bearer." With this information, explain why Rotifera is a good phylum name for the organisms it describes.

Multiple Choice

6. A pseudocoelom is located (a) in the endoderm (b) in the ectoderm (c) between the mesoderm and the ectoderm (d) between the endoderm and the mesoderm.
7. A planarian uses its pharynx to help it (a) move (b) reproduce (c) feed (d) respond to light.
8. Blood flukes of the genus *Schistosoma* reproduce sexually (a) in water (b) inside a snail (c) inside a cow's intestine (d) inside a human's blood vessels.
9. A tapeworm uses its scolex to (a) attach itself to its host (b) force food into its mouth (c) reproduce (d) eliminate excess water.
10. Most rotifers (a) are parasitic (b) live in the soil (c) feed with the help of cilia (d) have a gastrovascular cavity.
11. The animal shown in the photograph below is a (a) flatworm (b) roundworm (c) tapeworm (d) rotifer.

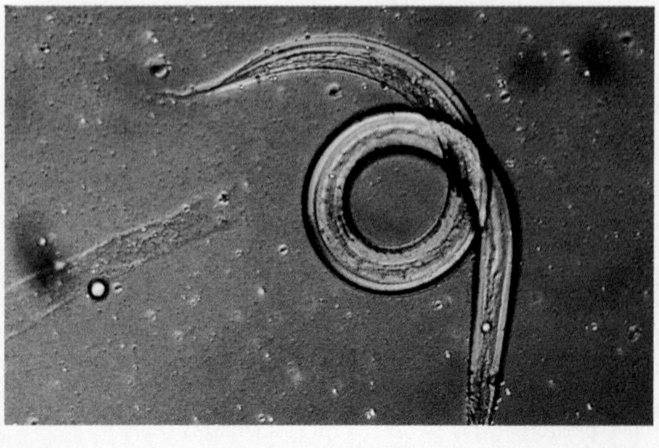

12. A characteristic that differentiates roundworms from flatworms is (a) the presence of a mesoderm (b) the presence of an anus (c) a bilaterally symmetrical body plan (d) cephalization.
13. Flame cells (a) collect excess water (b) respond to light (c) produce eggs (d) produce sperm.
14. In a fluke, the long, coiled tube that stores eggs until they are ready to be released is the (a) excretory tubule (b) mastax (c) tegument (d) uterus.
15. Humans become infected with tapeworms when they (a) eat undercooked meat containing tapeworm cysts (b) drink water containing tapeworm eggs (c) walk barefoot on contaminated soil (d) wade in contaminated water.

Short Answer

16. Explain how a flatworm can survive without a circulatory system or a respiratory system.
17. Describe the nervous system of a planarian.
18. How do planarians reproduce?
19. Arrange the following steps in the schistosome life cycle in the correct order, beginning with the first step after sexual reproduction.
    a. Tailed larvae swim through the water.
    b. Ciliated larvae swim through the water.
    c. Larvae penetrate the skin of a human.
    d. Larvae reproduce asexually in the intermediate host.
    e. Larvae burrow into a snail.
    f. Larvae enter a blood vessel and develop into adults.
    g. Fertilized eggs leave the primary host in the feces or urine.
20. What are some adaptations of flukes to a parasitic way of life?
21. How are tapeworms specialized for a parasitic way of life?
22. What are some features shared by roundworms and rotifers?
23. What is the difference between a primary host and an intermediate host?
24. Describe the life cycle of the roundworm *Ascaris.*
25. How can people avoid becoming infected with the beef tapeworm?

CRITICAL THINKING

1. When would asexual reproduction be advantageous to a free-living flatworm?
2. Parasites are host-specific. This means that they must find the correct host species in order to survive. Suggest a mechanism whereby a parasite could recognize its correct host species.
3. The Aswan High Dam across the Nile River in Egypt was completed in 1970. The dam was built to increase the supply of irrigation water, control major flooding, and provide a source of hydroelectric power. Since the dam was built, however, there has been an increase in the incidence of schistosomiasis in the region. Why do you think this has happened?
4. Why do biologists use the term *eyespot* instead of *eye* to refer to the light-sensitive structures at the anterior end of a planarian?
5. Look again at the diagram of a rotifer in Figure 36-10. Notice the two extensions, called spurs, at the rotifer's posterior end. Suggest a possible function for the spurs.
6. Hookworm infections are extremely common in China, where rice is grown in paddies that are periodically flooded. Considering what you know about how hookworms invade the body, why do you think hookworm infections are so common in this part of the world?
7. A person infected with a tapeworm may show symptoms such as tiredness, loss of weight, and anemia, which could indicate any number of diseases. How might a doctor be certain that the symptoms are caused by a tapeworm?
8. Some rotifers can survive being dried out for as long as four years. When they are placed in water again, they revive. For what kind of environment might this characteristic be adaptive?
9. The two curves below show the number of parasite eggs released each day by a person infected with both *Schistosoma* and *Ascaris* parasites. One curve represents *Schistosoma* eggs, and the other curve represents *Ascaris* eggs. At the time indicated by the arrow, the person was given a medicine that kills adult parasites in the digestive tract. Which curve corresponds to *Schistosoma* eggs? Explain your reasoning.

Egg Production in Parasites

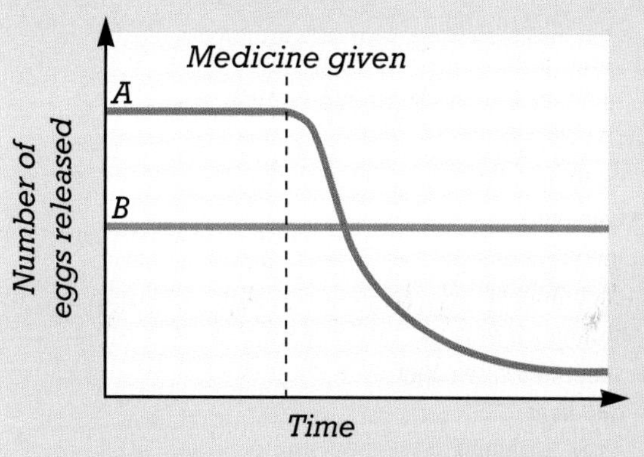

EXTENSION

1. Read "Wonderful Worms" in *New Scientist,* August 7, 1999, on page 4. Explain why people living in developed countries, who are free of intestinal parasites, are more likely to suffer from inflammatory bowel disease.
2. Collect a sample of moist soil, or soil and sand from the bottom of a pond. Using a dissecting microscope or a large magnifier, look for roundworms or rotifers, and observe their behavior.
3. Collect a sample of pond water, or place dried grass and other plants in a container of water for a few days. With the use of a microscope, examine the water for rotifers. (If you can't find any rotifers, your teacher may be able to obtain a culture of rotifers from a biological supply house.) Describe the feeding behavior of the rotifers you find.

ANSWERS

CONTINUED

ing and mating with another individual of the same species.
2. Any answers that suggest parasites use chemical or tactile receptors should be accepted.
3. The lake behind the dam and the many new irrigation canals provide favorable habitats for snails, which serve as intermediate hosts to schistosomes.
4. The term *eyespot* suggests that these structures sense light intensity and direction but do not form visual images.
5. The spurs anchor the rotifer to the substrate.
6. Hookworm eggs hatch in warm, damp soil, such as that found in and around rice paddies. Because the field workers often go barefoot, they are more likely to become infected with hookworm larvae, which enter their host by boring through the feet.
7. The doctor could look for proglottids or tapeworm eggs in the person's feces.
8. This characteristic would be adaptive in pools of water that periodically dry up.
9. Curve B corresponds to *Schistosoma* eggs. Since adult schistosomes live inside blood vessels, they would not be killed by the medicine and therefore would continue to produce eggs.

EXTENSION

1. Inflammatory bowel disease is caused by an overactive immune system. Intestinal parasites calm the immune system.
2. Caution students to handle the soil with tweezers or a probe instead of their hands.
3. Point out to students the sweeping action of the double rings of cilia around the rotifer's head, which explains why rotifers are called "wheel bearers." Students should be able to observe food particles being swept into the pharynx and see the mastax in action.

CHAPTER 36
INVESTIGATION

TIME REQUIRED
One 50-minute class period

SAFETY PRECAUTIONS
Caution students to use care when handling glass culture dishes and test tubes because they break easily.

QUICK REFERENCES
Lab Preparation Notes are found on page 707B. *Holt Biosources* provides a Teaching Resources CD-ROM that contains "Using Gowin's Vee in the Lab" and "Scoring Rubrics."

PROCEDURAL TIP
You may purchase planarians from a biological supply company, or if there is a pond or stream nearby, you may collect them yourself from the undersurface of submerged rocks and logs.

ANSWERS TO BACKGROUND
1. Most turbellarians are nonparasitic (like *Dugesia*) and live in the ocean (unlike *Dugesia,* which lives in fresh water). They glide on a layer of mucus, propelled by cilia.
2. Members of the phylum Platyhelminthes have bilateral symmetry, a gastrovascular cavity, and a solid body that is made of three tissue layers (ectoderm, mesoderm, and endoderm) and that does not have a hollow cavity between the endoderm and the mesoderm.
3. Cephalization refers to the concentration of sense organs and nerve cells at the anterior end of the body. This arrangement permits more-complicated sense organs and brains.
4. A planarian senses light with two eyespots located at the anterior end of its body.

CHAPTER 36 INVESTIGATION

Observing Flatworm Responses to Stimuli

OBJECTIVES
- Observe the feeding behavior of flatworms.
- Study the response of flatworms to different stimuli.

PROCESS SKILLS
- observing
- experimenting
- collecting data
- analyzing data

MATERIALS
- *Dugesia*
- culture dish
- stereomicroscope
- blunt probe
- test tube
- test-tube rack
- aluminum foil
- wax pencil
- medicine dropper
- cork stopper
- stopwatch or clock
- sheet of white paper
- raw liver
- small flashlight

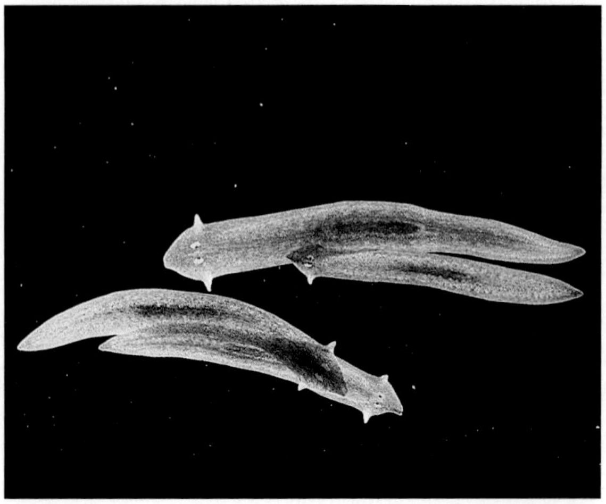

Background
1. *Dugesia,* a planarian, is a nonparasitic flatworm belonging to the class Turbellaria. What are some characteristics of this class?
2. What are some characteristics of the phylum Platyhelminthes?
3. What is meant by cephalization, and why is it an important evolutionary advance?
4. What structures does a planarian have that enable it to sense light, and where are these structures located?

PART A Observing a Flatworm
1. CAUTION **You will be working with a live animal. Be sure to treat it gently and to follow directions carefully.** Use a medicine dropper to transfer one flatworm to a small culture dish. Gently cover the flatworm with water from the culture jar. Why should you use water from the culture jar instead of tap water?
2. Examine the flatworm under the low-power setting of the stereomicroscope. Notice the shape of its body. What kind of symmetry does it have? What structural features demonstrate this symmetry? Does the flatworm appear to have a distinct head and tail? How can you tell?
3. Observe how the flatworm moves. Look carefully at the surface of the flatworm under high power. Can you see any structures that could account for its movement?

PART B Response to Touch
4. In your lab report, make a data table like the one shown. As you complete the following steps, record your observations of the flatworm in your data table.
5. CAUTION **Touch the flatworm gently.** Using a blunt probe, gently touch the posterior end of the flatworm. Notice its response. Now gently touch its anterior end. How does the response compare? What can you conclude from your observations?

PART C Feeding Behavior

6. Place a tiny piece of raw liver in the culture dish. Observe the flatworm for several minutes and describe its feeding response in your data table. Why is it important to use flatworms that have not been recently fed?
7. After several minutes, use the probe to gently turn the flatworm over. What do you observe? If you have time, watch as the flatworm eats the liver. Where do you think the undigested waste will come out?

PART D Response to Gravity

8. With a wax pencil, draw a line around the middle of a test tube.
9. Fill the test tube almost full with water from the culture jar. Then use a medicine dropper to transfer one flatworm to the test tube. Seal the test tube with a cork stopper.
10. Hold the test tube horizontally and move it slowly back and forth until the flatworm is centered on the line you drew.
11. To test whether the flatworm can sense gravity, place the test tube vertically in the test-tube rack. Which way should the flatworm move to show a positive response to gravity? Observe the flatworm for several minutes. Use a stopwatch to measure the amount of time the flatworm spends above the line and below the line. Record the times in your data table.

PART E Response to Light

12. Check the lighting in the room. The light must be low and even during this part of the investigation.
13. Using a piece of aluminum foil, make a cover that is big enough to fit over the bottom half of the test tube you used in Part D. Set the cover aside. Place the test tube horizontally on a white sheet of paper. Make sure the test tube is level. Why is it important for the test tube to be level in this experiment?
14. Position the flashlight so that it will shine directly on the test tube, but do not turn it on.
15. Wait for the flatworm to move to the center line. Then gently place the foil cover over the bottom half of the test tube and turn the flashlight on. Which way should the flatworm move to show a positive response to light? Observe the movements of the flatworm. Use the stopwatch to measure the amount of time the flatworm spends in each half of the test tube. Record the times in your data table.
16. Return the flatworm to the culture jar. Then clean up your materials and wash your hands before leaving the lab.

Analysis and Conclusions

1. What evidence does the flatworm show of cephalization?
2. The flatworm has an incomplete digestive system and no circulatory system. How do you think a flatworm's food gets to the cells after it is digested?
3. How does the flatworm respond to gravity? Is the response positive or negative?
4. How does the flatworm respond to light? Is the response positive or negative?
5. Are the flatworm's anterior and posterior ends equally sensitive to light?

Further Inquiry

Design an experiment to study the responses of flatworms to other stimuli, such as vibrations or sound.

OBSERVATIONS OF FLATWORM BEHAVIOR

Behavior	Observations
Response to touch with blunt probe	
Feeding behavior	
Response to gravity	
Response to light	

CHAPTER 36 INVESTIGATION

ANSWERS TO PARTS A–E

1. Tap water might not be at the right temperature or pH, and it might contain harmful dissolved substances, such as chlorine.
2. *Dugesia* has bilateral symmetry, as demonstrated by its paired anterior projections and eyespots. Its head and tail are distinct: the head has eyespots and is broader than the tail.
3. Cilia are responsible for the movement, although they may not be visible in this investigation.
5. There should be a stronger response at the anterior end, where the sense organs are concentrated.
6. Flatworms that have not been recently fed are more likely to eat.
7. The pharynx extrudes to ingest the liver. Undigested waste will leave through the pharynx.
11. The flatworm should move down in the tube.
13. The test tube should be level to eliminate the influence of gravity on the worm's movements.
15. The flatworm should move toward the lighted half of the test tube.

ANSWERS TO ANALYSIS AND CONCLUSIONS

1. Eyespots and lateral projections are at the anterior end. The flatworm leads with that end when it moves.
2. The gastrovascular cavity is so highly branched that it reaches nearly every cell of the body.
3. The responses of individual flatworms to gravity may vary.
4. The responses of individual flatworms to light may vary.
5. The anterior end is more sensitive to light.

FURTHER INQUIRY

Students might gently shake the dish, make a sudden noise, or feed the flatworm live protozoa.

PLANNING GUIDE
37

CHAPTER 37
MOLLUSKS AND ANNELIDS

Block	TOPICS	TEACHING RESOURCES	LABS, CLASSWORK, AND HOMEWORK
BLOCKS 1 & 2 90 minutes	**Introducing the Chapter, p. 724**	**ATE** Focus Concept, p. 724 **ATE** Assessing Prior Knowledge, p. 724 **ATE** Understanding the Visual, p. 724	■ Supplemental Reading Guide, *Journey to the Ants: A Story of Scientific Exploration*
	37-1 Mollusca, p. 725 Characteristics of Mollusks, p. 725 Body Plan of Mollusks, p. 726 Class Gastropoda, p. 727 Class Bivalvia, p. 728 Class Cephalopoda, p. 730	**ATE** Section Overview, p. 725 **ATE** Critical Thinking, p. 729 **ATE** Visual Link Figure 37-8, p. 730 141–142, 151, 159	**PE** **Quick Lab**, p. 730, Describing a Mollusk **Lab B24,** Snails ★ Study Guide, Section 37-1 **PE** Section Review, p. 731
BLOCK 3 45 minutes	**37-2 Annelida, p. 733** Characteristics and Classification of Annelids, p. 733 Class Oligochaeta, p. 733 Classes Polychaeta and Hirudinea, p. 736	**ATE** Section Overview, p. 733 **ATE** Visual Link Figure 37-11, p. 734 86, 136A, 152, 160	**PE** **Chapter 37 Investigation,** p. 740 **Lab B22,** Earthworm Dissection **Lab B23,** Live Earthworms ★ Study Guide, Section 37-2 **PE** Section Review, p. 736
BLOCKS 4 & 5 90 minutes	**REVIEW and ASSESSMENT** **PE** Chapter 37 Review, pp. 738–739 ■ Performance-Based Assessment—Exploring Invertebrates	Audio CD Program	**CHAPTER TESTING** ■ Chapter Test (blackline master) ▲ Test Generator ■ Assessment Item Listing ■ Scoring Rubrics and Classroom Management Checklists

MODERN BIOLOGY

KEY

- PE Pupil's Edition
- ATE Teacher's Edition
- ■ Active Reading Guide
- ★ Study Guide

HOLT BIOSOURCES

- Teaching Transparencies
- Laboratory Program

One-Stop Planner CD-ROM
Shows these resources within a customizable daily lesson plan:

- ■ Holt BioSources Teaching Resources
- ■ Active Reading Guide
- ★ Study Guide
- ▲ Test Generator

READING FOR CONTENT MASTERY

Preview: Have students preview the objectives, the introductory paragraph, topic sentences, and captions before reading the chapter.

- ATE Active Reading Technique: Brainstorming, p. 725
- ■ Active Reading Guide Worksheet 37-1
- PE Summary/Vocabulary, p. 737

- ■ Active Reading Guide Worksheet 37-2
- PE Summary/Vocabulary, p. 737

TECHNOLOGY AND INTERNET RESOURCES

Holt Biology Videodiscs Teacher's Correlation Guide, Lessons 37-1 through 37-2

Audio CD Program, Sections 37-1 through 37-2

internetconnect

SCILINKS NSTA

On-line Resources: www.scilinks.org
The following *sci*LINKS Internet resources can be found in the student text for this chapter:

Topics:
- Squids, p. 731
- Annelids, p. 734

go.hrw.com

On-line Resources: go.hrw.com
Visit the HRW Web site for a variety of resources related to this chapter. Just type in the keyword HM2 Chapter 37.

Smithsonian Institution®
Visit **www.si.edu/hrw** for additional on-line resources.

CNNfyi.com
Visit **www.cnnfyi.com** for late-breaking news and current event stories selected just for you.

LAB ACTIVITY PLANNING

CHAPTER 37 INVESTIGATION

Observing Earthworm Behavior, pp. 740–741

OVERVIEW
In this investigation, students will observe a live earthworm and will test how it responds to light, touch, and ammonia. Students will also observe how the earthworm's heart rate changes in response to varying body temperature.

SAFETY
Students will wear lab aprons and safety goggles during the lab. You might want to provide latex gloves for students. Instruct students to be careful when handling living organisms. Students must wash their hands after the lab.

PREPARATION
1. Prepare a 3 percent ammonia solution by adding 15 mL of a 10 percent ammonia solution to 35 mL of distilled water. Each lab group will need 5 mL of 3 percent ammonia solution.
2. Each lab group will need one earthworm, a shallow pan, a medicine dropper, a hand lens, two cotton swabs, a Petri dish, a thermometer, a 4 × 4 in. piece of black paper, and two plastic tubs for water baths. Each lab group will also need access to a fluorescent lamp and a stopwatch or clock with a second hand.
3. Each group will need access to hot and cold water. Set up a warm-water bath (30°C), and using ice cubes, set up a cold-water bath (10°C) before class.

Quick Labs

Describing a Mollusk, p. 730
Each student will need two or three bivalve shells, colored pencils, and paper for this activity.

UNDERSTANDING THE VISUAL

Tell students that the octopus has a true coelom, bilateral symmetry, a fairly complex nervous system, and the largest brain of any invertebrate. Ask students why a large brain might be an advantage for the octopus. (A large brain permits more complex decision making and enables the octopus to hunt food and defend itself more effectively.) Ask students how the tentacles of the octopus compare with those of cnidarians. (Both animals use their tentacles for feeding, but octopus tentacles have suction cups instead of stinging cells.)

FOCUS CONCEPT

Evolution

The evolution of animals with a true coelom represents a major advance in animal structure. A coelom provides space for circulatory and other organ systems and allows the muscles of the body wall to function separately from the muscles that surround the digestive tract.

ASSESSING PRIOR KNOWLEDGE

Review the following concepts.

Respiratory and Circulatory Systems: *Chapter 34*
Ask students to describe the function of gills and to name the type of environment in which they are used. Ask students to distinguish between closed and open circulatory systems.

Body Cavities: *Chapter 36*
Ask students to distinguish between acoelomates and pseudocoelomates. Ask students to name a phylum of acoelomate worms and a phylum of pseudocoelomate worms and to give two examples of worms in each phylum.

CHAPTER 37

MOLLUSKS AND ANNELIDS

This Caribbean octopus, Octopus briareus, *is an active predator with a complex brain.*

FOCUS CONCEPT: *Evolution*

As you read, notice the diversity in these two phyla of animals, which share the coelomate body plan and usually develop from trochophore larvae.

MOLLUSCA

Despite their very different appearances, invertebrates such as clams, snails, slugs, and octopuses belong to the same phylum, Mollusca. Members of this phylum are called **mollusks,** *a name that comes from the Latin* molluscus, *which means "soft." Although some mollusks have soft bodies, most have a hard shell that protects and conceals them.*

SECTION

37-1

OBJECTIVES

▲ Summarize the adaptive advantages of a true coelom.

● Identify two features shared by mollusks and annelids.

■ Describe the structure and function of the radula.

◆ Name the characteristics of four major classes of mollusks.

CHARACTERISTICS OF MOLLUSKS

The phylum Mollusca is a diverse group of more than 112,000 species. Among animals, only the phylum Arthropoda has more species. Some mollusks are sedentary filter feeders, while others are fast-moving predators with complex nervous systems.

Mollusks are among several phyla of animals known as coelomates. As you learned in Chapter 34, coelomates are so named because they have a true coelom, a hollow, fluid-filled cavity that is completely surrounded by mesoderm. Coelomates differ from *pseudo*coelomates, such as roundworms, which have a body cavity lined by mesoderm on the outside and endoderm on the inside.

A coelom has several advantages over a pseudocoelom. With a coelom, the muscles of the body wall are separated from those of the gut. Therefore, the body wall muscles can contract without hindering the movement of food through the gut. A coelom also provides a space where the circulatory system can transport blood without interference from other internal organs. The coelomate body plan is shared by annelids, which are discussed in the second half of this chapter, and by three major phyla of animals covered in Chapters 38–40: arthropods, echinoderms, and chordates, including humans.

Another feature that is shared by most aquatic mollusks and annelids is a larval stage of development called a **trochophore** (TRAHK-uh-FOHR), illustrated in Figure 37-1. In some species, the trochophore hatches from the egg case and exists as a free-swimming larva. Cilia on the surface of a free-swimming trochophore propel the larva through the water and draw food into its mouth. As free-swimming trochophores are carried by ocean currents and tides, they contribute to the dispersal of their species. The presence of a trochophore in mollusks and annelids suggests that these two groups of animals may have evolved from a common ancestor.

FIGURE 37-1

A trochophore is a larva that develops from the fertilized egg of most mollusks and annelids. Cilia at both ends and in the middle propel free-swimming trochophores through the water.

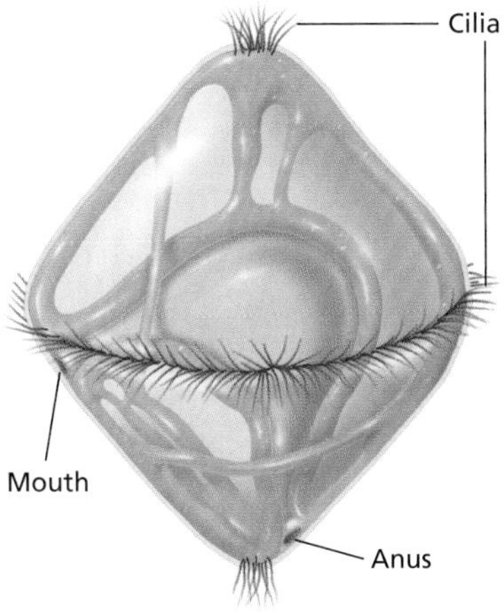

SECTION 37-1

SECTION OVERVIEW

Evolution

Section 37-1 describes the presence of a true coelom and the existence of a trochophore stage during development, two features that are shared by mollusks and annelids and that suggest that these groups of animals evolved from a common ancestor.

Stability and Homeostasis

Section 37-1 discusses the more-advanced organ systems of mollusks that give mollusks a greater capacity for maintaining homeostasis. With their large surface area, gills permit efficient exchange of oxygen and carbon dioxide between the body and the surrounding water. A larger nervous system, especially in the cephalopods, allows more-flexible behaviors and more-complex processing of sensory information.

ACTIVE READING

Technique: Brainstorming

Ask students to compare mollusks and annelids with humans. What body structures and systems do we have in common with them? Work with the class to develop a list of obvious similarities, and write the list on the board or overhead. After reading the chapter, add any items that might have been overlooked. *(Mollusks, annelids, and humans all have a respiratory system, a reproductive system, a circulatory system, a nervous system, a digestive system, an excretory system, and the presence of a coelom.)*

DEMONSTRATION

Obtain several large aquatic snails from a biological supply company or a local pet store, and put them in an aquarium whose sides are covered with a thick growth of algae, such as *Chlorella.* Have students use hand lenses to observe the action of the radula, the rasp-like organ that snails use to scrape up food. Explain how torsion twists most of a snail's body over its head and brings the outlets of its digestive, excretory, and reproductive systems to the front of the snail.

CULTURAL CONNECTION

Purple Dye

The Mediterranean rock whelk, *Murex brandaris,* was used by the Phoenicians and Romans to dye clothing purple. Because each whelk yielded just a small amount of dye, the dye was very expensive and only wealthy people could afford to have purple clothing. As a result, the color purple came to be associated with royalty.

RECENT RESEARCH

Medicine from Cone Shells

Scientists have been investigating cone shell venom, called omega-conotoxin, as a possible medicine for treating stroke victims. The venom acts by closing calcium ion channels in brain cells. During a stroke these channels open, allowing calcium to enter brain cells, killing many of them.

FIGURE 37-2

In the basic body plan of a mollusk, the body is divided into the head-foot and the visceral mass, which contains the internal organs. Covering the visceral mass is the mantle, which secretes the shell.

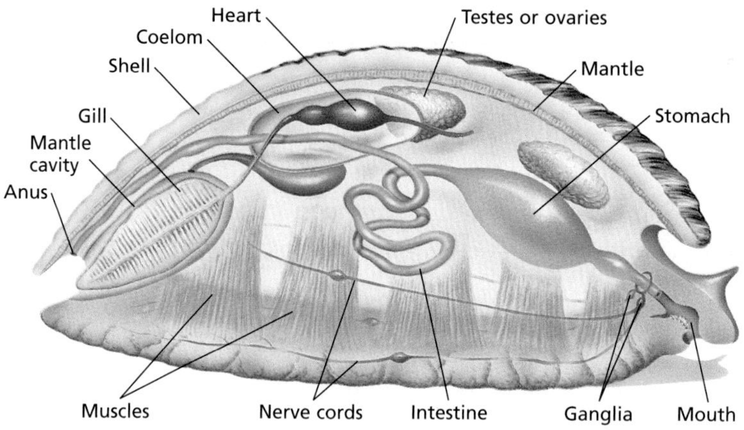

FIGURE 37-3

Inside the mouth (a), many mollusks have a radula, a band of tissue covered with teeth that can scrape food from other surfaces. The SEM in (b) shows the sharp edges of these teeth. (600×)

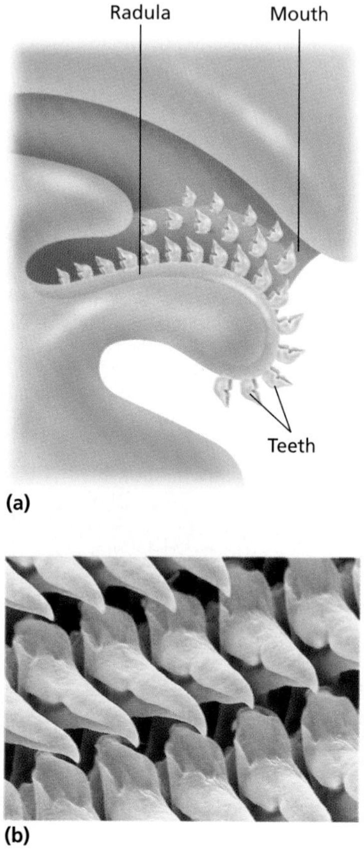

BODY PLAN OF MOLLUSKS

Figure 37-2 shows that the body of a mollusk is generally divided into two main regions: the head-foot and the visceral mass. As its name suggests, the **head-foot** consists of the head, which contains the mouth and a variety of sensory structures, and the foot, a large, muscular organ usually used for locomotion. Above the head-foot is the **visceral** (VIS-uhr-uhl) **mass,** which contains the heart and the organs of digestion, excretion, and reproduction. As you can see in Figure 37-2, the coelom is limited to a space around the heart. Covering the visceral mass is a layer of epidermis called the **mantle.**

In most mollusks, the mantle secretes one or more hard shells containing calcium carbonate. Although shells protect the soft bodies of mollusks from predators, they also reduce the surface area available for gas exchange. This disadvantage was overcome by the evolution of another structural adaptation, gills. Providing a large surface area in contact with a rich supply of blood, gills are specialized for the exchange of gases with water. Figure 37-2 also shows that the delicate gills of mollusks are protected within the **mantle cavity,** a space between the mantle and the visceral mass.

Like the animals you studied in Chapter 36, most mollusks are bilaterally symmetrical. This symmetry is apparent in the nervous system, which consists of paired clusters of nerve cells called **ganglia.** The ganglia are situated in the head-foot and visceral mass and are connected by two pairs of long nerve cords. Nerve cells in the ganglia control the muscles involved in locomotion and feeding and process sensory information from specialized cells that respond to light, touch, and chemicals in the environment.

The main feeding adaptation of many mollusks is the **radula** (RAD-yuh-luh). As Figure 37-3 shows, in most species the radula is a flexible, tongue-like strip of tissue covered with tough, abrasive teeth that point backward. Through evolution, the radula has become adapted for a variety of functions in different mollusks. Terrestrial snails use the radula to cut through the leaves of garden plants, while aquatic snails use it to scrape up algae or to drill

TABLE 37-1 Features of Three Classes of Mollusks

Feature	Gastropoda	Bivalvia	Cephalopoda
External shell	one (most species); none (slugs and nudibranchs)	two	none (except chambered nautilus)
Head	yes	no	yes
Radula	yes	no	yes
Locomotion	crawling (most)	sessile (most)	rapid swimming

holes in the shells of other mollusks. The cone shell has a harpoon-shaped radula that it uses to capture fish and inject venom.

Most biologists use structural differences to divide mollusks into eight classes. Three of these classes are discussed below: Gastropoda (gas-TRAHP-uh-duh), Bivalvia (bie-VALV-ee-uh), and Cephalopoda (SEF-uh-LAHP-uh-duh). Table 37-1 summarizes the major features of these three classes.

Word Roots and Origins

gastropod

from the Greek *gaster,* meaning "belly," and *podion,* meaning "foot"

CLASS GASTROPODA

The largest and most diverse class of mollusks is Gastropoda, whose members are called **gastropods** (GAS-truh-PAHDZ). Most of the 90,000 species of gastropods, including snails, abalones, and conchs, have a single shell. Others, such as slugs and nudibranchs, have no shell at all.

Gastropods undergo a process called **torsion** during larval development. During torsion, the visceral mass twists around 180 degrees in relation to the head. This twisting brings the mantle cavity, gills, and anus to the front of the animal, as shown in Figure 37-4. Because of torsion, a gastropod can withdraw its head into its mantle cavity when threatened. Coiling of the shell is unrelated to torsion.

Wavelike muscular contractions of the foot move gastropods smoothly over surfaces. You can see these contractions if you look closely at the underside of a snail or slug as it crawls across a window pane or the side of an aquarium.

Gastropods have an open circulatory system, meaning that the circulatory fluid, called **hemolymph,** does not remain entirely within vessels. Instead, it is collected from the gills or lungs, pumped through the heart, and released directly into spaces in the tissues. These fluid-filled spaces compose what is known as a **hemocoel** (HEE-muh-SEEL), or blood cavity. From the hemocoel, the hemolymph returns via the gills or lungs to the heart.

Snails

Snails are gastropods that live on land, in fresh water, and in the ocean. Two eyes at the end of delicate

FIGURE 37-4

In a gastropod, such as this snail, the mantle cavity, anus, and gills are near the head as a result of torsion during development.

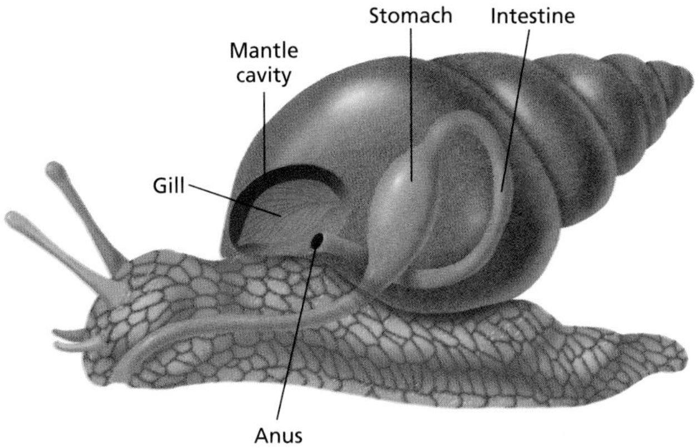

SECTION 37-1

RETEACHING ACTIVITY

Provide students with photographs of various types of mollusks in the classes Gastropoda, Bivalvia, and Cephalopoda, and ask students to sort the photographs by class. Have students describe the characteristics of mollusks in each class and explain which characteristics they considered in assigning each photograph to a class.

GIFTED ACTIVITY

Have students research one class of mollusks. Ask them to focus on specific features, such as the type of shell, the configuration of the digestive tract, the method of feeding, the form of locomotion, and the degree of cephalization. Ask them to develop a hypothetical evolutionary tree showing the possible relationships among orders in that class and the sequence in which the orders might have evolved. Suggest that they use Figure 38-3 as a model.

MAKING CONNECTIONS

Mathematics

The twelfth-century mathematician Fibonacci devised a number sequence by adding successive numbers together: 0, 1, 1, 2, 3, 5, 8, 13, 21, 34, etc. He and others since his time have found that many natural phenomena are organized according to this sequence. For instance, the diameter of the spirals of a snail shell adhere to the Fibonacci sequence when measured from the center of the shell outward.

CONTENT CONNECTION

Section 36-1: Blood Flukes

Larvae of the blood fluke *Schistosoma* use specific types of aquatic snails as intermediate hosts during a portion of their life cycle. Typical hosts are *Biomphalaria* in South America and the West Indies, *Bulinus* and *Physopsis* in Africa, and *Oncomelania* in East Asia.

TEACHING STRATEGY

Garden Slugs

Obtain several slugs from a local garden and bring them to class. Explain that gas exchange in these terrestrial gastropods occurs in a highly vascularized region of the mantle that functions as a lung. Ask students why gills would not be useful for gas exchange in these animals. (Gills are specialized for aquatic environments. They would collapse without the support of water, reducing the area available for gas exchange.) Have students locate the pneumostome, the opening to the lung on the right side of the slug's body. They can time its opening and closing to determine whether there is a correlation with the slug's activity level. (The pneumostome will open and close more frequently with increased activity.)

FIGURE 37-5

The extensions on the back of this horned nudibranch, *Hermissenda crassicornia,* provide a large surface area for gas exchange.

tentacles on the head help the snail locate food. If danger arises, the tentacles retract into the head. Aquatic snails respire through gills in the mantle cavity. In land snails, the mantle cavity acts as a modified lung that exchanges oxygen and carbon dioxide with the air. The thin membrane lining the mantle cavity must be kept moist to allow gases to diffuse through it. For this reason, land snails are most active when the air has a high moisture content. Snails survive dry periods by becoming inactive and retreating into their shells. They seal the opening to their shell with a mucous plug, which keeps them from drying out.

Land snails are hermaphrodites, but aquatic snails have separate sexes. In either case, the eggs are almost always fertilized internally when two individuals mate.

Other Gastropods

Slugs are terrestrial gastropods that look like snails without shells. Like land snails, slugs respire through the lining of their mantle cavity. They avoid desiccation by hiding in moist, shady places by day and feeding at night.

Nudibranchs, like the one shown in Figure 37-5, are marine gastropods that lack shells. *Nudibranch* means "naked gill," which refers to the fact that gas exchange occurs across the entire body surface of these animals. The surface of most nudibranchs is covered with numerous ruffles or delicate, fingerlike extensions that increase the total area available for gas exchange.

Some gastropods show unusual adaptations of the foot. In pteropods, or "sea butterflies," for example, the foot is modified into a winglike flap that is used for swimming rather than crawling.

CLASS BIVALVIA

FIGURE 37-6

The two valves that make up the hinged shell of a bivalve can separate, allowing water to circulate through the animal. Some bivalves, like this calico scallop, *Aequipecten viradians,* have a row of eyes near the outer margin of each valve.

Members of the class Bivalvia include aquatic mollusks, such as clams, oysters, and scallops. These mollusks are called **bivalves** because, as Figure 37-6 shows, their shell is divided into two halves, or valves, connected by a hinge. A bivalve can close its shell by contracting the powerful **adductor muscles** that are attached to the inside surface of each valve. You can see the adductor muscles in Figure 37-7. When the adductor muscles relax, the valves open.

Each valve consists of three layers that are secreted by the mantle. The thin outer layer protects the shell against acidic conditions in the water. The thick middle layer of calcium carbonate strengthens the shell. The smooth, iridescent inner layer protects the animal's soft body. If a grain of sand or other irritant gets inside the shell of a bivalve, the mantle coats it with the same material that lines the inner layer. Multiple applications of this material form a pearl.

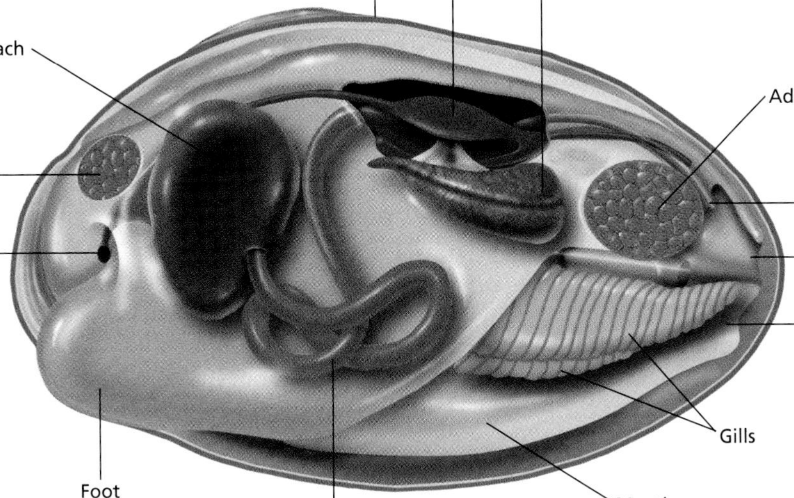

FIGURE 37-7

In this illustration, one valve has been omitted to show a clam's anatomy. The internal structure of a clam is typical of most bivalves.

In contrast with gastropods, which move about in search of food, most bivalves are sessile. Some species extend their muscular foot into the sand and fill the foot with hemolymph to form a hatchet-shaped anchor. The muscles of the foot then contract, pulling the animal down into the sand. As an adaptation for a sessile existence, bivalves usually are filter feeders. They are the only mollusks without a radula.

Bivalves lack a distinct head. Their nervous system consists of three pairs of ganglia: one pair near the mouth, another pair in the digestive system, and the third pair in the foot. The ganglia are connected by nerve cords. Nerve cells in the ganglia receive information from sensory cells in the edge of the mantle that respond to touch or to chemicals in the water. Some bivalves also have a row of small eyes along each mantle edge. Stimuli detected by these sensory structures can trigger nerve impulses that cause the foot to withdraw and the shell to close.

Clams

Clams are bivalves that live buried in mud or sand. The mantle cavity of a clam is sealed except for a pair of hollow, fleshy tubes called **siphons,** which you can see in Figure 37-7. Cilia beating on the gills set up a current of water that enters through the **incurrent siphon** and leaves through the **excurrent siphon.** As the water circulates inside the clam, the gills filter small organisms and organic debris from the water. The filtered material becomes trapped on the gills in a sticky mucus that moves in a continuous stream toward the mouth. Water passing over the gills also exchanges oxygen and carbon dioxide with the hemolymph.

Most species of clams have separate sexes. Marine clams reproduce by shedding sperm and eggs into the water, and fertilization occurs externally. The fertilized egg becomes a trochophore that eventually settles to the bottom and develops into an adult. In some

INCLUSION *ACTIVITY*

Have students use clay to build a model of a clam. Different colors of clay could be used for the shell, mantle, gills, siphons, adductor muscles, and digestive tract. When they have finished their model, have them use it to trace the path of water through a clam, and ask them to explain how the siphons and gills are used in gas exchange and feeding. Students could also use clay to make models of a snail and a cephalopod, using an invertebrate zoology textbook as a reference for internal structures.

CRITICAL THINKING

Salinity Tolerance

Large marine mollusks, such as helmet shells and clams, can usually tolerate a wide range of salinity in the water that surrounds them. What factors might cause the salinity to change in a tide pool or a lagoon inside a coral reef? (Evaporation of water during low tide would increase the salinity. Addition of fresh water from rivers and during periods of heavy rain would decrease the salinity.)

QUICK FACT

☑ Coquina clams on the Atlantic Coast of the United States actively seek large waves to carry them out to sea. They use their hairlike cilia to distinguish between large and small waves.

Quick Lab

Describing a Mollusk

Time Required 20 minutes

Procedural Tips Bags of assorted seashells can be obtained in bulk from biological supply houses or may be collected from seashores.

Answers to Analysis
Descriptions will vary. Some shells are smooth, and some are deeply ribbed. The hinge area may vary in location slightly to the right or left of the center edge. Some bivalves have very colorful patterns on their shells. The mantle is iridescent and very smooth.

VISUAL LINK

Figure 37-8

Like squids, cuttlefishes have a shell that is small and completely enclosed within the mantle. In octopuses, there is no shell at all. Ask students what problems these cephalopods would face if their body were covered by a hard shell. (The weight and inflexibility of the shell would restrict the mobility of these highly active predators.) Ask students how the chambered nautilus, which is a cephalopod with an external shell, overcomes the weight problem. (The nautilus shell contains gas-filled chambers that provide buoyancy.)

Quick Lab

Describing a Mollusk

Materials 2–3 bivalve shells, colored pencils, paper

Procedure

1. Draw a bivalve shell on a sheet of paper using your colored pencils.
2. Use Figure 37-7 to help you locate and label the adductor muscle scars, the mantle cavity, and the hinge area on the bivalve.

Analysis Describe the shell of the bivalve, including its color, its ridges, the appearance and texture of the mantle, and the location of the hinge area.

species, the adults may weigh 200 kg (440 lb) and be over 1 m (3.3 ft) across. In most freshwater clams, eggs are fertilized internally by sperm that enter through the incurrent siphon. The larvae that develop are discharged into the water through the excurrent siphon. If they contact a passing fish, they may live as parasites on its gills or skin for several weeks before settling to the bottom.

Other Bivalves

Oysters are bivalves that become permanently attached to a hard surface early in their development. Some are grown commercially as food or as sources of cultured pearls. Scallops can move through the water by repeatedly opening their valves and snapping them shut. This motion expels bursts of water, creating a form of jet propulsion. The teredo, or shipworm, is one of the few bivalves that does not filter-feed. Instead, it bores into driftwood or ship timbers and ingests the particles that are produced by the drilling. The wood cellulose is broken down by symbiotic bacteria that live in the shipworm's intestine.

CLASS CEPHALOPODA

Members of the class Cephalopoda include octopuses, squids, cuttlefishes, and chambered nautiluses. These marine mollusks are called **cephalopods** (SEF-uh-luh-PAHDZ), a term that means "head-foot." The name refers to the fact that a cephalopod's foot is concentrated in its head region. Cephalopods are specialized for a free-swimming, predatory existence. Extending from the head is a circle of tentacles, as you can see in Figure 37-8. The tentacles' powerful suction cups allow cephalopods to grasp objects and capture prey. Cephalopods kill and eat their prey with the help of a pair of jaws that resemble a parrot's beak.

The nervous system is more advanced in cephalopods than in any other group of mollusks. The cephalopod brain, which is the largest of any invertebrate brain, is divided into several lobes and contains millions of nerve cells. The large number of nerve cells enables cephalopods to process information in sophisticated ways. Octopuses, for example, can learn to solve simple problems, perform tasks, and discriminate between objects on the basis of their shape or texture. The sensory systems of cephalopods are also well developed. Most cephalopods have highly advanced eyes that are capable of forming images of objects. The tentacles contain numerous cells that sense chemicals in the water.

Cephalopods have a closed circulatory system, which means blood circulates entirely within a system of vessels. Closed circulatory systems transport fluid more rapidly than open circulatory systems do. Thus, nutrients, oxygen, and carbon dioxide are carried quickly through the body of these highly active animals. Cephalopods also have separate sexes. The male uses a special-

FIGURE 37-8

Cephalopods, like this cuttlefish, *Sepia latimanus,* have several long tentacles surrounding their mouth. The streamlined body of many cephalopods enables them to swim rapidly in pursuit of prey.

ized tentacle to transfer packets of sperm from his mantle cavity to the mantle cavity of the female, where fertilization occurs. The female lays a mass of fertilized eggs and guards the eggs until they hatch. Unlike other mollusks, cephalopods develop from an egg into a juvenile without becoming a trochophore.

Many cephalopods can release a cloud of ink into the water to temporarily distract predators. They also have pigment cells called **chromatophores** (kroh-MAT-uh-FOHRS), which are located in the outer layer of the mantle. Chromatophores can produce a sudden change in the color of a cephalopod, allowing the animal to blend in with its surroundings.

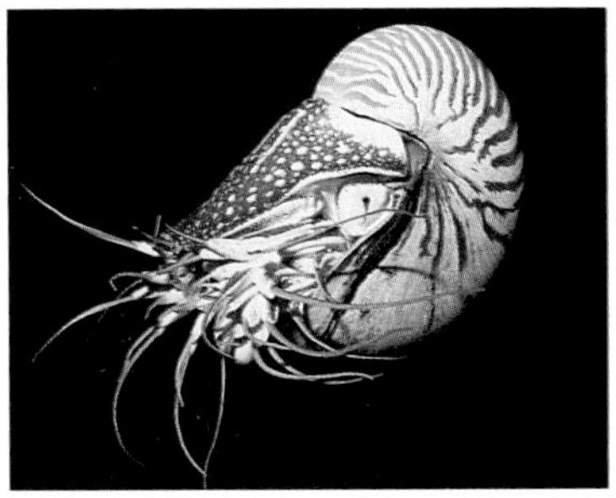

FIGURE 37-9
The chambered nautilus is the only cephalopod with an external shell.

Squids

Squids are cephalopods with ten tentacles. The longest two tentacles are used for capturing prey, and the other eight force the prey into the squid's mouth. The muscular mantle propels the squid swiftly through the water by pumping jets of water through an excurrent siphon. Most squids grow to about 30 cm (1 ft) in length, but a few species can be much longer. The giant squid, *Architeuthis,* may reach a length of 18 m (about 60 ft) and a weight of more than 3,300 kg (about 3.5 tons). It is the world's largest known invertebrate.

Octopuses and Chambered Nautiluses

Octopuses have eight tentacles and share many characteristics with squids, including their methods of escaping from predators. Instead of using jet propulsion to chase prey, however, octopuses are more likely to crawl along the bottom with their tentacles or lie in wait in caves and rock crevices. Octopuses seldom exceed 30 cm (1 ft) in diameter, although the giant Pacific octopus may grow to a diameter of 2.5 m (about 8 ft).

The chambered nautilus, shown in Figure 37-9, is the only existing cephalopod that has retained its external shell. The shell is coiled and divided into a series of gas-filled chambers separated by partitions. The soft body of the nautilus is confined to the outermost chamber. As the nautilus grows, it moves forward in its shell, makes a new partition, and fills the chamber behind the partition with gas. The gas makes the nautilus buoyant.

internetconnect

SCILINKS NSTA

TOPIC: Squids
GO TO: www.scilinks.org
KEYWORD: HM731

SECTION 37-1 REVIEW

1. What is one advantage of a true coelom over a pseudocoelom?
2. What is a trochophore? In what phyla of animals is it found?
3. What is a radula, and what is it used for?
4. Why are land snails more active when the air around them is moist?
5. In what ways is a squid adapted for a predatory way of life?
6. **CRITICAL THINKING** Suggest why an open circulatory system is sufficient to meet the needs of a gastropod.

QUICK FACT

The eyes of a giant squid are twice the size of a soccer ball.

ANSWERS TO SECTION 37-1 REVIEW

1. A true coelom separates the muscles of the body wall from those of the gut. It also provides space for the circulatory system to transport blood without interference from other internal organs.
2. A trochophore is a pear-shaped larva that constitutes the first stage of larval development in both mollusks and annelids.
3. A radula is a flexible, tongue-like strip of tissue used by mollusks to scrape food into the mouth, drill holes in the shells of prey, and harpoon prey.
4. In land snails, the mantle cavity acts as a modified lung that exchanges gases with the air. The thin membrane that lines the cavity must be kept moist to allow gases to diffuse through it. Therefore, land snails are most active when the air has a high moisture content.
5. Squids have tentacles and a beaklike structure for capturing and devouring prey, and a muscular mantle that propels them swiftly through the water.
6. Gastropods move by slowly crawling over surfaces. Their slow movements consume energy at a low rate, which means that their needs for nutrients and gas exchange are low. These needs can be met by an open circulatory system, which transports fluid more slowly than a closed circulatory system.

BACKGROUND

Leech is a term that applies to any annelid in the class Hirudinea. There are about 300 known species of leeches. Leeches are found mostly in fresh water or moist terrestrial environments and can reach lengths of 40 cm (16 in.).

Hirudo medicinalis, the medicinal leech described in this feature, feeds primarily on mammalian blood, but it also sucks blood from amphibians, reptiles, and fish. Within the mouth are three jaws with sharp teeth that make a Y-shaped incision in the skin of the host. Once engorged with blood, the leech detaches from its host and may not feed again for 18 months.

DISCUSSION

Guide the discussion by posing the following questions.

1. How are leeches used in surgery today? (Leeches are used to remove congested blood from areas where surgeons have reattached limbs, fingers, or toes and to prevent the formation of life-threatening blood clots in heart patients.)

2. What are some of the useful components of leech saliva? (Leech saliva contains an anticoagulant and an anesthetic.)

3. How is the leech's anticoagulant useful to the leech? (It prevents blood from clotting at the wound, allowing a continuous flow of blood into the leech.)

FURTHER READINGS

1. "Buggy Medicine," in *Boys' Life,* February 1996, p. 16. This article discusses how leeches and maggots have proven beneficial to medicine.

2. "The Worm Turns—into a Source of New Drugs," in *Science News,* March 9, 1996, p. 150. This article explains how hookworms synthesize safe anticoagulants.

Research Notes

Leeches: New Uses for an Old Remedy

Why are leeches called bloodsuckers? Do they really suck blood? Yes, they do, and although it may sound disgusting, their role as bloodsuckers could help save your life.

For centuries, leeches were a common tool of medical practice. In the second century A.D., the Greek physician Galen described the usefulness of leeches in removing blood from patients, a procedure called bloodletting. An excess of blood in the body was believed to be responsible for a variety of illnesses, from headaches and fevers to heart disease. Physicians regarded the leech's habit of feeding on blood as a simple way to remove this "bad blood" from a patient's body. Bloodletting remained very common in Europe through the early nineteenth century, and leeches were grown in ponds and harvested in large numbers for use in medicine. In fact, bloodletting was prescribed so often that physicians themselves were sometimes referred to as leeches. During the late 1800s, however, medical science discredited the idea that excess blood causes disease, and bloodletting fell out of favor.

Surprisingly, leeches are making a comeback in medicine, although with new purposes. One of these purposes is to increase the success rate of operations to reattach severed limbs, fingers, or toes. Such operations involve microsurgery, a process in which surgeons reconnect tendons, blood vessels, and nerves using tiny instruments and powerful microscopes. Along with today's sophisticated technology, some surgeons have added an unlikely tool—the leech.

During microsurgery, physicians can reconnect arteries but not small veins, which are more delicate. As a result, circulation in the reattached limb, finger, or toe is impaired, and the tissues may become congested with blood. If this happens, the tissues of the reattached part die, and the part cannot heal and rejoin the body. One solution to this problem is to place leeches on the reattached body part. There, they begin to suck out the accumulated blood, relieving congestion and allowing the tissues to remain healthy until the veins can grow back. At a cost of about $7.00 apiece, leeches are an inexpensive treatment for a serious problem.

Leeches have medical uses that go beyond their ability to remove blood. Scientists have known since the 1800s that leech saliva contains a powerful anticoagulant, a substance that inhibits blood clotting. The leech's anticoagulant, called hirudin, can cause four hours or more of steady bleeding. The word *hirudin* derives from the scientific name of the medicinal leech, *Hirudo medicinalis.* Today hirudin is made through genetic engineering, without the aid of leeches. It has proven useful in the treatment of some heart patients, particularly those who have had heart attacks, who suffer from angina, or who have undergone angioplasty, a procedure to open blocked arteries. One research study even indicated that hirudin may be effective against the spread of cancer.

The amazing uses that have been found for a substance in leech saliva are encouraging to medical researchers, who continue to explore how knowledge of invertebrate organisms can be beneficially combined with medical technology.

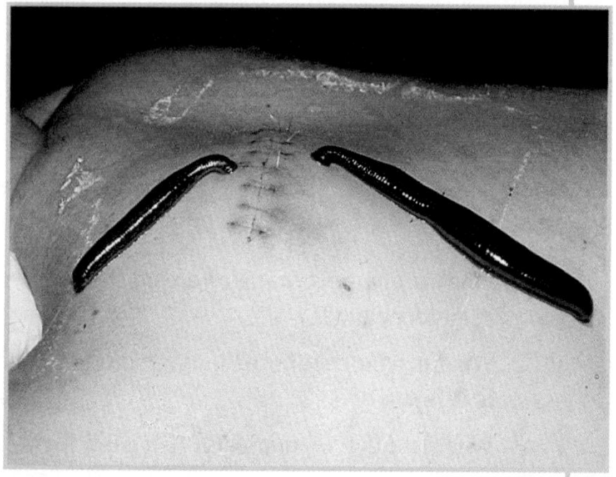

Leeches have been applied to this patient's sutures across his upper back to reduce congestion of blood. Each leech can remove up to 5 mL of blood.

ANNELIDA

Colorful feather-duster worms, common earthworms, and bloodsucking leeches are all members of the phylum Annelida (uh-NEL-uh-duh). An animal in this phylum is called an ***annelid*** *(AN-uh-LID), a term that means "little rings." The name refers to the many body segments that make an annelid look like it is composed of a series of rings.*

SECTION

37-2

OBJECTIVES

▲ List the advantages of body segmentation.

● Explain how earthworms move.

■ Describe the organ systems of earthworms.

◆ Distinguish between the three classes of annelids.

CHARACTERISTICS AND CLASSIFICATION OF ANNELIDS

The phylum Annelida consists of about 15,000 species of bilaterally symmetrical, segmented worms. Segmentation is the most distinctive feature of annelids. Like mollusks, annelids have a true coelom, but the coelom in annelids is divided into separate compartments by partitions. Division of the coelom represents an evolutionary advance over the earliest wormlike coelomates. In an undivided coelom, the force of muscle contraction in one part of the body is transmitted to other parts by the fluid in the coelom. A segmented coelom enables different parts of the body to contract or expand independently. In addition, duplication of some of the organ systems in each segment provides a form of insurance against injury. If one segment becomes disabled, the others can still function.

Most annelids have external bristles called **setae** (SEE-tee), and some have fleshy protrusions called **parapodia** (PAR-uh-POH-dee-uh). Both of these structures are visible in Figure 37-10. The number of setae and the presence or absence of parapodia provide the basis for dividing annelids into three classes: Oligochaeta (AHL-uh-goh-KEET-uh), Polychaeta (PAHL-ee-KEET-uh), and Hirudinea (HIR-yuh-DIN-ee-uh). All organ systems are well developed in most members of each class.

FIGURE 37-10

Numerous setae help this bristle worm move through its environment. The setae extend from fleshy flaps called parapodia. The bristle worm is a member of the class Polychaeta and genus *Hermodice*.

CLASS OLIGOCHAETA

Annelids of the class Oligochaeta generally live in the soil or in fresh water and have no parapodia. *Oligochaeta* means "few bristles," and as the name suggests, these annelids have a few setae on each segment. The most familiar member of the class Oligochaeta is the earthworm. As you read about the earthworm, look for adaptations that enable this animal to lead a burrowing life.

SECTION 37-2

SECTION OVERVIEW

Evolution

Section 37-2 describes the evolution of body segmentation, which in the annelids is represented by the division of the coelom into compartments separated by partitions. Segmentation affords a greater flexibility of movement and the potential for duplication of organ systems.

Matter, Energy, and Organization

Section 37-2 describes the organization of the earthworm's digestive system, which is divided into a series of specialized compartments with different but complementary functions.

▶ ENGAGE STUDENTS

Have students watch a video clip or laserdisc motion sequence showing the movement of earthworms and leeches. Explain how the circular and longitudinal muscles in an earthworm are coordinated as the animal moves. (A body segment elongates when the circular muscles in that segment contract, and it shortens when the longitudinal muscles contract.) Explain that leeches swim by alternately contracting longitudinal muscles in the dorsal and ventral sides of the body, producing a wave that passes along the body from anterior to posterior.

QUICK FACT

☑ An earthworm can pull 10 times its own weight.

RECENT RESEARCH

Leech Anticoagulant

Scientists have recently determined the three-dimensional structure of decorsin, one of three anticoagulants found in leech saliva. Decorsin inhibits the aggregation of blood platelets by binding to glycoproteins on their surface.

VISUAL LINK

Figure 37-11

This figure illustrates the internal structure of the anterior end of an earthworm. The body is divided by partitions (not shown) that confine the coelomic fluid to individual segments. However, the circulatory, digestive, and nervous systems penetrate the partitions and extend the length of the body. Posterior to the clitellum, the segments are virtually identical.

DEMONSTRATION

Bring an earthworm and a balloon to class. Place the earthworm in a large, shallow dish lined with wet paper towels. Fill the balloon with water and tie off the end. As the earthworm moves around in the dish, use the balloon to demonstrate what happens in one segment of the earthworm. Cause the balloon to elongate by squeezing the middle of the balloon with your hands, which represent the worm's circular muscles. Then release your grip to allow the balloon to return to its normal, rounded shape. Explain that the elastic balloon walls represent the worm's longitudinal muscles. Ask students to picture a chain of 100 water-filled balloons linked end to end, with balloons at one end of the chain being squeezed while those at the other end are shortening.

INCLUSION ACTIVITY

Have students imagine that they own an earthworm farm, and ask them to prepare a brochure advertising their earthworms to gardeners. The brochure should contain drawings of the internal and external structure of earthworms, a description of how earthworms feed, and an explanation of why earthworms are beneficial to gardeners.

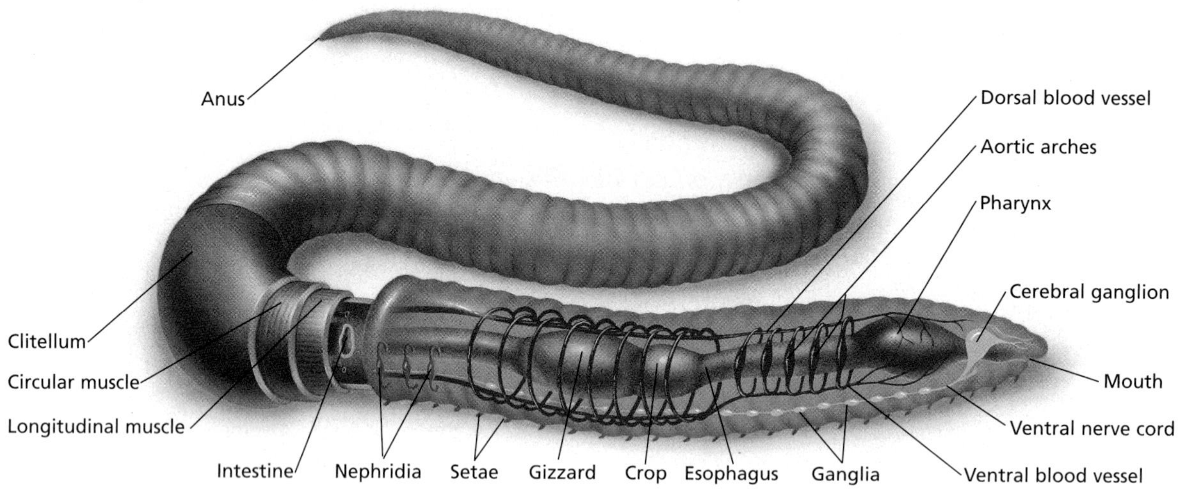

FIGURE 37-11

The segmentation of annelids, such as this earthworm, is visible both externally and internally. Some of the internal structures, such as ganglia and nephridia, are repeated in each segment.

Structure and Movement

An earthworm's body is divided into more than 100 segments, most of which are virtually identical. Figure 37-11 shows that circular and longitudinal muscles line the interior body wall of an earthworm. To move, the worm anchors some of the middle segments by their setae and contracts the circular muscles in front of those segments. Contraction of the circular muscles increases the pressure of the coelomic fluid in those segments. This increased pressure elongates the animal and pushes the anterior end forward. Setae in the anterior segments then grip the ground as the longitudinal muscles contract, pulling the posterior segments forward. This method of locomotion in earthworms is an example of the kind of movement made possible by segmentation.

Feeding and Digestion

Earthworms ingest soil as they burrow through it. Soil is sucked into the mouth by the muscular pharynx. The soil then passes through a tube called the **esophagus** (ee-SAHF-uh-guhs) to a temporary storage area known as the **crop.** From the crop, the soil moves to a thick, muscular part of the gut called the **gizzard.** Find these parts of the digestive tract in Figure 37-11. The gizzard grinds the soil, releasing and breaking up organic matter. As the soil passes through the long intestine, digested organic compounds and nutrients in the soil are absorbed by the blood. An infolding of the intestinal wall called the **typhlosole** (TIF-luh-SOHL) increases the surface area available for digestion and absorption. Undigested material is eliminated from the earthworm's body through the anus.

Earthworms play an important role in maintaining the fertility of soil. By decomposing dead leaves and other organic materials, earthworms help release nutrients into the soil. The burrows made by earthworms allow air to penetrate into the soil, bringing oxygen to plant roots and soil microorganisms. Earthworms also loosen the soil, making it easier for roots to grow and for water to seep in.

internetconnect

SCILINKS NSTA

TOPIC: Annelids
GO TO: www.scilinks.org
KEYWORD: HM734

Circulation

A closed circulatory system transports oxygen, carbon dioxide, nutrients, and wastes through the body of an earthworm. The blood travels toward the posterior end through a ventral blood vessel and then returns to the anterior end through a dorsal blood vessel. As you can see in Figure 37-11, five pairs of muscular tubes, the **aortic** (ay-OHR-tik) **arches,** link the dorsal and ventral blood vessels near the anterior end of the worm. Contractions of the ventral blood vessel and the aortic arches force blood through the circulatory system.

Respiration and Excretion

Earthworms have no specialized respiratory organs. Oxygen and carbon dioxide diffuse directly through the skin, which contains many small blood vessels. This exchange of gases can take place only if the skin is moist. Therefore, earthworms avoid dry ground and extreme heat. Secretions of mucus and the presence of a thin cuticle also help keep an earthworm's skin moist.

Earthworms eliminate cellular wastes and excess water through excretory tubules called **nephridia** (ne-FRID-ee-uh), some of which are shown in Figure 37-11. Each segment except the first three and the last one contains a pair of nephridia. As coelomic fluid passes through the nephridia, some of the water is reabsorbed by blood vessels. The remaining fluid and the wastes dissolved in it are released from the body through pores on the ventral surface.

Neural Control

The nervous system of an earthworm consists of a chain of ganglia connected by a ventral nerve cord. Most body segments contain a single ganglion. Nerves branching from each ganglion carry impulses to the muscles and from the sensory cells in that segment. In the most anterior segments, several ganglia are fused to form the cerebral ganglia, or brain, as you can see in Figure 37-11. One of the main functions of the cerebral ganglia is to process information from simple sensory structures that respond to light, touch, chemicals, moisture, temperature, and vibrations. Although these sensory structures are found in all segments, they are especially concentrated at the anterior end.

Reproduction

Earthworms are hermaphrodites, but an individual worm cannot fertilize its own eggs. Mating occurs when two earthworms press their ventral surfaces together with their anterior ends pointing in opposite directions. The two worms are held together by their setae and by a film of mucus secreted by each worm's **clitellum** (klie-TEL-uhm). The clitellum, also visible in Figure 37-11, is a thickened section of the body. Each earthworm injects sperm into the mucus. The sperm from each worm move through the mucus to the pouchlike **seminal receptacle** of the other, where they are stored. The worms then separate, and after several days the clitellum of each worm secretes a tube

Word Roots and Origins

nephridium

from the Greek *nephros,* meaning "kidney," and *idion,* meaning "small"

CONTENT CONNECTION

Section 5-1: Diffusion

By continuously moving blood beneath the surface of the skin, the earthworm's circulatory system maintains the concentration gradients of O_2 and CO_2 across the skin, allowing these gases to diffuse into and out of the body.

GIFTED ACTIVITY

Have students research and report on three annelid species other than earthworms and leeches. Possibilities include feather worms, peacock worms, fan worms, and the giant tubeworms found in deep-sea thermal vents. The report should explain how structural differences among different annelids relate to specific adaptive functions.

RETEACHING ACTIVITY

Ask students how earthworms and leeches differ. (Earthworms have setae and leeches do not. Leeches have anterior and posterior suckers and earthworms do not.) Have students explain why both animals are classified as annelids. (Both have a true coelom and a segmented body.)

TEACHING STRATEGY

Polychaetes

Show students photographs of several types of polychaetes to illustrate the diversity of forms in the largest class of annelids. Examples include sessile featherduster worms and Christmas-tree worms, burrowing clam worms, and carnivorous scale worms and fireworms. The photographs should clearly show the characteristic features of polychaetes: antennae, specialized mouthparts, and parapodia with numerous setae. Use the photographs to illustrate similarities and differences between the annelid classes.

SECTION 37-2

QUICK FACT

✔ The Amazonian leech, *Haementeria ghilianii,* may reach 30 cm in length.

ANSWERS TO SECTION 37-2 REVIEW

1. Body segmentation makes it possible for different parts of the body to contract or expand independently. In addition, duplication of organ systems in each segment provides insurance against injury.
2. Contraction of the circular muscles increases the pressure of the coelomic fluid, elongating the animal and pushing the anterior end forward. Contraction of the longitudinal muscles pulls the posterior segments forward.
3. Oxygen and carbon dioxide diffuse directly through the skin, which contains many small blood vessels.
4. Polychaetes have more setae than do earthworms, and the setae of polychaetes project from parapodia, which earthworms lack. Also, most polychaetes live in marine environments, whereas earthworms live in the soil.
5. Leeches have suckers that they use to attach to the skin of a host. Parasitic leeches secrete an anesthetic that prevents the host from feeling their presence and another substance that prevents blood from clotting.
6. Leeches are external parasites. They don't cycle between primary and intermediate hosts, and they remain with their host only while they are feeding. Tapeworms and liver flukes are internal parasites whose life cycles involve stages that alternate between primary and intermediate hosts. They remain with their hosts for much longer periods.

made of mucus and a tough carbohydrate known as **chitin** (KIE-tin). As this tube slides forward, it picks up the worm's eggs and the stored sperm from the other worm. Fertilization occurs inside the tube, which closes up to form a protective case. The young worms develop inside the case for 2–3 weeks before hatching.

CLASSES POLYCHAETA AND HIRUDINEA

About two-thirds of all annelids are members of the class Polychaeta and are called polychaetes. *Polychaeta* means "many bristles," which refers to the numerous setae that help polychaetes move. The setae project from parapodia, some of which function in gas exchange. Polychaetes differ from other annelids in that they have antennae and specialized mouthparts. They are also the only annelids that have a trochophore stage in development. Most polychaetes, like the one shown in Figure 37-10, live in marine habitats. Some are free-swimming predators that use their strong jaws to feed on small animals. Others feed on sediment as they burrow through it or use their tentacles to scour the ocean bottom for food.

Hirudinea is the smallest class of annelids, consisting of about 300 species of leeches. Most leeches live in calm bodies of fresh water, but some species live among moist vegetation on land. Leeches have no setae or parapodia. At each end of a leech's body is a sucker that can attach to surfaces. By attaching the anterior sucker and then pulling the rest of the body forward, leeches can crawl along solid objects. Aquatic leeches can also swim with an undulating movement of their body. Many leeches are carnivores that prey on small invertebrates, but some species, including the one shown in Figure 37-12, are parasites that suck blood from other animals. After attaching themselves to the skin of their host, parasitic leeches secrete an anaesthetic that prevents the host from feeling their presence. They also secrete a substance that prevents blood from clotting. If undisturbed, a leech can ingest 10 times its own weight in blood.

FIGURE 37-12

The leech *Haemadipsa* sp. is a parasite that sucks blood from animals, including humans. Other leeches are free-living carnivores.

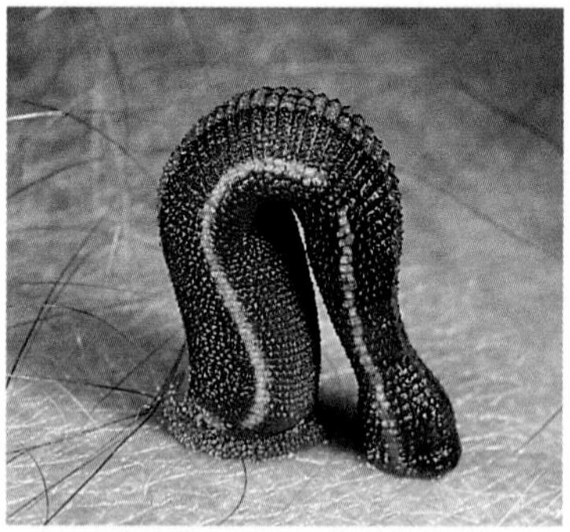

SECTION 37-2 REVIEW

1. What are the advantages of a segmented body?
2. How are an earthworm's circular and longitudinal muscles used in locomotion?
3. How does an earthworm exchange oxygen and carbon dioxide with its environment?
4. How do polychaetes differ from earthworms?
5. How are some leeches adapted to a parasitic lifestyle?
6. **CRITICAL THINKING** How is the form of parasitism that some leeches engage in different from that of a tapeworm or a liver fluke, which you read about in Chapter 36?

CHAPTER 37 REVIEW

SUMMARY/VOCABULARY

37-1

- Mollusks have a true coelom and usually develop from a pear-shaped larva called a trochophore. Their body is divided into the head-foot and the visceral mass, which contains the internal organs.
- Most mollusks have at least one shell, which is secreted by a layer of epidermis called the mantle. Aquatic mollusks have gills through which they exchange gases with water.
- The main feeding adaptation of most mollusks is the radula, a tongue-like structure that is modified in different species for scraping, drilling, or harpooning.
- Gastropods undergo a process called torsion, in which the visceral mass twists during larval development. Snails and most other gastropods have a single shell, while some gastropods, such as slugs and nudibranchs, lack shells. Gastropods move by means of wavelike, muscular contractions of the foot and have an open circulatory system.
- Bivalves have a shell that is divided into two valves, which they can pull together by contracting powerful adductor muscles. Bivalves lack a distinct head and have no radula. Most are sessile and filter food from the water. In clams, water enters through an incurrent siphon and exits through an excurrent siphon. Food is strained from the water as it passes through the gills.
- Cephalopods, including octopuses and squids, are free-swimming, predatory mollusks with numerous tentacles. They have an advanced nervous system with a large brain and well-developed sensory organs. Cephalopods have a closed circulatory system and do not pass through a trochophore stage during development.

Vocabulary

adductor muscle (728)	ganglion (726)	incurrent siphon (729)	siphon (729)
bivalve (728)	gastropod (727)	mantle (726)	torsion (727)
cephalopod (730)	head-foot (726)	mantle cavity (726)	trochophore (725)
chromatophore (731)	hemocoel (727)	mollusk (725)	visceral mass (726)
excurrent siphon (729)	hemolymph (727)	radula (726)	

37-2

- Annelids have a true coelom and a body that is divided into many segments. Most annelids have external bristles called setae, and some have fleshy protrusions called parapodia.
- Members of the class Oligochaeta generally live in the soil or in fresh water. They have no parapodia and relatively few setae.
- The most familiar member of the class Oligochaeta is the earthworm, which feeds on organic matter as it burrows through the soil. Earthworms have a closed circulatory system. They exchange gases through their skin and eliminate cellular wastes and excess water through excretory tubules called nephridia.
- Polychaetes have numerous setae that project from parapodia. They also have antennae and specialized mouthparts, and they pass through a trochophore stage during their development. Most polychaetes live in the ocean.
- Members of the class Hirudinea—leeches—live in fresh water or on land. They have no setae or parapodia. Many leeches are carnivores that prey on small invertebrates, but some are bloodsucking parasites.

Vocabulary

annelid (733)	clitellum (735)	gizzard (734)	seminal receptacle (735)
aortic arch (735)	crop (734)	nephridium (735)	seta (733)
chitin (736)	esophagus (734)	parapodium (733)	typhlosole (734)

CHAPTER 37 REVIEW ANSWERS

REVIEW

1. A coelom is a hollow, fluid-filled cavity that is completely surrounded by mesoderm. A pseudocoelom is a body cavity lined by mesoderm on the outside and endoderm on the inside.
2. In an open circulatory system, hemolymph leaves certain vessels and percolates through a system of spaces in the tissues. In a closed circulatory system, blood circulates entirely within a system of vessels.
3. Chromatophores are pigment cells. They can produce sudden color changes in cephalopods, allowing the animals to blend in with their surroundings.
4. The crop stores ingested soil. The gizzard grinds the soil, breaking up organic matter.
5. Nephridia eliminate cellular wastes and excess water from an earthworm's body.

6. c	**7.** a
8. d	**9.** c
10. c	**11.** a
12. b	**13.** c
14. d	**15.** a

16. The cilia on a free-swimming trochophore propel the trochophore through the water and draw food to its mouth.
17. The main parts are the head-foot, the visceral mass, and the mantle.
18. Torsion is a developmental process in which the visceral mass twists around 180 degrees in relation to the head. This twisting brings the mantle cavity to the front of the snail.
19. In most snails, the radula is a rough tongue-like structure used to cut through leaves, scrape up algae, or drill holes in the shells of other mollusks. In the cone shell, it is a poisonous stinger used to harpoon fish.
20. Gills have a large surface area that is in contact with a rich supply of blood. Thus, they are specialized for gas exchange.

21. Earthworms exchange oxygen and carbon dioxide by diffusion through their skin, and this exchange can take place only if the skin is moist.
22. The nervous system consists of a chain of ganglia connected by a ventral nerve cord. In most body segments, nerves extend from a single ganglion to the muscles and sensory cells in that segment. In the most anterior segments, several ganglia are fused to form the cerebral ganglia.
23. Polychaetes have more setae than do other annelids, and they are the only annelids with parapodia, antennae, specialized mouthparts, and a trochophore larval stage. Also, most polychaetes are marine, while most other annelids live in fresh water or on land.
24. The correct order of steps is e, d, b, c, a.
25. *A,* stomach; *B,* adductor muscle; *C,* excurrent siphon; *D,* gills; *E,* intestine; *F,* foot

Critical Thinking

1. Releasing sperm and eggs into the water makes the chance of fertilization very small, because they can be dispersed by currents and eaten by animals. To maximize the chance of fertilization, aquatic mollusks must release large numbers of sperm and eggs.
2. Both animals feed by extracting nutrients from large quantities of material—water in the case of the clam and soil in the case of the earthworm—as that material moves in a continuous stream through their bodies.
3. Oysters make pearls by covering sharp-edged particles of sand or rock with a smooth coating. By doing so, the oyster makes those particles less irritating.
4. The fact that the posterior half of an earthworm can make these movements indicates that the

Review

Vocabulary

1. How does a coelom differ from a pseudocoelom?
2. What is the difference between an open circulatory system and a closed circulatory system?
3. What are chromatophores, and of what value are they to a cephalopod?
4. What are the functions of the crop and the gizzard in an earthworm?
5. What is the function of the nephridia in an earthworm?

Multiple Choice

6. Mollusks and annelids share all of the following characteristics except (a) a coelom (b) a trochophore larva (c) segmentation (d) bilateral symmetry.
7. Gills are organs specialized for (a) gas exchange (b) movement (c) digestion (d) excretion.
8. Most bivalves are (a) predators (b) parasites (c) land dwellers (d) filter feeders.
9. The only mollusks with a closed circulatory system are (a) gastropods (b) bivalves (c) cephalopods (d) snails.
10. Terrestrial snails and slugs require an environment with a high moisture content in order to (a) reproduce (b) feed (c) respire (d) all of the above.
11. Slugs are members of the class (a) Gastropoda (b) Cephalopoda (c) Polychaeta (d) Oligochaeta.
12. Annelids are divided into three classes based partly on the number of their (a) segments (b) setae (c) nephridia (d) aortic arches.
13. Earthworms respire by means of (a) gills (b) lungs (c) diffusion across the skin (d) all of the above.
14. The movement of earthworms involves (a) pressure in the coelomic fluid (b) muscle contractions (c) traction provided by setae (d) all of the above.
15. Parapodia are a distinguishing characteristic of the class (a) Polychaeta (b) Oligochaeta (c) Hirudinea (d) Bivalvia.

Short Answer

16. What functions are performed by the cilia on a free-swimming trochophore?
17. What are the main parts in the basic body plan of a mollusk?
18. What is torsion? What effect does it have on the location of a snail's mantle cavity?
19. Describe how the radula is modified in different groups of mollusks.
20. How is the structure of a gill related to its function?
21. Why do earthworms require a moist environment?
22. Describe the organization of an earthworm's nervous system.
23. What characteristics distinguish polychaetes from other annelids?
24. Arrange the following steps in the locomotion of an earthworm in the correct order, beginning after the worm anchors some of its middle segments by their setae.
    a. The posterior segments are pulled forward.
    b. The anterior end moves forward as the worm elongates.
    c. Setae in the anterior segments grip the ground as the longitudinal muscles contract.
    d. The pressure of the coelomic fluid increases.
    e. The circular muscles contract.
25. Identify the structures labeled A through F in the diagram below.

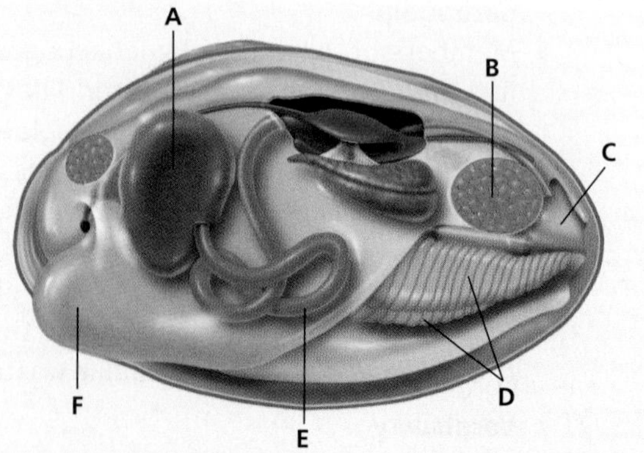

CRITICAL THINKING

1. Clams and other aquatic mollusks reproduce by releasing sperm and eggs into the water. How might this process affect the reproductive success of these mollusks? Would you expect aquatic mollusks to release many sperm and eggs or only a few?
2. Clams are aquatic and earthworms are terrestrial. Nevertheless, the feeding methods of clams and earthworms are basically similar. Explain how they are similar.
3. Humans value pearls for their luster and color, features that are of no significance to an oyster. Furthermore, making a pearl consumes resources that an oyster could use for other purposes, such as strengthening its shell. Given these facts, of what advantage is it to an oyster to manufacture a pearl?
4. For quite a while after an earthworm is cut in half, both halves will continue to move about, and both will retract if they are touched. What do these observations suggest about the role the brain has in coordinating these movements?
5. Land snails are hermaphrodites. Of what advantage is this characteristic to the land snail?
6. Many clams have very long incurrent and excurrent siphons. For example, the siphons of a clam called the geoduck, *Panope generosa,* may exceed one meter in length. What is the adaptive advantage of such long siphons? Keep in mind the habitat of most clams.
7. A mutation results in the birth of an earthworm that lacks moisture-sensing cells in its skin. Explain why this earthworm is less likely to survive than one with such sensory cells.
8. The graph below plots the movement of the anterior end of an earthworm over an interval of several seconds, as the worm crawled along a flat surface. Was the anterior end of the worm moving or stationary during the periods represented by the horizontal sections of the graph? Which of the earthworm's sets of muscles were contracting during the periods represented by the horizontal sections? Explain your answers.

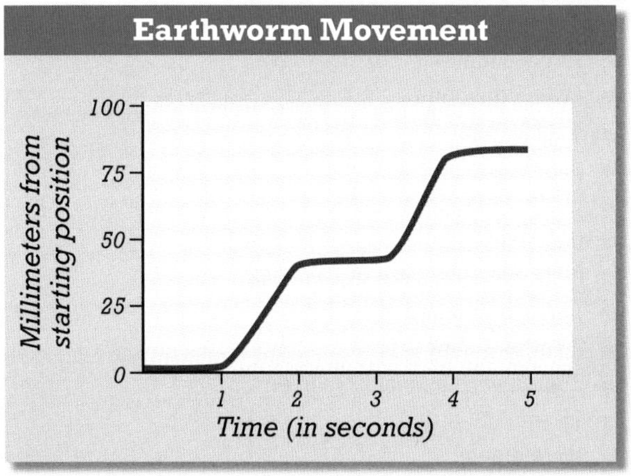

EXTENSION

1. Read "Sponging Off Mussels" in *National Wildlife,* February/March 2000, on page 10. Explain why the zebra mussel is an ecological pest. How does the native sponge in the Great Lakes help solve the problem?
2. Many people collect the shells of mollusks as a hobby. You may be able to see some of these shells in your school, in the homes of friends or relatives, or at a local museum. Using a book such as the *National Audubon Society Field Guide to North American Seashore Creatures,* identify five shells or pictures of shells. Draw each one, and under each drawing give the common name, the scientific name, the part of the world where it is found, and the size.
3. Research an annelid or mollusk species that was not covered in this chapter. Gather information on the anatomy, feeding, habitat, and reproduction of that species. How does the species you chose illustrate an evolutionary adaptation to the environment in which it lives?

ANSWERS

CONTINUED

brain is not necessary for coordinating them. The parts of the nervous system in the posterior half must be sufficient.
5. Hermaphroditism allows a land snail to mate with any other individual of its species, rather than with just half its population. This increases the snail's chance of reproducing, since snails move slowly and are often inactive.
6. Long siphons enable clams to bury themselves deep in the sand, where it is harder for predators to find them, while still obtaining oxygen and nutrients from the water.
7. Without moisture-sensing cells, an earthworm would be unable to distinguish damp soil from dry soil. In dry soil, the worm's skin would dry out, preventing O_2 and CO_2 from diffusing through it.
8. During the periods represented by the horizontal sections of the graph, the anterior end was stationary, and the longitudinal muscles in the posterior segments were contracting. The anterior end would be anchored by its setae as the longitudinal muscles pulled the posterior segments forward. Thus, the position of the anterior end wouldn't change during these periods.

EXTENSION

1. Zebra mussels clog pipes and water-treatment plants. The sponges smother and starve the mussels.
2. This is self-explanatory and easy to confirm by examining the drawing and comparing it with the specimen or the reference book the student is using.
3. Students should focus on the animal's class and highlight how its differences from the other classes affect how it interacts with its environment. You may suggest that students investigate cockles, limpets, conchs, mussels, and tubeworms, such as the Christmas-tree worm.

CHAPTER 37 INVESTIGATION

TIME REQUIRED

One 50-minute class period

SAFETY PRECAUTIONS

Remind students to wear safety goggles when they handle the ammonia solution.

QUICK REFERENCES

Lab Preparation Notes are found on page 723B.

Holt Biosources provides a Teaching Resources CD-ROM that contains "Using Gowin's Vee in the Lab" and "Scoring Rubrics."

PROCEDURAL TIPS

1. Live earthworms may be purchased from a biological supply company or bait shop or collected from local soil. Obtain the largest worms available.

2. Store the worms in the refrigerator in a covered container with a perforated lid. Add damp soil, sphagnum moss, or shredded newspapers to the container. After the lab, the worms may be returned to the soil.

ANSWERS TO BACKGROUND

1. Cephalization allows an earthworm to avoid harmful substances and seek optimal conditions.

2. Gases diffuse directly through an earthworm's skin.

ANSWERS TO PART A

3. The clitellum is closer to the anterior end. It secretes mucus that holds two worms together during mating, and it produces a protective case within which the fertilized eggs develop into young worms.

5. The setae are located on the ventral surface.

6. The earthworm has a closed circulatory system.

CHAPTER 37 INVESTIGATION

Observing Earthworm Behavior

OBJECTIVES

- Observe a live earthworm.
- Test how an earthworm responds to light, moisture, and ammonia.
- Test the effect of temperature on heart rate.

PROCESS SKILLS

- observing
- hypothesizing
- experimenting
- collecting data
- analyzing data

MATERIALS

- safety goggles
- live earthworm
- shallow pan
- paper towels
- medicine dropper
- hand lens
- black paper or piece of cardboard
- fluorescent lamp
- 2 cotton swabs
- 3% aqueous ammonia solution
- 15 cm Petri dish
- thermometer
- stopwatch or clock with second hand
- 2 plastic tubs for water baths
- warm tap water
- ice cubes

Background

1. How does an earthworm benefit from cephalization?
2. Describe how gases enter and exit an earthworm's body.

PART A Observing an Earthworm

1. CAUTION **You will be working with a live animal. Be sure to treat it gently and to follow directions carefully.** Place a moist paper towel in a pan, and place an earthworm on the paper towel. CAUTION **Rinse the earthworm frequently with water from a medicine dropper to prevent the worm from drying out and becoming lethargic.**
2. Observe the behavior of the earthworm for a few minutes. Identify the earthworm's anterior and posterior ends by watching it move in the pan. As the worm crawls around in the pan, it will lead with its anterior end.
3. Locate the earthworm's clitellum. Is the clitellum closer to the anterior end or the posterior end? What is the function of the clitellum?
4. Identify the earthworm's dorsal and ventral surfaces by gently rolling the worm over. The dorsal surface will be on top after the worm rights itself.
5. Pick up the earthworm and feel its skin with your fingers. One surface of the earthworm should feel slightly rougher than the other. The roughness is due to the hairlike setae that project from the earthworm's skin. On which surface are the setae located? Use a hand lens to examine the setae up close.
6. Return the earthworm to the pan, and use the hand lens to find a thick purple line running along the dorsal surface of the worm. This line is the dorsal blood vessel. Does the earthworm have an open or a closed circulatory system?
7. Draw a picture of the earthworm, and label its anterior and posterior ends, dorsal and ventral surfaces, clitellum, setae, and dorsal blood vessel.

PART B Earthworm Responses to Stimuli

8. In this part of the laboratory investigation, you will test the earthworm's responses to three different stimuli. With your lab partners, develop three separate hypotheses that describe an earthworm's responses to light, moisture, and a base. In your lab report, make a data table like the one on the next page.
9. To test the earthworm's response to light, cover half of the pan with black paper or cardboard. Check the lighting in the room. The light must be low and even during this test. Position the fluorescent lamp over the uncovered portion of the pan. Place the earthworm in the center of the pan and observe its movements. Record your observations in your data table.

OBSERVATIONS OF EARTHWORM BEHAVIOR

Behavior	Observations
Response to light	
Response to moisture	
Response to water on a swab	
Response to ammonia	

10. To test the earthworm's response to moisture, turn off the fluorescent lamp, move it away from the pan, and remove the paper covering half of the pan. Place a piece of dry paper towel on one side of the pan and a piece of wet paper towel on the other side of the pan. Lay the earthworm across the two paper towels. Observe the earthworm's response to the two environments, and record your observations in your data table.

11. To test the earthworm's response to ammonia, make sure the paper towels on both sides of the pan are wet. Moisten a cotton swab with water. Hold the cotton swab first near the earthworm's anterior end and then near its posterior end. Do not touch the earthworm with the swab. Record your observations in your data table.

12. CAUTION **Wear safety goggles at all times during the following procedure. If you get ammonia on your skin or clothing, wash it off at the sink while calling to your teacher. If you get ammonia in your eyes, immediately flush it out at the eyewash station while calling to your teacher.** Moisten a different cotton swab with ammonia solution and repeat step 11. Do not touch the earthworm with the swab or the ammonia solution. Record your observations in your data table.

PART C Effect of Temperature on Heart Rate

13. In this part of the laboratory investigation, you will examine how an earthworm's heart rate changes as its body temperature changes. Add enough tap water to a Petri dish to barely cover the bottom of the dish. Place an earthworm in the dish.

14. Using a hand lens, look for rhythmic contractions of the dorsal blood vessel. Each contraction represents a single heartbeat. Calculate the worm's heart rate by counting the number of contractions that occur in exactly one minute. This is easiest to do if one person counts contractions while another person watches a stopwatch or clock.

15. Place a thermometer next to the worm in the petri dish and measure the temperature. Record the worm's heart rate and the temperature in a table on the chalkboard.

16. Float the Petri dish containing the worm on top of either a warm-water bath or a cold-water bath. Place the thermometer next to the worm in the Petri dish, and watch the temperature until it reaches either 30°C (for the warm-water bath) or 10°C (for the cold-water bath).

17. Remove the Petri dish from the water bath and immediately begin counting heartbeats for exactly one minute. After one minute, measure the temperature in the dish again. Calculate the average temperature to the nearest degree. Record the worm's heart rate and the average temperature in the table on the chalkboard.

18. Using data from the whole class, graph heart rate as a function of temperature. Draw the best-fit curve through the points.

19. Return the earthworm to the container from which you obtained it. Clean up your materials and wash your hands before leaving the lab.

Analysis and Conclusions

1. State whether your hypotheses in Part B were supported by your observations. Explain.
2. What is the adaptive advantage of the earthworm's responses to light and moisture?
3. List variables that, if not controlled, might have affected the results in Part B.
4. Describe the relationship between the earthworm's heart rate and temperature as shown by your graph.

Further Inquiry

Design an experiment to determine which colors of light an earthworm is sensitive to or which areas on an earthworm are sensitive to light.

ANSWERS TO ANALYSIS AND CONCLUSIONS

1. Responses of the earthworms will vary, but generally earthworms prefer dark, moist environments and will move away from noxious substances such as ammonia.
2. The earthworm's responses prevent it from leaving the soil and drying out.
3. Variables include the intensity of the light, whether the pan was level, the moistness of the paper towels, and the closeness of the cotton swabs to the worm.
4. Heart rate should increase as temperature increases.

FURTHER INQUIRY

To test an earthworm's color-sensitivity, students could wrap translucent sheets of colored plastic around a fluorescent light. To identify the areas on an earthworm that are sensitive to light, students should work with a small spot of light, which they could produce by placing an index card with a small hole in it over the end of a flashlight.

CHAPTER 38
ARTHROPODS

	TOPICS	TEACHING RESOURCES	LABS, CLASSWORK, AND HOMEWORK
BLOCK 1 45 minutes	**Introducing the Chapter, p. 742**	ATE Focus Concept, p. 742 ATE Assessing Prior Knowledge, p. 742 ATE Understanding the Visual, p. 742	■ Supplemental Reading Guide, *Journey to the Ants: A Story of Scientific Exploration*
	38-1 Phylum Arthropoda, p. 743 Characteristics of Arthropods, p. 743 Molting, p. 744 Evolution and Classification, p. 744	ATE Section Overview, p. 743 ATE Visual Link Figure 38-3, p. 745	**Lab B20,** Life in a Pine Cone ★ Study Guide, Section 38-1 PE Section Review, p. 745
BLOCK 2 45 minutes	**38-2 Subphylum Crustacea, p. 746** Characteristics and Diversity of Crustaceans, p. 746 The Crayfish, p. 747	ATE Section Overview, p. 746 ATE Critical Thinking, p. 746 ATE Visual Link Figure 38-6, p. 747 161	PE **Quick Lab,** p. 748, Observing Crayfish Behavior **Lab B25,** Crayfish Dissection ★ Study Guide, Section 38-2 PE Section Review, p. 750
BLOCK 3 45 minutes	**38-3 Subphylum Chelicerata and Uniramia, p. 751** Class Arachnida, p. 751 Myriapods, p. 754	ATE Section Overview, p. 751 ATE Critical Thinking, p. 752 134A	PE **Chapter 38 Investigation,** p. 758 ★ Study Guide, Section 38-3 PE Section Review, p. 754
BLOCKS 4 & 5 90 minutes	**REVIEW and ASSESSMENT** PE Chapter 38 Review, pp. 756–757 ■ Performance-Based Assessment—Exploring Invertebrates	Audio CD Program	**CHAPTER TESTING** ■ Chapter Test (blackline master) ▲ Test Generator ■ Assessment Item Listing ■ Scoring Rubrics and Classroom Management Checklists

KEY

MODERN BIOLOGY

- PE Pupil's Edition
- ATE Teacher's Edition
- ■ Active Reading Guide
- ★ Study Guide

HOLT BIOSOURCES

- Teaching Transparencies
- Laboratory Program

One-Stop Planner CD-ROM
Shows these resources within a customizable daily lesson plan:

- ■ Holt BioSources Teaching Resources
- ■ Active Reading Guide
- ★ Study Guide
- ▲ Test Generator

READING FOR CONTENT MASTERY

Preview: Have students preview the objectives, the introductory paragraph, topic sentences, and captions before reading the chapter.

- **ATE** Active Reading Technique: Reading Effectively, p. 743
- ■ Active Reading Guide Worksheet 38-1
- **PE** Summary/Vocabulary, p. 755

- ■ Active Reading Guide Worksheet 38-2
- **PE** Summary/Vocabulary, p. 755

- ■ Active Reading Guide Worksheet 38-3
- **PE** Summary/Vocabulary, p. 755

TECHNOLOGY AND INTERNET RESOURCES

Holt Biology Videodiscs Teacher's Correlation Guide, Lessons 38-1 through 38-3

Audio CD Program, Sections 38-1 through 38-3

internetconnect

SCILINKS NSTA

On-line Resources: www.scilinks.org
The following *sci*LINKS Internet resources can be found in the student text for this chapter:

Topics:
- Arthropods, p. 744
- Crustaceans, p. 750
- Arachnida, p. 752

go.hrw.com

On-line Resources: go.hrw.com
Visit the HRW Web site for a variety of resources related to this chapter. Just type in the keyword HM2 Chapter 38.

Smithsonian Institution®

Visit **www.si.edu/hrw** for additional on-line resources.

CNNfyi.com

Visit **www.cnnfyi.com** for late-breaking news and current event stories selected just for you.

LAB ACTIVITY PLANNING

CHAPTER 38 INVESTIGATION

Behavior of Pill Bugs, pp. 758–759

OVERVIEW

Pill bugs are arthropods and are also called sow bugs, wood lice, or slaters. They are a group of crustaceans called isopods. In this investigation, students will observe the external anatomy of a pill bug. They will also observe behavioral responses of pill bugs to several different stimuli.

SAFETY

Instruct students to handle living organisms with care, to be careful using scissors, and to wash their hands after the lab.

PREPARATION

1. Pill bugs can be found outdoors under logs and rocks. Each lab group will need five pill bugs. Keep the pill bugs in a plastic container with crumpled, moist paper towels and a slice of potato until lab time.
2. Each lab group will need a medicine dropper, three sheets of filter paper, a Petri dish, a thin slice of potato, a 4 × 4 in. piece of aluminum foil, a bright lamp or flashlight, four 4 × 4 in. pieces of fabric (each with a different texture, such as wool, polyester, silk, and flannel), cellophane tape, and scissors.

Quick Labs

Observing Crayfish Behavior, p. 748

Each lab group will need a living crayfish in a container with water and a tapping instrument.

UNDERSTANDING THE VISUAL

One characteristic that has contributed to the success of arthropods is their exoskeleton. Ask students to look at the crustacean in the photograph and to think of some advantages and disadvantages of having an exoskeleton. (An exoskeleton offers protection from predators, disease, and desiccation in terrestrial environments. However, an arthropod must shed its exoskeleton to grow, and that makes it vulnerable to predators until the new exoskeleton hardens. Having an exoskeleton also limits the maximum size of arthropods.)

FOCUS CONCEPT

Evolution

The presence of a true coelom, jointed appendages, and an exoskeleton suggests that all arthropods evolved from a common ancestor. As arthropods evolved, many of their segments were fused into larger structures called tagmata and their appendages became specialized for functions such as feeding, locomotion, and defense.

ASSESSING PRIOR KNOWLEDGE

Review the following concepts.

Exoskeleton: *Chapter 34*
Ask students to distinguish between an endoskeleton and an exoskeleton.

Segmentation: *Chapter 37*
Ask students to name a phylum of worms that have segmented bodies. Ask students to explain the advantages of body segmentation.

CHAPTER 38

ARTHROPODS

The jointed appendages and hard exoskeleton of this flame lobster, Enoplometopus occidentalis, *are characteristic of arthropods.*

FOCUS CONCEPT: *Evolution*

As you read, note the ways in which the basic arthropod body plan has become modified for life in diverse environments.

PHYLUM ARTHROPODA

Three-fourths of all animal species belong to the phylum Arthropoda (ahr-THRAHP-uh-duh). This phylum contains a diverse assortment of bilaterally symmetrical coelomates, including lobsters, crabs, spiders, millipedes, centipedes, and insects. The characteristics of these animals have enabled them to adapt to almost every environment on Earth.

SECTION

38-1

OBJECTIVES

▲ Describe the distinguishing characteristics of arthropods.

● Relate the structure of the arthropod exoskeleton to its function.

■ Explain the process of molting in an arthropod.

◆ Name the four subphyla of the phylum Arthropoda, and describe the characteristics of each subphylum.

CHARACTERISTICS OF ARTHROPODS

Members of the phylum Arthropoda are called **arthropods** (AHR-thruh-PAHDS). Like the annelids you studied in Chapter 37, arthropods are segmented animals. In arthropods, however, the body segments bear jointed extensions called **appendages,** such as legs and antennae. In fact, *arthropod* means "jointed foot."

Another distinguishing feature of arthropods is their exoskeleton, which provides protection and support. As Figure 38-1 shows, the arthropod exoskeleton is made up of three layers that are secreted by the epidermis, which lies just beneath the layers. The waxy outer layer is composed of a mixture of protein and lipid. It repels water and helps prevent desiccation in terrestrial species. The middle layer, which provides the primary protection, is composed mainly of protein and chitin. In some arthropods, the middle layer is hardened by the addition of calcium carbonate. The inner layer also contains protein and chitin, but it is flexible at the joints, allowing arthropods to move freely. Muscles that attach to the inner layer on either side of the joints move the body segments relative to each other.

The arthropod body shows a high degree of cephalization. A variety of segmented appendages around the mouth serve as sensors and food handlers. Most arthropods have segmented antennae at the anterior end of the body that are specialized for sensing the environment and detecting chemicals. Most arthropods also have **compound eyes**—eyes composed of many individual light detectors, each with its own lens. In addition, many arthropods have simpler structures that sense light intensity. These sensory structures on the head send nerve impulses to the brain, which coordinates the animal's actions. As in annelids, impulses travel from the brain along a ventral nerve cord, which links ganglia in the other segments of the body. All arthropods have open circulatory systems.

FIGURE 38-1

The arthropod exoskeleton consists of three layers that cover the epidermis. Wax in the outer layer is secreted by wax glands. Sensory hairs projecting from the exoskeleton allow arthropods to respond to vibrations and chemicals in their environment.

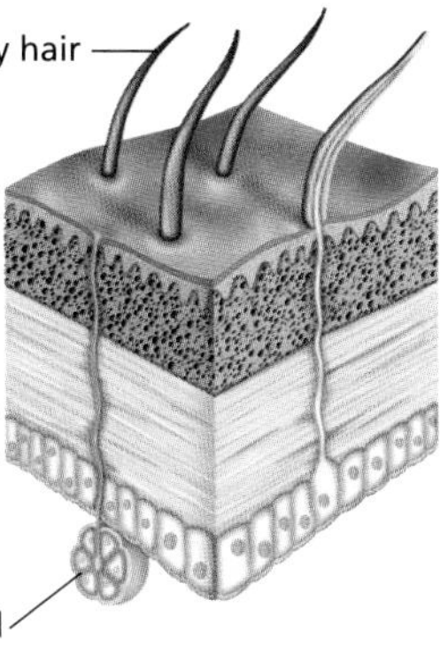

SECTION 38-1

SECTION OVERVIEW

Evolution

Section 38-1 discusses the common trends in arthropod evolution and briefly describes the differences between the four subphyla of arthropods.

ACTIVE READING

Technique: Reading Effectively

As students read through this chapter, have them make an outline of the material presented. You may want to review outlining methods, such as using Roman numerals, letters, and Arabic numerals. See the example below.

I.
  A.
    1.
    2.
  B.
    1.
II.

Review the outlines to be sure students are including the main ideas.

CONTENT CONNECTION

Section 3-2: Organic Compounds

The hydrophobic nature of lipids gives the outer layer of the exoskeleton its water repellency. Chitin, a major component of the middle and inner layers, is a leathery polysaccharide.

RETEACHING *ACTIVITY*

Present students with photographs of various types of arthropods and ask the students to name the subphylum to which each arthropod belongs. If key features, such as mouthparts and other appendages, are not visible in a photograph, add a legend to the photograph describing those features. Ask students to explain why they assigned each photograph to a particular subphylum.

TEACHING STRATEGY

Look for discarded arthropod exoskeletons around the school or at home, and bring them to class. Point out the area along the dorsal surface of the exoskeleton where the animal emerged during molting. Ask students to note that the exoskeleton covered the entire surface of the animal, including the legs, mouthparts, antennae (if present), and eyes. Have students observe the light weight of the exoskeleton, and explain that arthropods reabsorb part of the old exoskeleton before they molt. Supplement these observations with a time-lapse video clip of an arthropod molting.

GIFTED *ACTIVITY*

Have students investigate the hormones involved in molting in arthropods. These hormones include ecdysone, which stimulates molting; brain hormone, which triggers the release of ecdysone; and bursicon, which causes the new exoskeleton to harden. Ask students to focus on where these hormones are produced and what effects they have on arthropod tissues.

DEMONSTRATION

Have students examine a preserved horseshoe crab. Point out the two main body sections—the fused cephalothorax and the abdomen. The cephalothorax has seven pairs of legs, while the abdomen has six pairs of appendages used for swimming. Note that the abdomen is not curled beneath the body as it is in true crabs. Explain that whereas true crabs are in the subphylum Crustacea, horseshoe crabs are in the subphylum Chelicerata, along with spiders. Like spiders, horseshoe crabs lack antennae and have pincerlike mouthparts.

FIGURE 38-2

This green cicada, *Tibicen superbus*, is in the process of molting. The outer layer of its old exoskeleton appears brown.

MOLTING

Because an arthropod's skeleton lies on the outside of its body, an arthropod cannot grow without periodically shedding its exoskeleton. This process is called **molting.** Figure 38-2 shows an insect in the process of molting. An arthropod molts many times during its life, and with each molt it becomes larger.

In between molts, the tissues of an arthropod swell until they put a good deal of pressure on the exoskeleton. A hormone is then produced that induces molting. In response to this hormone, the cells of the epidermis secrete enzymes that digest the flexible inner layer of the exoskeleton. At the same time, the epidermis begins to synthesize a new exoskeleton, using much of the digested material. Eventually the outer layer of the old exoskeleton loosens, breaks along specific lines, and is shed. The new exoskeleton, which is flexible at first, stretches to fit the enlarged animal.

It takes a few days for the new exoskeleton to become as hard as the one it replaced. During this time, the animal is extremely vulnerable to predators and, in the case of terrestrial arthropods, susceptible to desiccation. For these reasons, arthropods usually remain in hiding from the time they begin to molt until their new exoskeleton has hardened. The "soft-shelled crabs" sold in some restaurants are crabs that have been caught immediately after molting, while their new exoskeleton is still flexible.

internet**connect**

SCILINKS NSTA

TOPIC: Arthropods
GO TO: www.scilinks.org
KEYWORD: HM744

EVOLUTION AND CLASSIFICATION

Animals with arthropod characteristics first appeared more than 600 million years ago. Because all arthropods have a true coelom, an exoskeleton, and jointed appendages, biologists have long inferred that they all evolved from a common ancestor.

The various groups of arthropods living today have undergone similar changes during evolution. For example, ancestral arthropods probably had one pair of appendages on every segment, but most living species have some segments that lack appendages. Ancestral arthropods also had bodies consisting of many segments that were nearly identical, but in most living species the segments are fused into a number of larger structures called **tagmata** (tag-MAHT-uh). The various tagmata are specialized to perform functions such as feeding, locomotion, and reproduction.

Arthropods are divided into four subphyla on the basis of differences in development and in the structure of mouthparts and other appendages. The possible evolutionary relationships among these subphyla are indicated in the phylogenetic tree shown in Figure 38-3.

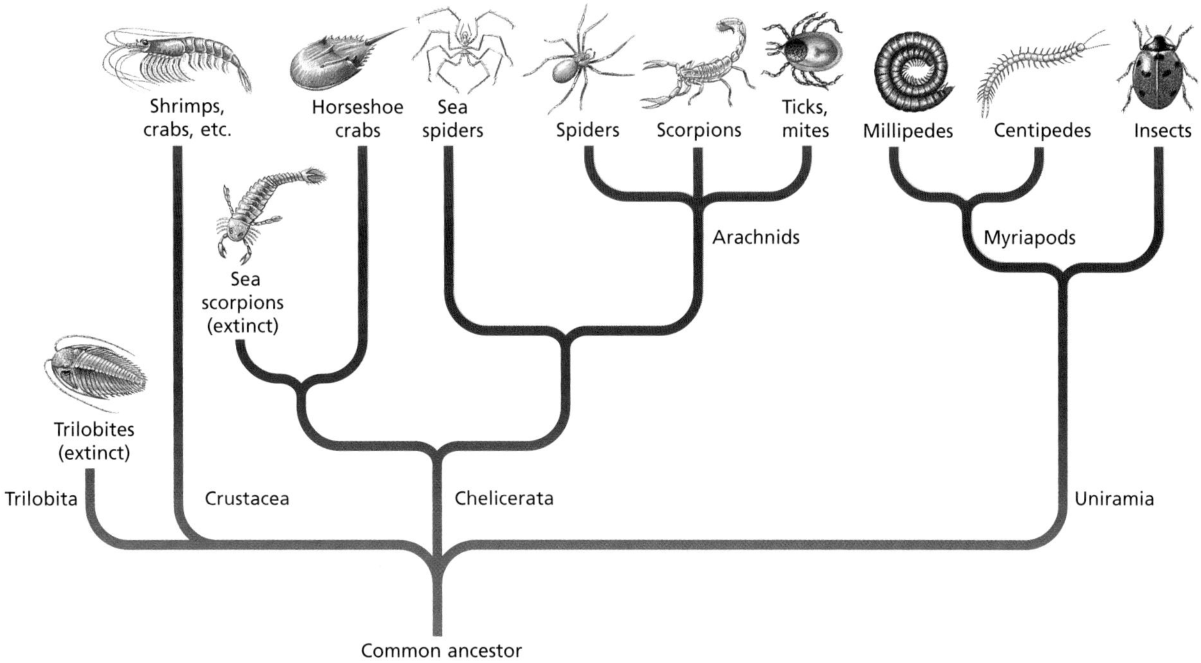

FIGURE 38-3

This phylogenetic tree shows the relationships between the four subphyla of arthropods: Trilobita, Crustacea, Chelicerata, and Uniramia.

- Trilobita (TRIE-luh-BIET-uh) includes extinct animals called trilobites, which had similar, paired appendages on each body segment.
- Crustacea (KRUHS-TAY-shuh) includes shrimps, lobsters, crabs, crayfish, barnacles, isopods, copepods, and water fleas. Members of this subphylum, known as **crustaceans,** have branched antennae and a pair of chewing mouthparts called **mandibles.**
- Chelicerata (kuh-LIS-uh-RAHT-uh) includes spiders, scorpions, mites, ticks, sea spiders, and horseshoe crabs. Members of this subphylum are distinguished from other arthropods by the absence of antennae and the presence of pincerlike mouthparts called **chelicerae** (kuh-LIS-uh-ree).
- Uniramia (YOO-nuh-RAY-mee-uh) includes centipedes, millipedes, and insects. The members of this subphylum also have antennae and mandibles, but their appendages are unbranched. *Uniramia* means "one branch." Uniramia is the only group that seems to have evolved on land.

Word Roots and Origins

chelicera

from the Greek *chele,* meaning "claw," and *keras,* meaning "horn"

SECTION 38-1 REVIEW

1. **What characteristics are shared by all arthropods?**
2. **How many layers are there in an arthropod's exoskeleton? What is the main function of each layer?**
3. **What is a compound eye?**
4. **What is molting?**
5. **What is the major structural difference between members of the subphyla Crustacea and Uniramia?**
6. **CRITICAL THINKING After its old exoskeleton has been shed but before the new one has hardened, an aquatic arthropod absorbs water and swells. What is the adaptive advantage of this behavior?**

SECTION 38-1

VISUAL LINK

Figure 38-3

Explain that organisms represented at the top level of this phylogenetic tree are still living, whereas those at lower levels (trilobites and sea scorpions) are extinct. Explain also that organisms within one branch of the tree are more closely related to each other than to organisms within other branches. For example, millipedes are more closely related to centipedes than to insects.

ANSWERS TO SECTION 38-1 REVIEW

1. All arthropods have a true coelom, a segmented body, jointed appendages, an exoskeleton, a high degree of cephalization, a ventral nerve cord, and an open circulatory system.
2. An arthropod's exoskeleton has three layers. The outer layer repels water and helps prevent desiccation in terrestrial species. The hard middle layer provides the primary protection. The inner layer provides flexibility at the joints, allowing arthropods to move freely.
3. A compound eye is an eye composed of many individual light detectors, each of which has its own lens.
4. Molting is the process by which an arthropod sheds its exoskeleton and develops a new one.
5. Crustaceans have branched appendages, whereas members of the subphylum Uniramia have unbranched appendages.
6. Swelling expands the new exoskeleton before it hardens, allowing room for the animal's growth before the next molt.

SECTION 38-2

SECTION OVERVIEW

Matter, Energy, and Organization

Section 38-2 discusses the organization of the crustacean body into digestive, respiratory, circulatory, excretory, and nervous systems, using the crayfish is a representative example.

▶ ENGAGE STUDENTS

Have students use a microscope to observe small crustaceans in water collected from a local pond. Ask them to look for organisms that seem to have complex body organs and resemble shrimp or crabs. Examples they might find are *Cyclops,* water fleas (*Daphnia*), and seed shrimp. Have students draw the crustaceans they find and use guide books to identify them.

CRITICAL THINKING

Hermit Crabs

Hermit crabs live with their abdomen coiled up inside a discarded snail shell. What problem does this lifestyle create as far as the crab's growth is concerned? (For a hermit crab to grow, it must find a larger snail shell to inhabit.)

SECTION 38-2

OBJECTIVES

▲ Describe the characteristics of crustaceans.

● Give examples of crustaceans that are adapted to marine, freshwater, and land environments.

■ Explain the functions of the appendages on a crayfish.

◆ Summarize digestion, respiration, circulation, and excretion in the crayfish.

SUBPHYLUM CRUSTACEA

The subphylum Crustacea contains approximately 40,000 species. Crustaceans are abundant in oceans, lakes, and rivers, and a few species are even found on land. Some crustaceans are sessile, while others move by walking on legs, swimming with paddle-like appendages, or drifting with the currents.

CHARACTERISTICS AND DIVERSITY OF CRUSTACEANS

Crustaceans are the only arthropods that have two pairs of appendages on their head that serve as feelers. Each of the other body segments generally has a pair of appendages, and at least some of those appendages are branched. Although some crustaceans have 60 or more body segments, most crustaceans have only 16–20 segments, which are fused into several tagmata. The exoskeletons of aquatic crustaceans, such as lobsters, often contain large amounts of calcium carbonate, making them extremely hard. Some small crustaceans exchange carbon dioxide and oxygen through the thin areas of their exoskeleton, but larger crustaceans respire with gills.

During the development of most crustaceans, the embryo becomes a free-swimming larva called a **nauplius** (NAH-plee-uhs), which looks quite different from the adults of its species. As you can see in Figure 38-4, a nauplius has three pairs of appendages and a single eye in the middle of its head. Through a series of molts, the nauplius eventually takes on the adult form.

FIGURE 38-4

The free-swimming nauplius larva is an early stage in the development of most crustaceans. It has one eye and three pairs of appendages.

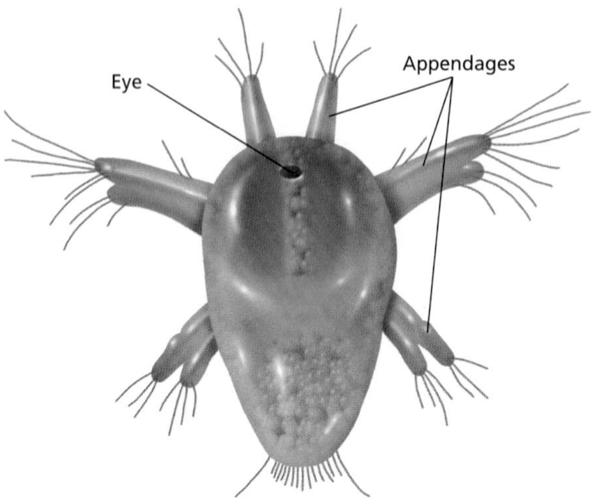

Crustaceans exist in a range of sizes, but most are small. For example, copepods, like the one shown in Figure 38-5a, are no larger than the comma in this sentence. Copepods are extremely abundant in some marine environments. In fact, they may be the most abundant animals in the world. Copepods constitute an important part of the ocean's **plankton,** the collection of small organisms that drift or swim weakly near the surface of a body of water. In freshwater environments, on the other hand, much of the plankton is composed of crustaceans known as water fleas, which are about the size of copepods. A common type of water flea, *Daphnia,* is illustrated in Figure 38-5b. At the other end of the crustacean size spectrum is the Japanese spider crab, shown in Figure 38-5c. With a leg span of 4 m (13 ft), it is the largest living arthropod.

(a)

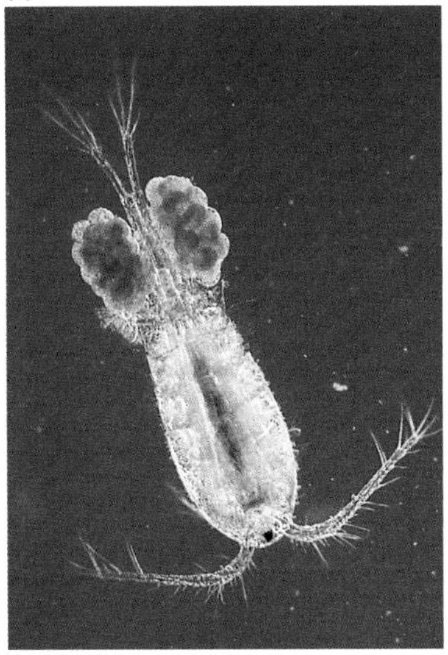

(b)

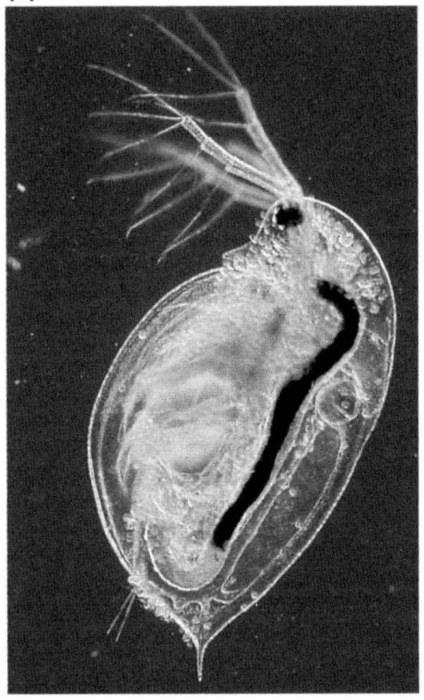

(c)

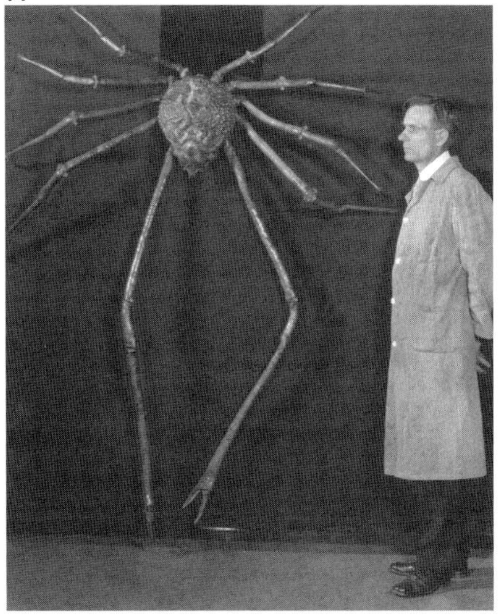

FIGURE 38-5

Crustaceans include tiny species such as the copepod *Cyclops* (a) and the water flea *Daphnia* (b), as well as giants like the Japanese spider crab, *Macrocheira kaempferi* (c).

Barnacles, like the one shown in Figure 38-6, are marine crustaceans that are adapted to a sessile lifestyle as adults. Free-swimming barnacle larvae attach themselves to rocks, piers, boats, sea turtles, whales, and just about any other surface. They then develop a very hard shell of calcium carbonate that completely encloses the body in most species. Their swimming appendages are replaced by six pairs of long legs called **cirri** (SIR-ie), each of which is covered with hairlike setae. The cirri extend through openings in the shell, sweeping small organisms and food particles from the water and directing them to the mouth.

Sow bugs and pill bugs are terrestrial members of a group of crustaceans called **isopods.** *Isopod* means "equal legs," which refers to the seven pairs of identical legs on these crustaceans. Terrestrial isopods can lose water quickly through their exoskeletons. Therefore, they live only in moist environments, such as those found under leaves and rocks, in crevices around garden beds, and in the spaces between house foundations and sidewalks. In addition, pill bugs are capable of rolling into a ball when disturbed or threatened with desiccation. Sow bugs and pill bugs generally feed on decaying vegetation, but they may also eat garden bulbs, vegetables, and fruits that lie on or in the soil.

FIGURE 38-6

Barnacles, like *Lepas anatifera*, are sessile marine crustaceans that filter food from the water with the help of their modified legs.

THE CRAYFISH

The crayfish is a freshwater crustacean that has been well studied because of its relatively large size and abundance. Crayfish are structurally similar to lobsters, their marine relatives. Both are called **decapods** (DEK-uh-PAHDS), a name that means "ten feet," because they have five pairs of legs. Shrimps and crabs are also decapods. The remainder of this section explores some of the details of crayfish structure and function.

QUICK FACT

☑ Shipbuilders estimate that a 12-month growth of barnacles on a ship weighing 10,000 metric tons may add another 30 metric tons to the weight of the ship.

TEACHING STRATEGY

Crab Sexual Dimorphism

Have students observe male and female crabs of any species. Then ask students to draw and describe any differences they see. Students should observe that in both males and females the abdomen is flattened and held beneath the cephalothorax. They should also note that the abdomen and telson are wider and shorter in females than in males. Females may also be carrying eggs attached to the abdomen. If fiddler crabs are used, students will notice that in males one cheliped is much larger than the other.

VISUAL LINK

Figure 38-6

At first glance, a barnacle may look like a mollusk. However, a barnacle's jointed cirri and internal anatomy indicate that it is a crustacean. Ask students to name some animals they have encountered in earlier chapters whose lifestyle and feeding habits resemble those of barnacles. (Such animals would be sessile or sedentary filter feeders, such as sponges and most bivalve mollusks.)

Quick Lab

Observing Crayfish Behavior

Time Required 15 minutes

Safety Remind students not to mistreat the crayfish. Crayfish have strong pincers and can inflict a painful pinch. Move the crayfish away from the students immediately after the conclusion of the activity to avoid safety problems to students or harm to the animals.

Procedural Tips Living crayfish may be obtained from a biological supply company. Have students work in groups. Supply each group with a large specimen bowl or container with one crayfish in water. A tapping instrument can be made by inserting a long metal or glass rod into the end of a spring-loaded lancet holder, which is available at any pharmacy.

Answers to Analysis A tap at the anterior end causes a tailflip that sends the animal straight back. A tap at the posterior end causes the animal to perform a forward somersault. These behavioral responses allow the animal to move suddenly away from the stimulus, which could be a predator.

RECENT RESEARCH

Crayfish Behavior

Recent research has shown that serotonin, a chemical produced by the crayfish nervous system, influences tailflip behavior in ways that depend on the social status of a crayfish. Serotonin apparently lowers the stimulus threshold needed to trigger a tailflip in dominant individuals but raises the threshold in subordinate individuals.

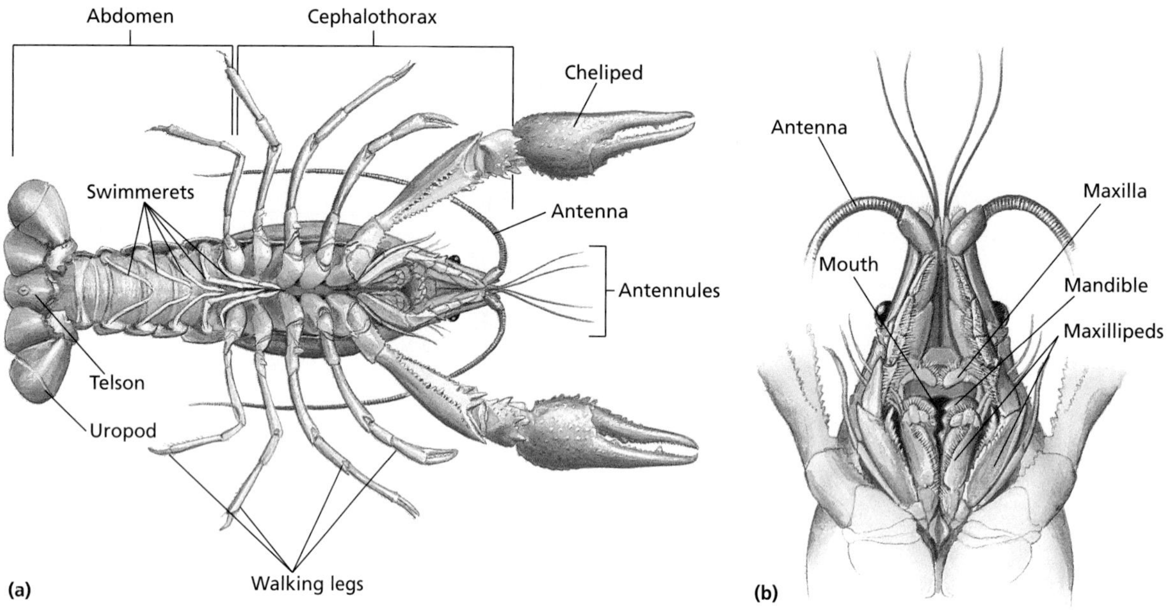

FIGURE 38-7

(a) Each of the 20 body segments of a crayfish except the telson bears a pair of appendages, which can be seen in this ventral view. (b) The appendages of the anterior cephalothorax are visible in this closer ventral view.

Quick Lab

Observing Crayfish Behavior

Materials crayfish in container with water, tapping instrument

Procedure

Gently tap the anterior (head) end of the crayfish, and record your observations. Tap the posterior (tail) end of the crayfish, and record your observations.

Analysis Describe each behavioral response you observed when you tapped the crayfish. How does each behavior aid the animal in its survival?

External Structure

The body of a crayfish is divided into two major sections: the cephalothorax (SEF-uh-luh-THOHR-aks) and the abdomen. The **cephalothorax,** in turn, consists of two tagmata: the head, which has five segments, and the **thorax,** which has eight segments and lies posterior to the head. The dorsal exoskeleton of the segments in the cephalothorax is fused into a single, tough covering known as the **carapace** (KAR-uh-PAYS). The **abdomen,** the tagma that lies posterior to the cephalothorax, is divided into seven segments. The seventh abdominal segment, called the **telson,** forms a flat paddle at the posterior end of the crayfish. Powerful muscles can bend the abdomen suddenly, propelling the animal rapidly backward in a movement referred to as a tailflip.

A pair of appendages is attached to each segment of the crayfish except the telson, as you can see in Figure 38-7. The **antennules** serve as feelers sensitive to touch, taste, and equilibrium. The long **antennae** are also feelers; they respond to touch and taste. Crayfish chew food with their mandibles and manipulate it with their two pairs of **maxillae** and three pairs of **maxillipeds** (mak-SIL-uh-PEDS). The posterior pair of maxillae also function in respiration and the maxillipeds are sensitive to touch and taste. The most anterior pair of appendages on the thorax, the **chelipeds** (KEE-luh-PEDS), end in large pincers used for capturing food and for defense. The four pairs of walking legs carry the crayfish over solid surfaces; the first two pairs end in small pincers that can grasp small objects. The **swimmerets,** which are attached to the anterior five abdominal segments, create water currents and function in reproduction. The **uropods** (YUR-uh-PAHDS), on the sixth abdominal segment, help propel the crayfish during tailflips. Table 38-1 summarizes the crayfish appendages and their functions.

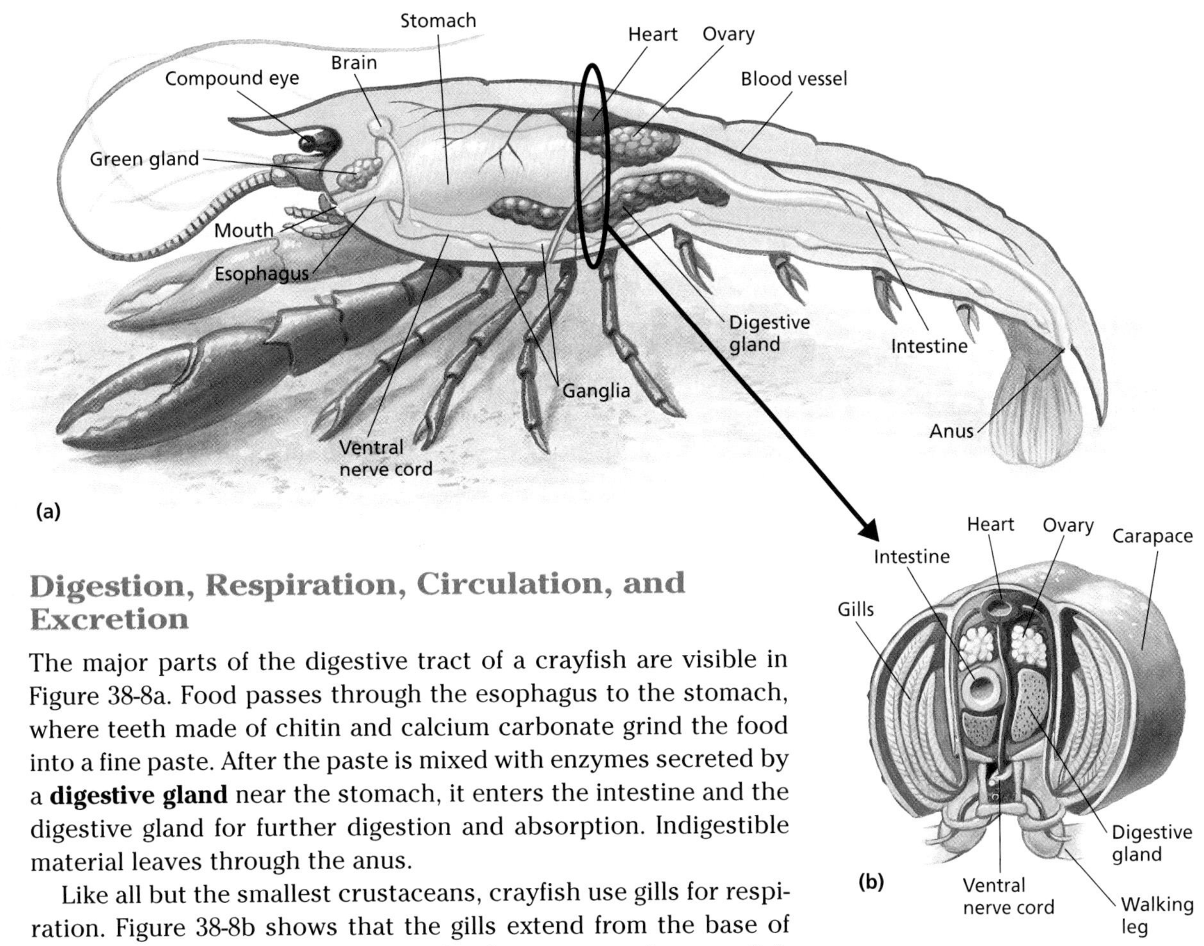

FIGURE 38-8

The major internal organs of a crayfish are seen in this cutaway side view (a) and cross section through the heart region (b).

Digestion, Respiration, Circulation, and Excretion

The major parts of the digestive tract of a crayfish are visible in Figure 38-8a. Food passes through the esophagus to the stomach, where teeth made of chitin and calcium carbonate grind the food into a fine paste. After the paste is mixed with enzymes secreted by a **digestive gland** near the stomach, it enters the intestine and the digestive gland for further digestion and absorption. Indigestible material leaves through the anus.

Like all but the smallest crustaceans, crayfish use gills for respiration. Figure 38-8b shows that the gills extend from the base of each walking leg into a chamber under the carapace. As a crayfish walks, its legs circulate water across its gills. Feathery branches on the posterior pair of maxillae also help direct water over the gills. Each gill is covered by an extension of the exoskeleton that is thin enough to permit gases to diffuse across the gill surface.

TABLE 38-1 ***Crayfish Appendages***

Appendage	Function
Antennule	touch, taste, equilibrium
Antenna	touch, taste
Mandible	chew food
Maxilla	manipulate food, draw water currents over gills
Maxilliped	touch, taste, manipulate food
Cheliped	capture food, defense
Walking leg	locomotion over solid surfaces
Swimmeret	create water currents, transfer sperm (males), carry eggs and young (females)
Uropod	propulsion during tailflips

QUICK FACT

✔ When ghost crabs are on the beach, they carry water with them inside their large, sealed gill chambers.

RETEACHING *ACTIVITY*

Have students compare and contrast the structure and function of a crayfish with the structure and function of an earthworm. Ask students to consider body segmentation, feeding, digestion, circulation, respiration, excretion, sensing of the environment, and neural control.

CONTENT CONNECTION

Section 5-1: Osmosis

Osmosis is the diffusion of water across a membrane. In brackish or dilute sea water, many crustaceans greatly increase their urine production to eliminate the water that enters their body through osmosis.

RECENT RESEARCH

Crab Shell Strength

Scientists have recently shown that the strength of a crab's exoskeleton, or shell, is related to its color. Dark parts of the shell are harder, less porous, and more resistant to cracking than lighter areas. The dark parts resemble a tough, synthetic polymer, while the lighter parts are more like a brittle ceramic.

SECTION 38-2

GIFTED ACTIVITY

Have students investigate how different varieties of aquatic arthropods maintain water balance. Students should include both euryhaline species, which can tolerate a wide range of salinity, and stenohaline species, which have a narrow salinity tolerance. Have students discuss the roles of gills, green glands, and other structures in this aspect of homeostasis.

ANSWERS TO SECTION 38-2 REVIEW

1. Crustaceans have two pairs of appendages on their head that serve as feelers, and they usually have a pair of appendages on most other body segments. At least some of the appendages are branched. Most crustaceans have gills, a nauplius larval stage, and 16 to 20 body segments. Aquatic crustaceans often incorporate large amounts of calcium carbonate into their exoskeletons.
2. Marine crustaceans include lobsters, crabs, shrimp, barnacles, and some copepods. Freshwater crustaceans include water fleas such as *Daphnia,* crayfish, and some copepods. Sow bugs and pill bugs live on land.
3. The mandibles are used for chewing food; the chelipeds are used for capturing food and for defense.
4. Gills are attached to the walking legs. When a crayfish walks, its legs circulate water across its gills.
5. The green glands eliminate excess water that enters a crayfish's body through osmosis.
6. A crayfish grows most rapidly during its first year of life, as indicated by the fact that it molts seven times in the first year.

internetconnect

SCILINKS NSTA
TOPIC: Crustaceans
GO TO: www.scilinks.org
KEYWORD: HM750

The main components of the crayfish's open circulatory system are shown in Figures 38-8a and 38-8b. The dorsal heart pumps hemolymph into several large vessels that carry it to different regions of the body. Hemolymph leaves the vessels and enters the hemocoel, bathing the various tissues. It then passes through the gills, where it exchanges carbon dioxide and oxygen with the water. From the gills, the hemolymph returns to the dorsal part of the crayfish and enters the heart.

As freshwater organisms, crayfish live in a hypotonic environment. Recall from Chapter 5 that a hypotonic environment is one in which the concentration of solute molecules is lower than that in the organism's cells. Therefore, water constantly enters the tissues of a crayfish by osmosis. This excess water is eliminated by excretory organs called **green glands,** which are visible in Figure 38-8a. The dilute fluid collected by the green glands leaves the body through a pore at the base of the antennae.

Neural Control

The nervous system of the crayfish is illustrated in Figure 38-8a. It is typical of arthropods and is similar to the nervous system of annelids. The crayfish brain consists of a pair of ganglia above the esophagus that receive nerve impulses from the eyes, antennules, and antennae. Two bundles of nerve fibers extend from the brain and pass around either side of the esophagus to a ganglion that controls the mandibles, maxillae, and maxillipeds. The ventral nerve cord runs posteriorly from this ganglion, connecting a series of ganglia that control the appendages and muscles in the segments of the thorax and abdomen.

Crayfish sense vibrations and chemicals in the water with thousands of small sensory hairs that project from the exoskeleton. Sensory hairs are visible in Figure 38-1. These sensory hairs are distributed over the entire body, but they are especially concentrated on the antennules, antennae, mouthparts, chelipeds, and telson. The compound eyes of a crayfish are set on two short, movable stalks. Each eye has more than 2,000 light-sensitive units with their own lenses. At the base of the antennules are organs that can detect the animal's orientation with respect to gravity.

SECTION 38-2 REVIEW

1. What characteristics are shared by most or all crustaceans?
2. Name one crustacean that lives in the ocean, one that lives in fresh water, and one that lives on land.
3. What are the functions of the mandibles and the chelipeds on a crayfish?
4. What structural adaptations of crayfish promote effective respiration in water?
5. What is the function of a crayfish's green glands?
6. **CRITICAL THINKING** In what year of its life would you expect a crayfish to grow most rapidly, given what you know about molting? Explain your reasoning.

SUBPHYLA CHELICERATA AND UNIRAMIA

SECTION

38-3

OBJECTIVES

▲ List the characteristics of the class Arachnida.

● Explain the adaptations spiders have for predatory life on land.

■ List the distinguishing characteristics of scorpions and of mites and ticks.

◆ Describe similarities and differences between millipedes and centipedes.

Unlike crustaceans, nearly all members of the subphyla Chelicerata and Uniramia are terrestrial. The major group in Chelicerata is the class Arachnida (uh-RAK-nuh-duh), which contains over 70,000 species. In Uniramia, members of the classes Diplopoda (di-PLAHP-uh-duh) and Chilopoda (ki-LAHP-uh-duh) have many body segments, and most segments have one or two pairs of legs. Thus, they are commonly called ***myriapods*** *(MIR-ee-uh-PAHDZ), which means "many feet."*

CLASS ARACHNIDA

Members of the class Arachnida, called **arachnids,** include spiders, scorpions, mites, and ticks. Like crayfish and other decapod crustaceans, arachnids have a body that is divided into a cephalothorax and an abdomen. The cephalothorax in arachnids usually bears six pairs of jointed appendages: one pair of chelicerae; one pair of **pedipalps,** which aid in holding food and chewing; and four pairs of walking legs.

Anatomy of a Spider

Spiders range in length from less than 0.5 mm to as large as 9 cm (3.5 in.) in some tropical tarantula species. As you can see in Figure 38-9, the body of a spider is constricted between the cephalothorax

FIGURE 38-9

The major internal organs of a spider are seen in this cutaway side view. The inset shows a closer view of a book lung, one of the spider's adaptations to life on land.

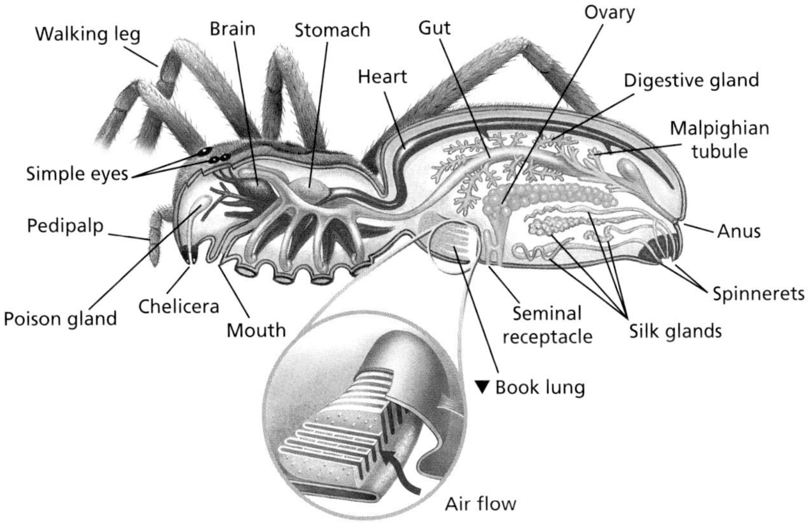

SECTION 38-3

SECTION OVERVIEW

Evolution

Section 38-3 describes the structural and functional adaptations of arachnids and myriapods that allow these two groups of arthropods to survive on land. The lack of segmental specialization in myriapods suggests that they may have been the first animals to evolve on land.

▶ ENGAGE STUDENTS

Collect several spiders of different species and bring them to class. Place them in Petri dishes covered with a damp paper towel. Have students observe each specimen through a dissecting microscope or hand lens and locate the cephalothorax, abdomen, chelicerae, pedipalps, walking legs, and spinnerets. Ask students to find and count the eyes. (Most spiders have eight simple eyes.) Point out that spiders have no antennae.

INCLUSION *ACTIVITY*

Give students a dozen or more plastic-foam balls of the same size and a box of toothpicks. Tell students that the balls represent arachnid and myriapod body segments and that the toothpicks represent appendages. The toothpicks can be used to link the balls together. Have students use the balls and toothpicks to make a model of each of the following: spider, scorpion, mite or tick, millipede, and centipede. Ask them to identify the appendages on each model.

CULTURAL CONNECTION

Insect Control

People in this country often regard centipedes with fear or disgust. In the tropics, however, many people allow centipedes to live in their houses to help control insect pests.

QUICK FACT

☑ If all the spiders in a single hectare (approximately 2 acres) of land were to spin their silk into one continuous strand for three days, that strand would be long enough to reach the moon.

TEACHING STRATEGY

Orb Weavers

Show a time-lapse videotape of a spider spinning an orb web. Emphasize that although spinning such a web may appear to be a very complicated process, it is a behavior that is programmed entirely by the spider's genes rather than learned. Explain that not all spiders make such geometrically regular webs; some spiders merely stretch a few strands of silk near their nest, while others spin no webs at all.

GIFTED *ACTIVITY*

Have students collect orb webs on black construction paper by bringing the paper in contact with the web and breaking the strands. Students can preserve the webs by covering them with tape or clear contact paper. Ask students whether the webs spun by the same spider on different days appear to be more similar to each other than to webs spun by different spiders of the same species.

CRITICAL THINKING

Spider Feeding

In what way are the feeding habits of web-spinning spiders and sponges similar? (Such spiders basically filter flying insects out of the air, much as sponges filter food out of the water that passes through their body.)

Word Roots and Origins

spiracle

from the Latin *spirare*, meaning "to breathe"

internet**connect**

SCLINKS. NSTA

TOPIC: Arachnida
GO TO: www.scilinks.org
KEYWORD: HM752

and the abdomen. The chelicerae are modified as fangs and are used to inject venom into prey. The venom is produced by poison glands in the cephalothorax and flows through ducts in the chelicerae to the tips of the fangs. Most spiders have eight simple eyes at the anterior end of the cephalothorax. Each simple eye has a single lens.

On the tip of the abdomen of many spiders are three pairs of organs called **spinnerets,** which are visible in Figure 38-9. Each spinneret is composed of hundreds of microscopic tubes that connect to silk glands in the abdomen. A protein-containing fluid produced in the silk glands hardens into threads as it is pulled from the spinnerets. Spiders use their silk threads to spin webs, build nests, and protect eggs. Some young spiders even move to new habitats when the wind pulls them through the air by their threads.

The nervous, digestive, and circulatory systems of spiders are similar to those of crustaceans. Because spiders are terrestrial, however, their respiratory system is quite different. In some spiders, respiration occurs in **book lungs,** paired sacs in the abdomen with many parallel folds that resemble the pages of a book. The folds in a book lung provide a large surface area for gas exchange. Other spiders have a system of tubes called **tracheae** (TRAY-kee-ee) that carry air directly to the tissues from openings in the exoskeleton known as **spiracles.** Some spiders have both book lungs and tracheae.

The excretory system of spiders is also modified for life on land. The main excretory organs, called **Malpighian** (MAL-PIG-ee-uhn) **tubules,** are hollow projections of the digestive tract that collect body fluids and wastes and carry them to the intestine. After most of the water is reabsorbed, the wastes leave the body in a nearly solid form with the feces. Thus, the Malpighian tubules help spiders conserve water in terrestrial environments. Some spiders also have **coxal glands,** organs that remove wastes and discharge them through openings at the base of some of the legs.

FIGURE 38-10

The black widow, *Latrodectus mactans* (a), and the brown recluse, *Loxosceles reclusa* (b), are the only two spiders in the United States whose venom is dangerous to humans. Note the red hourglass-shaped spot on the abdomen of the black widow and the dark violin-shaped marking on the cephalothorax of the brown recluse.

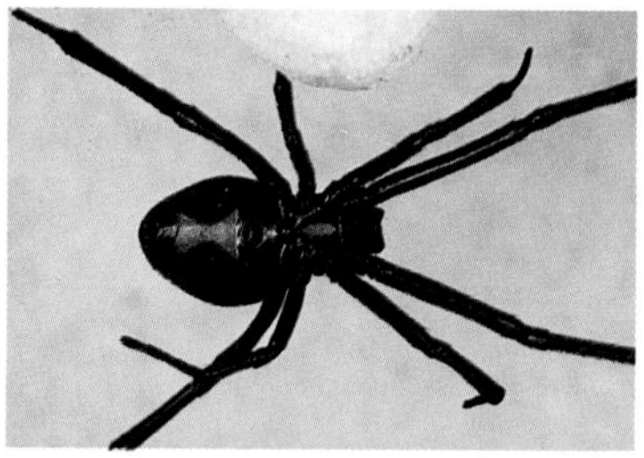

(a)

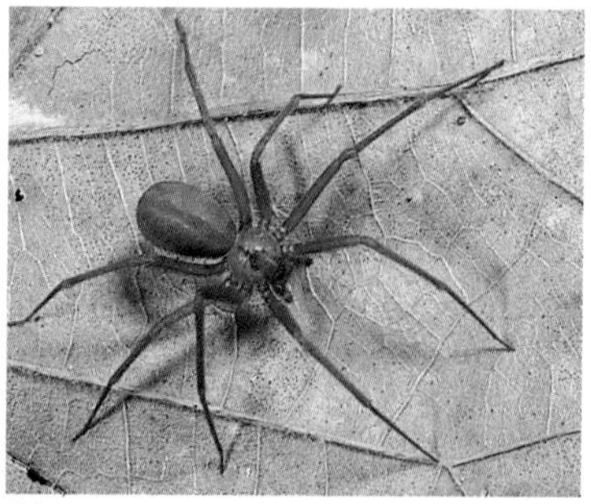

(b)

Life of a Spider

Spiders feed mainly on insects, although some can catch fish, frogs, and even birds. Different species of spiders are adapted to capture their prey in different ways. Some chase after prey, some hide beneath trapdoors waiting for prey to approach, and some snare prey in webs spun from silk. When an insect becomes trapped in the sticky web, the spider emerges from its hiding place near the edge of the web and paralyzes the insect with its venom. Many spiders also immobilize their prey by wrapping them in silk. They can then consume the body fluids of the prey at a later time.

Spider venom is usually harmless to humans, and most spiders bite only when they or their young are threatened. There are, however, two kinds of spiders in the United States whose bites can be serious and occasionally fatal to humans. They are the black widow and the brown recluse, shown in Figure 38-10. The black widow has a bright red or orange mark shaped like an hourglass on the ventral surface of its abdomen. Its venom attacks the nervous system. The brown recluse has a violin-shaped mark on the dorsal surface of its cephalothorax. Therefore, it is sometimes called the "violin spider." The venom of the brown recluse digests the tissues surrounding the bite.

A male spider is usually smaller than a female of the same species. When the male is mature, he transfers sperm to special sacs in the tips of his pedipalps. The sperm are then placed in the seminal receptacle of the female during mating. As soon as mating has occurred, the male darts away. If he is not quick enough, he may be eaten by the female. Eggs are fertilized as they pass out of the female into a silken case that she has spun. The female may carry the egg case with her or attach it to a web or plant. The young spiders hatch in about two weeks and undergo their first molt before leaving the case.

Scorpions

Scorpions, like the one shown in Figure 38-11, differ from spiders in two ways. Scorpions have large, pincerlike pedipalps, which they hold in a forward position. They also have a large stinger on the last segment of the abdomen, which is curled over the body. Scorpions usually hide during the day and hunt at night, mostly for insects and spiders. They seize prey with their pedipalps and inject venom into the prey with their stinger. Only a few species of scorpions have venom that can be fatal to humans. Most scorpions live in tropical or semitropical areas, but some are found in dry temperate or desert regions.

Mites and Ticks

Mites and ticks are the most abundant and most specialized arachnids. About 30,000 species have been identified, but the actual number of species may be much larger than that. Unlike spiders and scorpions, mites and ticks have a completely fused cephalothorax and abdomen, with no sign of separation between them.

Most mites are less than a millimeter in length, and some are small enough to live on particles of dust. They can be found in freshwater, marine, and terrestrial habitats. Many mites are entirely free living, while others are parasites during at least part of their life cycle. Spider mites parasitize fruit trees and many other agricultural crops by sucking the fluid from their leaves. The larvae of harvest mites—also known as chiggers—attach themselves to the skin of vertebrates, including humans. They break the skin with their chelicerae and feed on blood, causing swelling and itching. Other species of mites live on the bodies of chickens, dogs, cattle, humans, and other animals, where they feed on sloughing skin, hair, and feathers.

Ticks range in length from a few millimeters to 3 cm (a little over 1 in.). Many parasitic ticks, including the lone star tick shown in Figure 38-12, pierce their host's skin to feed on blood. In the process, they can transmit bacteria and other microorganisms that cause diseases such as Rocky Mountain spotted fever and Lyme disease.

FIGURE 38-11

Scorpions, like *Paruroctonus mesaensis,* have pincerlike pedipalps and a venomous stinger at the end of their abdomen.

FIGURE 38-12

The lone star tick, *Amblyomma americanum,* is an arachnid that parasitizes humans and other mammals.

SECTION 38-3

OVERCOMING MISCONCEPTIONS

Many people believe that all spiders are poisonous and dangerous to humans. Emphasize that in the United States only black widow and brown recluse spiders have venom that is harmful to humans and that these spiders bite people only when they are disturbed.

RECENT RESEARCH

Wheeling Spiders

Scientists have recently discovered how the wheel spider of the Namib Desert defends itself against its natural enemy, the pompilid wasp. To escape from the wasp, the wheel spider curls its legs like spokes and rolls rapidly downhill. The spider's habitat has steep, sandy slopes that are excellent for wheeling.

MAKING CONNECTIONS

Technology

Chemical companies are attempting to develop a synthetic spider silk that would be as strong as steel but much lighter. Fabric woven from the synthetic silk would then be used to make products such as parachutes and bulletproof vests.

RETEACHING *ACTIVITY*

Set up observation stations with preserved specimens of arachnids and myriapods, and ask students to classify the specimens into the following five categories: spider, scorpion, mite or tick, millipede, and centipede. Ask students to specify the criteria they used to classify each specimen. (Acceptable criteria should relate to the number and arrangement of body segments, number of legs per segment, and type of head appendages.)

DEMONSTRATION

Bring live millipedes and centipedes to class, and allow students to observe them. Ask students to describe how these two types of myriapods differ in their locomotion and how that difference is related to the feeding habits of the animals. (Millipedes are slow-moving herbivores; centipedes are fast-moving predators.)

ANSWERS TO SECTION 38-3 REVIEW

1. Arachnids have a body that is divided into a cephalothorax and an abdomen. The cephalothorax usually has six pairs of jointed appendages: chelicerae, pedipalps, and four pairs of walking legs.
2. Pedipalps are a pair of appendages on the cephalothorax of an arachnid. They aid in holding food and chewing, and in spiders they transfer sperm from the male to the female's seminal receptacle during mating.
3. The Malpighian tubules remove wastes from the body without excreting much water, thereby helping spiders conserve water in terrestrial environments where water may be scarce.
4. Scorpions have large pincerlike pedipalps and a stinger at the end of their abdomen, which is curled over the body.
5. Millipedes are herbivorous, have rounded bodies, and have two pairs of legs on each segment except the last two. Centipedes are carnivorous, have flattened bodies, and have one pair of legs on each segment except the first and last two.
6. Spiders feed on insects, which can damage crops or carry diseases.

(a)

(b)

FIGURE 38-13

(a) Millipedes, like *Narceus americanus,* have a rounded body and two pairs of legs on each body segment except the last two. (b) Centipedes, including *Scolopendra,* have a flattened body and one pair of legs on each body segment except the first and the last two.

MYRIAPODS

Myriapods—millipedes and centipedes—are members of the subphylum Uniramia. Early myriapods may have been the first animals to appear on land, and all present-day myriapods are terrestrial. Their bodies are highly segmented, and the segments bear appendages that show little variation. Like insects, myriapods have a single pair of unbranched antennae. However, myriapods do not have a waxy exoskeleton. They avoid desiccation by living in damp environments under leaves, rocks, and logs.

Class Diplopoda

The class Diplopoda consists of millipedes. Most millipedes have two pairs of legs on each body segment except the last two. The legs are well adapted for burrowing through humus and soil, but because they are short, millipedes move slowly. As you can see in Figure 38-13a, the bodies of millipedes are rounded.

Millipedes have short antennae and two groups of simple eyes on their head. They have poor vision, but a good sense of smell. Millipedes use their maxillae and mandibles to chew on plants or decaying plant matter in the soil. When threatened, they coil up and may secrete a noxious fluid that contains cyanide.

Class Chilopoda

Centipedes are members of the class Chilopoda. Centipedes may have as few as 15 or as many as 175 pairs of legs. In tropical regions centipedes reach lengths of 30 cm (12 in.). Their bodies are more flattened than those of millipedes, and their legs are longer relative to their body, as Figure 38-13b shows. Each body segment behind the head, except the first segment and the last two segments, has one pair of jointed legs. The appendages on the first segment are modified into a pair of poison claws. Long antennae and two clusters of simple eyes are located on the head.

Centipedes can move quickly in search of earthworms, insects, and other prey. They kill the prey with their poison claws and use their mandibles and maxillae to tear it apart. Most centipedes are not harmful to humans.

SECTION 38-3 REVIEW

1. What are the characteristics of arachnids?
2. Define *pedipalps,* and explain their function.
3. How is the excretory system of spiders adapted for life on land?
4. Name two ways in which scorpions differ from spiders.
5. How are millipedes and centipedes different?
6. CRITICAL THINKING How might the feeding habits of spiders benefit humans?

CHAPTER 38 REVIEW

SUMMARY/VOCABULARY

38-1

- Arthropods are segmented animals that have jointed appendages, an exoskeleton, a high degree of cephalization, a ventral nerve cord, and an open circulatory system.
- To grow, an arthropod must shed its exoskeleton periodically in a process called molting.
- The subphylum Trilobita includes extinct animals called trilobites.
- Members of the subphylum Crustacea have branched antennae and a pair of chewing mouthparts called mandibles.
- Members of the subphylum Chelicerata lack antennae and have pincerlike mouthparts called chelicerae.
- Members of the subphylum Uniramia have antennae, mandibles, and unbranched appendages.

Vocabulary

appendage (743)
arthropod (743)
chelicera (745)
compound eye (743)
crustacean (745)
mandible (745)
molting (744)
tagma (744)

38-2

- Crustaceans include shrimps, lobsters, crabs, crayfish, barnacles, isopods, copepods, and water fleas.
- Crustaceans have two pairs of feelers on their head and one pair of usually branched appendages on their other body segments. Their exoskeletons often contain large amounts of calcium carbonate.
- Most crustaceans are aquatic, respire with gills, and go through a nauplius larval stage during development.
- The body of a crayfish is divided into a cephalothorax and an abdomen. The appendages on each body segment are specialized for sensing, feeding, respiration, locomotion, or reproduction.

Vocabulary

abdomen (748)
antenna (748)
antennule (748)
carapace (748)
cephalothorax (747)
cheliped (748)
cirrus (747)
decapod (747)
digestive gland (749)
green gland (750)
isopod (747)
maxilla (748)
maxilliped (748)
nauplius (746)
plankton (746)
swimmeret (748)
telson (748)
thorax (748)
uropod (748)

38-3

- Arachnids include spiders, scorpions, mites, and ticks. Their bodies are divided into a cephalothorax and an abdomen, and they usually have six pairs of jointed appendages: one pair of chelicerae, one pair of pedipalps, and four pairs of walking legs.
- Spiders have eight simple eyes, and chelicerae modified as fangs that can inject venom. They produce silk threads that are used for prey capture and other functions. Spiders are terrestrial; they respire by means of book lungs, tracheae, or both, and they use Malpighian tubules to excrete wastes while conserving water.
- Scorpions have large, pincerlike pedipalps and a stinger on the last segment of the abdomen.
- Mites and ticks have a completely fused cephalothorax and abdomen. Many species are parasitic, and some spread diseases that affect humans.
- Myriapods include millipedes and centipedes. Millipedes have rounded bodies and two pairs of jointed legs on each body segment except the last two. Centipedes have flattened bodies and one pair of jointed legs on each body segment except the first and the last two.

Vocabulary

arachnid (751)
book lung (752)
coxal gland (752)
Malpighian tubule (752)
myriapod (751)
pedipalp (751)
spinneret (752)
spiracle (752)
trachea (752)

CHAPTER 38 REVIEW ANSWERS

REVIEW

1. The appendages are the antennules, antennae, mandibles, maxillae, maxillipeds, chelipeds, and walking legs.
2. The appendages are the swimmerets and uropods.
3. *Centipede* does not belong in the group, because centipedes are not members of the subphylum Chelicerata.
4. Tagmata are body structures in an arthropod that result from the fusion of two or more segments.
5. Chelicerae are pincerlike mouthparts, and mandibles are chewing mouthparts.

6. c
7. d
8. d
9. a
10. b
11. c
12. d
13. a
14. d
15. a

16. The correct order is c, b, d, a.
17. The criteria include the existence of a nauplius larval stage; the presence of antennae, chelicerae, and mandibles; and the presence of branched or unbranched appendages.
18. All three subphyla have experienced a loss of appendages on some segments and the fusion of certain segments into larger structures with specialized functions.
19. Crayfish respire through gills, whereas spiders respire by means of tracheae, book lungs, or both.
20. Spiders have chelicerae modified as fangs and venom to kill prey. Many spiders also use silk to build webs and entrap prey.
21. On the end of the spider's abdomen are three pairs of spinnerets, which are connected to silk glands in the abdomen. A protein-containing fluid produced in the silk glands hardens into threads as it is pulled from the spinnerets.
22. Parasitic ticks may transmit disease-causing microorganisms when the ticks bite their hosts to feed on blood.

23. Millipedes coil up and may secrete a noxious fluid.
24. Myriapods avoid desiccation through behavioral adaptations, such as living in damp environments under rocks and logs.
25. *A,* antennule; *B,* cheliped; *C,* walking leg; *D,* swimmeret; *E,* uropod; *F,* telson

CRITICAL THINKING

1. Most algae live in lighted areas. The eyespot of *Daphnia* senses where light is brightest and therefore where most algae are likely to be.
2. Barnacles have long appendages called cirri that sweep small organisms and food particles from the water into the barnacle's mouth. Barnacles have a very hard shell that completely encloses the body in most species.
3. Fusion of the exoskeleton strengthens a vulnerable joint where the head and thorax join. However, it also eliminates movement at this joint, limiting the flexibility of the animal.
4. Sensory hairs on the telson can pick up vibrations in the water that may signal the approach of a predator from the rear. Thus, they provide a defensive warning system for the posterior end of the animal.
5. The lobster's senses of taste and smell would enable the animal to find food and mates at night, when it is active and its vision is limited by darkness.
6. A thick, strong exoskeleton is better able to prevent an arthropod's body from collapsing under the pressure exerted by water.
7. Arthropods' segmented bodies and jointed appendages give these animals great flexibility. Specialization allows the appendages to be used for a variety of functions, including feeding, defense, and locomotion. The exoskeleton provides support and limits water loss. In some species,

REVIEW

Vocabulary

1. Name the appendages on the cephalothorax of a crayfish.
2. Name the appendages on the abdomen of a crayfish.
3. Choose the term that does not belong in the following group, and explain why it does not belong: scorpion, centipede, mite, spider.
4. What are tagmata?
5. Distinguish between chelicerae and mandibles of arthropods.

Multiple Choice

6. All arthropods have (a) a cephalothorax (b) spiracles (c) jointed appendages (d) antennae.
7. The exoskeleton of an arthropod (a) provides protection and support (b) plays a role in movement (c) contains chitin (d) all of the above.
8. Myriapods are members of the subphylum (a) Arachnida (b) Chelicerata (c) Chilopoda (d) Uniramia.
9. The major respiratory organs of crayfish are the (a) gills (b) lungs (c) tracheae (d) book lungs.
10. Compound eyes (a) have a single lens (b) are composed of many individual light detectors (c) are found in all arthropods except the crayfish (d) are located on the abdomen of scorpions.
11. Spiders feed mainly on (a) plants (b) decayed matter (c) insects (d) other spiders.
12. Mites and ticks differ from spiders by having (a) mandibles (b) a unique respiratory system (c) two pairs of antennae (d) a fused cephalothorax and abdomen.
13. Book lungs help spiders respire on land by (a) providing a large surface area for the exchange of gases (b) carrying air directly to tissues (c) both a and b (d) neither a nor b.
14. Centipedes and millipedes differ in (a) the way their bodies are shaped (b) the number of legs they have on each segment (c) their feeding habits (d) all of the above.
15. The subphylum of modern arthropods that is primarily aquatic is (a) Crustacea (b) Chelicerata (c) Trilobita (d) Uniramia.

Short Answer

16. Arrange the following steps in the molting of an arthropod in the correct order:
    a. The old exoskeleton loosens, breaks along specific lines, and is shed.
    b. Enzymes digest the inner layer of the exoskeleton.
    c. A hormone is produced that induces molting.
    d. The epidermis begins to synthesize a new exoskeleton.
17. What criteria do biologists use to classify arthropods into four subphyla?
18. What evolutionary trends have all three subphyla of modern arthropods undergone?
19. Contrast the process of respiration in crayfish and spiders.
20. In what ways are spiders adapted for a predatory way of life?
21. Describe how a spider produces silk threads.
22. How do parasitic ticks spread diseases?
23. What defensive behaviors are common in millipedes?
24. How do myriapods avoid desiccation in terrestrial environments?
25. Identify the structures labeled *A* through *F* in the diagram below.

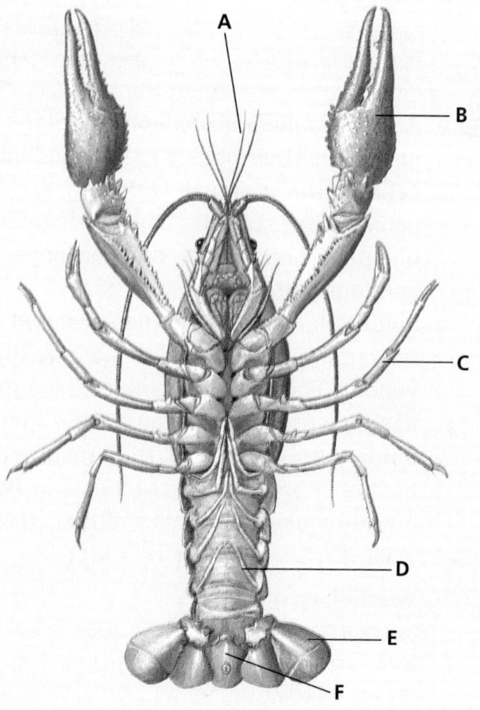

CRITICAL THINKING

1. The water flea *Daphnia* eats algae. It also has a prominent eyespot. How might the eyespot be connected with the ability of *Daphnia* to find food?
2. Barnacles are sessile crustaceans. What structural adaptation do barnacles have that enables them to compete with motile organisms for food? What structural adaptation do barnacles have that might protect them from predators?
3. The cephalothorax of a crayfish is covered by the carapace, a single, fused plate of exoskeleton. What are some possible advantages and disadvantages of this fused structure?
4. Like other arthropods, crayfish are cephalized, with a variety of specialized sensory structures on their head. However, crayfish also have a high concentration of sensory hairs on their telson. What might be the advantage of having so many sensory structures at the posterior end of the animal?
5. The American lobster, *Homarus americanus,* is a nocturnal organism. Marine biologists have discovered that the lobster's senses of taste and smell are over 1,000 times more powerful than those senses in humans. The lobster uses taste and smell both to search for food and to detect mates. What adaptive advantage would these highly developed senses provide for the lobster?
6. Aquatic arthropods, such as crabs and crayfish, typically have thicker, stronger exoskeletons than do terrestrial arthropods, such as spiders and insects. What advantage does a thick, strong exoskeleton provide in an aquatic environment?
7. Arthropods first invaded land about 400 million years ago. They have survived several mass extinctions in which many other kinds of organisms became extinct. What characteristics have enabled arthropods to thrive?
8. The graph below shows how the weight of a spider changed over the first 60 days of its life. What events occurred at the beginning of each sharp upward turn in the curve? Why did the spider's weight increase more gradually between these events?

Weight Gain by a Spider

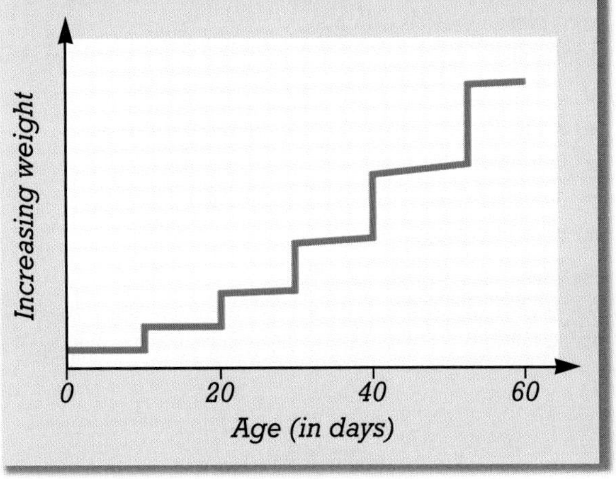

EXTENSION

1. Read "Having Mom for Dinner" in *Natural History,* April 1999, on page 21. What is the usual form of cannibalism in the animal world? What is the unusual form of cannibalism that is practiced by the spider *Amaurobius ferox*? How many eggs does the spider lay at one time? What does the mother spider do to protect herself from being eaten by her young?
2. Contact your local public health department and ask for information about arthropod-transmitted diseases that might be found in your area. If possible, find out how many cases of each disease have been reported in the last year. Report your findings to the class, or prepare a table summarizing your findings.
3. Do library research to discover the life history of a horseshoe crab, a sea spider, or a tick. Before beginning your research, examine Figure 38-3 for the phylogenetic relationships of these three types of arthropods.

Malpighian tubules also help conserve water. A high degree of cephalization enables arthropods to sense and respond to numerous stimuli in their environment.
8. The spider molted at the beginning of each upward turn in the curve. Its weight increased more gradually between molts because its rigid exoskeleton restricted the expansion of its body.

EXTENSION

1. Some adult insects, fish, and mammals occasionally eat their young. *Amaurobius ferox* feeds herself to her young hatchlings. The spider lays from 80 to 100 eggs at one time. When these eggs begin to hatch about 3 weeks later, the mother spider lays another batch of eggs on which the hatchlings begin to feed. Then the mother spider performs certain behaviors that invite her brood to swarm over her and eat her. She does nothing to fend off the attack.
2. Answers will vary, depending on the area. As an alternative activity, students may research Lyme disease or deer fever transmitted by ticks; the bubonic plague and other illnesses transmitted by fleas; or plant diseases caused by mites.
3. As Figure 38-3 shows, all of these animals are members of the subphylum Chelicerata. Students should note that horseshoe crabs are not crustaceans and sea spiders are not arachnids.

CHAPTER 38
INVESTIGATION

TIME REQUIRED

One 50-minute class period

SAFETY PRECAUTIONS

Caution students to use care when handling scissors.

QUICK REFERENCES

Lab Preparation Notes are found on page 741B.

Holt BioSources provides a Teaching Resources CD-ROM that contains "Using Gowin's Vee in the Lab" and "Scoring Rubrics."

PROCEDURAL TIPS

1. Pill bugs may be found outdoors under logs and rocks or purchased from WARD'S. Sow bugs may be used instead, although they will not curl up into a ball.
2. For these experiments, you may wish to make a chart on the chalkboard so that the results of the entire class (or more than one class) can be combined. The data can be quantified if students record the position of each pill bug at the end of each experiment.

ANSWERS TO BACKGROUND

1. Arthropods have jointed appendages and hard exoskeletons made of chitin. Crustaceans have two pairs of appendages on their head that serve as feelers and a pair of chewing mouthparts called mandibles, and most breathe with gills.
2. *Isopod* means "equal legs," and pill bugs, like other isopods, have seven pairs of identical legs.
3. Most crustaceans are aquatic, but pill bugs are terrestrial.

ANSWERS TO PARTS A–C

4. The pill bugs would be expected to move into the dark half of the dish.

CHAPTER 38 INVESTIGATION

Behavior of Pill Bugs

OBJECTIVES

- Review characteristics of the phylum Arthropoda and the subphylum Crustacea.
- Observe the external anatomy of a living terrestrial isopod.
- Investigate the behavior of terrestrial isopods.

PROCESS SKILLS

- observing
- hypothesizing
- experimenting

MATERIALS

- 5 live pill bugs or sow bugs for each pair of students
- 1 plastic medicine dropper
- water
- potato
- 3 sheets of filter paper cut to fit a Petri dish
- Petri dish with cover
- aluminum foil
- bright lamp or flashlight
- 4 fabrics of different texture
- cellophane tape
- scissors

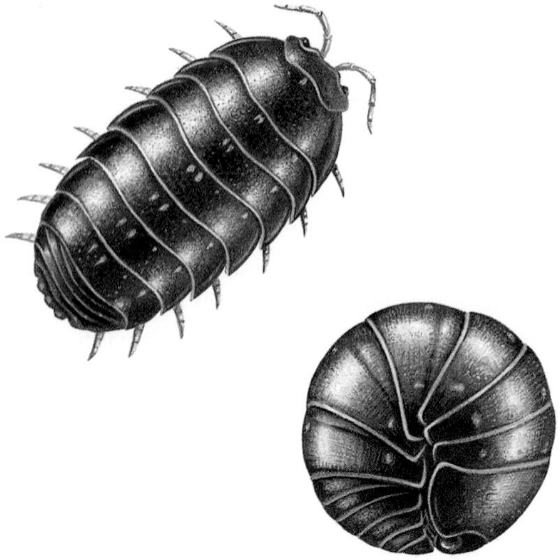

Background

1. What are the major characteristics of the members of the phylum Arthropoda and the subphylum Crustacea?
2. Why are pill bugs called isopods?
3. How are pill bugs different from most other crustaceans?

PART A Response to Light

1. Put several drops of water on a piece of filter paper until the paper becomes slightly moist. Place the filter paper in the bottom of a Petri dish. Cover half the bottom of the Petri dish with aluminum foil.
2. Check the lighting in the room. The light must be low and even during this part of the investigation.
3. CAUTION **You will be working with live animals. Be sure to treat them gently and to follow directions carefully.** Place five pill bugs in the center of the filter paper. Shine a lamp directly over the Petri dish so that half the filter paper is brightly illuminated and the other half is in darkness, shaded by the foil.
4. Based on your knowledge of the natural habitat of pill bugs, can you predict where they will go? Make a data table like the one shown, and record your prediction as well as the actual responses of the pill bugs.

PART B Response to Moisture

5. Cut a piece of filter paper in half. Moisten one of the halves with water and place it in the bottom of a Petri dish. Make sure that drops of water do not leak onto the bottom of the dish.
6. Place the dry half of the filter paper in the bottom of the Petri dish, leaving a 2 mm gap between it and the damp filter paper.
7. Place five pill bugs along the boundary between the wet and dry areas. Place the top on the dish.
8. To which side do you predict the pill bugs will move? Write your prediction in your data table. Observe the pill bugs for 3 to 5 minutes, and record your observations in your data table. Do your observations agree with your predictions?

PART C Response to Food

9. Again dampen a piece of filter paper and place it in the bottom of a Petri dish. Next place a thin slice of potato near the edge of the dish.
10. Place five pill bugs in the Petri dish opposite the potato slice, and place the lid on the dish.
11. Where do you predict the pill bugs will go? Write your prediction in your data table. Observe the pill bugs for 3 to 5 minutes, and record your observations in your data table. Do your observations agree with your predictions?

PART D Response to Surface Texture

12. Trace the outline of the bottom of a Petri dish on one of the fabrics. Cut the circle out of the fabric and fold it in half. Then cut along the fold to produce two half-circles.
13. Repeat step 12 using the other three fabrics. You should now have eight half-circles.
14. Tape together two half-circles, each of a different fabric. Place the two-fabric circle in the bottom of a Petri dish, tape side down, as shown in the figure below.
15. On a sheet of paper, draw the fabric circle and label the two types of fabric that make up the circle.

16. Place a pill bug in the center of the circle and observe its movements. One student should keep track of the amount of time the animal spends on each fabric. On the drawing that was made in step 15, the other student should draw the path the pill bug travels in the circle. After 5 minutes, stop your observations and record the amount of time the pill bug spent on each fabric.
17. Repeat steps 14–16 for two other pairs of fabrics.
18. Return the pill bugs to their container. Clean up your materials and wash your hands before leaving the lab.

Analysis and Conclusions

1. In Part A, why was the entire filter paper moistened?
2. In Part B, why was there a slight separation between the wet and dry halves of the filter paper?
3. In Part C, why was the entire filter paper moistened?
4. In Part D, which fabric did the pill bugs prefer? Describe the texture of that fabric.
5. How do the responses of pill bugs to light, moisture, and food in these experiments reflect adaptations to their natural surroundings?
6. How is being able to detect surface texture a good adaptation for pill bugs in their natural habitat?

Further Inquiry

1. Design an experiment to investigate the response of pill bugs to temperature. Think carefully about how you will construct your apparatus. Seek approval from your teacher before you actually conduct this experiment. How do you think the pill bugs will respond?
2. Design an experiment to investigate whether pill bugs have preferences for certain types of food.

OBSERVATION OF PILL-BUG BEHAVIOR

Stimulus	Prediction	Observation
Light		
Moisture		
Food		

8. The pill bugs would be expected to move into the wet area of the dish.
11. The pill bugs would be expected to cluster around the slice of potato.

ANSWERS TO ANALYSIS AND CONCLUSIONS

1. The independent variable in Part A was light. All other conditions should have been the same, so the entire filter paper was moistened.
2. Separating the wet and dry halves of the filter paper kept the moisture in half the dish.
3. The independent variable in Part C was food. If the filter paper were not moistened, it would have been impossible to know whether the pill bugs moved to the potato slice for moisture or for food.
4. Answers will vary. Pill bugs should prefer rough-textured fabrics.
5. Pill bugs seek damp, dark places where they can avoid desiccation and find food.
6. This ability may help pill bugs find substrates into which they can dig.

FURTHER INQUIRY

1. Students might place half of a larger dish on ice or shine a bright light on half of the dish. If they do the latter, the entire dish should be covered with aluminum foil so that temperature is the only variable. Students should predict that the pill bugs will move where the temperature is moderate.
2. Several Petri dishes could be prepared with damp filter paper covering the bottoms and different kinds of food on opposite sides of each dish. For a more controlled experiment, slices of potato could be soaked overnight in a strong sugar solution, salt solution, or water.

CHAPTER 39
INSECTS

	TOPICS	TEACHING RESOURCES	LABS, CLASSWORK, AND HOMEWORK
BLOCK 1 (45 minutes)	**Introducing the Chapter, p. 760**	ATE Focus Concept, p. 760 ATE Assessing Prior Knowledge, p. 760 ATE Understanding the Visual, p. 760	■ Supplemental Reading Guide, *Journey to the Ants: A Story of Scientific Exploration*
	39-1 The Insect World, p. 761 Characteristics and Classification of Insects, p. 761 The Grasshopper, p. 764 Development, p. 768 Defense, p. 770	ATE Section Overview, p. 761 ATE Critical Thinking, p. 767 ATE Visual Link Figure 39-3, p. 765 134A, 135A, 136–140	**Lab B20,** Life in a Pine Cone ★ Study Guide, Section 39-1 PE Section Review, p. 770
BLOCK 2 (45 minutes)	**39-2 Insect Behavior, p. 771** Communication, p. 771 Behavior in Honeybees, p. 772	ATE Section Overview, p. 771 ATE Visual Link Figure 39-10, p. 772	PE **Chapter 39 Investigation,** p. 778 PE **Quick Lab,** p. 773, Interpreting Nonverbal Communication **Lab A24,** Observing Insect Behavior **Lab C39,** Response in the Fruit Fly ★ Study Guide, Section 39-2 PE Section Review, p. 774
BLOCKS 3 & 4 (90 minutes)	**REVIEW and ASSESSMENT** PE Chapter 39 Review, pp. 776–777 ■ Performance-Based Assessment—Exploring Invertebrates	Audio CD Program	**CHAPTER TESTING** ■ Chapter Test (blackline master) ▲ Test Generator ■ Assessment Item Listing ■ Scoring Rubrics and Classroom Management Checklists

MODERN BIOLOGY — HOLT BIOSOURCES

KEY

PE Pupil's Edition
ATE Teacher's Edition
■ Active Reading Guide
★ Study Guide

Teaching Transparencies
Laboratory Program

One-Stop Planner CD-ROM
Shows these resources within a customizable daily lesson plan:
■ Holt BioSources Teaching Resources
■ Active Reading Guide
★ Study Guide
▲ Test Generator

READING FOR CONTENT MASTERY

Preview: Have students preview the objectives, the introductory paragraph, topic sentences, and captions before reading the chapter.

ATE Active Reading Technique: L.I.N.K., p. 761
■ Reading Strategy: L.I.N.K. Teacher Notes
■ Active Reading Guide Worksheet 39-1
PE Summary/Vocabulary, p. 775

■ Active Reading Guide Worksheet 39-2
PE Summary/Vocabulary, p. 775

TECHNOLOGY AND INTERNET RESOURCES

Holt Biology Videodiscs Teacher's Correlation Guide, Lessons 39-1 through 39-2

Audio CD Program, Sections 39-1 through 39-2

internetconnect

sciLINKS NSTA

On-line Resources: www.scilinks.org
The following *sci*LINKS Internet resources can be found in the student text for this chapter:

Topics:
- Grasshopper, p. 765
- Honeybees, p. 772

go.hrw.com

On-line Resources: go.hrw.com
Visit the HRW Web site for a variety of resources related to this chapter. Just type in the keyword HM2 Chapter 39.

Smithsonian Institution®
Visit **www.si.edu/hrw** for additional on-line resources.

CNNfyi.com
Visit **www.cnnfyi.com** for late-breaking news and current event stories selected just for you.

LAB ACTIVITY PLANNING

CHAPTER 39 INVESTIGATION

Anatomy of a Grasshopper, pp. 778–779

OVERVIEW
In this investigation, students will examine the external and internal anatomy of a grasshopper.

SAFETY
Students will wear disposable gloves, lab aprons, and safety goggles during the lab. Instruct students in the correct handling of all dissecting instruments. Remind students to keep their hands away from their faces while handling preserved specimens and to wash their hands after completing the lab.

PREPARATION
1. Purchase several *Lubber* grasshoppers from a biological supply house.
2. Each lab group will need one grasshopper, a dissection tray, fine dissection scissors, forceps, a blunt probe, a sharp probe, and several dissection pins.
3. Students will need access to a stereomicroscope.
4. If dissection is not practical for your classroom, alternative methods of studying these structures are available in models, filmstrips, videotapes, and computer simulations.

Quick Labs

Interpreting Nonverbal Communication, p. 773
Have students form groups of at least five or six. Each group will need pencils, paper, and several wrapped pieces of candy.

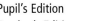

UNDERSTANDING THE VISUAL

Insects have successfully adapted to many kinds of habitats and exhibit a remarkable variety of adaptations. Ask students to look at the insect shown in the photograph and to suggest how its coloration might be adaptive. (The coloration might warn potential predators that the insect is poisonous or distasteful, allow the insect to blend in with a similarly colored part of its environment, or help in attracting a mate.)

FOCUS CONCEPT

Evolution

Insects first appeared about 400 million years ago. As they evolved, they retained the features that made their arthropod ancestors successful—a segmented body, jointed appendages, and an exoskeleton—and they added another important characteristic, the ability to fly.

ASSESSING PRIOR KNOWLEDGE

Review the following concepts.

Molting: *Chapter 38*
Ask students to define molting and to explain its importance to an arthropod. Ask students to describe the steps involved in molting.

Arthropod Classification: *Chapter 38*
Ask students to name the subphylum to which insects belong and to name two other arthropod classes in that subphylum. Ask students what characteristics are shared by the members of that subphylum.

CHAPTER 39

INSECTS

The leaf-footed bug, Diactor bilineatus, *is a colorful member of the extremely diverse world of insects.*

FOCUS CONCEPT: ***Evolution***

As you read, look for the structural, developmental, and behavioral adaptations that have made insects such a successful group of animals.

THE INSECT WORLD

Insects have thrived for more than 300 million years, since long before the rise and fall of the dinosaurs. The story of insects is one of great biological success through evolution and adaptation. Today, insects account for about three-fourths of all animal species on Earth.

SECTION

39-1

OBJECTIVES

▲ State the major characteristics of the class Insecta.

● Explain why insects are so successful.

■ List both harmful and beneficial effects of insects on human society.

◆ Describe the external structure and organ systems of a grasshopper.

▲ Explain incomplete and complete metamorphosis in insects.

CHARACTERISTICS AND CLASSIFICATION OF INSECTS

Many of the adaptations that have made insects successful are characteristics they share with other arthropods, such as a segmented body, jointed appendages, and an exoskeleton. Insects belong to the class Insecta in the subphylum Uniramia, which, as you may recall from Chapter 38, also includes millipedes and centipedes. The body of an insect is divided into three tagmata: the head, thorax, and abdomen. Like other members of the subphylum Uniramia, insects have mandibles and one pair of antennae on their head, and the antennae and other appendages are unbranched. The thorax has three pairs of jointed legs and, in many species, one or two pairs of wings. The abdomen is composed of 9 to 11 segments, and in adults it has neither wings nor legs.

Most insects are small. Among the smallest is the fairyfly, which is only 0.2 mm (0.008 in.) in length. Some insects are much larger. For example, the African Goliath beetle exceeds 10 cm (4 in.) in length, and the atlas moth has a wingspan of more than 25 cm (10 in.). These two giants of the insect world are shown in Figure 39-1.

FIGURE 39-1

Two of the largest insects are the African Goliath beetle, *Goliathus meleagris* (a), and the atlas moth, *Attacus atlas* (b).

(a)

(b)

SECTION 39-1

SECTION OVERVIEW

Evolution
Section 39-1 discusses the reasons why insects have evolved into one of the most successful groups of organisms on Earth. The physical and behavioral adaptations of insects allow them to live in nearly every terrestrial and freshwater habitat.

Reproduction and Inheritance
Section 39-1 describes two types of metamorphosis in insects. Metamorphosis is a part of the insect life cycle that contributes to the success of insects by reducing competition between larvae and adults and by helping insects survive harsh weather.

ACTIVE READING
Technique: L.I.N.K.

Write the word *insect* on the board, and have students **L**ist on their own paper all the words, phrases, and ideas that they associate with the word. After several minutes, ask the students to share one or two of their ideas as you record them. Facilitate a discussion in which students **I**nquire of each other or the teacher to clarify each of the ideas listed. After all discussion is completed, have students write **N**otes on paper about what they remember. Allow about 1 minute for students to write their notes. Have them look over their notes to see what they **K**now about insects based on experience and discussion.

SECTION 39-1

RECENT RESEARCH

Bees and Mistletoe

New Zealand's giant mistletoes produce large buds that cannot open without outside help. It had been thought that only a few species of native birds were capable of twisting off the ends of the buds, allowing the mistletoe to flower. Recently, however, scientists have discovered that a small bee of the genus *Hylaeus* can also open the buds. This finding argues against the idea that there is a one-to-one matching between a pollinator and its favorite blossom.

INCLUSION *ACTIVITY*

Have students design an educational poster to increase awareness of butterflies and the need to protect them. For example, students could focus on the monarch butterfly and draw a map showing the migration route of the monarch from the United States to central Mexico. The poster could also include illustrations of the plants butterflies feed on and pollinate. Have the students present their poster to the class.

QUICK FACT

☑ Wings are outgrowths of the insect's exoskeleton and are not modified limbs as are wings in bats and birds. Wings of many insects are made of a double layer of chitin with a supporting network of hollow veins filled with air or blood.

Most insects fold their wings over the abdomen when at rest. A few, such as the dragonfly, cannot fold their wings and must keep them outstretched at all times.

Wings may be thin and membranous, as in flies; thin and horny, forming a cover for hindwings, as in the front wings of beetles; covered with hair, as in caddis flies; or covered with fine, often highly pigmented scales, as in butterflies and moths.

The study of insects is called **entomology** (ENT-uh-MAHL-uh-jee), and the scientists who engage in it are known as entomologists. Entomologists classify insects into more than 30 orders based on characteristics such as the structure of mouthparts, number of wings, and type of development. Several of the more common insect orders are listed in Table 39-1.

TABLE 39-1 *Common Insect Orders*

Order	Number of species	Examples	Type of metamorphosis	Characteristics	Significance to humans
Thysanura ("bristled tail")	2,400	bristletails silverfish firebrats 	none	chewing mouthparts	feed on paste in wallpaper and starch in book bindings and labels
Anoplura ("unarmed tail")	2,400	sucking lice 	incomplete	wingless; piercing, sucking mouthparts	parasitize humans, other mammals, and birds; transmit diseases
Dermaptera ("skin wing")	1,000	earwigs 	incomplete	two pairs of wings; biting mouthparts; pincerlike appendages at tip of abdomen	damage plants; transmit diseases
Ephemeroptera ("for-a-day wing")	1,500	mayflies 	incomplete	membranous wings (triangular forewings); nonfunctioning mouthparts in adults	nymphs serve as food for fresh-water fish
Hemiptera ("half wing")	55,000	true bugs 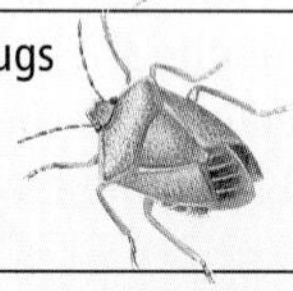	incomplete	two pairs of wings during part of life; piercing, sucking mouthparts	damage crops and garden plants
Homoptera ("like wing")	20,000	aphids mealy bugs cicadas 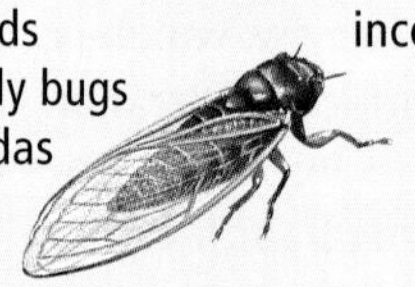	incomplete	membranous wings held like roof over body (some species wingless); piercing, sucking mouthparts	damage crops and garden plants
Isoptera ("equal wing")	2,000	termites 	incomplete	at times, two pairs of membranous wings; chewing mouthparts	decompose wood in buildings; recycle resources in forests
Odonata ("toothed")	5,000	dragonflies damselflies 	incomplete	two pairs of long, narrow, membranous wings; chewing mouthparts	destroy harmful insects; nymphs serve as food for freshwater fish
Orthoptera ("straight wing")	30,000	grasshoppers crickets katydids cockroaches 	incomplete	two pairs of straight wings; chewing mouthparts	damage crops, garden plants, and stored foods

The Success of Insects

Insects live almost everywhere in the world except in the deep ocean. Water striders glide on the surface of oceans and lakes, beetles inhabit the hottest deserts, and snow fleas survive on permanent glaciers. Entomologists have described and classified more than 700,000 insect species, or about three times as many species

Order	Number of species	Examples	Type of metamorphosis	Characteristics	Significance to humans
Coleoptera ("sheathed wing")	500,000	weevils ladybugs beetles 	complete	hard forewings, membranous hind wings; chewing mouthparts	destroy crops; prey on other insects
Diptera ("two wing")	80,000	mosquitoes flies gnats 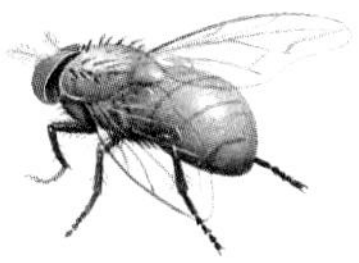	complete	one pair of wings (hind pair reduced to knobs); sucking, piercing, or lapping mouthparts	carry diseases; destroy crops; pollinate flowers; act as decomposers
Hymenoptera ("membrane wing")	90,000	bees wasps ants 	complete	two pairs of membranous wings (some species wingless); biting, sucking or lapping mouthparts; many have constriction between thorax and abdomen; some species social	pollinate flowers; make honey; destroy harmful insects
Lepidoptera ("scaled wing")	140,000	butterflies moths 	complete	large, scaled wings; chewing mouthparts in larvae, siphoning mouthparts in adults	pollinate flowers; larvae and pupae produce silk; larvae damage clothing and crops
Neuroptera ("nerve wing")	4,600	dobsonflies lacewings ant lions 	complete	two pairs of membranous wings; sucking or chewing mouthparts in larvae, chewing mouthparts in adults	destroy harmful insects; larvae serve as food for freshwater fish
Siphonaptera ("tubed, wingless")	1,200	fleas 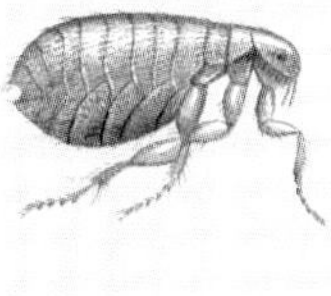	complete	wingless as adults; chewing mouthparts in larvae, sucking mouthparts in adults	parasitize birds and mammals; carry diseases

SECTION 39-1

OVERCOMING MISCONCEPTIONS

Locust is a name frequently applied to cicadas. In fact, cicadas are members of the order Homoptera, while locusts are a type of grasshopper and belong to the order Orthoptera.

QUICK FACT

☑ A single locust swarm may contain as many as 124 billion locusts and consume as much as 1,270,000 kg (1,400 tons) of food per day.

TEACHING STRATEGY

Orders of Insects

Have students play an insect identification game in order to help them distinguish characteristics of the different orders of insects. Divide the class into about five groups, and have each group choose an insect order from Table 39-1. Then have the other groups take turns asking one question at a time about the unknown insect order, such as "Does it have transparent wings?" The questions should be phrased so that they can be answered yes or no. The group that guesses the insect order should be allowed to take the next turn.

CULTURAL CONNECTION

Crickets in Japan

To the Japanese, crickets are considered pets and a source of musical and artistic pleasure. Favorite pet crickets are often given elaborate cages and porcelain water dishes. Japanese cricket owners even "tickle" their crickets with delicate hand-carved brushes to encourage them to sing.

GIFTED ACTIVITY

Have students prepare a report on a specific example of the role of insects in human history. Examples include fleas that spread the bubonic plague ("black death") of the Middle Ages and mosquitoes that transmitted yellow fever during the construction of the Panama Canal.

CULTURAL CONNECTION

Sacred Scarabs

Scarabs are dung beetles found in arid regions of the world. In ancient Egypt, the golden scarab was a symbol of the sun and was believed to be a form of the sun god on Earth. It was also a symbol of immortality, and its image was engraved on beads and amulets.

TEACHING STRATEGY

A World Without Insects

Have students describe what a world without insects might be like (no buzzing of bees, no mosquito bites, no insect pollinators, and so forth). Stress the importance of insects in ecosystems. While many insects may simply seem to be pests, such stereotypical images of them often mask a lack of understanding of their role in nature.

(a)

(b)

FIGURE 39-2

Some insects are harmful to humans, but most are beneficial. (a) Termites, such as *Reticulitermes flavipes,* can destroy a building by feeding on the wood. (b) The blister beetle, *Lutta fulvipennis,* cross-fertilizes plants by spreading pollen from flower to flower as it searches for nectar.

as exist in all other animal groups combined. Based on current knowledge, some entomologists believe that as many as 10 million insect species may exist. In terms of their widespread distribution and great abundance, insects are extremely successful.

One of the most important factors responsible for the remarkable success of insects is their ability to fly, which enables them to escape from predators and disperse rapidly into new environments. Like other arthropods, insects also benefit from having a light but sturdy exoskeleton and jointed appendages that perform a variety of functions. In addition, most insects are small, so several species can inhabit different local environments within an area without competing with one another for food or other resources. Finally, insects generally have very short life spans and produce large numbers of eggs. Therefore, natural selection can occur more quickly in insects than in organisms that take longer to reach maturity.

Insects and People

Since insects are so abundant, it is not surprising that they affect our lives in many ways. Some insects, such as grasshoppers, boll weevils, and corn earworms, compete with humans for food by eating crops. In fact, nearly every crop plant has some insect pest. Other insects spread diseases by biting humans or domesticated animals. Some fleas carry plague; female *Anopheles* mosquitoes transmit *Plasmodium,* the protozoan that causes malaria; and flies transmit the bacterium *Salmonella typhi,* which causes typhoid fever. Termites, shown in Figure 39-2a, attack the wood in buildings, and some moths consume wool clothing and carpets.

Despite the problems some insects cause, it would be a serious mistake to think that the world would be better off without any insects. Insects play vital roles in almost all terrestrial and freshwater environments. They serve as food for numerous species of fish, birds, and other animals. Many kinds of insects, such as the beetle shown in Figure 39-2b, are essential for the cross-pollination of plants. It is estimated that insects pollinate 40 percent of the world's flowering plants, including many of those cultivated as food for humans and livestock. Insects also manufacture a number of commercially valuable products, including honey, wax, silk, and shellac. We tend to think of termites as destructive pests because of their effects on buildings, but by feeding on decaying wood, they also help recycle nutrients needed to maintain a healthy forest. Other insects recycle the nutrients contained in animal carcasses.

THE GRASSHOPPER

In this section, the grasshopper will be used to demonstrate some of the details of insect structure and function. As you read, remember that these details are not shared by all insects. The diversity of the insect world is so great that no typical insect exists.

External Structure

The major features of an adult grasshopper's external structure are illustrated in Figure 39-3. The body of a grasshopper clearly shows the three insect tagmata. The most anterior tagma, the head, bears the mouthparts. It also has a pair of unbranched antennae as well as simple and compound eyes.

The middle tagma, the thorax, is divided into three parts: the prothorax, mesothorax, and metathorax. The **prothorax** attaches to the head and bears the first pair of walking legs. The **mesothorax** bears the forewings and the second pair of walking legs. The **metathorax** attaches to the abdomen and bears the hindwings and the large jumping legs. A springlike mechanism inside the jumping legs stores mechanical energy when the legs are flexed. Release of this mechanism causes the legs to extend suddenly, launching the grasshopper into the air and away from danger. A flexible joint at the base of each leg provides the legs with great freedom of motion. Spines and hooks on the legs enable the grasshopper to cling to branches and blades of grass.

The leathery forewings cover and protect the membranous hindwings when the grasshopper isn't flying. Although the forewings help the grasshopper glide during flight, the hindwings actually propel it through the air. The wings are powered by muscles attached to the inside of the exoskeleton in the thorax. Note that insect wings develop as outgrowths from the epidermal cells that produce the exoskeleton. Unlike the wings of birds and bats, the wings of insects did not evolve from legs.

The segments in the most posterior tagma, the abdomen, are composed of upper and lower plates that are joined by a tough but flexible sheet of exoskeleton. The same flexible sheet also connects the segments to one another. The exoskeleton is covered by a waxy cuticle that is secreted by the cells of the epidermis. The rigid exoskeleton supports the grasshopper's body, and the cuticle retards the loss of body water. Both structures are adaptations for a terrestrial life.

internetconnect

SCILINKS NSTA

TOPIC: Grasshoppers
GO TO: www.scilinks.org
KEYWORD: HM765

FIGURE 39-3

The external anatomy of a grasshopper shows features that are characteristic of most insects: a body consisting of a head, thorax, and abdomen; a pair of unbranched antennae; three pairs of jointed legs; and two pairs of wings.

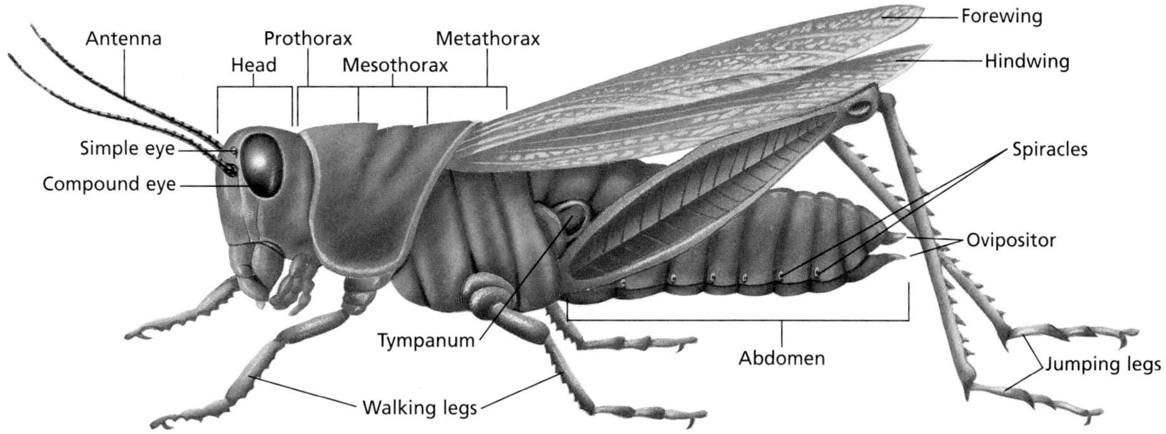

SECTION 39-1

TEACHING STRATEGY
Making Comparisons

Ask students why it is an advantage when discussing insect anatomy to describe a species of insect that is not highly specialized, rather than to make generalizations about all insects. (Great variety in the external anatomy of insects limits the generalizations that can be made to a very few, such as that insects have three body parts, one pair of antennae, and three pairs of jointed legs.) Ask students to give examples of the variety of insects' external characteristics. (Answers may include differences in color, shapes and sizes of wings, body shapes and sizes, and mouthparts.) Tell students that describing an insect, such as a grasshopper, provides a basis for making comparisons with other insects, rather than for making generalizations.

VISUAL LINK
Figure 39-3

Have students compare the external anatomy of the grasshopper, shown in Figure 39-3, with that of the crayfish, shown in Figure 38-7. Ask students to list the similarities and differences exhibited by these two arthropods. (Both animals have a segmented body divided into a head, a thorax, and an abdomen; both have jointed appendages and an exoskeleton. The head and thorax are separate in the grasshopper but fused in the crayfish. The grasshopper has three pairs of legs attached to the thorax; the crayfish has five. The grasshopper also has two pairs of wings arising from the thorax; the crayfish has none.)

TEACHING STRATEGY
Feeding Structures

Have students look at specimens or pictures of a butterfly, a mole cricket, and a praying mantis. Ask students to find structures on each insect that are likely to be used in feeding and to speculate on how they are used. (In butterflies, the maxillae are modified into a long, coiled proboscis that the butterfly uses to suck nectar from flowers. The mole cricket uses the shovel-like appendages on its forelegs to dig for the roots on which it feeds. The praying mantis uses the spines on its forelegs to grasp and immobilize other insects while it consumes them.)

DEMONSTRATION

Collect several mayfly or stonefly nymphs from beneath rocks in a local stream or pond and bring them to class. Place the nymphs in a shallow bowl of water, and place the bowl on an overhead projector. Turn on the projector and project the image for the class to see. Ask students to look for short, feathery gills that move rapidly back and forth on the legs, thorax, or abdomen of the nymphs. Ask students to explain the importance of gill movement. (Movement circulates oxygen-rich water over the gills.) Ask students to name the structures that will take over the function of the gills after metamorphosis (tracheae).

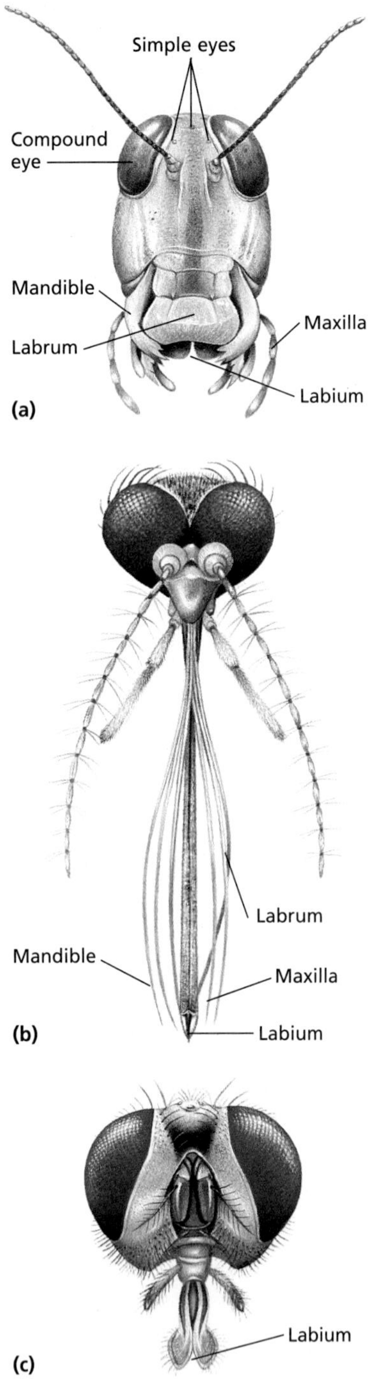

FIGURE 39-4

Insect mouthparts are adapted for different functions in different species. Mouthparts are used for biting and chewing in grasshoppers (a), piercing and sucking in mosquitoes (b), and sponging and lapping in houseflies (c).

Feeding and Digestion

Grasshoppers feed on plants. The mouthparts of grasshoppers, shown in Figure 39-4a, are modified for cutting and chewing leaves and blades of grass. The **labrum** and **labium** are mouthparts that function like upper and lower lips, respectively. They hold the food in position so that the sharp-edged mandibles can tear off edible bits. Behind the mandibles are the maxillae, which also help hold and cut the food.

The mouthparts of other insects are specialized for the types of food they eat, as you can see in Figures 39-4b and 39-4c. For example, mosquitoes have long, thin mouthparts that fit together to form a needle-like tube, which the females use to pierce the skin of a larger animal and suck up blood. The mouthparts of many flies, in contrast, are soft, spongelike lobes that soak up fruit juices and other liquids.

The structures that make up the digestive tract of a grasshopper are visible in Figure 39-5. Food that enters the mouth is moistened by saliva from the **salivary** (SAL-uh-VER-ee) **glands.** The moistened food then passes through the esophagus and into the crop for temporary storage. From the crop, food passes into the gizzard, where sharp, chitinous plates shred it. The shredded mass then enters a portion of the digestive tract called the **midgut.** There, the food is bathed in enzymes secreted by the **gastric ceca** (SEE-kuh), which are pockets that branch from the digestive tract. Nutrients are absorbed into the coelom through the wall of the midgut. Undigested matter travels into the posterior section of the digestive tract, the **hindgut,** and leaves the body through the anus.

Circulation, Respiration, and Excretion

Nutrients and other materials are transported through the body of a grasshopper by an open circulatory system like that of the crayfish. Hemolymph flows through a large dorsal vessel called the **aorta** (ay-OHR-tuh), which is shown in Figure 39-5. The muscular heart, which is located in the abdomen and thorax, pumps the hemolymph forward through the aorta and into the part of the coelom nearest the head. The hemolymph then percolates through the coelom toward the abdomen and reenters the heart through small pores along its length.

Unlike most other animals, insects do not use their circulatory system to transport oxygen and carbon dioxide. Instead, they exchange these gases with the environment by pumping air deep into their body through a complex network of air tubes called tracheae. Recall from Chapter 38 that tracheae are also used for this purpose in some spiders. In grasshoppers, air enters the tracheae through spiracles located on the sides of the thorax and abdomen. You can see the tracheae in Figure 39-5 and the spiracles in Figure 39-3. The ends of the tracheae branch near the cells of the body and are filled with fluid. Oxygen diffuses into the cells from this fluid while carbon dioxide diffuses in the reverse direction. Air can be pumped in and out of the tracheae by the movements of the abdomen and wings.

Another anatomical structure shared by insects and spiders is Malpighian tubules, which are excretory organs that collect water

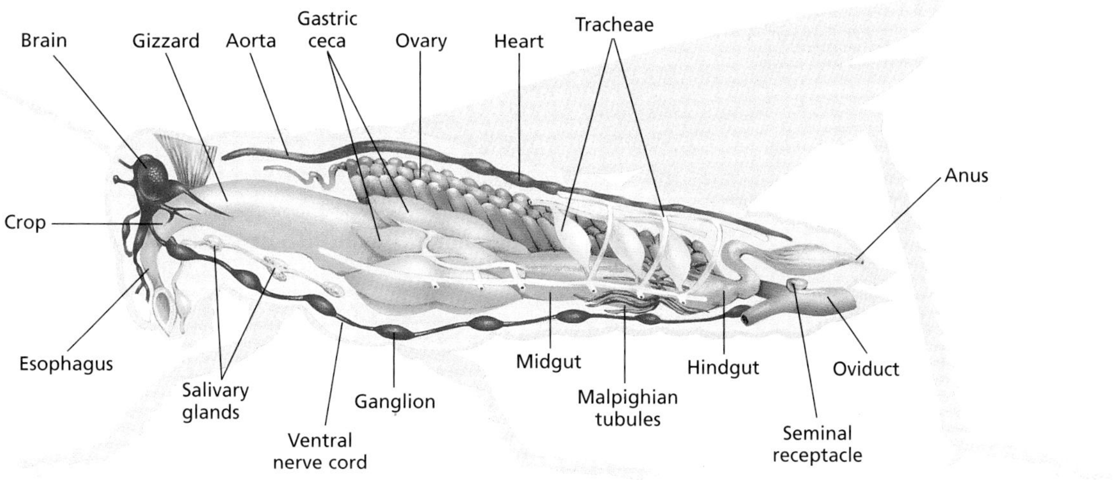

FIGURE 39-5

The major internal organs of a female grasshopper are seen in this cutaway side view.

and cellular wastes from the hemolymph. As Figure 39-5 shows, the Malpighian tubules in insects are attached to the digestive tract at the junction between the midgut and the hindgut. In insects that live in dry environments, the Malpighian tubules return most of the water to the hemolymph, producing a very concentrated mixture of wastes that is deposited in the hindgut and leaves the body with the feces. This is another method by which insects are adapted for life on land.

Neural Control

The grasshopper's central nervous system consists of a brain and a ventral nerve cord with ganglia located in each body segment. Nerves extend from the brain to the antennae, eyes, and other organs of the head. The antennae contain sensory structures that respond to both touch and smell. Look again at Figure 39-4a, and you can see that the three simple eyes are arranged in a row just above the base of the antennae. The simple eyes function merely to sense the intensity of light. Two bulging compound eyes, which are composed of hundreds of individual light detectors and lenses, allow the grasshopper to see in several directions at once. In addition to sensing light intensity, the compound eyes can detect movement and form images.

Other nerves extend from each of the ganglia to the muscles and sensory structures in the thorax and abdomen. One such structure is a sound-sensing organ called the **tympanum** (TIM-puh-nuhm). The tympanum is a large, oval membrane that covers an air-filled cavity on each side of the first abdominal segment. Sounds cause the tympanum to vibrate, and the vibrations are detected by nerve cells that line the cavity. Tympana are also found in many other insects that use sound in communication, such as crickets, katydids, and cicadas. In addition, sensory hairs like those described for crayfish in Chapter 38 are distributed over an insect's body. At the base of each hair is a nerve cell that is activated if the hair is touched or moved by vibration.

CRITICAL THINKING

Light Detection

If a grasshopper had a mutation that caused its compound eyes not to develop, how do you think that mutation would affect the grasshopper's ability to move from a dark to a light area? (That ability should be nearly normal, since a grasshopper can use its simple eyes to sense light intensity.)

RETEACHING *ACTIVITY*

Divide the class into two groups. Have one group make a detailed drawing of the external anatomy of a grasshopper, and have the other group do the same for the internal anatomy. Then have students from each group identify the major structures on the other group's drawing and explain the function of each structure.

GIFTED *ACTIVITY*

Have students write a short paper comparing incomplete metamorphosis with complete metamorphosis. Tell them to include advantages of complete metamorphosis and to provide an explanation of why metamorphosis is an adaptation that contributes to insects' success. Suggest that they consider why natural selection has led to so many insect species that undergo complete metamorphosis and so few that undergo incomplete metamorphosis.

Word Roots and Origins

ovipositor

from the Latin *ovum,* meaning "egg," and *positus,* meaning "to place"

Reproduction

Grasshoppers have separate sexes, as do all insects. During mating, the male deposits sperm into the female's seminal receptacle, where they are stored until the eggs are released by the ovaries. After release, the eggs are fertilized internally. The last segment of the female's abdomen forms a pointed organ called an **ovipositor** (OH-vuh-PAHZ-uht-uhr), which you can see in Figure 39-3. The female grasshopper uses her ovipositor to dig a hole in the soil, where she lays the fertilized eggs.

DEVELOPMENT

After hatching from the egg, a young insect must undergo several molts before it reaches its adult size and becomes sexually mature. Only silverfish and a few other insects go through this process without any change in body form. The majority of insects undergo some type of change in form as they develop into adults. This phenomenon of developmental change in form is called **metamorphosis** (MET-uh-MOHR-fuh-suhs). There are two main kinds of metamorphosis in insects: incomplete and complete.

Incomplete Metamorphosis

In **incomplete metamorphosis,** illustrated in Figure 39-6, a nymph hatches from an egg and gradually develops into an adult. A **nymph** is an immature form of an insect that looks somewhat like the adult, but it is smaller, and its wings and reproductive organs are undeveloped. The nymph molts several times. With each molt, the wings become larger and more fully formed. The final molt transforms the nymph into an adult that can reproduce and, in most species, fly. Insects that undergo incomplete metamorphosis include grasshoppers, mayflies, dragonflies, and termites. Several other examples are listed in Table 39-1 on pages 762–763.

FIGURE 39-6

In incomplete metamorphosis, shown here in a grasshopper, a nymph hatches from an egg and molts several times before becoming an adult. Nymphs resemble adults but are not sexually mature and lack functional wings.

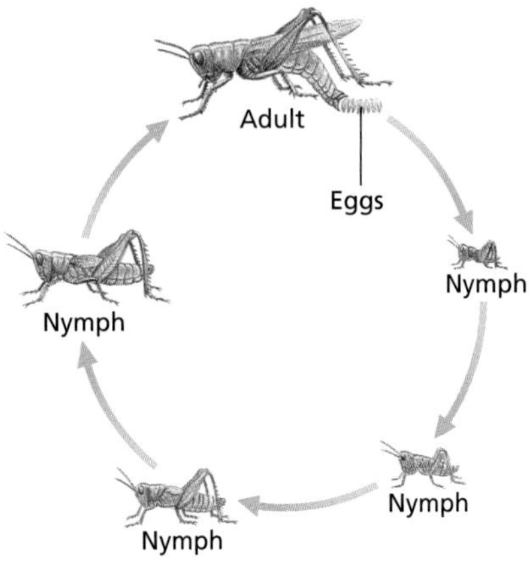

Complete Metamorphosis

In **complete metamorphosis,** an insect undergoes two stages of development between the egg and the adult. In both of those stages, the insect looks substantially different from its adult form. Figure 39-7 illustrates complete metamorphosis in the monarch butterfly. A wormlike larva, commonly called a caterpillar, hatches from the egg. The larva has three pairs of jointed legs on the thorax and several pairs of nonsegmented legs on the abdomen. The larva eats almost constantly, growing large on a diet of milkweed leaves. Thus, it is the larval stage of most insects that causes the most damage to plants.

The monarch larva molts several times as it grows. In the last larval stage, it develops bands of black, white, and yellow along its body. It continues to feed, but soon finds a sheltered spot and hangs upside down. Its body becomes shorter and thicker. Its exoskeleton then splits down the dorsal side, as shown in Figure 39-7, and falls off, revealing a green pupa. A **pupa** (PYOO-puh) is a stage of development in which an insect changes from a larva to an adult. The pupa of butterflies is enclosed in a protective case called a **chrysalis** (KRIS-uh-luhs). Moth pupae are enclosed in a case called a **cocoon.** Inside the pupa, the larval tissues break down, and groups of cells called imaginal disks develop into the wings and other tissues of the adult. When metamorphosis is complete, the pupa molts into a sexually mature, winged butterfly. Most insects go through complete metamorphosis. Table 39-1 lists several examples besides butterflies and moths, such as beetles, mosquitoes, and bees.

Importance of Metamorphosis

In a life cycle based on complete metamorphosis, the larval and adult stages often fulfill different functions, live in different habitats, and eat different foods. Therefore, the larvae and adults do not compete for space and food. For example, mosquito larvae live in fresh water and feed by filtering small food particles out of the water. When they become adults, the mosquitoes leave the water and feed on plant sap or the blood of terrestrial animals.

Metamorphosis also enhances insect survival by helping insects survive harsh weather. For instance, most butterflies and moths spend the winter as pupae encased in chrysalises or cocoons, which are often buried in the soil.

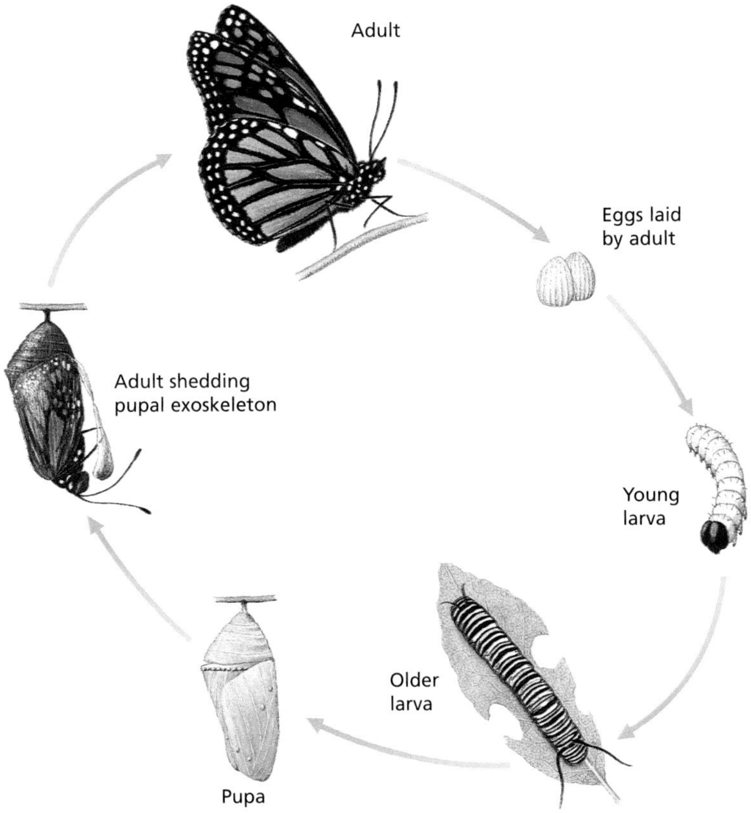

FIGURE 39-7

In complete metamorphosis, shown here in the monarch butterfly, a larva hatches from an egg and goes through several molts before becoming a pupa, which then develops into an adult. Neither the larva nor the pupa resembles the adult.

Eco Connection

Biological Control of Insects

Humans have been competing with insects for food since the invention of agriculture. To limit the damage that insects do to our food crops, we have developed a variety of poisons that kill a broad range of insects. However, these poisons have some serious drawbacks: they kill beneficial as well as harmful insects; they persist in the environment, accumulating in animals at higher levels in the food web; and they select for strains of insects that are resistant to the poisons. These drawbacks have led scientists to develop biological controls of insect pests.

One type of biological control is the use of natural predators or parasites that attack specific kinds of insects. For example, the bacterium *Bacillus thuringiensis* is used to control cabbage worms, tomato worms, and other moth larvae.

Biological control also includes methods that interfere with the reproduction of insects. In the sterile-male approach, for instance, large numbers of male insects sterilized by radiation are introduced into an area. The females lay eggs that never develop, so the next generation is smaller. Used over several generations, this technique can nearly eliminate some pest species in selected areas.

INCLUSION ACTIVITY

Show students photographs of different insects in various stages of complete or incomplete metamorphosis. Have each student draw and label the stages, and identify particular insect features.

CONTENT CONNECTION

Section 38-1: Molting

The hormone that triggers molting in insects also triggers molting in all other arthropods. In insects that undergo complete metamorphosis, another hormone controls whether each molt will produce a larva, a pupa, or an adult.

RECENT RESEARCH

Wasp-Corn Alliance

Scientists have recently discovered how corn plants defend themselves against one species of herbivorous insect. The beet armyworm caterpillar releases a chemical called volicitin when it feeds on corn seedlings. Volicitin causes the corn plant to emit volatile compounds that attract parasitic wasps, which are natural enemies of the caterpillar.

SECTION 39-1

QUICK FACT

☑ The rhinoceros beetle can support 100 times its own body weight.

ANSWERS TO SECTION 39-1 REVIEW

1. Characteristics include a segmented body that is divided into a head, thorax, and abdomen; mandibles; one pair of unbranched antennae; three pairs of jointed legs; and, in many species, one or two pairs of wings.
2. Adaptations include a segmented body, jointed appendages, exoskeleton, ability to fly, small size, short life span, life cycles that involve metamorphosis, and use of defenses such as camouflage, warning coloration, and mimicry.
3. Harmful effects include eating crops; spreading diseases; and destroying buildings, clothing, and carpets. Beneficial effects include serving as food for other animals; pollinating flowers; making honey, wax, silk, and shellac; and recycling nutrients.
4. The labrum helps hold food in position while the mandibles tear off pieces. The tympanum picks up sound vibrations, enabling the grasshopper to hear. The ovipositor is used by the female to dig a hole in the soil, into which eggs are laid.
5. In incomplete metamorphosis, a nymph hatches from an egg and gradually develops into an adult. The nymph looks somewhat like the adult. In complete metamorphosis, a wormlike larva hatches from an egg and eventually transforms into a pupa. The pupa then molts to produce an adult, which resembles neither the larva nor the pupa.
6. Viceroys would exhibit Batesian mimicry.

DEFENSE

Insects have many defensive adaptations that increase their chances for survival. Some adaptations provide a passive defense. One form of passive defense that is frequently used by insects is camouflage, which was discussed in Chapter 21. Camouflage enhances survival by making it difficult for predators to recognize an insect. Insects often resemble parts of the plants on which they feed or hunt for food. For example, many varieties of stick insects and mantises look so much like twigs or leaves that they are easy to overlook unless they move.

Other defensive adaptations of insects are more aggressive, such as the venomous stingers of female bees and wasps. One of the most elaborate adaptations is that of the bombardier beetle, which defends itself by spraying a hot stream of a noxious chemical. The beetle can even rotate an opening on its abdomen to aim the spray at an attacker.

Insects that defend themselves by being dangerous or poisonous or by tasting bad often have bold, bright color patterns that make them clearly recognizable and warn predators away. This type of coloration is known as **warning coloration.** In some cases, several dangerous or poisonous species have similar patterns of warning coloration. For example, many species of stinging bees and wasps display a pattern of black-and-yellow stripes. This adaptation, in which a member of one dangerous species mimics the warning coloration of another, is called **Müllerian** (myoo-LER-ee-uhn) **mimicry.** Figure 39-8 shows that the black-and-yellow stripes of bees and wasps are also shared by some species of flies, which lack stingers and are therefore harmless. Mimicry of this type, in which a harmless species mimics the warning coloration of a dangerous species, is called **Batesian** (BAYTZ-ee-uhn) **mimicry.** Both Müllerian and Batesian mimicry encourage predators to avoid all similarly marked species.

FIGURE 39-8

Batesian mimicry is shown by the harmless syrphid fly, *Arctophila* (a), which looks very similar to the stinging bumblebee, *Bombus* (b).

(a)

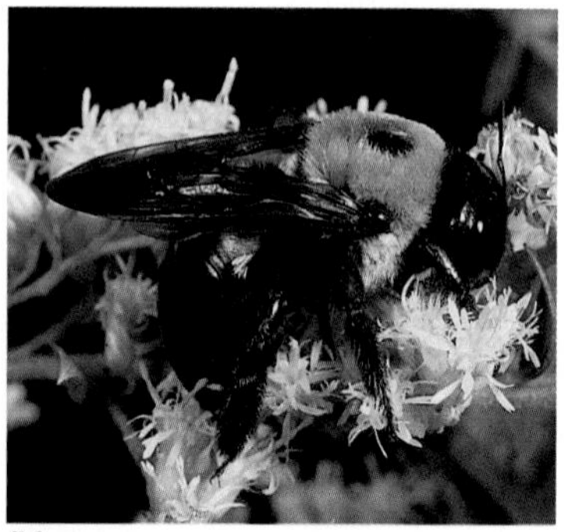

(b)

SECTION 39-1 REVIEW

1. What are the major characteristics of the class Insecta?
2. What are some adaptations that have allowed insects to become successful?
3. List two ways insects are harmful to society and two ways they are beneficial.
4. State the function of each of the following parts of a grasshopper: labrum, tympanum, ovipositor.
5. What are the differences between incomplete and complete metamorphosis?
6. CRITICAL THINKING The monarch butterfly and the viceroy butterfly have similar bright colors and markings, and birds generally avoid eating both butterflies. If monarchs were distasteful to birds but viceroys were not, what type of defensive mechanism would viceroys exhibit?

SECTION

39-2

OBJECTIVES

▲ Name three ways insects communicate, and give an example of each.

● Describe the social organization of honeybees.

■ Explain how honeybees communicate information about the location of food.

INSECT BEHAVIOR

One reason for the success of insects is their ability to engage in complex behaviors. This ability is made possible by insects' jointed appendages, elaborate sense organs, and relatively complex brains. Insects are capable of interpreting sensory information to escape from predators, find food and mates, and communicate with one another.

COMMUNICATION

One of the most common forms of communication among insects is chemical communication involving pheromones. A **pheromone** (FER-uh-MOHN) is a chemical released by an animal that affects the behavior or development of other members of the same species through the sense of smell or taste. Pheromones play a major role in the behavior patterns of many insects. For example, you may have noticed ants, like those in Figure 39-9, marching along a tightly defined route on the ground. The ants are following a trail, of pheromones left by the ants that preceded them. Such trails are often laid down by ants that have found a source of food as they make their way back to the nest. As other ants follow the trail, they too deposit pheromones, so the trail becomes stronger as more ants travel along it. Pheromones are also used by honeybees to identify their own hives and to recruit other members of the hive in attacking animals that threaten the hive.

Some insects secrete pheromones to attract mates. The female silkworm moth, for example, can attract males from several kilometers away by secreting less than 0.01 μg of a pheromone. Sensory hairs on the large antennae of a male moth make it exquisitely sensitive to the pheromone.

Many insects communicate through sound. Male crickets produce chirping sounds by rubbing a scraper located on one forewing against a vein on the other forewing. They make these sounds to attract females and to warn other males away from their territories. Each cricket species produces several calls that differ from those of other cricket species. In fact, because many species look similar, entomologists often use a cricket's calls rather than its appearance to identify its species. Mosquitoes communicate through sound, too. Males that are ready to mate fly directly to the buzzing

FIGURE 39-9

These leaf-cutter ants, *Atta colombica*, are following a pheromone trail as they carry sections of leaves to their underground nest.

SECTION 39-2

SECTION OVERVIEW

Interdependence of Organisms
Section 39-2 explains how insects communicate with members of their own species by releasing pheromones, producing sounds, and generating light. This section also describes the highly structured interdependence of individuals in a honeybee colony.

Reproduction and Inheritance
Section 39-2 discusses kin selection as a mechanism by which honeybees increase the propagation of their own genes by helping closely related individuals reproduce.

▶ ENGAGE STUDENTS

Do an experiment to demonstrate how ants use pheromones to mark trails. Lay a sheet of paper on the ground, with the top edge of the paper next to the entrance to an ant nest. Place a cookie or a piece of candy on the ground next to the bottom edge of the paper. About 10 to 15 minutes after the first ants find the food, remove the paper, shake off any ants that are attached to it, and place it on the opposite side of the nest. Watch and explain the behavior of the ants. (At first, many ants will follow the pheromone trail to the end of the paper, thus moving away from the food. After a few minutes, some ants will rediscover the food, and a new pheromone trail will be established.)

QUICK FACT

☑ The pheromone released by the female gypsy moth can trigger behavioral responses in male gypsy moths at concentrations as low as 1 molecule of pheromone per 10^{17} molecules of air.

DEMONSTRATION

Purchase some crickets from a local pet store and separate them into individual containers. By listening for chirping sounds coming from specific containers, you can determine which containers have male crickets, since only males chirp. In class, allow students to listen to the crickets. Then immobilize a male cricket with sticky wax, place it on the stage of a dissecting microscope, and look for the file-and-scraper mechanism that produces these sounds. The file is a row of teeth on one of the veins on the underside of each forewing. The scraper is a hard ridge of exoskeleton on the inner edge of each forewing.

TEACHING STRATEGY

Insect Communication

Ask students to observe live ants and to determine if the ants are communicating with some kind of body language. Have students list the behaviors that indicate communication between individuals. (Behaviors could include feeling each other with antennae or other body parts, running around in deliberate paths, and repeating an action.)

VISUAL LINK

Figure 39-10

The three types of individuals shown in the figure demonstrate the importance of both genetic and environmental influences on the development of organisms. Eggs that are not fertilized produce males, which develop into drones. Eggs that are fertilized produce females, which can develop into either queens or workers. Queens and workers differ in their structure, reproductive ability, and behavior, and these differences are due entirely to the royal jelly that the queen is fed as a larva.

sounds produced by females. A male senses the buzzing by means of sensory hairs on his antennae that vibrate only at the frequency produced by females of the same species.

Insects may also communicate by generating flashes of light. Fireflies, for example, use light to find mates. Males emit flashes in flight, and females flash back in response. Each species of firefly has its own pattern of flashes, which helps males find females of the same species.

BEHAVIOR IN HONEYBEES

internetconnect

SCILINKS NSTA

TOPIC: Honeybees
GO TO: www.scilinks.org
KEYWORD: HM772

Some insects, such as certain species of bees, wasps, ants, and termites, live in complex colonies. In these colonies, some individuals gather food, others protect the colony, and others reproduce. Insects that live in such colonies are called **social insects.** The division of labor among social insects creates great interdependence and a heightened need for communication. This section will look at the behavioral adaptations of one well-studied species of social insect, the honeybee.

As you read about the complex behaviors of honeybees, keep in mind that these behaviors are neither taught nor learned. Instead, they are genetically determined. That is, the behaviors of honeybees are programmed by instructions stored in their genes. Genetically determined behavior is called **innate behavior.**

FIGURE 39-10

In a honeybee colony, worker bees perform the work of the hive, while drones and the queen are involved exclusively with reproduction.

A honeybee colony consists of three distinct types of individuals, which are illustrated in Figure 39-10: worker bees, the queen bee, and drones. **Worker bees** are sterile females that make up the vast majority of the hive population, which may reach more than 80,000. The workers perform all the duties of the hive except reproduction. The **queen bee** is the only fertile female in the hive, and her only function is to reproduce. **Drones** are males that develop from unfertilized eggs. Their sole function is to deliver sperm to the queen. Their mouthparts are too short to obtain nectar from flowers, so the workers must feed them. The number of drones in the hive may reach a few hundred during the summer, but when the honey supply begins to run low, the workers kill the drones and clear them from the hive.

Worker Bees

Worker bees perform many functions at different times during their brief lifetime, which lasts about six weeks. After making the transition from pupa to adult, workers feed honey and pollen to the queen, drones, and larvae. During this stage the workers are called nurse bees. They secrete a high-protein substance known as **royal jelly,** which they feed to the queen and youngest larvae.

After about a week, worker bees stop producing royal jelly and begin to secrete wax, which they use to build and repair the honeycomb. During this stage they may also remove wastes from the hive, guard the hive, and fan their wings to circulate air through the hive.

The workers spend the last weeks of their life gathering nectar and pollen. A number of structural adaptations aid them in this work. Their mouthparts are specialized for lapping up nectar, and their legs have structures that serve as pollen packers, pollen baskets, and pollen combs.

The sterile workers do not use their ovipositors for egg laying. Instead, these structures are modified into barbed stingers that the workers use to protect the hive. When a worker bee stings an animal, the stinger and attached venom sac are left behind in the victim as the bee flies away. The worker, having lost part of its body and much of its hemolymph, dies a day or two later. Wasps also have stingers that are modified ovipositors. Unlike honeybees, they can sting many times because their stinger is not barbed.

The Queen Bee

The queen bee develops from an egg identical to those that develop into the workers. The differences between the queen and the workers result from the continuous diet of royal jelly that the queen is fed throughout her larval development. In addition, the queen herself secretes a pheromone called the **queen factor** that prevents other female larvae from developing into queens.

The queen's role is to reproduce. Within a few days after she completes metamorphosis and emerges as an adult, she flies out of the hive and mates in the air with one or more drones. During mating, millions of sperm are deposited in the queen's seminal receptacle, where they will remain for the five or more years of her life. Although the queen mates only once, she may lay as many as a million eggs each year.

When the hive becomes overcrowded, the queen stops producing the queen factor and leaves the hive. As she leaves, she secretes a swarming pheromone that induces about half of the workers in the hive to follow her and form a swarm. Eventually the swarm finds another location for a new hive. Meanwhile, in the old hive, the remaining workers begin feeding royal jelly to other larvae. When a new queen emerges, she produces the queen factor, and in response, the workers destroy the other developing queens. The new queen departs on a mating flight, and the cycle begins again.

The Dances of the Bees

When honeybees leave the hive and find a source of pollen and nectar, how do they communicate the location of this food source to other workers in the hive? An Austrian biologist, Karl von Frisch (1886–1982), spent 25 years answering this question. His careful experimentation earned him a Nobel Prize in 1973.

To study bees, von Frisch built a glass-walled hive and placed feeding stations stocked with sweetened water near the hive. He noted that "scout bees" returning from the feeding stations would perform a series of dancelike movements in the hive. As shown in Figure 39-11a, the scout bee would circle first to the right and then to the left, a behavior that von Frisch called the **round dance.** After

Quick Lab

Interpreting Nonverbal Communication

Materials pencil, paper, wrapped candy pieces

Procedure

1. Choose one member of your group to play the part of the "scout" bee. The others in the group will be the "worker" bees.
2. Your teacher will secretly tell the scout bee where a piece of candy is located. The scout bee will develop a method of nonverbal communication to let the worker bees know where the candy is hidden. The scout bee may not point to the candy. Use Figure 39-11 as a guide to developing your method of communication.
3. When the candy has been located, the scout will hide another piece of candy and select a new scout. The new scout will develop a different way to tell the group where the candy is located. Repeat the procedure until everyone in your group has been a scout.

Analysis

1. How effective were each of your scout's methods for showing the location of the candy?
2. Did the worker bees improve their ability to find the candy after several trials?
3. List and describe some types of nonverbal communication that humans use.

Quick Lab

Interpreting Nonverbal Communication

Time Required 30 minutes

Procedural Tips As an alternative to using candy, you might want to use a small prize or an object in the classroom for students to locate. Place the candy or object in unusual places in the classroom or outside the classroom, if possible. Inform the scout bee of the location of the candy. Instruct the scout not to speak or directly indicate the location of the candy in any way. Suggest that scouts mimic the round and waggle dances of the bees, shown in Figure 39-11.

Answers to Analysis Answers will vary according to each team's results. The questions are open-ended to stimulate discussion and inquiry. The team's ability to find the candy should improve after repeated trials because they will be learning and practicing nonverbal communication skills. Humans use many kinds of nonverbal communication methods. Students will list and describe hand gestures, such as a thumbs-up sign for congratulations or approval. Students should list and describe a sign language, such as American Sign Language. Students might also list flag signals, traffic signals and symbols, lighthouse and foghorn warnings, and Morse code.

CONTENT CONNECTION

Section 8-1: Chromosome Number

Queen bees and worker bees develop from fertilized eggs and thus are diploid, whereas drones develop from unfertilized eggs and are haploid.

SECTION 39-2

RECENT RESEARCH

Waggle Dances

Researchers have recently found that scout honeybees perform their dances at particular sites in the hive. If an established dance site is moved, the scout bees will seek out the site before beginning their dance. The researchers concluded that dance sites are labeled in some way.

ANSWERS TO SECTION 39-2 REVIEW

1. Methods of communication include pheromones, such as those that guide ants along trails, help honeybees identify their own hives, and attract mates in moths; sound, which attracts mates in crickets and mosquitoes; and light flashes, which attract mates in fireflies.
2. A pheromone is a chemical released by an animal that affects the behavior or development of other animals of the same species.
3. If the larva from a fertilized egg is fed royal jelly throughout its development, it will become a queen; if it is not, it will become a worker.
4. The queen leaves the hive and secretes a pheromone that causes many workers to leave with her to find a location for a new hive.
5. Scout bees perform a waggle dance, in which the direction of food is indicated by the angle of the straight run and the distance to the food is indicated by the duration of the dance and the number of waggles on each run.
6. Mating with drones from other hives provides a source of genetic diversity; drones from the queen's own hive share all of their genes with her, since they develop from unfertilized eggs laid by the queen herself.

(a) ROUND DANCE

(b) WAGGLE DANCE

FIGURE 39-11

Honeybees use two types of dances to convey information about food sources. (a) The round dance indicates that a food source is nearby but gives no information about its direction or distance. (b) The waggle dance is performed on a vertical surface in the hive when the food source is distant. The angle of the straight run part of the dance indicates the direction of food, while the duration of the dance and number of waggles indicate the food's distance from the hive.

many observations, von Frisch concluded that the round dance told other worker bees that a food source was near the hive, but it did not inform them of the exact location of the food.

Von Frisch also observed that when the food source was far from the hive, the scout bees would perform another type of dance on a vertical surface inside the hive. He called this dance the **waggle dance** because the scout bees waggled their abdomens from side to side. As you can see in Figure 39-11b, the pattern of the waggle dance is like a figure eight. The scout bee makes a circle in one direction, then a straight run while waggling her abdomen, and then another circle in the opposite direction from the first. Numerous experiments by von Frisch showed that the direction of food is indicated by the angle of the straight run on the vertical surface. Straight up, for example, indicates a direction toward the sun. The distance to the food source is indicated by the duration of the dance and the number of waggles on each run.

Altruistic Behavior

When worker bees sting an intruder to defend the colony, they cause their own deaths, as you learned earlier. This behavior is an example of **altruistic** (AL-troo-IS-tik) **behavior,** which is the aiding of other individuals at one's own risk or expense. The stinging of honeybees is an innate behavior. You might think that the genes directing this behavior would eventually be eliminated from the population, since dead bees can't reproduce. However, this does not happen.

To understand why evolution has selected for altruistic behavior in honeybees, you must focus not on the selection of individual bees, but rather on the selection of genes they possess. Remember that worker bees are sterile. Therefore, they cannot pass on their own genes by reproducing. However, they can pass on some of their genes by helping a closely related individual reproduce. By defending the colony, a worker bee increases the chances that the queen bee will survive. If the queen survives, she will produce more workers who will share many of the same genes. Thus, by behaving altruistically, a worker can cause more of her genes to be propagated in the population. This mechanism of increasing the propagation of one's own genes by helping a closely related individual reproduce is called **kin selection.**

SECTION 39-2 REVIEW

1. Name three ways insects communicate, and give an example of each.
2. What is a pheromone?
3. What determines whether a fertilized honeybee egg will develop into a worker or a queen?
4. How do honeybees behave when their hive is overcrowded?
5. How do honeybees convey information about the direction and distance of a food source that is far from the hive?
6. **CRITICAL THINKING** Current research indicates that a queen honeybee mates with drones from another hive rather than from her own hive. In what way might this behavior benefit the honeybee colony?

CHAPTER 39 REVIEW

SUMMARY/VOCABULARY

39-1

- The insect body is divided into three tagmata: the head has mandibles and one pair of unbranched antennae; the thorax has three pairs of jointed legs and, in many species, one or two pairs of wings; the abdomen has 9 to 11 segments but neither wings nor legs in adults.
- Insects live in almost every terrestrial and freshwater environment. Factors responsible for their success include their ability to fly, exoskeleton, jointed appendages, small size, and short life span.
- Insects negatively affect humans by competing for food, transmitting diseases, and destroying buildings and other manufactured products. However, insects are also beneficial: they serve as food for other animals, pollinate flowers, make valuable products such as honey, and recycle nutrients.
- The mouthparts of an insect are specialized for tearing and cutting solid food or for sucking or soaking up liquid food.
- Insects have an open circulatory system that transports nutrients through the body. Gas exchange occurs by means of air-filled tracheae that reach deep into the body. Malpighian tubules remove cellular wastes from the hemolymph while conserving water.
- Insect sensory structures include simple and compound eyes, sound-sensing tympana in some species, and sensory hairs on the antennae and other body parts.
- Most insects go through a metamorphosis. In incomplete metamorphosis, a nymph hatches from an egg and resembles the adult but has undeveloped reproductive organs and no wings. The nymph molts several times to become an adult. In complete metamorphosis, a wormlike larva hatches from an egg and molts several times before becoming a pupa. The pupa molts to produce the adult, which resembles neither the larva nor the pupa.
- Insects can defend themselves by using camouflage, stinging, or releasing noxious chemicals. Insects that are dangerous or taste bad often have warning coloration that makes them recognizable to predators. The warning coloration of a dangerous species may be mimicked by other species.

Vocabulary

aorta (766)
Batesian mimicry (770)
chrysalis (769)
cocoon (769)
complete metamorphosis (768)
entomology (762)
gastric cecum (766)
hindgut (766)
incomplete metamorphosis (768)
labium (766)
labrum (766)
mesothorax (765)
metamorphosis (768)
metathorax (765)
midgut (766)
Müllerian mimicry (770)
nymph (768)
ovipositor (768)
prothorax (765)
pupa (769)
salivary gland (766)
tympanum (767)
warning coloration (770)

39-2

- Insects communicate by releasing pheromones and by producing sounds and flashes of light.
- Honeybees live in complex colonies consisting mostly of sterile female workers that perform all duties except reproduction. Each colony also contains one fertile queen bee and a few hundred male drones.
- Honeybees communicate the direction and distance to food sources by performing dances inside the hive.
- In defending the colony, worker bees show altruistic behavior toward their close relatives in the colony. By doing so, they increase the propagation of their own genes.

Vocabulary

altruistic behavior (774)
drone (772)
innate behavior (772)
kin selection (774)
pheromone (771)
queen bee (772)
queen factor (773)
round dance (773)
royal jelly (772)
social insect (772)
waggle dance (774)
worker bee (772)

CHAPTER 39 REVIEW ANSWERS

REVIEW

1. Labium does not belong because it's a mouthpart rather than a developmental stage.
2. The spiracles admit air into the grasshopper's tracheae.
3. The three tagmata are the head, thorax, and abdomen.
4. An innate behavior is a behavior programmed by instructions stored in genes. Examples include the round and waggle dances of honeybees as well as any other behavior of insects.
5. A chrysalis is another name for the pupa of a butterfly.

6. d **7.** c
8. d **9.** b
10. a **11.** d
12. c **13.** a
14. c **15.** b

16. Small size allows several species to inhabit different environments within an area without competing with one another.
17. Characteristics include the kind of mouthparts, number of wings, and type of development.
18. Insects consume crops; spread disease; and damage buildings, clothing, and carpets.
19. A tympanum is an organ that is sensitive to sound vibrations. It is located on each side of the first abdominal segment.
20. Insect wings develop from cells that produce the exoskeleton, but they did not replace legs during evolution. The wings of birds and bats develop from structures that would have formed legs.
21. Camouflage allows animals to hide from predators by blending in with their surroundings. Warning coloration causes animals to stand out, making it more likely that predators will associate them with danger or an unpleasant taste.
22. Crickets produce sounds by rubbing a scraper on one forewing against a vein on the other forewing. These sounds are used

to attract mates and protect territories.

23. The three types are workers, the queen, and drones.

24. By defending the colony, a sterile worker bee increases the chances that the queen bee will survive and produce more workers, who will share many of the same genes. Thus, a worker can cause more of her genes to be propagated in the population. Some of these genes are responsible for programming the defensive behavior.

25. *A,* gastric ceca; *B,* tracheae; *C,* seminal receptacle; *D,* Malpighian tubules; *E,* ventral nerve cord; *F,* salivary glands.

Critical Thinking

1. Insects with genes that confer resistance to an insecticide survive and reproduce in large numbers. Thus, they proliferate through natural selection (although the insecticide may be synthetic). This process occurs rapidly because insects have short life spans and produce large numbers of eggs.

2. Both crustaceans and insects have exoskeletons. Large crustaceans live underwater, and the water helps support their weight, taking some of the load off their exoskeleton. The largest insects are terrestrial, and the ability of their exoskeleton to support their body weight puts a much lower limit on their size.

3. The circulatory system of a cephalopod carries oxygen to and carbon dioxide away from the animal's active cells. Insects use tracheae to transport these gases, so they can remain active with a less-efficient circulatory system.

4. If a worker bee dies from losing its stinger while defending the hive, it will have already completed most of its other functions. It would be a greater loss to the hive if death occurred early in the worker's life.

Review

Vocabulary

1. Choose the term that does not belong in the following group, and explain why it does not belong: egg, larva, pupa, labium.

2. What is the function of the spiracles on a grasshopper?

3. Name the three tagmata that make up the body of an insect.

4. Define *innate behavior,* and give an example.

5. What is a chrysalis?

Multiple Choice

6. Adaptations responsible for the success of insects include (a) jointed appendages (b) the ability to fly (c) short life spans (d) all of the above.

7. The mouthparts of a grasshopper are specialized for (a) sucking fluids (b) lapping up liquids (c) cutting and tearing (d) filter feeding.

8. A female grasshopper uses her ovipositor to (a) store sperm from a male (b) inject venom (c) hold food in position during feeding (d) dig holes in the soil and lay eggs.

9. An insect's legs and wings are attached to its (a) head (b) thorax (c) abdomen (d) labrum.

10. The function of the gastric ceca is to (a) secrete digestive enzymes (b) secrete saliva (c) store undigested food (d) remove cellular wastes.

11. The immature form of an insect that undergoes incomplete metamorphosis is called a (a) drone (b) pupa (c) caterpillar (d) nymph.

12. Metamorphosis benefits insect species by (a) ensuring that different developmental stages fulfill the same functions (b) increasing competition between larvae and adults (c) promoting survival during harsh weather (d) all of the above.

13. Bees perform the round dance to inform other bees of (a) the presence of food near the hive (b) the distance to a food source from the hive (c) the direction of a food source from the hive (d) the type of food they have located.

14. The substance called queen factor (a) causes female larvae to develop into queen bees (b) causes worker bees to form swarms (c) prevents female larvae from becoming sexually mature (d) attracts drones.

15. Altruistic behavior is (a) behavior that benefits oneself at the expense of others (b) the aiding of other individuals at one's own risk or expense (c) behavior that is always genetically determined (d) none of the above.

Short Answer

16. How has the small size of insects contributed to their success?

17. Name three characteristics that entomologists use to divide insects into orders.

18. What are three ways that insects cause harm to human society?

19. What is a tympanum, and where is it located on a grasshopper?

20. How is the origin of wings in insects different from that in birds and bats?

21. What is the difference between camouflage and warning coloration?

22. How do crickets produce sounds? What functions do these sounds serve?

23. What three types of individuals make up a honeybee society?

24. When a worker bee stings to defend the hive, it loses its stinger and dies. Explain how such behavior has been selected in honeybees.

25. Identify the structures labeled *A* through *F* in the diagram below.

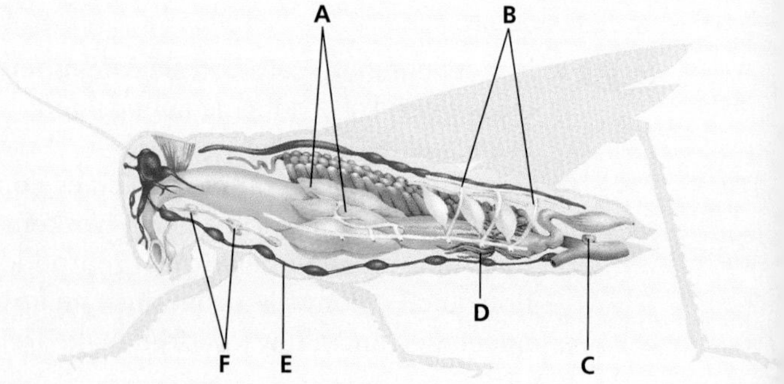

CRITICAL THINKING

1. Farmers who use an insecticide to control insect pests often find that strains of insects resistant to the insecticide rapidly appear. How do these insecticide-resistant strains develop? Why does this happen so quickly among insects?
2. Insects and crustaceans both belong to the phylum Arthropoda and share many characteristics. As you learned in Chapter 38, the largest crustacean, the Japanese spider crab, has a leg span of 4 m. In contrast, the largest insects, such as the atlas moth, have a wingspan of only about 25 cm. Why are the largest crustaceans so much bigger than the largest insects?
3. You learned in Chapter 37 that squids and other cephalopods have a closed circulatory system, which supports their active lifestyle by circulating blood quickly through their bodies. Many insects, such as dragonflies and bees, are also very active, but all insects have an open circulatory system. How can insects maintain active lifestyles while having an open circulatory system?
4. During their lives, worker bees move from first feeding and caring for larvae inside the hive to later gathering nectar and defending the hive. Why do you think this sequence is more advantageous for the hive than the reverse sequence might be?
5. What characteristics may have helped insects survive the major climatic changes that led to the extinction of the dinosaurs and many other species about 65 million years ago?
6. Worker honeybees guard the entrance to their hive, allowing nestmates to enter but turning away intruder bees. The graph below plots the percentage of bees turned away versus the genetic relatedness of the guards and the intruders. A genetic relatedness of 1.0 means that a guard and an intruder are genetically identical. Are guards more likely to turn away bees with a high or a low genetic relatedness? How might a guard determine the genetic relatedness of a bee that enters the hive?

Hive Guarding in Honeybees

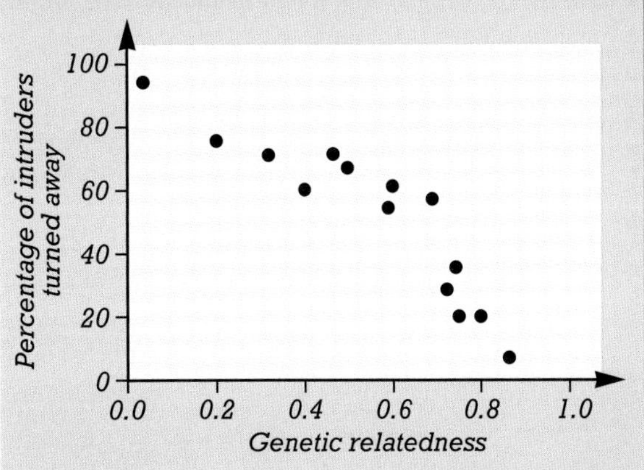

EXTENSION

1. Read “Pretty Poison” in *Audubon,* September/October 1999, on page 159. Explain how the monarch butterfly becomes toxic to bird predators. Describe the migration of the monarch butterfly.
2. Read “Child Care Among the Insects” in *Scientific American,* January 1999, on page 72. Why do some insects risk their own lives to protect their young? Describe how the male giant water bug takes care of his young.
3. Select any insect that lives in your area. Using a field guide, identify the insect. Make a drawing that shows the insect’s characteristics. Use your library and other sources to find information on the insect’s food sources, economic importance, and distribution.
4. Use your school library or public library to research the topic of sociobiology. Prepare a written report on what sociobiologists mean by *altruistic behavior* and how they explain such behavior in insect societies.

ANSWERS

CONTINUED

5. Characteristics include insects’ short life span, rapid reproduction with the production of many eggs, metamorphosis, and structural and behavioral adaptations. All of these characteristics allow insects to adapt rapidly to new and changing environments.
6. Guards are more likely to turn away bees with a low genetic relatedness. Guards could determine the genetic relatedness of other bees by comparing its similarity to themselves in terms of characteristics that are genetically determined, such as specific pheromones and behaviors.

EXTENSION

1. Monarch caterpillars feed almost exclusively on milkweed, which is poisonous to insects and birds. The adult monarch butterflies remain toxic even after metamorphosis. Monarchs migrate each October from Canada to central Mexico or coastal California. Very few make it back to Canada. They “breed” their way home to Canada as successive generations follow the north-moving direction of sprouting milkweed.
2. Some insect parents protect their young to ensure their survival in harsh conditions. Some lay eggs only once in a lifetime and have only one chance to reproduce. The female giant water bug glues her eggs on the male’s back where he keeps the eggs moist.
3. If the insect is a pest or feeds on crops, students could also write about methods of controlling the insect.
4. Direct students to include a comparison of altruistic behavior in insects and humans. Students can explore the altruistic behavior of ants through a computer program called SimAnt™.

CHAPTER 39 INVESTIGATION

TIME REQUIRED

One 50-minute class period

SAFETY PRECAUTIONS

Caution students to wear safety goggles and gloves when handling the preserved specimens. Instruct students to use care when handling the dissection scissors and sharp probe.

QUICK REFERENCES

Lab Preparation Notes are found on page 759B.

Holt BioSources provides a Teaching Resources CD-ROM that contains "Using Gowin's Vee in the Lab" and "Scoring Rubrics."

PROCEDURAL TIPS

1. Materials for this investigation can be purchased from WARD'S Natural Science Establishment, Inc.
2. Emphasize to students that dissection does not simply mean "cutting and slicing." Rather, it means making careful incisions to expose parts and then using a probe to separate organs from their coverings. Point out that the intent is to carefully unwrap the animal's structures without damaging them.
3. To make certain structures more visible, students may gently add enough water to the dissection tray to cover the specimen.
4. At the end of the investigation, wrap the remains of dissected grasshoppers in an old newspaper, place the bundle in a plastic bag, and tie the bag securely. Inform the school custodian of the bag's contents and hand the bag over personally for safe disposal.

ANSWERS TO BACKGROUND

1. Insects have a body divided into head, thorax, and abdomen;

CHAPTER 39 INVESTIGATION

Anatomy of a Grasshopper

OBJECTIVES

- Examine the external and internal anatomy of a grasshopper.
- Infer function from observation of structures.

PROCESS SKILLS

- relating structure and function
- observing

MATERIALS

- safety goggles
- gloves
- preserved grasshopper (1 for each student)
- dissection tray
- forceps
- fine dissection scissors, with pointed blades
- hand lens or dissecting microscope
- blunt probe
- sharp probe
- dissection pins

Background

1. List the distinguishing characteristics of insects.
2. In this investigation, you will dissect a grasshopper to observe its external and internal structure. To which order of insects do grasshoppers belong?
3. What characteristics of the grasshopper place it into this order?

PART A Observing the External Anatomy of the Grasshopper

1. CAUTION **Wear safety goggles and gloves during this investigation. Keep your hands away from your eyes and face when working with preserved specimens.** Using forceps, hold a preserved grasshopper under running water to gently but thoroughly remove excess preservative. Then place the grasshopper in a dissection tray.
2. Use a hand lens to observe the grasshopper's parts. While referring to Figure 39-3, identify the head, thorax, and abdomen. Note how the thorax and abdomen are divided into segments.
3. Use forceps to spread out and examine both pairs of wings. Notice that the forewings are narrow and the hindwings are wide. Observe how the hindwings fold fanlike against the body.
4. Observe the legs. Grasp one of each pair of legs and notice how the legs are divided into segments. Gently bend the legs to observe their normal range of motion.
5. Examine the 11 segments of the abdomen. On abdominal segment 1, find the tympanum. Then, along each side of abdominal segments 1–8, locate the spiracles, which look like small dots. Gently touch the abdomen with a blunt probe to find the flexible membrane that connects the segments to one another.
6. While referring to Figure 39-4a, examine the grasshopper's head. Find the two antennae, two compound eyes, and three simple eyes. Use a sharp probe to push apart the mouthparts. Locate and identify the mandibles, maxillae, labium, and labrum. Note that each maxilla has a segmented feeler called a palpus and that the labium has two palpi.

PART B Observing the Internal Anatomy of the Grasshopper

7. CAUTION **Dissecting instruments can cut you. Always cut in a direction away from your face and body.** Using scissors, snip off the grasshopper's legs, wings, and antennae at their bases. Pin the body to the dissection tray. Then use the scissors to make a shallow cut just above the spiracles through the exoskeleton along both sides of the thorax and abdomen.
8. As you remove the exoskeleton, look at its underside, where the heart and aorta may be attached. Also look for muscles attached to the inside of the exoskeleton. Find and remove any fatty tissue (which your teacher can help you identify) that may hide other organs.
9. If your grasshopper is a female, look for its ovaries, which may contain elongated eggs. Refer to Figure 39-5 to see what the ovaries look like. If your grasshopper is a male, examine the ovaries and eggs in the female grasshopper of another student. Remove the ovaries and eggs (if present) from one side of the abdomen to uncover the digestive tract.
10. Make a table in your lab report like the one shown. As you observe each of the structures listed in the table, fill in the function of that structure.
11. Referring to Figure 39-5, look for the organs of the digestive tract: esophagus, crop, gizzard, midgut, hindgut, and anus. Find the salivary glands and the gastric ceca, which are also parts of the digestive system.
12. On the surface of the midgut and hindgut, look for the Malpighian tubules, which are tiny tubes that connect to the digestive tract.
13. Locate and identify the brain, ventral nerve cord, and ganglia.
14. Carefully cut away and remove the organs of the digestive system to expose some parts of the respiratory system. Referring to Figure 39-5, locate and identify tracheae that run along the dorsal and ventral parts of the body, as well as other tracheae that connect them. The larger tracheae lead to spiracles in the abdomen. Look also for swollen tracheae, called air sacs, which increase the volume of air drawn into the abdomen when the grasshopper breathes.
15. Dispose of your specimen according to the directions from your teacher. Then clean up your materials and wash your hands before leaving the lab.

FUNCTION OF GRASSHOPPER STRUCTURES

Structure	Function
Esophagus	
Crop	
Gizzard	
Midgut	
Hindgut	
Salivary glands	
Gastric ceca	
Malpighian tubules	
Tracheae	
Spiracles	

Analysis and Conclusions

1. How do you think the membrane between segments helps the grasshopper in its movements?
2. How does the function of the stiff, leathery forewings differ from that of the more delicate hindwings?
3. Trace the path of food through the grasshopper's digestive tract.
4. To what system do the Malpighian tubules belong?
5. Why is the circulatory system of the grasshopper described as an open circulatory system?
6. Compared with invertebrates such as flatworms and earthworms, grasshoppers are highly responsive to environmental stimuli. What are some structural adaptations of the grasshopper that make this responsiveness possible?

Further Inquiry

1. Prepare an illustrated chart that compares and contrasts the characteristics of grasshoppers, beetles (order Coleoptera), and butterflies (order Lepidoptera). What trait of each kind of insect is reflected in the name of its order? Include other traits in your chart.
2. The fruit fly *Drosophila* is an insect that, unlike the grasshopper, undergoes complete metamorphosis. Research the life cycles of the grasshopper and the fruit fly, and make a chart that compares and contrasts their life cycles.

one pair of unbranched antennae; one pair of compound eyes; three pairs of jointed legs; and usually one or two pairs of wings.
2. Grasshoppers belong to the order Orthoptera.
3. Characteristics include two pairs of straight wings, chewing mouthparts, and incomplete metamorphosis.

ANSWERS TO ANALYSIS AND CONCLUSIONS

1. The membrane makes the body more flexible.
2. The forewings serve as protective covers for the hindwings, which are used for flying. The forewings also help the grasshopper glide during flight.
3. Food travels from the mouth through the esophagus, crop, gizzard, midgut, and hindgut. Wastes leave via the anus.
4. The Malpighian tubules belong to the excretory system.
5. The fluid (hemolymph) does not circulate entirely within a system of vessels.
6. The grasshopper's specialized sense organs, such as its antennae, simple and compound eyes, tympana, and sensory hairs, enable it to detect many different stimuli. Its segmented body, jointed legs, wings, and mouthparts allow it to respond to stimuli in a variety of ways.

FURTHER INQUIRY

1. The traits that are reflected in the name of each order are straight wings (Orthoptera), hard forewings (Coleoptera, "sheath-wing"), and scaled wings (Lepidoptera).
2. The life cycle of the grasshopper is illustrated in Figure 39-6 and described in the text. The life cycle of the fruit fly includes an egg, larva, pupa, and adult fly.

CHAPTER 40
ECHINODERMS AND INVERTEBRATE CHORDATES

	TOPICS	TEACHING RESOURCES	LABS, CLASSWORK, AND HOMEWORK
BLOCK 1 45 minutes	**Introducing the Chapter, p. 780**	ATE Focus Concept, p. 780 ATE Assessing Prior Knowledge, p. 780 ATE Understanding the Visual, p. 780	■ Supplemental Reading Guide, *Journey to the Ants: A Story of Scientific Exploration*
	40-1 Echinoderms and Invertebrate Chordates, p. 781 Characteristics, p. 781 Classification, p. 782 Structure and Function of Echinoderms, p. 785	ATE Section Overview, p. 781 ATE Critical Thinking, p. 786 ATE Visual Link Figure 40-7, p. 785 162	★ Study Guide, Section 40-1 PE Section Review, p. 787
BLOCK 2 45 minutes	**40-2 Invertebrate Chordates, p. 788** Characteristics, p. 788 Evolution and Classification, p. 789	ATE Section Overview, p. 788 ATE Visual Link Figure 40-11, p. 789 163	PE **Chapter 40 Investigation,** p. 794 PE **Quick Lab,** p. 789, Identifying Chordate Characteristics ★ Study Guide, Section 40-2 PE Section Review, p. 790
BLOCKS 3 & 4 90 minutes	**REVIEW and ASSESSMENT** PE Chapter 40 Review, pp. 792–793 ■ Performance-Based Assessment—Exploring Invertebrates	Audio CD Program	**CHAPTER TESTING** ■ Chapter Test (blackline master) ▲ Test Generator ■ Assessment Item Listing ■ Scoring Rubrics and Classroom Management Checklists

KEY

MODERN BIOLOGY

- **PE** Pupil's Edition
- **ATE** Teacher's Edition
- ■ Active Reading Guide
- ★ Study Guide

HOLT BIOSOURCES

- Teaching Transparencies
- Laboratory Program

One-Stop Planner CD-ROM
Shows these resources within a customizable daily lesson plan:

- ■ Holt BioSources Teaching Resources
- ■ Active Reading Guide
- ★ Study Guide
- ▲ Test Generator

READING FOR CONTENT MASTERY

Preview: Have students preview the objectives, the introductory paragraph, topic sentences, and captions before reading the chapter.

- **ATE** Active Reading Technique: K-W-L, p. 781
- ■ Reading Strategy: K-W-L Worksheet
- ■ Active Reading Guide Worksheet 40-1
- **PE** Summary/Vocabulary, p. 791

- **ATE** Active Reading Technique: K-W-L, p. 790
- ■ Reading Strategy: K-W-L Worksheet
- ■ Active Reading Guide Worksheet 40-2
- **PE** Summary/Vocabulary, p. 791

TECHNOLOGY AND INTERNET RESOURCES

Holt Biology Videodiscs Teacher's Correlation Guide, Lessons 40-1 through 40-2

Audio CD Program, Sections 40-1 through 40-2

internetconnect

sciLINKS NSTA

On-line Resources: www.scilinks.org
The following *sci*LINKS Internet resources can be found in the student text for this chapter:

Topics:
- Echinoidea, p. 783
- Asteroidea, p. 784
- Chordates, p. 788

go.hrw.com

On-line Resources: go.hrw.com
Visit the HRW Web site for a variety of resources related to this chapter. Just type in the keyword HM2 Chapter 40.

Smithsonian Institution®

Visit **www.si.edu/hrw** for additional on-line resources.

CNNfyi.com

Visit **www.cnnfyi.com** for late-breaking news and current event stories selected just for you.

LAB ACTIVITY PLANNING

CHAPTER 40 INVESTIGATION

Comparing Echinoderms, pp. 794–795

OVERVIEW
In this investigation, students will study the anatomy of two types of echinoderms. Students will compare similar structures in a sea star and a sea urchin.

SAFETY
Students will wear disposable gloves, lab aprons, and safety goggles during the lab. Bleach is a strong oxidant and irritant. Avoid contact with eyes, skin, and clothing. If bleach gets in a student's eye, flush the eye with water for 15 minutes and get immediate medical attention. Demonstrate the safe use of dissection instruments. Students must wash their hands after completing the lab.

PREPARATION
1. Purchase preserved sea stars and sea urchins from a biological supply house, such as WARD'S. Each lab group will need one specimen of each echinoderm.
2. Each lab group will need a test (the dried, outer covering) of a sea urchin, a dissection tray, fine dissection scissors, forceps, a blunt probe, a sharp probe, several dissection pins, a 100 mL beaker, a Petri dish, and 25 mL of bleach. Each lab group will also need access to a microscope or hand lens.

Quick Labs

Identifying Chordate Characteristics, p. 789
Each pair of students will need several colors of clay, toothpicks, and masking tape for this activity.

UNDERSTANDING THE VISUAL

Ask students what type of body symmetry is shown by the blue sea star in the photograph (radial symmetry). Ask them to name other radially symmetrical animals they have already studied (cnidarians and ctenophores). Explain that a sea star is an echinoderm. Ask students to describe the type of lifestyle they think echinoderms would have, based on what they know about other radially symmetrical animals. (Like cnidarians, most echinoderms are sessile or slow-moving.)

FOCUS CONCEPT

Evolution

Although echinoderms are radially symmetrical as adults, they are bilaterally symmetrical as larvae, suggesting that they evolved from bilaterally symmetrical ancestors. Because both echinoderms and chordates are deuterostomes, it is likely that they have a common ancestor.

ASSESSING PRIOR KNOWLEDGE

Review the following concepts.

Patterns of Development: *Chapter 34*
Ask students to define *protostome* and *deuterostome.* Ask students how cleavage patterns differ in protostomes and deuterostomes.

Chordates: *Chapter 34*
Ask students to define *notochord.* Ask students to explain the relationship between chordates and vertebrates.

CHAPTER 40

ECHINODERMS AND INVERTEBRATE CHORDATES

This blue sea star, Linckia laevigata, *shows the pentaradial symmetry of most echinoderms.*

FOCUS CONCEPT: *Evolution*

As you read, note the unique structural features of echinoderms and think about how echinoderms are similar to their closest relatives, the chordates.

ECHINODERMS

The phylum Echinodermata (ee-KIE-noh-duhr-MAH-tuh) is a group of invertebrates that includes sea stars, sand dollars, sea urchins, and sea cucumbers. The members of this phylum, called ***echinoderms,*** *inhabit marine environments ranging from shallow coastal waters to ocean trenches more than 10,000 m deep. They vary in diameter from 1 cm to 1 m and often are brilliantly colored.*

SECTION

40-1

OBJECTIVES

▲ Discuss the evolutionary origin of echinoderms.

● List the characteristics that distinguish echinoderms from other phyla.

■ Name representative species in each echinoderm class.

◆ Explain how the water-vascular system aids in movement and feeding.

▲ Describe sexual and asexual reproduction in the sea star.

CHARACTERISTICS

Echinoderms are radially symmetrical animals. Like cnidarians and ctenophores, which are also radially symmetrical, echinoderms have no head or any other sign of cephalization. Unlike cnidarians and ctenophores, however, adult echinoderms develop from bilaterally symmetrical larvae. A few examples of echinoderm larvae are illustrated in Figure 40-1. This feature of development indicates that echinoderms almost certainly evolved from bilaterally symmetrical ancestors.

The fossil record of echinoderms dates back to the Cambrian period, more than 500 million years ago. Early echinoderms from this period appear to have been sessile, and biologists believe these animals evolved radial symmetry as an adaptation to a sessile existence. Echinoderms later evolved the ability to move from place to place. Today the vast majority of echinoderm species can move by crawling slowly along the ocean bottom, and only about 80 species are sessile.

Echinoderms are deuterostomes, which makes them different from all of the other invertebrates you have studied so far. Recall from Chapter 34 that deuterostomes are coelomates whose embryos

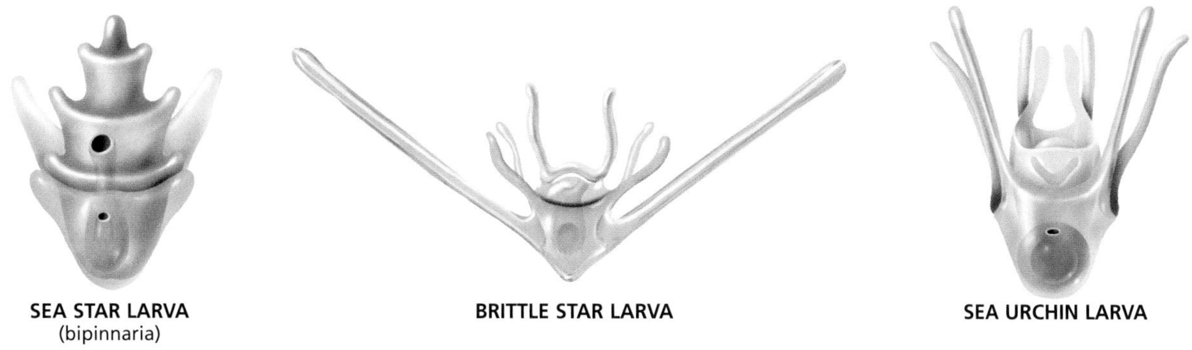

FIGURE 40-1

Notice the bilateral symmetry in these echinoderm larvae. The larvae develop into radially symmetrical adults.

SECTION 40-1

SECTION OVERVIEW

Evolution

Section 40-1 discusses the evidence that echinoderms evolved from bilaterally symmetrical animals and that they are more closely related to chordates than to other invertebrates.

Matter, Energy, and Organization

Section 40-1 describes the organization of body systems in the sea star, a representative echinoderm.

ACTIVE READING

Technique: K-W-L

Before beginning this chapter, divide the class into two groups. Have students in one group make a list of all the things they **Know** about echinoderms. The students in the second group should make a list of all the things they **Know** about chordates. When the groups finish their lists, have them write their entries on the board. Then have students make individual lists of things they **Want** to know about echinoderms and chordates. Have students save their lists for use later in the Active Reading activity on page 790.

CULTURAL CONNECTION

Echinoderm Cuisine

Trepang, made from the cured body wall of certain species of sea cucumbers, is considered a delicacy in some countries of East Asia. Being more than half protein, it is a highly nutritious food. Sea urchin gonads, either raw or roasted on the half shell, are another favorite food of people in these countries.

MAKING CONNECTIONS

Social Studies

The crown-of-thorns sea star, *Acanthaster planci,* feeds on coral polyps. The population of this sea star appears to be on the rise, posing a serious threat to Pacific coral reefs, including the Great Barrier Reef off the northeast coast of Australia. The Australian government has spent millions of dollars in various attempts to control this predator but so far has had little success. The incentive to win this battle is great, however, since the health of the fishing and tourist industries in many Pacific nations depends heavily on the survival of coral reefs.

have radial cleavage, whose anus forms near the blastopore, and whose mesoderm arises from outpockets of the endoderm. Because they develop as deuterostomes, echinoderms are more closely related to chordates, which are discussed in the second part of this chapter, than to other invertebrates.

Most echinoderms have a type of radial symmetry called **pentaradial symmetry,** in which the body parts extend from the center along five spokes. In addition to their pentaradial symmetry, echinoderms have three other major characteristics that are not shared by any other phylum. (1) They have an endoskeleton composed of calcium carbonate plates known as **ossicles.** The ossicles may be attached to spines or spicules that protrude through the skin. The name *echinoderm* means "spiny skin." (2) They have a **water-vascular system,** which is a network of water-filled canals inside their body. (3) They have many small, movable extensions of the water-vascular system called **tube feet,** which aid in movement, feeding, respiration, and excretion. The water-vascular system and tube feet will be discussed in more detail.

CLASSIFICATION

Taxonomists divide the 7,000 species of echinoderms into six classes, five of which will be described here: Crinoidea (kri-NOID-ee-uh), Asteroidea (AS-tuh-ROID-ee-uh), Ophiuroidea (OH-fee-yoor-OID-ee-uh), Echinoidea (EK-uh-NOID-ee-uh), and Holothuroidea (HOH-loh-thuh-ROID-ee-uh).

Class Crinoidea

Members of the class Crinoidea, called crinoids (KRI-NOIDS), include the sea lilies and feather stars, which are shown in Figure 40-2. The name *crinoid* means "lily-like." Sea lilies most closely resemble the fossils of ancestral echinoderms from the Cambrian period. They are sessile as adults, remaining attached to rocks or the sea bottom by means of a long stalk. Feather stars, in contrast, can swim or crawl as adults, although they may stay in one place for long periods. In

FIGURE 40-2

This sea lily, *Cenocrinus* (a), and these feather stars, *Oxycomanthus bennetti* (b), are members of the class Crinoidea. Notice their adaptations for filter feeding.

(a)

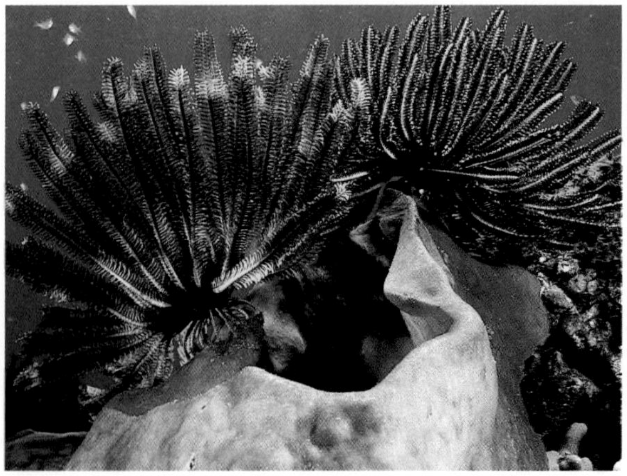

(b)

both types of crinoids, five arms extend from the body and branch to form many more arms—up to 200 in some feather star species. Sticky tube feet located at the end of each arm filter small organisms from the water. The tube feet also serve as a respiratory surface across which crinoids exchange oxygen and carbon dioxide with the water. Cilia on the arms transport trapped food to the crinoid's mouth at the base of the arms. The mouth faces up in crinoids, while in most other echinoderms the mouth faces toward the sea bottom.

Class Ophiuroidea

The 2,000 species of basket stars and brittle stars make up the largest echinoderm class, Ophiuroidea, which means "snake-tail." Members of this class are distinguished by their long, narrow arms, which allow them to move more quickly than other echinoderms. As you can see in Figure 40-3, the thin, flexible arms of basket stars branch repeatedly to form numerous coils that look like tentacles. Brittle stars, so named because parts of their arms break off easily, can regenerate missing parts.

Basket stars and brittle stars live primarily on the bottom of the ocean, often beneath stones or in the crevices and holes of coral reefs. They are so numerous in some locations that they cover the sea floor. Some species feed by raking in food with their arms or gathering it from the ocean bottom with their tube feet. Others trap suspended particles with their tube feet or with mucous strands located between their spines.

Class Echinoidea

The class Echinoidea consists of about 900 species of sea urchins and sand dollars. *Echinoidea* means "spinelike," a description that applies especially well to many of the sea urchins, such as the ones shown in Figure 40-4. In both sea urchins and sand dollars, the internal organs are enclosed within a compact, rigid endoskeleton called a **test.**

The spherical sea urchins are well adapted to life on hard sea bottoms. They use their tube feet for locomotion and feed by scraping algae from hard surfaces with the five teeth that surround their mouth. The teeth and the muscles that move them are part of a complex jawlike mechanism called Aristotle's lantern. The spines that protrude from the test may be short and flat, long and thin, or wedge-shaped, depending on the species. In some sea urchins, the spines are barbed, while in others, they are hollow and contain a venom that is dangerous to predators as well as swimmers.

Sand dollars live along seacoasts. As their name implies, they are usually found in sandy areas and have the flat, round shape of a silver dollar. Their shape is an adaptation for shallow burrowing. The short spines on a sand dollar are used in locomotion and burrowing, and they help clean the surface of the body. Sand dollars use their tube feet to capture food that settles on or passes over their body.

FIGURE 40-3
The basket star, *Oreaster reticulatus,* has long, flexible arms with many coiled branches.

internetconnect
SCILINKS NSTA
TOPIC: Echinoidea
GO TO: www.scilinks.org
KEYWORD: HM783

FIGURE 40-4
The long, sharp spines that cover these sea urchins, *Strongylocentrotus,* provide protection against most predators.

TEACHING STRATEGY
Echinoderm Spines

Allow students to examine live specimens or close-up photographs of sea urchins and sand dollars. Have students note the differences in the spines on these two members of the class Echinoidea. Ask students to name the functions of the spines on each animal. (Sea urchin spines are used mainly for defense; sand dollar spines are used for locomotion, burrowing, and cleaning the body surface.)

RECENT RESEARCH
Symbiosis in Echinoderms

An interesting symbiotic relationship has been found between sea urchins and a newly discovered species of brittle star. The urchin bores a hole in its rocky habitat, and the brittle star lives beneath the urchin at the bottom of the hole, where it is protected from waves and predators. As many as 10 brittle stars may live under each urchin.

RETEACHING ACTIVITY

Show photographs or preserved specimens of various species of echinoderms. Ask students to name the class that each species belongs to and to describe the feeding habits of that species. Ask students to name the features that place all of these species in the same phylum.

DEMONSTRATION

Students may erroneously believe that sea cucumbers are bilaterally symmetrical, since many sea cucumbers are elongated and appear to have anterior and posterior ends. Demonstrate the radial symmetry of sea cucumbers by rolling a lump of clay the size of your fist into a cylinder about 6–8 in. long. Stand the cylinder upright and etch five evenly spaced longitudinal lines in it from top to bottom. Tell students that in this orientation, a sea cucumber is similar to a stretched-out, upside-down sea urchin without its spines. Then lay the cylinder on its side and explain that this is the orientation most sea cucumbers have as they crawl along the ocean bottom.

GIFTED *ACTIVITY*

Have students investigate evisceration in sea cucumbers, a defensive behavior in which the animal forcefully ejects part of its internal organs when threatened. Ask students to find out which organs are ejected, how this behavior is adaptive to the sea cucumber, and how the sea cucumber compensates for the displaced organs.

RECENT RESEARCH

Sea Cucumber Microfibrils

Biologists have learned recently that the ability of sea cucumbers to change their body shape and tension is due to the microfibrils that exist in a special elastic tissue in the sea cucumbers' skin. The same type of microfibril is found in cnidarians, suggesting that it is evolutionarily very old.

FIGURE 40-5

Tentacles around the mouth of this sea cucumber collect food and bring it to the animal's mouth. Five rows of tube feet that run along the body are evidence of the sea cucumber's pentaradial symmetry.

internet**connect**

SCILINKS NSTA

TOPIC: Asteroidea
GO TO: www.scilinks.org
KEYWORD: HM784

Class Holothuroidea

Sea cucumbers belong to the class Holothuroidea. Most of these armless echinoderms live on the sea bottom, where they crawl or burrow into soft sediment. The ossicles that make up their endoskeleton are very small and are not connected to each other, so their bodies are soft. Modified tube feet form a fringe of tentacles around the mouth. When these tentacles are extended, as shown in Figure 40-5, they resemble the polyp form of some cnidarians. That explains the name of this class, which means "water polyp." A sea cucumber uses its tentacles to sweep up sediment and water. It then stuffs its tentacles into its mouth and cleans the food off them.

Class Asteroidea

The sea stars, or starfish, belong to the class Asteroidea, which means "starlike." Sea stars live in coastal waters all over the world. They exist in a variety of colors and shapes, two of which are shown in Figure 40-6. Sea stars are economically important because they prey on oysters, clams, and other organisms that humans use as food.

FIGURE 40-6

Sea stars are found on rocky coastlines worldwide. One of the more colorful varieties is the African sea star, *Protoreaster linckii* (a). The sunflower star, *Pycnopodia helianthoides* (b), can have up to two dozen arms.

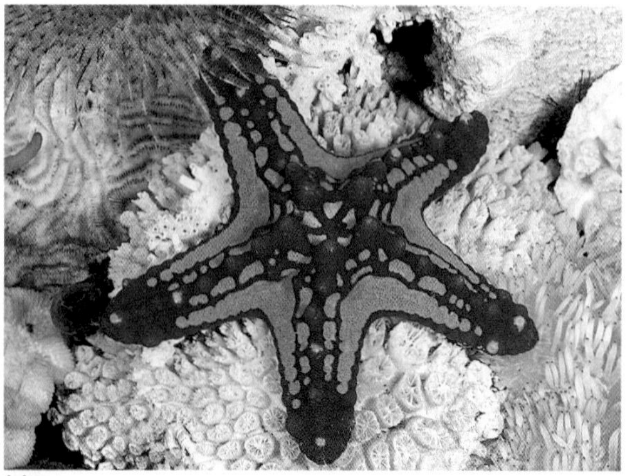

(a)

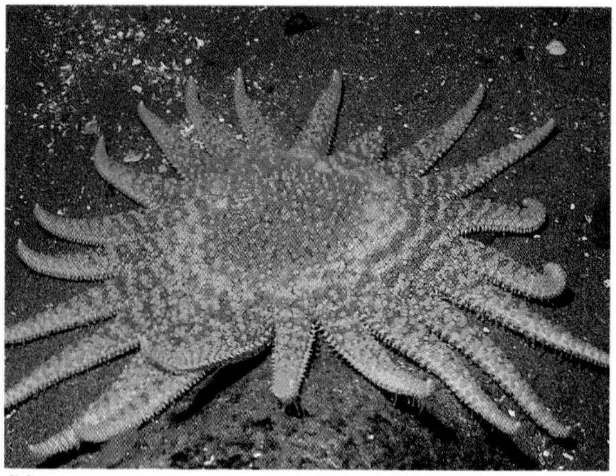

(b)

STRUCTURE AND FUNCTION OF ECHINODERMS

The sea star will be used to demonstrate some of the details of echinoderm structure and function. As you read about how echinoderms carry out life functions, consider how they differ from the other groups of invertebrates you have studied.

External Structure

As you can see in Figure 40-7, the body of a sea star is composed of several arms that extend from a central region. Sea stars typically have five arms, but in some species, such as the one shown in Figure 40-6b, there may be as many as 24. Two rows of tube feet run along the underside of each arm. The body is often flattened.

In echinoderms, the side of the body where the mouth is located is referred to as the **oral surface.** The side of the body that is opposite from the mouth is called the **aboral** (A-BOHR-uhl) **surface.** In sea stars, the oral surface is on the underside of the body.

The body of a sea star is usually covered with short spines that give the animal a rough texture. Surrounding each spine in many sea stars are numerous tiny pincers called **pedicellariae** (PED-uh-suh-LAR-ee-ee), which are shown in Figure 40-7. Pedicellariae help keep the body surface free of foreign objects, including algae and small animals that might grow on the sea star or damage its soft tissues. Pedicellariae are found in sea stars and some sea urchins.

FIGURE 40-7

Sea stars have a number of structural features that are unique to the phylum Echinodermata. Their pentaradial symmetry is indicated by their five arms, each of which contains a division of their internal organ systems. The water-vascular system consists of a network of canals connected to hundreds of tube feet. The inset shows that the sea star's exterior is dotted with short spines, pincerlike pedicellariae, and skin gills.

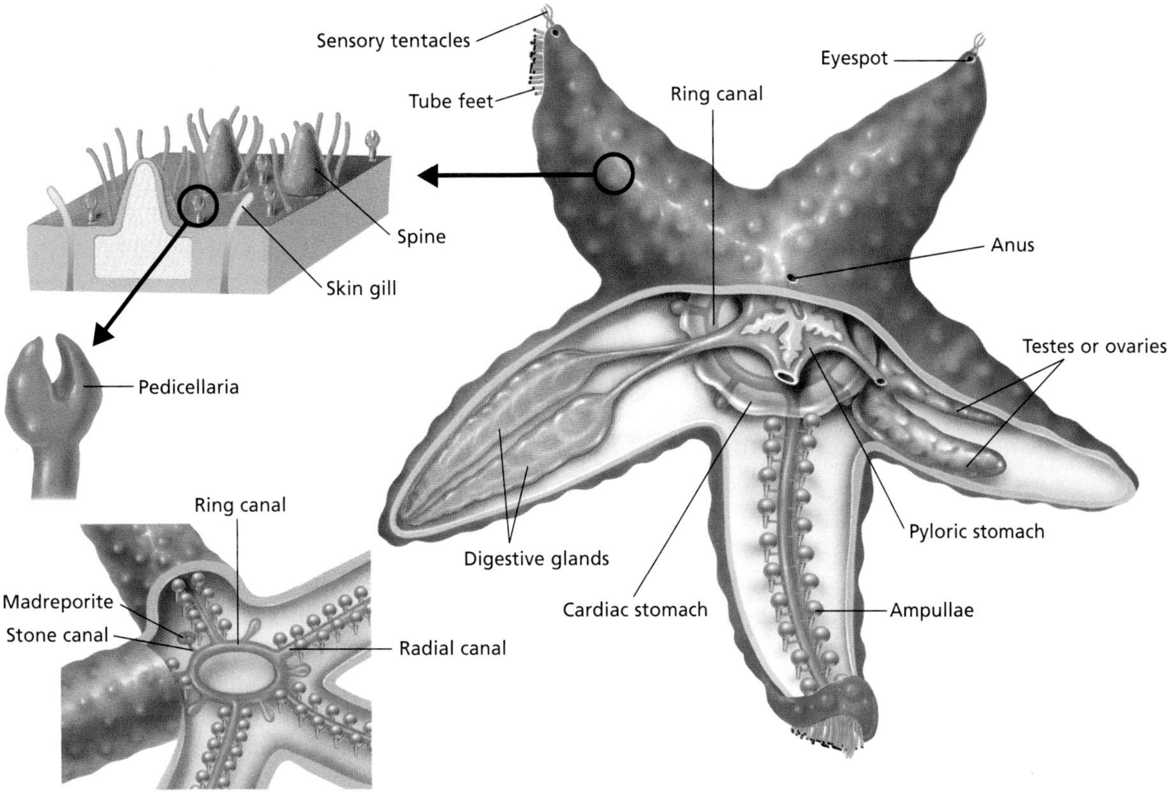

SECTION 40-1

INCLUSION ACTIVITY

Ask a student to stand up and identify his or her left, right, dorsal, and ventral sides. Then have five students form a circle by linking arms with their backs toward the center of the circle. Tell them they have formed a "human sea star." Ask the "sea star" to identify its left, right, dorsal, and ventral sides. (It cannot, because it has none.) Explain that in a radially symmetrical animal such as a sea star, these four terms do not apply. This point should be clear to students when they see that the left side of one student in the circle is next to the right side of another student, and that the dorsal sides of all students in the circle are confined to the interior of the circle.

OVERCOMING MISCONCEPTIONS

Sea stars are commonly called starfish, which may lead some students to conclude that sea stars are a type of fish. Emphasize that sea stars are not fish because fish are vertebrates with bilateral symmetry.

VISUAL LINK

Figure 40-7

Have students examine the sea star shown in the figure. Ask them what evidence they can find of pentaradial symmetry in the sea star's internal structure. (There are five sets each of digestive glands, reproductive organs, radial canals, and ampullae.) Point out how the ampullae, which lie inside the body, are connected to the tube feet, which protrude through the endoskeleton to the outside of the body. Use the inset in the figure to make sure students understand the structural and functional differences between spines, pedicellariae, and skin gills.

QUICK FACT

✔ Not all sea stars eject their stomach and predigest food outside their body. The sun star, which feeds on sea urchins, swallows its prey whole.

CONTENT CONNECTION

Section 37-1: Bivalve Anatomy

The muscles that are responsible for closing the shell of a bivalve mollusk are the adductor muscles. Although these muscles are strong enough to resist the pull of an attacking sea star initially, they fatigue like all muscles. When that happens, the sea star begins its meal.

CRITICAL THINKING

Echinoderm Respiration

All echinoderms require oxygen for respiration but lack a respiratory system. What features of the tube feet and skin gills make them good respiratory structures? (Their thin walls permit gases to diffuse between the coelom and the environment, and their elongated shape and great number create a large surface area for diffusion.)

Word Roots and Origins

ampulla

form the Latin *ampulla*, meaning "flask"

Water-Vascular System

The water-vascular system is a network of water-filled canals that are connected to the tube feet. Use Figure 40-7 to follow the path of water through the water-vascular system. Water enters the system through small pores in the **madreporite** (MAD-ruh-POHR-IET), a sievelike plate on the aboral surface. Water then passes down the **stone canal,** a tube that connects the madreporite to the **ring canal,** which encircles the mouth. Another tube, the **radial canal,** extends from the ring canal to the end of each arm. The radial canals carry water to the hundreds of hollow tube feet. Valves prevent water from flowing back into the radial canals from the tube feet.

The upper end of each tube foot is expanded to form a bulblike sac called an **ampulla** (AM-PYU-luh). Contraction of muscles surrounding the ampullae forces water into the tube feet, causing them to extend. Contraction of muscles lining the tube feet forces water back into the ampullae and shortens the tube feet. In this way, the sea star uses water pressure to extend and withdraw its tube feet. In many species, small muscles raise the center of each tube foot's disklike end, creating suction when the tube feet are pressed against a surface. These coordinated muscular contractions enable sea stars to climb slippery rocks and capture prey.

Feeding and Digestion

The sea star's mouth is connected by a short esophagus to the **cardiac stomach,** which the sea star can turn inside out through its mouth when it feeds. The cardiac stomach transfers food to the **pyloric stomach,** which connects to a pair of digestive glands in each arm. The cardiac stomach, pyloric stomach, and digestive glands break down food with the help of the enzymes they secrete. Nutrients are absorbed into the coelom through the walls of the digestive glands, and undigested material is expelled through the anus on the aboral surface.

Most sea stars are carnivorous, feeding on mollusks, worms, and other slow-moving animals. When a sea star captures a bivalve mollusk, such as a clam, it attaches its tube feet to both halves of the clamshell and exerts a steady pull, as Figure 40-8 shows. Eventually, the clam's muscles tire and the shell opens slightly. The sea star then inserts its cardiac stomach into the clam and digests the clam's soft tissues while they are still in the shell. The sea star then withdraws the stomach, containing the partially digested food, back into its body, where the digestive process is completed.

FIGURE 40-8

This sea star, *Asterias rubens,* is prying open the shell of a clam to feed on the clam's soft tissues.

Other Body Systems

Like other echinoderms, the sea star has no circulatory, excretory, or respiratory organ systems. Fluid in the coelom bathes the organs and distributes nutrients and oxygen. Gas exchange and waste excretion take place by diffusion through the thin walls of the tube feet and through the **skin gills,** hollow tubes that project from the coelom lining to the exterior. You can see the skin gills in the inset in Figure 40-7, on the previous page.

The nervous system in sea stars is primitive. Since echinoderms have no head, they also have no brain. The nervous system consists mainly of a **nerve ring** that circles the mouth and a **radial nerve** that runs from the nerve ring along the length of each arm. The nerve ring and radial nerves coordinate the movements of the tube feet. If the radial nerve in one arm is cut, the tube feet in that arm lose coordination. If the nerve ring is cut, the tube feet in all arms become uncoordinated and the sea star cannot move. Sea stars also have a nerve net near the body surface that controls the movements of the spines, pedicellariae, and skin gills. The end of each arm has an eyespot that responds to light and several tentacles that respond to touch. The tube feet also respond to touch, and other touch-sensitive and chemical-sensitive cells are scattered over the surface of the sea star's body.

Reproduction and Development

Most sea star species have separate sexes, as do most other echinoderms. Each arm of the sea star contains a pair of ovaries or testes. Females produce up to 200 million eggs in one year. Fertilization occurs externally, when the eggs and sperm are shed into the water. Each fertilized egg develops into a bilaterally symmetrical, free-swimming larva called a **bipinnaria** (BIE-pin-AR-ee-uh). After about two months, the larva settles to the bottom, and metamorphosis begins. During metamorphosis the larva develops into a pentaradially symmetrical adult.

Echinoderms have remarkable powers of regeneration. Sea stars can regenerate arms from the central region of their body, even if they lose all of their arms. The process of regeneration is very slow, taking as long as a year. Sea stars use their regenerative ability as a defensive mechanism, automatically shedding an arm at its base when the arm is captured by a predator. As you can see in Figure 40-9, some sea stars can even regenerate a complete, new individual from a detached arm, as long as the arm is attached to a portion of the central region. Certain species reproduce asexually by splitting their body through the central region. The two parts that are formed then regenerate the missing structures.

FIGURE 40-9

As long as a sea star retains part of its central region, it can regenerate any arms it loses. The sea star shown here, a member of the genus *Echinaster,* is regenerating five new arms.

SECTION 40-1 REVIEW

1. Why are echinoderms thought to have evolved from bilaterally symmetrical ancestors?
2. What are the characteristics that distinguish echinoderms from other phyla?
3. Name the representative species in each class of echinoderms.
4. Explain how a sea star extends and withdraws its tube feet.
5. How do sea stars reproduce asexually?
6. **CRITICAL THINKING** How would a sea star's ability to feed be affected if the sea star lost the water in its water-vascular system?

QUICK FACT

✔ A few sea star species brood their eggs by arching and wrapping their arms around them. Other species brood eggs in spine pouches on their arms.

ANSWERS TO SECTION 40-1 REVIEW

1. Although echinoderms are radially symmetrical as adults, they develop from bilaterally symmetrical larvae.
2. Distinguishing characteristics of echinoderms include pentaradial symmetry, an endoskeleton composed of ossicles, a water-vascular system, and tube feet.
3. class Crinoidea: sea lilies and feather stars; class Asteroidea: sea stars; class Ophiuroidea: basket stars and brittle stars; class Echinoidea: sea urchins and sand dollars; class Holothuroidea: sea cucumbers
4. Contraction of muscles surrounding the ampullae forces water into the tube feet, causing them to extend. Contraction of muscles lining the tube feet forces water back into the ampullae and makes the tube feet withdraw.
5. Sea stars reproduce asexually by splitting apart through the central region. Each of the two halves that are formed then regenerate the missing half.
6. The sea star would be unable to use water pressure to extend and withdraw its tube feet, and since the tube feet are used to capture prey, the sea star would not be able to feed.

SECTION 40-2

SECTION OVERVIEW

Evolution
Section 40-2 discusses two subphyla of animals in the phylum Chordata. One of these subphyla, Cephalochordata, contains animals that are thought to resemble the ancestral chordates from which vertebrates evolved.

Matter, Energy, and Organization
Section 40-2 describes the structural features—notochord, dorsal nerve cord, pharyngeal pouches, and postanal tail—that contribute to the organization of the chordate body.

▶ ENGAGE STUDENTS

Purchase some tunicates from a biological supply company, and have the students examine them. Ask the students if they can distinguish the incurrent siphon from the excurrent siphon. (Look for the direction of water currents near the siphons.) Ask them if they can see any similarities between these animals and any vertebrate they can think of. (No similarities should be visible.) Tell students that despite their appearance, tunicates are members of the phylum Chordata, which includes all vertebrates. Explain that phylogenetic relationships are not always obvious from the external appearance of adult animals; sometimes such relationships can be revealed only when the development of the animals is studied.

SECTION

40-2

OBJECTIVES

▲ List the major characteristics of chordates.

● Describe the structure of lancelets.

■ Describe the structure of tunicates.

INVERTEBRATE CHORDATES

The phylum Chordata (kohr-DAHT-uh) includes all of the vertebrates, or animals with backbones. It also includes two groups of animals that lack backbones and are therefore invertebrates. The development of chordates is similar to the development of echinoderms, suggesting that these two phyla descended from a common ancestor.

CHARACTERISTICS

All animals with a backbone are vertebrates, and they make up one of the subphyla in the phylum Chordata, whose members are called chordates. Chordates are so named because they have a notochord, a stiff but flexible rod of cells that runs the length of the body near the dorsal surface. The notochord is illustrated in Figure 40-10. The stiffness of the notochord provides a resistance against which the body muscles can exert force when they contract. The flexibility of the notochord allows the body to bend from side to side as well as up and down. Some kinds of chordates retain the notochord throughout their life. In most vertebrates, however, the notochord is present in embryos but becomes greatly reduced when the vertebral column, or backbone, develops. In adult mammals, the notochord persists only as small patches of tissue between the bones of the vertebral column.

You learned in Chapter 34 that in addition to a notochord, all chordates have the following three characteristics during some stage of their life: (1) a dorsal nerve cord, (2) pharyngeal pouches, and (3) a postanal tail. These characteristics are also illustrated in Figure 40-10. Unlike the ventral nerve cords of invertebrates such as

internet**connect**

SCILINKS NSTA
TOPIC: Chordates
GO TO: www.scilinks.org
KEYWORD: HM788

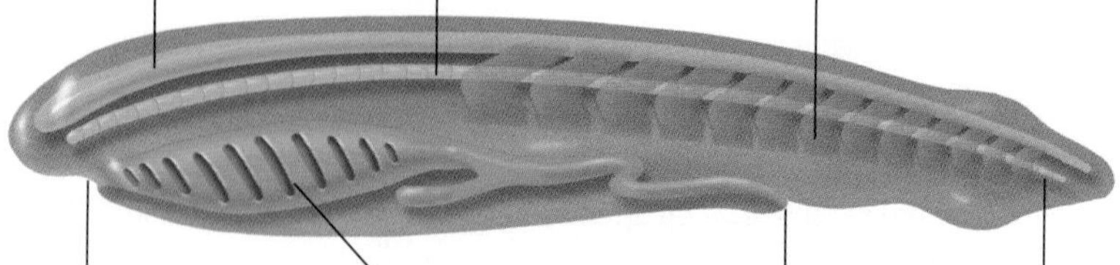

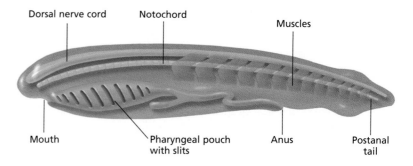

FIGURE 40-10
All chordates have a notochord, a dorsal nerve cord, pharyngeal pouches, and a postanal tail during at least some stage of their life.

annelids and arthropods, the dorsal nerve cord of a chordate is a hollow tube. In vertebrates, the anterior end of the nerve cord enlarges during development to form the brain, and the posterior end forms the spinal cord. The brain receives information from a variety of complex sensory organs, many of which are concentrated at the anterior end of the body. The pharyngeal pouches are outpockets in the pharynx, the portion of the digestive tract between the mouth and the esophagus. In aquatic chordates, the pharyngeal pouches became perforated by slits and evolved first into filter-feeding structures and later into gill chambers. In terrestrial chordates, the pouches evolved into a variety of structures, including the jaws, inner ear, and tonsils. The notochord or backbone extends into the postanal tail, and muscles in the tail can cause it to bend. The postanal tail provides much of the propulsion in many aquatic chordates. Invertebrates in other phyla lack this form of propulsion, because the anus, if present, is located at the end of the body.

EVOLUTION AND CLASSIFICATION

Like echinoderms, chordates are deuterostomes. This similarity provides evidence that echinoderms and chordates likely evolved from a common ancestor. The phylum Chordata is divided into three subphyla: Vertebrata, Cephalochordata (SEF-uh-loh-kohr-DAHT-uh), and Urochordata (YUR-uh-kohr-DAHT-uh). Members of the subphylum Vertebrata, the vertebrates, constitute more than 95 percent of all chordate species. They will be considered in detail in Chapters 41–45. Members of the other two subphyla live only in the ocean. They are the closest living relatives of the early animals from which all chordates evolved.

Subphylum Cephalochordata

The subphylum Cephalochordata contains about two dozen species of blade-shaped animals known as lancelets. Figure 40-11 shows that lancelets look much like the idealized chordate drawn in Figure 40-10. They retain their notochord, dorsal nerve cord,

Quick Lab

Identifying Chordate Characteristics

Materials several colors of clay, toothpicks, masking tape

Procedure Build clay models of a lancelet and an adult tunicate by using different colors of clay for the structures shown in Figures 40-11 and 40-12. Make flags using masking tape attached to toothpicks, and use them to identify any of the four major characteristics of chordates that are found in your models.

Analysis Which of the major characteristics of chordates are found in the lancelet? the adult tunicate? Which of the four characteristics is shared by both? Why is the tunicate classified as a chordate despite the lack of all four chordate characteristics?

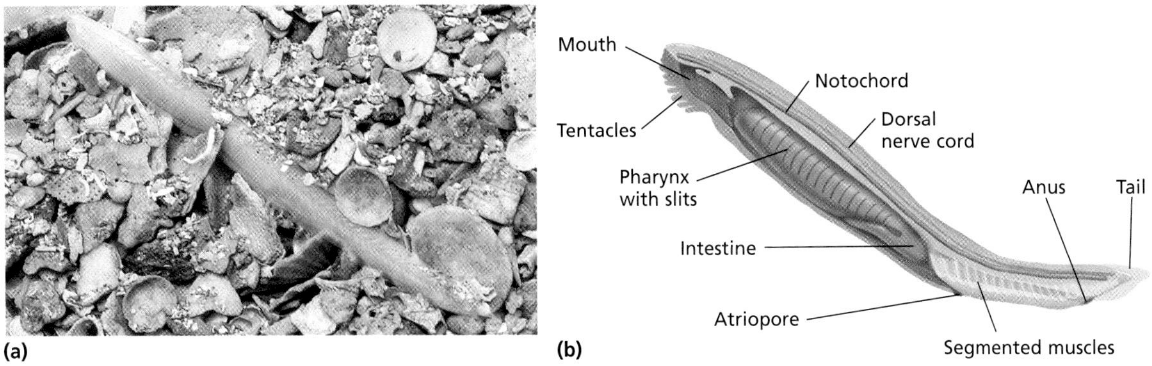

FIGURE 40-11

(a) The lancelet *Branchiostoma lanceolatum* lives with most of its body buried in the sand. (b) Even as adults, lancelets clearly show all four chordate characteristics.

Quick Lab

Identifying Chordate Characteristics

Time Required 20 minutes

Procedural Tips Have students work in pairs.

Answers to Analysis All four chordate characteristics are found in the lancelet including the notochord, dorsal nerve cord, pharyngeal pouches, and postanal tail. Adult tunicates have only one of the chordate characteristics—pharyngeal pouches. Both lancelets and tunicates possess pharyngeal pouches. Tunicates are classified as chordates because they possess all four chordate characteristics during their larval stage, although tunicates lose most of them during metamorphosis.

TERMINOLOGY NOTE

The genus of the lancelet used to be called *Amphioxus* but has been changed to *Branchiostoma.*

VISUAL LINK

Figure 40-11

Have students look at the lancelets shown in the photograph and diagram. Ask them how a lancelet is similar to a fish. (Both have a streamlined body divided into muscular segments.) Ask them how a lancelet differs from a fish. (A lancelet lacks a head, eyes, a vertebral column, and fins.)

SECTION 40-2

RECENT RESEARCH
Introduced Tunicates

A recent survey of tunicates along the western coast of Africa revealed two species that had never been found in the area before. Both species are thought to have been introduced to the area by passing ships because the tunicates can often be found attached to the hulls of ships.

ACTIVE READING
Technique: K-W-L

Tell students to read over the lists that they made using the Active Reading on page 781. Have students list what they have Learned. Conclude by asking students which questions are still unanswered and what new questions they have.

ANSWERS TO SECTION 40-2 REVIEW

1. The four major characteristics are a notochord, a dorsal nerve cord, pharyngeal pouches, and a postanal tail.
2. Both chordates and echinoderms are deuterostomes.
3. In vertebrates, the notochord usually is present only in embryonic stages. It is later replaced by the vertebral column.
4. Lancelets have segmented muscles that run the length of the body.
5. Adult tunicates retain an enlarged pharynx with slits.
6. Undigested wastes that are eliminated by the anus are carried away from the body by the water that leaves through the excurrent siphon.

pharyngeal pouches, and postanal tail throughout their life. Lancelets live in warm, shallow waters and use their muscular tail to wriggle backward into the sand. Only their anterior end protrudes from the sand. Lancelets use cilia to draw water into their pharynx through their mouth. Food particles in the water are trapped as the water passes through the numerous slits in the pharynx. The food enters the intestine to be digested, while the water leaves the body through an opening called the **atriopore** (AY-tree-oh-POHR).

Lancelets can swim weakly, powered by the coordinated contraction of muscles that run the length of their body. If you look closely at Figure 40-11, you can see that these muscles are arranged as a series of repeating segments. Body segmentation is another common feature of chordates. As you learned in Chapters 37 and 38, annelids and arthropods also have segmented bodies. However, animals in those phyla probably evolved body segmentation independently of chordates.

(a)

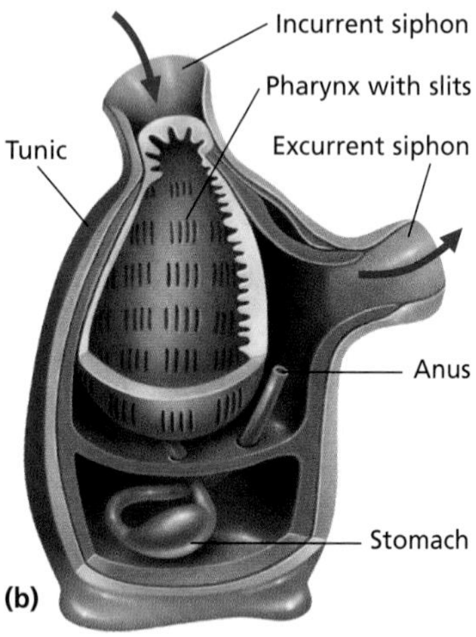

(b)

FIGURE 40-12

Most adult tunicates, such as *Polycarpa auranta* (a), are sessile filter feeders. A drawing of a tunicate's internal structure (b) shows its enlarged pharynx with slits, the only chordate characteristic retained by adult tunicates.

Subphylum Urochordata

The 2,000 species in the subphylum Urochordata are commonly called tunicates because their bodies are covered by a tough covering, or tunic. Tunicates are also called sea squirts because they squirt out a stream of water when touched. As adults, most tunicates are sessile, barrel-shaped animals that live on the sea bottom. They may be solitary or colonial. As you can see in Figure 40-12, tunicates are adapted for filter feeding. Propelled by the beating of cilia, water enters the body through an incurrent siphon, passes through slits in the pharynx, and exits through an excurrent siphon. Food that is filtered by the pharynx moves into the stomach. Undigested material leaves via the anus, which empties into the excurrent siphon. Tunicates are hermaphrodites. Sperm and eggs are released through the excurrent siphon into the surrounding water, where fertilization occurs.

Adult tunicates bear little resemblance to the idealized chordate shown in Figure 40-10. Although they do have a pouchlike pharynx with slits, they have no notochord, dorsal nerve cord, or postanal tail. Larval tunicates, however, possess all four chordate characteristics, but they lose most of them during metamorphosis.

SECTION 40-2 REVIEW

1. What are the four major characteristics of chordates?
2. Why are chordates and echinoderms thought to have evolved from a common ancestor?
3. How do vertebrates differ from the members of the other subphyla of chordates?
4. What evidence of body segmentation do lancelets display?
5. Which of the chordate characteristics do tunicates retain as adults?
6. CRITICAL THINKING In tunicates, the anus empties into the excurrent siphon. What is the advantage of this anatomical arrangement?

CHAPTER 40 REVIEW

SUMMARY/VOCABULARY

40-1

- Most echinoderms develop from free-swimming, bilaterally symmetrical larvae into bottom-dwelling adults with pentaradial symmetry.
- Echinoderms have an endoskeleton made of ossicles, which may be attached to spines or spicules that protrude through the skin. They also have a water-vascular system, which includes many movable extensions called tube feet.
- The class Crinoidea includes sea lilies and feather stars, which are filter feeders that catch small organisms with their sticky tube feet.
- The class Ophiuroidea consists of basket stars and brittle stars, fast-moving echinoderms with long, flexible arms.
- The class Echinoidea includes sea urchins and sand dollars, whose internal organs are enclosed inside a rigid endoskeleton called a test. Many sea urchins have long spines.
- The class Holothuroidea is made of sea cucumbers, armless echinoderms with soft bodies.
- The class Asteroidea consists of sea stars, which have from 5 to 24 arms. Two rows of tube feet run along the underside of each arm.
- The water-vascular system of a sea star consists of a network of canals that connect to bulblike ampullae. Contraction of muscles surrounding the ampullae extends the tube feet, while contraction of muscles lining the tube feet makes the tube feet retract.
- Sea stars can turn one of their stomachs inside out through their mouth to feed on prey they have captured. After the food is partially digested outside the body, it is brought inside, where digestion is completed.
- Sea stars lack circulatory, excretory, and respiratory organ systems, and they have no head or brain. They use skin gills for gas exchange and waste excretion. Most have separate sexes.

Vocabulary

aboral surface (785)
ampulla (786)
bipinnaria (787)
cardiac stomach (786)
echinoderm (781)
madreporite (786)
nerve ring (787)
oral surface (785)
ossicle (782)
pedicellaria (785)
pentaradial symmetry (782)
pyloric stomach (786)
radial canal (786)
radial nerve (787)
ring canal (786)
skin gill (786)
stone canal (786)
test (783)
tube foot (782)
water-vascular system (782)

40-2

- Chordates have a notochord, a stiff but flexible rod of cells that runs the length of the body near the dorsal surface. In one group of chordates, the vertebrates, the notochord is largely replaced by the vertebral column, or backbone.
- Other defining characteristics of chordates are a dorsal nerve cord, pharyngeal pouches, and a postanal tail. These characteristics are not present at all life stages in all chordates.
- Like echinoderms, chordates are deuterostomes, which suggests that echinoderms and chordates evolved from a common ancestor.
- Animals in the subphylum Cephalochordata are called lancelets. These blade-shaped animals live partially buried in the sand, but they can swim from place to place. They retain all of the major chordate characteristics throughout their life.
- Animals in the subphylum Urochordata are called tunicates. Tunicate larvae have all of the major chordate characteristics, but they lose most of them when they develop into adults. Most tunicate adults are sessile.

Vocabulary

atriopore (790)

CHAPTER 40 REVIEW ANSWERS

REVIEW

1. Pentaradial symmetry is a type of radial symmetry in which the body parts extend from the center along five spokes.
2. In an echinoderm, the oral surface is the side of the body where the mouth is located, and the aboral surface is the side of the body that is opposite from the mouth.
3. The endoskeleton is called a test.
4. Pedicellariae are tiny pincers surrounding the spines on the surface of a sea star (or other echinoderm).
5. A notochord is a stiff but flexible rod of cells that runs the length of the body of a chordate, near the dorsal surface. The dorsal nerve cord is the part of the nervous system that develops into the brain and spinal cord in a chordate.

6. a	**7.** d
8. b	**9.** c
10. b	**11.** a
12. d	**13.** c
14. a	**15.** b

16. Water enters the madreporite, goes down the stone canal to the ring canal, and enters the five radial canals that extend into each arm. The radial canals carry water to hundreds of paired, hollow tube feet.
17. Tube feet aid echinoderms in movement, feeding, respiration, and excretion.
18. Sea urchins are adapted for life on rocky bottoms because they are spherical in shape. Sand dollars are adapted for life in a sandy environment because they are flat and able to burrow.
19. Sea stars consume oysters and other mollusks that are harvested as food for humans.
20. A fringe of tentacles, actually modified tube feet, surrounds the mouth and sweeps up food. Sea cucumbers then stuff their tentacles into their mouth and clean the food off them.

21. The sea star attaches its tube feet to both halves of a clamshell, for example, and pulls until the shell opens slightly. The sea star then turns its cardiac stomach inside out through its mouth and digests the clam's soft parts while they are still in the shell. The sea star then withdraws its stomach, containing the partially digested food, and digestion is completed in the pyloric stomach and digestive glands.

22. In aquatic chordates, the pharyngeal pouches became perforated by slits and evolved first into filter-feeding structures and later into gill chambers. In terrestrial chordates, the pouches evolved into structures such as the jaws, inner ear, and tonsils.

23. The notochord provides a resistance against which the body muscles can exert force when they contract, but it also allows the body to bend. The postanal tail provides much of the propulsion in many aquatic chordates.

24. Their bodies are covered by a tough covering, or tunic.

25. *A*, madreporite; *B*, stone canal; *C*, ring canal; *D*, radial canal; *E*, cardiac stomach; *F*, digestive gland

CRITICAL THINKING

1. Swimming larvae allow these echinoderms to disperse widely. Thus, they may be able to spread out to other areas where there may be more food or fewer predators.

2. No, this was not a good method. A new sea star may grow from an arm if part of the central region is still attached. A better method might be to raise the oysters in cages covered with wire mesh, which would keep the sea stars out but allow smaller organisms that the oysters feed on to enter.

3. With their hard endoskeletons, echinoderms would have been

REVIEW

Vocabulary

1. What is pentaradial symmetry?
2. Distinguish between an oral surface and an aboral surface.
3. What term is used for the endoskeleton in members of the class Echinoidea?
4. What are pedicellariae?
5. Distinguish between a notochord and a dorsal nerve cord.

Multiple Choice

6. Scientists believe that the earliest adult echinoderms were (a) bilaterally symmetrical and sessile (b) bilaterally symmetrical and capable of moving about (c) radially symmetrical and sessile (d) radially symmetrical and capable of moving about.
7. Echinoderms lack all of the following organ systems except (a) a respiratory system (b) an excretory system (c) a circulatory system (d) a digestive system.
8. The class of echinoderms whose members most closely resemble the fossils of ancestral echinoderms is (a) Asteroidea (b) Crinoidea (c) Echinoidea (d) Holothuroidea.
9. The class Ophiuroidea consists of (a) sea cucumbers (b) sea stars (c) basket stars and brittle stars (d) sea urchins and sand dollars.
10. The endoskeleton of an echinoderm is composed of calcium carbonate plates referred to as (a) tests (b) ossicles (c) bipinnaria (d) pedicellariae.
11. In a sea star, gas exchange and excretion of wastes take place by diffusion through the walls of the (a) skin gills (b) atriopore (c) radial canals (d) pharynx.
12. When the muscles surrounding the ampullae of a sea star contract, the sea star's (a) stomach turns inside out (b) stomach retracts (c) tube feet retract (d) tube feet extend.
13. A sea star's nervous system does not include (a) a nerve ring (b) a nerve net (c) a brain (d) radial nerves.
14. Echinoderms and chordates are believed to have evolved from a common ancestor because both groups (a) are deuterostomes (b) are bilaterally symmetrical (c) have radially symmetrical larvae (d) have a postanal tail.
15. Adult tunicates have (a) a dorsal nerve cord (b) a pharynx with slits (c) a notochord (d) all of the above.

Short Answer

16. Trace the path of water through the water-vascular system of a sea star.
17. For what functions do echinoderms use their tube feet?
18. Contrast the ways that sea urchins and sand dollars are adapted to their environment.
19. Why are sea stars of economic importance to humans?
20. How does the sea cucumber transport food to its mouth?
21. Summarize the process of feeding and digestion in the sea star.
22. How have pharyngeal pouches become modified through evolution in aquatic chordates? How have they become modified in terrestrial chordates?
23. What is the significance of the notochord and the postanal tail to aquatic chordates?
24. Why are members of the subphylum Urochordata called tunicates?
25. Identify the structures labeled *A–F* in the diagram below.

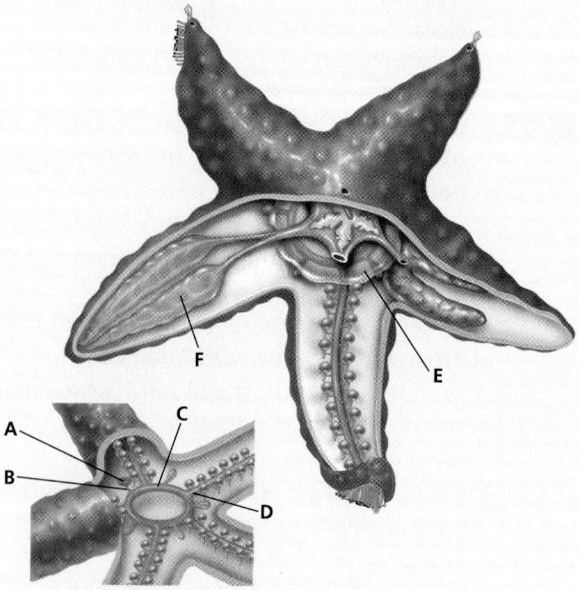

CRITICAL THINKING

1. Sea cucumbers and sea lilies are relatively sessile animals. Their larvae, however, are capable of swimming. What advantage do swimming larvae provide for these echinoderms?
2. Commercial oyster farmers used to remove sea stars from their oyster beds, chop them in half, and throw them back into the water. Was this a good way to protect the oysters from sea stars? If your answer is yes, explain why. If your answer is no, explain why not and describe a better method.
3. Scientists have found many echinoderm fossils from the Cambrian period, but they have found few fossils of other species that must have lived during the Cambrian period. What might explain the large number of fossilized echinoderms?
4. Sea urchin eggs, or roe, are highly prized in Japan for sushi. In trying to supply the Japanese market for this delicacy, divers have nearly wiped out sea urchin populations in some areas of California. What steps could be taken to reestablish sea urchin populations in those areas?
5. Great populations of plankton rise to the surface of the ocean at night and return to the ocean depths during the day. Basket stars are also active at night, when they uncoil their thin, flexible arms. During the day, basket stars curl up and become a compact mass. Why do you think basket stars uncoil their arms at night and coil up during the day?
6. Neoteny is a phenomenon in which larvae become sexually mature while retaining larval characteristics. The term has been used to help explain the evolution of vertebrates from ancestral chordates. Use what you know about tunicates to support this hypothesis.
7. The two curves below show relative changes in the populations of two mollusk species (*A* and *B*) in a coastal area over time. At the time indicated by the dotted line, a species of sea star that preys on both *A* and *B* was introduced to the area. Which mollusk species is the preferred prey of the sea stars? Why did the population of the other mollusk species also decline in size after the sea stars were introduced? Explain your answers.

Decline of Mollusk Populations

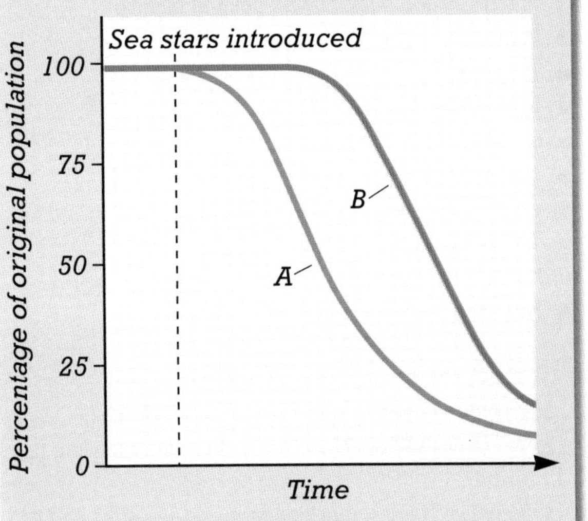

EXTENSION

1. Read "Universe in a Puddle" in *Life,* April 1998, on page 52. Describe some of the adaptations that tidal pool organisms have that help them survive the crashing and pounding of incoming tides.
2. Write a short paper on a species of echinoderms that is not discussed in the chapter. Focus on the particular adaptations that enable the species you have selected to survive in its environment.
3. Research several species of sea urchins that have very different types of spines. Why and how might these differences have evolved? Report on why you think sea urchin spines have different forms. Mention in your report the mechanisms that predators use to get past these spines.

more easily fossilized than soft-bodied animals.

4. Harvesting seasons could be set up, with limits on the number and size of sea urchins that could be collected. These limits would allow the animals to reproduce and increase their populations.

5. Basket stars are filter feeders, so they have much more food available to them at night. By curling up during the day, they make themselves less noticeable to predators.

6. Larval tunicates can swim, but adults are sessile. No vertebrates are sessile. If vertebrates evolved from tunicate-like ancestors, that might have happened when these ancestors became sexually mature as larvae. That would have led to chordates that can reproduce while still being able to swim.

7. Species *A* is the preferred prey, since its population began to decline much sooner after the sea stars were introduced than the population of species *B*. The population of species *B* also declined eventually, as species *A* became rarer and the sea stars were forced to prey on the less-preferred species *B*.

EXTENSION

1. Mussels anchor themselves to rocks. Barnacles glue themselves to a hard surface. Sea stars use tube feet to attach themselves securely between rocks.

2. The adaptations that are discussed should apply to the principles covered in the chapter.

3. Sea urchin spines may be long, short, blunt, sharp, colored, or black. Sea urchins' predators are as varied as their spines, which partly explains why a single type of spine isn't standard. Marine encyclopedias and books on marine invertebrates are a good starting place to identify these predators and their adaptations for eating urchins.

TIME REQUIRED

One 50-minute class period

SAFETY PRECAUTIONS

Students should wear safety goggles, gloves, and a lab apron. Caution students to keep their hands away from their eyes while handling the preserved specimens and to use care when working with the dissection scissors, sharp probe, and bleach.

QUICK REFERENCES

Lab Preparation Notes are found on page 779B.

Holt BioSources provides a Teaching Resources CD-ROM that contains "Using Gowin's Vee in the Lab" and "Scoring Rubrics."

PROCEDURAL TIP

Materials for this investigation can be purchased from WARD'S.

ANSWERS TO BACKGROUND

1. Major characteristics are pentaradial symmetry, an endoskeleton made of ossicles, a water-vascular system, and tube feet.
2. Spines give the phylum its name, which means "spiny skin."
3. Sea stars belong to the class Asteroidea; sea urchins belong to the class Echinoidea.
4. Pentaradial symmetry is a body symmetry in which parts extend from the center along five spokes.

ANSWERS TO PART A

2. The sea star's five arms are evidence of pentaradial symmetry.
4. The madreporite is a small spot on the aboral surface near the junction of two of the arms.
5. Spines are short, scattered over the surface, exposed, and fixed.
6. Pedicellariae are distributed between the spines.

Comparing Echinoderms

OBJECTIVES

- Compare the structure of two types of echinoderms.
- Infer function from the observation of structures.

PROCESS SKILLS

- using dissection instruments and techniques
- observing

MATERIALS

- preserved sea star
- preserved sea urchin
- sea urchin test
- dissection tray
- fine dissection scissors with pointed blades
- hand lens or dissecting microscope
- forceps
- blunt probe
- sharp probe
- dissection pins
- gloves (optional)
- 100 mL beaker
- household bleach
- Petri dish

Background

1. List the major characteristics of echinoderms.
2. Which feature of echinoderms gives their phylum its name?
3. To which classes of echinoderms do sea stars and sea urchins belong?
4. What is pentaradial symmetry?

PART A Observing the External Anatomy of a Sea Star

1. CAUTION **Put on safety goggles, gloves, and a lab apron.** Using forceps, hold a preserved sea star under running water to gently but thoroughly remove excess preservative. Then place the sea star in a dissection tray.
2. What evidence can you find that indicates that the sea star has pentaradial symmetry?
3. Make a table in your lab report like the one shown on the next page. As you observe each of the structures listed in the table, fill in the function of that structure.
4. While referring to Figure 40-7, find the madreporite on the aboral surface of your sea star. Record the madreporite's position and appearance.
5. Use a hand lens to examine the sea star's spines. Describe their size and shape. Are they distributed in any recognizable pattern on the surface of the sea star? Are they covered by tissue or exposed? Are they movable or fixed?
6. Are pedicellariae present? What is their location and arrangement? Draw one as it appears under the hand lens or dissecting microscope.
7. Use the dissecting microscope to look for small skin gills. If any are present, describe their location and structure.
8. Now examine the sea star's oral surface. Find the mouth, and describe its location and structure. Use forceps or a probe to gently move aside any soft tissues. What structures are found around the mouth?
9. Locate the tube feet. Describe their distribution. Using a dissecting microscope, observe and then draw a single tube foot.

PART B Observing the Internal Anatomy of a Sea Star

10. Using scissors and forceps, carefully cut the body wall away from the aboral surface of one of the sea star's arms. Start near the end of the arm and work toward the center. The internal organs may stick to the inside

of the body wall, so use a sharp probe to gently separate them from the body wall as you cut. Be careful not to damage the madreporite.

11. While referring to Figure 40-7, find the digestive glands in the arm you have opened. Describe their appearance. If you have dissected carefully, you should be able to find a short, branched tube that connects the digestive glands to the pyloric stomach. If you cannot find the digestive glands or this tube, repeat step 10 on one of the other arms and look for them there.
12. Cut the tube that connects the digestive glands to the pyloric stomach, and move the digestive glands out of the arm. Look for the testes or ovaries. If your specimen is an immature animal, these organs may be small and difficult to find.
13. Locate the two rows of ampullae that run the length of the arm. What is the relationship between the ampullae and the tube feet, which you observed on the oral surface?
14. Carefully remove the body wall from the aboral surface of the central region of the sea star. Try to avoid damaging the underlying structures. Locate the pyloric stomach and the cardiac stomach. How does a sea star use its cardiac stomach during feeding?
15. Remove the stomachs and find the canals of the water-vascular system: stone canal, ring canal, and radial canals. In which direction does water move through these canals?
16. CAUTION **Bleach is a highly corrosive agent. If you get it on your skin or clothing, wash it off at the sink while calling to your teacher. If you get it in your eyes, immediately flush it out at the eyewash station while calling to your teacher.** Cut a 1 cm cross section out of the middle of one of the arms. Using forceps, transfer the section to a small beaker containing enough bleach to cover it. The bleach will eat away the soft tissues, exposing the endoskeleton.
17. After about 10–15 minutes, use the forceps to carefully transfer the endoskeleton to a Petri dish containing tap water. Observe the endoskeleton under a dissecting microscope. Can you find individual ossicles?

FUNCTION OF SEA STAR STRUCTURES

Structure	Function
Madreporite	
Pedicellaria	
Skin gill	
Tube foot	
Digestive gland	
Ampulla	
Ossicle	

PART C Observing the Anatomy of a Sea Urchin

18. Using forceps, hold a preserved sea urchin under running water to remove excess preservative. Then place the sea star in a dissection tray. What evidence can you find that indicates that the sea urchin has pentaradial symmetry?
19. Observe the sea urchin's spines. Answer the same questions for the sea urchin's spines that you answered for the sea star's spines in step 5.
20. Examine the sea urchin's oral surface. Find the mouth and use a sharp probe to explore the structures around the mouth. How does the sea urchin's mouth differ from the sea star's?
21. Examine the sea urchin test. What might be the function of the rows of small pores on the test? What might be the function of the small bumps on the test?
22. Dispose of the specimens according to the directions from your teacher. Then clean up your materials and wash your hands before leaving the lab.

Analysis and Conclusions

1. What features are shared by sea stars and sea urchins?
2. What are some of the structural differences between sea stars and sea urchins?

Further Inquiry

Observe how living sea stars move in a saltwater aquarium. Add some live mussels, and observe the feeding behavior of the sea stars.

7. Skin gills are distributed between the spines but may not be visible in preserved specimens.
8. The mouth is in the center of the oral surface and is surrounded by five pairs of movable spines.
9. Tube feet are lined up in two rows running the length of each arm on the oral surface.

ANSWERS TO PART B

11. Digestive glands are long, paired organs in each arm that are connected by a duct to the pyloric stomach.
13. Each ampulla is connected to a tube foot.
14. A sea star turns its cardiac stomach inside out through its mouth when it feeds.
15. Water moves from the stone canal to the ring canal to the radial canals.
17. Individual ossicles should be visible when soft tissues are removed.

ANSWERS TO PART C

18. Differences in the lengths of the spines often follow a pentaradial pattern.
19. Spines are long, scattered over the surface, exposed, and movable.
20. The sea urchin's mouth contains five teeth connected to the plates of Aristotle's lantern.
21. Tube feet protruded through the pores. Spines were attached to the bumps.

ANSWERS TO ANALYSIS AND CONCLUSIONS

1. Shared features include pentaradial symmetry, spines, and tube feet.
2. Differences include the overall body shape, length and movability of spines, and location of tube feet.

Unit 9

VERTEBRATES

CHAPTERS

"Nature discloses the secrets of her past with the greatest reluctance. We paleontologists weave our tales from fossil fragments poorly preserved in incomplete sequences of sedimentary rocks. Most fossil mammals are known only from teeth—the hardest substance in our bodies—and a few scattered bones."

internet**connect**

SCI LINKS NSTA

National Science Teachers Association *sci*LINKS Internet resources are located throughout this unit.

Opossums are the only North American marsupial mammals.

This blue-spotted stingray, Taeniura lymna, is one of about 100 species of stingrays that belong in the class Chondrichthyes.

Antlers, such as those on this caribou, are bony outgrowths that are shed each winter.

These colorful rainbow lorikeets, also known as brush-tongued parrots, can be found in eastern Australia, where they feed on eucalyptus flowers.

Flap-necked chameleon

unit 9

CHAPTERS

41 *Fishes*

42 *Amphibians*

43 *Reptiles*

44 *Birds*

45 *Mammals*

VERTEBRATES

CONNECTING TO THE STANDARDS

The following chart shows the correlation of Unit 9 with the *National Science Education Standards* (grades 9–12). The items in each category are addressed in this unit and should be used to guide instruction. Annotated descriptions of the **Life Science Standards** are found below the chart. Consult the *National Science Education Standards* for detailed descriptions of each category.

UNIFYING CONCEPTS
- Systems, order, and organization
- Evolution and equilibrium
- Form and function

SCIENCE AS INQUIRY
- Abilities necessary to do scientific inquiry
- Understanding about scientific inquiry

SCIENCE AND TECHNOLOGY
- Abilities of technological design
- Understanding about science and technology

PHYSICAL SCIENCE
- Structure and properties of matter
- Interactions of energy and matter

LIFE SCIENCE

EARTH AND SPACE SCIENCE
- Energy in the Earth system
- Origin and evolution of the Earth system

SCIENCE IN PERSONAL AND SOCIAL PERSPECTIVES
- Natural resources
- Environmental quality
- Natural and human-induced hazards

HISTORY AND NATURE OF SCIENCE
- Nature of scientific knowledge
- Historial perspectives

ANNOTATED DESCRIPTIONS OF THE CORRELATED LIFE SCIENCE STANDARDS

Although all eight categories of the *National Science Education Standards* are important, the following descriptions summarize the **Life Science Standards** that specifically relate to Unit 9.

THE CELL
- Cells have particular structures that underlie their functions.
- Cells can differentiate. Complex multicellular organisms are formed from highly organized arrangements of differentiated cells.

BIOLOGICAL EVOLUTION
- Species evolve over time.
- The great diversity of organisms is the result of more than 3.5 billion years of evolution that has filled every available niche with life forms.
- Biological classifications are based on how organisms are related.

MATTER, ENERGY, AND ORGANIZATION IN LIVING SYSTEMS
- The complexity and organization within an organism result from the organism's need to obtain, transform, transport, release, and eliminate the matter and energy required to sustain the organism.

BEHAVIOR OF ORGANISMS
- Multicellular animals have nervous systems that generate behavior.
- Organisms have behavioral responses to internal changes and to external stimuli.
- Behaviors have evolved through natural selection.

TRENDS IN BIOLOGY

Cladistics

Birds are dinosaurs. A lungfish is more closely related to a cow than it is to a salmon. These are two of the surprising groupings that arise when the systematic method known as cladistics is applied to vertebrates. Cladistics, or phylogenetic systematics, is not new—the German entomologist Willi Hennig articulated its principles in 1950—but only in recent years has it gained widespread acceptance among biologists.

Cladistics comes from the Greek word for "branch," and the goal of a cladistic analysis is to reconstruct the branching sequence within a group, that is, the order in which the taxa diverged from a common ancestor. Cladists recognize branch points by identifying specific homologies known as derived characters, which are recently evolved features found in only some of the organisms in a given group. Because derived characters are distributed unevenly, they can serve as a basis for sorting the group members into subgroups. Older, more general homologies found in all of the organisms of the group are called ancestral characters. Since ancestral characters are present in all group members, they provide no information about the branching within the group and are ignored in cladistic analyses.

From the distribution of derived characters in the group, cladists are able to infer the evolutionary relationships among the group members, usually presenting the results in the form of a branching diagram called a cladogram (see the illustration at right). Each branch, or clade, contains taxa that share one or more derived characters. For example, the clade defined by the derived character of jaws, the Gnathostomes, includes all vertebrate groups except the jawless fishes. It is important to remember that all cladograms are provisional—they show hypotheses of evolutionary relationships that are subject to further testing and that can be disproved by additional evidence.

HOW TO READ A CLADOGRAM

A cladogram does not show ancestor-descendant relationships (which describe how one group gives rise to another). Instead, taxa are grouped by how recently they shared a common ancestor. Thus, placement of birds and crocodilians on the same branch does not mean that birds evolved from crocodilians, or vice versa, but rather that these two groups share a more-recent common ancestor with each other than they do with any other living vertebrate. The fact that cladistics ignores overall similarity accounts for some of the unfamiliar groupings, such as the lungfishes being placed with the tetrapods instead of with the bony fishes. Traditionally, systematists have classified lobe-finned fishes and ray-finned fishes together because of the presence of "fish" characters such as gills and fins in both groups. But cladists reject this classification because it is based on ancestral characters.

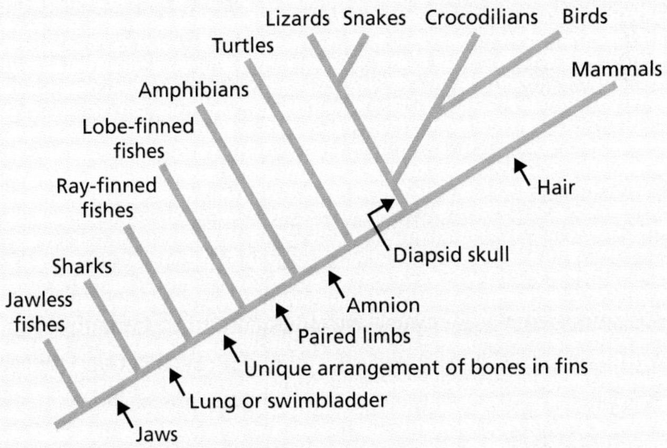

This cladogram of vertebrates shows some of the derived characters that define the major branches.

Additional Information

Gaffney, Eugene, Lowell Dingus, and Miranda Smith. "Why Cladistics?" *Natural History,* June 1995, 33–35.

Gould, Stephen Jay. "We Are All Monkeys' Uncles." *Natural History,* June 1992, 14–21. (Reprinted as "If Kings Can Be Hermits, Then We Are All Monkeys' Uncles." *Dinosaur in a Haystack.* New York: Harmony Books, 1996. 388–400.

Pough, F. Harvey, John Heiser, and William McFarland. *Vertebrate Life.* 4th ed. Upper Saddle River, NJ: Prentice Hall, 1996.

University of Arizona. Tree of Life page. World Wide Web: http://ag.arizona.edu/ENTO/tree/phylogeny.html

CHAPTER 41

FISHES

	TOPICS	TEACHING RESOURCES	LABS, CLASSWORK, AND HOMEWORK
BLOCK 1 45 minutes	**Introducing the Chapter, p. 798**	ATE Focus Concept, p. 798 ATE Assessing Prior Knowledge, p. 798 ATE Understanding the Visual, p. 798	■ Supplemental Reading Guide, *The Dinosaur Heresies*
	41-1 Introduction to Vertebrates, p. 799 Characteristics, p. 799 Evolution, p. 800	ATE Section Overview, p. 799 ATE Critical Thinking, p. 800 137A, 164, 168	PE **Quick Lab,** p. 800, Analyzing a Phylogenetic Tree **Lab A23,** Observing Some Major Animal Groups **Lab A26,** Vertebrate Skeletons ★ Study Guide, Section 41-1 PE Section Review, p. 801
BLOCK 2 45 minutes	**41-2 Jawless Fishes, Sharks, and Rays, p. 802** Life in the Water, p. 802 Class Agnatha, p. 803 Class Chondrichthyes, p. 804	ATE Section Overview, p. 802 ATE Critical Thinking, pp. 803, 804 ATE Visual Link Figure 41-6, p. 805 139A	PE **Quick Lab,** p. 804, Modeling a Shark Adaptation ★ Study Guide, Section 41-2 PE Section Review, p. 806
BLOCKS 3 & 4 90 minutes	**41-3 Bony Fishes, p. 807** Characteristics, p. 807 External Anatomy, p. 808 Internal Anatomy, p. 809 Reproduction, p. 812	ATE Section Overview, p. 807 ATE Critical Thinking, p. 808 ATE Visual Link Figure 41-8, p. 808 138A, 140A, 165–166, 172	PE **Chapter 41 Investigation,** p. 816 **Lab B26,** Perch Dissection **Lab A26,** Schooling Behavior in Fishes ★ Study Guide, Section 41-3 PE Section Review, p. 812
BLOCKS 5 & 6 90 minutes	**REVIEW and ASSESSMENT** PE Chapter 41 Review, pp. 814–815 ■ Performance-Based Assessment—The Challenge of Water Retention	Audio CD Program	**CHAPTER TESTING** ■ Chapter Test (blackline master) ▲ Test Generator ■ Assessment Item Listing ■ Scoring Rubrics and Classroom Management Checklists

KEY

MODERN BIOLOGY

- **PE** Pupil's Edition
- **ATE** Teacher's Edition
- ■ Active Reading Guide
- ★ Study Guide

HOLT BIOSOURCES

- Teaching Transparencies
- Laboratory Program

One-Stop Planner CD-ROM
Shows these resources within a customizable daily lesson plan:

- ■ Holt BioSources Teaching Resources
- ■ Active Reading Guide
- ★ Study Guide
- ▲ Test Generator

READING FOR CONTENT MASTERY

Preview: Have students preview the objectives, the introductory paragraph, topic sentences, and captions before reading the chapter.

- **ATE** Active Reading Technique: Reading Organizer, p. 799
- ■ Reading Strategy: Series-of-Events-Chain Worksheet
- ■ Active Reading Guide Worksheet 41-1
- **PE** Summary/Vocabulary, p. 813

- ■ Active Reading Guide Worksheet 41-2
- **PE** Summary/Vocabulary, p. 813

- **ATE** Active Reading Technique: K-W-L, p. 807
- ■ Reading Strategy: K-W-L Worksheet
- ■ Active Reading Guide Worksheet 41-3
- **PE** Summary/Vocabulary, p. 813

TECHNOLOGY AND INTERNET RESOURCES

CNN PRESENTS Video Segment 27 Fish Farming

Holt Biology Videodiscs Teacher's Correlation Guide, Lessons 41-1 through 41-3

Audio CD Program, Sections 41-1 through 41-3

Interactive Exploration: Human Biology, Evolution of the Heart

internetconnect

SciLINKS NSTA

On-line Resources: www.scilinks.org
The following *sci*LINKS Internet resources can be found in the student text for this chapter:

Topics:
- Fish, p. 799
- Anatomy of a Bony Fish, p. 809

go.hrw.com

On-line Resources: go.hrw.com
Visit the HRW Web site for a variety of resources related to this chapter. Just type in the keyword HM2 Chapter 41.

Smithsonian Institution®
Visit **www.si.edu/hrw** for additional on-line resources.

CNNfyi.com
Visit **www.cnnfyi.com** for late-breaking news and current event stories selected just for you.

LAB ACTIVITY PLANNING

CHAPTER 41 INVESTIGATION

Exploring the Fish Heart, pp. 816–817

OVERVIEW
This lab introduces students to the function of a fish heart. This lab also compares the function and efficiency of a fish heart with the function and efficiency of other vertebrate hearts.

PREPARATION
1. Access to the following materials is needed:
   - computer with CD-ROM drive
   - CD-ROM *Interactive Explorations in Biology: Human Biology*. The title of this program on the CD-ROM is "Evolution of the Heart."
2. Have students read the Objectives, Background, and Prelab Preparation sections before starting the lab.

Quick Labs

Analyzing a Phylogenetic Tree, p. 800
Students need only paper and pencils for this activity.

Modeling a Shark Adaptation, p. 804
Each student group will need 8 cm dialysis tubing, two 8 cm lengths of string, 100 mL salt solution (5 percent), a 250 mL beaker, 10 mL distilled water, a scale, and a graduated cylinder for this activity. You will need to prepare the salt solution ahead of time by diluting 5 g of salt with water to a final volume of 100 mL.

UNDERSTANDING THE VISUAL

Ask students to examine the whale shark and compare it to the diver shown above the whale shark. Students should immediately notice the difference in size. Point out that the whale shark is the largest shark species. Ask them to contrast the adaptations the shark has developed which the diver does not possess. (streamlined body, fins and a tail for swimming, and gills for breathing).

FOCUS CONCEPT

Structure and Function

All fishes use feathery respiratory structures called gills to remove oxygen from and release waste products into the water. Cartilaginous fishes and bony fishes have jaws, which they use to grasp and manipulate prey. Bony fishes use a structure called a swim bladder to regulate their vertical position in the water.

ASSESSING PRIOR KNOWLEDGE

Review the following concepts.

Evolution: *Chapter 15*
Ask students to explain the process of evolution by natural selection.

Classification: *Chapter 34*
Have students name the major chordate subphyla and the invertebrate phylum that probably shares a common ancestor with the chordates.

Diffusion: *Chapter 5*
Ask students to compare and contrast the terms *hypertonic, hypotonic,* and *isotonic.*

CHAPTER 41

FISHES

The whale shark lives in a saltwater environment, for which it has special organs and biochemical adaptations.

FOCUS CONCEPT: ***Structure and Function***

As you read the chapter, focus on the adaptations that allow fishes to survive and reproduce in aquatic environments.

INTRODUCTION TO VERTEBRATES

Although the vertebrates are not the largest or most abundant group of animals, they are the most familiar to us. This is partly because we are vertebrates. Vertebrates are an important part of our diet. In many parts of the world, vertebrates, not machines, perform the hard work of pulling plows and hauling heavy loads. We have also taken vertebrates into our homes as pets.

SECTION

41-1

OBJECTIVES

▲ Identify the distinguishing characteristics of vertebrates.

● List the seven major groups of vertebrates, and give an example of each.

■ Describe the characteristics of the early vertebrates.

◆ Explain the importance of jaws and paired fins for fishes.

CHARACTERISTICS

Vertebrates are one subphylum within the phylum Chordata. Like other chordates, vertebrates have, at some stage of life, a notochord, a dorsal hollow nerve cord, pharyngeal pouches, and a post-anal tail. Vertebrates are a distinct group because they have three characteristics that distinguish them from other chordates. First, vertebrates have vertebrae, bones or cartilage that surround and protect the dorsal nerve cord. The vertebrae form the **vertebral column,** or spine. Second, vertebrates have a **cranium,** or skull, that protects the brain. Third, all vertebrates have an endoskeleton composed of bone or cartilage.

Classification

Today there are about 45,000 species of vertebrates. They occupy all but the most extreme terrestrial habitats. More than 24,000 vertebrate species are fishes. Fishes are found in a wide range of water habitats. The major groups of vertebrates are summarized below.

- **Lampreys and Hagfishes** (class Agnatha)—These fishes have elongated, eel-like bodies, and they lack jaws, paired fins, and bone. There are about 80 species.
- **Sharks, Rays, and Skates** (class Chondrichthyes)—These predatory fishes have jaws and paired fins. Their skeleton is made of cartilage, not bone, and their skin is covered by a unique kind of scale. There are about 800 species.
- **Bony Fishes** (class Osteichthyes)—Most familiar fishes, such as guppies, salmon, bass, and catfish, are bony fishes. All have jaws, and most species have a skeleton composed of bone. This group comprises two main lineages: the ray-finned fishes and the lobe-finned fishes. There are over 23,000 species of bony fishes.

Word Roots and Origins

vertebra

from the Latin *vertebra*, meaning "a joint"

internetconnect

SCILINKS NSTA

TOPIC: Fish
GO TO: www.scilinks.org
KEYWORD: HM799

SECTION 41-1

SECTION OVERVIEW

Structure and Function
Section 41-1 describes the anatomical structures that characterize all vertebrates.

Evolution
Section 41-1 describes the classification of modern vertebrates into seven groups. It also discusses the evolution of vertebrates and jaws.

ACTIVE READING

Technique: Reading Organizer

As students read pp. 799–801, have them complete a series-of-events reading organizer to describe the adaptations in body structure for the vertebrate classes Agnatha, Chondrichthyes, and Osteichthyes. Organizers will vary but should summarize the key features of each group.

RECENT RESEARCH

The First Vertebrate

How vertebrates first evolved from invertebrates is a question that scientists have studied over for the past two centuries. Recent genetic evidence indicates that a doubling in the gene composition of an *Amphioxus*-like invertebrate may have produced the first vertebrate. The scientists think that this kind of sudden genetic expansion could have been caused by random mutation.

Quick Lab

Analyzing a Phylogenetic Tree

Time Required 20 minutes

Procedural Tips Have students work in pairs. Refer students to pp. 342–343 for a review of phylogenetic trees.

Answers to Analysis All vertebrates have a cranium, an endoskeleton, and vertebrae that form a spine. The development of bone leads to the class Osteichthyes. Mammals have hair and nurse their young with milk. Class Aves and class Reptilia share the most recent common ancestor.

CRITICAL THINKING

What Happened to the Jawless Fishes?

During the first 50 million years of vertebrate evolution, many kinds of jawless fishes thrived. Today, lampreys and hagfishes are the only jawless fishes that survive. Ask students to offer reasons for the low number of species that remain today based on what they know about the structure of jawless fishes. (Without paired fins, jawless fishes are not good swimmers, and without jaws, they are not efficient hunters. Once fishes with jaws and paired fins evolved, they soon became dominant.)

FIGURE 41-1
This phylogenetic tree shows the evolutionary relationships among vertebrates.

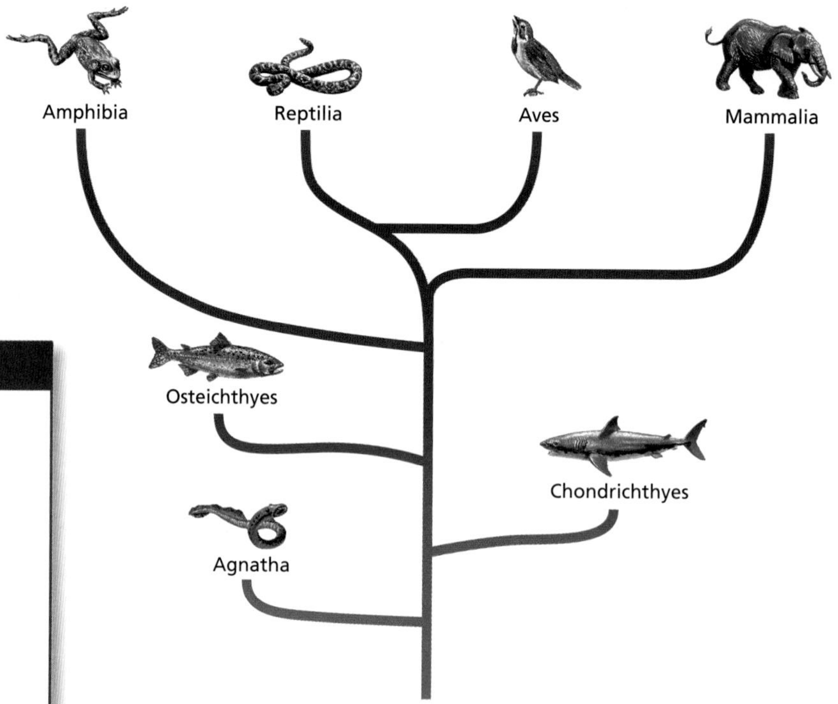

Quick Lab

Analyzing a Phylogenetic Tree

Materials paper, pencil

Procedure

1. Draw the phylogenetic tree shown on this page on your paper.
2. Using the information on pp. 799 and 800, determine the key characteristics that distinguish each vertebrate group. Indicate these evolutionary changes on the branches of the tree to make a diagram of the relationship that exists among vertebrates. Begin at the bottom of the tree with the key characteristics that distinguish vertebrates from other chordates.

Analysis What characteristics are shared by all vertebrates? What key characteristic separates the classes Chondrichthyes and Osteichthyes? What adaptations lead to the divergence of mammals? Which two groups of vertebrates share the most recent common ancestor?

- **Amphibians** (class Amphibia)—Frogs, toads, and salamanders belong to this group. Their skin is thin and is permeable to gases and water. Most species lay their eggs in water and pass through an aquatic larval stage.
- **Reptiles** (class Reptilia)—This group includes turtles, crocodiles, alligators, lizards, and snakes. The skin of reptiles is dry and scaly. The eggs of reptiles protect the embryo from drying out and can be laid on land. There are about 6,000 species.
- **Birds** (class Aves)—Birds are characterized by adaptations that enable flight, including feathers, hollow bones, and a unique respiratory system. There are over 10,000 species.
- **Mammals** (class Mammalia)—Humans, cats, mice, and horses are among the members of this group. All mammals have hair and nurse their young with milk. There are about 4,400 species. Figure 41-1 shows the relationships among the major groups of living vertebrates.

EVOLUTION

Most biologists think that vertebrates originated about 550 million years ago, shortly after the first chordates appear in the fossil record. The oldest known vertebrate fossils are those of jawless fishes. They appear in the fossil record starting a little over 500 million years ago. Most early jawless fishes did not have paired fins. Their bodies were covered with heavy, bony scales, but their skeletons were composed of cartilage. Figure 41-2 shows an artist's reconstruction of one of these fishes.

Jawless fishes were the only vertebrates for more than 50 million years, and they diversified into many evolutionary lines. By about 350 million years ago, most of these lines had become extinct. The survivors became the ancestors of today's jawless fishes (class Agnatha).

FIGURE 41-2

Early jawless fishes, such as this *Pharyngolepis,* lacked paired fins and probably fed on small invertebrates. Most species of early jawless fishes were less than 15 cm (6 in.) in length.

Origin of Jaws

About 440 million years ago, the first fishes with jaws and paired fins appeared. Paired fins increased their stability and maneuverability, and jaws allowed them to seize and manipulate prey. Jaws are thought to have evolved from the first pair of **gill arches,** the skeletal elements that support the pharynx. Figure 41-3 shows three possible stages in this transformation.

The first fishes to have paired fins and jaws were the acanthodians (AY-kan-THOH-dee-uhnz), or spiny fishes, in the class Acanthodii. Acanthodians became extinct about 270 million years ago. Modern fishes—the sharks and rays (class Chondrichthyes) and the bony fishes (class Osteichthyes)—make their first appearance in the fossil record about 400 million years ago.

FIGURE 41-3

Jaws are thought to have developed from the gill arches of early jawless fishes. These figures represent hypothesized stages of evolution.

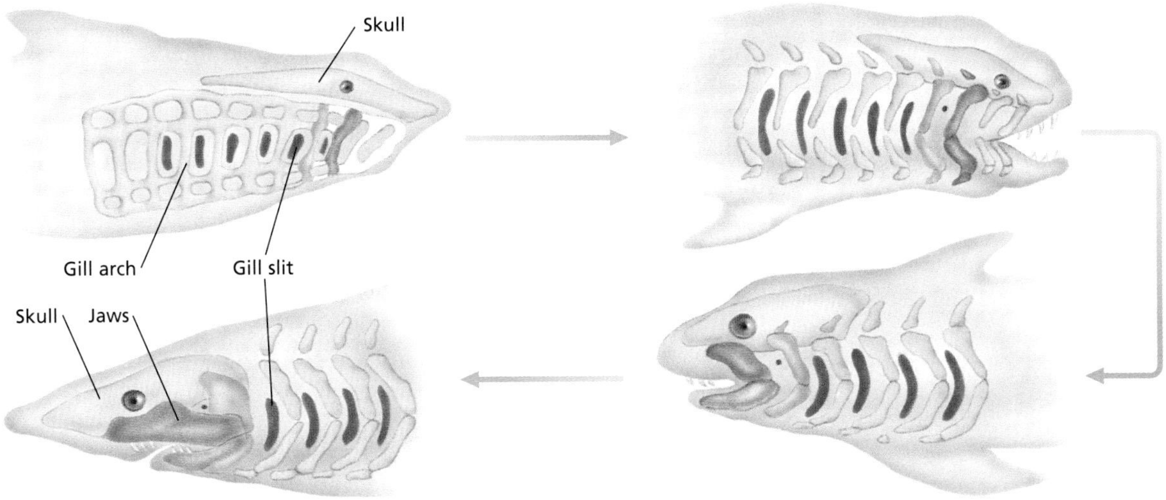

SECTION 41-1 REVIEW

1. What are three characteristics of vertebrates?
2. Which group of modern vertebrates contains the jawless fishes?
3. Name two differences between the first fishes and modern bony fishes.
4. Explain why scientists think that the early jawless fishes were probably awkward swimmers.
5. List the function of jaws.
6. **CRITICAL THINKING** Explain why scientists think that the vertebrates evolved from chordates in the sea.

TEACHING STRATEGY

Observing *Amphioxus*

Furnish students with slides of *Amphioxus* and hand lenses or stereoscopes. Have them observe and sketch this cephalochordate and label its muscle bundles, dorsal nerve cord, notochord, and pharyngeal gill slits. After students finish their observations, tell them that scientists think the first vertebrates probably evolved from a cephalochordate that resembled *Amphioxus.* Ask students to use their observations of *Amphioxus* to justify this reasoning. (Students can cite the fishlike appearance of *Amphioxus,* the presence of bilaterally symmetrical muscle mass, a tail fin, gill slits, and the dorsally located nerve cord.)

GIFTED *ACTIVITY*

Ask students to research and prepare written reports on one or more of the extinct groups of fishes, such as the ostracoderms, acanthodians, and placoderms. Students should describe the structural and behavioral characteristics of these fishes and offer reasons for their extinction.

ANSWERS TO SECTION 41-1 REVIEW

1. They have vertebrae, a cranium, and an endoskeleton of bone or cartilage.
2. class Agnatha
3. Early fishes did not have jaws, or a bony skeleton, and they were covered with bony plates.
4. because they had no paired fins, which stabilize fish during swimming
5. for feeding
6. Early chordate and vertebrate fossils come from marine sediment, and most of the closest relatives of vertebrates live in the sea.

SECTION OVERVIEW

Structure and Function
Section 41-2 describes how structures such as fins, gills, and lateral lines enable fishes to dominate aquatic environments.

Stability and Homeostasis
Section 41-2 explains how osmosis causes freshwater fishes to gain water and saltwater fishes to lose water. Various physiological mechanisms allow fishes to maintain a stable internal environment.

▶ ENGAGE STUDENTS

Fill three 500 mL beakers halfway with water. Drop a marble into one beaker, add a dropperful of cooking oil to the second beaker, and drop a small inflated plastic bag or balloon in the third beaker. Ask students to comment on what they observe. (The marble sinks, but the oil and bag float.) Tell students that sharks store a lot of oil in their bodies, which helps them stay buoyant in water. More-advanced fishes, such as bony fishes, contain gas-filled sacs that keep them from sinking to the bottom.

TEACHING STRATEGY

Freshwater and Saltwater Fishes

Have students put 10 mL of a 5 percent salt solution inside a piece of dialysis tubing and tie both ends of the tube with string. Students should then record the weight of the filled tube and place it in a beaker containing distilled water. After 10 minutes, have students remove the tube and reweigh it. (Explain that the hypertonic solution will tend to gain water from the hypotonic solution, so fish in a saltwater environment will lose water.)

SECTION

41-2

OBJECTIVES

▲ Identify three problems faced by all fishes.

● Describe the feeding behavior of lampreys and hagfishes.

■ Identify two characteristics of cartilaginous fishes.

◆ Describe how sharks detect prey.

▲ Describe reproduction in cartilaginous fishes, and contrast it with reproduction in jawless fishes.

JAWLESS FISHES, SHARKS, AND RAYS

The term fish *generally refers to three distinct classes of living vertebrates: Agnatha, jawless fishes; Chondrichthyes, cartilaginous fishes; and Osteichthyes, bony fishes. Fishes are the most numerous and widespread of all vertebrates.*

LIFE IN THE WATER

The body plan of a fish makes it well-suited to live in a water environment. A streamlined shape and a muscular tail enable most fishes to move rapidly through the water. Paired fins allow fishes to maneuver right or left, up or down, and backward or forward. Unpaired fins on the back and belly increase stability. In addition, most fishes secrete a mucus that reduces friction as they swim.

Most of the tissues in a fish's body are denser than water. By controlling the amount of gas in their bodies, many fishes can regulate their vertical position in the water. Some fishes also store lipids, which are less dense than water and therefore help them to float.

Fishes need to absorb oxygen and rid themselves of carbon dioxide. However, scales on fishes limit diffusion through the skin. Instead, most exchanges between water and blood take place across the membranes of gills—the internal respiratory organs of fishes.

Homeostasis

The concentration of solutes in a fish's body usually differs from the concentration of solutes in the water in which the fish swims. The body of a freshwater fish has a higher concentration of solutes than the surrounding water does, so the fish tends to gain water through osmosis and lose ions, such as sodium and chloride ions, through diffusion. Most saltwater fishes contain lower concentrations of solutes than their surroundings do. Thus, saltwater fishes tend to lose water and gain ions.

Like all organisms, fishes must also rid themselves of the waste products produced by metabolism. The kidneys and gills play important roles in maintaining homeostasis in the tissues and in getting rid of metabolic wastes. The kidneys filter the blood and help regulate the concentration of ions in the body. The gills release wastes, such as carbon dioxide and ammonia, and either absorb or release ions, depending on whether the fish lives in fresh water or in salt water.

Eco Connection

Hagfish Depletion

Hagfishes have recently become economically important. Most "eel-skin" products, such as wallets, are actually made from the tanned skin of hagfishes. The demand for these products is so high that hagfish populations in some parts of the world have been almost wiped out by overfishing.

Sensory Functions

A prominent adaptation present in nearly all fishes is the **lateral line system.** This system consists of a row of sensory structures that run the length of the fish's body on each side and that are connected by nerves to the brain. The lateral line system detects vibrations in the water, such as those caused by a fish swimming nearby. Most fishes also have highly developed senses of smell and sight. Hearing is an important sense for some fishes, and some species can detect electrical currents.

Word Roots and Origins

agnatha

from the Greek *gnathus,* meaning "jaws," and *a,* meaning "without"

CLASS AGNATHA

The only existing jawless fishes are the 80 species of hagfishes and lampreys that compose the class Agnatha. Figure 41-4 shows an example of both types of jawless fishes. Their skin has neither plates nor scales. Hagfishes and lampreys have an eel-like body, a cartilaginous skeleton, and unpaired fins. The notochord remains throughout life. Hagfishes have small eyes that are beneath the skin, while lampreys have large eyes. Hagfishes lack vertebrae and are unique among vertebrates in that their body fluids have nearly the same ion concentration as sea water. Hagfishes live only in the oceans. Many lampreys live permanently in fresh water, and all species reproduce in fresh water.

Hagfishes

Hagfishes are bottom dwellers in cold marine waters. They feed on small invertebrates or on dead and dying fish. Because the hagfish lacks jaws, it cannot bite, but within its mouth are two movable plates and a rough tonguelike structure that it uses to pinch off chunks of flesh. Hagfishes often burrow into the body of a dead fish through the gills, skin, or anus. Once inside, they eat the internal organs. Hagfishes can tie their bodies into knots to evade capture. They also secrete bad-tasting slime to discourage predators. Hagfishes usually remain hidden in mud burrows on the ocean floor.

Lampreys

About half the species of lampreys are free-living and do not feed as adults. The other half are parasites as adults and feed on the blood and body fluids of other fishes. Once a suitable host is located, a lamprey uses its disk-shaped mouth to attach to the host. Then it scrapes a hole in the host with its rough tongue and secretes a chemical that keeps the host's blood from clotting. The lamprey feeds on the blood and fluids that leak from the wound. After feeding, the lamprey drops off. The host may recover, bleed to death, or die from an infection.

Some lamprey species spend most of their adult lives in the ocean. Others live in rivers or lakes and never enter salt water. All lampreys breed in fresh water, usually choosing a shallow stream

FIGURE 41-4

Hagfishes (a) and lampreys (b) are modern jawless fishes. What feature can you see on the lamprey and hagfish that would prevent them from swimming as well as bony fishes and sharks?

(a)

(b)

CRITICAL THINKING

Feeding Adaptations

Hagfishes have tiny eyes buried beneath skin and lampreys have large, exposed eyes. Have students explain how each type of eye is an adaptation for feeding style. (Hagfishes are bottom-dwelling scavengers that burrow into the bodies of dying animals. They do not need well-developed eyes because light is scarce where they live. The skin protects their eyes as they feed. Parasitic lampreys swim in open water and need good vision to locate hosts.)

QUICK FACT

☑ Most fishes exchange gases by taking water into their mouths and passing it over their gills. Lampreys cannot exchange gases in this manner because their mouth is usually tightly clamped to a host. The gills of lampreys are housed inside muscular pouches that pump water through their own openings.

DEMONSTRATION

Lateral Line Mechanisms

Fill a shallow dish about halfway with water. Color the water with several drops of food dye. Cut a 3 in. piece of thin thread, and lay it over the side of the dish so that one end is floating on the surface of the water. Let students strike and listen to a tuning fork. Now strike the tuning fork and touch its vibrating end to the surface of the water close to but not touching the thread. Ask students to observe the end of the thread for movement. Tell students that the lateral lines of fishes contain long, hairlike structures that are attached to nerve cells. When sound vibrations in the water cause the hairs to move, the nerve cells send signals to the fish's brain.

Quick Lab

Modeling a Shark Adaptation

Time Required 20 minutes

Procedural Tips Prepare the salt solution ahead of time by dissolving 5 g of salt in 100 mL of tap water. Cut the string in 8 cm lengths. Have students soak the dialysis tubing in water for 30 seconds before using. Refer students to pp. 96–97 for a review of osmosis.

Answers to Analysis Students will observe a decrease in the weight of the tubing. The salt solution is hypertonic, and water is lost through osmosis. Sharks tend to lose water and gain ions because sharks contain lower concentrations of solutes than their surroundings. The kidneys and gills play a key role in maintaining homeostasis in sharks by regulating the concentration of ions in the body.

CRITICAL THINKING

Jaw Evolution

Tell students to use their understanding of evolution by natural selection and jaw development to offer a scenario that explains how certain early jawless fishes may have given rise to fishes with jaws. (Any answer in which students correctly cite the mechanism of natural selection should be accepted. One possible scenario is that some early jawless fishes had a more flexible first gill arch than others, allowing them to grab and hold on to food more easily than the others. This advantageous trait was selected for and eventually led to the development of a jawed mouth.)

Quick Lab

Modeling a Shark Adaptation

Materials 8 cm dialysis tubing, 8 cm length of string (2), 100 mL salt solution (5 percent), 250 mL beaker, 10 mL distilled water, scale, graduated cylinder

Procedure

1. Tightly tie one end of the dialysis tubing with string. Place 10 mL of distilled water inside the dialysis tubing, and tie the other end of the tube tightly with string.
2. Record the initial weight of the filled tube.
3. Add the filled dialysis tubing to a beaker filled with 100 mL of salt solution.
4. After 10 minutes, remove the tubing from the beaker, blot the outside, and reweigh it. Record your observations.

Analysis Explain the reason for the change in the weight of the tube, if any. Are sharks likely to lose or gain ions? What physical structures of a shark play a key role in maintaining ionic homeostasis in sharks?

FIGURE 41-5

These teethlike placoid scales are found on the skin of cartilaginous fishes. What advantage might they give a shark in swimming?

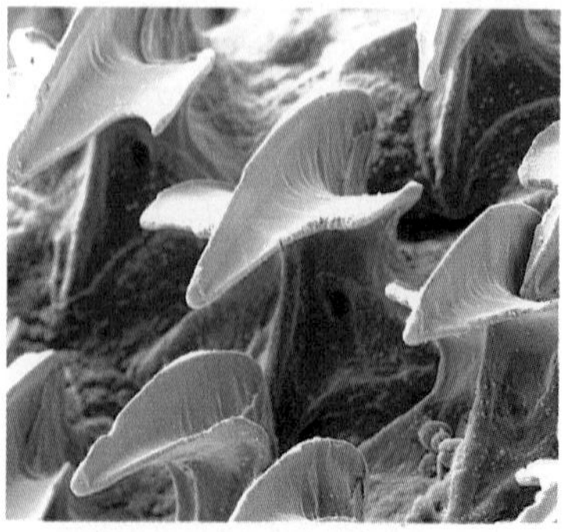

with a gravel bottom. They scrape out a small nest in the gravel, and the female releases eggs while the male releases sperm. Fertilization occurs outside the body of either parent. This type of fertilization is known as **external fertilization.** The eggs hatch into small larvae that closely resemble an amphioxus, or lancelet. The larvae eventually transform into adults.

CLASS CHONDRICHTHYES

Sharks, skates, and rays belong to the class Chondrichthyes. Because the fishes in this class have skeletons composed of cartilage, they are also called cartilaginous fishes. **Cartilage** is a flexible, lightweight material made of cells surrounded by tough fibers of protein. Sharks, skates, and rays differ from lampreys and hagfishes in that they have movable jaws, skeletons, and paired fins. Almost all of the approximately 800 species of sharks, skates, and rays live in salt water. All species are carnivores and some are scavengers. They eat many different kinds of foods, including fishes, seals, aquatic invertebrates, and plankton. The skin of cartilaginous fishes is covered with **placoid** (PLA-koyd) **scales**—small, toothlike spines that feel like sandpaper. Placoid scales, shown in Figure 41-5, probably reduce turbulence of the water flow and thus increase swimming efficiency.

Sharks

The largest sharks, the whale shark (up to 18 m, or 59 ft, long) and the basking shark (up to 15 m, or 49 ft, long), feed on plankton. Sharks swim in a side-to-side pattern created by the motion of their asymmetrical tail fin. Just behind a shark's head are paired **pectoral** (PEK-tuh-ruhl) **fins,** which jut out from the body like the wings of a plane. Figure 41-6 shows the external structure and internal anatomy of a shark.

The mouth of a typical shark has 6 to 20 rows of teeth that point inward. When a tooth in one of the front rows breaks or wears down, a replacement moves forward to take its place. One shark may use more than 20,000 teeth over its lifetime. The structure of each species' teeth is adapted to that species' feeding habits. Sharks that feed primarily on large fish or mammals have big, triangular teeth with sawlike edges that hook and tear flesh.

Sharks use several senses to locate prey. Their ability to detect chemicals—that is, their sense of smell—is particularly acute. Paired nostrils on the snout have specialized nerve cells that connect with the **olfactory** (awl-FAK-tuh-ree) **bulbs** of the brain, where the information from the nostrils is analyzed. Water entering the nostrils is continuously monitored for chemicals. Sharks also have a well-developed lateral line system. Their vision is keen, even at low levels of light. They are also extremely sensitive to electrical fields, such as those generated by the muscular contractions of animals.

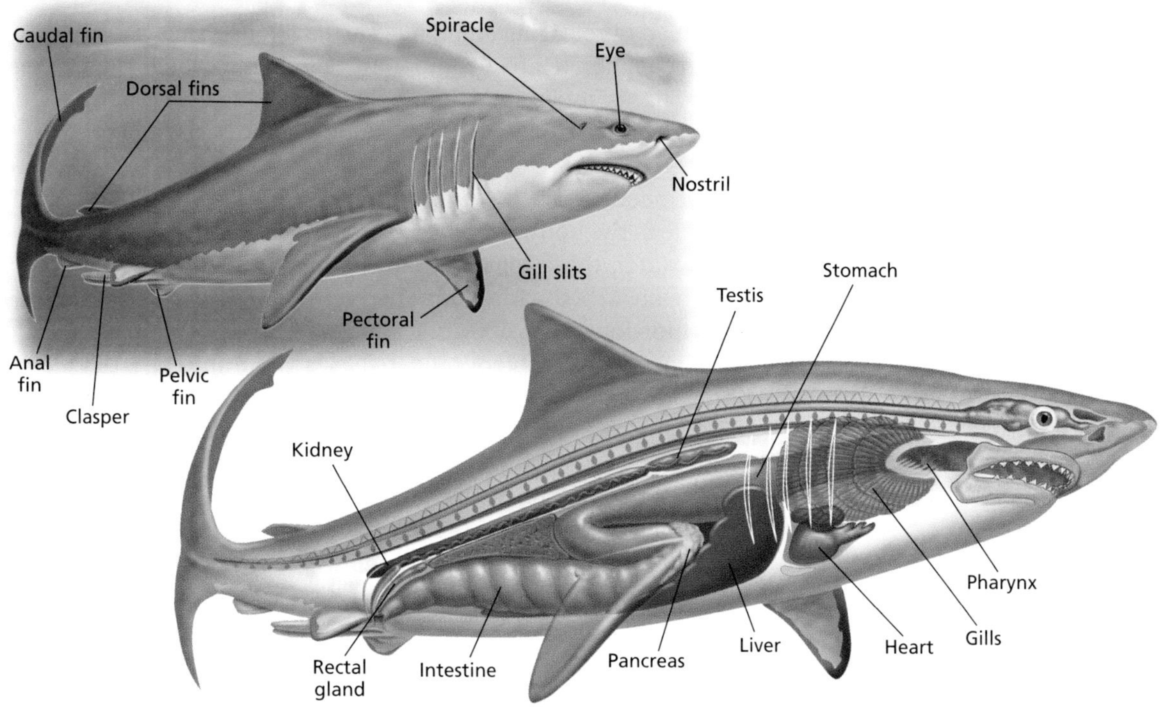

FIGURE 41-6

A streamlined shape allows the shark to slip through the water with little resistance. The shark's powerful caudal fin propels it forward, and the paired fins help steer and stabilize it. The shark's internal anatomy includes organs for digestion, reproduction, and maintaining homeostasis. Note how the rectal gland is connected to the lower part of the intestine. This makes the disposal of excess salts easier.

Sharks detect these fields by means of sensitive receptors located in small pits scattered over their head. Tests have shown that sharks can find and capture prey using only their sensitivity to electrical fields. Evaluating and integrating information from all these different senses requires a large and complex brain. Sharks have the largest brain—for their body size—of any fishes.

Rays and Skates

Rays and skates have flattened bodies with paired winglike pectoral fins and, in some species, whiplike tails. Rays have diamond- or disk-shaped bodies, while most skates have triangular bodies. Most rays and skates are less than 1 m (3.3 ft) long. Rays and skates are primarily bottom dwellers. Their flat shape and coloration camouflage them on the ocean floor. Most rays and skates feed on mollusks and crustaceans. Figure 41-7 shows an example of a ray.

Adaptations of Cartilaginous Fishes

In cartilaginous fishes, gas exchange occurs in the gills, which lie behind the head. Efficient gas exchange requires a continuous flow of water across the gills. Some fast-swimming sharks are able to push water through their mouth, over their gills, and out of their gill slits by swimming alone. However, most sharks and rays pump water over their gills by expanding and contracting their mouth cavity and pharynx. When lying on the bottom, rays and skates cannot bring in water through their mouth, which is located on the ventral surface of their body, so instead, they draw water in through their spiracles, which are two large openings on the top of the head, just behind the eyes.

FIGURE 41-7

This blue-spotted stingray, *Taeniura lymna,* is an example of a bottom dweller. This stingray was photographed in the Red Sea near Egypt.

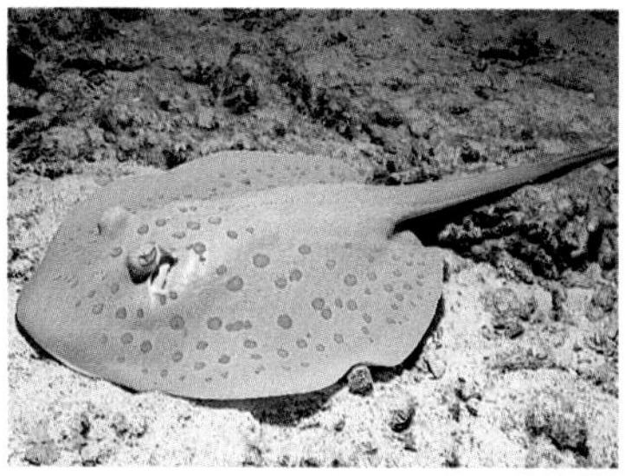

SECTION 41-2

VISUAL LINK

Figure 41-6

Refer students to Figure 41-6, and ask them to compare the shark body with the hagfish and lamprey bodies. Ask students what features make sharks better predators (jaws, teeth, powerful muscles to swim fast, and paired fins to control swimming).

RETEACHING *ACTIVITY*

Sketch the outline of a shark on the board. Make the sketch 5 or 6 ft long, a realistic size for a shark. Have the students use Figure 41-6 as a reference as they draw internal organs inside the outline. Have them label each organ and write its function on the board.

MAKING CONNECTIONS

Mathematics

The number of species of jawless, cartilaginous, and bony fishes has varied during the past 500 million years. Have students research the species diversity of these fish groups over this time period, and ask them to graphically represent what they find. (The greatest number of species of jawless fishes lived from about 500 million to 400 million years ago. Sharks and rays had two periods of species expansion. One period was about 375 million to 275 million years ago, and the other was about 200 million to 75 million years ago. Bony fishes expanded 400 million to 275 million years ago. Another expansion began 150 million years ago and continues today. Today, there are more species of bony fishes than ever.)

SECTION 41-2

INCLUSION ACTIVITY

Whale sharks can grow to almost 60 ft long. Show students a picture of a whale shark and challenge them to create a life-size outline of one on a playing field or gym floor. Tell them to use yarn to make the image. Have them begin the project by making a scaled drawing of a whale shark on a sheet of graph paper. If they allow 5 grid spaces on their paper to equal 3 ft, then their shark drawing will be 100 grid spaces long. Once the students agree on an accurate scaled drawing, give them skeins of yarn and measuring tapes, and let them work as a team to create their shark outline. After it is finished, have the students stand at different places along the outline to give them a better sense of the whale shark's immense size. Stand in the middle of the shark outline, and ask the students why there is no need to worry about being eaten by a whale shark. (Whale sharks eat only tiny plankton.)

ANSWERS TO SECTION 41-2 REVIEW

1. It detects vibrations in the water.
2. They attach to the body of a host fish, scrape a hole in the body, and drink the blood and other fluids of the host.
3. cartilage skeletons and placoid scales
4. sense of smell and electrical-field detection
5. Fertilization that occurs outside the body of either parent is known as external fertilization. Internal fertilization occurs within the body of the female.
6. Fishes that have lived a long time with lampreys have evolved ways to protect themselves. The Great Lakes fishes had not yet done so.

Instead of releasing ammonia, cartilaginous fishes use energy to convert ammonia into a compound called urea, which is much less toxic. Sharks retain large amounts of urea in their blood and tissues, thus raising the concentration of solutes in their body to at least the same level as that found in sea water. As a result, sharks do not need to drink. Because the concentration of sodium and chloride in the body of a shark is less than the concentration found in sea water, these ions still diffuse into the body across the gills and are absorbed with food. The **rectal gland,** located in the posterior portion of the intestine, removes excess sodium and chloride ions from the blood and releases them into the rectum for elimination.

Cartilaginous fishes maintain their position in the water in two ways. First, because a fish generates lift, or upward force, as it swims, it can remain at the same level in the water, counteracting the tendency to sink, as long as it keeps moving. Some open-water sharks rely entirely on this method, but it does use up a lot of energy. Second, many cartilaginous fishes store large amounts of low-density lipids, usually in the liver. A shark's lipid-filled liver may account for 25 percent of its mass. Lipids give sharks buoyancy by reducing the overall density of the body. Because continuous swimming is not necessary, less energy is used.

Reproduction in Cartilaginous Fishes

Cartilaginous fishes differ from jawless fishes in that fertilization occurs inside the body of the female. This type of fertilization is called **internal fertilization.** During mating, the male transfers sperm into the female's body with modified pelvic fins called **claspers.** In a few species of sharks and rays, the females lay large yolky eggs right after fertilization. The young develop within the egg, are nourished by the yolk, and hatch as miniature versions of the adults. The eggs of many species develop in the female's body, and the young are born live. In some of these species, the mother nourishes the developing sharks while they are in her body. No cartilaginous fishes provide parental care for their young after birth or hatching.

SECTION 41-2 REVIEW

1. What is the function of the lateral line in fishes?

2. How do parasitic lampreys feed?

3. Name two characteristics shared by all cartilaginous fishes.

4. A shark can locate and capture a fish buried in the gravel at the bottom of its tank. Describe at least two senses that the shark might be using to find the fish.

5. Define the terms *internal fertilization* and *external fertilization.*

6. CRITICAL THINKING When sea lampreys invaded the Great Lakes, they devastated populations of their hosts. In lakes where sea lampreys have lived for a long time, host populations have not declined. Using what you know about co-evolution, explain this difference.

BONY FISHES

Of the 24,000 known species of fishes, about 95 percent are in the class Osteichthyes—the bony fishes. Bony fishes account for most of the vertebrates living in fresh water and in salt water. In this section you will study some of the adaptations of this very large and successful group.

SECTION

41-3

OBJECTIVES

▲ List three characteristics of bony fishes.

● Distinguish between ray-finned fishes and lobe-finned fishes.

■ Trace the flow of blood through a fish's heart.

◆ Explain how gills function in respiration.

▲ Describe the function of the swim bladder.

● Compare and contrast reproduction in bony fishes with reproduction in cartilaginous fishes.

CHARACTERISTICS

The bony fishes are characterized by three key features:

- **Bone**—This material is typically harder and heavier than cartilage. The skeletons of most bony fishes contain bone.
- **Lungs or swim bladder**—Early bony fishes had **lungs,** internal respiratory organs in which gas is exchanged between the air and blood. Only a few species of bony fishes have lungs today. Most bony fishes have a **swim bladder,** a gas-filled sac that is used to control buoyancy. The swim bladder is thought to have evolved from the lungs of the early bony fishes.
- **Scales**—The body of a bony fish is usually covered with scales. Scales protect the fish and help reduce water resistance.

Biologists divide the bony fishes into two main groups: lobe-finned fishes and ray-finned fishes.

Lobe-Finned Fishes

The **lobe-finned fishes** have fleshy fins that are supported by a series of bones. Only seven species of lobe-finned fishes exist today, six species of lungfishes and one species of coelacanth. Lungfishes, which exchange gases through lungs and gills, live in shallow tropical ponds that periodically dry up. They burrow into the mud and cover themselves with mucus to stay moist until the pond refills. Lobe-finned fishes are important because extinct lobe-finned fishes are ancestors of amphibians and all other terrestrial vertebrates.

Ray-Finned Fishes

Ray-finned fishes have fins that are supported by long, segmented, flexible bony elements called rays. Rays probably evolved from scales. In contrast to lobe-finned fishes, ray-finned fishes do not have fins with a central bony axis. Ray-finned fishes are diverse in appearance, behavior, and habitat. Ray-finned fishes include most familiar fishes, such as eels, yellow perch, trout, salmon, guppies, bass, herring, and darters.

SECTION 41-3

SECTION OVERVIEW

Structure and Function

Section 41-3 describes the structure of a bony fish, including its scales, fins, skeleton, and internal organs. It also points out the structural differences between lobe-finned fishes and ray-finned fishes.

ACTIVE READING

Technique: K-W-L

Before students read this chapter, have them make short lists of all the things they already **K**now about fish. When students finish their lists, ask them to contribute their entries to a class list on the board or overhead. Then have students list things they **W**ant to know about fish. Have students save their lists for later use in the Active Reading activity on page 812.

GIFTED ACTIVITY

Have students use library references to research the coelacanths and write reports to present to the class. Tell them to examine different aspects of coelacanths, such as their evolution, natural history, and anatomy, or the circumstances of their rediscovery.

CONTENT CONNECTION

Section 31-3 Tree Rings

The growth rings in a tree and those on a fish's scale are caused by the same environmental phenomenon. In both cases, the seasonal variation in available nutrients causes the formation of the rings.

QUICK FACT

☑ Three types of bony-fish scales exist. Ctenoid scales are thin and have tiny hairlike structures projecting from their posterior edge. Cycloid scales are also thin, but their edges are smooth all the way around. Ganoid scales are thick and heavy because they are layered with an enamel-like substance.

CRITICAL THINKING

Fin Use

The posterior dorsal fin of many ray-finned fishes, such as perch, is supported by rigid spines. Considering the manner in which a predator might stalk a perch, explain why having a prickly dorsal fin is an adaptive advantage. (The backward-pointing spines of the perch can pierce the throat of a predator that may try to swallow the perch from the rear.)

OVERCOMING MISCONCEPTIONS

Flying fish cannot truly fly, but they are capable of gliding distances of up to 1,200 ft. A flying fish works up to its glorious leap by building up enormous swimming speed. When it finally breaks the water's surface and launches into the air, the fish extends its oversized pectoral fins and uses them as airfoils.

VISUAL LINK

Figure 41-8

Use this figure to point out that the shape of bony fishes is like that of sharks, streamlined to facilitate movement through water. Point out that even the scales are layered to decrease drag.

Word Roots and Origins

operculum

from the Latin *operculum*, meaning "cover"

EXTERNAL ANATOMY

Figure 41-8 shows the external anatomy of a yellow perch, a bony fish that is common in the Great Lakes and in other freshwater lakes of the eastern United States and Canada. The yellow perch, like all bony fishes, has distinct head, trunk, and tail regions. On each side of the head is the **operculum** (oh-PERK-yoo-LUHM), a hard plate that opens at the rear and covers and protects the gills.

Fins

The fins of the yellow perch are adapted for swimming and navigating through the water. The **caudal fin** extends from the tail. It moves from side to side and amplifies the swimming motion of the body. Two **dorsal fins,** one anterior and one posterior, and a ventral **anal fin** help keep the fish upright and moving in a straight line. The fish uses paired **pelvic fins** and pectoral fins to navigate, stop, move up and down, and even back up. The pelvic fins also orient the body when the fish is at rest. The fins are supported by either rays or spines. Rays are bony yet flexible, while spines are bony and rigid.

Skin

The skin of the yellow perch is covered with scales. Scales are thin, round disks of a bonelike material that grow from pockets in the skin. As Figure 41-8 shows, scales overlap like roof shingles. They all point toward the tail to minimize friction as the fish swims. Scales grow throughout the life of the fish, adjusting their growth pattern to the food supply. The scales grow quickly when food is abundant and slowly when it is scarce. In fishes that live in habitats with annual variations in food availability, the resulting growth rings give a good approximation of the fish's age.

FIGURE 41-8

The external features of the yellow perch, *Perca flavescens*, are representative of bony fishes. Note the growth rings on the scales shown in the inset. They indicate the fish's approximate age.

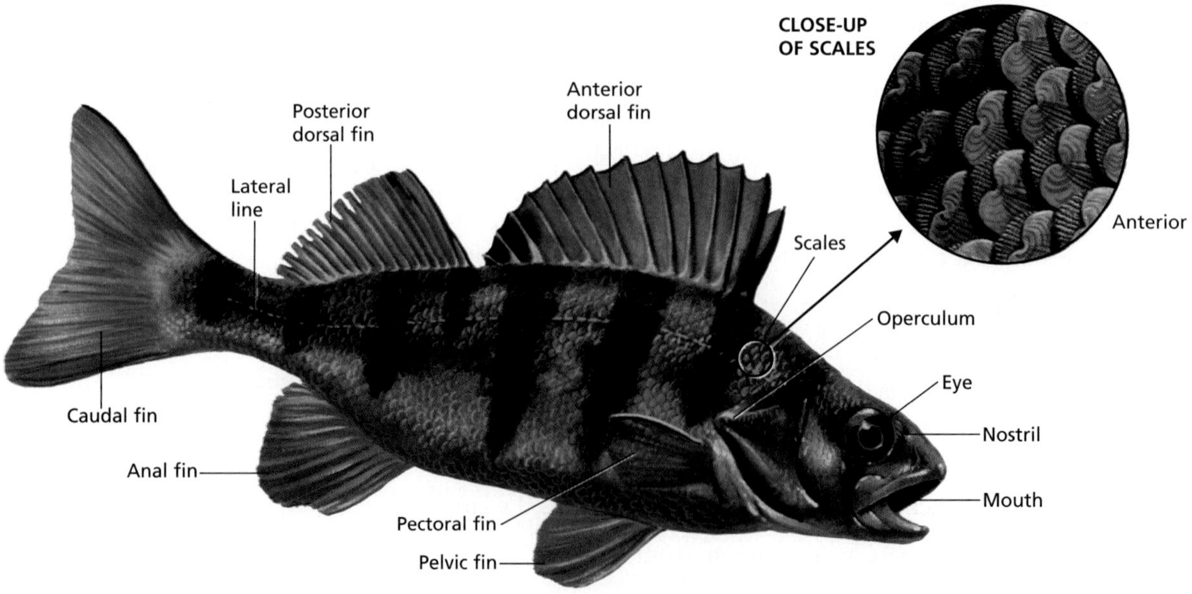

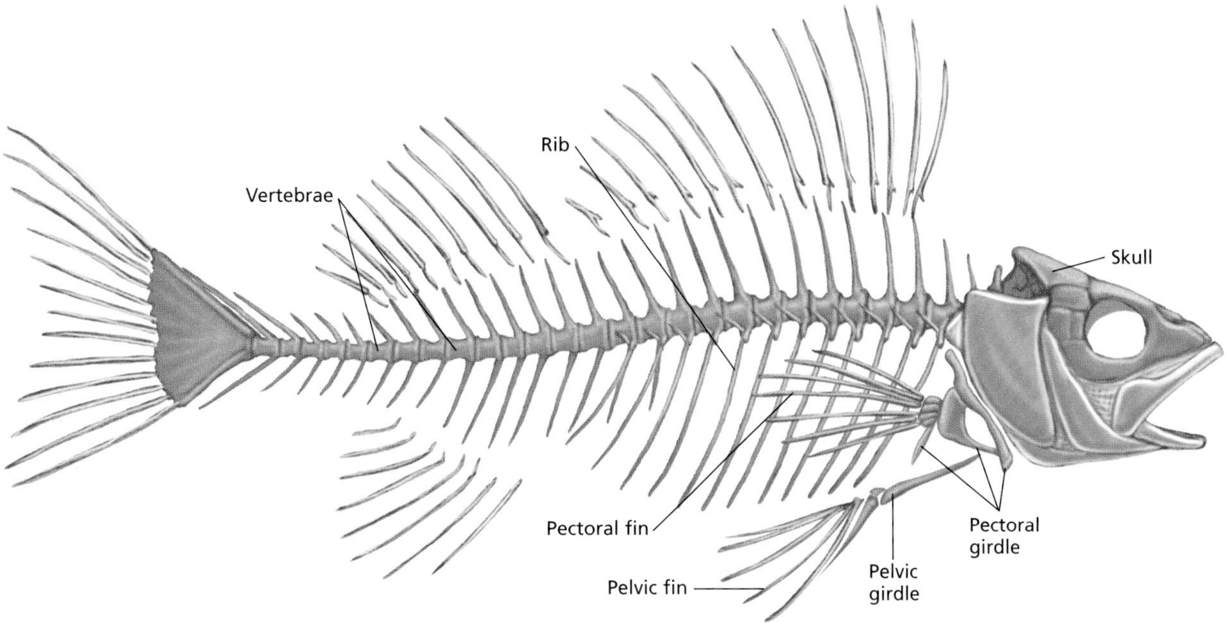

FIGURE 41-9

The skeleton of *Perca flavescens* is similar to that of other bony fishes. The general structure of the vertebrae, rib cage, and fins is found in many fishes.

INTERNAL ANATOMY

The major parts of a fish's skeleton, shown in Figure 41-9, are the skull, spinal column, pectoral girdle, pelvic girdle, and ribs. The spinal column is made up of many bones, called the vertebrae, with cartilage pads between each. The spinal column also partly encloses and protects the spinal cord. The **pectoral girdle** is the attachment point for the pectoral fins. The **pelvic girdle** is the attachment point for the pelvic fins. In a human skeleton, the pectoral girdle is the shoulder and its supporting bones, while the pelvic girdle is the hips. A fish's skull is composed of a large number of bones (far more than are in the human skull) and is capable of a wide range of movements. Figure 41-9 shows the skeleton of a bony fish.

Digestive System

Most bony fishes are generalized carnivores. The jaws of predatory fishes are lined with many sharp teeth that point inward to keep prey from escaping. Strong muscles operate the jaws, which are hinged to allow the mouth to open wide.

Figure 41-10 shows the internal anatomy of a bony fish. Food passes from the mouth into the pharynx, or throat cavity, and then moves through the **esophagus** to the stomach. The **stomach** secretes acid and digestive enzymes that begin to break down food. From the stomach, food passes into the **intestine,** where digestion is completed and nutrients are absorbed. The **liver,** located near the stomach, secretes **bile,** which helps break down fats. The **gallbladder** stores bile and releases it into the intestine. The **pancreas,** also located near the stomach, releases digestive enzymes into the intestine. The lining of the intestine is covered with fingerlike extensions called villi that increase the surface area for absorption of digested foods. Undigested material is then eliminated through the **anus.**

internetconnect

SCILINKS NSTA

TOPIC: Anatomy of a bony fish
GO TO: www.scilinks.org
KEYWORD: HM809

CULTURAL CONNECTION

Deadly Dinner

Japanese people have been eating pufferfish for hundreds of years. But a plate of *fugu* (the Japanese word for pufferfish) is no ordinary fish dinner. The organs of this bony fish contain tetrodotoxin, one the deadliest poisons on Earth. Someone who mistakenly ingests a bit of liver in poorly prepared fugu could die in a matter of minutes. So, those who dine on pufferfish are literally putting their lives in their chef's hands. It's no wonder that a chef must take intensive courses, endure long apprenticeships, and pass stringent exams to become a licensed preparer of *fugu.* Tetrodotoxins are also found in certain species of newts.

RECENT RESEARCH

Endothermic Fishes

Although most fishes are ectothermic, a few species, such as bluefin tuna and swordfish, are able to maintain their body temperatures higher than the temperature of the water that surrounds them. Genetic studies of these endotherms have revealed that the ability to maintain a high body temperature gives these fishes an adaptive advantage by allowing them to hunt in much colder waters than their competitors can.

DEMONSTRATION

Spinal Column Structure

Point out the spinal column on a model of a human skeleton. Explain to students that each vertebra is an individual bone. Show them where the cartilage pads and spinal cord were located.

TEACHING STRATEGY

Swim Bladder Function

Have students work in groups of two or three to perform the following exercise. Fill an aquarium with water. Have students use a rubber band to attach a series of weights to a table-tennis ball. Weights can be metal washers, coins, and modeling clay. Students should place the weighted ball in the aquarium and note whether it floats on the surface of the water, sinks, or floats beneath the surface of the water. Have students continue adjusting the weights until neutral buoyancy is achieved. Tell students to weigh all of the materials, including the rubber band and the table-tennis ball. Most of this weight reflects the weight of the water displaced by the air in the ball. Have the students discuss their results in class. Ask them how neutral buoyancy is adaptive for a fish. (A neutrally buoyant fish uses no energy to keep afloat. Some sharks, on the other hand, must constantly swim to avoid sinking.) Ask students what types of fish do not find neutral buoyancy a helpful adaptation. (Fishes such as flounders, which lack a swim bladder, are adapted to living on the bottom.)

INCLUSION *ACTIVITY*

Write the names of anatomical structures that are bold-faced throughout this section on small slips of paper (one name per slip). Put the slips of paper in a shoe box, shuffle them, and then have a student pick a slip from the box. Ask the student to read the name of the structure aloud and to describe its function. If he or she has difficulty, ask the rest of the class to help out with the answer. Let other students take turns picking a slip from the box.

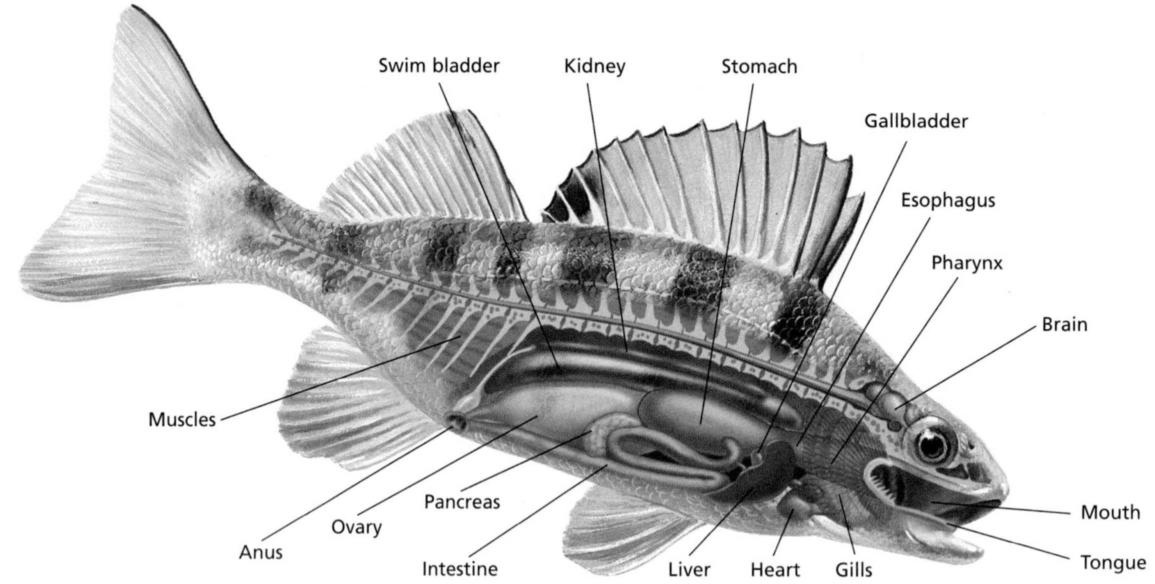

FIGURE 41-10

The internal anatomy of a bony fish, such as this perch, is a model for the arrangement of organs in all vertebrate descendants of fish. Food passes first from the mouth through the esophagus, then to the stomach and intestines. Finally, undigested waste is eliminated through the anus. Digestion of protein occurs in the stomach, and absorption of nutrients occurs in the intestine.

Circulatory System

The circulatory system of a fish delivers oxygen and nutrients to the cells of the body. It also transports wastes produced by metabolism—carbon dioxide and ammonia—to the gills and kidneys for elimination. The circulatory system consists of a heart, blood vessels, and blood. The heart pumps blood through **arteries** to small, thin-walled vessels, called **capillaries,** in the gills. There the blood picks up oxygen and releases carbon dioxide. From the gills, the blood then travels to the body tissues, where nutrients and wastes are exchanged. The blood returns to the heart through **veins.**

The heart of a bony fish has four chambers in a row, as you can see in Figure 41-11. Deoxygenated blood from the body empties into a collecting chamber called the **sinus venosus.** Next, blood moves into the larger **atrium.** Contraction of the atrium speeds up the blood and drives it into the muscular **ventricle,** the main pumping chamber of the heart. Contraction of the ventricle provides most of the force that drives the blood through the circulatory system. The final chamber is the **conus arteriosus.** It has an elastic wall and usually contains valves to prevent blood from flowing back into the ventricle. The conus arteriosus smooths the flow of blood from the heart.

FIGURE 41-11

A fish's heart is a series of four chambers that act in sequence to move blood through the body, transporting oxygen to the cells and wastes to organs for elimination. Note the thickness of the muscle in the ventricle. Why would the ventricle muscle be so much thicker than the other muscles of the heart?

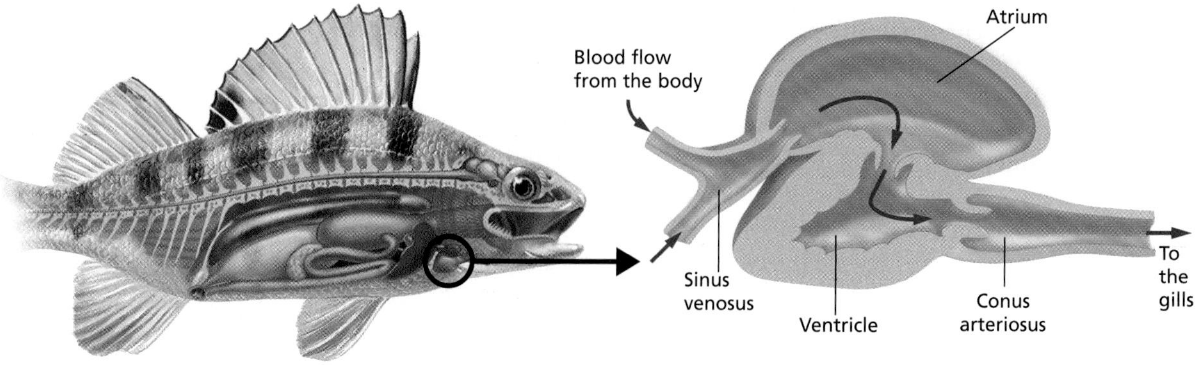

Respiratory and Excretory Systems

The large surface area of a fish's gills allows for rapid gas exchange. Gills are supported by four sets of curved bones on each side of the fish's head. Each gill has a double row of thin projections, called gill filaments. In most bony fishes, water is taken into the mouth and pumped over the gills, where it flows across the gill filaments before exiting behind the operculum. As you can see in Figure 41-12, water flows away from the head while blood flows toward the head. This arrangement is known as **countercurrent flow.** Countercurrent flow allows more oxygen to diffuse into the gills than would be possible if blood and water flowed in the same direction.

A fish's kidneys filter dissolved chemical wastes from the blood. The resulting solution, called **urine,** contains ammonia, ions such as sodium and chloride, and water. Urine is carried from the kidneys through a system of ducts to the **urinary bladder,** where it is stored and later expelled. By varying the amount of water and salts in the urine, the kidneys help regulate the water and the ion balance.

As blood flows through the gill filaments, ammonia generated by metabolism diffuses from the blood into the water passing over the gills and is removed from the body. The gills also regulate the concentration of ions in the body. Recall that saltwater fishes have lower ion concentrations than sea water has. Therefore, they lose water through osmosis and gain ions, such as sodium and chloride ions. Saltwater fishes make up for this water loss by excreting very little urine and by drinking sea water, but this increases their internal concentration of sodium and chloride ions. Both kinds of ions are actively transported out through the gills. Freshwater fishes tend to gain water and lose ions. They respond by producing large amounts of urine and actively transporting sodium and chloride ions in through the gills.

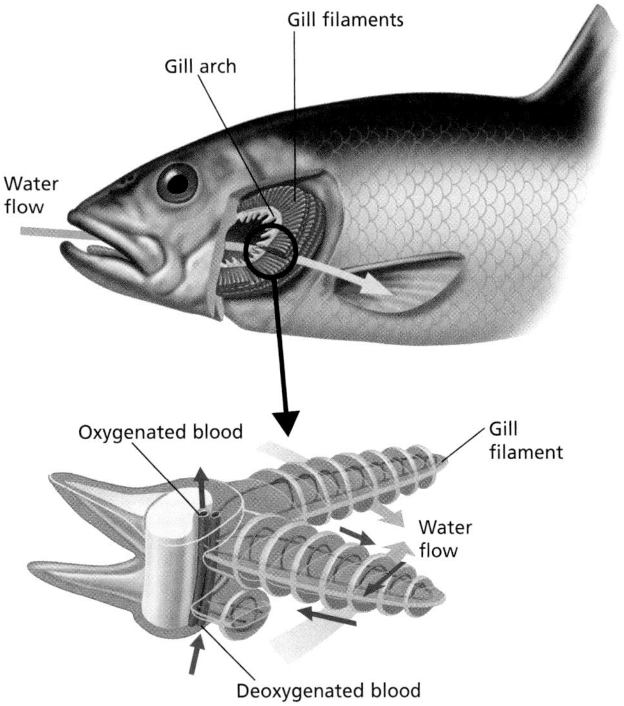

FIGURE 41-12

The gills are located directly behind the head and beneath the operculum. The gill filaments provide the organism with a large surface area, thus enabling gas exchange to occur quickly.

Swim Bladder

Most bony fishes have a swim bladder. This thin-walled sac in the abdominal cavity contains a mixture of oxygen, carbon dioxide, and nitrogen obtained from the bloodstream. Fish adjust their overall density by regulating the amount of gas in the swim bladder, enabling them to move up or down in the water.

Nervous System

The nervous system of a bony fish includes the brain and spinal cord, nerves that lead to and from all parts of the body, and various sensory organs. The major sensory organs are connected directly

TEACHING STRATEGY
Fish Breathing

Obtain several live goldfish, and put each fish in a clean 400 mL beaker half-filled with water. Let small groups of students observe a live goldfish breathing. Before students make their observations, have them mimic the way a fish breathes. Tell them to imagine that their hands are opercula. Have them cup each hand on either side of their neck and open their mouth. As they do this, tell them that this is analogous to what happens when water flows into a fish's mouth. Now have them close their mouth and flap open the thumb side of their hands. As they do this, tell them that this is analogous to what happens when the water inside the fish's mouth is forced over the gills and expelled from its body. Now have the students observe their fishes.

MAKING CONNECTIONS
Language Arts

Have students read one or more poems about fishes. Several can be found at the end of chapters in John Hersey's book *Blues* (Knopf, 1987). Other recommended poems include "The Fish," by Elizabeth Bishop; "The Lung Fish," by John Ciardi; "Pike," by Ted Hughes; and "The Red Mullet," by Robert Penn Warren. Before making assignments, check with your school's guidelines on literature. Have students write reports about the poems. Ask the following questions to focus students' thinking: How did the poet use the fish in the poem? How did the fish add to the meaning of the poem? Was the poet's description of the fish biologically accurate?

ACTIVE READING

Technique: K-W-L

Tell students to return to their lists of things they **W**ant to know about fish, which they made using the Active Reading exercise on page 807. Have them place check marks next to the questions that they are now able to answer. Students should finish by making a list of what they have **L**earned. Conclude by asking students which questions are still unanswered. Ask if they have any new questions.

ANSWERS TO SECTION 41-3 REVIEW

1. Bony fishes have a skeleton of bone, scales on the body, and a swim bladder or lungs.
2. The two kinds of living lobe-finned fishes are the coelacanth and the lungfish.
3. The ventricle pumps blood through the circulatory system.
4. Countercurrent flow occurs in the gills of fishes. Blood in the capillaries and water flow in opposite directions. This optimizes the diffusion of oxygen into the blood.
5. Many of the eggs and offspring die before reaching maturity. To make any contribution to the next generation, it is necessary to produce a large number of eggs, a few of which may survive.
6. Because swim bladders help fishes float, a bottom dweller would be disadvantaged if it did float.

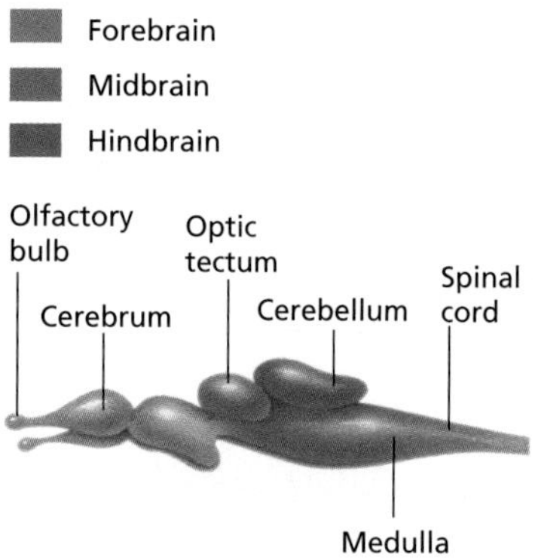

FIGURE 41-13

The fish brain, like the shark brain, has a well-developed medulla to coordinate muscle control.

with the brain via **cranial nerves.** The fish brain is illustrated in Figure 41-13. The most anterior part of the brain, the forebrain, contains the very prominent olfactory bulbs, which process information on smell. The forebrain, which includes the **cerebrum,** has areas that integrate sensory and other types of information from other areas of the brain. Behind the forebrain lies the midbrain, which is dominated by the **optic tectum.** The optic tectum receives and processes information from the fish's visual, auditory, and lateral line systems. The optic tectum is also involved in turning an animal's body toward or away from a stimulus. The most posterior division of the brain is the hindbrain, which contains the **cerebellum** (SER-uh-BEL-uhm) and the **medulla oblongata** (muh-DUL-uh AHB-lawn-GAHT-uh). The cerebellum helps coordinate motor output. The medulla oblongata helps control some body functions and acts as a relay station for stimuli from sensory receptors throughout the fish's body.

From the medulla oblongata, the spinal cord extends the length of the body and carries nerve impulses to and from the brain. **Spinal nerves** connect the spinal cord with the internal organs, muscles, and sense organs.

FIGURE 41-14

This male ringtailed cardinal fish, *Apogon aureus,* is carrying fertilized eggs in its mouth. This behavior lowers losses of eggs to predators and contributes to the success of the species.

REPRODUCTION

Eggs are produced by ovaries in the female, and sperm are produced by the testes in the male. Eggs and sperm are released through an opening behind the anus. Fertilization in most species takes place externally. Mortality among the eggs and young fishes is often very high. Many species of fishes lay large numbers of eggs to ensure that at least a few individuals survive to become adult fish.

Some bony fishes bear live young. Using a modified anal fin, the male inserts sperm into the female, and fertilization is internal. The female carries the eggs in her body until the young are born. Other species care for the young as shown in Figure 41-14.

The reproductive, or **spawning,** behavior of bony fishes varies widely. Some species build crude nests from plants, sticks, and shells. Many species migrate to warm, protected shallow water to spawn. For example, adult salmon migrate back to fresh water to spawn.

SECTION 41-3 REVIEW

1. **Describe two characteristics of bony fishes.**
2. **Identify the two kinds of lobe-finned fishes.**
3. **What is the function of the ventricle in the fish's heart?**
4. **What is countercurrent flow, and how does it relate to respiration in fishes?**
5. **Why is laying a large number of eggs an adaptation of some fishes?**
6. **CRITICAL THINKING Bottom-dwelling fish often lack a swim bladder. Explain the adaptive advantage of this.**

CHAPTER 41 REVIEW

SUMMARY/VOCABULARY

41-1
- Vertebrates are chordates and have a notochord, pharyngeal pouches, a dorsal nerve cord, and a post-anal tail at some stage of life.
- Vertebrates have vertebrae, a cranium, and an endoskeleton.
- Vertebrates are classified into seven classes: jawless fishes (class Agnatha); sharks and rays (class Chondrichthyes); bony fishes (class Osteichthyes); amphibians (class Amphibia); reptiles (class Reptilia); birds (class Aves); and mammals (class Mammalia).
- The oldest known vertebrates are jawless fishes.

Vocabulary

cranium (799)
gill arch (801)
vertebral column (799)

41-2
- Fishes have streamlined bodies and most have paired fins. Fishes float by means of a swim bladder or lipids in the body.
- Fishes use gills for respiration in water. The kidney and the gills help them maintain homeostasis.
- Living jawless fishes lack paired fins and jaws and have a notochord throughout life.
- Many lampreys are external parasites that feed on other fishes. Hagfishes feed on invertebrates or on dead or dying fishes.
- Sharks, rays, and skates are cartilaginous fishes.
- Sharks can smell, see, and detect electrical fields to detect prey. Rays and skates are flattened bottom-dwellers.

Vocabulary

cartilage (804)
clasper (806)
external fertilization (804)
internal fertilization (806)
lateral line system (803)
olfactory bulb (804)
pectoral fins (804)
placoid scale (804)
rectal gland (806)

41-3
- The characteristics of bony fishes are scales on the body, lungs or a swim bladder, and bone in the skeleton.
- Fishes use their fins for propulsion, maneuvering, and stability.
- The digestive system of a fish consists of a mouth, pharynx, esophagus, stomach, intestine, liver, gallbladder, pancreas, and anus.
- The fish heart has a sinus venosus, atrium, ventricle, and conus arteriosus.
- The main parts of a fish's brain are optic tectum, olfactory bulbs, the cerebrum, the medulla oblongata, and the cerebellum.
- Fertilization is external in most bony fishes and is internal in some species.

Vocabulary

anal fin (808)
anus (809)
artery (810)
atrium (810)
bile (809)
capillary (810)
caudal fin (808)
cerebellum (812)
cerebrum (812)
conus arteriosus (810)
countercurrent flow (811)
cranial nerve (812)
dorsal fin (808)
esophagus (809)
gallbladder (809)
intestine (809)
liver (809)
lobe-finned fish (807)
lung (807)
medulla oblongata (812)
operculum (808)
optic tectum (812)
pancreas (809)
pectoral girdle (809)
pelvic fin (808)
pelvic girdle (809)
ray-finned fish (807)
sinus venosus (810)
spawning (812)
spinal nerves (812)
stomach (809)
swim bladder (807)
urinary bladder (811)
urine (811)
vein (810)
ventricle (810)

CHAPTER 41 REVIEW ANSWERS

REVIEW

1. Gills are not part of the heart.
2. The operculum is not part of the brain.
3. All except the lung are characteristics of cartilaginous fishes.
4. The cranium is not part of the circulatory system.
5. The swim bladder is not part of the digestive system.

6. b **7.** a
8. c **9.** a
10. d **11.** c
12. c **13.** c
14. a **15.** d

16. Agnatha—lamprey, Chondricthyes—shark, Osteichthyes—bony fishes
17. Lungs enable lungfishes to breathe air when the oxygen concentration in water drops. Having gills enables lungfishes to stay underwater.
18. to reduce friction as the fish swims
19. Freshwater fishes produce large amounts of urine and absorb sodium and chloride ions through the gills. Saltwater fishes drink water, produce little urine, and excrete sodium and chloride.
20. It removes excess sodium and chloride ions from the blood.
21. sinus venosus →atrium→ ventricle → conus arteriosus
22. By keeping the eggs within her body, the female gives them protection from predators and environmental fluctuations.
23. This ensures that the shark always has a complete set of teeth. In humans, the first set is replaced by permanent teeth, which cannot be replaced if lost.
24. diffuse nitrogenous wastes, regulate sodium and chloride ion concentrations, diffuse oxygen into and carbon dioxide out of body
25. A–anterior dorsal fin, B–posterior dorsal fin, C–caudal fin, D–anal fin, E–pelvic fin, F–pectoral fin, G–operculum

CRITICAL THINKING

1. Adult tunicates are baglike filter-feeding organisms that spend their lives attached to some object. The only chordate characteristic they retain is gill slits, so they are unlikely ancestors of the vertebrates. Their larvae, however, are active swimmers that have all four vertebrate characteristics. A reasonable hypothesis of vertebrate origins is that vertebrates may have evolved from chordates that remained in the larval stage but could reproduce.
2. The spiral valve probably increases the surface area available for the absorption of nutrients.
3. The experiment does not exclude the possibility that sharks are finding their prey by smell. Although the insulating material would block electrical signals from the prey fish, it might also block the fish's odor. (In fact, Kalmijn conducted further experiments to demonstrate that sharks could detect the fish by its electrical discharges only.)
4. The body fluids of a freshwater fish have a higher concentration of ions than does fresh water. Thus, freshwater fishes gain water and rid themselves of this water by producing large amounts of urine. They do not need to drink. Saltwater fishes, by contrast, have lower ionic concentrations than sea water, so they lose water. They must drink to replace this water, and they conserve water by producing little urine.
5. It is likely that the bass would produce fewer eggs for two reasons. First, because it provides some protection for its eggs, their mortality likely will be lower. Fewer eggs will be lost and need to be replaced. Second, because the bass invests more in each egg, it does not have the resources to produce as many eggs.
6. It probably attracts desirable prey.

REVIEW

Vocabulary

In each set of terms below, choose the term that does not belong, and explain why it does not belong.

1. atrium, ventricle, sinus venosus, gill
2. cerebrum, cerebellum, operculum, optic tectum
3. internal fertilization, lung, placoid scale, rectal gland
4. cranium, vein, capillary, artery
5. stomach, liver, pancreas, swim bladder

Multiple Choice

6. Which of the following is *not* a characteristic of the first fishes? (a) bony plates on body (b) jaws (c) no paired fins (d) cartilaginous skeleton
7. The spinal column (a) protects the spinal cord (b) anchors limb muscles (c) protects the lateral line (d) collects sensory information.
8. One characteristic found in lampreys but not in hagfishes is (a) a notochord (b) paired fins (c) vertebrae (d) a cranium.
9. Which of the following is true of sharks and rays? (a) They have placoid scales. (b) Most species live in fresh water. (c) They have lungs. (d) They do not have a lateral line system.
10. Which of the following is not a way to control buoyancy? (a) the swim bladder (b) a fat-filled liver (c) continuous swimming (d) the rectal gland
11. The lateral line system (a) keeps fish moving in a straight line (b) initiates migration (c) detects vibrations (d) acts as camouflage.
12. Sharks use claspers to (a) increase maneuverability (b) startle other fish (c) transfer sperm while mating (d) hold on to prey while feeding.
13. The esophagus of a fish (a) creates buoyancy (b) fertilizes eggs (c) carries food from the mouth to the stomach (d) holds wastes moving from the stomach to the anus.
14. A fish's ventricle (a) pumps blood through the body (b) collects blood returning to the heart (c) facilitates gas exchange through diffusion (d) carries oxygen through the capillaries.
15. Bony fishes that live in fresh water (a) lose water (b) gain sodium and chloride ions (c) store urea (d) produce large amounts of urine.

Short Answer

16. Name the three classes of fishes, and give a representative of each.
17. What is the advantage to lungfishes of having both lungs and gills?
18. Why do all scales point toward a fish's tail?
19. Explain how bony fishes living in salt water and fresh water maintain their salt and water balance. How do sharks and rays prevent water loss?
20. What is the function of the rectal gland in sharks and rays?
21. Trace the flow of blood through the heart of a fish.
22. Many female sharks and rays retain the eggs inside their body, and some may even provide nutrition for the developing young. Explain why natural selection might have favored this behavior.
23. A shark replaces lost and worn teeth throughout its life. Why is this ability advantageous? How is this pattern of tooth replacement different from that of humans?
24. Describe as many functions of gills as possible. Be sure to include respiratory and excretory functions in your answer.
25. Identify each lettered structure in the figure below.

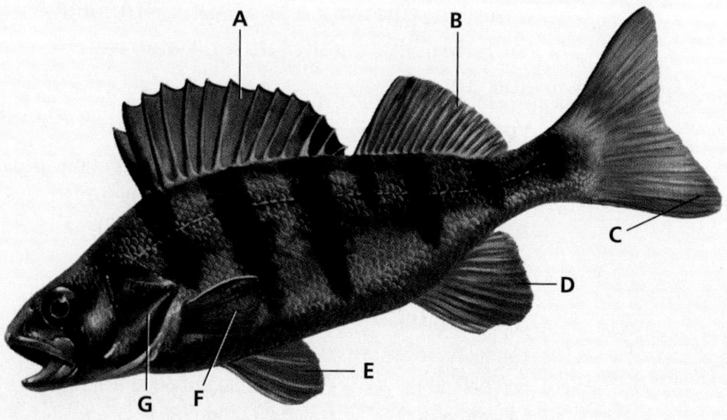

CRITICAL THINKING

1. Neoteny is a phenomenon in which larvae become sexually mature but still retain larval characteristics. The term has been used to explain in part the evolution of vertebrates from ancestral chordates. Use what you know about tunicates to test this hypothesis.
2. Sharks have a large corkscrew-shaped structure in the intestine called a spiral valve. How might this organ function in digestion?
3. In a famous set of experiments, A. J. Kalmijn studied the ability of sharks to find prey. He showed that a shark could locate and capture a stationary fish buried in the sand at the bottom of its tank. When the fish was enclosed in electrical insulation and then buried, however, the shark could not locate it. From this information alone, can you conclude that the shark is using only its electrical sense to locate the buried fish? Explain your answer.
4. Saltwater fishes drink more water and produce much less urine than freshwater fishes do. How do you account for this difference?
5. Cod and many other ocean fishes lay eggs near the surface of the water. The male largemouth bass scoops out a nest in a lake or river bottom and waits for a female to deposit her eggs. What hypothesis would you make regarding the relative number of cod and bass eggs? Explain your answer.
6. Many species of fishes that live deep in the ocean, where there is little or no light, are luminescent. What might be the advantages of such an adaptation?
7. Many fish are able to regulate their internal ion concentration as they go from a saltwater environment to a freshwater environment. Study the chart below, which shows the excreted ion concentration of a fish as it travels from one body of water to another. Is the fish traveling from fresh water to salt water or salt water to fresh water? Explain your reasoning.

Effect of Environment on Ion Excretion

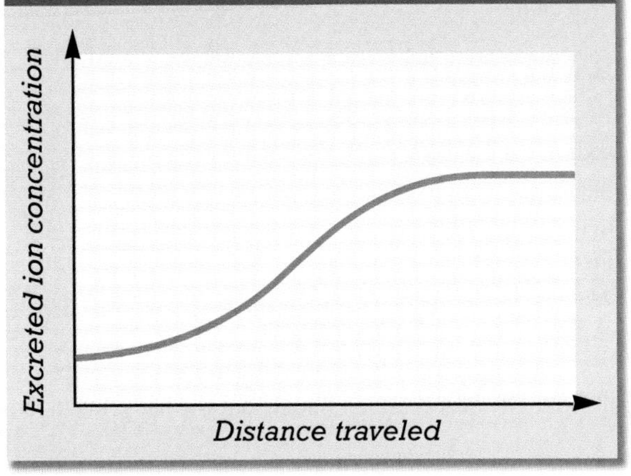

EXTENSION

1. Scientists are investigating an unexplained aspect of shark biology: sharks rarely get cancer. Use library resources or an on-line database to research this phenomenon. Gather information on how scientists are studying sharks' resistance to cancer. Also gather information on shark-cartilage dietary supplements. How has the popularity of shark-cartilage affected shark populations?
2. Read "Relax, It's Only a Piranha" in *Smithsonian,* July 1999, on page 42. Why does the author believe that the piranha has an undeserved and mistaken reputation for viciousness? How do piranhas usually feed? Describe the environmental conditions that led to the legendary voracious reputation that piranhas have.
3. Study the anatomy of any fish other than a yellow perch. Identify the operculum, vertebral column, and each fin. Label the internal organs. Compare the internal anatomy of your fish with that of the fish shown in Figure 41-10. What conclusions can you draw about habitat and behavior on the basis of the fish's structure?

ANSWERS

CONTINUED

7. Because the ion concentration of the excreta is increasing, the fish must be traveling from fresh water to salt water. In fresh water, it would not have to excrete many ions to maintain homeostasis; in a hypertonic environment, the fish would have to excrete more ions because more ions would diffuse into the fish.

EXTENSION

1. Scientists have recently discovered a substance in sharks that inhibits the growth of tumor cells. The craze for products containing shark cartilage is based on the idea that something in the cartilage is responsible for the low incidence of cancer among sharks. There is no scientific evidence for this claim, but the resulting demand for shark cartilage is threatening shark populations around the world.

2. The author of this article tried on several occasions to induce the piranha to attack but could not. He found that when nature is in balance, piranha tend to nip at the fins and scales of other fish. Their typical method is to sneak up on fish, nip, and flee. They seem to be extraordinarily shy. The only evidence found to support the swarming, devouring behavior piranhas have by reputation occurred in areas where the watering holes were becoming dryer, where competition was fierce because resources were scarce, or where fishermen had been cleaning their catch and throwing fish scraps into the water for a long time. Only in cases of rare accidents when humans had fallen into one of these areas did any harm come to people from piranhas.

3. Answers will vary, depending on the species chosen. Be sure that students relate the fish's anatomy to its habitat.

For an overview of this Interactive Exploration and preparatory information related to this activity, please see the CD-ROM User's Guide.

PREPARATION NOTES

Time required: 1 hour

PRELAB DISCUSSION

Have students answer the following questions:

1. What is a "one-cycle pump?" (A one-cycle heart pump pumps blood first to the gills then to the body, then back to the heart.)
2. What is the main purpose of the ventricle? (It is the main pumping chamber of the heart.)
3. Why would increasing the efficiency of the heart be advantageous to an organism? (More oxygen could be transported to the tissues for more-efficient metabolic processing of nutrients.)

ANSWERS TO PART A

2. The blood flows from the gills to the body, then back to the heart. Efficiency is 100 percent. Delivery rate is 6 mm Hg.
3. The separation is zero because the fish has a single-loop circulation, which means the ventricle does not need to pump two different populations of blood. A two-cycle heart without separation in the ventricle could have partial mixing of oxygenated and deoxygenated blood, which could decrease efficiency.

ANSWERS TO PART B

4. No, the shape is basically the same, but there is now a two-cycle route. The blood now flows first to the lungs, then to the heart, and then out to the body.
5. The efficiency rate has dropped, the delivery rate has increased. No, because no septum developed.

Exploring the Fish Heart

OBJECTIVES

- Simulate the action of a fish heart.
- Compare the structure and efficiency of a fish heart with those of the hearts of other vertebrates.

MATERIALS

- computer with CD-ROM drive
- CD-ROM *Interactive Explorations in Biology: Human Biology*

Background

This Interactive Investigation allows you to examine the action of a fish heart. Complex animals such as fishes require a uniform system of oxygen and nutrient distribution throughout their body. They also require a way to rid themselves of waste products, such as urea and carbon dioxide. The heart and the system of veins and arteries work together to provide these functions. The fish heart is a muscular tube with valves and chambers. It pumps blood to the gills and then throughout the body. The blood travels in a single loop, so that only deoxygenated blood enters the heart. Because the fish heart must pump blood through the fish's body in one cycle, it is less efficient than a human heart, which pumps blood first to the lungs, then to the heart, and finally to the rest of the body. Single-cycle pumping is considered to be less efficient than double-cycle pumping because it does not deliver as much oxygen and nutrients to the body's tissues.

Prelab Preparation

1. Start up the program Evolution of the Heart. Click the Topic Information button. Read the focus questions and review the three key concepts: the circulatory system, the heart, and the pulmonary vein.
2. Click the word *File* at the top left of the screen, and select Interactive Exploration Help. Listen to the instructions for operating the exploration. Click the Exploration button at the bottom right of the screen to begin the exploration. You will see an animated drawing like the one below.

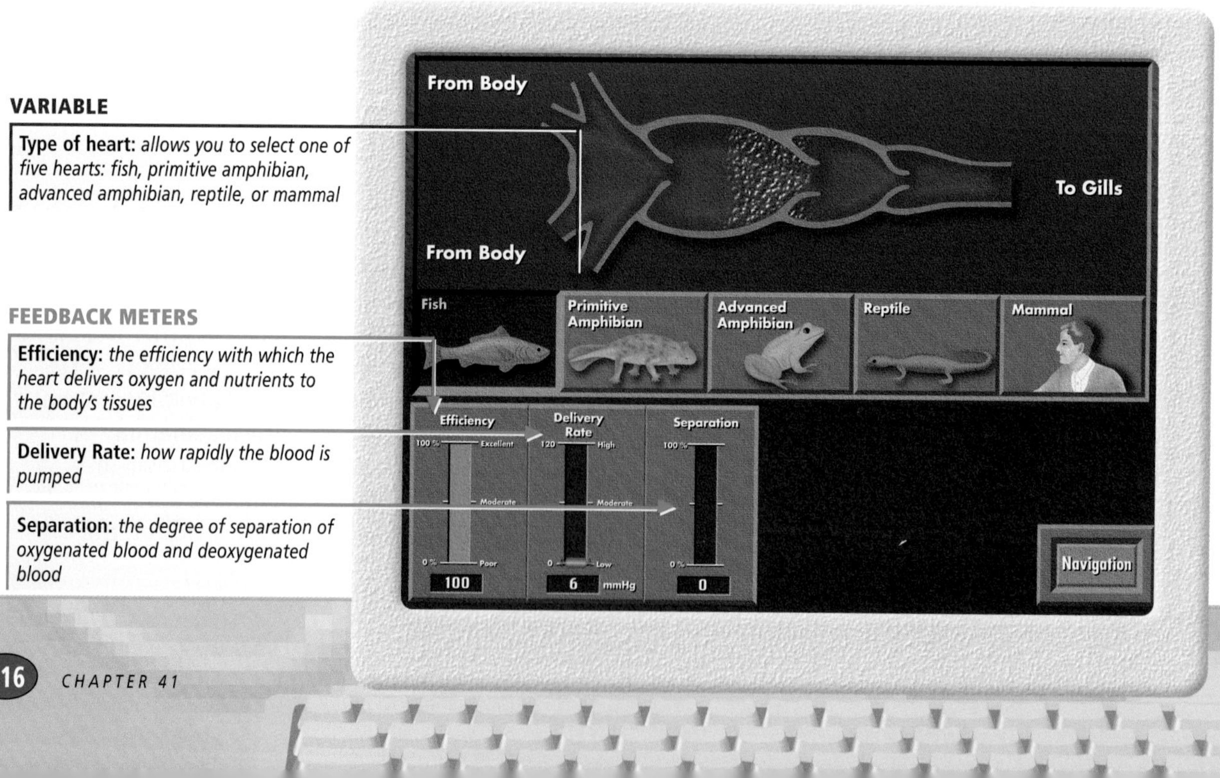

Procedure

Create a table like the one below for recording your data.

PART A The Fish Heart

First you will investigate the shape of the fish heart and its efficiency in pumping blood through the fish body.

1. Click the Interactive Explorations button (the one with the "magnifying glass" icon) at the top right of the screen. A fish heart is displayed with its efficiency, delivery rate, and separation in the feedback meters on the bottom left of the screen.
2. Note the route that blood travels through the fish heart. How does the blood complete the circuit? What is the efficiency of the fish heart? What is the delivery rate? Record this information in your data table.
3. Why is the separation in the feedback meter so low? Why does the lack of separation not matter in a single-cycle heart? Would this affect a two-cycle heart?

PART B The Primitive Amphibian Heart

4. Move the pointer to the primitive amphibian icon and click it. Has the shape of the heart changed significantly? How has the route of blood flow through the body changed?
5. How have the feedback meters changed? Did the amount of separation change from the fish heart?

PART C The Advanced Amphibian Heart

6. Move the pointer to the advanced amphibian icon and click it. What significant change occurs that separates it from the primitive amphibian heart?
7. Note the changes in the feedback meters. How did the change in the heart architecture change the readings on the feedback meters?

PART D The Reptile Heart

8. Move the pointer to the reptile icon and click it. What changes did you see take place?
9. Do the changes in the feedback meter indicate an improvement for the reptile's heart over the advanced amphibian's heart? Explain your answer.
10. Compare the reptile data with the fish heart data. Is the reptile heart better than the fish heart? Are the reptile data uniformly better than the fish heart data? Explain any inconsistencies.

PART E The Mammal Heart

11. Move the pointer to the mammal icon and click it. Record the feedback meter readings in the table you made. How do these changes compare with the readings from the other hearts?
12. What is the major advancement in the structure of the mammalian heart over the other animal hearts?

Analysis and Conclusions

1. What advantage does the fish's heart have over the reptilian and amphibian hearts? Explain the reason for the difference.
2. What advantage do the reptilian and amphibian hearts have over a fish's heart? How is this advantage improved on in the mammalian heart?

Further Investigation

Click the Navigation button and then click the Readings button. Click the "book" icon to the left of the article titled "The Beat Goes On." Prepare a report summarizing how successful scientists have been in producing artificial hearts.

COMPARISON OF VERTEBRATE HEARTS

Animal	Efficiency rate	Delivery rate	Separation

ANSWERS TO PART C

6. A partial septum now separates the ventricle.
7. The efficiency doubled, the delivery rate increased to 55 mm Hg, and the separation of the ventricle increased to 40 percent.

ANSWERS TO PART D

8. All the meters increased.
9. Yes, because more oxygen is being transported to the tissues.
10. The reptilian heart delivers blood to the tissues at a higher rate than the fish, but does not deliver oxygen as efficiently. The blood-delivery rate and percent separation of the amphibian heart are better than the fish's heart, but the efficiency of oxygen delivery is not. There is no mixing of oxygenated and deoxygenated blood in the fish. The amphibian heart allows some mixing; the oxygen-delivery efficiency is therefore only 75 percent.

ANSWERS TO PART E

11. The feedback-meter readings are higher than the readings of the other animal hearts.
12. The complete separation of the ventricle allows the two-cycle pump to be highly efficient.

ANALYSIS AND CONCLUSIONS

1. The fish heart is highly efficient at delivery of oxygen to a fish's body tissues. Because the amphibian and reptilian hearts are two-cycle pumps that partially mix oxygenated and deoxygenated blood they dilute oxygen delivery to the body tissues.
2. Both the reptilian and amphibian hearts deliver blood to the tissues at a higher rate than fish hearts do. The complete separation of the ventricle in the mammalian heart increases the efficiency of oxygen delivery to the body tissues.

CHAPTER 42
AMPHIBIANS

	TOPICS	TEACHING RESOURCES	LABS, CLASSWORK, AND HOMEWORK
BLOCK 1 45 minutes	**Introducing the Chapter, p. 818**	**ATE** Focus Concept, p. 818 **ATE** Assessing Prior Knowledge, p. 818 **ATE** Understanding the Visual, p. 818	■ Supplemental Reading Guide, *The Dinosaur Heresies*
	42-1 Origin and Evolution of Amphibians, p. 819 Adaptation to Land, p. 819 Modern Amphibians, p. 821	**ATE** Section Overview, p. 819 **ATE** Critical Thinking, p. 821 **ATE** Visual Link Figure 42-2, p. 820 167, 170	**PE** **Quick Lab,** p. 820, Comparing Fish and Amphibian Skin ★ Study Guide, Section 42-1 **PE** Section Review, p. 823
BLOCKS 2 & 3 90 minutes	**40-2 Characteristics of Amphibians, p. 824** External Covering, p. 824 Internal Anatomy, p. 824	**ATE** Section Overview, p. 824 **ATE** Critical Thinking, pp. 824, 827 **ATE** Visual Link Figure 42-8, p. 825 171–172, 180	**Lab A25,** Observing a Frog **Lab B28,** Frog Dissection ★ Study Guide, Section 42-2 **PE** Section Review, p. 829
BLOCK 4 45 minutes	**43-3 Reproduction in Amphibians, p. 830** Life Cycle, p. 830 Parental Care, p. 832	**ATE** Section Overview, p. 830 **ATE** Visual Link Figure 42-12, p. 831 169	**PE** **Chapter 42 Investigation,** p. 836 **Lab A34,** Culturing Frog Embryos ★ Study Guide, Section 42-3 **PE** Section Review, p. 832
BLOCKS 5 & 6 90 minutes	**REVIEW and ASSESSMENT** **PE** Chapter 42 Review, pp. 834–835 ■ Performance-Based Assessment—The Challenge of Water Retention	Audio CD Program	**CHAPTER TESTING** ■ Chapter Test (blackline master) ▲ Test Generator ■ Assessment Item Listing ■ Scoring Rubrics and Classroom Management Checklists

KEY

MODERN BIOLOGY

- **PE** Pupil's Edition
- **ATE** Teacher's Edition
- ■ Active Reading Guide
- ★ Study Guide

HOLT BIOSOURCES

- Teaching Transparencies
- Laboratory Program

One-Stop Planner CD-ROM
Shows these resources within a customizable daily lesson plan:

- ■ Holt BioSources Teaching Resources
- ■ Active Reading Guide
- ★ Study Guide
- ▲ Test Generator

READING FOR CONTENT MASTERY

Preview: Have students preview the objectives, the introductory paragraph, topic sentences, and captions before reading the chapter.

- **ATE** Active Reading Technique: Anticipation Guide, p. 819
- ■ Active Reading Guide Worksheet 42-1
- **PE** Summary/Vocabulary, p. 833

- ■ Active Reading Guide Worksheet 42-2
- **PE** Summary/Vocabulary, p. 833

- ■ Active Reading Guide Worksheet 42-3
- **PE** Summary/Vocabulary, p. 833

TECHNOLOGY AND INTERNET RESOURCES

CNN PRESENTS Video Segment 1 Frog Pollution

Holt Biology Videodiscs Teacher's Correlation Guide, Lessons 42-1 through 42-3

Audio CD Program, Sections 42-1 through 42-3

internetconnect

SCILINKS NSTA

On-line Resources:
www.scilinks.org
The following *sci*LINKS Internet resources can be found in the student text for this chapter:

Topics:
- Frogs, p. 821
- Salamanders, p. 822
- Life Cycle of Frogs, p. 830

go.hrw.com

On-line Resources:
go.hrw.com
Visit the HRW Web site for a variety of resources related to this chapter. Just type in the keyword HM2 Chapter 42.

Smithsonian Institution®
Visit **www.si.edu/hrw** for additional on-line resources.

CNNfyi.com
Visit **www.cnnfyi.com** for late-breaking news and current event stories selected just for you.

LAB ACTIVITY PLANNING

CHAPTER 42 INVESTIGATION

Observing Live Frogs, pp. 836–837

OVERVIEW
In this investigation, students study the reflexive behavior of frogs and the adaptive significance of these behaviors. Students will investigate and compare the anatomical differences that distinguish frog, fish, and salamander locomotion.

SAFETY
Remind students of the proper treatment and handling of laboratory animals. Remind students to wash their hands after handling live animals.

Release into the environment only native animals. Consult the *NABT Guidelines for the Use of Live Animals.*

PREPARATION
1. Feed frogs live crickets or other moving insects for several days before the lab. Set aside the frogs that do not feed, because they will not respond to stimuli in the experiment.
2. Do not feed frogs on the day of the lab. Because some frogs are more nervous than others, use frogs that respond and feed best if possible.

Quick Labs

Comparing Fish and Amphibian Skin, p. 820
Each student will wear disposable gloves, lab aprons, and safety goggles during this activity. Each lab group will also require paper, colored pencils, and living or preserved specimens of a fish and a frog.

UNDERSTANDING THE VISUAL

Tell students that some amphibians, like frogs, are adapted to live in both water and land environments. Ask students to identify the structural adaptations that enable a frog to live on land and in the water. (Legs and eardrums are adaptations for living on land. Webbed feet, and eyes and nostrils on top of the head are adaptations for living in the water.)

FOCUS CONCEPT

Structure and Function

Most amphibians live in water as larvae and on land as adults. Structural changes occur in the bodies of larvae to prepare them for life on land. For example, larvae breathe with gills. But as the larvae develop into terrestrial adults, the gills disappear and lungs form.

ASSESSING PRIOR KNOWLEDGE

Review the following concepts.

Vertebrates: *Chapter 41*
Draw the phylogenetic tree of vertebrates (see Figure 41-1) on the board but do not write in the class names. Have students provide the names (Agnatha, Chondrichthyes, Osteichthyes, Amphibia, Reptilia, Aves, and Mammalia) and give their proper location. Have a student circle *amphibians* on the tree.

Lobe-finned fishes: *Chapter 41*
Ask students to explain the evolutionary origin of the swim bladder in bony fishes. Then ask them to explain the evolutionary advantage of lungs in land-living organisms.

CHAPTER 42

AMPHIBIANS

Amphibians, such as this southern leopard frog (Rana pipiens)*, are thought to have been the first vertebrates on land. Many amphibians still live part or all of their life in water.*

FOCUS CONCEPT: ***Structure and Function***

As you read, note the adaptations of amphibians that enable them to live on land and in water.

ORIGIN AND EVOLUTION OF AMPHIBIANS

About 370 million years ago, the first amphibians evolved from lobe-finned bony fishes and became the first vertebrates to live on land. The name amphibian *comes from the Greek words meaning "double" and "life" and reflects the fact that many amphibians spend part of their life on land and part in water.*

SECTION 42-1

OBJECTIVES

▲ Describe two similarities between amphibians and lobe-finned fishes.

● List three characteristics of *Ichthyostega.*

■ List the major characteristics of living amphibians.

◆ Name the three orders of living amphibians, and give an example of each.

ADAPTATION TO LAND

According to one hypothesis, aquatic vertebrates first moved onto land as shallow pools of water began drying up, leaving these amphibian ancestors without water. Vertebrates adapted to land, according to this hypothesis, as a way to escape shrinking pools and move to those that still contained water. This hypothesis is no longer widely accepted because, as many critics have pointed out, it is unlikely that all the complex land-dwelling adaptations shown by amphibians would have evolved merely for short periods of overland travel. It is more likely that the ancestors of amphibians left the water to escape predation and competition and to gain access to the resources that were becoming abundant on land. At that time, oceans, lakes, rivers, and ponds supported a tremendous number and variety of fishes. Food and space were limited, and the numerous species of fishes competed intensely for them. On land, however, there were no vertebrates, and terrestrial invertebrates such as insects, a promising food source, were beginning to diversify.

Characteristics of Early Amphibians

Scientists have long recognized that amphibians evolved from lobe-finned fishes. The groups share many anatomical similarities, including features of the skull and vertebral column. Also, the bones in the fin of a lobe-finned fish are similar in shape and position to the bones in the limb of an amphibian. Figure 42-1 shows a crossopterygian (kraw-SEP-te-RIJ-ee-uhn), an extinct lobe-finned fish that is thought to be closely related to amphibians. This fish probably lived in shallow water and used its sturdy pelvic and pectoral fins to move along the bottom and to support its body while resting.

The oldest known amphibian fossils date from about 370 million years ago. *Ichthyostega* is the best known early amphibian. Like

FIGURE 42-1

Early lobe-finned fishes, such as this crossopterygian, are thought to be the immediate ancestors of the first land vertebrates.

SECTION 42-1

SECTION OVERVIEW

Evolution
Section 42-1 discusses the evolution of the first amphibians from early lobe-finned fishes. It describes the characteristics of *Ichthyostega,* an extinct early amphibian, as well as those of modern amphibians.

Structure and Function
Section 42-1 describes the characteristics that amphibians have evolved in response to their environment.

ACTIVE READING

Technique: Anticipation Guide
Before students read this chapter, write the following statements on the board or overhead:

1. Amphibians were the first vertebrates to adapt to living partially on land.

2. Most amphibians use gills, lungs, and skin in respiration.

Let students think a minute about the statements and decide if they agree with them. Have students read the chapter to see if their opinions are confirmed.

▶ ENGAGE STUDENTS

Group students into pairs. Ask each pair to provide a detailed drawing of what they think the first amphibians might have looked like. Tell them to use their imagination and the knowledge they already have about modern amphibians. Collect the drawings, and compare them with a picture of *Ichthyostega* in Section 42-1, the best-known early amphibian. Emphasize that modern amphibians are different from early amphibians.

SECTION 42-1

VISUAL LINK

Figure 42-2

Draw students' attention to the figure and point out how fishlike *Ichthyostega* appears. Ask students to imitate a fish swimming in water. Point out that they are using their forelimbs and hind limbs much as a fish or amphibian would.

Quick Lab

Comparing Fish and Amphibian Skin

Time Required 20 minutes

Safety Remind students to handle living animals gently. Have them wash their hands immediately after making their observations.

Procedural Tips Whole fish can be purchased from a seafood market. Keep fresh fish on ice. You may substitute a salamander for the frog. Frogs and salamanders can be purchased from a pet store.

Answers to Analysis The scales on the fish prevent oxygen from diffusing across the fish's skin. Student drawings should show oxygen molecules from the outside environment diffusing across the moist skin and into the blood vessels directly underneath the skin. Respiration through the skin is called cutaneous respiration. The excretory system helps to conserve water by transforming ammonia into less toxic urea, which can be excreted without using much water. Behaviors include burrowing in moist soil and being active at night.

FIGURE 42-2

Ichthyostega had well-developed limbs and is thought to have been a crawler on land, but it still had an overall fishlike body.

Quick Lab

Comparing Fish and Amphibian Skin

Materials disposable gloves, lab apron, safety goggles, paper, colored pencils, living or preserved specimens of a fish and a frog

Procedure

1. Put on your disposable gloves, lab apron, and safety goggles.
2. Touch and examine the skin of the specimens provided by your teacher. Record your observations. Handle living animals gently.
3. When you are finished with your observations, remove your disposable gloves, lab apron, and safety goggles. Wash your hands with soap and water.

Analysis Why can a frog use its skin as a respiratory membrane, while a fish cannot? Using your colored pencils, draw a diagram to illustrate how a frog's skin functions as a respiratory membrane. What is this type of respiration called? How does the excretory system help to maintain moisture by conserving water? What behaviors in amphibians enable them to maintain moist skin?

Ichthyostega, which is shown in Figure 42-2, all of the early amphibians had four strong limbs, which developed from the fins of their fish ancestors. The forelimbs of amphibians (and all other terrestrial vertebrates) are homologous to the pectoral fins of fishes, and the hind limbs are homologous to the pelvic fins. The early amphibians also breathed air with lungs. As you learned in Chapter 41, lungs arose early in the history of fishes and are found in the descendants of these early fishes—including terrestrial vertebrates.

Although the early amphibians showed several adaptations for life on land, such as sense organs for detecting airborne scents and sounds, they probably spent most of their time in the water. For example, *Ichthyostega* had a large tail fin and lateral-line canals on its head. Its teeth were large and sharp, indicating a diet of fish, not insects. In addition, some of the early amphibians appear to have had gills like those of fishes. An unusual characteristic of the early amphibians is the number of toes on their feet. Most present-day terrestrial vertebrates have five toes on each foot. However, *Ichthyostega* had seven toes on each hind foot (no fossils of its front feet have been discovered), while *Acanthostega,* another early amphibian, had eight toes per foot. Apparently, the five-toed characteristic had not yet occurred.

Diversification of Amphibians

During the late Devonian period and the Carboniferous period (360 million to 286 million years ago), amphibians split into two main evolutionary lines. One line included the ancestors of modern amphibians, and the other line included the ancestors of reptiles. Amphibians have been a diverse, widespread, and abundant group since this early diversification.

Today there are about 4,500 species of amphibians, belonging to three orders. The largest order, with more than 3,900 species, is Anura, which comprises the frogs and toads. The order Urodela contains about 400 species of salamanders. And the third order, Apoda, consists of about 160 species of caecilians, which are legless, wormlike, tropical amphibians. Figure 42-3 shows the phylogenetic relationships between these three groups.

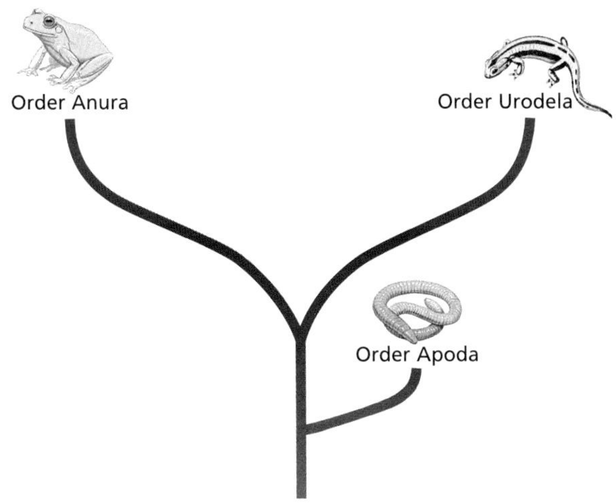

FIGURE 42-3

This phylogenetic tree represents the evolutionary relationships among the amphibians.

internet**connect**

SCI*LINKS*. NSTA
TOPIC: Frogs
GO TO: www.scilinks.org
KEYWORD: HM821

MODERN AMPHIBIANS

Modern amphibians share several key characteristics:

- Most change from an aquatic larval stage to a terrestrial adult form. This transformation is called metamorphosis.
- Most have moist, thin skin with no scales.
- Feet, if present, lack claws and often are webbed.
- Most use gills, lungs, and skin in respiration.
- Eggs lack multicellular membranes or shells. They are usually laid in water or in moist places and are usually fertilized externally.

Order Anura

Anurans (frogs and toads) are found worldwide except in polar climates and a few isolated oceanic islands. They live in a variety of habitats, from deserts and tundra to tropical rain forests. Figure 42-4 shows two examples of anurans. The term *toad* is commonly used for any anuran that has rough, bumpy skin, while the term *frog* commonly refers to anurans having smooth, moist skin. These

(a)

(b)

FIGURE 42-4

Anurans include toads and frogs such as the plains spadefoot toad (a), *Scaphiopus bombifrons,* which can be found throughout the United States, and the White's tree frog (b), *Litoria caerulea,* which is common in Australia.

SECTION 42-1

CRITICAL THINKING

Nose Breathing

Have students close their mouth and breathe in through their nose. Tell them all tetrapods (vertebrates with two pairs of limbs) and one group of fishes have internal nares that open to the outside. Have students name the group of fishes that have these structures and explain why they chose them. (Lobe-finned fishes have internal nares. Lobe-finned fishes are the logical choice because they gave rise to amphibians.)

MAKING CONNECTIONS

Social Studies

Have students examine a climate map of North America; they can be found in atlases and on wall charts. Ask students to hypothesize where the highest and lowest concentrations of amphibian species are found. Have students support their predictions. Then provide students with blank maps of North America, or have them trace a map from an atlas. Using a field guide to North American amphibians, have students shade the ranges of amphibian species on the blank maps. (Amphibian ranges are often included in field guides.) Ask students to compare actual amphibian ranges with their predictions and to explain why some areas provide more suitable amphibian habitats than others. (Weather variations can make an area unsuitable for amphibians.)

RECENT RESEARCH

The Tiniest Frog

Researchers recently discovered a new frog species of the genus *Eleutherodactylus* on the floor of a rain forest in Cuba. The little anuran, less than 10 mm long, is reported to be the smallest frog in the Northern Hemisphere.

SECTION 42-1

INCLUSION ACTIVITY

Before class begins, draw a 1 m (3 ft) long outline of *Ichthyostega* on a rectangular piece of poster board. Hang the poster on a wall in the classroom. Tell the students that this is how long a real *Ichthyostega* grew. Furnish students with stacks of old nature magazines (*Sierra, Smithsonian, Natural History, National Geographic*) that they can cut up. Tell them to cut out pictures of different types of modern amphibians. Have them place the pictures in a medium-sized cardboard box. Ask a student to select a picture from the box and name the amphibian's order (Anura, Urodela, or Apoda). Have the student attach the amphibian image inside the *Ichthyostega* outline using double-sided tape. Let other students take turns selecting pictures, classifying the organisms, and attaching the pictures to *Ichthyostega.*

TEACHING STRATEGY

Skin Breathing

Let the students touch and examine the skin of a live or preserved frog. Then have them observe the skin of a fish (you can obtain a few whole fishes from a seafood market). Ask why a fish cannot use its skin as a respiratory membrane but an amphibian can. (The scales prevent oxygen from diffusing across the fish's skin.) Ask students to illustrate how a frog's skin functions as a respiratory organ. (Students' drawings should show oxygen molecules from the outside environment diffusing across the moist skin and into blood vessels directly underneath.) CAUTION **Students should wash their hands after touching the fish or frog.**

Word Roots and Origins

Urodela

from the Latin *Ur,* meaning "tail," and the Greek *dēlos,* meaning "visible"

terms are general descriptions, however, and do not refer to any specific groups of anurans. Many anurans spend at least part of their life in water, and some species are permanently aquatic. However, many other species live and reproduce on land.

Anurans are characterized by a body adapted for jumping. Long, muscular legs provide power for the jump. The anuran body is compact, with a short, rigid spine and strong forelimbs that help absorb the shock of landing. The word *anuran* means "tailless" and reflects the fact that no adult anuran has a tail.

Adult anurans are carnivores that feed on any animal they can capture. Some frogs have a sticky tongue that can be extended to catch prey. Many species of anurans return to water to reproduce. In nearly all species, eggs are fertilized externally. The fertilized eggs hatch into swimming, tailed larvae called **tadpoles.**

internetconnect

SCILINKS NSTA

TOPIC: Salamanders
GO TO: www.scilinks.org
KEYWORD: HM822

Order Urodela

Salamanders have elongated bodies, long tails, and moist skin. Except for a few aquatic species, they have four limbs. The smallest salamanders are only a few centimeters long, while the largest reach lengths of 1.5 m (4.5 ft). Like anurans, salamander species range from fully aquatic to permanently terrestrial. Terrestrial salamanders usually live in moist places, such as under logs and stones. Larval and adult salamanders are carnivores. They are active mainly at night. Figure 42-5 shows two representative salamanders.

Most salamander species live in North America and Central America. There are very few species in Africa and South America, several species are found in Asia and in Europe, and there are no species found in Australia. With more than 300 species, the lungless salamanders (family Plethodontidae) are the largest group of salamanders. As their name suggests, these salamanders lack lungs. They absorb oxygen and release carbon dioxide through their skin.

Like most anurans, many salamanders lay their eggs in water, and the eggs hatch into swimming larval forms. Other species can reproduce in moist land environments. Eggs laid on land usually hatch into miniature adult salamanders and do not pass through a

(a)

(b)

FIGURE 42-5

The flatwoods salamander (a), *Ambystoma cingulatum,* and the spotted salamander (b), *Ambystoma maculatum,* are members of the order Urodela. *A. cingulatum,* which lives only in Florida, is an endangered species; *A. maculatum* can be found from eastern Canada to eastern Texas.

(a)

(b)

FIGURE 42-6

Caecilians, such as *Ichthyophis kohtaoensis* (a) and *Caecilia nigricans* (b), are primarily carnivores. They are burrowing amphibians that are usually blind, and a few species have scales embedded in their skin.

free-living larval stage. Most salamander species have a type of internal fertilization by which females pick up sperm packets deposited by males. In some terrestrial species, the female stays with the eggs until they hatch, which can take up to several weeks.

Order Apoda

The common name used to refer to members of the order Apoda is caecilian (see-SIL-yuhn). Caecilians are a highly specialized group of legless amphibians that resemble small snakes, as you can see in Figure 42-6. Caecilians average about 30 cm (12 in.) in length, but some species reach lengths of 1.5 m (4.5 ft). Because they have very small eyes that are located beneath the skin or even under bone, caecilians often are blind. Caecilians are rarely seen, and little is known about their ecology and behavior. Most species burrow in the soil, but some species are aquatic. All species have teeth in their jawbones that enable them to catch and consume prey. They eat worms and other invertebrates, which they detect by means of a chemosensory tentacle located on the side of their head. All species are thought to have internal fertilization. Some species lay eggs, which the female guards until they hatch. In a few species, the young are born alive. Caecilians live in tropical areas of Asia, Africa, and South America.

SECTION 42-1 REVIEW

1. **Explain the most likely reason that aquatic vertebrates evolved from an aquatic existence to a land-based existence.**
2. **What characteristics suggest that lobe-finned fishes are the ancestors of amphibians?**
3. **What features of *Ichthyostega* suggest that it spent most of its time in water?**
4. **What is metamorphosis?**
5. **Give two examples of modern amphibians, and name the order to which each belongs.**
6. **CRITICAL THINKING Fossils of *Ichthyostega* and other early amphibians do not indicate the presence of lungs. Explain why scientists think the early amphibians had lungs.**

SECTION 42-1

RETEACHING ACTIVITY

Have each student prepare a table listing characteristics of the three amphibian orders and showing which are adaptations for life on land or in water. Have them begin by drawing a table with three columns and three rows. Then tell them to write the following headings for each column: "Order," "Adaptations to Land," and "Adaptations to Water." Ask them to use information from their textbooks to fill in the table.

ANSWERS TO SECTION 42-1 REVIEW

1. The most likely reason aquatic vertebrates moved onto land was to escape predation and competition and to obtain access to newly developing resources on the land.
2. Lobe-finned fishes share some characteristics of the skull and vertebral column. The bones in the fins of a lobe-finned fish are similar in position and shape to the bones in the leg of an amphibian.
3. *Ichthyostega* had a large tail fin and a lateral line.
4. The transformation of an aquatic larvae into a terrestrial adult amphibian is called metamorphosis.
5. Answers will vary. Frogs and toads belong to the order Anura. Salamanders belong to the order Urodela, and caecilians belong to the order Apoda.
6. Lungs evolved in early fishes and are found in all their descendants—though sometimes in the modified form of a swim bladder—including lobe-finned fishes and modern amphibians.

SECTION 42-2

SECTION OVERVIEW

Structure and Function
Section 42-2 describes how the external and internal anatomy of an amphibian suits it for a semi-terrestrial lifestyle.

Matter, Energy, and Organization
Section 42-2 also describes the arrangement of the organ systems found in a frog.

▶ ENGAGE STUDENTS

Have students run in place until they begin to tire. Then ask them what two changes occurred in their bodies as a result of their running. (Breathing and heart rates increased.) Tell students that the heart and lungs work closely together in the bodies of terrestrial vertebrates. The lungs receive oxygen and pass it on to the heart, which then pumps the oxygen to the body's tissues. Tell students that this important heart-lung connection first appeared in amphibians.

CRITICAL THINKING

Health or Diversity
Tell students that a recent scientific study confirmed that California newts are being eliminated by mosquitofish, an introduced species. The mosquitofish were seeded into streams in Los Angeles County to control mosquitoes. The fish feed on the insect's larvae. Scientists now know that the fish also eat the larvae of the California newts, which also inhabit the streams. Streams that contain the fish have few if any newts. Ask students how they would feel about losing the newts if they lived in Los Angeles County. Remind students that mosquitoes can transmit human diseases, such as encephalitis. Then ask students if they think mosquitofish should continue to be used to control mosquitoes.

SECTION

42-2

OBJECTIVES

▲ Relate the structure of amphibian skin to the types of habitats in which amphibians can survive.

● Identify three adaptations for life on land shown by the skeleton of a frog.

■ Describe the pattern of blood flow through an amphibian's heart.

◆ Describe how a frog fills its lungs with air.

▲ Explain the function of each organ of an amphibian's digestive system.

● Compare the amphibian's nervous system with that of a bony fish.

CHARACTERISTICS OF AMPHIBIANS

As you have already seen, terrestrial vertebrates face challenges that are far different from those faced by aquatic vertebrates. In this section, you will learn about some of the ways amphibians meet the challenges of living on land.

EXTERNAL COVERING

The skin of an amphibian serves two important functions—respiration and protection. The skin is moist and permeable to gases and water, allowing rapid diffusion of carbon dioxide and water. Numerous **mucous glands** supply a lubricant that keeps the skin moist in air. This mucus is what makes a frog feel slimy. The skin also contains glands that secrete foul-tasting or poisonous substances that provide protection from predators.

However, the same features that allow efficient respiration also make amphibians vulnerable to dehydration, the loss of body water. Therefore, amphibians live mainly in wet or moist areas on land. Many species are active at night, when loss of water through evaporation is reduced. Although some species of frogs and toads survive in deserts, they spend most of their life in moist burrows deep in the soil. Only after heavy rains do these amphibians come to the surface to feed and reproduce.

INTERNAL ANATOMY

While water supports the body of an aquatic vertebrate against the force of gravity, terrestrial vertebrates must rely on the support of their strong internal skeleton. The vertebrae of the spine interlock and form a rigid structure that can bear the weight of the body. Strong limbs support the body during walking or standing. The forelimbs attach to the pectoral girdle (the shoulder and supporting bones), while the hind limbs attach to the pelvic girdle (the "hips"). The pectoral and pelvic girdles transfer the body's weight to the limbs. The cervical vertebra at the anterior end of the spine allows neck movement.

The frog skeleton in Figure 42-7 shows several specializations for absorbing the forces created by jumping and landing. In most terrestrial vertebrates, there are two bones in the lower hind limb, the tibia and the fibula, and two bones in the lower forelimb, the ulna and the radius. In frogs, the bones of the lower forelimb are fused into a single bone, the radio-ulna. The bones of the lower hind limb are fused into the tibiofibula. Frogs have few vertebrae, and the vertebrae at the posterior end of the spine are fused into a singe bone called the urostyle.

Heart and Circulatory System

The circulatory system of an amphibian, illustrated in Figure 42-8, is divided into two separate loops. The **pulmonary circulation** carries deoxygenated blood from the heart to the lungs, then returns the oxygenated blood to the heart. The **systemic circulation** carries oxygenated blood from the heart to the muscles and organs of the body and brings deoxygenated blood back to the heart. All other terrestrial vertebrates also have a "double-loop" circulatory pattern. This pattern of circulation provides a significant advantage over the "single-loop" circulation of a fish—faster blood flow to the body. In a fish, the blood loses some of its force as it passes through the narrow capillaries of the gills, and blood flow slows as a result. The lungs of an amphibian also contain narrow capillaries that slow blood flow. But after passing through the capillaries of the lung, blood returns to the heart to be pumped a second time before circulating to the body.

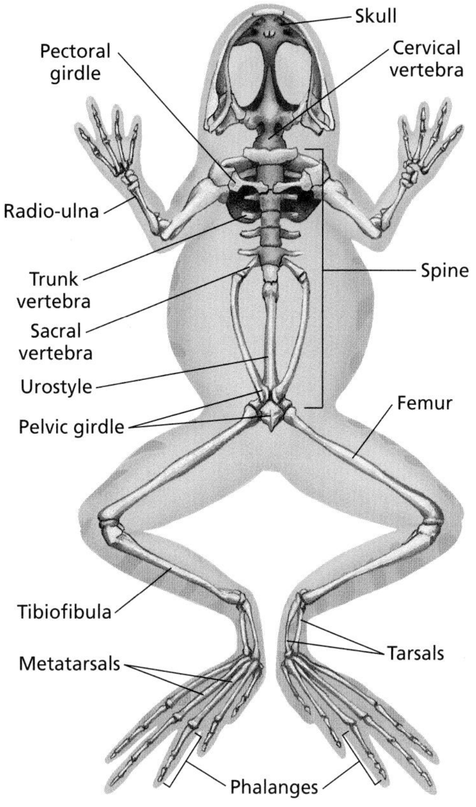

FIGURE 42-7

The skeleton of the frog absorbs shocks when the frog jumps. The urostyle and the two bones of the pelvic girdle extend outward to meet the sacral vertebra. This structure absorbs the shock for the long leg bones when the frog lands.

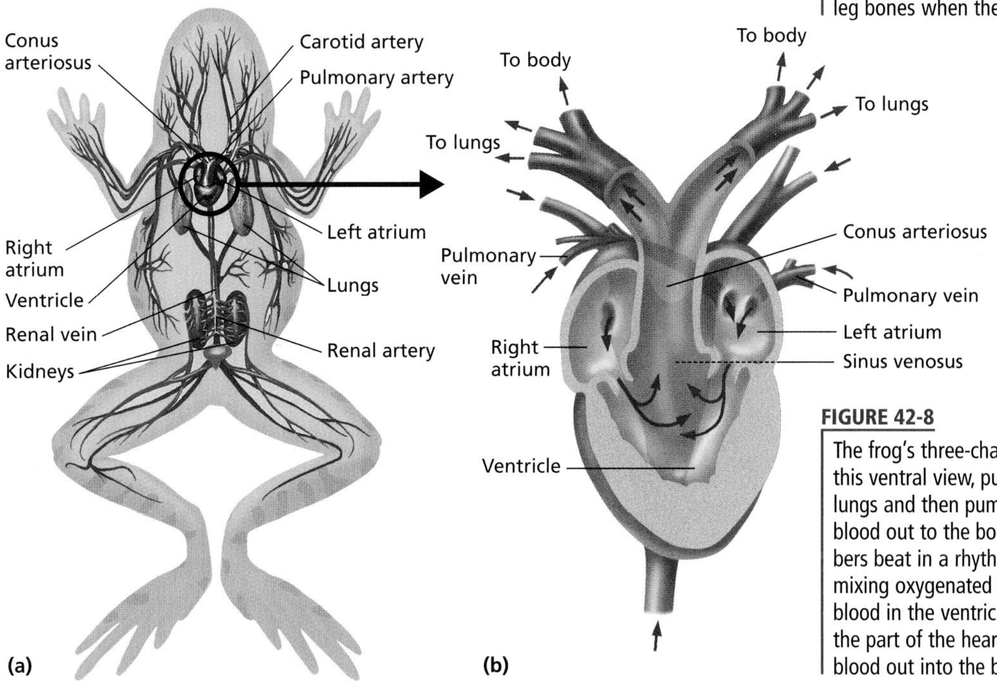

FIGURE 42-8

The frog's three-chambered heart, in this ventral view, pumps blood to the lungs and then pumps the oxygenated blood out to the body. The three chambers beat in a rhythm that minimizes mixing oxygenated and deoxygenated blood in the ventricle. The ventricle is the part of the heart that moves the blood out into the body and lungs.

OVERCOMING MISCONCEPTIONS

It is a common misconception that people can get warts from touching a toad. Although this is not true, you should wash your hands after handling a toad because toads release an irritant that can cause discomfort if it gets in your eyes or mouth.

TEACHING STRATEGY

Blood-Cell Differences

Provide students with microscopes, and let them compare frog blood with human blood (prepared slides of both types of blood can be obtained from a biological supply house). Do not allow students to prepare slides of their own blood for comparison. Tell students to view the cells under high power and to pay particular attention to the internal structure of the red blood cells. Have them sketch a few red cells from each organism. (Students will see that frog red blood cells have nuclei, whereas human red blood cells are anucleated.) Ask students to hypothesize a reason for the difference. (Humans require more energy, and therefore more oxygen, than amphibians. Anucleated red blood cells are able to carry more oxygen than nucleated red blood cells.)

VISUAL LINK

Figure 42-8

Point out to students that the amphibian heart does not mix oxygenated and deoxygenated blood very much. Ask students why mixing the blood would not be advantageous to the frog. (Mixing the blood would lower the amount of oxygen that is delivered to the tissues.)

RECENT RESEARCH

Vertebrate Limb Regeneration

Although it has been known since 1768 that some organisms can regenerate a lost limb, until recently scientists have not understood how this happens. Scientists at University College London have identified groups of cells on newts that are able to dedifferentiate, grow, and then redifferentiate into several cell types to produce a new limb. Although the molecular basis of this phenomenon is not yet understood, researchers hope that this knowledge can be applied to human medicine.

DEMONSTRATION

Inflate a sturdy elongated balloon until it is slightly less than full. Then tie the open end into a knot. Using a marker, draw a line around the middle of the balloon. On one side of the balloon, write the word *Mouth.* On the other side of the circle write the word *Lungs.* Now ask a student to assist you with the demonstration. Gently hold one end of the balloon in both hands, and have your assistant hold the other end the same way. The person holding the end labeled "Mouth" should then squeeze the balloon and force air into the end labeled "Lungs." Then the person holding the "Lungs" end should squeeze the balloon and force the air back into the "Mouth" end. Relate the movement of air in the balloon to the way an amphibian breathes.

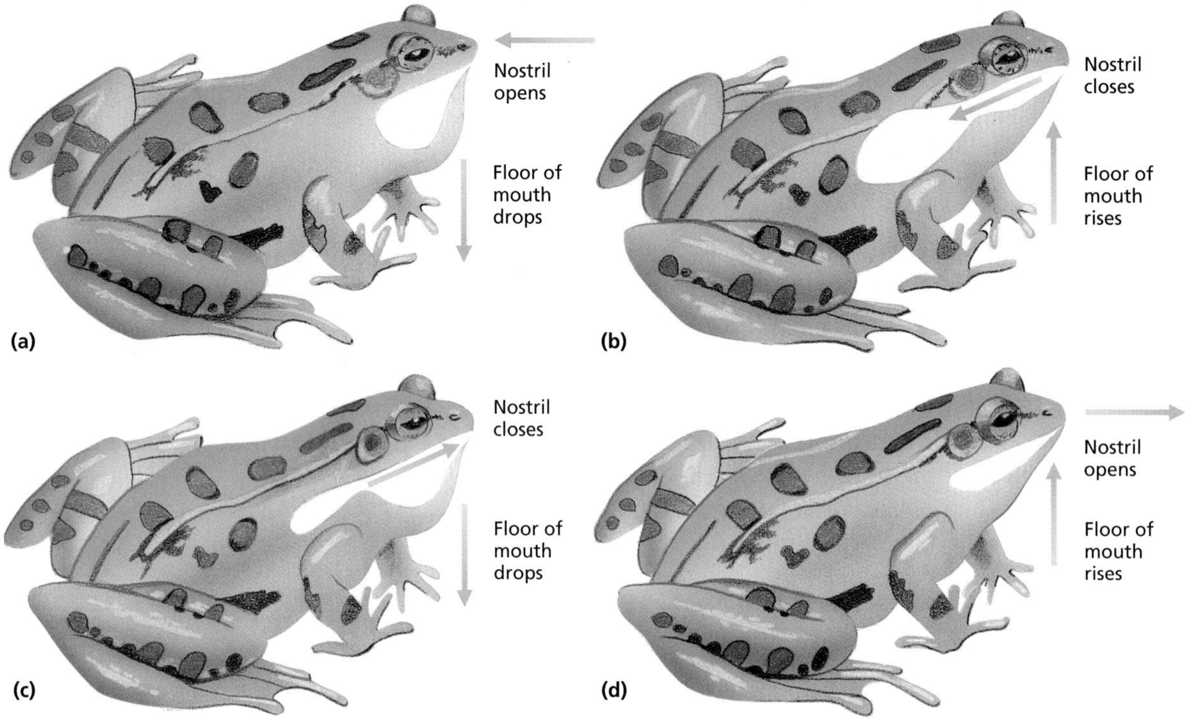

FIGURE 42-9

Frogs breathe by creating pressure that forces air into their lungs. When the floor of the frog's mouth drops, air capacity increases in the frog's mouth (a), and air rushes in. When the nostril is closed and the mouth floor rises, the air is forced into the lungs of the frog (b). The elasticity of the lungs and the use of muscles force the air back out (c). Then the nostril opens and the mouth floor rises again, forcing air out the nostril (d).

The three-chambered heart of an amphibian reflects the division of the circulatory system into pulmonary and systemic circulation. Deoxygenated blood from the body first enters the sinus venosus, a collecting chamber on the dorsal right side of the heart, indicated by the dashed line in Figure 42-8. From the sinus venosus, blood moves into the right atrium. Oxygenated blood from the lungs enters the left atrium, and contraction of the atria forces the deoxygenated and oxygenated blood into the single ventricle, the main pumping chamber of the heart. Although the ventricle is not divided, its spongy, irregular interior surface and the coordinated contractions of the atria keep the oxygenated and deoxygenated blood from mixing. Ventricular contraction expels both kinds of blood into the conus arteriosus, which directs deoxygenated blood to the lungs and oxygenated blood to the body. A valve within the conus arteriosus prevents mixing of oxygenated and deoxygenated blood.

Respiration

Larval amphibians respire, or exchange carbon dioxide and oxygen, through their gills and skin. Most adult amphibians lose their gills during metamorphosis, but they can respire in two ways: through the lungs and through the skin.

Respiration through the lungs is called **pulmonary respiration.** Amphibians ventilate their lungs with a unique mechanism that pumps air into the lungs; this is called positive-pressure breathing. For example, a frog breathes by changing the volume and pressure of air in its mouth while either opening or closing its nostrils, as shown in Figure 42-9. Both inhalation and exhalation involve a two-step process during which the floor of the frog's mouth is raised and lowered. The frog controls the direction of air flow by opening or closing its nostrils. Because amphibians have a small surface area in the lungs for gas exchange, respiration through the skin, or **cutaneous respiration,** is very important to most aquatic and terrestrial amphibians.

Digestive System

All adult amphibians are carnivorous. Because most amphibians are small, insects and other arthropods are their most commonly consumed prey. Larger amphibians sometimes eat mice, snakes, fish, and other amphibians. Many amphibian larvae, such as those of frogs, are herbivorous, feeding on algae or bacteria. The larvae of some species, such as those of salamanders, are carnivorous, and some feed on the larvae of other species.

The amphibian digestive system includes the pharynx, esophagus, stomach, liver, gallbladder, small intestine, large intestine, and cloaca. Figure 42-10 shows the digestive system of a frog. The elastic esophagus and stomach allow an amphibian to swallow large amounts of food. Once food reaches the stomach, tiny glands in the stomach walls secrete gastric juices that help break down, or digest, the food. A muscle called the pyloric sphincter at the lower end of the stomach relaxes, which allows digested food to move into the small intestine. The upper portion of the small intestine is called the **duodenum** (DOO-oh-DEE-nuhm). The coiled middle portion of the small intestine is the **ileum** (IL-ee-uhm). A membrane resembling plastic wrap, called the **mesentery,** holds the small intestine in place. Inside the small intestine, digestion is completed and the released nutrients pass through capillary walls into the bloodstream, which carries them to all parts of the body.

The lower end of the small intestine leads into the large intestine. Here indigestible wastes are collected and pushed by muscle action into a cavity called the cloaca (kloh-AY-kuh). Waste from the kidneys and urinary bladder, as well as either eggs or sperm from the gonads, also passes into the cloaca. Waste materials exit the body through the **vent.**

Other glands and organs aid in the digestion process. The liver produces bile, which is stored in the gallbladder. Bile helps break down fat into tiny globules that can be further digested and absorbed. A gland called the pancreas, located near the stomach, secretes enzymes that enter the small intestine and help break down food into products that can be absorbed by the blood.

FIGURE 42-10

The frog digestive system is shown in ventral view. Notice how the short small intestine is an adaptation for a carnivorous diet. Recall that the shark and perch also have short intestines.

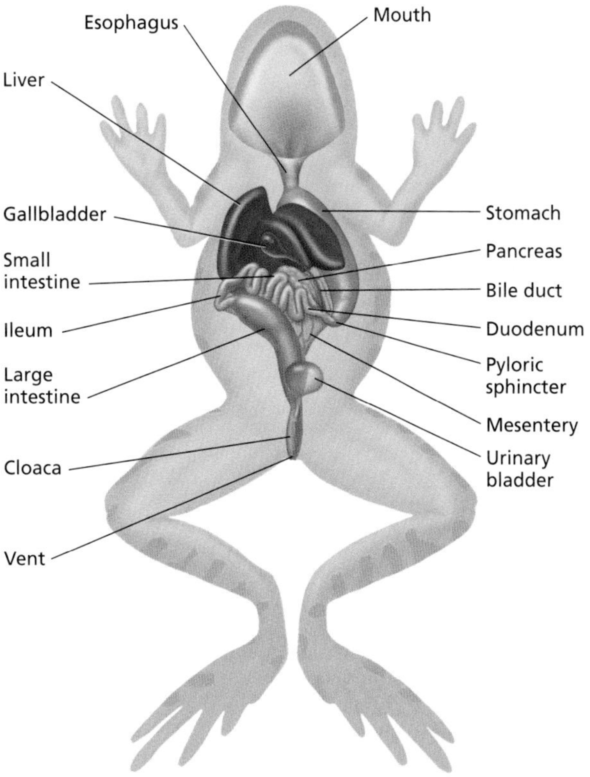

CONTENT CONNECTION

Section 7-2: Cell Respiration

The energy demands of land vertebrates require that their cells receive a steady supply of oxygen to act as a terminal electron acceptor. A double-loop circulatory system provides an efficient way of delivering oxygen to tissues.

QUICK FACT

☑ Frogs use their teeth only to hold food in place, not for chewing. Amphibians generally swallow their prey whole, while it's still alive.

CRITICAL THINKING

Digestive Differences

Tell students that food made from plants takes longer to digest than food made from animals because plants contain indigestible cellulose. Then ask students to consider the food eaten by tadpoles and adult frogs. Would they expect their digestive systems to be exactly the same? Why? (No. A tadpole's small intestine is longer and more coiled, allowing more time for the digestion of the plants and algae that the tadpole consumes.)

RETEACHING ACTIVITY

Divide the class into five groups and ask each group to stand at a different place near the board. Let each group represent a different frog organ system (circulatory, respiratory, digestive, excretory, or nervous). Ask the students to brainstorm about the structures and functions of the organs in their assigned system, and have them write their conclusions on the board. Let them use their notes and textbook. Ask the groups to rotate positions so that each group can work on a different system. Tell the students to add to or refine the information already posted. Continue the rotation until all the groups have had a chance to work on every system. Have students discuss the anatomy of the frog, and use the students' work as your guide.

GIFTED ACTIVITY

Ask students to choose an organ system in the frog and prepare a written report comparing and contrasting it with the human organ system. Tell students that their report should explain the similarities and differences they find. Remind them to think in terms of adaptations.

RETEACHING ACTIVITY

Have students make a chart with two columns. In the left column, have them list names of the major organ systems (circulatory, skeletal, respiratory, digestive, excretory, nervous, and sensory). Then ask students to join in a word-association exercise. As you say the name of each organ system, have students write in the right column words that they associate with it. (for example: Skeletal—bones, strength; Circulatory—blood, heart; etc.). Ask students to contribute their associations, and add them to a chart on the board.

Excretory System

The kidneys are the primary excretory organs. One kidney lies on either side of the spine against the dorsal body wall. The kidneys filter nitrogenous wastes from the blood. These wastes, flushed from the body with water, are known as urine. Urine flows from the kidneys to the cloaca through tiny tubes called urinary ducts. From the cloaca, it flows into the urinary bladder, which branches from the ventral wall of the cloaca. For many terrestrial amphibians, the urinary bladder serves as a water-storage organ. During dry periods, water can be reabsorbed from the urine in the bladder.

Like the larvae of fishes, most amphibian larvae excrete the nitrogen-containing wastes as ammonia. Because ammonia is very toxic, it must be removed from the body quickly or diluted with large amounts of water in the urine. To conserve water, adult amphibians transform ammonia into urea, a less-toxic substance that can be excreted without using as much water. Although this transformation uses energy, it helps save water. During metamorphosis, larval amphibians change from excreting ammonia to excreting urea.

Nervous System

Use the diagram in Figure 42-11 to find the main components of the amphibian nervous system. An amphibian's brain is about the same size as that of a similarly sized fish. The olfactory lobes, which are the center of the sense of smell, are larger in amphibians than in fish, and they lie at the anterior end of the brain. Notice that behind the

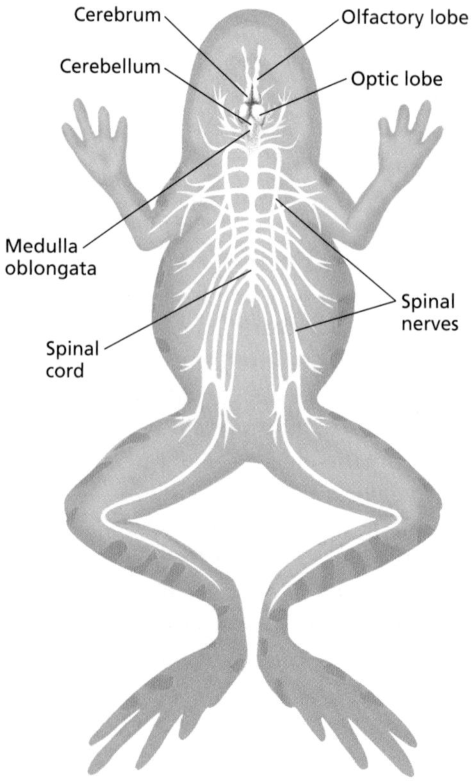

FIGURE 42-11

This diagram shows the frog's nervous system in ventral view. The brain of the frog is sufficiently developed to cope with both land and water environments.

olfactory lobes are the long lobes of the cerebrum, the area of the brain that integrates behavior and is responsible for learning. The optic lobes, which process information from the eyes, lie behind the cerebrum. The cerebellum, a small band of tissue that lies at a right angle to the long axis of the brain, is the center of muscular coordination and is not as well developed in amphibians as it is in other tetrapods. The medulla oblongata lies at the back of the brain and joins the spinal cord. It controls some organ functions, such as heart rate and respiration rate. There is continuous communication among most areas of the brain. Ten pairs of cranial nerves extend directly from the brain. The spinal cord conducts signals from all parts of the body to the brain and from the brain back to the body. Encased in protective bony vertebrae, the spinal cord extends down the back. As in fishes, the spinal nerves branch from the spinal cord to various parts of the body.

Sense Organs

Some sense organs work as well in air as in water, but others do not. For example, the lateral line system, used by fishes to detect disturbances in the water, works only in water. Thus, while larval amphibians have a lateral line, it is usually lost during metamorphosis. Only a few species of aquatic amphibians have a lateral line as adults. The senses of smell, sight, and hearing are well developed in most amphibians. All amphibians have eyes, and visual information is often important in hunting and in avoiding predators. The eyes are covered by a transparent, movable membrane called a **nictitating** (NIK-ti-tayt-eeng) **membrane.** Sound receptors are located in the inner ear, which is embedded within the skull. Sounds are transmitted to this organ by the **tympanic** (tim-PAN-ik) **membrane,** or eardrum, and the **columella** (CAHL-yoo-MEL-uh), a small bone that extends between the tympanic membrane and the inner ear. Sounds first strike the tympanic membrane, which is usually located on the side of the head, just behind the eye. Vibrations of the tympanic membrane cause small movements in the columella that are transmitted to the fluid-filled inner ear. In the inner ear, the sound vibrations are converted to nervous impulses by sensitive hair cells. These nervous impulses are then transmitted to the brain through a nerve.

Word Roots and Origins

nictitating

from the Latin *nictare*, meaning "to wink"

SECTION 42-2 REVIEW

1. Why do most amphibians live in wet or moist habitats?
2. Describe two features of a frog's skeleton that are adaptations for jumping.
3. Trace the flow of blood through a frog's heart.
4. Explain how a frog moves air into and out of its lungs.
5. What is the function of the large intestine in amphibians?
6. CRITICAL THINKING Which sense organ of a terrestrial amphibian resembles the fish's lateral line in function? Explain your answer.

CONTENT CONNECTION

Section 41-2: Lateral line

The lateral line of a fish functions similarly to the tympanic membrane of an amphibian. Both systems detect vibrations of water or air in their environment.

ANSWERS TO SECTION 42-2 REVIEW

1. An amphibian's skin must allow the rapid diffusion of gases, so the skin cannot be watertight. Thus, amphibians can lose body water to evaporation, and they must remain in wet or moist places.
2. The hind limbs are long, and the tibia and fibula in the lower leg are fused. The posterior vertebrae are fused. The radius and ulna are fused in the lower forelimb.
3. sinus venosus$\rightarrow$ right atrium $\rightarrow$ ventricle $\rightarrow$ conus arteriosus $\rightarrow$ lung$\rightarrow$ left atrium $\rightarrow$ ventricle $\rightarrow$ conus arteriosus $\rightarrow$ body
4. The frog draws air into its mouth by lowering the floor of its mouth cavity. The frog then pumps this air into its lungs by raising the floor of its mouth cavity and closing its nostrils. The muscles and elasticity of the lungs force air back out to the mouth, and the floor of the mouth drops. The nostril opens, and the floor of the mouth rises to force the air out.
5. The large intestine collects indigestible material and passes it to the cloaca for elimination.
6. The lateral line detects changes in water pressure, such as changes that result from the movement of a fish nearby. The ear of terrestrial vertebrates resembles the lateral line in function, because it detects airborne vibrations.

SECTION 42-3

SECTION OVERVIEW

Reproduction and Inheritance
Section 42-3 describes the life cycle of a frog, from fertilized egg to adult.

Structure and Function
Section 42-3 describes the structures that compose the reproductive systems in male and female frogs, and it explains how these systems operate.

Evolution
Section 42-3 discusses the importance of water in frog reproduction. This attachment to water links amphibians phylogenetically with fishes.

▶ ENGAGE STUDENTS

Give each student five jewelry beads from an arts and crafts store and a piece of string long enough to hold the beads. Ask the students to string their beads and knot each end of the string. Tell the students that the strings of beads represent amphibian eggs. Have them form the "eggs" into a clump in the middle of their palm. Add a couple of loose beads to each student's hand. Ask students what the loose beads represent (sperm). Ask students to name the process that occurs when a male frog releases sperm on a female frog's eggs (fertilization). Tell them that frogs typically lay their eggs in clumps. Have students dangle the string of beads by holding it by one end. Tell them that toads usually lay their eggs in strings.

QUICK FACT

☑ Some salamanders, such as the mudpuppy and the axolotl, are able to reproduce as larvae. This is known as neoteny. They keep their larval appearance throughout their life.

SECTION

42-3

OBJECTIVES

▲ Describe the life cycle of a frog.

● Describe the changes that occur during metamorphosis in frogs.

■ Identify two examples of parental care in amphibians.

REPRODUCTION IN AMPHIBIANS

One of the biggest differences between aquatic and terrestrial life-forms is their method of reproduction. Most amphibians depend on water for reproduction. They lay their eggs in water and spend the early part of their lives as aquatic larvae.

LIFE CYCLE

Consider the life cycle of a wood frog, illustrated in Figure 42-12. The reproductive system of the male frog includes two bean-shaped testes located near the kidneys. During the breeding season, sperm cells develop in the testes and pass through tubes to the kidneys and urinary ducts. During mating, sperm leave the body through the cloacal opening. In female frogs, a pair of large, lobed ovaries containing thousands of tiny immature eggs lie near the kidneys. During the breeding season, the eggs enlarge, mature, and burst through the thin ovarian walls into the body cavity. Cilia move the eggs forward into the funnel-like openings of the oviducts. As the eggs pass down the oviducts, they are coated with a protective jellylike material. The eggs exit by the cloaca to the external environment, where they are fertilized.

Courtship and Fertilization

In the first warm days of spring in the temperate zones, frogs emerge from hibernation. They migrate in great numbers to ponds and slow-moving streams. Males call to attract females of their own species and to warn off other males. Each species has its own mating call. The frog's croak is produced by air that is driven back and forth between the mouth and the lungs, vibrating the vocal folds. Male frogs have vocal sacs that amplify their calls. The female responds only to the call from a male of the same species.

When a female approaches, the male frog climbs onto her back. He grasps her firmly in an embrace called **amplexus** (am-PLEKS-UHS), which is shown in Figure 42-12a. The male clings to the female, sometimes for days, until she lays her eggs. When the female finally releases her eggs into the water, the male frog discharges his sperm over them, and direct external fertilization takes place. The frogs then separate and resume their solitary lives. Courtship behavior and fertilization often vary from one species to another.

internet**connect**

SCILINKS NSTA

TOPIC: Life cycle of frogs
GO TO: www.scilinks.org
KEYWORD: HM830

Word Roots and Origins

amplexus

from the Latin *amplexus*, meaning "embrace"

FIGURE 42-12

The life cycle of the wood frog, *Rana sylvatica,* begins with mating (a). The male frog holds the female frog in amplexus, and when she releases her eggs, he releases his sperm on the eggs. When the eggs hatch, a tadpole is released (b). One of the first developments of metamorphosis is the growth of hind legs (c). When the tadpole completes metamorphosis, a small adult (d) emerges from the water onto the land.

Metamorphosis

Within a few days of fertilization, the eggs hatch into tadpoles. A newly hatched tadpole lives off yolk stored in its body. It gradually grows larger and develops three pairs of gills. Eventually, the tadpole's mouth opens, allowing it to feed. The tadpole grows and slowly changes from an aquatic larva into an adult. This process of change is called metamorphosis. Legs grow from the body, and the tail and gills disappear. The mouth broadens, developing teeth and jaws, and the lungs become functional.

Biologists have long studied the process of metamorphosis and regeneration to learn what controls such dramatic physical changes. Increasing levels of a hormone called thyroxine, which circulates throughout the bloodstream, stimulate metamorphosis.

The life cycles of many amphibians are similar to that of the wood frog. But there are a variety of alternative reproductive patterns among amphibians. For example, many amphibians do not lay their eggs in water. They select a moist place on land, such as under a rock, inside a rotting log, or in a tree. One or both parents may even construct a nest for the eggs. A number of frog species make a nest of mucus, whipping it into a froth by kicking their hind legs. And not all amphibians undergo metamorphosis. Some salamanders remain in the larval stage for their entire life. Other amphibians bypass the free-living larval stage and hatch from the egg as a small version of the adult.

VISUAL LINK

Figure 42-12

Draw students' attention to Figure 42-12 and point out that amplexus is a behavior that makes amphibian reproduction more efficient than fish reproduction. Ask students to explain why this is so. (Amplexus in frogs places the male's sperm close to the female's eggs as they are being released. In fish mating, fertilization of the eggs is less certain.)

GIFTED ACTIVITY

Tell students that some amphibians *hibernate* and others *estivate.* Ask students to search library and on-line resources to distinguish these two processes. (*Estivate* means to be dormant in the summer, and *hibernate* means to be dormant in the winter.) Have them report on various amphibian species that express these behaviors and adaptive advantages conferred by each.

TEACHING STRATEGY

Observing Frog Eggs

Obtain live fertilized frog eggs and an aquarium from a biological supply house. Then organize students into small groups, and provide them with hand lenses or stereomicroscopes to observe the fertilized eggs. Have students view the eggs in a watch glass containing some aquarium water. Provide diagrams of the various stages of frog embryonic development, and encourage students to find similar stages in their live frog eggs. Place the eggs in the aquarium, and have the students periodically examine them. Remind students to wash their hands after handling the eggs.

SECTION 42-3

CULTURAL CONNECTION

Amphibian Cuisine

Although many people may know that French cuisine includes frog legs, many may not realize that in one region of Mexico, a certain salamander is considered edible. The axolotl, *Ambystoma mexicanum,* which grows to about 25 cm, is found in lakes surrounding Mexico City.

RETEACHING *ACTIVITY*

Have students prepare a circular time line of a frog's life, beginning and ending with the release of eggs. Ask students to divide the time line into aquatic and terrestrial stages and to identify and describe the important events that occur.

ANSWERS TO SECTION 42-3 REVIEW

1. The male frog's call attracts females and repels males.
2. During metamorphosis, the tail and gills disappear, and lungs and limbs grow.
3. Thyroxine is a hormone that stimulates the changes that occur during metamorphosis.
4. The female gastric-brooding frog swallows her eggs, which hatch and undergo metamorphosis in her stomach.
5. Not all amphibian species require standing water; some lay eggs in moist places on land or in a constructed nest.
6. The young are protected from predators and from some environmental changes during the egg, larval, and metamorphic stages. The female may reduce her likelihood of survival because she cannot eat while her young live inside her.

FIGURE 42-13

Rhinoderma darwinii male frogs exhibit parental care by holding the maturing eggs and larvae in their vocal sacs. This frog has already released his offspring, whose tails are still visible. Not all frogs express parental care in this way, and some frogs express no parental behavior at all.

PARENTAL CARE

Parental care is common among amphibians. Eggs and larvae are vulnerable to predators, but parental care helps increase the likelihood that some offspring will survive. Most often, one parent (often the male) remains with the eggs, guarding them from predators and keeping them moist until they hatch. The male Darwin's frog *(Rhinoderma darwinii)* takes the eggs into his vocal sacs, where they hatch and eventually undergo metamorphosis. The young frogs climb out of the vocal sacs and emerge from the male's mouth, as shown in Figure 42-13. Female gastric-brooding frogs of Australia swallow their eggs, which hatch and mature in the stomach. The eggs and tadpoles are not digested because the stomach stops producing acid and digestive enzymes until the young pass through metamorphosis and are released. There are two species of gastric-brooding frogs. Both appear to have become extinct within the last two decades. Females of some species of frogs, such as *Eleutherodactylus,* sit on their eggs until they hatch, not to provide warmth but to prevent the eggs from dessicating. The female normally lays the eggs in the leaves of trees or bushes, where they may dry up.

SECTION 42-3 REVIEW

1. What is the function of the male frog's call?

2. Describe the changes that occur during metamorphosis.

3. What role does thyroxine play in metamorphosis?

4. How does the female gastric-brooding frog care for her eggs?

5. Why is standing water not always required for amphibian reproduction?

6. CRITICAL THINKING How does the parental care of a female gastric-brooding frog help her offspring survive? How might her behavior reduce her likelihood of survival?

CHAPTER 42 REVIEW

SUMMARY/VOCABULARY

42-1

- Amphibians evolved from lobe-finned fishes about 370 million years ago.
- *Ichthyostega* and the other early amphibians had lungs and four legs. They were predominantly aquatic, they had a lateral line and a fishlike tail fin, and some had gills.
- Modern amphibians are divided into three orders: Anura (frogs and toads), Urodela (salamanders), and Apoda (caecilians).
- Anurans are found on all continents except Antarctica, and they have large hind legs for jumping and a strong skeleton.
- Salamanders have four legs and a tail. They are found mainly in North America and Central America.
- Caecilians are legless tropical amphibians that resemble earthworms.

Vocabulary

tadpole (822)

42-2

- Amphibians respire through their lungs and skin. Mucous glands produce slimy mucus that helps retain moisture.
- The skeleton of an amphibian supports the body against the pull of gravity. The spine and limbs are strong so that they can bear the body's weight.
- The amphibian pulmonary circuit carries blood between the heart and lungs. The systemic circuit carries blood to the body and returns it to the heart.
- The heart of an amphibian consists of the sinus venosus, two atria, one ventricle, and the conus arteriosus. Although the ventricle is not divided, deoxygenated and oxygenated blood mix little in the heart.
- Larval amphibians respire with gills and through their skin. Adult amphibians respire with the lungs and the skin. Amphibians pump air into their lungs by raising and lowering the floor of their mouth cavity.
- In amphibians, food passes through the mouth, esophagus, stomach, small intestine, large intestine, and cloaca. The kidneys remove wastes from the blood. Adult amphibians eliminate nitrogenous wastes as urea.
- The brain of a frog has large optic lobes that process visual information, large olfactory lobes that control the sense of smell, a cerebrum that integrates behavior and controls learning, a cerebellum that coordinates movement, and a medulla oblongata that controls the heart and respiration rate.

Vocabulary

columella (829)	ileum (827)	nictitating membrane (829)	systemic circulation (825)
cutaneous respiration (827)	mesentery (827)	pulmonary circulation (825)	tympanic membrane (829)
duodenum (827)	mucous gland (824)	pulmonary respiration (826)	vent (827)

42-3

- Most amphibians lay their eggs in water and have an aquatic larval stage.
- Male frogs call to attract females. Females of some species can distinguish the species of the caller and determine his size.
- The male frog grasps the female and fertilizes her eggs as they are released.
- The hormone thyroxine triggers metamorphosis. During metamorphosis, the tadpole loses its tail and gills and grows legs and lungs.
- Many amphibians show parental care, guarding their eggs and keeping them moist, and some take them into their body to develop.

Vocabulary

amplexus (830)

CHAPTER 42 REVIEW ANSWERS

REVIEW

1. Systemic circulation is the route of blood flow through the body, while pulmonary circulation refers to the circulation of the blood to the lungs.
2. Amplexus is the gripping of the female frog by the male frog during mating, and metamorphosis is the change amphibians undergo during maturation to an adult.
3. The ileum is the second segment of the small intestine. The cloaca is the terminal section of the large intestine and is where the reproductive and urinary systems release their products.
4. The conus arteriosus directs deoxygenated blood to the lungs and oxygenated blood to the body. The sinus venosus collects deoxygenated blood from the body and feeds it into the right atrium.
5. Pulmonary respiration is gas exchange that occurs in the lungs. Cutaneous respiration is gas exchange that occurs through the skin.

6. d **7.** b
8. c **9.** b
10. c **11.** d
12. a **13.** c
14. a **15.** c

16. Most amphibians have a larval aquatic stage that undergoes metamorphosis into a terrestrial adult; they have a moist, thin skin through which gas exchange occurs; they have clawless feet that are often webbed; they use lungs, gills, the skin; the eggs lack a shell or multicellular membranes, are usually fertilized externally, and are usually laid in water or moist places.
17. That amphibians evolved adaptations to land in order to move from one drying puddle to another is no longer accepted. It is considered unlikely that complex adaptations would have developed for short-term uses. Vertebrates probably moved onto land to avoid predators and

resource competition and to take advantage of developing resources on the land.

18. *Ichthyostega* had lungs and four strong limbs.

19. Frogs hide underground during dry times and emerge only after heavy rains.

20. They use cutaneous respiration.

21. Deoxygenated blood leaves the ventricle, goes through the conus arteriosus, travels to the lungs, and then moves into the left atrium. The blood moves into the ventricle, which pumps it into the conus arteriosus and out to the rest of the body. Blood returns to the heart through the sinus venosus and moves into the right atrium. The frog circulatory system is a double loop, but the fish system is a single loop.

22. They transmit sound from the surface of the animal's head to the inner ear.

23. The female can identify the male's species.

24. To save water, adults turn ammonia into less toxic urea, which can be excreted with less water. Larvae do not need to save water.

25. Oxygenated blood is found in the left atrium (C), the ventricle (D), and the conus arteriosus (B); deoxygenated blood is found in the right atrium (A), the sinus venosus (E), the conus arteriosus, and the ventricle.

Critical Thinking

1. Adult amphibians often live in different habitats than their larvae. Most larvae are aquatic, while most adults are at least partially terrestrial. The larvae often eat different foods than the adults. Many larvae are herbivorous, while adults are carnivorous. The separation between larvae and adults may reduce competition.

2. In both, a larva loses some structures and grows others.

Review

Vocabulary

Explain the difference between each pair of related terms.

1. systemic circulation, pulmonary circulation
2. amplexus, metamorphosis
3. ileum, cloaca
4. conus arteriosus, sinus venosus
5. pulmonary respiration, cutaneous respiration

Multiple Choice

6. Which of the following is not a characteristic of *Ichthyostega*? (a) tail fin (b) four limbs (c) lungs (d) pelvic fins
7. The forelimbs of vertebrates evolved from which structures in lobe-finned fishes? (a) pelvic fins (b) pectoral fins (c) pectoral girdle (d) anal fin
8. Amphibians must lay eggs in water primarily because the eggs (a) need oxygen from water (b) are not laid in nests (c) do not have multicellular membranes and a shell (d) need protection from predators.
9. Metamorphosis must take place before amphibians are able to (a) swim (b) live on land (c) feed themselves (d) respire with gills.
10. Salamanders differ from frogs in that they have (a) aquatic larvae (b) four limbs (c) a tail (d) moist skin.
11. The frog's ventricle pumps (a) only oxygenated blood (b) only deoxygenated blood (c) only blood returning from the lungs (d) both oxygenated and deoxygenated blood.
12. Adult amphibians release their nitrogenous wastes in the form of (a) urea (b) uric acid (c) guanine (d) ammonia.
13. Bile is a fluid that (a) aids in circulation (b) lubricates skin (c) breaks down fats into globules (d) aids in respiration.
14. The frog's tympanic membranes are (a) eardrums (b) mouth parts (c) eyelids (d) vocal folds.
15. Which of the following is true of reproduction in most frogs? (a) female calls to attract male (b) eggs surrounded by tough shell (c) fertilization is external (d) fertilization is internal

Short Answer

16. Identify and describe the major characteristics of amphibians.
17. Explain why scientists no longer favor one early theory of why vertebrates moved from water to land. What is the most likely reason to explain why vertebrates moved from water to land?
18. Identify two of *Ichthyostega's* adaptations for life on land.
19. Although frogs do not have watertight skin, some species can survive in deserts. Explain how frogs survive such dry conditions.
20. Salamanders in the family Plethodontidae have no lungs. Explain how these salamanders respire.
21. Outline the route of blood flow through the body of a frog, beginning with the ventricle. How does this route differ from the circulatory system of a fish?
22. What functions do the tympanic membrane and columella perform?
23. What kind of information can a female frog get from a male frog's call?
24. Why are the nitrogenous wastes of an adult amphibian and its larvae different?
25. Look at the diagram of a frog's heart shown below. Identify the structures indicated by the letters. In which structures would you find deoxygenated blood? In which structures would you find oxygenated blood?

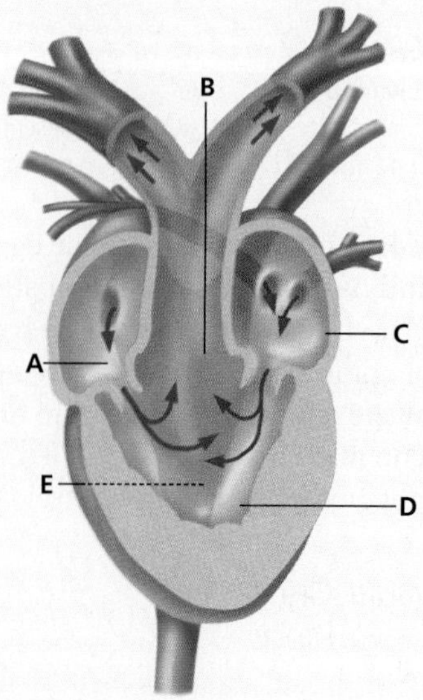

CRITICAL THINKING

1. There are usually many ecological differences between a frog and its tadpoles. Describe some of these differences. Explain why such differences might have been favored by natural selection.
2. As you learned in Chapter 39, insects also undergo metamorphosis. Compare metamorphosis in insects with metamorphosis in amphibians.
3. Charles Darwin noticed that frogs and toads are often absent from oceanic islands, such as the Galápagos Islands, even though they may be found on the nearby mainland. Darwin conducted some experiments that showed that frogs' eggs cannot tolerate exposure to salt water. What hypothesis do you think Darwin was trying to test? Explain your answer.
4. In the brains of amphibians, the largest parts are the olfactory lobes and the optic lobes, the centers of smell and sight. This is very important to amphibians in hunting prey and avoiding predators. In what other biological process is the capacity for hearing important? Explain why hearing must then be especially distinctive.
5. The female gastric-brooding frogs of Australia do not produce stomach acid or digestive enzymes while brooding their young in their stomachs until the tadpoles have completed metamorphosis and leave. If the mother frog does not eat during this period, from where does she get her energy? What other types of frogs must live off the same energy source during similar periods of fasting?
6. When tadpoles undergo metamorphosis, their bodies begin to produce an enzyme that converts ammonia into urea. The time that a tadpole takes to produce this enzyme varies among species. In the graph below, the rate of enzyme production in metamorphosis is shown for a species that inhabits a desertlike environment and a species that inhabits a forest environment. Which curve represents which frog? Explain your answer.

Enzyme Production in Two Frog Species

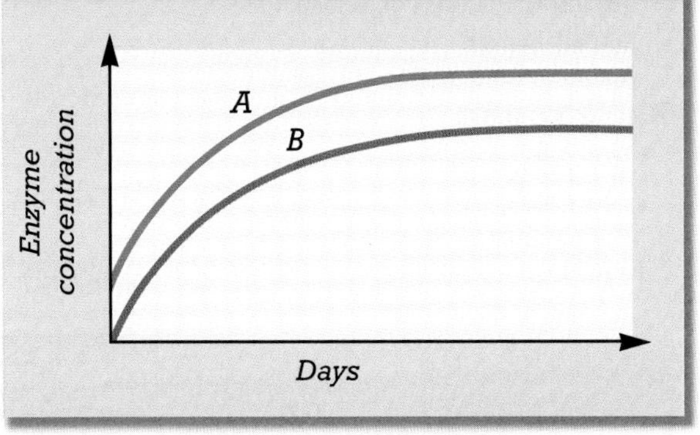

EXTENSION

1. Read "Having Their Toxins and Eating Them, Too" in *BioScience,* December 1999, on page 945. Answer the following question from the section titled "A Frog One Doesn't Kiss": Where does the family Dendrobatidae of poison-Dart frogs acquire the toxins found in their skin?
2. Research human uses of the skin secretions of frogs. Find out how South American Indians use the skin secretions of frogs in the family Dendrobatidae. What are some potential medical uses of skin secretions of frogs? Write a report describing what you have learned.
3. Read "Are Pathogens Felling Frogs?" in *Science,* April 30, 1999, on page 728. Frog deformities have been blamed on pollution in the past. What does new data from Australia now suggest may be the real cause of frog deformities and die-offs?

ANSWERS

CONTINUED

Metamorphosis is controlled by hormones in both. Insects become dormant during metamorphosis, which may reduce its vulnerability to predators, but amphibians do not.

3. He was trying to test the hypothesis that the islands separated from the mainland before frogs evolved. Salt water would keep recently evolved frogs off the island.

4. Females must be able to hear mating calls of males and know which calls are of her species.

5. She must obtain energy from stored fat. Desert-dwelling frogs that burrow into the earth to preserve their moisture have to fast until it rains.

6. A frog in a desert has little time to metamorphose, so it will produce the enzymes for making urea sooner than a frog living in a wet climate. Curve (A) represents the desert frog, and (B) represents the forest frog.

EXTENSION

1. Researchers have isolated over 400 toxic compounds from Dendrobatidae. It is believed that the frogs are acquiring these poisons from the arthropods they eat, such as ants, mites, beetles, and flies.

2. Toxic secretions are coated on arrow tips. Some secretions are antibiotic, and others may help neurological disorders.

3. Most epidemiologists think that a virulent new parasitic fungus, *Batrachochytrium dendrobatidis,* may be the key factor in the sudden decline of frogs around the world. This species of fungus may also be responsible for deformities in frogs, such as missing legs.

CHAPTER 42
INVESTIGATION

TIME REQUIRED

One 50-minute class period

SAFETY PRECAUTIONS

Because students will be working with a live frog, they should be reminded to be gentle with the animal and to follow your instructions carefully.

QUICK REFERENCES

Lab Preparation Notes are found on page 817B.

Holt BioSources provides a Teaching Resources CD-ROM that contains "Using Gowin's Vee in the Lab" and "Scoring Rubrics."

PROCEDURAL TIPS

1. Be sure that the terrarium glass is as clean as possible to allow students unimpeded observation of the frog.
2. If you have more than a few classes to teach, you may want to keep a fresh frog for each class.
3. If your class is large, you may want to have more than one terrarium or aquarium setup so that all students may clearly observe the frog.
4. Have students make their observation tables before beginning the lab.

ANSWERS TO BACKGROUND

1. *Amphibious* means "having two natures" which is reflected in beginning life in the water and continuing as an adult on land.
2. Adult amphibians vary in the amount of time they spend on land. Most amphibians lay their eggs in water, where the young hatch and develop into adults.
3. Amphibians are ectothermic vertebrates that have a three-chambered heart, breathe both through their skin and with lungs, and have scaleless skin through

CHAPTER 42 INVESTIGATION

Observing Live Frogs

OBJECTIVES

- Observe the behavior of a frog.
- Explain how a frog is adapted to life on land and in water.

PROCESS SKILLS

- observing
- relating structure to function
- recognizing relationships

MATERIALS

- live frog in a terrarium
- aquarium half-filled with dechlorinated water
- live insects (crickets or mealworms)
- 600 mL beaker

Background

1. What does *amphibious* mean?
2. Describe how amphibians live part of their life on land and part in water.
3. What are some major characteristics of amphibians?

PART A Observing the Frog in a Terrarium

1. Observe a live frog in a terrarium. Closely examine the external features of the frog. Make a drawing of the frog in your lab report. Label the eyes, nostrils, tympanic membranes, front legs, and hind legs. The tympanic membrane, or eardrum, is a disklike membrane behind each eye.
2. In your lab report, make a table similar to the one on the facing page to note all your observations of the frog in this investigation.
3. Watch the frog's movements as it breathes air with its lungs. Record your observations in your data table.
4. Look closely at the frog's eyes, and note their location. Examine the upper and lower eyelids as well as a third transparent eyelid called a nictitating membrane. The upper and lower eyelids do not move. The nictitating membrane moves upward over the eye. This eyelid protects the eye when the frog is underwater and keeps the eye moist when the frog is on land.
5. Study the frog's legs. Note in your data table the difference between the front and hind legs.
6. Place a live insect, such as a cricket or a mealworm, in the terrarium. Observe how the frog reacts.
7. Gently tap the side of the terrarium farthest from the frog, and observe the frog's response.

PART B Observing the Frog in an Aquarium

8. CAUTION **You will be working with a live animal. Handle it gently and follow instructions carefully.** Frogs are slippery. Do not allow the frog to injure itself by jumping from the lab bench to the floor. Place a 600 mL beaker in the terrarium. Carefully pick up the frog and examine its skin. How does it feel? The skin of a frog acts as a respiratory organ, exchanging oxygen and carbon dioxide with the air or water. A frog also takes in and loses water through its skin.
9. Place the frog in the beaker. Cover the beaker with your hand, and carry it to a freshwater aquarium. Tilt the beaker and gently submerge it beneath the surface of the water until the frog swims out of the beaker.

OBSERVATIONS OF A LIVE FROG

Characteristic	Observation
Breathing	
Eyes	
Legs	
Response to food	
Response to noise	
Skin texture	
Swimming behavior	
Skin coloration	

10. Watch the frog float and swim in the aquarium. How does the frog use its legs to swim? Notice the position of the frog's head.
11. As the frog swims, bend down and look up into the aquarium so that you can see the underside of the frog. Then look down on the frog from above. Compare the color on the dorsal and ventral sides of the frog. When you are finished observing the frog, your teacher will remove the frog from the aquarium.
12. Record your observations of the frog's skin texture, swimming behavior, and skin coloration in your data table.
13. Clean up your materials and wash your hands before leaving the lab.

Analysis and Conclusions

1. From the position of the frog's eyes, what can you infer about the frog's field of vision?
2. How does the position of the frog's eyes benefit the frog while it is swimming?
3. How does a frog hear?
4. How can a frog take in oxygen while it is swimming in water?
5. Why must a frog keep its skin moist while it is on land?
6. How are the hind legs of a frog adapted for life on land and in water?
7. What adaptive advantage do frogs have in showing different coloration on their dorsal and ventral sides?
8. What features provide evidence that an adult frog has an aquatic life and a terrestrial life?
9. What adaptations does the frog display in order to eat? What senses are involved in catching prey?
10. What movement does the frog make in order to breathe?

Further Inquiry

Observe other types of amphibians, or do research to find out how they are adapted to life on land and in water. How do the adaptations of other types of amphibians compare with those of the frog you observed in this investigation?

which they take in and lose water. Most amphibians also have appendages and the females lay eggs in water.

ANALYSIS AND CONCLUSIONS

1. A frog can see in almost all directions.
2. Having eyes on the top of its head enables a frog to see above the waterline while its body is submerged in water, thus allowing it to remain camouflaged.
3. Sound waves hit the tympanic membranes, causing them to vibrate. The vibrations send signals to the brain, allowing the frog to hear the sounds.
4. By keeping its nostrils above the waterline, the frog can breathe with its lungs while it is swimming.
5. Moisture is necessary for the diffusion of oxygen and carbon dioxide through the skin. It also keeps the frog from drying out.
6. The webs of the hind legs push back water as the frog swims. The large muscles of the hind legs allow the frog to jump on land.
7. The dark dorsal skin blends in with the color of the water as seen from above, making the frog less visible to predators above water. The light ventral skin blends in with the color of the sky above the water, making the frog less visible to predators under water.
8. aquatic life: webbed feet, eyes on top of head, breathes through skin; terrestrial life: jumping legs, breathes with lungs
9. Frogs possess a long, sticky tongue, which they can quickly project to catch prey. Frogs depend on sight and hearing in order to detect prey.
10. A frog must lower the floor of its mouth and open its nostrils in order to breathe.

PLANNING GUIDE 43

CHAPTER 43

REPTILES

	TOPICS	TEACHING RESOURCES	LABS, CLASSWORK, AND HOMEWORK
BLOCKS 1 & 2 90 minutes	**Introducing the Chapter,** p. 838	**ATE** Focus Concept, p. 838 **ATE** Assessing Prior Knowledge, p. 838 **ATE** Understanding the Visual, p. 838	■ Supplemental Reading Guide, *The Dinosaur Heresies*
	43-1 Origin and Evolution of Reptiles, p. 839 History of Reptiles, p. 839 Success of Reptiles, p. 841	**ATE** Section Overview, p. 839 137A, 141A, 142A, 167, 173, 179	**PE** **Quick Lab,** p. 842, Modeling an Amniotic Egg ★ Study Guide, Section 43-1 **PE** Section Review, p. 843
BLOCK 3 45 minutes	**43-2 Characteristics of Reptiles,** p. 845 Circulatory System, p. 845 Respiration, p. 846 Nervous System, p. 847 Thermoregulation, p. 848 Reproduction and Parental Care, p. 849	**ATE** Section Overview, p. 845 **ATE** Critical Thinking, p. 848 171	**Lab A22,** Comparing Animal Eggs **Lab A26,** Vertebrate Skeletons ★ Study Guide, Section 43-2 **PE** Section Review, p. 849
BLOCK 4 45 minutes	**43-3 Modern Reptiles,** p. 850 Order Chelonia, p. 850 Order Crocodilia, p. 851 Order Squamata, p. 852 Order Rhynchocephalia, p. 854	**ATE** Section Overview, p. 850 **ATE** Critical Thinking, p. 851 174	**PE** **Chapter 43 Investigation,** p. 858 **PE** **Quick Lab,** p. 853, Demonstrating Muscle Contractions ★ Study Guide, Section 43-3 **PE** Section Review, p. 854
BLOCKS 5 & 6 90 minutes	**REVIEW and ASSESSMENT** **PE** Chapter 43 Review, pp. 856–857 ■ Performance-Based Assessment—The Challenge of Water Retention	Audio CD Program	**CHAPTER TESTING** ■ Chapter Test (blackline master) ▲ Test Generator ■ Assessment Item Listing ■ Scoring Rubrics and Classroom Management Checklists

KEY

MODERN BIOLOGY

PE Pupil's Edition
ATE Teacher's Edition
■ Active Reading Guide
★ Study Guide

HOLT BIOSOURCES

Teaching Transparencies

Laboratory Program

One-Stop Planner CD-ROM
Shows these resources within a customizable daily lesson plan:

■ Holt BioSources Teaching Resources
■ Active Reading Guide
★ Study Guide
▲ Test Generator

READING FOR CONTENT MASTERY

Preview: Have students preview the objectives, the introductory paragraph, topic sentences, and captions before reading the chapter.

ATE Active Reading Technique: Discussion, p. 839
■ Active Reading Guide Worksheet 43-1
PE Summary/Vocabulary, p. 855

■ Active Reading Guide Worksheet 43-2
PE Summary/Vocabulary, p. 855

■ Active Reading Guide Worksheet 43-3
PE Summary/Vocabulary, p. 855

TECHNOLOGY AND INTERNET RESOURCES

CNN PRESENTS Video Segment 23 Mexico Sea Turtle

Holt Biology Videodiscs Teacher's Correlation Guide, Lessons 43-1 through 43-3

Audio CD Program, Sections 43-1 through 43-2

internet**connect**

SCI LINKS NSTA

On-line Resources: www.scilinks.org
The following *sci*LINKS Internet resources can be found in the student text for this chapter:

Topics:
- Amniotic egg, p. 843
- Reptiles, p. 845
- Lizards, p. 852

go.hrw.com

On-line Resources: go.hrw.com
Visit the HRW Web site for a variety of resources related to this chapter. Just type in the keyword HM2 Chapter 43.

Smithsonian Institution®
Visit **www.si.edu/hrw** for additional on-line resources.

CNNfyi.com
Visit **www.cnnfyi.com** for late-breaking news and current event stories selected just for you.

LAB ACTIVITY PLANNING

CHAPTER 43 INVESTIGATION

Observing Color Adaptation in Anoles, pp. 858–859

OVERVIEW
In this investigation, students will observe color change in the anole lizard in response to environmental color. Anoles may change color if they are frightened or deprived of food. Very cool temperatures can also produce a change to brown.

SAFETY
Anoles can run fast and are easily frightened. Remind students to handle anoles with care. If the anole is picked up by the tail, it may break off. Note that anoles that are no longer needed in the classroom make ideal pets at home.

PREPARATION
1. At least 1 hour before the lab exercise begins, place the anoles under fluorescent light in two terraria. Place half the anoles in a terrarium lined with black construction paper. These anoles should turn brown.
2. Place the remaining anoles in a terrarium lined with white construction paper. These anoles should turn green.
3. Provide a shallow dish of drinking water in each terrarium. Use jars large enough to allow the anoles to settle comfortably on the bottom.

Quick Labs

Modeling an Amniotic Egg, p. 842
Each student group will need a paper lunch bag, a large self-sealing plastic bag, a small self-sealing plastic bag, modeling clay, a yellow balloon, and a red balloon for this activity.

Demonstrating Muscle Contractions, p. 853
Each student group will need a plastic drinking straw and a small ball of modeling clay for this activity.

UNDERSTANDING THE VISUAL

Ask students if they have ever seen a frog or a salamander basking in the sun. Tell them that an amphibian exposed to the sun for too long would lose water through its thin skin and dehydrate. Ask students what prevents the marine iguanas, which are reptiles, from dehydrating. (Their thick, scaly skin keeps body fluids from escaping.)

FOCUS CONCEPT

Structure and Function

The unique design of a reptile enables it to spend its entire life cycle on land. Reptiles do not dehydrate because they have waterproof skin and because their kidneys produce a concentrated urine. Their lungs and heart can supply more oxygen than an amphibian's heart and lungs. Reptiles are able to produce waterproof amniotic eggs that are laid on land, and they do not need to return to the water to reproduce.

ASSESSING PRIOR KNOWLEDGE

Review the following concepts.

Circulation: *Chapters 41 and 42*
Review the structure of the heart and the pattern of blood circulation in fishes and amphibians. Sketch the circulatory patterns of a fish and of an amphibian on the board. Ask a volunteer to come to the board and describe the course that the blood takes in each.

Respiration: *Chapter 42*
Review breathing and lung structure in amphibians. Have students describe the pressure changes that occur in the body of a frog when it breathes.

CHAPTER 43

REPTILES

A marine iguana, Amblyrhynchus cristatus, *of the Galápagos Islands warms itself by basking on a rock. These are the only marine lizards in the world, and they feed exclusively on seaweed.*

FOCUS CONCEPT: *Structure and Function*

As you read the chapter, compare reptiles' structural and functional adaptations for living on land with those of amphibians.

ORIGIN AND EVOLUTION OF REPTILES

The reptiles (class Reptilia) appeared more than 300 million years ago. They are one of the largest and most evolutionarily successful groups of terrestrial vertebrates. In this chapter, you will study the diversity of reptiles and learn about some of the characteristics that make the reptiles a successful group.

SECTION

43-1

OBJECTIVES

▲ Identify and describe three groups of reptiles that lived during the Mesozoic era.

● Describe the asteroid-impact hypothesis to explain the extinction of the dinosaurs.

■ Describe the structure of the amniotic egg, and explain the functions of its parts.

◆ Contrast the skin of reptiles with that of amphibians.

HISTORY OF REPTILES

From studies of fossils and comparative anatomy, biologists infer that reptiles arose from amphibians. The oldest known fossils of reptiles were found in deposits from the early Carboniferous period (360 million to 286 million years ago) and are about 350 million years old. The earliest reptiles were small, four-legged vertebrates that resembled lizards and had teeth adapted for eating insects. The abundance of insects at the time may have been one reason the early reptiles flourished.

Age of Reptiles

The reptiles diversified rapidly, and by the Permian period (286 million to 245 million years ago) they had become the dominant land vertebrates. The Mesozoic era (245 million to 65 million years ago) is often called the Age of Reptiles because nearly all of the large vertebrates on Earth were reptiles during that time. On land, the most famous and spectacular reptiles, the **dinosaurs,** appeared and evolved into a great variety of forms during the Mesozoic era. Dinosaurs are known for the great size of some species. One of the largest dinosaurs, *Brachiosaurus,* measured 23 m (75.4 ft) long, stood 12 m (39.4 ft) tall, and weighed more than 77,000 kg (169,400 lb). In other words, *Brachiosaurus* was as long as a tennis court, as tall as a four-story building, and heavier than 10 elephants. Although the size of some dinosaurs has captured our imagination, many species were small, some no larger than a chicken.

Over 300 genera of dinosaurs have been identified. Their fossils have been discovered on all continents, even Antarctica, which had a much milder climate during the Mesozoic era than it does today. Dinosaurs were adapted to a wide range of environments and to different ways of life. Some reconstructions of dinosaurs are shown in Figure 43-1 on the following page.

Word Roots and Origins

dinosaur

from the Greek *deinos,* meaning "terrible," and the Greek *sauros,* meaning "lizard"

SECTION 43-1

SECTION OVERVIEW

Evolution
Section 43-1 discusses the evolution of reptiles from amphibian ancestors. It also describes early reptiles, such as dinosaurs, ichthyosaurs, plesiosaurs, and pterosaurs.

Structure and Function
Section 43-1 describes the structural adaptations that have allowed reptiles to become true land vertebrates. These adaptations include waterproof skin, efficient respiratory and excretory systems, and the ability to produce an amniotic egg.

ACTIVE READING

Technique: Discussion
Before students begin reading this chapter, have a volunteer explain how the vertebrates they have learned about so far have been dependent on water. Tell students that in this chapter they will learn about the first vertebrates to fully adapt to life on land—the reptiles.

TEACHING STRATEGY

Dinosaur Timeline

Tell students that they will construct a timeline emphasizing various reptiles from the Mesozoic era. It should contain both written and visual information. First assign groups of students a different kind of early reptile to research. Examples are plesiosaurs, ichythyosaurs, sauropods, theropods, and duck-billed dinosaurs. Tell each group to condense their research information so that it fits onto a 3 × 5 in. note card. Have them enhance the written information with drawings and clippings from magazine articles. After students complete the research, have them construct the timeline by drawing a time scale on butcher paper that runs along a wall in the classroom. Have the groups then use tape to attach their visual and written information to the timeline.

RECENT RESEARCH

The Biggest Mass Extinction

The extinctions that occurred at the end of the Cretaceous period eliminated about 60 percent of all land species, including the dinosaurs. But an even more catastrophic mass extinction occurred about 185 million years earlier, at the end of the Permian period. About 95 percent of Earth's species were wiped out, most of them invertebrates living in the sea. One recent theory attributes this extinction to an accumulation of carbon dioxide in the ocean caused by the decomposition of organic matter and uneven ocean water mixing. Scientists say the elevated CO_2 levels would have poisoned ocean organisms on a catastrophic scale.

FIGURE 43-1

Brachiosaurus, shown in (a), and related dinosaurs were herbivores that probably used their long necks to reach vegetation in the treetops. *Deinonychus,* shown in (b), and other carnivores walked on their hind legs and used sharp claws to rip apart prey. *Triceratops,* shown in (c), had body armor and horns to defend itself against meat-eating dinosaurs, but it was a plant eater. Scientists believe *Triceratops* may have roamed in great herds across western North America.

Reptilian success during the Mesozoic era was not limited to terrestrial habitats. Several groups of reptiles, including the ichthyosaurs and plesiosaurs, lived in the oceans. Ichthyosaurs, illustrated in 43-2b, were sleek aquatic reptiles that resembled modern bottlenose dolphins. Plesiosaurs had long, flexible necks and compact bodies. Mesozoic reptiles called pterosaurs, shown in Figure 43-2a, evolved the ability to fly. There are no flying reptiles today.

Extinction of Dinosaurs

Although the fossil record provides many clues about what dinosaurs were like, paleontologists who study dinosaurs still have many unanswered questions. For example, why did the dinosaurs become extinct 65 million years ago, at the end of the Cretaceous period? Many species of aquatic and terrestrial organisms besides the dinosaurs became extinct at this time.

Most scientists think that a catastrophic cosmic event was responsible for the mass extinction. Supporters of this hypothesis—called the **asteroid-impact hypothesis**—suggest that a huge asteroid hit the Earth, sending so much dust into the atmosphere that the amount of sunlight reaching the Earth's surface was greatly reduced. The reduced sunlight caused severe climatic changes that led to the mass extinction. According to this hypothesis, the dinosaurs would have become extinct very quickly, perhaps even within a few months.

The asteroid-impact hypothesis was first proposed in 1980 by Luis Alvarez, a Nobel Prize–winning physicist, and his son Walter, a geologist. They noted that sediments from the end of the Cretaceous period contain unusually high concentrations of iridium. Iridium is a metal that is very rare in the Earth's crust but more abundant in asteroids and other extraterrestrial bodies. Other scientists discovered that some sediments from this time contain quartz crystals that have been deformed by a powerful force, such as that resulting from the collision of a large asteroid with Earth. Further evidence for this hypothesis was provided by the discovery of the likely site of impact, a crater located on the Yucatan Peninsula in southern Mexico. The crater dates from the end of the Cretaceous period and is about 180 km (110 mi) across.

FIGURE 43-2

The smallest pterosaurs, such as the one in (a), were only the size of sparrows, while the largest were about the size of a small airplane, with wingspans of 12 m (about 39 ft). Like dolphins, ichthyosaurs, as in (b), were probably fast swimmers and fed on fish.

SUCCESS OF REPTILES

Representatives of the four modern orders of reptiles—turtles and tortoises, lizards and snakes, tuataras, and crocodilians—survived the mass extinction of the Cretaceous period. These four orders of reptiles have diversified to more than 6,000 species. Reptiles successfully occupy a variety of terrestrial and aquatic habitats on all continents except Antarctica. Figure 43-3 shows the evolutionary relationships between modern reptiles and birds.

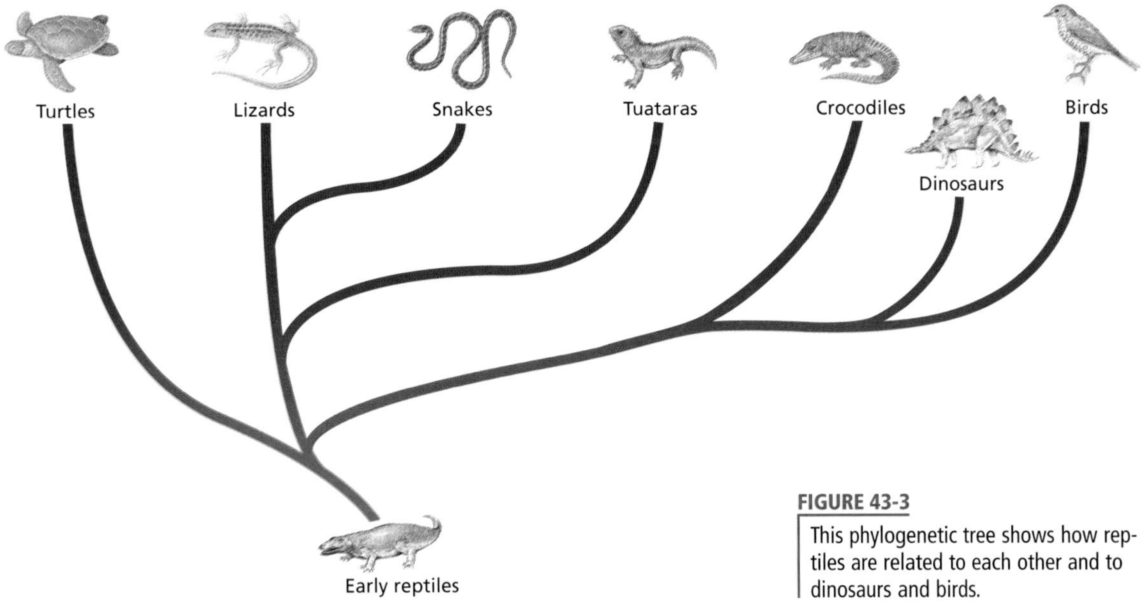

FIGURE 43-3

This phylogenetic tree shows how reptiles are related to each other and to dinosaurs and birds.

SECTION 43-1

RECENT RESEARCH

Ancient Flying Reptile

The first fossils collected of *Coelurosauravus jaekeli,* the earliest known flying vertebrate, were so poorly formed that its wings were not obvious. For years, scientists believed the reptile was a terrestrial species. But recently discovered fossils of the animal portray a crystal-clear image of its body. These fossils clearly show the presence of wings, and the wings are unique among those of other known living or extinct winged vertebrates. Instead of being supported by elements of the skeleton, the wings of *Coelurosauravus* were composed of skin that contained bony spines developed independently of the skeleton. *Coelurosauravus jaekeli* lived about 250 million years ago.

OVERCOMING MISCONCEPTIONS

Students may assume that pterosaurs are the ancestors of modern birds. Emphasize that the pterosaurs had died out by the end of the Mesozoic era and that their line was an evolutionary dead end. Most scientists think modern birds evolved from bipedal dinosaurs called theropods. The popular *Tyrannosaurus rex* was a member of this group.

SECTION 43-1

Quick Lab

Modeling an Amniotic Egg

Time Required 20 minutes

Procedural Tips Have students work in groups of two to four. Encourage students to devise their own models without having the information below. You may choose to write the following information on the board or overhead:

- paper bag = shell
- large plastic bag = chorion
- small plastic bag = amnion
- clay = embryo
- yellow balloon = yolk sac
- red balloon = allantois

Answers to Analysis The model demonstrates three layers of protection—the shell, the chorion, and the amnion. This adaptation allowed reptiles to live on land. The amniotic egg indicates that reptiles, birds, and mammals evolved from a common ancestor.

GIFTED *ACTIVITY*

Divide the class into groups of four. Divide each of these groups into two teams, and have them debate how the dinosaurs became extinct. One team will defend the sudden extinction that is supported by the asteroid-impact hypothesis. The other team will defend the idea of gradual extinction due to changing climates, drifting continents, or other factors. Both students on each team are responsible for researching the team position. Tell students to use facts, logic, and conviction to defend their position and to point out inconsistencies in their opponents' position. All students should have a chance to speak.

Quick Lab

Modeling an Amniotic Egg

Materials paper lunch bag; large, sealable plastic bag; small, sealable plastic bag; modeling clay; yellow balloon; red balloon

Procedure Design an amniotic egg model using the materials provided and the information in Figure 43-4. Use all the materials you have been provided.

Analysis According to your model, how many layers of protection surround the embryo? Why is this structure an important adaptation for reptiles? The amniotic egg links reptiles to what other two groups of organisms?

The Amniotic Egg

Although amphibians were the first vertebrates to successfully invade land, they could not make a full transition to terrestrial life. They still require water in their environment to reproduce. Reptiles are considered the first fully terrestrial vertebrates because they do not need to reproduce in water, as most amphibians do. Reptiles produce **amniotic eggs,** which encase the embryo in a secure, self-contained aquatic environment. Amniotic eggs provide more protection for the developing embryo than the jellylike eggs of amphibians.

Figure 43-4 shows the internal structure of the amniotic egg, including its four specialized membranes: the amnion, yolk sac, allantois, and chorion. The egg is named for the **amnion** (AM-nee-AHN), the thin membrane enclosing the fluid in which the embryo floats. The **yolk sac** encloses the yolk, a fat-rich food supply for the developing embryo. The **allantois** (uh-LAN-toh-wis) stores the nitrogenous wastes produced by the embryo. Because it contains many blood vessels, the allantois also serves as the embryo's "lung," exchanging carbon dioxide for oxygen from the environment. The **chorion** (KOR-ee-AHN) surrounds all the other membranes and helps protect the developing embryo. Protein and water needed by the embryo are contained in the **albumen** (al-BYOO-muhn). You are familiar with albumen as the egg white in a chicken's egg. In most reptiles, the tough outer shell provides protection from physical damage and limits the evaporation of water from the egg.

The amniotic egg first evolved in reptiles, but it also occurs in mammals and birds. The presence of this feature is strong evidence that reptiles, birds, and mammals evolved from a common ancestor. The eggs of some reptiles and nearly all mammals lack shells, and the embryo develops within the mother's body.

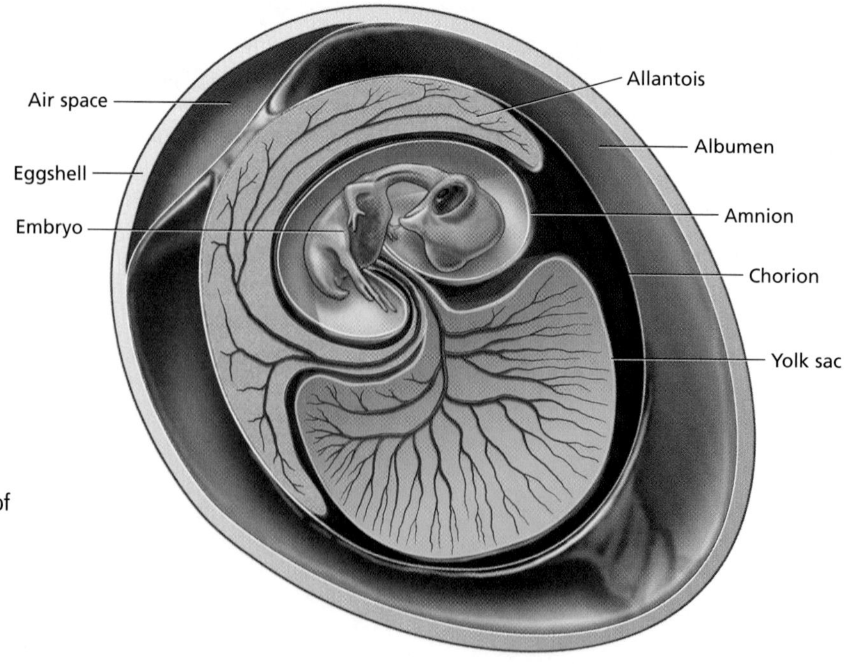

FIGURE 43-4

This amniotic egg at an early stage of development shows a chick embryo and the four major membranes. The porous shell allows the exchange of oxygen and carbon dioxide.

FIGURE 43-5

The skin of the spiny lizard *Sceloporus poinsetti* protects it from the rugged terrestrial environment and from water loss.

Watertight Skin

Because amphibians exchange gases through their skin, the skin must be moist and thin enough to allow rapid diffusion. A drawback of this kind of skin is that it cannot be watertight, and one problem amphibians face on land is the loss of body water through evaporation. Reptiles, like the lizard shown in Figure 43-5, are covered by a thick, dry, scaly skin that prevents water loss. This scaly covering develops as surface cells fill with **keratin,** the same protein that forms your fingernails and hair. Lipids and proteins in the skin help make the skin watertight. The tough skin of a reptile not only helps conserve body water but also protects the animal against infections, injuries, and the wear and tear associated with living in terrestrial environments.

Respiration and Excretion

All reptiles have lungs for gas exchange. Thus, all of the tissues involved in gas exchange are located inside the body, where they can be kept moist in even the driest environments. The excretory system of reptiles also helps them conserve body water. Snakes, lizards, and other land-dwelling reptiles excrete nitrogenous wastes in the form of uric acid. Uric acid is much less toxic than ammonia or urea. Thus, it requires little water for dilution, and reptiles lose only small amounts of water in their urine.

internet**connect**

SCILINKS NSTA

TOPIC: Amniotic egg
GO TO: www.scilinks.org
KEYWORD: HM843

SECTION 43-1 REVIEW

1. **Describe three kinds of Mesozoic reptiles, not including dinosaurs.**
2. **Why is the presence of iridium in sediments from the end of the Cretaceous period an important piece of evidence supporting the asteroid-impact hypothesis?**
3. **Explain the significance of the amniotic egg for reproduction on land.**
4. **Describe the functions of the chorion and the allantois in the amniotic egg.**
5. **What are two differences between the skin of an amphibian and the skin of a reptile?**
6. **CRITICAL THINKING If it were shown that dinosaurs disappeared slowly over many millions of years, would this contradict or support the asteroid-impact hypothesis? Explain your answer.**

INCLUSION ACTIVITY

Have pairs of students design and make posters of ancient reptiles. Tell them that each poster must have a picture of a dinosaur or some other ancient reptile on it and that it must list three adaptations that helped the animal survive in its environment. Have the students present their posters to the rest of the class. Provide dinosaur books, or let the students use the library to do research for their posters.

ANSWERS TO SECTION 43-1 REVIEW

1. Pterosaurs were flying reptiles of the Mesozoic period. Their wings were modified forelimbs. Plesiosaurs were long-necked aquatic reptiles that fed on fish. Ichthyosaurs were sleek fish eaters shaped like today's bottlenose dolphins.
2. Iridium is very rare on Earth and is much more common in asteroids and other extraterrestrial debris.
3. The amniotic egg protects the embryo from drying out by encasing it in an aquatic environment. The amniotic egg need not be laid in water or moist places.
4. The chorion and allantois control the exchange of gases with the environment and help retain water within the egg.
5. Amphibians have a thin, moist skin through which they exchange gases. Reptiles have a dry, scaly skin that is watertight.
6. The asteroid-impact hypothesis suggests a rapid die-off resulting from the after-effects of impact. Thus, evidence of a gradual decline would contradict the hypothesis.

BACKGROUND

This feature presents Robert Bakker's ideas about temperature regulation and the extinction of dinosaurs. *Warm-blooded* and *cold-blooded* are inaccurate terms. For example, some present-day "cold-blooded" lizards have a higher body temperature than some "warm-blooded" mammals. Accurate terms are *ectothermic* and *endothermic.* An ectothermic animal is one whose body temperature depends on gaining heat from the environment. An endothermic animal is one whose body temperature depends on internal heat production. Endothermy is an expensive metabolism because to produce constant heat, great amounts of food are needed.

READING FOR MEANING

Bakker uses the dinosaurs' extinction as proof that they were endothermic. He believes that species with a high metabolic rate are more vulnerable to extinction. Bakker also cites as proof the rapid evolutionary pace of the dinosaurs—even faster than modern-day reptiles.

READ FURTHER

The "best natural design for avoiding extinction" would be a slow metabolism, like that of the turtles and crocodiles. Species with low metabolic rates, such as the snapping turtle, monitor lizard, alligator, and crocodile, also have slow evolutionary rates. The type of animal most susceptible to extinction has a high metabolism and requires large amounts of food. This allows extinction of a species when major changes in the environment take place.

Literature & Life

Dinosaurs Are Extinct—But Why?

The following excerpt is from *The Dinosaur Heresies,* by Robert T. Bakker.

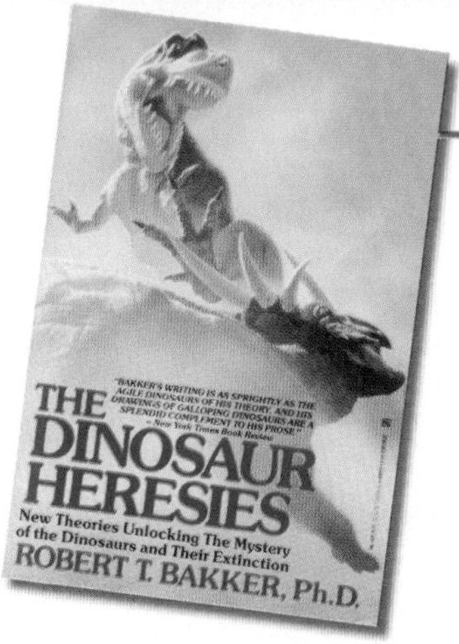

A very well-preserved segment of fossil history . . . permits a computation of how long the average species and genus of dinosaur lasted. And that can be compared with the rates of change computed for warm-blooded animals on the one hand and cold-blooded on the other. Rates of evolutionary change must somehow be linked to the metabolic rate of any given organism. When metabolism is very high, the high physical costs of its living must drive the animal to be an aggressive competitor or predator. And the rapid reproduction typical of warm-blooded animals tends to fill habitats to overcrowded levels much more quickly than the more leisurely breeding schedules characteristic of cold-bloods. An ecological community full of warm-blooded species is therefore a tough environment where the resident species jostle one another for food and water, breeding sites and burrows, all year round. In this sort of environment, the average species will not last long in terms of geological time before it is driven to extinction either by a new species, or by a combination of old species, or by an adverse change of climate.

Clams, by contrast, with their low metabolism, move around very little to accomplish the tasks of their adult lives. It is consequently not surprising that species of clams last much longer than warm-blooded species. Crocodiles and turtles are far more active than clams, but are still sluggish compared to the fully warm-blooded mammals. If dinosaurs resembled cold-blooded reptiles metabolically, their rate of evolutionary change would very nearly match that found in crocodiles or large turtles. But if the dinosaurs' metabolism was heated, then the average life span of one of their species or genera would have to be short, like a mammal's.

One of the best places to study the rate of evolutionary changes in dinosaurs is in the Late Cretaceous deltas of Wyoming, Montana, and Alberta. There the changes through the last ten million years of the dinosaurs' history can be followed. As might be expected, the turtles and crocodiles show very little change in these strata. The genera representing these cold-bloods hold on through formation after formation. But what about the dinosaurs here? Their evolutionary pace stands out as quite different. New species and genera kept appearing and eliminating older ones at quite a brisk tempo, geologically speaking. The average genus of dinosaur lasted for only a fraction of the time of that of the average crocodile. . . .

One part of the orthodox story does appear to be unassailable, an ineradicable fact safe from even the wildest heretic: Dinosaurs are indeed all extinct. The fact of their extinction is the cornerstone underlying the orthodox belief that dinosaurs were maladapted failures.

Dinosaurs are incontrovertibly dead. But that does not prove what orthodoxy believes about them. Paradoxically, the extinction of the dinosaurs is strong evidence that their biology was heated to levels far above those of typical reptiles. The basic principle is simple: The higher the metabolic needs of a group of species, the more vulnerable it is to sudden and catastrophic extinction.

Reading for Meaning

What is Bakker's explanation for the extinction of the dinosaurs?

Read Further

Bakker follows the text excerpted here by describing the best natural design for avoiding extinction. Explain what you think that design might be, and name some existing animals that are similar to that design.

Characteristics of Reptiles

Reptiles live in many different kinds of habitats and show a great deal of diversity in size and shape. Think of the differences between a snake and a turtle or between a lizard and a crocodile. In this section, we will look at some of the anatomical, physiological, and behavioral characteristics of reptiles.

SECTION

43-2

OBJECTIVES

▲ Describe the pattern of blood flow through the heart of a lizard.

● Compare the lungs of reptiles with those of amphibians.

■ Identify three senses of reptiles.

◆ Contrast endothermy with ectothermy.

▲ Explain the differences between oviparity, ovoviviparity, and viviparity.

CIRCULATORY SYSTEM

The circulatory system of a reptile, like those of all terrestrial vertebrates, is composed of two loops. The pulmonary loop carries deoxygenated blood from the heart to the lungs and returns oxygenated blood to the heart. The systemic loop transports oxygenated blood to the tissues of the body, where oxygen and nutrients are unloaded and carbon dioxide is picked up, and returns deoxygenated blood to the heart.

Heart Structure and Function

In lizards, snakes, tuataras, and turtles, the heart has two atria and a single ventricle partially divided by a wall of tissue called a **septum.** In crocodiles, there are two atria and two separate ventricles. The sinus venosus and the conus arteriosus, which are major structures in the heart of a fish, are much smaller in reptiles. In fact, the sinus venosus is absent in some species. When it is present, it collects blood from the body and channels it into the right atrium. The conus arteriosus forms the base of the three large arteries exiting from the reptilian heart.

Because the ventricle is not completely divided (except in crocodiles), it might seem that deoxygenated and oxygenated blood would mix. However, recent studies have shown that blood mixing does not occur when a reptile is active. Deoxygenated and oxygenated blood are kept separate during contraction of the heart by the actions of the heart valves and the movement of the septum and ventricular walls.

Pumping blood through lungs requires energy. Under some conditions, it is advantageous for a reptile to divert blood away from the lungs to conserve energy. For example, an inactive reptile needs so little oxygen that it may go a long time without breathing. Similarly, aquatic reptiles do not breathe while they are underwater.

internet**connect**

SCILINKS. NSTA

TOPIC: Reptiles
GO TO: www.scilinks.org
KEYWORD: HM845

SECTION 43-2

SECTION OVERVIEW

Structure and Function

Section 43-2 describes the organ systems of a reptile. This section discusses how the operation of each system is an adaptation for a reptile to live on land.

Reproduction and Inheritance

Section 43-2 describes the reproductive strategies in reptiles and discusses parenting behavior.

▶ ENGAGE STUDENTS

Ask students to stand, cross their arms in front of their body, and wrap their arms around their chest. Tell them to take a deep breath and then exhale. Ask them to describe what happened to their chest when they inhaled and exhaled. (Their chest expanded and contracted.) Tell them that reptiles were the first vertebrate group to breathe by expanding and contracting their chest. Contrast this method of breathing with the way amphibians breathe. (Amphibians inhale by trapping air in their mouth and squeezing it down into the lungs. To exhale, they contract their chest muscles and force the air back out.)

RECENT RESEARCH

Rattlesnake Muscles

Rattlesnakes have the quickest-moving skeletal muscles in the animal world. Their tail muscle allows them to shake their rattle 90 times each second. Most muscles would quickly tire from this kind of energetic activity, but the rattlesnake can shake its tail for long periods of time. The muscle's ability to use ATP in an efficient manner allows it to sustain a high level of activity.

GIFTED ACTIVITY

Have students search for a diagram that shows the internal anatomy of a reptile. Tell them it can be any kind of reptile and that they are to trace or draw the picture. Then have them write a report that describes the anatomical structures in their reptile and explains how they function. Tell students to organize the organs discussed in their report by system and to attach their drawing to the finished product.

INCLUSION ACTIVITY

Use string to create the schematic of the turtle heart shown in Figure 43-6 on the floor of the classroom. Make the replica large enough that students can walk through the chambers. Use another piece of string to fashion lungs on the floor, above the replica of the heart. With a third piece of string, form a circle to symbolize tissue below the heart. Tell the students to imagine that they are blood cells. Give each student a small square of red construction paper and a small square of blue construction paper. Tell them that when they are deoxygenated blood, they are to hold up the blue square and when they are oxygenated, they are to hold up the red square. Let students practice tracing the path of blood during double-loop circulation by marching through the setup on the floor.

FIGURE 43-6

The turtle's heart, shown in cross section, has a partially divided ventricle, unlike an amphibian's three-chambered heart or a crocodile's four-chambered heart. Because the flow of blood through a turtle's heart is asynchronous, deoxygenated blood and oxygenated blood pass through the upper part of the ventricle at different times and so do not mix.

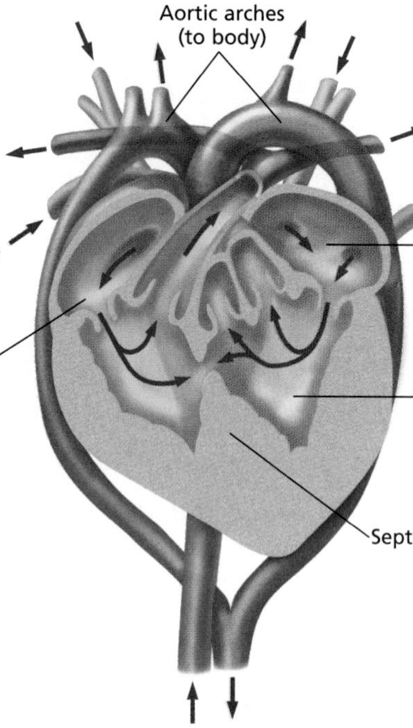

By constricting the pulmonary arteries, a reptile's blood flow through the heart can be redirected to send some deoxygenated blood back to the body instead of to the lungs. Bypassing the lungs may also help a reptile raise its body temperature quickly—warm blood from the skin can be directed to the organs deep within the body. The reptilian heart has a degree of circulatory flexibility that the hearts of birds and mammals do not. Instead of being a handicap, this flexibility is actually well suited to reptilian physiology and activity patterns. Figure 43-6 shows a schematic diagram of the heart of a turtle. Compare this diagram with the illustrations of the three-chambered heart of a frog (Figure 42-8) and the two-chambered heart of a fish (Figure 41-11).

RESPIRATION

The lungs of reptiles are large, and they are often divided internally into several chambers. The lining of the lungs may be folded into numerous small sacs called **alveoli.** Alveoli greatly increase the internal surface area of the lungs, thus increasing the amount of oxygen that can be absorbed. In most snakes, only the right lung actively functions. It is elongated and may be half as long as the body. The left lung is either reduced to a small nonfunctional sac or absent entirely.

A reptile fills its lungs by expanding its rib cage. This expansion reduces the pressure within the thorax and draws air into the lungs. When the ribs return to their resting position, pressure within the thorax increases and air is forced out of the lungs. Similar movements help you to breathe.

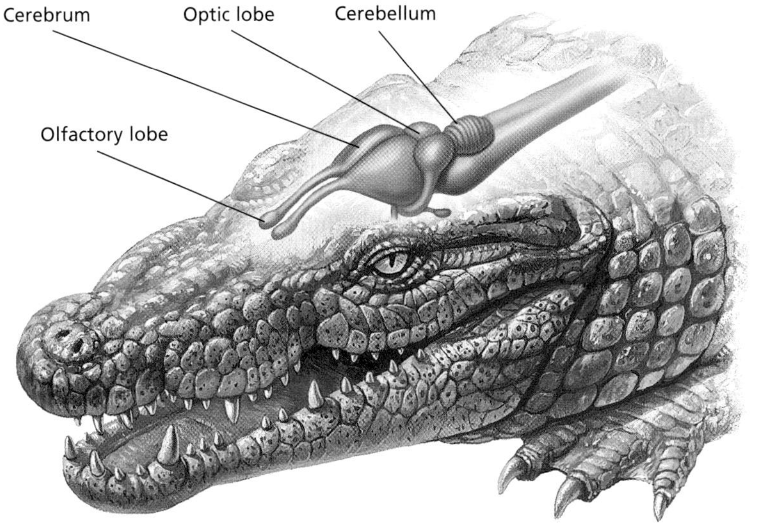

FIGURE 43-7

The crocodile's sense of smell is very important for its survival. The olfactory lobe of the reptile's brain, where the sense of smell is located, is highly developed.

NERVOUS SYSTEM

The brain of a reptile is about the same size as that of an amphibian of the same size. However, the reptilian cerebrum is much larger. This region of the brain is involved in controlling and integrating behavior. Because vision is an important sense for most reptiles, the optic lobes, which receive input from the eyes, are large. Figure 43-7 shows the structure of a crocodile's brain.

Most reptiles rely on their sense of sight to detect predators and prey. The eyes of reptiles are usually large, and many species have keen vision. Hearing is also an important sense. As in amphibians, sound waves first strike the tympanum, or eardrum, and are transmitted to the inner ear through the movements of a small bone called the columella. The inner ear contains the receptors for sound. Snakes lack a tympanum and are sensitive only to low-frequency sounds. They are able to detect ground vibrations, which are transmitted to the columella by the bones of the jaw.

Jacobson's organ is a specialized sense organ located in the roof of the mouth of reptiles. Jacobson's organ is sensitive to odors. Like the snake shown in Figure 43-8, reptiles use their tongue to gather chemicals from the environment. These chemicals are transferred to the Jacobson's organ when the tongue is drawn back into the mouth. Jacobson's organ is found in all reptiles except crocodiles and most turtles, but it is highly developed in lizards and snakes.

Pit vipers, such as rattlesnakes, copperheads, and water moccasins, are able to detect the heat given off by warm-bodied prey, such as mammals and birds. These snakes have one heat-sensitive pit below each eye, as shown in Figure 43-8. Input from these pits allows a snake to determine the direction of and distance to a warm object.

FIGURE 43-8

Some snakes have reduced senses of sight and hearing. They compensate with a sensitive forked tongue that is an organ of touch and smell. As the tongue darts in and out of the mouth, it picks up particles that are taken into the Jacobson's organ inside the snake's mouth, where even extremely low concentrations of odors can be detected.

TEACHING STRATEGY

Reptile Ears

Have pairs of students build a listening device out of two cans and a string. Let them try it out, and then ask them to explain how it is like the hearing apparatus of a reptile. (One can represents the tympanum, the string represents the columella, and the other can represents the inner ear. In a reptile, sound vibrations travel through these auditory structures.)

DEMONSTRATION

Feeding Behavior

Obtain a harmless, live snake and let students observe the way it feeds. CAUTION **Do not allow students to handle the snake. Live animals must be treated gently.** Small garter snakes will eat crickets. A bull snake prefers to eat mice. Have the students take notes about the behavior of the snake from the time the food is dropped into the cage or vivarium until it is swallowed. A hungry snake will continuously flick its tongue in and out as soon as food is made available. Ask students why the snake exhibits this behavior. (It is smelling the food with its Jacobson's organ.) If students look closely at the snake's mouth while the prey is in it, they may be able to see how the snake extends its windpipe to breathe during the swallowing process. A video may be used if a live snake is not available.

CRITICAL THINKING

A Sick Pet

Suppose a captive lizard is not given access to heat. It becomes lethargic, stops eating, and eventually dies. Use the above information to describe a series of events that could have caused the lizard's death. (The lizard could not warm itself up enough to feed. Weakened by a lack of nutrition, it eventually got sick and died.)

QUICK FACT

☑ Some early reptiles were masters of thermoregulation. *Dimetrodon,* a dinosaur from the early Permian period, had a high vertical fin along its back. Scientists think the fin was rich with blood vessels and was used by the dinosaur to absorb heat from the sun, much like a solar panel is used today. Refer the class to Figure 45-1.

TEACHING STRATEGY

Releasing Body Heat

Take the students outside and ask healthy volunteers to run a lap or two around a playing field. Then call them back together and have them hold their hands near their mouth so that they can feel the exhaled air. Ask them how the air feels (warm). Tell the students that by exhaling at an accelerated rate, they are able to release excess heat from their bodies. If any students are sweating, ask them the function of this process. (As the sweat evaporates, it carries heat away from the surface of their skin.) Tell the students that reptiles can't sweat. Ask them how an overheated reptile would probably cool itself down. (It would find a shady spot and remain inactive.)

THERMOREGULATION

The control of body temperature is known as **thermoregulation.** Vertebrates regulate their body temperature in two different ways. An **ectotherm** warms its body by absorbing heat from its surroundings. It has a slow metabolism that produces little heat. Reptiles, fishes, and amphibians are ectotherms. In contrast, **endotherms,** such as mammals and birds, have a rapid metabolism, which generates heat needed to warm the body. Most endotherms have insulation, such as hair, feathers, or fat, to retain the heat. The body temperatures of many aquatic ectotherms, such as fishes and amphibians, remain close to the temperature of their surroundings. When active, however, terrestrial ectotherms, such as lizards and snakes, usually keep their body temperatures about the same as the body temperatures of endotherms.

Reptiles regulate their body temperature by controlling how much heat they absorb. For example, when a lizard emerges from its nest after a cool night, its body temperature is low and must be raised before it can become active. The lizard warms itself by basking in the sun, as shown in Figure 43-9a. The lizard's warm blood is diverted from the skin to the interior of the body. The lizard continues to bask until it is warm enough to become active. As the graph in Figure 43-9b shows, a lizard can maintain its body temperature within a narrow range by using a variety of behaviors and body positions, despite variations in air temperature.

Advantages and Limitations of Ectothermy

Ectotherms require very little energy because their metabolism is very slow, and they also need only about one-tenth as much food as an endotherm of the same size. Ectotherms cannot live in very cold climates, and they can survive temperate climates only by becoming dormant during the coldest months. Furthermore, ectotherms can run or swim at maximum speed only for short periods of time. Ectothermic metabolism cannot provide enough energy for sustained exertion.

FIGURE 43-9

(a) A lizard regulates its body temperature throughout the day, basking to warm and seeking shade to prevent overheating. If its body temperature rises too high, the lizard may pant to accelerate heat loss. The graph in (b) shows an early-morning increase in the lizard's body temperature. The body temperature fluctuates only slightly during the remainder of the day despite wide fluctuations in ground temperature.

(a)

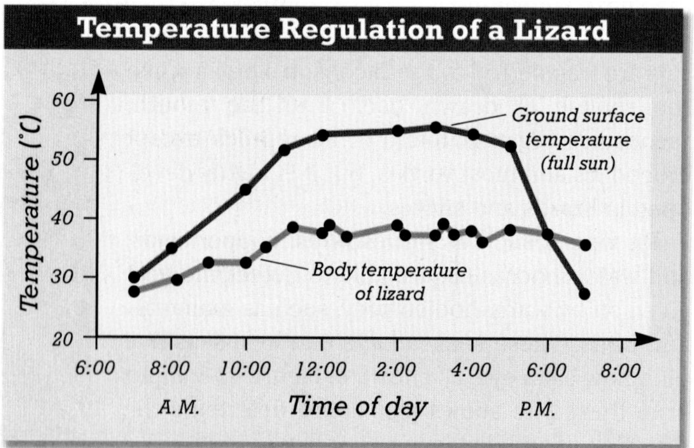

(b)

REPRODUCTION AND PARENTAL CARE

There are three patterns of reproduction among reptiles. The differences between these three patterns lie in how long the eggs remain within the female and in how she provides them with nutrition.

In **oviparity,** the female's reproductive tract encloses each egg in a tough protective shell. The female then deposits the eggs in a favorable place in the environment. Oviparity is characteristic of most reptiles, all birds, and three species of mammals.

Once released from the female's body, an egg is particularly vulnerable to predators and to environmental hazards. One way to reduce exposure to these hazards is to retain the eggs within the female's body for a time. This strategy is called **ovoviviparity.** The eggs may be laid shortly before hatching, or they may hatch within the female's body. The eggs absorb water and oxygen from the female, but they receive no nutrition other than the yolk.

In **viviparity,** a shell does not form around the egg, and the young are retained within the female's body until they are mature enough to be born. Nutrients and oxygen are transferred from mother to embryo through a structure called the **placenta.** The placenta forms from the membranes within the egg, and it brings blood vessels from the embryo near the vessels of the mother. Viviparity is the reproductive pattern shown by most mammals, but it is also found in a few species of lizards and snakes.

Many reptiles provide no care for their eggs or young. However, some species of lizards and snakes guard and warm the eggs until they hatch. Crocodiles and alligators provide the greatest amount of parental care. A female crocodilian, for example, builds a nest for her eggs. She remains nearby while the eggs incubate, guarding against nest-robbing predators. After the young hatch, they produce a high-pitched yelp, which summons the mother. She then breaks open the nest and carries the hatchlings to the water in her mouth, as shown in Figure 43-10. The mother crocodile may protect her young for a year or more.

Word Roots and Origins

ovoviviparous

from the Latin *ovum,* meaning "egg," the Latin *vivus,* meaning "alive," and the Latin *parere,* meaning "to bring forth"

FIGURE 43-10

Reptiles generally provide no care for their young, but the hatchlings are usually able to fend for themselves as soon as they emerge from the shell. Crocodiles and alligators, however, care for their young for up to two years. The female crocodile in the photograph is transporting her baby in her mouth.

SECTION 43-2 REVIEW

1. Describe how the heart of a lizard differs from the heart of a crocodile.
2. Explain how a reptile inflates its lungs.
3. What is the function of the Jacobson's organ?
4. Compare ectothermy with endothermy.
5. Contrast oviparity with viviparity.
6. **CRITICAL THINKING** Describe which of the three patterns of reproduction in reptiles best serves to protect the eggs from predators. Explain your answer.

SECTION 43-2

OVERCOMING MISCONCEPTIONS

Most people think that all snakes lay eggs, but not all modern reptiles reproduce on land and lay eggs. The sea snake is a totally aquatic reptile that bears live young in the open ocean.

RECENT RESEARCH

Turtle Behavior

Sea turtles migrate great distances across open oceans to get to beach breeding sites. These sea turtles use the same migratory pathways every breeding cycle. Today, every species of sea turtle is threatened by extinction. Mapping out their ocean highways may prove useful in preventing turtles from being caught in the nets of commercial fishermen.

ANSWERS TO SECTION 43-2 REVIEW

1. A lizard's heart has two atria and a partially divided ventricle. A crocodile's heart has two atria and two ventricles.
2. Reptiles fill their lungs by expanding the ribs. This draws air into the lungs.
3. The Jacobson's organ allows reptiles to analyze chemical odors picked up by their tongue.
4. Ectotherms warm their body by absorbing heat from their surroundings. Endotherms produce heat internally through metabolism.
5. Oviparous reptiles lay shelled eggs. Viviparous reptiles transfer nutrients and oxygen to their embryos and do not lay eggs.
6. Ovoviviparity reduces the time the eggs spend outside the mother's body and reduces the amount of time they are exposed to predators.

SECTION 43-3

SECTION OVERVIEW

Structure and Function
Section 43-3 describes the four orders of modern reptiles.

Evolution
Section 43-3 describes the adaptations each kind of reptile has for living in its environment.

▶ ENGAGE STUDENTS

Ask students who have kept reptiles as pets or who have encountered reptiles in their natural habitats to share what they have learned about reptiles from their experiences. Ask if any students have ever eaten snake or turtle meat. Ask them to describe the experience to the class.

GIFTED ACTIVITY

Ask students to research and report on a reptile in their state or in another part of the country that is listed as threatened or endangered. Tell students to include in their reports information about the animal's natural history, reasons for its dwindling numbers, and actions being taken to help the species.

SECTION

43-3

OBJECTIVES

▲ Compare the anatomy of turtles with that of other reptiles.

● Describe how crocodilians capture prey.

■ Explain two antipredator defenses of lizards.

◆ Describe two ways snakes subdue their prey.

▲ Identify two characteristics of tuataras.

MODERN REPTILES

Modern reptiles are classified into four orders: Chelonia, Crocodilia, Squamata, and Rhynchocephalia. As different as two species of reptile—such as a turtle and a snake—appear to be, all species of modern reptiles share the following characteristics: an amniotic egg; internal fertilization of eggs; dry, scaly skin; respiration through lungs; and ectothermic metabolism.

ORDER CHELONIA

The order Chelonia consists of about 250 species of turtles and tortoises. The term *tortoise* is generally reserved for the terrestrial members of the order, such as the Galápagos tortoise shown in Figure 43-11a. *Turtle* usually refers to chelonians that live in water, such as the green sea turtle shown in Figure 43-11b.

The earliest known turtle fossils, which are more than 200 million years old, show that ancient chelonians differed little from today's turtles and tortoises. This evolutionary stability may be the result of the continuous benefit of the basic turtle design—a body covered by a shell. The shell consists of fused bony plates. The **carapace** is the top, or dorsal, part of the shell, and the **plastron** is the lower, or ventral, portion. In most species, the vertebrae and ribs are fused to the inner surface of the carapace. Turtles are also distinctive in that the pelvic and pectoral girdles lie within the ribs instead of outside the ribs, as they do in all other terrestrial vertebrates. Unlike other reptiles, turtles have a sharp beak instead of teeth.

FIGURE 43-11
The Galápagos tortoise, *Geochelone gigantops*, shown in (a), is protected from predators by its high domed carapace. The green sea turtle, *Chelonia mydas*, shown in (b), is streamlined for life in the sea.

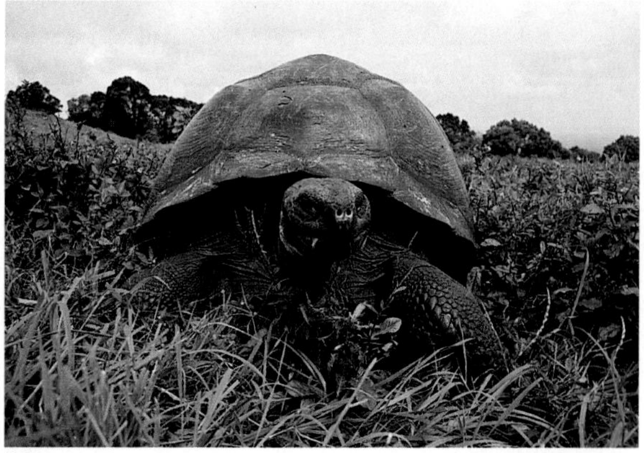

(a)

(b)

Turtles and tortoises live in a variety of habitats. Some species are permanently aquatic, some are permanently terrestrial, and some spend time both on land and in the water. The differing demands of these habitats are reflected in the shells and limbs of turtles. For example, water-dwelling turtles usually have a streamlined, disk-shaped shell that permits rapid turning in water, and their feet are webbed for swimming. Many tortoises have a domed carapace into which they can retract their head, legs, and tail as a means of protection from predators. Their limbs are sturdy and covered with thick scales. The limbs of marine turtles, which spend their entire lives in the ocean, have evolved into flippers for swimming and maneuvering.

Reproduction

All turtles and tortoises lay eggs. The female selects an appropriate site on land, scoops out a hole with her hind limbs, deposits the eggs, and covers the nest. She provides no further care for the eggs or the hatchlings. Marine turtles often migrate long distances to lay their eggs on the same beach where they hatched. For example, Atlantic green sea turtles travel from their feeding grounds off the coast of Brazil to Ascension Island in the South Atlantic—a distance of more than 2,000 km (1,242 mi). These turtles probably rely on several environmental cues, possibly even the Earth's magnetic field and the direction of currents, to find this tiny island.

ORDER CROCODILIA

The living reptiles most closely related to the dinosaurs are the crocodilians, order Crocodilia. This group is composed of about 21 species of large, heavy-bodied, aquatic reptiles. In addition to crocodiles and alligators, the order includes the caimans and the gavial. Figure 43-12 shows some examples of crocodilians.

Crocodilians live in many tropical and subtropical regions of the world. Alligators live in China and the southern United States. Caimans are native to Central America and South America, but they have been introduced into Florida.

FIGURE 43-12

Crocodiles, such as genus *Crocodylus,* shown in (a), are found in Africa, Asia, and the Americas. The gavial, *Gavialis gangeticus,* shown in (b), is a crocodilian with an extremely long and slender snout adapted for seizing and eating fish. Gavials live only in India and Burma.

(a)

(b)

DEMONSTRATION

Point out the relative positions of the rib cage, pectoral girdle, and pelvic girdle on a real or model human skeleton. Compare the positions of these structures in humans to their locations in a turtle's skeleton. (In humans, the pelvic and pectoral girdles lie outside the ribs. In turtles, these structures lie within the ribs.) Ask students to offer reasons for the unique design of the turtle's skeleton. (Having the pectoral girdle up and out of the way leaves more space inside the shell for the turtle to retract its head and legs.)

TEACHING STRATEGY

Reptiles Up-Close

Have groups of students observe different live or preserved reptiles. Ask them to sketch the organisms, note special characteristics they have, and explain how their adaptations help them survive. Provide reptile-identification books and taxonomic guides so students can classify the specimens. Tell them to be thorough in their classifications and to list every taxonomic level, kingdom through species, for each animal.

CRITICAL THINKING

Adaptations and Functions

Ask students how the nostrils and eyes of crocodiles and frogs are similar. (They are positioned on top of the head on each animal, allowing the animals to lie submerged below the surface of the water but still able to see and breathe.) Ask them if these adaptations are beneficial to each animal in the same way. (No, they make it easier for the frog to avoid predators and for the crocodile to find prey.)

SECTION 43-3

RECENT RESEARCH

Lizard Romance

Marine iguanas live on the rocky shores of the Galápagos archipelago. During their breeding season, the largest males bully the smaller ones and keep them from mating. Even when the smaller lizards do get a chance to mate, they are often interrupted and run off by the bigger lizards. Scientists have discovered that smaller males still have a good chance of fertilizing a female even if mating is prematurely interrupted. It seems that the smaller iguanas release sperm prior to coupling and store it in their cloaca. The stored sperm is quickly transferred into the female as soon as mounting occurs.

CULTURAL CONNECTION

Snake Tales

Snakes have been hallmarks of human mythology for thousands of years. Some cultures have held snakes in high regard, while others have despised and feared them. The Aztecs worshiped Quetzalcoatl, the mythical plumed serpent. In India, cobras were thought to be the reincarnated spirits of society's elite. Some cultures in Africa worshiped pythons. Killing one was considered a major crime. Western culture's largely negative view of snakes probably derives at least in part from its Judeo-Christian heritage. The story of the garden of Eden and the role the serpent played in man's fall from grace has not been friendly to snakes.

internetconnect

SciLINKS NSTA

TOPIC: Lizards
GO TO: www.scilinks.org
KEYWORD: HM852

All crocodilians are carnivorous. They feed on fish and turtles, and on land animals that come to the water to feed or drink. Crocodilians capture their prey by lying in wait until an animal approaches and then attacking swiftly. A crocodilian can see and breathe while lying quietly submerged in water. A valve at the back of the throat prevents water from entering the air passage when a crocodilian feeds underwater.

ORDER SQUAMATA

The order Squamata consists of about 5,500 species of lizards and snakes. A distinguishing characteristic of this order is an upper jaw that is loosely joined to the skull. Squamates are the most structurally diverse of the living reptiles, and they are found worldwide.

Lizards

There are about 3,000 species of living lizards. Common lizards include iguanas, chameleons, and geckos. Lizards live on every continent except Antarctica. Figure 43-13 shows some examples of lizards. Most lizards prey on insects or on other small animals. A few of the larger species, such as the chuckwalla and desert iguana of the southwestern United States, feed on plants. The Komodo dragon feeds on prey as large as goats and deer. Only two species of lizards are venomous. They are the Gila monster of the southwestern United States and northern Mexico and the related beaded lizard of southern Mexico.

Most lizards rely on agility, speed, and camouflage to elude predators. If threatened by a predator, some lizards have the ability to detach their tail. This ability is called **autotomy.** The tail continues to twitch and squirm after it detaches, drawing the predator's attention while the lizard escapes. The lizard grows a new tail in several weeks to several months, depending on the species.

Most lizards are small, measuring less than 30 cm (12 in.) in length. The largest lizards belong to the monitor family (Varanidae). Like snakes, monitors have deeply forked tongues that pick up airborne particles and transfer them to the Jacobson's organ in the roof of the mouth.

FIGURE 43-13

The largest of all monitors is the Komodo dragon, *Varanus komodoensis,* of Indonesia, shown in (a). The Komodo dragon can grow to 3 m long (10 ft). A colorful gecko, genus *Phelsuma,* is shown in (b). The gecko has specialized structures on the pads of its fingers and toes that allow it to cling to almost any surface.

(a)

(b)

(a)

(b)

FIGURE 43-14

The Gaboon viper, *Bitis gabonica,* shown in (a), injects a toxic venom to kill its prey before it begins the process of swallowing. The boa constrictor, *Constrictor constrictor,* shown in (b), suffocates its prey.

Snakes

There are about 2,500 species of snakes, and like lizards, they are distributed worldwide. Figure 43-14 shows some examples of snakes. The most obvious characteristic of snakes is the lack of legs, which affects all other aspects of their biology. What was the selective pressure that caused snakes to evolve leglessness? One possibility is that the ancestors of snakes were terrestrial but lived in thick vegetation, where legs were a hindrance to rapid movement.

The graceful movements of snakes are made possible by their unique anatomy. A snake has a backbone of 100 to 400 vertebrae, and a pair of ribs are attached to each vertebra. These bones provide the framework for thousands of muscles. The muscles manipulate not only the skeleton but also the snake's skin, causing the overlapping scales to extend and contract.

Capturing and Consuming Prey

A snake may just seize and swallow its prey. However, many snakes employ one of two methods for killing: constriction or injection of venom. Snakes that are **constrictors** wrap their bodies around prey. A constrictor suffocates its prey by gradually increasing the tension in its coils, squeezing a little tighter each time the prey breathes out. This technique is used both by large snakes, such as boas, pythons, and anacondas, and by smaller snakes, such as gopher snakes and king snakes.

Some snakes inject their prey with a toxic venom in one of three different ways. The snakes with fangs in the back of the mouth, such as the boomslang and twig snakes of Africa, bite the prey and use grooved teeth in the back of the mouth to guide the venom into the puncture. Cobras, kraits, and coral snakes are elapids. **Elapid** snakes inject poisons through two small, fixed fangs in the front of the mouth. **Vipers** inject venom through large, mobile fangs in the front of the mouth. Rattlesnakes, copperheads, and water moccasins are examples of vipers. When a viper strikes, these hinged fangs swing forward from the roof of the mouth and inject venom more deeply than can the fangs of elapids.

Quick Lab

Demonstrating Muscle Contractions

Materials plastic drinking straw, small ball of modeling clay

Procedure Make a small ball of clay slightly smaller than the opening of the drinking straw. Find a way to get the ball of clay into the middle of the straw.

Analysis How is this model similar to the feeding mechanism of a snake? If you used a ball of clay that was larger than the opening of the drinking straw, what problems would you encounter? Why is the size of larger prey not a problem for snakes?

RECENT RESEARCH

Primitive Snake

After reevaluating a Cretaceous period lizard fossil, scientists are convinced that the creature is really a snake. *Pachyrhachis problematicus* has skull bones similar to those of a modern snake's as well as a long and slender body. But it also has a well-developed sacrum, pelvis, and hindlimbs. The reclassification of the specimen as a snake is a significant scientific event because it means that this snake/lizard is the ancestor to all other snakes.

Quick Lab

Demonstrating Muscle Contractions

Time Required 20 minutes

Procedural Tips You may choose to cut the drinking straws in half.

Answers to Analysis Students will discover that they must pinch the drinking straw above the ball of clay and push the clay toward the center of the drinking straw. The "pinch and push" method used to move the clay to the middle of the drinking straw is similar to the muscle contractions involved in the feeding mechanism of snakes. A larger ball of clay will not fit in the opening of the drinking straw. A snake is able to swallow prey larger than the diameter of its head because of loosely hinged jaws and flexible ligaments in the head.

SECTION 43-3

QUICK FACT

☑ All male reptiles except tuataras have a penis, which they use to deposit sperm into the female. Tuataras mate by joining their cloacal regions. The females lay eggs only once every four years, the longest reproductive cycle of any reptile.

RETEACHING *ACTIVITY*

Have students prepare a poster featuring the characteristics of a reptile that interests them. Tell them the poster should highlight the animal's adaptations for survival, including internal anatomy and behavioral adaptations. Ask students to share their posters with the rest of the class.

ANSWERS TO SECTION 43-3 REVIEW

1. Tuataras live on several small islands of New Zealand, and they feed on insects and other small animals.
2. A turtle's shell is composed of bony plates. The ribs and vertebrae of a turtle are fused to the interior surface of the shell. The upper part of the shell is called the carapace, and the lower part is called the plastron.
3. Crocodiles ambush their prey. Eyes high on the head and nostrils on top of the snout allow crocodiles to lie in wait in the water.
4. The ability to detach the tail is called autotomy. The tail distracts the predator while the lizard escapes.
5. Constrictors wrap their body in a coil around the prey and slowly tighten, killing the prey by suffocation.
6. Replacing the tail requires energy and materials, such as muscle, skin, and blood.

FIGURE 43-15

This series of photographs shows a snake, *Dasypeltis scabra,* swallowing a bird's egg. Prey is often larger than the diameter of the snake's head, so the process of swallowing can take an hour or more.

Once killed, the prey must be swallowed whole because a snake's curved, needlelike teeth are not suited for cutting or chewing. Several features of a snake's skull enable it to swallow an animal larger in diameter than its head, as shown in Figure 43-15. The upper and lower jaws are loosely hinged and move independently, and can open to an angle of 130 degrees. In addition, a snake's lower jaw, palate, and parts of its skull are joined by a flexible, elastic ligament that allows the snake's head to stretch around its prey.

ORDER RHYNCHOCEPHALIA

FIGURE 43-16

Unlike most reptiles, the endangered tuataras, such as the one shown, are most active at low temperatures.

The order Rhynchocephalia (RING-koe-suh-FAY-lee-uh) is an ancient one that contains only two living species, the tuataras of the genus *Sphenodon.* Today tuataras inhabit only a few small islands of New Zealand. The Maoris of New Zealand named the tuataras for the conspicuous spiny crest that runs down the animal's back, which you can see in Figure 43-16. The word *tuatara* means "spiny crest" in the Maori language. Tuataras resemble large lizards and grow to about 60 cm (24 in.) in length. They usually hide in a burrow during the day and feed on insects, worms, and other small animals at night.

Since arriving in New Zealand about 1,000 years ago, humans have radically changed the landscape and introduced predators such as rats and cats, which feed on tuataras and their eggs. As a result, tuataras have disappeared from most of their original range.

SECTION 43-3 REVIEW

1. **Where do tuataras live, and what do they feed on?**
2. **Describe the structure of a turtle's shell.**
3. **How does the position of a crocodile's eyes and nostrils relate to its hunting behavior?**
4. **What is autotomy? How does it help a lizard escape predators?**
5. **How does a constrictor kill its prey?**
6. **CRITICAL THINKING Describe the costs of autotomy for a lizard.**

CHAPTER 43 REVIEW

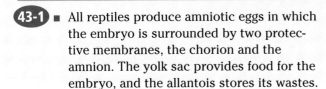

SUMMARY/VOCABULARY

43-1

- All reptiles produce amniotic eggs in which the embryo is surrounded by two protective membranes, the chorion and the amnion. The yolk sac provides food for the embryo, and the allantois stores its wastes.
- Reptiles conserve water by having a dry, scaly watertight skin and by eliminating their nitrogenous wastes as uric acid.
- During the Mesozoic period, most large vertebrates were reptiles. The dinosaurs lived at this time, as did the aquatic plesiosaurs and ichthyosaurs and the flying pterosaurs.
- Dinosaurs became extinct about 65 million years ago, at the end of the Cretaceous period. Most scientists think that the collision of a large asteroid with Earth caused this extinction.

Vocabulary

albumin (842)
allantois (842)
amnion (842)
amniotic egg (842)
asteroid-impact hypothesis (840)
chorion (842)
dinosaur (839)
keratin (843)
yolk sac (842)

43-2

- Lizards, turtles, snakes, and tuataras have a heart with two atria and one partially divided ventricle. Crocodiles have a four-chambered heart, as mammals do.
- Reptiles inflate their lungs by expanding the ribs, drawing in air by decreasing abdominal pressure.
- A reptile's brain is about the same size as the brain of an amphibian, but it has a much larger cerebrum.
- The Jacobson's organ detects chemicals picked up by the tongue.
- All living reptiles are ectotherms. Ectotherms warm their body mainly by absorbing heat from their surroundings.
- Reptiles can keep their body temperature fairly constant by moving into and out of the sun.
- Ectotherms require very little food, but they cannot live in cold climates or be active when temperatures are low.
- Many reptiles lay shelled eggs. This is called oviparity. Some species transfer nutrients and oxygen to the developing embryos through a placenta. This strategy is called viviparity. Some species retain the eggs inside the female's body. This is called ovoviviparity.

Vocabulary

alveolus (846)
ectotherm (848)
endotherm (848)
Jacobson's organ (847)
oviparity (849)
ovoviviparity (849)
placenta (849)
septum (845)
thermoregulation (848)
viviparity (849)

43-3

- Living reptiles are classified into four orders: Rhynchocephalia, Chelonia, Crocodilia, and Squamata.
- All turtles (order Chelonia) have a shell composed of bony plates. The vertebrae and ribs are fused to the interior surface of the shell. Turtles are mainly aquatic, while tortoises are terrestrial. All species lay eggs.
- Crocodilians (order Crocodilia) are large aquatic or semiaquatic carnivores. This group includes crocodiles, alligators, caimans, and the gavial.
- Lizards and snakes belong to the largest order of reptiles (order Squamata). Most lizards are small, agile, and feed on insects.
- Some snakes kill their prey by constriction, suffocating it with their coiled body. Some species kill by injecting venom. Snakes can swallow objects larger than their head because they have a very flexible skull.

Vocabulary

autotomy (852)
carapace (850)
constrictor (853)
elapid (853)
plastron (850)
viper (853)

CHAPTER 43 REVIEW
ANSWERS

REVIEW

1. Elapid snakes inject poisons through two small, fixed fangs in the front of the mouth. Vipers inject venom through large, mobile fangs at the front of the mouth. Cobras and coral snakes are elapids. Rattlesnakes and copperheads are vipers.
2. A septum partially divides the ventricle in the reptilian heart.
3. turtles
4. The placenta is associated with viviparity. It is the structure that allows viviparous animals to transfer nutrients and oxygen to their embryos.
5. *Ecto-* means "outside," *endo-* means "within," and *-therm* means "heat." Endotherms produce heat through metabolism; thus, they get their heat from within. Active, terrestrial ectotherms absorb much of their body heat from outside.

6. a **7.** b
8. c **9.** c
10. c **11.** a
12. b **13.** b
14. c **15.** b

16. The amniotic egg enabled organisms to reproduce in areas where there is no body of water. The amniotic egg encloses the embryo in its own aquatic environment.
17. A turtle's heart normally keeps deoxygenated and oxygenated blood separate. It can allow the two bloodstreams to mix if the turtle is resting, underwater, or trying to raise its body temperature.
18. The lizard can distract a predator by leaving its tail behind, making the predator believe it has caught its prey. Growing a new tail ensures that the lizard will be able to use that kind of escape another time.
19. When active, reptiles keep their body temperatures near those of warmblooded endotherms, so they are not coldblooded.

20. Ectothermy is very efficient in that very little food is needed. However, it limits where reptiles can live and when they can be active. Reptiles also cannot sustain strenuous activity.
21. Reptiles use the Jacobson's organ, which is located in the roof of the mouth. It is sensitive to odors collected by the tongue. Using pits located below the eye, some snakes can detect the heat given off by warm-bodied prey.
22. Female crocodilians build nests for their eggs and protect the nests during incubation. When the young hatch, the mothers carry them to the water in their mouths and may protect them for a year or more.
23. Some snakes guide venom into the prey through grooved teeth at the back of the mouth. Some species inject venom through small fixed fangs at the front of the mouth. Some species inject venom with large fangs that can be folded back against the roof of the mouth.
24. Humans have altered the landscape of New Zealand and introduced predators that feed on tuataras. As a result, tuataras have been driven from most of their original range.
25. A, the allantois, where O_2 and CO_2 are exchanged; B, the amnion, the membrane that encloses the fluid in which the embryo grows; C, the embryo, the new individual; D, the chorion, surrounds and protects all the other membranes; E, the yolk sac, encloses the protein and fat-rich food supply for the embryo.

CRITICAL THINKING

1. The turtle is digging a decoy hole that may distract the attention of predators from the nest containing the eggs.
2. Killing the prey prevents it from escaping and ensures that the snake won't be injured by its struggles.

REVIEW

Vocabulary

1. Briefly contrast elapids with vipers and give one example of each.
2. What is the function of the septum in the heart?
3. Which reptiles have a carapace and plastron?
4. The term *placenta* is most closely associated with which of the following terms: *oviparity, viviparity, ovoviviparity?* Explain your answer.
5. Use a dictionary to find the meanings of the following Greek roots: *endo-, ecto-, -therm.* Relate these roots to the meaning of the terms *ectotherm* and *endotherm.*

Multiple Choice

6. The membrane that encloses the fluid around a reptilian embryo is the (a) amnion (b) yolk sac (c) allantois (d) chorion.
7. The food supply within the egg is contained within the (a) amnion (b) yolk sac (c) allantois (d) chorion.
8. Which of the following is not true of dinosaurs?
   (a) They lived during the Mesozoic era.
   (b) They became extinct at the end of the Cretaceous period.
   (c) They were all large.
   (d) They lived on all continents.
9. Evidence for the asteroid-impact hypothesis of dinosaur extinction includes all of the following, except (a) sediments rich in iridium from the end of Cretaceous period (b) quartz crystals with blast damage (c) lack of dinosaur fossils from the early Permian period (d) remains of a large impact crater in Mexico.
10. Which of the following is true of the heart of a crocodile?
    (a) It has one ventricle.
    (b) Its atrium is partially divided by a septum.
    (c) It doesn't allow mixing of deoxygenated and oxygenated blood.
    (d) It only pumps deoxygenated blood.
11. The Jacobson's organ is most like our sense of (a) smell (b) hearing (c) sight (d) touch.
12. The two basic parts of a turtle's shell are the (a) septum and amnion (b) carapace and plastron (c) chorion and allantois (d) keratin and columella.
13. All of the following reptiles belong to the order Crocodilia, except (a) alligators (b) tuataras (c) caimans (d) gavial.
14. The ability to lose its tail and grow a new one helps a lizard (a) reduce its need for food (b) hide from predators (c) escape from predators (d) capture prey.
15. Long legless bodies may have arisen as an adaptation that helped snakes (a) absorb oxygen through their skin (b) move through thick vegetation (c) catch prey (d) swallow large animals.

Short Answer

16. What problem of life on land was solved by the evolution of the amniotic egg?
17. Describe how the structure of a turtle's heart allows for flexibility in blood circulation.
18. Explain the purpose of a lizard's ability to lose its tail and grow a new one.
19. Explain why it is inaccurate to call reptiles coldblooded.
20. What is the major benefit of ectothermy?
21. Describe two senses, other than vision and hearing, that reptiles use to find prey.
22. Describe the parental care of crocodilians.
23. Describe three ways snakes inject venom.
24. How has the settling of New Zealand by humans affected the tuataras?
25. The following diagram shows five parts of the amniotic egg, indicated by A, B, C, D, and E. Name and define the function of these parts.

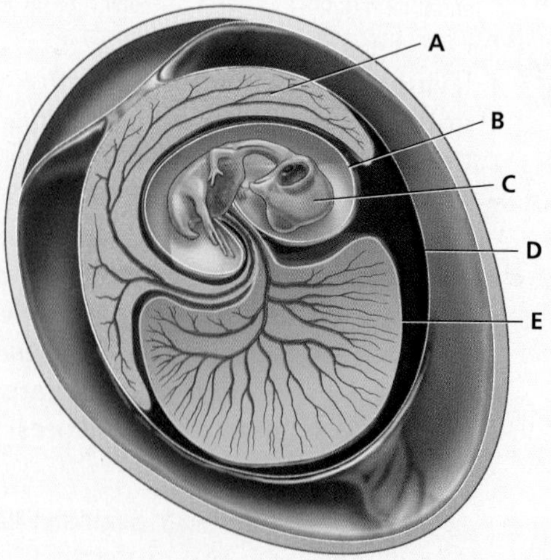

Critical Thinking

1. When a female leatherback turtle comes up on a beach to lay eggs, she first digs a deep hole, lays her eggs, and covers them with sand. Next she crawls about 100 m and digs another hole. This time she lays no eggs but just covers the hole with sand. Suggest a possible explanation for this behavior.
2. Why is it advantageous for a snake to kill its prey, either through constriction or venom, before trying to eat it.
3. The skin of a basking lizard is usually dark. As the lizard warms, the skin lightens. Suggest a functional explanation for this change. (Hint: Consider how this change might affect the lizard's absorption of heat.)
4. Fossil evidence collected in Alaska suggests that some dinosaurs were year-round residents of areas subject to freezing temperatures and long periods of darkness. Does this evidence of arctic dinosaurs support or contradict the hypothesis that the extinction of dinosaurs was due to the intense cold produced by a cloud of debris in the atmosphere? Explain your answer.
5. Many viviparous snakes and lizards live in cold climates. Why might viviparity be advantageous in such environments?
6. Luis and Walter Alvarez first proposed the asteroid-impact hypothesis after the discovery of abnormally high levels of iridium in sediments from the end of the Cretaceous period. According to the Alvarezes, what was the source of the iridium? High iridium levels were initially discovered at one site in Italy. Since then, high levels have been found at more than 100 sites around the world, all dating from the end of the Cretaceous period. Explain how the worldwide distribution of iridium is important evidence for the asteroid-impact hypothesis.
7. Examine the photo of the inside of a turtle's carapace. In cartoons, turtles often crawl out of their shell. Could a real turtle do this? What do you see in the photo that would indicate whether a real turtle could crawl out of its shell?

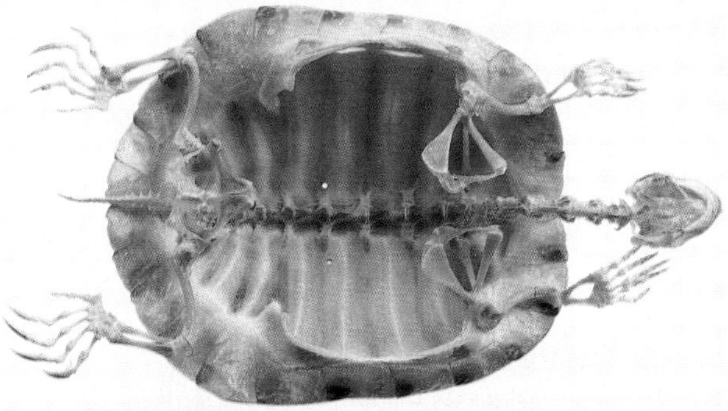

Extension

1. Read "A Dinosaur with Altitude" in *Time,* November 15, 1999, on page 96. Describe the dinosaur that paleontologists say may be the tallest dinosaur ever. How have they come to this conclusion?
2. Read "Snake Charmer" in *National Wildlife,* February–March 1999, on page 36. Explain why Dr. D. Bruce Means believes we should care about the fate of the eastern diamondback rattlesnake.
3. Use a library or on-line database to research an endangered species of reptile, such as Kemp's ridley turtle. Prepare a report describing the conditions that threaten the survival of the species and the efforts being made to save it (if any).
4. Observe the behavior of a small lizard kept in a terrarium. Keep a log of your observations, noting differences in behavior at various times of day and before and after feeding.

ANSWERS
CONTINUED

3. Dark objects absorb more heat. The lizard is dark when it first begins basking so that it can absorb as much heat as possible. As the lizard warms, it lightens, lessening its heat absorption.
4. It has been proposed that a cloud of debris in the atmosphere caused the extinction of the dinosaurs. Evidence of arctic dinosaurs would contradict the asteroid-impact hypothesis.
5. Viviparity reduces exposure to cold and allows eggs to incubate inside the mother's body.
6. The Alvarezes proposed that iridium came from an asteroid that struck Earth. If iridium was found in only one location, it might indicate a local event. Discoveries of iridium in sites worldwide indicate that the debris from the asteroid's impact spread over the globe, suggesting a very large catastrophe.
7. Turtles' ribs and other skeletal structures are attached to their shells, so turtles cannot crawl out of their shells.

Extension

1. *Sauroposeidon* lived 100 million years ago, was a vegetarian, and was similar in size and shape to *Apatosaurus.* Paleontologists think it could stand 60 ft tall, judging from the 5 ft vertebrae that were found.
2. Dr. Means states that the snake is a highly evolved life-form with complicated behaviors and unique attributes. The species contributes to biodiversity and nutrient cycling, and is *good* simply because it *exists.*
3. Answers should include the danger of commercial fishing and should point out that there are fewer than 1,000 left in the wild.
4. Answers will vary according to the species of lizard, feeding habits, temperature, light, and intruders.

CHAPTER 43 INVESTIGATION

TIME REQUIRED

Two 50-minute lab periods

SAFETY PRECAUTIONS

Remind students to handle anoles with care. They are easily frightened and run fast. If an anole is picked up by the tail, the tail may break off. Tell students that anoles that are no longer needed in the classroom make ideal pets at home. You may want to show students how to set up a vivarium for their anoles.

QUICK REFERENCES

Lab Preparation Notes are found on page 837B.

Holt Biosources provides a Teaching Resources CD-ROM that contains "Using Gowin's Vee in the Lab" and Scoring Rubrics."

PROCEDURAL TIPS

1. At least one hour before the lab exercise begins, place the anoles under fluorescent light in two terraria.
2. Place half the anoles in a terrarium lined with black construction paper. These anoles should turn brown.
3. Place the remaining anoles in a terrarium lined with white construction paper. These anoles should turn green.
4. Provide a shallow dish in each terrarium for drinking.
5. Use jars large enough to allow the anoles to settle comfortably on the bottom.
6. You may want to place anoles in jars prior to student arrival in class due to the difficulty of this task and the need to handle the anoles with great care.

CHAPTER 43 INVESTIGATION

Observing Color Adaptation in Anoles

OBJECTIVES

- Observe live anoles.
- Test whether background color stimulates color change in anoles.

PROCESS SKILLS

- observing
- hypothesizing
- experimenting
- organizing data
- analyzing data

MATERIALS

- glass-marking pencil
- 2 large clear jars with wide mouths and lids with air holes
- 2 live anoles
- 6 shades each of brown and green construction paper, ranging from light to dark (2 swatches of each shade)

Background

1. Anoles include 250–300 species of lizards in the genus *Anolis.*
2. Anoles can change color, ranging from brown to green, and are sometimes mottled.
3. Anoles live in shrubs, grasses, and trees. Describe some ways in which the ability to change color might be an advantage to anoles.
4. Light level, temperature, and other factors, such as whether the anole is frightened or whether it has eaten recently, can affect color. When anoles are frightened, they usually turn a dark grey or brown and are unlikely to respond to other stimuli.
5. Anoles can change color within a few minutes.

Procedure

1. Observe the anoles in two terraria, and discuss the purpose of this investigation with your partners. Develop a hypothesis that describes a relationship between anole skin color and background color. Write your hypothesis in your lab report.
2. Obtain swatches of construction paper in at least six different shades of green and brown. You will need two swatches of paper in each color.
3. Obtain two clear jars. Label one jar "Anole 1," and label the other jar "Anole 2."
4. CAUTION **You will be working with live animals. Handle them gently and follow instructions carefully.** Select two anoles of the same color from the terraria. Plan your actions and cooperate with a partner to transfer one anole into each labeled jar. Anoles will run fast and are easily frightened. Carefully pick them up and place the animals in separate jars. Do not pick up anoles by their tails. Grasp them gently behind the head. Quickly and carefully place a lid with air holes on each jar.
5. Gently place the jar with Anole 1 on a swatch of construction paper that most closely matches the anole's color. Try not to jostle the anole in the jar, and move the jar as little and as gently as possible. Repeat this procedure for Anole 2. Both anoles should closely match the color of the swatch.

6. When you have obtained and matched two anoles to closely matching colors, label the back of the pieces of paper "Initial Color of Anole 1" and "Initial Color of Anole 2," respectively. Replace the swatches underneath the jars after you have labeled them. The anoles should stay in their respective jars until the end of this investigation.
7. Using the given setup and the remaining swatches of colored construction paper, devise a control experiment to test whether background color stimulates color change in the anole.
8. In your lab report, list the independent variable and the dependent variables that you intend to use in your experiment. Describe how you will vary the independent variable and how you will measure changes in your dependent variable.
9. In your lab report, describe the control you will use in your experiment.
10. Create a data table similar to the one below to record your experimental observations for your lab report. For example, the table below is designed to record any change in anole skin color on four different background colors and the time it took for each change to take place. Design your data table to fit your own experiment. Remember to allow plenty of space to write your observations.
11. Have your experiment approved by your teacher before conducting it. As you conduct your experiment, be sure to record all of your data and observations in your lab report.
12. Attach your color swatches to your lab report, or include a color-coded key so that others reading your report will be able to understand how you measured initial color and color changes in your anoles. Be sure the color that most closely represents the initial color of both anoles is clearly indicated in your lab report.
13. Clean up your materials and wash your hands before leaving the lab.

Analysis and Conclusions

1. What effect, if any, did changes in the independent variable have on the dependent variable in your experiment?
2. Do your data support your hypothesis? Explain.
3. Can you think of any sources of error in your experiment?
4. Was your experiment a controlled experiment? If yes, describe your control and why you think a control is necessary for your experiment.
5. Were there any uncontrolled variables in your experiment, such as loud noises, bright light, or sudden movements, that could have affected your experiment? Describe how you might be able to improve your methods.

Further Inquiry

Design an experiment that tests the effects of temperature on anole skin color.

DATA TABLE *OBSERVING ANOLES*

	Color 1		Color 2		Color 3		Color 4	
	Change	**Time**	**Change**	**Time**	**Change**	**Time**	**Change**	**Time**
Anole 1								
Anole 2								

ANSWERS TO ANALYSIS AND CONCLUSIONS

1. Students should be able to state which color swatch caused a color change in the anole.
2. Students should recognize whether their data support their hypotheses.
3. Anoles may be reacting to stimuli other than the independent variable selected, such as fear, feeding patterns, or cool temperatures.
4. Students should design a control in which the anole remains on the original color swatch and does not change color throughout the experiment.
5. Students may list several uncontrolled variables that could have affected their results. They should suggest improving their methods by eliminating or controlling as many of the variables as possible.

FURTHER INQUIRY

Set upper and lower limits to the temperature students use in designing this experiment.

CHAPTER 44

BIRDS

	TOPICS	TEACHING RESOURCES	LABS, CLASSWORK, AND HOMEWORK
BLOCK 1 45 minutes	**Introducing the Chapter, p. 860**	**ATE** Focus Concept, p. 860 **ATE** Assessing Prior Knowledge, p. 860 **ATE** Understanding the Visual, p. 860	■ Supplemental Reading Guide, *The Dinosaur Heresies*
	44-1 Origin and Evolution of Birds, p. 861 Characteristics, p. 861 Evolution, p. 862	**ATE** Section Overview, p. 861 **ATE** Critical Thinking, pp. 862, 863 137A, 173, 60	★ Study Guide, Section 44-1 **PE** Section Review, p. 863
BLOCKS 2 & 3 90 minutes	**44-2 Characteristics of Birds, p. 864** Feathers, p. 864 Skeleton and Muscles, p. 865 Metabolism, p. 865 Reproduction, p. 868 Migration, p. 870	**ATE** Section Overview, p. 864 **ATE** Critical Thinking, pp. 865, 867 **ATE** Visual Link Figure 44-5, p. 866 143A, 171, 182–183	**PE** **Quick Lab,** p. 865, Comparing Wing Structures **Lab A22,** Comparing Animal Eggs **Lab A26,** Vertebrate Skeletons **Lab C22,** Examining Owl Pellets **Lab C40,** Conducting a Bird Survey ★ Study Guide, Section 44-2 **PE** Section Review, p. 870
BLOCK 4 45 minutes	**44-3 Classification, p. 871** Diversity, p. 871	**ATE** Section Overview, p. 871 **ATE** Critical Thinking, p. 872 **ATE** Visual Link Figure 44-11, p. 873 175–176	**PE** **Chapter 44 Investigation,** p. 878 ■ Occupational Applications Worksheets—Wildlife Biologist ★ Study Guide, Section 44-3 **PE** Section Review, p. 874
BLOCKS 5 & 6 90 minutes	**REVIEW and ASSESSMENT** **PE** Chapter 44 Review, pp. 876–877 ■ Performance-Based Assessment—The Challenge of Water Retention	Audio CD Program	**CHAPTER TESTING** ■ Chapter Test (blackline master) ▲ Test Generator ■ Assessment Item Listing ■ Scoring Rubrics and Classroom Management Checklists

MODERN BIOLOGY — HOLT BIOSOURCES

KEY

- PE Pupil's Edition
- ATE Teacher's Edition
- ■ Active Reading Guide
- ★ Study Guide
- Teaching Transparencies
- Laboratory Program

One-Stop Planner CD-ROM
Shows these resources within a customizable daily lesson plan:

- ■ Holt BioSources Teaching Resources
- ■ Active Reading Guide
- ★ Study Guide
- ▲ Test Generator

READING FOR CONTENT MASTERY

Preview: Have students preview the objectives, the introductory paragraph, topic sentences, and captions before reading the chapter.

ATE Active Reading Technique: Brainstorming, p. 861

■ Active Reading Guide Worksheet 44-1

PE Summary/Vocabulary, p. 875

■ Active Reading Guide Worksheet 44-2

PE Summary/Vocabulary, p. 875

■ Active Reading Guide Worksheet 44-3

PE Summary/Vocabulary, p. 875

TECHNOLOGY AND INTERNET RESOURCES

CNN PRESENTS Video Segment 26 For the Birds

Holt Biology Videodiscs Teacher's Correlation Guide, Lessons 44-1 through 44-3

Audio CD Program, Sections 44-1 through 44-3

internetconnect

sciLINKS NSTA

On-line Resources: www.scilinks.org
The following *sci*LINKS Internet resources can be found in the student text for this chapter:

Topics:
- Birds, p. 862
- Migration of birds, p. 870
- Classification of birds, p. 871

go.hrw.com

On-line Resources: go.hrw.com
Visit the HRW Web site for a variety of resources related to this chapter. Just type in the keyword HM2 Chapter 44.

Smithsonian Institution®

Visit **www.si.edu/hrw** for additional on-line resources.

CNNfyi.com

Visit **www.cnnfyi.com** for late-breaking news and current event stories selected just for you.

LAB ACTIVITY PLANNING

CHAPTER 44 INVESTIGATION

Comparing Feather Structure and Function, pp. 878–879

OVERVIEW
In this lab exercise, students compare the three basic types of bird feathers: quill, contour, and down. Students examine the central shaft and vane macroscopically and the barbs, barbules, and hooks under a microscope. Students compare these structures in the three feather types and speculate on how the structures contribute to the primary function of each type of feather.

SAFETY
Feathers collected from the field may contain disease-causing agents. If such feathers are used, remind students to wash their hands immediately after handling the feathers and to keep the feathers away from their faces.

PREPARATION
Prepare or purchase slides of contour feathers and slides of down feathers. Whole processed feathers may be obtained from a biological supply house. Feathers can also be collected in the field (pigeon feathers, for instance, can be found on most city streets). Make soap, water, and paper towels available so that students can wash after cleanup.

Quick Labs

Comparing Wing Structures, p. 865
Each student will need photocopies of different birds featuring their wings, and a ruler for this activity. Each sheet will be of a different bird with several different pictures of that bird's wings from several different angles. Have students hand these sheets in so that they can be reused.

UNDERSTANDING THE VISUAL

Ask students to study the owl closely. Ask them to name the one external characteristic that sets birds apart from other vertebrates. (Birds have feathers.) Tell them that some dinosaurs may also have had feathers, but no other living vertebrate possesses these structures. If students answer scales, remind them that the bodies of fishes and reptiles are also covered with scales.

FOCUS CONCEPT

Structure and Function

Many structures unique to birds are adaptations for flight. Their hollow bones and rigid skeleton make their body light yet strong. Efficient respiratory and circulatory systems supply body tissues with the oxygen needed for flight. Feathers streamline the body of a bird and allow lift to occur.

ASSESSING PRIOR KNOWLEDGE

Review the following concepts.

Cellular Respiration: *Chapter 7*
Ask students how birds generate the energy necessary to fly. Then ask what structures prevent heat from escaping from a bird's body on cold days.

Dinosaurs: *Chapter 43*
Ask students what the characteristics of dinosaurs were. Then ask them which of these characteristics can be found in birds today.

CHAPTER 44

BIRDS

Young birds, such as this owl, depend on their parents for food and protection.

FOCUS CONCEPT: *Structure and Function*
As you read, note the characteristics of birds that are adaptations for flight.

ORIGIN AND EVOLUTION OF BIRDS

Birds belong to the class Aves, which, with about 9,700 species, is the largest class of terrestrial vertebrates. Birds are also the most recently evolved group of vertebrates, having appeared only about 150 million years ago. Among living vertebrates, only birds and bats can fly. The bodies of birds are well adapted to flight.

SECTION

44-1

OBJECTIVES

▲ Identify and describe seven characteristics of birds.

● List three similarities between birds and dinosaurs.

■ Describe the characteristics of *Archaeopteryx.*

◆ Summarize the two main hypotheses for the evolution of flight.

CHARACTERISTICS

Birds are so distinctive that it is difficult to mistake one for any other kind of vertebrate. Seven important characteristics of birds are described below.

- **Feathers**—Feathers are unique to birds, and all birds have them. Like hair, feathers are composed mainly of the versatile protein keratin. Feathers are essential for flight, and they insulate a bird's body against heat loss.
- **Wings**—A bird's forelimbs are modified into a pair of wings. Feathers cover most of the surface area of the wing.
- **Lightweight, rigid skeleton**—The skeleton of a bird reflects the requirements of flight. Many of the bones are thin-walled and hollow, making them lighter than the bones of nonflying animals. Air sacs from the respiratory system penetrate some of the bones. Because many bones are fused, the skeleton is rigid and can resist the forces produced by the strong flight muscles.
- **Endothermic metabolism**—A bird's rapid metabolism supplies the energy needed for flight. Birds maintain a high body temperature of 40–41°C (104–106°F). The body temperature of humans, by contrast, is about 37°C, or 98.6°F.
- **Unique respiratory system**—A rapid metabolism requires an abundant supply of oxygen, and birds have the most efficient respiratory system of any terrestrial vertebrates. The lungs are connected to several sets of air sacs, an arrangement that ensures that oxygen-rich air is always in the lungs.
- **Beak**—No modern bird has teeth, but the jaws are covered by a tough, horny sheath called a beak.
- **Oviparity**—All birds lay amniotic eggs encased in a hard, calcium-containing shell. In most species, the eggs are incubated in a nest by one or both parents.

SECTION 44-1

SECTION OVERVIEW

Structure and Function
Section 44-1 summarizes the major characteristics of modern birds and explains how they contribute to a bird's ability to fly.

Evolution
Section 44-1 discusses the evidence that links modern birds with their dinosaur ancestors.

ACTIVE READING

Technique: Brainstorming
Before beginning this chapter, ask each student to write down five characteristics of birds. Then have the students read the first page of this chapter. They should formulate one "how" and one "why" question for each of the characteristics they have written down that are specific to flight. Have them brainstorm in groups to come up with possible answers to their questions. Each group should share its questions, answers, and reasons for its answers with the rest of the class. Have the class discuss some of the more interesting ideas after they have read Sections 44-1 and 44-2.

RECENT RESEARCH

A Clunky Flier
All modern birds have alulas, tufts of feathers on the first digit of each wing. Alulas allow birds to make smooth and accurate landings. Without alulas, a bird would drop quickly when braking. *Archaeopteryx,* which lacked alulas, probably broke its sudden fall by breaking into a full run upon landing.

CRITICAL THINKING

Sparse Fossil Record

Tell students that bird fossils are rarer than reptile or mammal fossils. Ask them why this is so. (The hollow skeletons of birds often disintegrate before fossilization can occur.)

QUICK FACT

☑ The common ancestors of birds, dinosaurs, crocodiles, and pterosaurs were reptiles called thecodontians. They first appeared over 250 million years ago.

GIFTED *ACTIVITY*

Ask students to search library and on-line resources and prepare a report on the events surrounding the discovery of the first *Archaeopteryx* fossil in 1861. Tell them to keep in mind that this reptile-bird "missing link" was discovered soon after Darwin published his theory of evolution.

TEACHING STRATEGY

Bone Density

Obtain chicken and beef bones from a supermarket. Clean the bones by boiling them in water and pulling off any remaining meat. Let them air-dry and store each one in a separate plastic bag. Give small groups of students a chicken bone and a beef bone and write the formula for density (density = mass/volume) on the board. Provide students with graduated cylinders and water, to determine the bones' volume by water displacement, and a balance, to determine their mass. Have students calculate the density of each bone with their data. Ask students which bone has the higher density (the beef bone). Have them carefully break open the bones with a hammer to compare their internal structures.

FIGURE 44-1

This phylogenetic tree shows the phylogenetic relationships between various terrestrial vertebrates. Note how birds, dinosaurs, and crocodiles evolved from a common ancestor, which split off very early from the other terrestrial vertebrates.

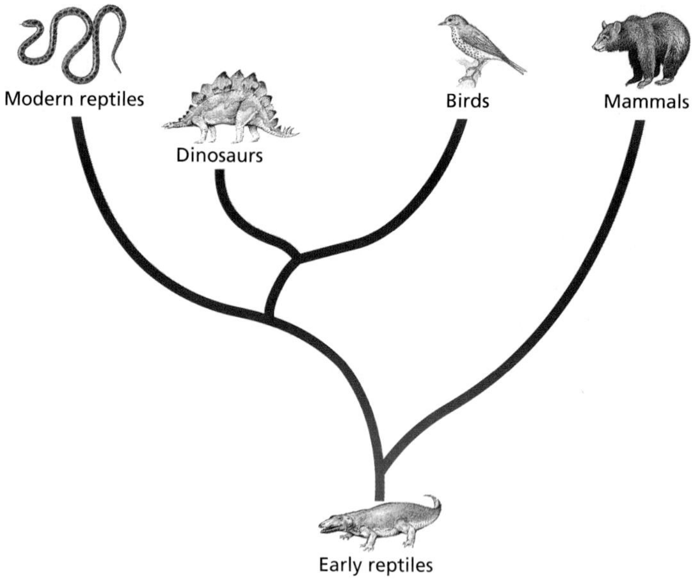

EVOLUTION

Scientists have long recognized the large number of similarities between birds and some dinosaurs. Three of these similarities include a flexible S-shaped neck, a unique ankle joint, and hollow bones. Birds are thought to have evolved from small, fast-running carnivorous dinosaurs during the Jurassic period (208–144 million years ago). Figure 44-1 shows the relationships between birds and other terrestrial vertebrates.

The oldest known bird fossils belong to the species *Archaeopteryx lithographica* and date from the late Jurassic period, about 150 million years ago. In the fossil in Figure 44-2a, the impressions of feathers are clearly visible. Feathers covered *Archaeopteryx*'s forelimbs, forming wings, and covered its body and tail as well. Like modern birds, *Archaeopteryx* had hollow bones and a **furcula** (FUR-kyuh-luh), the fused collarbones commonly called a wishbone. However, *Archaeopteryx* also had several characteristics of its dinosaur ancestors, including teeth, claws on its forelimbs, and a long, bony tail. Figure 44-2b shows an artist's conception of what an *Archaeopteryx* might have looked like.

Based on certain key similarities with modern birds, most scientists think that *Archaeopteryx* could fly. The furcula plays an important role in flight by helping to stabilize the shoulder joint.

Origin of Flight

The evolution of a flying animal from nonflying ancestors entails many changes in anatomy, physiology, and behavior. According to one hypothesis, the ancestors of birds were tree dwellers that ran along branches and occasionally jumped between branches and trees. Wings that allowed these animals to glide from tree to tree evolved. Once gliding was possible, the ability to fly by flapping the wings evolved. Another hypothesis draws on the fact that the

internetconnect

SCILINKS NSTA

TOPIC: Birds
GO TO: www.scilinks.org
KEYWORD: HM862

Word Roots and Origins

archaeopteryx

from the Greek *archaios*, meaning "ancient," and *pteryx*, meaning "wing"

(a)

(b)

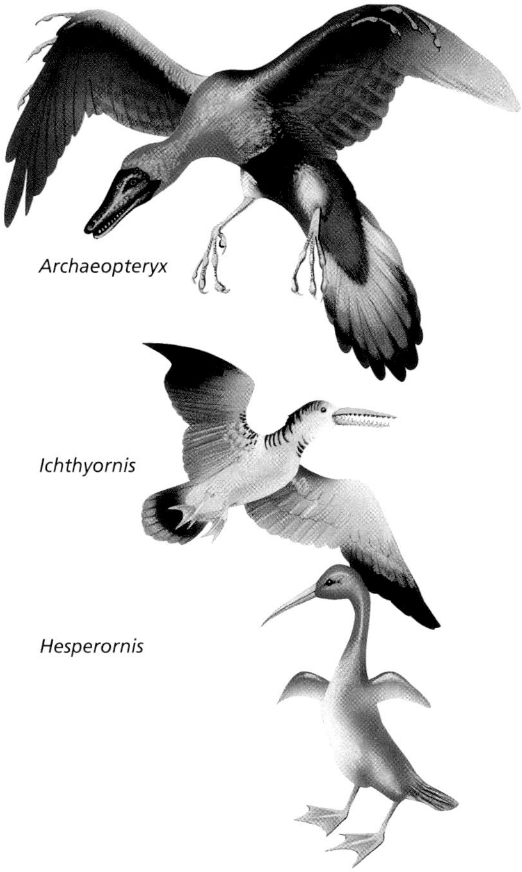

dinosaurs most closely related to birds were terrestrial and states that the evolution of birds must have occurred on the ground, not in the trees. Wings may have originally served to stabilize the animals as they leapt after prey. Or they may have been used for trapping or knocking down insect prey. Over generations, the wings became large enough to allow the animal to become airborne.

Evolution After *Archaeopteryx*

A number of recent discoveries show that by the early Cretaceous period (144–65 million years ago), birds had already begun diversifying. *Sinornis,* a 140-million-year-old specimen discovered in China in 1987, had some key features of modern birds, including a shortened, fused tail and a wrist joint that allowed the wings to be folded against the body. The diversification of birds continued throughout the Cretaceous period. Figure 44-2 shows three birds from the late Cretaceous period.

Only two of the modern orders of birds had appeared by the end of the Cretaceous period. Birds survived the asteroid impact that is thought to have wiped out the dinosaurs and underwent a dramatic and rapid radiation shortly afterward. By about 40 million years ago, most of the modern orders of birds had originated.

FIGURE 44-2

In this fossil of *Archaeopteryx lithographica* (a), one can see the link between birds and dinosaurs. These artist's renderings (b) of three extinct birds are based on fossil evidence. *Archaeopteryx* is the oldest bird; it still had claws on its wings. *Ichthyornis* had strongly developed wings and was about 21–26 cm (8–10 in.) in length. *Hesperornis* lacked a keel, had small wings, and was considered flightless, but its well-developed legs made it a strong swimmer.

SECTION 44-1 REVIEW

1. **List two unique features of a bird's skeleton.**

2. **What are two functions of feathers?**

3. **Describe two characteristics shared by *Archaeopteryx* and modern birds.**

4. **Name two differences between *Archaeopteryx* and modern birds.**

5. **Summarize the two major hypotheses for the evolution of flight.**

6. **CRITICAL THINKING Modern birds lack teeth. Form a hypothesis to explain why this characteristic might have evolved.**

CRITICAL THINKING

Surviving a Catastrophe

Ask students what characteristics may have helped birds survive the Cretaceous mass extinction. (They could fly far to find food and mates. Their ability to maintain a stable body temperature would have allowed them to survive climatic changes.)

RECENT RESEARCH

Cretaceous Extinction

A new study using DNA from modern birds suggests that the Cretaceous mass extinction may not have been as devastating to birds as once thought. The results indicate that as many as 22 different avian lineages may have survived the cataclysm.

ANSWERS TO SECTION 44-1 REVIEW

1. The skeleton has lightweight, hollow bones, some of which are penetrated by air sacs. Some of the bones are fused to make the skeleton rigid.
2. Feathers aid in flight and provide insulation.
3. Modern birds and *Archaeopteryx* share feathers, wings, hollow bones, and a furcula.
4. Unlike modern birds, *Archaeopteryx* had teeth, claws on its forelimbs, and a bony tail.
5. One hypothesis is that tree-dwelling dinosaurs first evolved the ability to glide, then the ability to fly. The second hypothesis is that flight evolved among ground-dwelling, running dinosaurs that used wings as stabilizers or to capture prey.
6. Answers will vary, but one possibility is that teeth were lost because they were heavy.

SECTION 44-2

SECTION OVERVIEW

Structure and Function
Section 44-2 discusses how feathers and the skeleton enable birds to achieve and maintain flight.

Stability and Homeostasis
Section 44-2 describes how efficient respiratory and digestive systems enable birds to maintain the high metabolic rate necessary to achieve flight.

▶ ENGAGE STUDENTS

Ask students to take out two sheets of notebook paper. Give them hole punchers, and have them use the devices to riddle one of the sheets with holes. Ask them to hold the perforated sheet with a hand at each end and move it through the air. Then have them move the unpunched sheet through the air in a similar manner. Ask students which sheet offered more resistance to the air. (The unpunched sheet offered greater resistance.) Show students a bird feather and create "holes" in its vanes by splitting apart the barbs in several locations along the feather's shaft. Tell them that a bird's wing feathers must resist the air in order for flight to occur. Add that birds spend a lot of their time realigning the barbs of their feathers to make sure they are "hole-free."

DEMONSTRATION

Obtain a spray bottle and a recently shed bird feather. Lightly mist the feather with water from the spray bottle. Gently place the feather in a plastic bag and pass it around so students can make observations. Ask them what they notice about the water on the feather. (The water beads up on the vanes of the feather.) Tell students that the oil that birds apply to their feathers during preening waterproofs them.

SECTION

44-2

OBJECTIVES

▲ Describe the structure of a contour feather.

● Identify two modifications for flight seen in a bird's skeletal system.

■ Contrast the function of the gizzard with that of the crop.

◆ Trace the movement of air through the respiratory system of a bird.

▲ Explain the differences between altricial and precocial young.

CHARACTERISTICS OF BIRDS

A number of unique anatomical, physiological, and behavioral adaptations enable birds to meet the aerodynamic requirements of flight. For example, natural selection has favored a lightweight body and powerful wing muscles that give birds their strength.

FEATHERS

Feathers are modified scales that serve two primary functions: providing lift for flight and conserving body heat. Soft, fluffy **down feathers** cover the body of nestling birds and provide an insulating undercoat in adults. **Contour feathers** give adult birds their streamlined shape and provide coloration and additional insulation. **Flight feathers** are specialized contour feathers on the wings and tail. Birds also have dust-filtering bristles near their nostrils.

The structure of a feather combines maximum strength with minimum weight. Feathers develop from tiny pits in the skin called **follicles.** A **shaft** emerges from the follicle, and two **vanes,** pictured in Figure 44-3a, develop on opposite sides. At maturity, each vane has many branches, called **barbs.** The barbs in turn have many projections, called **barbules,** equipped with microscopic hooks, as shown in Figure 44-3b. The hooks interlock and give the feather its sturdy but flexible shape.

Feathers need care. In a process called **preening,** birds use their beaks to rub their feathers with oil secreted by a **preen gland,** located at the base of the tail. Birds periodically molt, or shed, their feathers. Birds living in temperate climates usually replace their flight feathers (called major molt) during the late summer.

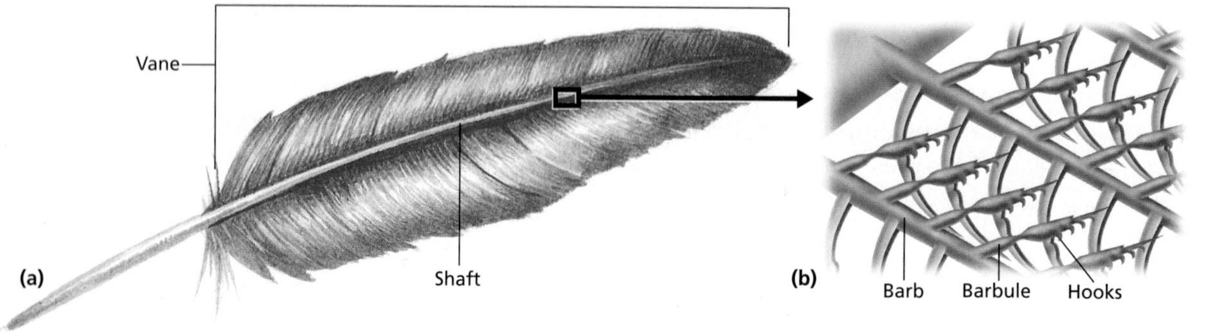

FIGURE 44-3
Bird feathers usually have a shaft, with two vanes growing out either side of the shaft. The vanes consist of barbs and barbules that interlock by means of hooks. Feathers are made of keratin, an insoluble protein that is highly resistant to enzyme digestion by bacteria. Keratin is also the protein that makes up fingernails, claws, hair, and scales.

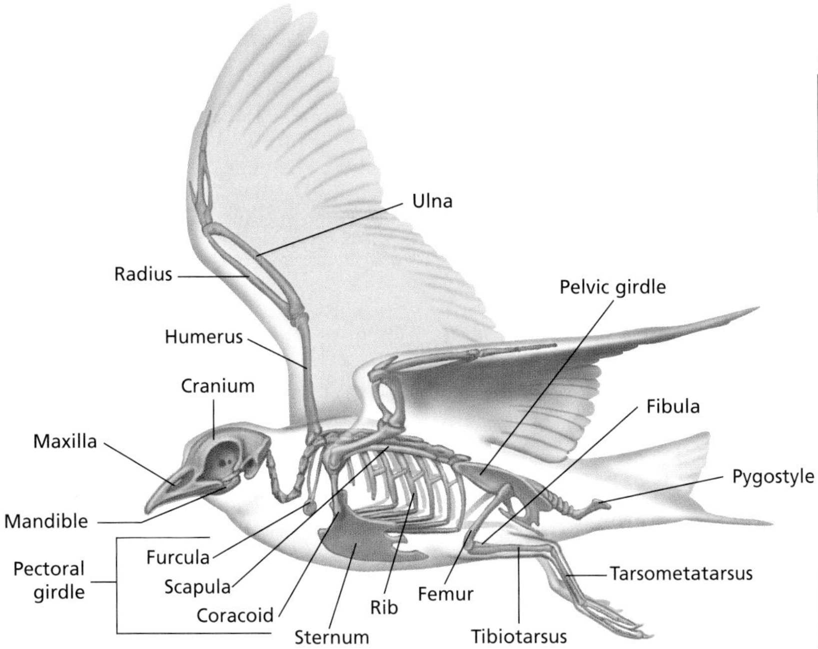

FIGURE 44-4

The avian skeleton is well adapted for flight. The bones are air-filled, making them light but strong. The skeleton is arranged to support the large muscles necessary for flight.

SKELETON AND MUSCLES

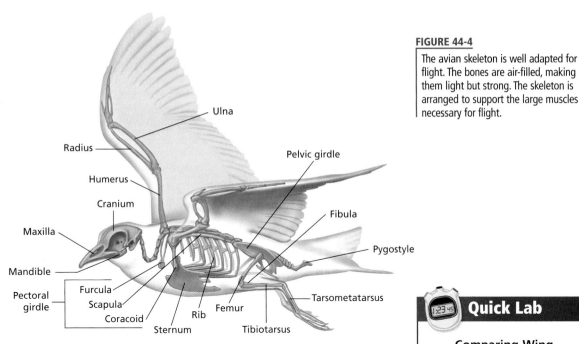

The avian skeleton combines lightness with strength. The bones are thin and hollow. Many bones are fused, so the skeleton is more rigid than the skeleton of a reptile or mammal. The rigid skeleton provides stability during flight. Note in Figure 44-4 the fused bones of the trunk and hip vertebrae and the pectoral and pelvic girdles. The large, keeled **sternum,** or breastbone, is an attachment point for flight muscles. The humerus, ulna, and radius, along with the pectoral girdle and the sternum, support the wing. The **pygostyle** (PIE-GUH-stiel), the terminal fused vertebrae of the spine, supports the tail feathers, which also play an important role in flight. The tail provides additional lift and aids in steering and braking.

Flight involves a series of complex wing movements, each one using a different set of muscles. On the downstroke the wings cut forward and downward through the air. During upstroke, they move upward and backward. These movements are made possible by large, powerful flight muscles. In some birds, flight muscles account for 50 percent of the body weight.

METABOLISM

Birds are endothermic; that is, they generate heat to warm the body internally. Rapid breathing and digestion of large quantities of food support the high metabolic rate necessary to generate this heat. Birds, unlike reptiles, cannot go for long periods without eating. To help conserve body heat they fluff out their feathers. Aquatic birds have a thin layer of fat that provides additional insulation.

Quick Lab

Comparing Wing Structures

Materials photocopies of different birds featuring their wings, ruler

Procedure Examine each sheet of birds and their wings. Compare the structure and shape of the wings. Measure the wingspan relative to the bird's body length. Record your observations.

Analysis Predict the type of habitat in which each bird lives. How does the shape of the wing relate to the bird's niche? Explain why the type of wings each bird has might make the bird unsuccessful if it were introduced into a much different environment.

Quick Lab

Comparing Wing Structures

Time Required 20 minutes

Procedural Tips Cut out several bird pictures from sources such as nature magazines or field guides. Get several angles of each bird with extended wings. Obtain pictures of eagles, peregrine falcons, penguins, and owls. Paste the various pictures of each bird on a sheet of paper, making one sheet for each bird type. Photocopy enough of the sheets for all the students in the class to have one of each.

Answers to Analysis Eagle: These birds live high in trees or mountain sides. They have wide wingspans for soaring in open spaces to find prey. Their tremendous wings would hinder them in a less open environment. Penguin: These birds live in an aquatic environment. They have short, stubby wings that have lost the ability to fly. Instead, their wings are used for swimming. Falcon: These birds are raptors that live in a forested environment. They have narrow, short wings adapted for enormous speed and maneuverability. Owl: These birds are stealth hunters that live in trees, barns, or anywhere there are small mammals for prey. Their wings have a medium span and are designed for maneuverability. They have "silent feathering" for quiet flight. Owls might not be competitive where there are other birds that are better adapted to the environment and other types of resources available.

CRITICAL THINKING

Death by Oil

Birds that get caught in oil spills often die of hypothermia (lowered body temperature). Ask students to explain how an oil-soaked bird becomes hypothermic. (When the feathers become soaked with oil, they can no longer effectively insulate the bird's body.)

SECTION 44-2

VISUAL LINK

Figure 44-5

Draw students' attention to the figure and point out that the bird stomach is divided into two parts, the proventriculus and the gizzard. Ask students why the phrase "scarce as hen's teeth" can be explained by the structure and function of the bird stomach. (There is no such thing as hen's teeth. The gizzard often contains stones to help grind food, so it serves the same function as teeth.)

CONTENT CONNECTION

Section 32-3: Seed Dispersal

Fruit-eating birds play a major role in scattering the seeds of plants. Their efficient digestive system can process large amounts of fruit, and the undigested seeds are frequently excreted as waste throughout the day.

INCLUSION *ACTIVITY*

Play this word-association game with students after they become familiar with the bird organ systems. Say a word related to one of the organ systems in a bird, then ask students to offer other words associated with the term. For example, if you say "proventriculus," student responses might include "crop," "digestion," "enzymes," or "pebbles." Write the responses on the board. After the list is complete, ask a volunteer to describe the structure and function of the organ system you named by using the terms offered by classmates.

FIGURE 44-5

A bird's digestive and excretory systems are adapted for the rapid processing of food and metabolic wastes. The high energy requirements of flight make this efficient system necessary. For example, the carnivorous shrike can digest a mouse in three hours.

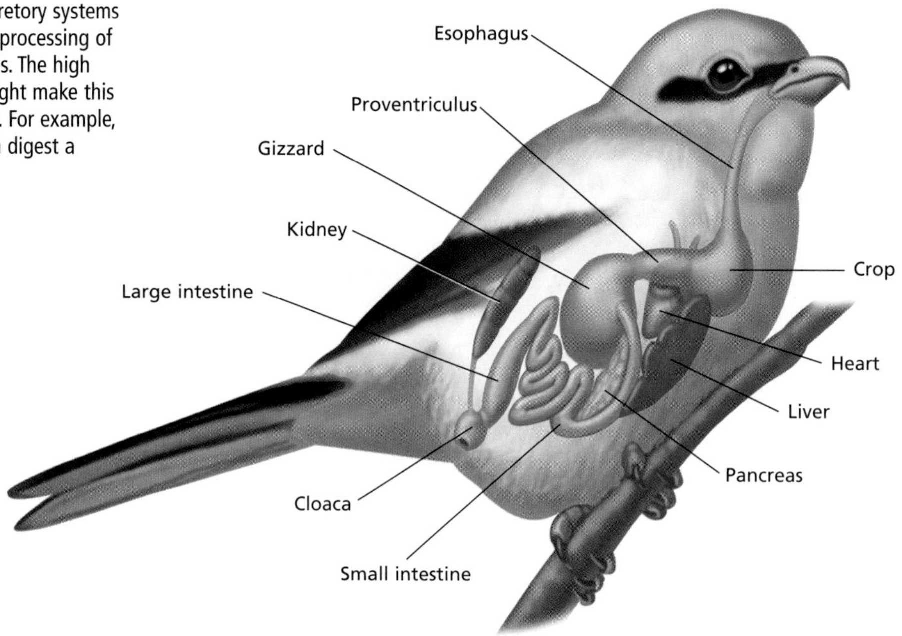

Word Roots and Origins

proventriculus

from the Greek *pro,* meaning "before," and the Latin *ventriculus,* meaning "stomach"

Digestive and Excretory Systems

The high amount of energy required to fly and regulate body heat is obtained by a quick and efficient digestive system, as illustrated in Figure 44-5. Because birds do not have teeth, they are not able to chew their food. Instead, food passes from the mouth cavity straight to the esophagus. An enlargement of the esophagus called the **crop** stores and moistens food. Food then passes to the two-part stomach. In the first chamber, the **proventriculus** (PROH-ven-TRIK-yuh-luhs), acid and digestive enzymes begin breaking down the food. Food then passes to the **gizzard,** the muscular portion of the stomach, which kneads and crushes the food. The gizzard often contains small stones that the bird has swallowed. These aid in the grinding process. Thus, the gizzard performs a function similar to that of teeth and jaws. Raptors, such as hawks and eagles, have modified systems for digesting protein. The crop and gizzard may be reduced or modified.

From the stomach, food passes into the small intestine. Here bile from the liver and enzymes made in the pancreas and intestine further break down the food. The nutrients are then absorbed into the bird's bloodstream. Passage of food through the digestive system of a bird is usually very rapid. For instance, a thrush can eat blackberries, digest them, and excrete the seeds 45 minutes later.

The avian excretory system is efficient and lightweight. Unlike other vertebrates, most birds do not store liquid waste in a urinary bladder. The two kidneys filter a nitrogenous waste called uric acid from the blood. Concentrated uric acid travels through ducts called ureters to the cloaca, where it mixes with undigested matter from the intestines and is then eliminated. Bird droppings are a mixture of feces and uric acid.

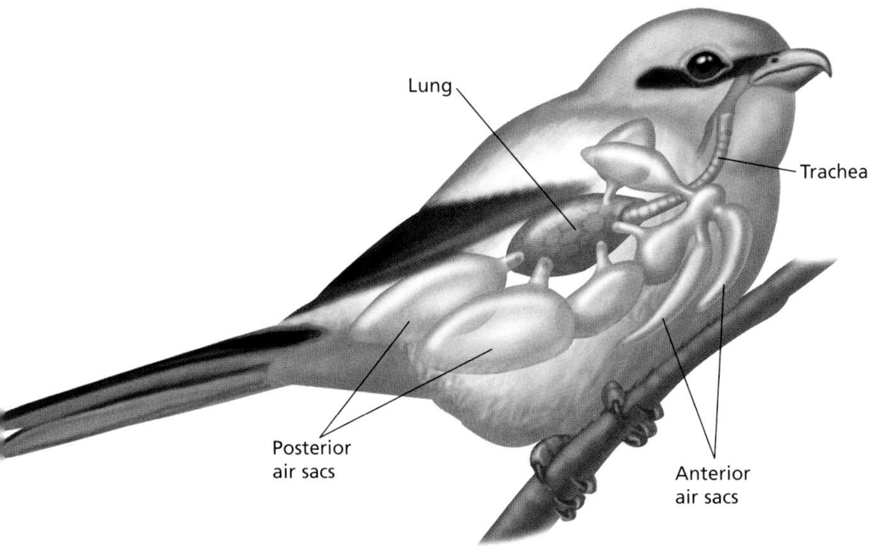

FIGURE 44-6

The unique architecture of the bird's respiratory system provides a virtual one-way flow of air. This highly efficient system of air flow allows birds to maintain the high metabolic rate necessary for flight. It also enables birds to function at high altitudes, where other animals would suffer from the lower levels of oxygen.

Respiratory System

The high metabolic rate of birds requires large amounts of oxygen. Yet some birds migrate thousands of miles at altitudes as high as 7,000 m (23,000 ft), where oxygen levels are very low. An elaborate and highly efficient respiratory system meets these oxygen needs. Air enters the bird's body through paired nostrils located near the base of the beak. The air passes down the trachea and enters the two primary bronchi. From the bronchi, some of the air moves to the lungs. However, about 75 percent of the air bypasses the lungs and flows directly to posterior air sacs, shown in Figure 44-6. Nine sacs extend from the lungs, in most birds, occupying a large portion of the bird's chest and abdominal cavity. These sacs also extend into some of the long bones. Thus, the air sacs not only function in respiration but also greatly reduce the bird's density.

Gas exchange does not occur in the air sacs. Their function is to store air. When the bird exhales, the carbon dioxide–rich air from its lungs is forced into the anterior air sacs, and the oxygen-rich air in the posterior air sacs is forced into the lungs. This way, the bird has oxygenated air in its lungs during both inhalation and exhalation.

Circulatory System

Like crocodiles and mammals, birds have a heart with two separate ventricles. Deoxygenated blood is always kept separate from oxygenated blood. In comparison with reptiles and most other vertebrates, most birds have a rapid heartbeat. A hummingbird's heart beats about 600 times a minute. An active chickadee's heart beats 1,000 times a minute. In contrast, the heart of the larger, less active ostrich averages 70 beats per minute, or about the same rate as a human heart. Avian red blood cells have nuclei.

CRITICAL THINKING

Avian Diagnosis

Tell students that a boy took his sick pet bird to a veterinarian. He told the vet that the bird would not fly when he took it out of its cage. After examining the bird, the vet told the boy that the bird has a severe respiratory infection. Ask students how this diagnosis explains the bird's inability to fly. (If its respiratory system is not functioning correctly, the bird cannot get sufficient oxygen to power its flight.)

TEACHING STRATEGY

Bird-Heart Model

Divide the class into groups of three or four. Give each group the following materials: a cardboard shoe box (with cover), an 11 × 17 in. sheet of cardboard or poster board, a blue marker, a red marker, four white straws, tape, and scissors. Also furnish students with reference books or a diagram that shows the external and internal structure of a bird's heart. Challenge the groups to build a model of a bird's heart out of the materials you have supplied. Tell them that the shoe box represents the outside of the heart. They are to use the cardboard and straws to make the inside chambers and the four main blood vessels of the heart. The markers are for color-coding the blood vessels. Tell students that red and blue signify vessels that contain oxygenated and deoxygenated blood, respectively. Finally, tell the groups to write a description of avian blood circulation on a 5 × 7 in. index card. Have them attach the card to the outside of the box's lid. Check each group's model to see how well it works as a prop for the written description.

SECTION 44-2

MAKING CONNECTIONS

Social Studies

Have students use library and online resources to research the uses and significance of feathers in Native American cultures. This could include their use in clothing and art, their ceremonial significance, their use as indicators of status within the tribe, or the significance of the species of birds from which feathers are obtained. Suggest that students obtain information about historical tribes or pre-Columbian civilizations. Have students present short oral reports based on their findings.

CULTURAL CONNECTION

Cultural Symbols

Birds have been important cultural symbols throughout the world. Eagles, for example, have been adopted as national symbols in the United States, Russia, Mexico, and Germany. The vulture was an important icon in ancient Egypt and the mythical phoenix of ancient Greece was also an important symbol.

TEACHING STRATEGY

Bird Songs

Tell students that many bird species sing to attract mates or establish territories. Obtain a recording of bird songs. Have groups of three or four students learn to identify a bird song. Provide them with bird identification books to research their bird. Have each group imitate the song and report about their bird to the rest of the class. Play the recording after students finish their presentation to see how closely their imitation sounded like the recorded song.

Nervous System and Sense Organs

Relative to their body size, birds have large brains. The most highly developed areas of the bird's brain are those that control flight-related functions, such as the cerebellum, which coordinates movement. The cerebrum is also large. It controls complex behavior patterns, such as navigation, mating, nest building, and caring for the young. The large optic lobes receive and interpret visual stimuli.

Keen vision is necessary for taking off, landing, spotting landmarks, hunting, and feeding. Most birds have good color vision that aids them in finding food. In most species, the eyes are large and are located near the sides of the head, giving the bird a wide field of vision. Birds that have eyes located near the front of the head have better binocular vision, meaning they can perceive depth in the area where the visual fields of the two eyes overlap.

Hearing is important to songbirds and to nocturnal species, such as owls, which rely on sounds to help them locate their prey. Though birds lack external ears, owls have feathers shaped to create a trough around their ear openings, which helps direct sound into the ear. The sense of smell is also developed in many birds.

REPRODUCTION

In the male bird, sperm is produced in two testes that lie anterior to the kidneys. Sperm passes through small tubes called **vasa deferentia** into the male's cloaca. During mating, the male presses his cloaca to the female's cloaca and releases sperm. Most females have a single ovary located on the left side of the body. The ovary releases eggs into a long funnel-shaped **oviduct,** where they are fertilized by sperm. Fertilized eggs move down the oviduct, where they receive a protective covering and a shell. The egg passes out of the oviduct and into the cloaca. From the cloaca, it is expelled from the bird.

Eco Connection

DDT and Bird Eggs

DDT is a pesticide that was widely used on crops until the 1970s. DDT was banned because of the harm it was causing to birds. DDT causes thinning of birds' egg shells, decreasing survival rates of the birds' offspring. This caused a significant drop in the numbers of raptors, brown pelicans, and other pelicans. The thinning of the eggs was so significant that even the weight of an incubating parent could crush the eggs. With the banning of DDT in the United States, populations of the affected birds have increased.

Nest Building and Parental Care

Birds usually lay their eggs in a nest. Nests hold the eggs, conceal young birds from predators, provide shelter from the elements, and sometimes serve to attract a mate. Most birds build nests in sheltered, well-hidden spots—from holes in the ground to treetops. Woodpeckers, for example, nest in a hole they have drilled in a tree. Orioles suspend their nests from branches, well beyond the reach of predators. And barn swallows build a saucer of mud on the beam of a building. Birds construct their nests from almost any available material. Twigs, grasses, feathers, and mud are the most common materials used.

One or both parents incubate, or warm, the eggs by sitting on them and covering them with a thickened, featherless patch of skin on the abdomen called a **brood patch.** Once the eggs hatch, the young usually receive extensive parental care. Birds have two methods of rearing young. Those that lay many eggs and incubate

Research Notes

A Hemispheric Strategy: A Single Vast Wildlife Reserve

The numbers of shorebirds—such as sandpipers, plovers, and curlews—are declining worldwide, but people in North America and elsewhere are working to preserve and protect these valuable species. To do this, researchers must first determine shorebird populations and map their migratory routes. To appreciate the magnitude of the job, you should know that more than 40 North American shorebird species breed in the Arctic and migrate to wintering sites in Central and South America, an annual round trip of 12,000 to 25,000 km.

Today the work of preserving shorebirds has been greatly advanced and simplified through the use of computer technology. One group that uses computers in this work is the International Shorebird Survey (ISS). Migrating shorebirds recognize no national boundaries, so any strategy for preserving them must be international. The International Shorebird Survey was established in 1974 and consists of both volunteers and professional researchers. The volunteers record information on populations, habitat characteristics, weather conditions, and human activity. Researchers use computers to identify relationships among these variables. Complex methods of analysis have revealed much about the migratory habits of shorebirds—their velocity, altitude, flight path, and gathering sites, for example. Computers have also helped researchers identify significant trends, such as a 70 percent to 80 percent drop in the population of sanderlings and short-billed dowitchers.

Another organization that uses computers in its work of preserving shorebirds is the Western Hemisphere Shorebird Reserve Network (WHSRN). As an international group of conservation organizations, the Western Hemisphere Shorebird Reserve Network has identified more than 90 sites where significant numbers of shorebirds breed, feed, and rest. The organization has proposed that these sites be recognized internationally as a single vast wildlife reserve whose boundaries are, in effect, defined by the migrating birds.

Shorebirds congregate in a few strategic sites, making them especially vulnerable to human activity. Replacing one beach or wetland with a housing development can have serious effects on the population of a species that depends on that site. However, humans must also develop a strategy for the survival of the congregation sites. The hemispheric wildlife reserve is the basis of such a strategy.

Shorebird populations are likely to need ongoing analysis, sometimes for new reasons. For example, possible consequences of global warming, such as changes in weather patterns and sea level, could affect shorebird habitats drastically. Thus computer technology will continue to facilitate the work of international conservation organizations and individuals who devote themselves to preserving and protecting shorebirds.

Delaware Bay, located between Delaware and New Jersey, is a staging site for shorebirds that migrate between South America and the Arctic. Up to 80 percent of the population of red knots feed and rest at this site. Preservation of the bay is critical to the species.

CHAPTER 44 Research Notes

BACKGROUND

International organizations are working to preserve bird populations through the protection of bird habitats. These efforts include determining bird population sizes, identifying habitats, mapping migration routes, and monitoring the weather.

DISCUSSION

Guide the discussion by posing the following questions.

1. What are researchers' first tasks in trying to preserve shorebirds? (Researchers must first determine shorebird populations and map their migratory routes.)

2. What are the names of two groups mentioned in this feature that use computers, and what do they do? (The International Shorebird Survey identifies relationships among shorebird populations, habitat characteristics, weather conditions, and human activity. The Western Hemisphere Shorebird Reserve Network identifies where shorebirds breed, feed, and rest.)

3. Why are shorebirds vulnerable to human activity? (Shorebirds are vulnerable because they congregate in a few strategic sites.)

FURTHER READINGS

1. "Bringing Back the Birds," *Science News*, August 17, 1996, pp. 108–109. This article describes efforts for improving and protecting bird habitats.

2. "Shorebird Patrol," *Nature Canada,* Winter 1996, pp. 10–12. This article describes one woman's efforts to save shorebirds in Canada from careless fishing practices.

SECTION 44-2

GIFTED ACTIVITY

Ask students to research a migratory bird species from the Northern Hemisphere, such as songbirds, waterfowl, and birds of prey. Have other students research a migrant from the Southern Hemisphere (mostly non-singing birds, such as cuckoos, nightjars, and swifts). Have them report on their bird's special adaptations and natural history. Reports should include information on where the bird summers and winters as well as on its breeding and nesting behaviors.

ANSWERS TO SECTION 44-2 REVIEW

1. The vanes are the flexible material on either side of the shaft. Barbs are the parallel branches that make up the vanes. Barbules are cross-pieces on the barbs that interlock with one another.
2. The keeled sternum is the attachment point for some of the flight muscles.
3. The gizzard, which often contains stones, pulverizes and grinds food.
4. The anterior and posterior air sacs store air but do not participate in gas exchange. The posterior air sacs receive oxygen-rich air and send it to the lungs. The anterior air sacs receive oxygen-depleted air leaving the lungs.
5. The nests of ground-dwelling birds are vulnerable to predators. Precocial young can leave the nest almost immediately and follow their mother, receiving her protection even when she is away from the nest.
6. Since many birds that nest in the United States winter in tropical areas, destruction of tropical rain forests reduces the populations of these birds, and some species may become extinct as a result.

FIGURE 44-7
This yellow warbler cares for its young by feeding and protecting them.

them for long periods hatch **precocial** (pree-KOH-shuhl) young. These young birds are active as soon as they hatch. The mother provides warmth and protection, but the newly hatched birds can walk, swim, and feed themselves. Ducks, quail, and other ground-nesting species produce precocial offspring. Birds that lay only a few eggs that hatch quickly produce **altricial** (al-TRI-shuhl) young. These birds hatch blind, naked, and helpless, as shown in Figure 44-7. They depend on both parents for several weeks. The parents keep the young warm and feed them. The young of woodpeckers, hawks, pigeons, parrots, warblers, and many aquatic birds are altricial.

MIGRATION

Each year, thousands of bird species exploit the spring and summer food resources of temperate regions. Then, when temperatures drop and the food supply dwindles, they travel to warmer climates. The seasonal movement of animals from one habitat to another habitat is called migration. Many of the birds that nest in the United States and Canada during the spring and summer fly south in the fall to spend the winter in Mexico, Central America, the Caribbean, or South America. For example, the blackburnian warbler is a small songbird that nests in forests of the northeastern United States and southern Canada but winters in Central and South America.

How do birds manage to navigate thousands of kilometers across varied terrains to the same spot year after year? **Ornithologists**—biologists who study birds—have learned that birds rely on a variety of cues to help them navigate. Some species monitor the position of the stars or the sun. Others rely on topographical landmarks, such as mountains. The Earth's magnetic field, changes in air pressure due to altitude, and low-frequency sounds may also provide information to migrating birds.

Many species migrate thousands of kilometers and must rely on their fat reserves in order to complete the journey. To prepare for their migration, some birds, such as blackpoll warblers, eat so much food before their journey that their weight nearly doubles.

internetconnect
SCILINKS NSTA
TOPIC: Migration of birds
GO TO: www.scilinks.org
KEYWORD: HM870

SECTION 44-2 REVIEW

1. Distinguish between vanes, barbs, and barbules.
2. What is the function of the keeled sternum?
3. In what way does the gizzard compensate for the lack of teeth in birds?
4. What are the functions of the anterior and posterior air sacs?
5. Ground-nesting birds often produce precocial young. How might this improve the odds of survival for the young?
6. **CRITICAL THINKING** How do you think the destruction of tropical rain forests affects populations of birds in the United States? Explain your answer.

CLASSIFICATION

Birds are the most widespread terrestrial vertebrates. Their ability to navigate over long distances and their many adaptations for flight enable them to migrate to and inhabit virtually any environment. Their anatomical diversity reflects the diversity of places they inhabit.

SECTION 44-3

OBJECTIVES

▲ Describe the relationship between beak shape and diet in birds.

● List 12 orders of living birds, and name an example of each order.

■ Describe the function of the syrinx.

DIVERSITY

By looking closely at a bird's beak and feet, you can infer many things about where it lives and how it feeds. Hawks and eagles have powerful beaks and clawed talons that help them capture and tear apart their prey. Swifts have a tiny beak that opens wide like a catcher's mitt to snare insects in midair. Because swifts spend most of their lives in flight, their feet are small and adapted for infrequent perching. The feet of flightless birds, on the other hand, are modified for walking and running. Some examples of the variety of bird beaks and feet are shown in Figure 44-8.

(a)

(b)

(c)

(d)

FIGURE 44-8

The cardinal (a), *Cardinalis cardinalis,* has a short, strong beak for cracking seeds and feet that enable it to perch on small tree branches. The kestrel (b), *Falco sparverius,* has a beak that enables it to tear flesh and talons that enable it to grip and kill prey. The calliope hummingbird (c), *Stellula calliope,* has a long, thin beak that enables it to extract nectar from flowers. The northern shoveler duck (d), *Anas clypeata,* has a flat beak that enables it to shovel mud while searching for food.

internetconnect

SCILINKS NSTA

TOPIC: Classification of birds
GO TO: www.scilinks.org
KEYWORD: HM871

SECTION 44-3

SECTION OVERVIEW

Structure and Function

Section 44-3 discusses the variation in beaks and feet among different birds and how these structures enable birds to gather food.

▶ ENGAGE STUDENTS

Ask students to contribute the names of as many birds as possible in 10 minutes. Write each name on the board as it is offered. (Accept all answers that are actually birds.) When time is up, ask students to suggest which birds on the list are likely to be closely related to each other. Have them reason on the basis of appearance, habitat, or feeding habits. Encourage interested students to consult a field guide or ornithology text to compare actual bird classifications.

RECENT RESEARCH

Bird-Friendly Coffee

Migratory songbird populations are steadily declining in the United States and Canada. Species such as thrushes, warblers, and orioles typically breed in the north in the summer and spend winter in the tropical forests of Mexico, Central America, and Caribbean islands. Scientists suspect that a switch to a new method of growing coffee plants may be contributing to songbird decline. Traditionally, coffee has been grown on shade plantations, where the coffee plants are grown under a canopy of planted trees. Over time, these plantations have become forest havens for wintering songbirds. Recently, coffee growers began planting new sun-tolerant coffee plants, which produce more coffee. As a result, the trees around the coffee plants are being cut down and, scientists fear, songbird winter homes are being lost.

GIFTED ACTIVITY

Have each student select an endangered bird species from a bird guide or textbook. Have each student give a five-minute oral report on the bird, its habitat, the reasons the bird is endangered, whether the bird will be able to survive, and what measures can be taken to protect it.

CRITICAL THINKING

Owl Pellets

Tell students that owls cough up pellets that contain the undigested bones of their prey. Ask them why mammalogists collect owl pellets. (By analyzing the contents of the pellets, mammalogists can determine the kinds of small mammals that live in an area.)

RETEACHING ACTIVITY

Have pairs of students prepare flash cards to help them learn the 12 major bird orders. On one side of each card, have them write the order name and sketch a representative bird from the order. Have them refer to bird field guides to make their drawings. On the other side of the card, they can list information about the bird in the order. Let students spend some time drilling each other using their cards, then orally quiz them to see how much they were able to retain.

CONTENT CONNECTION

Section 18-2: Molecular Taxonomy

Remind students that the sequence of nucleotides in a gene determines the amino-acid composition of the protein made from the gene. Then tell them that closely related organisms will have similar DNA, RNA, and proteins. Taxonomists compare molecules from different organisms to understand the evolutionary relationships.

FIGURE 44-9

This mute swan, *Cygnus olor,* is able to take off from water and fly at very high speeds despite its great weight. Weighing up to 23 kg (50 lb), mute swans are the heaviest flying birds. Swans are monogamous, meaning they mate for life. While the female incubates the eggs, the male helps guard the nest.

FIGURE 44-10

The parrot pictured below is a lesser sulfur-crested cockatoo, *Cacatua sulphurea.* Parrots range in length from 8 cm (3 in.) to over 91 cm (3 ft). The earliest fossils of parrots indicate that they have existed as a group for at least 20 million years. Like most parrots, these cockatoos nest in holes in trees, and they usually lay only two eggs per year. Because of their low reproductive rate and destruction of habitat, many species have become extinct or endangered.

Most taxonomists divide the 10,000 species of living birds into 29 orders. Taxonomists have traditionally used morphological evidence from beaks, feet, plumage, bone structure, and musculature to classify birds. Technological advances in the analysis of blood proteins, chromosomes, and DNA have also been used. Despite the introduction of these new methods, the relationships among the 29 orders of birds are still not well resolved. Following are 12 of the most familiar orders of living birds.

Order Anseriformes

Swans, geese, and ducks—commonly called waterfowl—belong to this order of 160 species. Found worldwide, members of this order are usually aquatic and have webbed feet for paddling and swimming. Waterfowl feed on a variety of aquatic and terrestrial foods, ranging from small invertebrates and fish to grass. The bill is typically flattened. The young are precocial, and parental care is usually provided by the female. A mute swan is shown in Figure 44-9.

Order Strigiformes

This order contains the owls, the nocturnal counterparts to the raptors. Owls are predators that have a sharp, curved beak and sharp talons or claws. As you can see in the figure on page 860, owls also have large, forward-facing eyes that provide improved vision at night. Owls rely on their keen sense of hearing to help locate prey in the dark. There are about 180 species of owls, and they are found throughout the world.

Order Apodiformes

Hummingbirds and swifts belong to this order. All of the roughly 420 species are small, fast-flying, nimble birds with tiny feet. Swifts pursue insects and capture them in flight. Hummingbirds, by contrast, feed on nectar, which they lap up with a very long tongue. The long, narrow bill of a hummingbird can reach deep into a flower to locate nectar. Swifts have a worldwide distribution, but hummingbirds live only in the Western Hemisphere.

Order Psittaciformes

This order includes the parrots and their relatives, the parakeets, budgerigars, cockatoos, and cockatiels. Most of the roughly 360 species in this order live in the tropics. Parrots are characterized by a strong, hooked beak that is often used for opening seeds or slicing fruits. Their upper mandible is hinged on the skull and movable. Unlike most birds, parrots have two toes that point forward and two toes that point toward the rear, an adaptation for perching and climbing. They are vocal birds, and many species gather in large, noisy flocks. Parrots have long been prized as pets because of their colorful plumage and intelligence and because some species can be taught to mimic human speech. However, excessive collecting for the pet trade and habitat destruction now threaten many parrot species with extinction. Figure 44-10 shows a cockatoo.

Order Piciformes

This diverse group of tree-dwelling birds contains woodpeckers, honeyguides, and toucans. All members of this order nest in tree cavities. Like parrots, they have two forward-pointing toes and two that point to the rear. There are about 350 species found throughout the world except in Australia. The diversity of foods consumed by these birds is reflected in the diversity of their bills. Woodpeckers, which drill holes into trees to capture insects, have strong, sharp, chisel-like bills. Toucans feed mainly on fruit, which they pluck with a long bill, as shown in Figure 44-11.

FIGURE 44-11

Toucans, such as this keel-billed toucan, *Ramphastos sulfuratus,* mate once per year, usually laying two to four eggs. The male and female toucans take turns sitting on the eggs. The eggs usually hatch after about 15 days of incubation.

Order Falconiformes

Known as raptors, the members of this order have a sharp, curved beak and sharp talons. Raptors include ospreys, hawks, falcons, vultures, and eagles. The 310 or so species of raptors are distributed throughout the world. Most species are diurnal (daytime) hunters with keen vision. Vultures, however, feed on dead animals and use their sense of smell to detect the odor of decomposing flesh.

Order Passeriformes

This large order contains about 5,700 species—more than half the total number of bird species—and includes most of the familiar North American birds. Robins, warblers, blue jays, and wrens are just some of the birds belonging to this group. In most birds, three toes point forward and one points backward. Passerines have this same arrangement of toes, but the rear toe is enlarged and particularly flexible to provide a better grip on branches. Passerines are sometimes called perching birds. Passerines feed on a variety of foods, including nectar, seeds, fruit, and insects.

Many passerines are called songbirds because the males produce long, elaborate, and melodious songs. Male birds sing to warn away other males and to attract females. The song is produced in the structure known as the **syrinx** (SIR-inks), which is located at the base of the bird's trachea. By regulating the flow of air through the syrinx, birds can generate songs of great range and complexity.

Order Columbiformes

This globally distributed group contains about 310 species of pigeons and doves. Figure 44-12 shows a mourning dove. These birds usually are plump-breasted and have relatively small heads; short necks, legs, and beaks; and short, slender bills. Most feed on fruit or grain. The crop, which in most other birds is used to store food, secretes a nutritious milklike fluid called **crop milk.** Both sexes produce crop milk to feed their young. Columbiform birds usually lay a clutch of two eggs, which hatch after a two-week incubation period. The young usually leave the nest two weeks after hatching. Another member of this order is the now-extinct dodo of Mauritius, an island in the Indian Ocean.

FIGURE 44-12

The adult mourning dove, *Zenaida macroura,* stands about 30 cm (12 in.) tall and nests in trees or bushes. Mourning doves breed throughout North America. They winter as far south as Panama.

VISUAL LINK

Figure 44-11

Have students examine the toucan in the figure and speculate about what adaptive advantage its coloring confers. (Scientists think the color scheme enables toucans to recognize members of their own species.) Then ask students what adaptive advantage a large bill provides. (Scientists think that the large bill frightens other birds that may represent a danger to them, such as hawks.)

TEACHING STRATEGY

Observing Wild Birds

Take students to an area near the school campus to observe birds, such as to a city park, nature area, or some other suitable site. Prepare them for bird-watching by furnishing them with field guides of birds that live in your area. Have students create blank check-off lists for the species they will most likely encounter. Ask them to bring binoculars from home if they have them, or supply one or two pairs for them to use. Have students check off and learn about the species they encounter.

QUICK FACT

☑ Birds with long bills do not find it difficult to preen hard-to-reach places, like the back of the head and neck. Most birds solve the problem by using their feet. Hummingbirds, for example, spend almost half their grooming time using their feet.

SECTION 44-3

INCLUSION ACTIVITY

Tell students that one result of habitat destruction is a reduction in available nesting sites for birds. Have students help wild birds by building houses for them. Provide students with cleaned quart or half-gallon cardboard milk cartons. Have students staple the top of the carton closed and cut out a round hole in one side about 3 in. down from the top. Tell students that the diameter of the hole will determine which species will use the house. Finally, have students attach a string to the top of the house and hang it outside. Encourage students to keep track of which bird species use their houses.

ANSWERS TO SECTION 44-3 REVIEW

1. Answers will vary. Students may describe any birds shown in the section.
2. eagle: Falconiformes; robin: Passeriformes; goose: Anseriformes; penguin: Sphenisciformes
3. Raptors and owls are predators that have sharp beaks and talons. Raptors are active during the day, and owls are active at night.
4. Crop milk is a nutritious fluid produced by the crop and used to feed the offspring. It is produced by pigeons and doves.
5. The syrinx is the sound-producing structure in birds.
6. Although penguins cannot fly, they swim with a similar motion and thus need a keeled sternum as an anchor for their muscles. Ostriches neither fly nor swim, and they do not have large flight muscles.

FIGURE 44-13
The great blue heron, *Ardea herodias,* uses its spearlike beak to stab fish, frogs, and other prey. Young herons must be taught how to hunt. Scientists have learned that young herons often miss their intended prey and must also learn what is and is not food.

Order Ciconiiformes

This group of long-necked, long-legged birds includes 120 species of herons, storks, ibises, and egrets. They tend to have a long, flexible neck and a long bill. Many are wading birds, and they feed on fish, frogs, and other small prey in shallow water. Some members of this order grow to be quite large. The marabou stork of Australia, for example, can be more than 1.5 m (59 in.) in height. Figure 44-13 shows a great blue heron, a common species in North America. The order has a worldwide distribution.

Order Galliformes

Members of this group, which includes turkeys, pheasants, chickens, grouse, and quails, are commonly called fowl. These terrestrial birds are usually plump-bodied and may have limited flying ability. Grains form a large part of the diet of many fowl, and all species have a large, strong gizzard. Some are also an important part of the human diet. The young are precocial. There are about 260 species distributed worldwide.

Order Sphenisciformes

Penguins are a unique group of flightless marine birds. All 17 species live in the Southern Hemisphere. The penguin's wedge-shaped wings have been modified into flippers, and the feet are webbed. Underwater, penguins flap their flippers to propel themselves forward—they "fly" through the water. Most species have a thick coat of insulating feathers and a layer of fat beneath the skin, enabling them to live in polar conditions. Penguins maintain this fat layer by consuming large quantities of fish and krill.

Word Roots and Origins

syrinx

from the Greek *syrinx,* meaning "reed" or "pipe"

Order Struthioniformes

Only the ostrich, the world's largest bird, belongs to this order. Ostriches can attain a height of nearly 3 m and weigh 150 kg. Ostriches cannot fly, but they are specialized as high-speed runners. Propelled by their long, strong legs, ostriches can reach speeds of 55 km per hour. Each foot has only two toes. Reduction in the number of toes is common in running animals—horses have just one toe per foot, for instance. Ostriches are native to Africa.

SECTION 44-3 REVIEW

1. Name two birds mentioned in this section, and describe how their beaks are suited to their way of life.

2. Identify the order to which each of the following birds belongs: eagle, robin, goose, penguin.

3. Name two similarities between raptors and owls. Name one difference.

4. What is crop milk? What types of birds produce it?

5. What is the function of the syrinx?

6. CRITICAL THINKING Penguins have a large, keeled sternum, but ostriches do not. Provide an explanation for this difference.

CHAPTER 44 REVIEW

SUMMARY/VOCABULARY

44-1

- Seven characteristics of birds are: feathers; wings; a lightweight, rigid flight skeleton; a respiratory system involving air sacs; endothermy; a beak instead of teeth; and oviparity.
- Birds evolved from small, carnivorous dinosaurs during the Jurassic period. The oldest known bird is *Archaeopteryx lithographica.*
- There are two hypotheses for the origin of flight in birds. One hypothesis is that flight began in the trees, with organisms progressing from jumping to gliding and flying. The second hypothesis is that birds began on the ground. Wings initially served as stabilizers or as "nets" for capturing insects. Feathers may have been used for heat regulation.
- Birds diversified after *Archaeopteryx,* a process that accelerated when the dinosaurs were wiped out.

Vocabulary

furcula (862)

44-2

- Feathers function in insulation and in flight. Down feathers provide insulation; contour feathers provide insulation and coloration and contribute most of the surface area of the wings and tail.
- Many of the bones in a bird's skeleton are hollow; many are also fused.
- The crop stores food. The two parts of the stomach are the proventriculus and the gizzard.
- Birds lack a urinary bladder and excrete their nitrogenous waste as uric acid.
- The lungs of a bird are connected to several air sacs that store air but do not participate in gas exchange.
- The cerebellum, cerebrum, and optic lobes of the bird's brain are large.
- All birds lay hard-shelled eggs. The eggs are usually laid in a nest and cared for by the parents.
- Many birds migrate using a variety of environmental cues to guide their migration.

Vocabulary

altricial (870)	down feather (864)	precocial (870)	shaft (864)
barb (864)	flight feather (864)	preen gland (864)	sternum (865)
barbule (864)	follicle (864)	preening (864)	vane (864)
brood patch (868)	gizzard (866)	proventriculus (866)	vasa deferentia (868)
contour feather (864)	ornithologist (870)	pygostyle (865)	
crop (866)	oviduct (868)		

44-3

- The feet and beak of a bird reflect its way of life.
- There are 29 orders of living birds, but scientists do not agree on the relationships among these orders.
- The majority of the world's birds belong to the order Passeriformes. Many are songbirds.
- Swifts and hummingbirds belong to the order Apodiformes. Parrots and relatives belong to the order Psittaciformes. Woodpeckers and toucans belong to the order Piciformes. Raptors belong to the order Falconiformes.
- Pigeons and doves belong to the order Columbiformes. Chickens and turkeys, belong to the order Galliformes. Owls are nocturnal hunters and belong to the order Strigiformes. Ducks, geese, and swans belong to the order Anseriformes.
- The herons and egrets belong to the order Ciconiiformes. Penguins belong to the Sphenisciformes, and the ostrich is in the order Struthioniformes.

Vocabulary

crop milk (873) syrinx (873)

CHAPTER 44 REVIEW
ANSWERS

REVIEW

1. *Altricial* refers to the naked, helpless young of some birds that hatch quickly. *Precocial* refers to young that are able to find food and walk soon after hatching. They require a long incubation period and are part of a larger clutch.

2. Barbs are the parallel filaments that make up the flexible part of a feather. They branch off the shaft of the feather. Barbules are extensions from barbs that interconnect with other barbules.

3. The furcula is the wishbone, or fused collarbone, and helps stabilize the shoulder joint. The sternum is the breastbone and serves as an attachment point for flight muscles.

4. The crop is an expanded section of the esophagus that stores and moistens food. The gizzard is the muscular part of the stomach that grinds food. It often contains stones that help grind food.

5. Migration is the seasonal movement from one habitat to another. Preening is grooming behavior by which a bird maintains its feathers.

6. c **7.** c
8. b **9.** b
10. d **11.** a
12. a **13.** b
14. a **15.** c

16. *Archaeopteryx* had several important flight characteristics of modern birds: hollow bones, a furcula, wings, and feathers.
17. *Archaeopteryx* had teeth, claws on its forelimbs, and a long, bony tail.
18. Down feathers are fluffy, insulating feathers. Contour feathers also serve as insulation, but they give a bird its coloration, and its streamlined shape.
19. The crop stores and moistens food prior to digestion in the proventriculus and gizzard.
20. Some air sacs store carbon-dioxide-rich air from the lungs and others store fresh air, which enables a one-way flow of air through the lungs. This keeps a steady supply of oxygen-rich air in the lungs whether the bird is inhaling or exhaling.
21. Binocular vision allows birds to perceive depth, which is important in flight and enables them to capture prey.
22. Birds use a variety of cues to navigate: the Earth's magnetic field, the positions of the sun and stars, landmarks, changes in air pressure, and low-frequency sound. All of these cues, except landmarks, would work over the ocean.
23. Owls and raptors are carnivores, but because they are active at different times of the day, they probably do not compete.
24. Birds migrate because of a seasonal change in the temperature or a change in the local food supply.
25. A—ulna, B—humerus, C—furcula, D—sternum, E—tibiotarsus, F—femur, G—pelvic girdle

REVIEW

Vocabulary

1. Differentiate between the terms *altricial* and *precocial.*
2. What are the differences between a barb and a barbule?
3. What are the differences between the furcula and the sternum?
4. Explain the difference between a crop and a gizzard.
5. Describe the differences between migration and preening.

Multiple Choice

6. Which of the following characteristics of *Archaeopteryx* is not shared by modern birds? (a) long tail (b) feathers (c) teeth (d) furcula
7. One similarity between birds and dinosaurs is (a) the presence of feathers (b) the lack of teeth (c) the structure of the ankle joint (d) size.
8. The function of the preen gland is to (a) produce digestive enzymes (b) produce an oily substance used to condition the feathers (c) release scents that help attract mates (d) control salt balance in the body.
9. The bone that supports the tail feathers is the (a) pelvic girdle (b) pygostyle (c) ulna (d) furcula.
10. Gas exchange in birds takes place in the (a) anterior air sacs (b) posterior air sacs (c) syrinx (d) lungs.
11. The structure that grinds food, aided by stones swallowed by the bird, is the (a) gizzard (b) proventriculus (c) crop (d) small intestine.
12. The excretory system of a bird (a) lacks a urinary bladder (b) excretes nitrogenous wastes as ammonia (c) stores large amounts of water (d) has only one kidney.
13. Birds that produce altricial young (a) abandon their newly hatched chicks immediately (b) lay only a few eggs at a time (c) usually build nests on the ground (d) incubate their eggs for a long period of time.
14. A cue that is believed to aid birds in navigation during their annual migrations is (a) the Earth's magnetic field (b) weather (c) cloud formations (d) phases of the moon.
15. Crop milk is produced by birds in the order (a) Ciconiiformes (b) Galliformes (c) Columbiformes (d) Falconiformes.

Short Answer

16. Summarize the evidence that *Archaeopteryx* could fly.
17. What characteristics did *Archaeopteryx* share with its dinosaur ancestors?
18. Contrast the function of down feathers with that of contour feathers.
19. What function does the crop serve in digestion?
20. How do air sacs increase respiratory efficiency in birds?
21. Why is binocular vision important to some birds?
22. Many birds migrate long distances over the ocean. What types of cues might these birds use to guide their movements?
23. Owls and raptors often coexist. Would you expect these two types of birds to compete? Explain your answer.
24. Explain why birds migrate.
25. Look at the diagram below of a bird's skeleton. Identify the following structures: pelvic girdle, furcula, sternum, femur, humerus, ulna, and tibiotarsus.

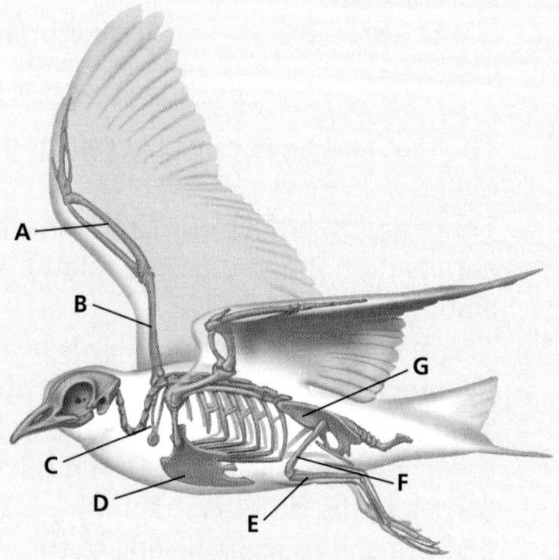

CRITICAL THINKING

1. What factors might explain why birds live in the Arctic but reptiles do not?
2. From studying the circulatory system of birds, you know that the right and left sides of a bird's heart are completely separated. This means that the oxygenated blood is never mixed with the deoxygenated blood. Why is this complete separation in the heart necessary?
3. Cowbirds lay their eggs in the nests of other birds. The young cowbirds hatch slightly earlier than the other birds do, and the cowbird hatchlings are slightly larger. Why might these distinctions be advantageous for the young cowbirds?
4. Although many species of temperate-zone birds migrate to the tropics to escape winter, some species remain behind. What benefits might these birds gain from not migrating?
5. From an evolutionary point of view, explain why the young of some birds, such as ducks and quail, are precocial.
6. The graph at right shows the results of an experiment on magpies in which the clutch size, that is, the number of eggs in the nest, was manipulated. Based on these data, determine whether the following statements are true or false.
   a. The greater the clutch size is, the greater the number of surviving offspring.
   b. More offspring died in nests containing eight eggs than in nests containing five eggs.
   c. Nests with nine eggs produced the fewest number of surviving offspring.

Effect of Varying Clutch Size

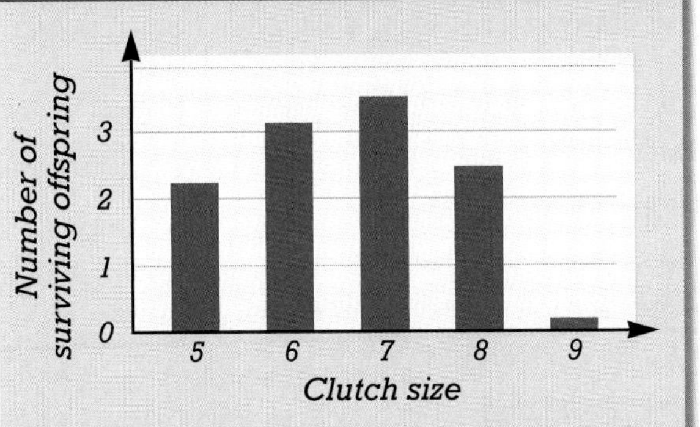

7. Drawing on data similar to that shown in the graph above, the ecologist David Lack argued that natural selection would favor intermediate clutch sizes. Explain the logic behind Lack's argument.

EXTENSION

1. Bird-watching is a fascinating and interesting hobby practiced by millions of Americans. From your library or bookstore, obtain a field guide to the birds in your area, and borrow a pair of binoculars. Take a long walk and try to locate as many birds as you can. Be sure to keep a notebook of the birds you find, including species, date sighted, location, and habits.
2. Many of the world's bird species have become extinct in the last 2,000 years. Three examples are the Carolina parakeet, the passenger pigeon, and the dodo. Using library resources or an on-line database, collect information on one of these extinct species. Find out as much as you can about its ecology and former distribution and about what caused it to become extinct. Write a report that summarizes your findings.
3. Read "Clown-Faced Hoarders" in *Audubon,* September–October 1999, on page 158. Describe the unique way acorn woodpeckers have of working together and storing food for the winter.

CRITICAL THINKING

1. Birds have insulation—feathers—and are endothermic. Reptiles lack both characteristics and depend on external heat to maintain their body temperature.
2. A bird's metabolism is so fast that oxygen must be continuously delivered to the tissues. Mixing of blood would dilute the oxygen content in the blood and hamper the bird's performance.
3. By hatching early, the cowbirds start to get food from the "parents" sooner. Because they are larger, they can dominate their nestmates in the competition for food.
4. Migration is expensive—it requires large amounts of food—and is dangerous. In addition, because so many birds migrate, birds that stay behind face less competition (although resources also are less abundant). And some birds don't migrate because they are physically unable to fly far enough.
5. Both types of birds build nests on the ground. If these birds had altricial young, they would likely be eaten by ground-based predators quickly. So, precocial young are selectively favored.
6. (a) false (b) true (c) true
7. Since intermediate clutch sizes yield the most surviving young, birds that lay medium-sized clutches will come to predominate in the population.

EXTENSION

1. Answers will vary.
2. Answers will vary, depending on the species chosen.
3. Acorn woodpeckers live in family groups who work together to protect their young, gather food, and defend their territory. They drill holes in trees or poles and stuff the holes with nuts. One tree may have 50,000 holes, each hole containing a nut.

CHAPTER 44
INVESTIGATION

TIME REQUIRED

One 50-minute class period

SAFETY PRECAUTIONS

Students should wash their hands with soap and water as soon as they are finished with the lab.

QUICK REFERENCES

Lab Preparation Notes are found on page 859B.

Holt BioSources provides a Teaching Resources CD-ROM that contains "Using Gowin's Vee in the Lab" and "Scoring Rubrics."

PROCEDURAL TIP

You may want to have students prepare their data collection tables ahead of time.

ANSWERS TO BACKGROUND

1. Birds have feathers, hollow bones, a furcula, a rigid skeleton, and a beak. They lay hard-shelled eggs, have wings, are endothermic, and have a unique respiratory system that keeps fresh air in their lungs while they are inhaling or exhaling.

2. Birds have feathers, hollow bones, a proventriculus and gizzard, a crop, and no teeth. Other vertebrates do not have feathers, hollow bones, a proventriculus, a gizzard, or a crop, but they do have teeth.

3. Feathers enable birds to fly and insulate them from heat loss.

CHAPTER 44 INVESTIGATION

Comparing Feather Structure and Function

OBJECTIVES

- Observe a quill feather, a contour feather, and a down feather.
- Compare the structure and function of different kinds of feathers.

PROCESS SKILLS

- observing
- relating structure to function
- comparing and contrasting

MATERIALS

- 1 quill feather
- 1 contour feather
- 1 down feather
- unlined paper
- prepared slide of a contour feather
- compound light microscope
- prepared slide of a down feather

Background

1. List several distinguishing characteristics of birds.
2. How do birds differ from other vertebrates?
3. What are the functions of feathers?

Procedure

1. In your lab report, make a table like the one on the next page. Record your observations of each kind of feather in your data table.
2. Examine a quill feather. Hold the base of the central shaft with one hand and gently bend the tip of the feather with your other hand. Be careful not to break the feather. Next hold the shaft and wave the feather in the air. Record your observations concerning the structure of the quill feather. Relate your observations to the feather's possible function.

Flight feather

3. Examine the vane of the feather. Does the vane appear to be a solid structure? Include a description of the quill feather's vane structure under "Structure of feather" in your data table.
4. Make a drawing of the quill feather. Label the shaft, vanes, and barbs. Compare your feather with the figure above.
5. Examine a contour feather. Make a sketch of the contour feather in your data table. Label the shaft, vanes, and barbs on your sketch. Does the feather resemble the one in the figure on the next page?
6. Describe the structure of the contour feather under "Structure of feather" in your data table.
7. Examine a prepared slide of a contour feather under low power. Note the smaller barbs, called barbules, extending from each of the barbs.
8. How might you observe the region between the barbs? Locate the tiny hooks at the end of each barbule. Note the arrangement of the hooks on adjacent barbs. Why do you think the hooks are so small? Make a labeled drawing of the hooks in your lab report.
9. Examine the down feather and sketch it in your data table. How does your down feather compare with the figure on the next page?
10. Describe the structure of the down feather in your data table. Do you notice a difference in the structure of the contour and down feathers?

COMPARISON OF FEATHERS

Type of feather	Sketch of feather	Structure of feather	Function of feather
Quill feather			
Contour feather			
Down feather			

Contour feather

Down feather

11. Examine the prepared slide of the down feather under low power. Locate the barbs and barbules. Switch your microscope to high power, and make a labeled drawing of the down feather in your lab report. Does it resemble the one in the figure above?

12. Clean up your materials and wash your hands before leaving the lab.

Analysis and Conclusions

1. What is the function of the shaft? What is the function of the vanes and barbs?
2. How do hooks increase the strength and air resistance of a feather?
3. How is the structure of the quill feather related to its function of aiding flight?
4. Based on your observations, why might down feathers be more effective at keeping a bird warm than the other two feather types you observed?
5. Based on your observations of the structure and function of the feathers, how do you expect to see these feathers oriented on a bird? Explain how orientation of the feather affects the function of the feather.
6. What evolutionary pressure(s) would have caused the evolution of these different types of feathers?

Further Inquiry

Each of the feather types you have examined has a specific structure and function. Review your observations, and try to think of features that account for the efficiency of the three types of feathers.

ANALYSIS AND CONCLUSIONS

1. The shaft is the main central structure of the feather and is where the barbs are attached. The barbs are the side branches that extend from the shaft. A vane is a group of barbs on one side of the feather shaft.
2. Hooks interlock on the feather and keep spaces from developing between the barbs. By keeping the area between the barbs unbroken, the feather resists the force of the wind by diverting the air around the feather rather than through it.
3. The barbs and barbules help maintain the quill feathers' shape and strength. The shape and strength of the quill feathers enable a bird to catch the air, thus enabling the bird to fly.
4. Down feathers have greater loft than quill and contour feathers do, so they supply more insulation. This increased insulation helps keep the bird warm.
5. Students should state that the feathers would be oriented so that they point backward relative to the desired direction of travel.
6. Cold weather could have prompted the evolution of down feathers to enable birds to stay warm. Answers may vary. Accept any answer that cites natural selection as a reason for the evolution of contour and quill feathers.

CHAPTER 45
MAMMALS

Block	Time	TOPICS	TEACHING RESOURCES	LABS, CLASSWORK, AND HOMEWORK
BLOCK 1	45 minutes	**Introducing the Chapter, p. 880**	**ATE** Focus Concept, p. 880 **ATE** Assessing Prior Knowledge, p. 880 **ATE** Understanding the Visual, p. 880	■ Supplemental Reading Guide, *The Dinosaur Heresies*
		45-1 Origin and Evolution of Mammals, p. 881 Key Characteristics, p. 881 Ancestors of Mammals, p. 882	**ATE** Section Overview, p. 881 **ATE** Visual Link Figure 45-3, p. 883 134	**Lab A23,** Observing Some Major Animal Groups ★ Study Guide, Section 45-1 **PE** Section Review, p. 884
BLOCKS 2 & 3	90 minutes	**45-2 Characteristics of Mammals, p. 885** Endothermy, p. 885 Nervous System and Sense Organs, p. 887 Reproduction, p. 888	**ATE** Section Overview, p. 885 **ATE** Critical Thinking, p. 885 **ATE** Visual Link Figure 45-5, p. 886 144A, 181, 183	**Lab A26,** Vertebrate Skeletons **Lab B29,** Fetal Pig Dissection ■ Supplemental Reading Guide, *Through a Window: My Thirty Years with the Chimpanzees of Gombe* ★ Study Guide, Section 45-2 **PE** Section Review, p. 888
BLOCK 4	45 minutes	**45-3 Mammalian Classification, p. 890** Order Monotremata, p. 890 Order Marsupalia, p. 891 Placental Mammals, p. 892	**ATE** Section Overview, p. 890 **ATE** Critical Thinking, p. 891 **ATE** Visual Link Figure 45-13, p. 894 177–178	**PE** **Chapter 45 Investigation,** p. 900 **PE** **Quick Lab,** p. 892, Comparing Gestation Periods ■ Occupational Applications Worksheets—Veterinary Technician ★ Study Guide, Section 45-3 **PE** Section Review, p. 896
BLOCKS 5 & 6	90 minutes	**REVIEW and ASSESSMENT** **PE** Chapter 45 Review, pp. 898–899 ■ Performance-Based Assessment—The Challenge of Water Retention	Audio CD Program	**CHAPTER TESTING** ■ Chapter Test (blackline master) ▲ Test Generator ■ Assessment Item Listing ■ Scoring Rubrics and Classroom Management Checklists

KEY

MODERN BIOLOGY

PE Pupil's Edition
ATE Teacher's Edition
■ Active Reading Guide
★ Study Guide

HOLT BIOSOURCES

Teaching Transparencies

Laboratory Program

One-Stop Planner CD-ROM
Shows these resources within a customizable daily lesson plan:

■ Holt BioSources Teaching Resources
■ Active Reading Guide
★ Study Guide
▲ Test Generator

READING FOR CONTENT MASTERY

Preview: Have students preview the objectives, the introductory paragraph, topic sentences, and captions before reading the chapter.

ATE Active Reading Technique: K-W-L, p. 881
■ Reading Strategy: K-W-L Worksheet
■ Active Reading Guide Worksheet 45-1
PE Summary/Vocabulary, p. 897

■ Active Reading Guide Worksheet 45-2
PE Summary/Vocabulary, p. 897

ATE Active Reading Technique: K-W-L, p. 896
■ Active Reading Guide Worksheet 45-3
PE Summary/Vocabulary, p. 897

TECHNOLOGY AND INTERNET RESOURCES

Holt Biology Videodiscs Teacher's Correlation Guide, Lessons 45-1 through 45-3

Audio CD Program, Sections 45-1 through 45-3

internetconnect

sciLINKS NSTA

On-line Resources: www.scilinks.org
The following *sci*LINKS Internet resources can be found in the student text for this chapter:

Topics:
- Mammals, p. 881
- Therapsids, p. 882
- Placental mammals, p. 896

go.hrw.com

On-line Resources: go.hrw.com
Visit the HRW Web site for a variety of resources related to this chapter. Just type in the keyword HM2 Chapter 45.

Smithsonian Institution®

Visit **www.si.edu/hrw** for additional on-line resources.

CNNfyi.com

Visit **www.cnnfyi.com** for late-breaking news and current event stories selected just for you.

LAB ACTIVITY PLANNING

CHAPTER 45 INVESTIGATION

Mammalian Characteristics, pp. 900–901

OVERVIEW
This laboratory lets students observe some differences between mammals and other vertebrates. It also lets them make observations on their own bodies. Depending on availability, a variety of materials can be used to demonstrate specific features and stimulate thought.

SAFETY
Students should handle slides and laboratory specimens with care. They should also wash their hands before and after counting their teeth.

PREPARATION
Hand lenses, microscopes, microscope slides of mammalian skin, and vertebrate skulls used in this lab are available from biological supply houses. Field guides may be obtained from a bookstore. Provide enough materials for half the class. While half the students perform part C, have the other students perform the other parts of the lab.

Quick Labs

Comparing Gestation Periods, p. 892
Students need only paper and pencils for this activity.

UNDERSTANDING THE VISUAL

Have students examine the opossums. Point out that they are mammals and have many of the traits unique to mammals, including endothermy, hair, mammary glands, a completely divided heart, a single jawbone, and specialized teeth.

FOCUS CONCEPT

Stability and Homeostasis

Mammals are composed of trillions of cells, each of which requires a nearly constant supply of nutrients, oxygen, water, and other substances. The mammalian body has many organs and organ systems that carry out specialized activities in a coordinated and highly regulated manner. Maintaining this level of organization requires a lot of energy.

ASSESSING PRIOR KNOWLEDGE

Review the following concepts.

Aerobic Respiration: *Chapter 7*
Ask students what the reactants and products of aerobic respiration are. Then ask them what form most of the energy not converted into ATP is given off in.

Mammalian Evolution: *Chapter 15*
Ask students what evidence we have that mammals evolved from a common ancestor. Then ask them by what process all of the species of mammals now living on Earth evolve.

CHAPTER 45

MAMMALS

These young opossums depend on their mother to provide them with nourishment from her mammary glands—a unique mammalian trait.

FOCUS CONCEPT: ***Stability and Homeostasis***

As you read the chapter, note the ways mammals meet the demands of their rapid metabolism.

ORIGIN AND EVOLUTION OF MAMMALS

There are about 4,400 species in the class Mammalia. Mammals are classified into more than 20 orders, one of which includes humans. Mammals live on every continent and in every ocean. Some mammals have the ability to fly, while others are exclusively aquatic.

SECTION 45-1

OBJECTIVES

▲ Name and describe six characteristics of mammals.

● Describe the characteristics of the early synapsids.

■ Summarize the importance of therapsids in mammalian evolution.

◆ Describe how dinosaurs affected the evolution of mammals.

KEY CHARACTERISTICS

Six important characteristics of mammals are listed below.

- **Endothermy**—Like birds, mammals produce body heat internally through metabolism. Mammals keep their body temperature high and nearly constant by controlling their metabolism and regulating the loss of heat through the body surface.
- **Hair**—All mammals have hair. The main function of hair is to insulate the body against heat loss. Most mammals (humans and whales are obvious exceptions) are covered with a thick coat of hair. Hair color also serves to camouflage a mammal from predators or a predator from being seen by its prey.
- **Completely divided heart**—Like crocodiles and birds, mammals have a four-chambered heart with two completely separate ventricles. Separate ventricles keep deoxygenated blood from diluting oxygenated blood and allow more efficient pumping of blood through both circuits of the circulatory system.
- **Milk**—Female mammals produce milk to feed their offspring. Milk is a nutritious fluid that contains fats, protein, and sugars. It is produced by the **mammary glands,** which are modified sweat glands located on the thorax or abdomen.
- **Single jawbone**—The lower jaw of a reptile is composed of several bones, but the lower jaw of a mammal is composed of a single bone. This characteristic is particularly important for identifying mammalian fossils because many of the other characteristics of mammals, such as hair and mammary glands, do not fossilize or leave traces on the skeleton.
- **Specialized teeth**—Teeth in different parts of a mammal's jaws are modified for different functions. Those at the front of the jaw are used for biting, cutting, or seizing prey, while those along the sides of the jaw are used for crushing, grinding, or slicing. The teeth of most reptiles are uniform in size and shape, regardless of where they are located in the mouth.

internetconnect

SCILINKS NSTA

TOPIC: Mammals
GO TO: www.scilinks.org
KEYWORD: HM881

SECTION 45-1

SECTION OVERVIEW

Evolution

Section 45-1 explores the evolution of mammals, from the reptilian synapsids that lived more than 300 million years ago to the three major mammalian groups—monotremes, marsupials, and placentals.

Structure and Function

Section 45-1 discusses structures unique to mammals and the advantages they provide mammals.

ACTIVE READING

Technique: K-W-L

Before students read this chapter, have them make short lists of all the things they already **K**now (or think they know) about mammals. Ask them to contribute their entries to a group list on the board or overhead. Then have students list things they **W**ant to know about mammals. Have students save their lists for later use in the Active Reading activity on page 896.

▶ ENGAGE STUDENTS

Using photographs from nature books and magazines, show students a variety of mammals in their habitats. Examples include dolphins, elephants, mice, tigers, anteaters, kangaroos, monkeys, bats, armadillos, deer, seals, beavers, and wolves. Ask students to infer what visible characteristic enables each mammal to survive in its habitat. Have students give reasons for their inferences (for example, bat wings for flight, dolphin flippers for swimming, tiger stripes for camouflage, and armadillo armor for protection).

GIFTED ACTIVITY

Have students use library and online resources to research the origin of endothermy. (Originally dinosaurs were thought to be ectotherms. In the late 1980s, scientists presented arguments to support the idea that dinosaurs were endotherms. Other scientists think dinosaurs had a physiology between endothermy and ectothermy.) Have students hypothesize about whether endothermy arose once, in a reptilian ancestor, or twice, in the mammals and the birds.

RECENT RESEARCH

Early Mammalian Teeth

Scientists studying early mammalian teeth believe that a dental feature that first appeared about 65 million years ago was key to the success of the mammals. This feature is known as the "hypocone," a fourth cusp that squared off upper molars. This adaptation made the mammalian mouth much better suited to grinding plant material. The presence of the hypocone is thought to be directly related to the explosive radiation of mammalian species that occurred 40 million years ago.

INCLUSION ACTIVITY

Have students prepare posters to illustrate mammalian characteristics. Provide students with old nature magazines, such as *National Geographic, National Wildlife,* and *Sierra,* that they can cut pictures out of. Have students label the photographs with the characteristic(s) that the photograph is meant to illustrate. Have students present their poster to the class and briefly talk about the characteristic(s) illustrated on their poster. Examples could include a polar bear (endothermy), a sheep (hair), a lion (teeth), an ape with her young (viviparity), and piglets suckling (mammary gland).

FIGURE 45-1

This *Dimetrodon* is an example of a synapsid. Note the sail-like structure on its back. Scientists think that it was probably filled with blood vessels and that it acted as a temperature-regulating device that could absorb heat during the day. *Dimetrodon* could grow up to 3.5 m (11 ft) in length.

ANCESTORS OF MAMMALS

The ancestors of mammals first appeared on Earth more than 300 million years ago. At that time, a major evolutionary split occurred in the terrestrial vertebrates, producing two groups of animals. Members of the first group gave rise to dinosaurs, birds, and all the living reptiles. Members of the second group, known as **synapsids,** include mammals and their closest relatives. Mammals are the only surviving synapsids. Early synapsids can be distinguished by the structure of their skull. There is a single opening in the outer layer of the skull just behind the eye socket. This same type of skull is found in all later synapsids, including mammals, although often in a highly modified form.

The first synapsids were only about 50 cm (20 in.) in length and resembled modern lizards. By the early Permian period (286–245 million years ago), an assortment of large synapsids, some of which reached 4 m (13 ft) in length and weighed more than 200 kg (440 lb), had appeared. Figure 45-1 shows a reconstruction of a carnivorous synapsid known as *Dimetrodon.* In most reptiles, teeth throughout the jaw are uniform in size and shape. But in many of the early carnivorous synapsids, including *Dimetrodon,* the teeth are specialized to a small degree, with long stabbing teeth in the front of the mouth and smaller teeth toward the back.

internet**connect**

SCILINKS NSTA

TOPIC: Therapsids
GO TO: www.scilinks.org
KEYWORD: HM882

Therapsids

Among the synapsids that appeared later in the Permian period were the **therapsids,** the group that gave rise to mammals. Figure 45-2 shows a rendering of a therapsid. Therapsids were the most

FIGURE 45-2

Lycaenops was a carnivorous therapsid. Its legs supported its body more directly than did those of the early synapsids. The repositioning of the legs also improved its speed, making it a better predator.

abundant terrestrial vertebrates during the late Permian period. They survived through the Triassic period (245–208 million years ago) and into the Jurassic period (208–144 million years ago).

The evolution of mammals from therapsids involved changes in anatomy, physiology, ecology, and behavior. The rich and unusually complete fossil record of therapsids preserves many transitional forms that bridge the gap between therapsids and mammals. By studying these fossils, scientists can follow the anatomical changes that occurred during this transition and infer when some of the physiological, ecological, and behavioral changes must have happened.

Many of the features we associate with mammals evolved first among therapsids. For example, many therapsids had teeth differing in size, shape, and function, depending on their position in the jaw. The limbs of mammals are positioned directly beneath the body, as were the limbs of many therapsids. Some therapsids were probably endothermic, and some may have had hair.

Early Mammals

The first mammals and the first dinosaurs appeared at about the same time, during the Triassic period, and the dinosaurs had an important influence on the evolution of mammals for more than 150 million years. Figure 45-3 shows a reconstruction of an early mammal. The early mammals were about 10 cm (4 in.) long and

FIGURE 45-3

The first mammals, such as this *Eozostrodon,* had to compete with the larger, dominant dinosaurs. These first mammals functioned at night, when dinosaurs were probably less active.

TEACHING STRATEGY

Comparing Shared Characteristics

Provide students with pictures of a therapsid, an early mammal, reptiles (such as a dinosaur and a modern lizard), and modern mammals (such as a rat or human). Ask students to identify characteristics that the therapsid shared with reptiles and those that were different. (Shared characteristics in most include ectothermy and lack of hair. Differing characteristics include specialized teeth and limbs positioned beneath the body of therapsids, but not of reptiles.) Then ask students to identify similarities and differences between the early mammals and modern mammals. (Shared characteristics include endothermy, hair, specialized teeth, and a four-chambered heart. Differing characteristics include size and diet.)

QUICK FACT

☑ During the evolution of mammals from reptiles, the lower jawbones of reptiles evolved to become bones in the middle ear, leaving a single jawbone.

VISUAL LINK

Figure 45-3

Have students examine the *Eozostrodon.* Ask them what characteristics the first mammals shared with their reptilian ancestors (both were vertebrates). Ask students what characteristics the first mammals had that their reptilian ancestors did not have (specialized teeth, hair, endothermy, mammary glands, small size, arboreal habitats, nocturnal behavior, and an insectivorous diet).

SECTION 45-1

RETEACHING *ACTIVITY*

Have students redraw the phylogenetic tree of vertebrates shown in Chapter 41. Ask them to add to the tree information they have learned from this chapter about the details of mammalian evolution. (This should include a synapsid as a common ancestor of both reptiles and mammals, a therapsid as a more recent common ancestor of both reptiles and mammals, and an early mammalian ancestor.)

ANSWERS TO SECTION 45-1 REVIEW

1. Hair insulates the body helping to retain body heat.
2. The jaw of a mammal is made of one bone instead of several, and it contains specialized teeth of different sizes and shapes.
3. They both have the synapsid type of skull, in which there is one opening in the side of the skull behind the eye socket.
4. Therapsids had legs positioned underneath the body, specialized teeth, and a synapsid skull. They also may have had hair and been endothermic.
5. Mammals were probably nocturnal, avoiding competition from and predation by dinosaurs.
6. The fragile skeletons of small mammals are hard to find, and this might be leading scientists to underestimate the abundance of early mammals. Their apparent rarity might be a consequence of infrequent preservation and discovery.

Word Roots and Origins

monotreme

from the Greek *monos,* meaning "single," and *trema,* meaning "hole"

20–30 g (0.5 oz) in mass, or about the size of a mouse. Fossil skeletons show that early mammals had large eye sockets, which suggests that these creatures probably were active at night. Their teeth indicate that they probably fed on insects. By hiding during the day, mammals would have avoided predation by dinosaurs, and by feeding on insects, they would not have competed with dinosaurs for food. Although there is no direct evidence of mammary-gland development in early mammals, scientists infer from similarities of mammary tissue among different mammalian orders that mammary glands were present as early as the late Triassic period.

During this time, dinosaurs filled the majority of terrestrial habitats. Mammals remained small and probably were neither abundant nor diverse. By the end of the Jurassic period, at least five orders of mammals had evolved. And by the middle of the Cretaceous period (144–65 million years ago), the three main groups to which modern mammals belong had appeared. The first group, the **monotremes,** are oviparous, that is, they lay eggs. The second group is the **marsupials.** Marsupials are viviparous, which means they give birth to live young, but the period of development within the mother is short. **Placental mammals,** the third group, are also viviparous, but the young are born at a much later stage of development than are marsupial offspring. As they develop inside the mother's uterus, placental young are nourished through a structure called the placenta.

Diversification of Mammals

The dinosaurs became extinct about 65 million years ago, at the end of the Cretaceous period. This opened up many new habitats and resources to mammals, and mammals took over many of the ecological roles previously fulfilled by dinosaurs. For example, during the Jurassic and Cretaceous periods, most of the large terrestrial carnivores and herbivores were dinosaurs. Today, nearly all large terrestrial carnivores (lions, tigers, and wolves, for example) and herbivores (elephants, giraffes, cattle, and horses, for example) are mammals.

SECTION 45-1 REVIEW

1. How does hair help a mammal maintain its body temperature?
2. Identify two differences between the jaw of a mammal and the jaw of a reptile.
3. What is one characteristic shared by mammals and the early synapsids?
4. List two characteristics of therapsids.
5. Why was it advantageous for the early mammals to be nocturnal?
6. **CRITICAL THINKING** Scientists think that mammals were rare during the time of the dinosaurs because there are few fossils of mammals from that time period. However, the early mammals were very small, and their fossils are very fragile. How might these factors influence estimates of the abundance of early mammals?

CHARACTERISTICS OF MAMMALS

Mammals inhabit a wide range of habitats on land and in the water. They can survive in some of the coldest and hottest environments on Earth. Adaptations have enabled mammals to inhabit such diverse environments.

SECTION

45-2

OBJECTIVES

▲ Describe the advantage of endothermy in mammals.

● Describe two features of the mammalian respiratory and circulatory systems that help sustain a rapid metabolism.

■ Describe two mammalian adaptations for digesting plants.

◆ Compare and contrast reproduction in monotremes, marsupials, and placental mammals.

ENDOTHERMY

Endotherms generate heat internally by breaking down food, while ectotherms absorb most of their heat from their surroundings. Endothermy determines many of the capabilities of mammals. For instance, mammals can live in cold climates and be active. Perhaps more important, their rapid metabolism provides mammals with the energy to perform strenuous activities for long periods of time. As a general rule, ectotherms are not capable of sustained activity because they tire quickly.

The organ systems of mammals also differ from the organ systems of ectotherms. Figure 45-4 shows the internal anatomy of a typical mammal.

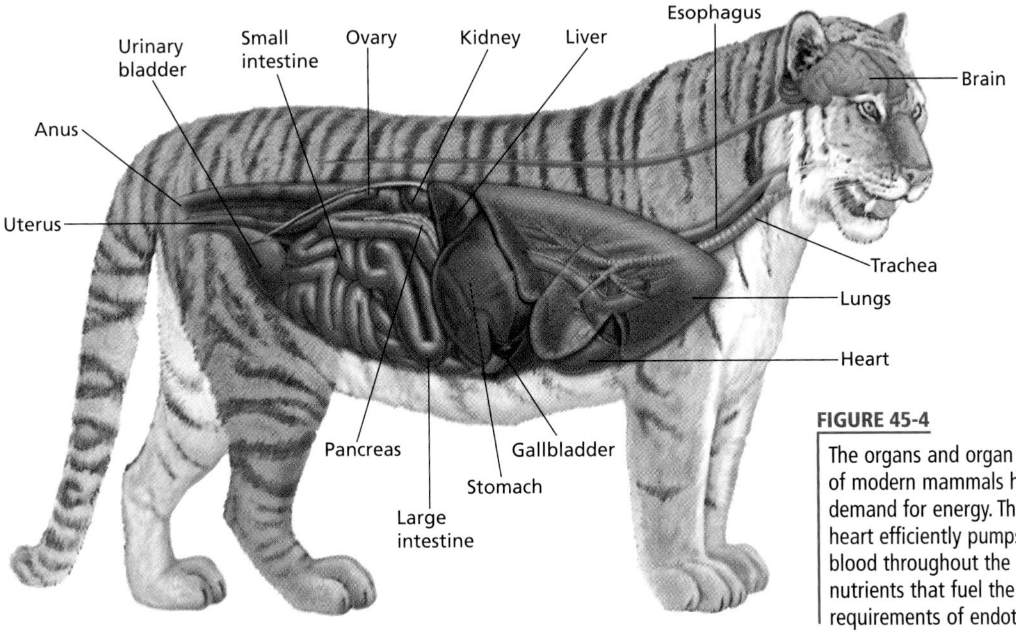

FIGURE 45-4

The organs and organ systems of modern mammals have a high demand for energy. The mammalian heart efficiently pumps oxygenated blood throughout the body, delivering nutrients that fuel the high-energy requirements of endothermy.

SECTION 45-2

SECTION OVERVIEW

Stability and Homeostasis
Section 45-2 discusses the various adaptations of mammals that enable them to maintain endothermy.

Reproduction and Inheritance
Section 45-2 discusses the various ways that mammals produce offspring and care for them.

▶ ENGAGE STUDENTS

Bring to class a small live mammal, such as a rat or a hamster, in a cage containing a wheel, food, and water. CAUTION **Do not allow students to handle the animal.** Because most small mammals are nocturnal, you may have to take the animal out of the cage to get it to do anything. Have students observe the mammal and make notes on activities that provide information about its physiology and the environment in which it lives. (The animal eats a lot of food, breathes rapidly, curls up to keep warm when it sleeps, exercises, urinates and defecates its wastes, has a rapidly beating heart, and is keenly attentive and responsive to its environment.) Ask students how they would expect the activities of a nonmammalian vertebrate to differ. (It would eat less, be less active, warm itself in the sun, and breathe less rapidly.)

CRITICAL THINKING

Metabolic Rate and Size

Ask students which has the higher metabolic rate, small or large mammals. Then ask them to explain why. (Small mammals have a higher metabolic rate because their bodies have a higher surface-to-volume ratio and therefore lose heat at a higher rate.)

VISUAL LINK

Figure 45-5

Have students examine Figure 45-5. Trace the pathways of blood through the heart, and explain to them how the four-chambered mammalian heart keeps oxygenated blood completely separate from deoxygenated blood. Tell students that deoxygenated blood returns to the heart from body tissues. It enters the right side and is pumped from the right ventricle to the lungs, where it is oxygenated. Then tell them that oxygenated blood is returned to the heart, enters the left side, and is pumped from the left ventricle to the body tissues. Emphasize that the left and right ventricles are completely separate.

QUICK FACT

☑ High fevers can be harmful to mammals, but recent evidence suggests that moderate fevers may accelerate the body's destruction of invading microorganisms and lead to recovery.

TEACHING STRATEGY

Comparing Mammalian Teeth

Provide students with unlabeled mammalian skulls or unlabeled pictures of mammalian skulls with teeth showing. (Skulls may be bought from a biological supply company.) Include skulls of carnivores, such as a cat, a hyena, or a shrew; herbivores, such as a rabbit or a pig; and omnivores, such as a rat or a dog. Point out the single jawbone on skulls that have jawbones. Have students note the presence or absence, number, and length of incisors in the front, which are used for biting; canines on the side, which are used for tearing flesh; and molars in the back, which are used for crushing or grinding plant material or flesh. Ask students to infer each animal's diet and feeding behavior from its teeth.

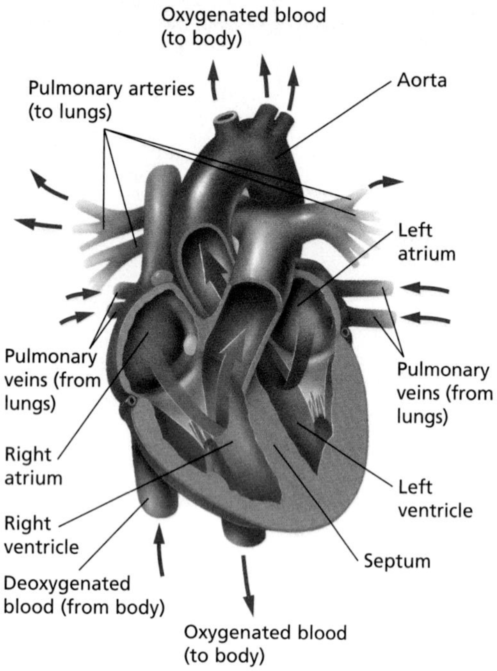

FIGURE 45-5

The red arrows show the path of oxygenated blood through the heart, and the blue arrows show the path of deoxygenated blood. As you can see, the ventricle is completely divided by a septum. As a result, the blood pumped to the body contains a higher percentage of oxygen than results from blood pumped by a reptilian or fish heart.

Because of its faster metabolism, a mammal needs more oxygen and food than does a reptile of the same size. Heat is constantly escaping from a mammal's body through its skin and exhaled air. Because producing heat requires so much energy, mammals need to limit their loss of body heat to the environment. Hair and fat layers serve as insulation for most mammals. For example, mammals that live in very cold climates usually have a very thick coat of hair. Large tropical mammals, such as elephants, have very little hair.

Circulatory System

The structure of the mammalian heart helps ensure the efficient flow of blood throughout the body. As you can see in Figure 45-5, the heart has two atria and two completely separated ventricles. Note that oxygenated and deoxygenated blood never mix. Recall that in the heart of a lizard or turtle, deoxygenated blood and oxygenated blood can mix when the animal is inactive. However, mammals have such a high demand for oxygen that they cannot tolerate the dilution of oxygenated blood by deoxygenated blood.

Respiratory System

The respiratory system of a mammal is adapted for efficient gas exchange. The lungs are large and contain millions of alveoli, the small chambers in which gas exchange occurs. Compared with the lung of a reptile, the lung of a mammal has a much larger surface area available for gas exchange. For example, the internal surface area of a pair of human lungs is about 140 m^2 (1,500 ft^2). Another adaptation that contributes to the efficiency of respiration is the **diaphragm,** a sheet of muscle below the rib cage. Contraction of the diaphragm during inhalation helps draw air into the lungs.

Feeding and Digestion

For most mammals, the breakdown of food begins with chewing in the mouth. Other vertebrates simply swallow their food whole or in large pieces. Chewing speeds up digestion by breaking food into small pieces that can be more easily digested by enzymes. Most mammals have specialized teeth for gripping and holding food, and for chewing. Chisel-like **incisors** cut. Pointed **canines** grip, puncture, and tear. **Premolars** shear, shred, cut, or grind. **Molars** grind, crush, or cut. Variations in the size and shape of teeth among different mammalian species reflect differences in diet. For instance, notice the differences between the teeth of a carnivore and those of a herbivore, as shown in Figure 45-6.

Baleen whales, such as the blue whale, do not have teeth. Instead they have **baleen,** thin plates of fingernail-like material that hang from the roof of the mouth. As a baleen whale swims, it gulps huge quantities of water. Then it closes its mouth and pushes the water through the baleen. Shrimp and other small invertebrates are trapped in the baleen and then swallowed.

Adaptations for Digesting Plants

Plants can be difficult to digest because they contain large amounts of cellulose, which is a polymer of glucose, and no vertebrates produce enzymes to break down cellulose. However, some mammals can digest cellulose with the aid of microorganisms. Cows, sheep, goats, giraffes, and many other hoofed mammals have a large stomach with four chambers. The first chamber, known as the **rumen** (ROO-muhn), contains symbiotic bacteria and other microorganisms. After plant material is chewed and swallowed, it enters the rumen. Microorganisms then begin to break down the cellulose. The food is further digested in the rumen, then regurgitated, chewed again, and swallowed again (a cow chewing its cud is actually rechewing partially digested food). Food may be regurgitated and swallowed several times. Microorganisms that live in the rumen break down cellulose into small molecules that are eventually absorbed into the animal's bloodstream when the food reaches the small intestine.

In horses, rodents, rabbits, and elephants, microorganisms that live in the **cecum** (SEE-kuhm) complete digestion of the food. The cecum is a large sac that branches from the small intestine and acts as a fermentation chamber. Food passes through the stomach and small intestine before entering the cecum. The human appendix is a vestigial cecum. Mammals with a cecum do not chew cud.

NERVOUS SYSTEM AND SENSE ORGANS

The brain of a mammal is about 15 times heavier than the brain of a similarly sized fish, amphibian, or reptile. Enlargement of one part of the brain, the cerebrum, accounts for most of this size increase. In mammals, the cerebrum is the largest part of the brain. The cerebrum's surface is usually folded and wrinkled, which greatly increases its surface area without increasing its volume. The cerebrum evaluates input from the sense organs, controls movement, and initiates and regulates behavior. It is also involved in memory and learning. Humans have the highest brain-to-body-size ratio.

Like other terrestrial vertebrates, mammals depend on five major senses: vision, hearing, smell, touch, and taste. The importance of each sense often depends on the mammal's environment. For instance, most bats, which are active at night, rely on sound rather than vision for navigating and finding food. Using a process called **echolocation** (EK-oh-loh-KAY-shuhn), these bats emit high-frequency sound waves, which bounce off objects, and then they analyze the returning echoes to determine the size, distance, direction, and speed of the objects.

FIGURE 45-6

Carnivores, such as this lion, have large, sharp incisor and canine teeth that are able to cut and tear flesh. Herbivores, such as the zebra, have flat teeth that are useful for grinding grass and grain, which are the staples of their diet.

Word Roots and Origins

cecum

from the Latin *intestinum caecum,* meaning "blind intestine"

INCLUSION ACTIVITY

Ask students to choose one herbivorous and one carnivorous mammal and to draw pictures of them on poster board. Have students list under each picture the adaptations that make each mammal well adapted to its environment. (Students should list the organ systems described in this chapter and the differences in tooth types and digestive systems of the herbivore and the carnivore.)

RETEACHING ACTIVITY

Ask students to infer how a mammal would be affected if one of the highly adapted systems described in this section were like that of a reptile instead of a mammal. (For example, a mammal with a three-chambered heart would not get sufficient oxygen to its body tissue and would therefore be less active and less able to maintain its body temperature.) Ask them if the animal could still survive in its environment. (Answers will vary, depending on the animal and the system chosen.)

CONTENT CONNECTION

Section 5-1: Diffusion

Oxygen and carbon dioxide waste products move across cell membranes in the lungs and body tissues by diffusion, which requires only that there be a concentration gradient across the membrane. Water moves into cells by osmosis; this requires that the solute concentration inside the cell be different from that outside the cell. Nutrient products of digestion move across body cell membranes by either facilitated diffusion, which requires a diffusion gradient and a transport protein, or active transport, which does not require a diffusion gradient but does require a transport protein and energy input.

RECENT RESEARCH

Menstruation

A scientist has challenged the current hypothesis that menstruation in female mammals evolved as a way of eliminating harmful microorganisms from the reproductive tract. She hypothesizes that menstruation eliminates energy-consumptive tissue from the body when it is not needed (when the one egg that matures per month is unfertilized).

ANSWERS TO SECTION 45-2 REVIEW

1. A mammal needs more food to fuel its rapid metabolism.
2. The mammal's heart has two ventricles completely divided from each other, instead of one partially divided ventricle, as in a lizard. The conus arteriosus is completely absent in the mammal, but it remains as a vestige in the lizard.
3. The rumen contains microorganisms that break down the cellulose from eaten plants into small molecules that can be absorbed by the mammal.
4. Bats rely on hearing to find their way through their environment, and large ears help them hear better by capturing more sound.
5. Monotremes lay eggs, while marsupials give birth to live young. Marsupial and monotreme young are only partially developed, and the mothers of both suckle their young.
6. The relationship is mutualistic. The microorganisms help the cow by digesting cellulose. They benefit by getting a warm place to live and a supply of food.

FIGURE 45-7

Marsupials, such as this young kangaroo, are born soon after conception and must crawl into their mother's pouch for the rest of their development. The level of development is so low that many marsupials are born before separation of the heart ventricle is complete. Other organs not yet completed include lungs, kidneys, and major nerves.

REPRODUCTION

Each main group of mammals—monotremes, marsupials, and placental mammals—has a unique reproductive pattern. A female monotreme typically lays one or two large eggs that have thin leathery shells and incubates them with her body heat. The developing embryo is nourished by the yolk within the egg. At hatching, a monotreme is very small and only partially developed. Its mother protects it and feeds it milk from her mammary glands until it is ready to survive on its own.

Marsupials are characterized by a very short period of development within the mother's uterus. Newborn marsupials emerge from the uterus after they have completed enough development to be able to survive outside the uterus. The newborn offspring of a kangaroo, for example, is only 2 to 3 cm (1 in.) long. Figure 45-7 shows a marsupial soon after birth. The newborn crawls into the mother's pouch, located on her abdomen, and attaches to a nipple. The newborn's development and growth then continue in the pouch.

Placental mammals, the group that includes cats, dogs, humans, and most other familiar mammals, give birth to their young after a much longer period of development. During this period, the mother provides nourishment and oxygen to her developing offspring through the placenta. The placenta begins to form shortly after fertilization, when the fertilized egg attaches to the lining of the uterus. Extensions from the chorion, the outer membrane surrounding the egg, grow into the lining of the uterus, and blood vessels from the uterus surround these extensions. When completely formed, the placenta contains tissue and blood vessels from both the mother and the offspring. Nutrients and oxygen diffuse from the mother's blood into the blood of the offspring, and carbon dioxide and other wastes diffuse from the offspring into the mother's blood.

Newborn mammals are nourished with milk from the mother's mammary glands. Young mammals are dependent on the mother for food and care.

SECTION 45-2 REVIEW

1. **A mammal eats about 10 times as much food as a lizard of the same size. Explain this difference.**
2. **How does the heart of a mammal differ from the heart of a lizard?**
3. **What is the function of the rumen in hoofed mammals?**
4. **Bats, which are active only at night, often have very large ears. Why might large ears be advantageous to a bat?**
5. **Identify one difference and one similarity between the reproductive patterns of monotremes and marsupials.**
6. **CRITICAL THINKING What sort of ecological relationship exists between a cow and the microorganisms living in its rumen? Explain your answer.**

Literature & Life

Gorillas in the Mist

This excerpt is from *Gorillas in the Mist,* by Dian Fossey.

During the early days of the study at Kabara, it was difficult to establish contacts because the gorillas were not habituated or accustomed to my presence and usually fled on seeing me. I could often choose between two different kinds of contacts: obscured, when the gorillas didn't know I was watching them, or open contacts, when they were aware of my presence.

Obscured contacts were especially valuable in revealing behavior that otherwise would have been inhibited by my presence. The drawback to this method was that it contributed nothing toward the habituation process. Open contacts, however, slowly helped me win the animals' acceptance. This was especially true when I learned that imitation of some of their ordinary activities such as scratching and feeding or copying their contentment vocalizations tended to put the animals at ease more rapidly than if I simply looked at them through binoculars while taking notes. I always wrapped vines around the binoculars in an attempt to disguise the potentially threatening glass eyes from the shy animals. With gorillas, as is often the case with humans, direct staring constitutes a threat.

Not only was it necessary to get the gorillas accustomed to the blue-jeaned creature who had become a part of their daily lives, it was also very necessary for me to know and recognize the particular animals of each group as the amazing individuals they were. Just as George Schaller had done some seven and a half years before me, I relied heavily upon "noseprints" for identification purposes. There is a tendency for the gorillas of each group to resemble one another, especially within matrilineal lines. As no two humans have exactly the same fingerprints, no two gorillas have the same "noseprint"—the shape of the nostrils and the outstanding troughs seen on the bridges of their noses. Since the gorillas initially were unhabituated, I had to use binoculars, but even from a distance I could quickly make sketches of noseprints seen on the more curious group members peeking back at me from partially hidden positions in the dense vegetation. These sketches proved invaluable at a time when close-up photography was out of the question. Also, I would have needed a third hand in order to manage a camera, binoculars, and note taking, not to mention carrying on with the imitative routine of feeding, scratching, and vocalizing needed to relax the gorillas as well as to arouse their curiosity.

Reading for Meaning

What do you think Fossey means when she says the gorillas "were not habituated"?

Read Further

Gorillas in the Mist tells about Dian Fossey's experiences during 13 years she spent observing African mountain gorillas in their natural habitat. How did her growing understanding of the gorillas help her become almost a part of their community?

BACKGROUND

This feature presents an excerpt from a book by primatologist Dian Fossey. The mountain gorilla, *Gorilla gorilla beringei,* is endangered. Its forest habitat has been destroyed through increased farming, grazing, and lumbering. The estimated population of mountain gorillas is 500 to 1,000. Dian Fossey studied occupational therapy in college. After graduating in 1954, she worked at a children's hospital. She became interested in doing research on apes after reading about them. In 1963, she met Louis Leakey (see **Great Discovery** in Chapter 17) in East Africa. He made it possible for her to stay in Africa and study gorillas. In 1967, she established a research center in the Republic of Congo (formerly Zaire) but was forced to leave because of political instability. She fled to Rwanda, where she established the Karisoke Research Center in the Virunga Mountains. As a result of her work, Fossey earned a Ph.D. from Cambridge University in 1974 and subsequently taught at Cornell University, in New York. She was murdered at her Virunga Mountains campsite in 1985.

READING FOR MEANING

She means that the gorillas were not accustomed to her presence in their group. As a result, the gorillas fled on seeing her.

READ FURTHER

She felt that her understanding and imitation of some of their ordinary behaviors (such as feeding, scratching, and vocalizing) helped to habituate the animals to her presence. It was easier for her to recognize each gorilla as a unique individual when she became "a part of their daily lives." Recognizing each gorilla helped her understand the behavior of the whole group.

SECTION 45-3

SECTION OVERVIEW

Structure and Function
Section 45-3 describes the various structures of mammals and how they enable mammals to function in their environment.

Evolution
Section 45-3 discusses how geographic isolation led to the evolution of marsupials.

▶ ENGAGE STUDENTS

Obtain a wall-sized physical map of the world that students can write on, and mount it on a wall in the classroom. Provide students with pictures of a variety of mammals from different habitats, including monotremes, marsupials, and placental mammals. Have each student choose a mammal, take the picture, and use library or on-line resources to learn about its geographic range. Have students shade in their animal's range on the world map and attach the picture, name, and order of the mammal on or near its range area. (Give each student a different-colored pencil.) Discuss with students the diverse types of mammals and the habitats in which they live. Remind students that endothermic animals can regulate their body temperature and thus live in a wide variety of climates and habitats.

CONTENT CONNECTION

Section 18-1: Taxonomy

All mammals are in the class Mammalia, chordates are in the phylum Chordata, and all animals are in the kingdom Animalia.

SECTION 45-3

OBJECTIVES

▲ Name the orders of mammals, and give an example of each.

● Describe two differences between marsupials and monotremes.

■ Distinguish between artiodactyls and perissodactyls.

◆ Identify two orders whose members are aquatic, and describe one adaptation for aquatic life shown by each order.

MAMMALIAN CLASSIFICATION

The roughly 4,400 species of mammals are divided into more than 20 orders. One order contains the monotremes, one order contains marsupials, and the remaining orders contain placental mammals.

ORDER MONOTREMATA

Monotremes are the only egg-laying mammals. Just three species exist, one platypus and two anteaters, or echidna, and they have a very restricted distribution. Monotremes live only in Australia and New Guinea. Figure 45-8 shows an example of a platypus and an echidna. Monotremes are considered to be the most primitive of all the mammals, as is illustrated in the phylogenetic tree in Figure 45-9.

The duckbill platypus is adapted to life in water. It has waterproof fur, webbed feet, and a flattened tail that aids in swimming. Its most distinctive characteristic is its flat, sensitive, rubbery muzzle, which it uses to root for worms, crayfish, and other invertebrates in soft mud. The female platypus digs a den in the bank of a river or stream to lay her eggs. She curls around the eggs to protect and incubate them. After hatching, the newborns lick milk from nippleless mammary glands on the mother's abdomen.

Echidnas are terrestrial. They have a coat of protective spines and a long snout for probing into anthills and termite nests. Echidnas incubate their eggs in a pouch on the belly.

FIGURE 45-8
The duckbill platypus (a) and the spiny anteater (b) are two of the three species in the order Monotremata. The platypus's scientific name is *Ornithorhynchus anatinus,* and the anteater's scientific name is *Tachyglossus aculeatus.*

(a)

(b)

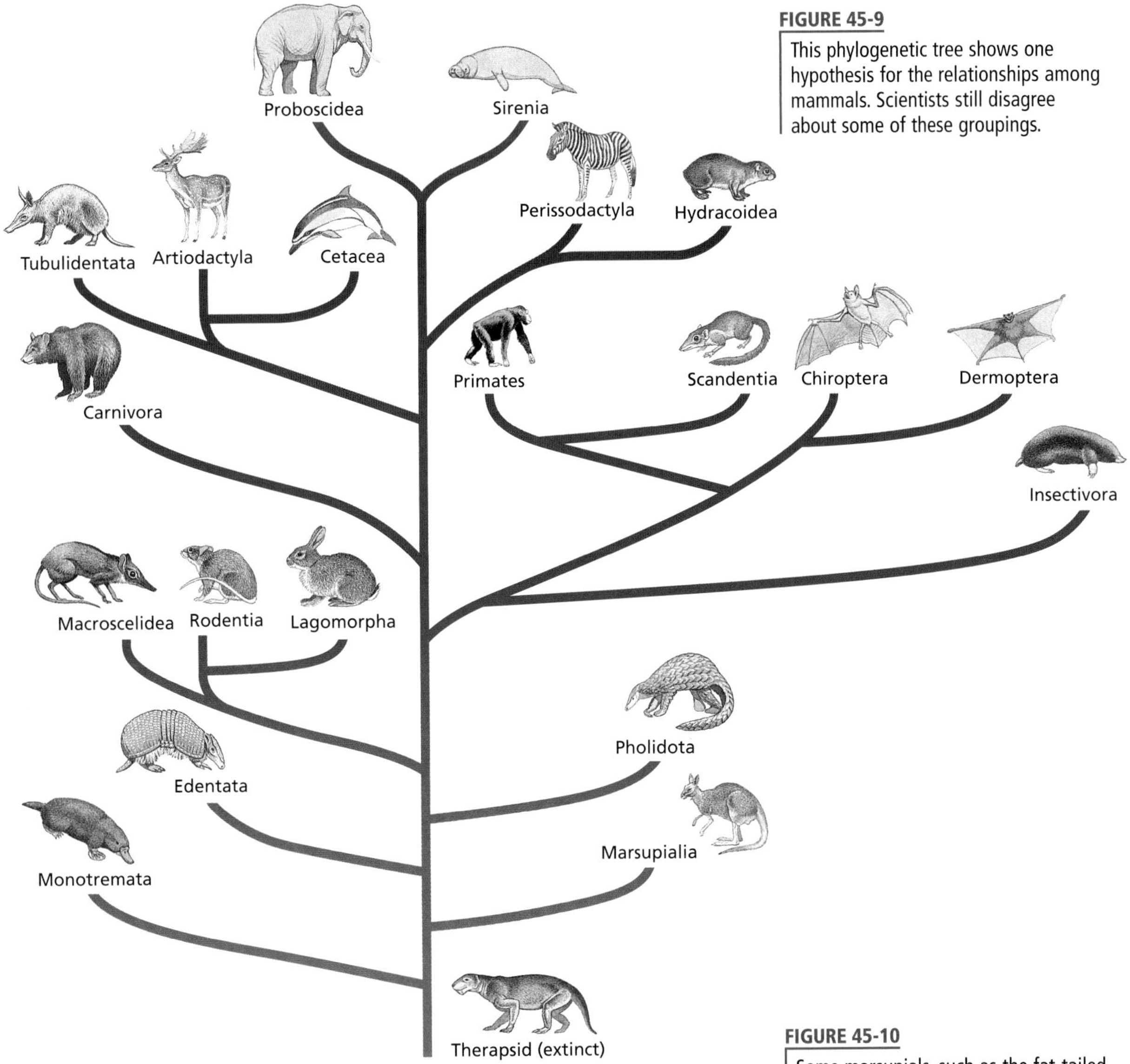

FIGURE 45-9

This phylogenetic tree shows one hypothesis for the relationships among mammals. Scientists still disagree about some of these groupings.

ORDER MARSUPIALIA

About 280 marsupial species exist today, inhabiting Australia, New Guinea, and the Americas. The United States has only one marsupial species, the Virginia opossum. Figure 45-10 shows an example of a marsupial.

These unique mammals evolved during a long period of isolation. Recall that Australia and New Guinea drifted away from the other continents more than 40 million years ago. At that time, placental mammals were apparently rare in the region. According to the fossil record, marsupials once dominated South America; they were gradually displaced by placental mammals, and many species became extinct.

FIGURE 45-10

Some marsupials, such as the fat-tailed dunnart, eat insects and small vertebrates. The marsupial line split from the placental-mammal line during the late Jurassic or early Cretaceous periods.

DEMONSTRATION

To demonstrate convergent evolution in mammals, provide students with pictures of pairs of mammals. Examples include the wombat and the groundhog, the flying phalanger and the flying squirrel, and the cuscus and the tamarin monkey. Ask students to examine the pictures and describe adaptations common to members of each pair (for example, the ability of both the flying phalanger and the flying squirrel to glide through the air). Point out that although the members of each pair are similar in many ways, they are only distantly related because their similarities result from convergent evolution.

CRITICAL THINKING

Early Competition

Ask students to explain what evolutionary mechanism permitted marsupials to succeed in Australia but not in the rest of the world. (When Australia drifted away from the other continents 40 million years ago, placental mammals were apparently rare so marsupials faced no competition from them there. However, marsupials that remained on the American continents were apparently unable to successfully compete with them. As a result, marsupials are seen less frequently than placental animals on these continents.)

Quick Lab

Comparing Gestation Periods

Time Required 20 minutes

Procedural Tips Have each student work independently and then compare his or her answers with another student's work. Each student may rearrange their table differently. As a class, discuss the advantages and disadvantages of each new arrangement.

Answers to Analysis

1. In general, the longer the gestation period, the fewer offspring are born.
2. Student hypotheses will vary but could include complexity or size related to the number of offspring.
3. The table could be rearranged to show the gestation period or number of offspring per pregnancy in decreasing or increasing numbers.

INCLUSION *ACTIVITY*

Have each student make a table dividing the placental mammal orders into those primarily active on land, in water, and in air. (Land orders include Insectivora, Rodentia, Lagomorpha, Edentata, Carnivora, Artiodactyla, Perissodactyla, and Proboscidea; water orders include Cetacea and Sirenia; and the only flying order is Chiroptera.) Challenge students to find examples of species that are exceptions to this division. (Hippopotamuses are artiodactyls and are primarily active in water; sea otters are carnivores and are primarily aquatic.)

FIGURE 45-11

These porcupines are members of the order Rodentia. The range of the North American porcupine, *Erithizon dorsatum,* on the left, extends from Canada to northern Mexico. The Brazilian porcupine, *Coendou prehensilis,* on the right, has a prehensile tail that enables it to maintain a firm grip on tree branches.

Quick Lab

Comparing Gestation Periods

Materials paper, pencil

Procedure

1. Make a table of gestation periods for different mammals. Make three columns labeled "Mammal," "Gestation period," and "Offspring per pregnancy."
2. Fill in your table with the following data:

Bat, 210 days, 1 offspring

Gerbil, 19–21 days, 4–7 offspring

Horse, 332–342 days, 1 offspring

Monkey, 226–232 days, 1 offspring

Squirrel, 44 days, 3 offspring

Rabbit, 31 days, 3–6 offspring

Whale, 420–430 days, 1 offspring

Wolf, 63 days, 4–5 offspring

Analysis

1. State a generalization about the relationship of the length of the gestation period to the number of offspring per pregnancy.
2. Propose a hypothesis to explain this relationship.
3. Rearrange your table, if necessary, to show this relationship more clearly.

PLACENTAL MAMMALS

Placental mammals are a diverse group that comprises at least 18 orders. Ninety-five percent of all mammal species are placental mammals. They live on land, in water, and in the air.

Order Rodentia

Rodentia, the largest mammalian order, includes more than 1,800 species, or about 40 percent of all placental mammals. Rodents flourish on every continent except Antarctica and are adapted to a wide range of habitats. Squirrels, marmots, chipmunks, gophers, muskrats, mice, rats, and porcupines are rodents. Rodents have only two pairs of incisors, which continue to grow as long as the rodent lives. Rodent incisors are sharp, an adaptation for the rodent's diet of hard seeds, twigs, roots, and bark. As a rodent gnaws, the back surface of the tooth wears away faster than the front surface, maintaining the tooth's chisel-like edge. Rodents also have a very high reproductive capacity. Figure 45-11 shows two examples of rodents.

Order Edentata

The order Edentata is made up of about 30 living species, including anteaters, armadillos, and sloths. These mammals are found in southern North America, Central America, and South America. The name of this order means "toothless." However, only anteaters are completely toothless; armadillos and sloths have teeth, although the teeth are peglike and lack enamel. Most edentates feed on insects, which they capture with a long, sticky tongue. Their powerful front paws have large, sharp claws, which they use to rip open anthills and termite nests. Armadillos supplement their insect diet with small reptiles, frogs, mollusks, and carrion. Sloths, on the other hand, are herbivores; their continuously growing teeth are an adaptation for grinding plants.

Order Lagomorpha

About 70 species make up the order Lagomorpha, which includes rabbits, hares, and pikas. Lagomorphs are found worldwide. They differ from rodents in that they have a double row of upper incisors, with two large front teeth backed by two smaller ones. Lagomorph teeth continue to grow throughout the animal's lifetime. Such teeth are an adaptation to the lagomorph's herbivorous diet.

Order Insectivora

The order Insectivora consists of about 390 species of shrews, hedgehogs, and moles. Insectivores are usually small animals with a high metabolic rate. Insectivores are found in North America, Africa, Europe, and Asia. Most have long, pointed noses that enable them to probe in the soil for insects, worms, and other invertebrates. They have teeth adapted for grasping and piercing their prey.

Insectivores are adapted to life on the ground, in trees, in water, and underground. Shrews, for example, feed on the surface of the ground, catching insects with their sharp teeth. Moles, on the other hand, burrow underground. They have stout limbs for digging, small eyes, and no external ears.

Order Primates

Recall from Chapter 17 that humans belong to the order Primates. The 235 species of living primates are classified as either prosimians, which include lemurs, tarsiers, and lorises, or anthropoids, which include monkeys, apes, and humans. Most primates are omnivores and have teeth suited for a varied diet. Primates have particularly large brains, which make possible the complex behaviors characteristic of this group.

Primates vary greatly in size and weight. The smallest primates weigh tens of grams, while the largest primate, the gorilla, *Gorilla gorilla,* can weigh 140–180 kg (300–400 lb). Most primates have two forward-facing eyes, a feature that enables depth perception. All primates have grasping hands and, with one notable exception, grasping feet, and many primates also have a grasping tail. The grasping feet, hands, and tail are excellent adaptations for life in the trees. Because of the wide range of body sizes and adaptations, primates live in a variety of habitats. The rhesus monkey, *Macaca mulatta,* is a tree-living species that is native to India. The gorilla, which is too large to climb trees, lives on the ground in the mountainous areas of Africa.

Order Chiroptera

This order includes bats, the only mammals that can fly. There are more than 900 species of bats, and they live throughout the world, except in polar environments. A bat's wing is a modified front limb with a skin membrane that stretches between extremely long finger bones to the hind limb. A bat's wingspan can measure up to 1.5 m (4.5 ft). The bat's clawed thumb sticks out from the top edge of the wing. Bats use their thumbs for walking, climbing, and grasping. Figure 45-12 shows an example of a bat.

Most bats use echolocation for navigation and have small eyes and large ears. Most feed on insects. However, some tropical bats feed on fruit or flower nectar and do not use echolocation. These bats, sometimes called "flying foxes," have large eyes and a keen sense of smell.

FIGURE 45-12

Peter's epauletted fruit bat is a member of the order Chiroptera, the only flying mammals. The physics of the bat's wing in flight gives the bat more lift in relation to its body weight than most birds have. This enables bats to remain airborne at slower speeds than birds are capable of.

MAKING CONNECTIONS

Language Arts

Have each student choose a mammal and write an essay that portrays a day in the animal's life from the animal's perspective. Students should include information about whether the animal is diurnal or nocturnal, what food it harvests or hunts and how it does so, what predators it must protect itself from and how it does so, whether it is a social or solitary animal, what kind of behaviors it engages in, and what kind of den or nest it lives in, if any. Encourage students to illustrate their essays with drawings.

RETEACHING *ACTIVITY*

Have students keep a record of every species of placental mammal they see, both domestic and wild, during an assigned week. Ask students to prepare a list of these mammals, grouped according to order. Remind students that humans are placental mammals.

SECTION 45-3

VISUAL LINK

Figure 45-13

Have students examine Figure 45-13 and compare it with Figure 45-14. Ask them to identify the characteristics that artiodactyls share with perissodactyls. (They are herbivorous, hooved, quadrupedal, terrestrial, and they are fast runners.) Then ask students to identify the characteristics that artiodactyls do not share with perissodactyls. (Artiodactyls have an even number of toes, and perissodactyls have an odd number of toes; artiodactyls have a rumen for digesting plant material, and perissodactyls have a cecum; and artiodactyls are native to every continent except Antarctica and Australia, and most perissodactyls are native only to Africa and Asia.)

CULTURAL CONNECTION

Forbidden Meat

Tell students that mammals often serve as sources of protein for humans. Then tell them that different cultures have different attitudes about the kinds of mammals that may be eaten. Ask students to name some foods that are forbidden or discouraged in certain cultures. (Hindus forbid eating beef, Muslims and some Jews forbid eating pork, and many western cultures do not eat cats, dogs, or horses.)

QUICK FACT

☑ While many mammals migrate to obtain food or avoid harsh weather, some hibernate. A hibernating mammal cannot be awakened quickly, and its body temperature approaches that of the surrounding air. It does not eat, but lives off the fat in its body. Examples of hibernating mammals include certain bats, echidnas, chipmunks, and woodchucks. Most scientists do not consider bears to be true hibernators because their body temperature falls only slightly.

FIGURE 45-13

This caribou is a member of the order Artiodactyla. Artiodactyls are native to all continents except Australia and Antarctica. The first domesticated artiodactyls were probably sheep and goats.

FIGURE 45-14

Although this tapir, *Tapirus bairdi*, looks like a pig, it is a member of the order Perissodactyla, making it more closely related to the horse and rhinoceros. This is another example of convergent evolution.

Order Carnivora

The 274 living species of the order Carnivora are distributed worldwide. Dogs, cats, raccoons, bears, hyenas, otters, seals, and sea lions are some well-known carnivores. Most of the species eat meat, which explains the name of the order. Carnivores generally have long canine teeth, strong jaws, and clawed toes, which they use for seizing and holding prey. Most have keen senses of sight and smell, which aid in hunting. Skeletal adaptations, such as long limbs, enable terrestrial carnivores to run quickly. Aquatic carnivores, such as sea lions, seals, and walruses, have streamlined bodies for efficient swimming and are known as **pinnipeds.** Pinnipeds propel themselves through the water and steer with their flattened, paddlelike forelimbs and hind limbs. Although pinnipeds spend much of their time in the sea feeding, they return to land to sleep and to give birth. Some scientists think that pinnipeds should be in their own order—Pinnipedia.

Pinnipeds are generally larger than land carnivores, weighing 90–3,600 kg (198–7,920 lb). Pinnipeds spend a large part of their life in cold water, so their large size serves to help maintain endothermy. Perhaps one of the most impressive feats of pinnipeds is their diving ability. Pinnipeds generally can dive to depths of 100 m (330 ft), and some species can dive to 400 m (1,313 ft). Most pinnipeds can remain underwater for up to five minutes, and some species can remain submerged for up to an hour.

Order Artiodactyla

Ungulates (UHN-gyoo-luhts) are mammals with hooves. Ungulates with an even number of toes make up the order Artiodactyla. Deer, elk, bison, moose, cattle, sheep, goats, pigs, and camels are artiodactyls (AHRT-ee-oh-DAK-tuhls). About 210 species belong to this order. Artiodactyls are native to every continent except Antarctica and Australia. They are fast runners and use speed as their major defense.

Most artiodactyls are herbivores, although pigs are omnivores. Their molars tend to be large and flat, for grinding plant material. As you learned in the last section, artiodactyls have a storage chamber in the stomach, called the rumen, in which bacteria and other microorganisms break down cellulose. This mutualistic relationship with microorganisms enables artiodactyls to extract nutrients efficiently from plants. Figure 45-13 shows an example of an artiodactyl.

Order Perissodactyla

Ungulates with an odd number of toes make up the order Perissodactyla, which includes about 17 living species. Horses, zebras, rhinoceroses, and tapirs are examples of perissodactyls (PER-is-oh-DAK-tuhls). Most species are native to Africa and Asia. However, some species of tapirs live in Central and South America. Figure 45-14 shows a tapir.

Like artiodactyls, perissodactyls have adaptations that help them digest the cellulose found in plant material. Members of this

FIGURE 45-15

Most cetaceans, like this killer whale, *Orcinus orca,* live in the oceans, but some species of dolphins live in freshwater lakes and rivers.

order have a cecum, instead of a rumen, in which microorganisms break down cellulose, releasing nutrients that the perissodactyl can absorb.

Order Cetacea

The order Cetacea includes 90 species of whales, dolphins, and porpoises. Cetaceans (se-TAY-shuhns) are distributed worldwide. Cetaceans have fishlike bodies with forelimbs modified as flippers. They lack hind limbs but have broad, flat tails that help propel them through the water. Cetaceans breathe through blowholes located on the top of the head. They are completely hairless, except for a few bristles on the snout. A thick layer of blubber below the skin provides insulation. Like bats, cetaceans navigate and find prey by using echolocation.

Cetaceans are exclusively aquatic and even give birth underwater, but they evolved from land-dwelling mammals. Cetaceans are divided into two groups: toothed whales and baleen whales. Toothed whales include sperm whales, beluga whales, narwhals, killer whales, dolphins, and porpoises. They can have from 1 to 100 teeth, and they prey on fish, squid, seals, and other whales. Figure 45-15 shows a killer whale. Baleen whales, such as blue whales, lack teeth and filter invertebrates from the water with a mesh of baleen that hangs from the roof of the mouth.

Word Roots and Origins

ungulate

from the Latin *ungula,* meaning "hoof"

Order Sirenia

The order Sirenia is made up of four species of manatees and dugongs. These large herbivores inhabit tropical seas, estuaries, and rivers. Their front limbs are flippers modified for swimming. Like whales (order Cetacea), sirenians lack hind limbs but have a flattened tail for propulsion. Although manatees and dugongs superficially resemble whales, they are more closely related to elephants. The similarities between whales and sirenians came about through convergent evolution. Figure 45-16 shows the only sirenian found in North America, the manatee.

FIGURE 45-16

This manatee, *Trichechus manatus,* or sea cow, is one of three species of manatee. Sirenian fossils that date to 50 million years ago have been found.

GIFTED ACTIVITY

Have students prepare a report on the intelligence of dolphins. Reports should focus on the fact that these mammals are surpassed in cerebrum-to-body-mass ratio only by chimpanzees and humans and that they are able to communicate with one another.

RECENT RESEARCH

Freshwater Whales

Scientists at Northeastern Ohio University have discovered that an early whale species, *Ambulocetus natans,* apparently lived part of its life in fresh water and part in salt water, where its fossils have been found. Scientists were able to determine this by measuring the amount of radioactive oxygen (^{18}O) in the fossils. Sea water has a higher concentration of ^{18}O than does fresh water. The earliest fossils show that 52 million years ago the whales drank fresh water, not salt water as do modern whales. Whales, which evolved from four-legged land-dwelling mammals, adapted to saltwater dwelling in only 4 million years, a very short period of time on the geologic scale.

OVERCOMING MISCONCEPTIONS

It is commonly believed that elephants are the largest mammals alive today. In fact, many whales are much larger than the largest elephants.

ACTIVE READING

Technique: K-W-L

Have students return to their lists of things they **W**ant to know about mammals, which they made as they began reading the chapter. (See Active Reading, page 881.) Have them place check marks next to questions that they are able to answer. Students should finish by making a list of what they have **L**earned. Have students identify unanswered questions, as well as new questions they may have.

ANSWERS TO SECTION 45-3 REVIEW

1. Students should pick any five orders from the 18 listed in this section.
2. Only Australia and New Guinea have both marsupials and monotremes.
3. The incisors of rodents grow continuously throughout the life of the animal.
4. Artiodactyls have an even number of toes and digest cellulose in the rumen. Perissodactyls have an odd number of toes and digest cellulose in the cecum.
5. Toothed whales have teeth for seizing fish or other mammals. Baleen whales capture their food by straining it from the water through a mesh of baleen in their mouth.
6. The reversed orientation probably prevents the pouch from filling with dirt while the mother burrows.

TABLE 45-1 ***Minor Orders of Mammals***

Order	Characteristic	Example
Macroscelidea	ground-dwelling insectivores with long, flexible snouts; 15 species found only in Africa	elephant shrew
Pholidota	insectivores with protective scales composed of fused hair; resemble reptiles; found in Africa and southern Asia	pangolin
Tubulidentata	nearly hairless insectivores with piglike bodies and long snouts; found in southern Africa	aardvark
Scandentia	squirrel-like omnivores that live on ground and in trees; feed on fruit and small animals; found in tropical Asia	tree shrew
Dermoptera	only two species exist; glide in air using a thin membrane stretched between their limbs; found only in parts of Asia	flying lemur
Hyracoidea	small rabbitlike herbivores; 7 species found only in Africa	hyrax

Order Proboscidea

Members of the order Proboscidea are characterized by a boneless, trunked nose, or proboscis. Only two species of this order exist today, the Asian elephant and the African elephant. Mammoths are an extinct member of this order. Elephants are the largest land animals alive today. The African elephant, the larger of the two species, can reach 6,000 kg (13,200 lb). To sustain such a large body, an elephant must feed on plants for up to 18 hours a day. Elephants use the trunk to gather leaves from high branches. They have modified incisors, called tusks, for digging up roots and for stripping bark from branches. The large, jagged molars, which can grow up to 30 cm (1 ft) long, are used for grinding plant material.

Elephants have one of the longest gestation periods of any animal. A female calf takes 20 months to develop, and a male calf takes 22 months to develop. Female elephants can continue to have calves until the age of 70, and elephants can live to be 80 years old.

The 12 orders just described include most of the familiar placental mammals. The six remaining orders contain just 1 percent of the mammalian species. These orders are summarized in Table 45-1.

internet**connect**

SCILINKS NSTA

TOPIC: Placental mammals
GO TO: www.scilinks.org
KEYWORD: HM896

SECTION 45-3 REVIEW

1. **List any five orders of placental mammals, and name an example of each order.**
2. **Where would you go to find both monotremes and marsupials?**
3. **What is unusual about the incisors of rodents?**
4. **Name two differences between artiodactyls and perissodactyls.**
5. **What are some feeding adaptations of whales?**
6. **CRITICAL THINKING The pouch of the marsupial mole opens toward the rear of the body. How would such a pouch aid in the survival of this burrowing animal's young?**

CHAPTER 45 REVIEW

SUMMARY/VOCABULARY

45-1

- Six key characteristics of mammals are endothermy; a fully divided heart; hair; milk production by females; a single jawbone; and complex, diverse teeth.
- Mammals belong to a group called synapsids. Synapsids have a skull with one opening behind the eye socket.
- Synapsids called therapsids gave rise to mammals. Therapsids had complex teeth and legs positioned beneath their body.
- Mammals first appeared about 225 million years ago. During the time of the dinosaurs they remained small and lacked diversity. After the extinction of dinosaurs, mammals evolved rapidly.

Vocabulary

mammary gland (881)
marsupial (884)
monotreme (884)
placental mammal (884)
synapsid (882)
therapsid (882)

45-2

- Mammals are endothermic.
- Endothermy expands the range of habitats mammals can occupy and facilitates sustained, strenuous activity. It also requires abundant supplies of food and oxygen.
- The heart of a mammal has two atria and two ventricles.
- Mammals have large lungs with a large internal surface area.
- Mammals, unlike most vertebrates, chew their food to begin its breakdown. Mammals have incisors, canines, premolars, and molars.
- Some hoofed mammals digest cellulose with the aid of microorganisms that live in the rumen. Elephants, rodents, and horses digest cellulose with the aid of microorganisms living in the cecum.
- Mammals have much larger brains than most other vertebrates.
- Monotremes lay eggs. Marsupials give birth to partially developed young that continue development in the mother's pouch. Placental mammals give birth to fully developed young.

Vocabulary

baleen (886)
canine (886)
cecum (887)
diaphragm (886)
echolocation (887)
incisor (886)
molar (886)
premolar (886)
rumen (887)

45-3

- There are at least 20 orders of mammals. Eighteen contain placental mammals. The other orders are monotremes and marsupials.
- Edentates—the sloths, anteaters, and armadillos—have either small teeth or no teeth. Rodents have two continuously growing incisors. Lagomorphs, such as rabbits, have four continuously growing incisors. Insectivores, such as shrews, are small and are adapted for eating invertebrates.
- Primates, including apes, monkeys, and humans, have very large brains and complex behaviors. Bats (order Chiroptera) are the only flying mammals. Carnivores, such as lions, are meat eaters. Deer, cattle, and sheep (order Artiodactyla) are hoofed mammals with an even number of toes. Hoofed mammals with an odd number of toes, such as rhinoceroses, belong to the order Perissodactyla. Whales and dolphins (order Cetacea) are exclusively aquatic and lack hind limbs. Elephants (order Proboscidea) have a long, prehensile nose called a trunk. Manatees and dugongs (order Sirenia) are large aquatic herbivores.

Vocabulary

pinniped (894)
ungulate (894)

CHAPTER 45 REVIEW ANSWERS

REVIEW

1. The rumen is a chamber of the stomach and is found in artiodactyls. The cecum is a sac protruding from the small intestine in rodents, elephants, and perissodactyls.
2. The ventricle is the heart chamber that pumps blood to either the lungs or the body. The diaphragm is a sheet of muscle below the rib cage that helps draw air into the lungs.
3. Baleen is the meshlike material in the mouth of baleen whales. It is used to filter small invertebrates from the water. Incisors are chisel-like teeth in mammals that cut food.
4. Monotremes are mammals that lay eggs. Placental mammals give birth to live young and nourish their unborn offspring through the placenta.
5. Canines are sharp teeth that stab or grip. Molars crush, grind, and cut food.

6. c **7.** c
8. b **9.** a
10. d **11.** a
12. b **13.** a
14. c **15.** c

16. Answers may include hair, mammary glands, a single jawbone, endothermy, diverse teeth, and a completely divided heart.
17. Endothermy expands the geographical range of mammals and makes it possible for them to be active when it is cold. It also increases mammals' endurance for strenuous activity.
18. Separated ventricles prevent mixing of deoxygenated and oxygenated blood. The efficient delivery of oxygenated blood maintains the high metabolic rate, which sustains endothermy.
19. Chewing food speeds digestion by breaking the food into pieces that digestive enzymes can more easily digest. Different teeth perform different chewing actions: cutting, gripping, puncturing, tearing, shredding, or grinding.

20. Some mammals have a rumen that contains cellulose-digesting microorganisms. Other mammals have a cecum that contains cellulose-digesting microorganisms.

21. Bats emit high-pitched sounds and then listen to the echoes and analyze them. This process allows them to navigate through their environment and locate prey. If a bat's ears are blocked, its source of information about its surroundings is cut off.

22. Monotremes and reptiles lay eggs, and the embryo is nourished by the yolk within the egg. Reptiles usually do not incubate their eggs and do not suckle the offspring after hatching, but monotremes do.

23. Whales have a thick layer of fat under the skin.

24. Rodents have two continuously growing incisors. Lagomorphs have four such teeth.

25. The mammal skull (B) has differentiated teeth and a single jawbone, but the reptile jaw (A) is made of several fused bones and has only one type of tooth.

Critical Thinking

1. Although mammals and dinosaurs appeared at about the same time, dinosaurs underwent rapid radiation. For 150 million years, the dinosaurs were diverse and the mammals remained small and gave rise to few species. Only after the dinosaurs' extinction did the mammals undergo large radiation.

2. As an animal becomes larger, its surface area does not increase as quickly as its volumez . For animals living in cold climates, this is an advantage because heat is lost through the surface of the body. Therefore, a larger animal has relatively less surface area through which to lose heat.

3. The short time between mating and birth means that many generations can be produced in a short

Review

Vocabulary

Explain the differences in meaning between each pair of terms.

1. rumen, cecum
2. ventricle, diaphragm
3. baleen, incisor
4. monotreme, placental mammal
5. canine, molar

Multiple Choice

6. Which of the following characteristics is present in all synapsids? (a) hair (b) placenta (c) skull with one opening behind the eye socket (d) endothermic metabolism
7. Which of the following is not true of therapsids? (a) may have been endothermic (b) gave rise to mammals (c) died out after dinosaurs (d) legs were positioned beneath the body
8. Evidence that early mammals were active at night includes the fossils of mammals with (a) a single jawbone (b) large eye sockets (c) a four-chambered heart (d) mammary glands.
9. After the dinosaurs became extinct, mammal species flourished because (a) many new habitats were suddenly available to them (b) the climate became warmer and wetter (c) the reptiles had undergone adaptive radiation (d) the reptiles had become endothermic.
10. The diaphragm allows mammals to (a) carry the young inside the uterus (b) have a divided ventricle (c) provide nourishment for their young (d) breathe efficiently.
11. Which of the following is not part of the mammalian heart? (a) conus arteriosus (b) septum (c) ventricle (d) atrium
12. Marsupials differ from monotremes in that marsupials (a) lay eggs (b) carry developing young in a pouch (c) live only in Australia (d) nourish their newborns with milk from mammary glands.
13. Which of these mammals is a marsupial? (a) kangaroo (b) duckbill platypus (c) lion (d) echidna
14. The order Chiroptera is made up of (a) toothless mammals (b) marine mammals (c) flying mammals (d) arboreal mammals.
15. Baleen whales feed on (a) fish (b) marine mammals (c) invertebrates (d) algae.

Short Answer

16. Name and describe two characteristics common to all mammals.
17. Describe two benefits of endothermy for mammals.
18. The ventricles of the mammalian heart are completely separated. How does this help a mammal maintain a constant body temperature?
19. Describe how specialized teeth aid mammals in digesting food.
20. Mammals cannot produce enzymes that break down cellulose, yet many mammals eat plants. Describe two mammalian adaptations for digesting plants.
21. In a famous study conducted 200 years ago, the Italian scientist Lazzaro Spallanzani showed that a blinded bat could still fly and capture insects. A bat whose ears had been plugged with wax could neither fly nor hunt. Explain Spallanzani's observations.
22. How is reproduction in reptiles and monotremes similar? How is it different?
23. Whales have only a few hairs on their snout, yet they are able to live in the coldest oceans. How do they retain their body heat in such environments?
24. How do the teeth of rodents differ from those of lagomorphs?
25. Examine the two skulls below. Which is mammalian? Justify your answer.

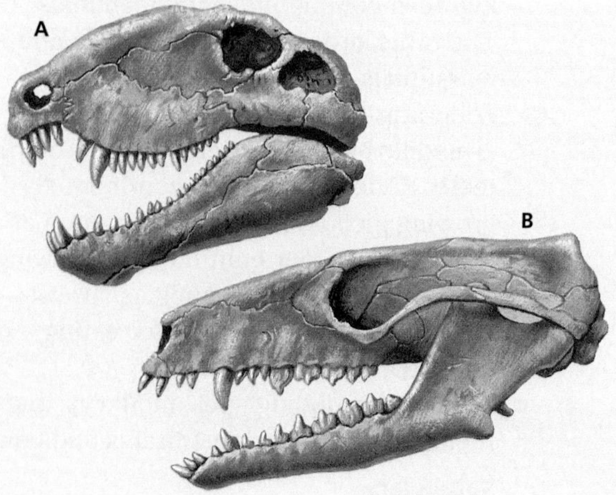

CRITICAL THINKING

1. Evaluate this statement by the paleontologist Robert Bakker: "Dinosaurs suppressed the evolutionary potential of mammals, not the other way around."
2. Mammals that live in very cold environments are usually larger than members of the same species that live in warmer climates. Propose a functional explanation for this. (Hint: Consider the effect that increasing size has on volume and surface area.)
3. Mice give birth only 21 days after mating. Why might this characteristic make mice ideal laboratory animals for experiments dealing with mammalian development and heredity?
4. Sloths are arboreal edentates that spend most of their lives hanging upside down from tree branches in the tropical forests of Central and South America. Most sloths have green algae growing in tiny pits in their hair. How might this be an advantage for the sloth? What advantage might the green algae gain by colonizing the sloth's fur?
5. In recent years, surgeons have tried transplanting baboon and pig hearts into humans. Explain why surgeons tried these hearts rather than a turtle's heart.
6. When performing surgery on animals such as cows, veterinarians try not to puncture the stomach because of the danger of infection. Explain what is meant by this statement. Why would it be unwise to completely sterilize the inside and outside of the animal?
7. The metabolic rates of two groups of sheep were measured as the amount of oxygen consumed per hour. One group was sheared before the experiment and the other was not. Explain the results shown below. How do these results support the theory that hair is an evolutionary advantage for endotherms?

Sheep-Shearing Experiment

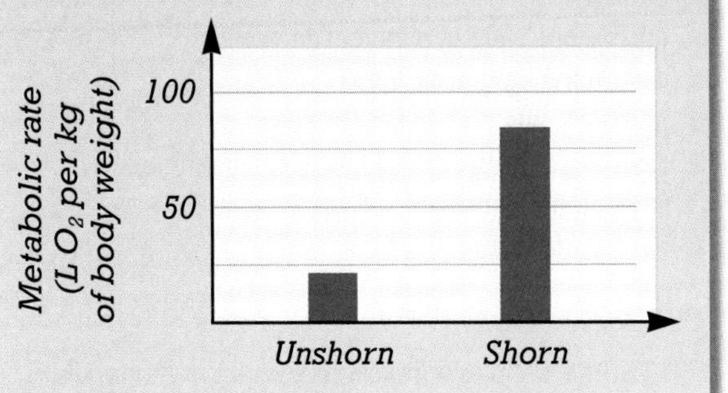

EXTENSION

1. Choose a nondomesticated mammal species, and write a report on its classification; habitat; morphology; life cycle; and adaptations for feeding, defense, and reproduction. Discuss its evolution if the information is available. You may wish to visit a zoo to observe the mammal. If possible, include labeled drawings or photographs of the mammal.
2. Read "Mind of a Dog" in *New Scientist*, March 4, 2000, on page 22. How long does dog mitochondrial DNA evidence suggest that humans have lived with dogs? Why does the author think dogs are particularly well suited to living with human society?
3. Obtain a live mammal, such as a mouse, hamster, or gerbil, from a pet store or your school. Get instructions on the care and feeding of the mammal. Then observe the mammal for at least 3 weeks. Take notes on its feeding and sleeping behavior. Share what you learn with your class.

ANSWERS

CONTINUED

time. It also means that all the steps of development can be observed within a relatively short time.

4. The sloth might benefit because the algae might give it some camouflage. Algae also might gain protection from their predators. They might also absorb nutrients from the sloth's hair.

5. Surgeons used mammal hearts because they are four-chambered, with a completely separated ventricle. This design would enable the human recipient to obtain enough oxygen to maintain normal human metabolic levels.

6. If the stomach is punctured, the microorganisms inside the digestive tract could infect the body cavity. This is harmful and potentially deadly to the animal. It is unwise to completely sterilize an animal because the microorganisms in the digestive tract help digest the food the animal eats.

7. The unshorn sheep did not have to use as much energy to maintain their body temperature as did the shorn sheep. Hair helps mammals maintain their body temperature by preventing body heat from radiating away.

EXTENSION

1. Answers will vary.

2. Dogs split with wolves and therefore became domesticated about 135,000 years ago. The author believes that humans and dogs evolved together and share certain patterns of thought. The author states that dogs are interested in the emotional content of the human mind and can learn and follow the rules of human society.

3. Answers will vary, but students should identify and describe as many behaviors as possible.

CHAPTER 45
INVESTIGATION

TIME REQUIRED

One 50-minute class period

QUICK REFERENCES

Lab Preparation Notes are found on page 879B.

Holt BioSources provides a Teaching Resources CD-ROM that contains "Using Gowin's Vee in the Lab" and "Scoring Rubrics."

PROCEDURAL TIPS

1. To save time, have half the students perform Part C while the other students perform the other parts of the lab.
2. Provide students with field guides to mammals so they can compare their lab findings with what others have discovered about mammals.
3. Have students prepare their data collection tables ahead of time.

ANSWERS TO BACKGROUND

1. Mammals have hair, have a completely divided four-chambered heart, produce milk, have a single jawbone, are endothermic, and have specialized teeth.
2. Endothermy is the ability of an animal to produce sufficient heat to maintain its own body temperature.

CHAPTER 45 INVESTIGATION

Mammalian Characteristics

OBJECTIVES

- Observe examples of mammals.
- Examine the distinguishing characteristics of mammals.

PROCESS SKILLS

- observing
- inferring

MATERIALS

- hand lens or stereomicroscope
- microscope slide of mammalian skin
- compound light microscope
- mirror
- selection of vertebrate skulls (some mammalian, some nonmammalian)
- field guide to mammals

Background

1. List the distinguishing characteristics of mammals.
2. Define the term *endothermy.*
3. Mammalian skin is characterized by cutaneous glands, such as sebaceous glands and sweat glands, that develop as ingrowths from the epidermis into the dermis.

PART A Mammalian Hair and Skin

1. Use a hand lens to examine several areas of your skin that appear to be hairless. Record your observations in your lab report.
2. Compare the amount of hair on humans with the amount on other mammals that you have seen or read about. What role does hair or fur play in endothermy? What other roles does hair (or whiskers) play in mammals?
3. Examine a slide of mammalian skin under low power. Notice the glands in the skin.
4. Identify the sebaceous glands and the sweat glands in the skin. Sweat glands are found only in mammals, but some mammals do not have them or have few of them. What mechanism for cooling might these other animals have? Which other glands are unique to mammals?

PART B Mammalian Reproduction

5. Look at the photographs of mammals on this page. What characteristics do these animals share?
6. Name the two kinds of mammals represented in the photographs on this page.

PART C Mammalian Mouth and Teeth

7. Use a mirror to look in your mouth, and identify the four kinds of mammalian teeth. Count how many of each you have on one side of your lower jaw.
8. Look at the skulls of several mammals. Identify the four kinds of teeth in each skull, and count them as you counted your own. How are the four types of teeth different from yours?
9. Look at the skulls of several nonmammalian vertebrates. Describe the teeth in each one, and compare them with mammalian teeth.
10. Breathe with your mouth closed. Do you feel a flow of air into your mouth? You have a hard palate (the roof of your mouth) that separates your mouth from your nose.
11. Look again at the different skulls. In which vertebrates do you see a hard palate? What is an advantage of having a hard palate?
12. Compare the jaws of the mammalian skulls with those of the nonmammalian skulls. Notice how the upper jawbone and the lower jawbone connect in each skull. Is there a similarity in the mammalian jaws that distinguishes them from the nonmammalian jaws? Explain.
13. Create a data table, similar to the model below, to record your observations for your lab report. For example, the table below is designed to record observations of differences that you will find among the animal skulls. Remember to allow plenty of space to record your observations.

PART D Vertebrate Diversity

14. Use a field guide to find out more about the following mammal orders: Cetacea, Edentata, Philodota, and Chiroptera. Answer the following questions about these mammals in your lab report:
    a. Cetaceans, such as whales and dolphins, are marine mammals. Except for a few bristles, cetaceans are hairless. Why do you think cetaceans are classified as mammals? How do they survive without hair or fur?
    b. Some mammals—including some members of Edentata (anteaters and armadillos) and Philodota (pangolins)—lack teeth. What characteristics do these animals share with other mammals?
    c. Like many birds, chiropterans (bats) have wings, fly, and are endotherms. What characteristics distinguish these mammals from birds?

Analysis and Conclusions

1. List the characteristics that distinguish mammals from other vertebrates.
2. List several characteristics you observed that most mammals share.
3. Birds are also endotherms. What structure in birds serves the same function as hair in mammals? Explain.
4. Compare the data you collected on the teeth from different animal skulls with the diet of each of those animals. How does the type of teeth that they have help with the particular diet that each animal has?

Further Inquiry

Find out how mammalian brains are different from the brains of other vertebrates. What adaptive advantage might these differences provide mammals?

OBSERVATIONS OF ANIMAL SKULLS

Animal	Mammal?	Number of incisors	Number of canines	Number of premolars	Number of molars	Hard palate?	Jaw

CHAPTER 45 INVESTIGATION

ANSWERS TO ANALYSIS AND CONCLUSIONS

1. Mammals produce milk and have a single jawbone, hair, and specialized teeth.
2. Answers will vary but should include hair, sebaceous glands, production of milk, single jawbone, and specialized teeth.
3. Feathers serve the same function as hair in mammals. Feathers not only enable birds to fly but also serve to insulate birds from the cold.
4. Answers will vary according to the types of skulls students are presented with. Answers should reflect that flat teeth are used to grind and crush and that incisors and canines are used to tear, grab, and cut meat.

Unit 16

HUMAN BIOLOGY

CHAPTERS

internet**connect**

SCILINKS NSTA

National Science Teachers Association *sci*LINKS Internet resources are located throughout this unit.

“The human body is marvelous. It can move freely, act deliberately, and survive under the most variable conditions. Its construction is complex and its requirements many.”

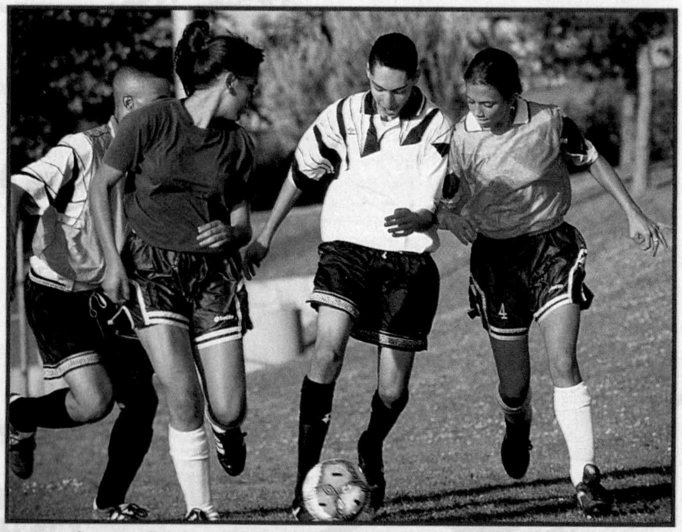

Near-perfect coordination of the many organ systems enables humans to play soccer and carry out daily activities.

At six weeks old, this developing human embryo, right, weighs less than 1 g. By eight weeks, all of the major organ systems will be recognizable.

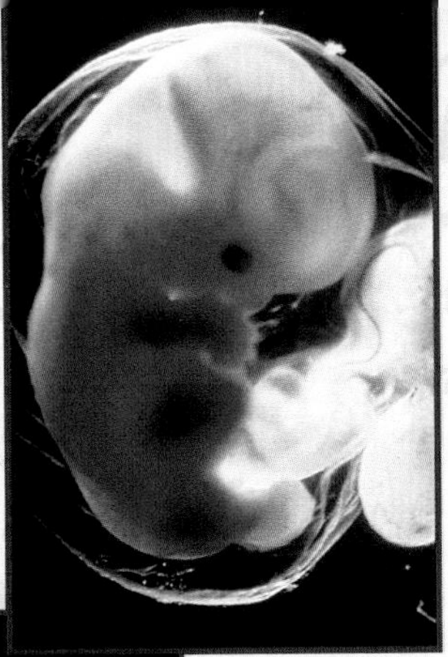

This X ray of a child's hand, below, reveals the hand's many bones.

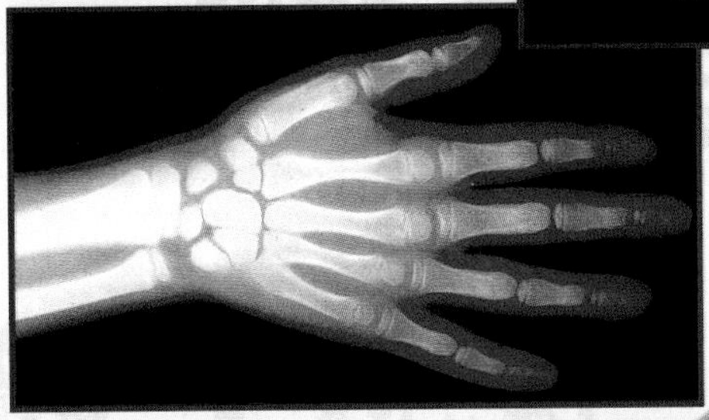

This researcher, like other scientists around the world, spends many hours in the lab each day searching for safe drugs that can be used to treat human ailments.

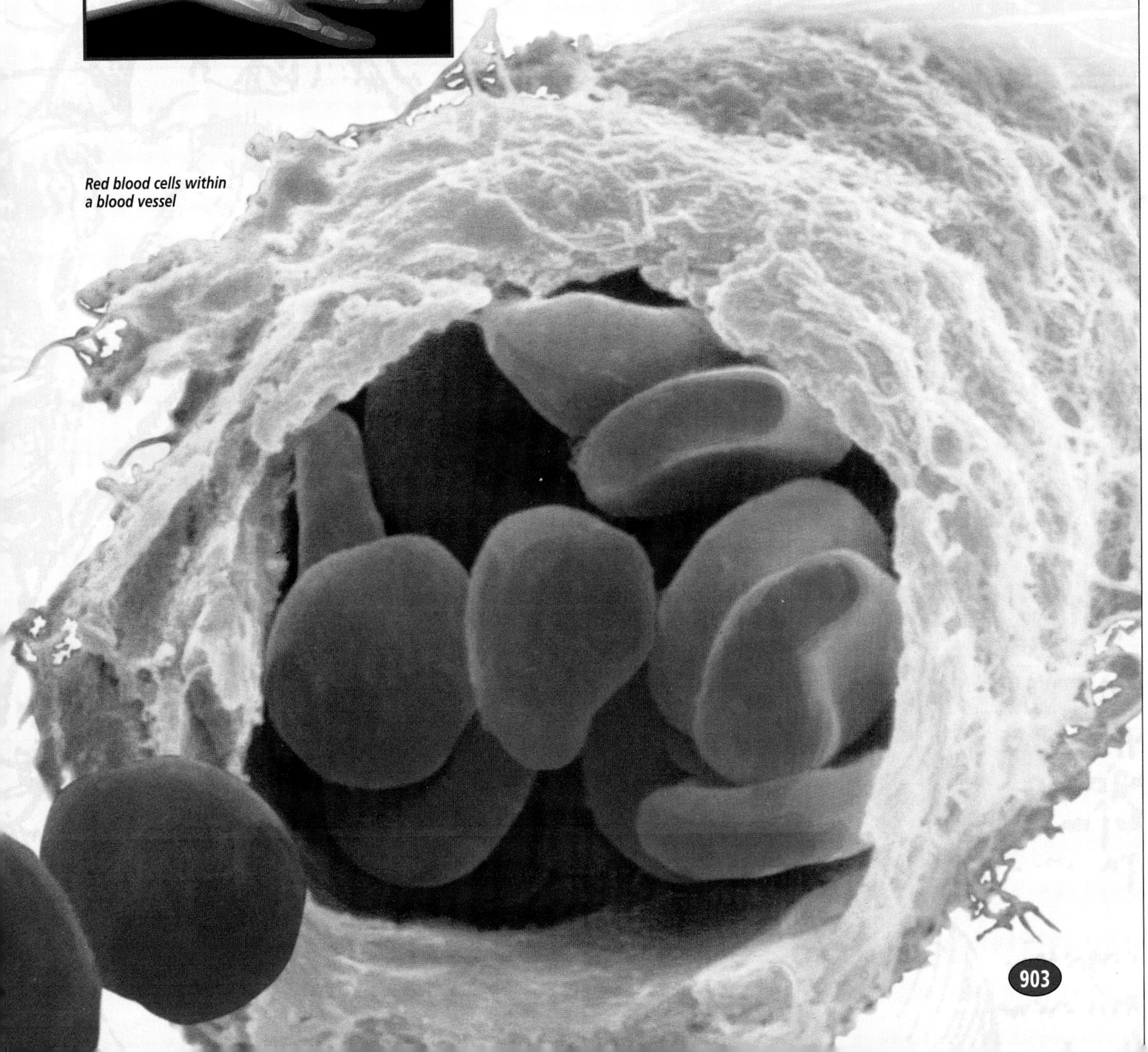

Red blood cells within a blood vessel

unit 10

CHAPTERS

46 *Skeletal, Muscular, and Integumentary Systems*

47 *Circulatory and Respiratory Systems*

48 *Infectious Diseases and the Immune System*

49 *Digestive and Excretory Systems*

50 *Nervous System and Sense Organs*

51 *Endocrine System*

52 *Reproductive System*

53 *Drugs*

HUMAN BIOLOGY

CONNECTING TO THE STANDARDS

The following chart shows the correlation of Unit 10 with the *National Science Education Standards* (grades 9–12). The items in each category are addressed in this unit and should be used to guide instruction. Annotated descriptions of the **Life Science Standards** are found below the chart. Consult the *National Science Education Standards* for detailed descriptions of each category.

UNIFYING CONCEPTS

- Systems, order, and organization
- Evidence, models, and explanation
- Form and function

SCIENCE AS INQUIRY

- Abilities necessary to do scientific inquiry
- Understanding about scientific inquiry

SCIENCE AND TECHNOLOGY

- Abilities of technological design
- Understanding about science and technology

PHYSICAL SCIENCE

- Chemical reactions
- Motions and forces
- Interactions of energy and matter

LIFE SCIENCE

EARTH AND SPACE SCIENCE

- Origin and evolution of the Earth system
- Energy in the Earth system

SCIENCE IN PERSONAL AND SOCIAL PERSPECTIVES

- Personal/community health
- Population growth
- Natural/human hazards

HISTORY AND NATURE OF SCIENCE

- Science as a human endeavor
- Nature of scientific knowledge
- Historical perspectives

ANNOTATED DESCRIPTIONS OF THE CORRELATED LIFE SCIENCE STANDARDS

Although all eight categories of the *National Science Education Standards* are important, the following descriptions summarize the **Life Science Standards** that specifically relate to Unit 10.

THE CELL

- Cells have particular structures that underlie their functions.
- Most cell functions involve chemical reactions.
- Cell functions are regulated.

MATTER, ENERGY, AND ORGANIZATION IN LIVING SYSTEMS

- The complexity and organization within an organism result from the organism's need to obtain, transform, transport, release, and eliminate the matter and energy required to sustain the organism.
- Living systems require a continuous input of energy to maintain their chemical and physical organizations.

BEHAVIOR OF ORGANISMS

- Multicellular animals have nervous systems that generate behavior.
- Organisms have behavioral responses to internal changes and to external stimuli. Responses to external stimuli can result from interactions with our own species and others and with environmental changes. Behavior must be flexible enough to deal with uncertainty and change.
- Behavioral biology has implications for humans because it provides links to psychology, sociology, and anthropology.

TRENDS IN BIOLOGY

Can We Cure AIDS?

For the first time, AIDS researchers have made significant advances in the fight against HIV. The clinical results from combining several antiviral drugs are so encouraging that scientists predict these drugs may be able to control HIV infection, transforming it into a chronic, manageable illness. Equally important, some people are genetically resistant to HIV, and an understanding of the molecular basis for this resistance may lead to improved treatments and means of prevention.

NEW THERAPIES

For many years, scientists have known that the amount of virus in the blood peaks shortly after infection and then falls to a low level, where it remains during the long, asymptomatic period that precedes the onset of AIDS. Why do viral levels stay so low for so long? Scientists had believed that after losing a skirmish with the immune system, the virus retreated into the cells of the body and entered a dormant state. They have discovered, however, that the asymptomatic period is not the result of a viral dormancy but rather of an intense struggle between HIV and the immune system. While HIV cranks out billions of new viral particles each day, the immune system destroys nearly all of these new viruses, keeping viral concentrations low. In a related discovery, scientists learned that the higher the viral level during the asymptomatic period, the faster the patient progresses to AIDS. Taken together, these discoveries suggest that suppressing viral replication might delay the onset of AIDS.

Since 1995, scientists have been treating people in various stages of HIV infection with a "cocktail" containing three drugs. Two of the drugs, AZT and 3TC, have been in use for years; these drugs interfere with the viral enzyme reverse transcriptase. The third drug is a protease inhibitor, which is a new class of drug that prevents the assembly of viral particles. So far, the results of this combination have been spectacular. In newly infected patients and in patients who are just beginning to show AIDS symptoms, the treatment reduces the viral concentration to below detectable levels. The treatment has also been credited with reducing the death rate among patients with advanced AIDS by 50 percent. Some scientists predict that early treatment may even eliminate the virus in some people, but it is too early to confirm this prediction.

RESISTANCE TO HIV

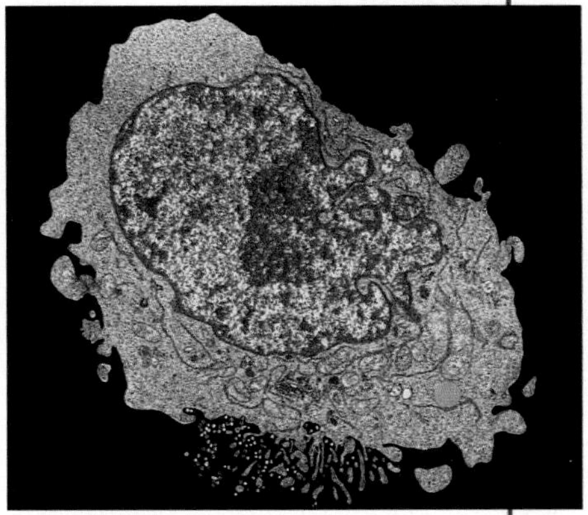

New therapies are allowing physicians to suppress the replication of HIV.

The second advance in fighting HIV came from a better understanding of how the virus infects cells. For more than a decade, scientists have known that HIV needs to bind to two receptors in order to enter a cell. One of these receptors, known as CD4, was identified in the mid-1980s. The second receptor, however, eluded scientists until 1996. That year, several research teams announced the discovery of a number of second receptors—there are at least four, and different strains of the virus seem to use different receptors. The most important second receptor discovered so far is CCR5 because there is a mutant gene for this protein. People who are homozygous for this mutant form of the gene seem to be resistant to HIV infection—they do not make a functional version of CCR5, thus HIV cannot invade their cells.

Additional Information

Cohen, John. "Exploiting the HIV-Chemokine Nexus." *Science,* February 28, 1997, 1261–1264.

Radetsky, Peter. "Immune to a Plague." *Discover,* June 1997, 61–67.

Richardson, Sarah. "Crushing HIV." *Discover,* January 1997, 28–29.

Richardson, Sarah. "The Second Key." *Discover,* January 1997, 29–30.

Centers for Disease Control and Prevention. HIV/AIDS Surveillance Report. World Wide Web: http://www. cdc.gov/nchstp/hiv_aids/stats/hivsur82.pdf

PLANNING GUIDE 46

CHAPTER 46

SKELETAL, MUSCULAR, AND INTEGUMENTARY SYSTEMS

	TOPICS	TEACHING RESOURCES	LABS, CLASSWORK, AND HOMEWORK
BLOCK 1 45 minutes	**Introducing the Chapter, p. 904**	**ATE** Focus Concept, p. 904 **ATE** Assessing Prior Knowledge, p. 904 **ATE** Understanding the Visual, p. 904	■ Occupational Applications Worksheets—Nurse Practitioner
	46-1 The Human Body Plan, p. 905 Body Tissues, p. 905 Organs and Organ Systems, p. 907 Body Cavities, p. 908	**ATE** Section Overview, p. 905 **ATE** Visual Link Table 46-1, p. 909 Figure 46-2, p. 908 190, 192–193	★ Study Guide, Section 46-1 **PE** Section Review, p. 908
BLOCKS 2 & 3 90 minutes	**46-2 Skeletal System, p. 909** The Skeleton, p. 909 Bone Structure, p. 910 Bone Development, p. 912 Joints, p. 913	**ATE** Section Overview, p. 909 **ATE** Critical Thinking, p. 910 **ATE** Visual Link Figure 46-3, p. 909 Figure 46-5, p. 911 Figure 46-7, p. 913 184, 184A, 185–186, 191, 198	**Lab A26,** Vertebrate Skeletons **Lab A27,** Comparing Skeletal Joints ■ Problem-Solving Worksheets—Determining Percent Composition ★ Study Guide, Section 46-2 **PE** Section Review, p. 914
BLOCKS 4 & 5 90 minutes	**46-3 The Muscular System, p. 915** Muscle Types, p. 915 Muscle Structure, p. 916 Muscular Movement of Bones, p. 919 Muscle Fatigue, p. 920	**ATE** Section Overview, p. 915 **ATE** Critical Thinking, pp. 916, 919 **ATE** Visual Link Figure 46-11, p. 917 190, 192–193	**Lab C41,** Evaluating Muscle Exhaustion ■ Occupational Applications Worksheets—Physical Therapist and Medical Sonographer ★ Study Guide, Section 46-3 **PE** Section Review, p. 920
BLOCK 6 45 minutes	**46-4 Integumentary System, p. 922** Skin, p. 922	**ATE** Section Overview, p. 922 **ATE** Critical Thinking, p. 923 188–189	**PE** **Chapter 46 Investigation,** p. 928 **Lab B30,** Touch Receptors in the Skin **Lab C45,** Screening Sunscreens ★ Study Guide, Section 46-4 **PE** Section Review, p. 924
BLOCKS 7 & 8 90 minutes	**REVIEW and ASSESSMENT** **PE** Chapter 46 Review, pp. 926–927 ■ Performance-Based Assessment—Testing Human Digestive Enzymes	Audio CD Program	**CHAPTER TESTING** ■ Chapter Test (blackline master) ▲ Test Generator ■ Assessment Item Listing ■ Scoring Rubrics and Classroom Management Checklists

KEY

MODERN BIOLOGY

PE Pupil's Edition
ATE Teacher's Edition
■ Active Reading Guide
★ Study Guide

HOLT BIOSOURCES

Teaching Transparencies
Laboratory Program

One-Stop Planner CD-ROM
Shows these resources within a customizable daily lesson plan:
■ Holt BioSources Teaching Resources
■ Active Reading Guide
★ Study Guide
▲ Test Generator

READING FOR CONTENT MASTERY

Preview: Have students preview the objectives, the introductory paragraph, topic sentences, and captions before reading the chapter.

ATE Active Reading Technique: K-W-L, p. 905
■ Reading Strategy: K-W-L Worksheet
■ Active Reading Guide Worksheet 46-1
PE Summary/Vocabulary, p. 925

■ Active Reading Guide Worksheet 46-2
PE Summary/Vocabulary, p. 925

■ Active Reading Guide Worksheet 46-3
PE Summary/Vocabulary, p. 925

ATE Active Reading Technique: K-W-L, p. 924
■ Reading Strategy: K-W-L Worksheet
■ Active Reading Guide Worksheet 46-4
PE Summary/Vocabulary, p. 925

TECHNOLOGY AND INTERNET RESOURCES

CNN PRESENTS Video Segment 30 Tanning Effects

Holt Biology Videodiscs Teacher's Correlation Guide, Lessons 46-1 through 46-4

Audio CD Program, Sections 46-1 through 46-4

internetconnect

SCILINKS NSTA

On-line Resources: www.scilinks.org
The following *sci*LINKS Internet resources can be found in the student text for this chapter:

Topics:
- Tissues, p. 905
- Bones and joints, p. 912
- Muscle fatigue, p. 920

go.hrw.com

On-line Resources: go.hrw.com
Visit the HRW Web site for a variety of resources related to this chapter. Just type in the keyword HM2 Chapter 46.

Smithsonian Institution®
Visit **www.si.edu/hrw** for additional on-line resources.

CNNfyi.com
Visit **www.cnnfyi.com** for late-breaking news and current event stories selected just for you.

LAB ACTIVITY PLANNING

CHAPTER 46 INVESTIGATION

Dehydrating and Demineralizing Bone, pp. 928–929

OVERVIEW
In this laboratory, students determine the amount of water and minerals in bone and the effect of the loss of these compounds on the physical properties of the bone. After dehydration, the bone will weigh less. Students should be able to determine the percentage of moisture in their bone sample. Following demineralization, the bone may melt if it is heated in an oven. Air-drying in a dessicator for 2 days may be necessary.

SAFETY
Students will wear disposable gloves, lab aprons, and safety goggles during the lab. If acid is splashed on skin or in the eyes, flush the area with water and seek medical attention. Students should also wash their hands thoroughly with soap and water after handling the chicken bones because of the danger of salmonella. All surfaces touched by the chicken should be washed with soap and water. Remind students to use tongs to handle hot samples.

PREPARATION
1. Obtain two fresh, raw-chicken leg bones for each lab group, and keep them refrigerated until ready for use.
2. Strip the bones of any meat with a knife and wash off any remaining flesh with water. Do not allow students to perform this procedure.
3. Although raw chicken flesh may be contaminated with Salmonella bacteria, cooking the bones may alter the outcome of the lab.

UNDERSTANDING THE VISUAL

Explain to students that this photograph shows an image produced by X rays. Ask if students know what an X ray is. (An X ray is an electromagnetic wave of very short wavelength.) Explain that an X-ray image of a human body is a shadowy negative of its structures. Dense structures, such as bone, absorb the X rays and appear as light areas on the film. Hollow, air-containing organs, as well as fat, absorb X rays to a lesser extent and appear as darkened areas.

FOCUS CONCEPT

Structure and Function

In this chapter students will learn that the relationship between structure and function is evident at the cellular, tissue, and organ levels. An example at the cellular level is muscle cell structure. An example at the tissue level is bone tissue. An example at the organ level is skin.

ASSESSING PRIOR KNOWLEDGE

Review the following concepts.

Protein Structure: *Chapter 10*
Have students review the structure of proteins.

Cell Structure: *Chapter 4*
Ask students to describe the basic structure of an animal cell.

Cell Division: *Chapter 8*
Have students outline the events of cell division.

CHAPTER 46

SKELETAL, MUSCULAR, AND INTEGUMENTARY SYSTEMS

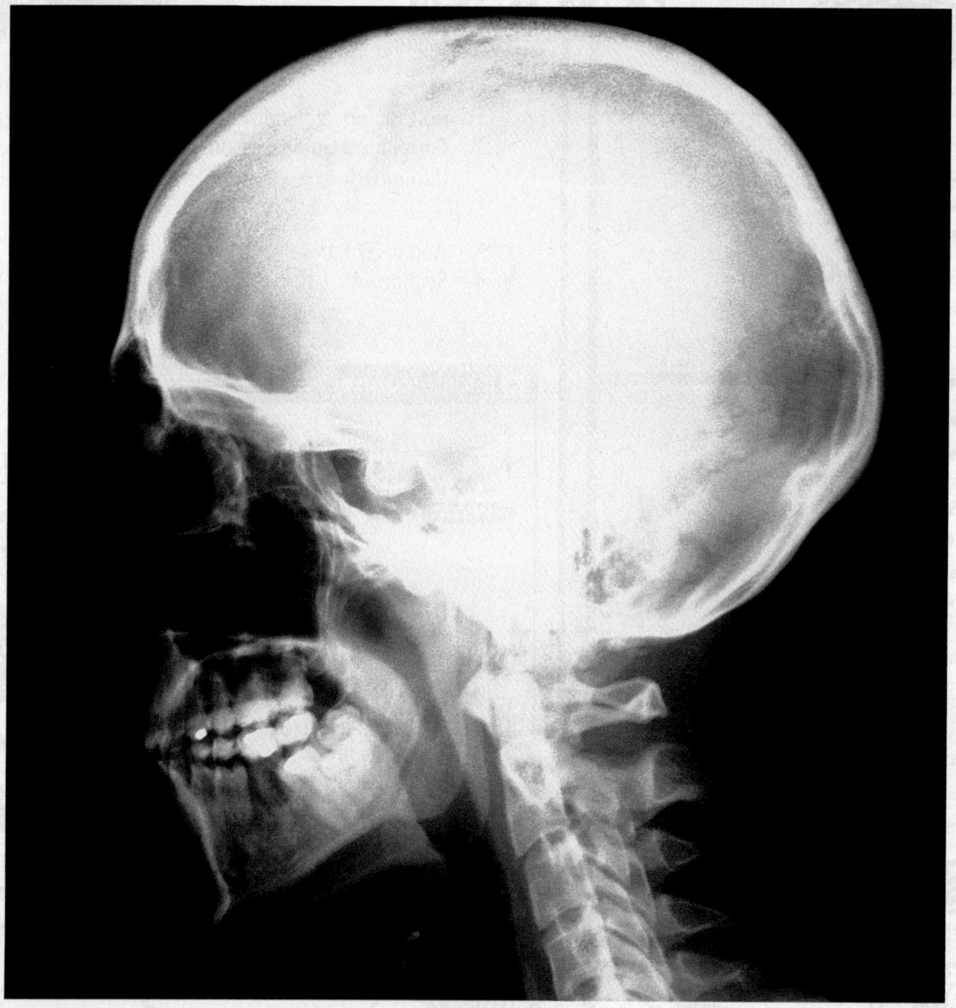

This X ray shows a color-enhanced image of the human skull, mandible, teeth, and neck.

FOCUS CONCEPT: ***Structure and Function***

As you read about bones, muscles, and skin, note how their structures are related to their functions in the human body.

THE HUMAN BODY PLAN

The human body begins to take shape during the earliest stages of embryonic development. While the embryo is a tiny hollow ball of dividing cells, it begins forming the tissues and organs that compose the human body. By the end of its third week, the human embryo has bilateral symmetry and is developing vertebrate characteristics that will support an upright body position.

SECTION

46-1

OBJECTIVES

▲

List and describe the four types of tissues that make up the human body.

●

Explain how tissues, organs, and organ systems are organized.

■

Summarize the functions of the primary organ systems in the human body.

◆

Name and locate four human body cavities, and describe the organs that each contains.

BODY TISSUES

In Chapter 4 you learned that a tissue is a collection of cells that are similar in structure and that work together to perform a particular function. The human body has four main types of tissues: muscle, nervous, epithelial, and connective.

Muscle Tissue

Muscle tissue is composed of cells that can contract. Every function that muscle tissue performs—from creating a facial expression to keeping the eyes in focus—is carried out by groups of muscle cells that contract in a coordinated fashion. The human body has three types of muscle tissue: skeletal, smooth, and cardiac. **Skeletal muscle** moves the bones in your trunk, limbs, and face. **Smooth muscle** handles body functions that you cannot control consciously, such as the movement of food through your digestive system. **Cardiac muscle,** found in your heart, pumps blood through your body. Figure 46-1a, on the following page, shows cells of skeletal muscle tissue.

Nervous Tissue

Nervous tissue contains cells that receive and transmit messages in the form of electrical impulses. These cells, called **neurons** (NOO-rahnz), are specialized to send and receive messages from muscles, glands, and other neurons throughout the body. Nervous tissue makes up your brain, spinal cord, and nerves. It is also found in parts of sensory organs, such as the retina in your eye. Nervous tissue provides sensation of the internal and external environment, and it integrates sensory information. Coordination of voluntary and involuntary activities and regulation of some body processes are also accomplished by nervous tissue. Figure 46-1b, on the following page, shows cells of nervous tissue.

internetconnect

SCILINKS NSTA

TOPIC: Tissues
GO TO: www.scilinks.org
KEYWORD: HM905

SECTION 46-1

SECTION OVERVIEW

Structure and Function
Section 46-1 describes how the body's cells, tissues, and organs interact to carry out life's functions. The human body has numerous cell types, four main tissue types, a variety of organs, and 11 organ systems, which all interact to carry out life functions.

Stability and Homeostasis
This section also emphasizes the role skin plays in maintaining homeostasis. Skin helps regulate body temperature and protects against pathogens.

ACTIVE READING

Technique: K-W-L
Before students begin this chapter, divide the class into three groups. Have one group list all the things they **K**now about bones; have the second group list all the things they **K**now about muscles; and have the third group list all the things they **K**now about skin. Ask students to write their lists on the board or the overhead. Then have students list things they **W**ant to know about bones, muscles, and skin. Have them save their lists for use in the Active Reading activity on page 924.

TEACHING STRATEGY

Analyzing Movement
Demonstrate a simple movement, such as bending and straightening your elbow. Ask students to identify the bones and muscles involved in this movement and to explain the mechanics. (The contraction of the muscles of the upper arm moves the bones of the forearm.) Emphasize the ways bones and muscles work together to perform this movement. Then point out that the elasticity of the skin covering the bones and muscles is essential for this and every movement of the body.

OVERCOMING MISCONCEPTIONS

The word *tissue* is used in everyday language, although not in the strict sense used in this section. Ask students to brainstorm and compile a list of body tissues as used in the everyday sense and write their suggestions on the board. (Students might name muscle, fat, skin, bone, and the brain.) Next ask students to identify each of these tissues as one of the four main types. (For example, muscle is made of muscle tissue, fat is made of connective tissue, etc.) Explain that organs, such as skin, bone, and the brain, cannot be identified by just one type of tissue because an organ is a combination of different tissues.

RECENT RESEARCH

Cartilage

Cartilage is a form of connective tissue. Like all cells, chondrocytes (cartilage cells) age and seem to respond to a programmed biological clock. Materials produced by aging chondrocytes are distinctly different from materials produced by young cells. In elderly people, these differences may account for some forms of osteoarthritis, in which the cartilage becomes thin and less resilient. The aging-chondrocyte hypothesis raises the possibility that some forms of osteoarthritis could be prevented or reversed by transplanting sheets of chondrocytes to joint surfaces. Such grafts are being tested for their effectiveness in resurfacing knee joints.

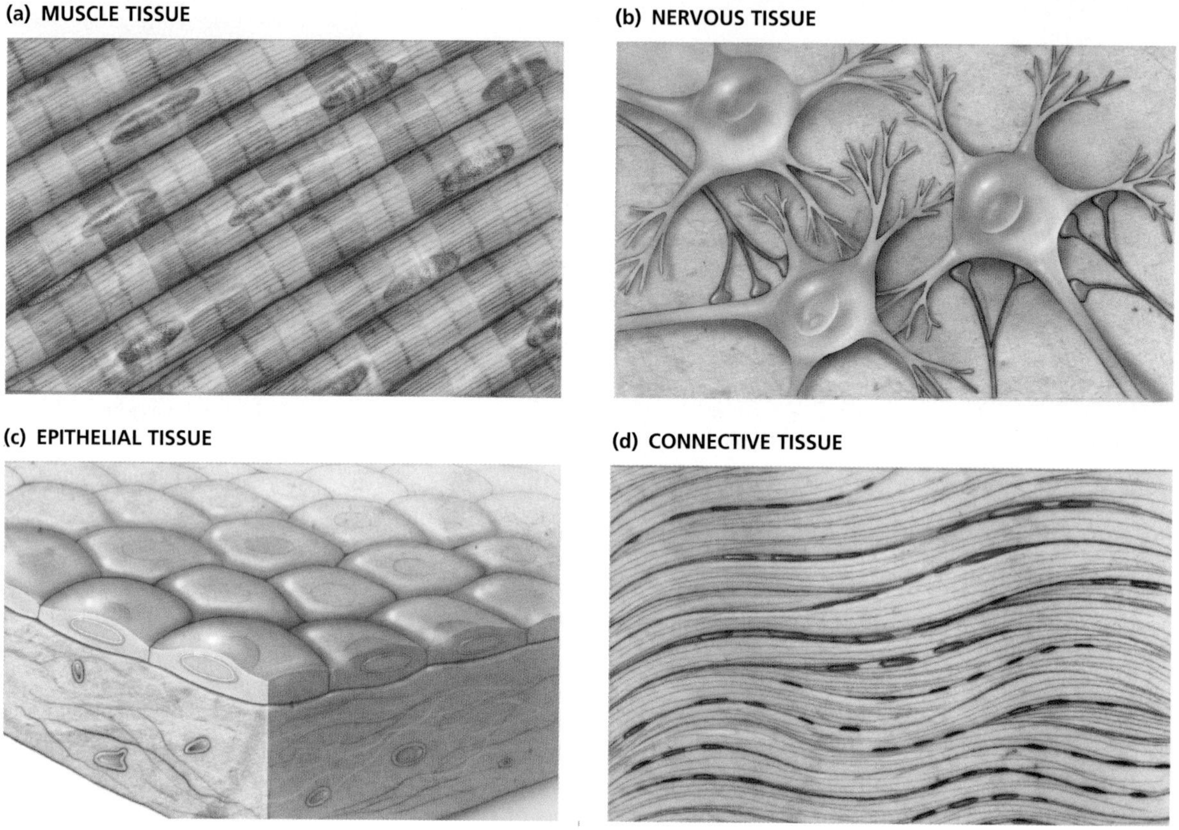

FIGURE 46-1

These four drawings show representative cells of the four main types of tissues in the human body: (a) muscle tissue, (b) nervous tissue, (c) epithelial tissue, and (d) connective tissue.

Epithelial Tissue

Epithelial (ep-uh-THEE-lee-uhl) **tissue** consists of layers of cells that line or cover all internal and external body surfaces. Each epithelial layer is formed from cells that are tightly bound together, providing a protective barrier for these surfaces. Epithelial tissue is found in various thicknesses and arrangements, depending on where it is located. For example, the epithelial tissue that lines blood vessels is a single layer of flattened cells through which substances can easily pass, but the epithelial tissue that lines the windpipe consists of a layer of cilia-bearing cells and mucus-secreting cells that act together to trap inhaled particles. The most easily observed epithelial tissue, the body's outer layer of skin, consists of sheets of dead, flattened cells that cover and protect the underlying living layer of skin. Figure 46-1c shows cells of epithelial tissue.

Connective Tissue

Connective tissue binds, supports, and protects structures in the body. Connective tissues are the most abundant and diverse of the four types of tissue, and include bone, cartilage, tendons, fat, blood, and lymph. These tissues are characterized by cells that are embedded in large amounts of an intercellular substance called **matrix.** Matrix can be solid, semisolid, or liquid. Bone cells are surrounded by a hard, crystalline matrix containing the mineral calcium. The cells in cartilage, tendons, and fat are surrounded by a semisolid fibrous matrix. Blood and lymph cells are suspended in a liquid matrix. Figure 46-1d shows cells of connective tissue.

ORGANS AND ORGAN SYSTEMS

An organ consists of various tissues that work together to carry out a specific function. The stomach, a saclike organ in which food is mixed with digestive enzymes, is composed of the four types of tissues. A single organ, such as the stomach, usually does not function in isolation. Rather, groups of organs interact in an organ system. For example, in the digestive system, the stomach, small intestine, liver, and pancreas all work together to break down food into molecules the body can use for energy. Table 46-1 lists the body's organ systems and names their major structures and functions. As you study the table, think about the ways in which the different organ systems work together to function in an efficient, integrated manner.

TABLE 46-1 ***Summary of Organ Systems***

System	Major structures	Functions
Skeletal	bones	provides structure; supports and protects internal organs
Muscular	muscles (skeletal, cardiac, and smooth)	provides structure; supports and moves trunk and limbs; moves substances through body
Integumentary	skin, hair, nails	protects against pathogens; helps regulate body temperature
Cardiovascular	heart, blood vessels, blood	transports nutrients and wastes to and from all body tissues
Respiratory	air passages, lungs	carries air into and out of lungs, where gases (oxygen and carbon dioxide) are exchanged
Immune	lymph nodes and vessels, white blood cells	provides protection against infection and disease
Digestive	mouth, esophagus, stomach, liver, pancreas, small and large intestines	stores and digests food; absorbs nutrients; eliminates waste
Excretory	kidneys, ureters, bladder, urethra, skin, lungs	eliminates waste; maintains water and chemical balance
Nervous	brain, spinal cord, nerves, sense organs, receptors	controls and coordinates body movements and senses; controls consciousness and creativity; helps monitor and maintain other body systems
Endocrine	glands (such as adrenal, thyroid, and pancreas), hypothalamus	maintains homeostasis; regulates metabolism, water and mineral balance, growth and sexual development, and reproduction
Reproductive	ovaries, uterus, mammary glands (in females), testes (in males)	produces ova and milk in females, sperm in males, and offspring after fertilization

VISUAL LINK
Table 46-1

Have students make a table that organizes the tissue types, using Table 46-1 as a model. Have students list organs according to their primary tissue type. (Hint: have students consider the organ's function as they determine the proper tissue category.)

TEACHING STRATEGY
Organ Systems

Have students work in pairs to review the major structures and functions of the 11 organ systems. Suggest that they discuss each system and then quiz each other about the material. For example, one student might ask the other what system is responsible for the regulation of homeostasis (endocrine). Circulate around the room, asking students questions about the major structures and functions of the organ systems.

CULTURAL CONNECTION
Bone China

Porcelain was first made in China during the Tang dynasty (618–907). For centuries Chinese porcelain was considered the finest in the world. It was not until the 1700s that Europe began producing its own porcelain. Around 1750, the English found a new way of making porcelain—with bone ash. Bone china is a form of porcelain made from burned animal bones. Bone ash is mixed with kaolin, a white clay found in Jiangxi Province, China, and petuntse, a type of mineral found only in China. The bone ash increases the procelain's translucence. Today England produces most of the world's bone china.

VISUAL LINK

Figure 46-2

Have students distinguish between dorsal (meaning "back") and ventral (meaning "belly"). Make sure that students can distinguish between the dorsal and ventral cavities shown in Figure 46-2 and that they understand that this is a longitudinal view not cross sectional.

ANSWERS TO SECTION 46-1 REVIEW

1. muscle tissue, such as the skeletal muscles that bend the arm; nervous tissue, such as the spinal cord; epithelial tissue, such as the lining of the digestive tract; and connective tissue, such as blood
2. Muscle tissue is able to contract, thereby moving limbs and pushing substances through the body; nervous tissue is specialized for sending and receiving messages in the form of electrical impulses.
3. They are all connective tissues, which are composed of cells embedded in a large amount of an intercellular substance called matrix.
4. Tissues form organs. Organs that interact form organ systems.
5. The thoracic cavity, located in the upper part of the trunk, contains the heart, lungs, trachea, and esophagus. The abdominal cavity, located in the lower part of the trunk, houses the organs that constitute most of the digestive system. A dome-shaped muscle called the diaphragm separates these two main cavities.
6. The skeletal, muscular, and nervous systems interact to move the body through the water. The respiratory system and the circulatory system work to deliver oxygen quickly to the muscles and other organs in the body.

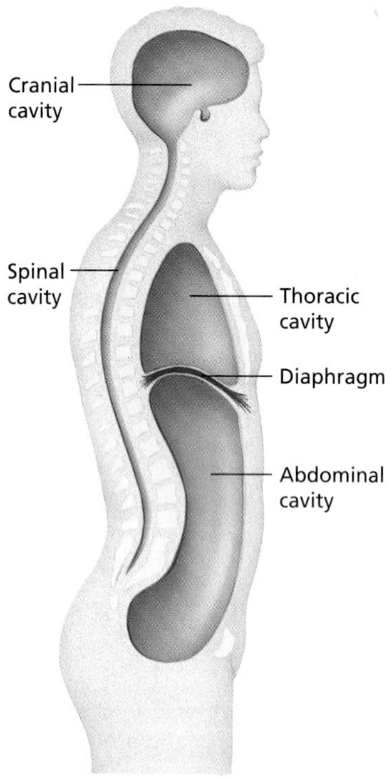

FIGURE 46-2

The human body has four main cavities that house and protect delicate internal organs.

Integration of Organ Systems

An even higher level of organization is the integration of organ systems. Each organ system has organs associated with it according to the organ's primary function. However, the boundaries are not always well defined. For example, nearly all of the juices produced by the pancreas are designed to aid in digestion. But because the pancreas produces vitally important hormones, it is also considered a component of the endocrine system. Each organ system carries out its own specific function, but for the organism to survive, the organ systems must work together. For example, nutrients from the digestive system are distributed by the circulatory system. The efficiency of the circulatory system depends on nutrients from the digestive system and oxygen from the respiratory system. An organism, whether it is a worm, a bird, or a human, is much more than an assembly of tissues. It is an integrated, whole being.

BODY CAVITIES

Many organs and organ systems in the human body are housed in compartments called body cavities. These cavities protect delicate internal organs from injuries and from the daily wear and tear of walking, jumping, or running. The body cavities also permit organs such as the lungs, the urinary bladder, and the stomach to expand and contract while remaining securely supported. As shown in Figure 46-2, the human body has four main body cavities, each of which contains one or more organs. The **cranial cavity** encases the brain. The **spinal cavity,** extending from the cranial cavity to the base of the spine, surrounds the spinal cord.

The two main cavities in the trunk of the human body are separated by a wall of muscle called the **diaphragm** (DIE-uh-FRAM). The upper compartment, or **thoracic** (thoh-RAS-ik) **cavity,** contains the heart, the esophagus, and the organs of the respiratory system—the lungs, trachea, and bronchii. The lower compartment, or **abdominal** (ab-DAHM-uh-nuhl) **cavity,** contains organs of the digestive, reproductive, and excretory systems.

SECTION 46-1 REVIEW

1. **Name the four types of tissues in the human body, and give an example of each.**
2. **How do muscle tissue and nervous tissue differ?**
3. **How are bone, cartilage, tendons, fat, blood, and lymph tissues structurally similar?**
4. **How are tissues, organs, and organ systems organized in the body?**
5. **Locate the thoracic and abdominal cavities, and describe the organs each cavity contains. What structure separates these cavities from each other?**
6. **CRITICAL THINKING Describe how the skeletal, muscular, nervous, respiratory, and circulatory systems function in a person swimming laps in a pool.**

SKELETAL SYSTEM

The adult human body consists of approximately 206 bones, which are organized into an internal framework called the **skeleton.** *Because the human skeleton is an internal structure, biologists refer to it as an endoskeleton. The variation in size and shape among the bones that make up the skeleton reflects their different roles in the body.*

THE SKELETON

As shown in Figure 46-3, the human skeleton is composed of two parts—the axial skeleton and the appendicular skeleton. The bones of the skull, ribs, spine, and sternum form the **axial skeleton.** The bones of the arms and legs, along with the scapula, clavicle, and pelvis, make up the **appendicular** (AP-uhn-DIK-yuh-luhr) **skeleton.**

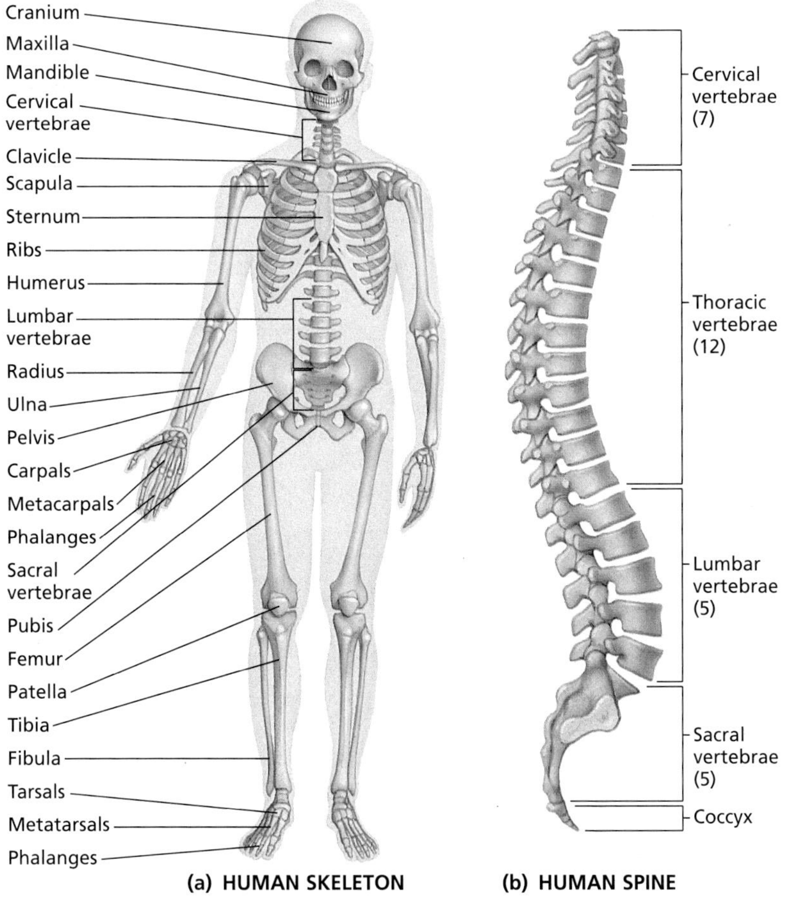

FIGURE 46-3

(a) The skeleton is the framework on which the rest of the body is built. (b) Like a stack of segmented tubes, the vertebrae of the spinal column offer strong protection for the spinal cord, yet they permit the trunk to bend and twist.

SECTION 46-2

OBJECTIVES

▲ Distinguish between the axial skeleton and the appendicular skeleton.

● Explain the function and structure of bones.

■ Summarize how bones develop and elongate.

◆ List three types of joints, and give an example of each.

▲ Describe a common disorder that affects the skeletal system.

SECTION 46-2

SECTION OVERVIEW

Stability and Homeostasis
Section 46-2 describes bones and how they form a system that supports and protects the body's organs. The skeletal system is composed of living material that interacts with other body systems, including the muscular, circulatory, and immune systems.

Cell Structure and Function
Cell structure and function is implicit in the discussion of how bones grow and develop throughout life. Bones are living material that grow and develop during a person's life.

▶ ENGAGE STUDENTS

Some students have probably experienced an injury to the skeletal system, such as a broken bone or an injured joint. Ask students who have had such an injury to share their experiences with the class. Have them point out the particular bone or joint that was injured (naming it, if possible), explain the cause of the injury, and describe how it was treated. If the injury was to a part of the axial skeleton, ask the student to speculate on what organs might have been injured if there had been no bones to protect them.

VISUAL LINK

Figure 46-3

Have students use Figure 46-3 to learn the names of the major bones. Whenever possible, have students relate the scientific name of a bone to its common name. For example, ask for the scientific name of the jawbone (mandible). Have students identify whether each bone is part of the axial or the appendicular skeleton.

SECTION 46-2

INCLUSION *ACTIVITY*

Have students work in pairs to construct a cutaway model of a long bone, showing the bone's inner anatomy. Students can use plastic wrap for periosteum, clay for compact bone, sponge or foam rubber for spongy bone, petroleum jelly for bone marrow, and red string for blood vessels. Have students form the clay first and then add the other materials. Students can work from Figure 46-4. Ask that each pair label their model with the proper terms, using toothpicks and paper flags.

CRITICAL THINKING

Bone Structure

Emphasize that bones consist of both organic and inorganic materials. Compact bone is solid because calcium and other minerals are deposited in it. Explain that when a bone is soaked in a vinegar solution for several days, calcium is dissolved from the bone. Ask students to design an experiment showing that calcium makes bones hard. (Break an animal bone in half and keep one half as a control. Soak the other half in vinegar or another dilute acid for several days. Then test each for hardness by attempting to bend the bones. The soaked half should be flexible, proving that calcium makes the bone hard.)

QUICK FACT

☑ Osteoporosis is a common and serious bone disease. It is characterized by excessive loss of calcified matrix and collagenous fibers from bone. The name *osteoporosis* means "condition of bone pores," referring to the holes or pores formed as bone tissue is lost.

The bones that make up the skeleton function in a variety of ways. Bones provide a rigid framework against which muscles can pull, give shape and structure to the body, and support and protect delicate internal organs. Notice, for example, that the ribs curve to form a cage that contains the heart and lungs. Similarly, bones in the skull form the cranium, a dome-shaped case that protects the brain. Bones also store minerals, such as calcium and phosphorus, which play vital roles in important metabolic processes. In addition, the internal portion of many bones produces red blood cells and certain types of white blood cells.

BONE STRUCTURE

Despite their number and size, bones make up less than 20 percent of the body's mass. The reason for their having relatively little mass can be better understood by looking at bone structure. Bones are not dry, rigid structures, as they may appear in a museum exhibit. They are moist, living tissues. As shown in Figure 46-4, a long bone consists of a porous central canal surrounded by a ring of dense material. The bone's surface is covered by a tough membrane called the **periosteum** (PER-ee-AHS-tee-uhm). This membrane contains a network of blood vessels, which supply nutrients, and nerves, which signal pain.

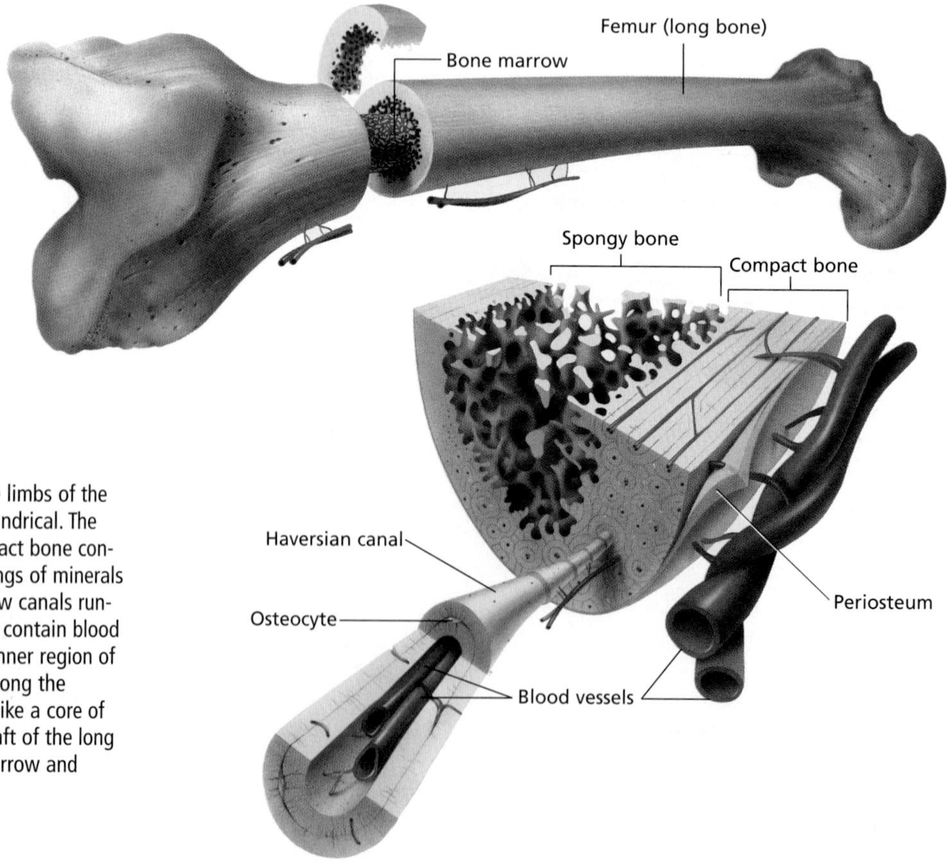

FIGURE 46-4

Long bones, found in the limbs of the body, are hollow and cylindrical. The outer shell of hard compact bone consists of closely packed rings of minerals and protein fibers. Narrow canals running through these rings contain blood vessels and nerves. The inner region of spongy bone stretches along the length of the long bone like a core of rigid lace. The central shaft of the long bone contains yellow marrow and blood vessels.

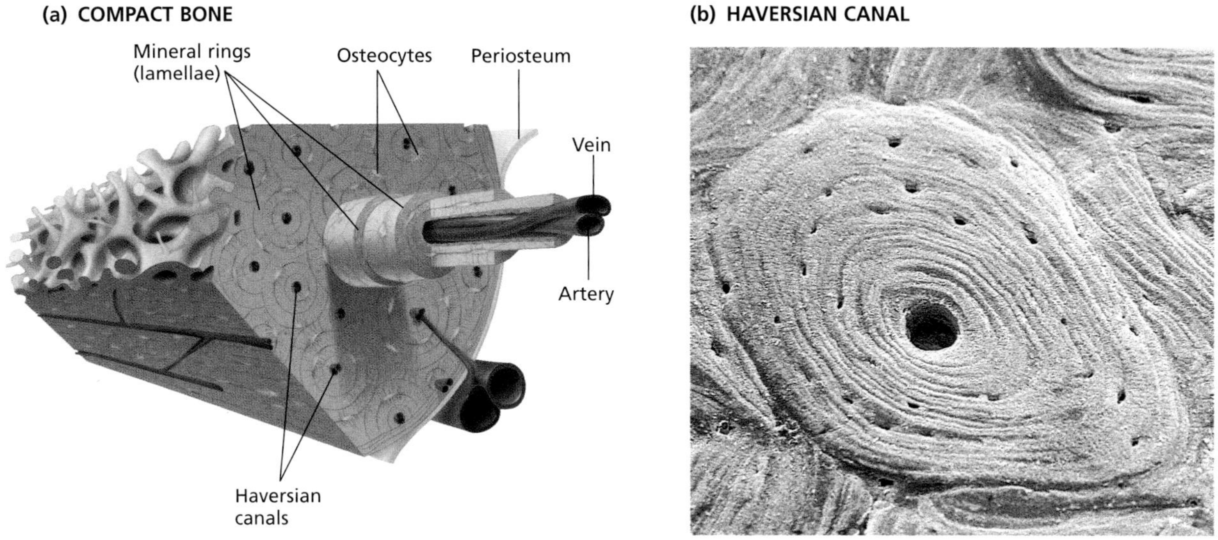

FIGURE 46-5

The cross section in (a) shows the internal structure of compact bone. A micrograph of a Haversian canal (380×) surrounded by lamellae in compact bone is shown in (b).

Under the periosteum is a hard material called **compact bone.** A thick layer of compact bone enables the shaft of the long bone to endure the large amount of stress it receives upon impact with a solid object. In the cross section shown in Figure 46-5a, notice that compact bone is composed of cylinders of mineral crystals and protein fibers called lamellae. In the center of each cylinder is a narrow channel called a **Haversian** (huh-VER-shuhn) **canal,** as shown in Figure 46-5b. Blood vessels run through interconnected Haversian canals, creating a network that carries nourishment to the living bone tissue. Several layers of protein fibers wrap around each Haversian canal. Embedded within the gaps between the protein layers are living bone cells called **osteocytes** (AHS-tee-uh-SIETS).

Inside the compact bone is a network of connective tissue called **spongy bone.** Although its name suggests that it is soft, this tissue is hard and strong. As shown in Figure 46-4, spongy bone has a latticework structure that consists of bony spikes. It is arranged along points of pressure or stress, making bones both light and strong.

Many bones also contain a soft tissue called **bone marrow,** which can be either red or yellow. Red bone marrow—found in spongy bone, the ends of long bones, ribs, vertebrae, the sternum, and the pelvis—produces red blood cells and certain types of white blood cells. Yellow bone marrow fills the shafts of long bones. It consists mostly of fat cells and serves as an energy reserve. It can also be converted to red bone marrow and produce blood cells when severe blood loss occurs.

Injury and Repair

Despite their strength, bones will crack or even break if they are subjected to extreme loads, sudden impacts, or stresses from unusual directions. The crack or break is referred to as a **fracture.** If circulation is maintained and the periosteum survives, healing will occur even if the damage to the bone is severe.

Eco Connection

Bones of Lead

Millions of Americans have been exposed to lead in the environment. Following exposure to lead, the kidneys excrete most of the metal. But 7 to 10 percent of the remaining lead in the body is stored in bone and can stay there for a lifetime. The rapid bone uptake of lead acts as a detoxifying mechanism. But lead is not permanently locked in bone. As people age, bone degeneration may occur, releasing lead into the bloodstream. Even very small concentrations of lead in the bloodstream can cause damage to kidneys, which in turn can cause high blood pressure.

The United States has outlawed the addition of lead to gasoline, water pipes, and paint. As a result, people who are now under age 25 may not accumulate as much lead in their bones as people from earlier generations.

VISUAL LINK

Figure 46-5

Point out the position of the Haversian canals in the compact bone. Have students describe the function of these canals. (Haversian canals create a network that carries nourishment to living bone tissue.)

RECENT RESEARCH

Bone Repair

"Sea bone" is a synthetic bone material made from coral. The coral is treated with heat to kill the living coral cells and convert its mineral structure (of calcium carbonates) to hydroxyapatite, the salt in human bone. The brittle coral graft is then carved to the desired shape, coated with naturally occurring bone-growth-enhancing proteins, and implanted. Typically 50 percent of sea-bone graft channels are filled with human bone in just three months.

QUICK FACT

☑ Bone tissue is renewed continuously. Every week, we recycle 5 percent to 7 percent of our bone mass, and as much as half a gram of calcium may enter or leave the adult skeleton each day.

GIFTED ACTIVITY

Have students research and report on Wolff's law (which states that a bone grows or remodels in response to the forces or stresses placed on it). Students may consider the effects of weightlessness (bone loss), weight lifting (the bones of weight lifters have enormous thickenings at the insertion sites of the most used muscles), or ballet dancing (the foot bones gradually thicken in response to the intense pressure of dancing *en pointe*).

DEMONSTRATION

Invite the school nurse to your classroom, and bring a first-aid manual and a supply of cloth bandages and splints to class. With the aid of the nurse, demonstrate how to wrap injuries to bones and joints. Divide the class into groups and have students practice these techniques on each other. If possible, obtain a series of X rays (from a hospital or a physician) that show the healing of a broken bone and the eight different types of fractures.

RECENT RESEARCH

Ultrasound and Bones

Ultrasound, one of the imaging techniques employed by physicians, speeds bone repair. Daily treatments reduce the healing time of broken arms and shinbones by 25 percent to 35 percent.

TEACHING STRATEGY

Hormones and Bones

During childhood the secretion of growth hormone is essential for bone growth. Discuss what happens when there is an excess or deficiency of this hormone. (Hypersecretion of growth hormone in children results in gigantism. Deficiencies of growth hormone or thyroid hormone result in dwarfism.)

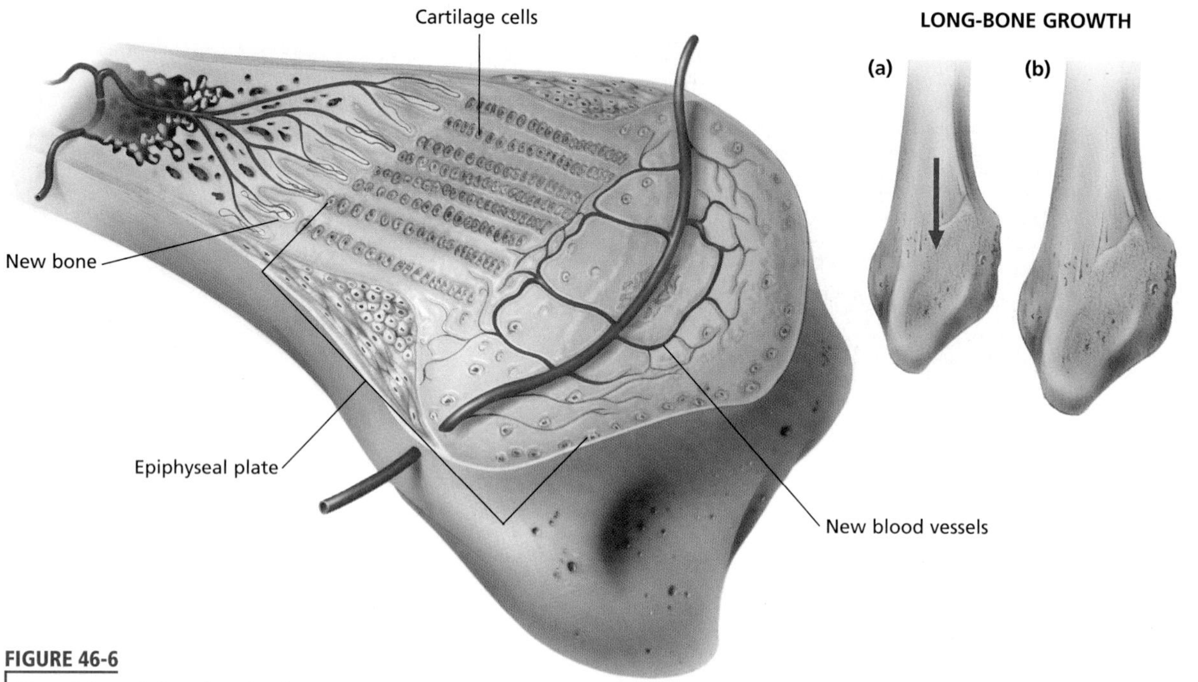

FIGURE 46-6

The epiphyseal plate, found at the ends of long bones, such as the femur shown above, is the site of bone elongation. This region is rich with cartilage cells, which divide, enlarge, and push older cells toward the middle of the bone shaft. As older cells move back, they are replaced by new bone cells, forming new regions of bone. A long bone (a) will grow in length, circumference, and density in this manner, as shown in (b).

internetconnect

SCILINKS NSTA
TOPIC: Bones and joints
GO TO: www.scilinks.org
KEYWORD: HM912

BONE DEVELOPMENT

Most bones develop from cartilage, a tough but flexible connective tissue. In the second month of fetal development, much of the skeleton is made of cartilage. During the third month, osteocytes begin to develop and release minerals that lodge in the spaces between the cartilage cells, turning the cartilage to bone. The process by which cartilage slowly hardens into bone as a result of the deposition of minerals is called **ossification** (AHS-uh-fuh-KAY-shuhn). Most fetal cartilage is eventually replaced by bone. However, some cartilage remains, lending flexibility to the areas between bones, at the end of the nose, in the outer ear, and along the inside of the trachea.

A few bones, such as some parts of the skull, develop directly into hard bone without forming cartilage first. In these cases, the osteocytes are initially scattered randomly throughout the embryonic connective tissue but soon fuse into layers and become flat plates of bone. In the skull, suture lines can be seen where the plates of bone meet.

Bone Elongation

Bones continue to develop after a person's birth. Between early childhood and late adolescence, bone cells gradually replace the cartilage in long bones of limbs, such as the arms and legs. Bone elongation takes place near the ends of long bones in an area known as the **epiphyseal** (EP-uh-FIZ-ee-uhl) **plate.** As shown in Figure 46-6, the epiphyseal plate is composed of cartilage cells that divide and form columns, pushing older cells toward the middle of the bone. As these older cells die, they are replaced by new bone cells.

Growth continues until bone has replaced all the cartilage in the epiphyseal plate. At this point, bones no longer elongate, and a person is considered to have reached full height.

JOINTS

The place where two bones meet is known as a **joint.** Three kinds of joints are found in the human body—fixed, semimovable, and movable. Examples of these joints are shown in Figure 46-7.

Joint Function

Fixed joints prevent movement. They are found in the skull, where they securely connect the bony plates and permit no movement of those bones. A small amount of connective tissue in a fixed joint helps absorb impact to prevent the bones from breaking.

Semimovable joints permit limited movement. For example, semimovable joints hold the bones of the vertebral column in place and allow the body to bend and twist. The vertebrae of the spine are separated by disks of cartilaginous tissue. These tough, springy disks compress and absorb shocks that could damage the fragile spinal cord. Semimovable joints are also found in the rib cage, where long strands of cartilage connect the upper seven pairs of ribs to the sternum, allowing the chest to expand during breathing.

Most of the joints in the body are **movable joints.** These joints enable the body to perform a wide range of movements and activities. Movable joints include hinge, ball-and-socket, pivot, saddle, and gliding joints. An example of a **hinge joint** is found in the elbow, which allows you to move your forearm forward and backward, like a hinged door. An example of a **ball-and-socket joint** is the shoulder joint, which enables you to move your arm up, down, forward, and backward, as well as to rotate it in a complete circle. The joint formed by the top two vertebrae of your spine is an example of a **pivot joint;** it allows you to turn your head from side to side, as when shaking your head "no." The **saddle joint,** found at the base of each thumb, allows you to rotate your thumbs and helps you grasp objects with your hand. Finally, **gliding joints** allow bones to slide over one another. Examples are the joints between the small bones of your foot, which allow your foot to flex when you walk.

Joint Structure

Joints, such as the knee, are often subjected to a great deal of pressure and stress, but their structure is well suited to meet these demands. As in all movable joints, the parts of the bones that come in contact with each other are covered with cartilage, which protects the bones' surface from friction. Tough bands of connective tissue, called **ligaments,** hold the bones of the joint in place. The surfaces of the joints that are subjected to a great deal of pressure are lined with tissue that secretes a lubricating substance called

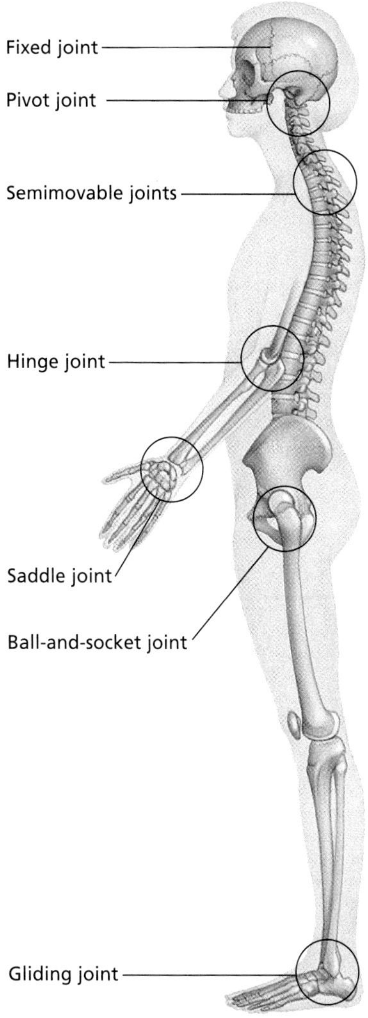

FIGURE 46-7

In addition to fixed joints and semimovable joints, the human body has five types of movable joints: pivot, hinge, saddle, ball-and-socket, and gliding.

VISUAL LINK
Figure 46-7

Using Figure 46-7 as a reference, have students identify the movable joints and classify each as one of the five types on a model or chart of the human skeleton. Write the five types of joints on the board, and have students list the specific examples under these headings. (Under ball-and-socket, for example, students should be able to name the shoulder and hip joints.)

TEACHING STRATEGY
Movable Joints

Have students work in groups of five to study the types of movable joints. Each group member should choose a different joint type and find common objects that are similar to it. For example, a door hinge is similar to a hinge joint, and a swivel is like a ball joint. Have the students bring the objects—or pictures of them—to class. Then each student should demonstrate how the objects are examples of the type of joint. Encourage groups to evaluate the appropriateness of each object.

RECENT RESEARCH
The Bionic Man

The history of artificial joints is less than 50 years old. In 1963, an English orthopedic surgeon revolutionized arthritic hip therapy. He created a vitallium metal ball on a stem in a cup-shaped polyethylene plastic socket anchored to the pelvis by methyl methacryulate cement. The prosthesis allows an older person perhaps another 10 years of mobility.

QUICK FACT

☑ Ligaments connect bones to bones. Tendons connect muscles to bones.

SECTION 46-2

RETEACHING *ACTIVITY*

Give each student a copy of an unlabeled drawing of the human skeleton. Ask students to label it with as many applicable terms as they can find in the section.

ANSWERS TO SECTION 46-2 REVIEW

1. The axial skeleton is made up of the skull, ribs, spine, and sternum. The appendicular skeleton consists of the arm and leg bones, scapula, clavicle, and pelvis.
2. Bone supports muscles and organs, gives shape and structure to the body, protects internal organs, stores minerals, and produces red and white blood cells.
3. Compact bone consists of a dense layer of minerals, protein fibers, and osteocytes, which form a durable, tubular structure. Spongy bone is made up of porous, bony spikes, which support points of stress and makes bone light in weight.
4. Most bones ossify in the fetal skeleton. Some bones continue to ossify until bone elongation ceases.
5. Fixed joints, such as those that connect the bones of the skull, are immovable. Semimovable joints, such as those that connect vertebrae, permit limited movement, such as bending and twisting. Movable joints, such as elbows, shoulders, hips, and knees, permit a wide range of motion.
6. A cartilaginous skeleton allows greater flexibility during development and birth.

FIGURE 46-8

The knee is a movable joint formed by the ends of the femur and the fibula. Many cordlike ligaments stabilize the joint, especially during movement. Pads of cartilage protect the ends of bones and act as shock absorbers. Like many joints in the body, the knee is a synovial joint. It contains membranes that secrete synovial fluid, which lubricates and nourishes the tissues inside the joint.

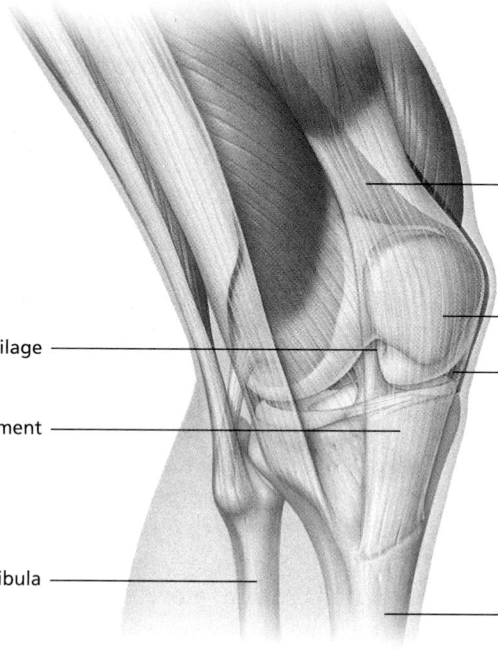

synovial (sih-NOH-vee-uhl) **fluid.** Synovial fluid helps protect the ends of bones from damage by friction. Figure 46-8 shows the internal structures of the synovial knee joint.

Sometimes these protective structures are not enough to prevent a joint from becoming injured. Of all the joints in the body, the knee joint is the most susceptible to injury because it carries the body's weight and relies on many ligaments for stability.

The term *arthritis* is used to describe several types of disorders that cause painful, swollen joints. There are two forms of arthritis that affect joints. **Rheumatoid arthritis** develops when the immune system begins to attack body tissues. The joints become inflamed, swollen, stiff, and deformed. **Osteoarthritis** is a degenerative joint disease in which the cartilage covering the surface of bone becomes thinner and rougher. As a result, bone surfaces rub against each other, causing severe discomfort.

SECTION 46-2 REVIEW

1. **Name the two main parts of the human skeleton, and list the bones that form them.**
2. **What are five functions of bones?**
3. **How do compact bone and spongy bone differ in structure and function?**
4. **When do most of the bones in the body begin to ossify? When does this process end?**
5. **What are the three major types of joints in the human body? Describe the function of each type and give an example.**
6. **CRITICAL THINKING What is the advantage of a cartilaginous skeleton during prenatal development?**

MUSCULAR SYSTEM

Muscles make up the bulk of the body and account for about one-third of its weight. Their ability to contract not only enables the body to move, but also provides the force that pushes substances, such as blood and food, through the body. Without the muscular system, none of the other organ systems would be able to function.

SECTION

46-3

OBJECTIVES

▲ Distinguish between the three types of muscle tissues.

● Describe the structure of skeletal muscle fibers.

■ Explain how skeletal muscles contract.

◆ Explain how muscles move bones.

MUSCLE TYPES

A muscle is tissue that can contract in a coordinated fashion and includes muscle tissue, blood vessels, nerves, and connective tissue. Recall that the human body has three types of muscle tissues: skeletal, smooth, and cardiac.

Skeletal muscle is responsible for moving parts of the body, such as the limbs, trunk, and face. Skeletal muscle tissue is made up of elongated cells called **muscle fibers.** Each muscle fiber contains many nuclei and is crossed by light and dark stripes, called **striations,** as shown on the following page in Figure 46-10a. Skeletal muscle fibers are grouped into dense bundles called **fascicles.** A

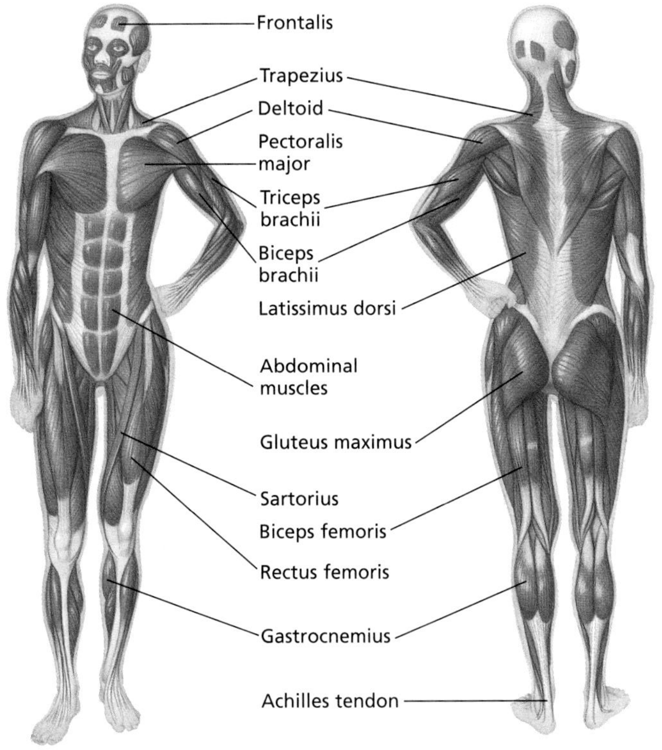

FIGURE 46-9

Skeletal muscle tissue is shown in these diagrams of some of the major muscles in the human body.

SECTION OVERVIEW

Stability and Homeostasis

Section 46-3 describes how movement depends on the proper functioning of the entire muscular system. The muscular system also interacts with other body systems. For example, the heart muscle pumps the blood required by all cells for oxygenation.

Cell Structure and Function

Section 46-3 describes the structure and function of muscle cells in the process of contraction.

▶ ENGAGE STUDENTS

Ask students to repeatedly open and close their hands. Challenge them to identify the muscles that contracted to perform these actions. (The hand is largely controlled by muscles in the forearm.) Explain that nerves stimulated the muscles, causing chemical reactions that contracted them. Ask students what shape the hand takes when the muscles are most relaxed. (The fingers curl.) Point out that straightening the fingers takes as much muscle action as closing them into a fist.

MAKING CONNECTIONS

Fine Arts

During the Renaissance, physicians began to study the ancient authors and to dissect cadavers. Their work impressed Leonardo da Vinci, who came to believe that the study of anatomy was essential for a painter. His system of anatomical drawing, which included cross-sectional representation, set the standard for centuries. Have students research Leonardo da Vinci, with an emphasis on his drawings of the human figure.

SECTION 46-3

TEACHING STRATEGY
Muscle Types

Challenge students to classify muscles in various ways. Ask which of the muscle types are striated and which are not. (Skeletal and cardiac muscles are striated; smooth muscle is not.) Ask students which types of muscle are voluntary and which are involuntary. (Skeletal muscle is voluntary; cardiac and smooth muscle are involuntary.) Finally, draw a table on the board with headings of the three muscle types. Ask students to name muscles that fall under each heading.

TEACHING STRATEGY
Keeping Muscles Healthy

Visit a local training facility or the high school weight-training room so students can see the types of equipment used to develop muscles. Point out that most muscle enlargement occurs when muscle cells increase in size, not through production of new muscle cells. Request an explanation of how different machines and regimens are designed to develop different muscles.

CRITICAL THINKING
Anabolic Steroids

Anabolic steroids are drugs that mimic the male hormone testosterone. While anabolic steroids do build up muscle mass, the side effects can be devastating. Though prolonged steroid use is hazardous to almost all body systems, the impairment of the cardiovascular system, including extremely high blood pressure and atherosclerosis, is the most dangerous. Have students research the uses—legal and illegal—and effects of anabolic steroids and present their findings to the class.

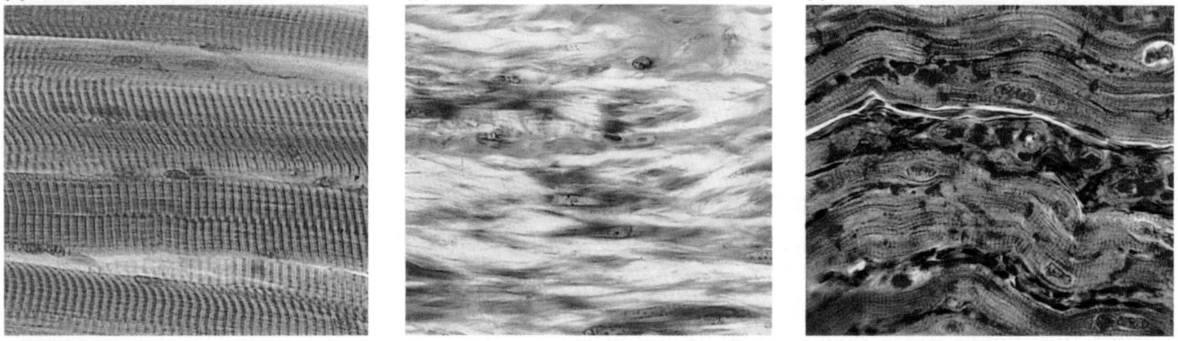

FIGURE 46-10

These light micrographs show the three types of muscle tissue. Skeletal muscle tissue (a) has a striped appearance when viewed under a microscope (430×). Smooth muscle tissue (b) is found in the digestive tract, the uterus, the bladder, and the blood vessels (400×). Cardiac muscle tissue (c) is found only in the heart (270×). Its interconnected fibers allow impulses to spread rapidly.

group of fascicles are bound together by connective tissue to form a muscle. Because their contractions can usually be consciously controlled, skeletal muscles are described as **voluntary muscles.**

Smooth muscle forms the muscle layers found in the walls of the stomach, intestines, blood vessels, and other internal organs. Individual smooth muscle cells are spindle-shaped, have a single nucleus, and interlace to form sheets of smooth muscle tissue, as shown in Figure 46-10b. Notice that smooth muscle lacks the striations found in skeletal muscle tissue. Smooth muscle fibers are surrounded by connective tissue, but the connective tissue does not unite to form tendons as it does in skeletal muscles. Because most of its movements cannot be consciously controlled, smooth muscle is referred to as **involuntary muscle.**

Cardiac muscle, shown in Figure 46-10c, makes up the walls of the heart. Cardiac muscle shares some characteristics with both skeletal muscle and smooth muscle. As with skeletal muscle, cardiac muscle tissue is striated; as with smooth muscle, it is involuntary and each cell has one nucleus. A bundle of specialized muscle cells in the upper part of the heart sends electrical signals through cardiac muscle tissue, causing the heart to rhythmically contract and pump blood throughout the body.

MUSCLE STRUCTURE

A muscle fiber is a single, multinucleated muscle cell. A muscle may be made up of hundreds or even thousands of muscle fibers, depending on the muscle's size. Although muscle fibers make up most of the muscle tissue, a large amount of connective tissue, blood vessels, and nerves are also present. Like all body cells, muscle cells are soft and easy to injure. Connective tissue covers and supports each muscle fiber and reinforces the muscle as a whole.

The health of a muscle depends on a sufficient nerve and blood supply. Each skeletal muscle fiber has a nerve ending that controls its activity. Active muscles use a lot of energy and therefore require a continuous supply of oxygen and nutrients, which are supplied by arteries. Muscles produce large amounts of metabolic waste that must be removed through veins.

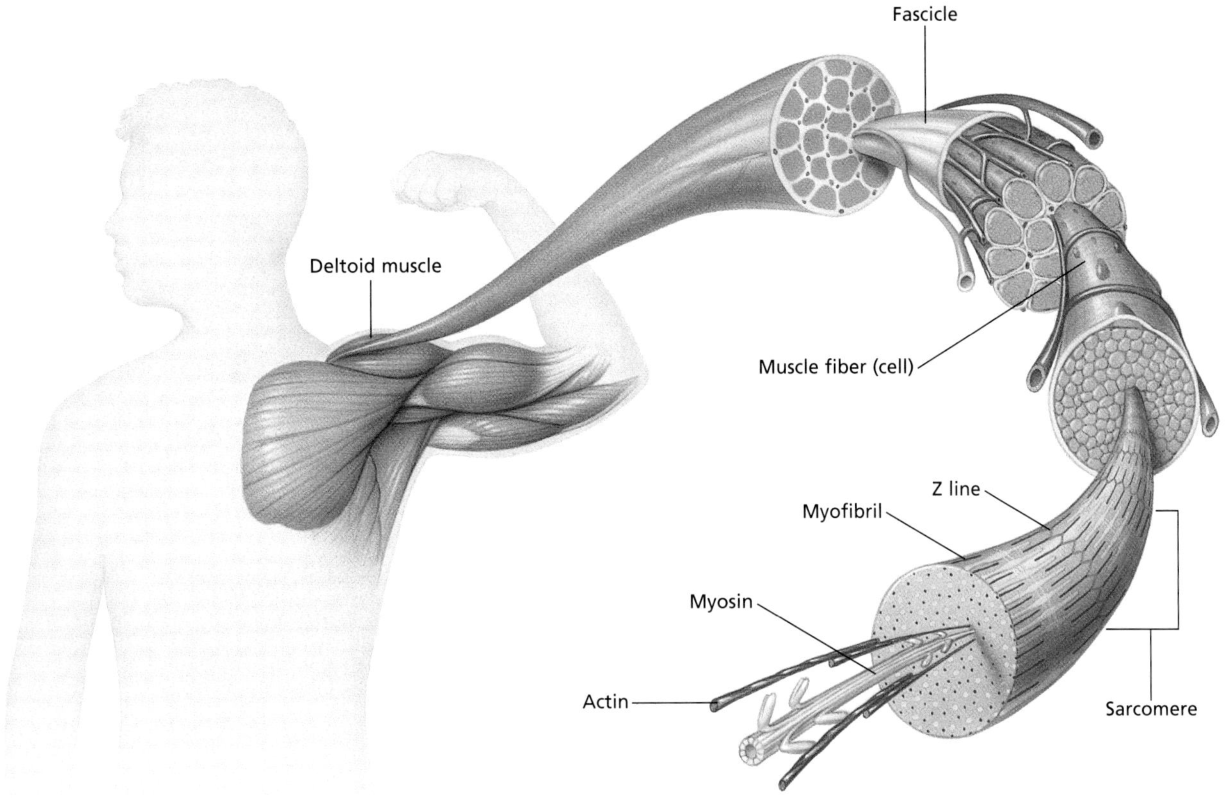

FIGURE 46-11

Skeletal muscles consist of densely packed groups of elongated cells, called fascicles, that are held together by connective tissue. Muscle fibers consist of protein filaments called myofibrils. Two types of filaments are found in muscle fibers—actin and myosin. The complementary structures of actin and myosin interact to contract and relax muscles.

A muscle fiber, such as the one shown in Figure 46-11, consists of bundles of threadlike structures called **myofibrils** (MIE-oh-FIE-bruhlz). Each myofibril is made up of two types of protein filaments—thick ones and thin ones. Thick filaments are made of the protein **myosin** (MIE-uh-suhn), and thin filaments are made of the protein **actin.** Myosin and actin filaments are arranged to form an overlapping pattern, which gives muscle tissue its striated appearance. Thin actin filaments are anchored at their endpoints to a structure called the **Z line.** The region from one Z line to the next is called a **sarcomere** (SAHR-kuh-MIR).

Muscular Contraction

The sarcomere is the functional unit of muscle contraction. When a muscle contracts, myosin filaments and actin filaments interact to shorten the length of a sarcomere. Myosin filaments have extensions shaped like oval "heads." Actin filaments look like a twisted strand of beads. When a nerve impulse stimulates a muscle fiber to contract, the heads at the end of the myosin filaments attach to points between the beads of the actin filaments. The myosin heads then bend inward, pulling the actin filaments with them. The myosin heads then let go, bend back into their original position, attach to a new point on the actin filament, and pull again. This action shortens the length of the

VISUAL LINK

Figure 46-11

Point out that a muscle fiber is like a rope made of many smaller fibers. Ask students to name these smaller fibers and the units that compose them (myofibrils, sarcomeres). What marks the boundaries of sarcomeres? (Z lines) Show students that sarcomere fibers contract as myosin filaments pull actin filaments together to make a muscle contract.

RETEACHING *ACTIVITY*

Bring to class photos of athletes in action, and ask students to identify the muscles the athletes are using.

GIFTED *ACTIVITY*

Have students research sports injuries, such as a torn rotator cuff or a pulled hamstring, and then determine the muscles affected, the cause of the injury, and the treatment. Have students share their findings with the class.

RECENT RESEARCH

Headaches

Researchers have found that people who have learned how to relax their trapezius muscles are able to cut the duration and intensity of their tension headaches in half. (Locate the trapezius muscles in Figure 46-9.) The experimental subjects used biofeedback techniques to relax the muscles. By practicing relaxation techniques when they were not in pain, they were also able to reduce the frequency of headaches. This group was compared with a group using temple massage and another using whole-body relaxation techniques.

QUICK FACTS

☑ The amount of work a muscle does is reflected in changes in the muscle itself. When used actively or strenuously, muscles may increase in size or strength and become more efficient and fatigue-resistant. On the other hand, muscle inactivity, whatever the cause, always leads to muscle weakness and wasting.

☑ Rigor mortis illustrates the fact that cross-bridge detachment, the detachment of the myosin heads from the actin filaments, is driven by ATP. Most muscles begin to stiffen three to four hours after death. They are stiffest after about 12 hours, and then they gradually relax over the next 48 to 60 hours. Because dying cells are unable to exclude calcium (which is in higher concentration in the extracellular fluid), the calcium influx promotes myosin cross-bridge binding. However, shortly after breathing stops, ATP synthesis ceases and cross-bridge detachment is impossible. Actin and myosin become irreversibly cross-linked, producing the stiffness of dead muscle. When actin and myosin break down, rigor mortis disappears.

TEACHING STRATEGY

Origin and Insertion

Have students refer to a muscle model or chart to draw a few major muscles of the human body, such as the biceps femoris, rectus femoris, deltoid, and pectoralis major. Have the students label the drawings and indicate the origin and insertion point of each muscle.

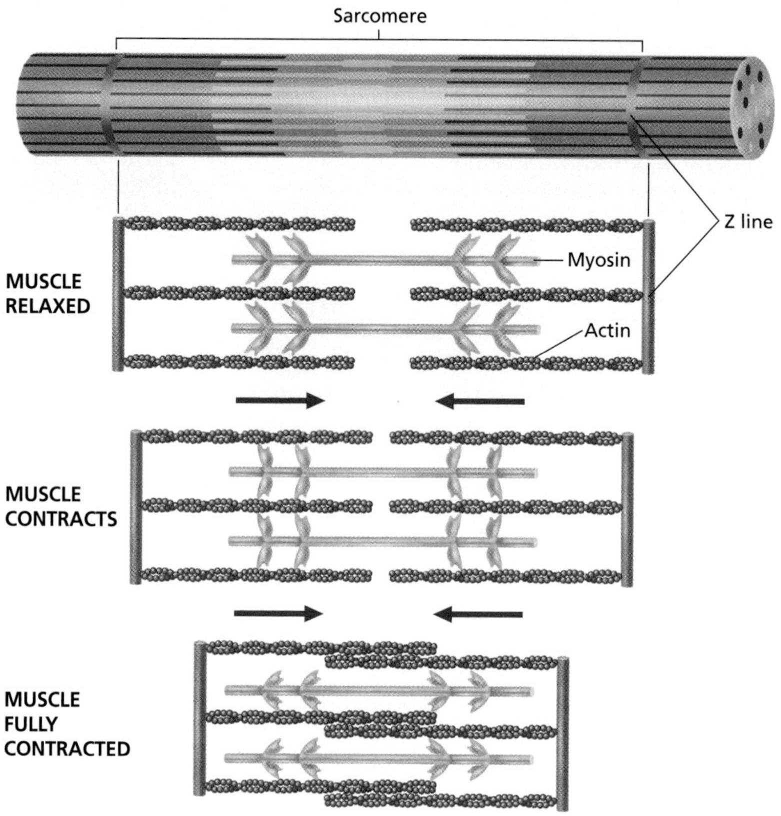

FIGURE 46-12

In a relaxed muscle, the actin and myosin filaments overlap. During a muscle contraction, the filaments slide past each other and the zone of overlap increases. As a result, the length of the sarcomere shortens.

sarcomere. The synchronized shortening of sarcomeres along the full length of a muscle fiber causes the whole fiber, and hence the muscle, to contract. Figure 46-12 shows the structures of the sarcomere.

Like all cellular activities, muscle contraction requires energy, which is supplied by ATP. This energy is used to detach the myosin heads from the actin filaments. Because myosin heads must attach and detach a number of times during a single muscle contraction, muscle cells must have a continuous supply of ATP. Without ATP, the myosin heads would remain attached to the actin filaments, keeping a muscle permanently contracted.

Muscle contraction is an all-or-none response—either the fibers contract or they remain relaxed. How, then, are you able to contract your muscles tightly enough to lift a dumbbell or gently enough to lift a pen? The force of a muscle contraction is determined by the number of muscle fibers that are stimulated. As more fibers are activated, the force of the contraction increases.

Some muscles, such as the muscles that hold the body in an upright position and maintain posture, are nearly always at least partially contracted. However, prolonged contraction may cause the muscle to use up all its available energy, become overly tired, and begin to cramp. To prevent this from happening, the nervous system automatically stimulates different muscle fibers alternately in the muscle, allowing some to be at rest while others are working to maintain the body's posture.

MUSCULAR MOVEMENT OF BONES

Generally, skeletal muscles are attached to one end of a bone, stretch across a joint, and are fastened to the end of another bone. Muscles are attached to the outer membrane of bone, either directly or by a tough fibrous cord of connective tissue called a **tendon.** For example, as shown in Figure 46-13, one end of the large biceps muscle in the arm is connected by tendons to the radius and ulna in the forearm, while the other end of the muscle is connected to the scapula in the shoulder. When the biceps muscle contracts, the forearm flexes upward while the scapula remains stationary. The point where the muscle attaches to the stationary bone—in this case, the scapula—is called the **origin.** The point where the muscle attaches to the moving bone—in this case the bones in the forearm—is called the **insertion.**

Most skeletal muscles are arranged in opposing pairs. One muscle in a pair moves a limb in one direction; the other muscle moves it in the opposite direction. Muscles move bones by pulling them, not by pushing them. For example, when the biceps muscle contracts, the elbow bends. The biceps muscle is known as a **flexor,** a muscle that bends a joint. Contraction of the triceps muscle in the upper arm straightens the limb. The triceps muscle is an example of an **extensor,** a muscle that straightens a joint. To bring about a smooth movement, one muscle in a pair must relax while the opposing muscle contracts.

FIGURE 46-13

Skeletal muscles, such as the biceps and triceps muscles in the upper arm, are connected to bones by tendons. (a) When the biceps muscle contracts, the elbow bends. (b) When the triceps muscle contracts, the elbow straightens.

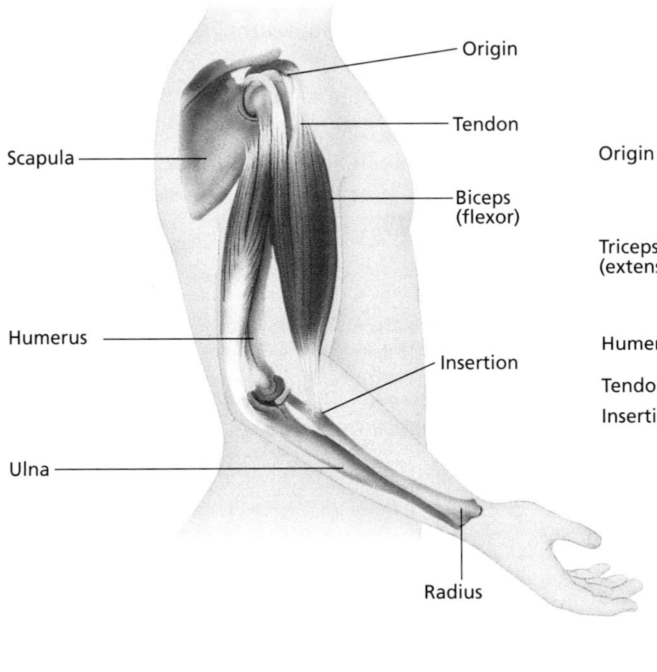

(a) BICEPS MUSCLE

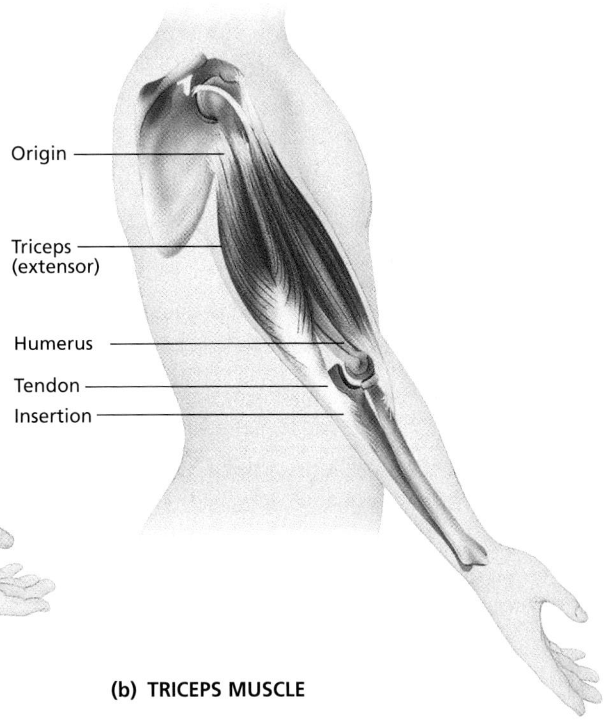

(b) TRICEPS MUSCLE

TEACHING STRATEGY
Flexors and Extensors

Two muscles named according to their actions are flexors (they decrease the angle of a joint) and extensors (they increase the angle of a joint). Have pairs of students use a human model or chart with muscles listed to identify some flexors and extensors. Examples in the arm are the triceps (extensors) and biceps (flexors), and examples in the leg are the biceps femoris (flexors) and rectus femoris (extensors). Ask students how the biceps brachii functions opposite to the biceps femoris. (Each muscle extends over the joint in an opposite direction.)

CRITICAL THINKING
Hyperventilation

Sometimes swimmers voluntarily hyperventilate so that they can hold their breath longer during competition. This is a very dangerous practice. The oxygen content of blood drops below 60 percent of normal during regular breath-holding. Carbon dioxide levels continue to rise when the breath is held, and breathing becomes unavoidable. However, strenuous hyperventilation can lower the P_{O_2} so much that a lag period occurs before it rebounds enough to stimulate respiration. This lag period may allow oxygen levels to fall well below 50 mm Hg, causing a swimmer to black out before the urge to breathe takes over. Ask students to research this phenomenon and to consider other situations (such as running the 100 yd sprint in 12 sec) in which the body experiences an oxygen debt.

SECTION 46-3

RETEACHING ACTIVITY

Have students name the parts of the body, and ask them to tell what types of muscle controls each part. For example, ask what type of muscle is found in the esophagus (smooth muscle).

ANSWERS TO SECTION 46-3 REVIEW

1. The human body has three types of muscle tissues: skeletal muscle, which moves limbs; smooth muscle, which lines the digestive system and blood vessels; and cardiac muscle, which pumps blood through the body.
2. Smooth muscle cannot be consciously controlled.
3. The muscle fiber is a long cell found in muscle tissue. It consists of thick and thin filaments that overlap, creating a striped appearance.
4. Actin is a protein found in thin filaments; myosin is a protein found in thick filaments. They overlap each other and form units called sarcomeres. During a muscle contraction, myosin heads attach to actin molecules and bend inward, pulling the actin filament with them. This action shortens the length of a sarcomere and the entire muscle fiber.
5. Muscles are arranged in opposing pairs. When one member of a pair contracts, a limb bends. When the opposing member of the pair contracts, the limb straightens.
6. The body stops producing ATP, which is needed to disengage the myosin and actin so the muscle can relax.

internet**connect**

SCILINKS NSTA

TOPIC: Muscle fatigue
GO TO: www.scilinks.org
KEYWORD: HM920

MUSCLE FATIGUE

Muscle cells store glycogen, which is used as a source of energy after blood-delivered glucose is exhausted. The breakdown of glycogen releases large amounts of energy, but sometimes even those reserves are used up, especially in athletes who participate in sports that require sustained exertion. During prolonged and vigorous exertion, fat molecules are utilized for energy. Fat molecules contain a greater concentration of potential energy than any other molecule in the body. When energy availability fails to keep pace with its use, muscle fatigue sets in and controlled muscle activity ceases, even though the muscle may still receive nerve stimulation to move. **Muscle fatigue** is the physiological inability of a muscle to contract. Muscle fatigue is a result of a relative depletion of ATP. When ATP is absent, a state of continuous contraction occurs. An example of depletion of ATP is when a marathon runner collapses during a race, suffering from severe muscle cramps.

FIGURE 46-14

These athletes are in the process of repaying their oxygen debts. Oxygen debt occurs frequently after strenuous, sustained exertion.

Oxygen Debt

Oxygen is used during cellular respiration in the synthesis of ATP. Large amounts of oxygen are needed to maintain the rate of maximum ATP production required to sustain strenuous exercise. However, after several minutes of exertion, the circulatory system and the respiratory system are not able to bring in enough oxygen to meet the demands of energy production. Oxygen levels in the body become depleted. This temporary lack of oxygen availability is called **oxygen debt.** Oxygen debt leads to an accumulation of metabolic wastes in the muscle fibers. The presence of this waste produces the soreness you may experience after prolonged exercise. Oxygen debt is what causes an athlete to spend a minute or two in rapid, deep breathing after strenuous exercise, as the athletes shown in Figure 46-14 are doing. The oxygen debt is repaid quickly as additional oxygen becomes available, but muscle soreness may persist until all of the metabolic wastes that have accumulated in the muscle fibers are carried away or converted.

SECTION 46-3 REVIEW

1. What are the three main types of muscle tissues found in the body, and what are their functions?
2. Why is smooth muscle referred to as involuntary muscle?
3. What is a muscle fiber? Why does it appear striated?
4. What are actin and myosin, and how are they organized in a muscle? How do they interact during a muscle contraction?
5. How do muscle pairs work together to move a limb?
6. CRITICAL THINKING Rigor mortis is a condition in which all the body muscles become rigid shortly after a person dies. Why does rigor mortis develop?

Research Notes

Looking Inside the Human Body

In 1895, the development of X-ray equipment provided physicians a way to look inside the body at images of dense tissue, such as bones. Modern imaging techniques rely on computers. For example, computerized tomography (CT) uses a focused beam of low-dose X rays to obtain cross-sectional images of structures in the body. Tomography is the X-ray technique used to image only a specific "slice" or plane of tissue. Computerized tomography can differentiate tissues of various densities. Another technology, magnetic resonance imaging (MRI), creates images of soft tissues by using radio waves emitted by the nuclei of hydrogen atoms activated by a magnetic scanner.

Still another and newer imaging technology is positron emission technology, or PET. Positrons are positively charged particles that result from the disintegration of radioisotopes. Michael M. TerPogossian and his colleagues at Johns Hopkins suggested using short-lived radioisotopes in medical research in 1966. They then developed scanners to detect the positrons released by radioisotopes that had been injected into a patient's bloodstream. As electronic and computer technology improved, biomedical engineers redesigned scanning equipment to create three-dimensional images from the positron emissions. Positron emission technology is often used in evaluating brain metabolism. It is also used to map areas of the brain involved in memory, sensation, perception, speech, and information processing. In addition, positron emission technology provides clues to the physiological and neurochemical basis of medical psychiatric disorders, such as depression.

A new holographic imaging system combines images obtained by computerized tomography or magnetic resonance scanners and displays an accurate three-dimensional image of the anatomical structures being viewed. A holograph is a method of photography that uses laser light. A three-dimensional image is produced when the holograph is viewed under visible light. The transparent but solid-seeming image floats in front of the holographic film and can be moved around or reversed so that it can be studied from all sides. Computer point-and-click methods allow viewers to focus on specific areas of the image. Physicians can make surgical plans by studying the actual appearance of a patient's organs, as shown in the 3-D PET scan of a brain that highlights the verbal center in (a). Compare the 3-D PET scan with the X ray of the same part of the body in (b). The X ray may not be as helpful as the 3-D PET scan is when a physician must diagnose an illness or injury of the brain.

Some of the newest techniques are interactive. On the Internet, for example, medical students and researchers can view a "Visible Man" made up of thousands of computerized tomography cross sections. They can use the images to test theories or study the relationships of various anatomical structures. An explosion in technology has provided medicine with new techniques, such as virtual reality. Virtual reality is a computer simulation of a system that allows the user to perform operations on the system and that shows the effects of those operations. Doctors are beginning to use virtual reality to navigate through a patient's computerized tomography images when planning for surgery or other treatments.

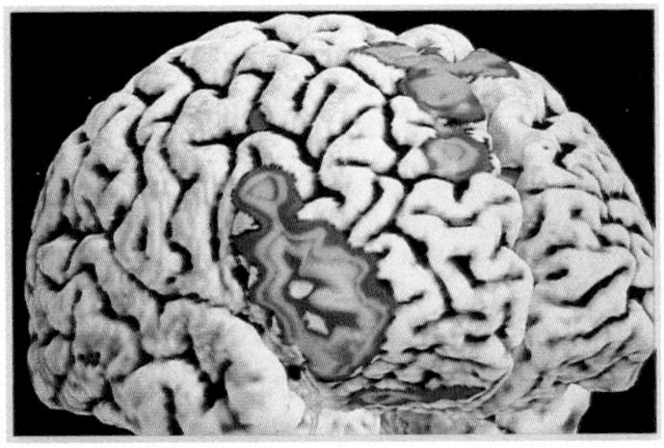

(a) 3-D PET SCAN

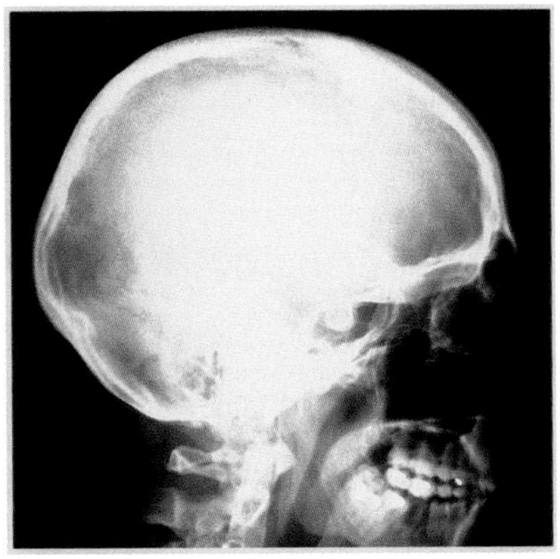

(b) X RAY

CHAPTER 46
Research Notes

BACKGROUND

Tomography is any of several techniques for making X-ray pictures of a plane section of a solid object by obscuring the images of other planes. The word comes from the Greek words *tomos,* which means "to cut," and *graphein,* which means "to draw or write." Computerized tomography (CT), or computerized axial tomography (CAT), has become a widely used diagnostic tool in medicine. CT is the preferred technique for assessing stroke and abdominal tumors. Magnetic resonance imaging (MRI) provides anatomical and chemical information and is especially useful in examining the brain, spinal cord, bones, urinary bladder, pelvic organs, and cartilage and ligaments in joints. Positron emission tomography (PET) is a very sensitive procedure for detecting blood flow in the brain and heart. It can be used for diagnosing stroke and neurological diseases, such as multiple sclerosis and epilepsy.

DISCUSSION

1. How does magnetic resonance imaging (MRI) provide information about the body's internal tissues? (MRI creates images activated by a magnetic scanner.)
2. How is positron emission tomography (PET) useful in brain research? (Positron emission tomography is used in evaluating brain metabolism and to map areas of the brain involved in memory, perception, speech, and information processing.)

FURTHER READING

Have students read "Sculptures of Light," by Corinna Wu, in *Science News,* October 26, 1996, Vol. 150, page 270. This article describes the production of three-dimensional, holographic images from traditional CT scans.

SECTION 46-4

SECTION OVERVIEW

Stability and Homeostasis
Section 46-4 describes how the integumentary system relates to and coordinates its functions with the rest of the body. For example, sweating helps keep body temperature optimal for metabolism and other biochemical functions.

▶ ENGAGE STUDENTS

Give students hand lenses to examine their skin. Ask them what structures they observe (pores and hairs). Ask them what passes through pores and the openings around hairs to the skin's surface (sweat and oil). Ask students how their skin looks (rough and leathery). Challenge students to infer the skin's functions from their observations (protection, fluid retention, waste elimination, temperature regulation).

TEACHING STRATEGY

Skin Cancer
Have students observe prepared slides of normal human skin and various forms of skin cancer under a microscope. Have available the photographs of skin cancers published by the American Cancer Society or by the American Medical Association. Have students draw both the microscopic and macroscopic differences between normal and cancerous skin. Lead a discussion on the causes of skin cancer (heredity and ultraviolet radiation).

SECTION

46-4

OBJECTIVES

▲ Describe the functions of the skin.

● Distinguish between the two layers that form the skin.

■ Compare the structure of hair with that of nails.

◆ Identify two types of glands found in the skin, and describe their functions.

INTEGUMENTARY SYSTEM

The integumentary system, consisting of the skin, hair, and nails, acts as a barrier to protect the body from the outside world. It also functions to retain body fluids, protect against disease, eliminate waste products, and regulate body temperature.

SKIN

The skin is one of the human body's largest organs. Subjected to a lifetime of wear and tear, the layers of skin are capable of repairing themselves. Skin contains sensory devices that monitor the external environment, and mechanisms that rid the body of wastes. The skin is composed of two layers—the epidermis and the dermis.

Epidermis

The **epidermis,** or outer layer of skin, is composed of many sheets of flattened, scaly epithelial cells. Its layers are made of mostly dead cells. These cells are exposed to the dangers of the external environment. Scraped or rubbed away on a daily basis, they are replaced by new cells made in the rapidly dividing lower layers. The cells of the epidermis are filled with a protein called **keratin,** which gives skin its rough, leathery texture and its waterproof quality.

There is a great variety in skin color among humans. The color of skin is mainly determined by a brown pigment called **melanin** (MEL-uh-nin), which is produced by cells in the lower layers of the epidermis. Melanin absorbs harmful ultraviolet radiation. The amount of melanin produced in skin depends on two factors: heredity and the length of time the skin is exposed to ultraviolet radiation. Increased amounts of melanin in a person's skin occurs in response to injury of the skin by ultraviolet radiation. All people, but especially people with light skin, need to minimize exposure to the sun and protect themselves from its ultraviolet radiation, which can damage the DNA in skin cells and lead to deadly forms of skin cancer.

Dermis

The **dermis,** the inner layer of skin, is composed of living cells and many kinds of specialized structures, such as sensory neurons, blood vessels, muscle fibers, hair follicles, and glands. Sensory neurons make it possible for you to sense many kinds of conditions and signals from the environment, such as heat and pressure. Blood vessels provide nourishment to the living cells and help regulate body

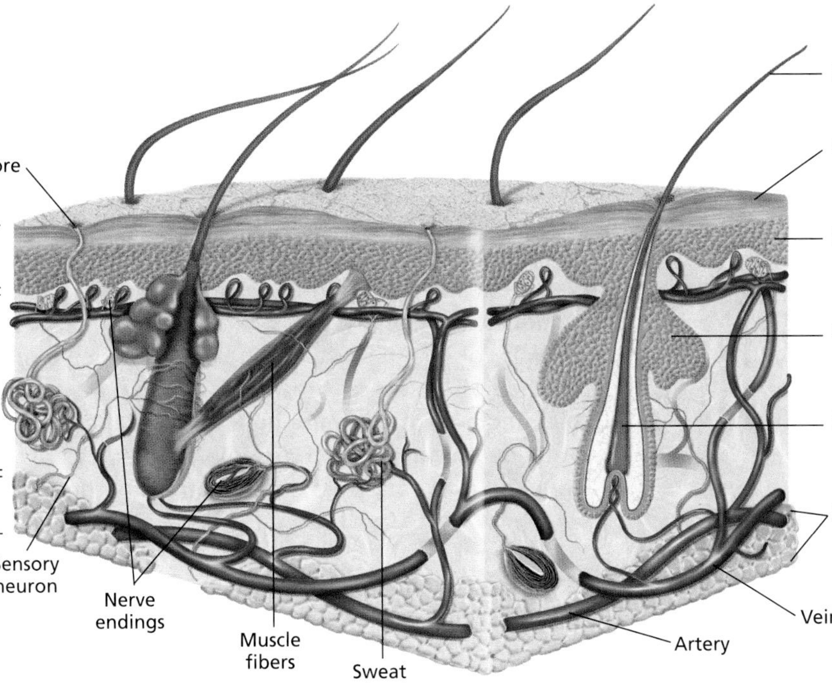

FIGURE 46-15

Skin is composed of two layers: the epidermis and the dermis. The epidermis consists of dead, flattened cells that are shed and replaced every day. The dermis contains a variety of specialized structures that carry out essential life processes, such as protecting the body from infectious diseases and regulating body temperature.

temperature. Tiny muscle fibers attached to hair follicles contract and pull hair upright when you are cold or afraid, producing what are commonly called goose bumps. Glands produce sweat, which helps cool your body, and oil, which helps soften your skin. A layer of fat cells lies below the dermis. These cells act as energy reserves; add a protective, shock-absorbing layer; and insulate the body against heat loss. Study the structures of the skin in Figure 46-15.

Nails and Hair

Nails, which protect the ends of the fingers and toes, form from nail roots under skin folds at their base and sides. As new cells form, the nail grows longer. Like hair, nails are composed primarily of keratin. The nail body is about 0.5 mm (0.02 in.) thick, and growth continues throughout life. Nails grow at about 1 mm (0.04 in.) per week. Nails rest on a bed of tissue filled with blood vessels, giving the nails a pinkish color. The structure of a fingernail can be seen in Figure 46-16.

Changes in the shape, structure, and appearance of the nails may be an indicator of a disease somewhere in the body. They may turn yellow in patients with chronic respiratory disorders, or they may grow concave in certain blood disorders.

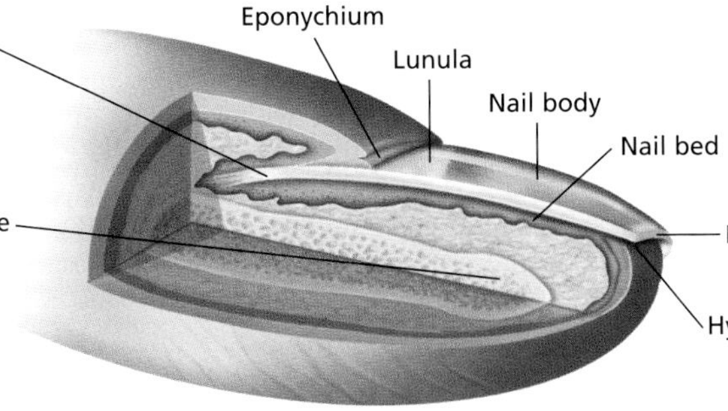

FIGURE 46-16

This illustration of the structure of a fingernail shows that the nail root, from which the nail is constantly regenerated, is protected well beneath the surface of the finger, next to the bone of the fingertip.

CRITICAL THINKING

Hair Thinning

Hair thinning, or alopecia, can be induced by a number of factors, such as acutely high fever, surgery, severe emotional trauma, certain drugs, protein-deficient diets, and lactation. In all of these cases, hair grows back if the cause of the thinning is removed or corrected. However, hair loss due to prolonged physical trauma, excessive radiation, or genetic factors is permanent. Have students research the different causes of alopecia and the various attempts at cures (drugs inhibiting testosterone production, minoxidil, hair implantation).

DEMONSTRATION

Three pigments contribute to skin color: melanin, carotene, and hemoglobin. Using photographs, diagrams, or student help, illustrate the following variations in human skin color:

1. Melanin varies from yellow to reddish brown to black. Differences in skin color are caused by the amount and type of melanin.
2. Freckles and moles are accumulations of melanin in one spot.
3. Tans are caused by the triggering of melanin production by sunlight.
4. Lack of oxygenated hemoglobin causes cyanosis, a bluish cast to the skin of fair-skinned people.
5. Pallor, or blanching, is caused by anemia or emotional stress.
6. Redness may indicate blushing, fever, inflammation, or allergy.

SECTION 46-4

ACTIVE READING
Technique: K-W-L
Tell students to return to their lists of things they **W**ant to know about bones, muscles, and skin. (Refer to the Active Reading activity on page 905.) Have them place check marks next to the questions that they are now able to answer. Students should finish by making a list of what they have **L**earned. Conclude by asking students which questions are still unanswered. Ask if they have any new questions.

ANSWERS TO SECTION 46-4 REVIEW

1. The epidermis protects the body from the scrapes and friction encountered in everyday life. The dermis contains the living cells of the skin.
2. Melanin is a brown pigment produced to protect the skin from harmful ultraviolet radiation. All colors of skin and especially light-colored skin can be easily injured by the sun's radiation during sunbathing.
3. The dermis contains a variety of structures, such as sensory neurons, blood vessels, and glands, that enable the body to respond to conditions in the internal and external environment.
4. Hair and nails grow from a bed of active cells and are made of keratin.
5. Sweat glands secrete water, which helps cool the body. Oil glands secrete sebum, which helps soften and waterproof the skin.
6. Third-degree burns leave the body susceptible to infections. The skin is no longer capable of releasing water, salts, and urea, and it is unable to regulate body temperature.

FIGURE 46-17
Skin acts as a temperature-controlling device. It contains millions of sweat glands that secrete microscopic droplets of water. The water droplets help cool the body when its temperature rises, such as after a rigorous workout.

Hair, which protects and insulates the body, is produced by a cluster of cells at the base of deep dermal pits called **hair follicles.** The shaft of the hair that extends beyond the skin is composed of dead, keratin-filled cells that overlap like roof shingles. Glands associated with hair follicles secrete oil, which prevents hair from drying out and breaking off. Most individual hairs grow for several years and then fall out. However, hair on the scalp can grow continuously for many years.

Hair color is the result of the presence of the pigment melanin in the hair shaft. Black, brown, and yellow variants of melanin combine in an almost infinite number of proportions to determine an individual's hair color. Hair color is influenced by hereditary factors and may vary over different areas of the body.

Glands

The skin contains **exocrine glands,** glands that release secretions through ducts. The main exocrine glands of the skin are the sweat glands and the oil glands.

The skin functions as an excretory organ by releasing excess water, salts, and urea through the **sweat glands.** By releasing excess water, the skin also helps regulate body temperature. When the body's temperature rises, circulation increases, and the skin becomes warm and flushed, as shown in Figure 46-17. The sweat glands then release sweat. As the water in sweat evaporates, the skin is cooled.

Oil glands, found in large numbers on the face and scalp, secrete a fatty substance known as **sebum.** Oil glands are usually connected by tiny ducts to hair follicles. Sebum coats the surface of the skin and the shafts of hairs, preventing excess water loss and lubricating and softening the skin and hair. Sebum is also mildly toxic to some bacteria. If the ducts of oil glands become clogged with excessive amounts of sebum, dead cells, and bacteria, the skin disorder **acne** can result. The production of sebum is controlled by hormones. During adolescence, high levels of sex hormones increase the activity of the skin's oil glands, often resulting in skin eruptions called acne. Although it is difficult to prevent, acne can sometimes be controlled with meticulous skin care.

SECTION 46-4 REVIEW

1. What are the names and functions of the two layers of skin?
2. What is melanin? Why can sunbathing be considered dangerous?
3. Why is the dermis considered the living layer of skin?
4. In what two ways are hair and nails similar?
5. What are the functions of the two types of exocrine glands found in the dermis?
6. CRITICAL THINKING Why are third-degree burns—which destroy the epidermis and dermis of the skin—over large regions of the body often fatal?

CHAPTER 46 REVIEW

SUMMARY/VOCABULARY

46-1

- A tissue is a collection of cells that work together to perform a particular function.
- The human body has four main types of tissue: muscle, nervous, epithelial, and connective.
- An organ consists of various tissues that work together to carry out a specific function.
- An organ system is a group of organs interacting to perform a life process.
- Many organs and organ systems in the human body are housed in body cavities.

Vocabulary

abdominal cavity (908)
cardiac muscle (905)
connective tissue (906)
cranial cavity (908)
diaphragm (908)
epithelial tissue (906)
matrix (906)
muscle tissue (905)
neuron (905)
nervous tissue (905)
skeletal muscle (905)
smooth muscle (905)
spinal cavity (908)
thoracic cavity (908)

46-2

- The human skeleton is composed of the axial skeleton (skull, ribs, spine, and sternum) and the appendicular skeleton (arms and legs, scapula, clavicle, and pelvis).
- Bones support muscles and organs, give structure to the body, protect internal organs, store minerals, and manufacture blood cells.
- Bones are made up of minerals, protein fibers, and cells called osteocytes. Most consist of compact and spongy bone and may contain bone marrow.
- Most bones develop from cartilage through a process called ossification.
- Bone elongation occurs near the ends of long bones, at the epiphyseal plate.
- The human body has three types of joints—fixed, semimovable, and movable.

Vocabulary

appendicular skeleton (909)
axial skeleton (909)
ball-and-socket joint (913)
bone marrow (911)
compact bone (911)
epiphyseal plate (912)
fixed joint (913)
fracture (911)
gliding joint (913)
Haversian canal (911)
hinge joint (913)
joint (913)
ligament (913)
movable joint (913)
ossification (911)
osteoarthritis (914)
osteocyte (911)
periosteum (910)
pivot joint (913)
rheumatoid arthritis (914)
saddle joint (913)
semimovable joint (913)
skeleton (909)
spongy bone (911)
synovial fluid (914)

46-3

- The human body has three types of muscle tissues—skeletal muscle, smooth muscle, and cardiac muscle.
- Skeletal muscles consist of groups of muscle fibers that contain threadlike myofibrils. Each myofibril is made up of two types of protein filaments, thin actin filaments and thick myosin filaments.
- A sarcomere is the fundamental unit of a muscle contraction. During a muscle contraction, myosin and actin filaments interact to shorten the length of a sarcomere.
- Most skeletal muscles are arranged in opposing pairs.

Vocabulary

actin (917)
extensor (919)
fascicle (915)
flexor (919)
insertion (919)
involuntary muscle (916)
muscle fatigue (920)
muscle fiber (915)
myofibril (917)
myosin (917)
origin (919)
oxygen debt (920)
sarcomere (917)
striation (915)
tendon (919)
voluntary muscle (916)
Z line (917)

CHAPTER 46 REVIEW ANSWERS

REVIEW

1. Epithelial tissue consists of layers of cells that line internal and external body surfaces.
2. Compact bone is a hard material composed of rings of mineral crystals, protein fibers, and osteocytes. Spongy bone is a porous network of tissue.
3. A sarcomere, the functional unit of muscle contraction, is composed of actin filaments and myosin filaments. Actin filaments are attached to Z lines, which define the boundaries of a sarcomere.
4. Skeletal muscle is striated, voluntary muscle attached to bone. Smooth muscle is involuntary muscle found in the walls of the digestive system and blood vessels. Cardiac muscle is striated involuntary muscle found in the walls of the heart.
5. The epidermis, the outer layer of skin, is made up primarily of dead cells. The dermis, the inner layer of skin, is composed of living cells.

6. c **7.** b
8. a **9.** c
10. d **11.** b
12. d **13.** c
14. d **15.** a

16. The structure is a sarcomere, the functional unit of a muscle contraction. A sarcomere comprises the distance between two Z lines. It shortens during a muscle contraction.
17. Cells form tissues, which work together in organs. Different organs interact in an organ system.
18. A body cavity is a compartment that contains organs. The abdominal cavity contains the organs of the digestive, reproductive, and excretory systems.
19. The skeletal system supports muscles and organs, gives the body shape and structure, protects delicate internal organs, and stores important minerals, such as calcium and phosphorus. The

ANSWERS

CONTINUED

internal portion of long bones produces red blood cells and certain types of white blood cells.

20. Haversian canals are interconnected and provide pathways for blood vessels and nerves to reach osteocytes.

21. Red bone marrow is soft tissue found in the ends of long bones. Red bone marrow produces red blood cells and certain types of white blood cells.

22. Rheumatoid arthritis develops when the immune system begins attacking body tissues. As a result, joints become swollen, deformed, and sore.

23. Tendons connect muscles to bones. Ligaments connect bones to bones.

24. Sebum, which is produced in oil glands, keeps skin and hair soft.

25. Melanin is a brown pigment produced by cells in the dermis. It absorbs the ultraviolet radiation in sunlight.

CRITICAL THINKING

1. Ossification continues after horses are born, and thus the bones of young horses are less able to withstand stress than are those of adult horses.

2. Red blood cells are transported to other parts of the body through the bloodstream, which enters the bones through the Haversian canals.

3. The larger, oval-shaped pelvis of the mother gives the baby more room to pass through the birth canal. The plasticity of the baby's skull bones allows the head to be compressed when passing through the birth canal.

4. Blood vessels need to constrict in only one direction to push blood through the body. The stomach and small intestines constrict in two directions to break up food and move it along the digestive system.

5. Washing too frequently can dry out the skin and remove the fatty

CHAPTER 46 REVIEW

46-4

- Skin, hair, and nails act as barriers that protect the body from the environment.
- Skin is composed of two layers: the epidermis (composed of dead, keratin-filled cells) and the dermis (composed of living cells and a variety of structures).
- Hair and nails are composed of the protein keratin; they grow from a bed of rapidly dividing cells.
- Sweat glands produce sweat, which evaporates and helps cool the body. Oil glands secrete sebum, which helps soften the skin and prevents hair from drying out.

Vocabulary

acne (924)	exocrine gland (924)	melanin (922)	sebum (924)
dermis (922)	hair follicle (924)	oil gland (924)	sweat gland (924)
epidermis (922)	keratin (922)		

REVIEW

Vocabulary

1. What is epithelial tissue?
2. Distinguish between compact bone and spongy bone.
3. Describe the components of a sarcomere.
4. Describe how skeletal muscle, smooth muscle, and cardiac muscle differ.
5. What is the difference between the epidermis and the dermis?

Multiple Choice

6. The thoracic cavity contains the (a) brain (b) organs of the digestive system (c) organs of the respiratory system (d) spine.
7. The cells of connective tissue are embedded in a substance called (a) marrow (b) matrix (c) synovial fluid (d) keratin.
8. The periosteum is a membrane that (a) covers the bone (b) produces red blood cells (c) contains marrow (d) protect the ends of long bones.
9. During ossification (a) cartilage replaces bone cells (b) bones become more porous (c) bone replaces cartilage (d) marrow is produced.
10. Hinge joints allow (a) gliding movement (b) circular movement (c) no movement (d) back-and-forth movement.
11. Cardiac muscle is (a) voluntary muscle (b) involuntary muscle (c) a type of smooth muscle (d) found in the lungs.
12. Actin and myosin are types of protein found in (a) cartilage (b) bone cells (c) hair and nails (d) muscle cells.
13. Red bone marrow (a) acts as an energy reserve (b) cushions the ends of bones (c) produces red and white blood cells (d) contains melanin.
14. The bones of a joint are held in place by (a) tendons (b) the epiphyseal plate (c) sarcomeres (d) ligaments.
15. The dermis (a) contains nerves and blood vessels (b) contains keratin (c) is the top layer of skin (d) is made up of dead cells.

Short Answer

16. Identify the structure in the figure below. Describe the components of this structure, and explain how it changes during a muscle contraction.

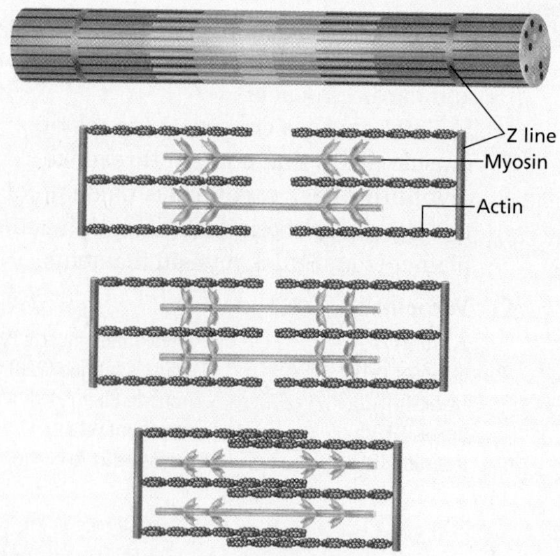

17. What is the relationship between cells and organ systems?
18. What is a body cavity? What organs are found in the abdominal cavity?
19. What are five functions of the skeletal system?
20. Explain the role Haversian canals play in compact bone.
21. What is red bone marrow? Where is it produced and what is its function?
22. Describe the cause and symptoms of the disease rheumatoid arthritis.
23. Compare the functions of tendons and ligaments.
24. What substance prevents the hair and skin from drying out? Where is this substance produced?
25. What is melanin? What is its role in the body?

CRITICAL THINKING

1. Young thoroughbred horses that are raced too early in life have an increased risk of breaking the bones in their legs. What can you infer about the process of ossification in horses?
2. Red bone marrow produces red blood cells. How are these cells transported to the rest of the body?
3. During a normal birth, a baby passes through the mother's pelvis. A woman's pelvis has a larger diameter and is more oval-shaped than a man's pelvis. In addition, a newborn's skull bones are not completely ossified. How are these skeletal properties advantageous to the birthing process?
4. The walls of blood vessels are encircled by a single layer of smooth muscle. The walls of the stomach and small intestine have a layer of circular smooth muscle and a layer of longitudinal smooth muscle. How do these muscle arrangements reflect the functions of each structure?
5. Oil glands secrete an oily substance that helps keep the skin soft and flexible. They also secrete fatty acids, which help kill bacteria. How can their function be affected if you wash your skin too frequently?
6. Examine the drawing of epithelial cells below. The flat epithelial cells of the skin overlap each other much like shingles on a roof do. How does this arrangement enable these cells to perform their protective function?

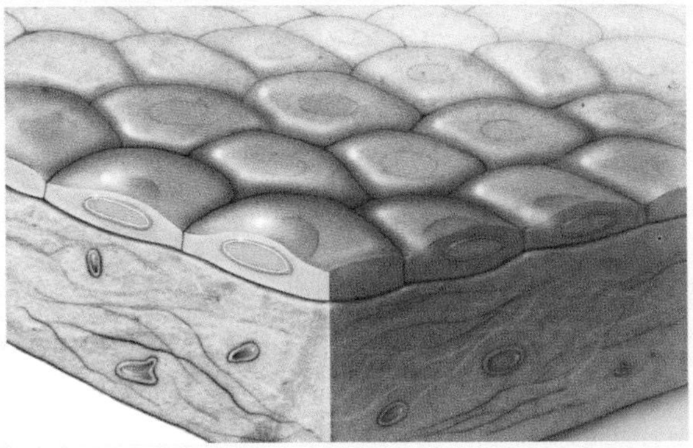

EXTENSION

1. The knee is a joint covered by the patella, or kneecap. Find out what function the kneecap serves. Research and write a report about an athlete who has suffered an injury to the patella. Report the cause of and treatment for the injury.
2. Read "Musical Muscles" in *Discover*, August 1999, on page 25. What are vocal cords? How do vocal cords make sound? Describe the damage that singers often do to their vocal cords. What is the condition called, and how is it treated?
3. Choose a form of exercise, such as running, cycling, or weight training. Write a report about the types of skills that are necessary for the exercise and the long-term effects it has on the body.

ANSWERS

CONTINUED

acids, thus reducing the skin's natural defense against bacteria.

6. The overlapping arrangement of cells helps eliminate cracks through which water and other substances can pass.

EXTENSION

1. The kneecap allows the hinge joint to open to 180 degrees. Professional athletes who have sustained serious knee injuries include former football player Joe Namath and tennis player Billie Jean King.

2. Vocal cords are two flaps of muscle that open and close quickly and function as valves. These flaps of muscle do not allow anything but air to enter the lungs, and they aid in the coughing up of phlegm. The vocal cords make sound like the open end of a blown-up balloon, which buzzes rapidly if pulled tight so that air can escape slowly. Singers often strain their voices to make sounds that are not natural to humans. Over time this straining can produce calluslike nodes on the cords that make the voice sound course. Vocal nodes are frequently treated with steroids.

3. Answers will vary. Encourage students to explore a variety of exercises.

CHAPTER 46 INVESTIGATION

TIME REQUIRED

Two 50-minute class periods, five to seven days apart

SAFETY PRECAUTIONS

Remind students to wear safety goggles, a lab apron, and gloves during all parts of this investigation. Discuss all safety symbols and caution statements with students. Review the safety rules for working with acids in the laboratory.

QUICK REFERENCES

Lab Preparation Notes are found on page 903B.

Holt BioSources provides a Teaching Resources CD-ROM that contains "Using Gowin's Vee in the Lab" and "Scoring Rubrics."

PROCEDURAL TIPS

1. Depending on the thickness of the bone, demineralizing may take from five to seven days. Students should check their progress daily.
2. The part of a bone that touches the glass of a beaker will not decalcify. Line the beaker with gauze to help the bones to decalcify more evenly.
3. You may want to have half of your class do Part A and the other half do Part B.
4. Remind students to always begin on low power when focusing the microscope.

ANSWERS TO ANALYSIS AND CONCLUSIONS

1. Water loss caused bone to become more brittle. Mineral loss caused bone to become more flexible.
2. Demineralizing bone requires that it soak in HCl for several days. HCl contains water that may be absorbed by the bone. This

CHAPTER 46 INVESTIGATION

Dehydrating and Demineralizing Bone

OBJECTIVES

- Determine the amount of water and minerals in bone.
- Identify structures in bone cells.

PROCESS SKILLS

- observing
- identifying
- calculating

MATERIALS

- safety goggles
- lab apron
- gloves
- bones (2)
- drying oven
- balance
- specimen tag
- tongs
- 250 mL beaker
- wax pencil
- gauze, circular piece
- hot pad
- 500 mL beakers (2)
- 300 mL of 1 M hydrochloric acid
- glass plate or parafilm
- prepared bone slides
- compound microscope
- lens paper
- resealable plastic bag
- permanent marker

Background

1. Dehydration is the process of removing the water from a substance.
2. Demineralization is the process of removing the minerals from a substance.

TABLE A *DEHYDRATION OF BONE*

Mass before drying	Mass after drying	Percentage of bone mass lost
	Answers will vary.	

PART A Dehydrating a Bone

1. In your lab report, prepare a data table similar to Table A.
2. Put on safety goggles, a lab apron, and gloves. Wear this protective gear during all parts of this investigation.
3. Obtain a bone from your teacher. Test the flexibility of the bone by trying to bend and twist it.
4. Place the bone on a balance. Measure the mass of the bone to the nearest 0.1 g, and record it in your data table. Then use a permanent marker to write the initials of each member of your group on a specimen tag, and tie the tag to the bone.
5. Place the bone in a drying oven at 100°C for 30 minutes. While the bone is in the oven, complete Part C.
6. CAUTION **Do not touch hot objects with your bare hands. Use insulated gloves and tongs as appropriate.** Using tongs, remove the bone from the oven and place it on a heatproof pad to cool for 10 minutes.
7. Use tongs to place the cooled bone on the balance. Measure the mass of the bone to the nearest 0.1 g, and record it in your data table.
8. Use the equation below to calculate the percentage of the bone's mass that was lost during heating.

Percentage mass lost =

$$\frac{\text{mass before heating} - \text{mass after heating}}{\text{mass before heating}} \times 100$$

PART B Demineralizing a Bone

9. In your lab report, prepare a data table similar to Table B.
10. Obtain a second bone from your teacher. Test the flexibility of the bone by trying to bend and twist it.

TABLE B DEMINERALIZATION OF A BONE

Mass before demineralizing	Mass after demineralizing and drying	Percentage of bone mass lost
	Answers will vary.	

11. Place the bone on a balance. Measure the mass of the bone, and record it in your data table.

12. CAUTION **Glassware is fragile. Notify your teacher promptly of any broken glass or cuts. Do not clean up broken glass or spills unless your teacher tells you to do so.** Using a wax pencil, label a 500 mL beaker "1 M HCl." Also label the beaker with the initials of all group members. Place a piece of gauze in the bottom of the beaker.

13. CAUTION **If you get an acid on your skin or clothing, wash it off at the sink immediately while calling to your teacher.** Place the bone on top of the gauze in the beaker, and add enough 1 M HCl to cover the bone. Use a glass plate or parafilm to cover the beaker.

14. Place the beaker under a fume hood, and allow the bone to soak in the acid until it softens and becomes spongy. This should take 5–7 days. Periodically use tongs to test the hardness of the bone. *Note: Do not touch the bone with your fingers while it is soaking in acid. Rinse the tongs with water thoroughly each time you finish testing the bone.*

15. When the bone becomes spongy, use tongs to carefully remove it from the beaker, and rinse it under running water for two minutes.

16. After the bone has been thoroughly rinsed, test the bone for hardness by twisting and bending it with your fingers. *Note: Be sure you are wearing gloves.*

17. Then use a permanent marker to write the initials of each member of your group on a specimen tag, and tie the tag to the bone. Place the bone in a drying oven at 100°C for 30 minutes.

18. CAUTION **Do not touch hot objects with your bare hands. Use insulated gloves and tongs as appropriate.** Using tongs, remove the bone from the oven and place it on a heat-proof pad to cool for 10 minutes.

19. Use tongs to place the cooled bone on the balance. Measure the mass of the bone to the nearest 0.1 g, and record it in your data table.

20. Use the equation below to calculate the percentage of the bone's mass that was lost through demineralization and dehydration.

Percentage of mass lost =

$$\frac{\text{mass before demineralizing} - \text{mass after demineralizing and drying}}{\text{mass before demineralizing}} \times 100$$

PART C Observing Prepared Slides of Bone

21. Using a compound light microscope, focus on a prepared slide of bone using low power, and then switch to high power. Locate a Haversian canal, the darkly stained circle in the center of a set of lamellae. Find the darkly stained osteocytes between the lamellae.

22. In your lab report, draw and label the following bone structures: Haversian canal, lamella, osteocyte.

Analysis and Conclusions

1. What effect did water loss have on the bone? What effect did mineral loss have on the bone?
2. Why did you have to dehydrate the bone before measuring its mass in Part B?
3. What percentage of bone is water? What percentage of bone is mineral?
4. If you were to prepare a slide using the dehydrated and demineralized bone, what do you think it would look like?
5. What happened when the demineralized bone was dried? Why do you think this happened?
6. If a person's diet lacked calcium, how could this affect his or her bones?

Further Inquiry

Research the differences in the amount and distribution of compact bone and spongy bone among bones from various parts of the human skeleton. Discover whether there are any other differences in the internal structure of different types of bones.

water must be removed before massing.
3. Answers will vary depending on the size of the bone and length of drying time. To calculate the percentage of water lost in dehydration, use the equation in step 8. To calculate the percentage of mineral lost, use the equation in step 20.
4. A slide of a completely demineralized and dehydrated bone would have a collapsed appearance and would show no distinguishing characteristics such as Haversian canals or osteocytes.
5. The bone sample probably fell apart because there was nothing left to hold it together.
6. A diet lacking calcium would result in bones that lack mineral density. This lack of mineral density may cause reduced growth, bone deformities, or weak bone.

PLANNING GUIDE 47

CHAPTER 47
CIRCULATORY AND RESPIRATORY SYSTEMS

	TOPICS	TEACHING RESOURCES	LABS, CLASSWORK, AND HOMEWORK
BLOCKS 1 & 2 90 minutes	**Introducing the Chapter, p. 930**	ATE Focus Concept, p. 930 ATE Assessing Prior Knowledge, p. 930 ATE Understanding the Visual, p. 930	■ Occupational Applications Worksheets—Emergency Medical Technician
	47-1 The Circulatory System, p. 931 The Heart, p. 931 Patterns of Circulation, p. 935 Lymphatic System, p. 937	ATE Section Overview, p. 931 ATE Critical Thinking, p. 934 ATE Visual Link Figure 47-2, p. 932 Figure 47-3, p. 933 Figure 47-4, p. 935 185A, 186A, 210–214	PE **Quick Lab**, p. 932, Determining Heart Rate ■ Problem-Solving Worksheets—Computing Rates and Heart Efficiency ★ Study Guide, Section 47-1 PE Section Review, p. 937
BLOCKS 3 & 4 90 minutes	**47-2 Blood, p. 939** Composition of Blood, p. 939 Blood Types, p. 941	ATE Section Overview, p. 939 ATE Critical Thinking, p. 941 ATE Visual Link Figure 47-11, p. 939 Figure 47-12, p. 940 187A, 188A, 189A	PE **Quick Lab**, p. 943, Identifying Offspring **Lab C42,** Blood Typing **Lab C43,** Blood Typing—Whodunit? **Lab C44,** Blood Typing—Pregnancy and Hemolytic Disease ■ Occupational Applications Worksheets—Blood-Bank Technologist ★ Study Guide, Section 47-2 PE Section Review, p. 943
BLOCK 5 45 minutes	**47-3 The Respiratory System, p. 944** The Lungs, p. 944 Gas Exchange and Transport, p. 946 Mechanism of Breathing, p. 947	ATE Section Overview, p. 944 ATE Critical Thinking, p. 946 ATE Visual Link Figure 47-16, p. 945 181, 215, 217	PE **Chapter 47 Investigation,** p. 952 **Lab A31,** Determining Lung Capacity **Lab E6,** Hemoglobin ■ Occupational Application Worksheets—Respiratory Therapist ■ Problem-Solving Worksheets—Pressure Gradients in Breathing and Calculating Respiratory Volume ★ Study Guide, Section 47-3 PE Section Review, p. 948
BLOCKS 6 & 7 90 minutes	**REVIEW and ASSESSMENT** PE Chapter 47 Review, pp. 950–951 ■ Performance-Based Assessment—Testing Human Digestive Enzymes	Audio CD Program	**CHAPTER TESTING** ■ Chapter Test (blackline master) ▲ Test Generator ■ Assessment Item Listing ■ Scoring Rubrics and Classroom Management Checklists

MODERN BIOLOGY | HOLT BIOSOURCES | One-Stop Planner CD-ROM

KEY

PE Pupil's Edition
ATE Teacher's Edition
■ Active Reading Guide
★ Study Guide

Teaching Transparencies

Laboratory Program

One-Stop Planner CD-ROM
Shows these resources within a customizable daily lesson plan:

- ■ Holt BioSources Teaching Resources
- ■ Active Reading Guide
- ★ Study Guide
- ▲ Test Generator

READING FOR CONTENT MASTERY

Preview: Have students preview the objectives, the introductory paragraph, topic sentences, and captions before reading the chapter.

ATE Active Reading Technique: Paired Reading, p. 931
■ Active Reading Guide Worksheet 47-1
PE Summary/Vocabulary, p. 949

■ Active Reading Guide Worksheet 47-2
PE Summary/Vocabulary, p. 949

■ Active Reading Guide Worksheet 47-3
PE Summary/Vocabulary, p. 949

TECHNOLOGY AND INTERNET RESOURCES

CNN PRESENTS Video Segment 31
High Blood Pressure

Holt Biology Videodiscs Teacher's Correlation Guide, Lessons 47-1 through 47-3

Audio CD Program, Sections 47-1 through 47-3

internetconnect

sciLINKS NSTA
On-line Resources: www.scilinks.org
The following *sci*LINKS Internet resources can be found in the student text for this chapter:

Topics:
- Blood types, p. 941
- Respiratory system, p. 944

go.hrw.com
On-line Resources: go.hrw.com
Visit the HRW Web site for a variety of resources related to this chapter. Just type in the keyword HM2 Chapter 47.

Smithsonian Institution®
Visit **www.si.edu/hrw** for additional on-line resources.

CNNfyi.com
Visit **www.cnnfyi.com** for late-breaking news and current event stories selected just for you.

LAB ACTIVITY PLANNING

CHAPTER 47 INVESTIGATION

Tidal Volume, Expiration Volume, and CO_2 Production, pp. 952–953

OVERVIEW
In this investigation students determine lung capacities and the effect of exercise on breathing rate and CO_2 production rate. Students form a hypothesis to explain the effect of exercise on breathing rate and CO_2 production. Students collect and analyze data to determine whether their data support their hypotheses.

SAFETY
Caution students with health problems not to perform vigorous exercise. Do not allow them to continue if they feel pain or become dizzy or tired. Caution students to breathe normally between steps to avoid hyperventilation. Students should wear safety goggles during the lab. Bromothymol blue may cause skin irritation. If it splashes in the eyes, flush with water and seek medical attention.

PREPARATION
1. Spirometers are available from biological supply companies, such as WARD'S. Provide enough clean spirometer mouthpieces for each participant. Provide one spirometer per lab group.
2. A 0.1 percent bromothymol indicator solution may be purchased from a biological supply company, such as WARD'S. Each lab group will need two 100 mL Erlenmeyer flasks with 100 mL of bromothymol blue in each flask.

Quick Labs

Determining Heart Rate, p. 932
Each student pair needs a stopwatch or a clock with a second hand for this activity.

Identifying Offspring, p. 943
Students need only paper and pencils for this activity.

UNDERSTANDING THE VISUAL

Explain to students that the photograph shows the smallest unit of the lung. These units are not just empty pockets of air. They contain cells that secrete fluids to keep the inner surfaces of the alveoli moist, and they contain macrophages that help fight bacteria, viruses, and foreign particles that are inhaled into the lungs. Point out that there is a vast network of capillaries that lie near the alveoli and facilitate the exchange of gases. Have students identify the red blood cells present in the photograph. Explain that they are inside the alveoli because this is a slide prepared from biopsied tissue. Red blood cells are not found inside alveoli in healthy, living lungs. Discuss the interrelatedness of the cardiovascular system and the respiratory system.

FOCUS CONCEPT

Structure and Function

As you read, note how the structural features of the organs in the circulatory and respiratory systems are related to the functions of transport and exchange of materials.

ASSESSING PRIOR KNOWLEDGE

Review the following concepts.

Cell Structure and Function: *Chapter 5*
Review cellular transport and exchange. Ask students to describe the direction that substances diffuse across a semipermeable membrane. Ask them if this is active or passive transport.

Matter, Energy, and Organization: *Chapter 46*
Review the organization of the body into tissues, organs, and organ systems. Have students list as many organ systems as they can and name the function of each system.

CHAPTER 47

CIRCULATORY AND RESPIRATORY SYSTEMS

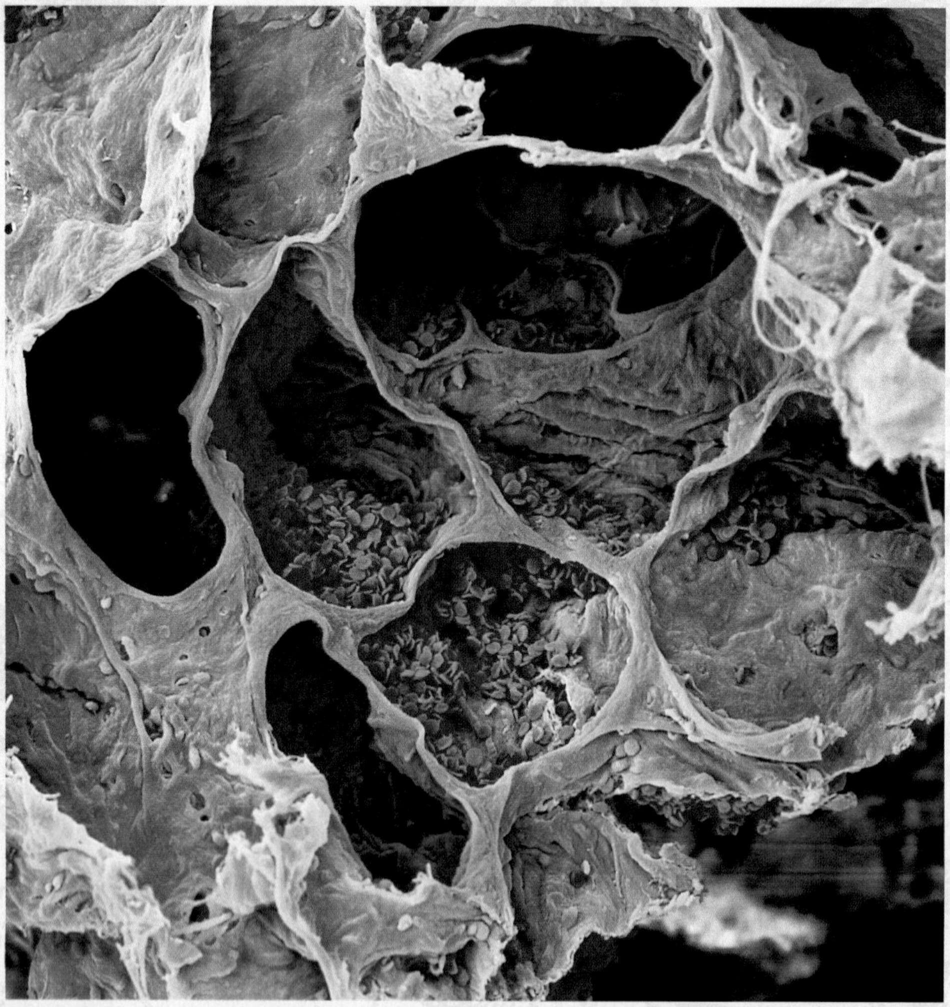

This photograph shows the air sacs of a human lung. (SEM 780×)

FOCUS CONCEPT: ***Structure and Function***

As you read, note how the structural features of the organs in the circulatory and respiratory systems are related to the transport and exchange of materials.

THE CIRCULATORY SYSTEM

Most of the cells in the human body are not in direct contact with the external environment. The circulatory system acts as a transport service for these cells. Two fluids move through the circulatory system: blood and lymph. The blood, heart, and blood vessels form the ***cardiovascular system.*** *The lymph, lymph nodes, and lymph vessels form the* ***lymphatic system.*** *The cardiovascular system and lymphatic system collectively make up the* ***circulatory system.***

SECTION

47-1

OBJECTIVES

▲ Describe the structure and function of the human heart.

● Trace the flow of blood through the heart and body.

■ Distinguish between arteries, veins, and capillaries in terms of their structure and function.

◆ Distinguish between pulmonary circulation and systemic circulation.

▲ Describe the functions of the lymphatic system.

THE HEART

The central organ of the cardiovascular system is the heart, the muscular organ that pumps blood through an intricate network of blood vessels. The heart beats more than 2.5 billion times in an average life span. Yet this organ that does so much work is slightly larger than a fist. The heart lies within the thoracic (chest) cavity, behind the sternum (breastbone) and between the two lungs. A tough, saclike membrane called the **pericardium** surrounds the heart and secretes a fluid that reduces friction as the heart beats.

Notice in Figure 47-1 that a **septum** vertically divides the heart into two sides. The right side pumps blood to the lungs, and the left side pumps blood to the other parts of the body. Each side of the heart is divided into an upper and lower chamber. Each upper chamber is called an **atrium,** and each lower chamber is called a **ventricle.**

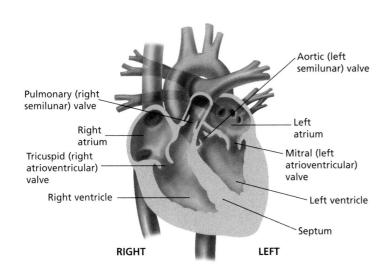

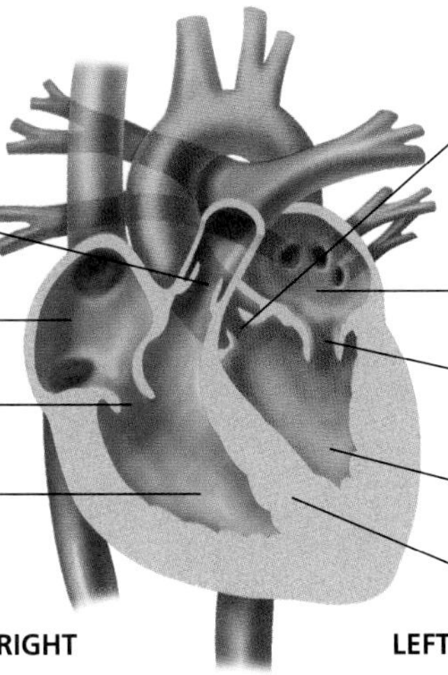

FIGURE 47-1

The septum prevents mixing of blood from the two sides of the heart, and the valves ensure that blood flows in only one direction.

SECTION 47-1

SECTION OVERVIEW

Structure and Function
Section 47-1 describes the structure of the heart and blood vessels. Structure and function is emphasized in the explanation of the subsystems of the circulatory system, each of which interacts with other body systems.

Stability and Homeostasis
Section 47-1 discusses the structures that promote stability in the cardiovascular system, such as valves to ensure unidirectional blood flow and maintenance of a stable blood pressure.

ACTIVE READING

Technique: Paired Reading
Pair each student with a partner. Have students identify one or more paragraphs or pages containing topics or concepts they find difficult to understand. One student should read the paragraph aloud, and then the other student should repeat the information in his or her own words. Both students should agree on the meaning of each paragraph.

RECENT RESEARCH

Gender Differences in Heart Disease
The first signs of heart disease are likely to be different in women than in men. In almost 67 percent of men, the first symptom is a heart attack. In 56 percent of women, the first sign is angina—fleeting, sharp, squeezing chest pains.

SECTION 47-1

Quick Lab

Determining Heart Rate

Time Required 15 minutes

Procedural Tips Have students work in pairs. Remind students to remain seated, standing, or lying down for a few minutes before their partner measures their pulse.

Answers to Analysis The pulse is caused by blood starting and stopping as it rushes through the arteries. The pulse changes in each position, depending on the force necessary to move the blood. When an individual stands upright, the heartbeat increases to push the blood through the body while overcoming the force exerted by gravity. When an individual lies down, the force is reduced and the heart beats more slowly.

VISUAL LINK

Figure 47-2

Tell students that the color of blood varies with the amount of oxygen it contains, from scarlet (oxygenated) to deep red (deoxygenated). Deoxygenated blood is traditionally depicted in blue to differentiate veins from arteries in illustrations. Blood vessels that are close to the surface of the skin appear blue because of the light-refractive properties of the vessel wall and the skin. Explain also that "deoxygenated" blood still contains up to 75 percent of its original oxygen in a human at rest.

Quick Lab

Determining Heart Rate

Materials stopwatch or clock with second hand

Procedure

1. Have your partner find the pulse in your wrist and count your heartbeats for 15 seconds while you are seated. Calculate your resting heart rate in beats per minute.
2. Have your partner count your heartbeats for 15 seconds while you are standing. Calculate your heart rate in beats per minute.
3. Have your partner count your heartbeats for 15 seconds while you are lying down. Calculate your heart rate in beats per minute.

Analysis What causes your pulse? What causes the change in your heart rate in each position?

A one-way valve separates each atrium from the ventricle beneath it. These valves are called **atrioventricular** (AY-tree-oh-ven-TRIH-kyuh-luhr) **(AV) valves.** They consist of flaps of tissue that open in only one direction. The AV valve on the right side is called the **tricuspid valve.** The **mitral valve,** also called the bicuspid valve, is on the left. As the ventricles pump, blood pressure closes the AV valves, preventing blood from flowing backward from the ventricles to the atria.

From the ventricles, blood flows out of the heart into large vessels. A **semilunar** (semee-LOON-uhr) **(SL) valve** separates the ventricles from these large vessels on each side of the heart. The SL valve on the right side is known as the **pulmonary valve,** and the SL valve on the left side is known as the **aortic valve.** The SL valves prevent blood from flowing back into the ventricles when the heart relaxes.

Circulation in the Heart

Refer to Figure 47-2 to trace the path of the blood as it circulates through the heart. Blood returning to the heart from parts of the body other than the lungs has a high concentration of carbon dioxide and a low concentration of oxygen. This deoxygenated blood enters the right atrium. Notice in Figure 47-2 that the flow of blood on the right side of the heart is illustrated with a blue arrow representing deoxygenated blood, which has a deep bluish-red color.

The right atrium pumps the deoxygenated blood into the right ventricle. The muscles of the right ventricle contract and force the blood into the pulmonary arteries, which lead to the lungs. In the lungs, the carbon dioxide diffuses out of the blood and oxygen diffuses into the blood. The oxygenated blood returns to the left

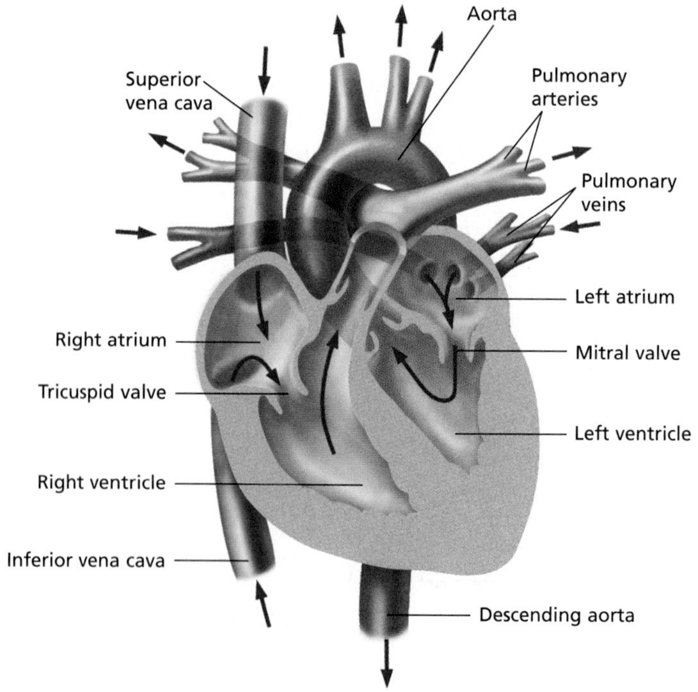

FIGURE 47-2

Trace the path of blood through the heart. Which side of the heart contains deoxygenated blood? Notice that illustrations of a heart are drawn as if the heart were in a person facing you, that is, the left side of the heart is shown on the right as you face it, and the right side of the heart is on the left as you face it.

atrium of the heart. Notice in Figure 47-2 that the flow of blood on the left side of the heart is illustrated with a red arrow representing oxygenated blood, which has a bright red color.

The oxygenated blood is then pumped into the left ventricle. Contraction of the muscular walls of the left ventricle force the blood into a large blood vessel called the **aorta.** From the aorta, blood is transported to all parts of the body except the lungs. The left ventricle is the thickest chamber of the heart because it has to do the most work to pump blood to all parts of the body.

Control of the Heartbeat

The heart consists of muscle cells that contract in waves. When the first group of cells are stimulated, they in turn stimulate neighboring cells. Those cells then stimulate more cells. This chain reaction continues until all the cells contract. The wave of activity spreads in such a way that the atria and the ventricles contract in a steady rhythm. The first group of heart-muscle cells that get stimulated lie in an area of the heart known as the sinoatrial node, shown in Figure 47-3.

The **sinoatrial** (SIEN-oh-AY-tree-uhl) **(SA) node** is a group of specialized heart-muscle cells located in the right atrium. These muscle cells spontaneously initiate their own electrical impulse and contract. The SA node is often called the **pacemaker** because it regulates the rate of contraction of the entire heart. The electrical impulse initiated by the SA node subsequently reaches another special area of the heart, known as the **atrioventricular (AV) node.** The AV node is located in the septum between the atria, as shown in Figure 47-3. The AV node relays the electrical impulse to the muscle cells that make up the ventricles. As a result, the ventricles contract a fraction of a second after the atria, completing one full heartbeat. In an average adult at rest, the heart beats about 70 times each minute.

A heartbeat has two phases. Phase one, called **systole** (SIS-tohl), occurs when the ventricles contract, closing the AV valves and opening the SL valves to pump blood into the two major vessels that exit the heart. Phase two, called **diastole** (DIE-a-stohl), occurs when the ventricles relax, allowing the back pressure of the blood to close the SL valves and opening the AV valves. The closing of these two heart valves results in the characteristic *lubb dup* sound we call a heartbeat. If one of the valves fails to close properly, some blood may flow backward, creating a different sound, which is known as a heart murmur.

Blood Vessels

The circulatory system is known as a closed system because the blood is contained within either the heart or the blood vessels at all times. This type of system differs from an open system, in which blood leaves the vessels and circulates within tissues throughout the organism's body. The blood vessels that are part of the closed circulatory system of humans form a vast network to help keep the blood flowing in one direction.

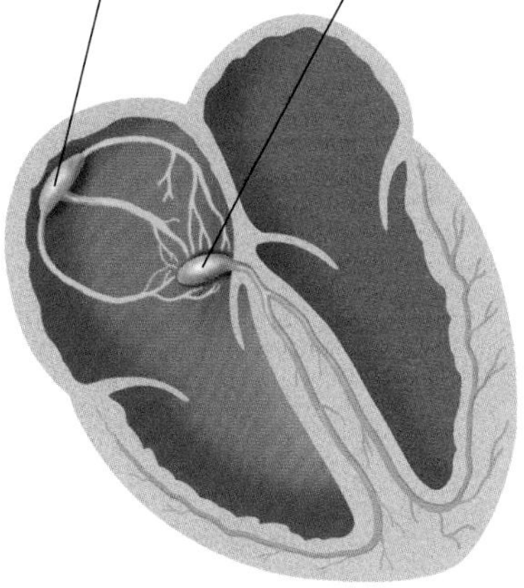

FIGURE 47-3

Two areas of specialized tissue, known as nodes, control the heartbeat. A person whose SA node is defective can have an operation to implant an artificial pacemaker. An artificial pacemaker can also help a defective AV node. Why is the term *artificial pacemaker* appropriate?

SECTION 47-1

VISUAL LINK
Figure 47-3

Tell students that if the heart is damaged so that no electrical impulses can be conducted from the atria to the ventricles, a fixed-rate pacemaker, set to deliver an impulse at a constant rate, is usually implanted. In hearts in which only a fraction of the atrial impulses reach the ventricles, a demand pacemaker, which delivers impulses only when the heart is not transmitting on its own, is usually implanted.

TEACHING STRATEGY
Pacemakers

Have students research artificial pacemakers and their use in patients with heart disease. Have students find out what types of heart patients are candidates for pacemaker implants. Students can make drawings of pacemakers connected to hearts and write explanations of how they work. Have students share their findings with the class.

TEACHING STRATEGY
Heartbeat

Obtain a stethoscope, and allow students to use it to listen to their heartbeat. Ask them to explain what causes the "lubb dup" sound. (The "lubb" sound is caused by the closing of the AV valves while the ventricles contract and force blood through the SL valves. The "dup" sound is caused by the snapping closed of the SL valves as blood flows into the ventricles from the atria.) Help students understand that although the heartbeat is described for convenience as occurring first on the right side and then on the left, the movement of blood occurs on both sides simultaneously.

TEACHING STRATEGY

Blood Pressure

Provide students with a sphygmomanometer (blood pressure cuff), alcohol swabs, and a stethoscope. Have the students work in pairs to obtain radial-artery blood-pressure readings. Students should clean the earpieces of the stethoscope with the alcohol swabs and completely deflate the blood-pressure cuff before using it. A partially inflated cuff can cause errors in measurement. Have the students take turns measuring their blood pressure, following the instructions with the cuff.

QUICK FACT

✔ One indicator of congestive heart failure is blood pressure that is consistently different in one arm than in the other arm.

CRITICAL THINKING

Hypertension

Chronic hypertension is a common and dangerous disease that warns of increased peripheral resistance. About 30 percent of people over 50 are hypertensive. Have students research and write a report on hypertension. Have them address the factors of diet, obesity, age, race, heredity, stress, and smoking. Have students determine how (or if) these causes of hypertension can be controlled and compare cases of hypertension caused by these factors with the 10 percent of hypertension cases that are caused by identifiable disorders, such as excessive renin secretion by the kidneys, arteriosclerosis, and endocrine disorders such as hyperthyroidism and Cushing's disease.

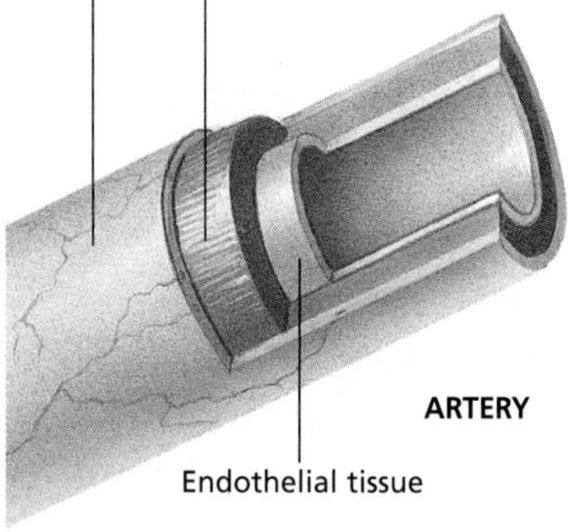

FIGURE 47-4
Notice the thick muscular layer of an artery. The layers of the artery wall are separated by elastic tissue. This tissue provides strength, preventing systolic pressure from bursting the artery.

FIGURE 47-5
The diameter of a capillary is so small that red blood cells must move single file through these vessels, as shown in this photograph (1,200×). All exchange of nutrients and waste between blood and cells occurs across the thin walls of the capillaries.

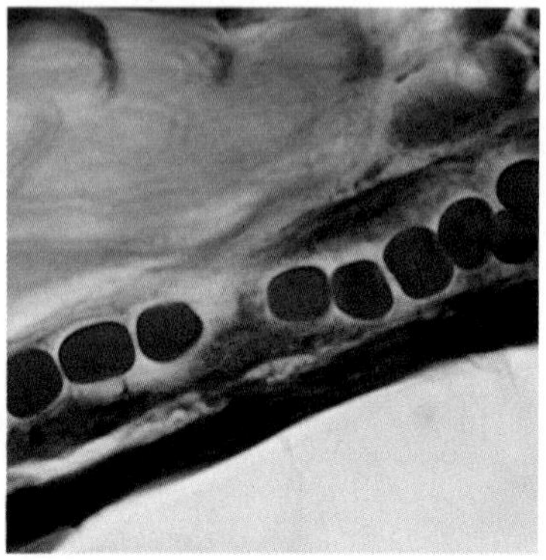

Arteries and Blood Pressure

Each beat of the heart forces blood through blood vessels. The large, muscular vessels that carry blood away from the heart are called **arteries.** As shown in Figure 47-4, the thick walls of the arteries have three layers: an inner endothelial layer, a middle layer of smooth muscle, and an outer layer of connective tissue. This structure gives arteries a combination of strength and elasticity, which allows them to stretch as pressurized blood enters from the heart. You can feel this stretching of arteries—it is your pulse.

Contraction of the heart propels the blood through the arteries with considerable force. The force that blood exerts against the walls of a blood vessel is known as **blood pressure.** Blood pressure is always highest in the two main arteries that leave the heart. It is usually measured in the artery that supplies blood to the arm. To measure blood pressure, a trained person inflates a cuff that is placed around a person's arm, temporarily stopping the flow of blood through the artery. Connected to the cuff is a gauge containing a column of mercury (Hg) that rises as the pressure in the cuff increases. The trained person then releases the air in the cuff slowly while listening to the artery with a stethoscope and watching the column of mercury. The first sounds of blood passing through the artery mean that the ventricles have pumped with enough force to momentarily overcome the pressure exerted by the cuff. This is known as the **systolic pressure,** or the pressure of the blood when the ventricles contract. In a normal adult, the systolic pressure is about 120 mm of Hg for males and 110 mm of Hg for females. Continuing to release the air in the cuff, the trained person next listens for the disappearance of sound, which indicates a steady flow of blood through the artery in the arm. At this point, the pressure of the blood is sufficient to keep the arteries open constantly even with the ventricles relaxed. This is known as **diastolic pressure.** In a normal adult, the diastolic pressure is about 80 mm of Hg for males and 70 mm of Hg for females.

High blood pressure, or **hypertension,** is a leading cause of death in many countries. Blood pressure that is higher than normal places a strain on the walls of the arteries and increases the chance that a vessel will burst. Because people with chronic high blood pressure often display no external symptoms, hypertension is sometimes referred to as the silent killer.

Capillaries and Veins

Recall that when the left ventricle contracts, it forces blood into the aorta, the body's largest artery. From the aorta, blood travels through a network of smaller arteries, which in turn divide and form even smaller vessels, called **arterioles.** The arterioles branch into a network of tiny vessels, called **capillaries.** A capillary is shown in Figure 47-5.

The network formed by capillaries is so extensive that all of the approximately 100 trillion cells in the body lie within about 125 μm of a capillary. This close association between capillaries and cells allows for rapid exchange of materials. Capillary walls are only one cell thick; gases and nutrients can diffuse through these thin walls. Wherever the concentration of oxygen or nutrients is higher in the blood than in the surrounding cells, the substance diffuses from the blood into the cells. Wherever the concentrations of carbon dioxide and wastes are higher in the cells than in the blood, these substances diffuse from the cells into the blood.

Blood flows through capillaries that merge to form larger vessels called **venules** (VEN-yoolz). Several venules in turn unite to form a **vein,** a large blood vessel that carries blood to the heart. Veins returning deoxygenated blood from the lower parts of the body merge to form the **inferior vena cava.** Veins returning deoxygenated blood from the upper parts of the body merge to form the **superior vena cava.** Refer back to Figure 47-2 and locate the inferior vena cava and the superior vena cava.

As you can see in Figure 47-6, although the walls of the veins are composed of three layers, like those of the arteries, they are thinner and less muscular. By the time blood reaches the veins, it is under much less pressure than it was in the arteries. With less pressure being exerted in the veins, the blood could flow backward and disrupt the pattern of circulation. To prevent that, valves in the veins help keep the blood flowing in one direction. Many veins pass through skeletal muscle. When these muscles contract, they are able to squeeze the blood through the veins. When these muscles relax, the valves can close, thus preventing the blood from flowing backward. Figure 47-7 shows the structure of a valve in a vein.

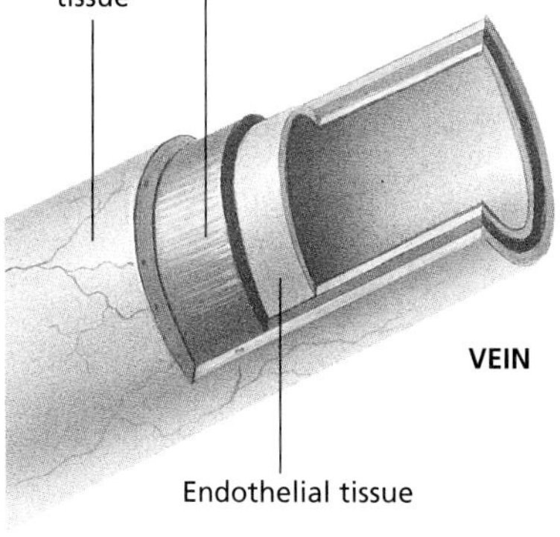

FIGURE 47-6

Like an artery, a vein has three layers. Notice the outer layer of connective tissue, the middle layer of smooth muscle, and the inner layer of endothelial tissue. Compare this figure with Figure 47-4. How do the three layers in a vein compare with those in an artery?

PATTERNS OF CIRCULATION

The English scientist William Harvey (1578–1657) first showed that the heart and the blood vessels form one continuous, closed system of circulation. He also reasoned that this system consists of two primary subsystems: **pulmonary circulation,** in which the blood travels between the heart and lungs, and **systemic circulation,** in which the blood travels between the heart and all other body tissues.

Pulmonary Circulation

Deoxygenated blood returning from all parts of the body except the lungs enters the right atrium, where it is then pumped into the right ventricle. When the right ventricle contracts, the deoxygenated blood is sent through the pulmonary artery to the lungs. The pulmonary artery is the only artery that carries deoxygenated blood. The pulmonary artery branches into two smaller arteries, with one artery going to each lung. These arteries branch into arterioles and then into capillaries in the lungs.

In the lungs, carbon dioxide diffuses out of the capillaries and oxygen diffuses into the capillaries. The oxygenated blood then

FIGURE 47-7

This figure shows the structure of the valves in veins. Many veins have valves to keep the blood flowing in one direction. If these valves fail to close properly, some blood can leak backward and expand a weak area of the vein. This results in a condition known as varicose veins.

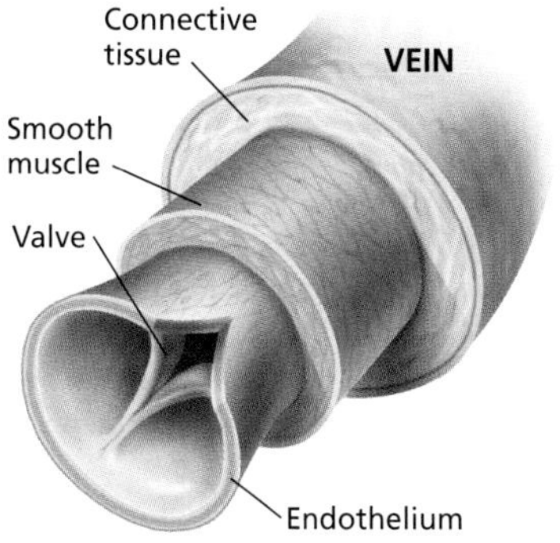

SECTION 47-1

VISUAL LINK
Figure 47-7

Tell students that varicose veins are veins that have become dilated because of weakened valves. Dilated veins in the anal area are called hemorrhoids.

RECENT RESEARCH
Atherosclerosis

Atherosclerosis, or clogged arteries, indirectly causes the death of half of all people in the Western world. Traditionally, coronary bypass surgery, in which a vein or small artery from the leg is removed and implanted in the heart to restore circulation, has been the treatment of choice. One new technique that is noninvasive is called enhanced external counterpulsation. For one hour daily for seven to eight weeks, inflatable belts are wrapped around the patient's lower body. They inflate and deflate in time with the heartbeat, forcing blood from the lower limbs to the chest during diastole. It is thought that the procedure enlarges existing vessels that have become narrowed and encourages formation of new channels to bypass the diseased vessels.

TERMINOLOGY NOTE

Some students may have heard the term *arteriosclerosis* but not the new term, introduced on page 936, *atherosclerosis.* Help them understand the difference by explaining that arteriosclerosis is a disease of the arteries characterized by thickening, loss of elasticity, and calcification of arterial walls, resulting in a decreased blood supply especially in the extremities and the cerebrum. Atherosclerosis is a form of arteriosclerosis in which the narrowing of the arteries is caused specifically by deposits of cholesterol, lipoid substances, and lipophages.

SECTION 47-1

TEACHING STRATEGY

Pulmonary Circulation

Provide students with a diagram of the human circulatory system. Have them use red and blue pencils or markers to trace the pathway of blood through the heart, lungs, and major pathways in the trunk and legs by adding arrows to the diagram. They should use red arrows for oxygen-rich blood and blue arrows for oxygen-poor blood. Have the students label the circulation pathways "pulmonary," "cardiac," and "systemic."

MAKING CONNECTIONS

Social Studies

Scientists' ideas about the composition, function, and circulation of blood have changed drastically since ancient times. Have students research some of the predominant theories once held about blood and the circulatory system, and the significant discoveries that changed or modified these theories. Some students may want to explore the ideas of Hippocrates, who lived about 400 B.C., that blood was one of the four humors, or fluids, of the body and relate this theory to the practice of bloodletting. Other students, focusing on William Harvey's discoveries, may want to find out about the ways in which scientists of the 1500s and 1600s thought that blood moved through the body.

INCLUSION ACTIVITY

Set up microscope stations with slides of arteries, veins, and capillaries. Have students observe each kind of vessel. Next have students construct models of the vessels. They can use materials such as modeling clay, plastic foam, fabric scraps, cardboard tubes from paper rolls, colored markers, and watercolors.

FIGURE 47-8

The pulmonary circulation between the heart and the lungs involves the pulmonary arteries and the pulmonary veins. Deoxygenated blood flows from the right side of the heart to the lungs. Oxygenated blood is returned to the left side of the heart from the lungs. This is the opposite of systemic and coronary blood flow, in which oxygen-rich blood flows from the heart and oxygen-poor blood is returned to the heart.

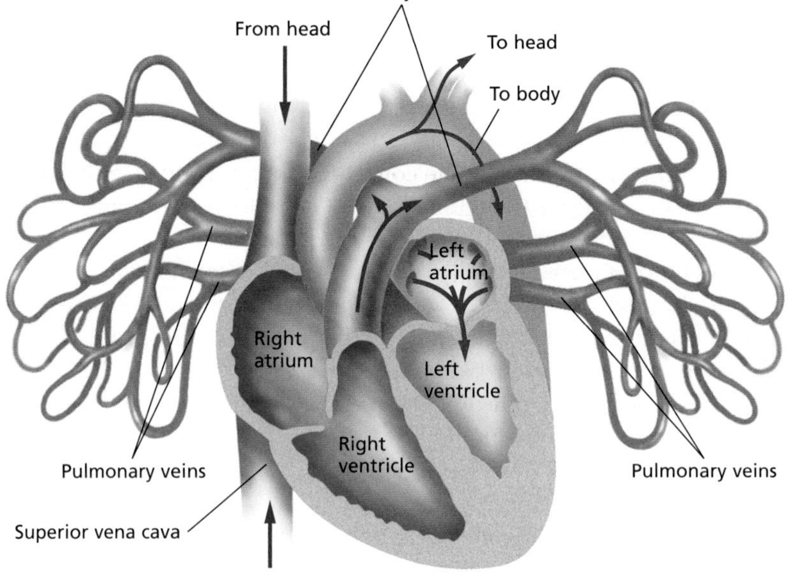

flows into venules, which merge into the **pulmonary veins** that lead to the left atrium of the heart. The pulmonary veins are the only veins that carry oxygenated blood. From the left atrium, blood is pumped into the left ventricle and then to the body through the aorta, supplying the cells with oxygen. In Figure 47-8, trace the path blood takes as it passes through pulmonary circulation.

FIGURE 47-9

Notice three subsystems of systemic circulation. Other subsystems transport blood between the heart and the head, arms, organs, and legs.

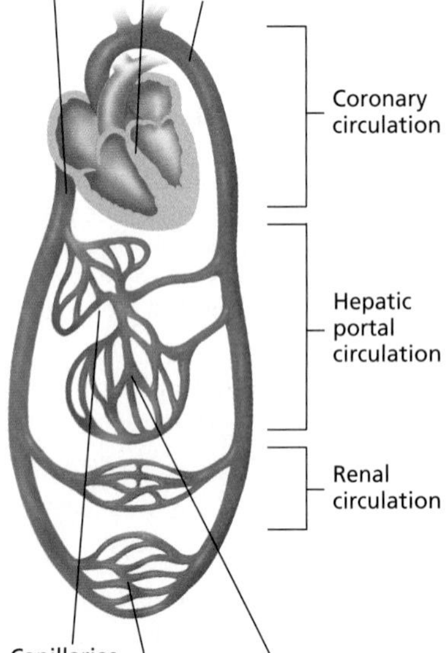

Systemic Circulation

Systemic circulation is the movement of blood between the heart and all parts of the body except the lungs. Trace the path blood follows in systemic circulation in Figure 47-9. Notice that oxygenated blood is pumped out of the left ventricle and into the aorta. From the aorta, blood flows into other subsystems of systemic circulation.

Coronary circulation is one subsystem of systemic circulation that supplies blood to the heart itself. The heart muscle is thick, and oxygen and nutrients must be supplied to each cell. If the blood supply to the heart is reduced or cut off, muscle cells will die. This can happen when an artery is blocked by a blood clot or by **atherosclerosis** (ATH-uhr-oh-skler-OH-sis), a disease characterized by the buildup of fatty materials on the interior walls of the coronary artery. If either type of blockage reduces the flow of blood to the heart muscle cells, a heart attack will result.

Renal circulation, another subsystem of systemic circulation, supplies blood to the kidneys. Nearly one-fourth of the blood that is pumped into the aorta by the left ventricle flows to the kidneys. The kidneys filter waste from the blood.

Hepatic portal circulation is a subsystem of systemic circulation. Nutrients are picked up by capillaries in the small intestine and are transported by the blood to the liver. Excess nutrients are stored in the liver for future needs. The liver receives oxygenated blood from a large artery that branches from the aorta.

LYMPHATIC SYSTEM

In addition to the cardiovascular system, the circulatory system also includes the lymphatic system. One function of the lymphatic system is to return fluids that have collected in the tissues to the bloodstream. Fluids diffuse through the capillary walls just as oxygen and nutrients do. Some of these fluids pass into cells, some return to the capillaries, and some remain in the intercellular spaces.

Excess fluid in the tissues moves into the tiny vessels of the lymphatic system; this fluid is called **lymph.** Lymph vessels merge to form larger vessels. The lymph vessels are similar in structure to capillaries, and the larger lymph vessels are similar in structure to veins. However, an important difference exists between blood vessels and lymph vessels. As you learned earlier, blood vessels form a complete circuit so that blood passes from the heart to all parts of the body and then back again to the heart. In contrast, lymph vessels form a one-way system that returns fluids collected in the tissues back to the bloodstream. In addition, the lymphatic system has no pump like the heart. Like the blood in veins, lymph must be moved through the vessels by the squeezing of skeletal muscles. Like veins, the larger lymph vessels have valves to prevent the fluid from moving backward.

Notice in Figure 47-10 that lymph vessels form a vast network that extends throughout the body. The lymph that travels in these vessels is a transparent yellowish fluid, much like the liquid part of the blood. As the lymph travels through these vessels on its way to the heart, it passes through small organs known as lymph nodes. Notice in Figure 47-10 that lymph nodes are like beads on a string. These nodes filter the lymph as it passes, trapping foreign particles, microorganisms, and other tissue debris. Lymph nodes also store **lymphocytes,** white blood cells that are specialized to fight disease. When a person has an infection, the nodes may become inflamed, swollen, and tender because of the increased number of lymphocytes.

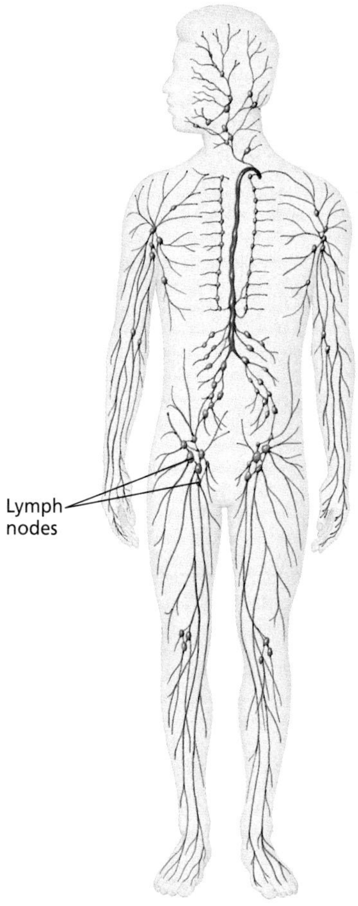

FIGURE 47-10

Like the cardiovascular system, the lymphatic system forms a vast network. Concentrated in certain regions of this network are lymph nodes that contain some of the disease-fighting cells of the immune system.

SECTION 47-1 REVIEW

1. **Identify the structure that prevents blood from mixing between the left and right sides of the heart. Explain what prevents blood from flowing from the ventricles backward into the atria.**
2. **Identify the structure that controls the heartbeat, and describe the process by which it regulates the heartbeat.**
3. **Outline the path that blood follows in pulmonary circulation.**
4. **Compare the lymphatic system with the cardiovascular system.**
5. **In which blood vessels would you expect to find the lowest average blood pressure? Explain your answer.**
6. **CRITICAL THINKING Some babies are born with a hole in the septum between the two atria. Based on what you know about blood flow through the heart, explain why this condition would be harmful to the baby.**

SECTION 47-1

RECENT RESEARCH

Lymphedema

Lymphedema (severe, localized swelling) is caused by anything that prevents the normal return of lymph to the blood, such as blockage of the lymph vessels and nodes by tumors, injury, or cancer surgery in which lymph nodes are removed. While there is no cure, recent research indicates that intensive physical therapy, in which the muscles are encouraged to pump the lymph back to the circulatory system, can help alleviate the problem.

ANSWERS TO SECTION 47-1 REVIEW

1. the septum; the AV valves
2. The SA node initiates a heartbeat by stimulating muscle cells in the atria. The impulse then passes to the AV node, which stimulates the muscle cells of the ventricles.
3. Blood takes the following path: right ventricle→pulmonary arteries→lungs→pulmonary veins→left atrium.
4. Both consist of vessels that transport a fluid. The veins in both have valves. The lymphatic system is one-way and does not have a pump.
5. The inferior vena cava and the superior vena cava have the lowest blood pressure because they are nearest the heart in terms of blood return and farthest from the heart in terms of the heart's pumping function.
6. The blood on the right side of the heart (deoxygenated blood) would mix with blood on the left side (oxygenated blood). As a result, cells would not get enough oxygen.

BACKGROUND

William Harvey depended on his own observations and experiments for proof of his claims about the heart. Before Harvey, it was assumed that the blood was sucked up by the heart during its expansion and flowed away with contraction. Harvey showed that the heart expands passively and contracts actively. He also showed that blood exits the heart in arteries and returns to the heart in veins. Harvey was born in 1578 in Kent, England, and was the oldest of nine children. He received his first university degree at age 19 at Cambridge. He continued his studies at the University of Padua in Padua, Italy, where his teacher encouraged him to study and compare the anatomy of all manner of animals, from insects to humans. Harvey returned to England, and in 1604 he became fully licensed by the College of Physicians in London. He was made physician to King James I and later physician to Charles I. Harvey died in 1657.

READING FOR MEANING

Student answers will vary. An example of a rewritten passage follows:

This evidence indicates that the motion of the heart contracts in the direction of the cardiac muscle fibers and develops tension as a result of the contraction.

READ FURTHER

Possible answers include X rays, CAT scans, MRIs, and thermal scans, which are used to make images of tissues to determine abnormalities that cause illness.

Literature & Life

On the Motion of the Heart

This excerpt is from William Harvey's anatomical essay *On the Motion of the Heart and Blood in Animals.* It was first presented in 1628.

The study of medicine in Harvey's time normally consisted of studying the works of ancient Greek doctors and philosophers, not direct observation. Harvey's work as a physician and anatomist was revolutionary because he actually observed the action of the heart and the movement of blood in live animals.

When the chest of a living animal is laid open and the capsule that immediately surrounds the heart is slit up or removed, the organ is seen now to move, now to be at rest; there is a time when it moves, and a time when it is motionless.

These things are more obvious in the colder animals, such as toads, frogs, serpents, small fishes, crabs, shrimps, snails, and shell-fish. They also become more distinct in warm-blooded animals, such as the dog and hog, if they be attentively noted when the heart begins to flag, to move more slowly, and, as it were, to die: the movements then become slower and rarer, the pauses longer, by which it is made much more easy to perceive and unravel what the motions really are, and how they are performed. In the pause, as in death, the heart is soft, flaccid, exhausted, lying, as it were, at rest.

In the motion, and interval in which this is accomplished, three principal circumstances are to be noted:

1. That the heart is erected, and rises upwards to a point, so that at this time it strikes against the breast and the pulse is felt externally.

2. That it is everywhere contracted, but more especially towards the sides so that it looks narrower, relatively longer, more drawn together. The heart of an eel taken out of the body of the animal and placed upon the table or the hand, shows these particulars; but the same things are manifest in the hearts of all small fishes and of those colder animals where the organ is more conical or elongated.

3. The heart being grasped in the hand, is felt to become harder during its action. Now this hardness proceeds from tension, precisely as when the forearm is grasped, its tendons are perceived to become tense and resilient when the fingers are moved.

4. It may further be observed in fishes, and the colder blooded animals, such as frogs, serpents, etc., that the heart, when it moves, becomes of a paler color, when quiescent of a deeper blood-red color.

From these particulars it appears evident to me that the motion of the heart consists in a certain universal tension—both contraction in the line of its fibres, and constriction in every sense.

Reading for Meaning

Because *On the Motion of the Heart and Blood in Animals* was written in Latin in the seventeenth century and was translated into English in 1910, the language is very old-fashioned. Rewrite the last paragraph of this passage, paraphrasing it in simple, modern language.

Read Further

In Harvey's time, knowledge about an animal's internal anatomy came from direct observation. Observations of human anatomy and physiology were rare because the observation techniques were injurious and often fatal. What equipment and techniques are used today in modern hospitals to make observations of human anatomy and physiology that are not invasive or harmful to the patient? How are these techniques used in the diagnosis of illness or disease?

BLOOD

Blood is a liquid connective tissue that constitutes the transport medium of the cardiovascular system. The two main functions of the blood are to transport nutrients and oxygen to the cells and to carry carbon dioxide and other waste materials away from the cells. Blood also transfers heat to the body surface and plays a major role in defending the body against disease.

SECTION

47-2

OBJECTIVES

▲ List the components of blood.

● Distinguish between red blood cells, white blood cells, and platelets in terms of their structure and function.

■ Summarize the process of blood clotting.

◆ Explain what determines the compatibility of blood types for transfusion.

COMPOSITION OF BLOOD

Blood is composed of a liquid medium and blood solids. Blood solids consist of red blood cells, white blood cells, and platelets. The liquid makes up about 55 percent of the blood, and blood solids make up the remaining 45 percent. A healthy adult has about 4 to 5 L of blood in his or her body.

Plasma

Plasma, the liquid medium, is a sticky, straw-colored fluid that is about 90 percent water. Cells receive nourishment from dissolved substances carried in the plasma. These substances, which may include vitamins, minerals, amino acids, and glucose, are absorbed from the digestive system and transported to the cells. Plasma also carries hormones and brings wastes from the cells to the kidneys or the lungs to be removed from the body.

Proteins are carried in the plasma and have various functions. Some of the proteins in the plasma are essential for the formation of blood clots. Another protein, called albumin, plays an important role in the regulation of osmotic pressure between plasma and blood cells and between plasma and tissues. Other proteins, called antibodies, help the body fight disease.

Red Blood Cells

Red blood cells, or **erythrocytes** (uh-RITH-ruh-siets), shown in Figure 47-11, transport oxygen to cells in all parts of the body. Red blood cells are formed in the red marrow of bones. Red blood cells synthesize large amounts of an iron-containing protein called **hemoglobin.** Hemoglobin is the molecule that actually transports oxygen and, to a lesser degree, carbon dioxide. During the formation of a red blood cell, its cell nucleus and organelles disintegrate. The mature red blood cell becomes little more than a membrane sac containing hemoglobin.

FIGURE 47-11

Notice that a mature red blood cell (RBC) is disk-shaped and is concave on both sides (5,250×). A red blood cell is little more than a cell membrane filled with hemoglobin. How is this structure related to its function?

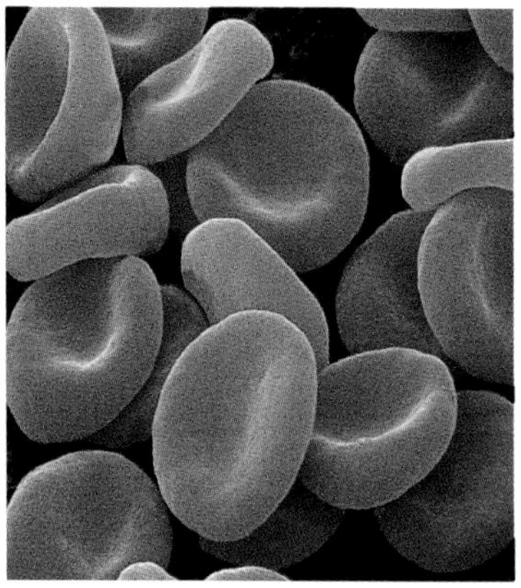

SECTION 47-2

SECTION OVERVIEW

Structure and Function
Section 47-2 describes the characteristics, composition, and function of the blood components—plasma, erythrocytes, leukocytes, and platelets.

Stability and Homeostasis
Section 47-2 discusses blood typing and compatibility of blood types. This section also discusses how blood clots prevent excess blood loss.

▶ ENGAGE STUDENTS

Ask students if they know their blood type. Many people do not know unless they have been in an accident, have undergone surgery, or have donated blood. Ask students why knowing blood types in such situations is important. (Blood of different types is not compatible, except in the case of type O donors or type AB recipients. Blood transfusions are often necessary in cases of severe blood loss. Blood-typing information is important in making a match between donated blood and its recipient.)

VISUAL LINK

Figures 47-11 and 47-12

Have students examine the red and white blood cells in Figures 47-11 and 47-12 and note the differences in size and shape. Ask students how these differences enable each type of cell to carry out its function. (The shape of the red blood cells provides greater surface area for gas diffusion, and their small size enables them to squeeze through tiny capillaries. The shape of the phagocyte enables it to engulf other particles, and its size ensures that it can defend the body against a large number of invaders.)

QUICK FACT

☑ Anemia, which literally means "lacking blood," is a condition in which the blood has an abnormally low oxygen-carrying capacity. It is a symptom of a disorder, such as a nutritional deficiency or sickle-cell anemia, rather than a disease in itself.

TEACHING STRATEGY

Blood

Provide pairs of students with microscopes and prepared, sealed slides of human blood. Caution students not to collect their own blood. Have them observe the cells on the slide and then identify, draw, and label the different cells they observe. Have available for the class a labeled chart or model showing a typical red blood cell and the different white blood cells and platelets. Use the text figures on pages 939, 940 and 941, or obtain charts from a biological supply house.

Because red blood cells lack nuclei, they cannot divide and they have a limited survival period, usually 120 to 130 days. Of the more than 30 trillion red blood cells circulating throughout the body at one time, 2 million disintegrate every second. To replace them, new ones form at the same rate in the red marrow of bones. Some parts of the disintegrated red blood cells are recycled. For example, the iron portion of the hemoglobin molecule is carried in the blood to the marrow, where it is reused in new red blood cells.

White Blood Cells

White blood cells, or **leukocytes** (LOO-kuh-siets), help defend the body against disease. They are formed in the red marrow, the lymph nodes, and the spleen. White blood cells are larger than red blood cells and significantly less plentiful. Each cubic millimeter of blood normally contains about 4 million red blood cells and 7,000 white blood cells. White blood cells can squeeze their way through openings in the walls of blood vessels and into the intercellular fluid. In that way, white blood cells can reach the site of infection and help destroy invading microorganisms.

Notice in Figure 47-12 that a white blood cell has a very different structure from that of a red blood cell. For instance, a white blood cell may be irregularly shaped and may have a rough outer surface. There are other differences between red blood cells and white blood cells as well. In contrast with the short-lived red blood cells, white blood cells may function for years. And while there is only one type of red blood cell, there are several types of white blood cells.

The white blood cell shown in Figure 47-12 is the type of white blood cell known as a **phagocyte** (FA-guh-siets). Phagocytes are cells that engulf invading microorganisms. Locate the microorganisms that are being engulfed by the phagocyte in Figure 47-12. Another type of white blood cell produces **antibodies.** Antibodies are proteins that help destroy substances, such as bacteria and viruses, that enter the body and can cause disease. When a person has an infection, the number of white blood cells can double.

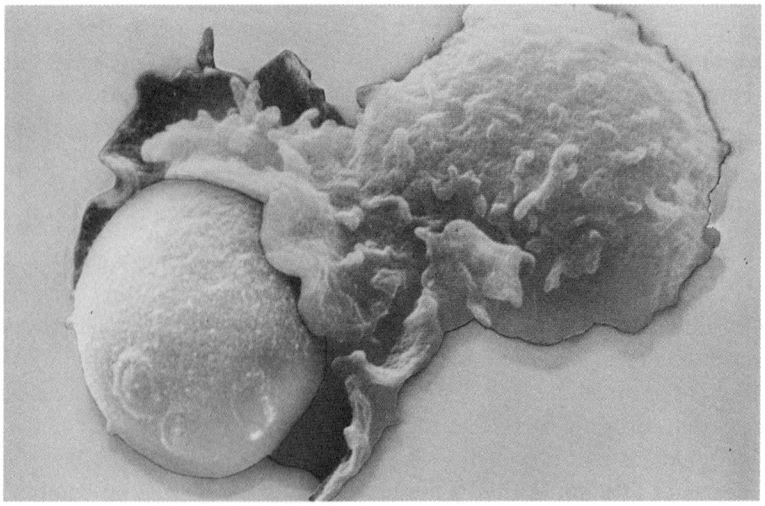

FIGURE 47-12

Some white blood cells, like the phagocyte shown in blue, engulf and destroy invading microorganisms.

(a) (b)

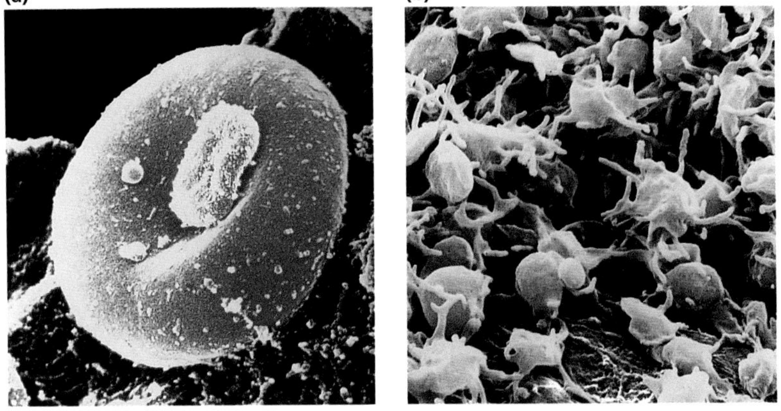

FIGURE 47-13

Inactive platelets, such as the yellow object shown in (a), derive their name from the fact that they look like little plates (7,410×). Platelets are colorless and contain chemicals that are involved in the clotting process. (b) The platelets change shape during the clotting process. When activated, the platelets settle and spread on the substrate (4,900×).

Platelets

Platelets are essential to the formation of a blood clot. A blood clot is a mass of interwoven fibers and blood cells that prevents excess loss of blood from a wound. Platelets are not whole cells. They are fragments of very large cells that were formed in the bone marrow. As you can see in Figure 47-13a, platelets get their name from their platelike structure. Platelets lack a nucleus and have a life span of 7 to 11 days. A cubic micrometer of blood may contain as many as half a million platelets.

When a blood vessel tears or rips, platelets congregate at the damaged site, sticking together and forming a small plug. The vessel constricts, slowing blood flow to the area. Then special clotting factors are released from the platelets. These factors begin a series of chemical reactions that occur at the site of the bleeding. The last step in this series brings about the production of a protein called **fibrin.** Fibrin molecules consist of long, sticky chains. As you can see in Figure 47-14, these chains form a net that traps red blood cells, and the mass of fibrin and red blood cells hardens into a clot, or scab.

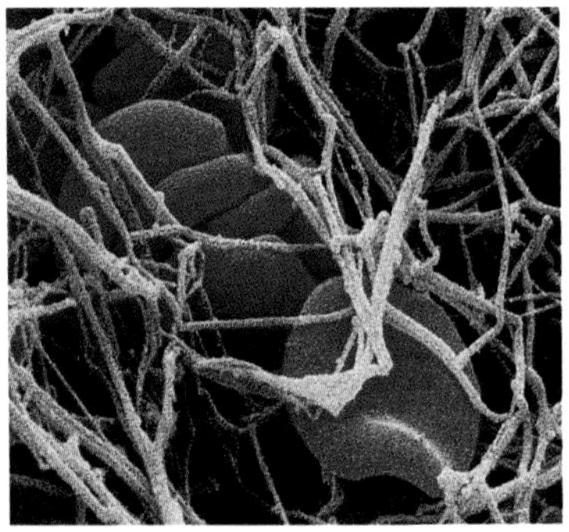

FIGURE 47-14

Long, sticky threads of fibrin trap blood cells and other materials to form a network to stop bleeding. This is comparable to placing a small piece of tissue paper over a cut. The fibers in the tissue paper act like fibrin to form a network to stop the bleeding. (5,000×)

BLOOD TYPES

Blood type is determined by the type of antigen present on the surface of the red blood cells. An **antigen** is a protein or carbohydrate that acts as a signal, enabling the body to recognize foreign substances that enter the body. Antigens that are normally present in a person's body provoke no response. However, when foreign antigens enter the body, cells respond by producing antibodies. In fact, the word *antigen* is an abbreviation for "antibody-generating substance."

Blood from the human population is classified into four groups, based on the antigens on the surface of the red blood cell. Physicians learned of the differences by observing what happened when blood samples from different patients were mixed. In the early 1900s, Karl Landsteiner wanted to find out why some blood transfusions caused no problems, while others led to complications and even death for the recipient. Using blood taken from his laboratory

internetconnect

SCILINKS NSTA

TOPIC: Blood types
GO TO: www.scilinks.org
KEYWORD: HM941

CRITICAL THINKING

Clots

Have students study the illustrations of clot formation on page 942. Then ask students to consider a scab on a cut. Ask them if there is a difference between a clot and a scab. (A scab is basically a dried clot consisting of interwoven threads of fibrin and trapped red blood cells.) Ask students what would happen if a scab did not form on even a very small cut. (Bleeding would not stop and death could occur.) Tell students that hemophilia is a disorder in which there is a failure in one of the steps of clot formation.

RECENT RESEARCH

Artificial Blood

Four blood substitutes are currently approved by the FDA for experimental use. "Fluosol" is a substance related to the nonstick coating used in cookware. It serves as a dissolving medium for oxygen. "Chemically altered hemoglobin" gives up more oxygen to tissues than regular hemoglobin. However, the risk of disease transmission is high. "Neohemocytes" are hemoglobin molecules packaged in fat bubbles. The cells are about one-twelfth the size of normal erythrocytes. "Hemopure" is a natural substance that contains purified hemoglobin extracted from cattle blood.

TEACHING STRATEGY

Blood Transfusion Mismatch

When mismatched blood is transfused into a person, a transfusion reaction occurs. Kidney failure is often the result of this reaction because the kidney tubules become clogged. Ask students to research and report on possible ways to treat transfusion-mismatch reactions. (Treatment includes infusing the patient with alkaline fluids to dilute and dissolve the hemoglobin and wash it out of the body, and administering diuretics and having the patient drink large amounts of liquids to increase urine output.)

QUICK FACT

☑ Because of the prevalence of HIV, hepatitis B, and other bloodborne pathogens, many people who undergo elective surgery now donate their own blood for use during their surgery.

RECENT RESEARCH

Nitric Oxide

Dr. Jonathan Stamler and his colleagues at Duke University Medical Center have discovered that blood delivers nitric oxide (NO), a noxious gas in the atmosphere, as well as oxygen (O_2). Their findings show that NO combined with hemoglobin is a major regulator of gas exchange in the circulatory system. The researchers measured blood flow and O_2 concentration in several regions of rats' brains while the rats breathed air with varying O_2 levels. They showed that hemoglobin releases NO in the small arteries, dilating the vessels and increasing blood flow and O_2 delivery. When O_2 levels in tissues were high, hemoglobin halted NO release, constricting the vessel. This information could be of importance in the treatment of heart disease, strokes, and sickle cell anemia.

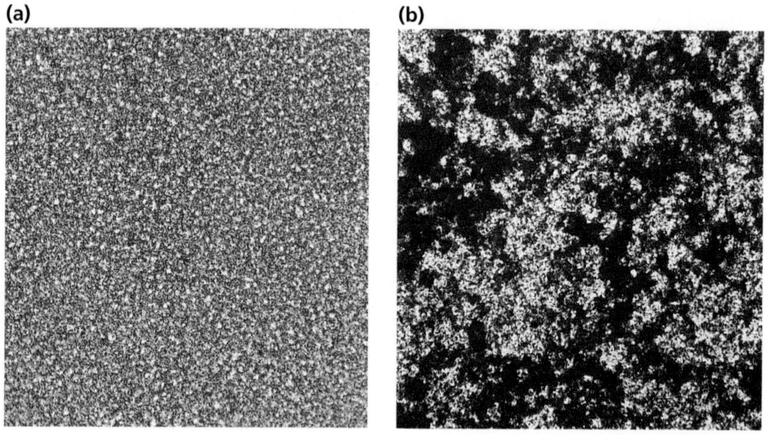

FIGURE 47-15

Notice that there is no agglutination of red blood cells in the slide in (a), where blood samples from two people with the same blood type were mixed (20×). Compare this with the slide in (b), where blood samples from two people with different blood types were mixed (20×).

workers, Landsteiner made observations similar to those you see in Figure 47-15. He noticed that mixing blood samples from two people sometimes resulted in the cells clumping together, or agglutinating. However, at other times no clumping or agglutination occurred when blood samples were mixed. Landsteiner reasoned that clumping occurred when blood samples of two different blood types were mixed.

When samples of two different blood types are mixed together, reactions occur between the antigens on the red blood cells and the antibodies in the plasma, causing the cells to agglutinate. When samples of the same blood type are mixed, no reaction occurs and the blood cells do not agglutinate. Landsteiner's observations led to the classification of human blood by blood types. Blood typing involves identifying the antigens in a sample. Three of the most important human antigens are called A, B, and Rh. The A-B-O system of blood typing, described below, is based on the A and B antigens.

A-B-O System

The A-B-O system is a means of classifying blood by the antigens located on the surface of the red blood cells and the antibodies circulating in the plasma. As shown in Table 47-1, an individual's red blood cells may carry an A antigen, a B antigen, both A and B antigens, or no antigen at all. These antigen patterns are called blood types A, B, AB, and O, respectively.

Notice in Table 47-1 that an individual with type A blood also has anti-B antibodies against type B blood. If type B blood is given to a recipient with type A blood, the recipient's anti-B antibodies will

TABLE 47-1 *Blood Types, Antigens, and Antibodies*

Blood types	Antigen on the red blood cells	Antibodies in the plasma	Can get blood from	Can give blood to
A	A	anti-B	O, A	A, AB
B	B	anti-A	O, B	B, AB
AB	A and B	none	A, B, AB, O	AB
O	none	anti-A, anti-B	O	A, B, AB, O

react with the B antigens on the donated red blood cells and the blood will agglutinate. In addition, the donor's type B blood has anti-A antibodies. Their presence will compound the antigen-antibody reaction in the recipient. The net result will be agglutinated blood that will block the flow of blood through the vessels. For this reason, transfusion recipients must receive blood that is compatible with their own. Based on the information in Table 47-1, why can a person with type AB blood receive blood from any of the four types?

Rh System

An antigen that is sometimes present on the surface of red blood cells is the **Rh factor,** named after the rhesus monkey in which it was first discovered. Eighty-five percent of the United States' population is Rh-positive (Rh^+), meaning that Rh antigens are present. People who do not have Rh antigens are called Rh-negative (Rh^-).

If an Rh^- person receives a transfusion of blood that has Rh^+ antigens, antibodies will react with the antigen and agglutination will occur. The most serious problem with Rh incompatibility occurs during pregnancy. If the mother is Rh^- and the father is Rh^+, the child may inherit the dominant Rh^+ allele from the father. The blood supplies of the mother and the fetus are separated during pregnancy, but during delivery, a small amount of the fetus's Rh^+ blood may reach the mother's bloodstream. If this happens, the mother will develop antibodies to the Rh factor. If a second Rh^+ child is conceived later, the mother's antibodies can cross the placenta and attack the blood of the fetus. This condition is called erythroblastosis fetalis. The fetus may die as a result of this condition, or if the child is born alive, he or she may need an immediate transfusion of Rh^+ blood.

To prevent this condition, an Rh^- mother of an Rh^+ child can be given antibodies to destroy any Rh^+ cells that have entered her bloodstream from the fetus. These antibodies must be administered to the mother within three days after the birth of her first Rh^+ child. By destroying any Rh^+ cells in her bloodstream, any danger to a second child is prevented because the mother will not make any antibodies against the blood cells of the Rh^+ fetus.

Quick Lab

Identifying Offspring

Materials pencil, paper

Procedure Two babies are believed to have been swapped at birth in error. Blood samples were taken from each of the parents and babies. The following results were obtained from the blood samples:

Family 1: mother, type B; father, type O; baby, type A

Family 2: mother, type O; father, type A; baby, type O

Design a chart or data table that correctly pairs the biological parents with their baby.

Analysis Are the babies with the correct biological parents? How do you know?

SECTION 47-2 REVIEW

1. What is plasma? Name at least one major function of plasma.
2. Distinguish between the three solid components of the blood.
3. Identify the stages and structures involved in the clotting process.
4. Explain why a pregnant woman should know her blood type and the blood type of her baby's father.
5. Which blood type, in terms of the A-B-O and Rh antigens, can be donated to all others? Why?
6. CRITICAL THINKING Why do you think some people turn pale when they are frightened?

Quick Lab

Identifying Offspring

Time Required 20 minutes

Procedural Tips Remind students that types A and B are dominant over O but that neither type A nor type B is dominant over the other. A type A phenotype could be an AA or AO genotype; a type B phenotype could be a BB or BO genotype; a type AB phenotype must be an AB genotype; and a type O phenotype is an OO genotype. Students might create charts similar to Punnett squares.

Answers to Analysis The babies are not with the correct biological parents. A man with type O blood could not be the father of a baby with type A blood if the mother is type B. The true father of the baby in Family 1 must be a type A or AB because the baby must have received an A allele from its father.

ANSWERS TO SECTION 47-2 REVIEW

1. Plasma is the liquid part of blood; it transports CO_2.
2. Red blood cells transport hemoglobin. White blood cells help the body fight disease. Platelets help form blood clots.
3. A blood clot is initiated by platelets, which stimulate the formation of fibrin. The long strands of fibrin trap red blood cells to form a mass that blocks the loss of blood from a wound.
4. The knowledge would help her determine ahead of time whether her baby's blood may be incompatible with her own.
5. Type O Rh-negative blood can be donated because there are no antigens in the blood.
6. Blood flow to the skin is reduced, and blood flow to muscles and glands is increased. Thus, a person may appear pale when frightened.

SECTION 47-3

SECTION OVERVIEW

Matter, Energy, and Organization
Section 47-3 describes the organs and organization of the respiratory system.

Interdependence of Organisms
Section 47-3 discusses the ways that the skeletal, muscular, and respiratory systems interact during breathing, as well as the relationship between the circulatory and respiratory systems in oxygen and carbon dioxide exchange.

▶ ENGAGE STUDENTS

Have students count the number of breaths they take in 15 seconds and multiply by 4 to find the number of breaths per minute. Ask students to explain why there are differences in breathing rates among members of the class. (Most people breathe about 16 to 24 times per minute, but rates differ depending on activity level and the resulting need for oxygen. The level of physical conditioning can also affect breathing rates.)

MAKING CONNECTIONS

Technology
Take students on a field trip to a hospital or medical school and arrange for a doctor, nurse, or technician to explain how a modern respirator works and how the iron lung (negative-pressure respirator) worked. Alternatively, invite a professional to the classroom to demonstrate the technology involved in respirators. Often, large hospitals and medical schools still have iron-lung machines in storage to use for demonstration. Ask the professional to compare the two modes of assisted breathing and then to compare each to the functioning of the human lung.

SECTION

47-3

OBJECTIVES

▲ Trace the passage of air from the environment to the bloodstream.

● Describe how gases are exchanged in the lungs.

■ Contrast the ways that oxygen and carbon dioxide are transported in the bloodstream.

◆ Summarize the skeletal and muscular changes that occur during breathing.

▲ Describe how the rate of breathing is controlled.

internet**connect**

SCILINKS NSTA

TOPIC: Respiratory system
GO TO: www.scilinks.org
KEYWORD: HM944

THE RESPIRATORY SYSTEM

You have read how the blood transports oxygen from the lungs to cells and carries carbon dioxide from the cells to the lungs. It is the function of the respiratory system to transport gases to and from the cardiovascular system. The respiratory system involves both external respiration and internal respiration.

External respiration *is the exchange of gases between the atmosphere and the blood.* ***Internal respiration*** *is the exchange of gases between the blood and the cells of the body. In Chapter 7, you learned how aerobic respiration involves the use of oxygen to break down glucose in the cell. In this section, you will examine the structures and mechanisms that carry oxygen to the cells for use in aerobic respiration and that eliminate the carbon dioxide that is produced by the same process.*

THE LUNGS

The lungs are the site of gas exchange between the atmosphere and the blood. Notice in Figure 47-16 that the right lung has three divisions, or lobes. It is slightly heavier than the two-lobed left lung. The lungs are located inside the thoracic cavity, bounded by the rib cage and the diaphragm. Lining the entire cavity and encasing the lungs are pleura, membranes that secrete a slippery fluid that decreases friction from the movement of the lungs during breathing.

The Passage of Air

Refer to Figure 47-16 to trace the path air follows from the atmosphere to the capillaries in the lungs. External respiration begins at the mouth and at the nose. Air filters through the small hairs of the nose and passes into the nasal cavity, located above the roof of the mouth. In the nasal cavity, mucous membranes warm and moisten the air, which helps prevent damage to the delicate tissues that form the respiratory system. The walls of the nasal cavity are also lined with cilia. These cilia trap particles that are inhaled and are eventually swept into the throat, where they are swallowed.

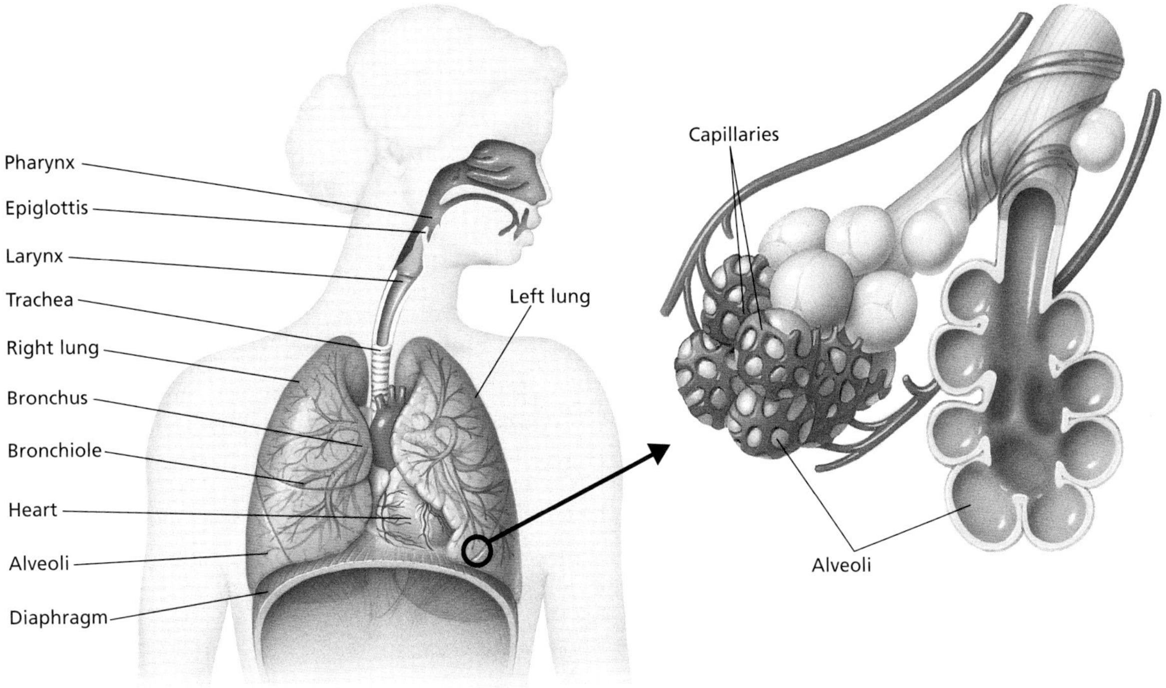

FIGURE 47-16

Trace the passage of air from the atmosphere to the lungs. Oxygen in the air finally reaches the alveoli, the functional units of the respiratory system. All exchange of gases between the respiratory system and the circulatory system occurs in the alveoli.

The moistened, filtered air then moves into the pharynx (FER-inks), a tube at the back of the nasal cavities and the mouth. The pharynx contains passageways for both food and air. When food is swallowed, a flap of cartilage, called the **epiglottis,** presses down and covers the opening to the air passage. When air is being taken in, the epiglottis is in an upright position, allowing air to pass into a cartilaginous tube called the **trachea** (TRAY-kee-uh). The trachea is about 10 to 12 cm long and has walls lined with ciliated cells that trap inhaled particles. The cilia sweep the particles and mucus away from the lungs toward the throat.

At the upper end of the trachea is the **larynx** (LER-inks). Sounds are produced when air is forced past two ligaments—the vocal cords—that stretch across the larynx. The pitch and volume of the sound produced varies with the amount of tension on the vocal cords and on the amount of air being forced past them.

The trachea then branches into two **bronchi** (BRAHN-kie), each of which leads to a lung. The walls of the bronchi consist of smooth muscle and cartilage and are lined with cilia and mucus. Within the lungs, the bronchi branch into smaller and smaller tubes. The smallest of these tubes are known as **bronchioles,** which are also lined with cilia and mucus. Eventually the bronchioles end in clusters of tiny air sacs called **alveoli** (al-VEE-oh-LIE). A network of capillaries surround each alveolus, as you can see in the detailed view shown in Figure 47-16. All exchange of gases in the lungs occurs in the alveoli. To facilitate this exchange, the surface area of the lungs is enormous. Each lung contains nearly 300 million alveoli and has a total surface area of 70 m^2—about 40 times the surface area of the skin.

INCLUSION ACTIVITY

Show students an anatomical model of the respiratory system, and provide them with a set of self-sticking labels that name its different structures. Review with students the path that air follows as it enters the respiratory system. As you name the different structures, have students attach the labels to the correct structures on the model. Allow students to remove the labels and practice reapplying them until they are familiar with the different structures.

VISUAL LINK

Figure 47-16

Help students locate the lungs, diaphragm, pharynx, trachea, larynx, and bronchi in Figure 47-16. Ask students how these structures contribute to the function of the respiratory system.

MAKING CONNECTIONS

Mathematics

To demonstrate the surface area of alveoli, have students calculate the total surface area of a pad of paper, including the areas of the top and bottom and the thickness of the pad (not including the cardboard). Tell students that the pad can be compared to one alveolus. Then arrange the individual sheets side by side and calculate their total surface area. The thickness of a single sheet can be measured by dividing the thickness of the pad (minus the cardboard) by the number of sheets. Explain that if the alveoli could be flattened and laid side by side like the paper, the students could see the tremendous surface area available in the lungs for gas exchange.

CRITICAL THINKING

Gas Exchange

Give each student a drinking straw and a test tube half full of limewater. Explain that limewater becomes cloudy when CO_2 is added to it. Ask students to inhale and hold their breath for 2 seconds and then exhale gently through the straw into the limewater. Have students repeat the process with a fresh test tube, but this time have them hold their breath for 10 seconds before exhaling. Ask them how the results differ and why. (The limewater was cloudier the second time because more CO_2 was present. Even while the breath is being held, the cells continue to use O_2 and give off CO_2 in respiration, and the CO_2 builds up in the lungs.)

QUICK FACT

☑ As altitude increases, the gases that make up the atmosphere, including oxygen, become less concentrated. When people who are used to breathing air at sea level visit a place at an altitude of 2,000 m, they sometimes experience shortness of breath, fatigue, nausea, and blackouts. Ask students how it is possible for people to live at altitudes of more than 2,500 m. (Over a period of weeks, our bodies are able to compensate for oxygen deprivation by producing more red blood cells, a process called erythropoiesis.)

GAS EXCHANGE AND TRANSPORT

In the lungs, gases are exchanged between the alveoli and the blood in the capillaries. Oxygen to be transported throughout the body moves into the bloodstream, and carbon dioxide to be eliminated from the body moves into the alveoli.

Gas Exchange in the Lungs

Figure 47-17 illustrates the direction in which oxygen and carbon dioxide move in the alveoli. When air moves into the lungs, the oxygen in the air crosses the thin alveolar membranes as well as the capillary walls and enters the blood. Carbon dioxide moves in the opposite direction, crossing the capillary walls and thin alveolar membranes and entering the alveoli.

Air moving into the alveoli is rich in oxygen and contains little carbon dioxide. In contrast, blood in the capillaries surrounding the alveoli is low in oxygen and contains high levels of carbon dioxide. Thus, concentration gradients for both oxygen and carbon dioxide exist. Remember from Chapter 5 that substances diffuse from an area of higher concentration to an area of lower concentration. Consequently, oxygen diffuses from the alveoli into the blood, and carbon dioxide diffuses from the blood into the alveoli. The enormous surface area of the alveoli increases the rate of diffusion of these two gases.

Hemoglobin and Gas Exchange

When oxygen diffuses into the blood, only a small amount dissolves in the plasma. Most of the oxygen—97 percent—moves into the red blood cells, where it combines with hemoglobin. Each hemoglobin molecule contains four iron atoms. Each iron atom can

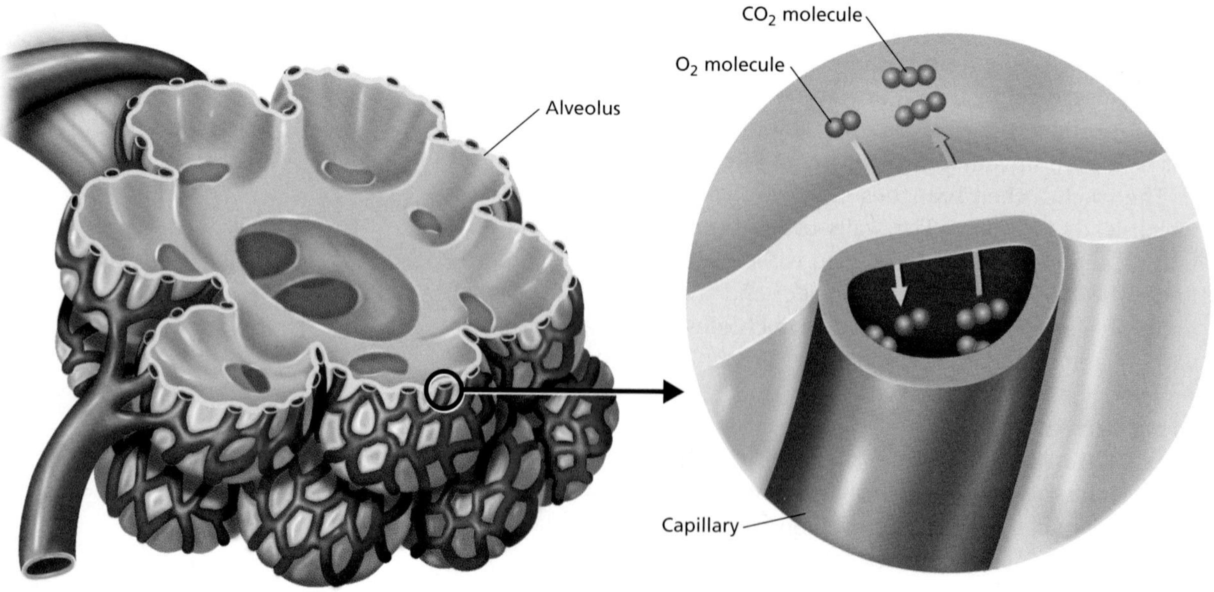

FIGURE 47-17

Because of concentration gradients, oxygen and carbon dioxide diffuse across the alveoli and capillary walls.

bind to one oxygen molecule. Thus, one hemoglobin molecule can carry up to four molecules of oxygen. When oxygenated blood reaches cells, the oxygen concentration is higher in the blood than in the cells. Thus, oxygen is released from the hemoglobin and diffuses out of the capillaries and into the surrounding cells.

Because the concentration of carbon dioxide is higher in the cells, it diffuses out of the cells and into the blood. Only about 8 percent of the carbon dioxide dissolves in the plasma. Approximately 25 percent binds to hemoglobin. The remaining 67 percent reacts with water in the plasma to form carbonic acid. In turn, the carbonic acid disassociates into bicarbonate ions and a proton. Thus, most of the carbon dioxide travels in the blood as bicarbonate ions. When the blood reaches the lungs, the series of reactions is reversed. Bicarbonate ions combine with a proton to form carbonic acid, which in turn forms carbon dioxide and water. The carbon dioxide diffuses out of the capillaries into the alveoli and is exhaled into the atmosphere.

MECHANISM OF BREATHING

Breathing is the process of moving air into and out of the lungs. **Inspiration,** shown in Figure 47-18a, is the process of taking air into the lungs. When you take a deep breath, your chest expands as muscles contract to move the ribs up and outward. At the same time, your diaphragm, a large skeletal muscle, flattens and pushes

FIGURE 47-18

The diaphragm, a large skeletal muscle that separates the thoracic cavity from the abdominal cavity, and the muscles between the ribs control the movement of the thoracic cavity during breathing. If these muscles were paralyzed, then inspiration and expiration would not occur.

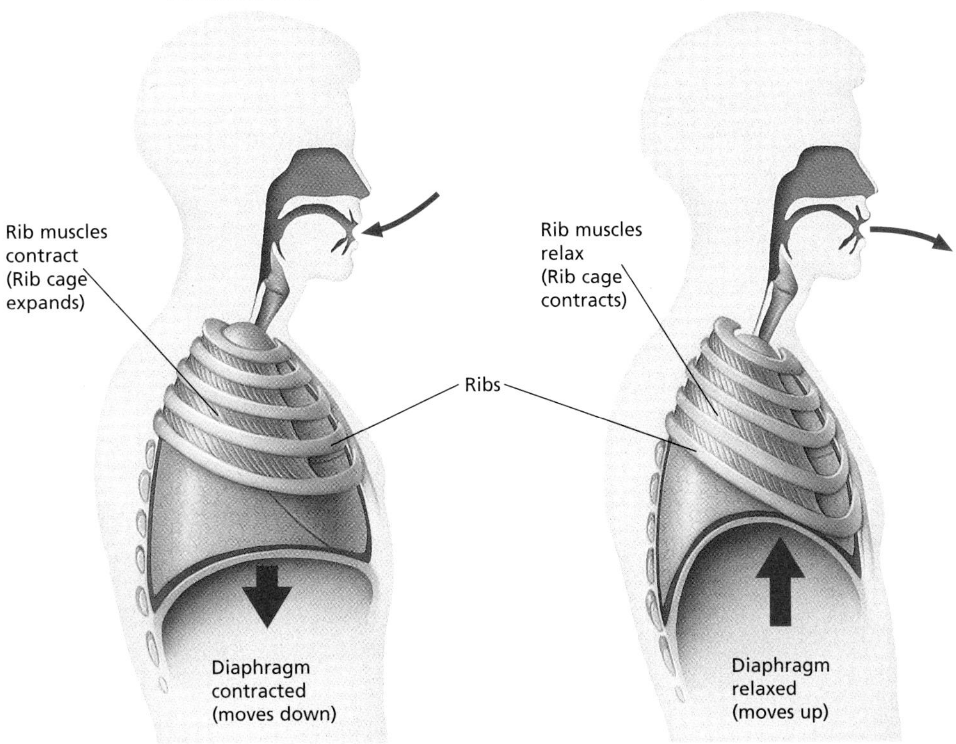

DEMONSTRATION

Demonstrate to students how the mechanism of breathing works by making a model of a lung and diaphragm. Fasten a small, deflated balloon to a short glass tube, and put the tube through a one-holed stopper. Use the stopper to plug the mouth of a bell jar, allowing the balloon to hang inside the jar. Next, with a rubber band, fasten a thin sheet of rubber to the open end of the jar. Pull the rubber bottom down and let students observe and explain the results. (The balloon inflates. When the rubber sheet was pulled down, the volume in the jar increased and the air pressure decreased, forcing air into the balloon.)

DEMONSTRATION

Have a professional show students how to perform the Heimlich maneuver. Have the professional gently demonstrate the maneuver on a volunteer. Have students discuss how the maneuver works. (The Heimlich maneuver forces air out of the lungs, dislodging a foreign object that may be stuck in the trachea.)

TEACHING STRATEGY

Pathological Lung Conditions

Provide pairs of students with a microscope and prepared slides of normal lung tissue and lung tissue exhibiting bronchitis, pneumonia, emphysema, and lung cancer. Have students observe the differences in the slides and state the physical changes each disease has caused in lung tissue. (Students should see the congestion of the alveoli in bronchitis and pneumonia, the hardening of cell walls in emphysema, and the tumor growth in cancer.)

SECTION 47-3

GIFTED ACTIVITY

Have students read the surgeon general's warnings on smoking and articles on smoking issues. Have them research current reports on the FDA's stand on smoking. Have the students form small groups to discuss the issues, and then have group spokespersons compare group findings in a classroom forum.

ANSWERS TO SECTION 47-3 REVIEW

1. The mouth, nose, nasal cavity, and pharynx are the structures that filter, warm, moisten, and direct air before it enters into the lungs.
2. Oxygen diffuses from an area of higher concentration to an area of lower concentration.
3. Carbon dioxide is mainly transported as bicarbonate (HCO_3^-) ions.
4. An increased level of CO_2 in the blood signals the brain to increase the rates of both inspiration and expiration. This helps bring more O_2 into the lungs and helps to quickly reduce the level of CO_2 in the blood.
5. Having the gas-exchange organ inside of the body and protected by the rib cage reduces the possibility of injury and desiccation. The organ must be moist so that the diffusion of gases can occur.
6. A single-celled organism does not need a respiratory system because the entire surface of the cell is exposed to its external environment. This exposure allows diffusion and osmosis to supply oxygen and nutrients and remove wastes from the cell.

down on the abdomen. Muscles in the abdominal wall in turn relax. This action provides room for the flattened diaphragm.

When the diaphragm flattens and the ribs are lifted up and out, the volume of the thoracic cavity increases. An increased volume reduces the air pressure within the cavity. At this point, the air pressure inside the thoracic cavity is lower than the air pressure outside the body. As a result, air from the atmosphere moves into the lungs.

During **expiration,** the reverse process takes place, as you can see in Figure 47-18b. As the diaphragm and rib muscles relax, the elastic tissues of the lungs recoil, deflating the lungs. The size of the thoracic cavity decreases. Because the volume is smaller, the air pressure inside the cavity becomes greater than the air pressure outside the body. This pressure difference forces air out of the lungs until the pressures are again equal.

Regulation of Breathing

The rate at which oxygen is used depends on the activity of the cells. The greater their activity, the more oxygen they need and the faster the body needs to breathe. The slower their activity, the slower the body needs to breathe.

The rate of breathing is controlled by the brain and brain stem, which monitors the concentration of carbon dioxide in the blood. As activity increases, high levels of carbon dioxide in the blood stimulate nerve cells in the brain. The brain stem in turn stimulates the diaphragm to increase the breathing rate and depth. When the carbon dioxide concentration in the blood returns to lower levels, the sensors in the brain send a message to the respiratory muscles to return to a slower breathing rate. All this is controlled subconsciously by control centers in the brain. However, a person can temporarily override the respiratory control system at any time, holding his or her breath until losing consciousness. Then the brain stem takes control, and normal breathing resumes. This mechanism allows humans to swim underwater for short periods and to sleep without concern for breathing.

SECTION 47-3 REVIEW

1. What structures of the respiratory system prepare air for entry into the lungs?
2. Why does oxygen diffuse from the alveoli into capillaries and then into cells in the body?
3. Describe the main way carbon dioxide is transported in the blood.
4. How does an increased carbon dioxide concentration in the blood affect expiration and inspiration?
5. What is the adaptive value of having the organ of gas exchange—in this case, the alveolus—inside the body?
6. **CRITICAL THINKING** Why does a single-celled organism not need a respiratory system?

CHAPTER 47 REVIEW

SUMMARY/VOCABULARY

47-1

- The circulatory system consists of the cardiovascular system and the lymphatic system.
- The human heart is located in the thoracic cavity and has two atria and two ventricles.
- Heartbeat is initiated by the sinoatrial (SA) node, also known as the pacemaker. A heartbeat has two phases: systole and diastole.
- Blood vessels include arteries, capillaries, and veins. Arteries are thick, muscular vessels that transport blood away from the heart. Arteries branch into smaller vessels known as arterioles and capillaries.
- Capillaries merge to form venules, which then collect into veins. Veins return blood to the heart.
- Pulmonary circulation involves blood flow between the heart and the lungs.
- Systemic circulation includes the heart, the kidneys, the liver, and all other organs, including skin and muscle.
- The lymphatic system returns intercellular fluid to the heart. Fluid in the lymphatic system is called lymph.

Vocabulary

aorta (933)
aortic valve (932)
arteriole (934)
artery (934)
atherosclerosis (936)
atrioventricular (AV) node (933)
atrioventricular valve (932)
atrium (931)
blood pressure (934)
capillary (934)
cardiovascular system (931)
circulatory system (931)
coronary circulation (936)
diastole (933)
diastolic pressure (934)
hepatic portal circulation (936)
hypertension (934)
inferior vena cava (935)
lymph (937)
lymphatic system (931)
lymphocyte (937)
mitral valve (932)
pacemaker (933)
pericardium (931)
pulmonary circulation (935)
pulmonary valve (932)
pulmonary vein (936)
renal circulation (936)
semilunar valve (932)
septum (931)
sinoatrial (SA) node (933)
superior vena cava (935)
systemic circulation (935)
systole (933)
systolic pressure (934)
tricuspid valve (932)
vein (935)
ventricle (931)
venule (935)

47-2

- Blood is composed of plasma, red blood cells, white blood cells, and platelets. Red blood cells transport oxygen. White blood cells help defend the body against disease. Platelets help form blood clots.
- Human blood can be grouped into four types: A, B, AB, and O. In addition, blood is either Rh-positive or Rh-negative.

Vocabulary

antibody (940)
antigen (941)
blood type (941)
erythrocyte (939)
fibrin (941)
hemoglobin (939)
leukocyte (940)
phagocyte (940)
plasma (939)
platelet (941)
Rh factor (943)

47-3

- Oxygen enters the bloodstream through the lungs.
- The epiglottis prevents food from entering the trachea during swallowing. The larynx contains the vocal cords.
- Oxygen and carbon dioxide are exchanged between the alveoli and blood and between blood and the cells.
- Nearly all the oxygen in the body is transported by hemoglobin. The level of carbon dioxide in the blood determines the rate of breathing, which is controlled by the brain.
- During inspiration, the thoracic cavity expands, pulling air into the lungs. During expiration, the thoracic cavity gets smaller, forcing air out of the lungs.

Vocabulary

alveolus (945)
bronchiole (945)
bronchus (945)
epiglottis (945)
expiration (948)
external respiration (944)
inspiration (947)
internal respiration (944)
larynx (945)
trachea (945)

CHAPTER 47 REVIEW ANSWERS

REVIEW

1. Systolic pressure is caused by the contractions of the ventricles. Diastolic pressure is maintained by the arterial walls when the ventricles relax.
2. Both transport blood away from the heart. The pulmonary artery carries deoxygenated blood, while the aorta transports oxygenated blood.
3. Both transport blood to the heart. The pulmonary vein carries oxygenated blood, while the inferior vena cava transports deoxygenated blood.
4. They are all vessels that transport blood.
5. Red blood cells, white blood cells, and platelets are the solid components of blood.

6. c **7.** d
8. a **9.** b
10. a **11.** d
12. b **13.** b
14. a **15.** d

16. Blood enters the right atrium, travels to the right ventricle, goes to the lungs, returns to the heart through the left atrium, goes to the left ventricle, and is pumped out the aorta. The pulmonary vein carries oxygenated blood.
17. The lymphatic system helps return intercellular fluid back to the bloodstream and is also part of the immune system.
18. Red blood cells lack nuclei and cannot repair themselves.
19. The parents should not be concerned about their child's Rh status. The only situation that is potentially dangerous is when the child inherits the father's Rh^+ blood type and the mother is Rh^-. The mother may develop antibodies that could be harmful to future children.
20. They are produced by air passing across the vocal cords in the larynx.
21. High levels of carbon dioxide in the blood cause the breathing rate to increase.

CHAPTER 47 REVIEW ANSWERS

CONTINUED

22. Unlike red blood cells, white blood cells have nuclei, come in several different types, and can function for years.
23. During inspiration, the diaphragm flattens and the muscles pull the ribs up and out, increasing the volume of the thoracic cavity. During expiration, the diaphragm and muscles relax, lowering the ribs and decreasing the volume of the thoracic cavity.
24. Both are extremely thin (single-cell thickness) to facilitate diffusion of materials across their surface.
25. A, aorta; B, superior vena cava; C, pulmonary valve; D, right atrium; E, tricuspid valve; F, right ventricle; G, inferior vena cava; H, septum; I, left ventricle; J, mitral valve; K, aortic valve; L, left atrium; M, pulmonary vein; N, pulmonary artery.

CRITICAL THINKING

1. The fewer red blood cells there are, the less hemoglobin there is. The smaller the amount of hemoglobin, the smaller the amount of oxygen that can be transported. A smaller amount of oxygen means a decreased level of aerobic respiration and less energy (ATP).
2. Polio can result in paralysis of the diaphragm and the muscles that control the rib cage. If these muscles cannot move, then the person cannot breathe.
3. The accident victim's loss of blood volume and the loss of body fluid can cause the victim to go into shock. Transfusing plasma replenishes the fluid, helps stabilize the patient, and can be done quickly without taking time to type. The small number of antibodies the plasma contains become harmlessly diluted in the patient's own blood.
4. Lymph must be moved through the vessels by the squeezing of skeletal muscles.
5. 2,759,400,000 beats in 75 years at 70 beats per minute. An

REVIEW

Vocabulary

1. Distinguish between systolic pressure and diastolic pressure.
2. What do the pulmonary arteries and the aorta have in common? How are they different?
3. What do the pulmonary veins and the inferior vena cava have in common? How are they different?
4. What do arteries, veins, and capillaries have in common?
5. Identify the solid components of the blood.

Multiple Choice

6. The wall that divides the heart vertically is the (a) ventricle (b) pericardium (c) septum (d) atrium.
7. During systole, blood moves from the (a) ventricles to the atria (b) atria to the veins (c) atria to the ventricles (d) ventricles to the arteries.
8. Pulmonary circulation involves movement of blood between the heart and the (a) lungs (b) brain (c) liver (d) kidneys.
9. One of the functions of the lymphatic system is that it (a) interacts with the respiratory system (b) helps the body fight infection (c) consists of a series of two-way vessels (d) transports intercellular fluid away from the heart.
10. One function of plasma is to (a) carry substances that nourish cells (b) aid in the formation of blood clots (c) carry the majority of the oxygen supply of the blood (d) defend against disease.
11. The function of fibrin is to (a) transport oxygen (b) destroy invading microorganisms (c) stimulate the production of antibodies (d) help form a blood clot.
12. A person who has no antigens present on the red blood cells has blood type (a) AB Rh^+ (b) O Rh^- (c) A Rh^+ (d) O Rh^+.
13. During internal respiration, gases are (a) exchanged between the atmosphere and the blood (b) exchanged between the blood and the cells (c) produced by the heart (d) warmed and moistened.
14. Gases diffuse (a) from an area of high concentration to an area of low concentration (b) from an area of low concentration to an area of high concentration (c) directly from the cells to the air passages (d) from the alveoli to the cells.
15. Cilia (a) move air molecules (b) moisten the air passages (c) sweep foreign particles into the stomach (d) sweep particles out of the air passages.

Short Answer

16. Describe the route of blood through the heart. Include circulation through the lungs, and specify whether the artery or the vein carries oxygenated blood.
17. What are two major roles of the lymphatic system?
18. What structure do red blood cells lack that limits their life span?
19. A child is about to be born to parents who are both Rh^-. Given their Rh status, what concerns might the parents have about the health of their child ? Explain your answer.
20. How are human vocal sounds produced?
21. What is one factor that stimulates the brain stem to increase the breathing rate?
22. Describe three differences between white blood cells and red blood cells.
23. Describe the movement of the diaphragm and the rib muscles during inspiration and expiration.
24. How are the structures of alveoli and capillaries related to their function?
25. List the part of the heart denoted by each letter in the diagram below.

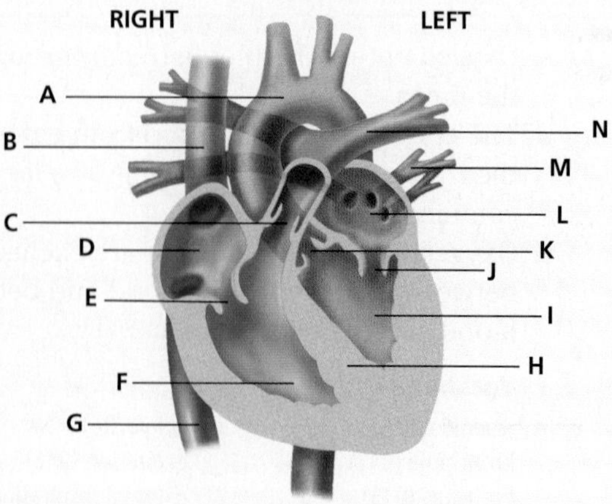

CRITICAL THINKING

1. A person with anemia has too few red blood cells. The most common symptom is a lack of energy. Why would anemia cause this symptom?
2. Polio is a disease that paralyzes muscles by affecting the nerves that make them move. Before the polio vaccine was developed, many people who had polio died because they could not breathe. Some of the survivors had to be placed in an "iron lung" that breathed for them. From what you know about the respiratory system, explain why people stricken with polio could no longer breathe on their own.
3. Even a small increase or decrease in blood volume has an effect on blood pressure. When an accident victim suffers significant blood loss, the person is transfused with plasma rather than whole blood. Why is plasma effective in meeting the immediate threat to life?
4. Explain how the lymphatic system moves lymph through the body without the aid of a pumping organ like that of the cardiovascular system.
5. Calculate the number of times a person's heart will beat if the person lives 75 years. Assume that the average heart beats 70 times per minute. Assuming that the heart of an overweight person beats an additional 10 beats per minute, explain why being overweight can put additional strain on the heart.
6. One function of the cardiovascular system is to help maintain a uniform body temperature. Explain how the constant circulation of blood throughout the body can accomplish this task.
7. Copy the blood-type table below on a sheet of paper. Fill in the missing information for each type.

Blood Types, Antigens, and Antibodies

Blood types	Antigen on the red blood cells	Antibodies in the plasma	Can get blood from	Can give blood to
A		anti-B	O, A	A, AB
B	B		O, B	B, AB
AB	A and B	none	A, B, AB, O	
O	none	anti-A, anti-B		A, B, AB, O

EXTENSION

1. Read "The Beat Goes On" in *Popular Science*, March 2000, on page 62. Describe the port-access mitral valve repair/replacement method of heart surgery. What is the major advantage of this kind of surgery?
2. Using the library or on-line references, research work that has been done about developing and implanting an artificial heart. Write a paper about the operation, the likely candidates for the operation, the number of people who have received an artificial heart, and some of the outcomes of these operations.
3. Carbon monoxide is an odorless gas that is extremely poisonous. Find out how this gas affects the respiratory and circulatory systems. What symptoms would a victim of carbon monoxide inhalation exhibit?
4. Using the library or an on-line data base, research a respiratory ailment, such as cystic fibrosis or emphysema. Write a report that includes information on the people who are most likely to get the disease. What are the causes, and what are the typical outcomes?

overweight person's heart will beat 394,200,000 more times in the same length of time at 80 beats per minute, which means much more work over a life of 75 years.
6. Blood circulates very close to every cell in the body, absorbing heat where the body is warmer than the blood and releasing heat where the body is cooler than the blood. Thus, blood distributes heat throughout the body, contributing to a uniform body temperature.
7. Type A antigen = A; Type B antibodies = anti-A; Type AB can only donate blood to AB; Type O can only receive blood from O.

EXTENSION

1. Port-access heart surgery is done through a very small incision. Specialized instruments and endoscopic monitors are used to view the heart. The surgery is minimally invasive, which allows the patient to have a rapid recovery period.
2. Have students consider how an artificial heart, especially Jarvik-7, works. Questions to explore can include how long patients have lived after transplants, what the quality of their life was, and whether students would consider having the surgery under similar circumstances.
3. Carbon monoxide has a greater affinity for hemoglobin than oxygen does. The binding of carbon monoxide to hemoglobin results in a bright red color. Thus, a victim of carbon monoxide inhalation would be flushed and short of breath.
4. Cystic fibrosis is a genetic disorder resulting in highly viscous lung fluids that clog air passages. Emphysema is a condition in which the alveolar walls become inelastic and permanently inflated. It is usually caused by years of breathing industrial waste gases or cigarette smoking.

CHAPTER 47
INVESTIGATION

TIME REQUIRED

One 50-minute lab period

SAFETY PRECAUTIONS

Remind students to wear safety goggles for Part B. Avoid skin and eye contact with the bromothymol indicator solution.

QUICK REFERENCES

Lab Preparation Notes are found on page 929B.

Holt BioSources provides a Teaching Resources CD-ROM that contains "Using Gowin's Vee in the Lab" and "Scoring Rubrics."

PROCEDURAL TIPS

1. To save time, you may want to have half of the students perform Part A while the other half perform Part B, and then switch.
2. To prevent the embarrassment of students who are not comfortable exercising in front of their peers, you may wish to ask for volunteers for Part B. These students should not have health conditions that would preclude sustained vigorous exercise.
3. Review the use of a spirometer and the terms *tidal volume, vital capacity*, and *expiratory reserve volume.*
4. You may wish to have students record their data in a class data table on the board. They need not identify themselves by name, only as male or female, athlete or nonathlete, smoker or nonsmoker. Calculate mathematical means for males, females, smokers, nonsmokers, and so on.

ANSWER TO BACKGROUND

4. Factors that may influence lung capacity include genetic makeup, smoking, health problems, and physical fitness.

CHAPTER 47 INVESTIGATION

Tidal Volume, Expiration Volume, and CO_2 Production

OBJECTIVES

- Use indirect measurement to determine lung capacity.
- Determine the effect of exercise on breathing rate and CO_2 production.

PROCESS SKILLS

- measuring
- hypothesizing
- collecting data
- analyzing data
- experimenting

MATERIALS

- safety goggles
- 1 L bromothymol indicator solution
- drinking straws
- 100 mL Erlenmeyer flasks, 2 per group
- 100 mL graduated cylinders
- marker
- plastic wrap
- spirometer
- stopwatch or clock with second hand

Background

1. A spirometer is an instrument used to measure the volume of air a person can breathe.
2. Examine the diagram of a spirometer on the right. Compare the diagram with the spirometer you will be using to complete this investigation. Note that the marking pen creates a line that can be compared with the scale on the left side to measure liters of air.
3. Tidal volume is the volume of air you inhale or exhale during a single, normal breath.
4. Lung capacity is the available volume of your lungs. Total lung capacity is 5 to 6 L. What factors might increase or reduce lung capacity?
5. Expiratory reserve volume is the amount of air remaining in the lungs after a normal exhalation.
6. Vital capacity is the maximum amount of air that can possibly be inhaled or exhaled.
7. Carbon dioxide is soluble in water. You can determine the relative amount of CO_2 in your breath at rest and after exercise by using an indicator to react with the CO_2. Higher CO_2 levels will react with the indicator solution faster.

PART A Tidal Volume, Expiratory Volume, and Vital Capacity

1. Make a data table in your notebook like one shown on the next page.
2. Place the tube of the spirometer near your mouth, and inhale a normal breath. Hold your nose, then exhale a normal breath into the spirometer and take a reading. Record your data in the table.
3. Measure your expiratory reserve volume by first breathing a normal breath and exhaling normally. Then put the spirometer tube to your mouth as you forcefully exhale whatever air is left in your lungs. Be sure to force out as much air as possible. Record your data in the table.

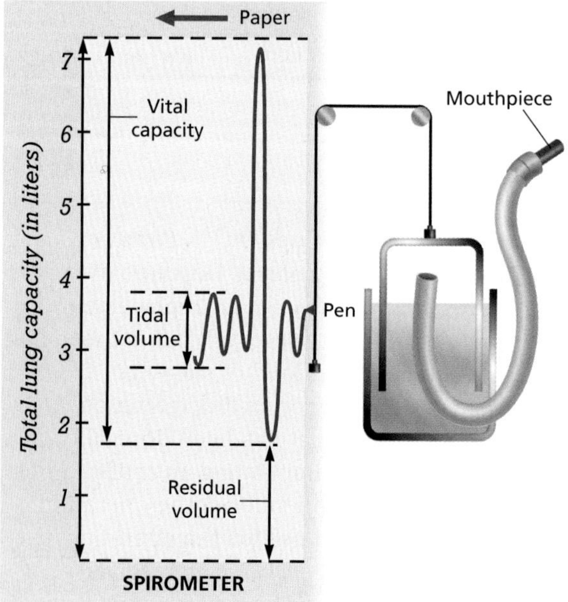

PART A *TIDAL VOLUME, EXPIRATORY VOLUME, AND VITAL CAPACITY*

	Average for young adult males	Average for young adult females	Average for athletes	Your readings
Tidal volume	500 mL	375 mL to 400 mL	650 mL to 700 mL – males 487 mL to 560 mL – females	
Expiratory reserve volume	100 mL	75 mL to 80 mL	130 mL to 140 mL – males 97 mL to 112 mL – females	
Vital capacity	4,600 mL	3,450 mL to 3,680 mL	5,980 mL to 6,440 mL – males 4,485 mL to 5,152 mL – females	

4. The table includes values for young adult males. The average volume for young adult females is 20–25 percent lower than that of males. Calculate the average volumes for young adult females. Athletes can have volumes that are 30–40 percent greater than the average for their gender. Calculate the average volume for an athlete.

PART B Breathing Rate and CO_2 Production

5. Discuss the purpose of this part of the investigation with your partners. You will use bromothymol blue as an indicator of the CO_2 you bubble into each flask. Develop a hypothesis that describes a relationship between air volume exhaled during rest or exercise and the volume of CO_2 exhaled.

6. CAUTION **Wear safety goggles at all times during this procedure. If you get the indicator solution on your skin or clothing, wash it off at the sink while calling to your teacher. If you get the indicator solution in your eyes, immediately flush it out at the eyewash station while calling to your teacher.**

7. Label the two flasks as 1 and 2.

8. Add 100 mL of indicator solution to each flask. Cover the mouth of each flask with plastic wrap.

9. Remove the plastic wrap from flask 1. Blow gently through one straw into flask 1 until the solution turns a yellowish color, exhaling slowly so that the solution does not bubble up. Be careful not to inhale the solution or get it in your mouth.

10. Record on your Part B Data Table the time in seconds that it took to see a color change in flask 1.

11. Exercise by jogging in place or doing jumping jacks for 2 min. Immediately blow gently through a new straw into flask 2 until the solution becomes the same yellowish color as the solution in flask 1.

12. In your Part B Data Table, record the amount of time in seconds that it took to get the same yellow color in flask 2 as you got in flask 1.

13. Calculate the difference in the amount of time it took to see a color change in the two flasks. What can you infer about the amount of CO_2 you exhaled before and after exercise?

14. Clean up your materials. Pour the solutions down the sink, and rinse the sink thoroughly with water. Wash your hands before leaving the lab.

Analysis and Conclusions

1. How did your tidal volume compare with that of your classmates?
2. What are the independent and dependent variables in Part B of the Investigation? How did you vary the independent variable and measure changes in the dependent variable?
3. Why were the flasks covered with plastic wrap?
4. Do your data support your hypothesis from Part B? Explain your answers.
5. How do you know whether you produced more carbon dioxide before or after you exercised? Support your answer with evidence from this lab.
6. What were some of the possible sources of error in your experiment?

PART B *DATA TABLE*

Time for color change in flask 1	
Time for color change in flask 2	
Difference in time between flask 1 and flask 2	

Further Inquiry

Design an experiment to determine whether exercise affects heart rate in the same way it affects breathing rate and tidal volume.

CHAPTER 47 INVESTIGATION

ANSWERS TO ANALYSIS AND CONCLUSIONS

1. Students should answer that their tidal volume was higher, lower, or in the middle of the ranked values of all their classmates.
2. The independent variable is exercise and the dependent variables are any measured changes in tidal volume and breathing rate. At one time the subject will exercise; at another time the subject will not. Changes in the dependent variable will be measured in milliliters or liters and breaths per unit of time. The controls are the resting tidal volume and resting breathing rate.
3. The plastic wrap keeps the flask contents from being contaminated by CO_2 in the air.
4. Students should recognize whether their data support their hypotheses. Breathing rates and tidal volumes should both increase.
5. The bromothymol blue indicator solution turns a yellowish color in the presence of CO_2. The faster the color changes, the higher the level of CO_2.
6. Sources of possible error include individual genetic differences; differences in level of fitness; differences in size, health, and gender; altitude to which a person is acclimated; error in measuring time and volume or in calculating; and letting air escape when exhaling.

PLANNING GUIDE 48

CHAPTER 48

INFECTIOUS DISEASES AND THE IMMUNE SYSTEM

	TOPICS	TEACHING RESOURCES	LABS, CLASSWORK, AND HOMEWORK
BLOCKS 1 & 2 90 minutes	**Introducing the Chapter, p. 954**	**ATE** Focus Concept, p. 954 **ATE** Assessing Prior Knowledge, p. 954 **ATE** Understanding the Visual, p. 954	■ Supplemental Reading Guide, *Microbe Hunters*
	48-1 Nonspecific Defenses, p. 955 Disease Transmission, p. 955 Skin and Mucous Membranes, p. 956 The Inflammatory Response, p. 957 Other Nonspecific Defenses, p. 958	**ATE** Section Overview, p. 955 **ATE** Critical Thinking, p. 956 **ATE** Visual Link Figure 48-1, p. 955 235	★ Study Guide, Section 48-1 **PE** Section Review, p. 958
BLOCKS 3 & 4 90 minutes	**48-2 Specific Defenses: The Immune System, p. 959** The Immune System, p. 959 Recognizing Pathogens, p. 960 Immune Response, p. 961 Immunity and Vaccination, p. 964 Allergies, p. 964 Autoimmune Diseases, p. 965	**ATE** Section Overview, p. 959 **ATE** Critical Thinking, pp. 960, 964 **ATE** Visual Link Figure 48-6, p. 961 Figure 48-7, p. 962 Figure 48-8, p. 963 202A, 236–237	**PE** **Quick Lab,** p. 964, Organizing the Immune Response **Lab B32,** Antigen-Antibody Interaction **Lab C42,** Blood Typing **Lab C43,** Blood Typing—Whodunit? **Lab C44,** Blood Typing—Pregnancy and Hemolytic Disease ★ Study Guide, Section 48-2 **PE** Section Review, p. 965
BLOCK 5 45 minutes	**48-3 AIDS, p. 968** HIV and AIDS, p. 968	**ATE** Section Overview, p. 968 203A, 238	**PE** **Chapter 48 Investigation,** p. 974 **Lab B33,** Transmission of a Communicable Disease ★ Study Guide, Section 48-3 **PE** Section Review, p. 970
BLOCKS 6 & 7 90 minutes	**REVIEW and ASSESSMENT** **PE** Chapter 48 Review, pp. 972–973 ■ Performance-Based Assessment—Testing Human Digestive Enzymes	Audio CD Program	**CHAPTER TESTING** ■ Chapter Test (blackline master) ▲ Test Generator ■ Assessment Item Listing ■ Scoring Rubrics and Classroom Management Checklists

KEY

MODERN BIOLOGY
- **PE** Pupil's Edition
- **ATE** Teacher's Edition
- ■ Active Reading Guide
- ★ Study Guide

HOLT BIOSOURCES
- Teaching Transparencies
- Laboratory Program

One-Stop Planner CD-ROM
Shows these resources within a customizable daily lesson plan:
- ■ Holt BioSources Teaching Resources
- ■ Active Reading Guide
- ★ Study Guide
- ▲ Test Generator

READING FOR CONTENT MASTERY

Preview: Have students preview the objectives, the introductory paragraph, topic sentences, and captions before reading the chapter.

ATE Active Reading Technique: Reading Effectively, p. 955

■ Active Reading Guide Worksheet 48-1

PE Summary/Vocabulary, p. 971

■ Active Reading Guide Worksheet 48-2

PE Summary/Vocabulary, p. 971

■ Active Reading Guide Worksheet 48-3

PE Summary/Vocabulary, p. 971

TECHNOLOGY AND INTERNET RESOURCES

CNN PRESENTS Video Segment 25 Asthma Roaches

Holt Biology Videodiscs Teacher's Correlation Guide, Lessons 48-1 through 48-3

Audio CD Program, Sections 48-1 through 48-3

internetconnect

sciLINKS NSTA

On-line Resources: www.scilinks.org
The following *sci*LINKS Internet resources can be found in the student text for this chapter:

Topics:
- Infectious diseases, p. 956
- Autoimmune diseases, p. 965
- AIDS, p. 969

go.hrw.com

On-line Resources: go.hrw.com
Visit the HRW Web site for a variety of resources related to this chapter. Just type in the keyword HM2 Chapter 48.

Smithsonian Institution®

Visit **www.si.edu/hrw** for additional on-line resources.

CNNfyi.com

Visit **www.cnnfyi.com** for late-breaking news and current event stories selected just for you.

LAB ACTIVITY PLANNING

CHAPTER 48 INVESTIGATION

Simulating Disease Transmission, pp. 974–975

OVERVIEW

Students will simulate the transmission of disease and will use the data they collect to determine who was the original source of the "outbreak." This lab will illustrate to students how easily an infectious disease can spread and will encourage them to use deductive reasoning to trace the source of infection.

SAFETY

Students will wear disposable gloves, lab aprons, and safety goggles during the lab. If the chemical solutions contaminate skin, eyes, or clothing, students should flush the affected area with water.

PREPARATION

1. To prepare unknown solutions, prepare stock dropper bottles of distilled water for half the students and dropper bottles containing 10 percent ascorbic acid for the rest of the students. Both solutions will be clear. Ascorbic acid may be obtained from a biological supply house, such as WARD'S. A 10 percent ascorbic acid solution may be prepared by dissolving 10 g of ascorbic acid in 50 mL of water and diluting to 100 mL with water.
2. Indophenol indicator solution may be purchased from a biological supply house.
3. Each student will need a dropper bottle of unknown solution, a large test tube, and indophenol indicator solution.

Quick Labs

Organizing the Immune Response, p. 964

Students will need only paper and pencils for this activity.

UNDERSTANDING THE VISUAL

Point out to students that natural killer cells (shown here in gold) are one of several classes of defensive cells in the body. Natural killer cells locate and destroy cancer cells and cells infected by viruses. Ask students what will happen to the cancer cell (shown here in red) after its membrane has been punctured. Also ask students to explain how killing virus-infected cells would help eliminate a virus from the body.

FOCUS CONCEPT

Stability and Homeostasis
In a healthy individual, cells, tissues, organs, and organ systems work together to maintain homeostasis. Failures of homeostasis often lead to disease, but they can also improve our understanding of the regulatory systems of the body.

ASSESSING PRIOR KNOWLEDGE

Review the following concepts.

Cell Structure and Function: *Chapter 4*
Ask students to describe the differences between human cells and bacterial cells. Have students name some of the functions performed by all cells.

Pathogens: *Chapters 24, 25, and 26*
Ask students to list several differences between viruses and bacteria. Have students explain how viruses replicate. Have students list as many disease-causing protozoans as they can.

CHAPTER 48

INFECTIOUS DISEASES AND THE IMMUNE SYSTEM

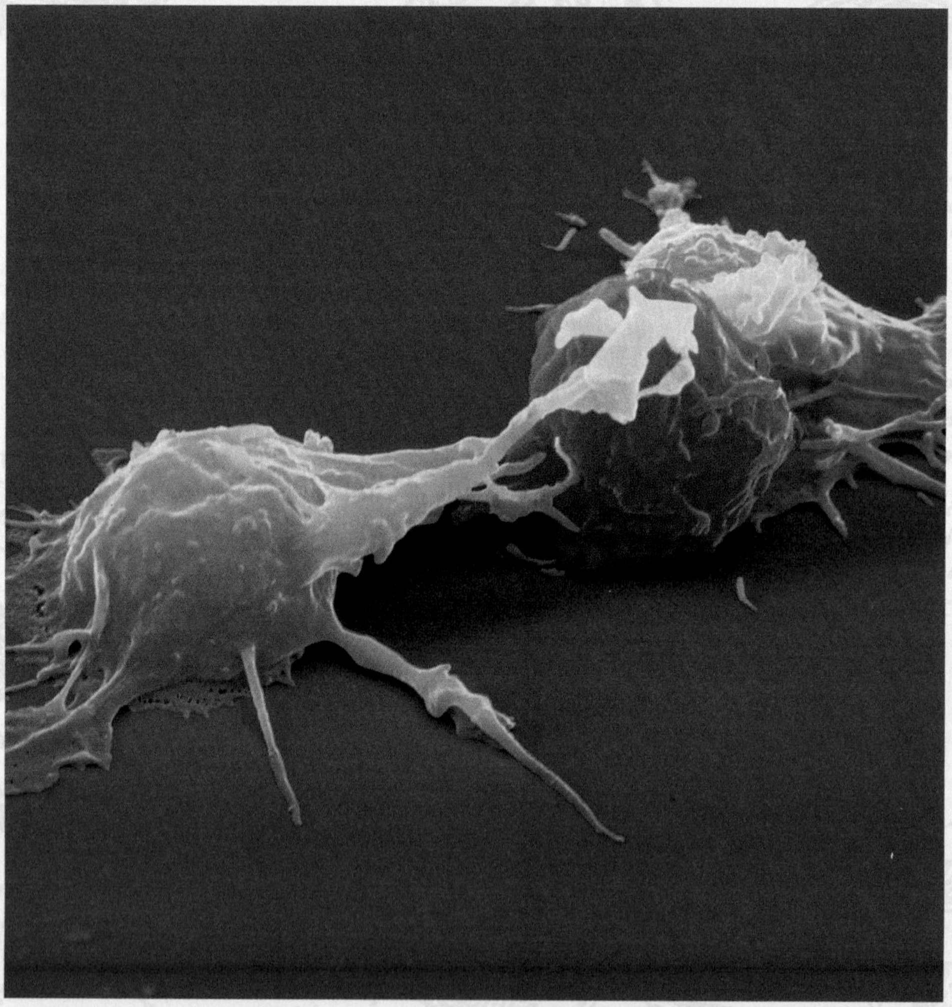

Two natural killer cells, which are defensive cells in the body, are attacking a cancer cell (shown in red). They will kill the cancer cell by puncturing its membrane (SEM 14,900×).

FOCUS CONCEPT: ***Stability and Homeostasis***
As you read, identify ways to prevent, diagnose, treat, and cure diseases.

NONSPECIFIC DEFENSES

*The human body is continuously exposed to pathogens, such as viruses and bacteria. When one of these pathogens enters the body and begins to multiply, it causes an **infectious disease.** In this section, you will examine the body's nonspecific defenses, which are the first lines of protection against invading pathogens.*

DISEASE TRANSMISSION

Robert Koch (1843–1910), a German physician, was the first scientist to establish a step-by-step procedure for identifying the particular pathogen responsible for a disease. In the 1870s, Koch was studying anthrax, a disease of cattle that can spread to humans. Koch observed that cattle with anthrax had swarms of bacteria in their blood. He hypothesized that these bacteria caused anthrax.

To test his hypothesis, Koch isolated bacteria from a cow with anthrax and grew the bacteria in a laboratory culture. Then he inoculated a healthy cow with bacteria from the culture. This cow developed anthrax, and Koch found that its blood contained the same kind of bacteria as the blood of the first cow. He concluded that these bacteria caused anthrax. The steps that Koch developed for determining the cause of a given disease, known as **Koch's postulates,** are listed in Figure 48-1.

Scientists have used Koch's postulates to identify thousands of pathogens. Human diseases are caused by bacteria, viruses, protists, fungi, and invertebrates. Each kind of pathogen affects the body differently. Among humans, pathogens can spread in five

Koch's Postulates
1. The suspected pathogen must occur in the body of an animal with the disease and not occur in the body of a healthy animal.
2. The suspected pathogen should be isolated and grown in a laboratory culture.
3. If a healthy animal is inoculated with this culture, the animal should develop the disease.
4. The pathogen from the second animal should be isolated and grown in the laboratory. It should be the same as the pathogen isolated from the first animal.

SECTION 48-1

OBJECTIVES

▲ Summarize Koch's postulates for identifying a disease-causing agent.

● Describe how the skin and mucous membranes protect the body against pathogens.

■ Describe the steps of the inflammatory response.

◆ Identify the white blood cells involved in a nonspecific response, and describe their functions.

▲ Explain the functions of interferon and fever.

FIGURE 48-1

The bacterium that causes anthrax, *Bacillus anthracis* (SEM 6,700×), was identified as a pathogen by means of Koch's postulates.

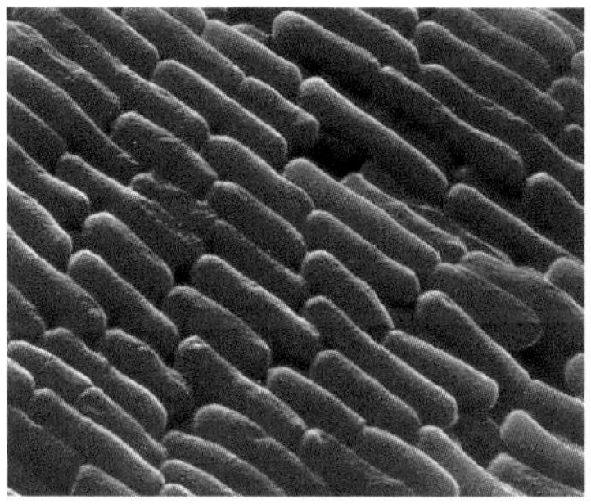

SECTION 48-1

SECTION OVERVIEW

Stability and Homeostasis

Section 48-1 describes Koch's postulates, the protocol that scientists use to determine the cause of a disease. The section also explains the nonspecific defenses that allow the body to maintain homeostasis in the face of invasion by pathogens.

ACTIVE READING

Technique: Reading Effectively

Select one paragraph for students to read. Have them write one sentence to summarize the main point of that paragraph. Repeat this activity in each section of this chapter.

DEMONSTRATION

Show pictures of various pathogens to students. Ask them to categorize the pathogens as one of five types: bacteria, viruses, protozoa, fungi, or invertebrates. Ask them to list some of the ways pathogens cause disease (cell destruction, toxin production, stealing nutrients from the body).

VISUAL LINK

Figure 48-1

Refer students to the list of Koch's postulates. Ask what process skills Koch used to determine the cause of a disease (observing, collecting data, hypothesizing, predicting, experimenting, and analyzing data).

SECTION 48-1

TEACHING STRATEGY
Colds

Check students' prior knowledge of infectious diseases by asking them what causes a cold (various viruses). Ask students what they should do to avoid catching a cold (avoid close contact with people with colds and wash their hands frequently). Have students discuss why these methods are not effective. (It is difficult to avoid exposure to people who have colds, especially at school or in other public places.)

QUICK FACT

☑ The skin is the largest and one of the most important organs of the body. In one square inch of skin there are 500 sweat glands, over 1,000 nerve endings, nearly 100 oil glands, 150 sensors for pressure, 75 sensors for heat, and 10 sensors for cold.

CRITICAL THINKING
Skin Bacteria

Ask students to identify which of the ecological relationships they studied in Chapter 22 (mutualism, commensalism, parasitism, predation, or competition) best describes the interaction between humans and the bacteria living on their skin. Have students justify their answers. (Some bacteria are parasites that can cause infections, while others are mutualists because they help suppress harmful bacteria.)

CONTENT CONNECTION
Section 46-3: Structure of the Skin

Remind students that the outer layer of the skin is known as the epidermis. The outer layer of the epidermis is composed of dead cells that are continuously being shed or worn away.

TABLE 48-1 *Some Human Diseases and Their Means of Transmission*

Disease	Pathogen	Principal means of transmission
AIDS	HIV (human immunodeficiency virus)	sexual intercourse, contaminated syringes
Common cold	any of 100 viruses	air currents, person-to-person contact
Malaria	protozoan parasites of genus *Plasmodium*	mosquitoes
Measles (rubeola)	paramyxovirus	air currents, person-to-person contact
Botulism	*Clostridium botulinum* (bacterium)	contaminated food

internetconnect
SCILINKS NSTA
TOPIC: Infectious diseases
GO TO: www.scilinks.org
KEYWORD: HM956

main ways—through air, food, water, person-to-person contact, and the bites of animals. Some pathogens are transmitted in only one way, while others may be transmitted in several ways. Table 48-1 lists some human pathogens, the diseases they cause, and their means of transmission.

SKIN AND MUCOUS MEMBRANES

The body's nonspecific defenses protect the body against any pathogen, regardless of its identity. Most pathogens must enter the body to cause disease. The skin, with its tough keratin shield, serves as a physical barrier to pathogens—as long as it remains intact. Any break in the skin may permit pathogens to enter the body. Extensive damage to the skin can be life threatening. A victim of severe burns, for example, is especially vulnerable to infections. Such a person must be thoroughly bandaged and treated with antibiotics until the skin has healed.

The skin continuously releases sweat, oils, and waxes, all of which contain chemicals that are toxic to many bacteria and fungi. Sweat, for example, contains the enzyme lysozyme, which can destroy the cell walls of bacteria. However, if all of the bacteria living on the skin were removed, harmful infections of the skin could occur because some of these bacteria inhibit the growth of pathogens.

Mucous membranes are epithelial tissues that protect the interior surfaces of the body that may be exposed to pathogens. Mucous membranes line the respiratory and digestive systems, the urethra, and the vagina. Mucous membranes serve as a barrier and secrete **mucus,** a sticky fluid that traps pathogens. The mucous membranes of the respiratory tract are covered with many beating cilia, as shown in Figure 48-2. These cilia constantly sweep mucus and pathogens up to the pharynx, where they are swallowed and forced into the stomach. Most swallowed pathogens are destroyed in the stomach by strong acids secreted by the stomach lining.

FIGURE 48-2
The passages of the respiratory system are lined with cells covered with beating cilia. Pathogens that become trapped in mucus secreted by these cells are swept upward, away from the lungs (5,325×).

THE INFLAMMATORY RESPONSE

Any pathogen that penetrates the skin or a mucous membrane will stimulate another nonspecific defense mechanism called the **inflammatory response,** which is summarized in Figure 48-3. Whenever an injury occurs, such as a cut or a splinter, some damaged cells release chemical messengers. One kind of chemical messenger is called histamine (HIS-tuh-MEEN). **Histamine** increases blood flow to the injured area and increases the permeability of the surrounding capillaries. As a result, fluids and white blood cells pass through capillary walls to the injured area. The changes caused by histamine result in redness, swelling, warmth, and pain—the familiar symptoms of inflammation.

If blood vessels have been damaged by the injury, platelets and clotting proteins initiate the blood-clotting process, sealing off the surrounding tissues and preventing pathogens from invading the rest of the body. To combat the pathogens that may have already entered through the wound, the body relies on white blood cells.

The chemical signals released by injured cells attract white blood cells to the site of the injury. Pus, the thick, whitish or yellowish fluid that often accumulates in wounds, contains dead pathogens and white blood cells. The presence of pus is a sign that white blood cells have arrived at the injury and are fighting pathogens.

Among the white blood cells that are involved in an inflammatory response are phagocytes. Phagocytes, as you learned in Chapter 5, are white blood cells that engulf pathogens by phagocytosis. The most common type of phagocyte, constituting 50 to 70 percent of the white blood cells in the body, is the **neutrophil** (NOO-truh-fil). Neutrophils circulate freely through blood vessels, and they

Eco Connection

Agriculture and the Origin of Human Diseases

Although some diseases are as old as our species, others have a more recent origin. One key development that changed the nature of human diseases was the beginning of farming and herding about 10,000 years ago. When humans began to keep large herds of domesticated animals, such as cattle and sheep, they were exposed to the pathogens of these animals. Some of these pathogens apparently began infecting humans. Measles, tuberculosis, smallpox, influenza, and pertussis (whooping cough) are among the diseases that appear to have been transmitted to humans from domesticated animals.

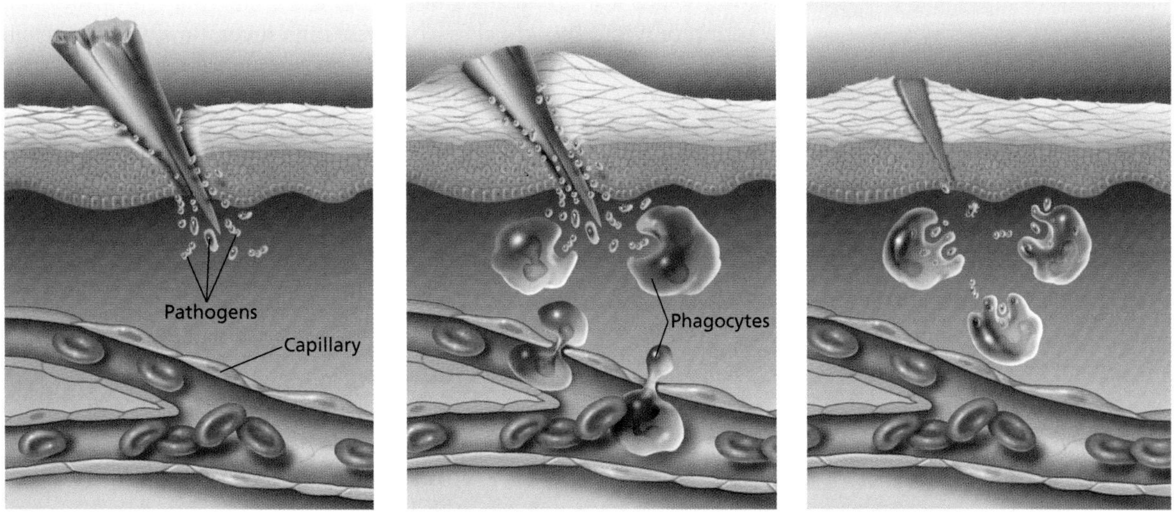

(a) (b) (c)

FIGURE 48-3

A small cut in the skin triggers an inflammatory response. (a) Injured cells release chemical alarm signals as pathogens enter through the cut. (b) In response, nearby capillaries swell and become leakier. The area around the wound swells and becomes warm. Phagocytes arrive to attack the pathogens. (c) Phagocytes destroy the pathogens, and the cut begins to heal.

TEACHING STRATEGY
Antihistamines

Antihistamines, such as epinephrine (adrenaline), reduce swelling. Histamines, such as norepinephrine, cause swelling. Ask students why the body has the ability to produce both kinds of molecules (so that it can respond to different situations).

DEMONSTRATION

Have students view a prepared slide of human blood through a microscope. Have them draw all the different types of cells they see. Review with them the differences between white and red blood cells, and have them locate both kinds of blood cells on their slide.

TEACHING STRATEGY
White-Blood-Cell Counts

Explain that low or high white-blood-cell counts often indicate disease. For example, neutrophils rise in number after microbes enter the body. Therefore, a high neutrophil count can signify an infection. Low lymphocyte counts are indicative of AIDS or other immune-suppression diseases.

INCLUSION *ACTIVITY*

To help students visually relate the vocabulary terms in this section to the functions of the nonspecific defenses, have them make posters illustrating nonspecific defenses. Encourage students to show how the nonspecific defenses work to deter pathogenic organisms. Make sure students label their posters using the vocabulary terms.

QUICK FACT

☑ The worldwide flu epidemic of 1918–1919 is the deadliest on record—21 million people died.

SECTION 48-1

TEACHING STRATEGY

Body Temperature

Tell students to take and record their temperature just before going to bed and when they get up in the morning. Have them take their temperature in class, then average all the temperatures taken at a given time. Is there a pattern? (Temperatures should be lowest in the early morning and highest at night.) Ask students whether the higher temperature at night is a fever. (No, body temperature fluctuates slightly throughout the day, mainly due to variation in the metabolic rate, which causes variation in the amount of body heat produced.)

ANSWERS TO SECTION 48-1 REVIEW

1. Koch isolated the bacteria from the blood of one cow, grew them in a laboratory culture, and injected them into a second cow, which subsequently developed anthrax. The bacteria in the second cow's blood were the same as those in the first cow.
2. The skin releases oils, waxes, and sweat that make its surface inhospitable to some pathogens.
3. Fluid leaks into the injury, causing swelling, and phagocytes can slip through the capillary walls to reach the site of injury.
4. Natural killer cells attack only infected cells and cancer cells, while macrophages attack pathogens and consume cellular debris.
5. Interferon is released by virally infected cells and increases the resistance of surrounding cells to infection.
6. Koch's third postulate cannot be applied to humans. It is unethical (and illegal) to infect a person with a pathogen that can cause serious disease.

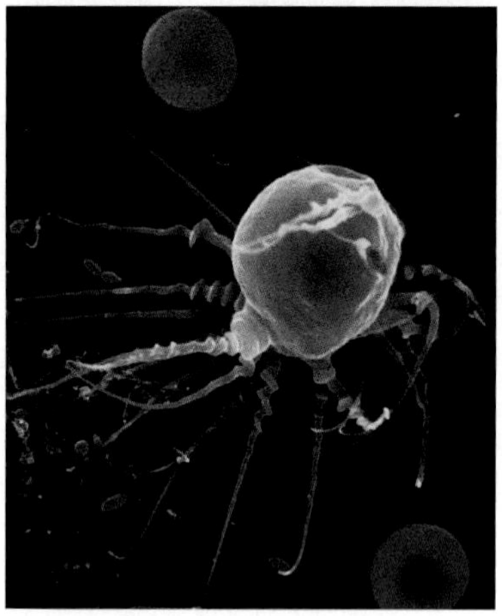

FIGURE 48-4
This macrophage is using cytoplasmic extensions to capture bacteria (shown in purple). (SEM 17,400×)

can squeeze between the cells in the wall of a capillary to reach the site of an infection. Once there, neutrophils engulf and destroy any pathogens they encounter. Another type of phagocyte is the **macrophage** (MA-kroh-FAYJ), shown in Figure 48-4. Macrophages consume and destroy any pathogens they encounter. They also rid the body of worn-out cells and cellular debris. Some macrophages are stationed in the tissues of the body, awaiting pathogens, while others move through the tissues and seek out pathogens.

Natural killer cells are large white blood cells that, unlike phagocytes, attack the cells that have been infected by pathogens, not the pathogens themselves. Natural killer cells are particularly effective in killing cancer cells and cells infected with a virus. A natural killer cell punctures the cell membrane of its target cell, allowing water to rush into the cell, causing the cell to burst.

Word Roots and Origins

macrophage

from the Greek *makros,* meaning "long," and *phagein,* meaning "to eat"

OTHER NONSPECIFIC DEFENSES

Interferon (in-tuhr-FEER-AHN) is a protein that inhibits the reproduction of viruses. It is produced in small amounts by cells that are infected by a virus. Interferon stimulates neighboring cells to produce an enzyme that inhibits the synthesis of viral proteins. This prevents viruses from reproducing within these cells. Scientists are investigating whether interferon can be used to treat viral diseases and cancer.

Fever is an elevation in body temperature above the normal 37°C (99°F). Fever is not a disease; it is a sign that the body is responding to an infection. Some pathogens trigger fever, as do chemical signals released by macrophages. Research suggests that a moderate fever stimulates the body's defense mechanisms. It suppresses the growth of some bacteria and may promote the action of white blood cells. However, high fevers are harmful. High temperatures can inactivate cellular enzymes in the body. In general, body temperatures greater than 39°C (103°F) are considered dangerous, and those greater than 41°C (105°F) are often fatal.

SECTION 48-1 REVIEW

1. How did Koch test his hypothesis about the cause of anthrax?
2. What chemical defenses does the skin use against pathogens?
3. What role does increased permeability of capillaries play in the inflammatory response?
4. How do natural killer cells differ from macrophages?
5. What is the function of interferon?
6. **CRITICAL THINKING** In many cases, not all of Koch's postulates can be applied to determine the cause of a human disease. Explain why.

SPECIFIC DEFENSES: THE IMMUNE SYSTEM

SECTION

48-2

OBJECTIVES

▲ Identify and describe the components of the immune system.

● Explain the functions of the three kinds of T cells.

■ Describe the actions of B cells in an immune response.

◆ Explain how a vaccine works.

▲ Contrast *allergy* with *autoimmune disease.*

Although the nonspecific defenses usually keep pathogens from entering and becoming established in the body, pathogens occasionally break through these defenses and begin to multiply. In response, the body's specific defenses are called into action. Unlike the nonspecific defenses, the specific defenses act against one particular pathogen.

THE IMMUNE SYSTEM

The body's specific defenses are part of the immune system, one of the major organ systems you studied in Chapter 46. The **immune system** has the job of fighting off invading pathogens and preventing the growth and spread of cancers. The immune system consists of several organs and the white blood cells found in these organs, as well as white blood cells in the blood and lymph. The organs of the immune system are scattered throughout the body; they include the bone marrow, thymus, lymph nodes, tonsils, adenoids, and spleen.

Each organ of the immune system plays a different role in defending the body against pathogens. Bone marrow, the soft material found inside long bones, such as the femur, manufactures the billions of new white blood cells needed by the body every day. Some newly produced white blood cells remain in the bone marrow to mature and specialize, while others travel to the **thymus** (THIE-muhs), a gland in the upper part of the chest just above the heart, to mature. Lymph nodes are located throughout the body along the vessels of the lymphatic system and contain large numbers of white blood cells. Lymph nodes filter pathogens from the lymph and expose them to white blood cells. The **spleen,** a fist-sized organ located just behind the stomach, filters pathogens from the blood. It is stocked with white blood cells that respond to the trapped pathogens. Figure 48-5, on the next page, shows the organs of the immune system.

The white blood cells of the immune system are known as lymphocytes (LIM-foh-SIETZ). As their name implies, these white blood cells accumulate in the lymph and lymph nodes, but lymphocytes are also found in the spleen and blood. There are two main types of lymphocytes: B cells and T cells. **B cells** are produced in the bone marrow and complete their development there. **T cells** are also produced in the bone marrow, but they mature after traveling to the thymus.

SECTION 48-2

SECTION OVERVIEW

Structure and Function
Section 48-2 describes the body's specific defenses, which come into action if pathogens penetrate the body's outer lines of defense. Each response from the specific defenses is directed against a particular pathogen. Section 48-2 also discusses malfunctions of the immune system, such as allergies and autoimmune diseases.

▶ ENGAGE STUDENTS

Ask students to recall what vaccinations they have had in their lives. If necessary, have them ask their parents or guardians for a list of their vaccinations. (Most have had polio, diphtheria, tetanus, pertussis (whooping cough), mumps, and rubella vaccinations.) Then ask students if any have had additional vaccinations for foreign travel or local situations. (Some may have had vaccinations against yellow fever, cholera, hepatitis, and typhoid fever.) Remind students that a vaccination stimulates the immune system to create defenses against a particular pathogen. Students will see in this section that vaccines rely on two properties of the immune system: specificity and "memory."

RETEACHING *ACTIVITY*

Refer students to Figure 48-5. Have students make a chart listing the major organs of the immune system, their locations, and their functions.

TEACHING STRATEGY

Tonsils

Students are probably familiar with tonsils, masses of lymphoid tissue located under the mucous membranes in the mouth and the back of the throat. Inform students that there are three sets of tonsils arranged in a protective ring. They are the palatine tonsils, which are located on each side of the throat; the pharyngeal tonsils, or adenoids, which are located near the posterior opening of the nasal cavity; and the lingual tonsils, which are located near the base of the tongue.

CRITICAL THINKING

Tonsillectomies

Until a few years ago, tonsillectomies were routine for young children with chronic throat infections (tonsillitis), and the tonsillectomies were usually complete, involving all three sets of tonsils. Today tonsillectomies usually are performed only if antibiotic therapy is unsuccessful or if swelling impairs breathing. Also, tonsillectomies now usually involve only one set of tonsils. Ask students to speculate about why there has been this change in procedure. (Students' answers will vary but should include the idea that today we know that tonsils play an important role in the body's defenses. Therefore, we are healthier if our tonsils stay intact.)

FIGURE 48-5

The organs of the immune system produce lymphocytes and filter pathogens from the blood and lymph.

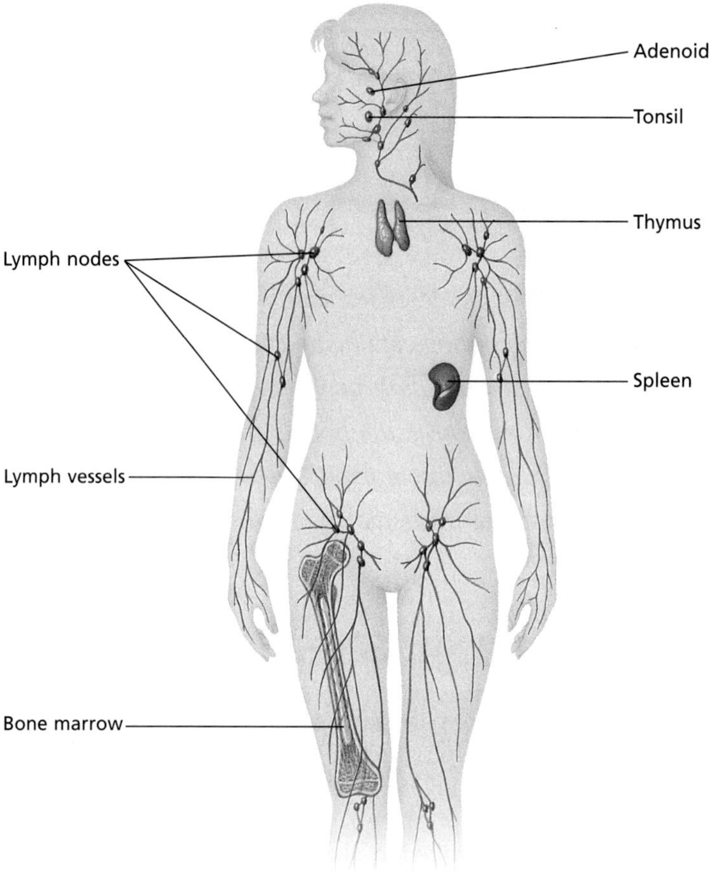

Word Roots and Origins

antigen

From the Greek *anti-*, meaning "against," and *-genes*, meaning "born"

RECOGNIZING PATHOGENS

Lymphocytes are the body's specific defenses. When the body is invaded by a pathogen, lymphocytes launch an attack known as an **immune response** to eliminate the pathogen. In order to respond to a pathogen, however, lymphocytes must be able to recognize it as a foreign invader and distinguish it from the cells of the body. Any substance that the immune system recognizes as a potential pathogen and that provokes an immune response is known as an **antigen** (AN-tuh-jin). A wide variety of substances can be antigens, including pathogens or parts of pathogens, bacterial toxins, insect venom, and pollen.

How do lymphocytes identify antigens? Every lymphocyte has receptor proteins scattered over the surface of its cell membrane. These receptor proteins recognize and bind to antigens that match their particular three-dimensional shape, as shown in Figure 48-6. The surface of a bacterial cell, for instance, can be covered with many different kinds of molecules—including proteins and polysaccharides—each of which can function as an antigen and cause lymphocytes to react. All of the receptors on an individual lymphocyte are the same shape and thus bind to the same antigen.

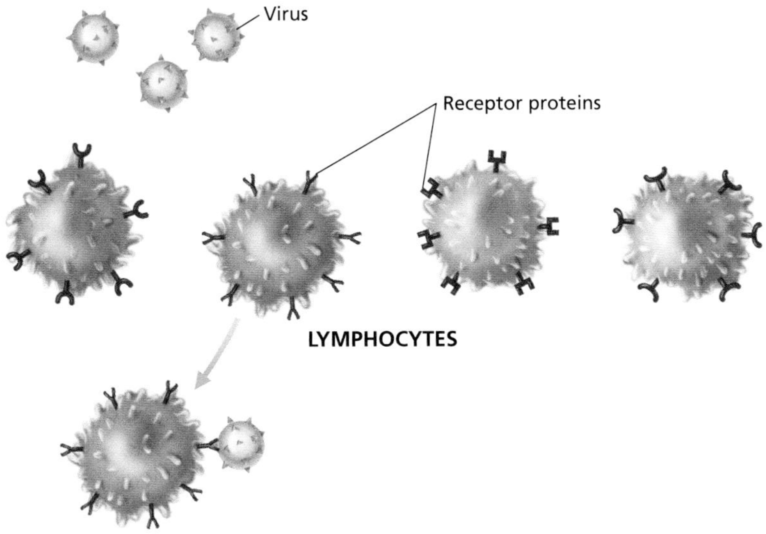

FIGURE 48-6

The receptor proteins on the surface of lymphocytes have a complex, three-dimensional structure. The receptors can fit onto, and thus bind, molecules on pathogens that have a complementary shape.

The body can defend itself against an enormous number of different pathogens because the immune system makes millions of different kinds of lymphocytes, each carrying uniquely shaped receptors. The specificity of the immune system is due to the specificity of the antigen receptors on the lymphocytes. For example, when a cold virus invades the body, lymphocytes whose receptor molecules match the antigens of the cold virus are activated to counterattack. Lymphocytes with different-shaped receptors, such as those that recognize an influenza (flu) virus, do not respond.

IMMUNE RESPONSE

An immune response is a two-pronged assault on a pathogen. One arm of the immune response, the **cell-mediated immune response,** involves T cells. The other arm involves mainly B cells and is called the **humoral** (HYOO-muh-ruhl) **immune response.** Both parts of the immune response are controlled by a type of T cell called a **helper T cell.**

The first step in an immune response occurs when a macrophage engulfs and destroys a pathogen. The macrophage then displays fragments of the pathogen's antigens on the surface of its own cell membrane. When a helper T cell with a receptor matching this antigen encounters the macrophage, the macrophage releases a cytokine called interleukin-1 (IN-tuhr-LOO-kin), which in turn triggers the helper T cell to release a second cytokine, called interleukin-2.

Cell-Mediated Immune Response

Interleukin-2 stimulates the helper T cells and two other types of T cells—cytotoxic T cells and suppressor T cells—to rapidly divide. **Cytotoxic** (SIE-toh-TAHK-sik) **T cells** combat the pathogen by destroying any of the body's cells that have been infected by the

VISUAL LINK

Figure 48-6

Use the figure to emphasize one of the immune system's key features—its specificity. Each time a pathogen enters the body, only a few of the millions of subclasses of lymphocytes will be called into action—those with antigen receptors that match the pathogen's antigens. Also point out to students that some pathogens exploit the immune system's specificity. By rapidly changing their antigens, such pathogens stay one step ahead of the immune system. HIV, the influenza virus, and the protists that cause African sleeping sickness are examples of pathogens with this ability.

TEACHING STRATEGY

Lymphocytes

To help students keep track of all the cells involved in an immune response, have them each make a table with the following headings: Type of Cell, Function, Interacts With. Have students complete their table as they read the section.

TEACHING STRATEGY

Cytotoxic T Cells Versus Natural Killer Cells

Cytotoxic T cells and natural killer cells have the same function—to attack and destroy virus-infected cells and cancer cells. Ask students to identify the key difference between these two kinds of cells. (Natural killer cells, as part of the nonspecific defenses, do not target specific antigens and will destroy any "foreign" cell they come across. Each cytotoxic T cell, by contrast, is programmed to destroy only cells with antigens that match its receptors.)

VISUAL LINK

Figure 48-7

Emphasize to students the central role of helper T cells in controlling the cell-mediated immune response. Cytokines released by stimulated helper T cells "recruit" cytotoxic T cells and trigger the division of suppressor T cells that will later shut down the immune response. Ask students whether a cell-mediated immune response can occur without helper T cells (no).

MAKING CONNECTIONS

Social Studies

Diseases have had a significant effect on history. When Europeans conquered the Americas beginning in the late 1400s, they introduced their animals, plants, and epidemic diseases. By some estimates, 95 percent of the population of the Americas perished within 100 years of Columbus's arrival; most were killed by diseases brought from Europe, such as smallpox, measles, and influenza. Why were the residents of the Americas so much more susceptible to these diseases than were Europeans? Coevolution. Because these diseases had existed in Europe for centuries, the Europeans had evolved some degree of resistance. The residents of the Americas, by contrast, had never been exposed to measles, smallpox, or influenza, and few people had any resistance.

CULTURAL CONNECTION

Cell Research

The first Japanese scientist to win the Nobel Prize in medicine and physiology, Dr. Susumu Tonegawa, has been studying B cells and T cells since the 1970s. He became famous for demonstrating that B cells are capable of producing a large variety of antibodies (between 10^6 and 10^9 forms) by genetic rearrangement.

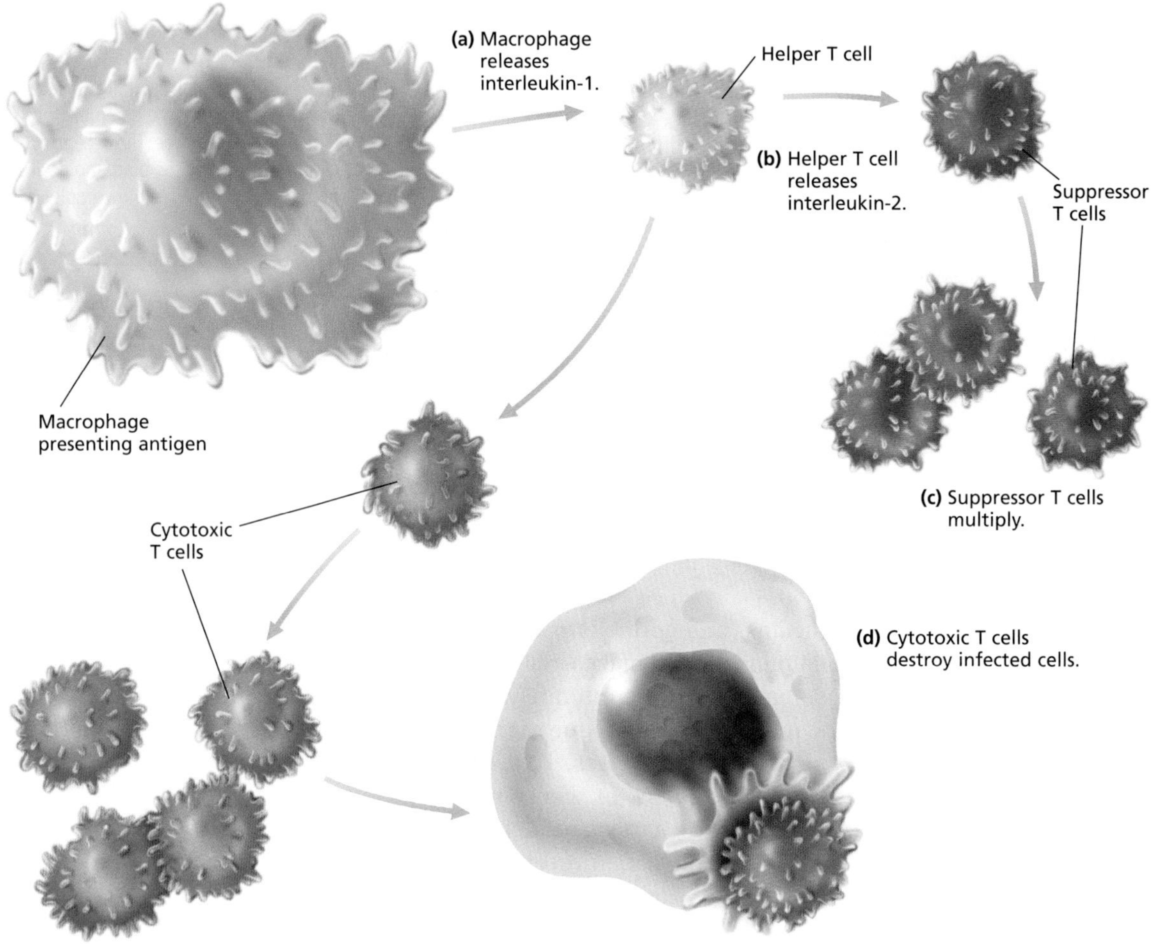

FIGURE 48-7

The cell-mediated immune response involves macrophages, helper T cells, cytotoxic T cells, and suppressor T cells. (a) The macrophage releases interleukin-1, which stimulates the helper T cell. (b) The helper T cell then releases interleukin-2. In response, cytotoxic T cells and suppressor T cells begin to divide. (c) Suppressor T cells divide slowly. Their role is to shut down the immune response. (d) Cytotoxic T cells destroy infected cells by puncturing their cell membranes.

pathogen. Cells that have been invaded by a pathogen are recognizable because they usually have some of the pathogen's antigens on their surface. Cytotoxic T cells attach to these antigens. Like natural killer cells, cytotoxic T cells kill by puncturing the cell membrane of their target. Cytotoxic T cells can also kill cancer cells and attack parasites and foreign tissues, such as those received during an organ transplant. **Suppressor T cells** help to shut down the immune response after the pathogen has been cleared from the body. Figure 48-7 illustrates the cell-mediated immune response.

Humoral Immune Response

Interleukin-2 and antigen presentation by a macrophage or T cell stimulate B cells to divide and differentiate into plasma cells. **Plasma cells** are highly specialized cells that produce defensive proteins and secrete them into the blood. These defensive proteins are identical to the plasma cell's antigen receptors and are known as **antibodies.** Antibodies are Y-shaped molecules. The two arms of each Y are identical, and they recognize and attach to the same antigen. One plasma cell can make up to 30,000 antibody molecules per second.

Antibodies bind to specific pathogens but do not destroy them directly. Instead, they either inactivate the pathogen or trigger its destruction by the nonspecific defenses. For example, by attaching

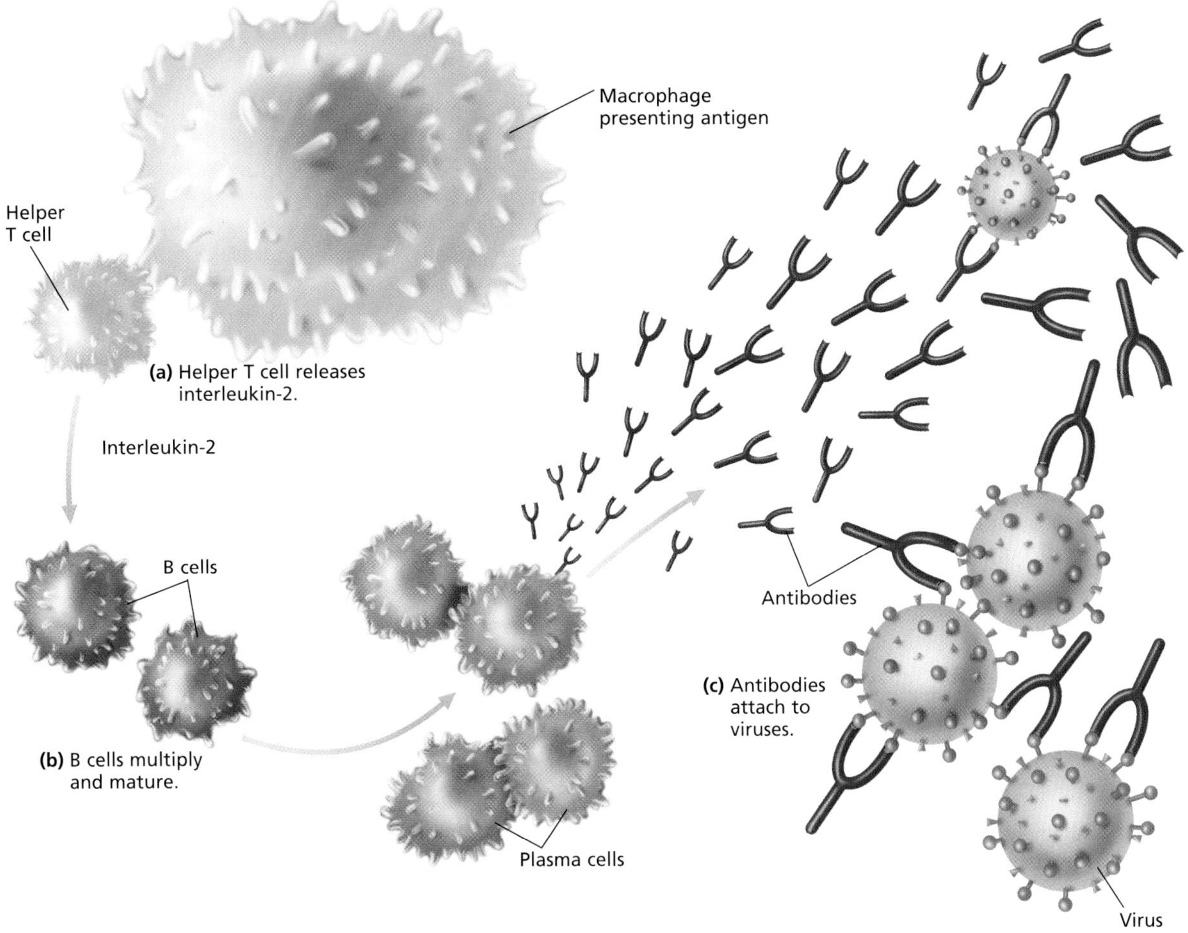

to the surface proteins of a virus, antibodies prevent the virus from entering a cell, thereby blocking its reproduction. Antibodies also cause pathogens to clump together, which facilitates phagocytosis by macrophages. Figure 48-8 summarizes the humoral immune response.

FIGURE 48-8

The humoral immune response involves the activation of B cells and the production of antibodies. (a) After encountering a macrophage, a helper T cell releases interleukin-2. This stimulates rapid division by B cells. (b) After dividing for five days, mature B cells differentiate into antibody-secreting plasma cells. (c) Antibodies bind to free antigen particles and cause antigens such as bacteria and viruses to clump.

Primary and Secondary Immune Responses

After an infection has been overcome, the immune response is shut down and most of the participating cells die. However, some B cells and T cells remain, often throughout the individual's life, as memory cells. **Memory cells** are the body's long-term protection against reinfection by a specific pathogen. Because of memory cells, you cannot get most diseases more than once. If you are exposed to a pathogen a second time, memory cells immediately recognize it and begin to divide rapidly. They eliminate the pathogen before it can produce disease. The first time the body encounters an antigen, the immune response is called a **primary immune response.** The response of memory cells to a subsequent infection by the same pathogen is called a **secondary immune response.** As you can see in Figure 48-9 on the next page, a secondary immune response is much faster and more powerful, producing many more antibodies. Keep in mind that memory cells only protect against a pathogen that the immune system has already encountered.

VISUAL LINK
Figure 48-8

Have students compare Figure 48-8 with Figure 48-7 and list as many similarities and differences as possible between the cell-mediated and humoral immune responses. Emphasize that helper T cells are also central to the humoral immune response. Inform students that HIV leads to the destruction of large numbers of helper T cells. Ask students to predict how a decline in T cell numbers would affect the immune response of an HIV-infected person and how the person's health might in turn be affected. (Without helper T cells, neither arm of the immune response can get started, and the body cannot mount a counterattack against pathogens. Thus, people infected with HIV are likely to be more susceptible to infections and cancer.)

RETEACHING *ACTIVITY*

Have students draw a flowchart that depicts the events of the humoral immune response, beginning with the production of B cells in the bone marrow and continuing through the differentiation of plasma cells and the formation of memory cells. Then have students make a similar diagram of the cell-mediated immune response, again beginning with the production of T lymphocytes and concluding with the proliferation of the different classes of T cells and the formation of memory cells. You may want to allow students time to share their charts with one another.

SECTION 48-2

GIFTED ACTIVITY

Among infectious diseases, pneumonia is the leading killer in North America, accounting for 28 of every 100,000 deaths. Pneumonia can be caused by viruses or bacteria. A vaccine has recently been developed to help protect the most vulnerable people—primarily the elderly and those with suppressed immune systems—from the most common kinds of pneumonia. But because of adverse side effects, doctors cannot repeat the vaccine for several years after inoculation. Have students research the use of pneumovaccines and write a report discussing the benefits and disadvantages of this vaccine.

Quick Lab

Organizing the Immune Response

Time Required 15 minutes

Procedural Tips Have students work in pairs. Refer them to Figures 48-7 and 48-8. Completed diagrams or flowcharts should show the four steps indicated in Figure 48-7, in their correct sequence.

Answers to Analysis Helper T cells control the immune response. The cell-mediated response attempts to destroy the pathogen. The humoral response leads toward production of antibodies.

TEACHING STRATEGY

Allergies

Poll the class to find out who has allergies or knows someone who does. On the board, make a list of the allergens that affect class members. Ask students to describe the symptoms of their allergies. Encourage students to describe how allergies affect their lives, and ask allergy sufferers how they are being treated for the allergy.

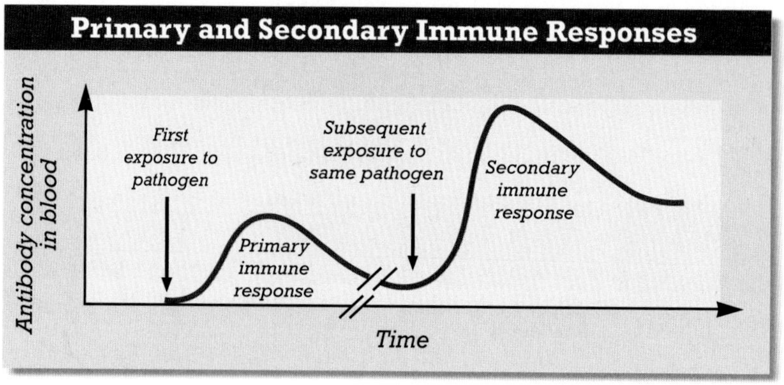

FIGURE 48-9

Compare the production of antibodies during the primary and secondary immune responses.

IMMUNITY AND VACCINATION

A person who is resistant to a specific pathogen is said to have **immunity** to it. One way to acquire immunity is to be infected by the pathogen and to survive the disease it causes. Another, safer way is through vaccination. Vaccines contain pathogens or toxins that have been modified so that they can no longer cause disease. Vaccines produce immunity because they contain antigens that stimulate an immune response, resulting in the production of memory cells. Thus, a vaccine prepares the immune system to fight off a pathogen should it ever appear in its disease-causing form.

Among the diseases that have been controlled through the use of vaccines are polio, measles, mumps, tetanus, and diphtheria. An intensive worldwide vaccination campaign has eliminated smallpox. Sometimes the protection provided by vaccination wears off over time. That's why doctors recommend periodic booster shots to restore immunity against some diseases, such as tetanus and polio.

Quick Lab

Organizing the Immune Response

Materials paper, pencil

Procedure Create a diagram or a flowchart that outlines the steps involved in a cell-mediated immune response. Label the cells and the steps.

Analysis What are helper T cells? How is a cell-mediated response different from a humoral response?

ALLERGIES

Sometimes the immune system reacts to otherwise harmless antigens in ways that can be harmful. This inappropriate reaction to a harmless substance is called **allergy.** Antigens that can trigger allergic reactions include pollen, animal dander (flakes of skin), dust mites, food, and fungal spores. Allergic reactions are characterized by watery eyes, wheezing, and sneezing. Many of the symptoms of allergy result from the release of histamine by cells that are exposed to the antigen. Drugs called antihistamines help counteract the effects of histamine and can relieve the symptoms of allergies. For most people, allergies are irritating and inconvenient but not life threatening. However, some people experience severe reactions that can be fatal.

TABLE 48-2 *Autoimmune Diseases and Their Target Tissues*

Disease	Tissues affected	Symptoms
Systemic lupus erythematosus	connective tissue throughout the body	facial skin rash, painful joints, fever, fatigue, kidney problems, weight loss
Type 1 diabetes	insulin-producing cells in pancreas	excessive urine production, blurred vision, weight loss, fatigue, irritability
Graves' disease	thyroid	weakness, irritability, heat intolerance, increased sweating, weight loss, insomnia
Rheumatoid arthritis	joints	painful, crippling inflammation of the joints

AUTOIMMUNE DISEASES

Lymphocytes that recognize and react to the body's own cells are usually eliminated during development, before they become functional. This prevents an attack directed at the body's own tissues. However, in rare cases the immune system does respond to the body's own cells, attacking them as if they were pathogens. Such a disorder is called an **autoimmune** (AW-toh-i-MYOON) **disease.** For example, multiple sclerosis is an autoimmune disease of the nervous system that affects mainly young adults. In this disease, T cells attack and slowly destroy the insulating material that covers nerves. Although the severity of multiple sclerosis varies from individual to individual, the damage may progress to a point at which nerve transmission is interrupted. In severe cases, paralysis, blindness, and even death can result. Scientists are still searching for the causes of multiple sclerosis and other autoimmune diseases. Table 48-2 lists several autoimmune diseases and describes their effects on the body.

internet**connect**

SCILINKS NSTA

TOPIC: Autoimmune diseases
GO TO: www.scilinks.org
KEYWORD: HM965

SECTION 48-2 REVIEW

1. **Describe the functions of the spleen and of the bone marrow.**
2. **Contrast the functions of helper T cells with those of cytotoxic T cells.**
3. **What is the role of B cells in an immune response? How do B cells depend on T cells?**
4. **Explain how a vaccine stimulates immunity to a disease.**
5. **Name one similarity and one difference between autoimmune diseases and allergies.**
6. **CRITICAL THINKING A friend tells you that because he has just recovered from a cold, he now cannot get the flu. Is your friend right? Explain your answer.**

SECTION 48-2

ANSWERS TO SECTION 48-2 REVIEW

1. The spleen filters pathogens from the blood and exposes them to attacks by white blood cells. The bone marrow produces new white blood cells and serves as the site of maturation for B cells.
2. Helper T cells activate the production of cytotoxic T cells, suppressor T cells, and B cells. Cytotoxic T cells attack and destroy infected body cells.
3. When stimulated by helper T cells, B cells divide and differentiate into plasma cells, which produce large amounts of the defensive proteins known as antibodies.
4. Vaccines expose the body to the antigens of a pathogen, thus stimulating a primary response from the immune system. This also stimulates the production of memory cells.
5. Both allergies and autoimmune diseases are attacks on an inappropriate target. An allergy is an attack on a harmless substance, such as plant pollen, while an autoimmune disease is an attack on the body's cells.
6. He is wrong. His immune system will be primed to fight off that particular cold virus if he is ever exposed to it again, but he has not developed immunity against the influenza virus.

BACKGROUND

Edward Jenner was born in 1749 in Gloucestershire, England. Edward Jenner's father, a clergyman, died when Edward Jenner was five years old. An older brother, also a clergyman, raised him from that point. Edward Jenner's love of nature and interest in science led to an apprenticeship with a local surgeon. From the age of 13 to the age of 21, he studied medicine and surgery. After his apprenticeship, he went to London and began medical studies at St. George's Hospital. In 1773, after two years in London, he returned to his country home and began a medical practice. Following the work described in the feature, Jenner tried in 1797 to have a short paper describing his results published by the Royal Society. However, the Royal Society rejected the paper. Jenner continued his work and in 1798 published a book that described his work.

Although vaccination is a simple procedure, there were many initial complications. Jenner had recommended that certain precautions be taken; they were ignored by many of the people who practiced vaccination. Despite all the problems, the procedure quickly proved its worth, and others tried to take credit for Jenner's ideas. Jenner saw vaccination become accepted throughout the world. After reading in 1799 about Jenner's vaccination procedures, an American physician and scientist named Benjamin Waterhouse became an advocate and pioneer in smallpox vaccination in America. Waterhouse was among the first faculty members of Harvard Medical School.

GREAT DISCOVERIES

Development of a Vaccine

HISTORICAL PERSPECTIVE

The vaccinations you received as a child were unknown in the 1700s. At that time, the diseases that are controlled today plagued children and adults, as they had for centuries. The odds were great that one or more children in a family would not live to be 10 years old. Today, however, in developed countries, it is rare for a child to die of any disease before the age of 10. A major reason is the use of vaccines.

A Dreaded Disease

Before the 1800s, a disease known as smallpox often reached epidemic proportions throughout the world. Thousands of people died of smallpox annually, and thousands of others were disfigured for life by deep scars.

It had long been observed that people who survived smallpox were thereafter immune to it. Thus the practice of inoculating a person with a mild form of smallpox had developed. But there were grave risks. When the disease was contracted in this way, there was still a risk that the symptoms would become severe, and many died as a result.

One Man's Plan

Edward Jenner, a British physician, was investigating cowpox. He had heard that milkmaids seemed to be immune to smallpox and that because of their contact with cows, they often contracted cowpox, a relatively harmless disease. It was also widely known at the time that milkmaids who had cowpox were immune to smallpox. Jenner saw a

Edward Jenner

connection between these observations and came to the conclusion that there was a safer way to immunize against smallpox—by using cowpox. All Jenner needed was an opportunity to test his hypothesis.

In 1796, a milkmaid named Sarah Nelmes contracted cowpox, and Jenner had his chance. He took matter from a cowpox sore on Sarah's hand and injected it into James Phipps, an 8-year-old boy. Two months later, Jenner inoculated the boy with material taken from a sore of a smallpox patient. James remained healthy. In fact, James never got smallpox, even after Jenner repeatedly inoculated him with smallpox material. Jenner named the procedure *vaccination,* after the Latin word *vacca,* which means "cow." (Note that Jenner's methods were acceptable by the ethical standards of his time, but they would not be permitted today.)

In London, Jenner sought volunteers for further tests, but no one was willing to take the risk. Jenner tried to convince the Royal Society, the foremost scientific society of his day, to publish his results. When the group refused, he published, at his own expense, a 64-page book, *An Inquiry into the Causes and Effects of the Variolae Vaccinae, a Disease Known by the Name of Cow Pox,* in which he described his work and its results:

> *I have proceeded in an inquiry founded on experiment and I shall continue to prosecute this inquiry, encouraged by the pleasing hope of its becoming essentially beneficial to mankind.*

Roadblocks and Rewards

Although other doctors had verified Jenner's results, the new procedure was not initially accepted by the medical community. Also, some of the few physicians who did use the vaccine used cowpox material contaminated with smallpox viruses, which confused the results. Jenner was able to prove that those experiments were flawed and thus gained acceptance for his conclusions. He worked actively for vaccination, and soon the procedure was used worldwide.

Much still remained to be done; many doctors who vaccinated for smallpox used improper methods. Still, the importance of Jenner's work became evident as the death rate from smallpox fell, and Jenner began to receive honors and awards. A Royal Jennerian Society was created to ensure that the vaccination process was used throughout London. Parliament twice voted to give Jenner monetary rewards, and statues were erected in his honor in London and Gloucester.

A Medical Success Story

For almost a hundred years, the use of cowpox matter to create immunity against smallpox was the only vaccination procedure. It wasn't until 1881 that a vaccine for another disease was produced. The French scientist Louis Pasteur was able to grow a weakened form of the cholera bacteria, and he inoculated chickens with these bacteria, immunizing them against the disease.

The vaccination for smallpox has been so successful that it is no longer necessary. The last smallpox case was reported in 1977 in Somalia. The disease is now declared eradicated worldwide.

Today's Search

Most children in the United States are routinely vaccinated against numerous diseases at an early age, so few American lives are cut short by measles, diphtheria, tetanus, or polio. But problems remain. For a number of deadly diseases, such as AIDS, malaria, schistosomiasis, and African sleeping sickness, no vaccine has yet been produced. For other diseases, such as influenza, vaccination provides only short-term immunity, and frequent re-vaccination is required. However, new vaccines made using DNA technology may soon be introduced into medical practice and may help to solve these problems.

Most of today's vaccines work by introducing into the patient weakened or dead bacteria or viruses that stimulate an immune response. New approaches, which immunize by using a DNA carrier, may lead to the prevention of diseases such as influenza, hepatitis, malaria, and AIDS. These DNA vaccines are easily stored, which would greatly facilitate their delivery to developing countries, bringing us even closer to a day when other deadly diseases will follow smallpox into extinction.

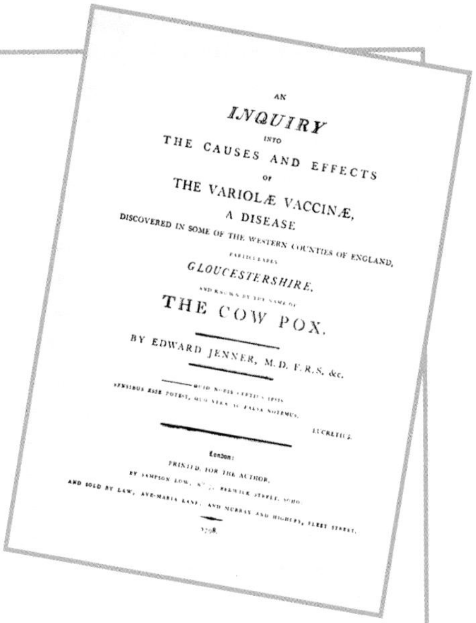
AN
INQUIRY
INTO
THE CAUSES AND EFFECTS
OF
THE VARIOLÆ VACCINÆ,
A DISEASE
DISCOVERED IN SOME OF THE WESTERN COUNTIES OF ENGLAND,
GLOUCESTERSHIRE,
THE COW POX.
BY EDWARD JENNER, M.D. F.R.S. &c.
LUCRETIUS.
London:
PRINTED, FOR THE AUTHOR,
1798.

Edward Jenner described the results of his work in the Inquiry, *published at his own expense in 1798.*

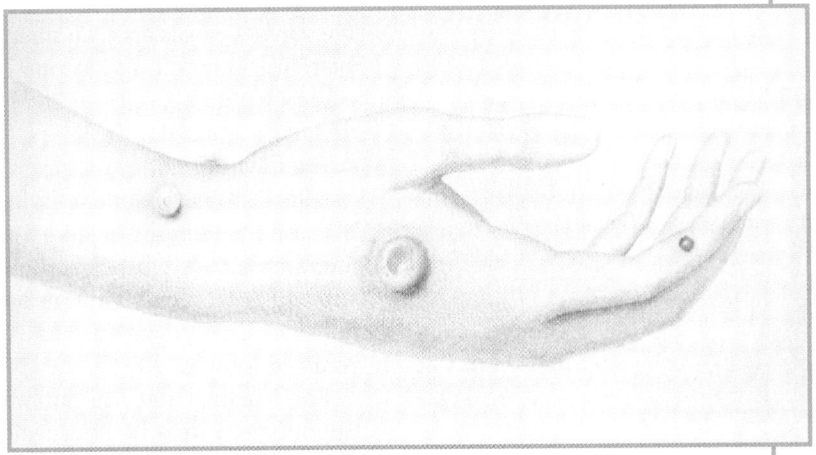

Jenner used matter taken from cowpox lesions like these as a vaccine against smallpox. He included this drawing of Sarah Nelmes's hand in the Inquiry.

DISCUSSION

Guide the discussion by posing the following questions:

1. What information did Jenner have that suggested that people could be protected from smallpox through inoculations? (It was well known at the time that once people had contracted smallpox, they were thereafter immune to it. Also, Jenner had heard that milkmaids who had been exposed to cowpox seemed to be immune to smallpox.)

2. What problem did Jenner face in getting the medical community to accept his ideas? (Although other doctors had verified Jenner's results, the new vaccination was not readily accepted by credible members of the medical community. Also, many doctors who vaccinated for smallpox used improper methods. One physician used cowpox material contaminated with smallpox, thereby distorting the results.)

FURTHER READINGS

1. "Swallowing *Shigella*," in *Science News,* May 11, 1996, pages 302–304. This article asks the question, "Can bacteria that cause food poisoning deliver oral DNA vaccines?" The radical new form of immunization using DNA vaccines may one day allow us to swallow a pill containing *Shigella flexneri* to ward off infectious diseases ranging from tuberculosis to AIDS.

2. "A pox on smallpox," in *U.S. News & World Report,* February 5, 1996, page 14. This article discusses the WHO governing board's recommendation that the only two known stocks of the smallpox virus be destroyed by 1999.

SECTION 48-3

SECTION OVERVIEW

Stability and Homeostasis
Section 48-3 describes how HIV causes AIDS, how the body responds to HIV, how the virus is transmitted, and how scientists are seeking to create vaccines and treatments for HIV.

▶ ENGAGE STUDENTS

Ask students to describe some ways in which AIDS has affected our society. (Hundreds of thousands of people have died. People should now be aware that there can be grave consequences to sexual encounters and the use of intravenous drugs. People also have concerns about receiving blood transfusions.) Ask students whether they believe that physicians or dentists who are infected with HIV should continue practicing and, if so, whether they should tell their patients. (Allow for students' personal opinions.) Stress that effective controls against the transmission of HIV to or from a health professional are proper hygiene (wearing gloves and eye protection) and the sterilization of instruments.

RETEACHING *ACTIVITY*

Review with students the role of helper T cells in the immune response (Section 48-2). Ask students to list the other cells of the immune cell that are affected if all helper T cells are killed (B cells, cytotoxic T cells, suppressor T cells).

DEMONSTRATION

Invite a physician, school nurse, or other health-care professional to class to discuss the latest information on HIV and AIDS. Ask the speaker to talk about how AIDS has challenged the health-care system.

SECTION 48-3

OBJECTIVES

▲ Describe the course of HIV infection.

● Identify four ways HIV is transmitted.

■ Describe how HIV's rate of evolution affects the development of vaccines and treatments.

AIDS

The immune system normally provides very effective protection against infectious diseases. Its importance to our health is dramatically illustrated by the diseases in which the immune system malfunctions. The most deadly of these diseases is AIDS, or acquired immunodeficiency syndrome. AIDS was first recognized as a disease in 1981, and since then it has killed more than 300,000 Americans.

HIV AND AIDS

AIDS is a disease in which the immune system loses its ability to fight off pathogens and cancers. As you learned in Chapter 24, AIDS is caused by the human immunodeficiency virus, or HIV, a type of retrovirus. Recall that retroviruses contain RNA, not DNA, as their genetic material. Like other viruses, HIV cannot reproduce outside a cell. HIV can invade several kinds of cells, including macrophages, but its main targets are helper T cells. HIV enters a cell by binding to a receptor called CD4 and one other receptor on the cell's membrane. An infected helper T cell manufactures large numbers of new virus particles, as shown in Figure 48-10.

FIGURE 48-10
This image shows particles of HIV escaping through the membrane of an infected T cell. (TEM 117,000×)

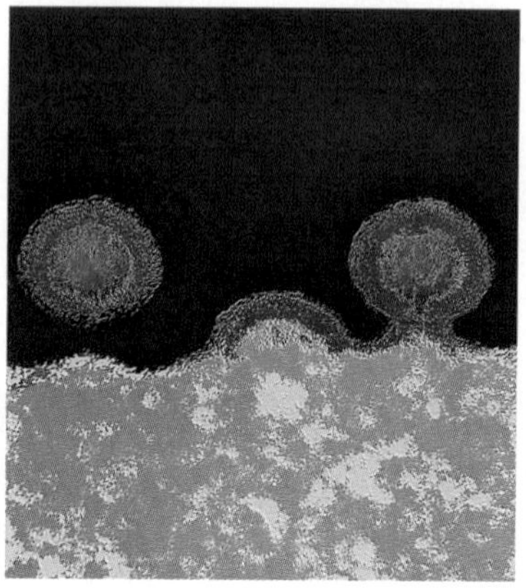

Course of the Disease

HIV begins to reproduce within the body shortly after infection. The presence of large numbers of viruses stimulates the immune system to launch a vigorous attack. At this point, which usually occurs within a month or so of infection, the individual often experiences a short flulike illness, with fever, fatigue, body aches, and swollen lymph nodes.

This first battle is the beginning of a long-term struggle between HIV and the immune system. HIV continues to replicate rapidly, but the immune system keeps the virus in check, and the infected person usually feels well and appears healthy. This stage of HIV infection can last for as little as two years, but typically lasts for 10 years or longer.

For reasons that are not well understood, HIV eventually gets the upper hand in its battle with the immune system. The number of helper T cells in the body begins to decline gradually, as shown in Figure 48-11.

The reduction in helper T cells is disastrous for the infected person. Helper T cells play a crucial role as the "commanders" of the

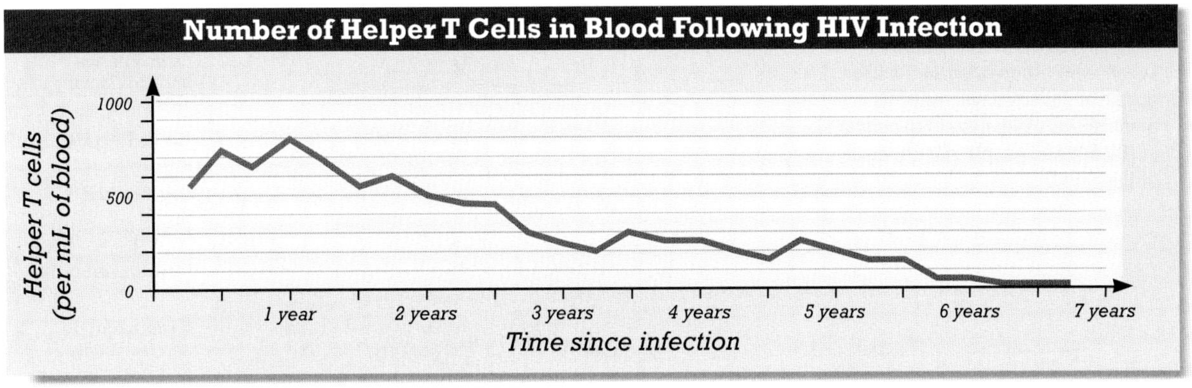

FIGURE 48-11

Once HIV begins to destroy the immune system, the number of helper T cells in the blood falls dramatically.

immune response, activating both cytotoxic T cells and B cells. As the number of helper T cells falls, the immune system becomes weaker, and the infected individual becomes vulnerable to a variety of cancers and diseases that a healthy immune system would normally defeat. These diseases are called **opportunistic infections,** and they usually strike only people with weakened immune systems. When the number of helper T cells in the blood falls below 200/mL (a normal amount is 600/mL–700/mL), the individual is said to have AIDS. AIDS is the last stage of HIV infection. Few individuals live more than two years after being diagnosed with AIDS, usually dying from opportunistic infections or cancer.

Transmission

HIV is transmitted by the transfer of body fluids containing HIV or HIV-infected cells. The most common means of infection is sexual intercourse with an infected person. HIV-infected cells are found in semen and vaginal secretions, and infection can occur through vaginal, oral, or anal intercourse. Use of a latex condom during sex greatly reduces the likelihood of transmission but does not eliminate it completely. It is important to remember that people who have been infected with HIV but have not developed AIDS can still transmit the virus.

The second most common means of infection is through the use of syringes and hypodermic needles that have been contaminated with blood containing HIV. People who inject intravenous drugs, such as heroin, and who share syringes or hypodermic needles with others are at very high risk of infection. During the early years of the AIDS epidemic, a number of individuals were infected by receiving HIV-contaminated blood transfusions. Donated blood is now tested for the presence of antibodies to HIV, and people who may have been exposed to the virus are discouraged from donating, so the likelihood of being infected through a transfusion is very low.

You cannot be infected with HIV through casual contact, such as by shaking hands with an infected person or by handling an object used by an infected person. HIV is apparently not transmitted through the air, in water, or on toilet seats. There is no evidence that it can be spread through the bites of arthropods, such as mosquitoes, fleas, and ticks.

internetconnect

SCILINKS NSTA

TOPIC: AIDS
GO TO: www.scilinks.org
KEYWORD: HM969

TEACHING STRATEGY
Opportunistic Infections

Inform students that two of the most common opportunistic infections among AIDS patients are pneumocystis pneumonia, caused by the fungus *Pneumocystis carinii,* and Kaposi's sarcoma, a cancer that is caused by a virus. Tuberculosis is also widespread among AIDS patients. Ask students to explain why these diseases are rare in people who do not have AIDS. (A healthy immune system is able to suppress the pathogens responsible for these diseases.)

TEACHING STRATEGY
The Course of HIV Infection

Ask students to chart the course of HIV infection on a time line, noting the symptoms associated with each stage of the disease.

CONTENT CONNECTION
Chapter 25: Retroviruses

Remind students that HIV is a retrovirus. Like other retroviruses, HIV has RNA as its genetic material and reproduces using an enzyme called reverse transcriptase, which makes a DNA version of the virus's genome. During replication, retroviruses undergo many mutations because they do not possess the controls necessary to ensure faithful genomic duplication.

RECENT RESEARCH

Locking Out HIV

Scientists have long known that to enter a cell and begin replicating, HIV must bind to two protein receptors on the cell's membrane. One of these receptors, a protein known as CD4, was identified not long after the discovery of HIV, but the second receptor eluded scientists until 1996. Called CCR5, the protein serves as an antenna for receiving messages from other cells. Moreover, a rare mutant CCR5 gene codes for a protein that HIV cannot bind to; in effect, the mutant protein locks HIV out of the cell. People with two versions of this mutant gene thus seem to be resistant to infection by HIV.

ANSWERS TO SECTION 48-3 REVIEW

1. HIV can infect macrophages and helper T cells.
2. The immune system attacks the virus and reduces its concentration to low levels, but it may not eliminate the virus.
3. Donated blood is tested for antibodies to HIV. Potential donors who are likely to be infected are discouraged from giving blood.
4. HIV is apparently not transmitted through casual contact or by insects.
5. The rapid rate of mutation means that the virus's surface proteins, which serve as antigens, are constantly changing shape. This variability has thus far made it impossible to develop an effective vaccine.
6. HIV infection happens when the virus has entered the body and begun reproducing. AIDS is the last stage of infection, when the immune system has been severely impaired.

FIGURE 48-12

According to the World Health Organization, more than 34.3 million people are infected with HIV. This graph shows estimates of the number of HIV-infected people in each region of the world.

Worldwide Distribution of HIV Infections

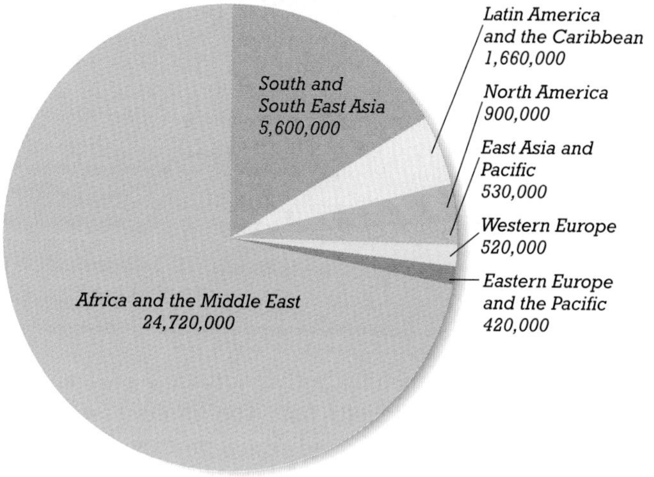

Although most early cases of AIDS were diagnosed in the United States and Western Europe, AIDS is now a worldwide problem. Figure 48-12 shows the estimated numbers of HIV-infected people in different parts of the world.

Treatments and Vaccines

Scientists trying to create vaccines for HIV must contend with its very rapid rate of evolution. The genes that code for the virus's surface proteins mutate frequently. As a result, new variants of the virus with slightly different surface proteins are constantly appearing. To produce effective immunity, a vaccine against HIV must stimulate the immune system to respond to many variants of the virus. Although several vaccines against HIV are under development or being tested, none have yet proven effective.

HIV quickly becomes resistant to drugs used against it. To avoid this problem, scientists have begun treating patients with three antiviral drugs. Even if the virus evolves resistance to one drug, it is unlikely to become simultaneously resistant to three drugs. Scientists have already observed some encouraging results from this treatment. However, it is important to note that there is currently no cure for HIV infection.

SECTION 48-3 REVIEW

1. **Name two kinds of cells that HIV can infect.**
2. **How does the immune system respond to the first appearance of HIV in the body?**
3. **Why is the risk of becoming infected with HIV through a blood transfusion extremely low?**
4. **Name two ways that HIV apparently cannot be transmitted.**
5. **How does HIV's rapid rate of evolution affect the development of a vaccine?**
6. **CRITICAL THINKING Explain the difference between HIV infection and AIDS.**

CHAPTER 48 REVIEW

SUMMARY/VOCABULARY

48-1

- Robert Koch developed four basic steps, known as Koch's postulates, for identifying the particular pathogen responsible for a disease.
- The skin is a nonspecific defense that helps keep pathogens out of the body. Mucous membranes line vulnerable interior surfaces of the body and serve as a barrier to pathogens.
- A break in the skin will trigger another nonspecific defense, called the inflammatory response, characterized by swelling, redness, raised temperature, and pain.
- Neutrophils and macrophages are phagocytes that engulf and destroy pathogens and cellular debris. Natural killer cells destroy infected cells.

Vocabulary

fever (958)
histamine (957)
infectious disease (955)
inflammatory response (957)
interferon (958)
Koch's postulates (955)
macrophage (958)
mucous membrane (956)
mucus (956)
natural killer cell (958)
neutrophil (957)

48-2

- The immune system consists of the spleen, tonsils, adenoids, lymph nodes, thymus, bone marrow, and white blood cells called lymphocytes.
- An antigen is any substance that can stimulate a response from the immune system. Lymphocytes have receptor proteins on their cell membrane that allow them to recognize antigens.
- The immune system's reaction to a pathogen is called an immune response. The immune response is controlled by helper T cells.
- Helper T cells activate the division of B cells to give rise to plasma cells, which produce and secrete defensive proteins called antibodies.
- An allergy is an extreme reaction by the immune system to a harmless antigen. An autoimmune disease is an attack on the cells of the body by the immune system.

Vocabulary

allergy (964)
antibody (962)
antigen (960)
autoimmune disease (965)
B cell (959)
cell-mediated immune response (961)
cytotoxic T cell (961)
helper T cell (961)
humoral immune response (961)
immune response (960)
immune system (959)
immunity (964)
memory cell (963)
plasma cell (962)
primary immune response (963)
secondary immune response (963)
spleen (959)
suppressor T cell (962)
T cell (959)
thymus (959)

48-3

- Acquired immunodeficiency syndrome (AIDS) is caused by the human immunodeficiency virus (HIV). HIV enters and reproduces inside helper T cells.
- HIV and the immune system fight a long battle, which the immune system eventually loses.
- Most AIDS patients die of opportunistic diseases or cancers.
- HIV is transmitted mainly through sexual intercourse and the use of HIV-contaminated hypodermic needles or syringes.

Vocabulary

opportunistic infection (969)

CHAPTER 48 REVIEW ANSWERS

REVIEW

1. A primary immune response is the immune system's initial reaction to a pathogen. A secondary immune response is a reaction to a subsequent exposure to the same pathogen.
2. Interferon increases the resistance of cells to viral infection.
3. Neutrophils are white blood cells that ingest pathogens. B cells are lymphocytes that differentiate into antibody-producing plasma cells.
4. *Opportunistic* means tending to take advantage of a situation. Pathogens that cause opportunistic infections take advantage of the immune system's weakness to invade the body.
5. T cells activate B cells to differentiate into plasma cells, which produce antibodies against an antigen.

6. a **7.** d
8. c **9.** a
10. a **11.** b
12. b **13.** b
14. d **15.** d

16. The suspected pathogen, which must be present in sick individuals but not healthy ones, must be isolated, grown in culture, and injected into a new host. If that new host contracts the disease, the pathogen should be isolated and shown to be identical to the original pathogen.
17. (a) macrophage
(b) This is an antibody, which immobilizes pathogens or marks them for destruction.
(c) humoral immune response
18. Mucous membranes, like the skin, serve as a barrier to pathogens. The skin and the mucous membranes also use chemical defenses against pathogens. Unlike the skin, the mucous membranes line the internal surfaces of the body.
19. Injured cells release chemical messengers that cause dilation of nearby capillaries and an increase in blood flow to the injured area. Fluid and white blood cells move

through the walls of the capillaries into the injury. White blood cells attack pathogens.

20. The thymus is a "nursery" in which T cells mature.

21. Helper T cells that have been stimulated by a macrophage release interleukin-2, which stimulates B cells to differentiate into antibody-producing plasma cells. It also stimulates division by cytotoxic T cells, which attack infected cells.

22. Antibodies are proteins that bind to specific antigens, inactivating them or marking them for destruction by macrophages.

23. Memory cells are specialized T and B cells that remain in the body after a primary immune response. They respond to the same pathogen if it reenters the body.

24. neutrophils, macrophages, and natural killer cells

25. The antigens that are present on the virus's surface continually change, making them unrecognizable to the immune system.

CRITICAL THINKING

1. High temperature from a fever may kill infectious pathogens and stimulate macrophages; reducing fever makes this defense ineffective. High fevers can be harmful or fatal, so they should be treated.

2. The smallpox virus must be evolving slowly enough so that it does not significantly change the shape of its antigens within the lifetime of an individual.

3. Cytotoxic T cells only attack pathogens they recognize as foreign, so the cancer cells must have abnormal proteins on their surface.

4. Students should choose to fund Proposal 2 because HIV must bind to CD4 to enter a cell. A drug that blocks CD4 might interrupt the virus's reproductive cycle. It is likely that a drug that blocks protein synthesis would have adverse side effects.

REVIEW

Vocabulary

1. Distinguish between a primary immune response and a secondary immune response.
2. Describe the function of interferon.
3. Contrast a neutrophil with a B cell.
4. Using a dictionary, find the meaning of the word *opportunistic.* Relate the meaning of this word to the action of opportunistic infections.
5. Explain the relationship between T cells and B cells.

Multiple Choice

6. Robert Koch (a) established procedures for identifying the cause of disease (b) identified specific neurotoxins (c) discovered interferon (d) pioneered the process of vaccination.
7. Pathogens can be (a) bacteria (b) viruses (c) invertebrates (d) any of the above.
8. Nonspecific defenses include the (a) primary immune response (b) secondary immune response (c) inflammatory response (d) T cells.
9. The skin acts as a defense against infection by (a) forming a physical barrier to pathogens (b) engulfing and digesting pathogens (c) forming blood clots (d) producing antibodies.
10. The primary immune response involves (a) the recognition of antigens (b) release of antibodies by macrophages (c) a general reaction to all pathogens (d) the action of cilia lining the air passages.
11. Antibodies are produced by (a) helper T cells (b) plasma cells (c) suppressor T cells (d) macrophages.
12. Cytotoxic T cells (a) are directly stimulated by macrophages (b) attack infected cells (c) mature in the bone marrow (d) help shut down the immune response.
13. Which of the following is *not* true of autoimmune diseases?
(a) An example is Type 1 diabetes.
(b) They are a type of cancer.
(c) They target the body's cells.
(d) They can be fatal.
14. HIV is transmitted through (a) infected food or water (b) sexual contact that involves the exchange of body fluids (c) intravenous injections with contaminated syringes (d) both b and c.
15. HIV (a) infects only macrophages (b) has DNA as its genetic material (c) can be eliminated with antibiotics (d) replicates in helper T cells.

Short Answer

16. What steps must be followed to prove that a particular pathogen is responsible for a disease?
17. Look at the diagram shown below.
    a. Identify the cell labeled *1.*
    b. What is the function of the structure labeled *2*?
    c. What process is illustrated by this diagram?

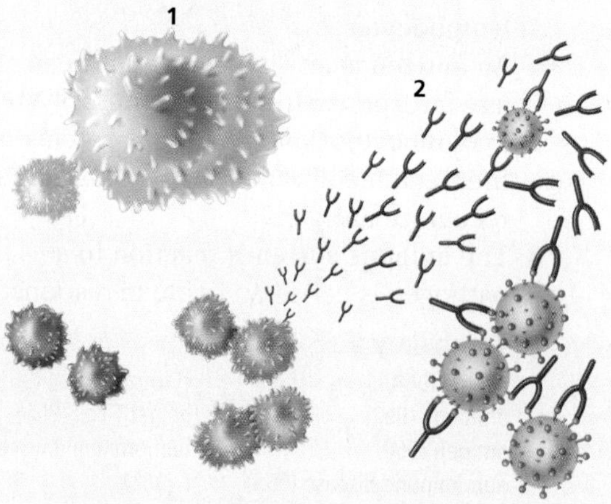

18. How does the function of the mucous membranes compare with that of the skin?
19. Describe the steps of an inflammatory response.
20. What is the function of the thymus?
21. Describe the primary immune response.
22. What functions do antibodies serve?
23. What are memory cells, and what is their role in providing immunity against disease?
24. What white blood cells are involved in the inflammatory response?
25. What problem have scientists encountered in trying to develop a vaccine against HIV?

CRITICAL THINKING

1. Many people take fever-reducing drugs as soon as their temperature exceeds 37°C (99°F). Why might it not be a good idea to immediately reduce a fever with drugs? What are the benefits of taking fever-reducing drugs?
2. Scientists created an effective vaccine for smallpox but have not been able to do so for AIDS. What does this suggest about the rate of evolution of the smallpox virus?
3. Cytotoxic T cells attack and destroy some kinds of cancer cells. What can you conclude about the surface proteins of these cancer cells?
4. A government agency is reviewing two proposals for HIV research, but it can fund only one. Suppose you are asked to provide input. Which proposal would you recommend that the agency fund? You should consider not only the likely effectiveness of the treatment but also the likely side effects. Explain how you made your choice. **Proposal 1**: Develop a drug that interferes with protein synthesis. **Proposal 2:** Develop a substance that binds to CD4 receptors on helper T cells.
5. Look at the graph shown below. It shows the amount of HIV in the blood of an infected person. Answer the following questions about the graph:
   a. What caused the peak in viral concentration at point *a*?
   b. Why did the level of virus drop between points *a* and *b*?
   c. Describe what is happening to both the virus and the immune system at points *c* and *d*.

Virus Concentration in HIV Infection

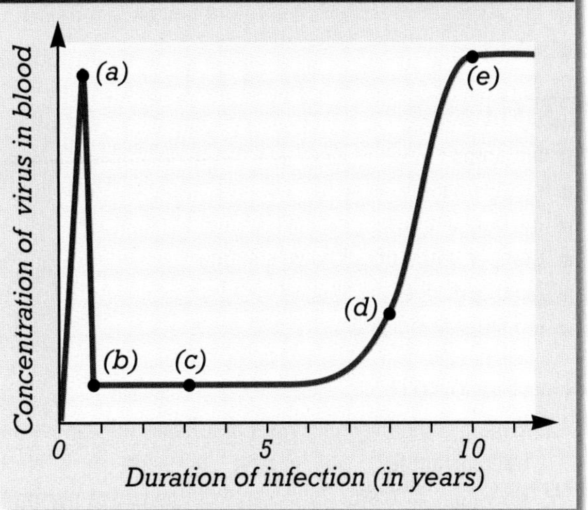

EXTENSION

1. Read "Granting Immunity" in *Scientific American*, March 2000, on page 15. Why are some parents becoming uncooperative about obtaining vaccines for their children? At what age should a person have the chicken pox vaccine?
2. Use the *Physicians' Desk Reference* (PDR) at your local library to research the large numbers of specialized antibiotics made by drug companies. To summarize what you have learned, make a large chart on poster board using the following headings: brand name of the drug, generic name of the drug, uses of the drug, company that manufactures the drug, and disease the drug controls. Select five or six antibiotics, and fill in the appropriate information on the chart.
3. During the early stages of the AIDS epidemic, many hemophiliacs became infected with HIV after using infected clotting factor. Using library resources or an on-line database, research this aspect of the AIDS epidemic. Why do hemophiliacs need clotting factor? How was it produced, and how did this lead to contamination with HIV? What changes have been made in the production of clotting factor to reduce the risk of infection?

ANSWERS

CONTINUED

5. At point *a,* the virus has just begun to reproduce within the body. Between points *a* and *b,* the immune system drives down the concentration of viruses. At point *c,* the immune system is still keeping the virus in control. By point *d,* the virus has begun to overwhelm the immune system.

EXTENSION

1. Some parents and others are becoming concerned without real cause that the vaccines might cause the diseases that they are supposed to prevent. Sometimes religious convictions prevent children from receiving vaccines. The chart on page 16 of the article points out that the chicken pox vaccine should be given to children between 9 and 21 months of age.
2. Answers will vary, depending on the antibiotics selected.
3. Hemophiliacs lack a blood-clotting protein. Treatment included injections of clotting proteins, or factors, isolated from the blood of nonhemophiliacs. Isolation of sufficient amounts of clotting factor required that blood from many donors be pooled. Some of the blood used was contaminated with HIV. Safety precautions include heating donated blood, heating to inactivate the virus, and using genetically engineered clotting factor.

CHAPTER 48 INVESTIGATION

TIME REQUIRED
About 45 minutes

SAFETY PRECAUTIONS
Caution students to not allow solutions to touch their skin or clothing. If this happens, flush the area with water. Students should wear a lab apron, goggles, and disposable gloves throughout the investigation. After completing the investigation, students should rinse the test tubes well.

QUICK REFERENCES
Lab Preparation notes are found on page 953B.

Holt BioSources provides a Teaching Resources CD-ROM that contains "Using Gowin's Vee in the Lab" and "Scoring Rubrics."

PROCEDURAL NOTES
1. Before students begin, remind them of safety practices, and emphasize that they should not allow solutions to touch their skin or clothes.
2. If the procedure is followed precisely, the route of transmission can be easily traced.

ANSWERS TO BACKGROUND
1. The five ways human diseases are transmitted are through air, food, water, person-to-person contact, and invertebrates.
2. The viruses that cause colds and flu are spread by tiny droplets of saliva and mucus expelled into the air by sneezing or coughing.
3. B cells are activated and begin to produce antibodies that prevent the virus from entering cells. Cytotoxic T cells destroy virus-infected cells that produce new viruses.
4. AIDS is a deadly disease that has spread rapidly throughout the world in the past several decades.

CHAPTER 48 INVESTIGATION

Simulating Disease Transmission

OBJECTIVES
- Simulate the transmission of a disease.
- Determine the original carrier of the disease.

PROCESS SKILLS
- organizing data
- analyzing data
- identifying
- modeling

MATERIALS
- lab apron
- safety goggles
- disposable gloves
- dropper bottle of unknown solution
- large test tube
- indophenol indicator

Background
1. What are the five main ways that human diseases can be transmitted?
2. How does a cold or flu spread from person to person?
3. How does the body fight invading viruses?
4. Why has the transmission of HIV become a great concern worldwide?
5. Why is a person with AIDS less able to combat infections than a person who does not have AIDS?

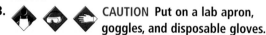

PART A Simulating the Transmission of a Disease
1. This investigation will involve the class in a simulation of disease transmission. After the simulation, you will try to identify the original infected person in the closed class population.
2. In your lab report, construct a data table similar to Table A.
3. CAUTION **Put on a lab apron, goggles, and disposable gloves.**
4. CAUTION **If you get any solution used in this investigation on your skin or clothing, wash it off at the sink while calling to your teacher. If you get any solution used in this investigation in your eyes, immediately flush your eyes with water at the eyewash station while calling to your teacher.** You have been given a dropper bottle of unknown solution and a clean test tube. The solution in the dropper bottle represents the pathogens that you carry. Handle the unknown solution with care because it is not simply water.
5. When your teacher says to begin, transfer three dropperfuls of your solution to your clean test tube. Then replace the lid on the dropper bottle and do not reopen it until Part B of this investigation.
6. Select one person to be your partner. Let one partner pour the contents of his or her test tube into the other partner's test tube. Then pour half the solution back into the first test tube. You and your partner now share pathogens of any possible transmittable disease that either of you might have had. Record the name of your first partner (Round 1) in your data table in your lab report.
7. For Round 2, wait for your teacher's signal, and then find a different partner and exchange solutions in the same manner as you did in step 6. Record the name of your second partner (Round 2) in your lab report. Do not exchange solutions with the same person more than once. Repeat this procedure again for Round 3.
8. After all rounds are finished, your instructor will ask you to add one dropperful of indophenol indicator to your test tube to see if the fluids in your test tube have become infected. Infected solutions will be colorless or light pink. All uninfected solutions will appear blue. Record the outcome of your tests in your lab report.

TABLE A *LIST OF PARTNERS' NAMES*

Round number	Partner's name
1	
2	
3	

TABLE B *PATH OF DISEASE TRANSMISSION*

Name of infected person	Names of infected person's partners		
	Round 1	Round 2	Round 3

PART B Tracing the Source of the Disease

9. If you are an infected person, give your name to your teacher. As names of infected people are written on the chalkboard or on the overhead projector, record them in your lab report in a table similar to Table B shown above.
10. Try to trace the original source of the infection, then determine the transmission route of the disease. In your table, cross out the names of all the uninfected partners in Rounds 1, 2, and 3. There should be only two people in Round 1 who were infected. One of these people was the original carrier.
11. Draw a diagram that shows the transmission route of the disease through all three rounds. Your diagram may look something like the chart below. Include your diagram in your lab report.

Transmission Route

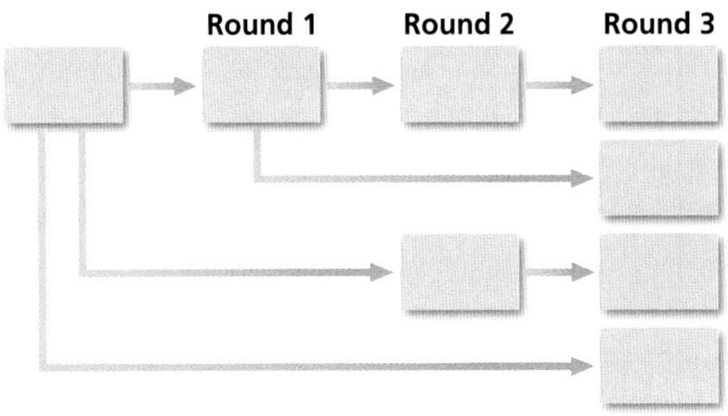

12. In your diagram, insert the names of the two people in Round 1 who were infected and the names of their partners in Rounds 2 and 3.
13. To test whether a person was the original disease carrier, pour a sample from his or her dropper bottle into a clean test tube and add indophenol indicator.
14. Clean up your materials and wash your hands before leaving the lab.

Analysis and Conclusions

1. What might the clear fluid in each student's dropper bottle represent?
2. Does the simulated disease have any apparent symptoms?
3. What chemical is added to the test tubes when the rounds are completed?
4. What color indicates a positive result?
5. What color indicates a negative result?
6. Who was the original disease carrier?
7. After the three rounds, how many students were infected? Express this as a percentage of the number of students in the class.
8. If an epidemic occurred in your community, how might public-health officials work to stop the spread of the disease?

Further Inquiry

A public-health official is sent to investigate an outbreak of a new disease. Devise an experiment to allow the official to determine whether the disease has been caused by the passing of pathogens from person to person or by environmental conditions.

CHAPTER 48
INVESTIGATION

Millions of people have died from AIDS, and millions more are infected with HIV and will die unless a cure is discovered.
5. HIV leads to the death of many helper T cells, without which cytotoxic T cells and B cells cannot be activated to fight pathogens.

ANSWERS TO ANALYSIS AND CONCLUSIONS

1. aerosol droplets from a cough or sneeze, blood, or other bodily fluids
2. No. The fluid is clear, so infected individuals cannot be identified without testing.
3. indophenol
4. clear to pink
5. blue
6. Answers will vary.
7. Answers will vary, depending on class size.
8. They might post signs, alert the media, notify schools and other places where many people congregate each day, and encourage people with symptoms to stay home. They might also begin a vaccination or treatment program.

CHAPTER 49
DIGESTIVE AND EXCRETORY SYSTEMS

	TOPICS	TEACHING RESOURCES	LABS, CLASSWORK, AND HOMEWORK
BLOCKS 1 & 2 90 minutes	**Introducing the Chapter, p. 976**	**ATE** Focus Concept, p. 976 **ATE** Assessing Prior Knowledge, p. 976 **ATE** Understanding the Visual, p. 976	■ Occupational Applications Worksheets—Dietitian
	49-1 Nutrients, p. 977 Six Basic Food Ingredients, p. 977	**ATE** Section Overview, p. 977 **ATE** Critical Thinking, p. 979 **ATE** Visual Link Figure 49-3, p. 979 Table 49-1, p. 980 Table 49-2, p. 981 190A, 191A, 192A, 222	**Lab C46,** Identifying Food Nutrients **Lab E7,** Diet and Weight Loss ■ Problem-Solving Worksheets—Using Food Labels to Calculate Percentage of Nutrients and Calories ★ Study Guide, Section 49-1 **PE** Section Review, p. 982
BLOCKS 3 & 4 90 minutes	**49-2 Digestive System, p. 983** The Gastrointestinal Tract, p. 983 Chemical Digestion, p. 986 Large Intestine, p. 989	**ATE** Section Overview, p. 983 **ATE** Visual Link Figure 49-4, p. 983 Figure 49-5, p. 984 Figure 49-13, p. 988 Figure 49-14, p. 988 193A, 216, 219	★ Study Guide, Section 49-2 **PE** Section Review, p. 989
BLOCKS 5 & 6 90 minutes	**49-3 Urinary System, p. 991** Kidneys, p. 991	**ATE** Section Overview, p. 991 **ATE** Visual Link Figure 49-17, p. 992 Figure 49-19, p. 994 218, 220–221	**PE** **Chapter 49 Investigation,** p. 1000 **PE** **Quick Lab,** p. 995, Analyzing Kidney Filtration **Lab C48,** Urinalysis Testing ★ Study Guide, Section 49-3 **PE** Section Review, p. 996
BLOCKS 7 & 8 90 minutes	**REVIEW and ASSESSMENT** **PE** Chapter 49 Review, pp. 998–999 ■ Performance-Based Assessment—Testing Human Digestive Enzymes	Audio CD Program	**CHAPTER TESTING** ■ Chapter Test (blackline master) ▲ Test Generator ■ Assessment Item Listing ■ Scoring Rubrics and Classroom Management Checklists

KEY

MODERN BIOLOGY
- PE Pupil's Edition
- ATE Teacher's Edition
- ■ Active Reading Guide
- ★ Study Guide

HOLT BIOSOURCES
- Teaching Transparencies
- Laboratory Program

One-Stop Planner CD-ROM
Shows these resources within a customizable daily lesson plan:
- ■ Holt BioSources Teaching Resources
- ■ Active Reading Guide
- ★ Study Guide
- ▲ Test Generator

READING FOR CONTENT MASTERY

Preview: Have students preview the objectives, the introductory paragraph, topic sentences, and captions before reading the chapter.

- **ATE** Active Reading Technique: Discussion, p. 977
- ■ Active Reading Guide Worksheet 49-1
- **PE** Summary/Vocabulary, p. 997

- ■ Active Reading Guide Worksheet 49-2
- **PE** Summary/Vocabulary, p. 997

- ■ Active Reading Guide Worksheet 49-3
- **PE** Summary/Vocabulary, p. 997

TECHNOLOGY AND INTERNET RESOURCES

CNN PRESENTS Video Segment 32 Going Vegetarian

Holt Biology Videodiscs Teacher's Correlation Guide, Lessons 49-1 through 49-3

Audio CD Program, Sections 49-1 through 49-3

internetconnect

SCI LINKS NSTA

On-line Resources: www.scilinks.org
The following *sci*LINKS Internet resources can be found in the student text for this chapter:

Topics:
- Nutrients, p. 977
- Chemical digestion, p. 986
- Urinary system, p. 992
- Excretory system, p. 996

go.hrw.com

On-line Resources: go.hrw.com
Visit the HRW Web site for a variety of resources related to this chapter. Just type in the keyword HM2 Chapter 49.

Smithsonian Institution®
Visit **www.si.edu/hrw** for additional on-line resources.

CNNfyi.com
Visit **www.cnnfyi.com** for late-breaking news and current event stories selected just for you.

LAB ACTIVITY PLANNING

CHAPTER 49 INVESTIGATION

Modeling Human Digestion, pp. 1000–1001

OVERVIEW
In this lab students learn what components must be present in the digestive fluid to cause the breakdown of protein. They observe that each component alone cannot promote digestion and that pepsin requires an acidic environment to be chemically active.

SAFETY
Students will wear a lab apron and safety goggles. If clothing, skin, or eyes become contaminated by the pepsin or HCl, flush with water and seek medical attention.

PREPARATION
1. A 0.2 percent hydrochloric acid solution can be made by diluting 2 mL of concentrated HCl in 1 L of distilled water.
2. A 1 percent sodium bicarbonate solution can be prepared by adding 10 g of sodium bicarbonate to 1 L of distilled water.
3. A 1 percent pepsin solution can be prepared by adding 1 g of pepsin to 100 mL of distilled water.
4. Each group will need 10 mL of water, 20 mL of 1 percent pepsin, 15 mL of 0.2 percent HCl, and 5 mL of 1 percent sodium bicarbonate solution. Each group will also need 30 g of cooked egg white, five pieces of blue litmus paper, five pieces of red litmus paper, five test tubes, and a test–tube rack.

Quick Labs

Analyzing Kidney Filtration, p. 995
Each student group will need disposable gloves, lab aprons, safety goggles, gelatin, flour, glucose, test tubes, filter, a beaker, biuret and Benedict's solutions, IKI solution, and pipets.

UNDERSTANDING THE VISUAL

The photograph shows a glomerulus of the human kidney. The ball of capillaries is covered with specialized cells, called podocytes, that interdigitate, forming slits. The porous capillaries and the slitted podocytes collectively function as a filter. The filter is permeable to small molecules but not to blood cells or proteins. Blood pressure forces water and small solutes through the filter and into the lumen of the Bowman's capsule, where the filtrate enters the nephron.

FOCUS CONCEPT

Structure and Function

Body structures all work together to perform the functions of a whole organism. Emphasize how the digestive and excretory systems work together to carry out physiological functions.

ASSESSING PRIOR KNOWLEDGE

Review the following concepts.

The Muscular System:
Chapter 46
Review with students how the stomach's muscle configuration allows it to perform mechanical digestion as well as chemical digestion. You may want to point out that the structure of the small intestine is specialized for chemical digestion and absorption.

The Circulatory System:
Chapter 47
Help students understand that the digestive and excretory systems work together with the circulatory system. Ask students to speculate about how blood is related to digestion and excretion.

CHAPTER 49

DIGESTIVE AND EXCRETORY SYSTEMS

This is a scanning electron micrograph of a filtration membrane in the human kidney. (SEM 3060×)

FOCUS CONCEPT: ***Structure and Function***

As you read, notice how the structure of each digestive and excretory organ relates to its function.

NUTRIENTS

Carrots, fish, eggs, hamburgers, blackberries, cow's milk—the human body is able to convert each of these foods into nutrients that body cells need to function, grow, and replicate. In this section you will learn what nutrients the human body needs and how it uses those nutrients to carry out life processes.

SECTION

49-1

OBJECTIVES

▲ List the four organic nutrients needed by the human body.

● Identify foods containing each of the organic nutrients.

■ Explain the importance of inorganic mineral nutrients.

◆ Summarize the functions that the six nutrients perform in the body.

▲ Explain why water is a vital nutrient.

SIX BASIC FOOD INGREDIENTS

All of the different foods in the world contain at least one of six basic ingredients: carbohydrates, proteins, lipids, vitamins, minerals, and water. These ingredients, called **nutrients,** are the chemical substances necessary for organisms to grow and function properly. Four of these nutrients—carbohydrates, proteins, fats, and vitamins—are organic compounds because they contain the elements carbon, hydrogen, and oxygen. The two remaining nutrients—minerals and water—are inorganic compounds.

Few foods contain all six nutrients. In fact, most foods contain a concentration of just one or two. Nutritionists classify foods into four groups—meat, milk, fruits and vegetables, and breads and cereals—based on nutrient similarity.

Carbohydrates

Carbohydrates are organic compounds composed of carbon, hydrogen, and oxygen in a ratio of about two hydrogen atoms to one oxygen atom and one carbon atom. Carbohydrates are broken down in aerobic respiration to provide most of the body's energy. Although proteins and fats also supply energy, the body most easily uses the energy provided by carbohydrates. Carbohydrates contain sugars that are quickly converted into the usable energy ATP, while proteins and fats must go through many chemical processes before the body can obtain energy from them.

The fructose and glucose (also known as dextrose) in fruit and honey are simple sugars, or monosaccharides. These sugars can be absorbed directly into the bloodstream and made available to cells for use in aerobic respiration. Sucrose (cane sugar), maltose, and lactose (milk sugar) are disaccharides. Disaccharides are sugars that consist of two chemically linked monosaccharides. These disaccharides must be converted into monosaccharides before they can be used by the body for energy. Disaccharides are split into two

internet**connect**

SCILINKS NSTA

TOPIC: Nutrients
GO TO: www.scilinks.org
KEYWORD: HM977

SECTION 49-1

SECTION OVERVIEW

Stability and Homeostasis
Section 49-1 describes nutrients as chemical substances that all organisms require in order to grow and function properly. The stability of living systems can be threatened by malnutrition.

Matter, Energy, and Organization
Section 49-1 reviews biochemical concepts that were first introduced in Chapter 3 and discusses the composition and function of carbohydrates, proteins, fats, vitamins, minerals, and water.

ACTIVE READING

Technique: Discussion
Before students begin reading the chapter, have each student list the components of a typical evening meal. Then have several volunteers write their menus on the board. Have students discuss which of the menus are healthy and why. Challenge students to explain what each food provides that the body requires. Tell students that in this chapter they will learn about nutrients and their importance to good health.

TEACHING STRATEGY

Potatoes
Have students carefully examine a picture of a baked potato and a picture of French fries, and ask them to describe differences between the two foods. (Unlike the baked potato, French fries are peeled, strip-sliced, fried in fat, and salted.) Ask students what the health implications are of a diet high in fat and salt. (Saturated fats may increase cholesterol levels. Over a long period of time, high salt intake can contribute to hypertension.)

SECTION 49-1

OVERCOMING MISCONCEPTIONS

A common misconception is that athletes need to eat more protein to improve their athletic performance and maintain their greater muscle mass. Actually, a diet rich in complex carbohydrates is much more effective in sustaining intense muscle activity than a diet high in protein.

INCLUSION *ACTIVITY*

Ask students what special foods they enjoy during holidays or festivals, or provide pictures of dishes from magazines or cookbooks. Have them identify the food groups and the nutrients found in each food. For example, a bean-and-rice dish contains complex carbohydrates and proteins as well as vitamins and minerals. List the dishes and the nutrients they contain on the board. Have students make posters illustrating each dish and the basic nutrients it contains. Have students use Tables 49-1 and 49-2 to analyze the vitamin and mineral content of the foods. Remind students to include water in their surveys of nutrients.

TEACHING STRATEGY

Vegetarian Meals

Design several vegetarian meals that are each missing an essential amino acid. Challenge students to supply the missing nutrient by adding the complementary food to each meal. Have them use Figure 49-2 to help them make their choices. For example, a meal of rice with tomatoes, cheese, and corn tortillas will have all the essential amino acids when beans are added.

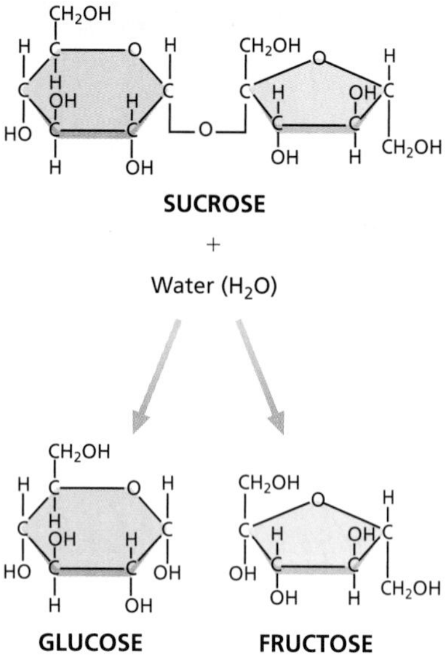

FIGURE 49-1

The hydrolysis, or digestion, of a disaccharide, such as sucrose, requires water and an enzyme. When sucrose is digested, two monosaccharides are formed—glucose and fructose. These monosaccharides are then transported through cell membranes to be used by cells.

monosaccharides in a process called hydrolysis. Figure 49-1 shows how sucrose is hydrolyzed to produce glucose and fructose.

The carbohydrates that require the longest digestion time are polysaccharides. Polysaccharides are complex molecules that consist of many monosaccharides bonded together. The starch found in rice, corn, and many other grains and vegetables is a polysaccharide made up of long chains of glucose molecules. During digestion, the body chemically breaks down these long chains into individual glucose units, which then can be absorbed by the blood and carried to the tissues.

Many foods we get from plants contain cellulose, a polysaccharide that the body cannot break down into individual component sugars. Cellulose, the substance that forms the walls of plant cells, is nevertheless an extremely important part of the human diet. Although indigestible, it provides fiber that aids in human digestion. Cellulose stimulates contractions of the smooth muscles that line the organs of the digestive system. These contractions help move the food along.

Proteins

Proteins are the major structural and functional material of body cells. Proteins from food help the body to grow and to repair tissues. Proteins consist of long chains of amino acids. The human body uses about 20 kinds of amino acids to construct the proteins it needs. The body manufactures many of these amino acids, but it cannot produce all of them in the quantities that it needs. Amino acids that the body produces are called nonessential amino acids because you do not have to get them from your food. Amino acids that must be obtained from food are called essential amino acids. Ten amino acids are essential to children and teenagers, while only eight are essential to adults. The two additional amino acids needed by children are involved in growth.

Most of the foods we get from plants do not contain all the essential amino acids. Eating certain combinations of two or more plant products, such as those shown in Figure 49-2, can supply all the essential amino acids. Vegetarians who do not eat animal

FIGURE 49-2

The combination of legumes with seeds or grains and the combination of grains with milk products furnish all the essential amino acids.

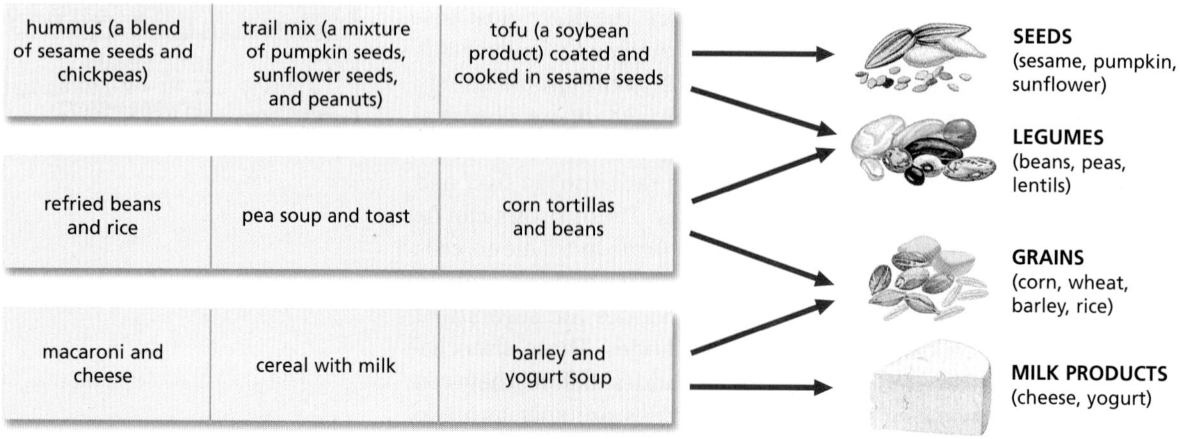

products get their proteins by eating combinations of two or more plant products or by combining seeds or grains with legumes. Most animal products, such as eggs, milk, fish, poultry, and beef, contain all the essential amino acids.

Lipids

Lipids are organic compounds that contain more carbon and hydrogen atoms than oxygen atoms. Lipids include triglycerides, commonly known as fats. Fats are organic molecules that the body uses to obtain energy and to build cell membranes and other cell parts. The body stores excess fat from the diet in special tissues under the skin and around the kidneys and liver. Excess carbohydrates may be converted to fat for storage.

Why does the human body store fat? The answer is that stored fats are beneficial unless they are excessive. A light layer of body fat beneath the skin provides insulation in cold weather. Fat surrounding vulnerable organs, such as the kidneys and liver, acts as protective padding. And most important, fat reserves are a source of energy. While you cannot live off just your body fat, the body can use its fat for energy, especially when carbohydrates are unavailable.

To use fats, the body must first break down each fat molecule into glycerol and fatty acids. The glycerol molecule is the same in all fats, but the fatty acids differ in both structure and composition. The body converts some fatty acids to other fatty acids, depending on which one the body needs at the time.

Scientists classify fats as saturated or unsaturated, based on structural differences in their fatty acids. A saturated fatty acid has all its carbon atoms connected by single bonds and thus contains as many hydrogen atoms as possible. An unsaturated fatty acid has at least one double bond between carbon atoms. If there are two or more double bonds, the fatty acid is called polyunsaturated. How many double bonds are shown in the structure of a fatty acid in Figure 49-3? In general, animal fats are saturated and plant oils are unsaturated. However, some vegetable oils, such as palm oil and coconut oil, are composed primarily of saturated fats.

FIGURE 49-3

The structure of linoleic acid, a fatty acid in margarine, is shown in this figure. Is this fatty acid saturated, unsaturated, or polyunsaturated?

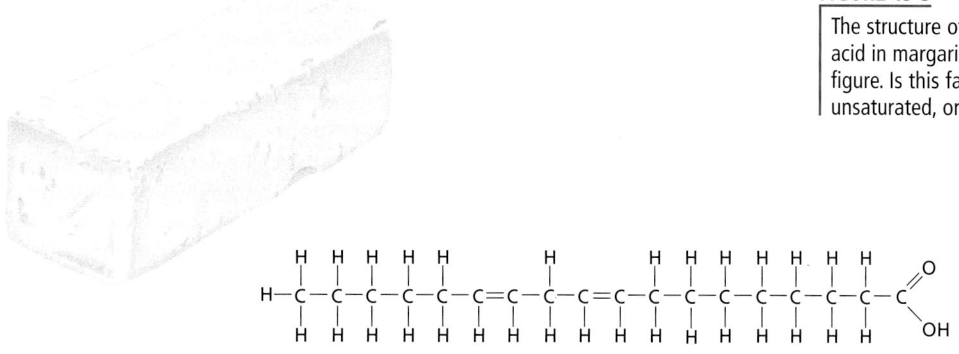

SECTION 49-1

CRITICAL THINKING

Vegetarian Diets

Have students use library resources or an on-line database to research the different kinds of vegetarian diets. (Vegetarians dine mainly on grains, vegetables, nuts, seeds, and fruits. There are three main vegetarian eating patterns. Vegans avoid meat, poultry, seafood, eggs, dairy products, and even foods containing animal products, such as honey. Lacto-vegetarians eat dairy products, but not meat, poultry, seafood, or eggs. Lacto-ovo vegetarians exclude meat, poultry, and seafood but eat eggs and dairy products. This is the most popular form of vegetarianism among Americans.)

QUICK FACT

☑ Cholesterol is carried in the bloodstream in two forms, as low density lipoprotein (LDL) and as high density lipoprotein (HDL). LDLs carry dietary and newly synthesized cholesterol to various organs of the body, whereas HDLs, because of their structure, seem to carry cholesterol to the liver to be broken down. When there is an excess of LDLs over HDLs due to high intake of saturated fats, the excess cholesterol may be deposited in the blood vessels. Strenuous exercise seems to increase the blood HDL-to-LDL ratio, whereas smoking decreases it.

VISUAL LINK

Figure 49-3

Point out to students that the linoleic acid in margarine is polyunsaturated and contains three double bonds.

SECTION 49-1

VISUAL LINK

Table 49-1 and Table 49-2

Pass a bottle of vitamin tablets around the class. Have students examine the label and identify the vitamins that are listed on the label but not in the tables. List these additional vitamins on the board. Have students make Table 49-1 and Table 49-2 more complete by checking references to determine the sources, functions, and deficiency information for each of the vitamins listed on the board.

OVERCOMING MISCONCEPTIONS

Some people have the idea that "If a little is good, then a whole lot ought to be better." Unfortunately, this is not the case with vitamins. For example, severe overdoses of vitamin A can cause enlargement of the liver and spleen. Overdoses of vitamin D can cause kidney damage and calcification of soft tissues. While vitamins A and D are both fat-soluble, making them more difficult for the body to eliminate, even the water-soluble vitamins have some negative effects if overused. Massive doses of vitamin C can cause kidney stones and enhance blood coagulation, and megadoses of niacin, one of the B vitamins, can cause liver damage and gout.

QUICK FACT

☑ A vegetarian diet reduces dietary intake of cholesterol, fat, and animal protein, but it may result in low intakes of certain nutrients. Vegetarians must take special care to get enough calcium, vitamin D, iron, vitamin B_{12}, and protein.

TABLE 49-1 ***Food Sources of Vitamins***

Vitamins	Best sources	Essential for	Deficiency diseases and symptoms
Vitamin A (carotene; fat soluble)	fish-liver oils, liver and kidney, green and yellow vegetables, yellow fruit, tomatoes, butter, egg yolk	growth, health of the eyes, and functioning of the cells of the skin and mucous membranes	retarded growth, night blindness, susceptibility to infections, changes in skin, defective tooth formation
Vitamin B_1 (thiamin; water soluble)	meat, soybeans, milk, whole grains, legumes	growth; carbohydrate metabolism; functioning of the heart, nerves, muscles	beriberi—loss of appetite and weight, nerve disorders, and faulty digestion
Vitamin B_2 (riboflavin; water soluble)	meat, fowl, soybeans, milk, green vegetables, eggs, yeast	growth, health of the skin and mouth, carbohydrate metabolism, functioning of the eyes, red blood cell formation	retarded growth, dimness of vision, inflammation of the tongue, premature aging, intolerance to light
Vitamin B_3 (niacin; water soluble)	meat, fowl, fish, peanut butter, potatoes, whole grains, tomatoes, leafy vegetables	growth; carbohydrate metabolism; functioning of the stomach, intestines, and nervous system	pellagra—smoothness of the tongue, skin eruptions, digestive disturbances, and mental disorders
Vitamin B_6 (pyridoxine; water soluble)	whole grains, liver, fish	coenzymes for metabolic reactions	dermatitis, nervous disorders
Vitamin B_{12} (cyanocobalamin; water soluble)	green vegetables, liver	preventing pernicious anemia	a reduction in number of red blood cells
Vitamin C (ascorbic acid; water soluble)	fruit (especially citrus), tomatoes, leafy vegetables	growth, strength of the blood vessels, development of teeth, health of gums	scurvy—sore gums, hemorrhages around the bones, and tendency to bruise easily
Vitamin D (calciferol; fat soluble)	fish-liver oil, liver, fortified milk, eggs, irradiated foods	growth, calcium and phosphorus metabolism, bones and teeth	rickets—soft bones, poor development of teeth, and dental decay
Vitamin E (tocopherol; fat soluble)	wheat-germ oil, leafy vegetables, milk, butter	normal reproduction	anemia in newborns
Vitamin K (naphthoquinone; fat soluble)	green vegetables, soybean oil, tomatoes	normal clotting of the blood, liver functions	hemorrhages

Vitamins

Vitamins are complex organic molecules that serve as coenzymes. Vitamins activate the enzymes and help them function. For example, vitamin B_1, or thiamin, is essential to the functioning of two enzymes that catalyze the first step in aerobic respiration. Because vitamins generally cannot be synthesized by the body, a diet should include the proper daily amounts of all vitamins. Like enzymes, coenzymes can be reused many times. Thus only small quantities of vitamins are needed in the diet. Table 49-1 summarizes the sources of the vitamins and their functions.

Vitamins dissolve in either water or fat. The fat-soluble vitamins include vitamins A, D, E, and K. The water-soluble vitamins are vitamins C and the B vitamins. Fat-soluble vitamins are absorbed and stored like fats. As with fats, the body can amass a reserve of fat-soluble vitamins. Unpleasant physical symptoms and even death can result from storing too much or having too little of a particular vitamin. Continual large doses of vitamin A, for example, can result in severe nausea and a yellow skin color. Because the body cannot store water-soluble vitamins, it excretes surplus amounts in urine.

The only vitamin that the body can synthesize in large quantities is vitamin D. This synthesis involves the conversion of cholesterol to vitamin D by intestinal enzymes and sunlight. People who do not spend a lot of time in the sun can get their vitamin D from food. With the exception of vitamin D, human beings must obtain the vitamins they need from food sources and vitamin supplements.

Minerals

Minerals are inorganic substances required for the normal functioning of the body. Some minerals, such as calcium, magnesium, and iron, are drawn from the soil and become part of plants. Animals that feed on plants extract the minerals and incorporate them into their bodies. Table 49-2 lists the primary sources and functions of a few of the minerals considered most essential to human beings. Iron, for example, is necessary for the formation of red blood cells, while potassium maintains the body's acid-base balance and aids in growth. Both are found in certain fruits and vegetables; iron is also found in meats. Iodine—found in seafood, water, and iodized salt—is needed for hormone production by the thyroid gland. Minerals are excreted through the skin in perspiration and through the kidneys in urine.

TABLE 49-2 ***Food Sources of Minerals***

Minerals	Source	Essential for
Calcium salts	milk, whole-grain cereals, vegetables, meats	deposition in bones and teeth; functioning of heart, muscles, and nerves
Iodine	seafoods, water, iodized salt	thyroid gland secretion
Iron salts	leafy vegetables, liver, meats, raisins, prunes	formation of red blood cells
Magnesium salts	vegetables	muscle and nerve action
Phosphorus salts	milk, whole-grain cereals, vegetables, meats	deposition in bones and teeth; formation of ATP and nucleic acid
Potassium salts	vegetables, citrus fruits, bananas, apricots	maintaining acid-base balance; growth; nerve action
Sodium salts	table salt, vegetables	blood and other body tissues; nerve action

TEACHING STRATEGY

Calcium and Vitamin D

Ask students what the predominant mineral in milk is (calcium). Point out that vitamin D is added to milk. Then ask them why. (Accept any answers that tie vitamin D and calcium together.) Tell students that vitamin D is needed for the absorption of calcium from the intestines. Explain that rickets, a disease that affects growing bones and is caused by a vitamin D deficiency, used to be common in the United States. In the 1930s the federal government mandated the enrichment of milk with vitamin D. Today the disease is virtually unknown in the United States.

MAKING CONNECTIONS

Social Studies

Have students each research the typical diet of people in another country. Countries to choose from could include Japan, India, Israel, Egypt, Nigeria, Italy, Greece, Germany, Mexico, and Russia. Ask students to investigate how the diet delivers the six nutrients discussed in this section. Also ask students to compare the diet with a typical American diet in terms of fat intake and cholesterol levels. Have students determine what factors might influence the composition of their country's diet.

SECTION 49-1

QUICK FACT

☑ Losses of water and salt (NaCl), often caused by profuse and prolonged sweating, may cause painful spasms of the skeletal muscles called heat cramps. This situation can be corrected by ingesting fluids.

ANSWERS TO SECTION 49-1 REVIEW

1. Nutrients are chemical substances that organisms need in order to grow and function properly. Carbohydrates, proteins, fats, and vitamins are organic nutrients because they contain carbon, hydrogen, and oxygen. Water and minerals do not, so they are inorganic nutrients.
2. Proteins help the body grow, help repair tissues, and serve as enzymes.
3. Body fat provides insulation, protects organs, and serves as an energy source.
4. Most chemical reactions that maintain life can take place only in water. Water is the medium in which waste materials are dissolved and carried away from body tissues and in which nutrients are carried throughout the body. Water also helps regulate body temperature.
5. Cellulose provides bulk material to stimulate muscular contractions in the large intestine, which in turn moves indigestible material out of the body as feces.
6. This vegetarian would suffer from deficiencies of vitamins, minerals, proteins, and fats.

TABLE 49-3 *Sources of Water Balance in Humans*

Source of water	Water gain (mL/day)	Water loss (mL/day)
Ingested in liquid	1,500 (60%)	
Ingested in food	750 (30%)	
Derived from metabolism	250 (10%)	
Evaporation		900 (36%)
Urine		1,500 (60%)
Feces		100 (4%)
Total gain	2,500 (100%)	
Total loss		2,500 (100%)

Water

Water accounts for over half of your body weight. Most of the reactions that maintain life can take place only in water. Water makes up over 90 percent of the fluid part of the blood, which carries essential nutrients to all parts of the body. It is also the medium in which waste products are dissolved and carried away from body tissues.

Water also helps regulate body temperature. It absorbs the heat released in cellular reactions and distributes the heat throughout the body. When the body needs to cool, perspiration—a water-based substance—evaporates from the skin, and heat is drawn away from the body.

Usually the water lost through your skin and kidneys is easily replaced by drinking water or consuming moist foods. If excess water is lost and not replenished, water moves from intercellular spaces to the blood by osmosis. Eventually, water will be drawn from the cells themselves. As a cell loses water, the cytoplasm becomes more concentrated until, finally, the cell can no longer function. This condition is referred to as **dehydration.** Humans can die if they lose as much as 12 percent of their body water. Table 49-3 summarizes sources of water gain and loss for the average person in the United States. The statistics vary according to the person's location and level of activity.

SECTION 49-1 REVIEW

1. **Differentiate between organic nutrients and inorganic nutrients.**
2. **What are the primary functions performed by proteins?**
3. **Describe three benefits of a normal distribution of body fat.**
4. **Explain why water is a vital nutrient.**
5. **Explain the function of indigestible cellulose in digestion.**
6. **CRITICAL THINKING What do you think the consequences might be of a diet of only water, brown rice, and fruits?**

DIGESTIVE SYSTEM

Before your body can use the nutrients in the food you consume, the nutrients must be broken down physically and chemically. This process of breaking down food into molecules the body can use is called ***digestion.***

SECTION

49-2

OBJECTIVES

▲ List the major organs of the digestive system.

● Distinguish between mechanical digestion and chemical digestion.

■ Relate the structure of each digestive organ to its function in mechanical digestion.

◆ Identify the source of each major digestive enzyme, and describe the function of the enzyme.

▲ Summarize the process of absorption in both the small and large intestine.

THE GASTROINTESTINAL TRACT

Digestion occurs in the **gastrointestinal tract,** or digestive tract, which begins at the mouth and winds through the body to the anus. The gastrointestinal tract, shown in Figure 49-4, is a long, winding tube that is divided into several distinct organs. These organs, such as the stomach and small intestine, carry out the digestive process. Located along the gastrointestinal tract are other organs that aid in digestion. These organs, such as the liver and pancreas, are not part of the gastrointestinal tract, but they deliver secretions into the tract through ducts. As you read, locate each of the structures of the digestive system in Figure 49-4.

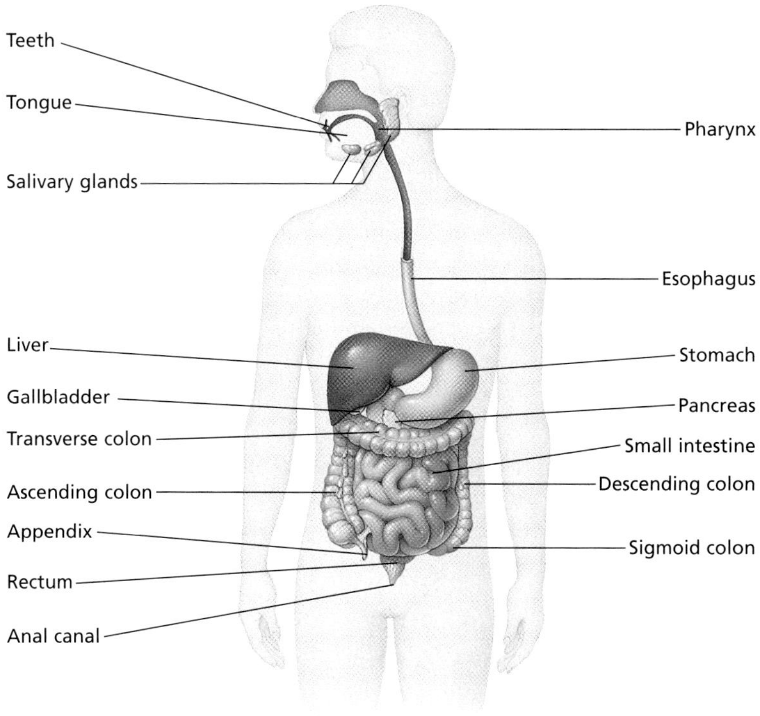

FIGURE 49-4

Notice that the gastrointestinal tract is basically a long tube with an opening at each end. Although food never passes through the liver and pancreas, these two organs are part of the digestive system because they secrete enzymes that break down food.

SECTION 49-2

SECTION OVERVIEW

Stability and Homeostasis
Section 49-2 describes the organs and structures of the alimentary canal. Students will learn how the organs of the alimentary canal digest and absorb nutrients.

Matter, Energy, and Organization
Section 49-2 discusses the coordinated structures and secretions of mechanical digestion and chemical digestion and summarizes absorption.

▶ ENGAGE STUDENTS

Use three identical pieces of hard candy to show the effectiveness of mechanical digestion. Drop one piece of candy into a clear glass of water. With a hammer, break a second piece into small bits and put them all into a second clear glass of water. Put a third piece of candy in an empty glass. Ask students to predict the results. The next day, have students explain the difference between the two jars. (The broken candy has all dissolved because of increased surface area.) Relate this to mechanical digestion by asking students what breaking the hard candy with the hammer represents. Ask students what the water represents (chemical digestion). Ask them what putting a piece of candy in an empty glass represents (a control).

VISUAL LINK

Figure 49-4
Point out that the illustration shows the complete alimentary canal. Ask a student volunteer to name the organs through which food passes (mouth, esophagus, stomach, small intestine, large intestine).

VISUAL LINK

Figure 49-5

Mumps, a common children's disease, is an inflammation of the parotid glands caused by the mumps virus (myxovirus), which spreads from person to person in saliva. Have students locate the parotid glands in Figure 49-5 and explain why people with mumps say that it hurts to open their mouth or chew. Besides the discomfort caused by the enlargement of the parotid glands, other signs and symptoms of mumps are fever and pain when swallowing acidic foods, such as sour pickles and grapefruit juice. When adult males get this disease, there is a 25 percent chance that their testes will become infected, which can lead to sterility.

QUICK FACT

☑ Halitosis (bad breath) results when decomposing food particles accumulate in the mouth, allowing bacteria to flourish. Saliva aids in washing away the food particles. Therefore, people who suffer from any disease that inhibits saliva secretion usually have problems with dental caries and bad breath, as well as difficulty talking, swallowing, and eating.

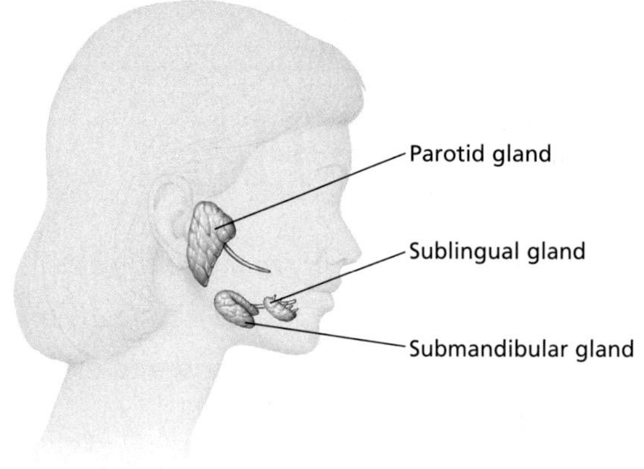

FIGURE 49-5

Saliva is produced by three sets of glands located near the mouth. The set closest to the ear is the target of the virus that causes mumps.

Mouth

When you take a bite of food, you begin the mechanical phase of digestion. In the mechanical phase, the body physically breaks down chunks of food into small particles. Mechanical digestion increases the surface area of food on which digestive enzymes can act. **Incisors**—sharp, flat front teeth—cut the food. Then the broad, flat surfaces of **molars,** or back teeth, grind it up. The tongue helps keep the food between the chewing surfaces of the upper and lower teeth by manipulating it against the **hard palate,** the bony, membrane-covered roof of the mouth. This structure is different from the **soft palate,** an area located just behind the hard palate. The soft palate is made of folded membranes and separates the mouth cavity from the nasal cavity.

While the mechanical phase of digestion is occurring, the chemical phase of digestion also takes place. Chemical digestion involves a change in the chemical nature of the nutrients. Preparations for chemical digestion begin even before the first bite of food is taken. The mouth starts to water—that is, the salivary glands increase their production of **saliva** (suh-LIE-vuh), a mixture of water, mucus, and a digestive enzyme called salivary amylase. Besides the many tiny salivary glands located in the lining of the mouth, there are three more pairs of larger salivary glands. Locate the three sets of salivary glands in Figure 49-5.

The mucus in the saliva softens and lubricates food and helps hold the food together. The salivary amylase begins the chemical digestion of carbohydrates by breaking down some starch into the disaccharide maltose.

FIGURE 49-6

The pharynx is the only passage shared by the digestive and respiratory systems. Notice how the epiglottis can close off the trachea so that food can pass only down the esophagus.

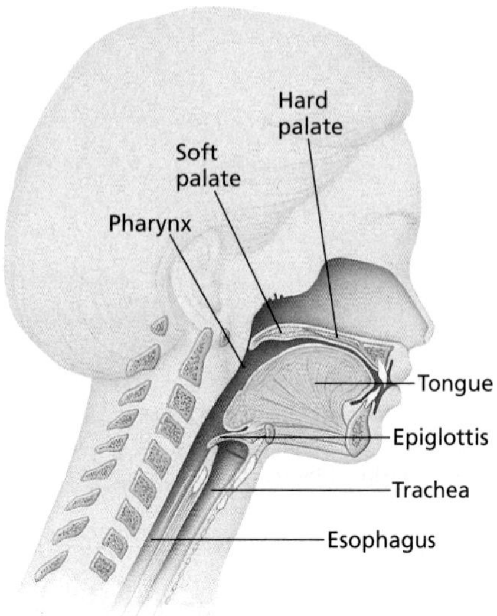

Esophagus

After food has been thoroughly chewed, moistened, and rolled into a **bolus,** or ball, it is forced into the pharynx by swallowing action. The pharynx, an open area that begins at the back of the mouth, serves as a passageway for both air and food. As Figure 49-6 shows, during swallowing, a flap of tissue called the **epiglottis** (ep-uh-GLAHT-is) prevents food from entering the trachea, or windpipe. Instead, the bolus passes into the esophagus, a muscular tube approximately 25 cm long that connects the pharynx with the stomach. The esophagus has two muscle layers: a circular layer that wraps around the esophagus and a longitudinal layer that runs the length of the tube. As you can see in Figure 49-7, alternating contractions of these muscle layers push the bolus through the esophagus and into the stomach. This series of rhythmic muscular contractions and relaxations is called **peristalsis.**

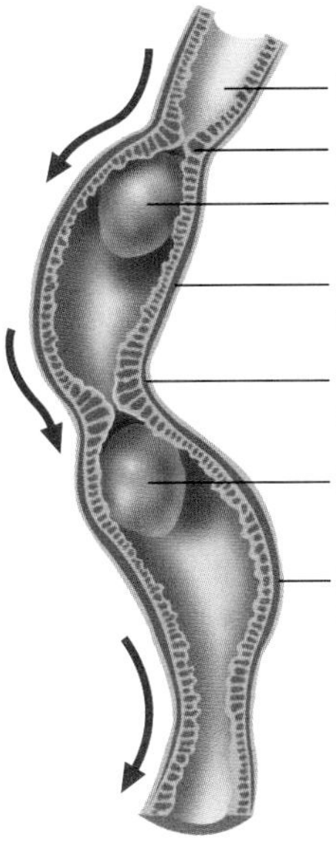

FIGURE 49-7

Peristalsis is so efficient at moving materials down the esophagus that you can drink while standing on your head. The smooth muscles move the water "up" the esophagus, against the force of gravity.

Stomach

The **stomach,** an organ involved in both mechanical and chemical digestion, is located in the upper left side of the abdominal cavity, just below the diaphragm. It is an elastic bag that is J-shaped when full and that lies in folds when empty. You have probably heard your stomach "growl" when it has been empty for some time. These sounds are made by the contraction of smooth muscles that line the stomach.

The walls of the stomach have several layers of smooth muscle. As you can see in Figure 49-8, there are three layers of muscle—a circular layer, a longitudinal layer, and a diagonal layer. Together, these muscles twist and turn the stomach. When nothing is present in the stomach, the growling sound is the result. When food is present, the muscles churn the contents of the stomach. This churning helps the stomach carry out mechanical digestion.

The inner lining of the stomach is a thick, wrinkled mucous membrane composed of epithelial cells. This membrane is dotted with small openings called gastric pits. **Gastric pits,** which are shown in Figure 49-9, are the open ends of gastric glands that release secretions into the stomach. Some of these glands secrete mucus, some secrete digestive enzymes, and still others secrete hydrochloric acid. The mixture of these secretions forms the acidic digestive fluid.

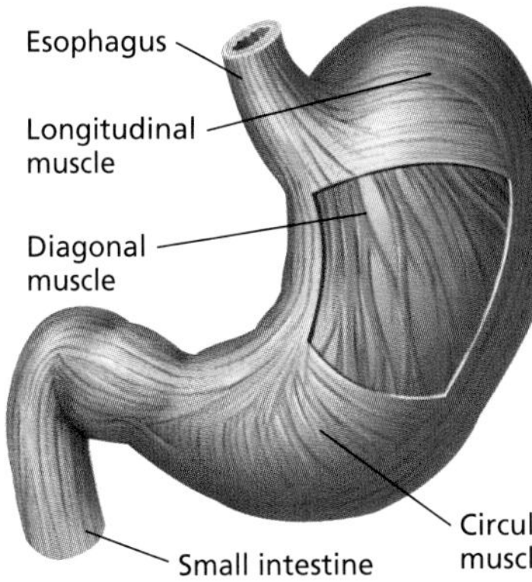

FIGURE 49-8

Each of the muscle layers of the stomach is oriented in a different direction. This allows the stomach to twist and turn in a variety of ways to thoroughly mix food.

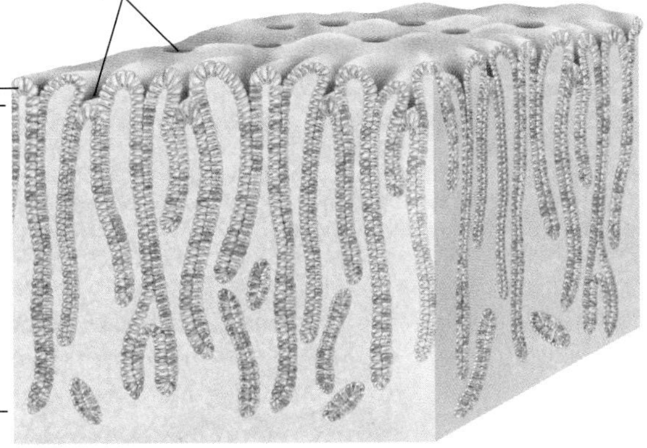

FIGURE 49-9

Because of the hydrochloric acid that the gastric glands secrete, the pH of the stomach is normally between 1.5 and 2.5, making it by far the most acidic environment in the body. Mucous cells lining the stomach wall protect the organ from damage.

TEACHING STRATEGY

Microscopic View of Digestive Structures

Provide pairs of students with microscopes, and have students observe, identify, and draw representative cells and tissue structures using prepared digestive-organ slides of the stomach, the small intestine, the liver, and the pancreas. Discuss how the structures relate to function, such as how the distribution of villi in the small intestine relates to nutrient absorption.

QUICK FACT

☑ Heartburn is the sharp, burning pain that occurs when gastric juice, which is very acidic, is regurgitated into the esophagus, an occurrence called reflux. Symptoms are so similar to a heart attack that many first-time sufferers of heartburn are rushed to the emergency room of a hospital. Heartburn is common with overeating in people who are obese or pregnant. People with hiatal hernias, a condition in which the stomach protrudes slightly above the diaphragm, can have heartburn regularly because gastric juices can reflux at any time.

DEMONSTRATION

Inflate a long, narrow balloon so that there is a ball of air at one end. Ask a student volunteer to demonstrate how peristalsis works. (The student should cut off the air flow at one end with the fingers of one hand and push with successive "waves" of movement using the thumb and forefinger of the other hand.)

TEACHING STRATEGY

Digestive System Enzymes

To help students comprehend the sources and functions of the various digestive enzymes, have them make a table with the following headings down the left side: Enzyme, Source, Site of enzymatic action, and Nutrient acted upon. Across the top of the table, have students write the following headings: Mouth, Stomach, Small intestine, Pancreas, and Large intestine. Then have students supply the information to complete the table.

RECENT RESEARCH

Miracle Cure

Researchers have found that there is a dietary cure for some young patients who suffer from 50 to 100 epileptic seizures per day for unknown reasons. These young patients are put on a high-fat, low-protein diet, such as one that includes whipping cream instead of milk. This throws their bodies into ketosis, a state in which ketone bodies are formed due to the rapid breakdown of fat. If these patients stay on such a diet for one or two years, the freedom from epileptic seizures can be permanent. Researchers are not sure why this dietary treatment is effective.

CHEMICAL DIGESTION

internetconnect

SCiLINKS NSTA

TOPIC: Chemical digestion
GO TO: www.scilinks.org
KEYWORD: HM986

Gastric fluid carries out chemical digestion in the stomach. An inactive stomach secretion called pepsinogen is converted into a digestive enzyme called **pepsin** at a low pH. Chemical digestion starts in the stomach when pepsin splits complex protein molecules into shorter chains of amino acids called peptides. The presence of hydrochloric acid in the stomach not only ensures the low pH needed for the pepsinogen to transform into pepsin, but also dissolves minerals and kills bacteria that enter the stomach along with food.

The mucus secreted in the stomach is vital to the survival of this organ. Mucus forms a coating that protects the lining from hydrochloric acid and prevents pepsin from digesting the proteins that make up the stomach tissue. In some people, the mucous coating of the stomach tissue breaks down, allowing digestive enzymes to eat through part of the stomach lining. The lesion, or sore, that results is called an **ulcer.** Scientists have recently discovered that the breakdown of the mucous layer is often caused by bacteria. These bacteria secrete toxins that destroy the epithelial cells that form the mucous layer.

Formation of Chyme

The **cardiac sphincter** (SFINK-tuhr) is a circular muscle located between the esophagus and the stomach. Food enters the stomach when the cardiac sphincter opens. After the food enters the stomach, the cardiac sphincter closes to prevent the food from reentering the esophagus. Food usually remains in the stomach for three to four hours. During this time, muscle contractions in the stomach churn the contents, breaking up food particles and mixing them with gastric fluid. This process forms a mixture called **chyme** (KIEM), a pastelike substance containing various nutrients. Chyme usually contains fats, sugars, starches, vitamins, minerals, peptides, and proteins that were not broken down by the pepsin.

Peristalsis forces chyme out of the stomach and into the small intestine. The **pyloric** (pie-LOR-ik) **sphincter,** a circular muscle between the stomach and the small intestine, regulates the flow of chyme. Each time the pyloric sphincter opens, about 5 to 15 mL (about 0.2 to 0.5 oz) of chyme moves into the small intestine, where it mixes with secretions from the liver and pancreas.

FIGURE 49-10

The liver is the body's largest internal organ, weighing about 1.5 kg (3 lb). If a small portion is surgically removed because of disease or injury, the liver regenerates the missing section.

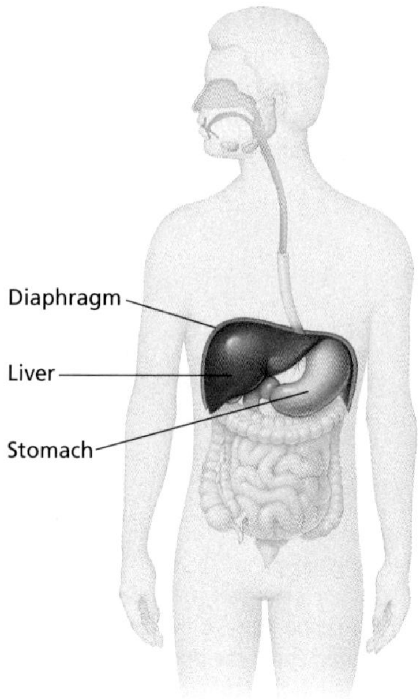

Liver

The **liver** is a large organ located to the right of the stomach and in the upper right area of the abdominal cavity, just below the diaphragm, as shown in Figure 49-10. The liver performs numerous functions in the body, including storing glycogen and breaking down toxic substances, such as alcohol. The liver also secretes bile, which is vital in the digestion of fats. Though it is not a digestive enzyme, bile breaks fat globules into small droplets, forming a

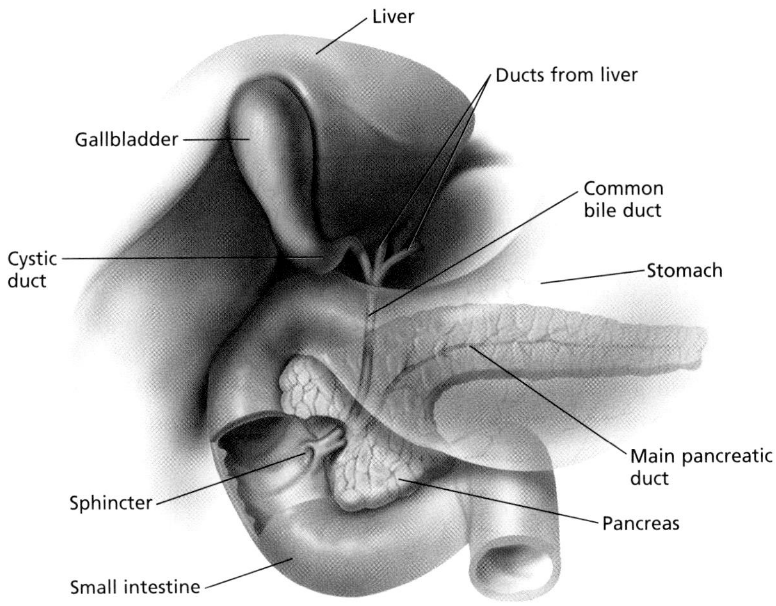

FIGURE 49-11

Cholesterol deposits known as gallstones can form in the ducts leading from the liver and gallbladder to the small intestine. If the gallstones interfere with the flow of bile, they must be removed, along with the gallbladder in most cases.

milky fluid in which fats are suspended. This process exposes a greater surface area of fats to the action of digestive enzymes and prevents small fat droplets from rejoining into large globules.

The bile secreted by the liver passes through a Y-shaped duct, as shown in Figure 49-11. The bile travels down one branch of the Y, the hepatic duct, and then up the other branch, the cystic duct, to the **gallbladder,** a saclike organ that stores and concentrates bile. When chyme is present in the small intestine, the gallbladder releases bile through the common bile duct into the small intestine.

Pancreas

As you can see in Figure 49-12, the pancreas is an organ that lies behind the stomach, against the back wall of the abdominal cavity. As part of the digestive system, the pancreas secretes pancreatic fluid, which contains digestive enzymes that help complete the breakdown of nutrients in the chyme. This pancreatic fluid enters the small intestine through the pancreatic duct, which joins the common bile duct just before it enters the intestine.

Pancreatic fluid contains sodium bicarbonate, which changes the pH of the chyme from an acid to a base. Many enzymes in the pancreatic fluid are activated by the higher pH. These enzymes hydrolyze disaccharides into monosaccharides, fats into fatty acids and glycerol, and proteins into amino acids.

Small Intestine

If you could stretch the small intestine to its full length, you would find that it is nearly 7 m (about 21 ft) long. The duodenum, the first section of this coiled tube, makes up only the first 25 cm (about

FIGURE 49-12

Most of the pancreas looks like the salivary glands of the mouth region. Like the salivary glands, the pancreas secretes amylases to digest carbohydrates. Other digestive enzymes are secreted by the pancreas to break down nutrients in the chyme.

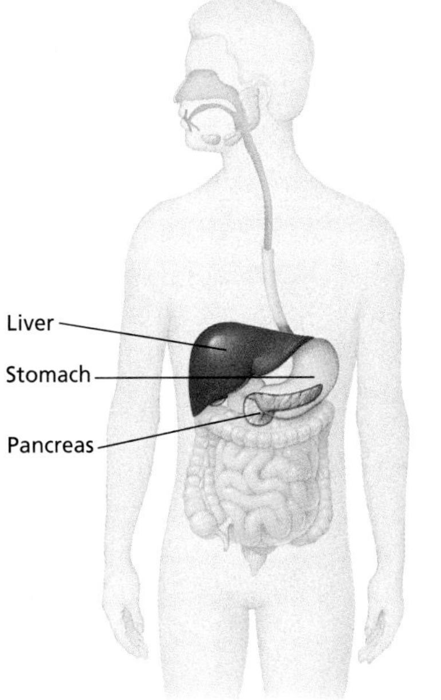

QUICK FACT

☑ The liver is the largest gland in the body.

GIFTED ACTIVITY

The liver is one of the most biochemically complex organs in the body. It processes nearly every class of nutrient and plays a major role in regulating plasma cholesterol levels. No mechanical device can substitute for a liver. Ask students to research and write a paper on the major metabolic functions of the liver.
(1. It packages fatty acids into forms that can be stored or transported.
2. It synthesizes plasma proteins.
3. It forms nonessential amino acids and converts the ammonia resulting from their deamination to urea.
4. It stores glucose as glycogen and regulates blood-sugar homeostasis.
5. It stores certain vitamins.
6. It conserves iron salvaged from worn-out red blood cells.
7. It degrades hormones.
8. It detoxifies substances such as alcohol and drugs.)

TEACHING STRATEGY

Diarrhea

When the intestine rushes its contents through without sufficient time to absorb the water from the wastes, diarrhea results. Irritations from drugs, parasites or bacteria, and stress can all cause diarrhea. Prolonged diarrhea, such as occurs with amebic dysentery and cholera, can cause lethal dehydration and salt imbalance. Ask students why children and infants are especially vulnerable to dehydration. (Their tissues cannot store as much water as the tissues of adults.)

TEACHING STRATEGY
Eating Disorders

Anorexia nervosa and bulimia most often affect females of high school or college age.

People who suffer from anorexia nervosa deliberately starve themselves and undertake strenuous exercise in an attempt to gain an "ideal" body image. However, the anorexic never can attain that ideal, because a poor self-image is often the true problem.

Those who suffer from bulimia binge on huge quantities of food and then purge themselves through vomiting, the use of huge doses of laxatives, or both. A bulimic may look healthy, but the acid from constant vomiting damages their esophagus and teeth, and many become laxative dependent. They may also suffer from life-threatening electrolyte (salt) imbalances and internal hemorrhages. Have students use library resources or an on-line database to investigate eating disorders, the treatments prescribed, and the average duration of treatment. (Treatments for both conditions are prolonged, often involving hospitalization during which new behavior patterns are taught. Treatment may also include drug and psychological therapy.)

VISUAL LINK
Figure 49-13 and Figure 49-14

Ask students to identify the structures shown in Figures 49-13 and 49-14. Ask them how villi are related to the function of the small intestine. (The major function of the small intestine is to absorb nutrients from the chyme; the villi give the organ more surface area to carry out that function.)

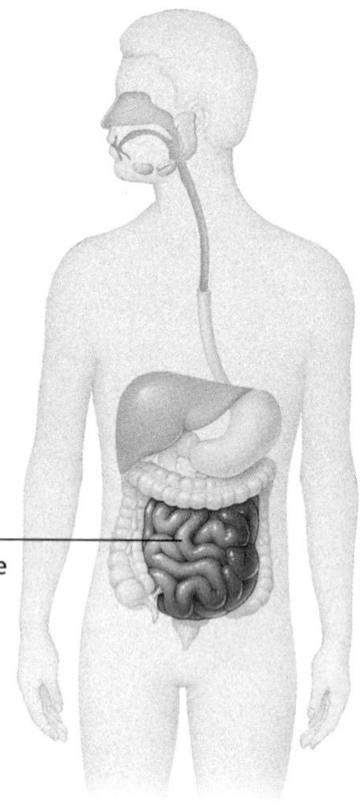

FIGURE 49-13

Although the small intestine is nearly 7 m long, only the first 25 cm is involved in digesting food. The rest is involved in the absorption of the digested products, including monosaccharides, amino acids, glycerol, and fatty acids.

10 in.) of that length. The **jejunum** (jee-JOO-nuhm), the middle section, is about 2.5 m (about 8 ft) long. The ileum, which makes up the remaining portion of the small intestine, is approximately 4 m (about 13 ft) in length. As you can see in Figure 49-13, the entire length of the small intestine lies coiled in the abdominal cavity.

The secretions from the liver and pancreas enter the duodenum, where they continue the chemical digestion of chyme. When the secretions from the liver and pancreas, along with the chyme, enter the duodenum, they trigger intestinal mucous glands to release large quantities of mucus. This mucus protects the intestinal wall from protein-digesting enzymes and the acidic chyme. Glands in the mucous lining of the small intestine release enzymes that complete digestion by breaking down peptides into amino acids, disaccharides into monosaccharides, and fats into glycerol and fatty acids.

Absorption

The end products of digestion—amino acids, monosaccharides, glycerol, and fatty acids—are absorbed into the circulatory system through blood and lymph vessels in the lining of the small intestine. The structure of this lining provides a huge surface area for absorption to take place. In **absorption,** the end products of digestion are transferred into the circulatory system. The highly folded lining of the small intestine is covered with millions of fingerlike projections called **villi,** which are shown in Figure 49-14. The cells covering the villi, in turn, have extensions on their cell membranes called **microvilli.** The folds, villi, and microvilli give the small intestine a surface area of about 250 m^2 (about 2,685 ft^2), or roughly the area of a tennis court. Nutrients are absorbed through this surface by means of diffusion and active transport.

Inside each of the villi are capillaries and tiny lymph vessels called **lacteals** (lak-TEE-uhls). Locate the lacteals in Figure 49-14.

FIGURE 49-14

Villi, as shown in the SEM (137×) and the diagram, expand the surface area of the small intestine to allow greater absorption of nutrients.

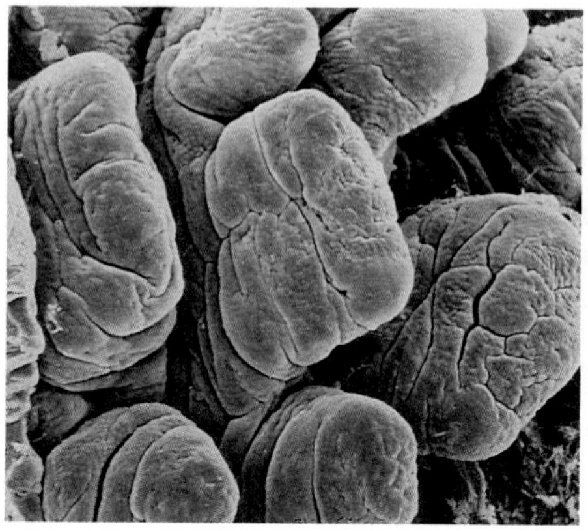

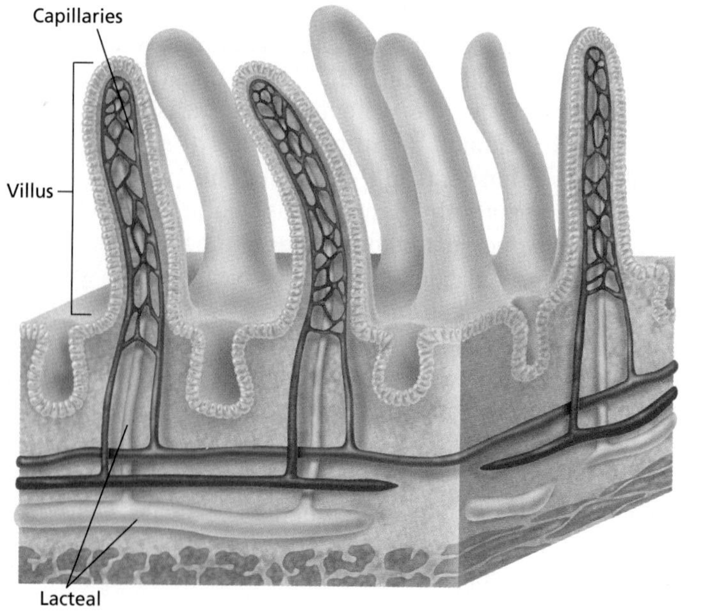

Glycerol and fatty acids enter the lacteals, which carry them through the lymph vessels and eventually to the bloodstream through lymphatic vessels near the heart. Amino acids and monosaccharides enter the capillaries and are carried to the liver. The liver neutralizes many toxic substances in the blood and removes excess glucose, converting it to glycogen for storage. The filtered blood then carries the nutrients to all parts of the body.

LARGE INTESTINE

After absorption in the small intestine is complete, peristalsis moves the remaining material on to the large intestine. The **large intestine,** or **colon,** is the final organ of digestion. Study Figure 49-15 to identify the four major parts of the colon: ascending colon, transverse colon, descending colon, and sigmoid colon. The sigmoid colon leads into the very short, final portions of the large intestine called the rectum and the anal canal.

Most of the absorption of nutrients and water is completed in the small intestine. About 9 L (about 9.5 qt) of water enters the small intestine daily, but only 0.5 L (about 0.53 qt) of water is present in the material that enters the large intestine. In the large intestine, only minerals and vitamins produced by bacteria that live in the colon, as well as most of the remainder of the water, are absorbed. Slow contractions move the material in the large intestine toward the rectum. Distension of the colon initiates reflex contractions which move the material out of the body. As this matter moves through the intestine, the absorption of water solidifies the mass. The solidified material is called **feces.**

As the fecal matter solidifies, cells lining the large intestine secrete mucus to lubricate the intestinal wall. This lubrication makes the passing of the feces less abrasive. Mucus also binds together the fecal matter, which is then eliminated through the anus.

FIGURE 49-15

This X ray shows the large intestine, or colon. Identify the four major regions of the colon in the X ray.

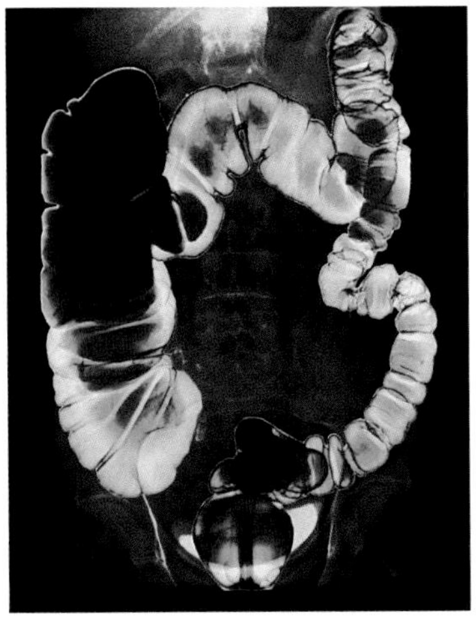

SECTION 49-2 REVIEW

1. Describe the gastrointestinal tract.
2. Name one structure involved in the mechanical phase of digestion, and explain its function in mechanical digestion.
3. Trace the route of bile, noting where it is produced, where it is stored, and how it functions in digestion.
4. Explain how the structure of the small intestine is related to the function of absorption.
5. Compare the process of chemical digestion with the process of mechanical digestion, and list the organs involved in each.
6. **CRITICAL THINKING** Which of the six basic nutrients might a person need to restrict after an operation to remove the gallbladder? Explain your answer.

RETEACHING *ACTIVITY*

Describe functions of the digestive system to the class, and ask volunteers to explain where and how each function is performed.

ANSWERS TO SECTION 49-2 REVIEW

1. The gastrointestinal tract is a long tube that begins at the mouth and ends at the anus. It involves organs such as the pancreas and liver, which produce chemicals that aid in digestion.
2. Any of the following are acceptable: the teeth, which cut and grind food; the tongue, which keeps food between the teeth; or the stomach, which churns the food.
3. Bile, which is produced in the pancreas and stored in the gallbladder, breaks fat globules into small droplets, exposing a greater surface area to the action of digestive enzymes.
4. The folds, villi, and microvilli provide a huge surface area for absorption to take place.
5. Chemical digestion involves the acids, bases, and enzymes produced by the liver and pancreas that chemically break down food into molecules the body can use. Mechanical digestion is the physical breaking apart of food particles by the teeth, tongue, and stomach to expose a larger surface area for the chemicals of digestion to act on.
6. Normally, the gallbladder stores bile, which is released for the digestion of fats (lipids). If this storage organ is removed, a steady flow of bile enters the digestive system. This may result in an insufficient quantity of bile to digest a meal high in fats. Thus, people who no longer have a gallbladder may be more comfortable with a low-fat diet.

BACKGROUND

The word *botulism* comes from the Latin word *botulus,* meaning "sausage." It was so named because 13 people in a small town in Germany died after they shared an uncooked smoked sausage. Botulin is the toxin secreted by the bacteria *Clostridium botulinum* and is one of the most potent poisons known. It causes the respiratory muscles to become paralyzed, resulting in suffocation. If the patient can be kept alive through tube-feeding and the use of an artificial respirator, paralyzed muscles can recover.

READING FOR MEANING

When humans first began to preserve foods using methods such as canning, which removes oxygen from the food, the spores of the germs were able to mature and multiply in the food, producing botulin. *Clostridium botulinum* can multiply and produce toxin only in the absence of oxygen.

Roueché conveys that the botulinus organism is present in abundance almost everywhere.

Prompt deoxidation means preserving foods rapidly before contamination can occur.

READ FURTHER

Sanitary working conditions and practices by food handlers result in less contamination and food poisoning. Also, there are new techniques for food preservation, such as sterilization through ionizing radiation, freeze-drying, and new chemical additives.

Infant botulism is the most common form of botulism in the United States. It is thought that 5 percent of sudden infant death syndrome (SIDS) deaths are related to botulism.

Literature & Life

Food Poisoning

The following excerpt is from "Family Reunion," a chapter in *Eleven Blue Men and Other Narratives of Medical Detection,* by Berton Roueché, published in 1954.

Botulism is a true but atypical form of food poisoning. Its methodical approach, its excessive lethality, and the predominantly neural cast of its clinical features all are unpleasantly peculiar. Even its history is unusual. Ordinary food poisoning, in common with many other ailments, is probably as old as mankind. Its beginnings go back to the first butcher with an infected finger, the first cook with a streptococcic cough, the first imprudent assumption of the first man rash enough to eat mushrooms.

Botulism is of far less fundamental origin. It is, in fact, one of the very few diseases for whose existence man has nobody to blame but himself. Like carbon-monoxide poisoning, and alcoholism, and the bends, botulism is essentially, if inadvertently, a product of human ingenuity. *Clostridium botulinum,* though plethorically abundant throughout the world, is not among man's natural antagonists. The organism is incapable of establishing itself in any living plant or animal. Its home is soil and earthy dust, its food is inanimate matter, and although it is able to exist in a dormant, sporal state almost indefinitely in almost any environment, it can mature and multiply and manufacture its vigorous venom only in the total absence of oxygen. Because of these physiological quirks, the toxin of *Clostridium botulinum,* under normal conditions, is safely out of human reach. It is dissipated deep in the earth. Exactly when botulism seized its first victim is unknown, but it could hardly have been more than eight or ten thousand years ago, when man ceased to subsist exclusively on fresh food. Freshly gathered food, along with everything else on the face of the earth, is exposed to the intrusion of dust-borne botulinus spores, but it is also exposed to the spore-stunting sweep of air. Botulism came into being when man made the otherwise triumphant discovery that prompt deoxidation would make numerous foods more or less permanently resistant to decay.

Reading for Meaning

How is botulism a result of human ingenuity?

Interpret the writer's meaning when he says that *Clostridium botulinum* is "plethorically abundant."

In the last paragraph, this excerpt explains how the botulism toxin first came into contact with humans. Explain what the writer means by "prompt deoxidation" of food.

Read Further

In an article more recent than this one, what new information might be included about how current methods of food handling help prevent food poisoning?

Use your school library or an online database to investigate infant botulism. How does infant botulism relate to sudden infant death syndrome (SIDS)?

SECTION 49-3

URINARY SYSTEM

The body must rid itself of the waste products of cellular activity. The process of removing metabolic wastes, called **excretion,** *is just as vital as digestion in maintaining the body's internal environment. Thus, the urinary system not only excretes wastes but also helps maintain homeostasis by regulating the content of water and other substances in the blood.*

OBJECTIVES

▲ Define the term *excretion,* and list the functions of each of the major excretory organs.

● Identify the major parts of the kidney.

■ Relate the structure of a nephron to its function.

◆ Explain how the processes of filtration, reabsorption, and tubular secretion help maintain homeostasis.

▲ Name the main parts of the urinary system.

KIDNEYS

The main waste products that the body must eliminate are carbon dioxide, from cellular respiration, and nitrogenous compounds, from the breakdown of proteins. The lungs excrete most of the carbon dioxide, and nitrogenous wastes are eliminated by the kidneys. The excretion of water is necessary to dissolve wastes and is closely regulated by the kidneys, the main organs of the urinary system.

Humans have two kidneys, bean-shaped excretory organs each about the size of a clenched fist. The kidneys are located in the small of the back, one behind the stomach and the other behind the liver. Together they regulate the chemical composition of the blood.

Structure

Figure 49-16 shows the three main parts of the kidney. The **cortex,** the outermost portion of the kidney, makes up about a third of the

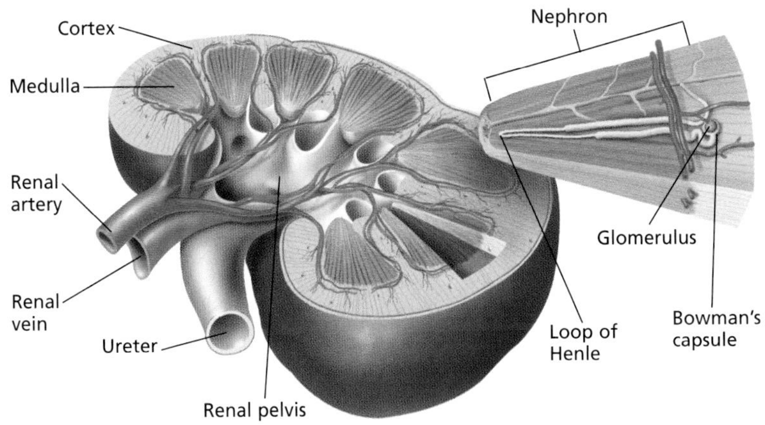

FIGURE 49-16

The outer region of the kidney, the cortex, contains structures that filter blood brought by the renal artery. The inner region, or medulla, consists of structures that carry urine, which empties into the funnel-shaped renal pelvis. The renal vein transports the filtered blood back to the heart.

SECTION 49-3

SECTION OVERVIEW

Stability and Homeostasis
Section 49-3 describes the excretory system. Students will learn about the structure and function of the kidneys, lungs, and skin as excretory organs.

Cell Structure and Function
Section 49-3 discusses how metabolic wastes are filtered from the blood and excreted by the body. This section also describes the difference between excretion and elimination.

▶ ENGAGE STUDENTS

Ask students if they ever feel bloated when they eat a lot of salty potato chips or other salty snacks. Tell students that this is a temporary retention of fluid in the body. Ask them why this happens. (The increased salt causes a cellular sodium imbalance. Fluid is retained to maintain osmotic equilibrium.) Ask what this equilibrium is called (homeostasis).

TEACHING STRATEGY

The Kidney

Provide pairs of students with microscopes and prepared slides of kidney cells in cross-section. Have students refer to Figure 49-16 and observe, identify, draw, and label the structures of the kidney.

QUICK FACT

☑ Kidney stones are formations of crystals of calcium, magnesium, or uric acid salts in the kidney. This relatively rare occurrence causes excruciating pain as the body attempts to pass the sharp crystals through the renal pelvis and ureter. In the past, surgery was the only remedy, but a new treatment uses sound waves to shatter the crystals into tiny, nonobstructive pieces.

TEACHING STRATEGY

The Excretory System

Have students locate the major organs of the excretory system on a human anatomical model. Point out the renal artery and renal vein, and ask how the blood moving through these two vessels differs. (The blood entering the kidney through the renal artery contains waste products of cellular activity and toxins that will be filtered out by the kidney before the blood leaves through the renal vein.) Ask students what the filtered wastes form (urine). Have students trace the path of urine from the kidneys through the ureter and urinary bladder to the urethra.

QUICK FACT

☑ Hemorrhagic shock is caused by a sudden loss of a large volume of blood, such as an accident victim might suffer. When a part of the body is seriously injured, circulation becomes constricted as the body attempts to retain body fluids. While shock is a reflexive reaction to preserve blood volume and pressure, it can be irreversible and fatal if it is prolonged. In the kidneys, both the renal vein and renal artery are affected. Because waste products are retained in the kidneys, prolonged shock may result in severe renal-tubule damage.

VISUAL LINK

Figure 49-17

Ask students to trace with their fingers the direction blood moves from the renal artery into the nephrons. Have students trace the pathway of the blood up the artery to the glomerulus. Ask students what happens to the blood in the glomerulus. (Fluid moves out of the blood and into the Bowman's capsule.) Have students turn to Figure 49-18 and name the process that occurs in the glomerulus (filtration).

internetconnect

SCILINKS NSTA

TOPIC: Urinary system
GO TO: www.scilinks.org
KEYWORD: HM992

kidney's tissue mass. The **medulla** is the inner two-thirds of the kidney. The **renal pelvis** is a funnel-shaped structure in the center of the kidney. Also notice in Figure 49-16 that blood enters the kidney through a renal artery and leaves through a renal vein. The renal artery transports nutrients and wastes to the kidneys. The nutrients are used by kidney cells to carry out their life processes. One such process is the removal of wastes brought by the renal artery.

The most common mammalian metabolic waste is **urea** (yoo-REE-uh), a nitrogenous product made by the liver. Nitrogenous wastes are initially brought to the liver as **ammonia,** a chemical compound of nitrogen so toxic that it could not remain long in the body without harming cells. The liver removes ammonia from the blood and converts it into the less harmful substance urea. The urea enters the bloodstream and is then removed by the kidneys.

Nephrons

The substances removed from the blood by the kidneys—toxins, urea, water, and mineral salts—form an amber-colored liquid called **urine.** Urine is made in structures called **nephrons** (NEF-rahns), the functional units of the kidney.

Take a close look at the structure of the nephron, shown in Figure 49-17. Each kidney consists of more than a million nephrons. If they were stretched out, the nephrons from both kidneys would extend for 80 km (50 mi). As you read about the structure of a nephron, locate each part in Figure 49-17.

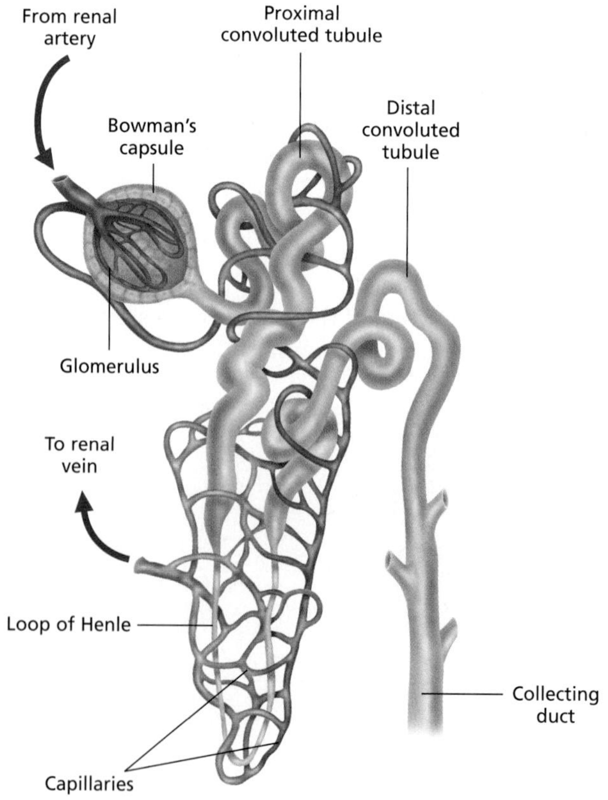

FIGURE 49-17

Notice the close association between a nephron of the kidney and capillaries of the circulatory system. Initially, fluid passes from the glomerulus, which is made of a capillary network, into a Bowman's capsule of the nephron. As the fluid travels through the nephron, nutrients that passed into the Bowman's capsule are removed and returned to the bloodstream. All that normally remains in the nephron are waste products, which form the urine that passes out of the kidney.

Each nephron has a cup-shaped structure, called a **Bowman's capsule,** that encloses a bed of capillaries. This capillary bed, called a **glomerulus** (glo-MER-yoo-luhs), receives blood from the renal artery. High pressure forces fluids from the blood through the capillary walls and into the Bowman's capsule. The material filtered from the blood then flows through the **renal tubule,** a long tube with permeable walls. The renal tubule consists of three parts: the proximal convoluted tubule, the loop of Henle, and the distal convoluted tubule. Blood remaining in the glomerulus then flows through a network of capillaries that wraps around these three parts of the renal tubule. The long and winding course of both the renal tubule and the surrounding capillaries provides a large surface area for the exchange of materials.

As the filtrate flows through a nephron, its composition is modified by the exchange of materials among the renal tubule, the capillaries, and the extracellular fluid. Various types of exchanges take place in the different parts of the renal tubule. To understand how the structure of each part of the nephron is related to its function, we will examine the three major processes that take place in the nephron: filtration, reabsorption, and secretion. Figure 49-18 shows the site of each of these processes in the nephron.

Filtration

Materials from the blood are forced out of the glomerulus and into the Bowman's capsule during a process called **filtration.** Blood in the glomerulus is under relatively high pressure. This pressure forces water, urea, glucose, vitamins, and salts through the thin capillary walls of the glomerulus and into the Bowman's capsule. About one-fifth of the fluid portion of the blood filters into the Bowman's capsule. The rest remains in the capillaries, along with proteins and cells that are too large to pass through the capillary walls. In a healthy kidney, the filtrate—the fluid that enters the nephron—does not contain large protein molecules.

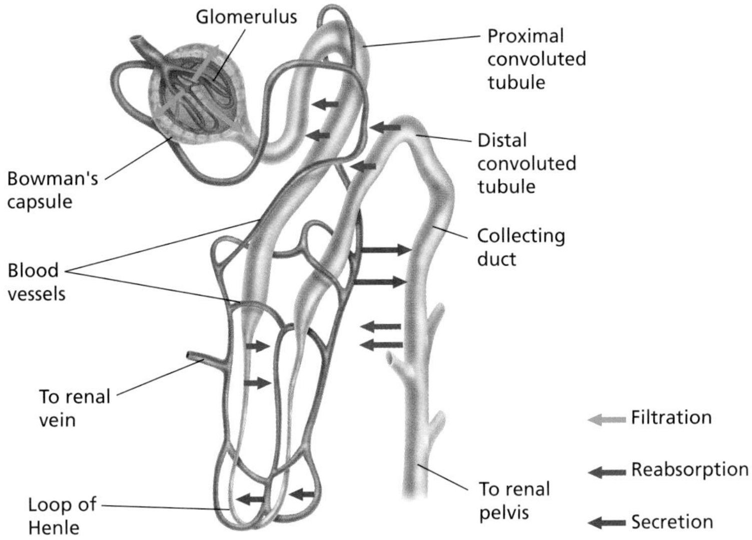

FIGURE 49-18
Color-coded arrows indicate where in the nephron filtration, reabsorption, and secretion occur.

Word Roots and Origins

glomerulus

from the French *glomérule,* meaning "a compact cluster"

MAKING CONNECTIONS

Mathematics

Glomerular filtration normally occurs at the rate of 125 mL per minute. Have students perform the following simple calculations to help them visualize how enormous this volume is: 125 mL/min × 60 min/hr = 7,500 mL of glomerular filtrate per hour. 7,500 mL/hr × 24 hr/day = 180,000 mL, or 180 L (about 47.5 gal), of fluid filtered out of glomerular blood per day. No one excretes 180 L of urine per day. Ask students why people do not excrete a vast amount of fluid. (Most of the fluid that leaves the blood by glomerular filtration returns to the blood by reabsorption.)

QUICK FACT

☑ More than 100,000 people in the United States suffer from kidney disease. Some of them will undergo hemodialysis in order to rid their body of nitrogenous wastes. A person on dialysis, as the process is commonly known, is connected to a stationary machine called a hemodialyzer for about four hours every other day. The machine filters out the wastes through a dialysis membrane and removes excess fluid from the body. A semipermanent tube called a catheter, which is implanted in a patient's artery, is connected to the hemodialyzer. Blood flows through the catheter to the hemodialyzer, is cleansed, and is then returned to the patient.

CONTENT CONNECTION

Section 5-1: Osmosis

Review the concept of osmosis with students, and ask them to define the terms *hypertonic, hypotonic,* and *isotonic.* Also correlate Figure 49-19 with Figure 49-18. Explain to students that Figure 49-19 demonstrates a process that is occurring in one portion of a nephron, which is the collecting duct and the capillaries that wrap around the renal tubule.

INCLUSION *ACTIVITY*

Have students make a sequential diagram of the events of glomerular filtration to help them understand how kidneys function.

QUICK FACT

☑ As food-conscious as Americans are, they often pay little attention to their fluid intake. In fact, people can survive without food much longer than they can survive without water. On average, an adult needs 1.5–3 L of water per day, some of it being supplied by moist food. Any combination of foods and beverages that provides the needed amount of water is acceptable, but most nutritionists recommend 6–8 glasses of liquid a day over and above the amount contained in food.

VISUAL LINK

Figure 49-19

Tell students that the drawing of the nephron is simplified. Tell them that the nephron *in vivo,* including the proximal convoluted tubule, loop of Henle, and distal convoluted tubule, is long and coiled. Have them point to the proximal convoluted tubule, and remind them that most reabsorption occurs there.

Eco Connection

Kidneys and Pollution

According to the U.S. Environmental Protection Agency, indoor areas, where we spend up to 90 percent of our time, contain substances that may be hazardous to our health. Because of their function in excretion, kidneys often are exposed to hazardous chemicals that have entered the body through the lungs, skin, or gastrointestinal tract. Household substances that, in concentration, can damage kidneys include paint, varnishes, furniture oils, glues, aerosol sprays, air fresheners, and lead.

Many factors in our environment are difficult to control, but the elimination of pollutants from our indoor living areas is fairly simple. The four steps listed below may help eliminate indoor pollutants.

1. Identify sources of pollutants in your home.
2. Eliminate the sources, if possible.
3. Seal off those sources that cannot be eliminated.
4. Ventilate to evacuate pollutants and bring in fresh air.

Reabsorption and Secretion

The body needs to retain many of the substances that were removed from the blood by filtration. Thus, as the filtrate flows through the renal tubule, these materials return to the blood by being selectively transported through the walls of the renal tubule and entering the surrounding capillaries. This process is called **reabsorption.** Most reabsorption occurs in the proximal convoluted tubule. In this region, about 75 percent of the water in the filtrate returns to the capillaries by osmosis. Glucose and minerals, such as sodium, potassium, and calcium, are returned to the blood by active transport. Some additional reabsorption occurs in the distal convoluted tubule.

When the filtrate reaches the distal convoluted tubule, some substances pass from the blood into the filtrate through a process called **secretion.** These substances include wastes and toxic materials. The pH of the blood is adjusted by hydrogen ions that are secreted from the blood into the filtrate.

Formation of Urine

The fluid and wastes that remain in the distal convoluted tubule form urine. The urine from several renal tubules flows into a collecting duct. Notice in Figure 49-19 that the urine is further concentrated in the collecting duct by the osmosis of water through the wall of the duct. This process allows the body to conserve water. In fact, osmosis in the collecting duct, together with reabsorption in other parts of the tubule, returns to the blood about 99 of every 100 mL (about 3.4 oz) of water in the filtrate.

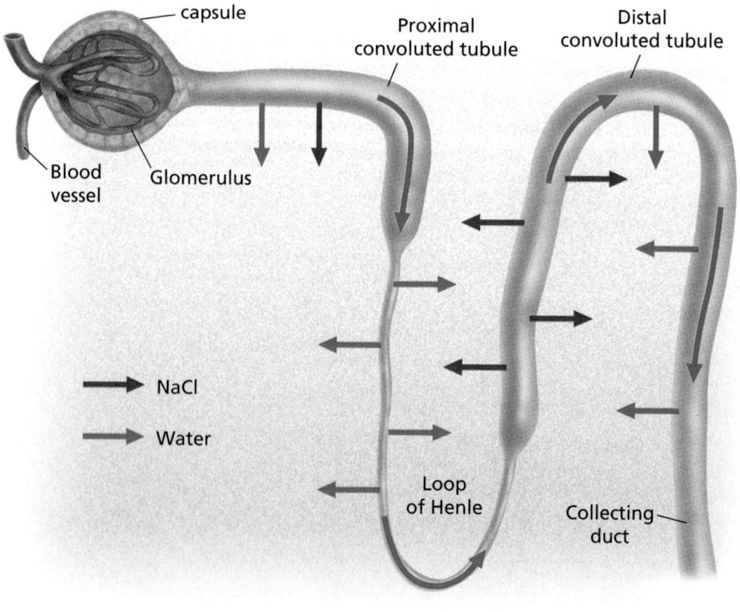

FIGURE 49-19

The sodium chloride that is actively transported out of the loop of Henle makes the extracellular environment surrounding the collecting duct hypertonic. Thus, water moves out of the collecting duct by osmosis into this hypertonic environment, increasing the concentration of urine.

The Loop of Henle

The function of the **loop of Henle** (HEN-lee) is closely related to that of the collecting duct. Water moves out of the collecting duct because the concentration of sodium chloride is higher in the fluid surrounding the collecting duct than it is in the fluid inside the collecting duct. This high concentration of sodium chloride is created and maintained by the loop of Henle. Notice in Figure 49-19 that cells in the wall of the loop actively transport negatively charged chloride ions from the filtrate to the fluid between the loops and the collecting duct. Positively charged sodium ions follow the chloride ions into the fluid. This ensures that the sodium chloride concentration of the fluid between the loops and the collecting duct remains high and thus promotes the reabsorption of water from the collecting duct.

Elimination of Urine

Urine from the collecting ducts flows through the renal pelvis and into a narrow tube called a **ureter** (yoo-REET-uhr). A ureter leads from each kidney to the **urinary bladder,** a muscular sac that stores urine. Muscular contractions of the bladder force urine out of the body through a tube called the **urethra** (yoo-REE-thruh). Locate the ureters, urinary bladder, and urethra in Figure 49-20.

At least 500 mL (17 oz) of urine must be eliminated every day because this amount of fluid is needed to remove potentially toxic materials from the body and to maintain homeostasis. A normal adult eliminates from 1.5 L (1.6 qt) to 2.3 L (2.4 qt) of urine a day, depending on the amount of water taken in and the amount of water lost through respiration and perspiration.

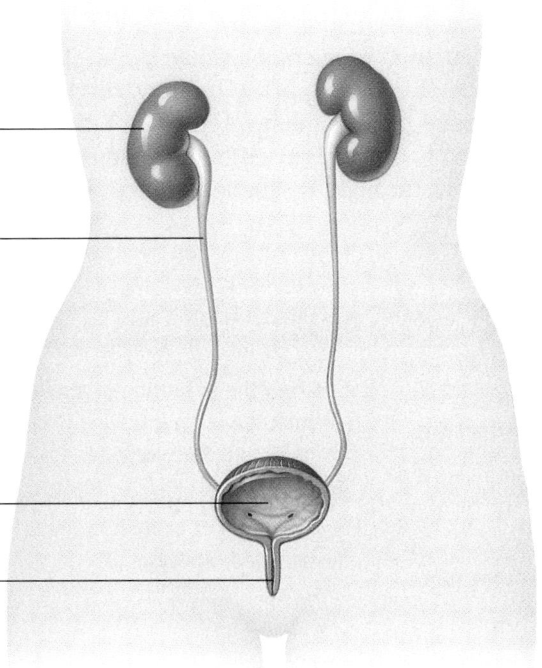

FIGURE 49-20

Urine trickles from each kidney through a ureter to the urinary bladder, where it is stored until it is eliminated from the body through the urethra.

Quick Lab

Analyzing Kidney Filtration

Materials disposable gloves, lab apron, and safety goggles, 20 mL of test solution, 3 test tubes, filter, beaker, 15 drops each of biuret and Benedict's solution, 2 drops IKI solution, 3 pipets, wax marker pen

Procedure

1. Put on your gloves, lab apron, and safety goggles.
2. Put 15 drops of the test solution into each of the test tubes. Label the test tubes "Protein," "Starch," and "Glucose."
3. Add 15 drops of biuret solution to the test tube labeled "Protein." Record your observations.
4. Add 15 drops of Benedict's solution to the test tube labeled "Glucose." Record your observations.
5. Add two drops of IKI solution to the test tube labeled "Starch." Record your observations.
6. Discard the tested solutions, and rinse your test tubes as your teacher directs.
7. Pour the remaining test solution through a filter into a beaker. Using the test solution from the beaker, repeat steps 3–5.

Analysis Which compounds passed through the filter paper? If some did not, explain why. How does the filtration of this activity resemble the activity of the kidney?

Quick Lab

Analyzing Kidney Filtration

Time Required 30 minutes

Procedural Tips To make 100 mL of test solution, add 1 g each of flour, gelatin, and glucose to 97 mL of cool water just before the activity (Do not use warm or hot water.) Each group will need 20 mL of solution to test. (Apple juice may be substituted for glucose. Add 1 g each of flour and gelatin to 98 mL of apple juice.) For filter paper, you may use a thick coffee filter like the kind used in drip coffee makers. Benedict's solution will need to be heated in a water bath prior to use by students.

Disposal Biuret, Benedict's, and IKI solutions are poisons and strong irritants. Have students empty test tubes in three large beakers labeled the same way as the students' test tubes. Test the solutions with litmus paper, and titrate to neutrality if necessary. Place the beakers in the sink, and run water gently into the beakers to overflowing for 10 minutes.

Answers to Analysis Smaller molecules, such as glucose, can pass through the filter. The starch and the gelatin should not pass through the filter. Some gelatin may dissolve and give a weak positive result for protein. Large molecules, such as proteins and complex carbohydrates, are not filtered out of the blood, but smaller molecules, such as glucose and amino acids, are removed. Glucose and other small molecules are reabsorbed by the kidney tubules.

QUICK FACT

✔ Tell students that only mammals and birds are able to produce urine more concentrated than the fluids that bathe the body tissues. Ask them what structure in the kidney allows this concentration to take place (the loop of Henle).

SECTION 49-3

RETEACHING ACTIVITY

Ask students to summarize the function of the urinary system in a paragraph or two, using their own words. Tell them not to look in their textbook or at their notes for information. Tell them to use as many science terms as possible in their summary. They may want to begin by making a list of the terms that relate to the urinary system and the formation of urine.

ANSWERS TO SECTION 49-3 REVIEW

1. Excretion is the process of removing metabolic wastes from the body.
2. The lungs excrete carbon dioxide and water vapor through respiration. The skin excretes water and excess minerals through perspiration.
3. The proximal convoluted tubule and the distal convoluted tubule are long, winding parts of the renal tubule that provide a large surface area for the exchange of materials between the tubule and the surrounding capillaries. The loop of Henle is the part of the renal tubule that maintains a high sodium chloride concentration, creating a concentration gradient that promotes the reabsorption of water from the collecting duct.
4. During filtration, the blood loses material, whereas during reabsorption, the blood regains materials.
5. Excretion is the process of removing metabolic wastes. The wastes that pass through the large intestine are the remnants of indigestible materials and not the products of a metabolic process.
6. In a healthy kidney, blood proteins never leave the blood to enter the tubules of the kidney. The urine should contain no proteins.

TABLE 49-4 *Waste Substances and Excretion Sites*

Substance excreted	Excreting organ(s)
Nitrogenous waste	kidneys, skin (a small amount in sweat)
Water	kidneys, skin, lungs
Salts	kidneys, skin (in sweat)
Carbon dioxide	lungs
Spices	lungs, kidneys

internet**connect**

SCILINKS NSTA
TOPIC: Excretory system
GO TO: www.scilinks.org
KEYWORD: HM996

The Excretory System

The kidneys are the primary organs of the excretory system. They play a vital role in maintaining the homeostasis of body fluids. While kidney failure is an immediate, life-threatening situation, there are other organs that are involved in the excretion of metabolic waste.

As you know, the lungs are the primary site of carbon dioxide excretion. The lungs carry out detoxification, altering harmful substances so that they are not poisonous. The lungs are also responsible for the excretion of the volatile substances in onions, garlic, and other strong spices. These strong spices are frequently detectable on a person's breath.

While the kidneys control the salt composition of the blood, some salt is excreted through perspiration. A person working in extreme heat may lose water through perspiration at the rate of 1 L per hour. This loss of water represents a loss of about 10 to 30 g of salt per day.

Table 49-4 summarizes some waste substances and the organ(s) that excrete them. Notice that undigested food is not in the table of substances excreted by the body. Undigested food is not excreted in the scientific sense; it is eliminated, meaning it is expelled as feces from the body without ever passing through a membrane or being subjected to metabolic processes. The term *excretion* is correctly used only to indicate those substances that must pass through a membrane to leave the body.

SECTION 49-3 REVIEW

1. **Define and describe *excretion*.**
2. **Explain the function of an organ other than the kidney that is involved in excretion.**
3. **Name and describe each structure of the renal tubule.**
4. **Contrast the function of filtration with the function of reabsorption.**
5. **Given the definition of excretion, why do you think the large intestine is not classified as a major excretory organ?**
6. **CRITICAL THINKING Explain why a high concentration of protein in the urine may indicate damaged kidneys.**

CHAPTER 49 REVIEW

SUMMARY/VOCABULARY

49-1

- The human body needs six nutrients—carbohydrates, proteins, lipids, vitamins, minerals, and water—to grow and function.
- Carbohydrates provide most of the body's energy. Monosaccharides are quickly processed by the body. Cellulose cannot be digested but is needed for fiber.
- Proteins help the body grow and repair tissues. Amino acids must be obtained from foods.
- Lipids are used to build cell membranes.
- Vitamins act as coenzymes. The body can synthesize vitamin D.
- Minerals are inorganic substances that are needed in small amounts.
- Water helps regulate body temperature and transports nutrients and wastes.

Vocabulary

dehydration (982)	mineral (981)	nutrient (977)	vitamin (980)

49-2

- Mechanical digestion involves the breaking of food into smaller particles. Chemical digestion involves changing the chemical nature of the food substance.
- The mouth, teeth, and tongue carry out mechanical digestion. Chemical digestion of carbohydrates occurs in the mouth.
- The esophagus is a passageway through which food passes by peristalsis.
- The stomach has layers of muscles that churn the food to assist in mechanical digestion. Pepsin in the stomach begins the chemical digestion of proteins.
- Bile assists in the mechanical digestion of lipids. Enzymes secreted by the pancreas complete the digestion of the chyme.
- The digested nutrients are absorbed through the villi of the small intestine.
- The large intestine absorbs water from the undigested mass. The undigested mass is eliminated as feces through the anus.

Vocabulary

absorption (988)	feces (989)	jejunum (988)	peristalsis (985)
bolus (984)	gallbladder (987)	lacteal (988)	pyloric sphincter (986)
cardiac sphincter (986)	gastric fluid (986)	large intestine (989)	saliva (984)
chyme (986)	gastric pit (985)	liver (986)	soft palate (984)
colon (989)	gastrointestinal tract (983)	microvillus (988)	stomach (985)
digestion (983)	hard palate (984)	molar (984)	ulcer (986)
epiglottis (984)	incisor (984)	pepsin (986)	villus (988)

49-3

- The kidneys are the main organs of the urinary system. Reabsorption and secretion occur in the proximal and distal convoluted tubules and the loop of Henle. Nutrients are removed from the nephron and returned to the blood. The filtrate is called urine.
- The urine passes through a ureter and is stored in the urinary bladder until it is eliminated through the urethra.

Vocabulary

ammonia (992)	glomerulus (993)	renal pelvis (992)	ureter (995)
Bowman's capsule (993)	loop of Henle (995)	renal tubule (993)	urethra (995)
cortex (991)	medulla (992)	secretion (994)	urinary bladder (995)
excretion (991)	nephron (992)	urea (992)	urine (992)
filtration (993)	reabsorption (994)		

REVIEW

1. Carbohydrates, proteins, and fats all contain carbon, hydrogen, and oxygen and therefore are organic nutrients. A mineral is an inorganic nutrient, so it does not belong in a group of organic nutrients.
2. The pharynx, epiglottis, and esophagus are all structures of the upper digestive tract. A bolus is the ball of material that passes through the gastrointestinal tract, so it does not belong with terms naming the structures found along the tract.
3. The cardiac sphincter helps prevent food from leaving the stomach and reentering the esophagus. Gastric pits secrete digestive fluids into the stomach. The pyloric sphincter regulates the flow of material from the stomach into the small intestine. The medulla names a part of the kidney, so it does not belong with terms associated with the stomach.
4. It is important to distinguish between absorption of nutrients in the digestive system and reabsorption that occurs in the kidneys. Absorption is a function that occurs in digestive organs, so it does not belong with functions performed by the kidneys.
5. The ileum is a structure of the small intestine, so it does not belong in a listing of structures of the kidney.

6. c **7.** d
8. c **9.** b
10. b **11.** a
12. c **13.** b
14. d **15.** c

16. Consuming saturated fats contributes to an increase in the level of cholesterol in the body, leading to an increased risk of heart disease.
17. Cholesterol is converted to vitamin D by intestinal enzymes and sunlight.

18. Bile functions in mechanical digestion by breaking down fat globules into smaller droplets.
19. Both the pancreas and the liver play major roles in other organ systems. They do not lie in the gastrointestinal tract but rather release their secretions into it.
20. The muscles of the stomach twist and churn the food, breaking it apart further and mixing it with gastric fluid.
21. Ammonia is so toxic that it would destroy cells if left in the body. The liver converts ammonia to urea, a less-harmful substance.
22. The water-soluble vitamins are vitamin C and the B vitamins. Amounts in excess of those needed by the body are excreted by the kidneys. Fat-soluble vitamins (A, D, E, and K) are stored in fat in excess, and may reach toxic levels.
23. Filtration, reabsorption, and secretion are performed in the nephron and maintain homeostasis.
24. Bowman's capsule, the glomerulus, the proximal convoluted tubule, the loop of Henle, and the distal convoluted tubule are the major structures of a nephron.
25. A—cortex, B—medulla, C—renal pelvis, D—ureter, E—loop of Henle, F—glomerulus, G—nephron, H—Bowman's capsule

CRITICAL THINKING

1. Few plant foods provide all the essential amino acids. Thus, kwashiorkor is a protein-deficiency disease.
2. The problem lies in the small intestine, where lactase is secreted to hydrolyze lactose.
3. If the large intestine did not absorb water, the person would have severe diarrhea and could die from dehydration.
4. The machine must be able to take the unfiltered blood away from the body, filter the blood, regulate fluid balance and concentration, and then return the

REVIEW

Vocabulary

For each set of terms below, choose the one that does not belong, and explain why it does not belong.

1. carbohydrate, protein, fat, mineral
2. pharynx, epiglottis, bolus, esophagus
3. cardiac sphincter, gastric pits, medulla, pyloric sphincter
4. absorption, filtration, secretion, reabsorption
5. nephron, ileum, glomerulus, renal tubule

Multiple Choice

6. The primary function of carbohydrates is to (a) break down molecules (b) aid in digestion (c) supply the body with energy (d) regulate the flow of chyme.
7. Cellulose (a) builds body tissue (b) is a monosaccharide (c) is used for energy (d) aids in digestion.
8. Proteins consist of (a) catalysts (b) enzymes (c) amino acids (d) polysaccharides.
9. The body needs vitamins because they (a) supply it with energy (b) serve as coenzymes (c) function as enzymes (d) act as hormones.
10. Dehydration is best prevented by (a) inhaling air (b) drinking water (c) not drinking water (d) perspiring.
11. The epiglottis is important because it (a) prevents food from going down the trachea (b) separates the pharynx from the nasal cavity (c) is the passage through which food travels to the stomach (d) regulates the flow of chyme.
12. The gallbladder (a) creates bile (b) stores urine (c) stores, concentrates, and secretes bile (d) is made up of nephrons.
13. During absorption, the lacteals absorb (a) glycogen (b) glycerol and fatty acids (c) amino acids and monosaccharides (d) lactose.
14. Organs involved in the excretion process include the (a) kidneys and stomach (b) liver and pancreas (c) nephron and glomerulus (d) kidneys and lungs.
15. During secretion in the kidney, substances move from (a) filtrate to blood (b) blood to blood (c) blood to filtrate (d) filtrate to filtrate.

Short Answer

16. Why can consuming a great deal of saturated fat be harmful?
17. Briefly explain the synthesis of vitamin D in the body.
18. What role does bile play in digestion?
19. In what two ways do the liver and pancreas differ from other digestive organs?
20. How is mechanical digestion carried out by the stomach?
21. Why does the body convert ammonia into urea?
22. Explain the difference between water-soluble and fat-soluble vitamins.
23. What processes occur in the nephron that maintain homeostasis?
24. List the structures of a nephron.
25. Examine the diagram of the kidney. Name each of the indicated structures.

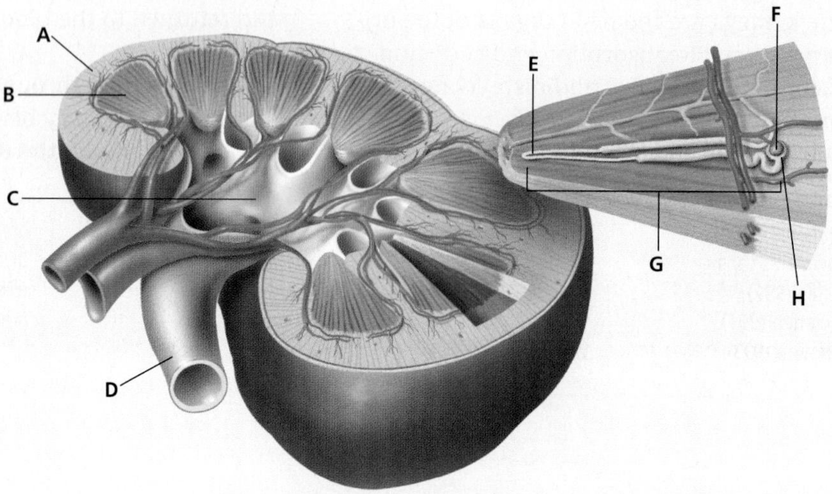

CRITICAL THINKING

1. In some countries many children suffer from a type of malnutrition called kwashiorkor. They have swollen stomachs and become increasingly thin until they die. Even when given rice and water, these children still die. What type of nutritional deficiency might these children have?
2. Some people cannot drink milk because they are unable to digest lactose, the sugar in milk. Doctors think this inability involves specific areas of the digestive system. What do you think these areas are?
3. Why is it important that the large intestine reabsorb water and not eliminate it?
4. When the kidneys stop functioning, urea builds up in the blood and poisons the body. A person with kidney failure will eventually die if the urea is not somehow removed. For the urea to be removed, the patient must be attached to an artificial kidney, also called a dialysis machine. Using your understanding of how a normal kidney functions, suggest a design for the major components of a dialysis machine.
5. A person has a small intestine that has villi but lacks microvilli. Would you expect this person to be underweight or overweight? Explain your answer.
6. The loop of Henle functions to conserve water by reabsorbing it. Its length varies among mammal species. Would you expect the loop of Henle of an animal like the beaver, which lives in a watery environment, to be longer or shorter than that found in humans? Explain your answer.
7. Look at the pictures of the teeth of different animals. What can you tell about the human diet by comparing the teeth of humans with those of the other animals shown here?

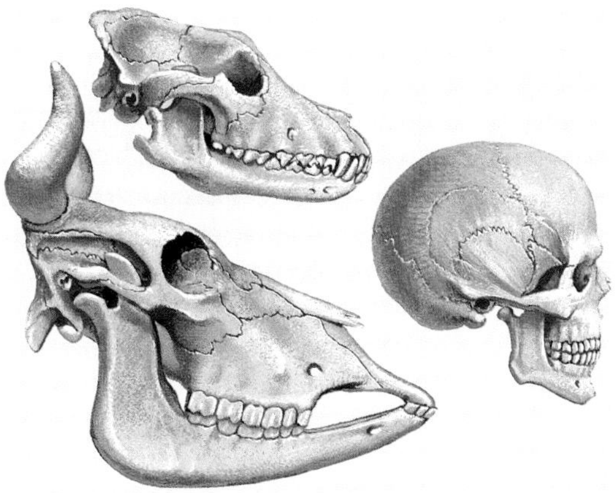

EXTENSION

1. Bring the label from a cereal box to class. Compare the "percent daily value" of vitamins and minerals on your label with those on labels brought in by other students.
2. Use the library and on-line sources for information about diets that claim to reduce the chances of a heart attack and cancer. Write a brief report summarizing a few of these articles.
3. Read "Fire in the Belly" in *Time*, April 26, 1999, on page 108. What is heartburn? What does heartburn have to do with the heart? What are the most common foods that cause heartburn?
4. Use library references to research kidney diseases and disorders. Write and illustrate a report focusing on the causes of these illnesses, their symptoms, and their treatment.
5. Look at your library's reference books that describe the cultures and customs of other nations. Find information relating to eating habits for at least five different countries. How do these customs differ from those in America? Investigate at least one nation or area where food is scarce. What are some of the causes and consequences of widespread food shortages?

ANSWERS

CONTINUED

filtered blood back to the body.

5. The person would be underweight. The reduced surface area of the intestinal lining would result in the inefficient absorption of food molecules.

6. The loop of Henle is shorter in beavers than in humans. Because beavers live in a watery environment, their need to conserve water is not as critical as it is in humans.

7. The sharp teeth in one animal is an adaptation for biting and tearing. The flat teeth in the other animal is an adaptation for chewing. Human teeth are adapted for both functions.

EXTENSION

1. Encourage students to bring in a variety of labels and place the data on the board.

2. Encourage students to locate information about high-fiber diets and low-fat diets.

3. Heartburn is a condition caused by the relaxation of the ring of muscle surrounding the entrance to the stomach. This relaxation allows acid to bubble up into the esophagus. Heartburn has nothing to do with the heart. It is most commonly caused by chocolate, mints, caffeine, and alcohol.

4. Students may include in their reports pyelonephritis, the most common kidney disease; glomerulonephritis, an inflammation caused by antibodies; kidney stones, caused by calcium deposits; and uremia, caused by wastes in the blood. They may also include information about hemodialysis and kidney transplants.

5. Students should include foods that are not typical components of diets in the United States, such as cassava, uncommon tropical fruits and berries, meats such as iguana and boar, and unusual seafood. Food scarcity can have several causes, such as political upheaval, overpopulation, and poverty.

CHAPTER 49 INVESTIGATION

TIME REQUIRED

Two 50-minute class periods

SAFETY PRECAUTIONS

Remind students to wear safety goggles, a lab apron, and protective gloves while handling hydrochloric acid solution.

QUICK REFERENCES

Lab Preparation Notes are found on page 975B.

Holt BioSources provides a Teaching Resources CD-ROM that contains "Using Gowin's Vee in the Lab" and "Scoring Rubrics."

PROCEDURAL TIPS

1. A 0.2 percent solution of hydrochloric acid can be made by diluting 2 mL of concentrated HCl to a final volume of 1 L with distilled water.
2. Use distilled water to dilute 10 g of sodium bicarbonate to a final volume of 1 L and 1 g of pepsin to a final volume of 100 mL. Commercial solutions are also available from biological supply houses.
3. Each test tube will hold 10 mL of fluid. Calculate the volume you will need for your class size.
4. Use 10 mL graduated cylinders if available. Decide on the best way to have students rinse their graduated cylinders when changing solutions.
5. If you want students to make their observations in 24 hours rather than 48 hours, provide a 37°C incubator for their test tubes.

ANSWERS TO BACKGROUND

1. Solid chunks of food are ground by the teeth, digested by enzymes in the mouth, stomach, and small intestine, and dissolved.
2. Pepsin digests protein into peptides.

Modeling Human Digestion

OBJECTIVES

- Test a model of digestion in the human stomach.

PROCESS SKILLS

- modeling
- observing
- predicting
- inferring

MATERIALS

- safety goggles
- lab apron
- glass-marking pencil
- 5 test tubes with stoppers
- test-tube rack
- scalpel
- cooked egg white
- balance
- 10 mL graduated cylinder
- 1% pepsin solution
- 0.2% hydrochloric acid
- 1% sodium bicarbonate
- distilled water
- red and blue litmus paper
- lined paper
- disposable gloves

Background

1. How is food changed from the chunks you chew with your teeth to the chyme absorbed in your small intestine?
2. What type of organic compound does the enzyme pepsin digest?

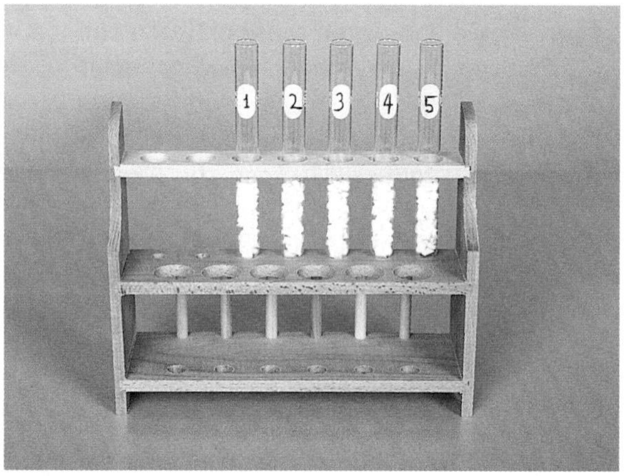

PART A Setting Up

1. Label five test tubes 1, 2, 3, 4, and 5, and place them in a test-tube rack.
2. CAUTION **Always cut in a direction away from your body.** Use a scalpel to cut a firm, cooked egg white into fine pieces.
3. Using the balance, measure and place equal amounts (about 6 g) of the fine egg white sample into each test tube, as shown in the illustration above.
4. CAUTION **Put on safety goggles and a lab apron. If you get hydrochloric acid solution on your skin or clothes, wash it off at the sink while calling to your teacher. If you get any solutions in this investigation in your eyes, immediately flush them out at the eyewash station while calling to your teacher.** Use a clean graduated cylinder to add the solutions listed below to the test tubes. Rinse the cylinder between additions so that you do not contaminate the samples.
   - test tube 1—10 mL of water
   - test tube 2—10 mL of pepsin solution
   - test tube 3—10 mL of hydrochloric acid
   - test tube 4—5 mL of pepsin solution and 5 mL of sodium bicarbonate solution
   - test tube 5—5 mL of pepsin and 5 mL of hydrochloric acid

DEGREE OF DIGESTION OF EGG WHITE UNDER VARYING CONDITIONS

Test tube number	Contents	pH	Degree of digestion
1	egg white 10 mL water		
2	egg white 10 mL 1% pepsin solution		
3	egg white 10 mL 0.2% hydrochloric acid solution		
4	egg white 5 mL 1% pepsin solution 5 mL 1% sodium bicarbonate solution		
5	egg white 5 mL 1% pepsin solution 5 mL 0.2% hydrochloric acid solution		

5. Stopper and gently shake each test tube.
6. In your lab report, make a data table like the one shown above. Fill in the first column with the test-tube numbers. Under the heading "Contents," fill in a description of the contents of each test tube, as shown.
7. Predict which test tube will show the most digestion after 48 hours. Explain your reasoning.
8. Label your test-tube rack with your initials. Store the test-tube rack for 48 hours at room temperature. Leave a note on the rack cautioning others not to spill the acids or bases.
9. Clean up your lab materials and wash your hands before leaving the lab.

PART B Recording the Results

10. After 48 hours, measure the pH of each solution with red and blue litmus paper. Record your results in the data table you created in your lab report.
11. Look for the egg white in each test tube. In your data table, describe the degree to which the egg white has broken down and dissolved in each test tube.
12. Clean up your lab materials and wash your hands before leaving the lab.

Analysis and Conclusions

1. What conditions caused the greatest digestion of cooked egg white?
2. Which test tube best modeled the chemical composition in the human stomach?
3. What information do test tubes 1, 2, and 3 give you? What do they control?
4. Compare test tubes 4 and 5. What can you conclude about the effects of the chemical environment on the activity of pepsin?
5. List some other foods that pepsin is likely to digest.
6. Do you think that pepsin would digest butter? Explain your answer.

Further Inquiry

Design an experiment to test the digestion of a food containing carbohydrates, such as a potato or an apple.

ANSWERS TO ANALYSIS AND CONCLUSIONS

1. The enzyme pepsin plus acid caused the greatest digestion of cooked egg white.
2. Based on the pupil's text, students should infer that test-tube 5, containing pepsin and hydrochloric acid, is the closest model.
3. Neither water, pepsin, nor acid alone would be expected to digest much protein. These test tubes control for the digestive action of each chemical or solution alone.
4. Pepsin requires an acidic environment for optimum digestive action.
5. Accept all answers that are lists of protein foods. Pepsin digests only protein.
6. No; however, accept all speculative answers, because students have no experimental evidence that enzymes are specific. Pancreatic lipase, not pepsin, will digest butter.

FURTHER INQUIRY

Design an experiment that is similar to this lab using chopped potato or apple plus saliva or pancreatic extract (as a source for amylases).

CHAPTER 50
Nervous System and Sense Organs

	TOPICS	TEACHING RESOURCES	LABS, CLASSWORK, AND HOMEWORK
BLOCKS 1 & 2 90 minutes	**Introducing the Chapter, p. 1002**	ATE Focus Concept, p. 1002 ATE Assessing Prior Knowledge, p. 1002 ATE Understanding the Visual, p. 1002	
	50-1 Central Nervous System, p. 1003 Organization, p. 1003 Brain, p. 1004	ATE Section Overview, p. 1003 ATE Visual Link Figure 50-3, p. 1005 197A, 199	★ Study Guide, Section 50-1 PE Section Review, p. 1008
BLOCK 3 45 minutes	**50-2 Peripheral Nervous System, p. 1009** Sensory Division, p. 1009 Motor Division, p. 1009	ATE Section Overview, p. 1009 ATE Visual Link Figure 50-6, p. 1010 196–197	**Lab A28,** Bias and Experimentation ★ Study Guide, Section 50-2 PE Section Review, p. 1011
BLOCK 4 45 minutes	**50-3 Transmission of Nerve Impulses, p. 1012** Neuron Structure, p. 1012 Neuron Function, p. 1013	ATE Section Overview, p. 1009 ATE Visual Link Figure 50-9, p. 1014 Figure 50-10, p. 1016 194–195	★ Study Guide, Section 50-3 PE Section Review, p. 1016
BLOCKS 5 & 6 90 minutes	**50-4 Sensory Systems, p. 1017** Receptors and Sense Organs, p. 1017 Hearing and Balance, p. 1018 Vision, p. 1020 Taste and Smell, p. 1021 Other Senses, p. 1022	ATE Section Overview, p. 1017 ATE Critical Thinking, p. 1021 ATE Visual Link Figure 50-11, p. 1018 Figure 50-13, p. 1020 200–203	PE **Chapter 50 Investigation,** p. 1026 PE **Quick Lab,** p. 1020, Observing a Lens **Lab B30,** Touch Receptors in the Skin **Lab B31,** Exploring Vision ★ Study Guide, Section 50-4 PE Section Review, p. 1022
BLOCKS 7 & 8 90 minutes	**REVIEW and ASSESSMENT** PE Chapter 50 Review, pp. 1024–1025 ■ Performance-Based Assessment—Testing Human Digestive Enzymes	Audio CD Program	**CHAPTER TESTING** ■ Chapter Test (blackline master) ▲ Test Generator ■ Assessment Item Listing ■ Scoring Rubrics and Classroom Management Checklists

KEY

MODERN BIOLOGY

- PE Pupil's Edition
- ATE Teacher's Edition
- ■ Active Reading Guide
- ★ Study Guide

HOLT BIOSOURCES

- Teaching Transparencies
- Laboratory Program

One-Stop Planner CD-ROM
Shows these resources within a customizable daily lesson plan:

- ■ Holt BioSources Teaching Resources
- ■ Active Reading Guide
- ★ Study Guide
- ▲ Test Generator

READING FOR CONTENT MASTERY

Preview: Have students preview the objectives, the introductory paragraph, topic sentences, and captions before reading the chapter.

- **ATE** Active Reading Technique: Anticipation Guide, p. 1003
- ■ Active Reading Guide Worksheet 50-1
- **PE** Summary/Vocabulary, p. 1023

- ■ Active Reading Guide Worksheet 50-2
- **PE** Summary/Vocabulary, p. 1023

- ■ Active Reading Guide Worksheet 50-3
- **PE** Summary/Vocabulary, p. 1023

- ■ Active Reading Guide Worksheet 50-4
- **PE** Summary/Vocabulary, p. 1023

TECHNOLOGY AND INTERNET RESOURCES

CNN PRESENTS Video Segment 34
Stroke Brain Repair

Holt Biology Videodiscs Teacher's Correlation Guide, Lessons 50-1 through 50-4

Audio CD Program, Sections 50-1 through 50-4

internetconnect

sciLINKS NSTA

On-line Resources: www.scilinks.org
The following *sci*LINKS Internet resources can be found in the student text for this chapter:

Topics:
- Central nervous system, p. 1005
- Peripheral nervous system, p. 1010
- The senses, p. 1021

go.hrw.com

On-line Resources: go.hrw.com
Visit the HRW Web site for a variety of resources related to this chapter. Just type in the keyword HM2 Chapter 50.

Smithsonian Institution®
Visit **www.si.edu/hrw** for additional on-line resources.

CNNfyi.com
Visit **www.cnnfyi.com** for late-breaking news and current event stories selected just for you.

LAB ACTIVITY PLANNING

CHAPTER 50 INVESTIGATION

Sheep's Eye Dissection, pp. 1026–1027

OVERVIEW
This lab allows students to examine in detail the optic structures (the iris and the lens) and the neural component (the retina) of the eye. Students will also identify structural features of the eye. Students should compare the sheep's eye with the human eye shown in Figure 50-13 on p. 1020.

SAFETY
Students will wear disposable gloves, lab aprons, and safety goggles during the lab. Formaldehyde, a preservative, is an irritant. If it gets on students' skin, have them wash the area with soap and water. If it gets in students' eyes, flush the eyes with water for 15 minutes and get medical attention. Do not let students clean up spills. Wrap the dissected specimens in old newspapers and secure and seal them in a large plastic bag. Inform the school custodian of the bag's contents, and hand the bag over personally for its safe disposal.

PREPARATION
1. Obtain preserved sheeps' eyes from a biological supply house, such as WARD'S.
2. Students will need a dissecting pan, a scalpel, fine scissors, a sheep's eye, tweezers, and a blunt probe.

Quick Labs

Observing a Lens, p. 1020
Students will need a beaker, water, newspaper print, and four drops of cooking oil for this activity.

UNDERSTANDING THE VISUAL

Magnetic resonance imaging (MRI) provides a detailed image of the body's internal structures, and is particularly useful for imaging the brain. For MRI, a subject is positioned within a strong magnetic field and then a radio wave is passed through the body. In response to this, some atomic nuclei in the body emit signals of their own as their protons absorb, then release, the energy from the radio wave. The MRI receiver is tuned to detect the signal emitted by hydrogen atoms, and it can quantify the proportion of hydrogen (and thus water) in tissues by assessing the strength of the radio signal. (Different types of tissue have different water content. Tissue with high water content, such as the wrinkled cortex of the brain—seen in the image to the right—produces stronger signals than tissue with low water content, such as dense bone.) A computer integrates data from different positions of the magnetic field relative to the body, and produces a clear image of internal features.

FOCUS CONCEPT

Stability and Homeostasis

The nervous system integrates the functions of all body systems. It enables behavior, homeostasis, learning, and memory, and it allows the body to respond to environmental stimuli.

ASSESSING PRIOR KNOWLEDGE

Review the following concepts.

Homeostasis and Transport: *Chapter 5*

Ask students to recall that the selective permeability of cells depends on transport mechanisms and cell membrane structure and function. Ask students why selective permeability varies with different cell types throughout the body.

CHAPTER 50

NERVOUS SYSTEM AND SENSE ORGANS

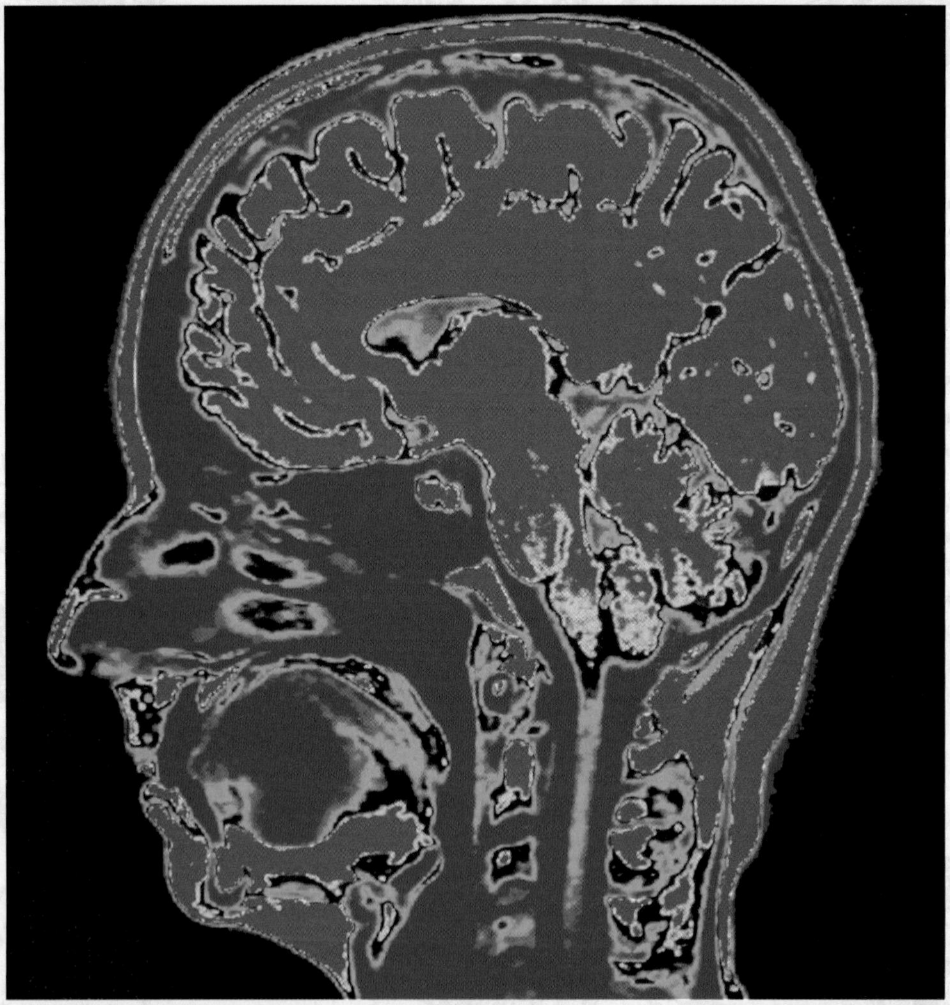

This color-enhanced magnetic-resonance imaging (MRI) of the head shows a living brain and spinal cord in cross section.

FOCUS CONCEPT: ***Stability and Homeostasis***

As you read, notice how the nervous system functions to ensure that other body systems work together in an efficient and coordinated manner.

CENTRAL NERVOUS SYSTEM

Mental and physical activity and many aspects of homeostasis are controlled by the ***nervous system,*** *a complex network of cells that communicate with one another. Within this communications network, a carefully organized division of labor exists so that each component of the nervous system operates effectively. As a result, a football player can weave through opposing tacklers, an architect can create an original design, and you can read and understand the words on this page.*

SECTION 50-1

OBJECTIVES

▲ Identify the two main organs of the central nervous system.

● Summarize the functions of the cerebrum, brain stem, and cerebellum.

■ Describe how the central nervous system is protected from injury.

◆ Describe the structure of the spinal cord.

▲ Distinguish between sensory receptors, motor neurons, and interneurons.

ORGANIZATION

The nervous system is composed of neurons, specialized cells that transmit information throughout the body. A neuron, such as the one shown in Figure 50-1, demonstrates the relationship between structure and function in living systems. The **axon,** a threadlike projection of the cell body, enables the neuron to transmit signals rapidly over relatively long distances in the body. Other structures projecting from the cell body allow the neuron to receive signals from other neurons. You will examine the structure and function of neurons more closely later in this chapter.

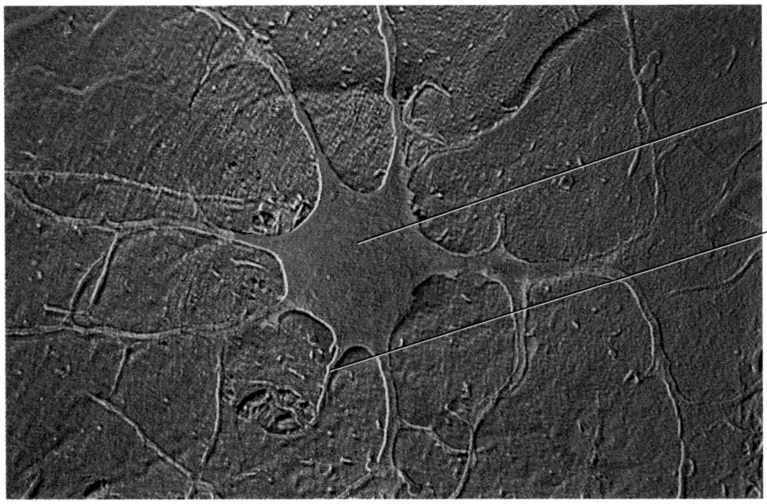

FIGURE 50-1

Neurons transmit information throughout the body in a combination of chemical and electrical signals. This stellate neuron is named for its starlike shape (SEM, 26,000×).

SECTION 50-1

SECTION OVERVIEW

Cell Structure and Function
Section 50-1 describes the general organization of the nervous system, which consists of billions of neurons that transmit information throughout the body. The nervous system is divided into the central nervous system (CNS) and the peripheral nervous system (PNS). This section lists the major structures of the CNS and their basic functions.

Stability and Homeostasis
Section 50-1 also emphasizes that the brain is responsible for maintaining homeostasis by controlling and integrating the body's systems.

ACTIVE READING
Technique: Anticipation Guide

Before students begin this chapter, write the following statements on the board or overhead:

- The human brain is responsible for processing thoughts, feelings, memories, and emotions.
- Electricity is the force that stimulates neurons, which stimulate muscle cells.
- Seeing, hearing, tasting, smelling, and touching are all controlled by the nervous system.

Give students time to form opinions about each statement. Ask volunteers who have different views about each statement to share the reasons for their opinions with the class. Then have students read the chapter to see if their opinions are confirmed or changed in any way.

MAKING CONNECTIONS

Technology

Ask students to compare the nervous system to a computer. (The brain is like the computer memory bank, central processing unit, and software; nerves are like wire circuits; and sensory receptors are like the keys of the keyboard. In both the nervous system and the computer, information travels in the form of electrical impulses. Electrical signals are analogous to nerve impulses.) Then ask them what primary functions of the brain cannot be duplicated by a computer. (Computers are incapable of thought, reason, emotion, and intuition.)

RECENT RESEARCH

Memory Formation

Many studies have attempted to shed light on how memories are formed. The predominant structure involved in the formation of memories is the hippocampus, located in the medial temporal lobe of the brain. Studies have shown that some people with damage to the hippocampus suffer from the inability to remember events that occur after the brain is damaged. This indicates that new episodic (event) memories are formed in the hippocampus. However, damage to the hippocampus generally does not affect memory of events that occurred prior to trauma. This indicates that the sites of long-term memory storage are located elsewhere in the brain. Many researchers claim that memories are formed by synaptic plasticity (modification) in the hippocampus and other brain structures. Synaptic plasticity enables long-term potentiation (strengthening) and long-term depression (weakening) of specific synapses, depending on their level of activity. Most researchers agree that synaptic plasticity enables the brain to "hard wire" distinct neural pathways that store memories.

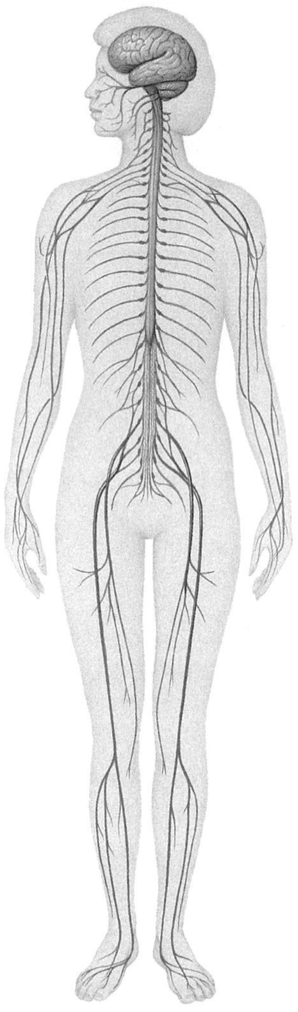

FIGURE 50-2

The central nervous system includes the brain and spinal cord, shown in orange. The peripheral nervous system, shown in violet, includes all other nervous tissue in the body.

Word Roots and Origins

corpus callosum

from the Latin *corpus*, meaning "body," and *callosus*, meaning "thick and hard"

The nervous system includes two major divisions: the central nervous system and the peripheral nervous system. The brain and spinal cord make up the **central nervous system,** shown in orange in Figure 50-2. The spinal cord carries messages from the body to the brain, where they are analyzed and interpreted. Response messages are then passed from the brain through the spinal cord and to the rest of the body.

The **peripheral nervous system,** shown in violet in Figure 50-2, consists of the neurons that are not included in the brain and spinal cord. Some peripheral neurons collect information from the body and transmit it *toward* the central nervous system. These are called **afferent neurons.** Other peripheral neurons transmit information *away* from the central nervous system. These are called **efferent neurons.**

BRAIN

The human brain is responsible for overseeing the daily operations of the human body and for interpreting the vast amount of information it receives. The adult human brain weighs an average of 1.4 kg, or about 2 percent of total body weight. Despite its relatively small mass, the brain contains approximately 100 billion neurons. Functioning as a unit, these neurons make up the most complex and highly organized structure on Earth. Because of the brain's complex nature, much remains to be learned about how it functions, but scientists have uncovered a good deal of information about the brain's many functions.

You may realize that the brain is responsible for many of the qualities that make each individual unique—thoughts, feelings, emotions, talents, memories, and the ability to process information. It is important to understand, however, that much of the brain is dedicated to running the body. For example, the brain is responsible for maintaining homeostasis by controlling and integrating the various systems that make up the body. Through painstaking trial-and-error study of the brain, scientists have established how and where various functions are localized in the brain.

Cerebrum

The largest portion of the human brain consists of the cerebrum (SER-uh-bruhm). The **cerebrum,** which is easily identified by its highly folded outer layer, is composed of two **cerebral hemispheres,** as shown in Figure 50-3a. The cerebral hemispheres are connected by the **corpus callosum** (KOR-puhs kuh-LOH-suhm), a heavy band of the axons of many neurons. The corpus callosum lies deep in the central groove that separates the right hemisphere from the left hemisphere. Other deep grooves separate each hemisphere into four lobes: the frontal, parietal (puh-RIE-uh-tuhl), temporal, and occipital (ahk-SIP-i-tuhl) lobes, as shown in Figure 50-3b.

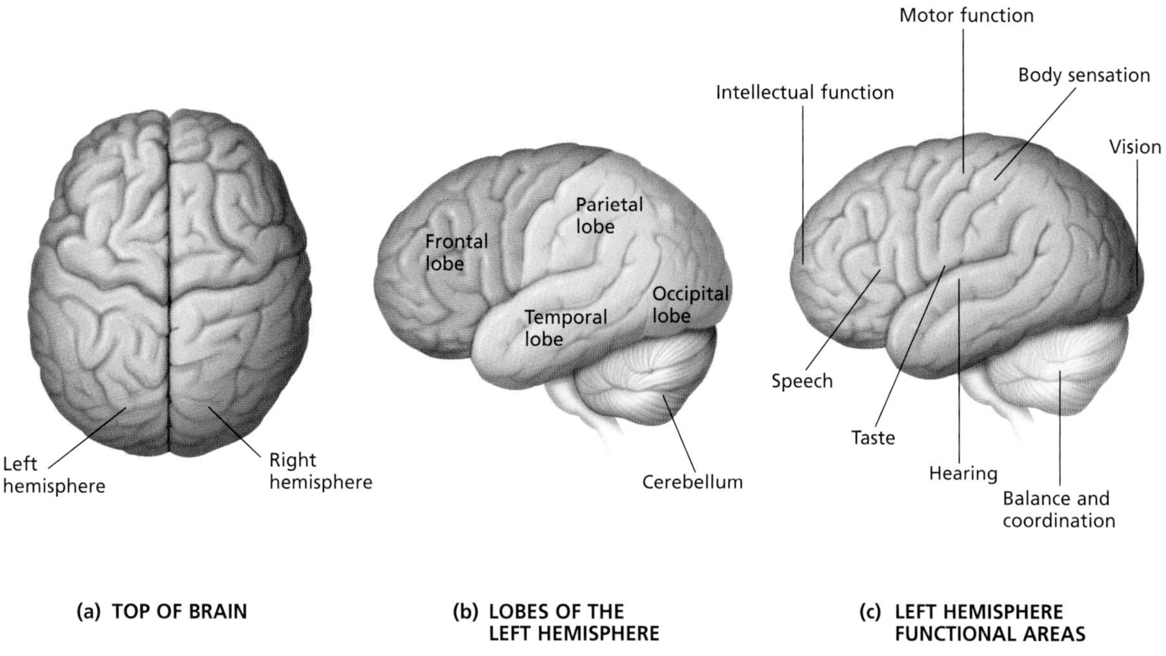

FIGURE 50-3

(a) A view of the top of the brain shows the left and right cerebral hemispheres. (b) Each cerebral hemisphere of the brain has four lobes. (c) Control centers for various functions are located in different areas of the brain.

The folded outer layer of the cerebral hemispheres is called the **cerebral cortex.** In humans, the cerebral cortex is a sheet of neurons about 2 to 4 mm (0.75 to 1.5 in.) thick. Scientists estimate that it contains nearly 10 percent, or about 10 billion, of the brain's total number of neurons. The surface area of the cerebral cortex is too large to lie smoothly over the lower structure of the brain. The folds, or convolutions, accommodate the surface area of the cortex within the limits of the interior of the skull.

The cerebral cortex is very important in sensory processing and in motor responses. As shown in Figure 50-3c, separate areas of the cerebral cortex control auditory, visual, body sensation, and motor processing. For example, an area of the cortex in the occipital lobe, at the rear of the brain, is essential for processing visual information. The area of the cortex that interprets touch information from all parts of the body lies in a band across the top of the brain, in the parietal lobe. Some functions are not localized in a symmetrical way in the brain. In right-handed individuals, cortical areas and other brain centers involved in speech and language production are located primarily in the left hemisphere. Brain centers involved in processing spatial information and in certain kinds of reasoning are located primarily in the right hemisphere. In left-handers, there is some variation in the location of functions.

Below the wrinkled surface of the cerebral cortex lies the **white matter,** which is composed of the axons of cortical neurons. These axons link specific regions of the cortex with each other and with other neural centers. There is a great degree of crossover of axons in the spinal cord and brain. That is, many impulses originating in the right half of the body are processed in the left half of the brain, and vice versa.

internetconnect

SCILINKS NSTA

TOPIC: Central nervous system
GO TO: www.scilinks.org
KEYWORD: HM1005

VISUAL LINK
Figure 50-3

This figure shows the basic organization of the cerebrum into four lobes: frontal, parietal, temporal, and occipital. These lobes function in association and motor planning, spatial location, object recognition and hearing, and visual processing, respectively. Explain to students that body parts are variably represented in the cerebral cortex, based on the complexity of their function. For example, the digits are extremely sensitive to touch, and their fine, detailed movements require precise contraction of many different muscles. Thus, the digits are overrepresented in the cortex relative to other body parts because they require many more sensory and motor neurons to enable their complex function.

CULTURAL CONNECTION

Gulf War Syndrome

Recent neuropsychological studies have suggested that illnesses reported by some veterans of the Gulf War are due to brain and nerve damage rather than to stress-related complications. Researchers at the University of Texas Southwestern Medical Center compared results of neuropsychological tests of healthy veterans with those of ill veterans. Results indicate that exposure to toxic chemical nerve agents, anti-nerve-gas pills, and flea collars may have caused deficits in thought, attention, concentration, balance, coordination, learning, memory, and limb sensation.

RECENT RESEARCH

Parkinson's Disease

Parkinson's disease mostly affects people over age 50, causing the progressive degeneration of neurons in a structure of the brain stem called the substantia nigra. Neurons of the substantia nigra use the neurotransmitter dopamine to innervate neurons that extend to regions of the frontal lobe which are involved in the planning of movements. When substantia nigra neurons release dopamine, they stimulate other neurons that in turn signal neurons in the frontal lobe to initiate voluntary movements.

Symptoms of Parkinson's disease include immobility, slowness of movement, tremors, stiffness, and muscle rigidity. These symptoms can be treated with L-dopa, a precursor of dopamine synthesis. Although L-dopa stimulates substantia nigra neurons to produce more dopamine, the drug becomes less effective as these neurons die. Now other drugs that mimic the action of dopamine or inhibit its breakdown are used to treat symptoms. Recent studies have suggested that oxygen free radicals contribute to the deterioration of cell membranes, proteins, and mitochondrial DNA in dopamine-producing neurons. This evidence has led to new treatments. For example, antioxidant drugs can minimize production of nitric oxide, a free radical, by cells in the substantia nigra.

GIFTED *ACTIVITY*

Have students research the causes, effects, and treatments of Parkinson's disease. The article titled "Understanding Parkinson's Disease," in the January 1997 issue of *Scientific American* (pp. 52–59), provides good background information on recent studies of the disorder.

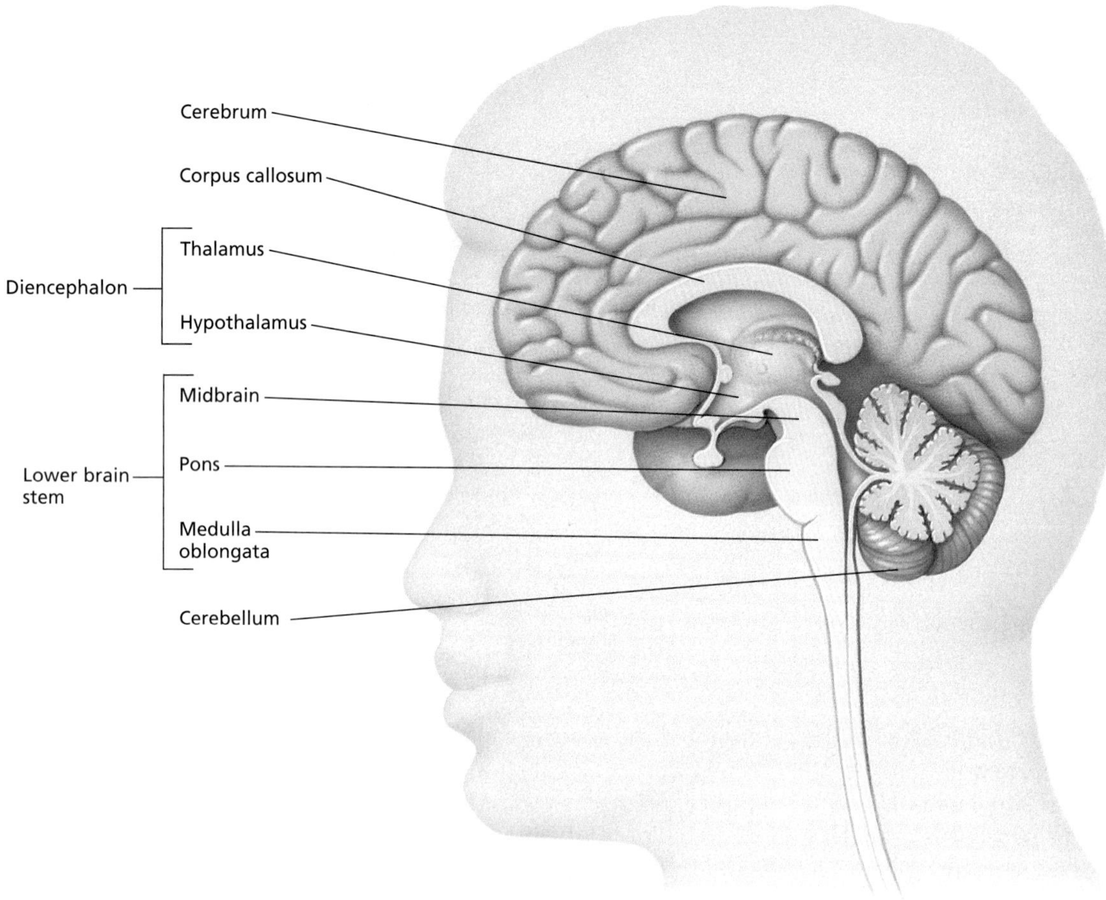

FIGURE 50-4

A side view through the center of the brain shows the right hemisphere. The wrinkled cerebral cortex is visible along the front, rear, and top of the brain, where it rolls into the deep groove separating the two hemispheres. The structures lying below the cerebrum are shown in cross section.

Upper Brain Stem—Diencephalon

Below the cerebrum, the **brain stem** links the cerebrum with the spinal cord. The upper part of the brain stem, the **diencephalon** (DIE-uhn-SEF-uh-lahn), contains important relay centers for information entering and exiting the brain. The upper relay center, the **thalamus** (THAL-uh-muhs), shown in Figure 50-4, directs most incoming sensory signals to the proper region of the cerebral cortex. The hypothalamus, located just below the thalamus, is linked to several other parts of the brain. The hypothalamus is very important in maintaining homeostasis, and it both directly and indirectly controls much of the body's hormone production.

Parts of the diencephalon and the cerebrum are included in an important group of connected brain centers called the **limbic system.** The limbic system includes the thalamus, the hypothalamus, some deeper parts of the cerebral cortex, and centers in the temporal lobes. The limbic system plays an important role in emotion, memory, and motivation, among other things. Scientists have tried to determine the functions of the different parts of the limbic system by observing problems associated with damage to its various regions. For example, people with damage to certain limbic areas may display inappropriate emotional responses, such as rage in response to a trivial, irritating stimulus.

Lower Brain Stem

Below the diencephalon, the brain stem narrows. This region of the brain stem has three main divisions: the midbrain, the pons, and the medulla oblongata.

The **midbrain** is an important relay center for visual and auditory information. The **pons** serves as a relay center between the neurons of the cerebral hemispheres and those of the cerebellum. The **medulla oblongata** (mi-DUHL-uh AHB-lahn-GAHT-uh) contains neurons that serve as both a relay center and a control center. Also within the medulla oblongata are centers that control various homeostatic activities, including heart rate and respiration rate.

Lying throughout the brain stem is a diffuse network of neurons called the **reticular formation.** The reticular formation helps to control respiration and circulation and serves as a filtering system for incoming sensory signals. The reticular formation separates signals that demand attention from those that are unimportant. Some of this function of the reticular formation is modified by learning. For example, a person can learn to sleep through the noise of a radio or television but awaken at the sound of a doorbell.

Cerebellum

Just below the occipital lobes of the cerebral hemispheres lies the **cerebellum** (SER-uh-BEL-uhm), a region of the brain that plays a vital role in the coordination of muscle action, particularly the timing of muscle contractions. As Figure 50-4 shows, the surface of the cerebellum is very highly folded, giving it a striated appearance. The cerebellum receives sensory impulses from muscles, tendons, joints, eyes, and ears, as well as input from other brain centers. It processes information about body position and controls posture by keeping skeletal muscles in a constant state of partial contraction. The cerebellum coordinates rapid and ongoing movements. It acts together with brain stem motor centers and with the section of cerebral cortex that is responsible for motor responses. Nerve impulses are sent down through the spinal cord and stimulate or inhibit the contraction of skeletal muscles.

Protection

The delicate neurons of the brain and spinal cord are surrounded by three protective layers called the **meninges** (muh-NIN-JEEZ). The outer layer, the **dura mater** (DU-ruh MAYT-uhr), consists of connective tissue, blood vessels, and neurons. The middle layer, the **arachnoid** (uh-RAK-NOYD) **layer,** is elastic and weblike. The thin inner layer, the **pia mater** (PIE-uh MAYT-uhr), adheres to the brain and spinal cord and contains many blood vessels and neurons.

A clear liquid called **cerebrospinal** (SER-uh-BROH-SPIE-NUHL) **fluid** provides a cushion that protects the brain and spinal cord from injury. Cerebrospinal fluid separates the middle and inner meninges and fills four interconnected **ventricles,** or cavities, in the brain. Within the ventricles, cerebrospinal fluid acts as a transport medium for substances that are important to brain function.

OVERCOMING MISCONCEPTIONS

The mental illness schizophrenia is often confused with multiple-personality disorder. In general, schizophrenia is characterized by irrational and illogical thoughts, moods, and behaviors; delusions and distortions of reality; and hallucinations. Other symptoms include indifference, emotional unresponsiveness, social withdrawal, and lack of motivation. Both the frontal cortex and the limbic system appear to be involved in the symptoms of schizophrenia.

Although the cause of schizophrenia is not known, some of its symptoms can be treated with drugs that interfere with the action of the neurotransmitter dopamine, which in excess, can produce psychotic symptoms. The effectiveness of dopamine-blocking drugs in some people with schizophrenia indicates that the causes of the disorder are likely biochemical, and genetic, as well as environmental.

GIFTED *ACTIVITY*

Have students research the use of gene therapy to treat degenerative nervous system disorders like Alzheimer's disease, Parkinson's disease, and amyotrophic lateral sclerosis (Lou Gehrig's disease), and to treat trauma caused by stroke. Students should find experimental evidence that viruses can be used to transport genetically engineered substances to damaged neurons, where they can stimulate production of proteins and other materials that can help prolong the life of the damaged cells. A good source is the article titled "Gene Therapy for the Nervous System," in *Scientific American,* June 1997 (pp. 116–120).

RETEACHING *ACTIVITY*

Have groups of four students make a diagram of the brain and a chart or diagram that illustrates which parts of the body are controlled by the cerebrum, the cerebellum, the diencephalon, and the brain stem. Students should choose one of these regions and describe its function. For example, the cerebellum coordinates and smooths ongoing movements and helps maintain posture and balance.

ANSWERS TO SECTION 50-1 REVIEW

1. The brain and spinal cord make up the central nervous system.
2. The cerebrum controls motor and sensory activities. The cerebellum coordinates movements and helps maintain balance and posture. The brain stem relays information to and from the cerebrum and helps regulate sensory and body systems.
3. Sensory (afferent) receptors carry signals from receptors to the spinal cord through dorsal roots. Motor (efferent) neurons carry signals from the spinal cord to muscles and glands through the ventral roots.
4. The skull protects the brain, and the bony vertebrae protect the spinal cord. Both organs are covered by the meninges and cushioned by cerebrospinal fluid.
5. Gray matter consists of cell bodies of neurons, whereas white matter consists of axons of neurons.
6. By observing which functions are impaired, a doctor can determine which area of the brain is affected. For example, inability to speak normally indicates damage to the speech area in the left cerebral hemisphere.

FIGURE 50-5

The spinal cord, shown in cross section, carries information toward and away from the brain. Sensory information from the body enters the spinal cord through the dorsal roots. Instructions to the body's many motor neurons exit the spinal cord through the ventral roots.

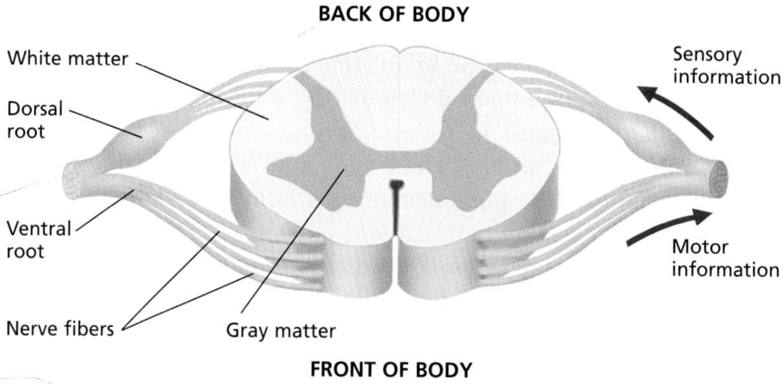

Spinal Cord

The **spinal cord** is a column of nerve tissue that starts in the medulla oblongata and runs down through the vertebral column. The spinal cord is composed of an outer sheath of white matter. As it does in the brain, the white matter contains primarily the axons of neurons. The white matter surrounds a rigid inner core of gray matter, which is composed of the cell bodies of neurons.

Thirty-one pairs of spinal nerves, part of the peripheral nervous system, originate in the spinal cord and branch out to both sides of the body. Recall that neurons have long axons that enable them to transmit signals rapidly. When bundled together, these axons form a **nerve.** Each spinal nerve consists of a dorsal root and a ventral root. The dorsal roots, shown in Figure 50-5, contain neurons that carry signals to the central nervous system from various kinds of sensory receptor neurons. A **sensory receptor** is a neuron that is specialized to detect a stimulus, such as pressure or heat.

The ventral roots, also shown in Figure 50-5, contain the axons of **motor neurons,** which are neurons that contact and carry information to muscles and glands. Within the spinal cord and elsewhere in the body are **interneurons,** which are neurons that connect other neurons to each other.

SECTION 50-1 REVIEW

1. **Name the two organs that make up the central nervous system.**
2. **How do the functions of the cerebral hemispheres, cerebellum, and brain stem differ?**
3. **Explain the difference between sensory receptors and motor neurons.**
4. **How are the brain and spinal cord protected from injury?**
5. **What are the gray matter and white matter in the spinal cord composed of?**
6. **CRITICAL THINKING Strokes result in the death of neurons in the brain. How can a doctor tell what areas of the brain have been affected in a person who has had a stroke?**

PERIPHERAL NERVOUS SYSTEM

The workings of your brain may be largely a mystery to you. You are undoubtedly more familiar with the workings of your peripheral nervous system. In the peripheral nervous system, 12 pairs of cranial nerves connect the brain with the head and neck and 31 pairs of spinal nerves connect the central nervous system with the rest of the body.

SENSORY DIVISION

The **sensory division** of the peripheral nervous system is composed of sensory receptors and the interneurons that connect them to the central nervous system. Sensory receptors acquire information from the external and internal environments of the body. Spinal and cranial nerves enable the flow of sensory information to the central nervous system. You will learn more about the processing of sensory information in Section 50-4.

MOTOR DIVISION

The **motor division** of the peripheral nervous system allows the body to react to the sensory information. The motor division is composed of two independent systems—the somatic nervous system and the autonomic nervous system.

Somatic Nervous System

The **somatic nervous system** of the motor division consists of motor neurons that control the movement of skeletal muscles. The somatic system is said to be voluntary—that is, skeletal muscles can be moved at will. The somatic system can also operate automatically, as it does when you maintain your balance.

One important function of the peripheral nervous system is the relay of signals in **reflexes,** which are involuntary and often self-protective movements. The patellar reflex, which depends on one of the simplest neural circuits in the human body, is shown in Figure 50-6.

SECTION

50-2

OBJECTIVES

▲ Name the two divisions of the peripheral nervous system and describe their function.

● Distinguish between the somatic nervous system and the autonomic nervous system.

■ Distinguish between the sympathetic division and the parasympathetic division.

◆ Describe a spinal reflex.

Word Roots and Origins

somatic

from the Greek *somatikos,* meaning "of the body"

SECTION 50-2

SECTION OVERVIEW

Stability and Homeostasis

Section 50-2 describes the sensory division of the peripheral nervous system, as well as the two systems of the motor division of the PNS—the somatic nervous system and the autonomic nervous system. These systems are involved in muscular and visceral control that are essential for the body's response to environmental stimuli and maintenance of homeostasis.

▶ ENGAGE STUDENTS

Sit on a desk or table so that your legs hang freely. Demonstrate the patellar (knee-jerk) reflex by tapping the tendon just below the kneecap with the side of your hand or with a medical reflex hammer (if available). Then, a minute or so later, have a student yell "go!" as your signal to extend the same leg. (This response is much slower.) Ask students to explain the time difference between the two responses, both of which result in extension of the same leg. (The first response is involuntary, a spinal reflex that reaches only the spinal cord and involves only a few synapses and neurons. The second response is voluntary, and it takes much longer because it involves the brain's reaction to the word "go" before the leg is extended.)

VISUAL LINK

Figure 50-6

Ask students how spinal reflexes are different from voluntary movement. (Spinal reflexes are involuntary and extremely rapid.) Ask students how reflexes are beneficial. (Reflexes generally occur to protect the body.) In the patellar reflex, stretching of the tendon below the knee stimulates a sensory neuron in the quadriceps, which sends a signal to the spinal cord. This sensory signal results in the excitation of motor neurons that stimulate the quadriceps, which are the leg extensor muscles; and in the inhibition of motor neurons that stimulate the hamstrings, which are the antagonistic leg-flexor muscles.

QUICK FACT

☑ Doctors often prescribe beta blockers to treat conditions of irregular heart beat and high blood pressure. When released by the sympathetic nervous system, the neurotransmitter norepinephrine binds to beta receptors on the heart and causes the heart to contract forcefully and rapidly, increasing heart rate and blood pressure. Beta blockers interfere with these norepinephrine receptors, antagonizing norepinephrine action.

FIGURE 50-6

In the patellar reflex, a sensory neuron (red) with a receptor that detects stretch in the quadriceps muscle sends signals to the spinal cord. The axon of the sensory neuron is split. One branch (green) stimulates motor neurons in the quadriceps, causing the muscle to contract and extend the leg. The other branch (blue) stimulates an interneuron which inhibits motor neurons in the hamstrings.

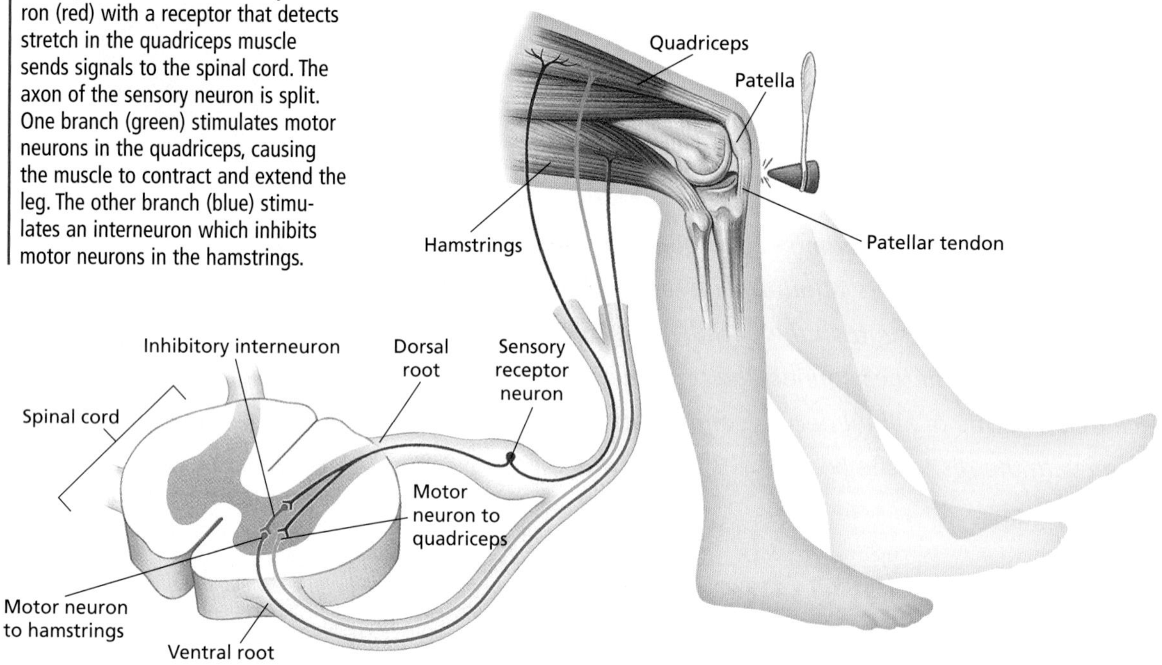

When the tendon below the patella is tapped sharply, a sensory receptor, shown in red, in the attached quadriceps muscle sends an impulse to the spinal cord. This impulse activates a motor neuron, shown in green, that leads back to the quadriceps muscle, causing the muscle to contract. The impulse also activates an interneuron, shown in blue, that has an inhibitory, or calming, effect on the motor neurons of the hamstrings in the lower thigh. The contraction of the quadriceps coupled with the relaxation of the hamstrings extends the lower leg. This type of reflex is a true **spinal reflex;** that is, it involves only neurons in the body and spinal cord, and completely bypasses the brain. Many of the body's involuntary actions, however, originate in a second subdivision of the motor division of the peripheral nervous system—the autonomic nervous system.

Autonomic Nervous System

The **autonomic nervous system** of the motor division consists of nerves that control the body's internal conditions by affecting smooth muscles, both in blood vessels and in organs. The main function of the autonomic nervous system is the control of respiration, heartbeat, and other functions involved in homeostasis. Stimulation and inhibition of body systems are the responsibility of two different subdivisions of the autonomic system—the sympathetic division and the parasympathetic division—as shown in Figure 50-7.

A major function of the **sympathetic division** is the shunting of blood from one part of the body to another. The sympathetic division of the autonomic nervous system is activated by conditions

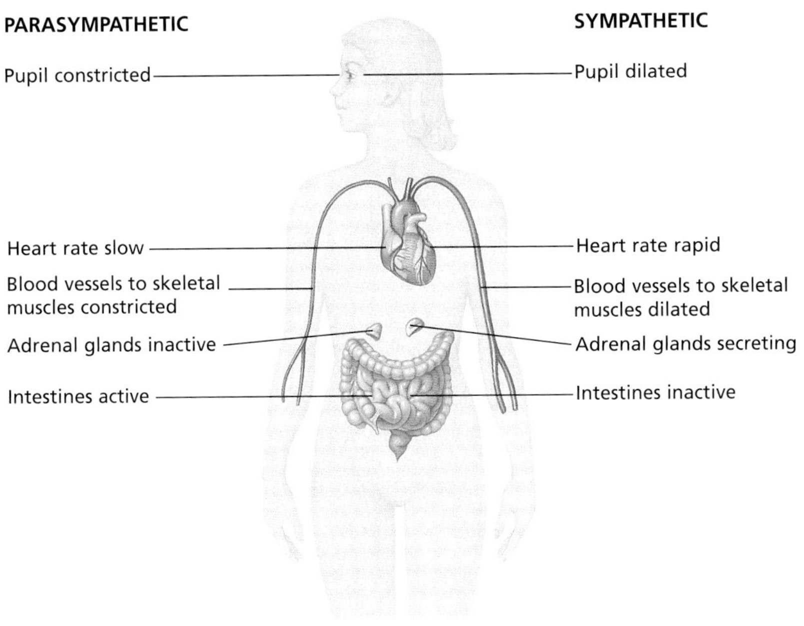

FIGURE 50-7

In the autonomic nervous system, the sympathetic division readies the body to respond to stress or danger. Activation of the parasympathetic division conserves energy and restores homeostasis.

of physical or emotional stress. For example, the threat of a physical attack would cause your sympathetic division to redirect the flow of blood away from your digestive organs and toward your heart and skeletal muscles. This is often called the "fight-or-flight" response. As Figure 50-7 shows, the sympathetic division stimulates activities that consume large amounts of energy.

The **parasympathetic division** of the autonomic nervous system controls the internal environment during routine conditions. After a threat of danger has passed, nerves from the parasympathetic division signal organs to revert to normal levels of activity. Blood flow to the heart and skeletal muscles is decreased and their function slows. The action of the parasympathetic division induces the body to conserve energy. Under normal conditions, both systems usually are activated to some degree. The balance of actions of the sympathetic division and the parasympathetic division of the autonomic nervous system help the body maintain homeostasis.

SECTION 50-2 REVIEW

1. **Name the two main divisions of the peripheral nervous system.**

2. **Name two types of neurons in the peripheral nervous system.**

3. **Name the parts of the nervous system that are involved in a spinal reflex.**

4. **How do the somatic nervous system and the autonomic nervous system differ?**

5. **Which division of the autonomic nervous system is involved in "fight-or-flight" reactions in response to danger?**

6. **CRITICAL THINKING** **Most organs in the body are stimulated by both the sympathetic division and the parasympathetic division of the autonomic nervous system. Explain how this helps maintain homeostasis.**

QUICK FACT

✔ Amyotrophic lateral sclerosis (ALS), or Lou Gehrig's disease, causes the degeneration of motor neurons. As motor neurons die, the muscles that they innervate atrophy, resulting in weakness, paralysis, and eventually death. Death occurs within two to five years and usually coincides with the failure of motor neurons that innervate muscles that control breathing. The onset of ALS in most victims occurs after age 40. Although the cause of ALS is not certain, nerve and muscle degeneration are slowed by a drug called riluzole, which blocks the release of the neurotransmitter glutamate from neurons. However, riluzole does not alter the course of the disease.

ANSWERS TO SECTION 50-2 REVIEW

1. The two main divisions of the peripheral nervous system are the sensory division and the motor division.
2. Sensory neurons and motor neurons are two kinds of neurons in the peripheral nervous system.
3. Sensory receptors, sensory and motor neurons, and interneurons of the spinal cord are involved in a spinal reflex.
4. The somatic nervous system controls both voluntary behaviors and reflexes, whereas the autonomic nervous system controls involuntary activities of body organs.
5. The sympathetic nervous system regulates the fight-or-flight response to danger.
6. Stimulation of body organs by the sympathetic division is counteracted by the steady state that is achieved by the parasympathetic division.

SECTION 50-3

SECTION OVERVIEW

Stability and Homeostasis
Section 50-3 describes the conduction of electrical impulses through neurons as action potentials. It explains the electrical and chemical changes that occur across the cell membrane as a neuron is excited from resting potential to action potential. This section also describes the role of neurotransmitters in chemical synaptic transmission.

▶ ENGAGE STUDENTS

Wire a small lamp to a 6 V battery and demonstrate how the lamp lights when the circuit is closed. Ask students where the energy to light the lamp comes from. (It comes from potential energy, which is expressed in volts.) Tell students that the potential energy of the battery is produced by the separation of positively and negatively charged particles. The potential energy (membrane potential) across the membrane of a neuron at rest (about –70 mV) enables the conduction of action potentials. Membrane potential is determined by the state of voltage-dependent protein channels (channels that are open or closed depending on the membrane potential) in the cell membrane of a neuron. Ask students to suggest how relative concentrations of ions inside and outside the cell membrane determine the membrane potential. (These channels are selectively permeable to ions—including sodium, potassium, and calcium [positively charged] and chloride [negatively charged]—which affect the membrane potential.)

SECTION

50-3

OBJECTIVES

▲
Describe the structure of a neuron.

●
Summarize the electrical and chemical conditions of resting potential.

■
Describe the electrical and chemical changes that occur during an action potential.

◆
Describe the role of neurotransmitters in transmitting a signal across a synapse.

TRANSMISSION OF NERVE IMPULSES

The ability of the nervous system to monitor and respond to the environment, both external and internal, depends on the transmission of signals within a neuron and from one neuron to another. The transmission of a signal along the axon of a neuron is known as an ***action potential.*** *The transmission of action potentials depends on electrochemical energy—electrical energy that is generated as charged chemical substances move into and out of neurons.*

NEURON STRUCTURE

Figure 50-8 shows the morphology of a typical neuron. The nucleus of a neuron, along with most of its organelles, is located in the cell body. Extending out from the cell body in different directions are membrane-covered extensions called dendrites. **Dendrites** receive action potentials from other neurons. The axon is a long, membrane-bound projection that transmits the action potentials away from the cell body. A neuron may have a single axon or branching axons that contact several other neurons. The end of the axon is called the **axon terminal.** It may lie near a muscle, a gland, or the dendrite or cell body of another neuron.

The axons of most neurons are covered with a lipid layer known as the **myelin** (MIE-uh-lin) **sheath.** The myelin sheath both insulates the axon much like the rubber coating of an electrical cord and speeds up transmission of action potentials through the axon. In the peripheral nervous system, myelin is produced by cells called **Schwann cells,** which surround the axon. Gaps in the myelin sheath along the length of the axon are known as the **nodes of Ranvier** (RAHN-vee-ay).

As Figure 50-8 shows, neurons do not touch each other. Instead a small gap, called a **synaptic cleft,** is present between the end of the axon of one neuron and the dendrite or cell body of another neuron. In most neurons, electrical activity in the neuron causes the release of chemicals into the synaptic cleft. These chemicals, called **neurotransmitters,** in turn elicit electrical activity in a second neuron. Thus, the signalling activity of the nervous system is composed of electrical activity *within* neurons and chemical flow

Word Roots and Origins

synapse

from the Greek *synaptein,* meaning "to fasten together"

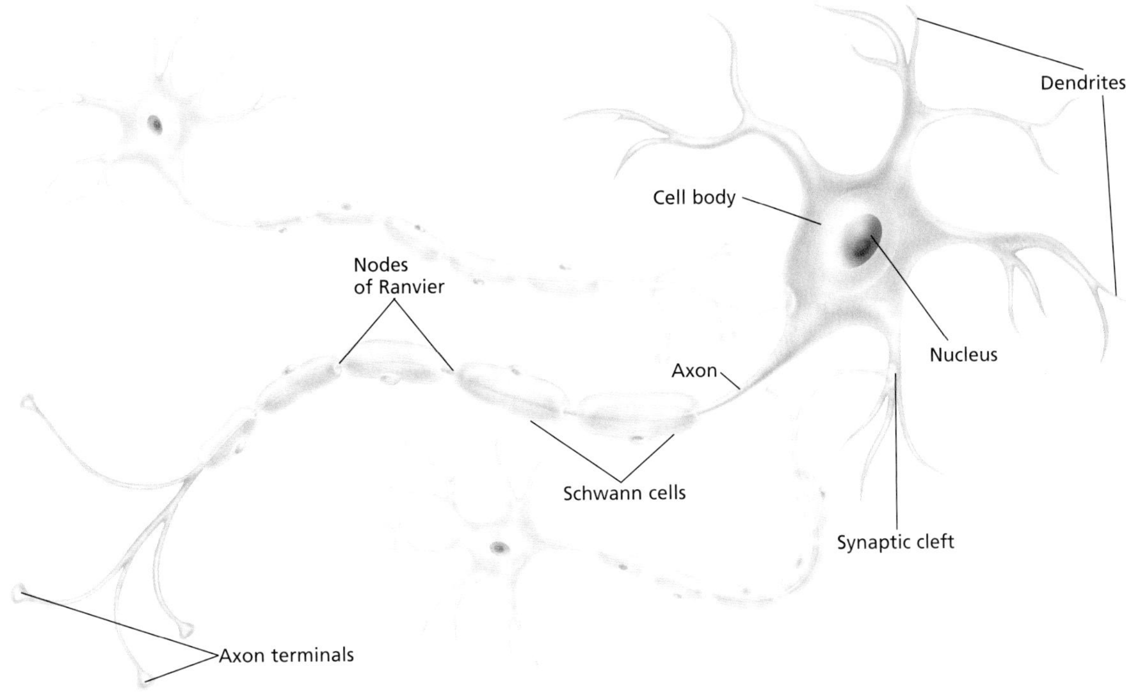

between neurons. Neurons form numerous interconnections that in turn form circuits in the same way that telephone lines interconnect to form circuits. These neural circuits form complex networks throughout the body, just as telephone circuits form networks that extend throughout the world.

FIGURE 50-8
Neurons are separated from other neurons by small gaps. Some neurons have many other neurons contacting their multiple dendrites. A long axon, which carries action potentials, extends from the cell body of the neuron. Most axons are covered by a fatty insulating layer called the myelin sheath.

NEURON FUNCTION

Nearly 200 years ago, scientists discovered that when an electrical current is passed through a muscle removed from a dead animal, the muscle will contract just as it did in life. Scientists have learned that electricity stimulates neurons that affect muscle cells. Neuron function is dependent on electrical activity.

Neurons have an electrical charge different from the extracellular fluid that surrounds them. A difference in the electrical charge between two locations is called a **potential.** In neurons, potentials are produced by a complex interplay of several different ions. The movement of these ions is affected by their ability to pass through the cell membrane, their concentrations inside and outside of the cell, and their charge.

Resting Potential

A neuron is at rest when it is not receiving or transmitting a signal. In a neuron at rest, the concentration of large, negatively-charged proteins and potassium, K^+, ions is greater inside the cell than outside. In contrast, the concentration of sodium, Na^+, ions is greater outside the cell than inside. This imbalance of Na^+ and K^+ ions is, in

CONTENT CONNECTION

Section 5-1: Transport
Ask students to review passive diffusion, facilitated diffusion, active transport, and the sodium-potassium pump. Explain to students that the membrane potential of neurons depends on relative concentrations of K^+ and Na^+ inside and outside the cell.

TEACHING STRATEGY

Conduction Velocity
Tell students that myelinated axons conduct much faster than unmyelinated axons. Ask students to suggest other characteristics of neurons that might affect conduction velocity. (Conduction velocity increases with increasing axon diameter.) To demonstrate the relationship between conduction velocity and axon diameter, have students watch you perform a quick qualitative experiment. Obtain two rubber hoses of different diameter and two beakers of equal size. Fasten each hose to a faucet, and place the free end of each hose in a beaker. Turn both faucets on simultaneously and at equal pressure. After 5 seconds, shut off both faucets. Ask students why more water flows through the larger hose. (The larger hose has a greater diameter, and it therefore has a greater cross-sectional area, so it carries more water than the smaller hose.) Explain that the flow rate of water is analogous to the conduction velocity of an action potential. Charged particles in large axons have more space in which to move than charged particles in small axons, which encounter more resistance to flow. Thus, large, myelinated axons are the fastest conductors, while small, unmyelinated axons are the slowest conductors.

VISUAL LINK

Figure 50-9

This figure shows an action potential. Stimulation of dendrites momentarily causes the inside of the cell to become less negatively charged. If the threshold for an action potential (about –60 mV) is reached, voltage-dependent Na^+ channels open, and Na^+ rushes into the cell. This rapid influx of Na^+ causes a reversal of polarity, raising the membrane potential from –70 mV to about +40 mV. This occurs over a small region in which positive charge moves like a wave down the axon. Ahead of and behind this region, the membrane potential is negative. Voltage-dependent Na^+ channels are located throughout the length of the axon. Thus, Na^+ that enters the cell causes adjacent Na^+ channels to open, and because opposite charges attract, the positive charge conducts down the axon. Shortly after Na^+ channels open, voltage-dependent K^+ channels open. This causes K^+ to flow out of the cell, causing the membrane to return to the resting potential.

INCLUSION *ACTIVITY*

Have students form a circle and hold hands. Have each person in the circle represent a neuron; the left hand represents dendrites, the person's body represents the cell body, and the right hand represents the axon. Initiate a nerve impulse by squeezing the hand of the student to your right. Instruct students to pass the "nerve impulse" around the circle to the right. Each student should call out the words *dendrites, cell body,* and *axons* in the correct order as the impulse moves along. Have students note the time it takes to go around the circle. Tell students that the conduction velocity of an action potential can exceed 20 m/s.

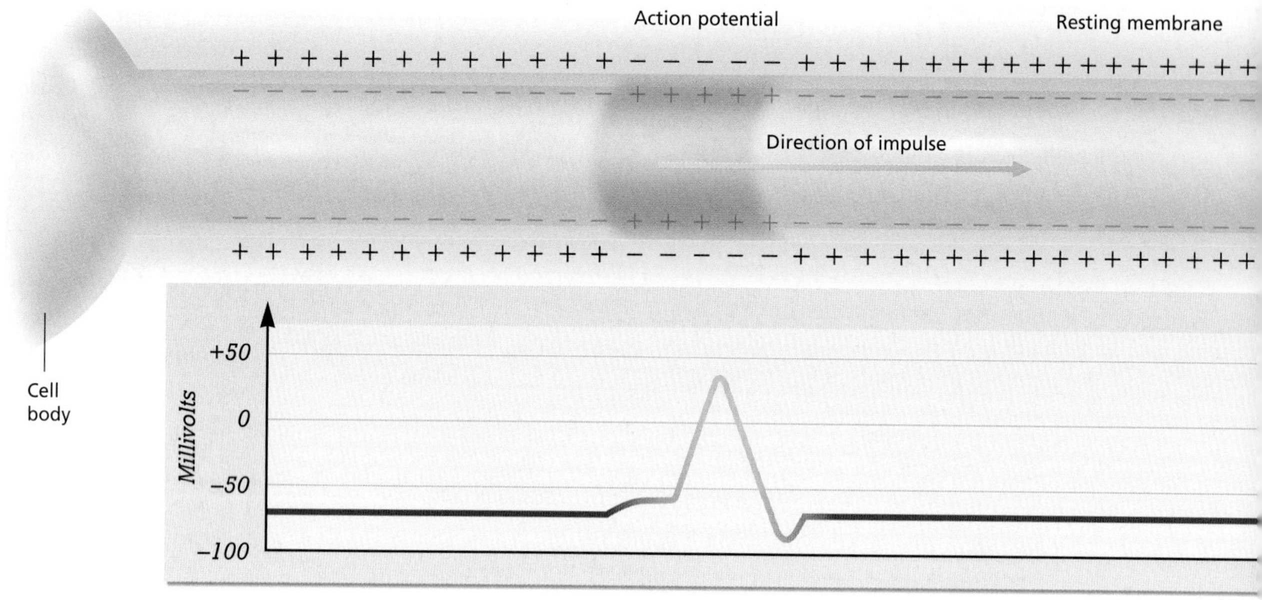

FIGURE 50-9

In the resting state, the interior of a neuron is negatively charged with respect to the extracellular fluid. The passage of an action potential over the membrane of the axon reverses this polarity, making the interior of the axon positively charged for a brief time. A graph shows the change in voltage during an action potential. The charge of the interior of the neuron rapidly switches from negative to positive, and then immediately reverses, the interior regaining its negative charge.

part, due to the action of the sodium-potassium pump. Recall from Chapter 5 that the sodium-potassium pump actively moves Na^+ out of cells while moving K^+ in.

The cell membrane is permeable to some ions but not to others. Na^+ ions do not readily diffuse through the membrane, and they accumulate outside the cell. The negatively charged proteins remain inside because they are too large to exit. K^+ ions, however, pass freely through the membrane and tend to diffuse out of the cell, down their concentration gradient. This exit of positively charged potassium ions, coupled with the retention of negatively charged protein ions, eventually causes the interior of the neuron to become negatively charged with respect to the exterior. This charge difference is called the **resting potential** of the membrane. In animal neurons, the resting potential is about –70 millivolts.

Action Potential

When a dendrite or the cell body of a neuron is stimulated, a sudden change occurs in the permeability of its cell membrane. At the point where it is stimulated, the cell membrane becomes permeable to Na^+ ions. The rush of Na^+ ions into the cell opens voltage gated channels in the membrane that allow even more Na^+ ions to diffuse rapidly from the outside of the membrane to the inside of the neuron. Remember from Chapter 5 that gated channels control the passage of ions through a cell membrane. As a result of the inward diffusion of Na^+ ions, the interior of the neuron's cell body becomes more positively charged than the outer surface. The interior, once negatively charged, is now positively charged. The exterior, once positively charged, is now negatively charged with respect to the interior. This reversal of polarity across the membrane begins an action potential. The action potential starts at the point where the cell body of the neuron joins the axon.

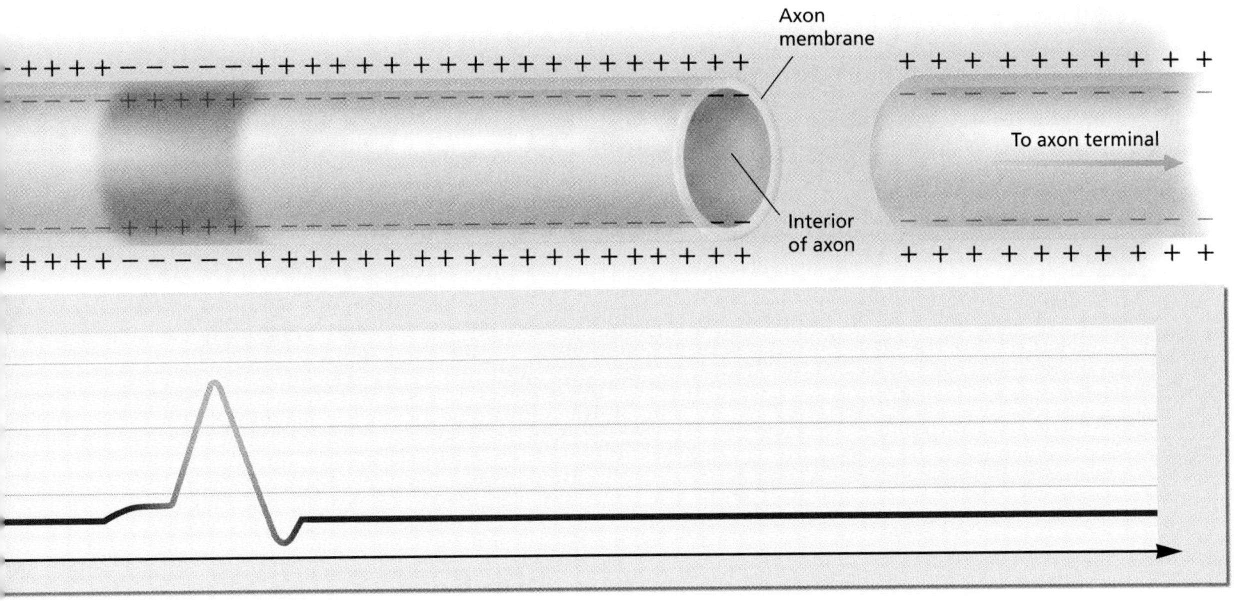

Voltage gated channels exist along the entire length of the axon. As the first small segment of the axon becomes positively charged, the rise in voltage opens channels in the adjacent segment of axon membrane. As before, Na^+ ions enter, driving the voltage in a positive direction and opening channels in the next segment of the axon. In this way, a wave of positive charge passes down the membrane of the axon like a flame burning down a length of rope. Axon potentials travel in one direction only—away from the cell body, where they begin, and toward the axon terminal.

Shortly after they open, the gated channels for Na^+ ions close. Then voltage gated channels for K^+ ions open. The result is an abrupt outward flow of K^+ ions. The outer surface again becomes positively charged, and the inner surface regains its negative charge. This signals the end of the action potential. The neuron cannot generate another action potential until the resting membrane potential is restored. This period, during which the neuron cannot fire, is called the **refractory period.**

After the action potential, the concentration of Na^+ inside the cell is abnormally high, and the concentration of K^+ inside the cell is abnormally low. The sodium-potassium pump reestablishes the original concentrations of Na^+ ions and K^+ ions on both sides of the membrane. Na^+ ions are actively pumped out across the cell membrane, while K^+ ions are actively pumped in across the membrane. Thus, the original ion concentrations are restored, and the neuron is ready for the next action potential. This restoration of resting potential comes at a price. Recall from Chapter 5 that the sodium-potassium pump requires energy in order to function. Neurons need a continuous supply of ATP to keep the sodium-potassium pump operating. In fact, neurons consume a great deal of energy in the body.

QUICK FACT

☑ Action potentials proceed in one direction: away from the cell body and toward the axon terminal. When the membrane potential becomes positive, voltage-dependent Na^+ channels close, and they cannot open again until the membrane potential becomes negative. The time during which Na^+ channels cannot open is called the refractory period. An action potential moves in one direction because of the refractory period.

VISUAL LINK

Figure 50-10 (p. 1016)

This figure illustrates the release of neurotransmitter from the presynaptic axon terminal. When an action potential reaches the axon terminal, vesicles fuse with the cell membrane. Neurotransmitter molecules are released, and they bind to receptor proteins on a postsynaptic neuron. Postsynaptic receptors often are chemical-gated ion channels that open when they are coupled with neurotransmitter molecules. Depending on the charge of the ions to which receptor proteins are permeable, neurotransmitters can have one of two general effects on a postsynaptic neuron: excitation or inhibition. Ask students what happens if a neurotransmitter binds to receptors that pass negatively charged ions, such as chloride (Cl^-), into the cell. (The effect is inhibitory, causing the membrane potential of the postsynaptic neuron to become more negative. Inhibition suppresses activity of the postsynaptic neuron.) Then ask students what happens if receptors pass positively charged ions, such as Na^+, into the cell. (The effect is excitatory, causing the membrane potential to become less negative. If the membrane potential reaches the threshold for an action potential, the postsynaptic neuron will fire.)

QUICK FACT

☑ Multiple sclerosis (MS) is a disorder in which myelin sheaths in the central nervous system gradually degrade. The loss of myelin leads to substantial deficiencies in nerve impulse conduction, causing problems with vision, muscle control, and coordination.

ANSWERS TO SECTION 50-3 REVIEW

1. A "typical" neuron consists of receptive dendrites, a cell body, a conducting axon, and its transmitting axon terminal. Some axons are myelinated to expedite the conduction of action potentials.
2. The resting potential (about –70 mV) is the charge across the cell membrane that exists when the neuron is not conducting an action potential.
3. Signals are chemically transmitted across synapses between neurons by neurotransmitters.
4. Neurons use a tremendous amount of ATP to drive sodium-potassium pumps, which help recharge a neuron after an action potential to its resting potential.
5. Neurotransmitters can either excite or inhibit a postsynaptic neuron.
6. A neuron with several dendrites can receive input from a number of different axons, allowing many presynaptic neurons to converge on one postsynaptic neuron.

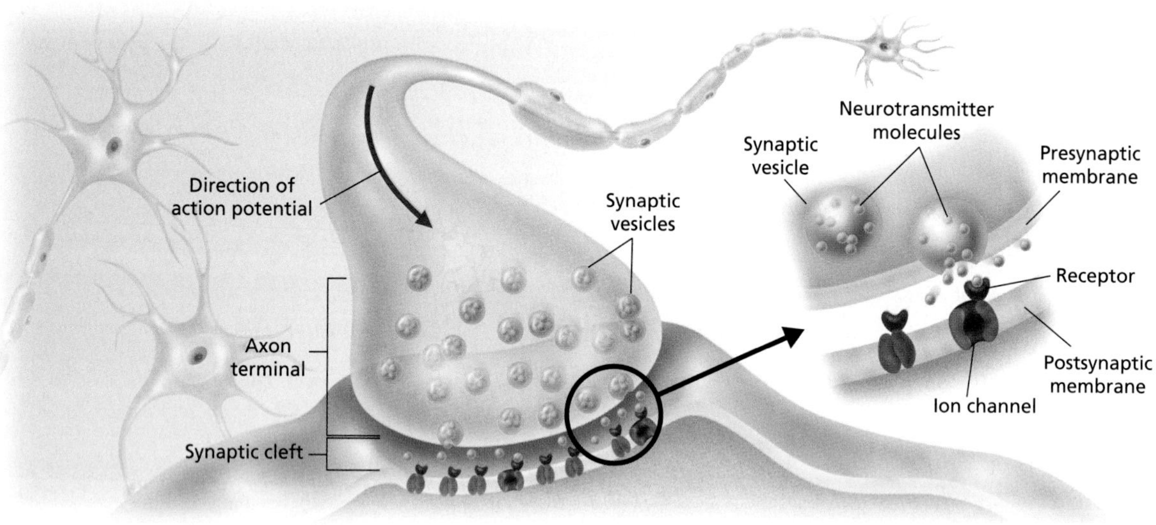

FIGURE 50-10

Neurotransmitter molecules are released into the synaptic cleft by the presynaptic neuron. These molecules bind at receptors in the postsynaptic membrane, opening ion channels. Positive ions entering through these channels cause the resting membrane potential of the postsynaptic neuron to rise. If it rises sufficiently, the postsynaptic neuron will generate an action potential, continuing the signal.

Neurotransmitter Function

When an action potential reaches the axon terminal, vesicles that are stored in the axon terminal and that contain neurotransmitters fuse with the presynaptic membrane. As Figure 50-10 shows, the fusion releases neurotransmitter molecules into the synaptic cleft. These molecules diffuse across the short distance of the synaptic cleft and bind to receptor molecules embedded in the postsynaptic membrane.

The interaction of neurotransmitter molecules and receptor molecules changes the permeability of the postsynaptic membrane by affecting chemically-gated ion channels. The opening of Na^+ channels in the postsynaptic membrane causes the neuron to become more positive in charge. If this positive change of the membrane potential is great enough, a new action potential is generated in the receiving neuron, effectively continuing the electrical signal.

On the other hand, if a sufficient number of Na^+ channels in the postsynaptic membrane fail to open, or if other channels open, allowing the rush of negatively charged ions in, the membrane potential of the receiving neuron will not rise appreciably or will become more negative. No new action potential will be generated in the receiving neuron, and the nervous signal will terminate.

SECTION 50-3 REVIEW

1. Describe the structure of a neuron.

2. What is the resting potential of the membrane?

3. How is a signal in the nervous system transmitted between adjacent neurons?

4. Why does the nervous system consume a large amount of energy?

5. Describe two possible effects that neurotransmitters may have at a synapse.

6. CRITICAL THINKING What functional advantage does a neuron with several dendrites have over a neuron with only one dendrite?

SENSORY SYSTEMS

Human experience is affected by both internal and external stimuli. Humans are able to distinguish among many different types of stimuli by means of a highly developed system of sense organs. Sensory systems represent an integration of the functions of the peripheral nervous system and the central nervous system. The sensory division of the peripheral nervous system gathers information about the body's internal conditions and the external environment. Sensory systems translate light, sound, temperature, and other aspects of the environment to electrical signals and transmit these signals, in the form of action potentials, to the central nervous system, where they are interpreted.

SECTION

50-4

OBJECTIVES

▲ List five types of sensory receptors.

● Describe the structure of the eye and the roles of rods and cones in vision.

■ Identify the parts of the ear responsible for hearing and for maintaining balance.

◆ Explain how taste and smell are detected.

RECEPTORS AND SENSE ORGANS

A sensory receptor is a neuron that is specialized to detect a stimulus. There are many kinds of sensory receptors, and they can be categorized on the basis of the types of stimuli they respond to.

- *Mechanoreceptors* respond to movement, pressure, and tension.
- *Photoreceptors* respond to variations in light.
- *Chemoreceptors* respond to chemicals.
- *Thermoreceptors* respond to changes in temperature.
- *Pain receptors* respond to tissue damage.

Sensory receptors are found in high concentrations in the **sense organs**—the eyes, ears, nose, mouth, and skin—and in fewer numbers, throughout the rest of the body. When a particular sense organ receives appropriate stimulation, its sensory receptors convert the stimulus into electrical signals that are sent to specific regions of the brain. The action potentials generated by the different sense organs are electrically similar, but the regions of the brain where they are interpreted vary according to the type of stimulus.

As you learned in Section 50-1, the brain has a specific region for each sense. Thus, signals received by the auditory region of the temporal lobe may be interpreted as sounds regardless of whether a sound wave was the original stimulus.

SECTION 50-4

SECTION OVERVIEW

Stability and Homeostasis

Section 50-4 describes the sensory systems, which enable perception of environmental stimuli. This section categorizes five primary types of sensory receptors, and it summarizes the role of these receptors in vision, hearing, balance, taste, and smell.

▶ ENGAGE STUDENTS

The area of the retina where the optic nerve exits and blood vessels enter the eye is called the optic disk. The optic disk has no photoreceptors, producing a "blind spot" in the visual field. To demonstrate the blind spot, provide each student with a plain white index card. Have students draw an *X* on the left side of the card and a filled-in square on the right side of the card. The drawings should be even with one another and equidistant from the edges of the card. Instruct students to hold the cards about one foot in front of their face. Have them close the left eye and look at the *X* with the right eye. Have students move the card around in front of their face until the square "disappears." At this point, the image of the square falls on the optic disk, or "blind spot," of the right eye. Explain to students that the visual cortex of the brain fills in the gap in visual perception that is produced by the optic disk.

SECTION 50-4

GIFTED ACTIVITY

Have students research and report various techniques that help scientists gather evidence about how the brain works. A good source is "Unlocking the Secrets of the Brain," parts I and II, in *Bioscience*, Volume 47, Numbers 6 and 7. Students should find information on lesion studies; studies of the effects of drugs on neurotransmitter action and behavior; and brain-imaging techniques, including X ray, computerized tomography (CT), positron emission tomography (PET), electroencephalogram (EEG), and magnetic resonance imaging (MRI).

VISUAL LINK

Figure 50-11

This figure illustrates the structure of the ear. Have students trace the path of sound waves through the ear. Sound waves vibrate the eardrum, which vibrates the bones of the middle ear. The stirrup vibrates the oval window, which pushes on cochlear fluid. High-frequency sound waves stimulate the end of the basilar membrane that is closest to the oval window, whereas low-frequency vibrations stimulate the far end. Movement of the basilar membrane causes hair cells in the organ of Corti to bend. Bending of hair cells triggers neurotransmitter release. Neurotransmitters activate spiral ganglion cells, whose axons make up the auditory nerve. The auditory nerve extends through the brain stem to the thalamus, which relays the incoming signals to the temporal lobe. Emphasize to students that perception of tonal qualities, such as pitch (frequency) and volume (amplitude), depends on stimulation of specific regions of the basilar membrane.

Word Roots and Origins

tympanic

from the Greek *tympanon,* meaning "drum"

HEARING AND BALANCE

The ear is specialized for two functions: detecting sound and maintaining balance. Sound is transmitted as vibrations of a substance; for humans, usually the substance is air. Sound is directed by the fleshy structure of the external ear into the ear itself, where the vibrations are translated into action potentials.

As shown in Figure 50-11, the **auditory canal** connects the external ear with the **tympanic** (tim-PAN-ik) **membrane,** also called the eardrum. Vibrations in the air passing through the auditory canal cause the tympanic membrane to vibrate. Air pressure in the chamber beyond the tympanic membrane, the middle ear, can be regulated and balanced by the amount of air allowed to enter the middle ear through the Eustachian tube. The **Eustachian** (yoo-STAY-kee-uhn) **tube** is an opening to the throat that enables you to equalize the pressure on both sides of your tympanic membrane whenever you experience a sudden change in atmospheric pressure, such as when you take off in an airplane. By swallowing, you allow air to enter the middle ear through the Eustachian tube. This air is under the same pressure as the air pressing against the tympanic membrane on the exterior side. Thus, the air pressure is equalized and you no longer feel pressure on the tympanic membranes in your ears.

Vibrations of the tympanic membrane are transmitted to three small bones of the middle ear: the hammer, the anvil, and the stirrup. The stirrup transfers the vibrations to a membrane called the **oval window,** which separates the middle ear from the inner ear. The inner ear contains the **cochlea** (KAHK-lee-uh), a coiled tube consisting of three fluid-filled chambers that are separated by membranes. The middle chamber contains the **organ of Corti,** which is the organ of

FIGURE 50-11

Sound waves, which are vibrations in the air, cause the tympanic membrane to move back and forth. This motion causes the small bones of the middle ear to move as well, transferring vibrations to the oval window. Mechanoreceptors in the inner ear translate these vibrations to action potentials, which travel through the auditory nerve to auditory processing centers in the brain.

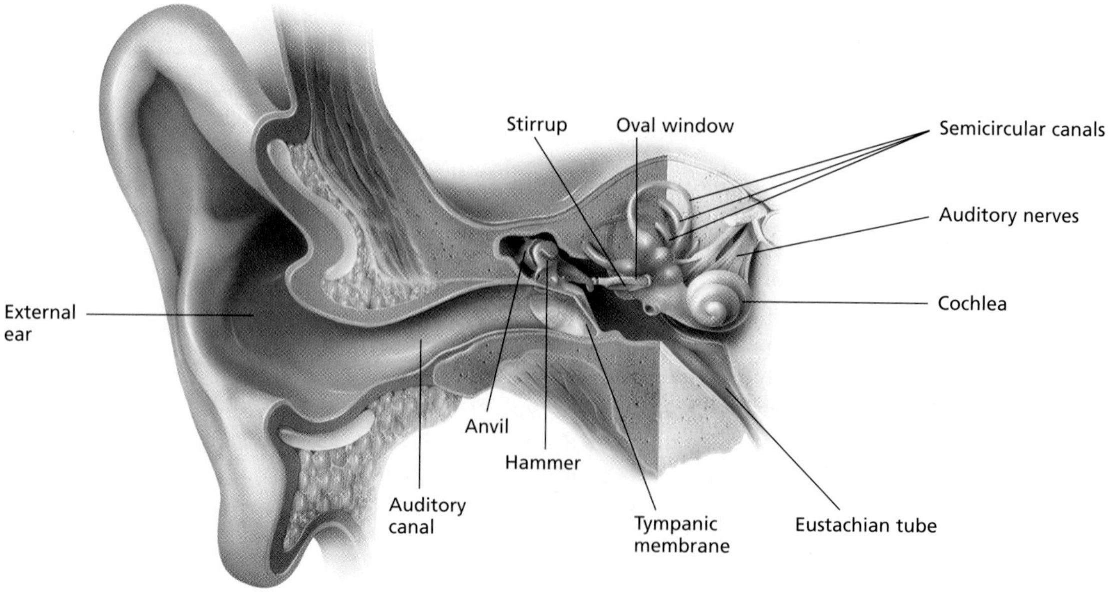

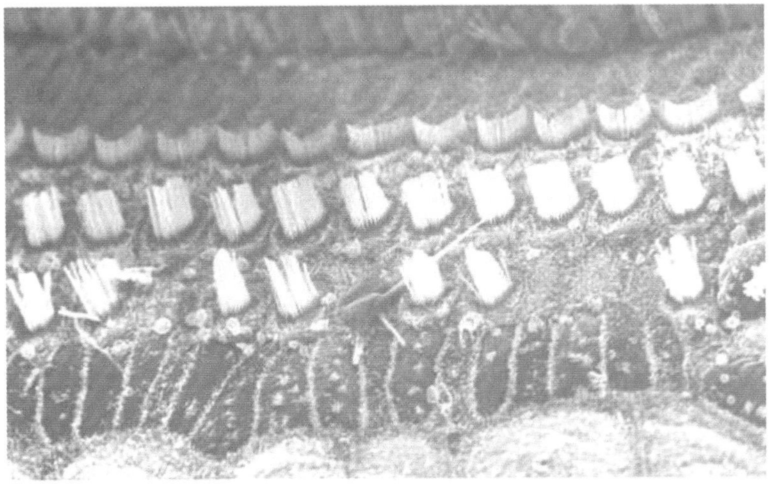

FIGURE 50-12

Hair cells are arranged in orderly rows in the organ of Corti in the cochlea. In this damaged section, the hair cells in the top and center rows are relatively intact. Many of the hair cells of the bottom row, however, are frayed and bent. Four clusters of hair cells (the third, sixth, ninth, and tenth from the left) have been completely destroyed.

hearing. The organ of Corti rests on the bottom membrane in the cochlea and contains sensory receptors known as hair cells. Vibrations of fluid in the cochlea move the bottom membrane and cause these hair cells to bend against a second membrane, which covers them like a roof. This movement stimulates the hair cells to produce action potentials. These action potentials travel along the auditory nerve to the auditory region of the brain stem, then to the thalamus, and finally to the auditory cortex, where they are interpreted as sound.

Different pitches stimulate different parts of the cochlea. Normally, the human ear can detect sounds ranging from 16 to 20,000 cycles per second. If the human ear could hear sounds lower than 16 cycles per second, it would pick up sounds generated by the flow of blood through the vessels and the movement of food through the digestive system.

The hair cells that line the cochlea are a delicate and vulnerable part of the ear. Repeated or sustained exposure to loud noise destroys the neurons of the organ of Corti. Once destroyed, the hair cells are not replaced, and the sound frequencies interpreted by them are no longer heard. Figure 50-12 shows rows of hair cells in a damaged section of the organ of Corti. Damage is most apparent in the bottom row of hair cells. Hair cells that respond to high frequency sound are very vulnerable to destruction, and loss of these neurons typically produces difficulty understanding human voices. Much of this type of permanent hearing loss is avoidable by reducing exposure to loud noises in the environment, such as industrial and machinery noise, gunfire, and loud music.

Balance is maintained with the help of mechanoreceptors in the three **semicircular canals** of the inner ear, shown in Figure 50-11. These canals are filled with fluid, and their interiors are lined with hair cells that have tiny particles of calcium carbonate on top of them. When the head moves, the hair cells are bent by the action of gravity on the calcium carbonate particles. The brain decodes the degree and direction of bend of the hair cells, interpreting the motion and the orientation of the head in space.

DEMONSTRATION

If possible, obtain a tuning fork or another noise-making device that can be moved easily, such as a bell. Have a student volunteer sit on a table or desk and close his or her eyes. Strike the tuning fork, and, while it is vibrating, move it around the student's head. Ask the student to guess the location of the sound source at various points above and beside the head. Then have the student plug one ear and try to guess the location. (The student should have trouble locating sound from the sides of the head and less trouble with sounds located behind, above, and in front of the head.) Ask students to suggest reasons for the differences in sound localization ability. (For example, sound waves coming from the left side reach the left ear first and the right ear a few milliseconds later. This delay makes sound localization difficult. The head also casts a sound "shadow" over the right ear, causing sound that reaches the right ear to be less intense than sound that reaches the left ear. This difference in sound intensity between ears also complicates sound localization.)

QUICK FACT

☑ Eye lenses are elastic, so they can change shape to adjust to changes in distance from a viewed object, a process called accommodation. Many vision disorders result from insufficient accommodation. If the eyeball is too long, light rays converge in front of the retina, causing myopia, or nearsightedness. Myopia is corrected by placing an artificial concave lens in front of the eye. If the eyeball is too short, light rays focus behind the retina, causing hyperopia, or farsightedness. Hyperopia is corrected by an artificial convex lens.

Quick Lab

Observing a Lens

Time Required 20 minutes

Preparation Tips If a glass is used instead of a beaker, it must be flat on the bottom and clear. Use newspaper with very small print. Have the students look through the sides of the glass for step 1. Then place the newspaper on the desk, and place the glass over the newspaper for step 2. An extension to this activity might be to have students add milk to the water and investigate an eye disease (cataracts) that the effect models.

Answers to Analysis The light waves were condensed as they passed through the water, which causes the print to look larger and more in focus. This activity illustrates the effect of the difference in density of the cornea, the fluid inside the eye, and the lens. The oil models either the cornea or the lens. The oil also demonstrates a difference in density, which would cause the print to appear larger.

VISUAL LINK

Figure 50-13

This figure shows the basic structure of the eye. Ask students how the cornea and lens focus images on the retina. (They both refract incoming light rays toward the center of the retina.) Note the suspensory ligaments that are attached to both the lens and the ciliary muscles. Ask students to suggest how the lens accommodates changes in focal length. (When an object is viewed from far away, ciliary muscles are at rest, and suspensory ligaments are tense, flattening the lens. When close objects are viewed, ciliary muscles contract and suspensory ligaments relax, rounding the lens.)

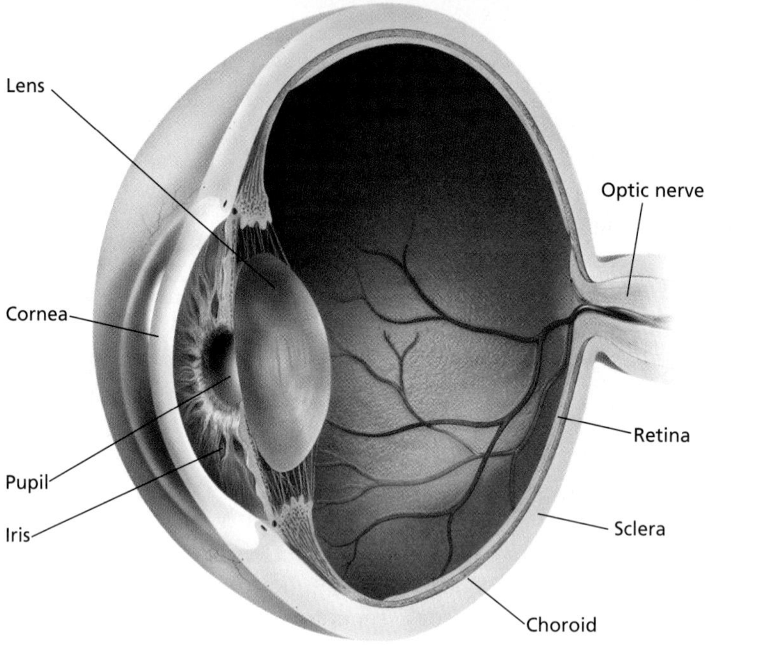

FIGURE 50-13

Light entering the eye travels through the cornea, pupil, and the clear lens to the retina, which contains millions of photoreceptors. Activation of these specialized sensory receptors sends a signal through the optic nerve to the optic centers of the brain—first to the thalamus and eventually to the visual cortex in the occipital lobe.

Quick Lab

Observing a Lens

Materials beaker, water, piece of newspaper, 4 drops cooking oil

Procedure

1. Observe the newspaper through the sides of an empty beaker. Fill the beaker with water, and observe the newspaper through the water. Note any difference in print size.
2. Add four drops of oil to the top of the water. Place the newspaper under the glass, and observe the print size through the oil drops and the water.

Analysis Infer why the print size changes when the newspaper is viewed through water. What structure of the eye does the oil on the water represent?

VISION

The eyes are specialized organs that function by receiving light and transmitting signals to visual processing areas of the brain. The eye is basically a hollow sphere filled with a clear fluid. The structures of the eye, shown in Figure 50-13, act together to focus light on the **retina,** the light-sensitive inner layer of the eye.

Light passes first through a clear, protective layer called the **cornea.** Light then passes through the **pupil,** the opening to the interior of the eye. The pupil becomes larger when light is dim and smaller when light is bright. These involuntary responses are controlled by muscles in the pigmented **iris** that surrounds the pupil.

After light passes through the pupil, it travels through a convex crystalline structure called the **lens.** Attached to the lens are muscles that adjust the lens's shape to bend the rays of the incoming light. This bending focuses the complete image formed by the light onto the retina.

Lying deep within the retina are rods and cones, photoreceptors that translate light energy into electrical signals that can be interpreted by the brain. There are nearly 125 million **rods** in a single retina. Rods contain rhodopsin, a light-sensitive pigment that allows the rods to respond to dim light. The 7 million **cones** in the retina are stimulated by bright light. The cones initiate the production of sharp images and respond to different colors. Humans have three kinds of cones. Each type of cone contains a pigment that absorbs different wavelengths of light. When the signals from these three kinds of cones are integrated, a person is able to see all the colors in the visible spectrum. Colorblindness, which is the inability to distinguish certain colors, is caused by a chemical disorder in the cones.

Each photoreceptor responds to light from a single location in the visual field. Signals from the stimulated photoreceptors in the deepest layer of the retina travel to neurons on the surface of the retina. From these neurons, millions of axons, which form the optic nerve, exit the eye. The **optic nerve** carries visual information in the form of action potentials from the retina to the thalamus. Visual information is ultimately processed in the cortex of the occipital lobe.

internetconnect

SCI LINKS NSTA

TOPIC: The senses
GO TO: www.scilinks.org
KEYWORD: HM1021

TASTE AND SMELL

Specialized chemoreceptors allow humans to perceive variations in tastes and odors. The chemoreceptors for taste are clustered in the **taste buds.** Most of the 10,000 taste buds are embedded between bumps called **papillae** (puh-PIL-ee) on the tongue; additional taste buds are found on the roof of the mouth and in the throat.

Chemicals from food dissolved in saliva enter a taste bud through a small opening and stimulate a signal in the neurons that line the inner surface of the taste buds. As Figure 50-14 shows, taste signals travel through a relay in the brain stem, then to the thalamus and finally to the cortex, where they are interpreted.

Other chemicals in the environment are perceived by receptors in the nasal passages. Specialized chemoreceptors called **olfactory** (ahl-FAK-tuh-ree) **receptors** are located in the olfactory epithelium of the nasal passage, as shown in Figure 50-14. These cells lie within the mucous lining of the epithelium. The binding of molecules in the air to specific receptor molecules in the olfactory receptors

FIGURE 50-14

Taste and smell are chemical senses. Sensory receptors in the mouth and nasal passage bind molecules from the environment, initiating neural signals that travel to the brain.

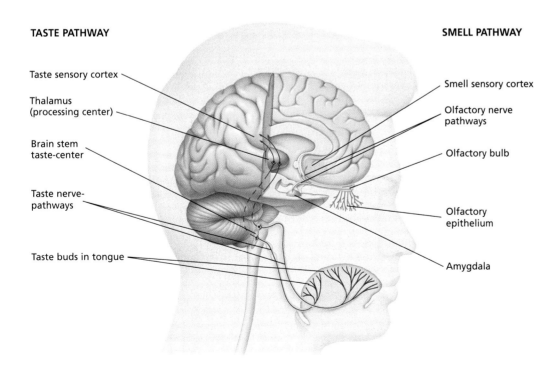

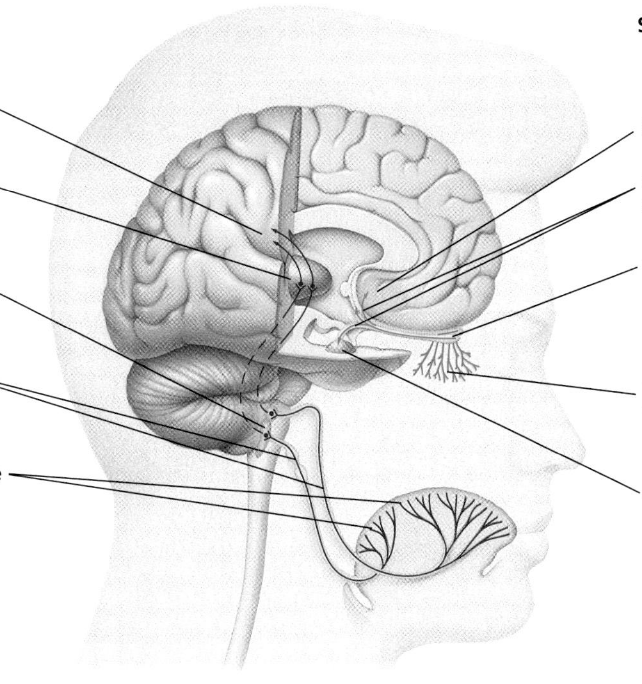

TEACHING STRATEGY
Binocular Vision

Neurons in the visual cortex respond to different regions of the retinas. Most neurons of the visual cortex respond to regions of either the left or the right retina. Explain to students that binocular vision is possible because some neurons in the visual cortex respond only to regions of both retinas. To demonstrate binocular vision, have pairs of students face each other standing about two feet apart. Instruct one student to extend his or her arm with the index finger extended upward. Tell the other student to cover one eye with one hand and use the index finger of the other hand to touch the partner's fingertip. Ask students why distance is difficult to judge with only one eye. (When students cover one eye, they disable binocular neurons in the visual cortex that enable depth perception.)

CRITICAL THINKING
Visual Development

At birth, axons projecting from both retinas to the thalamus, and from the thalamus to the visual cortex, overlap. After several weeks, however, axons from both eyes segregate, forming inputs from either the left or the right eye. Ask students to suggest how this "refining" of visual connections occurs during development. (The development of the visual system depends on the strengthening and weakening of specific synapses. This is called synaptic plasticity, or modification.)

SECTION 50-4

RETEACHING ACTIVITY

Have students make a table of the sense organs, showing the types of receptors found in each organ and the types of stimuli to which each organ responds. For example, mechanoreceptive hair cells in the organ of Corti respond to bending against the tectoral or "roof" membrane.

ANSWERS TO SECTION 50-4 REVIEW

1. Chemoreceptors detect chemicals, whereas mechanoreceptors detect movement, pressure, and tension.
2. Sound waves enter the auditory canal and cause the tympanic membrane to vibrate, which moves bones in the middle ear. The stirrup causes the oval window to vibrate, which transfers vibrations to cochlear fluid. Cochlear fluid stimulates the basilar membrane, which stimulates hair cells of the organ of Corti according to the sound frequency.
3. Rods and cones are the two types of photoreceptors in the retina. Photoreceptors convert light energy into electrical energy that can be used by the nervous system. Three different types of cones are stimulated by bright light and respond to colors. Much more sensitive than cones, rods are active primarily in dim light.
4. Both senses use chemoreceptors to detect sensory stimuli.
5. Humans use their hands to gather information about the environment and to perform tasks. Pain receptors help protect the hands.
6. The body is mapped in the nervous system according to location and function. Because the legs are represented in the lower spinal cord, injury to the lower spinal cord affects the functioning of the legs.

stimulates the receptors. Signals from olfactory receptors travel to the olfactory bulb, a structure of the limbic system, then to olfactory areas of the cortex and to the **amygdala,** another limbic structure important in emotion, memory, and eating behavior.

OTHER SENSES

Mechanoreceptors located throughout the skin make it possible to sense touch, pressure, and tension. In humans, the receptors for touch are concentrated in the face, tongue, and fingertips. Body hair also plays an important role in the ability to sense touch. Large numbers of mechanoreceptors are found in the skin at the base of the hair follicles.

Two types of specialized thermoreceptors monitor temperature. Cold receptors are most sensitive to temperatures below 20°C. Heat receptors respond to temperatures between about 30°C and 45°C.

Pain receptors are sensory neurons located in the base of the epidermis and throughout the interior of the body. Pain receptors are stimulated by mechanical, thermal, electrical, or chemical energy. The type and number of pain receptors vary at different locations throughout the body. For example, the hands and mouth have a high concentration of pain receptors.

Sensory input from the surface of the body goes to the spinal cord in a very orderly way, as shown in Figure 50-15. Sensory input from the shoulders enters upper dorsal root sections of the spinal cord, while sensory input from the lower body enters the dorsal roots of the lower spinal cord. Damage to a specific section of the spinal cord results in sensory problems limited to a well-defined area of the body. This is an example of how function is mapped throughout the nervous system. In the brain, specific areas of the sensory cortex and motor cortex likewise correspond to specific body parts, and damage to specific areas of the cortex results in isolated problems, such as the loss of sensation in part of a hand.

FIGURE 50-15

Specific dorsal roots of the spinal cord carry sensory information from certain parts of the skin and body.

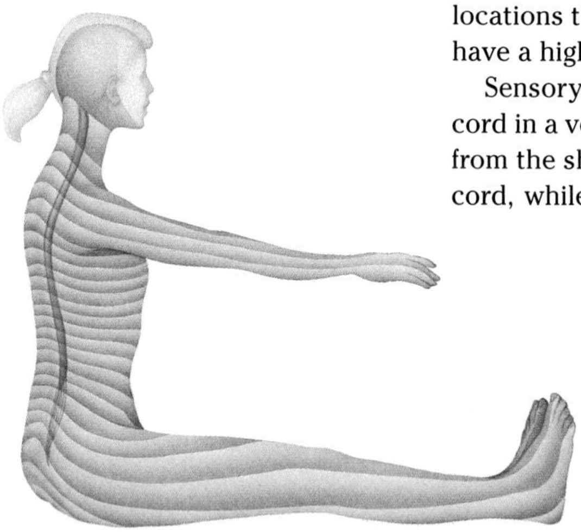

SECTION 50-4 REVIEW

1. Distinguish between chemoreceptors and mechanoreceptors.
2. How are sound vibrations transmitted through the ear?
3. Explain the role of rods and cones in the perception of images by humans.
4. What mechanisms do the senses of taste and smell have in common?
5. What is the importance of the high concentration of pain receptors in the hands?
6. **CRITICAL THINKING** Why might an injury to the lower spinal cord cause a loss of sensation in the legs?

CHAPTER 50 REVIEW

SUMMARY/VOCABULARY

50-1

- The central nervous system (CNS) includes the brain and spinal cord.
- Main divisions of the brain include the cerebrum and the brain stem—including the diencephalon—and the cerebellum. The cerebral cortex forms the outer layer of the cerebrum.
- The cerebellum coordinates muscle action and helps the body maintain its balance.
- The lower brain stem has three main sections—the midbrain, the pons, and the medulla oblongata.
- Thirty-one pairs of spinal nerves each consist of a dorsal root containing sensory neurons and a ventral root containing motor neurons.

Vocabulary

afferent neuron (1004)
arachnoid layer (1007)
axon (1003)
brain stem (1006)
central nervous system (1004)
cerebellum (1007)
cerebral cortex (1005)
cerebral hemispheres (1004)
cerebrospinal fluid (1007)
cerebrum (1004)
corpus callosum (1004)
diencephalon (1006)
dura mater (1007)
efferent neuron (1004)
interneuron (1008)
limbic system (1006)
medulla oblongata (1007)
meninges (1007)
midbrain (1007)
motor neuron (1008)
nerve (1008)
peripheral nervous system (1004)
pia mater (1007)
pons (1007)
reticular formation (1007)
sensory receptor (1008)
spinal cord (1008)
thalamus (1006)
ventricle (1007)
white matter (1005)

50-2

- The peripheral nervous system (PNS) links the central nervous system and the rest of the body. The PNS is composed of a sensory division and a motor division.
- The motor division of the PNS is divided into the somatic nervous system and the autonomic nervous system.
- The somatic nervous system controls skeletal muscles and is under voluntary control.
- In a reflex, a signal is routed from sensory neurons, through the spinal cord and out through motor neurons.
- In autonomic control of internal conditions, the sympathetic division and the parasympathetic division have largely opposite effects.

Vocabulary

autonomic nervous system (1010)
motor division (1009)
parasympathetic division (1011)
reflex (1009)
sensory division (1009)
somatic nervous system (1009)
spinal reflex (1010)
sympathetic division (1010)

50-3

- A neuron consists of a cell body, dendrites, and an axon. The transmission of a signal through a neuron is called an action potential.
- In a neuron at rest, K^+ ions are concentrated inside the membrane while Na^+ ions are concentrated outside the cell membrane. The inside of the cell membrane has a negative charge relative to the outside.
- During an action potential, the polarity of the membrane is reversed briefly as Na^+ ions diffuse into the neuron and K^+ ions diffuse out of the neuron through gated channels.
- When an action potential reaches the presynaptic membrane of the next neuron, neurotransmitter molecules diffuse across the synapse, where they can either stimulate or inhibit the start of an action potential in the postsynaptic neuron.

Vocabulary

action potential (1012)
axon terminal (1012)
dendrite (1012)
myelin sheath (1012)
neurotransmitter (1012)
node of Ranvier (1012)
potential (1013)
refractory period (1015)
resting potential (1014)
Schwann cell (1012)
synaptic cleft (1012)

CHAPTER 50 REVIEW ANSWERS

REVIEW

1. The autonomic nervous system is part of the motor division of the peripheral nervous system, and it contains two divisions: the sympathetic and parasympathetic divisions. Each division of the autonomic system helps maintain homeostasis by controlling blood vessels and visceral organs.
2. The hypothalamus is located below the thalamus.
3. The resting potential is the electrical charge across the cell membrane of a neuron that is prepared to conduct an impulse. During an action potential, the membrane potential momentarily becomes positive, and then the resting potential is restored.
4. The pons serves as a "bridge" that connects the cerebellum to the cerebrum. The pons relays information to and from the cerebellum.
5. Afferent neurons carry information toward the central nervous system, whereas efferent neurons transmit signals away from it.

6. a **7.** b
8. c **9.** a
10. c **11.** b
12. b **13.** d
14. d **15.** b

16. *A*, axon terminal; *B*, Schwann cell (or myelin sheath); *C*, nodes of Ranvier; *D*, axon; *E*, dendrite
17. The cerebrospinal fluid protects the brain and spinal cord, and it serves as a medium through which nutrients and other substances are transported to the CNS.
18. Afferent neurons transmit signals to the spinal cord and brain. Efferent neurons transmit signals away from the central nervous system. Interneurons connect afferent and efferent neurons.
19. The autonomic nervous system controls blood vessels and body organs to regulate visceral systems and to help maintain homeostasis.

ANSWERS

CONTINUED

20. The midbrain relays visual and auditory information. The pons connects the cerebellum to the cerebrum and the spinal cord. The medulla oblongata relays information from various sensory systems to and from the brain and the spinal cord, and it controls some homeostatic activities.
21. At the resting potential, sodium ions do not readily cross the cell membrane of a neuron.
22. The sodium-potassium pump reestablishes the original concentrations of Na^+ and K^+ ions on both sides of the cell membrane, restoring the membrane potential.
23. An action potential triggers the release of neurotransmitter molecules.
24. Dendrites are the receptive endings of neurons. Dendritic membranes contain receptor proteins that bind to specific neurotransmitter molecules.
25. Hair cells in the semicircular canals of the inner ear send information about head orientation and body position to the brain.
26. Rods and cones are the two types of photoreceptors in the retina. Rods respond to dim light, while cones respond to bright light and color.

CRITICAL THINKING

1. Repeated, rapid impulses from the brain cause intense contractions of skeletal muscles, causing rigidity and convulsions.
2. There are several possible explanations for synesthesia. It is possible that inputs of sensory neurons to the thalamus are incorrect. Perhaps neurons that extend from the thalamus to the cerebral cortex are misdirected, causing an improper relay of information from various sensory systems to the cortex. For example, the thalamus of a person who associates musical notes with colors could be misdirecting sensory

50-4

- Sensory receptors include mechanoreceptors, photoreceptors, chemoreceptors, thermoreceptors, and pain receptors.
- The eye receives light through the pupil. The lens focuses this light on the retina, which is composed of photoreceptors. The optic nerve that exits at the back of the retina carries a signal to the brain.
- In the ear, vibrations pass through the auditory canal, the tympanic membrane, the bones of the middle ear, the oval window, and the cochlea of the inner ear. Hair cells in the organ of Corti produce signals that travel through the auditory nerve to the brain.
- Hair cells in the semicircular canals of the inner ear monitor the body's position in space.
- Stimulation of neurons that line the inner surface of a taste bud initiates signals that travel to the brain, where they are interpreted as taste. Likewise, olfactory receptors in the nasal passage transmit signals to the brain, where they are interpreted as odor.

Vocabulary

amygdala (1022)	Eustachian tube (1018)	organ of Corti (1019)	rod (1020)
auditory canal (1018)	iris (1020)	oval window (1018)	semicircular canal (1019)
cochlea (1018)	lens (1020)	papilla (1021)	sense organ (1017)
cone (1020)	olfactory receptor (1021)	pupil (1020)	taste bud (1021)
cornea (1020)	optic nerve (1021)	retina (1020)	tympanic membrane (1018)

REVIEW

Vocabulary

1. Explain the relationship between the autonomic nervous system, the sympathetic division, and the parasympathetic division.
2. Where is the hypothalamus located with respect to the thalamus?
3. Explain the relationship between the resting potential of the membrane and the action potential.
4. The word *pons* means "bridge." How is this related to the function of the pons in the brain?
5. What do the terms *afferent* and *efferent* mean with respect to neurons?

Multiple Choice

6. The central nervous system consists of (a) the brain and spinal cord (b) 31 pairs of spinal nerves (c) the cranial nerves (d) the peripheral nervous system.
7. The cerebral cortex (a) is located deep in the brain (b) is the folded outer covering of the brain (c) is the lobed, highly folded structure located at the back of the brain (d) contains the reticular formation.
8. The thalamus (a) lies below the medulla oblongata in the brain stem (b) controls homeostasis (c) directs sensory information to the proper regions of the cerebral cortex (d) is important in maintaining balance.
9. The cerebellum is important in (a) coordinating motor responses and maintaining posture (b) controlling hormone levels and maintaining homeostasis (c) protecting the brain and spine (d) processing olfactory and taste information.
10. In a spinal reflex, the signal travels (a) immediately to the brain (b) to the spinal cord and then to the brain (c) to the spinal cord and out to a muscle (d) directly to a sense organ.
11. When a neuron is at resting potential, (a) the outside is negatively charged (b) the inside is negatively charged (c) both sides are equally charged (d) the polarity across the membrane reverses.
12. Neurotransmitters (a) diffuse across the membrane, carrying charge with them (b) diffuse across the synaptic cleft and open channels on the postsynaptic neuron (c) are transported across the synaptic cleft and close channels in the postsynaptic neuron (d) do not work at the synaptic cleft.
13. Sensory receptors (a) link interneurons with motor neurons (b) are found only in the spine (c) are found only in the brain (d) respond to stimuli.

14. Photoreceptors are stimulated by (a) heat (b) pressure (c) chemicals (d) light.
15. Mechanoreceptors are stimulated by (a) heat (b) pressure (c) chemicals (d) light.

Short Answer

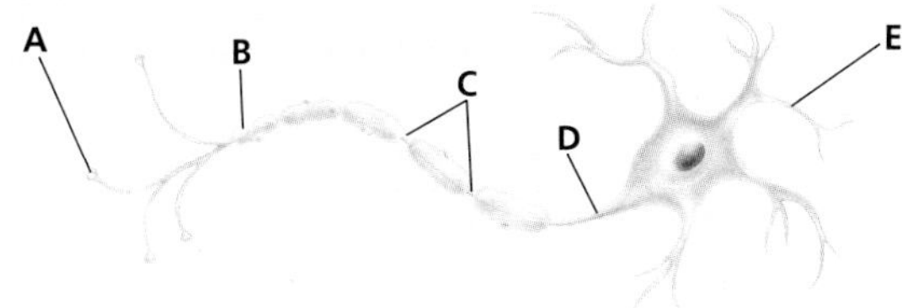

16. List the part of the neuron denoted by each letter in the diagram above.
17. What are two functions of cerebrospinal fluid?
18. What is the relationship between afferent neurons, interneurons, and efferent neurons?
19. How does the autonomic nervous system work to maintain homeostasis?
20. What are the roles of the three sections of the lower brain stem?
21. At resting potential, what ions do not readily cross the cell membrane of the neuron?
22. What role does the sodium-potassium pump play in the restoration of the membrane potential?
23. What event triggers release of neurotransmitter molecules?
24. What are dendrites and what function do they play in signaling in neurons?
25. What is the role of the ear in maintaining balance?
26. What are rods and cones, and how do their functions differ?

CRITICAL THINKING

1. Epilepsy affects one of every 200 Americans. Brain neurons normally produce small bursts of action potentials in varying patterns. During an epileptic seizure, large numbers of brain neurons send rapid bursts of action potentials simultaneously. The body of an individual having a seizure may grow rigid and jerk or convulse. From what you know about the brain's control of muscles and posture, how might you explain these symptoms?
2. Synesthesia is a puzzling phenomenon in which one type of sensory input is interpreted by the brain as another type. For example, a person hearing music might associate certain notes with certain colors. What may be happening in the central nervous system to produce this effect?
3. An imbalance of electrolytes, the ion-containing fluids of the body, can impair transmission of nerve signals. Why might this be so?
4. Look at the diagram below of the brain of a fish. How does the cerebrum differ from that of a human? The fish's brain has large olfactory bulbs. What does this tell you about the relative importance of the sense of smell to the fish?

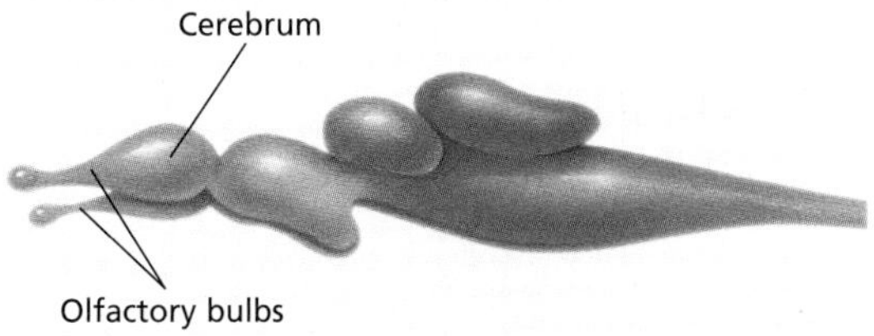

EXTENSION

1. Read "Why We Cry" in *Current Science,* February 11, 2000, on page 8. Describe the three kinds of tears that humans produce. Do any other animals cry tears? Explain the following comment by the writer: "Contrary to common belief, recent studies show that people don't weep while they're feeling a strong emotion."
2. Research the way general anesthetics work. Then write a short report in which you discuss a specific anesthetic, the effect of the anesthetic on the brain, and the effect of body weight and the passage of time on the action of an anesthetic.

inputs from the auditory nerve to the visual cortex.

3. Potassium and sodium ions are necessary for the conduction of action potentials and, therefore, neurotransmitter release. Thus, electrolyte imbalance can affect nerve impulse transmission.

4. Unlike the human cerebrum, the fish's cerebrum is small and its surface is not folded. The fish's large olfactory bulb indicates that it relies heavily on the sense of smell.

EXTENSION

1. Basal tears are those that slosh over our eyes every time we blink. These tears are lubricating tears that keep our eyes clean. Reflex tears are those that flood and wash our eyes in response to an environmental irritant, such as onion juice. Psychic tears are those that we cry in response to an emotion. All animals that live on land and have eyes produce basal and reflex tears. Humans are the only animals that cry psychic tears, or weep. The writer of the article states that we cry only *after* experiencing a strong emotion. Crying is controlled by the parasympathetic nervous system, which restores and maintains stability in the body and helps us recover from emotion.

2. In their research, students should investigate the effects of various anesthetics on nerve-impulse conduction and synaptic transmission.

CHAPTER 50 INVESTIGATION

TIME REQUIRED

One 50-minute class period

SAFETY PRECAUTIONS

Students should wear a lab apron, safety goggles, and safety gloves at all times during this investigation. Tell students to handle the sharp and pointed instruments—including the scalpel, tweezers, scissors, and the probe—carefully. At the end of the investigation, instruct students to place the sheep's eyes in a plastic biohazard bag. Secure the bag, and hand it over personally to the school custodian for safe disposal. Also, place all used scalpel blades in a biohazard sharps container for proper disposal.

QUICK REFERENCES

Lab Preparation notes are found on page 1001B.

Holt BioSources provides a Teaching Resources CD-ROM that contains "Using Gowin's Vee in the Lab" and "Scoring Rubrics."

PROCEDURAL TIP

During the dissection, have students compare the sheep's eye to the diagram of the human eye shown in Figure 50-13.

ANSWERS TO BACKGROUND

2. The pupil controls how much light enters the eye. In diffuse or dim light, the pupils dilate so that sufficient light reaches the retina. In bright light, the pupils constrict so that too much light does not enter the eye.
3. The sensory receptors in the retina are photoreceptors. The two different types of photoreceptors in the retina are called rods and cones.
6. The optic disk is the region where neurons from the retina exit and blood vessels enter

CHAPTER 50 INVESTIGATION

Sheep's Eye Dissection

OBJECTIVES

- Describe the main external and internal structures of a sheep's eye.
- Name the various structures associated with sight.

PROCESS SKILLS

- observing
- identifying
- comparing and contrasting

MATERIALS

- safety goggles
- disposable gloves
- lab apron
- preserved sheep's eye
- dissection tray
- scalpel
- tweezers
- fine scissors
- blunt probe

Background

1. The sheep's eye is very similar to the human eye, as shown in the diagram below. The wall of the eyeball is made up of three layers. The outer layer, the sclera, is a tough tissue that forms the white of the eye. At the front of the eye, the sclera becomes thin and transparent to form the cornea.

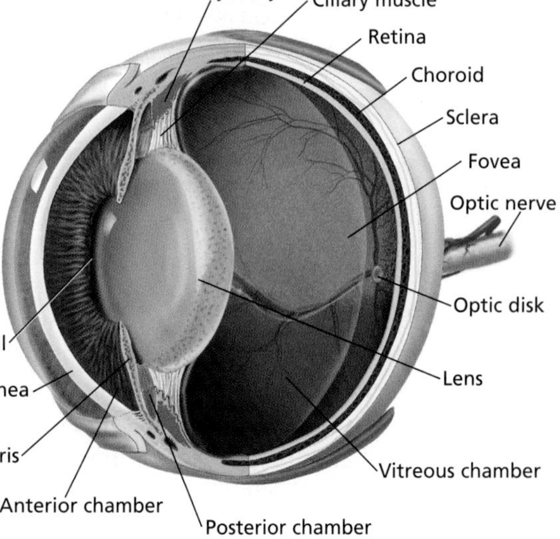

2. The middle layer, the choroid, is dark and rich with blood vessels. At the front of the eye behind the cornea, the choroid is modified into the iris and the ciliary body. The pigment in the doughnut-shaped iris determines the color of the eye. The opening in the center of the iris is the pupil. What is the function of the pupil?
3. The inner layer, the retina, is sensitive to light. What are the sensory receptor neurons in the retina called?
4. Directly behind the iris is the elastic, transparent lens. The suspensory ligament attaches the ciliary muscle, the main part of the ciliary body, to the lens. This ring of muscle changes the shape of the lens to focus on near and far objects.
5. The lens and its suspensory ligament divide the eye into two chambers. The large vitreous chamber extends from the retina to the lens and ligaments. It is filled with a gelatinous mass, the vitreous humor. This vitreous humor helps to maintain the shape of the eye and hold the retina in place. The second chamber, which extends from the iris to the cornea, is subdivided into two parts, the anterior and posterior chambers. The anterior chamber extends from the cornea to the iris; the posterior chamber from the iris to the suspensory ligament. Both chambers contain aqueous humor, a watery substance that bathes the front part of the eye.
6. In the retina, the nerves from the rods and cones bundle together at the optic nerve. The region of the retina where its nerve fibers and blood vessels enter the optic nerve is the small optic disk, which contains no rods or cones. Why is the optic disk also called the blind spot?
7. A very short distance away is a yellowish spot, the fovea. It is the site of sharpest vision because it has a high concentration of photoreceptors.

Procedure

1. Put on safety goggles, gloves, and a lab apron.
2. Locate the six main muscles on the outside of the eye, as shown in the diagram below. These

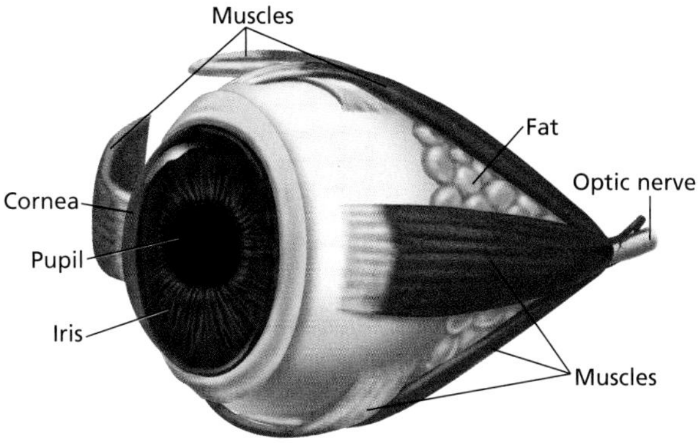

muscles move the eye in various directions. Use a scalpel to carefully cut the muscles near the eye. This will expose the sclera.
3. Observe the fatty tissue that cushions the eye in its socket, especially around the optic nerve. This fatty tissue helps to prevent shock. With a tweezer and scalpel, remove the fatty tissue. This will expose the optic nerve more fully.
4. Using a scalpel, carefully cut the sclera about 1 cm behind the cornea. Using fine scissors, extend the cut to make a flap that you can lift, as shown in the diagram below.

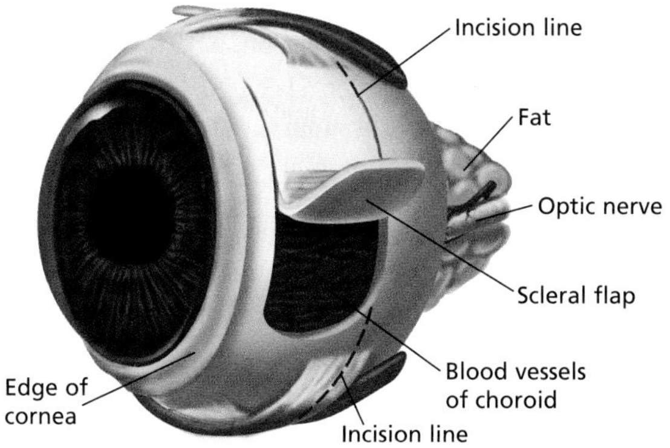

5. With forceps, carefully remove the sclera in this area and observe the dark choroid layer immediately below the sclera.
6. Next, use a scalpel to make an incision through the eye. Following along the incision you made in step 4, cut almost completely around the eye. You have separated the eye into an anterior and a posterior portion.
7. In the posterior section, observe the whitish retina. It is probably shriveled and may have fallen into the vitreous chamber.
8. In the anterior section, use a blunt probe to expose the lens. In a preserved eye, the lens is no longer clear.
9. In the anterior section, also locate the ciliary muscle and as many other structures as possible.
10. When you have finished your dissection, remove the specimen from the dissecting tray. Dispose of your materials according to the directions from your teacher.
11. Clean up your work area and wash your hands before leaving the lab.

Analysis and Conclusions

1. How is the lens different from the other structures of the eye?
2. How do the nature of the sclera and choroid change as they reach the front part of the eye?
3. How does the vitreous chamber differ from the anterior and posterior chambers?
4. What would be the result of the cornea becoming cloudy during an animal's lifetime?
5. Sometimes the retina of the eye becomes detached. Why is it important that the retina be reattached, if possible?

Further Inquiry

You see light if the vision center of your cerebrum is stimulated electrically. Scientists are trying to find out which brain cells are involved in forming images in our brains. How might scientists someday use this knowledge to help blind people see?

through the back of the eye. Because there are no photoreceptors in the region of the optic disk, there is no visual stimulation there. Thus, the optic disk is called the blind spot.

ANSWERS TO ANALYSIS AND CONCLUSIONS

1. The lens is quite hard and transparent, yet it is elastic enough to be stretched by suspensory ligaments.
2. At the front of the eye, the sclera becomes thin and transparent to form the cornea, and the choroid is modified into the iris and the ciliary body.
3. The large vitreous chamber contains extremely viscous fluid, whereas the smaller anterior and posterior chambers contain a watery fluid.
4. A cloudy cornea would prevent some light rays from reaching the retina, causing dark spots in the visual field.
5. The retina contains photoreceptors, which convert light energy to electrical energy, and retinal ganglion cells, whose axons exit the eye together as the optic nerve. If the retina became detached, light would hit the back of the eye unevenly, and some regions of the retina would be deprived of light. Also, many cells in the retina could perish if the retina became detached. If the retina were not reattached, serious vision disorders, even blindness, could result.

CHAPTER 51
ENDOCRINE SYSTEM

	TOPICS	TEACHING RESOURCES	LABS, CLASSWORK, AND HOMEWORK
BLOCKS 1 & 2 90 minutes	**Introducing the Chapter, p. 1028**	**ATE** Focus Concept, p. 1028 **ATE** Assessing Prior Knowledge, p. 1028 **ATE** Understanding the Visual, p. 1028	
	51-1 Hormones, p. 1029 Types of Glands, p. 1029 Types of Hormones, p. 1029 Hormone Action, p. 1030 Prostaglandins, p. 1031	**ATE** Section Overview, p. 1029 **ATE** Critical Thinking, p. 1031 **ATE** Visual Link Figure 51-1, p. 1030 198A, 204, 206–207, 229–231	**Lab A29,** Graphing Growth Rate Data ★ Study Guide, Section 51-1 **PE** Section Review, p. 1031
BLOCKS 3 & 4 90 minutes	**51-2 Endocrine Glands, p. 1032** Pituitary Gland, p. 1032 Thyroid Gland, p. 1034 Adrenal Glands, p. 1034 Gonads, p. 1035 Pancreas, p. 1036 Other Endocrine Glands, p. 1037	**ATE** Section Overview, p. 1032 **ATE** Visual Link Figure 51-4, p. 1033 194A, 195A, 205, 209	**PE** **Quick Lab,** p. 1034, Observing Solubilities ★ Study Guide, Section 51-2 **PE** Section Review, p. 1037
BLOCK 5 45 minutes	**51-3 Feedback Mechanisms, p. 1038** Homeostasis, p. 1038 Negative Feedback Mechanisms, p. 1039	**ATE** Section Overview, p. 1038 **ATE** Visual Link Figure 51-11, p. 1039 199A, 200A, 208	**PE** **Chapter 51 Investigation,** p. 1044 ★ Study Guide, Section 51-3 **PE** Section Review, p. 1040
BLOCKS 6 & 7 90 minutes	**REVIEW and ASSESSMENT** **PE** Chapter 51 Review, pp. 1042–1043 ■ Performance-Based Assessment—Testing Human Digestive Enzymes	Audio CD Program	**CHAPTER TESTING** ■ Chapter Test (blackline master) ▲ Test Generator ■ Assessment Item Listing ■ Scoring Rubrics and Classroom Management Checklists

KEY

MODERN BIOLOGY

PE Pupil's Edition
ATE Teacher's Edition
■ Active Reading Guide
★ Study Guide

HOLT BIOSOURCES

Teaching Transparencies
Laboratory Program

One-Stop Planner CD-ROM
Shows these resources within a customizable daily lesson plan:
■ Holt BioSources Teaching Resources
■ Active Reading Guide
★ Study Guide
▲ Test Generator

READING FOR CONTENT MASTERY

Preview: Have students preview the objectives, the introductory paragraph, topic sentences, and captions before reading the chapter.

ATE Active Reading Technique: Reading Effectively, p. 1029
■ Active Reading Guide Worksheet 51-1
PE Summary/Vocabulary, p. 1041

■ Active Reading Guide Worksheet 51-2
PE Summary/Vocabulary, p. 1041

■ Active Reading Guide Worksheet 51-3
PE Summary/Vocabulary, p. 1041

TECHNOLOGY AND INTERNET RESOURCES

CNN PRESENTS Video Segment 35 Obesity Hormone

Holt Biology Videodiscs Teacher's Correlation Guide, Lessons 51-1 through 51-3

Audio CD Program, Sections 51-1 through 51-3

internet**connect**

sciLINKS NSTA
On-line Resources: www.scilinks.org
The following *sci*LINKS Internet resources can be found in the student text for this chapter:

Topics:
- Glands, p. 1029
- Hormones, p. 1030
- Homeostasis, p. 1038

go.hrw.com
On-line Resources: go.hrw.com
Visit the HRW Web site for a variety of resources related to this chapter. Just type in the keyword HM2 Chapter 51.

Smithsonian Institution®
Visit **www.si.edu/hrw** for additional on-line resources.

CNNfyi.com
Visit **www.cnnfyi.com** for late-breaking news and current event stories selected just for you.

LAB ACTIVITY PLANNING

CHAPTER 51 INVESTIGATION

Observing the Effects of Thyroxine on Frog Metamorphosis, pp. 1044–1045

OVERVIEW
In this lab, students will study the effects of the hormone thyroxine on the development of tadpoles. Thyroxine affects the transition from gills to lungs, the decrease in tail size, and the growth of limbs. Thyroxine increases cellular metabolic rates and thereby increases the rate of metamorphosis.

SAFETY
Students should wear disposable gloves and safety goggles at all times during this lab. Tell students that thyroxine is highly toxic and can be absorbed through the skin. See safety information on page 1044.

PREPARATION
1. See p. 1044 for a list of materials needed for this lab.
2. Tadpoles can be ordered from any biological supply company. However, they are not always available year round, so be sure to order them well in advance.
3. To prepare the 0.01 percent thyroxine solution, dissolve 10mg crystalline thryoxine in 5 mL of 1 percent sodium hydroxide solution. Sodium hydroxide solution can be prepared by dissolving 0.5 g of NaOH pellets in 50 mL water. Dilute this solution by adding distilled water to a final volume of 1 L. Keep the thyroxine solution refrigerated.
4. If pond water is not available, set out open containers of water the night before the lab session. Commercially prepared baby food is a good source of strained spinach.

Quick Labs

Observing Solubilities, p. 1034
Each student group will need 100 mL beakers, water, gelatin, cooking oil, vitamin A or E capsules, a dissecting pin, and a measuring spoon for this activity.

UNDERSTANDING THE VISUAL

This photograph shows molecules of the neurotransmitter dopamine. Like many amino acid–based hormones, dopamine acts as a first messenger by binding to specific receptors on a postsynaptic neuron and activating a second messenger. In the endocrine system, dopamine released from the hypothalamus into the anterior pituitary inhibits the release of the hormone prolactin from the anterior lobe of the pituitary gland.

FOCUS CONCEPT

Stability and Homeostasis

The primary role of the endocrine system is to help maintain homeostasis. The endocrine system and the nervous system interact through the hypothalamus and pituitary gland, which together regulate production and secretion of hormones in many other endocrine glands. Endocrine glands secrete hormones into the bloodstream to influence the activity of distant target cells. The endocrine system uses cyclical feedback mechanisms to regulate hormone concentrations in the blood.

ASSESSING PRIOR KNOWLEDGE

Review the following concepts.

Homeostasis and Transport:
Chapter 5
Ask students to name the two main components of cell membranes. Ask them how membrane structure determines a cell's selective permeability to different materials. Ask students how concentration gradients influence the passage of materials across cell membranes. Ask them how homeostasis on a cellular level depends on the movement of substances into and out of cells.

CHAPTER 51

ENDOCRINE SYSTEM

The neurotransmitter dopamine is one of many amino acid–based compounds that act as chemical messengers in the body.

FOCUS CONCEPT: ***Stability and Homeostasis***

As you read, examine the various ways that the endocrine system helps the body maintain homeostasis and stability.

HORMONES

*The **endocrine** (EN-duh-KRIN) **system** consists of glands that transmit chemical messengers throughout the body. These chemical messengers, called **hormones** (HOHR-MOHNZ), circulate in the bloodstream and affect many types of body cells. In this section, you will learn about hormones and mechanisms of their action.*

TYPES OF GLANDS

A gland is an organ that consists of cells that secrete materials into other regions of the body. The body contains two types of glands: exocrine glands and endocrine glands. **Exocrine** (EK-suh-KRIN) **glands** secrete nonhormonal chemicals into ducts, which transport the chemicals to specific locations inside and outside the body. Sweat glands, mucous glands, salivary glands, and other digestive glands are examples of exocrine glands. **Endocrine glands** are ductless glands that are located throughout the body. Endocrine glands secrete hormones into the bloodstream through the fluid that surrounds their cells. You will learn more about endocrine glands in Section 51-2.

TYPES OF HORMONES

Hormones are compounds that are secreted in small amounts into the bloodstream and that influence the activity of distant cells. Hormones diffuse into the blood, which is their vehicle for transportation to various cells throughout the body. Hormones can be grouped into two general categories based on their structure: amino acid–based hormones and steroid, or lipid, hormones. **Amino acid–based hormones,** such as epinephrine (adrenaline), include proteins, peptides, amino acids, and other forms that are derived from amino acids. **Steroid** (STER-OYD) **hormones,** such as estrogen and testosterone, are lipids that the body synthesizes from cholesterol.

These two general classes of hormones differ significantly in their physical, and therefore chemical, properties. Thus, their mechanisms of action on body cells are also different. As you will see, mechanisms of hormone action depend on the way that hormones interact with cells that they affect.

SECTION

51-1

OBJECTIVES

▲ Compare exocrine glands with endocrine glands.

● Contrast amino acid–based hormones with steroid hormones.

■ Describe two ways that hormones affect their target cells.

◆ Distinguish hormones from prostaglandins.

internet**connect**

SCILINKS NSTA
TOPIC: Glands
GO TO: www.scilinks.org
KEYWORD: HM1029

Word Roots and Origins

hormone

from the Greek *hormon,* meaning "to excite"

SECTION 51-1

SECTION OVERVIEW

Stability and Homeostasis
Section 51-1 describes how endocrine glands are different from exocrine glands in that they secrete hormones directly into the bloodstream. This section defines amino acid–based hormones and steroid hormones, and it describes differences in the action of both types of hormones on their target cells. This section also describes prostaglandins.

ACTIVE READING

Technique: Reading Effectively
As students read through this chapter, have them make an outline of the material presented. You may want to review outlining methods, such as using Roman numerals, letters, and numbers. Have students develop an outline using the format shown as an example below.

I.
  A.
    1.
      a.
    2.
      a.

Go over the outlines to be sure students are developing the main ideas.

SECTION 51-1

TEACHING STRATEGY
Hormone Action

Ask students to recall how enzymes work as catalysts. (Enzymes bind to specific substrate molecules, lowering the activation energy of the substrates.) Ask them how hormones affect their target cells. (Hormones bind to specific receptor proteins on or inside their target cells.) Tell students that, like an enzyme-substrate complex, a hormone-receptor complex is similar to a "lock-and-key" mechanism. If the key (hormone) does not fit the lock (receptor protein), then there is no effect on the target cell.

TERMINOLOGY NOTES

The term *amino acid–based hormone* replaces *protein hormone.* The new term more accurately describes this group of hormones, which includes amino acids, proteins, peptides, and amines. Also, the term *steroid hormone* replaces *lipid hormone.* Steroids are a type of lipid, and not all steroids are hormones.

The term *cyclic AMP* stands for 3′,5′-cyclic adenosine monophosphate.

VISUAL LINK
Figure 51-1

This figure illustrates how an amino acid–based hormone binds to receptor proteins on the exterior surface of a target cell. A hormone-receptor complex indirectly activates c-AMP, a second messenger that activates other enzymes, which in turn activate several substrate proteins. Substrate proteins can cause changes in the cell's activity. For example, epinephrine causes reactions that lead to the breakdown of glycogen into glucose within its target cells.

internet**connect**

SCILINKS NSTA

TOPIC: Hormones
GO TO: www.scilinks.org
KEYWORD: HM1030

HORMONE ACTION

The body produces many hormones, and each hormone affects only specific cells, called **target cells.** Target cells have receptors that recognize and bind to specific hormones. **Receptors** are proteins that are located both inside the cytoplasm and on the surface of a target cell. When a hormone binds to a receptor on its target cell, it triggers events that lead to changes within the cell.

Amino Acid–Based Hormones

Because most amino acid–based hormones cannot diffuse passively across the membranes of their target cells, a two-messenger system is commonly required for the action of most of these hormones. Amino acid–based hormones identify their target cells by their attraction to receptor proteins that are embedded in the target cell membrane. The hormone acts as a **first messenger** by binding to a specific receptor on the surface of its target cell. This forms a **hormone-receptor complex,** which activates a second messenger located inside the target cell. A **second messenger** relays and amplifies the hormone signal. Figure 51-1 shows how, in many cases, the hormone-receptor complex indirectly activates an enzyme that converts molecules of ATP to cyclic AMP (c-AMP) inside the cell. Cyclic AMP acts as a second messenger by indirectly activating other enzymes and proteins in the target cell. Thus, c-AMP initiates a chain of biochemical events that leads to functional changes within the target cell.

Steroid Hormones

Steroid hormones do not act through cell surface receptors. Instead, they diffuse through the membranes of their target cells and bind to receptors in the cytoplasm. The newly formed hormone-receptor complexes cause the cells to activate existing enzymes or to initiate synthesis of new enzymes or proteins. Figure 51-2 illustrates how a hormone-receptor complex binds to DNA inside the nucleus of a target cell. Once bound to DNA, the hormone-receptor complex activates transcription of mRNA. By activating mRNA transcription, steroid hormones stimulate production of new proteins, which cause changes in the target cell.

FIGURE 51-1

(a) An amino acid–based hormone acts as a first messenger by binding to receptor proteins located on the target cell membrane. (b) The hormone-receptor complex indirectly activates an enzyme that converts ATP to cyclic AMP. (c) Cyclic AMP indirectly activates other enzymes that cause changes in the target cell.

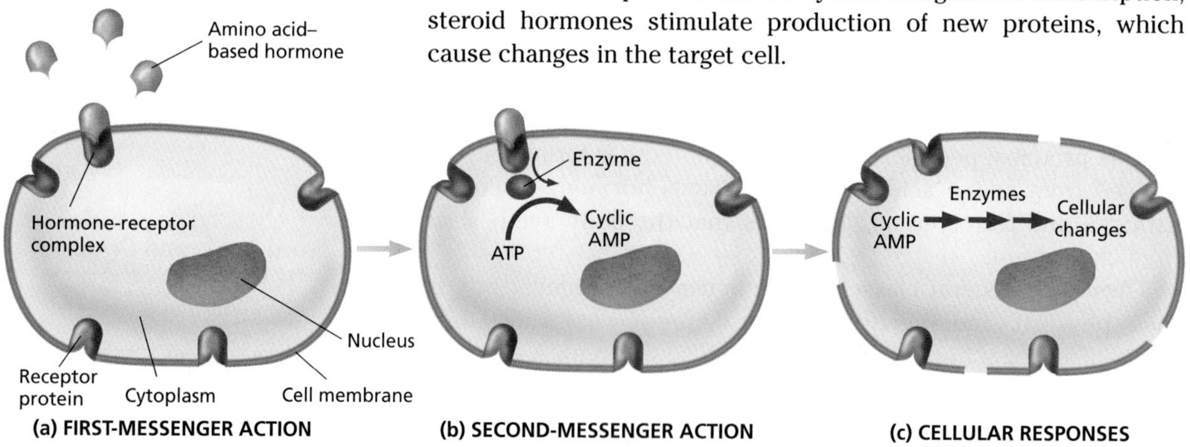

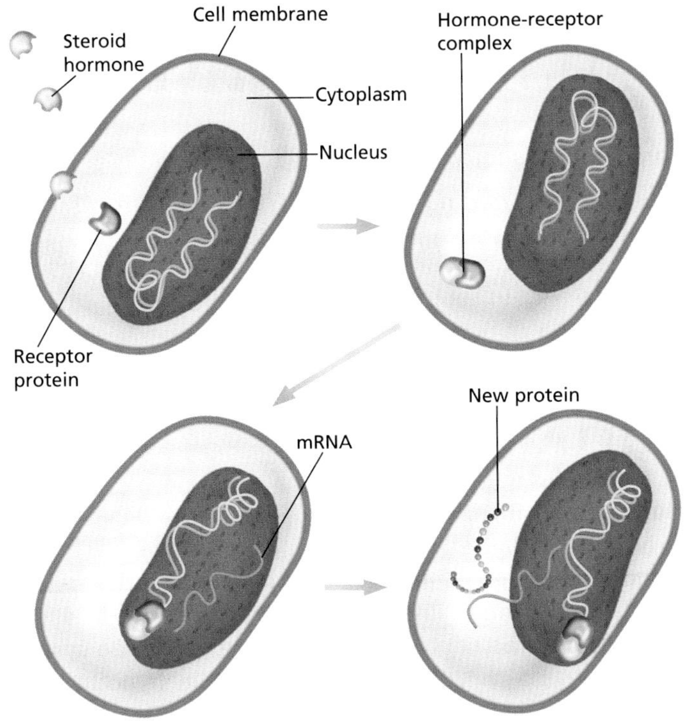

FIGURE 51-2

(a) Steroid hormones diffuse through the target cell membrane. (b) Once inside the target cell, the hormone binds to a receptor protein. (c) The newly formed hormone-receptor complex enters the nucleus of the target cell and binds to DNA, activating mRNA transcription. (d) Genes are activated, mRNA is transcribed, and new proteins are synthesized.

PROSTAGLANDINS

Prostaglandins (PRAHS-tuh-GLAN-dinz), a group of hormonelike lipids, also regulate cell activities. Unlike hormones, prostaglandins are not produced by specific endocrine glands. Instead, these chemicals are produced in small quantities by many cells throughout the body. Rather than being transported through the blood to distant regions of the body, prostaglandins act locally. Their effects include relaxation of smooth muscles that line air passageways and blood vessels, regulation of blood pressure, contraction of the intestinal walls and the uterus, and stimulation of the body's inflammatory response to infection.

SECTION 51-1 REVIEW

1. Compare exocrine glands with endocrine glands.
2. What are hormones, and what do they do?
3. Explain the differences in the way amino acid–based hormones and steroid hormones affect their target cells.
4. How are hormones transported throughout the body?
5. Compare prostaglandins with hormones.
6. CRITICAL THINKING Why are nonendocrine messengers not considered to be hormones?

SECTION 51-1

CRITICAL THINKING

Prostaglandins

Tell students that prostaglandins stimulate contraction of uterine muscles during labor and promote blood clotting, which helps minimize hemorrhages. Ask students how drugs that inhibit prostaglandins, such as aspirin and ibuprofen, can affect childbirth. (Labor may be delayed, and large amounts of blood could be lost, possibly threatening the mother's life.)

ANSWERS TO SECTION 51-1 REVIEW

1. Exocrine glands secrete nonhormonal chemicals into ducts. Endocrine glands secrete hormones directly into the bloodstream.
2. Hormones are chemicals that are secreted into the bloodstream and that influence the activity of distant target cells.
3. An amino acid–based hormone cannot pass directly through the cell membrane. It uses a second messenger to relay its signals to its target cells. A steroid hormone diffuses through the membrane of its target cell and binds to a receptor protein inside the cell. The hormone-receptor complex enters the nucleus and activates mRNA transcription.
4. Hormones are transported in the bloodstream.
5. Unlike hormones, prostaglandins are produced by many cells throughout the body, and they act locally.
6. Unlike hormones, nonendocrine chemical messengers do not affect distant cells.

SECTION 51-2

SECTION OVERVIEW

Stability and Homeostasis
Section 51-2 describes the function of each endocrine gland. It describes the hormones that are produced and secreted by endocrine glands and their role in regulating metabolic activities and maintaining homeostasis. This section emphasizes that the hypothalamus, a structure of the nervous system, and the pituitary gland produce and secrete hormones that affect the activity of many other glands.

▶ ENGAGE STUDENTS

Ask students to record their heart rate and breathing rate. Sometime later, expose them to a stressful situation by announcing a pop quiz that tests their knowledge of the endocrine system and the nervous system. Immediately after the quiz, have students record their heart and breathing rates. (Students should observe an increase in both.) Then tell students that the quiz does not count toward the course grade. Just before the class period ends, have students record their heart and breathing rates again. Ask students to account for the differences in rate. (Stressful situations cause the adrenal glands to secrete the hormones epinephrine [adrenaline] and norepinephrine [noradrenaline] into the bloodstream. The effects include increases in heart rate, breathing rate, and blood pressure, and dilation of bronchial tubes and the pupils.)

TERMINOLOGY NOTE

The term *antidiuretic hormone (ADH)* replaces the term *vasopressin.* ADH is a more common and widely used name for vasopressin, a hormone secreted by the posterior pituitary.

SECTION 51-2

OBJECTIVES

▲ List the major endocrine glands and hormones found in the body.

● Discuss the relationship between the hypothalamus and the pituitary gland.

■ Describe the function of each endocrine gland.

ENDOCRINE GLANDS

Endocrine glands, shown in Figure 51-3, are located throughout the body and regulate many of its vital processes. This section discusses the hormones that each endocrine gland produces and their effects on the body.

PITUITARY GLAND

The **pituitary** (puh-TOO-uh-TER-ee) **gland,** shown in Figure 51-4, secretes hormones that affect other glands and organs. The two lobes of the pituitary gland are regulated by the hypothalamus, a structure of the nervous system, also shown in Figure 51-4. The hypothalamus links the nervous system to the endocrine system. **Neurosecretory** (NOO-roh-SEE-kret-ohr-ee) **cells** of the hypothalamus produce hormones that either are stored in the pituitary gland or regulate the pituitary's activity. The hypothalamus and the pituitary gland are the primary regulators of the endocrine system.

Posterior Pituitary

Two amino acid–based hormones—oxytocin and antidiuretic hormone—are produced by neurosecretory cells whose axons extend into the posterior pituitary lobe, as shown in Figure 51-4. These hormones are transported down these axons into the posterior pituitary, where they are stored for eventual release into the bloodstream. Signaling through these neurons, the hypothalamus also regulates the secretion of these hormones from the posterior pituitary.

Oxytocin (AHK-see-TOH-sin) stimulates contraction of the uterus—a structure of the female reproductive system—during childbirth and flow of breast milk from the mammary glands during nursing. **Antidiuretic** (AN-tie-die-yoo-RET-tik) **hormone (ADH)** helps regulate the concentration of solutes in the blood by controlling the amount of water excreted by the kidneys. When the concentration of solutes in the blood increases, the hypothalamus signals the posterior pituitary to secrete ADH. ADH causes tubules in the kidneys to reabsorb water into the blood. Thus, the kidneys produce urine with a high solute concentration. Hypothalamic receptors detect the subsequent decrease in the concentration of solutes in the blood and stop signaling the posterior pituitary to release ADH.

Anterior Pituitary

Neurosecretory cells also produce and secrete **releasing hormones,** which stimulate endocrine cells of the anterior pituitary

FIGURE 51-3
Endocrine glands are located throughout the body. All of these glands contain cells that secrete hormones. The hypothalamus is a part of the nervous system that regulates the pituitary gland.

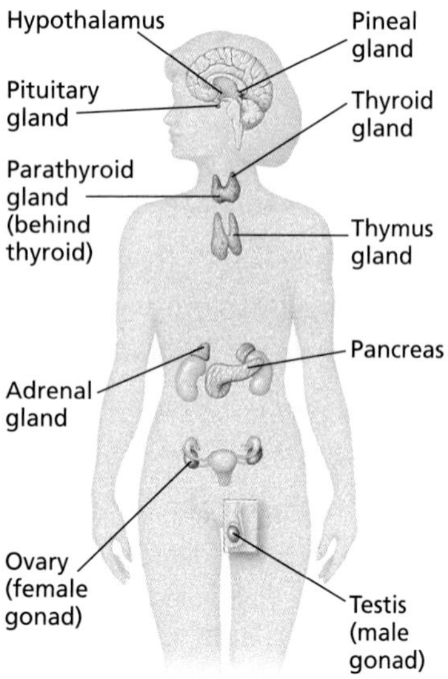

lobe to produce and secrete hormones. Other hypothalamic neurosecretory cells produce **release-inhibiting hormones,** which inhibit production and secretion of anterior-pituitary hormones. Releasing hormones and release-inhibiting hormones are produced in response to various stimuli that are processed by the nervous system. There is at least one releasing hormone for each anterior-pituitary hormone. As shown in Figure 51-4, a specialized system of blood vessels connects the hypothalamus with the anterior pituitary. Neurosecretory cells release hormones into the anterior pituitary through blood vessels.

Some anterior pituitary hormones, such as prolactin and growth hormone, are regulated by the hypothalamus through both a releasing hormone and a release-inhibiting hormone. **Growth hormone (GH)** controls skeletal and muscular growth. **Prolactin** (proh-LAK-tin) **(PRL)** stimulates and sustains production of breast milk during lactation. PRL-releasing hormone stimulates PRL production and secretion, whereas PRL release-inhibiting hormone stops PRL secretion. Releasing hormones also regulate production and secretion of other hormones—luteinizing hormone, follicle-stimulating hormone, adrenocorticotropic hormone, and thyroid-stimulating hormone—in the anterior pituitary that stimulate other endocrine glands. Table 51-1 lists eight hormones that are secreted by the pituitary gland.

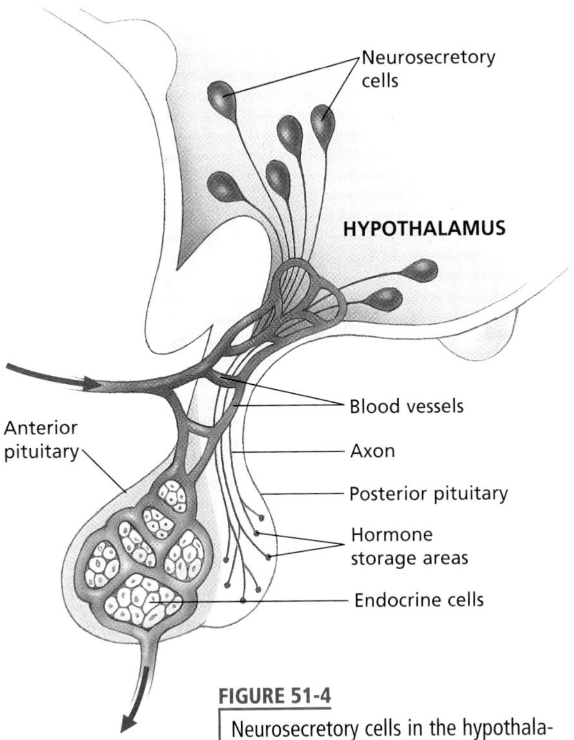

FIGURE 51-4
Neurosecretory cells in the hypothalamus produce hormones that affect the pituitary gland. The hypothalamus regulates the posterior pituitary through axons and the anterior pituitary through blood vessels. (Blood vessels in the posterior pituitary have been omitted to show axon projections.)

TABLE 51-1 *Hormones Secreted by the Pituitary Gland*

Hormone	Target	Function
Adrenocorticotropic hormone (ACTH)	adrenal cortex	stimulates secretion of cortisol and aldosterone by the adrenal cortex
Antidiuretic hormone (ADH)	kidney tubules	stimulates reabsorption of water by kidneys, reducing the concentration of solutes in the blood
Follicle-stimulating hormone (FSH)	ovaries in females; testes in males	stimulates egg production in females; stimulates sperm production in males
Growth hormone (GH)	muscle and bone	regulates development of muscles and bones
Luteinizing hormone (LH)	ovaries in females; testes in males	stimulates progesterone and estrogen production; initiates ovulation in females; stimulates testosterone production
Oxytocin	mammary glands and uterine muscles	initiates uterine contractions during childbirth; stimulates flow of milk from breasts during lactation
Prolactin (PRL)	mammary glands	stimulates milk production in breasts during lactation
Thyroid-stimulating hormone (TSH)	thyroid gland	regulates secretion of the thyroid hormones—thyroxine and triiodothyronine

VISUAL LINK
Figure 51-4

This figure illustrates how the hypothalamus and the pituitary gland are connected. Ask students which lobe of the pituitary is directly connected to the hypothalamus (the posterior lobe). Ask them how this relates to the function of the posterior pituitary. (It enables the hypothalamus to transport hormones directly into the posterior pituitary, where they are stored for eventual release.) Ask students what the anterior pituitary contains. (It contains blood vessels and endocrine cells.) Ask them how this relates to the function of the anterior pituitary. (The hypothalamus secretes releasing hormones and release-inhibiting hormones into the anterior pituitary through blood vessels. Endocrine cells in the anterior pituitary secrete hormones into the bloodstream.) Emphasize to students that the hypothalamus is part of the nervous system but that it connects to the endocrine system through the pituitary gland. Ask students why the hypothalamus is important to endocrine-system function. (The hypothalamus controls the release of all hormones from both lobes of the pituitary gland, and it links the nervous system to the endocrine system.)

QUICK FACT

☑ Two hormones that are secreted by the anterior pituitary—prolactin (PRL) and growth hormone (GH)—affect target cells of nonendocrine organs. The other four anterior pituitary hormones —follicle-stimulating hormone (FSH), luteinizing hormone (LH), thyroid-stimulating hormone (TSH), and adrenocorticotropic hormone (ACTH)—are tropic hormones. *Tropic hormones* affect target cells in other endocrine glands.

CONTENT CONNECTION

Section 5-1: Transport

Ask students to recall the discussions of membrane structure in Chapter 4 and diffusion in Chapter 5. Ask them how the composition of a cell membrane determines its selective permeability to materials. (Cell membranes consist of a hydrophobic, fatty phospholipid bilayer and hydrophilic proteins.) Ask students how cell membrane properties determine hormone action. (Lipid-soluble molecules, such as steroid hormones, can diffuse through the cell membrane. Hydrophilic molecules, such as amino acid–based hormones, rely on membrane proteins and second messengers to carry out their action.)

Quick Lab

Observing Solubilities

Time Required 20 minutes

Procedural Tips Have students work in small groups. If measuring spoons are not available, students can use their fingers to place a few pinches of the gelatin into the beakers. If small beakers are not available, test tubes can be substituted, but reduce the amount of gelatin by half.

Disposal Solutions can be poured down the drain. The empty gelatin capsules from the vitamins should be discarded in the trash.

Answers to Analysis The vitamins are fat soluble. The gelatin is water soluble. Because cell membranes are made of fatty phospholipids, fat-soluble substances can pass through the membranes, but water-soluble compounds cannot. This is why amino acid–based hormones remain outside their target cells while steroid hormones can enter their target cells.

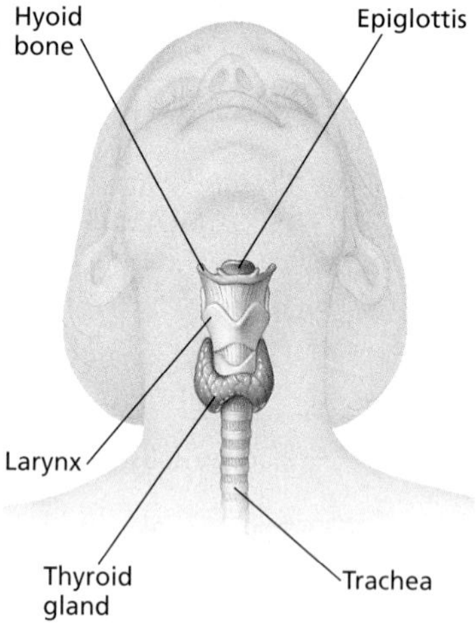

FIGURE 51-5

The thyroid gland is located under the larynx and on the trachea.

Quick Lab

Observing Solubilities

Materials 100 mL beakers (4), water, gelatin, cooking oil, vitamin A or E capsules, dissecting pin, measuring spoon

Procedure

1. Put 75 mL of water into a beaker. Place 2.5 g of gelatin (protein) into the beaker, and swirl the solution. Does the gelatin dissolve? Record your observations.
2. Put 75 mL of oil into a clean beaker. Repeat the procedure in step 2 using oil instead of water.
3. Repeat steps 1 and 2 using the contents of a vitamin capsule (fat) instead of gelatin.

Analysis Which substance is fat soluble? Which substance is water soluble? What type of substance can diffuse through a cell membrane? Relate the solubilities of hormones to whether or not they enter their target cells or work outside them.

THYROID GLAND

The two lobes of the **thyroid** (THIE-ROYD) **gland** are located near the lower part of the larynx, as shown in Figure 51-5. **Thyroid-stimulating hormone (TSH)** regulates the thyroid gland. Release of TSH from the anterior pituitary is regulated by both a releasing hormone and a release-inhibiting hormone. When stimulated by TSH, the thyroid gland produces and secretes the hormones **thyroxine** (thie-RAHK-sin) and **triiodothyronine** (TRIE-ie-oh-DOH-THIE-roh-NEEN). Both of these hormones are derived from the same amino acid and are synthesized with iodine atoms. The thyroid hormones help maintain normal heart rate, blood pressure, and body temperature. They stimulate enzymes that are associated with glucose oxidation and oxygen consumption, generating heat and increasing cellular metabolic rates. They also promote carbohydrate usage over fat usage for energy.

The thyroid gland is important to human development. It also produces **calcitonin** (KAL-sih-TOH-nin), a hormone that stimulates the transfer of calcium ions from blood to bone, where they can be used to generate bone tissue.

Abnormal thyroid activity can be detrimental to the body's metabolism. Overproduction of the thyroid hormones is called **hyperthyroidism** (HIE-puhr-THIE-royd-izm). Symptoms of hyperthyroidism include overactivity; weight loss; and high blood pressure, heart rate, and body temperature. Hyperthyroidism can be treated with medication or by surgical removal of part of the thyroid gland. Thyroid-hormone deficiency is known as **hypothyroidism** (HIE-poh-THIE-royd-izm). Symptoms of hypothyroidism include growth retardation, lethargy, weight gain, and low heart rate and body temperature. It can also cause **cretinism** (KREET-uhn-IZ-uhm), a form of mental retardation, during fetal and childhood development. If hypothyroidism is caused by iodine deficiency, then **goiter** (GOY-tuhr), or swelling of the thyroid gland, results. Hypothyroidism can be treated with supplementary thyroxine.

ADRENAL GLANDS

One **adrenal** (uh-DREE-nuhl) **gland** is located above each kidney, as shown in Figure 51-6. Each adrenal gland has an inner core, the medulla, and an outer layer, the cortex. The medulla and cortex function as separate endocrine glands. Medullary hormone secretion is controlled by the nervous system, whereas the anterior pituitary regulates cortical hormone secretion.

Adrenal Medulla

The adrenal medulla produces two amino acid–based hormones: **epinephrine** (EP-i-NEF-rin) and **norepinephrine** (NOHR-ep-i-NEF-rin) **(NE),** also known as adrenaline (uh-DREN-uh-lin) and noradrenaline (NOR-uh-DREN-uh-lin), respectively. These hormones orchestrate the nervous system's reaction to stress and its "fight-or-flight" response to danger. When a person is stressed, the medulla secretes epinephrine and NE into the bloodstream. These hormones cause the liver to break down glycogen into glucose, raising the level of glucose, which is oxidized for additional energy, in the blood. The results include enlargement of the bronchial tubes, dilation of the pupils, and an increase in heart rate. As the heart beats faster, surface blood vessels constrict, blood pressure rises, and more blood circulates to the muscles, brain, and heart.

Adrenal Cortex

The adrenal cortex responds to **adrenocorticotropic** (uh-DRE-noh-KOHR-ti-koh-TROH-pik) **hormone (ACTH),** which is secreted by the anterior pituitary. Stress causes the hypothalamus to secrete ACTH-releasing hormone. ACTH then stimulates the adrenal cortex to produce the steroid hormones **cortisol** (KOHRT-uh-SAWL), which regulates metabolism of carbohydrates and proteins, and **aldosterone** (al-DAHS-tuh-ROHN), which helps maintain the salt-and-water balance in the body by affecting the kidneys.

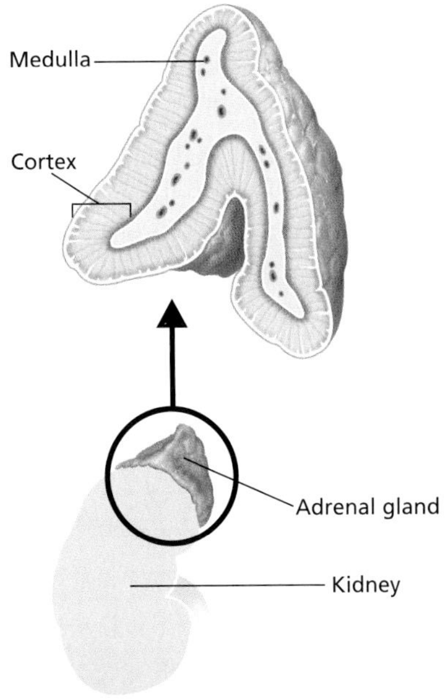

FIGURE 51-6

The adrenal glands, located above each kidney, consist of an inner medulla and an outer cortex. Epinephrine and norepinephrine are produced in the medulla, while cortisol and aldosterone are produced in the cortex.

GONADS

Gonads—the ovaries in females and the testes in males—are gamete-producing organs that also produce a group of steroid sex hormones. **Sex hormones** regulate body changes that begin at puberty. **Puberty** (PYOO-buhr-tee) is the adolescent stage during which the sex organs mature and secondary sex characteristics, such as facial hair, appear. During puberty in males, sperm production begins, the voice deepens, the chest broadens, and hair grows on the body and face. In females, the menstrual cycle begins, the breasts grow, and the hips widen. When secreted by the anterior pituitary, **luteinizing** (LOO-tee-in-IZE-ing) **hormone (LH)** and **follicle-stimulating** (FOL-uh-kuhl) **hormone (FSH)** stimulate secretion of sex hormones from the gonads.

In females, LH and FSH stimulate secretion of **estrogen** (ES-truh-jen) and **progesterone** (proh-JES-tuh-ROHN) from the ovaries. In preparation for a possible pregnancy, these sex hormones cause the monthly release of an egg by an ovary and buildup of the uterine lining. Estrogen also regulates female secondary sex characteristics. In males, LH stimulates the testes to secrete a group of sex hormones called **androgens** (AN-druh-jenz). **Testosterone** (tes-TAHS-tuh-ROHN) is an androgen that regulates male secondary sex characteristics. Along with FSH, testosterone also stimulates sperm production. You will learn more about the gonads and the sex hormones in Chapter 52.

TERMINOLOGY NOTE

The terms *epinephrine* and *norepinephrine* replace *adrenaline* and *noradrenaline.* The former two terms are now commonly used in the scientific and medical communities, whereas the latter two terms, although still valid, have become obsolete.

DEMONSTRATION

To demonstrate the "fight-or-flight" response to stress or danger, make a very loud, startling noise by activating an alarm, blowing a coach's whistle, or honking an air horn at the beginning of the class period, when students are settling in. Ask students to describe their reactions to the noise. (Responses might include perspiration, nervousness, fear, surprise, stress, concern, and tension.) Ask them to explain the body's reaction to the alarming stimulus. (Loud, unexpected noises usually signal a potentially dangerous situation, triggering the body's "fight-or-flight" response to danger.) Ask students how the "fight-or-flight" response works. (The adrenal glands release epinephrine and norepinephrine into the bloodstream. Like glucagon, these hormones stimulate liver cells to release stored glucose, which supplies additional energy. Heart and breathing rates increase, and blood is diverted to skeletal muscles and the brain.) The adrenal hormones prolong and intensify changes caused by stimulation of the sympathetic division of the autonomic nervous system. This is a good example of the interaction of the endocrine system and the nervous system.

SECTION 51-2

MAKING CONNECTIONS

Social Studies

Many athletes use anabolic-steroid hormones to enhance their performance. Simply referred to as *steroids,* anabolic-steroid hormones are synthesized as an artificial version of the androgen testosterone, a steroid hormone primarily responsible for the development of secondary sex characteristics and other physical changes, including muscle development, that occur in males during puberty. Anabolic steroids stimulate tissue buildup; some athletes, both male and female, take them orally or by injection to increase muscle mass, speed, and strength. Many anabolic-steroid users exhibit manic-depressive behavior, and some become aggressive and violent. Ask students why some athletes continue to use these drugs even though they are harmful. (Athletes in the United States, in particular, have the potential to make a lot of money with their skills. Thus, some of them view anabolic-steroid use as critical to their chances for success.) Ask students to research long-term effects of anabolic-steroid use. (For example, abuse of these artificial hormones can lead to sterility, liver cancer, stunted growth, altered concentrations of "good" cholesterol in the blood, and, in males, deformation of the testes.)

QUICK FACT

☑ Abuse of anabolic-steroid hormones disrupts the normal negative feedback mechanisms that regulate hormone concentrations throughout the body. This can lead to tissue damage, sterility, behavioral disorders, and life-threatening metabolic problems.

FIGURE 51-7

A cross section of pancreatic tissue shows the islets of Langerhans (lightly colored region). These endocrine cells are surrounded by exocrine cells that produce digestive fluids. (LM 315×)

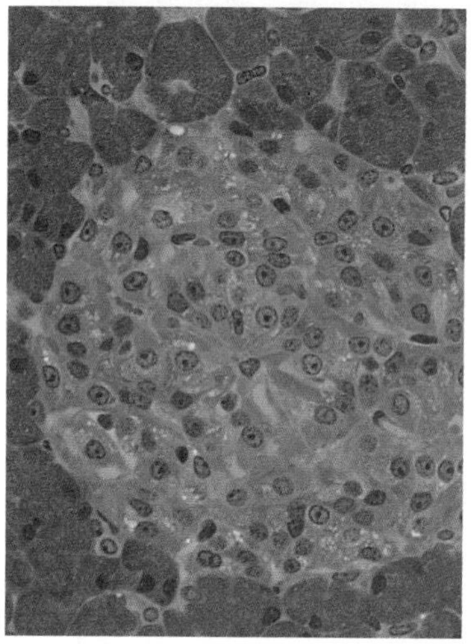

PANCREAS

The pancreas mostly contains exocrine cells, but specialized cells in the pancreas called the **islets of Langerhans** (LANG-uhr-HANZ) function together as an endocrine gland. Shown in Figure 51-7, these endocrine cells secrete two amino acid–based hormones that regulate the level of sugar in the blood. **Insulin** (IN-suh-lin) lowers the blood sugar level by stimulating body cells, especially muscles, to absorb glucose. In contrast, **glucagon** (GLOO-kuh-gahn) stimulates release of glucose into the bloodstream by liver cells.

Insulin deficiency causes **diabetes mellitus** (die-uh-BEET-eez muh-LIET-uhs), a condition of abnormally high blood glucose concentration. Type I diabetes is a severe childhood disorder in which insulin-producing islet cells die. Type I generally is treated with daily injections of insulin into the blood, and sometimes is treated with islet cell transplant. Type II diabetes usually occurs after age 40, and it is more common and less severe than type I. Type II is caused by insufficient insulin or unresponsive target cell receptors. Although type II is hereditary, its onset correlates with obesity, and type II often can be controlled through exercise and diet. In people with diabetes, excess glucose inhibits water reabsorption by the kidneys, producing large amounts of urine. Thus, dehydration and kidney damage can result. Lack of insulin can also cause nausea and rapid breathing, possibly leading to oxygen deficiency, circulatory and nervous system failure, diabetic coma, or even death.

TABLE 51-2 ***Summary of Endocrine Glands and Their Functions***

Glands	Hormone	Function
Adrenal glands		
Cortex	aldosterone	maintains salt-and-water balance
	cortisol	regulates carbohydrate and protein metabolism
Medulla	epinephrine, norepinephrine	initiate body's response to stress and the "fight-or-flight" response to danger
Ovaries	estrogen	regulates female secondary sex characteristics
	progesterone	maintains growth of uterine lining
Pancreas (islets of Langerhans)	glucagon	stimulates release of glucose
	insulin	stimulates absorption of glucose
Parathyroid glands	parathyroid hormone	increases blood calcium concentration
Pineal gland	melatonin	regulates sleep patterns
Pituitary gland	see Table 51-1	see Table 51-1
Testes	androgens (testosterone)	regulate male secondary sex characteristics
Thymus gland	thymosin	stimulates T-cell formation
Thyroid gland	thyroxine, triiodothyronine	increase cellular metabolic rates

Excessive insulin causes **hypoglycemia** (HIE-poh-glie-SEE-mee-uh), a disorder in which glucose is stored, rather than being properly delivered to body cells. This leads to a lowered blood glucose concentration and subsequent release of glucagon and epinephrine. Symptoms of hypoglycemia include lethargy, dizziness, nervousness, overactivity, and in extreme cases, unconsciousness and death.

OTHER ENDOCRINE GLANDS

There are several other glands in the endocrine system, including the thymus gland, the pineal gland, and the parathyroid glands. There are also specialized digestive endocrine cells. The endocrine glands and their functions are listed in Table 51-2.

The **thymus** (THIE-muhs) **gland,** located beneath the sternum and between the lungs, consists mostly of T-cells and plays a role in the development of the immune system. The thymus gland secretes **thymosin** (THIE-moh-sin), an amino acid–based hormone that stimulates formation of T-cells, which help defend the body from pathogens.

The **pineal** (PIEN-ee-uhl) **gland** is located near the base of the brain, as shown in Figure 51-8. It secretes the hormone melatonin. **Melatonin** (mel-uh-TOH-nin) concentrations increase sharply at night and decrease dramatically during the day. This cyclic release of melatonin indicates that it helps regulate sleep patterns.

As Figure 51-9 shows, the four **parathyroid glands** are embedded in the back of the thyroid gland, two in each lobe. These glands secrete **parathyroid hormone,** which increases the concentration of calcium ions in the blood. A proper balance of calcium ions is necessary for normal bone growth, muscle tone, and neural activity.

Endocrine cells within the walls of some digestive organs also secrete a variety of hormones that help digest food. When food is eaten, endocrine cells in the stomach lining secrete **gastrin** (GAS-trin), a hormone that stimulates other stomach cells to release digestive enzymes and hydrochloric acid. Endocrine cells of the small intestine release **secretin** (si-KREE-tin), a hormone that stimulates the release of various digestive fluids from the pancreas and bile from the liver.

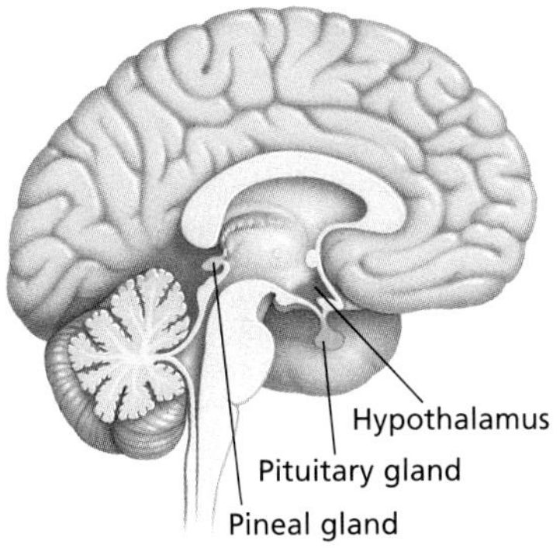

FIGURE 51-8
The pineal gland, located near the base of the brain, secretes the hormone melatonin at night.

FIGURE 51-9
The four parathyroid glands are embedded in the dorsal side of the thyroid gland. They secrete a hormone that regulates the concentration of calcium ions in the blood.

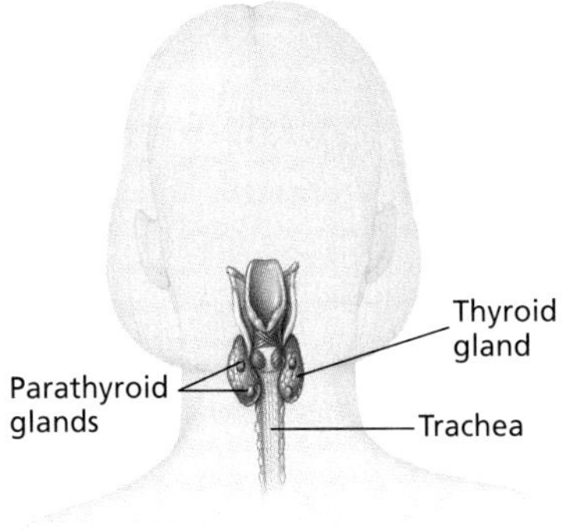

SECTION 51-2 REVIEW

1. List six endocrine glands and their functions.
2. What is a releasing hormone?
3. Describe the effects of two releasing hormones on the pituitary gland.
4. Explain why the pineal gland is considered the "biological clock" of the body and why the pituitary gland is considered the "master gland" of the body.
5. Describe the malfunctions that cause Type I and Type II diabetes.
6. CRITICAL THINKING Why might overactive parathyroid glands cause bone problems?

RETEACHING *ACTIVITY*

Using Table 51-1 and Table 51-2, review with students the actions of all hormones secreted by each endocrine gland, including antagonistic hormones. Ask students to give examples of hormone types (for example, epinephrine is an amino acid–based hormone).

ANSWERS TO SECTION 51-2 REVIEW

1. Answers will vary but may include the adrenals, the thyroid gland, the pituitary gland, the thymus gland, the pineal gland, and the gonads.
2. A releasing hormone is a chemical that is secreted by the hypothalamus into the anterior pituitary. Releasing hormones stimulate production and secretion of hormones from the anterior pituitary.
3. Answers will vary but may include PRL-releasing hormone and TSH-releasing hormone, which stimulate secretion of PRL and TSH, respectively, from the anterior pituitary.
4. Secretion of melatonin from the pineal gland increases at night and decreases during the day. This suggests that the pineal gland regulates sleep patterns. The pituitary gland regulates the activity of many other endocrine glands. Thus, it is known as the master gland.
5. Type I is caused by degeneration of insulin-producing islet cells, whereas type II is caused by insufficient insulin or lack of functional insulin receptors.
6. Overactive parathyroid glands secrete too much parathyroid hormone, which increases the concentration of calcium ions in the blood by stimulating bone cells to release calcium. Thus, overactive parathyroids can cause softening and deformation of bones.

SECTION 51-3

SECTION OVERVIEW

Stability and Homeostasis
Section 51-3 describes how hormone production is regulated by feedback mechanisms. It explains how both negative and positive feedback control hormone production and secretion. This section includes examples of how negative feedback mechanisms regulate the concentration of hormones in the blood.

▶ ENGAGE STUDENTS

Ask students to imagine or recall waiting for the school bus on a cold morning. Ask students to suggest how the body adjusts to the outside temperature. (The hypothalamus monitors body temperature and triggers a negative feedback mechanism to adjust to external temperature.) Ask students how the hypothalamus initiates the adjustment. (It produces and secretes TSH-releasing hormone, which causes the anterior pituitary to release TSH. TSH in turn stimulates secretion of the thyroid hormones, which generate body heat by increasing cellular metabolic rates.) Ask students what happens after the body adjusts to the low temperature. (The hypothalamus stops secreting TSH-releasing hormone and begins secreting TSH release-inhibiting hormone, decreasing thyroid activity.) Emphasize that the endocrine system uses negative feedback to regulate the concentration of the thyroid hormones (and, therefore, body temperature) within a certain range.

SECTION

51-3

OBJECTIVES

▲ Define the terms *feedback mechanism* and *antagonistic hormones.*

● Distinguish positive feedback from negative feedback.

■ Explain the role of negative feedback mechanisms in maintaining homeostasis.

◆ Give examples of negative feedback in the endocrine system.

FEEDBACK MECHANISMS

The endocrine system uses feedback mechanisms to respond and adjust to changes that occur in and outside the body. In a ***feedback mechanism,*** *the last step in a series of events controls the first step.*

HOMEOSTASIS

Recall that homeostasis is defined as a stable internal environment. The endocrine system plays an important role in the maintenance of homeostasis because it affects the activities of cells, tissues, and organs throughout the body. For example, glucagon and insulin together maintain a balanced blood glucose concentration. These hormones are considered **antagonistic hormones** because their actions have opposite effects. Figure 51-10 illustrates how an increase in glucose concentration following glucagon secretion is counteracted by insulin secretion.

To maintain homeostasis, hormone secretion must be tightly regulated. One example of hormone regulation was discussed in Section 51-2: ADH secretion is controlled by hypothalamic receptors that detect the concentration of solutes in the blood. More-common types of endocrine control are **feedback mechanisms,** which are illustrated in Figure 51-11. Most hormone systems use **negative feedback,** in which release of an initial hormone stimulates release or production of other hormones or substances that subsequently inhibit further release of the initial hormone. In **positive feedback,** release of an initial hormone stimulates release or production of other hormones or substances, which stimulate further release of the initial hormone. For example, although LH regulates estrogen production by the ovaries, increased estrogen concentrations stimulate a surge in LH secretion prior to ovulation.

internetconnect

SCILINKS NSTA

TOPIC: Homeostasis
GO TO: www.scilinks.org
KEYWORD: HM1038

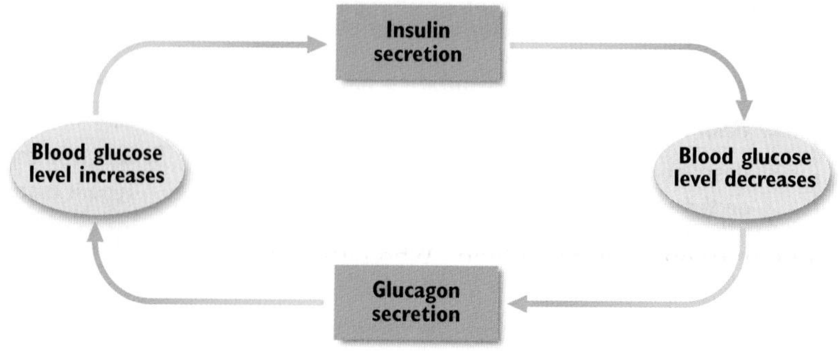

FIGURE 51-10

Working in opposition, glucagon and insulin maintain a balanced blood glucose concentration. These antagonistic hormones oppositely affect the amount of glucose in the blood.

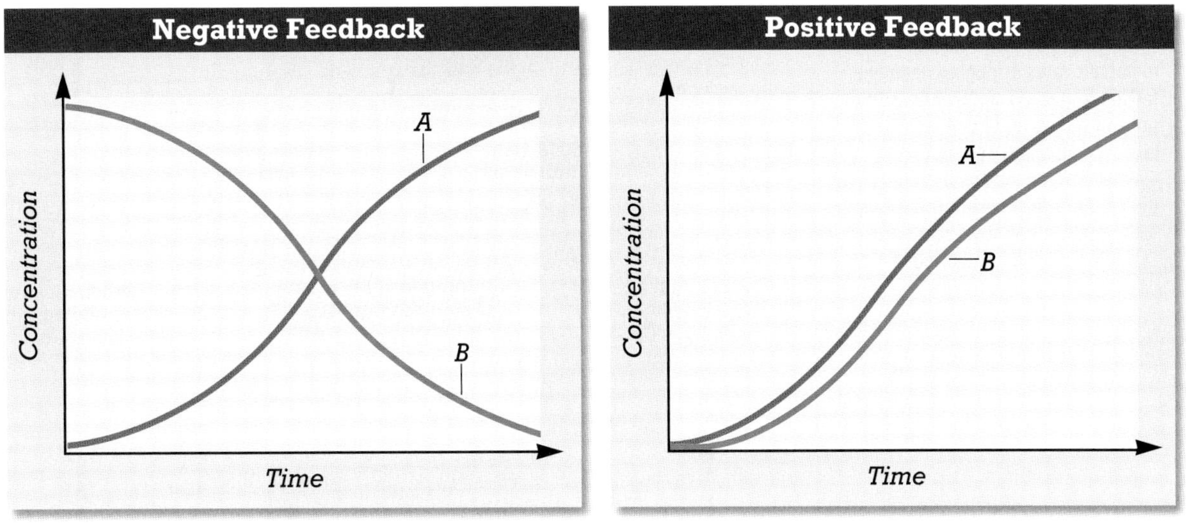

FIGURE 51-11

In negative feedback, a secondary substance (A) inhibits production of its initial stimulating substance (B). In positive feedback, a secondary substance (A) stimulates production of its initial stimulating substance (B).

NEGATIVE FEEDBACK MECHANISMS

Negative feedback mechanisms in the body involve interactions of the nervous, endocrine, and circulatory systems. In negative feedback, the final step in a series of events inhibits the initial signal in the series. To understand negative feedback, consider a home heating system that is controlled by a thermostat, as illustrated in Figure 51-12. When the room temperature drops below a set point, the thermostat activates the heater to produce heat. When room temperature returns to the set point, the thermostat shuts off the heater.

Negative feedback mechanisms help maintain hormone concentrations at a certain range. A good example of negative feedback in the endocrine system is the hypothalamus–anterior-pituitary–testis system discussed in Section 51-2. The hypothalamus secretes LH-releasing hormone, which stimulates secretion of LH from the anterior pituitary. LH is released into the bloodstream and is transported throughout the body. LH binds to its target cells in the testes, forming hormone-receptor complexes that stimulate c-AMP production, which leads to testosterone secretion into the bloodstream. Testosterone binds to its target cells, some of which are the hypothalamic neurosecretory cells that produce LH-releasing hormone. If the testosterone concentration is higher than normal, then secretion of LH-releasing hormone will be inhibited.

Another example of negative feedback in the endocrine system is the regulation of the concentration of the thyroid hormones in the blood. When the hypothalamus detects low concentrations of thyroxine and triiodothyronine, it secretes TSH-releasing hormone into the anterior

FIGURE 51-12

Negative feedback mechanisms inhibit the original signal as its products build up.

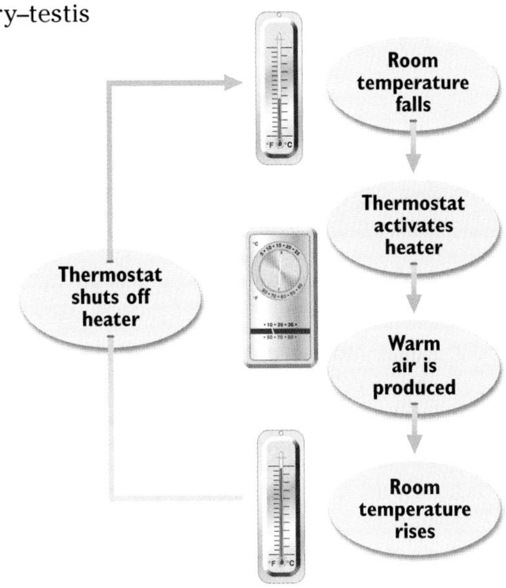

VISUAL LINK

Figure 51-11

In the graph labeled "Negative Feedback," substance *B* is the initial step in the series of events that causes production of substance *A*. Ask students what happens as the concentration of *A* increases. (*A* inhibits *B*, and the concentration of *B* decreases.) In the graph labeled "Positive Feedback," initial production of *B* causes production of *A*. Ask students what happens now as the concentration of *A* increases. (*A* stimulates *B*, and the concentrations of both *A* and *B* continue to increase steadily.)

RECENT RESEARCH

Melatonin

In humans, the hormone melatonin is secreted by the pineal gland at night and is inhibited during the day, suggesting that melatonin plays a role in the sleep patterns of humans. A recent study by researchers at the University of Virginia found that hamster retinas produce melatonin on a similar time-dependent schedule, providing evidence that melatonin regulates the daily function of eyes. They also found that the cyclical production of melatonin in the eyes could be altered by shining light on the retinas of the hamsters at various times. This study provides substantial evidence that melatonin regulates the body's daily cycles and that the eyes and the brain communicate information about melatonin production.

SECTION 51-3

INCLUSION *ACTIVITY*

Divide the class into groups of three, and assign each group a major endocrine gland. Ask group members to discuss what problems might occur if the gland is overactive or underactive. Have students contribute their conclusions to a class discussion on hyperglandular and hypoglandular conditions.

ANSWERS TO SECTION 51-3 REVIEW

1. Insulin and glucagon are antagonistic hormones because their actions have opposite effects.

2. In a positive feedback mechanism, insulin secretion would cause more insulin to be secreted, causing the concentration of glucose in the blood to decrease steadily.

3. The endocrine system uses feedback mechanisms to regulate the concentration of hormones within a certain range so that a steady state is maintained.

4. In a positive feedback mechanism, the thermostat would continue to signal the furnace to produce heat. Thus, the room temperature would increase steadily.

5. Because the thyroid hormones are synthesized from iodine atoms, insufficient dietary iodine could cause a deficiency of thyroid hormones, which would disable the negative feedback mechanism that regulates thyroid activity.

6. Thyroxine signals the hypothalamus to stop secreting TSH-releasing hormone and to begin secreting TSH release-inhibiting hormone. Thus, if too much thyroxine were administered to a person with hypothyroidism, thyroid activity could be further suppressed.

FIGURE 51-13

The thyroid hormones regulate cellular metabolic rates through a negative feedback mechanism. Low concentrations of the thyroid hormones stimulate production and secretion of TSH-releasing hormone from the hypothalamus; high concentrations inhibit TSH-releasing hormone but stimulate TSH release-inhibiting hormone.

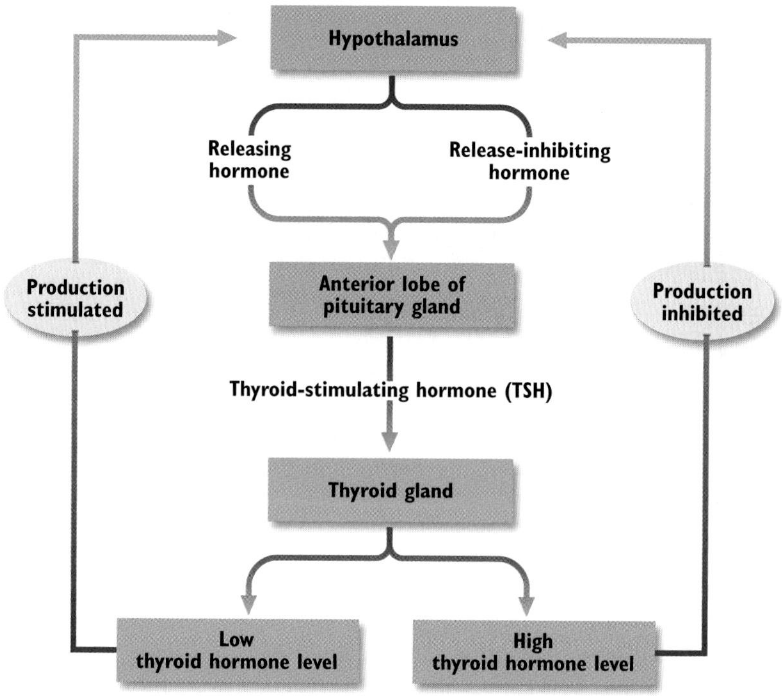

pituitary, as Figure 51-13 illustrates. The anterior pituitary then secretes TSH into the bloodstream. TSH stimulates the thyroid gland to secrete thyroxine and triiodothyronine, whose target cells include hypothalamic neurosecretory cells that produce TSH-releasing hormone and TSH release-inhibiting hormone. When the hypothalamus detects increasing concentrations of the thyroid hormones in the blood, it stops secreting TSH-releasing hormone and begins to secrete TSH release-inhibiting hormone into the anterior pituitary. Thus, the anterior pituitary no longer secretes TSH, and the thyroid gland no longer secretes thyroxine and triiodothyronine. As a result, the concentration of the thyroid hormones in the blood decreases. When the concentration again drops to a certain point, the cycle is repeated.

SECTION 51-3 REVIEW

1. **Explain why glucagon and insulin are antagonistic hormones.**
2. **Explain what might happen to the level of glucose circulating in the blood if the production of insulin were controlled by a positive, rather than a negative, feedback mechanism.**
3. **How do feedback mechanisms help maintain homeostasis?**
4. **What would happen if a thermostat operated under a positive feedback mechanism?**
5. **How would lack of dietary iodine affect the negative feedback mechanism that controls the level of the thyroid hormones?**
6. **CRITICAL THINKING Why is it important to carefully control the dose of thyroxine administered to a person with hypothyroidism?**

CHAPTER 51 REVIEW

SUMMARY/VOCABULARY

51-1

- Endocrine glands produce and secrete hormones that affect distant target cells.
- Amino acid–based hormones bind to receptors on their target cells, activating a second messenger.
- Steroid hormones diffuse through the membranes of their target cells and bind to cytoplasmic receptors.
- Prostaglandins are hormonelike lipids that affect nearby cells.

Vocabulary

amino acid–based hormone (1029)
endocrine gland (1029)
endocrine system (1029)
exocrine gland (1029)
first messenger (1030)
hormone-receptor complex (1030)
hormone (1029)
prostaglandin (1031)
receptor (1030)
second messenger (1030)
steroid hormone (1029)
target cell (1030)

51-2

- The hypothalamus links the nervous system to the endocrine system by regulating the pituitary gland.
- Releasing hormones and release-inhibiting hormones secreted by the hypothalamus regulate the anterior-pituitary hormones.
- The thyroid gland secretes hormones that regulate metabolism.
- The adrenal glands help regulate metabolism and the body's responses to stress and danger.
- The gonads secrete sex hormones that are responsible for body changes that begin at puberty.
- The islets of Langerhans of the pancreas secrete hormones that regulate the blood glucose level.
- Other endocrine glands include the thymus gland, the pineal gland, the parathyroid glands, and endocrine cells of the digestive system.

Vocabulary

adrenal gland (1034)
adrenocorticotropic hormone (1035)
aldosterone (1035)
androgens (1035)
antidiuretic hormone (1032)
calcitonin (1034)
cortisol (1035)
cretinism (1034)
diabetes mellitus (1036)
epinephrine (1034)
estrogen (1035)
follicle-stimulating hormone (1035)
gastrin (1037)
glucagon (1036)
goiter (1034)
gonads (1035)
growth hormone (1033)
hyperthyroidism (1034)
hypoglycemia (1037)
hypothyroidism (1034)
insulin (1036)
islets of Langerhans (1036)
luteinizing hormone (1035)
melatonin (1037)
neurosecretory cell (1032)
norepinephrine (1034)
oxytocin (1032)
parathyroid gland (1037)
parathyroid hormone (1037)
pineal gland (1037)
pituitary gland (1032)
progesterone (1035)
prolactin (1033)
puberty (1035)
release-inhibiting hormone (1033)
releasing hormone (1032)
secretin (1037)
sex hormones (1035)
testosterone (1035)
thymosin (1037)
thymus gland (1037)
thyroid gland (1034)
thyroid-stimulating hormone (1034)
thyroxine (1034)
triiodothyronine (1034)

51-3

- Feedback mechanisms help maintain homeostasis.
- In negative feedback, the final step in a series inhibits the first step.
- In positive feedback, the final step in a series stimulates the first step.

Vocabulary

antagonistic hormone (1038)
feedback mechanism (1038)
negative feedback (1038)
positive feedback (1038)

CHAPTER 51 REVIEW ANSWERS

REVIEW

1. prostaglandin; it is not a hormone
2. epinephrine; it is not secreted by the pituitary gland
3. parathyroid hormone; it is not secreted by the adrenal glands
4. aldosterone; it is not secreted by the gonads
5. liver; the others are endocrine glands

6. b **7.** a
8. c **9.** b
10. d **11.** d
12. a **13.** a
14. a **15.** b

16. A steroid hormone diffuses through the cell membrane and binds to a receptor protein in the cytoplasm. The hormone-receptor complex enters the nucleus and activates mRNA transcription.
17. Positive feedback would cause hormone levels to increase continuously; a steady state of hormone concentrations could not be maintained.
18. More water would be lost in urine, and the concentration of solutes in the blood would increase.
19. Excess glucose inhibits water reabsorption by the kidneys, causing frequent urination.
20. Gastrin stimulates stomach cells to secrete digestive enzymes and HCl.
21. Both the nervous system and the endocrine system help maintain homeostasis and use chemical messengers to carry out their functions.
22. Thymosin, secreted by the thymus gland, stimulates the development of T-cells, which aid the body's immune response.
23. For example, calcitonin and parathyroid hormone are antagonistic hormones.
24. In males, LH stimulates testosterone secretion from the testes. In females, LH and FSH stimulate estrogen release from the ovaries. Estrogen and

testosterone regulate the secondary sex characteristics.

25. This figure illustrates how the concentration of testosterone in the blood is regulated by a negative feedback mechanism. If the testosterone concentration is low, the hypothalamus secretes LH-releasing hormone, which stimulates LH secretion from the anterior pituitary. LH then stimulates the testes to secrete testosterone into the bloodstream. If the testosterone concentration is high, testosterone inhibits production and secretion of LH-releasing hormone.

CRITICAL THINKING

1. Damage to the pituitary gland will subsequently affect the other endocrine glands and organs that the pituitary regulates.

2. Because of glucose's role in aerobic respiration and ATP production, precise regulation of the concentration of glucose in the blood is critical to maintaining homeostasis. Thus, glucose production is regulated by several hormones. For example, if only one hormone secreted by one endocrine gland were to regulate glucose, damage to that endocrine gland could be detrimental.

3. Gigantism is caused by hypersecretion of growth hormone (GH) by the anterior pituitary, whereas pituitary dwarfism is caused by GH deficiency.

4. Many other conditions, such as stress, the menstrual cycle, diet, and disease, can produce symptoms that are similar to those of diabetes.

5. As shown in the graph, the glucose level increases after eating. This increase causes islet cells of the pancreas to secrete insulin, which returns the glucose level to normal.

6. Insulin and glucagon are responsible for changes in blood glucose level.

REVIEW

Vocabulary

For each set of terms below, choose the one that does not belong and then explain why it does not belong.

1. insulin, prostaglandin, thyroxine, GH
2. oxytocin, prolactin, epinephrine, ADH
3. cortisol, parathyroid hormone, epinephrine, aldosterone
4. aldosterone, estrogen, progesterone, testosterone
5. thyroid gland, liver, adrenal glands, pineal gland

Multiple Choice

6. The endocrine system (a) affects only the nervous system (b) helps maintain homeostasis (c) affects only the reproductive system (d) primarily uses positive feedback.
7. In a negative feedback system, the end product (a) inhibits the first step (b) inhibits the last step (c) stimulates the first step (d) stimulates the last step.
8. Steroid hormones (a) bind to receptors on a target cell (b) convert ATP to cyclic AMP (c) diffuse through the target cell membrane (d) activate a second messenger.
9. Cyclic AMP is produced by a cell in response to (a) estrogen (b) amino acid–based hormones (c) testosterone (d) steroid hormones.
10. Which of the following organs does not contain endocrine cells? (a) stomach (b) pancreas (c) small intestine (d) liver
11. Diabetes mellitus can be caused by (a) insufficient insulin (b) lack of insulin receptors (c) obesity (d) all of the above.
12. The interaction of the hypothalamus and the pituitary gland demonstrates (a) the relationship between the nervous system and the endocrine system (b) the relationship between the endocrine system and the digestive system (c) that energy from the hypothalamus transports glucose (d) the involvement of the pituitary gland in prostaglandin production.
13. Prostaglandins are (a) lipids (b) hormones (c) endocrine glands (d) proteins.
14. Oxytocin (a) is secreted by the posterior pituitary (b) controls the amount of water excreted in the urine (c) increases the calcium ion concentration in the blood (d) regulates secondary sex characteristics.
15. Which of the following is both an exocrine gland and an endocrine gland? (a) the thyroid gland (b) the pancreas (c) the thymus gland (d) the pituitary gland

Short Answer

16. Describe the interaction of a steroid hormone and its target cell.
17. Why is positive feedback not an efficient way to control hormone levels?
18. What would be the consequence if the pituitary gland secreted too little ADH?
19. Explain why frequent urination is a symptom of diabetes.
20. How does gastrin help digestion?
21. Describe two ways in which the endocrine system and the nervous system are similar.
22. How does the endocrine system help the immune system?
23. Name two antagonistic hormones besides insulin and glucagon.
24. How do LH and FSH help regulate secondary sex characteristics in males and females?
25. Identify and describe the type of feedback mechanism operating in the diagram shown below.

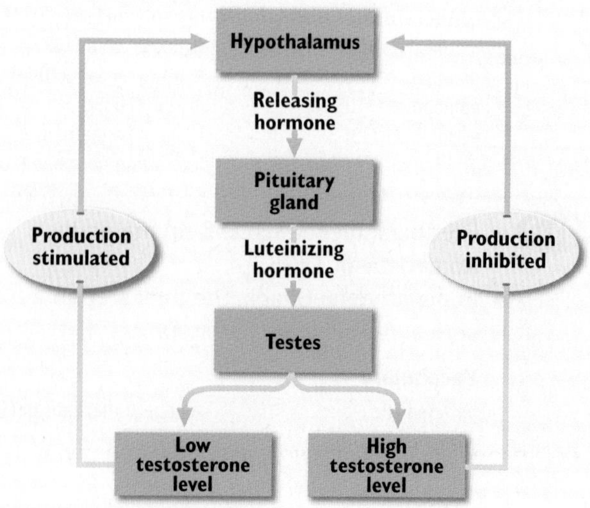

CRITICAL THINKING

1. Why might damage to the pituitary gland be considered far more serious than damage to one of the other endocrine glands?
2. A number of different hormones secreted by various endocrine glands help regulate the blood glucose concentration. Based on your knowledge of the importance of glucose, hypothesize why glucose should be controlled by several hormones rather than just one hormone.
3. Severe structural abnormalities result from gigantism, a condition of extremely rapid growth, and from pituitary dwarfism, a condition of slowed growth. Based on your knowledge of the endocrine system, suggest what causes these disorders.
4. Suppose that a friend tells you that he or she has recently experienced some of the warning signs of diabetes mellitus. What other conditions could cause symptoms that are similar to those of diabetes?
5. The graph shown at right illustrates the change in the blood glucose level before and after eating. According to the graph, what happens after food is eaten?
6. Which hormones are primarily responsible for the changes in blood glucose level shown in the graph?
7. Referring to the graph, identify and explain the type of feedback mechanism that enables the body to adjust the blood glucose level after eating.

Blood Glucose Level

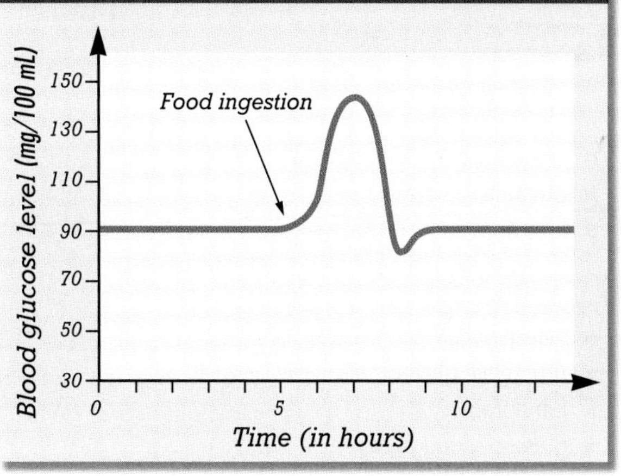

EXTENSION

1. The connection between melatonin and sleep patterns has kindled interest in this hormone. Check the library and on-line references for information about other uses that have been proposed for melatonin.
2. Read "Hormone Helps Ring Internal Alarm Clock" in *Science News*, January 9, 1999, on page 22. Describe the hormones that control waking from sleep. What did the researchers conclude was the primary factor that set the neurological timer to prepare the body to awaken?
3. Read "Is Leptin a 'Thrifty' Hormone in Muscle and Fat?" in *Science*, March 10, 2000, on page 1739. Explain the "thrifty genotype" proposition as an evolutionary mechanism that prepares a body for times of famine. How does this hypothesis explain why many people who live in developed countries struggle with obesity?
4. Check the library and on-line references for information about disorders of the endocrine system, such as Addison's disease, diabetes mellitus, Cushing's syndrome, and pituitary gigantism and dwarfism. Describe the symptoms and treatments for these and other disorders of the endocrine system.
5. Read "Hormone Recall" in *New Scientist*, September 18, 1999, on page 12. Investigate evidence presented in the article that estrogen is involved in memory and cognitive circuits in the brain. Also investigate the evidence that hormone replacement therapy protects the brains of postmenopausal women against diseases such as Alzheimer's.

ANSWERS

CONTINUED

7. After food is eaten, a negative feedback mechanism readjusts the glucose level (about 90 mg/100 mL). When food is eaten, the glucose concentration increases due to sugars in the food. When the glucose level reaches a certain point (about 150 mg/100 mL), insulin is released, causing body cells to absorb glucose. When the glucose level falls below the normal range, glucagon is released, causing the liver to release glucose and inhibiting insulin secretion. The glucose level then returns to the normal range.

EXTENSION

1. Melatonin has been proposed as a panacea for "reversing" the effects of aging and for treating lethargy and insomnia.
2. Cortisol, adrenocorticotropin (ACTH), and corticotropin-releasing hormones react to and prepare the body for stressful events, including waking from sleep. Levels of ACTH and cortisol increase sharply before waking. The expectation of waking at a certain time caused the amount of ACTH to rise just before the expected time of waking.
3. The "thrifty genotype" hypothesis holds that there is a mechanism that allows animals and people to hoard fat in times of abundance for use during times of famine. Most people who live in developed countries always have an abundance of food and never experience famine.
4. Students should find that most endocrine disorders are caused by hormone imbalances; hyposecretion or hypersecretion of hormones by endocrine glands; target cell unresponsiveness or deficiency; or a combination of these or other factors. Most treatments entail hormone replacement.
5. Students should cite evidence that women deprived of estrogen suffer memory lapses that vanish when the hormone is replaced.

CHAPTER 51 INVESTIGATION

TIME REQUIRED

One 50-minute class period to set up and one 15-minute observation period once a week for three weeks

SAFETY PRECAUTIONS

Safety goggles and gloves should be worn at all times. Instruct students to be careful when handling thyroxine and to wash out with water any thyroxine that gets on the skin or clothing or in the eyes. Remind students that tadpoles are living animals and that they should be handled with care.

QUICK REFERENCES

Lab Preparation Notes are found on page 1027B.

Holt Biosources provides a Teaching Resources CD-ROM that contains "Using Gowin's Vee in the Lab" and "Scoring Rubrics."

PROCEDURAL TIP

This investigation works well when students work in groups of three.

ANSWERS TO BACKGROUND

1. A hormone is a chemical that is secreted into the bloodstream by endocrine cells and that affects distant target cells.
2. Metamorphosis is the changing of an organism from an immature stage to a mature stage.
3. Frog eggs (laid in water) hatch into larvae (tadpoles), which develop hind legs and then front legs. As their tails shrink and their legs develop, tadpoles develop lungs and then leave the water to continue growth as a frog.

CHAPTER 51 INVESTIGATION

Observing the Effects of Thyroxine on Frog Metamorphosis

OBJECTIVES

- Observe the effects of the hormone thyroxine on the development of tadpoles.

PROCESS SKILLS

- observing
- measuring
- comparing and contrasting
- organizing data
- inferring

MATERIALS

- safety goggles
- protective gloves
- lab apron
- glass-marking pencil
- six 600 mL beakers
- pond water
- 10 mL graduated cylinder
- 0.01% thyroxine solution
- strained spinach
- graph paper marked in 1 mm squares
- Petri dish
- small fish net
- 9 tadpoles with budding hind legs
- 3 pencils in different colors

Background

1. What is a hormone?
2. What is metamorphosis?
3. Describe the stages of frog development.
4. What are the effects of thyroxine in humans?
5. What effects do you predict the hormone thyroxine will have on tadpole growth and development?

PART A Setting Up the Experiment

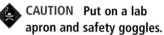

1. In your lab report, make a data table similar to the one shown on the facing page.
2. Use a glass-marking pen to label three beakers A, B, and C. Also write your initials on the beakers.
3. Add 500 mL of pond water to each beaker.
4. **CAUTION Put on a lab apron and safety goggles. If you get thyroxine on your skin or clothing, rinse with water while calling to your teacher. If you get thyroxine in your eyes, immediately flush them with water at the eyewash station while calling to your teacher.**
5. Use a graduated cylinder to measure 10 mL of thyroxine solution. Add the 10 mL of solution to beaker A. Measure and add 5 mL of thyroxine solution to beaker B.
6. Add equal amounts (about 1 mL) of strained spinach to beakers A, B, and C.
7. Place a sheet of graph paper, ruled side up, under a Petri dish.
8. **CAUTION You will be working with live animals. Handle them with care and follow directions carefully.** Catch a tadpole with a fish net and place the tadpole in the Petri dish. Measure the tadpole's total length, tail length, and body length in millimeters by counting the number of squares that it covers on the graph paper. Then place the tadpole in beaker A.

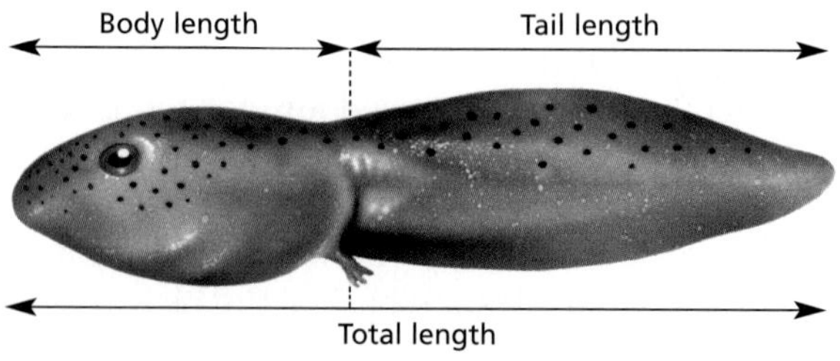

9. Repeat step 8 with two more tadpoles. Average the total length, tail length, and body length of the three tadpoles that you placed in beaker A. In your data table, record the average measurements under the column labeled "Beaker A."

10. Repeat step 8 with three more tadpoles, this time placing the tadpoles in beaker B. Average the total length, tail length, and body length of the three tadpoles. Record your average measurements in your data table under "Beaker B."
11. Repeat step 8 with three more tadpoles. Place these three tadpoles in beaker C. Average the total length, tail length, and body length of the three tadpoles. Record your average measurements in your data table under "Beaker C." You should have measured a total of nine tadpoles and placed three in each beaker.
12. Clean up your materials and wash your hands before leaving the lab.

Part B Observing the Effects of Thyroxine on Tadpoles

13. Feed the tadpoles 1 mL of spinach every other day. Be careful not to overfeed the tadpoles. Change the water in the beakers every 4 days, adding thyroxine solution in the original amounts to beakers A and B. Label the beakers, and do not mix up the tadpoles during the water changes.
14. Measure the tadpoles once a week for 3 weeks and average the length of the tadpoles in each beaker. Record the average lengths in your data table.
15. Calculate the average growth-per-week for each set of three tadpoles. For example, the average growth in total length during the second week is equal to the average total length at the end of week 2 minus the average total length at the end of week 1. Record your values in the appropriate spaces of your data table.
16. Graph your data using a different colored pencil for the tadpoles in each beaker. Label the horizontal axis "Time in weeks," and label the vertical axis "Length in centimeters." You should have three graphs. Each graph should include the changes in average total length, average tail length, and average body length for one group of tadpoles.
17. Clean up your materials and wash your hands before leaving the lab.

Analysis and Conclusions

1. Did this investigation include a control group? If so, describe it. If not, suggest a possible control that you could have used.
2. Why are three tadpoles used for each beaker rather than just one?
3. According to the data that you collected in this investigation, what is the effect of thyroxine on tadpole metamorphosis?
4. Which concentration of thyroxine solution caused the greatest visible change in the tadpoles?
5. How do average body length and tail length change during metamorphosis?

Further Inquiry

Iodine is needed to produce thyroxine. Design—but do not conduct—an experiment that shows the effect of adding iodine to water that contains tadpoles. NOTE: Iodine is a highly poisonous substance.

MEASUREMENT OF TADPOLE GROWTH

	Beaker A			Beaker B			Beaker C		
	Average total length	**Average tail length**	**Average body length**	**Average total length**	**Average tail length**	**Average body length**	**Average total length**	**Average tail length**	**Average body length**
Initial									
End of week 1									
End of week 2									
End of week 3									

CHAPTER 51 INVESTIGATION

4. In humans, thyroxine is secreted by the thyroid gland. It increases cellular metabolic rates and helps maintain normal blood pressure, heart rate, and body temperature. Abnormal thyroxine levels can be detrimental to the body's metabolism.
5. Thyroxine should increase the rate of tadpole growth and development.

ANSWERS TO ANALYSIS AND CONCLUSIONS

1. The tadpoles in water without thyroxine serve as a control group to which the treated tadpoles can be compared.
2. Using several tadpoles yields more accurate results by preventing a single abnormal tadpole from distorting data.
3. Thyroxine increases the rate of tadpole growth and metamorphosis.
4. The strongest concentration causes the greatest visible change in the tadpoles.
5. As the body gets longer, the tail gets shorter.

CHAPTER 52

REPRODUCTIVE SYSTEM

Block	TOPICS	TEACHING RESOURCES	LABS, CLASSWORK, AND HOMEWORK
BLOCK 1 45 minutes	**Introducing the Chapter, p. 1046**	ATE Focus Concept, p. 1046 ATE Assessing Prior Knowledge, p. 1046 ATE Understanding the Visual, p. 1046	■ Occupational Applications Worksheets—Medical Sonographer and Nurse Practitioner
	52-1 Male Reproductive System, p. 1047 Formation of Sperm, p. 1047 Male Reproductive Structures, p. 1048	ATE Section Overview, p. 1047 ATE Visual Link Figure 52-6, p. 1052 224–225, 232	★ Study Guide, Section 52-1 PE Section Review, p. 1049
BLOCKS 2 & 3 90 minutes	**52-2 Female Reproductive System, p. 1050** Formation of Eggs, p. 1050 Female Reproductive Structures, p. 1051 Preparation for Pregnancy, p. 1052	ATE Section Overview, p. 1050 ATE Visual Link Figure 52-6, p. 1052 223, 226	★ Study Guide, Section 52-2 PE Section Review, p. 1053
BLOCKS 4 & 5 90 minutes	**52-3 Gestation, p. 1054** Fertilization, p. 1054 Pregnancy, p. 1055	ATE Section Overview, p. 1054 ATE Visual Link Figure 52-8, p. 1055 196A, 227–228	PE **Chapter 52 Investigation,** p. 1062 PE **Quick Lab,** p. 1055, Summarizing Vocabulary **Lab A34,** Culturing Frog Embryos **Lab B34,** Embryonic Development ★ Study Guide, Section 52-3 PE Section Review, p. 1058
BLOCKS 6 & 7 90 minutes	**REVIEW and ASSESSMENT** PE Chapter 52 Review, pp. 1060–1061 ■ Performance-Based Assessment—Testing Human Digestive Enzymes	Audio CD Program	**CHAPTER TESTING** ■ Chapter Test (blackline master) ▲ Test Generator ■ Assessment Item Listing ■ Scoring Rubrics and Classroom Management Checklists

MODERN BIOLOGY — **HOLT BIOSOURCES** — **One-Stop Planner CD-ROM**

KEY

- PE Pupil's Edition
- ATE Teacher's Edition
- ■ Active Reading Guide
- ★ Study Guide
- Teaching Transparencies
- Laboratory Program

One-Stop Planner CD-ROM
Shows these resources within a customizable daily lesson plan:
- ■ Holt BioSources Teaching Resources
- ■ Active Reading Guide
- ★ Study Guide
- ▲ Test Generator

READING FOR CONTENT MASTERY

Preview: Have students preview the objectives, the introductory paragraph, topic sentences, and captions before reading the chapter.

- ATE Active Reading Technique: K-W-L, p. 1047
- ■ Reading Strategy: K-W-L Worksheet
- ■ Active Reading Guide Worksheet 52-1
- PE Summary/Vocabulary, p. 1059

- ■ Active Reading Guide Worksheet 52-2
- PE Summary/Vocabulary, p. 1059

- ATE Active Reading Technique: K-W-L, p. 1055
- ■ Reading Strategy: K-W-L Worksheet
- ■ Active Reading Guide Worksheet 52-3
- PE Summary/Vocabulary, p. 1059

TECHNOLOGY AND INTERNET RESOURCES

Holt Biology Videodiscs Teacher's Correlation Guide, Lessons 52-1 through 52-3

Audio CD Program, Sections 52-1 through 52-3

internetconnect

SCILINKS NSTA

On-line Resources: www.scilinks.org
The following *sci*LINKS Internet resources can be found in the student text for this chapter:

Topics:
- Male reproductive system, p. 1048
- Female reproductive system, p. 1053

go.hrw.com

On-line Resources: go.hrw.com
Visit the HRW Web site for a variety of resources related to this chapter. Just type in the keyword HM2 Chapter 52.

Smithsonian Institution®

Visit **www.si.edu/hrw** for additional on-line resources.

CNNfyi.com

Visit **www.cnnfyi.com** for late-breaking news and current event stories selected just for you.

LAB ACTIVITY PLANNING

CHAPTER 52 INVESTIGATION

Observing Embryonic Development, pp. 1062–1063

OVERVIEW
This lab will allow students to observe the early stages of echinoderm embryonic development. Mammalian embryonic development is difficult to study because the zygote is so small. Thus, it is necessary to study an animal whose early development is in many ways similar to that of a mammal. Students will describe changes that occur during early development and will compare the stages of embryonic development. Stages of development that will be examined include the 2-, 4-, 8-, 16-, 32-, and 64-cell stages, the blastula, the early gastrula, the middle gastrula, the late gastrula, and a young sea-star larva.

SAFETY
Caution students to handle glass slides carefully.

PREPARATION
1. Obtain prepared slides of sea-star development from a biological supply house, such as WARD'S.
2. Provide compound microscopes for student use.

Quick Labs

Summarizing Vocabulary, p. 1055

Students will need paper, pencils, and dictionaries for this activity.

UNDERSTANDING THE VISUAL

This photograph shows a fetus. Ask students where the fetus develops. (It develops in the uterus.) Have students note the umbilical cord and the placenta. Ask students to suggest how oxygen and nutrients pass between the mother and the fetus. (They diffuse through the lining of the placenta and are transported in the bloodstream through fetal arteries and veins that make up the umbilical cord.)

FOCUS CONCEPT

Structure and Function

The function of sexual reproduction is to produce a new organism by combining the hereditary information of a male and a female. Hormones regulate the male and female reproductive systems. Gamete formation, menstruation, development, and the regulation of hormone production are examples of the relationship between structure and function in the reproductive system.

ASSESSING PRIOR KNOWLEDGE

Review the following concepts.

Cell Reproduction: *Chapter 8*
Ask students to distinguish mitosis from meiosis. Then ask them why meiosis is necessary for reproduction. Ask students to suggest how the haploid (1*n*) number, achieved during meiosis, is restored to the diploid (2*n*) number through sexual reproduction.

Endocrine System: *Chapter 51*
Ask students to define positive and negative feedback. Then ask them why feedback mechanisms are crucial to hormone regulation. Ask students to identify some of the hormones that regulate the cyclic changes of the female reproductive system and that help maintain the steady state of the male reproductive system.

CHAPTER 52

REPRODUCTIVE SYSTEM

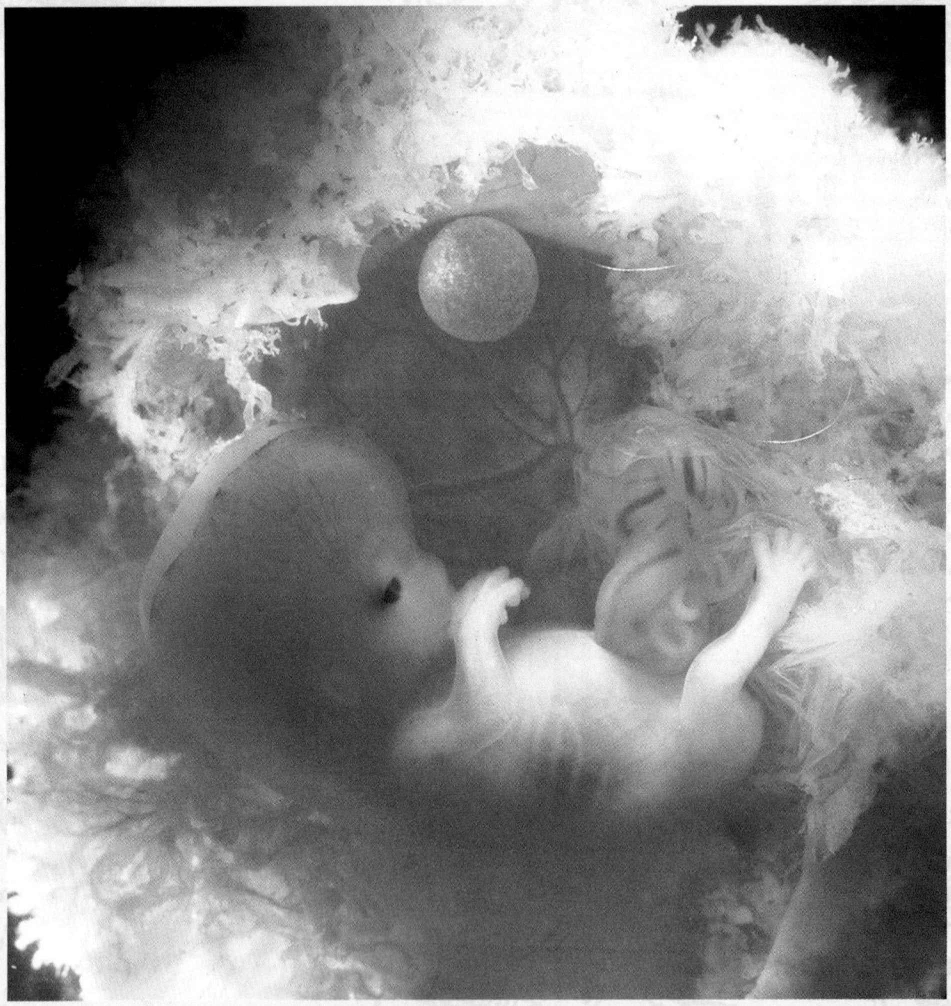

This is a photograph of a fetus. Notice the umbilical cord and the placenta, through which oxygen and nutrients are passed from the mother to the fetus.

FOCUS CONCEPT: *Structure and Function*

As you read this chapter, notice how the structures of the male and female reproductive systems are adapted for fertilization and development.

MALE REPRODUCTIVE SYSTEM

Recall from Chapter 51 that the gonads—testes and ovaries—are endocrine glands that secrete sex hormones. However, the primary function of the gonads is not to produce hormones but to produce and store gametes—sperm and eggs. Other organs in the male reproductive system prepare sperm for the possible fertilization of an egg.

FORMATION OF SPERM

Males begin to produce sperm during puberty, the adolescent stage of development when changes in the body make reproduction possible. At this time, the concentration of the hormone testosterone is high enough to stimulate sperm production. Recall from Chapter 51 that testosterone is the main androgen (male sex hormone) produced by the testes. The **testes** (TES-teez) are the gamete-producing organs of the male reproductive system. A male will continue to produce sperm as long as his testosterone level is high enough—usually for most of his life.

In Chapter 8, you learned that the formation of gametes in humans involves the process of meiosis. Recall that meiosis results in a reduction of the number of chromosomes from the diploid ($2n$) number to the haploid ($1n$) number. As cells that produce sperm within the testes undergo meiosis, their chromosome number drops from 46 to 23. Four sperm cells result from each cell that begins meiosis. The immature sperm that result from meiosis then undergo significant changes. These changes prepare the sperm for passage through the female reproductive system.

The structure of a mature sperm is shown in Figure 52-1. Notice that a mature sperm consists of three regions—a head, a midpiece, and a tail. The tip of the head region contains enzymes. During fertilization, these enzymes help the sperm penetrate the protective layers that surround an egg cell. Also located in the head region are the 23 chromosomes that will be delivered to the egg. The midpiece is packed with mitochondria. Remember from Chapter 6 that mitochondria are the sites of aerobic respiration and ATP production. These mitochondria supply the energy that is required for sperm to reach an egg. The tail consists of a single, powerful flagellum that propels the sperm.

SECTION 52-1

OBJECTIVES

▲ Describe the structure of a human sperm.

● Identify the major parts of the male reproductive system.

■ Describe the function of each part of the male reproductive system.

◆ Trace the path that sperm follow in leaving the body.

FIGURE 52-1

A mature sperm is an elongated cell with three distinct parts, all of which are enclosed by a cell membrane.

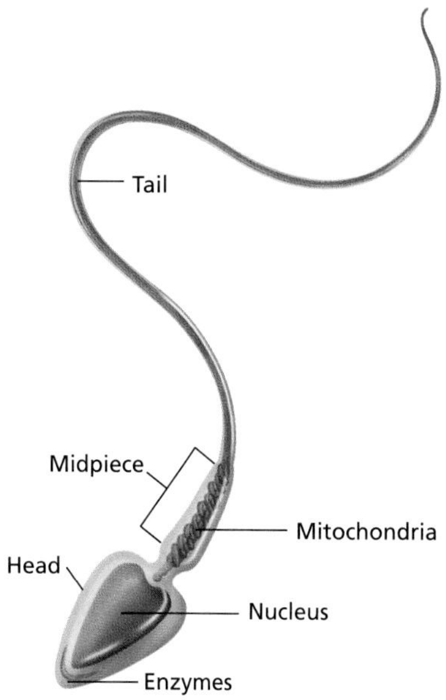

SECTION 52-1

SECTION OVERVIEW

Cell Structure and Function
Section 52-1 describes spermatogenesis and the male reproductive organs. It explains how sperm is delivered into the female reproductive system.

Reproduction and Inheritance
Section 52-1 also explains that spermatogenesis involves changes that enable fertilization.

ACTIVE READING

Technique: K-W-L
Before students begin this chapter, have them write short lists of all the things they **Know** about reproduction and development. Write some of these ideas on the board or overhead. Then have students list things they **Want** to know about reproduction and development. Have students save their lists for use later in the Active Reading activity on page 1055.

CULTURAL CONNECTION

Sperm Production
A 40 percent decline in the average sperm count of men has been documented over the last 40 years. Some researchers blame environmental toxins, such as drugs, lead, and pesticides, for the decline in male fertility. Others claim that men who wear tight pants may decrease sperm production because the pants hold the testes too close to the body; the resultant increase in temperature causes sperm to die.

TEACHING STRATEGY
Microscopic Anatomy

Provide pairs of students with a microscope and prepared slides of epididymis and penis cross sections. Slides can be purchased from a biological supply house, such as WARD'S. Have students identify, draw, and label the tubules of the coiled epididymis (at low magnification) and sperm within the tubules (at high magnification). Have students identify, draw, and label the spongy tissue and the urethra of the penis. Ask students to suggest the function of these structures. (The tissue fills with blood to make the penis hard for intercourse. The urethra passes both urine and sperm through the penis.) Provide a chart or model of the male reproductive system for student reference.

QUICK FACT

☑ Hypertrophy of the prostate gland, a troublesome condition commonly seen in elderly men, constricts the urethra, making urination difficult.

RETEACHING *ACTIVITY*

Using Figure 52-3, have students trace the path of sperm through the male reproductive system. (Sperm move from the seminiferous tubules to the epididymis, through the vas deferens and urethra, and from the penis.) Have students identify the hormones and accessory structures that are involved with each step of spermatogenesis and delivery of sperm to the female reproductive system.

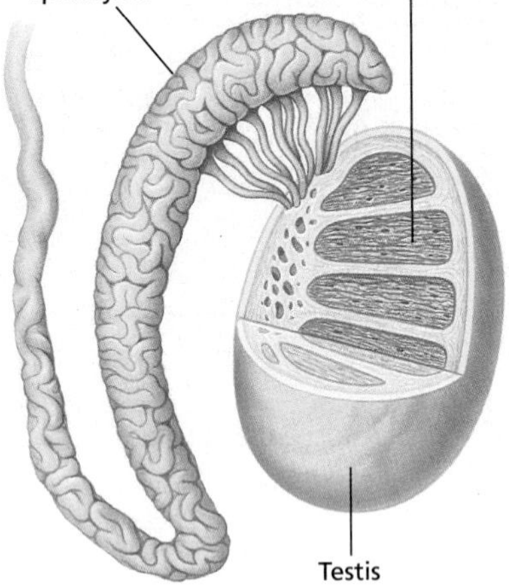

FIGURE 52-2

Sperm are formed continuously within the seminiferous tubules, which make up the bulk of each testis. Cells in the tissue that surrounds the seminiferous tubules secrete testosterone, which stimulates the production of sperm. Before leaving the body, sperm mature and are stored in each epididymis.

internet**connect**

SCILINKS NSTA
TOPIC: Male reproductive system
GO TO: www.scilinks.org
KEYWORD: HM1048

MALE REPRODUCTIVE STRUCTURES

The male reproductive system contains two egg-shaped testes. Each testis, which is about 4 cm (1.5 in.) long and 2.5 cm (1 in.) in diameter, has about 250 compartments. As shown in Figure 52-2, these compartments contain many tightly coiled tubules, called **seminiferous** (SEM-uh-NIF-uhr-uhs) **tubules.** Each seminiferous tubule is approximately 80 cm (32 in.) long. If all of the tubules in both testes were stretched out end to end, they would extend about 500 m (1,640 ft). Sperm form through meiosis in the specialized lining of this extensive network of tubules.

The testes develop within the abdominal cavity. Before a male is born, the testes leave this cavity and descend into an external sac called the **scrotum** (SKROHT-uhm). The temperature within the scrotum is about 3°C cooler than the temperature inside the abdomen. The slightly cooler temperature of the scrotum is necessary for the development of normal sperm.

Path of Sperm Through the Male Body

Mature sperm move through and past several other male reproductive structures, some of which further prepare the sperm for a possible journey through the female reproductive system. Figure 52-3 shows the path taken by sperm as they exit the body.

Sperm move from the seminiferous tubules to the **epididymis** (EP-uh-DID-i-mis), a long, coiled tubule that is closely attached to each testis. Within each epididymis, a sperm matures and gains the ability to swim as its flagellum completes development. Although most sperm remain stored in each epididymis, some leave the epididymis and pass through the **vas deferens** (vas DEF-uh-RENZ), a duct that extends from the epididymis. Smooth muscles that line each vas deferens contract to help move sperm along as they exit the body. Each vas deferens enters the abdominal cavity, where it loops around the urinary bladder and merges with the urethra. Recall from Chapter 49 that the urethra is the duct through which urine exits the urinary bladder. Thus, in a male, both urine and sperm exit the body through the urethra.

In the urethra, sperm mix with fluids that are secreted by three exocrine glands—the seminal vesicles, bulbourethral glands, and the prostate gland. Recall from Chapter 51 that exocrine glands deliver their products through ducts. Ducts that extend from the seminal vesicles, bulbourethral glands, and the prostate gland connect with the urethra. These glands secrete fluids that nourish and protect the sperm. Together, sperm and these secretions form a fluid called **semen** (SEE-muhn). Semen has a high concentration of fructose. Sperm break down fructose by aerobic respiration to obtain energy that they need for movement. To increase sperm survival, semen also contains alkaline fluids that help neutralize the acidic environment of the female's vagina. To help sperm move

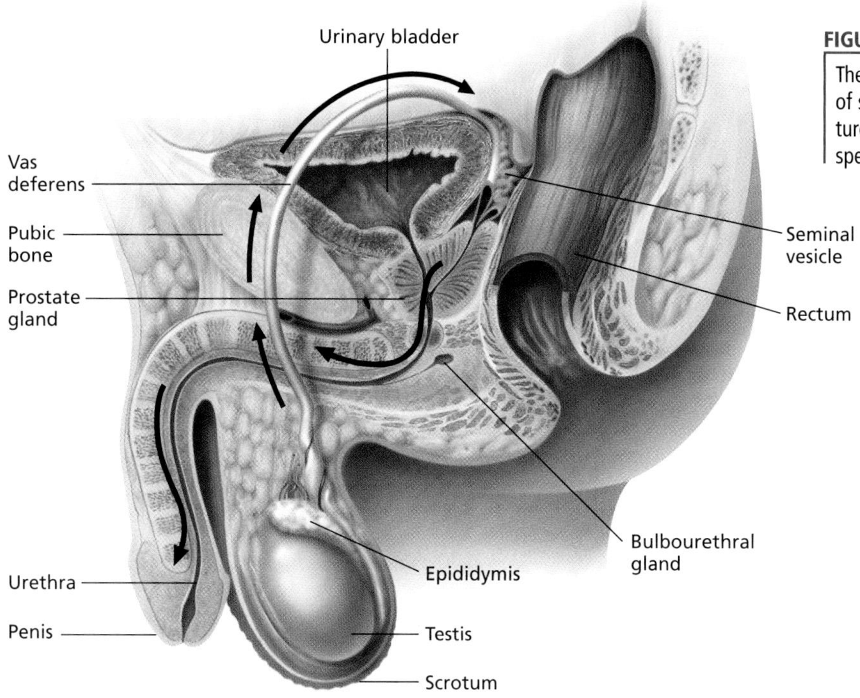

FIGURE 52-3

The male reproductive system consists of several internal and external structures. Arrows indicate the path taken by sperm as they leave the body.

through the female reproductive system, semen also contains prostaglandins that stimulate contractions of smooth muscles that line the female reproductive tract.

Delivery of Sperm

The urethra passes through the **penis,** the organ that deposits sperm in the female reproductive system. When a male becomes sexually aroused, spongy tissue in the penis fills with blood. This causes the penis to become erect, enabling it to deposit sperm. Semen is forcefully expelled from the penis by contractions of the smooth muscles that line the urethra. This process is called **ejaculation** (ee-JAK-yoo-LAY-shun). Each ejaculation expels 3–4 mL (0.10–0.14 fl oz) of semen. Sperm make up only 10 percent of this volume. Fluids that are secreted by the seminal vesicles, bulbourethral glands, and the prostate gland make up the other 90 percent. Even so, 300–400 million sperm leave the male body during a single ejaculation. Still, only a few sperm reach the site of fertilization. Most sperm are killed by the acidic environment of the female reproductive tract.

SECTION 52-1 REVIEW

1. **What parts make up a mature sperm? What does each part contain?**
2. **Which structures in a male produce fluids that mix with sperm to form semen?**
3. **What is the function of the testes? the epididymis?**
4. **Describe the composition of semen.**
5. **Trace the path that sperm take in exiting the body.**
6. **CRITICAL THINKING Why are so many sperm produced by the male reproductive system?**

RECENT RESEARCH

Prostate Cancer

Researchers at the Kettering Cancer Center in New York City are investigating whether diet can delay the development of prostate cancer. They found that mice on low-fat diets are less likely to grow cancerous prostate tumors than mice on normal diets. Other evidence suggests that excessive fat intake increases testosterone levels. This apparently stimulates the growth of prostate tumors. These studies suggest that prostate cancer can be deterred by a diet of low-fat foods, such as tomato and soy products.

ANSWERS TO SECTION 52-1 REVIEW

1. The head, the midpiece, and the tail make up a mature sperm. The head contains enzymes and the haploid (1*n*) number of chromosomes. The midpiece contains mitochondria. The tail consists of a flagellum.
2. Three exocrine glands—seminal vesicles, bulbourethral glands, and the prostate gland—produce fluids that mix with sperm to form semen.
3. The testes produce testosterone, the hormone that stimulates sperm production. Sperm mature in the epididymis.
4. Semen is about 10 percent sperm and 90 percent fluids that are secreted by exocrine glands.
5. Sperm move from the seminiferous tubules to the epididymis, through the vas deferens and urethra, and from the penis.
6. Because most sperm die in the female reproductive tract, many sperm are needed to increase the likelihood of fertilization.

SECTION 52-2

SECTION OVERVIEW

Cell Structure and Function
Section 52-2 describes how each structure of the female reproductive system functions during the menstrual cycle. It describes the four phases of the menstrual cycle, during which the body prepares for a possible pregnancy. This section also explains how the menstrual cycle is regulated by both negative and positive feedback mechanisms that are regulated by hormones.

Reproduction and Inheritance
Section 52-2 also contrasts egg formation with spermatogenesis.

▶ ENGAGE STUDENTS

On an overhead projector, crack a chicken egg into a Petri dish so that students can observe the egg. Ask students to identify the egg white, or albumen, and the yolk. Ask a volunteer student to locate the embryo in the egg. The student might point to the dot on the surface of the yolk. This dot is the actual egg, or ovum, and it is usually not fertilized in store-bought eggs. If it were fertilized, students could see the embryo under a dissecting microscope. (Chicken embryos begin their development while the egg travels through the egg canal.) Ask students to describe how a human egg differs from a chicken egg. (The human egg does not have the supplementary structures—albumen as a shock absorber or yolk as a nutrient supplier—although it develops a yolk sac during pregnancy. Also, the human egg is protected inside the mother, whereas the chicken egg is protected by a shell.)

SECTION

52-2

OBJECTIVES

▲ Compare eggs with sperm.

● Identify the major parts of the female reproductive system.

■ Describe the function of each part of the female reproductive system.

◆ Describe the menstrual cycle and explain how it is regulated.

FEMALE REPRODUCTIVE SYSTEM

Like the testes, the female gonads—ovaries—are endocrine glands that produce gametes. The female reproductive system prepares the female gametes—eggs—for possible fertilization. It also contains structures that enable fertilization to occur and that house and nourish a developing baby.

FORMATION OF EGGS

A female is born with more than 400,000 eggs in her **ovaries** (OH-vuh-reez), the gamete-producing organs of the female reproductive system. When a female is born, her eggs are immature and cannot be fertilized. Typically, a female will release 300–400 mature eggs during her lifetime, averaging one egg about every 28 days from the time she reaches puberty to about age 50. Thus, fewer than 1 percent of a female's eggs reach maturity.

Like sperm formation, the formation of eggs involves meiosis. Unlike in sperm production—in which four sperm result from each cell that begins meiosis—only one egg results from each cell that undergoes meiosis. All immature eggs begin meiosis, but the process is stalled in prophase I until the female reaches puberty, when levels of the sex hormones become high enough to complete the maturation of the eggs. Regulated by these hormones, 10–20 immature eggs resume meiosis every 28 days. However, typically only one mature egg is released each time. A mature egg, or **ovum** (OH-vuhm), shown in Figure 52-4, is about 75,000 times larger than a sperm and is visible to the unaided eye. Meiosis II is not completed unless a sperm fuses with the egg. If the ovum is fertilized, it completes the final meiotic division. Only one cell retains most of the cytoplasm, which provides nutrients that are needed for the egg's survival through the early stages of development. Deprived of cytoplasm, the other three meiotic cells die.

FIGURE 52-4
This ovum is being approached by a single sperm. Notice the tremendous size difference between the egg and the sperm. (SEM 1225×)

FEMALE REPRODUCTIVE STRUCTURES

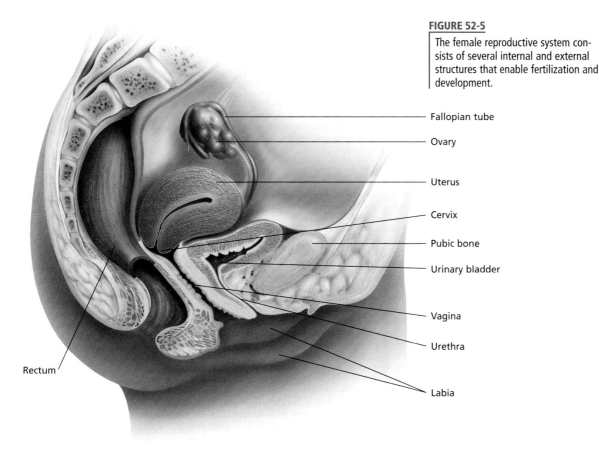

The female reproductive system contains two almond-shaped ovaries that are located in the lower abdomen. Eggs mature near the surface of the ovaries, which are about 3.5 cm (1.4 in.) long and 2 cm (0.8 in.) in diameter. An ovum is released into the abdominal cavity, where it is swept by cilia into the opening of a nearby **fallopian** (fuh-LOH-pee-uhn) **tube,** or uterine tube. This narrow passageway leads to the uterus, as shown in Figure 52-5. The **uterus** (YOOT-uhr-uhs) is a hollow, muscular organ about the size of a fist. If an egg is fertilized, it will develop in the uterus. The lower entrance to the uterus is called the **cervix** (SUHR-viks). A sphincter muscle in the cervix controls the opening to the uterus. Leading from the cervix to the outside of the body is a muscular tube called the **vagina** (vuh-JIE-nuh). The vagina receives sperm from the penis; it is also the channel through which a baby passes during childbirth. The external structures of the female reproductive system are collectively called the **vulva** (VUHL-vuh). The vulva includes the **labia** (LAY-bee-uh), folds of skin and mucous membranes that cover and protect the opening to the female reproductive system.

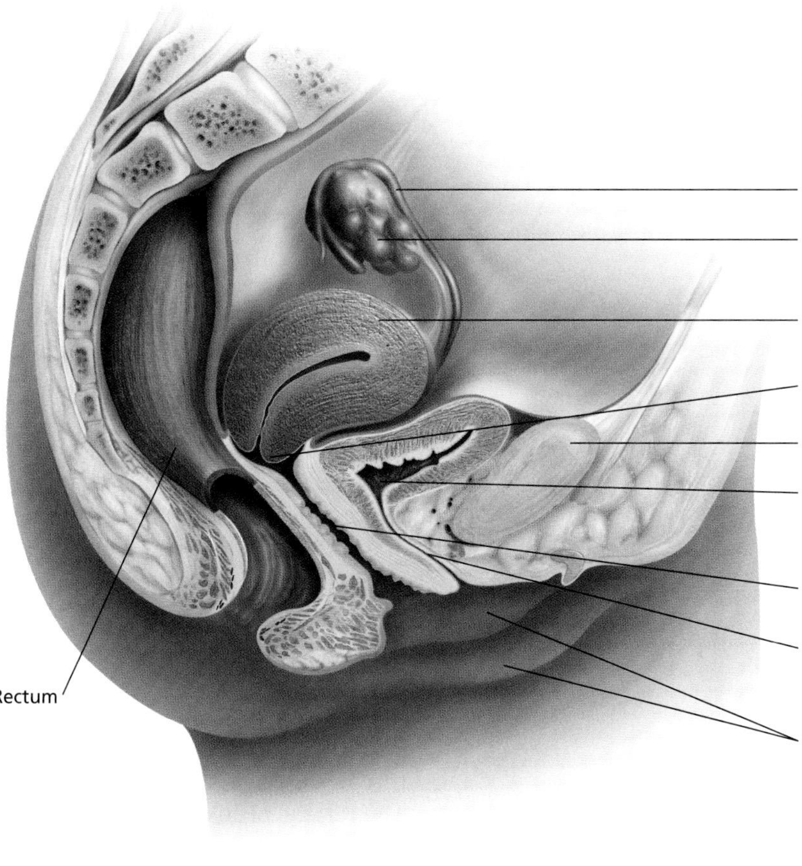

FIGURE 52-5

The female reproductive system consists of several internal and external structures that enable fertilization and development.

SECTION 52-2

TEACHING STRATEGY

Human Egg Production

Provide pairs of students with a microscope and prepared slides of a human ovary cross section and uterine endometrium during the four phases of the menstrual cycle. Provide either a wall chart or a model of the female reproductive system for reference. Have students identify, draw, and label the following structures: the germinal epithelium (the outermost layer of the ovary); the primary follicle (several cuboidal follicle cells surrounding a developing ovum); secondary, or growing, follicles (several larger cell layers with fluid accumulation and central cavity formation); the follicle; and the corpus luteum (the solid, granular structure that develops from the ovulated follicle). Have students observe each phase of the uterine lining and compare the relative thickness and vascularity in each stage. Supply students with photomicrographs from a histology atlas so that they can compare them with the slides.

CONTENT CONNECTION

Section 8-3: Gametes

The difference in the size of eggs and sperm is a result of the nature of egg and sperm formation. During egg formation, unequal divisions of the cytoplasm follow meiosis I and meiosis II; this produces one large egg and three tiny cells that degenerate. During sperm formation, equal divisions of the cytoplasm follow meiosis I and meiosis II; this yields four equal-sized sperm cells. In humans, meiosis II occurs at fertilization.

RETEACHING ACTIVITY

Have students design a table in which they list the major structures and functions of the male and female reproductive systems.

QUICK FACT

☑ Throughout history, the perception of menstruating women has varied among cultures. Both shame and fear have led to the segregation of menstruating women from society. Attitudes toward menstruating women are changing in the United States. For example, until 1972, televised advertisements of female sanitary products were considered taboo.

VISUAL LINK

Figure 52-6

This figure illustrates the four stages of the menstrual cycle. Review the events of each phase with students. Ask students to explain how hormones regulate the cycle. (FSH stimulates follicle growth. Follicle cells secrete estrogen, which stimulates release of LH. LH initiates ovulation, which leads to the luteal phase. The corpus luteum secretes progesterone and estrogen, causing FSH and LH levels to fall. If the egg is not fertilized, the corpus luteum stops producing estrogen and progesterone.) Ask students to describe how the egg and the reproductive organs change during the cycle. (The follicle develops to nourish the immature egg, the corpus luteum grows, and the uterine lining thickens.) Emphasize that feedback mechanisms regulate hormone levels.

Word Roots and Origins

menstrual

from the Latin *mensis,* meaning "month"

PREPARATION FOR PREGNANCY

Each month, the female reproductive system prepares for a possible pregnancy by undergoing a series of changes called the **menstrual** (MEN-struhl) **cycle.** For most women, the menstrual cycle lasts about 28 days. During this time, an egg matures and enters a fallopian tube, where it is able to fuse with a sperm. If the egg does not fuse with a sperm, the egg degenerates. The menstrual cycle has four stages: the follicular phase, ovulation, the luteal phase, and menstruation. These stages are regulated by hormones secreted by the endocrine system. Figure 52-6 summarizes the stages of the menstrual cycle.

FIGURE 52-6

During the 28-day menstrual cycle, an egg matures and is released by an ovary, and the uterus prepares for a possible pregnancy. The events of the menstrual cycle are regulated by hormones that are produced by the anterior pituitary and the ovaries.

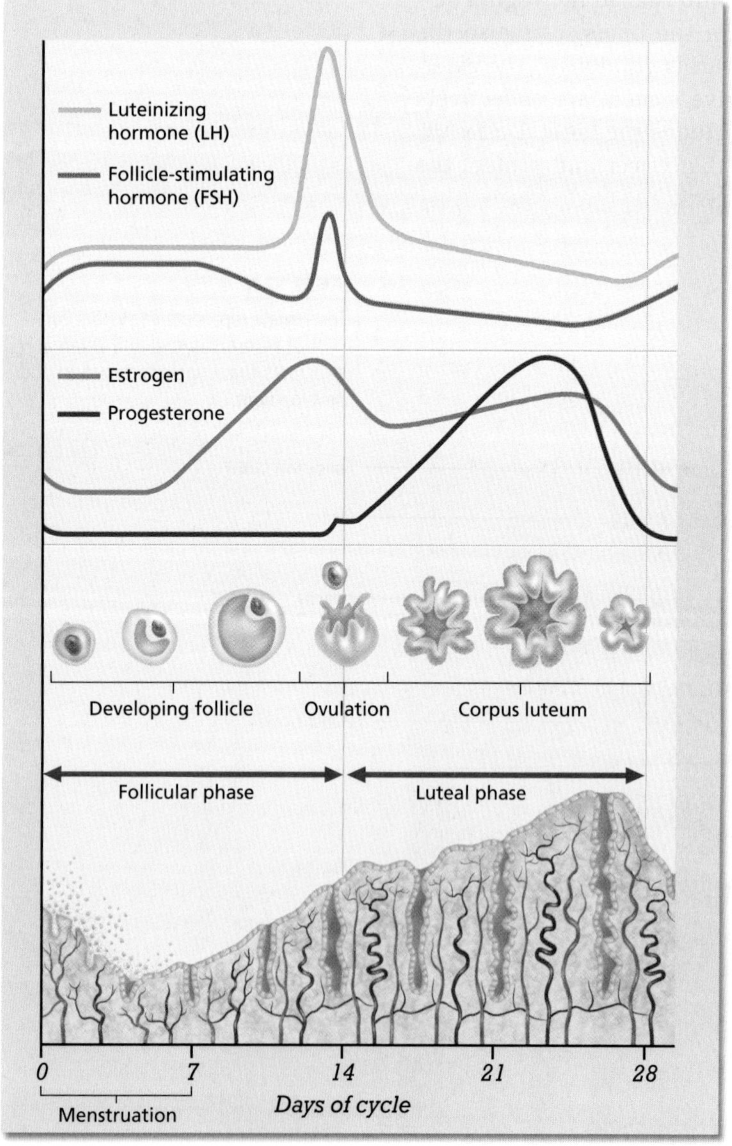

Follicular Phase

An immature egg cell completes its first meiotic division during the **follicular** (fuh-LIK-yoo-luhr) **phase.** This phase begins when the hypothalamus secretes a releasing hormone that stimulates the anterior lobe of the pituitary gland to secrete follicle-stimulating hormone (FSH). FSH stimulates cell division in a **follicle,** a layer of cells that surrounds an immature egg. Follicle cells supply nutrients to the egg. They also secrete estrogen, which stimulates mitotic divisions of cells in the lining of the uterus, causing the lining to thicken. The follicular phase lasts approximately 14 days. During this time, the estrogen level in the blood continues to rise until it reaches a peak and the egg moves to the surface of the ovary. The elevated estrogen level stimulates the anterior pituitary to secrete luteinizing hormone (LH), which initiates the next stage of the menstrual cycle (a positive feedback mechanism).

Ovulation

The sharp rise in the level of LH that occurs midway through the menstrual cycle causes the follicle to rupture and release its egg at that time. The release of an egg from a ruptured follicle is called **ovulation** (AHV-yoo-LAYshuhn). Following ovulation, an egg is swept into a fallopian tube, where it travels toward the uterus awaiting fertilization. The egg has enough stored nutrients to survive about 48 hours.

Luteal Phase

The cells of the ruptured follicle grow larger and fill the cavity, forming a new structure called a **corpus luteum** (KOHR-puhs LOOT-ee-uhm). Thus, this stage of the menstrual cycle is called the **luteal** (LOOT-ee-uhl) **phase.** The corpus luteum begins to secrete large amounts of progesterone and estrogen. Progesterone stimulates growth of blood vessels and storage of fluids and nutrients in the lining of the uterus. This causes the uterine lining to become even thicker. In addition, increased levels of estrogen and progesterone cause the pituitary gland to stop secreting LH and FSH (a negative feedback mechanism). The luteal phase lasts about 14 days. During this time, estrogen and progesterone levels in the blood rise, while the FSH and LH levels drop.

Menstruation

If an egg is fertilized, the resulting zygote attaches to the lining of the uterus, where it will develop for the next nine months. A hormone that is produced early in pregnancy stimulates the corpus luteum to continue producing estrogen and progesterone, and the thickened lining of the uterus is maintained. If the egg is not fertilized, the corpus luteum stops producing sex hormones. Without estrogen and progesterone to maintain the thickened uterine lining, the lining begins to slough off. In the last stage of the menstrual cycle, called **menstruation** (men-STRAY-shuhn), the lining of the uterus and blood from ruptured blood vessels are discharged through the vagina. Menstruation lasts about five days at the beginning of the follicular phase.

Menstruation continues in most women until about age 50. By then, most of a woman's follicles have either matured and ruptured or degenerated. Without follicles, the ovaries cannot secrete enough estrogen and progesterone to continue the menstrual cycle, and menstruation ceases. This stage is called **menopause** (MEN-uh-PAHZ).

internetconnect

SCILINKS NSTA

TOPIC: Female reproductive system
GO TO: www.scilinks.org
KEYWORD: HM1053

SECTION 52-2 REVIEW

1. What are the main female reproductive organs?
2. How are eggs and sperm similar? How are they different?
3. Where does fertilization occur?
4. What is ovulation? When does it occur?
5. How do high levels of estrogen and progesterone during the luteal phase of the menstrual cycle affect the uterus?
6. CRITICAL THINKING What might happen if more than one egg were simultaneously released from the ovaries?

GIFTED ACTIVITY

Have students research and report on the causes, symptoms, treatments, and prevention of premenstrual syndrome (PMS). Students likely will find biopsychological evidence for the symptoms of PMS. For example, recent studies have suggested that changes in the level of the neurotransmitter serotonin in the brain might contribute to the irritability, mood changes, and stress that are characteristic of PMS. There is also evidence that diet, exercise, and use of fluoxetine (in severe cases), the popular antidepressant that selectively inhibits reuptake of serotonin in the central nervous system, all might help alleviate PMS. Eating carbohydrates, for instance, stimulates release of serotonin in the brain.

ANSWERS TO SECTION 52-2 REVIEW

1. The ovaries, fallopian tubes, the uterus, the cervix, the vagina, and the vulva are the primary female reproductive organs.
2. Both are gametes that contain the haploid (1*n*) number of chromosomes. The egg is much larger than the sperm.
3. Fertilization usually occurs in a fallopian tube, but it can also occur in the uterus.
4. Ovulation is the release of an egg from a ruptured follicle. It occurs midway through the menstrual cycle.
5. These hormones cause the uterine lining to thicken.
6. More than one egg might be fertilized, causing twins, triplets, or more.

SECTION 52-3

SECTION OVERVIEW

Reproduction and Inheritance
Section 52-3 explains fertilization and the three stages of pregnancy. It describes how fusion of a sperm's nucleus and an egg's nucleus leads to the formation of a zygote, in which the diploid (2*n*) number is restored. It also describes how the fertilized egg divides and grows in a pattern of cellular differentiation. Finally, this section describes childbirth.

▶ ENGAGE STUDENTS

From the *Holt BioSources Teaching Transparencies,* photocopy Blackline Master 44A, "Comparing Vertebrate Embryo Development," as either a transparency or a handout. This figure shows various animal embryos at different stages of development. Display the figure to students, and have them compare the different animal species. Ask students to identify the human embryo. (At early stages of development, this will be difficult, if not impossible.) Ask students to consider the similarities of the various species in the early stages of development and the differences between them in the later stages. Remind them of the embryological evidence for evolution that was discussed in Chapter 15. Ask students what they can infer from this evidence. (Humans share a common ancestry with other animals.)

SECTION

52-3

OBJECTIVES

▲ Describe the processes of fertilization, cleavage, and implantation.

● Summarize the changes that take place during the development of an embryo.

■ List the three stages of pregnancy, and describe each stage.

GESTATION

A new individual is produced when a sperm fertilizes an egg, resulting in the formation of a zygote. During the nine-month period of ***gestation*** *(jes-TAY-shuhn), a series of changes transforms a single cell into a complex organism made of trillions of cells—a human.*

FERTILIZATION

Recall that with one ejaculation, a male releases hundreds of millions of sperm into the vagina of a female. Once sperm are released, they swim through the vagina, cervix, and uterus, and, finally, up the fallopian tubes. If ovulation occurs anytime from 72 hours before to 48 hours after ejaculation, sperm may encounter an egg in one of the fallopian tubes. Fertilization occurs when a sperm and an egg fuse and form a zygote.

An egg in a fallopian tube is encased in a jellylike substance and surrounded by a layer of cells from the follicle of the ovary. As shown in Figure 52-7, several sperm may attach to an egg and attempt to penetrate its outer layers. Recall from Section 52-1 that the head of a sperm contains digestive enzymes. These enzymes break down an egg's outer layers and enable the cell membrane that surrounds the head of the sperm to fuse with the egg's cell membrane. The sperm's nucleus and midpiece then enter the cytoplasm of the egg. The tail of the sperm remains outside the egg. Usually only one sperm is successful in penetrating an egg. Changes that occur in an egg's cell membrane after a sperm enters the egg help keep other sperm from penetrating the egg.

FIGURE 52-7
Several sperm surround this ovum, but only one will be able to fertilize it. (SEM 1165×)

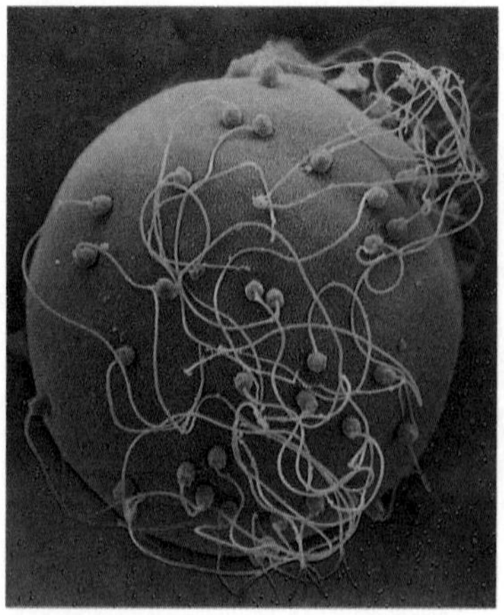

After a sperm enters an egg, its nucleus fuses with the egg's nucleus. The diploid cell that results from this fusion is called a zygote. Recall that each gamete contains 23 chromosomes, the haploid (1*n*) number. Thus, fusion of a sperm nucleus and an egg nucleus causes a zygote to have 46 chromosomes, the diploid (2*n*) number.

Cleavage and Implantation

Immediately following fertilization and while still in the fallopian tube, the zygote begins a series of mitotic divisions known as cleavage. The resulting cells do not increase in size during these cell divisions. Cleavage produces a ball of cells called a **morula** (MOHR-yoo-luh), which is not much larger than the zygote. Cells of the morula divide and release a fluid, resulting in a blastocyst. A **blastocyst** (BLAS-toh-SIST) is a ball of cells with a large, fluid-filled cavity.

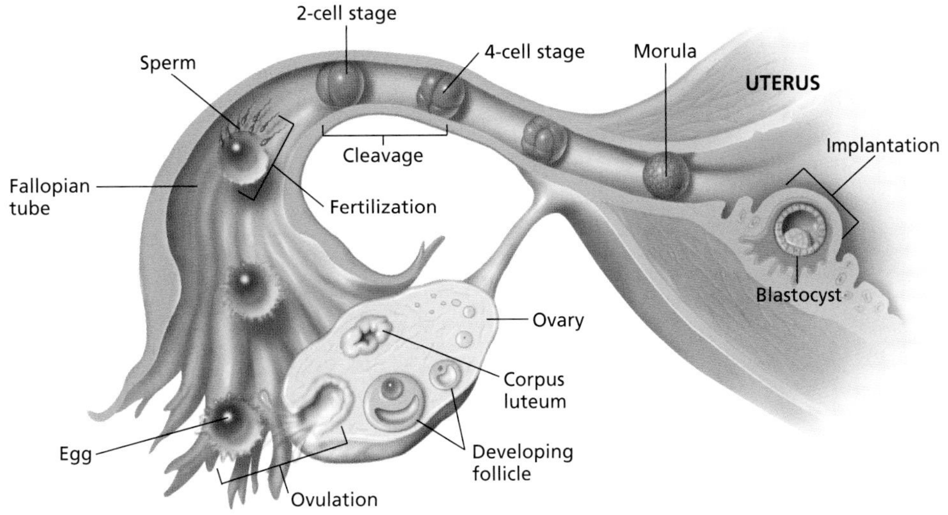

FIGURE 52-8

The earliest stages of development occur within a fallopian tube as a zygote travels toward the uterus.

As shown in Figure 52-8, the morula has become a blastocyst by the time it reaches the uterus. In the uterus, the blastocyst attaches to the thickened uterine lining. The blastocyst then releases an enzyme that breaks down the epithelial tissue that lines the uterus and burrows into the thickened lining. The process in which the blastocyst burrows and embeds itself into the lining of the uterus is called **implantation** (IM-plan-TAY-shun). Pregnancy begins at implantation, which occurs about six days after fertilization.

PREGNANCY

After implantation, the blastocyst slowly takes on the recognizable features of a human infant. This nine-month period of development is called gestation, or **pregnancy.** Pregnancy is divided into three equal periods, or **trimesters.** Significant changes occur during each trimester.

First Trimester

The most dramatic changes in human development take place during the first trimester. Throughout the first two to three weeks following fertilization, a developing human embryo resembles the embryos of other animals. The embryo develops from the mass of cells on the inner surface of the blastocyst. At first, all of the cells in the mass look alike. But the cells soon reorganize, first into two and then into three distinct types of cells, forming the primary germ layers: the ectoderm, mesoderm, and endoderm. Different parts of the body develop from each of the primary germ layers. The nervous system and the epidermis of the skin develop from the ectoderm. Muscle tissue, connective tissue, and organs develop from the mesoderm. The linings of the digestive, respiratory, and urinary tracts develop from endoderm, as do glands, such as the liver, pancreas, thyroid, and thymus.

Quick Lab

Summarizing Vocabulary

Materials pencil, paper, dictionary

Procedure Write and define the following list of words: *ovary, ovum, follicle, gestation, morula, blastocyst, amnion, chorion, umbilical, uterus, corpus,* and *luteum.*

Analysis Do any of the meanings of the words surprise you? Explain. How does knowing the roots and meanings of the words help you remember them?

VISUAL LINK

Figure 52-8

Explain to students that a blastocyst occasionally embeds and develops within the fallopian tube—a circumstance called ectopic pregnancy. Such a pregnancy is extremely painful and potentially fatal to the mother because heavy bleeding generally occurs as the embryo grows. The embryo cannot survive in the fallopian tube, and it must be removed surgically.

ACTIVE READING

Technique: K-W-L

Tell students to return to the list that they created using the Active Reading on page 1047. Have them check off the questions that they can now answer. Students should then make a list of what they have Learned. Have students research their unanswered questions.

Quick Lab

Summarizing Vocabulary

Time Required 20 minutes

Procedural Tips Have students work in pairs. Students could perform this activity in competitive teams. You might provide a small prize to the winners (pencils, candy, etc.).

Answers to Analysis Words are defined in students' books and its glossary. The word meanings are as follows: *ovum,* egg; *follicle,* small bag; *gestation,* to bear or carry; *morula,* mulberry; *blastocyst,* sprout or bud with sac or pouch; *amnion,* lamb; *chorion,* yarn; *umbilical,* navel; *uterus,* belly; *corpus,* body; and *luteum,* golden yellow tinged with green. The relevance of several of the words to their respective structures is obscure, such as *morula* or *amnion.* The meaning of the word in Latin or Greek often describes the structure, such as *corpus luteum.*

QUICK FACT

☑ Pregnant women can transmit HIV to their children through blood either before or during birth.

DEMONSTRATION

Use photographs and illustrations to demonstrate which body tissues develop from each of the primary germ layers—ectoderm, mesoderm, and endoderm. Call out the names of various body tissues and organs, and ask students to identify the germ layer from which each develops. (For example, skin arises from the ectoderm, muscles from the mesoderm, and lungs from the endoderm.)

MAKING CONNECTIONS

Language Arts

The word *hysterectomy* originated with two Greek words: *hyster,* meaning "uterus," and *ektome,* meaning "to cut out." By definition, a hysterectomy is the surgical removal of all or part of the uterus.

RETEACHING *ACTIVITY*

Divide the class into groups. Ask each group to make a poster that illustrates either fertilization, embryonic development, or fetal development. Allow each group time to display and discuss its poster with the rest of the class.

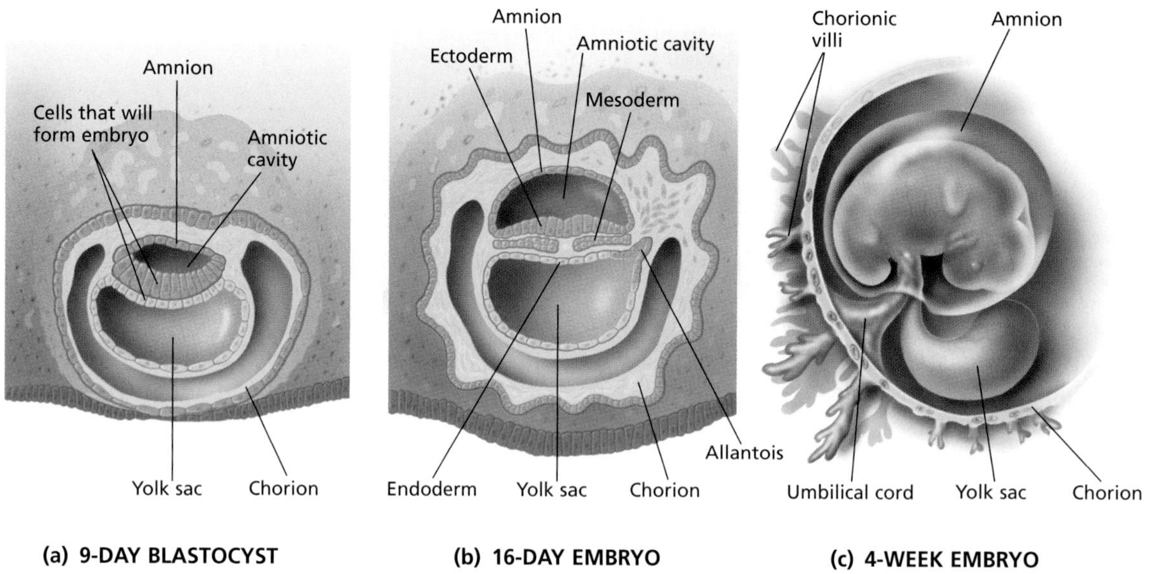

FIGURE 52-9

(a) An embryo develops from the mass of cells on one side of a blastocyst. (b) The primary germ layers develop by the third week of pregnancy, and the four embryonic membranes form. (c) By the end of the first month of pregnancy, all of the embryonic membranes are formed.

Four membranes that aid the development of the embryo also form during the first trimester. One of these membranes, called the amnion (AM-nee-uhn), forms the fluid-filled **amniotic** (AM-nee-AHT-ik) **sac,** which surrounds the developing embryo. The fluid in the amniotic sac cushions the embryo from injury and keeps it moist. A second membrane forms the yolk sac. Although it does not contain yolk, the yolk sac is an important structure because it is where the first blood cells and reproductive cells originate. A third membrane, called the allantois (uh-LAN-toh-is), forms near the yolk sac. The fourth membrane, called the chorion (KOHR-ee-AHN), surrounds all of the other membranes. As shown in Figure 52-9, one side of the chorion forms many small, fingerlike projections called **chorionic villi** (KOHR-ee-AHN-ik VIL-IE), which extend into the uterine lining. Blood vessels that form within the chorionic villi originate in the allantois.

Together, chorionic villi and the portion of the uterine lining that they invade form a close-knit structure called the placenta. A human placenta is about 15–20 cm (6–8 in.) in diameter and about 2.5 cm (1 in.) thick. Nutrients, gases, pathogens, drugs, and other substances pass from the mother to the fetus through the placenta. The developing fetus is attached to the placenta by the **umbilical** (uhm-BIL-i-kuhl) **cord,** which contains arteries and veins that carry blood between the fetus and the placenta. As Figure 52-10 shows, blood from the mother and fetus never mixes. Instead, materials are exchanged by diffusion.

By this time, had the egg not been fertilized, the corpus luteum would have stopped producing sex hormones, and menstruation would have followed. A developing placenta begins to secrete a hormone called **human chorionic gonadotropin (HCG)** early in the second week after fertilization. In the early stages of pregnancy, HCG stimulates the corpus luteum to continue producing sex hormones, and thus the uterine lining and the embryo are retained. As the placenta grows, it begins to secrete large amounts of progesterone and estrogen, which take over maintenance of the uterine lining from hormones that were produced by the corpus luteum. Continued production of estrogen and progesterone throughout pregnancy prevents the release of FSH and LH, and eggs are not ovulated.

The brain, the spinal cord, and the rest of the nervous system begin to form in the third week. The heart begins to beat at 21 days, and it develops a smooth, rhythmic beat at 28 days. By the fifth week, human features, including arms, legs, eyes, ears, and the digestive system have begun to develop. At six weeks, the fingers and toes form, and the brain shows signs of activity. The embryo also begins to move, although it is so small that the mother cannot feel it. From eight weeks until birth, the developing child is called a **fetus** (FEET-uhs). The fetus is only about 5 cm (2 in.) long when the first trimester ends, but all of its organ systems have begun to form.

Second Trimester

During the second trimester, the mother's abdomen begins to swell as her uterus enlarges. The fetus's heartbeat can be heard, its skeleton begins to form, and a layer of soft hair grows over its skin. At this time, the fetus also begins to wake and sleep. The mother may feel the fetus move about. The fetus swallows, hiccups, sucks its thumb, and makes a fist. It also kicks its feet and curls its toes. By the end of the second trimester, the fetus is about 32 cm (12.6 in.) long, and its eyes are open.

Third Trimester

In the third trimester, the fetus grows quickly and undergoes changes that will enable it to survive outside the mother. The fetus can see light and darkness through the mother's abdominal wall,

FIGURE 52-10

About two weeks after fertilization, chorionic villi form and the placenta begins to develop. The placenta and umbilical cord function as the lifeline between an embryo and its mother.

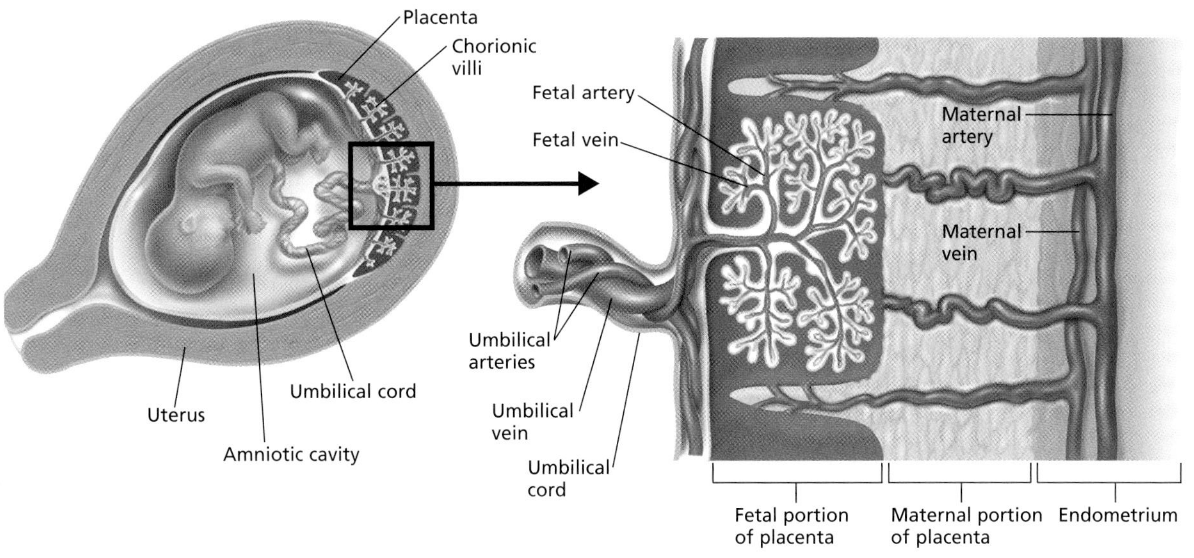

TERMINOLOGY NOTE

In humans, gestation lasts about 280 days, the duration from the last menstrual period to birth. The *embryo* stage of human development begins in the third week after fertilization, after implantation has occurred. Prior to implantation, the developing human is called a *blastocyst.* After eight weeks, it is called a *fetus.* The term *pregnancy* refers to all of the events from implantation to birth.

QUICK FACT

☑ The average size of an animal is directly related to the duration of its gestation period. For example, the average time of gestation is 16 days for hamsters, 63 days for cats, 114 days for pigs, 425 days for giraffes, and 645 days for elephants.

RECENT RESEARCH

Smoking and Pregnancy

Researchers at the University of California at San Diego have suggested that fetuses and infants exposed to smoke are more likely to die from Sudden Infant Death Syndrome (SIDS). This concurs with other evidence that both a mother's smoking during pregnancy and a child's exposure to secondhand smoke after birth detrimentally affect the child's lungs. Other studies indicate that the fetuses of smokers are more likely to develop birth defects, such as cleft lip or cleft palate, than those of nonsmokers. Further evidence suggests that smoking causes a decrease in the mother's supply of vitamin C, an important nutrient for the fetus.

SECTION 52-3

GIFTED ACTIVITY

Ask students to research the effects of alcohol, cocaine, and other drugs on development. Students will find that all drugs are potentially harmful to fetuses and babies. Fetal exposure to alcohol that is ingested by the mother causes fetal alcohol syndrome (FAS), which is characterized by severe birth defects, including a flat face, small, wide-set eyes, and an abnormally small body. Severe FAS can cause brain damage—causing balance and coordination deficiencies and mental retardation—and even death. FAS generally results in learning disabilities and behavioral abnormalities, such as hyperactivity. Babies that were exposed to cocaine *in utero* can also develop behavioral disorders and learning disabilities. Other drugs, including some prescribed painkillers, narcotics, and antibiotics, can produce harmful side effects during pregnancy as well.

ANSWERS TO SECTION 52-3 REVIEW

1. During implantation, the blastocyst embeds itself in the uterine lining.
2. The blastocyst is a sphere of cells with a large, fluid-filled cavity at its center.
3. The fetus is nourished through the placenta, which passes nutrients from the mother to the fetus.
4. In the first trimester, the primary germ layers arise; the placenta attaches to the fetus and secretes HCG so that the uterine lining is maintained.
5. In the third trimester, the fetus's sensory systems become active. Fat deposits enable the fetus to maintain its body temperature.
6. Substances in the maternal bloodstream pass through the placenta to the fetus.

and it can react to music and loud sounds. During the last half of this trimester, the fetus develops fat deposits under its skin. These fat deposits, which make the fetus look rounded and less wrinkled, insulate the body so that it can maintain a steady body temperature.

Birth

Birth occurs about 270 days after fertilization. Prostaglandins that are produced by the fetal membranes, along with hormones that are produced by the fetus and the mother, initiate childbirth. High levels of estrogen, prostaglandins, and oxytocin, a hormone that is secreted by the pituitary gland, cause the smooth muscles that line the uterus to contract. The amniotic sac breaks, and its fluid flows out through the vagina. This is called "breaking water." Muscles in the cervix and the vagina relax, enabling the cervix and vagina to enlarge and allowing the fetus to pass through them. The muscular contractions and other events that lead up to childbirth are called **labor.**

During childbirth, strong contractions of the uterus push a fetus out through the cervix, through the vagina, and from the body, as shown in Figure 52-11. The placenta, the amnion, and the uterine lining, collectively called **afterbirth,** are expelled from the mother's body about 10 minutes after the baby is born. Following birth, the newborn baby's lungs expand for the first time as the baby cries and begins to breathe on its own. At this time, the umbilical cord is tied and cut. The umbilical arteries and veins close off within 30 minutes after birth. This and other changes in the baby's blood vessels lead to completion of cardiopulmonary and renal circulation, allowing the baby to function independent of the mother. The newborn baby's respiratory and excretory systems soon become fully functional.

FIGURE 52-11

During childbirth, the fetus passes through the greatly enlarged cervix and vagina.

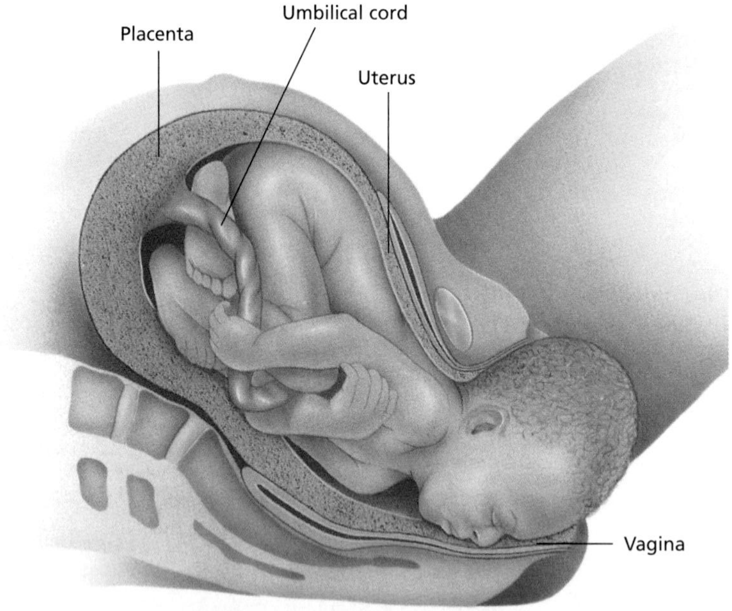

SECTION 52-3 REVIEW

1. What is the process of implantation?
2. What is a blastocyst?
3. How is a fetus nourished during development?
4. Summarize the events of human development that occur during the first trimester of pregnancy.
5. What changes occur in a fetus during the third trimester of pregnancy?
6. CRITICAL THINKING Why is it important for a pregnant woman to eat healthy foods, to avoid unhealthy substances, and to follow her doctor's advice?

CHAPTER 52 REVIEW

SUMMARY/VOCABULARY

52-1

- Sperm form in the seminiferous tubules of the testes. Meiosis reduces the number of chromosomes in sperm to 23.
- A mature sperm consists of a head, which contains the nucleus and chromosomes; a midpiece that contains mitochondria; and a tail, which consists of a flagellum.
- The testes are contained in the scrotum, where the cooler temperature allows normal sperm development.
- Sperm take the following path to exit the body: seminiferous tubules of the testes → epididymis → vas deferens → urethra.
- Fluids that are secreted by various exocrine glands are mixed with sperm to produce semen.

Vocabulary

ejaculation (1049)
epididymis (1048)
penis (1049)
scrotum (1048)
semen (1048)
seminiferous tubules (1048)
testes (1047)
vas deferens (1048)

52-2

- Eggs form in the ovaries. Meiosis reduces the chromosome number in eggs to 23.
- A female is born with all of the eggs that she will ever produce.
- Starting at puberty, one egg matures (completes meiosis) approximately every 28 days during the menstrual cycle.
- The menstrual cycle consists of four phases: follicular phase, ovulation, luteal phase, and menstruation.
- In the follicular phase, FSH causes a follicle to grow in an ovary. Estrogen produced by the follicle causes an egg to mature and stimulates the buildup of the uterine lining.
- Ovulation occurs midway through the menstrual cycle, when LH causes the follicle to rupture and release its egg.
- In the luteal phase, the follicle becomes a corpus luteum. The corpus luteum secretes progesterone, which stimulates further buildup of the uterine lining.
- Menstruation, the discharge of the uterine lining, occurs when a corpus luteum stops secreting progesterone and estrogen.

Vocabulary

cervix (1051)
corpus luteum (1053)
fallopian tube (1051)
follicle (1052)
follicular phase (1052)
labia (1051)
luteal phase (1053)
menopause (1053)
menstrual cycle (1052)
menstruation (1053)
ovary (1050)
ovulation (1052)
ovum (1050)
uterus (1051)
vagina (1051)
vulva (1051)

52-3

- Fertilization occurs in a fallopian tube.
- Pregnancy begins when a blastocyst implants itself in the lining of the uterus.
- The three primary germ layers—the ectoderm, mesoderm, and endoderm—form early in embryonic development.
- Four membranes—the amnion, yolk sac, allantois, and chorion—also form early in embryonic development.
- Part of the chorion and part of the lining of the uterus form the placenta, through which nutrients, gases, and other substances pass by diffusion to the fetus from the mother.
- From the eighth week until birth, a developing human is known as a fetus.
- During childbirth, contractions of the uterus initiated by prostaglandins and oxytocin push the baby from the mother's body through the vagina.

Vocabulary

afterbirth (1058)
amniotic sac (1056)
blastocyst (1054)
chorionic villi (1056)
fetus (1057)
gestation (1054)
human chorionic gonadotropin (1056)
implantation (1055)
labor (1058)
morula (1054)
pregnancy (1055)
trimester (1055)
umbilical cord (1056)

CHAPTER 52 REVIEW ANSWERS

REVIEW

1. Testes and ovaries are the gamete-producing organs of the male and the female, respectively.
2. Sperm are the male gametes, whereas semen, which includes sperm, is the fluid vehicle of the sperm.
3. Menstruation is the final phase of the menstrual cycle during which the uterine lining is flushed out. Menstruation stops at menopause.
4. The chorionic villi and the uterine lining form the placenta.
5. Until eight weeks after implantation, the developing child is an embryo. From eight weeks until birth, it is a fetus.

6. d **7.** a
8. b **9.** c
10. c **11.** d
12. b **13.** a
14. b **15.** d

16. A sperm has a large head that contains its nucleus (with its 23 chromosomes), a midpiece filled with mitochondria, and a flagellum.
17. Semen forms when fluids secreted by three exocrine glands—seminal vesicles, bulbourethral glands, and the prostate gland—mix with sperm.

18. Both involve meiosis so that the mature gamete contains the haploid (1*n*) number of chromosomes.
19. (a) The period labeled *A* represents the follicular phase. (b) The period labeled *B* represents the luteal phase. (c) Ovulation occurs midway through the menstrual cycle after LH secretion. (d) Fertilization is likely to occur just after ovulation. (e) The estrogen level is highest during the follicular phase just prior to ovulation. (f) The progesterone level falls just before menstruation.
20. Eggs are produced in the ovaries.
21. If implantation occurs, menstruation does not occur. The uterine lining cannot be discharged because it is the site where the blastocyst is implanted.
22. When a sperm attaches to an egg, digestive enzymes in its head break down the egg's outer layers. The two cell membranes fuse, and the sperm's nucleus enters the egg.
23. Nutrients and oxygen diffuse from the maternal blood to the fetal blood through the placenta.
24. The fluid-filled amnion protects the embryo from injury.
25. The cervix relaxes and enlarges during childbirth.

REVIEW

Vocabulary

1. Name the male and female gamete-producing organs.
2. What is the difference between sperm and semen?
3. Explain the relationships between the following: the menstrual cycle, menstruation, and menopause.
4. Which two structures form the placenta?
5. What is the difference between an embryo and a fetus?

Multiple Choice

6. The correct pathway for sperm is from the (a) testes to the vas deferens to the epididymis (b) epididymis to the urethra to the vas deferens (c) urethra to the vas deferens to the testes (d) testes to the epididymis to the vas deferens.
7. One similarity between a sperm and an egg is their (a) chromosome number (b) size (c) ability to move quickly (d) role in producing hormones.
8. Which of the following is not a function of the uterus? (a) house the developing fetus (b) help produce eggs (c) provide protection for an embryo (d) aid the birth of the baby.
9. Follicle-stimulating hormone (a) promotes contractions of the uterine muscles (b) is secreted by the follicle (c) is secreted by the pituitary gland (d) stimulates the development of the placenta.
10. During menstruation (a) the egg moves into the fallopian tube (b) the corpus luteum develops (c) the uterine lining is discharged (d) the egg matures.
11. Fertilization takes place in the (a) vagina (b) cervix (c) uterus (d) fallopian tube.
12. The embryonic membranes that help form the placenta and umbilical cord are (a) the amnion and yolk sac (b) the chorion and allantois (c) the chorion and yolk sac (d) the amnion and chorion.
13. Which of the following organs develops from ectoderm? (a) brain (b) liver (c) heart (d) lungs
14. The placenta is the (a) site of yolk development (b) site of nutrient and gas exchange between the mother and the fetus (c) structure that protects the fetus from injury (d) tip of the uterus.
15. By the end of the first trimester (a) the fetus can suck its thumb (b) the brain of the fetus is fully developed (c) the fetus uses its lungs to breathe (d) all of the organs of the fetus have begun to form.

Short Answer

16. Describe the structure of a mature human sperm cell.
17. How is semen formed?
18. How are sperm formation and egg formation similar?
19. Use the graph shown below to answer the following questions:
(a) What phase of the menstrual cycle is occurring during the period labeled *A*?
(b) What phase of the menstrual cycle is occurring at the period labeled *B*?
(c) When during the menstrual cycle does ovulation usually occur?
(d) When during the menstrual cycle is fertilization most likely to occur?
(e) When during the menstrual cycle is the estrogen level highest?
(f) When during the menstrual cycle does the progesterone level start to fall?

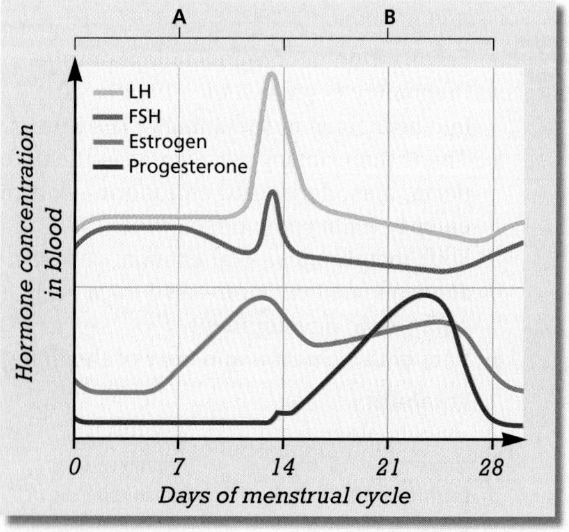

20. Where are eggs produced?
21. Which part of the menstrual cycle does not occur if implantation occurs? Why is this important?
22. How does a sperm penetrate an egg during fertilization?
23. How does the developing fetus receive nourishment?
24. What role does the amnion play in human development?
25. What changes does the cervix undergo during childbirth?

CRITICAL THINKING

1. Why are sperm able to survive as long as they do in the female, even though they have very little cytoplasm to provide nutrients?
2. What do you think might happen if more than one sperm were able to penetrate the cell membrane of an egg?
3. A human female produces on average only one mature egg every 28 days. In contrast, a female salmon lays 50 million eggs at each spawning. Hypothesize why there is such a great difference in egg production between the two species.
4. Women who smoke tobacco, drink alcoholic beverages, or consume other drugs or harmful substances during pregnancy give birth to infants who are often addicted to drugs, have severe birth defects, or develop learning disabilities and behavioral disorders. Why is this so?
5. About 20 years ago, a large number of women who took a tranquilizer called thalidomide early in pregnancy gave birth to babies with serious limb defects. Other mothers who took the drug later on in pregnancy had normal children. What does this tell you about the sequence of development in a fetus?
6. What is the fetus doing in the photograph below? Suggest an adaptive advantage for this activity.

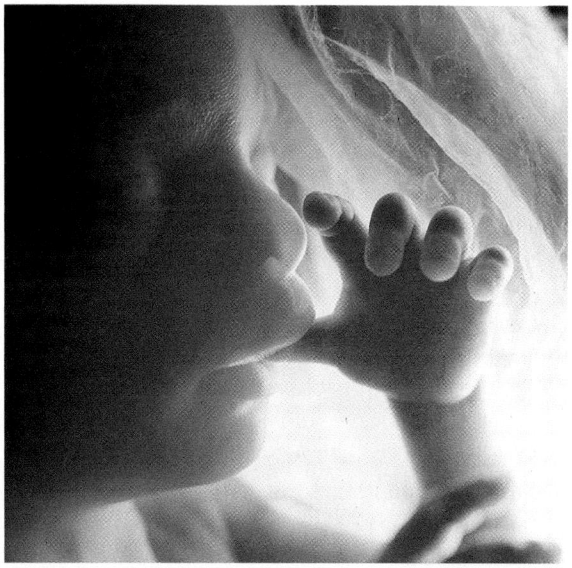

EXTENSION

1. Concerned about the rapid increase in the world's population, several countries are attempting to reach zero population growth (ZPG). These countries have encouraged women to have fewer children or have instituted laws that limit the number of children that each woman may have. Check on-line references to identify one such country, and prepare a report summarizing what measures the country has taken to attain ZPG.
2. Read "Why We Don't Lay Eggs" in *New Scientist,* June 12, 1999, on page 12. Prepare a report on the origin of ERVs in the human placenta. Include in your report how ERVs are involved in protecting the fetus.
3. Research some of the physical and hormonal causes of infertility. Write a report describing the conditions, and explain how medical science is helping couples overcome infertility.

ANSWERS

CRITICAL THINKING

1. The seminal fluids secreted by exocrine glands nourish and protect the sperm.
2. Normal development depends on a diploid ($2n$) set of chromosomes. If more than one sperm were to penetrate the egg, the additional chromosomes would restrict development.
3. Most of the eggs produced by a salmon die in water before fertilization. Even many fertilized eggs die. A single fertilized egg of a human has an excellent chance of survival.
4. Drugs and other harmful substances that are taken by pregnant women are passed from the mother to the fetus through the placenta.
5. Limbs develop during the first trimester.
6. The fetus is sucking its thumb. This behavior facilitates the sucking response required to feed after birth.

EXTENSION

1. India, for example, has tried to limit the number of children that a woman can have by encouraging contraception.
2. ERVs are pathogens that became trapped in the DNA of human ancestors some 120 million years ago. ERVs, in hypothesis, help the placenta to form a barrier between mother and fetus. One of the two primary ways ERVs protect the fetus is by stopping other pathogens from reaching the fetus by occupying the pathogen's receptor sites on cells. Another protective service is that ERVs deactivate some of the mother's immune cells that would otherwise attack the fetus.
3. The article titled "A Pill to Treat Impotence?" in *Time,* May 20, 1996, page 54, discusses several methods used to treat infertility.

CHAPTER 52 INVESTIGATION

TIME REQUIRED

40 minutes

SAFETY PRECAUTIONS

Tell students not to scrape the microscope objectives on the slides. If a slide breaks, remind students to be careful of its sharp edges.

QUICK REFERENCES

Lab Preparation notes are found on page 1045B.

Holt BioSources provides a Teaching Resources CD-ROM that contains "Using Gowin's Vee in the Lab" and "Scoring Rubrics."

PROCEDURAL TIPS

1. Remind students that their drawings should be informative rather than artistic.
2. If possible, have students compare the stages of sea-star development to photographs of the mammalian stages.

ANSWER TO BACKGROUND

1. A fertilized egg becomes a completely developed organism through cell division and differentiation.

ANSWERS TO ANALYSIS AND CONCLUSIONS

1. The zygote and the blastula are the same size. By the time the embryo develops into a gastrula, it is larger than the zygote.
2. When the embryo reaches the gastrula stage, some cells look different from others.
3. Until the gastrula stage, all cells are roughly spherical. After that, some become flattened, elongated, or irregularly shaped. From the first cleavage until the gastrula stage, the cells get progressively smaller. After that, some cells begin to grow.

CHAPTER 52 INVESTIGATION

Observing Embryonic Development

OBJECTIVES

- Identify the stages of early animal development.
- Describe the changes that occur during early development.
- Compare the stages of human embryonic development with those of echinoderm embryonic development.

PROCESS SKILLS

- observing
- comparing and contrasting
- making drawings
- drawing conclusions

MATERIALS

- prepared slides of sea-star development, including
  unfertilized egg
  zygote
  2-cell stage
  4-cell stage
  8-cell stage
  16-cell stage
  32-cell stage
  64-cell stage
  blastula
  early gastrula
  middle gastrula
  late gastrula
  young sea-star larva
- compound light microscope
- paper and pencil

Background

1. Most members of the animal kingdom (including sea stars and humans) begin life as a single cell—the fertilized egg, or zygote. How does a fertilized egg become a completely developed organism?
2. The early stages of development are quite similar in different species. Cleavage follows fertilization. During cleavage, the zygote divides many times without growing. The new cells migrate and form a hollow ball of cells called a blastula. The cells then begin to organize into the three primary germ layers: endoderm, mesoderm, and ectoderm. During this process, the developing organism is called a gastrula.
3. The early stages of mammalian development are difficult to study because mammalian eggs are tiny and are not produced in great numbers. In addition, mammalian embryos develop within the mother's body. In the laboratory, it is difficult to replicate the internal conditions of the mother's body. Because the early stages of echinoderm development are similar to those of human development, and because echinoderm development is easier to study in the laboratory than human development, you will observe the early developmental stages of an echinoderm—the sea star—in this investigation.

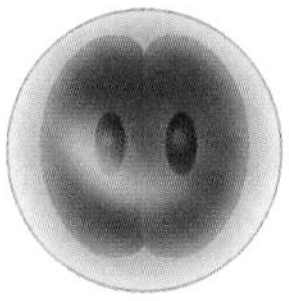
2-cell stage

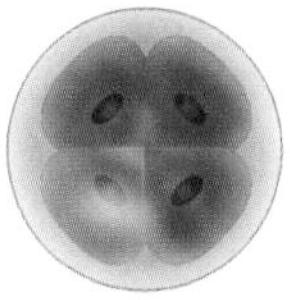
4-cell stage

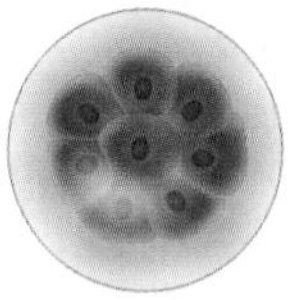
8-cell stage

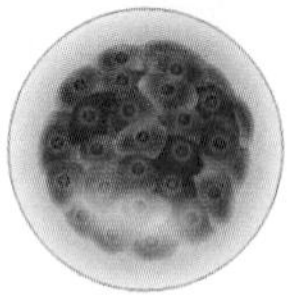
64-cell stage

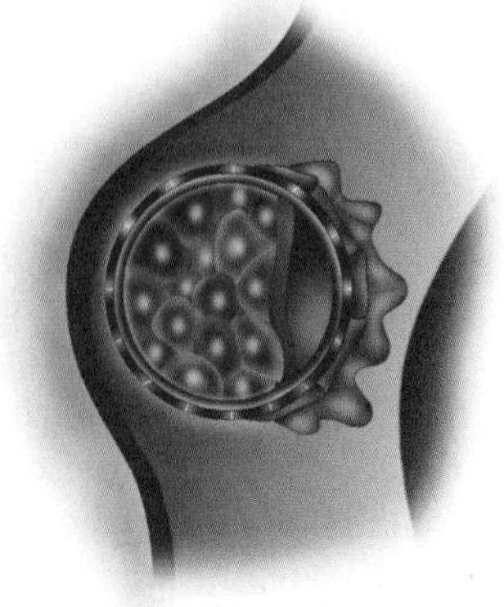
Blastocyst

4. As development continues, the cells continue to specialize as they become part of specific tissues and complex structures. Ectoderm forms the epidermis and nerve tissue. Mesoderm forms muscle, connective tissue, and vascular organs. Endoderm forms the lining of the digestive, urinary, and respiratory tracts.
5. Similarities and differences in early stages of development reflect evolutionary relationships between species.

Procedure

1. Obtain a set of prepared slides that show sea-star eggs at different stages of development. Choose slides labeled unfertilized egg, zygote, 2-cell stage, 4-cell stage, 8-cell stage, 16-cell stage, 32-cell stage, 64-cell stage, blastula, early gastrula, middle gastrula, late gastrula, and young sea-star larva. (Note: *Blastula* is the general term for the embryonic stage that results from cleavage. In mammals, a blastocyst is a modified form of the blastula.)
2. Examine each slide using a compound light microscope. Using the microscope's low-power objective first, focus on one good example of the developmental stage listed on the slide's label. Then switch to the high-power objective, and focus on the image with the fine adjustment.
3. In your lab report, draw a diagram of each developmental stage that you examine (in chronological order). Label each diagram with the name of the stage it represents and the magnification used. Record your observations as soon as they are made. Do not redraw your diagrams. Draw only what you see; lab drawings do not need to be artistic or elaborate. They should be well organized and include specific details.
4. Compare your diagrams with the diagrams of human embryonic stages shown at left.
5. Clean up your materials and wash your hands before leaving the lab.

Analysis and Conclusions

1. Compare the size of the sea-star zygote with that of the blastula. At what stage does the embryo become larger than the zygote?
2. At what stage do all of the cells in the embryo not look exactly like each other?
3. How do cell shape and size change during successive stages of development?
4. Do the cell nuclei stay the same size, get larger, or get smaller as the stages progress?
5. Compare the number of chromosomes in a fertilized sea-star egg with the number of chromosomes in one cell of each of the following phases: 2-cell stage, blastula, gastrula, and adult stage.
6. From your observations of changes in cellular organization, why do you think the blastocoel (the space in the center of the hollow sphere of cells of a blastula) is important during embryonic development?
7. Label the endoderm and ectoderm in your drawing of the late gastrula stage. What do these two tissue types eventually develop into?
8. How are the symmetries of a sea-star embryo and a sea-star larva different from the symmetry of an adult sea star? Would you expect to see a similar change in human development?
9. What must happen to the sea-star gastrula before it becomes a mature sea star?
10. How do your drawings of sea-star embryonic development compare with those of human embryonic development? Based on your observations, in what ways do you think sea-star embryos could be used to study early human development?
11. Describe one way that the cleavage of echinoderms and mammals is alike.
12. Describe two ways that the cleavage of echinoderms and mammals is different.
13. Why are sea-star eggs a good choice for the study of embryonic development in humans?

Further Inquiry

Using the procedure that you followed in this investigation, compare embryonic development in other organisms with embryonic development in sea stars. Which types of organisms would you expect to develop similarly to sea stars? Which types of organisms would you expect to develop differently from sea stars?

4. The nuclei remain the same size as the stages progress.
5. The chromosome number is the same in each cell during embryonic development.
6. A blastocoel enables invagination (folding) to occur during gastrulation, when the three primary germ layers form.
7. Refer to the figure for labels. The endoderm forms the lining of the gut. The ectoderm forms the epidermis and the nervous tissue.
8. Both sea-star embryos and larvae have bilateral symmetry, whereas the adult has radial symmetry. No; both human embryos and adults have bilateral symmetry.
9. Before the gastrula becomes a mature sea star, it develops into a larva, which has bilateral symmetry. Because the adult has radial symmetry, a repositioning of body organs must occur to conform to the radial symmetry of the adult.
10. Sea-star embryos can be used as a model of early human development because they exhibit similar stages of cell division and differentiation.
11. The cleavage in both echinoderms and mammals results in equal-sized cells.
12. In echinoderms, the entire zygote cleaves at regular intervals. The first two divisions are meridonal (from pole to pole), and the third division is equatorial. Subsequent divisions alternate between meridonal and equatorial. In mammals, the zygote divides asymmetrically and more slowly than the echinoderm. The first cleavage is meridonal, yet incomplete. In the second cleavage, one cell divides meridonally, while the other divides equatorially.
13. Sea-star eggs are a good choice for the study of early embryonic development because they are large, numerous, and easy to culture, and because their development is similar to that of mammals.

CHAPTER 53

DRUGS

	TOPICS	TEACHING RESOURCES	LABS, CLASSWORK, AND HOMEWORK
BLOCK 1 45 minutes	**Introducing the Chapter, p. 1064**	**ATE** Focus Concept, p. 1064 **ATE** Assessing Prior Knowledge, p. 1064 **ATE** Understanding the Visual, p. 1064	■ Occupational Applications Worksheets—Pharmacist
	53-1 Role of Drugs, p. 1065 Drug Use and Administration, p. 1065 Medical Use, p. 1066	**ATE** Section Overview, p. 1065	★ Study Guide, Section 53-1 **PE** Section Review, p. 1067
BLOCK 2 45 minutes	**53-2 Social Drugs, p. 1065** Tobacco, p. 1068 Alcohol, p. 1070	**ATE** Section Overview, p. 1068 **ATE** Critical Thinking, p. 1071	**PE** **Quick Lab,** p. 1071, Graphing Tobacco Use **Lab A31,** Determining Lung Capacity ■ Occupational Applications Worksheets—Respiratory Therapist ★ Study Guide, Section 53-2 **PE** Section Review, p. 1072
BLOCKS 3 & 4 90 minutes	**53-3 Abuse of Drugs, p. 1073** Drug Addiction, p. 1073 Drugs of Abuse, p. 1075	**ATE** Section Overview, p. 1073 **ATE** Visual Link Figure 53-6, p. 1074 201A, 233–234	**PE** **Chapter 53 Investigation,** p. 1080 ■ Occupational Applications Worksheets—Substance-Abuse Counselor ■ Occupational Applications Worksheets—Forensic Toxicologist ★ Study Guide, Section 53-3 **PE** Section Review, p. 1076
BLOCKS 5 & 6 90 minutes	**REVIEW and ASSESSMENT** **PE** Chapter 53 Review, pp. 1078–1079 ■ Performance-Based Assessment—Testing Human Digestive Enzymes	Audio CD Program	**CHAPTER TESTING** ■ Chapter Test (blackline master) ▲ Test Generator ■ Assessment Item Listing ■ Scoring Rubrics and Classroom Management Checklists

MODERN BIOLOGY — **HOLT BIOSOURCES** — **One-Stop Planner CD-ROM**

KEY

- PE Pupil's Edition
- ATE Teacher's Edition
- ■ Active Reading Guide
- ★ Study Guide
- Teaching Transparencies
- Laboratory Program

One-Stop Planner CD-ROM
Shows these resources within a customizable daily lesson plan:
- ■ Holt BioSources Teaching Resources
- ■ Active Reading Guide
- ★ Study Guide
- ▲ Test Generator

READING FOR CONTENT MASTERY

Preview: Have students preview the objectives, the introductory paragraph, topic sentences, and captions before reading the chapter.

- **ATE** Active Reading Technique: Reader Response Logs, p. 1065
- ■ Active Reading Guide Worksheet 53-1
- **PE** Summary/Vocabulary, p. 1077

- ■ Active Reading Guide Worksheet 53-2
- **PE** Summary/Vocabulary, p. 1077

- ■ Active Reading Guide Worksheet 53-3
- **PE** Summary/Vocabulary, p. 1077

TECHNOLOGY AND INTERNET RESOURCES

Holt Biology Videodiscs Teacher's Correlation Guide, Lessons 53-1 through 53-3

Audio CD Program, Sections 53-1 through 53-3

Interactive Exploration: Human Biology, Drug Addiction

internetconnect

SCILINKS NSTA

On-line Resources: www.scilinks.org
The following *sci*LINKS Internet resources can be found in the student text for this chapter:

Topics:
- Drugs, p. 1065

go.hrw.com

On-line Resources: go.hrw.com
Visit the HRW Web site for a variety of resources related to this chapter. Just type in the keyword HM2 Chapter 53.

Smithsonian Institution®

Visit **www.si.edu/hrw** for additional on-line resources.

CNNfyi.com

Visit **www.cnnfyi.com** for late-breaking news and current event stories selected just for you.

LAB ACTIVITY PLANNING

Chapter 53 Investigation

Exploring a Model of Cocaine Addiction, pp. 1080–1081

OVERVIEW

Students will examine the molecular mechanisms involved in cocaine addiction. The program begins with an examination of how a normal synapse works when no cocaine is present. The program monitors the number of receptors on a postsynaptic neuron of a synapse in the brain's pleasure center and the number of times the neuron fires. As the simulation begins, students will observe several meters that record nerve firing rate (pleasure), the number of receptors (addiction), drug-use frequency, and drug dosage. Students should conclude that there is no safe dose of cocaine; its use always results in a decrease in the number of postsynaptic dopamine receptors. Because of this decrease in receptors, the dose of cocaine required to achieve the desired effect increases, which produces addiction.

PREPARATION

1. Access to the following materials is needed:
   - computer with CD-ROM drive
   - CD-ROM *Interactive Explorations in Biology: Human Biology.* The title of this program on the CD-ROM is "Drug Addiction."
2. Have students read the Objectives, Background, and Prelab Preparation sections before starting the lab.

Quick Labs

Graphing Tobacco Use, p. 1071

Each pair of students will need graph paper, poster paper, pencils, colored markers or colored pencils, and a ruler for this activity.

UNDERSTANDING THE VISUAL

This is a scanning electron micrograph (SEM) of a cirrhotic liver. *Cirrhosis* is chronic inflammation of the liver that is usually caused by alcoholism, overuse of certain drugs, or chronic hepatitis. Cirrhosis is characterized by an excessive buildup of scar tissue (the light brown areas in the micrograph) around healthy liver cells (the darker brown areas). Scar tissue interferes with the liver cells' ability to filter blood. Cirrhosis severely impairs liver function and can be fatal.

FOCUS CONCEPT

Stability and Homeostasis

Drugs affect the body's ability to maintain homeostasis by altering the actions of molecules, cells, tissues, organs, and organ systems. Some drugs help maintain stability, and others alter normal homeostatic mechanisms. If drugs are used improperly or excessively, they can permanently damage the body.

ASSESSING PRIOR KNOWLEDGE

Review the following concepts.

Cell Structure and Function:
Chapters 4 and 5
Ask students to recall that cells transport substances across the cell membrane. Ask them how a cell responds to relative concentrations of substances inside and outside of its membrane.

Chemical Messengers:
Chapters 50–52
Ask students to recall that chemicals in the body, including hormones and neurotransmitters, act as messengers that enable communication between cells. Ask students how chemical messengers bind to receptor proteins that are located on and within the membranes of their target cells.

CHAPTER 53

DRUGS

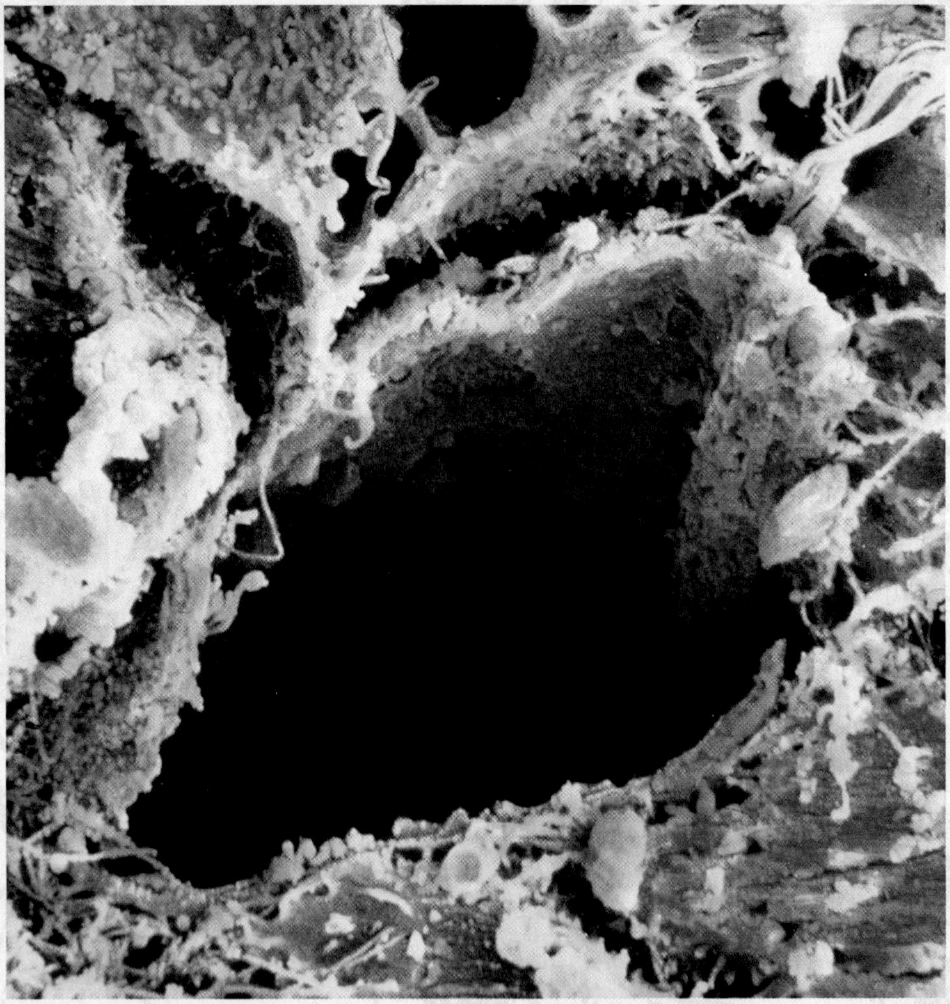

This color-enhanced micrograph (SEM, 6000×) shows a liver affected by the disease cirrhosis. The deep blue central cavity is a blood vessel. Fibers and clumps of scar tissue surrounding the blood vessel have replaced much of the liver's normal cellular architecture.

FOCUS CONCEPT: ***Stability and Homeostasis***

As you read, note how certain drugs affect the body's ability to maintain homeostasis.

ROLE OF DRUGS

As biologists explore the structure and function of the various systems that make up the human body, they discover the complex series of chemical reactions that form the basis of each life process. Understanding the chemistry of the body has enabled scientists to alter it by administering naturally occurring or laboratory-synthesized chemicals called drugs. *The wise use of drugs can alleviate suffering by curing or controlling many kinds of diseases. But the unwise use of drugs can cause great harm because many drugs taken for nonmedical purposes have the potential to destroy human lives.*

SECTION 53-1

OBJECTIVES

▲ Define the term *drug.*

● Name four ways that a drug can be administered.

■ List five common types of prescription and nonprescription drugs.

◆ Describe the potential consequences of a drug interaction.

▲ Describe the consequences of the overuse of a drug.

DRUG USE AND ADMINISTRATION

A **drug** is a chemical compound that affects the structure of a body part or the functioning of a biological process. There are thousands of different drugs, and they affect many different parts of the body. Some drugs are synthesized in laboratories, while other drugs are derived from natural sources. Many drugs are refined from plant or animal tissue. For example, *Rauwolfia serpentina* is a plant native to India and neighboring countries. Ancient Hindu writings describe use of the plant's roots as a remedy for snakebites, insomnia, and psychotic behavior. The modern drug reserpine, which is used to control blood pressure, is extracted from the roots of *Rauwolfia serpentina.*

A drug can be introduced into the body in a variety of ways. Drugs taken orally, in pill or liquid form, are absorbed by the small intestine and then enter the bloodstream. Drugs can also be injected directly into the bloodstream or muscle tissue or under the skin. Other methods of administering drugs include inhaling the drug, placing the drug under the tongue, inserting it in the rectum in the form of a suppository, or applying the drug to the skin.

The method of administering a drug depends on its chemical nature and the organ to be treated. For example, because insulin is a protein, if it were taken orally, it would be digested in the stomach and would therefore be useless. Thus, insulin is injected under the skin, where it can diffuse into the bloodstream for transport throughout the body. In contrast, drugs used to treat respiratory problems are often administered by inhalation so that they can reach their destination quickly.

internet**connect**

SCILINKS NSTA

TOPIC: Drugs
GO TO: www.scilinks.org
KEYWORD: HM1065

SECTION 53-1

SECTION OVERVIEW

Stability and Homeostasis
Section 53-1 describes how both prescription and nonprescription drugs are administered for medicinal purposes. It explains that drug interactions and overuse can be harmful to the body. This section also explains how drugs are synthesized from various sources.

ACTIVE READING

Technique: Reader Response Logs
Students may already hold opinions about drugs. So that students can make personal responses to the concepts presented in this chapter, have them make a Reader Response Log. Ask students to divide their paper in half lengthwise. On the left side of the paper, have them copy a word, phrase, or passage from the text. On the right side, have them write their reactions, thoughts, or questions about the entries on the left side of the paper.

QUICK FACT

☑ Nonprescription drugs such as ibuprofen and naproxen sodium are widely used as alternatives to aspirin for pain relief and for the reduction of fever and inflammation. These drugs inhibit the synthesis and release of prostaglandins. Their side effects include irritation of the stomach lining, disturbed blood platelet function, and allergic reaction.

MAKING CONNECTIONS

Social Studies

Americans are obsessed with being slim and are anxious about being overweight, yet more than 50 percent of Americans are clinically obese and continue to overeat. Drug companies try to mass-market new quick-weight-loss drugs, which many people take to accompany (or, in many cases, to avoid) proper diet and exercise. In April 1996, the Food and Drug Administration approved dexfenfluramine as a prescription drug to treat obesity. It was hailed as a "miracle cure" for exceedingly overweight people.

Dexfenfluramine promotes the action of the neurotransmitter serotonin in the brain. Serotonin release is associated with feelings of satiation and satisfaction. Relatively minor side effects of dexfenfluramine include diarrhea, fatigue, and dry mouth. The drug was removed from the market, however, because it is suspected of causing physical dependence, damage to the brain, damage to blood vessels in the heart and lungs, and mood disorders. Many people improperly use weight-loss drugs for cosmetic purposes, which can result in dangerous consequences. Drugs like dexfenfluramine are powerful social and medical phenomena, and dexfenfluramine, like many other drugs, has the potential to be overprescribed and abused.

TERMINOLOGY NOTE

The term *nonprescription drug* replaces *over-the-counter drug.*

MEDICAL USE

Drugs classified as **prescription drugs** cannot be obtained unless they are prescribed by a physician or dentist. The physician or dentist specifies the **dose,** that is, the amount of the drug to be taken and the number of times per day that the drug should be taken. Thousands of prescription drugs are currently available to treat a variety of diseases and disorders.

Nonprescription drugs, also called over-the-counter (OTC) drugs, can be purchased without a prescription and are available at pharmacies and supermarkets. A quick look at the shelves in a supermarket will show the variety of nonprescription drugs that are available. Such drugs are used to relieve pain, treat minor skin infections, reduce stomach acidity, and alleviate symptoms of a cold or flu.

The labels on nonprescription drugs contain information regarding correct usage. These drugs are safe for most people when taken in the recommended dose but can be dangerous if too much is taken. More than 100 children in the United States die each year from accidental overdoses of aspirin, and even children's vitamins are harmful in large doses. Certain precautions may be stated on the drug's label. For example, aspirin labels caution against giving aspirin to children or teenagers who have flulike symptoms or chickenpox because of the possibility of their developing Reye's syndrome, an often fatal condition. Pregnant women and nursing mothers should consult their physician before using any drug, even nonprescription drugs, because many drugs can cross the placenta or enter a mother's milk and can affect a fetus or a young child.

TABLE 53-1 ***Common Types of Prescription and Nonprescription Drugs***

Drug type	Medical use	Mode of action
Antacids, acid reducers	heartburn, indigestion, ulcer	neutralize excess stomach acid or slow its production
Antibiotics	bacterial infection	kill or prevent growth of infectious bacteria
Antidepressants	mental depression	increase activity in some neurons in the brain
Anti-inflammatory and non-narcotic pain medications	arthritis; headache; injury to bone, joint, or muscle	reduce inflammatory response (swelling, redness, pain) to injury; relieve pain
Antihistamines	allergy, cold, influenza	reduce allergic response (excess fluid in tissue, contraction of bronchi, dilation of capillaries)
Blood pressure medications	hypertension	dilate blood vessels or cause excretion of water, which reduces blood volume
Decongestants	asthma, cold, influenza	dilate nasal passages and bronchi
Hormone replacements	diabetes mellitus, hypothyroidism, post-menopausal hormone loss	restore normal levels of hormones

Drug Interactions

Table 53-1 lists several common types of drugs. You may have taken one or more of these kinds of drugs. Have you ever taken more than one drug at the same time? Many people do, particularly the elderly and those with chronic diseases. For example, a person with a heart ailment might take a drug to reduce blood clotting, a drug to reduce high cholesterol levels, and a drug to regulate heart function. Although simultaneous use of more than one drug is common, some combinations of drugs are inappropriate because of the chemical interactions between them. In such cases, one drug might add to the effect of the other drug, resulting in symptoms of overdose, or one drug might cancel the effectiveness of the other drug. For example, the commonly prescribed antibiotic tetracycline is rendered inactive in the presence of calcium. Therefore, taking tetracycline within an hour of taking a chewable antacid containing calcium will impair the drug's effectiveness. Similarly, some antibiotics interfere with the action of birth-control pills.

In some cases, drug interactions can lead to more-serious problems and can even cause death. A commonly used antihistamine combined with an often-prescribed antifungal drug can cause an irregular heartbeat and even a heart attack. A physician or pharmacist must closely monitor anyone who is taking more than one drug simultaneously. And a prescribing physician must be informed of the name and dose of *any* drugs—prescription or nonprescription—that a patient is taking or has taken in the past few weeks.

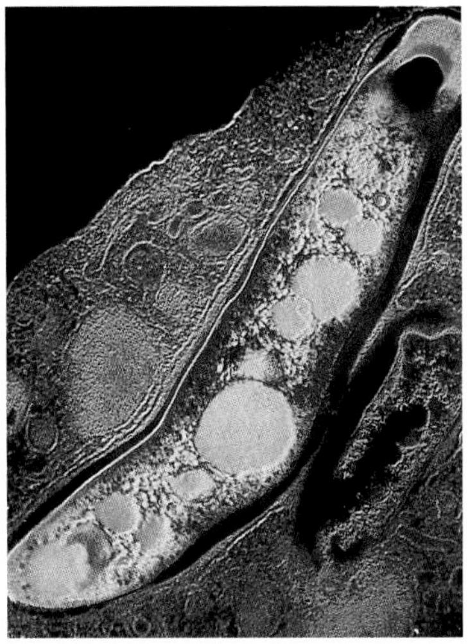

FIGURE 53-1

Some strains of the common pathogen *Mycobacterium tuberculosis* are resistant to antibiotics.

Drug Overuse

Overuse of both prescription drugs and nonprescription drugs can lead to serious problems. Continual use of acetaminophen, a common nonprescription pain medication, can damage the liver. Large doses of magnesium-containing antacids can interfere with the functioning of the heart and kidneys. Anyone who is taking any drug for an extended period of time should consult a doctor.

The problems of drug overuse are not confined to damage to the body. The overuse and misuse of antibiotics have resulted in strains of bacteria that are resistant to these drugs. For example, some strains of the organism, shown in Figure 53-1, that causes the lung disease tuberculosis, are resistant to all antibiotics currently in use.

SECTION 53-1 REVIEW

1. What is a drug?

2. Describe four ways that a drug can be administered.

3. Explain why doctors ask patients to list all the medications they are taking.

4. Name five commonly used types of drugs.

5. What problem has the overuse of antibiotics produced?

6. **CRITICAL THINKING** Colds, flu, and 90 percent of sore throats are caused by viruses. Why would taking an antibiotic be useless in these cases?

GIFTED ACTIVITY

Some doctors recommend that adults take aspirin regularly to help prevent heart attacks, strokes, and other disorders that involve blood clots. Have students research the advantages and disadvantages of taking aspirin regularly. Students should cite that aspirin relieves pain, reduces fever and inflammation, and "thins" the blood by interfering with blood platelets and preventing blood clots from forming. Aspirin is considered a relatively safe drug for adults (children who have a fever should not be given aspirin) when it is taken as directed. Disadvantages are that it can irritate the stomach lining and cause stomach bleeding. Also, some people are allergic to aspirin.

ANSWERS TO SECTION 53-1 REVIEW

1. A drug is a chemical compound that is introduced into the body and that affects either the structure of a body part or the functioning of a biological process.
2. For example, a drug can be administered orally, topically, by injection, or by inhalation.
3. Doctors will then be aware of any possible drug interactions.
4. Examples of five commonly used types of drugs are those used to treat pain, infection, cold symptoms, heart ailments, and diabetes.
5. Some strains of bacteria have become resistant to antibiotics.
6. Antibiotics are specifically used to eliminate infection-causing bacteria from the body. Thus, they are ineffective against viruses.

SECTION 53-2

SECTION OVERVIEW

Stability and Homeostasis
Section 53-2 describes how nicotine from tobacco and alcohol in beverages affect the user. It explains that use of tobacco can cause cancer and can damage many other body tissues. It describes both short-term and long-term effects of alcohol on a drinker. This section also describes how alcohol and nicotine consumed by pregnant women affect their unborn children.

▶ ENGAGE STUDENTS

Bring in a newspaper from a Monday or the day after a holiday weekend. Show students the stories about accidents in which the use of alcohol or other drugs was implicated. (Note: the newspaper should be from another city or state to prevent the possibility that students might know the people involved in the accidents.) Ask students why drunken driving is so dangerous. (When people drink alcohol, their judgment, motor control, and sensory perception are impaired, and their reaction time increases.) Ask students if they know what the legal level of intoxication is. (In many states, it is a blood alcohol concentration of 0.10.) Emphasize that about half of all fatal automobile accidents that involve young people are caused by drunken drivers. Note that drunken drivers often cause accidents that kill other people and that many drunken drivers serve time in prison for charges of DWI (driving while intoxicated) or DUI (driving under the influence). Ask students to discuss the effectiveness of laws that regulate alcohol consumption, such as set legal drinking ages and blood alcohol concentration standards. Ask them whether the laws should be more strict or more lenient. Use Table 53-2 for reference.

SECTION

53-2

OBJECTIVES

▲ Describe the effects of nicotine on the body.

● List three diseases that can result from smoking or chewing tobacco.

■ Explain what can happen to the fetus of a pregnant woman who smokes.

◆ Distinguish between a stimulant and a depressant.

▲ Define *blood alcohol concentration,* and name three factors that affect it.

● Explain what can happen as a result of excessive alcohol consumption.

SOCIAL DRUGS

Two drugs that are often used by people in a social context are nicotine, found in all tobacco products, and alcohol. The leaves of the tobacco plant, Nicotiana tabacum, *are dried and crushed, then smoked in cigarettes, cigars, and pipes. Tobacco is also chewed and snuffed. Ethanol—the alcohol found in beer, wine, liquor, and certain other beverages—comes from the anaerobic respiration of sugars in fruits and grains. Both tobacco and alcohol have gained wide social acceptance, but that acceptance has come at a great price.*

TOBACCO

Nicotine, the major drug found in tobacco, is a stimulant. A **stimulant** is a drug that increases the activity of the central nervous system. When tobacco is inhaled, nicotine is absorbed into the bloodstream through the lining of the mouth and through the lungs. Nicotine is quickly transported throughout the body, penetrating the brain, all other organs, and, in pregnant women, the fetus.

Effects of Tobacco

Nicotine increases blood pressure and heart rate while decreasing the oxygen supply to body tissues and the blood supply to the hands and feet. Nicotine is an addictive drug. In **addiction,** the user of a drug becomes dependent on the drug and cannot function comfortably without it. Nicotine is a poison—60 mg of nicotine is a lethal dose for an adult. But nicotine is not the only poison found in tobacco. There are more than 2,000 potentially toxic chemical compounds produced when tobacco is burned. Collectively, these are called tars.

Tars are complex mixtures of chemicals and smoke particles produced by burning tobacco. Tars paralyze the cilia that line the air passages. Remember from Chapter 47 that these cilia move particles out of the air passages and protect the passages from disease-causing microorganisms. Tars irritate the nose, throat, trachea, and bronchial tubes, causing sore throat and coughing. Eventually, tars settle in the lungs. The result is a reduction in breathing capacity and increased susceptibility to infections. The lungs of a smoker look much different from those of a nonsmoker due to the accumulation of tars, as shown in Figure 53-2.

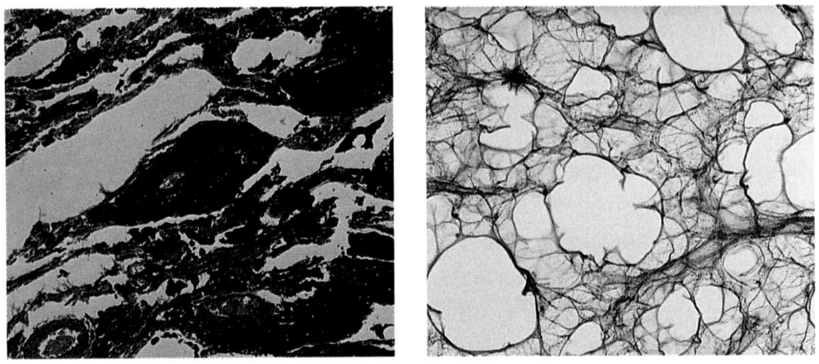

FIGURE 53-2
The blackened appearance of a smoker's lung is the result of years of tar accumulation. Contrast the smoker's lung (left) with the healthy lung (right).

Hazards of Long-Term Tobacco Use

Figure 53-3 illustrates the hazards of tobacco use. Notice how many parts of the body can be affected by tobacco. As it circulates in the bloodstream, nicotine can affect the performance of nearly every system in the body. The system that is most affected is the cardiovascular system. Approximately 25 percent of all heart attacks are associated with the use of tobacco.

The causal relationship between smoking and lung cancer is well established. As you read in Chapter 11, lung cancer is one of the deadliest forms of cancer. More than 90 percent of all lung cancer deaths can be attributed to smoking. Many smokers contract **chronic bronchitis** (brahn-KIET-is), an inflammation of the bronchi and bronchioles, or **emphysema** (EM-fuh-SEE-muh), a degenerative lung disease. In emphysema, the alveoli lose their elasticity and eventually rupture, reducing the surface area available for gas exchange in the lungs.

Users of smokeless tobacco, such as chewing tobacco and snuff, have a higher rate of lip, gum, and mouth cancer than people who

FIGURE 53-3
The health risks associated with smoking extend to many parts of the body.

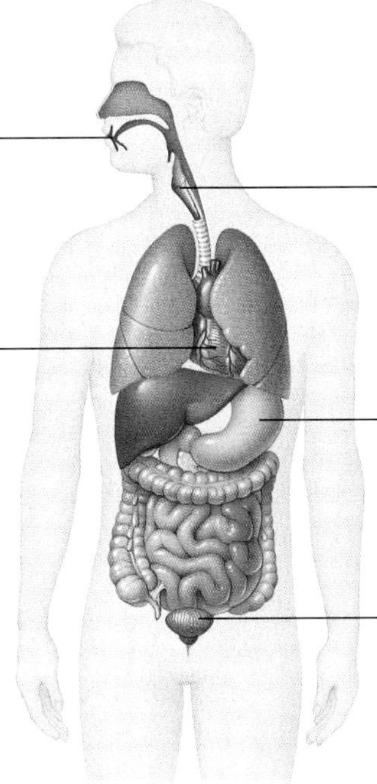

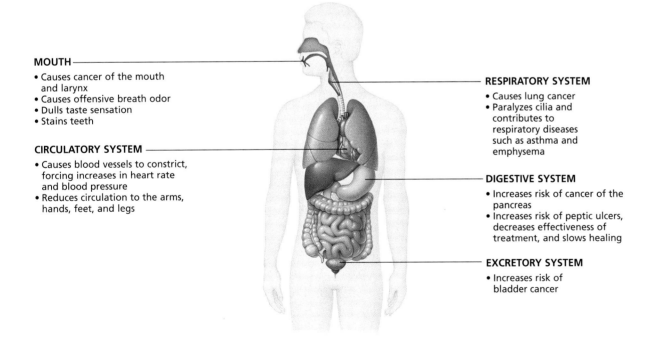

CONTENT CONNECTION

Section 51-1: Chemical Messengers

Ask students to recall that chemical messengers affect their target cells in different ways. Ask students how amino acid–based hormones and some neurotransmitters, such as norepinephrine, affect their target cells. (They bind to receptor proteins on their target cells.) Tell students that drugs often affect the interactions of naturally occurring chemical messengers and their target cells. Tell them that drugs that mimic or enhance the action of chemical messengers are called *agonists.* Conversely, drugs that interfere with or inhibit the action of chemical messengers are called *antagonists.*

QUICK FACT

☑ As the text states, nicotine is extremely toxic and poisonous; a dose of about 60 mg is fatal to adults. Symptoms of nicotine poisoning include decreased blood pressure, irregular heart beat, nausea, abdominal pain, chills, sweating, diarrhea, dizziness, confusion, weakness, excessive salivation, blurred vision, and vomiting. Death can result after collapse, convulsions, and respiratory failure.

INCLUSION *ACTIVITY*

Ask students why alcohol and tobacco are considered drugs. (They alter the biological processes of the central nervous system and many other organs throughout the body.) Have students think of their own definitions of alcohol and tobacco and share them with the class. (Alcohol is any beverage that contains ethanol, a central nervous system depressant. Tobacco contains nicotine—a central nervous system stimulant—and tars.)

TEACHING STRATEGY

Drug Advertisements

Have students cut out advertisements of various tobacco and alcohol products from magazines. Have them note the Surgeon General's warnings on these advertisements. Ask students how these advertisements are effective regardless of the warning labels. (In their advertisements, many tobacco and alcohol companies feature pictures of glamorous people using their products in interesting social settings.) Ask students to suggest why people smoke and drink in spite of the well-known risks and how advertisements encourage such behavior. (Answers will vary. Students may suggest that tobacco and alcohol companies are probably aware that nicotine and ethanol, the primary drugs found in their products, are addictive. Thus, the aim of their advertisements is to encourage people to begin using their products by portraying attractive people engaged in fun activities.)

OVERCOMING MISCONCEPTIONS

A common misconception is that smoking marijuana is less harmful than smoking cigarettes. All smoke is harmful because it contains tars and carbon monoxide, which contribute to respiratory problems. Carbon monoxide remains in the blood for hours, reducing its oxygen-carrying capacity. Tars in marijuana smoke and tobacco smoke are carcinogenic. Smoking marijuana increases both heart rate and blood pressure and can also cause infertility.

do not use smokeless tobacco. Tobacco-induced cancers can also occur far from the respiratory system. Smoking is associated with cancers of the bladder and pancreas and may be associated with other types of cancer as well. Cancers associated with smoking likely result from the effects of tars on body tissues.

The consequences of smoking during pregnancy are serious. Pregnant women who smoke are twice as likely as nonsmoking mothers to suffer miscarriages. Their babies tend to have lower birth weights than babies of nonsmokers and are twice as likely to die in the first few months of life. The infants of nursing mothers who smoke may have concentrations of nicotine in their blood as high as those of their mothers.

Tobacco even affects nonsmokers. Tobacco smoke released by cigarettes or exhaled by a smoker is called **secondhand smoke.** Laboratory studies have shown that people exposed to secondhand smoke are at risk for the same diseases as people who smoke. Researchers have recently found that exposure to secondhand smoke increases the risk of sudden infant death syndrome in infants up to the age of 12 months.

ALCOHOL

Ethanol is the alcohol found in beer, wine, and mixed drinks. Alcohol is generally a **depressant,** a drug that decreases the activity of the central nervous system. Following consumption, alcohol is immediately absorbed by the stomach and intestines, where it enters the bloodstream and is transported to the brain and other organs.

FIGURE 53-4

A Breathalyzer is an instrument that estimates blood alcohol concentration from alcohol vapors released by the lungs.

Effects of Alcohol

Alcohol increases circulation to the skin, resulting in decreased blood flow to internal organs and a net drop in body temperature. Alcohol increases excretion of water by the kidneys and can result in dehydration if the fluid is not replaced. The greatest and most immediate effects of alcohol, however, occur in the brain.

Alcohol affects different classes of neurons in the brain at different times as a person drinks. This accounts for the wide spectrum of behaviors associated with alcohol consumption. Behavioral excitability and an increased respiration rate are often seen immediately after a person begins to drink. As drinking continues, judgment and coordination are significantly impaired, speech becomes slurred, and reaction time lengthens. Respiration rate slows after the initial increase, and high doses of alcohol can cause death by respiratory failure.

The severity of all of these effects depends largely on **blood alcohol concentration (BAC),** a measurement of the amount of alcohol in the blood. As shown in Figure 53-4, BAC can be determined by a breath test, which measures alcohol vapors given off by the lungs. Table 53-2 illustrates the relationship between body

TABLE 53-2 *Blood Alcohol Concentration (in mg of alcohol per mL of blood) and Its Effects*

Drinks per hour	Body weight in pounds								Result
	100	120	140	160	180	200	220	240	
1	0.04	0.03	0.03	0.02	0.02	0.02	0.02	0.02	impairment
2	0.06	0.06	0.05	0.05	0.04	0.04	0.03	0.03	impairment
3	0.11	0.09	0.08	0.07	0.06	0.06	0.05	0.05	intoxication
4	0.15	0.12	0.11	0.09	0.08	0.08	0.07	0.06	intoxication
5	0.19	0.16	0.13	0.12	0.11	0.09	0.09	0.08	intoxication
6	0.23	0.19	0.16	0.14	0.13	0.11	0.10	0.09	intoxication
7	0.26	0.22	0.19	0.16	0.15	0.13	0.12	0.11	severe intoxication
8	0.30	0.25	0.21	0.19	0.17	0.15	0.14	0.13	severe intoxication

One drink is 1 oz of 80 proof (40%) liquor, 12 oz of 4.5% beer, or 4 oz of 12% wine.

weight, rate of alcohol consumption, and BAC values. A BAC level of 0.30 or greater produces unconsciousness, and a BAC of 0.50 can be fatal. The BAC can be affected by factors other than body weight, such as gender—BAC rises more rapidly in female drinkers—and the ratio of lean body mass to fat. Food in the stomach slows the absorption of alcohol, so a person who drinks on an empty stomach will have a higher BAC than a person of similar weight who drinks the same amount but has recently eaten.

Drivers who have consumed alcohol are much more likely to be involved in motor-vehicle accidents than are sober drivers. In fact, alcohol is a factor in nearly 50 percent of all fatal crashes involving young people, and alcohol-related motor-vehicle accidents are the leading cause of death among people 15 to 24 years of age. You can see in Table 53-3 that many tragedies stem from the impaired judgment that invariably results from drinking.

Like other drugs, alcohol can contribute to serious and even fatal drug interactions. For example, when alcohol is combined with another depressant, such as a tranquilizer or a narcotic analgesic, the cumulative effects of the two drugs can slow the cardiac and respiratory systems to the point of causing death.

Quick Lab

Graphing Tobacco Use

Materials graph paper, poster board, pencil, colored markers or pencils, ruler

Procedure

1. Examine the following table, which lists the percentage of students who said on national surveys that they smoked.

Grade	1991	1994	1998
8th	14.3%	18.4%	19.1%
10th	20.9%	25.4%	27.6%
12th	28.3%	31.2%	35.1%

2. Using a different color for each grade, graph each grade on the same set of axes. Put the year along the *x*-axis and the percentage along the *y*-axis.
3. Using the data in the graph and the table, design a poster to help prevent a teenager from smoking.

Analysis Analyze the resulting graph. What trends do you notice?

TABLE 53-3 *Alcohol-Related Tragedies*

Alcohol is a factor in the following:
• 70 percent of all murders
• more than 50 percent of all suicides
• more than 50 percent of all drownings
• more than 50 percent of all fire-related deaths
• 50 percent of all arrests
• nearly 50 percent of all fatal car accidents involving young people

Quick Lab

Graphing Tobacco Use

Time Required 20 minutes

Procedural Tips Have students work in pairs. When the posters are done, place them around the school as an anti-smoking campaign.

Answers to Analysis The trend for all grade levels is toward an increase in smoking from 1991 through 1998. There is another trend toward an increase in the percentage of students who smoke as they go from 8th grade to 12th grade.

TEACHING STRATEGY

Alcoholic Beverages

Tell students that 12 oz of beer, 4 oz of wine, and 1 oz of liquor contain about the same amount of ethanol. Ask students why liquor affects the body more quickly and potently than beer or wine. (Although wine, beer, and liquor contain the same amount of ethanol, their ethanol concentrations are different.) Tell students that drinking a shot of liquor is equivalent to chugging a beer.

CRITICAL THINKING

Diuretics

Based on their knowledge of endocrine hormones, ask students to suggest how consumption of diuretics, such as alcohol and caffeine results in excretion of copious amounts of urine. (Alcohol inhibits the release of antidiuretic hormone. This causes the kidneys to reabsorb water from the blood. The subsequent increase in the permeability of the kidney to water causes more water to be excreted in the urine. Caffeine is circulated through the bloodstream to the kidneys, where it inhibits the reabsorption of solutes and water. Thus, caffeine also increases urination.)

RETEACHING *ACTIVITY*

Have groups of students make a table of different kinds of drugs that they can think of (not only commonly abused drugs), their desired effects, and their short-term and long-term effects. Have them categorize the drugs as medical or social, legal or illegal, and nonprescription or prescription. Their tables should include stimulants, depressants, narcotics, hallucinogens, inhalants, hormones, analgesics, antibiotics, antihistamines, and appetite suppressants. Ask students to share their tables with the rest of the class. Students should find that most drugs have more side effects than desired effects.

ANSWERS TO SECTION 53-2 REVIEW

1. Nicotine is a stimulant that causes an increase in heart rate and blood pressure and a decrease in blood circulation to the hands and feet.
2. Tars paralyze cilia in air passages that help remove disease-causing materials.
3. Blood alcohol concentration is influenced by body weight, gender, the amount of food in the stomach, the rate of consumption, and the ratio of lean body mass to fat.
4. A stimulant drug increases the activity of the central nervous system, while a depressant drug decreases the activity of the central nervous system.
5. Fetal alcohol syndrome may result in low birth weight, below-average body length, stunted growth, heart defects, problems with coordination and balance, and mental retardation.
6. Alcohol affects different classes of neurons in the brain at different times while a person is drinking.

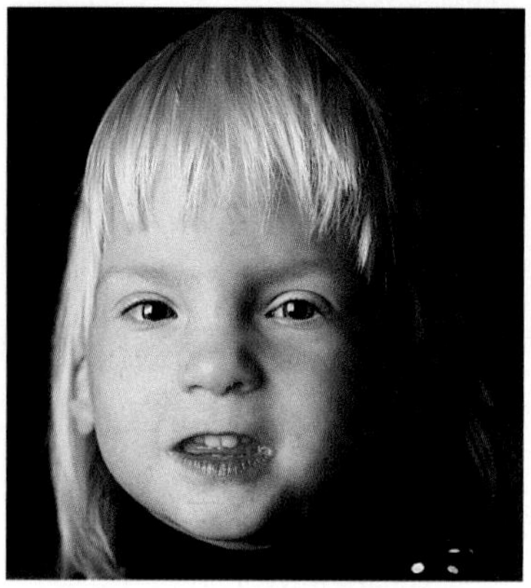

FIGURE 53-5

This child shows some of the abnormalities associated with fetal alcohol syndrome. The forehead is narrow, the wide-set eyes have a shallow crease in the upper eyelid, the upper lip is thin, and the span from the upper lip to the nose is long and lacks a central groove.

Hazards of Long-Term Alcohol Use

People who consume alcohol in large amounts over long periods of time are more likely than nonusers to suffer from high blood pressure and some forms of heart disease. Excessive alcohol use can irritate the lining of the stomach. The most serious result of long-term or inappropriate use of alcohol is **alcoholism,** the addiction to alcohol.

In alcoholics, alcohol takes the place of nutritious foods, causing metabolic disorders. Long-term use of alcohol is associated with the degeneration of neurons in the peripheral nervous system as well as irreversible damage to the brain. Chronic alcohol use forces the liver to use alcohol as an energy source. Eventually, liver cells become unable to function properly, and they accumulate fat deposits. As a result, the liver enlarges. This **fatty liver** is a condition that occurs in 75 percent of all alcoholics. If the drinker abstains from alcohol, the liver can return to normal functioning. If drinking continues, however, the drinker may develop **alcoholic hepatitis** (HEP-uh-TIET-is), an inflammation of the liver, or even **cirrhosis** (suh-ROH-sis), a condition in which normal liver tissue is replaced by scar tissue. Locate the scar tissue in the liver shown in the first illustration in this chapter, on page 1064. Cirrhosis causes severe impairment of liver function and is often fatal.

Like many other drugs, alcohol passes through the placenta and thus can affect a fetus. Children born to women who drink during pregnancy, such as the child shown in Figure 53-5, may suffer from fetal alcohol syndrome. **Fetal alcohol syndrome (FAS)** is a cluster of physical and mental disabilities associated with exposure of the developing fetus to alcohol. Babies that exhibit FAS are small and have an abnormally formed head. They lack normal motor coordination, grow slowly, and may be mentally retarded. Mild degrees of FAS in children often go undiagnosed and may exhibit themselves as behavioral, intellectual, or motor problems.

The severity of FAS seems to be related to the amount of alcohol a pregnant woman drinks and when, during her pregnancy, drinking occurs. There is no established safe level of alcohol intake during pregnancy; researchers think that fetal brain damage occurs at very low levels of alcohol exposure. Moreover, the fetus may be vulnerable to serious damage very early in development, even before a woman realizes that she is pregnant.

Word Roots and Origins

cirrhosis

from the Greek *kirrhos,* meaning "tawny-colored"

SECTION 53-2 REVIEW

1. **In what ways does nicotine affect the body?**
2. **How do tars contribute to a smoker's increased susceptibility to colds and infections?**
3. **What factors influence blood alcohol concentration in a drinker?**
4. **How does a stimulant drug differ from a depressant drug?**
5. **What are some of the effects of fetal alcohol syndrome?**
6. **CRITICAL THINKING Why does alcohol, a single drug, have a wide range of effects on a user?**

Abuse of Drugs

When used correctly, both prescription and nonprescription drugs can be beneficial. However, many drugs are abused, that is, taken for nonmedical reasons. Drug abuse includes taking a drug in doses greater than prescribed or for a longer time than recommended, as well as taking a drug that is obtained illegally or without a doctor's prescription.

DRUG ADDICTION

Addiction to a drug involves physiological changes in neurons, especially those in the brain. Studies have shown that physical addiction results in changes in body and brain chemistry that in turn create a demand for a steady or increasingly large supply of the drug. The drive to consume greater and greater amounts of a drug is all too familiar to alcoholics and abusers of many other drugs.

Tolerance

As a person becomes addicted to a drug, his or her body may become less responsive to the drug. The addict has developed **tolerance,** which means that increasingly large amounts of the drug are needed to attain the sensation previously achieved with a smaller dose. This rise in the **effective dose,** the dose that results in the desired feeling, is a deadly situation for users of some drugs, particularly depressant drugs. As tolerance increases, the addict must take higher and higher doses of the drug, approaching the lethal dose. The **lethal dose** of a drug is the amount of drug that will cause the user to die, and it does not change as tolerance builds.

If the drug supply is cut off, the addict will go through **withdrawal,** a physical and mental response to the lack of the drug. The symptoms of withdrawal vary, depending on the drug being used and the duration of use. Symptoms may include nausea, headache, insomnia, breathing difficulties, depression, mental instability, and seizures. Withdrawal from addiction to some depressant drugs, such as alcohol and barbiturates, can be life threatening. Addicts undergoing withdrawal are often hospitalized so that their responses can be monitored. Although drug users develop an emotional dependence on a drug, tolerance and withdrawal are physiological effects of drug use. Figure 53-6 shows what happens in neurons of a person addicted to cocaine. The excitatory effect sought by users of cocaine is probably due to the drug's action on neurons that use the neurotransmitters **dopamine** and norepinephrine, NE.

SECTION 53-3

OBJECTIVES

▲ Define the term *tolerance.*

● Explain the physical basis of drug addiction.

■ Describe the process of withdrawal.

◆ Identify five types of psychoactive drugs.

▲ List one example of each type of drug, and describe how each affects the body.

Word Roots and Origins

tolerance

from the Latin *tolerare,* meaning "to bear"

SECTION 53-3

SECTION OVERVIEW

Stability and Homeostasis
Section 53-3 uses cocaine as a model to describe how drug addiction results from physiological changes in the nervous system. This section also describes the effects on the body of commonly abused psychoactive drugs.

▶ ENGAGE STUDENTS

Ask students to suggest why someone would use drugs other than those that are taken for medical or health reasons. (Answers will vary, but students might comment on many sociological factors of drug abuse and addiction. For example, some people begin using drugs as an escape mechanism to avoid dealing with problems, or because they desire to belong to a certain peer group.) Ask students if they know the meanings of *tolerance* and *withdrawal.* (Tolerance is the physiological basis of addiction; as their tolerance for a drug increases, drug addicts must take increasing amounts of the drug to achieve the desired effect. Withdrawal is a set of symptoms that drug addicts experience when they stop taking drugs.) Tell students that addicts often continue to use drugs to avoid painful withdrawal symptoms. Emphasize to students that people who use drugs as a problem-solving device only compound their problems by introducing addiction into their lives; the costs of drug abuse far outweigh any foreseen benefits. Have the class collaborate to make a list of the harmful effects of drugs. Emphasize that drug addiction can cause serious health problems and even death.

VISUAL LINK
Figure 53-6

Ask students how cocaine affects neurons in the central nervous system. (Cocaine blocks transporter proteins that enable the reuptake of dopamine and norepinephrine into presynaptic neurons. By inhibiting reuptake, cocaine causes excess neurotransmitter to remain in the synapses, prolonging and intensifying excitation of postsynaptic neurons.) Tell students that the popular prescription antidepressant fluoxetine works by a similar mechanism to block the reuptake of the neurotransmitter serotonin. Explain to students that reuptake inhibition is not the only mechanism of drug action in the central nervous system. For example, morphine and other opiates mimic the action of enkephalins, the body's natural painkilling neurotransmitters. Opiates bind selectively to receptors in the central nervous system and inhibit the release of neurotransmitters that relay pain signals to the brain.

TEACHING STRATEGY
Drug Addiction

Lead a class discussion about drug addiction. Ask students to suggest reasons why addiction to psychoactive drugs is so powerful. (Most addictive drugs affect neurons in the central nervous system that are part of a neural "reward pathway," which reinforces behaviors that make the person feel "good." These neurons stimulate the limbic system and the hypothalamus, which regulate emotions and behaviors associated with pleasurable stimuli. Excessive stimulation of the limbic system caused by cocaine results in a craving for the drug.)

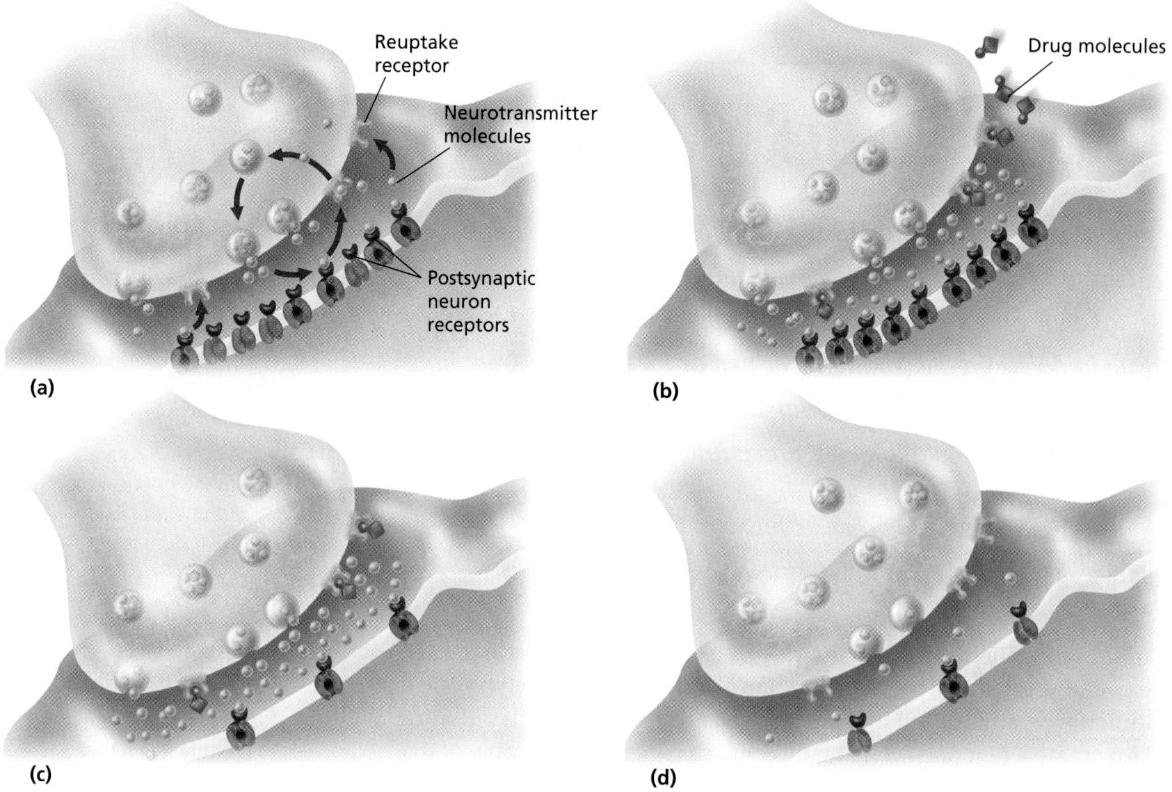

FIGURE 53-6

(a) At a normal NE synapse, neurotransmitter molecules are removed by reuptake receptors. (b) Cocaine and certain other drugs bind at reuptake receptors, blocking them. Neurotransmitter molecules remain in the synaptic cleft. (c) The overstimulated postsynaptic neuron reduces its number of receptors over time to normalize firing. (d) When the drug is removed, the number of neurotransmitter molecules drops to normal—a level too low to stimulate the altered postsynaptic neuron normally.

Neural Changes

Recall from Chapter 50 that when an axon terminal receives a signal—an action potential—it releases molecules of neurotransmitter into the synaptic cleft. In dopamine and NE neurons, and in many other kinds of neurons, neurotransmitter molecules are removed from the synaptic cleft when they bind to reuptake receptors on the presynaptic (neurotransmitter-releasing) neuron, as shown in Figure 53-6a. **Reuptake receptors** are molecular transporters that move neurotransmitter molecules back into the presynaptic neuron. Cocaine acts by binding at these presynaptic reuptake receptors, stopping their function. As a result, a large number of neurotransmitter molecules collect in the synaptic cleft, as shown in Figure 53-6b. These excess neurotransmitter molecules in the synapse excite the postsynaptic neuron to an abnormal degree, increasing firing rate. This excitatory response in dopamine and NE neurons in the brain provides the sensation sought by the drug user.

In response to this chronic surplus of neurotransmitter, the postsynaptic (neurotransmitter-receiving) neurons reduce their number of neurotransmitter receptors over time, as shown in Figure 53-6c. This response is the body's attempt to restore homeostasis in these neurons. As you can see in Figure 53-6d, when the drug is removed, the number of neurotransmitter molecules in the synaptic cleft decreases. (They are taken up by the presynaptic neuron, as is normal.) However, this now-normal level of neurotransmitter is not sufficient to stimulate the postsynaptic neuron, with its now-reduced number of receptors, and transmission of the signal is disrupted.

DRUGS OF ABUSE

Drugs that affect the functioning of the central nervous system are known as **psychoactive drugs.** Psychoactive drugs are often addictive, and possession of many types of psychoactive drugs is illegal. Table 53-4 lists the broad categories of frequently abused drugs; all groups but one, the steroids, are primarily psychoactive.

TABLE 53-4 *Common Drugs of Abuse*

Drug	Short-term effects	Dangers of use
Depressants • Alcohol • Barbiturates • Tranquilizers • Nonbarbiturate sleeping drugs	temporary sense of well-being, impaired judgment, impaired sensory perception, loss of motor control, confusion, sedation	emotional depression, liver damage, brain and peripheral nerve damage, respiratory failure (Withdrawal from alcohol or barbiturates can cause death; withdrawal should be medically supervised.)
Narcotics • Codeine • Heroin • Morphine • Synthetic opiates	temporary feeling of euphoria, impaired reflexes, impaired sensory perception, sedation	coma, respiratory failure
Stimulants • Amphetamines • Caffeine • Cocaine/crack cocaine	temporary feeling of exhilaration and energy, distorted thoughts, irritability, anxiety, restlessness, insomnia, elevated blood pressure, increased heart rate	sleep disturbances, irregular heartbeat; with amphetamines and cocaine, paranoia, delusions, hallucinations, loss of coordination, permanent brain damage, respiratory paralysis, cardiac arrest
Hallucinogens • Amphetamine-like drugs • Ecstasy (MDMA) • Marijuana/hashish • LSD • Mescaline/peyote • PCP (phencyclidine) • Psilocybe mushrooms	temporary dreamlike sense of detachment from surroundings, sensory distortion, hallucinations, delusions, anxiety, slurred speech, numbness, bizarre or violent behavior	mental depression, "flashbacks," paranoia, genetic damage; with marijuana, amotivational syndrome
Inhalants • Aerosol propellants • Alkyl nitrates • Anesthetic gases • Solvents in glue, paint, correction fluid, marking pens, gasoline additives	disorientation, confusion, memory loss; with anesthetic gases, sedation	severe permanent brain damage, hearing loss, limb spasms, bone marrow damage, liver and kidney damage, cardiac arrest; with anesthetic gases, respiratory failure
Anabolic steroids	increase of muscle mass, fluid retention, insomnia, depression	cancer, liver disease, stunted growth, heart disease, heart attack

RECENT RESEARCH

Cocaine Immunotherapy

Cocaine causes overstimulation of neurons in the reward pathway of the central nervous system. Cocaine blocks the reuptake of the neurotransmitter dopamine by interfering with proteins that transport dopamine back into presynaptic neurons after its release into the synaptic cleft. Cocaine addiction is difficult to treat because any drug that is used to block cocaine at the synapse also interferes with reuptake, producing essentially the same effects that cocaine does.

Researchers at Columbia University have devised a method that uses catalytic antibodies to eliminate cocaine in the blood before it reaches the central nervous system. In laboratory animals, they stimulated production of enzymatic antibodies that inactivate cocaine by cutting the cocaine molecule at specific chemical bond sites. To produce these antibodies, they introduced into mice a synthetic molecule that is chemically analogous to cocaine. They found that cells in the blood produced antibodies to the analog. The researchers produced nine different catalytic antibodies to cocaine, and now they are investigating ways to administer sufficient antibodies to humans in order to combat the large doses of cocaine normally consumed by addicts.

QUICK FACT

☑ The central nervous system contains naturally occurring receptors for opiates and 9-tetrahydrocannabinol (THC), the active ingredient in marijuana. Enkephalins are neurotransmitters that bind to opiate receptors, blocking neural signals that relay pain stimuli to the brain. THC targets cannabinoid receptors in the brain that are normally activated by two naturally occurring chemicals called anandamide and 2-AG.

SECTION 53-3

QUICK FACT

☑ Morphine, an opiate, cannot be used to anesthetize patients before surgery, even though it is an effective painkiller. (The body has two responses to pain: initial and long-term. Although morphine blocks prolonged pain sensation, it does not intercept new or changed pain responses. Thus, if morphine were used as a surgical anesthetic, the patient could feel pain from each incision or movement made by the surgeon.)

ANSWERS TO SECTION 53-3 REVIEW

1. Examples of psychoactive drugs include stimulants, depressants, hallucinogens, narcotics, and inhalants.
2. Tolerance is a physiological decrease in the body's responsiveness to a drug that causes a user to take more of the drug to achieve the desired effects.
3. In drug addicts, withdrawal is characterized by harsh symptoms, including nausea, seizures, and mental instability.
4. Cocaine is a powerful stimulant that results in the continuous stimulation of norepinephrine-producing neurons in the brain.
5. Drug addiction causes physiological and structural changes in the central nervous system. Drug addicts may also develop an emotional dependence on the drug, feeling as if they "need" it. In this manner, drug addiction can cause both physical and psychological problems.
6. As tolerance increases, the effective dose approaches the lethal dose, which causes death.

FIGURE 53-7

Several narcotic drugs are made from the resin of the opium poppy.

Marijuana

One of the hallucinogens listed in Table 53-4 is marijuana, which comes from the flowers, leaves, and seeds of *Cannabis sativa,* the hemp plant. The active ingredient in marijuana is 9-tetrahydrocannabinol, or THC. When smoked, or when heated and then eaten, marijuana produces feelings of disorientation in space and time. Continual use of marijuana can lead to addiction and a suppressed activity level referred to as **amotivational syndrome.**

Cocaine

One of the illegal stimulants listed in Table 53-4 is cocaine. Cocaine is a white powder that is extracted from the leaves of the coca plant, *Erythroxylon coca.* Cocaine is generally inhaled, although it can be injected. Crack cocaine is a smokable form of the drug. Cocaine is a powerful stimulant that raises heart rate, blood pressure, and body temperature. A single dose of cocaine can cause a heart attack. Cocaine is one of the most addictive drugs known.

Opium

Figure 53-7 shows the poppy plant, *Papaver somniferum,* from which the resinous extract opium is obtained. Opium is used to make morphine and codeine, prescription drugs used to alleviate pain. Heroin, an illegal drug, is a semisynthetic, more potent form of morphine. Morphine, codeine, heroin, and similar synthetic drugs are known as narcotics. A **narcotic** is a pain-relieving drug that also induces sedation and sleep. Tolerance and addiction to narcotics occur rapidly.

Narcotics work by mimicking natural painkillers that are produced by the body. Natural painkillers include a class of neurotransmitters known as **enkephalins** (en-KEF-uh-linz). When enkephalins bind to receptor proteins in spinal neurons, they block action potentials from reaching the brain, where they would be interpreted as pain. Narcotics bind to the same receptor sites that enkephalins bind to, triggering and amplifying the body's own pain-blocking system.

At times, morphine is used under medical supervision when the intensity or duration of pain is too much for a person to bear. However, keep in mind that pain plays an important role in the body. Pain makes you aware of some damage that has occurred to a body part or of some malfunction that has occurred in a physiological process. In effect, pain can tell you that an organ system is not operating at its normal level of performance—something that is critical for good health.

SECTION 53-3 REVIEW

1. Name five types of psychoactive drugs.
2. What is tolerance to a drug?
3. How does withdrawal affect the body?
4. Explain how cocaine affects NE neurons.
5. Why is drug addiction considered a physical problem in addition to being a psychological problem?
6. CRITICAL THINKING Why are long-term drug users, who have built a large degree of tolerance, endangered by their effective dose?

CHAPTER 53 REVIEW

SUMMARY/VOCABULARY

53-1

- A drug is a chemical compound that affects either the structure of a body part or the functioning of a physiological process. A drug can be introduced into the body orally, by injection, by being placed under the tongue, through inhalation, or by application to the skin.
- Prescription drugs are prescribed by a doctor or dentist. Nonprescription drugs can be purchased without a prescription.
- Taking more than one drug at the same time can cause a harmful drug interaction.
- Overuse of prescription drugs can lead to serious problems. Overuse of antibiotics has fostered the evolution of increasing numbers of resistant bacteria.

Vocabulary

dose (1066)
drug (1065)
nonprescription drug (1066)
prescription drug (1066)

53-2

- Tobacco contains more than 2,000 potentially toxic chemical compounds, including nicotine and tars. Nicotine is an addictive stimulant drug that increases the heart rate and decreases blood flow to the hands and feet.
- Tars paralyze the cilia lining the respiratory passages.
- Health problems associated with tobacco use include lung, lip, mouth, gum, bladder, and pancreatic cancers; chronic bronchitis; emphysema; heart attacks; and health problems in babies of smokers.
- Alcohol is a depressant drug that affects multiple organ systems. Alcohol interferes with normal brain function. Impairment caused by alcohol depends on the amount consumed; the person's body weight, gender, rate of consumption, and ability to metabolize alcohol; and the amount of food in the person's stomach.
- Health problems associated with prolonged, excessive alcohol consumption include high blood pressure, heart disease, fatty liver, alcoholic hepatitis, cirrhosis, and fetal alcohol syndrome. Alcohol can lead to the addiction known as alcoholism.

Vocabulary

addiction (1068)
alcoholic hepatitis (1072)
alcoholism (1072)
blood alcohol concentration (1070)
chronic bronchitis (1069)
cirrhosis (1072)
depressant (1070)
emphysema (1069)
fatty liver (1072)
fetal alcohol syndrome (1072)
nicotine (1068)
secondhand smoke (1070)
stimulant (1068)
tar (1068)

53-3

- Drug abuse can lead to addiction, in which the body undergoes physiological changes that result in a need for the drug.
- Drug addiction results in tolerance as the addict's body becomes less responsive to the drug. During withdrawal, an addict will experience both physical and mental reactions to the lack of the drug.
- Cocaine is an extremely addictive stimulant drug that causes repeated stimulation of NE neurons. Crack cocaine is a smokable form of the drug.
- Marijuana is a hallucinogen that produces feelings of disorientation in space and time. Prolonged use of marijuana can lead to amotivational syndrome.
- Morphine, codeine, and heroin are narcotics derived from opium. Narcotics are powerful painkillers that mimic the body's natural painkillers.

Vocabulary

amotivational syndrome (1076)
dopamine (1073)
effective dose (1073)
enkephalin (1076)
lethal dose (1073)
narcotic (1076)
psychoactive drug (1075)
reuptake receptor (1074)
tolerance (1073)
withdrawal (1073)

CHAPTER 53 REVIEW ANSWERS

REVIEW

1. A drug is a chemical compound that affects the structure of a body part or the functioning of a biological process.
2. Chronic bronchitis and emphysema are caused by tobacco.
3. Cirrhosis and alcoholic hepatitis are caused by alcohol.
4. Blood alcohol concentration is a measurement of the level of alcohol in the blood. Factors that affect blood alcohol concentration are gender, weight, rate of consumption, ability to metabolize alcohol, and the amount of food in the stomach.
5. A person who has developed an addiction needs a drug to function comfortably. An addict will develop a tolerance for the drug, reducing its effectiveness. If the addict stops using the drug, then withdrawal will occur.

6. d **7.** c
8. a **9.** b
10. c **11.** d
12. a **13.** a
14. b **15.** c

16. A nonprescription drug could be dangerous if taken in doses higher than those recommended, or for a longer period than is recommended, or if it interacts in a harmful way with another drug.
17. Secondhand smoke can have the same respiratory and cardiovascular effects as firsthand smoke. It also increases the risk of sudden infant death syndrome.
18. The mother is twice as likely to suffer a miscarriage or to give birth to low-weight babies who are twice as likely to die within the first few months after birth.
19. Excessive use of alcohol causes the liver cells to use alcohol as an energy source. Eventually fat deposits build up around the cells and they become unable to function, leading to fatty liver and cirrhosis.
20. Coordination and vision become impaired, speech may

become slurred, and reaction time increases.

21. The long-term effects of alcohol use may include heart disease, high blood pressure, fatty liver, cirrhosis, and alcoholism.

22. If a person continues to drink excessive amounts of alcohol, a fatty liver can eventually become cirrhotic.

23. Psychoactive drugs are chemicals that affect the functioning of the central nervous system.

24. A reuptake-inhibiting drug, such as cocaine, has blocked the proteins that transport neurotransmitter molecules back into the presynaptic terminal after they have been released into the synapse. Thus, the neurotransmitter molecules remain in the synapse.

25. The overstimulated postsynaptic neuron has reduced its number of neurotransmitter receptors. The body has developed a tolerance, the physiological basis of addiction, to the reuptake-inhibiting drug.

CRITICAL THINKING

1. Among the early effects of alcohol are increased feelings of relaxation and lightheadedness. People in this stage of intoxication appear carefree and happy.

2. If use of an antibiotic is discontinued early, not all bacteria will be eliminated. Thus, the immune system will be overwhelmed as bacteria reproduce. Also, bacteria that survive in the presence of an antibiotic can produce new bacteria that are resistant to the drug.

3. The addiction of a newborn to a drug is the result of the mother's drug-taking. Drugs consumed by the mother diffuse through the placenta into the fetus's bloodstream. Thus, infants can become addicted to drugs *in utero*.

4. Long-term addicts of some drugs increase their tolerance almost to the point that the

REVIEW

Vocabulary

1. Define the term *drug.*
2. Identify two respiratory diseases caused by tobacco.
3. Identify two diseases caused by alcohol.
4. What is blood alcohol concentration, and what are some factors that affect it?
5. Describe the relationship between addiction, tolerance, and withdrawal.

Multiple Choice

6. Drugs can be administered (a) orally (b) by injection (c) by inhalation (d) all of the above.
7. Tars (a) cause an increase in heart rate (b) are stimulants (c) paralyze cilia (d) are neurotransmitters.
8. Emphysema is (a) a degenerative lung disease (b) a form of cancer (c) an inflammation of the throat (d) a result of chewing tobacco.
9. Alcohol is a (a) stimulant (b) depressant (c) narcotic (d) steroid.
10. BAC specifically measures (a) the rate at which a person has been drinking (b) a person's genetic ability to metabolize alcohol (c) the amount of alcohol in a person's blood (d) the time at which a person was drinking.
11. Fetal alcohol syndrome can cause (a) low birth weight (b) heart defects (c) learning difficulties (d) all of the above.
12. Enkephalins are (a) natural painkillers (b) illegal drugs (c) stimulants (d) synthetic narcotics.
13. Addiction (a) occurs when the body becomes dependent on a drug (b) occurs when the supply of a drug is stopped (c) is easily reversed when the supply of a drug is stopped (d) is caused only by illegal drugs.
14. Depressants include (a) cocaine (b) barbiturates (c) nicotine (d) amphetamines.
15. Narcotics cause (a) an increased heart rate (b) increased alertness (c) decreased alertness (d) lung damage.

Short Answer

16. Describe a situation in which a nonprescription drug could be dangerous.
17. Describe the harmful effects of secondhand smoke.
18. What are some of the harmful effects of smoking during pregnancy?
19. Describe the specific effects of alcohol on the cells of the liver.
20. In what ways are mental capabilities diminished as BAC increases?
21. What are some potential effects of long-term, excessive alcohol use?
22. What is the relationship between fatty liver and cirrhosis?
23. What are psychoactive drugs?
24. In the figure below, explain what has just occurred in (a).
25. In the figure below, explain what has just occurred in (b).

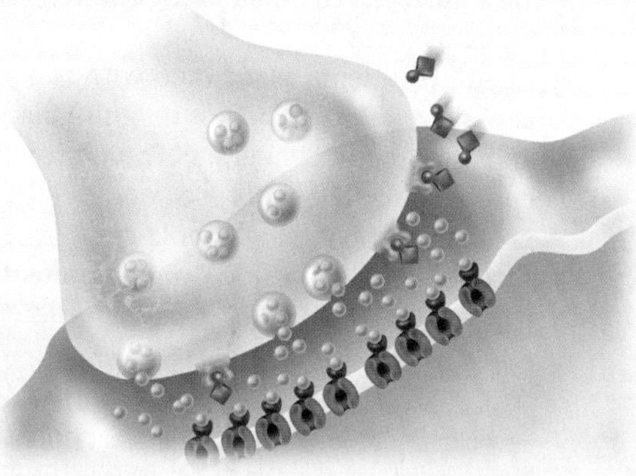

(a)

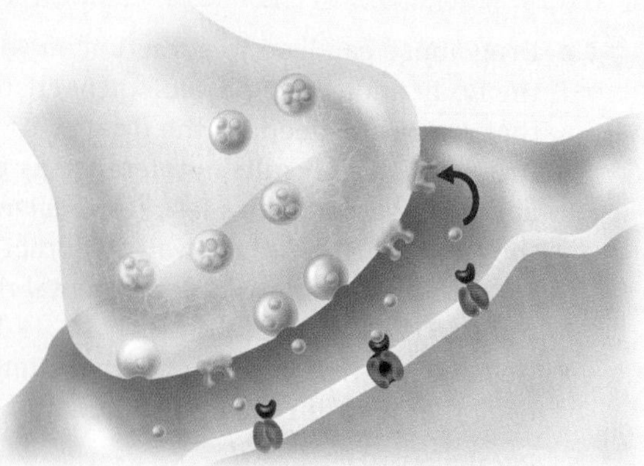

(b)

CRITICAL THINKING

1. Alcohol is primarily a depressant drug. How do you explain the fact that consumption of alcohol sometimes results in a carefree attitude and a feeling of elation?
2. If you are given a 10-day course of an antibiotic for a bacterial infection and you take the medicine for only five days, how might the infectious bacteria respond?
3. Some infants are born addicted to a drug. Explain how this is possible.
4. Why is tolerance a deadly problem for long-term addicts of some drugs?
5. The table below shows typical blood alcohol concentration values for individuals of different weights. Use the table to answer the following.
   a. What is the blood alcohol concentration of a 140 lb person who has consumed three drinks in one hour?
   b. How many drinks would a 200 lb person have to consume in an hour to equal the BAC of a 140 lb person who consumed three drinks in an hour?
   c. At what weight (among those shown on the table) does a person's BAC rise most rapidly?

Blood Alcohol Concentration (in mg of alcohol per mL of blood) and Its Effects

Drinks per hour	Body weight in pounds								Results
	100	120	140	160	180	200	220	240	
1	0.04	0.03	0.03	0.02	0.02	0.02	0.02	0.02	impairment
2	0.06	0.06	0.05	0.05	0.04	0.04	0.03	0.03	impairment
3	0.11	0.09	0.08	0.07	0.06	0.06	0.05	0.05	intoxication
4	0.15	0.12	0.11	0.09	0.08	0.08	0.07	0.06	intoxication
5	0.19	0.16	0.13	0.12	0.11	0.09	0.09	0.08	intoxication
6	0.23	0.19	0.16	0.14	0.13	0.11	0.10	0.09	intoxication
7	0.26	0.22	0.19	0.16	0.15	0.13	0.12	0.11	severe intoxication
8	0.30	0.25	0.21	0.19	0.17	0.15	0.14	0.13	severe intoxication

EXTENSION

1. In the 1960s, many pregnant women in Europe took the prescription drug thalidomide. Many of the women gave birth to children with severe deformities. In the United States this drug was not available because it had not undergone the lengthy testing required for approval. Find out what is required before a drug is made available to the public in the United States.
2. Investigate and report on the resources available in your local community for treating addiction to both social drugs and illegal drugs.
3. Read "Of Two Minds About Marijuana" in *Discover*, November 1999, on page 26. Why, according to the article, do schizophrenics have a higher rate of marijuana abuse than the general population? Explain why researchers say marijuana is not suitable to treat brain disorders.

effective dose matches the lethal dose. Because drugs are consumed over a long time period, tissue damage can also be fatal.
5. a. The blood alcohol concentration is 0.08. **b.** The 200 lb person would have to consume four drinks. **c.** The blood alcohol concentration of a 100 lb person increases most rapidly.

EXTENSION

1. Students should obtain information about how the Food and Drug Administration determines whether a drug is safe for public use. Encourage them to explore the issue of increased public pressure to make drugs that treat AIDS and other terminal diseases available as quickly as possible.
2. Have students look in the Yellow Pages under "Alcoholism" and "Drug Abuse" or speak with the school's health teacher or nurse.
3. Many schizophrenics say that the abuse of marijuana eases their psychotic symptoms. Marijuana has a broad, nonspecific sedative effect on the brain's dopamine levels. Researchers are looking for a drug that will enhance the body's natural chemical pathways to treat diseases such as schizophrenia, autism, and Huntington's disease.

CHAPTER 53
INTERACTIVE EXPLORATION

For an overview of this Interactive Exploration and preparatory information related to this activity, please see the CD User's Guides.

PREPARATION NOTES

Time Required: One 50-minute class period

PRELAB DISCUSSION

Before students begin the lab, they should be able to answer the following questions:

1. How are signals transmitted *between* neurons? (Chemical neurotransmitters relay signals between neurons.)

2. What events occur in chemical synaptic transmission? (Neurotransmitter is released from a presynaptic neuron and binds to receptor proteins on a postsynaptic neuron, causing a change in its electrical activity.)

PROCEDURAL TIPS

1. Make sure that students are familiar with instructions for this activity before they begin.

2. The animation depicts a schematic cross section of a synapse in the limbic system. Molecules of the neurotransmitter dopamine, which is released from a presynaptic neuron, diffuse across the synaptic cleft and bind to receptor proteins on a postsynaptic neuron. Dopamine stimulates a neural reward pathway, resulting in a pleasurable feeling.

3. Molecules of cocaine inhibit the reuptake of dopamine by binding to the transporter proteins—the reuptake receptor—that transport dopamine back into the presynaptic neuron.

(Note: In this animation, the reuptake-receptor transporter molecules are depicted as free-floating so that students can clearly see their function. They are actually embedded in the presynaptic membrane.)

CHAPTER 53
INTERACTIVE EXPLORATION

Exploring a Model of Cocaine Addiction

OBJECTIVES

- Simulate the effects of cocaine use on a synapse in the brain.
- Observe the effect of cocaine use on the number of receptors in a synapse.

MATERIALS

- computer with CD-ROM drive
- CD-ROM *Interactive Explorations in Biology: Human Biology*

Background

This interactive exploration enables you to learn about the physical basis of drug addiction by exploring the consequences of exposure of a neuron to cocaine. The exploration presents an animated diagram of a single neural synapse in a "pleasure center" in the human brain. You will explore the consequences of introducing cocaine into a normal neural synapse.

Prelab Preparation

1. Load and start the program Drug Addiction. Click the Topic Information button on the Navigation Palette. Read the focus questions, and review these concepts: The Synapse, Neurotransmitters, Transporters, Neuromodulators, Desensitization, and Addiction.

2. Now click the word *File* at the top left of the screen, and select Interactive Exploration Help. Listen to the instructions that explain the operation of the exploration. Click the Exploration button at the bottom right of the screen to begin the exploration. You will see an animated diagram like the one below.

Procedure

PART A Normal Synapse

First you will investigate how the synapse works when no cocaine is present.

1. Before beginning the simulation, record the number of receptors in the post-synaptic neuron. As the simulation runs, observe what happens to the Pleasure

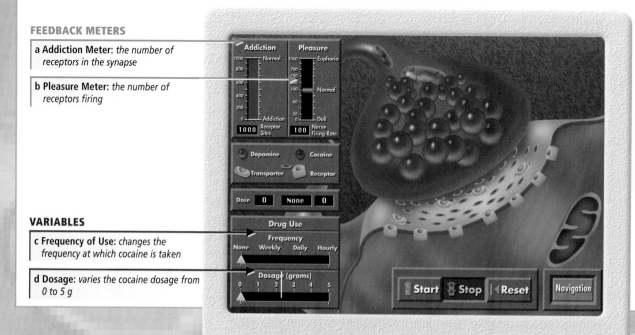

FEEDBACK METERS

a Addiction Meter: *the number of receptors in the synapse*

b Pleasure Meter: *the number of receptors firing*

VARIABLES

c Frequency of Use: *changes the frequency at which cocaine is taken*

d Dosage: *varies the cocaine dosage from 0 to 5 g*

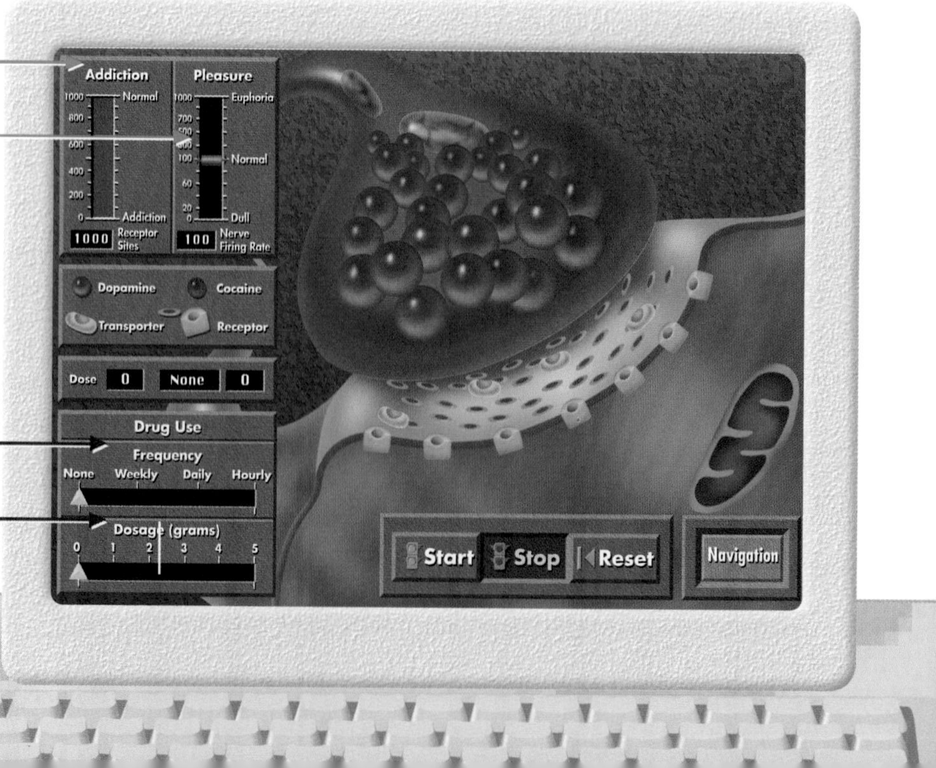

Meter and the Addiction Meter. Record the highest and lowest levels reached on the Pleasure Meter.

2. Click the Start button to begin the simulation. Observe what happens when neurotransmitter molecules (the blue spheres) are released into the synapse. For the purpose of this demonstration, each contact with a receptor on the other side of the synapse "fires" the postsynaptic neuron, producing a sensation of pleasure. Note carefully that in the normal synapse, the neurotransmitter molecules are recycled by the presynaptic neuron so that they do not accumulate in the synapse.
3. Allow the simulation to run for 3 to 4 minutes, noting any changes in the number of receptors (recorded by the Addiction Meter) and in the postsynaptic neuron's firing rate (recorded by the Pleasure Meter). End the simulation by clicking the Stop button.
4. Record the number of receptors in the postsynaptic neuron at the end of the simulation. Did the number of receptors change during the simulation? Explain why or why not.

PART B Data Collection

Now you will explore the effects of cocaine on the synapse.

5. Click the Reset button to clear the simulation.
6. Slide the Dosage indicator to 1 g and the Frequency of Use indicator to Weekly.
7. On a separate sheet of paper, create a six-column table like the one shown below.
8. Record the number of postsynaptic receptors in your table in the column labeled Addiction Meter: Start. Record the postsynaptic neuron firing rate in the column labeled Pleasure Meter: Start.
9. Click the Start button. For each dose of cocaine, you will see a burst of red spheres, representing cocaine, enter the synapse. What happens to the red spheres after they enter the synapse?
10. Based on what happens to the red spheres, explain why cocaine induces pleasurable sensations.
11. The simulation will run until five doses have been taken, which should require about 5 minutes. At the end of the simulation, record in your table the number of receptors and the postsynaptic neuron's firing rate.
12. How does the number of receptors in the synapse change with repeated cocaine doses?
13. Is this dose large and frequent enough to cause addiction? Explain your answer.
14. What was the highest level reached by the Pleasure Meter? Compare this level with the highest level reached without cocaine use, and account for the difference.
15. Explain how the simulation you carried out in Part A serves as a control for this simulation.
16. Click the Reset button to clear the simulation. Set the Frequency of Use Indicator at Weekly, but slide the Dosage indicator to 2 g. Repeat steps 8–14. How do the results of this simulation differ from those obtained with a 1 g dose?

Analysis and Conclusions

1. Based on your results, can you conclude that cocaine can be taken at a safe dose, at which addiction does not occur? Explain your answer.
2. Explain why people who are addicted to cocaine require higher and higher doses to produce pleasure.

Further Investigation

Selective serotonin reuptake inhibitors (SSRIs) constitute a relatively new class of antidepressants. Read about SSRIs, and write a short description of how they work; include names of specific drugs and the subclasses of serotonin receptors they target.

TABLE A *CHANGES IN RECEPTOR NUMBER AND FIRING RATE AFTER COCAINE USE*

	Addiction Meter: Start	Addiction Meter: End	Change in receptor number	Pleasure Meter: Start	Pleasure Meter: End	Change in firing rate
No drug (control)						
1 g cocaine/week						
2 g cocaine/week						

CHAPTER 53
INTERACTIVE EXPLORATION

ANSWERS TO PART A

4. The number of receptors should not change. The synapse loses receptors only when cocaine is used.

ANSWERS TO PART B

9. The red spheres bind to yellow transporter molecules that remove dopamine from the synapse.
10. Because dopamine remains in the synapse, it continues to stimulate the postsynaptic neuron.
12. The number of receptors decreases.
13. Yes, because as postsynaptic neurons are overstimulated by dopamine, the number of dopamine receptors decreases. Addiction results because more cocaine is needed to achieve the desired effect.
14. It should reach 190. The highest level reached without cocaine should be 100. Cocaine prolongs and intensifies dopamine action.
15. Part A demonstrates "normal" synaptic transmission without cocaine.
16. The Pleasure Meter should reach a much higher value (300), and the number of postsynaptic receptors should decrease to a lower value (500).

ANSWERS TO ANALYSIS AND CONCLUSIONS

1. There is no safe dose of cocaine; the lowest dose taken at the lowest frequency still results in a decrease of postsynaptic dopamine receptors, which in turn leads to addiction.
2. As the number of postsynaptic dopamine receptors decreases, more cocaine is required to achieve the desired effect.

SAFE LABORATORY PRACTICES

It is your responsibility to protect yourself and other students by conducting yourself in a safe manner while in the laboratory. You will avoid accidents in the laboratory by following directions, handling materials carefully, and taking your work seriously. Read the following general safety guidelines before attempting to work in the laboratory. Make sure that you understand all safety guidelines before entering the laboratory. If necessary, ask your teacher for clarification of laboratory rules and procedures.

General Guidelines for Laboratory Safety

Do not perform experiments not specifically assigned by your teacher. Do not attempt any laboratory procedure without your teacher's direction, and do not work in the laboratory by yourself.

Familiarize yourself with the investigation and all safety precautions before entering the lab. Be aware of the potential hazards of the required materials and procedures. Ask your teacher to explain any parts of an investigation that you do not understand before you begin.

Before beginning work, tie back long hair, roll up loose sleeves, and put on any required personal protective equipment as directed by your teacher. Avoid or confine loose clothing that could knock things over, catch on fire, or absorb chemical solutions. Nylon and polyester fabrics burn and melt more readily than do cotton fabrics. Do not wear open-toed shoes, sandals, or canvas shoes in the laboratory.

Always wear a lab apron and safety goggles. Wear this equipment even if you are not working on an experiment at the time. Laboratories contain chemicals that can damage your clothing, skin, and eyes. If your safety goggles cloud up or are uncomfortable, ask your teacher for help. Lengthening the strap slightly, washing the goggles with soap and warm water, or using an anti-fog spray may help alleviate the problems.

Do not wear contact lenses in the lab. Even if you are wearing safety goggles, chemicals could get between contact lenses and your eyes and cause irreparable eye damage. If your doctor requires that you wear contact lenses instead of glasses, then you should wear eye-cup safety goggles—similar to goggles that are worn for underwater swimming—in the lab. Ask your doctor or your teacher how to use eye-cup safety goggles to protect your eyes.

Know the location of all safety and emergency equipment used in the laboratory. Ask your teacher where the nearest eyewash stations, safety blankets, safety shower, fire extinguisher, first aid kit, and chemical spill kit are located.

Immediately report any accident, incident, or hazard—no matter how trivial—to your teacher. Any incident involving bleeding, burns, fainting, chemical exposure, or ingestion should also be reported immediately to the school nurse or to a physician.

In case of fire, alert the teacher and leave the laboratory. Standard school fire safety procedures should be followed.

Do not fool around in the lab. Take your lab work seriously and behave appropriately in the laboratory. Be aware of your classmates' safety as well as your own at all times.

Do not have or consume food or drink in the laboratory. Do not store food in the laboratory. Keep your hands away from your face. Do not apply cosmetics in the laboratory. Some hair-care products and nail polish are highly flammable.

Keep your work area neat and uncluttered. Have only books and other materials that are needed to conduct the experiment in the laboratory.

Clean your work area at the conclusion of each lab period as your teacher directs. Broken glass, chemicals, and other laboratory waste products should be disposed of in separate special containers. Dispose of waste materials as directed by your teacher.

Wash your hands with soap and water after each lab period. Also wash your hands before each lab period to avoid contamination.

Key to Safety Symbols and Their Precautions

Before you begin working in the laboratory, familiarize yourself with the following safety symbols that are used in this textbook and the guidelines that you should follow when you see these symbols.

EYE SAFETY

- **Wear approved safety goggles as directed.** Goggles should always be worn in the laboratory, especially when you are working with a chemical or a solution, a heat source, or a mechanical device.
- **In case of eye contact, do the following:** Go to an eyewash station immediately, and flush your eyes (including under the eyelids) with running water for at least 15 minutes. Hold your eyelids open with your thumb and fingers, and roll your eyeball around. While doing this, have another student notify your teacher.
- **Do not wear contact lenses in the laboratory.** Chemicals can be drawn up under the contact lens and into the eye. If you must wear contacts, tell your teacher. You must also wear approved eye-cup safety goggles to protect your eyes.
- **Do not look directly at the sun through any optical device or lens system or reflect direct sunlight to illuminate a microscope.** Such actions concentrate light rays to an intensity that can severely burn your retina, possibly causing blindness.

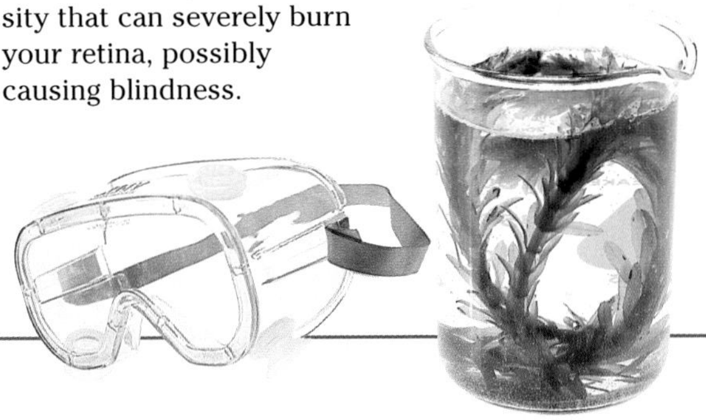

HAND SAFETY

- **Do not cut objects while holding them in your hand.** Dissect specimens in a dissecting tray.
- **Wear protective gloves when working with an open flame, chemicals, solutions, or wild or unknown plants.**

SAFETY WITH GASES

- **Do not inhale any gas or vapor unless directed to do so by your teacher.** Do not breathe pure gases.
- **Handle materials prone to emit vapors or gases in a well-ventilated area.** This work should be done in an approved chemical fume hood.

SHARP-OBJECT SAFETY

- **Use extreme care when handling all sharp and pointed instruments, such as scalpels, sharp probes, and knives.**
- **Do not use double-edged razor blades in the laboratory.**
- **Do not cut objects while holding them in your hand.** Cut objects on a suitable work surface. Always cut in a direction away from your body.

1083

CLOTHING PROTECTION

- **Wear an apron or laboratory coat at all times in the laboratory to prevent chemicals or solutions from contacting skin or clothes.**

ANIMAL CARE AND SAFETY

- **Do not touch or approach wild animals.** When working in the field, be aware of poisonous or dangerous animals in the area.
- **Always obtain your teacher's permission before bringing any animal (including pets) into the school building.**
- **Handle animals only as your teacher directs.** Mishandling or abusing any animal will not be tolerated.

HEATING SAFETY

- **Be aware of any source of flames, sparks, or heat (open flames, electric heating coils, hot plates, etc.) before working with flammable liquids or gases.**
- **When heating chemicals or reagents in a test tube, do not point the test tube toward anyone.**
- **Avoid using open flames.** If possible, work only with hot plates that have an "On-Off" switch and an indicator light. Do not leave hot plates unattended. Do not use alcohol lamps. Turn off hot plates and open flames when they are not in use.
- **Know the location of laboratory fire extinguishers and fire-safety blankets.**
- **Use tongs or other appropriate, insulated holders when heating objects.** Heated objects often do not appear to be hot. Thus, do not pick up an object with your hand if it could be warm.
- **Keep flammable substances away from heat, flames, and other ignition sources.**

HYGIENIC CARE

- **Keep your hands away from your face and mouth while working in the laboratory.**
- **Wash your hands thoroughly before leaving the laboratory.**
- **Remove contaminated clothing immediately.** If you spill caustic substances on your skin or clothing, use the safety shower or a faucet to rinse. Remove affected clothing while under the shower, and call to your teacher. (It may be temporarily embarrassing to remove clothing in front of your classmates, but failing to rinse a chemical off your skin could cause permanent damage.)
- **Launder contaminated clothing separately.**
- **Use the proper technique demonstrated by your teacher when handling bacteria or similar microorganisms.** Do not open Petri dishes to observe or count bacterial colonies.

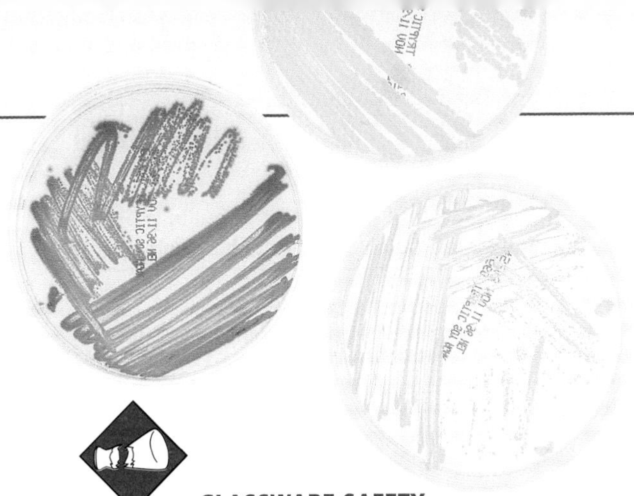

GLASSWARE SAFETY

- **Inspect glassware before use; do not use chipped or cracked glassware.** Use heat-resistant glassware for heating materials or storing hot liquids.
- **Do not attempt to insert glass tubing into a rubber stopper without specific instruction from your teacher.**
- **Immediately notify your teacher if a piece of glassware breaks. Do not attempt to clean up broken glass.**

PROPER WASTE DISPOSAL

- **Clean and decontaminate all work surfaces and personal protective equipment after each lab period as directed by your teacher.**
- **Dispose of all sharp objects (such as broken glass) and other contaminated materials (biological or chemical) in special containers as directed by your teacher.**

ELECTRICAL SAFETY

- **Do not use equipment with frayed electrical cords or loose plugs.**
- **Fasten electrical cords to work surfaces with tape.** This will prevent tripping and will ensure that equipment cannot fall off the table.
- **Do not use electrical equipment near water or with wet hands or clothing.**
- **Hold the rubber cord when you plug in or unplug equipment. Do not touch the metal prongs of the plug, and do not unplug equipment by pulling on the cord.**

PLANT SAFETY

- **Do not ingest any plant part used in the laboratory (especially commercially sold seeds).** Do not touch any sap or plant juice directly. Always wear gloves.
- **Wear disposable polyethylene gloves when handling any wild plant.**
- **Wash hands thoroughly after handling any plant or plant part (particularly seeds).** Avoid touching your face and eyes.
- **Do not inhale or expose yourself to the smoke of any burning plant.** Smoke contains irritants that can cause inflammation in the throat and lungs.
- **Do not pick wildflowers or other wild plants unless otherwise directed by your teacher.**

CHEMICAL SAFETY

- **Always wear appropriate protective equipment.** Eye goggles, gloves, and a lab apron or lab coat should always be worn when working with any chemical or chemical solution.
- **Do not taste, touch, or smell any substance or bring it close to your eyes, unless specifically instructed to do so by your teacher.** If you are directed by your teacher to note the odor of a substance, do so by waving the fumes toward you with your hand. Do not pipette any substance by mouth; use a suction bulb as directed by your teacher.
- **Always handle chemicals and solutions with care.** Check the labels on bottles, and observe safety procedures. Do not return unused chemicals or solutions to their original containers. Return unused reagent bottles or containers to your teacher.
- **Do not mix any chemicals unless specifically instructed to do so by your teacher.** Two harmless chemicals can be poisonous if combined.
- **Do not pour water into a strong acid or strong base.** The mixture can produce heat and splatter.
- **Report any spill immediately to your teacher.** Do not clean up spills yourself unless your teacher instructs you otherwise.

1085

MEASUREMENT

Scientists throughout the world use the Système International d'Unités, or International System of Measurements. It is usually referred to simply as SI. You will often use SI units to express the measurements that you will make in the laboratory. Most measurements in this book are expressed in SI units.

SI is a decimal system; that is, all relationships between SI units are based on powers of 10. Most units have a prefix that indicates the relationship of that unit to a base unit. For example, the SI base unit for length is the meter. A meter equals 100 centimeters (cm), or 1,000 millimeters (mm). A meter also equals 0.001 kilometers (km). Table 1 summarizes the prefixes and abbreviations that are commonly used to represent SI units.

TABLE 1 SI PREFIXES

Prefix	Abbreviation	Factor of base unit
giga	G	1,000,000,000
mega	M	1,000,000
kilo	k	1,000
hecto	h	100
deka	da	10
deci	d	0.1
centi	c	0.01
milli	m	0.001
micro	µ	0.000001
nano	n	0.000000001

SI Units

BASE UNITS

Seven fundamental quantities are represented by base units in SI. These base units include familiar quantities, such as length, mass, and time. These base quantities, their units, and their abbreviations are given in Table 2.

TABLE 2 SI BASE UNITS

Base quantity	Unit	Abbreviation
Length	meter	m
Mass	kilogram	kg
Time	second	s
Electric current	ampere	A
Thermodynamic temperature	kelvin	K
Amount of substance	mole	mol
Luminous intensity	candela	cd

DERIVED UNITS

The base units in Table 2 cannot be used to express measurements such as the surface area of a wildlife preserve or the speed of a running cheetah. Therefore, other important quantities, such as area, volume, and velocity, are expressed in derived units. Derived units are mathematical combinations of one or more base units.

TABLE 3 EXAMPLES OF SI DERIVED UNITS

Derived quantity	Unit	Abbreviation
Area	square meter	m^2
Volume	cubic meter	m^3
Mass density	kilogram per cubic meter	kg/m^3
Specific volume	cubic meter per kilogram	m^3/kg
Velocity	meter per second	m/s
Celsius temperature	degree Celsius	°C

Table 3 lists some derived units that you may encounter in your study of biology.

Like base units, derived units also can be expressed using prefixes. In the lab, you will often express volume measurements in cubic centimeters (cm^3). Graduated cylinders are calibrated in milliliters or cubic centimeters. A square meter equals 10,000 square centimeters (cm^2). Large area measurements are often expressed in hectares (ha). A hectare equals 10,000 square meters (m^2).

UNITS ACCEPTED FOR USE WITH SI

Certain units of measure that are not SI units are still acceptable for use with SI units. They are the units of time—minute, hour, and day; a unit of volume—liter; and a unit of mass—metric ton. These units are listed in Table 4.

TABLE 4 EXAMPLES OF UNITS ACCEPTED FOR USE WITH SI

Unit	Abbreviation	Value in SI units
Minute	min	1 min = 60 s
Hour	h	1 h = 3,600 s 1 h = 60 min
Day	d	1 d = 24 h
Liter	L	1 L = 0.001 m^3
Metric ton	t	1 t = 1,000 kg

Equivalent Measurements and Conversions

Conversion between SI units requires a conversion factor. For example, to convert from meters to centimeters, you need to know the relationship between meters and centimeters.

1 cm = 0.01 m or 1 m = 100 cm

If you needed to convert a lab measurement of 15.5 centimeters to meters, you could do either of the following:

$$15.5 \text{ cm} \times \frac{1 \text{ m}}{100 \text{ cm}} = 0.155 \text{ m}$$

or

$$15.5 \text{ cm} \times \frac{0.01 \text{ m}}{1 \text{ cm}} = 0.155 \text{ m}$$

The following are some measurement equivalents for length, area, mass, and volume.

LENGTH

1 kilometer (km) = 1,000 m
1 meter (m) = base unit of length
1 centimeter (cm) = 0.01 m
1 millimeter (mm) = 0.001 m
1 micrometer (μm) = 0.000001 m

AREA

square kilometer (km^2) = 100 ha
1 hectare (ha) = 10,000 m^2
1 square meter (m^2) = 10,000 cm^2
1 square centimeter (cm^2) = 100 mm^2

MASS

1 kilogram (kg) = base unit of mass
1kg = 1,000 grams (g)
1 gram (g) = 0.001 kg
1 milligram (mg) = 0.001 g
1 microgram (μg) = 0.000001 g

LIQUID VOLUME

1 kiloliter (kL) = 1,000 L
1 liter (L) = base unit of liquid volume
1 milliliter (mL) = 0.001 L
1 mL = 1 cm^3

NOTE: When measuring liquid volume in a graduated cylinder, read the measurement at the bottom of the meniscus, or curve.

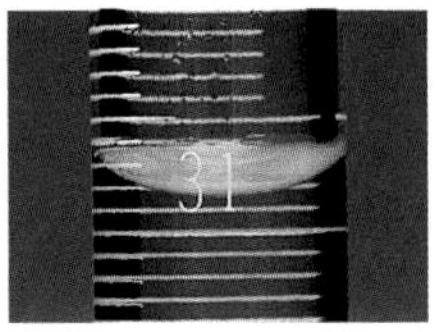

TEMPERATURE

In this textbook, the Celsius scale is used to express temperature measurements. In Celsius, 0° is the freezing point of water, and 100° is the boiling point of water. You can use the scale shown below to convert between the Celsius scale and the Fahrenheit scale.

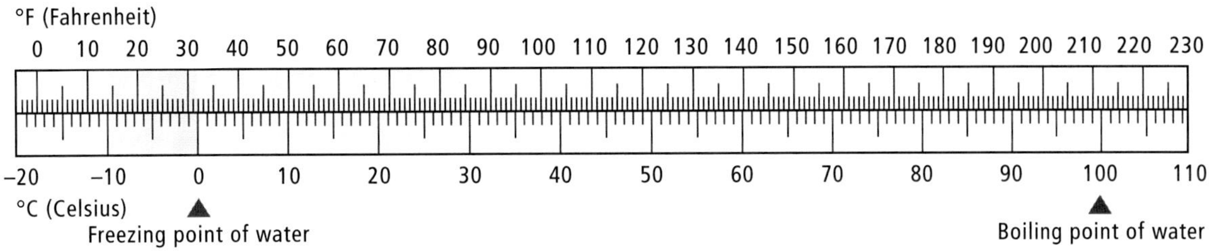

1087

USING A COMPOUND LIGHT MICROSCOPE

Parts of the Compound Light Microscope

- The **eyepiece** magnifies the image, usually 10×.
- The **low-power objective** further magnifies the image, up to 4×.
- The **high-power objectives** further magnify the image, from 10× to 43×.
- The **nosepiece** holds the objectives and can be turned to change from one objective to another.
- The **body tube** maintains the correct distance between the eyepiece and the objectives. This is usually about 25 cm (10 in.), the normal distance for reading and viewing objects with the naked eye.
- The **coarse adjustment** moves the body tube up and down in large increments to allow gross positioning and focusing of the objective lens.
- The **fine adjustment** moves the body tube slightly to bring the image into sharp focus.
- The **stage** supports a slide.
- The **stage clips** secure the slide in position for viewing.
- The **diaphragm** (not labeled), located under the stage, controls the amount of light that is allowed to pass through the object being viewed.
- The **light source** provides light for viewing the image. It can be either a light reflected with a mirror or an incandescent light from a small lamp. NEVER use reflected direct sunlight as a light source.
- The **arm** supports the body tube.
- The **base** supports the microscope.

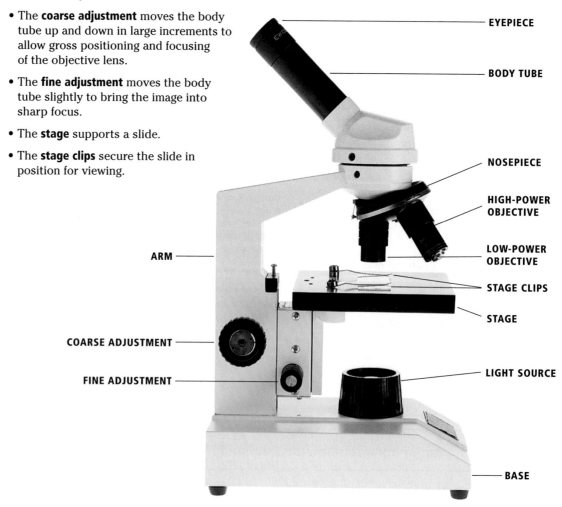

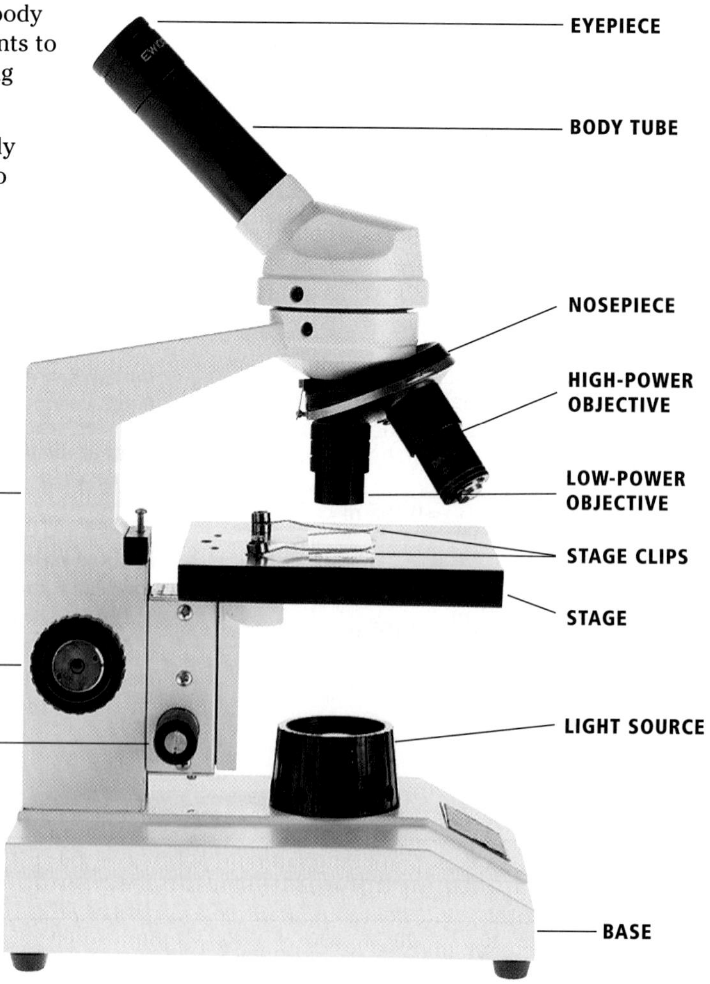

Proper Handling and Use of the Compound Light Microscope

1. Carry the microscope to your lab table using both hands, one supporting the base and the other holding the arm of the microscope. Hold the microscope close to your body.
2. Place the microscope on the lab table at least 5 cm (2 in.) from the edge of the table.
3. Check to see what type of light source the microscope has. If the microscope has a lamp, plug it in, making sure that the cord is out of the way. If the microscope has a mirror, adjust it to reflect light through the hole in the stage.

   CAUTION **If your microscope has a mirror, do not use direct sunlight as a light source. Using direct sunlight can damage your eyes.**
4. Adjust the revolving nosepiece so that the low-power objective is in line with the body tube.
5. Place a prepared slide over the hole in the stage, and secure the slide with the stage clips.
6. Look through the eyepiece, and move the diaphragm to adjust the amount of light that passes through the specimen.
7. Now look at the stage at eye level, and slowly turn the coarse adjustment to raise the stage until the objective almost touches the slide. Do not allow the objective to touch the slide.
8. While looking through the eyepiece, turn the coarse adjustment to lower the stage until the image is in focus. Never focus objectives downward. Use the fine adjustment to attain a sharply focused image. Keep both eyes open while viewing a slide.
9. Make sure that the image is exactly in the center of your field of vision. Then switch to the high-power objective. Focus the image with the fine adjustment. Never use the coarse adjustment at high power.
10. When you are finished using the microscope, remove the slide. Clean the eyepiece and objectives with lens paper, and return the microscope to its storage area.

Procedure for Making a Wet Mount

1. Use lens paper to clean a glass slide and coverslip.
2. Place the specimen that you wish to observe in the center of the slide.
3. Using a medicine dropper, place one drop of water on the specimen.
4. Position the coverslip so that it is at the edge of the drop of water and at a 45° angle to the slide. Make sure that the water runs along the edge of the coverslip.
5. Lower the coverslip slowly to avoid trapping air bubbles.
6. If a stain or solution will be added to a wet mount, place a drop of the staining solution on the microscope slide along one side of the coverslip. Place a small piece of paper towel on the opposite side of the coverslip.
7. As the water evaporates from the slide, add another drop of water by placing the tip of the medicine dropper next to the edge of the coverslip, just as you would if you were adding stains or solutions to a wet mount. If you have added too much water, remove the excess by using the corner of a paper towel as a blotter. Do not lift the coverslip to add or remove water.

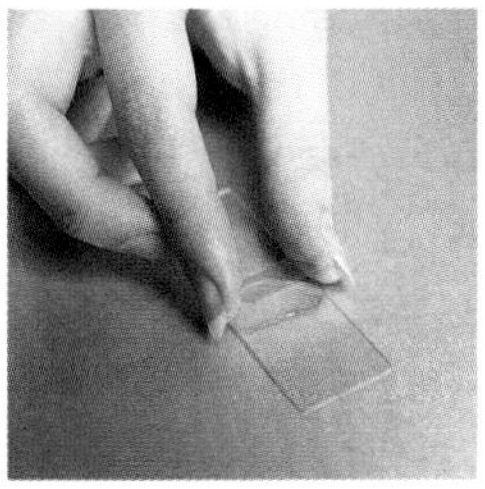

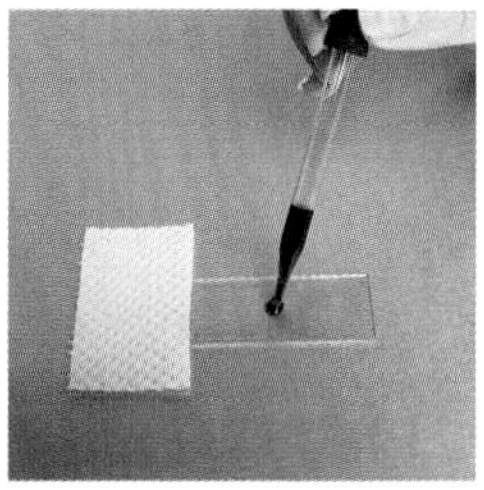

INTERACTIVE EXPLORATIONS

Imagine that you could watch what happens within a plant cell during photosynthesis or that you could observe how drug use changes the human nervous system. You cannot do these things in a conventional laboratory exercise, but you can in an Interactive Exploration. By using the *Interactive Explorations in Biology* CD-ROMs, you can explore biological processes that are important to both human biology and cell biology. The instructions on these two pages describe how to use the CD-ROMs.

Using the CD-ROM *Interactive Explorations in Biology: Human Biology*

1. Obtain the CD-ROM *Interactive Explorations in Biology: Human Biology* from your teacher. Load the disc into the CD-ROM drive as directed by your teacher.
2. An icon for the CD-ROM will appear on the screen. Click it, and then click the icon labeled "Start Explorations."
3. After a few moments, the title screen will appear. Click the button labeled "Menu" at the bottom right of the screen.
4. The computer will then display the menu screen, which lists all the Interactive Explorations on the disc. Select the Interactive Exploration that you would like to run, and click on its title.
5. After a few moments, the computer will display a screen containing a column of buttons along the right edge. This column is the Navigation Palette, and the buttons it contains enable you to access the various features found in each Interactive Exploration. To begin, click the top button of the column. This button displays a magnifying glass and is labeled "Interactive Investigation."
6. If you have questions about an Interactive Exploration, you can play the instructions for operating the Interactive Exploration by clicking the File menu at the top of the screen and selecting "Interactive Exploration Help." You may also refer to the User's Guide for additional information or help.

1090

Using the CD-ROM *Interactive Explorations in Biology: Cell Biology and Genetics*

1. Obtain the CD-ROM *Interactive Explorations in Biology: Cell Biology and Genetics* from your teacher. Load the disc into the CD-ROM drive as directed by your teacher.
2. An icon for the CD-ROM will appear on the screen. Click it, and then click the icon labeled "Start Explorations."
3. After a few moments, the title screen will appear. Click the button labeled "Menu" at the bottom right of the screen.
4. The computer will then display the menu screen, which lists all the Interactive Explorations on the disk. Select the Interactive Exploration that you would like to run, and click on its title.
5. After a few moments, the computer will display the Interactive Exploration and play a short narration that describes the problem you will be studying. (To shut off this narration, select "Sound Off" from the Sound menu at the top right of the screen.) After this brief narration has played, the Interactive Exploration is ready to run.
6. If you have questions about an Interactive Exploration, you can play the instructions for operating the Interactive Exploration by clicking the Help menu at the top of the screen, and selecting "How to Use This Exploration." Clicking the Help button on the Navigation Palette will call up instructions on how to use the features of the Navigation Palette. You may also refer to the User's Guide.

FEEDBACK METER: Displays a response to changes in variables

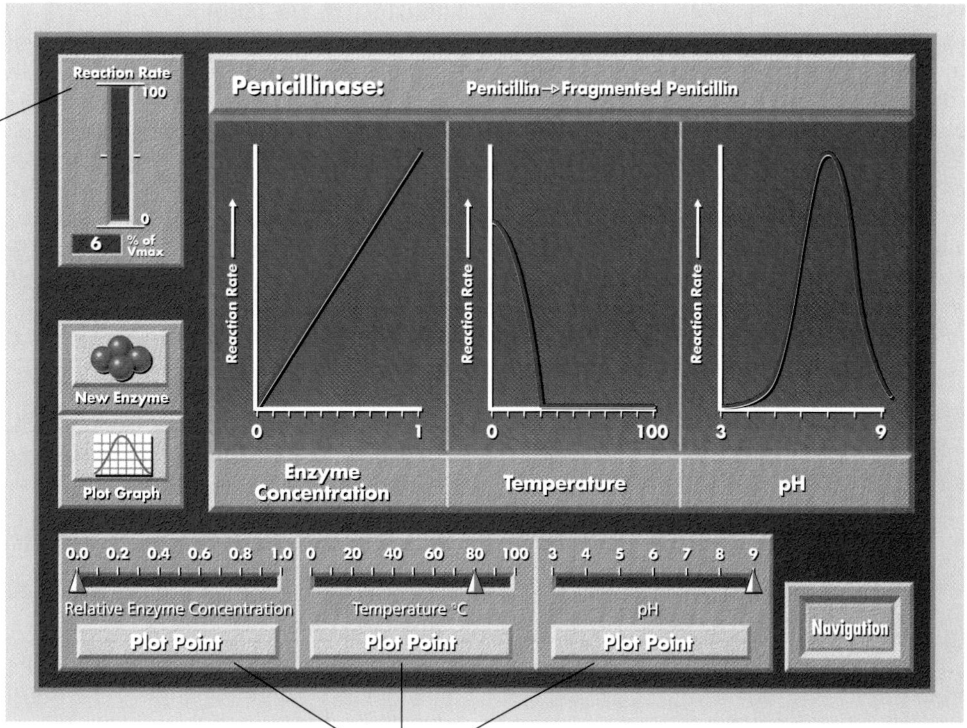

VARIABLES: Conditions that you can change in the Exploration

1091

SIX-KINGDOM SYSTEM FOR THE CLASSIFICATION OF ORGANISMS

The classification of organisms in this textbook is based on the six-kingdom system described in Chapter 18. Not all groups are shown, and numbers of species are approximate. Also, since the division of the kingdoms Eubacteria and Archaebacteria into phyla is controversial, only broad, generally recognized groups are presented for these two kingdoms. As you read about the groups below that comprise the six kingdoms, bear in mind that the classification of organisms changes as new information about the evolutionary relationships among organisms is uncovered.

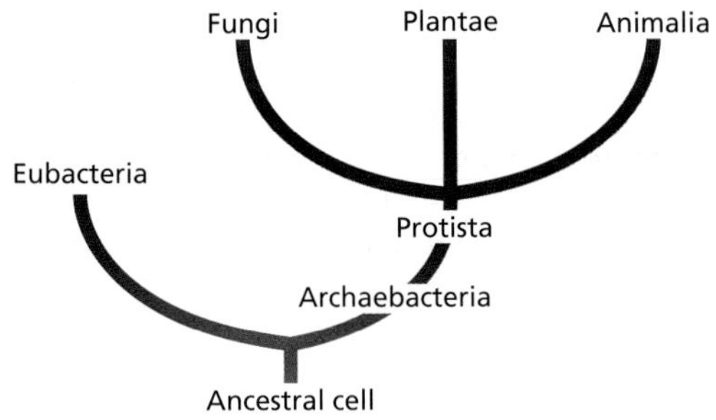

KINGDOM EUBACTERIA

Typically unicellular; prokaryotic: without membrane-bound organelles or a nucleus; nutrition mainly by absorption, but some are photosynthetic or chemosynthetic; includes anaerobic and aerobic species; reproduction usually by fission or budding; about 5,000 species have been recognized, but there are undoubtedly many thousand more

CYANOBACTERIA
Photosynthetic bacteria surrounded by a gooey, deeply pigmented covering; photosynthetic mechanism is very similar to that in plants; some species can fix nitrogen; had an important effect on life's evolution because photosynthesis by early cyanobacteria raised the oxygen content of the atmosphere; common on land and in the ocean; some early species were probably ancestors of chloroplasts in some kinds of protists: *Anabaena, Oscillatoria, Spirulina*

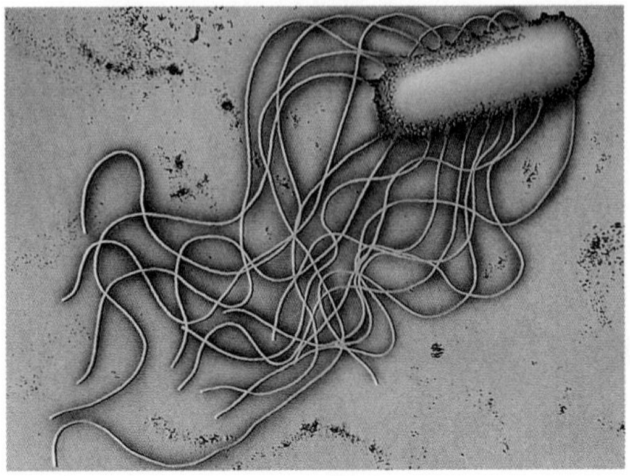

Salmonella

SPIROCHETES
Long, spiral cells with flagella originating at each end; aerobic and anaerobic heterotrophs; move with corkscrew-like motion; parasitic, symbiotic, or free-living; responsible for several serious diseases, including syphilis (*Treponema pallidum*) and Lyme disease (*Borrelia burgdorferi*)

PROTEOTIC BACTERIA
A large, metabolically and ecologically diverse group of Gram-negative bacteria; includes anaerobes and aerobes, heterotrophs and autotrophs; free-living, symbiotic, and parasitic species; many species are of great ecological or medical significance; examples include nitrogen-fixing *Azotobacter* and *Rhizobium*, pathogens such as *Rickettsia rickettsii* (Rocky Mountain spotted fever), *Salmonella typhi* (typhoid), and *Vibrio cholerae* (cholera), and intestinal symbionts such as *Escherichia coli*

GRAM-POSITIVE BACTERIA
A group of predominantly Gram-positive bacteria; heterotrophic; anaerobic and aerobic species; many can form endospores to resist unfavorable environmental conditions: *Streptococcus, Staphylococcus, Lactobacillus, Bacillus, Clostridium*

CHLAMYDIAE
A small group of obligate intracellular parasites that infect endothermic hosts; heterotrophic; cell walls do not have peptidoglycan; *Chlamydia trachomatis* causes chlamydia, one of the most common sexually transmitted diseases in humans

KINGDOM ARCHAEBACTERIA

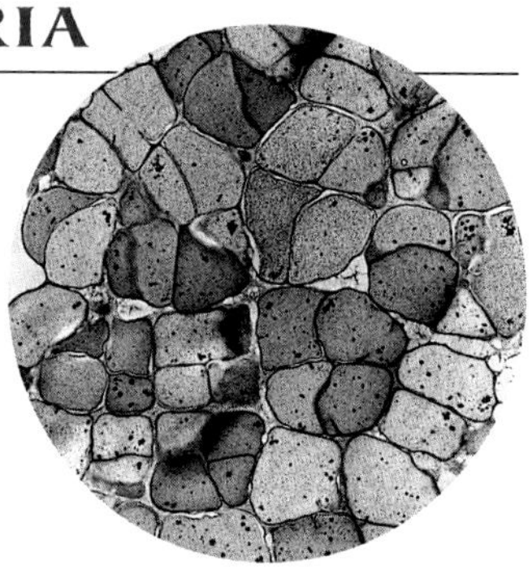
Methanosarcina

Anaerobic and aerobic prokaryotes that are biochemically distinct from eubacteria; many species are adapted to environments with extreme temperatures, acidity, or salt content; differ from eubacteria in structure of cell membrane and cell wall; RNA polymerase and ribosomal RNA sequences are similar to those in eukaryotes, suggesting that archaebacteria are more closely related to eukaryotes than to eubacteria; undergo asexual reproduction only; fewer than 100 named species, divided into three broad groups

METHANOGENS
Anaerobic methane producers; most species use carbon dioxide as a carbon source; found in soil, swamps, and the digestive tract of animals, particularly grazing mammals such as cattle; also inhabit the human large intestine; produce nearly 2 trillion kg (2 billion tons) of methane gas annually: *Methanococcus, Methanosarcina*

THERMOACIDOPHILES
Inhabit very hot environments that are often very acidic; some species can tolerate temperatures of 110°C (230°F); require sulfur; nearly all are anaerobes: *Sulfolobus, Pyrodictium*

EXTREME HALOPHILES
Live in environments with very high salt content, including the Dead Sea and the Great Salt Lake; nearly all are aerobic; all are Gram-negative: *Heliobacterium, Halococcus*

KINGDOM PROTISTA

A catchall kingdom for the eukaryotes that are not plants, fungi, or animals; the most structurally diverse kingdom; unicellular and multicellular representatives; all have a membrane-bound nucleus; nearly all have chromosomes, mitochondria, and internal compartments; many have chloroplasts; most have cell walls; reproduce sexually and asexually; more than 60,000 living species

PHYLUM SARCODINA
Heterotrophic, unicellular protists that move and feed with flexible cytoplasmic extensions called pseudopods; aquatic (fresh and salt water) and terrestrial forms; most are free-living, but a few are parasitic; undergo asexual and sexual reproduction: amoebas, radiolarians, foraminiferans (*Entamoeba, Amoeba*); several hundred living species

PHYLUM CILIOPHORA
Very complex single cells; heterotrophic; have rows of cilia and two types of cell nuclei: ciliates (*Didinium, Paramecium, Stentor, Vorticella*); about 8,000 species

PHYLUM ZOOMASTIGINA
Mostly unicellular; heterotrophic; all have at least one flagellum: zoomastigotes (*Giardia, Leishmania, Trichonympha, Trypanosoma*); about 2,500 species

PHYLUM SPOROZOA
Unicellular, heterotrophic, spore-forming parasites of animals; have complex life cycles with asexual and sexual stages; nonmobile as adults; *Plasmodium* is responsible for malaria, which kills more than 2 million people each year: sporozoans (*Plasmodium, Toxoplasma*); about 6,000 species

PHYLUM MYXOMYCOTA
Heterotrophic; stream along as a multinucleate mass of cytoplasm; when dry or starving, can give rise to spores that start a new individual in a more favorable environment: plasmodial slime molds (*Physarum*); about 450 species

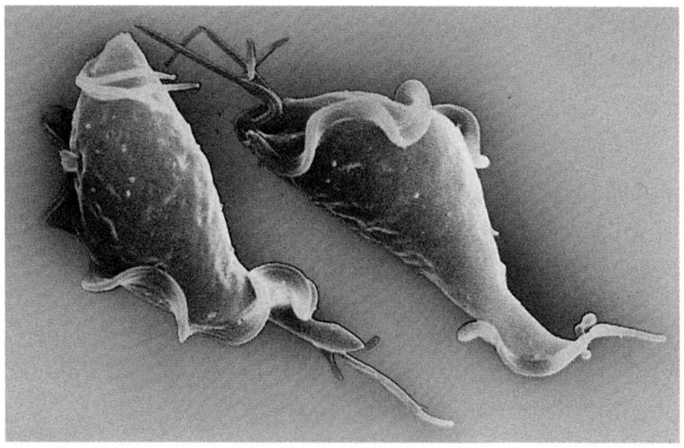
Giardia

1093

PHYLUM ACRASIOMYCOTA
Heterotrophic, amoeba-shaped cells that aggregate into a moving mass called a slug when deprived of food; cells within the slug retain their membranes and do not fuse; a slug produces spores that form new amoebas elsewhere: cellular slime molds (*Dictyostelium*); about 65 species

PHYLUM OOMYCOTA
Heterotrophic, unicellular parasites or decomposers; cell walls composed of cellulose, not chitin as in fungi; asexual and sexual reproduction: water molds, white rusts, downy mildews (*Phytophthora, Plasmopara*); about 580 species

PHYLUM CHYTRIDIOMYCOTA
Funguslike, heterotrophic decomposers and parasites with cell walls of chitin; most are unicellular; gametes and zoospores have a single, posterior flagellum; live in water and the soil: chytrids (*Allomyces*); about 750 species (some scientists place this group within the kingdom Fungi)

PHYLUM CHLOROPHYTA
Unicellular, colonial, and multicellular algae; all are photosynthetic; have chlorophylls *a* and *b;* use starch to store food; contain chloroplasts very similar to those of plants; most scientists think that plants descended from this group: green algae (*Chlamydomonas, Chlorella, Oedogonium, Spyrogyra, Ulva, Volvox*); about 7,000 species

PHYLUM PHAEOPHYTA
Multicellular and photosynthetic; nearly all are marine; contain chlorophylls *a* and *c* and fucoxanthin, the source of their brownish color; store food as laminarin: brown algae (*Fucus, Laminaria, Postelsia, Sargassum*); about 1,500 species

PHYLUM RHODOPHYTA
Almost all are multicellular; all are photosynthetic; most are marine; contain chlorophyll *a* and phycobilins; chloroplasts probably evolved from symbiotic cyanobacteria: red algae (*Porphyra*); about 4,000 species

PHYLUM BACILLARIOPHYTA
Unicellular photosynthetic inhabitants of fresh and salt water; secrete a unique shell made of opaline silica that resembles a box with a lid; chloroplasts resemble those of brown algae; contain chlorophylls *a* and *c* and fucoxanthin; important component of plankton: diatoms; more than 11,500 species

PHYLUM DINOFLAGELLATA
Unicellular and aquatic; heterotrophic and autotrophic species; mostly marine; body is enclosed within two cellulose plates; most have two flagella; contain chlorophylls *a* and *c* and carotenoids: dinoflagellates (*Gonyaulax, Noctiluca*); about 1,100 species

PHYLUM CHRYSOPHYTA
Photosynthetic, usually aquatic, and often colonial; two flagella at the end of each cell; golden brown color imparted by carotenoid pigments; can form durable cysts to survive adverse environmental conditions: golden algae; about 850 species

PHYLUM EUGLENOPHYTA
Unicellular; cells lack a cell wall and are surrounded by a flexible protein covering called pellicle; both photosynthetic and heterotrophic species; asexual reproduction; most live in fresh water; chloroplasts are similar to those of green algae and are thought to have evolved from the same symbiotic bacteria: euglenoids (*Euglena*); about 1,000 species

KINGDOM FUNGI

Eukaryotic heterotrophs with nutrition by absorption; all but yeasts are multicellular; nearly all are terrestrial; body is typically composed of filaments (called hyphae) and is multinucleate, with incomplete divisions (called septa) between cells; cell walls are made of chitin; asexual or sexual reproduction; about 100,000 species

PHYLUM ASCOMYCOTA
Hyphae usually have perforated septa; fusion of hyphae leads to formation of densely interwoven mass that contains characteristic microscopic reproductive structures called asci; terrestrial, marine, and freshwater species: brewer's and baker's yeasts, molds, morels, truffles (*Neurospora, Saccharomyces, Candida)*; more than 60,000 species

PHYLUM BASIDIOMYCOTA

Hyphae usually have incomplete septa; reproduction is typically sexual; fusion of hyphae leads to the formation of a densely interwoven reproductive structure (mushroom) with characteristic microscopic, club-shaped structures called basidia; terrestrial: mushrooms, toadstools, shelf fungi, rusts, smuts; about 25,000 species

PHYLUM ZYGOMYCOTA

Usually lack septa; fusion of hyphae leads directly to formation of a zygote, which divides by meiosis when it germinates; terrestrial or parasitic: bread molds (*Pilobolus, Rhizopus*); about 660 species

Amanita muscaria

KINGDOM PLANTAE

Multicellular, eukaryotic, autotrophic, terrestrial organisms with tissues and organs; cell walls contain cellulose; chlorophylls *a* and *b* are present and are localized in plastids; all have alternation of generations between diploid sporophyte and haploid gametophyte; about 270,000 species

PHYLUM BRYOPHYTA

Small and usually found in moist environments; most lack vascular tissues; gametophyte is the dominant generation; gametophyte lacks roots, stems, and leaves and is anchored to a substrate by projections called rhizoids: mosses (*Sphagnum*); about 10,000 species

PHYLUM HEPATOPHYTA

Gametophyte is dominant, usually small, and grows close to the ground; gametophyte lacks stomata, vascular tissue, roots, stems, and leaves; sporophyte is inconspicuous and dependent on the gametophyte: liverworts (*Marchantia*); about 6,500 species

PHYLUM ANTHOCEROPHYTA

Gametophyte is the dominant generation; usually small and flat; stoma are present on sporophyte; lack vascular tissue, roots, stems, and leaves; contain a single chloroplast per cell: hornworts (*Anthoceros*); about 100 species

PHYLUM PSILOTOPHYTA

Vascular system present; seedless; no roots or leaves; gametophyte is small and independent; sporophyte is dominant and has small leaves: whisk ferns (*Psilotum*); about 10 species

PHYLUM LYCOPHYTA

Vascular system present; sporophyte is dominant and mosslike; has roots, stems, and leaves; gametophyte is small and independent; seedless: club mosses (*Lycopodium, Selaginella*); about 1,000 species

PHYLUM SPHENOPHYTA

Vascular system present; gametophyte is small and independent; sporophyte is dominant; roots are present; seedless: horsetails (*Equisetum*); 15 species

PHYLUM PTEROPHYTA

Vascular system present; gametophyte is small and independent; sporophyte generation is dominant; has roots, stems, and leaves; seedless: ferns (*Salvinia*); about 12,000 species

PHYLUM CYCADOPHYTA

Palmlike gymnosperms; vascular system present; male and female cones produced on different trees; naked seeds; sporophyte is dominant; tropical and subtropical: cycads (*Cycas*); about 100 species

1095

PHYLUM CONIFEROPHYTA
Gymnosperms that produce cones; vascular system lacks vessels; leaves are usually needles or scales; typically evergreen; sporophyte is dominant; ovules are exposed at time of pollination; pollen is dispersed by wind: pines, spruces, firs, redwoods, cypresses, yews; about 550 species

PHYLUM GNETOPHYTA
Specialized gymnosperms; vascular system contains water-conducting vessels; seeds are naked; sporophyte dominant: gnetophytes (*Ephedra, Welwitschia*); about 70 species

PHYLUM GINKGOPHYTA
Deciduous, gymnosperm tree; vascular system is present; has fanlike leaves; sporophyte is dominant; produces conelike male reproductive structures and uncovered seeds on different individuals: the ginkgo, *Ginkgo biloba*, is the only living species

PHYLUM ANTHOPHYTA
Angiosperms; vascular system present; sporophyte is dominant, ovules are fully enclosed by ovary, gametophyte greatly reduced after fertilization, ovary and seed mature to become fruit; flowers are reproductive structures: flowering plants (*Aster, Prunus, Quercus, Zea*); about 240,000 species

Redwoods

Class Monocotyledones
Embryo has one cotyledon; flower parts in threes; leaf veins are parallel; vascular bundles are scattered through stem tissue: grasses, sedges, lilies, irises, palms, orchids; around 70,000 species

Class Dicotyledones
Embryo has two cotyledons; flower parts in fours or fives; leaves have netlike veins; vascular bundles in orderly arrangement in stems: roses, maples, elms, cactuses; about 170,000 species

KINGDOM ANIMALIA

Multicellular, eukaryotic, heterotrophic organisms; nutrition mainly by ingestion; most have specialized tissues, and many have organs and organ systems; no cell walls or chloroplasts; sexual reproduction predominates; both aquatic and terrestrial forms; more than 1 million species

PHYLUM PORIFERA
Typically asymmetrical; lack tissues and organs; body wall consists of two cell layers penetrated by numerous pores; internal cavity is lined with unique food-filtering cells called choanocytes; adults are sessile; mostly marine; sexual and asexual reproduction: sponges; about 10,000 species

PHYLUM CNIDARIA
Radially symmetrical and gelatinous; most have distinct tissues; baglike body of two cell layers; marine and freshwater species: hydras, jellyfishes, corals, sea anemones (*Hydra, Obelia*); about 10,000 species

Class Hydrozoa
Most have both polyp and medusa stages in life cycle; usually colonial: hydras, Portuguese men-of-war; about 3,700 species

Class Scyphozoa
Exclusively marine; medusa stage is usually dominant: jellyfishes; about 200 species

Class Anthozoa
Marine; solitary or colonial; medusa stage absent: sea anemones, corals, sea fans; about 6,100 species

PHYLUM CTENOPHORA
Transparent, gelatinous marine animals resembling jellyfishes; radially symmetrical; propelled by cilia: comb jellies; about 100 species

PHYLUM PLATYHELMINTHES
Body flat and ribbonlike, without true segments; bilaterally symmetrical acoelomates; organs present; three germ layers: flatworms (*Dugesia, Planaria, Schistosoma*) more than 18,000 species

Class Turbellaria
Mostly free-living aquatic or terrestrial forms: planarians (*Dugesia*); about 4,500 species

Class Trematoda
Internal parasites with mouth at anterior end; often have complex life cycle with alternation of hosts: human blood flukes (*Schistosoma*), human liver fluke (*Chlonorchis sinensis*); about 9,000 species

Class Cestoda
Extremely specialized internal parasites; hooked scolex for attaching to host: tapeworms (*Taenia*); about 5,000 species

PHYLUM ROTIFERA
Small, wormlike or spherical animals; bilaterally symmetrical and transparent; pseudocoelomates; almost all live in fresh water: rotifers; about 1,750 species

PHYLUM NEMATODA
Typically tiny, parasitic, unsegmented worms; body slender and elongated; pseudocoelomates with one-way gut; includes important human parasites such as *Ascaris,* pinworms, hookworms, *Trichinella*, and *Wuchereria*: roundworms; more than 80,000 species

PHYLUM MOLLUSCA
Soft-bodied animals with a true coelom; usually two-part body plan consisting of head-foot and visceral mass; visceral mass is covered by mantle; protostomes; most have a unique rasping tongue called a radula; terrestrial, freshwater, and marine: clams, snails, octopuses, squids, mussels, slugs; more than 110,000 species

Class Cephalopoda
Foot modified into tentacles; closed circulatory system: squids, octopuses, nautiluses; over 600 species

Class Bivalvia
Two shells connected by a hinge; no radula; large, wedge-shaped foot: clams, oysters, scallops; more than 10,000 species

Class Gastropoda
Visceral mass twisted during development; head, distinct eyes, and tentacles usually present: snails, slugs, whelks; about 90,000 species

Class Polyplacophora
Elongated body and reduced head: chitons; about 600 species

PHYLUM ANNELIDA
Serially segmented worms; bilaterally symmetrical; protostomes: annelids; about 15,000 species

Class Polychaeta
Fleshy outgrowths called parapodia extend from segments; marine; many bristles: sandworms

Class Oligochaeta
Head not well developed; no parapodia; terrestrial and freshwater forms: earthworms

Class Hirudinea
Body flattened; no parapodia; usually suckers at both ends; many are external parasites: leeches

PHYLUM ARTHROPODA
Segmented bodies with paired, jointed appendages; bilaterally symmetrical coelomates; chitinous exoskeleton; protostomes; aerial, terrestrial, and aquatic forms: arthropods; about 1 million species

Subphylum Chelicerata
Distinguished by absence of antennae and presence of chelicerae; all appendages are unbranched; four pairs of walking legs; body has two regions (cephalothorax and abdomen); predominantly terrestrial

Class Merostomata
Cephalothorax covered by protective "shell"; sharp spike on tail: horseshoe crabs (*Limulus*); 4 species

Class Pycnogonida
Small, marine predators or parasites; usually four pairs of legs but sometimes five or six pairs: sea spiders; about 1,000 species

Class Arachnida
Terrestrial; book lungs and tracheae for respiration: spiders, scorpions, ticks, mites; about 70,000 species

Lone star tick

1097

Subphylum Crustacea
Two pairs of antennae; chewing mandibles; appendages with two branches; predominantly aquatic: crayfish, lobsters, crabs, shrimps, sowbugs; about 40,000 species

Subphylum Uniramia
Have antennae, mandibles, and unbranched appendages

Class Chilopoda
Body flattened and consisting of 15 to 170 or more segments; one pair of legs attached to each segment; carnivorous: centipedes; about 2,500 species

Class Diplopoda
Elongated body of 15 to 200 segments; two pairs of legs per segment; primarily herbivorous: millipedes; about 10,000 species

Class Insecta
Body has three regions—head, thorax, and abdomen; three pairs of legs, all attached to thorax; usually two pairs of wings: insects; about 750,000 described species, but millions more may exist

Order Thysanura: silverfishes
Order Ephemeroptera: mayflies
Order Odonata: dragonflies, damsel flies
Order Orthoptera: grasshoppers, cockroaches, walking sticks, praying mantises, crickets
Order Isoptera: termites
Order Dermaptera: earwigs
Order Anoplura: sucking lice
Order Hemiptera: true bugs—water striders, water boatmen, back swimmers, bedbugs, squash bugs, stink bugs, assassin bugs
Order Homoptera: cicadas, aphids, leaf hoppers, scale insects
Order Neuroptera: ant lions, lacewings
Order Coleoptera: beetles—ladybugs, fireflies, boll weevils
Order Lepidoptera: butterflies and moths
Order Diptera: flies, mosquitoes, gnats, midges
Order Siphonaptera: fleas
Order Hymenoptera: bees, ants, wasps, hornets

PHYLUM ECHINODERMATA
Deuterostomes; adults are radially symmetrical with five-part body plan; most forms have a water vascular system with tube feet for locomotion; marine: echinoderms; about 7,000 species

Class Crinoidea
Mouth faces upward and is surrounded by many arms: sea lilies, feather stars; about 600 species

Class Asteroidea
Body usually has five arms and double rows of tube feet on each arm; mouth directed downward: sea stars; about 1,500 species

Class Ophiuroidea
Usually has five slender, delicate arms or rays: brittle stars, basket stars; about 2,000 species

Class Echinoidea
Body spherical, oval, or disk-shaped; arms lacking but five-part body plan still apparent: sea urchins, sand dollars; about 900 species

Class Holothuroidea
Elongated, thickened body with tentacles around the mouth: sea cucumbers; about 1,500 species

PHYLUM HEMICHORDATA
Wormlike relatives of chordates; deuterostomes; body in three regions: acorn worms; about 90 species

PHYLUM CHORDATA
Bilaterally symmetrical; deuterostomes; coelom present; have a notochord, a dorsal nerve cord, pharyngeal slits, and a tail at some stage of life; aquatic and terrestrial; more than 47,000 species

Subphylum Urochordata
Body with saclike covering, or tunic, in adults; larvae are free-swimming and have nerve cord and notochord; marine: tunicates; about 2,000 species

Subphylum Cephalochordata
Small and fishlike with a permanent notochord; filter feeders: lancelets (*Branchiostoma*); about 25 species

Subphylum Vertebrata
Most of the notochord is replaced by a spinal column composed of vertebrae that protect the dorsal nerve cord; recognizable head containing a brain: vertebrates; about 45,000 species

Class Agnatha
Freshwater or marine eel-like fishes without jaws, scales, or paired fins; cartilaginous skeleton: lampreys, hagfishes; about 80 species

Class Chondrichthyes
Fishes with jaws and paired fins; gills present; no swim bladder; cartilaginous skeleton: sharks, rays, skates; about 800 species

Blue-spotted stingray

Class Osteichthyes

Freshwater and marine fishes with gills attached to gill arch; jaws and paired fins; bony skeleton; most have swim bladder: bony fishes; about 23,000 species

Class Amphibia

Freshwater or terrestrial; gills present at some stage; skin is often slimy and lacking scales; eggs typically laid in water and fertilized externally: amphibians; about 4,500 species

Order Apoda: caecilians
Order Urodela: salamanders and newts
Order Anura: frogs and toads

Class Reptilia

Terrestrial or semiaquatic vertebrates; breathe with lungs at all stages; body covered by scales; most species lay amniotic eggs covered with a protective shell; fertilization internal: reptiles; about 6,000 species

Order Rhynchocephalia: tuataras
Order Chelonia: turtles and tortoises
Order Crocodilia: alligators, crocodiles, gavials, caimans
Order Squamata: lizards and snakes

Class Aves

Body covered with feathers; forelimbs modified into wings; four-chambered heart; endothermic; lay shelled, amniotic eggs: birds; about 10,000 species

Order Gaviiformes: loons
Order Pelecaniformes: pelicans, cormorants, ganets
Order Ciconiiformes: herons, bitterns, egrets, storks, spoonbills, ibises
Order Anseriformes: ducks, geese, swans
Order Falconiformes: hawks, falcons, eagles, kites, vultures
Order Galliformes: pheasants, turkeys, quails, partridges, grouse
Order Gruiformes: cranes, coots, gallinules, rails
Order Charadriiformes: snipes, sandpipers, plovers, gulls, terns, auks, puffins, ibises
Order Procellariiformes: albatrosses, petrels
Order Columbiformes: pigeons and doves
Order Psittaciformes: parrots, parakeets, macaws, cockatoos
Order Cuculiformes: cuckoos, roadrunners
Order Strigiformes: owls
Order Caprimulgiformes: goatsuckers, whippoorwills, nighthawks
Order Apodiformes: swifts, hummingbirds
Order Coraciiformes: kingfishers
Order Sphenisciformes: penguins
Order Piciformes: woodpeckers, sapsuckers, flickers, toucans, honeyguides
Order Passeriformes: robins, bluebirds, sparrows, warblers, thrushes
Order Struthioniformes: ostriches

Class Mammalia

Hair on at least part of body; young nourished with milk secreted by mammary glands; endothermic; breathe with lungs: mammals; about 4,400 species

Order Monotremata: duckbill platypuses and spiny anteaters
Order Marsupialia: opossums, kangaroos, koalas, wallabies
Order Insectivora: moles and shrews
Order Chiroptera: bats
Order Edentata: armadillos, sloths, anteaters
Order Pholidota: pangolins
Order Rodentia: squirrels, woodchucks, mice, rats, muskrats, beavers
Order Lagomorpha: rabbits, hares, pikas
Order Carnivora: bears, weasels, mink, otters, skunks, lions, tigers, wolves, seals
Order Cetacea: whales, porpoises, dolphins
Order Sirenia: sea cows, dugongs, manatees
Order Proboscidea: elephants
Order Pinnipedia: seals, sea lions, walruses
Order Perissodactyla: tapirs, rhinoceroses, horses, zebras
Order Artiodactyla: hippopotamuses, camels, llamas, deer, giraffes, cattle, sheep, goats
Order Primates: monkeys, lemurs, gibbons, orangutans, gorillas, chimpanzees, humans
Order Macroscelidea: elephant shrews
Order Scandentia: tree shrews
Order Hyracoidea: hyraxes
Order Dermoptera: flying lemurs
Order Tubulidentata: aardvark

1099

GLOSSARY

Phonetic Key

Capital letters indicate primary stress. Small caps indicate secondary stress.

Sound symbol	Key word(s)	Phonetic respelling
a	map	MAP
ay	face day	FAYS DAY
ah	father cot	FAH-thuhr KAHT
aw	caught law	KAWT LAW
ee	eat ski	EET SKEE
e	wet rare	WET RER
oy	boy foil	BOY FOYL
ow	out now	OWT NOW
oo	shoot suit	SHOOT SOOT
u	book put	BUK PUT
uh	sun cut	SUHN KUHT
i	lip	LIP
ie	tide sigh	TIED SIE
oh	over coat	OH-vuhr KOHT
yoo	yule globule	YOOL GLAH-byool
yu	cure	KYUR
uhr	paper fern	PAY-puhr FUHRN
k	card kite	KAHRD KIET
s	cell sit	SEL SIT
y	yes	YES
j	job	JAHB
g	got	GAHT
zh	pleasure	PLE-zhuhr

Each page number provided indicates where the word is defined and not necessarily where it is first used. Remember to refer to the index for more page numbers if further investigation is needed.

A

abdomen a posterior segment of an animal that usually houses the organs of digestion and excretion (748)
abiotic factor any nonliving component of an ecosystem (368)
aboral surface the surface opposite the mouth on an echinoderm (785)
abscisic acid a hormone in plants that helps regulate the growth of buds and the germination of seeds (651)
abscission zone the area at the base of a leaf petiole where the leaf breaks off from the stem (651)
absorption in digestion, the movement of nutrients to the circulatory system (988)
accessory pigment a pigment that absorbs light energy and transfers energy to chlorophyll *a* in photosynthesis (113)
acclimation the process of an organism's adjustment to an abiotic factor (370)
acetyl coenzyme A (acetyl CoA) the compound that reacts with oxaloacetic acid in the first step of the Krebs cycle (134)
acid any substance that increases the concentration of hydrogen (H^+) ions when added to a water solution (41)
acne an inflammatory disorder of the sebaceous glands characterized by skin lesions in the form of pimples, blackheads, and sometimes cysts (924)
acoelomate an animal with no coelom, or body cavity (686)
acquired immune deficiency syndrome (AIDS) a group of diseases resulting from a viral disruption of the immune system (968)
actin one of the two protein filaments in a muscle cell that function in contraction (79)
actinomycete a member of a class of bacteria characterized as rod-shaped monerans that form branched filaments (472)
activation energy amount of energy required for a chemical reaction to start and to continue on its own (36)
active transport the movement of a substance across a cell membrane against a concentration gradient; requires the cell to expend energy (101)
acyclovir an antiviral drug (498)
adapt in populations, to change genetically over generations to become more suited to the environment (288)
adaptation an inherited trait that increases an organism's chance of survival in a particular environment (288)
adaptive radiation an evolutionary pattern in which many species evolve from a single ancestral species (292)
addiction a dependency on a drug (1068)
adductor muscle the muscle with which a bivalve closes its shell (728)
adenine a nitrogen-containing base that is a component of a nucleotide (185)
adenosine diphosphate (ADP) a substance involved in energy metabolism formed by the breakdown of adenosine triphosphate (116)
adenosine triphosphate (ATP) a molecule present in all living cells and acting as an energy source for metabolic processes (54)
adhesion the attractive force between unlike substances (51)
adrenal gland an endocrine gland located on the top of a kidney (1034)
adrenaline epinephrine; a hormone secreted by the adrenal medulla that raises the level of glucose in the blood, increases the heartbeat, and raises blood pressure (1034)
adrenocorticotropic hormone a hormone that stimulates the adrenal cortex (1035)
aerobic respiration the process in which cells make ATP by breaking down organic compounds, with oxygen as the final electron acceptor (133)
afferent neuron a neuron that conducts impulses toward the central nervous system (1008)
aflatoxin a toxic fungal poison that causes liver cancer; found as contaminants in peanuts and corn; produced by species of *Aspergillus* (551)
afterbirth the remains of the placenta and the amnion, expelled from the mother's body following birth (1058)
agar a gel-like base for culturing microbes; extracted from the cell walls of red algae (531)
age structure the distribution of individuals among different ages in a population (382)
alcoholic fermentation the process by which pyruvic acid is converted to ethyl alcohol; the anaerobic action of yeast on sugars (129)
alcoholism the disease of addiction to ethanol (1072)
aldosterone a hormone that helps maintain water and salt balance (1035)
algae autotrophic protists (5)
allantois in amniotic eggs, the membranous sac that contains many blood vessels; in humans, a membrane surrounding the embryo that becomes the umbilical cord (842)
allele an alternative form of a gene (169)
allele frequency the percentage of an allele in a gene pool (300)
allergen a usually harmless antigen in the environment that is capable of inducing an allergic reaction (964)
allergy immune reaction to an allergen (964)
alternation of generations a sexual life

cycle in plants and algae involving two or more phases (581)
altricial referring to birds that, at hatching, are immature and in need of parental care (870)
altruistic behavior sacrifice by one individual that results in a benefit for another individual (774)
alveolus one of the tiny air sacs found in the lungs (945)
amebic dysentery a sometimes fatal disease caused by an amoeba that enters the body in contaminated food or water (513)
amebocyte in sponges, an amoebalike cell that moves through the body cells, supplying nutrients, removing wastes, and transporting sperm during sexual reproduction (694)
amino acid a carboxylic acid with an amino group; one of 20 monomers that form proteins (56)
amino acid–based hormones a group of hormones that includes proteins, peptides, amino acids, and other forms derived from amino acids (1029)
ammonification in the nitrogen cycle, the formation of ammonia compounds (423)
amniocentesis a procedure used in fetal diagnosis in which fetal cells are removed from the amniotic fluid (232)
amnion one of four membranes of an amniotic egg; surrounds and protects the embryo (842)
amniotic egg the four-membrane egg of a terrestrial vertebrate (842)
ampulla a bulblike sac at the base of the tube foot of an echinoderm that functions in movement (786)
anal pore in protozoa, an opening from which wastes are eliminated (514)
analogous structure in evolution, structures in more than one organism that have similar appearance and function, but different embryological origin (290)
anaphase a phase of mitosis and meiosis in which the chromosomes separate (150)
androgen a hormone secreted by the testes that controls secondary sex characteristics (1035)
angiosperm a flowering plant (581)
annelid a bilaterally symmetrical, segmented worm, such as an earthworm or a leech (733)
annual ring a yearly growth ring in a woody plant (611)
antenna a sensory appendage on the head of an arthropod (748)
anterior the front end of a bilaterally symmetrical organism (670)
anther the microsporangium of an angiosperm in which pollen grains are produced (166)
antheridium in plants, algae, and fungi a reproductive structure that produces gametes by mitosis (625)
anthropoid primate one of a subgroup of primates that includes monkeys and apes (320)
antibiotic a chemical that can inhibit the growth of some bacteria (472)
antibody a protein produced by B cells that binds to antigens (940)
anticodon a region of tRNA consisting of three bases complementary to the codon of mRNA (195)
antidiuretic hormone a hormone that helps regulate solute concentration in the blood (1033)
antigen a substance that stimulates an immune response (960)
anus the posterior opening of the digestive tract (809)
aorta the largest artery in the human body; carries blood from the left ventricle to systemic circulation (932)
aortic valve the semilunar valve on the left side of the heart (932)
aphotic zone the ocean layer that receives no light (431)
apical dominance a plant growth pattern in which branches near the shoot tip are shorter than branches farther from the shoot tip (649)
apical meristem the growing region at the tips of stems and roots in plants (602)
appendage any complex part or organ extending from the body (743)
appendicular skeleton in vertebrates, the bones that form the limbs (909)
aqueous solution a solution in which water is the solvent (40)
arachnid an arthropod with four pairs of walking legs; a spider, scorpion, mite, or tick (751)
arachnoid membrane the middle layer of the three sacs that surround the brain and spinal cord (1007)
archaebacterium a prokaryotic organism distinguished from other prokaryotes by the composition of the cell membranes and walls (467)
archegonium in seedless plants, a reproductive structure that produces a single egg by mitosis (625)
arteriole a branch of an artery that gives rise to capillaries (934)
artery a vessel that carries blood away from the heart to the body's organs (934)
arthropod segmented animal with jointed appendages and an exoskeleton (742)
artificial selection breeding of organisms by humans for specific phenotypic characteristics (292)
ascocarp the fruiting body of an ascomycete (548)
ascogonium a gamete-producing structure in ascomycetes (548)
ascospore one of eight haploid cells in an ascus (548)
ascus a spore sac that forms on the surface of an ascocarp (548)
asexual reproduction the production of offspring that does not involve the union of gametes (156)
asteroid-impact hypothesis the hypothesis that the extinction of the dinosaurs was caused by the impact of a huge asteroid (840)
asthma a chronic respiratory condition characterized by recurring attacks of wheezing, coughing, and labored breathing (1066)
atherosclerosis a disease characterized by the buildup of fatty materials on the interior walls of the arteries (936)
atom the simplest particle of an element that retains all the properties of that element (32)
atomic number the number of protons in an atom (32)
ATP synthase an enzyme that catalyzes the synthesis of ATP from ADP and phosphate (116)
atrioventricular node a group of nerves in the heart that functions in the heartbeat (933)
atrioventricular valve a one-way valve separating each atrium from the ventricle beneath it (931)
atrium an anterior chamber of the heart (931)
attenuated strain refers to a preparation of a virus for a vaccine in which the virus is incapable of causing disease under normal circumstances (497)
auditory canal the tube through which air enters the ear (1018)
australopithecine early hominid from the genus *Australopithecus* (325)
autoimmune disease a disorder in which the immune system attacks the organism's own cells (965)
autonomic nervous system a division of the peripheral nervous system that controls the involuntary actions that regulate the body's internal environment (1010)
autosome a chromosome that is not a sex chromosome (146)
autotomy the ability of an organism to deliberately drop a body part, most often to escape a predator (852)
autotroph an organism that uses energy to synthesize organic molecules from inorganic substances (10)
auxin a plant hormone that regulates cell elongation (647)
axial skeleton the backbone, skull, and associated bones of vertebrates (909)
axon in a neuron, an elongated extension that carries impulses away from the cell body (1003)
azidothymidine (AZT) an antiviral drug that inhibits the reverse transcriptase of retroviruses, such as HIV (498)

B

B cell a lymphocyte that produces antibodies in response to antigens (959)
bacillus a rod-shaped bacterium (470)
bacteriophage a virus that infects bacteria (491)
baleen one of the keratinous plates that function to filter food from water in some whales (886)
ball-and-socket joint allows movement in all directions, as in the shoulder (913)
barb a branch of a vane in the feather of a bird (864)
barbule a branch of a barb in the feather of a bird (864)
base any substance that increases the concentration of hydroxide (OH^-) ions when added to a water solution (42)
base unit one of seven fundamental units of SI measurement that describe length, mass, time, and other

1101

quantities (23)
base-pairing rule the rule stating that in DNA, cytosine pairs with guanine and adenine pairs with thymine and in RNA, adenine pairs with uracil (187)
basidiocarp the fruiting body of a basidiomycete (547)
basidium a specialized club-shaped reproductive structure that forms on the gills of mushrooms (547)
benign tumor an abnormal but non-threatening cell mass (211)
benthic zone the ocean bottom (431)
biennial a plant having a two-year life cycle (658)
bilateral symmetry in animals, a body plan in which the left and right sides mirror each other (670)
bile a yellowish fluid secreted by the liver that functions as a fat emulsifier in digestion (809)
binary fission an asexual cell division of prokaryotes that produces identical offspring (148)
binomial nomenclature a system of naming organisms that uses the genus name and a species identifier (339)
biochemical pathway a series of chemical reactions in which the product of one reaction is consumed in the next reaction (111)
biodiversity the number and variety of organisms in a given area during a specific period of time (448)
biogenesis the theory that living organisms come only from other living organisms (261)
biogeochemical cycle the process by which materials necessary for organisms are circulated through the environment (420)
biogeography study of the geographical distribution of fossils and living organisms (281)
biological species concept the principle that defines a species as those organisms that can produce offspring together (310)
bioluminescence the production of light by means of a chemical reaction in an organism (702)
biomass the dry weight of organic material in an ecosystem (415)
biome a geographic area characterized by specific kinds of plants and animals (424)
biosphere the area on and around Earth where life exists (363)
biotic factor a living component of an ecosystem (368)
bipedalism the ability to walk upright on two legs (319)
bipinnaria the larva of echinoderms (787)
birth rate the number of births occurring in a period of time (381)
bivalve an aquatic mollusk with a shell divided into two halves connected by a hinge, such as a clam, oyster, or scallop (728)
blade the broad, flat portion of a typical leaf (616)
blastocoel the central cavity of a blastula (683)
blastopore a depression formed when cells of the blastula move inward (344)
blastula a hollow ball of cells formed when a zygote undergoes repeated cycles of cell division (344)
blight a disease of plants characterized by quickly developing decay and discoloring of leaves, stems, and flowers (535)
blood alcohol concentration (BAC) a measurement of the amount of alcohol in a person's blood (1071)
blood pressure the force that blood exerts against the walls of a blood vessel (934)
blood type a specific characteristic of the blood of an individual; A, B, AB, or O, depending on the type of antigen present on the surface of the red blood cell (941)
bolting the rapid lengthening of internodes caused by gibberellic acid (658)
bone marrow the soft tissue in the center and ends of long bones where blood cells are produced (911)
book lung in the abdomen of an arachnid, an organ for gas exchange with parallel folds that resembles the pages of a book (752)
botany the scientific study of plants (561)
Bowman's capsule cup-shaped structure of the nephron of a kidney which encloses the glomerulus and in which filtration takes place (991)
brain stem the region of the brain that lies posterior to the cerebrum and that connects the cerebrum to the spinal cord (1007)
broad-spectrum antibiotic an antibiotic that is able to affect a wide variety of organisms (479)
bronchiole a small tube that branches from the bronchi within the lungs (945)
bronchus one of the two branches of the trachea that enter the lungs (945)
bryophyte a plant that has no vascular tissue and does not form true roots, stems, and leaves (583)
bud a structure that develops on the stem at the point of attachment of each leaf (609)
bud scale a modified leaf that forms a protective covering for a bud until it opens (609)
budding in fungi, a form of asexual reproduction in which a part of a cell pinches off to produce an offspring cell (544)
buffer chemical that neutralizes small amounts of acids or bases added to a solution (42)

C

calcitonin a hormone that stimulates removal of calcium from the blood (1036)
Calvin cycle a biochemical pathway of photosynthesis in which CO_2 is converted into carbohydrate (117)
CAM crassulacean acid metabolism; a biochemical pathway in certain plants in which CO_2 is incorporated into organic acids at night and released for fixation in the Calvin cycle during the day (119)
cancer the uncontrolled growth of cells (211)
canopy layers of treetops that shade the forest floor (429)
capillarity the reaction of a liquid surface with a solid; capillarity allows water to creep up the interior of a narrow vessel (51)
capillary the smallest vessel in the blood vessel network (810)
capsid the protein covering a virus (488)
capsule in mosses, a sporangium that produces spores; in bacteria, a protective layer of polysaccharides around the cell wall (475)
carageenan a substance found in the cell walls of red algae that is used commercially as a smoothing agent (531)
carapace a tough covering over the cephalothorax of some crustaceans; the dorsal part of a turtle's shell (748, 850)
carbohydrate an organic compound present in the cells of all living things and a major organic nutrient for humans (55)
carbon cycle process in which carbon is cycled through the biosphere (421)
carbon fixation the incorporation of carbon dioxide into organic compounds (117)
carcinogen a cancer-causing substance (212)
carcinoma a malignant tumor that grows in body tissues (211)
cardiac muscle the involuntary muscle of the heart (905)
cardiac sphincter a circular muscle located between the esophagus and the stomach (986)
cardiac stomach in an echinoderm, the stomach closer to the mouth (786)
cardiovascular system the blood, the heart, and the blood vessels (931)
carnivore a consumer that eats other consumers (416)
carotenoid a light-absorbing compound that functions as an accessory pigment in photosynthesis (113)
carrier protein a protein that transports specific substances across a biological membrane (99)
carrying capacity the number of individuals of a species that an ecosystem is capable of supporting (385)
cartilage a strong, flexible connective tissue (804)
cast a type of fossil formed when sediments fill in the cavity left by a decomposing organism (279)
catalyst a chemical that reduces the amount of activation energy needed for a reaction but is not a reactant (36)
caudal fin in fishes, a fin extending from the tail that moves from side to side and amplifies the swimming motions (808)
cause and effect a relationship between two variables in which a change in one variable leads to a change in the other (444)
cecum a sac usually found at the beginning of the large intestine (887)
cell a membrane-bound structure that is

the basic unit of life (11)
cell cycle the events of cell division; includes interphase, mitosis, and cytokinesis (149)
cell differentiation the change in morphology, physiology, or function of a cell in relation to its neighboring cells (209)
cell junction connection between cells that holds them together as a unit (667)
cell-mediated response part of an immune response involving a T-cell attack on an antigen (961)
cell membrane the lipid bilayer that forms the outer boundary of a cell (59)
cell plate a membrane that divides newly forming plant cells following mitosis (151)
cell theory the theory that all living things are made up of cells, that cells are the basic units of organisms, and that cells come only from existing cells (69)
cell wall a rigid structure that surrounds the cells of plants, fungi, many protists, and most bacteria (82, 468, 543)
cellular respiration the process in which cells make ATP by breaking down organic compounds (127)
cellular slime mold individual haploid cells that move about like amoebas; member of the phylum Acrasiomycota (534)
central nervous system the brain and the spinal cord (1003)
centriole a structure that appears during mitosis in animal cells (150)
centromere a region of the chromosome where the two sister chromatids are held together and which is the site of attachment of the chromosome to the spindle fibers during mitosis (146)
centrosome a dark body containing a centriole in animal cells but not in plant cells; spindle fibers radiate from the centrosome in preparation for mitosis (150)
cephalization concentration of nerve tissue and sensory organs at the anterior end of an organism (670)
cephalopod a free-swimming, predatory mollusk with a circle of tentacles extending from the head; an octopus, squid, cuttlefish, or chambered nautilus (730)
cephalothorax in arachnids and some crustaceans, a body part formed by the fusion of the head with the thorax (748)
cerebellum a posterior portion of the brain that controls movement and muscle coordination (1007)
cerebral cortex the folded outer layer of the cerebrum that controls motor and sensory activities (1005)
cerebral ganglion one of a pair of nerve-cell clusters that serve as a brain at the anterior end of some invertebrates (771)
cerebrospinal fluid a watery substance that provides a cushion that protects the brain and spinal cord (1007)
cerebrum the anterior portion of the brain where higher brain functions occur (1004)
character displacement evolution of anatomical differences that reduce competition between similar species (401)
chelicera a pincerlike mouthpart of some arthropods (745)
cheliped in arthropods, a claw used to capture food and for defense (748)
chemical bond a chemical attachment between atoms (33)
chemical reaction the process of breaking chemical bonds, forming new bonds, or both (33)
chemiosmosis a process in chloroplasts and mitochondria in which the movement of protons down their concentration gradient across a membrane is coupled to the synthesis of ATP (116)
chemoautotroph an organism that synthesizes organic compounds using chemicals instead of light (473)
chemosynthesis the production of carbohydrates through the use of energy from inorganic molecules instead of light (271)
chemotropism plant growth in response to a chemical (654)
chitin a carbohydrate that forms part of the arthropod exoskeleton (736)
chlorofluorocarbons (CFCs) a class of industrial chemicals found to be destroying the ozone layer in the upper atmosphere (443)
chlorophyll a class of light-absorbing pigments used in photosynthesis (113)
chloroplast a plastid containing chlorophyll; the site of photosynthesis (83)
chordate an animal that, at some stage of its life cycle, has a dorsal nerve cord, a notochord, and pharangeal pouches, such as mammals, birds, reptiles, amphibians, fish, and some marine lower forms (788)
chorion the outer membrane surrounding an embryo (842)
chorionic villi fingerlike projections of the chorion that extend into the uterine lining (1056)
chorionic villi sampling a procedure involving the analysis of the chorionic villi to diagnose fetal genotypes (232)
chromatid one of two identical parts of a chromosome (146)
chromatin the DNA and proteins in the nucleus of a nondividing cell (80)
chromosome DNA and protein in a coiled, rod-shaped form that occurs during cell division (81)
chromosome map a diagram of allele positions on a chromosome (224)
chronic bronchitis an inflammation of the bronchi and bronchioles (1069)
chrysalis the outer covering of a butterfly pupa (769)
chyme the mixture formed in the stomach from digested food particles and gastric fluid (986)
chytrid an aquatic protist characterized by gametes and zoospores with a single, posterior flagellum (536)
cilium a short, hairlike organelle that extends from a cell and functions in locomotion or in the movement of substances across the cell surface (80)
circulatory system the system that distributes oxygen and nutrients to cells in all parts of the body (675)
cirrhosis a condition in which normal liver tissues are replaced by scar tissue (1072)
citric acid a six-carbon compound formed in the Krebs cycle (134)
cladistics a system of phylogenetic classification using shared derived characters and the recency of ancestry is the sole criterion for grouping taxa (345)
clasper in fishes, a structure that transfers sperm into the female's body (806)
class in taxonomy, a group of related orders (338)
cleavage the divisions of the zygote immediately following fertilization (683)
cleavage furrow the area of the cell membrane that pinches in and eventually separates the dividing cell (151)
clitellum a noticeable swelling around the sex organs of an earthworm during reproduction (735)
cloaca in animals, such as birds, amphibians, and reptiles, a common chamber into which the digestive, reproductive, and excretory systems enter (718)
clone an offspring produced by asexual production (639)
cloning vector a carrier used to clone a gene and transfer it from one organism to another (240)
closed circulatory system a system in which blood is contained within vessels as in the human circulatory system (675)
cnidarian a phylum of animals with radially symmetrical bodies, a saclike internal cavity, tentacles, and nematocysts, such as jellyfish, hydras, sea anemones, and corals (696)
cnidocyte a stinging cell of a cnidarian (697)
coacervate a cell-like droplet formed from dissimilar substances (268)
coccus sphere-shaped bacterium (470)
cochlea a coiled tube in the inner ear that is filled with fluid and lined with hair-like cells that translates the mechanical energy from sound into nerve impulses (1018)
cocoon the outer covering of a moth pupa; a chrysalis (769)
codominance an inheritance relationship in which neither of two alleles of the same gene totally masks the other (176)
codon a group of three sequential nitrogen bases of an mRNA molecule (193)
coelom a body cavity formed within the mesoderm (685)
coenocytic referring to filaments without internal cross walls (543)
coevolution the mutual evolution of two different species interacting with each other (291)
cohesion the attraction of like molecules to each other (51)
cohesion-tension theory states that water is able to move up the stem xylem due to the strong attraction of water molecules to each other (613)
collar cell a cell lining the inside of sponges that circulates water; also called a choanocyte (693)

1103

collenchyma plant tissue made up of elongated cells with unevenly thickened, flexible walls (599)
colloblast a cell on the tentacles of a ctenophore that secretes a sticky substance that binds to the organism's prey (702)
colon the large intestine (989)
colonial algae algae that have a structure that consists of groups of cells behaving in a coordinated manner (526)
colonial organism a collection of genetically identical cells that live together in a closely connected group (87)
columella a bone that transmits sounds from the eardrum to the inner ear (829)
commensalism an ecological relationship in which one organism benefits and the other neither benefits nor is harmed (402)
community all the populations in one area (363)
compact bone the layer of bone just beneath the periosteum that gives the bone its strength (911)
companion cell a specialized plant cell that helps control the movement of sugars through sieve tubes (601)
competitive exclusion the local extinction of a species due to competition (400)
complete dominance an inheritance relationship in which one allele is completely dominant over the other; both Bb and BB have the same phenotype (176)
compound eye an eye composed of many light detectors, each with its own lens (743)
compound leaf a type of leaf in which the blade is divided into leaflets (616)
compound light microscope an instrument that magnifies small objects so they can be seen easily using two or more lenses (21)
concentration a measurement of the amount of solute dissolved in a fixed amount of solvent (40)
concentration gradient the difference in concentration of a substance across space (95)
condensation reaction a chemical reaction, also called dehydration synthesis, in which one molecule of water is produced (53)
cone a photoreceptor within the retina that can produce sharp images and distinguish colors; in gymnosperms, a seed-bearing structure (588, 1020)
conformer an organism that does not regulate its internal environment (370)
conidium a spore produced during asexual reproduction in ascomycetes (544)
conjugation the union of two protists to exchange genetic material (477)
conjugation bridge in certain algae and fungi, a passageway for the transfer of the genetic information from one organism to another (477)
connective tissue body tissue that supports and holds body structures together (906)
conservation biology a branch of biology concerned with preserving biodiversity in natural areas (452)
constrictor snake that kills its prey by constriction and suffocation (853)
consumer a heterotroph that obtains energy from organic molecules made by other organisms (416)
contour feather a feather that provides coloration, insulation, and a streamlined shape to adult birds (864)
contractile vacuole an organelle in protists that expels water (97)
control group in an experiment, a group or individual that serves as a standard of comparison with another group or individual to which it is identical except for one factor (17)
controlled experiment a test of variables using a comparison of a control group with an experimental group (17)
conus arteriosus in fishes and frogs, the final chamber of the heart (810)
convection cell a regional pattern of rising and falling air (441)
convergent evolution the process by which unrelated species become more similar as they adapt to the same kind of environment (292)
coral reef a rocklike formation in warm, shallow seas composed of the skeletons of corals (432)
cork cambium the lateral meristem of a plant; produces cork (602)
cornea protective layer that covers the iris and pupil of the eye (1020)
coronary circulation a subsystem of systemic circulation that supplies blood to the heart (936)
corpus callosum the structure formed by nerve fibers connecting the cerebral hemispheres of the brain (1004)
corpus luteum the structure that forms from the ruptured follicle in the ovary at ovulation (1053)
correlation a relationship between two variables in which both variables change together (444)
cortex in plants, a mature ground tissue located just inside the epidermis; in animals, the outermost portion of an organ such as the kidney (605, 991)
cortisol a hormone that regulates certain phases of carbohydrate and protein metabolism (1035)
cotyledon a seed leaf in a plant embryo (592)
countercurrent flow in gills of fishes, an arrangement whereby water flows away from the head and blood flows toward the head (811)
covalent bond a bond that forms when two atoms share one or more pairs of electrons (33)
coxal gland in some arachnids, an organ that removes wastes and discharges them at the base of the legs (752)
C_3 plant a plant that fixes carbon exclusively through the Calvin cycle, named for the three-carbon compound that is initially formed (119)
C_4 plant a plant that incorporates CO_2 into four-carbon compounds (119)
cranial cavity the area in which the brain rests (908)
cranial nerve one of the 12 nerves of the head that relay information between the brain and the muscular and sensory structures in the head (811)
cretinism a form of mental retardation caused by hypothyroidism (1034)
crista a fold of the inner membrane of mitochondria (77)
crop in earthworms, a structure that stores soil; in birds, a structure that stores food (734, 766, 866)
crop milk a nutritious milklike fluid secreted by the crop of pigeons and doves to feed their young (873)
cross-pollination a reproductive process in which pollen from one plant is transferred to the stigma of another plant (166)
crossing-over the exchange of genes by reciprocal segments of homologous chromosomes during meiosis (153)
crustacean an arthropod with mandibles and branched appendages; examples include shrimps, barnacles, crabs, and pill bugs (746)
crustose a type of lichen that grows as a layer on rocks and trees (549)
cultivar a combination of the words *cultivated* and *variety* for a food plant that has a specific distinguishing characteristic, such as Thompson Seedless grapes (561)
cutaneous respiration in some animals, the exchange of gases through the skin (827)
cuticle in plants and insects, a waxy covering that prevents desiccation; in parasitic roundworms, a noncellular layer that protects the worm from the host's digestive and immune systems (600, 715)
cyanobacteria photosynthetic unicellular prokaryotes (271)
cyst a resistant, thick-walled structure that encloses and protects a dormant organism (714)
cytokine a chemical signal between cells of the immune system (957)
cytokinesis the division of the cytoplasm of one cell into two new cells (149)
cytokinins plant hormones that promote cell division (651)
cytolysis the bursting of a cell (99)
cytoplasm the region of a cell between the cell membrane and the nucleus (75)
cytoplasmic streaming the circular motion of the cytoplasm (512)
cytosine a nitrogen-containing base; a pyrimidine of DNA and RNA (185)
cytoskeleton a network of long protein strands in the cytosol that helps maintain the shape and size of a eukaryotic cell (79)
cytosol the gelatinlike aqueous fluid that bathes the organelles on the inside of the cell membrane (75)
cytotoxic T cell a type of T cell that destroys infected body cells (961)

D

day-neutral plant a plant unaffected by day length (656)
death rate or **mortality rate** the number

of deaths occurring in a period of time (381)

debt-for-nature swap a conservation strategy in which a developing country takes steps to preserve its biodiversity in exchange for a reduction in its foreign debt (450)

decapod a crustacean with five pairs of legs, such as shrimps, crayfish, and lobsters (747)

deciduous referring to trees that lose their leaves at the end of the growing season (589)

decomposer an organism that obtains nutrients from dead plants and animals (416)

dehydration the condition resulting from excessive water loss (982)

deletion a mutation in which a segment of DNA breaks off of a chromosome (225)

denitrification a final step in the nitrogen cycle, during which nitrogen gas is returned to the atmosphere (423)

density-dependent factor a variable related to the density of a population that affects population size (386)

density-independent factor a variable that affects population size regardless of population density (386)

dental formula the number, kind, and placement of teeth that are specific for a particular kind of organism (320)

deoxyribonucleic acid (DNA) a double-helix-shaped nucleic acid (7)

dependent variable the responding variable in an experiment (18)

depressant a drug that decreases the activity of the central nervous system (1070)

derived character a feature that evolved only within the group under consideration, such as the feathers of birds (345)

dermal tissue system one type of plant tissue system, which forms the outer covering of plants (600)

desert an area where rainfall averages less than 25 cm per year (416)

determinate cleavage a process in which each cell develops into a specific part of the gastrula (684)

detritivore a consumer that feeds on dead plants and animals (416)

deuterostome an organism in which the blastopore develops into the anus and the coelom arises by enterocoely and whose embryo has indeterminate cleavage (684)

development the process by which an organism grows (12)

diabetes mellitus a disorder of glucose metabolism caused by insufficient insulin (60)

diaphragm a sheet of muscle below the rib cage that functions in inspiration and expiration (886)

diastole a phase in the heartbeat during which the ventricles relax and the blood flows in from the atria (933)

diastolic pressure the pressure maintained in the blood when the ventricles are relaxed (934)

diatom an abundant component of phytoplankton; member of phylum Bacillariophyta (532)

dicot an angiosperm with two cotyledons, net venation, and flower parts in fours or fives (580)

diencephalon in the brain, an area below the cerebral hemisphere that contains the thalamus and the hypothalamus (1006)

differentiation a process in which the cells of a multicellular individual become specialized during development (209)

diffusion the process by which molecules move from an area of greater concentration to an area of lesser concentration (95)

digestion breaking down food into usable molecules (983)

dihybrid cross a cross between individuals that involves the pairing of contrasting traits (177)

dimorphism the existence of distinct, genetically determined forms of the same species, such as male and female forms (544)

dinosaur one of a great variety of terrestrial reptilian forms during the Mesozoic era 245 million to 65 million years ago (839)

dipeptide a formation from two amino acids bonded together by means of a condensation reaction (57)

diploid a cell that contains both chromosomes of a homologous pair (147)

disaccharide a double sugar formed from two monosaccharides (56)

dispersion the spatial distribution of individuals in a population (380)

disruptive selection a type of natural selection in which individuals with two extreme forms of a trait have an advantage (307)

dissociation the separating of a molecule into simpler molecules, atoms, radicals, or ions (41)

divergent evolution the process of two or more related species becoming more and more dissimilar (292)

division in taxonomy, a grouping of similar classes of plants (338)

DNA hereditary information in the form of a large molecule called deoxyribonucleic acid (7)

DNA fingerprint a pattern of bands made up of specific fragments from an individual's DNA (243)

DNA polymerase an enzyme that binds to the separated strands of DNA and assembles each strand's complement in replication (188)

domain in the three-domain system of classification, one of three broad groups that all living things fall into on the basis of rRNA analysis (350)

dominant referring to an allele that masks the presence of another allele for the same characteristic (168)

donor gene a specific gene that is isolated from another organism and spliced into a plasmid (240)

dormancy a state of decreased metabolism (370)

dorsal the top of a bilaterally symmetrical animal (670)

dorsal nerve cord a neural tube dorsal to the notochord (673)

double fertilization in plants, the process in which two types of cell fusion take place in the embryo sac (633)

down feather one of the soft, fluffy feathers that cover the body of nesting birds and help insulate adult birds (864)

Down syndrome a disorder caused by an extra twenty-first chromosome and characterized by a number of physical and mental abnormalities (231)

drone a male bee that develops from an unfertilized egg and functions only to fertilize eggs from the queen (772)

drug any chemical taken into the body that alters the normal processes of either the mind or the body (1065)

Duchenne muscular dystrophy a form of muscular dystrophy that weakens and progressively destroys muscle tissue (229)

duodenum the first section of the small intestine (827)

dura mater the outermost of the three sacs that enclose the brain and spinal cord (1007)

E

echinoderm a radially symmetrical marine invertebrate with an endoskeleton, a water-vascular system, and tube feet, such as a sea star, a sea urchin, or a sea cucumber (781)

echolocation a method of navigation similar to sonar in which the sound produced bounces off an object used by bats and cetaceans (887)

ecology the study of the relationship between organisms and their environment (359)

ecosystem all the biotic and abiotic components of an environment (363)

ecotourism a form of tourism that helps local people make money from an intact ecosystem; tourists pay for nature guides, food, and lodging to tour the ecosystem (451)

ectoderm the outermost of the three germ layers of the gastrula that develops into the epidermis and epidermal tissues, the nervous system, external sense organs, and the mucous membranes lining the mouth and anus (683)

ectoparasite a parasite that lives on a host but does not enter the host's body (399)

ectoplasm the region in the cytoplasm located directly inside the cell membranes (512)

ectothermic referring to an animal whose body temperature is determined by the environment (848)

efferent neuron a neuron that conducts impulses away from the central nervous system (1008)

ejaculation contractions of the smooth muscles surrounding the urethra by which semen is forcefully expelled (1049)

elapid a venomous snake with two small, fixed fangs in the front of the mouth (853)

El Niño a disruption of the normal air and water circulation patterns in the Pacific Ocean, leading to unusual weather in

1105

many parts of the world (442)

electron a particle with a negative electric charge that orbits the nucleus of an atom (32)

electron microscope an instrument that uses a beam of electrons rather than a beam of light to enlarge the image of an extremely small object so that it can be seen (22)

electron transport chain molecules in the thylakoid membrane or inner mitochondrial membrane that uses some of the energy in electrons to pump protons across the membrane (115)

element a substance that ordinarily cannot be broken down chemically to form simpler kinds of matter (31)

elephantiasis a condition of swollen lymphatic vessels in the limbs caused by a parasitic filarial worm (718)

embryo sac in plants, a megagametophyte containing seven cells and eight nuclei (630)

emigration the movement of individuals out of a population (304)

emphysema a degenerative lung disease (1069)

endocrine gland a ductless gland that secretes hormones into the blood (1029)

endocrine system a system of glands that transmit chemical messages throughout the body (1029)

endocytosis the process by which a cell surrounds and engulfs substances (103)

endoderm the innermost of the three germ layers of the gastrula and develops into the epithelium of the pharynx, respiratory tract, digestive tract, bladder, and urethra (683)

endodermis in plants, a specialized layer of cells that regulates substances entering the center of the root (605)

endoparasite a parasite that lives inside the host's body (399)

endoplasm the region in the cytoplasm found in the interior of the cell (512)

endoplasmic reticulum a system of membranous tubules and sacs in eukaryotic cells that functions as a path along which molecules move from one part of the cell to another (77)

endoskeleton an internal skeleton (909)

endosperm a tissue in angiosperms that provides food for the developing embryo (579)

endospore a dormant bacterial cell enclosed by a tough coating (475)

endosymbiosis a mutually beneficial relationship between one organism and another that lives within it; the hypothesis of endosymbiosis relating to the evolution of eukaryotes that holds that chloroplasts and mitochondria evolved from endosymbiotic bacteria (132)

endotherm an animal that generates its own body heat through metabolism (848)

endotoxin a substance that causes a poison reaction; a compound that makes up part of the cell wall of Gram-negative bacteria (479)

enhancer a region adjacent to a eukaryotic gene that must be activated if the gene is to be expressed (208)

enkephalin a neurotransmitter that blocks pain messages to the brain (1076)

enteric bacteria Gram-negative heterotrophic bacteria that inhabit animal intestinal tracts (473)

enterocoely the process of mesoderm formation in deuterostomes in which the coelom forms in folded mesoderm (685)

entomology the study of insects (762)

environmental science a field of study that uses biological principles to look at the relationships between humans and the Earth (440)

enzyme a catalyst, usually a protein, in living systems (37)

ephedrine a decongestant drug derived from *Ephedra,* a genus of desert shrubs (590)

epidermis a layer of cells that forms a continuous sheet over the outer surface of a plant or animal (696)

epididymis coiled tube on each testis where sperm complete their development (1048)

epiglottis a flap of tissue that covers the trachea (945)

epiphyseal plate the site of bone elongation at the ends of long bones (912)

epiphyte a plant that grows on other plants (429)

epithelial tissue tissue composed of cells that create a solid sheet that covers a body part (906)

equilibrium a state that exists when the concentration of a substance is the same throughout a space (95)

erosion the loss of soil that is caused by wind or water (822)

erythrocyte a red blood cell (939)

esophagus a tube connecting the mouth or pharynx to the stomach or crop (734)

estrogen hormone that helps regulate the menstrual cycle (1035)

estuary an aquatic biome found where freshwater streams and rivers flow into the sea; where the tides meet a river current (433)

ethephon a synthetic chemical that breaks down to release ethylene gas; used to ripen fruit (650)

ethylene a gaseous hormone produced by various parts of plants and released into the air (650)

eubacteria the lineage of prokaryotes that includes all contemporary bacteria except archaebacteria (348)

euchromatin the uncoiled form of chromatin (206)

euglenoid flagellated unicellular algae ; many are photosynthetic (533)

eukaryote a cell that contains a nucleus and membrane-bound organelles (72)

eustachian tube a canal that connects the middle ear to the mouth cavity (1018)

eutrophication the increase of nutrients in an environment (472)

evenness the number of individual organisms that belong to each species in a given area (449)

evolution all the changes that have formed life on Earth from its earliest beginnings to the diversity that characterizes it today (8)

excretion the process of removing metabolic wastes (991)

excurrent siphon a tube through which water exits the mantle cavity of a bivalve (729)

exergonic reaction a chemical reaction that involves a net release of free energy (36)

exocrine gland a gland that secretes nonhormonal chemicals through a duct (1029)

exocytosis a process in which a vesicle inside a cell fuses with the cell membrane and releases its contents to the external environment (104)

exon coding region of a eukaryotic gene (206)

exoskeleton the hard external covering of some invertebrates that provides protection and support (675)

exotoxin a toxic protein secreted by pathogenic bacteria (478)

expiration the process in which air is forced out of the lungs (948)

exponential growth a model of population growth in which the birth and death rates are constant (384)

extensor a muscle that straightens a joint (919)

external fertilization the union of gametes outside the bodies of the parents, as in many fishes and amphibians (804)

external respiration the exchange of gases between the atmosphere and the blood (948)

extinction the dying out of a species (281)

extreme halophile an archaebacterium that lives in very high salt concentrations (468)

eyespot a localized region of pigment in some invertebrates and protozoa that detects changes in the quantity and quality of light (508)

F

facilitated diffusion a process in which substances move down their concentration gradient across the cell membrane with the assistance of carrier proteins (99)

facultative anaerobe an organism that can metabolize with or without oxygen (476)

FAD flavin adenine dinucleotide; a compound that accepts electrons during redox reactions (134)

fallopian tube a tube through which eggs move from the ovary to the uterus (1051)

family in taxonomy, a group of related genera (338)

fascicle a dense bundle of skeletal muscle fibers (916)

fatty acid a monomer that is part of most lipids (58)

feces waste materials that pass through the anus or cloacal opening (989)

feedback mechanism (positive or negative) a series of events in which the last step controls the first (1038)

fermentation a process in which cells make a limited amount of ATP by converting glucose into another organic compound, such as lactic acid or ethyl alcohol, in the absence of oxygen (129)
fertilizer a compound that provides plants with essential mineral nutrients like nitrogen and phosphorus (564)
fetal alcohol syndrome a condition characterized by growth and developmental defects affecting some children born to women who drink alcohol during pregnancy (1072)
fetus referring to a developing human from the end of the eighth week until birth (1057)
F_1 generation the offspring of cross-pollinated P_1 generation plants (166)
F_2 generation the offspring of self-pollinated F_1 generation plants (166)
fibrin the long, sticky threads that function in blood clotting (941)
fibrous root system a mat of threadlike roots that spread out below the surface of the soil to maximize exposure to water and minerals and to anchor the plant firmly to the soil (603)
filament in flowers, the structure that supports the anther (629)
filamentous alga an alga that has a slender, rod-shaped thallus composed of rows of cells joined end to end (526)
filarial worm one of a variety of parasitic roundworms that cause elephantiasis in humans and heartworm disease in dogs (717)
filter feeding the method by which a sessile organism screens food out of the surrounding water (694)
filtration the process in the kidney in which materials from the blood are forced out of the glomerulus into Bowman's capsule (993)
first messenger a hormone that activates a second messenger inside its target cell (1030)
fission a type of asexual reproduction in which the organism splits in two and regenerates the parts missing on each new half to produce two whole organisms (711)
fitness a measurement of the ability of a species to respond to the pressures of natural selection; the ability of individuals to survive to propagate their genes (288)
fixed joint a place where two bones meet but movement does not occur, as in the skull (913)
flagellum a hairlike structure made up of microtubules that function in locomotion (80)
flame cell in a flatworm, a cell that collects excess body water and transfers it to the excretory system(710)
flatworm the common name for any of the dorso ventrally flattened worms of the phylum Platyhelminthes, such as a planarian, fluke, and a tapeworm (709)
flavin adenine dinucleotide see FAD
flexor a muscle that bends a joint (919)
flight feather a specialized contour feather on the wing or tail of a bird (864)
fluid mosaic model a model of cell membrane structure representing the dynamic nature of membrane lipids and proteins (75)
fluke a leaf-shaped parasitic flatworm (711)
flyway the route followed by migratory birds (453)
foliose lichen a type of lichen that lives on soil surfaces, where it forms a mat-like growth (549)
follicle a pit in the skin in which hair or a feather develops; a structure surrounding an immature egg in an ovary (864)
follicle stimulating hormone (FSH) a gonadotropin that stimulates sperm production and causes cells around an immature egg to form a follicle (1035)
follicular phase the stage in which an immature egg completes its first meiotic division (1052)
food chain pathway beginning with producers along which energy is transferred from trophic level to trophic level (417)
food web the interconnected food chains in an ecosystem (417)
foraminifera an ancient group of shelled sarcodines found in oceans (513)
fossil the remains or traces of a once-living organism (279)
fracture a crack or break in a bone (911)
fragmentation in fungi, a form of asexual reproduction in which individual cells are released that act like spores (544)
frame shift mutation a mutation that results in the misreading of the code during translation because of a change in the reading frame (226)
free energy the energy in a system available for work (35)
frond the mature leaf of a fern (588)
fruit a mature plant ovary; a modified flower part that encloses a seed or seeds (563)
fruiting body spore-bearing structure of a slime mold during reproduction (534)
fruticose a shrublike lichen (549)
fucoxanthin a carotenoid pigment that is one of the primary pigments found in diatoms and brown algae (531)
functional group a structural building block that determines the characteristics of a chemical compound (53)
fundamental niche the full range of conditions that a species can tolerate and resources it can use (371)
fungi imperfecti or **deuteromycota** a group of fungi whose sexual phase has not been discovered (549)
furculum the fused collarbones in a bird's skeleton; the "wishbone" (862)

G

gallbladder the saclike organ in vertebrates that is connected to the liver that stores bile (809)
gametangium in algae and fungi, a gamete chamber (525)
gamete a reproductive cell (153)
gametocyte an undifferentiated cell that develops into a gamete (517)
gametophyte in plants and algae, a multicellular structure that produces gametes (529)
ganglion a mass of nerve cells (726)
gasohol an alternative fuel made from alcohol from fermented grains mixed with gasoline (567)
gastric cecum a pocket that branches from the digestive tract and secretes enzymes (766)
gastric fluid a liquid secreted by gastric glands in the stomach (986)
gastric pit the open end of gastric glands through which secretions are released into the stomach (985)
gastrin a hormone that stimulates the stomach to secrete digestive enzymes (1037)
gastrodermis in cnidarians, the layer of cells surrounding the gastrovascular cavity (696)
gastrointestinal tract the digestive tract from the mouth to the anus (983)
gastropod a mollusk that has one or no shell and moves by gliding on a muscular foot; examples include snails, slugs, and nudibranchs (727)
gastrovascular cavity a digestive chamber with a single opening found in cnidarians, ctenophores, and flatworms (696)
gastrula the cup-shaped embryo formed as the blastula folds inward (683)
gel electrophoresis technique used to separate nucleic acids or proteins by size and charge (244)
gemmule an asexual reproductive structure produced by some freshwater sponges (695)
gene a segment of DNA that contains coding for a polypeptide or protein; a unit of hereditary information (7)
gene clone an exact copy of a gene (240)
gene expression the transcription of DNA into RNA and then into proteins (202)
gene flow the movement of genes into or out of a population (305)
gene pool all the genes for all of the traits in a population (300)
gene therapy treatment of a genetic disorder by introducing a healthy gene into a cell (246)
generalist a species with a broad niche that can tolerate a wide range of conditions and can use a variety of resources (372)
generative cell in a pollen grain, the cell that forms two sperm (632)
genetic code triplets of nucleotides in mRNA that determine the sequence of amino acids in protein (193)
genetic counseling the process of informing a couple about their genetic makeup, which has the potential to affect their offspring (231)
genetic disorder a disease that has a genetic basis (228)
genetic diversity the amount of genetic variation within a population or species (449)
genetic drift a shift in allele frequencies in a population due to chance (305)
genetic engineering a form of applied genetics in which scientists directly manipulate genes (238)
genetic marker a short section of DNA

1107

that indicates the presence of an allele that codes for a trait (228)
genetic screening an examination of a person's genetic makeup (231)
genetics the field of biology devoted to understanding how traits are passed from parents to offspring (164)
genome the complete genetic material contained in an individual (203)
genomic library a set of thousands of DNA pieces cut by a specific restriction enzyme (241)
genotype the genetic makeup of an organism (172)
genotypic ratio the probable ratio of genotypes produced by a cross (175)
genus in taxonomy, a group of similar species (338)
geographic isolation the physical separation of populations (310)
germ cell mutation a change in the DNA of a sex cell (224)
germ layer a specific layer of cells in an embryo from which specific organ systems are derived (671)
germinate in plants, sprout; the seed embryo begins to grow (588)
gestation the length of a normal pregnancy (1054)
giardiasis an illness characterized by severe diarrhea and intestinal cramps caused by a protozoan parasite (507)
gibberellin a plant hormone that, among other actions, stimulates bolting and cessation of dormancy (650)
gill in mushrooms, one of the radiating rows on the underside of the cap; in animals, an organ specialized for the exchange of gases with water (802)
gizzard a muscular region of the digestive tract in birds that crushes food (734, 866)
gliding joint a joint that allows bones to slide over one another, as in foot bones that flex in walking (913)
glomerulus a cluster of capillaries that receives blood from the renal artery and that is enclosed in Bowman's capsule (993)
glucagon a hormone that raises the blood glucose level (1036)
glycolysis a pathway in which glucose is oxidized to pyruvic acid (127)
glycoprotein a molecule of protein with attached sugar chains (489)
Golgi apparatus a system of membranes in eukaryotic cells that modifies proteins for export by the cell (78)
gonad the organ that produces gametes and some sex hormones (1035)
G_0 phase a phase of the cell cycle in which the cell is not dividing and its DNA is not replicating (149)
G_1 phase the first period of interphase, in which the cell doubles in size (149)
G_2 phase the final period of interphase, in which the cell undergoes rapid growth and prepares mitosis (149)
grafting a technique in which a portion of one plant is inserted into and grows with the root or shoot of another (640)
Gram-negative bacterium a bacterium that stains pink with Gram stain and that usually has an outer covering on its cell wall (471)
Gram-positive bacterium a bacterium that stains purple with Gram stain and that usually lacks an outer covering on its cell wall (471)
Gram stain a series of dyes that stain bacteria either purple or pink according to the chemistry of the bacterial cell wall (471)
granum a stack of thylakoids in a chloroplast (112)
gravitropism a plant growth response to gravity (654)
great ape a gibbon, orangutan, gorilla, or a chimpanzee (320)
green gland an excretory organ of some crustaceans that helps regulate the solute concentration of the hemolymph (750)
greenhouse effect warming of the Earth due to the insulating effect of gases in the atmosphere, such as carbon dioxide and water vapor (361)
gross primary productivity the rate at which energy is produced in an ecosystem (415)
ground water water found in the soil or in underground formations of porous rock (421)
ground-tissue system one type of plant-tissue system, made up of parenchyma, collenchyma, and sclerenchyma (601)
growth factor protein that regulates the rate and sequence of the events of cell division (212)
growth hormone a hormone secreted by the anterior lobe of the pituitary gland that regulates skeletal and muscular growth (1033)
growth rate the amount by which a population's size changes in a given time (383)
growth regulator a hormone that stimulates or inhibits plant growth (647)
growth retardant a chemical that retards plant growth (651)
guanine a nitrogen-containing base that is a component of a nucleotide (185)
guard cell one of two cells bordering a stoma, each of which has flexible cell walls that regulate gas and water exchange (618)
gullet the structure that forms food vacuoles that circulate throughout the cytoplasm of the *Paramecium* (514)
gut digestive tract (676)
gymnosperm a seed plant whose seeds are not enclosed by an ovary (581)

H

habitat the physical area in which an organism lives (368)
half-life the period of time in which half of a radioactive substance decays (265)
haploid having only one chromosome of each homologous pair (147)
hard palate the bony roof of the mouth (984)
Hardy-Weinberg principle of genetic equilibrium the stability of gene frequencies across generations (302)
Haversian canal a channel containing nerves and blood vessels making up compact bone (911)
hay fever a pollen allergy that results in sneezing, a running nose, and watering eyes (572)
head-foot one of the two regions of the mollusk body consisting of the head and the foot (726)
heartwood the darker wood in the center of a tree (611)
heartworm disease a disease of the heart and large arteries in dogs caused by a filarial worm (718)
helicase an enzyme that separates DNA strands before replication (188)
helper T cell a type of T cell that coordinates the immune response (961)
hemocoel the hemolymph-filled space or body cavity some invertebrates (727)
hemoglobin the oxygen-carrying pigment in red blood cells (939)
hemolymph the fluid that is circulated through the body of an animal with an open circulatory system (766)
hemophilia a trait in which the blood lacks a protein that is essential for clotting (229)
hepatic portal circulation the system of blood movement through the digestive tract and liver (936)
herb a flowering, vascular seed plant that lacks a woody stem; a plant part used to add flavor to food (563)
herbaceous plant an annual plant that is nonwoody (580)
herbivore a consumer that eats primary producers (398)
heredity the transmission of traits from parents to their offspring (165)
heterocyst a cell of cyanophytes that fixes atmospheric nitrogen (472)
heterospory a phenomenon in which two types of spores are produced by an individual plant (627)
heterotroph an organism that obtains organic food molecules by eating other organisms or their by-products (10)
heterozygous referring to a gene pair in which the two alleles do not code for the same trait (173)
hilum a scar along the edge of a plant seed marking where the seed was attached to the ovary wall (636)
hindgut the colon and rectum of some invertebrates (766)
hinge joint allows movement back and forth like a hinged door, as in the elbow (913)
histamine a kind of cytokine that causes an inflammatory response to an injury resulting in redness, swelling, warmth, and pain (957)
histone a protein molecule that DNA wraps around during chromosome formation (145)
holdfast the part of an algal thallus that anchors it to a substrate (529)
homeobox a specific DNA sequence within a homeotic gene that regulates patterns of development (209)
homeostasis the stable internal conditions of a living thing (6)
homeotic gene a gene that controls the development of a specific adult structure (209)
hominid a subgroup of primates that

includes modern humans and their bipedal ancestors (319)
homologous chromosome one of a pair of morphologically similar chromosomes (146)
homologous features similar features that originated in a shared ancestor (289)
homospory a phenomenon in which all spores look alike and produce similar gametophytes (626)
homozygous referring to a gene pair in which the two alleles code for the same trait (173)
hookworm a parasitic roundworm that feeds on its host's blood by cutting into the intestinal wall (716)
hormone a chemical secreted by an endocrine gland that specifically influences the activity of distant cells (1029)
host an organism that supports a parasite (399)
human chorionic gonadotropin (HCG) a hormone secreted by the placenta that stimulates the corpus luteum (1056)
Human Genome Project worldwide scientific collaboration to determine the nucleotide sequence of the entire human genome (246)
human immunodeficiency virus (HIV) the virus that causes AIDS (488)
humoral immune response an arm of the immune response that involves mainly B cells and antibodies (961)
hunter-gatherer lifestyle a nomadic way of life in which food is found by hunting animals and gathering uncultivated plants, fruits, and shellfish (388)
Huntington's disease a human genetic disorder caused by a dominant allele resulting in involuntary movements, mental deterioration, and eventual death (228)
hydrogen bond a weak chemical bond between the hydrogen atom in one molecule and a negatively-charged region of another molecule (50)
hydrolysis the splitting of a molecule through reaction with water (54)
hydrophilic referring to the molecular attraction to water (58)
hydrophobic referring to the molecular repulsion of water (58)
hypertension high blood pressure (934)
hyperthyroidism overproduction of thyroid hormones (1034)
hypertonic describing a solution whose solute concentration is higher than that inside a cell (96)
hypha the vegetative filament of a fungus (543)
hypocotyl a stemlike area in plants between the cotyledons and radicle (636)
hypoglycemia overproduction of insulin (1036)
hypothesis a statement that can be tested experimentally (17)
hypothyroidism thyroid hormone deficiency (1034)
hypotonic describing a solution whose solute concentration is lower than that inside a cell (96)

I

icosahedron a polyhedron or a capsid of a virus with 20 triangular faces (489)
ileum the middle portion of the small intestine (827)
immigration the movement of individuals into a population (366)
immune response a reaction of the body against a foreign substance (960)
immune system a body system that has the job of fighting off invading pathogens and preventing the growth and spread of cancers (959)
immunity resistance to a specific pathogen (964)
implantation the process by which the newly fertilized egg in the blastocyst stage imbeds itself into the lining of the uterus (1055)
inactivated vaccine a vaccine that has been treated so that its component microorganisms no longer have the ability to cause disease (497)
inbreeding mating with relatives (387)
incisor a chisel-like tooth used for biting and cutting (984)
incomplete dominance an inheritance relationship that occurs when both alleles influence the phenotype (176)
incurrent siphon a tube through which water enters the mantle cavity of a bivalve (729)
independent assortment during meiosis, the random distribution of genes from different chromosomes to the gametes (154)
independent variable an experimentally manipulated variable (18)
indeterminate cleavage a pattern of development in which the functional destiny of each cell is not determined early in the development of the embryo (685)
indoleacetic acid (IAA) an auxin produced in actively growing regions of plants (647)
inducer in the *lac* operon, the molecule that binds to repressor molecules and induces gene expression (205)
infectious disease a disease caused by a pathogen that can be transmitted from one individual to another (496)
inferior vena cava the large vein carrying blood from the lower part of the body (935)
inflammatory response a response in which white blood cells engulf foreign substances and body temperature rises (957)
insertion the attachment point of a tendon to a moving bone (919)
inspiration the process of taking air into the lungs (947)
insulin a hormone that lowers the blood glucose level (1036)
integral protein a protein imbedded in the bilayer of the cell membrane (74)
integument in plants, one or two outer layers of the ovule; in animals, the outer covering (628)
intercalary meristem a region of plant tissue that allow grass leaves to quickly regrow (958)
interferon a protein produced by the body that inhibits the reproduction of viruses (958)
intermediate host the host from which the larvae of a parasite derive their nourishment (713)
internal fertilization fertilization in which sperm fuses with an egg inside the female's body (806)
internal respiration the exchange of gases between the blood and the cells of the body (944)
interneuron a neuron that connects with another neuron (1008)
internode one of the different segments of a stem (609)
interphase a period of cell growth and development that precedes eukaryotic cell nuclear division (149)
intertidal zone an area along ocean shorelines that is repeatedly covered and uncovered by ocean tides (431)
intestine the part of the body where digestion is completed and nutrients are absorbed (809)
intron a section of a structural gene that does not code for an amino acid and is excised before translation (206)
inversion a mutation that occurs when a chromosome piece breaks off and reattaches in reverse orientation (225)
invertebrate an animal without a backbone (667)
involuntary muscle a muscle whose movement cannot be consciously controlled (916)
ion an atom or a compound with a net electrical charge (34)
ion channel a membrane protein that provides a passageway across the cell membrane through which an ion can diffuse (100)
ionic bond a bond formed by electrical attraction between two oppositely charged ions (34)
islets of Langerhans specialized cells in the pancreas that secrete two hormones, insulin and glucagon (1036)
isomer one of two or more compounds that differ in structure but not in molecular composition (55)
isopod a crustacean with seven pairs of identical legs; examples include sow bugs and pill bugs (745)
isotonic describing a solution whose solute concentration equals that inside a cell (97)
isotope one of multiple forms of a chemical element that has a different number of neutrons than other atoms of that element (264)

J

Jacobson's organ one of two pits in the mouth of a snake that contain nerves which are extremely sensitive to odors (847)
jejunum the middle section of the small intestine (988)
joint the place where two bones meet (913)

K

karyotype a picture of an individual's chromosomes (147)
keratin a protein that forms hair, bird feathers, human fingernails, and the horny scales of reptiles (843)
kidney an excretory organ in which wastes are filtered from the blood (678)
kilocalorie a unit of energy equal to 1,000 calories; the amount of heat needed to raise the temperature of 1 kg of water by 1° C (130)
kin selection a mechanism for increasing the propagation of one's own genes by helping a closely related individual reproduce (774)
kinetochore a disk-shaped protein found in the centromere region of a chromosome and attaches the chromosomes to the mitotic spindle (150)
kinetochore fiber a spindle fiber that extends from the centrosome at one pole of the cell to a chromatid during mitosis (150)
kingdom in taxonomy, a group of similar phyla or divisions (338)
Koch's postulates the steps that Robert Koch, a Nobel Prize–winner in 1905 for his work on tuberculosis, formulated for determining the cause of a given disease (955)
Krebs cycle a series of biochemical reactions that release carbon dioxide and result in the formation of ATP (133)

L

labium an insect mouthpart that functions like a lower lip (766)
labor the muscular contractions of the uterus and other events that lead to birth (1058)
labrum an insect mouthpart that functions like an upper lip (766)
***lac* operon** DNA segment that codes for the regulatory enzymes for lactose metabolism in *E. coli* (203)
lacteal a tiny lymph vessel (989)
lactic acid fermentation the process by which pyruvic acid is converted to lactic acid (129)
laminarin a type of polysaccharide with glucose units linked differently from those in starch (531)
large intestine an organ of digestion in which water is reabsorbed (989)
larva an immature form of an organism that is morphologically distinct from that of an adult (695)
larynx the voice box that houses the vocal cords (945)
lateral line system a row of sensory structures that run the length of the body of a fish (803)
lateral meristem in plants, a growing region that increases the diameter of roots and stems (602)
law of independent assortment law stating that pairs of genes separate independently of one another in meiosis (168)
law of segregation law stating that pairs of genes separate in meiosis and each gamete receives one gene of a pair (168)
law of superposition a law that states that successive layers of rock or soil were deposited on top of one another and the lowest level is oldest (280)
layering the process of causing roots to form on a stem (640)
leaflet one of the divisions of a blade (616)
legume any plant of the family Leguminosae, such as a bean, a pea, or a lentil (412)
lens a convex crystalline structure in the eye that focuses light on the retina (1020)
leukemia a progressive, malignant disease of the blood-forming organs (211)
leukocyte a white blood cell (940)
lichen a symbiotic association between fungi and green algae (549)
life expectancy how long, on average, an individual is expected to live (381)
ligament a connective tissue in a joint (913)
light reactions the initial reactions in photosynthesis, including the absorption of light by photosystems I and II, the passage of electrons along the electron transport chains, the production of NADPH and O_2, and the synthesis of ATP through chemiosmosis (112)
limbic system a group of clusters of neurons in the brain that helps regulate the emotions (1006)
limiting factor biotic or abiotic factor that restrains the growth of a population (385)
linkage group the group of genes, located on the same chromosome, that are usually inherited together (222)
lipid a kind of organic compound that is insoluble in water, such as fats and steroids (58)
liver organ that performs vital metabolic processes including filtration of blood, secretion of bile, and conversion of sugars into glycogen (986)
lobe-finned fish a fish having paddlelike fins with fleshy bases (807)
logistic growth a model of population growth in which birth and death rates vary with population size and are equal at the carrying capacity (385)
long-day plant a plant that flowers when exposed to day lengths longer than the plant's critical length (656)
loop of Henle in the kidney, the part of the nephron that maintains a high concentration of sodium chloride in the surrounding fluid (993)
lung the respiratory organ where gas exchange occurs (678)
luteal phase the menstrual stage in which the corpus luteum develops (1053)
luteinizing hormone (LH) a hormone that stimulates ovulation and androgen secretion by the corpus luteum (1035)
lymph the fluid of the lymphatic system (937)
lymphatic system a series of one-way vessels that carry intercellular fluid from tissues to the bloodstream (931)
lymphocyte a type of white blood cell occurring in two primary forms, T cells and B cells (959)
lymphoma a tumor in the tissues that form blood cells (211)
lysis the disintegration of a cell by disruption of the plasma membrane (493)
lysosome an organelle containing digestive enzymes, existing primarily in animal cells (79)
lytic cycle the replication process of viruses that results in the destruction of the host cell (492)

M

macromolecule a very large organic molecule composed of many smaller molecules (53)
macronucleus in *Paramecia,* the larger of the two types of nuclei (514)
macronutrient an element used by plant cells in relatively large amounts (606)
macrophage a large white blood cell that engulfs pathogens and cellular debris (961)
madreporite a sievelike plate on the aboral surface of an echinoderm through which water enters the water-vascular system (786)
malignant tumor a cancerous mass of cells (211)
Malpighian tubule the main excretory organ of most insects and some myriapods and arachnids (752)
mammary gland one of the milk-secreting glands of mammals (881)
mandible a movable mouthpart that usually functions in chewing (745)
mantle the epidermal layer of mollusks (726)
mantle cavity a space between the mantle and the visceral mass in mollusks (726)
map unit a unit in chromosome mapping equal to a 1 percent occurrence of crossing-over (224)
marsupial a mammal that carries its young in a pouch (884)
mass a fundamental property of an object generally regarded as equivalent to the amount of matter in the object (31)
mass extinction one of the brief periods of time during which large numbers of species disappeared (281)
mass number the sum of the protons and the neutrons in an atom (265)
mastax a muscular organ that breaks up food in rotifers (718)
matter anything that occupies space and has mass (10)
maxilla a head appendage in arthropods that is used to manipulate food (748)
maxilliped a thoracic appendage in arthropods that is used to manipulate food and to detect touch and chemicals (748)
medulla the innermost portion of an organ such as the kidney (992)
medulla oblongata in fish, the posterior brain lobes that regulate the internal organs; in humans, a part of the brain stem (812, 1007)
medusa a free-swimming, bell-shaped stage in the life cycle of a cnidarian (696)

megagametophyte female gametophyte (628)
megasporangium structure that produces megaspores (628)
megaspore a spore produced by meiosis in a megasporangium (627)
megaspore mother cell a diploid cell in the ovule that undergoes meiosis to produce four haploid megaspores (630)
meiosis the process of nuclear division that reduces the number of chromosomes in a cell by half (148)
melatonin a hormone that probably regulates sleep patterns (1037)
memory cell a specialized white blood cell that protects the body from reinfection by a specific pathogen (963)
meninges the three protective membranes that surround the brain and spinal cord (1007)
menopause the cessation of the menstrual cycle (1053)
menstrual cycle the monthly changes in the female reproductive system (1052)
menstruation the release of the uterine lining, blood, and an unfertilized egg at the end of the menstrual cycle (1053)
meristem the growing region in plants (602)
merozoite an asexually produced stage in the life cycle of some parasitic sporozoan protozoans that disperse and infect additional cells (517)
mesentery a fanlike membrane that holds the digestive organs in place (827)
mesoderm a layer of cells in the gastrula that gives rise to muscles and to interior body linings (683)
mesoglea in cnidarians, the jellylike material located between the epidermis and the gastrodermis (696)
mesophyll in leaves, the tissue where photosynthesis occurs (616)
mesothorax the middle segment of an insect's thorax (765)
messenger RNA (mRNA) the type of RNA that carries genetic information from the nucleus to the ribosomes (191)
metabolism the sum of all chemical processes in living things (10)
metamorphosis a major change in form that occurs as a larval animal develops into an adult (768)
metaphase the second phase of mitosis, during which all the chromosomes move to the cell's equator (150)
metastasis the spread of cancer cells beyond their original site (211)
metathorax the most posterior segment of an insect's thorax (765)
methanogen a bacterium that produces methane during anaerobic respiration (468)
microfilament a polymer chain of the protein actin; the smallest strand in the cytoskeleton (79)
micronucleus in *Paramecia*, the smaller of the two nuclei; involved in sexual reproduction (515)
micronutrient an element used by plants in relatively small amounts (606)
microsphere a type of microscopic droplet enclosed by a membrane composed of organic molecules (268)
microsporangium structure that produces microspores (628)
microspore a spore produced by meiosis in a microsporangium (627)
microspore mother cell diploid cell in the pollen sac that undergoes meiosis to produce four haploid microspores (631)
microtubule a hollow tube of protein that constitutes the largest strand in the cytoskeleton (79)
microvillus one of the extensions of membranes covering the villi (688)
midbrain the section of the brain stem, which acts as a relay between the cerebrum and cerebellum (709)
midgut the stomachlike region of some invertebrates (766)
migratory bird a bird that overwinters in the tropics and then travels to temperate areas to breed and raise its offspring (453)
mimicry a defense in which one organism resembles another that is dangerous or poisonous (398)
mineral an inorganic solid formed in the Earth (981)
mitochondrial matrix the space inside the inner membrane of a mitochondrion (133)
mitochondrion the organelle that is the site of aerobic respiration in eukaryotic cells (76)
mitosis eukaryotic nuclear division(148)
mitotic spindle the array of spindle fibers that serve to divide the chromatids during nuclear division (150)
mitral valve bicuspid valve; the atrioventricular valve on the left side of the heart (932)
molar a flattened tooth used to grind and crush (984)
mold a type of fossil formed from an impression of the shape or tracks of an organism; in fungi, a mass of tangled filaments of cells, such as the fungus that grows on bread (279)
molecular genetics the study of the structure and function of chromosomes and genes (169)
molecule a chemical structure composed of one or more atoms (34)
mollusk an invertebrate with a soft, unsegmented body often enclosed in a hard shell containing calcium carbonate; examples include snails, clams, and squids (725)
molting the process of shedding an exoskeleton or other outer layer on an animal (744)
monocot an angiosperm with a single cotyledon, parallel veins, and flower parts in multiples (592)
monohybrid cross a cross between individuals that involves one pair of contrasting traits (174)
monomer a repeated, single-molecule unit in a polymer (53)
monosaccharide a simple sugar such as fructose or glucose (55)
monosomy a condition in a diploid cell in which one chromosome of one pair is missing as a result of nondisjunction during meiosis (231)
monotreme egg-laying mammal (884)
morphogenesis the change in form of an organism resulting from cell differentiation (209)
morphology the study of the internal and external structure and form of an organism (309)
morula a ball of cells produced by cleavage of the zygote (1054)
motor division a division of the peripheral nervous system that allows the body to react to the brain's interpretation of sensory information (1009)
motor neuron a neuron that carries impulses from the spinal cord to muscle cells (1004)
mouth pore the structure at the oral groove into which the *Paramecium* sweeps food (514)
movable joint enables the skeleton to move where bones come together to perform a wide range of activities (913)
M phase mitosis; the phase of cell division in which the nucleus divides (148)
mucous membrane the mucus-secreting epithelial tissue lining of the interior surfaces of the body (956)
mucus a thick, slimy fluid produced by the mucous membranes (956)
multicellular organism a living thing that consists of more than one cell (6)
multiple fission a form of cell division that results in a number of identical individuals (507)
muscle fatigue the physiological inability of a muscle to contract (920)
muscle fiber a skeletal muscle cell that runs the length of the muscle and contains many nuclei (916)
muscle tissue the body tissue that enables movement (905)
mutagen an environmental factor that damages DNA (212)
mutation a change in DNA (189)
mutualism a form of symbiosis in which both organisms benefit from living together (402)
mycelium a mat of interwoven hyphae (543)
mycology the study of fungi (543)
mycorrhiza a symbiotic association between fungi and plant roots (549)
myofibril a contracting thread in a skeletal muscle (917)
myosin the thick protein filament of a sarcomere (917)

N

NAD⁺ nicotinamide adenine dinucleotide; an organic molecule that accepts electrons during redox reactions (128)
NADP⁺ nicotinamide adenine dinucleotide phosphate; an organic molecule that accepts electrons during redox reactions (115)
naphthalene acetic acid (NAA) synthetic auxin used to promote root formation on stem and leaf cuttings (648)
narcotic a drug derived from opium that relieves pain and induces sleep (1076)
nastic movement a type of plant movement that is independent of the direction of the stimuli (654)
natural killer cell a large white blood cell that attacks pathogen-infected cells

1111

and cancer cells (958)

natural selection the process by which organisms with favorable variations reproduce at higher rates than those without such variations (284)

nauplius the free-swimming larva of most crustaceans (746)

nematocyst a stinging organelle in the cnidocytes of cnidarians (697)

nephridium tubule through which some invertebrates eliminate wastes (735)

nephron the functional unit of the kidney (992)

neritic zone the ocean subarea over the continental shelf (431)

nerve a strand of axons through which impulses travel (1008)

nerve net a diffuse web of interconnected nerve cells in the nervous system of cnidarians (697)

nerve ring a nerve that encircles the central region of an echinoderm (787)

nervous tissue a type of animal tissue that conducts electrical impulses (905)

net primary productivity the rate at which biomass accumulates in an ecosystem (415)

net venation in plant leaves, the repeated branching of veins to form a network of smaller veins (592)

neuron a cell that transmits electrical signals (668)

neurosecretory cells cells of the hypothalamus that produce and secrete hormones (1032)

neutron a subatomic particle with no electric charge (32)

neutrophil a large leukocyte containing a lobed nucleus and many cytoplasmic granules (957)

niche the way of life of a species (371)

nicotine a toxic, addictive alkaloid derived from tobacco and one of the major contributors to the harmful effects of smoking; used as an insecticide (1068)

nictitating membrane a third membrane that can cover the eyeball (829)

nitrification the process in the nitrogen cycle by which nitrites and nitrates are produced (423)

nitrogen-containing base a DNA nucleotide containing nitrogen, a sugar molecule, and a phosphate group (185)

nitrogen cycle the pathway that nitrogen follows within an ecosystem (422)

nitrogen fixation the process by which gaseous nitrogen in the air is converted into ammonia (251)

nitrogen-fixing bacteria bacteria that live in the roots of legumes and convert atmospheric nitrogen into ammonia (423)

node the end of the internode where one or more leaves are attached (609)

nondisjunction the failure of homologous chromosomes to separate during meiosis or the failure of sister chromatids to separate during mitosis (225)

nonutilitarian value the intrinsic value of biodiversity apart from any human use or economic value (451)

nonvascular plant a plant that lacks vascular tissue and true roots, stems, and leaves (581)

norepinephrine (noradrenaline) a chemical that is a hormone in the endocrine system and is a neurotransmitter in the nervous system (1035)

notochord a flexible rod of tissue found in the dorsal part of chordates (673)

nuclear envelope a double membrane that surrounds the nucleus of a eukaryotic cell (80)

nuclear matrix the nuclear skeleton, a shape-maintaining protein (80)

nuclear pore a small hole in the nuclear envelope through which substances pass between the nucleus and the cytoplasm (81)

nucleic acid an organic molecule, DNA or RNA, that stores and carries important information for cell function (59)

nucleolus the structure in which ribosomes are synthesized and partially assembled; found in most nuclei (81)

nucleotide a monomer of DNA and RNA, consisting of a nitrogen base, a sugar, and a phosphate group (59)

nucleus in an atom, the core of protons and neutrons; in a eukaryotic cell, the organelle that contains most of the DNA and directs most of the cell's activities (32)

nutrient a chemical substance an organism needs to grow and function properly (977)

nyctinastic movement plant movement in response to the daily cycle of light and temperature (655)

nymph an immature form of an insect (768)

O

obligate aerobe an organism that requires oxygen (476)

obligate anaerobe an organism that requires the absence of oxygen (476)

obligate intracellular parasite a parasite that requires a host cell in order to reproduce (491)

oceanic zone one of two subareas of the pelagic zone (432)

ocular lens in a microscope, the lens in the eyepiece (21)

olfactory lobe the brain region that interprets smell (812)

olfactory nerve the nerve that carries impulses from olfactory receptors in the nasal passage to the brain (721)

omnivore an animal that eats both plants and animals (416)

oncogene a gene that induces cancer or other uncontrolled cell proliferation; a mutated or activated proto-oncogene that is associated with the development of tumor cells (213)

oogenesis the production of mature egg cells (155)

oogonium a unicellular gametangium that produces eggs (528)

open circulatory system a system in which the circulatory fluid is not contained entirely within vessels (675)

operator a sequence of DNA to which a repressor binds to prevent mRNA synthesis from the adjacent gene; characteristically composed of one or more palindromic sequences (204)

operculum in fish, a hard plate attached to each side of the head that covers gills and is open at the rear (808)

operon a unit of DNA common in bacteria and phages consisting of clusters of regulated genes (204)

opportunistic infection an infection that does not normally cause disease but becomes pathogenic if the patient's immune system is weakened (969)

optic nerve the nerve that transmits signals from the retina of the eye to the brain (1021)

optic tectum an area of the fish brain that receives and processes information from the fish's visual, auditory, and lateral line systems (812)

oral groove a funnel-like structure used by ciliates for feeding (514)

oral surface the surface of an echinoderm where the mouth is located (785)

order in taxonomy, a group of similar families (338)

organ several types of body tissues that together perform a function (86)

organelle one of several formed bodies with a specialized function that is suspended in the cytoplasm and found in eukaryotic cells (71)

organ of Corti in the ear, the middle fluid-filled chamber of the cochlea; the organ of hearing (1019)

organ system a group of organs that interact to perform a set of related tasks (8)

organic compound a compound that is derived from living things and contains carbon (52)

origin attachment point of a tendon to a stationary bone (919)

osculum an opening at the top of a sponge through which water leaves the sponge (693)

osmosis the diffusion of water across a selectively permeable membrane (96)

ossicle one of the small calcium carbonate plates that make up the endoskeleton of an echinoderm (782)

ossification process by which cartilage is converted to bone (912)

osteoarthritis degenerative joint disease (914)

osteocyte a living bone cell (911)

oval window the membrane that separates the middle ear from the inner ear (1018)

ovary an egg-producing gonad of a female (1050)

oviparous referring to organisms that produce eggs that hatch outside the body of the mother (849)

ovipositor in many female insects, a structure at the end of the abdomen used for laying eggs (768)

ovoviviparous referring to organisms that produce eggs that hatch inside the body of the mother (849)

ovulation menstrual stage in which the ovarian follicle ruptures and releases an egg (1052)

ovule the megasporangium of a flowering plant (630)

ovum a female gamete or an egg (1050)

oxaloacetic acid a four-carbon compound that combines with acetyl CoA in the Krebs cycle to produce citric acid (134)
oxidation reaction a chemical reaction in which a reactant loses one or more electrons, becoming more positive in charge (37)
oxygen debt a temporary lack of oxygen availability due to exertion (920)
oxytocin a hormone that stimulates uterine contractions during childbirth (1032)
ozone a molecule composed of three atoms of oxygen; ozone in the upper atmosphere shields the Earth from ultraviolet radiation (271)

P

P_1 a strain obtained through self-pollination (166)
pacemaker the sinoatrial node that regulates the contraction of the heart (933)
palisade mesophyll in plants, a layer of mesophyll directly beneath the upper epidermis where most photosynthesis takes place (616)
pancreas the organ that lies behind the stomach and produces and secretes insulin, glucagon, and digestive enzymes (809)
papilla bumps of tissue on the tongue, between which are embedded taste buds (1021)
parallel venation in plants, the arrangement of several main veins parallel to each other (616)
parapodium one of the fleshy appendages on each side of most segments in a polychaete annelid (733)
parasite an organism that obtains its nutrition at the expense of another (399)
parathormone a hormone that regulates the levels of calcium ions and phosphate ions in the blood (1037)
parathyroid gland a gland that secretes hormones that regulate calcium and phosphate ions in the blood (1037)
parenchyma in plants, the living tissue of the ground system, consisting of cells with thin cell walls (599)
passive transport the movement of substances across a cell membrane without the use of cell energy (95)
pathogen an organism that causes disease (250)
pathology the scientific study of disease (804)
pectoral fin a fin located on the side of a fish body just behind the head (804)
pedicellaria one of many pincerlike structures on the surface of certain echinoderms (785)
pedigree a diagram of the genetic history of an individual; can show how a trait is inherited over several generations (227)
pedipalp in arachnids, an appendage that aids in chewing (753)
pellicle a rigid protein covering of some unicellular organisms (514)
pelvic fin in fishes, one of the paired fins positioned in the ventral, posterior portion of the body (808)
pelvis the hipbones (319)
penicillin antibiotic drug that combats bacteria by interfering with bacterial cell-wall synthesis (479)
pentaradial symmetry body symmetry in which body parts extend from the center along five spokes (782)
pepsin digestive enzyme of the stomach (986)
peptide bond a covalent bond between two amino acids (57)
peptidoglycan a protein-carbohydrate compound found in the cell walls of bacteria (468)
pericardium the tough saclike membrane that surrounds the heart (931)
pericycle in plants, the outermost cylinder of cells in the central vascular area (606)
periosteum the tough outer membrane of a bone; the point where tendons attach muscles to bones (910)
peripheral nervous system the network of nerves that connects the central nervous system with all parts of the body (1004)
peripheral protein a protein attached to the interior or exterior surface of the cell membrane (74)
peristalsis a series of rhythmic muscular contractions that force food through the digestive system (985)
permafrost in the tundra, a permanently frozen layer of ground (425)
pesticide a chemical used to control insects (564)
petal one of the leaf-shaped parts of the corolla of a flower, usually brightly colored and conspicuous (629)
petiole in leaves, the structure that attaches blade and stem (616)
PGA phosphoglycerate; a three-carbon molecule formed in the first step of the Calvin cycle (117)
PGAL glyceraldehyde phosphate; a three-carbon molecule formed in the second step of the Calvin cycle that can leave the cycle and be used to make other organic compounds (118)
phagocyte a cell that engages in phagocytosis (104)
phagocytosis a type of endocytosis in which a cell engulfs large particles or whole cells (103)
pharyngeal pouch a small outpocket of the anterior gut in a chordate (673)
pharynx in flatworms, a muscular tube that leads to the gastrovascular cavity; in animals with a digestive tract, the part of the tract between the mouth and the esophagus (710)
phenotype the external appearance of an organism that is determined by the individual's genotype (172)
phenotype frequency a ratio stating the number of times a specific phenotype occurs in a population in a single generation (175)
phenylketonuria (PKU) a genetic disorder in which the body cannot metabolize phenylalanine (232)
pheromone a chemical released by an animal that affects the behavior or development of other animals (771)
phloem in plants, the specialized tissue in the vascular system that transports dissolved sugars (580)
phospholipid a complex lipid having two fatty acids joined by a molecule of glycerol (58)
photic zone the layer of the ocean that receives light (431)
photoautotroph an organism that uses sunlight as an energy source (476)
photoperiodism a plant response to changes in day length (656)
photosynthesis the conversion of light energy into chemical energy stored in organic compounds (10, 111)
photosystem in plants, a unit of several hundred chlorophyll molecules and carotenoid pigment molecules in the thylakoid membrane (114)
phototropism plant growth in response to light (652)
pH scale a numeric range that quantifies the relative concentrations of hydronium ions and hydroxide ions in a solution (42)
phycobilin a pigment in red algae capable of absorbing the wavelengths of extremely dim light, allowing red algae to exist in very deep water (531)
phylogeny the evolutionary history of a species or taxonomic group (339)
phylum in taxonomy, a group of similar classes (338)
phytochrome a pigment in plants that monitors changes in day length (657)
phytoplankton a photosynthetic, aquatic microorganism (525)
pia mater the innermost layer of the three sacs that surround the brain and spinal cord (1007)
pigment in plants, a compound that absorbs light and imparts color (112)
pilus an appendage that bacteria use to attach to objects (475)
pineal gland endocrine gland that secretes melatonin (1037)
pinocytosis a type of endocytosis in which a cell engulfs solutes or fluids (103)
pinworm roundworm parasite that lives in the lower intestine of humans but causes no serious disease (717)
pioneer species the first species to colonize a new habitat (406)
pistil a plant structure formed of fused carpels (629)
pith the plant tissue located in the center of the stem (606)
pituitary gland the endocrine gland suspended directly below the brain that secretes several different regulatory hormones (1032)
pivot joint allows movement from side to side and up and down, as in the top two vertebrae of the spine (913)
placenta the organ that nourishes unborn young (884)
placental mammal an animal that nourishes its young through a placenta inside the mother's body (884)
placoid scale a toothlike spine found on shark skin (804)
plankton the abundant small plant and animal organisms that drift or swim weakly near the surface of a body of

1113

water (432)
plant growth regulator a hormone that affects growth (421)
planula a free-swimming, ciliated larva of cnidarians (700)
plasma the liquid component of blood (939)
plasma cell a differentiated B cell that produces antibodies and secretes them into the blood (962)
plasmid a single ring of extrachromosomal DNA in bacteria (240)
plasmodial slime mold a member of phylum Myxomycota; a multinucleate mass of cytoplasm that engulfs food by phagocytosis (534)
plasmodium the multinucleate cytoplasm of a slime mold surrounded by a membrane that moves as a mass (517)
plasmolysis the shrinking or wilting of a walled cell in a hypertonic environment (98)
plastid an organelle of plant cells that contains starch, fats, or pigments (83)
plastron the ventral portion of a turtle shell (850)
platelet a partial cell needed to form blood clots (941)
plumule in plants, the structure composed of the epicotyl plus any embryonic leaves (636)
point mutation the change of a single nitrogen-containing base within a codon (225)
polar body one of two small cells produced and discarded during each of the two meiotic divisions that yield the haploid egg (156)
polar compound a compound with one side having a negative charge and the other side a positive charge (49)
polar nucleus in plants, a nucleus that migrates to the center of the cell during embryo-sac formation (630)
pollen grain the microgametophyte of seed plants (628)
pollen tube a structure that grows from a pollen grain, down which the sperm travels to the eggs (628)
pollination the transfer of pollen from an anther to a stigma of a flower of the same species (166)
pollinator an animal that carries pollen, such as insects, birds, or bats (402)
pollution an alteration of the natural environment producing a condition that is harmful to living organisms (825)
polygenic trait a trait controlled by multiple genes (229)
polymer a compound consisting of repeated linked monomers (53)
polymerase chain reaction (PCR) technique used to quickly make many copies of selected segments of DNA (245)
polyp a sessile, vase-shaped stage in the life cycle of a cnidarian (696)
polypeptide a long chain of several amino acids (57)
polysaccharide a complex carbohydrate composed of three or more monosaccharides (56)
pons a segment of the brain stem (1007)
population all the members of a species that live in the same area and make up a breeding group (363)
population density the number of individuals in a population in a given area at a specific time (380)
population genetics the study of evolution from a genetic point of view (299)
posterior the back or dorsal side of an organism (670)
postzygotic isolation a barrier to successful breeding that occurs after fertilization, such as the production of nonviable or sterile offspring (311)
power of magnification in a microscope, the factor of enlargement (22)
precocial referring to birds that, when hatched, are physically active and relatively mature (870)
preen gland the oil-secreting gland of a bird (864)
preening in birds, the act of grooming their feathers with oil secreted by the preen gland (864)
prehensile grasping, as in a primate's hand, foot, or tail (320)
premolar a tooth used to shear, shred, and grind (886)
pre-mRNA a form of mRNA that contains introns and exons (207)
pressure-flow hypothesis an explanation of the movement of sugars in the phloem of plants (612)
prezygotic isolation a barrier to successful breeding that occurs before fertilization, such as differences in mating time or behavior (311)
primary electron acceptor a molecule in the thylakoid membrane that accepts the electrons that chlorophyll *a* loses in the light reactions (115)
primary growth the tissue produced by apical meristems in plants (602)
primary host the host from which the adults of a parasite derive their nourishment and in which sexual reproduction occurs (712)
primary immune response an immunological response that occurs when an antigen is encountered for the first time (963)
primary succession the process of sequential replacement of populations in an area that has not previously supported life (406)
primate a member of the order Primates, the mammalian order that includes humans (320)
primer an artificially made single-stranded sequence of DNA required for the initiation of replication (245)
prion a glycoprotein particle implicated in diseases with long incubation periods (490)
probability the number of times an event is expected to happen divided by the number of opportunities for an event to happen (173)
probe a radioactive segment of DNA used in DNA fingerprinting (244)
producer an autotrophic organism that captures energy to make organic molecules (415)
product a compound formed by a chemical reaction (36)
progesterone a steroid hormone secreted by the corpus luteum of the ovary that stimulates the uterine lining to prepare for the implantation of a fertilized egg; produced by the placenta during pregnancy (1035)
proglottid one of the many body sections of a tapeworm; contains reproductive organs (713)
prokaryote a unicellular organism that lacks a nucleus and membrane-bound organelles (72)
prolactin a hormone that stimulates and maintains milk production (1033)
promoter a nucleotide sequence on a DNA molecule that, when attached to an RNA polymerase molecule, will initiate transcription of a specific structural gene (192)
prophage an intracellular bacteriophage that is harmless to the host cell (494)
prophase the first stage of mitosis and meiosis, characterized by condensation of chromosomes (150)
prosimian primate a suborder of modern primates that includes lemurs, lorises, and tarsiers (320)
prostaglandin a hormonelike chemical messenger that functions as a cell regulator (1031)
protease inhibitor an antiviral drug that interferes with the synthesis of viral capsids during viral replication; in combination with AZT has shown some promise in treating patients infected with HIV (498)
protein an organic compound composed of one or more chains of polypeptides, which in turn are formed from amino acids (56)
protein synthesis the formation of proteins using information coded on DNA and carried by RNA (193)
prothorax the most anterior segment of an insect's thorax (765)
proton a subatomic particle with a positive charge (32)
proto-oncogene oncogene precursor that controls a cell's growth and differentiation (214)
protostome an organism in which the blastopore develops into the mouth and the coelom arises by schizocoely and whose embryo has determinate cleavage (684)
protozoan a single-celled, eukaryotic protist that is able to move independently (84)
proventriculus the first of two chambers in the stomach of a bird (866)
provirus a viral DNA molecule produced by reverse transcriptase (495)
pseudocoelom in animals, a cavity between the mesoderm and the endoderm (686)
pseudoplasmodium a group of cells that act together as a unit to form a sporangium (534)
pseudopodium a retractile, temporary cytoplasmic extension that functions in movement in certain ameboid organisms (512)
psychoactivc drug a chemical compound that affects the central nervous system (1075)
puberty a stage of human life when the

1114

sex organs mature and secondary sex characteristics appear (1035)

pulmonary artery the artery that carries deoxygenated blood from the heart to the lungs (935)

pulmonary circulation the movement of blood between the heart and the lungs (935)

pulmonary respiration a type of respiration involving lungs, in which air is inhaled and exhaled by changing the volume and pressure of air in the mouth (826)

pulmonary vein the vein that carries oxygenated blood from the lungs to the heart (936)

punctuated equilibrium a theory that speciation occurs during brief periods of rapid genetic change (312)

Punnett square a model used to establish the probabilities of the results of a genetic cross (174)

pupa a stage in which an insect that goes through complete metamorphosis changes from a larva to an adult (769)

pupil the opening in the eye that allows light to enter (1020)

purine an organic molecule that has a double ring of carbon and nitrogen atoms (186)

pyloric sphincter the valve that separates the stomach from the small intestine (986)

pyloric stomach in a sea star, the stomach that is connected to the digestive glands (786)

pyrenoid a small protein structure found within the chloroplasts of certain algae that stores starch (525)

pyrimidine an organic molecule that has a single ring of carbon and nitrogen atoms (186)

pyruvic acid a three-carbon molecule that is the end product of glycolysis (128)

Q

quadrupedal able to walk on four limbs (324)

queen bee the only fertile female in a hive of honeybees (772)

queen factor a pheromone secreted by a queen bee that prevents other female larvae from developing into fertile females (773)

quinine a drug used to treat malaria (399)

R

radial canal the portion of a sea star's water-vascular system that runs the length of each arm (786)

radial cleavage in deuterostomes, the division of cells parallel to or at right angles to the polar axis of a fertilized egg (684)

radial nerve a nerve that runs along the length of each arm of a sea star (787)

radial symmetry in animals, an arrangement of body parts around a central axis (670)

radicle in plants, the embryonic root (636)

radioactive dating a method of determining the age of an object by measuring the amount of a specific radioactive isotope it contains (264)

radioactive decay the release of particles or radiant energy by the unstable nuclei of some isotopes (265)

radioactive isotope an isotope with an unstable nucleus (265)

radula a rough, tonguelike structure used in feeding by many mollusks (726)

ray-finned fish a fish with fins supported by several flexible bones rather than a single bony axis (807)

reabsorption a process in the kidney by which materials return to the blood from the nephron (994)

reactant a compound or atom involved in a chemical reaction (36)

realized niche the range of resources and conditions a species actually uses or can tolerate at optimal efficiency; smaller than fundamental niche (372)

receptacle the swollen tip of a branch that is the base of a flower (629)

receptor site the area of a cell membrane where antigen attachment takes place (492)

recessive referring to an allele that is masked by the presence of another allele for the same characteristic (168)

recombinant DNA a DNA segment from at least two different organisms (241)

rectal gland in fishes, a gland that removes excess sodium and chloride ions from the blood and releases them into the rectum for elimination (806)

red tide discoloration of sections of the ocean caused by a population explosion of certain dinoflagellates; contains a poisonous alkaloid produced by the dinoflagellates (528)

redox reaction a reduction-oxidation reaction in which electrons are transferred between atoms (37)

reduction reaction a chemical reaction in which a reactant gains one or more electrons, becoming more negative in charge (37)

regeneration the regrowth of missing body parts (695)

regulator an organism that uses energy to control its internal environment (370)

regulator gene a genetic unit that regulates or suppresses the activity of one or more structural genes (205)

releasing hormones hormones secreted by the hypothalamus that stimulate the anterior pituitary (1033)

release-inhibiting hormone hormones secreted by the hypothalamus that inhibit the anterior pituitary (1033)

renal circulation a subsystem of the systemic circulation that involves the movement of blood through the kidneys (936)

renal pelvis the central structure of the kidney (992)

renal tubule a long tube with permeable walls in a nephron (993)

replication the process by which DNA copies itself during interphase (188)

replication fork the points at which the DNA strands separate during replication (188)

repressor in the *lac* operon, a protein that inhibits an operator and thus stops gene expression (204)

reproduction the production of new offspring (7)

reproductive isolation the inability of formerly interbreeding organisms to produce offspring (311)

resolution the power of a microscope to show detail (21)

resource partitioning among similar species, a pattern of resource use in which species reduce their use of shared resources (401)

restoration biology the science of reversing major changes in ecosystems and replacing missing ecosystem components (452)

restriction enzyme a protein that recognizes specific sequences in a DNA molecule and cuts it into pieces (239)

reticular formation a network of nerve fibers running through the brain stem and the thalamus (1007)

retina the light-sensitive inner layer of the eye (1020)

retrovirus a virus, such as HIV, that contains RNA and reverse transcriptase, (490)

reverse transcriptase an enzyme that makes DNA from an RNA template (490)

RFLP analysis restriction fragment length polymorphism; the method for preparing a DNA fingerprint (244)

Rh factor a type of antigen found in the blood that may cause complications during some pregnancies (943)

rheumatoid arthritis an immune-system disorder causing painful joints (914)

rhizoid a rootlike structure attaching a gametophyte to soil, rock, or tree bark; a rootlike structure that anchors fungal hyphae to a solid substance (546)

rhizome the underground stem of a fern sporophyte (588)

ribonucleic acid (RNA) a nucleic acid composed of a single strand and distinguished from DNA by containing ribose and uracil (59)

ribose the five-carbon sugar in RNA (190)

ribosomal RNA (rRNA) the type of RNA found in a ribosome (191)

ribosome an organelle that functions in the synthesis of proteins (77)

ribozyme a type of RNA that is able to act as an enzyme (270)

ring canal the portion of a sea star's water-vascular system that encircles the mouth (786)

RNA polymerase the primary transcription enzyme (192)

root cap a shield covering the apical meristem (604)

root crop a plant grown for the nutrients it stores in its roots or modified underground stems (563)

root hair an extension of the epidermal cells (604)

rotifer a small, aquatic invertebrate with a crown of cilia surrounding its mouth (718)

rough endoplasmic reticulum the portion of the endoplasmic reticulum that

1115

contains attached ribosomes (76)
roundworm a worm in the phylum Nematoda with a long, slender body that tapers at both ends, such as hookworms and pinworms (715)
royal jelly a high-protein substance secreted by worker bees and fed to the queen and the youngest larvae (772)
RuBP ribulose bisphosphate; a five-carbon carbohydrate that combines with CO_2 to form two molecules of PGA in the first step of the Calvin cycle (117)
rumen the storage chamber in the stomach of artiodactyls, such as the antelope and the bison (887)

S

saddle joint type of joint found at the base of each thumb; allows grasping and rotation (913)
saliva a watery fluid secreted into the mouth; moistens food and begins digestion (984)
saprophyte an organism that feeds on dead organic material (476)
sapwood lighter-colored wood near the outside of a tree trunk active in transporting sap; tissue of the secondary xylem system (611)
sarcoma a tumor growing in bone or muscle tissue (211)
sarcomere the basic contractile unit of skeletal and cardiac muscle that is the portion of the myofibril between two adjacent Z lines (917)
saturated solution a solution in which no more solute can dissolve (40)
savanna subtropical or tropical grassland with scattered trees and shrubs (428)
scanning electron microscope a microscope that produces an enlarged image of the surface of an object with a beam of electrons rather than light (22)
schistosomiasis a disease caused by a parasitic blood fluke of the genus *Schistosoma* (713)
schizocoely the method of coelom formation in protostomes that involves splitting the embryonic mesoderm into two layers (685)
sclerenchyma the tissue of the ground system that supports and strengthens the plant (600)
scolex a knob-shaped organ bearing hooks and suckers; lies at the anterior end of a tapeworm (713)
scrotum an external sac of skin that contains the testes (1048)
second messenger a chemical signal that relays a message from a cell's surface to its interior (1030)
secondary compound a poisonous, irritating, or bad-tasting substance synthesized by plants as a defense mechanism (398)
secondary growth the tissue produced by lateral meristems of plants (602)
secondary immune response the response to a subsequent infection by the same pathogen (963)
secondary succession the sequential replacement of populations in a disrupted habitat (406)
sediment dust, sand, or mud deposited over time by wind or water (279)
seed a plant embryo enclosed in a protective coat (579)
seed coat the protective structure of a plant seed (636)
seed plant a plant that produces seeds for reproduction (581)
segmentation division of a body into units (675)
selectively permeable membrane a membrane that keeps out some molecules but allows others to pass through (73)
self-pollination pollination involving the same flower, flowers on the same plant, or two genetically identical plants (166)
semen a fluid formed from sperm and various exocrine secretions (1048)
semicircular canal a structure in the inner ear that helps maintain balance (1019)
semilunar valve valve that separates the ventricles from the large vessels on each side of the heart (932)
semimovable joint a joint that allows limited movement, as in vertebrae (913)
seminal receptacle a structure on a female or hermaphrodite that receives sperm (735)
seminiferous tubules the location of sperm formation in the testes (1048)
sense organ an organ that contains sensory neurons which detect external stimuli (1017)
sensory division a division of the peripheral nervous system that acquires information from the external and internal environments of the body (1009)
sensory neuron a neuron that transmits impulses from sensory receptors to the central nervous system (1004)
sensory receptor a neuron that detects a stimulus (1009)
sepal a modified leaf that protects a young flower (629)
septum a wall dividing hyphae into cells; a thick wall that divides an organ, such as the heart (931)
sessile referring to an organism that attaches firmly to a surface and does not move (694)
seta one of the external bristles on annelids (733)
sex chromosome a chromosome that determines sex (146)
sex-influenced trait a trait that is influenced by the presence of male or female sex hormones (229)
sex-linkage the presence of a gene on a sex chromosome (222)
sexual reproduction the production of offspring from the combination of genetic material from two parent organisms (156)
sexual selection the preferential choice of a mate based on a specific phenotypic trait (308)
shaft the center of a bird feather (864)
short-day plant a plant that flowers when exposed to day lengths shorter than the plant's critical length (656)
sieve plate a region at the end of a sieve tube member through which compounds move from cell to cell (601)
sieve tube structure formed by stacked sieve tube members (601)
sieve tube member the conducting cell of angiosperm phloem (601)
simple leaf a leaf with only one blade (616)
single-allele trait a trait controlled by a single allele (228)
sink the place where a plant stores or uses sugars (613)
sinoatrial node the nerve tissue in the right atrium that regulates the contraction rate of the heart (933)
sinus venosus in fishes and amphibians, a collecting chamber for deoxygenated blood (810)
siphon on a clam, a tube through which water enters or leaves the body (729)
SI Système International d'Unités (International System of Measurements) the scientific standard of measurement which employs a set of units that describe length, mass, time and other attributes of matter (23)
skeletal muscle voluntary muscle, attached to bones by tendons, that moves parts of the body (905)
skeleton the bones of an animal body that form its internal framework (909)
skin gill one of many hollow tubes that project from the surface of a sea star, through which gas exchange and excretion take place (785)
smooth endoplasmic reticulum the portion of the endoplasmic reticulum that lacks attached ribosomes (76)
smooth muscle elongated, spindle-shaped involuntary muscles (except cardiac muscles) (905)
sodium-potassium pump a carrier protein that actively transports K^+ ions into and Na^+ ions out of cells (101)
soft palate a soft area made of folded membranes that separates the mouth cavity from the nasal cavity (984)
solar tracking the phototropism of leaves or flowers as they follow the sun's movement across the sky (653)
solute a substance dissolved in a solution (40)
solution a mixture in which one or more substances are uniformly dissolved in another substance (40)
solvent in a solution, the substance in which a solute is dissolved (40)
somatic mutation a mutation that occurs in a body cell (224)
somatic nervous system the division of the peripheral nervous system that controls skeletal muscles (1009)
sorus a cluster of sporangia in ferns (627)
source the place where sugars are made in a plant (613)
spawning the egg-laying behavior of some bony fishes (812)
specialist a species with a narrow niche that can tolerate a narrow range of conditions and can use only a few specific resources (372)
specialization the differentiation of a cell for a particular function (667)
speciation the formation of a new species (309)

species a group of organisms of a single type that are capable of producing fertile offspring in the natural environment (338)
species-area effect a pattern of species distribution in which larger areas contain more species than smaller areas do (404)
species diversity an index combining the number and relative abundance of different species in a community (403)
species identifier the second part of a species name; humans are known by genus name *Homo* and by species identifier *sapiens*; specific epithet (339)
species name the two-part scientific name of a species composed of the genus and the species identifier (339)
species richness the number of different species in a community (403)
spermatid in meiosis, a haploid cell that develops into a mature sperm cell (156)
spermatogenesis the production of sperm cells (155)
S phase the second period of interphase during which replication of DNA occurs (132)
spicule one of the small, spike-shaped particles of calcium carbonate or silicon dioxide that make up the skeleton of some sponges (693)
spinal cavity the area that surrounds and protects the spinal cord (908)
spinal cord a column of nerve tissue running through the vertebral column (1008)
spinal nerve a nerve connection between the spinal cord and an organ or muscle (812)
spinal reflex a muscular reaction to a stimulus that is initiated by the neurons of the spinal cord instead of the brain (1009)
spindle fiber one of the microtubules that extend across a dividing eukaryotic cell; assists in the movement of chromosomes (150)
spinneret the structure through which fluid from the silk gland of spiders passes (752)
spiracle an external opening of a trachea in an insect, myriapod, or spider (752)
spiral cleavage in protostomes, the division of cells in a spiral arrangement (684)
spirillum spiral-shaped eubacteria (470)
spirochete Gram-negative, spiral-shaped, heterotrophic bacterium (472)
spleen the largest lymphatic organ in the body; serves as a blood reservoir, disintegrates old red blood cells, and produces lymphocytes and plasmids (959)
sponge a sessile, aquatic invertebrate in the phylum Porifera with no true tissues or organs (693)
spongin the network of protein fibers making up the simple skeleton of some sponges (693)
spongy bone the lacy network of connective tissue in the center of a bone (911)
spongy mesophyll a mesophyll of irregularly shaped cells surrounded by air spaces that allow oxygen, carbon dioxide, and water to diffuse into and out of the leaf (616)
spontaneous generation an early and now disproved hypothesis that living organisms develop from nonliving material (261)
sporangiophore in some fungi, an upright hypha that produces sporangia (544)
sporangiospore in fungi, a specialized hypha supporting a sporangium (544)
sporangium a spore-bearing sac in plants, algae, and fungi (544)
spore a resistant reproductive cell of some organisms, such as certain plants, protists, and fungi (535)
sporophyte one of the diploid generation of land plants that produce spores (581)
sporozoite a sporozoan that has been released from the oocyst and is ready to penetrate a new host cell (517)
springwood in woody plants, secondary xylem with cells that are wide and thin walled; the first wood of a growing season (611)
stabilizing selection a type of natural selection in which the average form of a trait causes an organism to have an advantage in reproduction (306)
stamen the structure in the flower of plants that produces pollen (629)
staphylococcus eubacterium occurring in grapelike clusters (470)
start codon the codon AUG which engages a ribosome to start translating an mRNA molecule (194)
steroid a lipid in which the molecule is composed of four carbon rings (74)
steroid hormones a class of hormones synthesized from cholesterol (1030)
sticky end the single-stranded end of a double-stranded DNA fragment that can form base pairs with another sticky end of a DNA fragment cut by the same enzyme (240)
stigma an expanded apex of a pistil, supported by the style; the part of the pistil that receives pollen (166)
stimulant a drug that increases the activity of the central nervous system (1068)
stipe in some algae, the stemlike region of a thallus (531)
stolon the horizontal hypha of some fungi (546)
stoma one of many small pores usually located on the underside of a leaf; aids in gas exchange (119)
stomach an organ that prepares food for absorption by both physical and chemical activities (985)
stone canal part of the water-vascular system in sea stars that connects the madreporite with the ring canal (786)
stop codon causes the ribosome to stop translating mRNA; UAA, UAG, UGA (194)
stratum in geology, a layer of soil or rock in a cross section of the Earth (280)
streptococcus Gram-positive eubacterium occurring in chains (470)
strobilus a conelike structure of sporangia-bearing leaves (587)
stroma in plants, the solution that surrounds the thylakoids in a chloroplast (112)
structural gene a gene that codes for a product, such as an enzyme, protein, or RNA, rather than serving as a regulator (204)
style in plants, a stalklike structure of the ovary (629)
subspecies varieties of animal species that occur in different geographical areas (339)
substitution a point mutation in which one nucleotide in a codon is replaced with a different nucleotide (225)
substrate a part, substance, or element that lies beneath and supports another; the reactant in any enzyme-catalyzed reaction (57)
succession the predictable, sequential replacement of populations in an ecosystem (406)
sulfa drug an antibiotic drug that combats bacteria by inhibiting bacterial cell metabolism (479)
summerwood in woody plants, secondary xylem cells that are small and thick walled and that are produced in hot, dry summers (611)
superior vena cava the large vein carrying blood from the upper part of the body (935)
suppressor T cell a type of T cell that shuts down the immune response after the pathogen has been eliminated (961)
survivorship curve the graph of a species' mortality-rate data (382)
swim bladder in bony fishes, a gas-filled sac that is used to control bouyancy (807)
swimmeret one of 10 bilaterally paired appendages on the abdomen of a crayfish (748)
swimmer's itch a condition characterized by minor skin irritation and swelling, caused by a small, brown fluke that usually lives in fresh water (713)
symbiosis the relationship between different species living in close association with one another (397)
sympathetic division the part of the autonomic nervous system that controls internal organs during conditions of high stress or increased activity (1010)
synapsis the pairing of homologous chromosomes during meiosis (153)
synovial joint a joint filled with synovial fluid (914)
syrinx the vocal organ of birds (873)
systematics systematic taxonomy; taxonomic organization of living things in the context of evolution (342)
systemic circulation the movement of blood between the heart and all parts of the body except the lungs (935)
systole the phase of the heartbeat in which the ventricles contract and force blood into the arteries (933)
systolic pressure pressure of the blood when the ventricles contract (934)

T

tagma in arthropods, a structure composed of several segments that have fused to perform specific functions (744)
taiga forested biome characterized by

1117

cone-bearing evergreen trees (426)
tapeworm a parasitic flatworm that lives in the vertebrate intestine (713)
taproot the organ that results when the primary root grows downward and becomes the largest root (603)
tar a sticky mixture of chemicals and particles from tobacco (1068)
target cell the specific cell affected by a hormone (1029)
taste bud a structure in which chemoreceptors for taste are clustered (1021)
taxonomy the science of grouping organisms according to their presumed evolutionary relationships (337)
T cell a thymus-derived lymphocyte that helps stimulate an immune response against an antigen and can attack certain antigens (499)
tegument a continuous sheet of fused cells that covers the external surface of a fluke or tapeworm and protects the worm from the host's digestive and immune systems (712)
telophase the final stage of mitosis during which a nuclear membrane forms around each set of new chromosomes (151)
telson a flat, paddlelike projection at the posterior end of many crustaceans (748)
temperate referring to a nonvirulent virus, rarely causing disease; moderate, not subject to prolonged extremes of hot or cold (493)
temperate deciduous forest a forest characterized by trees that shed their leaves in the fall (426)
tendon a tough, solid band of connective tissue that supports and connects body parts (919)
tendril a specialized leaf found in many vines that takes on a coiled appearance (615)
tentacle a flexible appendage with which an animal feels its environment or grasps objects (696)
termination signal a specific sequence of nucleotides that marks the end of a gene in eukaryotes (192)
test the shell, or hardened covering, of some invertebrates (783)
test cross the crossing of an individual of unknown genotype with a homozygous recessive individual to determine the unknown genotype (175)
testis the gamete-producing gonad of a male (1047)
testosterone a hormone that regulates male secondary sex characteristics and the production of sperm cells (1035)
tetracycline an antibiotic drug that combats bacteria by interfering with bacterial protein synthesis (479)
tetrad a group of two homologous chromosomes during meiosis (153)
thalamus the brain structure that directs incoming sensory impulses to the proper region of the cerebral cortex (1006)
thallus the body of an alga (525)
theory a broad and comprehensive statement of what is believed to be true, supported by considerable experimental evidence resulting from many tests of related hypotheses (19)
thermoacidophile an archaebacterium that lives only in hot, acid places (468)
thermoregulation control of body temperature (848)
thigmotropism a plant's growth response to touching a solid object, as in the tendrils of vines that coil when they touch an object (653)
thoracic cavity the upper ventral cavity of the human body, containing the heart, lungs, and esophagus (908)
thorax in insects and some crustaceans, the part of the body between the head and the abdomen; in mammals, the part of the body between the neck and the abdomen (748)
thylakoid a flattened, membranous sac inside a chloroplast; contains most of the components involved in the light reactions of photosynthesis (83)
thymine a nitrogen-containing base, one component of a nucleotide (185)
thymosin a hormone produced in the thymus gland that stimulates the differentiation of T cells (1037)
thymus gland endocrine gland involved in immune-system development (1037)
thyroid gland the endocrine gland that secretes thyroxine (1034)
thyroxine a thyroid hormone that increases metabolism (1034)
tissue in most multicellular organisms, a group of similar cells that carry out a common function (86)
tissue culture the growing of living cells in a controlled medium (640)
tolerance a lessening response to a drug as a result of physical addiction (1073)
tolerance curve a graph of an organism's tolerance to a range of an environmental variable (369)
topsoil the uppermost layer of soil (822)
torpor a state of stupor (556)
torsion a process occurring during the larval development of gastropods in which the visceral mass twists 180° in relation to the head (727)
toxin a chemical substance, usually biological in origin, that is harmful to the normal functioning of cells (478)
toxoplasmosis a disease caused by a sporozoan that can cause serious harm to fetuses and newborns and can cause flulike symptoms in adults (517)
trachea in insects, myriapods, or spiders, one of a network of air tubes; in vertebrates, the tube that connects the pharynx to the lungs (752, 945)
tracheid a long, thick wall with tapering ends, forming part of the xylem tissue of plants (601)
trait in genetics, a category within which alternate characteristics, such as height and eye color, can be observed (165)
transcription the process in which RNA is made from DNA (191)
transcription factor one of the additional proteins bound to enhancers and RNA polymerase that regulate transcription (208)
transduction in viruses, the process in which genetic material is transferred from one cell to another (477)
transfer RNA (tRNA) the type of RNA that carries amino acids from the cytoplasm to the ribosomes (191)
transformation a process of direct transfer of nucleic acid from one bacterium to another (477)
transgenic organism a host organism that has received recombinant DNA (242)
translation the process of converting the genetic code in RNA into the amino acid sequence that makes up a protein (194)
translocation a mutation in which a broken piece of chromosome attaches to a nonhomologous chromosome; movement of organic molecules in plant tissues (225)
transmission electron microscope a microscope that transmits a beam of electrons rather than light through a thinly sliced specimen (22)
transpiration the evaporation of water through stomata (421)
trichinosis a disease caused by a parasitic roundworm of the genus *Trichinella;* characterized by muscle pain and stiffness (717)
triglyceride a lipid made of three fatty-acid molecules and one glycerol molecule (58)
trimester one of three divisions of pregnancy (1055)
trisomy a chromosomal anomaly in which an individual has an extra chromosome in any of the chromosome pairs (231)
trisomy 21 a human congenital disorder caused by trisomy of chromosome 21 due to the failure of the sister chromatids to separate during mitosis or the failure of homologous chromosomes to separate during meiosis (see Down syndrome) (231)
trochophore a free-swimming, ciliated larva of most mollusks and some annelids (725)
trophic level a feeding level in an ecosystem (417)
tropical rain forest a biome near the equator characterized by large amounts of rain and sunlight (429)
tropism plant movement toward or away from an environmental stimulus (652)
trypanosomiasis any of the forms of sleeping sickness or Chagas' disease (516)
tube cell the cell of a pollen grain that contains the tube nucleus (631)
tube foot one of many small, flexible, fluid-filled tubes that project from the body of an echinoderm and are used in locomotion, feeding, gas exchange, and excretion (782)
tumor an abnormal mass of cells (211)
tumor-suppressor gene a gene that suppresses tumor formation but when mutated causes a loss in cell function, resulting in tumor formation (213)
tundra a biome of low-growing vegetation that forms a continuous belt across North America, Europe, and Asia (425)
turgor pressure water pressure within a plant cell (98)
tympanic membrane the eardrum (1018)

tympanum a sound-sensing membrane on each side of the abdomen of an insect (767)
typhlosole in an earthworm, an infolding of the intestinal wall (734)

U

umbilical cord blood vessels that connect the placenta to the fetus (1056)
ungulate a hoofed mammal (894)
unicellular gametangium a single-celled gamete holder (321)
unicellular organism a living thing composed of one cell (6)
uniformitarianism a principle that states that the geological structure of Earth resulted from cycles of observable processes and that these processes operate continuously (284)
upwelling cold water rising from the ocean floor bringing organic material and nutrients (442)
uracil a nitrogen-containing base found in RNA (190)
urban ecology a new environmental field wherein people are challenged to increase biodiversity even in the most heavily developed areas (458)
urea product of the liver's conversion of ammonia that is eliminated by the kidneys (992)
ureter a narrow tube through which urine flows from the renal pelvis to the urinary bladder (995)
urethra the tube through which urine flows from the urinary bladder out of the body (995)
urinary bladder a muscular sac that stores urine (995)
urine a liquid containing nitrogenous wastes that have been removed from the blood (995)
uropod the most posterior appendage on many crustaceans (748)
uterus a female reproductive structure where eggs are stored or fertilized or where development of the young occurs (1051)
utilitarian value a way of relating the importance of biodiversity in terms of the economic benefits biodiversity provides to humans (451)

V

vaccine a solution containing a harmless version of a virus, bacterium, or toxin that causes an immune response when introduced to the body (250)
vacuole a fluid-filled organelle that stores enzymes or metabolic wastes in a plant cell (82)
vane the projection on either side of a bird's feather (864)
variety a further division or subset of species (339)
vascular cambium the lateral meristem in a plant that produces additional vascular tissue (602)
vascular plant a plant that has xylem and phloem (581)
vascular tissue system the system that includes the xylem and phloem tissues of plants (601)
vas deferens a duct through which sperm leave the epididymis (1048)
vasopressin a hormone that helps control the amount of water excreted by the kidneys (1033)
vegetable food derived from any part of soft plants (563)
vegetative propagation asexual reproduction in plants (639)
vein a vessel carrying blood to the heart (934)
venation the arrangement of veins in a leaf (616)
vent in amphibians and reptiles, the opening through which waste and eggs or sperm exit the body (827)
ventral the bottom of a bilaterally symmetrical organism (670)
ventricle a lower, and the most muscular, chamber of the heart (931)
venule a small vessel in the network of veins (935)
vernalization the requirement that some seeds or spores be exposed to cold before they can germinate (658)
vertebra one of the repeating bony units of the backbone (677)
vertebrate an animal with a backbone (667)
vesicle a membrane-bound sac in a eukaryotic cell that contains materials involved in endocytosis, exocytosis, or transport within the cell (103)
vessel in plants, a structure formed by stacked vessel elements; in animals, a tubular conducting structure for blood or other body fluid (601)
vessel element an elongated, water-conducting plant cell wall with slanting ends and holes through which water can pass (601)
vestigial referring to a functionless structure that was functional in an ancestral species (290)
villus one of the small projections lining the small intestine (988)
viper venomous snake with two large, mobile fangs at the front of the mouth (853)
viroid a short, single strand of RNA that causes disease in plants (490)
virology the study of viruses (487)
virulent disease-causing and highly infectious (492)
virus a nonliving infectious particle composed of a nucleic acid and a protein coat (486)
visceral mass the central region of a mollusk, located between the head-foot and the mantle (726)
visible spectrum the portion of sunlight perceived by the human eye as various colors (112)
vitamin complex organic molecule that serves as a coenzyme; one of the basic six nutrients (980)
viviparous referring to organisms that carry and nourish the young inside the body before birth (849)
voluntary muscle a muscle whose movement can be consciously controlled (916)
vulva the external part of the female reproductive system (1051)

W

warning coloration a bold, bright color pattern that makes dangerous, poisonous, or bad-tasting animals clearly recognizable and warns predators away (770)
water cycle the movement of water between the ground, the atmosphere, and bodies of water (420)
water mold a funguslike organism composed of branching filaments of cells, found in freshwater systems (535)
water-vascular system a network of water-filled canals in echinoderms (782)
wavelength the distance between crests in a wave (113)
withdrawal a reaction to the lack of a drug (1073)
woody tissue in plants, tissue that is formed from several layers of xylem, usually concentrated in the center of the stem (580)
worker bee a sterile female bee that makes up the majority of the beehive population (772)

X

xylem the vascular tissue in plants that transfers water and minerals from the roots to the leaves (580)
X-linked gene a gene found on the X chromosome (222)

Y

yeast unicellular fungus whose colonies resemble those of bacteria; known as the microorganisms that make bread rise (543)
Y-linked gene a gene found on the Y chromosome (222)
yolk sac in amniote eggs, the membrane that encloses the yolk and supplies the embryo with food (842)

Z

Z line boundary of a sarcomere; point of anchor for thin actin filaments (917)
zooplankton microscopic animal organisms that drift in bodies of water worldwide; represent a basic level of feeding relationships (507)
zoospore a flagellated spore (527)
zygosporangium in fungi, fused gametangia in conjugation (547)
zygospore the protective structure that results when the wall surrounding a zygote thickens (527)
zygote the diploid cell that results from the fusion of gametes (1054)

1119

INDEX

Page numbers preceded by the letter *t* refer to tables.
Page numbers in **boldface** refer to figures.

A

B

1121

C

1122

D

E

1126

1127

M

1129

P

1131

Q

R

S

1133

1135

V

W

X

Y

Z

1136

CREDITS

ILLUSTRATION

Abbreviation Code
AT = Alexander & Turner; SB = Sally J. Bensusen/ Visual Science Studio; KC = Kip Carter; BH = Barbara Hoopes-Ambler; RH = Robert Hynes; BJ = Bridgette James; JK = John Karapelou; KK = Keith Kasnot; GK = George Kelvin; CK = Christy Krames; JL = Joe LeMonnier; MCA = Morgan-Cain & Associates; PG = Precision Graphics; SR = Steve Roberts; WAA = Wildlife Art Agency; MW = Michael Woods; SW = Sarah Woodward

All Safety Symbols were designed by Michael Helfenbein/Evolution Design.

xi KK; **xiv** MW; **Chapter 1: 16, 21, 26** MCA; **Chapter 2: 31–37, 42, 44, 46** MCA; **Chapter 3: 49–50, 52–56** MCA; **57** (t) MCA; (b) Robert Margulies/MMA; **58–59, 62** MCA; **Chapter 4: 70, 74–79, 83** MCA; **86** JK; **91** MCA; **Chapter 5: 95, 97, 99, 102–104, 106–108** MCA; **Chapter 6: 112–117, 120, 122–124** MCA; **Chapter 7: 127–129, 133–137, 140–141** MCA; **Chapter 8: 145** PG/MCA; **148** T Narashima; **149–151** MCA; **154** (t) MCA; (b) PG; **155** (t) MCA; (b) PG; **159** MCA; **Chapter 9: 166** MW; **167** PG/(flowers) MW; **169** MCA; **174** (t) (spot) MW; (graphics) MCA; **174** (b) (spot) BH; (graphics) MCA; **175** (spot) BH; (graphics) MCA; **176–178** (spots) PG; (graphics) MCA; **180** MCA; **Chapter 10: 185–186, 188, 190–191** MCA; **194** GK; **195, 199** MCA; **Chapter 11: 204–205, 207–208** MCA; **210** (spots) SR/WAA; (graphics) MCA; **214, 216–217, 219** MCA; **Chapter 12: 221** (spots) SR/WAA; (graphics) MCA; **222** MCA; **223** (t) (spots) SR/WAA; (graphics) MCA; **223** (b) JL; **224** (spots) SR/WAA; (graphics) MCA; **225–227, 231** MCA; **232** CK; **235–236** MCA; **Chapter 13: 239–241, 244–245** MCA; **250** SD/David Fischer; **255** MCA; **Chapter 14: 261–263, 265, 267** Mellander; **269, 272** MCA; **274** Mellander; **275** MCA; **Chapter 15: 281** MW; **285** (map) MCA/Mountain High Map Resources; **286** Doug Schneider; **287** (spots) MW, (graphics) MCA; **289** JK; **291** SW; **282** BH; **294** (spots) MW; (graphics) MCA; **295–296** MCA; **Chapter 16: 299, 301** (spots) PG; (graphics) MCA; **305** MCA; **307** (spots) PG; (graphics) MCA; **311** MCA; **312** (spots) MW, (graphics) MCA; **309** Doug Schneider; **314** PG; **315** MCA; **Chapter 17: 321** JK; **326** MCA; **328** Chris Forsey; **330** MCA; **332** David Fischer; **333–335** MCA; **Chapter 18: 338** MCA; **343** (spots) BH; (graphics) MCA; **344** SW; **346** (spots) BH; (graphics) MCA; **350, 352** MCA; **353** BH; **354** Pond & Giles; **Chapter 19; 361** MCA; **362, 364** BH; **367, 369–370** MCA; **371** (t) MW; (b) MCA; **372, 374–375** (MCA); **Chapter 20: 381–382, 384–386, 389, 392–394** MCA; **Chapter 21: 400** MCA; **401** RH; **404, 410** MCA; **411** Cecile Duray-Bito; **Chapter 22: 416** MCA; **417** (spots) MW; (graphics) MCA; **418** (t) (spots) MW; (graphics) MCA; **418** (b) PG; **420–422** RH, (graphics) MCA; **424** MCA; **431** JL; **437** (spots) MW, (graphics) MCA; **Chapter 23: 441–442, 444** David Uhl Illustration; **445** MCA; **448** BJ/WAA; **449** BH; **450, 453–454, 457, 460–461** MCA; **Chapter 24: 471, 482** MCA; **Chapter 25: 488** MCA; **489** MCA; **491** MCA; **492** GK; **493** MCA; **494** (t) GK, (b) MCA; **502** GK; **503** MCA; **Chapter 26: 510** MCA; **513** Claire Booth/CBMI; **514–515** MCA; **518** SB; **Chapter 27: 528** (t) Henry Hill/John Edwards & Associates; (b) MCA; **529** Andrew Grivas; **533, 538** MCA; **539** Andrew Grivas; **Chapter 28: 543** MCA; **546–547** MW; (zooms, graphics) MCA; **548** MCA; **554** MW; (zoom) MCA; **555** MCA; **Chapter 29: 574** Ian Milnes/WAA; **Chapter 30: 579** MW; **581** (spots) Cy Baker/WAA; (graphics) MCA; **582** (spots) MW; (graphics) MCA; **590** Stansbury, Ronsaville, Wood, Inc.; **592** Cy Baker/WAA; (graphics) MCA; **594** MCA; **Chapter 31: 599, 601** MCA; **609** Stansbury, Ronsaville, Wood, Inc.; **612, 617** MCA; **Chapter 32: 625–627, 629–631, 633, 635** SW; **636** (l) (c) SW; (r) MW; **638** SW; **Chapter 33: 647** RH, **657** (spots) RH; (graphics) MCA; **Chapter 34: 669** AT; **671** (tl) RH; (tc) MCA; (tr) BJ/WAA; **671** (b) AT; **672** (spots) BH; (graphics) MCA; **676** BH; **677, 679** KC; **682** SW; **683** (t) SW, (b) SW/MCA; **685–686, 689** MCA; **690–691** CK; **Chapter 35: 694–696** SW; **697, 699** MCA; **700** SW; **704–705** MCA; **Chapter 36: 710** AT; **712** David DeGasperis; **713** MCA; **714** JK; **718** AT; **721** MCA; **Chapter 37: 725** AT; **726** (t) SW; (c) MCA; **727, 729** AT; **734** Cynthia Turner; **738** AT; **739** MCA; **Chapter 38: 743** MCA; **745** (spots) SR/WAA; (graphics) MCA; **746** AT; **748** SB; **749,** SR/WAA Walter Stuart; **756** SB; **757** MCA; **758** SR/WAA; **Chapter 39: 762–763** (spots) SR/WAA; **765** PG; **766** SR/WAA; **767** PG; **768** SR/WAA; **769** BJ/WAA; **772** PG; **774** BJ/WAA; **776** PG; **777** MCA; **Chapter 40: 781, 785, 788–790, 792** AT; **793** MCA; **Chapter 41: 800** (spots) BH; (graphics) MCA; **801** (t) BH; (b) Laurie O'Keefe; **805** KC; **808** BH; **809, 810** (t) (br) KC; **811** KC; **812** Todd Buck/HS; **814** BH; **815–816** MCA; **Chapter 42: 819** PG; **820** Peg Gerrity/HS; **821** (spots) MW; (graphics) MCA; **825** PG, (silhouettes & heart) KC; **826** PG; **827** KC; **828** PG; (silhouette) KC; **834** KC; **835** MCA; **Chapter 43: 840** BH; **841** (t) BH; (b) (dinosaurs) Chris Forsey; (animals) Pond & Giles; (graphics) MCA; **842, 846** KC; **847** (head) BH; (brain) KC; **848** MCA; **856** KC; **Chapter 44: 862** (dinosaurs) Chris Forsey; (animals) Pond & Giles; (graphics) MCA; **863** PG; **864** (l) RH; (r) MCA; **865–867** KC; **876** KC; **877** MCA; **Chapter 45: 882** BH; **883** Howard Friedman; **885** KC; **886** E. Alexander/AT; **891** (spots) MW; (graphics) MCA; **898** BH; **899** MCA; **Chapter 46: 906** KK; **908** KK; (silhouette) JK; **909** JK; **910, 911** (l) **912** KK; **913–915, 917** JK; **918** MCA; **919** KK, (silhouette) JK; **923** (t) MCA; (b) CK; **926** MCA; **927** CK; **Chapter 47: 931–932** KC; **933** CK; **934, 935** (t) Walter Stuart; **935** (b) **936** CK; **937, 945** JK; **946** MCA; **947** JK; **950** KC; **952** MCA; **Chapter 48: 957** CK; **960** JK; **961–964, 969–970, 972–973** MCA; **Chapter 49: 978** (t) (bl) MCA; (br) MW; **979** (spot) MW, (graphics) MCA; **983–984** JK; **985** (t) CK; (c) KK/CK; (b) MCA; **986** JK; **987** (t) KK/CK; (b) JK; **988** (t) JK; (r) MCA; **991** KK/CK; **992–994** MCA; **995** JK; **998** KK/CK; **999** BH; **Chapter 50: 1004** JK; **1005** CK; **1006** JK; **1008** CK; **1010** MCA; **1011** JK; **1013, 1014–1015, 1016** MCA; **1018, 1020** KK; **1021–1022** JK; **1025** (t) MCA; (b) Todd Buck; **1026–1027** KC; **Chapter 51: 1030–1031** MCA; **1032** JK; **1033** Tonya Hines; **1034** JK; **1035** CK; **1037** JK; **1038–1040, 1043** MCA; **1044** AT; **Chapter 52: 1047–1048** MCA; **1049, 1051** KK; **1052** MCA; **1055–1057** CK; **1058** C. Turner/AT; **1060, 1062** MCA; **Chapter 53: 1069** JK; **1074, 1078, 1080** MCA; **1091, 1092** MCA

PHOTOGRAPHY

Abbreviation Code
AA = Animals, Animals; BPA = Biophoto Associates; BPS = Biological Photo Service; CB = Corbis-Bettmann; ES = Earth Scenes; FH = Fran Heyl Associates; GH = Grant Heilman Photography; HRW = Holt, Rinehart and Winston staff photograph; JS/FS = Jeff Smith/FOTOSMITH; MP = Minden Pictures; NASC = National Audubon Society Collection; OSF = Oxford Scientific Films; PA = Peter Arnold, Inc.; PH = Phototake NYC; PR = Photo Researchers; RLM = Robert & Linda Mitchell; SPL = Science Photo Library; SS = Science Source; TSI = Tony Stone Images; VU = Visuals Unlimited

Cover: © Manfred Danegger/Okrpai/Photo Researchers, Inc. © Michael Fogden/DRK Photo; **iv** (t) © Tim Fitzharris; (b) © Flip Nicklin/MP; v © BPA/SS/PR; **vi** © Mark Joseph/ TSI; **vii** AA/© Anup & Manuj Shah; **viii** © Oliver Meckes/MPI-Túbingen; **ix** (t) © John Kaprielian/ PR; (b) © Neil McDaniel/PR; **x** © Kevin Schafer/Martha Hill; **xi** © U.H.B. Trust/TSI; **xiii** © Cornell University Photography; **Unit 1: 2** © E.R. Degginger; **3** (tl) © Tony Stone Imaging/TSI; (tc) © Luiz C. Marigo/PA; (tr) © Frans Lanting/MP; (b) © Tim Davis/TSI; **Chapter 1: 4** © Art Wolfe; **5** © Philip Sze/VU; **6** © Carolina Biological Supply/PH; **7** (t) © Dan Guravich/NASC/PR; (b) © Pr. Stanley Cohen/SPL/PR; **8** (l) © Jim Brandenburg/MP; (r) © Steve Kaufman/PA; **9** © Gunter Ziesler/PA; **10** © Alfred Pasieka/PA; **11** © Runk/Schoenberger/GH; **12** (t) © Stephen J. Krasemann/NASC/PR; (b) © Dr. K.S. Kim/PA; **13** © A. Kerstitch/VU; **15** © Sinclair Stammers/ SPL/PR; **16** © Barry Dowsett/SPL/PR; **19** © Agence France Presse/CB; © Will & Demi McIntyre/PR; **22** (t) © E. White/VU; (b) © Jerome Paulin/VU; **27** (t) © Oliver Meckes/SS/PR; (b) © CNRI/SPL/PR; **28** © JS/FS; **Chapter 2: 30** © Alfred Pasieka/PA; **38** (bkgnd) © Superstock/ Mark Romine; courtesy of the American Red Cross; **39** (t) courtesy of the American Red Cross; (b) © Larry Mulvehill/SS/PR; **41** © Harvey Lloyd/PA; **Chapter 3: 48** © Fred Bavendam/TSI; **51** © William Hopkins; **60** (bkgnd) Geoff Tompkison/SPL/PR; © UPI/CB; **Unit 2: 66–67** (bkgnd) © BPA/PR; **66** © FH; **67** (white blood cells) © Don Fawcett/E. Shelton/PR; (t) © Barry L. Runk/Rannels/GH; (cr) © FH; (br) © Don W. Fawcett/ VU; **Chapter 4: 68** © Dr. Gopal Murti/PH; **70** ©Barry L. Runk/Rannels/GH; **71** (tl) © Dr. David Scott/CNRI/PH; (tc) © David Scharf/PA; (tr) © Meckes/Ottawa/SS/PR; (b) © Dr. Patricia Schultz/PA; **72** © John Cardmore/ BPS/TSI; **77** © Don W. Fawcett/VU; **78** © R. Bolender–D. Fawcett/VU; **79** (t) © R. Bolender–D. Fawcett/VU; (b) © Pr. Jan de May/PH; **80** (t, bl, br) © David M. Phillips/ VU; **81** © Don Fawcett/VU; **82** (t) © BPA/SS/PR; (b) © Dr. Jeremy Burgess/SPL/SS/PR; **83** PR; **84** (bkgnd) © Superstock/Mark Romine; © CB; **85** (l) RLM; (r) © K. Talaro/VU; **88** © Roland Burke/PA; **90** © Don W. Fawcett/VU; **92** © Sergio Purtell/FOCA; **93** © Runk/ Schoenberger/GH; **Chapter 5: 94** © FH; **97** © M. Abbey/ VU; **98** (t) © Runk/Schoenberger/GH; (b) © David M. Phillips/VU; **104** © Prof. Birgit H. Satir; **Chapter 6: 110** © Arthur C. Smith III/GH; **119** © Dr. Jeremy Burgess/SS/ PR; **Chapter 7: 126** © Frans Lanting/MP; **130** © John Colwell/GH; **132** (t) From THE LIVES OF CELLS by Lewis Thomas. © Used by permission of Bantam Books, a division of Bantam Doubleday Dell Publishing Group, Inc.; (b) © BPA/SS/PR; **142** © WARD'S Natural Science; **Chapter 8: 144** © CNRI/SPL/PR; **146** © Gunther F. Bahr/AFIP/TSI; **147** © CNRI/PH; **151** (t) © David M. Phillips/VU; (b) © R. Calentine/VU; **152** (bkgnd) Geoff Tompkison/SPL/PR; © CNRI/PH; **156** © Jerome Wexler/PR; **158** © David M. Phillips/VU; **160** © John D. Cunningham/VU; **Unit 3: 162** © Schafer & Hill/TSI; **163** (t) © Mark Joseph/TSI; (c) © David Scharf; (b) © James H. Robinson; **Chapter 9: 164** © James H. Robinson; **170** (bkgnd) © Superstock/Mark Romine; © AP/Wide World Photos; **171** © Gregory G. Dimijian/PR; **172** (l) © Jane Grushow/GH; (r) © Christian Grzimek, OKAPIA/PR; **176** © Larry Lefever/GH; **Chapter 10: 184** © Tony Stone Imaging/TSI; **187** © Rosalind Franklin/SS/ PR; **198** © Dr. Gopal Murti/PH; **200-202** JS/FS; **Chapter 11: 202** © David Scharf; **206** © Professor Oscar Miller/ SPL/PR; **210** © Oliver Meckes/PR; **211** (l) © Bob Thomason/TSI; (r) © Martin M. Rotker/SS/PR; **212** © Tom Tracy/TSI; **213** (bkgnd) Geoff Tompkison/SPL/PR; James King-Holmes/SPL/PR; **Chapter 12: 220** © BPA/ SS/PR; **232** © CNRI/PH; **234** © Gopal Murti/CNRI/PH; **Chapter 13: 238** © Agricultural Research Service, USDA; **243** © Mark Joseph/TSI; **247** from GENETHICS by David Suzuki and Peter Knudston © 1989 by New Data Enterprises and Peter Knudston. Reprinted by permission of Harvard University Press; **248** © University of Florida; **251** AA/© Zig Leszczynski; **252** © Agricultural Research Service, USDA; **254** © Courtesy of Cellmark Diagnostics, Germantown, Md; **256** © Sergio Purtell/FOCA; **Unit 4: 258–259** (bkgnd) Iconographic Encyclopedia of Science, Art and Literature (1851); **258** (l) © Tui DeRoy/Bruce Coleman, Inc.; (r) © CB; **259** (tl) © Nanci Kahn/Institute of Human Origins; (tr) Smithsonion Institution (b) © Cabisco/VU; **Chapter 14: 260** © Douglas Peebles; **266** © Kenneth Garrett; **268** © Sidney Fox/VU; **271** (t) © Henry C. Aldrich, U. of Florida; (b) © Stanley Awramik/ BPS/TSI; **276** © WARD'S Natural Science; **Chapter 15: 278** © Tui DeRoy/Bruce Coleman, Inc.; **279** © RLM; **282** (bkgnd) Geoff Tompkison/SPL/PR; **284** © Bridgeman/ Art Resource, NY; **290** (t) AA/© Joe McDonald; (b) © Runk/Schoenberger/GH; **291** © Stephen Dalton/PR; **292** AA/© Fritz Prenzel; **Chapter 16: 298** © Art Wolfe/TSI; **300** © HRW photo/Sam Dudgeon; **302** © Art Wolfe/TSI; **303** (t) The cover from THE FLAMINGO'S SMILE: Reflections in Natural History by Stephen Jay Gould. © 1985 by Stephen Jay Gould. Reprinted by permission of W. W. Norton & Company, Inc.; (b) © Paula Lerner/ Woodfin Camp & Associates; **305** AA/© Ralph A. Reinhold; **308** © Art Wolfe/TSI; **310** © Tom McHugh, Steinhart Aquarium/PR; **311** (t) AA/© Zig Leszczynski; (b) © Barry L. Runk/GH; **Chapter 17: 318** © Enrico Ferorelli/Institute of Human Origins; **320** (t) © Frans Lanting/MP; (b) © Adam Jones/PR; **322** (bkgnd) © Superstock/Mark Romine; © Des & Jen Bartlett/ National Geographic Image Collection; **323** (t) © E. R.

TE CREDITS

ILLUSTRATION
6T MW; **14T** MW; **67B, 163B, 259B, 357B** MCA

PHOTOGRAPHY
Cover: © Manfred Danegger/Okrpai/Photo Researchers, Inc. **1T** © © Manfred Danegger/ Okrpai/Photo Researchers, Inc.; **3T** © Michael Fogden/ DKR Photo; **4T** © BPA/SS/PR; **5T** (t) © Runk/Schoenberger/GH; (b) Mark Joseph/ TSI; **6T** AA/© Anup & Manuj Shah; **7T** (t) AA/© Raymond A. Mendez; (b) Oliver Mekes/ MPI-Túbingin; **8T** © John Kaprielian/PR; **9T** (t) © Fred Bavendam; (b) © Kevin Schafer/ Martha Hill; **10T** © U.H.B. Trust; **11T** © Cornell University Photography; **12T** (t) JS/FS; (b) SP/FOCA; **13T** SP/FOCA; **14T** (b) Reprinted with permission from National Science Education Standards. Copyright 1996 by the National Academy of Sciences. Courtesy of the National Academy Press, Washington, D.C.; **18T, 19T** HRW; **29T** © SP/FOCA; **36T** SP/FOCA; **37T** HRW; **49T, 50T** © JS/FS; **51T** SP/FOCA; **3B** © Paul Chesley/TSI; **559B** © Jacques Jangoux/TSI; **665B:** James G. Gehling/© 1997. Reprinted with permission of Discover Magazine; **903B** © Hans Gelderblom/TSI

Degginger; (b) © Des & Jen Bartlett/National Geographic Image Collection; **325** (t) © John Reader/ SPL/PR; (b) © Cabisco/VU; **328** © E. R. Degginger; **Chapter 18: 336** © Nigel J. Dennis/PR; **338** (l) © Tom Tietz/TSI; (c) © E.R. Degginger; (r) © Ed Reschke/PA; **339** © Carolina Biological Supply Company/PH; **340** (bkgnd) © Superstock/Mark Romine; © CB; **341** © Lefever/Grushow/GH; **345** (l) © Gerard Lacz/PA; (tr, br) © Leonard Lessin/PA; **348** (t) ES/© Patti Murray; (b) © Chuck Davis/TSI; **349** (t) © RLM; (b) © Frans Lanting/MP; **Unit 5: 356** © Fred Bavendam; **357** (tl) © AA/© Tom Edwards; (tr) © DRA/Still Pictures/PA; (bl) AA/© Doug Wechsler; (br) © Art Wolfe/TSI; **Chapter 19: 358** © DRA/Still Pictures/PA; **360** (t) © Phil Degginger/ Color-Pic, Inc.; (bl, br) © Bishop Museum; **361** © Goddard Space Flight Center; **366** (bkgnd) © Superstock/Mark Romine; (l) © Princeton University; (r) © UPI/CB; **369** (l, r) ES/© E.R. Degginger; **Chapter 20: 378** © Flip Nicklin/MP; **380** © Mitsuaki Iwago/MP; **381** (l) AA/© Carl Roessler; (r) © Robert W. Hernandez/ PR; **386** © Tom J. Ulrich/VU; **387** AA/© Anup & Manuj Shah; **394** (l, r) © Stephen Kron/University of Chicago; **Chapter 21: 396** © Franklin J. Viola; **397** AA/© C.W. Schwartz; **398** (l) AA/© Doug Wechsler; (r) AA/© John Netherton; **399** (l) © Suzanne L. Collins & Joseph T. Collins/NASC/PR; (r) © E.R. Degginger; **400** (t) © Otto Hahn/PA; (b) © C. James Webb/PH; **402** © Merlin D. Tuttle/Bat Conservation International/NASC/PR; **405** © Terry Donnelly/TSI; **407** (tl) © Ken M. Johns/NASC/PR; (tr) © Glenn M. Oliver/VU; (b) © E.R. Degginger; **408** (t) © Kirtley Perkins/VU; (c, b) © John Sohlden/VU; **413** (t) © Runk/Schoenberger/GH; (b) RLM; **Chapter 22: 414** © Art Wolfe/TSI; **425** © Arthur C. Smith III/GH; **426** © E.R. Degginger; **427** (t) © Art Wolfe/TSI; (b) © Jim Brandenburg/MP; **428** (t) © Philip and Karen Smith/ TSI; (b) © Art Wolfe/TSI; **429** (t) © Luiz C. Marigo/PA; (b) © Ray Pfortner/PA; **430** (t) Reprinted by permission of Random House, Inc./Marston Bates; (b) © Mark Moffett/MP; **432** (t) © Norbert Wu/PA; (b) © Fred Bavendam; **433** (t) © Kim R. Reisenbichler for MBARI; (b) AA/© C.C. Lockwood; **434** © Luiz C. Marigo/PA; **437** © Kim R. Reisenbichler for MBARI; **Chapter 23: 440** © Art Wolfe/TSI; **443** © NASA/Goddard Space Flight Center; **447** (bkgnd) Geoff Tompkison/SPL/PR; © NORUT Information Technology, Earth Observation Group, Tromsø, Norway; **453** © Anthony Mercieca/PR; **455** (t) © Stephen J. Krasemann /DRK PHOTO; (b) © Art Wolfe/TSI; **456** (l, r) © Jeff & Alexa Henry/PA; **457** © Jeff Greenberg/PR; **463** JS/FS; **Unit 6: 464–465** (bkgnd) © BPA/PR; (amoeba) © M. Abbey/VU; **464** © Manfred Danegger/OKAPIA/PR; **465** (tl) © FH; (tr) © 1997 Kent Wood; (br) © Hank Morgan/PR; **Chapter 24: 466** © CNRI/SPL; **468** © E.R. Degginger; **469** (bkgnd) Geoff Tompkison/SPL/PR; © NSF Oasis/Mo Yung Productions; **470** (l) © FH; (c) © G. Shih-R. Kessel/VU; (r) © David M. Phillips/VU; **476** © Chris Bjornberg/PR; **480** © Runk/Schoenberger/GH; **483** © A.M. Siegelman/ VU; **484** JS/FS; **Chapter 25: 486** © Hans Gelderblom /TSI; **490** © Agricultural Research Service/USDA; **491** © Oliver Meckes/MPI-Tú bingen; **496** © SIU/VU; **497** (t) © Centers for Disease Control; (b) © Barts Medical Library/PH; **498** (t) © Yoav Levy/PH; (b) © K. Doyle/ VU; **499** © Hans Gelderblom/TSI; **504** JS/FS; **505** © Norm Thomas/PR; **506** © Eric Grave/PH; **507** © M. Abbey/VU; **508** (tl) © Robert Brons/BPS; (tr) © Omikron/SS/PR; (bl) © David M. Phillips/SS/PR; (br) © M. Abbey/VU; **510** (bkgnd) © Superstock/Mark Romine; **511** © Dr. Kwang W. Jeon; **512** © M. Abbey/VU; **513** (t) © Robert Brons/BPS; (b) © Dennis Kunkel/PH; **514** © Michael Abbey/PR; **516** (t) © Professors P.M. Motta & F.M. Magliocca/SPL/PR; (b) © C. James Webb/ PH; **517** © Moredun Animal Health Ltd./SPL/PR; **520** © Robert Brons/BPS; **521** © PH; **Chapter 27: 524** © Gregory Ochocki/PR; **526** (tl) © Microfield Scientific Ltd./SPL/PR; (tc) © Cabisco/VU; (tr) © T.E. Adams/VU; (b) AA/© Doug Wechsler ; **530** (l) © RLM; (r) ES/© Doug Wechsler; **531** (t) © P. & J. Clement/PR; (b) AA/© Anne Wertheim; **532** (t) © 1997 Kent Wood; (b) © Roland Birke/PA; **534** © Carolina Biological Supply/PH; **535** (t) © L. West/PR; (b) © FH; **536** (t) © American Photopathological Society; (b) © Herb Charles Ohlmeyer/FH; **540** © Herb Charles Ohlmeyer/FH; **541** © RLM; **Chapter 28: 542** © Manfred Danegger/ OKAPIA/PR; **544** (t) © University of Texas Medical Mycology Research Center; (bl) © J. Forsdyke/SPL/PR; (bc) © RLM; (br) © Dr. Dennis Kunkel/PH; **548** © Rod Planck/PR; **549** (t) © Runk/Schoenberger/GH; (b) © Grant Heilman/GH; **550** © SPL/PR; **551** © E. R. Degginger; **552** (t) © Francois Ducasse/PR; (b) © Dr. Paul A. Zahl/PR; **556–557** JS/FS; **Unit 7: 558** © Pal Hermansen/TSI; **559** (orchid) © Kurt Coste; (tl) © Frans Lanting/MP; (tr) © William E. Ferguson; (br) © RLM; **Chapter 29: 560** © Yann Layma/TSI; **561** © TSI; **563** (t) © Peter Holden/VU; (b) © Michel Viard/PA; **565** (tl) © E.R. Degginger; (tr) © Jim Strawser/GH; (b) © RLM; **567** © H. Richard Johnston/TSI; **568** (bkgnd) © Superstock/Mark Romine; © National Portrait Gallery, Smithsonian Institution/Art Resource, NY; **569** © AP/ Wide World Photos; **570** © Frans Lanting/MP; **571** (t) © Bob Gibbons/NASC/PR; (b) © RLM; **572** © Martha Cooper/PA; **576** © WARD'S Natural Science; **Chapter 30: 578** © Michael Sewell/PA; **583** © Ed Reschke/PA; **584** (tl) © L. West/PR; (tr) © RLM; (b) © Runk/ Schoenberger/GH; **585** (t) Cover of SILENT SPRING. © 1962 by Rachel Carson, renewed 1990 by Roger Christie. Reprinted by permission of Houghton Mifflin Company. All rights reserved; (b) © The Granger Collection, New York; **586** © RLM; **587** © Runk/ Schoenberger/GH; **588** (t) © Tom Dietrich/TSI; (b) © Tom Till/TSI; **589** (t) © Lefever/Grushow/GH; (b) © E. Webber/VU; **590** © D. Cavagnaro/VU; **591** (t) © RLM; (b) © Frans Lanting/MP; **595** © Willard Clay/TSI; **596** (t) © Pat Anderson/VU; (b) © David Sieren/VU; **597** (t) © Runk/Schoenberger/GH; (b) © E.R. Degginger; **Chapter 31: 598** © Dr. Jeremy Burgess/SPL/PR; **603** (l) © Dwight R. Kuhn; (r) © R. Calentine/VU; **604** (tl) © Jane Grushow/GH; (tr) © Brian Rogers/VU; (b) © Runk/Rannels/GH; **605** (t) © RLM; (bl) © Stan Elems/VU; (br) © Runk/Schoenberger/GH; **606** © Dwight R. Kuhn; **608** (t, bl) © William E. Ferguson; (br) © E.R. Degginger/PR; **610** (t) © Carolina Biological Supply/PH; (b) © Ken Wagner/PH; **611** (l) © Runk/ Schoenberger/GH; (r) © Albert Copley/VU; **614** © Cornell University Photography; **615** (l) ES/© Donald Specker; (c) © E.R. Degginger; (r) ES/© John Pontier; **616** (t) © John Kaprielian/PR; (c) © E.R. Degginger; (b) JS/FS; **617** © Michael Fogden/DRK PHOTO; **618** (l, r) © P. Dayanandan/PR; **620** © Runk/Schoenberger/GH; **621** ES/© John Gerlach; **622** © Ed Reschke/PA; **623** (t) © Carolina Biological Supply/PH; (c) © PH; (b) © RLM; **Chapter 32: 624** © Dr. Jeremy Burgess/SPL/PR; **628** © RLM; **632** © Michael Fogden/DRK Photo; **634** © Dick Canby/DRK PHOTO; 637 © Runk/Schoenberger/GH; **639** © Runk/Schoenberger/GH; **640** © Rosenfeld Images Ltd./SPL/PR; **642** © ES/Richard Shiell; **643** © Runk/Schoenberger/GH; **644** © JS/FS; **Chapter 33: 646** © Jack Dykinga; **649** © Stefan Eberhard/FH; **650** (t) © E.R. Degginger (b) © Abbot Laboratories; **652** © Cathlyn Melloan/TSI; **653** © David Newman/VU; **654** (t) © Runk/Schoenberger/GH; (bl, br) © Christi Carter/ GH; **655** (l, r) © Tom McHugh/PR; **658** © Pal Hermansen/TSI; **660** © John D. Cunningham/VU; **661** © David Newman/VU; **Unit 8: 664–665** (bkgnd) © BPA/PR; **664** © Kevin Schafer; **665** (ants) © Mark Moffett/MP; (tl) © E.R. Degginger; (tc) © CNRI/SPL/PR; (tr) AA/© Raymond A. Mendez; **Chapter 34: 666** © Kevin Schafer; **668** © Merlin D. Tuttle, Bat Conservation International/PR; **673** © Fred Bruemmer/DRK PHOTO; **674** © S.J. Krasemann/PA; **675** © Phil Degginger/TSI; **677** (l) © Stephen J. Krasemann/DRK PHOTO; (r) © Stephen J. Krasemann/PR; **680** (bkgnd) © Superstock/ Mark Romine; © The Granger Collection, New York; **681** © Runk/Rannels/GH; **Chapter 35: 692** AA/© Joyce Burek; **698** (l) © Robert Brons/BPS/TSI; (r) © Runk/ Schoenberger/GH; **699** © Runk/Schoenberger/GH; **701** (t) © Neil McDaniel/PR; (c) © Fred Bavendam; (b) © Stuart Westmorland/TSI; **702** © Andrew J. Martinez/PR; **706** © Carolina Biological Supply/PH; **Chapter 36: 708** © RLM; **709** © Milton Love/PA; **711** © CNRI/SPL/PR; **716** (t) © C. James Webb/PH; (b) © Jan Callagan/PH; **717** © Andrew Syred/SPL/PR; **720** © A.M. Siegelman/ VU; **722** © T.E. Adams/VU; **Chapter 37: 724** © Franklin J. Viola; **726** © Andrew Syred/SS/PR; **728** (t) © Hal Beral/VU; (b) © David & Hayes Norris/PR; **730** © Andrew Martinez/PR; **731** © AA/Rudie Kuiter/OSF; **732** (bkgnd) Geoff Tompkison/SPL/PR; © St. Bartholomew's Hospital/SPL/PR; **733** © Marty Snyderman/VU; **736** © RLM; **Chapter 38: 742** AA/© E.R. Degginger; **744** RLM; **747** (tl) AA/© OSF; (tc) © M.I. Walker/PR; (tr) Neg. No. 312007 (Photo by I. Dutchner) Courtesy Dept. of Library Services, American Museum of Natural History; (b) © RLM; **752** (t) © RLM; (b) © Sturgis McKeever/PR; **753** (t, b) © RLM; **754** (t) © Thomas Gula/VU; (b) © Edward S. Ross/PR; **759** JS/FS; **Chapter 39: 760** AA/© Raymond A. Mendez; **761** (l) © WHM Bildarchiv/PA; (r) AA/© E.R. Degginger; **764** (t, b) © RLM; **770** (t) © James C. Cokendolpher/FH; (b) AA/© E.R. Degginger; **771** © Tom McHugh/PR; **778** © Stephen Dalton/PR; **Chapter 40: 780** © Fred Bavendam; **782** (l) F. Stuart Westmorland/PR; (r) © Fred Bavendam; **783** (t) AA/© Capt. Clay H. Wiseman; (b) © Fred Bavendam; **784** (t) AA/© Steven David Miller; (bl) © E.R. Degginger; (br) © Daniel W. Gotshall/VU; **786** © Fred Winner/Jacana/PR; **787** © Fred McConnaughey/PR; **789** AA/© G.I. Bernard; **790** © Fred Bavendam; **794** © Pat O'Hara/TSI; **Unit 9: 796** © Art Wolfe; **797** (chameleon) © Frans Lanting/MP; (tl) © Fred Bavendum; (c) © S.J. Krasemann/PA; (tr) © Art Wolfe/TSI; **Chapter 41: 798** © Stuart Westmorland/TSI; **803** (t) © William E. Ferguson; (b) © Berthoule-Scott/PR; **804** © Meckes/Ottawa/PR; **805** © Fred Bavendam; **812** © Fred Bavendam; **Chapter 42: 818** © William J. Weber/VU; **821** (l, r) © RLM; **822** (l, r) © Suzanne L. & Joseph T. Collins/PR; **823** (l) AA/© Zig Leszczynski; (r) AA/© Juan M. Renjifo; **831** © Dwight R. Kuhn; **832** © Michael Fogden/DRK Photo; **836** © Runk/Schoenberger/GH; **837** © Dwight R. Kuhn; **Chapter 43: 838** © Kevin Schafer/Martha Hill; **843** AA/ © Zig Leszcynski; **844** Cover from THE DINOSAUR HERESIES by Robert T. Bakker. Copyright © 1986 by Robert T. Bakker. Reprinted by permission of William Morrow & Company, Inc.; **847** AA/© Zig Leszczynski; **848** © RLM; **849** AA/© Roger de la Harpe; **850** (l) AA/© Joe McDonald; (r) AA/© Victoria McCormick; **851** (l) © Beth Davidow/VU; (r) © E.R. Degginger; **852** (l) AA/© Michael Dick; (r) AA/© Zig Leszczynski; **853** (l) © E.R. Degginger; (r) AA/© Breck P. Kent; **854** (tl, tc, tr) © Michael Fogden/DRK Photo; (b) © Tom J. Ulrich/VU; **857** © Sergio Purtell/FOCA/ HRW; **858** © JS/FS; **Chapter 44: 860** © Art Wolfe/TSI; **863** © James L. Amos/PR; **869** (bkgnd) Geoff Tompkison/SPL/PR; © Fred Bruemmer/ DRK Photo; **870** AA/© E.R. Degginger; **871** (tl) AA/© Tom Edwards; (tr) © Gary Meslaros/VU; (bl) © Neal & Mary Mishler/TSI; (br) AA/© John Gerlach; **872** (t) © Carl R. Sams II/PA; (b) AA/© Tony Tilford; **873** (t) © Frans Lanting/MP; (b) © S. Maslowski/VU; **874** © Joe McDonald/VU; **878–879** © WARD'S Natural Science; **Chapter 45: 880** © Art Wolfe; **887** (t) AA/© Hamman/ Heldring; (b) © Art Wolfe/TSI; **888** © Mitsuaki Iwago/ MP; **889** (t) Reprinted by permission Houghton Mifflin Company; (b) © Peter Veit/DRK Photo; **890–891** © Tom McHugh/PR; **892** (l) © S.J. Krasemann/PA; (r) © Luiz C. Marigo/PA; **893** © Art Wolfe/TSI; **894** (t) © S.J. Krasemann; (b) AA/© Ken Cole; **895** (t) © Fred Felleman/TSI; (b) © Douglas Faulkner/PR; **900** (l) © Phil Degginger/Color-Pic, Inc.; (r) © David Austen/TSI; **Unit 10: 902** © David Young Wolff/TSI; **903** (blood cells) Professors P.M. Motta & S. Correr/SPL/PR; (tl) © Image Shop/PR; (tc) © Nestle/ Petit Format/PR; (tr) Terry Vine/TSI; **Chapter 46: 904** © U.H.B. Trust/TSI; **911** © Andrew Syred; **916** (l) © John D. Cunningham/ VU; (c) © David M. Phillips/VU; (r) © Eric Grave/SS/PR; **920** © David Madison; **921** (bkgnd) Geoff Tompkison/ SPL/PR; (t) CNRI/PH; (b) © U.H.B. Trust/TSI; **924** © David Madison; **Chapter 47: 930** © Oliver Meckes/PR; **934** © Ed Reschke/PA; **938** © St. Bartholomew's Hospital/SPL/PR; **939** © Andrew Syred; **940** © Biology Media/PR; **941** (l) © A.B. Dowsett/SPL/PR; (c) © NIBSC/ SPL/PR; (r) © Andrew Syred; **945** (l, r) © Runk/ Schoenberger/GH; **Chapter 48: 954** © Meckes/ Ottawa/PR; **955** © CNRI/SPL/PR; **956** © Prof. P. Motta/Dept. of Anatomy/University "La Sapienza," Rome/SPL/PR; **958** © Dr. Dennis Kunkel/PH; **966** © CB; **967** (t) © The Granger Collection, New York; (b) CB; **968** © NIBSC/SPL/PR; **Chapter 49: 976** © John Kennedy/BPS/TSI; **988** © Prof. P. Motta/Dept. of Anatomy/University "La Sapienza," Rome/SPL/PR; **989** © Scott Camazine/PR; **990** © AP/Wide World Photos; **1000** © JS/FS; **Chapter 50: 1002** © Comp-Unique/ Strange/CMSP; **1003** © Dr. David Scott/PH; **1019** Courtesy of Prof. Dr. G. Reiss, Dept. of Cell Biology/ Electron Microscopy, Hnaover Medical School, Germany; **Chapter 51: 1028** © Herb Charles Ohlmeyer/ FH; **1036** © Carolina Biological Supply/PH; **Chapter 52: 1046** © Lennart Nilson/Bonnier Alba AB, from A Child is born (Dell Publishing Co.); **1050, 1054** © David M. Phillips/PR; **1061** © Lennart Nilson/Bonnier Alba AB, from A Child is born (Dell Publishing Co.); **Chapter 53: 1064** © Professors P. Motta & T. Fujita/University "La Sapienza," Rome/SPL/PR; **1067** © Institut Pasteur/ CNRI/PH; **1069** (l) © Manfred Kage/PA; (r) © Manfred Kage/PA; **1070** © JS/FS; **1072** © George Steinmetz; **1076** © E.R. Degginger; **1082** HRW; **1083** SP/FOCA; **1084** HRW; **1085** SP/FOCA; **1087** Charles D. Winters; **1088, 1089, 1090** SP/FOCA; **1091** HRW; **1092** © Chris Bjornberg/PR, **1093** (t) © Henry C. Aldrich, U. of Florida; (b) © David M. Phillips/SS/PR; **1095** © Manfred Danegger/OKAPIA/ PR; **1096** © H. Richard Johnston/ TSI, **1097** © RLM; **1099** © Fred Bavendam

1138